GOODE'S
WORLD ATLAS
Twenty-Third Edition

Editor

Christopher J. Sutton, Ph.D.
Western Illinois University

Editorial Advisor

Howard Veregin, PhD
University of Wisconsin-Madison

RAND McNALLY

Executive Editor: *Christian Botting*
Content Producer: *Anton Yakovlev*
Executive Marketing Manager, Product: *Neena Bali*
Senior Marketing Manager, Field: *Mary Salzman*
Senior Project Manager (International Mapping): *Kevin Lear*
Lead Cartographer (International Mapping): *Marissa Wood*
Cartographer (International Mapping): *Kim Clark*
Cartographer (International Mapping): *Vickie Taylor*
Map Editor (International Mapping): *Robert Harding*
Full-Service Project Management (SPi Global): *Patty Donovan*
Composition (SPi Global): *Steve Martel*

Cover Credit: Left half: Blue Marble: Next Generation monthly composite image for August. Blue Marble Next Generation images are derived from MODIS data at a resolution of 500 meters. MODIS (Moderate Resolution Imaging Spectroradiometer) sensors on board the Terra and Aqua Satellites provide global coverage every one to two days in 36 spectral bands. Source: NASA's *Earth Observatory* (http://neo.sci.gsfc.nasa.gov/).
Right half: Earth at Night composite assembled from data acquired by the Suomi National Polar-orbiting Partnership (Suomi NPP) satellite using the "day-night band" of the Visible Infrared Imaging Radiometer Suite (VIIRS). VIIRS detects light in a range of wavelengths from green to near-infrared and uses filtering techniques to observe dim signals such as gas flares, auroras, wildfires, city lights, and reflected moonlight. Source: NASA's *Earth Observatory* (http://neo.sci.gsfc.nasa.gov/).

Section Opener graphics: Blue Marble Next Generation, available from NASA's *Earth Observatory* (http://neo.sci.gsfc.nasa.gov) and cross-blended hypsometric tints from Natural Earth (http://naturalearthdata.org).

Library of Congress Control Number: 2016945027

ISBN-10: 0-13-386464-2
ISBN-13: 978-0-13-386464-9

10 9 8 7 6 5 4 3 2 1

Goode's Atlas is named for John Paul Goode, who created the atlas and served as its editor for many editions. Goode was one of the first U.S. academic cartographers. He was born in rural Minnesota in 1862, received his bachelor's degree from the University of Minnesota in 1889, and earned his doctorate in economic geography from the University of Pennsylvania in 1903. He spent much of his professional career at the University of Chicago. Among his many accomplishments he is perhaps best known for the development of the Interrupted Homolosine projection, which he first presented at the Association of American Geographers meeting in 1923, and which has been used extensively in *Goode's Atlas* and in many other geographic publications to the present day.

The Homolosine is a composite of two projections, the Mollweide (Homolographic) and the Sinusoidal. Goode interrupted the Homolosine over the oceans to minimize distortion of shapes over continental land masses. Lines of latitude on the Homolosine are straight lines, to facilitate analysis of comparative latitudes. Also, the projection is equal area. Goode was a strong proponent of equal area projections and an equally strong opponent of the Mercator projection, widely used in the early part of the 20th century for world maps. As Goode stated in the introduction to the 1st edition of the atlas (1923, p. x), the distortion of area on the Mercator projection is so extreme that "it becomes pedagogically a crime to use Mercator's map" for studies of areal distributions such as population density, rainfall, or sizes of countries.

Under Goode's editorship the atlas doubled in size. The 1st edition of *Goode's School Atlas* contained 96 pages of maps. The 4th edition (1932), the last edition that Goode would edit before his death, contained 174 pages of maps. Goode introduced many of the thematic map topics that are still found in the atlas today, including world economic maps of agricultural commodities, minerals, energy, and international trade. These topics reflect Goode's interest and training in economic geography.

Goode remained the only name on *Goode's School Atlas* until the 8th edition (1949), on which Edward B. Espenshade, Jr., was credited with numerous updates and revisions. Espenshade was then named editor for the 9th edition (1953). Espenshade was one of Goode's students and spent his academic career at Northwestern University in Evanston, Illinois. The 9th edition was significant in many respects. It boasted a new title, *Goode's World Atlas,* and contained many of the features of the modern atlas.
In particular, Espenshade made extensive use of maps compiled by experts in specific subdisciplines of geography. Examples include natural vegetation by A. W. Küchler, physiography by Erwin Raisz, climate regions by Glenn Trewartha, and agricultural regions by

John Paul Goode

Derwent Whittlesey. By relying on the research of these and other scholars, Espenshade was able to incorporate the latest advances in the study of geographical phenomena. Espenshade also oversaw the creation of a new reference map series, which included hand-drawn shaded relief for the first time in the atlas. These reference maps were introduced in the 11th edition (1960).

Joel L. Morrison, then at the University of Wisconsin, joined Espenshade as associate editor on the 14th edition (1974). Morrison, who had a distinguished career in academia and the federal government, was affiliated with the atlas through the 19th edition (1995). In the 1970s and 1980s the atlas saw numerous innovations, including the introduction of ocean floor shaded relief maps, reference maps of major world cities, a continent environments map series, and the first use of cartograms.

The 19th edition was Espenshade's last as editor. On that edition, John C. Hudson assumed the role of associate editor. Hudson, a distinguished academic geographer at Northwestern University, then took on the role of editor for the 20th edition. Hudson introduced many new thematic maps, including world ecoregions, origins of plants, refugees, conflicts, and oceanic environments.

Howard Veregin of the University of Minnesota succeeded Hudson as editor for the 21st edition. Veregin then moved to Rand McNally to serve as director of geographic information services and, in that capacity, edited the 22nd edition. Under Veregin's stewardship, the atlas became all-digital, with most maps produced using geographic information systems (GIS) technology, including numerous new thematic maps.

Christopher Sutton was named editor for the 23rd edition. Sutton was a member of the geography faculty at Western Illinois University. Sutton expanded the Atlas, introducing more than sixty new world thematic and regional reference maps, and an updated design.

Throughout its history *Goode's Atlas* has adapted to changes in cartographic technology, map design, and geographic curricula. However, it has always maintained the pedagogical foundation that John Paul Goode established in the 1st edition in 1923. It should be seen first and foremost as a work of scholarship, incorporating the latest insights into geographical research and knowledge. It is also a fascinating portrait of almost nine decades of evolution in geography and cartography.

Robert B. McMaster, Ph.D.
Susanna A. McMaster, Ph.D.
University of Minnesota

Interest in geography has increased dramatically in the last few decades.

Perhaps it is because of efforts by those who teach geography or study it. Perhaps, because of instant global communications and the Internet, we're all more aware of global events. Maybe it's globalization, or recent wars, or global terrorism. Perhaps, because of environmental concerns, we feel a responsibility to better understand and manage Earth and its resources.

Whatever the reasons, this renewed interest in geography is serious. Billions of dollars are being spent every year collecting geographic data. Globally, tens of thousands of organizations of all kinds — government, business, academic, non-profit — have recognized that many of the problems they face must be understood geographically.

In emergencies, government agencies at every level need geographic information about the hurricanes, wildfires, earthquakes, tsunamis, storm surges, and floods they must respond to. They also need to know about the geography of political trouble spots, famines, droughts, terrorism, the narcotics trade, energy resources, shipping, war fighting, and a long list of other topics.

But spending for geographic information goes far beyond governments. Geographic information is also essential to understanding commerce, business, history, military campaigns, migrations, exploration, evangelization, cultural diffusion, origins of civilizations, distribution of organisms and their ecology, agriculture, climate change, natural resources, transportation patterns, productivity, epidemiology, conservation, election results, and many, many other topics. Organizations throughout the world recognize this, pay for geographic information, and hire people to manage and analyze it for them.

Geography matters! The community of people who rely on it is growing every day.

Geography today is a multidisciplinary science. Our ability to collect geographic data scientifically is exploding and we have begun to acquire the methods needed to effectively manage this data explosion for the entire planet. This includes new tools like satellite images of Earth, geographic information system (GIS) software to process and display enormous volumes of data, and global positioning system (GPS) devices that can accurately determine locations anywhere on Earth.

Goode's World Atlas is part of this revolutionary growth. The atlas has long been a staple of the college classroom, educating students about important geographic issues of the day. The current 23rd edition, which you hold in your hands, makes extensive use of digital geographic information of the kind I refer to above. It focuses on important contemporary issues like globalization, global climate change, food security, and environmental degradation. It uses GIS to integrate information and render it for cartographic display.

Goode's World Atlas helps us understand our world and our place in it. It helps us interpret stories in the news, understand international conflicts, evaluate foreign competition for jobs, make informed decisions about free trade or immigration, respond to changing oil prices or possible climate change, and think through complex domestic and foreign policy issues.

Goode's World Atlas is an essential guidebook to the new geography, helping us sort through and decipher patterns in a flood of geographic data. It helps us make sense of these data, and it provides authoritative cartographic interpretations of complex geographic issues.

I have studied and worked with geographic information for more than forty years. My experiences have given me insights into what our world was and is, and what we can make of it in the future. I think that geography can give us new eyes with which to see the world, so that — as the poet remarked — after all our traveling we return home and see it for the first time. I believe that *Goode's World Atlas* is an invaluable component of this learning process. The Atlas continues to evolve and adapt, but remains rooted in its original function — helping us develop geographic understanding and knowledge as a way to make sense of our world.

Jack Dangermond
President, ESRI

Goode's World Atlas, 23rd edition

The 23rd edition of *Goode's World Atlas* maintains the tradition of excellence introduced in Professor J. Paul Goode's first edition of the *Atlas* in 1923.

The 23rd edition includes more than 60 new reference and thematic maps. New regional maps include the Arctic Islands and Greenland in the North America section and Turkey in Asia. New geology maps are now included for the world as well as for each of the continents. Additional city maps have been added to North America, reflecting population growth in the southern and southwestern United States. Also new to the 23rd edition are maps of Europe's microstates and Africa's island countries.

The 23rd edition features more than thirty new thematic maps, including several new maps that expand on the Earth's oceanic environments, including: tropical cyclones, benthic biomass, oceanic dead zones, coral reef distribution, tsunamis, and areas vulnerable to sea level rise. Additionally, more than a dozen new maps focus on population characteristics and changes, megacities, politics, labor, and migration.

The 23rd edition also features several design enhancements to facilitate navigation. Each section of the atlas has a new introductory section that presents a brief overview of the region as well as a "mini-index" of the maps contained within the section. Each section now also features a colored heading, with the charts that accompany the world thematic maps keyed to these colors.

All prior thematic maps are updated with the latest available data and to reflect the many changes to place names that have occurred since the 22nd edition.

Finally, the 23rd edition is now offered in a number of flexible formats for teachers and students, with eBook and custom library versions of the atlas available from Pearson:

www.pearsonhighered.com

www.pearsoncustomlibrary.com

Any project such as this is a considerable undertaking, which requires a skilled team of researchers, cartographers, and publishing professionals. The production team at International Mapping—Kevin Lear, Marissa Wood, Kim Clark, Vickie Taylor, and Robert Harding—worked tirelessly on the 23rd Edition's numerous revisions and new artwork to ensure the success of this update. The dedicated professionals at Pearson, and their partnership with Rand McNally, allowed the 23rd edition to become a reality. I am particularly indebted to Christian Botting, Geosciences Executive Editor at Pearson, whose vision to provide a comprehensive update to the *Atlas* brought this project to the fore and whose stewardship allows students and "armchair geographers" to view our world in a way that only an atlas can provide.

Finally, I would like to acknowledge the input of Howard Veregin, editor of the 21st and 22nd Editions. His insights were invaluable, particularly those related to content new to the 23rd Edition.

It has been more than ninety years since the first edition of J. Paul Goode's first atlas. Over the decades, the atlas has changed with the times while maintaining an unparalleled reputation of meticulous research and the dedication to mapping excellence provided by Rand McNally's outstanding cartographers. *Goode's World Atlas* has long been a widely-used educational tool and the 23rd edition continues this tradition.

Christopher J. Sutton, Ph.D., Editor
Western Illinois University

Introduction

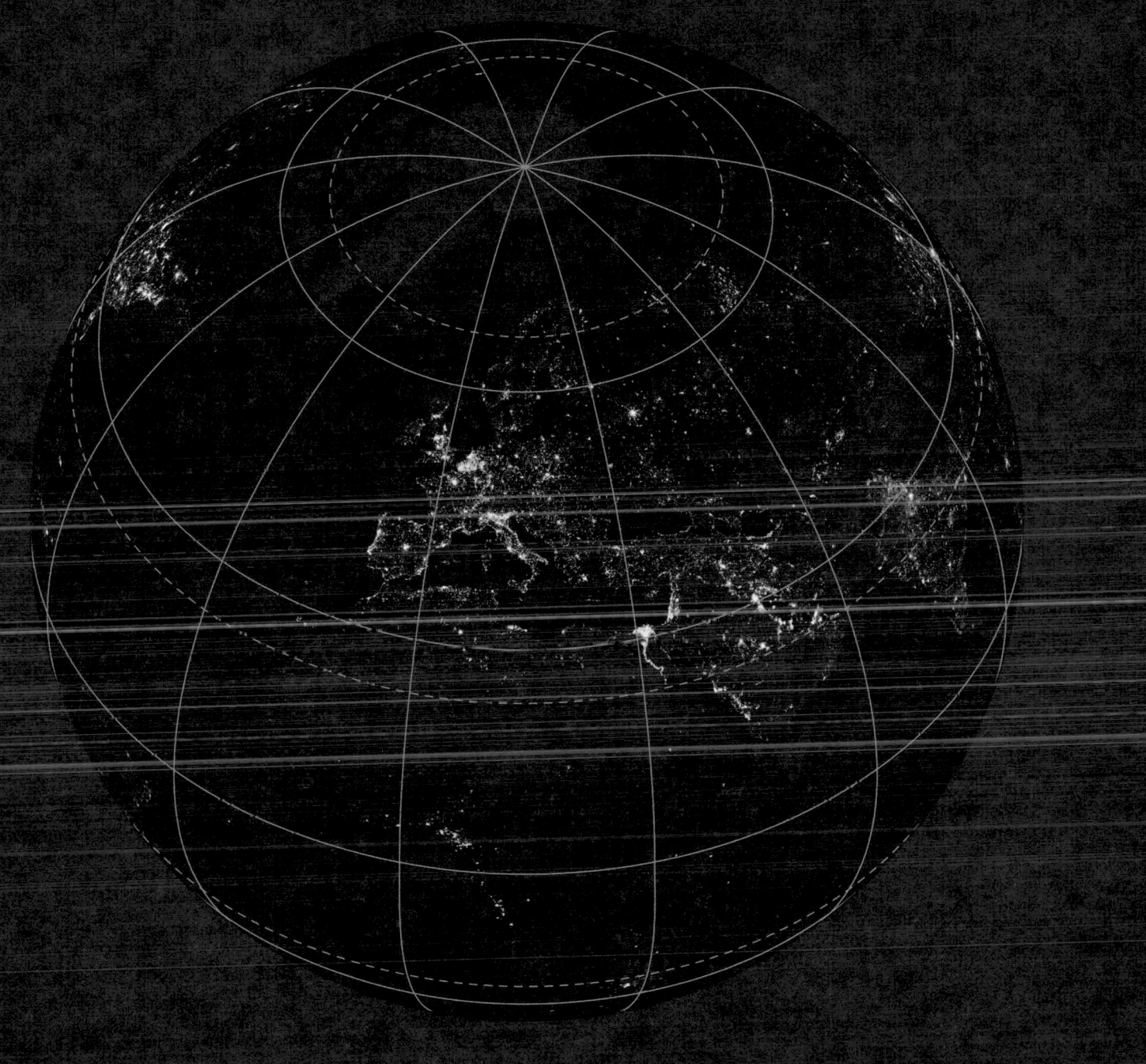

From the first printing of his atlas in 1923, J. Paul Goode sought to provide the geographically curious a tool to explore Earth's complex landscapes and the relationships between people and the environment. The 23rd Edition continues this tradition with more than 300 pages of richly detailed reference maps, informative human and environmental thematic maps, charts and graphs, and a comprehensive pronouncing index.

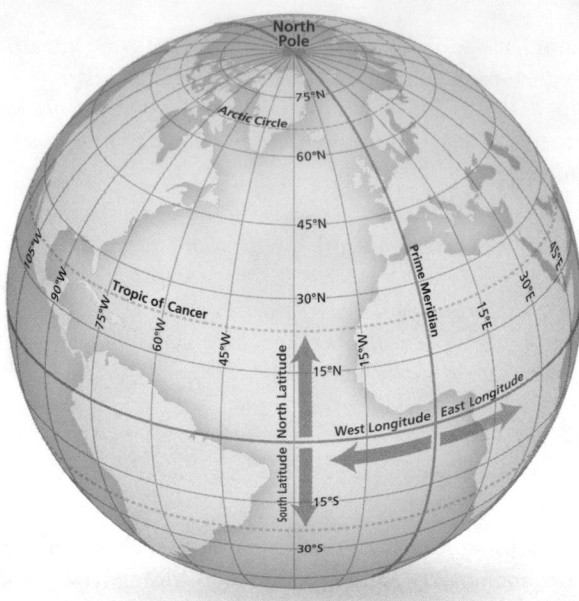

View of Earth centered on 30° N, 30° W

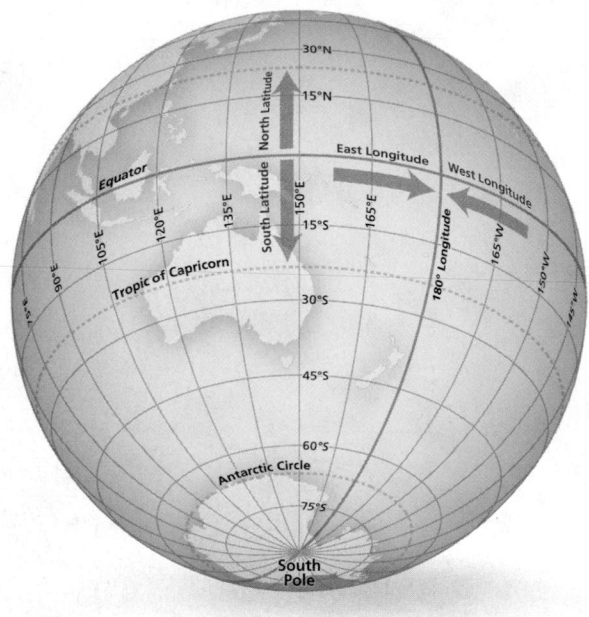

View of Earth centered on 30° S, 150° E

Basic Earth Properties

Earth is essentially spherical in shape. The North and South Poles are aligned with Earth's axis of rotation. The Equator is equidistant between the Poles and divides Earth into northern and southern hemispheres.

Latitude and longitude identify the locations of features on Earth's surface. Latitude is the angle north or south of the Equator. Longitude is the angle east or west of the Prime (Greenwich) Meridian. A meridian is a line of longitude extending from the North Pole to the South Pole. The Prime Meridian is the meridian passing through the Royal Observatory in Greenwich, England. This location for the Prime Meridian was adopted at the International Meridian Conference in Washington, D.C., in 1884.

Latitude and longitude are usually given in degrees, minutes and seconds. There are 60 minutes in a degree and 60 seconds in a minute. The symbols °, ' and " represent degrees, minutes and seconds, respectively. For latitude the symbols N and S indicate degrees north or south of the Equator. The latitude of the Equator is 0° and the latitudes of the North and South Poles are 90° N and 90° S. For longitude the symbols E and W indicate degrees east or west of the Prime Meridian. Longitude ranges from 0° at the Prime Meridian to 180° E or W. The meridian at 180° E is the same as the meridian at 180° W, and this meridian is the approximate location of the International Date Line. This meridian and the Prime Meridian divide Earth into eastern and western hemispheres.

A latitude-longitude coordinate pair defines the location of a feature on Earth. As an example, the Rand McNally building in Skokie, Illinois, has the coordinates 42° 3' 37" N, 87° 45' 39" W. Often the number of seconds are omitted from the coordinates if a high level of precision is not required.

Lines of latitude are also known as parallels. Two parallels of special importance are the Tropic of Cancer and the Tropic of Capricorn, at approximately 23° 30' N and S respectively. This angle coincides with the inclination of Earth's axis relative to its orbital plane around the Sun. The Tropics are the lines of latitude where the noon sun is directly overhead on the solstices. Two other important parallels are the Arctic Circle and the Antarctic Circle, at approximately 66° 30' N and S respectively. These lines mark the most northerly and southerly points at which the Sun can be seen on the solstices.

The Geographic Grid

The geographic grid is the grid of latitude-longitude lines on Earth. The following are some important characteristics of the grid.

Lines of longitude (meridians) are equal in length and meet at the Poles.

Lines of latitude (parallels) are parallel to each other and equally spaced along meridians.

The length of parallels decreases as one gets closer to the Poles. For example, the length of the parallel at 60° latitude is one-half the length of the Equator.

Meridians get closer together with increasing distance from the Equator, and finally converge at the Poles.

Parallels and meridians meet at right angles.

Cartography and Geospatial Technology

Geography's subject matter includes the people, landforms, climate, and other physical and human phenomena that make up Earth's environments and give unique character to different places. Geographers construct maps to visualize how these phenomena vary over geographic space. Maps help geographers understand and explain phenomena and their interactions.

The art and science of mapmaking is known as cartography. Although maps were once drawn by hand, they are now usually created using digital technology. This technology includes GIS (geographic information systems), as well as GPS (global positioning system) and remote sensing. Collectively these geospatial technologies are used to both produce as well as consume maps.

GIS is a specialized type of software that enables the integration, processing, analysis, and display of digital geographic information. It combines an underlying database with spatial analysis tools and cartographic rendering capabilities. First developed in the 1960s, GIS has evolved rapidly in the last decade with advances in computer processing power and the increased availability of digital geographic information. Applications of GIS have also diversified rapidly as more users have recognized its utility for solving geographic problems.

In cartography, GIS has redefined how maps are made. Since the underlying data for a map are stored in a database, a map is just one of the many possible data representations. Many different maps can be produced from one database, based on different permutations of attributes, for different map scales, geographic areas, or time periods, and using different map treatments. GIS also enhances the efficiency of map production. For example, map symbology can be driven off stored attributes, selection and generalization can be conducted using defined rules, map text can be placed automatically, and index creation can be automated. All of the maps in this edition of *Goode's World Atlas* are digitally produced, and the vast majority have been created using GIS software.

GIS also greatly enhances the ability to integrate data from a variety of sources and process these data for specific mapping purposes. In this sense GIS is closely related to other geospatial technologies such as GPS and remote sensing. GPS is a satellite-based system for capturing precise information about locations on Earth's surface. Originally developed by and for the military, GPS is now the underlying technology behind personal navigation devices and location-based services. GPS has revolutionized the field of surveying and is used in a wide variety of fields where accurate coordinates are needed.

Remote sensing refers to the collection of data about Earth from satellites and aircraft. Advances in remote sensing have greatly magnified the volume and types of geographic data available for mapping. Many of the maps in this atlas are derived from remote sensing imagery, including the maps of land cover, gravity, sea level change, sea ice, and forest loss. Modern remote sensing systems are designed to focus on specific portions of the electromagnetic spectrum, some of which cannot be seen by the human eye. This capability allows very specific geographic phenomena to be imaged and analyzed.

The diagram below illustrates how geospatial technology can be used to map and monitor changes in the global environment. This edition of *Goode's World Atlas* reflects this growing awareness and capability by incorporating the latest digital data sources wherever possible.

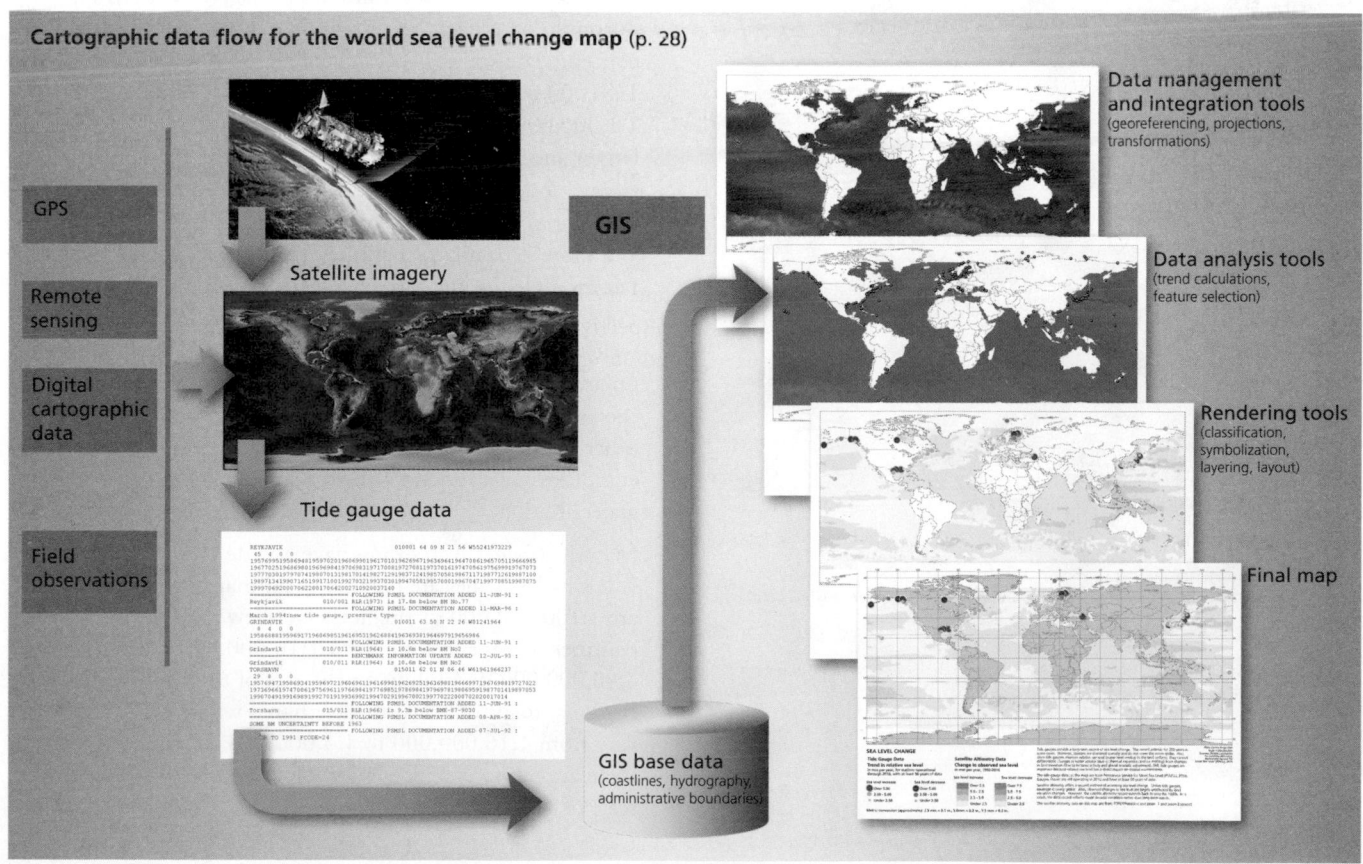

Cartographic data flow for the world sea level change map (p. 28)

GPS

Remote sensing

Digital cartographic data

Field observations

Satellite imagery

Tide gauge data

GIS

GIS base data
(coastlines, hydrography, administrative boundaries)

Data management and integration tools
(georeferencing, projections, transformations)

Data analysis tools
(trend calculations, feature selection)

Rendering tools
(classification, symbolization, layering, layout)

Final map

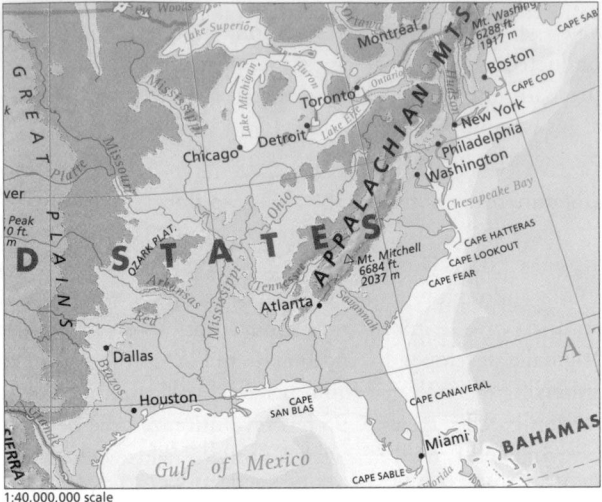

1:40,000,000 scale

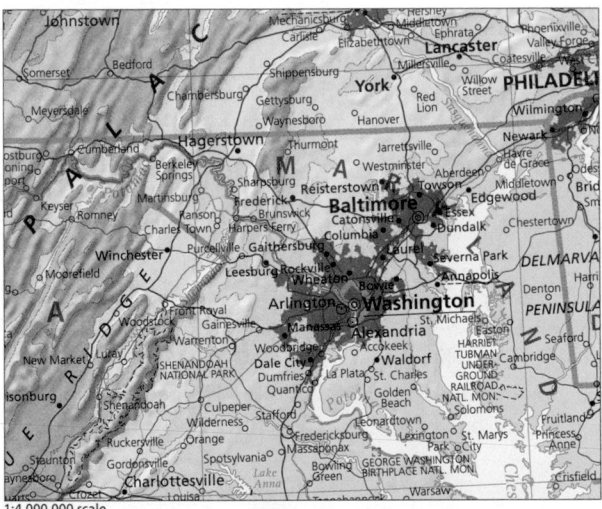

1:4,000,000 scale

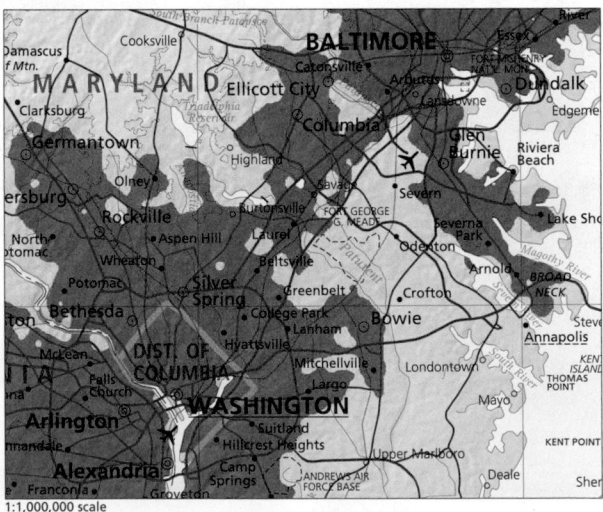

1:1,000,000 scale

Map Scale

Map scale is the ratio of distance on a map to distance on Earth's surface. For example, if two towns on a map are separated by a distance of 1 inch, and these towns are actually 1 mile apart, then the scale of the map is 1 inch to 1 mile.

The statement "1 inch to 1 mile" is a verbal scale. Verbal scales are simple and intuitive, but it can be difficult to compare verbal scales for different maps on which different linear units are used, such as kilometers instead of miles. A more flexible way to express map scale is the representative fraction. To construct a representative fraction, the numerator and denominator are first converted to the same units. For example, since there are 63,360 inches in a mile, the verbal scale "1 inch to 1 mile" can be expressed as "1 inch to 63,360 inches". Next the unit names are dropped and the scale is expressed as a ratio, in this case 1:63,360. This means that 1 linear unit on the map represents 63,360 linear units on Earth, whether those units are inches, miles, kilometers, or some other unit of measurement.

Map scale can also be represented in graphical form. Many maps contain a graphic scale (or bar scale) showing real-world units such as miles or kilometers. The bar scale is usually subdivided to allow easy calculation of distance on the map. However, using a bar scale to measure distance can result in significant errors, especially on small-scale maps covering large areas. This is due to the distortion of distances on the map, as discussed in the map projection section below.

Map scale determines the amount of detail that can be portrayed on a map. The maps on this page illustrate this concept. The scale of these maps increases from 1:40,000,000 (top map) to 1:4,000,000 (middle map) to 1:1,000,000 (bottom map). On small-scale maps, only the largest and most important features can be shown, such as large cities, major rivers and lakes, and international boundaries. Features on small-scale maps are also smaller and more generalized than they are on larger-scale maps. For example, on the top map (smallest scale), Washington, D.C., appears as a small dot. On the middle map (larger scale), it is represented by a red blob indicating the built-up area of Washington. The bottom map (largest scale) shows additional detail that could not be shown on the other maps. This change in map content and feature complexity as a function of map scale is known as map generalization.

Maps in *Goode's World Atlas* have a wide range of scales. The smallest scales are for the world maps, where scales are 1:100,000,000 or smaller. Overview maps of the continents range in scale from 1:16,000,000 to 1:40,000,000 depending on the size of the continent. Regional maps of areas smaller than a whole continent vary from 1:16,000,000 to 1:4,000,000. In addition there are numerous inset maps of cities and islands at a scale of 1:1,000,000.

Map Projections

A map projection is a geometric representation of Earth's surface on a flat surface. Since Earth is roughly spherical, a map projection is needed to produce any flat map, whether a page in this atlas or a computer-generated map of driving directions on www.RandMcNally.com. Hundreds of projections have been developed since the dawn of cartography. A limitation of all of these projections is that they introduce geometric distortion. Some projections distort shape, others distort area, and all distort distance to some degree.

In order to choose an appropriate projection for a particular map, cartographers must pay careful attention to the properties that are distorted and the properties that are preserved by the projection. If shape is preserved, the projection is "conformal." On conformal projections the shapes of geographic features agree with their shapes on Earth. However, a limitation of conformal projections is that they necessarily distort area. This means that the sizes of the geographic features on the map will not be directly comparable. Some will be too large and others too small.

If areas are correctly represented, the projection is "equal area." On equal area projections the sizes of features on the map are directly comparable and in correct proportion to their sizes on Earth. However, in order to achieve this effect, equal area projections distort shape. No projection can preserve both shape and area simultaneously. Some projections preserve neither shape nor area, but instead balance shape and area distortion, creating a compromise projection.

The term "equidistant" is often used for projections that preserve distance. However this can be misleading since distance can only be preserved selectively, such as along specific meridians or parallels. No projection correctly preserves distance in all directions at all locations. Since distance is closely related to scale, one implication is that map scale is often only approximate and may not apply to the entire coverage area of a map. This problem is especially acute for small-scale maps covering large areas.

The projection selected for a particular map depends on the relative importance of different types of distortion, which in turn depends on the purpose of the map. For example, world maps showing phenomena that vary with area, such as population density, often use an equal area projection to give an accurate depiction of the importance of each region.

Map projections are created using mathematical procedures. To illustrate the general principles of projections without using mathematics, we can view a projection as the geometric transfer of information from a globe to a flat projection surface, such as a sheet of paper. If we allow the paper to be rolled in different ways, we can derive three basic types of map projections called cylindrical, conic, and azimuthal.

For cylindrical projections, the sheet of paper is rolled into a tube and wrapped around the globe so that it is tangent (touching) along a circle such as the Equator (see figure below). Information from the globe is transferred to the tube, and the tube is then unrolled to produce the final flat map.

Conic projections use a cone rather than a cylinder. The figure shows the cone tangent to the globe along a line of latitude with the apex of the cone over the North Pole. The line of tangency is called the standard parallel of the projection. Azimuthal projections use a flat projection surface that is tangent to the globe at a single point, such as the North Pole (see figure below).

In general, map distortion increases with distance away from the point or line of tangency. This is why maps of equatorial, mid-latitude, and polar regions often use cylindrical, conic and azimuthal projections, respectively.

The projection surface model is useful for illustrating how projections are developed. However, each of the three projection surfaces actually represents scores of individual projections. There are, for example, many projections with the term "cylindrical" in the name, each of which has the same basic rectangular shape, but different spacings of parallels and meridians.

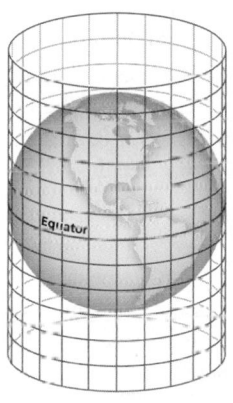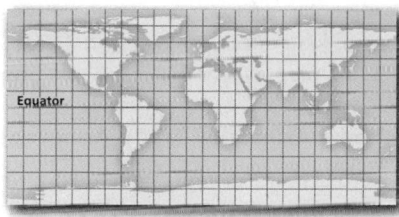

Cylindrical Projection

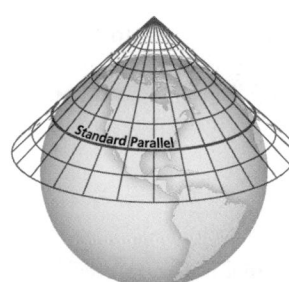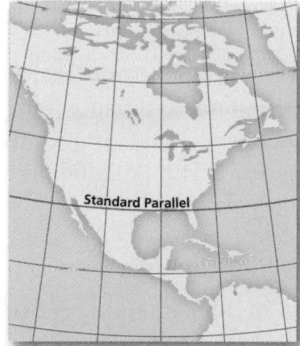

Conic Projection

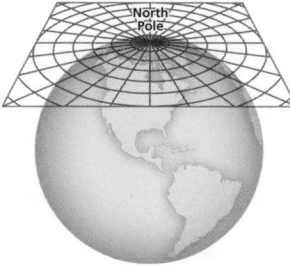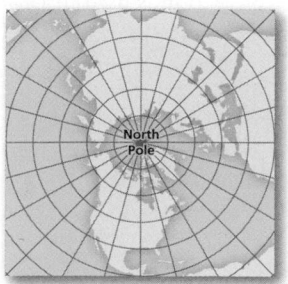

Azimuthal Projection

Map Projections Used in *Goode's World Atlas*

Of the hundreds of projections that have been developed, only a fraction are in everyday use. The main projections used in *Goode's World Atlas* are described below.

Lambert Conformal Conic Projection

On this conic projection, spacing between parallels increases with distance away from the standard parallel, which allows the geometric property of shape (but not area) to be preserved. The projection is named after Johann Lambert, an 18th century mathematician who developed some of the most important projections in use today. It became widely used in the United States in the 20th century following its adoption for many state mapping programs. This projection is used extensively in *Goode's World Atlas* for larger-scale reference maps.

Albers Equal Area Conic Projection

On this conic projection, spacing between parallels decreases with distance away from the standard parallel, which allows the geometric property of area (but not shape) to be preserved. The projection is named after Heinrich Albers, who developed it in 1805. It became widely used in the 20th century, when the United States Coast and Geodetic Survey made it a standard for equal area maps of the United States. This projection is used in *Goode's World Atlas* for continent thematic maps where the equal area property is important.

Lambert Azimuthal Equal Area Projection

On this azimuthal projection, area is preserved, but at the expense of significant shape distortion as distance from the point of tangency increases. This projection is most appropriate for areas of roughly circular shape. This projection, like the Lambert Conformal Conic, is named after Johann Lambert. It is used in *Goode's World Atlas* for smaller-scale reference maps.

Stereographic Projection

On this azimuthal projection, shape is preserved, but distortion of area becomes significant as distance from the point of tangency increases. As a result, this projection is often used for areas that are roughly circular in shape. This projection is used in *Goode's World Atlas* for maps of the polar regions.

Miller Cylindrical Projection

This cylindrical projection is neither conformal nor equal area. However, it is a useful compromise projection to show Earth in a simple, rectangular form. One problem is that polar areas exhibit significant exaggeration of area, a problem common to many cylindrical projections. The projection is named after Osborn Miller, director of the American Geographical Society, who developed it in 1942. The projection is used in *Goode's World Atlas* for many of the world climate maps.

Lambert Conformal Conic Projection

Albers Equal Area Conic Projection

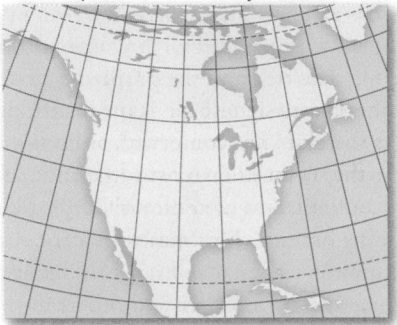

Lambert Azimuthal Equal Area Projection

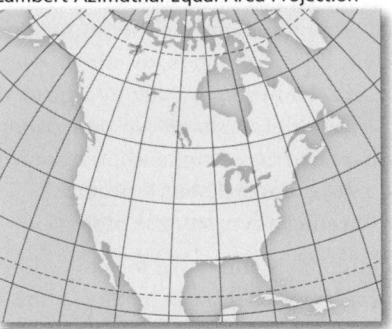

Stereographic Projection

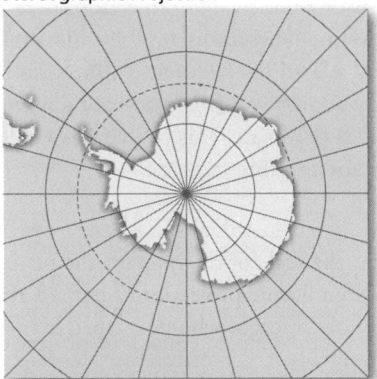

Miller Cylindrical Projection

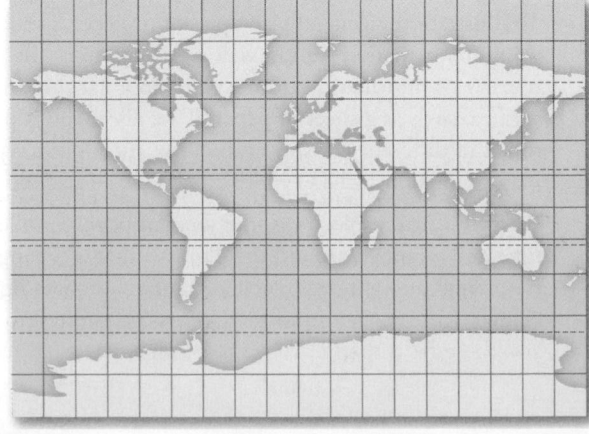

Plate Carrée Projection

This cylindrical projection is neither conformal nor equal area. Its main utility lies in the fact that it shows lines of latitude as evenly spaced lines on the map. This allows for effective thematic map display for phenomena that are measured at regular intervals of latitude. The projection is used in *Goode's World Atlas* for world climate change maps.

Sinusoidal Projection

The straight, evenly-spaced parallels on this pseudocylindrical projection resemble the parallels on cylindrical projections. Unlike cylindrical projections, however, meridians are curved and converge at the poles. This causes significant shape distortion in polar regions. The projection is therefore not conformal, although it is equal area. The Sinusoidal is the oldest-known pseudocylindrical projection, dating to the 16th century. It is not used extensively in *Goode's World Atlas*. However, along with the Mollweide projection, it is the basis for the Goode's Interrupted Homolosine projection described below.

Mollweide Projection

The Mollweide (or Homolographic) projection resembles the Sinusoidal but has less shape distortion in polar areas due to its elliptical form. Like the Sinusoidal projection, it is equal area but not conformal. It is one of several pseudocylindrical projections developed in the 19th century, and is named after Karl Mollweide, an astronomer and mathematician, who developed it in 1805. It is not used extensively in *Goode's World Atlas*. However, along with the Sinusoidal projection, it is the basis for the Goode's Interrupted Homolosine projection described below.

Goode's Interrupted Homolosine Equal Area Projection

This projection is a fusion of the Sinusoidal projection between 40° 44' N and S, and the Mollweide projection between these parallels and the Poles. The projection is equal area but not conformal. The unique appearance of the projection is due to the introduction of discontinuities in oceanic regions, the goal of which is to reduce distortion for continental land masses. A condensed version of the projection also exists in which the Atlantic Ocean is compressed to help maximize the scale of the map on the page. The Goode's Interrupted Homolosine projection is named after J. Paul Goode of the University of Chicago, who developed it in 1923. Goode was an advocate of interrupted projections and, as editor of Goode's School Atlas, promoted their use in education. This projection is used extensively in *Goode's World Atlas* for world thematic maps.

Robinson Projection

This pseudocylindrical projection resembles the Mollweide projection except that polar regions are flattened and stretched out. While neither conformal nor equal area, the Robinson projection manages to balance shape and area distortion in an effective way. The projection was developed in 1963 by Arthur Robinson of the University of Wisconsin, at the request of Rand McNally. The Robinson projection is widely used in *Goode's World Atlas* for world thematic maps where the interrupted nature of the Goode's Homolosine projection would be inappropriate.

Plate Carrée Projection

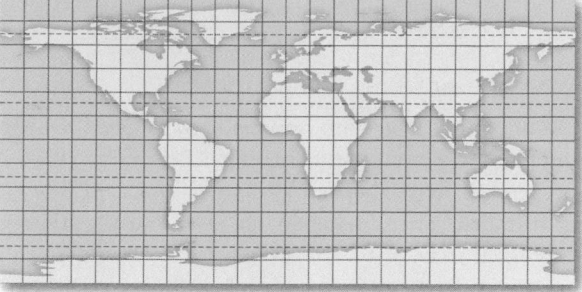

Sinusoidal Projection

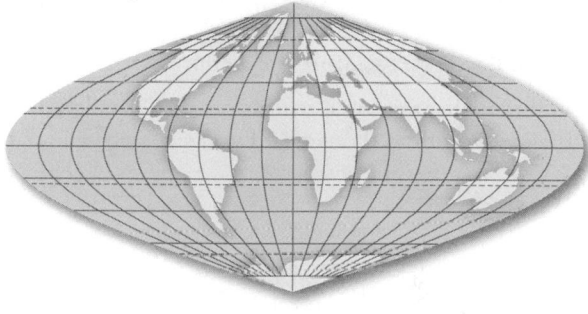

Mollweide Projection

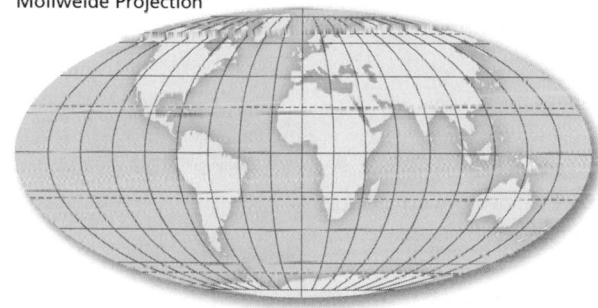

Goode's Interrupted Homolosine Equal Area Projection

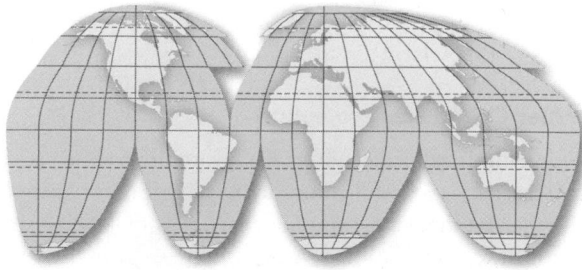

Robinson Projection

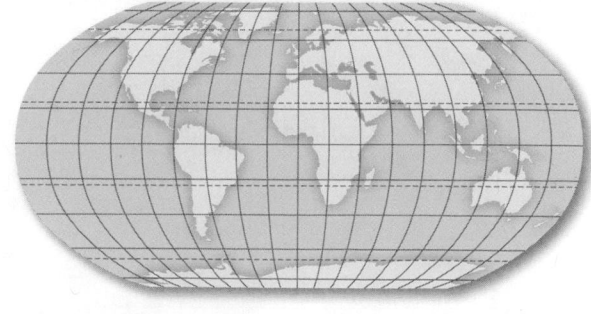

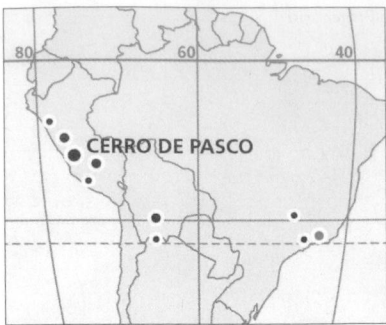

Point symbol map: Detail of Zinc and Coltan (p. 74)

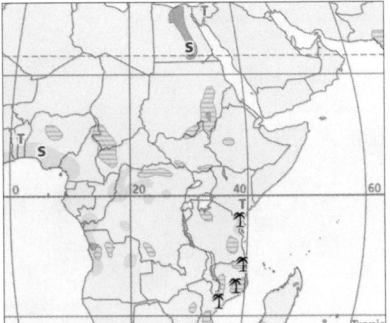

Area symbol map: Detail of Vegetable Oils (p. 69)

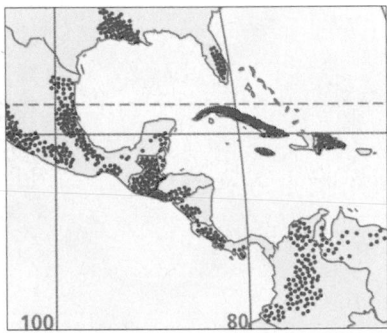

Dot map: Detail of Sugar, Spices (p. 67)

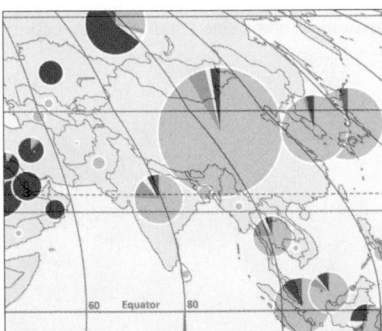

Proportional symbol map: Detail of Exports (p. 82)

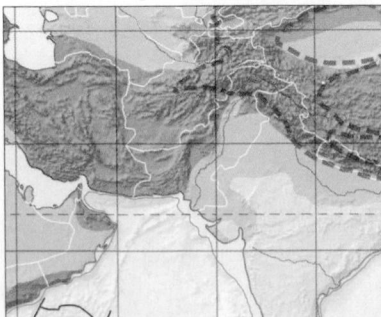

Area class map: Detail of Landforms (pp. 12-13)

Thematic Map Types in *Goode's World Atlas*

Thematic maps depict a single theme such as population density, agricultural productivity or annual precipitation. The selected theme is presented on a base of locational information, such as coastlines, country boundaries, and major drainage features.

Goode's World Atlas contains many different types of thematic maps. The characteristics of each are summarized below.

Point Symbol Maps

Point symbol maps are perhaps the simplest type of thematic map. They show features that occur at discrete locations. Examples include earthquakes, nuclear power plants, and mineral-producing areas. The Zinc and Coltan map (p. 74) is an example of a point symbol map. Different colors represent the two different materials, and various symbol sizes show relative importance.

Area Symbol Maps

Area symbol maps are useful for delineating regions of interest. For example, the Vegetable Oils maps (p. 69) show major oil-producing regions in different colors. Some point symbols also appear on this map, for less extensive oil crops.

Dot Maps

Dot maps show a distribution using a pattern of dots, where each dot represents a certain quantity or amount. For example, on the Sugar and Spice map (p. 67), each dot represents 20,000 metric tons of sugar produced. The different dot colors represent different sources of sugar (cane vs. beet). Dot maps are an effective way of representing the variable density of geographic phenomena.

Proportional Symbol Maps

Proportional symbol maps portray numeric quantities, such as total toxic chemical releases per state, or the total value of agricultural goods produced by country. The symbols on these maps — usually circles — are drawn such that the size the symbol is proportional to the value at that location. For example, the Exports map (p. 82) shows the value of goods exported by each country in the world, in billions of U.S. dollars. Proportional symbols are frequently subdivided based on the percentage of individual components making up the total. The Exports map uses wedges of different color to show the percentages of various types of exports, such as manufactured articles and raw materials.

Area Class Maps

Area class maps divide Earth into zones based on categories of a particular geographic phenomenon. For example, the Landforms map (pp. 12-13) divides Earth into seven unique structural regions based on landform type and origin. Other examples of area class maps in *Goode's World Atlas* include soil taxonomy (pp. 40-41), terrestrial biomes (pp. 46-47), and natural vegetation (pp. 36-37).

Flow Line Maps

Flow line maps show flows between locations. Usually the thickness of the flow lines is proportional to the volume of the flow. Flows may be physical commodities like petroleum or less tangible quantities like information. The flow lines on the Communication Network Infrastructure map (pp. 90-91) represent bandwidth usage in gigabits per second. Note that the locations of flow lines may not accurately represent the actual physical route.

Choropleth Maps

Choropleth maps apply distinctive colors to predefined areas, such as counties or states, to represent different quantities in each area. The quantities shown are usually rates, percentages, or densities. For example, the Birth Rate map (p. 51) shows the annual number of births per one thousand people for each country.

Isoline Maps

Isoline maps are used to portray quantities that vary continuously over space. These maps are frequently used for climate variables such as precipitation and temperature. For example, the January Temperature map for the North Polar region (p. 23) contains isolines at intervals of 5° C. Colors are also used to assist map interpretation.

Grid-Based Maps

Grid-based maps rely on data points occurring at regular intervals in a two dimensional grid. Some grid-based maps are actually digital images, analogous to the pictures captured by digital cameras. These maps are created from a very fine grid of cells called pixels, each of which is assigned a color that corresponds to a specific value or range of values. The population density maps in this atlas (pp. 46-47, for example) are examples of this type. Other grid-based maps are based on data integrated over a coarser grid, such as the map showing temperature change for 5-degree grid cells (p. 28) and the tornado map showing the frequency of tornadoes within 1-degree grid cells (p. 101). Grid-based mapping is increasingly being used to map environmental phenomena observable from remote sensing systems.

Cartograms

Cartograms are maps on which shapes and areas have been deliberately distorted. The cartograms in this atlas adjust the size of each country proportional to the population of the country. This means that the countries with the largest areas are those with the largest populations, regardless of actual country area. Because border relationships are maintained, the resizing of the countries can also result in considerable shape distortion. Cartograms make explicit the relationship between the mapped variable and the size of the affected population. As an example, consider the HIV Infection cartogram (p. 58). Both Chad and Nigeria have relatively high rates of HIV infection, but Nigeria is much larger than Chad on the Cartogram because Nigeria's population is much larger. This informs the cartogram reader that the population affected by HIV is much larger in Nigeria.

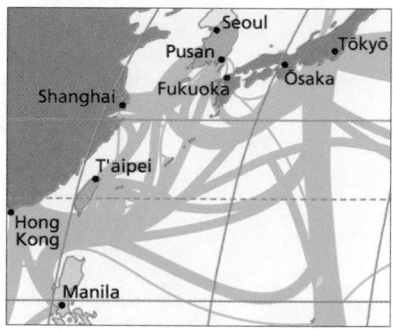

Flow line map: Detail of Communication Network Infrastructure (pp. 90-91)

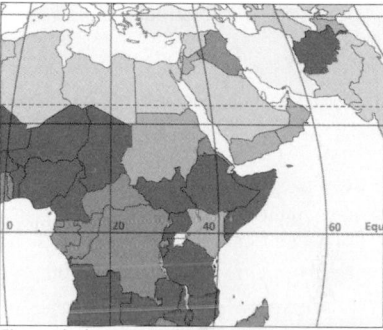

Choropleth map: Detail of Birth Rate (p. 51)

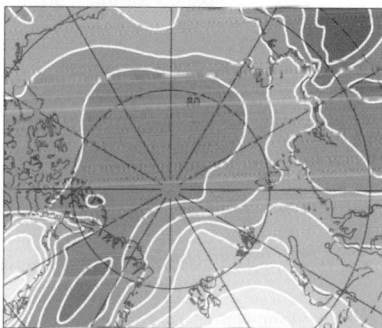

Isoline map: Detail of January Temperature (p. 23)

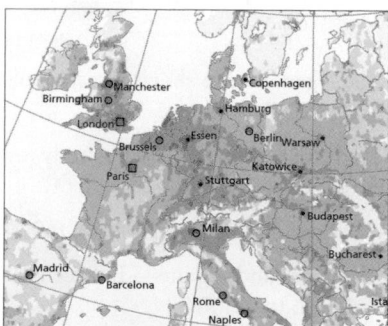

Grid-based map: Detail of Population Density (pp. 46-47)

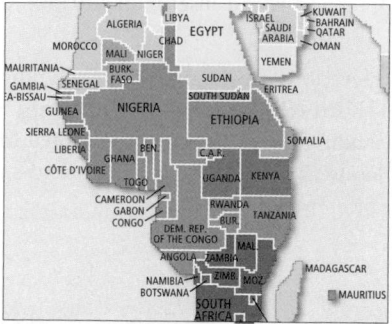

Cartogram: Detail of HIV Infection (p. 58)

Political Boundaries

- - - — ━━━ International

━━ ━ Disputed or Unrecognized

━━ ━ ━━ Secondary (State, Provincial, etc.)

─ ─ ─ ─ International Boundary over Water

─ ─ · ─ Secondary Boundary over Water

⬚ Park, Indian Reservation, Area of Interest

🌳 Urbanized Area

Populated Places

TŌKYŌ   National Capital

Boise   Secondary Capital

1:2,000,000, 1:1,000,000 and 1:500,000 Inset Maps

◉   1,000,000 and over

◎   100,000 to 1,000,000

☉   50,000 to 100,000

•   10,000 to 50,000

○   Under 10,000

▫   Neighborhood, Section of City

Other Reference Maps

◉   1,000,000 and over

◎   250,000 to 1,000,000

☉   100,000 to 250,000

•   25,000 to 100,000

○   Under 25,000

Note: Type size indicates the relative importance of the city. On the continent physical maps, city populations and relative importance are not differentiated.

Cultural Features

🌊 Dam

▫ Point of Interest

∴ Ruins

PALESTINE Cultural or Historic Region

Transportation

——— Major Road

——— Minor Road

——— Railroad

✈ Airport

Land Features

△ Peak, Spot Height

≍ Pass

Sand

Contours

Elevation

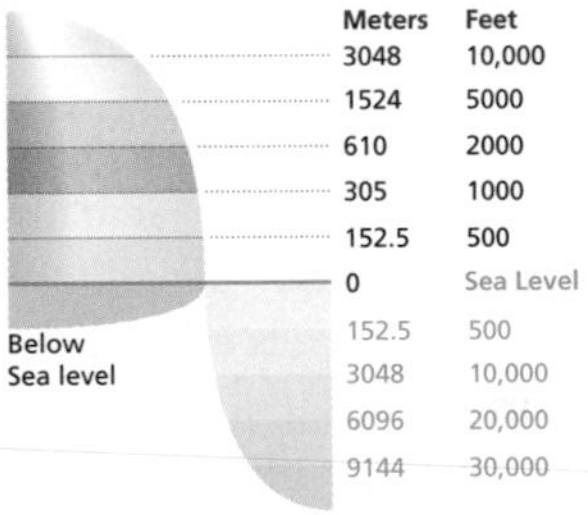

Meters	Feet
3048	10,000
1524	5000
610	2000
305	1000
152.5	500
0	Sea Level
152.5	500
3048	10,000
6096	20,000
9144	30,000

Below Sea level

Note: The 500 foot contour is not shown on the small-scale oceans and polar regions maps.

Lakes and Reservoirs

Fresh Water

Fresh Water: Intermittent

Dry Lake

Salt Water

Salt Water: Intermittent

Other Water features

Swamp

Glacier

Ice Cap

River

Intermittent River

· · · · · · Canal (navigable)

Falls

• · • Springs

∿∿∿ Reef

———→ Warm Ocean Current

———→ Cold Ocean Current

The legend above shows the symbols used for reference maps in *Goode's World Atlas*.

To portray relative areas correctly, uniform map scales have been used wherever possible:

Continents – Between 1:16,000,000 and 1:40,000,000

Countries and regions – Between 1:4,000,000 and 1:16,000,000

World, polar areas and oceans – 1:40,000,000 and smaller

City and island inset maps – 1:500,000, 1:1,000,000 and 1:2,000,000

Elevations on the maps are shown using a combination of shaded relief and hypsometric tints. Shaded relief (or hill-shading) gives a three-dimensional impression of the landscape, while hypsometric tints show elevation ranges in different colors.

The choice of names for mapped features is complicated by the fact that a variety of languages and alphabets are used throughout the world. A local-names policy is used in *Goode's World Atlas* for populated places and local physical features. For some major features, an English form of the name is used with the local name, e.g., Vienna (Wien) and Naples (Napoli). In countries where more than one official language is used, names are given in the dominant local language. For large physical features spanning international borders, the conventional English form of the name is used. In cases where a non-Roman alphabet is used, names have been transliterated according to accepted practice.

Selected features are also listed in the Index, which includes a pronunciation guide. A list of foreign geographic terms is provided in the Glossary at the end of the atlas.

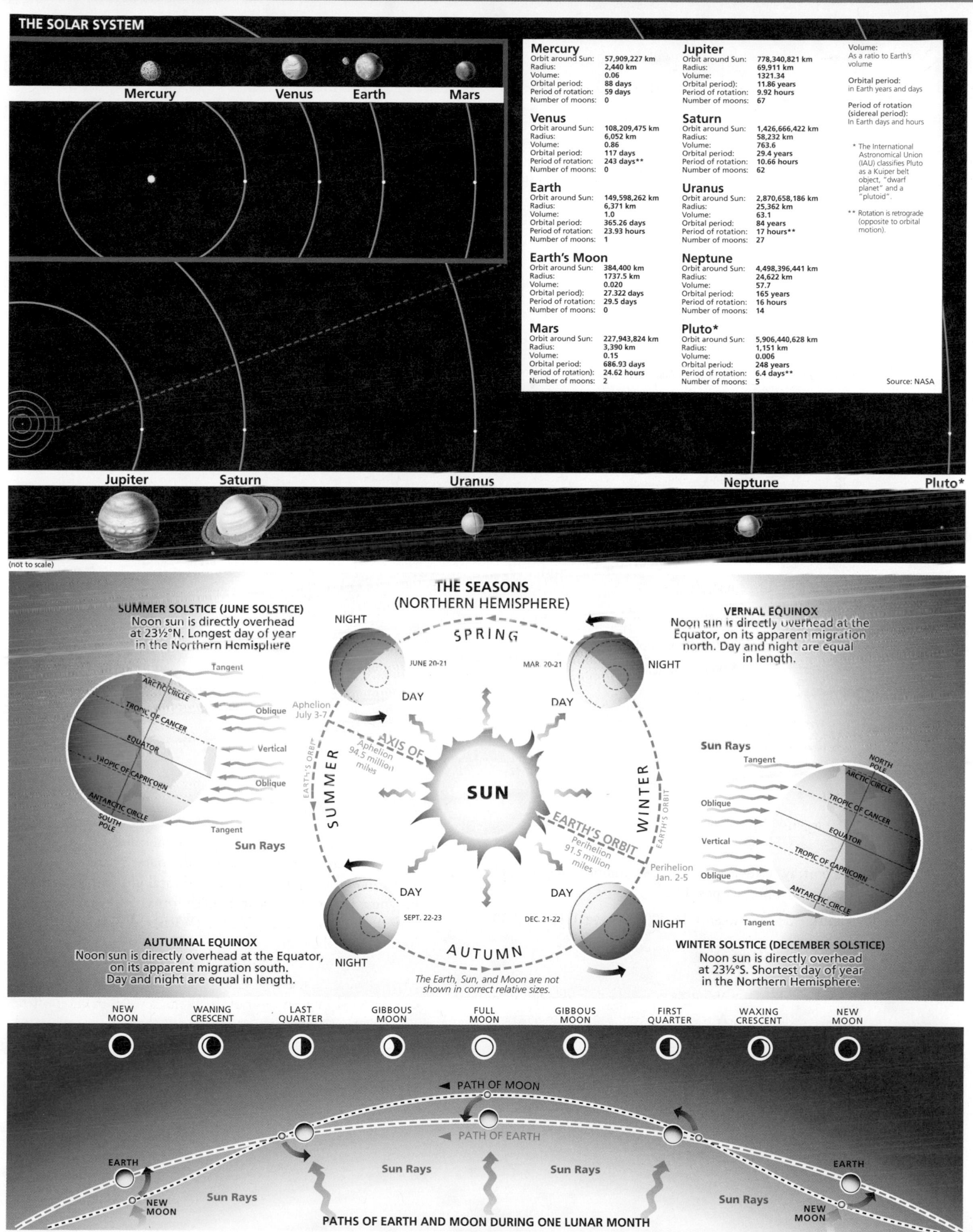

The World

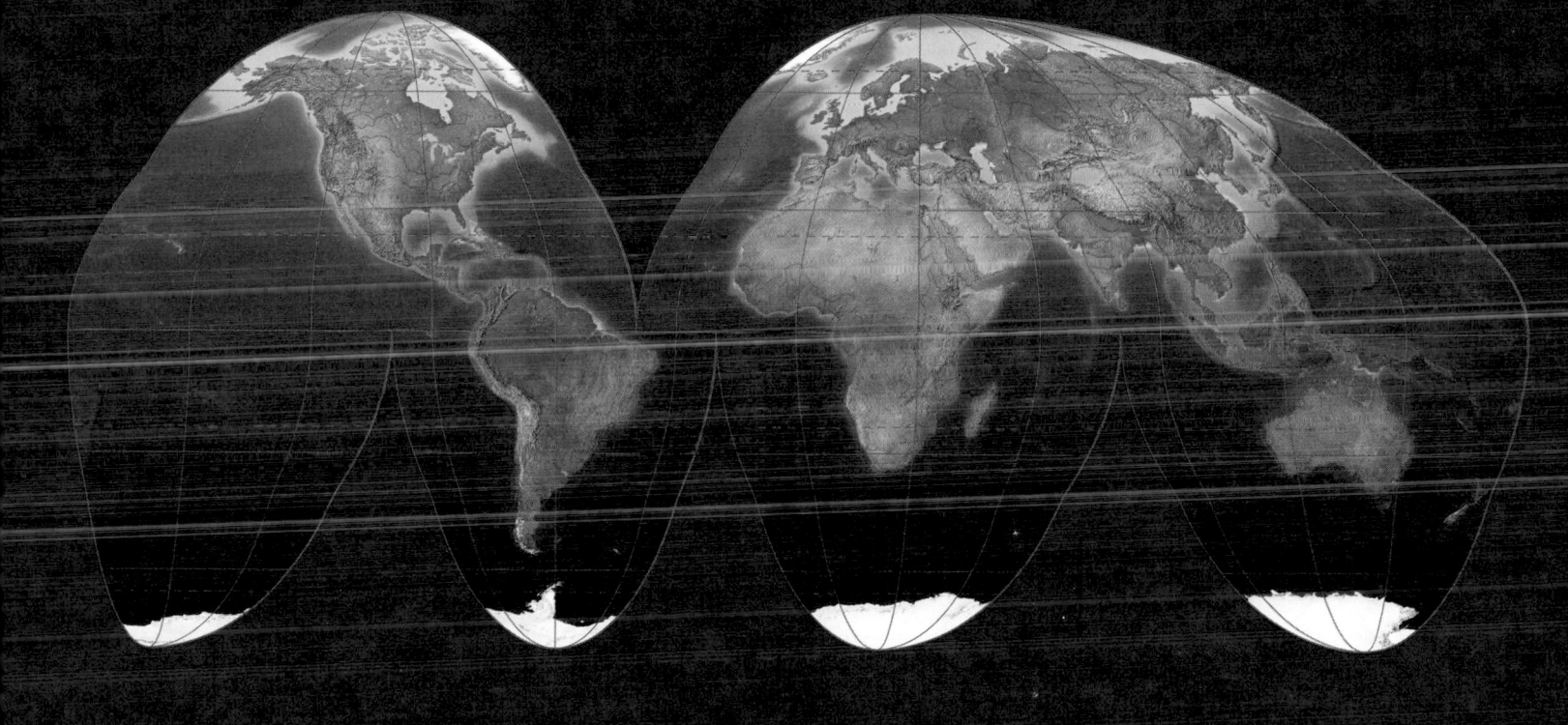

Like an enormous puzzle, the world is comprised of massive tectonic plates, whose slow movement builds mountains, creates oceans, spawns volcanoes, and unleashes earthquakes. A thin envelope of gases cloaks the planet, allowing for a great diversity of plant and animal life. Humans have indelibly imprinted the planet with cities and roads, the introduction of agriculture into arid regions, and the cutting and burning of forests. Humans are defined by cultural characteristics like language and religion, and organize ourselves into countries and alliances. We are diverse and distinct, and yet global communications and transportation systems link and intertwine people, governments, and economies like never before. As the human population continues to grow—we now exceed seven billion people on Earth—our connections with the environment and each other are increasingly complex.

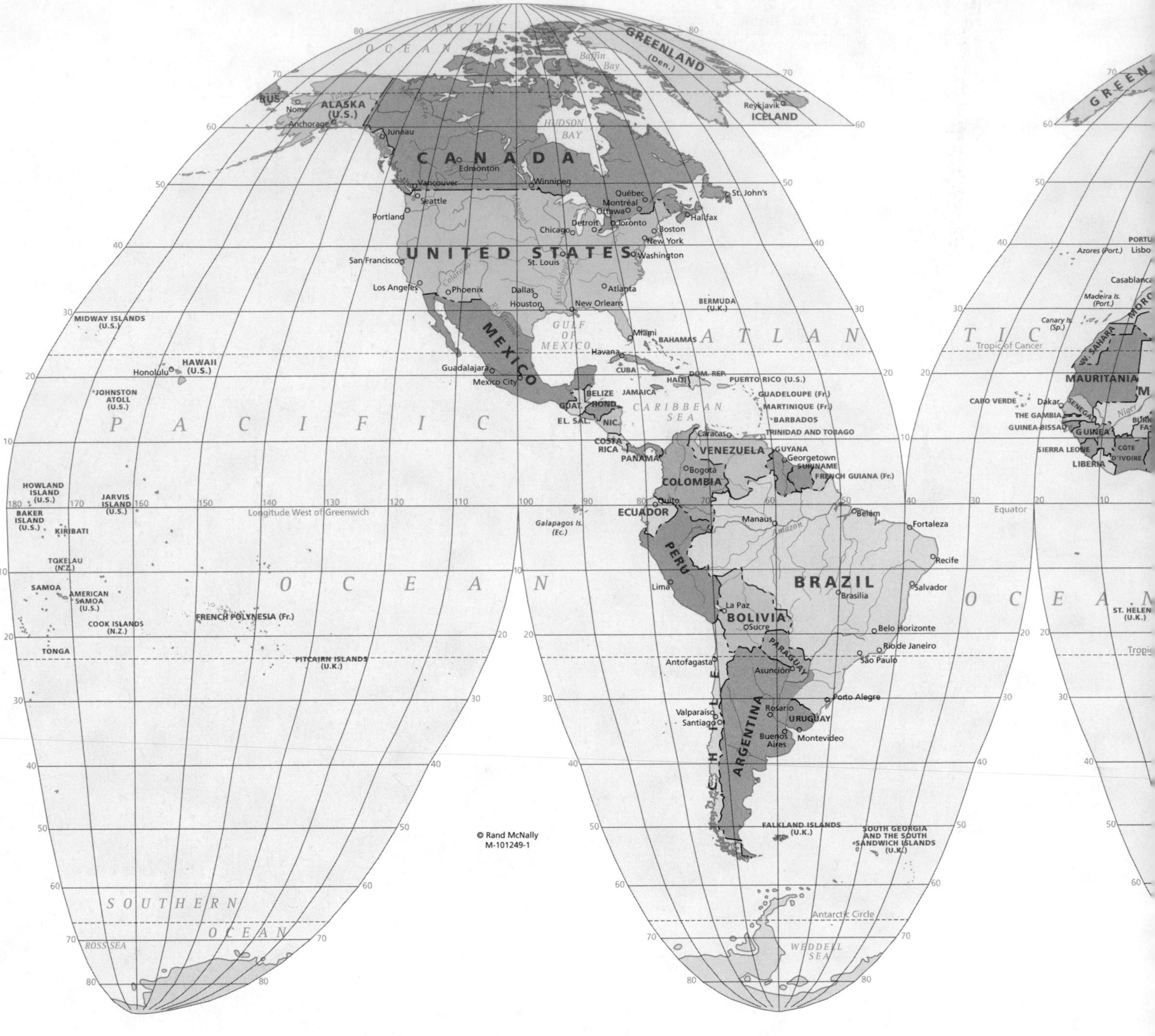

ARCTIC OCEAN — GREENLAND (Den.) — Baffin Bay — Reykjavik ICELAND — ALASKA (U.S.) — Nome — Anchorage — Juneau — HUDSON BAY — CANADA — Edmonton — Vancouver — Winnipeg — Québec — St. John's — Seattle — Portland — Montréal — Ottawa — Halifax — Chicago — Detroit — Toronto — Boston — UNITED STATES — San Francisco — St. Louis — New York — Washington — Los Angeles — Phoenix — Dallas — Atlanta — Houston — New Orleans — BERMUDA (U.K.) — Miami — BAHAMAS — ATLANTIC — MEXICO — GULF OF MEXICO — Havana — CUBA — Guadalajara — Mexico City — HAITI — DOM. REP. — PUERTO RICO (U.S.) — GUADELOUPE (Fr.) — MARTINIQUE (Fr.) — BELIZE — HOND. — JAMAICA — CARIBBEAN SEA — BARBADOS — GUAT. — EL SAL. — NIC. — TRINIDAD AND TOBAGO — COSTA RICA — PANAMA — Caracas — VENEZUELA — GUYANA — Georgetown — SURINAME — FRENCH GUIANA (Fr.) — Bogotá — COLOMBIA — ECUADOR — Quito — Galapagos Is. (Ec.) — Manaus — Belém — Fortaleza — PERU — Lima — BRAZIL — Brasília — Recife — Salvador — La Paz — BOLIVIA — Sucre — Belo Horizonte — Rio de Janeiro — Antofagasta — PARAGUAY — São Paulo — Asunción — Valparaíso — Santiago — Rosario — URUGUAY — Porto Alegre — ARGENTINA — Buenos Aires — Montevideo — FALKLAND ISLANDS (U.K.) — SOUTH GEORGIA AND THE SOUTH SANDWICH ISLANDS (U.K.) — PACIFIC OCEAN — MIDWAY ISLANDS (U.S.) — HAWAII (U.S.) — Honolulu — JOHNSTON ATOLL (U.S.) — HOWLAND ISLAND (U.S.) — BAKER ISLAND (U.S.) — JARVIS ISLAND (U.S.) — KIRIBATI — TOKELAU (N.Z.) — SAMOA — AMERICAN SAMOA (U.S.) — COOK ISLANDS (N.Z.) — FRENCH POLYNESIA (Fr.) — TONGA — PITCAIRN ISLANDS (U.K.) — SOUTHERN OCEAN — ROSS SEA — WEDDELL SEA — Antarctic Circle — Longitude West of Greenwich — Equator — Tropic of Cancer — © Rand McNally M-101249-1 — AZORES (Port.) — Lisbon — Casablanca — Madeira Is. (Port.) — Canary Is. (Sp.) — MOROCCO — W. SAHARA — MAURITANIA — CABO VERDE — Dakar — SENEGAL — THE GAMBIA — Niger — GUINEA-BISSAU — GUINEA — BURKINA FASO — SIERRA LEONE — LIBERIA — CÔTE D'IVOIRE — GREENLAND — PORTUGAL

Comparative Land Areas
Includes land and inland water. Numbers indicate thousands of square kilometers.

ASIA 44,900

CHINA	INDIA	KAZAKHSTAN	SAUDI ARABIA	INDONESIA	IRAN	MONGOLIA	PAKISTAN	TURKEY	MYANMAR	OTHER ASIA	RUSSIA (ASIA)
9,557	3,166	2,717	2,150	1,904	1,648	1,567	880	784	677	2,775	13,120

EUROPE 9,900

RUSSIA (EUROPE)	UKRAINE	FRANCE	OTHER EUROPE
3,955	604	540	8,756

AFRICA 30,300

ALGERIA	D.R. OF CONGO	SUDAN	LIBYA	CHAD	NIGER	ANGOLA	MALI	S. AFRICA	ETHIOPIA	MAURITANIA	EGYPT	TANZANIA	NIGERIA	NAMIBIA
2,382	2,345	1,861	1,760	1,284	1,267	1,247	1,240	1,219	1,104	1,031	1,001	945	924	873

Comparative Populations
Estimated population as of June 1, 2015. Numbers indicate millions of people.

ASIA 4,385.1

CHINA	INDIA	INDONESIA	PAKISTAN	BANGLADESH	JAPAN	PHILIPPINES	VIETNAM	TURKEY
1,361.5	1,257.1	256.0	199.1	169.0	126.9	109.6	94.3	82

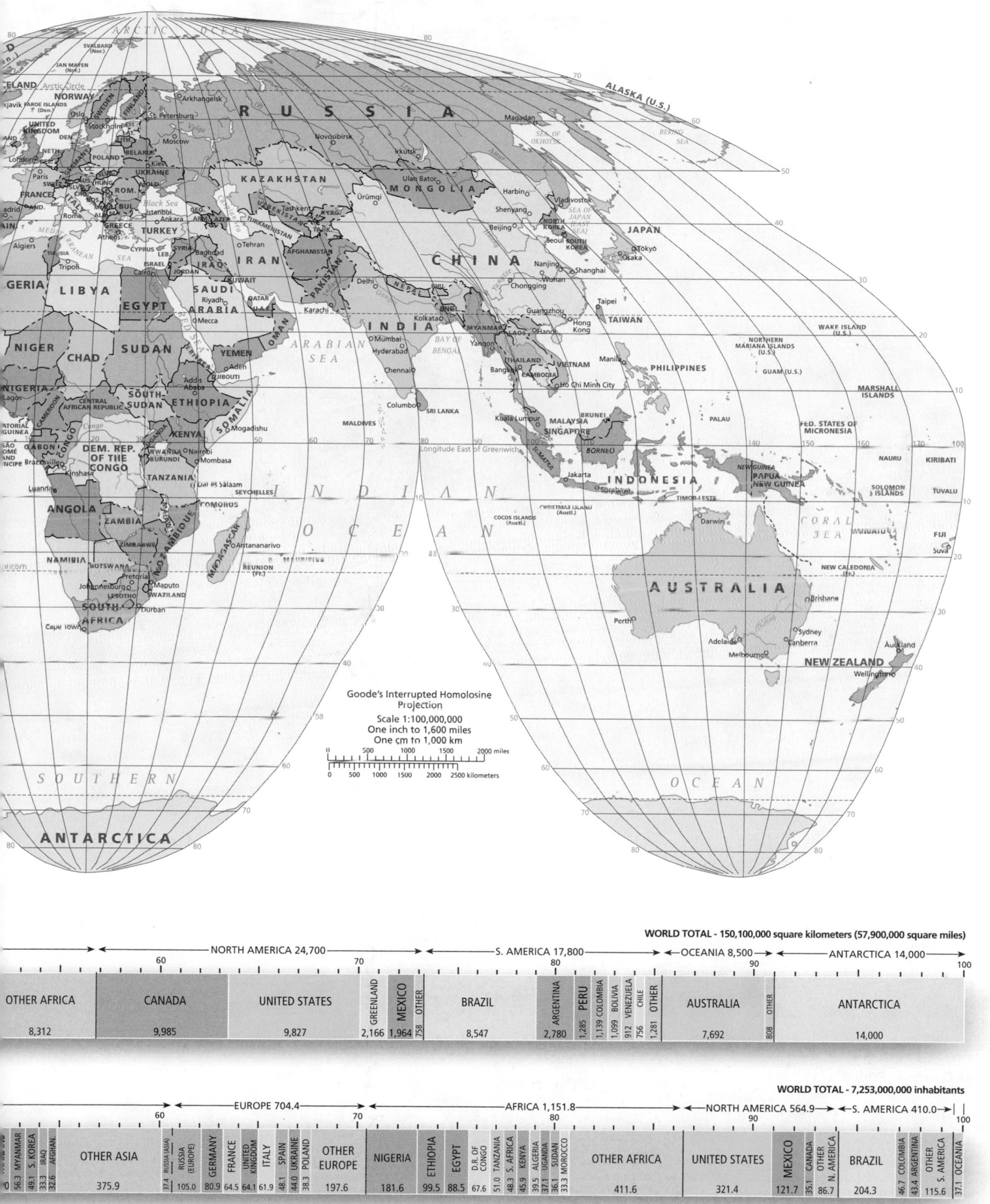

Goode's Interrupted Homolosine
Projection

Scale 1:100,000,000
One inch to 1,600 miles
One cm to 1,000 km

500   1000   1500   2000 miles

0   500   1000   1500   2000   2500 kilometers

WORLD TOTAL - 150,100,000 square kilometers (57,900,000 square miles)

	NORTH AMERICA 24,700					S. AMERICA 17,800							OCEANIA 8,500		ANTARCTICA 14,000

OTHER AFRICA	CANADA	UNITED STATES	GREENLAND	MEXICO	OTHER	BRAZIL	ARGENTINA	PERU	COLOMBIA	BOLIVIA	VENEZUELA	CHILE	OTHER	AUSTRALIA	OTHER	ANTARCTICA
8,312	9,985	9,827	2,166	1,964	758	8,547	2,780	1,285	1,139	1,099	912	756	1,281	7,692	808	14,000

WORLD TOTAL - 7,253,000,000 inhabitants

					EUROPE 704.4									AFRICA 1,151.8												NORTH AMERICA 564.9			S. AMERICA 410.0		

	MYANMAR	S. KOREA	IRAQ	AFGHAN.	OTHER ASIA	RUSSIA (ASIA)	RUSSIA (EUROPE)	GERMANY	FRANCE	UNITED KINGDOM	ITALY	SPAIN	UKRAINE	POLAND	OTHER EUROPE	NIGERIA	ETHIOPIA	EGYPT	D.R. OF CONGO	TANZANIA	S. AFRICA	KENYA	UGANDA	ALGERIA	SUDAN	MOROCCO	OTHER AFRICA	UNITED STATES	MEXICO	CANADA	OTHER N. AMERICA	BRAZIL	COLOMBIA	ARGENTINA	OTHER S. AMERICA	OCEANIA
56.3	49.1	33.3	32.6		375.9	37.4	105.0	80.9	64.5	64.1	61.9	48.1	44.0	38.3	197.6	181.6	99.5	88.5	67.6	51.0	48.3	45.9	39.5	37.1	36.1	33.3	411.6	321.4	121.7	35.1	86.7	204.3	46.7	43.4	115.6	37.1

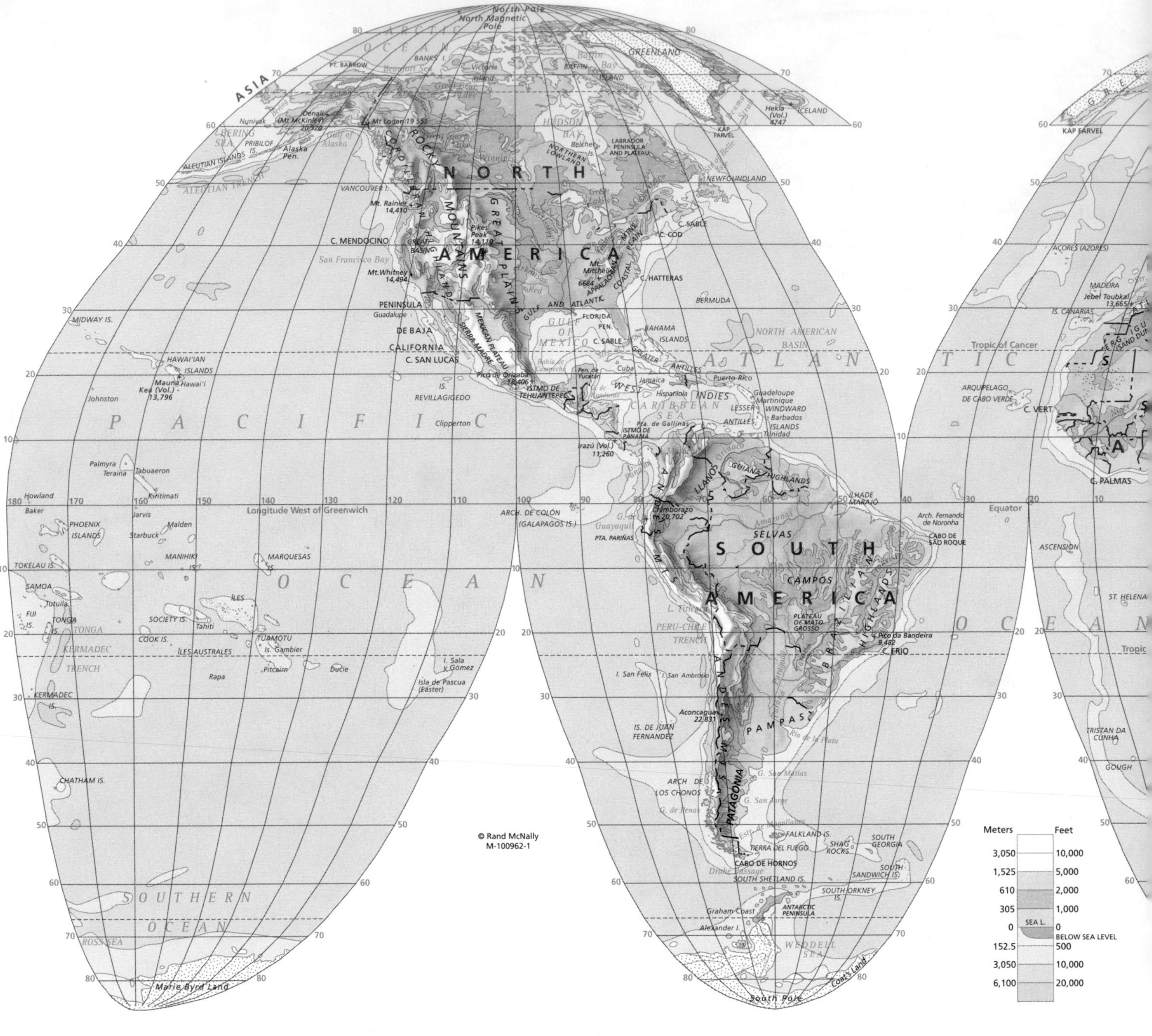

North Pole
North Magnetic Pole

ARCTIC OCEAN

ASIA

PT. BARROW
Denali (Mt. McKinley) 20,328
Mt Logan 19,551
BANKS I.
Victoria Island
Baffin Island
Baffin Bay
GREENLAND
Hekla (Vol.) 5247
ICELAND

NORTH AMERICA

Nunivak
Gulf of Alaska
Alaska Pen.
PRIBILOF IS.
BERING SEA
ALEUTIAN ISLANDS
ALEUTIAN TRENCH
ROCKY MOUNTAINS
HUDSON BAY
Belcher Is.
NORTHERN LOWLAND
LABRADOR PENINSULA AND PLATEAU
KAP FARVEL
NEWFOUNDLAND
KAP FARVEL

VANCOUVER I.
Mt. Rainier 14,410
C. MENDOCINO
Pikes Peak 14,110
San Francisco Bay
Mt. Whitney 14,494
GREAT PLAINS
Mt. Mitchell 6684
APPALACHIAN
C. SABLE
C. COD
C. HATTERAS
BERMUDA
AÇORES (AZORES)

PENINSULA DE BAJA CALIFORNIA
Guadalupe
MIDWAY IS.
SIERRA MADRE
MEXICAN PLATEAU
GULF OF MEXICO
FLORIDA PEN.
C. SABLE
BAHAMA ISLANDS
NORTH AMERICAN BASIN
Tropic of Cancer
MADEIRA
Jebel Toubkal 13,665
IS. CANARIAS

C. SAN LUCAS
Pico de Orizaba 18,406
ISTMO DE TEHUANTEPEC
Pen. de Yucatán
Cuba
GREATER ANTILLES
Jamaica
Hispaniola
Puerto Rico
WEST INDIES
LESSER ANTILLES
Guadeloupe
Martinique
WINDWARD ISLANDS
Barbados
Trinidad
CARIBBEAN SEA
ATLANTIC

HAWAI'IAN ISLANDS
Mauna Kea (Vol.) Hawai'i 13,796
Johnston
REVILLAGIGEDO IS.
Clipperton
Irazú (Vol.) 11,260
ISTMO DE PANAMÁ
Pta. de Gallinas
GUIANA HIGHLANDS
ILHA DE MARAJÓ
Arch. Fernando de Noronha
CABO DE SÃO ROQUE
ASCENSION
C. VERT

PALMYRA
Teraina
Tabuaeron
Kiritimati
Chimborazo 20,702
Guayaquil
PTA. PARIÑAS
SELVAS
BRAZILIAN HIGHLANDS
Equator
ARQUIPÉLAGO DE CABO VERDE
C. PALMAS

PACIFIC OCEAN

Howland
Baker
PHOENIX ISLANDS
Jarvis
Malden
Starbuck
MANIHIKI IS.
MARQUESAS
SOUTH AMERICA
CAMPOS
L. Titicaca
PLATEAU DE MATO GROSSO
Pico da Bandeira 9,469
C. FRIO
ST. HELENA

TOKELAU IS.
SAMOA
Tutuila
FIJI IS.
TONGA IS.
SOCIETY IS.
Tahiti
COOK IS.
ÎLES AUSTRALES
TUAMOTU Is. Gambier
KERMADEC IS.
Rapa
Pitcairn
Ducie
I. Sala y Gómez
Isla de Pascua (Easter)
PERU-CHILE TRENCH
PAMPAS
Rio de la Plata
Tropic

Aconcagua 22,831
I. San Félix I. San Ambrosio
IS. DE JUAN FERNÁNDEZ
ANDES MTS.

KERMADEC TRENCH
CHATHAM IS.
ARCH. DE LOS CHONOS
G. de Penas
PATAGONIA
G. San Matias
G. San Jorge

Longitude West of Greenwich
ARCH. DE COLÓN (GALÁPAGOS IS.)

© Rand McNally
M-100962-1

SOUTHERN OCEAN
ROSS SEA
Marie Byrd Land
South Pole
FALKLAND IS.
Estr. de Magallanes
TIERRA DEL FUEGO
CABO DE HORNOS
Drake Passage
SOUTH SHETLAND IS.
ANTARCTIC PENINSULA
Graham Coast
Alexander I.
SHAG ROCKS
SOUTH GEORGIA
SOUTH ORKNEY
SOUTH SANDWICH IS.
WEDDELL SEA
Coats Land
TRISTAN DA CUNHA
GOUGH

Meters		Feet
3,050		10,000
1,525		5,000
610		2,000
305		1,000
0	SEA L.	0
		BELOW SEA LEVEL
152.5		500
3,050		10,000
6,100		20,000

Land Elevations in Profile

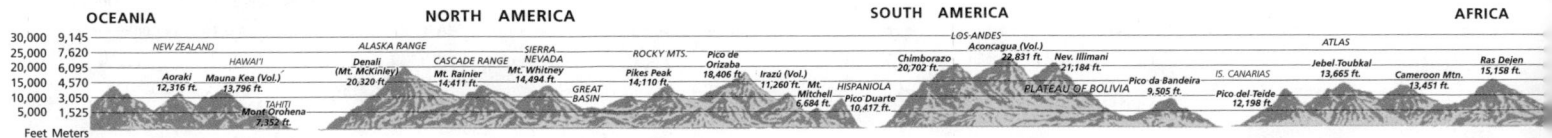

OCEANIA — NORTH AMERICA — SOUTH AMERICA — AFRICA

NEW ZEALAND
HAWAI'I
Aoraki 12,316 ft.
Mauna Kea (Vol.) 13,796 ft.
TAHITI
Mont Orohena 7,352 ft.
ALASKA RANGE
Denali (Mt. McKinley) 20,320 ft.
CASCADE RANGE
Mt. Rainier 14,411 ft.
SIERRA NEVADA
Mt. Whitney 14,494 ft.
GREAT BASIN
ROCKY MTS.
Pikes Peak 14,110 ft.
Pico de Orizaba 18,406 ft.
Irazú (Vol.) 11,260 ft.
HISPANIOLA
Mitchell 6,684 ft.
Pico Duarte 10,417 ft.
LOS ANDES
Chimborazo 20,702 ft.
Aconcagua (Vol.) 22,831 ft.
Nev. Illimani 21,184 ft.
FLATEAU OF BOLIVIA
Pico da Bandeira 9,505 ft.
IS. CANARIAS
Pico del Teide 12,198 ft.
ATLAS
Jebel Toubkal 13,665 ft.
Cameroon Mtn. 13,451 ft.
Ras Dejen 15,158 ft.

30,000 9,145
25,000 7,620
20,000 6,095
15,000 4,570
10,000 3,050
5,000 1,525
Feet Meters

Ocean Depths in Profile

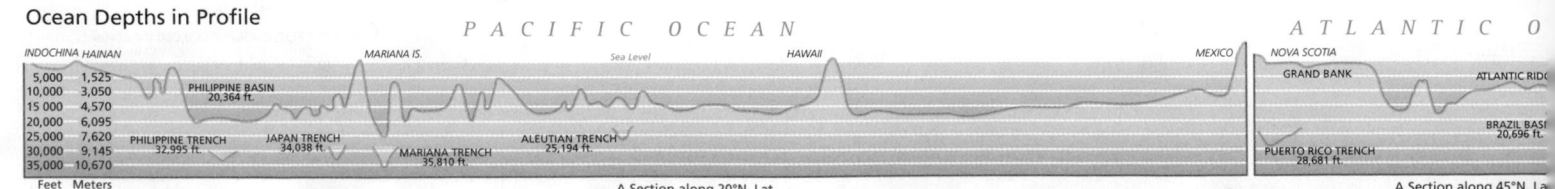

PACIFIC OCEAN — ATLANTIC O[CEAN]

INDOCHINA HAINAN
MARIANA IS.
Sea Level
HAWAII
MEXICO
NOVA SCOTIA
GRAND BANK
ATLANTIC RIDGE

PHILIPPINE BASIN 20,364 ft.
BRAZIL BASIN 20,696 ft.

PHILIPPINE TRENCH 32,995 ft.
JAPAN TRENCH 34,038 ft.
MARIANA TRENCH 35,810 ft.
ALEUTIAN TRENCH 25,194 ft.
PUERTO RICO TRENCH 28,681 ft.

5,000 1,525
10,000 3,050
15,000 4,570
20,000 6,095
25,000 7,620
30,000 9,145
35,000 10,670
Feet Meters

A Section along 20°N. Lat.

A Section along 45°N. La[t.]

PLATE TECTONIC BOUNDARIES

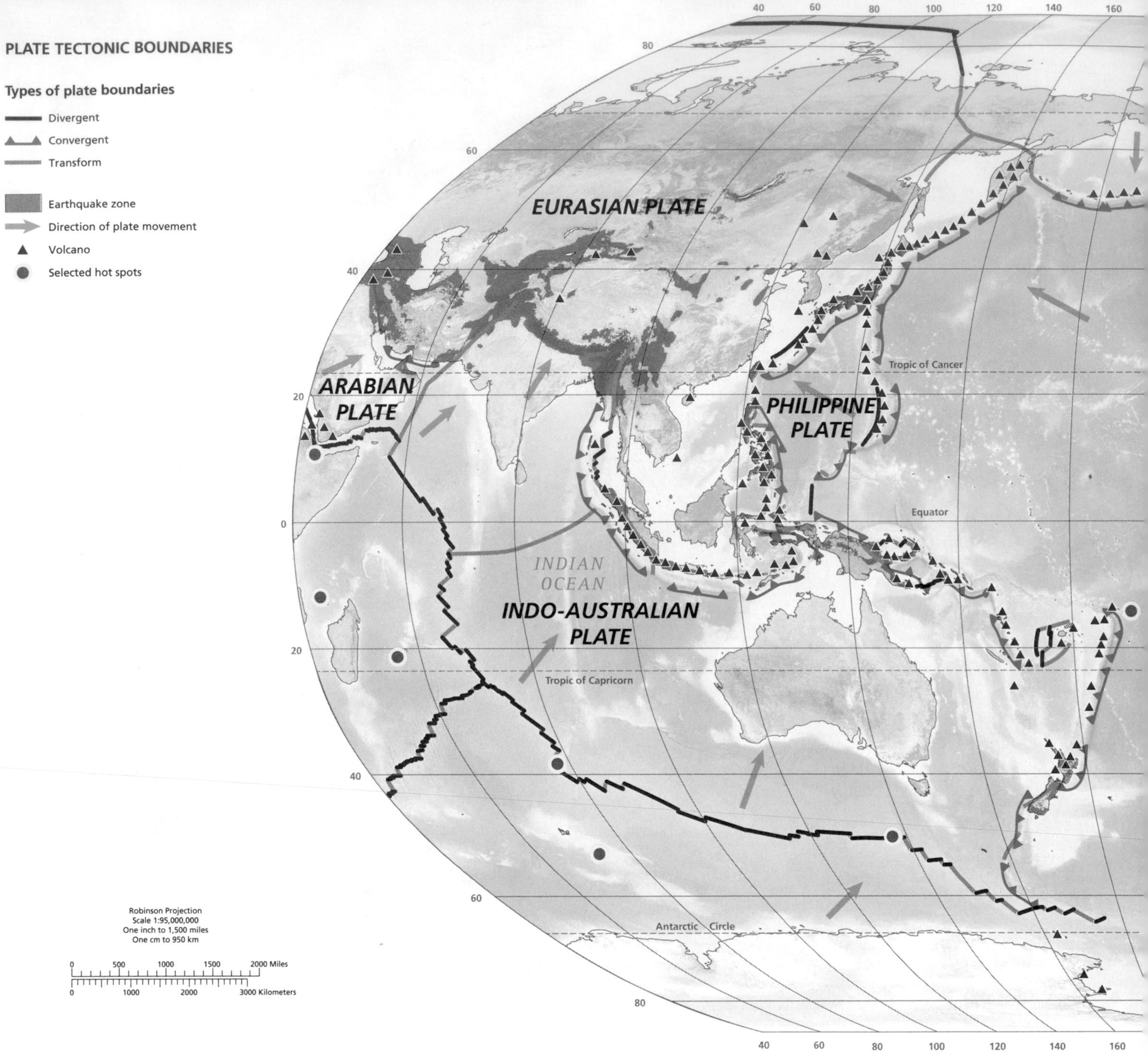

Types of plate boundaries

— Divergent

▲▲▲ Convergent

— Transform

▨ Earthquake zone

➡ Direction of plate movement

▲ Volcano

● Selected hot spots

EURASIAN PLATE

ARABIAN PLATE

PHILIPPINE PLATE

Tropic of Cancer

Equator

INDIAN OCEAN

INDO-AUSTRALIAN PLATE

Tropic of Capricorn

Antarctic Circle

Robinson Projection
Scale 1:95,000,000
One inch to 1,500 miles
One cm to 950 km

```
0    500   1000   1500   2000 Miles
0      1000     2000      3000 Kilometers
```

Plate tectonic theory describes the motions of the lithosphere, the outer surface of which forms the Earth's crust. The theory originated with scientist Alfred Wegener's work on continental drift in the early part of the 20th century. According to plate tectonic theory, the lithosphere is composed of distinct plates that move relative to each other as a result of convection currents deep within the Earth's mantle. The largest of these plates and their movements are shown on the map above.

There are three main types of plate boundaries.

Divergent plate boundaries occur where two adjacent plates move away from each other. As the plates separate, upwelling magma from the mantle solidifies, and new crust is formed. These boundaries frequently make up oceanic ridge zones, such as the Mid-Atlantic Ridge. This spreading explains why North and South America have separated from Eurasia and Africa over time. The Mid-Atlantic Ridge is actually part of a much larger subaqueous divergent boundary system that encircles the Earth.

Convergent plate boundaries occur where two adjacent plates collide with one another. When two continental plates collide, the resulting compression of lithospheric material causes large mountain ranges to form. The Himalayas, for example, were formed by the collision of the Eurasian and Indo-Australian Plates.

In other cases one plate is forced (subducted) under the other and the lithospheric material from the descending plate is recycled within the mantle. These areas are called subduction zones.

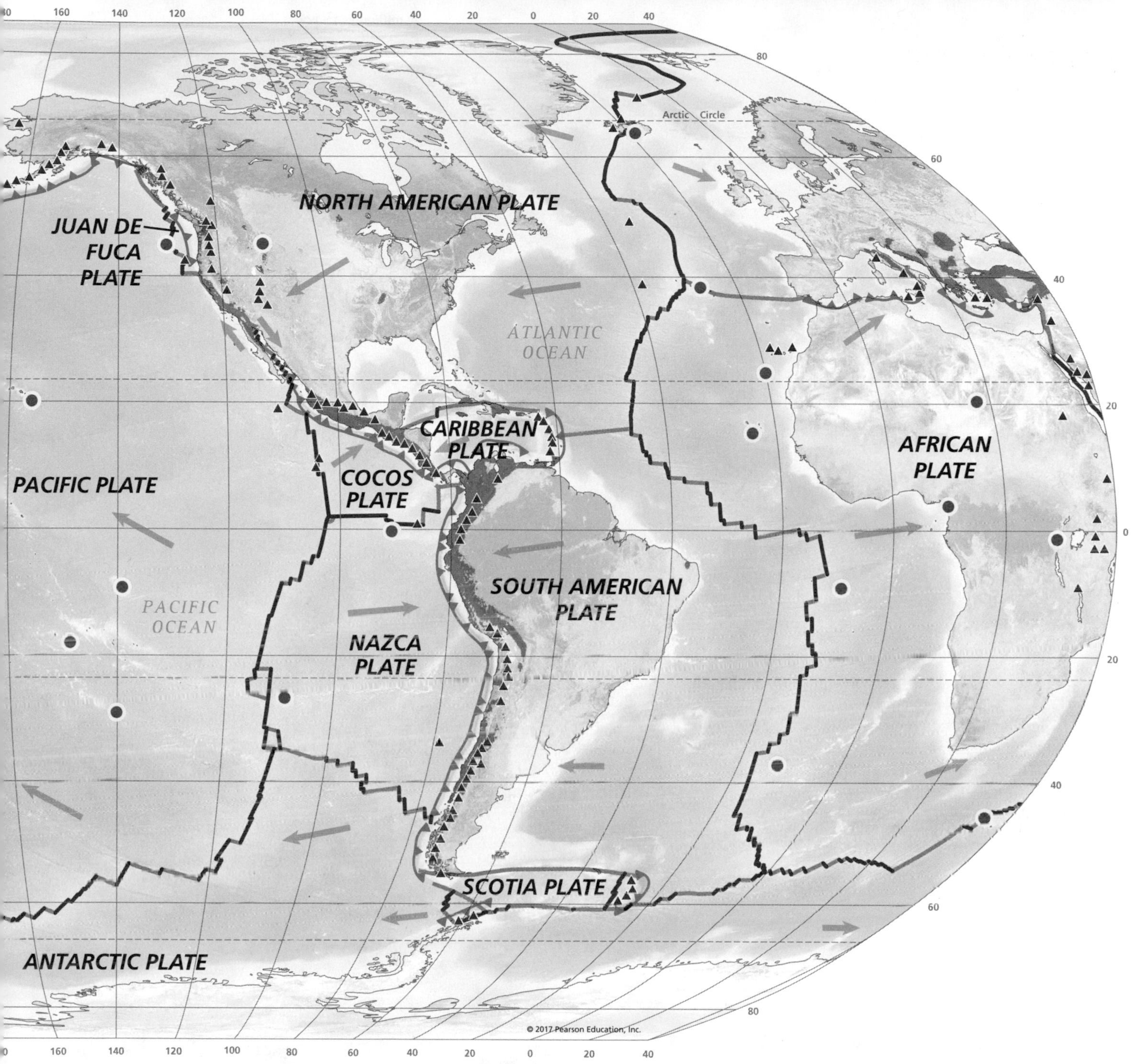

JUAN DE
FUCA
PLATE

NORTH AMERICAN PLATE

Arctic Circle

ATLANTIC
OCEAN

CARIBBEAN
PLATE

COCOS
PLATE

AFRICAN
PLATE

PACIFIC PLATE

SOUTH AMERICAN
PLATE

PACIFIC
OCEAN

NAZCA
PLATE

SCOTIA PLATE

ANTARCTIC PLATE

© 2017 Pearson Education, Inc.

Subduction zones occur when a continental plate collides with an oceanic plate. An example occurs along the west coast of South America where the Nazca Plate is being subducted under the South American Plate, creating the long, deep Peru-Chile trench and the Andes mountain chain. This area is part of a much larger ring of convergent plate boundaries circling the Pacific and known as the Ring of Fire. Volcanoes and earthquakes are common features in this region.

Subduction zones can also occur when two oceanic plates collide. Intense volcanic activity in these areas eventually results in the formation of long, volcanic island chains. The Aleutian Islands of Alaska are one example.

Transform boundaries occur when two plates slide laterally past each other with no divergence or convergence. Commonly they offset the active spreading ridges of divergent boundaries on the ocean floor. The San Andreas fault zone of California is an example of a terrestrial transform boundary.

Volcanoes and earthquakes do not occur only at plate boundaries. At certain isolated hot spots, upwelling magma rises to the surface to create tall volcanoes. Over time, as the plate moves, long islands chains are formed. The Hawai'ian Islands are one such example.

The rate of movement of tectonic plates is very slow, on the order of several centimeters per year. Over geological time, these small movements accumulate and cause fragmentation and reformation of continental land masses. The process is still underway, which implies that the arrangement of the continents millions of years from now will be quite different from what it is today.

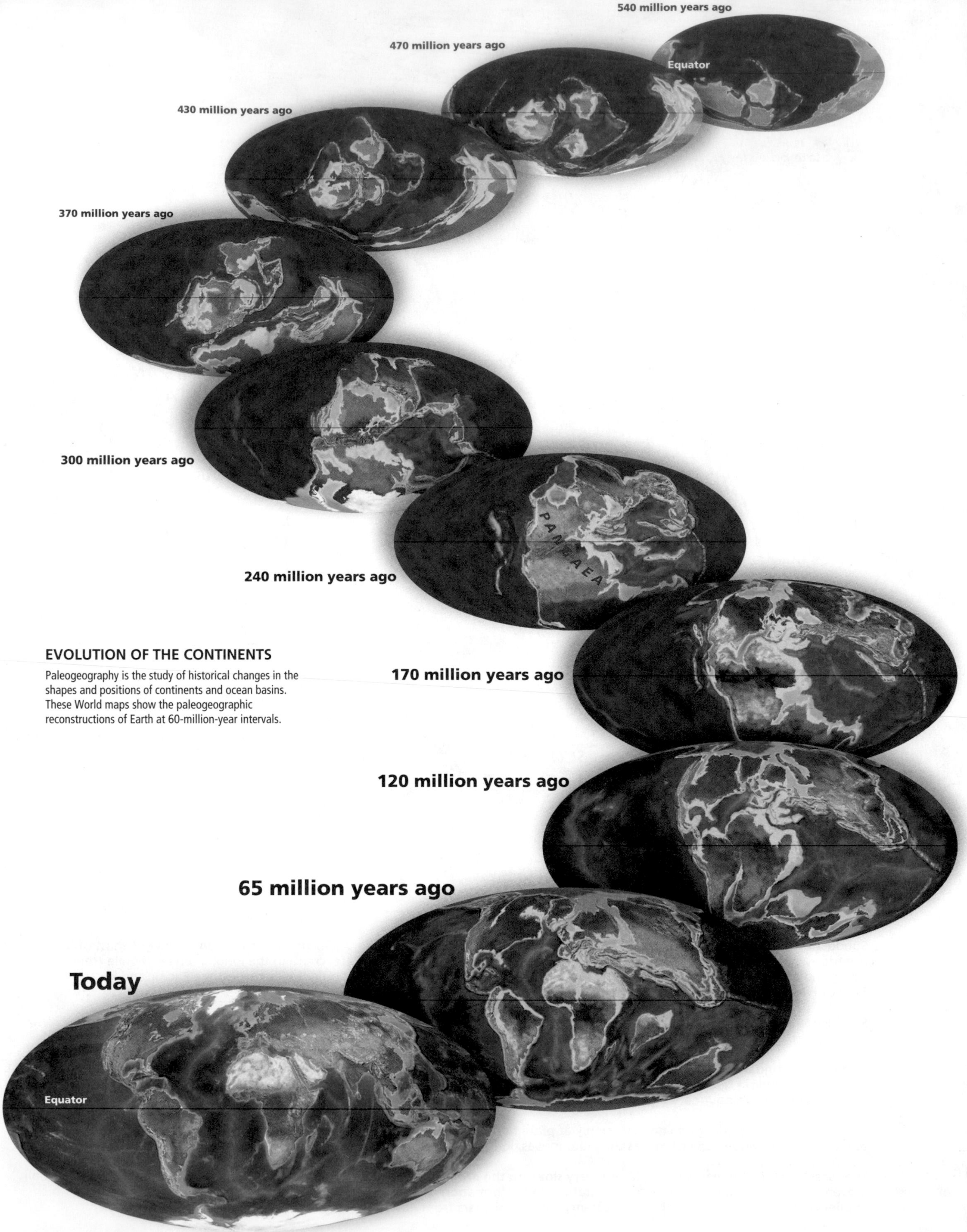

540 million years ago

Equator

470 million years ago

430 million years ago

370 million years ago

300 million years ago

240 million years ago

PANGAEA

170 million years ago

EVOLUTION OF THE CONTINENTS

Paleogeography is the study of historical changes in the shapes and positions of continents and ocean basins. These World maps show the paleogeographic reconstructions of Earth at 60-million-year intervals.

120 million years ago

65 million years ago

Today

Equator

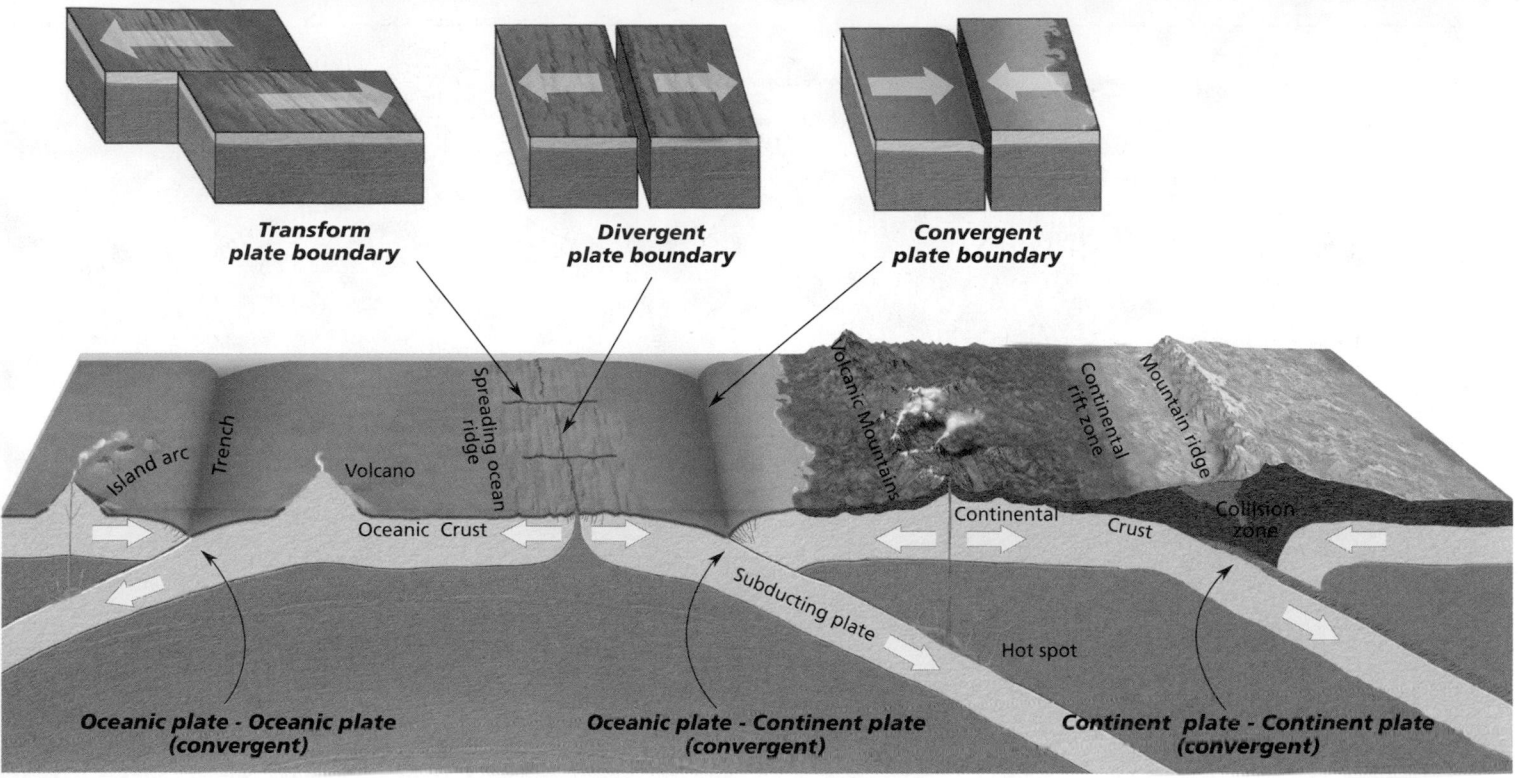

Transform plate boundary

Divergent plate boundary

Convergent plate boundary

Spreading ocean ridge

Volcanic Mountains

Continental rift zone

Mountain ridge

Island arc

Trench

Volcano

Oceanic Crust

Continental Crust

Collision zone

Subducting plate

Hot spot

Oceanic plate - Oceanic plate (convergent)

Oceanic plate - Continent plate (convergent)

Continent plate - Continent plate (convergent)

TECTONIC PLATE MOVEMENT

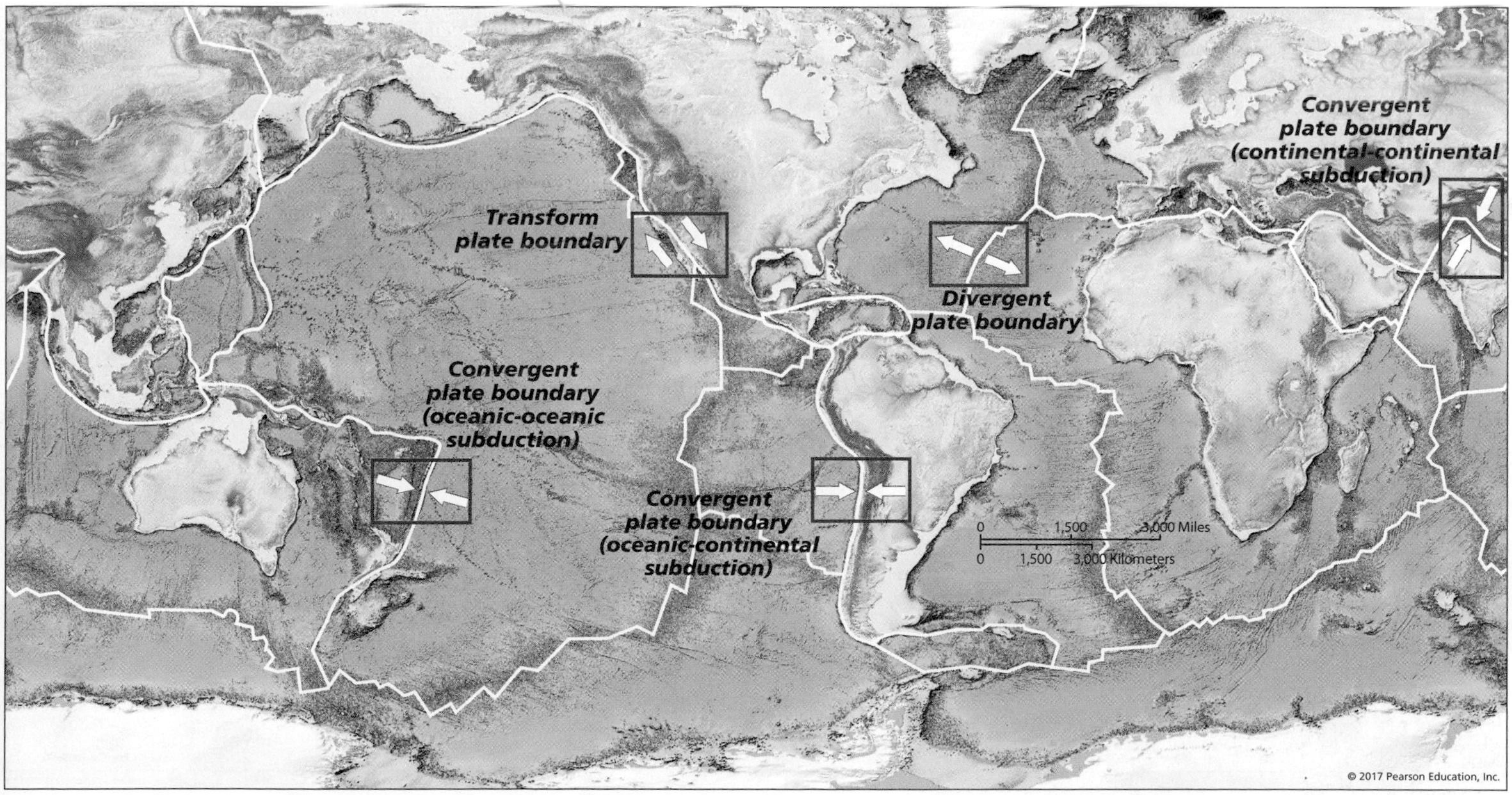

Transform plate boundary

Convergent plate boundary (continental-continental subduction)

Divergent plate boundary

Convergent plate boundary (oceanic-oceanic subduction)

Convergent plate boundary (oceanic-continental subduction)

0   1,500   3,000 Miles
0   1,500   3,000 Kilometers

© 2017 Pearson Education, Inc.

TECTONIC PLATE BOUNDARIES

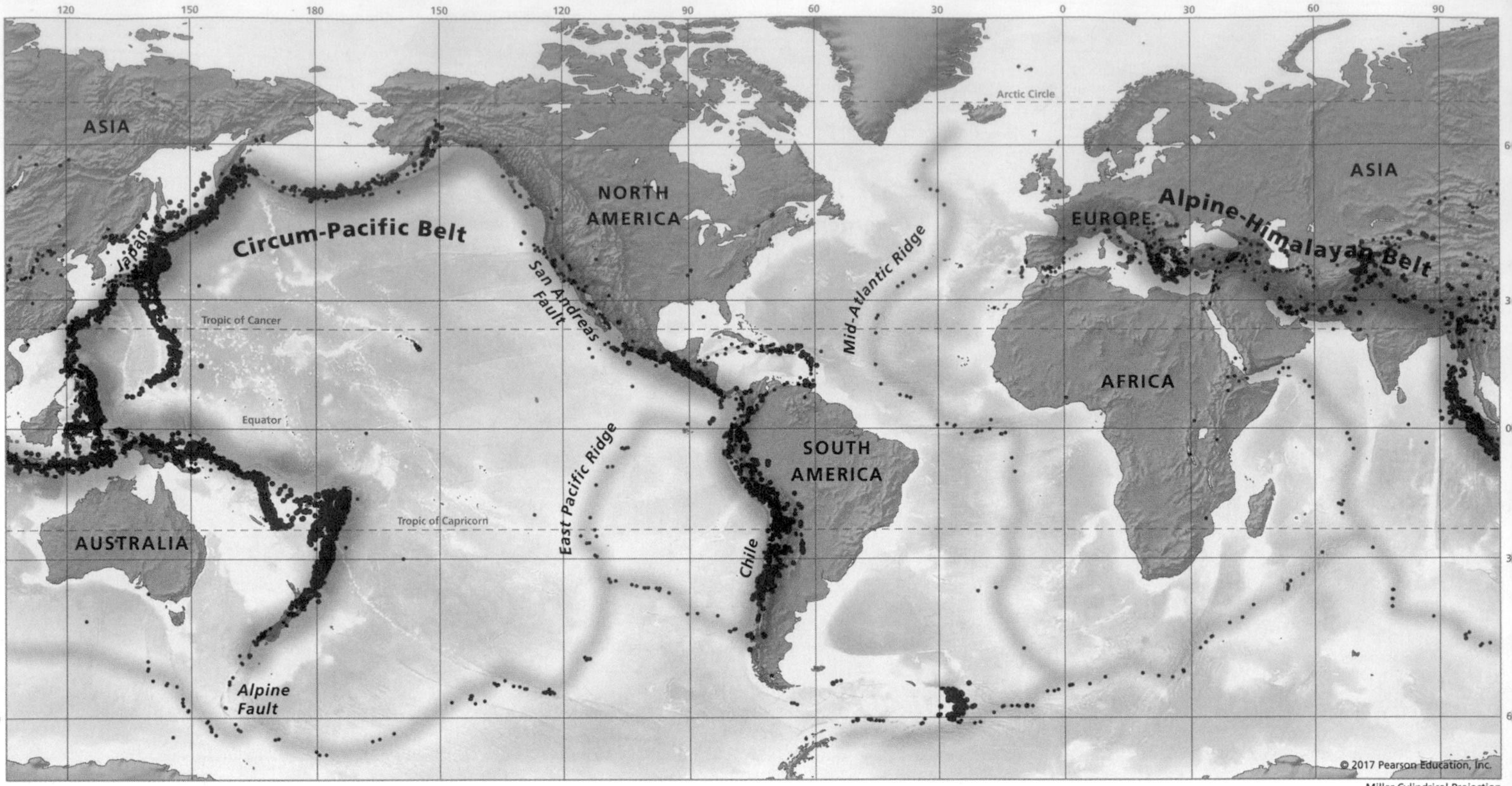

EARTHQUAKE BELTS

- • Recorded earthquakes

- Earthquake belts

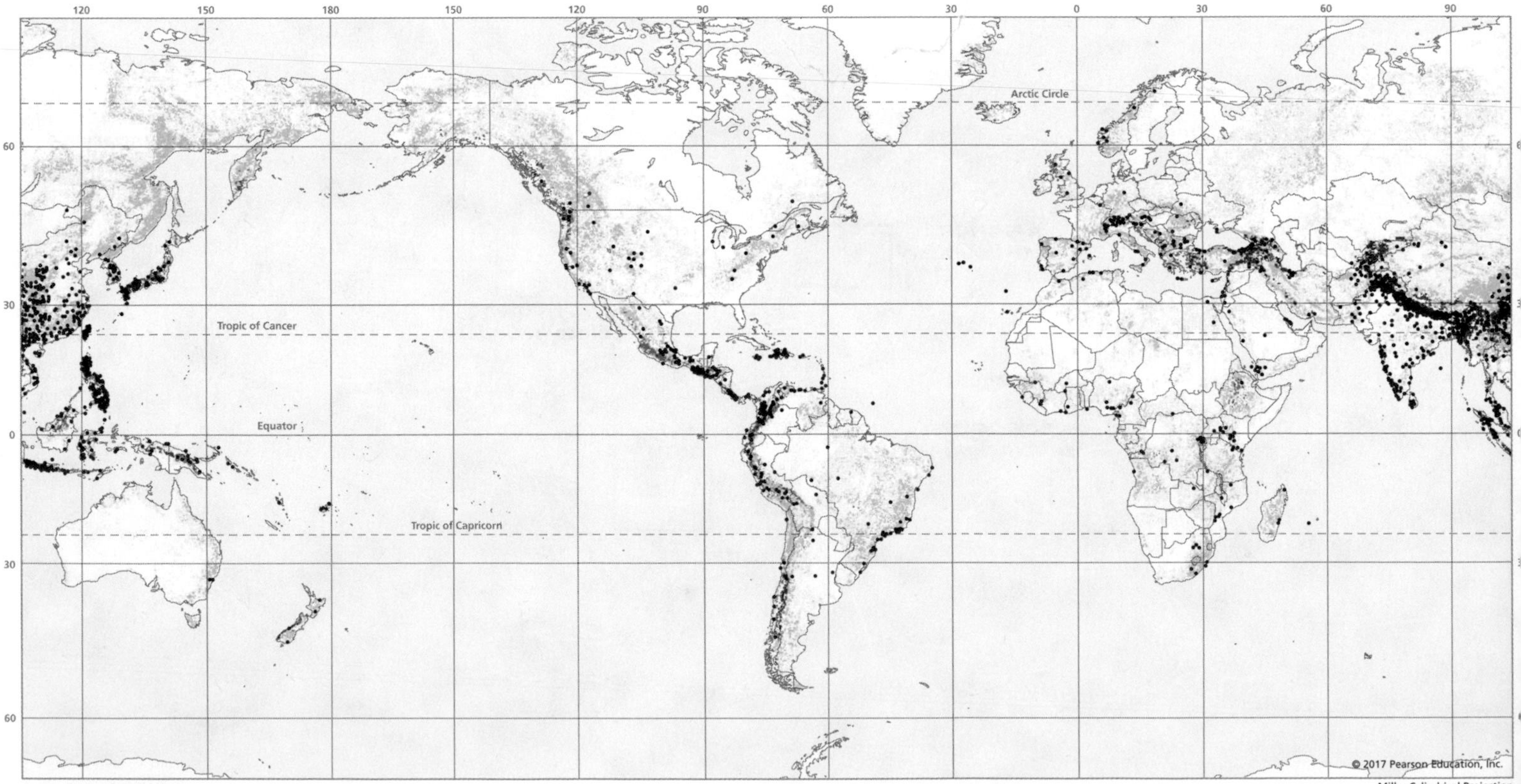

Miller Cylindrical Projection
Scale 1:200,000,000

LANDSLIDE RISK

- • Fatal landslide (2004–2010)

Slight   Moderate   Severe

Miller Cylindrical Projection
Scale 1:200,000,000
Sources: Global Risk Data Platform/
Durham Fatal Landslide Database

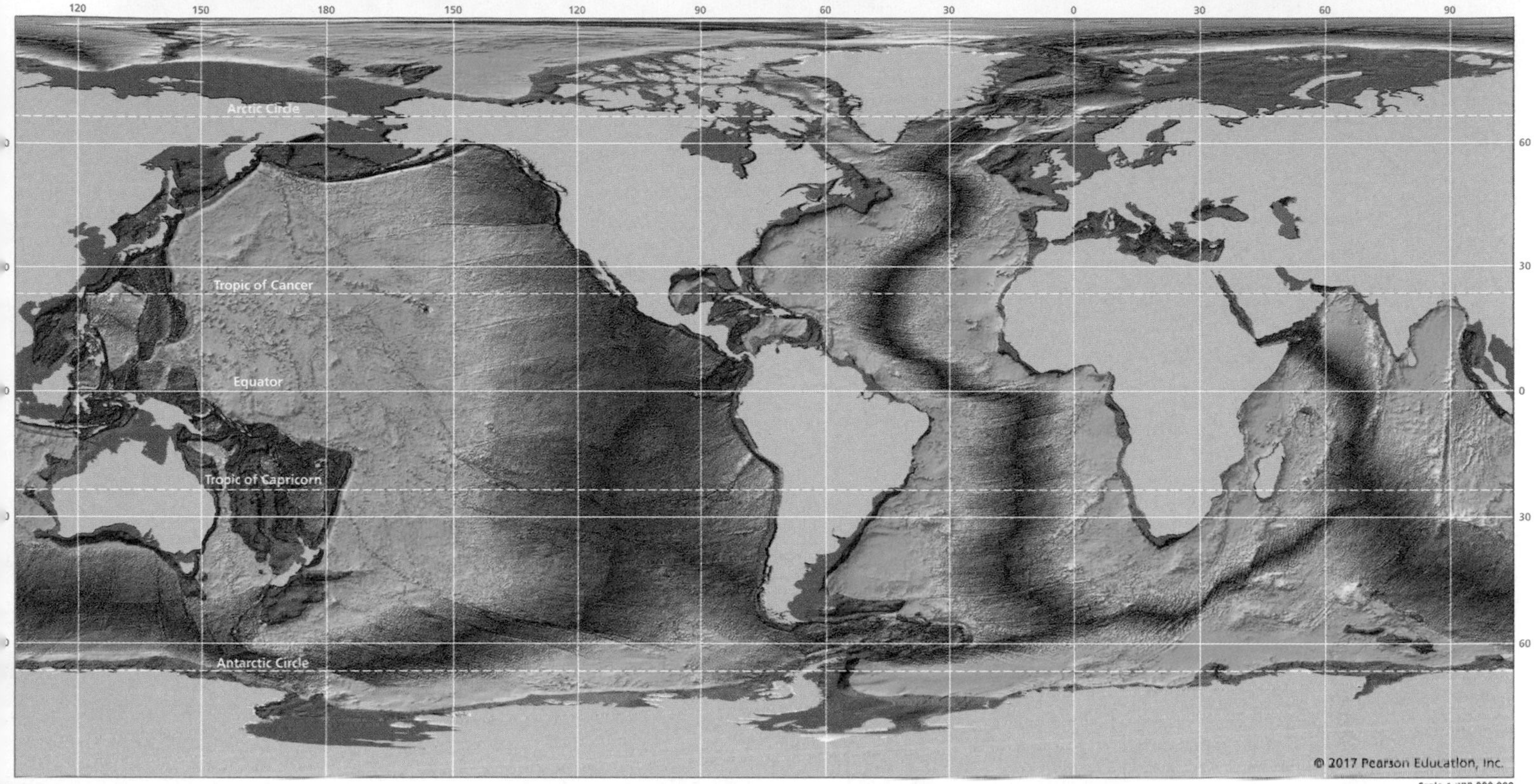

AGE OF OCEAN FLOOR

© 2017 Pearson Education, Inc.

Scale 1:100,000,000
Source: Mr. Elliot Lim, CIRES & NOAA/NCEI

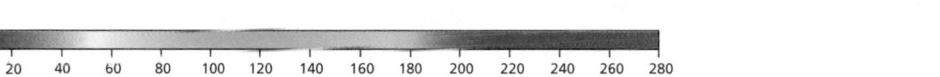

0   20   40   60   80   100   120   140   160   180   200   220   240   260   280

Age (millions of years ago)

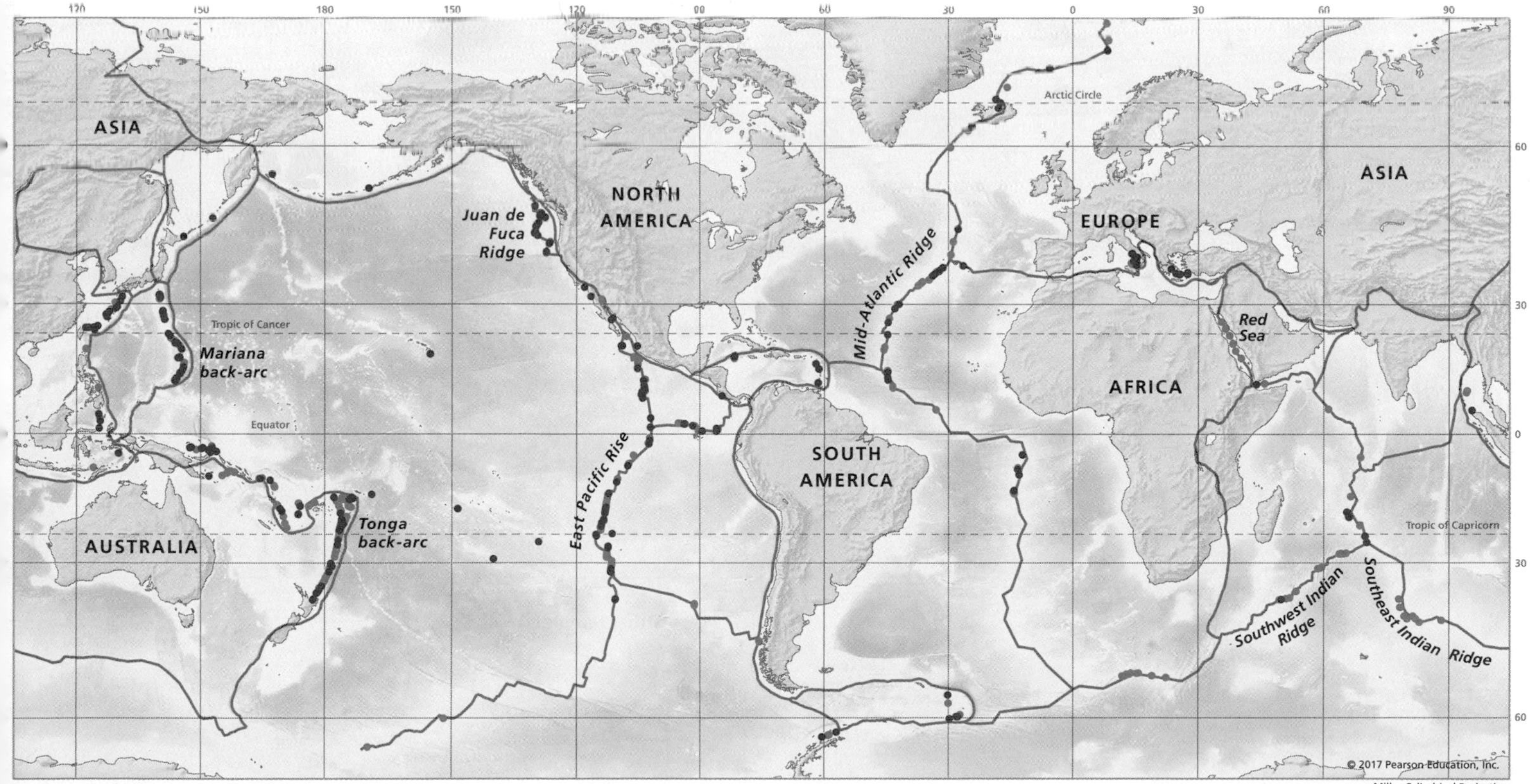

ASIA

NORTH
AMERICA

Juan de
Fuca
Ridge

Mariana
back-arc

Tropic of Cancer

Equator

AUSTRALIA

Tonga
back-arc

East Pacific Rise

SOUTH
AMERICA

Mid-Atlantic Ridge

Arctic Circle

EUROPE

ASIA

Red
Sea

AFRICA

Tropic of Capricorn

Southwest Indian Ridge

Southeast Indian Ridge

© 2017 Pearson Education, Inc.

Miller Cylindrical Projection
Scale 1:200,000,000

HYDROTHERMAL VENT FIELDS

- • Active vent
- • Unconfirmed vent
- —— Tectonic plate boundary

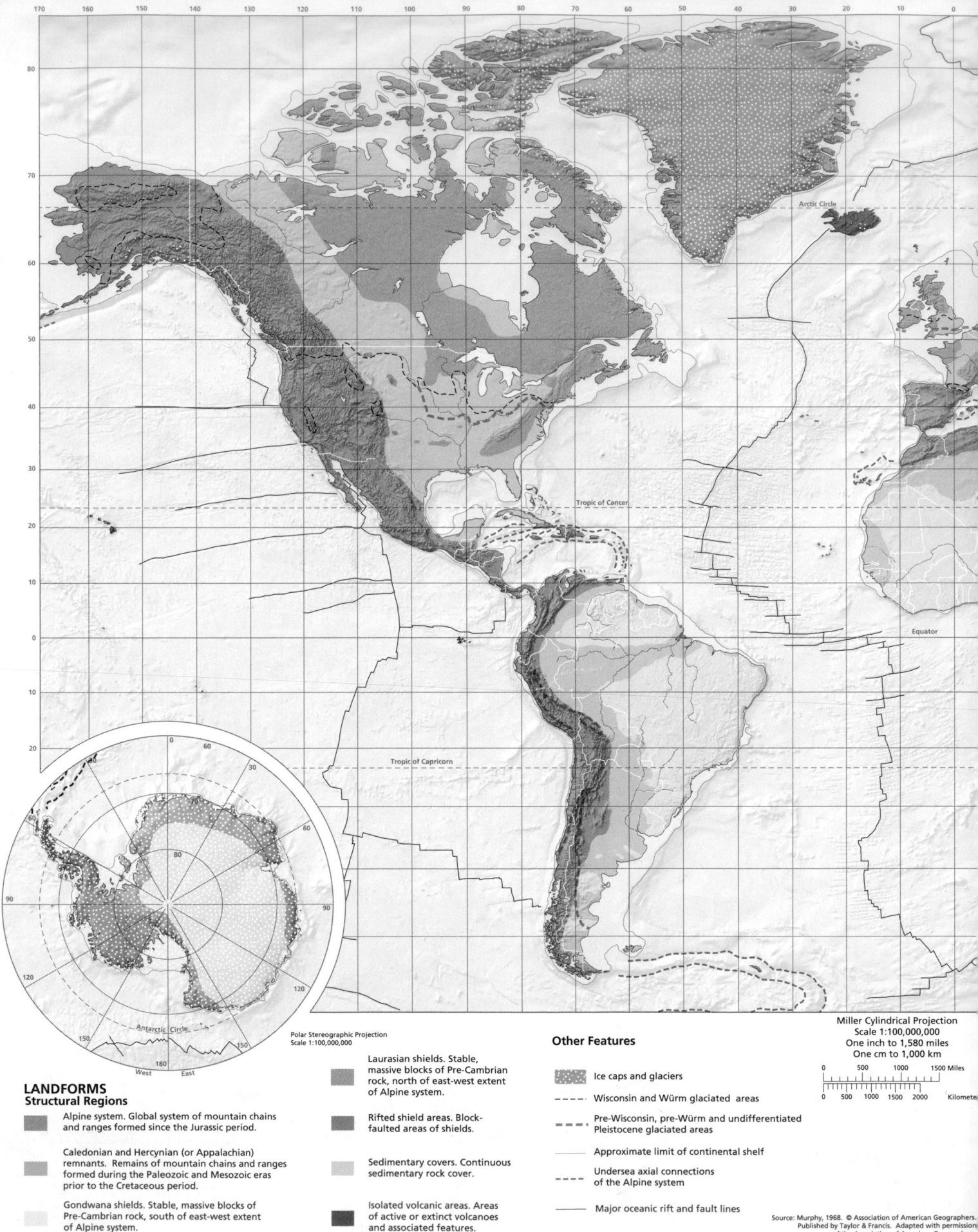

Miller Cylindrical Projection
Scale 1:100,000,000
One inch to 1,580 miles
One cm to 1,000 km

0   500   1000   1500 Miles

0   500   1000   1500   2000   Kilometer

Polar Stereographic Projection
Scale 1:100,000,000

Other Features

Ice caps and glaciers

- - - - Wisconsin and Würm glaciated areas

▬ ▬ ▬ Pre-Wisconsin, pre-Würm and undifferentiated
Pleistocene glaciated areas

──────── Approximate limit of continental shelf

▬ ▬ ▬ Undersea axial connections
of the Alpine system

──────── Major oceanic rift and fault lines

LANDFORMS
Structural Regions

Alpine system. Global system of mountain chains
and ranges formed since the Jurassic period.

**Caledonian and Hercynian (or Appalachian)
remnants.** Remains of mountain chains and ranges
formed during the Paleozoic and Mesozoic eras
prior to the Cretaceous period.

Gondwana shields. Stable, massive blocks of
Pre-Cambrian rock, south of east-west extent
of Alpine system.

Laurasian shields. Stable,
massive blocks of Pre-Cambrian
rock, north of east-west extent
of Alpine system.

Rifted shield areas. Block-
faulted areas of shields.

Sedimentary covers. Continuous
sedimentary rock cover.

Isolated volcanic areas. Areas
of active or extinct volcanoes
and associated features.

Source: Murphy, 1968. © Association of American Geographers.
Published by Taylor & Francis. Adapted with permission
of the Association of American Geographers.

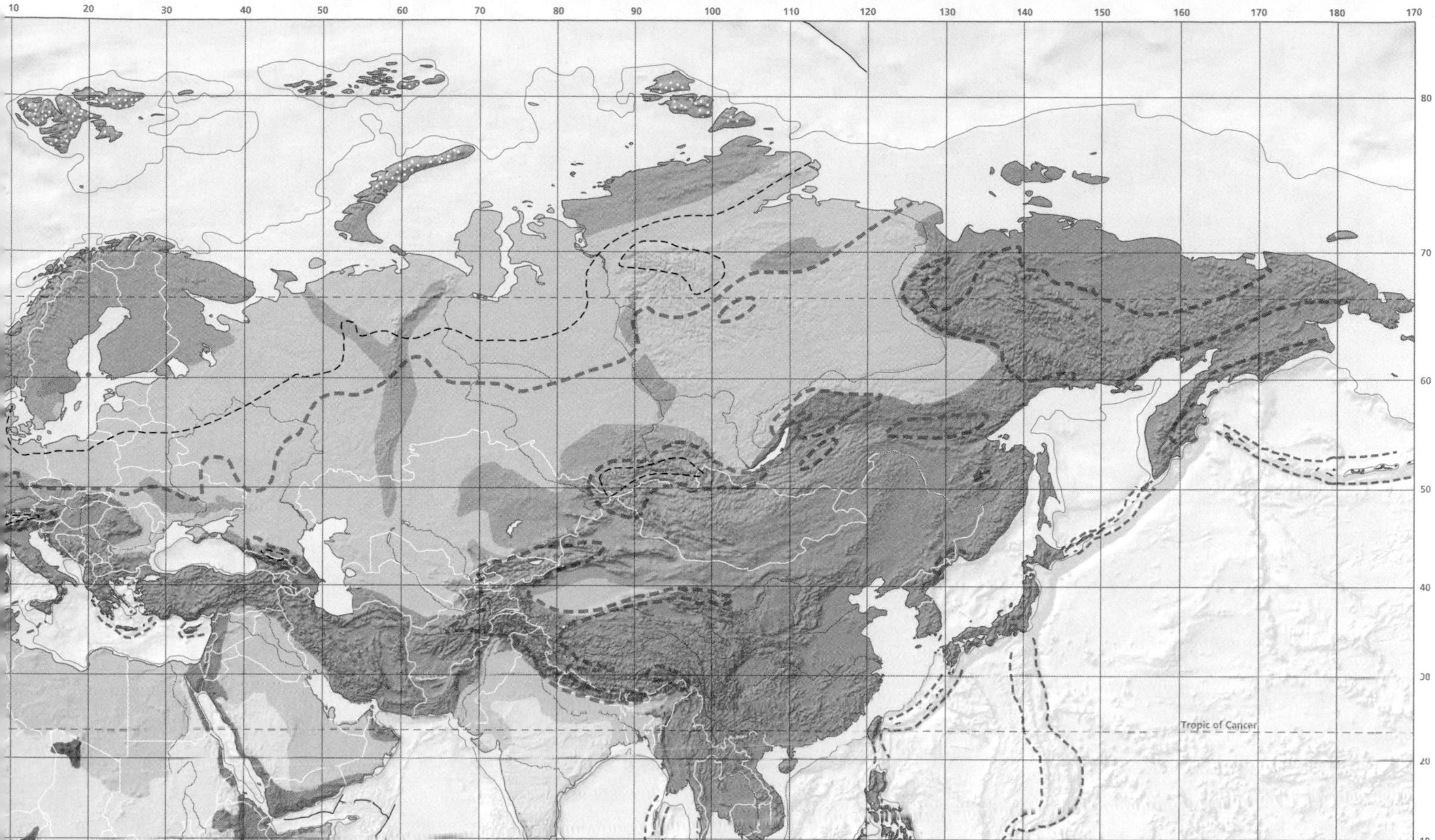

Mollweide Projection
Scale 1:275,000,000
Source: Tapley et al., 2005

A-101928-1

© Rand McNally

GRAVITY ANOMALY

mGal 75 50   25   0   -25   -50 -75

This map is based on the GGM02C gravity anomaly model. This model was derived from over a year of GRACE (Gravity Recovery And Climate Experiment) satellite data coupled with terrestrially-based gravity observations. The gravity anomaly is the difference between observed gravity and the standard gravity on a reference ellipsoid. It is measured in milligals (mGal), defined as 10^{-5} m/s^2 or approximately one millionth of the standard acceleration on the Earth's surface. Values above zero have higher than standard gravity, and vice versa.

Tropic of Cancer

Equator

Tropic of Capricorn

Barents Sea

Norwegian Basin

Arctic Circle

NORWAY SWEDEN FINLAND

RUSSIA

Ob'

Yenisey

Lena

Magadan

60°

North Sea

DEN. Baltic Sea

Moscow

EST. LAT. LITH. BELARUS

POLAND

GERMANY

EUROPE

AUS.

UKRAINE

Volga

KAZAKHSTAN

Aral Sea

Balqash Köli

MONGOLIA

Ozero Baykal

Okhotsk Basin

Sea of Okhotsk

SAKHALIN

Japan Basin

Amur

ROMANIA

Black Sea

Caspian Sea

UZBEKISTAN

KYRGYZSTAN

ASIA

NORTH KOREA

Japan Trench

Kuril Trench

North Pac

45°

Istanbul

GREECE TURKEY

Mediterranean Sea

SYRIA

IRAQ

TURKMENISTAN

TAJIKISTAN

Tehran

AFGHANISTAN

IRAN

CHINA

Beijing

Huang

Seoul SOUTH KOREA

Yellow Sea

JAPAN

Tokyo

Shanghai

Izu Trench

Ba

Mid-Pac

30°

Cairo

EGYPT

Red Sea

Nile

SAUDI ARABIA

PAKISTAN

Indus

NEPAL

Delhi

Ganges

BNGL.

Kolkata

DHAKA

East China Sea

Taipei

Ryukyu Trench

Tropic of Cancer

Karachi

Dhaka

MYANMAR

Hong Kong

TAIWAN

Philippine Sea

NORTHERN MARIANA ISLANDS (U.S.)

YEMEN

OMAN

Arabian Sea

INDIA

Mumbai

LAOS

South China Sea

Philippine Basin

Kyushu-Palau Ridge

Mariana Trench

East Mariana Basin

15°

SUDAN

ERITREA

Arabian Basin

Carlsberg Ridge

Chennai

ANDAMAN ISLANDS (India)

Bangkok

THAILAND

CAMB.

VIETNAM

South China Basin

Ho Chi Minh City

Manila

PHILIPPINES

Philippine Trench

PALAU

CAROLINE ISLANDS

Eauripik Rise

FEDERATED STATE OF MICRONESIA

AFRICA

SOUTH SUDAN

ETHIOPIA

SOMALIA

MALDIVES

MID-INDIAN RIDGE

Chagos-Laccadive Plateau

NICOBAR ISLANDS (India)

Andaman Basin

SRI LANKA

Sunda Shelf

BRUNEI

Sulu Basin

West Caroline Basin

East Caroline Basin

MELAN

0°

Equator

UGANDA KENYA

Nairobi

Lake Victoria

Somali Basin

SEYCHELLES

Mascarene Plateau

INDONESIA

SINGAPORE

MALAYSIA

BORNEO

SULAWESI

NEW GUINEA

Bismarck Sea

Solomon Basin

TANZANIA

COMOROS

Mascarene Basin

INDIAN

Mid-Indian Basin

Ninety-East Ridge

COCOS ISLANDS (Austl.)

CHRISTMAS ISLAND (Austl.)

Jakarta

JAVA

Java Trench

TIMOR LESTE

Darwin

Arafura Shelf

PAPUA NEW GUINEA

15°

MALAWI

ZAMBIA

ZIMBABWE

MOZAMBIQUE

Mozambique Channel

MADAGASCAR

REUNION (Fr.)

MAURITIUS

OCEAN

Wharton Basin

North Australian Basin

Gulf of Carpentaria

Coral Sea Basin

Coral Sea

SOUTH AFRICA

Johannesburg

Mozambique Plateau

Madagascar Basin

MID-INDIAN RIDGE

Southwest Indian Ridge

Broken Ridge

Perth Basin

Perth

AUSTRALIA

Darling

Brisba

30°

Tropic of Capricorn

Madagascar Plateau

ÎLE AMSTERDAM (Fr.)

ÎLE ST. PAUL (Fr.)

Great Australian Bight

South Australian Basin

Sydney

Melbourne

Tasm Se

Agulhas Basin

Crozet Basin

Southeast Indian Ridge

South Tasman Rise

TASMANIA

Tasma Basi

45°

PRINCE EDWARD ISLANDS (S. Afr.)

ÎLES CROZET (Fr.)

ÎLES KERGUELEN (Fr.)

Kerguelen Plateau

Atlantic - Indian Ridge

HEARD ISLAND (Austl.)

South Indian Basin

arie R.

60°

Atlantic - Indian Basin

South Indian Basin

South Magnetic Pole

SOUTHERN

Antarctic Circle

ANTARCTICA

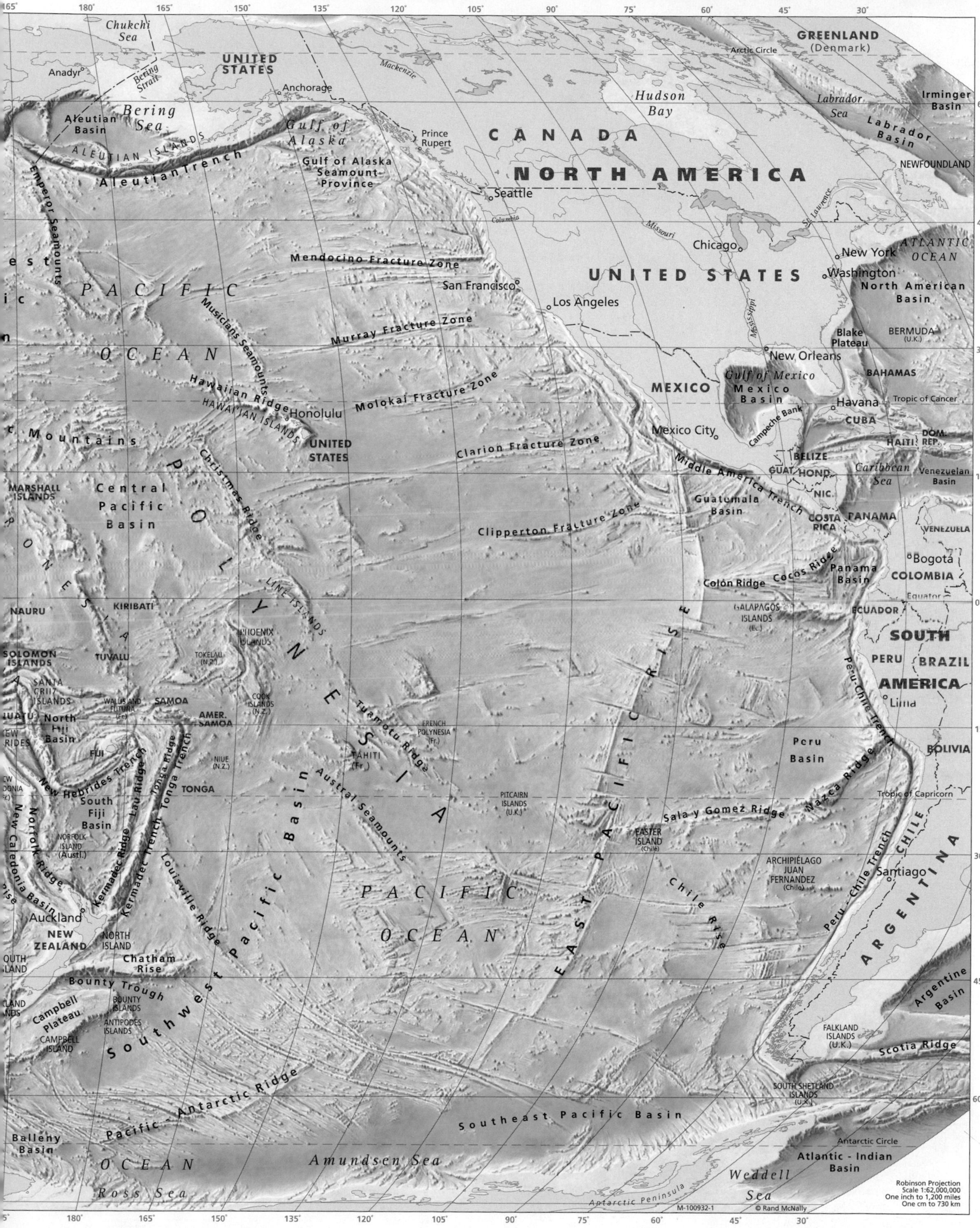

Chukchi Sea
180° 165° 150° 135° 120° 105° 90° 75° 60° 45° 30°

Anadyr

Bering Strait

UNITED STATES

Mackenzie

Arctic Circle

GREENLAND (Denmark)

Anchorage

Aleutian Basin

Bering Sea

ALEUTIAN ISLANDS

Aleutian Trench

Emperor Seamounts

Gulf of Alaska

Gulf of Alaska Seamount Province

Prince Rupert

Hudson Bay

CANADA

NORTH AMERICA

Irminger Basin

Labrador Sea

Labrador Basin

NEWFOUNDLAND

60°

Seattle

Columbia

45°

Missouri

St. Lawrence

PACIFIC

Mendocino Fracture Zone

San Francisco

Los Angeles

Chicago

UNITED STATES

New York

Washington

ATLANTIC OCEAN

North American Basin

Musicians Seamounts

OCEAN

Murray Fracture Zone

Mississippi

New Orleans

BERMUDA (U.K.)

Blake Plateau

30°

Mountains

Hawaiian Ridge

HAWAIIAN ISLANDS

Honolulu

Molokai Fracture Zone

UNITED STATES

MEXICO

Gulf of Mexico

Mexico Basin

Campeche Bank

Havana

CUBA

BAHAMAS

Tropic of Cancer

Mexico City

Clarion Fracture Zone

Middle America Trench

BELIZE

GUAT. HOND.

NIC.

Caribbean Sea

HAITI DOM. REP.

Venezuelan Basin

15°

MARSHALL ISLANDS

Central Pacific Basin

Christmas Ridge

MICRONESIA

POLYNESIA

Clipperton Fracture Zone

Guatemala Basin

Colón Ridge

Cocos Ridge

COSTA RICA

PANAMA

Panama Basin

VENEZUELA

Bogotá

COLOMBIA

Equator

0°

NAURU

KIRIBATI

LINE ISLANDS

PHOENIX ISLANDS

GALÁPAGOS ISLANDS (Ec.)

ECUADOR

SOLOMON ISLANDS

TUVALU

TOKELAU (N.Z.)

SOUTH

PERU

BRAZIL

SANTA CRUZ ISLANDS

WALLIS AND FUTUNA (Fr.)

SAMOA

COOK ISLANDS (N.Z.)

AMERICA

Lima

VANUATU

North Fiji Basin

AMER. SAMOA

Tuamotu Ridge

FRENCH POLYNESIA (Fr.)

Peru Basin

15°

NEW HEBRIDES

FIJI

NIUE (N.Z.)

TAHITI (Fr.)

Peru-Chile Trench

BOLIVIA

NEW CALEDONIA

South Fiji Basin

Lau Ridge

Kermadec Ridge

Tonga Ridge

TONGA

Austral Seamounts

PITCAIRN ISLANDS (U.K.)

Sala y Gomez Ridge

Nazca Ridge

Tropic of Capricorn

Norfolk Basin

New Caledonia Ridge

Kermadec Trench

Tonga Trench

Louisville Ridge

Southwest Pacific Basin

POLYNESIA

EASTER ISLAND (Chile)

EAST PACIFIC RISE

ARCHIPIÉLAGO JUAN FERNÁNDEZ (Chile)

Santiago

CHILE

30°

Auckland

NEW ZEALAND

NORTH ISLAND

NORFOLK ISLAND (Austl.)

PACIFIC

Chile Rise

Peru-Chile Trench

ARGENTINA

SOUTH ISLAND

Chatham Rise

OCEAN

45°

Bounty Trough

Campbell Plateau

BOUNTY ISLANDS

ANTIPODES ISLANDS

CAMPBELL ISLAND

Southwest Pacific Basin

FALKLAND ISLANDS (U.K.)

Argentine Basin

CHATHAM ISLANDS

SOUTH SHETLAND ISLANDS (U.K.)

Scotia Ridge

60°

Balleny Basin

Pacific - Antarctic Ridge

Southeast Pacific Basin

Antarctic Circle

Atlantic - Indian Basin

OCEAN

Ross Sea

Amundsen Sea

Antarctic Peninsula

Weddell Sea

180° 165° 150° 135° 120° 105° 90° 75° 60° 45° 30°

M-100932-1

© Rand McNally

Robinson Projection
Scale 1:62,000,000
One inch to 1,200 miles
One cm to 730 km

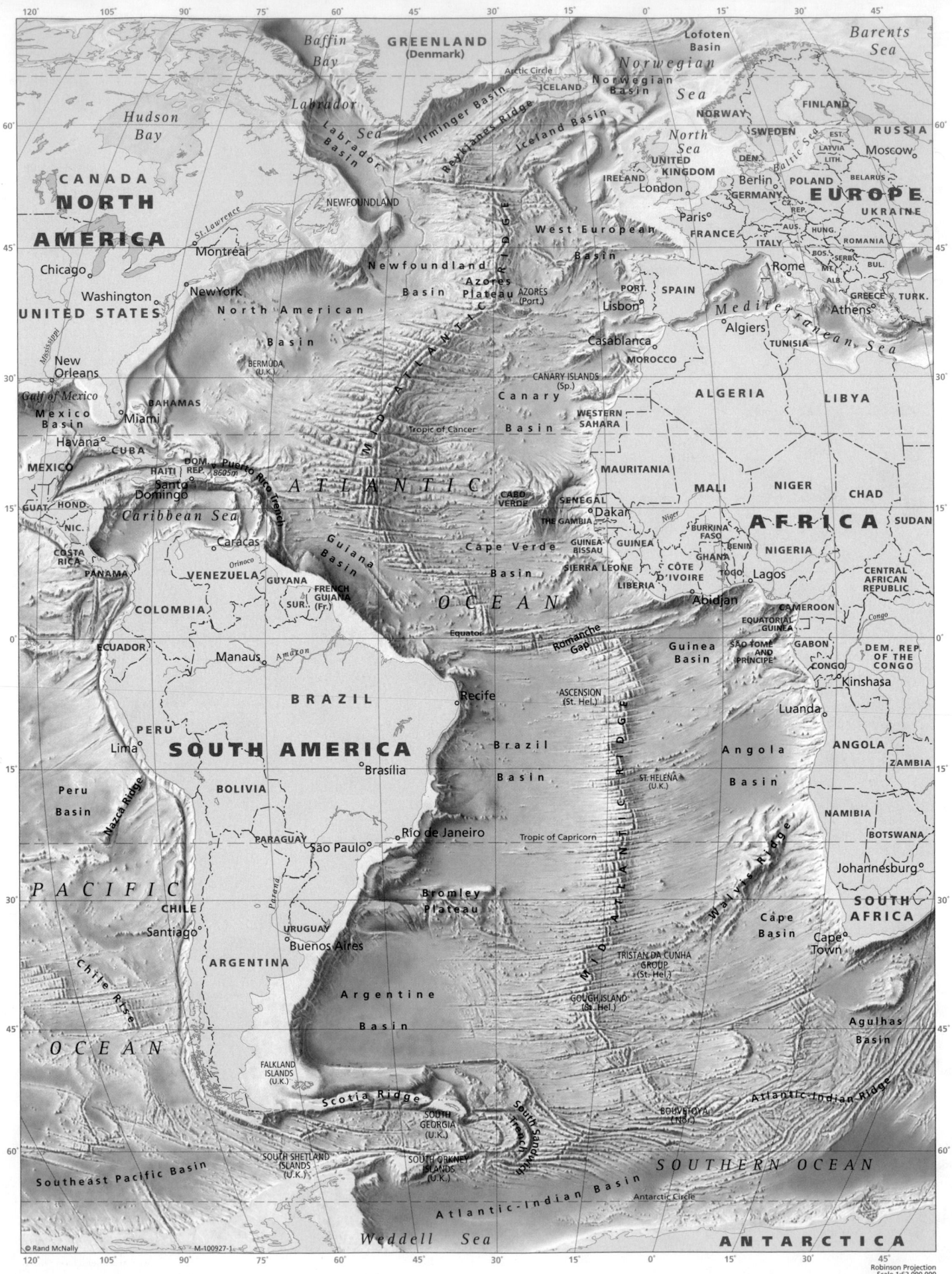

© Rand McNally   M-100927-1

Robinson Projection
Scale 1:62,000,000
One inch to 1,200 miles
One cm to 730 km

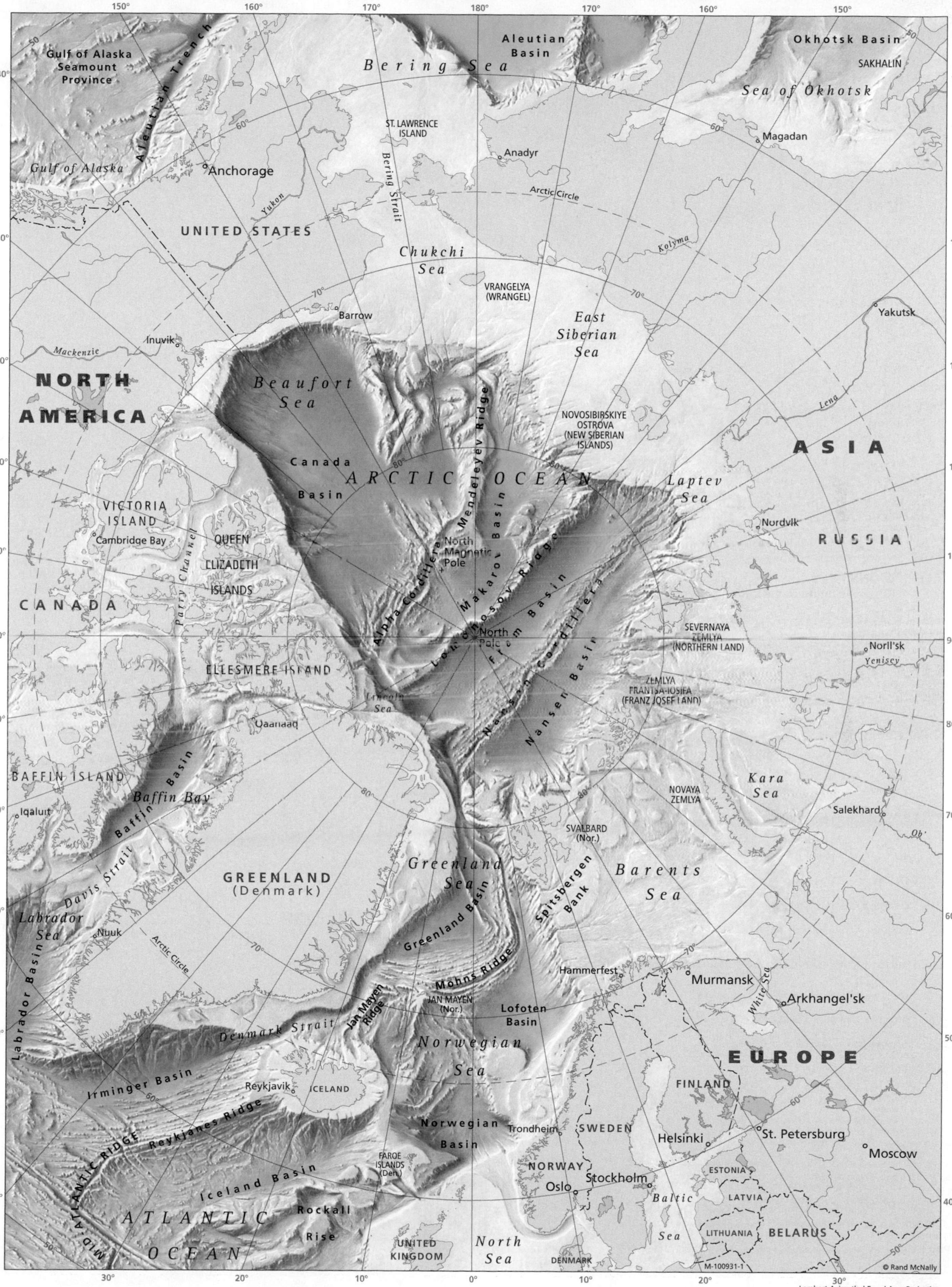

Gulf of Alaska Seamount Province

Aleutian Trench

Bering Sea

Aleutian Basin

Okhotsk Basin

SAKHALIN

Sea of Okhotsk

ST. LAWRENCE ISLAND

Anadyr

Magadan

Arctic Circle

Gulf of Alaska

Anchorage

Bering Strait

Yukon

Kolyma

Yakutsk

UNITED STATES

Chukchi Sea

VRANGELYA (WRANGEL)

East Siberian Sea

Lena

Barrow

Inuvik

Mackenzie

Beaufort Sea

Mendeleyev Ridge

NOVOSIBIRSKIYE OSTROVA (NEW SIBERIAN ISLANDS)

ASIA

NORTH AMERICA

Canada Basin

ARCTIC OCEAN

Makarov Basin

Laptev Sea

Nordvik

RUSSIA

VICTORIA ISLAND

Cambridge Bay

QUEEN ELIZABETH ISLANDS

North Magnetic Pole

Alpha Cordillera

Lomonosov Ridge

Makarov Ridge

Fram Basin

Amundsen Basin

SEVERNAYA ZEMLYA (NORTHERN LAND)

Noril'sk

CANADA

Patry Channel

North Pole

Nansen Cordillera

Nansen Basin

Yenisey

ELLESMERE ISLAND

Lincoln Sea

ZEMLYA FRANTSA-IOSIFA (FRANZ JOSEF LAND)

Qaanaaq

Kara Sea

BAFFIN ISLAND

Baffin Basin

Baffin Bay

NOVAYA ZEMLYA

Salekhard

Iqaluit

Baffin

Ob'

Davis Strait

SVALBARD (Nor.)

GREENLAND (Denmark)

Greenland Sea

Spitsbergen Bank

Barents Sea

Labrador Sea

Labrador Basin

Nuuk

Greenland Basin

Mohns Ridge

Hammerfest

Murmansk

White Sea

Arkhangel'sk

Arctic Circle

JAN MAYEN (Nor.)

Jan Mayen Ridge

Lofoten Basin

EUROPE

Denmark Strait

Norwegian Sea

FINLAND

Irminger Basin

Reykjavik

ICELAND

Norwegian Basin

Trondheim

SWEDEN

Helsinki

St. Petersburg

MID-ATLANTIC RIDGE

Reykjanes Ridge

NORWAY

Oslo

Stockholm

ESTONIA

Moscow

Iceland Basin

Rockall Rise

Baltic Sea

LATVIA

ATLANTIC OCEAN

FAROE ISLANDS (Den.)

North Sea

UNITED KINGDOM

DENMARK

LITHUANIA

BELARUS

M-100931-1

© Rand McNally

Lambert Azimuthal Equal Area Projection
Scale 1:30,000,000
One inch to 500 miles
One cm to 300 cm

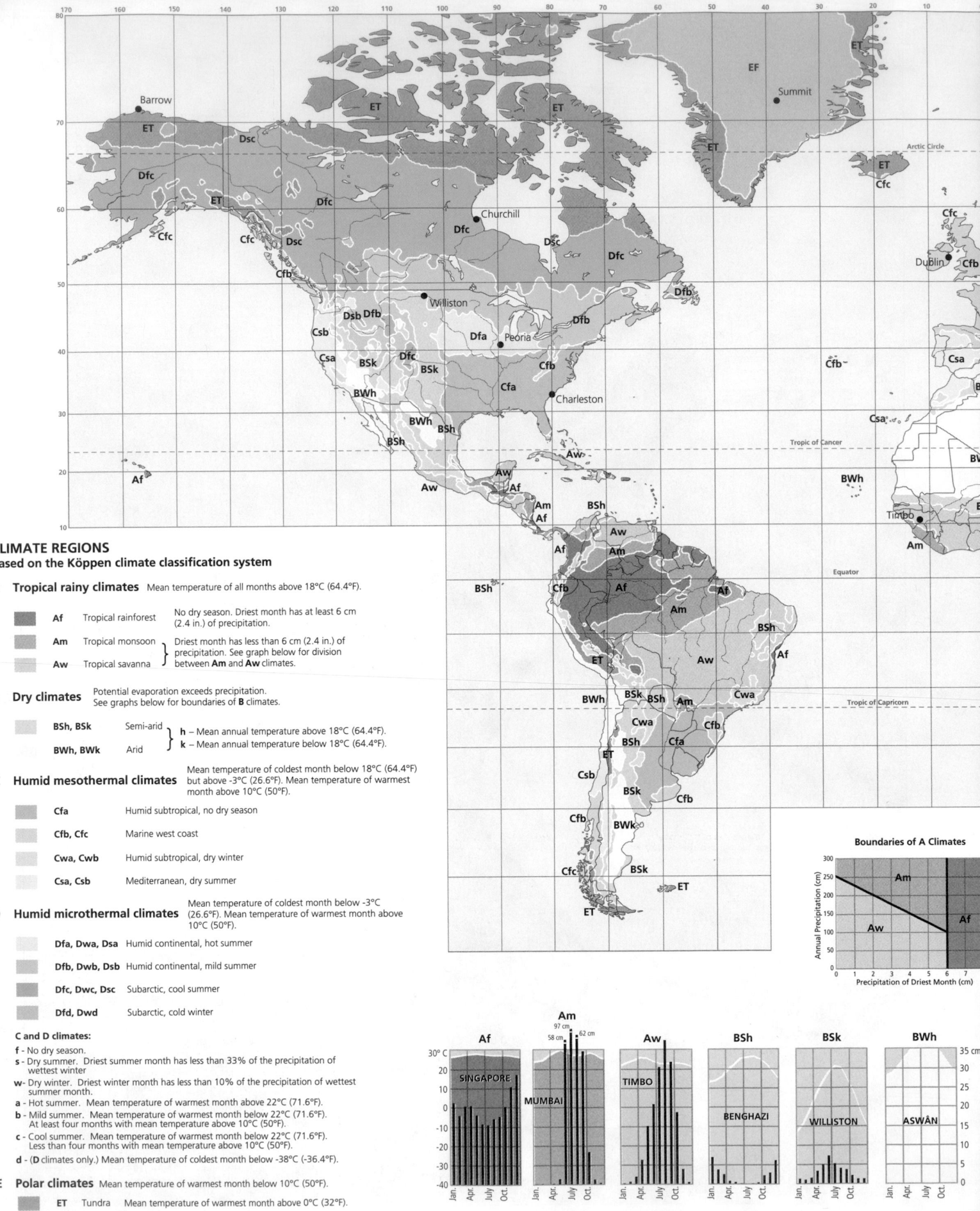

18

CLIMATE REGIONS
Based on the Köppen climate classification system

A   Tropical rainy climates   Mean temperature of all months above 18°C (64.4°F).

	Af	Tropical rainforest	No dry season. Driest month has at least 6 cm (2.4 in.) of precipitation.
	Am	Tropical monsoon	Driest month has less than 6 cm (2.4 in.) of precipitation. See graph below for division between **Am** and **Aw** climates.
	Aw	Tropical savanna	

B   Dry climates   Potential evaporation exceeds precipitation. See graphs below for boundaries of **B** climates.

| | BSh, BSk | Semi-arid | **h** – Mean annual temperature above 18°C (64.4°F). |
| | BWh, BWk | Arid | **k** – Mean annual temperature below 18°C (64.4°F). |

C   Humid mesothermal climates   Mean temperature of coldest month below 18°C (64.4°F) but above -3°C (26.6°F). Mean temperature of warmest month above 10°C (50°F).

	Cfa	Humid subtropical, no dry season
	Cfb, Cfc	Marine west coast
	Cwa, Cwb	Humid subtropical, dry winter
	Csa, Csb	Mediterranean, dry summer

D   Humid microthermal climates   Mean temperature of coldest month below -3°C (26.6°F). Mean temperature of warmest month above 10°C (50°F).

	Dfa, Dwa, Dsa	Humid continental, hot summer
	Dfb, Dwb, Dsb	Humid continental, mild summer
	Dfc, Dwc, Dsc	Subarctic, cool summer
	Dfd, Dwd	Subarctic, cold winter

C and D climates:
f - No dry season.
s - Dry summer. Driest summer month has less than 33% of the precipitation of wettest winter
w- Dry winter. Driest winter month has less than 10% of the precipitation of wettest summer month.
a - Hot summer. Mean temperature of warmest month above 22°C (71.6°F).
b - Mild summer. Mean temperature of warmest month below 22°C (71.6°F). At least four months with mean temperature above 10°C (50°F).
c - Cool summer. Mean temperature of warmest month below 22°C (71.6°F). Less than four months with mean temperature above 10°C (50°F).
d - (D climates only.) Mean temperature of coldest month below -38°C (-36.4°F).

E   Polar climates   Mean temperature of warmest month below 10°C (50°F).

| | ET | Tundra | Mean temperature of warmest month above 0°C (32°F). |
| | EF | Icecap | Mean temperature of all months below 0°C (32°F). |

Boundaries of A Climates

Am   97 cm   58 cm   62 cm

Af   SINGAPORE
MUMBAI

Aw   TIMBO

BSh   BENGHAZI

BSk   WILLISTON

BWh   ASWÂN

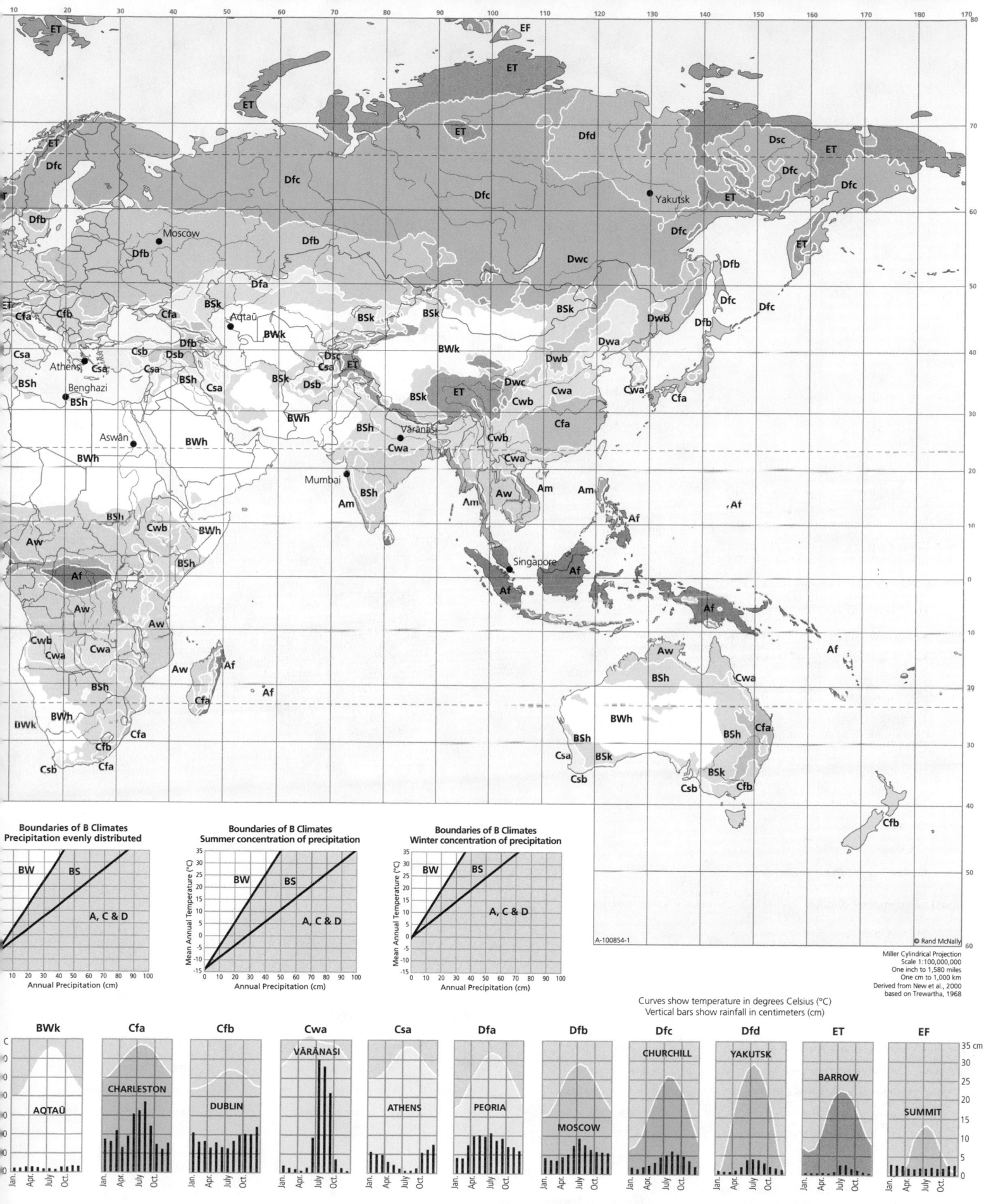

A-100854-1

Miller Cylindrical Projection
Scale 1:100,000,000
One inch to 1,580 miles
One cm to 1,000 km
Derived from New et al., 2000
based on Trewartha, 1968

Boundaries of B Climates
Precipitation evenly distributed

BW   BS

A, C & D

Annual Precipitation (cm)
10 20 30 40 50 60 70 80 90 100

Boundaries of B Climates
Summer concentration of precipitation

Mean Annual Temperature (°C)
35 30 25 20 15 10 5 0 -5 -10 -15

BW   BS

A, C & D

Annual Precipitation (cm)
0 10 20 30 40 50 60 70 80 90 100

Boundaries of B Climates
Winter concentration of precipitation

Mean Annual Temperature (°C)
35 30 25 20 15 10 5 0 -5 -10 -15

BW   BS

A, C & D

Annual Precipitation (cm)
0 10 20 30 40 50 60 70 80 90 100

Curves show temperature in degrees Celsius (°C)
Vertical bars show rainfall in centimeters (cm)

BWk — AQTAŪ
Cfa — CHARLESTON
Cfb — DUBLIN
Cwa — VĀRĀNASI
Csa — ATHENS
Dfa — PEORIA
Dfb — MOSCOW
Dfc — CHURCHILL
Dfd — YAKUTSK
ET — BARROW
EF — SUMMIT

Jan. Apr. July Oct.

35 cm
30
25
20
15
10
5

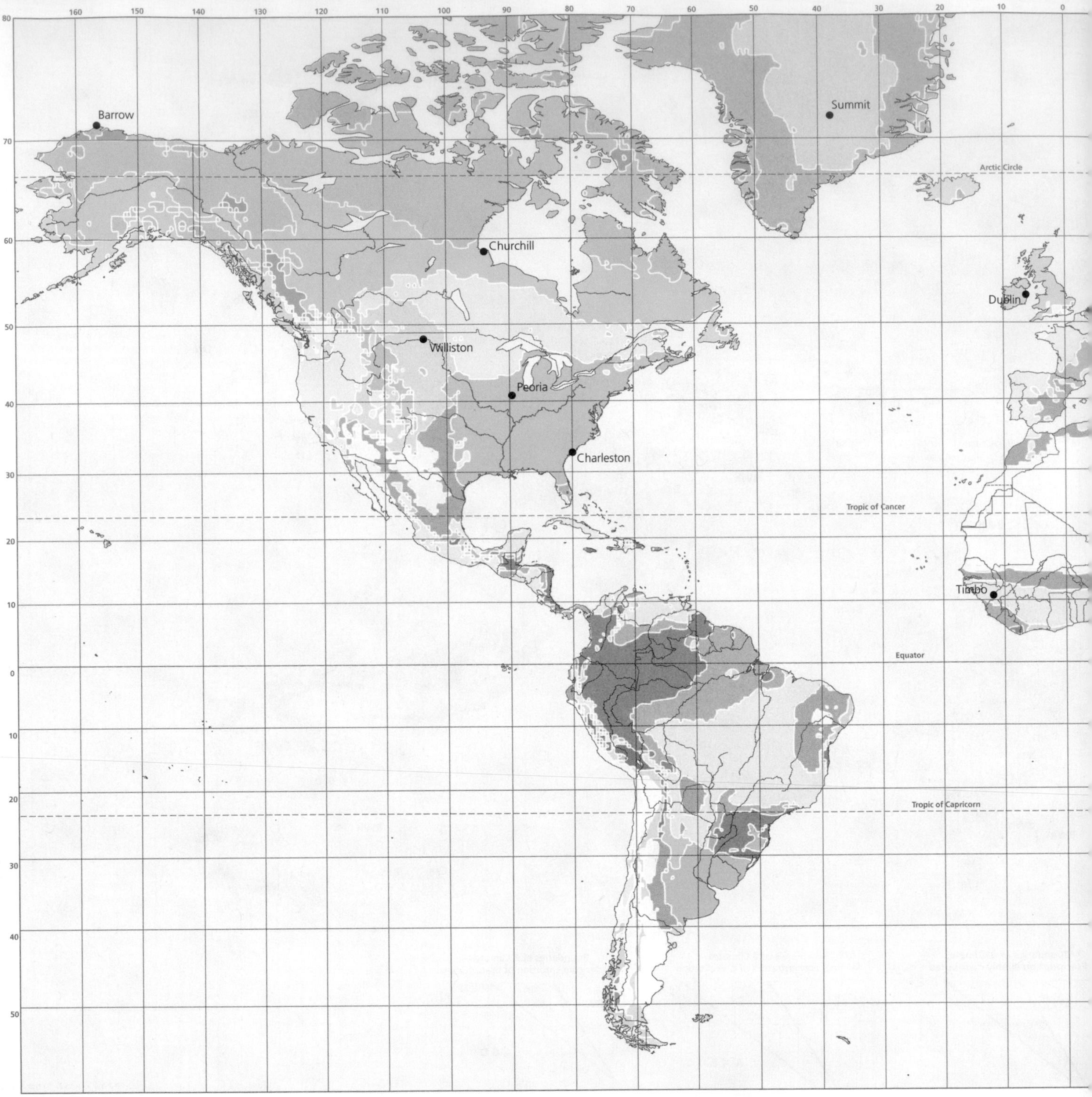

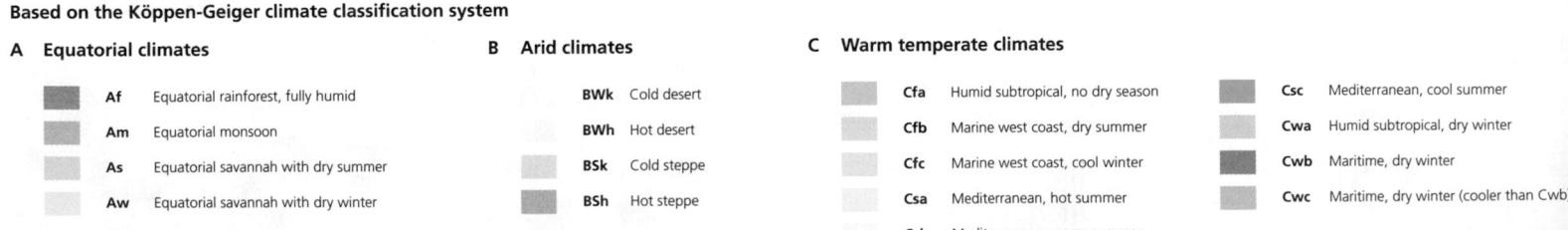

FUTURE CLIMATE REGIONS

Based on the Köppen-Geiger climate classification system

A   Equatorial climates

Af	Equatorial rainforest, fully humid
Am	Equatorial monsoon
As	Equatorial savannah with dry summer
Aw	Equatorial savannah with dry winter

B   Arid climates

BWk	Cold desert
BWh	Hot desert
BSk	Cold steppe
BSh	Hot steppe

C   Warm temperate climates

Cfa	Humid subtropical, no dry season
Cfb	Marine west coast, dry summer
Cfc	Marine west coast, cool winter
Csa	Mediterranean, hot summer
Csb	Mediterranean, warm summer

Csc	Mediterranean, cool summer
Cwa	Humid subtropical, dry winter
Cwb	Maritime, dry winter
Cwc	Maritime, dry winter (cooler than Cwb)

World Map of Köppen-Geiger climate classification projected with Tyndall temperature and precipitation data for the period 2076–2100, A1FI emission scenario

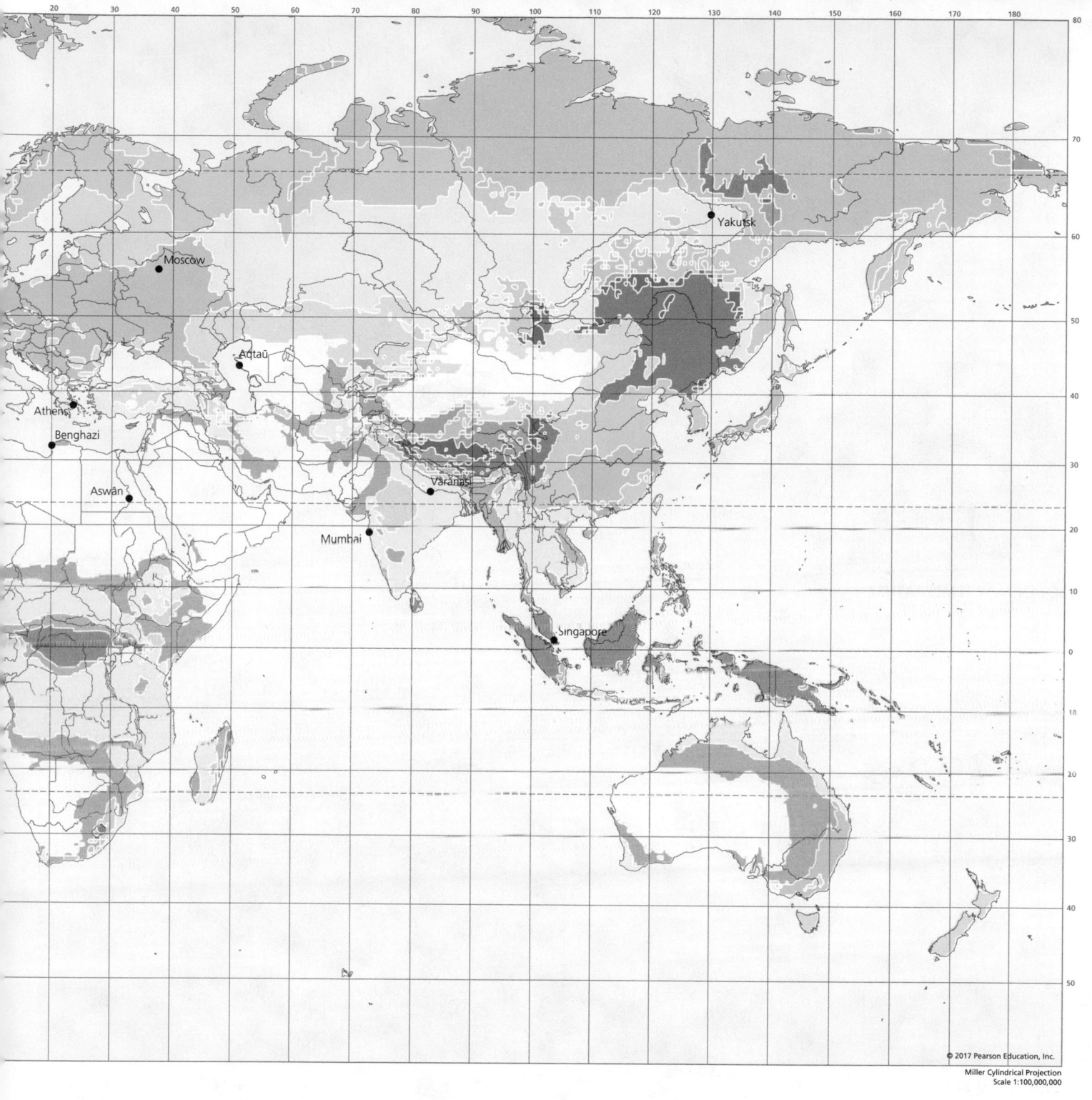

20 30 40 50 60 70 80 90 100 110 120 130 140 150 160 170 180

Moscow
Aqtaū
Athens
Benghazi
Aswān
Varanasi
Mumbai
Singapore
Yakutsk

© 2017 Pearson Education, Inc.

Miller Cylindrical Projection
Scale 1:100,000,000

D Snow climates

Dfa	Humid continental, hot summer
Dfb	Humid continental, warm summer
Dfc	Humid continental, cool summer
Dfd	Humid continental, very cold winter
Dsa	Continental, hot/dry summer
Dsb	Continental, warm/dry summer

Dsc	Continental, cool/dry summer
Dsd	Continental, dry summer/cold winter
Dwa	Continental, dry winter/hot summer
Dwb	Continental, dry winter/warm summer
Dwc	Continental, dry winter/cool summer
Dwd	Continental, dry/very cold winter

E Polar climates

EFrost
ETundra

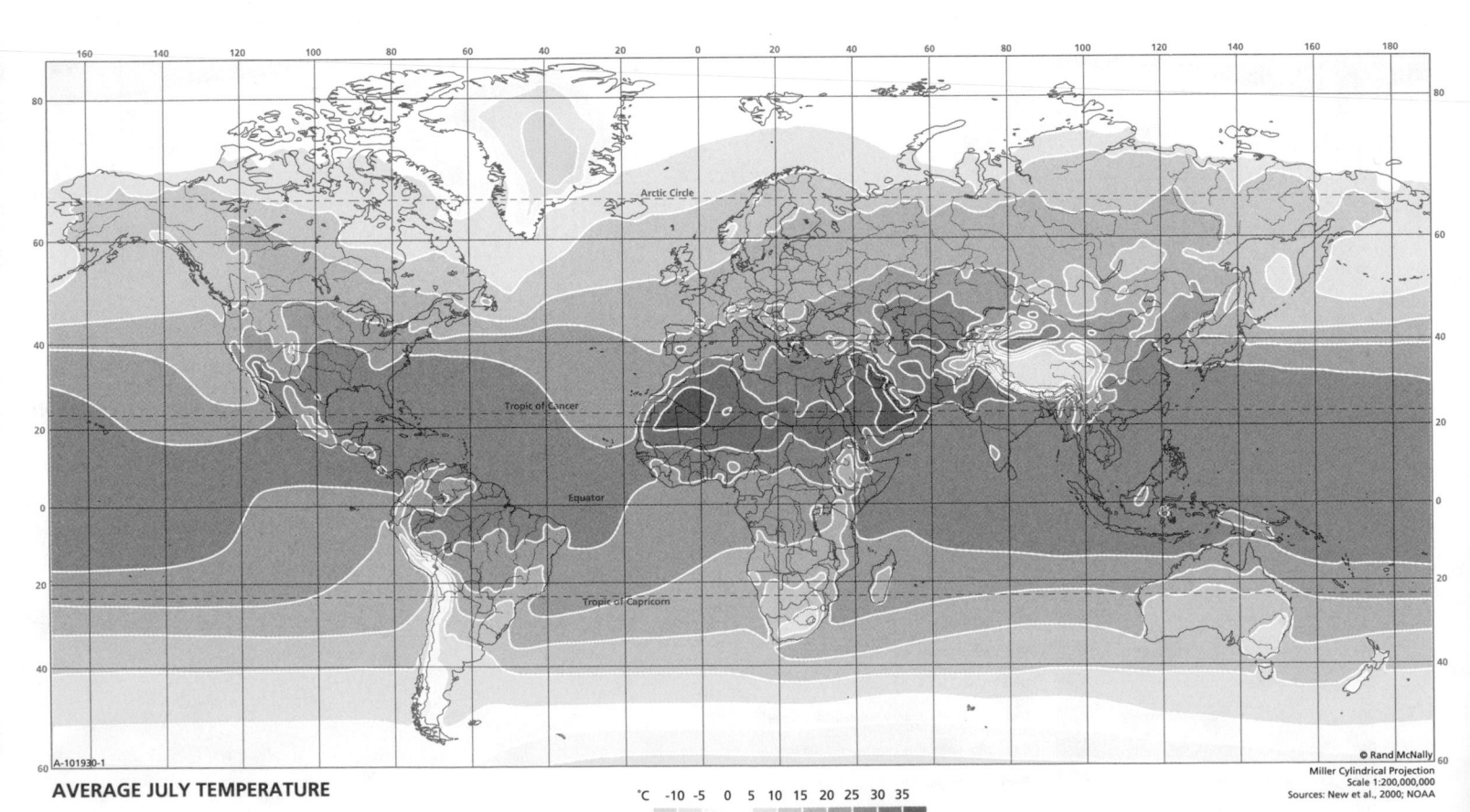

°C -45 -40 -35 -30 -25 -20 -15 -10 -5 0 5 10 15 20 25 30

°F -49 -40 -31 -22 -13 -4 5 14 23 32 41 50 59 68 77 86

AVERAGE JANUARY TEMPERATURE

Miller Cylindrical Projection
Scale 1:200,000,000
Sources: New et al., 2000; NOAA

© Rand McNally

A-101929-1

°C   -10 -5 0 5 10 15 20 25 30 35

°F   14 23 32 41 50 59 68 77 86 95

AVERAGE JULY TEMPERATURE

Miller Cylindrical Projection
Scale 1:200,000,000
Sources: New et al., 2000; NOAA

© Rand McNally

A-101930-1

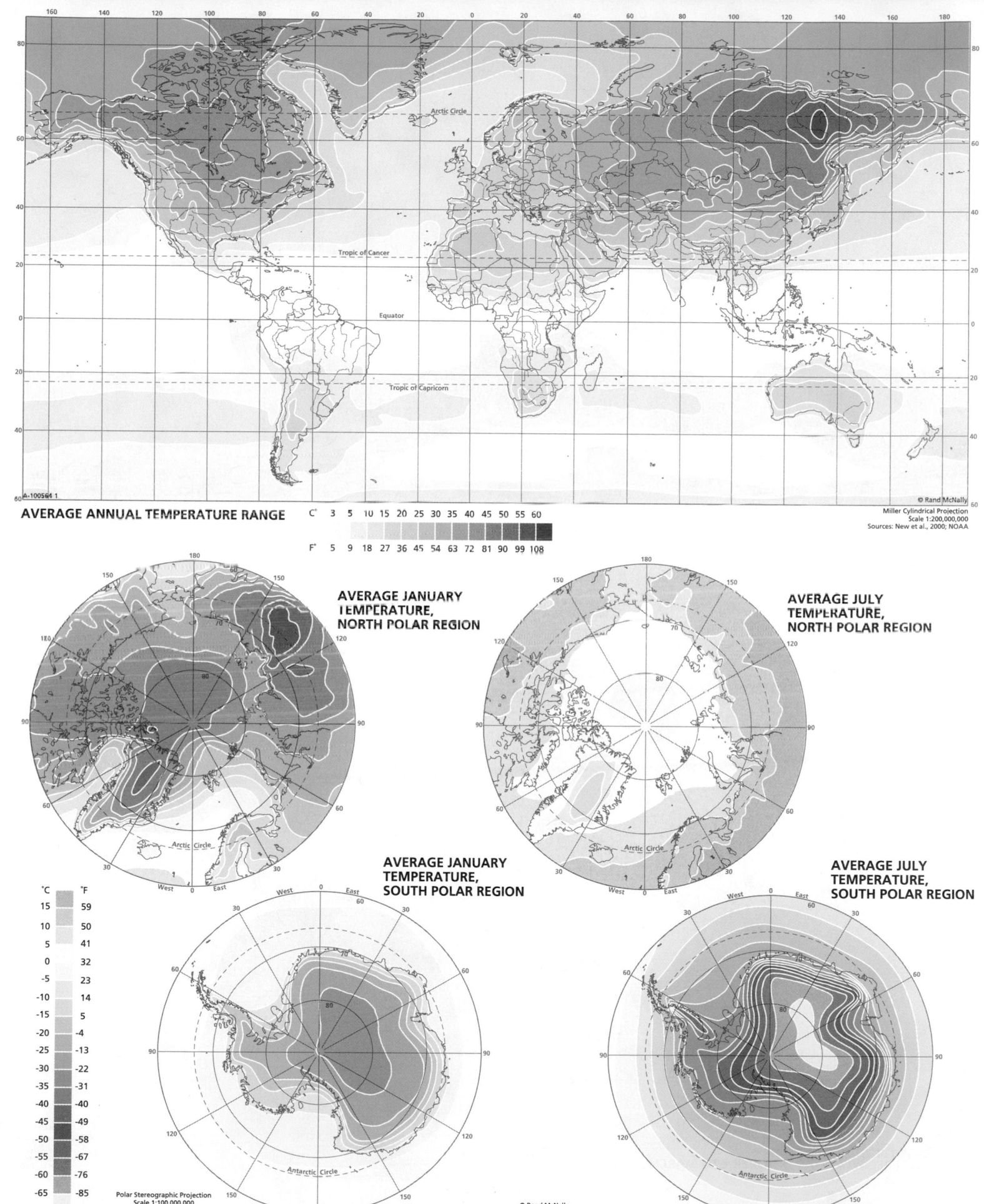

AVERAGE ANNUAL TEMPERATURE RANGE

C° 3 5 10 15 20 25 30 35 40 45 50 55 60

F° 5 9 18 27 36 45 54 63 72 81 90 99 108

Miller Cylindrical Projection
Scale 1:200,000,000
Sources: New et al., 2000; NOAA

© Rand McNally

AVERAGE JANUARY TEMPERATURE, NORTH POLAR REGION

AVERAGE JULY TEMPERATURE, NORTH POLAR REGION

AVERAGE JANUARY TEMPERATURE, SOUTH POLAR REGION

AVERAGE JULY TEMPERATURE, SOUTH POLAR REGION

°C	°F
15	59
10	50
5	41
0	32
-5	23
-10	14
-15	5
-20	-4
-25	-13
-30	-22
-35	-31
-40	-40
-45	-49
-50	-58
-55	-67
-60	-76
-65	-85

Polar Stereographic Projection
Scale 1:100,000,000
Sources: New et al., 2000; NOAA

© Rand McNally
A-101983-1

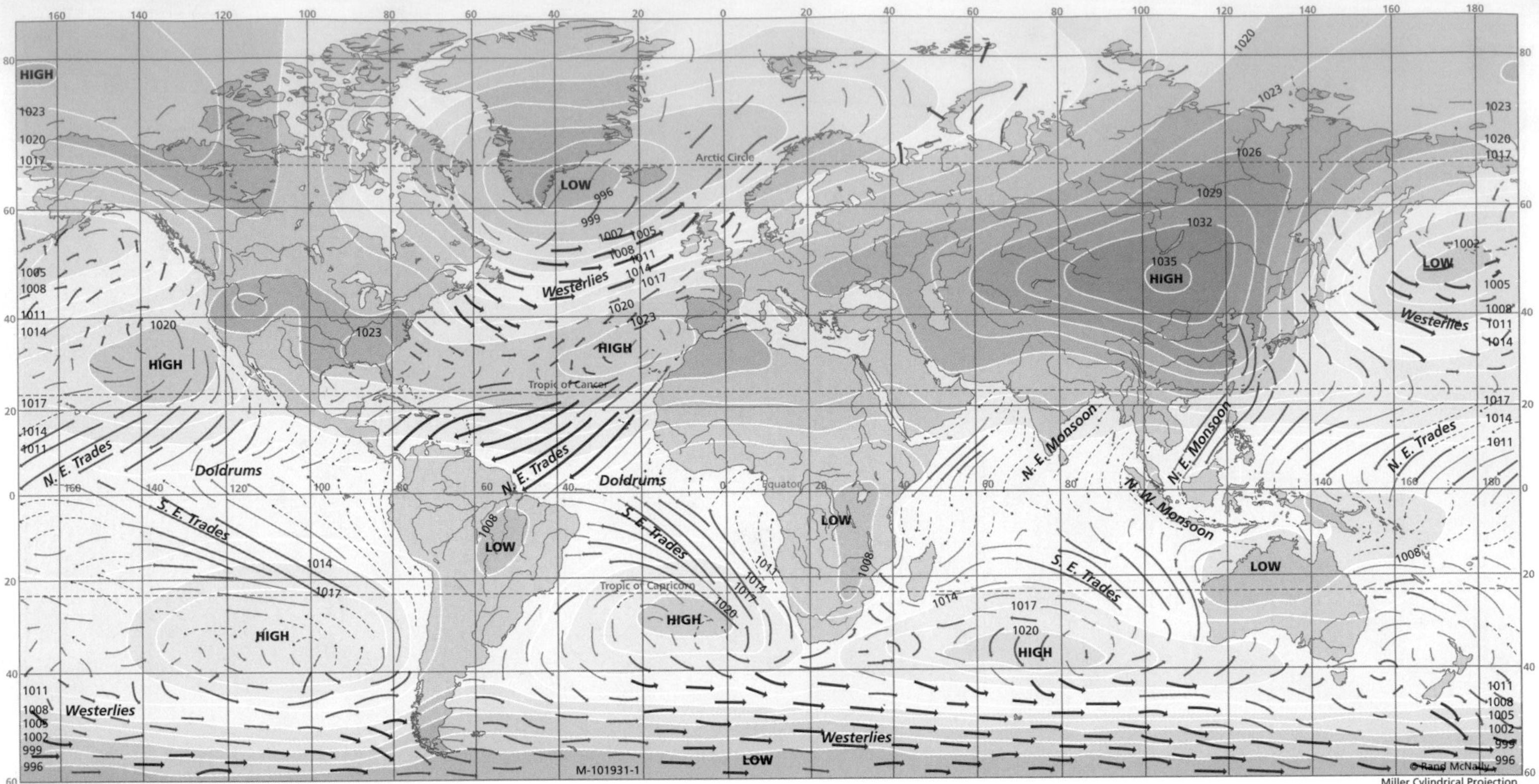

JANUARY PRESSURE AND PREDOMINANT WINDS

Atmospheric Pressure
in millibars (mb)

Normal sea-level pressure (1013.25 mb)

1032 1026 1020 1014 1008 1002 996

Isobars on map at intervals of 3 millibars

Wind Speed

Kilometers per hour (kph)	Miles per hour (mph)
0-16	0-10
19-24	10-15
24-40	15-25
Over 40	Over 25

Direction of arrow indicates dominant wind direction.
Length of arrow indicates steadiness of wind.

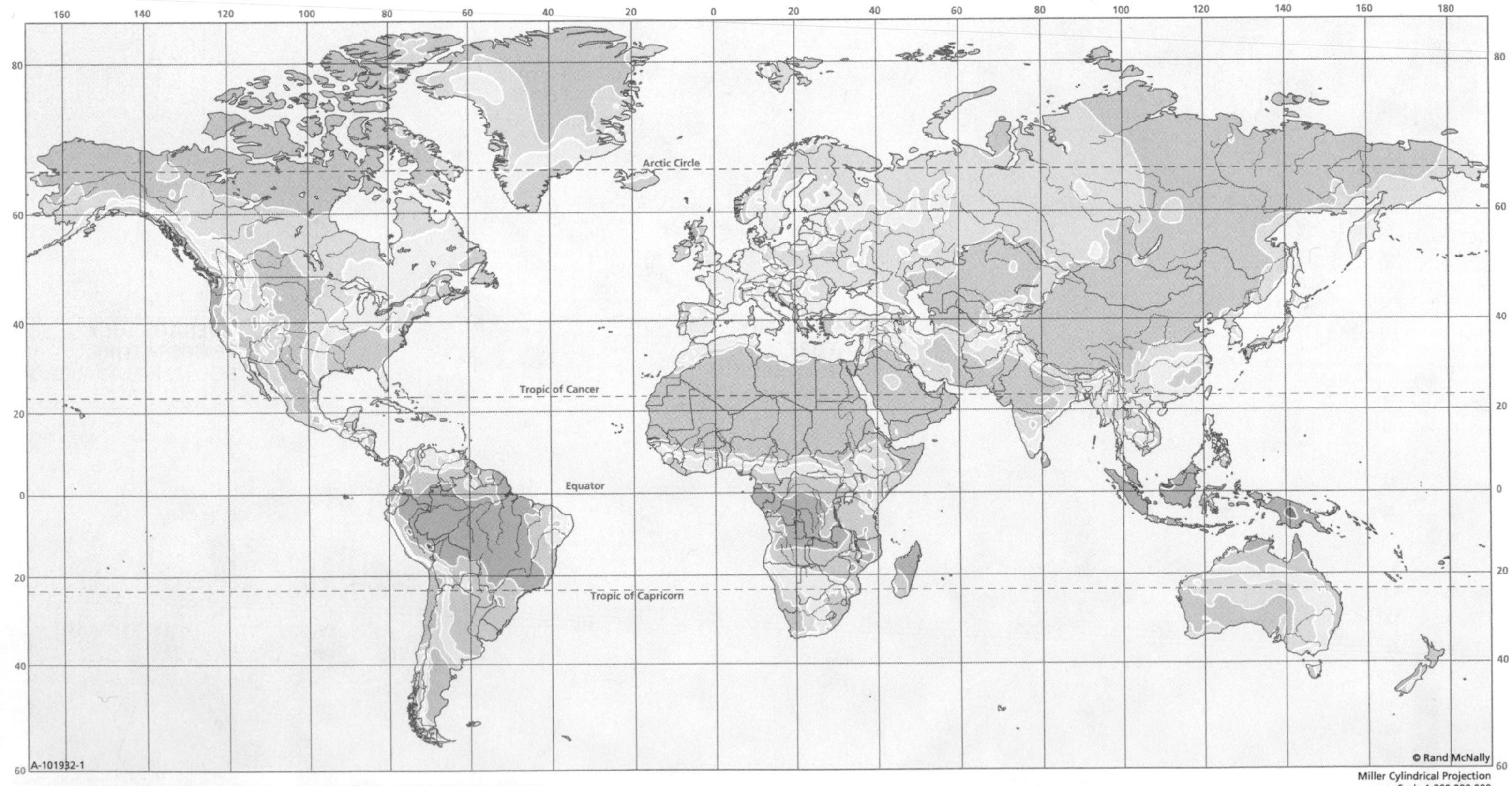

Miller Cylindrical Projection
Scale 1:200,000,000

AVERAGE PRECIPITATION - OCTOBER 1 TO MARCH 31

12.5 25 50 100 200 Centimeters

5 10 20 40 80 Inches

Miller Cylindrical Projection
Scale 1:200,000,000
Source: New et al., 2000

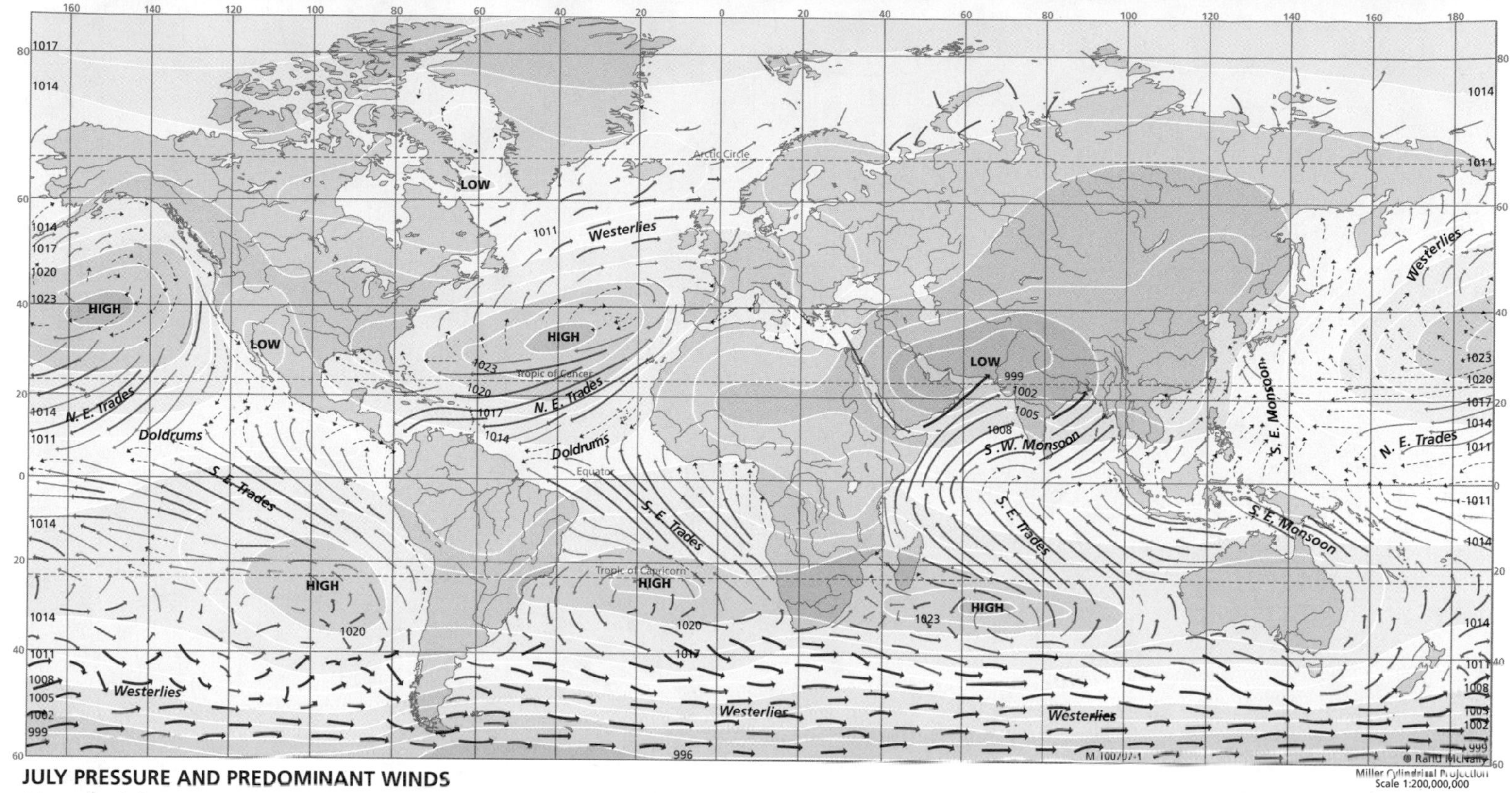

JULY PRESSURE AND PREDOMINANT WINDS

Text labels on map:

1017
1014

LOW

1014
1017
1020
1023 HIGH

LOW

Westerlies

1011

HIGH

1023 Tropic of Cancer
1020
999
1017
1008
1014

N. E. Trades

Doldrums

Doldrums
Equator

LOW 999
1002
1005

S. W. Monsoon

S. E. Monsoon

N. E. Trades

Westerlies

Westerlies

N. E. Trades

S. E. Trades

S. E. Trades

S. E. Monsoon

1014
1011

S. E. Trades

HIGH

Tropic of Capricorn

HIGH
1020
1011

HIGH
1023

1014

1014
1011
1008
1005
1002
999

Westerlies

Westerlies

Westerlies

996

Arctic Circle

© Rand McNally

M 100707-1

Miller Cylindrical Projection
Scale 1:200,000,000

Atmospheric Pressure
in millibars (mb)

Normal sea-level pressure (1013.25 mb)

| 1026 | 1020 | 1014 | 1008 | 1002 | 996 |

Isobars on map at intervals of 3 millibars

Wind Speed

Kilometers per hour (kph)	Miles per hour (mph)
0-16	0-10
16-24	10-15
24-40	15-25
Over 40	Over 25

Direction of arrow indicates dominant wind direction.
Length of arrow indicates steadiness of wind.

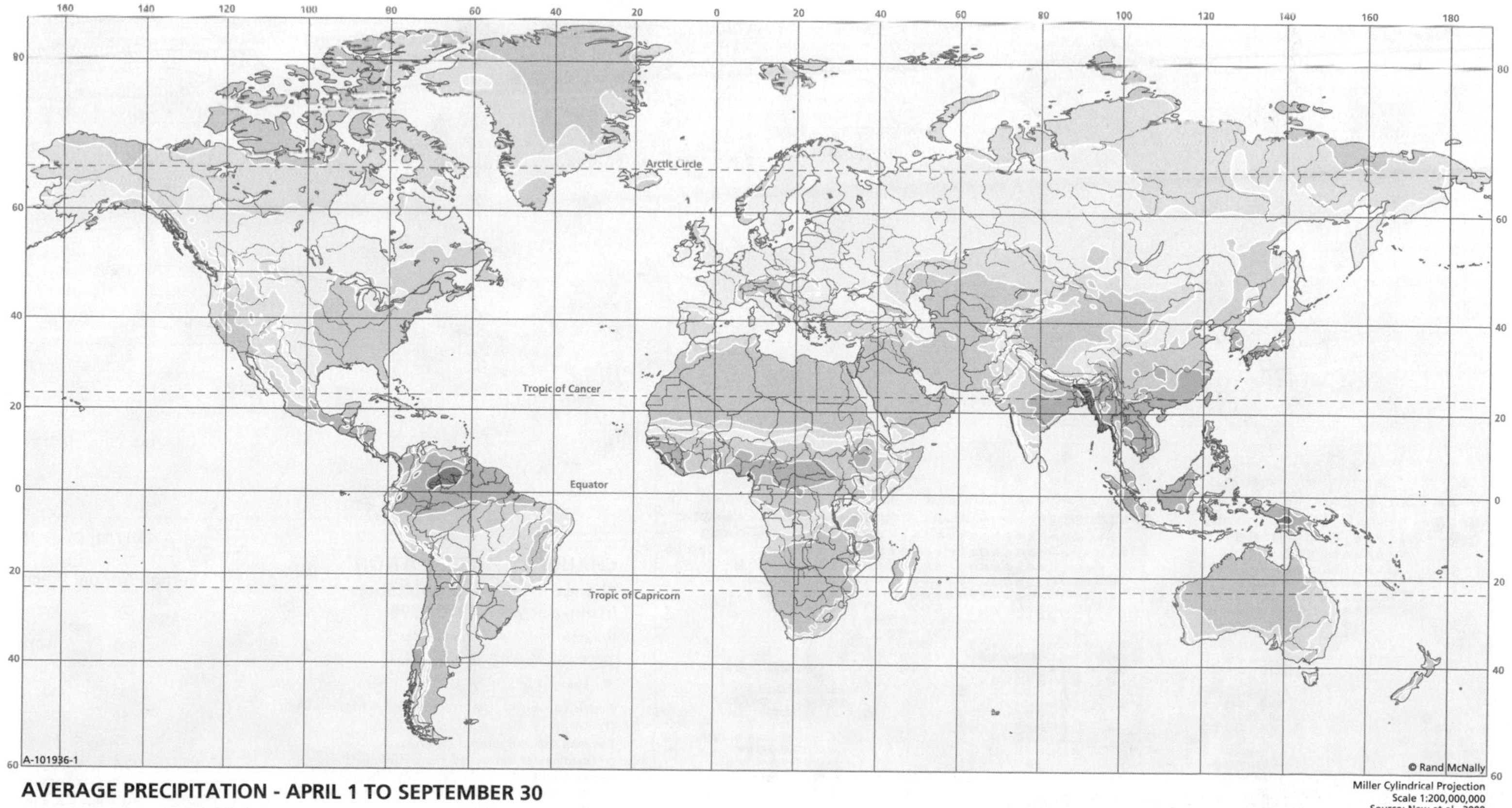

AVERAGE PRECIPITATION - APRIL 1 TO SEPTEMBER 30

Map labels: Arctic Circle, Tropic of Cancer, Equator, Tropic of Capricorn

A-101936-1

© Rand McNally

Miller Cylindrical Projection
Scale 1:200,000,000
Source: New et al., 2000

| 12.5 | 25 | 50 | 100 | 200 | Centimeters |
| 5 | 10 | 20 | 40 | 80 | Inches |

26

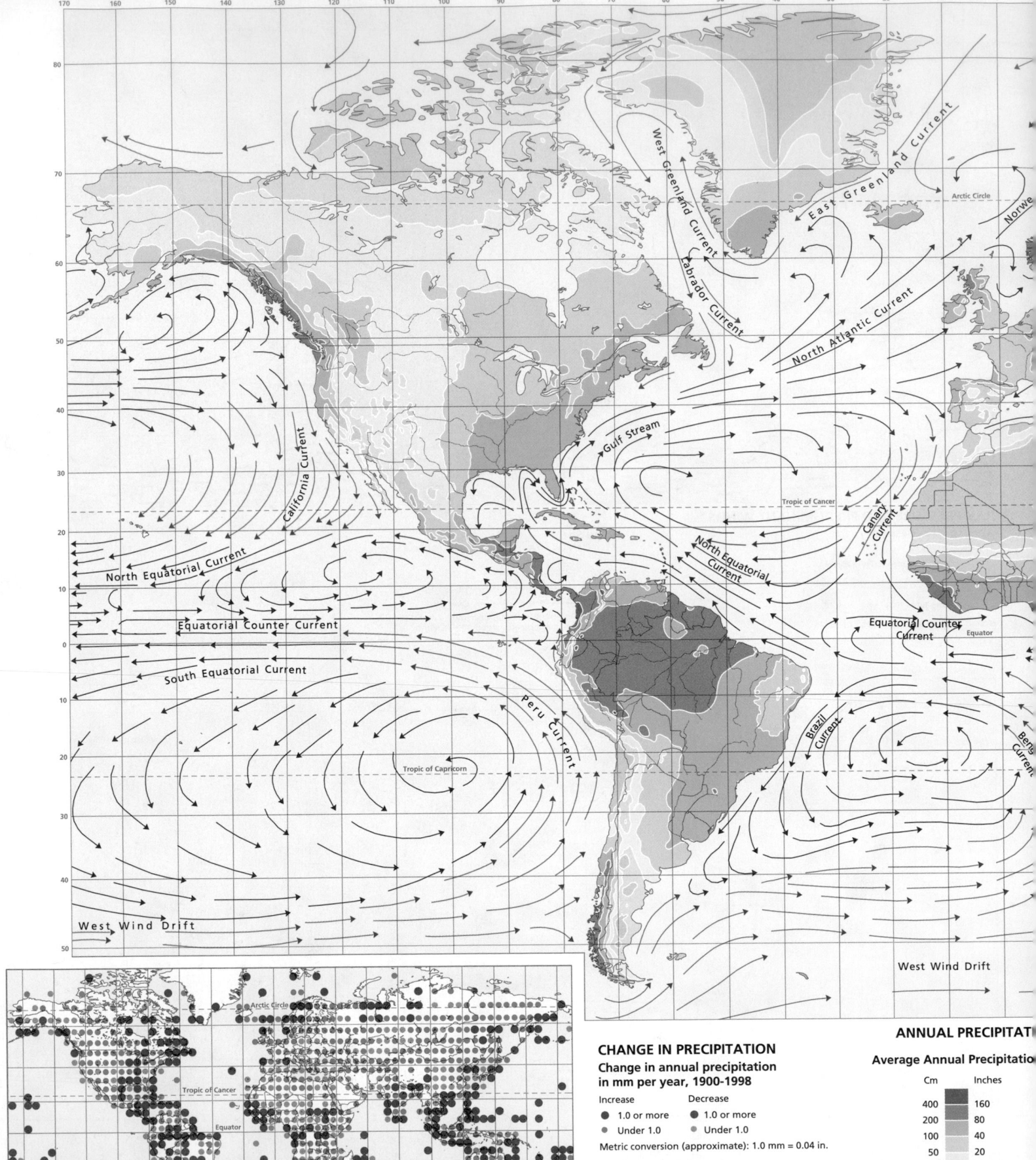

CHANGE IN PRECIPITATION

Change in annual precipitation in mm per year, 1900-1998

Increase
- ● 1.0 or more
- ● Under 1.0

Decrease
- ● 1.0 or more
- ● Under 1.0

Metric conversion (approximate): 1.0 mm = 0.04 in.

This map shows the trend in annual precipitation for the period 1900-1998. Each symbol on the map is a 5-degree by 5-degree grid cell. The trend for each cell was computed by fitting a regression line to the data.

Derived from Hulme, 1998

ANNUAL PRECIPITAT

Average Annual Precipitatio

Cm	Inches
400	160
200	80
100	40
50	20
25	10
12.5	5

Source: New et al., 2000

West Greenland Current
East Greenland Current
Labrador Current
Norwe
Arctic Circle
North Atlantic Current
Gulf Stream
California Current
Canary Current
Tropic of Cancer
North Equatorial Current
North Equatorial Current
Equatorial Counter Current
Equatorial Counter Current
Equator
South Equatorial Current
Peru Current
Brazil Current
Beng Current
Tropic of Capricorn
West Wind Drift
West Wind Drift

Miller Cylindrical Projection
Scale 1:350,000,000

A-101935-1

© Rand McNally

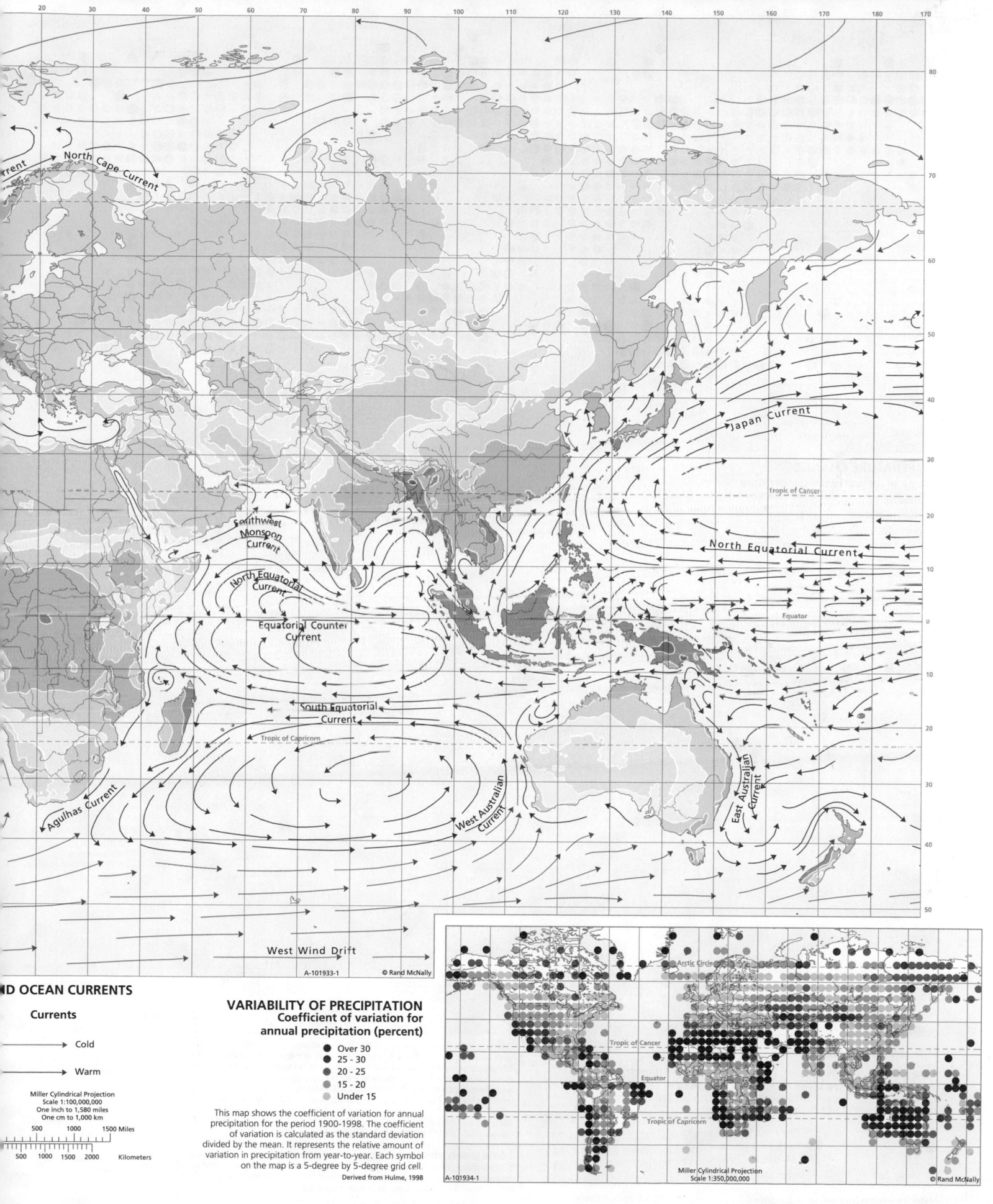

North Cape Current

North Cape Current

Japan Current

Tropic of Cancer

North Equatorial Current

Southwest
Monsoon
Current

North Equatorial
Current

Equatorial Counter
Current

Equator

South Equatorial
Current

Tropic of Capricorn

West Australian
Current

East Australian
Current

Agulhas Current

West Wind Drift

A-101933-1   © Rand McNally

D OCEAN CURRENTS

Currents

→ Cold

→ Warm

Miller Cylindrical Projection
Scale 1:100,000,000
One inch to 1,580 miles
One cm to 1,000 km

500   1000   1500 Miles

500   1000   1500   2000   Kilometers

VARIABILITY OF PRECIPITATION
Coefficient of variation for
annual precipitation (percent)

- ● Over 30
- ● 25 - 30
- ● 20 - 25
- ● 15 - 20
- ● Under 15

This map shows the coefficient of variation for annual
precipitation for the period 1900-1998. The coefficient
of variation is calculated as the standard deviation
divided by the mean. It represents the relative amount of
variation in precipitation from year-to-year. Each symbol
on the map is a 5-degree by 5-degree grid cell.

Derived from Hulme, 1998

Arctic Circle

Tropic of Cancer

Equator

Tropic of Capricorn

A-101934-1   Miller Cylindrical Projection
Scale 1:350,000,000   © Rand McNally

© 2017 Pearson Education, Inc.

Plate Carrée Projection
Scale 1:200,000,000
Source: ClimateDataGuide, NCAR

TEMPERATURE CHANGE

Change of annual mean temperature
in Celsius degrees (C°) between 1901:1920 and 1991:2010

Temperature increase	Temperature decrease
● Over 1.00	● Over 0.50
● 0.50 - 1.00	● Under 0.50
● Under 0.50	

Temperature conversion (approximate): 0.1 C° = 0.18 F°; 0.2 C° = 0.36 F°

This map is derived from the HadCRUT4 dataset. This data takes the average temperature for the ten-year period 1901-1920 and the average temperature for the period 1991-2010. This determines whether the temperature has increased or decreased (in degrees Celsius) in the last 90 years. The global average is 0.8 degrees.

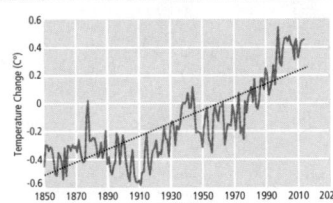

Average Annual Global Temperature Trend, 1850-2014

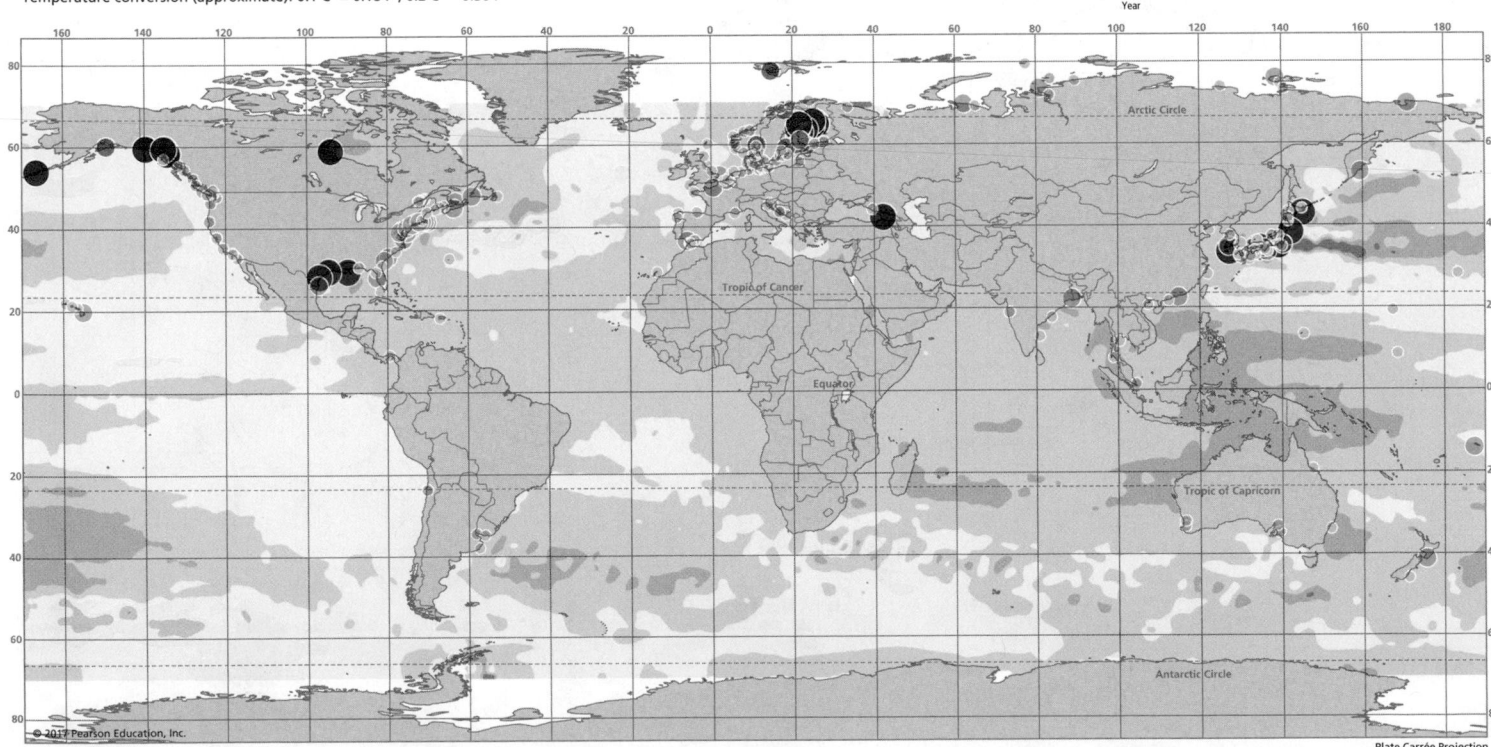

© 2017 Pearson Education, Inc.

Plate Carrée Projection
Scale 1:200,000,000
Sources: NOAA Laboratory
for Satellite Altimetry;
Permanent Service for
Mean Sea Level (PSMSL), 2016

SEA LEVEL CHANGE

Tide Gauge Data
Trend in relative sea level
in mm per year, for stations operational through 2010, with at least 50 years of data

Sea level increase	Sea level decrease
● Over 5.00	● Over 5.00
● 2.50 - 5.00	● 2.50 - 5.00
● Under 2.50	● Under 2.50

Satellite Altimetry Data
Change in observed sea level
in mm per year, 1992-2016

Sea level increase	Sea level decrease
Over 7.5	Over 7.5
5.0 - 7.5	5.0 - 7.5
2.5 - 5.0	2.5 - 5.0
Under 2.5	Under 2.5

Metric conversion (approximate): 2.5 mm = 0.1 in.; 5.0mm = 0.2 in.; 7.5 mm = 0.3 in.

Tide gauges provide a long-term record of sea level change. The record extends for 200 years in some cases. However, stations are clustered spatially and do not cover the entire globe. Also, since tide gauges measure relative sea level (water level relative to the land surface), they cannot differentiate changes in water volume (due to thermal expansion and ice melting) from changes in land elevation (due to tectonic activity and glacial isostatic adjustment). Still, tide gauges are important because relative sea level has a direct impact on coastal environments.

The tide gauge data on this map are from Permanent Service for Mean Sea Level (PSMSL), 2016. Gauges shown are still operating in 2010 and have at least 50 years of data.

Satellite altimetry offers a second method of assessing sea level change. Unlike tide gauges, coverage is nearly global. Also, observed changes in sea level are largely unaffected by land elevation changes. However, the satellite altimetry record extends back to only the 1990s. As a result, the data record reflects major decadal variations rather than long-term trends.

The satellite altimetry data on this map are from TOPEX/Poseidon and Jason-,1 and Jason-2 sensors.

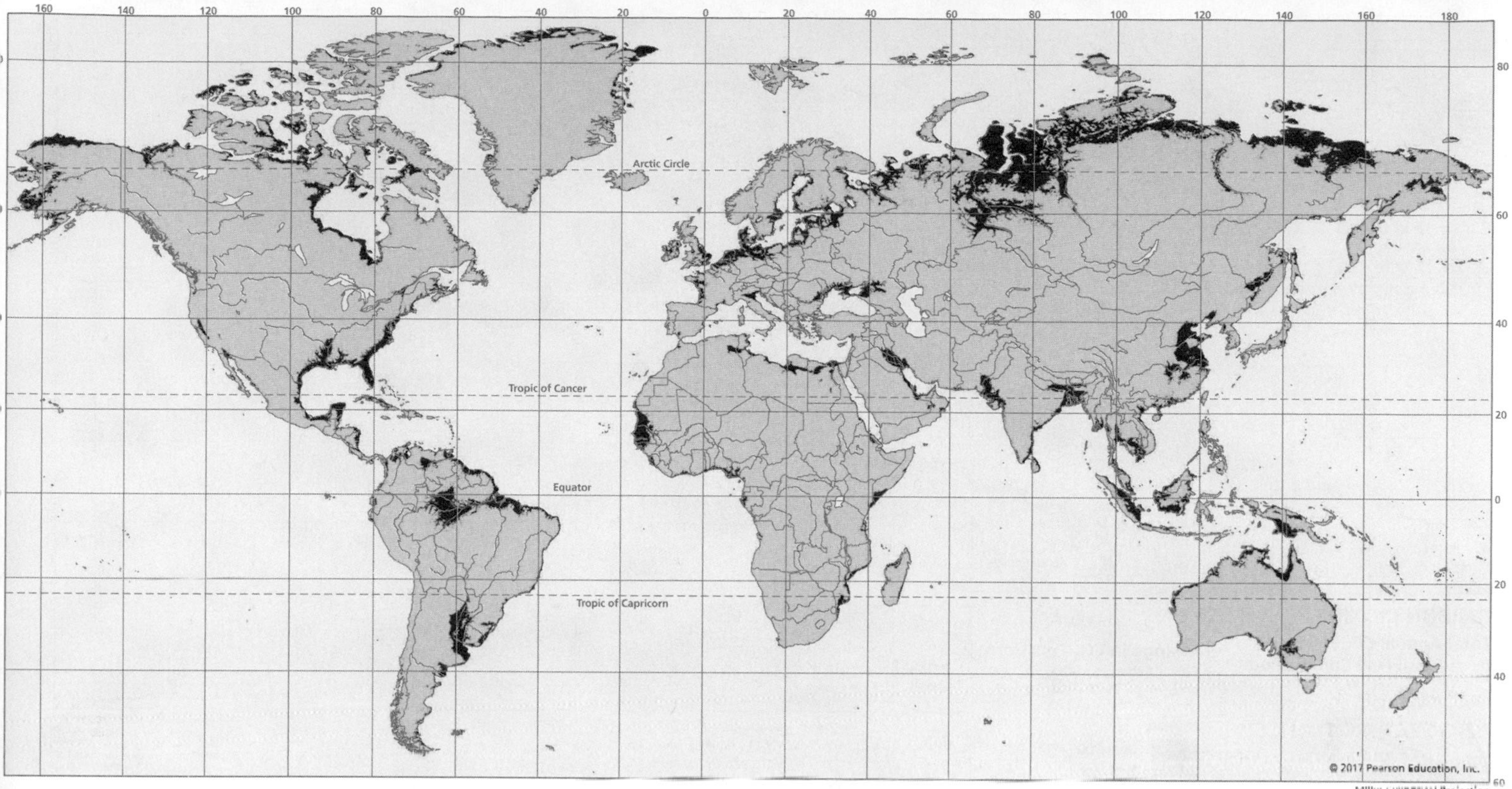

AREAS VULNERABLE TO SEA LEVEL RISE

Warming of Earth's atmosphere is causes ocean levels to rise. This map reflects areas vulnerable to a one-meter rise in sea level. Coastal lowlands are particularly vulnerable as are inland areas along major rivers that emptying into the ocean.

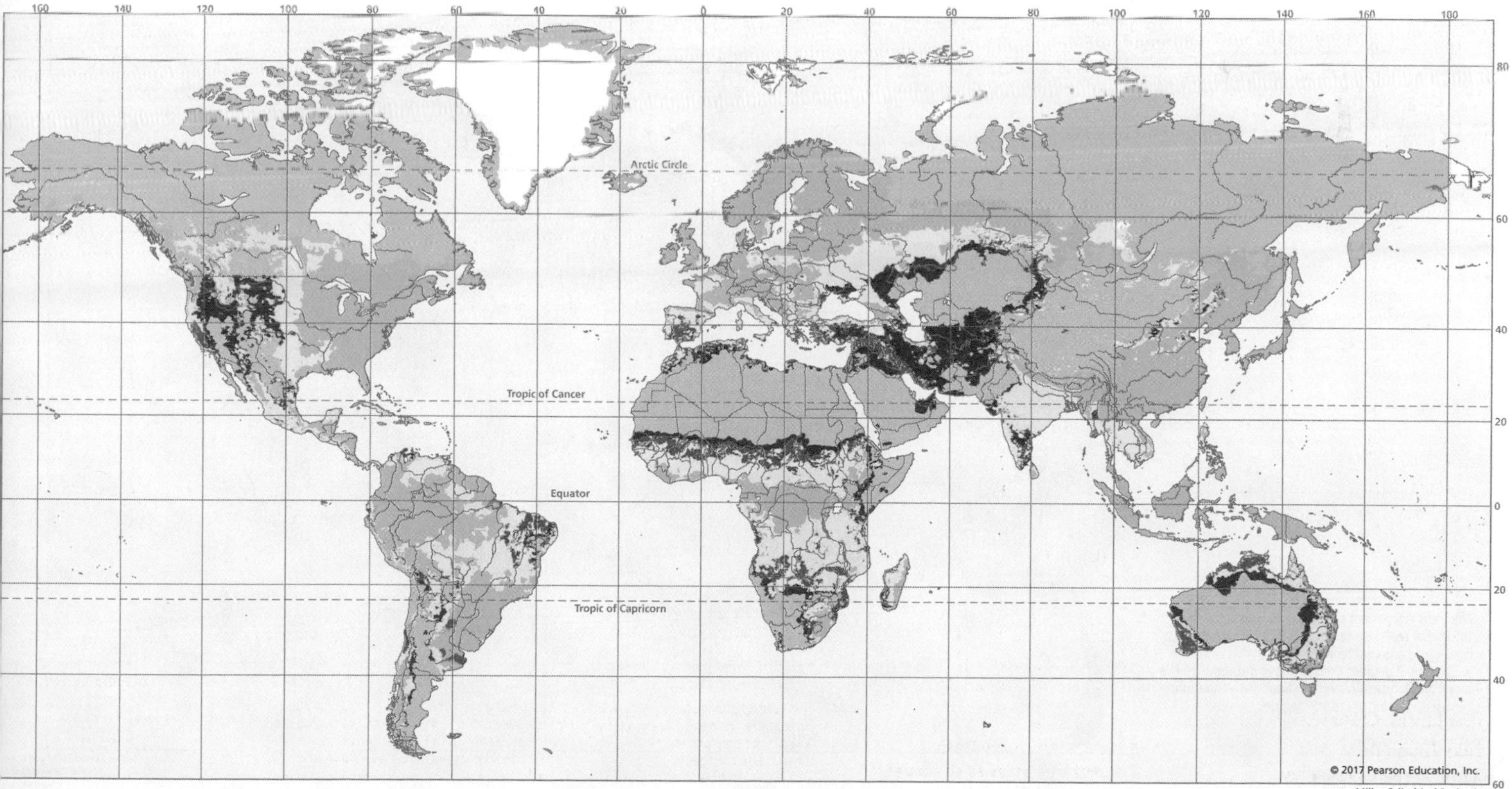

© 2017 Pearson Education, Inc.
Miller Cylindrical Projection
Scale 1:200,000,000

DESERTIFICATION VULNERABILITY

Vulnerability
- Low
- Moderate
- High
- Very high

Other regions
- Dry
- Cold
- Humid/not vulnerable
- Ice/glacier

The transformation of and area to arid conditions, with loss of vegetation and waterbodies, is desertification. Many factors can cause desertification including climate shifts, droughts, overgrazing by livestock and land degradation by human activities including deforestation and agriculture. Semiarid locations at the margins of deserts are particularly vulnerable to desertification.

© 2017 Pearson Education, Inc.
Miller Cylindrical Projection
Scale 1:200,000,000
Source: USDA

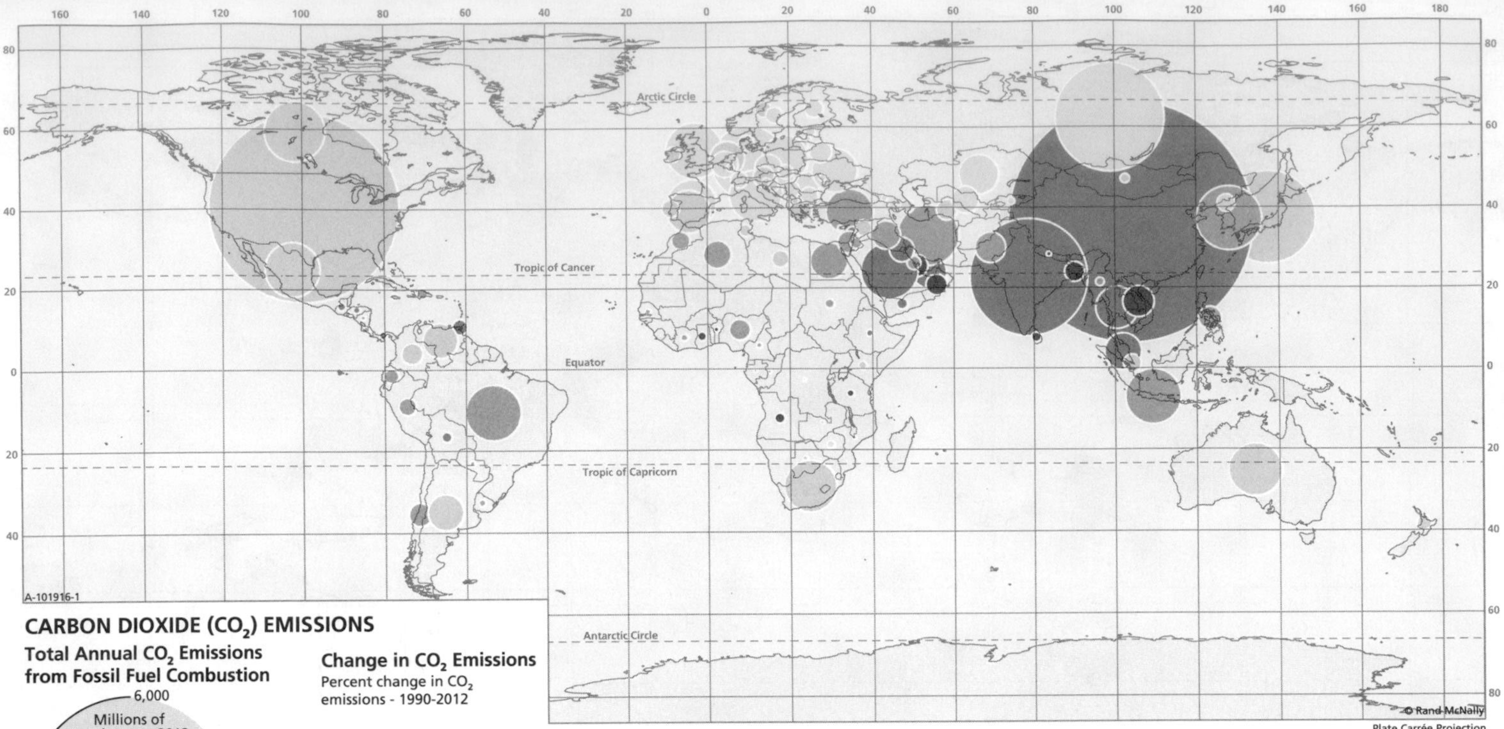

CARBON DIOXIDE (CO₂) EMISSIONS

Total Annual CO₂ Emissions from Fossil Fuel Combustion

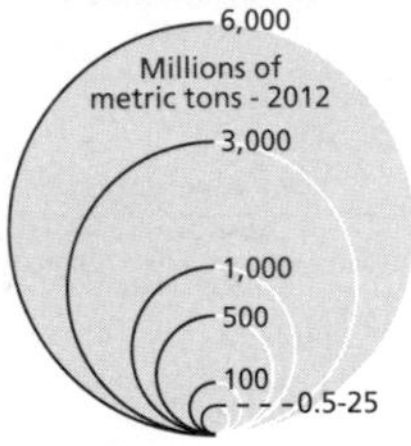

6,000

Millions of metric tons - 2012

3,000

1,000

500

100
0.5-25

Countries with emissions below 500,000 metric tons are not shown.

Change in CO₂ Emissions

Percent change in CO₂ emissions - 1990-2012

Increase

Over 300

200 - 300

100 - 199

Under 100

Decrease

Under 100

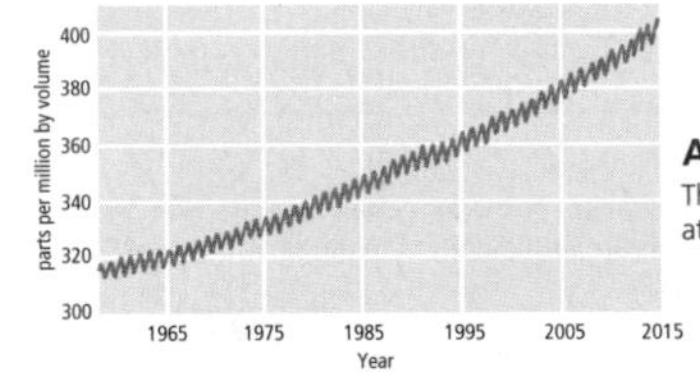

A-101916-1

Plate Carrée Projection
Scale 1:200,000,000
Sources: Energy Information
Administration; Scripps Institution of
Oceanography's The Keeling Curve

© Rand McNally

Atmospheric CO₂ Concentrations - 1958-2015

This graph (the "Keeling Curve") shows the rising level of CO₂ in the atmosphere, as well as the seasonal pattern of CO₂ uptake by plants.

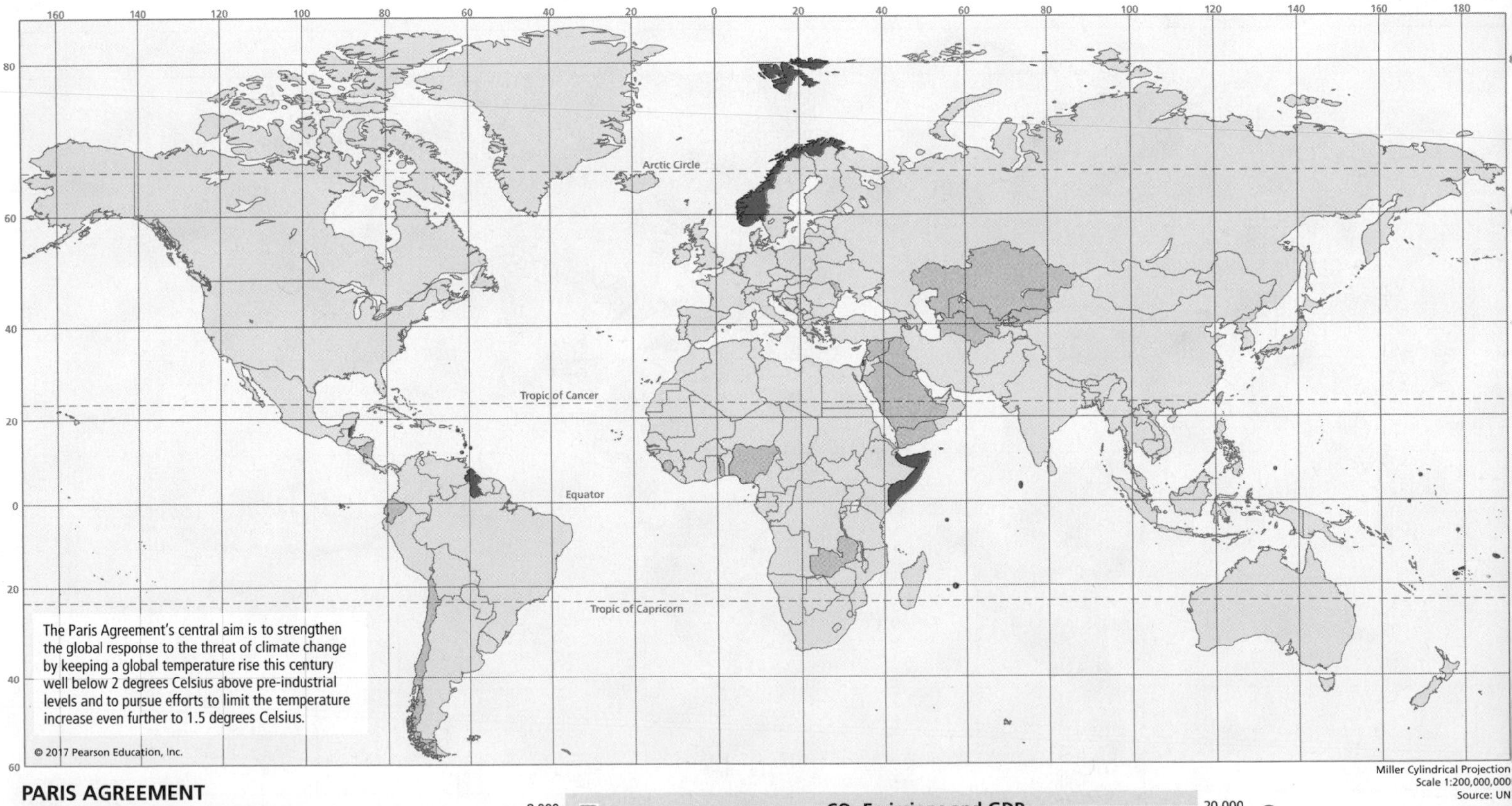

(graph: parts per million by volume, 300–400, vs Year 1965–2015)

The Paris Agreement's central aim is to strengthen the global response to the threat of climate change by keeping a global temperature rise this century well below 2 degrees Celsius above pre-industrial levels and to pursue efforts to limit the temperature increase even further to 1.5 degrees Celsius.

© 2017 Pearson Education, Inc.

Miller Cylindrical Projection
Scale 1:200,000,000
Source: UN

PARIS AGREEMENT

United Nations Framework Convention on Climate Change, 12 December 2015

Participant's action as of 22 April 2016

Signature

Ratification

Countries that have not signed or ratified the agreement

No data

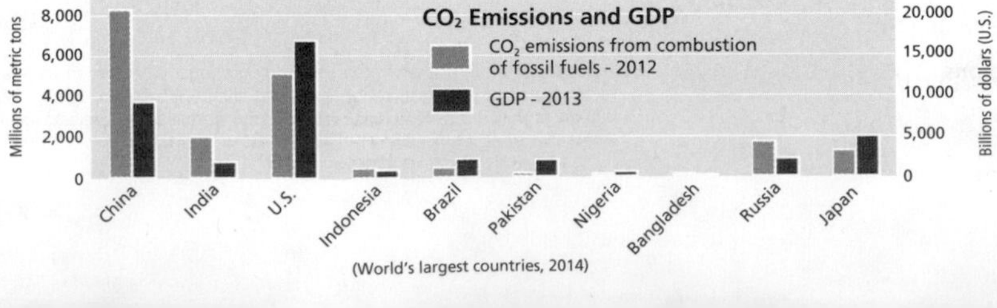

CO₂ Emissions and GDP

CO₂ emissions from combustion of fossil fuels - 2012

GDP - 2013

(bar chart: Millions of metric tons, 0–8,000; Billions of dollars (U.S.), 0–20,000; countries: China, India, U.S., Indonesia, Brazil, Pakistan, Nigeria, Bangladesh, Russia, Japan)

(World's largest countries, 2014)

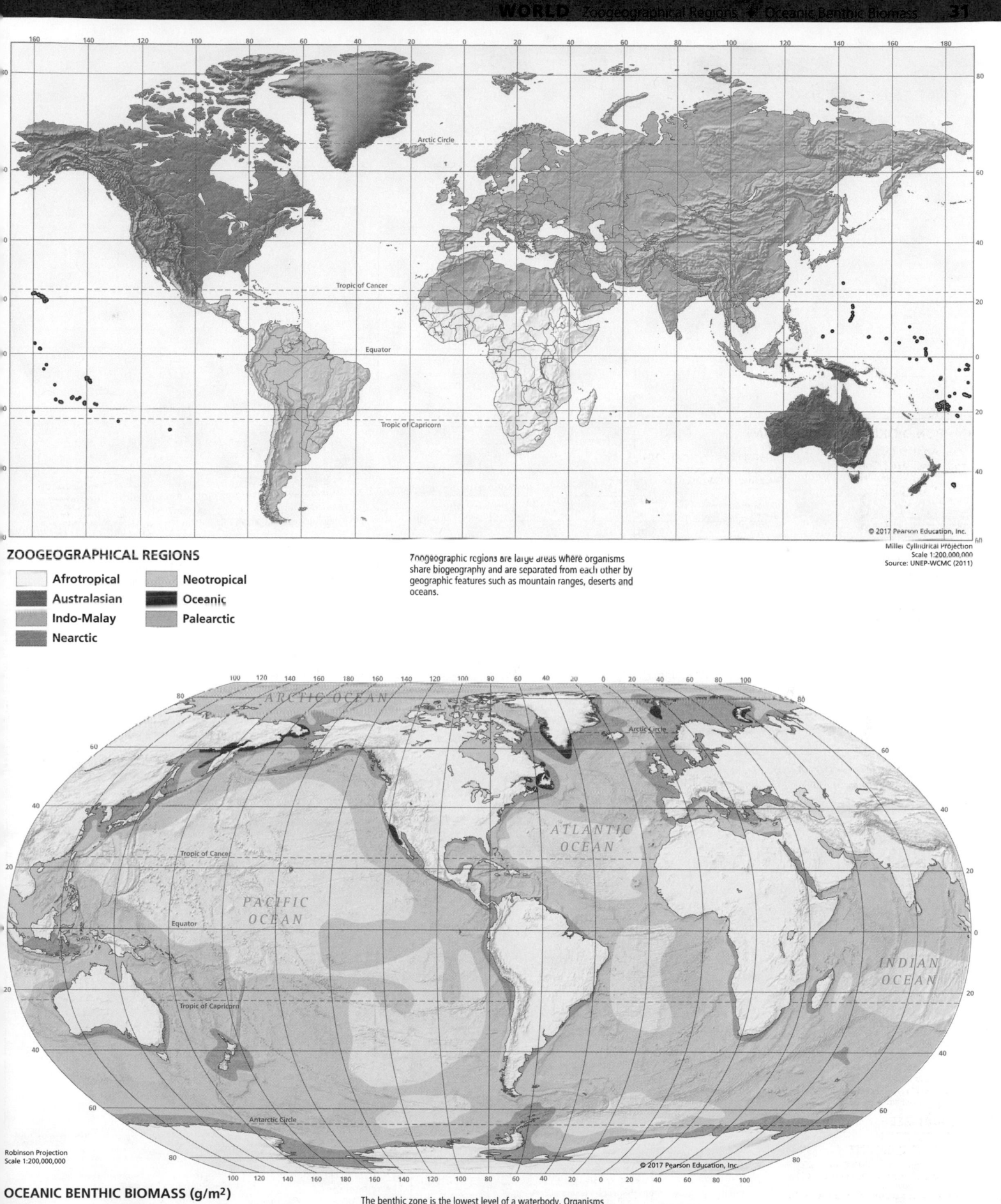

ZOOGEOGRAPHICAL REGIONS

- Afrotropical
- Australasian
- Indo-Malay
- Nearctic
- Neotropical
- Oceanic
- Palearctic

Zoogeographic regions are large areas where organisms share biogeography and are separated from each other by geographic features such as mountain ranges, deserts and oceans.

Miller Cylindrical Projection
Scale 1:200,000,000
Source: UNEP-WCMC (2011)

© 2017 Pearson Education, Inc.

Robinson Projection
Scale 1:200,000,000

© 2017 Pearson Education, Inc.

OCEANIC BENTHIC BIOMASS (g/m²)

- <0.1
- 0.1–10
- 10–300
- >300

The benthic zone is the lowest level of a waterbody. Organisms living in this zone rely on organic matter from higher in the water. Benthic organisms are more plentiful in shallower coastal waters where dead food material is more plentiful.

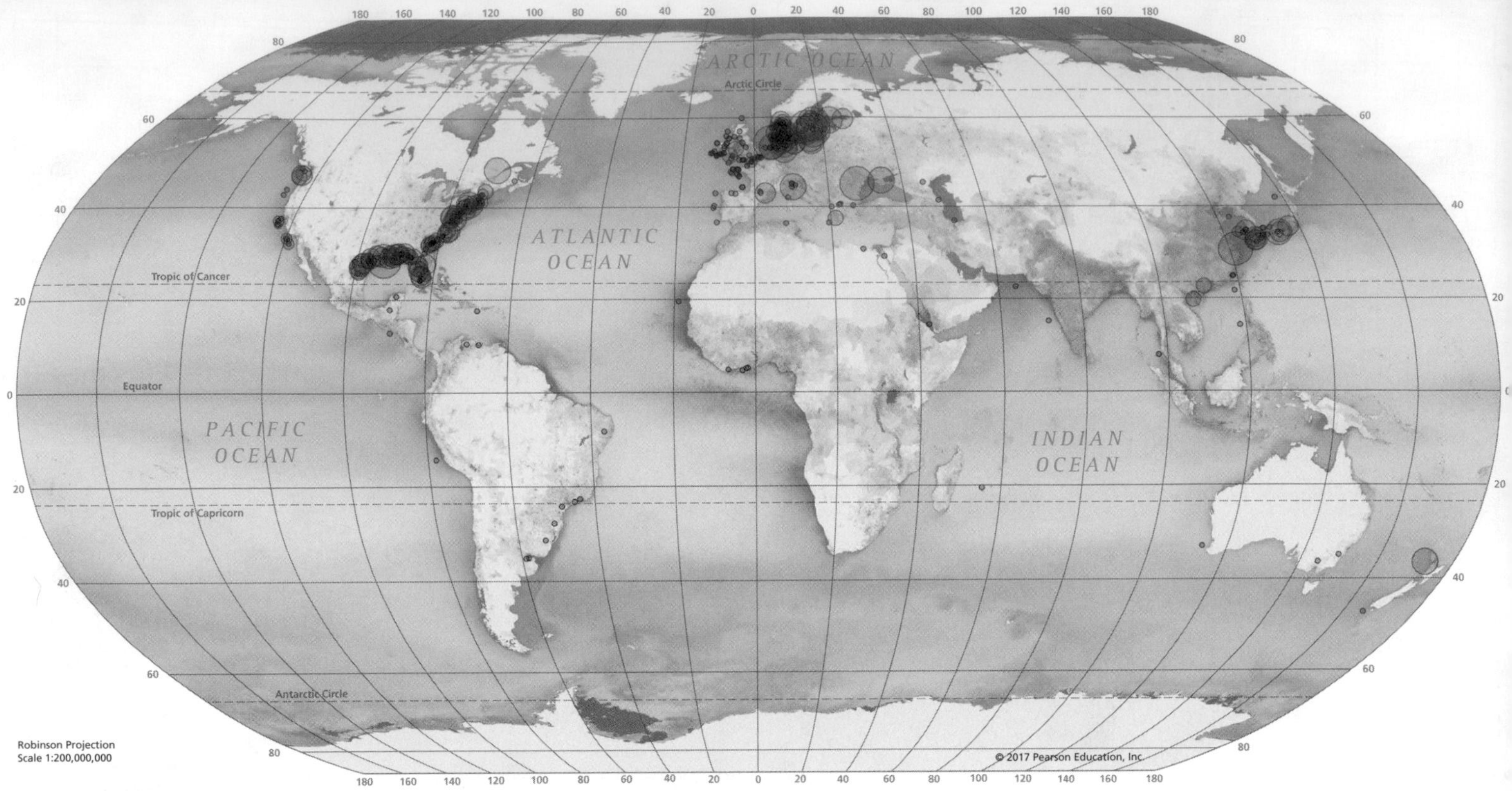

ARCTIC OCEAN

Arctic Circle

ATLANTIC OCEAN

PACIFIC OCEAN

INDIAN OCEAN

Tropic of Cancer

Equator

Tropic of Capricorn

Antarctic Circle

Robinson Projection
Scale 1:200,000,000

© 2017 Pearson Education, Inc.

WORLDWIDE DEAD ZONES

Dead zones are areas of oxygen-poor water near the mouths of major rivers. Delivery by rivers of excess nutrients to the ocean results in algae blooms. When these blooms die and decompose, they rob the water of oxygen and marine organisms suffocate.

Particulate organic carbon (mg/m^3)

10   20   50   100   200   500   1,000

Population density (persons/km^2)

1   10   100   1,000   10k   100k

Dead zone size (km^2)

unkown   0.1   1   10   100   1k   10k

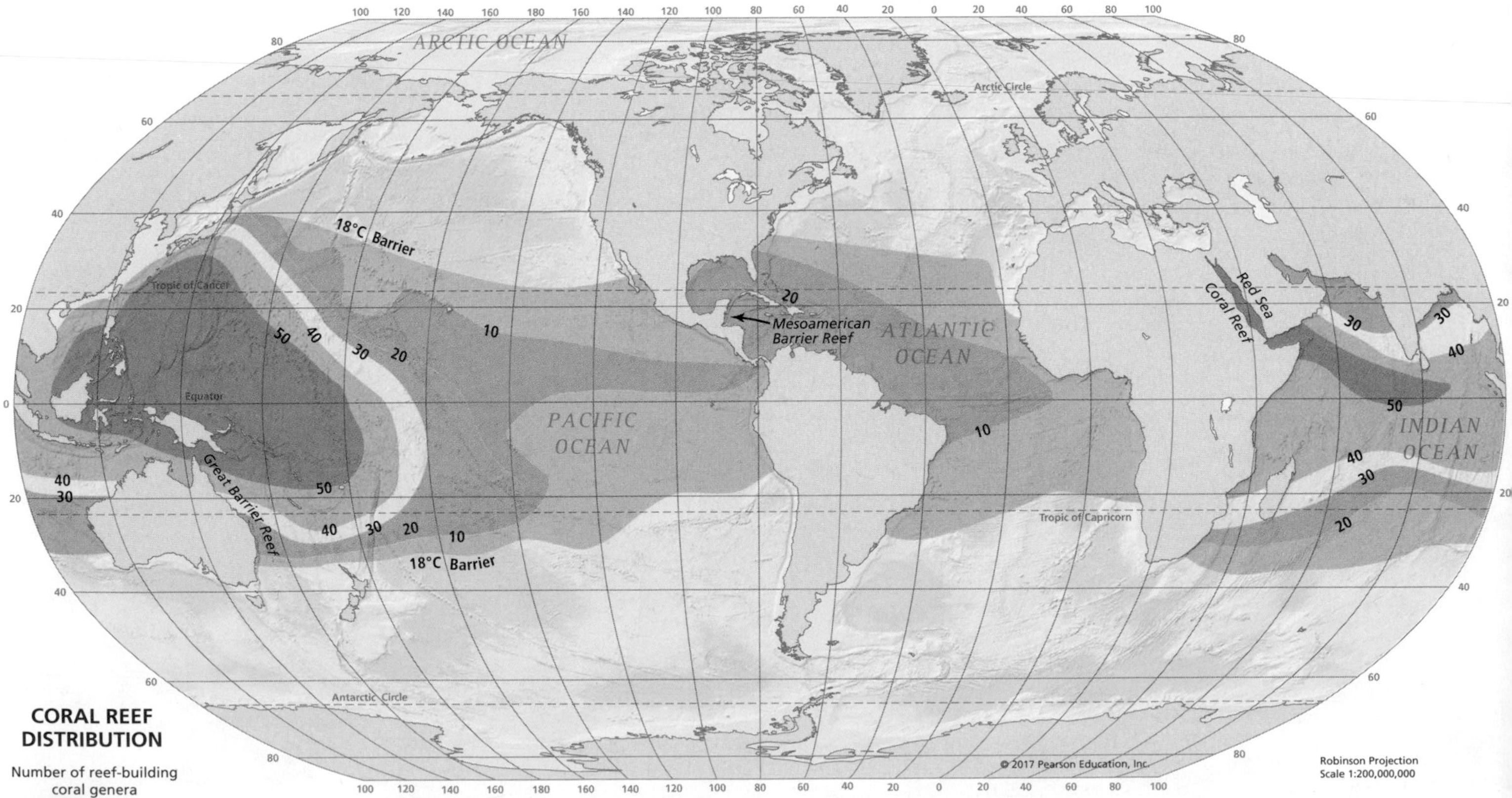

ARCTIC OCEAN

Arctic Circle

18°C Barrier

Tropic of Cancer

Mesoamerican Barrier Reef

ATLANTIC OCEAN

Red Sea Coral Reef

50   40   30   20

Equator

Great Barrier Reef

PACIFIC OCEAN

INDIAN OCEAN

40   30

40   30   20   10

18°C Barrier

Tropic of Capricorn

20

Antarctic Circle

© 2017 Pearson Education, Inc.

Robinson Projection
Scale 1:200,000,000

CORAL REEF DISTRIBUTION

Number of reef-building coral genera

- Above 50
- 40–50
- 30–39
- 20–29
- 10–19
- 0–9

Corals only grow within the tropics where average annual water temperature is above 18°C (64°F).

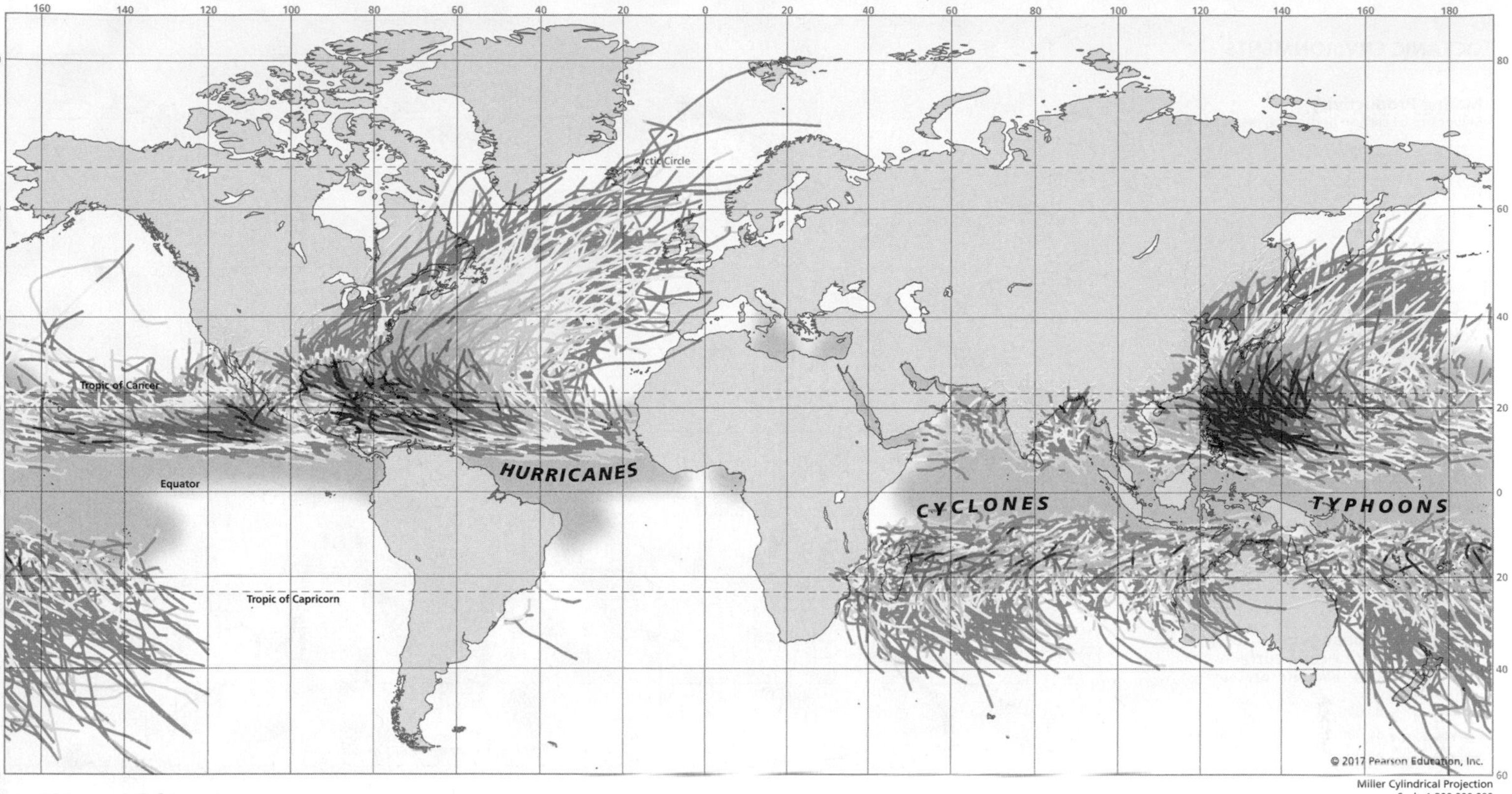

HISTORICAL TROPICAL CYCLONE TRACKS

Miller Cylindrical Projection
Scale 1:200,000,000
Source: NOAA

Water warm enough to produce tropical cyclones.

Names of tropical cyclones vary in each ocean basin.

Saffir-Simpson hurricane wind scale

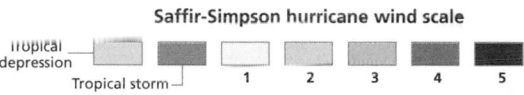

Tropical depression — Tropical storm — 1 2 3 4 5

The Saffir-Simpson hurricane wind scale classifies tropical storm and hurricane intensity using sustained wind speeds. While the scale is used to describe storms originating in the Atlantic Ocean and northern Pacific Ocean, it is used here for all tropical storms and hurricanes, regardless of origin.

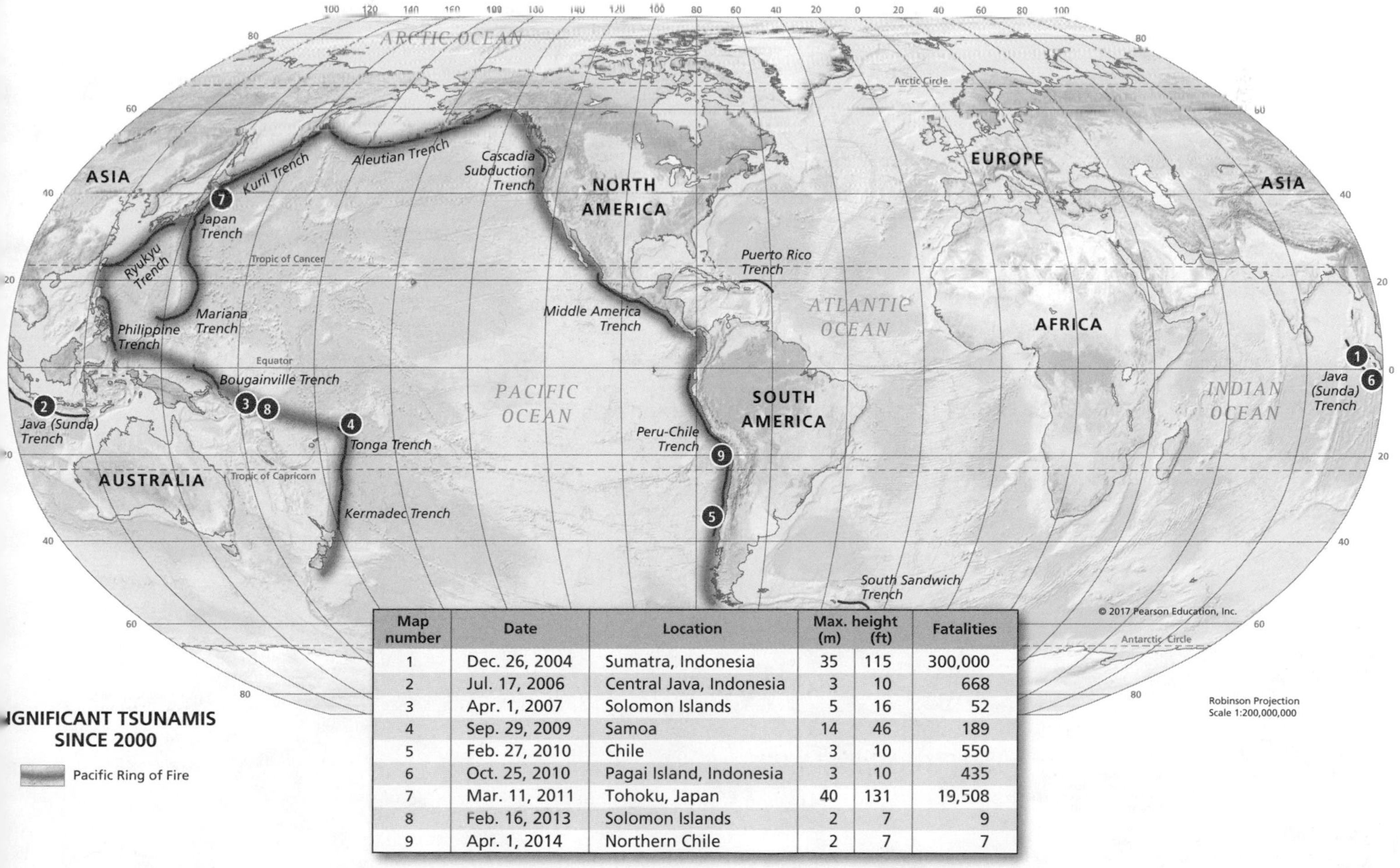

© 2017 Pearson Education, Inc.

Robinson Projection
Scale 1:200,000,000

SIGNIFICANT TSUNAMIS SINCE 2000

Pacific Ring of Fire

Map number	Date	Location	Max. height (m)	(ft)	Fatalities
1	Dec. 26, 2004	Sumatra, Indonesia	35	115	300,000
2	Jul. 17, 2006	Central Java, Indonesia	3	10	668
3	Apr. 1, 2007	Solomon Islands	5	16	52
4	Sep. 29, 2009	Samoa	14	46	189
5	Feb. 27, 2010	Chile	3	10	550
6	Oct. 25, 2010	Pagai Island, Indonesia	3	10	435
7	Mar. 11, 2011	Tohoku, Japan	40	131	19,508
8	Feb. 16, 2013	Solomon Islands	2	7	9
9	Apr. 1, 2014	Northern Chile	2	7	7

OCEANIC ENVIRONMENTS

Marine Productivity
Milligrams of carbon per square meter per day

- Over 500
- 250-500
- 150-250
- 100-150
- Under 100

Velocity of current
Nautical miles per day

→ Over 36
→ 24 - 36
→ 12 - 24
→ Under 12

〰 Areas of upwelling cold water

〰 Average limits of sea ice or drift ice

🪸 Coral reefs

**Atmospheric heat gain (or loss)
by contact with ocean surface**
Calories per square centimeter per year

- + 80,000
- + 60,000
- + 40,000
- 0
- - 40,000
- - 60,000

Robinson Projection
Scale 1:95,000,000
One inch to 1,500 miles
One cm to 950 km

0 500 1000 1500 2000 Miles
0 1000 2000 3000 Kilometers

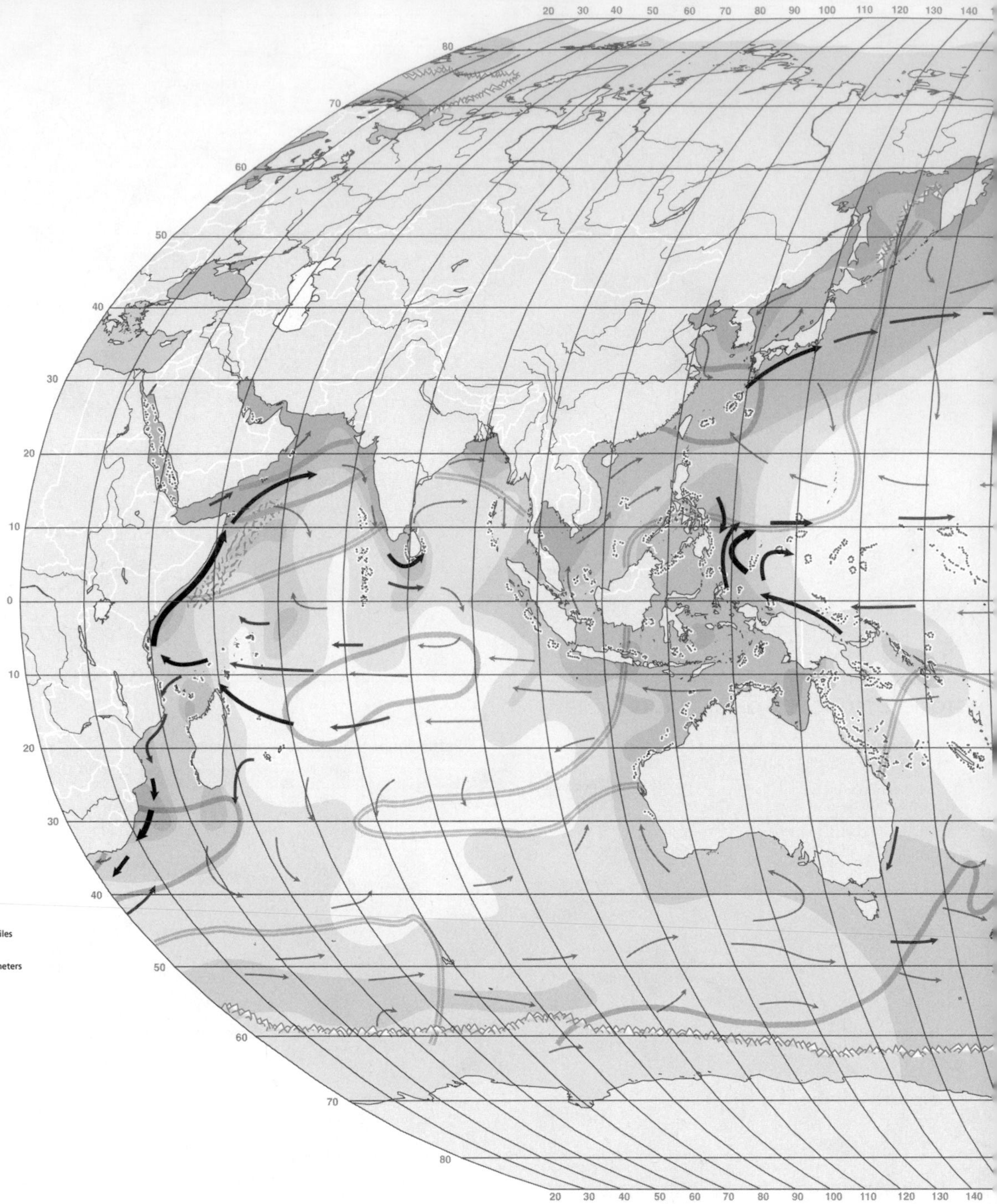

CHANGE IN ARCTIC SEA ICE EXTENT

☐ Monthly sea ice extent

〰 Median monthly sea ice extent, 1981-2010

▨ Ice sheets, ice caps, and glaciers

March, 2014

September, 2014

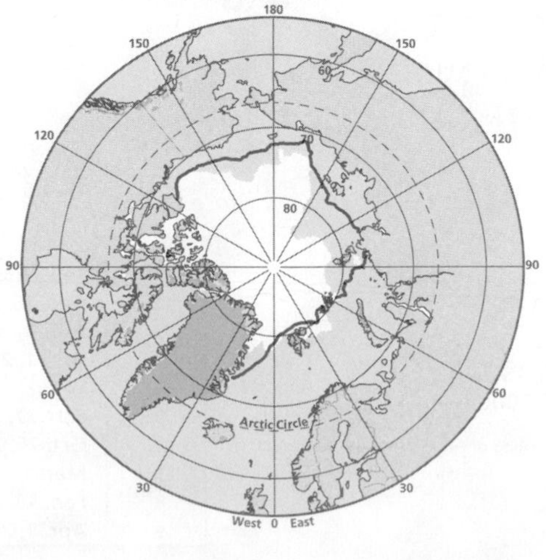

Polar Stereographic Projection
Scale 1:140,000,000
Source: National Snow and Ice Data Center
© Rand McNally A-101961-1

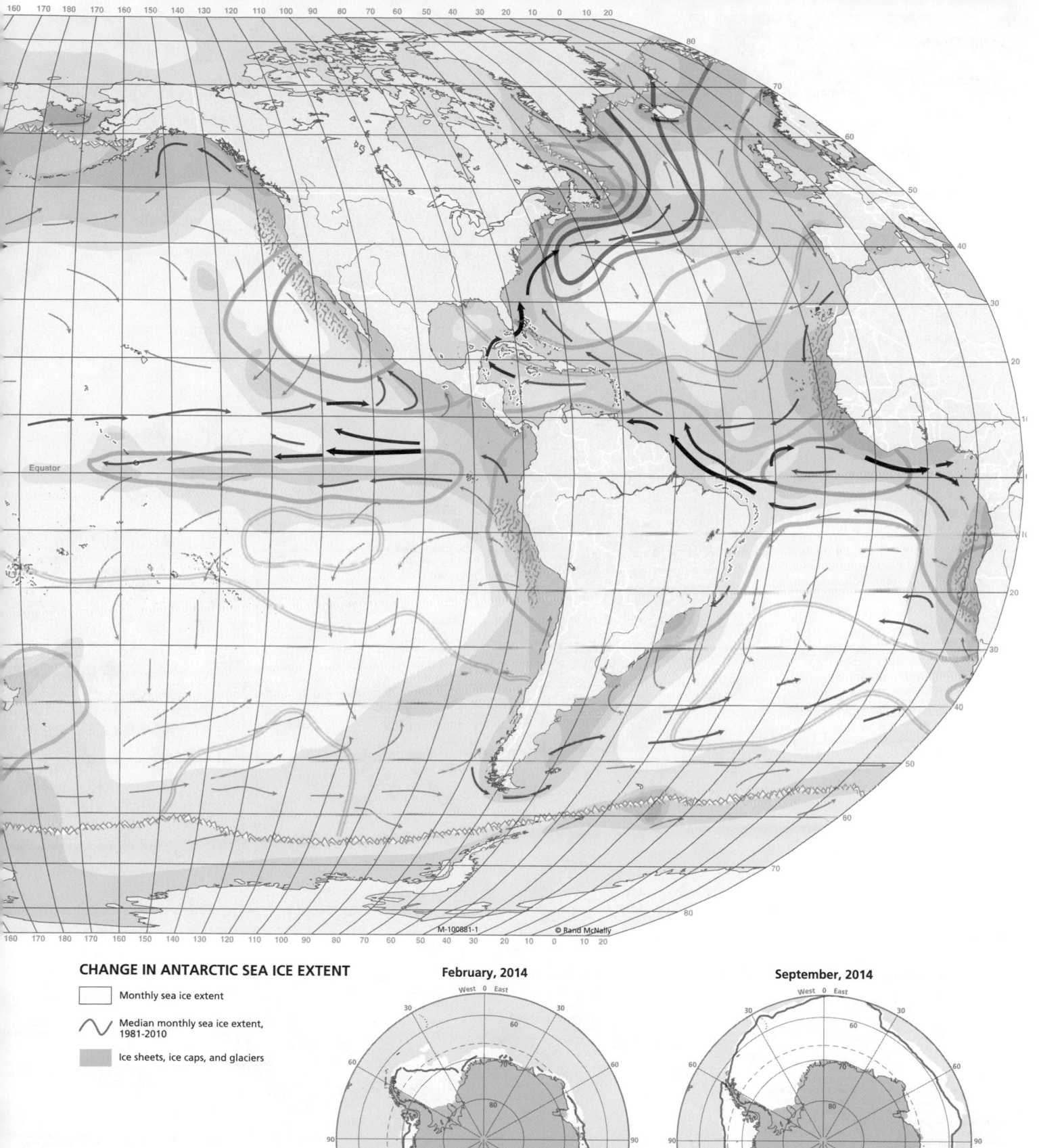

CHANGE IN ANTARCTIC SEA ICE EXTENT

☐ Monthly sea ice extent

〰 Median monthly sea ice extent, 1981-2010

▨ Ice sheets, ice caps, and glaciers

February, 2014

September, 2014

Polar Stereographic Projection
Scale 1:140,000,000
Source: National Snow and Ice Data Center
© Rand McNally A-101962-1

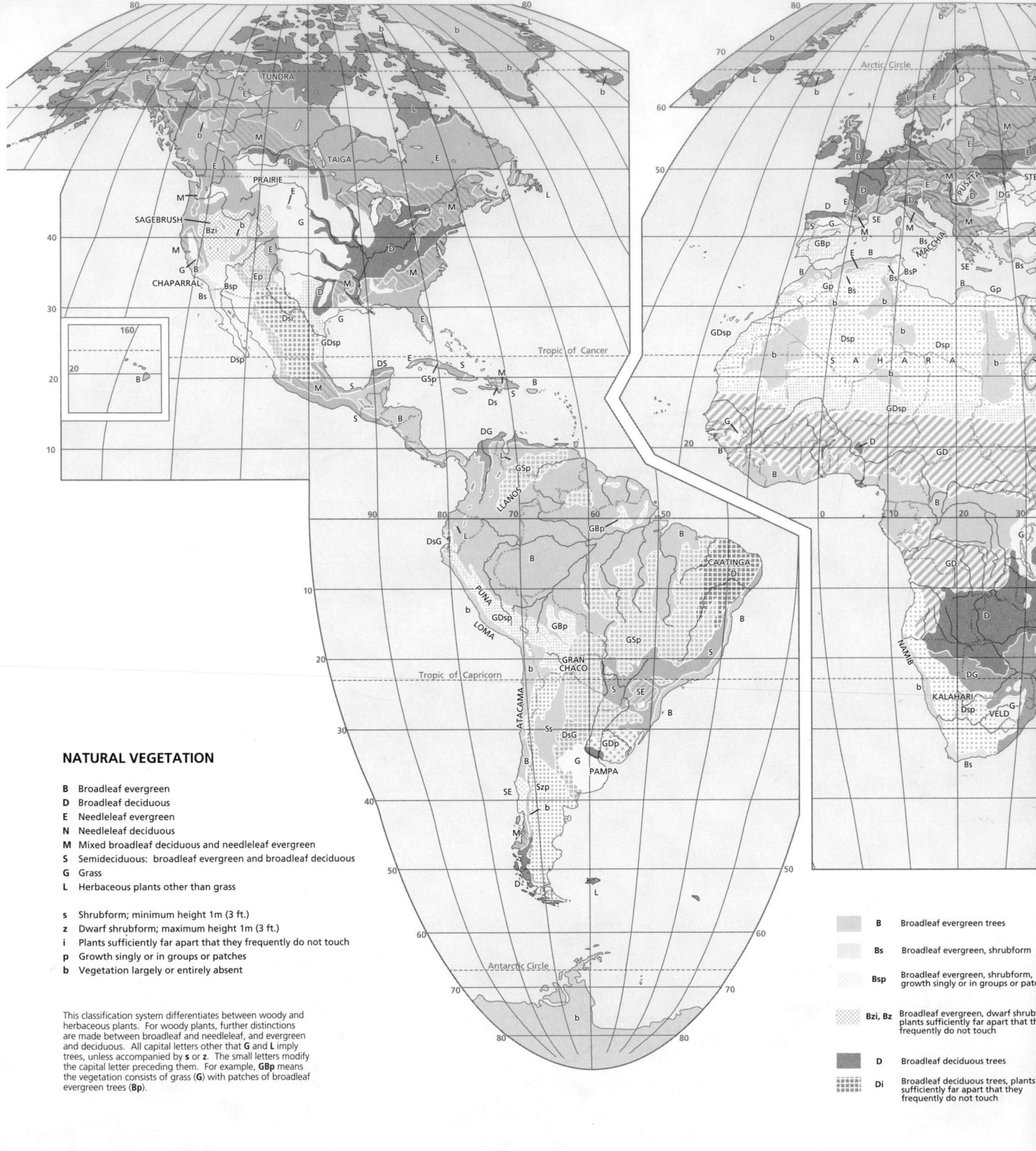

NATURAL VEGETATION

- **B** Broadleaf evergreen
- **D** Broadleaf deciduous
- **E** Needleleaf evergreen
- **N** Needleleaf deciduous
- **M** Mixed broadleaf deciduous and needleleaf evergreen
- **S** Semideciduous: broadleaf evergreen and broadleaf deciduous
- **G** Grass
- **L** Herbaceous plants other than grass

- **s** Shrubform; minimum height 1m (3 ft.)
- **z** Dwarf shrubform; maximum height 1m (3 ft.)
- **i** Plants sufficiently far apart that they frequently do not touch
- **p** Growth singly or in groups or patches
- **b** Vegetation largely or entirely absent

This classification system differentiates between woody and herbaceous plants. For woody plants, further distinctions are made between broadleaf and needleleaf, and evergreen and deciduous. All capital letters other that **G** and **L** imply trees, unless accompanied by **s** or **z**. The small letters modify the capital letter preceding them. For example, **GBp** means the vegetation consists of grass (**G**) with patches of broadleaf evergreen trees (**Bp**).

	B	Broadleaf evergreen trees
	Bs	Broadleaf evergreen, shrubform
	Bsp	Broadleaf evergreen, shrubform, growth singly or in groups or patch
	Bzi, Bz	Broadleaf evergreen, dwarf shrubfe plants sufficiently far apart that the frequently do not touch
	D	Broadleaf deciduous trees
	Di	Broadleaf deciduous trees, plants sufficiently far apart that they frequently do not touch

Map labels: TUNDRA, TAIGA, PRAIRIE, SAGEBRUSH, CHAPARRAL, SAHARA, PUSZTA, STEP, MACCHIA, Arctic Circle, Tropic of Cancer, Tropic of Capricorn, Antarctic Circle, LLANOS, PUNA, LOMA, GRAN CHACO, ATACAMA, PAMPA, CAATINGA, NAMIB, KALAHARI, VELD

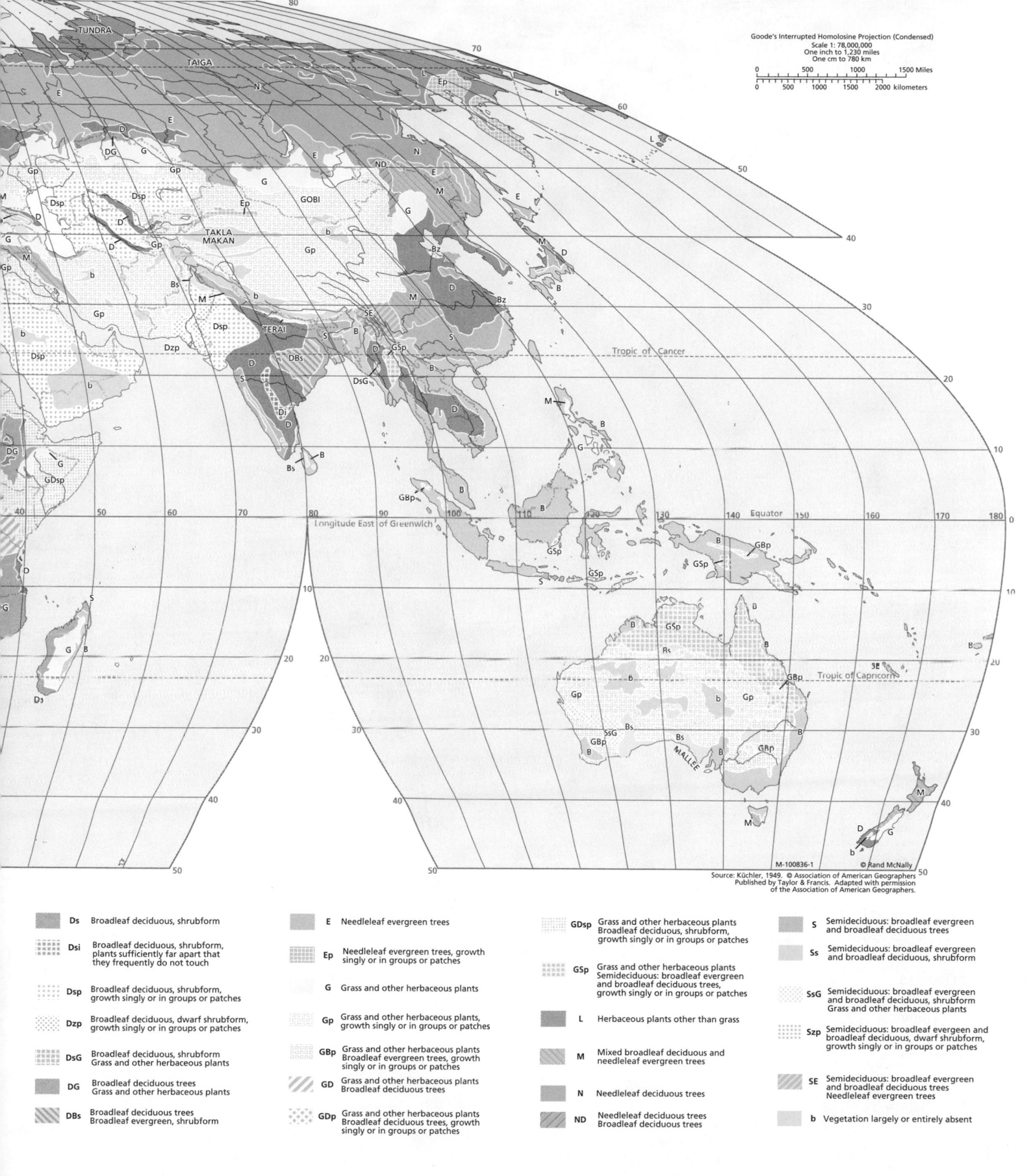

Goode's Interrupted Homolosine Projection (Condensed)
Scale 1: 78,000,000
One inch to 1,230 miles
One cm to 780 km

Source: Küchler, 1949. © Association of American Geographers
Published by Taylor & Francis. Adapted with permission
of the Association of American Geographers.

M-100836-1   © Rand McNally

Ds	Broadleaf deciduous, shrubform	
Dsi	Broadleaf deciduous, shrubform, plants sufficiently far apart that they frequently do not touch	
Dsp	Broadleaf deciduous, shrubform, growth singly or in groups or patches	
Dzp	Broadleaf deciduous, dwarf shrubform, growth singly or in groups or patches	
DsG	Broadleaf deciduous, shrubform Grass and other herbaceous plants	
DG	Broadleaf deciduous trees Grass and other herbaceous plants	
DBs	Broadleaf deciduous trees Broadleaf evergreen, shrubform	

E	Needleleaf evergreen trees	
Ep	Needleleaf evergreen trees, growth singly or in groups or patches	
G	Grass and other herbaceous plants	
Gp	Grass and other herbaceous plants, growth singly or in groups or patches	
GBp	Grass and other herbaceous plants Broadleaf evergreen trees, growth singly or in groups or patches	
GD	Grass and other herbaceous plants Broadleaf deciduous trees	
GDp	Grass and other herbaceous plants Broadleaf deciduous trees, growth singly or in groups or patches	

GDsp	Grass and other herbaceous plants Broadleaf deciduous, shrubform, growth singly or in groups or patches	
GSp	Grass and other herbaceous plants Semideciduous: broadleaf evergreen and broadleaf deciduous trees, growth singly or in groups or patches	
L	Herbaceous plants other than grass	
M	Mixed broadleaf deciduous and needleleaf evergreen trees	
N	Needleleaf deciduous trees	
ND	Needleleaf deciduous trees Broadleaf deciduous trees	

S	Semideciduous: broadleaf evergreen and broadleaf deciduous trees	
Ss	Semideciduous: broadleaf evergreen and broadleaf deciduous, shrubform	
SsG	Semideciduous: broadleaf evergreen and broadleaf deciduous, shrubform Grass and other herbaceous plants	
Szp	Semideciduous: broadleaf evergreen and broadleaf deciduous, dwarf shrubform, growth singly or in groups or patches	
SE	Semideciduous: broadleaf evergreen and broadleaf deciduous trees Needleleaf evergreen trees	
b	Vegetation largely or entirely absent	

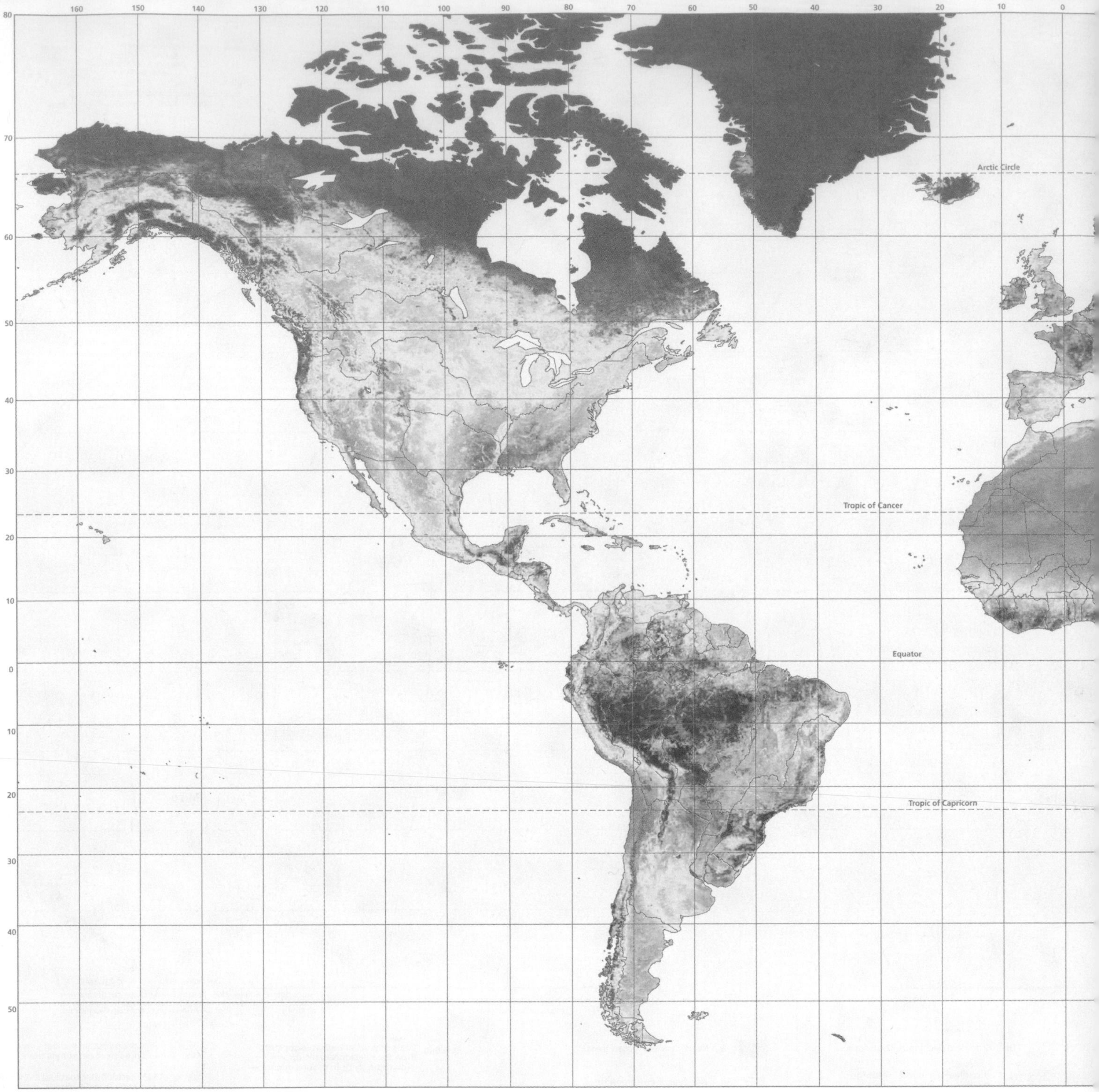

160 150 140 130 120 110 100 90 80 70 60 50 40 30 20 10 0

80

70

Arctic Circle

60

50

40

30

Tropic of Cancer

20

10

Equator

0

10

20

Tropic of Capricorn

30

40

50

GLOBAL GREENNESS

Based on the Normalized Difference Vegetation Index (NDVI)

-.1

.4

.9

Higher values (dark greens) show land areas with plenty of leafy green vegetation, such as the Amazon Rainforest.

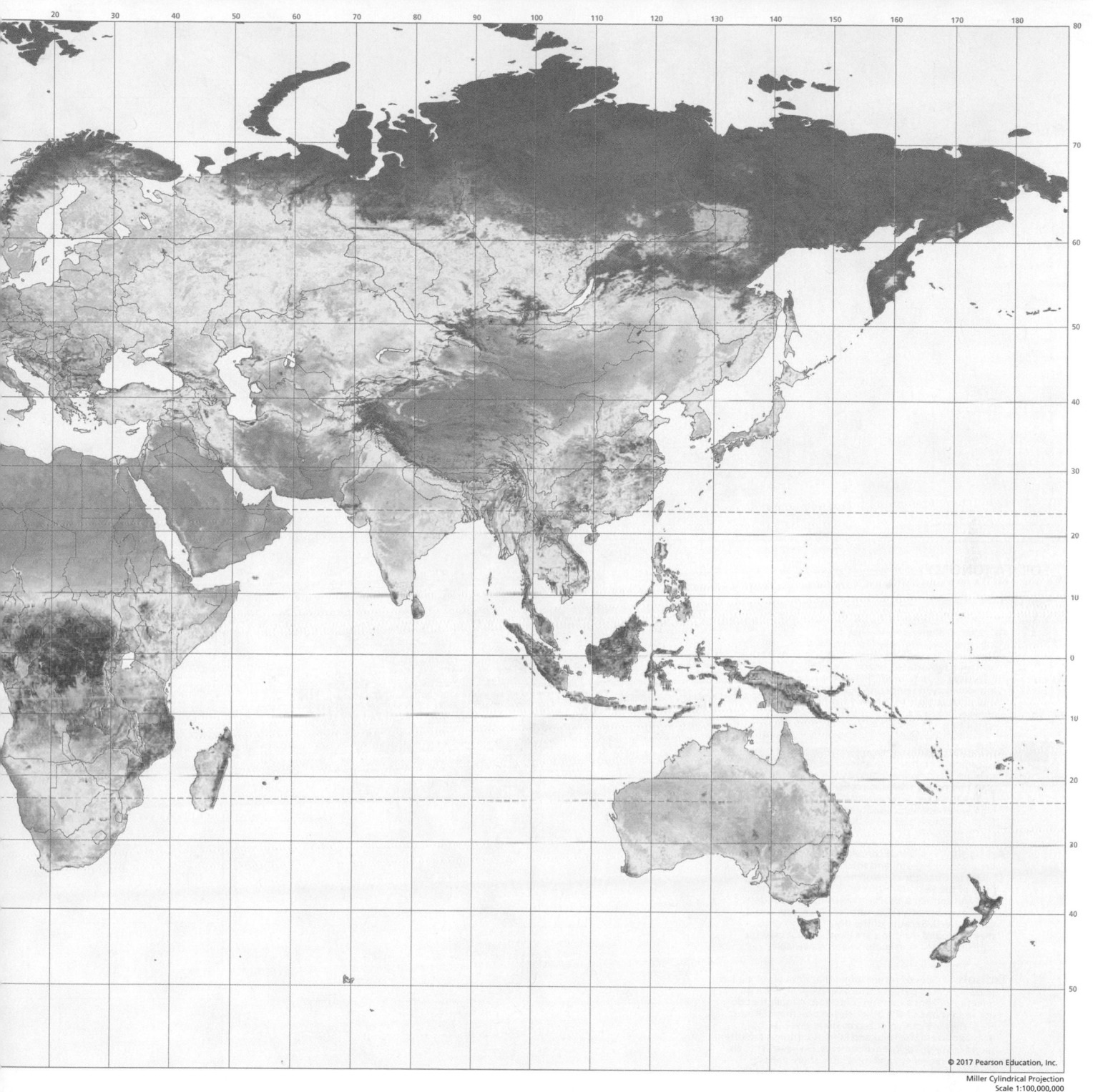

Miller Cylindrical Projection
Scale 1:100,000,000
Source: NASA

NDVI April, 2016 based on observations from the
Moderate Resolution Imaging Spectroradiometer (MODIS) on NASA's Terra satellite.

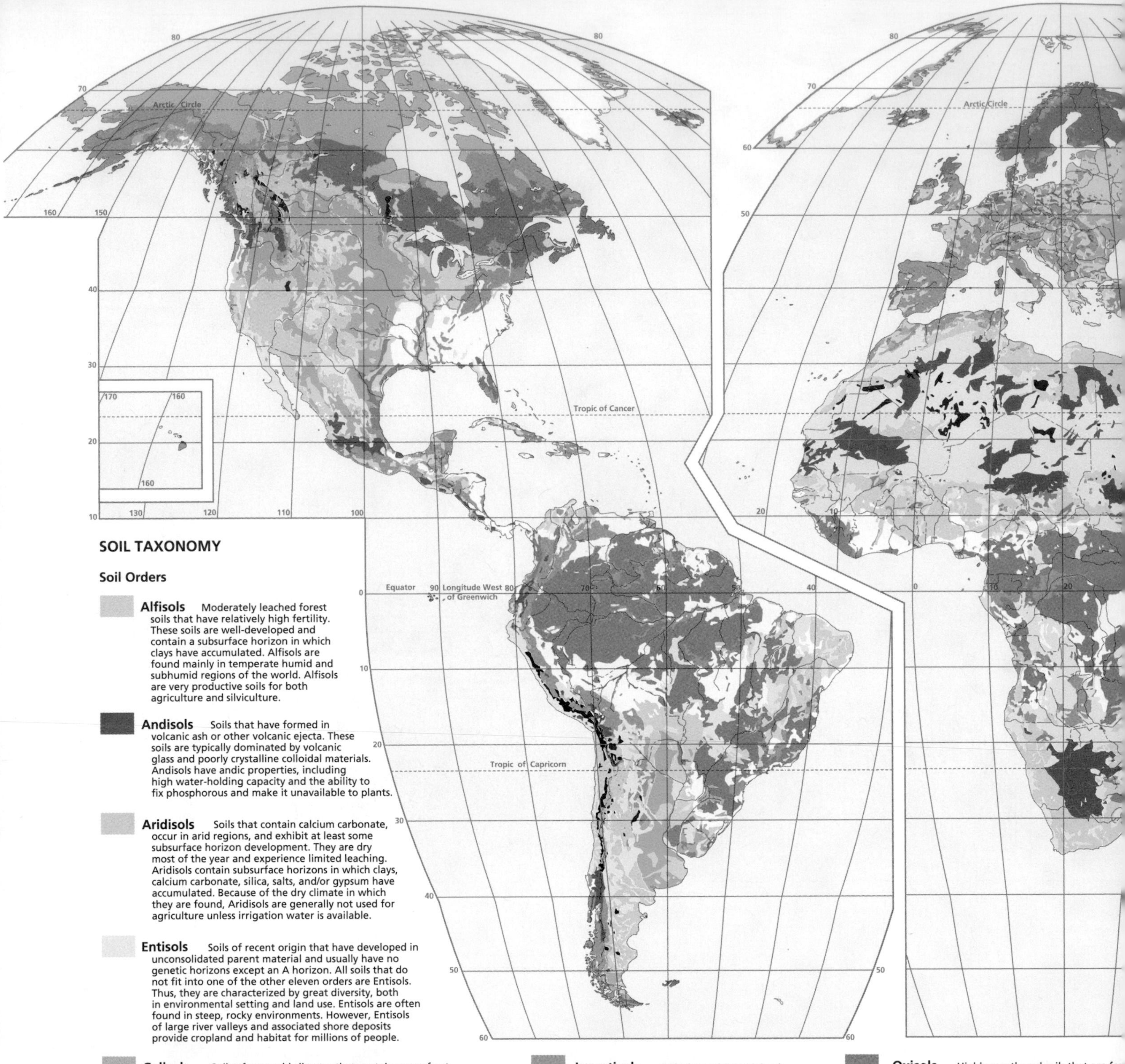

SOIL TAXONOMY

Soil Orders

Alfisols   Moderately leached forest soils that have relatively high fertility. These soils are well-developed and contain a subsurface horizon in which clays have accumulated. Alfisols are found mainly in temperate humid and subhumid regions of the world. Alfisols are very productive soils for both agriculture and silviculture.

Andisols   Soils that have formed in volcanic ash or other volcanic ejecta. These soils are typically dominated by volcanic glass and poorly crystalline colloidal materials. Andisols have andic properties, including high water-holding capacity and the ability to fix phosphorous and make it unavailable to plants.

Aridisols   Soils that contain calcium carbonate, occur in arid regions, and exhibit at least some subsurface horizon development. They are dry most of the year and experience limited leaching. Aridisols contain subsurface horizons in which clays, calcium carbonate, silica, salts, and/or gypsum have accumulated. Because of the dry climate in which they are found, Aridisols are generally not used for agriculture unless irrigation water is available.

Entisols   Soils of recent origin that have developed in unconsolidated parent material and usually have no genetic horizons except an A horizon. All soils that do not fit into one of the other eleven orders are Entisols. Thus, they are characterized by great diversity, both in environmental setting and land use. Entisols are often found in steep, rocky environments. However, Entisols of large river valleys and associated shore deposits provide cropland and habitat for millions of people.

Gelisols   Soils of very cold climates that contain permafrost within 2 m (6.5 ft.) of the surface. These soils are limited to the high-latitude polar regions and high mountain elevations. Gelisols show relatively little morphological development. Low soil temperatures cause soil-forming processes such as decomposition of organic materials to proceed slowly. As a result, Gelisols store large quantities of organic carbon. Because of the extreme environment in which they are found, Gelisols support only a small fraction of the world's population. The frozen condition of Gelisol landscapes makes them sensitive to human activities.

Histosols   Soils that are composed mainly of organic materials. They contain at least 20 to 30 percent organic matter by weight and are more than 40 cm (15.75 in.) thick. Most Histosols form in settings such as wetlands where restricted drainage inhibits the decomposition of plant and animal remains, allowing these organic materials to accumulate over time. As a result, Histosols are ecologically important because of the large quantities of carbon they contain. Histosols are often referred to as peats and mucks and are mined for fuel and horticultural products.

Inceptisols   Soils that exhibit minimal horizon development. They are more developed than Entisols, but still lack the features that are characteristic of other soil orders. Inceptisols are widely distributed and occur under a wide range of ecological settings. They are often found on fairly steep slopes, young geomorphic surfaces, and on resistant parent materials. Land use varies considerably with Inceptisols.

Mollisols   Soils of grassland ecosystems. These soils are characterized by a thick, dark surface horizon that results from the long-term addition of organic materials derived from plant roots. Mollisols primarily occur in the mid-latitudes and are extensive in prairie regions. Mollisols are among some of the most important and productive agricultural soils in the world.

Oxisols   Highly-weathered soils that are fou primarily in the intertropical regions of the wo These soils contain few weatherable minerals are often rich in iron and aluminum oxide min Most Oxisols have extremely low native fertili resulting from very low nutrient reserves, high phosphorus retention by oxide minerals, and l cation exchange capacity. Oxisols can be quite productive with inputs of lime and fertilizers.

Spodosols   Acid soils characterized by a subsurface accumulation of humus that is complexed with aluminum and iron. Spodoso often occur under coniferous forest in cool, moist climates. Because they are naturally infertile, Spodosols require additions of lime in order to be productive agriculturally.

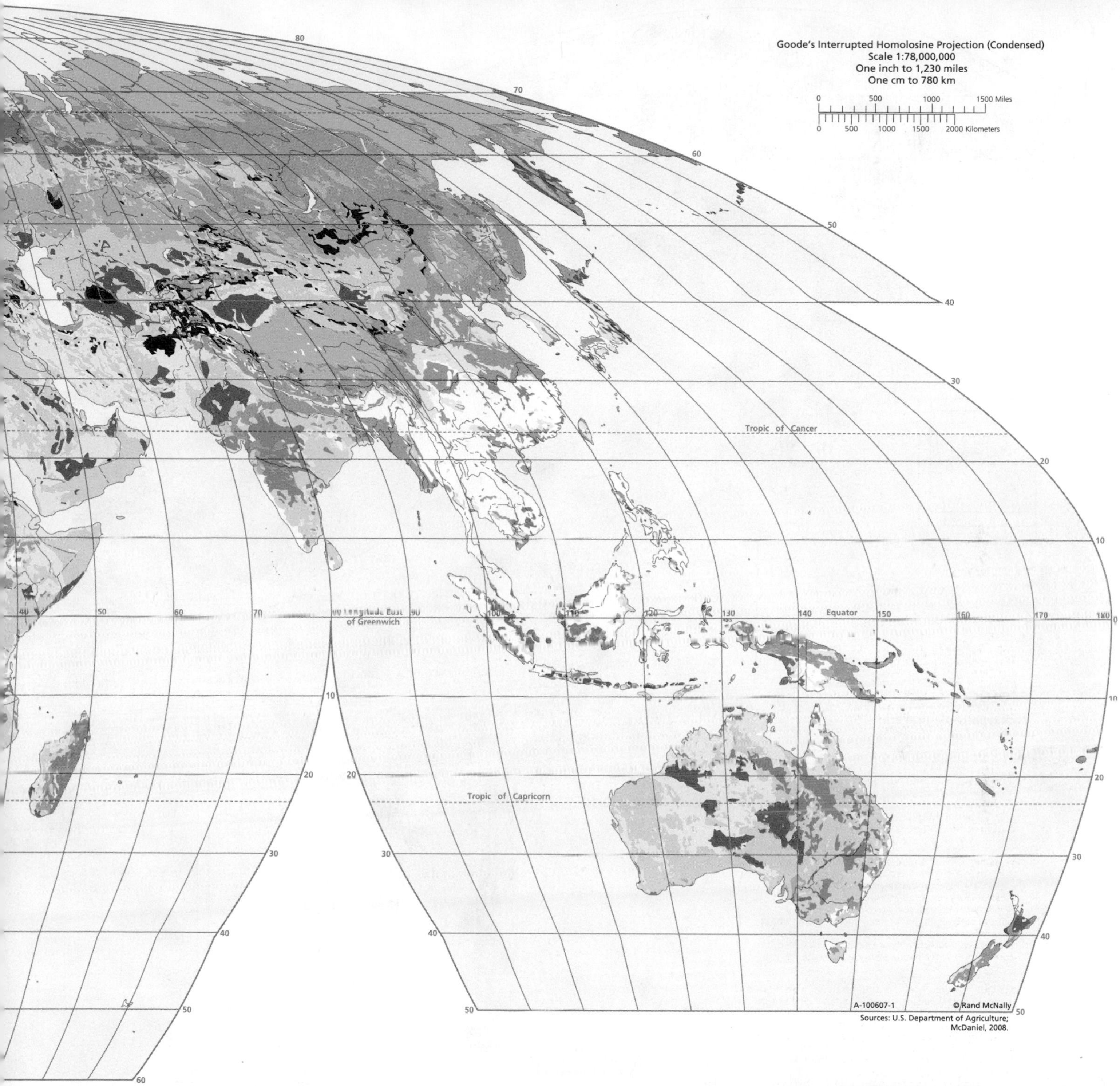

Goode's Interrupted Homolosine Projection (Condensed)
Scale 1:78,000,000
One inch to 1,230 miles
One cm to 780 km

0   500   1000   1500 Miles

0   500   1000   1500   2000 Kilometers

70

80

60

50

40

30

Tropic of Cancer

20

10

longitude East of Greenwich

Equator

40   50   60   70   80   90   100   110   120   130   140   150   160   170   180   0

10

Tropic of Capricorn

20

30

40

50

60

A-100607-1   © Rand McNally

Sources: U.S. Department of Agriculture;
McDaniel, 2008.

Ultisols   Strongly leached, acid forest soils with relatively low native fertility. They are found primarily in humid temperate and tropical areas of the world, typically on older, stable landscapes where intense weathering of primary minerals has occurred. Because of the favorable climate regimes in which they are typically found, Ultisols often support productive forests. However, high acidity and limited availability of nutrients makes them poorly suited to agriculture without the use of fertilizer and lime.

Vertisols   Clay-rich soils that shrink and swell with changes in moisture content. This shrink-swell action creates serious engineering problems and generally prevents formation of distinct, well-developed horizons.

Ice/Glacier

Rocky land

Salt

Shifting sands

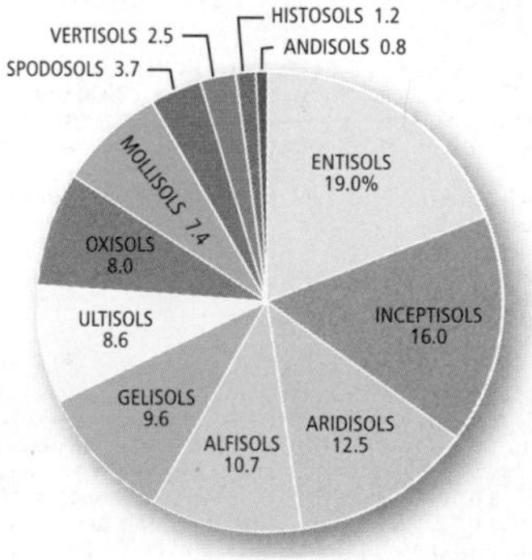

VERTISOLS 2.5
SPODOSOLS 3.7
HISTOSOLS 1.2
ANDISOLS 0.8
MOLLISOLS 7.4
OXISOLS 8.0
ULTISOLS 8.6
GELISOLS 9.6
ALFISOLS 10.7
ARIDISOLS 12.5
INCEPTISOLS 16.0
ENTISOLS 19.0%

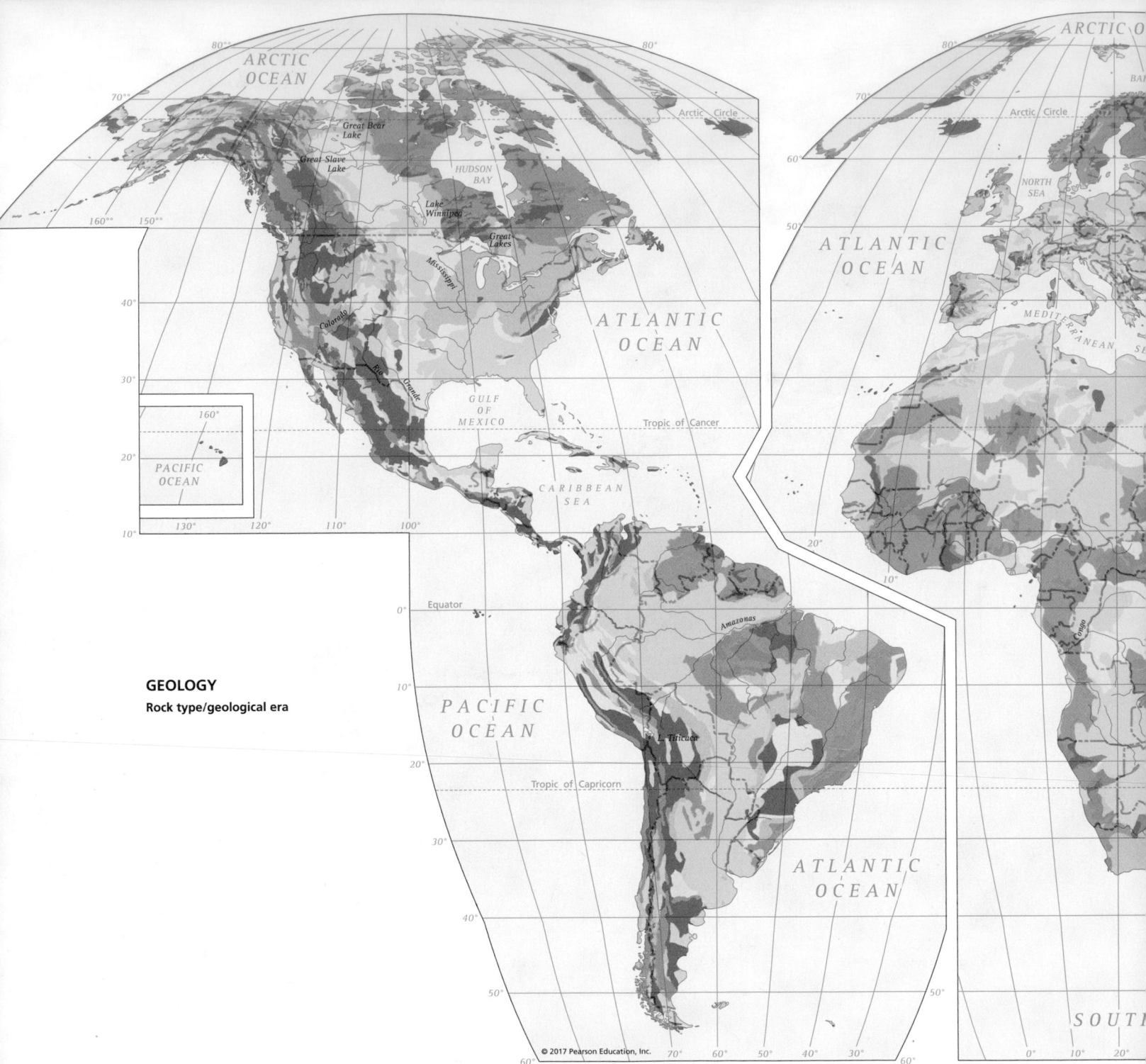

GEOLOGY

Rock type/geological era

Note: Areas classified as sedimentary also
include some sedimentary/volcanic areas.

© 2017 Pearson Education, Inc.

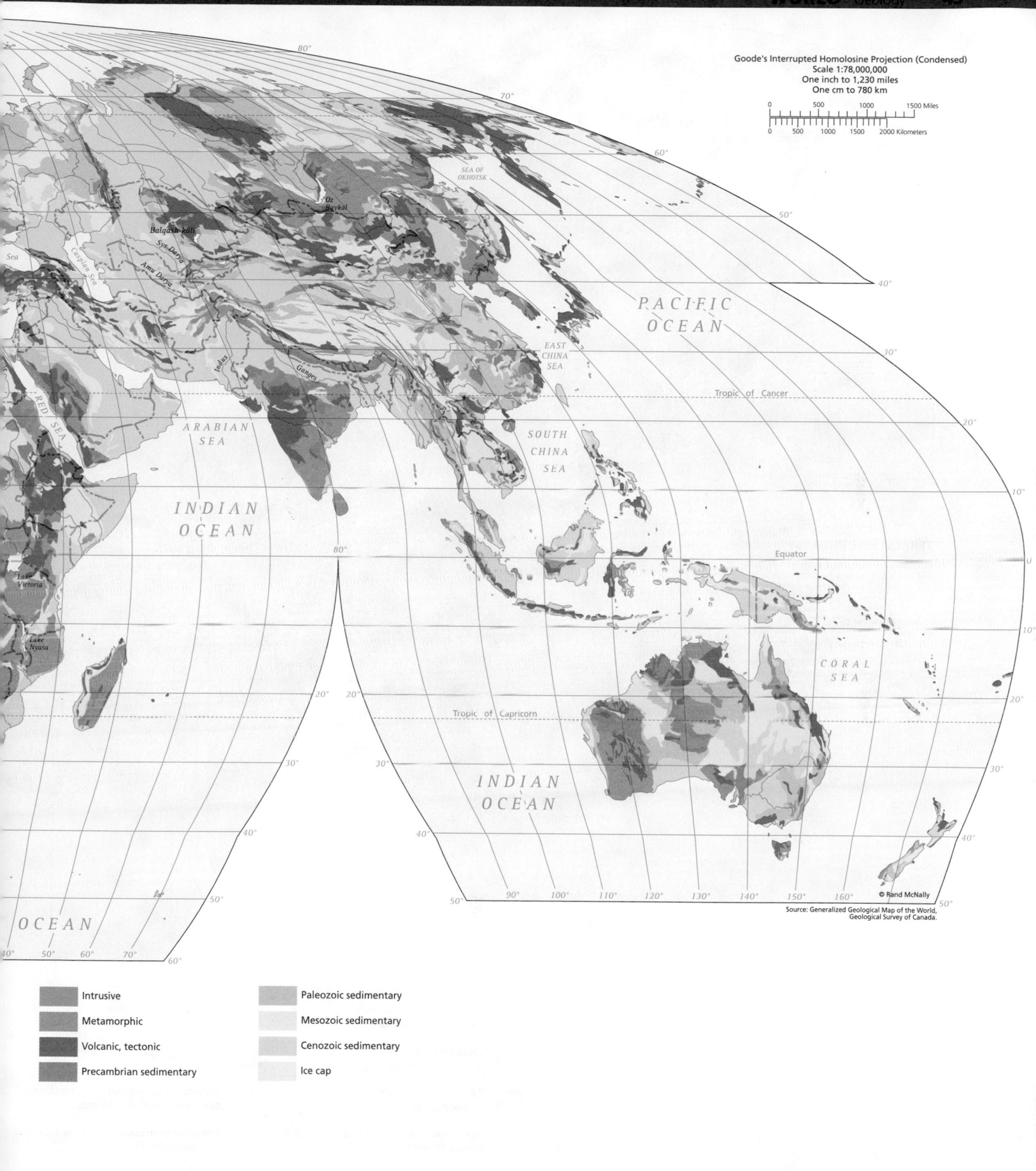

Goode's Interrupted Homolosine Projection (Condensed)
Scale 1:78,000,000
One inch to 1,230 miles
One cm to 780 km

	Intrusive		Paleozoic sedimentary
	Metamorphic		Mesozoic sedimentary
	Volcanic, tectonic		Cenozoic sedimentary
	Precambrian sedimentary		Ice cap

Source: Generalized Geological Map of the World,
Geological Survey of Canada.

© Rand McNally

TERRESTRIAL BIOMES

Terrestrial biomes are large geographic regions within which living organisms exhibit similar adaptations to environmental and climatic conditions. Biomes are a broad classification of the earth's ecosystems, and may be further subdivided into ecoregions. Ecoregions are geographic areas with distinct assemblages of natural communities of plant and animal species. The World Wildlife Fund's terrestrial ecoregions database, from which this map was derived, contains fourteen biomes and 867 ecoregions.

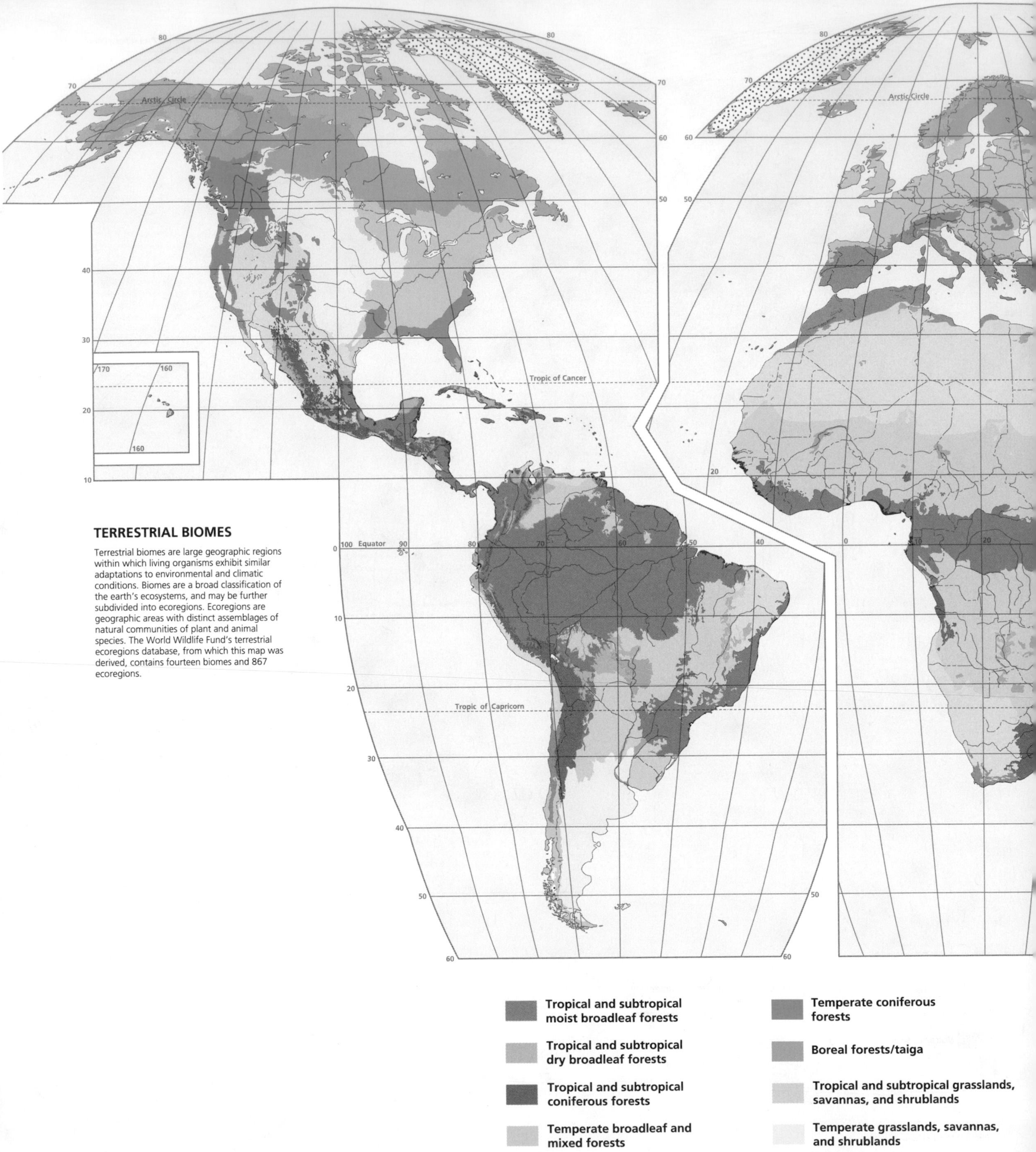

Arctic Circle
Tropic of Cancer
Equator
Tropic of Capricorn
Tropical and subtropical moist broadleaf forests
Tropical and subtropical dry broadleaf forests
Tropical and subtropical coniferous forests
Temperate broadleaf and mixed forests
Temperate coniferous forests
Boreal forests/taiga
Tropical and subtropical grasslands, savannas, and shrublands
Temperate grasslands, savannas, and shrublands

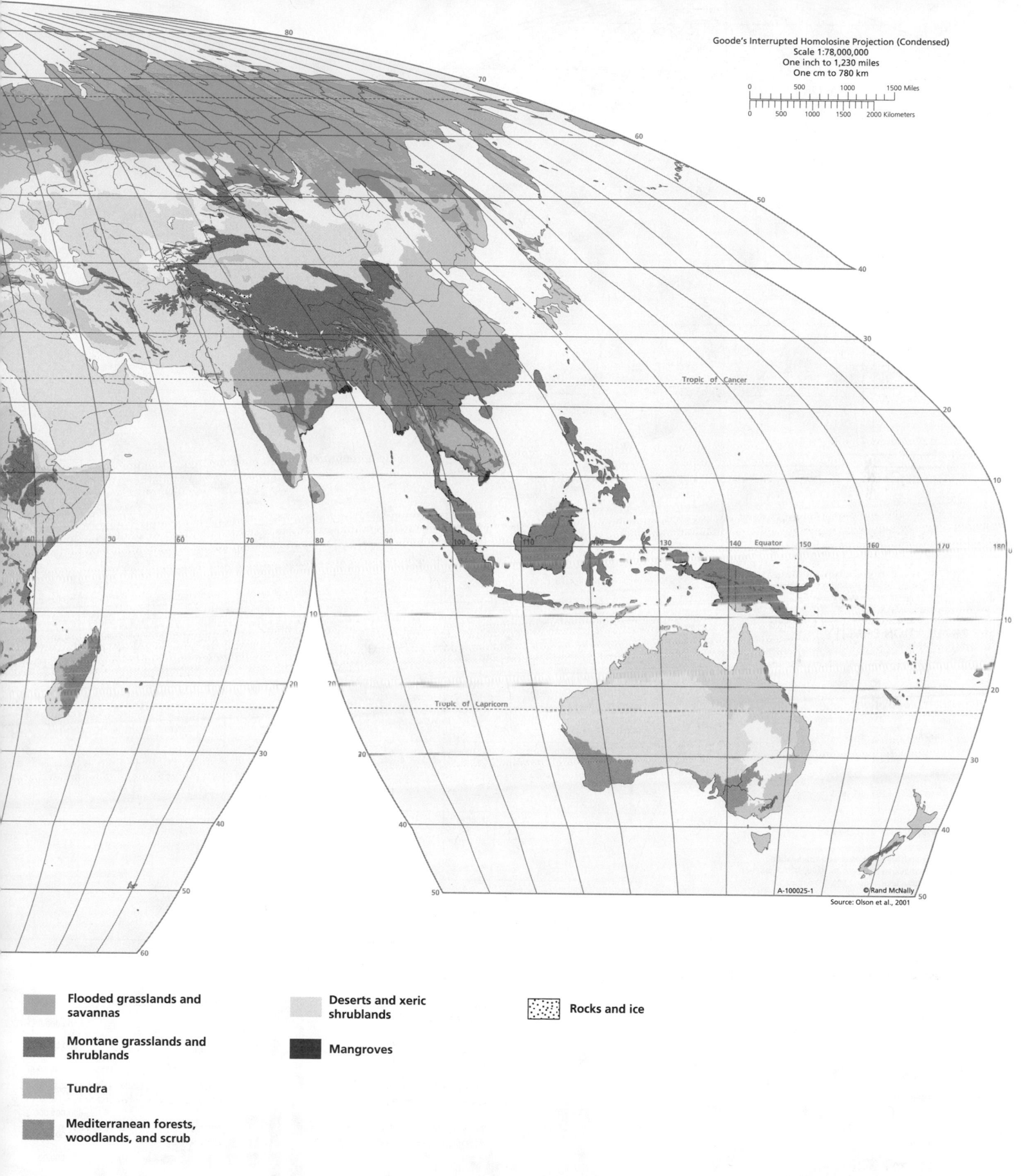

Goode's Interrupted Homolosine Projection (Condensed)
Scale 1:78,000,000
One inch to 1,230 miles
One cm to 780 km

A-100025-1 ©Rand McNally
Source: Olson et al., 2001

Flooded grasslands and savannas

Montane grasslands and shrublands

Tundra

Mediterranean forests, woodlands, and scrub

Deserts and xeric shrublands

Mangroves

Rocks and ice

80
70
Arctic Circle
60

Seattle
Toronto
Montreal
Chicago
Detroit
Boston
New York
Washington
Philadelphia
San Francisco
40
Phoenix
Dallas
Atlanta
Los Angeles
San Diego
Houston
Monterrey
30
Miami
Tropic of Cancer
20
Guadalajara
Mexico City

170   160

160

10

Medellín
Bogotá

100   Equator   90   Longitude West 80
of Greenwich
70   60   50   40

0

Fortaleza

Lima   Recife

10

Brasília   Salvador

Belo Horizonte

Campinas   Rio de Janeiro
São Paulo
20
Tropic of Capricorn
Curitiba

Porto Alegre

Santiago
30

Buenos Aires

40

50

60
Sources: U.S. Census Bureau; U.S. Department of Energy; United Nations
60

St. Petersburg
Moscow
60
Copenhagen
Hamburg   Berlin   Warsaw
London   Essen   Katowice
Paris   Stuttgart   Donets'
Milan   Budapest
50
Bucharest
Madrid   Barcelona   Rome
Istanbul   Ankara
Athens   İzmir   Ale
Casablanca
Tel Aviv-Yaf
Alexandria
Cairo
30

Khartoum
20
Dakar
Kano
Ibadan
Abidjan   Lagos
Yaoundé
10

Kinshasa
Luanda

10

Johann
20

Cape Town
30

POPULATION DENSITY

Population

per sq. km		per sq. mile
Over 500		Over 1,250
100 - 500		250 - 1,250
25 - 100		62.5 - 250
10 - 25		25 - 62.5
1 - 10		2.5 - 25
Under 1		Under 2.5

□ Metropolitan area over 10,000,000 population
O Metropolitan area 3,000,000 to 10,000,000 population
• Metropolitan area under 3,000,000 population

Largest Countries of the World 1950, 2015, 2050

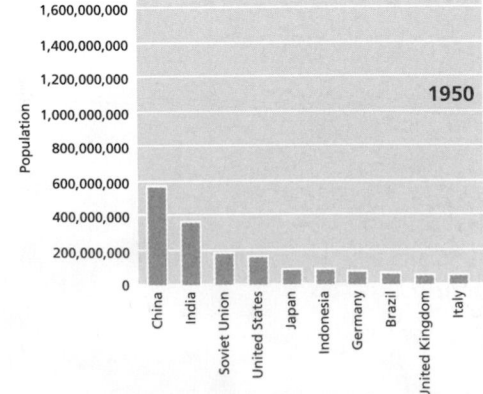

1950

Population

China
India
Soviet Union
United States
Japan
Indonesia
Germany
Brazil
United Kingdom
Italy

1,600,000,000
1,400,000,000
1,200,000,000
1,000,000,000
800,000,000
600,000,000
400,000,000
200,000,000
0

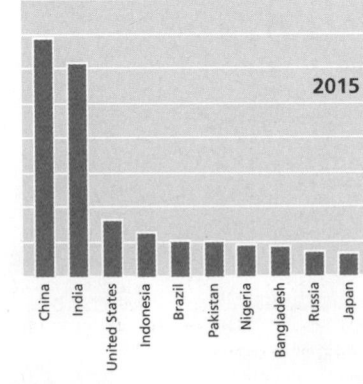

2015

China
India
United States
Indonesia
Brazil
Pakistan
Nigeria
Bangladesh
Russia
Japan

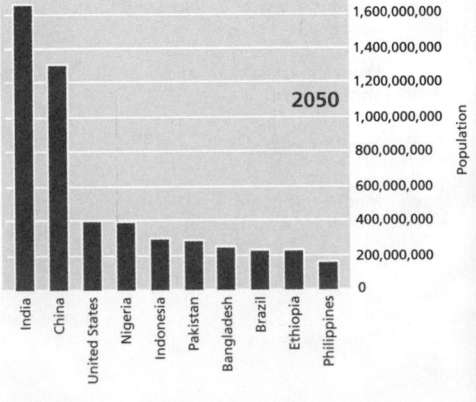

2050

India
China
United States
Nigeria
Indonesia
Pakistan
Bangladesh
Brazil
Ethiopia
Philippines

1,600,000,000
1,400,000,000
1,200,000,000
1,000,000,000
800,000,000
600,000,000
400,000,000
200,000,000
0

Population

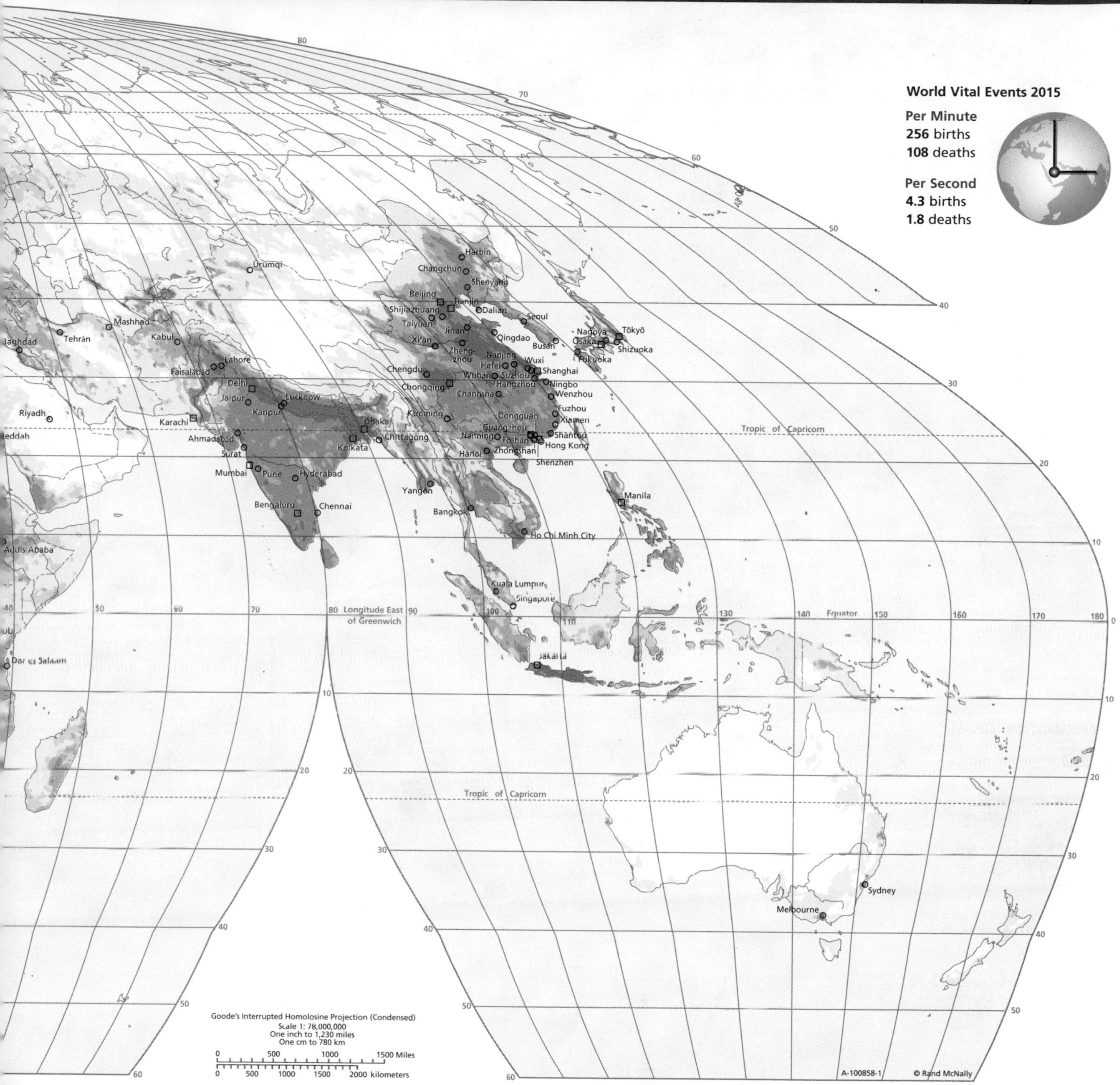

World Vital Events 2015

Per Minute
256 births
108 deaths

Per Second
4.3 births
1.8 deaths

Baghdad
Tehrān
Mashhad
Kabul
Riyadh
Jeddah
Addis Ababa
Dar es Salaam
Ürümqi
Faisalabad
Lahore
Karachi
Delhi
Jaipur
Kanpur
Lucknow
Ahmadabad
Surat
Mumbai
Pune
Hyderabad
Bengalūru
Chennai
Kolkata
Dhaka
Chittagong
Yangon
Bangkok
Ho Chi Minh City
Kuala Lumpur
Singapore
Jakarta
Hanoi
Shenzhen
Hong Kong
Zhongshan
Foshan
Shantou
Nanning
Guangzhou
Dongguan
Xiamen
Fuzhou
Wenzhou
Ningbo
Hangzhou
Shanghai
Suzhou
Wuxi
Wuhan
Hefei
Nanjing
Changsha
Chongqing
Kunming
Chengdu
Zheng-zhou
Xi'an
Jinan
Qingdao
Busan
Osaka
Nagoya
Fukuoka
Shizuoka
Tōkyō
Seoul
Dalian
Taiyuan
Shijiazhuang
Beijing
Tianjin
Shenyang
Changchun
Harbin
Manila
Sydney
Melbourne

Tropic of Capricorn
Longitude East of Greenwich
Equator
Tropic of Capricorn

Goode's Interrupted Homolosine Projection (Condensed)
Scale 1: 78,000,000
One inch to 1,230 miles
One cm to 780 km

500 | 1000 | 1500 Miles
0 | 500 | 1000 | 1500 | 2000 kilometers

A-100858-1   © Rand McNally

Age and Sex Composition 2015

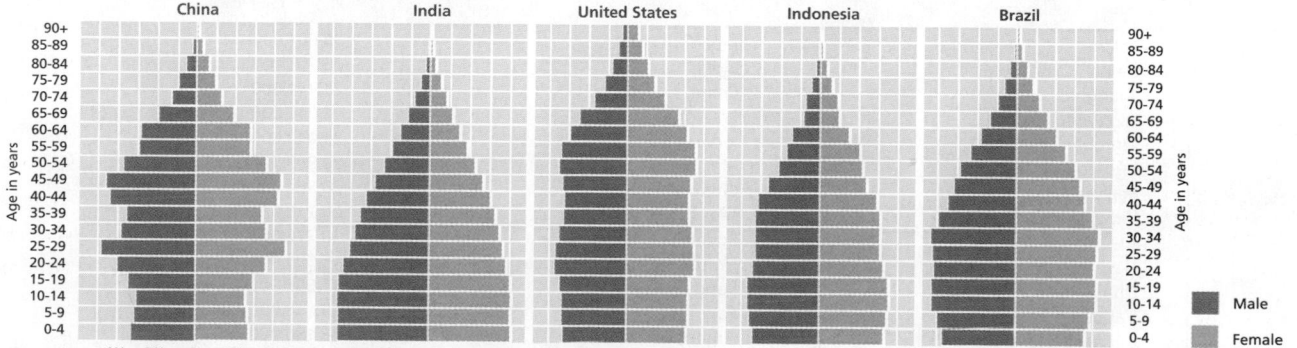

China · India · United States · Indonesia · Brazil

Age in years

90+, 85-89, 80-84, 75-79, 70-74, 65-69, 60-64, 55-59, 50-54, 45-49, 40-44, 35-39, 30-34, 25-29, 20-24, 15-19, 10-14, 5-9, 0-4

4% 2% 0 2% 4%

Percent of total population

■ Male   ■ Female

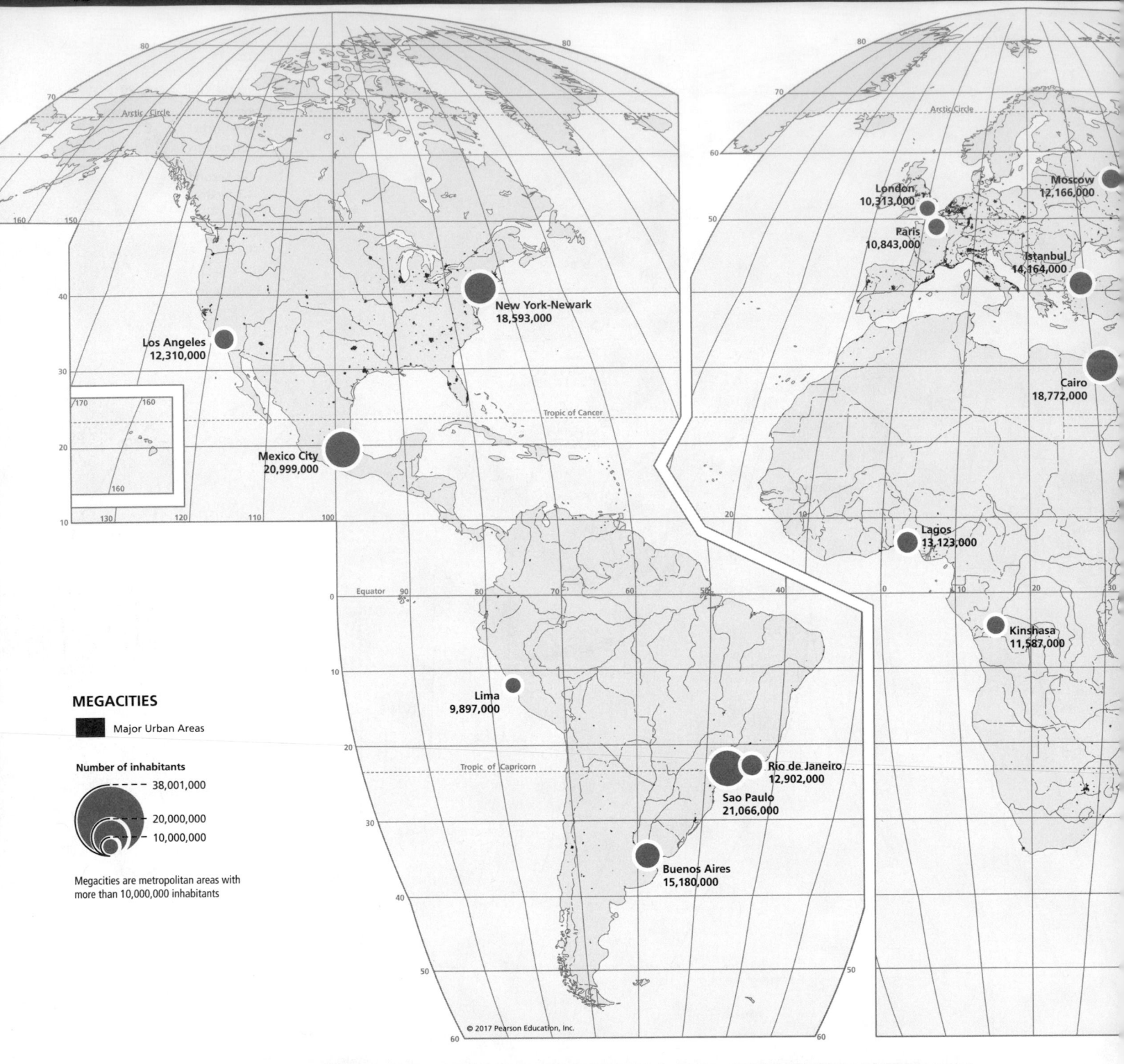

MEGACITIES

Major Urban Areas

Number of inhabitants

38,001,000

20,000,000

10,000,000

Megacities are metropolitan areas with more than 10,000,000 inhabitants

Los Angeles
12,310,000

New York-Newark
18,593,000

Mexico City
20,999,000

Lima
9,897,000

Rio de Janeiro
12,902,000

Sao Paulo
21,066,000

Buenos Aires
15,180,000

London
10,313,000

Paris
10,843,000

Moscow
12,166,000

Istanbul
14,164,000

Cairo
18,772,000

Lagos
13,123,000

Kinshasa
11,587,000

© 2017 Pearson Education, Inc.

Urban Agglomeration	Country	Population (2014)
Tokyo	Japan	38,001,000
Delhi	India	25,703,000
Shanghai	China	23,741,000
São Paulo	Brazil	21,066,000
Mumbai (Bombay)	India	21,043,000
Mexico City	Mexico	20,999,000
Beijing	China	20,384,000
Kinki M.M.A. (Osaka)	Japan	20,238,000
Al-Qahirah (Cairo)	Egypt	18,772,000
New York-Newark	United States of America	18,593,000
Dhaka	Bangladesh	17,598,000
Karachi	Pakistan	16,618,000
Buenos Aires	Argentina	15,180,000
Kolkata (Calcutta)	India	14,865,000
Istanbul	Turkey	14,164,000

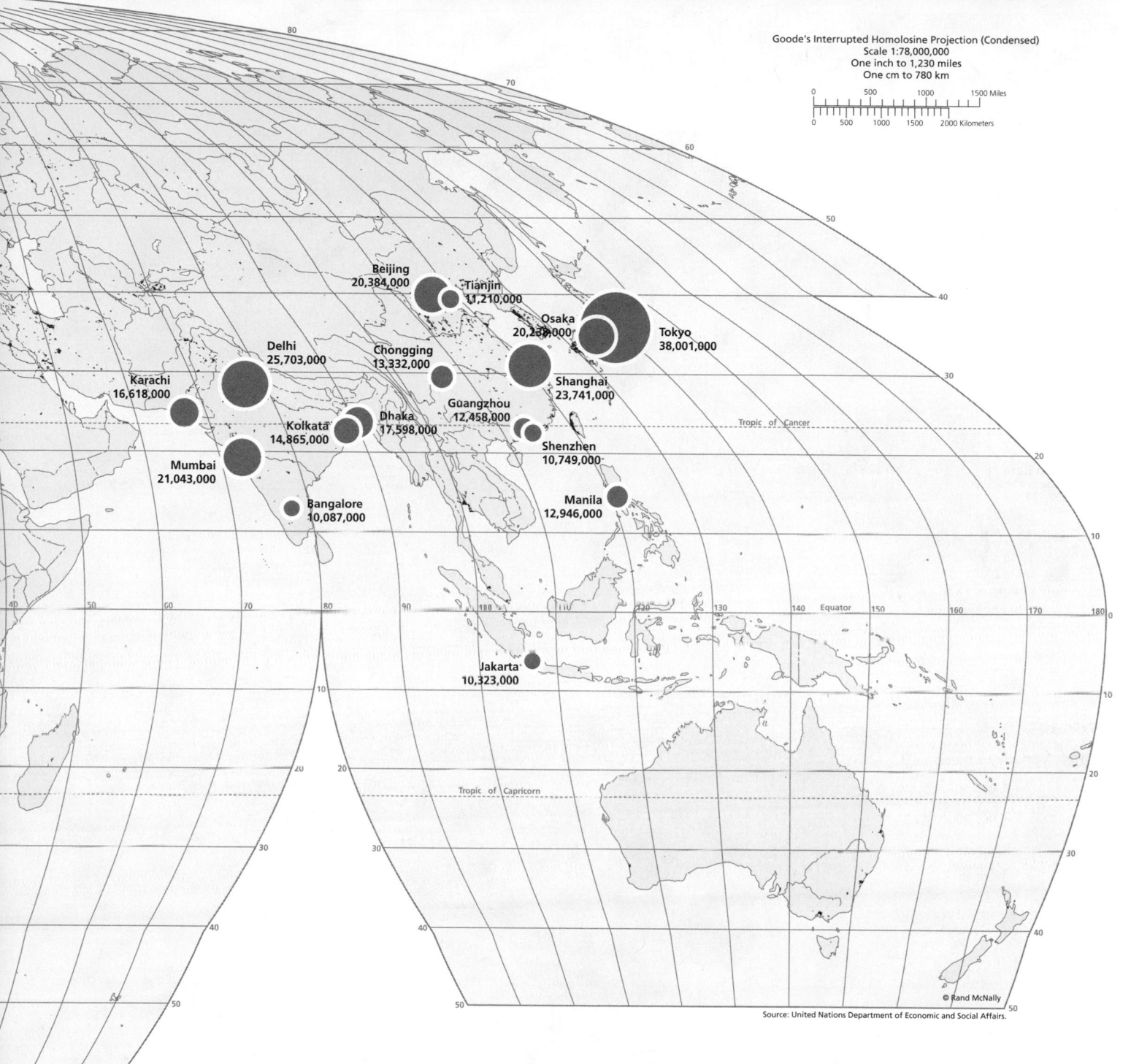

Goode's Interrupted Homolosine Projection (Condensed)
Scale 1:78,000,000
One inch to 1,230 miles
One cm to 780 km

Beijing
20,384,000

Tianjin
11,210,000

Osaka
20,238,000

Tokyo
38,001,000

Delhi
25,703,000

Chongqing
13,332,000

Shanghai
23,741,000

Karachi
16,618,000

Guangzhou
12,458,000

Kolkata
14,865,000

Dhaka
17,598,000

Tropic of Cancer

Shenzhen
10,749,000

Mumbai
21,043,000

Bangalore
10,087,000

Manila
12,946,000

Equator

Jakarta
10,323,000

Tropic of Capricorn

© Rand McNally

Source: United Nations Department of Economic and Social Affairs.

Urban Agglomeration	Country	Population (2014)
Chongqing	China	13,332,000
Lagos	Nigeria	13,123,000
Manila	Philippines	12,946,000
Rio de Janeiro	Brazil	12,902,000
Guangzhou, Guangdong	China	12,458,000
Los Angeles-Long Beach-Santa Ana	United States of America	12,310,000
Moscow	Russian Federation	12,166,000
Kinshasa	Democratic Republic of the Congo	11,587,000
Tianjin	China	11,210,000
Paris	France	10,843,000
Shenzhen	China	10,749,000
Jakarta	Indonesia	10,323,000
London	United Kingdom	10,313,000
Bangalore	India	10,087,000
Lima	Peru	9,897,000

POPULATION GROWTH

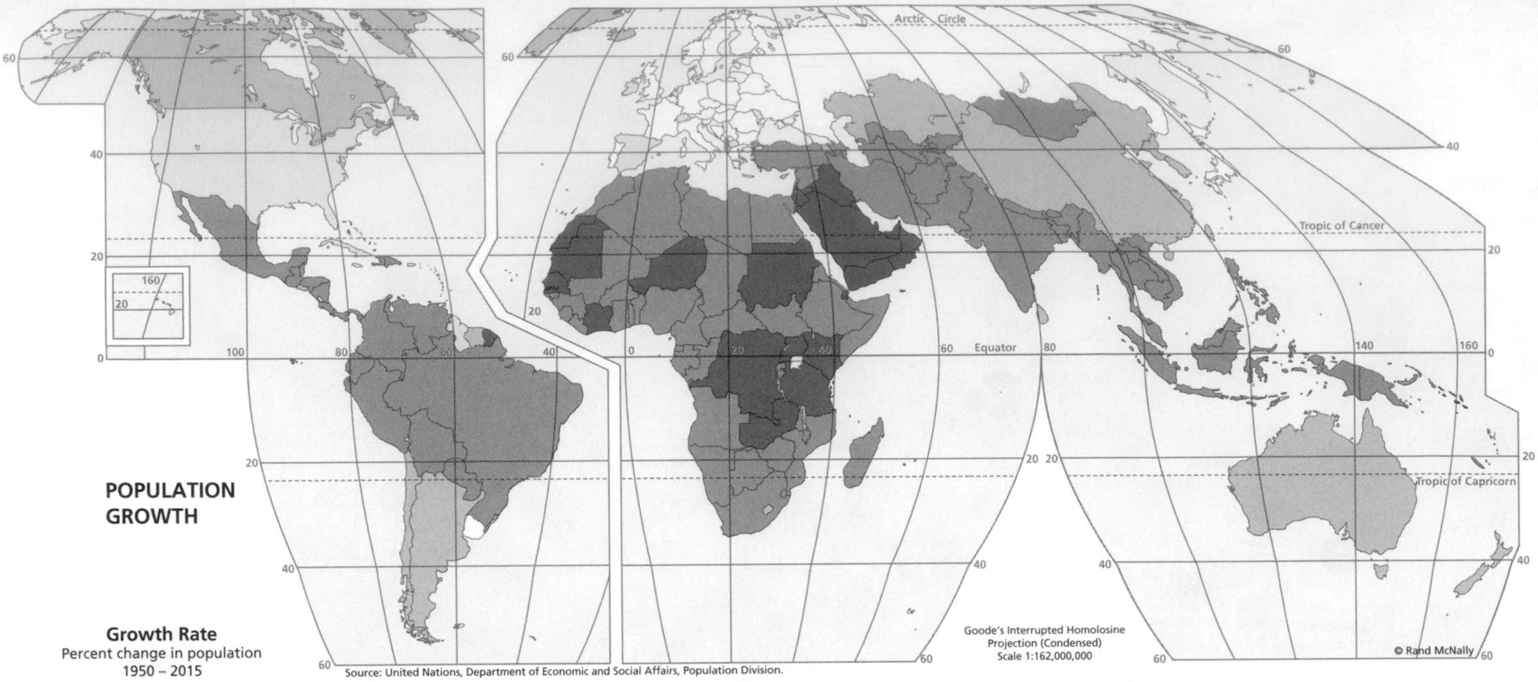
Arctic Circle
Tropic of Cancer
Equator
Tropic of Capricorn
Goode's Interrupted Homolosine
Projection (Condensed)
Scale 1:162,000,000
© Rand McNally

Source: United Nations, Department of Economic and Social Affairs, Population Division.

Growth Rate
Percent change in population
1950 – 2015

- Under 0.6%
- 0.6 - 1.0%
- 1.1 - 2.0% (World 1.9%)
- 2.1 - 5.0%
- Over 5.0%
- Data not available

FERTILITY RATE

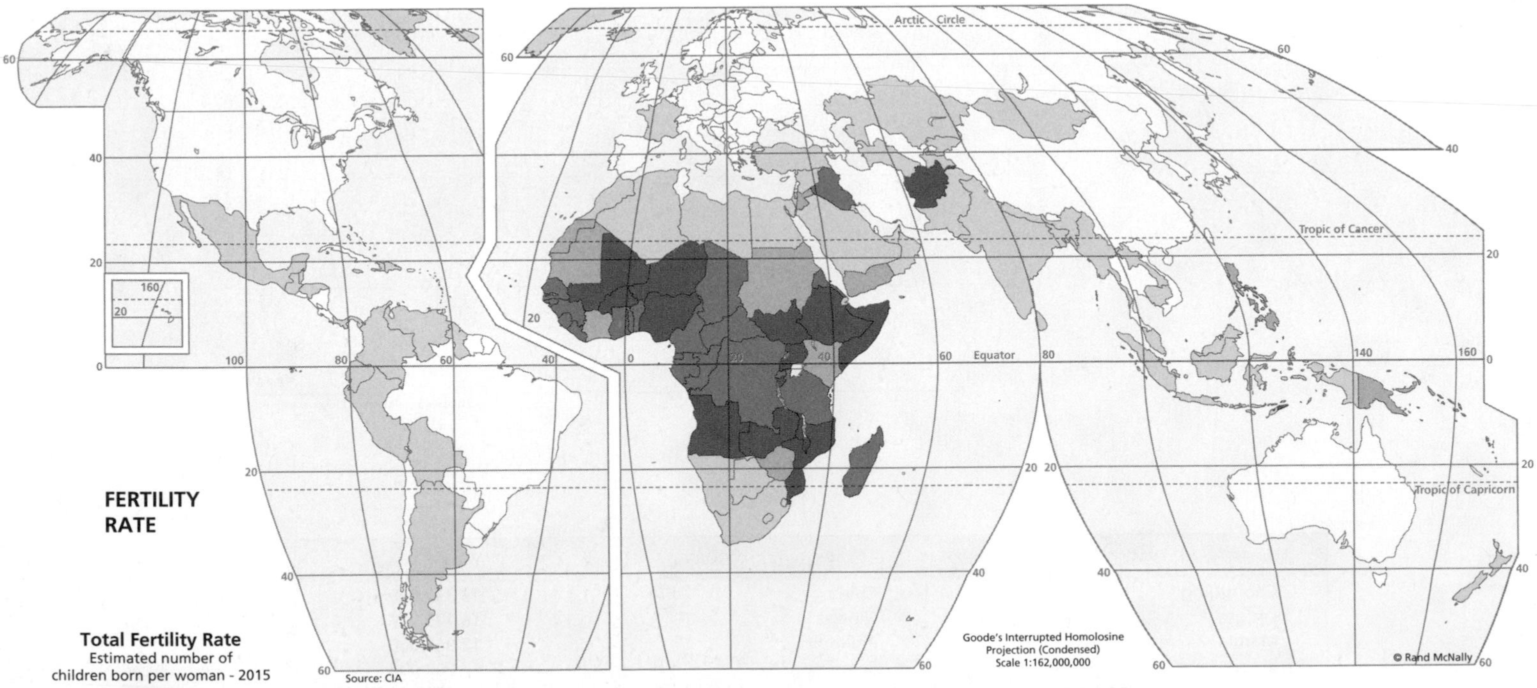
Arctic Circle
Tropic of Cancer
Equator
Tropic of Capricorn
Goode's Interrupted Homolosine
Projection (Condensed)
Scale 1:162,000,000
© Rand McNally

Source: CIA

Total Fertility Rate
Estimated number of
children born per woman - 2015

- Over 5.00
- 4.01 - 5.00
- 3.01 - 4.00
- 2.00 - 3.00 (World Avg. 2.42)
- Under 2.00
- Data not available

Total Fertility Rate (TFR) is the average number of children that would be born per woman if all women lived to the end of their childbearing years and bore children according to a given fertility rate at each age.
A rate of two children per woman is considered the replacement rate for a population, resulting in relative stability in terms of total numbers.

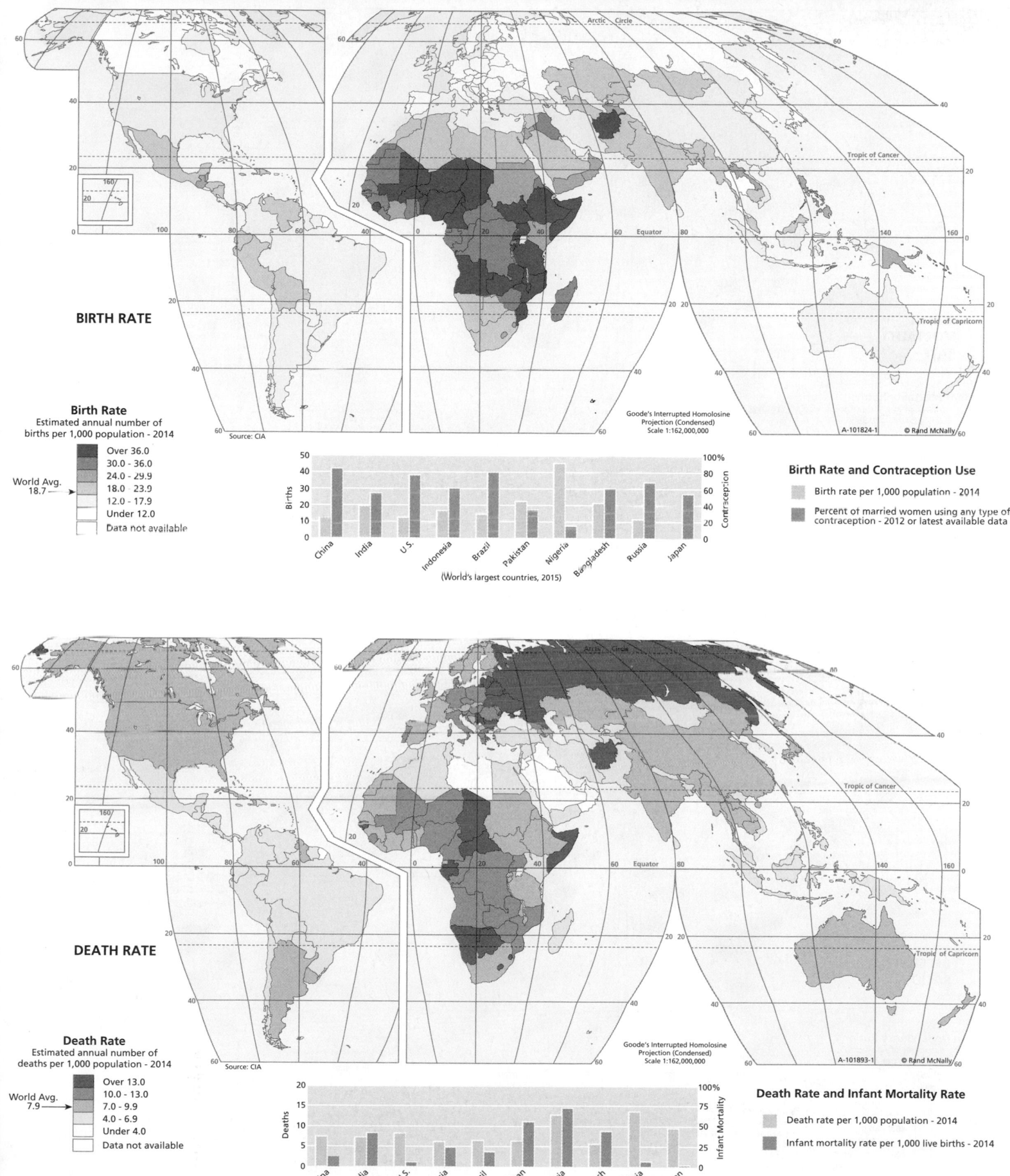

BIRTH RATE

Birth Rate
Estimated annual number of
births per 1,000 population - 2014

Source: CIA

Goode's Interrupted Homolosine
Projection (Condensed)
Scale 1:162,000,000

A-101824-1   © Rand McNally

World Avg.
18.7 →

- Over 36.0
- 30.0 - 36.0
- 24.0 - 29.9
- 18.0 - 23.9
- 12.0 - 17.9
- Under 12.0
- Data not available

Birth Rate and Contraception Use

- Birth rate per 1,000 population - 2014
- Percent of married women using any type of contraception - 2012 or latest available data

(World's largest countries, 2015)

China, India, U.S., Indonesia, Brazil, Pakistan, Nigeria, Bangladesh, Russia, Japan

DEATH RATE

Death Rate
Estimated annual number of
deaths per 1,000 population - 2014

Source: CIA

Goode's Interrupted Homolosine
Projection (Condensed)
Scale 1:162,000,000

A-101893-1   © Rand McNally

World Avg.
7.9 →

- Over 13.0
- 10.0 - 13.0
- 7.0 - 9.9
- 4.0 - 6.9
- Under 4.0
- Data not available

Death Rate and Infant Mortality Rate

- Death rate per 1,000 population - 2014
- Infant mortality rate per 1,000 live births - 2014

(World's largest countries, 2015)

China, India, U.S., Indonesia, Brazil, Pakistan, Nigeria, Bangladesh, Russia, Japan

INFANT MORTALITY RATE

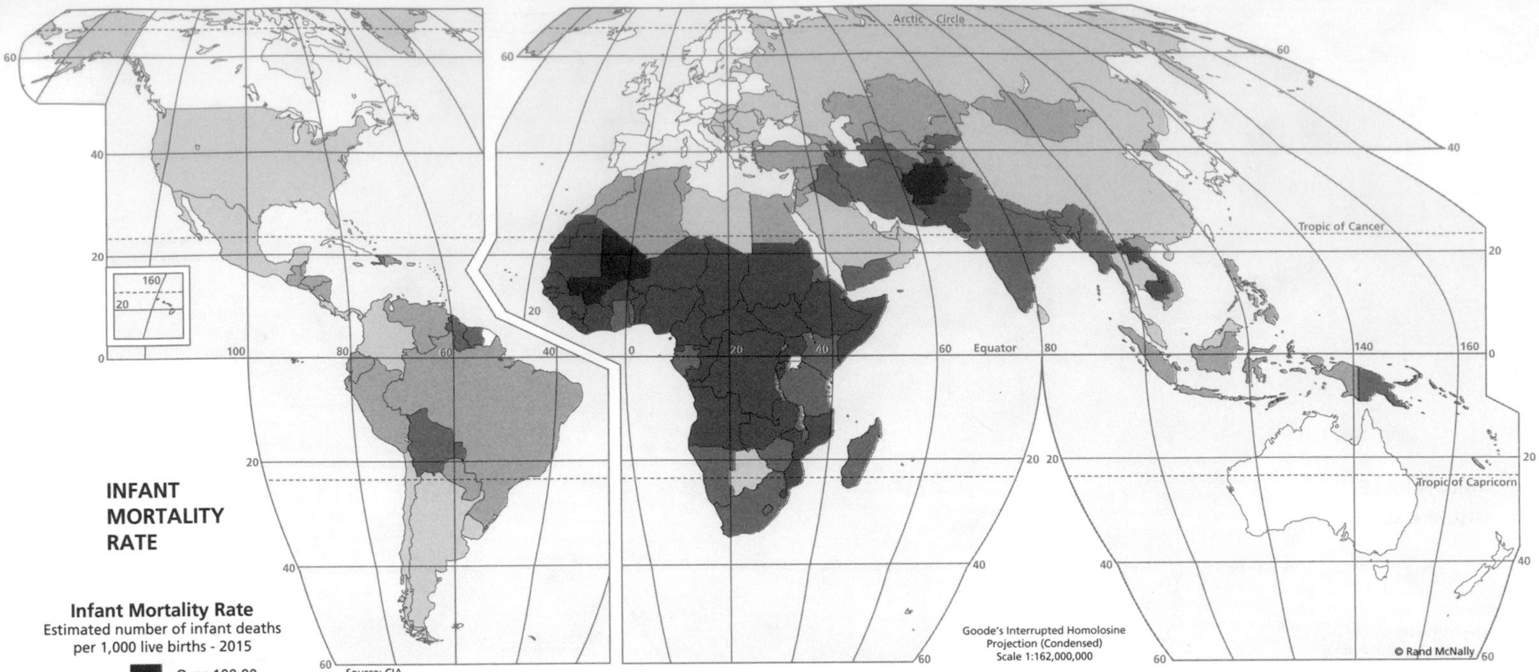
Arctic Circle
Tropic of Cancer
Equator
Tropic of Capricorn
Goode's Interrupted Homolosine
Projection (Condensed)
Scale 1:162,000,000
Source: CIA
© Rand McNally

Infant Mortality Rate
Estimated number of infant deaths per 1,000 live births - 2015

- Over 100.00
- 50.01 - 100.00
- 25.01 - 50.00 (World Avg. 32.00)
- 15.01 - 25.00
- 5.00 - 15.00
- Under 5.00
- Data not available

Infant Mortality Rate is the number of deaths of infants under one year old per 1,000 live births in the same year.

DEPENDENCY RATIO

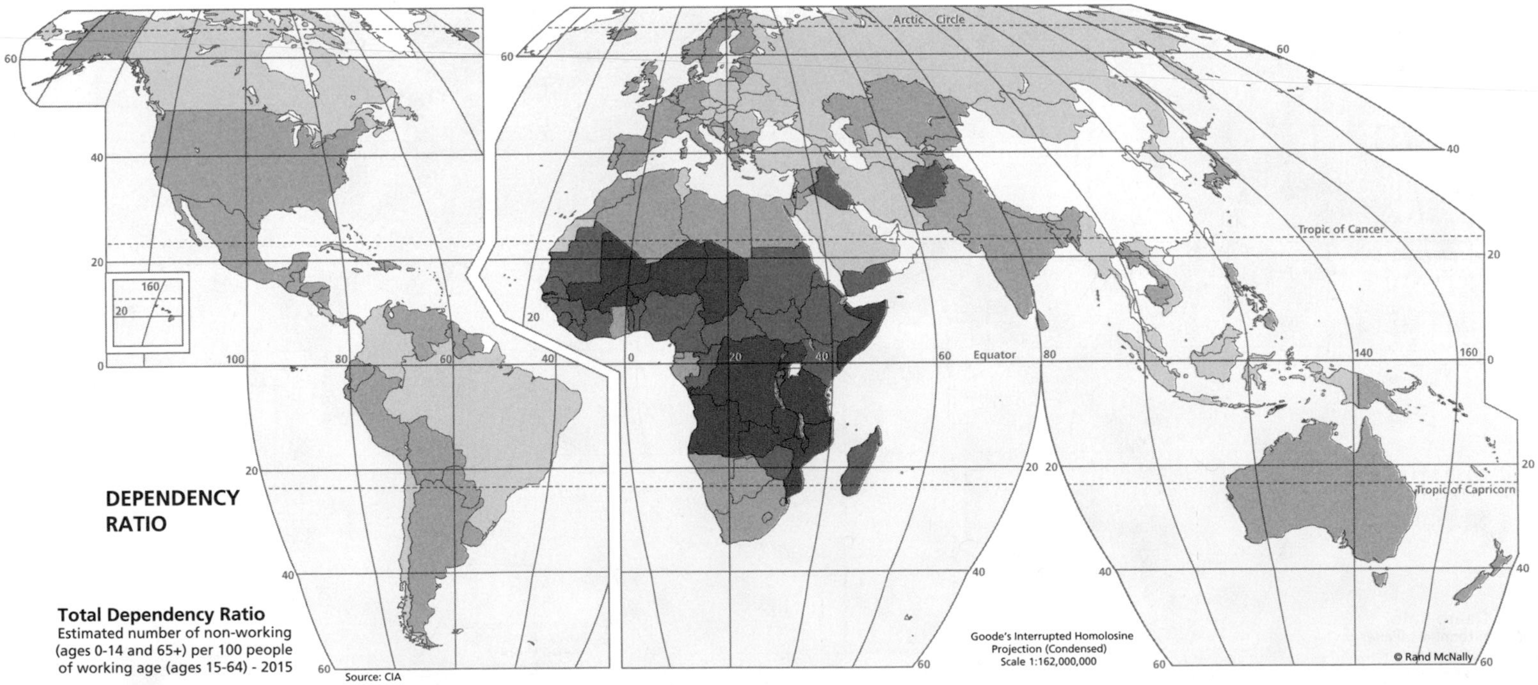
Arctic Circle
Tropic of Cancer
Equator
Tropic of Capricorn
Goode's Interrupted Homolosine
Projection (Condensed)
Scale 1:162,000,000
Source: CIA
© Rand McNally

Total Dependency Ratio
Estimated number of non-working (ages 0-14 and 65+) per 100 people of working age (ages 15-64) - 2015

- Over 90.0
- 75.1 - 90.0
- 50.1 - 75.0 (World Avg. 52.3)
- 40.0 - 50.0
- Under 40.0
- Data not available

Dependency ratios are a measure of the age structure of a population. They relate the number of individuals that are likely to be economically "dependent" on the support of others.

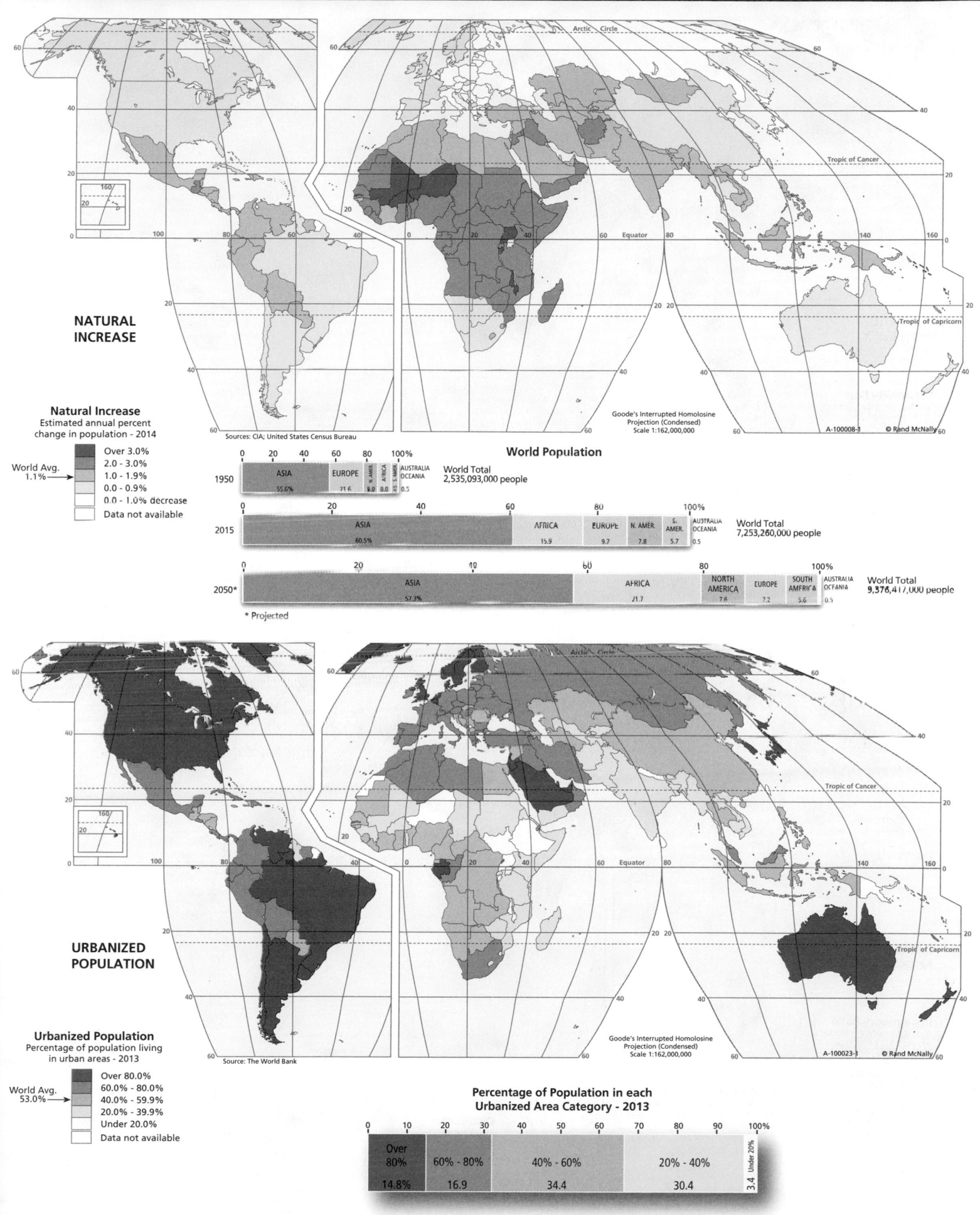

NATURAL INCREASE

Natural Increase
Estimated annual percent change in population - 2014

World Avg.
1.1%

- Over 3.0%
- 2.0 - 3.0%
- 1.0 - 1.9%
- 0.0 - 0.9%
- 0.0 - 1.0% decrease
- Data not available

Sources: CIA; United States Census Bureau

Goode's Interrupted Homolosine Projection (Condensed)
Scale 1:162,000,000

A-100008-1   © Rand McNally

World Population

1950 ASIA 55.6% EUROPE 21.6 N. AMER. AFRICA S. AMER. AUSTRALIA OCEANIA 0.5
World Total 2,535,093,000 people

2015 ASIA 60.5% AFRICA 15.9 EUROPE 9.7 N. AMER. 7.8 S. AMER. 5.7 AUSTRALIA OCEANIA 0.5
World Total 7,253,260,000 people

2050* ASIA 57.3% AFRICA 21.7 NORTH AMERICA 7.6 EUROPE 7.2 SOUTH AMERICA 5.6 AUSTRALIA OCEANIA 0.5
World Total 9,376,417,000 people

* Projected

URBANIZED POPULATION

Urbanized Population
Percentage of population living in urban areas - 2013

World Avg.
53.0%

- Over 80.0%
- 60.0% - 80.0%
- 40.0% - 59.9%
- 20.0% - 39.9%
- Under 20.0%
- Data not available

Source: The World Bank

Goode's Interrupted Homolosine Projection (Condensed)
Scale 1:162,000,000

A-100023-1   © Rand McNally

Percentage of Population in each Urbanized Area Category - 2013

Over 80%	60% - 80%	40% - 60%	20% - 40%	Under 20%
14.8%	16.9	34.4	30.4	3.4

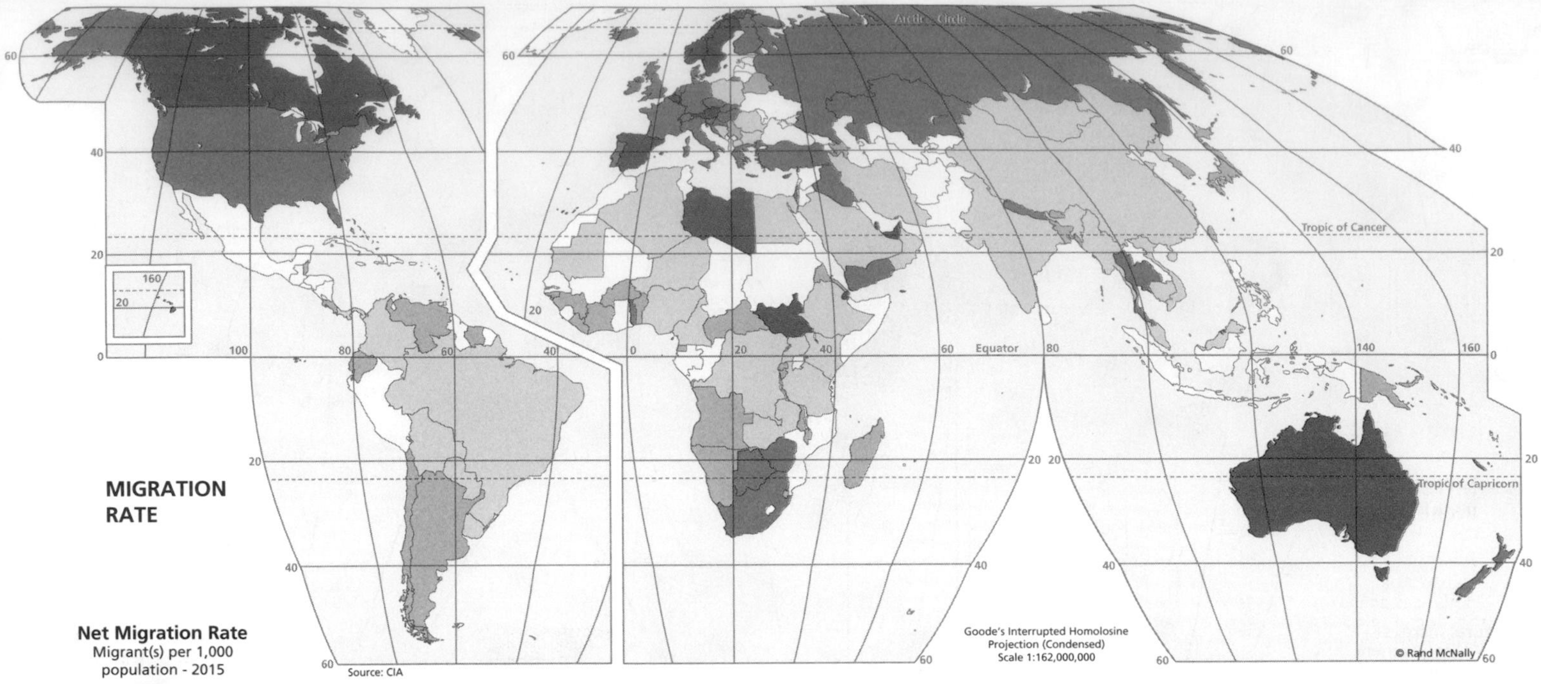
Arctic Circle
Tropic of Cancer
Equator
Tropic of Capricorn
MIGRATION RATE
Goode's Interrupted Homolosine Projection (Condensed) Scale 1:162,000,000
© Rand McNally
Source: CIA

Net Migration Rate
Migrant(s) per 1,000 population - 2015

- Over 5.0
- 1.0 to 4.9
- 0.0 to 0.9
- −0.9 to −0.1
- −4.9 to −1.0
- Under −5.0
- Data not available

Net migration rate compares the difference between the number of persons entering and leaving a country during the year per 1,000 persons (based on midyear population).

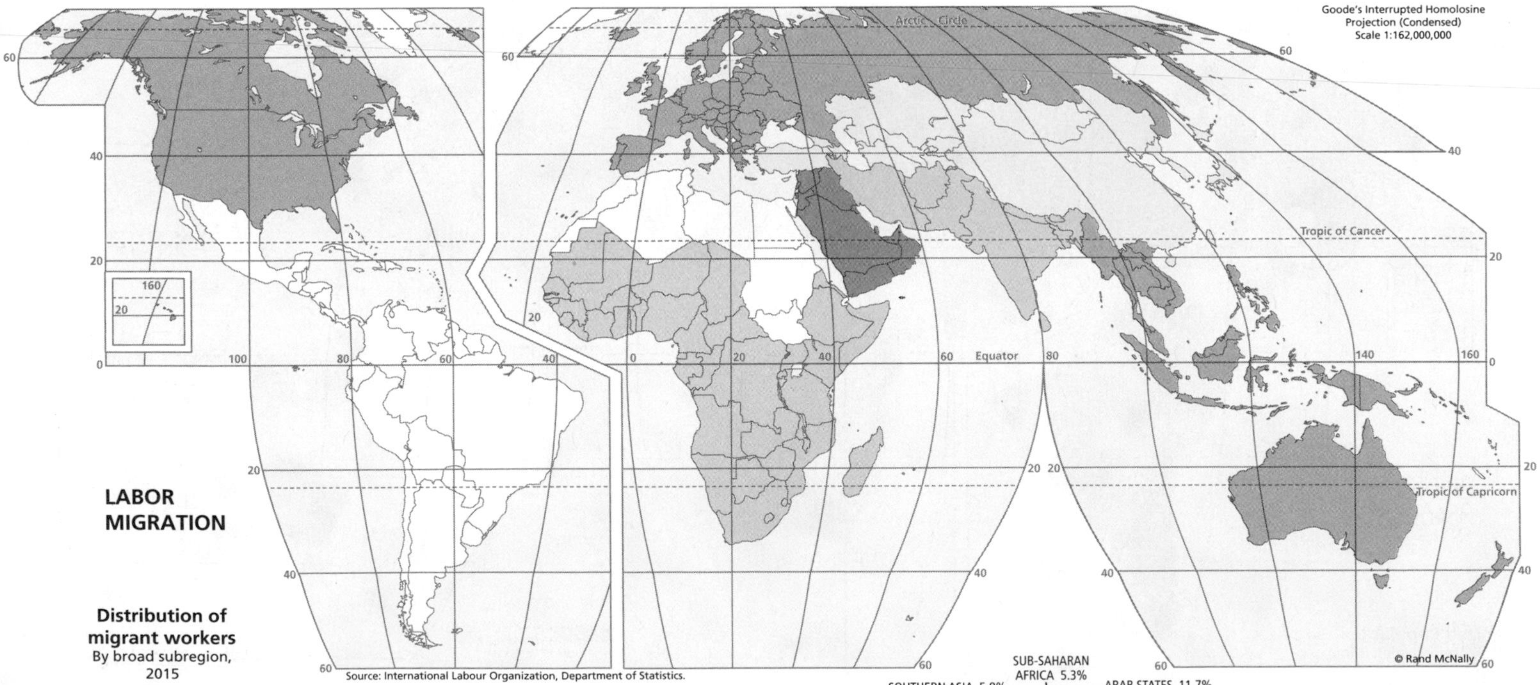
Goode's Interrupted Homolosine Projection (Condensed) Scale 1:162,000,000
Arctic Circle
Tropic of Cancer
Equator
Tropic of Capricorn
LABOR MIGRATION
© Rand McNally
Source: International Labour Organization, Department of Statistics.

Distribution of migrant workers
By broad subregion, 2015

- Under 3.0%
- 3.1 - 5.0%
- 5.1 - 7.0%
- 7.1 - 10.0%
- 10.1 - 20.0%
- Over 20.0%
- Data not available

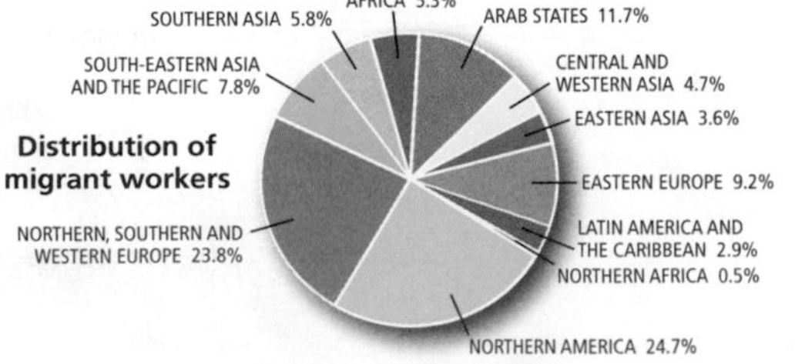
Distribution of migrant workers
SUB-SAHARAN AFRICA 5.3%
ARAB STATES 11.7%
CENTRAL AND WESTERN ASIA 4.7%
EASTERN ASIA 3.6%
EASTERN EUROPE 9.2%
LATIN AMERICA AND THE CARIBBEAN 2.9%
NORTHERN AFRICA 0.5%
NORTHERN AMERICA 24.7%
NORTHERN, SOUTHERN AND WESTERN EUROPE 23.8%
SOUTH-EASTERN ASIA AND THE PACIFIC 7.8%
SOUTHERN ASIA 5.8%

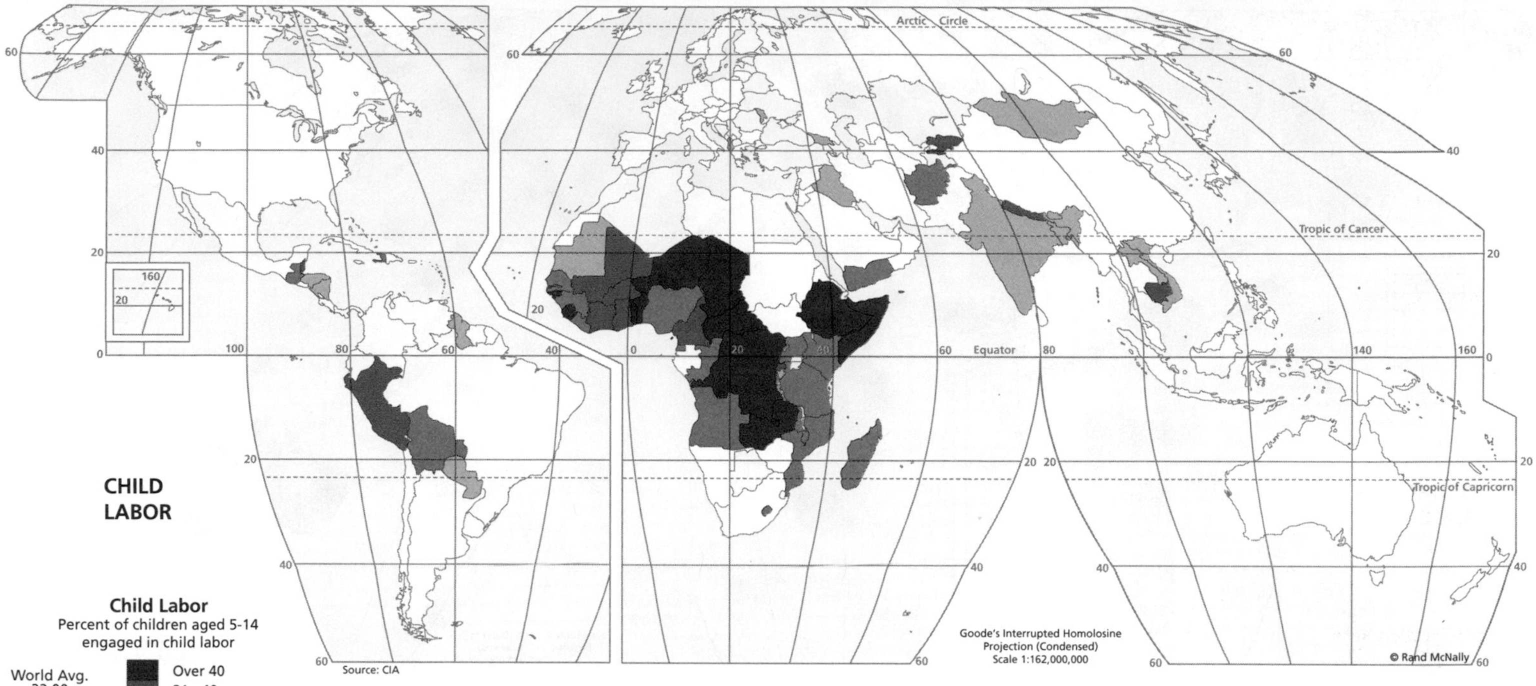

CHILD LABOR

Child Labor
Percent of children aged 5-14
engaged in child labor

World Avg.
32.00 →

- Over 40
- 31 - 40
- 21 - 30
- 11 - 20
- 1 - 10
- Data not available

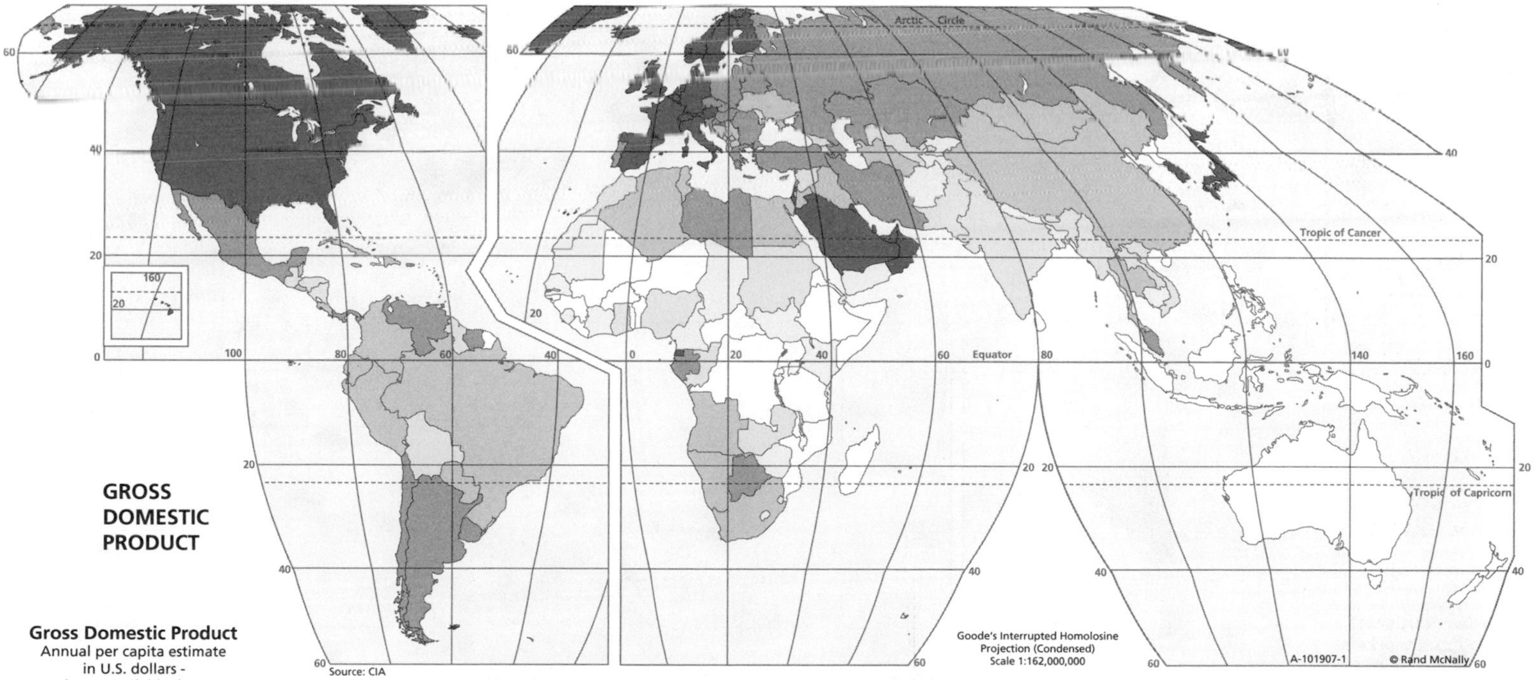

Source: CIA

Goode's Interrupted Homolosine
Projection (Condensed)
Scale 1:162,000,000

© Rand McNally

GROSS DOMESTIC PRODUCT

Gross Domestic Product
Annual per capita estimate
in U.S. dollars -
latest available data

World Avg.
$16,100 →

- Over $32,000
- $16,000 - $32,000
- $8,000 - $15,999
- $4,000 - $7,999
- $2,000 - $3,999
- Under $2,000
- Data not available

Source: CIA

Goode's Interrupted Homolosine
Projection (Condensed)
Scale 1:162,000,000

A-101907-1   © Rand McNally

Percentage of World Population in each Per Capita GDP Category

0	10	20	30	40	50	60	70	80	90	100%

Over $32,000	$16,000-$32,000	$8,000-$15,999	$4,000 - $7,999	$2,000-$3,999	Under $2,000
14.5%	10.3%	32.8%	29.1%	6.2%	6.9%

SCHOOL LIFE EXPECTANCY

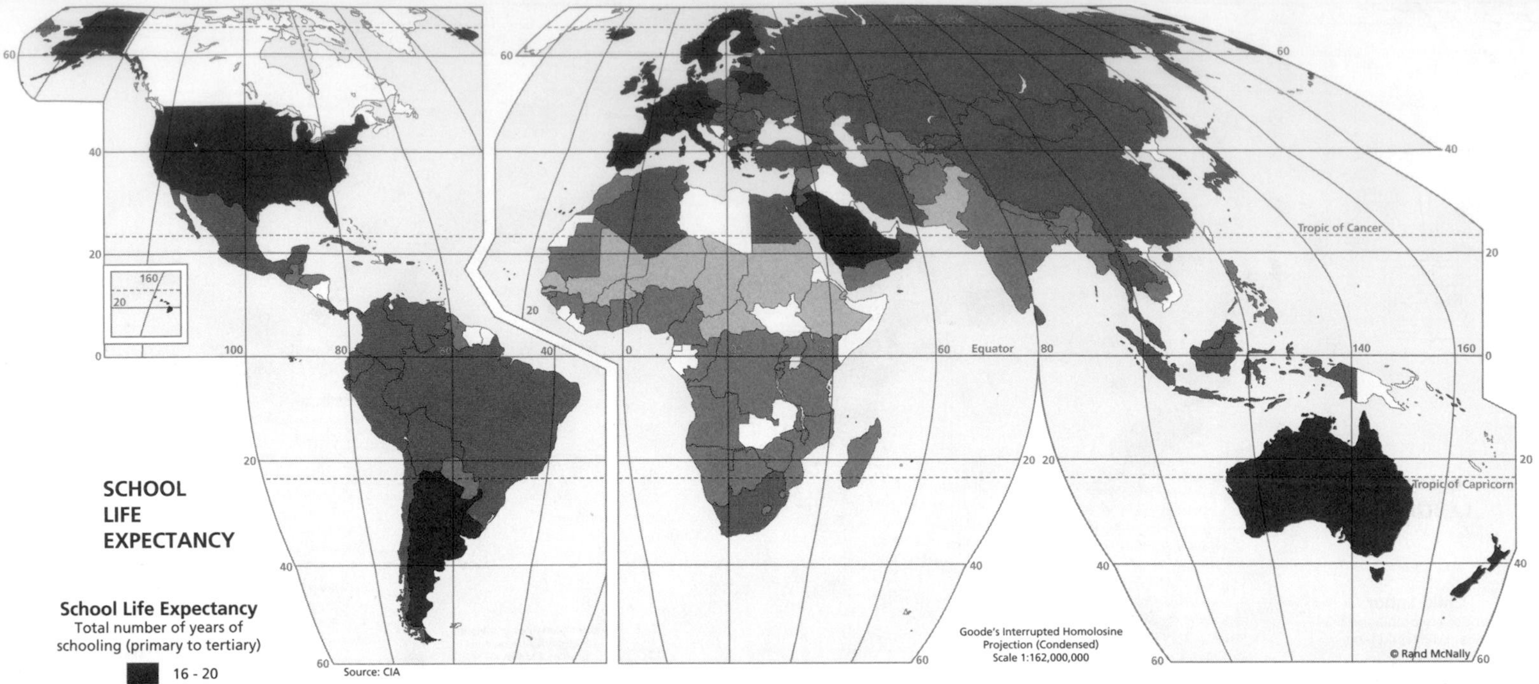
Tropic of Cancer
Equator
Tropic of Capricorn
Source: CIA
Goode's Interrupted Homolosine
Projection (Condensed)
Scale 1:162,000,000
© Rand McNally

School Life Expectancy
Total number of years of schooling (primary to tertiary)

World Avg. 12 →

- 16 - 20
- 13 - 15
- 9 - 12
- 5 - 8
- Data not available

LITERACY

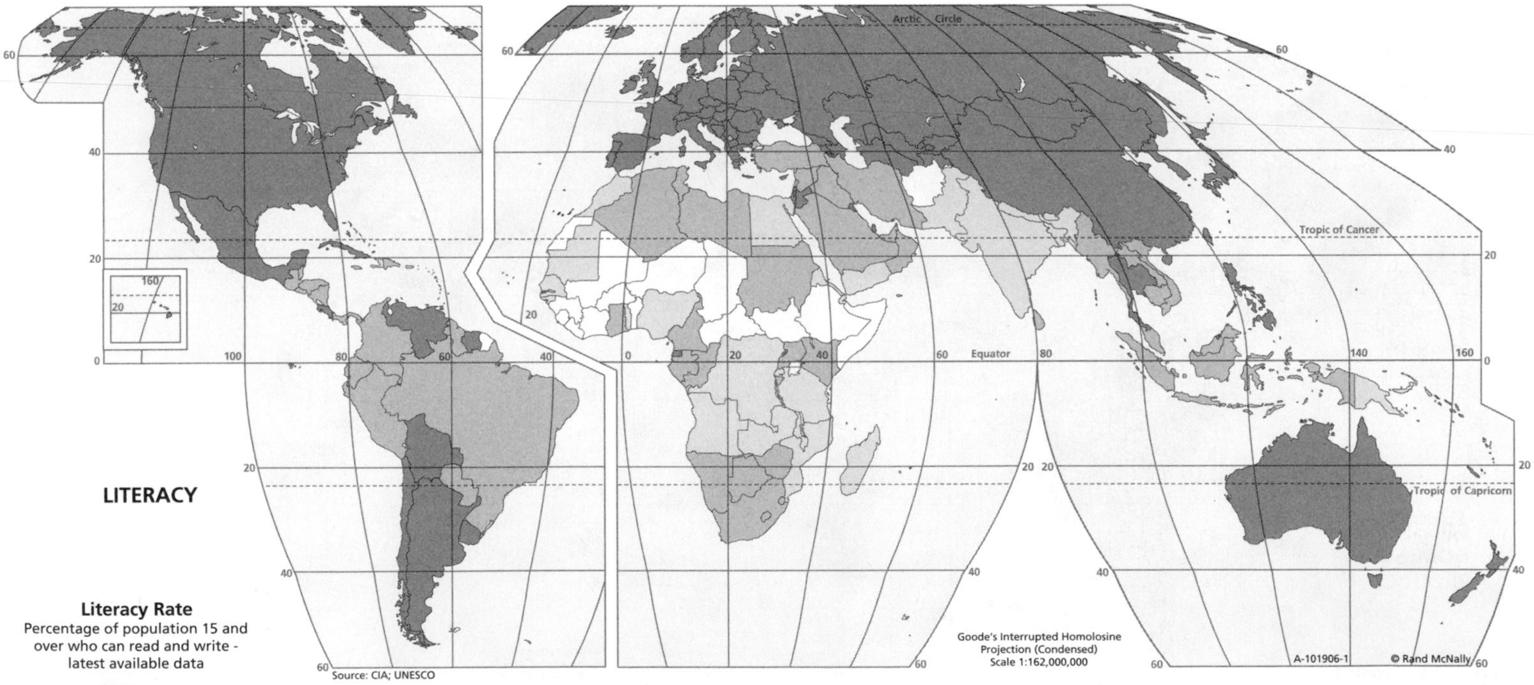
Arctic Circle
Tropic of Cancer
Equator
Tropic of Capricorn
Source: CIA; UNESCO
Goode's Interrupted Homolosine
Projection (Condensed)
Scale 1:162,000,000
A-101906-1
© Rand McNally

Literacy Rate
Percentage of population 15 and over who can read and write - latest available data

World Avg. 86% →

- Over 95%
- 75 - 95%
- 50 - 74%
- Under 50%
- Data not available

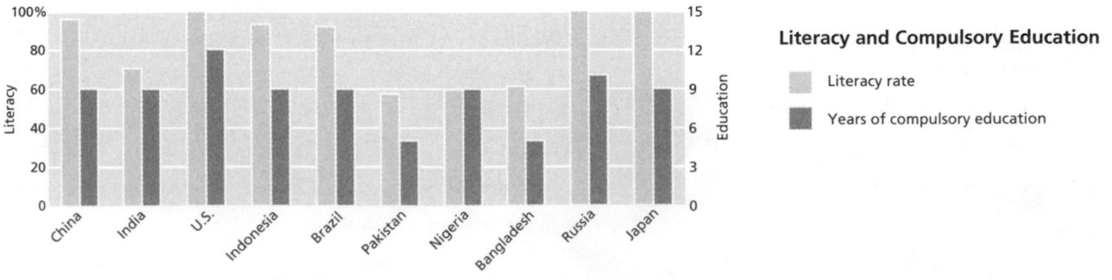
Literacy and Compulsory Education
Literacy rate
Years of compulsory education
Literacy
Education
100%
80
60
40
20
0
15
12
9
6
3
China
India
U.S.
Indonesia
Brazil
Pakistan
Nigeria
Bangladesh
Russia
Japan

(World's largest countries, 2015)

Arctic Circle
Tropic of Cancer
Equator
Tropic of Capricorn
E
F
S
S, E
P
A
G
R
H
B
C
Goode's Interrupted Homolosine Projection (Condensed) Scale 1:162,000,000
M-102080-2
© Rand McNally

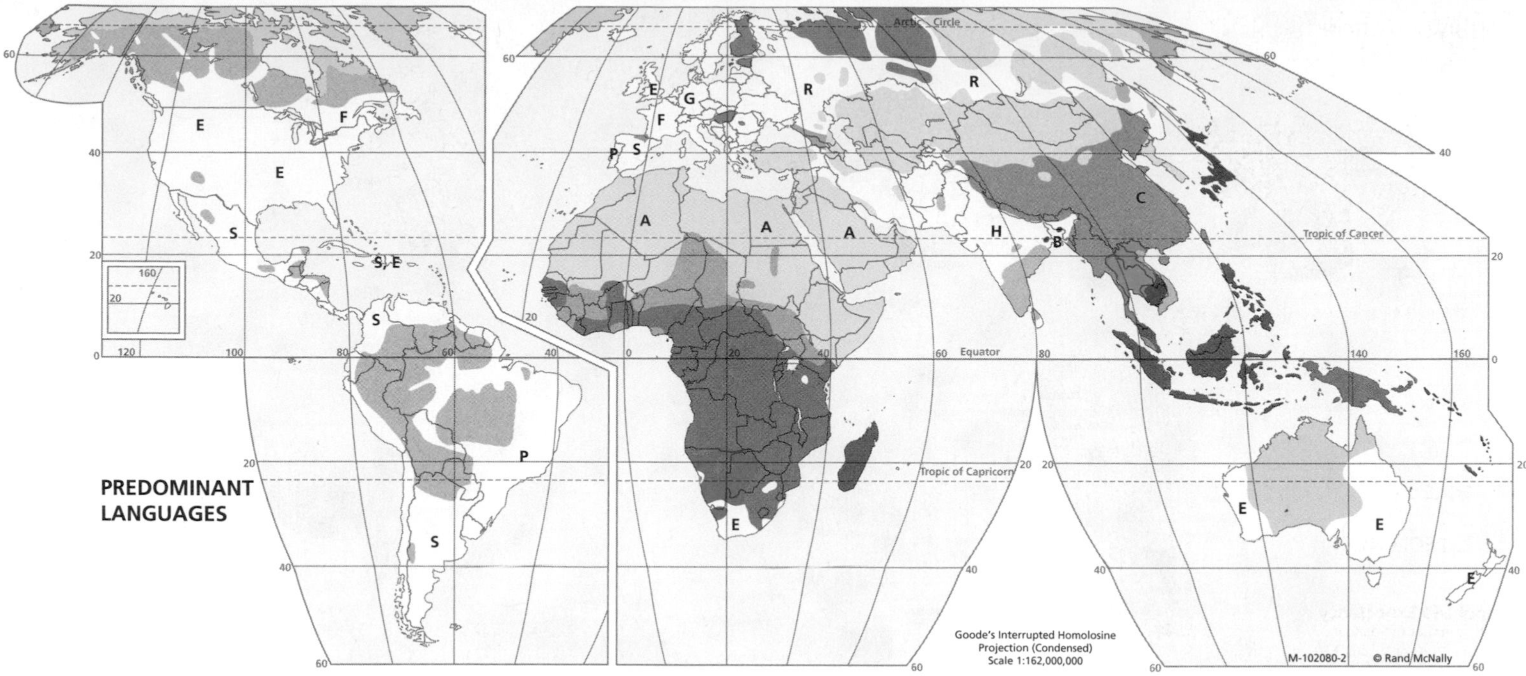

PREDOMINANT LANGUAGES

Language Families

- 1 - Afro-Asiatic
- 2 - Altaic
- 3 - American Indian
- 4 - Australian Aborigine
- 5 - Austro Asiatic
- 6 - Austronesian
- 7 - Basque
- 8 - Dravidian
- 9 - Eskimo-Aleut
- 10 - Indo-European
- 11 - Japanese
- 12 - Khoisan
- 13 - Korean
- 14 - Niger-Congo
- 15 - Nilo-Saharan
- 16 - North Caucasian
- 17 - Sino-Tibetan
- 18 - Tai-Kadai
- 19 - Trans New Guinea
- 20 - Uralic
- 21 - Vietnamese
- 22 - Yeniseian

Major Languages

- A - Arabic
- B - Bengali
- C - Chinese
- E - English
- F - French
- G - German
- H - Hindi
- P - Portuguese
- R - Russian
- S - Spanish

Arctic Circle
Tropic of Cancer
Equator
Tropic of Capricorn
Goode's Interrupted Homolosine Projection (Condensed) Scale 1:162,000,000
M-102081-2
© Rand McNally

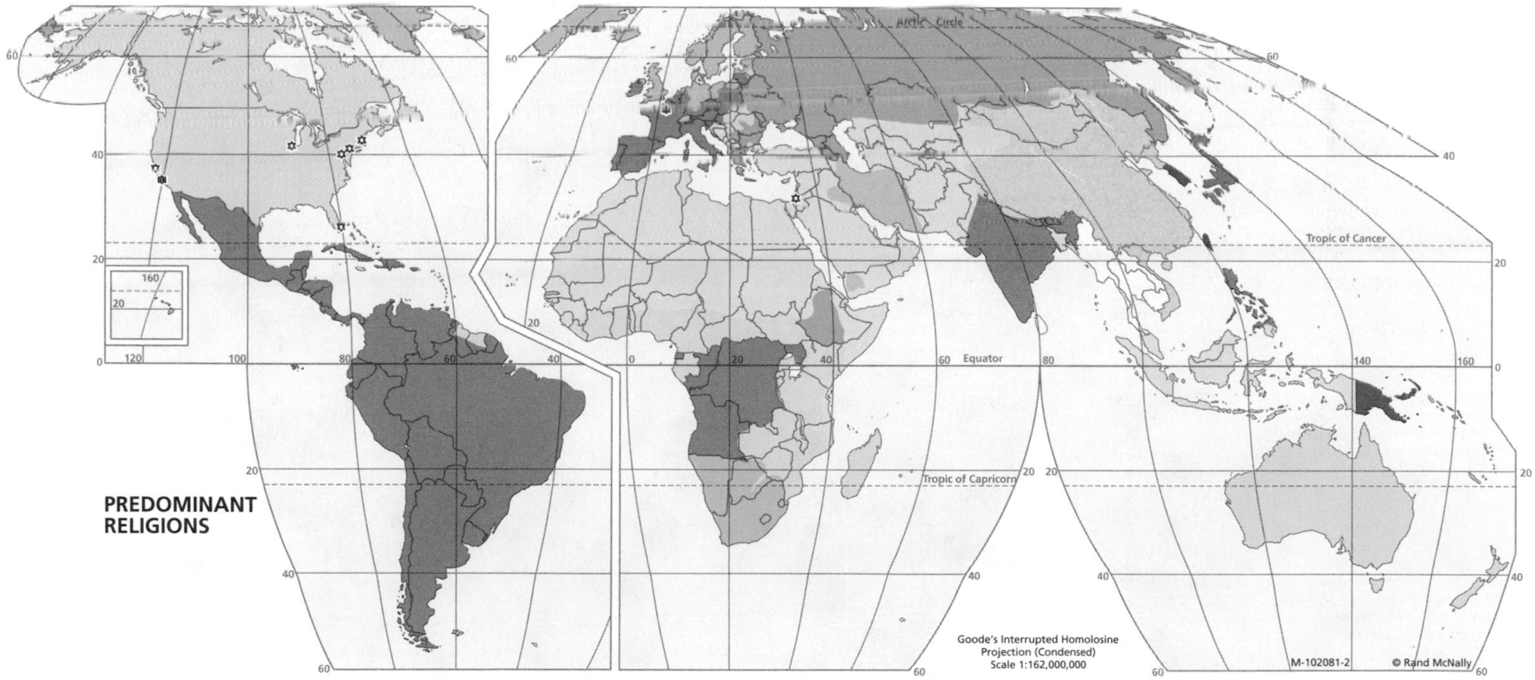

PREDOMINANT RELIGIONS

Christianity

- Roman Catholic
- Eastern Churches (Orthodox, Armenian, and Coptic)
- Protestant
- Sect not differentiated

Islam

- Sunni Muslim
- Shia Muslim

Buddhism

- Southern Buddhism
- Lamaism (Northern Buddhism)

- Hinduism
- ✡ Judaism (mainly in cities)
- Religions of Japan
- Mixed
- None / Unorganized

HIV INFECTION

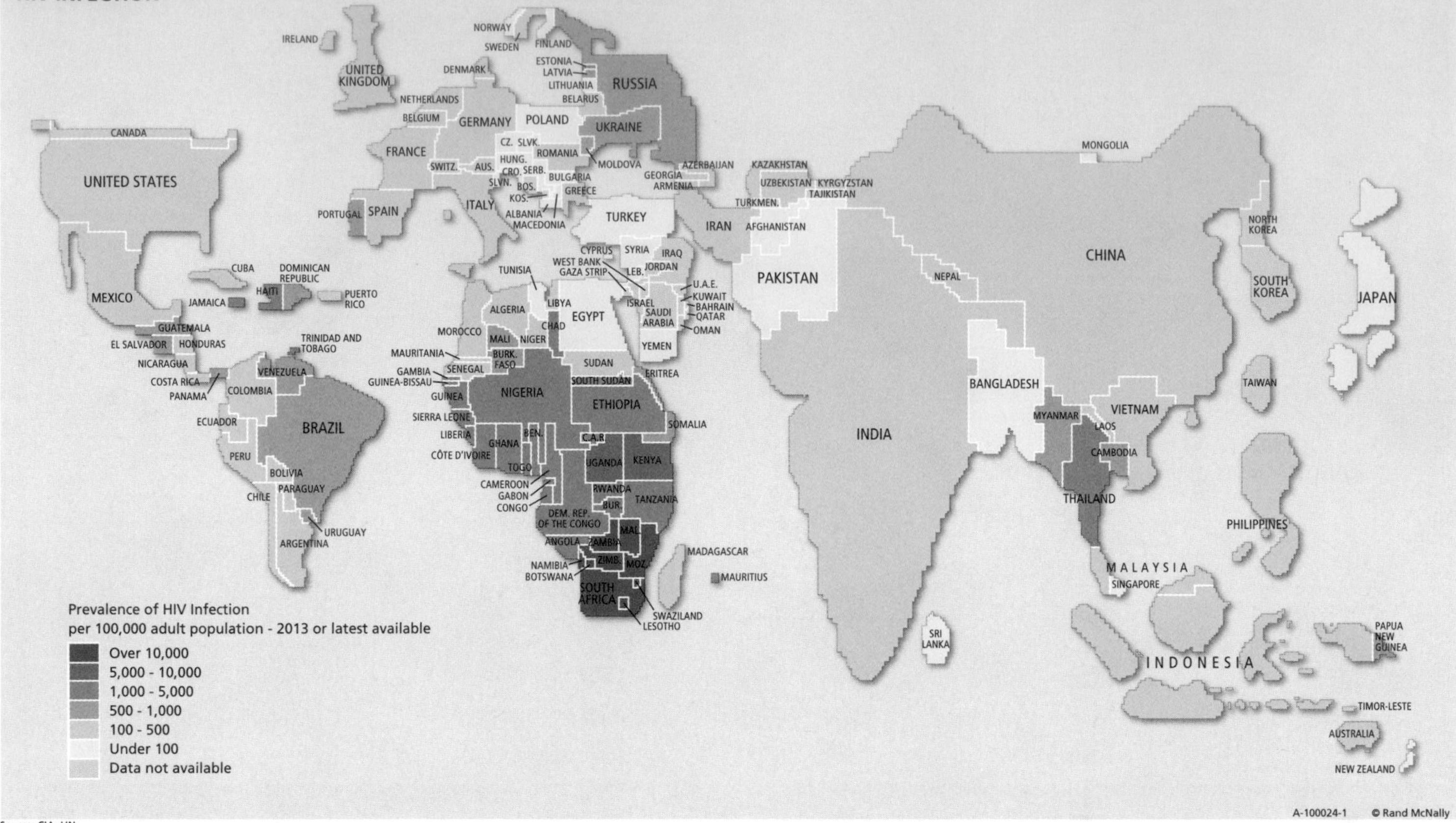

IRELAND NORWAY SWEDEN FINLAND ESTONIA LATVIA LITHUANIA BELARUS RUSSIA MONGOLIA
UNITED KINGDOM DENMARK NETHERLANDS BELGIUM GERMANY POLAND UKRAINE KAZAKHSTAN
CANADA FRANCE SWITZ. AUS. CZ. SLVK. HUNG. ROMANIA MOLDOVA GEORGIA AZERBAIJAN UZBEKISTAN KYRGYZSTAN NORTH KOREA
UNITED STATES SLVN. BOS. CRO. SERB. BULGARIA ARMENIA TAJIKISTAN CHINA
PORTUGAL SPAIN ITALY KOS. GREECE TURKEY TURKMEN. SOUTH KOREA JAPAN
ALBANIA MACEDONIA CYPRUS SYRIA IRAQ IRAN AFGHANISTAN
CUBA DOMINICAN REPUBLIC PUERTO RICO WEST BANK GAZA STRIP JORDAN LEB. PAKISTAN NEPAL NORTH KOREA
MEXICO JAMAICA HAITI TUNISIA ISRAEL KUWAIT BAHRAIN QATAR OMAN TAIWAN
GUATEMALA MOROCCO ALGERIA LIBYA EGYPT SAUDI ARABIA U.A.E. BANGLADESH MYANMAR VIETNAM
EL SALVADOR HONDURAS MAURITANIA MALI NIGER CHAD SUDAN YEMEN ERITREA INDIA LAOS
NICARAGUA VENEZUELA SENEGAL BURK. FASO SOUTH SUDAN ETHIOPIA SOMALIA CAMBODIA
COSTA RICA PANAMA COLOMBIA GAMBIA GUINEA-BISSAU GUINEA NIGERIA THAILAND PHILIPPINES
ECUADOR SIERRA LEONE LIBERIA CÔTE D'IVOIRE GHANA TOGO BEN. C.A.R. UGANDA KENYA
PERU BRAZIL CAMEROON GABON CONGO RWANDA BUR. TANZANIA MADAGASCAR MAURITIUS MALAYSIA SINGAPORE
BOLIVIA DEM. REP. OF THE CONGO MAL. MOZ. SRI LANKA INDONESIA PAPUA NEW GUINEA
PARAGUAY ANGOLA ZAMBIA ZIMB. NAMIBIA BOTSWANA SOUTH AFRICA SWAZILAND LESOTHO TIMOR-LESTE
CHILE URUGUAY ARGENTINA AUSTRALIA NEW ZEALAND

Prevalence of HIV Infection
per 100,000 adult population - 2013 or latest available

- Over 10,000
- 5,000 - 10,000
- 1,000 - 5,000
- 500 - 1,000
- 100 - 500
- Under 100
- Data not available

Source: CIA; UN

A-100024-1　© Rand McNally

Size of each country is proportional to its population

☐ = 25,000,000 people

Countries with populations under 1,000,000 are not shown.

TUBERCULOSIS

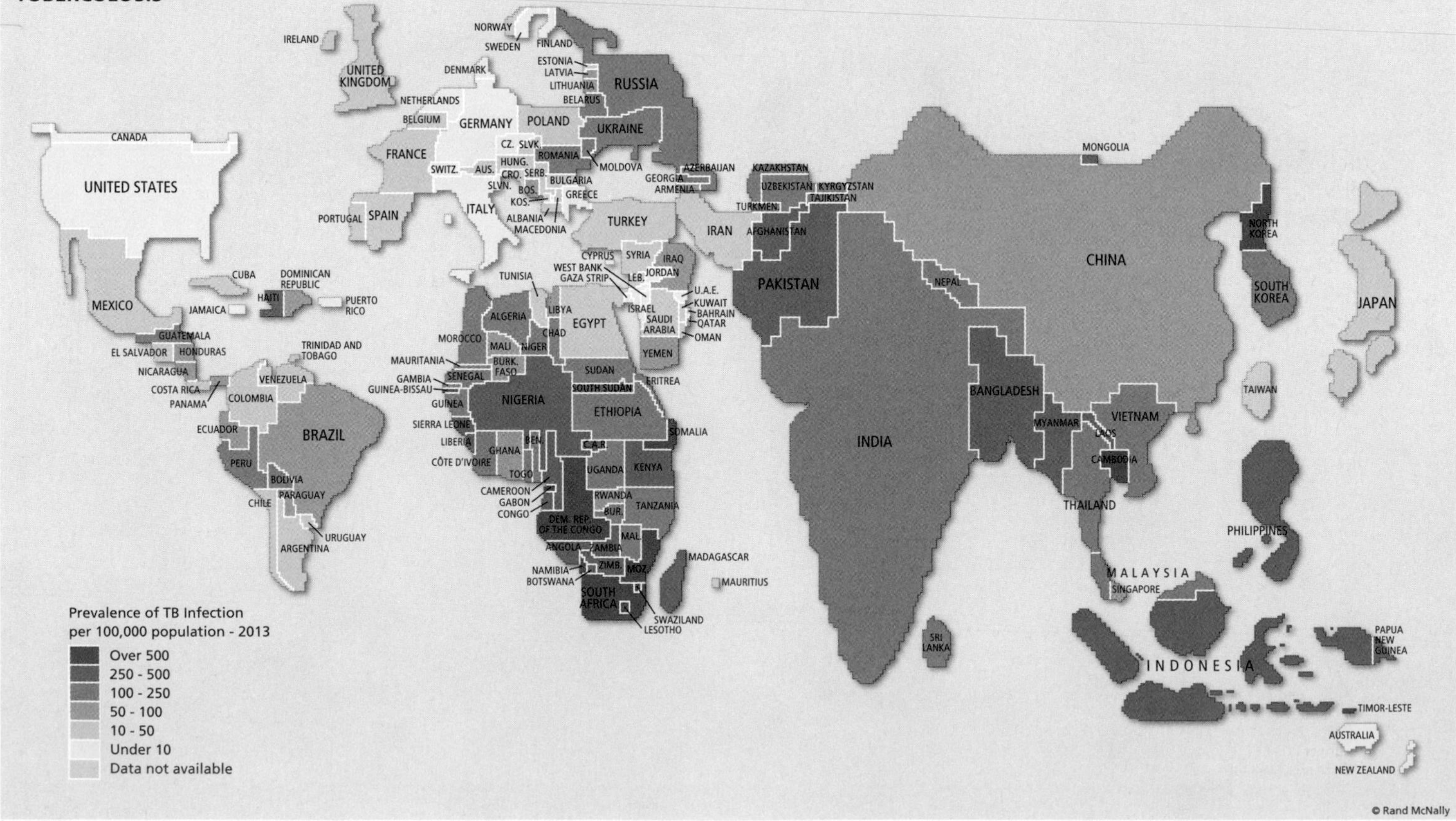

IRELAND NORWAY SWEDEN FINLAND ESTONIA LATVIA LITHUANIA BELARUS RUSSIA MONGOLIA
UNITED KINGDOM DENMARK NETHERLANDS BELGIUM GERMANY POLAND UKRAINE
CANADA FRANCE SWITZ. AUS. CZ. SLVK. HUNG. ROMANIA MOLDOVA GEORGIA AZERBAIJAN KAZAKHSTAN UZBEKISTAN KYRGYZSTAN NORTH KOREA
UNITED STATES SLVN. BOS. CRO. SERB. BULGARIA ARMENIA TAJIKISTAN CHINA
PORTUGAL SPAIN ITALY KOS. GREECE TURKEY TURKMEN. SOUTH KOREA JAPAN
ALBANIA MACEDONIA CYPRUS SYRIA IRAQ IRAN AFGHANISTAN
CUBA DOMINICAN REPUBLIC PUERTO RICO WEST BANK GAZA STRIP JORDAN LEB. PAKISTAN NEPAL
MEXICO JAMAICA HAITI TUNISIA ISRAEL KUWAIT BAHRAIN QATAR OMAN TAIWAN
GUATEMALA MOROCCO ALGERIA LIBYA EGYPT SAUDI ARABIA U.A.E. YEMEN BANGLADESH MYANMAR VIETNAM
EL SALVADOR HONDURAS MAURITANIA MALI NIGER CHAD SUDAN ERITREA INDIA LAOS
NICARAGUA VENEZUELA SENEGAL BURK. FASO SOUTH SUDAN ETHIOPIA SOMALIA CAMBODIA
COSTA RICA PANAMA COLOMBIA GAMBIA GUINEA-BISSAU GUINEA NIGERIA THAILAND PHILIPPINES
ECUADOR SIERRA LEONE LIBERIA CÔTE D'IVOIRE GHANA TOGO C.A.R. UGANDA KENYA
PERU BRAZIL CAMEROON GABON CONGO RWANDA BUR. TANZANIA MADAGASCAR MAURITIUS MALAYSIA SINGAPORE
BOLIVIA DEM. REP. OF THE CONGO MAL. MOZ. SRI LANKA INDONESIA PAPUA NEW GUINEA
PARAGUAY ANGOLA ZAMBIA ZIMB. NAMIBIA BOTSWANA SOUTH AFRICA SWAZILAND LESOTHO TIMOR-LESTE
CHILE URUGUAY ARGENTINA AUSTRALIA NEW ZEALAND

Prevalence of TB Infection
per 100,000 population - 2013

- Over 500
- 250 - 500
- 100 - 250
- 50 - 100
- 10 - 50
- Under 10
- Data not available

© Rand McNally

Source: WHO

MALARIA

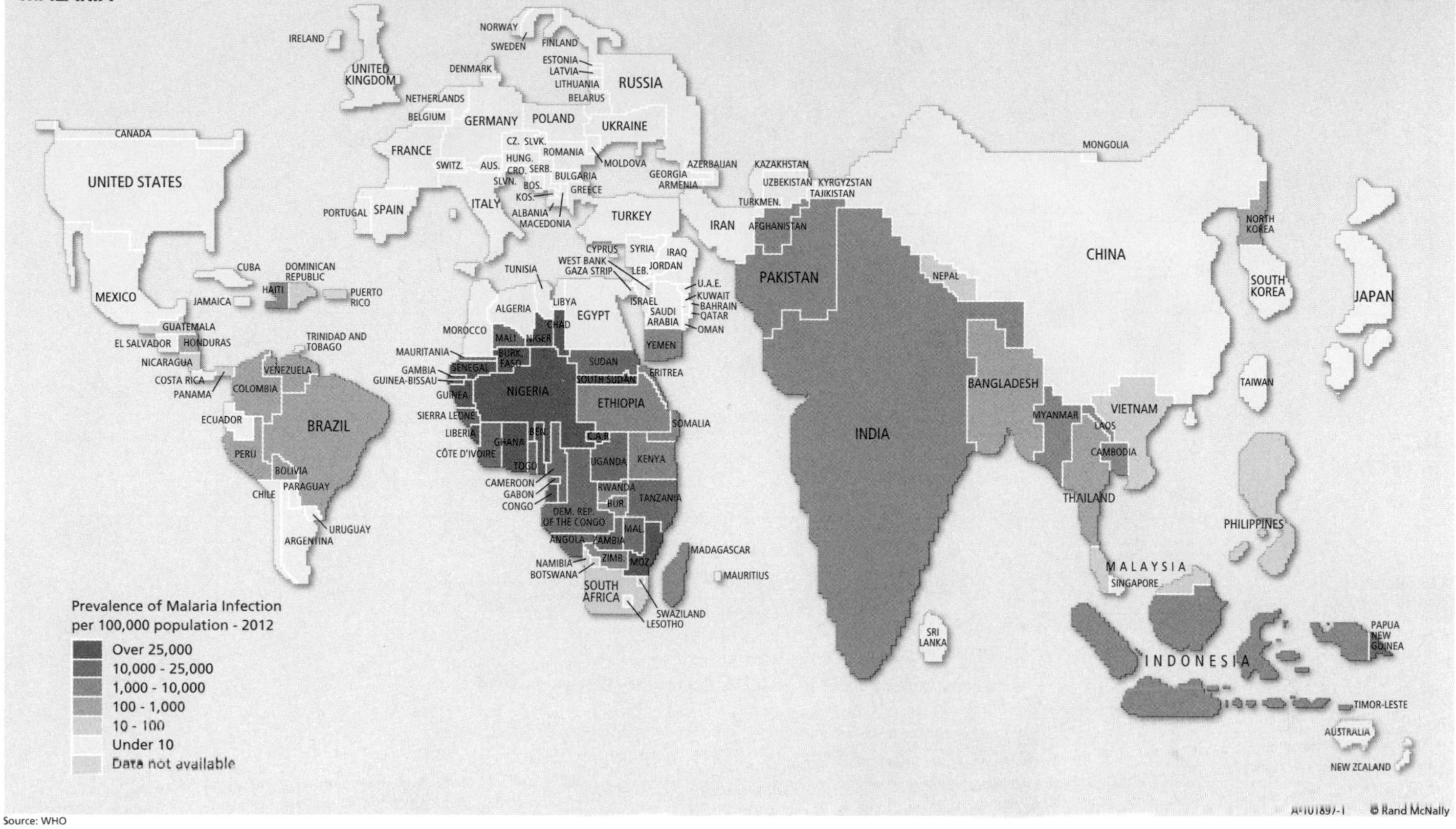

NORWAY, SWEDEN, FINLAND, IRELAND, UNITED KINGDOM, DENMARK, NETHERLANDS, BELGIUM, GERMANY, POLAND, ESTONIA, LATVIA, LITHUANIA, BELARUS, RUSSIA, UKRAINE, MONGOLIA, CANADA, FRANCE, SWITZ., AUS., HUNG., CRO. SERB., SLVN. BOS., KOS., ROMANIA, MOLDOVA, AZERBAIJAN, KAZAKHSTAN, UZBEKISTAN, KYRGYZSTAN, TAJIKISTAN, TURKMEN., UNITED STATES, PORTUGAL, SPAIN, ITALY, ALBANIA, MACEDONIA, BULGARIA, GREECE, GEORGIA, ARMENIA, TURKEY, IRAN, AFGHANISTAN, CHINA, NORTH KOREA, SOUTH KOREA, JAPAN, PAKISTAN, NEPAL, CYPRUS, SYRIA, IRAQ, JORDAN, WEST BANK, GAZA STRIP, LEB., ISRAEL, U.A.E., KUWAIT, BAHRAIN, QATAR, OMAN, CUBA, DOMINICAN REPUBLIC, HAITI, PUERTO RICO, MEXICO, JAMAICA, TUNISIA, ALGERIA, LIBYA, EGYPT, SAUDI ARABIA, YEMEN, GUATEMALA, EL SALVADOR, HONDURAS, TRINIDAD AND TOBAGO, MOROCCO, MALI, NIGER, CHAD, SUDAN, ERITREA, NICARAGUA, COSTA RICA, PANAMA, VENEZUELA, COLOMBIA, MAURITANIA, GAMBIA, SENEGAL, GUINEA-BISSAU, BURK. FASO, NIGERIA, SOUTH SUDAN, ETHIOPIA, SOMALIA, BANGLADESH, MYANMAR, LAOS, VIETNAM, TAIWAN, ECUADOR, BRAZIL, GUINEA, SIERRA LEONE, LIBERIA, CÔTE D'IVOIRE, GHANA, BEN., TOGO, C.A.R., UGANDA, KENYA, INDIA, THAILAND, CAMBODIA, PHILIPPINES, PERU, BOLIVIA, PARAGUAY, CHILE, URUGUAY, ARGENTINA, CAMEROON, GABON, CONGO, DEM. REP. OF THE CONGO, RWANDA, BUR., TANZANIA, MAL., ANGOLA, ZAMBIA, MADAGASCAR, MAURITIUS, MALAYSIA, SINGAPORE, NAMIBIA, BOTSWANA, ZIMB., MOZ., SOUTH AFRICA, SWAZILAND, LESOTHO, SRI LANKA, INDONESIA, PAPUA NEW GUINEA, TIMOR-LESTE, AUSTRALIA, NEW ZEALAND

Prevalence of Malaria Infection
per 100,000 population - 2012

- Over 25,000
- 10,000 - 25,000
- 1,000 - 10,000
- 100 - 1,000
- 10 - 100
- Under 10
- Data not available

Source: WHO

A-101897-1   © Rand McNally

The maps on these two pages are called **cartograms**. On these cartograms, the size of each country is proportional to its total population. This means that the countries with the largest areas are those with the largest populations. The shapes of countries must be distorted in order to achieve this proportional representation. One advantage of these cartograms is that they reveal the relationship between the mapped variable and the affected population. Consider the example of Chad and Nigeria. Both have relatively high rates of HIV infection (between 1,000 and 5,000 cases per 100,000 population). But Nigeria is much larger than Chad on the cartogram, which informs the reader that the population affected by HIV is much larger in Nigeria.

PHYSICIANS

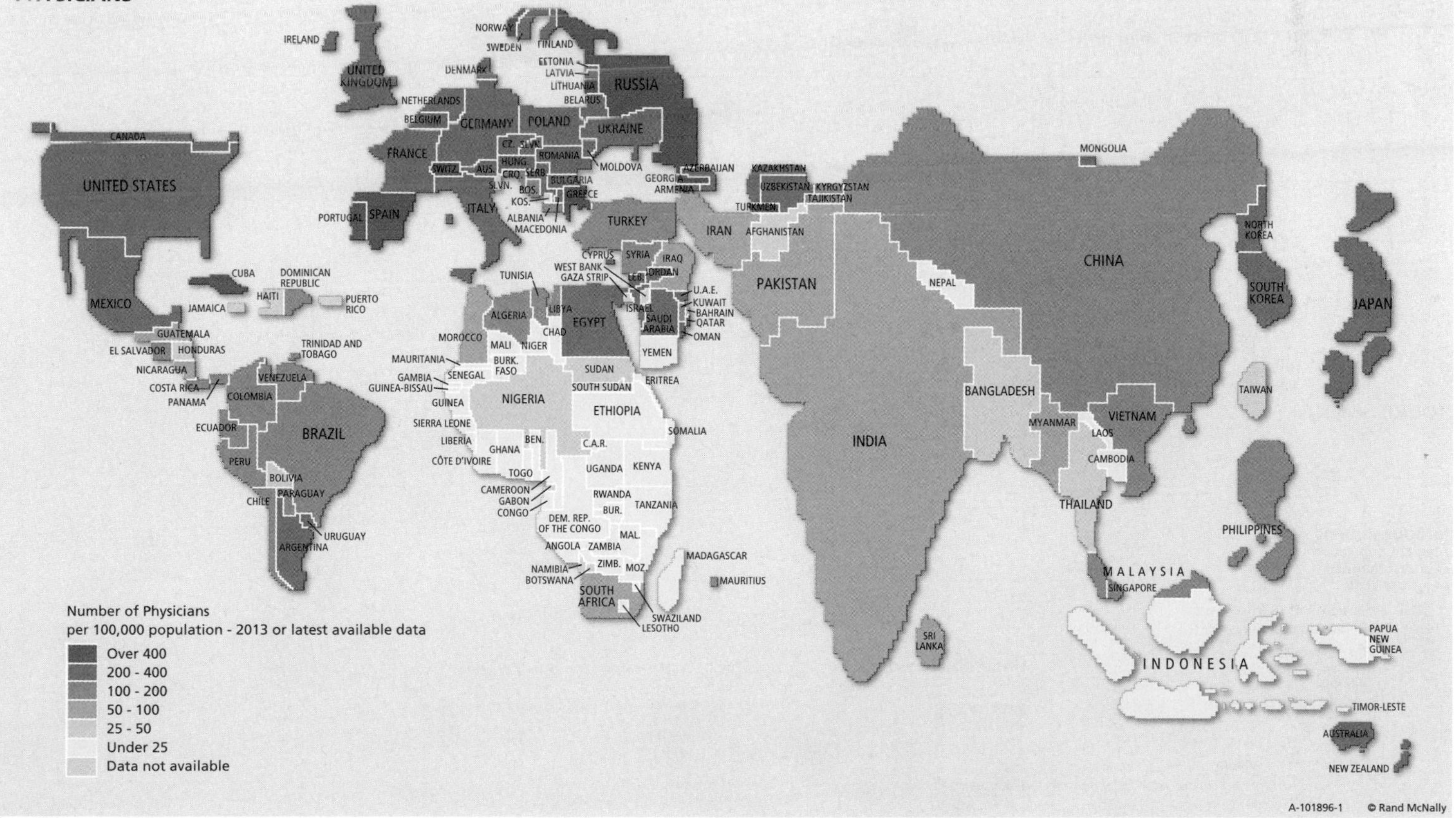

Number of Physicians
per 100,000 population - 2013 or latest available data

- Over 400
- 200 - 400
- 100 - 200
- 50 - 100
- 25 - 50
- Under 25
- Data not available

Source: WHO

A-101896-1   © Rand McNally

LIFE EXPECTANCY

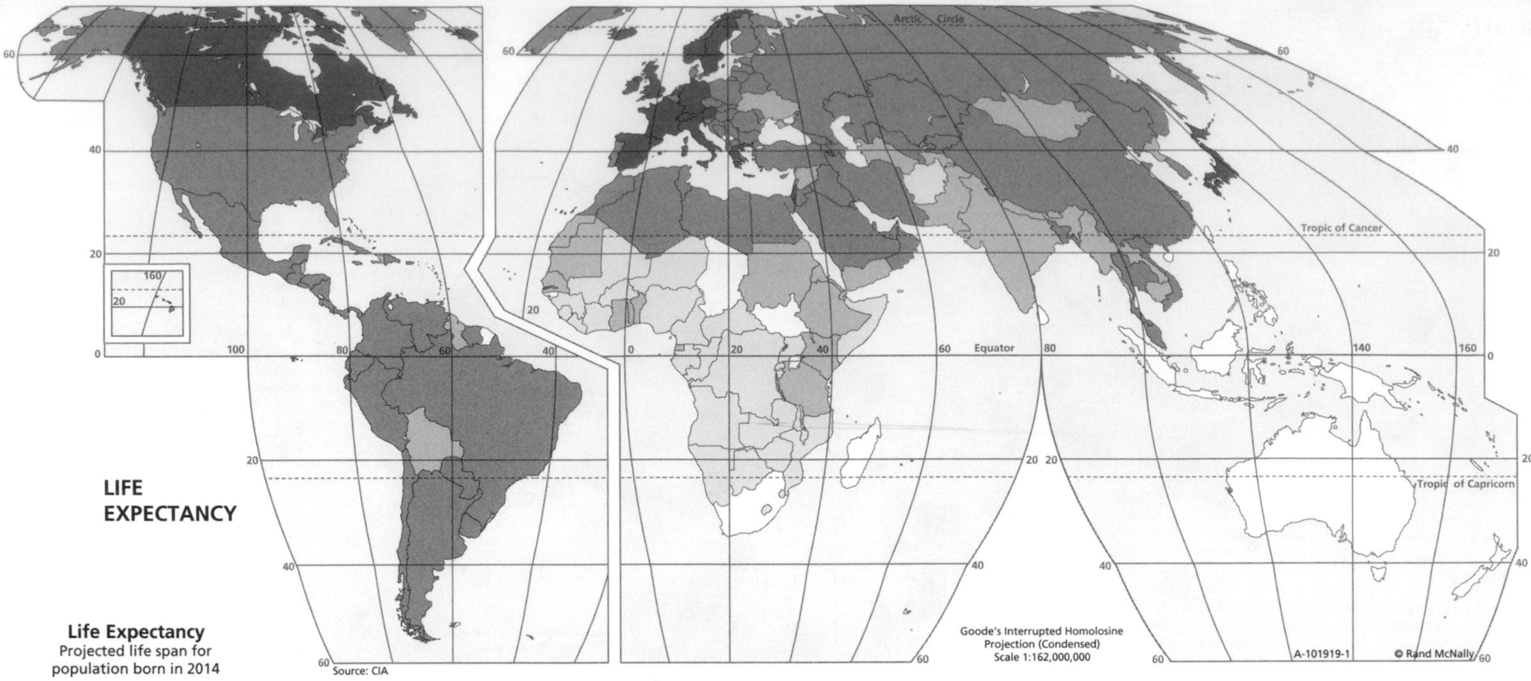
Arctic Circle
Tropic of Cancer
Equator
Tropic of Capricorn
Goode's Interrupted Homolosine Projection (Condensed) Scale 1:162,000,000
A-101919-1
© Rand McNally
Source: CIA

Life Expectancy
Projected life span for population born in 2014

- Over 80
- 70 - 80
- 60 - 69
- 50 - 59
- Under 50
- Data not available

World Avg. 68 →

Percentage of Births in each Life Expectancy Category - 2014

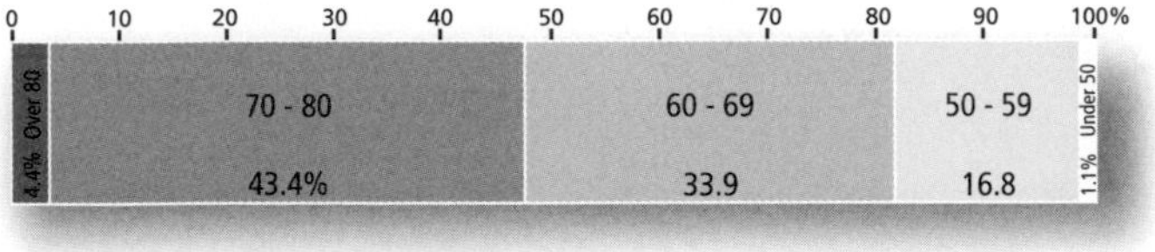
0
10
20
30
40
50
60
70
80
90
100%
4.4% Over 80
70 - 80
43.4%
60 - 69
33.9
50 - 59
16.8
1.1% Under 50

UNDERNOURISHMENT

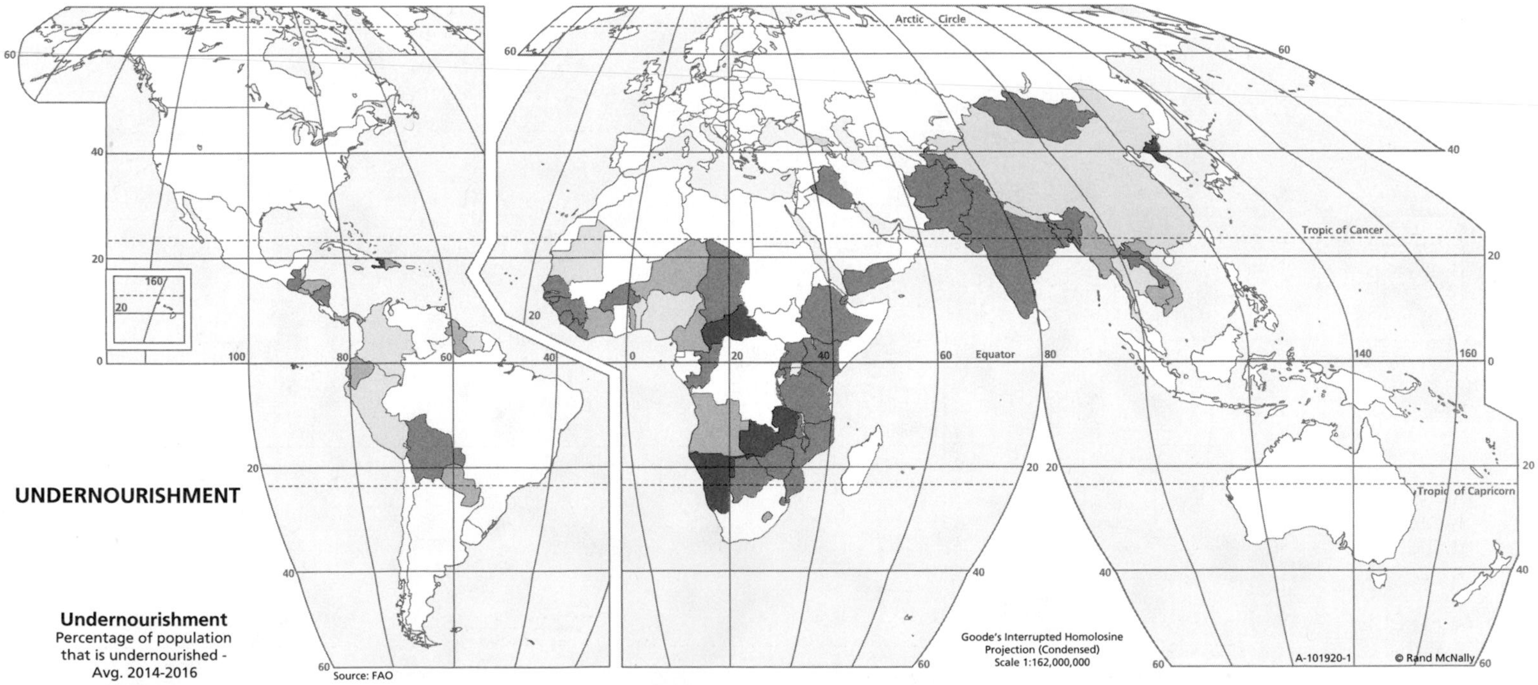
Arctic Circle
Tropic of Cancer
Equator
Tropic of Capricorn
Goode's Interrupted Homolosine Projection (Condensed) Scale 1:162,000,000
A-101920-1
© Rand McNally
Source: FAO

Undernourishment
Percentage of population that is undernourished - Avg. 2014-2016

- Over 40%
- 15% - 40%
- 10% - 14%
- 5% - 9%
- Under 5%
- Data not available

Undernourished People

World Total - 794,600,000 people - Avg. 2014-2016

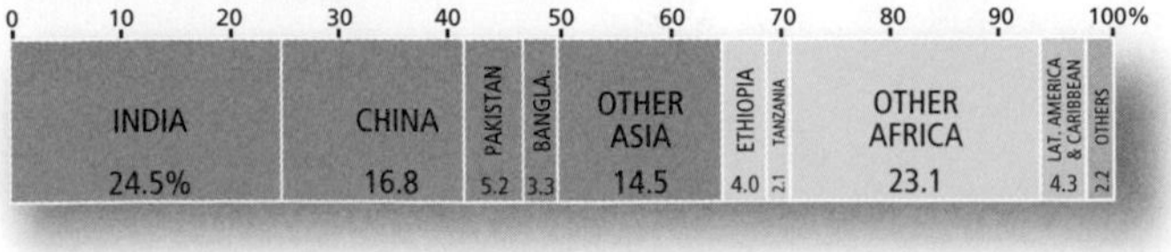
0
10
20
30
40
50
60
70
80
90
100%
INDIA
24.5%
CHINA
16.8
PAKISTAN
5.2
BANGLA.
3.3
OTHER ASIA
14.5
ETHIOPIA
4.0
TANZANIA
2.1
OTHER AFRICA
23.1
LAT. AMERICA & CARIBBEAN
4.3
OTHERS
2.2

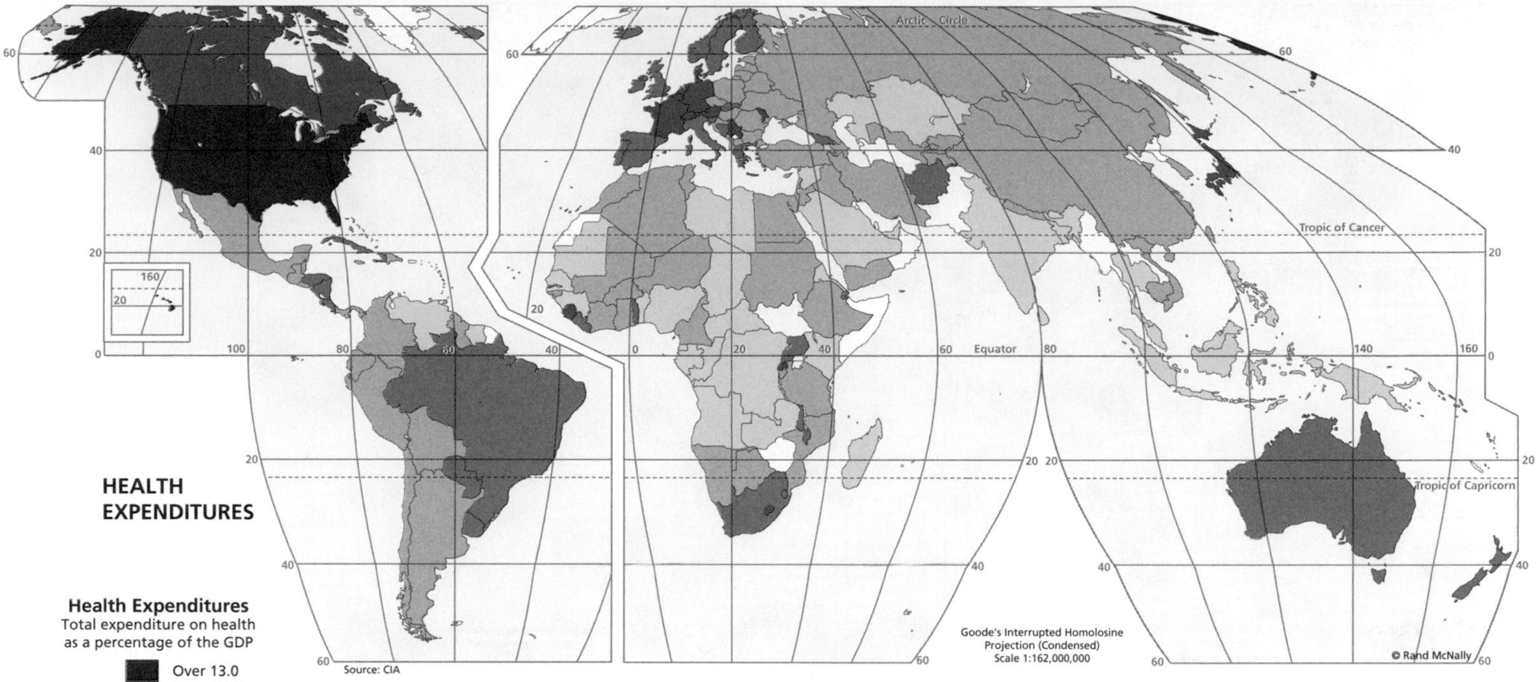
Arctic Circle
Tropic of Cancer
Equator
Tropic of Capricorn
Goode's Interrupted Homolosine
Projection (Condensed)
Scale 1:162,000,000
© Rand McNally
Source: CIA

HEALTH EXPENDITURES

Health Expenditures
Total expenditure on health as a percentage of the GDP

- Over 13.0
- 13.0 - 10.1
- 8.1 - 10.0
- 5.1 - 8.0
- 3.0 - 5.0
- Less than 3.0
- Data not available

Health Expenditures are broadly defined as activities performed either by institutions or individuals through the application of medical, paramedical, and/or nursing knowledge and technology, the primary purpose of which is to promote, restore, or maintain health.

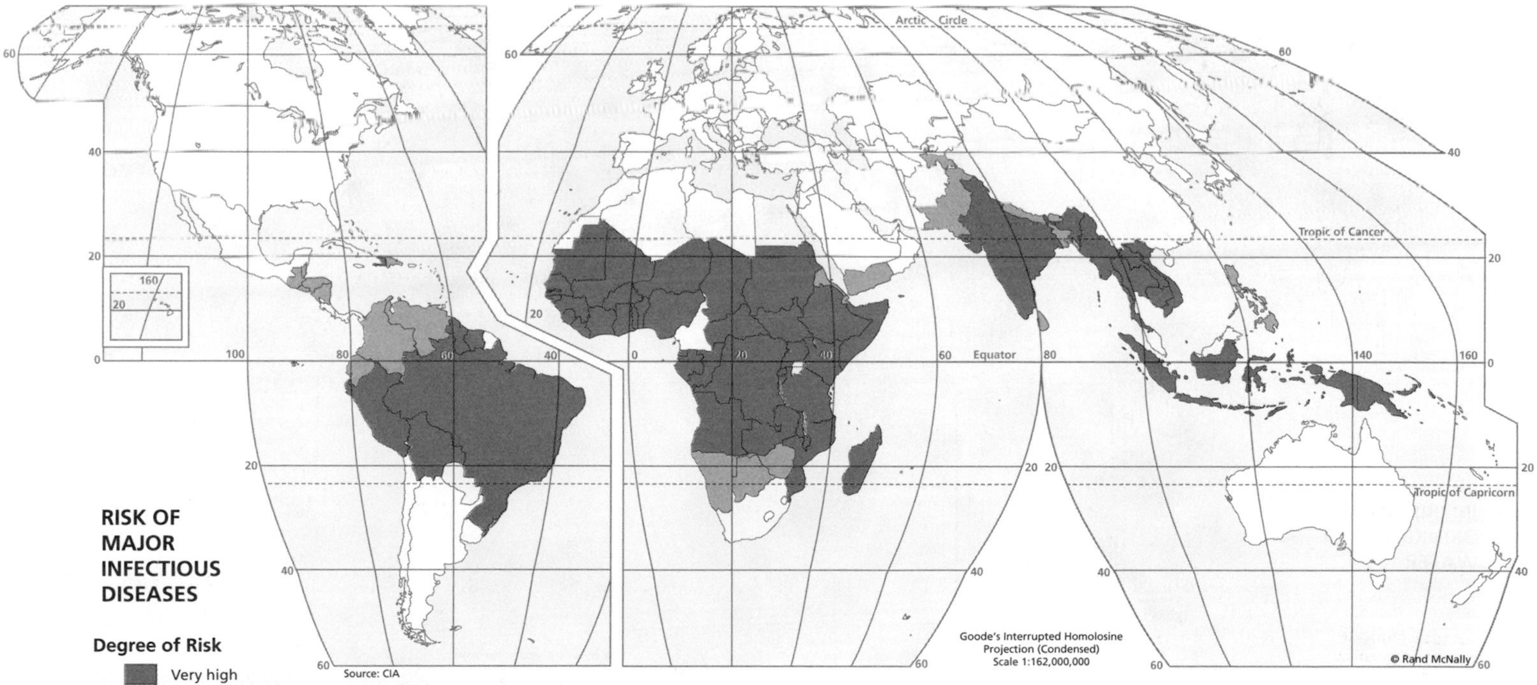
Arctic Circle
Tropic of Cancer
Equator
Tropic of Capricorn
Goode's Interrupted Homolosine
Projection (Condensed)
Scale 1:162,000,000
© Rand McNally
Source: CIA

RISK OF MAJOR INFECTIOUS DISEASES

Degree of Risk

- Very high
- High
- Intermediate
- Data not available

Major Infectious Diseases likely to be encountered in countries where the risk of such diseases is assessed to be very high as compared to the United States.

The degree of risk is assessed by considering the foreign nature of these infectious diseases, their severity, and the probability of being affected by the diseases present.

FOOD AID

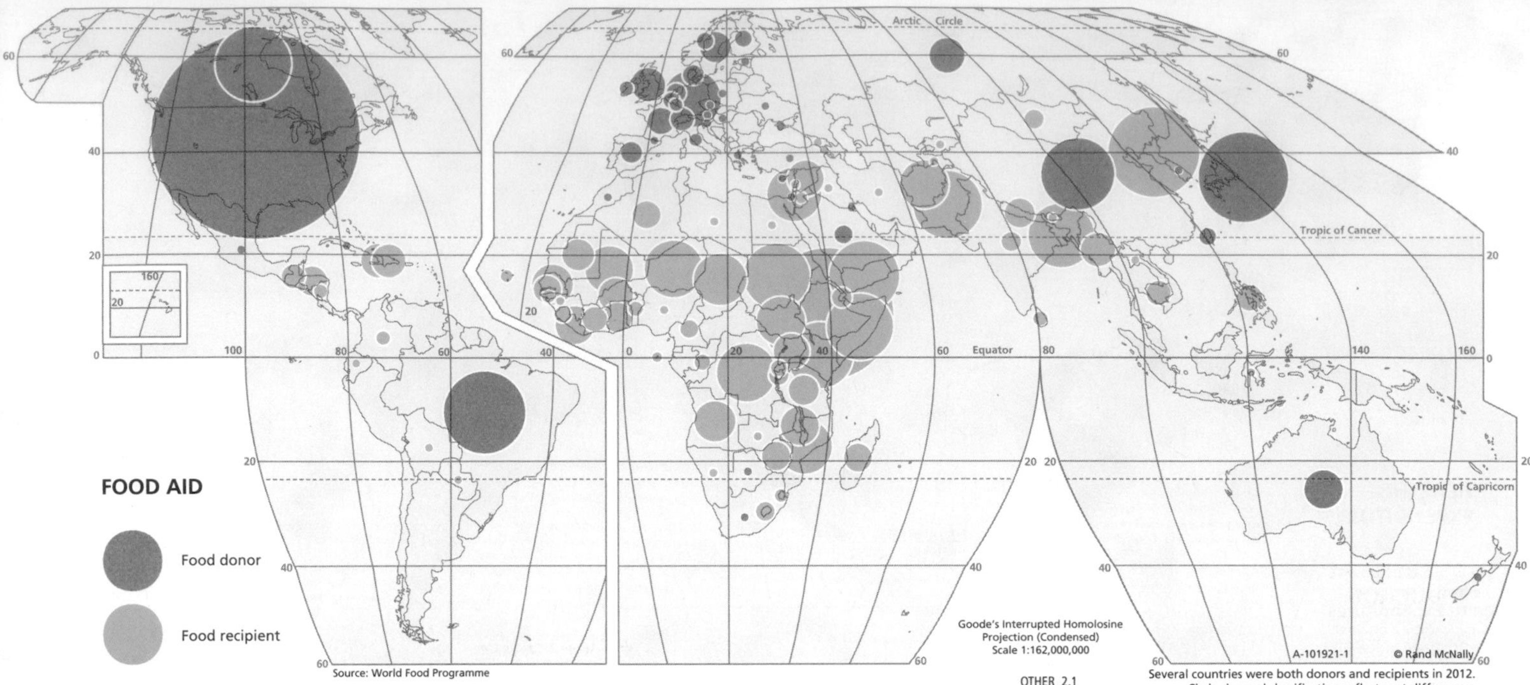

Food donor

Food recipient

Source: World Food Programme

Goode's Interrupted Homolosine
Projection (Condensed)
Scale 1:162,000,000

A-101921-1   © Rand McNally

Several countries were both donors and recipients in 2012.
Circle size and classification reflects net differences.

An additional 820,000 metric tons were donated by
international agencies, NGOs, non-profits,
and public and private organizations.

**Amount of Food
Donated or Received**
in metric tons - 2012

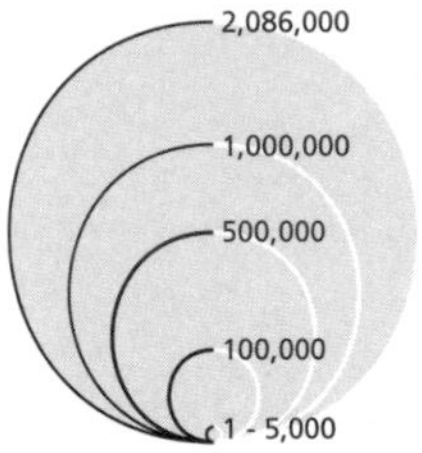

2,086,000

1,000,000

500,000

100,000

1 - 5,000

**Aid by
Type of Food**

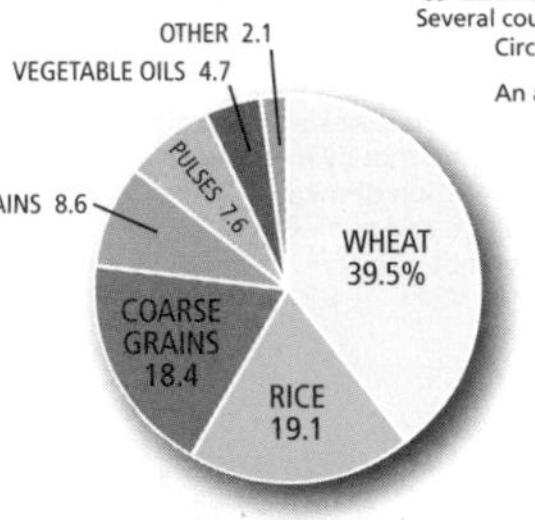

OTHER 2.1
VEGETABLE OILS 4.7
PULSES 7.6
BLENDED GRAINS 8.6
WHEAT 39.5%
COARSE GRAINS 18.4
RICE 19.1

IMPROVED DRINKING WATER

Source: The World Bank

Goode's Interrupted Homolosine
Projection (Condensed)
Scale 1:162,000,000

A-101922-1   © Rand McNally

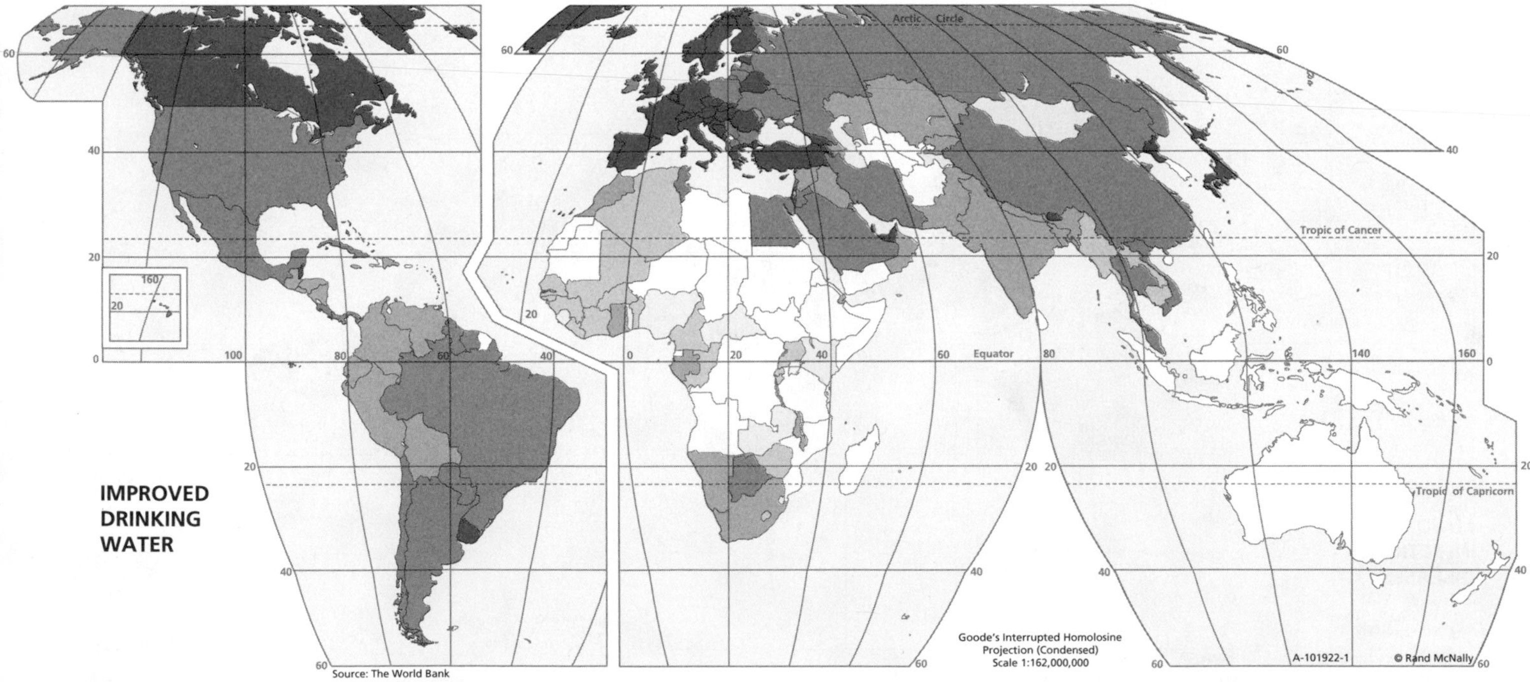

Drinking Water
Percentage of population
with sustainable access to
improved drinking water - 2015

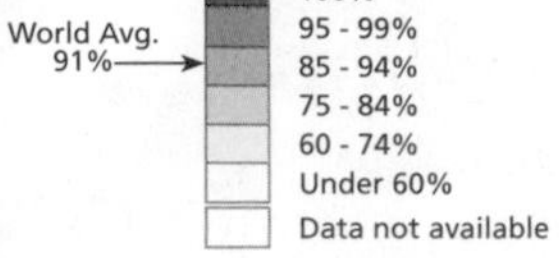

World Avg.
91%

- 100%
- 95 - 99%
- 85 - 94%
- 75 - 84%
- 60 - 74%
- Under 60%
- Data not available

**Percentage of World Population in
Each Drinking Water Access Category - 2015**

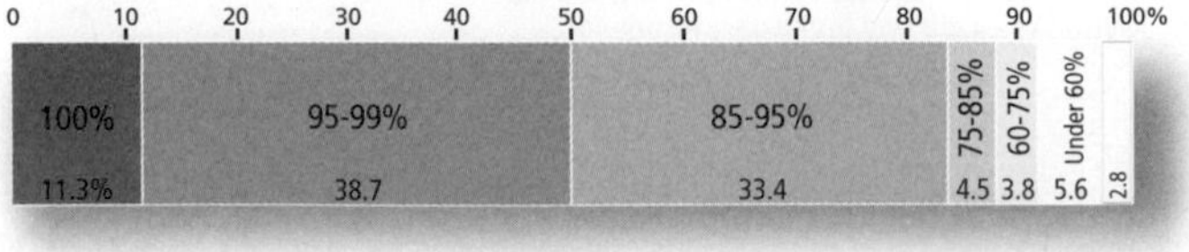

0	10	20	30	40	50	60	70	80	90	100%

100%	95-99%	85-95%	75-85%	60-75%	Under 60%
11.3%	38.7	33.4	4.5	3.8	5.6

2.8

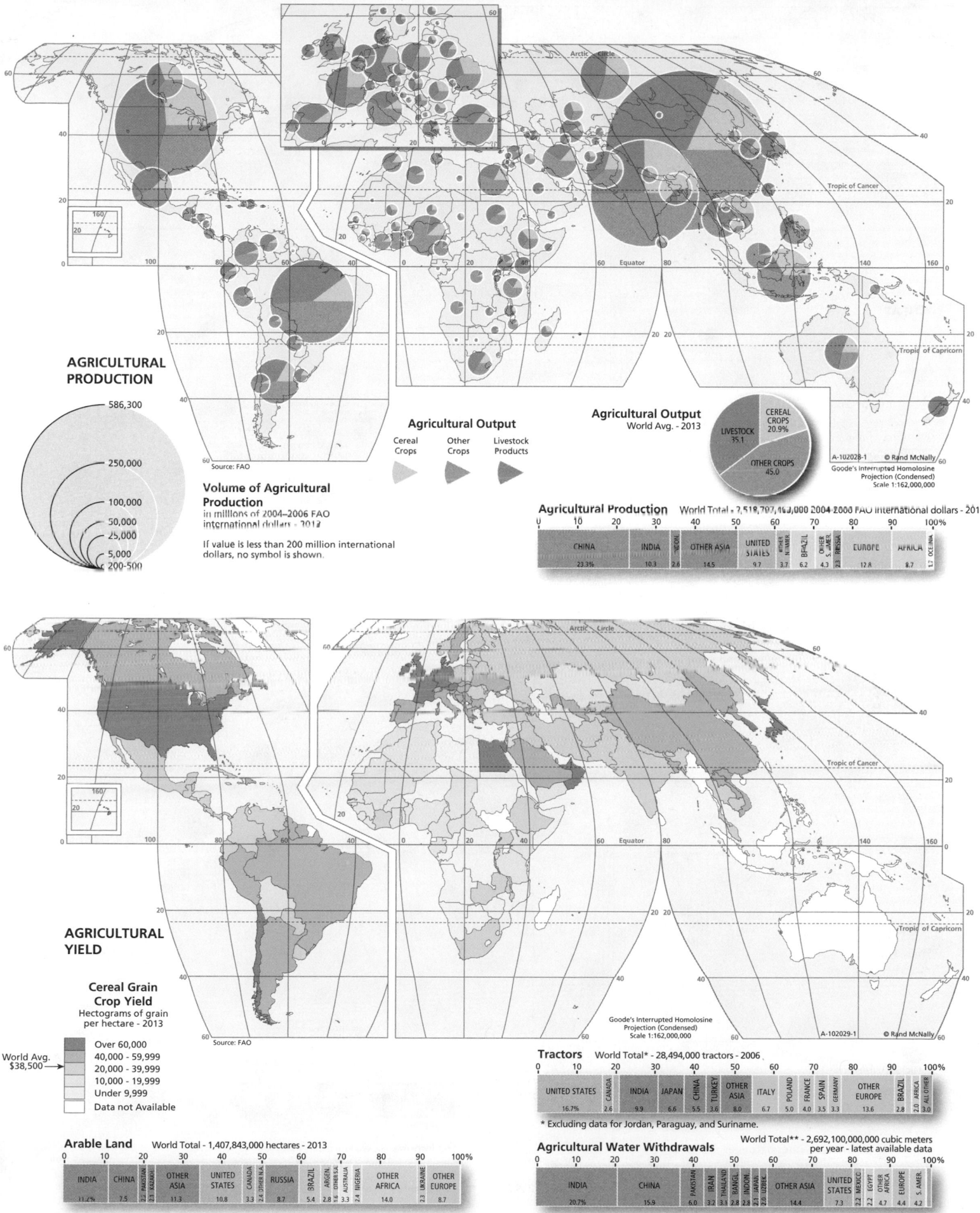

AGRICULTURAL PRODUCTION

586,300
250,000
100,000
50,000
25,000
5,000
200-500

Source: FAO

Volume of Agricultural Production
in millions of 2004–2006 FAO international dollars - 2013

If value is less than 200 million international dollars, no symbol is shown.

Agricultural Output
Cereal Crops — Other Crops — Livestock Products

Agricultural Output
World Avg. - 2013

CEREAL CROPS 20.9%
LIVESTOCK 35.1
OTHER CROPS 45.0

A-102028-1   © Rand McNally
Goode's Interrupted Homolosine
Projection (Condensed)
Scale 1:162,000,000

Agricultural Production   World Total - 7,518,707,163,000 2004-2006 FAO international dollars - 2013

0	10	20	30	40	50	60	70	80	90	100%

CHINA	INDIA	MEXI.	OTHER ASIA	UNITED STATES	OTHER N. AMER.	BRAZIL	OTHER S. AMER.	RUSSIA	EUROPE	AFRICA	OCEANIA
23.3%	10.3	2.6	14.5	9.7	3.7	6.2	4.3	2.3	12.8	8.7	1.7

AGRICULTURAL YIELD

Cereal Grain Crop Yield
Hectograms of grain per hectare - 2013

World Avg. $38,500

Over 60,000
40,000 - 59,999
20,000 - 39,999
10,000 - 19,999
Under 9,999
Data not Available

Source: FAO

Goode's Interrupted Homolosine
Projection (Condensed)
Scale 1:162,000,000

A-102029-1   © Rand McNally

Tractors   World Total* - 28,494,000 tractors - 2006

0	10	20	30	40	50	60	70	80	90	100%

UNITED STATES	CANADA	INDIA	JAPAN	CHINA	TURKEY	OTHER ASIA	ITALY	POLAND	FRANCE	SPAIN	GERMANY	OTHER EUROPE	BRAZIL	AFRICA	ALL OTHER
16.7%	2.6	9.9	6.6	5.5	3.6	8.0	6.7	4.0	3.5	3.3		13.6	2.8	2.0	3.0

* Excluding data for Jordan, Paraguay, and Suriname.

Arable Land   World Total - 1,407,843,000 hectares - 2013

0	10	20	30	40	50	60	70	80	90	100%

INDIA	CHINA	PAKISTAN	OTHER ASIA	UNITED STATES	CANADA	RUSSIA	BRAZIL	ARGEN.	OTHER S.A.	AUSTRALIA	NIGERIA	OTHER AFRICA	UKRAINE	OTHER EUROPE
11.2%	7.5	2.2	11.3	10.8	3.3	8.7	5.4	2.8	1.6	3.3	2.4	14.0	2.3	8.7

Agricultural Water Withdrawals   World Total** - 2,692,100,000,000 cubic meters per year - latest available data

0	10	20	30	40	50	60	70	80	90	100%

INDIA	CHINA	PAKISTAN	IRAN	THAILAND	BANGL.	INDON.	JAPAN	UZBEK.	OTHER ASIA	UNITED STATES	MEXICO	OTHER AFRICA	EUROPE	S. AMER.
20.7%	15.9	6.0	3.2	3.1	2.8	2.8	2.3		14.4	7.3	2.2	4.7	4.4	4.2

** Excluding data for Slovakia and the former Yugoslav republics.

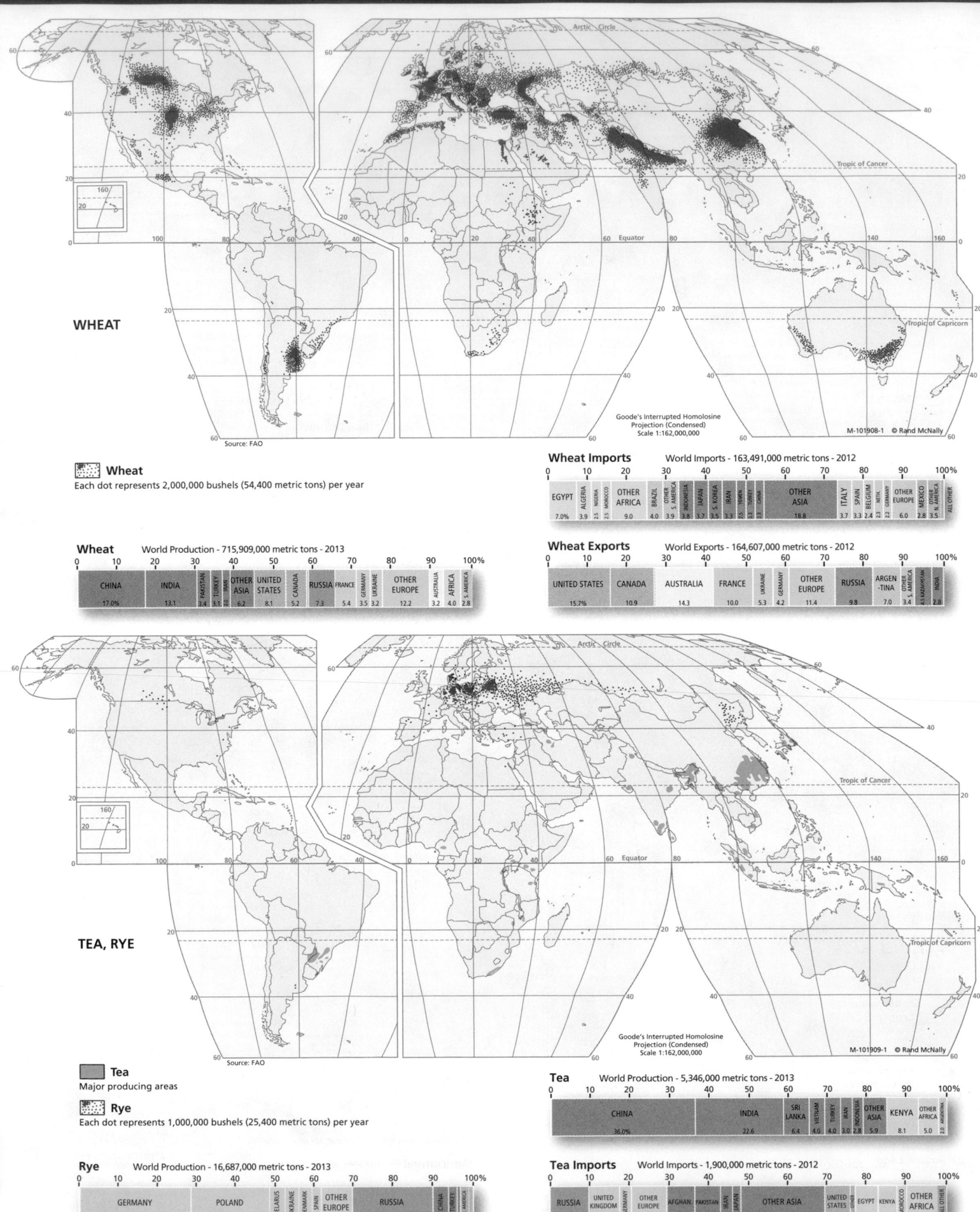

WHEAT

Source: FAO

Goode's Interrupted Homolosine
Projection (Condensed)
Scale 1:162,000,000

M-101908-1   © Rand McNally

Wheat
Each dot represents 2,000,000 bushels (54,400 metric tons) per year

Wheat Imports   World Imports - 163,491,000 metric tons - 2012

0	10				20		30				40				50	60		70		80		90			100%
EGYPT	ALGERIA	NIGERIA	MOROCCO	OTHER AFRICA	BRAZIL	OTHER S. AMERICA	INDONESIA	JAPAN	S. KOREA	IRAN	TURKEY	CHINA	OTHER ASIA	ITALY	SPAIN	BELGIUM	NETH.	GERMANY	OTHER EUROPE	MEXICO	OTHER N. AMERICA	ALL OTHER
7.0%	3.9	2.5	2.5	9.0	4.0	3.9	3.8	3.5	3.5	2.5	2.5	2.3	18.8	3.7	3.3	2.4	2.3	2.2	6.0	2.8	3.5	

Wheat   World Production - 715,909,000 metric tons - 2013

0	10	20			30		40			50	60		70		80	90			100%
CHINA	INDIA	PAKISTAN	IRAN	TURKEY	OTHER ASIA	UNITED STATES	CANADA	RUSSIA	FRANCE	GERMANY	UKRAINE	OTHER EUROPE	AUSTRALIA	AFRICA	S. AMERICA				
17.0%	13.1	3.4	2.0	1.1	6.2	8.1	5.2	7.3	5.4	3.5	3.2	12.2	3.2	4.0	2.8				

Wheat Exports   World Exports - 164,607,000 metric tons - 2012

0	10		20		30		40		50		60		70		80	90			100%
UNITED STATES	CANADA	AUSTRALIA	FRANCE	UKRAINE	GERMANY	OTHER EUROPE	RUSSIA	ARGEN-TINA	OTHER S. AMERICA	KAZAKSTAN	INDIA								
15.7%	10.9	14.3	10.0	5.3	4.2	11.4	9.8	7.0	3.4	4.3	2.8								

TEA, RYE

Source: FAO

Goode's Interrupted Homolosine
Projection (Condensed)
Scale 1:162,000,000

M-101909-1   © Rand McNally

Tea
Major producing areas

Rye
Each dot represents 1,000,000 bushels (25,400 metric tons) per year

Tea   World Production - 5,346,000 metric tons - 2013

0	10	20			30	40		50		60	70		80		90		100%
CHINA	INDIA	SRI LANKA	VIETNAM	TURKEY	IRAN	INDONESIA	OTHER ASIA	KENYA	OTHER AFRICA	ARGENTINA							
36.0%	22.6	6.4	4.0	4.0	3.0	2.8	5.9	8.1	5.0	2.0							

Rye   World Production - 16,687,000 metric tons - 2013

0	10	20		30	40		50		60		70	80		90		100%
GERMANY	POLAND	BELARUS	UKRAINE	DENMARK	SPAIN	OTHER EUROPE	RUSSIA	CHINA	TURKEY	N. AMERICA						
28.1%	20.1	3.9	3.8	3.2	2.3	8.0	20.1	3.9	2.4	1.2						

Tea Imports   World Imports - 1,900,000 metric tons - 2012

0	10		20		30			40	50		60		70	80		90			100%
RUSSIA	UNITED KINGDOM	GERMANY	OTHER EUROPE	AFGHAN.	PAKISTAN	IRAN	JAPAN	OTHER ASIA	UNITED STATES	CANADA	EGYPT	KENYA	MOROCCO	OTHER AFRICA	ALL OTHER				
9.5%	7.6	3.0	8.9	7.1	6.4	3.3	2.0	21.3	6.6	2.2	5.8	4.6	2.8	7.6	2.3				

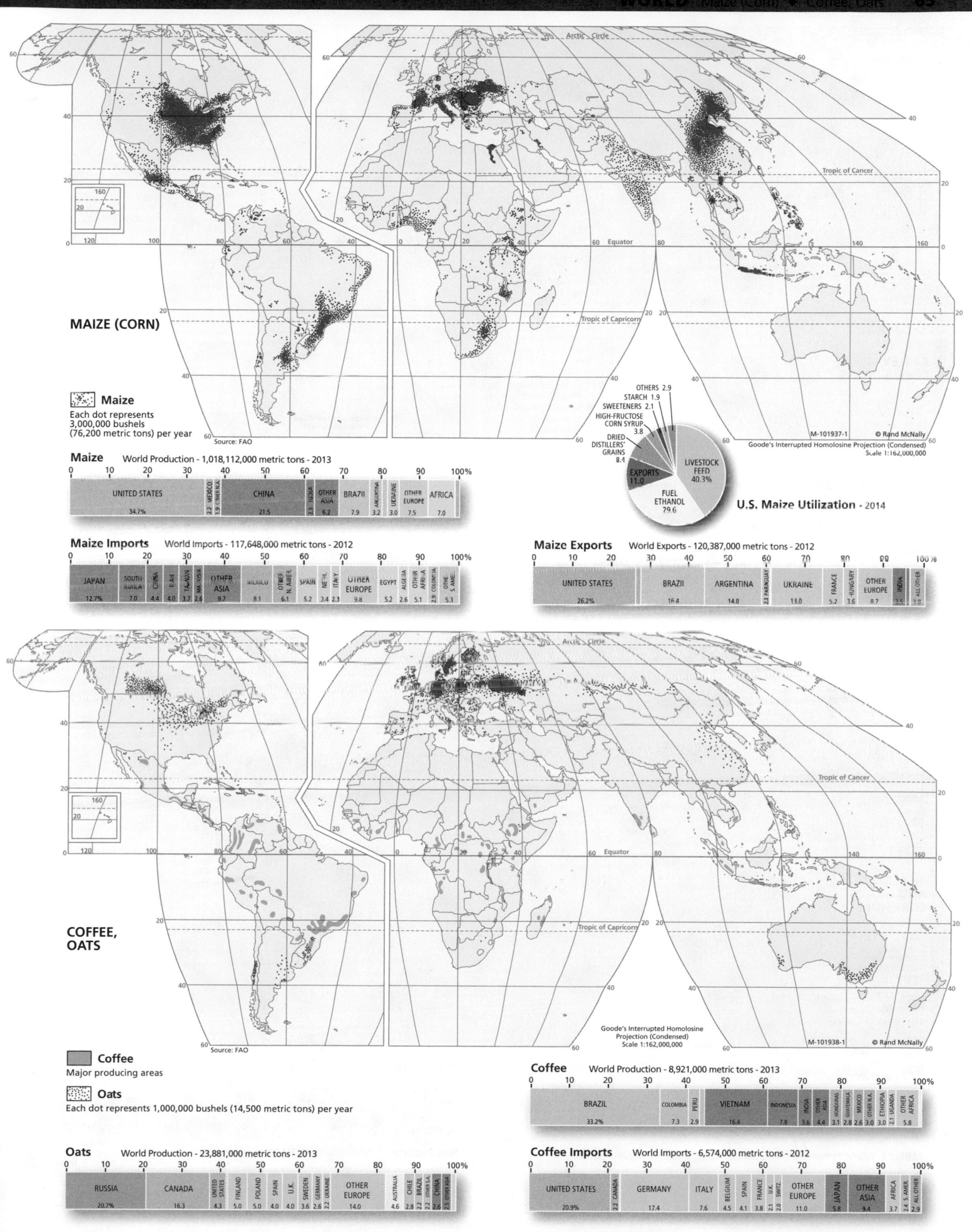

MAIZE (CORN)

Maize
Each dot represents
3,000,000 bushels
(76,200 metric tons) per year

Source: FAO

Maize World Production - 1,018,112,000 metric tons - 2013

	0	10	20	30	40	50	60	70	80	90	100%

| UNITED STATES 34.7% | MEXICO 2.2 | OTHER N.A. 1.9 | CHINA 21.5 | INDIA 2.3 | OTHER ASIA 6.2 | BRAZIL 7.9 | ARGENTINA 3.2 | UKRAINE 3.0 | OTHER EUROPE 7.5 | AFRICA 7.0 |

Maize Imports World Imports - 117,648,000 metric tons - 2012

	0	10	20	30	40	50	60	70	80	90	100%

| JAPAN 12.7% | SOUTH KOREA 7.0 | CHINA 4.4 | IRAN 4.0 | TAIWAN 3.7 | MALAYSIA 2.6 | OTHER ASIA 9.7 | MEXICO 8.1 | OTHER N. AMER. 6.1 | SPAIN 5.2 | NETH. 3.4 | ITALY 2.3 | OTHER EUROPE 9.8 | EGYPT 5.2 | ALGERIA 2.6 | OTHER AFRICA 5.1 | COLOMBIA 2.9 | OTHER S. AMER. 5.3 |

Maize Exports World Exports - 120,387,000 metric tons - 2012

	0	10	20	30	40	50	60	70	80	90	100%

| UNITED STATES 26.2% | BRAZIL 16.4 | ARGENTINA 14.0 | PARAGUAY 2.3 | UKRAINE 13.0 | FRANCE 5.2 | HUNGARY 3.6 | OTHER EUROPE 8.7 | INDIA 2.5 | ALL OTHER |

U.S. Maize Utilization - 2014

OTHERS 2.9
STARCH 1.9
SWEETENERS 2.1
HIGH-FRUCTOSE CORN SYRUP 3.8
DRIED DISTILLERS' GRAINS 8.4
EXPORTS 11.0
LIVESTOCK FEED 40.3%
FUEL ETHANOL 29.6

M-101937-1   © Rand McNally
Goode's Interrupted Homolosine Projection (Condensed)
Scale 1:162,000,000

COFFEE, OATS

Coffee
Major producing areas

Oats
Each dot represents 1,000,000 bushels (14,500 metric tons) per year

Source: FAO

Goode's Interrupted Homolosine Projection (Condensed)
Scale 1:162,000,000
M-101938-1   © Rand McNally

Coffee World Production - 8,921,000 metric tons - 2013

	0	10	20	30	40	50	60	70	80	90	100%

| BRAZIL 33.2% | COLOMBIA 7.3 | PERU 2.9 | VIETNAM 16.4 | INDONESIA 7.8 | INDIA 3.6 | OTHER ASIA 4.4 | HONDURAS 3.1 | GUATEMALA 2.8 | MEXICO 2.6 | OTHER N.A. 3.0 | ETHIOPIA 3.0 | UGANDA 2.1 | OTHER AFRICA 5.8 |

Oats World Production - 23,881,000 metric tons - 2013

	0	10	20	30	40	50	60	70	80	90	100%

| RUSSIA 20.7% | CANADA 16.3 | UNITED STATES 4.3 | FINLAND 5.0 | POLAND 5.0 | SPAIN 4.0 | U.K. 4.0 | SWEDEN 3.6 | GERMANY 2.6 | UKRAINE 2.2 | OTHER EUROPE 14.0 | AUSTRALIA 4.6 | CHILE 2.8 | BRAZIL 2.2 | OTHER S.A. 2.2 | CHINA 2.5 | OTHER ASIA 2.6 |

Coffee Imports World Imports - 6,574,000 metric tons - 2012

	0	10	20	30	40	50	60	70	80	90	100%

| UNITED STATES 20.9% | CANADA 2.2 | GERMANY 17.4 | ITALY 7.6 | BELGIUM 4.5 | SPAIN 4.1 | FRANCE 3.8 | U.K. 2.1 | SWITZ. 2.0 | OTHER EUROPE 11.0 | JAPAN 5.8 | OTHER ASIA 9.4 | AFRICA 3.7 | S. AMER. 2.4 | ALL OTHER 2.9 |

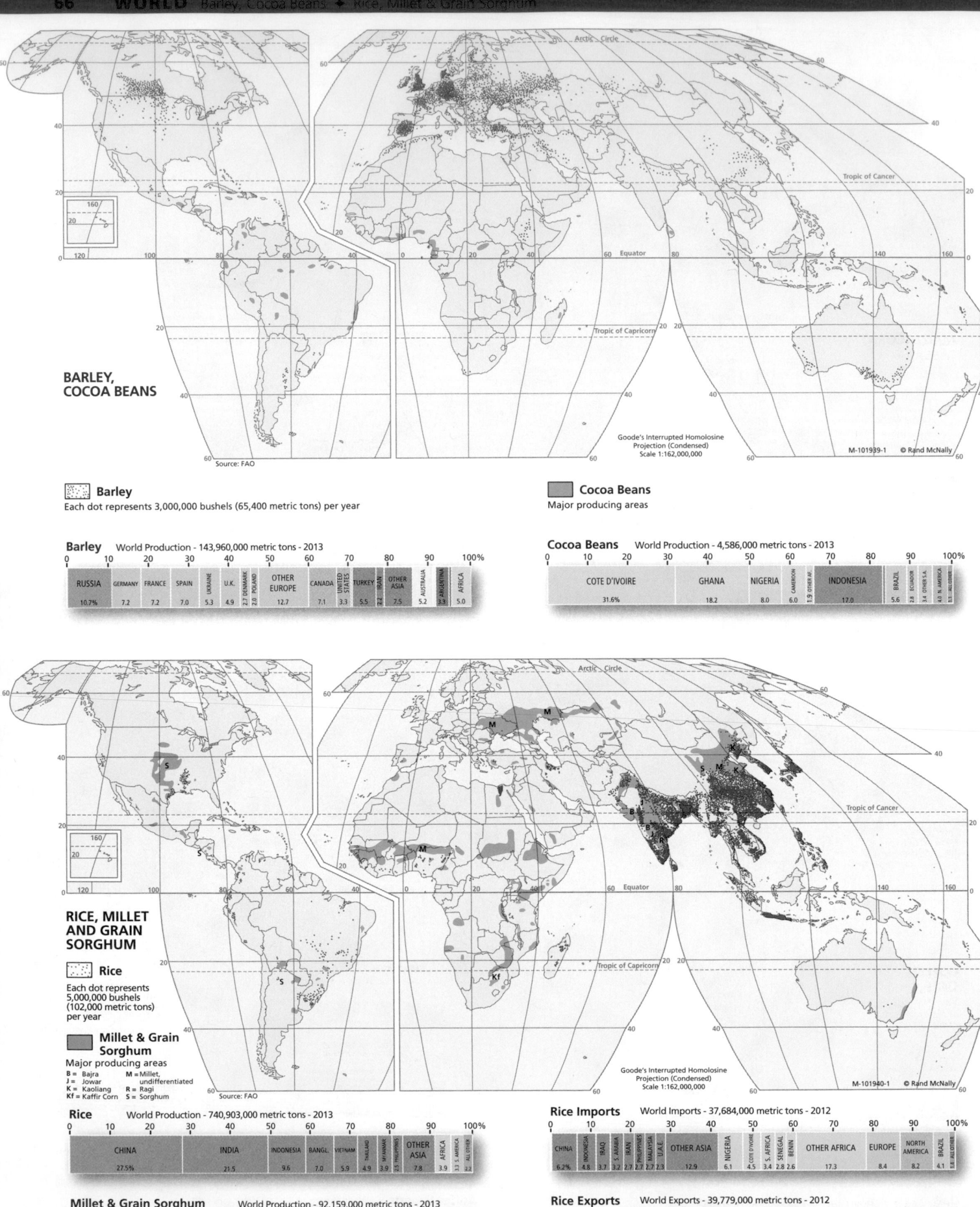

BARLEY, COCOA BEANS

Barley
Each dot represents 3,000,000 bushels (65,400 metric tons) per year

Cocoa Beans
Major producing areas

Goode's Interrupted Homolosine Projection (Condensed)
Scale 1:162,000,000
M-101939-1 © Rand McNally

Source: FAO

Barley World Production - 143,960,000 metric tons - 2013

	0	10	20	30	40	50	60	70	80	90	100%

RUSSIA	GERMANY	FRANCE	SPAIN	UKRAINE	U.K.	DENMARK	POLAND	OTHER EUROPE	CANADA	UNITED STATES	TURKEY	IRAN	OTHER ASIA	AUSTRALIA	ARGENTINA	AFRICA
10.7%	7.2	7.2	7.0	5.3	4.9	2.7	2.0	12.7	7.1	3.3	5.5	2.2	7.5	5.2	3.3	5.0

Cocoa Beans World Production - 4,586,000 metric tons - 2013

	0	10	20	30	40	50	60	70	80	90	100%

COTE D'IVOIRE	GHANA	NIGERIA	CAMEROON	OTHER AF.	INDONESIA	BRAZIL	ECUADOR	OTHER S.A.	N. AMERICA	ALL OTHER
31.6%	18.2	8.0	6.0	1.9	17.0	5.6	2.8	3.4	4.0	1.5

RICE, MILLET AND GRAIN SORGHUM

Rice
Each dot represents 5,000,000 bushels (102,000 metric tons) per year

Millet & Grain Sorghum
Major producing areas
B = Bajra
J = Jowar
K = Kaoliang
Kf = Kaffir Corn
M = Millet, undifferentiated
R = Ragi
S = Sorghum

Goode's Interrupted Homolosine Projection (Condensed)
Scale 1:162,000,000
M-101940-1 © Rand McNally

Source: FAO

Rice World Production - 740,903,000 metric tons - 2013

	0	10	20	30	40	50	60	70	80	90	100%

CHINA	INDIA	INDONESIA	BANGL.	VIETNAM	THAILAND	MYANMAR	PHILIPPINES	OTHER ASIA	AFRICA	S. AMERICA	ALL OTHER
27.5%	21.5	9.6	7.0	5.9	4.9	3.9	2.5	7.8	3.9	3.3	1.5

Millet & Grain Sorghum World Production - 92,159,000 metric tons - 2013

	0	10	20	30	40	50	60	70	80	90	100%

INDIA	CHINA	OTHER	NIGERIA	SUDAN	ETHIOPIA	NIGER	BURKINA F.	MALI	OTHER AFRICA	UNITED STATES	MEXICO	ARGENTINA	AUSTRALIA	ALL OTHER
17.6%	5.0	2.7	12.7	6.1	5.6	4.6	3.2	2.1	9.7	11.2	6.8	4.0	2.3	2.5

Rice Imports World Imports - 37,684,000 metric tons - 2012

	0	10	20	30	40	50	60	70	80	90	100%

CHINA	INDONESIA	IRAQ	S. ARABIA	IRAN	PHILIPPINES	MALAYSIA	U.A.E.	OTHER ASIA	NIGERIA	COTE D'IVOIRE	S. AFRICA	SENEGAL	BENIN	OTHER AFRICA	EUROPE	NORTH AMERICA	BRAZIL	ALL OTHER
6.2%	4.8	3.7	3.2	2.7	2.7	2.3		12.9	6.1	4.5	3.4	2.8	2.6	17.3	8.4	8.2	4.1	5.8

Rice Exports World Exports - 39,779,000 metric tons - 2012

	0	10	20	30	40	50	60	70	80	90	100%

INDIA	VIETNAM	THAILAND	PAKISTAN	OTHER ASIA	UNITED STATES	BRAZIL	URUGUAY	EUROPE	ALL OTHER
26.3%	20.1	16.9	8.6	4.1	8.2	2.8	2.7	4.6	3.1

POTATOES, CASSAVA

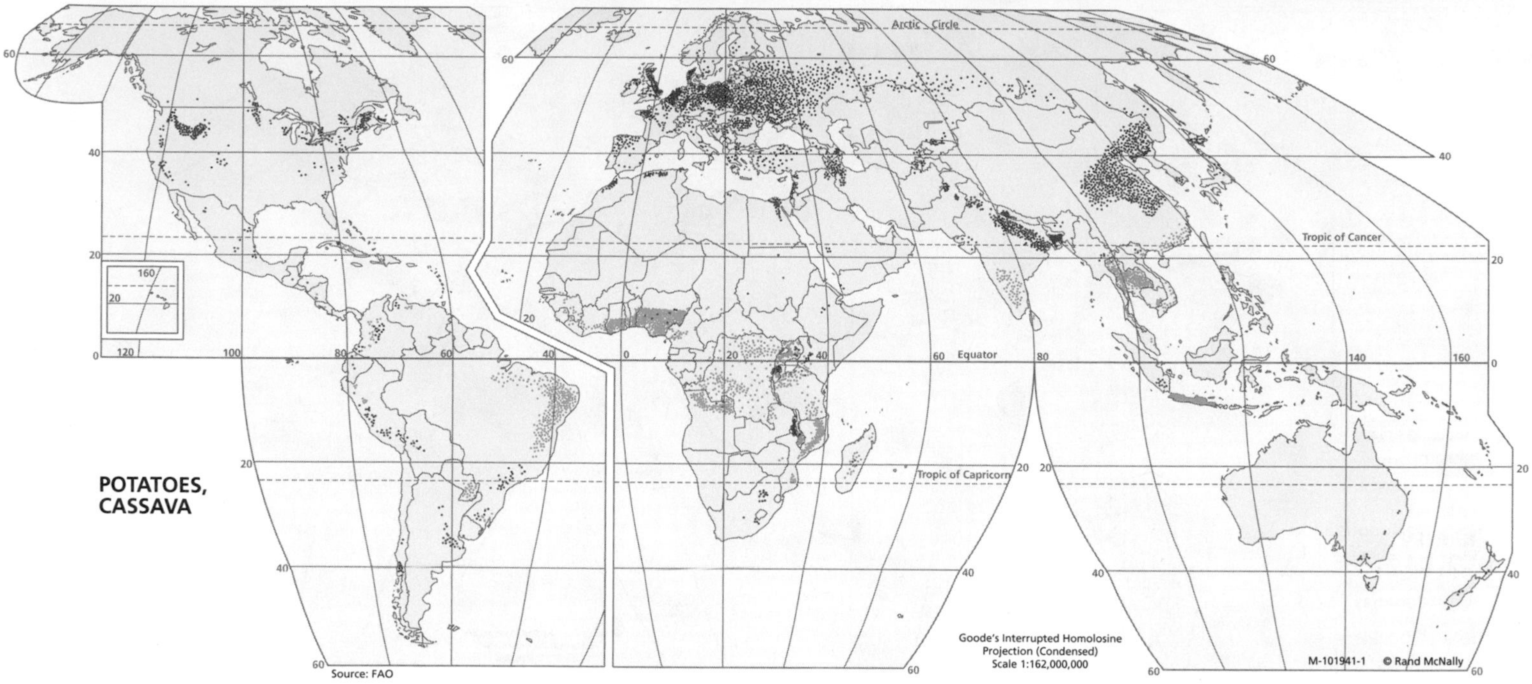

160
20
120
100
0

160
20

60
40
20
0
20
40
60

Arctic Circle

Tropic of Cancer

Equator

Tropic of Capricorn

Goode's Interrupted Homolosine
Projection (Condensed)
Scale 1:162,000,000

M-101941-1   © Rand McNally

Source: FAO

Potatoes
Each dot represents 100,000 metric tons average annual production

Cassava
Each dot represents 100,000 metric tons average annual production

Potatoes   World Production - 376,453,000 metric tons - 2013

	0	10	20	30	40	50	60	70	80	90	100%

| CHINA 25.5% | INDIA 12.0 | BANGLADESH 2.3 | OTHER ASIA 9.9 | RUSSIA 8.0 | UKRAINE 6.0 | GERMANY 2.8 | OTHER EUROPE 13.3 | UNITED STATES 5.3 | OTHER N.A. 1. | AFRICA 8.1 | SOUTH AMERICA 4.1 |

Cassava   World Production - 276,762,000 metric tons - 2013

	0	10	20	30	40	50	60	70	80	90	100%

| NIGERIA 19.2% | DEM REP OF THE CONGO 6.0 | ANGOLA 5.9 | GHANA 5.8 | MOZAM. 3.6 | OTHER AFRICA 16.6 | THAILAND 10.9 | INDONESIA 8.6 | VIET NAM 3.5 | CAMBODIA 2.9 | INDIA 2.6 | OTHER ASIA 3.3 | BRAZIL 7.8 | OTHER S.A. 2.6 |

SUGAR, SPICES

160
20
120
100
0

160
20

60
40
20
0
20
40
60

Arctic Circle

Tropic of Cancer

Equator

Tropic of Capricorn

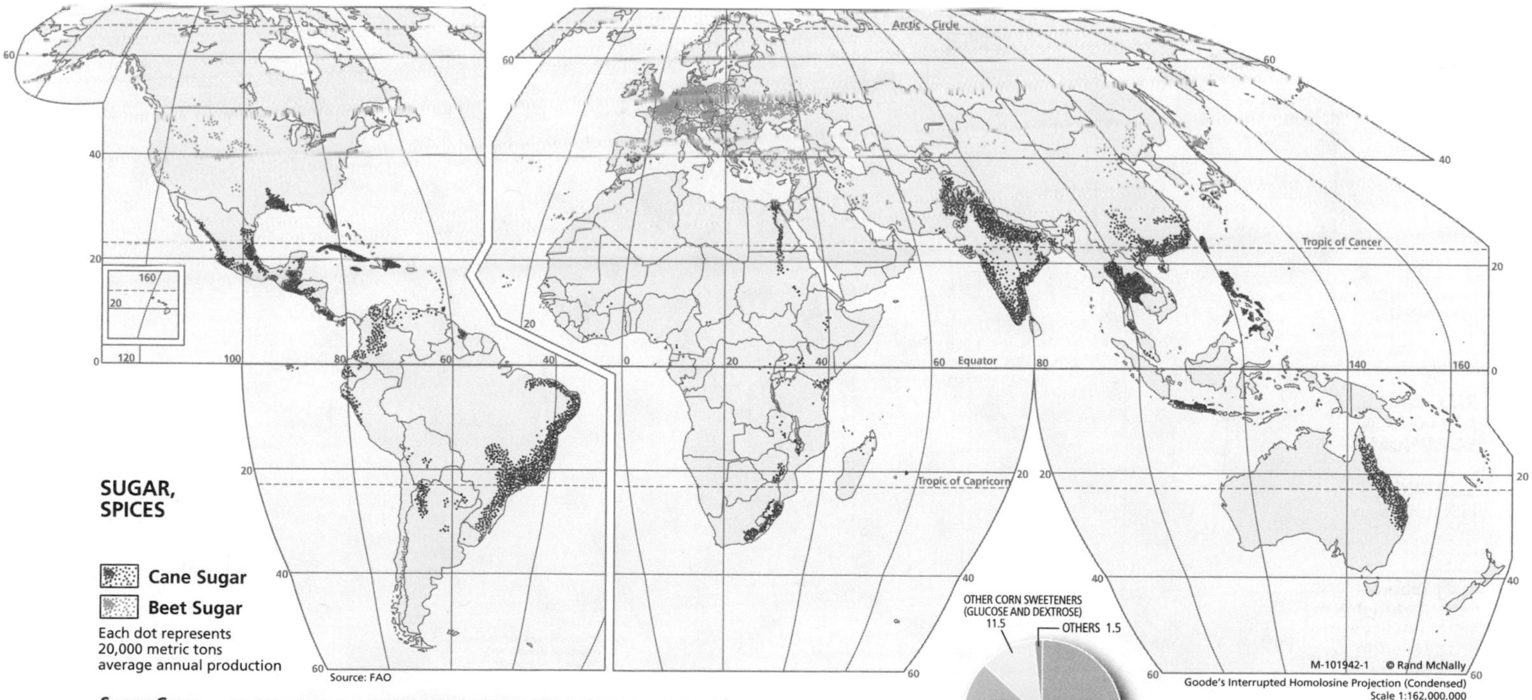

Cane Sugar

Beet Sugar

Each dot represents
20,000 metric tons
average annual production

Source: FAO

Goode's Interrupted Homolosine
Projection (Condensed)
Scale 1:162,000,000

M-101942-1   © Rand McNally

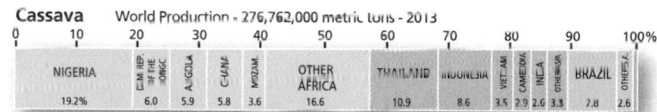

OTHER CORN SWEETENERS
(GLUCOSE AND DEXTROSE)
11.5

OTHERS 1.5

HIGH
FRUCTOSE
CORN
SWEETENERS
34.8

REFINED
SUGAR
52.1%

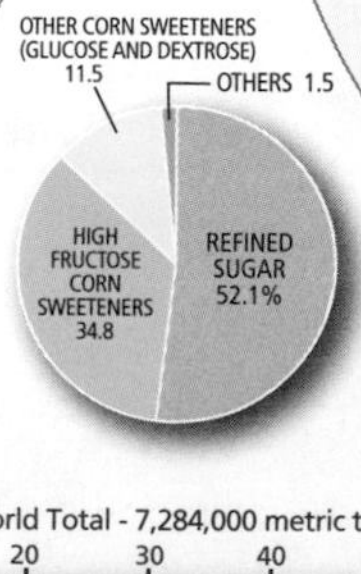

U.S. Sweetener Consumption Per Person
Total - 59.5 kilograms - 2014

Sugar Cane   World Production - 1,911,180,000 metric tons - 2013

	0	10	20	30	40	50	60	70	80	90	100%

| BRAZIL 40.2% | OTHER S.A. 5.3 | INDIA 17.9 | CHINA 6.7 | THAILAND 5.2 | PAKISTAN 3.3 | OTHER ASIA 6.0 | MEXICO 3.2 | OTHER N.A. 5.6 | AFRICA 5.1 | ALL OTHER 1.5 |

Sugar Beets   World Production - 246,522,000 metric tons - 2013

	0	10	20	30	40	50	60	70	80	90	100%

| RUSSIA 16.0% | FRANCE 13.6 | GERMANY 9.3 | UKRAINE 4.4 | POLAND 4.3 | U.K. 3.2 | NETH. 2.3 | OTHER EUROPE 14.9 | UNITED STATES 12.1 | TURKEY 6.7 | CHINA 3.8 | OTHER ASIA 3.5 | EGYPT 4.1 |

Spices   World Total - 7,284,000 metric tons - 2013

	0	10	20	30	40	50	60	70	80	90	100%

| INDIA 38.4% | CHINA 9.0 | INDONESIA 7.5 | NEPAL 5.6 | BANGL. 3.1 | VIETNAM 2.6 | TURKEY 2.3 | THAILAND 2.0 | OTHER ASIA 8.0 | BELGIUM 3.7 | OTHER EUROPE 5.3 | OTHER AFRICA 2.3 | NIGERIA 4.1 | CANADA 2.2 | OTHER N.A. 2.0 | ALL OTHER 2.6 |

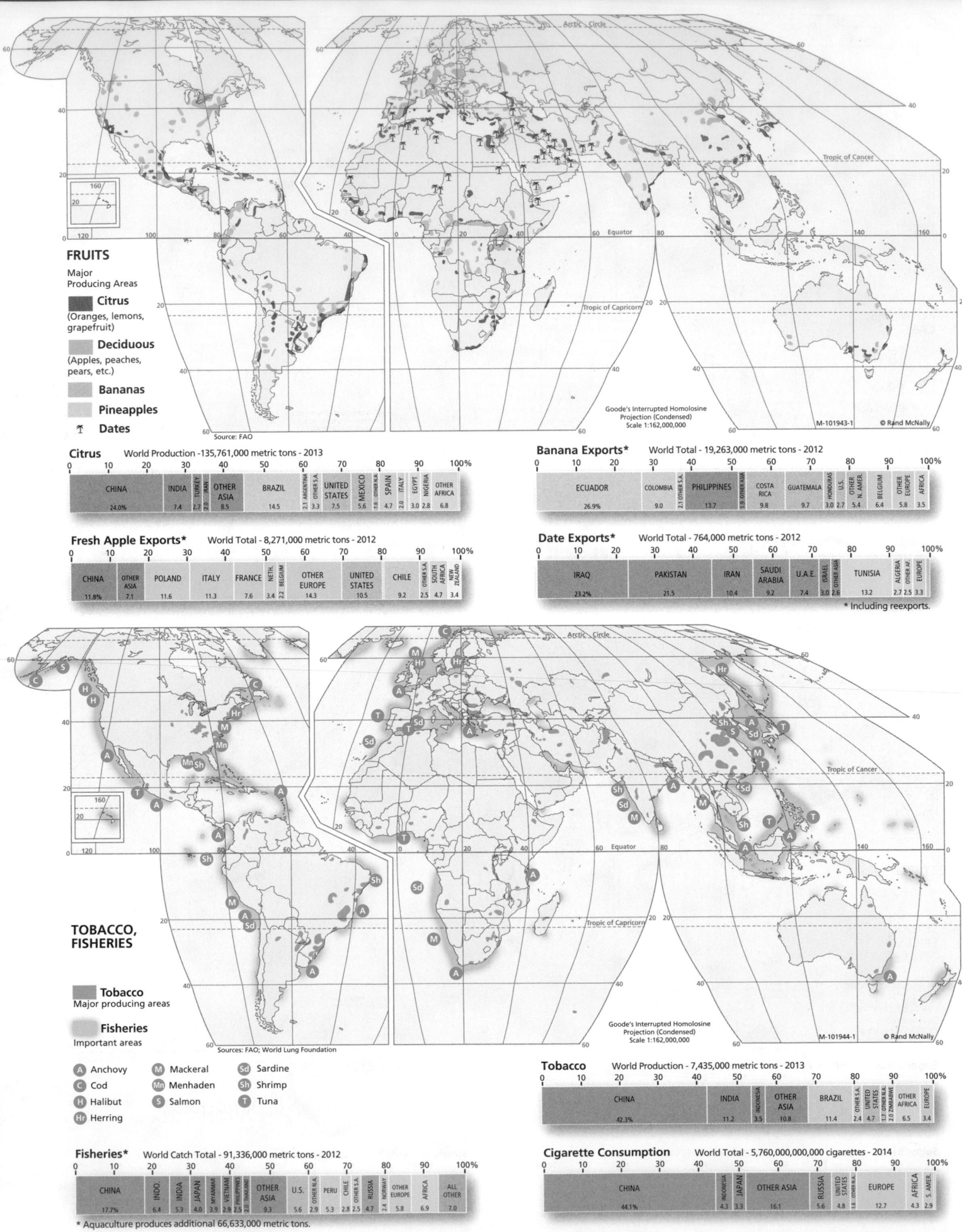

FRUITS

Major Producing Areas

Citrus
(Oranges, lemons, grapefruit)

Deciduous
(Apples, peaches, pears, etc.)

Bananas

Pineapples

☂ **Dates**

Source: FAO

Goode's Interrupted Homolosine Projection (Condensed)
Scale 1:162,000,000
M-101943-1 © Rand McNally

Citrus World Production -135,761,000 metric tons - 2013

CHINA 24.0%	INDIA 7.4	TURKEY 2.7	IRAN 2.0	OTHER ASIA 8.5	BRAZIL 14.5	ARGENTINA 2.1	OTHER S.A. 3.3	UNITED STATES 7.5	MEXICO 5.6	OTHER N.A. 1.6	SPAIN 4.7	ITALY 2.0	EGYPT 3.0	NIGERIA 2.8	OTHER AFRICA 6.8

Banana Exports* World Total - 19,263,000 metric tons - 2012

| ECUADOR 26.9% | COLOMBIA 9.0 | OTHER S.A. 2.1 | PHILIPPINES 13.7 | OTHER ASIA 3.9 | COSTA RICA 9.8 | GUATEMALA 9.7 | HONDURAS 3.0 | U.S. 2.7 | OTHER N. AMER. 5.4 | BELGIUM 6.4 | OTHER EUROPE 5.8 | AFRICA 3.5 |

Fresh Apple Exports* World Total - 8,271,000 metric tons - 2012

| CHINA 11.8% | OTHER ASIA 7.1 | POLAND 11.6 | ITALY 11.3 | FRANCE 7.6 | NETH. 2.2 | BELGIUM | OTHER EUROPE 14.3 | UNITED STATES 10.5 | CHILE 9.2 | OTHER S.A. 2.5 | SOUTH AFRICA 4.7 | NEW ZEALAND 3.4 |

Date Exports* World Total - 764,000 metric tons - 2012

| IRAQ 23.2% | PAKISTAN 21.5 | IRAN 10.4 | SAUDI ARABIA 9.2 | U.A.E. 7.4 | ISRAEL 3.0 | OTHER ASIA 2.6 | TUNISIA 13.2 | ALGERIA 2.7 | OTHER AF. 2.5 | EUROPE 3.3 |

* Including reexports.

TOBACCO, FISHERIES

Tobacco
Major producing areas

Fisheries
Important areas

Sources: FAO; World Lung Foundation

Goode's Interrupted Homolosine Projection (Condensed)
Scale 1:162,000,000
M-101944-1 © Rand McNally

Ⓐ Anchovy Ⓜ Mackeral Ⓢd Sardine
Ⓒ Cod Ⓜn Menhaden Ⓢh Shrimp
Ⓗ Halibut Ⓢ Salmon Ⓣ Tuna
Ⓗr Herring

Tobacco World Production - 7,435,000 metric tons - 2013

| CHINA 42.3% | INDIA 11.2 | INDONESIA 3.5 | OTHER ASIA 10.8 | BRAZIL 11.4 | OTHER S.A. 2.4 | UNITED STATES 4.7 | OTHER N.A. 1.2 | ZIMBABWE 2.0 | OTHER AFRICA 6.5 | EUROPE 3.4 |

Fisheries* World Catch Total - 91,336,000 metric tons - 2012

| CHINA 17.7% | INDO. 6.4 | INDIA 5.3 | JAPAN 4.0 | MYANMAR 3.9 | VIETNAM 2.9 | PHILIPPINES 2.3 | THAILAND | OTHER ASIA 9.3 | U.S. 5.6 | OTHER N.A. 2.9 | PERU 5.3 | CHILE 2.8 | OTHER S.A. 2.5 | RUSSIA 4.7 | OTHER EUROPE 5.8 | AFRICA 6.9 | ALL OTHER 7.0 |

* Aquaculture produces additional 66,633,000 metric tons.

Cigarette Consumption World Total - 5,760,000,000,000 cigarettes - 2014

| CHINA 44.1% | INDONESIA 4.3 | JAPAN 3.3 | OTHER ASIA 16.1 | RUSSIA 5.6 | UNITED STATES 4.8 | OTHER N.A. 1.6 | EUROPE 12.7 | AFRICA 4.3 | S. AMER. 2.9 |

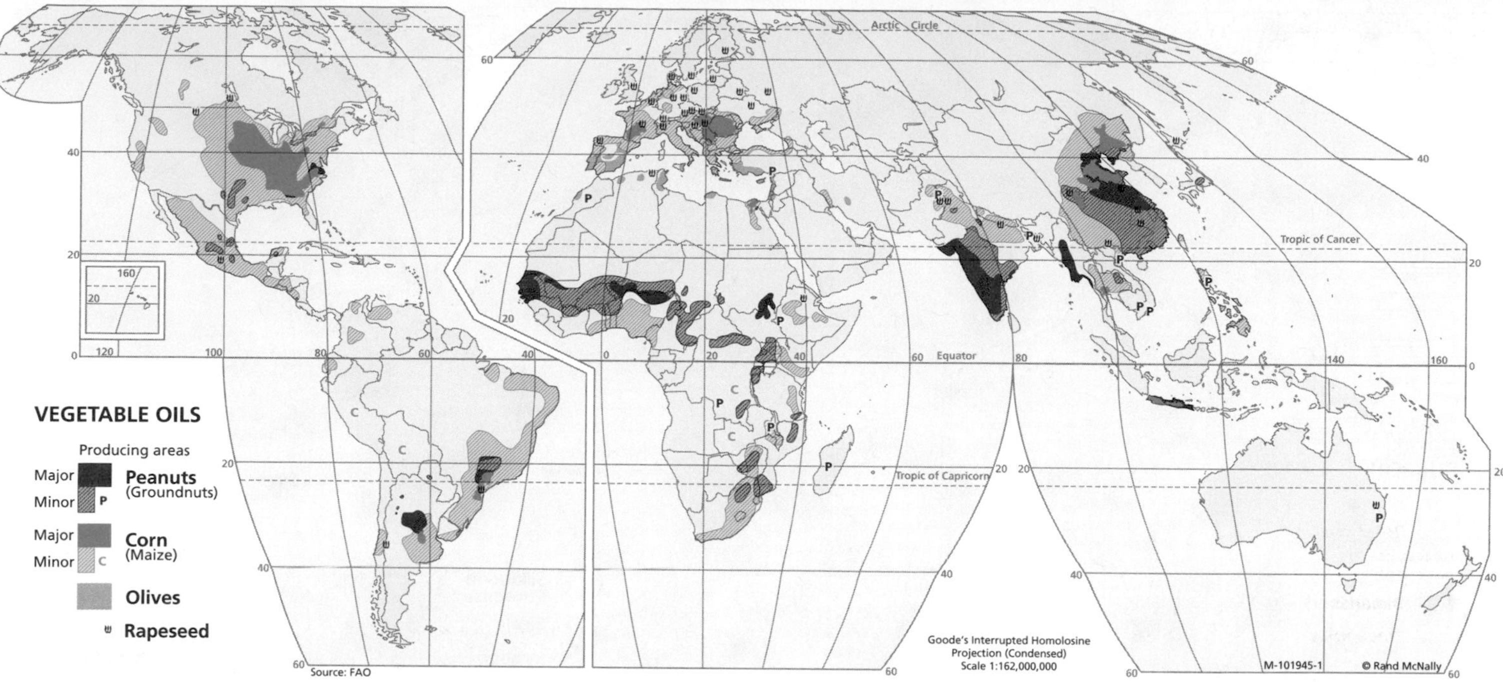

Arctic Circle

Tropic of Cancer

Equator

Tropic of Capricorn

Goode's Interrupted Homolosine
Projection (Condensed)
Scale 1:162,000,000

M-101945-1   © Rand McNally

VEGETABLE OILS

Producing areas

Major | **Peanuts** (Groundnuts)
Minor P

Major | **Corn** (Maize)
Minor C

Olives

Rapeseed

Source: FAO

Peanut Oil   World Production - 5,177,000 metric tons - 2013

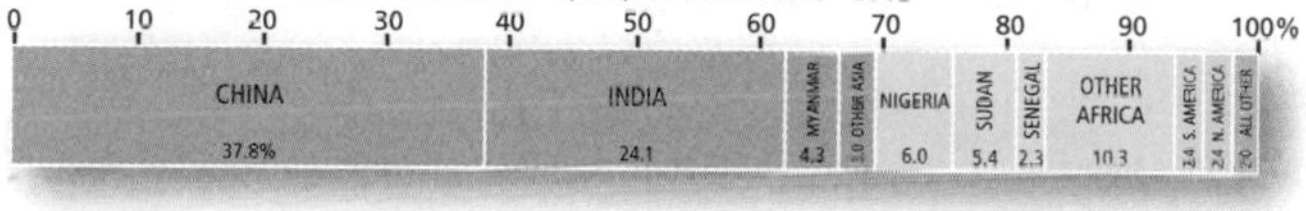

0	10	20	30	40	50	60	70	80	90	100%

CHINA 37.8% | INDIA 24.1 | MYANMAR 4.3 | OTHER ASIA 3.0 | NIGERIA 6.0 | SUDAN 5.4 | SENEGAL 2.3 | OTHER AFRICA 10.3 | S. AMERICA 2.4 | N. AMERICA 2.0 | ALL OTHER

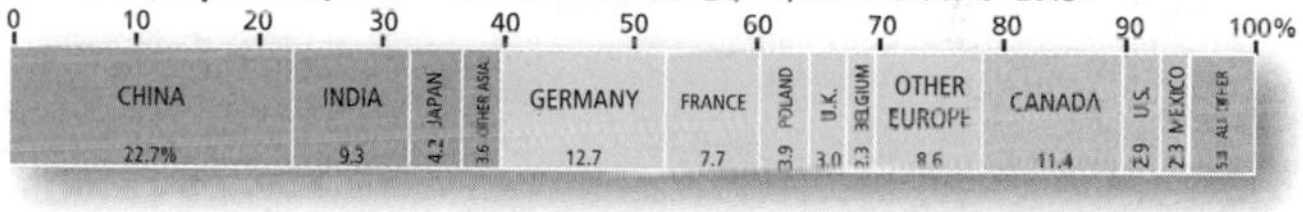

Corn Oil   World Production - 2,856,000 metric tons - 2013

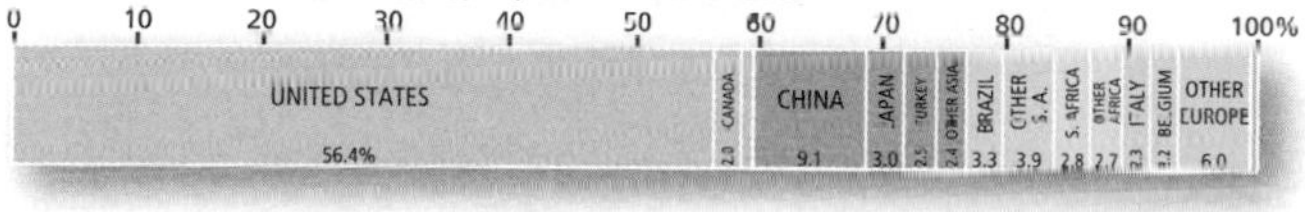

0	10	20	30	40	50	60	70	80	90	100%

UNITED STATES 56.4% | CANADA 2.8 | CHINA 9.1 | JAPAN 3.0 | TURKEY 2.3 | OTHER ASIA | BRAZIL 3.3 | OTHER S.A. 3.9 | S. AFRICA 2.8 | OTHER AFR 2.7 | ITALY 3.2 | OTHER EUROPE 6.0

Canola (Rapeseed) Oil   World Production - 24,688,000 metric tons - 2013

0	10	20	30	40	50	60	70	80	90	100%

CHINA 22.7% | INDIA 9.3 | JAPAN 4.2 | OTHER ASIA 3.6 | GERMANY 12.7 | FRANCE 7.7 | POLAND 3.9 | U.K. 3.0 | BELGIUM 2.3 | OTHER EUROPE 9.6 | CANADA 11.4 | U.S. 2.9 | MEXICO 2.3 | ALL OTHER 5.3

Olive Oil   World Production - 2,826,000 metric tons - 2013

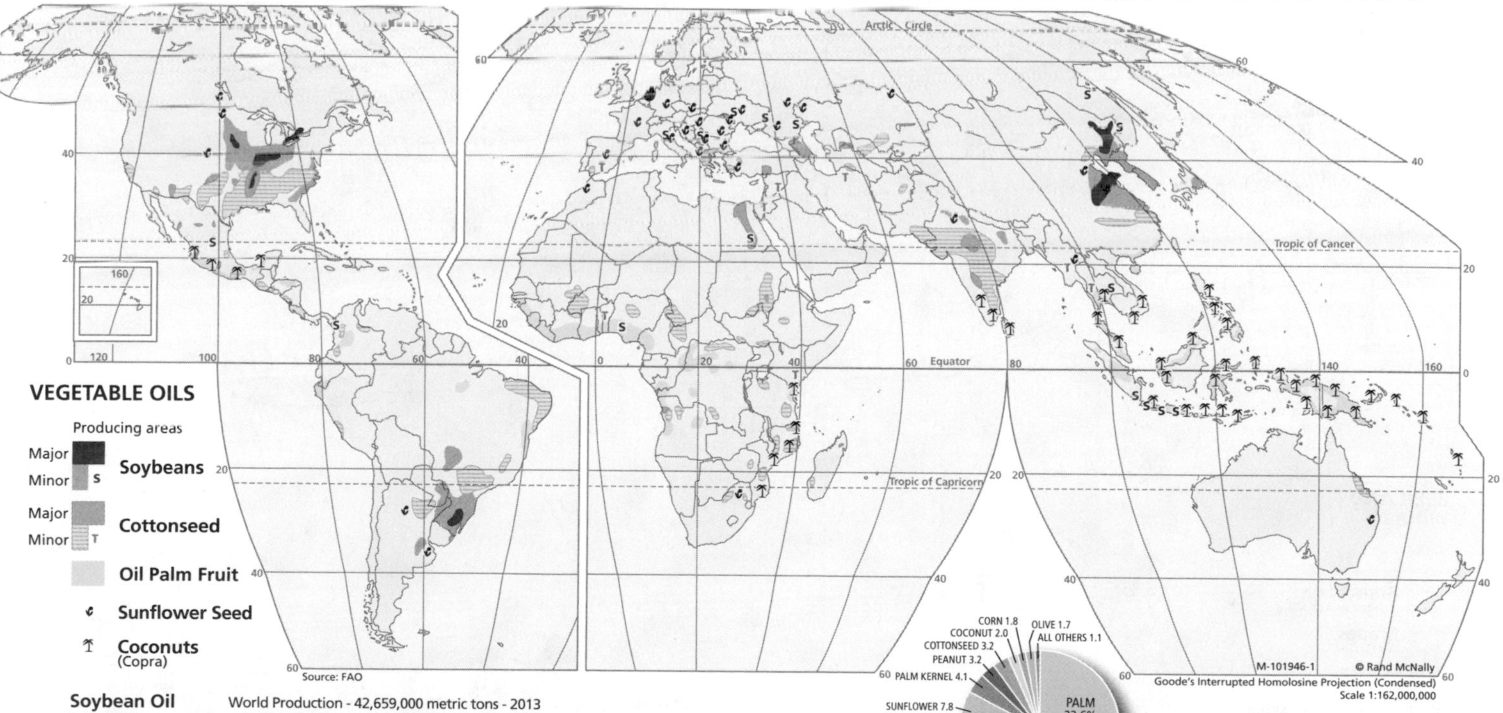

0	10	20	30	40	50	60	70	80	90	100%

SPAIN 19.3% | ITALY 15.6 | GREECE 10.0 | PORTUGAL 4.4 | TUNISIA 6.8 | MOROCCO 4.0 | ALGERIA 2.1 | TURKEY 6.0 | SYRIA 3.6 | S. OTHER

Arctic Circle

Tropic of Cancer

Equator

Tropic of Capricorn

Goode's Interrupted Homolosine Projection (Condensed)
Scale 1:162,000,000

M-101946-1   © Rand McNally

VEGETABLE OILS

Producing areas

Major | **Soybeans**
Minor S

Major | **Cottonseed**
Minor T

Oil Palm Fruit

Sunflower Seed

Coconuts (Copra)

Source: FAO

Soybean Oil   World Production - 42,659,000 metric tons - 2013

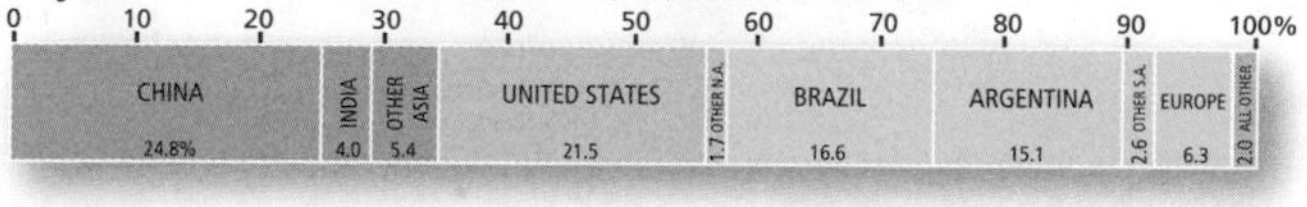

0	10	20	30	40	50	60	70	80	90	100%

CHINA 24.8% | INDIA 4.0 | OTHER ASIA 5.4 | UNITED STATES 21.5 | OTHER N.A. 1.7 | BRAZIL 16.6 | ARGENTINA 15.1 | OTHER S.A. 2.6 | EUROPE 6.3 | ALL OTHER 2.0

Palm Oil   World Production - 54,385,000 metric tons - 2013

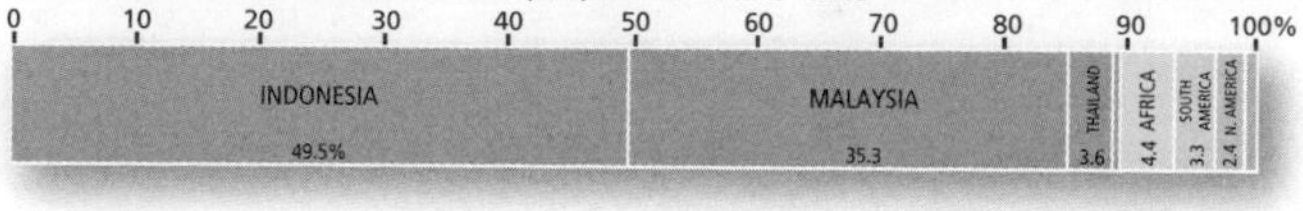

0	10	20	30	40	50	60	70	80	90	100%

INDONESIA 49.5% | MALAYSIA 35.3 | THAILAND 3.6 | AFRICA 4.4 | SOUTH AMER 3.3 | N. AMERICA 2.4 | ALL OTHER

Sunflower Oil   World Production - 12,590,918 metric tons - 2013

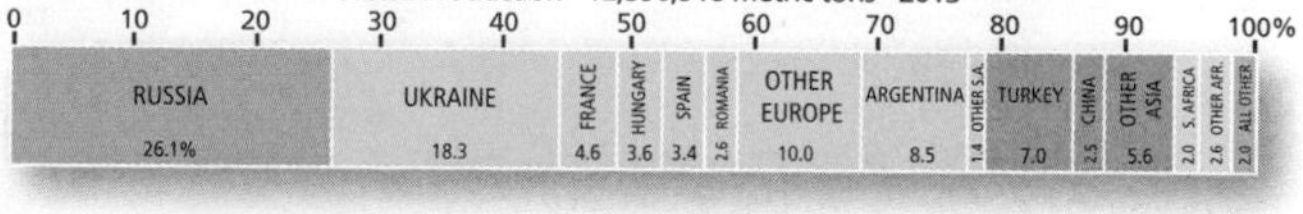

0	10	20	30	40	50	60	70	80	90	100%

RUSSIA 26.1% | UKRAINE 18.3 | FRANCE 4.6 | HUNGARY 3.6 | SPAIN 3.4 | ROMANIA 2.6 | OTHER EUROPE 10.0 | ARGENTINA 8.5 | TURKEY 7.0 | CHINA 1.4 | OTHER ASIA 5.6 | S. AFRICA 2.0 | OTHER AFR. 2.6 | ALL OTHER

Vegetable Oils
World Production - 162,008,000 metric tons - 2013

PALM 33.6%
SOYBEAN 26.3
CANOLA 15.2
SUNFLOWER 7.8
PALM KERNEL 4.1
PEANUT 3.2
COTTONSEED 3.2
COCONUT 2.0
CORN 1.8
OLIVE 1.7
ALL OTHERS 1.1

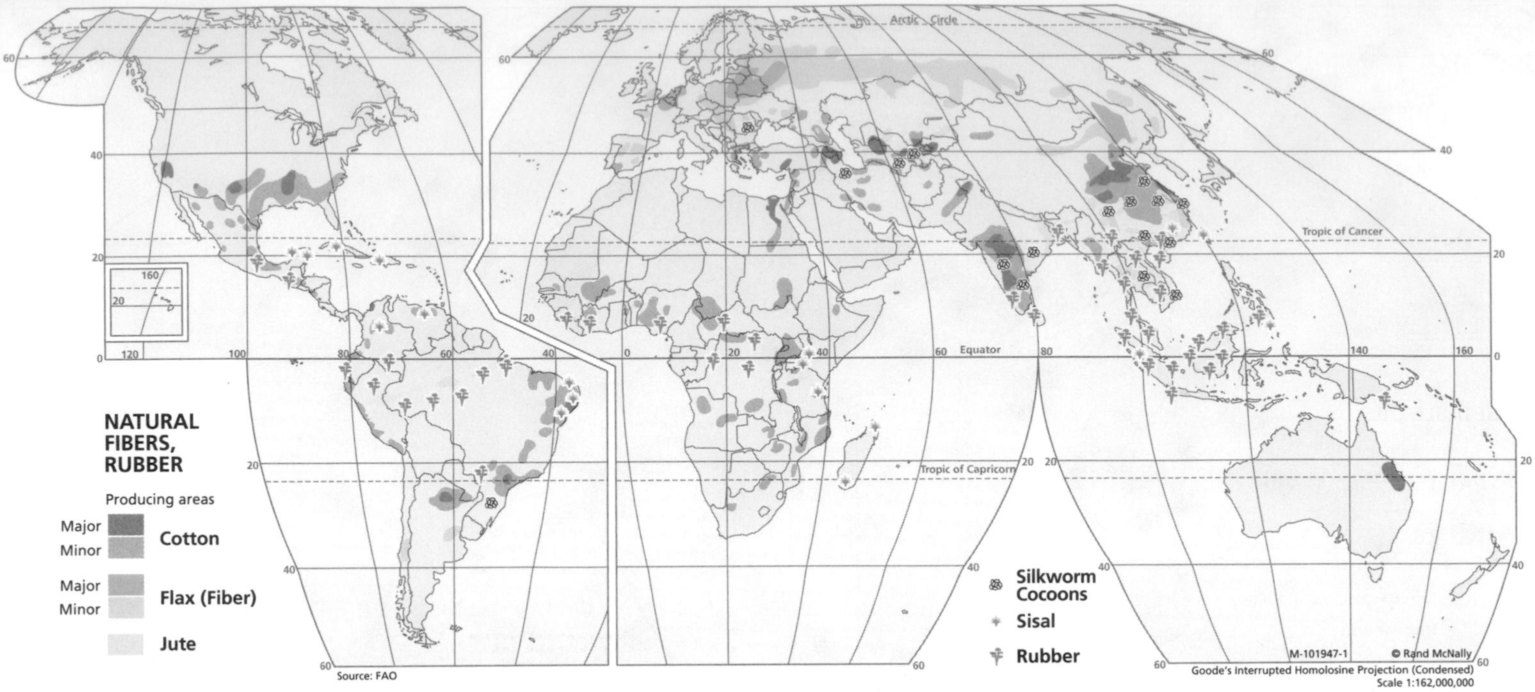

NATURAL FIBERS, RUBBER

Producing areas

Major / Minor **Cotton**

Major / Minor **Flax (Fiber)**

Jute

⊗ **Silkworm Cocoons**

✦ **Sisal**

✤ **Rubber**

Source: FAO

M-101947-1 © Rand McNally
Goode's Interrupted Homolosine Projection (Condensed)
Scale 1:162,000,000

Cotton (Lint) World Production - 24,544,000 metric tons - 2013

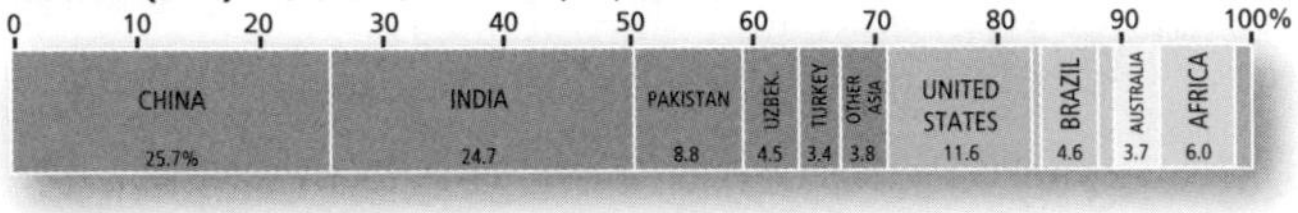

CHINA	INDIA	PAKISTAN	UZBEK.	TURKEY	OTHER ASIA	UNITED STATES	BRAZIL	AUSTRALIA	AFRICA
25.7%	24.7	8.8	4.5	3.4	3.8	11.6	4.6	3.7	6.0

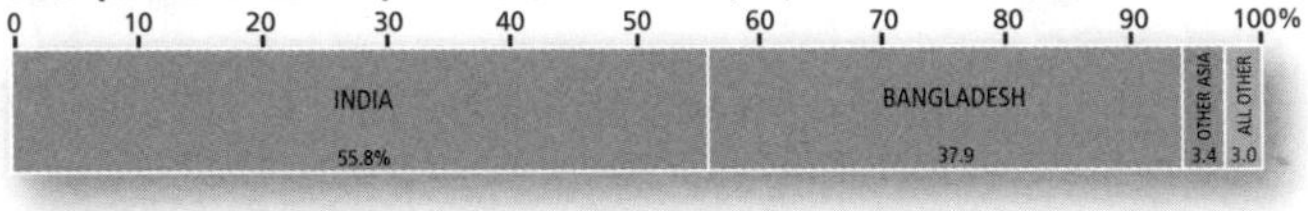

Silk (Raw) Production World Production - 168,000 metric tons - 2013

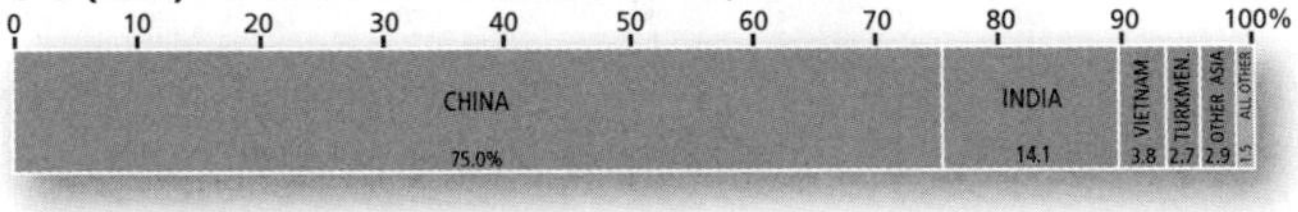

CHINA	INDIA	VIETNAM	TURKMEN.	OTHER ASIA	ALL OTHER
75.0%	14.1	3.8	2.7	2.9	1.5

Jute (and Substitutes) World Production - 3,680,000 metric tons - 2013

INDIA	BANGLADESH	OTHER ASIA	ALL OTHER
55.8%	37.9	3.4	3.0

Flax (Fiber and Tow) World Production - 303,000 metric tons - 2013

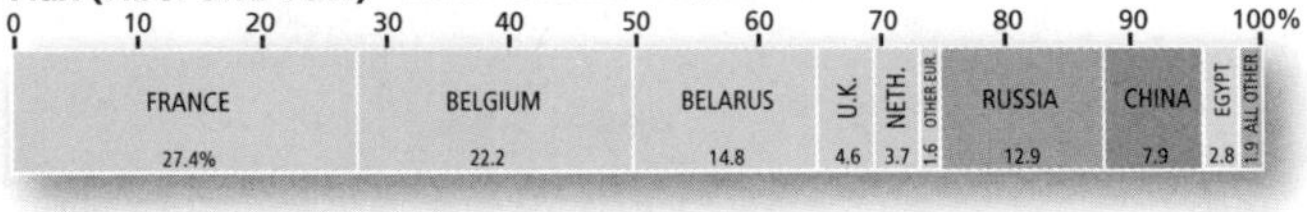

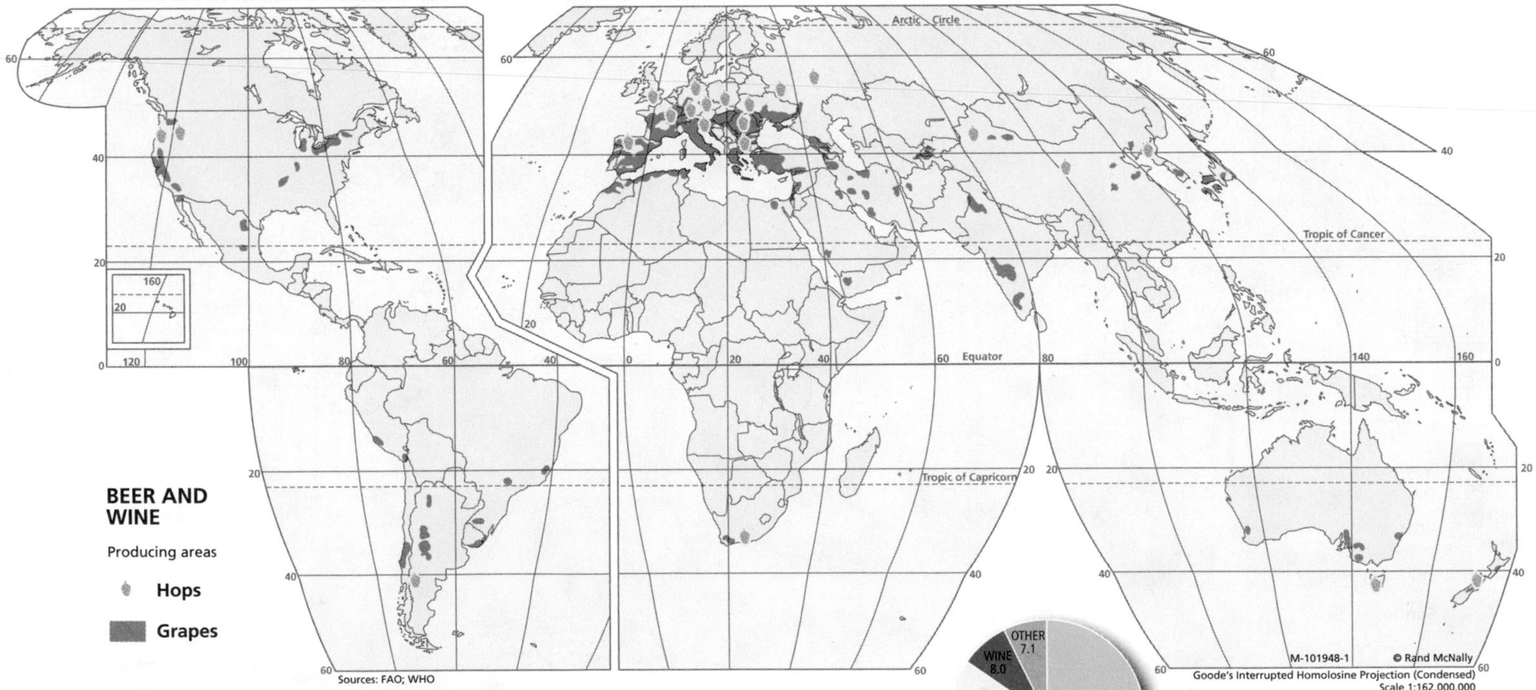

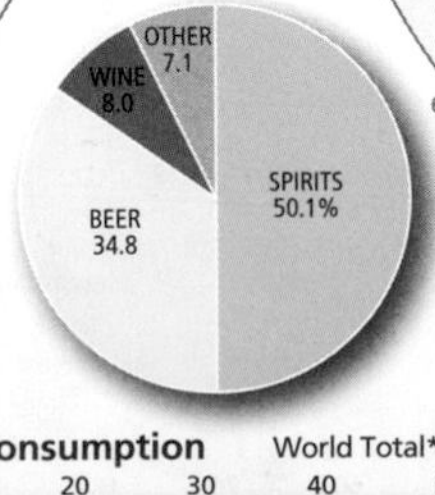

FRANCE	BELGIUM	BELARUS	U.K.	NETH.	OTHER EUR.	RUSSIA	CHINA	EGYPT	ALL OTHER
27.4%	22.2	14.8	4.6	3.7	1.6	12.9	7.9	2.8	1.9

BEER AND WINE

Producing areas

🌰 **Hops**

▇ **Grapes**

Sources: FAO; WHO

M-101948-1 © Rand McNally
Goode's Interrupted Homolosine Projection (Condensed)
Scale 1:162,000,000

Beer World Production - 189,097,000 metric tons - 2013

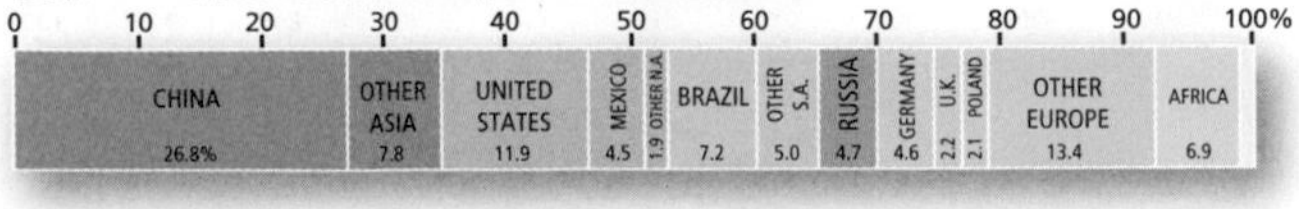

CHINA	OTHER ASIA	UNITED STATES	MEXICO	OTHER N.A.	BRAZIL	OTHER S.A.	RUSSIA	GERMANY	U.K.	POLAND	OTHER EUROPE	AFRICA
26.8%	7.8	11.9	4.5	1.9	7.2	5.0	4.6	4.7	2.2	2.1	13.4	6.9

Wine World Production - 27,422,000 metric tons - 2013

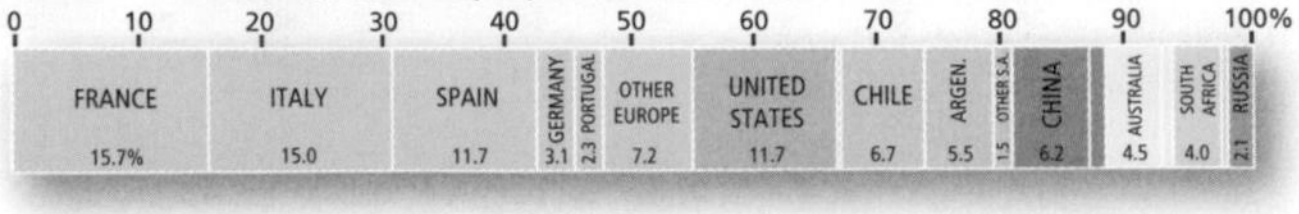

FRANCE	ITALY	SPAIN	GERMANY	PORTUGAL	OTHER EUROPE	UNITED STATES	CHILE	ARGEN.	OTHER S.A.	CHINA	AUSTRALIA	SOUTH AFRICA	RUSSIA
15.7%	15.0	11.7	3.1	2.3	7.2	11.7	6.7	5.5	1.5	6.2	4.5	4.0	2.1

World Consumption of Alcoholic Beverages

OTHER 7.1
WINE 8.0
BEER 34.8
SPIRITS 50.1%

Alcohol Consumption World Total* - 31,137,823,000 liters - 2010

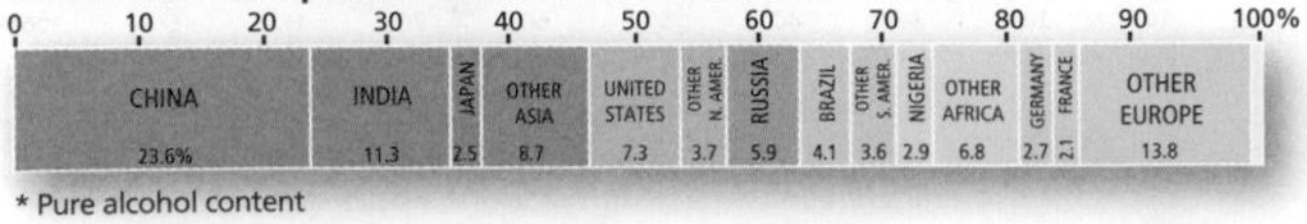

CHINA	INDIA	JAPAN	OTHER ASIA	UNITED STATES	OTHER N. AMER.	RUSSIA	BRAZIL	OTHER S. AMER.	NIGERIA	OTHER AFRICA	GERMANY	FRANCE	OTHER EUROPE
23.6%	11.3	2.5	8.7	7.3	3.7	5.9	4.1	3.6	2.9	6.8	2.7	2.1	13.8

* Pure alcohol content

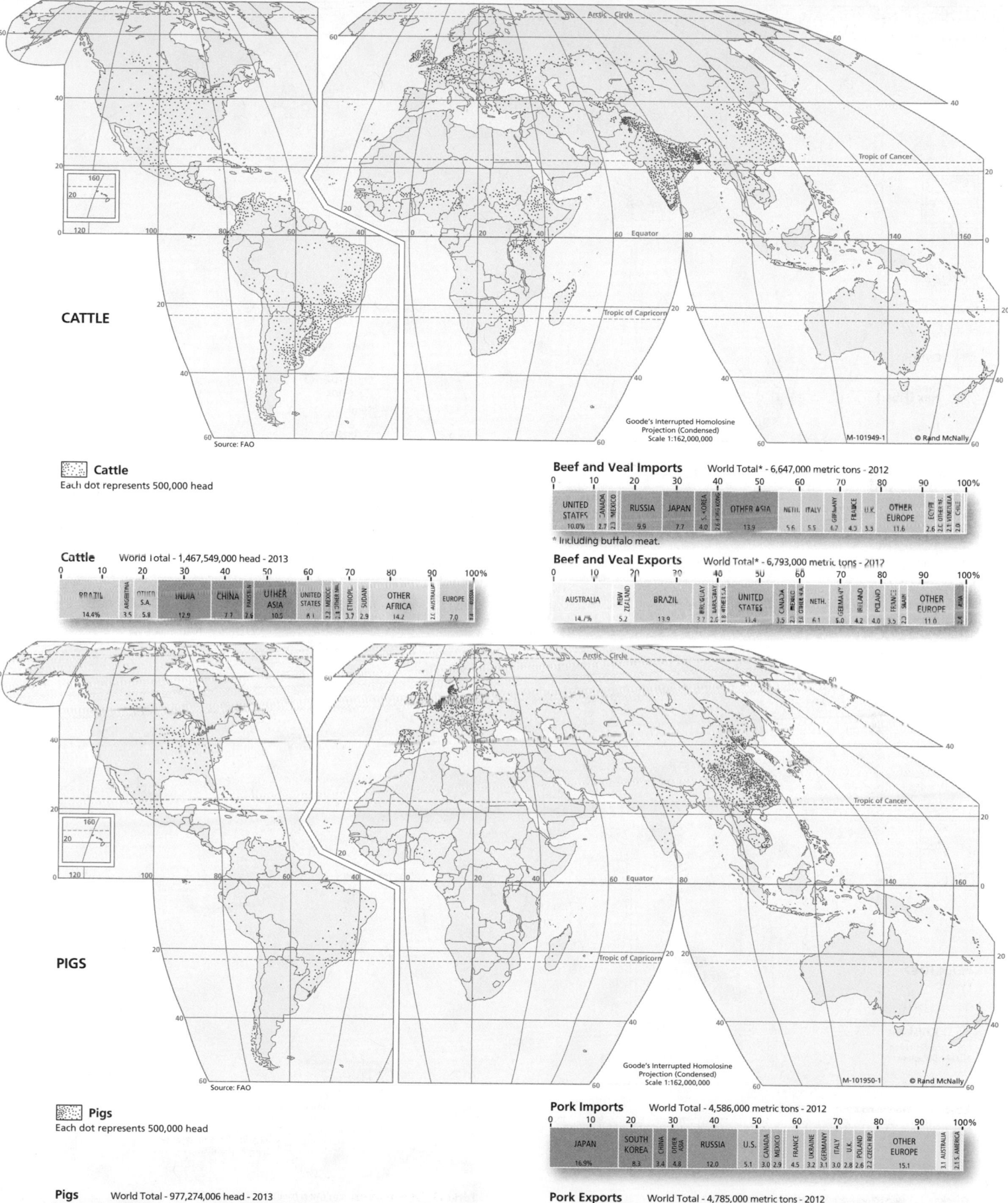

CATTLE

Goode's Interrupted Homolosine
Projection (Condensed)
Scale 1:162,000,000

Source: FAO

M-101949-1 © Rand McNally

Cattle
Each dot represents 500,000 head

Cattle   World Total - 1,467,549,000 head - 2013

BRAZIL 14.4%	ARGENTINA 3.5	OTHER S.A. 5.8	INDIA 12.9	CHINA 7.7	PAKISTAN 2.6	OTHER ASIA 10.5	UNITED STATES 6.1	MEXICO 2.2	CHAD 2.4	ETHIOPIA 3.7	SUDAN 2.9	OTHER AFRICA 14.2	AUSTRALIA 2.6 EUROPE 7.0 RUSSIA

Beef and Veal Imports   World Total* - 6,647,000 metric tons - 2012

UNITED STATES 10.0%	CANADA 2.3	MEXICO 2.7	RUSSIA 9.9	JAPAN 7.7	S. KOREA 2.6 HONG KONG	OTHER ASIA 13.9	NETH. 5.6	ITALY 5.5	GERMANY 4.7	FRANCE 4.3	U.K. 3.5	OTHER EUROPE 11.6	EGYPT 2.6 OTHER N.A. 2.2	VENEZUELA 2.1	CHILE 2.0

* Including buffalo meat.

Beef and Veal Exports   World Total* - 6,793,000 metric tons - 2012

AUSTRALIA 14.7%	NEW ZEALAND 5.2	BRAZIL 13.9	URUGUAY 3.7	PARAGUAY 2.6 OTHER S.A. 1.8	UNITED STATES 11.4	CANADA 2.5	MEXICO 2.0 OTHER N.A.	NETH. 6.1	GERMANY 5.0	IRELAND 4.2	POLAND 4.0	FRANCE 2.3 SPAIN	OTHER EUROPE 11.0	ASIA 2.6

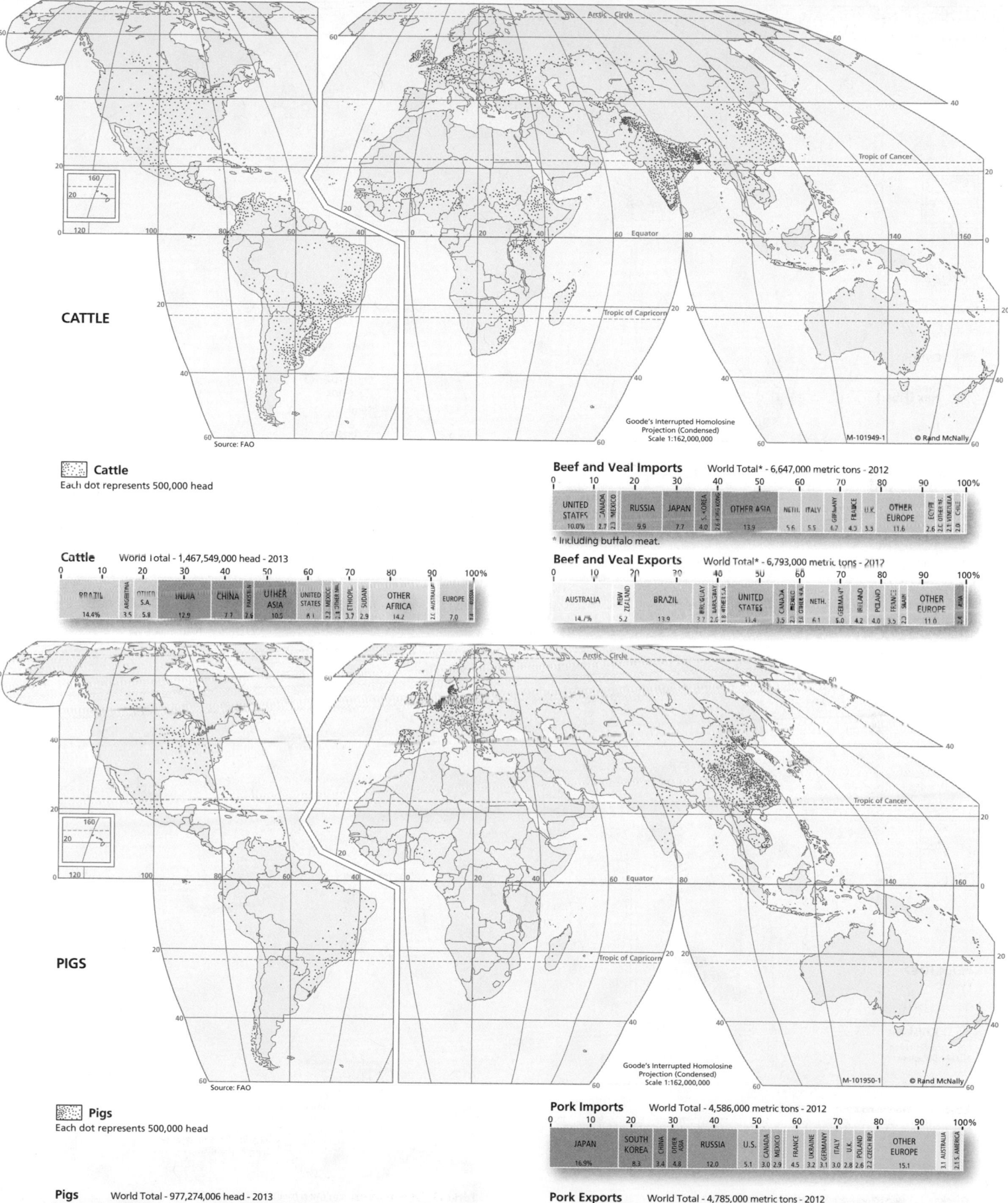

PIGS

Goode's Interrupted Homolosine
Projection (Condensed)
Scale 1:162,000,000

Source: FAO

M-101950-1 © Rand McNally

Pigs
Each dot represents 500,000 head

Pigs   World Total - 977,274,006 head - 2013

CHINA 48.7%	VIETNAM 2.7	OTHER ASIA 9.0	UNITED STATES 6.6	OTHER N.A. 3.8	BRAZIL 3.8	OTHER S.A. 2.8	GERMANY 2.6 SPAIN	OTHER EUROPE 11.5	AFRICA 3.7	ALL OTHER 2.4

Pork Imports   World Total - 4,586,000 metric tons - 2012

JAPAN 16.9%	SOUTH KOREA 8.3	CHINA 3.4	OTHER ASIA 4.8	RUSSIA 12.0	U.S. 5.1	CANADA 3.0	MEXICO 2.9	FRANCE 4.5	UKRAINE 3.2	GERMANY 3.1	ITALY 2.8	U.K. 2.6	POLAND 2.2 CZECH REP.	OTHER EUROPE 15.1	AUSTRALIA 3.1 S. AMERICA 2.1

Pork Exports   World Total - 4,785,000 metric tons - 2012

UNITED STATES 24.3%	CANADA 15.7	OTHER N.A. 1.5	GERMANY 15.5	SPAIN 7.8	DENMARK 7.5	NETH. 2.5 POLAND 2.3	BELGIUM 2.1	OTHER EUROPE 9.7	BRAZIL 8.7	OTHER S.A. 1.2

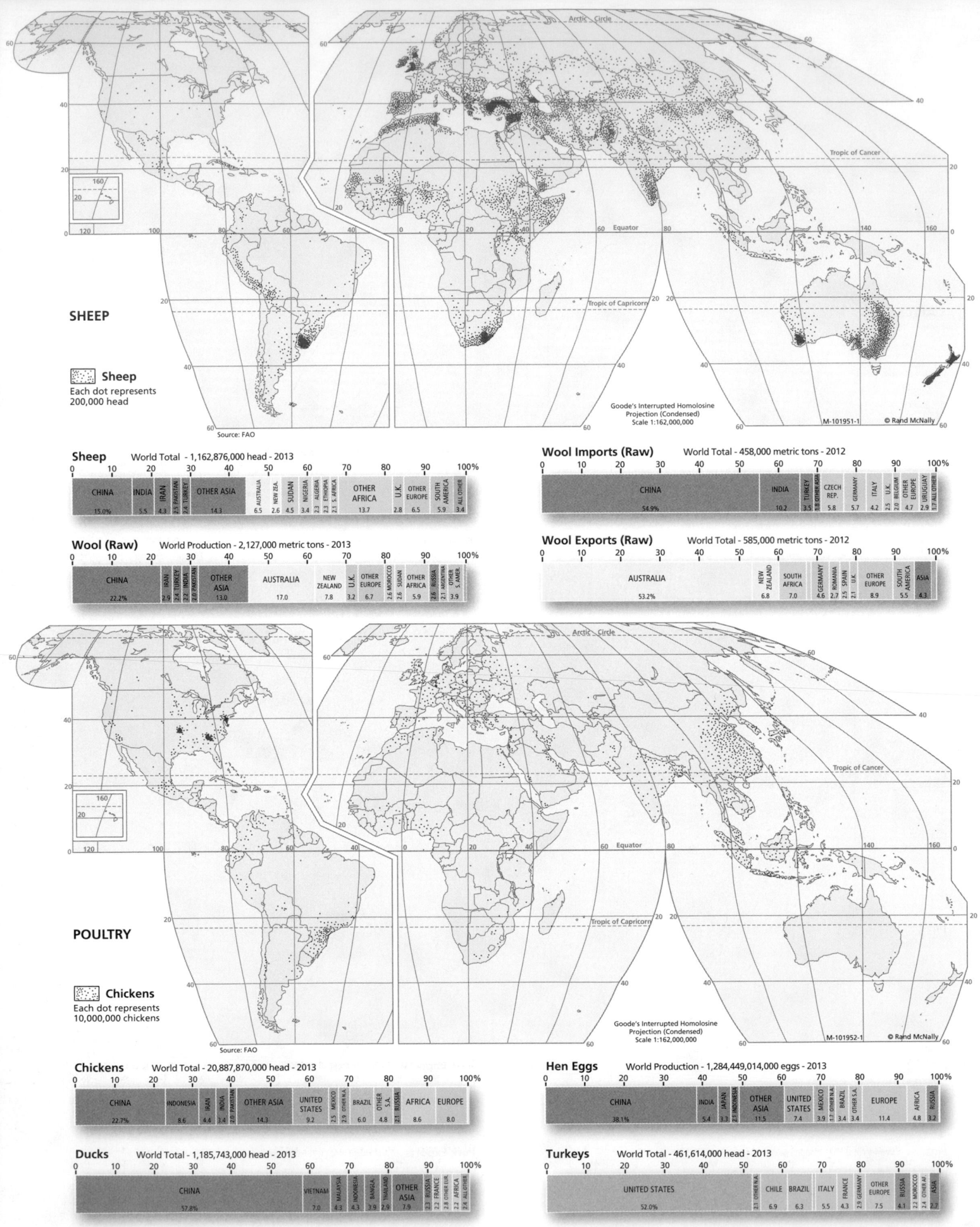

SHEEP

Sheep
Each dot represents
200,000 head

Source: FAO

Goode's Interrupted Homolosine
Projection (Condensed)
Scale 1:162,000,000

M-101951-1   © Rand McNally

Sheep   World Total - 1,162,876,000 head - 2013

CHINA 15.0%	INDIA 5.5	IRAN 4.3	PAKISTAN 2.5	TURKEY 7.4	OTHER ASIA 14.3	AUSTRALIA 6.5	NEW ZEA. 2.6	SUDAN 4.5	NIGERIA 3.4	ALGERIA 2.3 ETHIOPIA 2.3 S. AFRICA 2.1	OTHER AFRICA 13.7	U.K. 2.8	OTHER EUROPE 6.5	SOUTH AMERICA 5.9	ALL OTHER 3.4

Wool (Raw)   World Production - 2,127,000 metric tons - 2013

CHINA 22.2%	IRAN 2.9	TURKEY 2.4	INDIA 2.2	PAKISTAN 2.0	OTHER ASIA 13.0	AUSTRALIA 17.0	NEW ZEALAND 7.8	U.K. 3.2	OTHER EUROPE 6.7	MOROCCO 2.6	SUDAN 2.6	OTHER AFRICA 5.9	RUSSIA 2.6	ARGENTINA 2.1	OTHER S. AMER. 3.9

Wool Imports (Raw)   World Total - 458,000 metric tons - 2012

CHINA 54.9%	INDIA 10.2	TURKEY 3.5	OTHER ASIA 1.8	CZECH REP. 5.8	GERMANY 5.7	ITALY 4.2	U.K. 2.5	BELGIUM 2.0	OTHER EUROPE 4.7	URUGUAY 2.9	ALL OTHER 1.7

Wool Exports (Raw)   World Total - 585,000 metric tons - 2012

AUSTRALIA 53.2%	NEW ZEALAND 6.8	SOUTH AFRICA 7.0	GERMANY 4.6	ROMANIA 2.7	SPAIN 2.5	U.K. 2.1	OTHER EUROPE 8.9	SOUTH AMERICA 5.5	ASIA 4.3

POULTRY

Chickens
Each dot represents
10,000,000 chickens

Source: FAO

Goode's Interrupted Homolosine
Projection (Condensed)
Scale 1:162,000,000

M-101952-1   © Rand McNally

Chickens   World Total - 20,887,870,000 head - 2013

CHINA 22.7%	INDONESIA 8.6	IRAN 4.4	INDIA 3.4	PAKISTAN 2.0	OTHER ASIA 14.3	UNITED STATES 9.2	MEXICO 2.5	OTHER N.A. 2.9	BRAZIL 6.0	OTHER S.A. 4.8	RUSSIA 2.1	AFRICA 8.6	EUROPE 8.0

Hen Eggs   World Production - 1,284,449,014,000 eggs - 2013

CHINA 38.1%	INDIA 5.4	JAPAN 3.3	INDONESIA 2.3	OTHER ASIA 11.5	UNITED STATES 7.4	MEXICO 3.9	OTHER N.A. 1.7	BRAZIL 3.4	OTHER S.A. 3.4	EUROPE 11.4	AFRICA 4.8	RUSSIA 3.2

Ducks   World Total - 1,185,743,000 head - 2013

CHINA 57.8%	VIETNAM 7.0	MALAYSIA 4.3	INDONESIA 3.2	BANGLA. 2.8	THAILAND 2.9	OTHER ASIA 7.9	RUSSIA 2.3	FRANCE 2.2	OTHER EUR. 2.8	AFRICA 2.2	ALL OTHER 2.3

Turkeys   World Total - 461,614,000 head - 2013

UNITED STATES 52.0%	OTHER N.A. 2.1	CHILE 6.9	BRAZIL 6.3	ITALY 5.5	FRANCE 4.3	GERMANY 2.9	OTHER EUROPE 7.5	RUSSIA 4.1	MOROCCO 2.2	OTHER AF. 2.4	ASIA 2.7

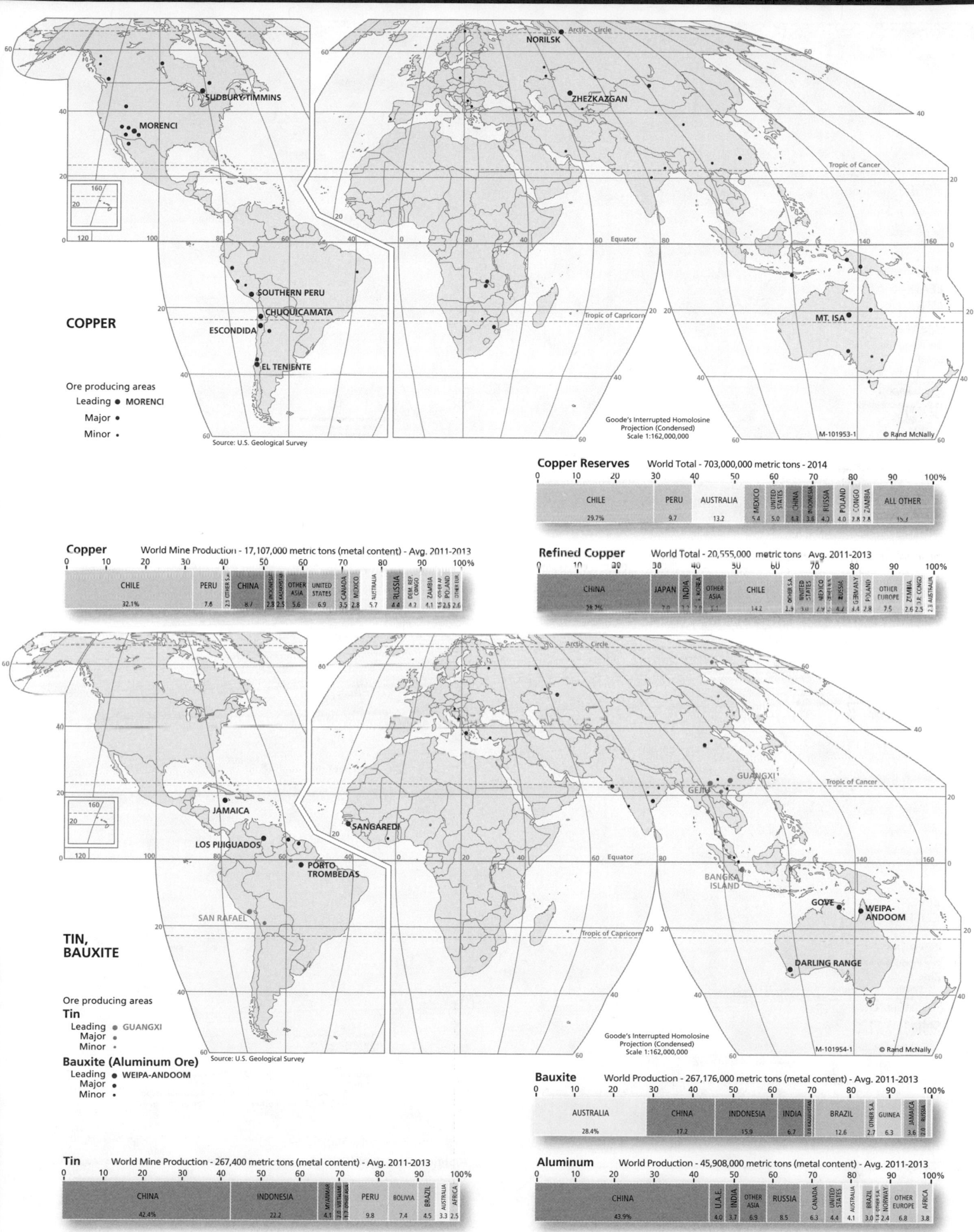

COPPER

Ore producing areas
Leading ● MORENCI
Major ●
Minor ·

Source: U.S. Geological Survey

Goode's Interrupted Homolosine Projection (Condensed)
Scale 1:162,000,000
M-101953-1   © Rand McNally

Map labels: NORILSK, ZHEZKAZGAN, SUDBURY-TIMMINS, MORENCI, SOUTHERN PERU, CHUQUICAMATA, ESCONDIDA, EL TENIENTE, MT. ISA

Copper Reserves   World Total - 703,000,000 metric tons - 2014

CHILE	PERU	AUSTRALIA	MEXICO	UNITED STATES	CHINA	INDONESIA	RUSSIA	POLAND	CONGO	ZAMBIA	ALL OTHER
29.7%	9.7	13.2	5.4	5.0	4.3	3.6	4.0	2.8	2.8		15.3

Copper   World Mine Production - 17,107,000 metric tons (metal content) - Avg. 2011-2013

CHILE	PERU	OTHER S.A.	CHINA	INDONESIA	KAZAKHSTAN	OTHER ASIA	UNITED STATES	CANADA	MEXICO	AUSTRALIA	RUSSIA	DEM. REP. CONGO	ZAMBIA	OTHER AF.	POLAND	OTHER EUR.
32.1%	7.6	2.1	8.7	2.8	2.5	5.6	6.9	3.5	2.8	5.7	4.4	4.2	4.1	3.6	2.6	2.6

Refined Copper   World Total - 20,555,000 metric tons   Avg. 2011-2013

CHINA	JAPAN	INDIA	KOREA	OTHER ASIA	CHILE	OTHER S.A.	UNITED STATES	MEXICO	RUSSIA	GERMANY	POLAND	OTHER EUROPE	ZAMBIA	D.R. CONGO	AUSTRALIA
28.7%	7.0	2.2			14.2	2.3	3.0	2.2	4.2	4.4	2.8	7.5	2.6	2.3	2.2

TIN, BAUXITE

Ore producing areas
Tin
Leading ● GUANGXI
Major ●
Minor ·

Bauxite (Aluminum Ore)
Leading ● WEIPA-ANDOOM
Major ●
Minor ·

Source: U.S. Geological Survey

Goode's Interrupted Homolosine Projection (Condensed)
Scale 1:162,000,000
M-101954-1   © Rand McNally

Map labels: GUANGXI, GEJIU, JAMAICA, LOS PIJIGUADOS, PORTO TROMBEDAS, SAN RAFAEL, SANGAREDI, BANGKA ISLAND, GOVE, WEIPA-ANDOOM, DARLING RANGE

Bauxite   World Production - 267,176,000 metric tons (metal content) - Avg. 2011-2013

AUSTRALIA	CHINA	INDONESIA	INDIA	OTHER ASIA	BRAZIL	OTHER S.A.	GUINEA	JAMAICA	RUSSIA
28.4%	17.2	15.9	6.7		12.6	2.7	6.3	3.6	2.0

Tin   World Mine Production - 267,400 metric tons (metal content) - Avg. 2011-2013

CHINA	INDONESIA	MYANMAR	VIETNAM	OTHER ASIA	PERU	BOLIVIA	BRAZIL	AUSTRALIA	AFRICA
42.4%	22.2	4.1	3.5		9.8	7.4	4.5	3.3	2.5

Aluminum   World Production - 45,908,000 metric tons (metal content) - Avg. 2011-2013

CHINA	U.A.E.	INDIA	OTHER ASIA	RUSSIA	CANADA	UNITED STATES	AUSTRALIA	BRAZIL	NORWAY	OTHER EUROPE	AFRICA
43.9%	4.0	3.7	6.9	8.5	6.3	4.4	4.1	3.0	2.4	6.8	3.8

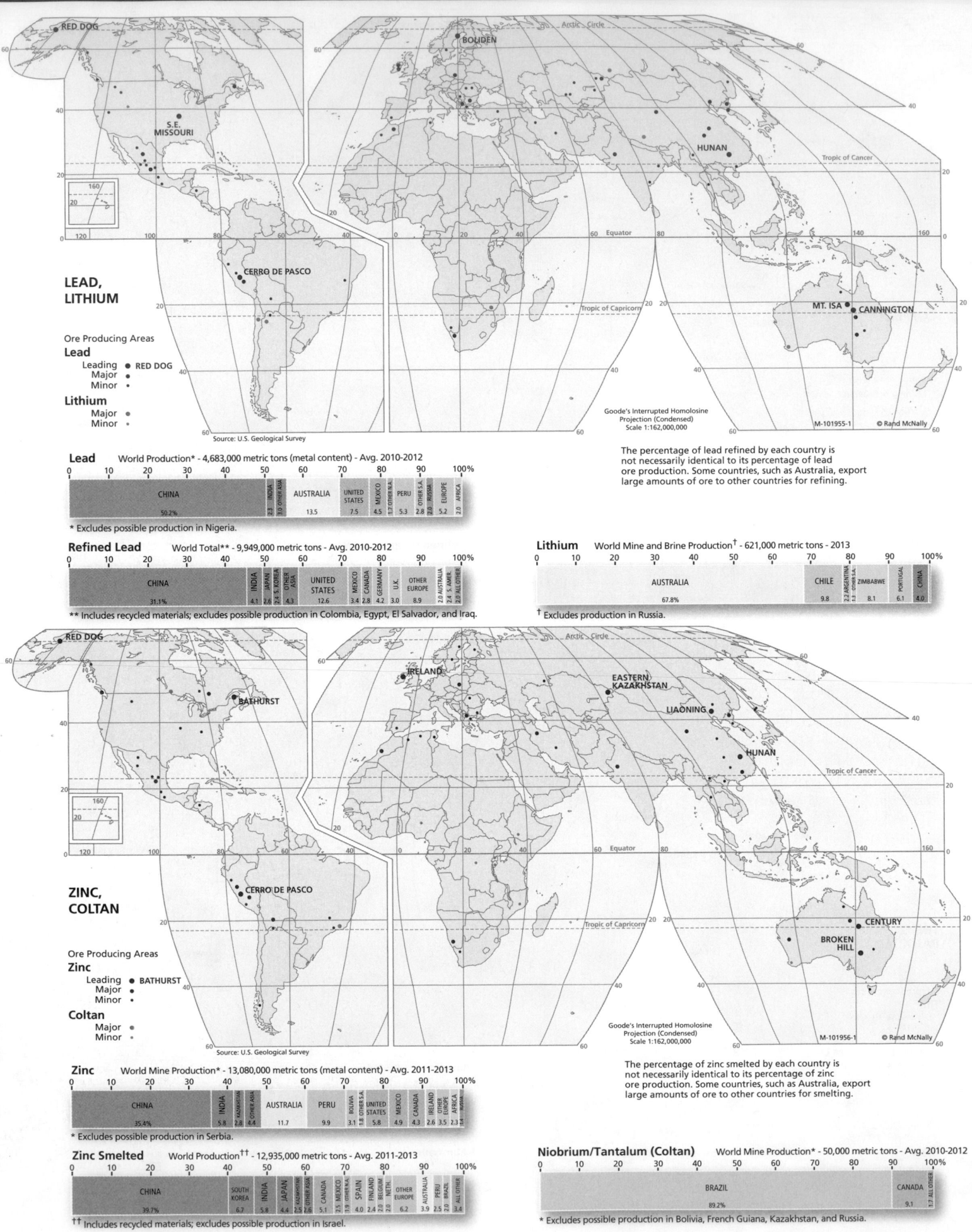

LEAD, LITHIUM

Ore Producing Areas

Lead
- Leading ● RED DOG
- Major ●
- Minor ·

Lithium
- Major ●
- Minor ·

Source: U.S. Geological Survey

Goode's Interrupted Homolosine
Projection (Condensed)
Scale 1:162,000,000

M-101955-1 © Rand McNally

The percentage of lead refined by each country is not necessarily identical to its percentage of lead ore production. Some countries, such as Australia, export large amounts of ore to other countries for refining.

Lead
World Production* - 4,683,000 metric tons (metal content) - Avg. 2010-2012

CHINA 50.2%	INDIA 2.3 / OTHER ASIA 3.0	AUSTRALIA 13.5	UNITED STATES 7.5	MEXICO 4.5 / OTHER N.A. 1.7	PERU 5.3	OTHER S.A. 2.8 / RUSSIA 2.0	EUROPE 2.6	AFRICA 2.0

* Excludes possible production in Nigeria.

Refined Lead
World Total** - 9,949,000 metric tons - Avg. 2010-2012

CHINA 31.1% | INDIA 4.1 | JAPAN 2.6 | S. KOREA 4.5 | OTHER ASIA 4.9 | UNITED STATES 12.6 | MEXICO 3.4 | CANADA 2.8 | GERMANY 4.2 | U.K. 3.0 | OTHER EUROPE 8.9 | AUSTRALIA 2.0 | S. AMER. 2.4 | ALL OTHER 4.0

** Includes recycled materials; excludes possible production in Colombia, Egypt, El Salvador, and Iraq.

Lithium
World Mine and Brine Production† - 621,000 metric tons - 2013

AUSTRALIA 67.8% | CHILE 9.8 | ARGENTINA 2.2 / OTHER S.A. 1.1 | ZIMBABWE 8.1 | PORTUGAL 6.1 | CHINA 4.0

† Excludes production in Russia.

ZINC, COLTAN

Ore Producing Areas

Zinc
- Leading ● BATHURST
- Major ●
- Minor ·

Coltan
- Major ●
- Minor ·

Source: U.S. Geological Survey

Goode's Interrupted Homolosine
Projection (Condensed)
Scale 1:162,000,000

M-101956-1 © Rand McNally

The percentage of zinc smelted by each country is not necessarily identical to its percentage of zinc ore production. Some countries, such as Australia, export large amounts of ore to other countries for smelting.

Zinc
World Mine Production* - 13,080,000 metric tons (metal content) - Avg. 2011-2013

CHINA 35.4% | INDIA 5.8 | KAZAKHSTAN 2.8 | OTHER ASIA 4.4 | AUSTRALIA 11.7 | PERU 9.9 | BOLIVIA 3.1 / OTHER S.A. 1.8 | UNITED STATES 5.8 | MEXICO 4.9 | CANADA 4.3 | IRELAND 2.6 | OTHER EUROPE 3.5 | AFRICA 2.3 | RUSSIA 1.4

* Excludes possible production in Serbia.

Zinc Smelted
World Production†† - 12,935,000 metric tons - Avg. 2011-2013

CHINA 39.7% | SOUTH KOREA 6.7 | INDIA 5.8 | JAPAN 4.4 | KAZAKHSTAN 2.6 / OTHER ASIA 1.9 | CANADA 5.1 | MEXICO 2.5 | SPAIN 4.0 | FINLAND 2.4 | BELGIUM 2.0 / NETH. 2.0 | OTHER EUROPE 6.2 | AUSTRALIA 3.9 | PERU 2.1 / BRAZIL 2.0 | ALL OTHER 3.4

†† Includes recycled materials; excludes possible production in Israel.

Niobrium/Tantalum (Coltan)
World Mine Production* - 50,000 metric tons - Avg. 2010-2012

BRAZIL 89.2% | CANADA 9.1 | ALL OTHER 1.7

* Excludes possible production in Bolivia, French Guiana, Kazakhstan, and Russia.

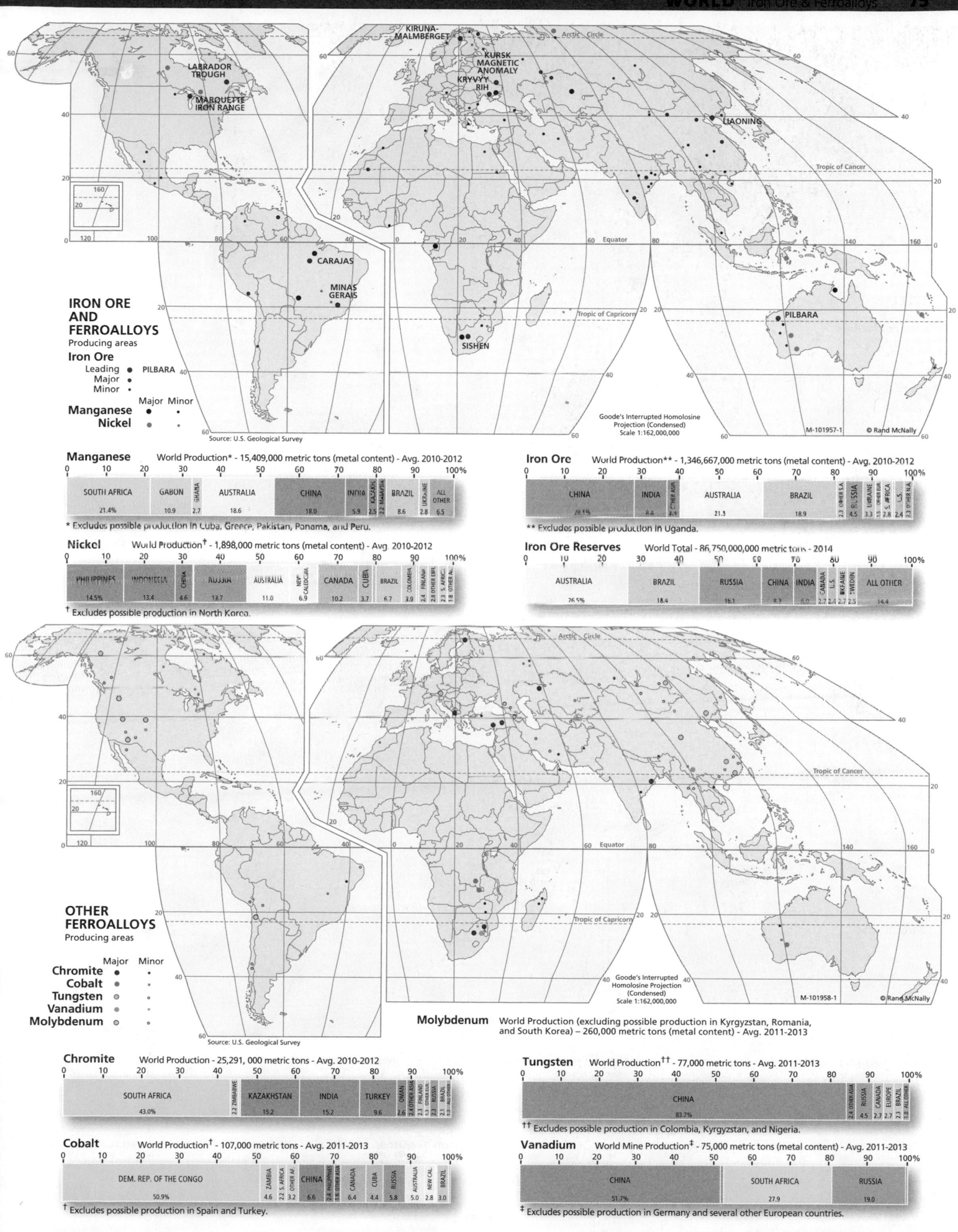

IRON ORE AND FERROALLOYS
Producing areas

Iron Ore
- Leading ● PILBARA
- Major ●
- Minor ·

	Major	Minor
Manganese	●	·
Nickel	●	·

Map labels: LABRADOR TROUGH, MARQUETTE IRON RANGE, KIRUNA-MALMBERGET, KURSK MAGNETIC ANOMALY, KRYVYY RIH, LIAONING, CARAJAS, MINAS GERAIS, SISHEN, PILBARA

Source: U.S. Geological Survey

Goode's Interrupted Homolosine Projection (Condensed)
Scale 1:162,000,000
M-101957-1
© Rand McNally

Manganese
World Production* - 15,409,000 metric tons (metal content) - Avg. 2010-2012

| SOUTH AFRICA 21.4% | GABON 10.9 | GHANA 2.7 | AUSTRALIA 18.6 | CHINA 18.0 | INDIA 5.9 | KAZAKH. 2.5 | MALAYSIA 2.2 | BRAZIL 8.6 | UKRAINE 2.8 | ALL OTHER 6.5 |

* Excludes possible production in Cuba, Greece, Pakistan, Panama, and Peru.

Nickel
World Production† - 1,898,000 metric tons (metal content) - Avg 2010-2012

| PHILIPPINES 14.5% | INDONESIA 13.4 | CHINA 4.6 | RUSSIA 13.7 | AUSTRALIA 11.0 | NEW CALEDONIA 6.9 | CANADA 10.2 | CUBA 3.7 | BRAZIL 6.7 | COLOMBIA 3.9 | OTHER EUR. 2.4 | OTHER ASIA 2.8 | S. AFRIC 2.3 | OTHER AF. 1.8 |

† Excludes possible production in North Korea.

Iron Ore
World Production** - 1,346,667,000 metric tons (metal content) - Avg. 2010-2012

| CHINA 28.1% | INDIA 8.8 | OTHER ASIA 3.9 | AUSTRALIA 21.3 | BRAZIL 18.9 | OTHER S.A. 2.3 | RUSSIA 4.5 | UKRAINE 3.3 | OTHER EUR. 1.5 | S. AFRICA 2.8 | U.S. 2.4 | OTHER N.A. 2.3 |

** Excludes possible production in Uganda.

Iron Ore Reserves
World Total - 86,750,000,000 metric tons - 2014

| AUSTRALIA 26.5% | BRAZIL 18.4 | RUSSIA 16.1 | CHINA 8.3 | INDIA 6.0 | CANADA 2.7 | U.S. 2.4 | MOZAMB. 2.7 | SWEDEN 2.5 | ALL OTHER 14.4 |

OTHER FERROALLOYS
Producing areas

	Major	Minor
Chromite	●	·
Cobalt	●	·
Tungsten	○	·
Vanadium	○	·
Molybdenum	○	·

Source: U.S. Geological Survey

Goode's Interrupted Homolosine Projection (Condensed)
Scale 1:162,000,000
M-101958-1
© Rand McNally

Molybdenum World Production (excluding possible production in Kyrgyzstan, Romania, and South Korea) – 260,000 metric tons (metal content) - Avg. 2011-2013

Chromite
World Production - 25,291,000 metric tons - Avg. 2010-2012

| SOUTH AFRICA 43.0% | ZIMBABWE 2.2 | KAZAKHSTAN 15.2 | INDIA 15.2 | TURKEY 9.6 | OMAN 2.6 | OTHER ASIA 2.3 | OTHER EUR. 1.8 | RUSSIA 2.3 | BRAZIL 2.1 | ALL OTHER 1.3 |

Cobalt
World Production† - 107,000 metric tons - Avg. 2011-2013

| DEM. REP. OF THE CONGO 50.9% | ZAMBIA 4.6 | S. AFRICA 2.2 | OTHER AF. 3.2 | CHINA 6.6 | PHILIPPINES 2.4 | OTHER ASIA 2.3 | CANADA 6.4 | CUBA 4.4 | RUSSIA 5.8 | AUSTRALIA 5.0 | NEW CAL. 2.8 | BRAZIL 3.0 |

† Excludes possible production in Spain and Turkey.

Tungsten
World Production†† - 77,000 metric tons - Avg. 2011-2013

| CHINA 83.7% | OTHER ASIA 2.4 | RUSSIA 4.5 | CANADA 2.7 | EUROPE 2.7 | BRAZIL 2.3 | ALL OTHER 1.8 |

†† Excludes possible production in Colombia, Kyrgyzstan, and Nigeria.

Vanadium
World Mine Production‡ - 75,000 metric tons (metal content) - Avg. 2011-2013

| CHINA 51.7% | SOUTH AFRICA 27.9 | RUSSIA 19.0 |

‡ Excludes possible production in Germany and several other European countries.

STEEL

Arctic Circle
Tropic of Cancer
Equator
Tropic of Capricorn
Goode's Interrupted Homolosine
Projection (Condensed)
Scale 1:162,000,000
A-101959-1
© Rand McNally
Sources: United Nations; U.S. Geological Survey

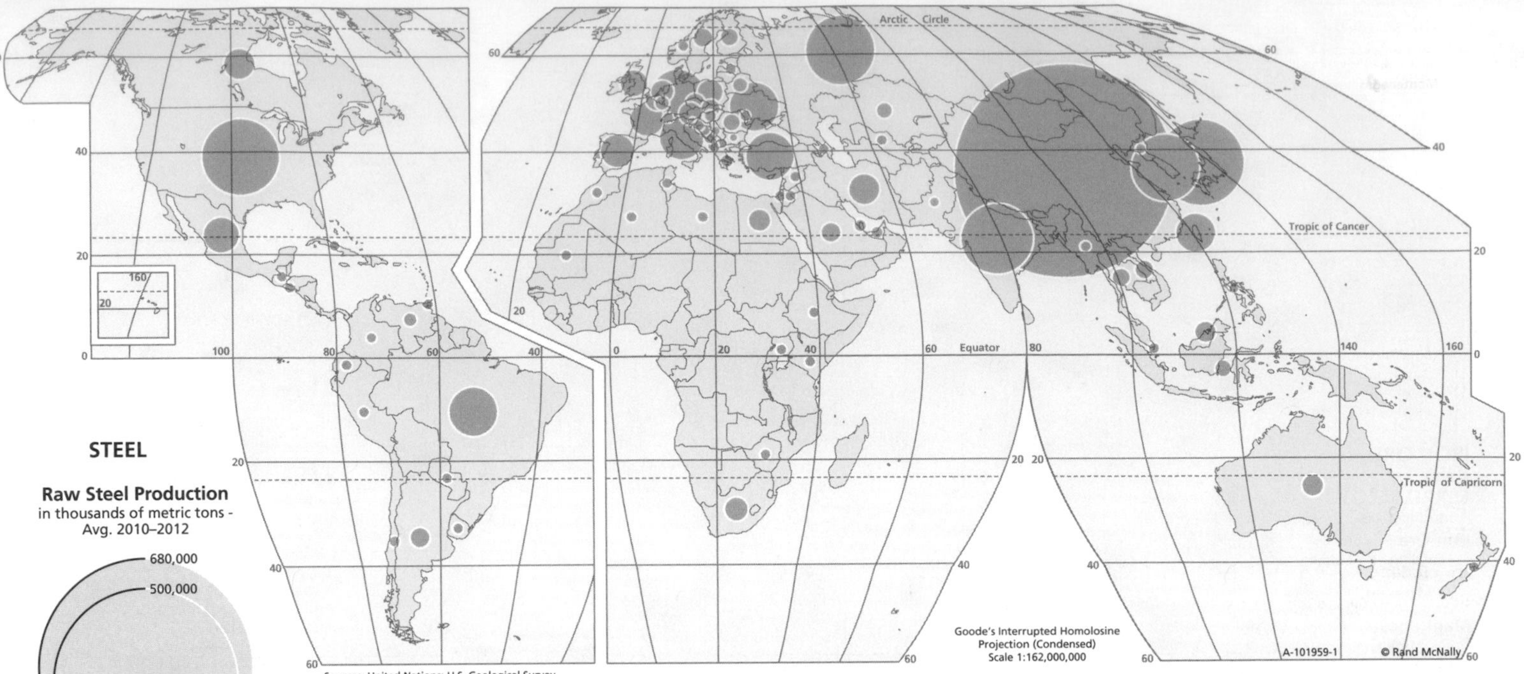

Raw Steel Production
in thousands of metric tons - Avg. 2010–2012

680,000
500,000
100,000
50,000
25,000
5,000
5 - 1,000

Iron and Steel Exports
World Total* - $389,713,000,000 (U.S. dollars) - 2014

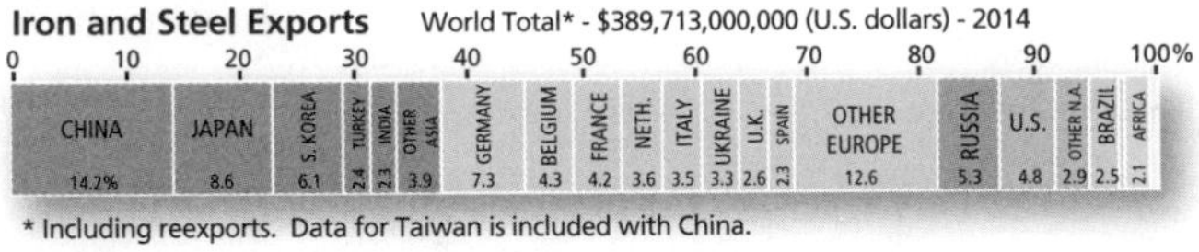

0 10 20 30 40 50 60 70 80 90 100%
CHINA 14.2%
JAPAN 8.6
S. KOREA 6.1
TURKEY 2.4
INDIA 2.3
OTHER ASIA 3.9
GERMANY 7.3
BELGIUM 4.3
FRANCE 4.2
NETH. 3.6
ITALY 3.5
UKRAINE 3.3
U.K. 2.6
SPAIN 2.3
OTHER EUROPE 12.6
RUSSIA 5.3
U.S. 4.8
OTHER N.A. 2.9
BRAZIL 2.5
AFRICA 2.1

* Including reexports. Data for Taiwan is included with China.

Steel Production Trends 1945-2014

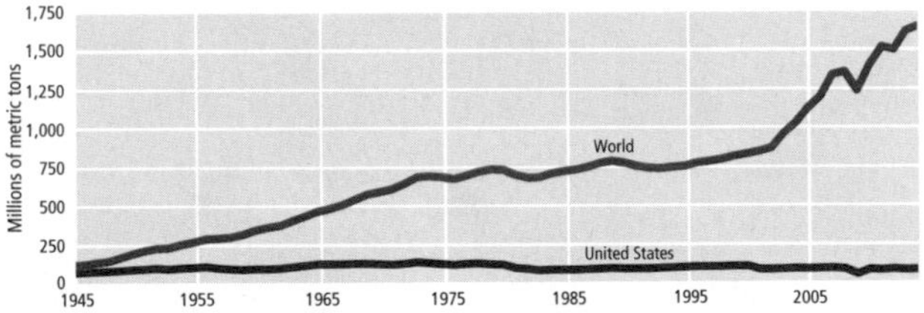

Millions of metric tons
1,750
1,500
1,250
1,000
750
500
250
0
World
United States
1945 1955 1965 1975 1985 1995 2005

PRECIOUS METALS

ESKAY CREEK
NORTHERN NEVADA
CENTRAL MEXICO
NORTHERN PERU
CENTRAL PERU
NORIL'SK
TRZEBINIA
MURUNTAU
BUSHVELD
WITWATERSRAND
CANNINGTON
WESTERN AUSTRALIA
Arctic Circle
Tropic of Cancer
Equator
Tropic of Capricorn
Goode's Interrupted Homolosine
Projection (Condensed)
Scale 1:162,000,000
M-101960-1
© Rand McNally
Source: U.S. Geological Survey

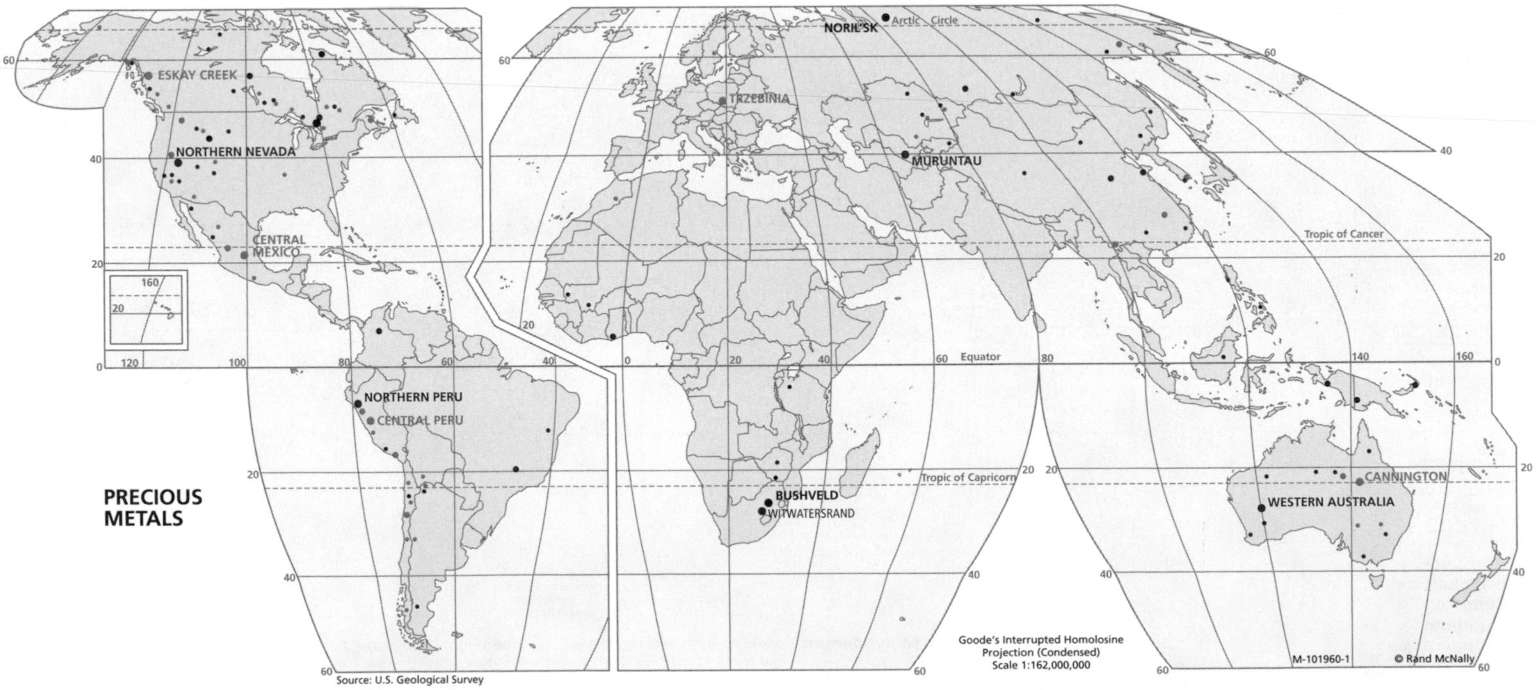

Gold Producing Areas
Leading ● MURUNTAU
Major •
Minor •

Silver Producing Areas
Leading ● CANNINGTON
Major •
Minor •

Platinum Producing Areas
Leading ● NORIL'SK
Major •
Minor •

Silver
World Production** - 24,500 metric tons - Avg. 2010-2012

0 10 20 30 40 50 60 70 80 90 100%
MEXICO 19.8%
UNITED STATES 4.7
CANADA 2.5
CHINA 15.1
KAZAKH. 2.9
OTHER ASIA 4.4
PERU 14.3
CHILE 5.1
BOLIVIA 5.0
ARGENTINA 3.1
AUSTRALIA 7.2
RUSSIA 5.7
POLAND 4.8
OTHER EUR. 2.3

**Excluding possible production in Botswana, Eritrea, Fiji, Georgia, Iran, Kyrgyzstan, and Zambia.

Gold
World Production† - 2,700 metric tons - Avg. 2011-2013

0 10 20 30 40 50 60 70 80 90 100%
CHINA 14.7%
UZBEKISTAN 3.5
INDONESIA 2.4
OTHER ASIA 5.8
AUSTRALIA 9.5
PAP. N.G. 2.1
UNITED STATES 8.6
CANADA 4.1
MEXICO 3.4
RUSSIA 8.0
SOUTH AFRICA 6.1
GHANA 3.2
OTHER AFRICA 10.1
PERU 5.9
BRAZIL 2.5
COLOMBIA 2.2
ARGENTINA 2.0
OTHER S.A. 3.3

† Excluding possible production in twelve additional countries.

Platinum-Group Metals
World Production* - 460 metric tons - Avg. 2011-2013

0 10 20 30 40 50 60 70 80 90 100%
SOUTH AFRICA 56.4%
ZIMBABWE 4.8
RUSSIA 26.4
CANADA 5.6
U.S. 3.5
JAPAN 2.0

* Excluding possible production in China, Indonesia, and Philippines.

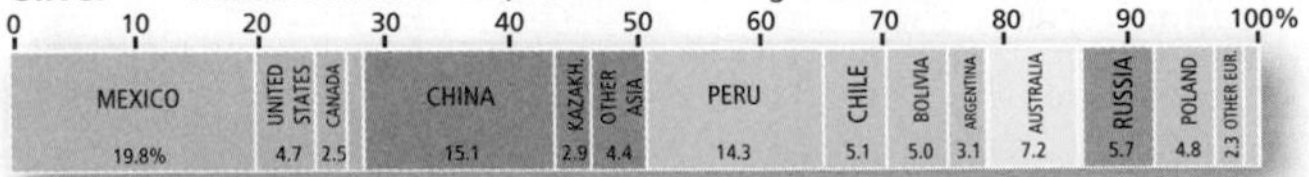

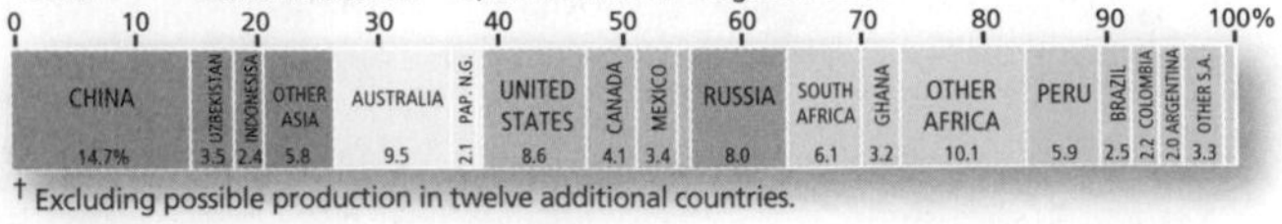

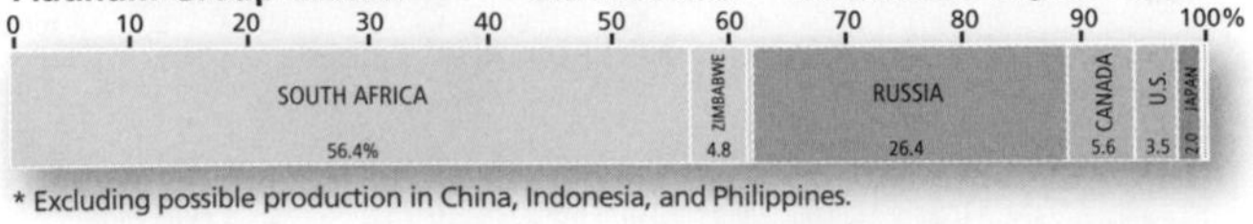

Taiwan figures are included with China.
Botswana, Lesotho, Namibia and Swaziland figures are included with South Africa.
Montenegro figures are included with Serbia.

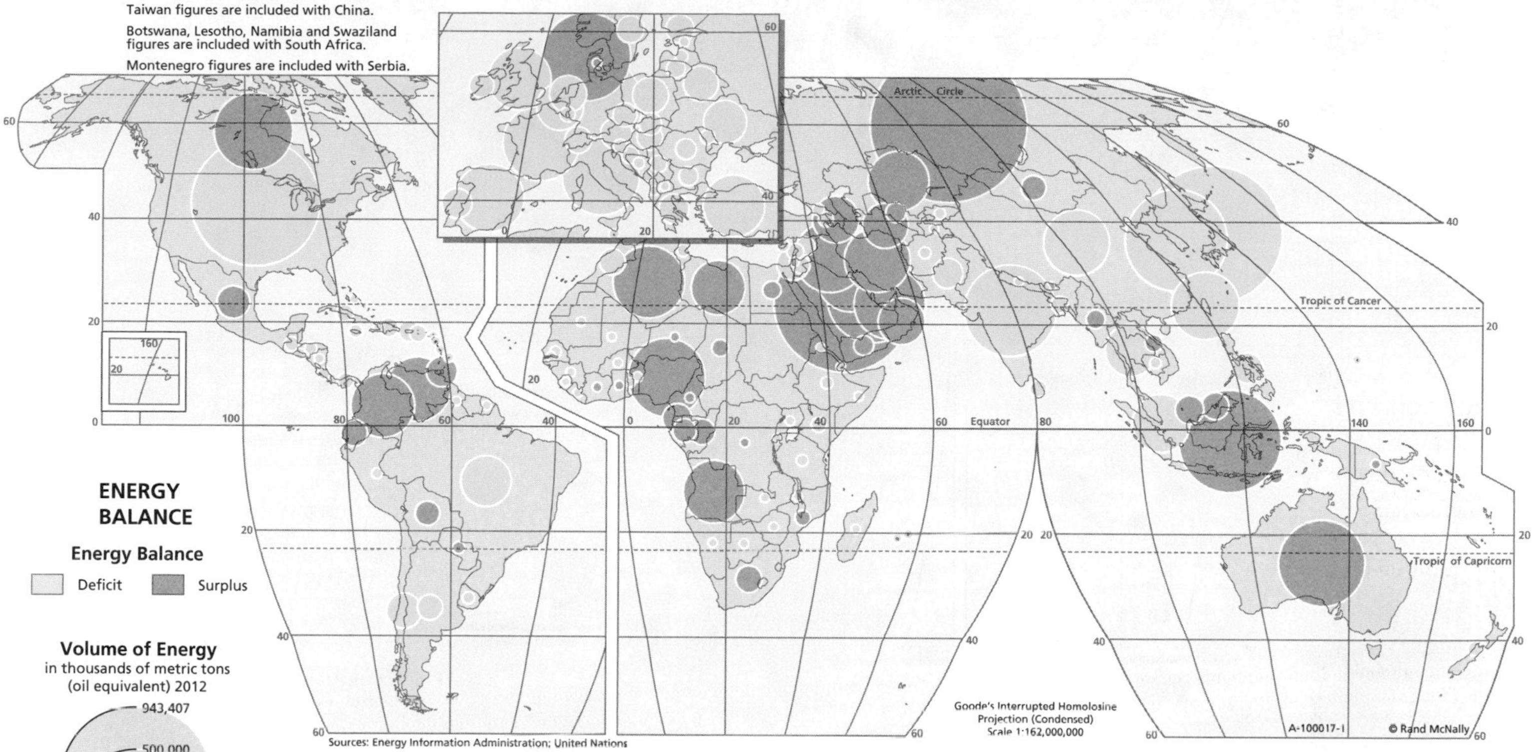

ENERGY BALANCE

Energy Balance

- Deficit
- Surplus

Volume of Energy
in thousands of metric tons
(oil equivalent) 2012

- 943,407
- 500,000
- 100,000
- 50,000
- 10,000
- 100 - 2,500

Sources: Energy Information Administration; United Nations

Arctic Circle

Tropic of Cancer

Equator

Tropic of Capricorn

Goode's Interrupted Homolosine
Projection (Condensed)
Scale 1:162,000,000

A-100017-1   © Rand McNally

Energy Consumption   World Total - 20,797,000,000 metric tons (oil equiv.) - 2012

CHINA	INDIA	JAPAN	S. KOREA	OTHER ASIA	UNITED STATES	CANADA	OTHER N.A.	RUSSIA	GERMANY	FRANCE	OTHER EUROPE	BRAZIL	OTHER S.A.	AFRICA	OCEANIA
20.1%	4.6	3.9	2.2	15.0	18.1	2.5	2.2	5.1	2.0		11.3	2.5	2.3	3.2	1.

Energy Consumption Trends 1980-2030

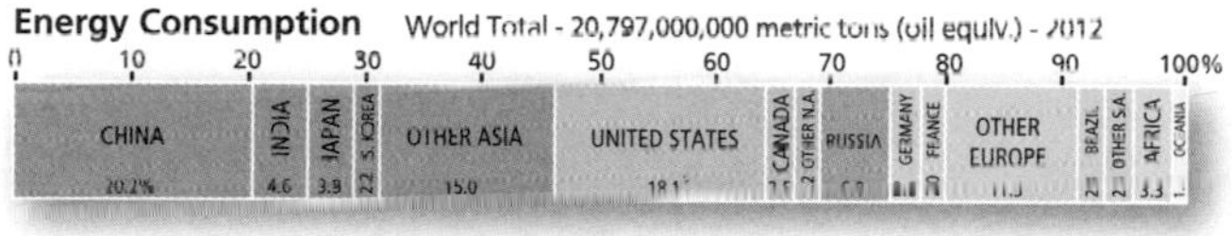

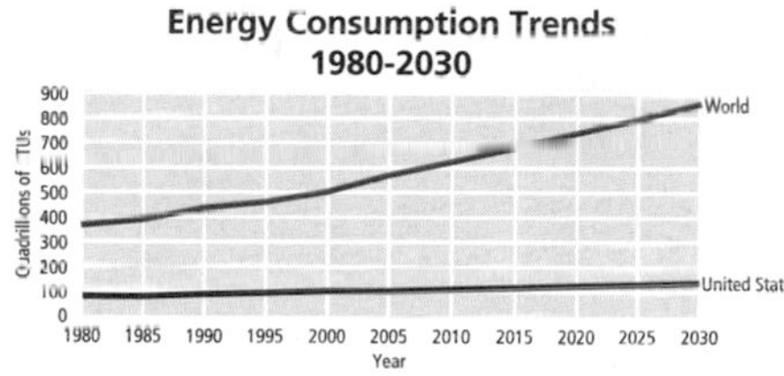

Sudan figures include South Sudan.

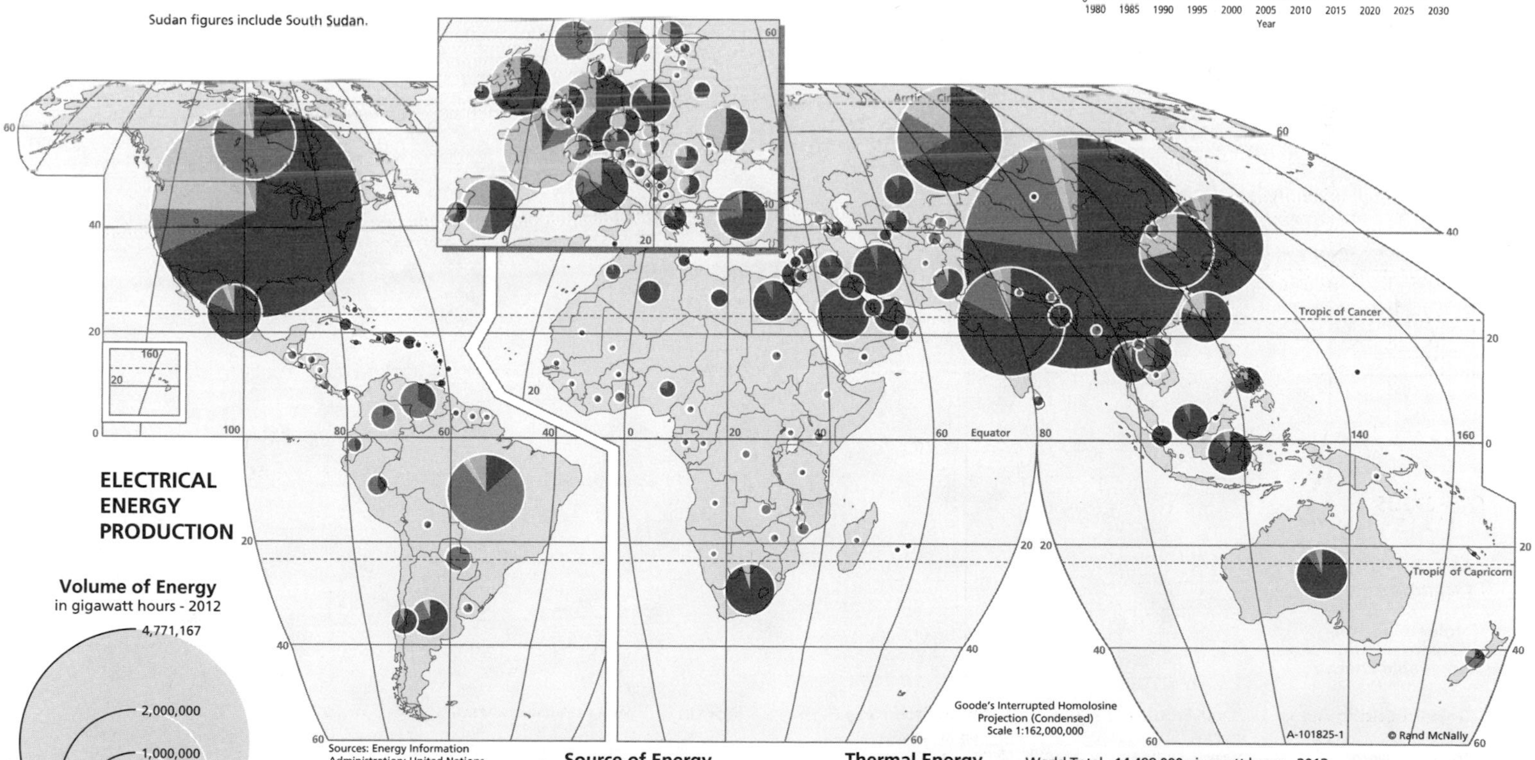

ELECTRICAL ENERGY PRODUCTION

Volume of Energy
in gigawatt hours - 2012

- 4,771,167
- 2,000,000
- 1,000,000
- 250,000
- 100,000
- 25,000
- 500 - 5,000

Sources: Energy Information
Administration; United Nations

No symbol is shown if production
is less than 500 gigawatt hours.

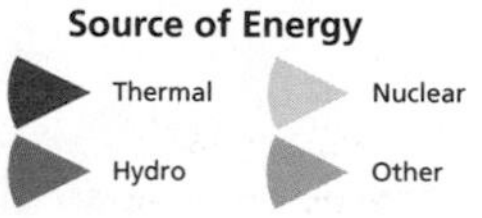

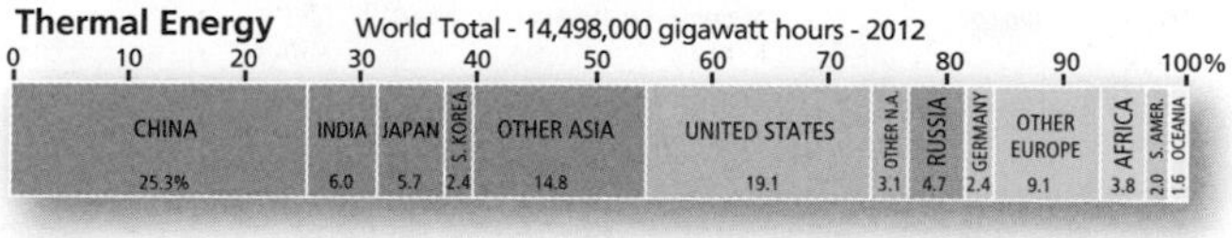

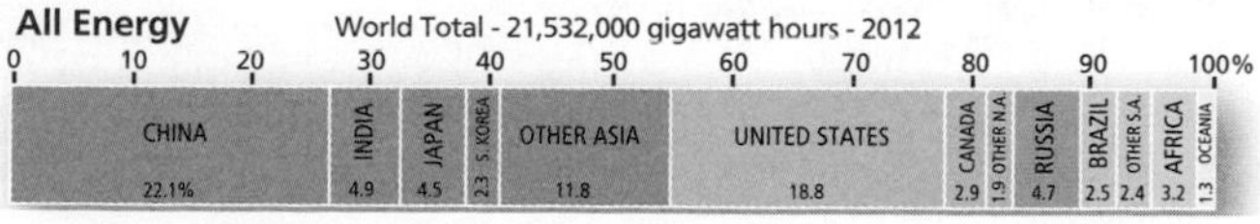

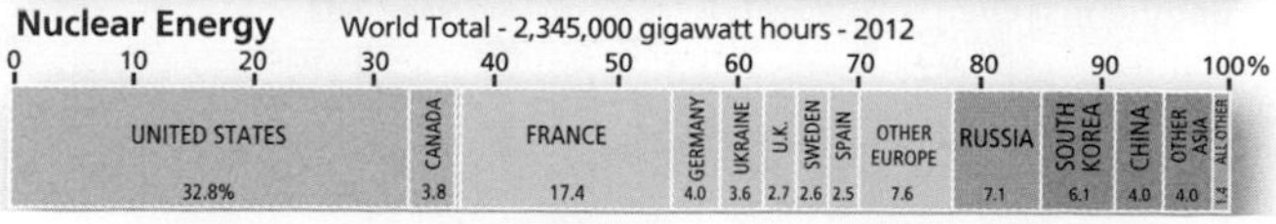

Source of Energy

- Thermal
- Hydro
- Nuclear
- Other

Arctic Circle

Tropic of Cancer

Equator

Tropic of Capricorn

Goode's Interrupted Homolosine
Projection (Condensed)
Scale 1:162,000,000

A-101825-1   © Rand McNally

Thermal Energy   World Total - 14,498,000 gigawatt hours - 2012

CHINA	INDIA	JAPAN	S. KOREA	OTHER ASIA	UNITED STATES	OTHER N.A.	RUSSIA	GERMANY	OTHER EUROPE	AFRICA	S. AMER.	OCEANIA
25.3%	6.0	5.7	2.4	14.8	19.1	3.1	4.7	2.4	9.1	3.8	2.0	1.6

All Energy   World Total - 21,532,000 gigawatt hours - 2012

CHINA	INDIA	JAPAN	S. KOREA	OTHER ASIA	UNITED STATES	CANADA	OTHER N.A.	RUSSIA	BRAZIL	OTHER S.A.	AFRICA	OCEANIA
22.1%	4.9	4.5	2.3	11.8	18.8	2.9	1.9	4.7	2.5	2.4	3.2	1.3

Nuclear Energy   World Total - 2,345,000 gigawatt hours - 2012

UNITED STATES	CANADA	FRANCE	GERMANY	UKRAINE	U.K.	SWEDEN	SPAIN	OTHER EUROPE	RUSSIA	SOUTH KOREA	CHINA	OTHER ASIA	ALL OTHER
32.8%	3.8	17.4	4.0	3.6	2.7	2.6	2.5	7.6	7.1	6.1	4.0	4.6	1.4

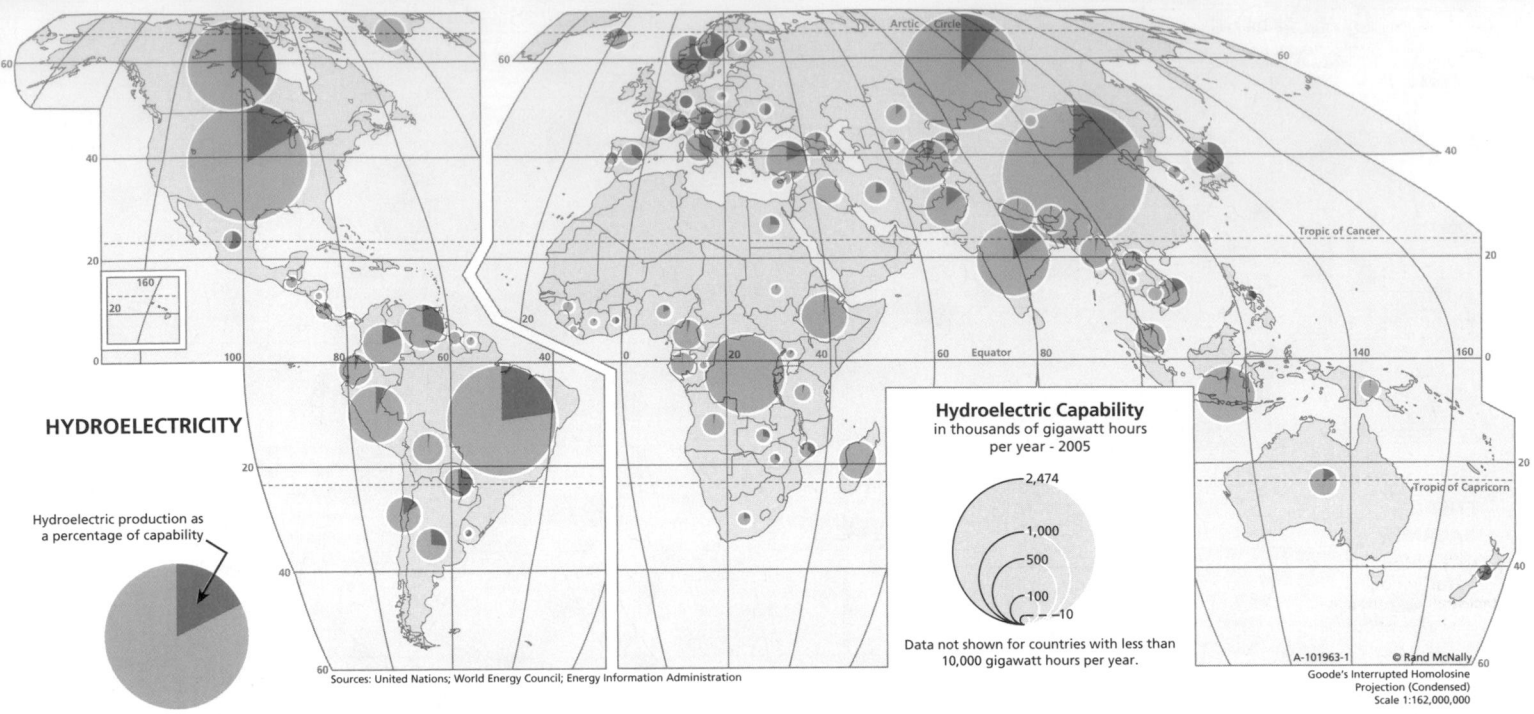

HYDROELECTRICITY

Hydroelectric production as
a percentage of capability

Hydroelectric Capability
in thousands of gigawatt hours
per year - 2005

2,474
1,000
500
100
10

Data not shown for countries with less than
10,000 gigawatt hours per year.

Sources: United Nations; World Energy Council; Energy Information Administration

A-101963-1 © Rand McNally

Goode's Interrupted Homolosine
Projection (Condensed)
Scale 1:162,000,000

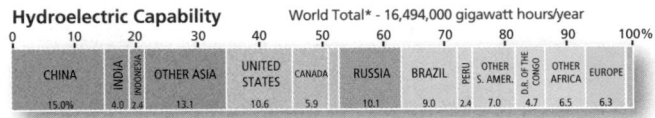

Hydroelectric Capability
World Total* - 16,494,000 gigawatt hours/year

CHINA	INDIA	INDONESIA	OTHER ASIA	UNITED STATES	CANADA	RUSSIA	BRAZIL	PERU	OTHER S. AMER.	D.R. OF THE CONGO	OTHER AFRICA	EUROPE
15.0%	4.0	2.4	13.1	10.6	5.9	10.1	9.0	2.4	7.0	4.7	6.5	6.3

* Technically exploitable capability.

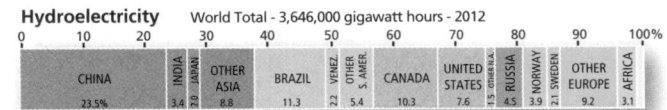

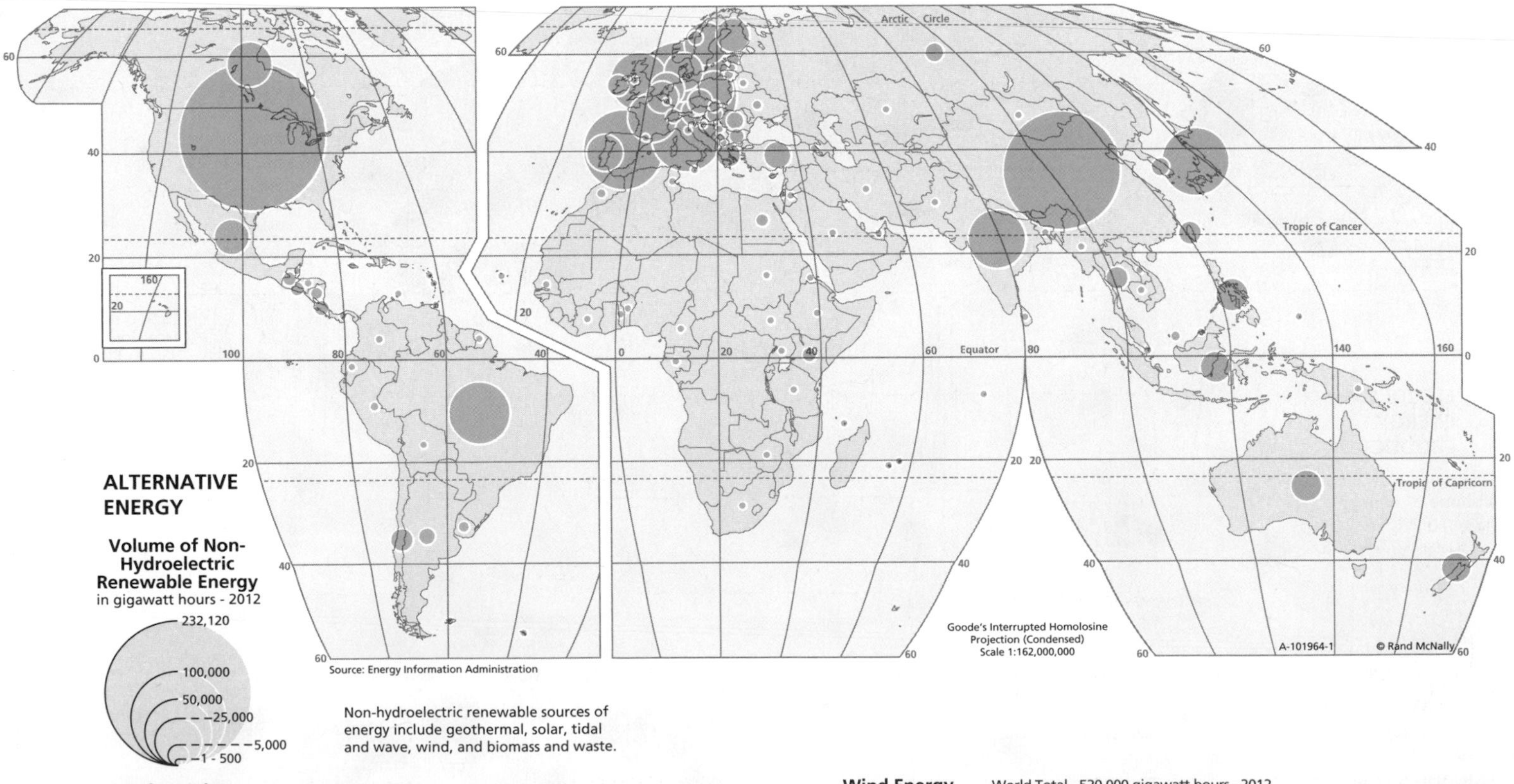

Hydroelectricity
World Total - 3,646,000 gigawatt hours - 2012

CHINA	INDIA	JAPAN	OTHER ASIA	BRAZIL	VENEZ.	OTHER S. AMER.	CANADA	UNITED STATES	OTHER N. A.	RUSSIA	NORWAY	SWEDEN	OTHER EUROPE	AFRICA
23.5%	3.4	2.0	8.8	11.3	2.2	5.4	10.3	7.6	1.5	4.5	3.9	2.1	9.2	3.1

ALTERNATIVE ENERGY

Volume of Non-Hydroelectric Renewable Energy
in gigawatt hours - 2012

232,120
100,000
50,000
25,000
5,000
1-500

Source: Energy Information Administration

Non-hydroelectric renewable sources of
energy include geothermal, solar, tidal
and wave, wind, and biomass and waste.

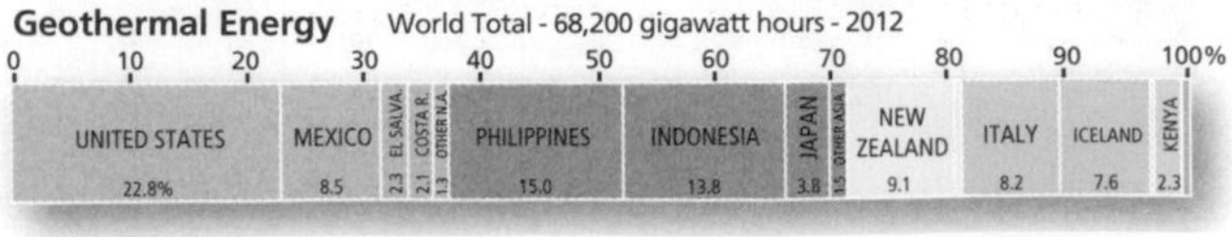

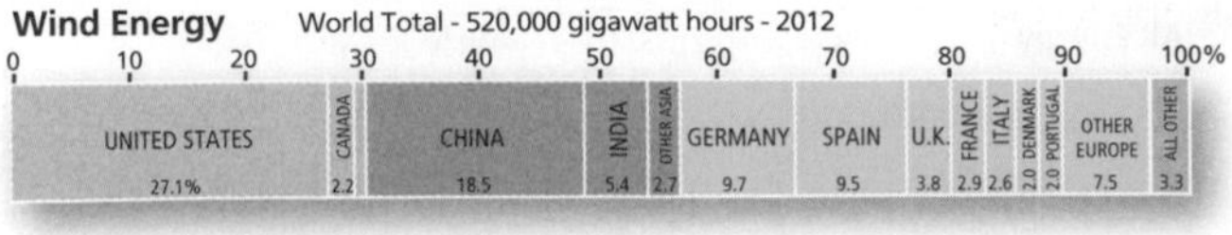

Goode's Interrupted Homolosine
Projection (Condensed)
Scale 1:162,000,000

A-101964-1 © Rand McNally

Geothermal Energy
World Total - 68,200 gigawatt hours - 2012

UNITED STATES	MEXICO	EL SALVA.	COSTA R.	OTHER N. A.	PHILIPPINES	INDONESIA	JAPAN	OTHER ASIA	NEW ZEALAND	ITALY	ICELAND	KENYA
22.8%	8.5	2.3	2.1	1.3	15.0	13.8	3.8	1.5	9.1	8.2	7.6	2.3

Wind Energy
World Total - 520,000 gigawatt hours - 2012

UNITED STATES	CANADA	CHINA	INDIA	GERMANY	SPAIN	U.K.	FRANCE	ITALY	DENMARK	PORTUGAL	OTHER EUROPE	ALL OTHER
27.1%	2.2	18.5	5.4	2.7	9.7	9.5	3.8	2.9	2.6	2.0	2.0	7.5	3.3

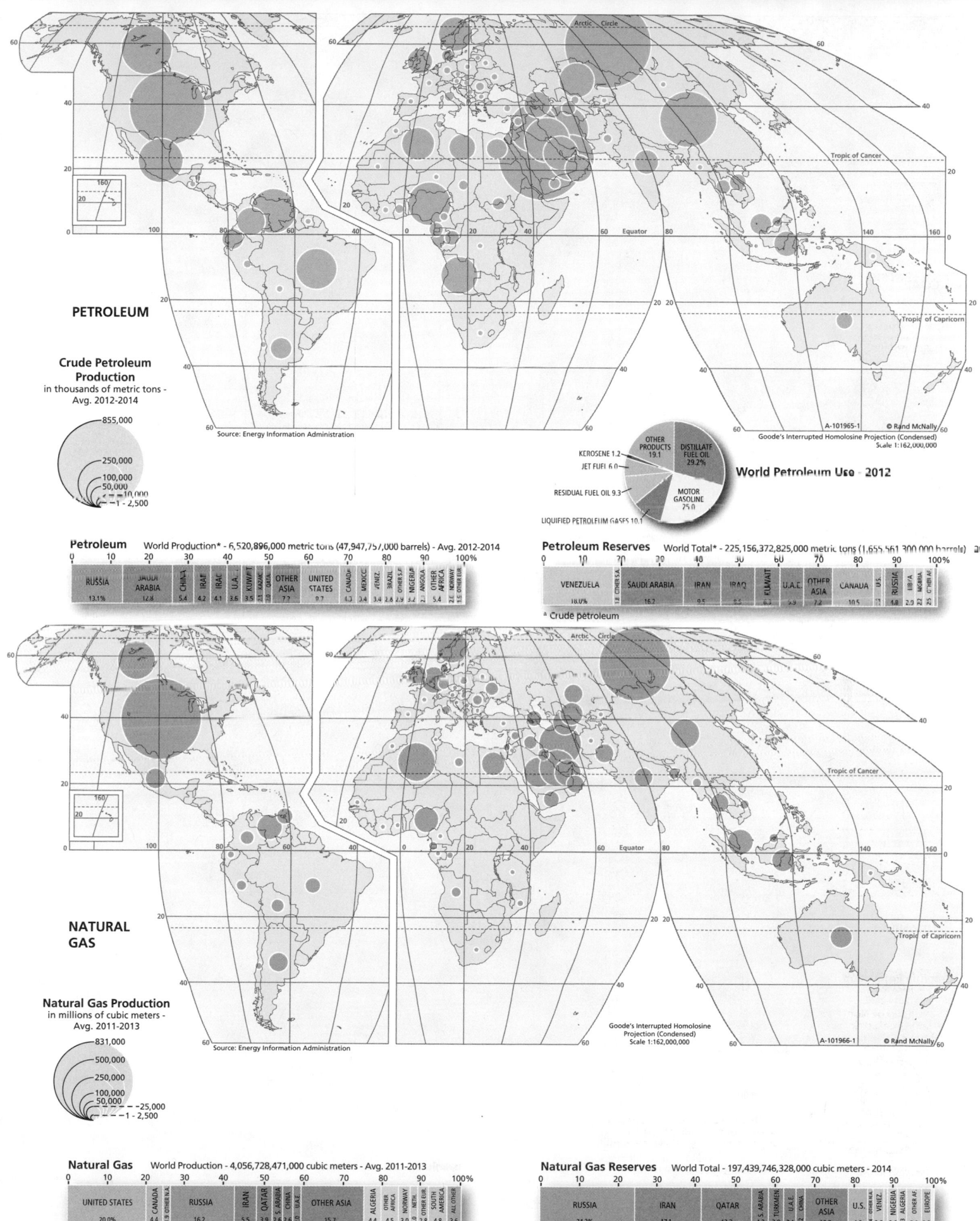

PETROLEUM

Crude Petroleum Production
in thousands of metric tons -
Avg. 2012-2014

- 855,000
- 250,000
- 100,000
- 50,000
- 10,000
- 1 - 2,500

Source: Energy Information Administration

Goode's Interrupted Homolosine Projection (Condensed)
Scale 1:162,000,000

A-101965-1   © Rand McNally

World Petroleum Use - 2012

- OTHER PRODUCTS 19.1
- DISTILLATE FUEL OIL 29.2%
- KEROSENE 1.2
- JET FUEL 6.0
- RESIDUAL FUEL OIL 9.3
- MOTOR GASOLINE 25.0
- LIQUIFIED PETROLEUM GASES 10.1

Petroleum World Production* - 6,520,896,000 metric tons (47,947,757,000 barrels) - Avg. 2012-2014

0	10	20	30	40	50	60	70	80	90	100%

RUSSIA	SAUDI ARABIA	CHINA	IRAN	IRAQ	U.A.E.	KUWAIT	KAZAK.	OTHER ASIA	UNITED STATES	CANADA	MEXICO	VENEZ.	BRAZIL	OTHER S.A.	NIGERIA	ANGOLA	OTHER AFRICA	NORWAY	OTHER EUR.
13.1%	12.8	5.4	4.2	4.1	3.6	3.5	2.7	7.2	9.7	5.5	3.3	4.0	3.4	2.8	2.9	2.3	5.4	2.0	1.5

Petroleum Reserves World Total* - 225,156,372,825,000 metric tons (1,655,561,300,000 barrels) 2011

0	10	20	30	40	50	60	70	80	90	100%

VENEZUELA	OTHER S.A.	SAUDI ARABIA	IRAN	IRAQ	KUWAIT	U.A.E.	OTHER ASIA	CANADA	U.S.	RUSSIA	LIB'A	NIGERIA	OTHER AF.
18.0%	1.8	16.2	9.5	8.5	6.3	5.9	7.2	10.5	1.7	4.8	2.9	2.2	2.5

* Crude petroleum

NATURAL GAS

Natural Gas Production
in millions of cubic meters -
Avg. 2011-2013

- 831,000
- 500,000
- 250,000
- 100,000
- 50,000
- 25,000
- 1 - 2,500

Source: Energy Information Administration

Goode's Interrupted Homolosine
Projection (Condensed)
Scale 1:162,000,000

A-101966-1   © Rand McNally

Natural Gas World Production - 4,056,728,471,000 cubic meters - Avg. 2011-2013

0	10	20	30	40	50	60	70	80	90	100%

UNITED STATES	CANADA	OTHER N.A.	RUSSIA	IRAN	QATAR	S. ARABIA	CHINA	U.A.E.	OTHER ASIA	ALGERIA	OTHER AFRICA	NORWAY	NETH.	OTHER EUR.	SOUTH AMERICA	ALL OTHER
20.0%	4.4	1.9	16.2	5.5	3.9	2.6	2.6	2.0	15.7	4.4	4.5	3.0	2.0	4.8	3.6	

Natural Gas Reserves World Total - 197,439,746,328,000 cubic meters - 2014

0	10	20	30	40	50	60	70	80	90	100%

RUSSIA	IRAN	QATAR	S. ARABIA	TURKMEN.	U.A.E.	CHINA	OTHER ASIA	U.S.	OTHER N.A.	VENEZ.	NIGERIA	ALGERIA	OTHER AF.	EUROPE
24.2%	17.1	12.7	4.2	3.8	3.1	10.8	4.9	1.7	2.8	2.6	2.3	3.8	2.5	

COAL

Coal Production
in thousands of metric tons -
Avg. 2010-2012

3,464,000

1,000,000

500,000

100,000

10,000

1 - 2,500

Source: Energy Information Administration

Goode's Interrupted Homolosine
Projection (Condensed)
Scale 1:162,000,000

A-101967-1   © Rand McNally

Coal   World Production* - 7,599,352,000 metric tons - Avg. 2010-2012

0	10	20	30	40	50	60	70	80	90	100%

CHINA	INDIA	INDON.	OTHER ASIA	UNITED STATES	AUSTRALIA	RUSSIA	S. AFRICA	GERMANY	OTHER EUROPE
45.6%	7.6	4.9	4.6	12.7	5.5	4.4	3.4	2.5	6.5

Anthracite and Bituminous: World Total - 6,788,057,000 metric tons

Coal Reserves   World Total* - 888,851,239,000 metric tons - 2011

0	10	20	30	40	50	60	70	80	90	100%

UNITED STATES	RUSSIA	CHINA	INDIA	KAZAKH.	INDONESIA	OTHER ASIA	AUSTRALIA	GERMANY	UKRAINE	OTHER EUROPE	S. AFRICA	S. AMER.
26.4%	17.7	12.9	6.8	3.8	3.2	2.4	8.6	4.6	3.8	3.7	3.4	1.8

Anthracite and Bituminous: World Total - 401,973,703,000 metric tons
* Includes anthracite, bituminous, and lignite coal.

URANIUM

Uranium Production
in metric tons -
Avg. 2010-2012

19,500

5,000

2,500

1,000

500

1 - 50

Source: United Nations

Goode's Interrupted Homolosine
Projection (Condensed)
Scale 1:162,000,000

A-101968-1   © Rand McNally

Uranium   World Production - 56,200 metric tons - Avg. 2010-2012

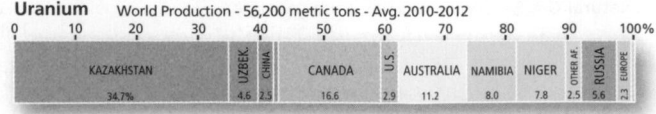

0	10	20	30	40	50	60	70	80	90	100%

KAZAKHSTAN	UZBEK	CHINA	CANADA	U.S.	AUSTRALIA	NAMIBIA	NIGER	OTHER AF	RUSSIA	EUROPE
34.7%	4.6	2.5	16.6	2.9	11.2	8.0	7.8	2.5	5.6	2.3

Uranium Reserves   World Total - 4,379,000 metric tons - 2011

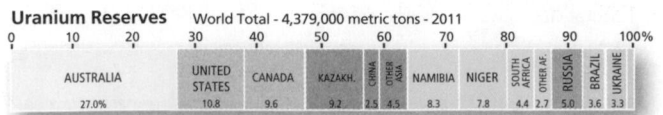

0	10	20	30	40	50	60	70	80	90	100%

AUSTRALIA	UNITED STATES	CANADA	KAZAKH.	CHINA	OTHER ASIA	NAMIBIA	NIGER	SOUTH AFRICA	OTHER AF	RUSSIA	BRAZIL	UKRAINE
27.0%	10.8	9.6	9.2	2.5	4.5	8.3	7.8	4.4	2.7	5.0	3.6	3.3

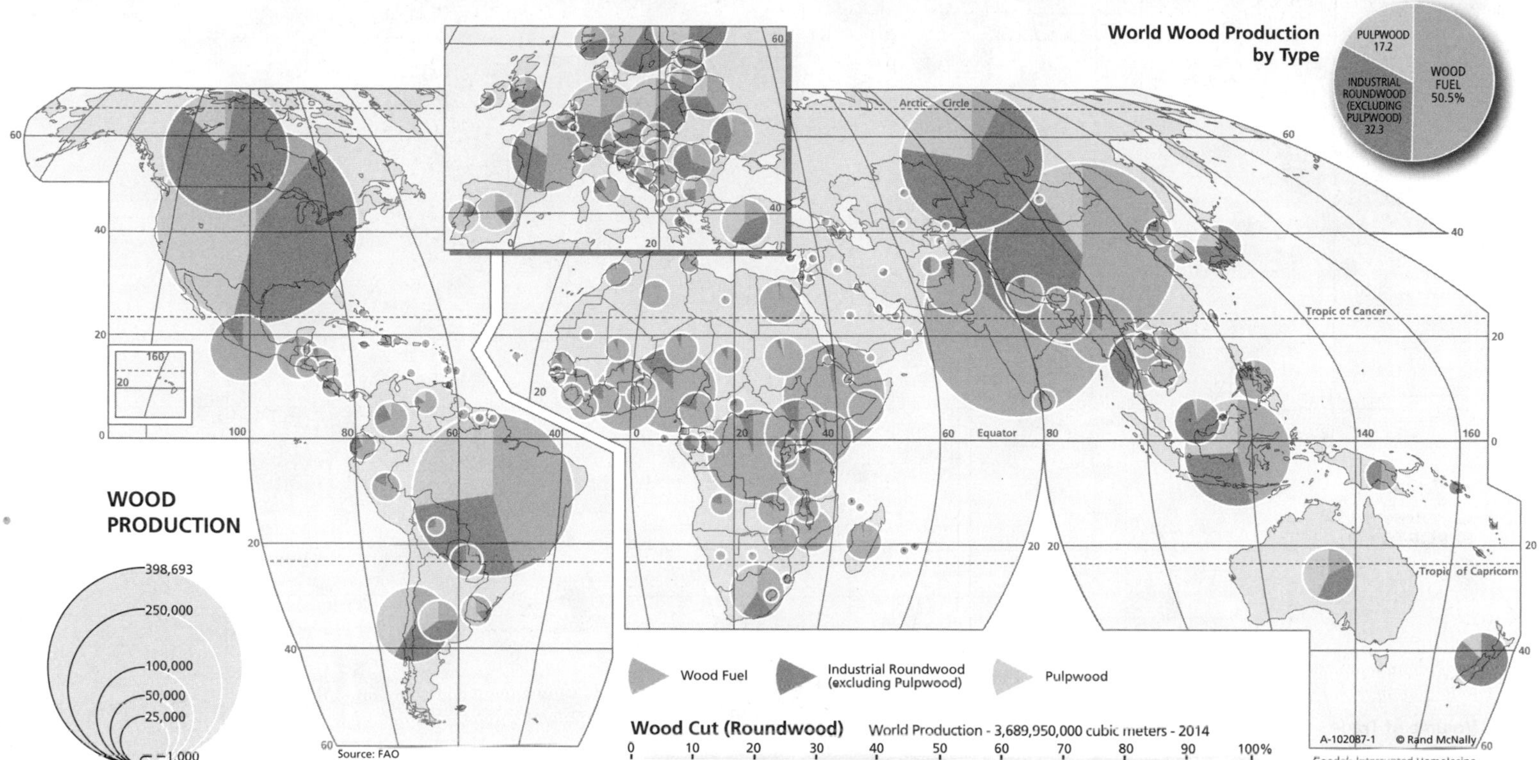

World Wood Production by Type

PULPWOOD 17.2

INDUSTRIAL ROUNDWOOD (EXCLUDING PULPWOOD) 32.3

WOOD FUEL 50.5%

WOOD PRODUCTION

398,693
250,000
100,000
50,000
25,000
1,000

160
20

Source: FAO

Wood Production (Roundwood)
in thousands of cubic meters - 2014

Wood Fuel

Industrial Roundwood (excluding Pulpwood)

Pulpwood

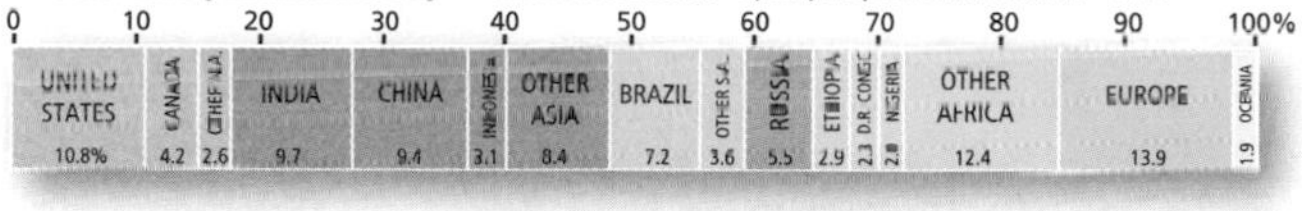

A-102087-1 © Rand McNally

Goode's Interrupted Homolosine Projection (Condensed)
Scale 1:162,000,000

Wood Cut (Roundwood)
World Production - 3,689,950,000 cubic meters - 2014

0	10	20	30	40	50	60	70	80	90	100%

UNITED STATES	CANADA	OTHER N.A.	INDIA	CHINA	INDONESIA	OTHER ASIA	BRAZIL	OTHER S.A.	RUSSIA	ETHIOPIA	D.R. CONGO	NIGERIA	OTHER AFRICA	EUROPE	OCEANIA
10.8%	4.2	2.6	9.7	9.4	3.1	8.4	7.2	3.6	5.5	2.9	2.1	2.8	12.4	13.9	1.9

Paper and Paperboard
World Total - 399,798,000 metric tons - 2014

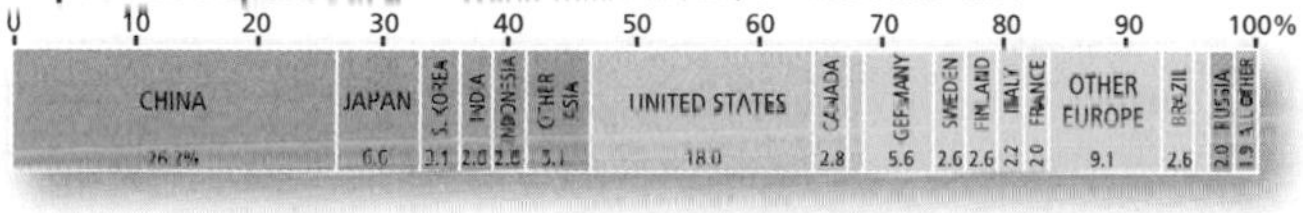

0	10	20	30	40	50	60	70	80	90	100%

CHINA	JAPAN	S. KOREA	INDIA	INDONESIA	OTHER ASIA	UNITED STATES	CANADA	GERMANY	SWEDEN	FINLAND	ITALY	FRANCE	OTHER EUROPE	BRAZIL	RUSSIA	ALL OTHER
26.2%	6.6	3.1	2.8	2.6	3.1	18.0	2.8	5.6	2.6	2.6	2.2	2.0	9.1	2.6	2.1	3.5

Recovered Paper
World Total - 221,216,000 metric tons - 2014

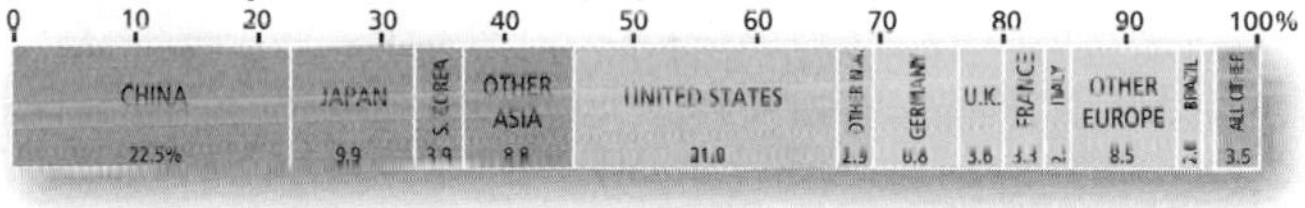

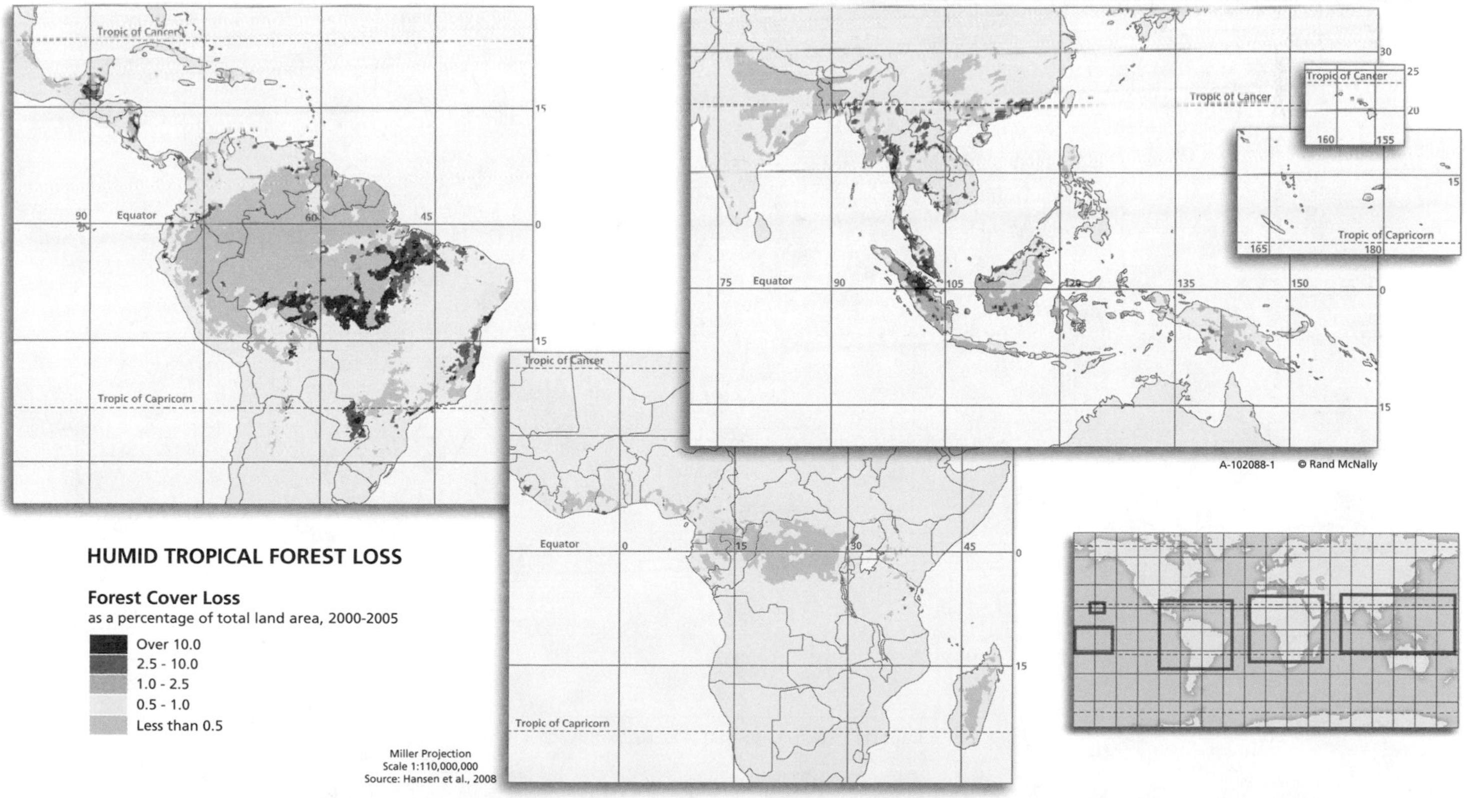

0	10	20	30	40	50	60	70	80	90	100%

CHINA	JAPAN	S. KOREA	OTHER ASIA	UNITED STATES	OTHER N.A.	GERMANY	U.K.	FRANCE	ITALY	OTHER EUROPE	BRAZIL	ALL OTHER
22.5%	9.9	2.9	8.8	31.0	1.9	6.8	3.6	3.1	2.1	8.5	2.6	3.5

HUMID TROPICAL FOREST LOSS

Forest Cover Loss
as a percentage of total land area, 2000-2005

- Over 10.0
- 2.5 - 10.0
- 1.0 - 2.5
- 0.5 - 1.0
- Less than 0.5

Miller Projection
Scale 1:110,000,000
Source: Hansen et al., 2008

A-102088-1 © Rand McNally

Humid Tropical Forest Area
World Total* - 11,487,357 square kilometers - 2000

0	10	20	30	40	50	60	70	80	90	100%

BRAZIL	PERU	COLOMBIA	VENEZUELA	BOLIVIA	OTHER S. AMER.	INDONESIA	CHINA	MYANMAR	INDIA	MALAYSIA	OTHER ASIA	D.R. OF CONGO	OTHER AFRICA	PAPUA NEW GUINEA	N. AMER.
31.4%	5.7	4.9	3.3	2.5	4.6	9.0	4.5	4.5	2.0	1.9	4.5	8.5	8.1	2.7	2.9

Humid Tropical Forest Cover Loss
World Total - 272,605 square kilometers - 2000-2005

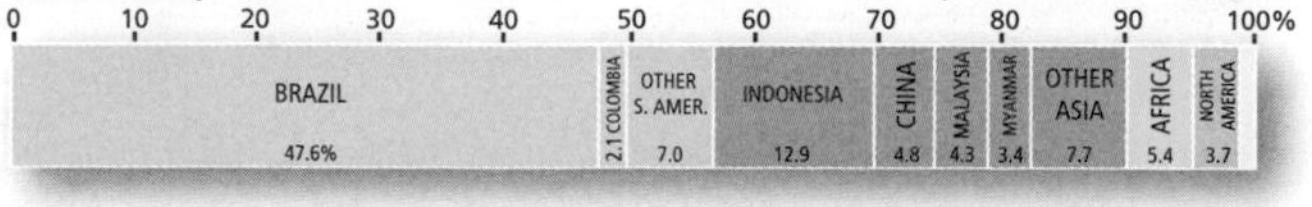

0	10	20	30	40	50	60	70	80	90	100%

BRAZIL	OTHER S. AMER.	COLOMBIA	INDONESIA	CHINA	MALAYSIA	MYANMAR	OTHER ASIA	AFRICA	NORTH AMERICA
47.6%	7.0	2.1	12.9	4.8	4.3	3.4	7.7	5.4	3.7

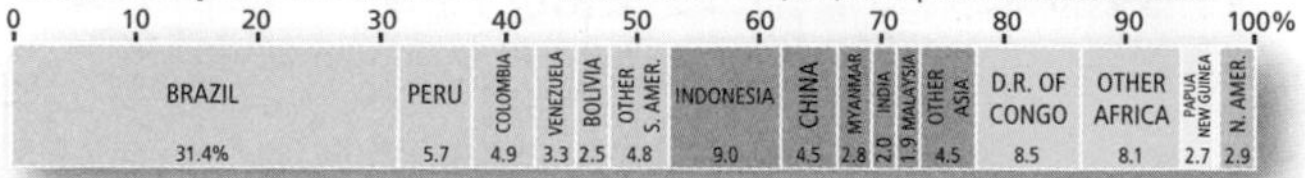

* Defined as areas with tree canopy cover of 25% or more.

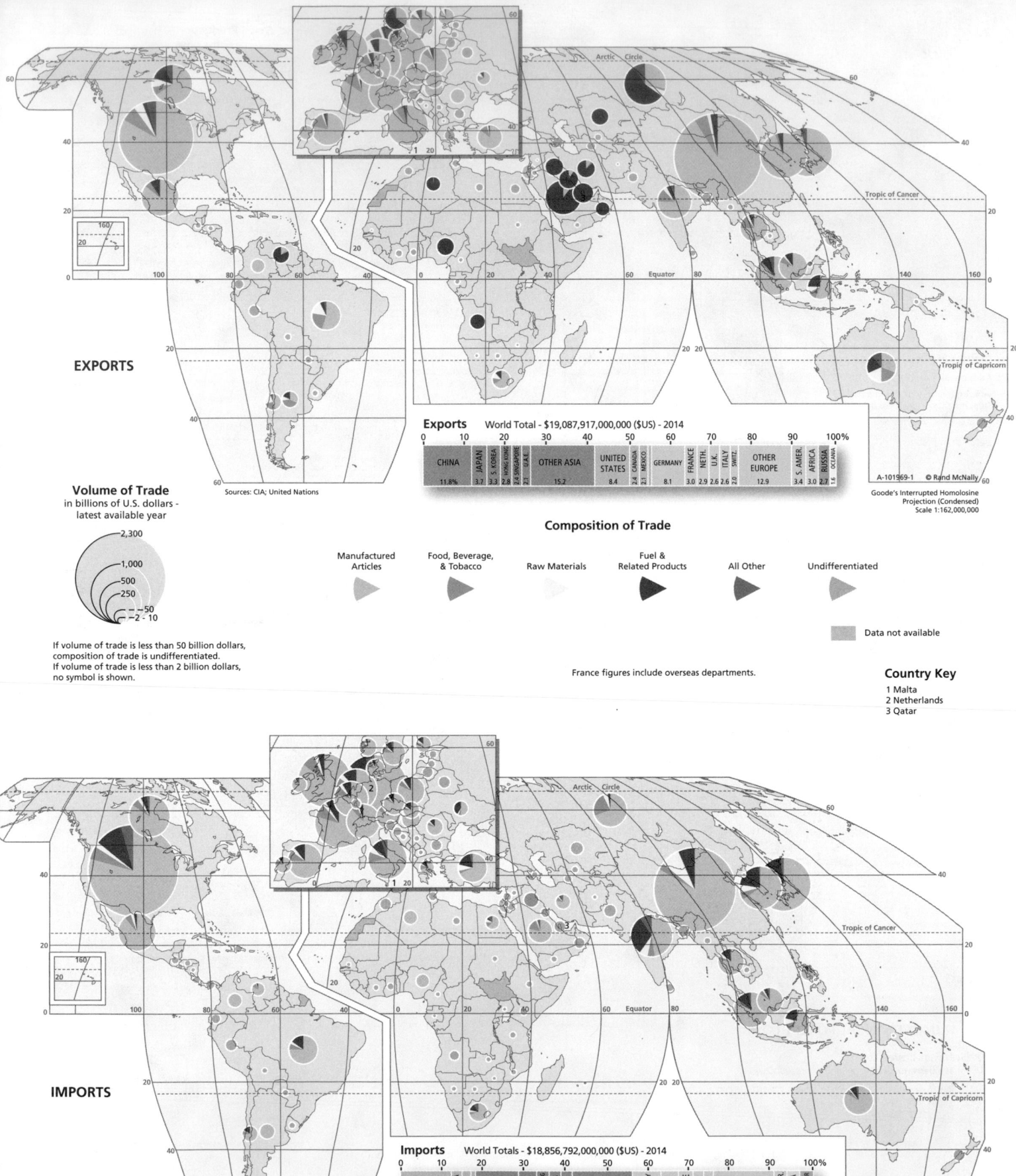

EXPORTS

Sources: CIA; United Nations

Volume of Trade
in billions of U.S. dollars -
latest available year

2,300
1,000
500
250
50
2 - 10

If volume of trade is less than 50 billion dollars,
composition of trade is undifferentiated.
If volume of trade is less than 2 billion dollars,
no symbol is shown.

Exports	World Total - $19,087,917,000,000 ($US) - 2014

0	10	20	30	40	50	60	70	80	90	100%

CHINA	JAPAN	S. KOREA	HONG KONG	SINGAPORE	U.A.E.	OTHER ASIA	UNITED STATES	CANADA	MEXICO	GERMANY	FRANCE	NETH.	U.K.	ITALY	SWITZ.	OTHER EUROPE	S. AMER.	AFRICA	RUSSIA	OCEANIA
11.8%	3.7	3.3	2.8	2.4	2.1	15.2	8.4	2.4	2.1	8.1	3.0	2.9	2.6	2.6	2.0	12.9	3.4	3.0	2.7	1.6

A-101969-1   © Rand McNally

Goode's Interrupted Homolosine
Projection (Condensed)
Scale 1:162,000,000

Composition of Trade

Manufactured
Articles

Food, Beverage,
& Tobacco

Raw Materials

Fuel &
Related Products

All Other

Undifferentiated

Data not available

France figures include overseas departments.

Country Key
1 Malta
2 Netherlands
3 Qatar

IMPORTS

Sources: CIA; United Nations

Imports	World Totals - $18,856,792,000,000 ($US) - 2014

0	10	20	30	40	50	60	70	80	90	100%

UNITED STATES	CANADA	MEXICO	CHINA	JAPAN	HONG KONG	S. KOREA	INDIA	OTHER ASIA	GERMANY	U.K.	FRANCE	NETH.	ITALY	OTHER EUROPE	S. AMER.	AFRICA	ALL OTHER
12.4%	2.6	2.2	10.4	4.3	2.8	2.7		14.6	7.0	4.3	3.6	2.6	2.4	14.8	3.1	3.1	3.3

A-101970-1   © Rand McNally

Goode's Interrupted Homolosine
Projection (Condensed)
Scale 1:162,000,000

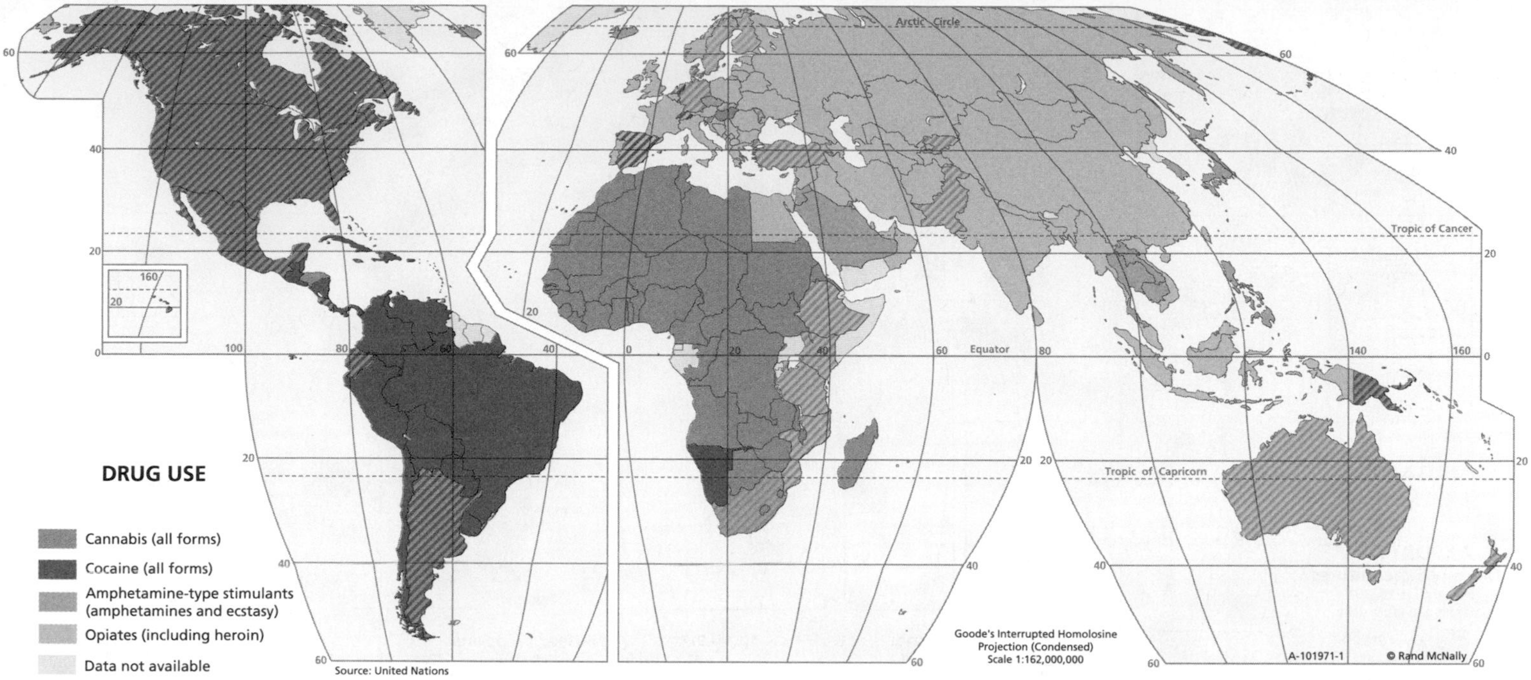

DRUG USE

- Cannabis (all forms)
- Cocaine (all forms)
- Amphetamine-type stimulants (amphetamines and ecstasy)
- Opiates (including heroin)
- Data not available

Circles show annual prevalence of drug use - latest available data.
Circle size is proportional to number of drug users per continent.

Country tints show primary drug or drugs of abuse based on drug treatment statistics - most current available year.

Other drugs, such as hallucinogens, depressants, and inhalants, are not included.

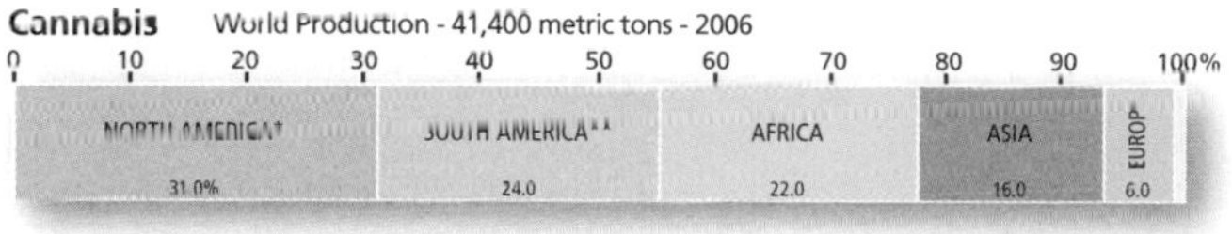

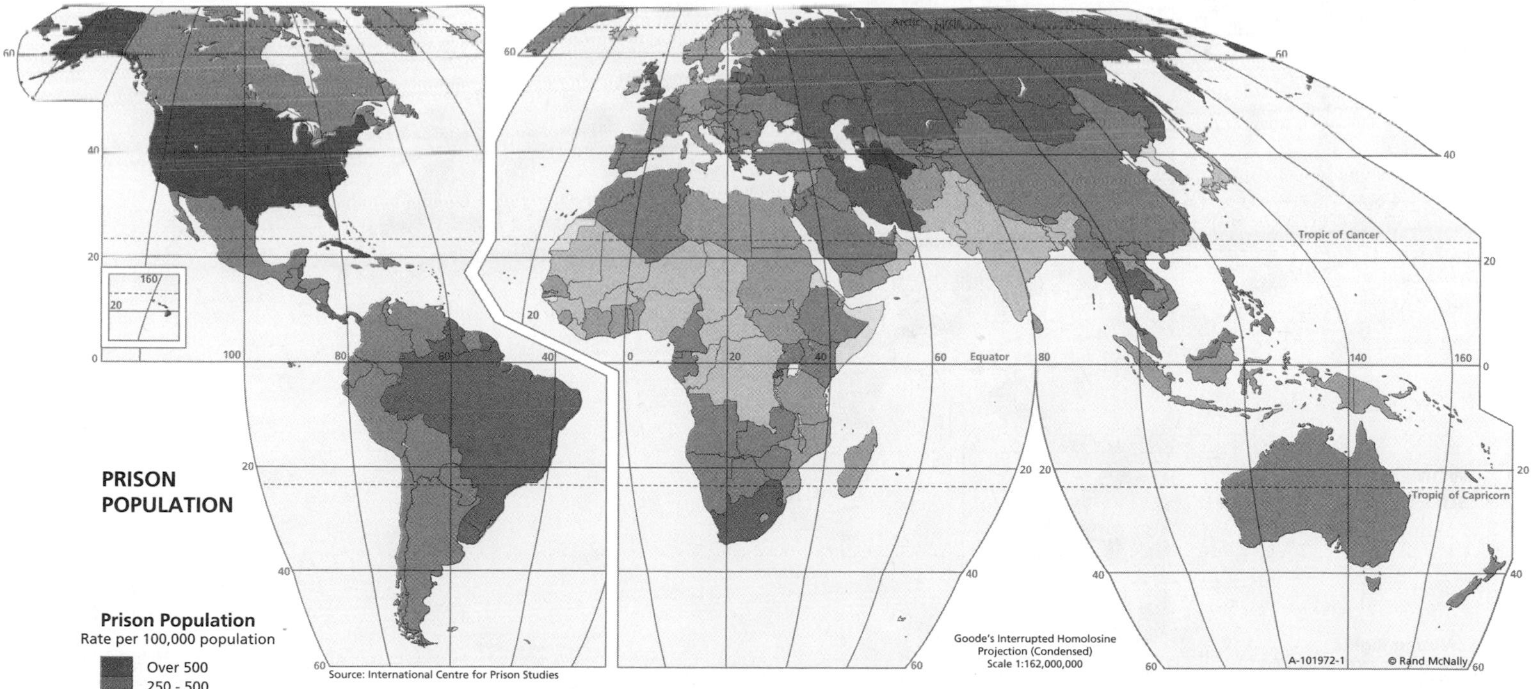

Source: United Nations

Goode's Interrupted Homolosine Projection (Condensed)
Scale 1:162,000,000
A-101971-1
© Rand McNally

Cannabis — World Production - 41,400 metric tons - 2006

0	10	20	30	40	50	60	70	80	90	100%
NORTH AMERICA‡ 31.0%			SOUTH AMERICA‡‡ 24.0		AFRICA 22.0		ASIA 16.0		EUROPE 6.0	

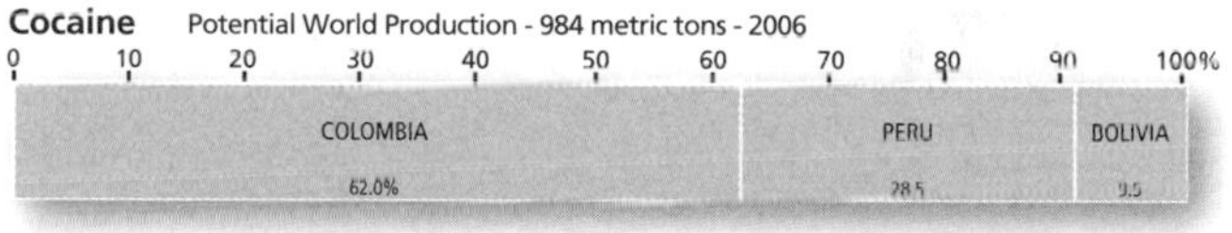

Cocaine — Potential World Production - 984 metric tons - 2006

0	10	20	30	40	50	60	70	80	90	100%
COLOMBIA 62.0%						PERU 28.5			BOLIVIA 9.5	

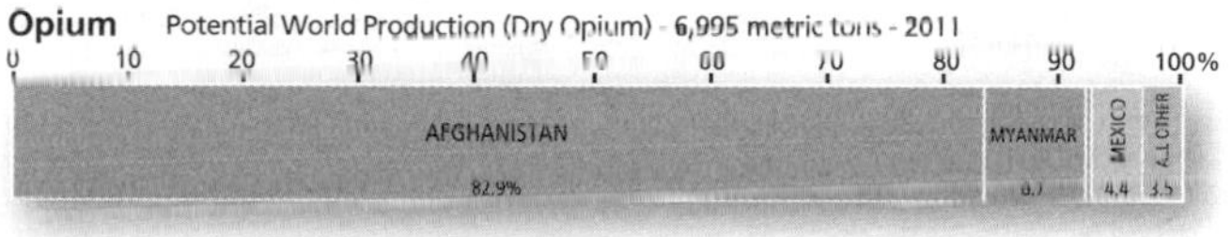

Opium — Potential World Production (Dry Opium) - 6,995 metric tons - 2011

0	10	20	30	40	50	60	70	80	90	100%
AFGHANISTAN 82.9%									MYANMAR 8.7	MEXICO 4.4	A.L OTHER 3.5

PRISON POPULATION

Prison Population
Rate per 100,000 population

- Over 500
- 250 - 500
- 100 - 249
- 50 - 99
- 0 - 49
- Data not available

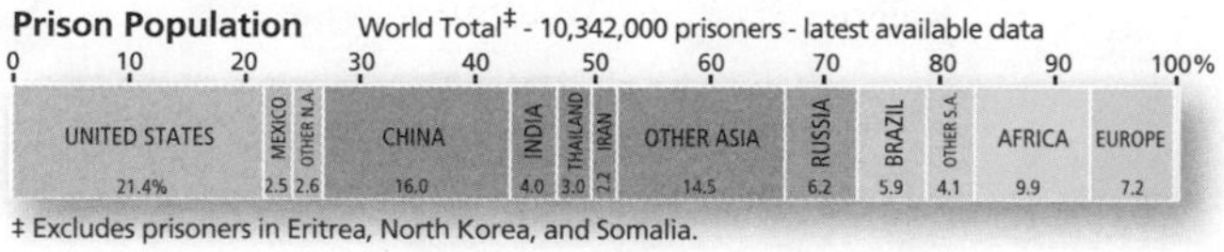

Source: International Centre for Prison Studies

Goode's Interrupted Homolosine Projection (Condensed)
Scale 1:162,000,000
A-101972-1
© Rand McNally

Prison Population — World Total‡ - 10,342,000 prisoners - latest available data

0	10	20	30	40	50	60	70	80	90	100%
UNITED STATES 21.4%		MEXICO 2.5	OTHER N.A. 2.6	CHINA 16.0		INDIA 4.0	THAILAND 3.0	IRAN 2.2	OTHER ASIA 14.5	RUSSIA 6.2	BRAZIL 5.9	OTHER S.A. 4.1	AFRICA 9.9	EUROPE 7.2

‡ Excludes prisoners in Eritrea, North Korea, and Somalia.

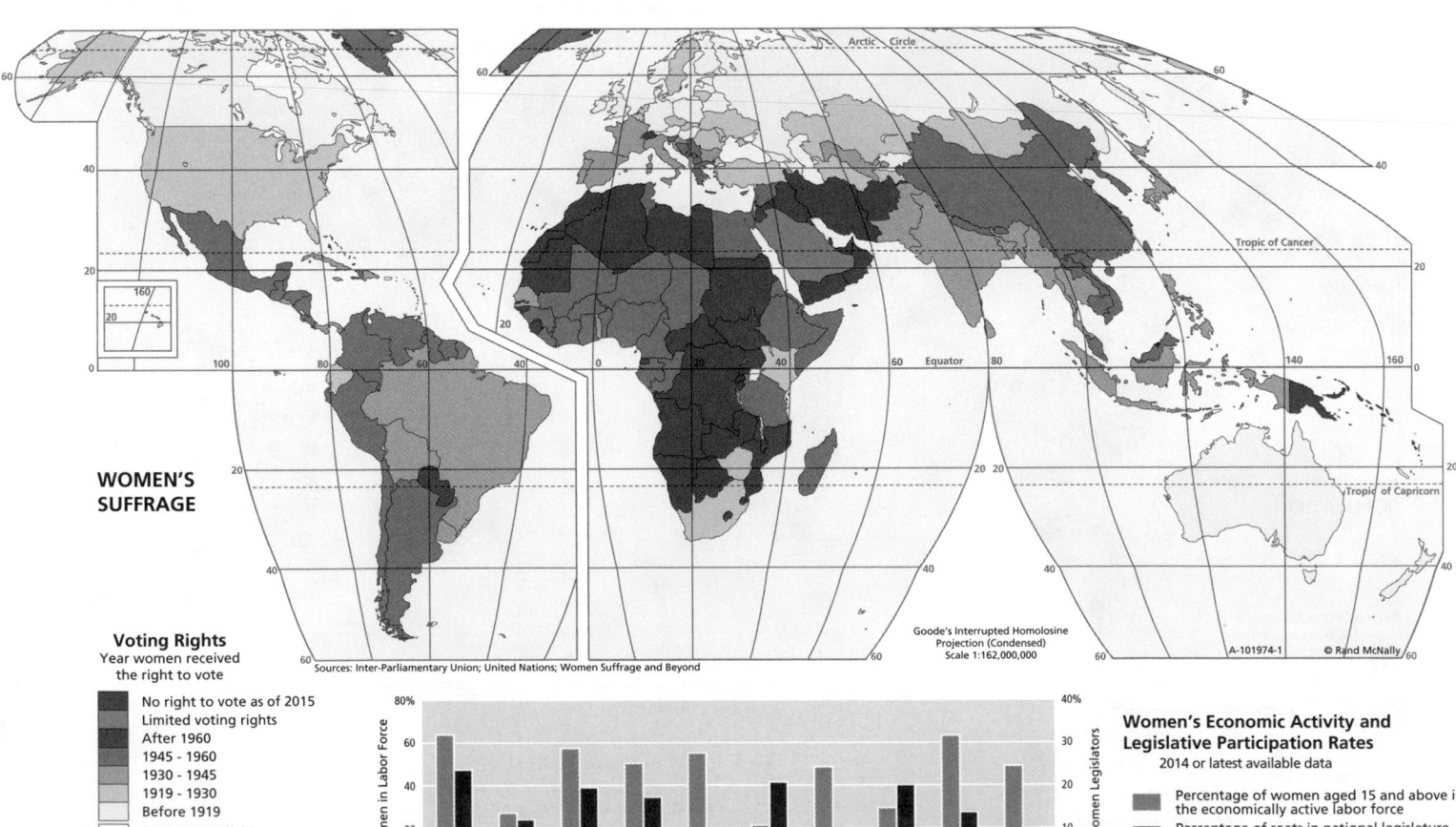

MILITARY EXPENDITURES

Military Expenditures
Percentage of GDP -
latest available data

- Over 5%
- 3.5% - 5%
- 2.5% - 3.4%
- World Avg. 2.4% → 1.5% - 2.4%
- 0 - 1.4%
- Data not available

Sources: CIA; Federation of American Scientists;
Stockholm International Peace Research Institute

Nuclear Weapons
World Total - 15,800 weapons - 2013

ISRAEL 0.5   PAKISTAN 0.8
UNITED KINGDOM 1.4   INDIA 0.7
CHINA 1.6   NORTH KOREA < 0.1
FRANCE 1.9

UNITED STATES 45.6   RUSSIA 47.5%

A-101973-1   © Rand McNally

Goode's Interrupted Homolosine
Projection (Condensed)
Scale 1:162,000,000

Armed Forces Personnel   World Total - 27,207,000 people - 2013

0	10	20	30	40	50	60	70	80	90	100%

CHINA	INDIA	N. KOREA	PAKISTAN	INDONESIA	S. KOREA	TURKEY	IRAN	OTHER ASIA	UNITED STATES	OTHER N.A.	RUSSIA	BRAZIL	OTHER S. AMER.	EUROPE	ALL OTHER
11.0%	10.1	5.1	3.5	2.5	2.4	2.3	2.1	18.9	5.3	2.4	4.6	2.6	4.3	9.4	1.4

Arms Exports   World Total - $51,360,000,000 (2014 $U.S.) - Avg. 2012 - 2014

0	10	20	30	40	50	60	70	80	90	100%

UNITED STATES	RUSSIA	CHINA	ISRAEL	OTHER ASIA	FRANCE	U.K.	GERMANY	UKRAINE	ITALY	SPAIN	NETH.	OTHER EUROPE
31.2%	26.8	5.7	2.5	1.8	5.4	4.8	3.9	3.3	3.0	2.8	2.0	4.7

WOMEN'S SUFFRAGE

Goode's Interrupted Homolosine
Projection (Condensed)
Scale 1:162,000,000

A-101974-1   © Rand McNally

Sources: Inter-Parliamentary Union; United Nations; Women Suffrage and Beyond

Voting Rights
Year women received
the right to vote

- No right to vote as of 2015
- Limited voting rights
- After 1960
- 1945 - 1960
- 1930 - 1945
- 1919 - 1930
- Before 1919
- Data not available

Neither women nor men are allowed
to vote in Brunei and Western Sahara.
Right for Saudi Arabian women to vote
in local elections gained in December 2015.

Women's Economic Activity and Legislative Participation Rates
2014 or latest available data

- Percentage of women aged 15 and above in the economically active labor force
- Percentage of seats in national legislature held by women

(World's largest countries, 2015)

China, India, U.S., Indonesia, Brazil, Pakistan, Nigeria, Bangladesh, Russia, Japan

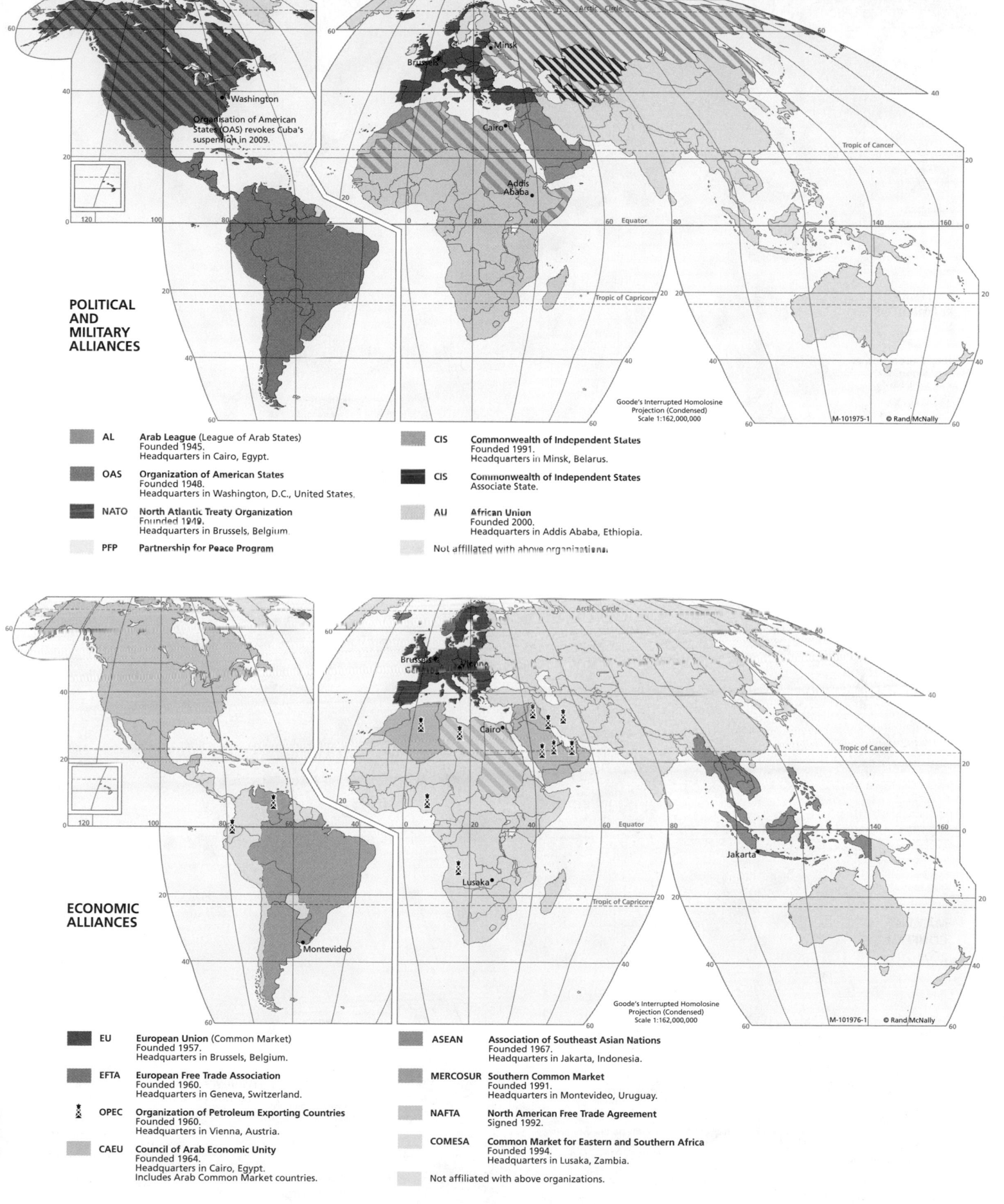

POLITICAL AND MILITARY ALLIANCES

Organisation of American States (OAS) revokes Cuba's suspension in 2009.

Goode's Interrupted Homolosine Projection (Condensed)
Scale 1:162,000,000

M-101975-1   © Rand McNally

	AL	**Arab League** (League of Arab States) Founded 1945. Headquarters in Cairo, Egypt.
	OAS	**Organization of American States** Founded 1948. Headquarters in Washington, D.C., United States.
	NATO	**North Atlantic Treaty Organization** Founded 1949. Headquarters in Brussels, Belgium.
	PFP	**Partnership for Peace Program**

	CIS	**Commonwealth of Independent States** Founded 1991. Headquarters in Minsk, Belarus.
	CIS	**Commonwealth of Independent States** Associate State.
	AU	**African Union** Founded 2000. Headquarters in Addis Ababa, Ethiopia.
		Not affiliated with above organizations.

ECONOMIC ALLIANCES

Goode's Interrupted Homolosine Projection (Condensed)
Scale 1:162,000,000

M-101976-1   © Rand McNally

	EU	**European Union** (Common Market) Founded 1957. Headquarters in Brussels, Belgium.
	EFTA	**European Free Trade Association** Founded 1960. Headquarters in Geneva, Switzerland.
	OPEC	**Organization of Petroleum Exporting Countries** Founded 1960. Headquarters in Vienna, Austria.
	CAEU	**Council of Arab Economic Unity** Founded 1964. Headquarters in Cairo, Egypt. Includes Arab Common Market countries.

	ASEAN	**Association of Southeast Asian Nations** Founded 1967. Headquarters in Jakarta, Indonesia.
	MERCOSUR	**Southern Common Market** Founded 1991. Headquarters in Montevideo, Uruguay.
	NAFTA	**North American Free Trade Agreement** Signed 1992.
	COMESA	**Common Market for Eastern and Southern Africa** Founded 1994. Headquarters in Lusaka, Zambia.

Not affiliated with above organizations.

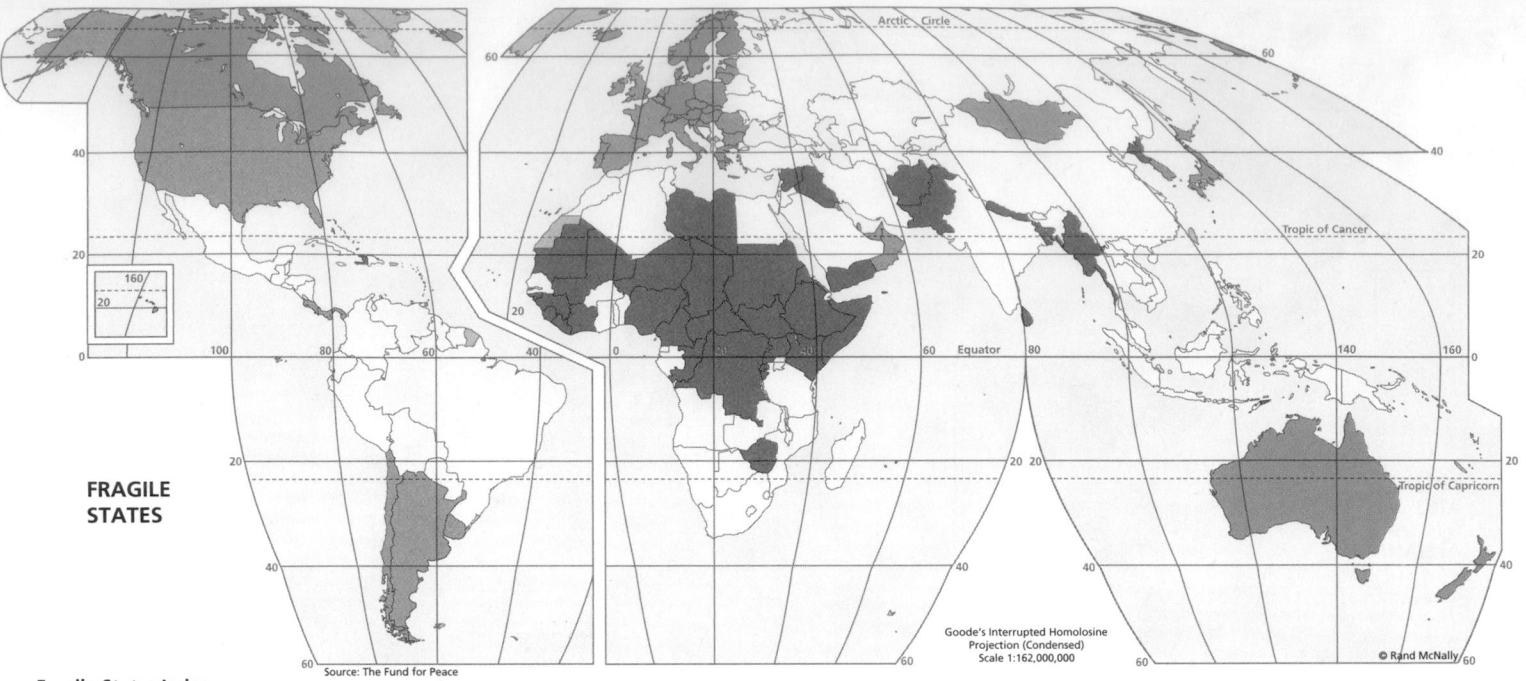

FRAGILE STATES

Fragile States Index
Fragility in the world - 2015

- Alert
- Warning
- Stable
- Sustainable
- Data not available

The Fragile States Index is based on twelve indicators measuring social conditions (demographic pressures, refugees and internally displaced persons, tension between groups, and human flight and brain drain), economic conditions (uneven economic development, poverty and economic decline), and political conditions (state legitimacy, provision of public services, human rights violations, security, factionalized elites and external intervention) and reflects a country's vulnerability to conflict or collapse.

Source: The Fund for Peace

Goode's Interrupted Homolosine
Projection (Condensed)
Scale 1:162,000,000

© Rand McNally

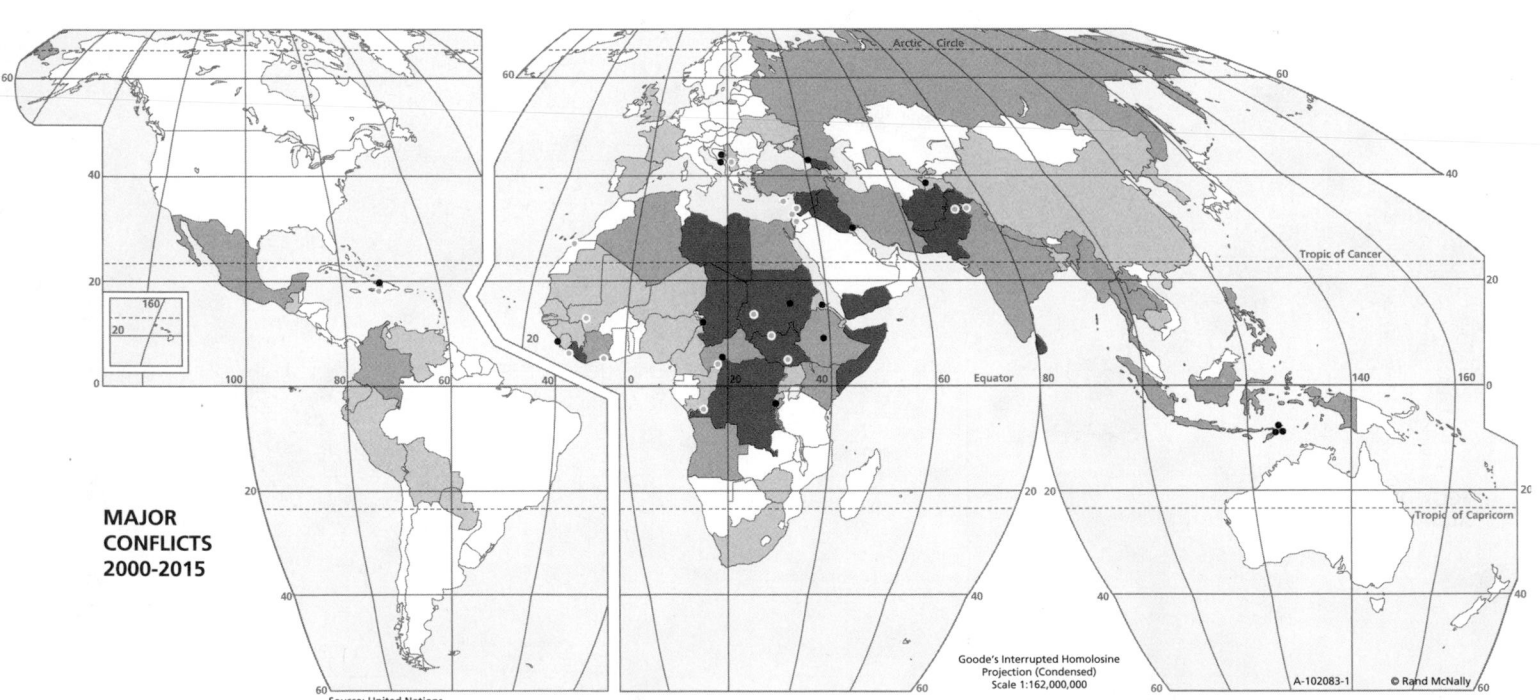

MAJOR CONFLICTS 2000-2015

Source: United Nations

Goode's Interrupted Homolosine
Projection (Condensed)
Scale 1:162,000,000

A-102083-1   © Rand McNally

- **Very Serious Conflict:** A sustained conflict in which organized, systematic, and continual violent force is used causing massive destruction.
- **Serious Conflict:** Severe crisis where organized violence is used regularly.
- **Hot Spot:** A tense situation in which at least one of the parties uses violence in sporadic incidents.

United Nations Peacekeeping Operations

- ○ Ongoing Peacekeeping Missions
- ● Completed Peacekeeping Missions

Arctic Circle
Tropic of Cancer
Equator
Tropic of Capricorn
© 2017 Pearson Education, Inc.

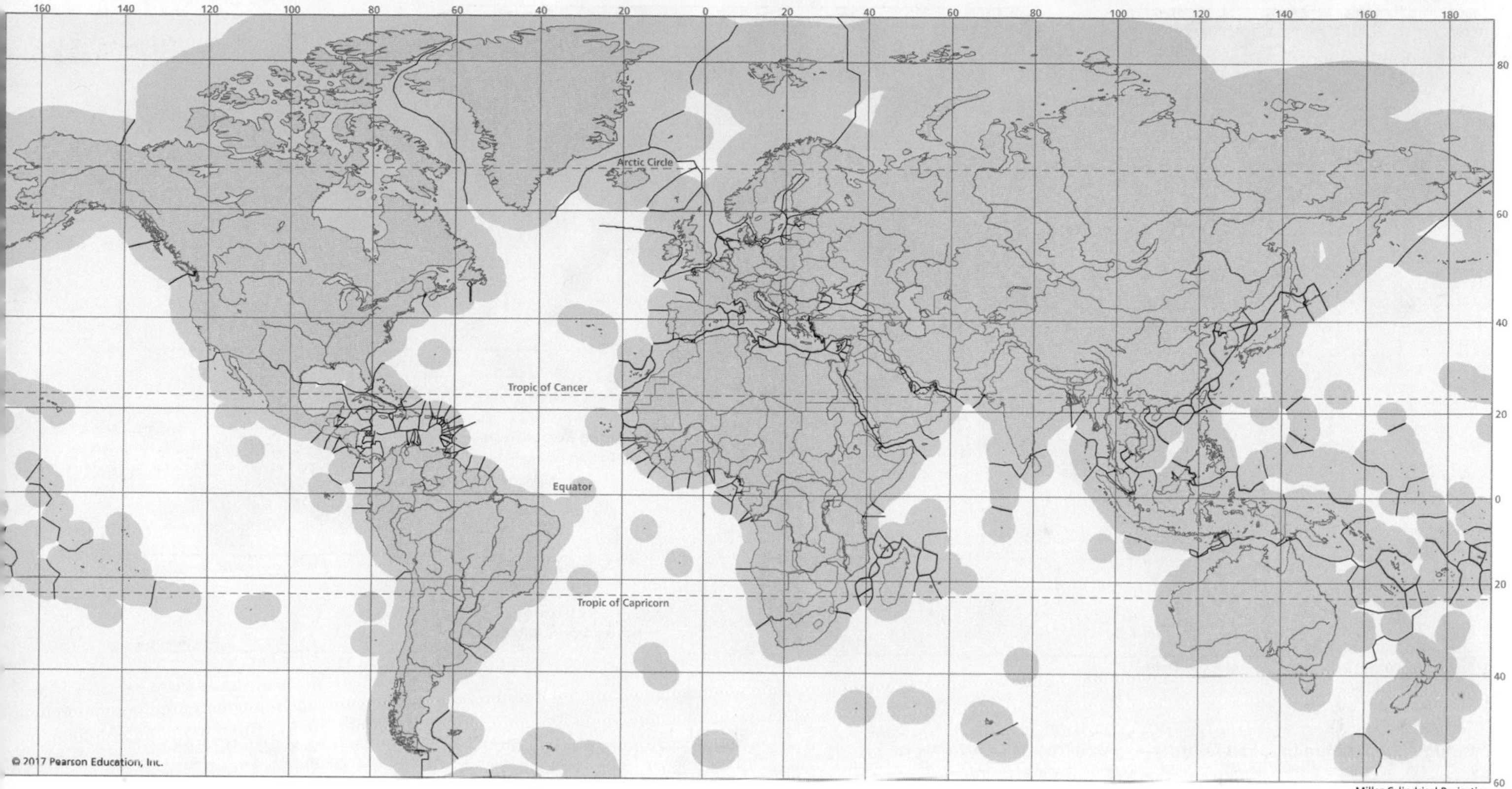

MARITIME BOUNDARIES

Miller Cylindrical Projection
Scale 1:200,000,000
Source: International Mapping

- Maritime boundary
- Equidistance line
- Exclusive Economic Zone (EEZ)

CHINA
TAIWAN
0 100 200 Miles
0 100 200 Kilometers
LAOS
THAILAND
VIETNAM
CAMBODIA
Vietnam claims the Paracel and Spratly Islands. Extent of their claims has not been delimited.
PARACEL ISLANDS
South China Sea
Chinese claim
Chinese claim
PHILIPPINES
Philippine claim
SPRATLY ISLANDS
Gulf of Thailand
claim
Malaysia
Brunei claim
MALAYSIA
BRUNEI
MALAYSIA
INDONESIA
© 2017 Pearson Education, Inc.

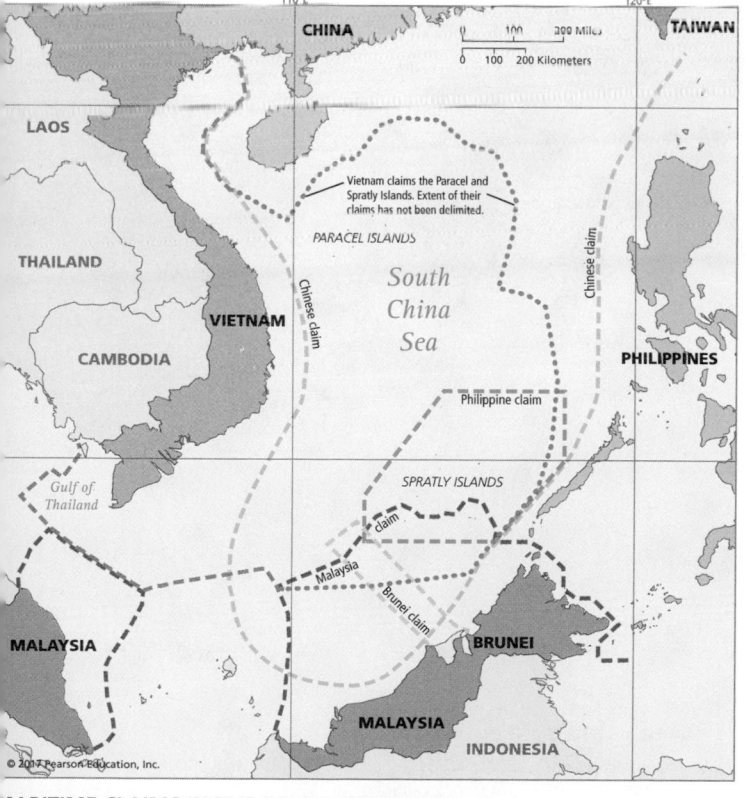

MARITIME CLAIMS IN THE SOUTH CHINA SEA

SOMALIA
Cadale
MOGADISHU
INDIAN OCEAN
Equator
KENYA
Kismaayo
Somalia's Outer Continental Shelf Claim
Kenya's Claim Line
Equidistance Line
Area of Overlapping Claims
Kenya's Outer Continental Shelf Claim
Mombasa
Kenya-Tanzania Maritime Boundary
PEMBA I.
ZANZIBAR I.
200 M limit
TANZANIA
0 50 100 Nautical Miles
0 50 100 Kilometers
© 2017 Pearson Education, Inc.

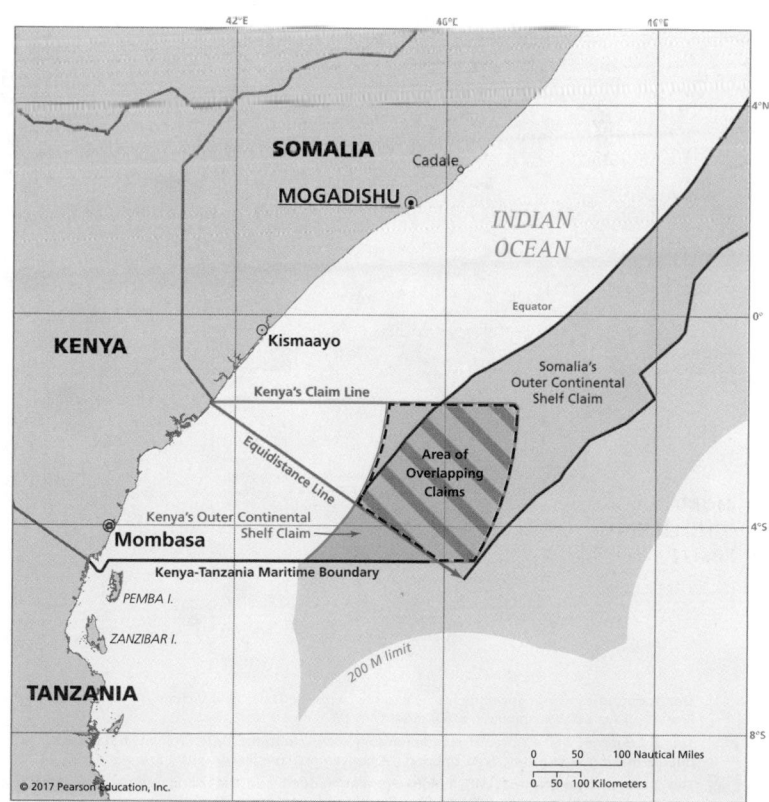

OUTER CONTINENTAL SHELF CLAIMS OF SOMALIA AND KENYA

WORLD REFUGEES

Refugee Population
by Host Country*

■	Over 250,000
■	100,000 - 250,000
■	10,000 - 99,999
■	Under 10,000

Source: United Nations

Refugee Population
by Country of Origin**

- 3,883,554
- 1,000,000
- 250,000
- 100,000
- 10,000 - 20,000

No symbol is shown for countries
with less than 10,000 refugees.

A-102082-1   © Rand McNally
Goode's Interrupted Homolosine
Projection (Condensed)
Scale 1:162,000,000

Refugee Population (by Host Country)*   World Total - 14,376,000 - 2014

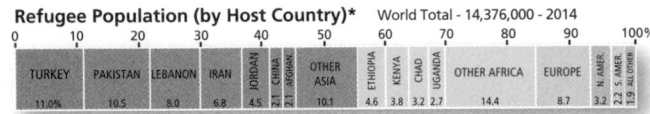

TURKEY	PAKISTAN	LEBANON	IRAN	JORDAN	CHINA	AFGHAN.	OTHER ASIA	ETHIOPIA	KENYA	CHAD	UGANDA	OTHER AFRICA	EUROPE	N. AMER.	S. AMER.	ALL OTHER
11.0%	10.5	8.0	6.8	4.5	2.1	2.1	10.1	4.6	3.8	3.2	2.7	14.4	8.7	3.2	2.2	1.9

* People who have come to this country from another country.

Refugee Population (by Country of Origin)**   World Total - 14,376,000 - 2014

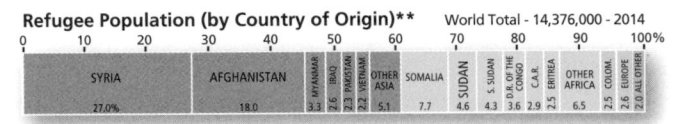

SYRIA	AFGHANISTAN	MYANMAR	IRAQ	PAKISTAN	VIETNAM	OTHER ASIA	SOMALIA	SUDAN	S. SUDAN	D.R. OF THE CONGO	C.A.R.	ERITREA	OTHER AFRICA	COLOM.	EUROPE	ALL OTHER
27.0%	18.0	3.3	2.6	2.3	2.2	5.1	7.7	4.6	4.3	3.6	2.9	2.5	6.5	2.5	2.6	2.0

** People who fled from this country.

HUMAN TRAFFICKING

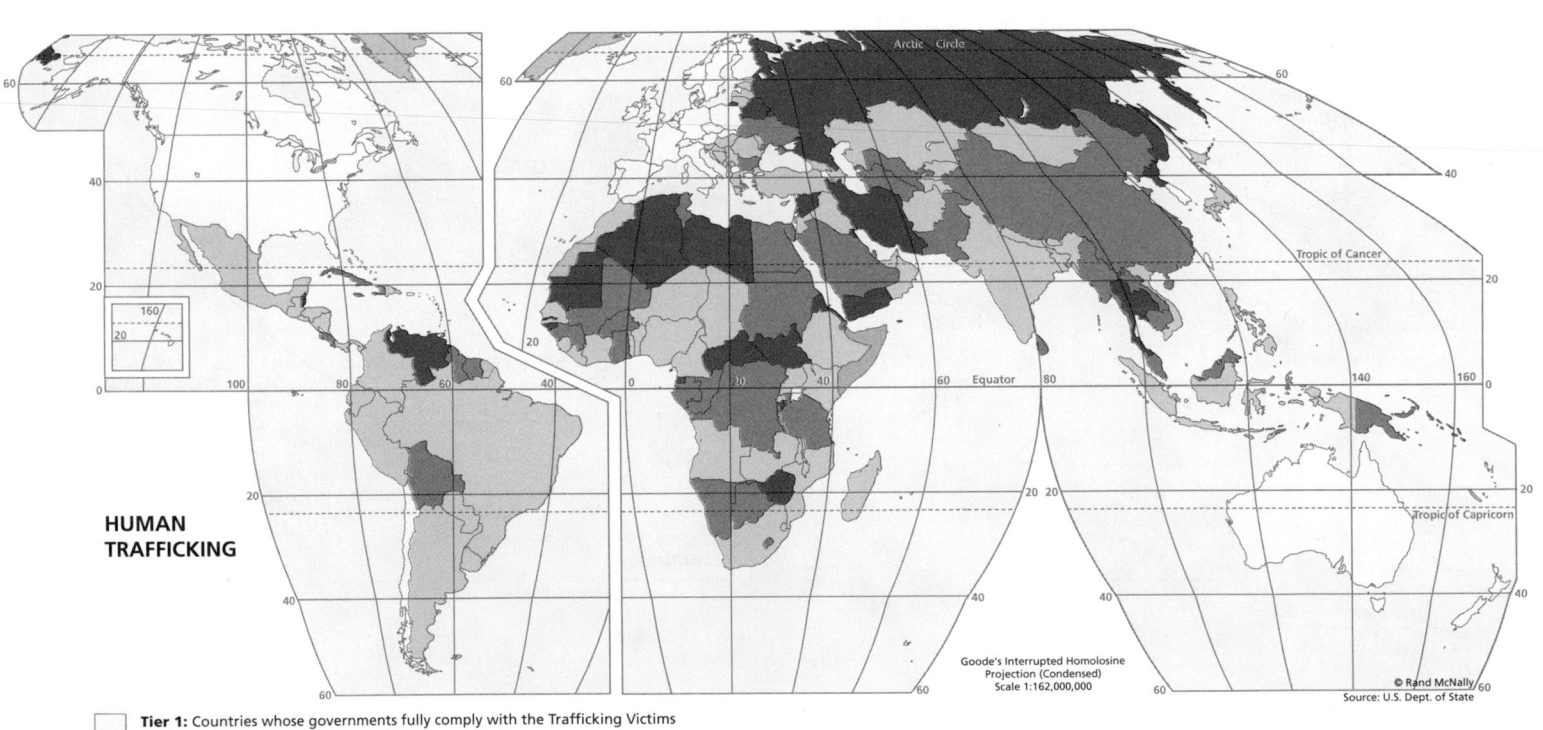

Goode's Interrupted Homolosine
Projection (Condensed)
Scale 1:162,000,000

© Rand McNally

Source: U.S. Dept. of State

☐	**Tier 1:** Countries whose governments fully comply with the Trafficking Victims Protection Act's (TVPA) minimum standards.
■	**Tier 2:** Countries whose governments do not fully comply with the TVPA's minimum standards, but are making significant efforts to bring themselves into compliance with those standards.
■	**Tier 2 Watch List:** Countries with a high or increasing number of trafficking victims whose governments do not fully comply with the TVPA and need to provide more evidence of their commitment to do so.
■	**Tier 3:** Countries whose governments do not fully comply with the minimum standards and are not making significant efforts to do so.
■	**Special Case**
☐	**No Tier Ranking**

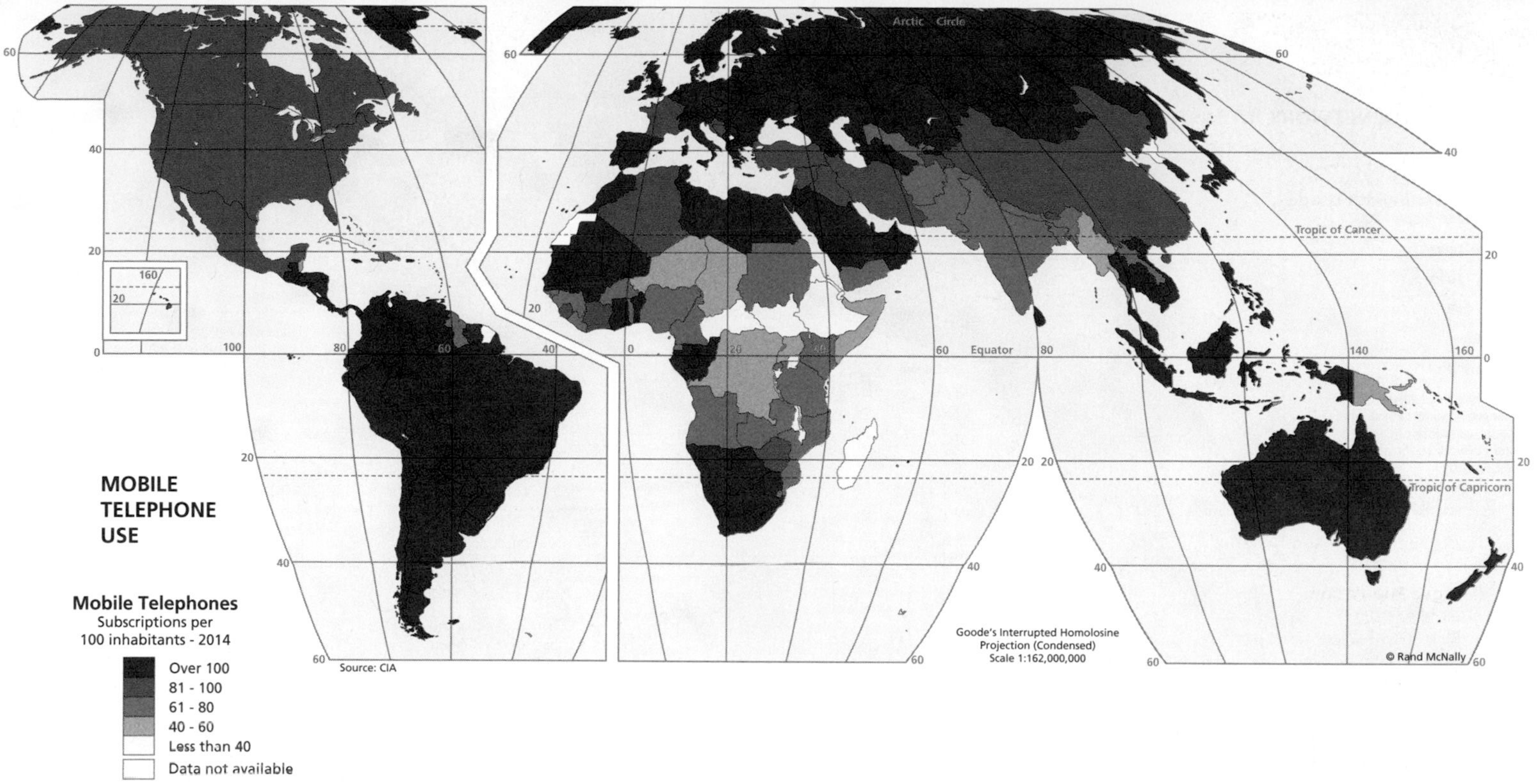
MOBILE TELEPHONE USE
Mobile Telephones
Subscriptions per 100 inhabitants - 2014
Over 100
81 - 100
61 - 80
40 - 60
Less than 40
Data not available
Arctic Circle
Tropic of Cancer
Equator
Tropic of Capricorn
Source: CIA
Goode's Interrupted Homolosine Projection (Condensed)
Scale 1:162,000,000
© Rand McNally

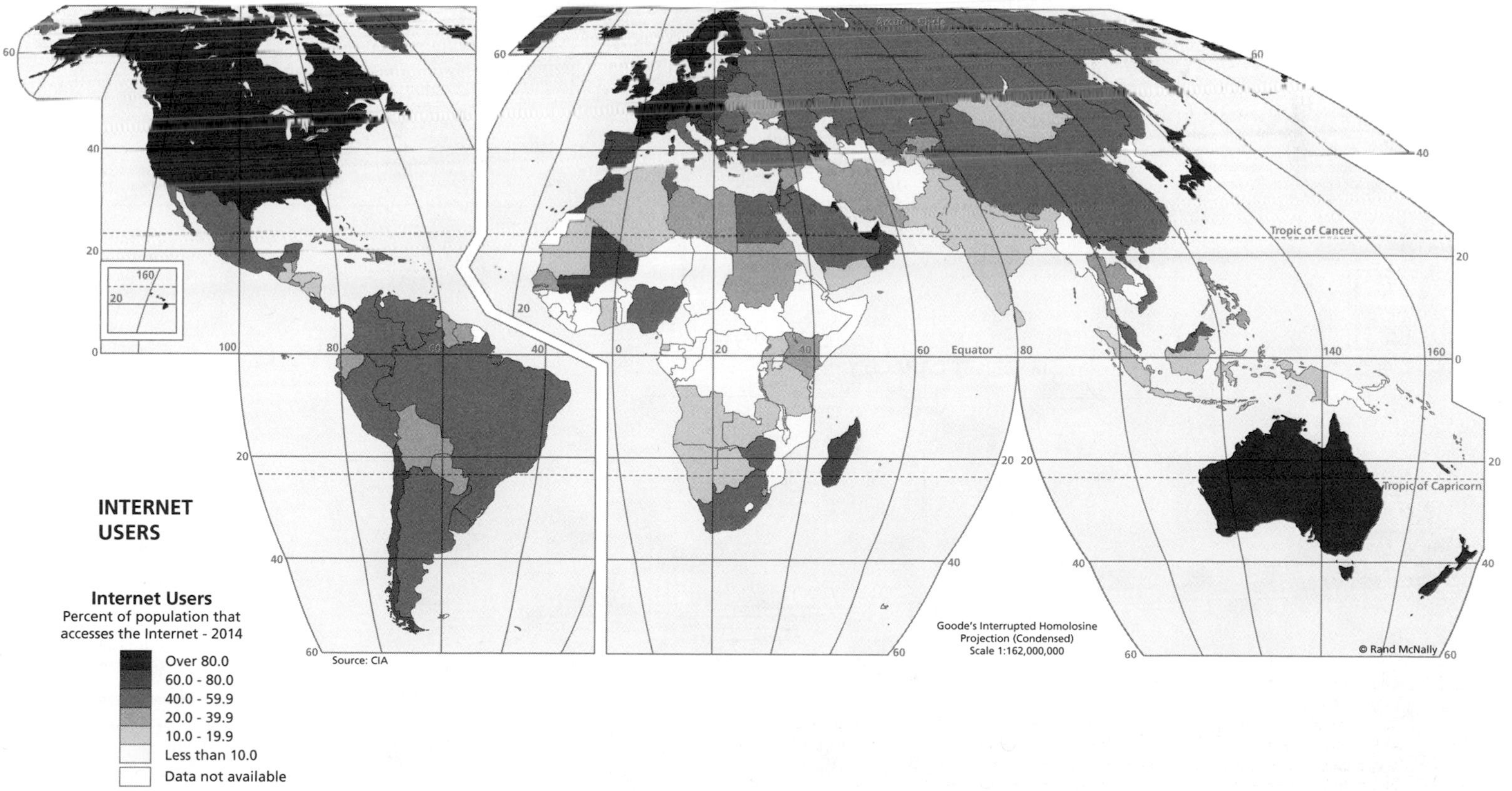
INTERNET USERS
Internet Users
Percent of population that accesses the Internet - 2014
Over 80.0
60.0 - 80.0
40.0 - 59.9
20.0 - 39.9
10.0 - 19.9
Less than 10.0
Data not available
Arctic Circle
Tropic of Cancer
Equator
Tropic of Capricorn
Source: CIA
Goode's Interrupted Homolosine Projection (Condensed)
Scale 1:162,000,000
© Rand McNally

COMMUNICATION NETWORK INFRASTRUCTURE

International Bandwidth Usage
Gigabits per second (Gbps) - 2007

- Over 1000
- 250 - 1000
- 50 - 250
- 1 - 50
- Less than 1

Capacity deployed by carriers, internet service providers (ISPs), and enterprises to carry internet, voice, and private network traffic across international borders.

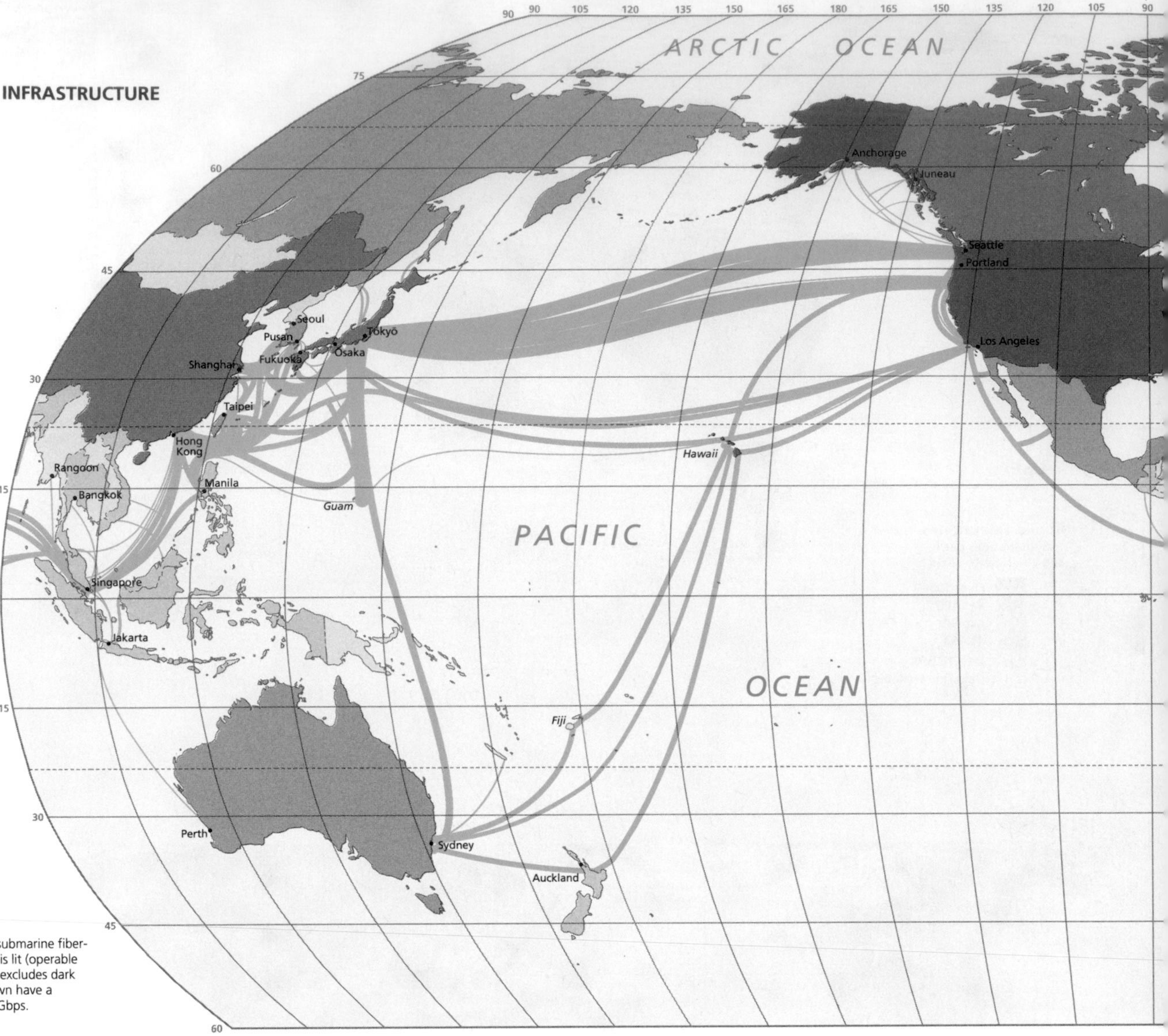

Submarine Cable Capacity
Lit capacity of submarine cables, in Gigabits per second (Gbps) - 2008

- Over 500
- 50 - 500
- 10 - 50

Line thickness is proportional to lit capacity of submarine fiber-optic cables. Lit capacity includes all cable that is lit (operable and capable of transmitting a light signal), but excludes dark fiber (inactive or inoperable cable). Cables shown have a maximum upgradeable capacity of at least 10 Gbps.

INTERNET CAPACITY

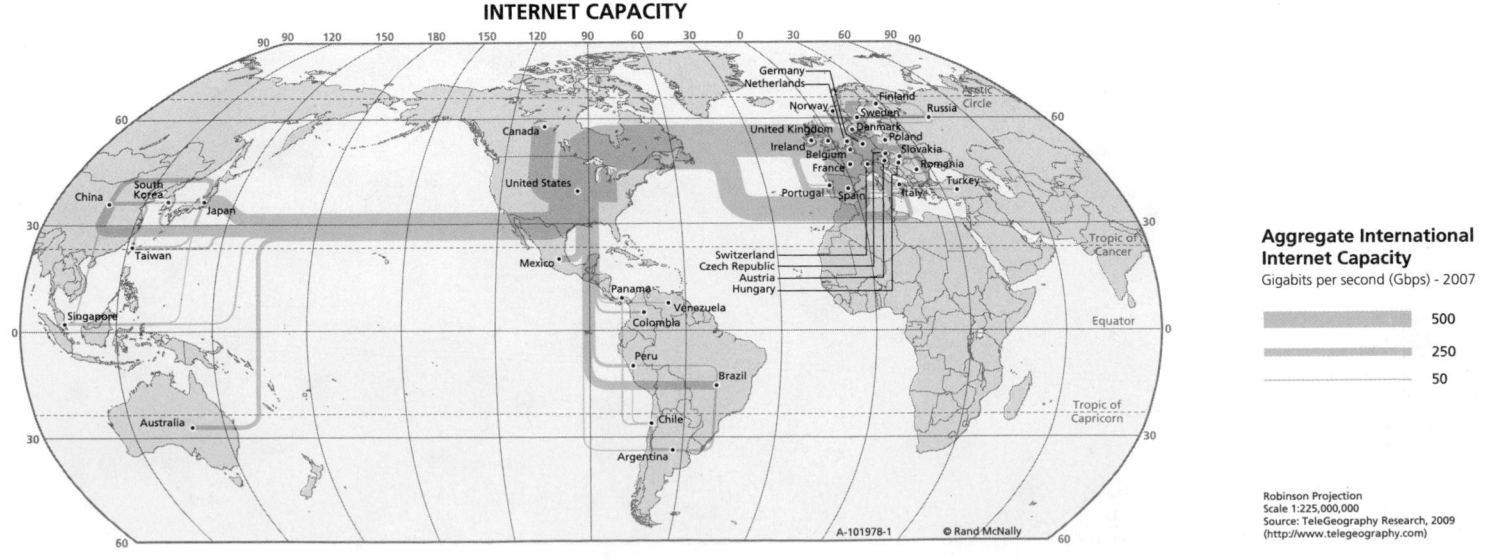

Aggregate International Internet Capacity
Gigabits per second (Gbps) - 2007

- 500
- 250
- 50

Robinson Projection
Scale 1:225,000,000
Source: TeleGeography Research, 2009
(http://www.telegeography.com)

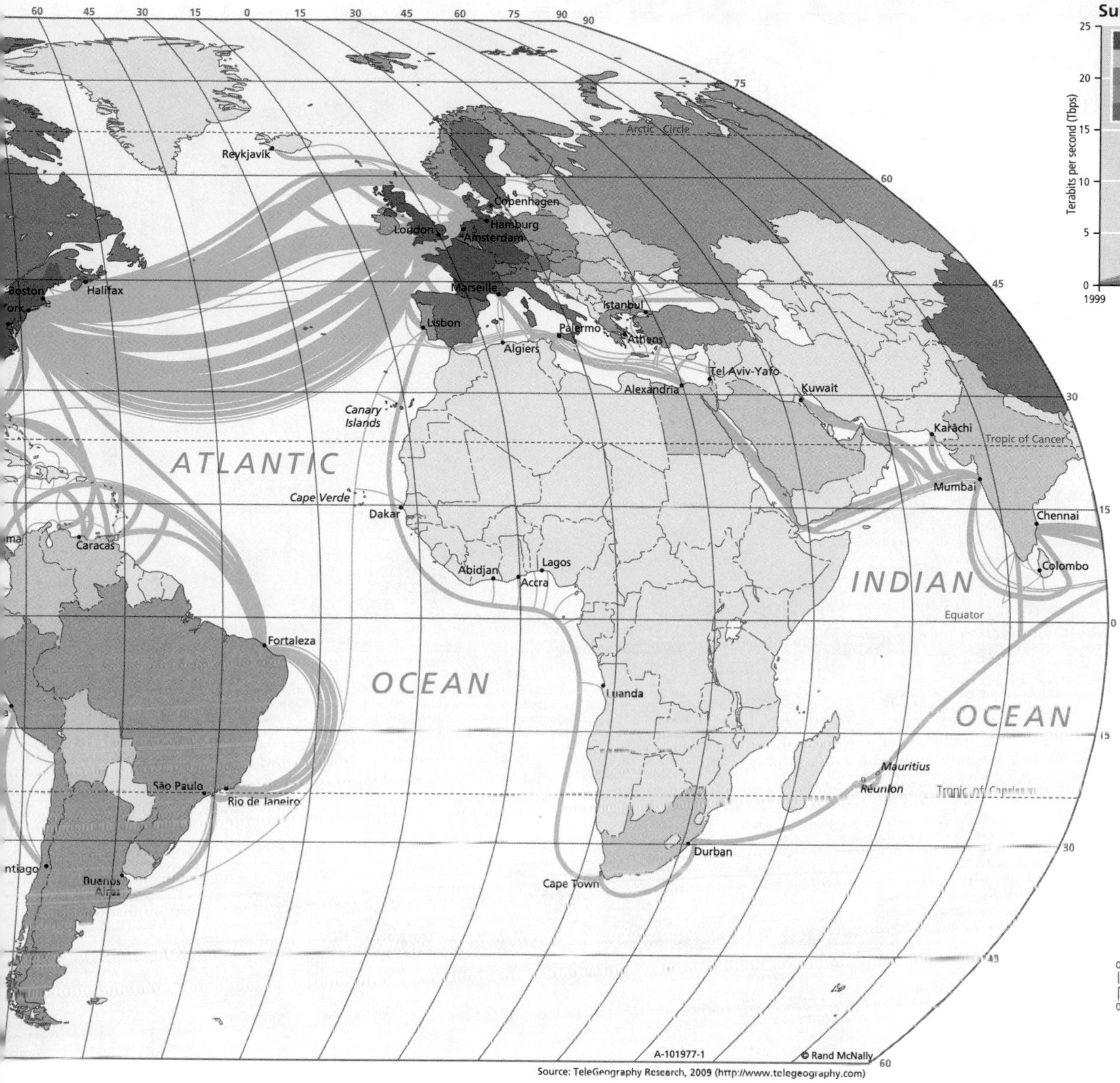

Submarine Cable Capacity by Route

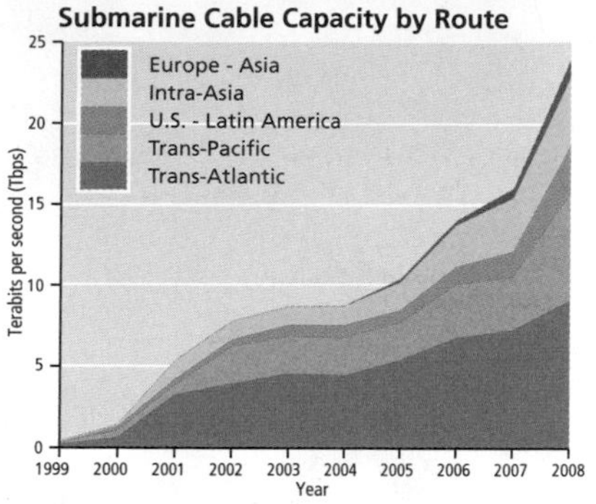

Legend:
- Europe - Asia
- Intra-Asia
- U.S. - Latin America
- Trans-Pacific
- Trans-Atlantic

Terabits per second (Tbps)

Years: 1999 2000 2001 2002 2003 2004 2005 2006 2007 2008

Figures denote lit capacity of submarine fiber-optic cables. Trans-Pacific capacity excludes cables linking the United States to Australia and New Zealand. Trans-Atlantic capacity excludes cables linking Europe to South America. Intra-Asia capacity includes cables with landings in both Hong Kong and Japan. Europe-Asia capacity reflects available capacity between the Middle East and Europe, and excludes Europe-Asia capacity routed via Russia or the United States.

Map labels: Reykjavik, Boston, Halifax, York, Copenhagen, London, Hamburg, Amsterdam, Marseille, Lisbon, Palermo, Istanbul, Athens, Algiers, Alexandria, Tel Aviv-Yafo, Kuwait, Karachi, Canary Islands, ATLANTIC, Cape Verde, Dakar, Caracas, Abidjan, Lagos, Accra, Mumbai, Chennai, Colombo, INDIAN, Fortaleza, OCEAN, Luanda, Equator, OCEAN, São Paulo, Rio de Janeiro, Mauritius, Réunion, Tropic of Capricorn, Santiago, Buenos Aires, Durban, Cape Town

Arctic Circle, Tropic of Cancer

Robinson Projection
Scale 1:100,000,000
One inch to 1,580 miles
One cm to 1,000 km

0 500 1000 1500 2000 Miles
0 500 1000 1500 2000 2500 Kilometers

A-101977-1 © Rand McNally

Source: TeleGeography Research, 2009 (http://www.telegeography.com)

SHIPPING LANES

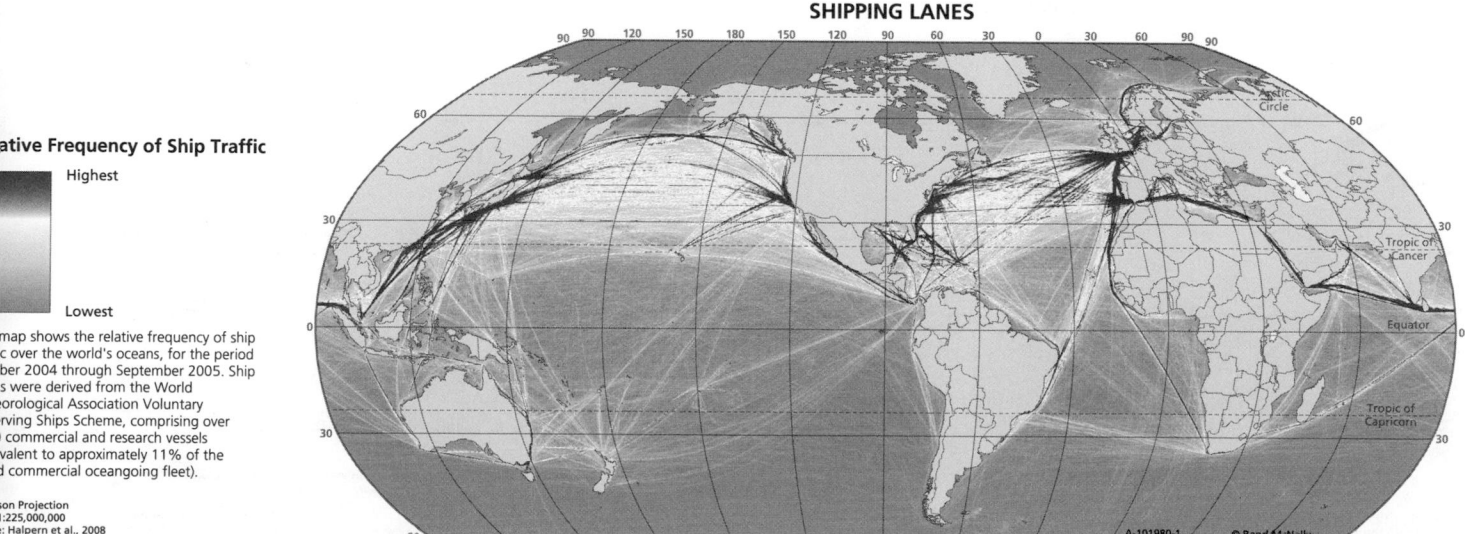

Relative Frequency of Ship Traffic

Highest

Lowest

map shows the relative frequency of ship
c over the world's oceans, for the period
ber 2004 through September 2005. Ship
s were derived from the World
eorological Association Voluntary
rving Ships Scheme, comprising over
commercial and research vessels
valent to approximately 11% of the
d commercial oceangoing fleet).

son Projection
:225,000,000
e: Halpern et al., 2008

A-101980-1 © Rand McNally

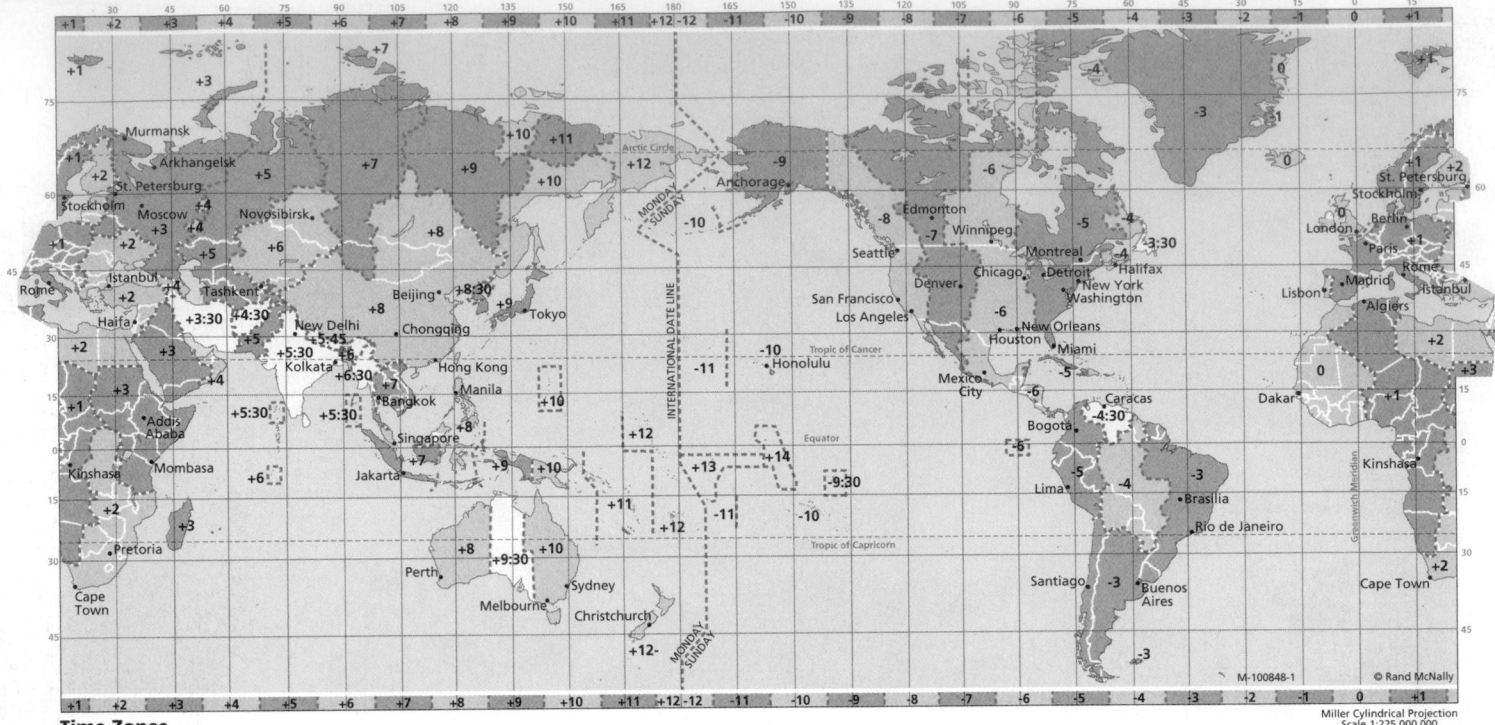

Time Zones

Coordinated Universal Time (UTC) is the standard for international time zones and the official reference for standard time across the world. Although UTC has officially replaced Greenwich Mean Time (GMT), both terms are widely employed and, in casual usage, are essentially synonymous. On the time zone map above, the numbers along the top and bottom edges indicate the time difference, in hours, from UTC. The first time zone, with a value of 0, is centered on the Prime Meridian running through Greenwich, England. To compute standard time at any location, add the value on the map to UTC at Greenwich. For example, Chicago is in time zone UTC -6, which means it is 6 hours earlier than UTC at Greenwich. This means that if it is noon at Greenwich then it is 6 a.m. in Chicago.

To ensure synchronization with the Sun's location, time zone boundaries should follow lines of longitude very precisely. However this is rarely the case, and most time zones boundaries are very irregular. They are often constrained to follow international or internal administrative boundaries, and may be shifted east or west for various reasons. Discontinuities sometimes exist where time changes by more than one hour across a zone boundary, and the UTC difference for some time zones is less than a full hour. To make matters even more complicated, these time zones do not account for Daylight Savings Time, which is observed in some jurisdictions for part of the year.

HOURS OF DAYLIGHT

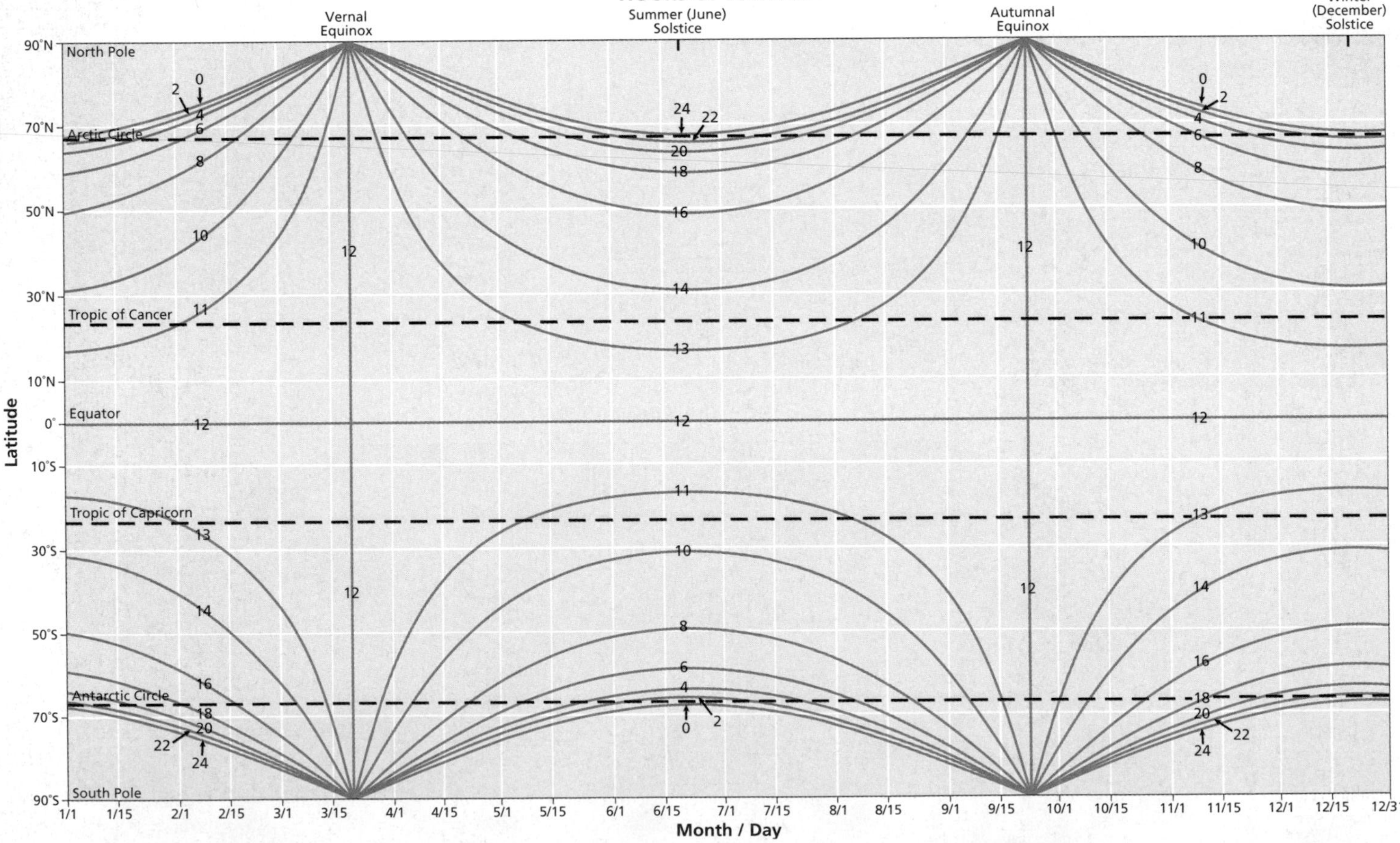

This graph shows hours of daylight at various latitudes for each day of the year. The following are some important patterns evident on the graph.
- The Equator experiences 12 hours of daylight every day of the year.
- Every point on the Earth experiences 12 hours of daylight at the vernal and autumnal equinoxes.
- The greater the distance from the Equator, the greater the variability in daylight length over the year.
- In the northern hemisphere, daylight length is greater than 12 hours between the vernal and autumnal equinoxes (the northern hemisphere summer) and less than 12 hours between the autumnal and vernal equinoxes (the northern hemisphere winter). The opposite pattern occurs in the southern hemisphere.

- Areas north of the Arctic Circle and south of the Antarctic Circle experience an annual pattern with periods of total darkness and periods of continuous daylight.

The data used to create this graph do not account for refraction of the Sun's rays by the Earth's atmosphere, which lengthens the daylight period slightly. The calculations are based on the center of the Sun, and do not account for the size of the solar disk, which also extends the daylight period by several minutes.

North America

NORTH AMERICA

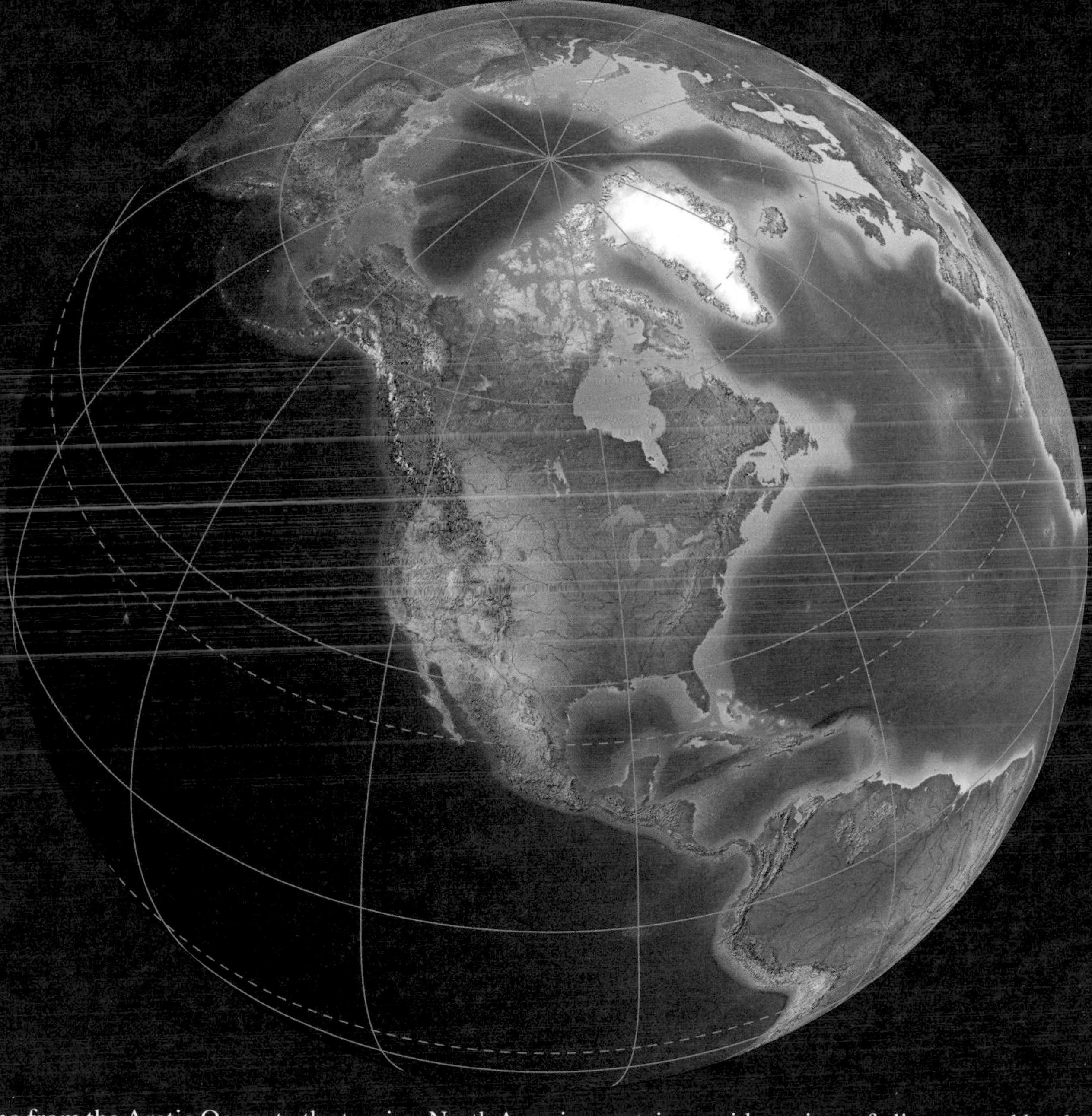

Stretching from the Arctic Ocean to the tropics, North America contains a wide variety of climates and landscapes. The western half of the continent is dominated by mountain chains stretching from Alaska to the isthmus of Central America. Some of the world's most productive agricultural land is found in North America, along with a vast northern boreal forest, large deserts, tropical and temperate rain forests, and the Great Lakes—the largest freshwater lake system in the world. The continent's cultural geography reflects a legacy of British, Spanish, and French colonial rule, and the millions of Africans forcibly relocated as slaves prior to the mid-1880s. Before Europeans arrived, North America was inhabited by a diversity of Native American cultures whose total population has been estimated to be in the tens of millions.

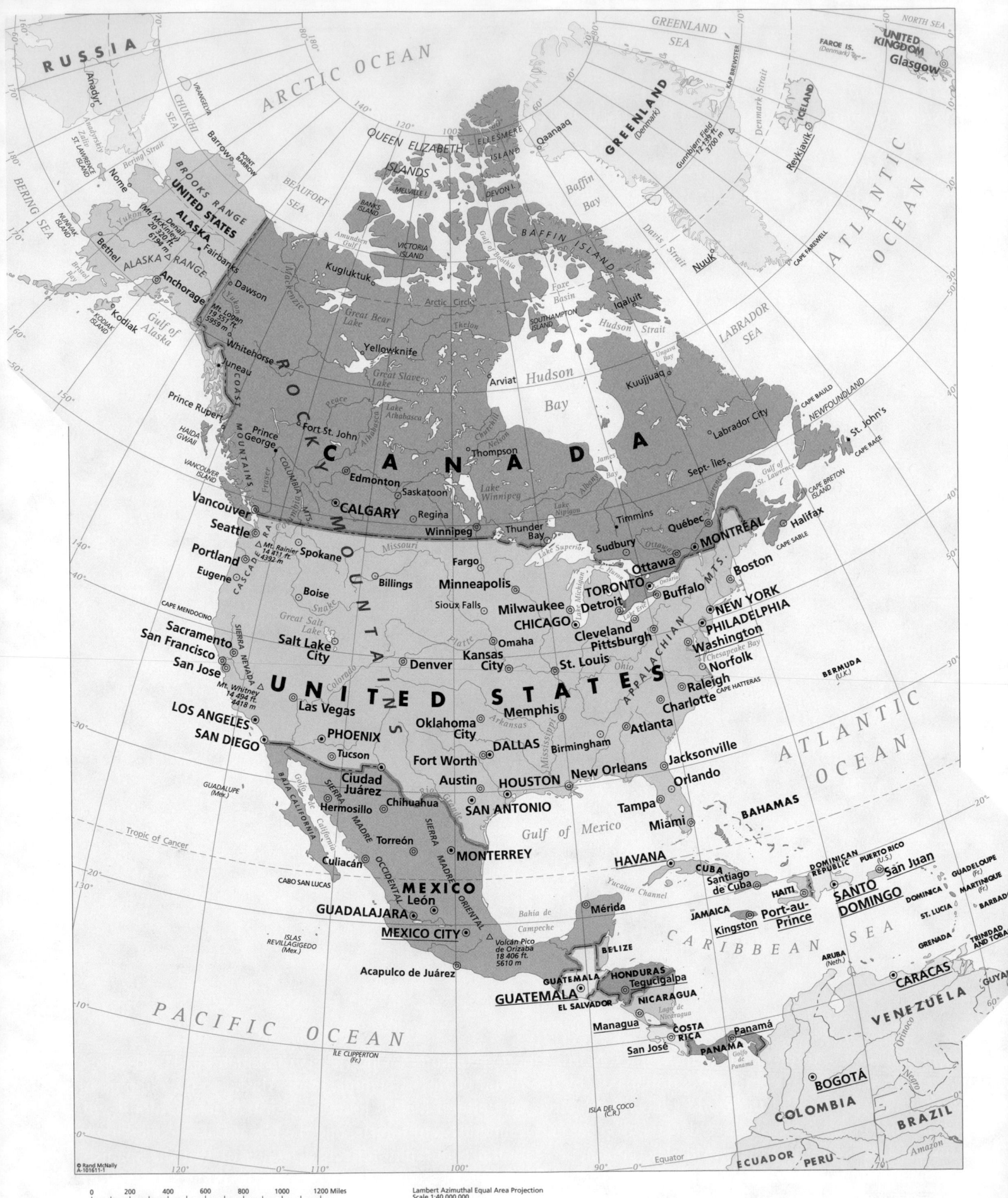

RUSSIA

ARCTIC OCEAN

GREENLAND SEA

NORTH SEA

UNITED KINGDOM
Glasgow

FAROE IS.
(Denmark)

ICELAND
Reykjavík

GREENLAND
(Denmark)

Gunnbjørn Field
12,139 ft.
3,700 m

Kap Brewster

Denmark Strait

ATLANTIC OCEAN

Anadyr
CHUKCHI SEA
Bering Strait
Nome
Barrow
POINT BARROW
BROOKS RANGE
BEAUFORT SEA

QUEEN ELIZABETH ISLANDS

ELLESMERE ISLAND

Qaanaaq

CAPE FAREWELL

Nuuk

UNITED STATES
ALASKA
Fairbanks
Denali
(Mt. McKinley)
20,320 ft.
6194 m
ALASKA RANGE
Anchorage
Dawson

MELVILLE ISLAND
DEVON I.
Baffin Bay

Iqaluit

Davis Strait

LABRADOR SEA

Bethel
Kodiak
KODIAK ISLAND
Gulf of Alaska

Whitehorse
Juneau
Yukon

VICTORIA ISLAND
Amundsen Gulf
Gulf of Boothia
Kugluktuk

BAFFIN ISLAND

Foxe Basin
SOUTHAMPTON ISLAND

CAPE BAULD
NEWFOUNDLAND
St. John's
CAPE RACE

Prince Rupert
HAIDA GWAII
VANCOUVER ISLAND

Mt. Logan
19,551 ft.
5959 m

Yellowknife
Great Bear Lake

Arctic Circle

Hudson Strait
Kuujjuaq
Ungava Bay

Labrador City

Sept-Îles

CAPE BRETON ISLAND

CANADA

Great Slave Lake

Arviat

Hudson Bay

James Bay

Gulf of St. Lawrence

Halifax

Prince George
Fort St. John

Peace

Athabasca

Lake Athabasca

Churchill

Nelson

Thompson

Timmins

Québec

MONTRÉAL

CAPE SABLE

Vancouver
Seattle
Portland
Eugene

ROCKY MOUNTAINS
COLUMBIA MTS.
Fraser
COAST RANGE

Edmonton
CALGARY
Saskatoon
Regina

Lake Winnipeg

Winnipeg
Thunder Bay

Sudbury

Ottawa
TORONTO

Lake Nipigon

Boston

Mt. Rainier
14,411 ft.
4392 m
CASCADE RANGE
Spokane

Missouri

Fargo

Lake Superior

Lake Michigan

Buffalo

NEW YORK
PHILADELPHIA

Sacramento
San Francisco
San Jose
CAPE MENDOCINO

Boise
SIERRA NEVADA
Great Salt Lake
Snake

Billings
Sioux Falls
Minneapolis
Milwaukee

Detroit

Lake Huron
Lake Erie
Lake Ontario

Cleveland
Pittsburgh

Washington
Norfolk

CHICAGO

APPALACHIAN MTS.

Chesapeake Bay

Salt Lake City

Omaha

Ohio

Raleigh

BERMUDA (U.K.)

Mt. Whitney
14,494 ft.
4418 m

Denver
Colorado
Platte

Kansas City

St. Louis

Charlotte

CAPE HATTERAS

LOS ANGELES
SAN DIEGO

Las Vegas

UNITED STATES

Memphis
Arkansas

Atlanta

ATLANTIC OCEAN

PHOENIX
Tucson

Oklahoma City

Mississippi

Birmingham

DALLAS
Fort Worth

Jacksonville

GUADALUPE (Mex.)

SIERRA MADRE OCCIDENTAL

Ciudad Juárez
Chihuahua
Hermosillo

Austin
HOUSTON

SAN ANTONIO

New Orleans

Orlando

Tampa
Miami

BAHAMAS

BAJA CALIFORNIA
Golfo de California

Torreón

MONTERREY

Gulf of Mexico

CABO SAN LUCAS

Culiacán

Tropic of Cancer

SIERRA MADRE ORIENTAL

HAVANA

CUBA
Santiago de Cuba

DOMINICAN REPUBLIC

PUERTO RICO (U.S.)
San Juan

GUADELOUPE (Fr.)

MEXICO
León

Yucatan Channel

HAITI
Port-au-Prince

SANTO DOMINGO

DOMINICA

MARTINIQUE (Fr.)

ISLAS REVILLAGIGEDO (Mex.)

GUADALAJARA

Mérida

JAMAICA
Kingston

ST. LUCIA

BARBADOS

MEXICO CITY

Bahía de Campeche

CARIBBEAN SEA

GRENADA

TRINIDAD AND TOBAGO

Volcán Pico de Orizaba
18,406 ft.
5610 m

BELIZE

ARUBA (Neth.)

Acapulco de Juárez

GUATEMALA
GUATEMALA

HONDURAS
Tegucigalpa

CARACAS

GUYANA

EL SALVADOR

NICARAGUA

Lago de Nicaragua

Managua

COSTA RICA
San José

PANAMÁ
PANAMA

Golfo de Panamá

VENEZUELA

ILE CLIPPERTON (Fr.)

PACIFIC OCEAN

BOGOTÁ

COLOMBIA

ISLA DEL COCO (C.R.)

Equator

ECUADOR
PERU

BRAZIL

Amazon
Negro
Orinoco

© Rand McNally
A-101611-1

| 0 | 200 | 400 | 600 | 800 | 1000 | 1200 Miles |

| 0 | 200 | 400 | 600 | 800 | 1000 | 1200 | 1400 | 1600 | 1800 | 2000 Kilometers |

Lambert Azimuthal Equal Area Projection
Scale 1:40,000,000
One inch to 640 miles
One cm to 400 km

RUSSIA

ARCTIC OCEAN

GREENLAND SEA

NORTH SEA

UNITED KINGDOM

Glasgow

FAROE IS. (Denmark)

ICELAND

Reykjavík

ATLANTIC OCEAN

GREENLAND (Denmark)

Gunnbjørn Field 12,139 ft. 3700 m

Mont Forel 11,024 ft. 3360 m

Denmark Strait

CAPE BREWSTER

QUEEN ELIZABETH ISLANDS

ELLESMERE ISLAND

KAP YORK

Baffin Bay

BAFFIN ISLAND

CAPE FAREWELL

Nuuk

Davis Strait

LABRADOR SEA

Hudson Bay

PÉNINSULE D'UNGAVA

Ungava Bay

NEWFOUNDLAND

St. John's

CAPE RACE

GRAND BANKS

CANADA

Calgary

Winnipeg

Lake Winnipeg

Lake Manitoba

BROOKS RANGE UNITED STATES

ALASKA

ALASKA RANGE

Mt. McKinley (Denali) 20,320 ft. 6194 m

Anchorage

COAST

Brønlund Peak 8510 ft. 2594 m

MACKENZIE MTS.

Great Bear Lake

Great Slave Lake

Lake Athabasca

Reindeer Lake

Arctic Circle

ROCKY MOUNTAINS

GREAT PLAINS

Montréal

Halifax

CAPE SABLE

Mt. Washington 6288 ft. 1917 m

Toronto

Boston

New York

Philadelphia

Washington

Detroit

Chicago

APPALACHIAN MTS.

CAPE COD

BERMUDA (U.K.)

Seattle

CASCADE RANGE

Mt. Rainier 14,411 ft. 4392 m

Mt. Hood 11,239 ft. 3426 m

San Francisco

Point Reyes

POINT CONCEPTION

Los Angeles

COAST RANGES

SIERRA NEVADA

GREAT BASIN

Mt. Whitney 14,494 ft. 4418 m

Mt. Elbert 14,433 ft. 4399 m

Pikes Peak 14,110 ft. 4301 m

Cloud Peak 13,167 ft. 4013 m

Gannett Peak 13,804 ft. 4207 m

Denver

UNITED STATES

Phoenix

Baldy Peak 11,403 ft. 3476 m

Dallas

Atlanta

Mt. Mitchell 6684 ft. 2037 m

CAPE HATTERAS

CAPE LOOKOUT

CAPE FEAR

ATLANTIC OCEAN

Houston

CAPE SAN BLAS

CAPE CANAVERAL

Chesapeake Bay

Tropic of Cancer

BAJA CALIFORNIA

SIERRA MADRE OCCIDENTAL

CHIHUAHUAN DESERT

SIERRA MADRE ORIENTAL

Monterrey

CABO SAN LUCAS

MEXICO

Guadalajara

Mexico City

SIERRA MADRE DEL SUR

Volcán Pico de Orizaba 18,406 ft. 5610 m

Gulf of Mexico

CAPE SABLE

Miami

Havana

CUBA

Straits of Florida

BAHAMAS

GREATER ANTILLES

HAITI

HISPANIOLA

DOMINICAN REPUBLIC

San Juan

PUERTO RICO (U.S.)

PUERTO RICO TRENCH

GUADELOUPE (Fr.)

MARTINIQUE (Fr.)

BARBADOS

JAMAICA

WEST INDIES

CARIBBEAN SEA

LESSER ANTILLES

TRINIDAD AND TOBAGO

YUCATAN PENINSULA

BELIZE

GUATEMALA

HONDURAS

Volcán Tajumulco 13,845 ft. 4220 m

EL SALVADOR

NICARAGUA

Managua

COSTA RICA

PANAMA

ISTMO DE PANAMA

Volcán Barú 11,401 ft. 3475 m

Pico Cristóbal Colón 18,947 ft. 5775 m

Caracas

VENEZUELA

LLANOS

GUYANA

Bogotá

COLOMBIA

ANDES

Nevado del Huila 18,865 ft. 5750 m

BRAZIL

Negro

Amazon

Orinoco

PACIFIC OCEAN

ÎLE CLIPPERTON (Fr.)

ISLAS REVILLAGIGEDO (Mex.)

GUADALUPE (Mex.)

ISLA DEL COCO (C.R.)

ECUADOR

PERU

Equator

| 0 | 200 | 400 | 600 | 800 | 1000 | 1200 Miles |

| 0 | 200 | 400 | 600 | 800 | 1000 | 1200 | 1400 | 1600 | 1800 | 2000 Kilometers |

Lambert Azimuthal Equal Area Projection
Scale 1:40,000,000
One inch to 640 miles
One cm to 400 km

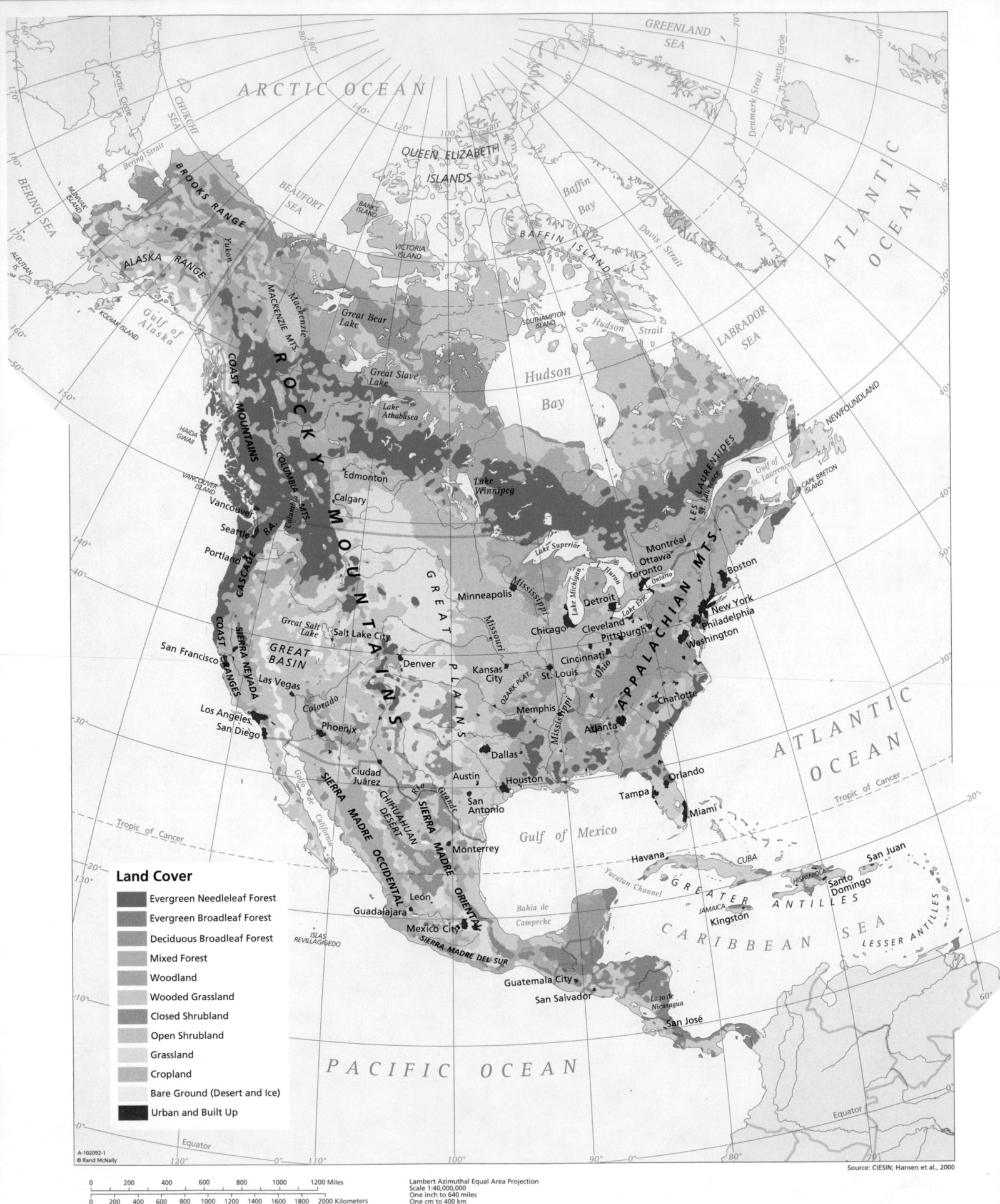

Land Cover

- Evergreen Needleleaf Forest
- Evergreen Broadleaf Forest
- Deciduous Broadleaf Forest
- Mixed Forest
- Woodland
- Wooded Grassland
- Closed Shrubland
- Open Shrubland
- Grassland
- Cropland
- Bare Ground (Desert and Ice)
- Urban and Built Up

A-102092-1
© Rand McNally

0 200 400 600 800 1000 1200 Miles
0 200 400 600 800 1000 1200 1400 1600 1800 2000 Kilometers

Lambert Azimuthal Equal Area Projection
Scale 1:40,000,000
One inch to 640 miles
One cm to 400 km

Source: CIESIN; Hansen et al., 2000

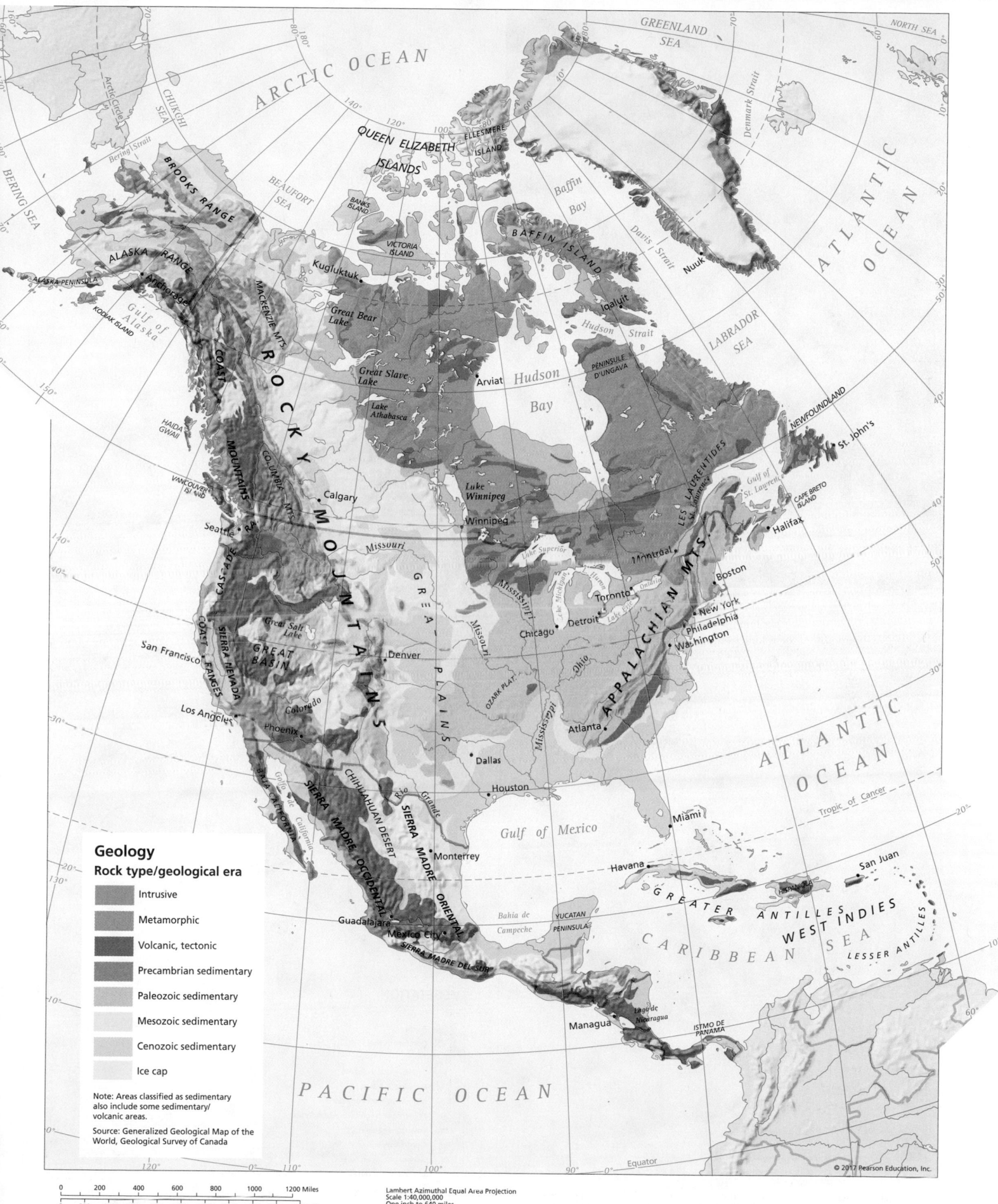

ARCTIC OCEAN

GREENLAND SEA

NORTH SEA

CHUKCHI SEA

BERING STRAIT

Bering Strait

BEAUFORT SEA

QUEEN ELIZABETH ISLANDS

ELLESMERE ISLAND

Baffin Bay

Denmark Strait

ATLANTIC OCEAN

Arctic Circle

BROOKS RANGE

BANKS ISLAND

VICTORIA ISLAND

BAFFIN ISLAND

Davis Strait

Nuuk

ALASKA RANGE

Kugluktuk

ALASKA PENINSULA

Archorage

KODIAK ISLAND

Gulf of Alaska

MACKENZIE MTS.

Great Bear Lake

Great Slave Lake

Iqaluit

Hudson Strait

LABRADOR SEA

PÉNINSULE D'UNGAVA

HAIDA GWAII

Lake Athabasca

Arviat

Hudson Bay

NEWFOUNDLAND

St. John's

VANCOUVER ISLAND

COLUMBIA Mts.

COAST MOUNTAINS

R O C K Y

Calgary

Lake Winnipeg

LES LAURENTIDES

Gulf of St. Lawrence

CAPE BRETON ISLAND

Seattle

CASCADE RANGE

M O U N T A I N S

Winnipeg

Missouri

Lake Superior

Halifax

Montréal

APPALACHIAN MTS.

Boston

G R E A T

Lake Michigan

Lake Huron

Lake Ontario

Toronto

Detroit

New York

SIERRA NEVADA

COAST RANGES

Missouri

Chicago

Lake Erie

Philadelphia

Washington

San Francisco

GREAT BASIN

Great Salt Lake

P L A I N S

Ohio

Denver

Colorado

OZARK PLAT.

Mississippi

Atlanta

Los Angeles

Phoenix

Dallas

ATLANTIC OCEAN

BAJA CALIFORNIA

Golfo de California

Rio Grande

Houston

Tropic of Cancer

SIERRA MADRE OCCIDENTAL

SIERRA MADRE

SIERRA MADRE ORIENTAL

Monterrey

Gulf of Mexico

Miami

Havana

San Juan

HISPANIOLA

CHIHUAHUAN DESERT

Guadalajara

Mexico City

Bahía de Campeche

YUCATAN PENINSULA

G R E A T E R   A N T I L L E S

WEST INDIES

LESSER ANTILLES

SIERRA MADRE DEL SUR

CARIBBEAN SEA

Managua

Lago de Nicaragua

ISTMO DE PANAMA

PACIFIC OCEAN

Geology
Rock type/geological era

- Intrusive
- Metamorphic
- Volcanic, tectonic
- Precambrian sedimentary
- Paleozoic sedimentary
- Mesozoic sedimentary
- Cenozoic sedimentary
- Ice cap

Note: Areas classified as sedimentary
also include some sedimentary/
volcanic areas.

Source: Generalized Geological Map of the
World, Geological Survey of Canada

© 2017 Pearson Education, Inc.

0	200	400	600	800	1000	1200 Miles

| 0 | 200 | 400 | 600 | 800 | 1000 | 1200 | 1400 | 1600 | 1800 | 2000 Kilometers |

Lambert Azimuthal Equal Area Projection
Scale 1:40,000,000
One inch to 640 miles
One cm to 400 km

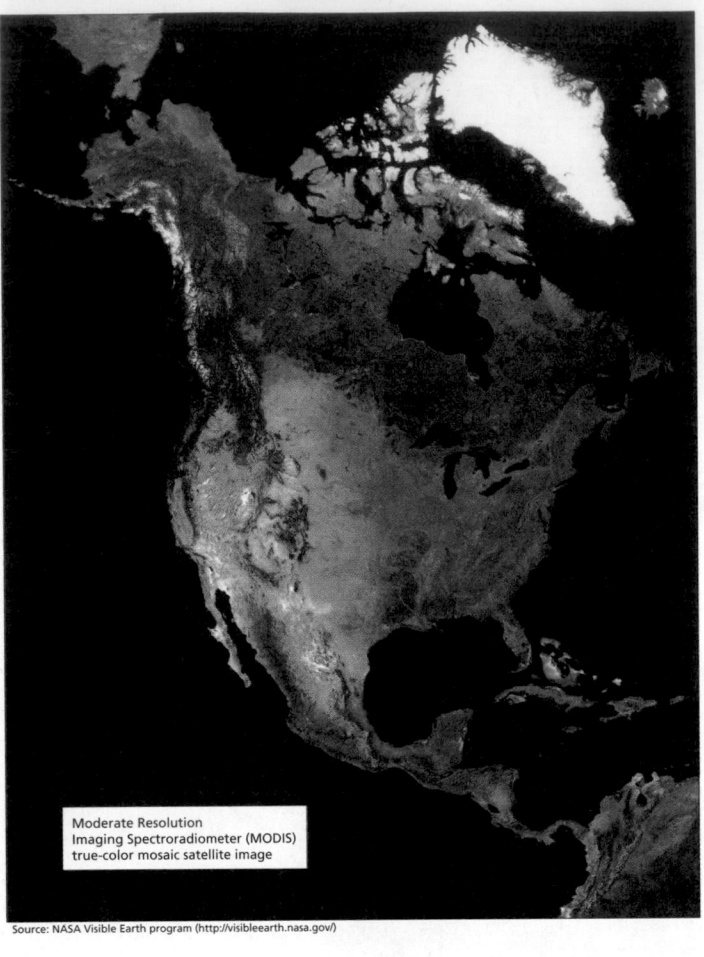

Moderate Resolution
Imaging Spectroradiometer (MODIS)
true-color mosaic satellite image

Source: NASA Visible Earth program (http://visibleearth.nasa.gov/)

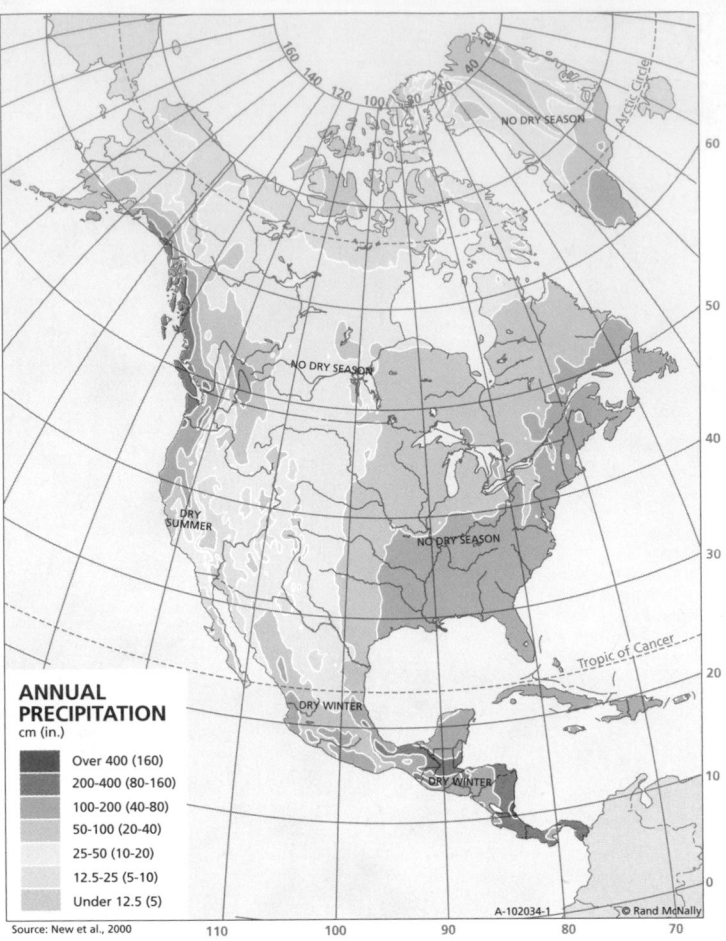

ANNUAL PRECIPITATION
cm (in.)

- Over 400 (160)
- 200-400 (80-160)
- 100-200 (40-80)
- 50-100 (20-40)
- 25-50 (10-20)
- 12.5-25 (5-10)
- Under 12.5 (5)

Source: New et al., 2000

A-102034-1   © Rand McNally

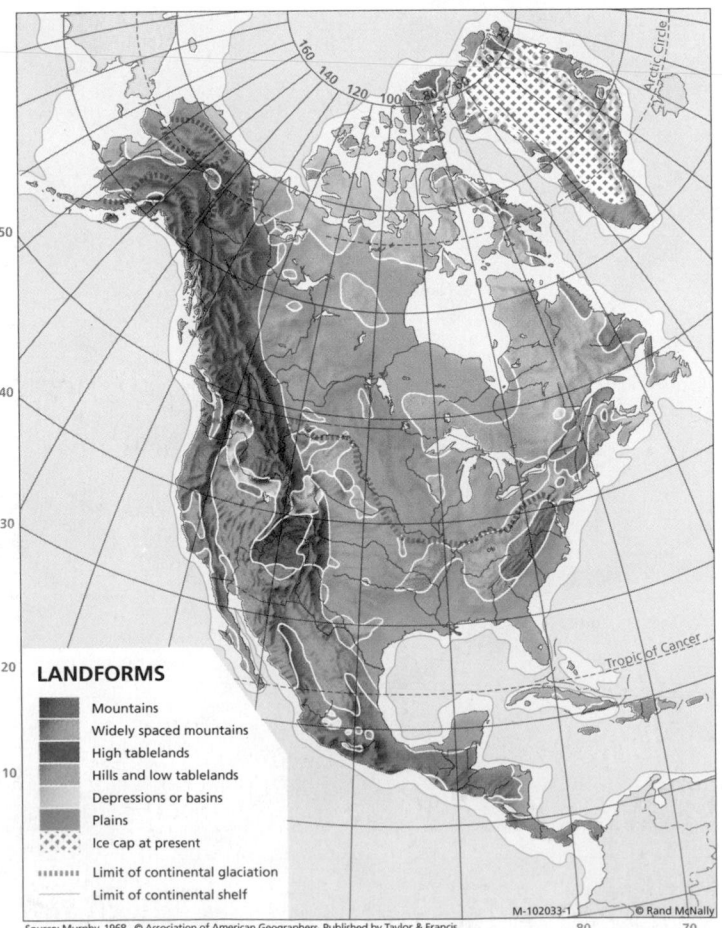

LANDFORMS

- Mountains
- Widely spaced mountains
- High tablelands
- Hills and low tablelands
- Depressions or basins
- Plains
- Ice cap at present

········· Limit of continental glaciation
———— Limit of continental shelf

Source: Murphy, 1968. © Association of American Geographers. Published by Taylor & Francis.
Adapted with permission of the Association of American Geographers.

M-102033-1   © Rand McNally

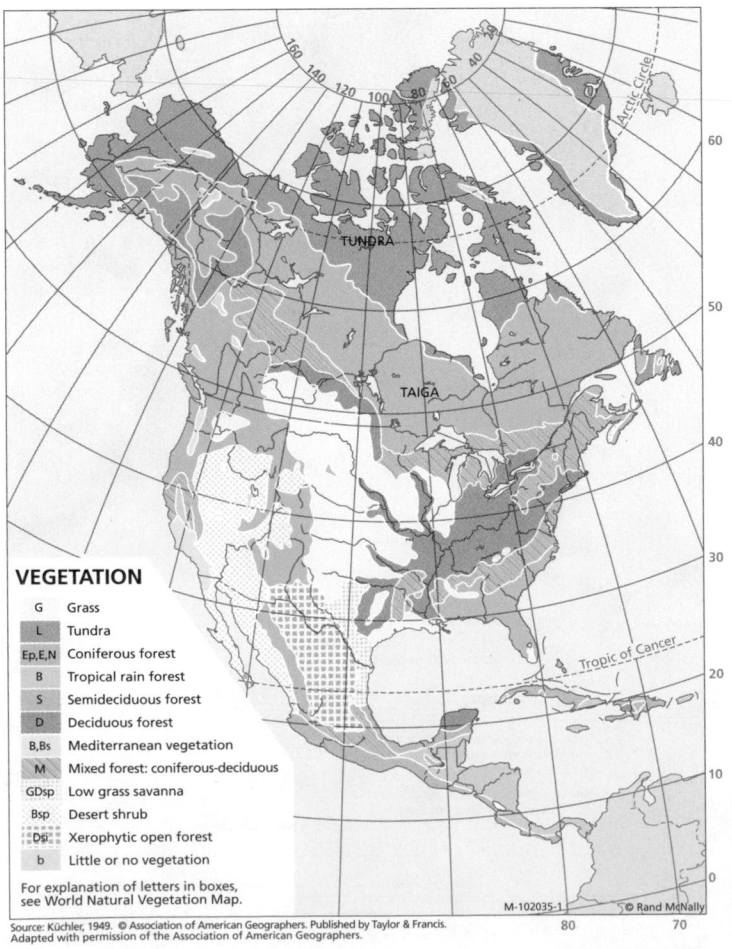

VEGETATION

G	Grass
L	Tundra
Ep,E,N	Coniferous forest
B	Tropical rain forest
S	Semideciduous forest
D	Deciduous forest
B,Bs	Mediterranean vegetation
M	Mixed forest: coniferous-deciduous
GDsp	Low grass savanna
Bsp	Desert shrub
Dsl	Xerophytic open forest
b	Little or no vegetation

For explanation of letters in boxes,
see World Natural Vegetation Map.

Source: Küchler, 1949. © Association of American Geographers. Published by Taylor & Francis.
Adapted with permission of the Association of American Geographers.

M-102035-1   © Rand McNally

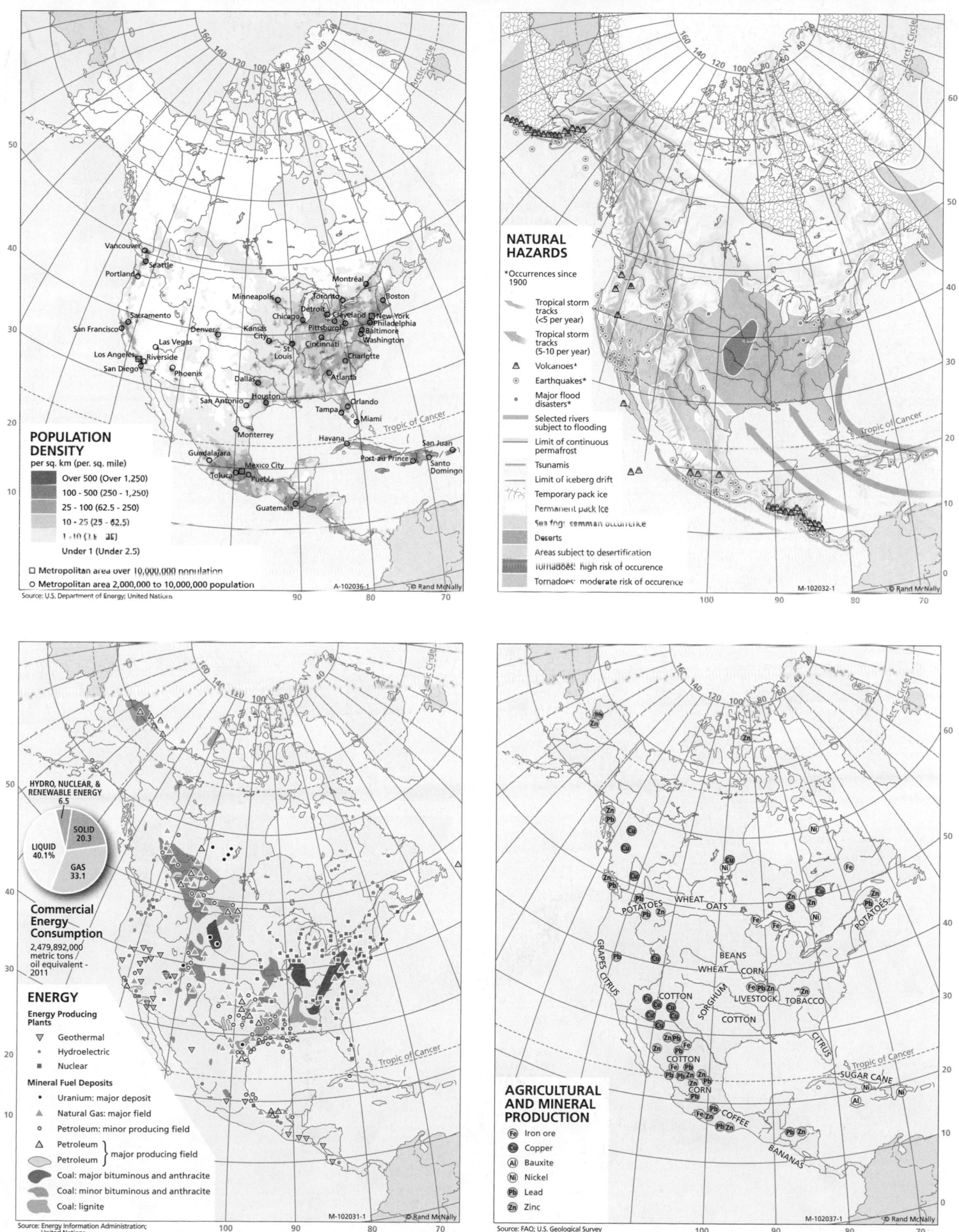

POPULATION DENSITY
per sq. km (per. sq. mile)

- Over 500 (Over 1,250)
- 100 - 500 (250 - 1,250)
- 25 - 100 (62.5 - 250)
- 10 - 25 (25 - 62.5)
- 1 - 10 (2.5 - 25)
- Under 1 (Under 2.5)

□ Metropolitan area over 10,000,000 population
○ Metropolitan area 2,000,000 to 10,000,000 population

Source: U.S. Department of Energy; United Nations

A-102036-1   © Rand McNally

NATURAL HAZARDS

*Occurrences since 1900

- Tropical storm tracks (<5 per year)
- Tropical storm tracks (5-10 per year)
- △ Volcanoes*
- ⊙ Earthquakes*
- • Major flood disasters*
- Selected rivers subject to flooding
- Limit of continuous permafrost
- Tsunamis
- Limit of iceberg drift
- Temporary pack ice
- Permanent pack ice
- Sea fog: common occurrence
- Deserts
- Areas subject to desertification
- Tornadoes: high risk of occurence
- Tornadoes: moderate risk of occurence

M-102032-1   © Rand McNally

Commercial Energy Consumption

2,479,892,000 metric tons / oil equivalent - 2011

HYDRO, NUCLEAR, & RENEWABLE ENERGY 6.5
SOLID 20.3
LIQUID 40.1%
GAS 33.1

ENERGY

Energy Producing Plants
- ▽ Geothermal
- • Hydroelectric
- ■ Nuclear

Mineral Fuel Deposits
- • Uranium: major deposit
- ▲ Natural Gas: major field
- ○ Petroleum: minor producing field
- △ Petroleum }
- ⬭ Petroleum } major producing field
- Coal: major bituminous and anthracite
- Coal: minor bituminous and anthracite
- Coal: lignite

Source: Energy Information Administration; United Nations

M-102031-1   © Rand McNally

AGRICULTURAL AND MINERAL PRODUCTION

- Fe Iron ore
- Cu Copper
- Al Bauxite
- Ni Nickel
- Pb Lead
- Zn Zinc

Source: FAO; U.S. Geological Survey

M-102037-1   © Rand McNally

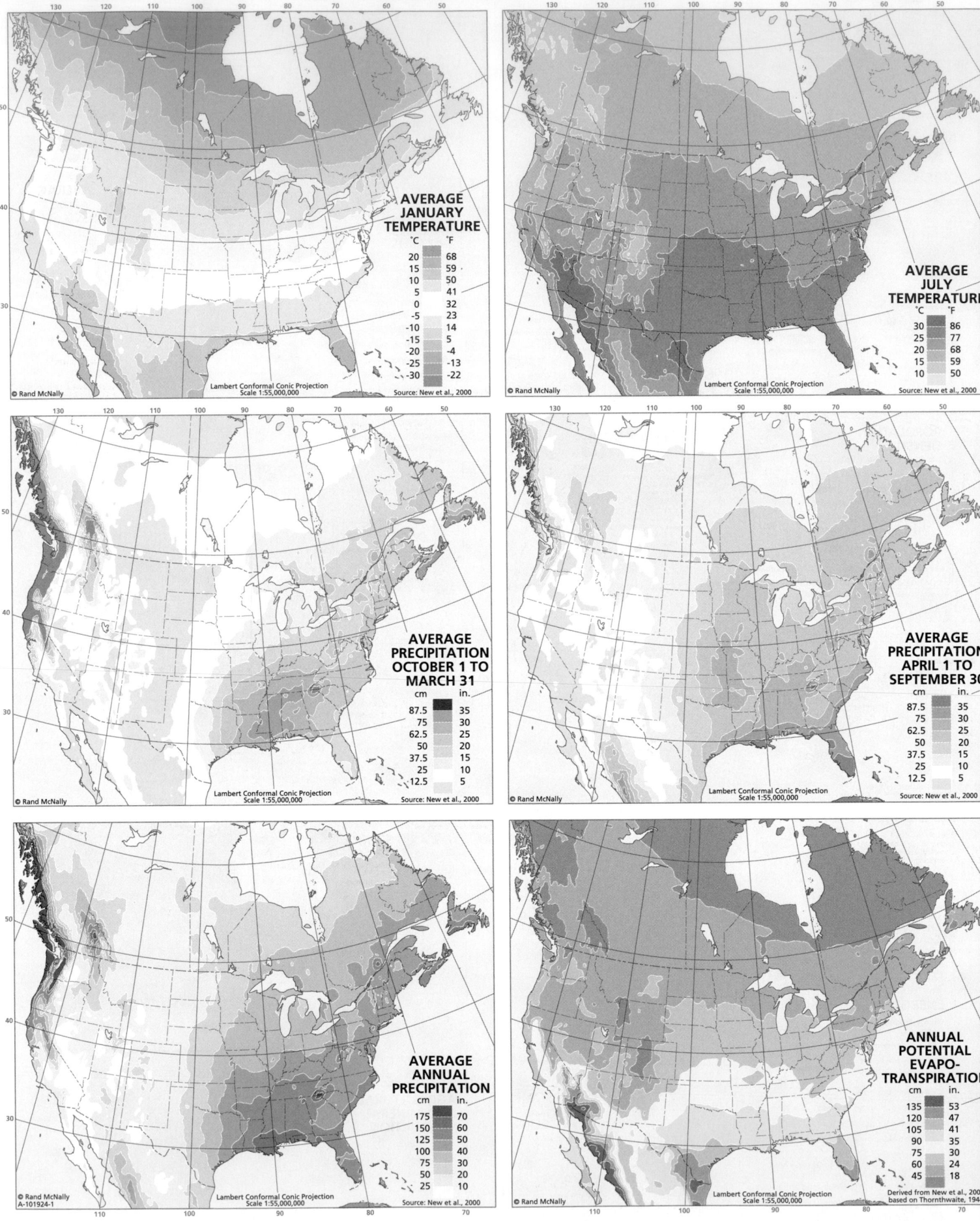

AVERAGE JANUARY TEMPERATURE

°C	°F
20	68
15	59
10	50
5	41
0	32
-5	23
-10	14
-15	5
-20	-4
-25	-13
-30	-22

Lambert Conformal Conic Projection
Scale 1:55,000,000
Source: New et al., 2000

AVERAGE JULY TEMPERATURE

°C	°F
30	86
25	77
20	68
15	59
10	50

Lambert Conformal Conic Projection
Scale 1:55,000,000
Source: New et al., 2000

AVERAGE PRECIPITATION OCTOBER 1 TO MARCH 31

cm	in.
87.5	35
75	30
62.5	25
50	20
37.5	15
25	10
12.5	5

Lambert Conformal Conic Projection
Scale 1:55,000,000
Source: New et al., 2000

AVERAGE PRECIPITATION APRIL 1 TO SEPTEMBER 30

cm	in.
87.5	35
75	30
62.5	25
50	20
37.5	15
25	10
12.5	5

Lambert Conformal Conic Projection
Scale 1:55,000,000
Source: New et al., 2000

AVERAGE ANNUAL PRECIPITATION

cm	in.
175	70
150	60
125	50
100	40
75	30
50	20
25	10

© Rand McNally
A-101924-1
Lambert Conformal Conic Projection
Scale 1:55,000,000
Source: New et al., 2000

ANNUAL POTENTIAL EVAPO-TRANSPIRATION

cm	in.
135	53
120	47
105	41
90	35
75	30
60	24
18	18

Lambert Conformal Conic Projection
Scale 1:55,000,000
Derived from New et al., 2000
based on Thornthwaite, 1944

© Rand McNally

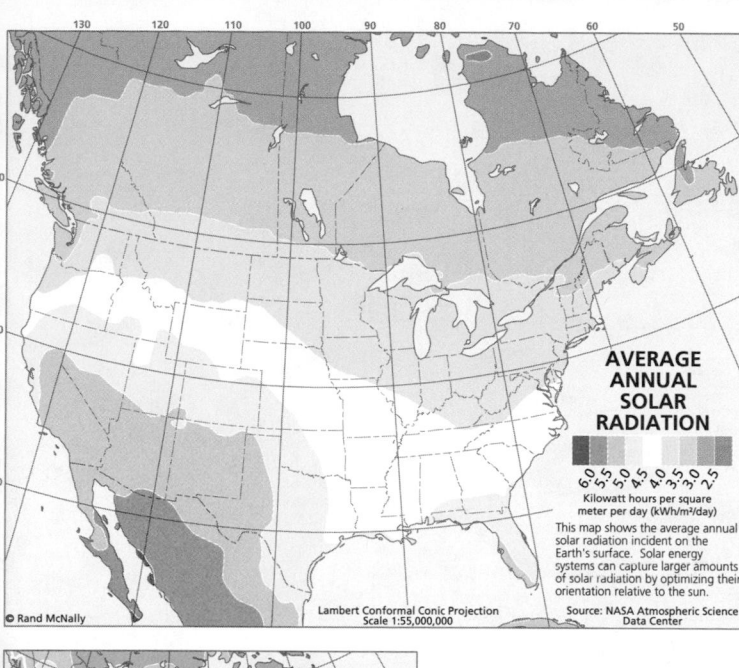

AVERAGE ANNUAL SOLAR RADIATION

6.0 5.5 5.0 4.5 4.0 3.5 3.0 2.5

Kilowatt hours per square meter per day (kWh/m²/day)

This map shows the average annual solar radiation incident on the Earth's surface. Solar energy systems can capture larger amounts of solar radiation by optimizing their orientation relative to the sun.

Source: NASA Atmospheric Science Data Center

© Rand McNally Lambert Conformal Conic Projection Scale 1:55,000,000

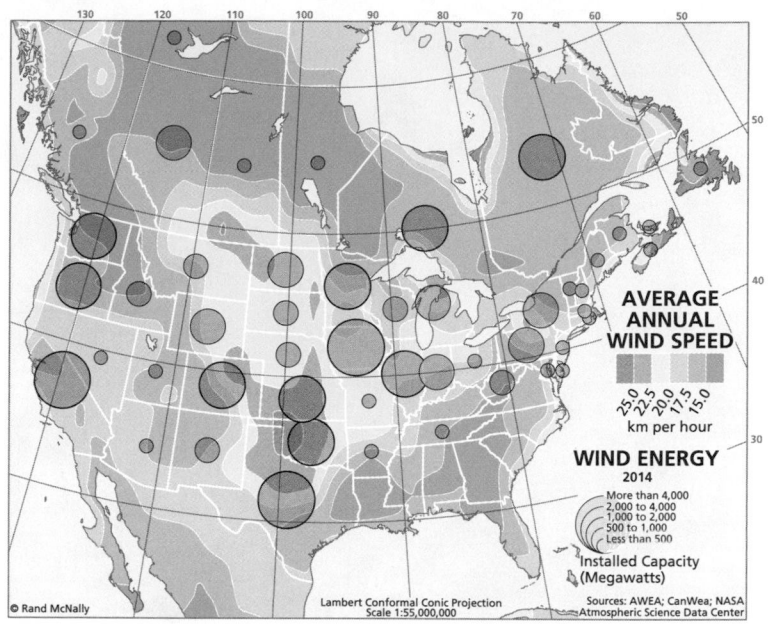

AVERAGE ANNUAL WIND SPEED

25.0 22.5 20.0 17.5 15.0
km per hour

WIND ENERGY
2014

More than 4,000
2,000 to 4,000
1,000 to 2,000
500 to 1,000
Less than 500

Installed Capacity (Megawatts)

© Rand McNally Lambert Conformal Conic Projection Scale 1:55,000,000

Sources: AWEA; CanWea; NASA Atmospheric Science Data Center

EL NIÑO CLIMATE ANOMALIES

These two maps show temperature and precipitation anomalies associated with the 1982-83 El Niño-Southern Oscillation (ENSO) event, one of strongest such events on record. The maps compare temperature and precipitation values for the 1982-83 winter season (October 1, 1982 through March 31, 1983) to winter averages for the 1961-90 baseline period. A positive anomaly indicates a higher than average temperature or precipitation value for 1982-83, while a negative anomaly indicates a lower than average value.

Derived from Brohan et al., 2006, and Hulme, 1998

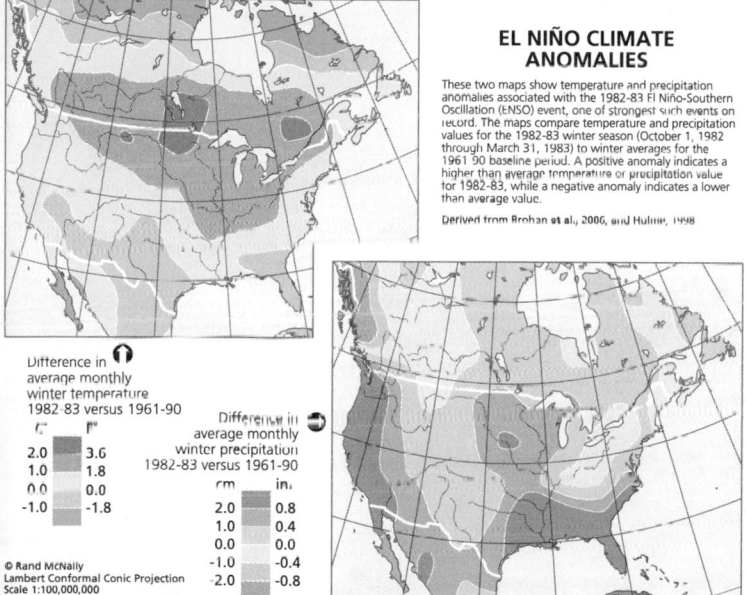

Difference in average monthly winter temperature 1982-83 versus 1961-90

C°	F°
2.0	3.6
1.0	1.8
0.0	0.0
-1.0	-1.8

Difference in average monthly winter precipitation 1982-83 versus 1961-90

cm	in.
2.0	0.8
1.0	0.4
0.0	0.0
-1.0	-0.4
-2.0	-0.8

© Rand McNally
Lambert Conformal Conic Projection
Scale 1:100,000,000

Negative anomaly in the Basin and Range is the result of hotter, less dense, and tectonically active crust.

The narrow positive anomaly than runs down the middle of the U.S. is the mid-continent rift, where dense rocks entered the crust more than a billion years ago.

Basin and Range Rocky Mountains Mid-continent rift Appalachian Mts.

The negative anomalies beneath the Rocky Mountains indicate that the crust has deep roots made up of less dense rock.

GRAVITY ANOMALY

-220 -200 -180 -160 -140 -120 -100 -80 -60 -40 -20 0
mGal

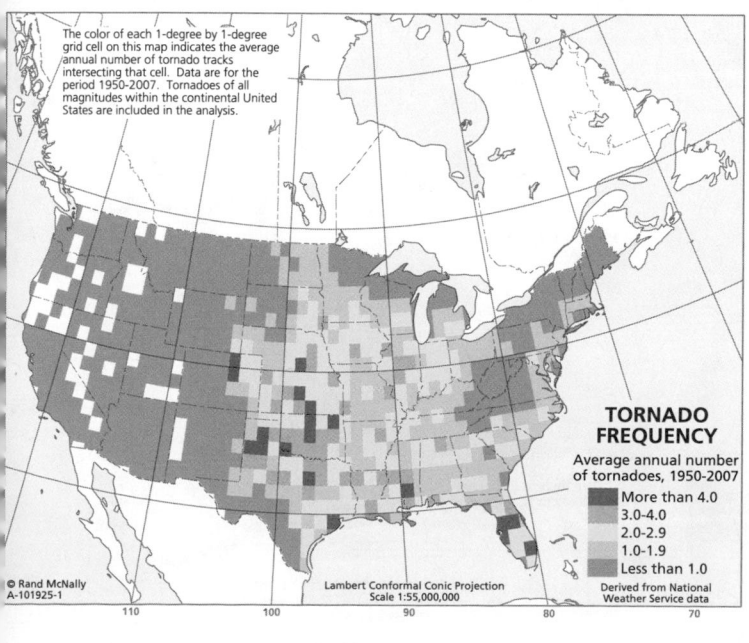

The color of each 1-degree by 1-degree grid cell on this map indicates the average annual number of tornado tracks intersecting that cell. Data are for the period 1950-2007. Tornadoes of all magnitudes within the continental United States are included in the analysis.

TORNADO FREQUENCY

Average annual number of tornadoes, 1950-2007

More than 4.0
3.0-4.0
2.0-2.9
1.0-1.9
Less than 1.0

© Rand McNally
A-101925-1

Lambert Conformal Conic Projection
Scale 1:55,000,000

Derived from National Weather Service data

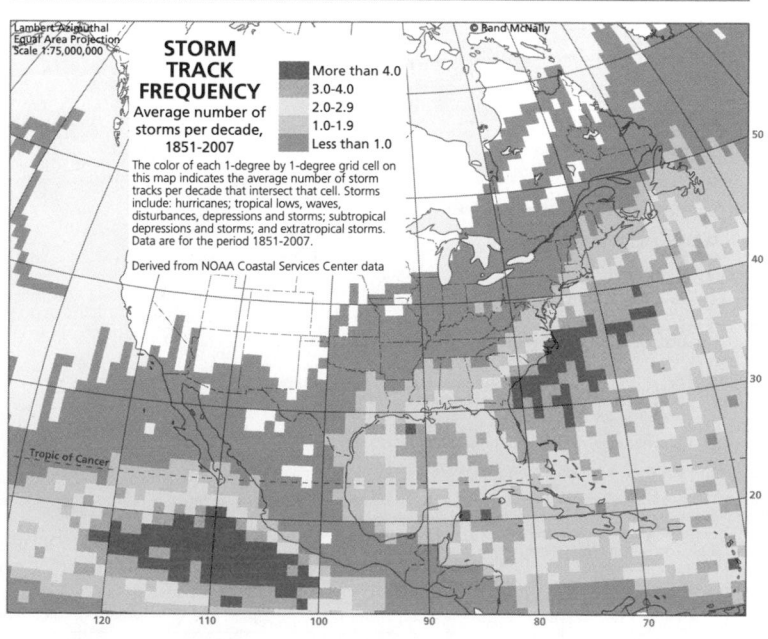

Lambert Azimuthal Equal Area Projection
Scale 1:75,000,000

© Rand McNally

STORM TRACK FREQUENCY

Average number of storms per decade, 1851-2007

More than 4.0
3.0-4.0
2.0-2.9
1.0-1.9
Less than 1.0

The color of each 1-degree by 1-degree grid cell on this map indicates the average number of storm tracks per decade that intersect that cell. Storms include: hurricanes; tropical lows; waves, disturbances; depressions and storms; subtropical depressions and storms; and extratropical storms. Data are for the period 1851-2007.

Derived from NOAA Coastal Services Center data

Tropic of Cancer

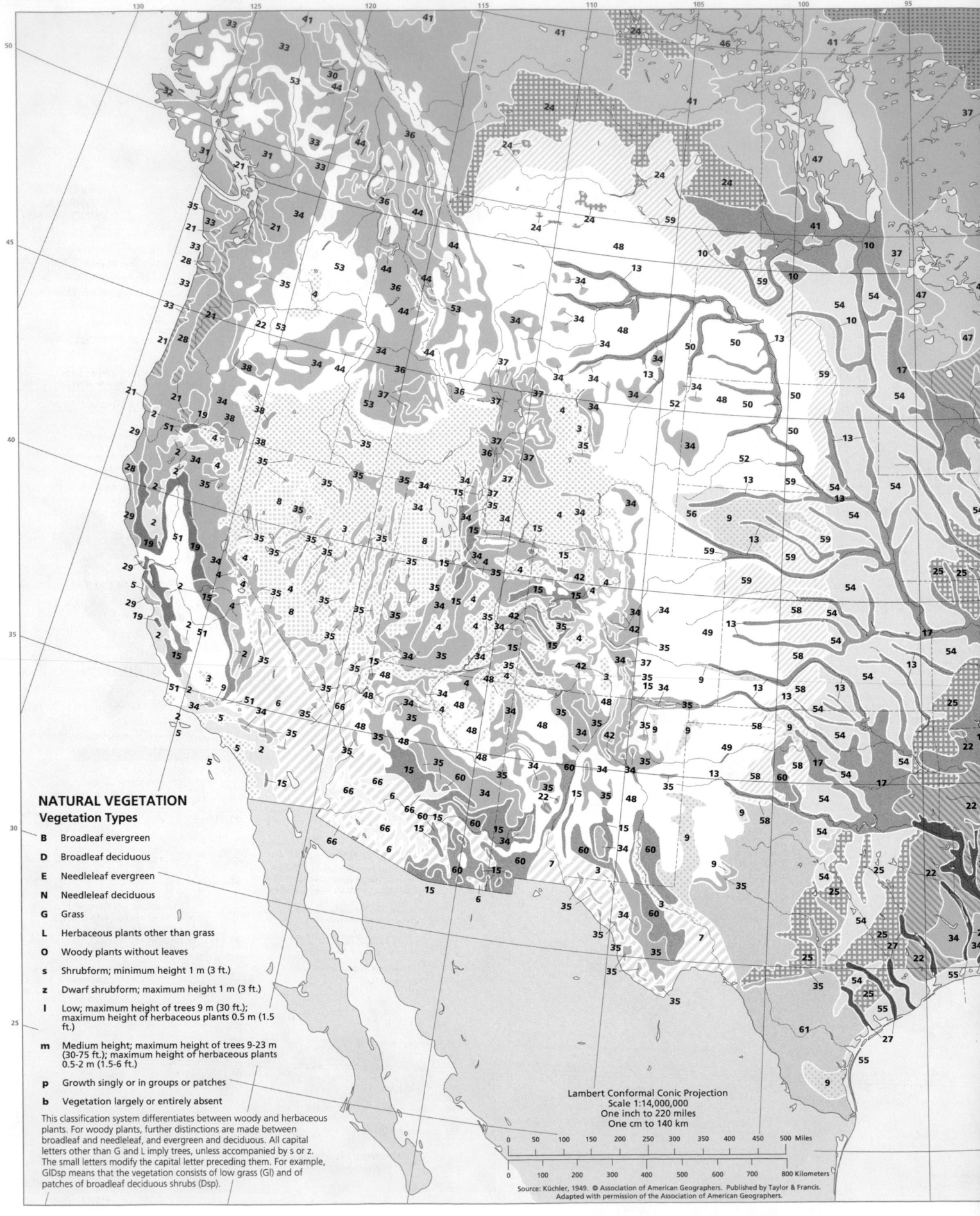

NATURAL VEGETATION
Vegetation Types

B Broadleaf evergreen

D Broadleaf deciduous

E Needleleaf evergreen

N Needleleaf deciduous

G Grass

L Herbaceous plants other than grass

O Woody plants without leaves

s Shrubform; minimum height 1 m (3 ft.)

z Dwarf shrubform; maximum height 1 m (3 ft.)

l Low; maximum height of trees 9 m (30 ft.); maximum height of herbaceous plants 0.5 m (1.5 ft.)

m Medium height; maximum height of trees 9-23 m (30-75 ft.); maximum height of herbaceous plants 0.5-2 m (1.5-6 ft.)

p Growth singly or in groups or patches

b Vegetation largely or entirely absent

This classification system differentiates between woody and herbaceous plants. For woody plants, further distinctions are made between broadleaf and needleleaf, and evergreen and deciduous. All capital letters other than G and L imply trees, unless accompanied by s or z. The small letters modify the capital letter preceding them. For example, GlDsp means that the vegetation consists of low grass (Gl) and of patches of broadleaf deciduous shrubs (Dsp).

Lambert Conformal Conic Projection
Scale 1:14,000,000
One inch to 220 miles
One cm to 140 km

0 50 100 150 200 250 300 350 400 450 500 Miles

0 100 200 300 400 500 600 700 800 Kilometers

Source: Küchler, 1949. © Association of American Geographers. Published by Taylor & Francis. Adapted with permission of the Association of American Geographers.

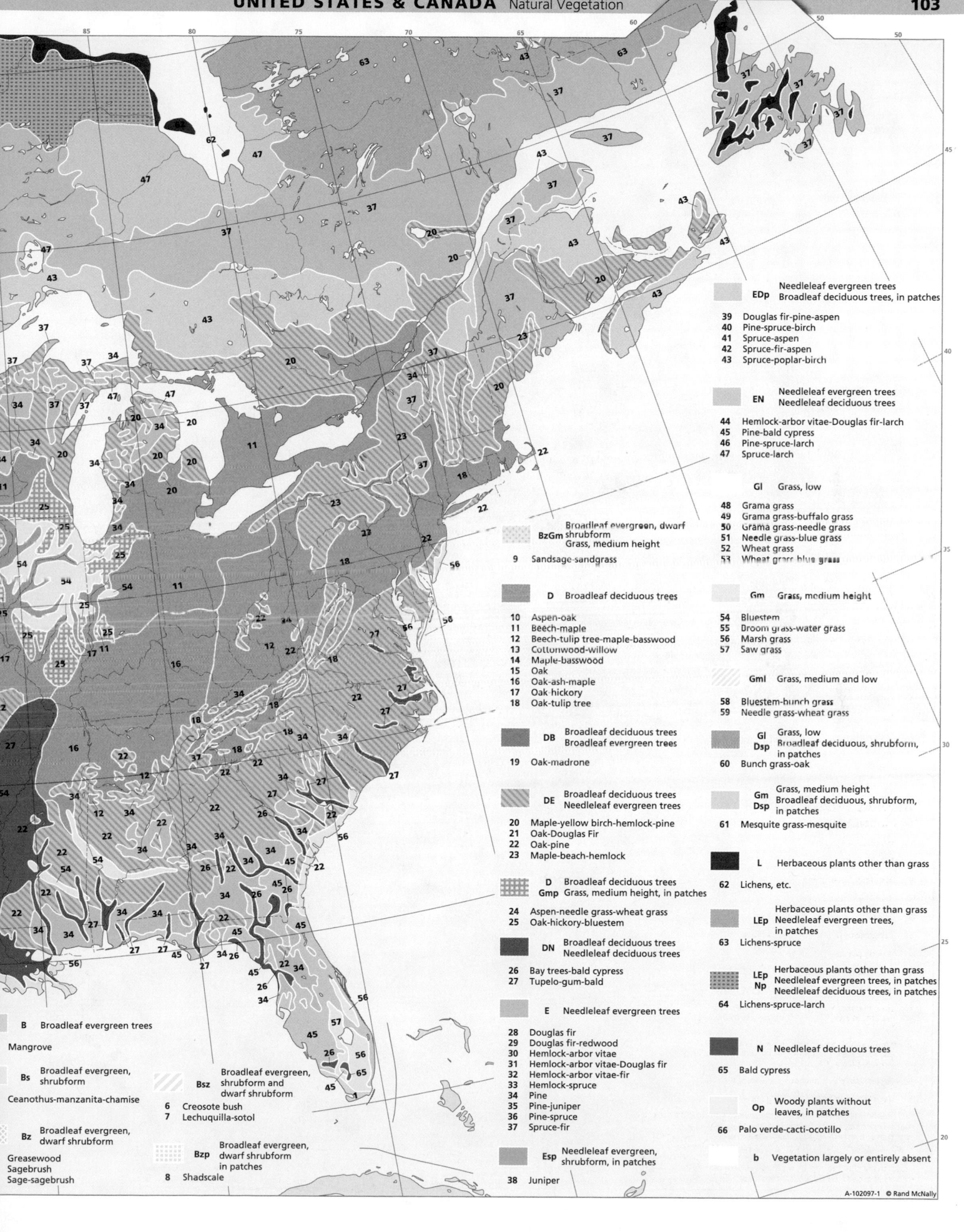

Legend

BzGm Broadleaf evergreen, dwarf shrubform / Grass, medium height
9 Sandsage-sandgrass

D Broadleaf deciduous trees
10 Aspen-oak
11 Beech-maple
12 Beech-tulip tree-maple-basswood
13 Cottonwood-willow
14 Maple-basswood
15 Oak
16 Oak-ash-maple
17 Oak-hickory
18 Oak-tulip tree

DB Broadleaf deciduous trees / Broadleaf evergreen trees
19 Oak-madrone

DE Broadleaf deciduous trees / Needleleaf evergreen trees
20 Maple-yellow birch-hemlock-pine
21 Oak-Douglas Fir
22 Oak-pine
23 Maple-beach-hemlock

D / Gmp Broadleaf deciduous trees / Grass, medium height, in patches
24 Aspen-needle grass-wheat grass
25 Oak-hickory-bluestem

DN Broadleaf deciduous trees / Needleleaf deciduous trees
26 Bay trees-bald cypress
27 Tupelo-gum-bald

E Needleleaf evergreen trees
28 Douglas fir
29 Douglas fir-redwood
30 Hemlock-arbor vitae
31 Hemlock-arbor vitae-Douglas fir
32 Hemlock-arbor vitae-fir
33 Hemlock-spruce
34 Pine
35 Pine-juniper
36 Pine-spruce
37 Spruce-fir

Esp Needleleaf evergreen, shrubform, in patches
38 Juniper

EDp Needleleaf evergreen trees / Broadleaf deciduous trees, in patches
39 Douglas fir-pine-aspen
40 Pine-spruce-birch
41 Spruce-aspen
42 Spruce-fir-aspen
43 Spruce-poplar-birch

EN Needleleaf evergreen trees / Needleleaf deciduous trees
44 Hemlock-arbor vitae-Douglas fir-larch
45 Pine-bald cypress
46 Pine-spruce-larch
47 Spruce-larch

Gl Grass, low
48 Grama grass
49 Grama grass-buffalo grass
50 Grama grass-needle grass
51 Needle grass-blue grass
52 Wheat grass
53 Wheat grass-blue grass

Gm Grass, medium height
54 Bluestem
55 Broom grass-water grass
56 Marsh grass
57 Saw grass

Gml Grass, medium and low
58 Bluestem-bunch grass
59 Needle grass-wheat grass

Gl / Dsp Grass, low / Broadleaf deciduous, shrubform, in patches
60 Bunch grass-oak

Gm / Dsp Grass, medium height / Broadleaf deciduous, shrubform, in patches
61 Mesquite grass-mesquite

L Herbaceous plants other than grass
62 Lichens, etc.

LEp Herbaceous plants other than grass / Needleleaf evergreen trees, in patches
63 Lichens-spruce

LEp / Np Herbaceous plants other than grass / Needleleaf evergreen trees, in patches / Needleleaf deciduous trees, in patches
64 Lichens-spruce-larch

N Needleleaf deciduous trees
65 Bald cypress

Op Woody plants without leaves, in patches
66 Palo verde-cacti-ocotillo

b Vegetation largely or entirely absent

B Broadleaf evergreen trees
Mangrove

Bs Broadleaf evergreen, shrubform
Ceanothus-manzanita-chamise

Bsz Broadleaf evergreen, shrubform and dwarf shrubform
6 Creosote bush
7 Lechuquilla-sotol

Bz Broadleaf evergreen, dwarf shrubform
Greasewood
Sagebrush
Sage-sagebrush

Bzp Broadleaf evergreen, dwarf shrubform, in patches
8 Shadscale

Lambert Conformal Conic Projection
Scale 1:15,000,000
One inch to 237 miles
One cm to 150 km

0   100   200   300 Miles

0   100   200   300   400 Kilometers

© Rand McNally

M-100990-1

AGRICULTURE

Dairying

Fruits and Vegetables

Wheat, Barley, and Oilseeds

Cash Corn and Soybeans

Tobacco

Cotton

Livestock and Feed Grains: Beef

Livestock and Feed Grains: Hogs

Livestock and Feed Grains: Poultry

Livestock and Feed Grains: Mixed

Specialty Crops (Peanuts, Potatoes, Rice, Sugar)

Western Livestock Ranching

Western Feedlots

Agriculture and Forestry

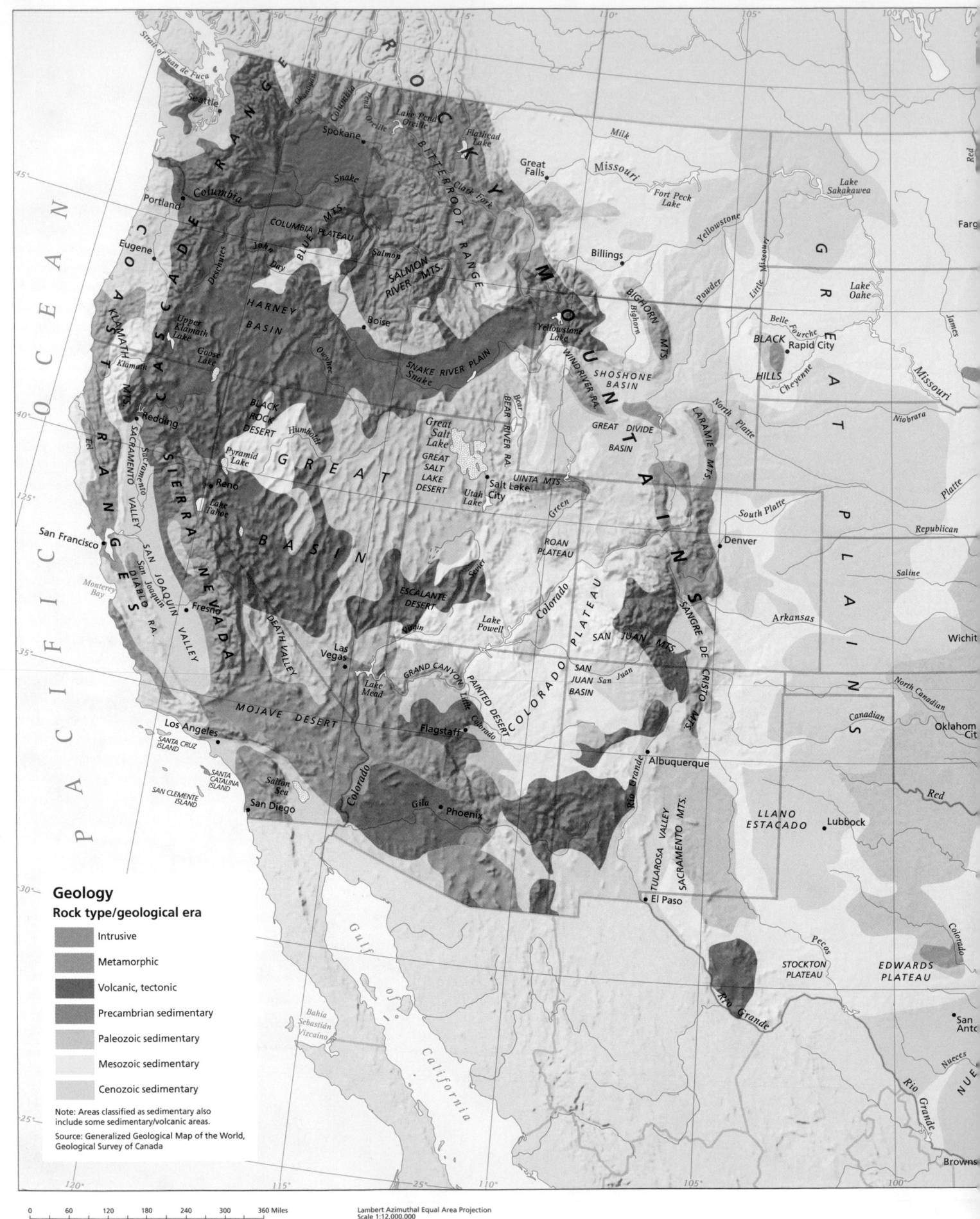

Geology

Rock type/geological era

Intrusive

Metamorphic

Volcanic, tectonic

Precambrian sedimentary

Paleozoic sedimentary

Mesozoic sedimentary

Cenozoic sedimentary

Note: Areas classified as sedimentary also
include some sedimentary/volcanic areas.

Source: Generalized Geological Map of the World,
Geological Survey of Canada

0   60   120   180   240   300   360 Miles

0   60   120 180 240 300 360 420 480 540 600 Kilometers

Lambert Azimuthal Equal Area Projection
Scale 1:12,000,000
One inch to 190 miles
One cm to 120 km

Lake of
the Woods
Rainy
ISLE
ROYALE
MICHIPICOTEN
ISLAND
Upper
Red Lake
ower
d Lake
LAKE SUPERIOR
Sault Ste. Marie
Leech
Lake
Duluth
KEWEENAW PEN.
MANITOULIN
ISLAND
Beaver
ISLAND
Georgian Bay
St. Lawrence
Lake
Champlain
ADIRONDACK
MTS.
Saint John
Bay of Fundy
Penobscot
Portland
Gulf of
Maine
Mille
Lacs Lake
Green Bay
DOOR PEN.
Saginaw Bay
LAKE HURON
LAKE ONTARIO
Connecticut
Boston
CAPE
COD
Minneapolis
Mississippi
St. Croix
LAKE MICHIGAN
Lake
Winnebago
Muskegon
Grand
Lake
St. Clair
Detroit
LAKE ERIE
Cleveland
Buffalo
Allegheny
CATSKILL
MTS.
Hartford
Hudson
New
York
MARTHA'S
VINEYARD
NANTUCKET
ISLAND
Minnesota
Milwaukee
Wisconsin
Chicago
Rock
Fox
Kankakee
Maumee
Pittsburgh
PLATEAU
Delaware
Philadelphia
LONG
ISLAND
ux Falls
Wabash
Columbus
Scioto
Ohio
Washington
Susquehanna
Potomac
DELMARVA PENINSULA
Delaware Bay
Omaha
Des
Moines
Des Moines
Mississippi
Illinois
Indianapolis
Cincinnati
ALLEGHENY
APPALACHIAN MOUNTAINS
Chesapeake Ba
Kanawha
James
Kansas
opeka
Kansas City
Missouri
St
Louis
Kaskaskia
Kentucky
Neu
Norfolk
MOUNTAINS
Lake of
the Ozarks
OZARK PLATEAU
Ohio
CUMBERLAND PLATEAU
Holston
BLUE RIDGE
Yadkin
Roanoke
HATTERAS
ISLAND
Raleigh
Neuse
Arkansas
White
Nashville
Cumberland
Tennessee
APPALACHIAN
PIEDMONT
Charlotte
Cape Fear
BOSTON MTS.
Memphis
Atlanta
J. Strom
Thurmond
Res.
Savannah
Great Pee Dee
OUACHITA MTS.
Little
Rock
Arkansas
Birmingham
Coosa
Flint
Lake
Marion
Charleston
Santee
Kansas
Ouachita
Tombigbee
Orange
HILTON HEAD ISLAND
Dallas
Sabine
Red
Alabama
Mobile
Chattahoochee
Ocmulgee
Altamaha
Trinity
Toledo
Bend
Reservoir
Neches
Mobile
Lake
Pontchartrain
Apalachicola
St. Marys
Jacksonville
Houston
New
Orleans
MISSISSIPPI
DELTA
Pearl
Suwannee
St. Johns
LAINS
GALVESTON
ISLAND
Yazoo
Mississippi
Brazos
Tampa
GULF OF MEXICO
Tampa Bay
Lake
Okeechobee
Miami
CAPE SABLE
KEY
WEST
KEY
LARGO
FLORIDA KEYS
Straits of Florida
Tropic of Cancer
ATLANTIC OCEAN

© 2017 Pearson Education, Inc.

PACIFIC RIM NATL. PARK
Vancouver
Victoria
MAKAH
QUINAULT
OLYMPIC N.P.
Olympic
Mt. Baker
Okanogan
NORTH CASCADES N.P.
Snoqualmie
Seattle
MT. RAINIER N.P.
Wenatchee
Spokane
COLVILLE
SPOKANE
Colville
Kaniksu
Coeur d'Alene
Kootenai
GLACIER N.P.
Flathead
Lewis and Clark
ROCKY BOYS
BLACKFEET
FORT BELKNAP
GRASSLANDS N.P.
C.F.B. Suffield
PIEGAN
WATERTON LAKES N.P.
BLOOD
Calgary
KOOTENAY N.P.
BANFF N.P.
BLACKFOOT
C.F.B. Shilo
Saskatoon
Regina
RIDING MOUNTAIN N.P.
TURTLE MOUNTAIN
Winnipeg

Portland
Mt. Hood
Siuslaw
GIFFORD PINCHOT
YAKAMA
HANFORD REACH N.M.
UMATILLA
Umatilla
WARM SPRINGS
HELLS CANYON N.R.A.
Wallowa
NEZ PERCE
Nez Perce
Clearwater
St. Joe
Coeur d'Alene
LOLO
FLATHEAD
Lewis and Clark
UPPER MISSOURI RIVER BREAKS N.M.
Charles M. Russell
FORT PECK
FORT BERTHOLD
THEODORE ROOSEVELT NATL. PARK
Bismarck
SPIRIT LAKE
Sheyenne

Siuslaw
Willamette
Ochoco
Deschutes
Winema
CRATER LAKE N.P.
Rogue River
Siskiyou
Umpqua
REDWOOD N.P.
Klamath
Modoc
Fremont
Sheldon
Malheur
Malheur
Payette
Boise
Boise
SAWTOOTH N.R.A.
Sawtooth
Salmon-Challis
Bitterroot
Beaverhead-Deerlodge
Butte
Helena
Gallatin
Lewis and Clark
POMPEYS PILLAR N.M.
NORTHERN CHEYENNE
BIGHORN CANYON N.R.A.
Custer
CROW
Custer
Grand River
STANDING ROCK
CHEYENNE RIVER
DEVILS TOWER N.M.
Black Hills
Rapid City
BADLANDS N.P.
Ft. Pierre
Buffalo Gap
CROW CREEK
PINE RIDGE
ROSEBUD
LOWER BRULE
Sioux Falls
YANKTON
Yankton

Shasta
Trinity
Six Rivers
Lassen
LASSEN VOLCANIC N.P.
Plumas
Tahoe
Reno
PYRAMID LAKE
Humboldt/Toiyabe
Hart Mountain
Humboldt/Toiyabe
HAGERMAN FOSSIL BEDS N.M.
CRATERS OF THE MOON N.M.
DUCK VALLEY
Caribou
FORT HALL
GRAND TETON N.P.
Teton
Targhee
YELLOWSTONE N.P.
Shoshone
Bighorn
WIND RIVER
Bridger
Medicine Bow
Oglala
Nebraska
AGATE FOSSIL BEDS N.M.
SCOTTS BLUFF N.M.
SANTEE
WINNEBAGO
OMAHA
Nebraska

POINT REYES N.S.
Sacramento
BERRYESSA SNOW MOUNTAIN N.M.
Mendocino
Stanislaus
San Francisco
San Jose
YOSEMITE N.P.
Sierra
Inyo
Humboldt/Toiyabe
Eldorado
Walker River
GOSHUTE
Utah Test & Training Range
Salt Lake City
Wasatch
Uinta
Ashley
FLAMING GORGE N.R.A.
Medicine Bow
FOSSIL BUTTE N.M.
Dugway Proving Ground
Manti-La-Sal
UINTAH AND OURAY
DINOSAUR N.M.
White River
Routt
Roosevelt
Pawnee
ROCKY MOUNTAIN N.P.
Arapaho
Denver
Cheyenne
Omaha

PINNACLES N.P.
Los Padres
Fort Hunter Liggett
KINGS CANYON N.P.
SEQUOIA N.P.
TULE RIVER
Sequoia
China Lake N.A.W.S.
DEATH VALLEY N.P.
Nevada National Security Site
Nellis Air Force Range
GREAT BASIN N.P.
Humboldt/Toiyabe
Fishlake
BASIN AND RANGE N.M.
Desert
Dixie
GRAND STAIRCASE-ESCALANTE N.M.
CAPITOL REEF N.P.
CANYONLANDS N.P.
ARCHES N.P.
Grand Mesa
Gunnison
Pike
San Isabel
Ft. Carson
Uncompahgre
CANYONS OF THE ANCIENTS N.M.
GREAT SAND DUNES N.P. AND PRES.
Rio Grande
Comanche
Cimarron
POTTAWATOMI
Ft. Riley
KICKA
Wichita

CARRIZO PLAIN N.M.
Ft. Irwin
Edwards A.F.B.
MOJAVE N. PRES.
Las Vegas
LAKE MEAD N.R.A.
HUALAPAI
GRAND CANYON-PARASHANT N.M.
Kaibab
GRAND CANYON N.P.
ZION N.P.
GLEN CANYON N.R.A.
NAVAJO N.M.
HOPI
NAVAJO
San Juan
SOUTHERN UTE
UTE MTN.
CANYON DE CHELLY N.M.
Santa Fe
Carson
JICARILLA APACHE
RIO GRANDE DEL NORTE N.M.
CAPULIN VOLCANO N.M.
Kiowa
Rita Blanca
Black Kettle
OSAG
Tulsa

Vandenberg A.F.B.
CHANNEL ISLANDS NATL. PARK
Los Angeles
Los Padres
Angeles
San Bernardino
Twentynine Palms M.C.A.G.C.C.
CHEMEHUEVI VALLEY
Prescott
Coconino
Sitgreaves
ZUNI
EL MALPAIS N.M.
Cibola
ACOMA
Albuquerque
FT. UNION N.M.
Lincoln
Comanche
Wichita

Camp Pendleton
Cleveland
San Diego
CABRILLO N.M.
JOSHUA TREE N.P.
Chocolate Mtn. Gunnery Range
COLORADO RIVER
Yuma Proving Ground
Kofa
Phoenix
Tonto
FORT APACHE
Apache
SAN CARLOS
Gila
Cibola
White Sands Missile Range
Lincoln
MESCALERO
Lincoln
Lubbock
Lyndon B. Johnson
Fort Worth

Goldwater A.F. Range
SONORAN DESERT
Cabeza Prieta
ORGAN PIPE CACTUS N.M.
SAGUARO N.P.
TOHONO O'ODHAM
Tucson
Coronado
ORGAN MOUNTAINS-DESERT PEAKS N.M.
WHITE SANDS N.M.
Fort Bliss
El Paso
GUADALUPE MOUNTAINS N.P.
CARLSBAD CAVERNS N.P.
Ft. Sill
CHICKASAW N.R.A.
Oklahoma City
WACO MAMMOTH N.M.
Ft. Hood
Austin
BIG BEND NATL. PARK
AMISTAD N.R.A.
San Antonio
Corpus Christi
PADRE ISL. N.S.
Dallas

Alaska inset:
NOATAK N. PRES.
GATES OF THE ARCTIC N.P. & PRES.
IVVAVIK N.P.
TUKTUT NOGAIT N.P.
CAPE KRUSENSTERN N.M.
KOBUK VALLEY N.P.
Arctic
VUNTUT N.P.
YUKON-CHARLEY RIVERS N. PRES.
NAATS'IHCH'OH NATIONAL PARK AND RESERVE
Selawik
WHITE MTS. N.R.A.
Kanuti
Yukon Flats
Arctic Circle
NAHANNI N.P.
BERING LAND BRIDGE N. PRES.
Koyukuk
Nowitna
Innoko
Fairbanks
Fort Greeley
Tetlin
Whitehorse
Yukon Delta
Togiak
WRANGELL-ST. ELIAS N.P. & PRES.
KLUANE N.P.
Anchorage
LAKE CLARK N.P. & PRES.
Kenai
Chugach
KENAI FORDS N.P.
Tongass
GLACIER BAY N.P. & PRES.
Juneau
Bécharof
Alaska Peninsula
KATMAI N.P. & PRES.
ANIAKCHAK N.M. & PRES.
Kodiak
Izembek
Alaska Peninsula
ADMIRALTY N.M.
MISTY FJORDS N.M.
Tongass
ANNETTE ISLAND

Hawaii inset:
Kilauea Point
Honolulu
HALEAKALĀ N.P.
Hakalau Forest
HAWAI'I VOLCANOES N.P.

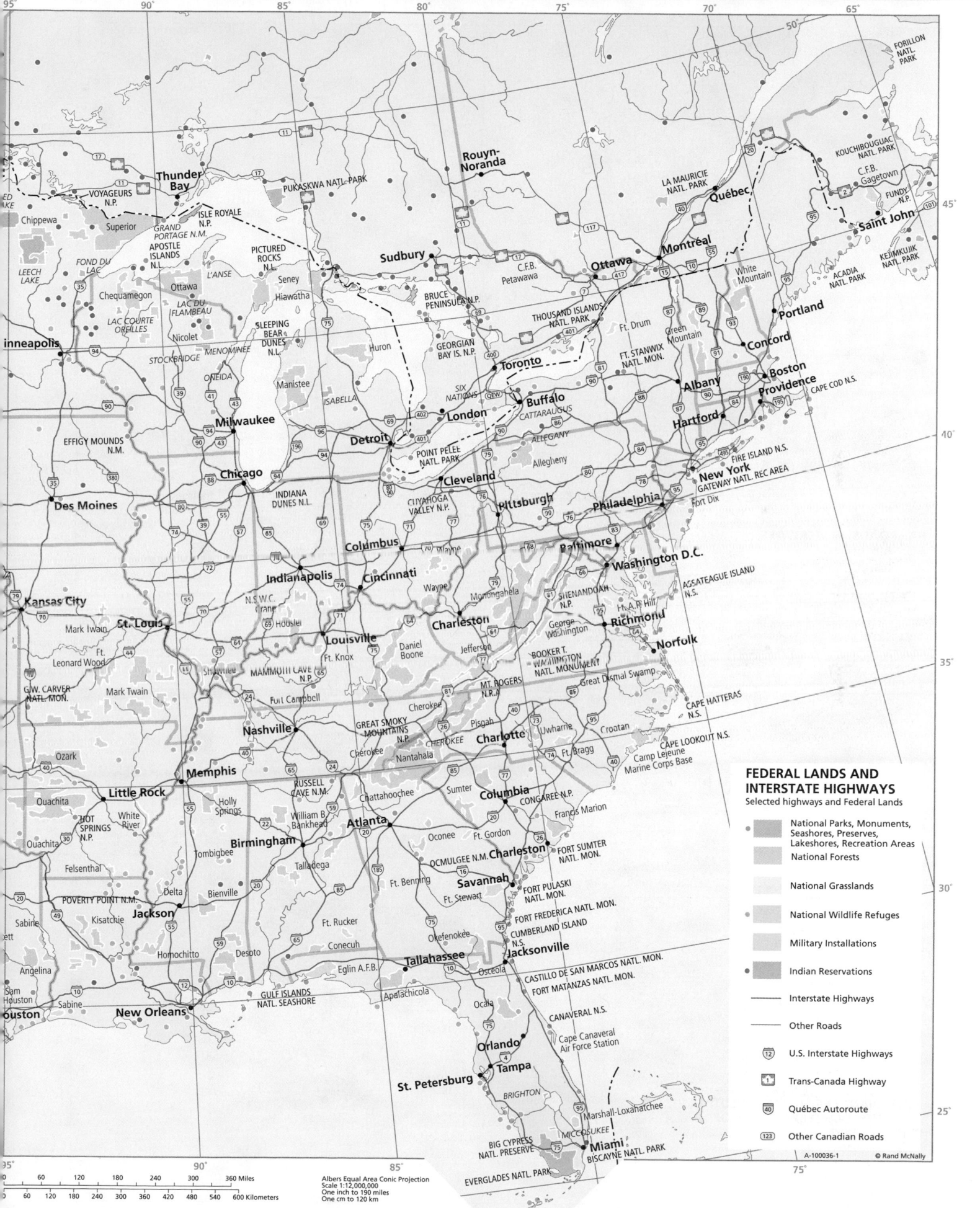

FORILLON NATL. PARK

KOUCHIBOUGUAC NATL. PARK

C.F.B. Gagetown

FUNDY N.P.

Saint John

KEJIMKUJIK NATL. PARK

ACADIA NATL. PARK

White Mountain

Portland

Concord

Boston

Providence

CAPE COD N.S.

Hartford

New York

GATEWAY NATL. REC. AREA

FIRE ISLAND N.S.

Albany

LA MAURICIE NATL. PARK

Québec

Montréal

Ottawa

C.F.B. Petawawa

Rouyn-Noranda

Thunder Bay

VOYAGEURS N.P.

GRAND PORTAGE N.M.

ISLE ROYALE N.P.

APOSTLE ISLANDS N.L.

PICTURED ROCKS N.L.

PUKASKWA NATL. PARK

Sudbury

BRUCE PENINSULA N.P.

GEORGIAN BAY IS. N.P.

Toronto

London

Buffalo

SIX NATIONS

CATTARAUGUS

ALLEGANY

POINT PELEE NATL. PARK

Detroit

Cleveland

Pittsburgh

Philadelphia

Fort Dix

Chippewa

Superior

Chequamegon

LEECH LAKE

FOND DU LAC

Ottawa

L'ANSE

Seney

Hiawatha

LAC COURTE OREILLES

LAC DU FLAMBEAU

Nicolet

Minneapolis

STOCKBRIDGE

MENOMINEE

ONEIDA

SLEEPING BEAR DUNES N.L.

Manistee

ISABELLA

Huron

Milwaukee

Chicago

EFFIGY MOUNDS N.M.

Des Moines

INDIANA DUNES N.L.

CUYAHOGA VALLEY N.P.

Columbus

Allegheny

Wayne

Wayne

Monongahela

Baltimore

Washington D.C.

ASSATEAGUE ISLAND N.S.

SHENANDOAH N.P.

H.A.P. Hill

George Washington

Richmond

Norfolk

Kansas City

St. Louis

N.S.W.C. Crane

Hoosier

Indianapolis

Cincinnati

Charleston

Jefferson

Louisville

Ft. Knox

Daniel Boone

Mark Twain

Ft. Leonard Wood

SHAWNEE

MAMMOTH CAVE N.P.

Fort Campbell

BOOKER T. WASHINGTON NATL. MONUMENT

Great Dismal Swamp

G.W. CARVER NATL. MON.

Mark Twain

Ozark

MT. ROGERS N.R.A.

CAPE HATTERAS N.S.

Cherokee

Nashville

GREAT SMOKY MOUNTAINS N.P.

Pisgah

Cherokee

Charlotte

Uwharrie

Croatan

CAPE LOOKOUT N.S.

Nantahala

Ft. Bragg

Camp Lejeune Marine Corps Base

Memphis

Little Rock

Ouachita

HOT SPRINGS N.P.

Ouachita

RUSSELL CAVE N.M.

Holly Springs

WILLIAM B. BANKHEAD

Columbia

CONGAREE N.P.

Francis Marion

White River

Chattahoochee

Sumter

Atlanta

Oconee

Ft. Gordon

Charleston

FORT SUMTER NATL. MON.

Birmingham

Talladega

OCMULGEE N.M.

Ft. Benning

FORT PULASKI NATL. MON.

Tombigbee

Savannah

Ft. Stewart

Felsenthal

POVERTY POINT N.M.

Delta

Bienville

Jackson

Kisatchie

Ft. Rucker

FORT FREDERICA NATL. MON.

CUMBERLAND ISLAND N.S.

Homochitto

Desoto

Conecuh

Okefenokee

Tallahassee

Jacksonville

CASTILLO DE SAN MARCOS NATL. MON.

FORT MATANZAS NATL. MON.

Eglin A.F.B.

Osceola

Sabine

Angelina

Sam Houston

Sabine

GULF ISLANDS NATL. SEASHORE

Apalachicola

Ocala

CANAVERAL N.S.

Houston

New Orleans

Orlando

Cape Canaveral Air Force Station

Tampa

St. Petersburg

BRIGHTON

Marshall-Loxahatchee

MICCOSUKEE

BIG CYPRESS NATL. PRESERVE

Miami

BISCAYNE NATL. PARK

EVERGLADES NATL. PARK

Ft. Drum

THOUSAND ISLANDS NATL. PARK

FT. STANWIX NATL. MON.

Green Mountain

FEDERAL LANDS AND INTERSTATE HIGHWAYS
Selected highways and Federal Lands

- National Parks, Monuments, Seashores, Preserves, Lakeshores, Recreation Areas
- National Forests
- National Grasslands
- National Wildlife Refuges
- Military Installations
- Indian Reservations
- ——— Interstate Highways
- ——— Other Roads
- U.S. Interstate Highways
- Trans-Canada Highway
- Québec Autoroute
- Other Canadian Roads

Albers Equal Area Conic Projection
Scale 1:12,000,000
One inch to 190 miles
One cm to 120 km

0 60 120 180 240 300 360 Miles
0 60 120 180 240 300 360 420 480 540 600 Kilometers

A-100036-1   © Rand McNally

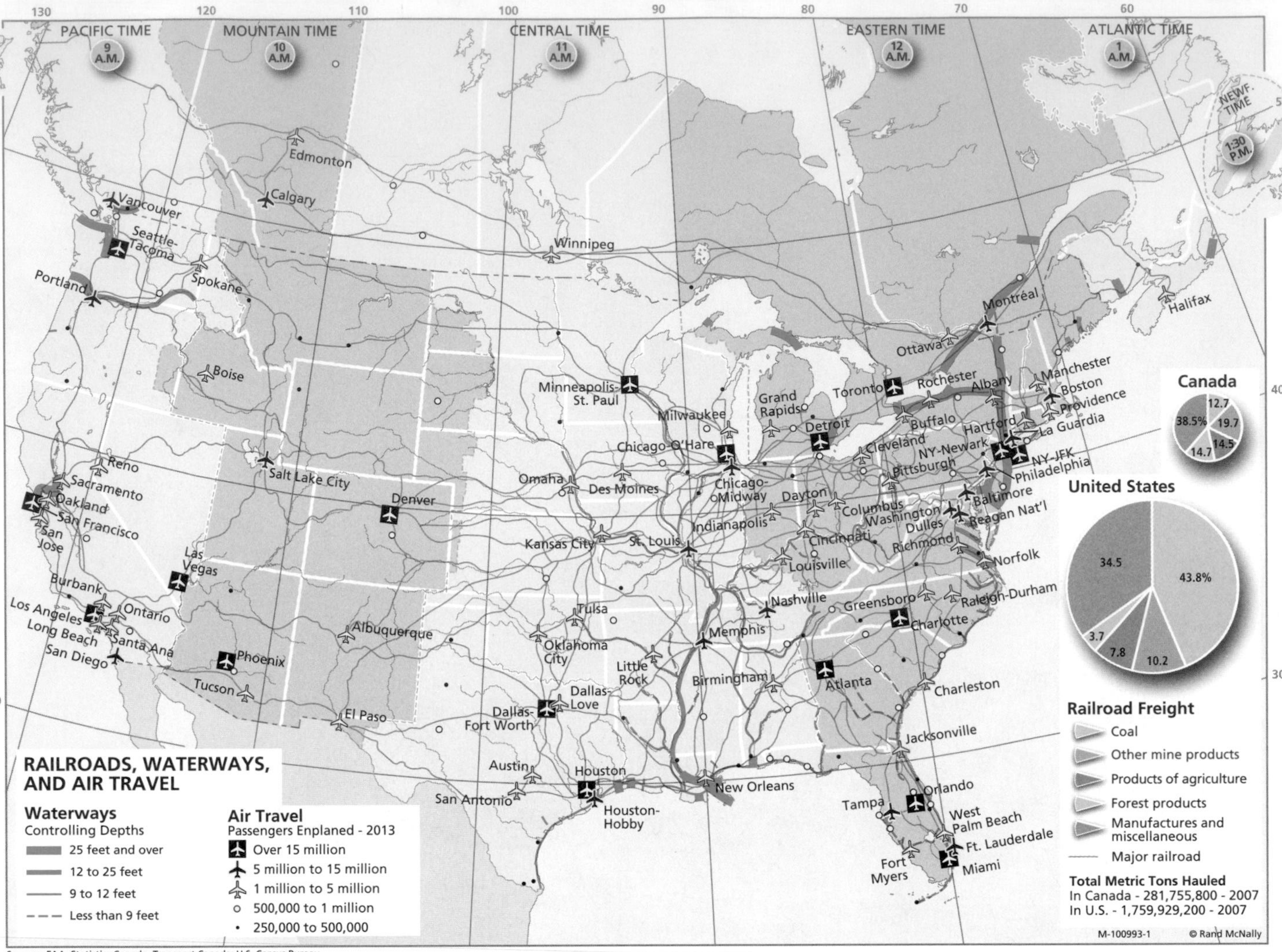

PACIFIC TIME MOUNTAIN TIME CENTRAL TIME EASTERN TIME ATLANTIC TIME
9 A.M. 10 A.M. 11 A.M. 12 A.M. 1 A.M.

NEWF. TIME
1:30 P.M.

RAILROADS, WATERWAYS, AND AIR TRAVEL

Waterways
Controlling Depths
- ▬ 25 feet and over
- ▬ 12 to 25 feet
- ▬ 9 to 12 feet
- ╌ Less than 9 feet

Air Travel
Passengers Enplaned - 2013
- ✈ Over 15 million
- ✈ 5 million to 15 million
- ✈ 1 million to 5 million
- ○ 500,000 to 1 million
- • 250,000 to 500,000

Canada
- 38.5%
- 12.7
- 19.7
- 14.5
- 14.7

United States
- 43.8%
- 34.5
- 10.2
- 7.8
- 3.7

Railroad Freight
- Coal
- Other mine products
- Products of agriculture
- Forest products
- Manufactures and miscellaneous
- ─ Major railroad

Total Metric Tons Hauled
In Canada - 281,755,800 - 2007
In U.S. - 1,759,929,200 - 2007

Sources: FAA; Statistics Canada; Transport Canada; U.S. Census Bureau

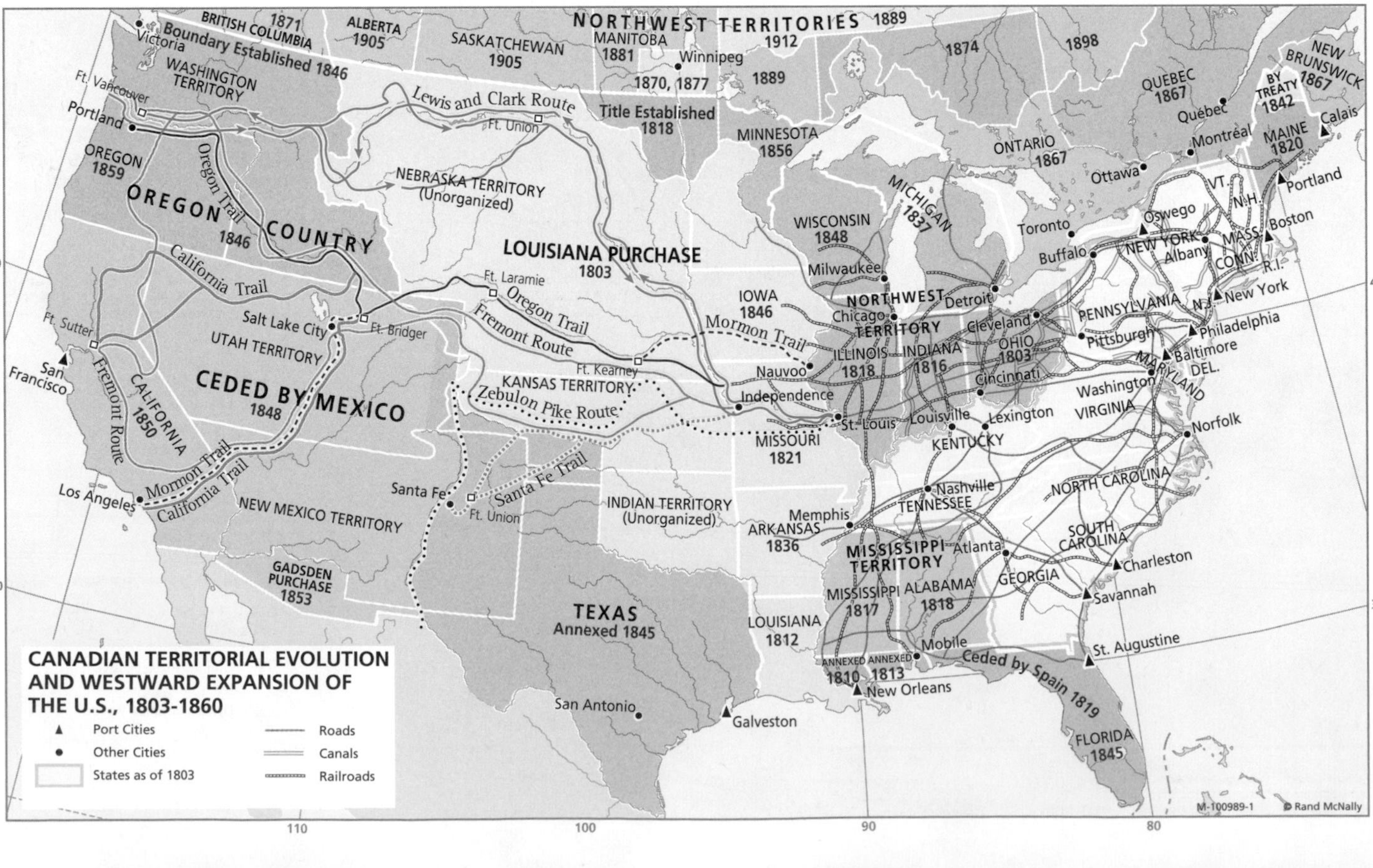

M-100993-1   © Rand McNally

CANADIAN TERRITORIAL EVOLUTION AND WESTWARD EXPANSION OF THE U.S., 1803-1860

- ▲ Port Cities
- • Other Cities
- ▭ States as of 1803
- ═ Roads
- ═ Canals
- ⋯ Railroads

M-100989-1   © Rand McNally

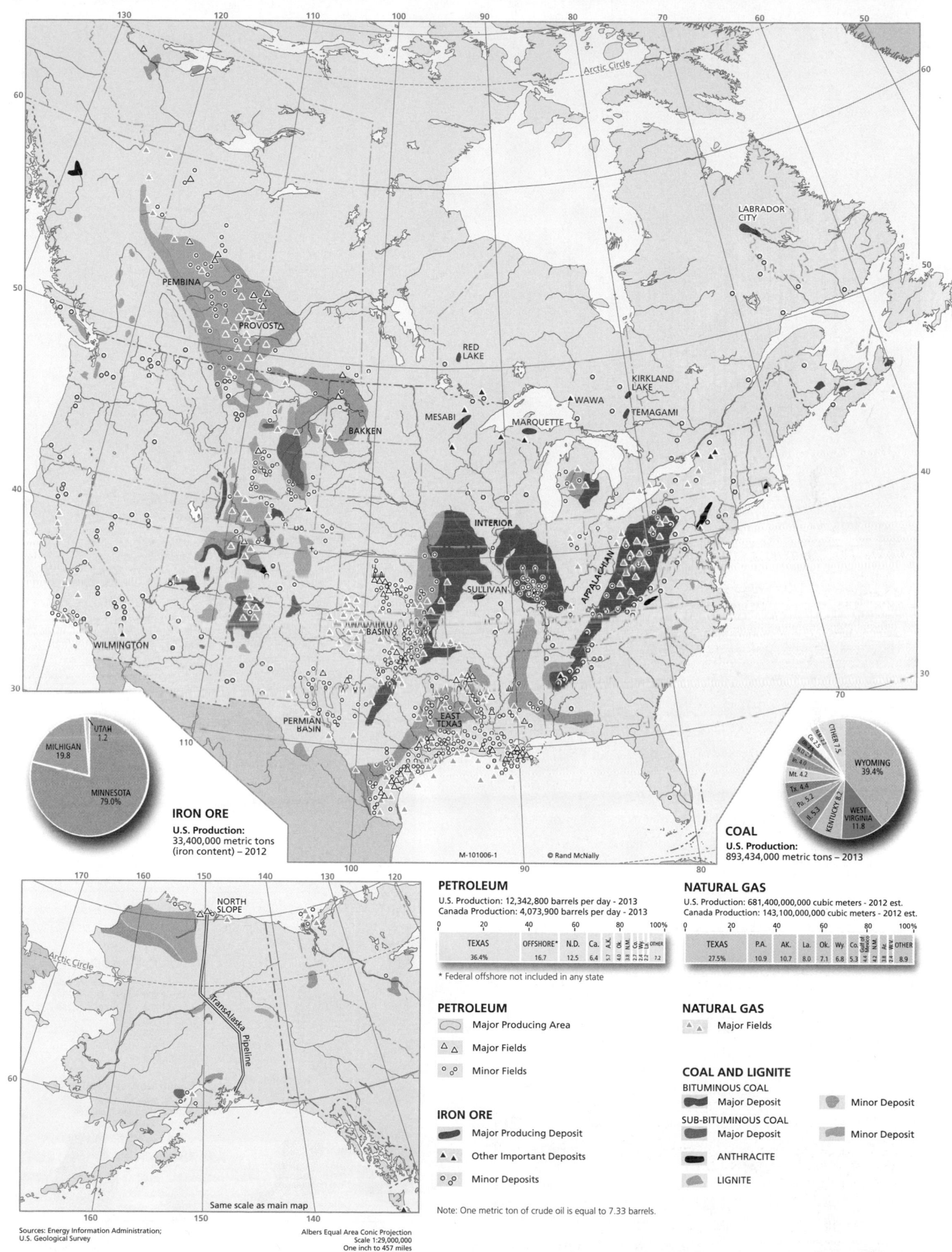

LABRADOR CITY

PEMBINA

PROVOSTA

RED LAKE

KIRKLAND LAKE

WAWA

MESABI

BAKKEN

MARQUETTE

TEMAGAMI

INTERIOR

SULLIVAN

APPALACHIAN

WILLISTON BASIN

WILMINGTON

PERMIAN BASIN

EAST TEXAS

IRON ORE

UTAH 1.2

MICHIGAN 19.8

MINNESOTA 79.0%

IRON ORE

U.S. Production:
33,400,000 metric tons
(iron content) – 2012

COAL

OTHER 7.5

IN 6.2
MD 2.5
Mt. 4.0
Mt. 4.2
Tx. 4.4
Pa. 5.2
Ill. 5.3
KENTUCKY 8.2
WEST VIRGINIA 11.8
WYOMING 39.4%

COAL

U.S. Production:
893,434,000 metric tons – 2013

M-101006-1   © Rand McNally

NORTH SLOPE

Arctic Circle

TransAlaska Pipeline

Same scale as main map

Sources: Energy Information Administration;
U.S. Geological Survey

Albers Equal Area Conic Projection
Scale 1:29,000,000
One inch to 457 miles
One cm to 290 km

PETROLEUM

U.S. Production: 12,342,800 barrels per day - 2013
Canada Production: 4,073,900 barrels per day - 2013

0	20	40	60	80						100%
TEXAS		OFFSHORE*	N.D.	Ca.	AK.	Ok.	N.M.	Co.	Wy.	Uk.	OTHER
36.4%		16.7	12.5	6.4	5.7	4.0	3.8	3.2	2.2	1.8	7.2

* Federal offshore not included in any state

NATURAL GAS

U.S. Production: 681,400,000,000 cubic meters - 2012 est.
Canada Production: 143,100,000,000 cubic meters - 2012 est.

0	20	40	60	80						100%
TEXAS		P.A.	AK.	La.	Ok.	Wy.	Co.	Gulf of Mexico	N.M.	Ar.	WV.	OTHER
27.5%		10.9	10.7	8.0	7.1	6.8	5.3	4.5	4.2	3.8	2.4	8.9

PETROLEUM

Major Producing Area

Major Fields

Minor Fields

IRON ORE

Major Producing Deposit

Other Important Deposits

Minor Deposits

NATURAL GAS

Major Fields

COAL AND LIGNITE

BITUMINOUS COAL

Major Deposit   Minor Deposit

SUB-BITUMINOUS COAL

Major Deposit   Minor Deposit

ANTHRACITE

LIGNITE

Note: One metric ton of crude oil is equal to 7.33 barrels.

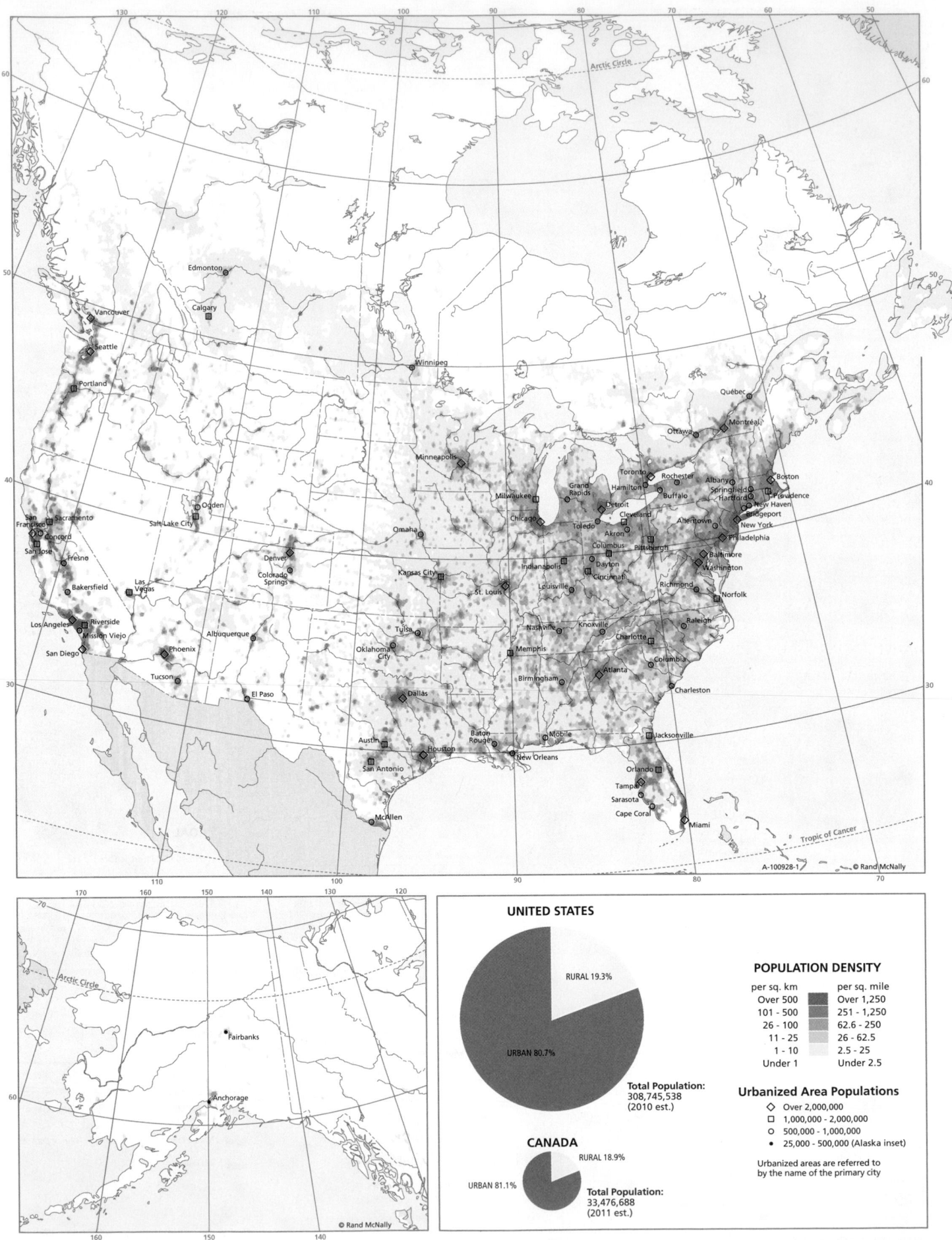

UNITED STATES

RURAL 19.3%

URBAN 80.7%

Total Population:
308,745,538
(2010 est.)

CANADA

RURAL 18.9%

URBAN 81.1%

Total Population:
33,476,688
(2011 est.)

POPULATION DENSITY

per sq. km	per sq. mile
Over 500	Over 1,250
101 - 500	251 - 1,250
26 - 100	62.6 - 250
11 - 25	26 - 62.5
1 - 10	2.5 - 25
Under 1	Under 2.5

Urbanized Area Populations

◇ Over 2,000,000
▢ 1,000,000 - 2,000,000
○ 500,000 - 1,000,000
• 25,000 - 500,000 (Alaska inset)

Urbanized areas are referred to
by the name of the primary city

Sources: Census of Canada; U.S. Census Bureau;
U.S. Department of Energy; United Nations

Albers Equal Area Conic Projection
Scale 1:29,000,000
One inch to 457 miles
One cm to 290 km

© Rand McNally

A-100928-1

WHITE POPULATION

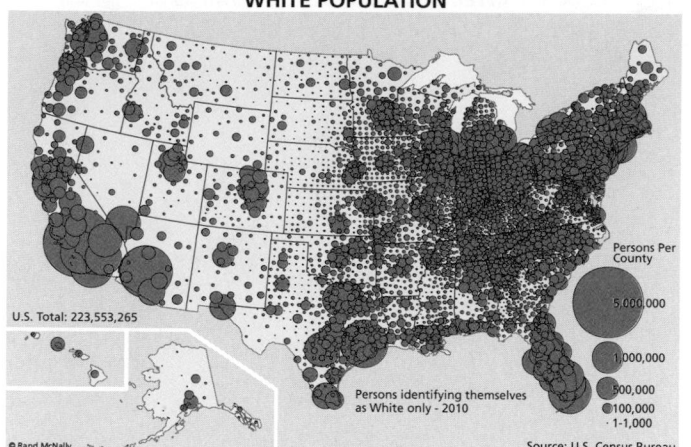

U.S. Total: 223,553,265

Persons Per County
5,000,000
1,000,000
500,000
100,000
1-1,000

Persons identifying themselves as White only - 2010

© Rand McNally

Source: U.S. Census Bureau

AFRICAN AMERICAN POPULATION

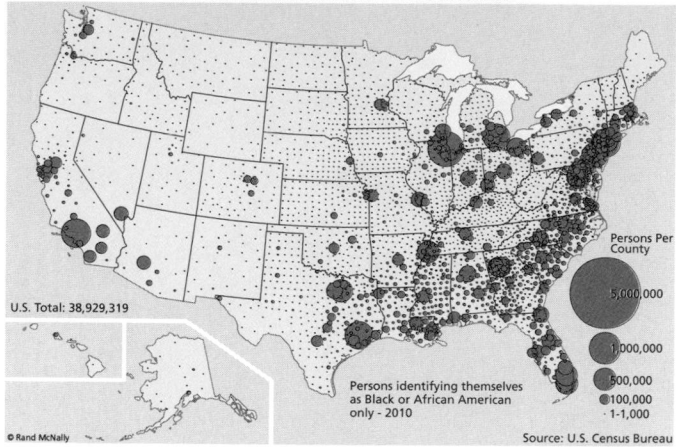

U.S. Total: 38,929,319

Persons Per County
5,000,000
1,000,000
500,000
100,000
1-1,000

Persons identifying themselves as Black or African American only - 2010

© Rand McNally

Source: U.S. Census Bureau

ASIAN POPULATION

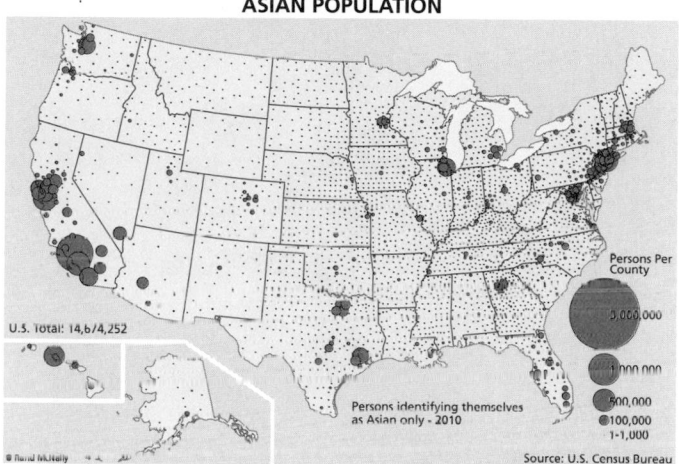

U.S. Total: 14,674,252

Persons Per County
5,000,000
1,000,000
500,000
100,000
1-1,000

Persons identifying themselves as Asian only - 2010

© Rand McNally

Source: U.S. Census Bureau

AMERICAN INDIAN AND ALASKA NATIVE POPULATION

U.S. Total: 2,932,248

Persons Per County
5,000,000
1,000,000
500,000
100,000
1-1,000

Persons identifying themselves as American Indian or Alaska Native only - 2010

© Rand McNally

Source: U.S. Census Bureau

NATIVE HAWAIIAN AND PACIFIC ISLANDER POPULATION

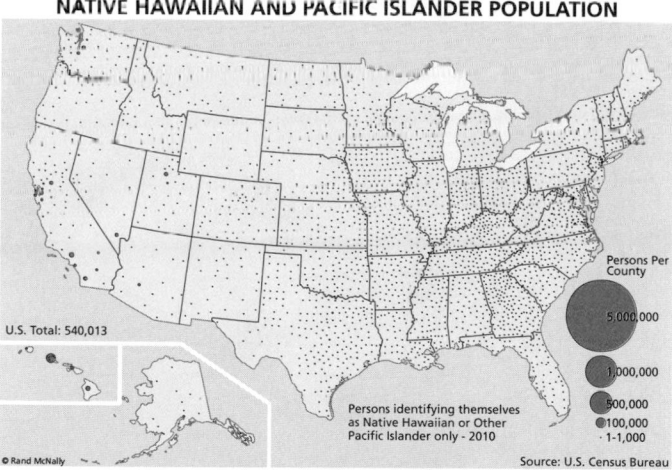

U.S. Total: 540,013

Persons Per County
5,000,000
1,000,000
500,000
100,000
1-1,000

Persons identifying themselves as Native Hawaiian or Other Pacific Islander only - 2010

© Rand McNally

Source: U.S. Census Bureau

SOME OTHER RACE

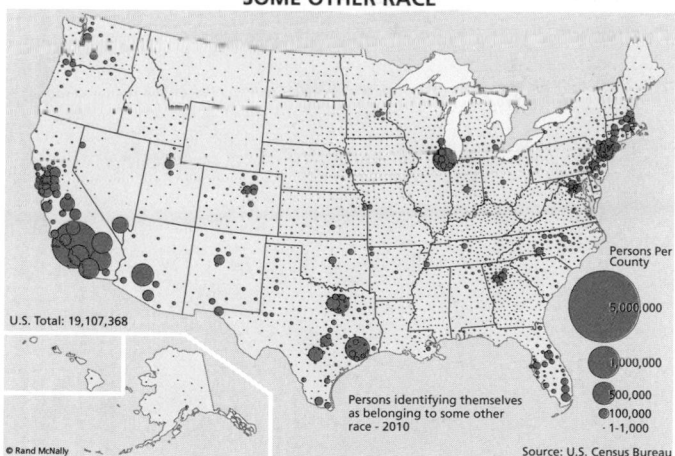

U.S. Total: 19,107,368

Persons Per County
5,000,000
1,000,000
500,000
100,000
1-1,000

Persons identifying themselves as belonging to some other race - 2010

© Rand McNally

Source: U.S. Census Bureau

TWO OR MORE RACES

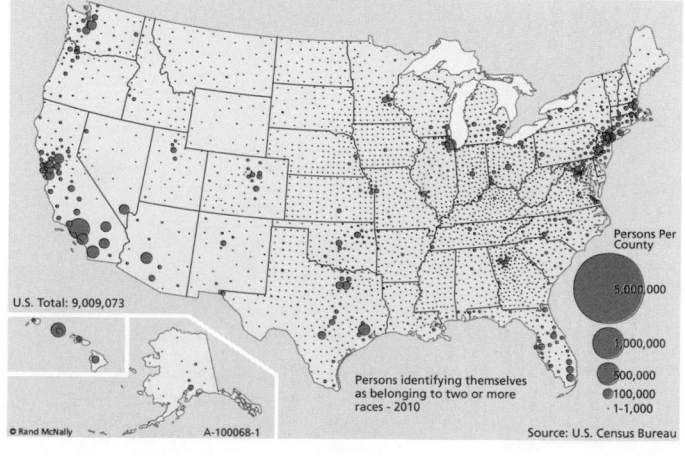

U.S. Total: 9,009,073

Persons Per County
5,000,000
1,000,000
500,000
100,000
1-1,000

Persons identifying themselves as belonging to two or more races - 2010

© Rand McNally   A-100068-1

Source: U.S. Census Bureau

HISPANIC POPULATION (ANY RACE)

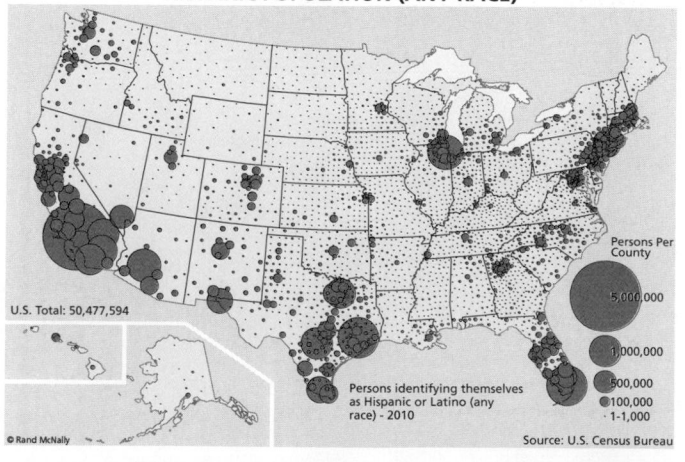

U.S. Total: 50,477,594

Persons Per County
5,000,000
1,000,000
500,000
100,000
1-1,000

Persons identifying themselves as Hispanic or Latino (any race) - 2010

© Rand McNally

Source: U.S. Census Bureau

POPULATION CHANGE

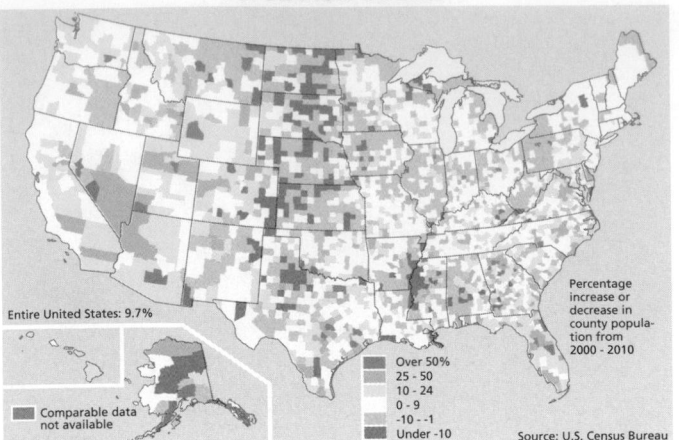

Entire United States: 9.7%

Comparable data not available

© Rand McNally

Percentage increase or decrease in county population from 2000 - 2010

Over 50%
25 - 50
10 - 24
0 - 9
-10 - -1
Under -10

Source: U.S. Census Bureau

INTER-STATE POPULATION SHIFTS

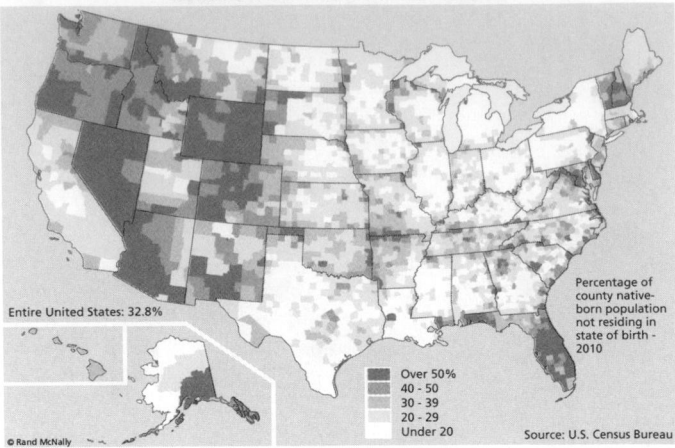

Entire United States: 32.8%

© Rand McNally

Percentage of county native-born population not residing in state of birth - 2010

Over 50%
40 - 50
30 - 39
20 - 29
Under 20

Source: U.S. Census Bureau

POPULATION UNDER 18

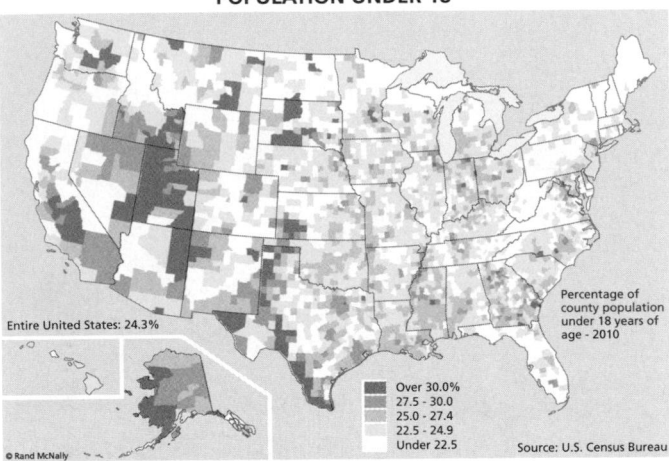

Entire United States: 24.3%

© Rand McNally

Percentage of county population under 18 years of age - 2010

Over 30.0%
27.5 - 30.0
25.0 - 27.4
22.5 - 24.9
Under 22.5

Source: U.S. Census Bureau

POPULATION 65 AND OVER

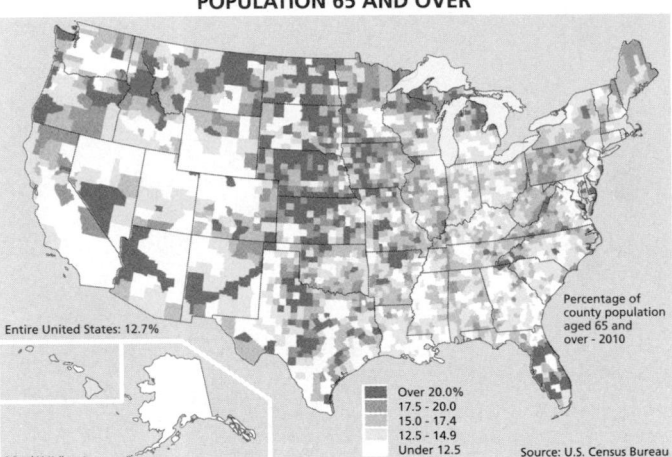

Entire United States: 12.7%

© Rand McNally

Percentage of county population aged 65 and over - 2010

Over 20.0%
17.5 - 20.0
15.0 - 17.4
12.5 - 14.9
Under 12.5

Source: U.S. Census Bureau

EDUCATIONAL ATTAINMENT RATE

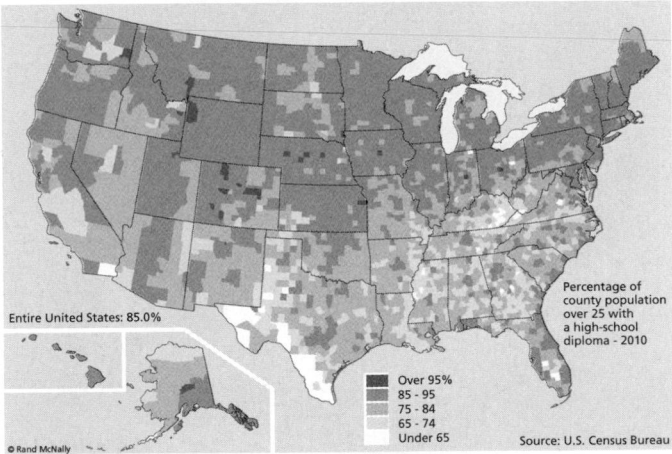

Entire United States: 85.0%

© Rand McNally

Percentage of county population over 25 with a high-school diploma - 2010

Over 95%
85 - 95
75 - 84
65 - 74
Under 65

Source: U.S. Census Bureau

COLLEGE ENROLLMENT RATE

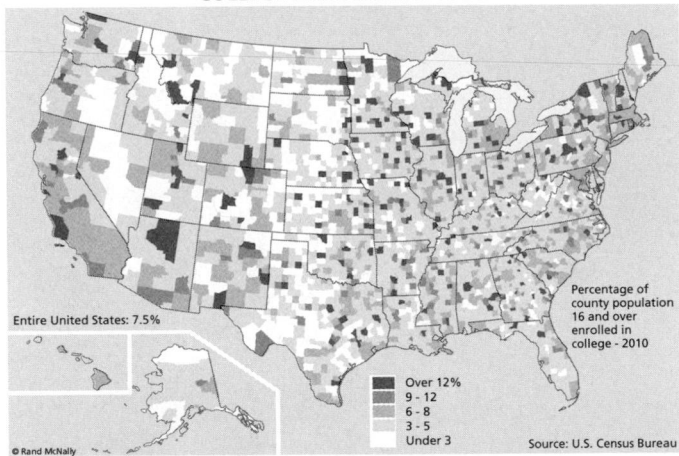

Entire United States: 7.5%

© Rand McNally

Percentage of county population 16 and over enrolled in college - 2010

Over 12%
9 - 12
6 - 8
3 - 5
Under 3

Source: U.S. Census Bureau

COMMUTING TIME

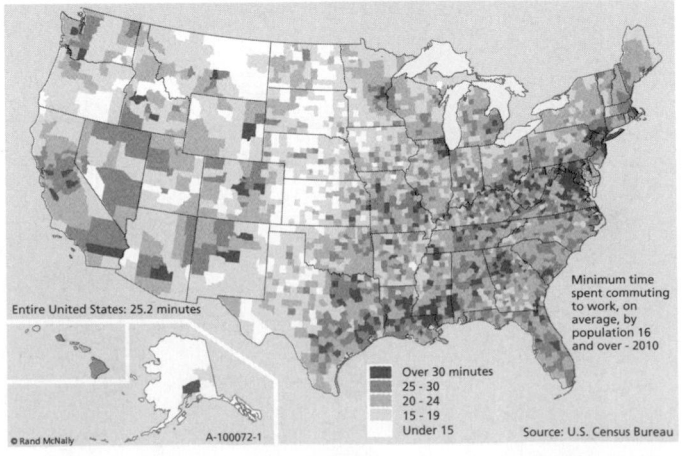

Entire United States: 25.2 minutes

© Rand McNally   A-100072-1

Minimum time spent commuting to work, on average, by population 16 and over - 2010

Over 30 minutes
25 - 30
20 - 24
15 - 19
Under 15

Source: U.S. Census Bureau

MEDIAN DECADE OF HOUSE CONSTRUCTION

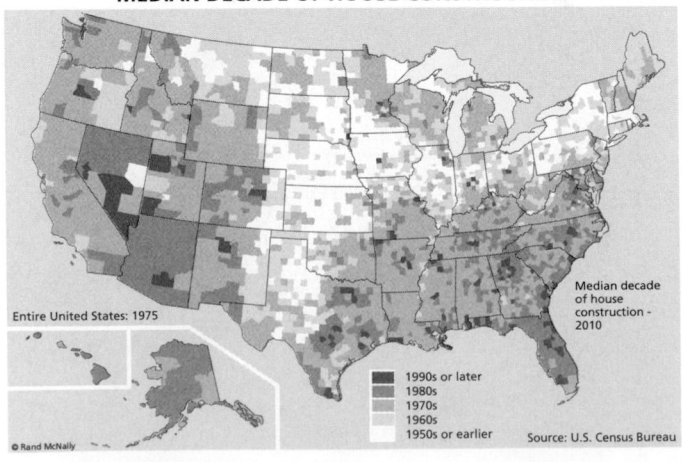

Entire United States: 1975

© Rand McNally

Median decade of house construction - 2010

1990s or later
1980s
1970s
1960s
1950s or earlier

Source: U.S. Census Bureau

WOMEN'S MEDIAN EARNINGS

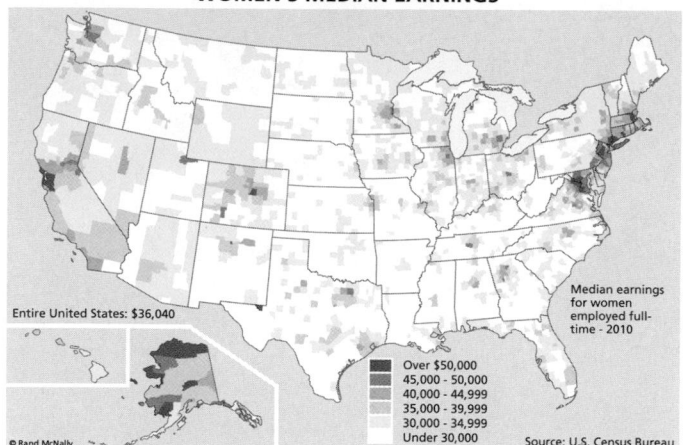

Entire United States: $36,040

Median earnings for women employed full-time - 2010

Over $50,000
45,000 - 50,000
40,000 - 44,999
35,000 - 39,999
30,000 - 34,999
Under 30,000

© Rand McNally

Source: U.S. Census Bureau

MEN'S MEDIAN EARNINGS

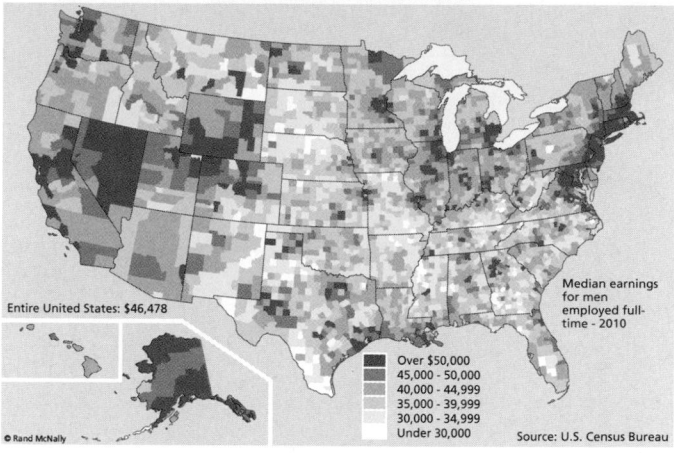

Entire United States: $46,478

Median earnings for men employed full-time - 2010

Over $50,000
45,000 - 50,000
40,000 - 44,999
35,000 - 39,999
30,000 - 34,999
Under 30,000

© Rand McNally

Source: U.S. Census Bureau

RATIO OF WOMEN'S TO MEN'S EARNINGS

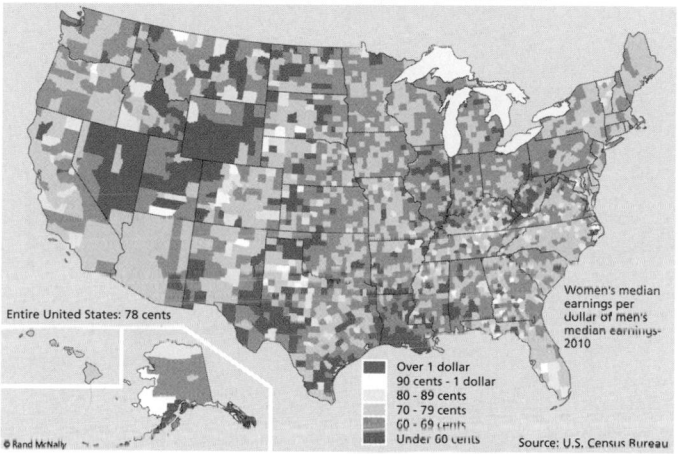

Entire United States: 78 cents

Women's median earnings per dollar of men's median earnings-2010

Over 1 dollar
90 cents - 1 dollar
80 - 89 cents
70 - 79 cents
60 - 69 cents
Under 60 cents

© Rand McNally

Source: U.S. Census Bureau

MEDIAN HOUSEHOLD INCOME

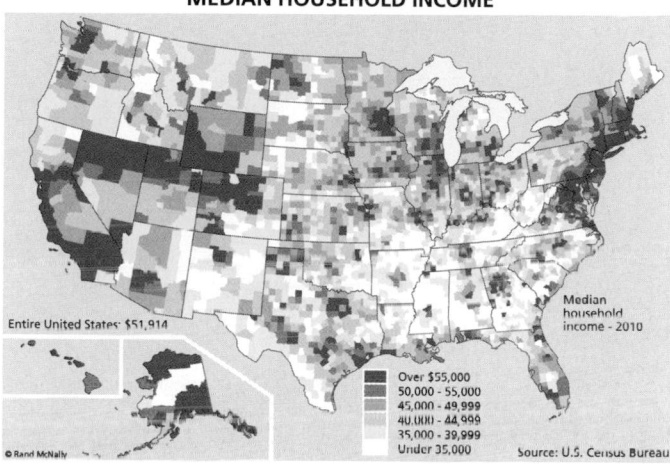

Entire United States: $51,914

Median household income - 2010

Over $55,000
50,000 - 55,000
45,000 - 49,999
40,000 - 44,999
35,000 - 39,999
Under 35,000

© Rand McNally

Source: U.S. Census Bureau

HOUSEHOLDS HEADED BY WOMEN

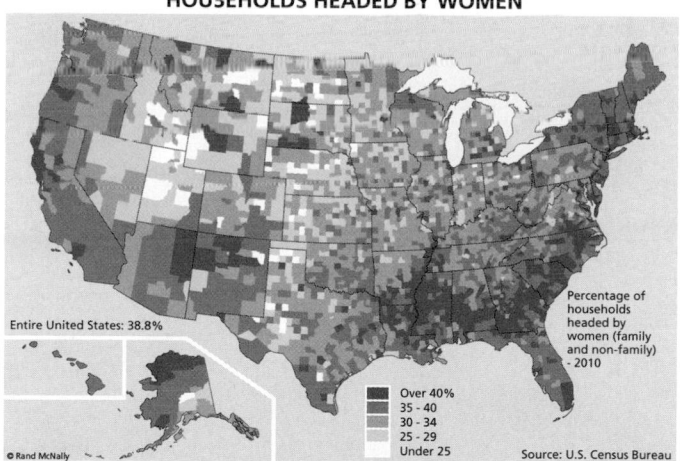

Entire United States: 38.8%

Percentage of households headed by women (family and non-family) - 2010

Over 40%
35 - 40
30 - 34
25 - 29
Under 25

© Rand McNally

Source: U.S. Census Bureau

CHILDREN LIVING IN POVERTY

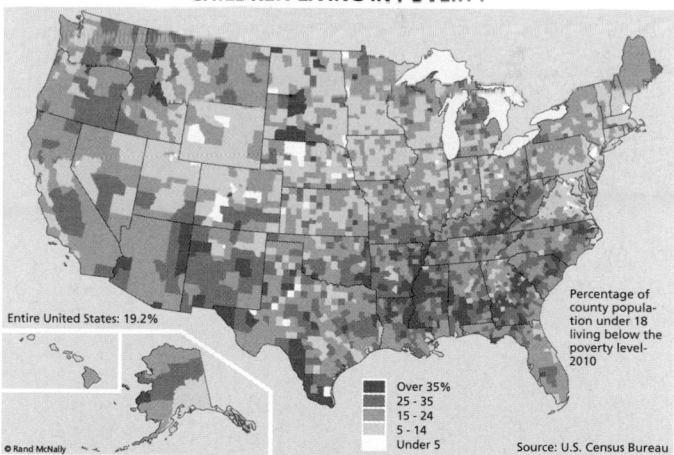

Entire United States: 19.2%

Percentage of county population under 18 living below the poverty level- 2010

Over 35%
25 - 35
15 - 24
5 - 14
Under 5

© Rand McNally

Source: U.S. Census Bureau

UNEMPLOYMENT RATE

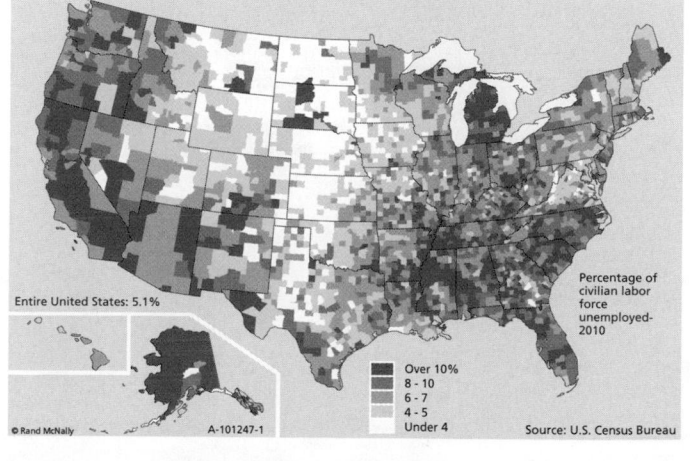

Entire United States: 5.1%

Percentage of civilian labor force unemployed- 2010

Over 10%
8 - 10
6 - 7
4 - 5
Under 4

© Rand McNally   A-101247-1

Source: U.S. Census Bureau

NON-ENGLISH SPEAKING HOUSEHOLDS

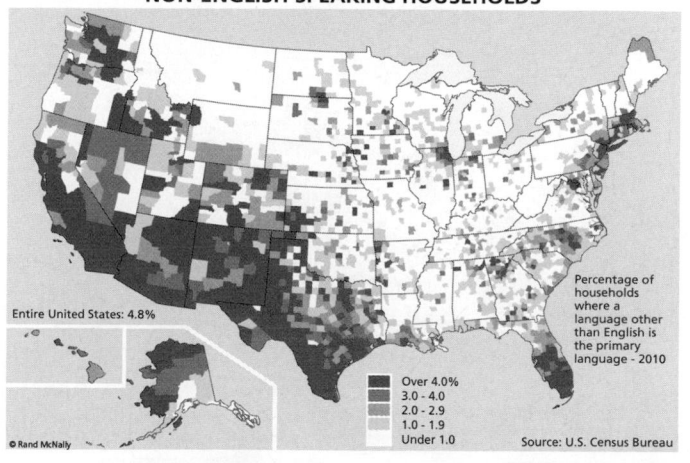

Entire United States: 4.8%

Percentage of households where a language other than English is the primary language - 2010

Over 4.0%
3.0 - 4.0
2.0 - 2.9
1.0 - 1.9
Under 1.0

© Rand McNally

Source: U.S. Census Bureau

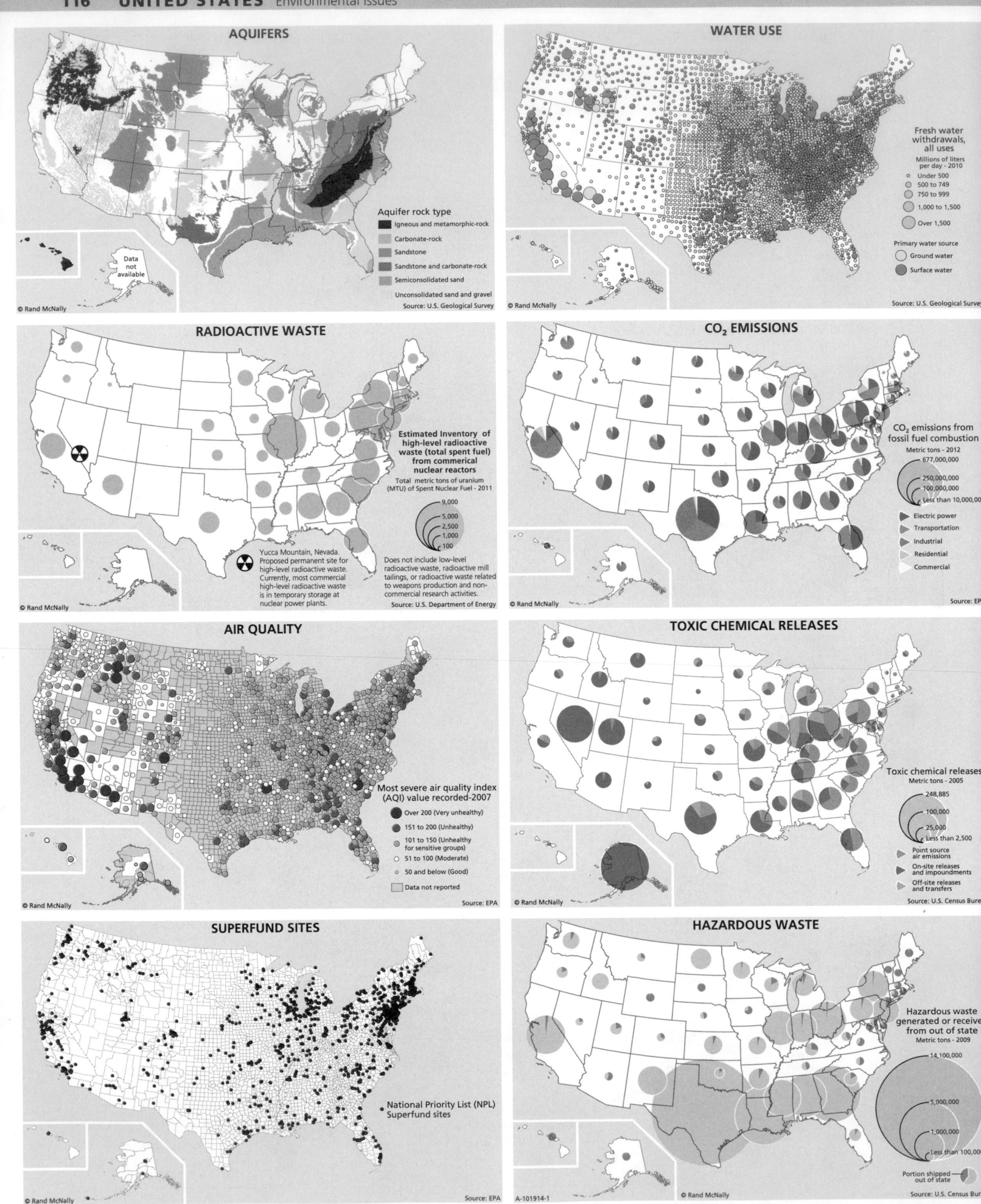

AQUIFERS

Aquifer rock type
- Igneous and metamorphic-rock
- Carbonate-rock
- Sandstone
- Sandstone and carbonate-rock
- Semiconsolidated sand
- Unconsolidated sand and gravel

Data not available

© Rand McNally

Source: U.S. Geological Survey

WATER USE

Fresh water withdrawals, all uses
Millions of liters per day - 2010
- Under 500
- 500 to 749
- 750 to 999
- 1,000 to 1,500
- Over 1,500

Primary water source
- Ground water
- Surface water

© Rand McNally

Source: U.S. Geological Survey

RADIOACTIVE WASTE

Estimated Inventory of high-level radioactive waste (total spent fuel) from commerical nuclear reactors
Total metric tons of uranium (MTU) of Spent Nuclear Fuel - 2011
- 9,000
- 5,000
- 2,500
- 1,000
- 100

Yucca Mountain, Nevada. Proposed permanent site for high-level radioactive waste. Currently, most commercial high-level radioactive waste is in temporary storage at nuclear power plants.

Does not include low-level radioactive waste, radioactive mill tailings, or radioactive waste related to weapons production and non-commercial research activities.

© Rand McNally

Source: U.S. Department of Energy

CO$_2$ EMISSIONS

CO$_2$ emissions from fossil fuel combustion
Metric tons - 2012
- 677,000,000
- 250,000,000
- 100,000,000
- Less than 10,000,000

- Electric power
- Transportation
- Industrial
- Residential
- Commercial

© Rand McNally

Source: EPA

AIR QUALITY

Most severe air quality index (AQI) value recorded-2007
- Over 200 (Very unhealthy)
- 151 to 200 (Unhealthy)
- 101 to 150 (Unhealthy for sensitive groups)
- 51 to 100 (Moderate)
- 50 and below (Good)
- Data not reported

© Rand McNally

Source: EPA

TOXIC CHEMICAL RELEASES

Toxic chemical releases
Metric tons - 2005
- 248,885
- 100,000
- 25,000
- Less than 2,500

- Point source air emissions
- On-site releases and impoundments
- Off-site releases and transfers

© Rand McNally

Source: U.S. Census Bureau

SUPERFUND SITES

- National Priority List (NPL) Superfund sites

© Rand McNally

A-101914-1

Source: EPA

HAZARDOUS WASTE

Hazardous waste generated or received from out of state
Metric tons - 2009
- 14,100,000
- 5,000,000
- 1,000,000
- Less than 100,000

- Portion shipped out of state

© Rand McNally

Source: U.S. Census Bureau

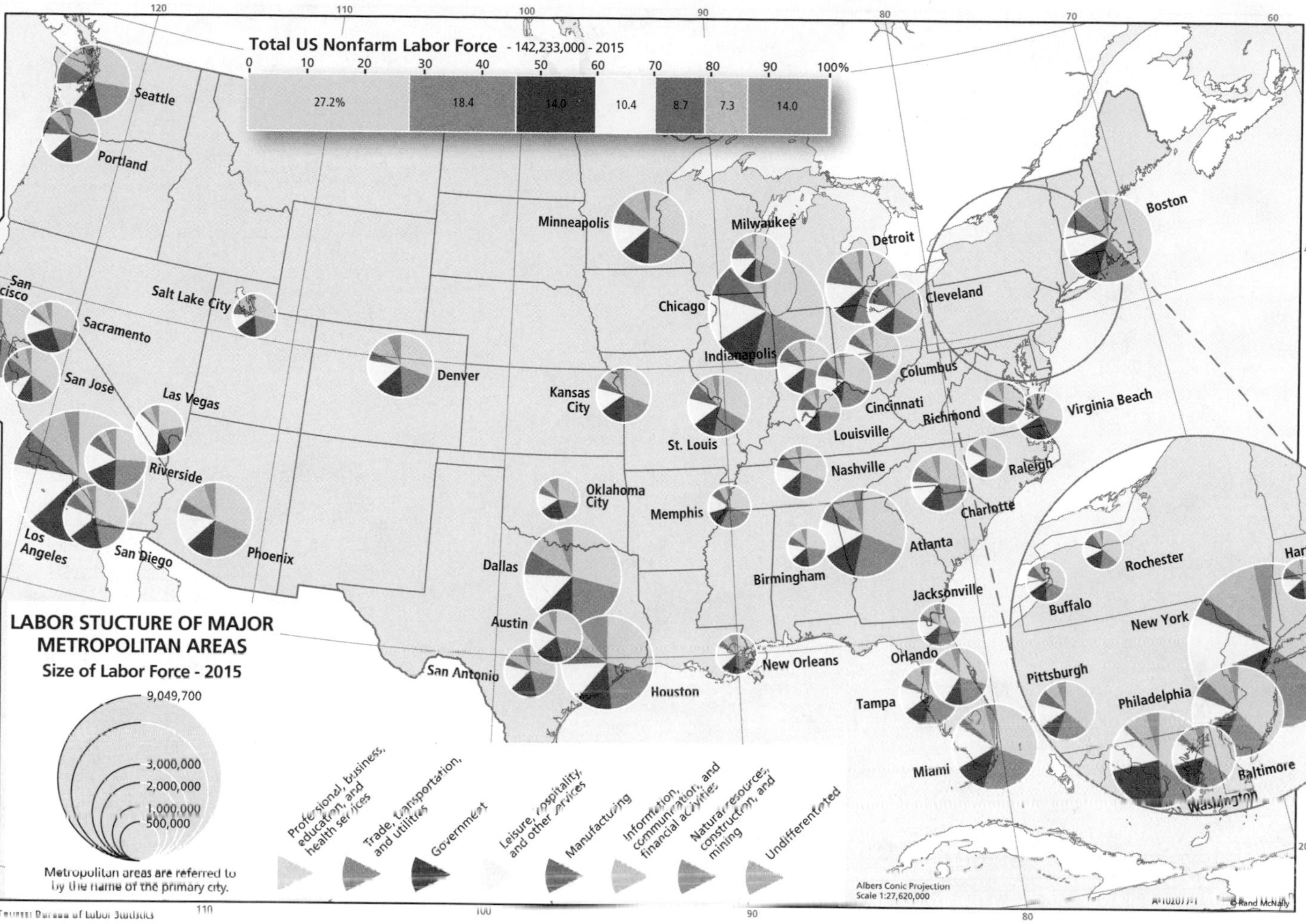

Total US Nonfarm Labor Force – 142,233,000 – 2015

27.2%	18.4	14.0	10.4	8.7	7.3	14.0

Seattle
Portland

San Francisco
Sacramento
San Jose
Salt Lake City
Las Vegas
Riverside
Los Angeles
San Diego
Phoenix

Denver

Minneapolis
Milwaukee
Detroit
Chicago
Cleveland
Indianapolis
Columbus
Kansas City
Cincinnati
Richmond
St. Louis
Louisville
Virginia Beach
Nashville
Raleigh
Oklahoma City
Memphis
Charlotte
Dallas
Atlanta
Austin
Birmingham
San Antonio
New Orleans
Houston
Jacksonville
Orlando
Tampa
Miami

Boston

Buffalo
Rochester
Hartford
New York
Providence
Pittsburgh
Philadelphia
Baltimore
Washington

LABOR STUCTURE OF MAJOR METROPOLITAN AREAS

Size of Labor Force - 2015

9,049,700
3,000,000
2,000,000
1,000,000
500,000

Metropolitan areas are referred to by the name of the primary city.

Professional, business, education, and health services
Trade, transportation, and utilities
Government
Leisure, hospitality, and other services
Manufacturing
Information, communication, and financial activities
Natural resource, construction, and mining
Undifferentiated

Source: Bureau of Labor Statistics

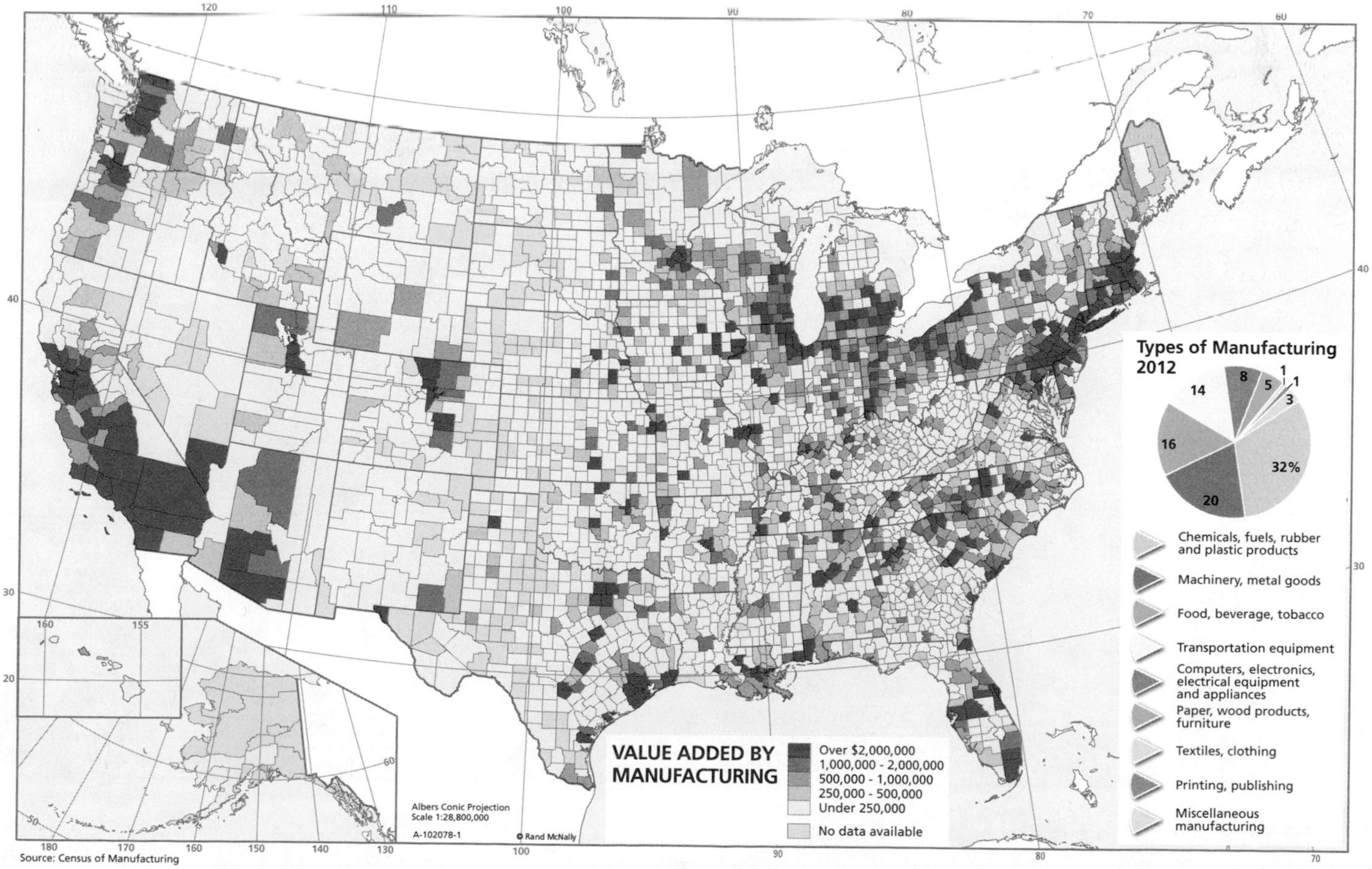

Albers Conic Projection
Scale 1:27,620,000

A-102077-1   © Rand McNally

VALUE ADDED BY MANUFACTURING

	Over $2,000,000
	1,000,000 - 2,000,000
	500,000 - 1,000,000
	250,000 - 500,000
	Under 250,000
	No data available

Albers Conic Projection
Scale 1:28,800,000
A-102078-1   © Rand McNally

Source: Census of Manufacturing

Types of Manufacturing 2012

8 5 1 1
14
3
16
32%
20

Chemicals, fuels, rubber and plastic products
Machinery, metal goods
Food, beverage, tobacco
Transportation equipment
Computers, electronics, electrical equipment and appliances
Paper, wood products, furniture
Textiles, clothing
Printing, publishing
Miscellaneous manufacturing

0   60   120   180   240   300   360 Miles
0 60 120 180 240 300 360 420 480 540 600 Kilometers

Lambert Azimuthal Equal Area Projection
Scale 1:12,000,000
One inch to 190 miles
One cm to 120 km

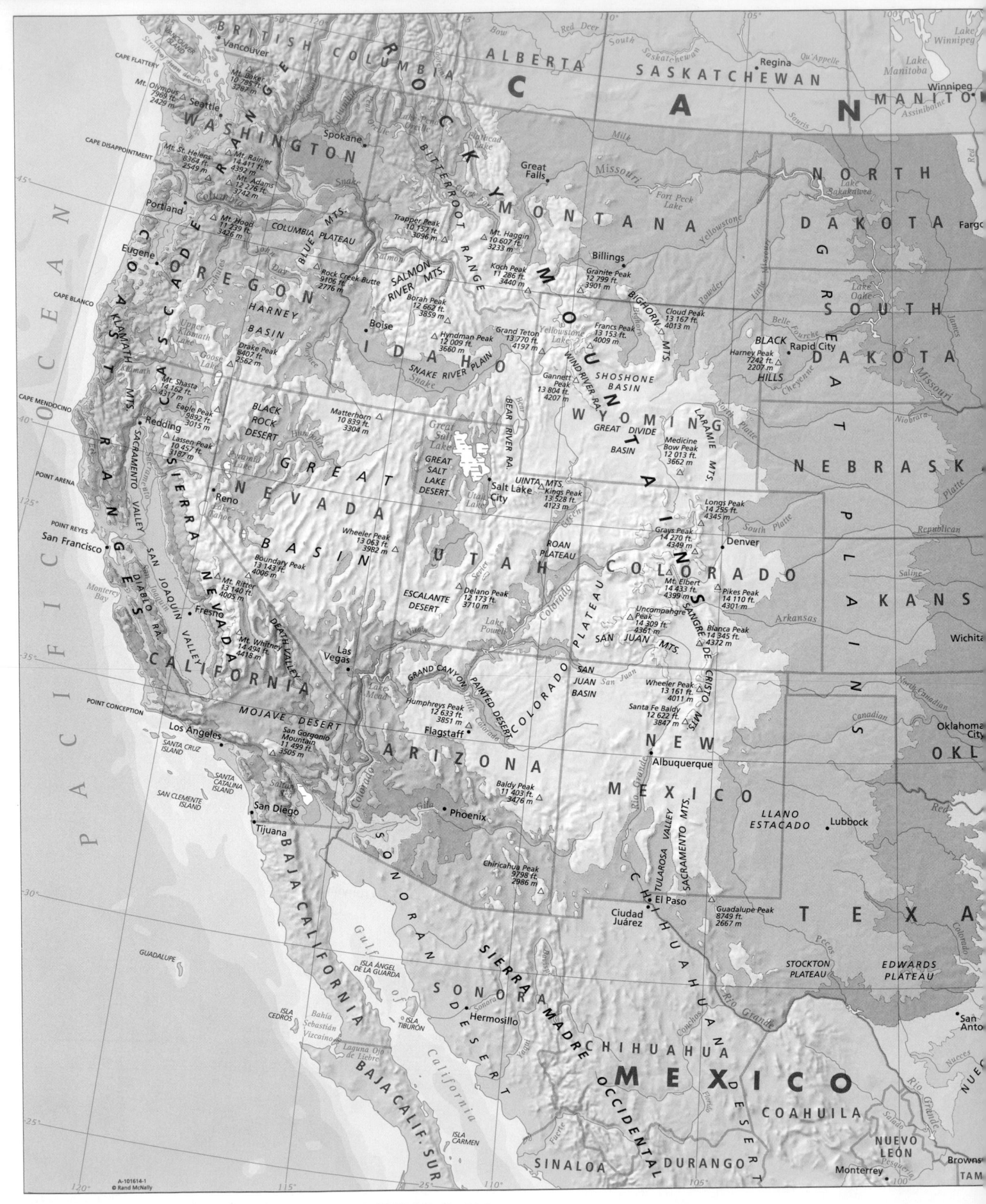

Lambert Azimuthal Equal Area Projection
Scale 1:12,000,000
One inch to 190 miles
One cm to 120 km

0 60 120 180 240 300 360 Miles

0 60 120 180 240 300 360 420 480 540 600 Kilometers

A-101614-1
© Rand McNally

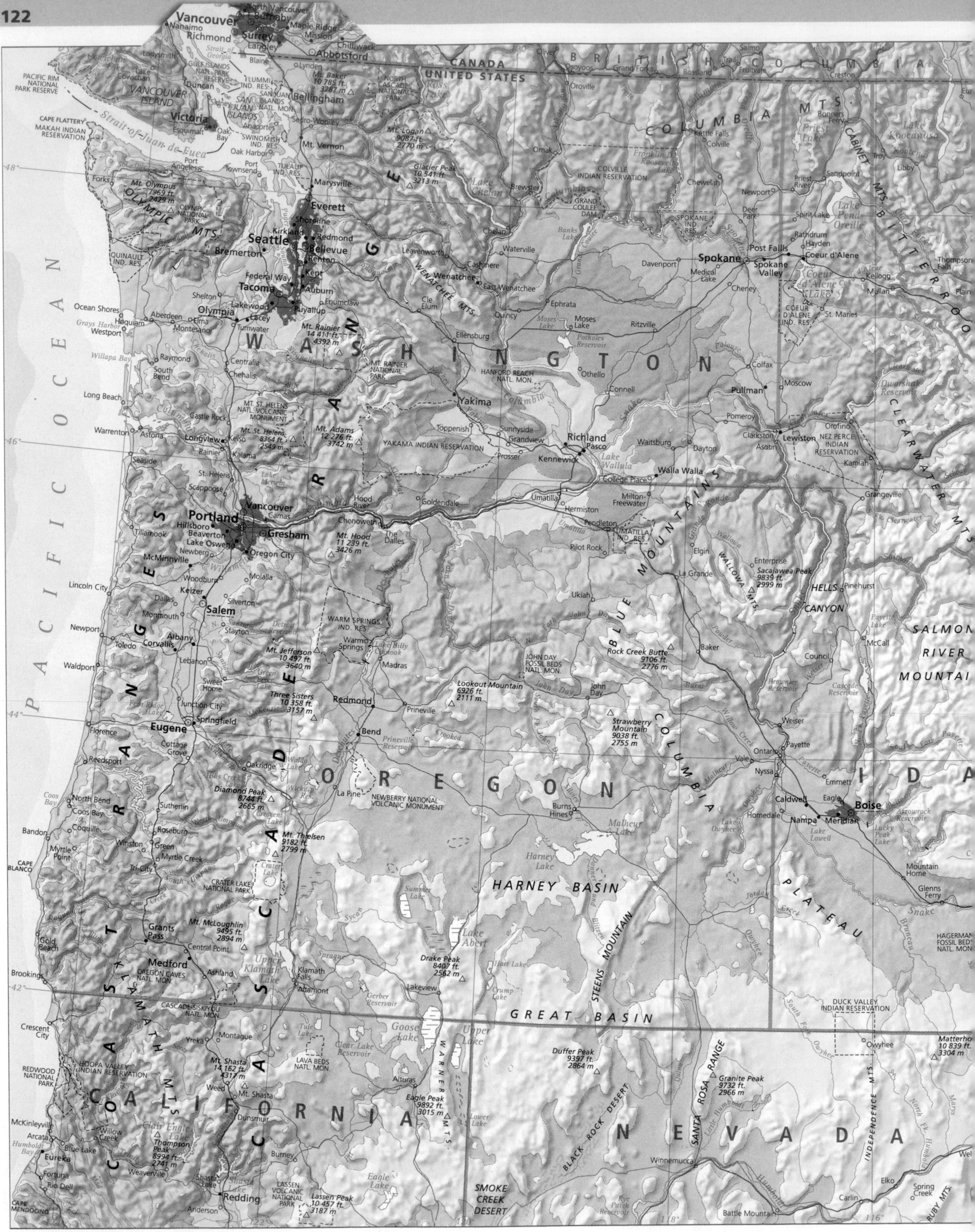

CANADA
UNITED STATES

BRITISH COLUMBIA

COLUMBIA MTS.

Vancouver
Burnaby
Nanaimo
Richmond
Surrey
Maple Ridge
Mission
Langley
Chilliwack
Abbotsford
Blaine
Lynden
Oliver
Osoyoos
Grand Forks
Trail
Rossland
Salmo
Fruitvale
Creston

PACIFIC RIM
NATIONAL
PARK RESERVE
VANCOUVER ISLAND
Ladysmith
Cowichan
Duncan
GULF ISLANDS NAT'L PARK RESERVE

CABINET MTS.
BITTERROO
Troy
Libby

CAPE FLATTERY
MAKAH INDIAN RESERVATION
Victoria
Esquimalt
Oak Bay
Sidney
SAN JUAN ISLANDS NATL. MON.
Anacortes
Sedro-Woolley
Mt. Baker
10 785 ft.
3287 m
NORTH CASCADES NATL. PARK
Oroville
Omak
Colville
COLUMBIA
Chewelah
Kettle Falls
Priest Lake
Bonners Ferry
Sandpoint
Kootenai
Kootenai

Port Angeles
Port Townsend
SWINOMISH IND. RES.
Mt. Vernon
LUMMI IND. RES.
TULALIP IND. RES.
Glacier Peak
10 541 ft.
3213 m
Lake Chelan
Brewster
COLVILLE INDIAN RESERVATION
Deer Park
SPOKANE IND. RES.
Newport
Spirit Lake
Rathdrum
Lake Pend Oreille

Mt. Olympus
7965 ft.
2429 m
OLYMPIC MTS.
OLYMPIC NATIONAL PARK
Forks
QUINAULT IND. RES.
Everett
Shoreline
Marysville
Kirkland
Redmond
Chelan
Waterville
Leavenworth
Cashmere
Ephrata
GRAND COULEE DAM
Banks Lake
Davenport
Medical Lake
Cheney
Spokane
Spokane Valley
Post Falls
Hayden
Coeur d'Alene
Kellogg
Mullan
Thompson Falls

Seattle
Bellevue
Bremerton
WENATCHEE MTS.
Wenatchee
East Wenatchee
Quincy
Moses Lake
Ritzville
Coeur d'Alene Lake
COEUR D'ALENE IND. RES.
St. Maries
Plummer

Federal Way
Tacoma
Lakewood
Kent
Auburn
Puyallup
Enumclaw
Cle Elum
Ellensburg
WASHINGTON
Othello
Connell
Pomeroy
Colfax
Moscow
Pullman
Dworshak Reservoir

Ocean Shores
Shelton
Olympia
Tumwater
Lacey
Mt. Rainier
14 411 ft.
4392 m
MT. RAINIER NATIONAL PARK
Yakima
HANFORD REACH NATL. MON.
Clarkston
Asotin
Lewiston
NEZ PERCE INDIAN RESERVATION
Kamiah
Orofino

Aberdeen
Hoquiam
Elma
Montesano
Westport
Grays Harbor
Centralia
Chehalis
MT. ST. HELENS NATL. VOLCANIC MONUMENT
Toppenish
Sunnyside
Grandview
Prosser
Richland
Pasco
Kennewick
Waitsburg
Dayton
Walla Walla
Grangeville
CLEARWATER MTS.

Willapa Bay
Raymond
South Bend
Long Beach
Mt. St. Helens
8364 ft.
2549 m
Mt. Adams
12 276 ft.
3742 m
YAKAMA INDIAN RESERVATION
Lake Wallula
College Place
Milton-Freewater

Warrenton
Astoria
Longview
Kelso
Kalama
Rainier
Goldendale
Umatilla
Hermiston
UMATILLA IND. RES.
Pendleton
Pilot Rock
Elgin
Enterprise
Sacajawea Peak
9839 ft.
2999 m
HELLS CANYON

Seaside
St. Helens
Scappoose
Vancouver
Camas
Hood River
Chenoweth
The Dalles
Mt. Hood
11 239 ft.
3426 m
Ukiah
La Grande
WALLOWA MTS.
Pinehurst

Tillamook
Portland
Hillsboro
Beaverton
Lake Oswego
Gresham
Oregon City
Newberg
WARM SPRINGS IND. RES.
Warm Springs
Madras
JOHN DAY FOSSIL BEDS NATL. MON.
Rock Creek Butte
9106 ft.
2776 m
Baker
Council
McCall
SALMON RIVER MOUNTAINS

McMinnville
Molalla
Woodburn
Keizer
Silverton
Salem
Stayton
Mt. Jefferson
10 497 ft.
3640 m
Detroit Reservoir
Lake Billy Chinook
Lookout Mountain
6926 ft.
2111 m
John Day
Strawberry Mountain
9038 ft.
2755 m
Weiser
Payette
IDAHO

Lincoln City
Dallas
Monmouth
Albany
Corvallis
Lebanon
Sweet Home
Three Sisters
10 358 ft.
3157 m
Redmond
Prineville
Ontario
Nyssa
Brownlee Reservoir

Newport
Toledo
Waldport
Junction City
Springfield
Bend
Prineville Reservoir
Vale

Florence
Eugene
Cottage Grove
Oakridge
OREGON
BLUE MOUNTAINS
COLUMBIA PLATEAU

Reedsport
Diamond Peak
8744 ft.
2665 m
La Pine
NEWBERRY NATIONAL VOLCANIC MONUMENT
Burns
Hines
Caldwell
Nampa
Boise
Meridian
Lucky Peak Lake

Coos Bay
North Bend
Coquille
Bandon
Sutherlin
Roseburg
Winston
Mt. Thielsen
9182 ft.
2799 m
Crescent
Malheur Lake
Homedale
Lake Lowell

CAPE BLANCO
Myrtle Point
Green
Myrtle Creek
CRATER LAKE NATIONAL PARK
Crater Lake
Harney Lake
Mountain Home
Glenns Ferry
Snake

Port Orford
Grants Pass
Mt. McLoughlin
9495 ft.
2894 m
Central Point
HARNEY BASIN
HAGERMAN FOSSIL BEDS NATL. MON.

Gold Beach
Medford
OREGON CAVES NATL. MON.
Ashland
Upper Klamath Lake
Summer Lake
STEENS MOUNTAIN

Brookings
KLAMATH MTS.
CASCADE-SISKIYOU NATL. MON.
Klamath Falls
Altamont
Lake Abert
Drake Peak
8407 ft.
2562 m
Hart Lake
Crump Lake

Crescent City
REDWOOD NATIONAL PARK
Montague
Yreka
Tule Lake
Clear Lake Reservoir
Gerber Reservoir
Lakeview
Goose Lake
Upper Lake
GREAT BASIN
DUCK VALLEY INDIAN RESERVATION
Owyhee

HOOPA VALLEY INDIAN RESERVATION
Mt. Shasta
14 162 ft.
4317 m
LAVA BEDS NATL. MON.
Alturas
WARNER MTS.
Duffer Peak
9397 ft.
2864 m
Matterho
10 839 ft.
3304 m

McKinleyville
Arcata
Blue Lake
Willow Creek
Weaverville
Thompson Peak
8994 ft.
2741 m
Weed
Mt. Shasta
Burney
Eagle Peak
9892 ft.
3015 m
Lower Lake
Eagle Lake
SANTA ROSA RANGE
Granite Peak
9732 ft.
2966 m
NEVADA
INDEPENDENCE MTS.

Eureka
Fortuna
Rio Dell
CAPE MENDOCINO
Anderson
Redding
LASSEN VOLCANIC NATIONAL PARK
Lassen Peak
10 457 ft.
3187 m
CALIFORNIA
COAST RANGES
CASCADE RANGE
BLACK ROCK DESERT
SMOKE CREEK DESERT
Winnemucca
Battle Mountain
Carlin
Elko
Spring Creek
RUBY MTS.

PACIFIC OCEAN

STRAIT OF JUAN DE FUCA

0   20   40   60   80   100   120 Miles
0   20   40   60   80   100   120   140   160   180   200 Kilometers

Lambert Conformal Conic Projection
Scale 1:4,000,000
One inch to 64 miles
One cm to 40 km

CANADA
UNITED STATES
SASKATCHEWAN
MANITOBA

MONTANA

NORTH DAKOTA

SOUTH DAKOTA

WYOMING

NEBRASKA

COLORADO

Lambert Conformal Conic Projection
Scale 1:4,000,000
One inch to 64 miles
One cm to 40 km

0   20   40   60   80   100   120 Miles
0   20   40   60   80   100   120   140   160   180   200 Kilometers

A-101620-1
© Rand McNally

Lambert Conformal Conic Projection
Scale 1:4,000,000
One inch to 64 miles
One cm to 40 km

Inset map a
Lambert Conformal Conic Projection
Scale 1:6,000,000
One inch to 96 miles
One cm to 60 km

PACIFIC
OCEAN

NEVADA

GREAT BASIN

CALIFORNIA

MOJAVE DESERT

DEVILS
PLAYGROUND

LOS ANGELES

CHANNEL ISLANDS

CHANNEL
ISLANDS
NATIONAL
PARK

SAN DIEGO

TIJUANA

BAJA
CALIFORNIA

Inset map a
Lambert Conformal Conic Projection
Scale 1:1,000,000
One inch to 16 miles
One cm to 10 km

| 0 | 20 | 40 | 60 | 80 | 100 | 120 Miles |

| 0 | 20 | 40 | 60 | 80 | 100 | 120 | 140 | 160 | 180 | 200 Kilometers |

Lambert Conformal Conic Projection
Scale 1:4,000,000
One inch to 64 miles
One cm to 40 km

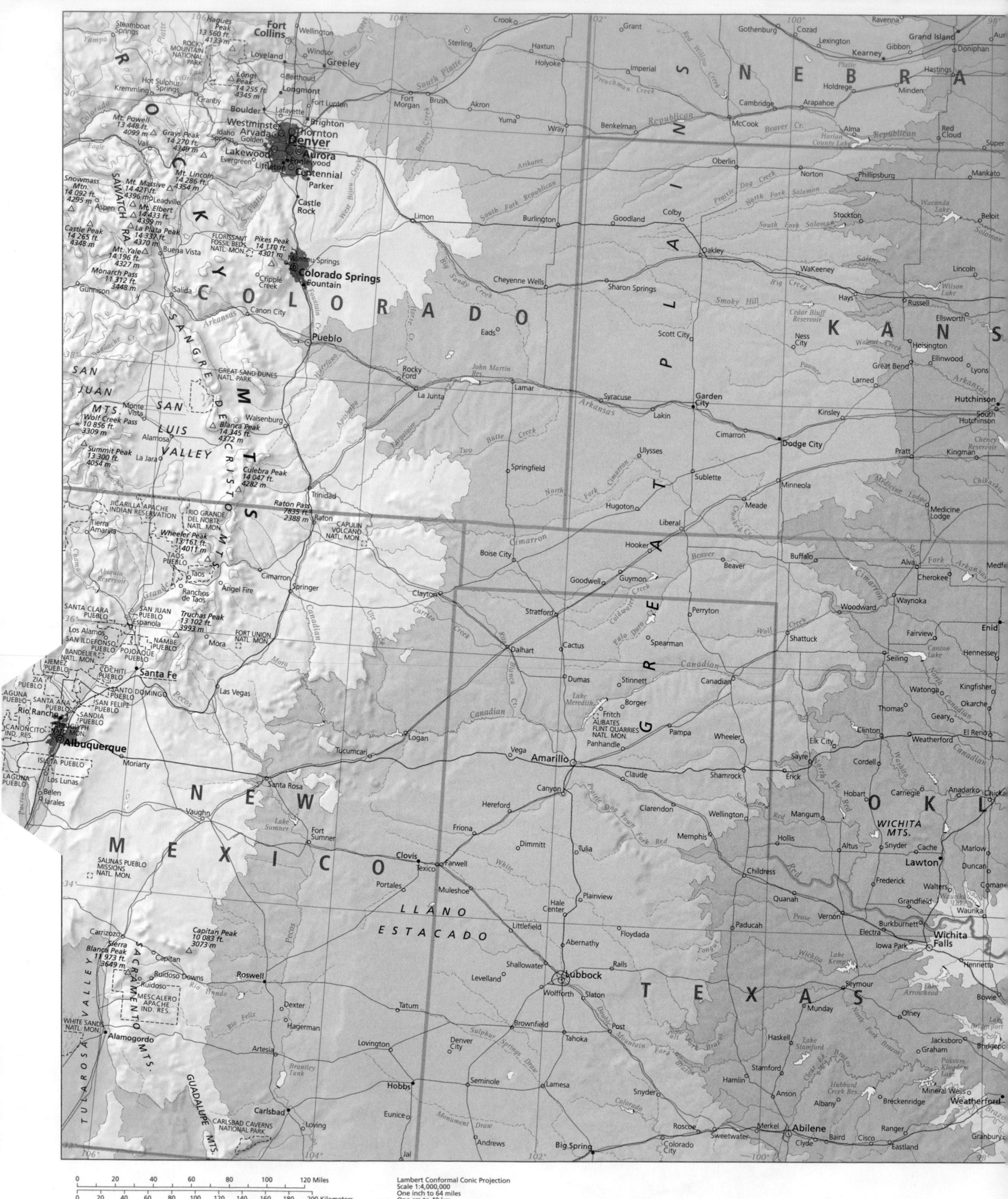

0   20   40   60   80   100   120 Miles
0   20   40   60   80   100   120   140   160   180   200 Kilometers

Lambert Conformal Conic Projection
Scale 1:4,000,000
One inch to 64 miles
One cm to 40 km

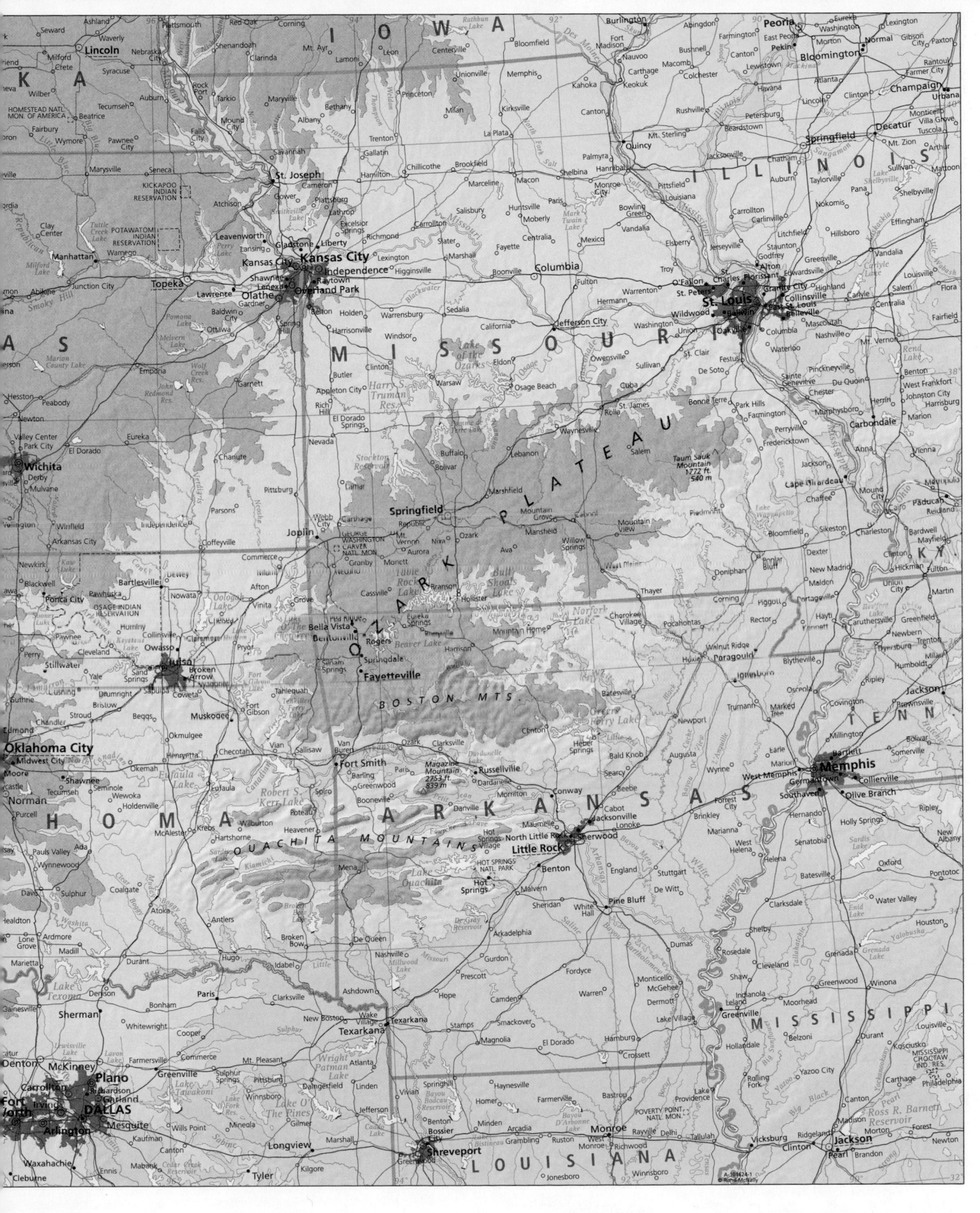

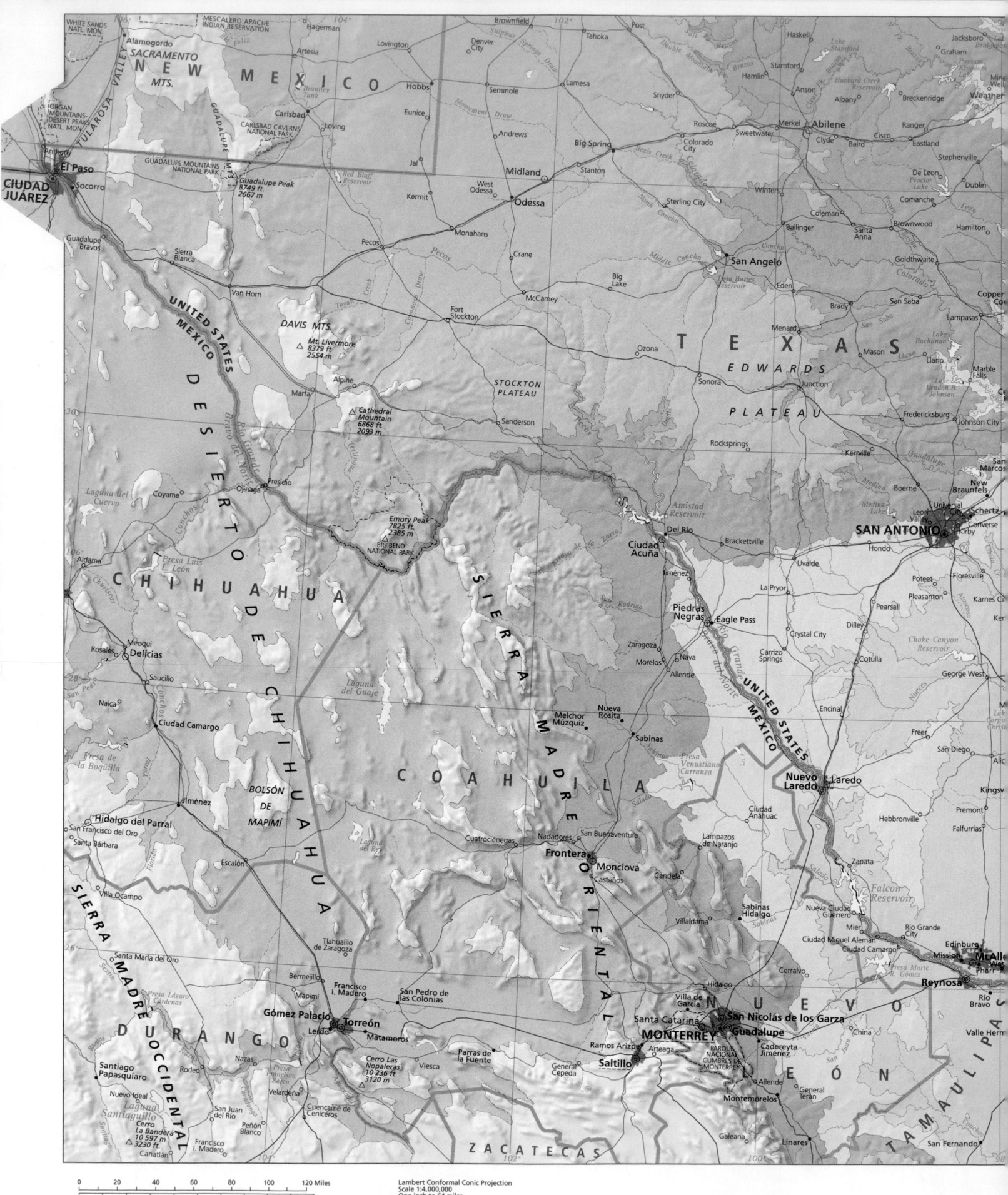

Inset map a
Lambert Conformal Conic Projection
Scale 1:1,000,000
One inch to 16 miles
One cm to 10 km

© Rand McNally
A-101626-1

0   20   40   60   80   100   120 Miles

0   20   40   60   80   100   120   140   160   180   200 Kilometers

Lambert Conformal Conic Projection
Scale 1:4,000,000
One inch to 64 miles
One cm to 40 km

Inset map a
Lambert Conformal Conic Projection
Scale 1:4,000,000
One inch to 64 miles
One cm to 40 km

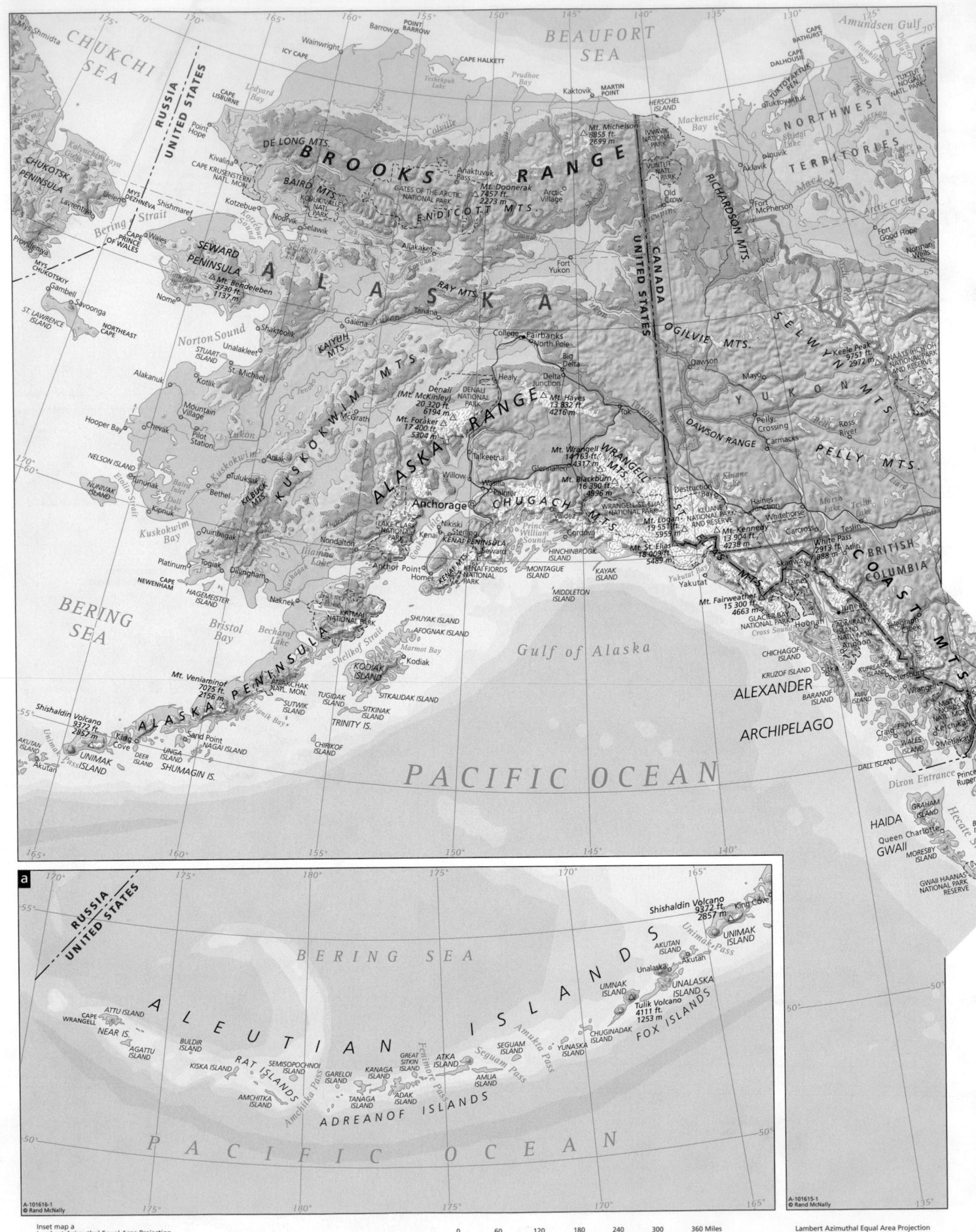

CHUKCHI SEA
RUSSIA
UNITED STATES
CHUKOTSKI PENINSULA
Mys Shmidta
Kolyuchinskaya Guba
Uelen
Mys Dezhneva
Laventiya
Providenya
Mys Chukotskiy
Gambell
Savoonga
ST LAWRENCE ISLAND
NORTHEAST CAPE

BEAUFORT SEA
Barrow
POINT BARROW
ICY CAPE
Wainwright
CAPE HALKETT
Prudhoe Bay
Kaktovik
MARTIN POINT
Mt. Michelson 8855 ft. 2699 m
HERSCHEL ISLAND
IVVAVIK NATIONAL PARK

AMUNDSEN GULF
CAPE BATHURST
CAPE DALHOUSIE
Franklin Bay
TUKTOYAKTUK PEN.
TUKTUT NOGAIT NATL. PARK
Tuktoyaktuk

NORTHWEST TERRITORIES

CAPE LISBURNE
Ledyard Bay
CAPE KRUSENSTERN NATL. MON.
Point Hope
Kivalina
Teshekpuk Lake
Colville
Anaktuvuk Pass
GATES OF THE ARCTIC NATIONAL PARK
KOBUK VALLEY NATL. PARK
BAIRD MTS.
DE LONG MTS.
BROOKS RANGE
Mt. Doonerak 7457 ft. 2273 m
ENDICOTT MTS.
Arctic Village
VUNTUT NATL. PARK
Old Crow
Arctic Circle
Inuvik
Aklavik
RICHARDSON MTS.
Fort McPherson
Fort Good Hope
Norman Wells
Mackenzie Bay

Kotzebue
Noorvik
Selawik
Kotzebue Sound
Allakaket
Fort Yukon
OGILVIE MTS.
Keele Peak 9751 ft 2972 m
NAATS IHCH'OH NATIONAL PARK AND RESERVE

Shishmaref
CAPE PRINCE OF WALES
Wales
Bering Strait
SEWARD PENINSULA
Mt. Bendeleben 3730 ft. 1137 m
RAY MTS.
Tanana
CANADA
UNITED STATES
Dawson
Mayo
SELWYN MTS.
YUKON MTS.
Ross River

Nome
Nome
Galena
College Fairbanks
North Pole
Big Delta
Pelly Crossing
Carmacks
PELLY MTS.
NORTON SOUND
Unalakleet
KAIYUH MTS.
Healy
Delta Junction
DAWSON RANGE
Alakanuk
STUART ISLAND
St. Michael
Shaktoolik
Denali (Mt. McKinley) 20,320 ft. 6194 m
DENALI NATIONAL PARK
Mt. Hayes 13 832 ft. 4216 m
Tok
Kotlik
Mountain Village
Chevak
Hooper Bay
Pilot Station
KUSKOKWIM MTS.
McGrath
Mt. Foraker 17 400 ft. 5304 m
Talkeetna
ALASKA RANGE
Mt. Wrangell 14 163 ft. 4317 m
WRANGELL MTS.
Destruction Bay
KLUANE NATIONAL PARK AND RESERVE
Whitehorse
Teslin Lake
Marsh Lake

NUNIVAK ISLAND
NELSON ISLAND
Bethel
KILBUCK MTS.
Aniak
Willow
Wasilla
Palmer
Mt. Blackburn 16 390 ft. 4996 m
CHUGACH MTS.
Glennallen
Mt. Logan 19 551 ft. 5959 m
Haines Junction
Carcross
White Pass 2913 ft 888 m
Skagway
Haines
BRITISH COLUMBIA

Tununak
Kuskokwim Bay
Quinhagak
Platinum
Togiak
Dillingham
Naknek
LAKE CLARK NATIONAL PARK
Nondalton
Anchorage
Nikiski
Sterling
KENAI PENINSULA
Kenai
Valdez
Cordova
Prince William Sound
WRANGELL-ST. ELIAS NATIONAL PARK
Mt. Kennedy 13 904 ft. 4238 m
Mt. St. Elias 18 009 ft. 5489 m
Yakutat Bay
Mt. Logan

BERING SEA
CAPE NEWENHAM
HAGEMEISTER ISLAND
Bristol Bay
Becharof Lake
KATMAI NATIONAL PARK
Anchor Point
Homer
Seward
KENAI FJORDS NATIONAL PARK
HINCHINBROOK ISLAND
MONTAGUE ISLAND
KAYAK ISLAND
MIDDLETON ISLAND
Mt. Fairweather 15 300 ft 4663 m
GLACIER BAY NATIONAL PARK
Cross Sound
Hoonah
Yakutat
Juneau

Mt. Veniaminof 7075 ft. 2156 m
ANIAKCHAK NATL. MON.
SHUYAK ISLAND
AFOGNAK ISLAND
Marmot Bay
Gulf of Alaska
ADMIRALTY ISLAND
ANGOON NATL. MON.
Angoon

ALASKA PENINSULA
Shishaldin Volcano 9372 ft. 2857 m
KODIAK ISLAND
Kodiak
CHICHAGOF ISLAND
KRUZOF ISLAND
BARANOF ISLAND
Petersburg
Wrangell
Sitka

King Cove
DEER ISLAND
Sand Point
NAGAI ISLAND
SHUMAGIN IS.
Ugnik Bay
SUTWIK ISLAND
TUGIDAK ISLAND
SITKALIDAK ISLAND
TRINITY IS.
CHIRIKOF ISLAND
ALEXANDER ARCHIPELAGO
BARANOF ISLAND
MISTY FJORD NATL. MON.
Craig
PRINCE OF WALES ISLAND
KETCHIKAN
Metlakatla
Stewart

AKUTAN ISLAND
UNIMAK ISLAND
Akutan
King Cove
Akutan Pass
Unimak Pass
DALL ISLAND
Dixon Entrance
Prince Rupert

HAIDA GWAII
GRAHAM ISLAND
Queen Charlotte
MORESBY ISLAND
BANKS ISLAND
Hecate Strait
GWAII HAANAS NATIONAL PARK RESERVE

PACIFIC OCEAN

RUSSIA
UNITED STATES
BERING SEA
Shishaldin Volcano 9372 ft. 2857 m
King Cove
AKUTAN ISLAND
Akutan
Unimak Pass
UNIMAK ISLAND
UMNAK ISLAND
Unalaska
UNALASKA ISLAND
Tulik Volcano 4111 ft. 1253 m
CHUGINADAK ISLAND
YUNASKA ISLAND
FOX ISLANDS

CAPE WRANGELL
ATTU ISLAND
NEAR IS.
AGATTU ISLAND
BULDIR ISLAND
SEMISOPOCHNOI ISLAND
KISKA ISLAND
RAT ISLANDS
Amchitka Pass
AMCHITKA ISLAND
GARELOI ISLAND
KANAGA ISLAND
TANAGA ISLAND
ADAK ISLAND
ADREANOF ISLANDS
ALEUTIAN ISLANDS
GREAT SITKIN ISLAND
ATKA ISLAND
Finmore Pass
SEGUAM ISLAND
Seguam Pass
AMLIA ISLAND
Amukta Pass

PACIFIC OCEAN

A-101616-1
© Rand McNally

Inset map a
Lambert Azimuthal Equal Area Projection
Scale 1:12,000,000
One inch to 190 miles
One cm to 120 km

0 60 120 180 240 300 360 Miles
0 60 120 180 240 300 360 420 480 540 600 Kilometers

A-101615-1
© Rand McNally

Lambert Azimuthal Equal Area Projection
Scale 1:12,000,000
One inch to 190 miles
One cm to 120 km

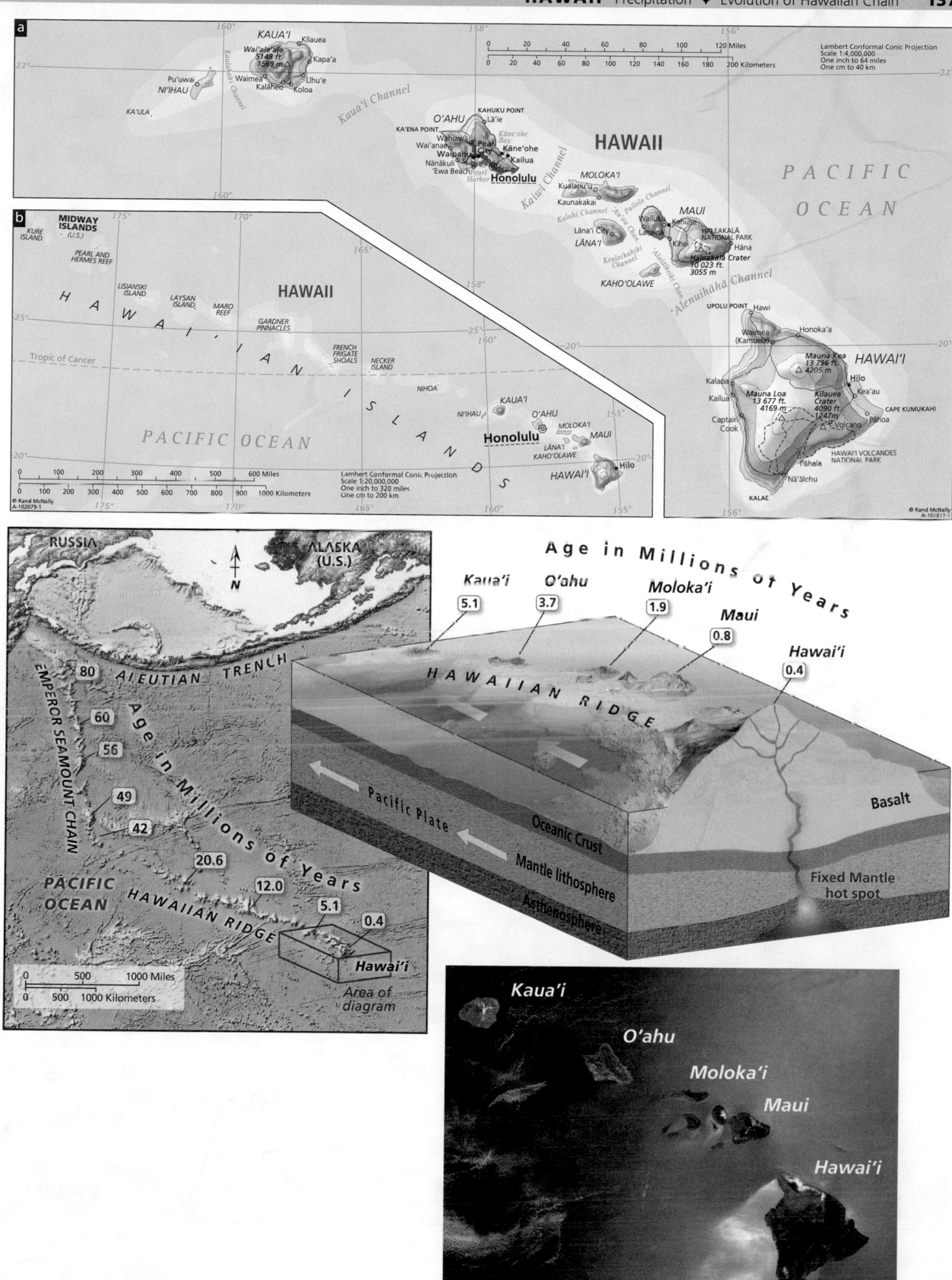

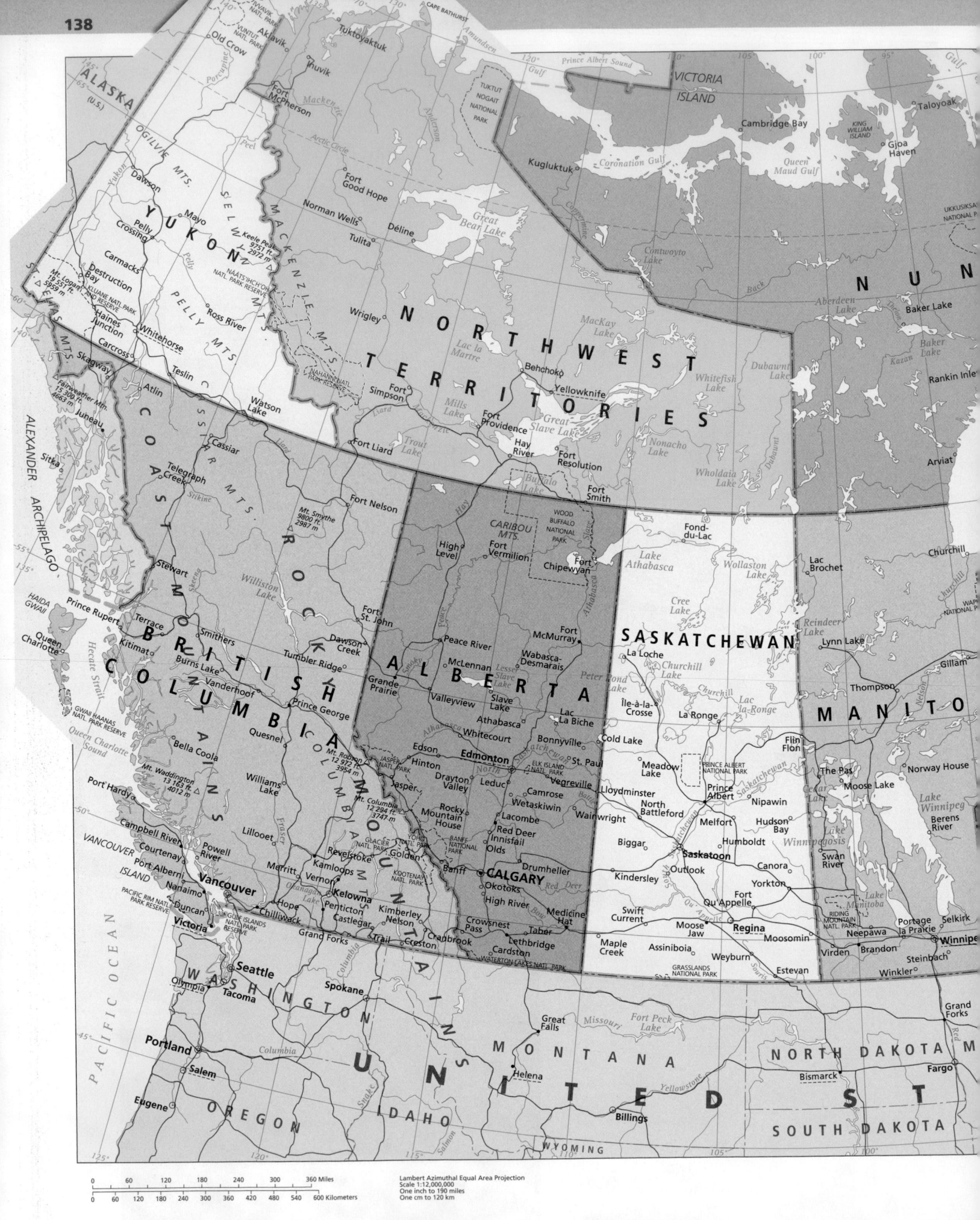

138

ALASKA (U.S.)

YUKON

NORTHWEST TERRITORIES

NUN

VICTORIA ISLAND

BRITISH COLUMBIA

ALBERTA

SASKATCHEWAN

MANITO

ROCKY MOUNTAINS

COLUMBIA MOUNTAINS

UNITED ST

WASHINGTON

OREGON

IDAHO

MONTANA

WYOMING

NORTH DAKOTA

SOUTH DAKOTA

M

PACIFIC OCEAN

ALEXANDER ARCHIPELAGO

COAST MTS.

OGILVIE MTS.

SELWYN MTS.

PELLY MTS.

MACKENZIE MTS.

CASSIAR MTS.

CARIBOU MTS.

VANCOUVER ISLAND

HAIDA GWAII

Old Crow
Aklavik
Inuvik
Tuktoyaktuk
CAPE BATHURST
Amundsen
Prince Albert Sound
Gulf
Cambridge Bay
KING WILLIAM ISLAND
Talooyak
Gjoa Haven
Kugluktuk
Coronation Gulf
Queen Maud Gulf
Fort McPherson
Norman Wells
Fort Good Hope
Déline
Tulita
Great Bear Lake
Coppermine
Contwoyto Lake
UKKUSIKSALIK NATIONAL PARK
Dawson
Mayo
Pelly Crossing
Carmacks
Keele Peak 9751 ft. 2972 m
NAATS'IHCH'OH NATL. PARK RESERVE
Wrigley
NORTHWEST TERRITORIES
MacKay Lake
Aberdeen Lake
Thelon
Baker Lake
Destruction Bay
Mt. Logan 19,551 ft. 5959 m
Mt. Fairweather 15,300 ft. 4663 m
KLUANE NATL. PARK RESERVE
Ross River
Fort Simpson
NAHANNI NATL PARK RESERVE
Mills Lake
Fort Providence
Behchokò
Yellowknife
Great Slave Lake
Fort Resolution
Whitefish Lake
Nonacho Lake
Dubawnt Lake
Dubawnt
Kazan
Baker Lake
Rankin Inlet
Haines Junction
Whitehorse
Carcross
Skagway
Teslin
Atlin
Watson Lake
Fort Liard
Trout Lake
Liard
Hay River
Fort Smith
Wholdaia Lake
Arviat
Juneau
Sitka
Telegraph Creek
Cassiar
Fort Nelson
Mt. Smythe 9800 ft. 2987 m
Hay
WOOD BUFFALO NATIONAL PARK
High Level
Fort Vermilion
Fort Chipewyan
Fort Smith
Buffalo Lake
Fond-du-Lac
Lac Brochet
Churchill
Stewart
Skeena
Fort St. John
Peace River
Fort McMurray
Lake Athabasca
Cree Lake
Wollaston Lake
La Loche
Reindeer Lake
Lynn Lake
Prince Rupert
Terrace
Kitimat
Smithers
Dawson Creek
Tumbler Ridge
McLennan
Wabasca-Desmarais
Lesser Slave Lake
Peter Pond Lake
Churchill
Lac la Ronge
La Ronge
Île-à-la-Crosse
Gillam
Queen Charlotte
Hecate Strait
Burns Lake
Vanderhoof
Prince George
Grande Prairie
Valleyview
Slave Lake
Athabasca
Lac la Biche
Bonnyville
Cold Lake
St. Paul
Flin Flon
The Pas
Thompson
Norway House
GWAII HAANAS NATL. PARK RESERVE
Quesnel
Bella Coola
Mt. Robson 12,972 ft. 3954 m
JASPER NATL. PARK
Edson
Hinton
Drayton Valley
Whitecourt
Edmonton
Leduc
ELK ISLAND NATL. PARK
Vegreville
Lloydminster
Meadow Lake
PRINCE ALBERT NATIONAL PARK
Prince Albert
Nipawin
Moose Lake
Lake Winnipeg
Berens River
Port Hardy
Mt. Waddington 13,163 ft. 4012 m
Williams Lake
Mt. Columbia 12,294 ft. 3747 m
Jasper
Rocky Mountain House
Camrose
Wetaskiwin
Wainwright
North Battleford
Biggar
Melfort
Humboldt
Hudson Bay
Swan River
Campbell River
Courtenay
Port Alberni
Nanaimo
Lilloet
Merritt
Revelstoke
Golden
GLACIER NATL. PARK
YOHO NATL. PARK
BANFF NATIONAL PARK
Lacombe
Red Deer
Innisfail
Olds
Saskatoon
Canora
Yorkton
Lake Winnipegosis
Powell River
Kamloops
Vernon
Kelowna
Okanagan Lake
KOOTENAY NATL. PARK
Banff
Drumheller
Outlook
Kindersley
Fort Qu'Appelle
RIDING MOUNTAIN NATIONAL PARK
Portage la Prairie
Neepawa
Selkirk
Winnipeg
VANCOUVER
Duncan
Hope
Penticton
Castlegar
Kimberley
Nelson
CALGARY
Okotoks
High River
Swift Current
Moose Jaw
Regina
Moosomin
Virden
Brandon
Steinbach
Victoria
Seattle
Olympia
Tacoma
PACIFIC RIM NATL. PARK RESERVE
GULF ISLANDS NATL. PARK RESERVE
Chilliwack
Grand Forks
Trail
Cranbrook
Creston
Crowsnest Pass
Taber
Lethbridge
Cardston
WATERTON LAKES NATL. PARK
Medicine Hat
Maple Creek
Assiniboia
Weyburn
Estevan
GRASSLANDS NATIONAL PARK
Winkler
Grand Forks
Portland
Salem
Eugene
Spokane
Great Falls
Helena
Billings
Bismarck
Fargo

Lambert Azimuthal Equal Area Projection
Scale 1:12,000,000
One inch to 190 miles
One cm to 120 km

0   60   120   180   240   300   360 Miles
0   60   120   180   240   300   360   420   480   540   600 Kilometers

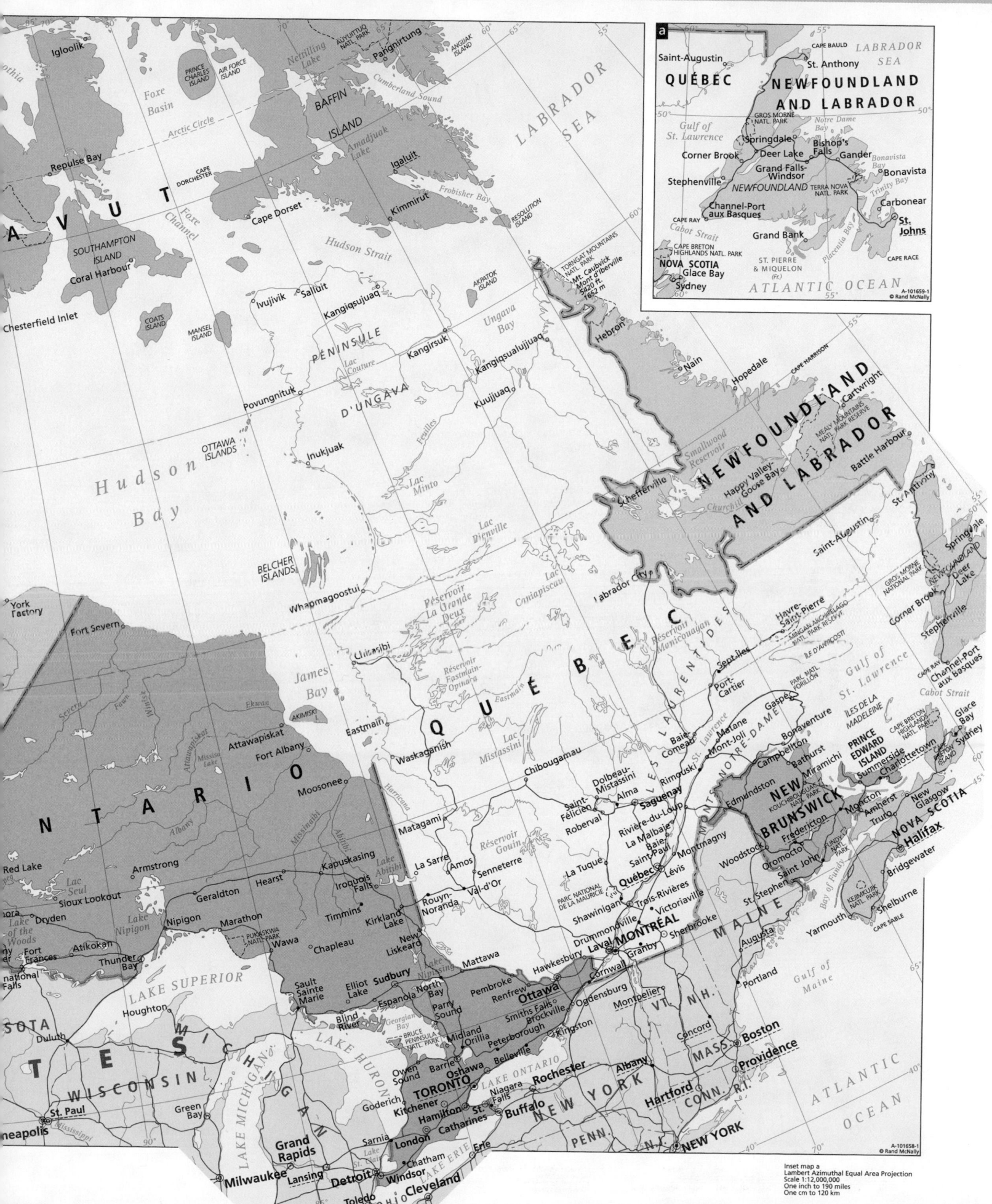

Inset map a

QUÉBEC

NEWFOUNDLAND AND LABRADOR

Saint-Augustin • CAPE BAULD • LABRADOR SEA
St. Anthony
Gulf of St. Lawrence • GROS MORNE NATL. PARK • Notre Dame Bay
Springdale • Bishop's Falls • Gander • Bonavista Bay
Corner Brook • Deer Lake • Bonavista
Stephenville • Grand Falls-Windsor • NEWFOUNDLAND • TERRA NOVA NATL. PARK • Trinity Bay
Channel-Port aux Basques • Carbonear
CAPE RAY • Cabot Strait • Grand Bank • St. John's
CAPE BRETON HIGHLANDS NATL. PARK • ST. PIERRE & MIQUELON (Fr.) • Placentia Bay • CAPE RACE
NOVA SCOTIA • Glace Bay
Sydney • ATLANTIC OCEAN
A-101659-1 © Rand McNally

Inset map a
Lambert Azimuthal Equal Area Projection
Scale 1:12,000,000
One inch to 190 miles
One cm to 120 km
A-101658-1 © Rand McNally

ALASKA (U.S.)

OGILVIE MTS.
Tombstone Mountain 7192 ft. 2192 m

RICHARDSON MTS.

Inuvik

TUKTOYAKTUK PENINSULA

CAPE DALHOUSIE

CAPE BATHURST

CAPE PARRY

Amundsen Gulf

CAPE BARING

Prince Albert Sound

VICTORIA ISLAND

Washburn Lake

CAPE FELIX

BOOTHIA PENINSULA

Gulf

Striding Lake

Mackenzie

Arctic Circle

WOLLASTON PENINSULA

Coronation Gulf

KENT PENINSULA

MELBOURNE ISLAND

Queen Maud Gulf

KING WILLIAM ISLAND

ADELAIDE PENINSULA

Cambridge Bay

MacAlpine Lake

N U N A

SELWYN MTS.

Nadaleen Mountain 7210 ft. 2210 m

MACKENZIE

Dawson

YUKON TERRITORY

Kluane Lake

St. LUCANIA 18 009 ft. 5489 m

Mt. Logan 19 551 ft. 5959 m

Mt. Kennedy 8 984 ft. 4238 m

St. ELIAS MTS.

Fairweather Mtn. 15 300 ft. 4663 m

Whitehorse

PELLY MTS.

Keele Peak 9751 ft. 2972 m

Ross Mountain 7887 ft. 2404 m

LOGAN MTS.

Mt. Sir James MacBrien 9062 ft. 2762 m

FRANKLIN MTS.

Mackenzie

Great Bear Lake

Horton

Aubry Lake

Colville Lake

N O R T H W E S T   T E R R I T O R I E S

Napaktulik Lake

Contwoyto Lake

Back

Garry Lake

Aberdeen Lake

Tehek Lake

Beechey Lake

MacKay Lake

Napaktulik Lake

COAST

BOUNDARY RANGES

Mt. Nesselrode 8104 ft. 2470 m

CHICHAGOF ISLAND

ADMIRALTY ISLAND

Mt. Sait 10 289 ft. 9136 m

BARANOF ISLAND

KUPREANOF ISLAND

ALEXANDER ARCHIPELAGO

PRINCE OF WALES ISLAND

CASSIAR MOUNTAINS

Teslin Lake

Sharktooth Mountain 8750 ft. 2667 m

SKEENA MTS.

OMINECA MOUNTAINS

Shelagyote Peak 8091 ft. 2466 m

Mt. Smythe 9800 ft. 2987 m

Bronlund Peak 8570 ft. 2594 m

Fort Nelson

CARIBOU MTS.

HORN PLATEAU

Yellowknife

Great Slave Lake

Buffalo Lake

Whitefish Lake

Nonacho Lake

Wholdaia Lake

Kasba Lake

Selwyn Lake

Nueltin Lake

Kaminak Lake

Kaminuriak Lake

Baker Lake

Mallery Lake

Dubawnt

Artillery Lake

Aylmer Lake

Lac la Martre

Trout Lake

Tathlina Lake

Bistcho Lake

Slave

Lake Claire

Lake Athabasca

Cree Lake

Wollaston Lake

Reindeer Lake

Tadoule Lake

Northern Indian Lake

Stephens Lake

Churchill

CHURCHILL

BIRCH MTS.

Athabasca

SASKATCHEWAN

M A N I T O

Thompson

CAPE KNOX

HAIDA GWAII

GRAHAM ISLAND

MORESBY ISLAND

Hecate Strait

CAPE ST. JAMES

Queen Charlotte Sound

CAPE KNOX

Prince Rupert

B R I T I S H   C O L U M B I A

Howson Peak 9009 ft. 2785 m

R O C K Y   M O U N T A I N S

Stuart Lake

Nechako Reservoir

Prince George

FRASER PLATEAU

CARIBOO MTS.

Good Hope Mountain 10 630 ft. 3240 m

Mt. Garibaldi 8786 ft. 2678 m

Okanagan

Fort St. John

Peace

Lesser Slave Lake

Smoky

Athabasca

A L B E R T A

Mt. Robson 12 972 ft. 3954 m

Mt. Columbia 12 294 ft. 3747 m

Mt. Assiniboine 11 870 ft. 3618 m

Edmonton

Battle

North Saskatchewan

Saskatoon

Lesser Slave Lake

Peter Pond Lake

Dore Lake

Lac la Ronge

Churchill

Montreal Lake

Deschambault Lake

Beaver

Churchill Lake

Highrock Lake

Flin Flon

Kisskittogisu Lake

Gods Lake

Island Lake

Lake Winnipeg

Cedar Lake

COLUMBIA MOUNTAINS

PRINCESS ROYAL ISLAND

Mt. Waddington 13 163 ft. 4017 m

PACIFIC RANGES

VANCOUVER ISLAND

CAPE COOK

CAPE FLATTERY

Mt. Olympus 7969 ft. 2429 m

Vancouver

Seattle

Mt. Baker 10 785 ft. 3278 m

Mt. Rainier 14 411 ft. 4392 m

Mt. Hood 11 239 ft. 3426 m

Portland

COAST RANGES

CASCADE RANGE

OREGON

WASHINGTON

Columbia

Shuswap Lake

Kootenay Lake

BITTERROOT RANGE

Salmon

I D A H O

Flathead Lake

Kootenay

Columbia

U N I T E D

M O N T A N A

Missouri

Fort Peck Lake

Milk

Yellowstone

Granite Peak 12 799 ft. 3901 m

South Saskatchewan

Red Deer

Bow

Calgary

Qu'Appelle

Regina

Lake Diefenbaker

Saskatchewan

Assiniboine

Souris

Lake Winnipegosis

Dauphin Lake

Lake Manitoba

Winnipeg

Winnipeg

Lake Sakakawea

Missouri

N O R T H   D A K O T A

S O U T H   D A K O T A

S T A

PACIFIC OCEAN

0   60   120   180   240   300   360 Miles

0   60   120   180   240   300   360   420   480   540   600 Kilometers

Lambert Azimuthal Equal Area Projection
Scale 1:12,000,000
One inch to 190 miles
One cm to 120 km

Inset map a
Lambert Conformal Conic Projection
Scale 1:1,000,000
One inch to 16 miles
One cm to 10 km

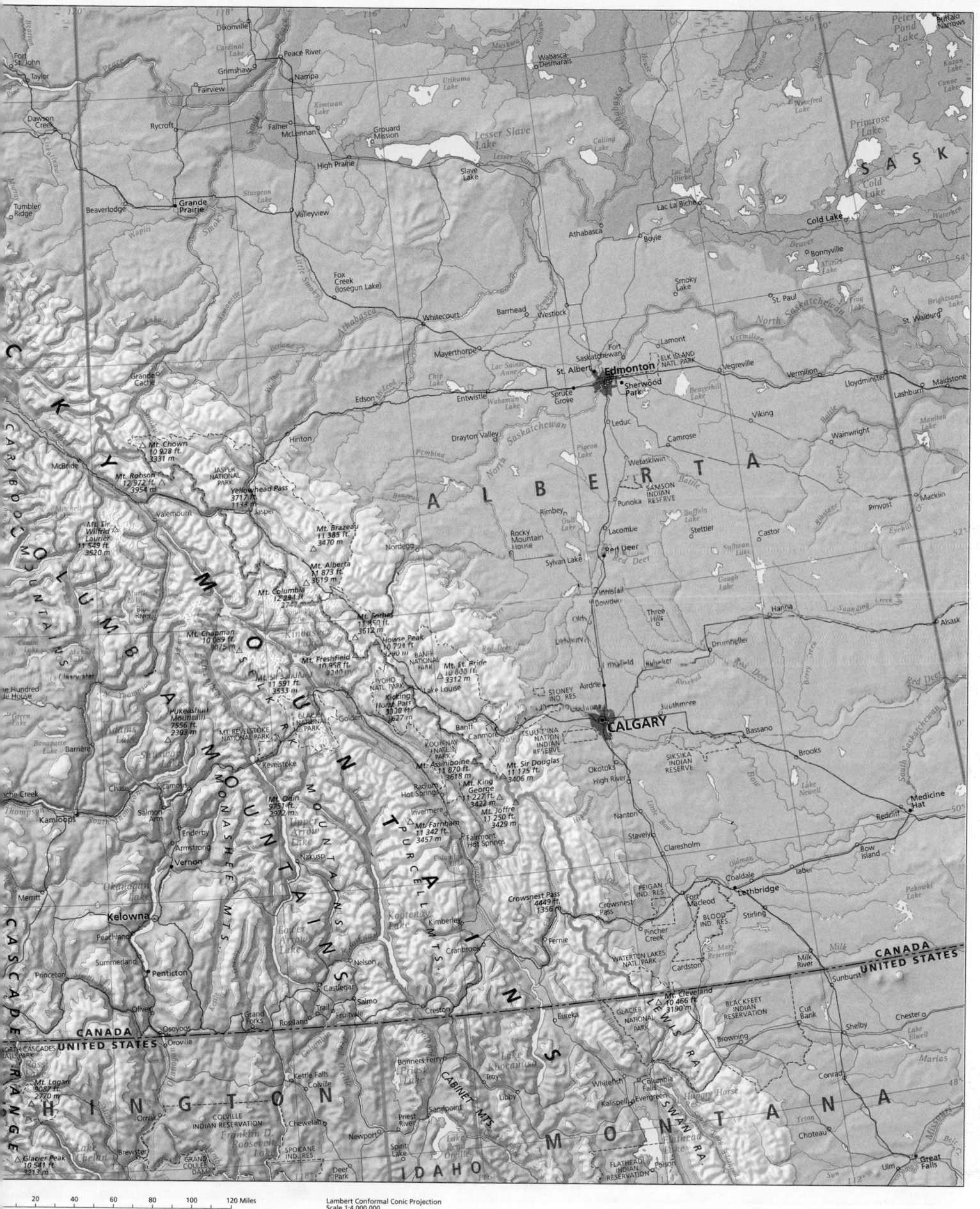

S A S K.

A L B E R T A

R O C K Y   M O U N T A I N S

C A R I B O O   M O U N T A I N S

C O L U M B I A   M O U N T A I N S

M O N A S H E E   M T S.

S E L K I R K   M O U N T A I N S

P U R C E L L   M T S.

L E W I S   R A N G E

C A S C A D E   R A N G E

W A S H I N G T O N

I D A H O

M O N T A N A

CANADA
UNITED STATES

Fort St. John
Taylor
Dawson Creek
Tumbler Ridge
Beaverlodge
Grande Prairie
Rycroft
Dixonville
Grimshaw
Peace River
Fairview
Nampa
Falher
McLennan
High Prairie
Grouard Mission
Slave Lake
Valleyview
Grande Cache
McBride
Fox Creek (Iosegun Lake)
Whitecourt
Barrhead
Westlock
Athabasca
Boyle
Smoky Lake
St. Paul
Lamont
Vegreville
Vermilion
Lloydminster
Lashburn
Maidstone
Cold Lake
Bonnyville
St. Walburg
Brightsand Lake

Mt. Chown
10 928 ft.
3331 m
Mt. Robson
12 972 ft.
3954 m
Jasper National Park
Yellowhead Pass
3717 ft.
1133 m
Jasper
Hinton
Edson
Mayerthorpe
Entwistle
Spruce Grove
Edmonton
St. Albert
Fort Saskatchewan
Sherwood Park
ELK ISLAND NATL. PARK

Mt. Sir Wilfrid Laurier
11 549 ft.
3520 m
Valemount
Blue River
Mt. Brazeau
11 385 ft.
3470 m
Mt. Alberta
11 873 ft.
3619 m
Mt. Columbia
12 294 ft.
3747 m
Nordegg
Drayton Valley
Leduc
Wetaskiwin
Camrose
Viking
Wainwright

Mt. Chapman
10 089 ft.
3075 m
Mt. Forbes
11 850 ft.
3612 m
Howse Peak
10 794 ft.
Mt. Freshfield
10 958 ft.
BANFF NATIONAL PARK
Mt. St. Bride
10 680 ft.
3312 m
Rocky Mountain House
Rimbey
Gull Lake
Punoka
Lacombe
Stettler
Castor
Provost
Macklin
SAMSON INDIAN RESERVE
Buffalo Lake

Mt. Sir Sandford
11 591 ft.
3533 m
Pukeashun Mountain
7556 ft.
2303 m
MT. REVELSTOKE NATIONAL PARK
GLACIER NATIONAL PARK
Golden
YOHO NATL. PARK
Kicking Horse Pass
5330 ft.
1627 m
Lake Louise
KOOTENAY NATL. PARK
Sylvan Lake
Red Deer
Innisfail
Dowson
Olds
Didsbury
Three Hills
Hanna
Alsask

Revelstoke
Sicamous
Chase
Salmon Arm
Enderby
Armstrong
Vernon
Mt. Odin
9751 ft.
2972 m
Upper Arrow Lake
Nakusp
Mt. Assiniboine
11 870 ft.
3618 m
Radium Hot Springs
Invermere
Mt. King George
11 227 ft.
3422 m
Mt. Joffre
11 250 ft.
3429 m
Mt. Farnham
11 342 ft.
3457 m
Fairmont Hot Springs
Banff
Canmore
Airdrie
CALGARY
TSUU T'INA NATION INDIAN RESERVE
STONEY IND. RES.
Cochrane
Southmore
Bassano
Brooks
SIKSIKA INDIAN RESERVE

Kamloops
Kelowna
Peachland
Summerland
Penticton
Merritt
Okanagan Lake
Lower Arrow Lake
Kootenay Lake
Kimberley
Cranbrook
Okotoks
High River
Nanton
Stavely
Claresholm
Medicine Hat
Redcliff
Bow Island
Lake Newell

Princeton
Oliver
Osoyoos
Oroville
Grand Forks
Rossland
Trail
Fruitvale
Salmo
Nelson
Castlegar
Creston
Crowsnest Pass
4449 ft.
1356 m
Crowsnest Pass
Fernie
Pincher Creek
Coaldale
Taber
Coalhurst
Fort Macleod
Lethbridge
Stirling
PEIGAN IND. RES.
BLOOD IND. RES.
Pakowki Lake

NORTH CASCADES NATL. PARK
Mt. Logan
9087 ft.
2770 m
Glacier Peak
10 541 ft.
3213 m
Brewster
COLVILLE INDIAN RESERVATION
GRAND COULEE DAM
Omak
Chewelah
Colville
Kettle Falls
Newport
Bonners Ferry
Priest River
Sandpoint
Libby
Troy
Eureka
WATERTON LAKES NATL. PARK
Cardston
GLACIER NATIONAL PARK
Mt. Cleveland
10 466 ft.
3190 m
Whitefish
Columbia Falls
Kalispell
Evergreen
BLACKFEET INDIAN RESERVATION
Browning
Cut Bank
Sunburst
Shelby
Chester
Conrad

SPOKANE IND. RES.
Deer Park
Spokane
Spirit Lake
Priest Lake
Lake Pend Oreille
CABINET MTS.
Lake Koocanusa
Hungry Horse
SWAN RA.
FLATHEAD INDIAN RESERVATION
Polson
Great Falls

CANADA
UNITED STATES
Milk River

20   40   60   80   100   120 Miles
20   40   60   80   100   120   140   160   180   200 Kilometers

Lambert Conformal Conic Projection
Scale 1:4,000,000
One inch to 64 miles
One cm to 40 km

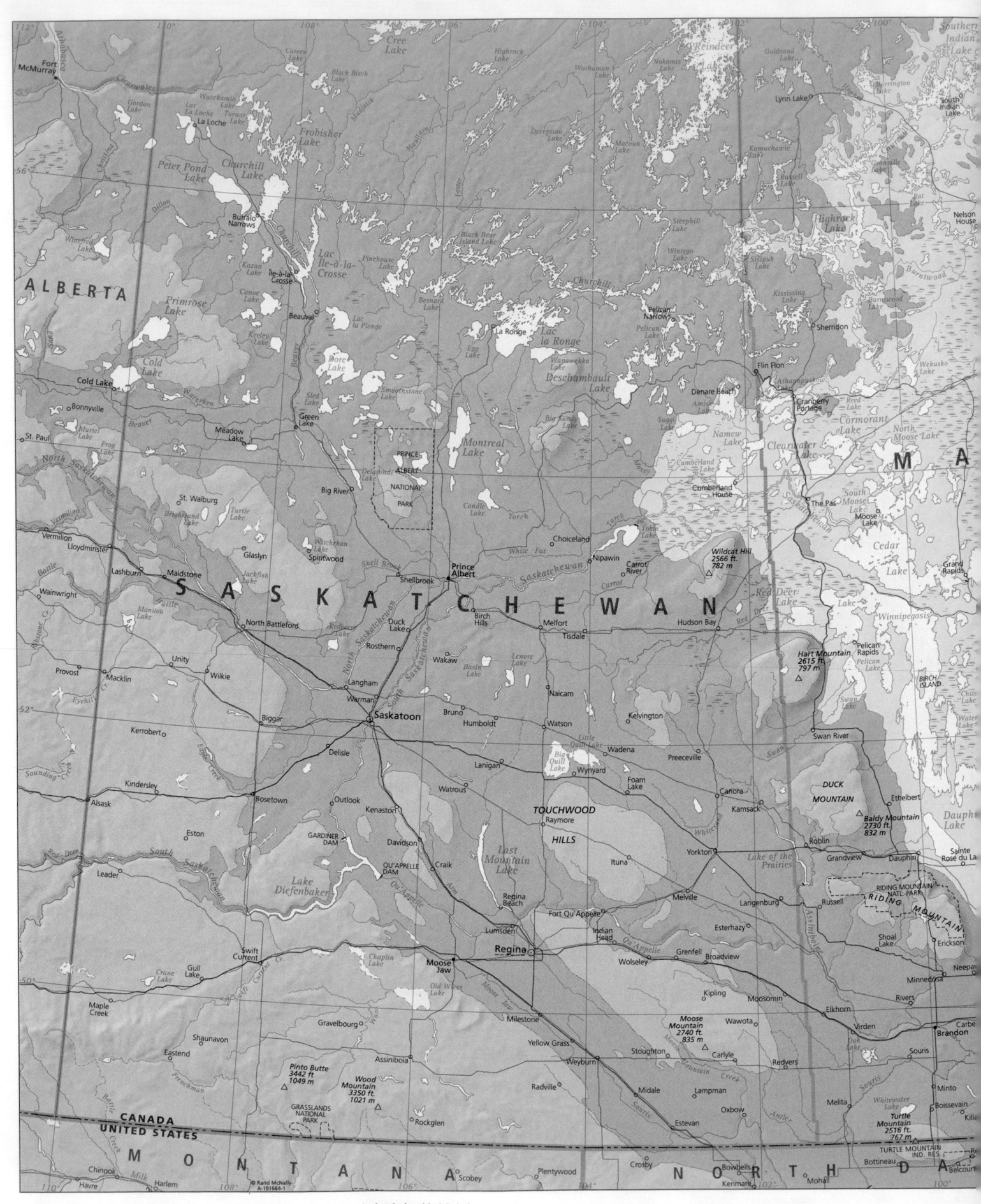

ALBERTA

SASKATCHEWAN

MANITOBA

Fort McMurray

Lynn Lake

Sherridon

Flin Flon

Denare Beach

Cranberry Portage

Cold Lake

Bonnyville

St. Paul

Muriel Lake

Vermilion

Lloydminster

St. Walburg

Glaslyn

Spiritwood

Big River

Green Lake

Meadow Lake

Beauval

Ile-à-la-Crosse

Buffalo Narrows

La Loche

Peter Pond Lake

Churchill Lake

Lac Ile-à-la-Crosse

Primrose Lake

Keeley Lake

Dore Lake

Sled Lake

Smoothstone Lake

PRINCE ALBERT NATIONAL PARK

Montreal Lake

La Ronge

Lac la Ronge

Pelican Narrows

Deschambault Lake

Cumberland House

The Pas

Moose Lake

South Moose Lake

Cedar Lake

Grand Rapids

Cormorant Lake

Clearwater Lake

Nelson House

Winnipegosis

Lake Winnipegosis

Wainwright

Provost

Macklin

Unity

Wilkie

North Battleford

Maidstone

Lashburn

Biggar

Kerrobert

Kindersley

Rosetown

Alsask

Eston

Leader

Maple Creek

Gull Lake

Swift Current

Eastend

Shaunavon

Gravelbourg

Assiniboia

Saskatoon

Delisle

Outlook

Kenaston

Davidson

Craik

GARDINER DAM

QU'APPELLE DAM

Lake Diefenbaker

Chaplin Lake

Moose Jaw

Rosthern

Warman

Langham

Wakaw

Bruno

Humboldt

Watson

Watrous

Lanigan

Last Mountain Lake

Regina Beach

Lumsden

Regina

Milestone

Yellow Grass

Radville

Rockglen

Weyburn

Stoughton

Midale

Lampman

Estevan

Oxbow

Birch Hills

Duck Lake

Shellbrook

Prince Albert

Melfort

Tisdale

Hudson Bay

Naicam

Kelvington

Wadena

Wynyard

Foam Lake

Big Quill Lake

Little Quill Lake

Raymore

TOUCHWOOD HILLS

Ituna

Fort Qu'Appelle

Indian Head

Wolseley

Grenfell

Broadview

Kipling

Carlyle

Redvers

Choiceland

Nipawin

Carrot River

Wildcat Hill 2566 ft. 782 m

Hart Mountain 2615 ft. 797 m

Preeceville

Canora

Kamsack

Yorkton

Melville

Langenburg

Esterhazy

Moosomin

Wawota

Moose Mountain 2740 ft. 835 m

DUCK MOUNTAIN

Baldy Mountain 2730 ft. 832 m

Roblin

Grandview

Russell

Shoal Lake

RIDING MOUNTAIN NATL. PARK

RIDING MOUNTAIN

Ethelbert

Dauphin

Dauphin Lake

Sainte Rose du La

Minnedosa

Neepawa

Rivers

Elkhorn

Virden

Brandon

Souris

Minto

Boissevain

Melita

Whitewater Lake

Turtle Mountain 2516 ft. 767 m

TURTLE MOUNTAIN IND. RES.

Bottineau

Belcourt

Pinto Butte 3442 ft 1049 m

Wood Mountain 3350 ft 1021 m

GRASSLANDS NATIONAL PARK

CANADA
UNITED STATES

MONTANA

NORTH DAKOTA

Havre

Chinook

Harlem

Scobey

Plentywood

Crosby

Bowbells

Kenmare

Mohall

Rand McNally
A-101564-1

Lambert Conformal Conic Projection
Scale 1:4,000,000
One inch to 64 miles
One cm to 40 km

0 20 40 60 80 100 120 Miles

0 20 40 60 80 100 120 140 160 180 200 Kilometers

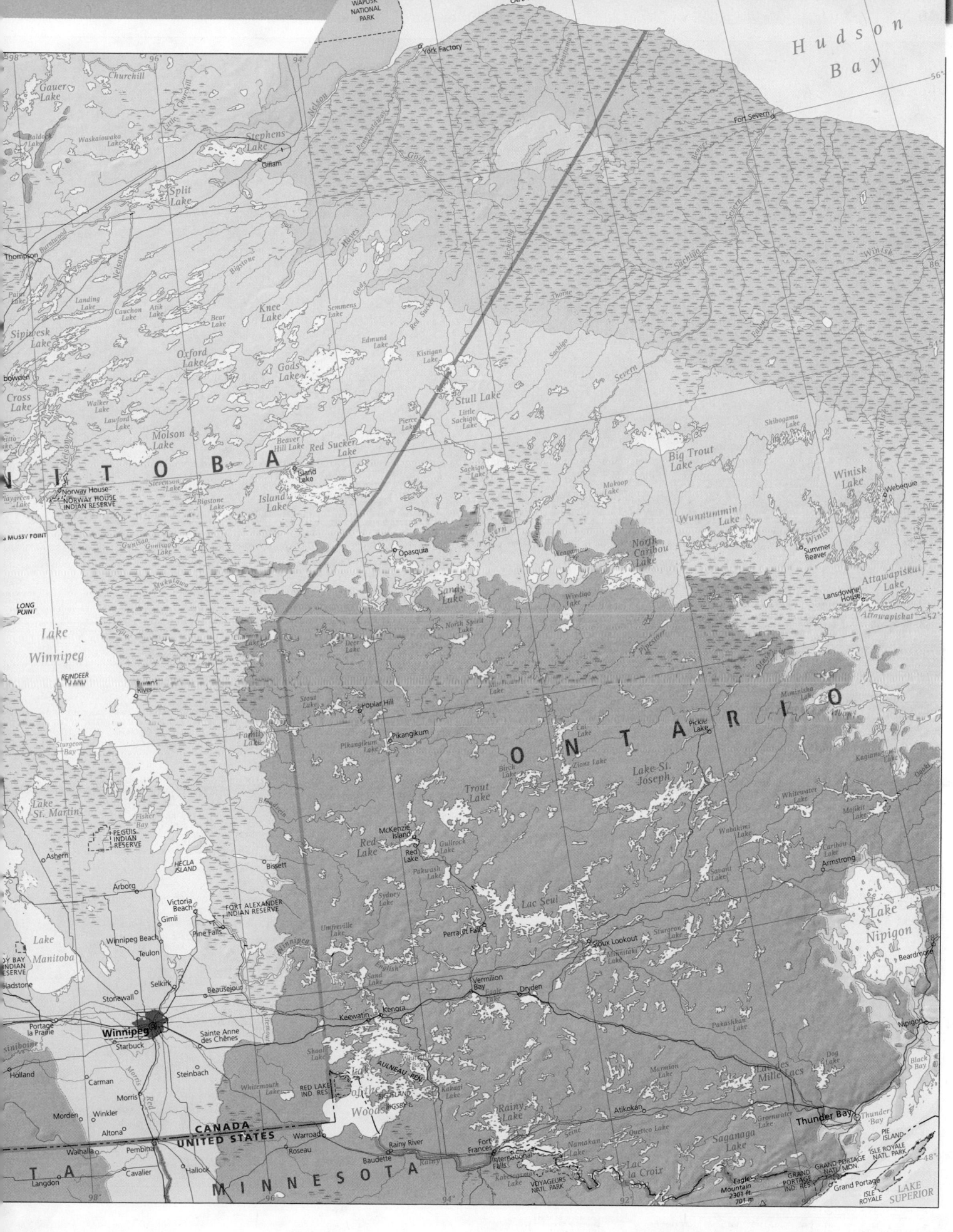

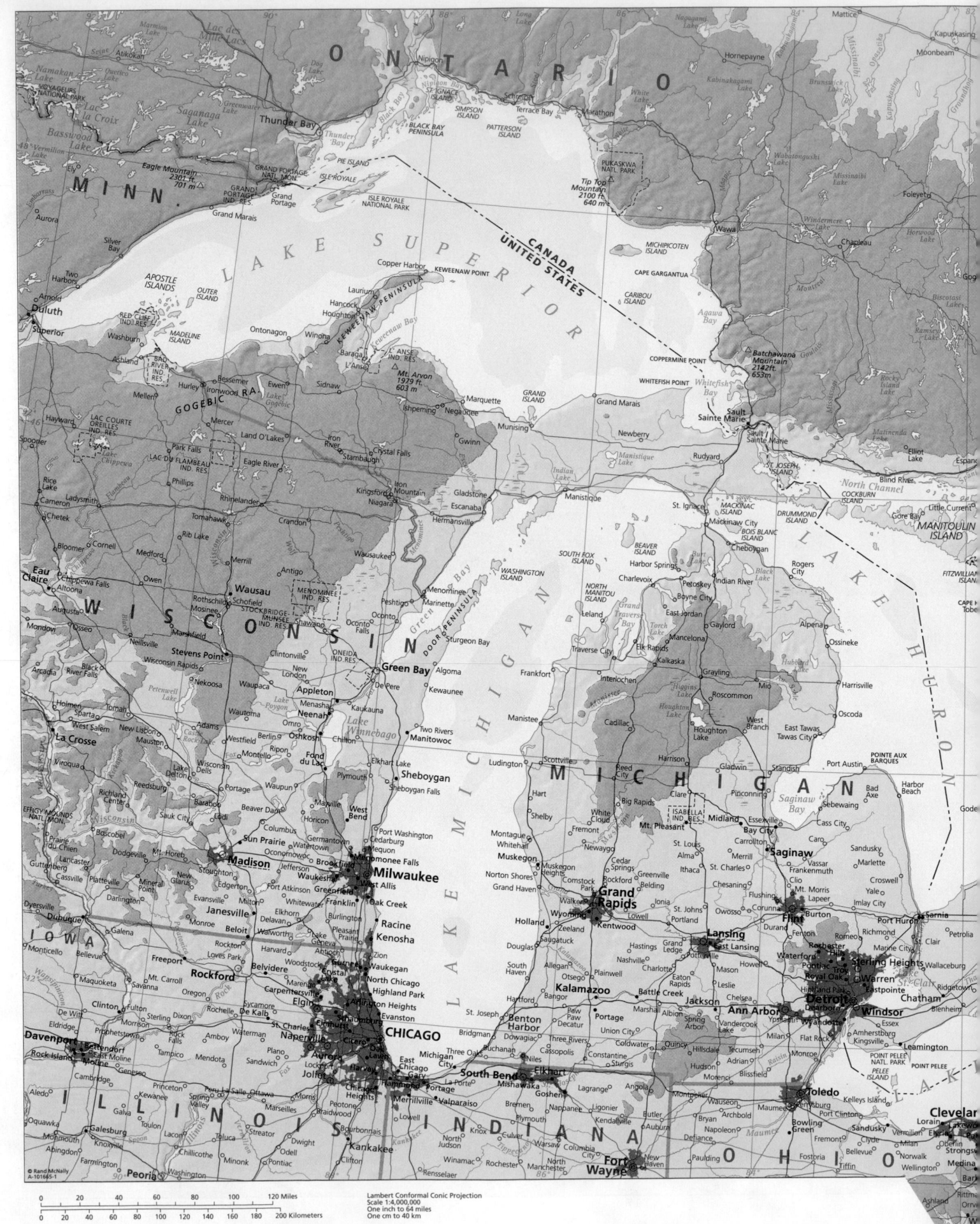

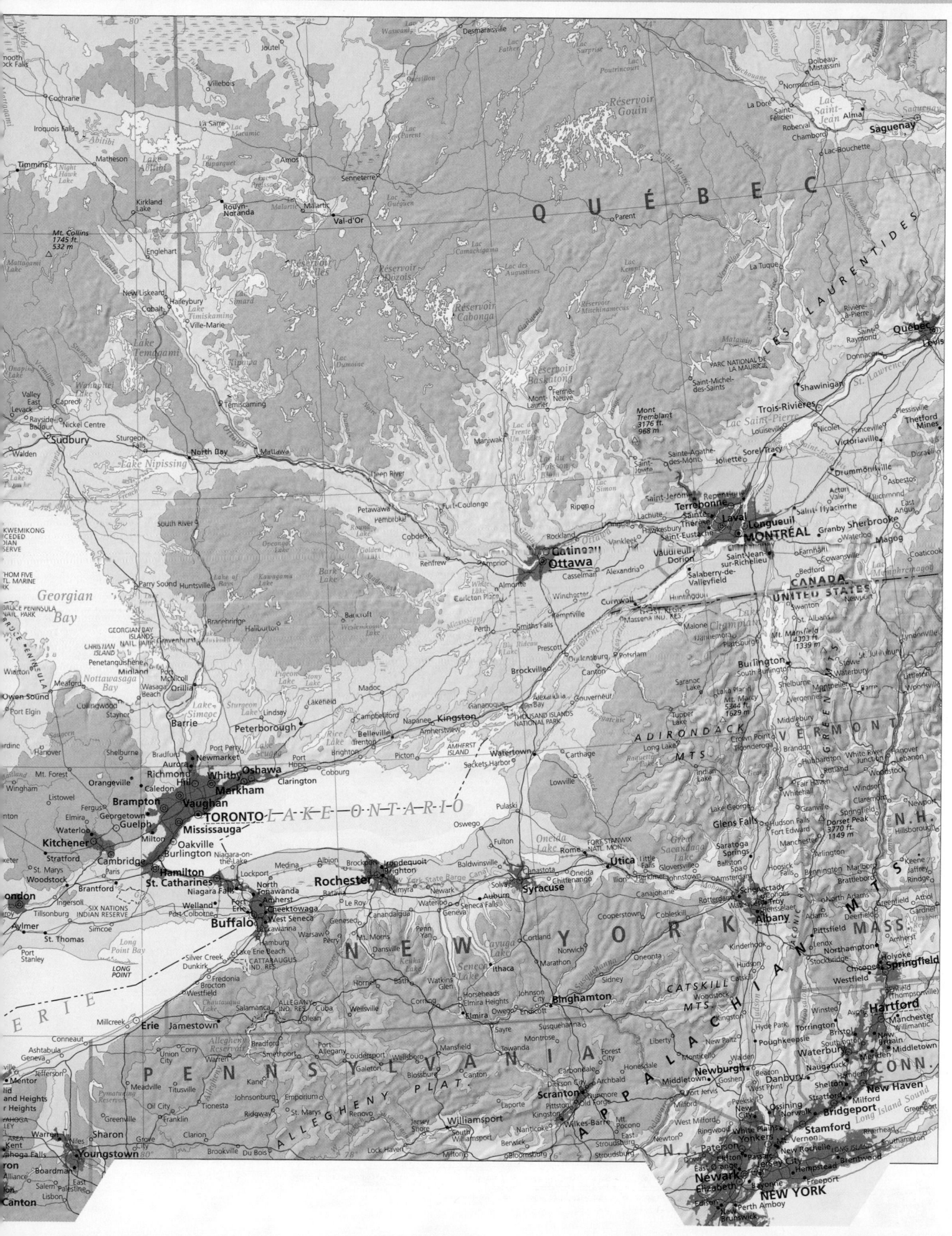

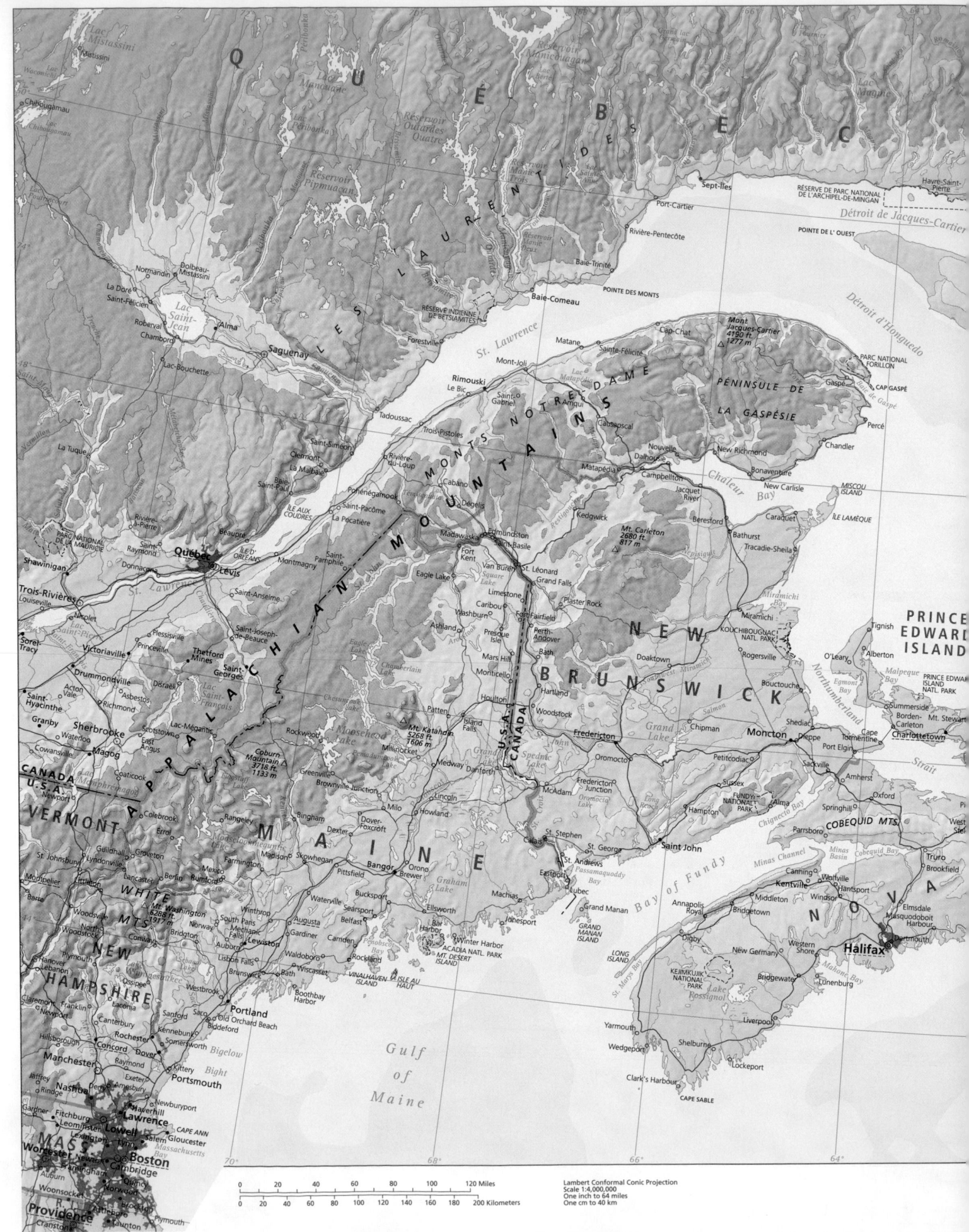

Red Bay
CAPE BAULD
Pistolet Bay
L'ANSE-AUX-MEADOWS
St. Anthony
Strait of Belle Isle
Saint-Augustin
Hare Bay
Mutton Bay
CAP DU GROS MÉCATINA
ST. JOHN ISLAND
Roddickton
LABRADOR
Île du Petit Mécatina
Port Saunders
GROAIS ISLAND
GREY ISLANDS
SEA
La Romaine
Blue Mountain 2129 ft. 649 m
BELL ISLAND
HORSE ISLANDS
Natashquan
CAP WHITTLE
CAPE ST. JOHN
La Scie
Notre Dame Bay
NEW WORLD ISLAND
FOGO ISLAND
GROS MORNE NATIONAL PARK
Fogo
Rocky Harbour
Gros Morne 2644 ft. 806 m
ÎLE D'ANTICOSTI
Durrell
Lewisporte
Hamilton Sound
Springdale
Robert's Arm
Carmanville
New-Wes-Valley
POINTE DE L'EST
Deer Lake
Sandy Lake
Botwood
Glenwood
CAPE FREELS
Bay of Islands
Lark Harbour
Pasadena
NEWFOUNDLAND
Badger
Bishop's Falls
Gander
Hare Bay
Corner Brook
Buchans
Grand Falls-Windsor
Gander Lake
Bonavista Bay
LONG POINT
Grand Lake
AND LABRADOR
Red Indian Lake
Bonavista
Gulf of
Port au Port Bay
Stephenville
Victoria Lake
NEWFOUNDLAND
Glovertown
Catalina
St. Lawrence
CAPE ST. GEORGE
St. George's Bay
TERRA NOVA NATIONAL PARK
RANDOM ISLAND
GRATES POINT
CAPE ANGUILLE
Jeddore Lake
Shoal Harbour
Trinity Bay
Bay de Verde
Meelpaeg Lake
CAPE RAY
Channel-Port aux Basques
Isle aux Morts
Burgeo
Pouch Cove
Carbonear
Harbour Grace
Torbay
Hermitage Bay
Harbour Breton
Belle Bay
Wabana
Bay Roberts
Brigus
ST. JOHN'S
CAPE SPEAR
BRUNETTE ISLAND
Fortune Bay
BURIN PENINSULA
Placentia Bay
AVALON PENINSULA
Witless Bay
MIQUELON
Fox Harbour
LA GROSSE ÎLE
ÎLE DE L'EST
Grand Bank
Marystown
ÎLE DU CAP AUX MEULES
ÎLES DE LA MADELEINE (Que.)
Cap-aux-Meules
Fortune
Burin
St. Mary's Bay
Cabot Strait
Saint Pierre
St. Lawrence
Branch
ÎLE DU HAVRE AUBERT
SAINT-PIERRE
CAPE ST. MARY'S
SAINT PIERRE & MIQUELON (Fr.)
St. Shotts
CAPE RACE
CAPE NORTH
Aspy Bay
Dingwall
CAPE BRETON HIGHLANDS NATL. PARK
Souris
St. Ann's Bay
Sydney Mines
New Waterford
Georgetown
North Sydney
Glace Bay
Hague
Sydney
SCATARIE ISLAND
Port Hood
NOVA SCOTIA
Louisbourg
Murray Harbour
St. Georges Bay
Bras d'Or Lake
St. Peters
CAPE BRETON ISLAND
Antigonish
ISLE MADAME
Glasgow
Port Hawkesbury
Chedabucto Bay
Canso
ATLANTIC OCEAN
SABLE ISLAND NATL. PARK RESERVE
SABLE ISLAND (N.S.)

Inset map a

a
LES LAURENTIDES
Saint-Joachim
Sainte-Anne-de-Beaupré
Beaupré
Lac-Beauport
L'Ange-Gardien
Château-Richer
ÎLE D'ORLÉANS
Saint-Raymond
Shannon
Charlesbourg
Boischatel
Beauport
Sainte-Pétronille
Sainte-Catherine-de-la-Jacques-Cartier
Val-Bélair
St. Lawrence
Loretteville
Vanier
QUÉBEC
L'Ancienne-Lorette
Beaumont
Lévis
Saint-Augustin-de-Desmaures
Sainte-Foy
Sillery
Saint-Basile
Pont-Rouge
Cap-Rouge
Saint-Romuald
Portneuf
Neuville
Saint-Jean-Chrysostome
Saint-Henri
Cap-Santé
Donnacona
St. Lawrence (Saint-Laurent)
Saint-Nicolas
Saint-Bélempteur
Saint-Anselme
Saint-Antoine-de-Tilly
Saint-Apollinaire
Saint-Agapit
Saint-Édouard-de-Lotbinière
Sainte-Croix
Laurier-Station
Saint-Flavien
Saint-Gilles
Dosquet
Saint-Bernard

Inset map a
Lambert Conformal Conic Projection
Scale 1:1,000,000
One inch to 16 miles
One cm to 10 km

ARCTIC OCEAN

+ North
Magnetic
Pole

ELLESMERE ISLAND

CAPE
COLUMBIA

CAPE

QUTTINIRPAAQ
NATL. PARK

Bardeau Pk.
8583 m

ALERT
POINT

Eureka

AXEL
HEIBERG
ISLAND

MEIGHEN
ISLAND

Peary Channel

CAPE
ISACHSEN

SVERDRUP ISLANDS

Ghse Fiord

Jones Sour

CAPE
VERA

GRINNELL
PENINSULA

Norwegian Bay

BORDEN
ISLAND

ELLEF
RINGNES
ISLAND

AMUND
RINGNES
ISLAND

BROCK
ISLAND

MACKENZIE
KING ISLAND

QUEEN ELIZABETH ISLANDS

Wainwright

Barrow

POINT
BARROW

BEAUFORT

SEA

PRINCE
PATRICK ISLAND

PARRY ISLANDS

Hazen Strait

DEVON
ISLAND

CORNWALLIS
ISLAND

CAPE
HALKETT

Meade

Teshekpuk
Lake

Prudhoe
Bay

Colville

CAPE
RUSSELL

MELVILLE ISLAND

BATHURST
ISLAND

WROTTESLEY

CAPE

Liddon Gulf

M'Clure Strait

CAPE
JAMES
ROSS

Resolute

Lancaster Sou

Kaktovik

MARTIN
POINT

Mount
Michelson
8855 ft
2699 m

DEMARCATION
POINT

Mackenzie
Bay

AULAVIK
NATIONAL
PARK

BANKS
ISLAND

RUSSELL
POINT

CAPE HAY

Viscount Melville Sound

Parry Channel

CAPE
CLARENCE

Fort Yukon

ALASKA
(U.S.)

IVVAVIK
NATL. PARK

VUNTUT
NATL. PARK

Aklavik

Tuktoyaktuk

CAPE
KELLETT

STAFANSSON
ISLAND

Peel Sound

SOMERSET
ISLAND

BRODEUR
PENINSULA

Tombstone Mountain
7192 ft 2192 m

OGILVIE MTS.

Dawson

RICHARDSON MTS.

TUKTOYAKTUK PENINSULA

Inuvik

Sitidgi
Lake

CAPE
DALHOUSIE

CAPE
BATHURST

PEEL
POINT

CAPE
LAMBTON

Prince of Wales Strait

PRINCE
OF WALES
ISLAND

M'Clintock Channel

CAPE
SWINBURNE

Gulf

Fort
McPherson

Mackenzie

Minto Inlet

Ulukhaktok

VICTORIA
ISLAND

BOOTHIA
PENINSULA

Nodaleen Mountain
7257 ft
2210 m

SELWYN MTS.

MACKENZIE MTS.

FRANKLIN MTS.

CAPE
PARRY

Paulatuk

CAPE
BARING

Prince Albert Sound

WOLLASTON
PENINSULA

Washburn
Lakes

CAPE FELIX

Taloyoak

CAPE
CHAPMAN

Keele Peak
9751 ft
2972 m

Aubry
Lake

TUKTUT
NOGAIT
NATL. PARK

Bluenose
Lake

Cambridge
Bay

Victoria Strait

KING
WILLIAM
ISLAND

Gjoa
Haven

Kugaaruk

Arctic Circle

Fort
Good
Hope

Great Bear
Lake

Kugluktuk

Coronation
Gulf

KENT
PENINSULA

MELBOURNE
ISLAND

Queen
Maud Gulf

ADELAIDE
PENINSULA

DAWSON RANGE

YUKON

PELLY MTS.

Mt. Sir
7088 ft
2404 m

Ross Mountain

LOGAN MTS.

NAATS'IHCH'OH
NATL. PARK RESERVE

Norman
Wells

Tulita

NORTHWEST
TERRITORIES

Déline

Napaktulik
Lake

MacAlpine
Lake

NUNA

Horton
Lake

Contwoyto
Lake

Back

Garry
Lake

UKKUSIKSALIK
NATL. PARK

Whitehorse

MT. St. James MacBri
2730 m

Lac la Martre

MacKay
Lake

Aylmer
Lake

Tehek
Lake

Ahetdech
Lake

Baker
Lake

0   60   120   180   240   300   360 Miles

0   60   120   180   240   300   360   420   480   540   600 Kilometers

Lambert Azimuthal Equal Area Projection
Scale 1:12,000,000
One inch to 190 miles
One cm to 120 km

GREENLAND SEA

JAN MAYEN

KAP EILER RASMUSSEN

KAP MORRIS JESUP

KAP BISMARK

KAP BREWSTER

Ittoqqortoormiit

Traill

SHANNON

Kangerlittivaq

NORTHEAST GREENLAND NATIONAL PARK

ICELAND

Hvannadalshnúkur 6952 ft, 2119 m

Hekla 4892 ft, 1491 m

FONTUR

Reykjavík

HORN

Denmark Strait

GREENLAND
(Denmark)

(KALAALLIT NUNAAT)

Gunnbjørn Field 12 139 ft, 3700 m

Mont Forel 11 024 ft, 3360 m

KAP GUSTAV HOLM

Tasiilaq

Haffner Bjerg 4865 ft, 1483 m

KAP ALEXANDER

Qaanaaq

KAP PARRY

KAP YORK

Qimusseriarsuaq

KAP SEDDON

Kullorsuaq

Upernavik

Tasiusaq

Uummannaq

Nuussuaq

KAP BROZTING

Ilulissat

Qasigiannguit

Qeqertarsuaq

Qeqertsuaq Tunua

Aasiaat

Qeqertarsuaq

Kangaatsiaq

Kangerlussuaq

CAPE DARKER

CAPE SHEPARD

CAPE LIVERPOOL

CAPE GRAHAM MOORE

SIRMILIK NAT'L PARK

Pond Inlet

Clyde River

Baffin Bay

Sisimiut

Maniitsoq

Nuuk

Paamiut

Narsarsuaq

Qaqortoq

Nanortalik

CAPE FAREWELL

Davis Strait

BAFFIN ISLAND

CAPE DYER

AUYUITTUQ NAT'L PARK

CUMBERLAND PENINSULA

Pangnirtung

ANGIJAK ISLAND

CAPE MERCY

Nettilling Lake

Cumberland Sound

LABRADOR SEA

Igloolik

ROWLEY ISLAND

Hall Beach

FOLEY ISLAND

PRINCE CHARLES ISLAND

AIR FORCE ISLAND

MELVILLE PENINSULA

Foxe Basin

CAPE WILSON

Arctic Circle

CAPE DOMINION

CAPE DORCHESTER

UT

Amadjuak Lake

HALL PENINSULA

Iqaluit

Frobisher Bay

FOXE PENINSULA

Cape Dorset

META INCOGNITA PENINSULA

RESOLUTION ISLAND

FAIR NESS

Kimmirut

BIG ISLAND

SALISBURY ISLAND

CAP DE NOUVELLE-FRANCE

Hudson Strait

Port Burwell

TORNGAT MOUNTAINS NATIONAL PARK

Mt. Caubvick Mont d'Iberville 5420 ft, 1652 m

SOUTHAMPTON ISLAND

Coral Harbour

BELL PEN.

NOTTINGHAM ISLAND

COATS ISLAND

CAPE KENDALL

Foxe Channel

Ivujivik

Salluit

AKPATOK ISLAND

CAP HOPES ADVANCE

Ungava Bay

ATLANTIC OCEAN

© 2017 Pearson Education, Inc.

Inset map a
Lambert Conformal Conic Projection
Scale 1:2,000,000
One inch to 32 miles
One cm to 20 km

BERMUDA
(U.K.)

ST. GEORGE'S
ISLAND
St. George
ST. DAVID'S
ISLAND

Harrington
Sound
Castle
Harbour

Hamilton
Flatts

Great
Sound

Town Hill
259 ft.
79 m

SOMERSET
ISLAND

HIGH POINT

ATLANTIC OCEAN

© Rand McNally
A-101670-1

W.VA.
VIRGINIA
TUCKY
Roanoke
Richmond
Norfolk
Virginia Beach
Chesapeake Bay
Knoxville
Mt. Mitchell
6684 ft.
2037 m
Raleigh
Chattanooga
Charlotte
NORTH CAROLINA
SEE
S
Atlanta
SOUTH
CAROLINA
Columbia
Fayetteville
CAPE LOOKOUT
CAPE HATTERAS
Columbus
GEORGIA
Charleston
CAPE FEAR
Wilmington
Montgomery
Savannah
Altamaha
Tallahassee
Jacksonville
St. Johns
FLORIDA

CAPE
BLAS
Orlando
CAPE CANAVERAL
St. Petersburg
Tampa
Lake
Okeechobee
West
Palm
Beach
Fort
Lauderdale
Miami
CAPE SABLE
Key West
Straits of Florida

ATLANTIC

OCEAN

Tropic of Cancer

BAHAMAS
Nassau
ELEUTHERA
ABACO
GRAND
BAHAMA
ANDROS
NEW
PROVIDENCE
CAT
ISLAND
San Salvador
MANGROVE
CAY
GREAT
EXUMA
LONG
ISLAND
CROOKED
ISLAND
HAVANA
Matanzas
Pinar del Río
Santa
Clara
Golfo de
Batabanó
Cienfuegos
CUBA
CAYO COCO
CAYO ROMANO
CAYO GUAJABA
GREAT
INAGUA
MAYAGUANA
ACKLINS
TURKS AND CAICOS
ISLANDS
(U.K.)
Grand Turk
CABO DE
SAN
ANTONIO
ISLA DE
LA JUVENTUD
Camagüey
Holguín
G
R
E
A
T
E
R
Bayamo
Manzanillo
CABO CRUZ
Santiago de
Cuba
6470 ft.
1972 m
Guantánamo
Cap-Haïtien
PUERTO RICO TRENCH
CAYMAN
ISLANDS
(U.K.)
George Town
GRAND
CAYMAN
Montego Bay
JAMAICA
Spanish Town
Kingston
Windward Passage
Gonaïves
HAITI
Port-au-Prince
Santiago de
los Caballeros
10,417 ft.
3175 m
HISPANIOLA
DOMINICAN
REPUBLIC
SANTO
DOMINGO
San Pedro
de Macorís
Ponce
San Juan
PUERTO
RICO
(U.S.)
VIRGIN
ISLANDS
(U.S.)
Charlotte
Amalie
ST. CROIX
(U.S.)
BRITISH VIRGIN
ISLANDS
(U.K.)
ANGUILLA
(U.K.)
ST. KITTS
AND NEVIS
ANTIGUA AND BARBUDA
MONTSERRAT
(U.K.)
GUADELOUPE
(Fr.)
Basse-Terre
GRANDE-TERRE
MARIE-GALANTE
L
E
E
W
A
R
D
I
S
L
A
N
D
S
DOMINICA
Roseau
MARTINIQUE
(Fr.)
Fort-de-France

BERMUDA
(U.K.)
Hamilton

A
N
T
I
L
L
E
S

WEST INDIES

CARIBBEAN SEA

LESSER

W
I
N
D
W
A
R
D

I
S
L
A
N
D
S

A
N
T
I
L
L
E
S

Castries
ST. LUCIA
Kingstown
ST. VINCENT AND
THE GRENADINES
BARBADOS
Bridgetown
GRENADA
TOBAGO
TRINIDAD
AND TOBAGO
Port of Spain
TRINIDAD

HONDURAS
CABO GRACIAS
A DIOS
Coco
icalpa
NICARAGUA
Lago de
Nicaragua
Bluefields
ISLA DE
PROVIDENCIA
(Col.)
ISLA DE
SAN ANDRÉS
(Col.)
COSTA
RICA
Volcán Irazú
11,260 ft.
3432 m
San José
intarenas
Puerto Limón
Volcán Barú
11,401 ft.
3475 m
Cerro Chirripó
12,530 ft.
3819 m
David
Santiago
ISTMO DE PANAMÁ
Colón
Panamá
Panama
Canal
Golfo de
Chiriquí
ISLA DE
COIBA
PEN. DE
AZUERO
PUNTA
MARIATO
Golfo de
Panamá

PUNTA GALLINAS
PEN. DE
LA GUAJIRA
Santa Marta
ARUBA
(Neth.)
Willemstad
CURAÇAO
(Neth.)
BONAIRE
(Neth.)
ISLA LA
ORCHILA
ISLA
BLANQUILLA
ISLA LA
TORTUGA
ISLA DE
MARGARITA
Carúpano
Golfo de
Paria

Barranquilla
Soledad
Pico Cristóbal Colón
18,947 ft.
5775 m
Cartagena
Punto Fijo
Golfo de
Venezuela
Puerto
Cabello
Cabimas
MARACAIBO
Valle de
la Pascua
CARACAS
Barcelona
Cumaná
Ciudad Guayana
Sincelejo
Magangué
Valera
Mérida
Lago de
Maracaibo
ACarigua
BARQUISIMETO
Acarigua
Calabozo
El Tigre
Ciudad
Bolívar
Monteria
Ocaña
Pico Bolívar
16,427 ft.
5007 m
San Fernando
de Apure
Cerro Mato
6312 ft.
1863 m
Georgetown
Cúcuta
San Cristóbal
Apure
Arauca
Orinoco
Mt. Roraima
9432 ft.
2875 m
GUYANA
Barrancabermeja
Bucaramanga
Puerto
Ayacucho
LLANOS
GUIANA
HIGHLANDS
BRAZIL
SURINAME
 OCCIDENTAL
MEDELLÍN
La Dorada
Sogamoso
Meta
Vichada
PAKARAIMA
Tunja
San Fernando
de Atabapo
Cerro Marahuaca
8461 ft.
2579 m
Boa Vista
Manizales
Nevado del Tolima
17,110 ft.
5215 m
Cartago
CORD.
Ibagué
Nevado del Huila
18,865 ft.
5750 m
BOGOTÁ
Villavicencio
COLOMBIA
Guaviare
CABO CORRIENTES
ISLA DE MALPELO
(Col.)
Buenaventura
Palmira
CALI
OCCIDENTAL
ORIENTAL
CORD.

VENEZUELA

80   160   240   320   400   480 Miles
80 160 240 320 400 480 560 640 720 800 Kilometers

Lambert Azimuthal Equal Area Projection
Scale 1:16,000,000
One inch to 256 miles
One cm to 160 km

Inset map b
Lambert Conformal Conic Projection
Scale 1:1,000,000
One inch to 16 miles
One cm to 10 km

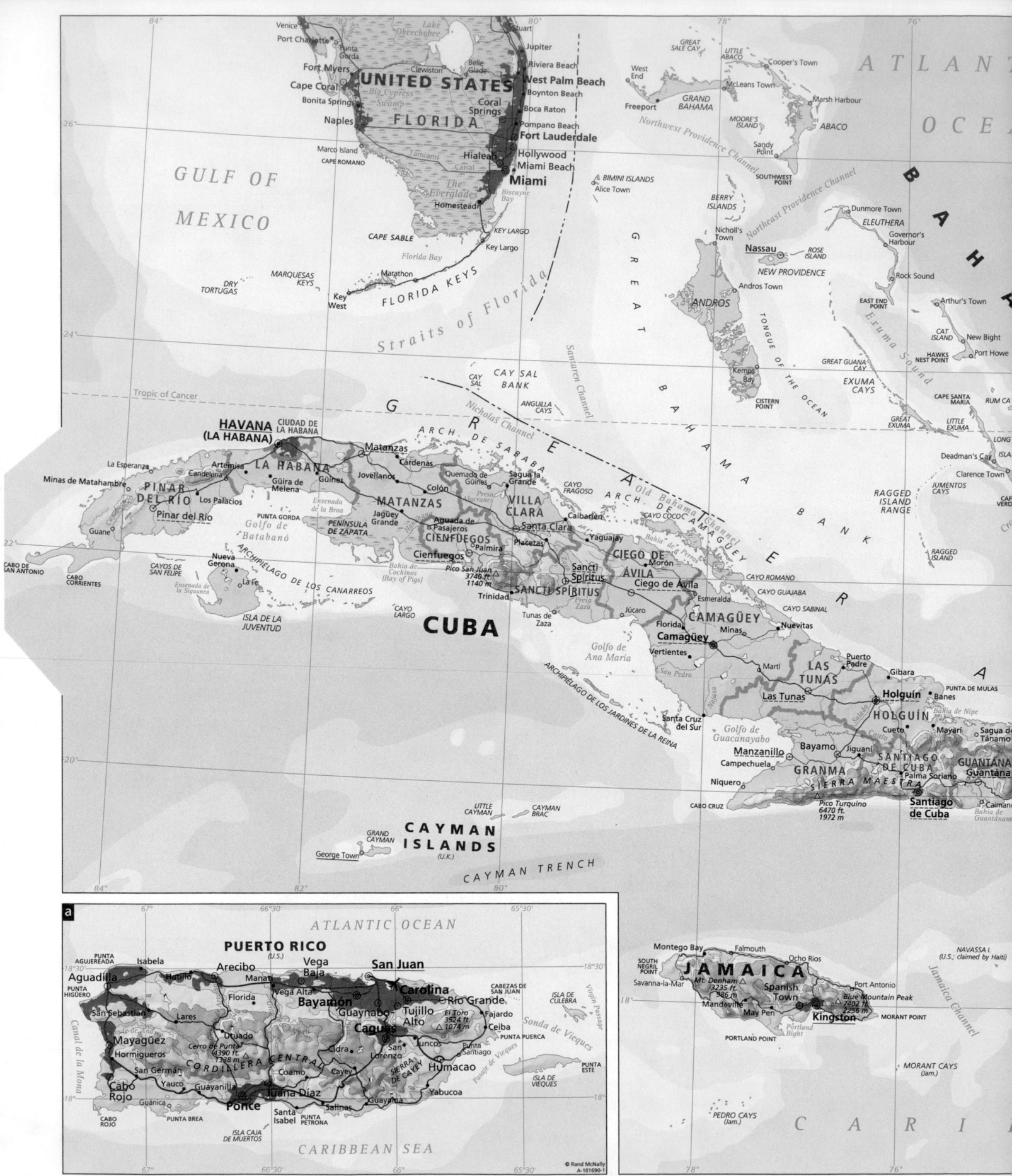

UNITED STATES

Venice
Port Charlotte
Punta Gorda
Fort Myers
Cape Coral
Bonita Springs
Naples
Marco Island
CAPE ROMANO

FLORIDA
Clewiston
Belle Glade
Big Cypress Swamp
Tamiami
The Everglades
Homestead
Biscayne Bay

Lake Okeechobee
Stuart
Jupiter
Riviera Beach
West Palm Beach
Boynton Beach
Boca Raton
Pompano Beach
Fort Lauderdale
Hollywood
Miami Beach
Hialeah
Miami
Coral Springs

West End
Freeport
GRAND BAHAMA
GREAT SALE CAY
LITTLE ABACO
Cooper's Town
McLeans Town
Marsh Harbour
ABACO
MOORE'S ISLAND
Sandy Point
SOUTHWEST POINT

ATLANTIC OCEAN

GULF OF MEXICO

CAPE SABLE
Florida Bay
Key Largo
Key Largo

MARQUESAS KEYS
DRY TORTUGAS
Marathon
Key West
FLORIDA KEYS

BIMINI ISLANDS
Alice Town
BERRY ISLANDS
Nicholl's Town
Nassau
ROSE ISLAND
NEW PROVIDENCE
Andros Town
ANDROS

Dunmore Town
Governor's Harbour
ELEUTHERA
Rock Sound
EAST END POINT
Arthur's Town
CAT ISLAND
New Bight
Port Howe
HAWKS NEST POINT

Straits of Florida

Santaren Channel

Tropic of Cancer

CAY SAL
CAY SAL BANK
ANGUILLA CAYS
Nicholas Channel

HAVANA (LA HABANA)
CIUDAD DE LA HABANA
La Esperanza
Artemisa
LA HABANA
Candelaria
Güira de Melena
Güines
Matanzas
Matanzas
Cárdenas
Jovellanos
Quemado de Güines
Sagua la Grande
ARCH. DE SABABA
CAYO FRAGOSO

Minas de Matahambre
PINAR DEL RIO
Los Palacios
Pinar del Rio
Guane

Colón
MATANZAS
Aguada de Pasajeros
Jagüey Grande
PUNTA GORDA
PENÍNSULA DE ZAPATA
Presa Macranes
Ensenada de la Broa
Golfo de Batabanó

CIENFUEGOS
Palmira
Santa Clara
VILLA CLARA
Placetas
Yaguajay
Caibarién
Bahía de Cádiz
CAYO COCOC
ARCH. DE CAMAGÜEY

Cienfuegos
Bahía de Cochinos (Bay of Pigs)
Pico San Juan 3740 ft. 1140 m
Sancti Spíritus
SANCTI SPÍRITUS
Trinidad
CIEGO DE ÁVILA
Morón
Ciego de Ávila
Esmeralda

La Fe
Nueva Gerona
CAYOS DE SAN FELIPE
ARCHIPIÉLAGO DE LOS CANARREOS
Ensenada de la Siguanea
ISLA DE LA JUVENTUD
CAYO LARGO

CUBA

Tunas de Zaza
Júcaro
Golfo de Ana Maria
San Pedro
CAMAGÜEY
Florida
Camagüey
Vertientes
Minas
Nuevitas
CAYO ROMANO
CAYO GUAJABA
CAYO SABINAL

ARCHIPIÉLAGO DE LOS JARDINES DE LA REINA

Santa Cruz del Sur
Golfo de Guacanayabo
Martí
LAS TUNAS
Puerto Padre
Las Tunas
Gibara
Holguín
Banes
PUNTA DE MULAS

Manzanillo
Campechuela
Niquero
GRANMA
Bayamo
Jiguaní
Cueto
Mayarí
Sagua de Tánamo
HOLGUÍN
Bahía de Nipe

SANTIAGO DE CUBA
Palma Soriano
SIERRA MAESTRA
GUANTÁNAMO
Guantána
CABO CRUZ
Pico Turquino 6470 ft. 1972 m
Santiago de Cuba
Bahía de Guantánamo
Caiman

LITTLE CAYMAN
CAYMAN BRAC
GRAND CAYMAN
CAYMAN ISLANDS (U.K.)
George Town

CAYMAN TRENCH

CABO DE SAN ANTONIO
CABO CORRIENTES

CAPE SANTA MARIA
RUM CAY
GREAT EXUMA
LITTLE EXUMA
Deadman's Cay
Clarence Town
JUMENTOS CAYS
RAGGED ISLAND RANGE
RAGGED ISLAND
Kemps Bay
CISTERN POINT
TONGUE OF THE OCEAN
EXUMA CAYS
Exuma Sound

Northwest Providence Channel
Northeast Providence Channel

Old Bahama Channel

GREAT BAHAMA BANK

Inset a — PUERTO RICO

ATLANTIC OCEAN

PUERTO RICO (U.S.)

PUNTA AGUJEREADA
PUNTA HIGÜERO
Isabela
Hatillo
Aguadilla
San Sebastián
Lares
Arecibo
Manati
Florida
Vega Baja
Vega Alta
Dorado
San Juan
Bayamón
Carolina
Guaynabo
Caguas
Río Grande
Tujillo Alto
CABEZAS DE SAN JUAN
ISLA DE CULEBRA
Fajardo
Ceiba
Juncos
PUNTA PUERCA
San Lorenzo
Cidra
Cayey
Juana Díaz
SIERRA DE CAYEY
Humacao
Santiago
PUNTA SANTIAGO
Yabucoa

Canal de la Mona
Mayagüez
Hormigueros
San Germán
Yauco
Guánica
Cabo Rojo
CABO ROJO
ISLA CAJA DE MUERTOS
Guayanilla
CORDILLERA CENTRAL
Cerro de Punta 4390 ft. 1338 m
El Toro 3524 ft. 1074 m
Coamo
Santa Isabel
Salinas
Guayama
Ponce
PUNTA BREA
PUNTA PETRONA

Sonda de Vieques
ISLA DE VIEQUES
PUNTA ESTE
Pasaje de Vieques
Virgin Passage

CARIBBEAN SEA

JAMAICA

Montego Bay
Falmouth
Ocho Rios
SOUTH NEGRIL POINT
Savanna-la-mar
JAMAICA
Mt. Denham 3235 ft. 986 m
Mandeville
May Pen
Spanish Town
Kingston
Port Antonio
Blue Mountain Peak 7402 ft. 2256 m
MORANT POINT
NAVASSA I. (U.S.) claimed by Haiti
PORTLAND POINT
Portland Bight
MORANT CAYS (Jam.)
Jamaica Channel

PEDRO CAYS (Jam.)

© Rand McNally
A-101690-1

Inset map a
Lambert Conformal Conic Projection
Scale 1:2,000,000
One inch to 32 miles
One cm to 20 km

Lambert Conformal Conic Projection
Scale 1:5,000,000
One inch to 80 miles
One cm to 50 km

Inset map b
Lambert Conformal Conic Projection
Scale 1:5,000,000
One inch to 80 miles
One cm to 50 km

0 40 80 120 160 200 240 Miles
0 40 80 120 160 200 240 280 320 360 400 Kilometers

Lambert Azimuthal Equal Area Projection
Scale 1:8,000,000
One inch to 128 miles
One cm to 80 km

UNITED STATES

Fort Worth
DALLAS
Cleburne
Abilene
Stamford
Sweetwater
San Angelo
Brownwood
Corsicana
Tyler
Jacksonville
Nacogdoches
Longview
Lake O the Pines
Ruston
Shreveport
Monroe
Tallulah
Vicksburg
Jackson
Brookhaven
Meridian
Troy
Ozark
Albany
Bainbridge
GEORGIA

T E X A S
Waco
Lufkin
Toledo Bend
Natchitoches
Black
Natchez
MISSISSIPPI
Hattiesburg
Laurel
McComb
ALABAMA
Andalusia
Dothan
Tallahassee

E D W A R D S
Sonora
P L A T E A U
Kerrville
Fredericksburg
Austin
Bryan
Huntsville
Conroe
LOUISIANA
Alexandria
De Ridder
Opelousas
Baton Rouge
Bogalusa
Hammond
Gulfport
Biloxi
Pascagoula
Mobile
Pensacola
FLORIDA
Fort Walton Beach
Panama City
CAPE SAN BLAS
ST. GEORGE ISLAND

San Marcos
SAN ANTONIO
Sugar Land
HOUSTON
Beaumont
Port Arthur
Baytown
Lake Charles
Lafayette
New Iberia
Morgan City
Houma
Thibodaux
New Orleans
Lake Pontchartrain
MARSH ISLAND
POINT AU FER ISLAND

Del Rio
Ciudad Acuña
Eagle Pass
Uvalde
Cotulla
Victoria
Beeville
Lake Jackson
Galveston
Texas City
Freeport
GALVESTON ISLAND

28°

Nuevo Laredo
Ciudad Anáhuac
Laredo
Kingsville
Zapata
Falcon Reservoir
Raymondville
MATAGORDA ISLAND
SAN JOSE ISLAND
Corpus Christi
PADRE ISLAND
Laguna Madre

26°

Sabinas Hidalgo
McAllen
Harlingen
Brownsville
Reynosa
Matamoros
Valle Hermoso
San Nicolás de los Garza
Guadalupe
MONTERREY
Saltillo
Montemorelos
Linares
San Fernando
Laguna Madre
BARRA DE LOS AMERICANOS

24°

NUEVO LEÓN

M E X I C O
Ciudad Victoria
TAMAULIPAS
Aldama
GULF OF MEXICO
Tropic of Cancer

Ciudad Mante
Cuauhtémoc
22°

SAN LUIS POTOSÍ
Ciudad Valles
Tampico
Pánuco
Laguna de Tamiahua
CABO ROJO
San Luis Potosí
Rioverde
Cerritos

GUANAJUATO
San Luis de la Paz
Tamazunchale
Tantoyuca
Tamiahua
Tuxpan de Rodríguez Cano
Río Lagartos
Panabá
CABO CATOCHE
Progreso
Tizimín
Cancún

Guanajuato
Ixmiquilpan
QUERÉTARO
Poza Rica de Hidalgo
Papantla de Olarte
EL TAJÍN
Hunucmá
Mérida
YUCATÁN
Temax
Playa del Carmen
Cozumel
ISLA COZUMEL

Irapuato
Salamanca
Querétaro
Celaya
Acámbaro
HIDALGO
Pachuca de Soto
Tulancingo
Teziutlán
Martínez de la Torre
Maxcanú
CHICHÉN ITZÁ
Ticul
Valladolid
Tulum
TULUM
20°

Morelia
MEXICO CITY
Ciudad Netzahualcóyotl
TLAXCALA
Xicohténcatl
Xalapa
Volcán Pico de Orizaba 18,406 ft. 5610 m
Veracruz
Dzitbalché
UXMAL
Tekax
Peto
YUCATÁN
Hopelchén
Campeche
Félipe Carrillo Puerto

Toluca de Lerdo
Nevado de Toluca 15,387 ft. 4690 m
MÉXICO
Cuernavaca
MORELOS
PUEBLA DE ZARAGOZA
PUEBLA
Orizaba
Córdoba
Tehuacán
Tierra Blanca
Alvarado
PUNTA ROCA PARTIDA
San Andrés Tuxtla
Bahía de Campeche
PUNTA MORRO
Seybaplaya
Champotón
CAMPECHE
PENÍNSULA
QUINTANA ROO
Chetumal
Corozal
CARIBBEAN SEA
Bahía Chetumal

Taxco de Alarcón
Iguala
GUERRERO
Huajuapan de León
Coatzacoalcos
Minatitlán
Villahermosa
Paraíso
Frontera
Ciudad del Carmen
Laguna de Términos
Escárcega
Sabancuy
Emiliano Zapata
AMBERGRIS CAY
Belize City
TURNEFFE ISLANDS
18°

SIERRA MADRE
Chilpancingo de los Bravo
Tlapa de Comonfort
Asunción Nochixtlán
Tuxtepec
Jonuta
Teapa
Tenosique
PALENQUE
TIKAL
San Pedro
La Libertad
San Benito
Belmopan
Dangriga
ISLAS DE LA BAHÍA (Hond.)
BELIZE

Acapulco de Juárez
San Marcos
Ometepec
Putla de Guerrero
Matías Romero
Oaxaca de Juárez
OAXACA
ISTMO DE
TEHUANTEPEC
Tuxtla Gutiérrez
San Cristóbal de las Casas
CHIAPAS
Comitán de Domínguez
San Luis
La Ceiba
Punta Gorda
PUNTA NEGRA
Golfo de Honduras
Puerto Cortés

Tecpan de Galeana
Santiago Pinotepa Nacional
Santiago Jamiltepec
Mihuatlán de Porfirio Díaz
Ixtepec
Juchitán de Zaragoza
Salina Cruz
Chahuites
Tonalá
Cintalapa
DEL SUR
Livingston
SIERRA MADRE
Puerto Barrios
San Pedro Sula
Tela
16°

MIDDLE AMERICA TRENCH
Puerto Escondido
PUNTA CORNETA
Puerto Ángel
Golfo de Tehuantepec
Pijijiapan
Mapastepec
Huixtla
Tapachula
Huehuetenango
Volcán Tajumulco 13,846 ft. 4220 m
Quetzaltenango
GUATEMALA
Mazatenango
Escuintla
Salamá
Cobán
GUATEMALA
Chiquimula
COPÁN
Cerro El Pital 8957 ft. 2730 m
EL SALVADOR
Cerro Las Minas 9347 ft. 2849 m
Zacapa
Santa Rosa de Copán
Chiquimulilla
Jutiapa
Jalapa
Santa Rita
Yoro
HONDURAS
Siguatepeque
Comayagua
Guaimaca
Tegucigalpa
Danlí

© Rand McNally
A-101672-1

PACIFIC

OCEAN

Tropic of Cancer

SINALOA

DURANGO

CHIHUAHUAN
DESERT

COAHUILA

ZACATECAS

SAN LUIS
POTOSÍ

NUEVA
LÉON

NAYARIT

SIERRA MADRE OCCIDENTAL

AGUASCALIENTES

GUANAJUATO

QUERÉ

JALISCO

COLIMA

MICHOACÁN

MÉ

SIERR

GUE

Mazatlán

Durango

Fresnillo

Zacatecas

San Luis
Potosí

Aguascalientes

León

Guadalajara

ZAPOPAN

Celaya

Querétaro

Morelia

Uruapan del Progreso

Manzanillo

Colima

Lázaro
Cárdenas

Zihuatanejo

ISLAS
TRES
MARÍAS

ISLA MARÍA
MADRE

ISLA MARÍA
MAGDALENA

ISLA MARÍA
CLEOFAS

EAST PACIFIC RISE

MIDDLE AMERICA TRENCH

© Rand McNally
A-101672-1

| 0 | 20 | 40 | 60 | 80 | 100 | 120 Miles |

| 0 | 20 | 40 | 60 | 80 | 100 | 120 | 140 | 160 | 180 | 200 Kilometers |

Lambert Conformal Conic Projection
Scale 1:4,000,000
One inch to 64 miles
One cm to 40 km

GULF OF
MEXICO

MEXICO BASIN

TAMAULIPAS

Tropic of Cancer

CAYO
NUEVO

CAMPECHE BANK

Laguna
Madre

BARRA DE LOS
AMERICANOS

BARRA SOTO
LA MARINA

Ascención
2499 ft.
200 m

Hidalgo

Santander
Jiménez

La Luz

Aramberri

Zaragoza   Ciudad
Victoria

Purificación

La Pesca

Soto la
Marina

Presa
Vicente
Guerrero

Abasolo

Las Guayabas

Jaumave

Llera de Canales

Tula

Xicoténcatl

González

Aldama

Ciudad Mante

Manuel

Cuauhtémoc

Antiguo
Morelos

Ebano

Ciudad
del Maíz

Pánuco

Ciudad
Valles

Cárdenas

Ciudad Madero
Tampico

Laguna del
Chairel

Laguna de San
Andrés

Laguna de
Tamiahua

San Ciro
e Acosta

Tamapatz

Tempoal de
Sánchez

Magozal

Ozuluama

CABO ROJO

ISLA DE
LOBOS

Jalpan de
Serra

Tamazunchale

Tantoyuca

Tamiahua

uisquiapan

Cerro Azul

Zimapán

Zacualtipán

Huichapan

Ixmiquilpan

Tuxpan de
Rodríguez Cano

Chicontepec
de Tejeda

Santiago de
la Peña

Huejutla
de Reyes

HIDALGO

Actopan

Poza Rica
de Hidalgo

Papantla
de Olarte

Gutiérrez
Zamora

EL TAJIN

Tihuatlán

Pachuca
de Soto

Tepeji de
Ocampo

Tulancingo

Huauchinango

Cuetzalan del
Progreso

Nautla

Zacatlán

Zacatlan

Martínez de la Torre

Misantla

Chignahuapan

Presa de
Huapango

Teziutlán

Santiago Tlaxhuala

San Juan Teotihuacan

TEOTIHUACAN

Perote

Xalapa

Tlalnepantla

Ciudad
Netzahualcóyotl

TLAXCALA

Apizaco

Cerro Cofre de Perote
13 944 ft.
4250 m

Libres

MEXICO CITY
(DE MÉXICO)

San Martín
Texmelucan

Tlaxcala de
Xicohténcatl

Vol. Iztaccíhuatl
17 159 ft.
5230 m

Toluca
de Lerdo

Nev. de Toluca
15 387 ft.
4690 m

Cholula
Rivadabia

PUEBLA DE ZARAGOZA

VERACRUZ

Veracruz

PUNTA ANTÓN LIZARDO

Bahía de
Campeche

DIST.
FED.

Amecameca
de Juárez

San
Andrés
Cholula

Tlapaca

Volcán Pico de Orizaba
18 406 ft.
5610 m

Córdoba

Cuernavaca

Yautepec

Volcán
Popocatépetl
17 930 ft.
5465 m

Tecamachalco

Orizaba

Presa Manuel
Ávila Camacho

Alvarado

Nuevo
Progreso

Emiliano
Zapata

Cuautla

MORELOS

Izúcar de
Matamoros

PUEBLA

Tlacotalpan

San Andrés
Tuxtla

Cerro Santa Marta
5577 ft.
1700 m

Laguna
Machona

Laguna
Mecoacán

Frontera

o de
rcón

Iguala

Chiautla
de Tapia

Tehuacán

Tierra
Blanca

Cosamaloapan
de Carpio

Paraíso

Comalcalco

Huitzuco
de los Figueroa

Acatlán
de Osorio

San Gabriel Chilac

Coatzacoalcos

Minatitlán

Laguna
del
Carmen

Cárdenas

TABASCO

Villahermosa

Balsas Sur

Mezcala

Petlalcingo

Huautla

Tuxtepec

Isla

Las Choapas

Macuspana

Huajuapan
de León

San Juan
Evangelista

Acayucan

Olinalá

Tamazulapan
del Progreso

Cuicatlán

Playa
Vicente

Teapa

Chilapa de
Álvarez

Tlapa de
Comonfort

Silacayoapan

Asunción
Nochixtlán

Ixtlán
de Juárez

Santiago
Choapan

Jesús
Carranza

Pichucalco

Chilpancingo
de los Bravo

Tlalixtaquilla

Santa María
Asunción Tlaxiaco

Huitzo

ISTMO DE

Petalcingo

Chilón

pulco
Juárez

Ayutla de
los Libres

Metlatonoc

Cerro Pedro de Olla
10 991 ft.
3350 m

Putla de
Guerrero

Oaxaca de
Juárez

OAXACA

TEHUANTEPEC

Matías
Romero

Simojovel

San Marcos

Malinaltepec

Ocotlán de
Morelos

SIERRA MADRE

San Cristóbal
de las Casas

Cruz Grande

Xochistlahuaca

Ometepec

Cerro Las Flores
7054 ft.
2150 m

Nejapa de Madero

Presa Benito
Juárez

Ixtepec

Juchitán de
Zaragoza

Santo Domingo
Zanatepec

Ocozocuautla

Cintalapa

Chiapa
de Corzo

Tuxtla
Gutiérrez

Cerro Las Bolones
9154 ft.
2790 m

CHIAPAS

Palos

Laguna
Chautengo

Cuajinicuilapa

Santiago
Pinotepa
Nacional

Santiago
Jamiltepec

Ejutla de Crespo

Mihuatlán de
Porfirio Díaz

Santo Domingo
Tehuantepec

Chahuites

Arriaga

Villa
Flores

Revolución
Mexicana

Las
Rosas

Salina Cruz

Laguna
Superior

Laguna
Inferior

Mar
Muerto

Tonalá

Cerro Tres Picos
8360 ft.
2550 m

Venustiano
Carranza

Río Grande

Santa María
Colotepec

San Pedro
Pochutla

Puerto
Arista

Presa de
Angostura

Lagunas de
Chacagua

Puerto
Escondido

Golfo de
Tehuantepec

PUNTA CORNETA Puerto
Ángel

Cerro el Triunfo
8038 ft.
2450 m

Pijijiapan

Mapastepec

CAMPECHE BANK

GULF OF MEXICO

PUNTA HOLOHIT
CABO CATOCHE
ISLA CONTOY
Laguna de Yalahau

Progreso
Dzemul
Chicxulub
Hunucmá
Dzilam González
Temax
Panabá
Tizimín
Motul de Felipe Carrillo Puerto
Espita
Río Lagartos
Kantunilkin
ISLA MUJERES
Isla Mujeres
Puerto Juárez
Cancún
PUNTA CANCÚN

Mérida
Celestún
Estero Celestún
Umán
Izamal
Tunkás
Dzitás
X-Can
Puerto Morelos

Maxcanú
YUCATÁN
CHICHÉN ITZÁ
Valladolid
Chichimilá
Chemax
COBÁ
Playa del Carmen

Halachó
Becal
Muna
Tekit
Chikindzonot
Tulum

Dzitbalché
Calkiní
UXMAL
Ticul
Okkutzcab
Peto
TULUM

Tenabo
Hecelchakán
Tekax
Tzucacab
Y U C A T A N

Cozumel
ISLA COZUMEL

Campeche
Bolonchén de Rejón
José María Morelos
Laguna Chichancanab
Felipe Carrillo Puerto
Bahía de la Ascensión

PUNTA MORRO
Seybaplaya
Hopelchén
Iturbide
Chunhuhux
QUINTANA ROO

Champotón
Dzibalchén
Bahía del Espíritu Santo

MEXICO
CAMPECHE
Nohbec

Sabancuy
Escárcega
CALAKMUL
BANCO CHINCHORRO

ISLA DEL CARMEN
Ciudad del Carmen
Laguna de Términos
Chetumal
Bahía Chetumal

Nuevo Progreso
Candelaria
Nicolás Bravo
Corozal

Frontera
Palizada
Caledonia
Xcalak

Paraíso
Laguna del Pom
Candelaria
Hondo
AMBERGRIS CAY

Comalcalco
Laguna del Este
Orange Walk
San Pedro

TABASCO
Jonuta
San Pedro Tabasco
Indian Church
CARIBBEAN SEA

Cárdenas
Villahermosa
Emiliano Zapata
Multé
Hill Bank
CAY CORKER

Macuspana
Laguna Chichil
Chuntuqui
Belize City
NORTHERN CAY

Teapa
PALENQUE
Palenque
Tenosique
Northern Lagoon
TURNEFFE ISLANDS

Pichucalco
Petalcingo
San Pedro
El Encanto
Belmopan
LONG CAY
HALF MOON CAY

Chilón
Piedras Negras
TIKAL
Ciudad Melchor de Menchos
Benque Viejo Del Carmen
Middlesex

Chicoasén
Simojovel
Lago Petén Itzá
Dangriga

Ocozocuautla
Ocosingo
San Benito
Flores
SOUTH WATER CAY

Tuxtla Gutiérrez
Chiapa de Corzo
San Cristóbal de las Casas
La Libertad
Sayaxché
Victoria Peak 3675 ft. 1120 m
LAUGHING BIRD CAY

Cerro Los Bolones 9154 ft. 1790 m
La Florida
Dolores
ISLA DE ROATÁN

CHIAPAS
Las Rosas
MAYA MOUNTAINS
RANGUANA CAY

Villa Flores
Venustiano Carranza
Las Margaritas
San Luis
PUNTA NEGRA
Roatán

Revolución Mexicana
Socoltenango
Comitán de Domínguez
Punta Gorda
ISLA DE UTILA
ISLAS DE LA BAHÍA

Presa de la Angostura
La Trinitaria
Gulf of Honduras
SAPODILLA CAYS
Utila

Pijijiapan
Las Delicias
Livingston
Bahía de Amatique
CABO TRES PUNTAS

Chicomuselo
Paso Hondo
Chisec
Puerto Cortés
Colorado
La Ceiba

Cerro El Triunfo 8038 ft. 2450 m
El Pacayal
Barillas
Chahal
Puerto Barrios
Baracoa
Tela
Pico Bonito 7989 ft. 2435 m

Mapastepec
Jacaltenango
Concepción Huista
El Estor
QUIRIGUÁ
Choloma
Santa Rita
Olanchito

Motozintla de Mendoza
Huehuetenango
Cobán
San Pedro Carchá
Panzós
San Pedro Sula
La Lima
El Progreso
Pico Pijol 7487 ft. 2282 m
Yoro

Volcán Tacaná 13 428 ft. 4093 m
San Andrés Sajcabajá
Salamá
San Jerónimo
Zacapa
COPÁN
Santa Bárbara
El Negrito
San Ignacio

Huixtla
San Pedro Sacatepéquez
Santa Cruz del Quiché
Chiquimula
Santa Rosa de Copán
Cerro Las Minas 9347 ft. 2849 m
Siguatepeque
Salamá

Tapachula
Volcán Tajumulco 13 845 ft. 4220 m
GUATEMALA
Jalapa
San Luis Jilotepeque
Comayagua
Guaimaca

Volcán Santa María 12 375 ft. 3772 m
Chichicastenango
HONDURAS
La Esperanza
San Ignacio

Puerto Madero
Lago de Atitlán
Chimaltenango
Antigua Guatemala
GUATEMALA
Jutiapa
Metapán
Cerro El Pital 8957 ft. 2730 m
La Paz
Montaña El Chile 7402 ft. 2256 m

Retalhuleu
Volcán de Fuego 12 346 ft. 3763 m
Villa Nueva
Escuintla
Cuilapa
Tegucigalpa
Yuscarán

Mazatenango
Barberena
Chalatenango
Danlí

Champerico
Chiquimulilla
Santa Ana
Cojutepeque
San Francisco
Sabanagrande

PACIFIC OCEAN
Puerto San José
Apopa
Soyapango
Nueva San Salvador
San Salvador
Volcán de San Vicente 7156 ft. 2181 m
San Miguel
San Marcos de Colón

MIDDLE AMERICA TRENCH
Sonsonate
Acajutla
PUNTA REMEDIOS
San Vicente
Usulután
La Unión
Golfo de Fonseca
Choluteca
El Corpus

EL SALVADOR
Bahía de Jiquilisco
Volcán Cosigüina 2818 ft. 859 m
NICARAGUA

© Rand McNally
A-101673-1

0 20 40 60 80 100 120 Miles
0 20 40 60 80 100 120 140 160 180 200 Kilometers

Lambert Conformal Conic Projection
Scale 1:4,000,000
One inch to 64 miles
One cm to 40 km

0   20   40   60   80   100   120 Miles

0   20   40   60   80   100   120   140   160   180   200 Kilometers

Lambert Conformal Conic Projection
Scale 1:4,000,000
One inch to 64 miles
One cm to 40 km

NICARAGUA

Laguna Páhara
Puerto Cabezas

CARIBBEAN SEA

ISLA DE PROVIDENCIA

SAN ANDRÉS Y PROVIDENCIA
(Colombia)

ISLA DE SAN ANDRÉS   San Andrés

CAYOS DEL ESTE SUDESTE

ISLAS DEL MAÍZ
(Nicaragua)

CAYOS DE ALBUQUERQUE

CARIBBEAN SEA

a

CARIBBEAN SEA

PUNTA MANZANILLO
Portobelo   Nombre de Dios   Palenque   Miramar
Playa Chiquita

Cerro Brujo 3212 ft. 979 m

María Chiquita   COLÓN

Coco Solo   Cativá   Salamanca   La Mesa
Colón   Puerto Pilón
Cristóbal   Rainbow City
GATÚN LOCKS   Río Rita   Gatuncillo   Lago Alajuela
Margarita
Nuevo Chagres   Gatún   Buena Vista
Palmas Bellas   Escobal   Chilibre
Lago Gatún   ISLA BARRO COLORADO   Panama Canal   Calzada Larga   Pacora
Lagarterita   Gamboa   Tocumen
Arenosa   Las Cumbres   Juan Díaz
Cerro Cama   Paraíso   Pedro Miguel
PEDRO MIGUEL LOCKS   San Miguelito
COCLÉ   Boca del Río Indio   MIRAFLORES LOCKS   Panamá
PANAMÁ   Diablo Heights   Balboa
La Zanguenga   Arraiján
Pueblo Nuevo   Nuevo Arraiján
Ciri Grande   La Chorrera   Vacamonte   Bahía de Panamá
Río Indio   Taboga
Lídice   Capira

©Rand McNally
A-101679-1

COSTA RICA

Puerto Limón
Vesta
PUNTA MONA
CORDILLERA DE TALAMANCA   ISLA COLÓN   ARCHIPIÉLAGO DE BOCAS DEL TORO
Cerro Kámuk 11 660 ft. 3554 m   Bocas del Toro   ISLA BASTIMENTOS
Buenos Aires   ISLA POPA
Almirante   ISLA ESCUDO DE VERAGUAS
Laguna de Chiriquí   PEN. VALIENTE   Golfo de los Mosquitos
Volcán Barú 11 401 ft. 3475 m   Chiriquí Grande
Volcán   Bajo Boquete
PEN. DE OSA   La Concepción   Dolega
Golfito   Gualaca   CORDILLERA CENTRAL   **PANAMA**   Santa Fé
Golfo Dulce   David   Cerro Santiago 6959 ft. 2121 m   El Valle
CABO MATAPALO   Puerto Armuelles   Pedregal   Las Lajas   Canazas   Peñonomé
Bahía de Charco Azul   ISLA SEVILLA   Remedios   Santa María   Natá   Antón   Río Hato
ISLA BOCA BRAVA   Las Palmas   Aguadulce
PUNTA BURICA   ISLA PARIDA   Golfo de Chiriquí   Soná   Santiago   Monagrillo
Montijo   Pesé   Chitré
ISLA DE COIBA   ISLA DE CEBACO   Las Tablas   Guararé   La Palma
Golfo de Montijo   PENÍNSULA DE AZUERO   Pedasí
ISLA JICARÓN   Tonosí   PUNTA MALA
PUNTA MARIATO

Puerto Limón
Panama Canal (Canal de Panamá)   Palmas Bellas
Colón   Cristóbal   ISTMO DE PANAMÁ   Golfo de San Blas   SERRANÍA DE SAN BLAS   Ticantiquí
El Porvenir
PUNTA MANZANILLO   Nombre de Dios   Portobelo
Lago Gatún   Gamboa   Chepo
Paraíso   Pacora   PUNTA MOSQUITO   Mansucum
San Miguelito   Lago Bayano
La Chorrera   Panamá   Chimán   SERRANÍA DEL DARIÉN   CABO TIBURÓN   PUNTA CARIBANA
Capira   Bahía de Panamá   Acandí   Golfo de Urabá
Bejuco   San Carlos   San Miguel   La Palma
San Miguel   ISLA DEL REY   El Real de Santa María   Turbo
ISLA SAN JOSÉ   Golfo de San Miguel   Garachiné   **ANTIOQUIA**
ARCHIPIÉLAGO DE LAS PERLAS   PUNTA GARACHINÉ   Chigorodó

Golfo de Panamá   Yaviza   Arrato   Ríosucio

Jaqué   Salaquí   **COLOMBIA**
CABO   Apartadó

CHOCÓ   ANT.

PUNTA MARZO   SERRANÍA DE BAUDÓ   Golfo de Cupica   Ensenada de Tribugá   Nuquí   Quibdó

PACIFIC OCEAN

PANAMA BASIN

©Rand McNally
A-101677-1

0 20 40 60 80 100 120 Miles
0 20 40 60 80 100 120 140 160 180 200 Kilometers

Lambert Conformal Conic Projection
Scale 1:4,000,000
One inch to 64 miles
One cm to 40 km

Inset map a
Lambert Conformal Conic Projection
Scale 1:1,000,000
One inch to 16 miles
One cm to 10 km

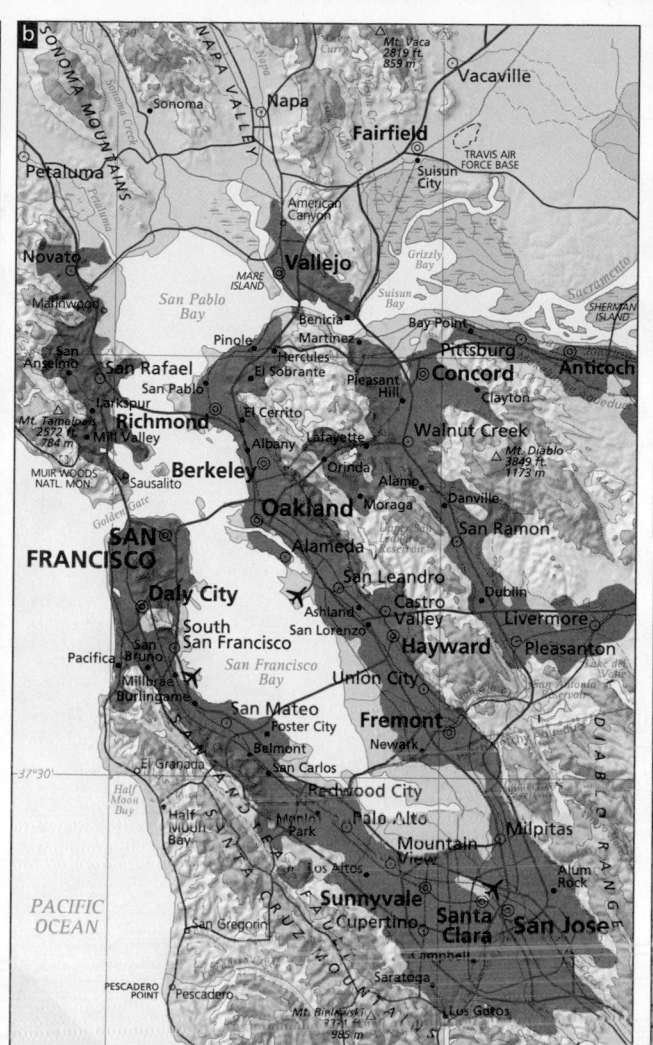
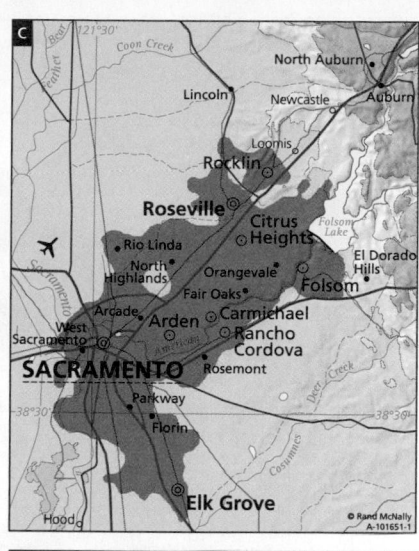

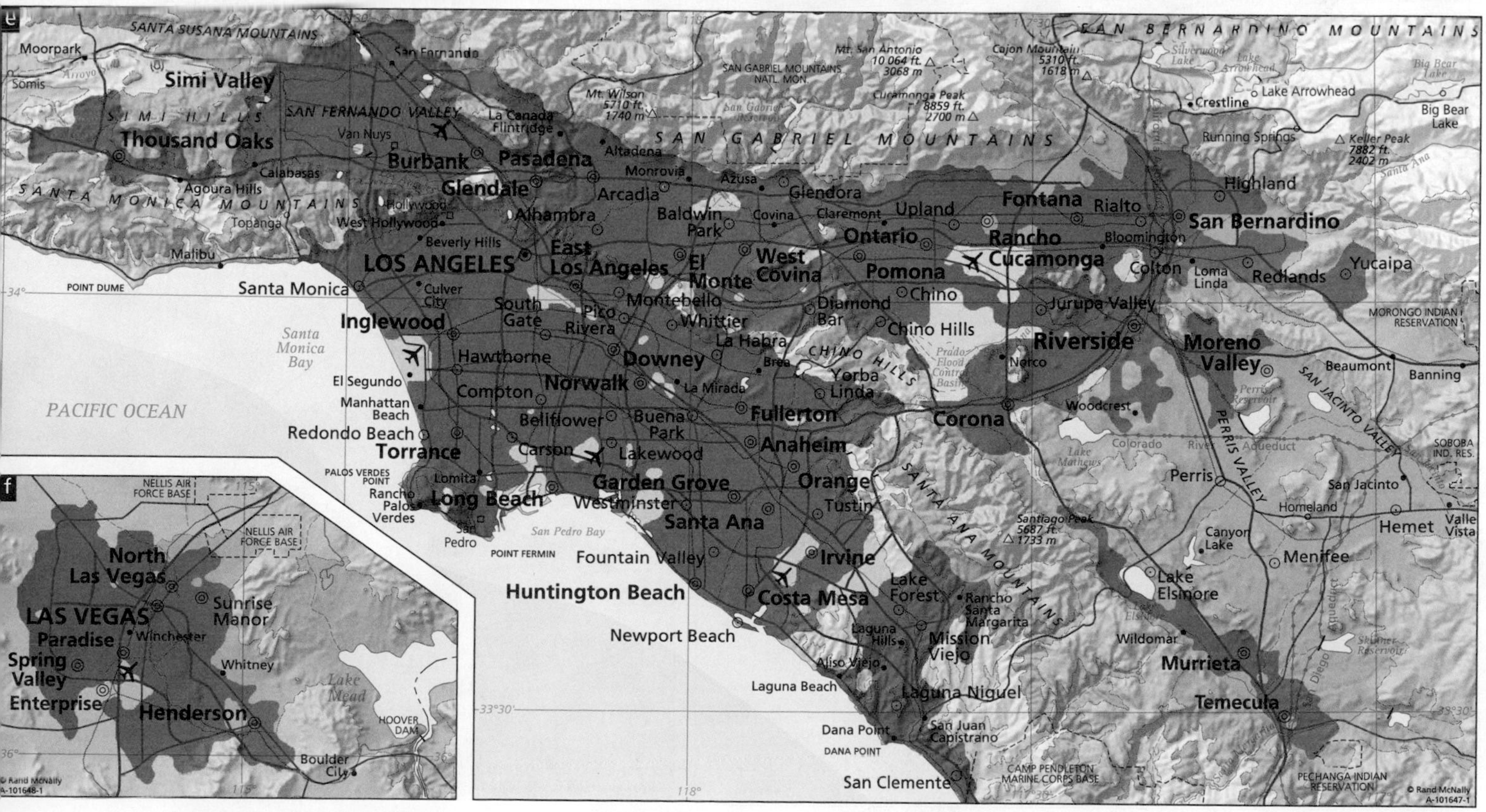

a

Lake Erie

CLEVELAND

Lorain
Sheffield Lake
Avon Lake
Bay Village
Rocky River
Lakewood
Euclid
Willowick
Willoughby
Willoughby Hills
Mentor
Eastlake
Mentor On The Lake
Fairport Harbor
Painesville
Chardon
Chesterland
Richmond Heights
Mayfield Heights
Cleveland Heights
Shaker Heights
Beachwood
South Russell
Chagrin Falls
Solon
Bedford
Maple Heights
Garfield Heights
Parma Heights
Parma
Brooklyn
Berea
Brook Park
North Olmsted
Westlake
Avon
Sheffield
North Ridgeville
Eaton Estates
Elyria
Amherst
Oberlin
Grafton
Lagrange
Wellington
West Salem
Lodi
Creston
Seville
Chippewa Lake
Medina
Copley
Fairlawn
Akron
Norton
Barberton
Clinton
Green
Greentown
Uniontown
Hartville
Randolph
Suffield
Mogadore
Tallmadge
Kent
Stow
Cuyahoga Falls
Bath
Richfield
Hinckley
Brecksville
North Royalton
Broadview Heights
Strongsville
Brunswick
Macedonia
Twinsburg
Aurora
Hiram
Mantua
Streetsboro
Hudson
Reminderville
Ravenna
Brimfield
Doylestown
Rittman
Wadsworth
Manchester
Canal Fulton
Massillon
Dalton
Apple Creek
Orrville
Smithville
Wooster
Madisonburg
Navarre
Brewster
North Industry
Perry Heights
CANTON
North Canton
Louisville
East Canton
Lake Cable
Richville
Bedford
CUYAHOGA VALLEY NATL. PARK
La Due Res.
Lake Rockwell
Mogador Res.
MOHICANVILLE DAM
© Rand McNally
A-101629-1

b

ATLANTA

White
Ball Ground
Canton
Cumming
Gainesville
Flowery Branch
Cartersville
Emerson
Acworth
Holly Springs
Milton
Buford
BUFORD DAM
Woodstock
Alpharetta
Sugar Hill
Suwanee
Kennesaw
Roswell
Johns Creek
Duluth
Auburn
Marietta
Sandy Springs
Dunwoody
Peachtree Corners
Norcross
Lawrenceville
Dallas
Fair Oaks
Powder Springs
Smyrna
Chamblee
Doraville
Brookhaven
Snellville
Loganville
Hiram
Mableton
Tucker
Austell
Decatur
Stone Mountain
Douglasville
ATLANTA
Redan
Lithonia
East Point
Pantersville
Conyers
Hapeville
College Park
Forest Park
Union City
Fairburn
Riverdale
Covington
Palmetto
Stockbridge
Jonesboro
Fayetteville
Hampton
McDonough
Sargent
Peachtree City
Newnan
Moreland
Senoia
Experiment
Jackson Lake
Allatoona Lake
Chattahoochee R.
Flint R.
Yellow R.
Alcovy R.
© Rand McNally
A-101631-1

c

BALTIMORE
WASHINGTON

MARYLAND
VIRGINIA
DIST. OF COLUMBIA

Union Bridge
Westminster
Hereford
Forest Hill
Woodsboro
New Windsor
Sandyville
Upperco
Fallston
Bel Air
Walkersville
Libertytown
Finksburg
Reisterstown
Cockeysville
Phoenix
Long Green
Pleasant Hill
Frederick
Middletown
Braddock Heights
New Market
Mount Airy
Sykesville
Owings Mills
Lutherville
Timonium
Towson
Parkville
Perry Hall
Joppatowne
Edgewood
Clarksburg
Damascus
Mount Airy
Randallstown
Pikesville
Overlea
Essex
Middle River
Dickerson
Point of Rocks
Sugar Loaf Mtn. 1282 ft. △ 391 m
Cooksville
BALTIMORE
Catonsville
Arbutus
Landsdowne
Dundalk
Edgemere
Germantown
Highland
Columbia
Ellicott City
Glen Burnie
Riviera Beach
Poolesville
Gaithersburg
Olney
Savage
Fort George G. Mead
Severn
Severna Park
Lake Shore
Rockville
Aspen Hill
Burtonsville
Laurel
Odenton
Arnold
BROAD NECK
North Potomac
Wheaton
Potomac
Beltsville
Silver Spring
Greenbelt
Crofton
Ashburn
Sterling
Herndon
Reston
Bethesda
College Park
Lanham
Bowie
Stevensville
Annapolis
KENT ISLAND
THOMAS POINT
McLean
Falls Church
Hyattsville
Mitchellville
Largo
Mayo
Eastern Bay
Chantilly
Vienna
ARLINGTON
WASHINGTON
Suitland
Hillcrest Heights
Upper Marlboro
Londontown
KENT POINT
Centreville
Fairfax
Annandale
Camp Springs
Wittman
Sherwood
Manassas Park
Burke
Franconia
Clinton
Rosaryville
Deale
Manassas
Springfield
Groveton
Mount Vernon
Fort Washington Forest
ANDREWS AIR FORCE BASE
Herring Bay
Tilghman
FORT BELVOIR
Lake Ridge
Woodbridge
Bryans Road
Accokeek
Owings
Chesapeake Beach
TILGHMAN ISLAND
Dale City
Dumfries
Potomac Heights
Waldorf
Huntingtown
QUANTICO MARINE CORPS BASE
Triangle
Quantico
Marbury
St. Charles
Prince Frederick
Lunga Reservoir
Golden Beach
La Plata
Hughesville
Taylors Island
Chesapeake Bay
Potomac R.
Patuxent R.
© Rand McNally
A-101630-1

d

PHILADELPHIA

PENNSYLVANIA
MARYLAND
DELAWARE

Lehighton
Palmerton
Columbia
BLUE MOUNTAIN
KITTATINNY MTN.
Pen Argyl
Bangor
Belvidere
Slatington
Walnutport
Nazareth
Northampton
Easton
Phillipsburg
Washi
Schnecksville
Catasauqua
Wilson
Alpha
Clinton
Whitehall
Bethlehem
Hamp
Allentown
Fountain Hill
Milford
Emmaus
Heilertown
Shoemakersville
Macungie
Coopersburg
Fleming
Kutztown
Frenchtown
Laureldale
Quakertown
Fleetwood
East Greenville
Perkasie
Reading
Boyertown
Red Hill
Shillington
Gilbertsville
Telford
Doylestown
Birdsboro
Stowe
Souderton
Lansdale
New Hope
Pottstown
Harleysville
Spring City
Royersford
Warminster
Honey Brook
Collegeville
Richboro
Phoenixville
Valley Forge
East Norriton
Willow Grove
Abington
Glenside
Exton
King of Prussia
Conshohocken
Cheltenham
Downingtown
Paoli
Coatesville
Newtown Square
Ardmore
Havertown
PHILADELPHIA
Parkesburg
West Chester
Broomall
Upper Darby
Pennsauken
Moore
Cherry Hi
Kennett Square
Darby
Camden
Haddonfield
Springdale
West Grove
Chester
Bellmawr
Stratford
Linde
Hockessin
Paulsboro
Woodbury
Gibbstown
Pine Hill
Wilmington
Elsmere
Claymont
Swedesboro
Mantua
Glassboro
Penns Grove
Pitman
Delaware R.
Schuylkill R.
Nockamixon Lake
Green Lane Reservoir
Spruce Run

```
0    5    10    15    20    25    30 Miles
0  5  10  15  20  25  30  35  40  45  50 Kilometers
```

Lambert Conformal Conic Projection
Scale 1:1,000,000
One inch to 16 miles
One cm to 10 km

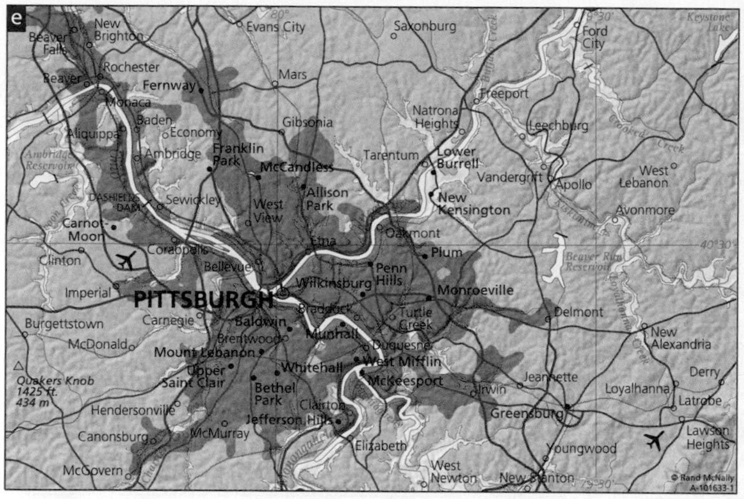

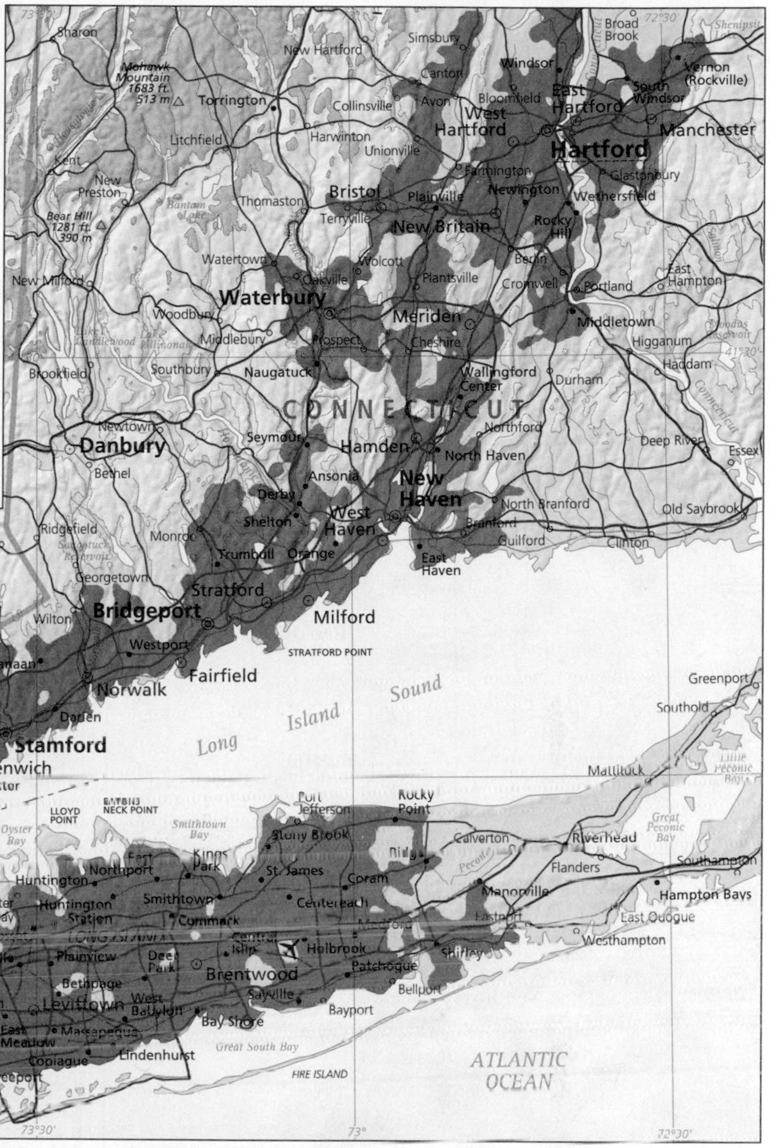

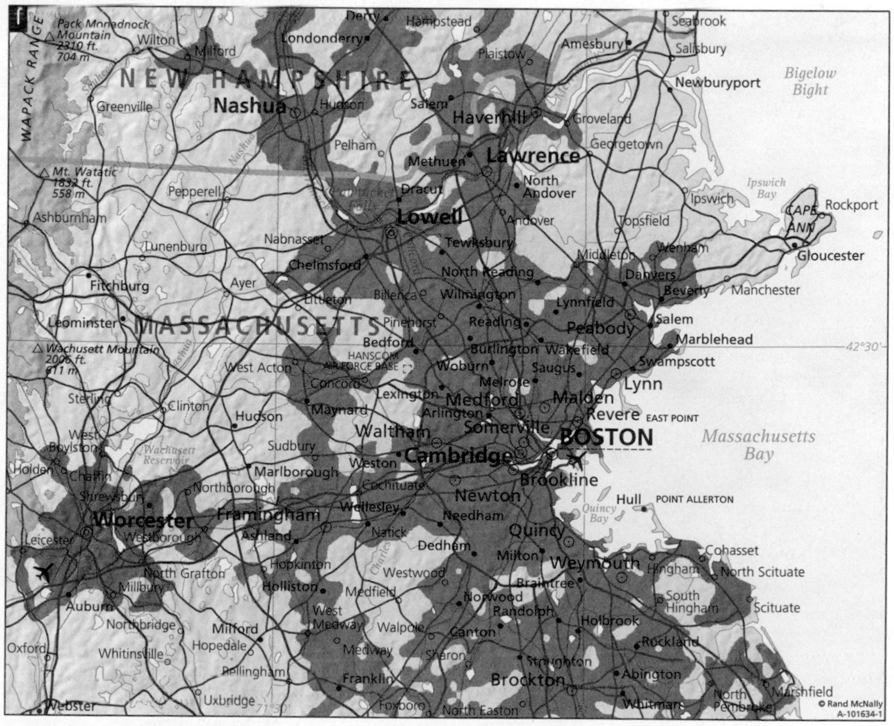

e

PITTSBURGH

Beaver Falls, New Brighton, Beaver, Rochester, Fernway, Evans City, Saxonburg, Ford City, Mars, Natrona Heights, Freeport, Leechburg, Vandergrift, Apollo, West Leechburg, Avonmore, Vandergrift, Tarentum, Lower Burrell, New Kensington, Penn Hills, Monroeville, Plum, Delmont, New Alexandria, Derry, Loyalhanna, Greensburg, Latrobe, Youngwood, Lawson Heights, Jeannette, Irwin, McKeesport, West Mifflin, Turtle Creek, Wilkinsburg, Braddock, Duquesne, Elizabeth, West Newton, New Eagle, McMurray, Canonsburg, McGovern, McDonald, Burgettstown, Hendersonville, Carnegie, Mount Lebanon, Upper Saint Clair, Bethel Park, Whitehall, Jefferson Hills, Clinton, Imperial, Carnot-Moon, Coraopolis, Sewickley, Belleview, West View, Allison Park, Oakmont, McCandless, Franklin Park, Economy, Baden, Ambridge, Aliquippa, Monaca, Gibsonia

Quakers Knob 1425 ft. 434 m

New Brunswick, North Brunswick, East Brunswick, Sayreville, Perth Amboy, Edison, Metuchen, Piscataway, Woodbridge, Carteret, Rahway, Linden, Elizabeth, Bayonne, Newark, Jersey City, **NEW YORK**, Yonkers, Paterson, Clifton, Passaic, Wayne, Parsippany-Troy Hills, Morristown, Madison, Summit, Union, Irvington, Cranford, Westfield, Plainfield, Bound Brook, Somerville, Manville, Hopewell, Princeton, Pennington, Lawrenceville, Ewing, Mercerville, **Trenton**, White Horse, Bordentown, Hightstown, Freehold, Englishtown, Matawan, Keansburg, Hazlet, Middletown, Red Bank, Eatontown, Long Branch, Asbury Park, Neptune, Spring Lake, Manasquan, Point Pleasant, Brick Township, Lakewood, Toms River, Beachwood, Forked River, Barnegat

NEW JERSEY

PINE BARRENS

McGuire Air Force Base, Fort Dix, Mount Holly, Browns Mills, Whiting, Bedford Lakes, New Egypt, Lakehurst Naval Air Engineering Station, Silverton, Gilford Park

CONNECTICUT

Hartford, West Hartford, East Hartford, Manchester, Bristol, New Britain, Rocky Hill, Newington, Wethersfield, Glastonbury, Farmington, Plainville, Southington, Bloomfield, Windsor, Windsor Locks, East Windsor, South Windsor, Vernon (Rockville), Simsbury, Avon, Canton, Collinsville, Unionville, Torrington, Litchfield, Harwinton, Thomaston, Terryville, Waterbury, Watertown, Oakville, Wolcott, Prospect, Naugatuck, Middlebury, Woodbury, Southbury, Brookfield, Newtown, Bethel, Danbury, Ridgefield, Seymour, Derby, Shelton, Ansonia, Orange, West Haven, New Haven, Hamden, North Haven, Meriden, Cheshire, Wallingford Center, Durham, Middletown, Cromwell, Portland, East Hampton, Haddam, Higganum, Deep River, Essex, Old Saybrook, Clinton, Guilford, Branford, North Branford, East Haven, Trumbull, Monroe, Stratford, Milford, Fairfield, Bridgeport, Norwalk, Westport, Darien, New Canaan, Stamford, Greenwich, Port Chester

Mohawk Mountain 1683 ft. 513 m
Bear Hill 1281 ft. 390 m

NEW YORK

Hudson, High Point 1803 ft. 550 m, Warwick, Sussex, Hamburg, Franklin, West Milford, Sparta, Ogdensburg, Bloomingdale, Pompton Lakes, Ringwood, Wanaque, Oakland, Wyckoff, Suffern, Spring Valley, New City, Nyack, Congers, Nanuet, Pearl River, Ramsey, Harriman, Greenwood Lake, Bear Mountain 1306 ft. 398 m, Stony Point, Haverstraw, Peekskill, Mohegan Lake, Yorktown Heights, South Salem, Bedford Hills, Croton-on-Hudson, Ossining, Chappaqua, Mount Kisco, Pleasantville, New Canaan, Wilton, Georgetown, Tarrytown, Dobbs Ferry, White Plains, Harrison, Rye, Mamaroneck, New Rochelle, Mount Vernon, Port Chester, Glen Cove, Oyster Bay, Huntington, Northport, Kings Park, Smithtown, Commack, Centereach, Coram, Port Jefferson, Rocky Point, Calverton, Riverhead, Flanders, Southampton, Westhampton, East Quogue, Hampton Bays, Manorville, Yaphank, Patchogue, Bellport, Sayville, Bayport, Bay Shore, Brentwood, West Babylon, Lindenhurst, Copiague, Amityville, Massapequa, Bethpage, Levittown, East Meadow, Hempstead, Hicksville, Mineola, Garden City, Floral Park, Valley Stream, Rockville Centre, Freeport, Long Beach, Woodmere, Great South Bay, Fire Island

ATLANTIC OCEAN

Long Island Sound

Rockaway Point, Sandy Hook, Highlands, Raritan Bay, Lloyd Point, Oyster Bay, Bay Ville, Smithtown Bay, Great Peconic Bay, Little Peconic Bay, Greenport, Southold, Mattituck, Stratford Point

f

MASSACHUSETTS

BOSTON, Cambridge, Brookline, Newton, Waltham, Somerville, Medford, Malden, Revere, Lynn, Quincy, Weymouth, Braintree, Randolph, Milton, Dedham, Needham, Wellesley, Natick, Framingham, Ashland, Hopkinton, Marlborough, Northborough, Westborough, Worcester, Auburn, Leicester, Shrewsbury, Hudson, Maynard, Concord, Lexington, Arlington, Bedford, Burlington, Woburn, Reading, Wakefield, Saugus, Melrose, Stoneham, Winchester, Lincoln, Sudbury, Weston, Waltham, Peabody, Danvers, Beverly, Salem, Marblehead, Swampscott, Lynnfield, North Reading, Wilmington, Tewksbury, Billerica, Chelmsford, Lowell, Lawrence, Methuen, Dracut, Andover, North Andover, Haverhill, Groveland, Georgetown, Topsfield, Ipswich, Rockport, Gloucester, Manchester, Wenham, Hull, Cohasset, Scituate, South Hingham, Hingham, Rockland, Abington, Brockton, Whitman, Holbrook, Stoughton, Canton, Norwood, Walpole, Medfield, Medway, Franklin, Foxboro, North Easton, Bellingham, Milford, Hopedale, Northbridge, Uxbridge, Whitinsville, Millbury, Oxford, Webster, Holliston, Medway

Wachusett Mountain 2006 ft. 611 m
Mt. Watatic 1832 ft. 558 m
Wachusett Reservoir
Hanscom Air Force Base

NEW HAMPSHIRE

Nashua, Hudson, Pelham, Salem, Derry, Londonderry, Hampstead, Plaistow, Atkinson, Windham, Wilton, Milford, Amherst, Greenville, Pepperell, Ashburnham, Fitchburg, Leominster, Gardner, Ashby, Townsend, Newburyport, Amesbury, Salisbury, Seabrook, Bigelow Bight, Ipswich Bay, Cape Ann

WAPACK RANGE
Pack Monadnock Mountain 2310 ft. 704 m

Massachusetts Bay
Point Allerton
East Point
Quincy Bay

© Rand McNally A-101632-1
© Rand McNally A-101633-1
© Rand McNally A-101634-1

0	5	10	15	20	25	30 Miles

| 0 | 5 | 10 | 15 | 20 | 25 | 30 | 35 | 40 | 45 | 50 Kilometers |

Lambert Conformal Conic Projection
Scale 1:1,000,000
One inch to 16 miles
One cm to 10 km

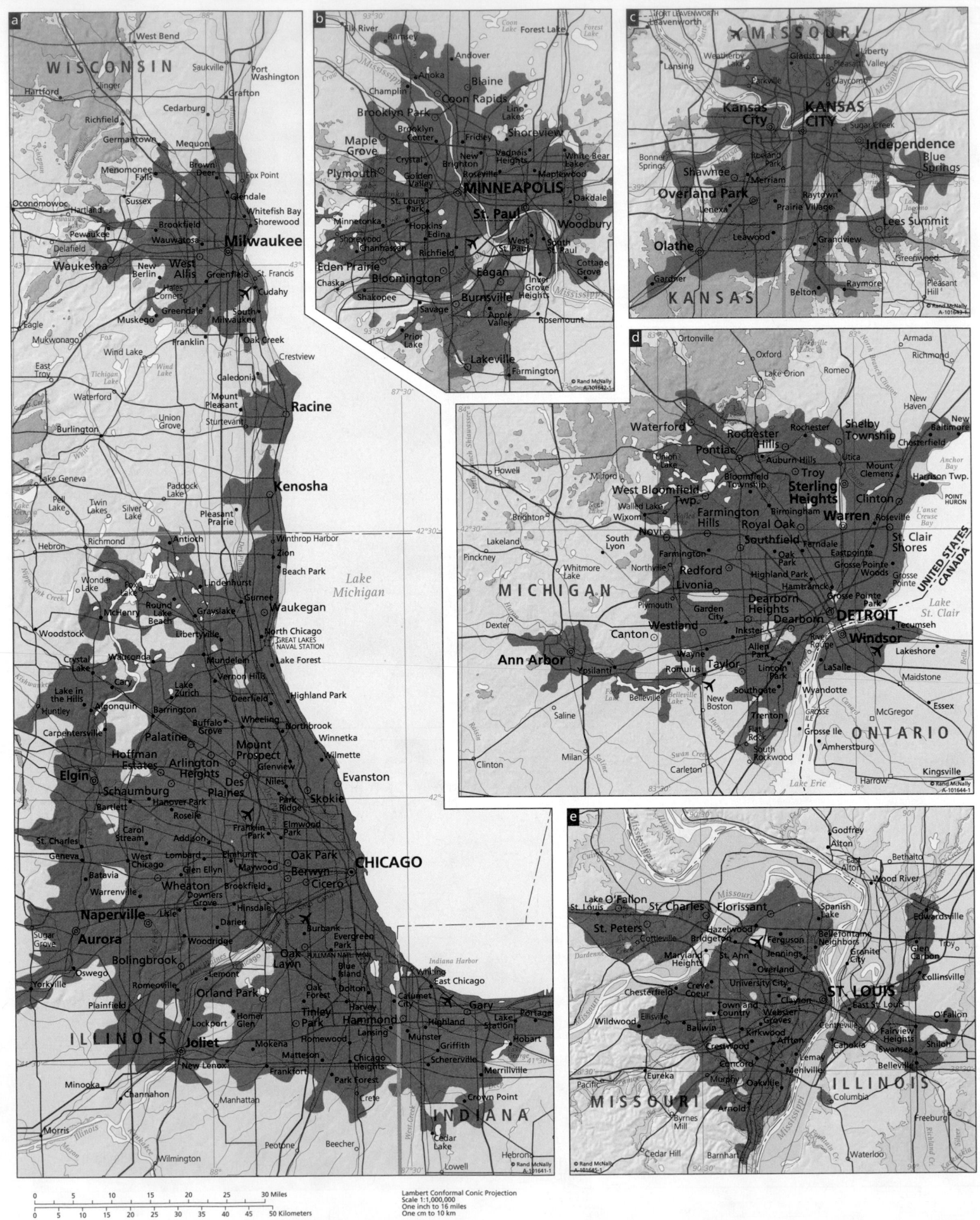

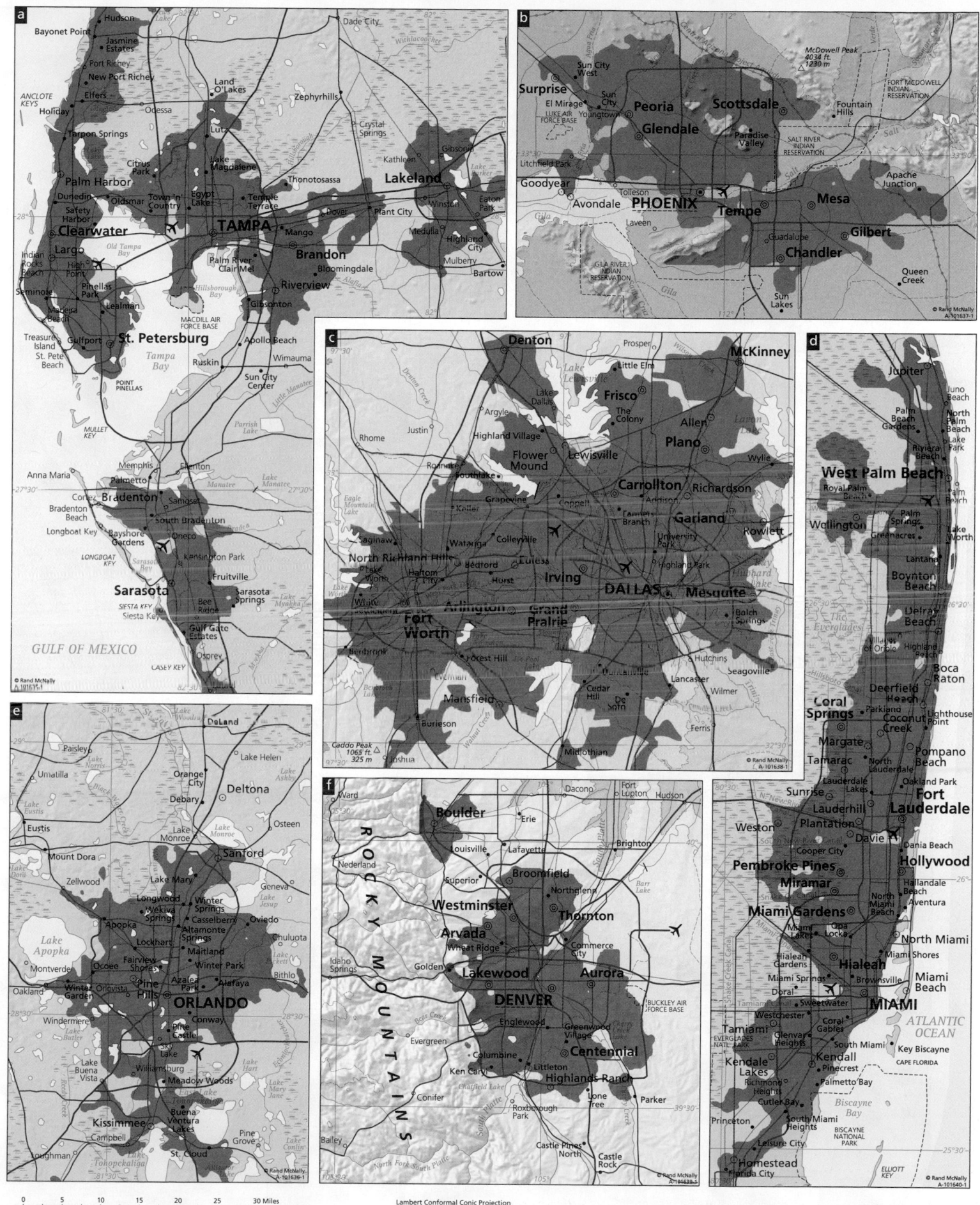

a

Bayonet Point
Hudson
Jasmine Estates
Port Richey
New Port Richey
Elfers
Holiday
Tarpon Springs
Palm Harbor
Dunedin
Oldsmar
Safety Harbor
Clearwater
Largo
High Point
Indian Rocks Beach
Seminole
Madeira Beach
Pinellas Park
Lealman
Treasure Island
Gulfport
St. Petersburg
St. Pete Beach
Indian Shores

ANCLOTE KEYS
Land O'Lakes
Odessa
Lutz
Citrus Park
Town 'n' Country
Egypt Lake
Lake Magdalene
Temple Terrace
TAMPA
Mango
Palm River-Clair Mel
Riverview
Gibsonton

Zephyrhills
Crystal Springs
Kathleen
Thonotosassa
Lakeland
Plant City
Winston
Eaton Park
Medulla
Highland City
Mulberry
Bartow

Dover
Brandon
Bloomingdale

POINT PINELLAS
Tampa Bay
Ruskin
Sun City Center
Wimauma
Apollo Beach

MULLET KEY
Memphis
Ellenton
Anna Maria
Palmetto
Cortez
Bradenton
Bradenton Beach
Longboat Key
Bayshore Gardens
South Bradenton
Oneco
LONGBOAT KEY
Sarasota
Kensington Park
Fruitville
Sarasota Springs
Bee Ridge
Gulf Gate Estates
SIESTA KEY
Siesta Key
Osprey
CASEY KEY

GULF OF MEXICO

© Rand McNally
A-101615-1

Old Tampa Bay
MACDILL AIR FORCE BASE
Hillsborough Bay
Lake Manatee
Parrish Lake
Little Manatee
Lake Myakka

b

Sun City West
Surprise
El Mirage
Youngtown
Luke Air Force Base
Litchfield Park
Goodyear
Avondale
Tolleson
Laveen
PHOENIX
Peoria
Glendale
Scottsdale
Paradise Valley
Tempe
Guadalupe
Chandler
Mesa
Gilbert
Apache Junction
Queen Creek
Sun Lakes

McDowell Peak
4034 ft.
1230 m
FORT McDOWELL INDIAN RESERVATION
Fountain Hills
SALT RIVER INDIAN RESERVATION
GILA RIVER INDIAN RESERVATION

© Rand McNally
A-101637-1

c

Denton
Prosper
McKinney
Little Elm
Frisco
Argyle
The Colony
Allen
Rhome
Justin
Highland Village
Plano
Roanoke
Flower Mound
Lewisville
Wylie
Southlake
Grapevine
Carrollton
Richardson
Addison
Keller
Farmers Branch
Garland
Saginaw
Watauga
Colleyville
University Park
Rowlett
North Richland Hills
Bedford
Euless
Highland Park
Haltom City
Fort Worth
Hurst
Irving
DALLAS
Mesquite
Arlington
Grand Prairie
Balch Springs
Benbrook
Forest Hill
Hutchins
Seagoville
Everman
Duncanville
Lancaster
Wilmer
Crowley
Mansfield
Cedar Hill
De Soto
Ferris
Burleson
Midlothian
Caddo Peak
1065 ft.
325 m
Joshua

Lake Lewisville
Lake Dallas
Lavon Lake
Eagle Mountain Lake
Lake Worth
White Rock Lake
Hubbard Lake
Joe Pool Lake

© Rand McNally
A-101638-1

d

Jupiter
Juno Beach
Palm Beach Gardens
North Palm Beach
Lake Park
West Palm Beach
Riviera Beach
Royal Palm
Palm Beach
Wellington
Palm Springs
Greenacres
Lake Worth
Lantana
Boynton Beach
Delray Beach
Highland Beach
The Everglades
Village of Golf
Boca Raton
Coral Springs
Deerfield Beach
Parkland
Coconut Creek
Lighthouse Point
Margate
Pompano Beach
Tamarac
North Lauderdale
Sunrise
Lauderdale Lakes
Oakland Park
Lauderhill
Fort Lauderdale
Weston
Plantation
Davie
Dania Beach
Cooper City
Hollywood
Pembroke Pines
Hallandale Beach
Miramar
Aventura
North Miami Beach
Miami Gardens
Opa-Locka
Miami Lakes
North Miami
Hialeah
Miami Shores
Hialeah Gardens
Brownsville
Miami Beach
Miami Springs
Doral
Sweetwater
MIAMI
Westchester
Coral Gables
Tamiami
Glenvar Heights
South Miami
Key Biscayne
CAPE FLORIDA
Kendale Lakes
Kendall
Pinecrest
Richmond Heights
Palmetto Bay
Cutler Bay
Princeton
South Miami Heights
Leisure City
Homestead
Florida City
BISCAYNE NATIONAL PARK
ELLIOTT KEY
EVERGLADES NATL. PARK

ATLANTIC OCEAN
Biscayne Bay

© Rand McNally
A-101640-1

e

DeLand
Paisley
Lake Helen
Umatilla
Orange City
Deltona
Debary
Eustis
Lake Monroe
Osteen
Mount Dora
Sanford
Zellwood
Lake Mary
Longwood
Geneva
Apopka
Wekiva Springs
Winter Springs
Casselberry
Oviedo
Chuluota
Lockhart
Altamonte Springs
Montverde
Fairview Shores
Maitland
Winter Park
Oakland
Winter Garden
Pine Hills
Azalea Park
Ocoee
Orlovista
Alafaya
Windermere
Conway
ORLANDO
Pine Castle
Lake Buena Vista
Sky Lake
Williamsburg
Meadow Woods
Buena Ventura Lakes
Kissimmee
Pine Grove
Campbell
St. Cloud
Loughman

Lake Apopka
Lake Woodruff
Lake Norris
Lake Eustis
Lake Dora
Lake Jesup
Lake Jessamine
Lake Butler
Lake Hart
Lake Mary Jane
Lake Tohopekaliga

© Rand McNally
A-101636-1

f

Ward
Dacono
Fort Lupton
Hudson
Boulder
Erie
Nederland
Louisville
Lafayette
Brighton
Superior
Broomfield
Northglenn
Barr Lake
Westminster
Thornton
Arvada
Wheat Ridge
Commerce City
Idaho Springs
Golden
Lakewood
Aurora
DENVER
BUCKLEY AIR FORCE BASE
Englewood
Evergreen
Greenwood Village
Columbine
Centennial
Ken Caryl
Littleton
Highlands Ranch
Conifer
Lone Tree
Parker
Bailey
Roxborough Park
Castle Pines North
Castle Rock

ROCKY MOUNTAINS
Bear Creek
South Platte
Chatfield Lake
North Fork South Platte

© Rand McNally
A-101639-1

Lambert Conformal Conic Projection
Scale 1:1,000,000
One inch to 16 miles
One cm to 10 km

0 5 10 15 20 25 30 Miles
0 5 10 15 20 25 30 35 40 45 50 Kilometers

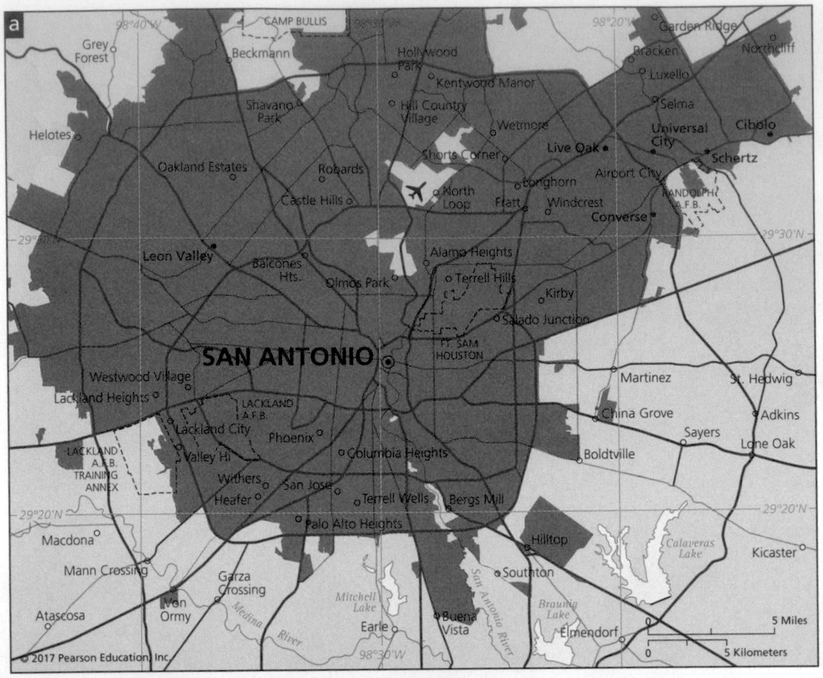

a — SAN ANTONIO

CAMP BULLIS
Grey Forest · Beckmann · Hollywood Park · Garden Ridge · Northcliff
Helotes · Shavano Park · Kentwood Manor · Bracken · Luxello · Selma
Hill Country Village · Wetmore · Universal City · Cibolo · Schertz
Oakland Estates · Shorts Corner · Live Oak · Airport City
Robards · Longhorn · Windcrest · RANDOLPH A.F.B.
Castle Hills · North Loop · Pratt · Converse
Leon Valley · Balcones Hts. · Alamo Heights
Olmos Park · Terrell Hills · Kirby · Salado Junction
SAN ANTONIO · FT. SAM HOUSTON
Westwood Village · Martinez · St. Hedwig
Lackland Heights · LACKLAND A.F.B. · China Grove · Adkins
LACKLAND A.F.B. TRAINING ANNEX · Lackland City · Phoenix · Sayers · Lone Oak
Valley Hi · Columbia Heights · Boldtville
Withers · San Jose · Terrell Wells · Bergs Mill
Heafer · Palo Alto Heights · Hilltop · Kicaster
Macdona · Garza Crossing · Southton · Calaveras Lake
Mann Crossing · Von Ormy · Earle · Buena Vista · Elmendorf
Atascosa · Mitchell Lake · Brauns Lake
San Antonio River · Medina River
5 Miles / 5 Kilometers
© 2017 Pearson Education, Inc.

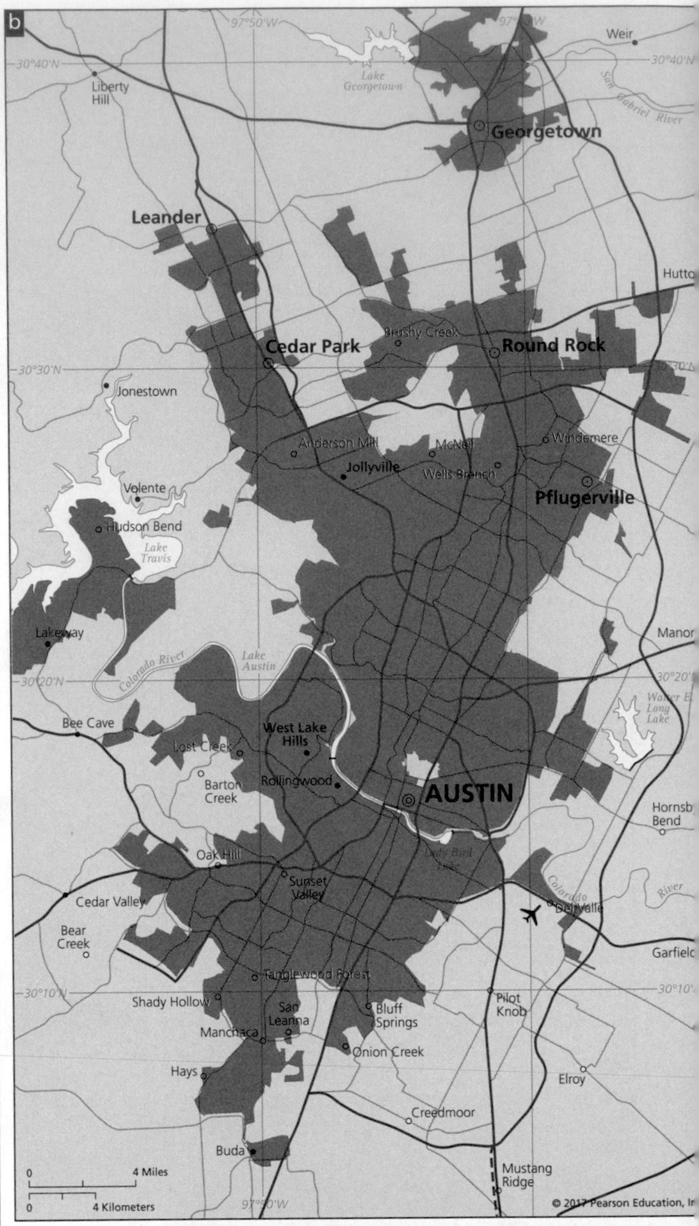

b — AUSTIN

Liberty Hill · Lake Georgetown · Weir · Georgetown
Leander · Hutto · Cedar Park · Brushy Creek · Round Rock
Jonestown · Anderson Mill · McNeil · Windemere · Pflugerville
Volente · Jollyville · Wells Branch
Hudson Bend · Lake Travis · Lake Austin
Lakeway · Colorado River · Manor
Bee Cave · Lost Creek · West Lake Hills · Long Lake
Cedar Valley · Barton Creek · Rollingwood · **AUSTIN** · Hornsby Bend
Bear Creek · Oak Hill · Sunset Valley · Garfield
Shady Hollow · Tanglewood Forest · San Leanna · Bluff Springs · Pilot Knob
Manchaca · Onion Creek · Elroy
Hays · Creedmoor
Buda · Mustang Ridge
4 Miles / 4 Kilometers
© 2017 Pearson Education, Inc.

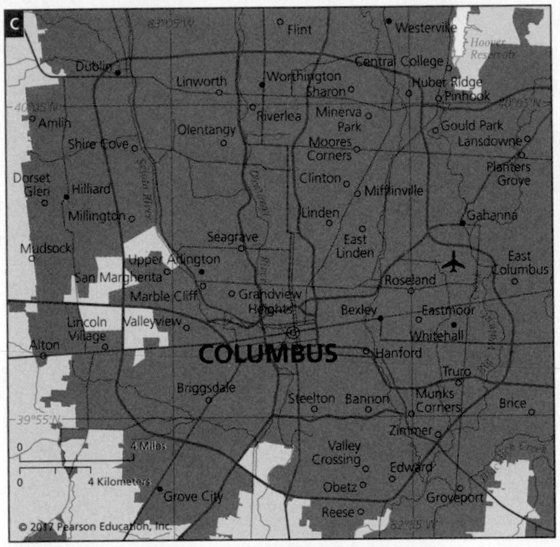

c — COLUMBUS

Flint · Westerville · Hoover Reservoir
Dublin · Central College · Hubet Ridge · Pinhook
Amlin · Linworth · Worthington · Sharon
Shire Cove · Olentangy · Riverlea · Minerva Park · Gould Park · Lansdowne
Dorset Glen · Hilliard · Moores Corners · Planters Grove
Millington · Clinton · Mifflinville
Mudsock · Seagrave · Linden · Gahanna
San Marghenta · Upper Arlington · East Linden · East Columbus
Marble Cliff · Grandview Heights · Roseland
Ailton · Lincoln Village · Bexley · Eastmoor
Valleyview · **COLUMBUS** · Whitehall
Hanford · Truro
Briggsdale · Steelton · Bannon · Munks Corners · Brice
Valley Crossing · Zimmer
Grove City · Obetz · Edward
Reese · Groveport
4 Miles / 4 Kilometers
© 2017 Pearson Education, Inc.

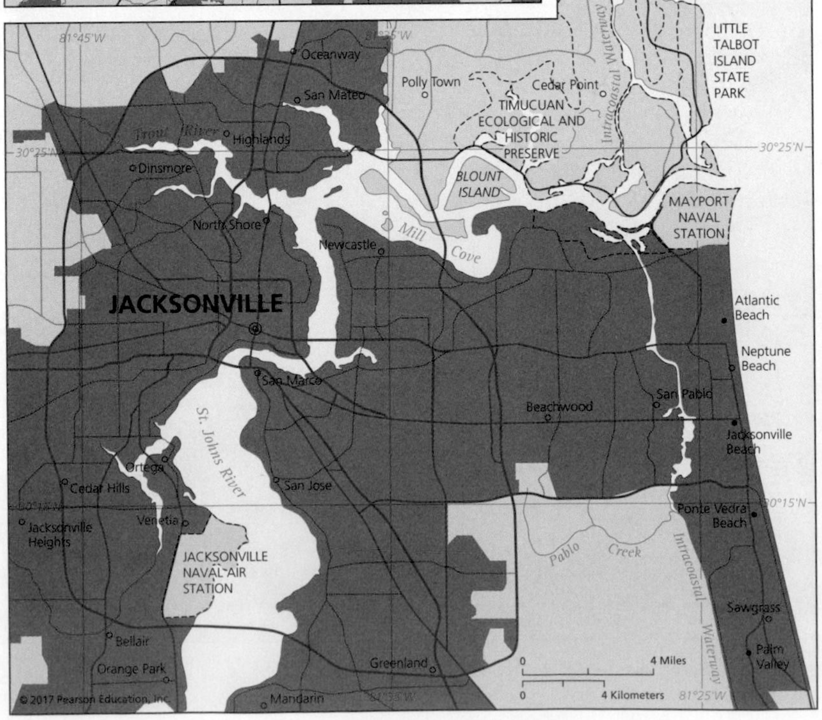

d — JACKSONVILLE / AMELIA ISLAND

ATLANTIC OCEAN
Fernandina Beach
AMELIA ISLAND
Amelia City
American Beach
Franklintown
St. Amelia River · Intracoastal Waterway
BIG TALBOT ISLAND S.P.
LITTLE TALBOT ISLAND STATE PARK
Oceanway
Polly Town · Cedar Point
San Mateo · TIMUCUAN ECOLOGICAL AND HISTORIC PRESERVE
Highlands · BLOUNT ISLAND · MAYPORT NAVAL STATION
Dinsmore
North Shore · Mill Cove
Newcastle · Atlantic Beach
JACKSONVILLE · Neptune Beach
San Marco · San Pablo
Beachwood · Jacksonville Beach
Ortega · St. Johns River · San Jose
Cedar Hills · Ponte Vedra Beach
Venetia · Intracoastal Waterway
Jacksonville Heights · JACKSONVILLE NAVAL AIR STATION
Bellair · Sawgrass
Orange Park · Greenland · Mandarin · Pablo Creek · Palm Valley
4 Miles / 4 Kilometers
© 2017 Pearson Education, Inc.

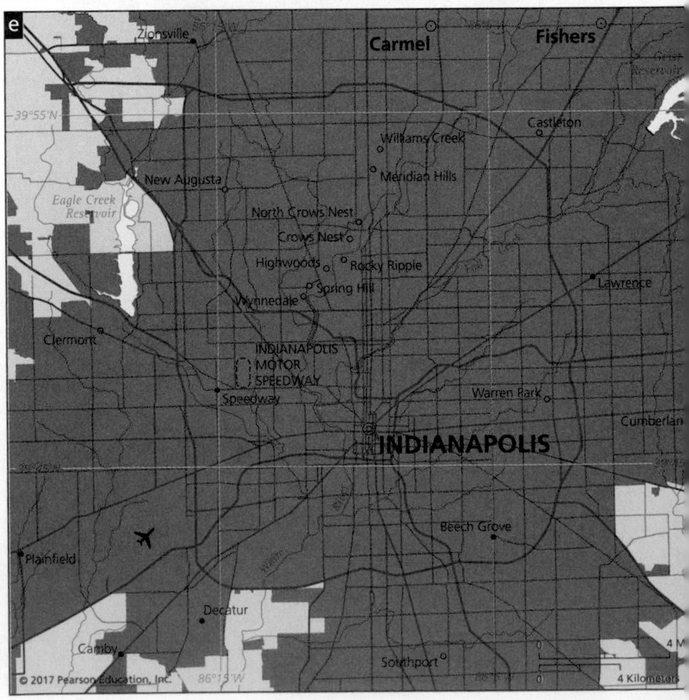

e — INDIANAPOLIS

Zionsville · Carmel · Fishers
Williams Creek · Castleton
New Augusta · Meridian Hills
Eagle Creek Reservoir · North Crows Nest · Lawrence
Crows Nest
Highwoods · Rocky Ripple
Wynnedale · Spring Hill
Clermont · INDIANAPOLIS MOTOR SPEEDWAY · Warren Park
Speedway · **INDIANAPOLIS** · Cumberland
Beech Grove
Plainfield
Decatur
Camby · Southport
4 Miles / 4 Kilometers
© 2017 Pearson Education, Inc.

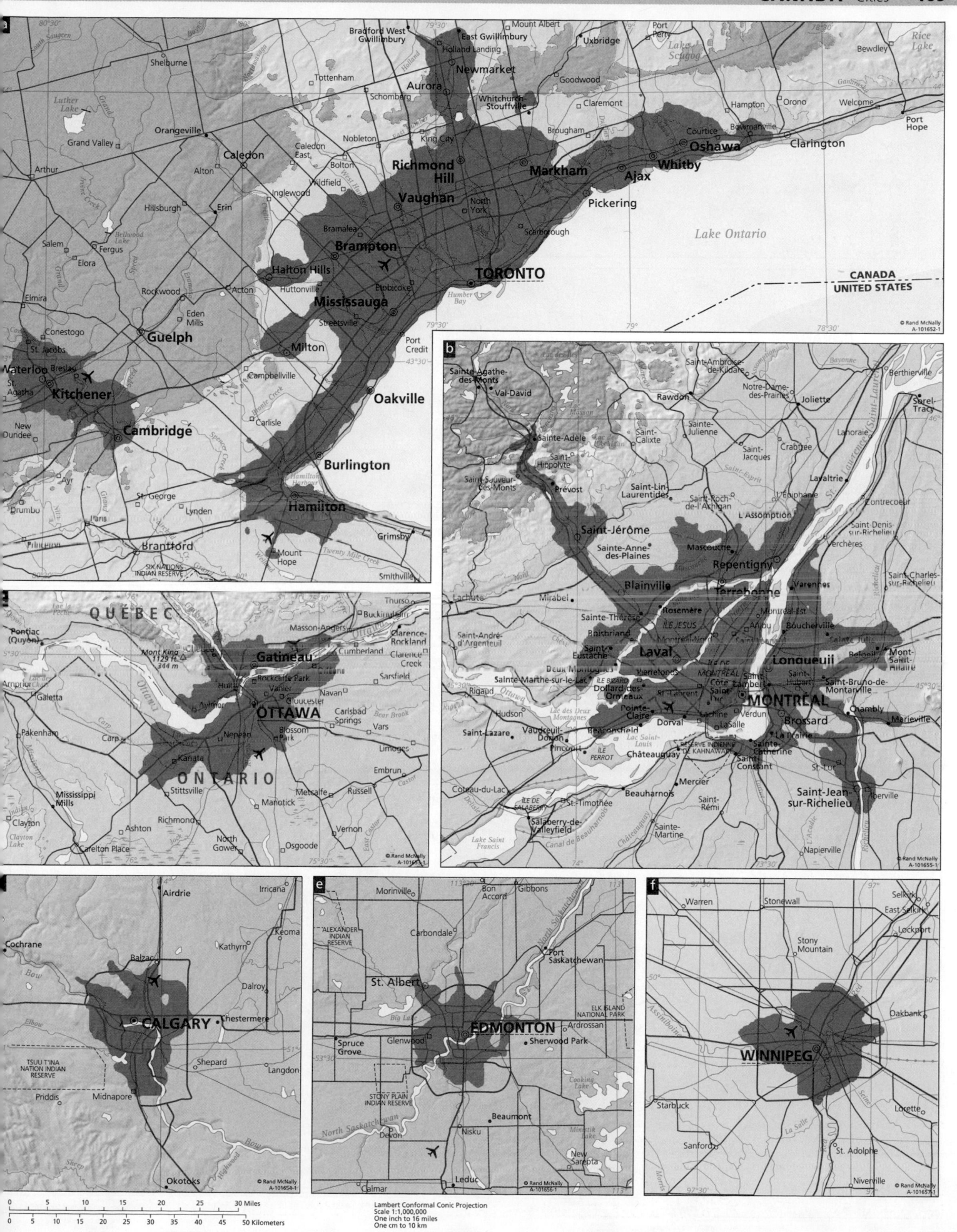

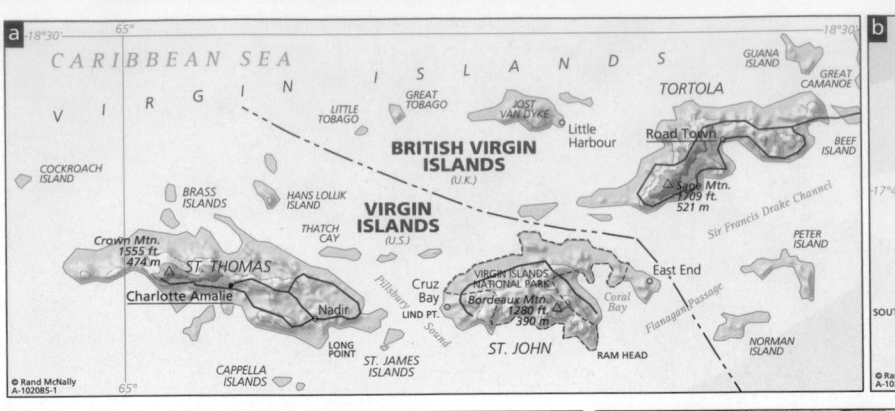
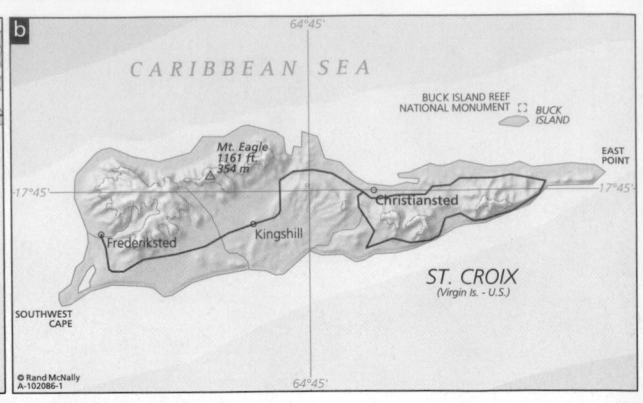
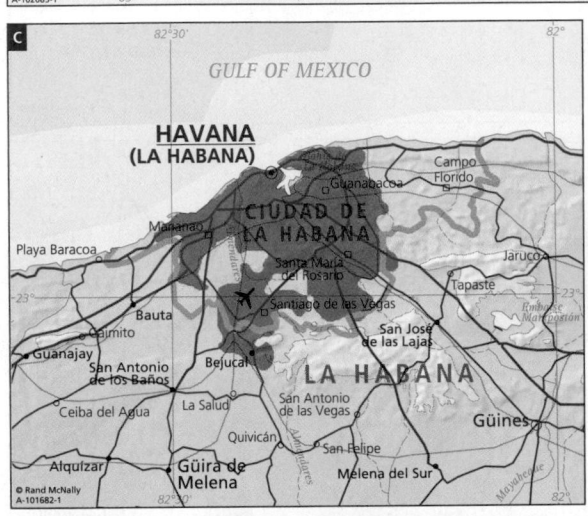

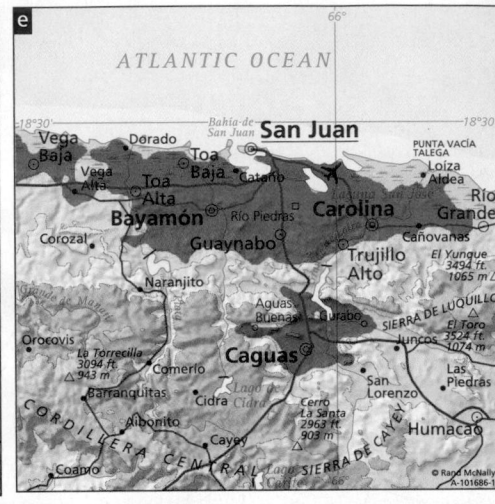
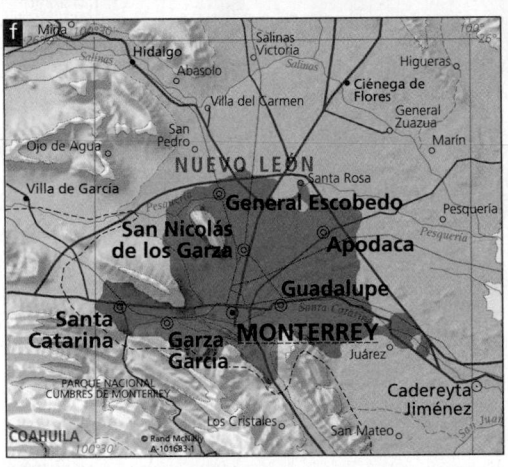

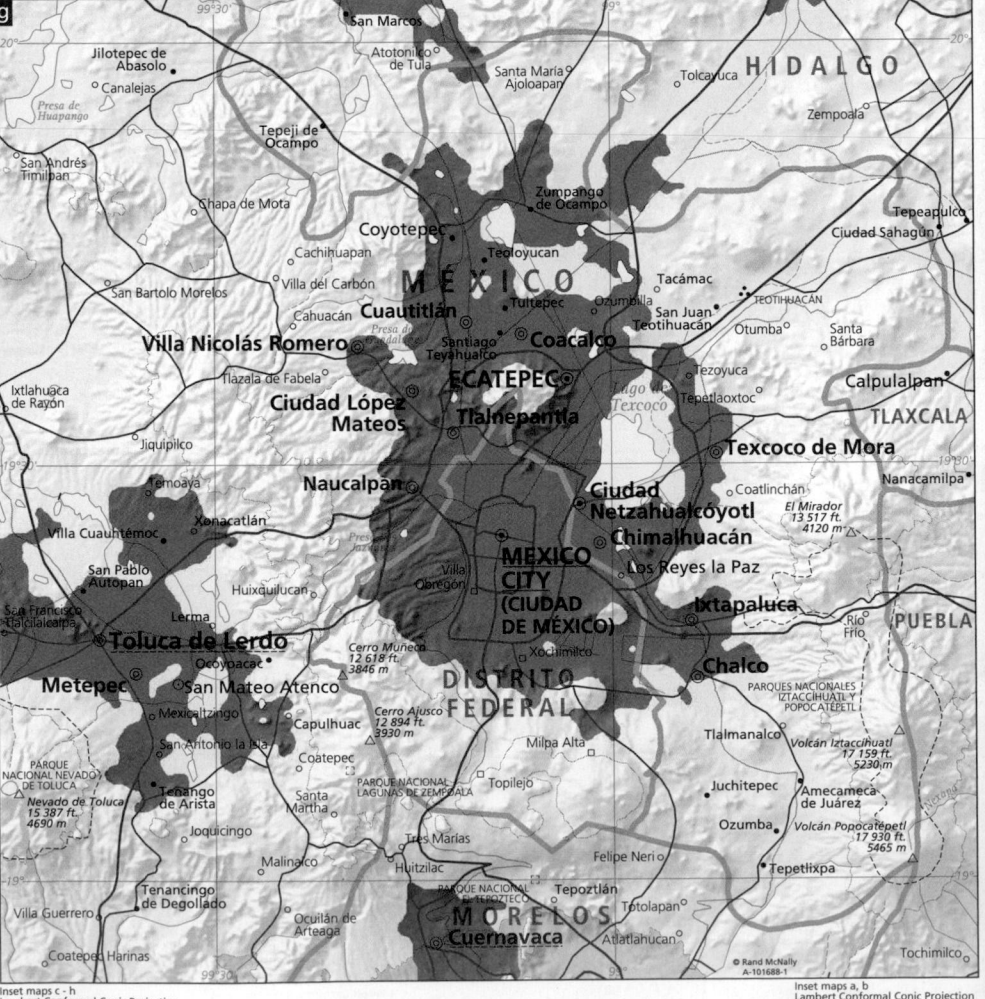

South America

SOUTH
AMERICA

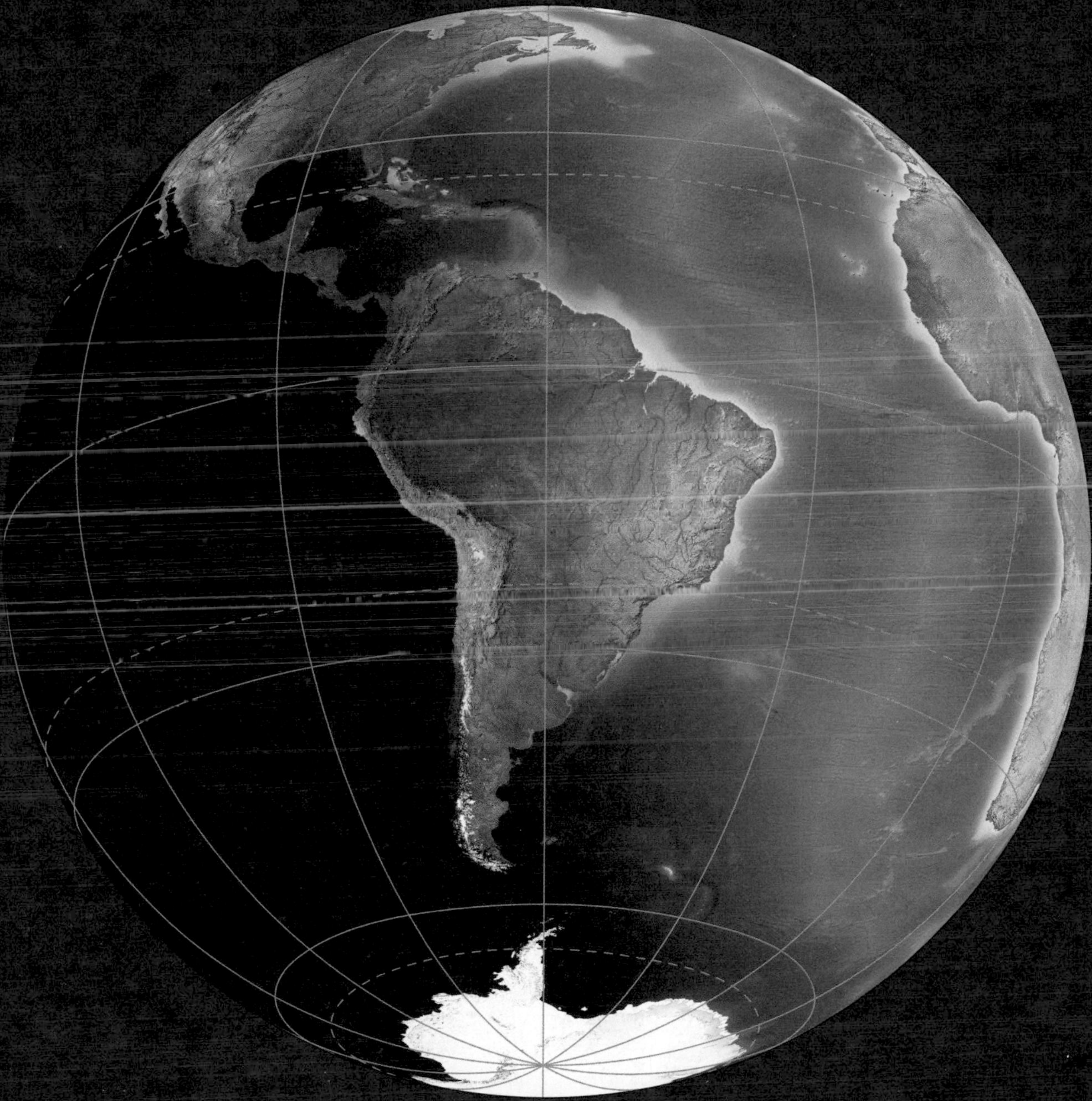

Stretching from the Caribbean Sea to the edge of the Southern Ocean, South America is home to a diverse range of landscapes. The Andes Mountains, which extend more than 7,000 kilometers (4,300 miles) along the western edge of the continent, form the longest continental mountain range on Earth. The Amazon River drains a vast basin containing Earth's largest and most biologically diverse rainforest and the Atacama Desert is the driest non-polar desert in the world. South America is a highly urbanized continent, with large cities located along coasts and in mountainous areas. Before colonial rule by the Spanish and Portuguese was established, South America was home to the Inca Empire, one of the most advanced civilizations in pre-Columbian America.

Gulf of Mexico

HAVANA *CUBA*

Mérida

JAMAICA
DOMINICAN
HAITI REPUBLIC **San**
Kingston Port-au- **Juan**
Prince **SANTO** PUERTO
DOMINGO RICO (U.S.)

GUADELOUPE (Fr.)
MARTINIQUE (Fr.)
ST. LUCIA BARBADOS

MEXICO

BELIZE

GUATEMALA HONDURAS
Guatemala
EL SALVADOR NICARAGUA

GRENADA

CARIBBEAN SEA

ARUBA (Neth.)

TRINIDAD AND
TOBAGO

ATLANTIC

Managua

Barranquilla **MARACAIBO** **CARACAS**

Cartagena
BARQUISIMETO Barcelona
Ciudad Guayana
Cúcuta San Cristóbal

OCEAN

COSTA
RICA Panama

PANAMA

ISLA DEL
COCO
(C.R.)

MEDELLÍN
Bucaramanga
Manizales **BOGOTÁ**

Buenaventura
CALI
COLOMBIA

Esmeraldas

QUITO
ECUADOR Chimborazo
20 702 ft.
6310 m
GUAYAQUIL
Cuenca
Loja Iquitos

Ciudad
Bolívar **GUYANA**
VENEZUELA **SURINAME**

Georgetown
Paramaribo
Cayenne
FRENCH CABO ORANGE
GUIANA
(Fr.)

Boa **Macapá**
Vista

Llanos

△ Pico da Neblina
9888 ft
3014 m

Equator

Amazon

S *E* *L* *V* *A* *S*

MANAUS Santarém **Belém** **São Luís**
Parnaíba
Fortaleza

Negro

Japurá
Putumayo

Chiclayo Cajamarca
Trujillo Pucallpa

Nevado Huascarán △
22 133 ft
6746 m

Rio
Branco **Porto Velho**

Marabá

Teresina

Marañón
Juruá
Purus
Madeira
Tapajós
Xingu
Tocantins
Araguaia
Parnaíba

Campina Natal
Grande João Pessoa
Caruaru **RECIFE**

PERU

Huancayo
LIMA Cusco

BOLIVIA
La Paz Trinidad
Cochabamba

BRAZIL

Juazeiro

São Francisco

Maceió
Aracaju
SALVADOR

Ica Puno
PUNTA CARRETA
Arequipa Oruro
**SANTA CRUZ
DE LA SIERRA**
Arica Potosí
Sucre
Iquique Tarija

Cuiabá **BRASÍLIA**

GOIÂNIA

Corumbá **BELO HORIZONTE**

Itabuna

Montes
Claros

A N D E S

Antofagasta

Tropic of Capricorn

ISLA SAN FÉLIX
(Chile)

PARAGUAY

GRAN Salta
CHACO

Asunción

San Miguel de Tucumán

Corrientes
Posadas

SÃO PAULO

Santos

Pico da Bandeira △
9505 ft
2897 m Vitória

Campos

CABO FRIO
RIO DE JANEIRO

Paraná

CURITIBA

ILHAS
MARTIN VAZ
(Braz.)

Santiago
del Estero

Santa
Maria

Florianópolis

PORTO ALEGRE

ARCHIPIÉLAGO DE COLÓN
(GALÁPAGOS ISLANDS)
(Ec.)

PACIFIC

Coquimbo

Cerro Aconcagua
22 831 ft
6959 m

CÓRDOBA
Mendoza
Valparaíso **Rosario**

A R G E N T I N A

Santa
Fe
Salto
Paysandú
Rio Grande
URUGUAY

Salado

Uruguay

OCEAN

ARCHIPIÉLAGO JUAN
FERNÁNDEZ
(Chile)

SANTIAGO
BUENOS AIRES
La Plata **MONTEVIDEO**

Pelotas

Concepción

Neuquén

Bahía
Blanca

PAMPA

CABO SAN ANTONIO

Mar del Plata

ATLANTIC

C H I L E

Valdivia

Negro

Osorno

Puerto Montt

ARCHIPIÉLAGO
DE LOS CHONOS

OCEAN

Golfo San
Jorge

Comodoro Rivadavia

Golfo San Matías

Monte San Valentín △
13 314 ft
4058 m

CABO TRES PUNTAS

P A T A G O N I A

**FALKLAND
ISLANDS**
(U.K.)

Río
Gallegos

Stanley

Punta Arenas

*SCOTIA
SEA*

**SOUTH GEORGIA
AND THE SOUTH
SANDWICH ISLANDS**
(U.K.)

CAPE HORN

Drake Passage

SOUTH
SHETLAND IS.
(U.K.)

SOUTH
ORKNEY IS.
(U.K.)

S O U T H E R N *O C E A N*

ANTARCTIC PENINSULA Antarctic Circle

© Rand McNally
A-101692-1

0 200 400 600 800 1000 1200 Miles
0 200 400 600 800 1000 1200 1400 1600 1800 2000 Kilometers

Lambert Azimuthal Equal Area Projection
Scale 1:40,000,000
One inch to 640 miles
One cm to 400 km

Gulf of Mexico
Havana
CUBA
GREATER
ANTILLES
HAITI
DOMINICAN
REPUBLIC
HISPANIOLA
PUERTO RICO TRENCH
Tropic of Cancer
YUCATAN
Yucatan Channel
PENINSULA
MEXICO
JAMAICA
PUERTO
RICO
(U.S.)
GUADELOUPE
(Fr.)
WEST INDIES
MARTINIQUE
(Fr.)
GUATEMALA
BELIZE
HONDURAS
EL SALVADOR
NICARAGUA
Lago de
Nicaragua
CARIBBEAN SEA
LESSER ANTILLES
BARBADOS

COSTA
RICA
Pico Cristóbal Colón
18,947 ft.
5775 m
Maracaibo
Pico Bolívar
16,427 ft.
5007 m
Caracas
TRINIDAD AND
TOBAGO
Orinoco

PANAMA
Golfo de
Panamá
ISTMO DE
PANAMÁ
Lago de
Maracaibo
VENEZUELA
GUYANA
Mt. Roraima
9432 ft.
2875 m
SURINAME
FRENCH
GUIANA
(Fr.)
Paramaribo
CABO ORANGE

ISLA DEL
COCO
(C.R.)
Nevado del Tolima
17,110 ft.
5215 m
Bogotá
LLANOS
Pico da Neblina
9888 ft.
3014 m
GUIANA HIGHLANDS

ISLA DE
MALPELO
(Col.)
Nevado del Huila
18,865 ft.
5750 m
COLOMBIA
Negro
Represa
Balbina
Equator

PUNTA GALERA
Cayambe
18,996 ft.
5790 m
Quito
ECUADOR
Chimborazo
20,702 ft.
6310 m
Iquitos
Napo
Putumayo
Japurá
Amazon
Manaus
Belém
Fortaleza
ILHA FERNANDO
DE NORONHA

ARCHIPIÉLAGO DE COLÓN
(GALAPAGOS ISLANDS)
(Ec.)
PUNTA PARIÑAS
Golfo de
Guayaquil
Marañón
S
E
L
V
A
S
Juruá
Purus
Madeira
Tapajós
Xingu
BRAZIL
CABO DE
SÃO ROQUE

Nevado Huascarán
22,133 ft.
6746 m
A
N
D
E
S
Porto
Velho
PONTA DO SEIXAS

PERU
Lima
PUNTA CARRETA
Lago
Titicaca
Madre de Dios
Beni
Nevado Illampu
21,066 ft.
6421 m
La Paz
Represa de
Sobradinho
Recife
São Francisco
Pico das Almas
6024 ft.
1835 m
Salvador
PLANALTO DO
MATO GROSSO
Brasília

Nevado Coropuna
20,086 ft.
6305 m
Nevado Sajama
21,463 ft.
6542 m
BOLIVIA
Guaporé
Grande
Pilcomayo
BRAZILIAN
HIGHLANDS
SERRA DE ESPINHAÇO

DESIERTO DE ATACAMA
PERU-CHILE TRENCH
ALTIPLANO
Volcán San Pedro
20,161 ft.
6145 m
CHACO
PARAGUAY
Belo Horizonte
Pico da Bandeira
9482 ft.
2097 m
ILHAS
MARTIN VAZ
(Braz.)

Tropic of Capricorn
ISLA SAN FÉLIX
(Chile)
Volcán Llullaillaco
22,110 ft.
6739 m
GRAN
CHACO
Asunción
Paraguay
São Paulo
Rio de Janeiro
CABO FRIO

Nevado Ojos del Salado
22,615 ft.
6893 m
CABO BASCUÑÁN
San Miguel
de Tucumán
Cerro General
Manuel Belgrano
20,505 ft.
6250 m
Córdoba
Salado
Paraná
Uruguay
Porto
Alegre
Lagoa
dos Patos

ARCHIPIÉLAGO
JUAN FERNÁNDEZ
(Chile)
Cerro Aconcagua
22,831 ft.
6959 m
Santiago
A
N
D
E
S
ARGENTINA
URUGUAY
Montevideo
Rio de
la Plata
Lagoa
dos Patos

PACIFIC
OCEAN
PUNTA LAVAPIÉ
CHILE
Buenos
Aires
PAMPA
CABO SAN ANTONIO
ATLANTIC
OCEAN

Monte Tronador
11,453 ft.
3491 m
Bahía Blanca
Colorado
Negro
Golfo San Matías

ISLA GRANDE
DE CHILOÉ
ARCHIPIÉLAGO
DE LOS CHONOS
PATAGONIA
CABO DOS BAHÍAS
Golfo
San Jorge
CABO TRES PUNTAS

Monte San Valentín
13,314 ft.
4058 m
ISLA WELLINGTON
FALKLAND
ISLANDS
(U.K.)
EAST
FALKLAND
WEST
FALKLAND

ISLA DESOLACIÓN
Punta Arenas
ISLA SANTA INÉS
TIERRA DEL FUEGO
CAPE DISAPPOINTMENT
SOUTH
GEORGIA

ISLA HOSTE
CAPE HORN
(CABO DE HORNOS)
SCOTIA
SEA
SOUTH GEORGIA
AND THE SOUTH
SANDWICH ISLANDS
(U.K.)
SOUTH
SANDWICH TRENCH

Drake Passage
SOUTH
ORKNEY IS.
(U.K.)
SOUTH
SANDWICH IS.

SOUTHERN
OCEAN
SOUTH
SHETLAND IS.
(U.K.)
ANTARCTIC PENINSULA
Antarctic Circle

© Rand McNally
A-101693-1

0 200 400 600 800 1000 1200 Miles
0 200 400 600 800 1000 1200 1400 1600 1800 2000 Kilometers

Lambert Azimuthal Equal Area Projection
Scale 1:40,000,000
One inch to 640 miles
One cm to 400 km

Land Cover

- Evergreen Needleleaf Forest
- Evergreen Broadleaf Forest
- Deciduous Broadleaf Forest
- Mixed Forest
- Woodland
- Wooded Grassland
- Closed Shrubland
- Open Shrubland
- Grassland
- Cropland
- Bare Ground (Desert and Ice)
- Urban and Built Up

CARIBBEAN SEA

ATLANTIC OCEAN

Barranquilla
Caracas
Orinoco
Medellín
LLANOS
Bogotá
GALAPAGOS ISLANDS
Quito
Negro
Japurá
Amazon
Manaus
Amazon
Fortaleza
Guayaquil
S   E   L   V   A   S
Purus
Madeira
Tapajós
Ucayali
Recife
Lima
CORDILLERA ORIENTAL
PLANALTO DO MATO GROSSO
Salvador
Lago Titicaca
La Paz
São Francisco
ALTIPLANO
Goiânia
Brasília
DESIERTO DE ATACAMA
Belo Horizonte
CHACO
Paraguay
Paraná
GRAN
Asunción
Rio de Janeiro
São Paulo
Salado
Curitiba
A N D E S
Córdoba
Paraná
Porto Alegre
Santiago
Bermejo
Buenos Aires
Montevideo
PAMPA

PACIFIC OCEAN

ATLANTIC OCEAN

FALKLAND ISLANDS

PATAGONIA

TIERRA DEL FUEGO

SCOTIA SEA

SOUTH GEORGIA

Drake Passage

SOUTH SHETLAND IS.

SOUTH ORKNEY IS.

SOUTH SANDWICH IS.

SOUTHERN   OCEAN

Antarctic Circle

Tropic of Cancer
Equator
Tropic of Capricorn

A-102093-1
©Rand McNally

Source: CIESIN; Hansen et al., 2000

0   200   400   600   800   1000   1200 Miles
0   200   400   600   800   1000   1200   1400   1600   1800   2000 Kilometers

Lambert Azimuthal Equal Area Projection
Scale 1:40,000,000
One inch to 640 miles
One cm to 400 km

Geology

Rock type/geological era

- Intrusive
- Metamorphic
- Volcanic, tectonic
- Precambrian sedimentary
- Paleozoic sedimentary
- Mesozoic sedimentary
- Cenozoic sedimentary

Note: Areas classified as sedimentary also include some sedimentary/volcanic areas.

Source: Generalized Geological Map of the World, Geological Survey of Canada

Lambert Azimuthal Equal Area Projection
Scale 1:40,000,000
One inch to 640 miles
One cm to 400 km

0 200 400 600 800 1000 1200 Miles
0 200 400 600 800 1000 1200 1400 1600 1800 2000 Kilometers

© 2017 Pearson Education, Inc.

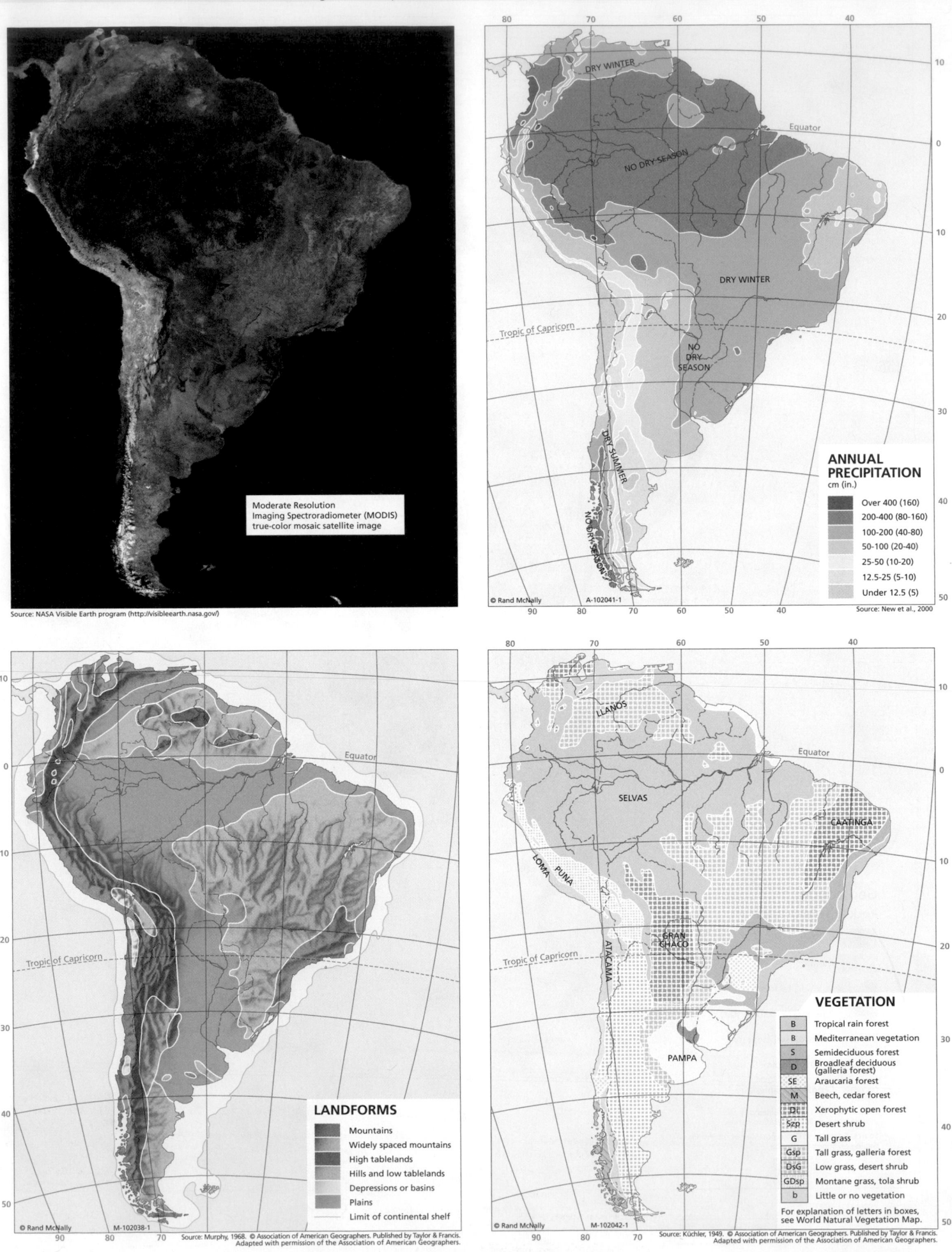

Moderate Resolution
Imaging Spectroradiometer (MODIS)
true-color mosaic satellite image

Source: NASA Visible Earth program (http://visibleearth.nasa.gov/)

DRY WINTER

Equator

NO DRY SEASON

DRY WINTER

Tropic of Capricorn

NO
DRY
SEASON

DRY SUMMER

NO DRY SEASON

**ANNUAL
PRECIPITATION**
cm (in.)

Over 400 (160)
200-400 (80-160)
100-200 (40-80)
50-100 (20-40)
25-50 (10-20)
12.5-25 (5-10)
Under 12.5 (5)

© Rand McNally   A-102041-1

Source: New et al., 2000

Equator

Tropic of Capricorn

LANDFORMS

Mountains
Widely spaced mountains
High tablelands
Hills and low tablelands
Depressions or basins
Plains
— Limit of continental shelf

© Rand McNally   M-102038-1

Source: Murphy, 1968. © Association of American Geographers. Published by Taylor & Francis.
Adapted with permission of the Association of American Geographers.

LLANOS

Equator

SELVAS

CAATINGA

LOMA
PUNA

ATACAMA

GRAN
CHACO

Tropic of Capricorn

PAMPA

VEGETATION

B	Tropical rain forest
B	Mediterranean vegetation
S	Semideciduous forest
D	Broadleaf deciduous (galleria forest)
SE	Araucaria forest
M	Beech, cedar forest
Dt	Xerophytic open forest
Szp	Desert shrub
G	Tall grass
Gsp	Tall grass, galleria forest
DsG	Low grass, desert shrub
GDsp	Montane grass, tola shrub
b	Little or no vegetation

For explanation of letters in boxes,
see World Natural Vegetation Map.

© Rand McNally   M-102042-1

Source: Küchler, 1949. © Association of American Geographers. Published by Taylor & Francis.
Adapted with permission of the Association of American Geographers.

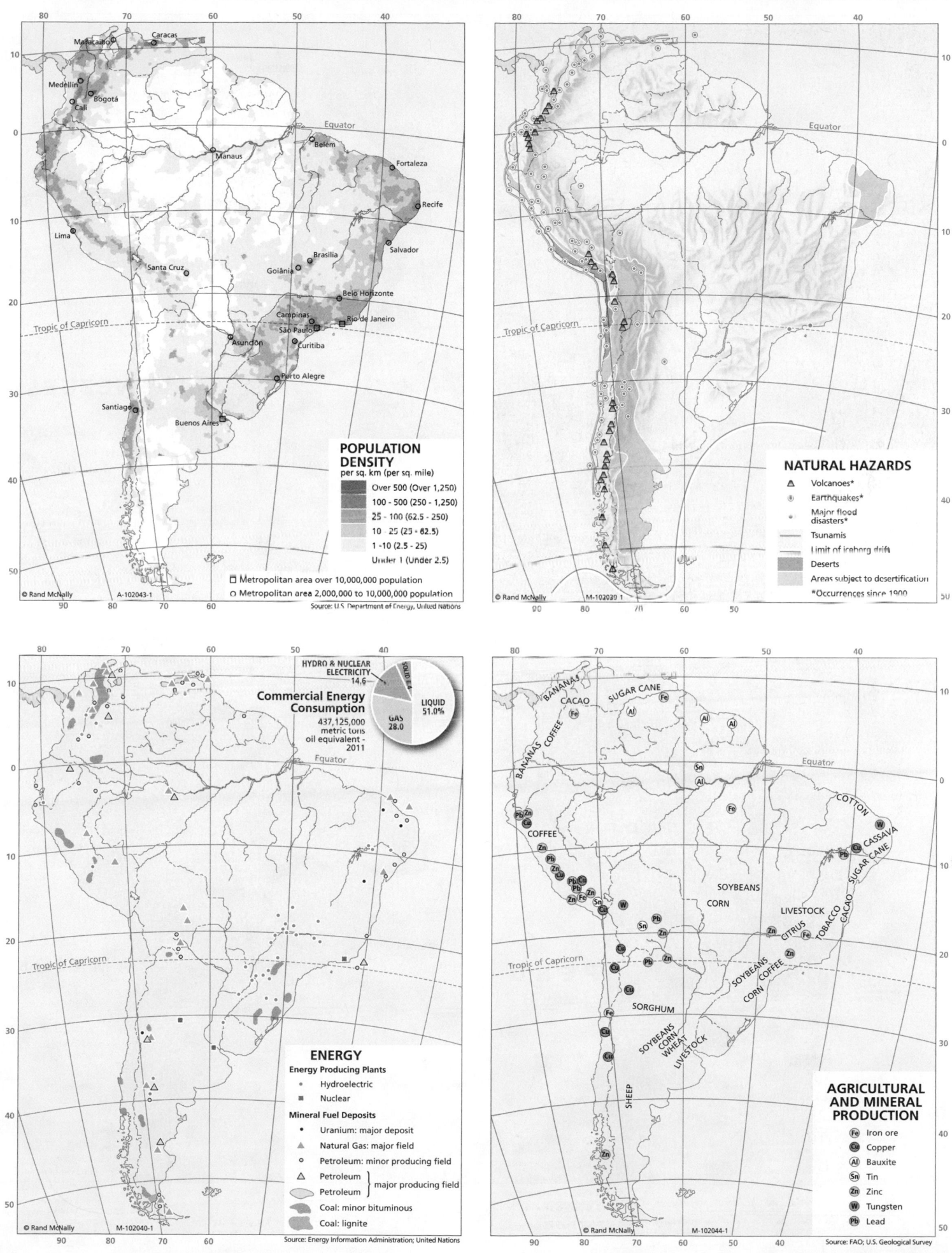

POPULATION DENSITY
per sq. km (per sq. mile)

- Over 500 (Over 1,250)
- 100 - 500 (250 - 1,250)
- 25 - 100 (62.5 - 250)
- 10 - 25 (25 - 62.5)
- 1 - 10 (2.5 - 25)
- Under 1 (Under 2.5)

☐ Metropolitan area over 10,000,000 population
○ Metropolitan area 2,000,000 to 10,000,000 population

© Rand McNally   A-102043-1
Source: U.S. Department of Energy, United Nations

Caracas
Maracaibo
Medellín
Bogotá
Cali
Manaus
Belém
Fortaleza
Recife
Lima
Salvador
Brasília
Santa Cruz
Goiânia
Belo Horizonte
Campinas
Rio de Janeiro
São Paulo
Asunción
Curitiba
Porto Alegre
Santiago
Buenos Aires

NATURAL HAZARDS

- △ Volcanoes*
- ⊕ Earthquakes*
- • Major flood disasters*
- Tsunamis
- Limit of iceberg drift
- Deserts
- Areas subject to desertification

*Occurrences since 1900

© Rand McNally   M-102030-1

ENERGY

Energy Producing Plants
- • Hydroelectric
- ■ Nuclear

Mineral Fuel Deposits
- • Uranium: major deposit
- ▲ Natural Gas: major field
- ○ Petroleum: minor producing field
- △ Petroleum } major producing field
- ⬭ Petroleum }
- Coal: minor bituminous
- Coal: lignite

© Rand McNally   M-102040-1
Source: Energy Information Administration; United Nations

Commercial Energy Consumption
437,125,000 metric tons oil equivalent - 2011

HYDRO & NUCLEAR ELECTRICITY 14.6
SOLID 6.4
LIQUID 51.0%
GAS 28.0

AGRICULTURAL AND MINERAL PRODUCTION

- Fe Iron ore
- Cu Copper
- Al Bauxite
- Sn Tin
- Zn Zinc
- W Tungsten
- Pb Lead

© Rand McNally   M-102044-1
Source: FAO; U.S. Geological Survey

BANANAS
CACAO
SUGAR CANE
BANANAS
COFFEE
COFFEE
COTTON
CASSAVA
SUGAR CANE
SOYBEANS
CORN
LIVESTOCK
CITRUS
TOBACCO
CACAO
SORGHUM
SOYBEANS
CORN
COFFEE
SOYBEANS
CORN
WHEAT
LIVESTOCK
SHEEP

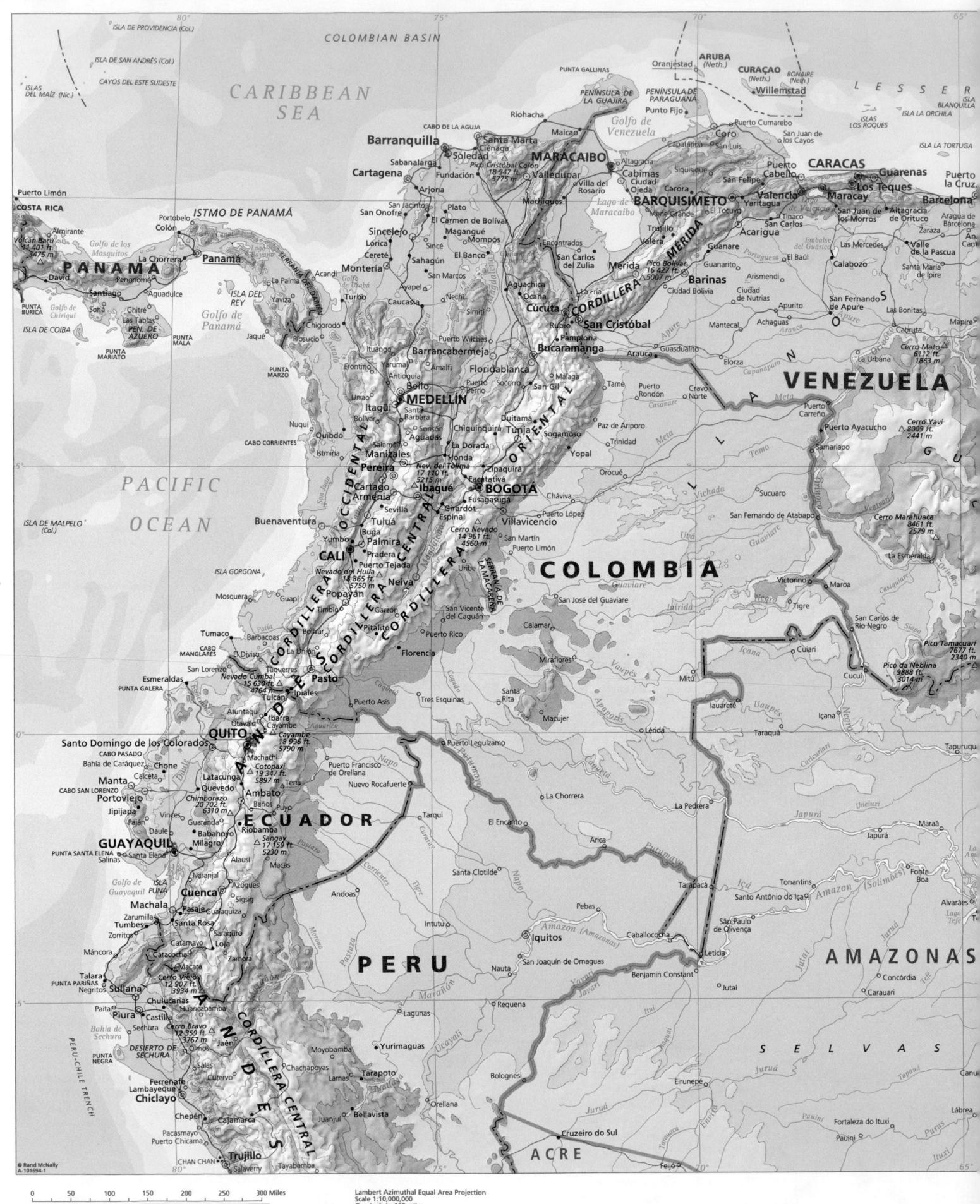

ISLA DE PROVIDENCIA (Col.)

COLOMBIAN BASIN

ISLA DE SAN ANDRÉS (Col.)

CAYOS DEL ESTE SUDESTE

CARIBBEAN SEA

ISLAS DEL MAÍZ (Nic.)

PUNTA GALLINAS

ARUBA (Neth.)
Oranjestad

CURAÇAO (Neth.)
Willemstad

BONAIRE (Neth.)

LESSER

ISLA BLANQUILLA

PENÍNSULA DE LA GUAJIRA

PENÍNSULA DE PARAGUANÁ
Punto Fijo

Golfo de Venezuela

ISLA LA ORCHILA

ISLAS LOS ROQUES

ISLA LA TORTUGA

Puerto Limón

COSTA RICA

Volcán Barú 11,401 ft. 3475 m
Almirante
La Chorrera
Colón
Portobelo

ISTMO DE PANAMÁ

Golfo de los Mosquitos

PANAMA

Panamá

Santiago
Penonomé
Aguadulce

Chitré
Las Tablas
PEN. DE AZUERO

Soná
Golfo de Chiriquí

ISLA DE COIBA

PUNTA BURICA
David

PUNTA MARIATO

PUNTA MALA

ISLA DEL REY

Golfo de Panamá

Jaqué

Yaviza

La Palma

SERRANÍA DEL DARIÉN

Acandí

Riohacha

CABO DE LA AGUJA

Barranquilla

Cartagena

Santa Marta
Ciénaga
Soledad
Sabanalarga

Pico Cristóbal Colón 18,947 ft. 5775 m

Maicao

Valledupar

Villa del Rosario

Machiques

Lago de Maracaibo

MARACAIBO

Altagracia
Ciudad Ojeda
Cabimas

Coro

Puerto Cumarebo

San Juan de los Cayos

Capatárida

San Luis

Siquisique

CARACAS

Puerto Cabelло
Valencia
San Felipe
Yaritagua

Guarenas
Los Teques
Maracay

Barcelona

Puerto la Cruz

Aragua de Barcelona

BARQUISIMETO

Carora

El Tocuyo

San Carlos

Trujillo
Valera

MÉRIDA

Mérida

Pico Bolívar 16,427 ft. 5007 m

Barinas

Acarigua

Guanare

Portuguesa

El Baúl

Las Mercedes

Calabozo

Valle de la Pasca

Santa María de Ipire

San Fernando de Apure

Las Bonitas

Cabruta

Mapire

San Jacinto
San Onofre
Arjona

Plato

El Carmen de Bolívar
Sincelejo
Magangué
Mompós

El Banco

Ocaña

Aguachica

La Fría

Encontrados

San Carlos del Zulia

Simití

Cúcuta

Rubio
San Cristóbal

Pamplona

Arauca

Elorza

La Urbana

Cerro Mato 6112 ft. 1863 m

VENEZUELA

Lorica
Cereté
Sahagún
Montería
San Marcos

Sincé

Nechí

Caucasia

Ayapel

Turbo

CABO CORRIENTES

Quibdó

Istmina

Nuquí

PACIFIC OCEAN

ISLA DE MALPELO (Col.)

Buenaventura

ISLA GORGONA

Mosquera
Guapi

Tumaco
Barbacoas

CABO MANGLARES

San Lorenzo

PUNTA GALERA

Esmeraldas

Santo Domingo de los Colorados

CABO PASADO

Bahía de Caráquez
Chone
Calceta

Manta
Portoviejo

CABO SAN LORENZO

Jipijapa

Pajám

QUITO

Machachi
Latacunga
Quevedo

Ambato
Baños
Puyo

Tena

Riobamba

ECUADOR

GUAYAQUIL

Milagro

PUNTA SANTA ELENA
Salinas
Santa Elena

Golfo de Guayaquil

ISLA PUNÁ

Daule
Babahoyo
Vinces

Naranjal

Azogues
Sigsig

Cuenca

Machala
Pasaje
Guayaquiza

Zarumilla
Tumbes
Zorritos

Máncora

Talara

PUNTA PARIÑAS
Negritos

Sullana

Paita

Piura
Castilla

Bahía de Sechura
Sechura

DESIERTO DE SECHURA

PUNTA NEGRA

PERÚ-CHILE TRENCH

Catamayo
Loja
Zamora

Saraguro

Macará
Catacocha

Jaén

Olmos

Salas

Ferreñafe
Lambayeque

Chiclayo

Cajamarca

Chepén

Puerto Chicama

CHAN CHAN

Trujillo
Salaverry

Ituango
Frontino
Yarumal
Antioquia
Amalfi

Bello
Santa Bárbara
Bolívar

MEDELLÍN

Itagüí

Sonsón
Aguadas

Manizales

Salamina
Pereira

Armenia
Cartago

Tuluá
Buga
Palmira

Yumbo

CALI

Pradera

Puerto Tejada

Nevado del Huila 18,865 ft. 5750 m

Popayán

Timbío

Bolívar

La Unión
Taqueres

Nevado Cumbal 15,630 ft. 4764 m

Pasto
Ipiales
Tulcán

Atuntaqui
Otavalo
Ibarra
Cayambe

Cayambe 18,996 ft. 5790 m

Cotopaxi 19,347 ft. 5897 m

Chimborazo 20,702 ft. 6310 m

Sangay 17,159 ft. 5230 m

Alausí
Macas

Guaranda

Barrancabermeja

Puerto Wilches

Bucaramanga

Floridablanca

San Gil

Málaga

Socorro

Puerto Berrío

Honda
La Dorada

Chiquinquirá

Tunja

Duitama
Sogamoso

Yopal

Paz de Ariporo

Trinidad

Orocué

BOGOTA

Zipaquirá
Facatativá
Fusagasugá
Girardot
Espinal

Nev. del Tolima 17,110 ft. 5215 m

Ibagué

Cerro Nevado 14,961 ft. 4560 m

Neiva

Garzón

Pitalito

San Vicente del Caguán

Florencia

Puerto Rico

Uribe

Puerto Limón

San Martín

Villavicencio

SERRANÍA DE LA MACARENA

Puerto López

Cháviva

Puerto Lleras

San José del Guaviare

Calamar

Miraflores

CORDILLERA OCCIDENTAL

CORDILLERA CENTRAL

CORDILLERA ORIENTAL

CORDILLERA DE LOS ANDES

ANDES

COLOMBIA

Guaviare

Inírida

Vaupés

San Fernando de Atabapo

Maroa

Victorino

Cerro Marahuaca 8461 ft. 2579 m

La Esmeralda

Cerro Yaví 8009 ft. 2441 m

Puerto Ayacucho

Samariapo

Puerto Carreño

Casanare

Meta

Puerto Rondón

Cravo Norte

Arauca

Guasdualito

Mantecal

Achaguas

Apurito

Apure

San Fernando de Apure

Arauca

Meta

Vichada

Tomo

Sucuaro

Guaviare

LLANOS

Cuiarí

Içana

Iauaretê

Taraquá

Mitú

Lérida

Taraquá

Puerto Leguízamo

Nuevo Rocafuerte

Puerto Francisco de Orellana

Napo

Tarqui

El Encanto

La Chorrera

La Pedrera

Arica

Pebas

Caballococha

Leticia

São Paulo de Olivença

Amazon (Solimões)

AMAZONAS

Tonantins

Santo Antônio do Içá

Fonte Boa

Alvarães

Lago Tefé

Maraã

Japurá

Uneiuxi

Puerto Asís

Tres Esquinas

Santa Rita

Macujer

Tarapacá

Ambato

Puyo

Tena

Macas

Moyobamba

Yurimaguas

Chachapoyas

Tarapoto

Lamas

Mendoza

Cutervo

Bellavista

Juanjuí

Celendín

Huancabamba

Chulucanas

Cerro Viejo 12,907 ft. 3934 m

Cerro Bravo 12,359 ft. 3767 m

CORDILLERA CENTRAL

ANDES

Iquitos

San Joaquín de Omaguas

Nauta

Requena

Lagunas

Bolognesi

Orellana

Juanjuí

Cruzeiro do Sul

ACRE

SELVAS

Jutaí

Concórdia

Carauari

Eirunepé

Fortaleza do Ituxi

Lábrea

PERU

Amazon (Amazonas)

Benjamin Constant

Mosquera

Santa Clotilde

Intuto

Andoas

Tarqui

Corrientes

Tigre

Pastaza

Marañón

Tapurucu

0	50	100	150	200	250	300 Miles
0	50	100	150 200 250 300	350	400	450	500 Kilometers

Lambert Azimuthal Equal Area Projection
Scale 1:10,000,000
One inch to 160 miles
One cm to 100 km

ATLANTIC OCEAN

St. VINCENT
Kingstown
ST. VINCENT
AND THE
GRENADINES

BARBADOS
Bridgetown

GRENADA
St. George's

A DE
RGARITA
La Asunción
Porlamar
Carúpano
PENÍNSULA
DE PARIA
Güiria
Golfo de
Paria

TOBAGO
Scarborough

Port of Spain   TRINIDAD
Arima   AND
TRINIDAD   TOBAGO
San Fernando

umaná
Carúpito

Maturín
Pedernales

Güsepín
José
Guanípa
re

Temblador
Tucupita

DELTA DEL ORINOCO

Barrancas

Morawhanna

Ciudad Guayana
Upata
Mabaruma

Ciudad
Bolívar
Ciudad Piar
Cerro Bolívar
2631 ft.
802 m
Paragua

Embalse
de Guri
Guasipati
El Callao
Tumeremo

Marlborough

Suddie

Georgetown

Cuyuni
El Dorado

Angel Falls
(Salto Ángel)
Luepa

Bartica
Hyde Park

New Amsterdam
Rockstone
Linden
Corriverton
Nieuw
Nickerie

Paramaribo
Groningen
Onverwacht

Moengo

Iracoubo
Sinnamary
Kourou

Saint Laurent
du Maroni

Cayenne
Rémiré

Auyán Tepuy
9678 ft
2950 m
LA GRAN
SABANA

Mt. Roraima
9432 ft
2875 m
Tumatumari

GUYANA

Issano

Brokopondo

W.J. van
Blommesten
Meer

SURINAME

WILHELMINA GEBERGTE
Juliana Top
4035 ft
1230 m

FRENCH
GUIANA
(France)

Saint-Élie

Saint-Georges
Oiapoque

CABO ORANGE

PAKARAIMA MTS.

NA HIGH

Conceição
do Mau

Lethem

LANDS

Vila Velha

Saül

Curianí
Calçoene

ILHA DE MARACÁ

Amapá

CABO NORTE
Sucuriju

Uraricoera

Boa Vista

Dadanawa

ACARAI MOUNTAINS

TUMUC-HUMAC MOUNTAINS

AMAPÁ

Serra do Navio

ROiRAIMA

Caracaraí

Fisherton

Porto Grande

Ferreira Gomes

ILHA JANAUCU

São José
de Anauá

Mucajaí

New

Peru du Oeste

Macapá
Porto Santana
Mazagão

ILHA CAVIANA DE FORA
ILHA MEXIANA

CABO
MAGUARI

Branco

Boiaçu

Trombetas

ILHA GRANDE
DO GURUPÁ

Boca do Jari

Itatupa

Soure

Baia de Marajó

Salinópolis

Negro

Barcelos

Jauari

Lago de
Erepecuru

Gurupá

São Miguel
dos Macacos

Anajás

Muaná

Murajá
Maracanã

Bragança
Capanema

Unini

Carvoeiro
Moura

Novo Airão

Uatumã

Oriximiná
Óbidos
Alenquer
Monte
Alegre

Prainha

Carrazedo
Porto de Moz

Breves

Curralinho

Belém

Igarapé-
Açu

Irituia

Represa
Balbina

Terra Santa
Faro

Juriti

Veiros

Portel

Pará

Cametá

São Domingos
do Capim

Acará

Itamataré

MANAUS
Itacoatiara

Urucará
Itapiranga

Parintins
Barreirinha
Ariaú

Santarém

Amazon (Amazonas)

Vitória

Juaba

Baião

Carapajó

Tomé-Açu

Codajás

Anamã
Manacapuru

Nova Olinda
do Norte
careiro

Maués

Tapajós

Altamira

ILHA TUPINAMBARANA
Paraná Urariá

Lago
Piorini

Camará
Beruri

Axinim
Canumã

Itaituba

PARÁ

Tucuruí

Jacundá

MARANHÃO

Coari

Aiapuá

Novo
Aripuanã

Borba

Xingu

Tocantins

Represa
de Tucuruí

Itupiranga

Açailândia

Abufari

Tapauá

Manicoré

Iriri

Marabá

Imperatriz

Amarante do
Maranhão

Marmelos

BRAZIL

Madeira

Aripuanã

Araguaína

Santa Isabel do Araguaia

Sítio
Novo

Humaitá

Prainha
Nova

Tapajós

SERRA DO CACHIMBO

SERRA DOS CARAJÁS

Carajás

Xambioá

Montes Altos

Tocantinópolis

Araguaína
Babaçulândia

Carolina

Riachão

Samaúna

MATO
GROSSO

Gradaús

São Félix do Xingu

TOCANTINS

Conceição do Araguaia

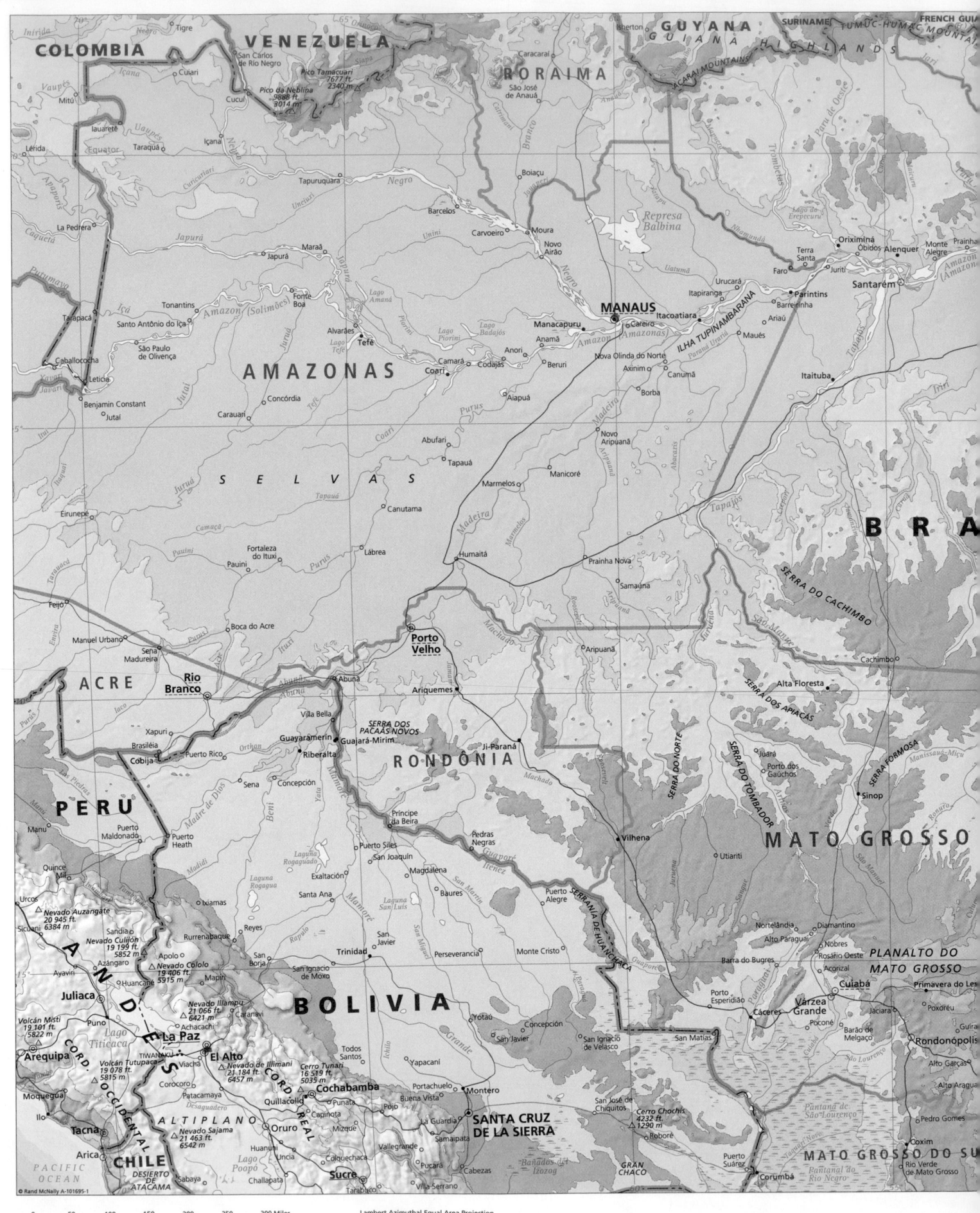

COLOMBIA

VENEZUELA

GUYANA

SURINAME TUMUC-HUMAC MOUNTAINS FRENCH GUIA

Inírida

Tigre

Negro

San Carlos de Río Negro

Pico Tamacuari 7677 ft. 2340 m

Siapa

Isherton

Guiana Highlands

ACARAI MOUNTAINS

Vaupés

Cuiari

Içana

Caracaraí

RORAIMA

Paru de Oeste

Mitú

Iauareté

Pico da Neblina 9888 ft. 3014 m

Cucuí

São José de Anauá

Branco

Trombetas

Lérida

Vaupés

Taraquá

Equator

Içana

Negro

Boiaçu

Anauá

Jari

Curicuriari

Tapuruquara

Unini

Carvoeiro

Moura

Jatapu

Represa Balbina

Catrimani

Lago do Erepecuru

Terra Santa

Oriximiná

Óbidos Alenquer

Monte Alegre

Prainha

La Pedrera

Caquetá

Içá

Maraã

Barcelos

Novo Airão

Faro

Urucará

Juruti

Amazon Amazonas

Santarém

Putumayo

Tonantins

Fonte Boa

Japurá

Lago Amanã

Negro

MANAUS

Itacoatiara

Itapiranga

Parintins

Barreirinha

Tarapacá

Santo Antônio do Içá

Alvarães

Tefé

Lago Tefé

Lago Badajós

Manacapuru

Careiro

Ariaú

Maués

Caballococha

São Paulo de Olivença

Leticia

Benjamin Constant

Jutaí

Javari

Jutaí

Concórdia

Camará

Lago Piorini

Coari

Codajás

Anamã

Anori

Beruri

Nova Olinda do Norte

Axinim

Canumã

ILHA TUPINAMBARANA

Paraná Urariá

Itaituba

Tapajós

AMAZONAS

Carauari

Abufari

Coari

Purus

Aiapuá

Novo Aripuanã

Borba

Madeira

Eirunepé

Juruá

SELVAS

Tapauá

Tapauá

Canutama

Jataí

Tapauá

Marmelos

Manicoré

Marmelos

Aripuanã

Abacaxis

B R A

SERRA DO CACHIMBO

Camaçã

Pauini

Fortaleza do Ituxi

Lábrea

Purus

Humaitá

Madeira

Prainha Nova

Samaúna

Roosevelt

Aripuanã

Juruena

Tapajós

Cachimbo

Feijó

Manuel Urbano

Iaco

Sena Madureira

Purus

Boca do Acre

Iquiri

Porto Velho

Machado

Jiparaná

SERRA DOS APIACÁS

Alta Floresta

ACRE

Rio Branco

Abuna

Abuna

Ariquemes

Aripuanã

SERRA DO NORTE

SERRA DO TOMBADOR

Juára

Xapuri

Brasiléia

Cobija

Puerto Rico

Vila Bella

Guayaramerín

SERRA DOS PACAÁS NOVOS

Guajará-Mirim

Machado

Ji-Paraná

Machado

Porto dos Gaúchos

SERRA FORMOSA

Manissaua-Miçu

Roturi

PERÚ

Orthon

Riberalta

RONDÔNIA

Sinop

Manu

Madre de Dios

Sena

Concepción

Beni

Mamoré

Príncipe da Beira

Pedras Negras

Vilhena

Utiariti

MATO GROSSO

Puerto Maldonado

Madidi

Yata

Puerto Siles

San Joaquín

Magdalena

Guaporé

Itenez

Juruena

São Manuel

Puerto Heath

Quince Mil

Urcos

Nevado Auzangate 20 945 ft. 6384 m

Sandia

Laguna Rogaguado

Exaltación

Baures

San Martín

Puerto Alegre

SERRANÍA DE HUANCHACA

Nortelândia

Diamantino

Sicuani

Nevado Culijón 19 199 ft. 5852 m

Ixiamas

Laguna Rogagua

Santa Ana

Guaporé

Alto Paraguai

Nobres

PLANALTO DO MATO GROSSO

Azángaro

Apolo

Nevado Cololo 19 406 ft. 5915 m

Rurrenabaque

Reyes

Laguna San Luis

San Martín

Monte Cristo

Paraguai

Rosário Oeste

Ayaviri

Huancané

Mapiri

San Borja

Trinidad

San Javier

Perseverancia

Barra do Bugres

Acorizal

Cuiabá

Primavera do Les

A N D E S

Juliaca

Nevado Illampu 21 066 ft. 6421 m

Caranavi

BOLIVIA

Itonamas

Frotáu

Concepción

San Javier

San Ignacio de Velasco

Porto Esperidião

Cáceres

Várzea Grande

Jaciara

Poxoréu

Volcán Misti 19 101 ft. 5822 m

Puno

Lago Titicaca

La Paz

El Alto

TIWANAKU

Viacha

San Ignacio de Moxo

Todos Santos

Mamoré

Grande

San José de Chiquitos

San Matías

San Lourenço

Rondonópolis

Arequipa

CORD. OCCIDENTAL

Nevado de Illimani 21 184 ft. 6457 m

Corocoro

Cochabamba

Yapacaní

Portachuelo

Montero

Buena Vista

Pojo

Barão de Melgaço

Alto Garças

Moquegua

CORD. REAL

Volcán Tutupaca 19 078 ft. 5815 m

Cerro Tunari 16 519 ft. 5035 m

Quillacollo

Punata

La Guardia

SANTA CRUZ DE LA SIERRA

Cerro Chochís 4232 ft. 1290 m

Roboré

Pantanal de São Lourenço

Alto Araguí

Ilo

Patacamaya

Capinota

Mizque

Samaipata

Pucará

Poconé

Pedro Gomes

Tacna

ALTIPLANO

Nevado Sajama 21 463 ft. 6542 m

Oruro

Huanuni

Vallegrande

San José de Chiquitos

Puerto Suárez

Coxim

Arica

CHILE

DESIERTO DE ATACAMA

Tabaya

Colquechaca

Lago Poopó

Uncía

Sucre

Cabezas

Bañados del Izozog

GRAN CHACO

Corumbá

Rio Verde de Mato Grosso

MATO GROSSO DO SU

Challapata

Villa Serrano

Tarabuco

PACIFIC OCEAN

© Rand McNally A-101695-1

PERU

Cailloma
Cotahuasi
Ayaviri
Azángaro Huancané
Nevado Coropuna
20 686 ft.
6305 m
Juliaca
Puno
Lago
Titicaca
TIWANAKU
La Paz
El Alto
Viacha
Nevado de Illimani
21 184 ft.
6457 m
Corocoro
Patacamaya

BOLIVIA

Yotaú
Concepción
San Ignacio
de Velasco
San Matías

MATO GROSSO

Porto Esperidião
Cáceres
Pocone
Cuiabá
Várzea
Grande

Mapiri
Caranavi
Achacachi
Nevado Illampu
21 066 ft.
6421 m
Volcán Tutupaca
19 101 ft.
5815 m
Moquegua
Tacna
Arica
Arequipa
Camaná
Mollendo
Ilo
Cerro Tunari
16 519 ft.
5035 m
Cochabamba
Quillacollo
Capinota
Punata
Pojo
Yapacaní
Buena Vista
Montero
Portachuelo

SANTA CRUZ DE LA SIERRA

San José
de Chiquitos
Corumbá
Porto
Esperança

MAT

Nevado Sajama
21 463 ft.
6542 m
Oruro
Huanuni
Uncia
Colquechaca
La Guardia
Samaipata
Vallegrande
Pucará
Cerro Chochis
4232 ft.
1290 m
Roboré
Puerto
Suárez
Bañados de
Isoso
Rio Verde
de Mato Grosso

Pantanal de
São Lourenço

A

Challapata
Sabaya
Chiapa
Lago
Poopó
Pisagua
Iquique
Pozo Almonte
Pintados
Huara

ALTIPLANO

CORDILLERA REAL

Sucre
Tarabuco
Villa Serrano
Lagunillas
Charagua
Camiri
Azurduy

Corumbá
Puerto
Bahía Negra

Miranda
Aquidauana
Bonito

SERRA DA BODOQUENA

N

Uyuni
Pulacayo
SALAR DE UYUNI
Rio Mulatos
Potosí
San Lucas
Camargo
Cotagaita

General Eugenio A. Garay
Fuerte Olimpo

Puerto
Murtinho
Bela Vista

Jardim

Dourados

D

Ollagüe
Volcán San Pedro
20 161 ft.
6145 m
Cerro López
19 452 ft.
5929 m
Atocha
Tupiza
Villa Abecia
San Lorenzo
Villa Montes
Yacuiba
Tarija
Villazón
La Quiaca

Mariscal Estigarribia
Puerto Pinasco

Pedro Juan
Caballero
Ponta P
Amambaí

E

Tocopilla
María Elena
Chuquicamata
Calama
Volcán Licancábur
19 409 ft.
5916 m
Sierra Gorda
Baquedano
SALAR DE ATACAMA

Aguaray
Tartagal
San Ramón de
la Nueva Orán
Pichanal

JUJUY

Cerro Coyaguaima
18 596 ft.
5668 m
Abra Pampa
Humahuaca
Susques

SALTA
Los Blancos

PARAGUAY

Concepción
Belén
Horqueta
Lima
Paranhos
Ygatimí

CHACO

FORMOSA
Las Lomitas
Pozo del Tigre
El Pintado

Comandante
Fontana
Pirané

Asunción
Villa Hayes
Caacupé
Paraguarí
Villarrica
Coronel
Oviedo
Ciuc
del E

S

PUNTA ANGAMOS
Mejillones
Tropic of Capricorn
Antofagasta
Paposo

San Antonio
de los Cobres
Volcán Llullaillaco
22 110 ft.
6739 m

PUNA
DE
Cachi
El Carril
Rosario de Lerma
Salta
San Salvador
de Jujuy
Libertador
General San Martín
San Pedro
Perico
Rivadavia
Apolinario Saravia
Las Lajitas
Joaquín V. González

Taco Pozo
Monte
Quemado
Castelli
Presidencia Roca

Palo Santo
Formosa

Pilar
San
Ignacio

San Juan
Bautista
San Pedro
de Ycuamandiyú
Yuty
Caazapá

Jardín América
Ign

PACIFIC OCEAN

PUNTA SAN PEDRO
Taltal
Catalina
Altamira
Chañaral

Volcán Antofalla
21 027 ft.
6409 m
Cerro Galán
19 396 ft.
5912 m
San Carlos
Cafayate
Angastaco
El Galpón
Rosario de
la Frontera
Metán

Presidencia Roque
Sáenz Peña
Campo Gallo
Pampa del
Infierno
Machagai
Makallé
Quimilí

Resistencia
Charadai

Corrientes
Empedrado
Ituzaingó
Posadas
Apóstoles

São
Gonza
São Borja

CHILE

ATACAMA

Nevado Ojos del Salado
22 615 ft.
6893 m
Copiapó
Caldera
PUNTA MEDIO
Carrizal Bajo

Cerro Pissis
22 241 ft.
6793 m
Tinogasta
Fiambalá
Andalgalá

TUCUMÁN
San Miguel
de Tucumán
Concepción
La Madrid
Bella Vista
Las Cejas

**SANTIAGO
DEL ESTERO**
Tintina
Suncho Corral
Pinedo
Villa
Ángela
Santa Sylvina
Villa Guillermina
Villa Ocampo
Intiyaco

Goya
Santo Tomé

CORRIENTES

S

Huasco
Freirina
Vallenar
El Tránsito
CABO BASCUÑAN
Domeyko

Cerro Bonete Grande
22 546 ft.
6872 m
Cerro General
Manuel Belgrano
20 505 ft.
6250 m
Chilecito
Aimogasta
Villa Mazán
Chumbicha
Recreo

**SALINAS DE
AMBARGASTA**
Frías
Colonia Dora
Herrera
Bandera
Selva
Tostado
Vera

Reconquista
Mercedes
La Cruz
Itaqui
Alegrete
Uruguaiana
Santi

La Rioja

LA RIOJA
Villa Unión
Guandacol
Patquía

San Francisco
del Chañar
Quilino
Deán Funes
Cruz del Eje
Chamical
La Falda
Jesús María

**SALINAS
GRANDES**
Villa Ojo
de Agua
San Cristóbal

San Justo
San Javier
Suardí
Sunchales

Curuzú Cuatiá
Esquina
Paso de
los Libres

COXILHA

Sauce
Monte
Caseros

Quaraí
Artigas
Baltasar
Brum
Rivera

Rosário do
Santan
Livram

A

La Serena
Coquimbo
PUNTA LENGUA
DE VACA
Tongoy
Ovalle

Cerro Las Tórtolas
20 735 ft.
6320 m
Rodeo
Tucunuco
Milagro
Desiderio Tello
Coquín

Rafaela
San Francisco

Córdoba
Río Segundo
Oliva
Villa María
Marcos
Juárez
Cañada
de Gómez
Casilda
Villa Constitución

Santa Fe
Paraná
Nogoyá
Diamante
Victoria
Gualeguay

**ENTRE
RÍOS**
Villaguay
Concepción
del Uruguay
Colón
Paso de
los Toros
Tacuarembó
Ansina

Salto
Concordia
Chajarí
Federal
La Paz

URUGUAY

N

Andacollo
Combarbalá
Monte
Patria
Huentelauquén
Illapel
Salamanca
Los Vilos

Cerro Mercedario
22 208 ft.
6769 m
Guandacol

SAN JUAN
Tamberías
San Juan
Caucete
Jáchal

D

Cruz del Eje

Bell Ville
Villa
Dolores
Quines
Río Cuarto
La Carlota

Rosario
San Lorenzo
San Nicolás

Gualeguaychú
Mercedes
Dolores
Durazno
Trinidad
José Batll
y Ordóñe
Florida
Mina

E

La Ligua
La Calera
Viña del Mar
Valparaíso
Quilpué
Santiago
San Antonio
San Bernardo

Volcán Aconcagua
22 837 ft.
6959 m
Cerro Tupungato
22 146 ft.
6750 m
San Felipe
Los Andes
Mendoza
San Martín
Godoy Cruz

Tunuyán
La Paz

SAN LUIS
San Luis
Villa
Mercedes
Beazley
Justo
Daract

CÓRDOBA
Sampacho

Venado Tuerto
Firmat
Pergamino
Colón
Zárate
Campana
BUENOS AIRES
General San Martín
Avellaneda
Lomas de
Zamora

La Plata
MONTEVIDE
Canelones
Las Piedras
Rio de la Plata

S

Rancagua
PUNTA TOPOCALMA
Pichilemu
San Fernando
Curicó
Licantén
Molina
Constitución
CABO CARRANZA
Cauquenes
Talca
Linares

Volcán Maipo
17 270 ft.
5264 m
Monte
Comán
San Rafael
General
Alvear
Malargüe
Volcán Nevado
12 575 ft.
3833 m
Telén

Bowen
Unión
Arizona

LA PAMPA

MENDOZA

Buena
Esperanza
Villa Valeria
Realicó
General
Villegas
Nueve de Julio
Pehuajó
Trenque Lauquen

Lincoln
Chivilcoy
Bragado
Veinticinco
de Mayo
Saladillo
General
Pico
Eduardo Castex

BUENOS AIRES
Las Flores
General Lavalle
Chascomús
Bahía
Samborombón
CABO SAN ANTO

Quirihue
Ñiquén
San Carlos
Coelemu
Chillán

Volcán Domuyo
15 427 ft.
4702 m
Bañados
del Atuel
Santa Isabel
Santa Rosa
Luan
Toro
Maza
Rivera
Guatraché
Carhué

Olavarría
Azul
Maipú
Tandil
Ayacucho
General Juan Madariaga

PERU-CHILE TRENCH

0   50   100   150   200   250   300 Miles
0   50   100   150   200   250   300   350   400   450   500 Kilometers

Lambert Azimuthal Equal Area Projection
Scale 1:10,000,000
One inch to 160 miles
One cm to 100 km

Primavera do Leste
General Carneiro
Poxoréu
Barra do Garças
ndonópolis
Aragarças
Jussara
Guiratinga
Baliza
Piranhas
Alto Garças
Caiapônia
 dro
Mineiros
omes
Alto Araguaia
Ribas do
Rio Pardo
Campo Grande
Bandeirantes
Camapuã
ROSSO DO SUL

Goianésia
Uruana
Ceres
Goianésia
Itapuranga
Itaberaí
Inhumas
GOIÂNIA
Silvânia
GOIÁS
Jandaia
Pontalina
Quirinópolis
Itumbiara
Cachoeira Alta
Santa Fé
do Sul
Paranaíba
Campo Verde
Três Lagoas

BRASÍLIA
Formosa
Cabeceiras
Cristalina
Luziânia
DISTRITO
FEDERAL
Unaí
Paracatu
Patos de
Minas
Ipameri
Catalão
Araguari
Uberlândia
Patrocínio
Ituiutaba
Uberaba
Araxá
Campina Verde
Iturama
Frutal
Conquista
Abaeté
Pompéu

Inset map a
Lambert Conformal Conic Projection
Scale 1:1,000,000
One inch to 16 miles
One cm to 10 km

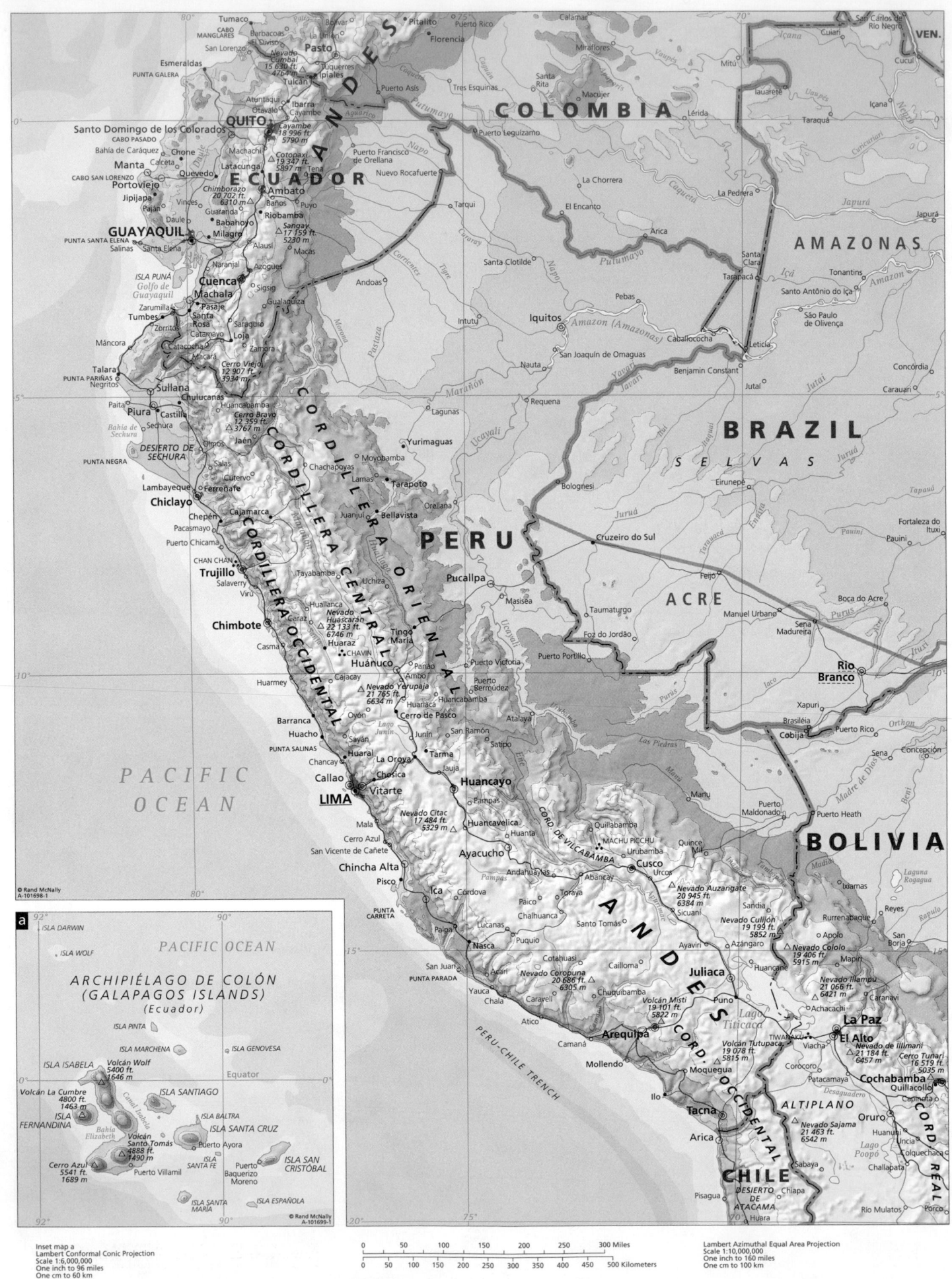

Inset map a
Lambert Conformal Conic Projection
Scale 1:6,000,000
One inch to 96 miles
One cm to 60 km

| 0 | 50 | 100 | 150 | 200 | 250 | 300 Miles |
| 0 | 50 | 100 | 150 | 200 | 250 | 300 | 350 | 400 | 450 | 500 Kilometers |

Lambert Azimuthal Equal Area Projection
Scale 1:10,000,000
One inch to 160 miles
One cm to 100 km

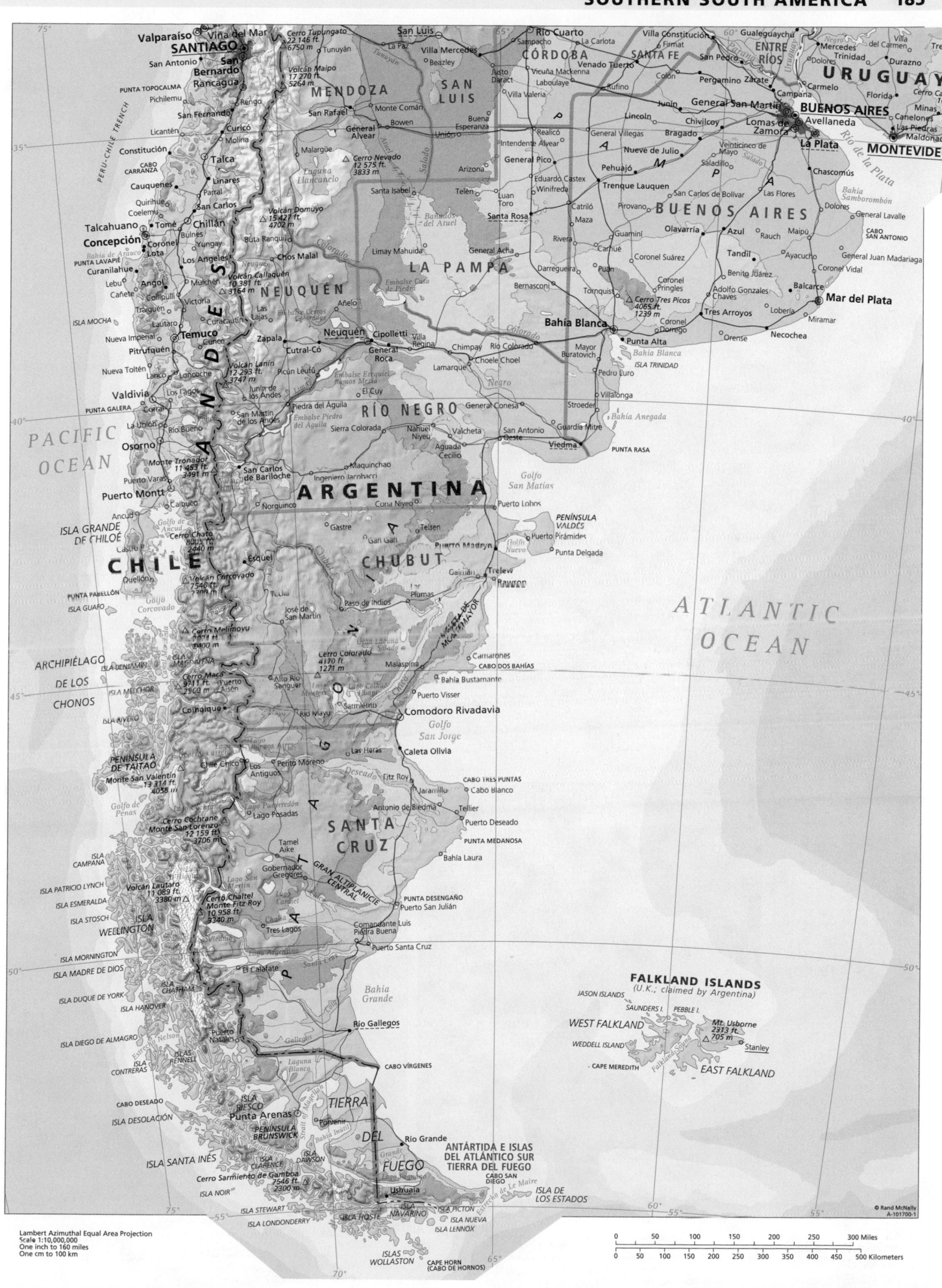

PACIFIC OCEAN

ATLANTIC OCEAN

URUGUAY

MONTEVIDEO

Valparaíso
Viña del Mar
SANTIAGO
San Antonio
San Bernardo
Rancagua
PUNTA TOPOCALMA
Pichilemu
San Fernando
Licantén
Constitución
CABO CARRANZA
Talca
Cauquenes
Quirihue
Coelemu
Talcahuano
Tomé
Concepción
Coronel
Lota
PUNTA LAVAPIÉ
Bahía de Arauco
Lebu
Cañete
Angol
Collipulli
ISLA MOCHA
Nueva Imperial
Victoria
Lautaro
Traiguén
Nueva Toltén
Pitrufquén
Lonco
Temuco
Valdivia
Punta Galera
Corral
La Unión
Río Bueno
Osorno
Puerto Varas
Puerto Montt
Calbuco
Ancud
ISLA GRANDE DE CHILOÉ
Castro
Quellón
ISLA GUAFO
Golfo Corcovado
PUNTA PABELLÓN
ARCHIPIÉLAGO DE LOS CHONOS
ISLA BENJAMIN
ISLA MELCHOR
ISLA NIVERO
PENÍNSULA DE TAITAO
ISLA CAMPANA
ISLA PATRICIO LYNCH
ISLA ESMERALDA
ISLA STOSCH
ISLA WELLINGTON
ISLA MORNINGTON
ISLA MADRE DE DIOS
ISLA DUQUE DE YORK
ISLA DIEGO DE ALMAGRO
CABO DESEADO
ISLA SANTA INÉS
ISLA NOIR
Punta Arenas
ISLA DESOLACIÓN
ISLA RIESCO
PENÍNSULA BRUNSWICK
ISLA CLARENCE
ISLA DAWSON
ISLA STEWART
ISLA LONDONDERRY
ISLA HOSTE
ISLA NAVARINO
ISLAS WOLLASTON
CAPE HORN (CABO DE HORNOS)

CHILE

ARGENTINA

MENDOZA
Volcán Maipo 17,270 ft. 5264 m
Cerro Tupungato 22,146 ft. 6750 m
Tunuyán
San Rafael
Malargüe
Cerro Nevado 12,575 ft. 3833 m
Volcán Domuyo 15,427 ft. 4702 m
Buta Ranquil
Chos Malal
Volcán Callaquén 10,381 ft. 3164 m
NEUQUÉN
Zapala
Cutral-Có
Neuquén
Cipolletti
Volcán Lanín 12,293 ft. 3747 m
General Roca
Junín de los Andes
San Martín de los Andes
Piedra del Águila
RÍO NEGRO
Valcheta
Maquinchao
Ingeniero Jacobacci
Norquinco
Cona Niyeu
Esquel
Gastre
Gan Gan
Telsen
CHUBUT
Volcán Corcovado 7546 ft. 2300 m
Tecka
José de San Martín
Paso de Indios
Las Plumas
Cerro Melimoyu 8011 ft. 2400 m
Cerro Chato 2440 m
Coihaique
Puerto Aisén
Cerro Maca 9711 ft. 2960 m
Alto Río Senguer
Sarmiento
Cerro Colorado 4170 m 1271 m
Gran Laguna Salada
Comodoro Rivadavia
Caleta Olivia
Las Heras
Perito Moreno
Chile Chico
Los Antiguos
Monte San Valentín 13,314 ft. 4058 m
Lago Posadas
Lago Pueyrredón
Fitz Roy
Jaramillo
Cerro Cochrane Monte San Lorenzo 12,159 ft. 3706 m
Tamel Aike
Gobernador Gregores
SANTA CRUZ
GRAN ALTIPLANICIE CENTRAL
Volcán Lautaro 11,089 ft. 3380 m
Cerro Chaltel Monte Fitz Roy 10,958 ft. 3340 m
Tres Lagos
El Calafate
Lago Cardiel
Comandante Luis Piedra Buena
Puerto Santa Cruz
Puerto San Julián
PUNTA DESENGAÑO
Antonio de Biedma
Tellier
Puerto Deseado
PUNTA MEDANOSA
Bahía Laura
ISLA CAMPANA
Puerto Natales
ISLAS CONTRERAS
ISLA HANOVER
CABO VÍRGENES
Río Gallegos
TIERRA DEL FUEGO
Porvenir
Río Grande
Ushuaia
CABO SAN DIEGO
ISLA DE LOS ESTADOS
ANTÁRTIDA E ISLAS DEL ATLÁNTICO SUR TIERRA DEL FUEGO
Cerro Sarmiento de Gamboa 7546 ft. 2300 m

San Luis
Tunuyán
Villa Mercedes
Beazley
SAN LUIS
Monte Comán
Bowen
General Alvear
Arizona
Santa Isabel
Telén
Luan Toro
LA PAMPA
General Acha
Santa Rosa
Maza
Catriló
Bernasconi
Guatraché

Río Cuarto
Sampacho
La Carlota
CÓRDOBA
Vicuña Mackenna
Venado Tuerto
Justo Daract
Villa Valeria
Realicó
Intendente Alvear
Eduardo Castex
Winifreda
General Pico
Pehuajó
Trenque Lauquen
Pirovano
Guaminí
Darregueira
Puan
Coronel Suárez
Tornquist
Cerro Tres Picos 4065 ft. 1239 m
Bahía Blanca
Punta Alta

Villa Constitución
Firmat
SANTA FE
San Pedro
Pergamino
Colón
Rufino
Lincoln
Junín
Nueve de Julio
Bragado
General Villegas
Chivilcoy
Saladillo
Veinticinco de Mayo
Bolívar
Olavarría
Azul
Rauch
Tandil
Benito Juárez
Adolfo Gonzales Chaves
Balcarce
Tres Arroyos
Coronel Dorrego
Coronel Pringles
BUENOS AIRES

Gualeguaychú
ENTRE RÍOS
Trinidad
Durazno
Treinta y Tres
Cerro Catedral 514 m
Minas
Florida
Canelones
Las Piedras
Rocha
Maldonado
Río de la Plata
Campana
Zárate
General San Martín
BUENOS AIRES
Avellaneda
Lomas de Zamora
La Plata
Chascomús
Las Flores
General Lavalle
CABO SAN ANTONIO
Maipú
Ayacucho
General Juan Madariaga
Coronel Vidal
Mar del Plata
Miramar
Necochea
Loberia
Orense

Bahía Blanca
Mayor Buratovich
ISLA TRINIDAD
Pedro Luro
Stroeder
Villalonga
Bahía Anegada
PUNTA RASA
Viedma
Guardia Mitre
Carmen de Patagones
San Antonio Oeste
Golfo San Matías
Puerto Lobos
PENÍNSULA VALDÉS
Puerto Pirámides
Puerto Madryn
Punta Delgada
Trelew
Rawson
Gaiman
Golfo Nuevo
Camarones
CABO DOS BAHÍAS
Bahía Bustamante
Puerto Visser
Golfo San Jorge
CABO TRES PUNTAS
Cabo Blanco

FALKLAND ISLANDS
(U.K.; claimed by Argentina)
JASON ISLANDS
SAUNDERS I.
PEBBLE I.
WEST FALKLAND
Mt. Usborne 2313 ft. 705 m
Stanley
WEDDELL ISLAND
CAPE MERIDITH
EAST FALKLAND
Falkland Sound

© Rand McNally
A-101700-1

Lambert Azimuthal Equal Area Projection
Scale 1:10,000,000
One inch to 160 miles
One cm to 100 km

0   50   100   150   200   250   300 Miles
0   50 100 150 200 250 300 350 400 450 500 Kilometers

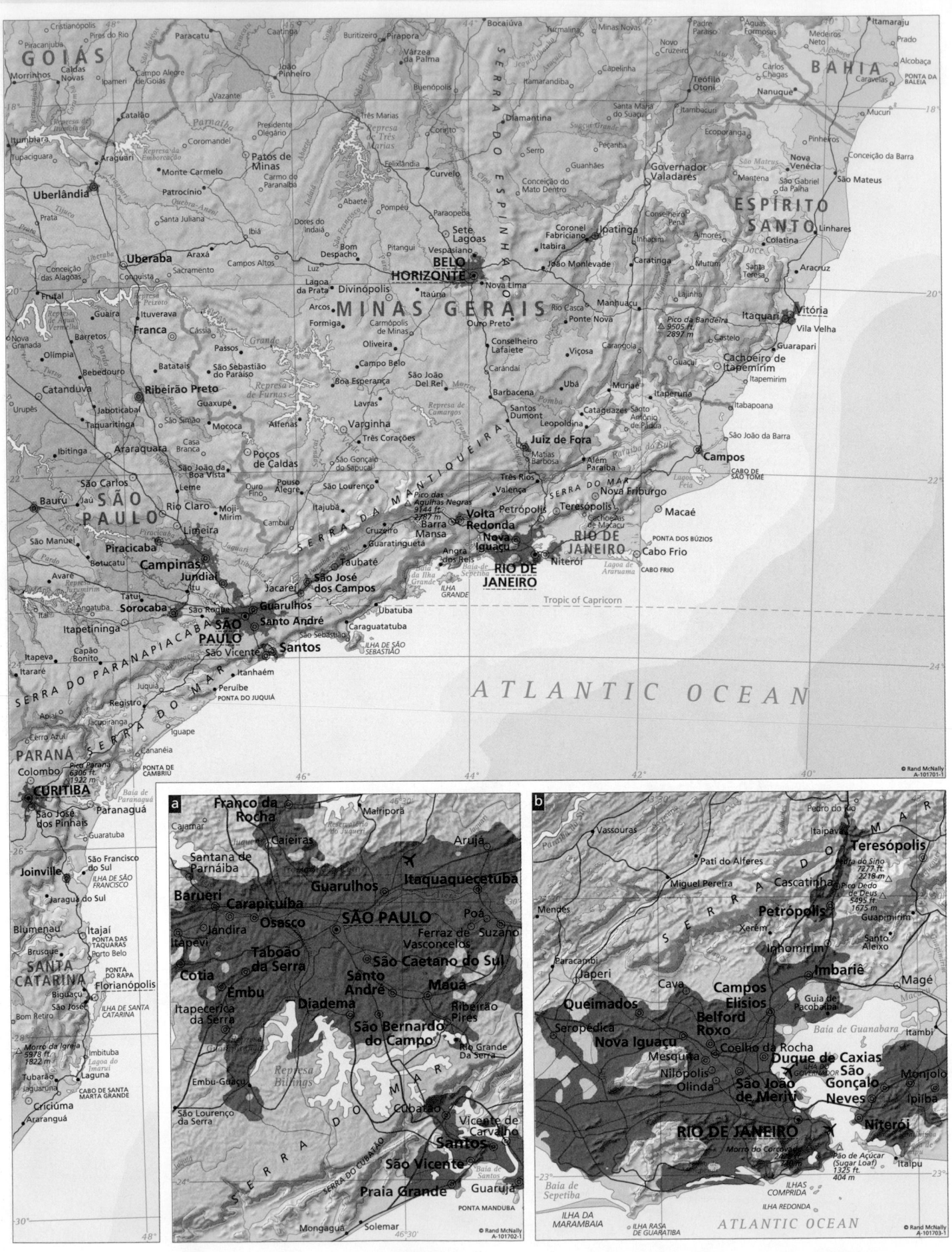

GOIÁS

Cristianópolis
Piracanjuba
Pires do Rio
Paracatu
Catinga
Buritizeiro
Pirapora
Bocaiúva
Turmalina
Minas Novas
Novo
Cruzeiro
Padre
Paraíso
Águas
Formosas
Medeiros
Neto
Itamaraju
Prado

Morrinhos
Caldas
Novas
Campo Alegre
de Goiás
João
Pinheiro
Presidente
Olegário
Várzea
da Palma
Capelinha
Teófilo
Otoni
Carlos
Chagas
Nanuque
Alcobaça
PONTA DA
BALEIA
Caravelas

Ipameri
Catalão
Coromandel
Patos de
Minas
Carmo do
Paranaíba
Diamantina
Santa Maria
do Suaçuí
Itambacuri
Ecoporanga
Conceição da Barra

Itumbiara
Araguari
Monte Carmelo
Santa Juliana
Ibiá
Dores do
Indaiá
Pompéu
Paraopeba
Conceição do
Mato Dentro
Serro
Guanhães
Peçanha
Pinheiros
BAHIA

Tupaciguara
Uberlândia
Patrocínio
Abaeté
Três Marias
Represa
de Três
Marias
Curvelo
Nova
Venécia
São Gabriel
da Palha
São Mateus

Conceição
das Alagoas
Uberaba
Araxá
Sacramento
Campos Altos
Sete
Lagoas
Vespasiano
Itabira
Coronel
Fabriciano
Ipatinga
Inhapim
Mantena
Martena
Colatina
Linhares

Prata
Santa Juliana
BELO
HORIZONTE
Nova Lima
João Monlevade
Caratinga
Mutum
Santa
Teresa
Aracruz

Frutal
Lagoa
da Prata
Divinópolis
Itaúna
Rio Casca
Manhuaçu
Pico da Bandeira
9505 ft
2897 m
Itaquari
Vitória
Vila Velha

MINAS GERAIS

ATLANTIC OCEAN

Tropic of Capricorn

© Rand McNally
A-101701-1

Inset a

Franco da
Rocha
Mairiporã
Arujá

Cajamar
Caieiras

Santana de
Parnaíba
Guarulhos
Itaquaquecetuba

Barueri
Carapicuíba
SÃO PAULO
Poá

Jandira
Osasco
Ferraz de
Vasconcelos
Suzano

Itapevi
Taboão
da Serra
São Caetano
do Sul
Mauá

Cotia
Embu
Santo
André
Ribeirão
Pires

Itapecerica
da Serra
Diadema
São Bernardo
do Campo
Rio Grande
Da Serra

Embu-Guaçu
Represa
Billings

São Lourenço
da Serra
Cubatão
Vicente de
Carvalho

Santos
São Vicente

Praia Grande
Guarujá

Mongaguá
Solemar
PONTA MANDUBA

© Rand McNally
A-101702-1

Inset b

Pedro do Rio
Itaipava
Teresópolis
Pico do Sino
7277 ft
2218 m

Vassouras
Patí do Alferes
Cascatinha
Pico Dedo
de Deus
5495 ft
1675 m

Miguel Pereira
Petrópolis
Guapimirim

Mendes
Xerém
Inhomirim
Santo
Aleixo

Paracambi
Japeri
Cava
Imbariê
Magé

Queimados
Campos
Elísios
Guia de
Pacobaíba

Seropédica
Belford
Roxo
Coelho da Rocha
Duque de Caxias
São
Gonçalo
Monjolo

Mesquita
Nova Iguaçu
Ipiíba

Nilópolis
Olinda
São João
de Meriti
Neves
Itambi

RIO DE JANEIRO
Niterói

Morro do Corcovado
Pão de Açúcar
(Sugar Loaf)
1325 ft
404 m
Italpu

Baía de
Sepetiba
ILHAS
COMPRIDA

ILHA DA
MARAMBAIA
ILHA RASA
DE GUARATIBA
ILHA REDONDA
ATLANTIC OCEAN

© Rand McNally
A-101703-1

Scale

0   30   60   90   120   150   180 Miles
0 30 60 90 120 150 180 210 240 270 300 Kilometers

Lambert Conformal Conic Projection
Scale 1:6,000,000
One inch to 96 miles
One cm to 60 km

Inset maps a, b
Lambert Conformal Conic Projection
Scale 1:1,000,000
One inch to 16 miles
One cm to 10 km

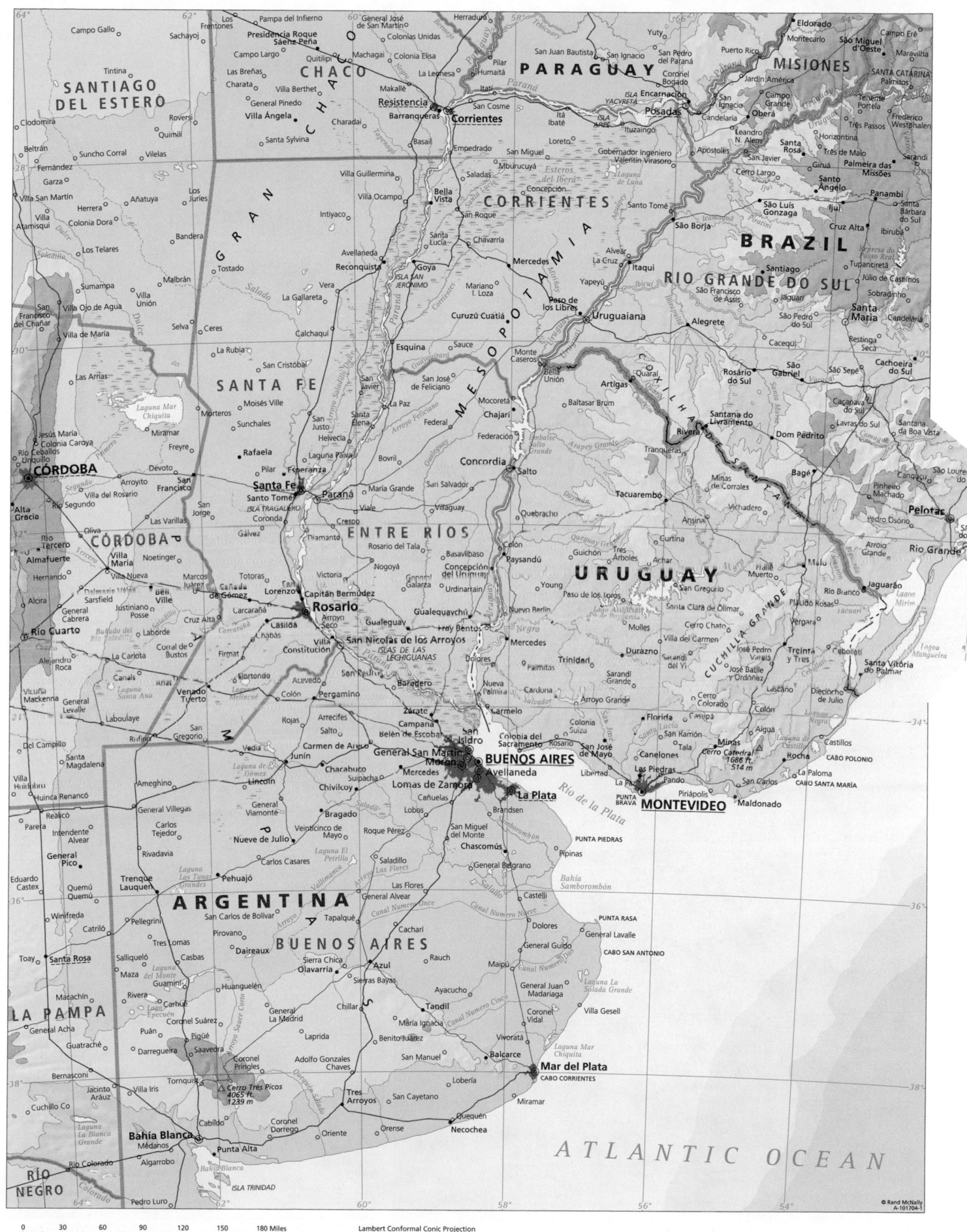

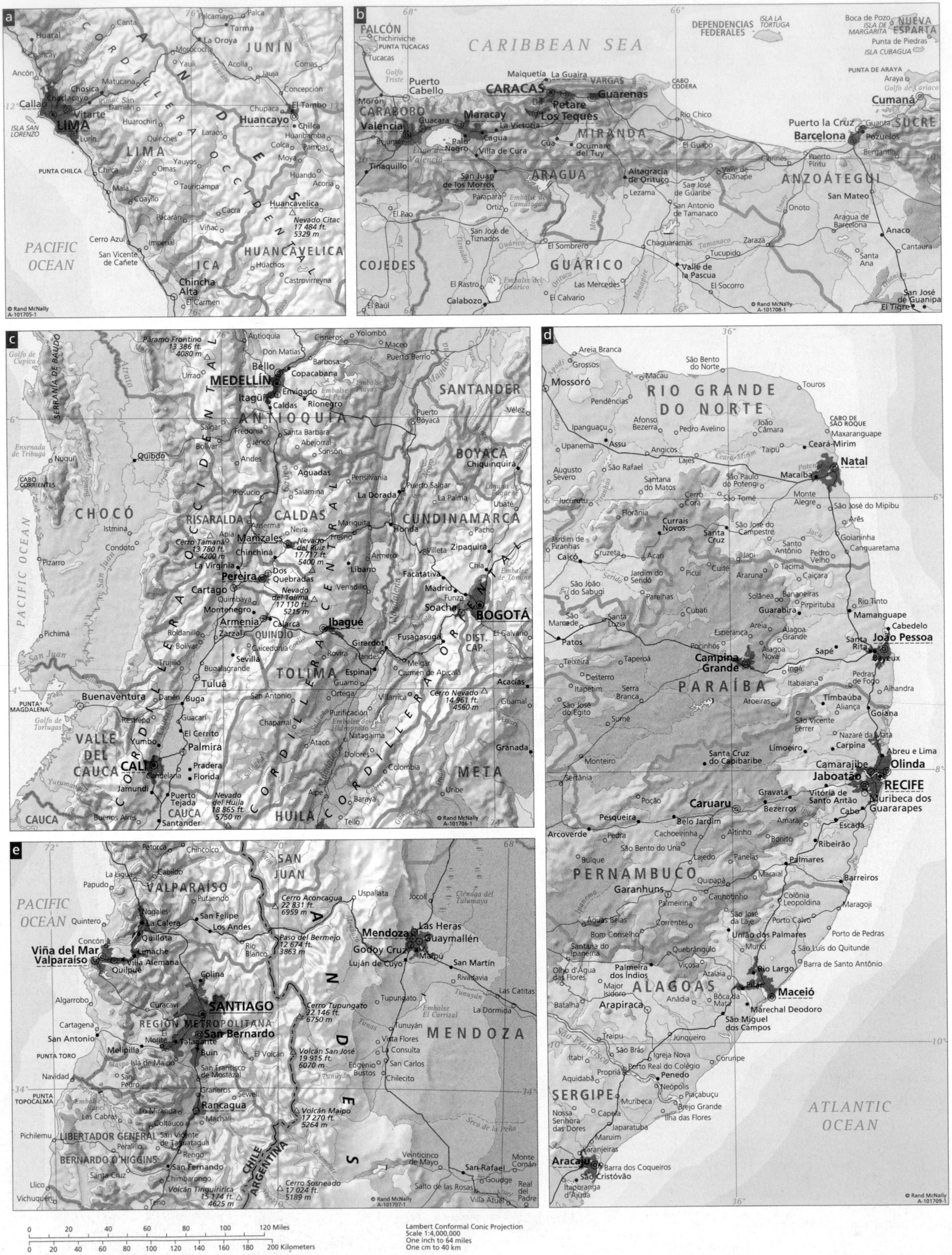

Map a

JUNÍN
La Oroya
Morococha
Tarma
Palcamayo · Palca
Huaral
Canta
Chosica
San
Damián
Acolla · Jauja
Concepción
Comas
Chupaca · El Tambo
Huancayo
Callao
LIMA
Vitarte
ISLA SAN
LORENZO
Lurín
LIMA
Huarochirí
Huamantanga
Larecos
Quinches
Moya
Huancavelica
Nevado Citac
17 484 ft.
5329 m
PACIFIC
OCEAN
ICA
PUNTA CHILCA
Chilca
Mala
Coayllo
Omas
Yauyos
Tauripampa
Cacra
HUANCAVELICA
Huachos
Castrovirreyna
Cerro Azul
Imperial
San Vicente
de Cañete
Viñac
Chincha
Alta
El Carmen
Pacarán

Map b

FALCÓN
Chichiriviche
PUNTA TUCACAS
Tucacas
Golfo
Triste
Puerto
Cabello
Morón
Maiquetía · La Guaira
CARACAS
VARGAS
Guarenas
CARIBBEAN SEA
DEPENDENCIAS
FEDERALES
ISLA LA
TORTUGA
Boca de Pozo
ISLA DE
MARGARITA
NUEVA
ESPARTA
Punta de Piedras
ISLA CUBAGUA
PUNTA DE ARAYA
Golfo de Cariaco
CABO
CODERA
Cumaná
SUCRE
CARABOBO
Valencia
Maracay
Guacara
Petare
Los Teques
MIRANDA
Río Chico
El Guapo
Puerto la Cruz
Barcelona
Guanta
Pozuelos
Bergantín
Lago
Negro
La Victoria
Bejuma
Villa de Cura
Cúa
Ocumare
del Tuy
Altagracia
de Orituco
Valle de
Guanape
Puerto
Píritu
Clarines
Tinaquillo
Lago de
Valencia
San Juan
de los Morros
ARAGUA
Parapara
Ortiz
Lezama
San José
de Guaribe
San Antonio
de Tamanaco
ANZOÁTEGUI
San Mateo
Aragua de
Barcelona
Anaco
Cantaura
Santa
Ana
COJEDES
El Pao
GUÁRICO
San José de
Tiznados
El Sombrero
Chaguaramas
Tucupido
Zaraza
Onoto
El Rastro
Las Mercedes
El Calvario
El Socorro
Valle de
la Pascua
San José
de Guanipa
El Tigre
Calabozo

Map c

Golfo de
Cupica
SERRANÍA DE BAUDÓ
Páramo Frontino
13 386 ft.
4080 m
Antioquia
Don Matías
Cisneros
Yolombó
Maceo
Urrao
Bello
Copacabana
Barbosa
Puerto Berrío
SANTANDER
MEDELLÍN
Itagüí
Envigado
Caldas
Rionegro
MAGDALENA
Embalse del
Peñol
ANTIOQUIA
Santa Bárbara
Fredonia
Abejorral
Sonsón
BOYACÁ
Vélez
Jericó
Andes
Aguadas
Pensilvania
Chiquinquirá
Ensenada
de Tribugá
Nuquí
Quibdó
Salamina
La Dorada
Puerto Salgar
La Palma
Ubaté
CHOCÓ
Istmina
Riosucio
RISARALDA
CALDAS
Neira
Mariquita
Fresno
Honda
CUNDINAMARCA
Pacho
CABO
CORRIENTES
Condoto
Apía
Cerro Tatamá
13 780 ft.
4200 m
Manizales
Chinchiná
Nevado del Ruiz
17 717 ft.
5400 m
Líbano
Armero
Villeta
Zipaquirá
Chía
Embalse
de Tominé
Pizarro
La Virginia
Pereira
Dos
Quebradas
Nevado
del Tolima
17 110 ft.
5215 m
Venadillo
Facatativá
Madrid
Funza
Soacha
BOGOTÁ
DIST.
CAP.
Pichimá
Cartago
Quimbaya
Montenegro
Calarcá
Ibagué
Girardot
Fusagasugá
El Colegio
El Galvano
Armenia
QUINDÍO
Sevilla
Caicedonia
Rovira
Flandes
Melgar
Carmen de Apicalá
Buenaventura
Darién
Buga
Guacarí
Tuluá
Bugalagrande
San Antonio
Ortega
Villarrica
Cerro Nevado
14 961 ft.
4560 m
Acacías
Granada
VALLE
DEL
CAUCA
Yumbo
Restrepo
El Cerrito
Palmira
TOLIMA
Chaparral
Ataco
Purificación
Natagaima
Dolores
Colombia
Baraya
META
Uribe
CALI
Candelaria
Pradera
Florida
Jamundí
Puerto
Tejada
Buenos Aires
CAUCA
Santander
Nevado del Huila
18 865 ft.
5750 m
HUILA
Aipe
Tello

Map d

Areia Branca
Grossos
Macau
São Bento
do Norte
Touros
RIO GRANDE
DO NORTE
Mossoró
Pendências
Afonso
Bezerra
Pedro Avelino
João
Câmara
CABO DE
SÃO ROQUE
Maxaranguape
Ipanguaçu
Upanema
Angicos
Lajes
Taipu
Ceará-Mirim
Macaíba
Natal
Augusto
Severo
São Rafael
Santana
do Matos
Cerro
Cora
São Tomé
São Paulo
do Potengí
Monte
Alegre
São José de Mipibu
Arês
Jucuruti
Florânia
Jardim de
Piranhas
Currais
Novos
Santa
Cruz
São José de
Campestre
Santo
Antônio
Pedro
Velho
Goianinha
Canguaretama
Caicó
Cruzeta
Acari
Picuí
Japi
Cuité
Solânea
Bananeiras
Pirpirituba
Rio Tinto
Guarabira
São João
do Sabugí
Parelhas
Cubati
Areia
Alagoa
Grande
Mamanguape
São
Mamede
Santa
Luzia
Esperança
Sapé
Santa
Rita
João Pessoa
Cabedelo
Bayeux
Patos
Teixeira
Desterro
Taperoá
Soledade
Ingá
Pedras
de Fogo
Alhandra
PARAÍBA
Itapetim
Serra
Branca
Itabaiana
Aroeiras
Timbaúba
Aliança
Goiana
Campina
Grande
Nazaré da
Mata
Monteiro
São José
do Egito
Sumé
São Vicente
Férrer
Carpina
Sertânia
Limoeiro
Camarajibe
Olinda
Jaboatão
RECIFE
Muribeca dos
Guararapes
PERNAMBUCO
Caruaru
Pesqueira
Poção
Belo Jardim
Bezerros
Gravatá
Vitória de
Santo Antão
Amaraji
Cabe
Arcoverde
Pedra
Cachoeirinha
Altinho
Bonito
Escada
Ribeirão
Palmares
Buíque
São Bento do Una
Lajedo
Panelas
Maraial
Barreiros
Garanhuns
Quipapá
Palmeirina
Canhotinho
Correntes
São José
da Laje
Porto Calvo
Porto de Pedras
Águas
Belas
Bom
Conselho
União dos Palmares
Murici
São Luís do Quitunde
ALAGOAS
Santana do
Ipanema
Olho d'Água
das Flores
Palmeira
dos Índios
Viçosa
Atalaia
Rio Largo
Barra de Santo Antônio
Major
Isidoro
Anadia
Maceió
Batalha
Boca da
Mata
Marechal Deodoro
Arapiraca
São Miguel
dos Campos
SERGIPE
Traipu
São
Francisco
Igreja Nova
Porto Real
do Colégio
Penedo
Corunpe
Aquidabã
Neópolis
Piaçabuçu
Propriá
Muribeca
Brejo
Grande
Itabi
Capela
Ilha das Flores
Nossa
Senhora
das Dores
Japaratuba
Aracaju
Barra dos Coqueiros
Maruim
Laranjeiras
São Cristóvão
Itaporanga
d'Ajuda
ATLANTIC
OCEAN

Map e

Petorca
Chincolco
La Ligua
Cabildo
SAN
JUAN
PACIFIC
OCEAN
Papudo
Putaendo
Uspallata
Ciénaga del
Tulumaya
Zapallar
VALPARAÍSO
Nogales
La Calera
San Felipe
Los Andes
Cerro Aconcagua
22 831 ft.
6959 m
Jocolí
Quintero
Quillota
Limache
Paso del Bermejo
12 674 ft.
3863 m
Las Heras
Mendoza
Guaymallén
Viña del Mar
Valparaíso
Villa Alemana
Quilpué
Río
Blanco
Godoy Cruz
Maipú
Concón
Curacaví
Colina
Luján de Cuyo
San Martín
Algarrobo
Rivadavia
SANTIAGO
REGIÓN
METROPOLITANA
San Bernardo
Cerro Tupungato
22 146 ft.
6750 m
Tupungato
Las Catitas
San Antonio
Monte
Talagante
Buin
Volcán San José
19 915 ft.
6070 m
Vista Flores
Tunuyán
Cartagena
Melipilla
El Volcán
San Carlos
Chilecito
PUNTA TORO
Isla de Maipo
San Francisco
de Mostazal
Eugenio
Bustos
MENDOZA
Navidad
San
Pedro
Graneros
Gewell
Tunuyán
Embalse
El Carrizal
Llico
PUNTA
TOPOCALMA
Machalí
Rancagua
Volcán Maipo
17 270 ft.
5264 m
La Dormida
Pichilemu
LIBERTADOR GENERAL
BERNARDO O'HIGGINS
Peralillo
San Vicente
de Taguatagua
San Fernando
Cerro Tupungato
Volcán Tinguiririca
15 174 ft.
4625 m
Santa
Cruz
Chimbarongo
Cerro Sosneado
17 024 ft.
5189 m
Veinticinco
de Mayo
San Rafael
Vichuquén
Teno
CHILE
ARGENTINA
Salto de las Rosas
Goudge
Monte
Comán
Real
del
Padre
Villa Atuel

Lambert Conformal Conic Projection
Scale 1:4,000,000
One inch to 64 miles
One cm to 40 km

0 20 40 60 80 100 120 Miles
0 20 40 60 80 100 120 140 160 180 200 Kilometers

© Rand McNally

Europe

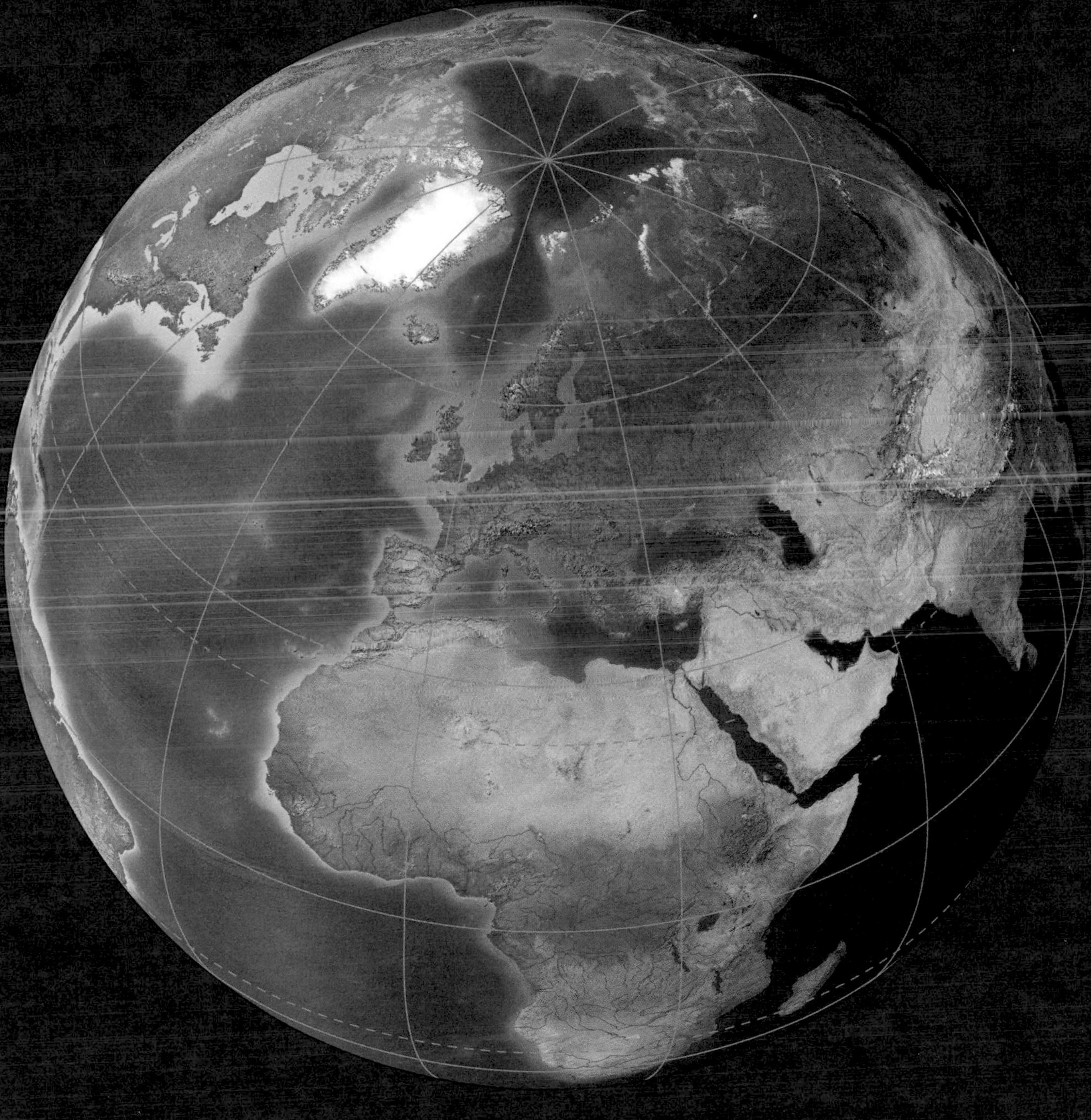

Small and densely populated, Europe is noted for the complexity of both its landscape and its human geography. Having a land area slightly larger than Australia, Europe is carved up into nearly fifty countries ranging from the world's smallest, Vatican City, to the world's largest, Russia, whose territory west of the Ural Mountains is included within the continent. Europe is dominated by temperate climate zones and fertile agricultural lands. The birthplace of democracy and the Industrial Revolution, Europe has been a major geopolitical and economic force at the global scale for centuries. For the better part of 400 years—from the 16th century to the 20th century—European colonial powers controlled large portions of territory throughout the world.

OSTROV
KOLGUYEV
MYS KANIN NOS
Naryan-Mar
Usinsk
Inta
Gora Narodnaya
6214ft
1894m
Pechora
Ponoy
Patity
Usa
Pechora
WHITE SEA
Severodvinsk
Arkhangel'sk
Ukhta
Segezha
Ozero
Vygozero
Syktyvkar
Nizhniy
Tagil
Tyumen'
OMSK
NOVOSIBIRSK
Barnaul
Ob'
Petrozavodsk
Kotlas
Solikamsk
Berezniki
Pervouralsk
YEKATERINBURG
CHELYABINSK
Astana
(Aqmola)
Semey
RUSSIA
Kirov
PERM'
Kungur
Zlatoust
Miass
Qaraghandy
Tikhvin
Cherepovets
Vologda
Glazov
Izhevsk
Magnitogorsk
Lake
Ladoga
Lake
Onega
Novgorod
Borovichi
Rybinsk
Yaroslavl
Ivanovo
Kostroma
Yoshkar-Ola
Naberezhnye
Chelny
UFA
Sterlitamak
Salavat
Orsk
KAZAKHSTAN
Vyshniy
Volochëk
Tver'
NIZHNIY
NOVGOROD
Cheboksary
KAZAN'
At'met'yevsk
Oktyabrskiy
Lake Balkhash
(Balqash köli)
Sergiyev Posad
Vladimir
Dzerzhinsk
Dimitrovgrad
SAMARA
Orenburg
Aqtöbe
Zhezqazghan
Vicebsk
MOSCOW
Podolsk
Serpukhov
Kolomna
Ryazan'
Murom
Saransk
Ulyanovsk
Tolyatti
Novokuybyshevsk
Smolensk
Kaluga
Novomoskovsk
Penza
Syzran
Oral
Aral Sea
Shymkent
TASHKENT
Mahilëu
Tula
Tambov
Saratov
Balakovo
Kamyshin
Bryansk
Orel
Elets
Lipetsk
Michurinsk
Engel's
Nyr Darya
Homel'
Kursk
Voronezh
Balashov
Volzhskiy
Atyrau
Nukus
UZBEKISTAN
Chernihiv
Staryy Oskol
KIEV
(KYÏV)
Sumy
KHARKIV
Belgorod
VOLGOGRAD
Samarqand
UKRAINE
Poltava
DNIPROPETROVS'K
Horlivka
Luhans'k
Astrahan
Atrak
TURKMENISTAN
MOLDOVA
Chişinău
Tiraspol
Zaporizhzhia
Mariupol'
DONETS'K
Taganrog
ROSTOV-
NA-DONU
Flista
Sarygamysh
köli
ODESA
Mykolaiv
Kherson
SEA OF
AZOV
Krasnodar
Armavir
Stavropol'
Kuma
CASPIAN
SEA
Aşgabat
Simferopol'
Kerch
Cherkessk
Pyatigorsk
Nalchik
Groznyy
Makhachkala
Derbent
Yalta
Novorossiysk
Maykop
Gora El'brus
18 510 ft
5642 m
Vladikavkaz
Sumqayit
BAKU
(BAKI)
MASHHAD
Constanța
BLACK
SEA
Sochi
CAUCASUS MOUNTAINS
GEORGIA
TBILISI
Gäncä
AZERBAIJAN
AFGHANISTAN
Varna
Burgas
ARMENIA
Yerevan
AZER.
TABRIZ
Rasht
İSTANBUL
TEHRAN
İZMIR
ANKARA
TURKEY
Van
Gölü
Daryācheh-ye
Orūmīyeh
Daryācheh-ye
Namak
IRAN
BURSA
Tuz
Gölü
MOSUL
ESFAHAN
Kermān
Antalya
ADANA
ALEPPO
(HALAB)
SYRIA
IRAQ
Kermānshāh
SHIRĀZ
RÓDOS
KÁRPATHOS
CYPRUS
Nicosia
LEBANON
Beirut
(Bayrūt)
DAMASCUS
(DIMASHQ)
BAGHDĀD
Euphrates
Tigris
Ahvāz
Basra
Bandar
'Abbas

0   80   160   240   320   400   480 Miles

0   80 160 240 320 400 480 560 640 720 800 Kilometers

Lambert Conformal Conic Projection
Scale 1:16,000,000
One inch to 256 miles
One cm to 160 km

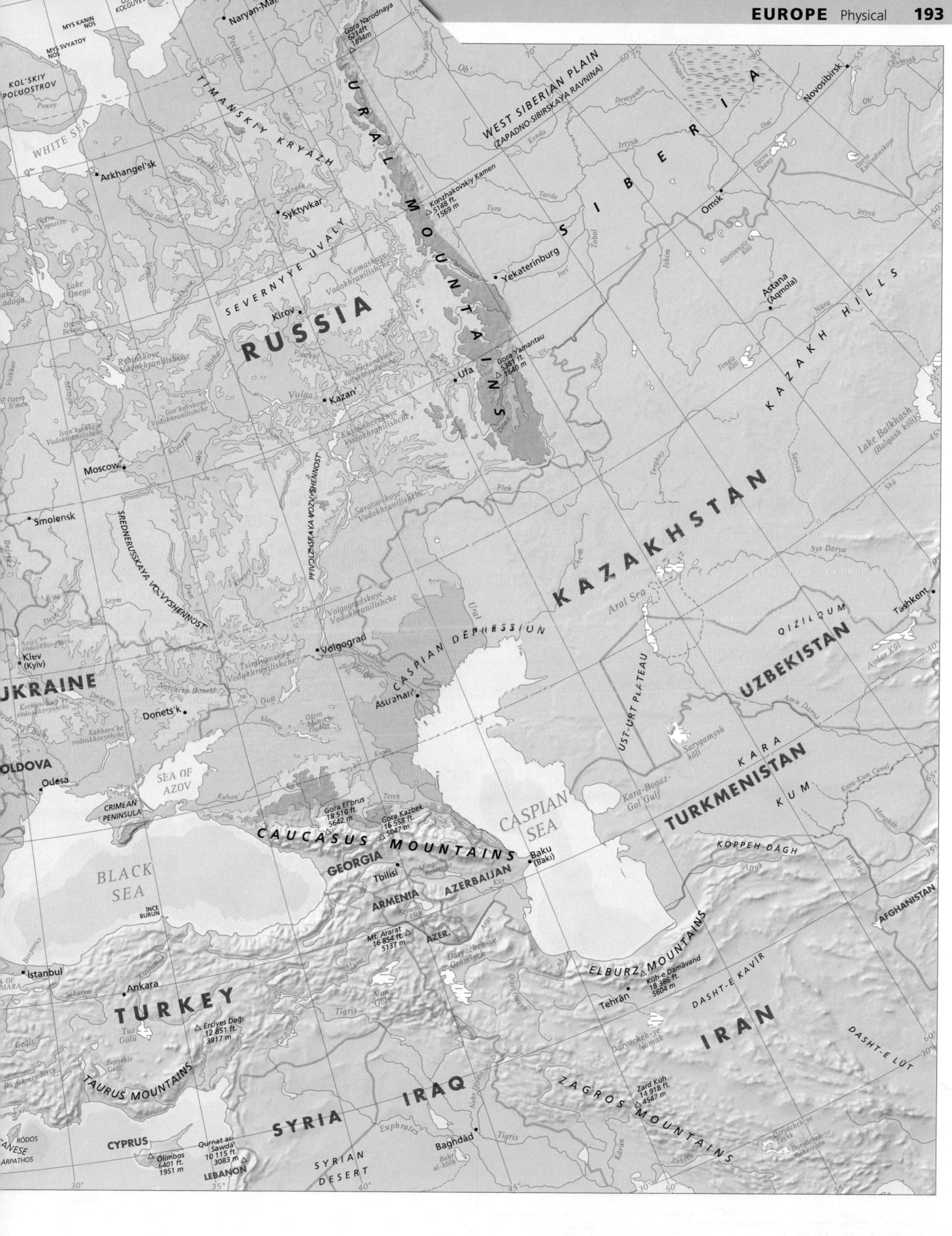

MYS KANIN NOS

KOL'SKIY POLUOSTROV

MYS SVYATOY NOS

OSTROV KOLGUYEV

• Naryan-Mar

Gora Narodnaya 6214ft 1894m

WHITE SEA

WEST SIBERIAN PLAIN (ZAPADNO-SIBIRSKAYA RAVNINA)

• Arkhangel'sk

U R A L

• Novosibirsk

S I B E R I A

Ponoy

TIMANSKIY KRYAZH

• Syktyvkar

M O U N T A I N S

Lake Ladoga

Ozero Vygozero

SEVERNYYE UVALY

Konzhakovskiy Kamen 5148 ft 1569 m

• Omsk

• Yekaterinburg

Ozero Chany

Lake Onega

Kirov •

R U S S I A

Astana (Aqmola) •

Rybinskoye Vodokhranilishche

Gora Yamantau 5381 ft 1640 m

K A Z A K H   H I L L S

Ozero Il'men

Volga

Vulga

Kazan •

Ufa •

Moscow •

Lake Balkhash (Balqash köli)

• Smolensk

SREDNERUSSKAYA VOZVYSHENNOST

PRIVOLZHSKAYA VOZVYSHENNOST

Saratovskoye Vodokhranilishche

K A Z A K H S T A N

Kiev (Kyiv) •

Volgograd •

CASPIAN DEPRESSION

Aral Sea

Tashkent •

U K R A I N E

Donets'k •

Astrahan' •

UST-URT PLATEAU

QIZILQUM

U Z B E K I S T A N

OLDOVA

Odesa •

SEA OF AZOV

Sarygamysh köli

Amu Darya

CRIMEAN PENINSULA

K A R A   K U M

BLACK SEA

Gora El'brus 18 510 ft 5642 m

Gora Kazbek 16 558 ft 5047 m

CASPIAN SEA

Kara-Bogaz-Gol Gulf

T U R K M E N I S T A N

Kara-Kum Canal

C A U C A S U S   M O U N T A I N S

Baku (Bakı) •

INCE BURUN

GEORGIA

Tbilisi •

KOPPEH DAGH

Mingechaur Reservoir

ARMENIA

AZERBAIJAN

• Istanbul

Mt. Ararat 16 854 ft 5137 m

AZER.

ELBURZ   MOUNTAINS

AFGHANISTAN

Ankara •

Daryächeh-ye Orümïyeh

Küh-e Damävand 18 386 ft 5604 m

T U R K E Y

Erciyes Dağı 12 851 ft 3917 m

DASHT-E KAVIR

Van Gölü

Tehrān •

Tuz Gölü

I R A N

TAURUS MOUNTAINS

Daryächeh-ye Namak

DASHT-E LUT

RODOS

Z A G R O S   M O U N T A I N S

Zard Küh 14 918 ft 4547 m

ARPATHOS

S Y R I A

I R A Q

CYPRUS

Ölimbos 6401 ft 1951 m

Qurnat as-Sawdä 10 115 ft 3083 m

Euphrates

Tigris

Baghdād •

LEBANON

SYRIAN DESERT

Bahr al-Milh

ATLANTIC
OCEAN

NORWEGIAN SEA

Arctic Circle

FAROE
ISLANDS

SHETLAND
ISLANDS

ORKNEY
ISLANDS

HEBRIDES

Trondheim

Gulf of Bothnia

Helsinki

Oslo

Stockholm

St. Petersb

Gulf of Finland

Göteborg

SAAREMAA

BALTIC

Glasgow

NORTH
SEA

Dublin

IRISH SEA

Leeds

Manchester

Birmingham

London

Amsterdam

Hamburg

Copenhagen

GOTLAND

Riga

Western Dvi

ÖLAND

BORNHOLM

Minsk

Berlin

Elbe

Warsaw

Bug

Brussels

Essen

Łódź

Rhine

Frankfurt
am Main

Oder

Katowice

English Channel

Paris

Seine

Mannheim

Prague

CARPATHIAN MOUNTAIN

Stuttgart

Danube

Dniester

Loire

Munich

Bay of
Biscay

Vienna

Danube

Budapest

Bordeaux

Lyon

Rhône

A L P S

MASSIF
CENTRAL

CORDILLERA CANTÁBRICA

Turin

Milan

TRANSYLVANIAN ALPS

Porto

PYRENEES

Ebro

Belgrade

Bucharest

Marseille

LIGURIAN
SEA

A P E N N I N E S

ADRIATIC SEA

Danube

Madrid

Tagus

Barcelona

CORSICA

Rome

Sofia

Lisbon

SIERRA MORENA

València

SARDINIA

Naples

Sevilla

BALEARIC ISLANDS

M E D I T E R R A N E A N

TYRRHENIAN
SEA

IONIAN
ISLANDS

IONIAN
SEA

PÍNDOS ÓROS

AEGEAN
SEA

Athens

CYCLAD

Strait of Gibraltar

SICILY

MALTA

S E A

SEA OF CR

CRE

0   80   160   240   320   400   480 Miles

0   80 160 240 320 400 480 560 640 720 800 Kilometers

Lambert Conformal Conic Projection
Scale 1:16,000,000
One inch to 256 miles
One cm to 160 km

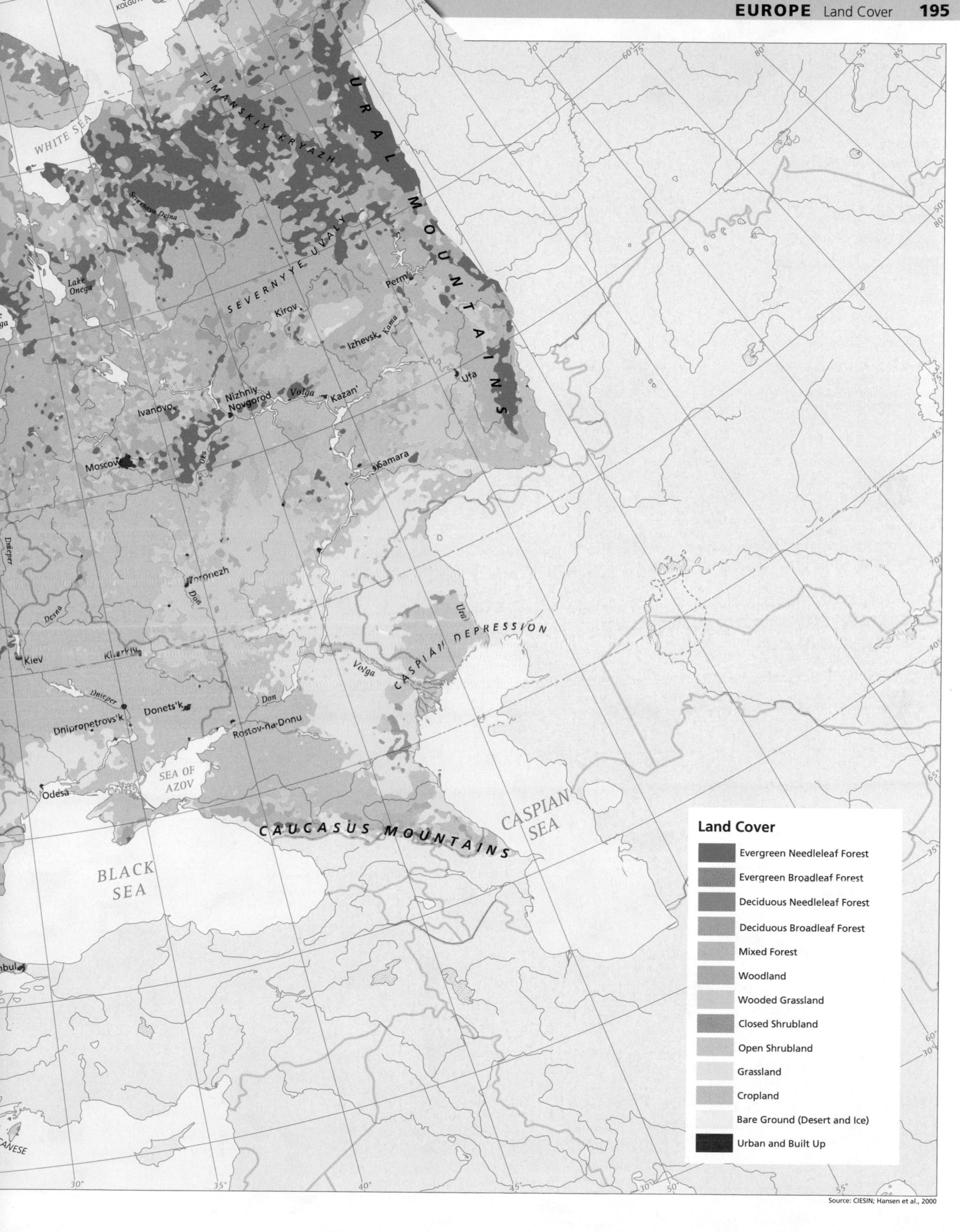

OSTROV
KOLGUYEV

URAL MOUNTAINS

TIMANSKIY KRYAZH

WHITE SEA

Severnaya Dvina

SEVERNYYE UVALY

Usa

Lake
Onega

Kirov

Perm

Izhevsk

Kama

Nizhniy
Novgorod

Volga

Kazan'

Ufa

Ivanovo

Oka

Moscow

Samara

Voronezh

Don

CASPIAN DEPRESSION

Ural

Desna

Dnieper

Volga

Kiev

Kharkiv

Dnipropetrovs'k

Donets'k

Don

Rostov-na-Donu

Odesa

SEA OF
AZOV

CAUCASUS MOUNTAINS

CASPIAN
SEA

BLACK
SEA

Land Cover

- Evergreen Needleleaf Forest
- Evergreen Broadleaf Forest
- Deciduous Needleleaf Forest
- Deciduous Broadleaf Forest
- Mixed Forest
- Woodland
- Wooded Grassland
- Closed Shrubland
- Open Shrubland
- Grassland
- Cropland
- Bare Ground (Desert and Ice)
- Urban and Built Up

Source: CIESIN; Hansen et al., 2000

Geology
Rock type/geological era

- Intrusive
- Metamorphic
- Volcanic, tectonic
- Precambrian sedimentary
- Paleozoic sedimentary
- Mesozoic sedimentary
- Cenozoic sedimentary

Note: Areas classified as sedimentary also
include some sedimentary/volcanic areas.

Source: Generalized Geological Map of the World,
Geological Survey of Canada

Reykjavík

Arctic Circle

NORWEGIAN SEA

Trondheim

FAROE
ISLANDS

SHETLAND
ISLANDS

HEBRIDES

ORKNEY
ISLANDS

Glasgow

Dublin

IRISH SEA

Leeds

Manchester

Birmingham

London

Amsterdam

English Channel

Brussels

Essen

Hamburg

Elbe

Berlin

NORTH
SEA

Oslo

Göteborg

Stockholm

Copenhagen

BORNHOLM

GOTLAND

ÖLAND

BALTIC

SEA

Gulf of Bothnia

Helsinki

Gulf of Finland

St. Petersb

SAAREMAA

Lake
Peip

Rīga

Western Dvi

Minsk

Warsaw

Bug

Łódź

Oder

Vistula

ATLANTIC

OCEAN

Bay of
Biscay

Bordeaux

Dordogne

MASSIF
CENTRALE

Rhône

Loire

Seine

Paris

Mannheim

Stuttgart

Frankfurt
am Main

Rhine

Munich

Danube

Prague

Vienna

Katowice

CARPATHIAN MOUNTAINS

Dniester

Danube

Budapest

Drava

Sava

Lyon

Turin

Milan

Po

ALPS

A P
E N N

LIGURIAN
SEA

Marseille

CORDILLERA CANTÁBRICA

PYRENEES

Porto

Duero

Ebro

Madrid

Tagus

Lisbon

SIERRA MORENA

Sevilla

València

Barcelona

BALEARIC ISLANDS

Strait of Gibraltar

CORSICA

SARDINIA

M E D I T E R R A N E A N   S E A

Rome

Naples

I N E S

ADRIATIC
SEA

TRANSYLVANIAN ALPS

Belgrade

Danube

Bucharest

Sofia

PINDOS OROS

AEGEAN
SEA

Athens

CYCLAD

SEA OF CR

CRE

IONIAN
SEA

IONIAN
ISLANDS

TYRRHENIAN
SEA

SICILY

MALTA

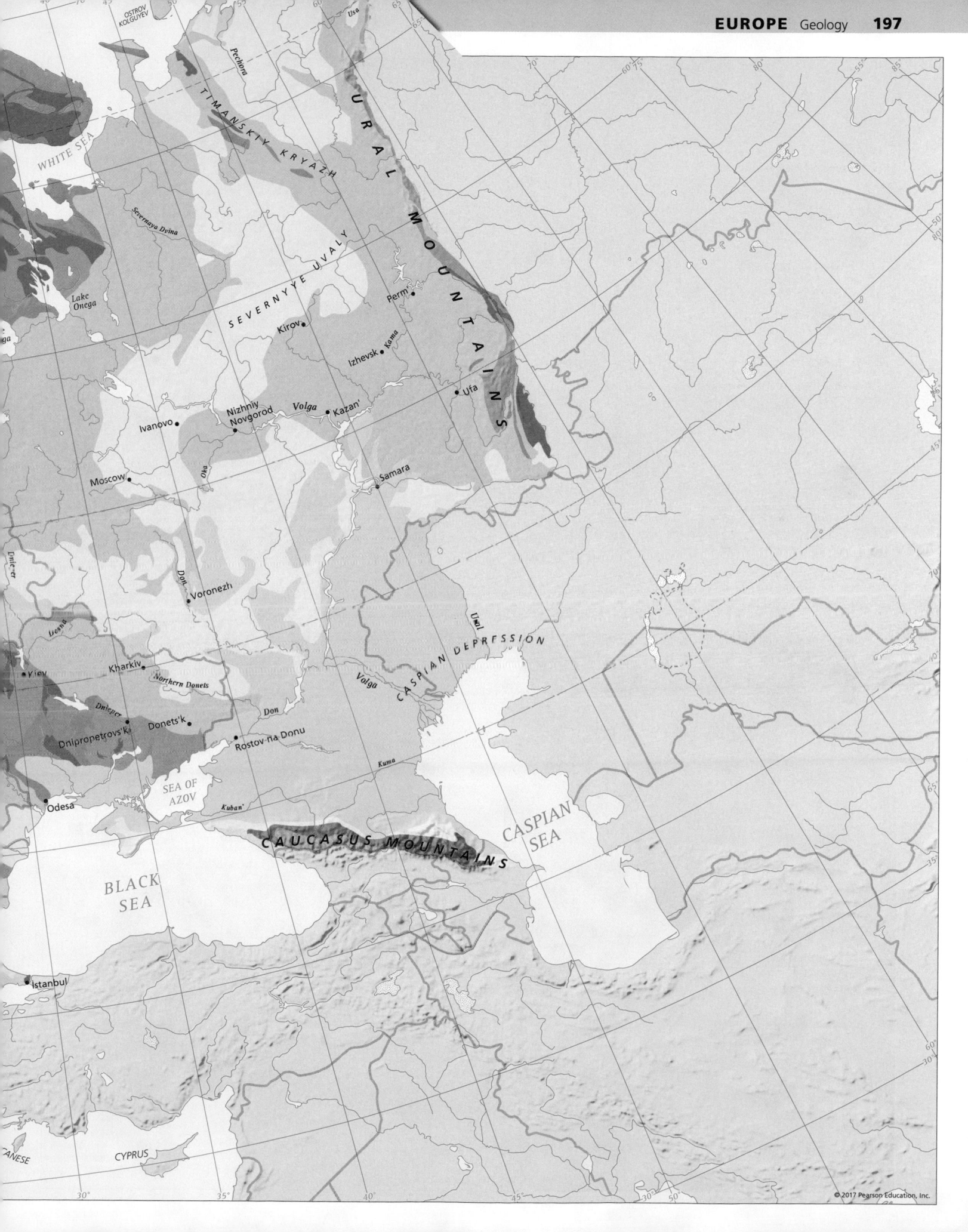

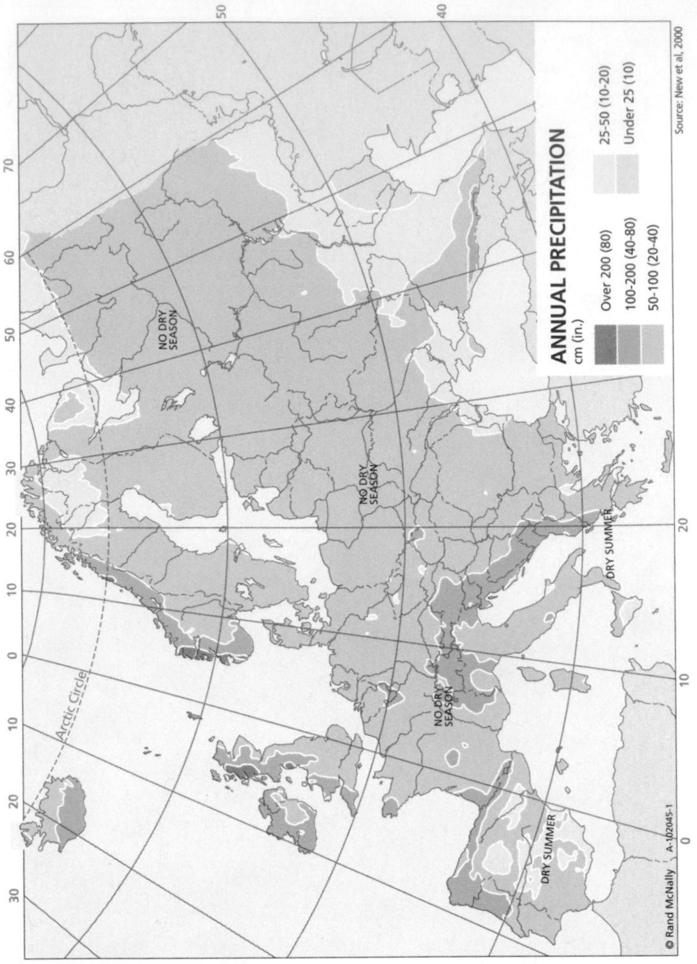

Moderate Resolution
Imaging Spectroradiometer (MODIS)
true-color mosaic satellite image

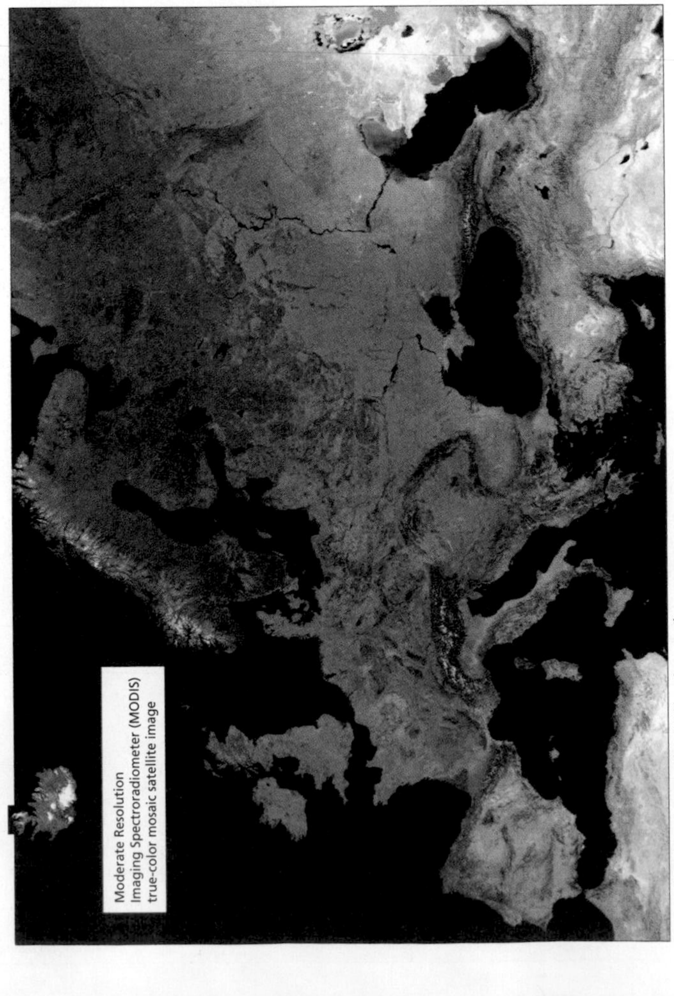

Source: NASA Visible Earth program (http://visibleearth.nasa.gov/)

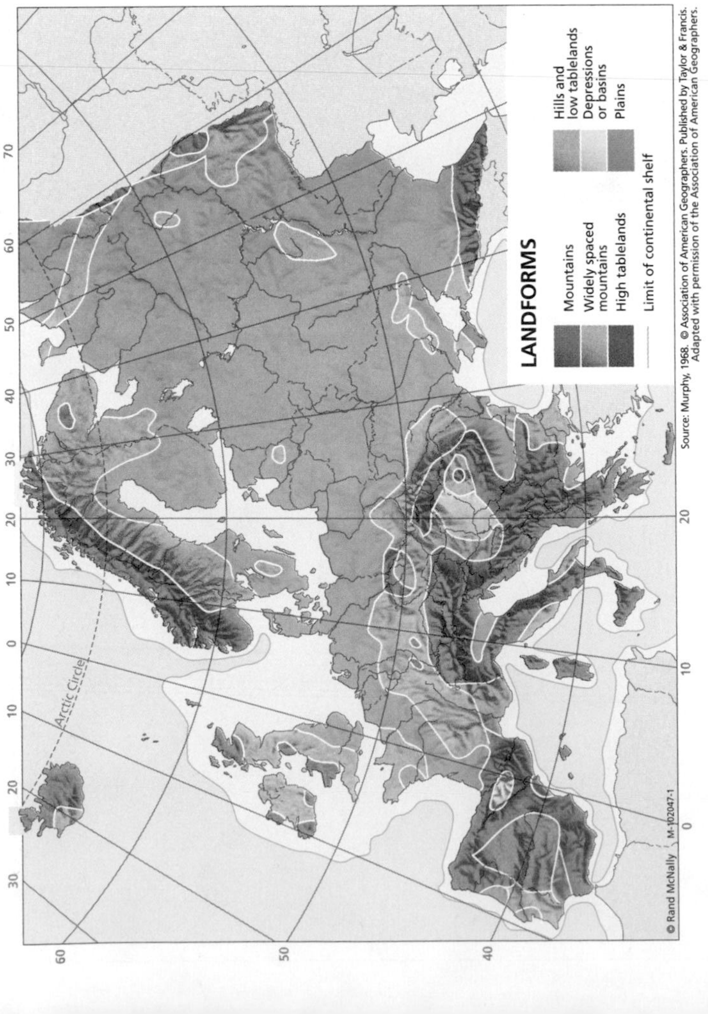

ANNUAL PRECIPITATION

cm (in.)

Over 200 (80)
100–200 (40–80)
50–100 (20–40)
25–50 (10–20)
Under 25 (10)

NO DRY SEASON
NO DRY SEASON
NO DRY SEASON
NO DRY SEASON
DRY SUMMER
DRY SUMMER
Arctic Circle

Source: New et al, 2000

© Rand McNally   A-\J02045-1

VEGETATION

E Coniferous forest
B,Bs Mediterranean vegetation
M Mixed forest; coniferous-deciduous
S Semideciduous forest
D Deciduous forest
DG Wooded steppe

G Grass (steppe)
Gp Short grass
Dsp Desert shrub
L Heath and moor
Alpine vegetation, tundra
b Little or no vegetation

For explanation of letters in boxes,
see World Natural Vegetation Map.

TAIGA
STEPPE
Arctic Circle

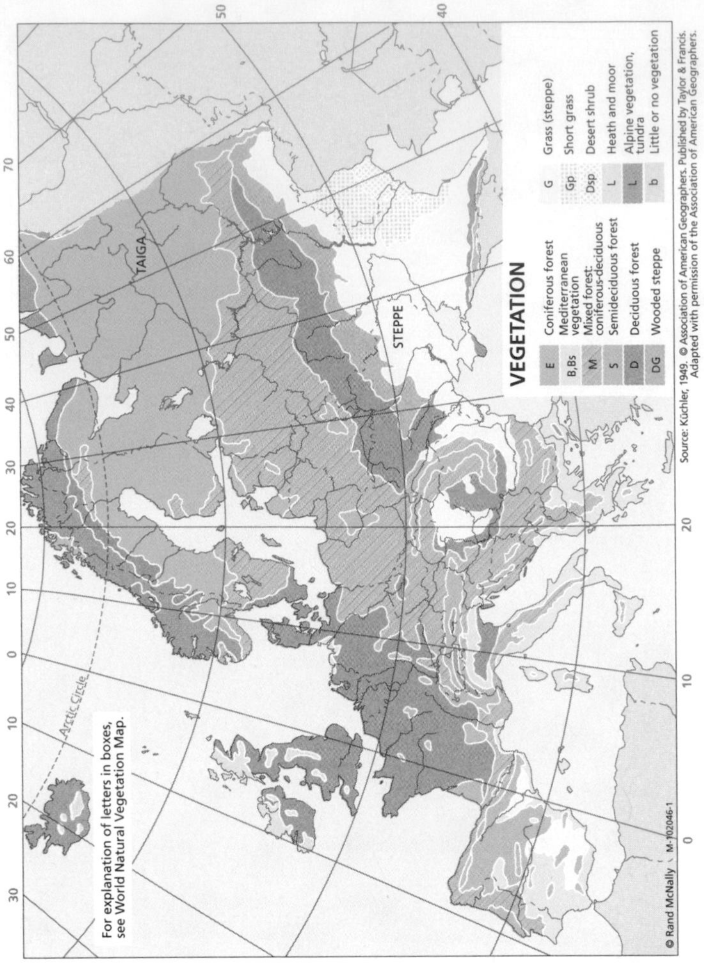

Source: Küchler, 1949. © Association of American Geographers. Published by Taylor & Francis.
Adapted with permission of the Association of American Geographers.

© Rand McNally   M-\J02046-1

LANDFORMS

Mountains
Widely spaced mountains
High tablelands
Limit of continental shelf

Hills and low tablelands
Depressions or basins
Plains

Arctic Circle

Source: Murphy, 1968. © Association of American Geographers. Published by Taylor & Francis.
Adapted with permission of the Association of American Geographers.

© Rand McNally   M-\J02047-1

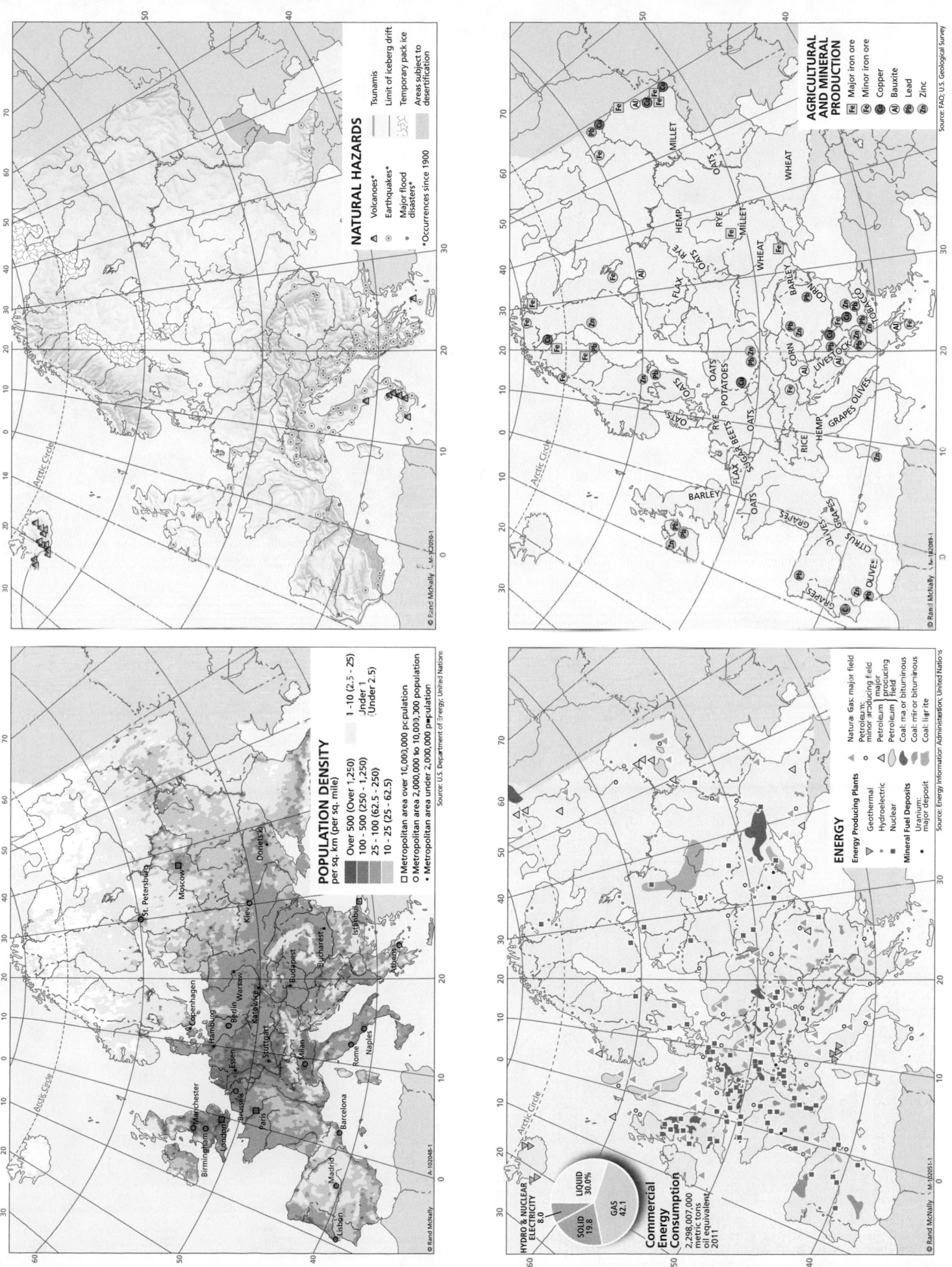

NATURAL HAZARDS

△ Volcanoes*
⊙ Earthquakes*
• Major flood disasters*

*Occurrences since 1900

Tsunamis

Limit of iceberg drift

Temporary pack ice

Areas subject to desertification

Source: U.S. Department of Energy; United Nations

AGRICULTURAL AND MINERAL PRODUCTION

Fe Major iron ore
Fe Minor iron ore
Cu Copper
Al Bauxite
Pb Lead
Zn Zinc

Source: FAO; U.S. Geological Survey

POPULATION DENSITY
per sq. km (per sq. mile)

Over 500 (Over 1,250)
100 - 500 (250 - 1,250)
25 - 100 (62.5 - 250)
10 - 25 (25 - 62.5)
1 - 10 (2.5 - 25)
Under 1 (Under 2.5)

□ Metropolitan area over 10,000,000 population
○ Metropolitan area 2,000,000 to 10,000,000 population
• Metropolitan area under 2,000,000 population

Source: U.S. Department of Energy; United Nations

ENERGY

Energy Producing Plants
▽ Geothermal
▲ Hydroelectric
△ Nuclear

Mineral Fuel Deposits
■ Uranium: major deposit

Natural Gas: major field

Petroleum: minor producing field
Petroleum: major producing field

Coal: major or bituminous
Coal: minor or bituminous
Coal: lignite

Source: Energy Information Administration; United Nations

Commercial Energy Consumption
2,298,007,000 metric tons oil equivalent
2011

HYDRO & NUCLEAR ELECTRICITY 8.0
SOLID 19.8
LIQUID 30.0%
GAS 42.1

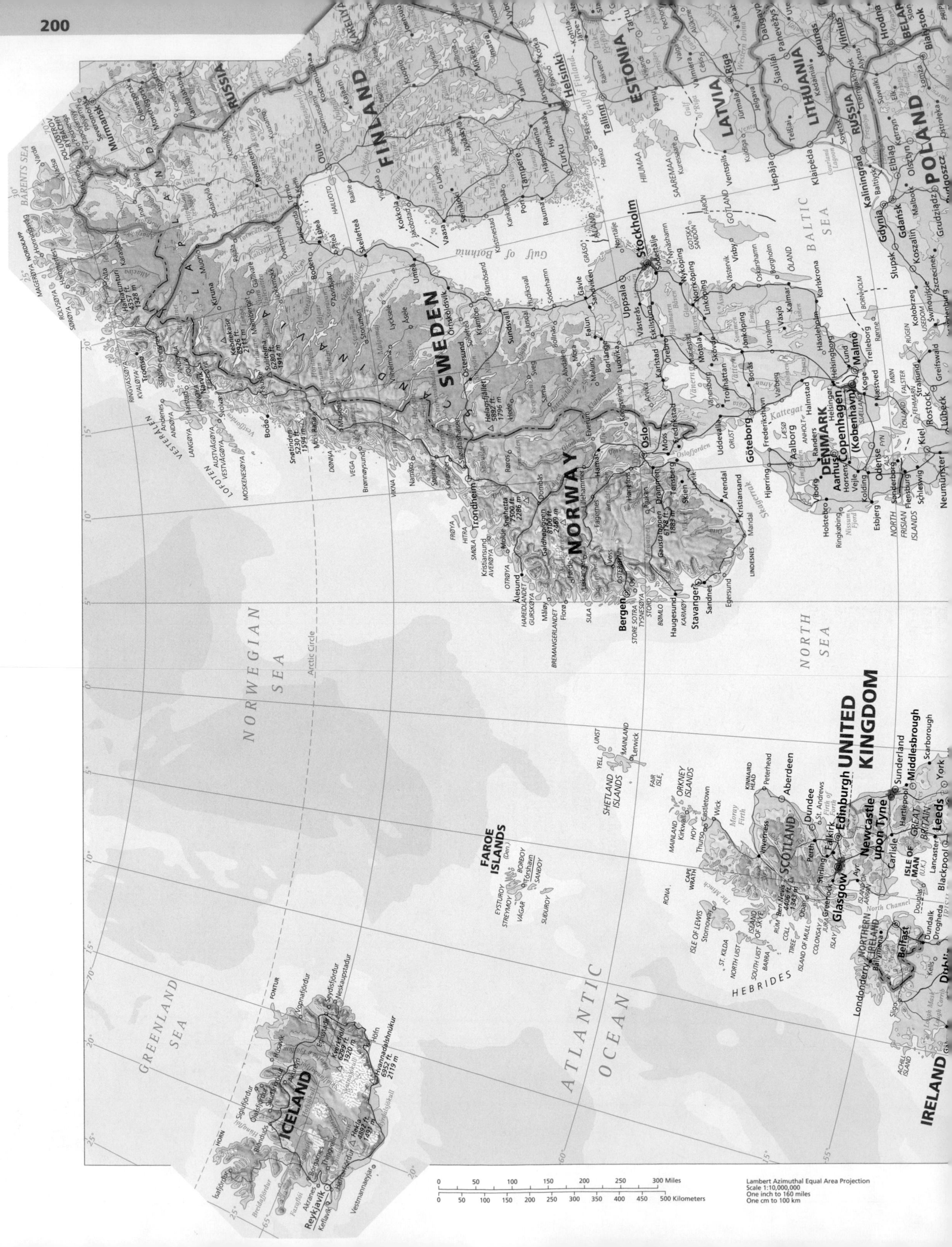

RUSSIA · FINLAND · SWEDEN · NORWAY · DENMARK · ESTONIA · LATVIA · LITHUANIA · POLAND · BELARUS

BARENTS SEA · Murmansk · KARELIYA · POLUOSTROV

Helsinki · Tampere · Turku · Tallinn · Riga · Vilnius · Kaliningrad · Gdańsk · Gdynia

NORWEGIAN SEA · Arctic Circle

Stockholm · Uppsala · Gulf of Bothnia · BALTIC SEA · GOTLAND

Oslo · Bergen · Stavanger · Trondheim · Ålesund · Kristiansand

DENMARK · Copenhagen (København) · Aarhus · Aalborg · Odense · Malmö · Göteborg

NORTH SEA · FAROE ISLANDS (Den.) · Tórshavn

SHETLAND ISLANDS · Lerwick · ORKNEY ISLANDS · Kirkwall

GREENLAND SEA

ICELAND · Reykjavík

ATLANTIC OCEAN

UNITED KINGDOM · SCOTLAND · Edinburgh · Glasgow · Aberdeen · Dundee · Inverness · Newcastle upon Tyne · Middlesbrough · Sunderland · Leeds · York · Blackpool · Lancaster · ISLE OF MAN (U.K.) · GREAT BRITAIN · HEBRIDES · ISLE OF LEWIS · ISLAND OF SKYE · ISLAND OF MULL

NORTHERN IRELAND · Belfast · Londonderry · IRELAND · Dublin

Lambert Azimuthal Equal Area Projection
Scale 1:10,000,000
One inch to 160 miles
One cm to 100 km

0 50 100 150 200 250 300 Miles
0 50 100 150 200 250 300 350 400 450 500 Kilometers

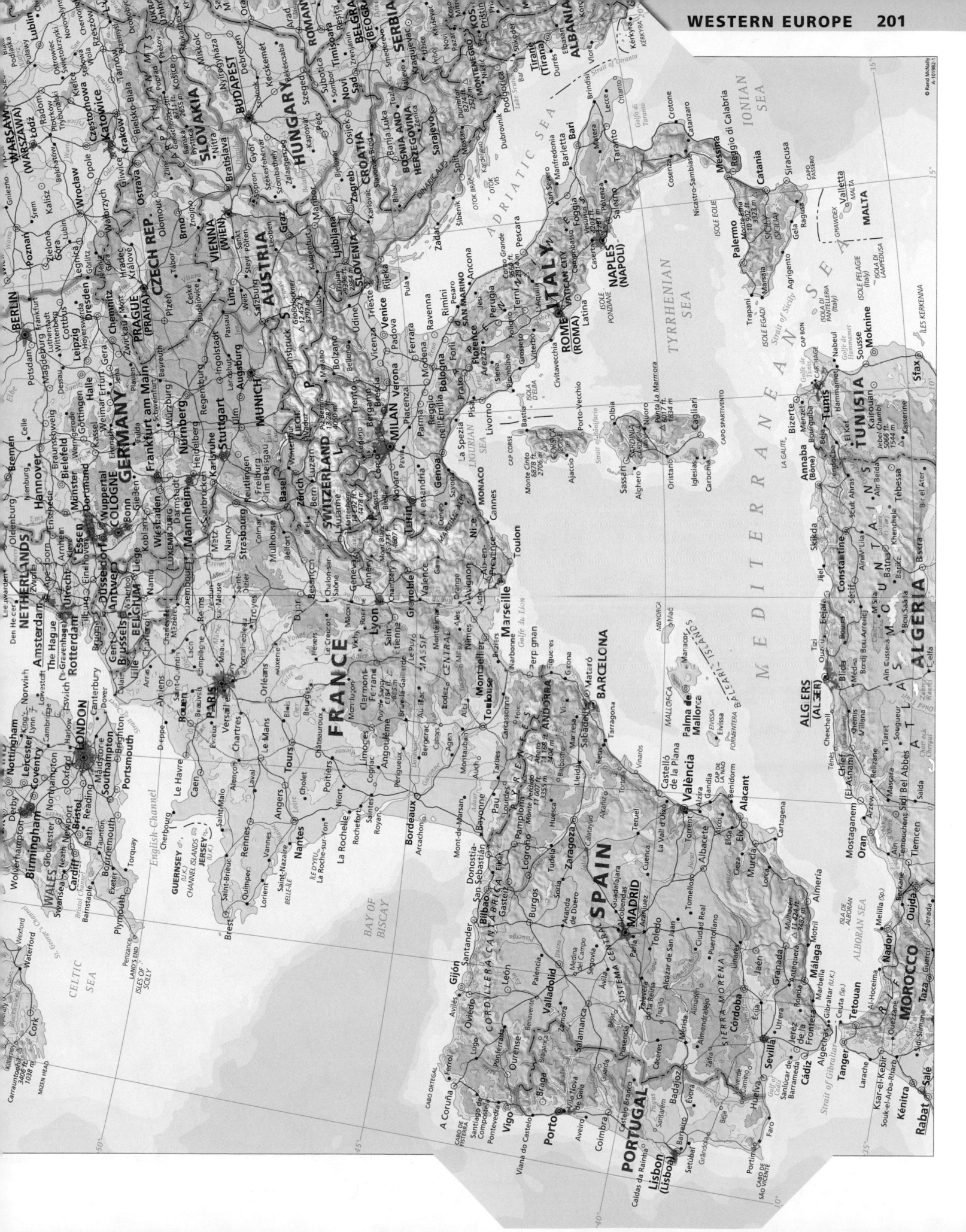

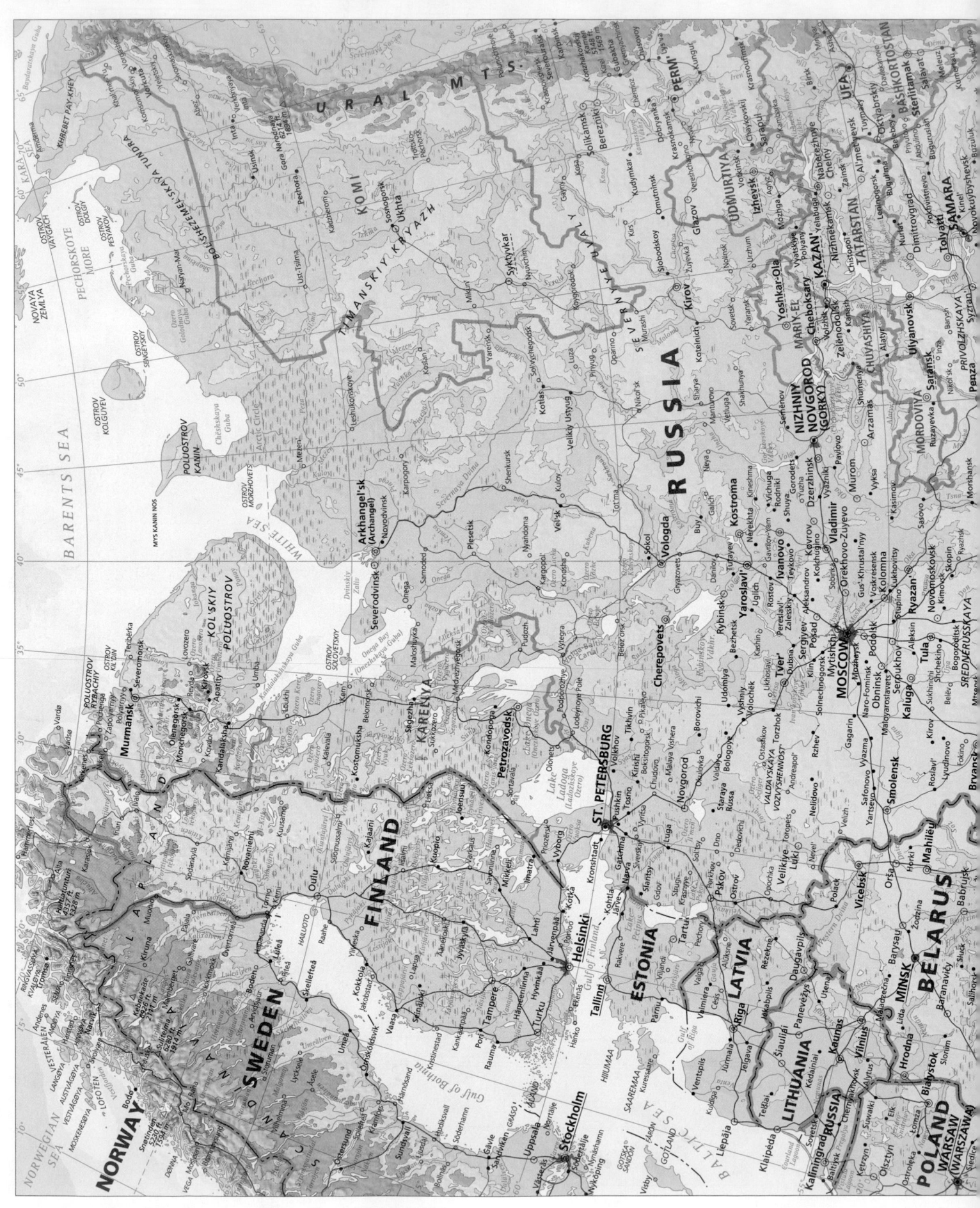

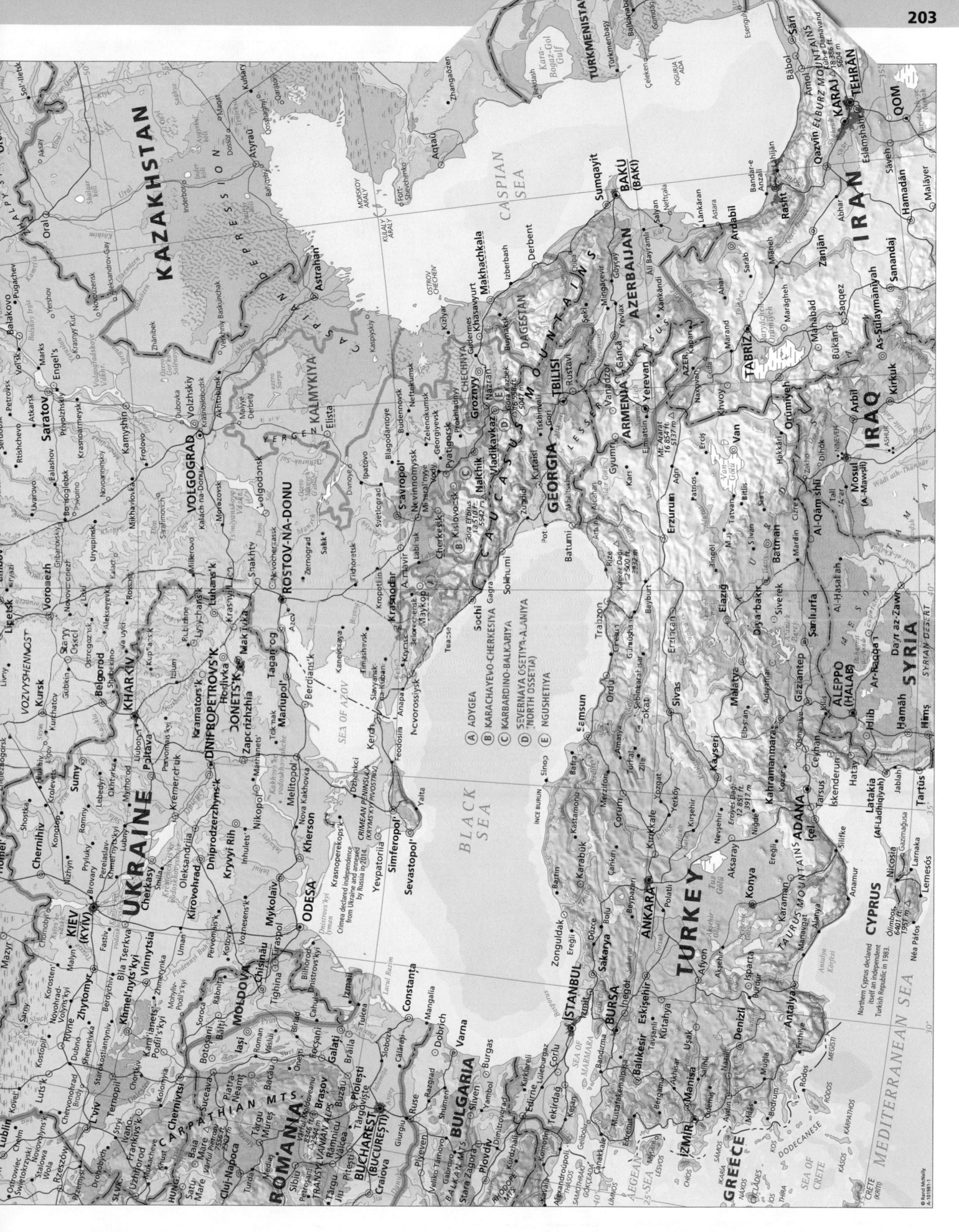

ATLANTIC OCEAN

BAY OF BISCAY

FRANCE

GERMANY

AUSTRIA

SWITZERLAND

SPAIN

PORTUGAL

ITALY

MOROCCO

ALGERIA

TUNISIA

MEDITERRANEAN SEA

TYRRHENIAN SEA

ALBORAN SEA

LIGURIAN SEA

ATLAS MOUNTAINS

GRAND ERG OCCIDENTAL

GRAND ERG ORIENTAL

TRIPOLITANIA

SAHARA

| 0 | 50 | 100 | 150 | 200 | 250 | 300 Miles |

| 0 | 50 | 100 | 150 | 200 | 250 | 300 | 350 | 400 | 450 | 500 Kilometers |

Lambert Azimuthal Equal Area Projection
Scale 1:10,000,000
One inch to 160 miles
One cm to 100 km

© Rand McNally
A-101877-1

a
GREENLAND SEA

Arctic Circle

Denmark Strait

HORN
RIFSTANGI
GRÍMSEY
Raufarhöfn
Kópasker
FONTUR
Þistilfjörður
Flateyri
Ísafjörður
Siglufjörður
Ólafsfjörður
Dalvík
Þórshöfn
Bakkaflói
Bakkafjörður
Skinnastaðir
Húsavík
Skagaströnd
Sauðárkrókur
Akureyri
Vopnafjörður
Vopnafjörður
Héraðsflói
Grímsstaðir
Borgarfjörður

Vatneyri
Hólmavík
Húnaflói
Herðubreið
5518 ft.
1682 m
Egilsstaðir
Seyðisfjörður
Neskaupstaður

BJARGTANGAR
Breiðafjörður
Askja
4954 ft.
1510 m
Eskifjörður
Reyðarfjörður
Búðir

ICELAND

Stykkishólmur
Búðardalur
Snæfell
6014 ft.
1833 m

SNÆFELLSNES
Hofsjökull
Kverkfjöll
6299 ft.
1920 m
Djúpivogur

Faxaflói
Langjökull
Vatnajökull
Höfn

Borgarnes
Grímsvötn
5640 ft.
1719 m
STOKKSNES

Akranes
Þingvellir
Hvítárvatn
Þórisvatn
Hvannadalshnúkur
6952 ft.
2119 m

Reykjavík
Hafnarfjörður
Kópavogur
Hveragerði
Hekla
4892 ft.
1491 m
Skarð
Kálfafell
Höfn

Keflavík
Selfoss
Kirkjubæjarklaustur
REYKJANES
Grindavík
Þorlákshöfn
Eyrarbakki
Hvolsvöllur
Ingólfshöfði

Vestmannaeyjar
HEIMAEY
Mýrdalsjökull

SURTSEY

ATLANTIC OCEAN

© Rand McNally
A-101769-1

b
NORWEGIAN SEA

Slættaratindur
2894 ft.
882 m
VIÐOY
FUGLOY
STREYMOY
SVÍNOY
VÁGAR
BORÐOY
EYSTUROY
MYKINES
Tórshavn
NÓLSOY

SANDOY
Húsavík
SKÚVOY
SUÐUROY

FAROE
ISLANDS
(Denmark)

ATLANTIC OCEAN

© Rand McNally
A-101763-1

c
SHETLAND
ISLANDS
(U.K.)
UNST
FETLAR
YELL
St. Magnus Bay
WHALSAY
Melby House
MAINLAND
ATLANTIC
Lerwick
BRESSAY
FOULA

OCEAN
SUMBURGH HEAD

FAIR ISLE

NORTH
RONALDSAY
WESTRAY
North
Sound
SANDAY
ROUSAY
STRONSAY
MAINLAND
SHAPINSAY
Stromness
Kirkwall
HOY
SOUTH RONALDSAY
ORKNEY
ISLANDS
(U.K.)
Burwick
Pentland Firth
DUNCANSBY HEAD
NORTH
Thurso
Castletown
SEA

© Rand McNally
A-101764-1

© Rand McNally
A-101759-1

Ireland / Northern Ireland area
Carloway
ISLE OF
LEWIS
Stornoway
SCARP
St. KILDA
HEBRIDES
TARANSAY
Tarbert
ENSAY
PABBAY
NORTH
UIST
Uig
ISLAND OF SKYE
RAASAY
Plockton
BENBECULA
Torridon
Kyle of
Lochalsh
SOUTH
UIST
Lochboisdale
ERISKAY
CANNA
SANDRAY
RUM
EIGG
Mallaig
BARRA
MINGULAY
SEA OF
THE HEBRIDES
COLL
TIREE
Tobermory
IONA
ISLE OF MULL
Firth of Lorn
SCARBA
COLONSAY
Lochgilphead
JURA
Sound of Jura
Port Askaig
ISLAY
Port Ellen
Campbeltown

MALIN
HEAD
Moville
RATHLIN
ISLAND
Ballycastle
Gaoth
Dobhair
An Earagail
2467 ft.
752 m
Buncrana
Londonderry
Coleraine
Ballymena
Larne
ÁRAINN
MHÓR
Letterkenny
Lifford
Strabane
Newtownabbey
CEANN ROS
EOGHAIN
Killybegs
NORTHERN
IRELAND
Antrim
Bangor
Donegal Bay
Dungannon
Belfast
Lisburn
Béal an
Mhuirthead
Killala
Ballina
Sligo
Lough
Erne
Enniskillen
Lough
Neagh
Lurgan
Armagh
Banbridge
Newry
Downpatrick
Newcastle
ACHILL
ISLAND
Bundoran
Monaghan
MOURNE
MTS.
Dundalk
Castlebar
Boyle
Carrick-on-
Shannon
Cavan
Dundalk Bay
CLARE ISLAND
Clew Bay
Westport
Swinford
Bailieborough
Ardee
Drogheda
INISHTURK
INISHBOFIN
Claremorris
Castlerea
Longford
Kells
Balbriggan
CONNEMARA
Ballinrobe
Roscommon
Lough Ree
Mullingar
Navan
Swords
Clifden
Tuam
Athlone
Edenderry
Maynooth
Dublin
(Baile Átha Cliath)
Galway
Ballinasloe
Royal Canal
Grand Canal
Dún Laoghaire
GARUMNA
Lough
Corrib
Clare
Galway Bay
Loughrea
Tullamore
Droichead Nua
Naas
Bray
INIS MÓR
INIS
MEÁIN
Birr
Roscrea
Athy
Wicklow
IRELAND
Ennis
Nenagh
Portlaoise
Carlow
Lugnaquillia
Mountain
3031 ft.
924 m
WICKLOW MTS.
LOOP
HEAD
Kilrush
Thurles
Kilkenny
Tullow
Gorey
Listowel
Newcastle
West
Tipperary
Cashel
Enniscorthy
AN BLASCAOD MÓR
An Daingean
Tralee
Rath Luirc
Carrick-
on-Suir
Clonmel
New Ross
Wexford
Rosslare
Dingle Bay
Castleisland
Mallow
Mitchelstown
Waterford
CARNSORE
POINT
VALENCIA ISLAND
Carrauntoohil
3406 ft.
1038 m
Killarney
Fermoy
Tramore
Dungarvan
Caherciveen
Kenmare
Macroom
Cork
Youghal
Passage
West
Cobh
DURSEY
ISLAND
Bantry
Clonakilty
Kinsale
Cork
Harbour
MIZEN
HEAD
CLEAR
ISLAND
Skibbereen
OLD HEAD
OF KINSALE

St. George's Channel
Fishguard
ST. DAVID'S
HEAD
St.
David's
Haverfordwest
Milford Haven
Pembroke

CELTIC
SEA

ATLANTIC
OCEAN

Camborne
Truro
ISLES OF
SCILLY
LAND'S END
Penzance
Falmouth
LIZARD
POINT

Scale bars
0 20 40 60 80 100 120 Miles
0 20 40 60 80 100 120 140 160 180 200 Kilometers

Lambert Conformal Conic Projection
Scale 1:4,000,000
One inch to 64 miles
One cm to 40 km

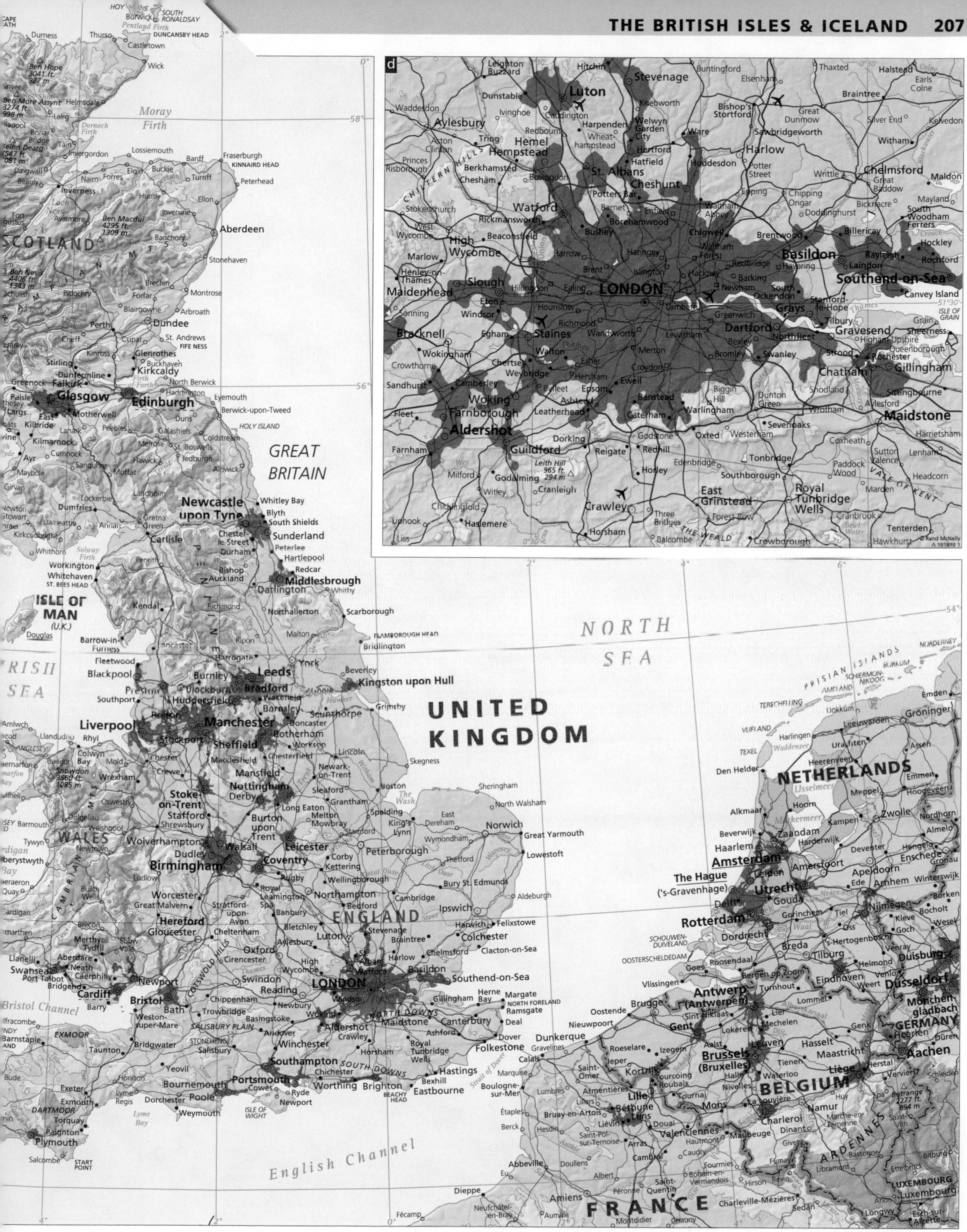

d

Inset map (London area)

Leighton Buzzard, Hitchin, Buntingford, Thaxted, Halstead, Earls Colne, Braintree, Silver End, Kelvedon, Witham, Chelmsford, Maldon, Luton, Stevenage, Knebworth, Bishop's Stortford, Great Dunmow, Writtle, Great Baddow, Mayland, South Woodham Ferrers, Dunstable, Ivinghoe, Redbourn, Caddington, Harpenden, Welwyn Garden City, Ware, Sawbridgeworth, Harlow, Hockley, Aylesbury, Waddesdon, Tring, Hemel Hempstead, Wheathampstead, Hatfield, Hoddesdon, Potter Street, Epping, Chipping Ongar, Doddinghurst, Bicknacre, Basildon, Rayleigh, Rochford, Aston Clinton, Berkhamsted, St. Albans, Cheshunt, Waltham Abbey, Brentwood, Billericay, Princes Risborough, Chesham, Bovingdon, Potters Bar, Barnet, Enfield, Southend-on-Sea, Stokenchurch, West Wycombe, Rickmansworth, Bushey, Borehamwood, Chigwell, Redbridge, Havering, Laindon, Canvey Island, High Wycombe, Beaconsfield, Watford, Harrow, Brent, Islington, Hackney, Barking, Newham, Grays, Tilbury, Gravesend, ISLE OF GRAIN, Marlow, Henley-on-Thames, Slough, Hillingdon, Ealing, LONDON, Greenwich, South Ockendon, Stanford-le-Hope, Grain, Sheerness, Queenborough, Maidenhead, Eton, Hounslow, Richmond, Lewisham, Bexley, Dartford, Northfleet, Strood, Rochester, Gillingham, Bracknell, Windsor, Staines, Wandsworth, Merton, Bromley, Swanley, Chatham, Sittingbourne, Aylesford, Maidstone, Sonning, Egham, Walton, Esher, Croydon, Biggin Hill, Dunton Green, Snodland, Wrotham, Wokingham, Chertsey, Weybridge, Byfleet, Epsom, Ewell, Banstead, Warlingham, Sevenoaks, Coxheath, Sutton Valence, Harrietsham, Crowthorne, Camberley, Hersham, Ashtead, Caterham, Oxted, Westerham, Paddock Wood, Lenham, Sandhurst, Woking, Leatherhead, Godstone, Redhill, Edenbridge, Tonbridge, Marden, Headcorn, Farnborough, Aldershot, Guildford, Dorking, Reigate, Royal Tunbridge Wells, Cranbrook, Tenterden, Farnham, Milford, Godalming, Leith Hill 965 ft 294 m, Horley, Southborough, East Grinstead, Hawkhurst, Witley, Cranleigh, Crawley, Forest Row, Bewl Water, Liphook, Chiddingfold, Three Bridges, Balcombe, Crowborough, Haslemere, Horsham, THE WEALD, VALE OF KENT, SOUTH DOWNS, Liss
© Rand McNally A 101840 1

Main map

SCOTLAND
Moray Firth
Ben Hope 3041 ft 927 m, Ben More Assynt 3274 ft 998 m, Beinn Dearg 3547 ft 1081 m, Ben Macdui 4295 ft 1309 m, Ben Nevis 4406 ft 1343 m
HOY, SOUTH RONALDSAY, Burwick, Pentland Firth, Duncansby Head, Durness, Thurso, Castletown, Wick, Helmsdale, Lairg, Dornoch Firth, Tain, Invergordon, Lossiemouth, Banff, Fraserburgh, KINNAIRD HEAD, Nairn, Forres, Elgin, Buckie, Turriff, Peterhead, Beauly, Inverness, Huntly, Ellon, Inverurie, Aberdeen, Aviemore, Banchory, Stonehaven, Pitlochry, Brechin, Montrose, Forfar, Arbroath, Blairgowrie, Dundee, Perth, Cupar, St. Andrews, FIFE NESS, Crieff, Kinross, Glenrothes, Buckhaven, Stirling, Dunfermline, Kirkcaldy, North Berwick, Firth of Forth, Greenock, Falkirk, Haddington, Paisley, Glasgow, Edinburgh, Eyemouth, East Kilbride, Motherwell, Peebles, Berwick-upon-Tweed, Largs, Lanark, Galashiels, Duns, Coldstream, HOLY ISLAND, Kilmarnock, Melrose, St. Boswells, Ayr, Cumnock, Hawick, Jedburgh, Alnwick, Maybole, Moffat, Langholm, GREAT BRITAIN, Lockerbie, Dumfries, Gretna Green, Annan, Newcastle upon Tyne, Whitley Bay, Kirkcudbright, Solway Firth, Carlisle, Chester-le-Street, Blyth, South Shields, Sunderland, Workington, Penrith, Bishop Auckland, Durham, Peterlee, Hartlepool, Whitehaven, ST. BEES HEAD, Darlington, Middlesbrough, Redcar, ISLE OF MAN (U.K.), Douglas, Kendal, Richmond, Northallerton, Whitby, Scarborough, Barrow-in-Furness, Ripon, Malton, Bridlington, FLAMBOROUGH HEAD, IRISH SEA, Lancaster, Harrogate, York, Beverley, Fleetwood, Burnley, Leeds, Kingston upon Hull, Blackpool, Preston, Blackburn, Bradford, Goole, Grimsby, Southport, Bolton, Huddersfield, Wakefield, Scunthorpe, Humber, NORTH SEA, Amlwch, Llandudno, Rhyl, Liverpool, Manchester, Stockport, Sheffield, Rotherham, Doncaster, Worksop, Lincoln, Skegness, ANGLESEY, Colwyn Bay, Mold, Chester, Macclesfield, Chesterfield, Caernarfon, Bangor, Wrexham, Crewe, Mansfield, Newark-on-Trent, Snowdon 3560 ft 1085 m, Oswestry, Stoke-on-Trent, Nottingham, Derby, Sleaford, Boston, Sheringham, Pwllheli, WALES, Stafford, Shrewsbury, Burton upon Trent, Long Eaton, Melton Mowbray, Spalding, The Wash, North Walsham, Tywyn, Newtown, Wolverhampton, Walsall, Leicester, Corby, King's Lynn, East Dereham, Norwich, Great Yarmouth, Aberystwyth, Builth Wells, Dudley, Birmingham, Coventry, Kettering, Peterborough, Wymondham, Lowestoft, CAMBRIAN MTS, Ludlow, Royal Leamington Spa, Rugby, Wellingborough, Northampton, Thetford, Bury St. Edmunds, Aldeburgh, Cardigan, Worcester, Stratford-upon-Avon, Banbury, Bedford, Cambridge, Ipswich, Stour, Felixstowe, Carmarthen, Brecon, Hereford, Great Malvern, Cheltenham, ENGLAND, Bletchley, Northampton, Harwich, Colchester, Clacton-on-Sea, Llanelli, Ebbw Vale, Gloucester, Cirencester, Oxford, Aylesbury, Luton, Stevenage, Braintree, Chelmsford, Merthyr Tydfil, Aberdare, COTSWOLD HILLS, High Wycombe, St. Albans, Harlow, Southend-on-Sea, Swansea, Neath, Caerphilly, Newport, Chippenham, Swindon, Reading, LONDON, Basildon, Gillingham, Herne Bay, Margate, Port Talbot, Bridgend, Cardiff, Barry, Bristol, Bath, Trowbridge, Newbury, Windsor, Woking, NORTH DOWNS, Maidstone, Canterbury, Ramsgate, NORTH FORELAND, Bristol Channel, Weston-super-Mare, Basingstoke, Andover, Aldershot, Crawley, Horsham, Royal Tunbridge Wells, Ashford, Deal, Dover, LUNDY, EXMOOR, Bridgwater, Salisbury, Winchester, SALISBURY PLAIN, STONEHENGE, Chichester, Hastings, Folkestone, Ilfracombe, Taunton, Yeovil, Southampton, Worthing, Brighton, Bexhill, Eastbourne, Barnstaple, Dorchester, SOUTH DOWNS, BEACHY HEAD, Bude, Exeter, Exmouth, Lyme Regis, Bournemouth, Poole, Portsmouth, Cowes, Ryde, Newport, DARTMOOR, Torquay, Paignton, Weymouth, ISLE OF WIGHT, Plymouth, Salcombe, START POINT, Lyme Bay, English Channel

NETHERLANDS
FRISIAN ISLANDS, NORDERNEY, BORKUM, SCHIERMONNIKOOG, AMELAND, TERSCHELLING, Emden, Groningen, VLIELAND, Dokkum, Leeuwarden, Assen, TEXEL, Harlingen, Drachten, Heerenveen, Meppel, Hoogeveen, Nordhorn, Den Helder, Waddenzee, IJsselmeer, Kampen, Zwolle, Almelo, Gronau, Alkmaar, Markermeer, Lelystad, Deventer, Enschede, Winterswijk, Beverwijk, Zaandam, Harderwijk, Apeldoorn, Arnhem, Bocholt, Haarlem, Amsterdam, Leiden, Amersfoort, Ede, Nijmegen, Kleve, Wesel, The Hague ('s-Gravenhage), Utrecht, Gouda, Tiel, Oss, 's-Hertogenbosch, Venray, Kevelaer, Delft, Rotterdam, Gorinchem, Waal, Duisburg, SCHOUWEN-DUIVELAND, Dordrecht, Breda, Tilburg, Helmond, Venlo, Weert, Düsseldorf, OOSTERSCHELDEDAM, Roosendaal, Bergen op Zoom, Eindhoven, München-Gladbach, GERMANY, Vlissingen, Goes, Turnhout, Lommel, Genk, Heerlen, Aachen, Oostende, Brugge, Antwerp (Antwerpen), Sint-Niklaas, Lier, Mechelen, Leuven, Hasselt, Maastricht, Herstal, Liège, Nieuwpoort, Gent, Lokeren, Aalst, Brussels (Bruxelles), Tienen, Verviers, Bitburg, Roeselare, Izegem, Kortrijk, Nivelles, Huy, Marche-en-Famenne, BELGIUM, Ieper, Roubaix, Lille, Tournai, Mons, La Louvière, Charleroi, Namur, Saint-Vith, Bastogne, Dunkerque, Gravelines, Calais, Armentières, Béthune, Lens, Douai, Valenciennes, Maubeuge, Dinant, Givet, ARDENNES, Boulogne-sur-Mer, Saint-Omer, Bruay-en-Artois, Liévin, Arras, Cambrai, Caudry, Fourmies, Hirson, Libramont, LUXEMBOURG, Luxembourg, Étaples, Hesdin, Saint-Pol-sur-Ternoise, Berck, Doullens, Albert, Bohain-en-Vermandois, Fumay, Revin, Arlon, Esch-sur-Alzette, Longwy, Dieppe, Eu, Abbeville, Amiens, Péronne, Saint-Quentin, Charleville-Mézières, Sedan, Bitburg, Fécamp, Neufchâtel-en-Bray, Aumale, Montdidier, Laon, FRANCE, Strait of Dover, Marquise, Pas de Calais, UNITED KINGDOM

Inset map d
Lambert Conformal Conic Projection
Scale 1:1,000,000
One inch to 16 miles
One cm to 10 km

NORWAY

SWEDEN

DENMARK

GERMANY

POLAND

NORTH SEA

BALTIC

Bergen

Oslo

Stockholm

Göteborg

Copenhagen
(København)

Malmö

Aalborg

Aarhus

HAMBURG

Gdynia

Gdańsk
(Danzig)

Skagerrak

Kattegat

0   20   40   60   80   100   120 Miles
0   20 40 60 80 100 120 140 160 180 200 Kilometers

Lambert Conformal Conic Projection
Scale 1:4,000,000
One inch to 64 miles
One cm to 40 km

FINLAND

Kristinestad
Haapamäki
Keuruu
Virrat
Jyväskylä
Mikkeli
Savonlinna
Kharlu
Ozero
Pryazh'
Merikarvia
Kankaanpää
Ikaalinen
Parkano
Mänttä
Joutsa
Otava
Parikkala
Lähdenpohj'ya
Pitkäranta
Salmi
KARELIYA
Mäntyluoto
Tampere
Kangasala
Jämsä
Syysmä
Mäntyharju
Valaam
OSTROV
VALAAM
OSTROV
MANTSINSARI
Il'inskiy
Vidlitsa
Pori
Nokia
Kuhmoinen
Heinola
Imatra
Elisenvaara
Priozërsk
Lake
Ladoga
(Ladozhskoye Ozero)
Olonets
Podporozh'ye
of Bothnia
Rauma
Säkylä
Pyhäjärvi
Lahti
Kuusankoski
Lappeenranta
Svetogorsk
Tesogorskiy
Kamennogorsk
Zaostrov'ye
Novaya
Ladoga
Svyas'stroy
Lodeynoye
Pole
Uusikaupunki
Laitila
Loimaa
Riihimäki
Hyvinkää
Myrskylä
Taavetti
Seleznëvo
Viborg
Sovetskiy
Primorsk
Vaskelovo
Volkhov
PYHÄMAA
Naantali
Paimio
Vihti
Korso
Porvoo
Kotka
Hamina
Vyborgskiy
Zaliv
Zelenogorsk
Sestroretsk
Vsevolozhsk
Tikhvin
Turku
Pargas
Salo
Lohja
Espoo
Vantaa
Kronshtadt
LENINGRADSKAYA
OBLAST'
ST. PETERSBURG
Naziya
Kirishi
Kukuy
Budogoshch'
Kimito
Kirkkonummi
Karis
Ekenäs
OSTROV
GOGLAND
OSTROV
MOSHCHNYY
Sosnovyy Bor
Lomonosov
Petrodvorets
Krasnoye Selo
Pushkin
Kolpino
Tosno
Nebolchi
Mariehamn
Hanko
Gulf of Finland
Narva
laht
Vistino
Kotly
Gatchina
Vyritsa
Lyuban
Malaya
Vishera
ÅLAND
Långnäs
NAISSAAR
Loksa
Kunda
Kohtla-
Järve
Sillamäe
Narva
Kingisepp
Slepino
Chudovo
ÅLAND
KÖKAR
SEA
Tallinn
Maardu
Kehra
Rakvere
Kiviõli
Nangorod
Siverskiy
Nesterkovo
Tesovo-
Netyl'skiy
Tesovskiy
Mata
Paldiski
Keila
Kohila
Rapla
Tapa
Tamsalu
Mustvee
Kozlov
Bereg
RUSSIA
Tolmachëvo
Luga
Novgorod
Krestsy
Viny
OSMUSSAAR
Risti
Marjamaa
Paide
Jõgeva
Kallaste
Lyady
Plyussa
Batetskiy
NOVGORODSKAYA
OBLAST'
VORMSI
Haapsalu
Vändra
Võhma
Põltsamaa
Lake
Peipus
Gdov
Yamm
Strugi
Krasnyye
Sol'tsy
Shimsk
Staraya
Russa
Parfino
Lychkovo
HIIUMAA
Kärdla
Soela väin
MUHU
Kuivastu
Pärnu
Jaagupi
Sindi
Viljandi
Abja-
Paluoja
Elva
Otepää
Põlva
Ostrov
Panina
Pechury
Pavy
Porkhov
Dno
Demyansk
GOTSKA
SANDÖN
Mustjala
SAAREMAA
Kuressaare
Kihelkonna
Salme
ABRUKA
SAAR
KIHNU
Kilingi-
Nõmme
Tõrva
Antsla
Võru
Pskov
Karamyshevo
Dedovichi
Zheluch'ye
Mārevo
FÅRÖN
RUHNU
SAAR
Ainaži
Jürjena
Valka
Valga
Paikuli
PSKOVSKAYA
OBLAST'
Khilm
Sopki
Kolka
KOLKASRAGS
Gulf
of Riga
Salacgrīva
Galgauska
Valmiera
Limbaži
Smiltene
Alūksne
Ostrov
Bezhely
Kruttsy
Chikhachevo
Krasnyy
Luch
Ploskosh'
VALDAYSKAYA
VOZVYSHENNOST'
Andreapol'
Toropets
TVERSKAYA
OBLAST'
Ventspils
Dundaga
Saulkrasti
Ēsis
Gulbene
Balvi
Vilaka
Pytalovo
Pushkinskiye
Gory
Novorzhev
Lokhnya
SEA
Talsi
Engure
Jaunpiebalga
LATVIA
Madona
Viļāni
Ludza
Opochka
Nasva
Novosokol'niki
Pustoshka
Velikiye
Luki
Zapadnaya
Dvina
Pavilosta
AKMEŅRAGS
Alsunga
Kuldīga
Tukums
Jūrmala
Riga
Ērgļi
Preiļi
Mozuli
Nevel
Plakstino
Zhizhitsa
Ostrov
Dvin'ye
ZSkrunda
Saldus
Jelgava
Aizkraukle
Plavinas
Jēkabpils
Rēzekne
Kuznetsovka
Idritsa
SMOLENSKAYA
OBLAST'
Liepāja
Durbe
Priekule
Auce
Bēne
Eleja
Bauska
Preili
Raznas
ezers
Asveja
Rayony
Zabor'ye
Ezjaryšča
Velizh
Selezni
Rucava
Mažeikiai
Joniškis
Birži
Nereta
Subate
Daugavpils
Druja
Verhnjadzvinsk
Navapolack
Polack
Suraž
Demidov
Skuodas
Telšiai
Kuršėnai
Pasvalys
Rokiškis
Zarasai
Druksiai
Brāslau
Dzisna
Obal'
Sumilina
Janavičy
Rudnya
Klaipėda
Plungė
Šiauliai
Radviliškis
Kupiškis
Anykščiai
Utena
Vidzy
Šarkauščyna
Disna
Vetryna
Vičebsk
Lëzna
Besankovičy
Baltušeysk
Gargždai
Rietavas
Kelmė
Panevėžys
Moletai
Ignalina
Varapayeva
VICEBSK
Usačy
Šianno
Orša
LITHUANIA
Šilalė
Skaudvilė
Raseiniai
Kėdainiai
Ukmergė
Svenčionėliai
Pastavy
Hlybokae
Lepel
Čašniki
Arehausk
Krasnyy
Courland
Lagoon
Šilutė
Tauragė
Jurbarkas
Vilkija
Jonava
Širvintos
Nemenčinė
Svir
Kryvičy
Mjadzel
Bjahoml
Novalukoml
Dubrouna
Svetlogorsk
Zelenogradsk
Sovetsk
Neman
Šakiai
Kaišiadorys
Vievis
Vilnius
Astravec
Vilejka
Smarhon
Maladzečna
Lahojsk
Baran'
Gulf of
Gdansk
Kaliningrad
Krasnoznamensk
Kazlu
Rūda
Garliava
Prienai
Lentvaris
Pleščanicy
Krupki
Horki
Svetlyy
Gvardeysk
Nesterov
Gusev
Vilkaviškis
Marijampole
Alytus
Šalčininkai
Varėna
Ašmjany
Smarhon
Radaškovičy
Talačyn
Baltiysk
KALININGRADSKAYA OBLAST'
Chernyakhovsk
Zheleznodorozhnyy
Ozërsk
Eišiškes
Radun
Višneva
Pjarša
MINSK
Berazino
Červen'
Sklou
Čavusy
RUSSIA
Polessk
Branievo
Bartoszyce
Ketrzyn
Giżycko
Suwałki
Augustów
Lazdijai
Druskininkai
Lida
Astryna
Dzjaržynsk
Neharelae
MAHILËU
Kirausk
Karma
Elblag
Pasłek
Lidzbark
Warmiński
Biskupiec
Elk
HRODNA
Hrodna
Ščučyn
Navahrudak
Stoŭbcy
Svislač
Bychau
Slauharad
Olsztyn
MAZURY
Rand McNally
A-101765-1
HOMEL

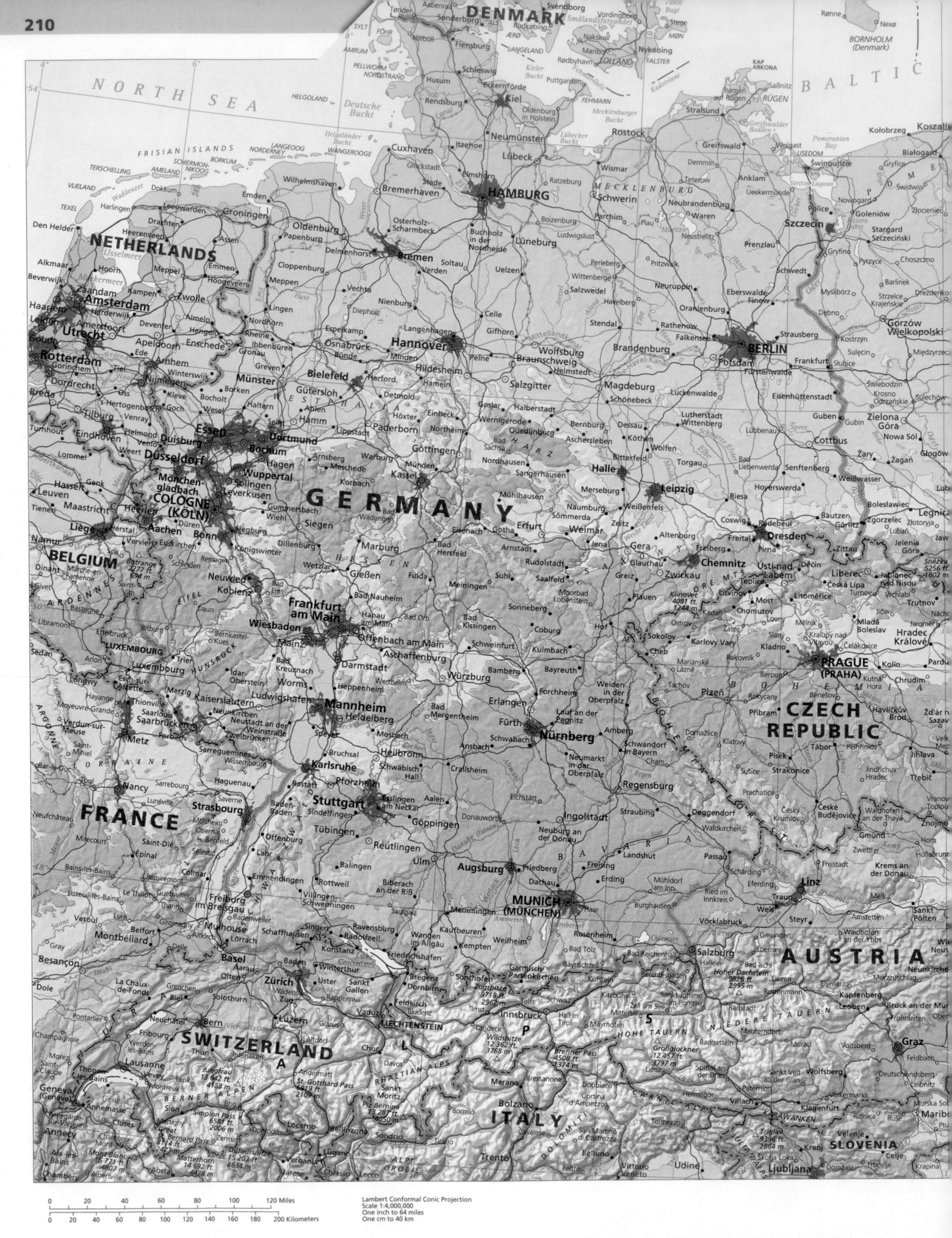

0   20   40   60   80   100   120 Miles

0 20 40 60 80 100 120 140 160 180 200 Kilometers

Lambert Conformal Conic Projection
Scale 1:4,000,000
One inch to 64 miles
One cm to 40 km

CELTIC SEA

UNITED KINGDOM

English Channel

BAY OF BISCAY

FRANCE

SPAIN

0   20   40   60   80   100   120 Miles
0   20   40   60   80 100 120 140 160 180 200 Kilometers

Lambert Conformal Conic Projection
Scale 1:4,000,000
One inch to 64 miles
One cm to 40 km

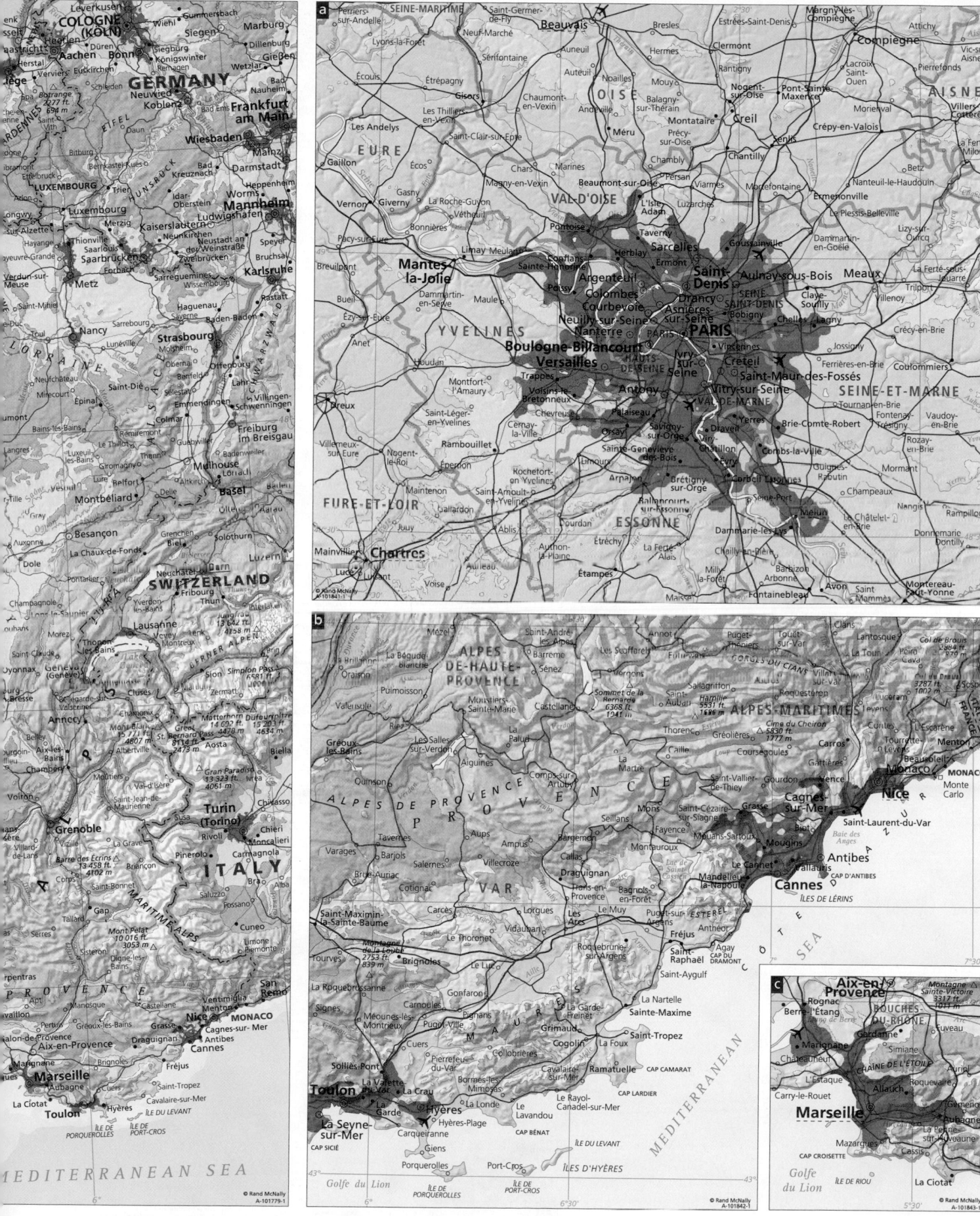

a

SEINE-MARITIME

Perriers-sur-Andelle · Saint-Germer-de-Fly · Beauvais · Bresles · Estrées-Saint-Denis · Margny-lès-Compiègne · Compiègne · Attichy

Lyons-la-Forêt · Neuf-Marché · Hermes · Clermont · Nogent-sur-Oise · Vic-sur-Aisne

Écouis · Étrépagny · Chaumont-en-Vexin · Auneuil · Noailles · Mouy · Balagny-sur-Thérain · Creil · Crépy-en-Valois · Pierrefonds · AISNE

OISE · Rantigny · Senlis · Morienval · Villers-Cotterêts

Gaillon · Les Thilliers-en-Vexin · Andelys · Montataire · Précy-sur-Oise · Nanteuil-le-Haudouin · La Ferté-Milon

EURE · Gasny · Chars · Marines · Persan · Viarmes · Montfermeil · Betz

Vernon · Giverny · La Roche-Guyon · Magny-en-Vexin · Beaumont-sur-Oise · Chantilly · Ermenonville

Bonnières · Vétheuil · VAL-D'OISE · L'Isle-Adam · Luzarches · Le Plessis-Belleville · Lizy-sur-Ourcq

Pacy-sur-Eure · Limay · Meulan · Pontoise · Taverny · Herblay · Sarcelles · Goussainville · Dammartin-en-Goële

Mantes-la-Jolie · Conflans-Sainte-Honorine · Ermont · Saint-Denis · Aulnay-sous-Bois · Meaux · La Ferté-sous-Jouarre · Trilport

Breuilpont · Dammartin-en-Serve · Maule · Argenteuil · Colombes · Drancy · SEINE-SAINT-DENIS · Claye-Souilly · Chelles · Lagny · Villenoy

Bueil · Ézy-sur-Eure · YVELINES · Poissy · Courbevoie · Asnières-sur-Seine · Bobigny · Crécy-en-Brie

Anet · Neuilly-sur-Seine · Nanterre · PARIS · Vincennes · Jossigny · Coulommiers

Houdan · Boulogne-Billancourt · HAUTS-DE-SEINE · Ivry-sur-Seine · Créteil · Saint-Maur-des-Fossés · Ferrières-en-Brie

Dreux · Montfort-l'Amaury · Versailles · Antony · VAL-DE-MARNE · Vitry-sur-Seine · Tournan-en-Brie · Vaudoy-en-Brie

Saint-Léger-en-Yvelines · Trappes · Voisins-le-Bretonneux · Palaiseau · Yerres · Brie-Comte-Robert · Fontenay-Trésigny

Villemeux-sur-Eure · Cernay-la-Ville · Orsay · Savigny-sur-Orge · Viry-Châtillon · Dravel · Combs-la-Ville · Rozay-en-Brie

Nogent-le-Roi · Rambouillet · Chevreuse · Sainte-Geneviève-des-Bois · Évry · Guignes · Mormant

EURE-ET-LOIR · Maintenon · Rochefort-en-Yvelines · Limours · Arpajon · Brétigny-sur-Orge · Corbeil-Essonnes · Champeaux · Nangis · Rampillon

Épernon · Ablis · Dourdan · Author-la-Plaine · ESSONNE · Ballancourt-sur-Essonne · Seine-Port · Melun · Le Châtelet-en-Brie · Donnemarie-Dontilly

Mainvilliers · Gallardon · Étréchy · La Ferté-Alais · Dammarie-les-Lys · Barbizon · Chailly-en-Bière

Chartres · Luce · Voise · Auneau · Milly-la-Forêt · Arbonne · Avon · Saint-Mammès · Montereau-Faut-Yonne

Jouy · Luisant · Étampes · Fontainebleau · Maisse

© Rand McNally
A-101841-1

b

Mézel · Saint-André-les-Alpes · Annot · Puget-Théniers · Touët-sur-Var · Clans · Lantosque · Col de Braus

ALPES-DE-HAUTE-PROVENCE · Barrême · Les Scaffarels · Villars-sur-Var · La Tour · Pairo · Cava

Oraison · La Bégude-blanche · Sénez · Saint-Auban · Sallagriffon · Roquestéron · GORGES DU CIANS · ALPES-MARITIMES · Sospel

Valensole · Moustiers-Sainte-Marie · Castellane · Sommet de la Bernarde 6368 ft. 1941 m · Hamille 5531 ft. 1686 m · Cime du Cheiron 5830 ft. 1777 m · Levens · Contes · Escarène

Gréoux-les-Bains · Riez · La Palud · Thorenc · Gréolières · Carros · Tourrette-Levens · Menton

Quinson · Les Salles-sur-Verdon · Aiguines · Caille · Coursegoules · Loup · Gattières · Beaulieu

ALPES DE PROVENCE · Comps-sur-Artuby · La Martre · Saint-Vallier-de-Thiey · Gourdon · Grasse · Cagnes-sur-Mer · Monaco · MONACO

PROVENCE · Mons · Saint-Cézaire-sur-Siagne · Mouans-Sartoux · Biot · Nice · Monte Carlo

Tavernes · Aups · Ampus · Bargemon · Fayence · Mougins · Antibes · Saint-Laurent-du-Var

Varages · Barjols · Salernes · Villecroze · Callas · Montauroux · Le Cannet · Vallauris · CAP D'ANTIBES

Cotignac · Bras · Bruc-Auriac · Draguignan · Bagnols-en-Forêt · Mandelieu-la-Napoule · Cannes · ÎLES DE LÉRINS

VAR · Carcès · Lorgues · Les Arcs · Le Muy · Puget-sur-Argens · ESTÉREL

Saint-Maximin-la-Sainte-Baume · Le Thoronet · Vidauban · Fréjus · Anthéor · CÔTE

Tourves · Montagne de la Loube 2753 ft. 839 m · Brignoles · Le Luc · Roquebrune-sur-Argens · Saint-Raphaël · CAP DU DRAMONT · Agay · SEA

La Roquebrussanne · Gonfaron · Saint-Aygulf

Signes · Méounes-lès-Montrieux · Carnoules · Pignans · La Garde-Freinet · La Nartelle · Sainte-Maxime

Puget-Ville · Grimaud · Saint-Tropez · CÔTE D'AZUR

Cuers · Pierrefeu-du-Var · Collobrières · Cogolin · La Foux · Ramatuelle · CAP CAMARAT · MEDITERRANEAN

Solliès-Pont · La Valette-du-Var · Bormes-les-Mimosas · Cavalaire-sur-Mer · CAP LARDIER

Toulon · La Crau · Le Londe · Le Rayol-Canadel-sur-Mer · CAP BÉNAT

La Seyne-sur-Mer · La Garde · Hyères · Hyères-Plage · Le Lavandou · ÎLE DU LEVANT

CAP SICIÉ · Carqueiranne · Giens · Porquerolles · Port-Cros · ÎLES D'HYÈRES

Golfe du Lion · ÎLE DE PORQUEROLLES · ÎLE DE PORT-CROS

© Rand McNally
A-101842-1

c

Aix-en-Provence · Montagne Sainte-Victoire 3317 ft. 1011 m

Rognac · BOUCHES-DU-RHÔNE

Berre-l'Étang · Gardanne · Fuveau

Étang de Berre · Simiane · Auriol

Marignane · Châteauneuf · CHAÎNE DE L'ÉTOILE · Roquevaire

L'Estaque · Allauch · Gémenos

Carry-le-Rouet · Marseille · La Penne-sur-Huveaune · Aubagne

Mazargues · La Ciotat

CAP CROISETTE · Golfe du Lion · ÎLE DE RIOU · Cassis

© Rand McNally
A-101843-1

Inset maps a - c
Lambert Conformal Conic Projection
Scale 1:1,000,000
One inch to 16 miles
One cm to 10 km

GERMANY · COLOGNE (KÖLN) · Frankfurt am Main · Wiesbaden · Mainz · Darmstadt · Mannheim · Ludwigshafen · Karlsruhe

LUXEMBOURG · Luxembourg · Trier · Saarbrücken · Strasbourg · Freiburg im Breisgau

LORRAINE · Metz · Nancy · Épinal · Mulhouse · Basel

SWITZERLAND · Neuchâtel · Bern · Fribourg · Luzern · Montbéliard · Besançon · La Chaux-de-Fonds · Dole

Lausanne · Genève (Geneva) · Annecy · Simplon Pass · Mont Blanc 15 771 ft. 4807 m · Monte Rosa 15 203 ft. 4634 m · Matterhorn 14 692 ft. 4478 m

Chambéry · Grenoble · Turin (Torino) · Gran Paradiso 13 323 ft. 4061 m · ITALY

Barre des Écrins 13 458 ft. 4102 m · Briançon · MARITIME ALPS · Mont Pelat 10 016 ft. 3053 m · San Remo

PROVENCE · Marseille · Aix-en-Provence · Nice · MONACO · Cannes · Toulon · La Ciotat · ÎLE DU LEVANT

MEDITERRANEAN SEA

© Rand McNally
A-101779-1

BAY OF BISCAY

ATLANTIC

OCEAN

PORTUGAL

SPAIN

MOROCCO

ALBORAN SEA

Gulf of Cadiz

Strait of Gibraltar

CORDILLERA CANTÁBRICA

SISTEMA CENTRAL

SIERRA MORENA

A Coruña
Ferrol
Ortigueira
Vivero
CABO ORTEGAL
Betanzos
Carballo
Laxe
CABO DE FISTERRA
Muros
Santiago de Compostela
Lugo
Vilalba
Vilagarcía de Arousa
O Grove
Marín
Pontevedra
Cangas
Vigo
Ourense
Verin
Viana do Castelo
Esposende
Póvoa de Varzim
Vila do Conde
Braga
Guimarães
Barcelos
Chaves
Bragança
Mirandela
Vila Real
Matosinhos
Porto
Vila Nova de Gaia
Espinho
Lamego
São João da Madeira
Ovar
Estarreja
Viseu
Aveiro
Ílhavo
Mira
Cantanhede
Guarda
CABO MONDEGO
Coimbra
Figueira da Foz
Covilhã
Estrela 6539 ft 1993 m
SIERRA DE GATA
Ciudad Rodrigo
Coria
Plasencia
Marinha Grande
Leiria
Batalha
Fátima
Serra
Castelo Branco
Nazaré
BERLENGA
CABO CARVOEIRO
Peniche
Caldas da Rainha
Tomar
Torres Novas
Abrantes
Nisa
Cáceres
Trujillo
Lourinhã
Santarém
Portalegre
San Vicente de Alcántara
SIERRA DE GUADALUPE
Torres Vedras
Ponte de Sor
Alburquerque
Mafra
Vila Franca de Xira
Coruche
Estremoz
Miajadas
Sintra
CABO DA ROCA
Lisbon (Lisboa)
Montijo
Vila Viçosa
Borba
Elvas
Badajoz
Mérida
Don Benito
Cascais
Barreiro
Setúbal
Montemor-o-Novo
Évora
Villanueva de la Serena
Herrera del Duque
Sesimbra
CABO ESPICHEL
Baía de Setúbal
Grândola
Beja
Moura
Zafra
Azuaga
Sines
Odemira
Castro Verde
Aracena
Monchique
Silves
Portimão
Lagos
CABO DE SÃO VICENTE
Sagres
Albufeira
Loulé
Faro
Olhão
Tavira
Vila Real de Santo António
Ayamonte
Isla Cristina
Lepe
Moguer
Huelva
Gibraleón
Bollullos Par del Condado
Valverde del Camino
Sanlúcar de Barrameda
Rota
El Puerto de Santa María
Jerez de la Frontera
Cádiz
San Fernando
Chiclana de la Frontera
Barbate
CABO TRAFALGAR
Algeciras
Tarifa
Gibraltar (U.K.)
La Línea de la Concepción
CAP SPARTEL
Tanger
Ceuta (Sp.)
Fnideq
Tetouan
Asilah
Larache
Ksar-el-Kebir
Chechaouen

Gijón
Oviedo
Mieres
Castropol
Avilés
CABO DE PEÑAS
Langreo
Cangas de Narcea
Ponferrada
MONTES DE LEÓN
Astorga
La Bañeza
León
Benavente
Zamora
Medina de Rioseco
Palencia
Santander
Llanes
San Vicente de la Barquera
Laredo
Torrelavega
Reinosa
Peña Prieta 8320 ft 2536 m
Guardo
Carrión
Bilbao
Santoña
Castro-Urdiales
Bermeo
Portugalete
Barakaldo
Durango
Bergara
Eibar
Donostia-San Sebastián
Irún
Rentería
Biarritz
Bayonne
Anglet
Saint-Jean-de-Luz
Gasteiz
Miranda de Ebro
Haro
Pamplona
Estella
Tafalla
Logroño
Calahorra
Arnedo
Alfaro
Ejea de los Caballeros
Tudela
Burgos
Aranda de Duero
Salas de los Infantes
Soria
Tarazona
Valladolid
Tordesillas
Cuéllar
Almazán
Zaragoza
Calatayud
Medina del Campo
Arévalo
Segovia
San Ildefonso o la Granja
Pico de Peñalara 7972 ft 2430 m
Sigüenza
Molina de Aragón
Calamocha
Monreal del Campo
Salamanca
Peñaranda de Bracamonte
Ávila
Colmenar Viejo
Alcobendas
Guadalajara
Béjar
San Martín de Valdeiglesias
Pozuelo de Alarcón
Alcalá de Henares
Jaraíz de la Vera
Talavera de la Reina
Leganés
MADRID
Getafe
Parla
Navalmoral de la Mata
Torrijos
Toledo
Aranjuez
Tarancón
Cuenca
Javalambre 6627 ft 2020 m
Navahermosa
Mora
Corral de Almaguer
MONTES DE TOLEDO
Madridejos
Villacañas
Alcázar de San Juan
NEW CASTILE
LA MANCHA
Tarazona de la Mancha
Utiel
Requena
Villarrobledo
La Roda
Albacete
Almansa
Yecla
Jumilla
Ciudad Real
Daimiel
Manzanares
La Solana
Tomelloso
Socuéllamos
Almagro
Valdepeñas
El Bonillo
Puertollano
Almadén
Cabeza del Buey
Zalamea de la Serena
Alcaraz
Hellín
Cieza
Peñarroya-Pueblonuevo
Pozoblanco
La Carolina
Orcera
SIERRA DE SEGURA
Molina de Segura
Fuente de Cantos
Bailén
Linares
Úbeda
Baeza
Sagra 7815 ft 2382 m
Caravaca de la Cruz
Mula
Alcantarilla
Córdoba
Andújar
Jaén
Iódar
Huéscar
Totana
Lorca
Palma del Río
Lora del Río
Montilla
Baena
Martos
Baza
Vera
Mojácar
Carmona
Écija
Aguilar
Cabra
Priego de Córdoba
Alcalá la Real
Mazarrón
Golfo de Mazarrón
Camas
Sevilla
Alcalá de Guadaira
Puente Genil
Lucena
Guadix
Águilas
Dos Hermanas
Arahal
Osuna
Lora
Santa Fe
Granada
CORDILLERA PENIBÉTICA
Los Palacios y Villafranca
Utrera
Morón de la Frontera
Campillos
Antequera
Loja
SIERRA NEVADA Mulhacén 11 424 ft 3482 m
Almería
CABO DE GATA
Las Cabezas de San Juan
Lebrija
Villamartín
Olvera
Vélez-Málaga
Nerja
Almuñécar
Motril
Adra
Berja
Golfo de Almería
Ubrique
Ronda
Coín
Alhaurín el Grande
Málaga
Torremolinos
COSTA DEL SOL
Marbella
Estepona
PUNTA DE ALMINA
Smir-Restinga
Dar Chaoui
Bou-Hamed
Al-Hoceima
Baie d'Al Hoceima
CAP DES TROIS FOURCHES
Melilla (Sp.)
ISLAS CHAFARINAS (Sp.)
Nador
ISLA DE ALBORÁN (Sp.)

© Rand McNally
A-101778-1

0 20 40 60 80 100 120 Miles
0 20 40 60 80 100 120 140 160 180 200 Kilometers

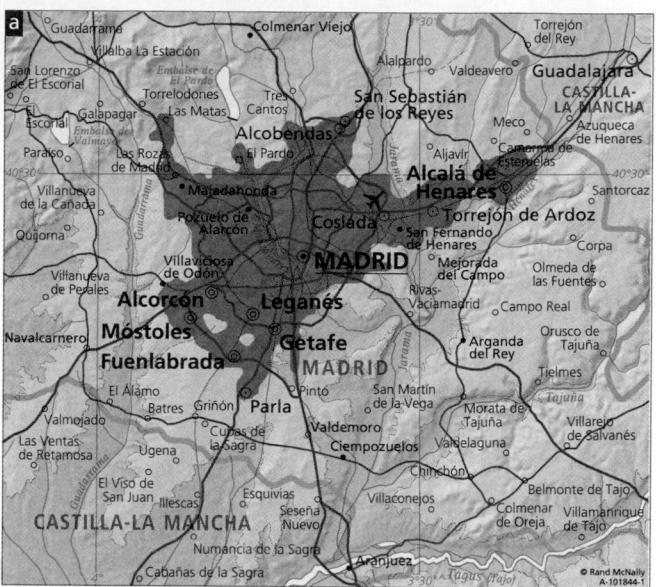

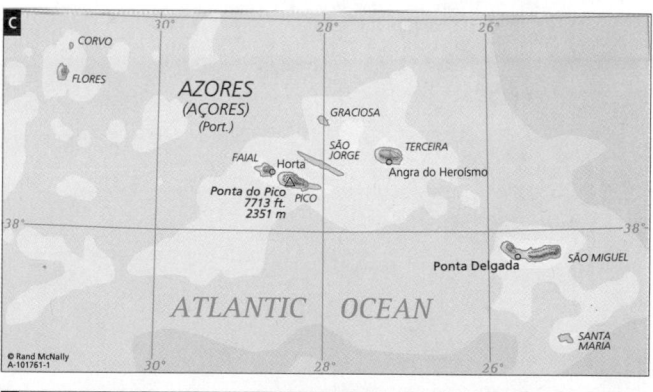

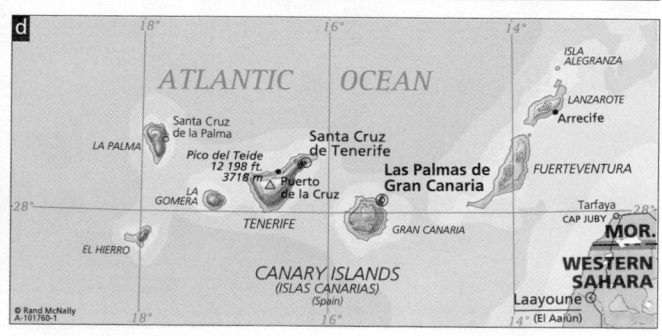

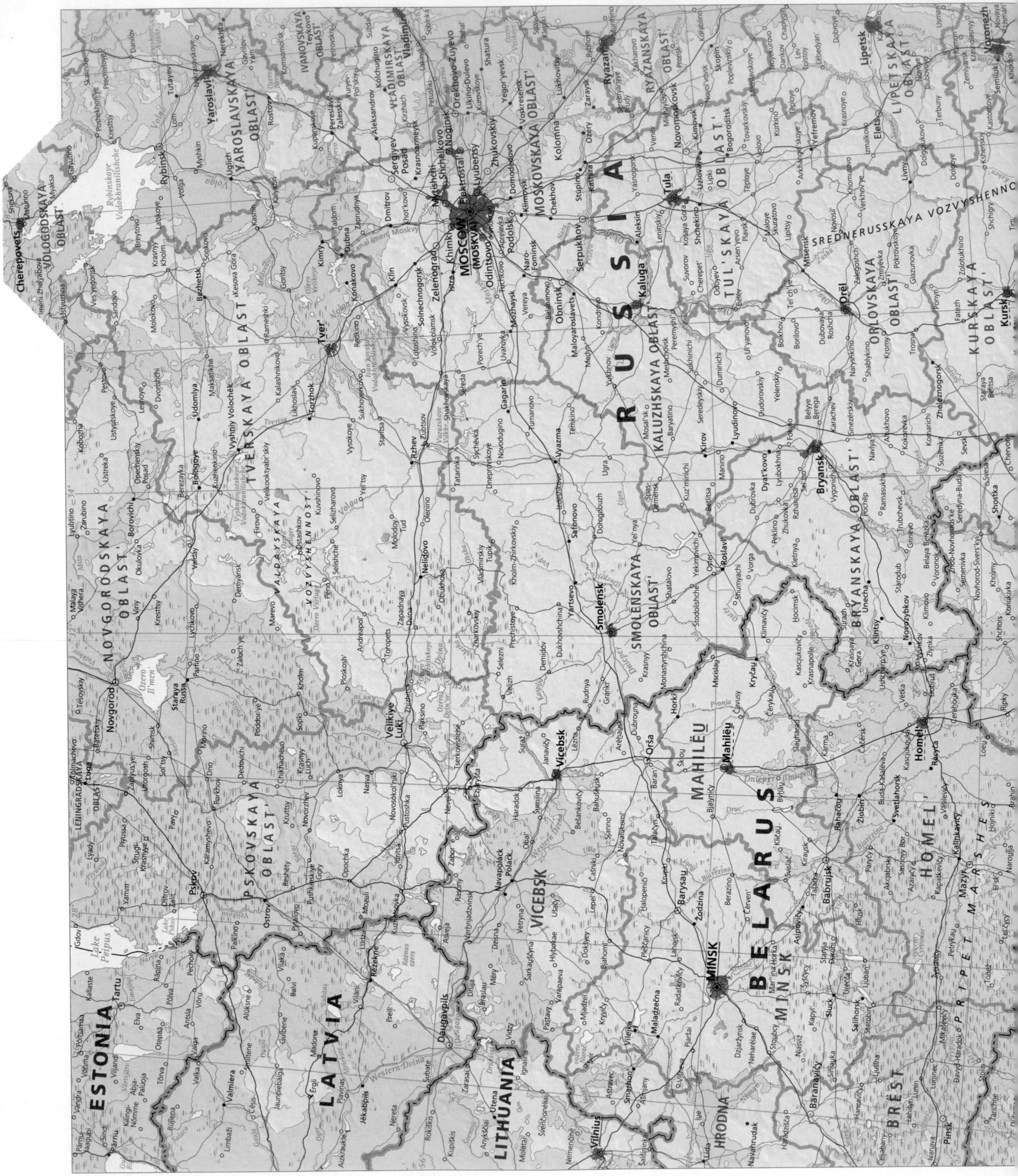

RUSSIA

BELARUS

ESTONIA

LATVIA

LITHUANIA

VOLOGODSKAYA OBLAST'

YAROSLAVSKAYA OBLAST'

IVANOVSKAYA OBLAST'

VLADIMIRSKAYA OBLAST'

MOSKOVSKAYA OBLAST'

RYAZANSKAYA OBLAST'

LIPETSKAYA OBLAST'

TUL'SKAYA OBLAST'

ORLOVSKAYA OBLAST'

KURSKAYA OBLAST'

KALUZHSKAYA OBLAST'

BRYANSKAYA OBLAST'

SMOLENSKAYA OBLAST'

TVERSKAYA OBLAST'

NOVGORODSKAYA OBLAST'

PSKOVSKAYA OBLAST'

LENINGRADSKAYA OBLAST'

VICEBSK

MINSK

HOMEL

BREST

HRODNA

MAHILÉU

SREDNERUSSKAYA VOZVYSHENNOST'

VALDAYSKAYA VOZVYSHENNOST'

PRIPET MARSHES

MOSCOW / MOSKVA

Yaroslavl'

Cherepovets

Rybinsk

Ryazan'

Lipetsk

Voronezh

Kursk

Orël

Bryansk

Tula

Kaluga

Smolensk

Mahilëu

Homel'

Minsk

Vicebsk

Navapolack

Velikiye Luki

Tver'

Pskov

Novgorod

Daugavpils

Vilnius

Tartu

0 20 40 60 80 100 120 Miles

0 20 40 60 80 100 120 140 160 180 200 Kilometers

Lambert Conformal Conic Projection
Scale 1:4,000,000
One inch to 64 miles
One cm to 40 km

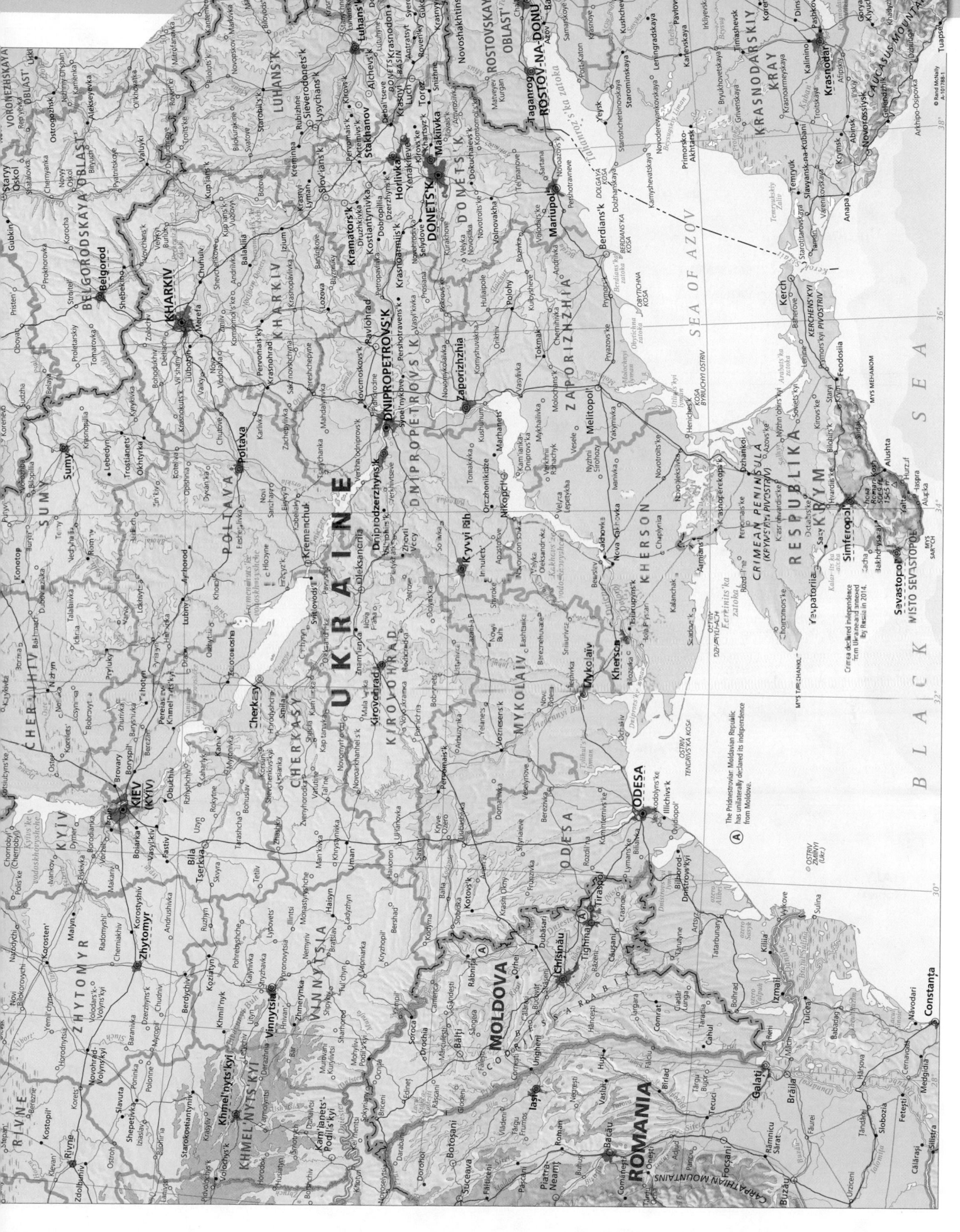

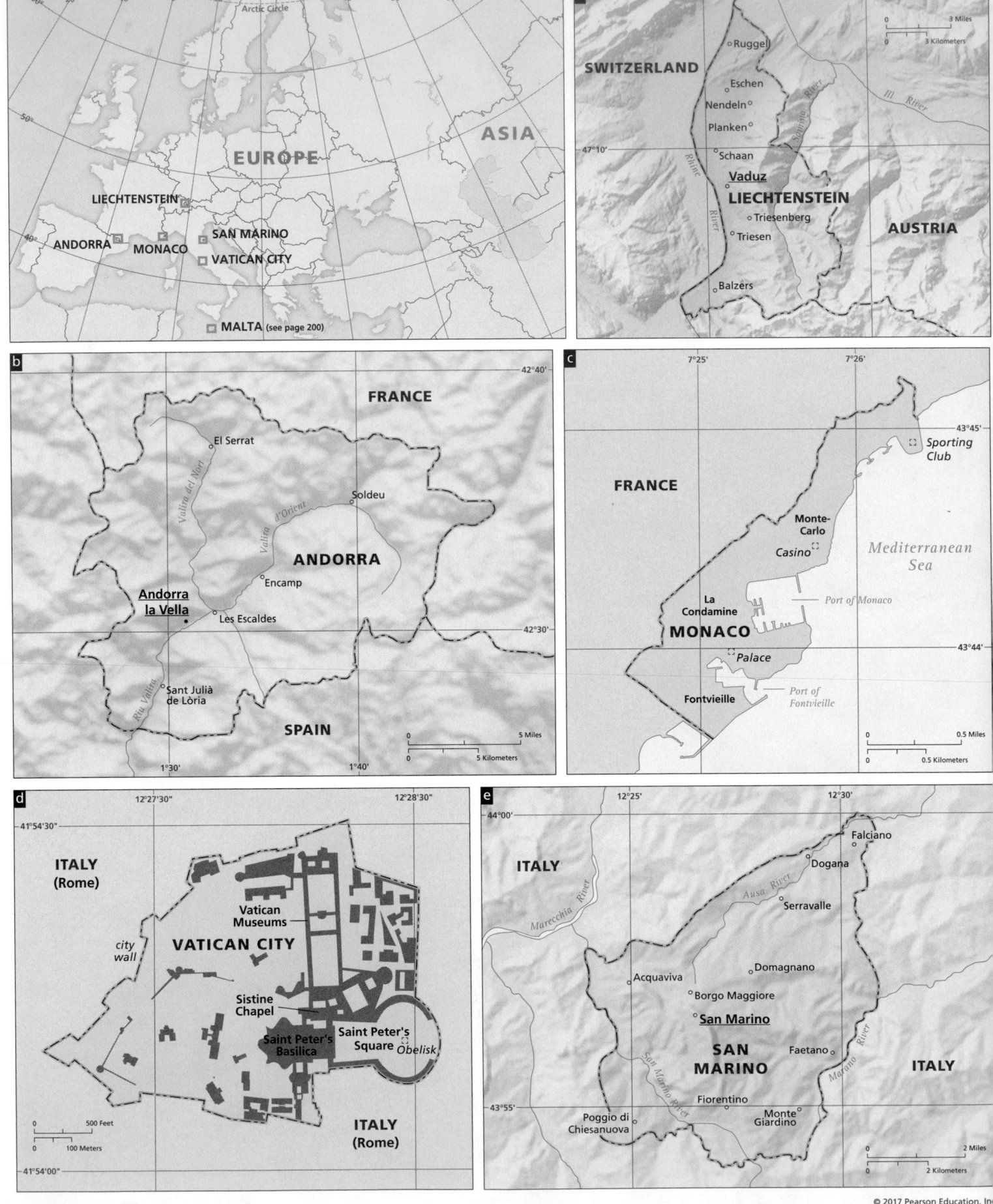

a

SWITZERLAND

Ruggel

Eschen

Nendeln

Planken

Schaan

Vaduz
LIECHTENSTEIN

Triesenberg

Triesen

AUSTRIA

Balzers

Rhine River

Samina River

Ill River

47°10'

9°30'   9°40'

0   3 Miles
0   3 Kilometers

ASIA

EUROPE

LIECHTENSTEIN

ANDORRA   MONACO

SAN MARINO

VATICAN CITY

☐ MALTA (see page 200)

Arctic Circle

---

b

FRANCE

El Serrat

Soldeu

ANDORRA

Encamp

**Andorra
la Vella**

Les Escaldes

Sant Julià
de Lòria

SPAIN

Valira del Nort

Valira d'Orient

Riu Valira

42°40'

42°30'

1°30'   1°40'

0   5 Miles
0   5 Kilometers

---

c

FRANCE

Monte-
Carlo

Casino

La
Condamine

MONACO

Palace

Fontvieille

Sporting
Club

*Mediterranean
Sea*

Port of Monaco

*Port of
Fontvieille*

43°45'

43°44'

7°25'   7°26'

0   0.5 Miles
0   0.5 Kilometers

---

d

ITALY
(Rome)

*city
wall*

VATICAN CITY

Vatican Museums

Sistine
Chapel

Saint Peter's
Basilica

Saint Peter's
Square

Obelisk

ITALY
(Rome)

41°54'30"

41°54'00"

12°27'30"   12°28'30"

0   500 Feet
0   100 Meters

---

e

ITALY

Falciano

Dogana

Serravalle

Domagnano

Acquaviva

Borgo Maggiore

San Marino

Faetano

SAN
MARINO

Fiorentino

Monte
Giardino

Poggio di
Chiesanuova

ITALY

Marecchia River

Ausa River

San Marino River

Marano River

44°00'

43°55'

12°25'   12°30'

0   2 Miles
0   2 Kilometers

Albers Equal Area Conic Projection

© 2017 Pearson Education, Inc

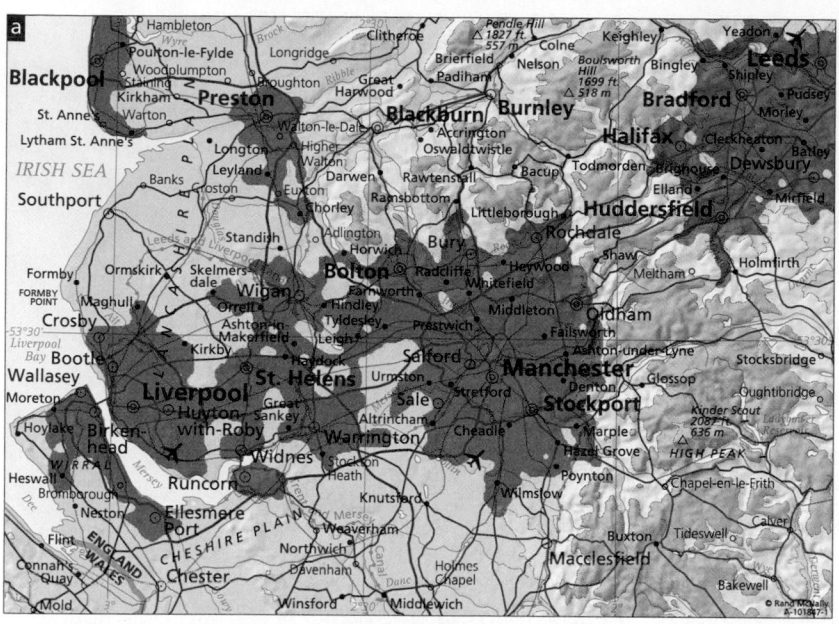

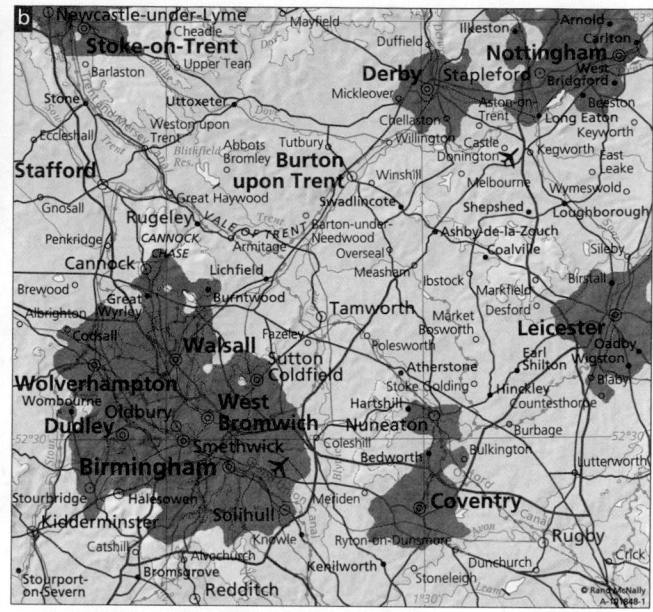

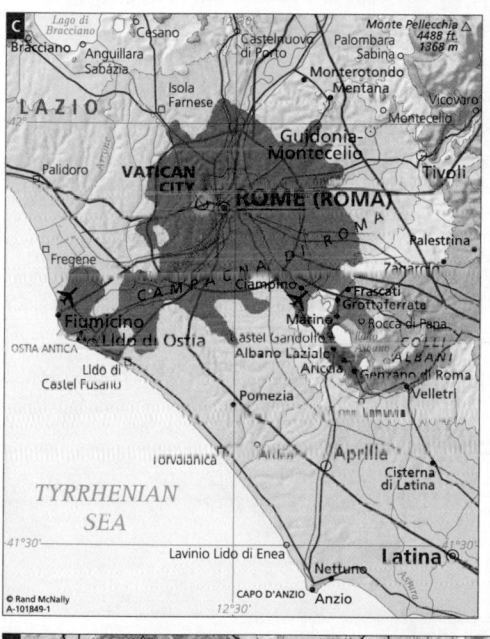

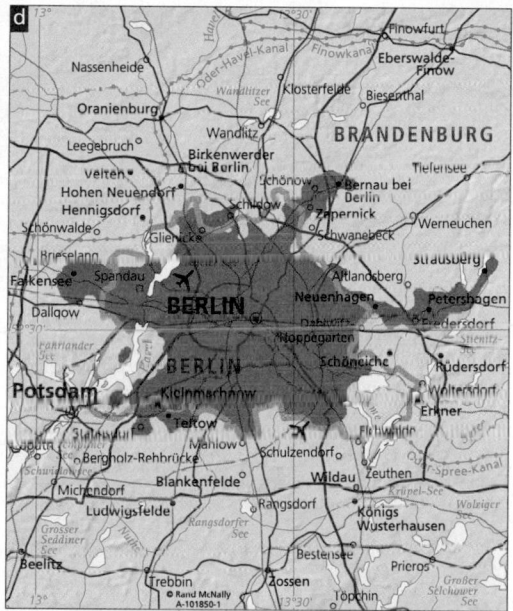

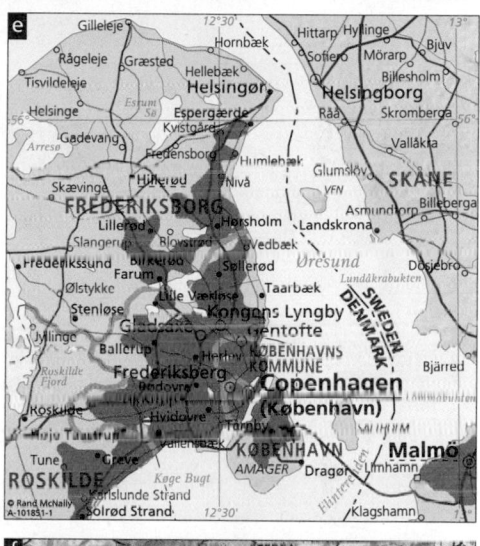

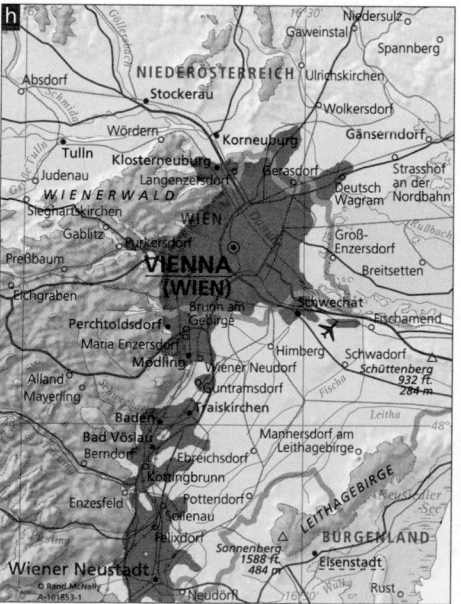

0   5   10   15   20   25   30 Miles

0   5   10   15   20   25   30   35   40   45   50 Kilometers

Lambert Conformal Conic Projection
Scale 1:1,000,000
One inch to 16 miles
One cm to 10 km

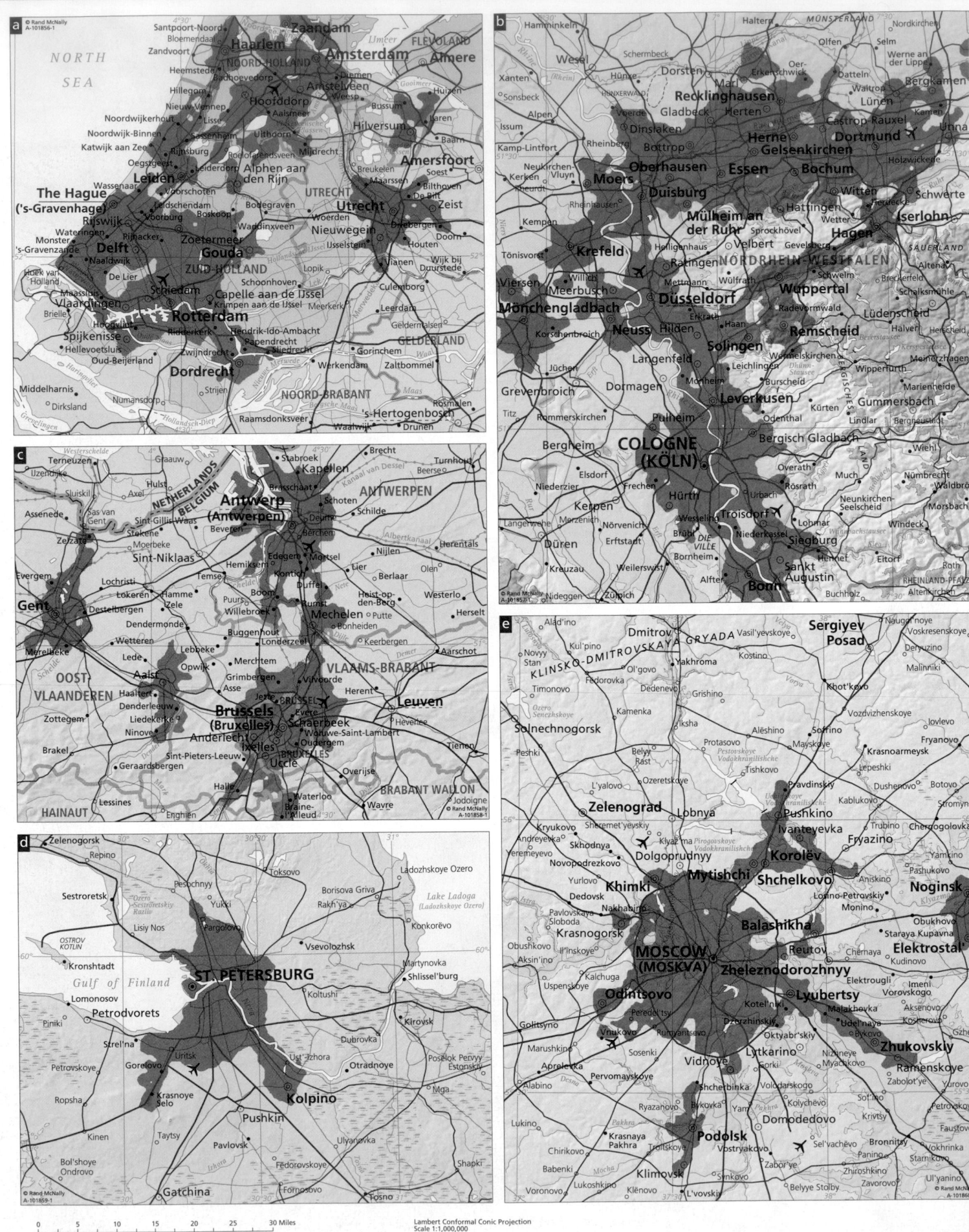

0   5   10   15   20   25   30 Miles
0 5 10 15 20 25 30 35 40 45 50 Kilometers

Lambert Conformal Conic Projection
Scale 1:1,000,000
One inch to 16 miles
One cm to 10 km

Asia

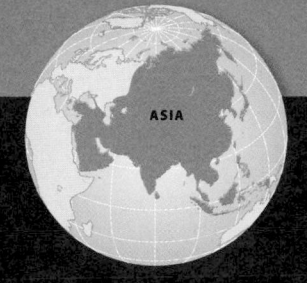

ASIA

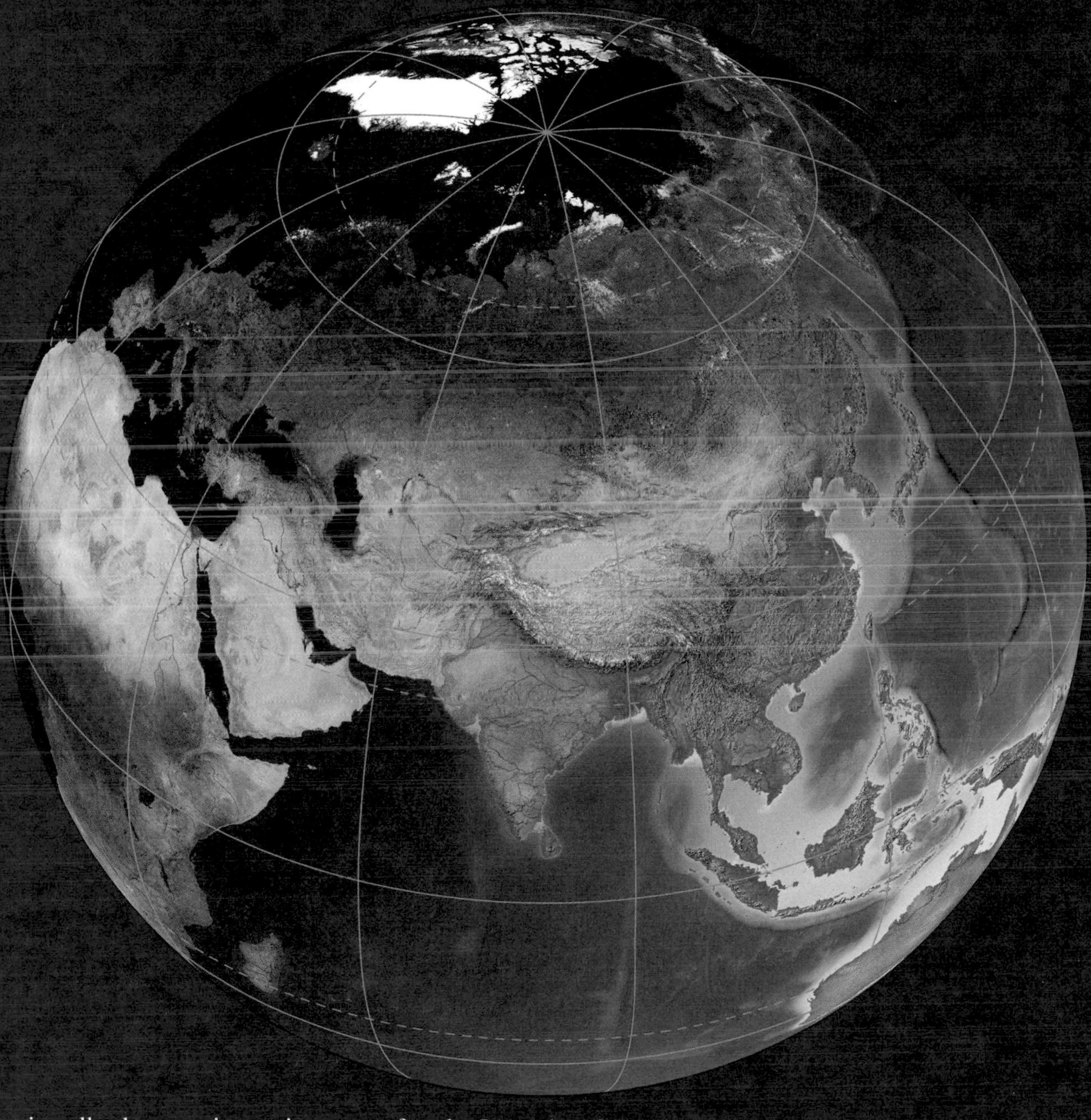

Asia outstrips all other continents in terms of scale. It reaches from the Arctic Ocean to the Southern Hemisphere and stretches almost halfway around Earth from east to west. As the largest and most populous continent, Asia contains nearly thirty percent of Earth's land area and sixty percent of its population. It is home to the world's highest mountains, its largest and deepest lakes, and some of its hottest, coldest, driest, and wettest places. The historic hearth of some of humankind's oldest civilizations, Asia is the birthplace of Hinduism, Buddhism, Judaism, Christianity, and Islam. Five of the world's seven most populous countries are in Asia, including China and India which both have populations in excess of one billion peop▮▮. Due to rapid urbanization over the last f▮▮ decades, Asia's population increasingly lives in urban areas, including a number of megacities each containing tens of millions of people.

ARCTIC OCEAN

ATLANTIC OCEAN

NORWEGIAN SEA

BARENTS SEA

SEVERNAYA ZEMLYA

FAROE ISLANDS (Den.)

SVALBARD (Norway)

FRANZ JOSEF LAND

AZORES (Port.)

IRELAND

UNITED KINGDOM

NORWAY

SWEDEN

FINLAND

OSTROV KOLGUYEV

NOVAYA ZEMLYA

KARA SEA

Khatang

LONDON

DENMARK

Stockholm

NETH.

BELG.

BERLIN

GERMANY

POLAND

BALTIC SEA

EST.

LATVIA

LITH.

ST. PETERSBURG

BELARUS

Arkhangel'sk

Pechora

Salekhard

Dudinka

Noril'sk

R U S

PORTUGAL

PARIS

FRANCE

SWITZ.

RHINE

AUSTRIA

CZECH REP.

SLVK.

KIEV

UKRAINE

MOSCOW

Oka

Gora Narodnaya 6214 ft. 1894 m

Khanty-Mansiysk

Nizhniy Tagil

YEKATERINBURG

Tyumen'

Irtysh

Tomsk

SPAIN

MADRID

ITALY

CROATIA

HUNGARY

BUDAPEST

ROMANIA

Volga

CHELYABINSK

OMSK

NOVOSIBIRSK

Magnitogorsk

Orsk

Astana

Barnaul

Novokuznec

MOROCCO

CASABLANCA

ALGIERS

SARDINIA

ROME

MONT.

BOS.

SERBIA

KOS.

ALB.

BUCHAREST

BULGARIA

Danube

Ural

Tobol

Qaraghandy

KAZAKHSTAN

Semey

Kyzy

ALGERIA

TUNISIA

SICILY

MEDITERRANEAN

MACE.

BLACK SEA

İSTANBUL

CAUCASUS MTS.

Gora El'brus 18 510 ft. 5642 m

Caspian Sea

Aral Sea

Zhezqazghan

Lake Balkhash

Balqash

ALTAY MTS.

GREECE

İZMİR

ANKARA

TURKEY

GEORGIA

Yerevan

ARM.

TBILISI

AZER.

BAKU

Nukus

UZBEKISTAN

ALMATY

TIEN SHAN

ÜRÜMC

CRETE

CYPRUS

Nicosia

ALEPPO

Beirut

SYRIA

TABRİZ

TASHKENT

Samarqand

Aşgabat

BISHKEK

KYRGYZSTAN

Kashi

Tarim

SEA

LEBANON

DAMASCUS

MASHHAD

TURKMENISTAN

Dushanbe

TAJIKISTAN

LIBYA

ISRAEL

Jerusalem

AMMAN

JORDAN

TEHRAN

BAGHDAD

IRAQ

Kühe Damvand 18 386 ft. 5604 m

Amu Darya

K2 28 250 ft. 8611 m

KUNLUN SHAN

CAIRO

EGYPT

ESFAHAN

IRAN

AFGHANISTAN

Kabul

Islamabad

Lhas

Tropic of Cancer

NIGER

Lake Nasser

Basra

Kuwait

KUWAIT

SHIRAZ

Quetta

LAHORE

DELHI

NEPAL

Mt. Everest 29 028 ft. 8848 m

Brahmap

BHUT

Thimphu

SAUDI

BAHRAIN

Persian Gulf

PAKISTAN

New Delhi

KATH-MANDU

Ganges

CHAD

Nile

RIYADH

QATAR

Doha

HYDERABAD

KANPUR

PATNA

DHAKA

Lake Chad

RED SEA

Medina

ARABIA

Abu Dhabi

U.A.E.

Muscat

HYDERABAD

KARACHI

RA'S AL-HADD

AHMADABAD

INDIA

KOLKATA (CALCUTTA)

BNG

SUDAN

JIDDAH

MECCA

OMAN

Narmada

NAGPUR

Khartoum

Chari

Blue Nile

White Nile

ERITREA

YEMEN

SANAA

ARABIAN SEA

MUMBAI (BOMBAY)

HYDERABAD

Godávari

Krishna

BAY OF BENGAL

CENTRAL AFRICAN REPUBLIC

SOUTH SUDAN

ETHIOPIA

Aden

DJIBOUTI

Gulf of Aden

SOCOTRA (Yemen)

BENGALURU (BANGALORE)

CHENNAI (MADRAS)

Kochi

CONGO

Congo

Ubanghi

DEMOCRATIC REPUBLIC OF THE CONGO

UGANDA

Lake Rudolf

SOMALIA

KENYA

CAPE COMORIN

SRI LANKA

Colombo

DONDRA HEAD

RWANDA

BURUNDI

Lake Victoria

NAIROBI

Equator

Male'

MALDIVES

Kasai

Lake Tanganyika

TANZANIA

DAR ES SALAAM

SEYCHELLES

CHAGOS ARCHIPELAGO (B.I.O.T.)

I N D I A N   O C E A N

ANGOLA

ZAMBIA

0 200 400 600 800 1000 1200 Miles

0 200 400 600 800 1000 1200 1400 1600 1800 2000 Kilometers

Lambert Azimuthal Equal Area Projection
Scale 1:40,000,000
One inch to 640 miles
One cm to 400 km

UNITED STATES

NEW SIBERIAN ISLANDS

OSTROV VRANGELYA

EAST SIBERIAN SEA

LAPTEV SEA

Tiksi

MYS DEZHNEVA

ST. LAWRENCE ISLAND

MYS NAVARIN

Arctic Circle

Cherskiy

BERING SEA

ALEUTIAN ISLANDS

Zhigansk

Gora Pobeda 10 325 ft. 3147 m

Anadyr'

CAPE OLYUTORSKIY

Yakutsk

Lensk

Aldan

Magadan

Vulkan Klyuchevskaya Sopka 15 584 ft. 4750 m

Petropavlovsk-Kamchatskiy

Lena

Vilyuy

MYS LOPATKA

SEA OF OKHOTSK

Irkutsk

Lake Baikal

Chita

Nikolayevsk-na-Amure

SAKHALIN

KURIL ISLANDS

OYARSK

Ulan-Ude

Kerulen

Komsomol'sk-na-Amure

Yuzhno-Sakhalinsk

MIDWAY ISLANDS (U.S.)

ULAANBAATAR

Khabarovsk

Amur

Ussuri

SAPPORO

HOKKAIDŌ

MONGOLIA

QIQIHAR

HARBIN

Vladivostok

GOBI DESERT

CHANGCHUN

SEA OF JAPAN (EAST SEA)

SENDAI

HONSHŪ

PACIFIC OCEAN

SHENYANG

NORTH KOREA

Tropic of Cancer

Yinchuan

BEIJING

P'YŎNGYANG

Fujisan 12 388 ft. 3776 m

TŌKYŌ

Huang

TIANJIN

SEOUL

NAGOYA

ŌSAKA

JAPAN

LANZHOU

TAIYUAN

JINAN

QINGDAO

SOUTH KOREA

BUSAN

SHIKOKU

WAKE ISLAND (U.S.)

CHINA

ZHENGZHOU

KYŪSHŪ

Kagoshima

XI'AN

NANJING

SHANGHAI

CHENGDU

WUHAN

EAST CHINA SEA

RYUKYU ISLANDS

BONIN ISLANDS (Japan)

Yangtze

CHONGQING

CHANGSHA

NANCHANG

Naha

OKINAWA-JIMA

VOLCANO ISLANDS (Japan)

GUIYANG

FUZHOU

TAIPEI

MICRONESIA

KUNMING

GUANGZHOU

TAIWAN

PHILIPPINE SEA

NORTHERN MARIANA ISLANDS (U.S.)

MARSHALL ISLANDS

Mandalay

Ha Noi

HONG KONG

MARIANA IS

Ha Noi

Haikou

SAIPAN

TINIAN

MYANMAR (BURMA)

LAOS

HAINAN DAO

LUZON

GUAM (U.S.)

GON OON)

Vientiane

Da Nang

ESCARPADA POINT

PHILIPPINES

POHNPEI

KOSRAE

THAILAND

VIETNAM

SOUTH

MANILA

MINDORO

SAMAR

CAROLINE ISLANDS

KIRIBATI

BANGKOK

CAMBODIA

CHINA

Cebu

FEDERATED STATES OF MICRONESIA

ANDAMAN ISLANDS (India)

PHNOM PENH

HO CHI MINH CITY (SAIGON)

SEA

NEGROS

SULU SEA

MINDANAO

DAVAO

PALAU

NAURU

NICOBAR ISLANDS (India)

MUI CA MAU

Gunong Kinabalu 13 455 ft. 4101 m

CELEBES SEA

ANDAMAN SEA

MALAYSIA

Bandar Seri Begawan

BRUNEI

MALAYSIA

Manado

TANJUNG D'URVILLE

NEW IRELAND

MELANESIA

MEDAN

KUALA LUMPUR

Jayapura

BISMARCK SEA

BOUGAINVILLE

CHOISEUL

SANTA ISABEL

SUMATRA

SINGAPORE

SINGAPORE

BORNEO

Balikpapan

SULAWESI (CELEBES)

MOLUCCAS

Puncak Jaya 16 503 ft. 5030 m

NEW BRITAIN

MALAITA

Padang

Gunung Kerinci 12 467 ft. 3800 m

PALEMBANG

Banjarmasin

NEW GUINEA

PAPUA NEW GUINEA

SANTA CRUZ IS.

Equator

JAVA SEA

MAKASSAR

BANDA SEA

SOLOMON SEA

SAN CRISTOBAL

SOLOMON ISLANDS

JAKARTA

SURABAYA

INDONESIA

TANJUNG VALS

Port Moresby

© Rand McNally A-101711-1

JAVA

BALI

FLORES SEA

Dili

TIMOR-LESTE

ARAFURA SEA

CHRISTMAS ISLAND (Austl.)

SUMBAWA

SUMBA

FLORES

TIMOR

MELVILLE ISLAND

TIMOR SEA

AUSTRALIA

ARCTIC OCEAN

ATLANTIC OCEAN

NORWEGIAN SEA

BARENTS SEA

KARA SEA

SVALBARD (Norway)

FRANZ JOSEF LAND

SEVERNAYA ZEMLYA

NOVAYA ZEMLYA

TAYMYR PENINSULA

FAROE ISLANDS (Den.)

MYS KANIN NOS

OSTROV KOLGUYEV

YAMAL PENINSULA

Noril'sk

NORWAY

SWEDEN

FINLAND

IRELAND

UNITED KINGDOM

London

DENMARK

BALTIC SEA

EST.

LATVIA

LITH.

Lake Ladoga

Lake Onega

St. Petersburg

R U S

WEST SIBERIAN PLAIN

Yekaterinburg

Novosibirsk

Moscow

SAYA

FRANCE

Paris

BELG.

NETH.

GERMANY

POLAND

BELARUS

UKRAINE

CZECH REP.

SLVK.

AUSTRIA HUNGARY

SWITZ.

Mont Blanc 15,771 ft. 4807 m

ALPS

SPAIN

Madrid

Mulhacén 11,424 ft. 3482 m

PYRENEES

PORTUGAL

CABO DA ROCA

SÃO CABO DE VICENTE

AZORES (Port.)

CORSICA

SARDINIA

ITALY

Rome

Vesuvius 4203 ft. 1281 m

CROATIA

BOS.

SERBIA

MONT.

KOS.

ALB.

MACE.

ROMANIA

MOLD.

BULGARIA

Mt. Olympus 9570 ft. 2917 m

GREECE

CRETE

Monte Etna 10,902 ft. 3323 m

MEDITERRANEAN SEA

Dnieper

Don

Volga

Oka

Kama

Sukhona

Pechora

Ural

URAL MTS.

Gora Narodnaya 6214 ft. 1894 m

Irtysh

Ob

Tobol

Ishim

Tura

Yenisey

KAZAKHSTAN

Qaraghandy

KAZAKH HILLS

Mt. Belukha 14,783 ft. 4506 m

ALTAY MTS.

Zhaysang köl

Lake Balkhash

Almaty

Ürümqi

KYRGYZSTAN

TIEN SHAN

Jengish Chokusu 24,406 ft. 7439 m

ALTUN SHA

Aral Sea

Syr Darya

Amu Darya

UZBEKISTAN

Tashkent

QIZILQUM

KARA KUM

UST-URT PLAT.

TURKMENISTAN

TAJIKISTAN

Pik Imeni Ismail Samani 24,590 ft. 7495 m

PAMIRS

K2 28,250 ft. 8611 m

KUNLUN SHAN

PLATEAU OF TIBET

TARIM PENDI

Tarim

HINDU KUSH

Kabul

AFGHANISTAN

Lahore

PAKISTAN

Annapurna 26,545 ft. 8091 m

Mt. Everest 29,028 ft. 8848 m

Lhasa

NEPAL

HIMALAY

BHUT

Delhi

Ganges

Brahmap

INDIA

Dhaka

Kolkata (Calcutta)

BNG

Narmada

Godávari

DECCAN

Mumbai (Bombay)

WESTERN GHATS

EASTERN GHATS

Krishna

BAY OF BENGA

BLACK SEA

Sea of Azov

CAUCASUS MTS.

Gora El'brus 18,510 ft. 5642 m

GEORGIA

ARM.

AZER.

Baku

CASPIAN DEPRESSION

CASPIAN SEA

TURKEY

Istanbul

Erciyes Dagi 12,851 ft. 3917 m

Mt. Ararat 16,854 ft. 5137 m

CYPRUS

GREECE

LEBANON

ISRAEL

SYRIA

Damascus

IRAQ

Baghdad

Tigris

Euphrates

JORDAN

SYRIAN DESERT

Kuh-e Damávand 18,386 ft. 5604 m

Tehrán

DASHT-E KAVIR

ZAGROS MOUNTAINS

IRAN

DASHT-E LUT

Daryácheh-ye Namak

KUWAIT

AN-NAFÚD

SAUDI ARABIA

Riyadh

ARABIAN PENINSULA

QATAR

U.A.E.

Jabal ash-Shám 9957 ft. 3035 m

Gulf of Oman

OMAN

RA'S AL-HADD

RUB' AL-KHALI

Persian Gulf

AL HIJAZ

ASIR

RED SEA

Jiddah

Cairo

EGYPT

LIBYA

Tropic of Cancer

Lake Nasser

Nile

SUDAN

Jabal Marrah 10,072 ft. 3070 m

CHAD

Lake Chad

NIGER

ALGERIA

TUNISIA

MOROCCO

ATLAS MOUNTAINS

Chott Melrhir

Chott ej Jerid

Khalíj Surt

ERITREA

Sanaa

YEMEN

RA'S FARTAK

DJIBOUTI

Gulf of Aden

SOCOTRA (Yemen)

GEES GWARDAFUY

ARABIAN SEA

Ras Dejen 15,155 ft. 4620 m

ETHIOPIA

Addis Ababa

SOUTH SUDAN

CENTRAL AFRICAN REPUBLIC

Mbomou

Uele

Congo

CONGO

DEMOCRATIC REPUBLIC OF THE CONGO

Margherita Peak 16,763 ft. 5109 m

Lake Albert

UGANDA

Lake Edward

RWANDA

BURUNDI

Lake Victoria

Lake Tanganyika

Lake Rudolf

Mt. Kenya 17,058 ft. 5199 m

KENYA

SOMALIA

Equator

Blue Nile

White Nile

Atbarah

Sobat

Kilimanjaro 19,340 ft. 5895 m

TANZANIA

Dar es Salaam

ANGOLA

ZAMBIA

Lake Mweru

Kasai

Kwango

LAKSHADWEEP

Ánai Mudi 8842 ft. 2695 m

Bengalúru (Bangalore)

CAPE COMORIN

MALDIVES

Colombo

SRI LANKA

Pidurutalagala 8281 ft. 2524 m

DONDRA HEAD

SEYCHELLES

CHAGOS ARCHIPELAGO (B.I.O.T.)

INDIAN OCEAN

0   200   400   600   800   1000   1200 Miles

0   200   400   600   800   1000   1200   1400   1600   1800   2000 Kilometers

Lambert Azimuthal Equal Area Projection
Scale 1:40,000,000
One inch to 640 miles
One cm to 400 km

UNITED STATES

NEW SIBERIAN ISLANDS

EAST SIBERIAN SEA

LAPTEV SEA

OSTROV WRANGELYA

CHUKOTSK PEN.

MYS DEZHNEVA

Gulf of Anadyr

MYS NAVARIN

BERING SEA

ALEUTIAN ISLANDS

ALEUTIAN TRENCH

CHERSKIY MTS.

Anadyr

Gora Pobeda 10,325 ft. 3147 m

Yakutsk

A S I A

VERKHOYANSK MTS.

Magadan

KAMCHATKA PENINSULA

KOMANDORSKI ISLANDS

CAPE WRANGELL

SIBERIA

STANOVOY MTS.

STANOVOY RANGE

Lake Baikal

Irkutsk

DZHUGDZHUR RANGE

Komsomol'sk-na-Amure

Vulkan Klyuchevskaya Sopka 15,584 ft. 4750 m

MYS YELIZAVETY

Petropavlovsk-Kamchatskiy

SEA OF OKHOTSK

MYS LOPATKA

KURIL ISLANDS

JAVA TRENCH

KURIL TRENCH

Ulaanbaatar

GREATER KHINGAN RANGE

SIKHOTE-ALIN'

SAKHALIN

Tatar Strait

MYS TERPENIYA

MONGOLIA

Harbin

NORTH KOREA

Sapporo

HOKKAIDO

JAPAN TRENCH

PACIFIC OCEAN

MIDWAY ISLANDS (U.S.)

HAWAI'IAN ISLANDS

NECKER RIDGE

GOBI DESERT

Beijing

Seoul

SOUTH KOREA

CHEJU-DO

SEA OF JAPAN (EAST SEA)

HONSHU

Tokyo

Fuji-san 12,388 ft. 3776 m

JAPAN

C H I N A

Xi'an

QIN LING

YELLOW SEA

SHIKOKU

KYUSHU

IZU-SHOTO

IZU TRENCH

Tropic of Cancer

Shanghai

Three Gorges Reservoir

JEJUDO

EAST CHINA SEA

BONIN IS. (Japan)

WAKE ISLAND (U.S.)

Chongqing

RYUKYU ISLANDS

OKINAWA-JIMA

VOLCANO ISLANDS (Japan)

MARIANA TRENCH

Taipei

TAIWAN

Yu Shan 13,114 ft. 3997 m

RYUKYU TRENCH

PHILIPPINE SEA

MARIANA ISLANDS

NORTHERN MARIANA ISLANDS (U.S.)

MICRONESIA

BIKINI

Hong Kong

Taiwan Strait

SAIPAN

ENEWETAK

KWAJALEIN

MARSHALL ISLANDS

MYANMAR (BURMA)

LAOS

HAINAN DAO

Gulf of Tonkin

SOUTH CHINA SEA

Luzon Strait

ESCARPADA POINT

LUZON

PHILIPPINES

Manila

SAMAR

TINIAN

GUAM (U.S.)

CHALLENGER DEEP

YAP

CAROLINE ISLANDS

CHUUK

POHNPEI

KOSRAE

MAJURO

KIRIBATI

Yangon (Rangoon)

THAILAND

Bangkok

CAMBODIA

VIETNAM

Tonle Sap

MINDORO

MINDORO

PANAY

PALAWAN

NEGROS

PHILIPPINE TRENCH

YAP TRENCH

FEDERATED STATES OF MICRONESIA

PALAU

TARAWA

ANDAMAN ISLANDS (India)

ANDAMAN SEA

Ho Chi Minh City (Saigon)

Gulf of Thailand

MUI CA MAU

Gunong Kinabalu 13,455 ft. 4101 m

SULU SEA

Mt. Apo 9692 ft. 2954 m

MINDANAO

TINACA POINT

NAURU

NICOBAR ISLANDS (India)

MALAY PENINSULA

BRUNEI

CELEBES SEA

HALMAHERA

PULAU WAIGEO

MANUS ISLAND

NEW HANOVER

NEW IRELAND

MELANESIA

PULAU SIMEULUE

Strait of Malacca

MALAYSIA

MALAYSIA

IRAN MTS.

Kapuas

BORNEO

Bukit Raya 7474 ft. 2278 m

SULAWESI (CELEBES)

Makassar Strait

MOLUCCA SEA

BURU

CERAM

MOLUCCAS

BIAK

TANJUNG D'URVILLE

Jayapura

PULAU YAPEN

Sepik

BISMARCK SEA

NEW BRITAIN

BOUGAINVILLE

CHOISEUL

SANTA ISABEL

SOLOMON ISLANDS

MALAITA

PULAU NIAS

SINGAPORE

Singapore

SUMATRA

GREATER SUNDA

TANJUNG PUTING

TANJUNG SELATAN

Makassar

BANDA SEA

PULAU BUTON

Puncak Jaya 16,503 ft. 5030 m

NEW GUINEA

Mt. Wilhelm 14,793 ft. 4509 m

NEW BRITAIN TRENCH

NEW GEORGIA

GUADALCANAL

SAN CRISTOBAL

PULAU SIBERUT

Gunung Kerinci 12,467 ft. 3800 m

ISLANDS

KEPULAUAN ARU

PAPUA NEW GUINEA

SOLOMON SEA

RENNELL

SAN CRISTOBAL

SANTA CRUZ IS.

Equator

JAVA SEA

I N D O N E S I A

PULAU WETAR

PULAU YAMDENA

TANJUNG VALS

Jakarta

JAVA

Gunung Semeru 12,060 ft. 3676 m

BALI

LOMBOK

SUMBAWA

LESSER SUNDA ISLANDS

FLORES SEA

FLORES

SUMBA

TIMOR-LESTE

TIMOR

MELVILLE ISLAND

ARAFURA SEA

CAPE YORK

CHRISTMAS ISLAND (Aust.)

JAVA TRENCH

TIMOR SEA

AUSTRALIA

CAPE ARNHEM

© Rand McNally A-101712-1

140° 80° 160° 180° 160° 150° 140° 170°

100° 110° 120° 130° 140° 150° 160° 170°

ARCTIC OCEAN

ATLANTIC OCEAN

BARENTS SEA

KARA SEA

SEVERNAYA ZEMLYA

URAL MOUNTAINS

WEST SIBERIAN PLAIN

Yenisey

Yekaterinburg

Chelyabinsk

CE

SAYA

ALTAY MTS.

Ural

Irtysh

MEDITERRANEAN SEA

BLACK SEA

CAUCASUS MTS.

CASPIAN SEA

Aral Sea

KAZAKH HILLS

Lake Balkhash

Istanbul

Ankara

UST-URT PLAT.

SYr DarYa

QIZILQUM

Almaty

TIEN SHAN

Tel Aviv-Yafo

Damascus

Tabrīz

Tehrān

Mashhad

Dushanbe

Tashkent

KARA KUM

Amu Darya

PAMIRS

TARIM PENDI

ALTUN SH

Euphrates

Tigris

Amman

Baghdād

Eṣfahān

DASHT-E KAVIR

DASHT-E LUT

HINDU KUSH

KUNLUN SHAN

PLATEAU OF TIBET

Tropic of Cancer

ZAGROS MTS.

AN-NAFŪD

Islāmābād

Lahore

Brāhma

H I M A L A

AL-HIJĀZ

RED SEA

Riyadh

Persian Gulf

Gulf of Oman

Indus

GREAT INDIAN DESERT

Delhi

Kānpur

Kathmandu

Ganges

Dhaka

Hyderābād

Karāchi

Ahmadābād

Indore

Kolkata

Jiddah

Mecca

ʿASĪR

RUB' AL-KHALI

DECCAN

Sanaa

Mumbai

WESTERN GHATS

Hyderābād

EASTERN GHATS

BAY O

BENGA

ARABIAN SEA

Gulf of Aden

Bengalūru

Chennai

Kochi

Colombo

INDIAN OCEAN

Equator

0   200   400   600   800   1000 1200 Miles

0   200 400 600 800 1000 1200 1400 1600 1800 2000 Kilometers

Lambert Azimuthal Equal Area Projection
Scale 1:40,000,000
One inch to 640 miles
One cm to 400 km

Land Cover

- Evergreen Needleleaf Forest
- Evergreen Broadleaf Forest
- Deciduous Needleleaf Forest
- Deciduous Broadleaf Forest
- Mixed Forest
- Woodland
- Wooded Grassland
- Closed Shrubland
- Open Shrubland
- Grassland
- Cropland
- Bare Ground (Desert and Ice)
- Urban and Built Up

Source: CIESIN; Hansen et al., 2000

Geology

Rock type/geological era

- Intrusive
- Metamorphic
- Volcanic, tectonic
- Precambrian sedimentary
- Paleozoic sedimentary
- Mesozoic sedimentary
- Cenozoic sedimentary

Note: Areas classified as sedimentary also include some sedimentary/volcanic areas.

Source: Generalized Geological Map of the World, Geological Survey of Canada

0 200 400 600 800 1000 1200 Miles
0 200 400 600 800 1000 1200 1400 1600 1800 2000 Kilometers

Lambert Azimuthal Equal Area Projection
Scale 1:40,000,000
One inch to 640 miles
One cm to 400 km

LAPTEV SEA

NEW SIBERIAN ISLANDS

EAST SIBERIAN SEA

OSTROV VRANGELYA

CHUKCHI PEN.

Gulf of Anadyr'

Anadyr'

BERING SEA

ALEUTIAN ISLANDS

0° 40'

Olenëk

Indigirka

Yana

CHERSKIY MTS.

Kolyma

Magadan

KAMCHATKA PENINSULA

VERKHOYANSK MTS.

Lena

Aldan

Petropavlovsk-Kamchatskiy

SEA OF OKHOTSK

Zhizhnyaya

Vilyuy

Yakutsk

SAKHALIN

KURIL ISLANDS

Tunguska

Lena

STANOVOY RANGE

YABLONOVYY RANGE

Angara

Kerulen

STANOVOY MTS.

Lake Baikal

Irkutsk

Ulaanbaatar

GREATER KHINGAN RANGE

Argun

Amur

Komsomol'sk-na-Amure

Songhua

Harbin

Ussuri

Amur

SIKHOTE-ALIN'

Sapporo

HOKKAIDŌ

SEA OF JAPAN (EAST SEA)

PACIFIC OCEAN

GOBI DESERT

Liao

Beijing

Seoul

HONSHŪ

Tōkyō

Tropic of Cancer

Huang

YELLOW SEA

JÉJUDO

KYŪSHŪ

SHIKOKU

Xi'an

QIN LING

Yalong

Three Gorges Reservoir

Yangtze

Chongqing

Han

Shanghai

EAST CHINA SEA

RYUKYU ISLANDS

Taipei

MARIANA ISLANDS

MICRONESIA

PHILIPPINE SEA

Salween

Xi

Hong Kong

HAINAN DAO

SOUTH CHINA SEA

LUZON

Manila

MINDORO

SAMAR

CAROLINE ISLANDS

Irrawaddy

Yangon (Rangoon)

Bangkok

ANDAMAN SLANDS

ANDAMAN SEA

Gulf of Thailand

Ho Chi Minh City

PALAWAN

PANAY

NEGROS

MINDANAO

NICOBAR ISLANDS

MALAY PENINSULA

Strait of Malacca

SUMATRA

Singapore

BORNEO

SULU SEA

CELEBES SEA

Makassar Strait

IRIAN MTS.

MOLUCCA SEA

MOLUCCAS

Jayapura

BISMARCK SEA

MELANESIA

GREATER SUNDA ISLANDS

SULAWESI (CELEBES)

Makassar

BANDA SEA

NEW GUINEA

SOLOMON SEA

Equator

Jakarta

JAVA

JAVA SEA

FLORES SEA

LESSER SUNDA ISLANDS

TIMOR

ARAFURA SEA

TIMOR SEA

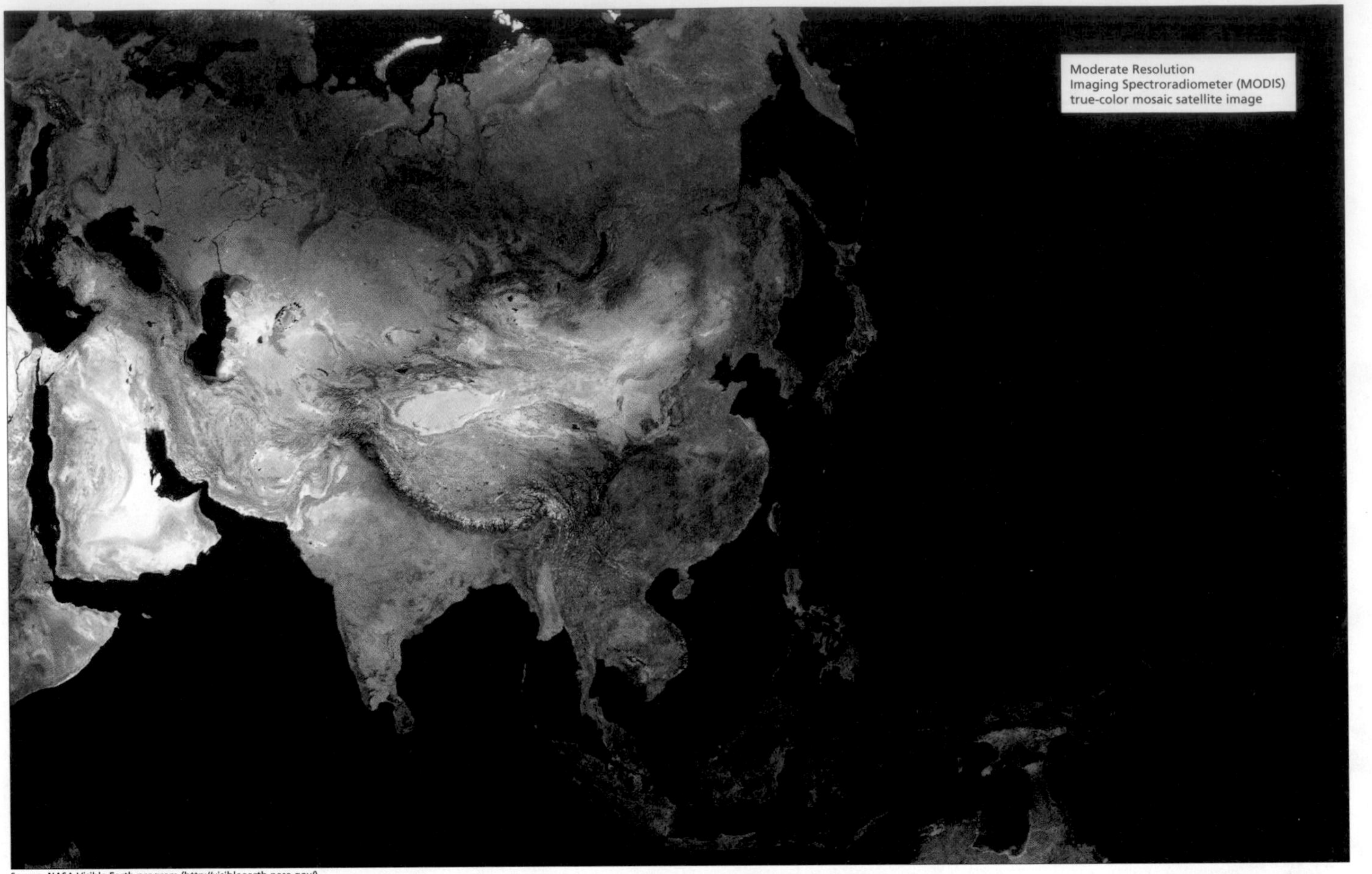
Moderate Resolution
Imaging Spectroradiometer (MODIS)
true-color mosaic satellite image

Source: NASA Visible Earth program (http://visibleearth.nasa.gov/)

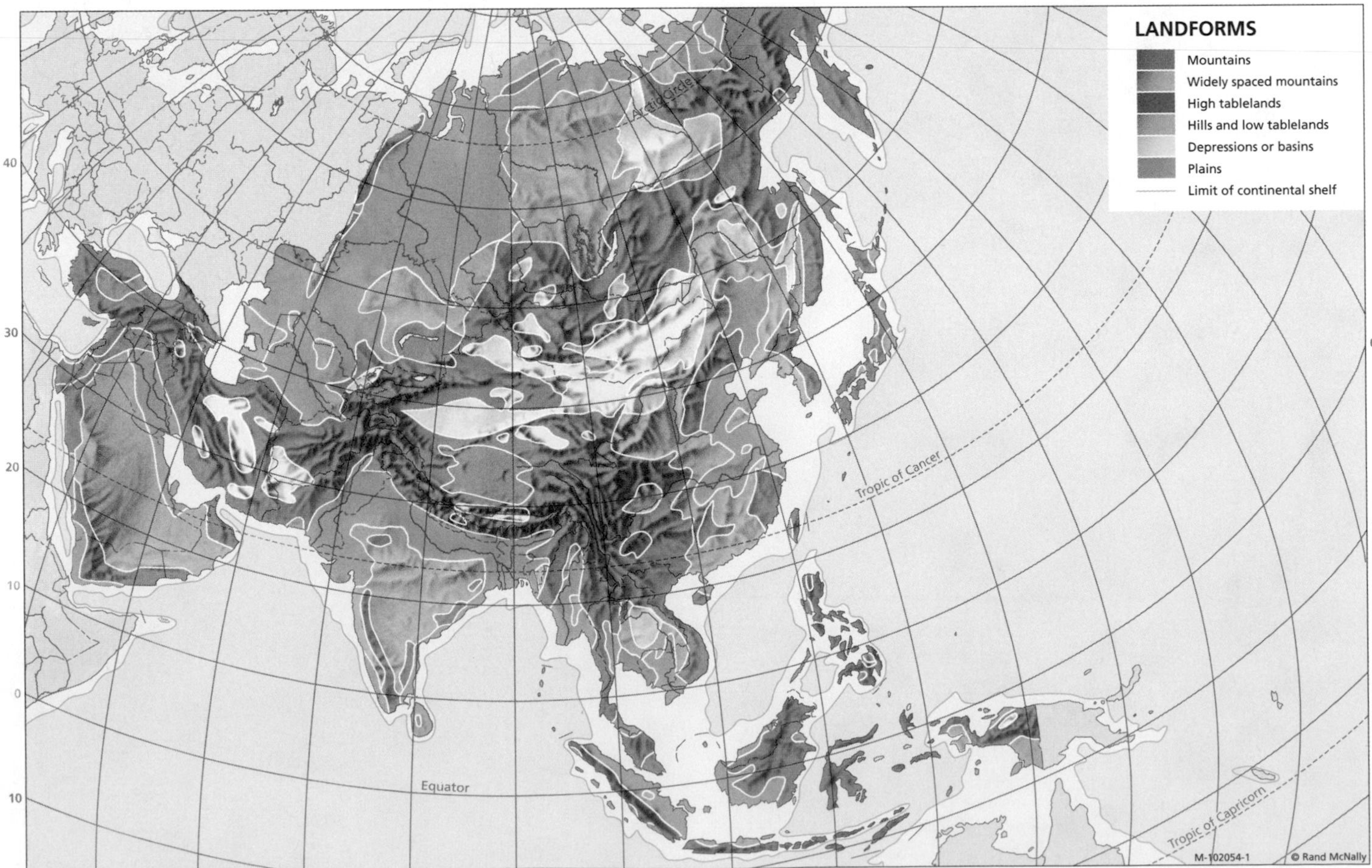
LANDFORMS
Mountains
Widely spaced mountains
High tablelands
Hills and low tablelands
Depressions or basins
Plains
Limit of continental shelf
Arctic Circle
Tropic of Cancer
Equator
Tropic of Capricorn
M-102054-1
© Rand McNally
40
30
20
10
0
10
0
10
20
30
90
100
110
120
130
140
150
160
170

Source: Murphy, 1968. © Association of American Geographers. Published by Taylor & Francis. Adapted with permission of the Association of American Geographers.

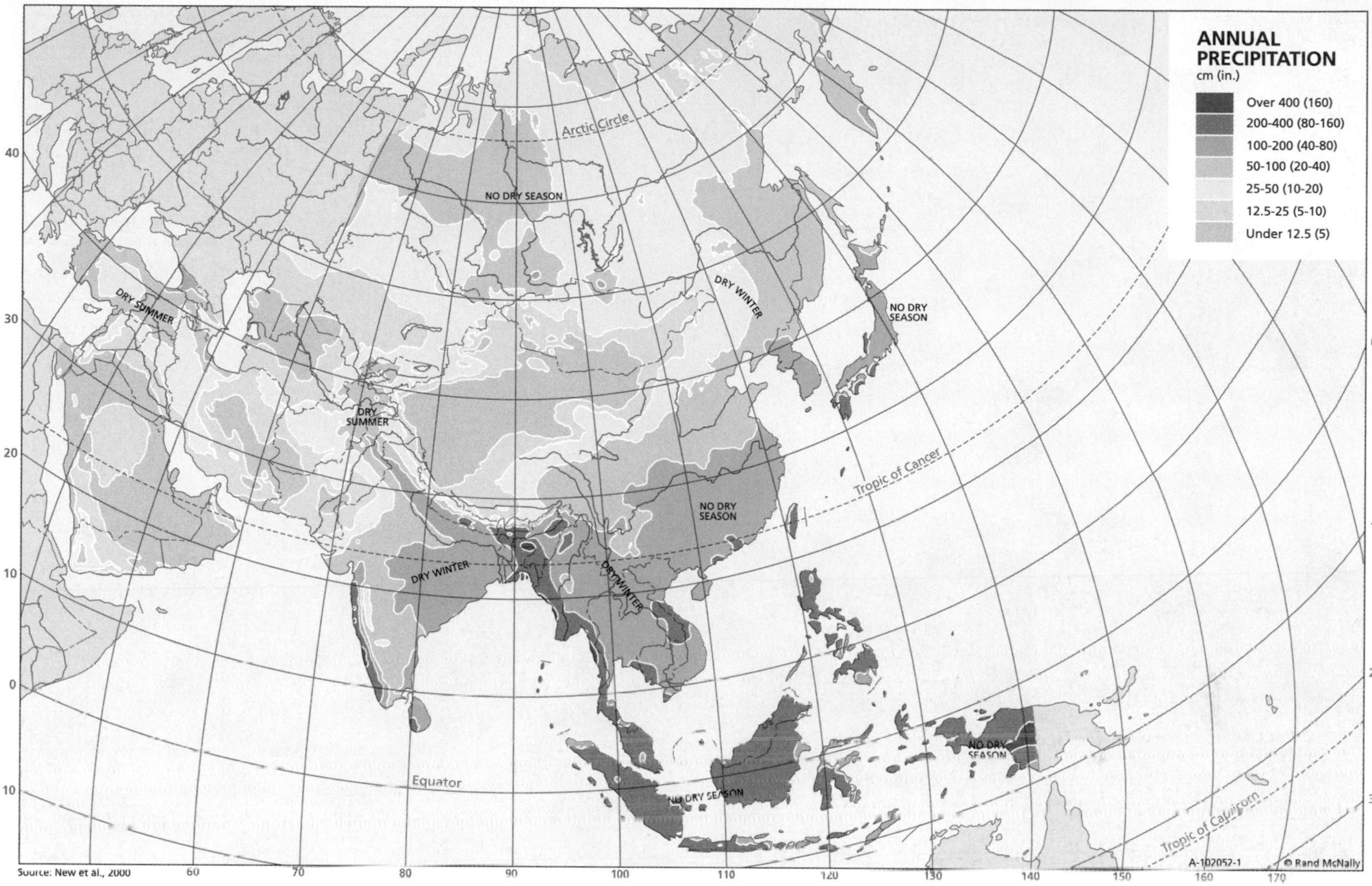

ANNUAL PRECIPITATION
cm (in.)

	Over 400 (160)
	200-400 (80-160)
	100-200 (40-80)
	50-100 (20-40)
	25-50 (10-20)
	12.5-25 (5-10)
	Under 12.5 (5)

NO DRY SEASON

DRY SUMMER

DRY WINTER

NO DRY SEASON

DRY SUMMER

NO DRY SEASON

DRY WINTER

DRY WINTER

Arctic Circle

Tropic of Cancer

Equator

Tropic of Capricorn

NO DRY SEASON

NO DRY SEASON

Source: New et al., 2000

A-102052-1 © Rand McNally

VEGETATION

	B	Tropical rain forest
		Subtropical rain forest
	B,Bs	Mediterranean vegetation
	S	Semideciduous mixed forest
DJRS,	D,Dh	Tropical dry deciduous forest
	ND-D	Temperate deciduous forest
	M,(SE)	Temperate mixed forest
	Ep,E,N	Coniferous forest
DsG,GBp,	GSp	Savanna (locally wooded)
	DG	Wooded steppe
	G	Grass (steppe)
	Gp	Short grass
Dzp,	Dzp	Desert shrub
	L	Tundra, alpine vegetation
	b	Little or no vegetation

For explanation of letters in boxes,
see World Natural Vegetation Map.

TAIGA

TAKLA MAKAN

GOBI

Arctic Circle

Tropic of Cancer

Equator

Tropic of Capricorn

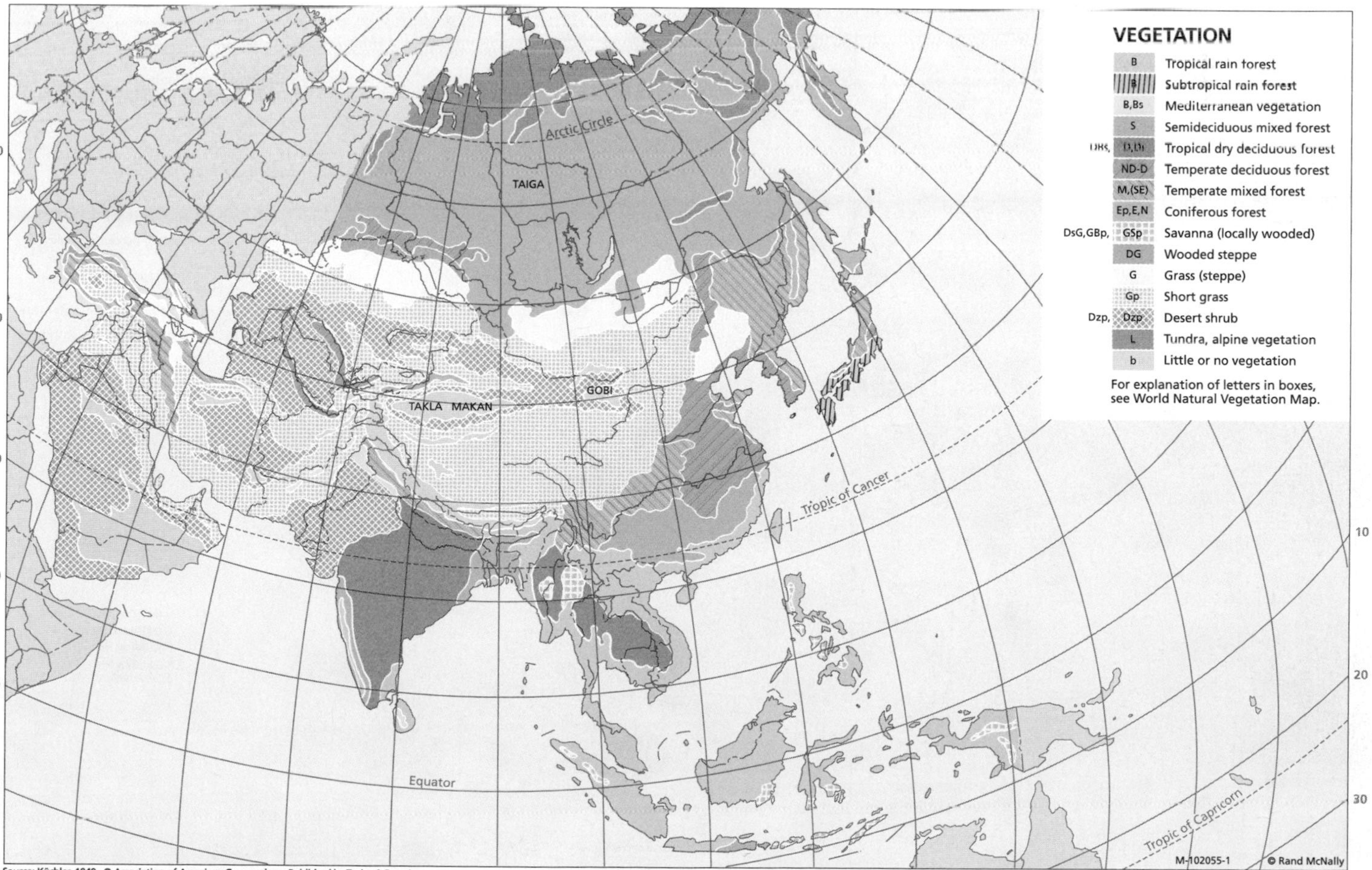

Source: Küchler, 1949. © Association of American Geographers. Published by Taylor & Francis.
Adapted with permission of the Association of American Geographers.

M-102055-1 © Rand McNally

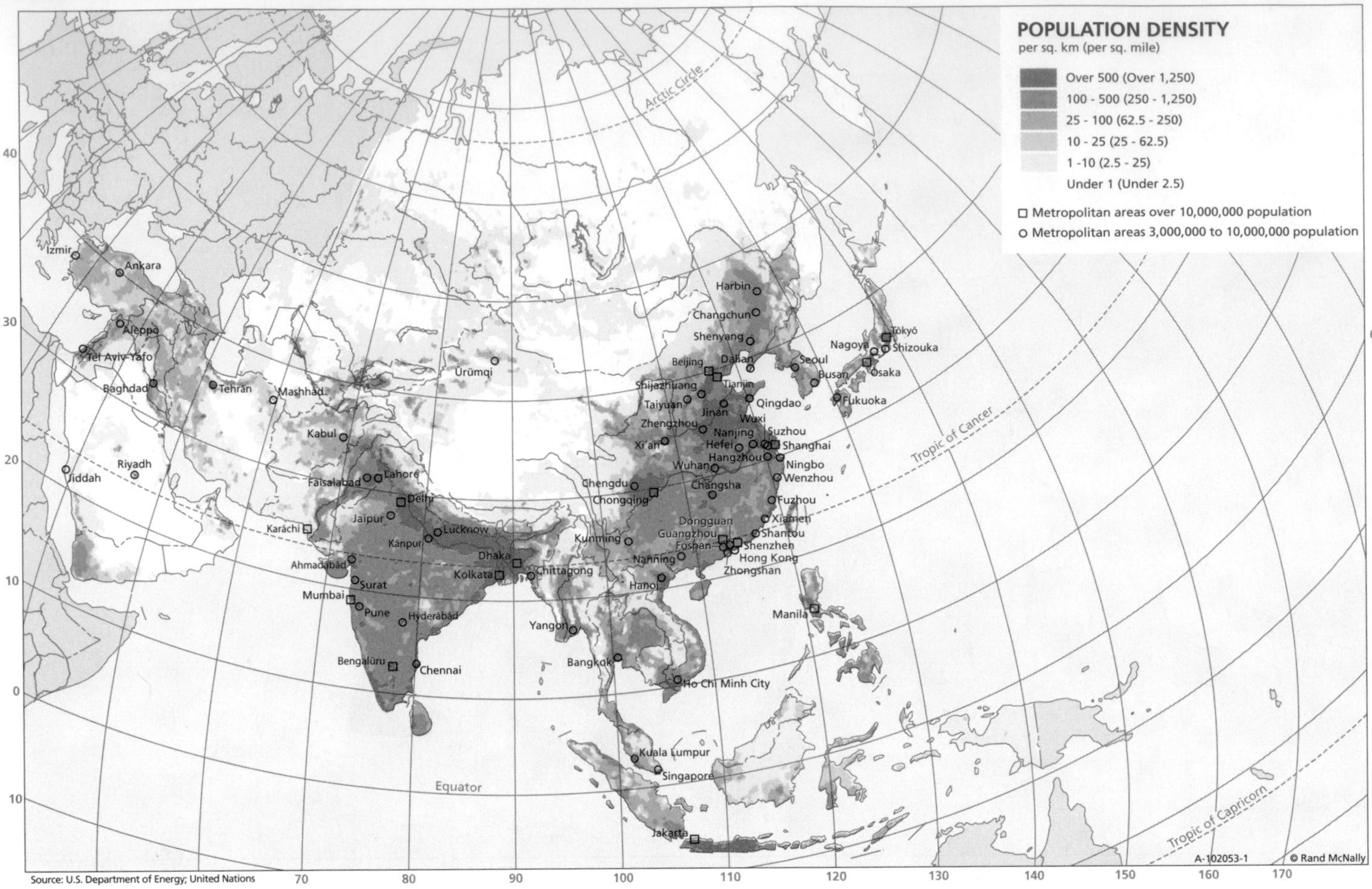

POPULATION DENSITY
per sq. km (per sq. mile)

Over 500 (Over 1,250)
100 - 500 (250 - 1,250)
25 - 100 (62.5 - 250)
10 - 25 (25 - 62.5)
1 -10 (2.5 - 25)
Under 1 (Under 2.5)

□ Metropolitan areas over 10,000,000 population
○ Metropolitan areas 3,000,000 to 10,000,000 population

Izmir
Ankara
Aleppo
Tel Aviv-Yafo
Baghdad
Tehran
Mashhad
Ürümqi
Riyadh
Jiddah
Kabul
Lahore
Faisalabad
Delhi
Karachi
Jaipur
Lucknow
Kanpur
Dhaka
Ahmadabad
Surat
Kolkata
Chittagong
Mumbai
Pune
Hyderabad
Yangon
Bengalūru
Chennai
Bangkok
Kuala Lumpur
Singapore
Jakarta

Harbin
Changchun
Shenyang
Beijing
Dalian
Seoul
Shijiazhuang
Tianjin
Busan
Taiyuan
Jinan
Qingdao
Zhengzhou
Nanjing
Wuxi
Xi'an
Hefei
Suzhou
Wuhan
Hangzhou
Shanghai
Chengdu
Changsha
Ningbo
Chongqing
Wenzhou
Kunming
Dongguan
Fuzhou
Guangzhou
Shantou
Xiamen
Nanning
Foshan
Shenzhen
Hong Kong
Zhongshan
Hanoi
Manila
Ho Chi Minh City

Nagoya
Tōkyō
Shizuoka
Osaka
Fukuoka

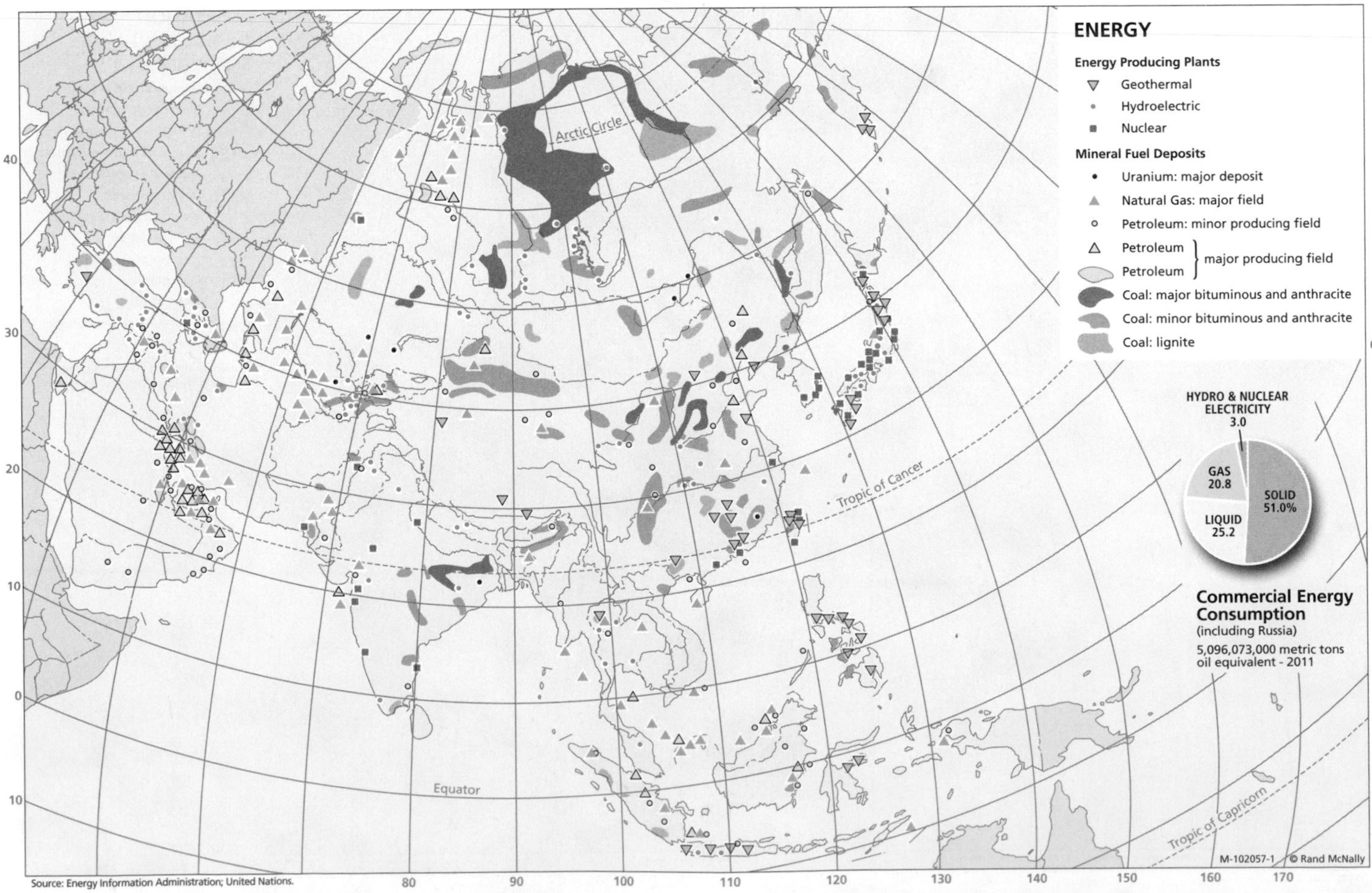

ENERGY

Energy Producing Plants
▽ Geothermal
• Hydroelectric
■ Nuclear

Mineral Fuel Deposits
• Uranium: major deposit
▲ Natural Gas: major field
○ Petroleum: minor producing field
△ Petroleum } major producing field
Petroleum }
Coal: major bituminous and anthracite
Coal: minor bituminous and anthracite
Coal: lignite

HYDRO & NUCLEAR
ELECTRICITY
3.0
GAS
20.8
SOLID
51.0%
LIQUID
25.2

**Commercial Energy
Consumption**
(including Russia)
5,096,073,000 metric tons
oil equivalent - 2011

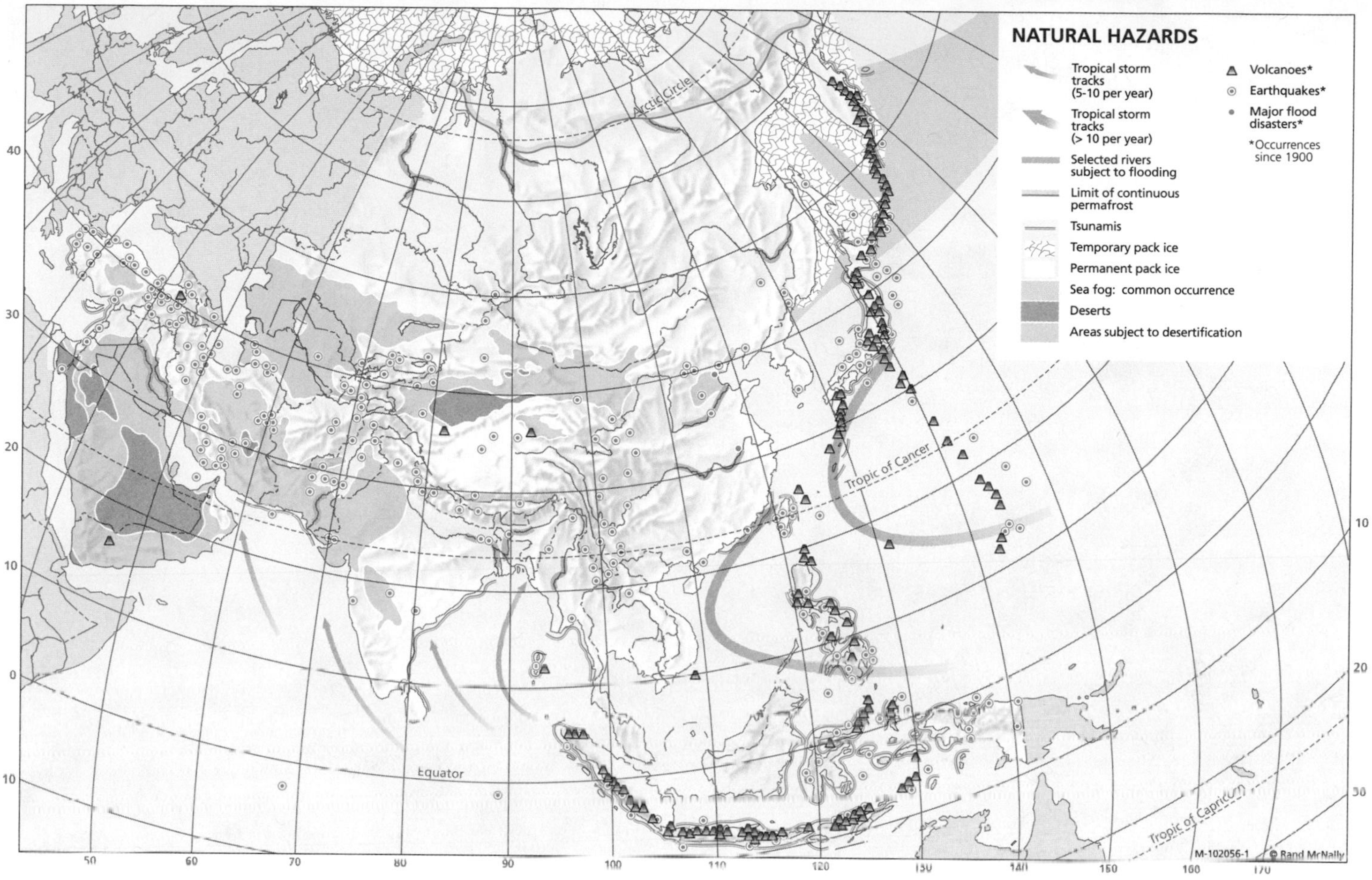

NATURAL HAZARDS

- Tropical storm tracks (5–10 per year)
- Tropical storm tracks (> 10 per year)
- Selected rivers subject to flooding
- Limit of continuous permafrost
- Tsunamis
- Temporary pack ice
- Permanent pack ice
- Sea fog: common occurrence
- Deserts
- Areas subject to desertification
- ▲ Volcanoes*
- ⊙ Earthquakes*
- ● Major flood disasters*

*Occurrences since 1900

Arctic Circle

Tropic of Cancer

Equator

Tropic of Capricorn

M-102056-1 © Rand McNally

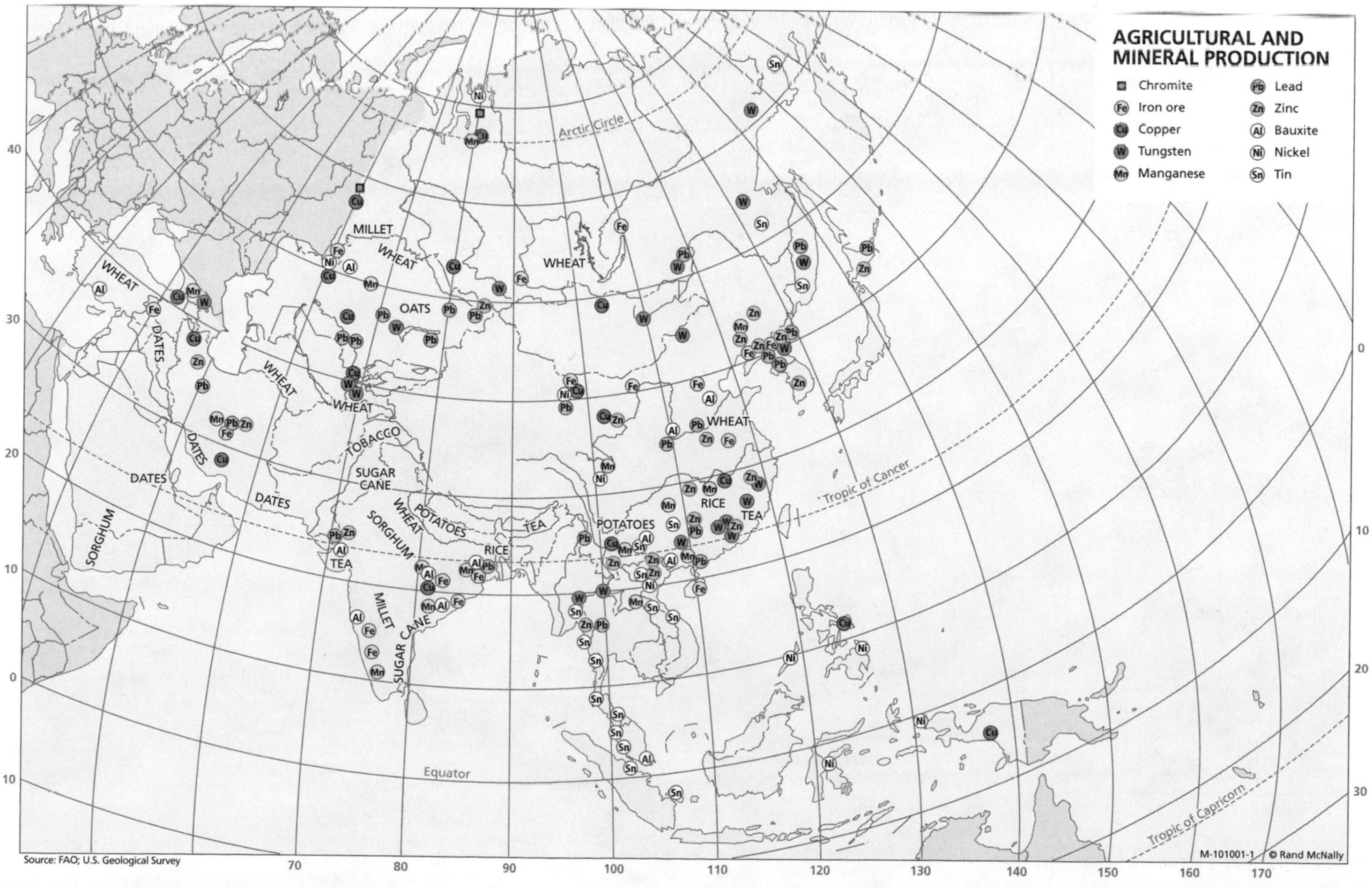

AGRICULTURAL AND MINERAL PRODUCTION

- ◼ Chromite
- Fe Iron ore
- Cu Copper
- W Tungsten
- Mn Manganese
- Pb Lead
- Zn Zinc
- Al Bauxite
- Ni Nickel
- Sn Tin

Arctic Circle

Tropic of Cancer

Equator

Tropic of Capricorn

Source: FAO; U.S. Geological Survey

M-101001-1 © Rand McNally

NORWEGIAN BASIN

NORWEGIAN SEA

ARCTIC

SPITSBERGEN

SVALBARD (Nor.)

FRANZ JOSEF LAND (ZEMLYA FRANTSA-IOSIFA)

OSTROV PIONER

OSTROV SHMIDTA

SEVERN

NORTH SEA

FAROE ISLANDS

ORKNEY IS. (U.K.)

SHETLAND IS. (U.K.)

Torshavn

NORWAY

SWEDEN

Trondheim

Narvik

Tromsø

Hammerfest

BARENTS TROUGH

BARENTS SEA

NOVAYA ZEMLYA

OSTROV VIZE

KARA SEA

OSTROV PIONER

Stavanger

Bergen

Oslo

Galdhøpiggen 8,100 ft 2469 m

Lofoten

Vesterålen

Bodø

Murmansk

KOLSKIY POLUOSTROV

Apatity

Kandalaksha

MYS KANIN NOS

OSTROV KOLGUYEV

Pechorskoye More

OSTROV VAYGACH

OSTROV BELYY

OSTROV SIBIRYAKOVA

Dikson

Kristiansand

Skagerrak

Göteborg

Stockholm

Gävle

Uppsala

Gulf of Bothnia

Luleå

Kemi

Oulu

Kajaani

WHITE SEA

Severodvinsk

Arkhangel'sk

Mezen

Naryan-Mar

Salekhard

POLUOSTROV YAMAL

Yar-Sale

Tazovskiy

Dudinka

Noril'sk

Ozero Pyasino

Pyasina

BALTIC SEA

Turku

Tampere

Helsinki

ESTONIA

Tallinn

Tartu

Narva

FINLAND

Lahti

St. Petersburg

Petrozavodsk

Vologda

Kotlas

Syktyvkar

Ukhta

Pechora

Vorkuta

Gora Narodnaya 6,214 ft 1894 m

Nadym

Arctic Circle

Ob'

Kazym

Tarko-Sale

Turukhansk

WEST SIBERIAN PLAIN

Yenisey

Klaipeda

Liepāja

Riga

LATVIA

LITHUANIA

Vilnius

Kaunas

Kaliningrad RUS.

POL.

Hrodna

BELARUS

Minsk

Vicebsk

Mahilëu

Daugavpils

Pskov

Novgorod

Borovichi

Tikhvin

Cherepovets

Rybinsk

Yaroslavl'

Kirov

Glazov

Berezniki

Khanty-Mansiysk

Surgut

Nizhnevartovsk

(ZAPADNO-SIBIRSKAYA

RUS

Pinsk

Babruisk

Homel

Smolensk

Kaluga

Moscow (Moskva)

Vladimir

Ivanovo

Kostroma

Nizhniy Novgorod

Cheboksary

Yoshkar-Ola

Izhevsk

Perm'

Serov

Tobol'sk

RAVNINA)

Ishim

Tara

Kolpashevo

Tomsk

Asino

Chernihiv

Bryansk

Orel

Tula

Ryazan'

Kolomna

Dzerzhinsk

Kazan'

Naberezhnye Chelny

Pervouralsk

Nizhniy Tagil

Yekaterinburg

Tyumen'

Kargasok

Kiev (Kyïv)

Sumy

Kursk

Elets

Tambov

Penza

Saransk

Ulyanovsk

Al'met'yevsk

Dimitrovgrad

Ufa

Zlatoust

Kurgan

Chelyabinsk

Petropavlovsk

Omsk

Kuybyshev

Yurga

Novosibirsk

Kemerovo

Anzhero-Sudzhensk

Ukraine

Poltava

Belgorod

Stary Oskol

Lipetsk

Voronezh

Syzran'

Tolyatti

Samara

Sterlitamak

Salavat

Magnitogorsk

Qostanay

Rudnyy

Zhetiqara

Kökshetaü

Shchüchinsk

Pavlodar

Leninsk-Kuznetskiy

Belovo

Prokopyevsk

Novokuznetsk

Abak

Kharkiv

Dnipropetrovs'k

Luhans'k

Saratov

Balakovo

Kamyshin

Oral

Orenburg

Orsk

Aqtöbe

Qostanay

Astana

Temirtaü

Ekibastuz

Barnaul

Bijsk

Gorno-Altaysk

Novosibirskoye Vdkhr.

Zaporizhzhia

Donets'k

Rostov-na-Donu

Volgograd

Astrahan

Atyraū

Arqalyq

Qaraghandy

Semey

Öskemen

Rubtsovsk

Ridder

Zyryanovsk

Mt. Belukha 14,783 ft 4506 m

Altay Mts.

Olgiy

Mariupol

Kerch

SEA OF AZOV

Elista

CASPIAN DEPRESSION

Aqtaū

Zhezqazghan

KAZAKH HILLS

Ayaköz

Krasnodar

Maykop

Sochi

Stavropol'

Cherkessk

Nalchik

Groznyy

Makhachkala

Derbent

CASPIAN SEA

Aral Sea

Bayqongyr (Baikonur)

Qyzylorda

Lake Balkhash (Balqash köli)

Balqash

Taldyqorghan

Karamay

Tacheng

MONG

BLACK SEA

CAUCASUS MTS.

Sukhumi

Batumi

Kutaisi

GEORGIA

TBILISI

Gyumri

ARMENIA

YEREVAN

Gäncä

Mingäçevir

Xankändi

AZERBAIJAN

Sumqayıt

BAKU (BAKI)

UST-URT PLATEAU

Nukus

Daşoguz

Urganch

UZBEKISTAN

QIZILQUM

Shymkent

Türkistan

Taraz

Bishkek

Balykchy

KYRGYZSTAN

TIEN SHAN

Almaty

Yining

Changji

ÜRÜMQI

Erzurum

Mt. Ararat 16,854 ft 5137 m

Van

TURKEY

TABRIZ

Orūmīyeh

Maragheh

Ardabil

Rasht

Zanjan

Qazvin

IRAN

Länkäran

KARA KUM

TURKMENISTAN

Balkanabat

Gyzylarbat

Türkmenbaşy

TASHKENT

Buxoro

Aşgabat

Türkmenabat

Qarshi

Samarqand

Dushanbe

TAJIKISTAN

Khujand

Namangan

Farg'ona

Andijon

Osh

Naryn

Karakol

Jengish Chokusu 24,406 ft 7439 m

Lake Issyk

Kashi

Aksu

Korla

TARIM PENDI

Pik Imeni Ismail Samani 24,590 ft 7495 m

PAMIRS

Shache

Yarkand

MASHHAD

Sabzevar

Quchan

Tejen

Mary

Kara-Kum Canal

Gonbad-e Kāvūs

Gorgan

AFGHAN.

IRAQ

Sanandaj

OCEAN

OSTROV KOMSOMOLETS
OSTROV OKTYABR'SKOY REVOLYUTSII
ZEMLYA OSTROV BOL'SHEVIK

MYS CHELYUSKIN
MYS MALYY TAYMYR

POLUOSTROV TAYMYR

SEVERO-SIBIRSKAYA NIZMENNOST'

GORY PUTORANA

LAPTEV SEA

NEW SIBERIAN ISLANDS (NOVOSIBIRSKIYE OSTROVA)

OSTROV BENNETTA
OSTROV FADDEYEVSKIY
OSTROV NOVAYA SIBIR
OSTROV BOL'SHOY LYAKHOVSKIY
OSTROV KOTEL'NYY
OSTROV BEL'KOVSKIY
OSTROV BOL'SHOY BEGICHEV
OSTROV STOLBOVOY

EAST SIBERIAN SEA

CHUKCHI SEA

OSTROV VRANGELYA

MYS SHMIDTA

Gulf of Anadyr

ST. LAWRENCE ISLAND (U.S.)

ST. MATTHEW ISLAND (U.S.)

BERINGIAN BASIN

ALEUTIAN SEA

KORYAKSKOYE NAGOR'YE

KHREBET CHERSKOGO

Gora Pobeda 10,325 ft. 3147 m

KAMCHATKA PENINSULA

SREDINNYY KHREBET

SEA OF OKHOTSK

OKHOTSK BASIN

KURIL ISLANDS

PACIFIC OCEAN

VERKHOYANSKIY KHREBET

Yakutsk

Mirnyy

ALDANSKOYE NAGOR'YE

STANOVOY KHREBET

STANOVOYE NAGOR'YE

Gora Skalistyy Golets 8094 ft. 2467 m

Tynda

SAYAN MTS.

Irkutsk
Ulan-Ude
Lake Baikal
YABLONOVYY KHREBET
Chita

KRASNOYARSK
Bratsk
Ust'-Ilimsk

GREATER KHINGAN RANGE (DA HINGGAN LING)

Blagoveshchensk

Komsomol'sk-na-Amure

Khabarovsk

SAKHALIN

Yuzhno-Sakhalinsk

OSTROV PARAMUSHIR

HOKKAIDO

Asahikawa
SAPPORO
Hakodate
Obihiro
Kushiro

Aomori

MONGOLIA

ULAANBAATAR
Darhan
Choybalsan

GOBI DESERT

Hami
BAOTOU
Hohhot
Datong

CHINA

QIQIHAR
HARBIN
CHANGCHUN
JILIN
SHENYANG
FUSHUN
ANSHAN
BEIJING
TIANJIN
TANGSHAN
DALIAN

NORTH KOREA
P'YONGYANG

SOUTH KOREA
SEOUL
DAEJEON
DAEGU
GWANGJU
BUSAN

SEA OF JAPAN (EAST SEA)

YELLOW SEA

JAPAN

HONSHU

TOKYO
YOKOHAMA
NAGOYA
KYOTO
KOBE
OSAKA
HIROSHIMA
FUKUOKA
SENDAI
Niigata
Nagano

SHIKOKU
KYUSHU

100   200   300   400   500   600 Miles
100 200 300 400 500 600 700 800 900 1000 Kilometers

Lambert Azimuthal Equal Area Projection
Scale 1:20,000,000
One inch to 320 miles
One cm to 200 km

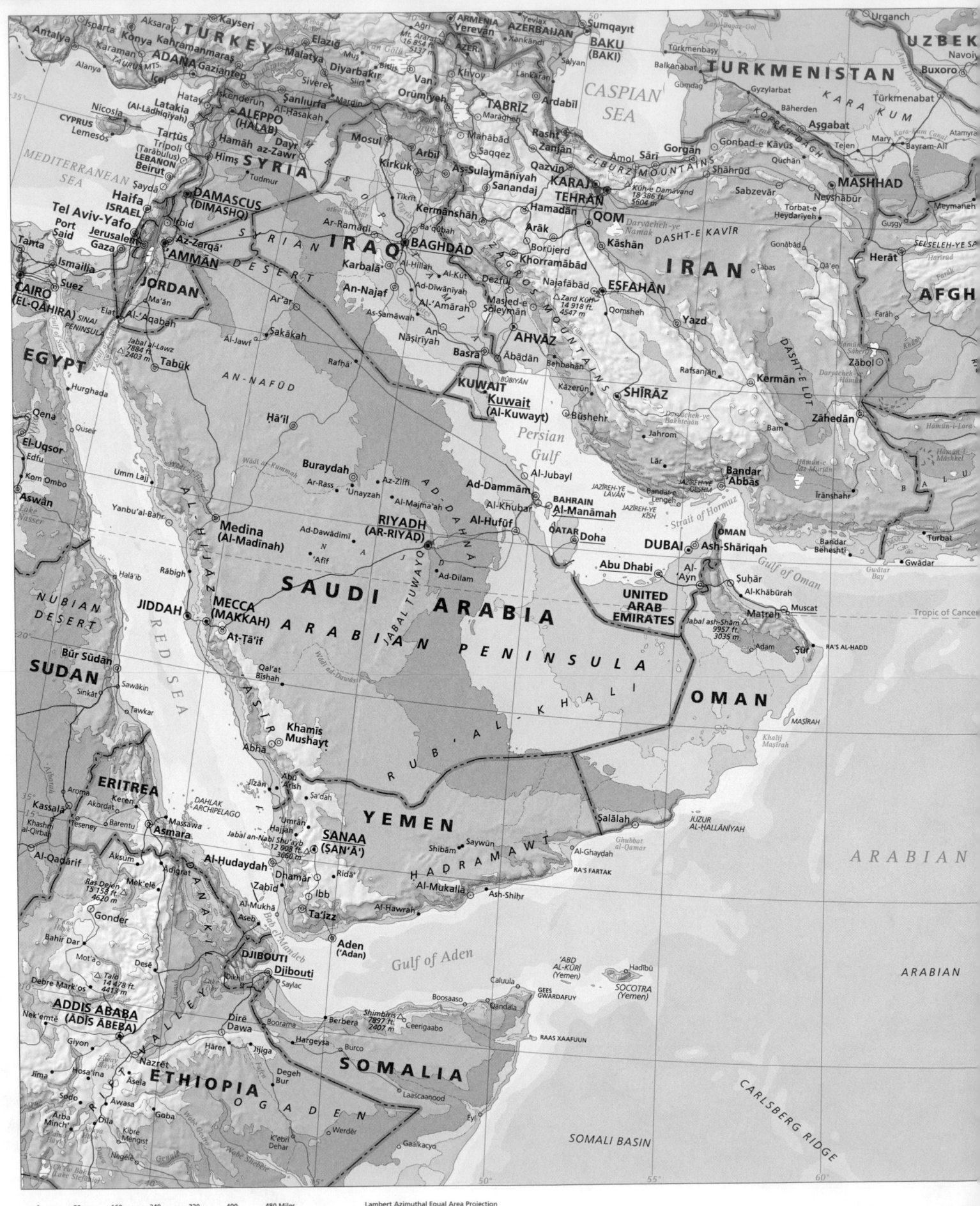

TURKEY
Antalya Isparta Aksaray Kayseri Kahramanmaras Ağrı ARMENIA AZERBAIJAN Sumqayıt
Konya Karaman Malatya Elazığ Muş Yerevan AZER. BAKU Türkmenbaşy Urganch
TAURUS MTS. İçel Siverek Diyarbakır Siirt Van Mt. Ararat Xankändi (BAKI) UZBEK
Alanya Nicosia Hatay İskenderun Şanlıurfa Mardin Orūmīyeh 16 854 ft. Khvoy Länkäran Salyan Balkanabat Buxoro
CYPRUS (Al-Lādhiqīyah) Al-Hasakah Mosul TABRĪZ Maraghen Ardabīl CASPIAN Gümdag TURKMENISTAN Navoiy
Lemesós Latakia Tartūs Dayr Kirkuk Sanandaj Saqqez Zanjān Rasht SEA Bäherden Aşgabat Türkmenabat
Tripoli az-Zawr Hamāh As-Sulaymānīyah Qazvīn Āmol Sārī Gorgān KARA Mary MASHHAD
(Tarābulus) LEBANON Ḥimṣ SYRIA Tudmur KARAJ Kūh-e Damāvand Shāhrūd KUM Quchān Neyshābūr Meymaneh
Beirut Şaydā DAMASCUS Ba'qūbah TEHRĀN 18 386 ft. Sabzevār Gonbad-e Kāvūs Güşgy Herāt
Haifa (DIMASHQ) An-Najaf BAGHDAD QOM 5604 m Torbat-e Gonābād AFGH
ISRAEL Tel Aviv-Yafo Az-Zarqā' Al-Hillah Al-Kūt Hamadān DASHT-E KAVĪR Heydariyeh
Port Jerusalem Gaza AMMĀN IRAQ Karbalā An-Najaf Arāk Qā'en
Said Al-'Aqabah JORDAN Al-Dīwānīyah Al-'Amārah Borūjerd Khorramābād Tabas
Tanta Ismailia Ma'ān Ar-Ramādī An-Nāṣirīyah Dezfūl Kāshān IRAN Farāh
CAIRO Suez Elat SYRIAN Ar'ar As-Samāwah Masjed-e ESFAHĀN Yazd
(EL-QĀHIRA) SINAI DESERT Sakākah Nāṣirīyah An Soleymān Qomsheh Dārreich-ye Zābol
EGYPT PENINSULA Jabal al-Lawz Al-Jawf Rafḥā' Basra Zard Kūh 14 918 ft. Nīrīz Dasht-e Rafsanjān SELSELEH-YE Hāmūn-i-Sabarī
Hurghada 7884 m Tabūk Ābādān Behbahān 4547 m Kerman LŪT
2403 m AN-NAFŪD Wādī ar-Rummah KUWAIT Kāzerūn SHĪRĀZ Zāhedān
Qena Qusier Umm Lajj Ḥā'il Kuwait Būbīyān Būshehr Jahrom Bam Irānshahr
El-Uqsor (Al-Kuwayt) Kerman Hāmūn-e DASHT-E Qā'en
Edfu Kom Ombo Yanbu'al-Bahr Buraydah Az-Ziffī Ad-Dammām JAZĪREH-YE Lār Bandar Jāz Mūrīān Turbat
Aswan Medina Ar-Rass Al-Majma'ah Al-Jubayl LĀVĀN Abbās B Gwādar
Lake (Al-Madīnah) 'Unayzah Al-Khubar BAHRAIN JAZĪREH-YE Strait of Hormuz Gwātar
Nasser RIYADH Al-Hufūf Al-Manāmah KĪSH OMAN Bandar Bay
Ad-Dawādimī (AR-RIYĀD) QATAR Beheshtī
Hala'ib Rābigh 'Afif Doha DUBAI Ash-Shāriqah
NUBIAN JIDDAH Abu Dhabi Al- Suḥār Tropic of Cancer
DESERT MECCA SAUDI ARABIA UNITED 'Ayn Al-Khābūrah
Bür Sūdān (MAKKAH) ARAB Jabal ash-Sham Matrah RA'S AL-HADD
SUDAN At-Tā'if ARABIAN PENINSULA EMIRATES 9957 ft. Muscat
Sinkāt Qal'at 3035 m Adam Şūr
Sawākin Bīshah Wādī ad-Dawāsir Aden OMAN MAṢĪRAH
Tawkar ASIR RUB' AL- KHALI Khalīj
Khamīs Maṣīrah
ERITREA Mushayt JUZUR
Kassala Keren Abha AL-HALLĀNĪYAH
Khashm Akordat DAHLAK 'Umrān Şa'dah Salālah
al-Qirbah Teseney Barentu ARCHIPELAGO Hajjah Ghubbat ARABIAN
Al-Qadārif Massawa Jabal an-Nabi Shu'ayb SANAA al-Qamar
Asmara 12 008 ft. (ŞAN'Ā') Shibām Saywūn RA'S FARTAK
Ras Dejen Al-Hudaydah 3660 m Al-Ghaydah
15 158 ft. Dhamār Ridā HADRAMAWT ARABIAN
4620 m Zabīd Ibb Al-Mukallā
Mek'elē Al-Mukhā YEMEN Ash-Shihr
Gonder Ta'izz Al-Hawrah
Bahir Dar Aden SOCOTRA
Debre Mark'os Dikhil ('Adan) Gulf of Aden (Yemen)
Mot'a Desē DJIBOUTI 'ABD Hadibū
ADDIS ABABA Djibouti Saylac AL-KŪRĪ
Nek'emtē (ĀDĪS ĀBEBA) Dirē Boorama (Yemen) Caluula
Giyon Dawa Berbera GEES Qandala
Nazrēt Hārēr Jijiga Boosaaso GWARDAFUY
Jima Hosa'ina Degeh SOMALIA RAAS XAAFUUN
Āsela Bur Shimbiris Ceerigaabo
Sodo Awasa Burco 7897 ft. CARLSBERG
Arba Dila Goba Hargeysa 2407 m
Minch' Kibre OGADEN Laascaanood RIDGE
Mengist ETHIOPIA Eyl
Negēle K'ebrī Werdēr SOMALI BASIN
Dehar Gaalkacyo

0 80 160 240 320 400 480 Miles
0 80 160 240 320 400 480 560 640 720 800 Kilometers

Lambert Azimuthal Equal Area Projection
Scale 1:16,000,000
One inch to 256 miles
One cm to 160 km

KAZAKH HILLS

RUSSIA

ALTAY MTS.

SAYAN MTS.

TYVA

BURYATIYA

Irkutsk

Angarsk

Öskemen

Ridder

Kyzyl

Shagonar

MONGOL ALTAYN NURUU

Uliastay

Ulaangom

Uvs Nuur

Darhan

Gusinoozërsk

HANGAYN NURUU

MONGOLIA

GOBI

Shymkent

Taraz

TASHKENT

Bishkek

ALMATY

Yining

Karamay

ÜRÜMQI

Hami

Dalandzadgad

UZBEKISTAN

KYRGYZSTAN

TIAN SHAN

Korla

TURFAN DEPRESSION

Barkol

TAJIKISTAN

Dushanbe

PAMIRS

Kashi

Aksu

Kuqa

Yanqi

Bosten Hu

Korla

NE

AFGHANISTAN

HINDU KUSH

KARAKORAM RANGE

XINJIANG

TARIM PENDI

TAKLA MAKAN DESERT

Hotan

Ruoqiang

ALTUN SHAN

QAIDAM PENDI

GREAT WALL

Yumen

QILIAN SHAN

Zhangye

Wuhai

PAKISTAN

Peshawar

RAWALPINDI

Srinagar

JAMMU AND KASHMIR

KUNLUN SHAN

XIZANG (TIBET)

PLATEAU OF TIBET

QINGHAI

Xining

LANZHOU

GANSU

NINGXIA

Yinchuan

Baiyin

LAHORE

FAISALABAD

LUDHIANA

AMRITSAR

HIMACHAL PRADESH

Shimla

Chandigarh

GANGDISE SHAN

TANGGULA SHAN

BAYAN HAR SHAN

CHINA

DELHI

New Delhi

MEERUT

UTTARAKHAND

Nagqu

Qamdo

Dêgê

Chengdu

SICHUAN

JAIPUR

AGRA

RAJASTHAN

NEPAL

HIMALAYA

KATHMANDU

Lhasa

NYAINQÊNTANGLHA SHAN

CHONGQING

INDIA

KANPUR

LUCKNOW

UTTAR PRADESH

BHUTAN

ARUNACHAL PRADESH

BHOPAL

MADHYA PRADESH

VARANASI (BENARES)

PATNA

BIHAR

ASSAM

NAGALAND

YUNNAN

NAGPUR

MAHARASHTRA

JHARKHAND

WEST BENGAL

DHAKA

BANGLADESH

MEGHALAYA

TRIPURA

MIZORAM

MYANMAR (BURMA)

KUNMING

GUIYANG

HYDERABAD

TELANGANA

ODISHA

KOLKATA (CALCUTTA)

CHITTAGONG

LAOS

ANDHRA PRADESH

EASTERN GHATS

BAY OF BENGAL

ARAKAN YOMA

THAILAND

VIETNAM

0   80   160   240   320   400   480 Miles
0   80   160   240   320   400   480   560   640   720   800 Kilometers

Lambert Azimuthal Equal Area Projection
Scale 1:16,000,000
One inch to 256 miles
One cm to 160 km

Lake Baikal (Ozero Baikal)
Ulan-Ude
Chita
Shilka
Nerchinsk
Mohe
Magdagachi
Guliah
Shimanovsk

YABLONOVYY KHREBET
Petrovsk-Zabaykal'skiy
Chikoy
Onon
Sherlovaya Gora
Borzya
Nahe
Bei'an
Heihe
Blagoveshchensk
Raychikhinsk
RUSSIA
Svobodnyy
Belogorsk
Khabarovsk
Birobidzhan

Manzhouli
Hailar
GREATER KHINGAN RANGE (DA HINGGAN LING)
MANCHURIA
Nunjiang
Yitulihe
Yi'an
Hailun
LESSER KHINGAN RANGE
Fujin
Litovko
Sovetskaya Gavan'
SAKHALIN
Makarov
Dolinsk
Yuzhno-Sakhalinsk
Korsakov
SEA OF OKHOTSK
OSTROV URUP

LAANBAATAR
Choybalsan
Zalantun
Kerulen
Arxan
Taonan
Zuoqi
QIQIHAR
Suihua
Nahe
HEILONGJIANG
Tieli
Hegang
Jiamusi
Shuangyashan
Boli
Mishan
Lesozavodsk
Dalnerechensk
SIKHOTE-ALIN
Vyazemskiy
Bikin
Wakkanai
RISHIRI-TO
Asahi-dake 7513 ft 2290 m
Kushiro
The islands known in Japan as the Northern Territories and in Russia as the Southern Kuril Islands are occupied by Russia and claimed by Japan.
OSTROV KUNASHIRI (KUNASHIR-TO)
MALAYA KURIL'SKAYA GRYADA (HABOMAI-SHOTO)

Buyant-Uhaa
Ulaan-Uul
Bayan Obo
Uliastai
Hailar
Hulan Nur
Arxan
HARBIN
Acheng
Shuangcheng
Anda
Mudanjiang
Spassk-Dal'niy
Arsen'yev
Kavalerovo
Dal'negorsk
Asahikawa
HOKKAIDO
SAPPORO
Tomakomai
Obihiro

A
Erenhot
Xilinhot
Tongliao
CHANGCHUN
Gongzhuling
Siping
Liaoyuan
Hunjiang
JILIN
Jilin
Nong'an
Dehui
Jiutai
Fuyu
Yanji
Huadian
Dunhua
Ch'ŏngjin
Ussuriysk
Artem
Vladivostok
Nakhodka
Aomori
Hachinohe
HAKODATE
Morioka
Akita

DESERT
Ulaan-Uul
Oagan Nur
Sonid Youqi
Linxi
Baicheng
Tongyu
Tieling
Kanggye
Hyesan
PAEKTU-SAN 9,003 ft 2,744 m
Hŭngdŏki-dong
SEA OF JAPAN (EAST SEA)
Sakata
Yamagata
Niigata
SADO
Fukushima
Koriyama
Iwaki

ONGOL
Bayan Obo
Wuyuan
Hohhot
Zhangjiakou
Chengde
Chifeng
Fuxin
LIAONING
SHENYANG
Benxi
ANSHAN
FUSHUN
Tonghua
NORTH KOREA
Hamhŭng
Kimch'aek
Nagaoka
Nagano
Utsunomiya
Mito
Toyama
Kanazawa
Maebashi
HONSHU
12,388 ft 3,776 m
Hamamatsu

BAOTOU
Jungar Qi
Huang (Yellow)
Datong
Beipiao
Chaoyang
Jinxi
Yingkou
Dandong
Sinŭiju
Wŏnsan
P'YŎNGYANG
Namp'o
Both North Korea and South Korea claim to be the sole legitimate government of Korea.
ULLEUNGDO (S. Kor.)
TŌKYŌ
YOKOHAMA
NAGOYA
KYOTO
KOBE
OSAKA

Jining
BEIJING
Qinhuangdao
Changli
Jinzhou
Lüshun
Kaesŏng
SEOUL
INCHEON
Cheongju
OKI-SHOTO
Tottori
Wakayama
MIYAKE-JIMA
HACHIJO-JIMA

Zhuozhou
TANGSHAN
TIANJIN
DALIAN
Bo Hai
Korea Bay
SOUTH KOREA
Pohang
Matsue
Okayama
Takamatsu
Matsuyama
HIROSHIMA
Kitakyūshū
AOGA-SHIMA

Baoding
HEBEI
Cangzhou
Weihai
Yantai
CHENGSHAN JIAO
DAEJEON
DAEGU
BUSAN
SHIKOKU
Kōchi
SHIJIAZHUANG
Dezhou
Lalyang
Jeonju
Gwangju
Korea Strait
Miyazaki

TAIYUAN
Yangquan
Yuci
JINAN
Zibo
Weifang
SHANDONG
QINGDAO
Mokpo
Matsuan
FUKUOKA
Saseho
Kumamoto
Kagoshima

SHANXI
Changzhi
Linfen
Handan
Taian
SHANDONG BANDAO
YELLOW SEA
Jejuju
JEJUDO
Nagasaki
YAKU-SHIMA
TANEGA-SHIMA

SHAANXI
Yan'an
Xinxiang
Jiaozuo
Jining
Yanzhou
Xuzhou
Lianyungang
Qingjiang
ZHENGZHOU
Kaifeng
Yancheng
SUWANOSE-JIMA
NAKANO-SHIMA
AMAMI-O-SHIMA
Naze

Tongchuan
Weinan
XI'AN
Luoyang
Pingdingshan
Xuchang
Suzhou
Fuyang
JIANGSU
Huaian
Xinghua
TOKUNO-SHIMA
OKINO-ERABU-SHIMA

QIN LING
Nanyang
HENAN
Bengbu
Yangzhou
Zhenjiang
CHANGXING DAO
EAST CHINA SEA
OKINAWA-JIMA (NANSEI-SHOTO)

NANJING
HEFEI
Changzhou
Wuxi
Naha
KUME-SHIMA
KITA-DAITŌ-JIMA
MINAMI-DAITŌ-JIMA (Japan)

A
HUBEI
Xiangfan
Xiaogan
Anlu
Lu'an
Wuhu
Suzhou
SHANGHAI
HANGZHOU
ZHOUSHAN DAO

Yichang
Enshi
Shashi
WUHAN
Huangshi
Anqing
Jiaxing
Shaoxing
NINGBO
Dongyang
Taizhou
OKINO-DAITŌ-JIMA (Japan)

Jinshi
Yueyang
Purqi
Jingdezhen
Jinhua
Quzhou
Lishui
Wenzhou
Tropic of Cancer

Changde
NANCHANG
Shangrao
ZHEJIANG

HUNAN
Yiyang
JIANGXI
Yichun
Fuzhou
Longquan
Shaowu

CHANGSHA
Xiangtan
Pingxiang
Ji'an
Jian'ou
Nanping

Shaoyang
Hongjiang
Jinggangshan
Yong'an
FUZHOU
MATSU TAO

Tongren
Hengyang
Yongzhou
Ganzhou
FUJIAN
Quanzhou
TAIPEI
Hsinchu
Taichung
Yu Shan 13,114 ft 3,997 m

Guilin
Chenzhou
NAN LING
Shaoguan
Zhangzhou
Xiamen
QUEMOY
Taiwan Strait
Chiai
TAIWAN
China and many other countries do not recognize the existence of Taiwan as a separate country.

Hechi
Liuzhou
Hexian
Meizhou
Chao'an
Tainan
IRIOMOTE-JIMA
ISHIGAKI-JIMA
Hirara
SAKISHIMA-SHOTO
RYUKYU ISLANDS
MIYAKO-SHOTO

GUANGXI
Wuzhou
GUANGDONG
GUANGZHOU
Qingtang
Shantou
KAOHSIUNG
OLUAN PI
PHILIPPINE SEA
OKINO-TORI-SHIMA (Japan)

Nanning
Yulin
Jiangmen
Foshan
Dongguan
Lufeng
Shenzhen
HONG KONG (XIANGGANG)
Macau (Aomen)

Zhanjiang
LEIZHOU BANDAO
PRATAS ISLAND (Occupied by Taiwan; claimed by China)
Basco
BATAN ISLANDS
Luzon Strait
PHILIPPINE BASIN

Beihai
Haikou
HAINAN
HAINAN DAO
Baoting
Wuzhi Shan 6,125 ft 1,867 m
SOUTH CHINA SEA
SOUTH CHINA BASIN
CAPE BOJEADOR
Laoag
ESCARPADA POINT
Balintang Channel
BABUYAN ISLANDS

Sanya
Vigan
PHILIPPINES
LUZON
Tuguegarao City

CHINA
YUNNAN
GUANGXI
GUANGDONG
HONG KONG (XIANGGANG)
KAOHSIUNG
TAIWAN

Nanning
Beihai
Zhanjiang
Macau (Aomen)
Shenzhen
Jiangmen
Guangzhou

MYANMAR (BURMA)
Mandalay
Maymyo
Mt. Victoria 10,016 ft. 3053 m
Paletwa
Myingyan
Meiktila
Taunggyi
Keng Tung
Kuang Xaignabouri
Louangphrabang
LAOS
Vientiane (Viangchan)
Fan Si Pan 10,312 ft. 3143 m
Ha Noi
Cao Bang
Yen Bai
Thai Nguyen
Hai Phong
Nam Dinh
Hon Gai
Thanh Hoa
Gulf of Tonkin
Vinh
HAINAN DAO
Dongfang
Haikou
Baoting
Wuzhi Shan 6125 ft. 1867 m
Sanya
Leizhou Bandao

Yenangyaung
Prome
Henzada
Tharrawaddy
Bago
Pha-an
Pinmana
Nay Pyi Taw
Chiang Mai
Chiang Rai
Lampang
Phitsanulok
Khon Kaen
Udon Thani
Savannakhet
Dong Hoi
Hue
Da Nang

YANGON (RANGOON)
Pathein
Mawlamyine
Ye
Nakhon Sawan
THAILAND
Nakhon Ratchasima
Ubon Ratchathani
Pakxé
INDOCHINA
VIETNAM
Play Ku
Quy Nhon

PARACEL ISLANDS (Occupied by China; claimed by Taiwan and Vietnam)

Dawei
Kadan Kyun
Mergui
Nakhon Pathom
Phra Nakhon Si Ayutthaya
Chon Buri
BANGKOK (KRUNG THEP)
Chanthaburi
Siĕmréab
Kâmpóng Thum
Stoeng Trêng
Krâchéh
Da Lat
Nha Trang

CAMBODIA
PHNOM PENH
Bătdâmbâng
Tonle Sap
KO CHANG
KO KUT

MUI KE GA
SOUTH CHINA BASIN

HO CHI MINH CITY (SAIGON)
Bien Hoa
Kâmpôt
Phan Thiet

ANDAMAN SEA
Chumphon
Kawthaung
ISTHMUS OF KRA
Rach Gia
Vung Tau
Can Tho
Bac Lieu

SOUTH CHINA SEA

SPRATLY ISLANDS (Claimed by China, Malaysia, the Philippines, Taiwan and Vietnam)

Surat Thani
Nakhon Si Thammarat
Trang
KO PHUKET
Thale Sap Songkhla
Songkhla
Hat Yai
Yala
Kota Bharu
PULAU LANGKAWI
Taiping
Ipoh
Gunong Tahan 7175 ft. 2187 m
Kuala Terengganu

MALAY PENINSULA

Banda Aceh
Lhokseumawe
Langsa
Alor Setar
George Town
PULAU PINANG
MALAYSIA
KUALA LUMPUR
Klang
Melaka
Batu Pahat
Johor Bahru
SINGAPORE
SINGAPORE

Binjai
MEDAN
Tebingtinggi
Pematangsiantar
Sibolga
Dumai
Pekanbaru
Bukittinggi
Padang
Gunung Kerinci 12,467 ft. 3800 m
Jambi
Tembilahan
PULAU NIAS
PULAU SIMEULUE
KEPULAUAN BANYAK
KEPULAUAN BATU
PULAU SIBERUT
PULAU SIPURA
PULAU PAGAI UTARA
PULAU PAGAI SELATAN
KEPULAUAN MENTAWAI

SUMATRA (SUMATERA)
PALEMBANG
Bengkulu
Gunung Dempo 10,364 ft. 3159 m
Kotabumi
Bandar Lampung
PEGUNUNGAN BARISAN

PULAU BANGKA
PULAU BELITUNG
KEPULAUAN LINGGA
PULAU SINGKEP
KEPULAUAN KARIMATA
Tanjungpandan
Ketapang

Singkawang
Pontianak
Sintang
BORNEO (KALIMANTAN)
PEG. SCHWANER
Bukit Raya 7474 ft. 2278 m
PEG. MULLER
Palangkaraya
Banjarmasin
Banjarbaru

NATUNA BESAR
KEPULAUAN ANAMBAS
PULAU TIOMAN
PULAU LAUT
SUNDA SHELF
TANJUNG DATU

Kuching
Sibu
Bintulu
Miri
BRUNEI
Bandar Seri Begawan
MALAYSIA
Gunong Murud 7946 ft. 2422 m
RAJANG MTS.
UPPER KAPUAS MTS.
Kota Kinabalu
Gunong Kinabalu 13,455 ft. 4101 m
Kudat
PULAU BANGGI
BALABAC ISLAND
Balabac Strait
Sandakan
Lahad Datu
Tawau
Tarakan
Samarinda
Balikpapan
Bontang
Palu
CELEBES SEA
SULU SEA
Zamboanga
Jolo
JOLO ISLAND
TAWITAWI ISLAND
BASILAN ISLAND

SULAWESI (CELEBES)
Bulu Rantekombola 11,335 ft. 3455 m
Majene
Parepare
Palopo
MAKASSAR (UJUNGPANDANG)
Watampone
Kendari
Teluk Bone
Teluk Tomini
Toli-toli
Luwuk
KEPULAUAN TOGIAN
PULAU PELENG

INDONESIA
GREATER SUNDA ISLANDS
JAVA SEA (LAUT JAWA)
TANJUNG PUTING
TANJUNG SELATAN
PULAU LAUT
KEPULAUAN LAUT KECIL
KEPULAUAN KANGEAN

JAKARTA
Bogor
Sukabumi
BANDUNG
Cirebon
SEMARANG
SURABAYA
Madura
Sumenep
Surakarta
Pasuruan
Kediri
Malang
Yogyakarta
Gunung Slamet 11,247 ft. 3428 m
Gunung Semeru 12,060 ft. 3676 m
Denpasar
JAVA (JAWA)
BALI SEA
Singaraja
Gunung Rinjani 12,224 ft. 3726 m
Mataram
LOMBOK
SUMBAWA
Sumbawa Besar
Raba
FLORES
Ende
FLORES SEA
LESSER SUNDA ISLANDS
Waikabubak
SUMBA
SAVU SEA

INDIAN OCEAN
JAVA TRENCH
PULAU ENGGANO
TANJUNG CINA
SELAT SUNDA
KEPULAUAN KARIMUNJAWA
PULAU BAWEAN
PULAU MASALEMBU BESAR
PULAU SELAYAR
PULAU KABAENA

COCOS ISLANDS (Austl.)
© Rand McNally A-101748-1
CHRISTMAS ISLAND (Austl.)
ASHMORE AND CARTIER ISLANDS (Austl.)

PRATAS ISLAND (Occupied by Taiwan; claimed by China)
Luzon Strait
BATAN ISLANDS
Basco
BABUYAN ISLAND
CAPE BOJEADOR
Laoag
Vigan
Tuguegarao City
Mt. Pulag 9626 ft. 2934 m
LUZON
SIERRA MADRE
Baguio
Dagupan
Angeles
San Fernando
Cabanatuan
QUEZON CITY
MANILA
Lucena
Naga
MINDORO
Calapan
TABLAS ISLAND
BUSUANGA ISLAND
CULION ISLAND
CUYO ISLANDS
Roxas
PANAY
Iloilo
PALAWAN
Puerto Princesa
Bacolod
NEGROS
Dumaguete
Pagadian
CELEBES BASIN

ARAKAN YOMA
RAMREE ISLAND
CHEDUBA ISLAND
PREPARIS ISLAND
COCO IS.
Coco Channel
NORTH ANDAMAN
MIDDLE ANDAMAN
SOUTH ANDAMAN
Port Blair
ANDAMAN ISLANDS
Ten Degree Channel
LITTLE ANDAMAN
CAR NICOBAR ISLAND
KATCHALL ISLAND
LITTLE NICOBAR
GREAT NICOBAR
NICOBAR ISLANDS
ANDAMAN AND NICOBAR ISLANDS (India)
Great Channel

Gulf of Martaban
MERGUI ARCHIPELAGO
Gulf of Thailand
DAO PHU QUOC
CON SON
MUI CA MAU

Strait of Malacca
Strait of Melaka

Equator
Tropic of Cancer

0 80 160 240 320 400 480 Miles
0 80 160 240 320 400 480 560 640 720 800 Kilometers

Lambert Azimuthal Equal Area Projection
Scale 1:16,000,000
One inch to 256 miles
One cm to 160 km

CHINA

FUJIAN

TAIWAN

FUZHOU

EAST
CHINA SEA

TAIPEI

Chilung

Taoyüan

Hsinchu

Taichung

Hualien

TAIWAN

Tainan

KAOHSIUNG

Fangshan

PHILIPPINE
SEA

SOUTH
CHINA SEA

Bashi Channel

China and many other countries do not recognize
the existence of Taiwan as a separate country.

PHILIPPINE
SEA

PHILIPPINE
BASIN

PHILIPPINES

DAVAO

PALAU

MOLUCCA
SEA

MOLUCCAS
(MALUKU)

CERAM SEA

BANDA SEA
(LAUT BANDA)

Ambon

WEST CAROLINE
BASIN

Equator

EAST CAROLINE
BASIN

Jayapura

BISMARCK
SEA

BISMARCK
ARCHIPELAGO

NEW GUINEA

Puncak Jaya
16 503 ft.
5030 m

PAPUA
NEW GUINEA

Mt. Wilhelm
14 793 ft.
4509 m

SOLOMON SEA

TIMOR-LESTE

Dili

ARAFURA SEA

CORAL SEA

TIMOR SEA

AUSTRALIA

Darwin

Port
Moresby

Mt. Victoria
13 238 ft.
4035 m

Inset map a
Lambert Conformal Conic Projection
Scale 1:4,000,000
One inch to 64 miles
One cm to 40 km

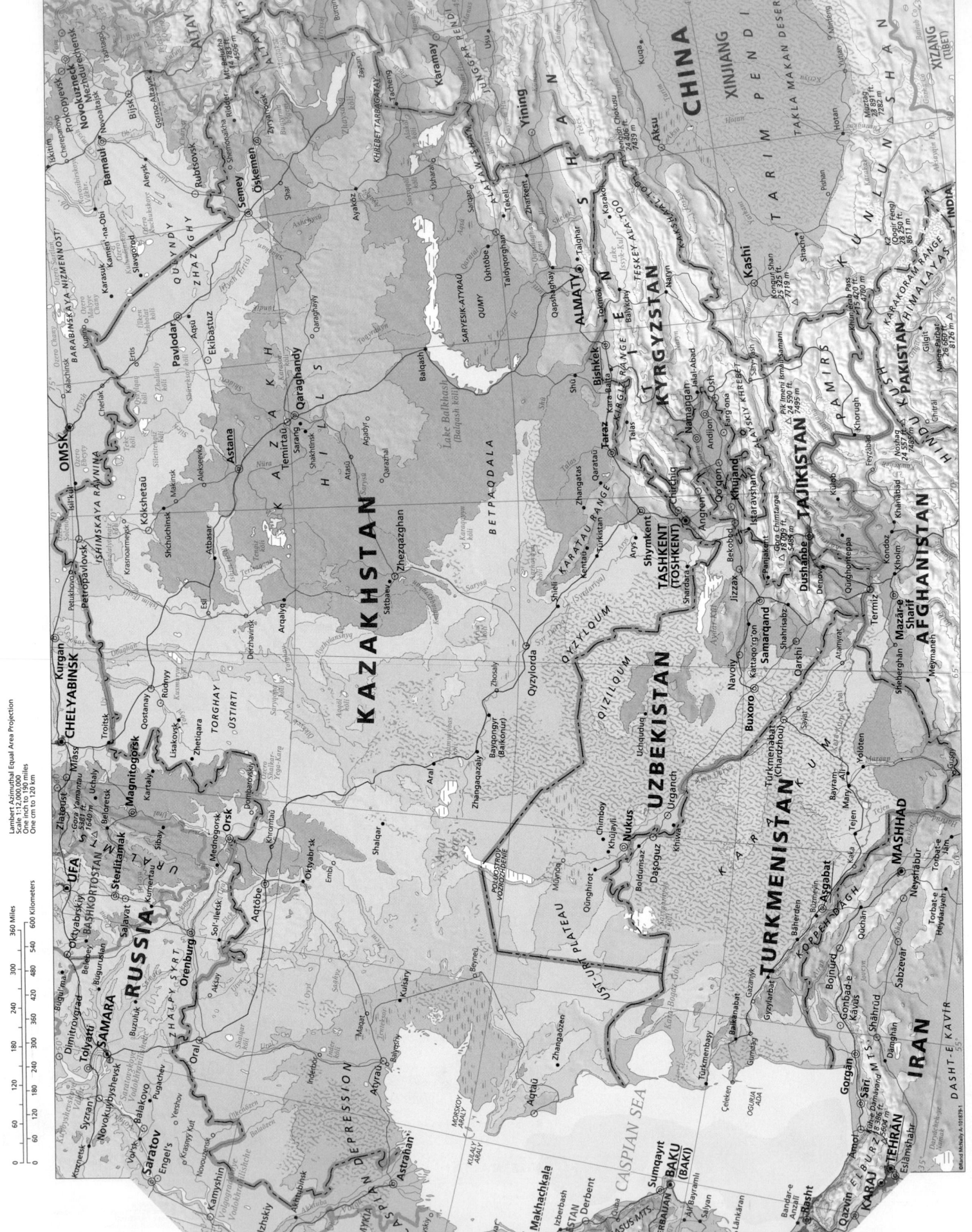

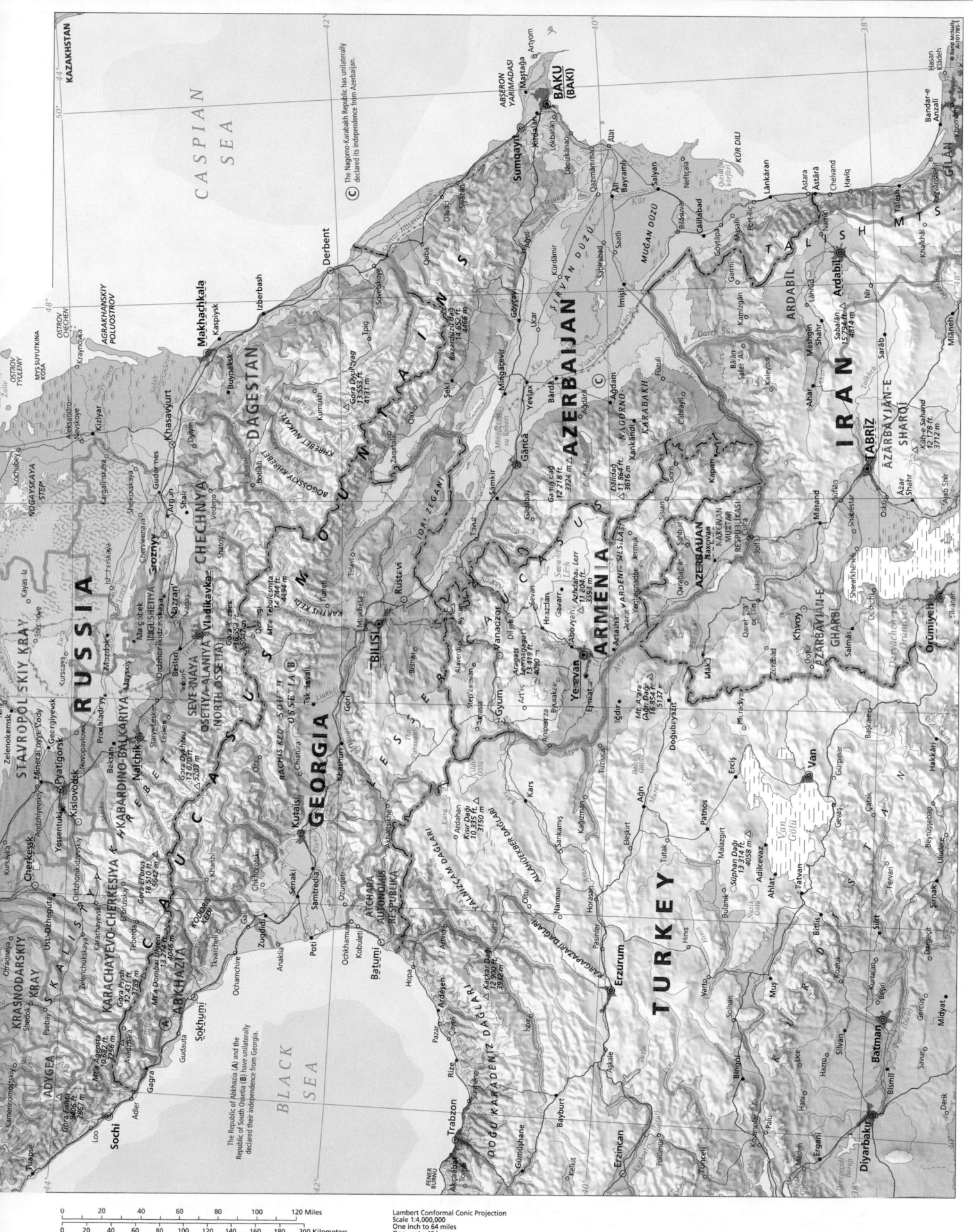

KAZAKHSTAN

C A S P I A N

S E A

ABŞERON
YARIMADASI
Maştağa
BAKU
(BAKI)

ⒸThe Nagorno-Karabakh Republic has unilaterally
declared its independence from Azerbaijan.

Sumqayıt

OSTROV
CHECHEN

OSTROV
TYULENIY

AGRAKHANSKIY
POLUOSTROV

MYS SUYUTKINA
KOSA

Makhachkala
Kaspiysk
Izberbash

D A G E S T A N

Derbent

KHREBET NUKATL

BOGOSSKIY KHREBET

A Z E R B A I J A N

SIRVAN DOZU

MUĞAN DOZU

Lankäran
Astara

RUSSIA

KABARDINO-BALKARIYA

KARACHAYEVO-CHERKESIYA

KRASNODARSKIY
KRAY

STAVROPOLSKIY KRAY

ADYGEA

Sochi

Groznyy

CHECHNYA

SEVERNAYA
OSETIYA-ALANIYA
NORTH OSSETIA

INGUSHETIYA

Vladikavkaz

C A U C A S U S   M O U N T A I N S

G R E A T E R   C A U C A S U S

Gäncä

NAGORNO-
KARABAKH

TBILISI

GEORGIA

SOUTH
OSSETIA Ⓑ

Rustavi

Kutaisi

Batumi

ABKHAZIA

L E S S E R   C A U C A S U S

ARMENIA

AZERBAIJAN
NAXÇIVAN
MUXTAR
RESPUBLIKASI

Yerevan

A T Ç A R A
AVTONOMIUS
RESPUBLIKA

The Republic of Abkhazia (A) and the
Republic of South Ossetia (B) have unilaterally
declared their independence from Georgia.

B L A C K
S E A

DOĞU KARADENİZ DAĞLARI

KARAÇALI DAĞLARI

T U R K E Y

Erzurum

Van
Gölü

Van

I R A N

TABRIZ

ARDABIL

ÂZARBÂYJÂN-E
SHARQI

ÂZARBÂYJÂN-E
GHARBI

Orūmīyeh

Diyarbakır

Batman

0   20   40   60   80   100   120 Miles

0   20   40   60   80   100   120   140   160   180   200 Kilometers

Lambert Conformal Conic Projection
Scale 1:4,000,000
One inch to 64 miles
One cm to 40 km

BLACK SEA

BULGARIA

GREECE

Edirne

İSTANBUL

SEA OF MARMARA
(MARMARA DENIZI)

BURSA

İZMIR

MANISA

ANKARA

Kütahya

Eskişehir

Uşak

Afyon

Konya

Denizli

Isparta

Antalya

Antalya
Körfezi

TAURUS MOUNTAINS

ADANA

MEDITERRANEAN SEA

Northern Cyprus declared
itself an independent
Turkish Republic in 1983.

CYPRUS

Nicosia

0 20 40 60 80 100 120 Miles

0 20 40 60 80 100 120 140 160 180 200 Kilometers

Lambert Conformal Conic Projection
Scale 1:4,000,000
One inch to 64 miles
One cm to 40 km

The Republic of Abkhazia (A) and the
Republic of South Ossetia (B) have unilaterally
declared their independence from Georgia.

RUSSIA

ABKHAZIA
GEORGIA
TBILISI
ARMENIA
YEREVAN
IRAN

TURKEY

Samsun
Trabzon
Erzurum
Sivas
Elazig
Malatya
Diyarbakır
Gaziantep
Şanlıurfa
Mardin
Van

SYRIA
ALEPPO
(HALAB)
Ar-Raqqah
Hamāh
Hims

IRAQ
Mosul
(Al-Mawşil)
Arbil
Kirkuk
(Karkūk)
Dayr az-Zawr

TURKEY

Şanlıurfa
Viranşehir
Derik

Alanya
Gazipaşa
Gülnar
Mut
Ermenek
Kuzucubelen
Tarsus
İçel
Tarsus
ADANA
İslâhiye
Gaziantep
Nizip
Oğuzeli
Birecik
Süruç
Akçakale
Ceylanpınar
Ra's al-'Ayn
Tall
Tamir

Antalya
Körfezi
Anamur
ANAMUR
BURNU
TAURUS MOUNTAIN
Silifke
Erdemli
Dörtyol
Kilis
Jarābulus
Dulq
Maghār
Al-Bāb
Manbij
A'zāz
Afrīn

Bozyazı
Anamur
İskenderun
İskenderun
Körfezi
Hatay
Reyhanlı
Al-Bāb
ALEPPO
(HALAB)
Maskanah
Buhayrat
al-Asad
Ar-Raqqah
Euphrates
Suwaydah

İNCEKUM
BURNU
Samandağı
AKINCI
BURUN
Seyvol
Dār
Ta'izzah
As-Safirah
Idlib
As-Safirah
SYRIA

Northern Cyprus declared
itself an independent
Turkish Republic in 1983.
Dipkarpaz
ZAFER BURNU
RA'S AL-BASĪT
Jisr ash-Shughūr
Ariha
Ma'arrat
an-Nu'mān
Al-Kawm
Dayr az-Zawr

KORUÇAM
BURNU
CYPRUS
Ziyamet
Latakia
(Al-Lādhiqīyah)
Al-Haffah
As-Saqlabīyah
'Āsī
Sūrān
At-Tayyibah
As-Sukhnah
Al-Mayādīn

Kölpos
Khrisokhoús
Güzelyurt
Körfezi
Girne
İskele
Gazimağusa
Körfezi
Yablah
Bāniyās
Hamāh
Salamīyah
Dabl

Polis
Güzelyurt
Strovolos
Nicosia
Gazimağusa
Tartūs
Tall
Kalakh
Hims
Shinshār
Furqlus
Tudmur
PALMYRA
As-Sukhnah

Néa Páfos
Olímbos
6401 ft.
1951 m
Yermasóya
Larnaka
AKROTÍRION
PIDÁLION
Tripoli
(Tarābulus)
Al-Batrūn
Qurnat
as-Sawdā'
10 115 ft.
3083 m
Al-Qusayr
Al-Hirmil
Al-Qaryatayn
Akāshāt

Lemesós
Akrotíri
AKROTÍRION
GÁTAS
Jubayl
BYBLOS
Bsharri
Sab 'Ābar
LEBANON

MEDITERRANEAN
SEA
Beirut
(Bayrūt)
Jūnīyah
Zahlan
B'abda
AFLEY
Ba'labakk
BAALBEK
An-Nabk
At-Tanf

Gaza Strip is administered by the Palestinian
Authority following unilateral withdrawal
by Israel in 2005.
Saydā
Az-
Zabadāni
Al-Qutayfah
Khān Abū
Shāmāt

West Bank is controlled by Israel and parts
are administered by the Palestinian Authority.
DAMASCUS
(DIMASHQ)
Dūmā
Jaramānah
Darayyā
Al-Kiswah

Golan Heights has been unilaterally annexed by Israel.
Tyre
(Sūr)
Iwayyā
Mt. Hermon
9232 ft.
2814 m
Qatana

Nahariyya
Har Meron
3963 ft.
1208 m
Zefat
Qiryat Shemona
Al-Qunaytirah
Al-Mismīyah
Lāhithah
At-Tanf
Wadi at-Wu'ayr

Haifa
(Hefa)
Akko
Teverya
GOLAN
HEIGHTS
Al-Qunaytirah
As-Suwaydā'
Jabal ad-Durūz
5909 ft.
1801 m
SYRIA

Tirat Karmel
Nazareth
(Nazerat)
Fiq
Dar'ā
DESERT

CAESAREA
Dor
'Afula
Bet She'an
Irbid
Al-Ramthā
Salkhad
Jabal 'Unayzah
3084 ft.
940 m

Hadera
Jahīn
Tūbas
'Ajlūn
Al-Mafraq
Mahattat
al-Hafif
AL-HAMĀD

ISRAEL
Netanya
Tūlkarm
Nābulus
Jarash
Mahattat
al-Hafif
Turayf

Herzliyya
Qalqilya
WEST
BANK
As-Salt
Az-Zarqā'
Ar-Rusayfah
AMMĀN
Azraq
ash-Shīshān
Al-Hadithah
Kāf

Tel Aviv-Yafo
Petah
Tiqwa
Arīha
(Jericho)
Madaba
Al-Jīzah
Al-Hadithah
Al-Jalāmīd

Rishon LeZiyyon
Rehovot
Jerusalem
(Yerushalayim)
Al-Khalīl
Dhībān
Turayf

Ashdod
Ashqelon
Bethlehem
(Bayt Lahm)
Qiryat
Gat
Al-Karak
WADI AS-SIRHAN
Al-'Īsāwīyah

GAZA STRIP
Gaza
(Ghazzah)
Be'er
Sheva
Yattah
MASADA
'Arad
Al-Mazra'ah
At-Tafilah
Sabkhat
Hazīwza
Sakākah

Port Said
(Būr Sa'īd)
Būr Fu'ād
Khān Yūnis
Rafah
Dimona
At-Qatrānah
Al-Jawf

El-Arish
HOLOT
HALUZA
Yeroham
Ash-Shawbak
Al-Jafr
QA' AL-JAFR
Al-Jalāmīd

Ismailia
El-Qantara
el-Sharqīya
Nizzana
Mizpé
Ramon
Al-Tafilah
Wadi Baīr
Qārah

EGYPT
El-Quseima
NEGEV
Gebel Yi'allaq
3589 ft.
1084 m
Ma'ān
PETRA
Wadi Hasā
SAUDI

Suez
(El-Suweis)
Bur Taufīq
El-Kuntilla
Ra's
an-Naqb
Al-Jafr
AL-BUSAYTA
AT-TAWIL

Nakhl
El-Thamad
Elat
Makhfar
al-Quwayrah
Jabal Ramm
5755 ft.
1754 m
Al-'Urayq

GEBEL EL-TĪH
SINAI PENINSULA
Al-'Aqabah
Al-Mudawwarah
AL-HUFRAH

Gulf of Suez
(Khalīg el-Suweis)
Abu
Zenīma
Ras el-Gineina
5335 ft.
1626 m
Nuweiba
Gulf of Aqaba
Haql
Al-Bad
Al-Bi'r

Mt. Sinai
(Gebel Mūsa)
7497 ft.
2285 m
Dahab
Jabal al-Lawz
7884 ft.
2403 m
Al-Qalibah

Gebel Abu
Khashaba
4797 ft.
1462 m
El-Tūr
Magna
Tabūk

Sharm
el-Sheikh
'Aynūnah
MIDYAN
Ash-Sharmah
Sakākah

TIRAN
SANAFIR
RED SEA
RĀS MOHAMMED

0   20   40   60   80   100   120 Miles
0 20 40 60 80 100 120 140 160 180 200 Kilometers

Lambert Conformal Conic Projection
Scale 1:4,000,000
One inch to 64 miles
One cm to 40 km

ARDABĪL

Rasht

TALISH MOUNTAINS

GĪLĀN

Ḩakkārī

Orūmīyeh

Ajab Shīr

Marāgheh

Mīāneh

Fowman

Roudbār

Şa'īdīyeh

Zanjān

Midyat

Dargeçit

Şirnak

Uludere

Şemdinli

Daryācheh-ye Orūmīyeh

Benāb

Malek Kandī

Mehrābād

Şīdān

Khorram Daraq

Abhar

Cizre

Zākho

Al-'Amādīyah

Silvāneh

Oshnavīyeh

ĀZARBĀYJĀN-E SHARQĪ

Mardin

İdil

DAHŪK

Dahūk

Naqadeh

Mīāndoāb

ĀZARBĀYJĀN-E GHARBĪ

Būkān

QAZVĪN

Īzīltepe

Nusaybin

Mālikīyah

Tall Kūşīk

Rawāndoz

Pīrān Shahr

ZANJĀN

Qeydār

Al-Qāmishlī

Sīnjār

Tall Kayf

NINEVEH

ARBĪL

Arbīl

Mahābād

Sardasht

Takāb

Harşīn

Al-Ḩasakah

Aqrah

Banah

DĪvāndarreh

Bījār

Razan

Nowbarān

ARABIA

Rawdah

MESOPOTAMIA

NĪNAWĀ

HATRA

Al-Hadr

ASHUR

Sharqāt

AT-TA'MĪM

Kirkuk (Karkūk)

As-Sulaymānīyah

AS-SULAYMĀNĪYAH

KORDESTĀN

Marīvān

Sanandaj

Hoseynābād

Dehgolān

Qorveh

Kabūdarāhang

Avaj

Mosul (Al-Mawşil)

NIMRUD

IRAN

HAMADĀN

Hamadān

QAZVĪN

MARKAZĪ

Taqtaq

Kifrī

Tozkhurmato

Halabjah

Pāveh

Ravānsar

Sonqor

Asadābād

Bāhār

Kangāvar

Tūysarkān

Malāyer

SALĀH AD-DĪN

Tikrīt

Sāmarrā'

Buhayrat ath-Tharthār

Balad

IRAQ

Khānaqīn

Qaşr-e Shīrīn

KERMĀNSHĀH

Kermānshāh

Kerend

Eslāmābād

Nahāvand

Sāmen

Oshtorīnān

Borūjerd

Deh Kord

Āstāneh

LORESTĀN

Hadīthah

Jadīdah

Khān al-Baghdādī

Hīt

As-Sa'dīyah

Jalūlā

Al-Miqdādīyah

Gīlān-e Gharb

Sūmār

Kūhdasht

Khorramābād

Oshtorān Kūh 14,009 ft. 4,331 m

ZAGROS MOUNTAINS

Do Kud

Ar-Ramādī

Al-Fallūjah

Al-Kāzimīyah

BAGHDAD

BAGHDĀD

DIYĀLĀ

Ba'qūbah

Mandalī

Īlām

Abdānān

ĪLĀM

KABĪR KŪH

AL-ANBĀR

Buhayrat al-Habbānīyah

Al-Mahmūdīyah

Al-Madā'in

CTESIPHON

Mehrān

Hoseynīyeh-ye Khodā-Dād

Dehlorān

Buhayrat al-Milh

Bahr al-Milh

Ar-Razzāzah

Al-Musayyib

Al-Hindīyah

BABYLON

BĀBIL

WĀSIT

Nu'mān

Shaykh Sa'd

'Alī al-Gharbī

Andīmeshk

Dezfūl

Shūsh

Shūshtar

Dār Khazīneh

Galvand

Shithāthah

Karbalā'

Al-Hillah

KARBALĀ'

Al-Kifl

Al-Kūfah

'Afak

Al-Hayy

MAYSĀN

Al-'Amārah

Al-Halfāyah

Susangerd

Hūzgān

KHŪZESTĀN

AHVĀZ

An-Najaf

Abū Şukhayr

Ash-Shāmīyah

Ad-Dīwānīyah

AL-QĀDISĪYAH

Hamzah

Ash-Shināfīyah

Ar-Rumaythah

Ar-Rifā'ī

Ash-Shatrah

Qal'at Şālih

Bostān

Qal'at Şālih

AN-NAJAF

As-Samāwah

DHĪ QĀR

Al-Bathā'

An-Nāṣirīyah

UR

Sūq ash-Shuyūkh

Al-Qurnah

Ash-Shuwayyib

Shādegān

Abādān

Ash-Shabakah

As-Salmān

AL-MUTHANNĀ

Kibāsī

Basra (Al-Başrah)

Az-Zubayr

Abū al-Khaşīb

Khorramshahr

Bandar-e Khomeynī

Badanah

Ar'ar

AL-ATHĀMĪN

AL-HAJARAH

Lawqah

Rafhā'

Şafwān

Umm Qaşr

Al-Fāw

AL-BAŞRAH

BŪBĪYĀN

RA'S AL-QULAY'AH

KUWAIT

FAYLAKAH

Kuwait (Al-Kuwayt)

Persian Gulf

Nişāb

Al-Jahrah

Aş-Şulaybīyah

South Khītān

Al-Fuhayhīl

AN-NAFŪD

Al-Hayyānīyah

Ünah

Al-Qaysūmah

AD-DIBDIBAH

Wafrah

Subahiya

Mīnā al-Ahmadī

Qalīb ash-Shuyūkh

RandMcNally A-101791-1

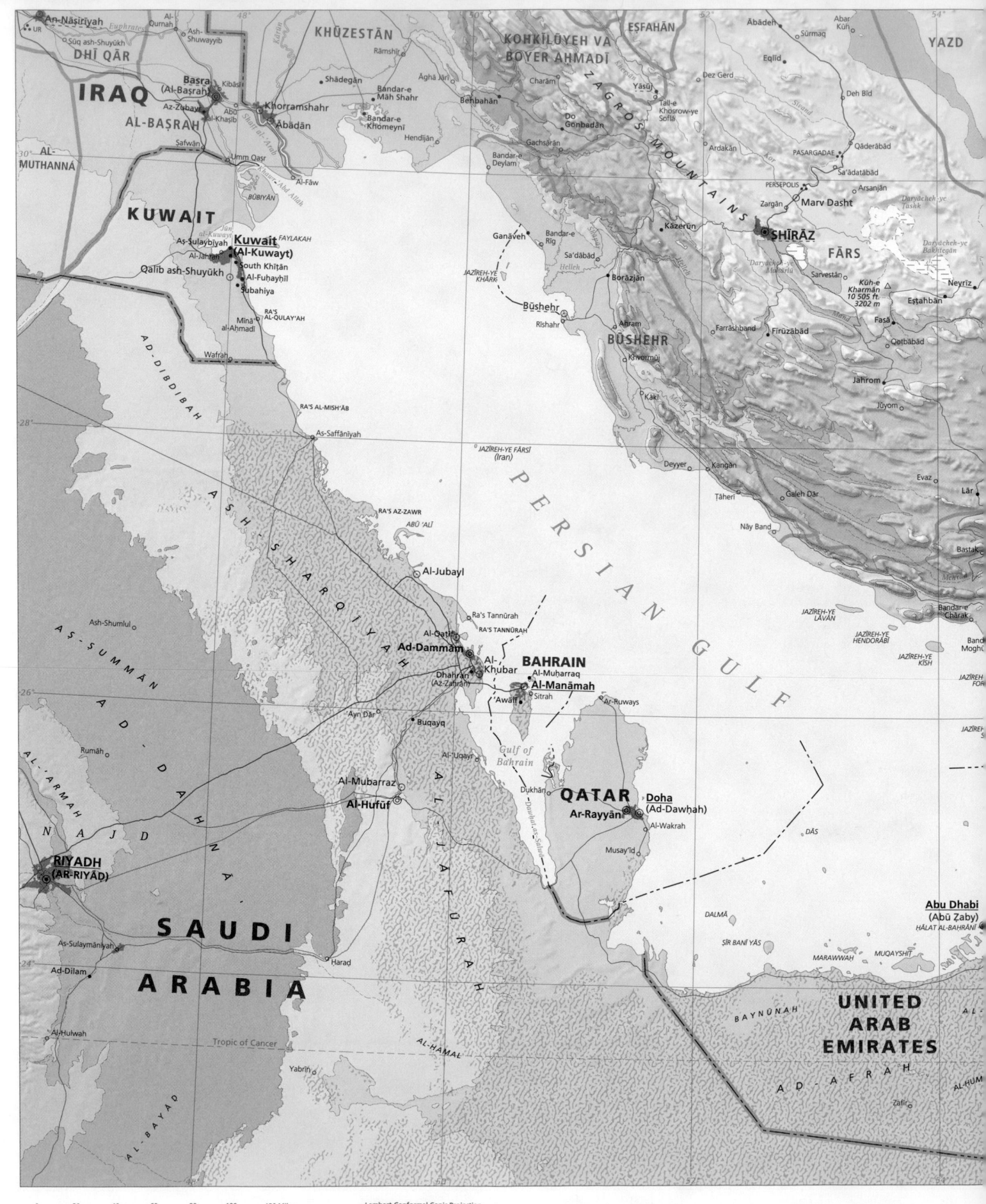

KHORĀSĀN-E
JONŪBĪ

Sefid
Ābeh

Zābol

DASHT-E
MĀRGOW

AFGHANISTAN

Lakar Kūh
9764 ft.
2976 m

Mohammadābād

Daryācheh-ye
Hāmūn

SĪSTĀN

Mirābād

Chār
Borjak

Rodbār

Helmand

Ānār

Zarand

Rāvar

Shahr-e
Bābak

Rafsanjān

Kermān

Bāghīn

Kūh-e Palvār
13 888 ft.
4233 m

Golbāf

GOWD-E ZEREH

Noṣratābād

Kūh-e Malek Sīāh
5390 ft.
1643 m

Zāhedān

Saindak

BALUCHISTĀN

IRAN

KERMĀN

Sīrjān

Kūh-e Laleh Zār
14 357 ft.
4376 m

Mashīz

Rāyen

Kūh-e Hazār
14 649 ft.
4465 m

Bāft

Bam

Fahraj

Mashki
Chāh

Mirjāveh

Lādiz

PAKISTAN

Nok
Kundi

Kūh-e Khabīr
12 612 ft.
3844 m

Esfandageh

Jīroft

Rīgān

Kūh-e Taftān
13 261 ft.
4042 m

Khāsh

BALUCHISTAN

Kūh-e
Qashqeh
9190 ft.
2801 m

Fūrg

Dowlatābād

Do Sārī

Halīl

Kūh-e Bazmān
11 453 ft.
3491 m

Bazmān

SĪSTĀN VA BALŪCHESTĀN

LĀRISTĀN

Sarā-ye
Ahmadī

Golāshkerd

Kahnūj

Bīzhanābād

Bampūr

Bampūr

Īrānshahr

Zāboli

Nāvar
Panāh

Sarāvān

Shamīl

Qal'eh-ye
Delīr Bālā

Manūjān

Hāmūn-e
Jaz Mūrīān

Saravān

Bandar
'Abbās

Mīnāb

Remeshk

Maskūtān

Sarbāz

JAZĪREH-YE
HORMOZ

Qeshm

Kūh-e Būnīken
7169 ft.
2185 m

Angohrān

Bent

Qaṣr-e
Qand

Ispikān

Randar-e
Lengeh

JAZĪREH-YE
LĀRAK

Strait of Hormuz

RA'S
SHARĪTAH

Nīkshahr

Mand

JAZĪREH-YE
QESHM

MŪSANDAM
PENINSULA

Gūr Kūh
6270 ft.
1911 m

Gāsrīk

Sūrak

Kārkīndar

Bāhū
Kalāt

Kohak

PAKISTAN

ABŪ
MŪSĀ

Jabal al-Harim
6847 ft.
2087 m

Al-Khaṣab

Govān

Ra's al-Khaymah

Jāsk

Hūmedān

Bandar
Beheshtī

Polānī

Dasht

Suntsar

Umm al-Qaywayn

Dadnah

RA'S-E
KŪH LAB

Gwātar
Bay

RĀS
NŪH

Ajmān

OMAN

Ash-Shāriqah

Bandar
Beheshtī

MAKRĀN   COAST

Jīwani

Gwādar

DUBAI
(DUBAYY)

OMAN

Al-Fujayrah

Kalbā'

Shināṣ

Al-Buraymī

Al-Buraymī

Ṣuḥār

Gulf of Oman

Al-'Ayn

Jabal Hafīt
3806 ft.
1160 m

Al-Qābil

Al-Ghurayfah

Al-Khābūrah

As-Suwayq

As-Sīb

Maṭrah

Muscat
(Masqaṭ)

ARABIAN

Dank

Maskin

Ar-Rustāq

Bawshar

Al-Amrat

RA'S ABŪ DĀ'ŪD

Tropic of Cancer

SEA

'Ibrī

Jabal ash-Shām
9957 ft.
3035 m

Samā'il

Sarūr

Qurayyāt

Ta'nam

Bahlā'

Nizwā

Izkī

Ibrā

Fins

Tīwī

OMAN

Al-Muḍaybī

Ṣūr

Al-Hadd

RA'S AL-HADD

© Rand McNally
A-101797-1

0   40   80   120   160   200   240 Miles

0   40   80 120 160 200 240 280 320 360 400 Kilometers

Lambert Azimuthal Equal Area Projection
Scale 1:8,000,000
One inch to 128 miles
One cm to 80 km

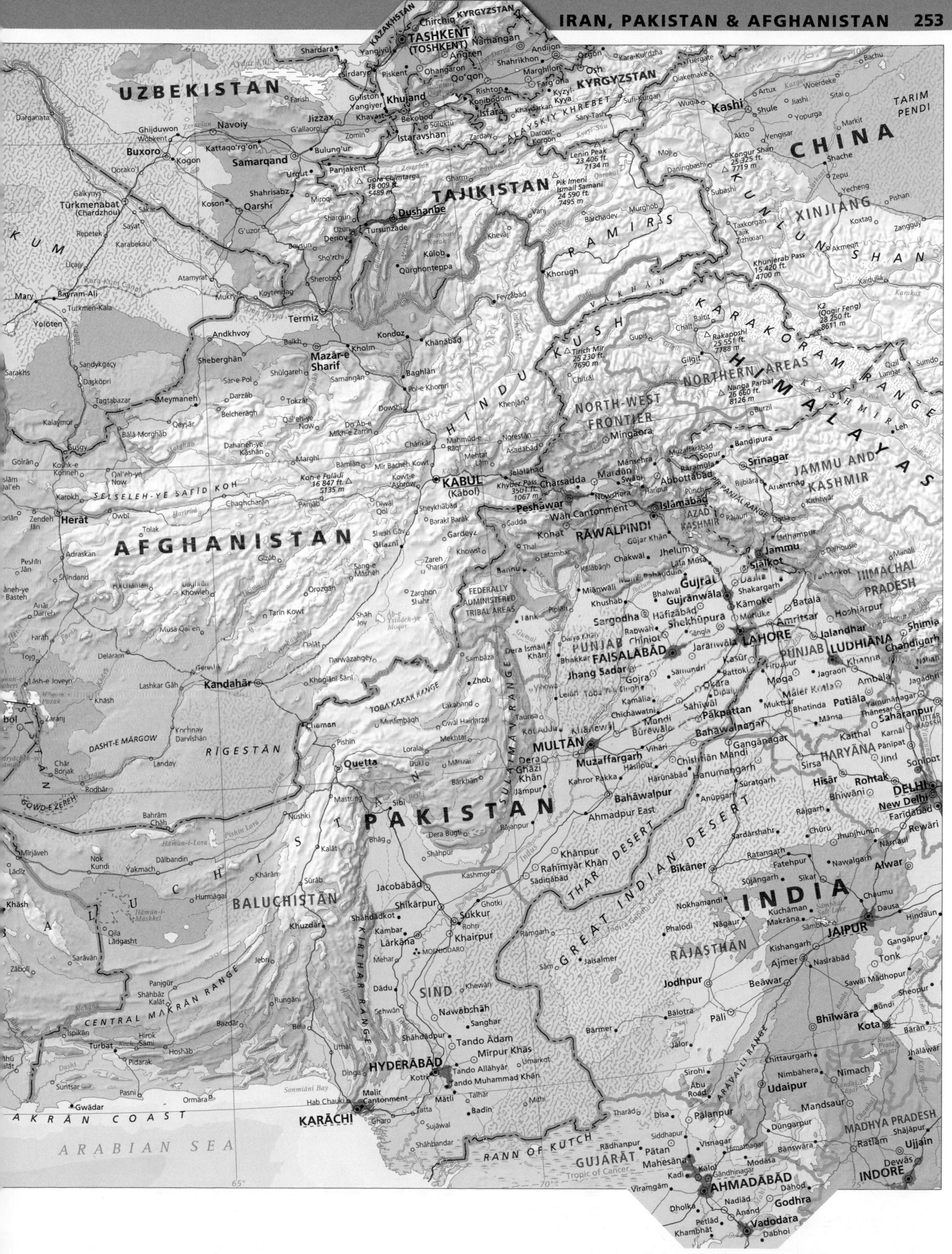

AFGHANISTAN

PAKISTAN

INDIA

UZBEKISTAN

TAJIKISTAN

KYRGYZSTAN

KAZAKHSTAN

CHINA

XINJIANG

Major cities and places

KABUL (Kābol)
KARACHI
Dushanbe
TASHKENT (TOSHKENT)
Peshāwar
RĀWALPINDI
Islāmābād
LAHORE
FAISALĀBĀD
MULTĀN
Quetta
Kandahār
Herāt
Mazār-e Sharif
HYDERĀBĀD
AHMADĀBĀD
DELHI
New Delhi
JAIPUR
Samarqand
Buxoro
Navoiy
Termiz
Kondoz
Khānābād
Baghlān

Regions and physical features

HINDU KUSH
KARAKORAM RANGE
HIMALAYAS
PAMIRS
KUNLUN SHAN
NORTHERN AREAS
JAMMU AND KASHMIR
NORTH-WEST FRONTIER
FEDERALLY ADMINISTERED TRIBAL AREAS
AZAD KASHMIR
PUNJAB
SIND
BALUCHISTĀN
RĀJASTHĀN
HARYĀNA
HIMACHAL PRADESH
MADHYA PRADESH
GUJARĀT
UTTAR PRADESH
TOBA KĀKAR RANGE
SULAIMĀN RANGE
CENTRAL MAKRĀN RANGE
KIRTHAR RANGE
ARAVALLI RANGE
SELSELEH-YE SAFID KOH
DASHT-E MĀRGOW
RĪGESTĀN
THAR DESERT
GREAT INDIAN DESERT
RANN OF KUTCH
MAKRĀN COAST
ARABIAN SEA
TARIM PENDI

Khyber Pass 3501 ft. 1067 m
Khunjerab Pass 15 420 ft. 4700 m
Lenin Peak 23 406 ft. 7134 m
Pik Imeni Ismail Samani 24 590 ft. 7495 m
K2 (Qogir Feng) 28 250 ft. 8611 m
Nanga Parbat 26 660 ft. 8126 m
Tirich Mīr 25 230 ft. 7690 m
Rakaposhi 25 551 ft. 7788 m
Koh-e Fūlādī 16 847 ft. 5135 m
Gora Chimtarga 18 009 ft. 5489 m

Tropic of Cancer

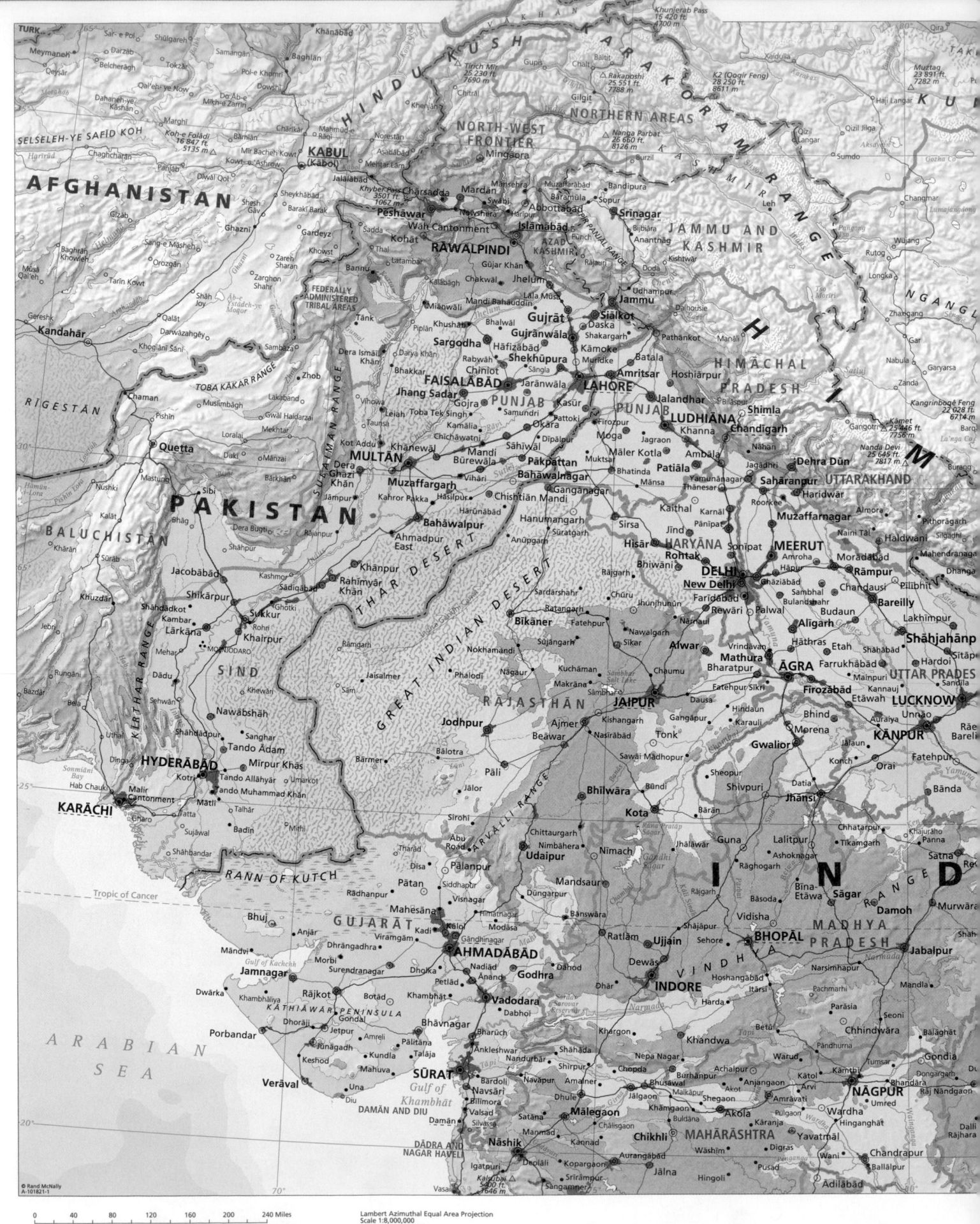

TAJIKISTAN

TURK.

Meymaneh
Sar-e Pol
Shūlgareh
Samangān
Khānābād
Baghlān

Qeysār
Ðarzāb
Belcheragh
Tokzar
Pol-e Khomrī

Maīmana
Mazār-e Sharīf
Samangān

HINDU KUSH
Tirich Mīr
25,230 ft.
7690 m

KARAKORAM
RANGE

Khunjerab Pass
15,420 m
4700 m

Muztag
23,891 ft.
7282 m

K2 (Qogir Feng)
28,250 ft.
8611 m

Rakaposhi
25,551 ft.
7788 m

NORTHERN AREAS

Gilgit

Nanga Parbat
26,660 ft.
8126 m

AFGHANISTAN

KABUL
(Kābol)

NORTH-WEST
FRONTIER

Mingaora

Srinagar

JAMMU AND
KASHMIR

Peshāwar
Islāmābād
RĀWALPINDI

AZAD
KASHMIR

PIR PANJAL RANGE

Jammu

HIMĀCHAL
PRADESH

HIMALAYA

FEDERALLY
ADMINISTERED
TRIBAL AREAS

Gujrāt
Siālkot

PUNJAB

Amritsar

Shimla

Chandīgarh

FAISALĀBĀD
LAHORE
Jalandhar

PUNJAB
LUDHIĀNA

Nandā Devi
25,645 ft.
7817 m

UTTARAKHAND

MULTĀN

Bahāwalpur

Sāhīwal

Patiāla

Saharanpur
Haridwār

Quetta

PAKISTAN

THAR DESERT

HARYĀNA

Meerut

DELHI
New Delhi

Moradābād
Rāmpur
Bareilly

BALUCHISTĀN

GREAT INDIAN DESERT

Bīkāner

SHĀHJAHĀNP.

Āgra
UTTAR PRADES

Jaisalmer

RĀJASTHĀN
JAIPUR

Mathura

Firozābād
LUCKNOW
KĀNPUR

Jodhpur
Ajmer

Gwalior

Jhānsi

HYDERĀBĀD

RANN OF KUTCH

KARĀCHI

Tropic of Cancer

ARAVALLI RANGE

Udaipur

Kota

VINDHYA RANGE

MADHYA
PRADESH

Jabalpur

GUJARĀT

AHMADĀBĀD

BHOPĀL

INDORE

ARABIAN
SEA

Gulf of
Khambhāt

SŪRAT

Nāgpur

MAHĀRĀSHTRA

Nāshik

Lambert Azimuthal Equal Area Projection
Scale 1:8,000,000
One inch to 128 miles
One cm to 80 km

0   40   80   120   160   200   240 Miles
0   40   80   120 160 200 240 280 320 360 400 Kilometers

© Rand McNally
A-101821-1

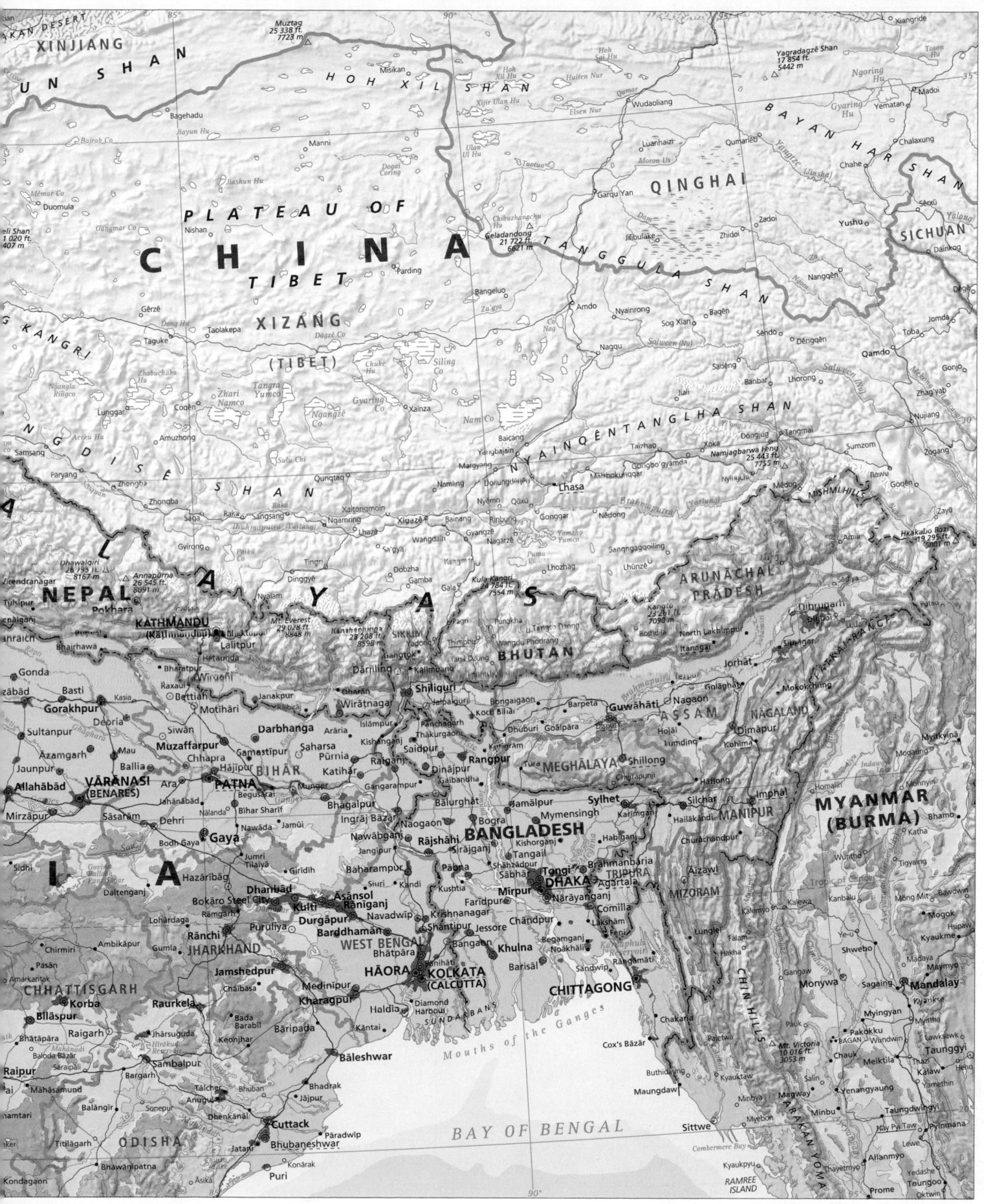

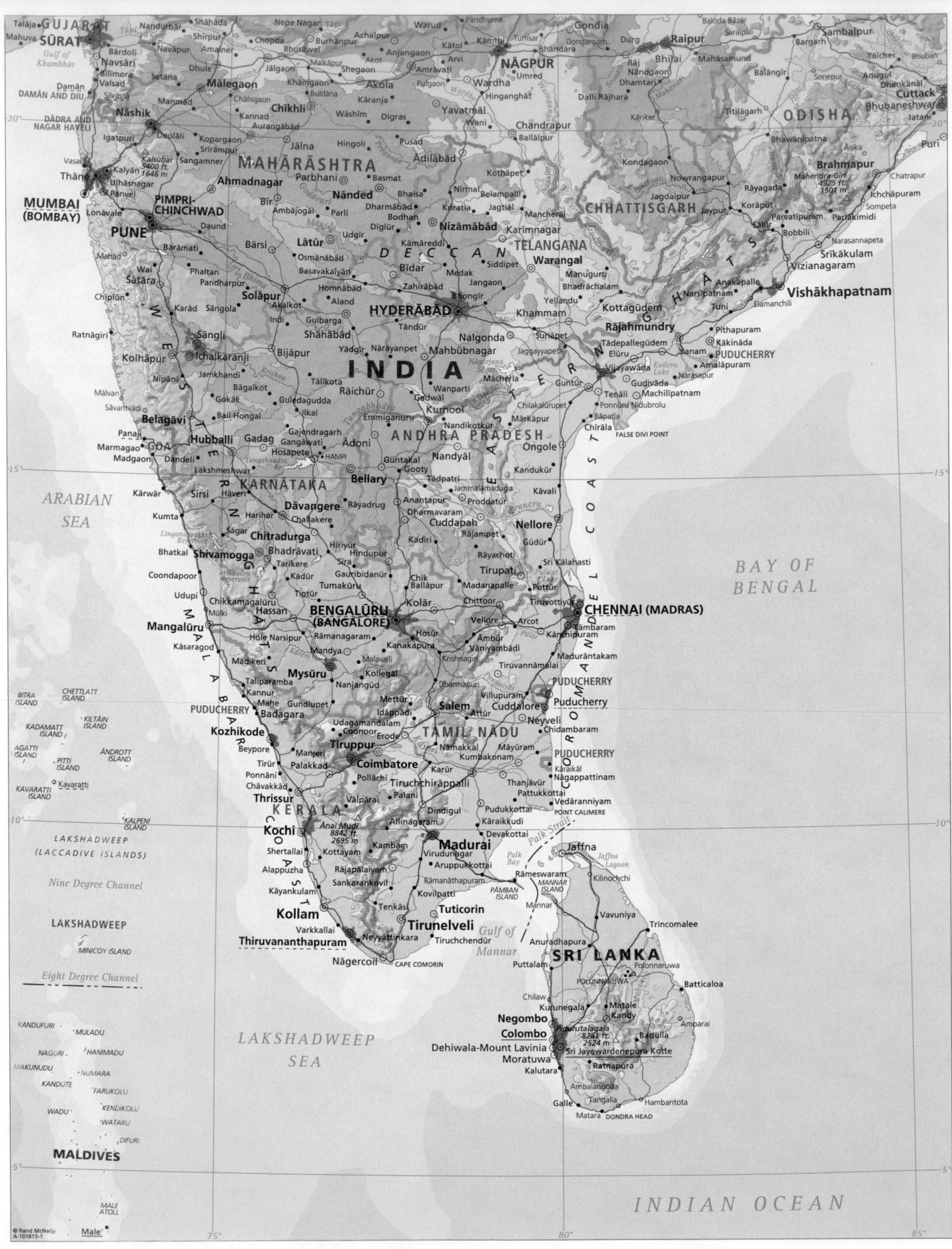

0 40 80 120 160 200 240 Miles

0 40 80 120 160 200 240 280 320 360 400 Kilometers

Lambert Azimuthal Equal Area Projection
Scale 1:8,000,000
One inch to 128 miles
One cm to 80 km

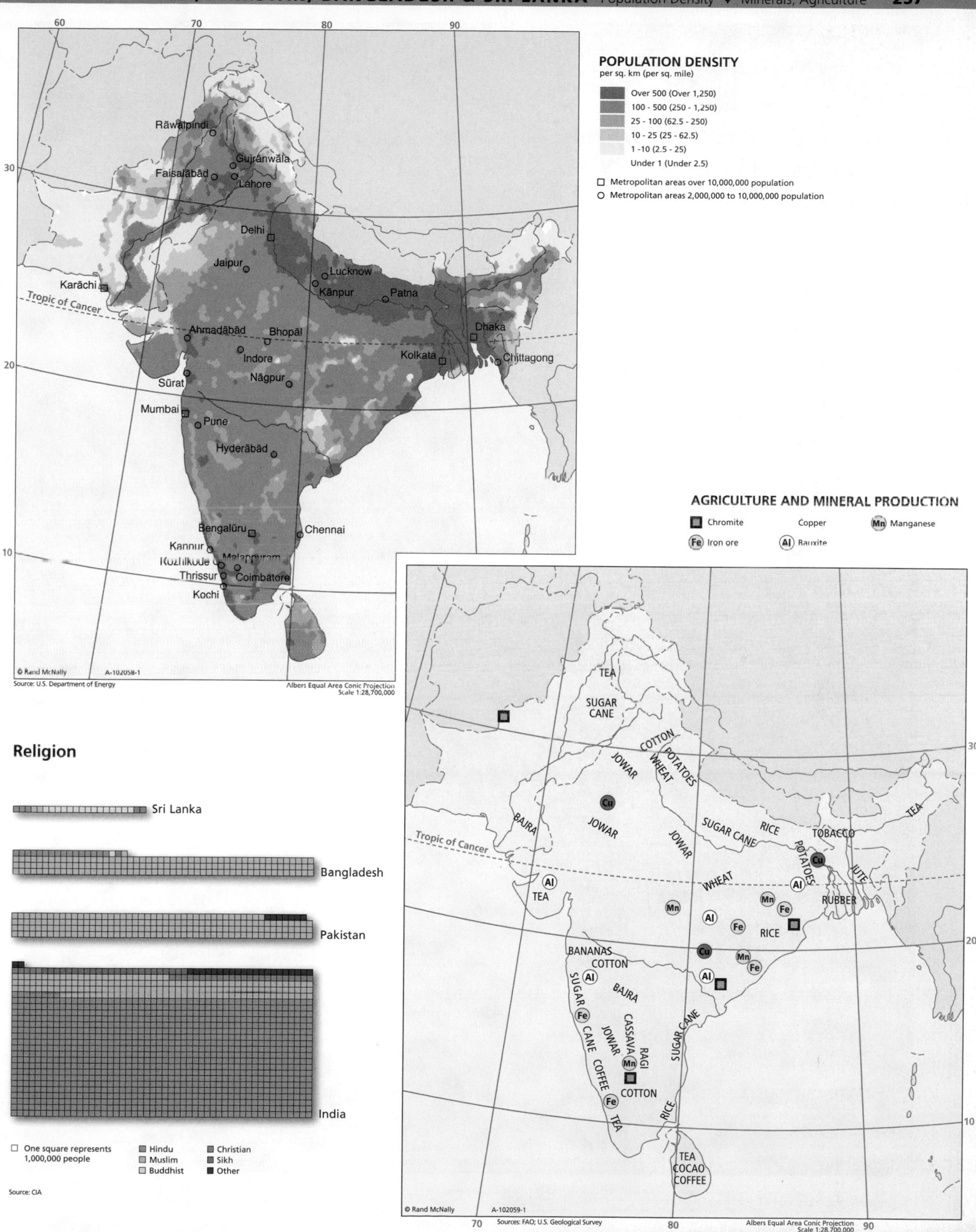

POPULATION DENSITY
per sq. km (per sq. mile)

- Over 500 (Over 1,250)
- 100 - 500 (250 - 1,250)
- 25 - 100 (62.5 - 250)
- 10 - 25 (25 - 62.5)
- 1 -10 (2.5 - 25)
- Under 1 (Under 2.5)

☐ Metropolitan areas over 10,000,000 population
○ Metropolitan areas 2,000,000 to 10,000,000 population

Tropic of Cancer

Rāwalpindi
Gujrānwāla
Faisalābād
Lahore
Delhi
Jaipur
Karāchi
Lucknow
Kānpur
Patna
Ahmadābād
Bhopāl
Dhaka
Indore
Kolkata
Chittagong
Sūrat
Nāgpur
Mumbai
Pune
Hyderābād
Bengalūru
Chennai
Kannur
Kozhikode
Malappuram
Thrissur
Coimbatore
Kochi

© Rand McNally   A-102058-1

Source: U.S. Department of Energy

Albers Equal Area Conic Projection
Scale 1:28,700,000

AGRICULTURE AND MINERAL PRODUCTION

☐ Chromite   Copper   (Mn) Manganese

(Fe) Iron ore   (Al) Bauxite

Religion

Sri Lanka

Bangladesh

Pakistan

India

☐ One square represents
1,000,000 people

- Hindu
- Muslim
- Buddhist
- Christian
- Sikh
- Other

Source: CIA

TEA
SUGAR CANE
COTTON
POTATOES
JOWAR
WHEAT
Cu
BAJRA
JOWAR
SUGAR CANE
RICE
TOBACCO
TEA
Tropic of Cancer
WHEAT
Cu
POTATOES
JUTE
Al
TEA
Mn
Al
Al
Mn
Fe
RUBBER
Mn
Fe
Al
Fe
RICE
BANANAS
COTTON
Cu
SUGAR
Al
Mn
Fe
CANE
Al
CASSAVA
RAGI
Fe
SUGAR CANE
JOWAR
COFFEE
Mn
COTTON
Fe
RICE
TEA
TEA
COCAO
COFFEE

© Rand McNally   A-102059-1

Sources: FAO; U.S. Geological Survey

Albers Equal Area Conic Projection
Scale 1:28,700,000

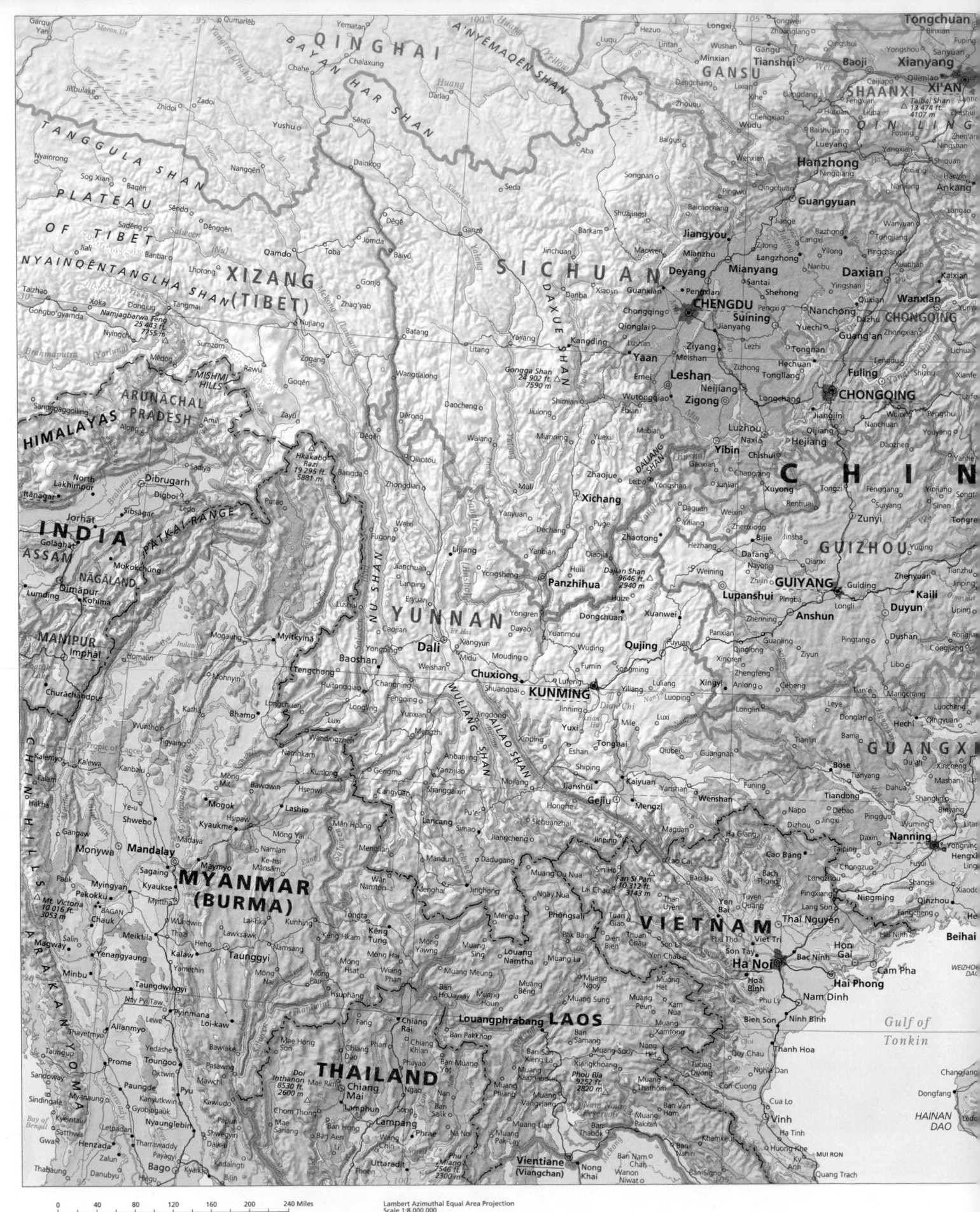

0   40   80   120   160   200   240 Miles
0 40 80 120 160 200 240 280 320 360 400 Kilometers

Lambert Azimuthal Equal Area Projection
Scale 1:8,000,000
One inch to 128 miles
One cm to 80 km

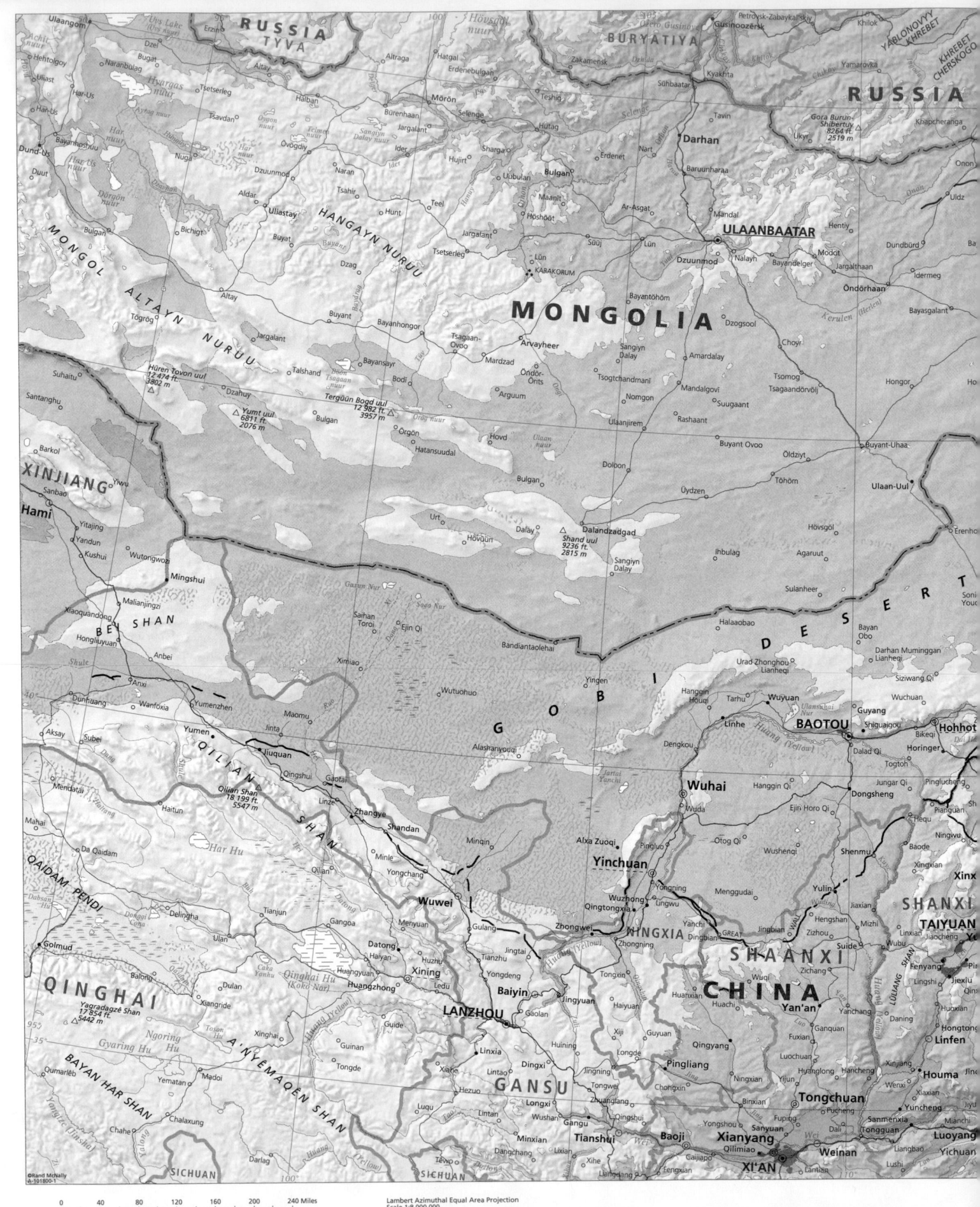

0	40	80	120	160	200	240 Miles

0	40	80	120	160	200	240	280	320	360	400 Kilometers

Lambert Azimuthal Equal Area Projection
Scale 1:8,000,000
One inch to 128 miles
One cm to 80 km

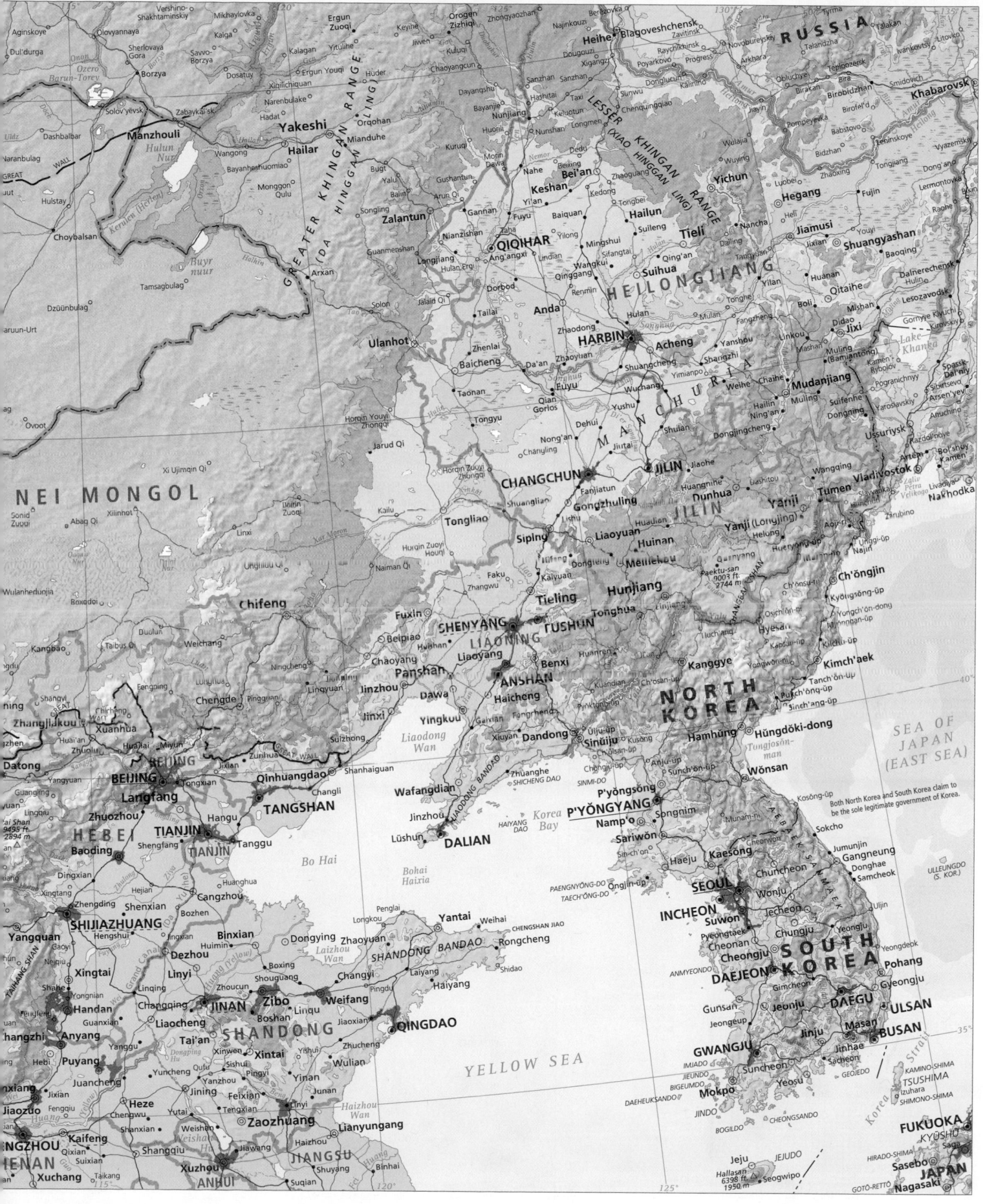

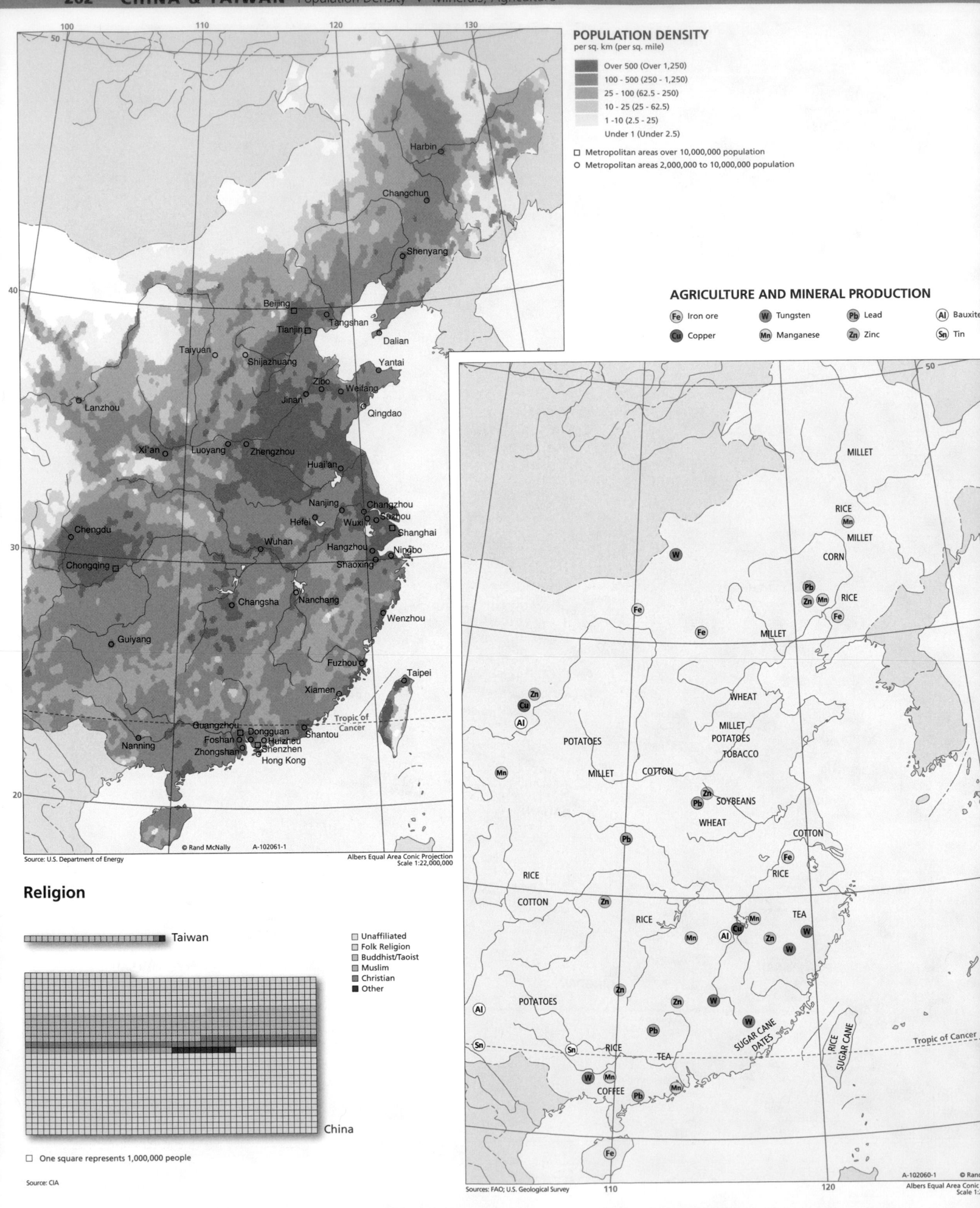

POPULATION DENSITY
per sq. km (per sq. mile)

- Over 500 (Over 1,250)
- 100 - 500 (250 - 1,250)
- 25 - 100 (62.5 - 250)
- 10 - 25 (25 - 62.5)
- 1 - 10 (2.5 - 25)
- Under 1 (Under 2.5)

☐ Metropolitan areas over 10,000,000 population
○ Metropolitan areas 2,000,000 to 10,000,000 population

AGRICULTURE AND MINERAL PRODUCTION

- Fe Iron ore
- Cu Copper
- W Tungsten
- Mn Manganese
- Pb Lead
- Zn Zinc
- Al Bauxite
- Sn Tin

Source: U.S. Department of Energy

© Rand McNally   A-102061-1

Albers Equal Area Conic Projection
Scale 1:22,000,000

Religion

Taiwan

China

- ☐ Unaffiliated
- ☐ Folk Religion
- ☐ Buddhist/Taoist
- ☐ Muslim
- ☐ Christian
- ■ Other

☐ One square represents 1,000,000 people

Source: CIA

Sources: FAO; U.S. Geological Survey

A-102060-1   © Rand

Albers Equal Area Conic P
Scale 1:22

RUSSIA

CHINA

JILIN

LIAONING

Tieling

SHENYANG FUSHUN

Benxi

Liaoyang

ANSHAN

Hunjiang

Tonghua

Kanggye

CHAGANG-DO

Dandong

Sinŭiju P'YONGAN-BUKTO

Kusŏng

HAMGYŎNG-BUKTO

Musan-up

Ch'ŏngjin

Kyŏngsŏng-up

Chuŭl-li

Kilchu-up

Kimch'aek

MUSU-DAN

HAMGYŎNG-NAMDO

Hamhŭng

Hŭngdŏki-dong

Paektu-san
9003 ft.
2744 m

YANGGANG-DO

Hyesan

Puksubaek-san
8274 ft.
2522 m

Kwanmo-bong
8337 ft.
2541 m

NAJIN-SŎNBONG-SI

NORTH
KOREA

P'YONGAN-NAMDO

P'yŏngsong

P'YŎNGYANG

P'YŎNGYANG-SI

NAMP'O-SI

Namp'o

Songnim

Sariwŏn

HWANGHAE-BUKTO

HWANGHAE-NAMDO

Haeju

Ongjin-up

Kaesŏng

Wŏnsan

KANGWON-DO

Both North Korea and South Korea claim to
be the sole legitimate government of Korea.

SEA OF
JAPAN
(EAST SEA)

Korea
Bay

Söjosŏn-man

Tongjosŏn-man

YELLOW
SEA

PAENGNYŎNG-DO

TAECH'ŎNG-DO

Chuncheon

Uijeongbu

SEOUL

INCHEON

Seongnam

Anyang

Suwon

Osan

GYEONGGI-DO

GANGWON-DO

Gyebangsan
5174 ft.
1577 m

Jumunjin

Gangneung

Donghae

Samcheok

ULLEUNGDO
(South Korea)

Gyeonggiman

Wonju

Yeongwol

Uljin

Pyeongtaek

Cheonan

Asan

Jecheon

Chungju

CHUNGCHEONGBUK-DO

SOUTH
KOREA

Yeongju

Yecheon

Andong

Cheongju

Cheongju

Taean

Seosan

Yesan

Hongseong

Jochiwon

Gongju

Sejong

DAEJEON

DAEJEON-GWANGYEOKSI

Yeongdong

Sangju

Uiseong

Yeongdeok

Boryeong

Gimcheon

Pohang

Buyeo

Gonggyeong

Gunsan

Iksan

Jeonju

Jeollabuk-do

Gimje

Jeongeup

Namwon

Gwangju
GWANGJU
GWANGYEOKSI

Gwangsan

Naju

Gwangju

Mokpo

JEOLLANAM-DO

Yeongam

Suncheon

Gwangyang

Yeosu

Goheung

Haenam

Daegu

DAEGU-GWANGYEOKSI

Gyeongsan

Gyeongju

Ulsan

ULSAN-GWANGYEOKSI

GYEONGSANGBUK-DO

GYEONGSANGNAM-DO

Miryang

Samnangjin

Jinju

Changwon

Masan

Gimhae

Jinhae

Busan

BUSAN-GWANGYEOKSI

Geoje

Namhaedo

Tongyeong

Sacheon

JAPAN

HONSHŪ

YAMAGUCHI

Shimonoseki

Kitakyūshū Ube

TSUSHIMA
(Japan)

KAMINO-SHIMA

SHIMONO-SHIMA

MI-SHIMA

0 20 40 60 80 100 120 Miles

0 20 40 60 80 100 120 140 160 180 200 Kilometers

Lambert Conformal Conic Projection
Scale 1:4,000,000
One inch to 64 miles
One cm to 40 km

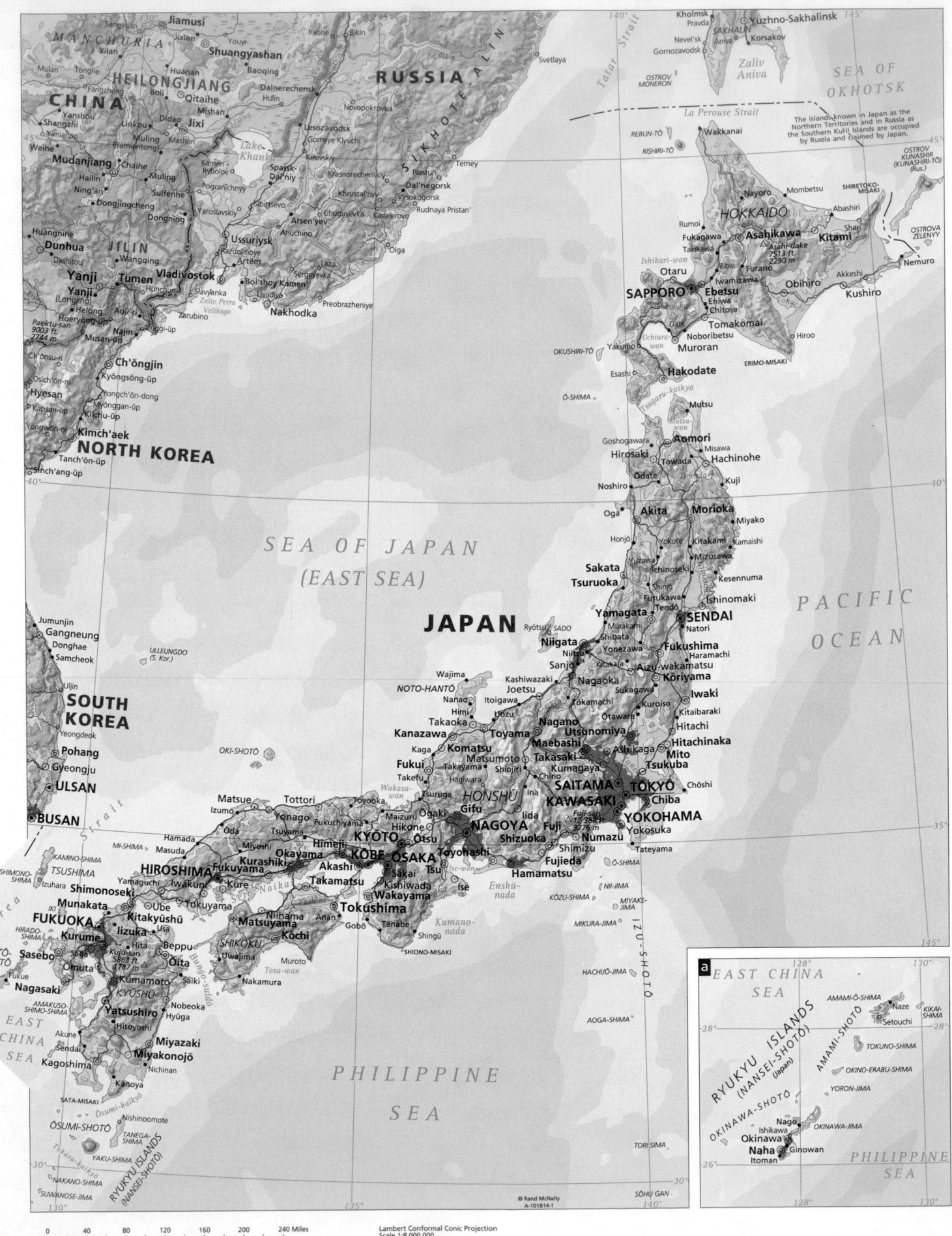

MANCHURIA

RUSSIA

SIKHOTE ALIN'

Jiamusi
Jixian
Youyi
Shuangyashan
Raohe
Bikin
Kholmsk
Pravda
Yuzhno-Sakhalinsk
Nevel'sk
Korsakov

HEILONGJIANG

Huanan
Baoqing
Gornozavodsk
Zaliv
Aniva

SEA OF
OKHOTSK

CHINA

Yilan
Mulan
Tonghe
Fangzheng
Mudan
Boli
Qitaihe
Svetlaya
OSTROV
MONERON
SAKHALIN

Shangzhi
Linkou
Didao
Mishan
Novopokrovka
La Perouse Strait

OSTROVA
ZELENYY

The islands known in Japan as the
Northern Territories and in Russia as
the Southern Kuril Islands are occupied
by Russia and claimed by Japan.

Yanshou
Mashan
Hulin
Lesozavodsk
Wakkanai
REBUN-TŌ
RISHIRI-TŌ

Mudanjiang
Chaihe
Kamen'-
Rybolov
Spassk-
Dal'niy
Kirovskiy

Hailin
Muling
Kirovskiy

Ning'an
Wangqing
Yarosklavskiy
Arsen'yev
Anuchino

Nayoro
Mombetsu
OSTROV
KUNASHIR
(KUNASHIRI-TŌ)
(Rus.)

Dunhua
Dongjingcheng
Dongning
Ussuriysk
Razdol'noye
Chuguyevka
Kavalerovo

JILIN
Rudnaya Pristan'
Rumoi
Fukagawa
Asahikawa
Asahi-dake
7513 ft.
2290 m
Kitami

Yanji
Tumen
Vladivostok
Bol'shoy Kamen'
Sergeyevka
Olga

Yanji
(Longjing)
Hunchun
Slavyanka
Lazo
Preobrazheniye

Helong
Zarubino
Nakhodka

Abashiri
HOKKAIDŌ

Otaru
SAPPORO
Ebetsu
Eniwa
Chitose
Obihiro

Takikawa
Iwamizawa
Bibai
Furano

SHIRETOKO-
MISAKI

OSTROVA
ZELENYY
Nemuro

Paektu-san
9003 ft
2744 m
Hoeryong-ŭp
Aoji-ri
Najin

Akkeshi
Kushiro

Ch'ŏnsu-ri
Musan-ŭp

Date
Tomakomai
Noboribetsu
Hiroo

NORTH KOREA

Ch'ŏngjin
Kyŏngsŏng-ŭp
Yongch'ŏn-dong

OKUSHIRI-TŌ
Yakumo
Uchiura-
wan
Muroran

Hyesan
Myŏnggan-ŭp
Kilchu-ŭp

Esashi
Hakodate
ERIMO-MISAKI

Kapsan-ŭp

Ō-SHIMA
Mutsu
Mutsu-
wan

Kimch'aek
Tanch'ŏn-ŭp

Goshogawara
Aomori
Misawa
Hachinohe

Sinch'ang-ŭp
40°
40°

Hirosaki
Towada
Kuji

SEA OF JAPAN
(EAST SEA)

Odate
Noshiro

Oga
Akita
Morioka
Miyako

Jumunjin
Gangneung
Honjō
Yokote
Kitakami
Kamaishi

Donghae
Samcheok

Sakata
Tsuruoka
Shinjō
Furukawa
Ishinomaki

ULLEUNGDO
(S. Kor.)

Yamagata
Murakami
Tendō
SENDAI

JAPAN
Ryōtsu
SADO
Niigata
Nihon
Shibata
Natori

SOUTH
KOREA

Uljin

Fukushima
Aizu-
wakamatsu
Haramachi

Sanjō
Kitakata
Kōriyama

Wajima
Kashiwazaki
Nagaoka
Sukagawa
Iwaki

Yeongdeok

NOTO-HANTŌ
Nanao
Joetsu
Tōkamachi
Kuroiso
Kitaibaraki

Pohang
Gyeongju

Itoigawa
Nagano
Utsunomiya
Hitachi

Himi
Uozu
Matsumoto
Otawara
Hitachinaka

ULSAN

Takaoka
Kanazawa
Toyama
Maebashi
Ashikaga
Mito

Kaga
Komatsu
Takasaki
Kumagaya
Tsukuba

Fukui
Takayama
Shiojiri
Chino
Ina
Fuji-san
12,388 ft.
3776 m

BUSAN
Takefu
Hagiwara
SAITAMA
TOKYO
Chōshi

Matsue
Tottori
Toyooka
Tsuruga
Gifu
Ōgaki
KAWASAKI
Chiba

Izumo
Fukuchiyama
Hikone
Nagoya
Fuji
YOKOHAMA
35°

Yonago
Maizuru
Ōtsu
Toyohashi
Shizuoka
Yokosuka

Hamada
Miyoshi
KYOTO
Tsu
Shimizu
Numazu

Masuda
Tsuyama
KOBE
OSAKA
Ise-wan
Tateyama

Yamaguchi
Okayama
Akashi
Sakai
Ise
Ō-SHIMA

HIROSHIMA
Fukuyama
Kishiwada
KŌZU-SHIMA
NII-JIMA

Iwakuni
Kure
Kurashiki
Wakayama
Enshū-
nada
MIYAKE-
JIMA

SHIMONO-
SEKI
Izuhara
TSUSHIMA
Takamatsu
Hamamatsu

Shimonoseki
Tokuyama
Naikai
Tokushima
Fujieda

KAMINO-
SHIMA
Ube
Kitakyūshū
Niihama
Kōchi
Anan
Kumano-
nada

IKI
Munakata
Iizuka
Matsuyama
Gobō
MIKURA-JIMA

FUKUOKA
Kurume
Usa
Tanabe
Shingū
SHIONO-MISAKI

HIRADO-
SHIMA
Saga
Hita
Beppu
Uwajima
Muroto
Tosa-wan
HACHIJŌ-JIMA

GOTŌ-
RETTŌ
Sasebo
Kuju-san
5863 ft
1787 m
Ōita
SHIKOKU
Nakamura

FUKUE-
JIMA
Ōmuta
Kumamoto
Saiki
Nobeoka
Hyūga

Nagasaki
Yatsushiro
Hitoyoshi
Kanoya

AMAKUSA-
SHIMO-SHIMA
Akune
Sendai
Miyazaki
Miyakonojō
Nichinan

EAST
CHINA
SEA

Izumi
KYŪSHŪ
Kagoshima
Kanoya
SATA-MISAKI
Nishinoomote

PHILIPPINE

SEA

ŌSUMI-SHOTŌ
TANEGA-
SHIMA
Tokara-kaikyō
YAKU-SHIMA
YAKU ISLANDS
(NANSEI-SHOTŌ)

NAKANO-SHIMA
SUWANOSE-JIMA
RYUKYU ISLANDS
(NANSEI-SHOTŌ)
SŌFU GAN
TORI SIMA

SŌHU GAN

© Rand McNally
A-101814-1

Lambert Conformal Conic Projection
Scale 1:8,000,000
One inch to 128 miles
One cm to 80 km

0   40   80   120   160   200   240 Miles
0 40 80 120 160 200 240 280 320 360 400 Kilometers

Inset (a)

EAST CHINA
SEA

RYUKYU ISLANDS
(NANSEI-SHOTŌ)

AMAMI-Ō-SHIMA
Naze
KIKAI-
SHIMA
Setouchi

AMAMI-SHOTŌ
TOKUNO-SHIMA
OKINO-ERABU-SHIMA
YORON-JIMA

OKINAWA-SHOTŌ
(Japan)
AMAMI-SHOTŌ

Nago
Ishikawa
OKINAWA-JIMA

Okinawa
Ginowan
Naha
Itoman

PHILIPPINE
SEA

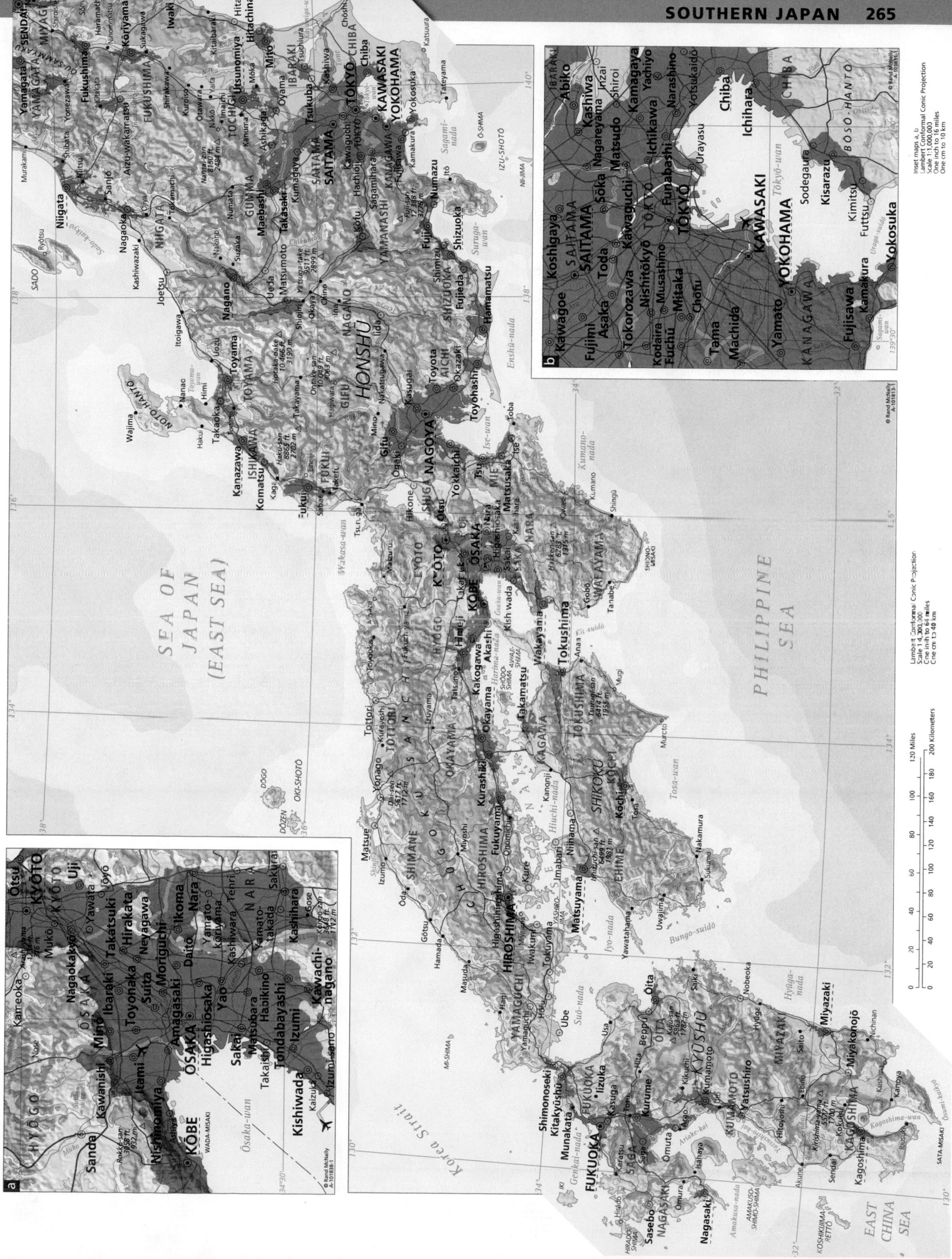

CHINA
GUANGXI
GUANGDONG
Nanning
Guixian
Yutian
Zhanjiang
Beihai
Haikou
HAINAN
HAINAN DAO
Sanya

SOUTH CHINA SEA

Gulf of Tonkin

VIETNAM
Da Nang
Hue
Huong Thuy
HÀ NỘI
Hai Phong
Nam Dinh
Ninh Binh
Thanh Hoa
Vinh

LAOS
Vientiane (Viangchan)
Louangphrabang

INDOCHINA

THAILAND
BANGKOK (KRUNG THEP)
Chon Buri
Nakhon Ratchasima
Udon Thani
Khon Kaen
Chiang Mai
Chiang Rai
Lampang
Phitsanulok

CAMBODIA
PHNOM PENH (Phnum Pénh)
Bătdâmbâng
Siĕmréab
Kâmpóng Saôm

HO CHI MINH CITY (SAIGON)
Biên Hoa
Vung Tau
Phan Thiet
Da Lat
Nha Trang

MYANMAR (BURMA)
YANGON (RANGOON)
Mandalay
Bago
Pathein
Moulmeingyun
Mawlamyine
Lashio
Meiktila
Taunggyi
Prome

CHIN HILLS
ARAKAN YOMA
INDIA
MIZORAM
BNGL
CHITTAGONG
Cox's Bazar

BAY OF BENGAL

ANDAMAN SEA

Gulf of Martaban

MERGUI ARCHIPELAGO

Gulf of Thailand

ANDAMAN AND NICOBAR ISLANDS (India)
Port Blair
NORTH ANDAMAN
MIDDLE ANDAMAN
SOUTH ANDAMAN
LITTLE ANDAMAN
CAR NICOBAR
COCO ISLANDS
PAGODA POINT
ISTHMUS OF KRA
BILAUKTAUNG RANGE
DAWNA RANGE
PHNOM DONGRAK RANGE
KRAVANH

0 40 80 120 160 200 240 Miles
0 40 80 120 160 200 240 280 320 360 400 Kilometers

Lambert Azimuthal Equal Area Projection
Scale 1:8,000,000
One inch to 128 miles
One cm to 80 km

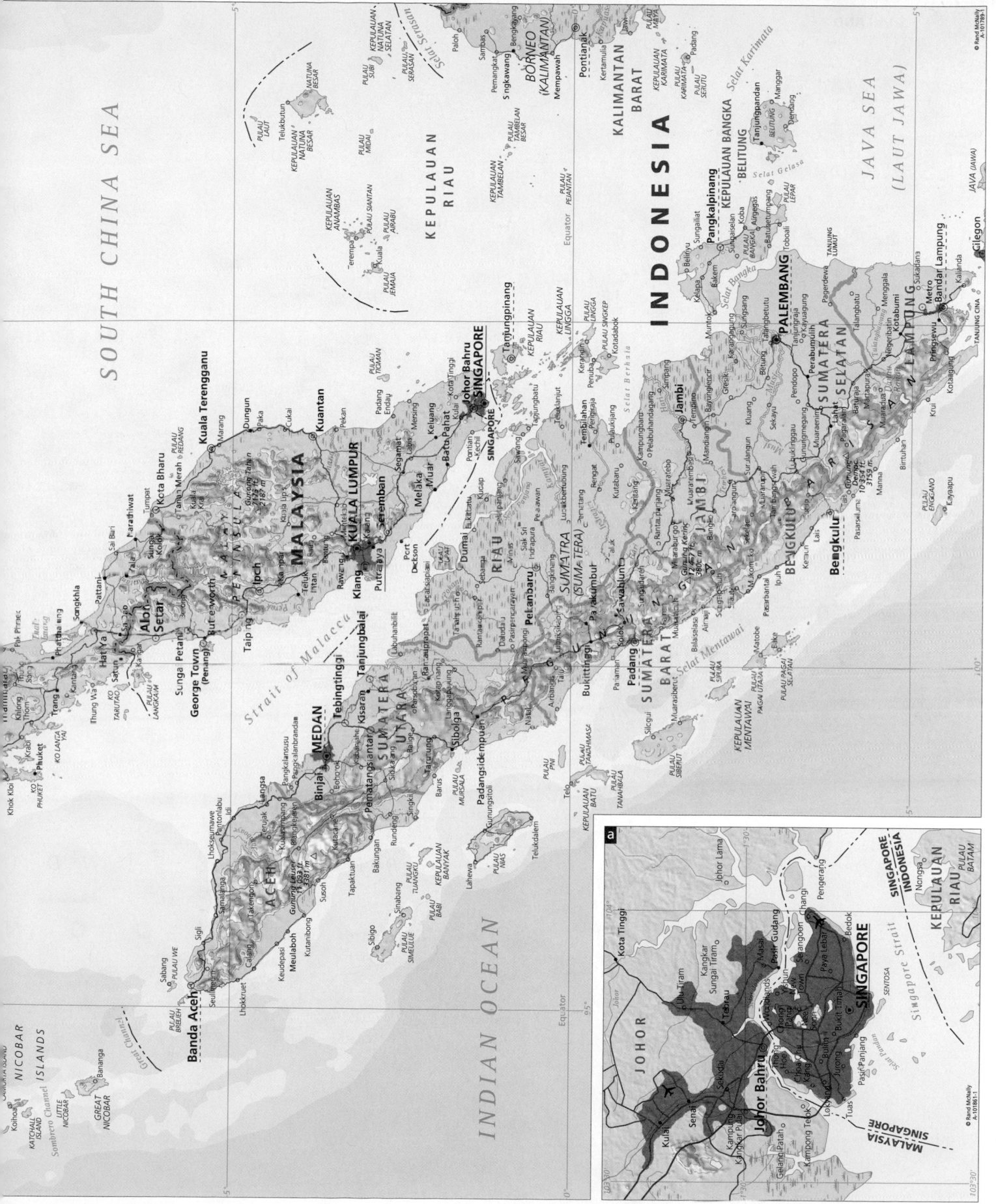

SOUTH CHINA SEA

BORNEO
(KALIMANTAN)

KALIMANTAN
BARAT

INDONESIA

KEPULAUAN
RIAU

KEPULAUAN BANGKA

JAVA SEA
(LAUT JAWA)

BELITUNG

JAVA (JAWA)

Pangkalpinang

PALEMBANG

SUMATERA
SELATAN

KUALA LUMPUR

MALAYSIA

Seremban

SINGAPORE

RIAU

SUMATERA
(SUMATERA)

JAMBI

Jambi

MEDAN

SUMATERA
UTARA

Padang

SUMATERA
BARAT

BENGKULU

Bengkulu

ACEH

Banda Aceh

INDIAN OCEAN

Strait of Malacca

NICOBAR
ISLANDS

LITTLE
NICOBAR

GREAT
NICOBAR

Equator

a

JOHOR

Kota Tinggi

SINGAPORE

MALAYSIA
SINGAPORE

SINGAPORE
INDONESIA

KEPULAUAN
RIAU

PULAU
BATAM

Johor Bahru

Singapore Strait

Inset map a
Lambert Conformal Conic Projection
Scale 1:1,000,000
One inch to 16 miles
One cm to 10 km

THAILAND

Thung Wa
Songkhla
Hat Yai
Pattani
Sa Dao
Sai Buri
Satun
Yala
Narathiwat
Kangar
Sungai Kolok
Tumpat
PULAU LANGKAWI
Alor Setar
Kota Bharu
Sungai Petani
Betong
Tanah Merah
PULAU REDANG
George Town (Penang)
Butterworth
Kuala Krai
Kuala Terengganu
Taiping
MALAY PENINSULA
Marang
Ipoh
Gunong Tahan 7175 ft 2187 m
Dungun
Kampar
Paka
Kuala Lipis
Cukai
Teluk Intan
Raub
Benting
Mentekab
Pekan
Kuantan
Rawang
MALAYSIA
Klang
KUALA LUMPUR
Putrajaya
Seremban
Padang Endau
PULAU TIOMAN
Tanjungbalai
Labuhanbilik
Port Dickson
Melaka
Segamat
Labis
Mersing
Bagansiapiapi
Muar
Keluang
Rantauprapat
Kotapinang
Batu Pahat
Kota Tinggi
SUMATERA UTARA
Tanahputih
Pontian Kechil
SINGAPORE
Langgapayung
Rantaukampar
Dumai
Bukitbatu
Kudap
Johor Bahru
SINGAPORE
Daludalu
Sebanga
RIAU
Selatpanjang
Tanjungpinang
Muarasipongi
Pasirpengaraian
Minas
Siak Sri Indrapura
KEPULAUAN RIAU
Pekanbaru
Bangkinang
Pelalawan
Lubukbertubung
Lubuksikaping
SUMATRA (SUMATERA)
Payakumbuh
Pematang
Teluklanjut
KEPULAUAN LINGGA
Equator
PULAU PEJANTAN
Bukittinggi
Taluk
Rengat
Perigiraja
Pariaman
Sawahlunto
Pulaukijang
PULAU LINGGA
Padang
Solok
Sungaidareh
Kelantang
Kotadabok
PULAU SINGKEP
Selat Berhala
Painan
SUMATERA BARAT
Muaralabuh
Rantaupanjang
Kampungbaru
di Pelabuhandagang
Simpang
Gunung Kerinci 12 467 ft 3800 m
Muaratebo
PULAU SIPURA
Sungaipenuh
Jambi
Balaiselasa
Airhaji
Sungaipenuh
Muaratembesi
Tempino
Bayunglencir
JAMBI
Mandiangin
KEPULAUAN MENTAWAI
Ipuh
Pasarbantal
Bangko
Sarolangun
Surulangun
Kluang
Gresik
Karangagung
Bake
Mukomuko
Sekelad
Muararupit
Sekayu
Betung
Talangbetutu
BENGKULU
Lais
Lubuklinggau
Pendopo
Muaraenim
PALEMBANG
Toboali
Curup
Gunungmegang
SUMATERA SELATAN
Bengkulu
Tais
Lahat
Perabumulih
Pagerdewa
Gunung Dempo 10 364 ft 3159 m
Batutaja
Kayuagung
Pasarseluma
Pagaralam
SUMATERA
Muaradua
Talangbatu
Manna
LAMPUNG
Menggala
Bintuhan
Negeribatin
Sukadana
Kotabumi
Danau Ranau
PULAU ENGGANO
Krui
Pringsewu
Metro
Kayaapu
Kotaagung
Bandar Lampung
Kalianda
TANJUNG CINA
KRAKATOA
Cilegon
Serang
JAKARTA
Karawang
PULAU PANAITAN
Labuhan
DEPOK
Cikampek
Indramayu
TANJUNG CANGKUANG
BANTEN
Bogor
Cianjur
Klangenang
Cirebon
Pelabuhanratu
Sukabumi
JAWA BARAT
Majalaya
Garut
Tegal
Pekalongan
Jampang-kulon
BANDUNG
Gunung Slamet 11 247 ft 3428 m
JAWA TENGAH
Purwokerto
Jampang-kulon
Tasikmalaya
Purworejo
Magelang
Sindangbarang
Karangnunggal
Cijulang
Cilacap
Yogyakarta
Pacitan
YOGYAKARTA

SOUTH CHINA SEA

PULAU LAUT
KEPULAUAN NATUNA BESAR
Telukbutun
Lutong
Miri
Niah
KEPULAUAN ANAMBAS
NATUNA BESAR
Terempa
PULAU SIANTAN
Bintulu
Tubau
Kuala
PULAU AIRABU
KEPULAUAN NATUNA SELATAN
PULAU BRUIT
Igan
Mukah
Belaga
PULAU SUBI
PULAU MIDAI
SARAWAK
Balingian
PULAU JEMAJA
SIBU
MALAYSIA
KEPULAUAN RIAU
PULAU SERASAN
TANJONG DATU
Paloh
Semantan
Kanowit
Kapit
Sibu
PULAU SERASAN
Selat Serasan
Sematan
UPPER KAPUAS MTS.
Betong
Pemangkat
Sambas
Bau
Serian
Sri Aman
Nangabadau
Nangaobat
KEPULAUAN TAMBELAN
Siluas
Kuching
Saratok
Semitau
Putussibau
PULAU TAMBELAN BESAR
Bengkayang
Gunung Niut 5581 ft 1701 m
Sanggau
Nangaraun
PEGUNUNGAN MÜLLER
Singkawang
Mualang
Longguntur
Kertamulia
Ngabang
Sosok
Sintang
Tanjungpinang
Mempawah
Pontianak
Meliau
PULAU MAYA
Teratak
KALIMANTAN BARAT
Nangapinoh
Rangantemia
Jawi
Kotabaharu
Bukit Raya 7474 ft 2278 m
Sungaipinang
PULAU MAYA
Telukbatang
Nangalangki
PEGUNUNGAN SCHWANER
Tewah
Kualakurun
Sukadana
Sandai
Nangatayap
Kualamanjual
KALIMANTAN TENGAH
KEPULAUAN KARIMATA
Ketapang
Panahan
Bakumpai
PULAU KARIMATA
Serengka
Mabau
Palangkaraya
PULAU SERUTU
Kualapesaguan
Sukaraja
Kotawaringin
Padang
Kendawangan
Telegapulang
Sampit
Mendawai
Teluk Sukadana
Pangkalanbuun
PULAU BAWAL
Matua
TANJUNG PUTING
Kumai

KEPULAUAN RIAU

KEPULAUAN BANGKA BELITUNG
Belinyu
Sungailiat
Kelapa
Muntok
Bakem
Pangkalpinang
Koba
Airgegas
Tanjungpandan
BELITUNG
Manggar
Batubetumpang
Dendang
PULAU LEPAR
PULAU GELASA

JAVA SEA
(LAUT JAWA)

GREATER SUNDA

KEPULAUAN KARIMUNJAWA
Tambak
PULAU BAWEAN

INDON

Jepara
Pati
Rembang
Kudus
Tuban
Bangkalan
Sumenep
MADURA
Cepu
Gresik
Pamekasan
Semarang
Salatiga
Surakarta
Jombang
JAWA TIMUR
SURABAYA
Boja
Madiun
Ponorogo
Kediri
Pasuruan
Blega
Kalianget
Probolinggo
Situbondo
Malang
Tulungagung
Lumajang
Jember
Blitar
Gunung Semeru 12 060 ft 3676 m
Gente

INDIAN OCEAN

JAVA (JAWA)

CHRISTMAS ISLAND (Austl.)
Settlement

0 40 80 120 160 200 240 Miles
0 40 80 120 160 200 240 280 320 360 400 Kilometers

Lambert Azimuthal Equal Area Projection
Scale 1:8,000,000
One inch to 128 miles
One cm to 80 km

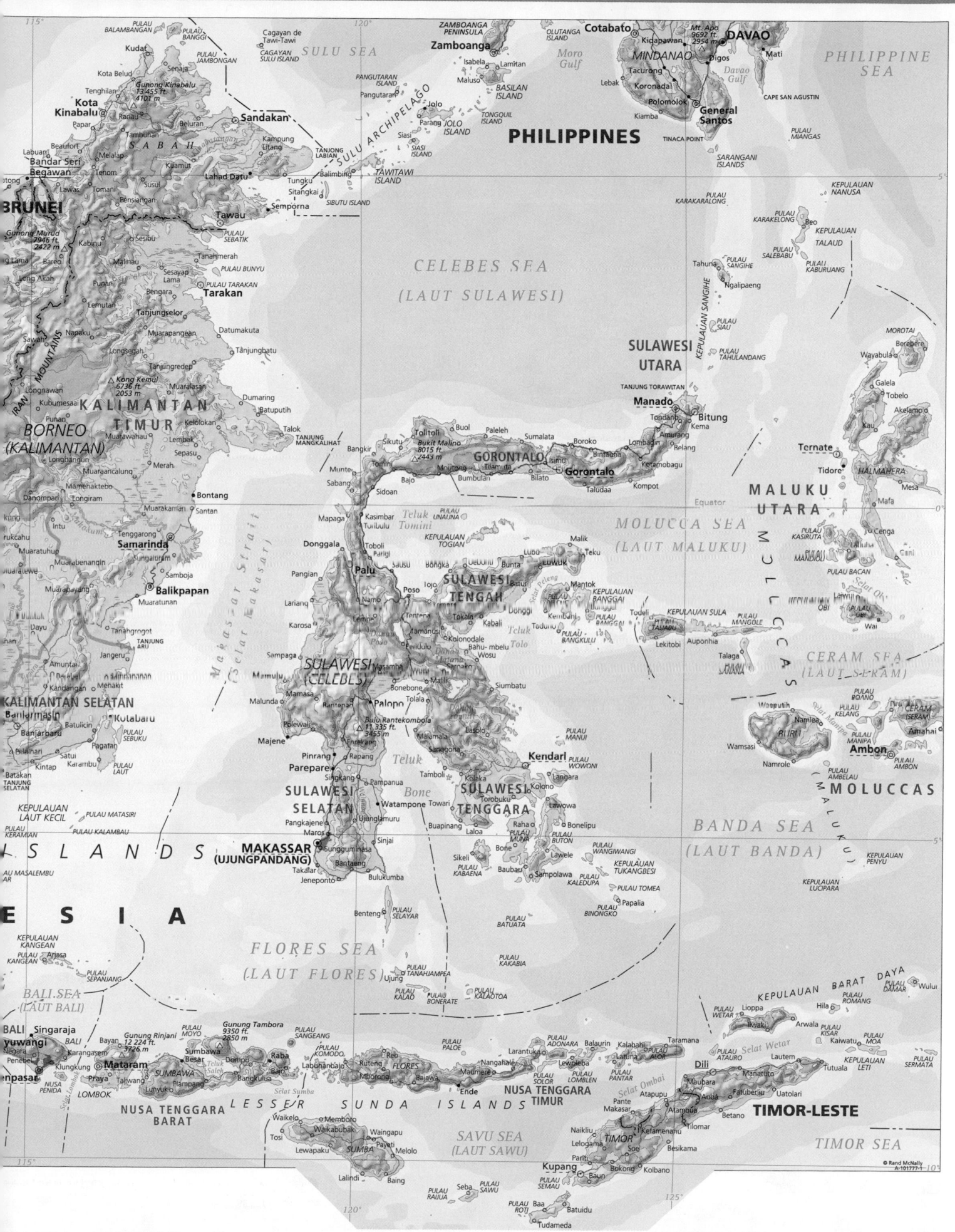

PHILIPPINES

Seas and Bodies of Water

SOUTH CHINA SEA
PHILIPPINE SEA
SULU SEA
CELEBES SEA
SIBUYAN SEA
VISAYAN SEA
SAMAR SEA
BOHOL SEA
Babuyan Channel
Luzon Strait
Bashi Channel
Balintang Channel
Lingayen Gulf
Baler Bay
Dingalan Bay
Lamon Bay
Tayabas Bay
Ragay Gulf
Panay Gulf
Moro Gulf
Iligan Bay
Illana Bay
Davao Gulf
Leyte Gulf
Palawan Passage
Balabac Strait
Mindoro Strait
Tablas Strait
Sibuyan Sea
Macajalar Bay
Sarangani Bay
Honda Bay
Manila Bay

Inset (a)

SOUTH CHINA SEA
Bashi Channel
AMIANAN ISLAND
ITBAYAT ISLAND
BATAN ISLANDS
Basco
BATAN ISLAND
Luzon Strait
Balintang Channel
BABUYAN ISLANDS
CALAYAN ISLAND
Calayan
DALUPIRI ISLAND
FUGA ISLAND
BABUYAN ISLAND
CAMIGUIN ISLAND
PHILIPPINE SEA
Babuyan Channel
PALAUI ISLAND
CAPE BOJEADOR
Pagudpud
Laoag
San Nicolas
Batac
LUZON
Mt. Sicapoo 7329 ft. 2234 m
Aparri
Gonzaga
ESCARPADA POINT
© Rand McNally A-101808-1

Luzon

Pagudpud
CAPE BOJEADOR
Laoag
San Nicolas
Batac
Mt. Sicapoo 7329 ft. 2234 m
Aparri
Gonzaga
PALAUI ISLAND
ESCARPADA POINT
Vigan
Bangued
Conner
Tuguegarao City
Ilagan
Tabuk
Lubuagan
Bontoc
COR. CENTRAL
SIERRA MADRE
Cagayan
San Fernando
La Trinidad
Baguio
Agno
Lingayen
Dagupan
Solano
Bayombong
Maddela
Santiago
Mt. Pulog 9626 ft. 2934 m
Mt. Palanan 3976 ft. 1212 m
CAPE SAN ILDEFONSO
San Carlos
Santa Cruz
Camiling
Iba
Palauig
Tarlac
Cabanatuan
San Jose
Baler
Guimba
Angeles
Mt. Pinatubo 5840 ft. 1780 m
San Fernando
Malolos
Olongapo
Oran
Meycauayan
MANILA
QUEZON CITY
Pasig
Cavite
BATAAN PENINSULA
Mariveles
Santa Cruz
San Pablo
Lucban
Balayan
Lipa
Batangas
Lucena
LUBANG ISLANDS
Lubang
LUBANG ISLAND
Paluan
Calapan

Mindoro / Central Islands

Mt. Halcon 8481 ft. 2585 m
MINDORO
Mamburao
Mt. Baco 8159 ft. 2487 m
MARINDUQUE
DUMALI POINT
Pinamalayan
Bongabong
San Jose
Pagsañgahan
BONDOC POINT
ILIN ISLAND
Romblon
ROMBLON I.
TABLAS ISLAND
Tablas Strait
GARABAO ISLAND
SIBUYAN ISLAND
BURIAS ISLAND
MASBATE
Masbate
Milagros
Aroroy
TICAO ISLAND
San Jacinto
Magallanes
Bulan

Bicol / Southern Luzon

Daet
Gumaca
Catanauan
Guinayangan
Naga
Pili
Iriga
Goa
Bato
Tabaco
Legaspi
Mayon Volcano 8077 ft. 2462 m
Sorsogon
Bulusan
CATANDUANES ISLAND
Virac
CAGRARAY ISLAND
BATAN ISLAND
RAPU RAPU ISLAND
San Miguel Bay
Lagonoy Gulf
BONDOC PENINSULA
Tayabas Bay

Palawan

PALAWAN
Puerto Princesa
Victoria Peaks 5607 ft. 1709 m
Mt. Mantalingajan 6841 ft. 2085 m
Marangas
Rio Tuba
Taytay
Caruray
Honda Bay
DUMARAN ISLAND
CAGAYAN ISLANDS
CALAMIAN GROUP
BUSUANGA ISLAND
CULION ISLAND
LINAPACAN ISLAND
Linapacan Strait
CUYO ISLANDS
CUYO ISLAND
AGUTAYA ISLAND
CABULAUAN ISLAND
BATAS ISLAND
LIBRO POINT
CALANDAGAN ISLAND
Coyo East Pass
Coyo West Pass
CAVILI ISLAND

Visayas

PANAY
Nabas
Kalibo
Roxas
Dumalag
Tibiao
Januiay
Silay
Iloilo
GUIMARAS ISLAND
San Jose
La Carlota
NEGROS
Bacolod
San Carlos
La Castellana
Binalbagan
Kabankalan
Sipalay
Bayawan
Tanjay
Dumaguete
Bonawon
SIQUIJOR ISLAND
Siquijor
Santander
PANGLAO ISLAND
Tagbilaran
BOHOL
Guindulman
CEBU
Toledo
Danao
Cebu
Mandaue
Lapu-Lapu
Talibon
Maasin
Sagay
Cadiz
Victorias
Sagay
Toboso
Bantayan
BANTAYAN ISLAND
CAMOTES ISLANDS
Camotes Sea
Bogo
Ormoc
LEYTE
Baybay
Burauen
MacArthur
Balangiga
Basey
Guiuan
HOMONHON ISLAND
Tacloban
Carigara
Villalon
SAMAR
Catarman
Calbayog
Catbalogan
Borongan
Llorente
Laoang
Gamay
BILIRAN ISLAND
Leyte Gulf
Maasin
Sogod
Hindang
Libagon
DINAGAT ISLAND
Dinagat
Surigao
SIARGAO ISLAND
BUCAS GRANDE ISLAND
PANAON ISLAND
Dinagat Sound
Surigao Strait
Jintotolo Channel
BANTAYAN ISLAND
Asid Gulf
SEMIRARA ISLAND
SIBAY ISLAND
CARABAO ISLAND

Mindanao

MINDANAO
Dipolog
Katipunan
Sindangan
Siocon
Siraway
ZAMBOANGA PENINSULA
SIBUGUEY PENINSULA
Zamboanga
Isabela
BASILAN ISLAND
Lamitan
Maluso
PILAS GROUP
PANGUTARAN GROUP
PANGUTARAN ISLAND
Pangutaran
SAMALES GROUP
JOLO GROUP
JOLO ISLAND
Jolo
Parang
TAPUL GROUP
PATA ISLAND
Siasi
SIASI ISLAND
TONGQUIL ISLAND
TAWITAWI ISLAND
Balimbing
Bongao
CAGAYAN SULU ISLAND
Cagayan de Tawi-Tawi
SULU ARCHIPELAGO
Oroquieta
Tudela
Ozamis
Tangub
Bonifacio
Pagadian
Malangas
Margosatubig
OLUTANGA ISLAND
Buenavista
Liloy
Iligan
Marawi
Mt. Kaatoan 9501 ft. 2896 m
Lake Sultan Alonto
Malaybalay
Valencia
Impasugong
Alubijid
Cagayan de Oro
Gingoog
Balingasag
Salay
Prosperidad
Bislig
Lianga
Lianga Bay
Tandag
Bhutan
Cabadbaran
Tabonga
Butuan
Caraga
Baganga
Lupon
Governor Generoso
CAPE SAN AGUSTIN
Mati
Tagum
Panabo
Babak
SAMAL ISLAND
Davao
Digos
Malita
Sultan Kudarat
Parang
Cotabato
Datu Piang
Talayan
Midsayap
Kabacan
Kidapawan
Mt. Apo 9692 ft. 2954 m
Isulan
Tacurong
Lebak
Koronadal
Mt. Busa 6334 ft. 2083 m
Kiamba
Palimbang
Polomolok
General Santos
Glan
Culaman
Jose Abad Santos
TINACA POINT
BALUT ISLAND
SARANGANI ISLAND
SARANGANI ISLANDS
PULAU MIANGAS
Sarangani Bay
Davao Gulf
Moro Gulf
Iligan Bay
Illana Bay
Macajalar Bay
Basilan Strait
Tolo-ong
Buluan

Malaysia / Borneo

MALAYSIA
BORNEO
Sandakan
Lahad Datu
Kunak
Lamag
Pintasan
Sukau
Kitagan
Beluran
Senaja
PULAU BALAMBANGAN
PULAU BANGGI
PULAU MALAWALI
PULAU JAMBONGAN
Telukan Labuk
TANJONG PISAU
Tanjong Labian
Kampung Litang
Segama
Sugut
North Balabac Strait
BUGSUK ISLAND
Balabac
BALABAC ISLAND
Balabac Strait
SIBUTU ISLAND
Sibutu Passage
Sitangkai
Bongao
Tungku

Other labels

SOUTH CHINA SEA
(Claimed by China, Taiwan and the Philippines)
SCARBOROUGH REEF
PHILIPPINE SEA
YOG POINT
POLILLO ISLAND
POLILLO ISLANDS
PATNANONGAN ISLAND
JOMALIG ISLAND
Polillo Strait
Burdeos
ALABAT ISLAND
Larap
CALAGUA ISLANDS

Scale / Projection

0 30 60 90 120 150 180 Miles
0 30 60 90 120 150 180 210 240 270 300 Kilometers

Lambert Conformal Conic Projection
Scale 1:6,000,000
One inch to 96 miles
One cm to 60 km

© Rand McNally A-101807-1

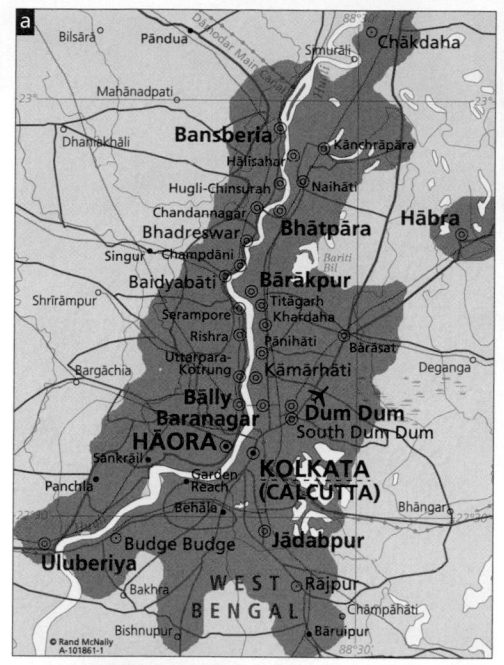

a

Bilsārā · Pāndua · Chākdaha
Mahānadpati · Simurāli
Dhanīākhāli · Bānsberia · Hālisahar · Kānchrāpāra
Hugli-Chinsurah · Naihāti
Chandannagar · Bhadreswar · Bhātpāra · Hābra
Singur · Champdāni · Bārākpur
Baidyabāti · Titāgarh · Khardaha
Serampore · Rishra · Pānihāti · Bārāsat
Uttarpara-Kotrung · Kāmārhāti · Deganga
Bārgāchia · Bally · Baranagar · Dum Dum · South Dum Dum
Sānkrāil · HĀORA
Panchla · Garden Reach · KOLKATA (CALCUTTA)
Behāla · Bhāngar
Uluberia · Budge Budge · Jādabpur
Bakhra · Rājpur
WEST BENGAL · Bishnupur · Bāruipur · Champāhāti

b

Kepong · Batu Caves
Sungai Buluh · Selapak · Kuala Ampang
KUALA LUMPUR · Meru · Ampang
Shah Alam · KUALA LUMPUR · Hulu Langat
Klang · Puchong · Petaling Jaya
Serdang · Kajang
Teluk Panglima Garang · Putrajaya · Semenyih
Jenjarum · Kampung Pulau Ibul · Beroga · Mantin
Chondoi · Kampung Dengkil · Bangi
Banting · NEGERI SEMBILAN
SELANGOR · Salak · Tiroi
Morib · Sepang · Seremban
Kampung Batu Laut · Kampung Janginl · Sungai Pelik · Kampoh
Tanah Merah · Mambau
Strait of Malacca · Kampung Cuah · Siliau
Port Dickson

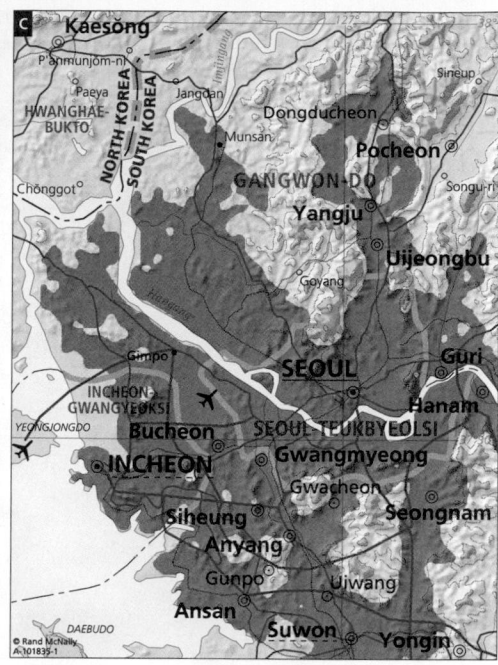

c

Kaesŏng · P'anmunjŏm-ni · Paeya · Jangdan · Sineup
HWANGHAE-BUKTO · NORTH KOREA · SOUTH KOREA · Songu-ri
Chŏnggot · Munsan · Dongducheon · Pocheon
GANGWON-DO · Yangju
Goyang · Uijeongbu · Guri
Gimpo · SEOUL · Hanam
INCHEON-GWANGYEOKSI · Bucheon · Gwangmyeong · Gwacheon
YEONGJONGDO · SEOUL-TEUKBYEOLSI
INCHEON · Siheung · Gunpo · Uiwang · Seongnam
Anyang · Ansan · Suwon · Yongin
DAEBUDO

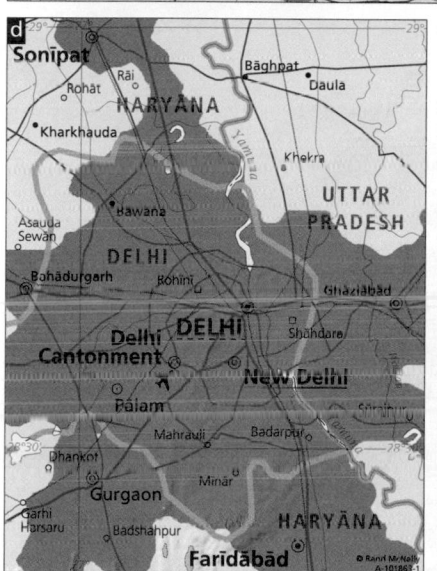

d

Sonīpat · Bāghpat · Daula
Rohat · Rāi
Kharkhauda · Khekra
HARYANA · UTTAR PRADESH
Asauda Sewān · Bawāna · Ghāziābād
Bahādurgarh · Rohini
DELHI · DELHI · Shāhdara
Delhi Cantonment
Pāam · NEW DELHI
Dhankot · Mahrauli · Badarpur
Gurgaon · Minār
Garhi Harsaru · Badshāhpur · HARYANA
Faridābād

e

FUKUEI CHIAO · Shihmen
Gangziping · Sanchih
EAST CHINA SEA · Tanshui · Chinshan · Wanli
Baoouotuo · Linkou · Chilung
Tayuan · Sangchungshih · Hsichih
Hsinchuang · TAIPEI
Taoyūan · Hsintien
Chungli · Sanhsia · Wulai
Pingchen · Tahsi
Longtan · Chiaohsi · Ilan
Kuanhsi · Chiaopan
Taman Shan 5190 ft 1582 m · Chuangwei
Kunghsi · Lotung
Sanhsing

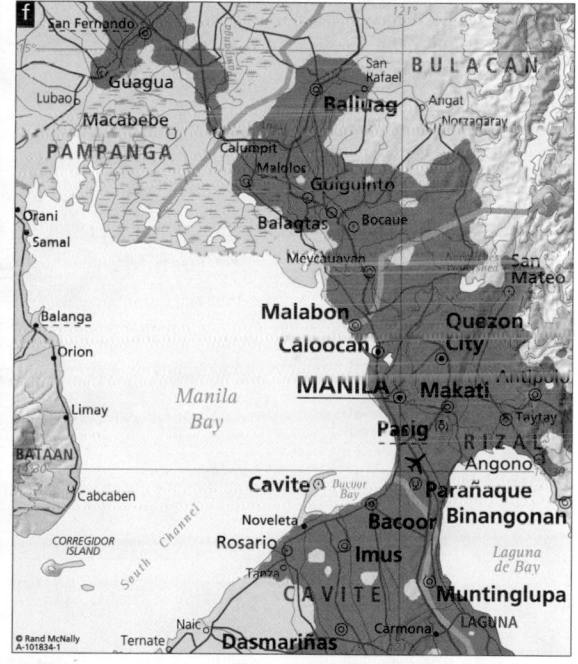

f

San Fernando · BULACAN
Lubao · San Rafael · Angat · Norzagaray
Guagua · Baliuag
Macabebe · Calumpit · Maigios · Guiguinto
PAMPANGA · Bocaue
Orani · Balagtas · San Mateo
Samal · Meycauayan
Balanga · Malabon · Quezon City
Orion · Caloocan · Antipolo
Limay · MANILA · Makati
BATAAN · Manila Bay · Pasig · RIZAL
Cavite · Taytay
Cabcaben · Noveleta · Parañaque · Angono
Rosario · Bacoor · Binangonan
CORREGIDOR ISLAND · Imus · Laguna de Bay
Naic · Tanza · CAVITE · Muntinlupa
Ternate · Carmona · LAGUNA
Dasmariñas

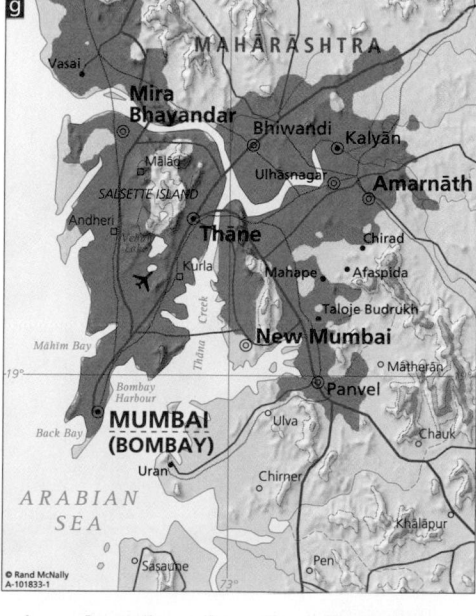

g

MAHĀRĀSHTRA
Vasai · Mira Bhayandar · Bhiwandi · Kalyān
Mālād · Ulhāsnagar · Amarnāth
SALSETTE ISLAND · Andheri · Thāne
Kurla · Chirad
Māhim Bay · Mahape · Afaspida
Taloje Budrukh
New Mumbai · Panvel
Bombay Harbour · Ulva · Chauk
Back Bay · MUMBAI (BOMBAY) · Uran · Māthcran
ARABIAN SEA · Sasaune · Pen

h

Rassām · Salih Hasan · Arab Yahūdah
Khalaf Laftah · Muhammad Bāqir · Hayy ath Thawrah
Al-Kāzimiyah · BAGHDAD
AL-ANBĀR · BAGHDAD
Abū Ghurayb · Hayy ad Durah
'Uwayrij · Midhat
Khamīs ash Shāhin · Majid Sha'lān · DIYĀLĀ
Abū Muhammad · Al Hājj Ja'far · Sālih Muhammad
Qal'at Hamīd al Habash · Al-Madā'in
BĀBIL · Zukaytūn

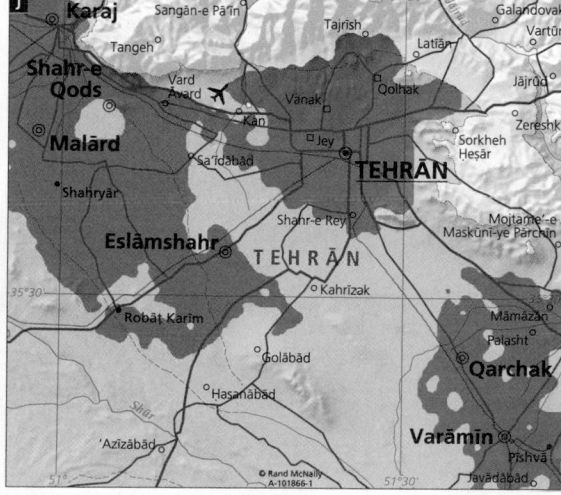

i

BALUCHISTĀN · Hab Chauki · Gadap · Malir Cantonment
North Karāchi
North Nazimabad · Malir Cantonment
Orangi · Pipnapur
Baldia · Malir Cantonment
Mauripur · KARĀCHI · SIND
Lyari · Jamshed
Kiamari · Landhi
Clifton · Korangi · Bin Qasim
ARABIAN SEA

j

Karaj · Sangān-e Pā'in · Tajrish · Galandovak · Vartūn
Tangeh · Latlān · Jājrūd
Shahr-e Qods · Vard Avard · Vanak · Qolhak · Zereshki
Malārd · Kan · Jey · Sorkheh Hesār
Sa'īdābād · TEHRĀN
Shahryār · Shahr-e Rey
Eslāmshahr · Mojtame-e Maskūni-ye Parchin
TEHRĀN · Kahrizak
Robāt Karim · Golābād · Māmāzon · Falasht
Hasanābad · Qarchak
'Azīzābād · Varāmīn · Pishvā · Javādābād

0 5 10 15 20 25 30 Miles
0 5 10 15 20 25 30 35 40 45 50 Kilometers

Lambert Conformal Conic Projection
Scale 1:1,000,000
One inch to 16 miles
One cm to 10 km

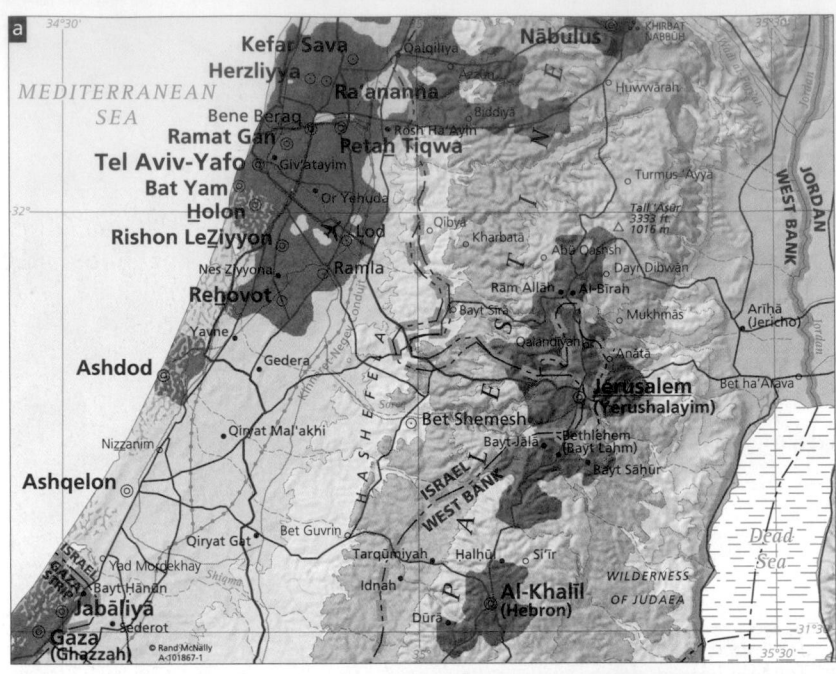

a

MEDITERRANEAN SEA

Kefar Sava
Herzliyya
Ra'ananna
Bene Beraq
Ramat Gan
Petah Tiqwa
Tel Aviv-Yafo
Bat Yam
Holon
Rishon LeZiyyon
Rehovot

Ashdod

Ashqelon

ISRAEL GAZA STRIP

Jabaliya
Gaza (Ghazzah)

Nabulus
Qalqilya
Biddya
Huwwarah
Turmus 'Ayya
Tall 'Asur 3333 ft. 1016 m

WEST BANK
JORDAN

Giv'atayim
Or Yehuda
Lod
Ramla
Nes Ziyyona
Yavne
Qiryat Mal'akhi
Gedera
Nizzanim
Bayt Hanan
Sederot

Rosh Ha'Ayin
Qibya
Kharbata
Ram Allah
Al-Birah
Mukhmas
Anata
Qalandiyah
Dayr Dibwan
Abu Dis

Jerusalem (Yerushalayim)
Bet Shemesh
Bethlehem (Bayt Lahm)
Bayt Jala
Bayt Sahur
Bet Guvrin
Qiryat Gat
Tarqumiyah
Halhul
Idnah
Dura
Al-Khalil (Hebron)

Bayt Sira
Ariha (Jericho)
Bet ha'Arava

WILDERNESS OF JUDAEA

Dead Sea

PALESTINE
ISRAEL
WEST BANK
HASHEFELA

© Rand McNally A-101867-1

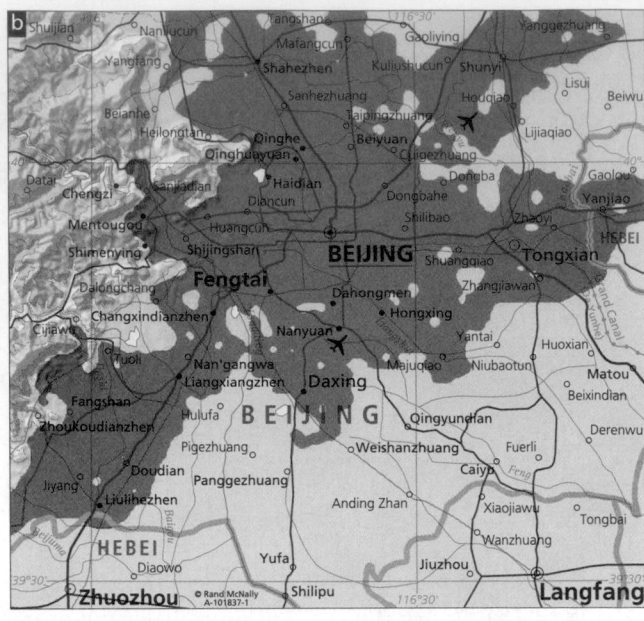

b

BEIJING
HEBEI

Shijiazhuang
Nanyucun
Mafangcun
Gooliying
Yanggezhuang

Shujiang
Yangfang
Shahezhen
Kuliushucun
Shunyi
Lisui
Beiwu

Beianhe
Heilongtan
Sanhezhuang
Taipingzhuang
Beiyuan
Liujiaqiao
Gaoluo

Datai
Chengzi
Qinghe
Haidian
Houjie
Donghe
Yanjiao

Mentougou
Huangcun
Diancun
Dongbahe

Shimenying
Shijingshan
BEIJING
Shuangqiao
TONGXIAN

Dalongchang
Tongxian
Hongxing
Yantai
Huoxian

Changxindianzhen
Fengtai
Dahongmen
Zhangjiawan
Matou
Beixindian

Fangshan
Tuoli
Nan'gangwa
Liangxiangzhen
Daxing
Qingyundian
Derenwu

Zhoukoudianzhen
Hulufa
Weishanzhuang
Fuerli

Pigezhuang
Doudian
Panggezhuang
Anding Zhan
Xiaojiawu
Tongbai

Jiyang
Liulihezhen
Wanzhuang

HEBEI
Yufa
Jiuzhou
Diaowo
Shilipu
ZHUOZHOU
Langfang

© Rand McNally A-101837-1

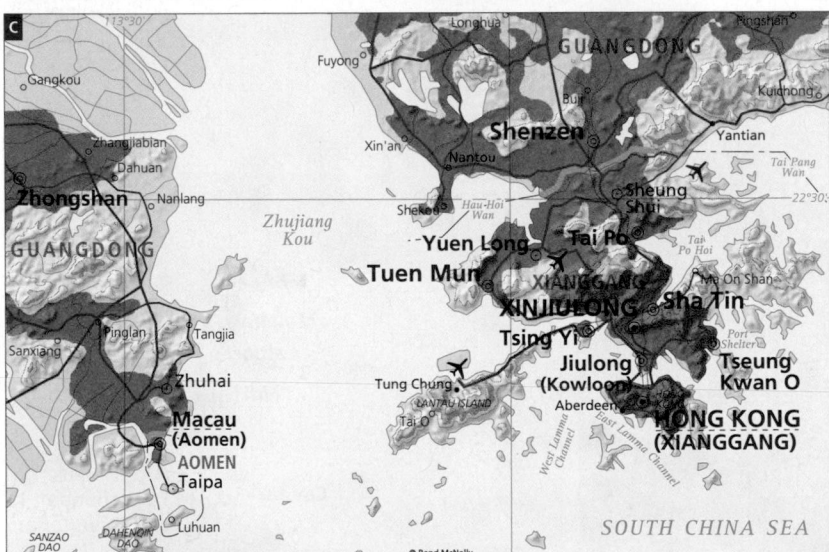

c

GUANGDONG

Longhua
Pingshan
Fuyong
Buji
Kuichong

Gangkou
Zhangjiablan
Dahuan
Xin'an
Nantou
SHENZEN
Yantian
Tai Pang Wan

Zhongshan
Nanlang
Shekou
Sheung Shui
Me On Shan

GUANGDONG
Pinglan
Sanxiang
Tangjia

Yuen Long
Tuen Mun
XIANGGANG
Tai Po
Sha Tin
Port Shelter

XINJIULONG
Tsing Yi
Jiulong (Kowloon)
Tseung Kwan O

Zhuhai
Tung Chung
Aberdeen
HONG KONG (XIANGGANG)

Macau (Aomen)
AOMEN
Taipa
LANTAU ISLAND
Tai O
Luhuan

Zhujiang Kou
Hau-Hoi Wan
Tai Po Hoi

West Lamma Channel
East Lamma Channel

SANZAO DAO
DAHENGQIN DAO

SOUTH CHINA SEA

© Rand McNally A-101839-1

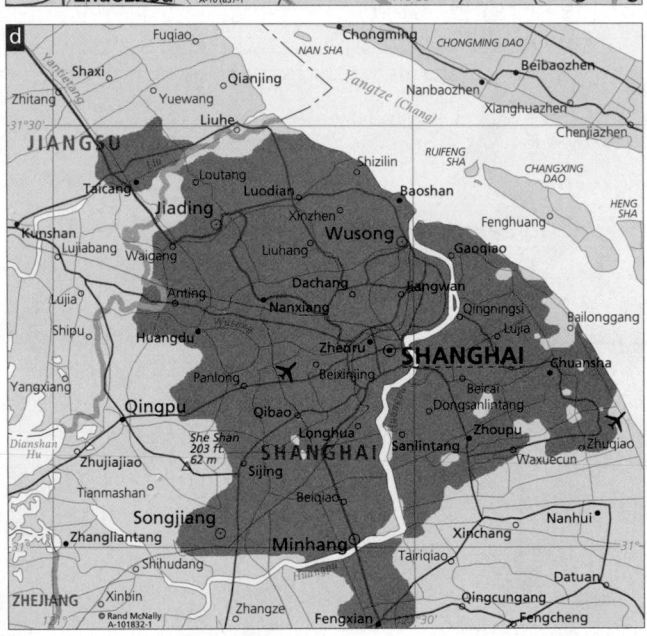

d

JIANGSU
SHANGHAI
ZHEJIANG

Fuqiao
Chongming
CHONGMING DAO
Beibaozhen

Yangtze (Chang)
Shaxi
Qianjing
Nanbaozhen

Zhitang
Yuewang
Xianghuazhen
Chenjiazhen

Taicang
Liuhe
Shizilin
RUIFENG SHA
CHANGXING DAO

Jiading
Loutang
Luodian
Baoshan
HENG SHA

Kunshan
Lujiabang
Liuhang
Xinzhen
Wusong
Fenghuang

Shipu
Waideng
Dachang
Jiangwan
Gaoqiao

Huangdu
Anting
Nanxiang
Qingningsi

Yangxiang
Zhenru
Beixinjing
Lujia
Bailonggang

Qingpu
Panlong
Qibao
SHANGHAI
Baicai
Dongsanlintang

Dianshan Hu
Zhujiajiao
She Shan 203 ft. 62 m
Sijing
Longhua
Sanlintang
Zhouqiao

Tianmashan
Beiqiao
SHANGHAI
Zhoupu
Chuansha

Songjiang
Zhangliantang
Shihudang
Minhang
Tairiqiao
Xinchang
Nanhui

Xinbin
Zhangze
Fengxian
Qingcungang
Datuan

Fengcheng

© Rand McNally A-101832-1

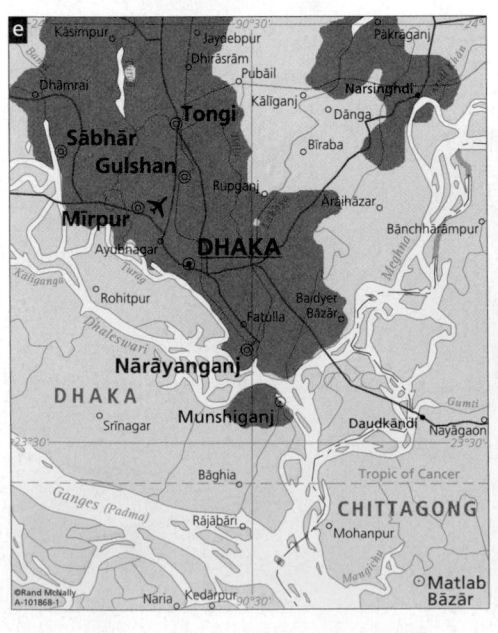

e

Kasimpur
Jaydebpur
Pakragonj
Dhirasram
Pubail

Dhamrai
Narsinghdi
Tongi
Kaliganj
Danga

Sabhar
Gulshan
Biraba
Bhanchharampur

Mirpur
Rupganj
Ayubnagar

DHAKA
Araihazar

Narayanganj
Baidyer Bazar
Fatulla

DHAKA
Rohitpur
Munshiganj
Srinagar
Daudkandi
Nayagaon

Baghia
Tropic of Cancer
Mohanpur
Ganges (Padma)
Rajabari
CHITTAGONG

Naria
Kedarpur
Matlab Bazar

© Rand McNally A-101868-1

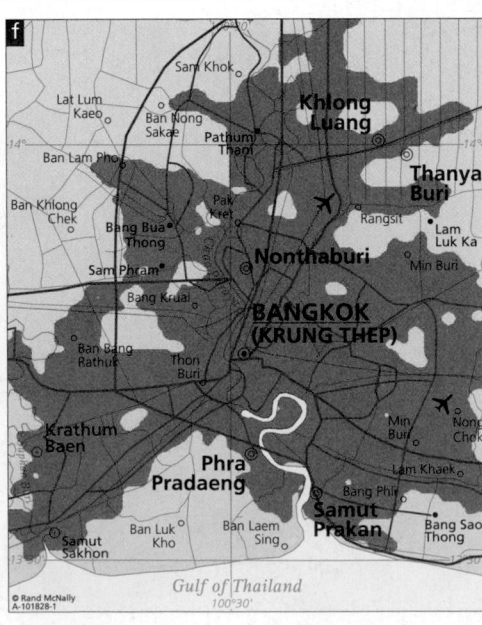

f

Sam Khok
Lat Lum Kaeo
Ban Nong Sakae
Khlong Luang

Ban Lam Pho
Pathum Thani
Thanya Buri

Ban Khlong Chek
Bang Bua Thong
Pak Kret
Rangsit
Lam Luk Ka

Sam Phran
Nonthaburi
Min Buri

Bang Kruai
Thon Buri
BANGKOK (KRUNG THEP)
Min Buri
Nong Chok

Ban Bang Rathut
Krathum Baen
Phra Pradaeng
Bang Phli
Lam Khaek

Ban Luk Kho
Samut Prakan
Bang Sao Thong

Samut Sakhon
Ban Laem Sing

Gulf of Thailand

© Rand McNally A-101828-1

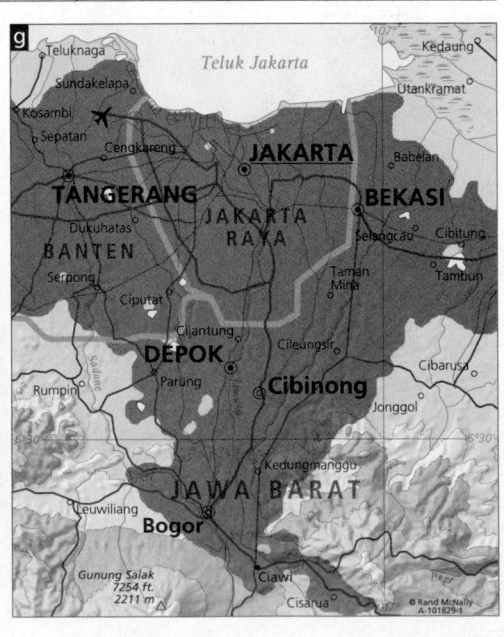

g

Teluknaga
Kedaung
Teluk Jakarta
Utankramat

Sundakelapa
JAKARTA
Babelan

Kosambi
Cengkareng
Sepatan
Bekasi

TANGERANG
JAKARTA RAYA
Cibitung

BANTEN
Dukuhatas
Selangca

Serpong
Taman Mini
Tambun

Ciputat
Cijantung
Cibinong

DEPOK
Parung
Cibinong

Rumpin
Cileungsir
Cibarusa

JAWA BARAT
Kedungmanggu
Jonggol

Leuwiliang
Bogor
Ciawi

Gunung Salak 7254 ft. 2211 m
Cisarua

© Rand McNally A-101829-1

0 5 10 15 20 25 30 Miles
0 5 10 15 20 25 30 35 40 45 50 Kilometers

Lambert Conformal Conic Projection
Scale 1:1,000,000
One inch to 16 miles
One cm to 10 km

Africa

AFRICA

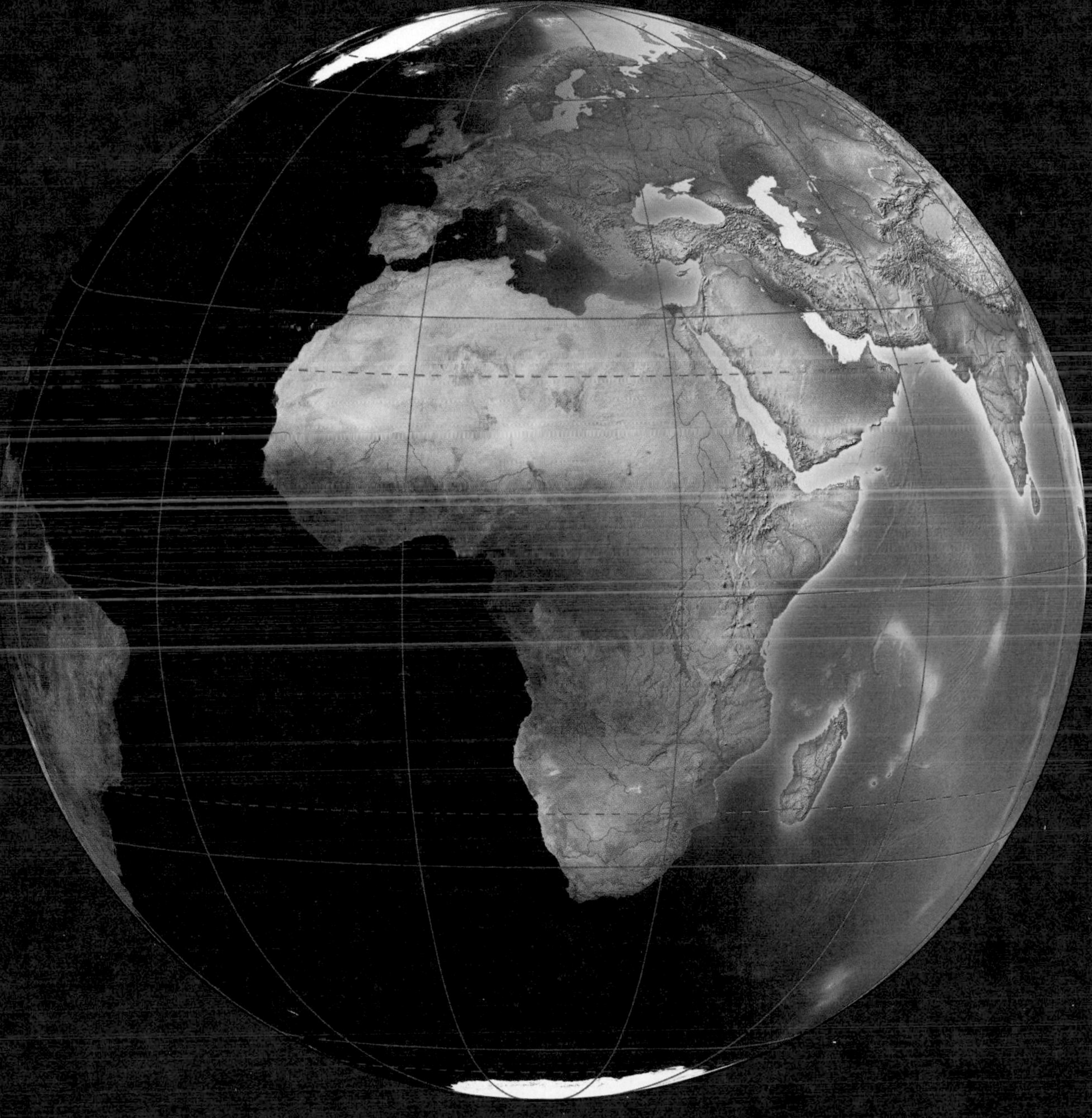

Straddling the equator, Africa is the birthplace of humankind and, among the continents, it is the second largest, the second most populous, and the fastest growing. There are more than fifty countries in Africa, many of which were under European colonial rule until the twentieth century. The Sahara Desert, which is roughly the same size as the United States, dominates North Africa. The Congo Basin, home to some of the largest areas of undisturbed tropical rainforest, dominates Central Africa. Africa's recent history includes disruptive civil wars and disease outbreaks, coupled with poverty, undernourishment, and low life expectancy. However, the continent is also rich in resources, rapidly urbanizing, and possesses some of the world's fastest growing economies.

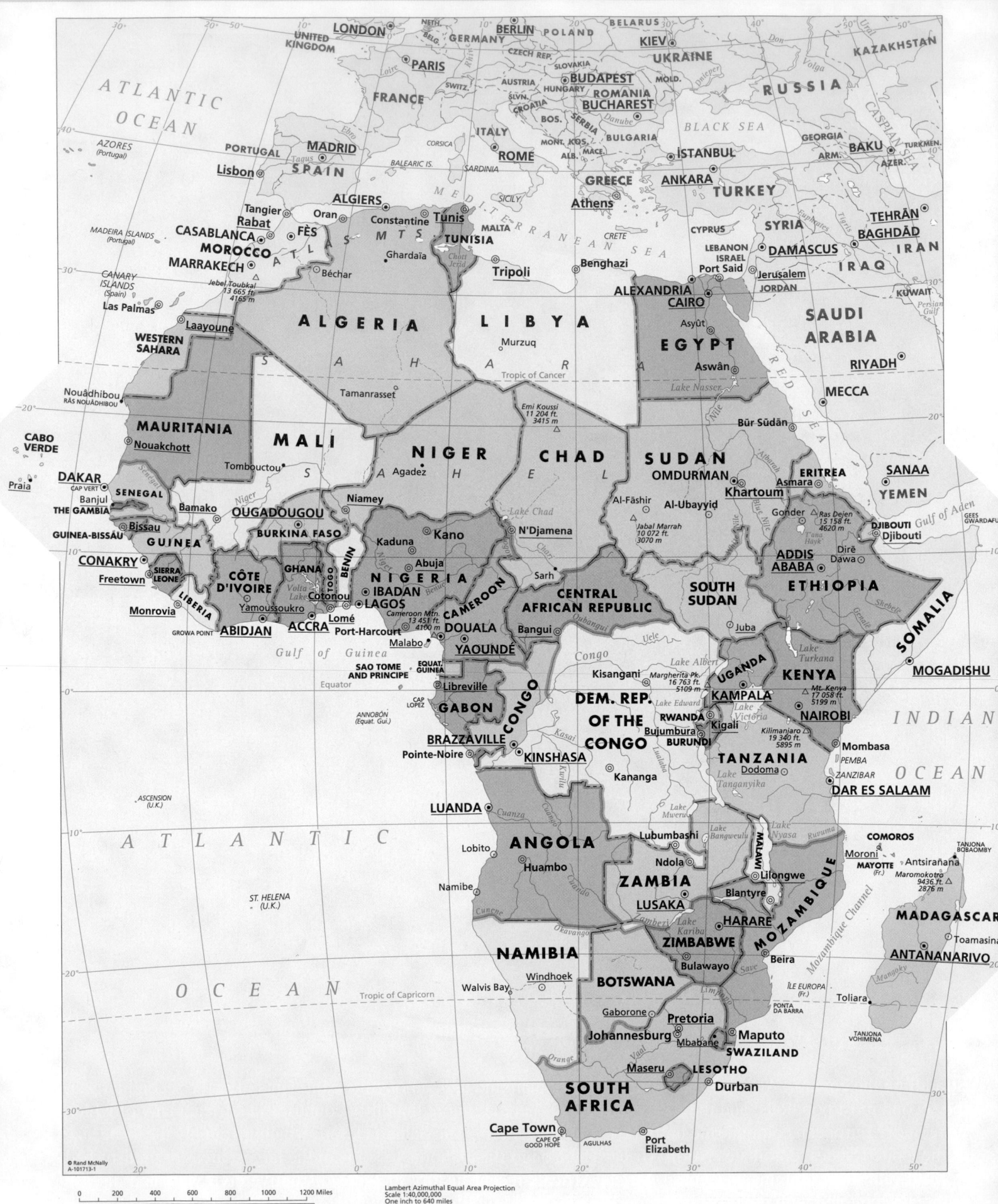

Lambert Azimuthal Equal Area Projection
Scale 1:40,000,000
One inch to 640 miles
One cm to 400 km

| 0 | 200 | 400 | 600 | 800 | 1000 | 1200 Miles |

| 0 | 200 | 400 | 600 | 800 | 1000 | 1200 | 1400 | 1600 | 1800 | 2000 Kilometers |

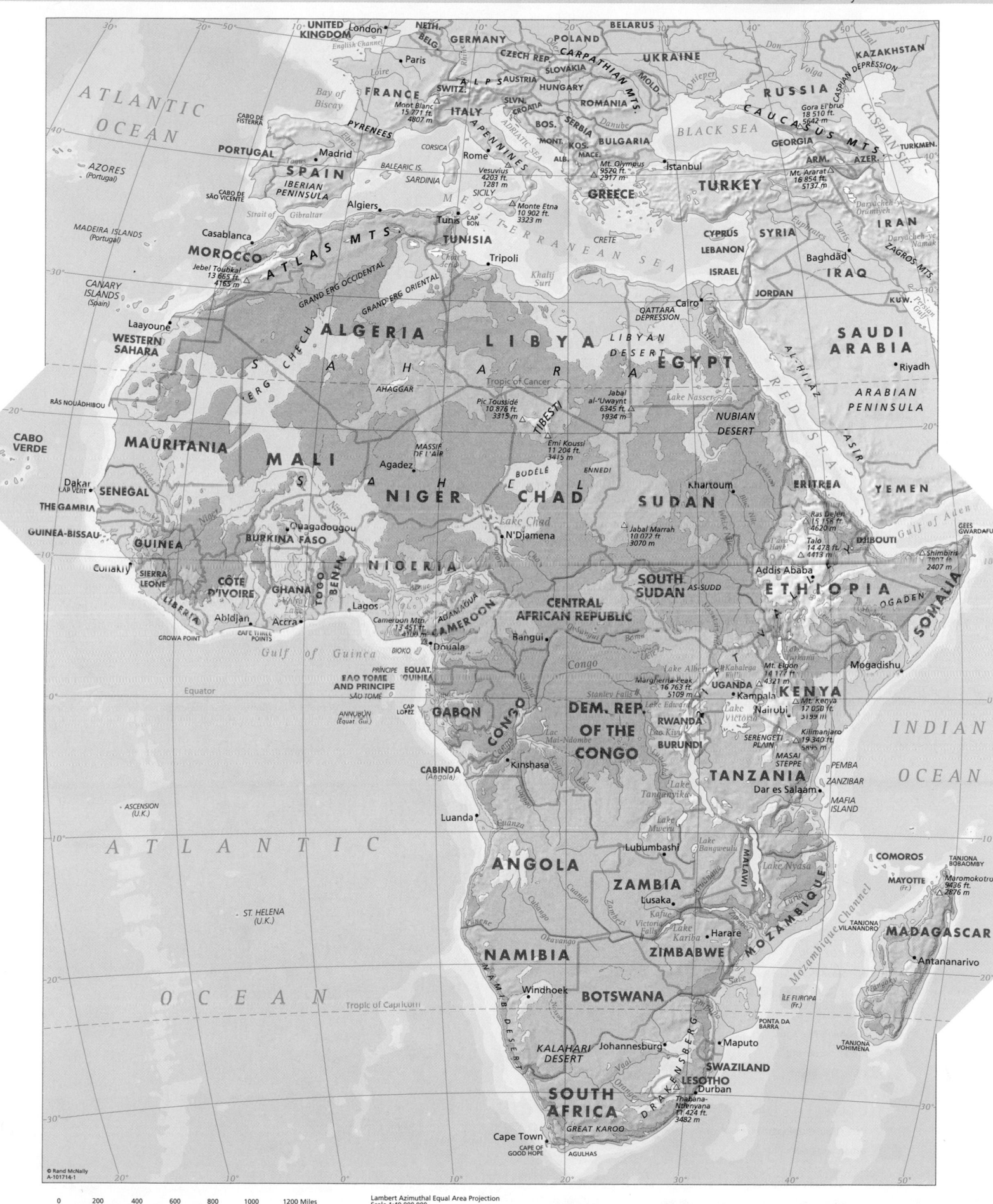

ATLANTIC OCEAN

UNITED KINGDOM London
NETH.
BELG.
GERMANY
POLAND
BELARUS

English Channel
Paris
FRANCE
CZECH REP.
SLOVAKIA
HUNGARY
CARPATHIAN MTS.
MOLD.
UKRAINE

Loire
ALPS
SWITZ.
AUSTRIA
SLVN.
CROATIA
ROMANIA

Mont Blanc 15,771 ft. 4807 m
ITALY
BOS.
SERBIA
BULGARIA

PORTUGAL

CABO DE FISTERRA

Madrid
SPAIN
IBERIAN PENINSULA

APENNINES
Rome
ADRIATIC SEA
MONT.
KOS.
MACE.
ALB.

Danube

RUSSIA
Volga
Don

KAZAKHSTAN

CASPIAN DEPRESSION

Ural

CAUCASUS MTS.
Gora El'brus 18,510 ft. 5642 m
CASPIAN SEA

GEORGIA

AZORES (Portugal)

CABO DE SÃO VICENTE

BALEARIC IS.
CORSICA
SARDINIA

Vesuvius 4203 ft. 1281 m
SICILY

Mt. Olympus 9570 ft. 2917 m
GREECE

Monte Etna 10,902 ft. 3323 m
CRETE
MEDITERRANEAN
SEA

Istanbul
TURKEY

ARM.
AZER.
TURKMEN.
Darýächeh-ye Orümiyeh

Mt. Ararat 16,854 ft. 5137 m
Daryächeh-ye Namak
ZAGROS MTS.

IRAN

MADEIRA ISLANDS (Portugal)

Casablanca
MOROCCO

Algiers
Tunis
CAP BON
TUNISIA

Tripoli

CYPRUS
SYRIA
LEBANON
Baghdad
IRAQ

Strait of Gibraltar

ATLAS MTS.

Jebel Toubkal 13,665 ft. 4165 m
Khalïj Surt

ISRAEL
JORDAN
KUW.
Persian Gulf

CANARY ISLANDS (Spain)

Laayoune

WESTERN SAHARA

GRAND ERG OCCIDENTAL
GRAND ERG ORIENTAL

ALGERIA
LIBYA

LIBYAN DESERT

EGYPT

Cairo
QATTARA DEPRESSION

SAUDI ARABIA

Riyadh

ARABIAN PENINSULA

RÅS NOUÂDHIBOU

S A H A R A

ERG CHECH
AHAGGAR
Tropic of Cancer

Pic Toussidé 10,876 ft. 3315 m
TIBESTI
Jabal al-'Uwaynt 6345 ft. 1934 m
Lake Nasser
NUBIAN DESERT

RED SEA
AL-HIJAZ
ASIR

CABO VERDE

MAURITANIA

MALI

MASSIF OF L'AIR
Agadez

Emi Koussi 11,204 ft. 3415 m

BUDÉLÉ
ENNEDI

YEMEN
Gulf of Aden
GEES GWARDAFUY

Dakar
CAP VERT
SENEGAL

THE GAMBIA

S A H E L

NIGER
CHAD

Lake Chad
Khartoum
SUDAN

Jabal Marrah 10,072 ft. 3070 m

ERITREA
Ras Dejen 15,158 ft. 4620 m
Talo 14,478 ft. 4413 m
DJIBOUTI
Shimbiris 7897 ft. 2407 m

GUINEA-BISSAU

GUINEA

Conakry
SIERRA LEONE

Ouagadougou
BURKINA FASO

N'Djamena

Niger
NIGERIA

Addis Ababa
ETHIOPIA
OGADEN

CÔTE D'IVOIRE
GHANA
TOGO
BENIN

Volta Lake

Lagos

SOUTH SUDAN
AS-SUDD

LIBERIA

Abidjan
Accra
CAPE THREE POINTS

Cameroon Mtn. 13,451 ft. 4110 m
BIOKO
Douala
CAMEROON
ADAMAOUA

CENTRAL AFRICAN REPUBLIC
Bangui

Ubangi

Lake Albert
Kabalega Falls
Mt. Eldon 14,177 ft. 4321 m
UGANDA
Kampala

Mt. Kenya 17,050 ft. 5199 m
KENYA
Nairobi

Mogadishu

SOMALIA

GROWA POINT

Gulf of Guinea
PRÍNCIPE
EQUAT. GUINEA
SÃO TOMÉ AND PRÍNCIPE
SÃO TOMÉ

CAP LOPEZ

ANNOBÓN (Equat. Gui.)

GABON
CONGO

Congo
Lac Mai-Ndombe

DEM. REP. OF THE CONGO

Lake Kivu
RWANDA
BURUNDI

Stanley Falls
Lake Edward

Lake Victoria

SERENGETI PLAIN

Kilimanjaro 19,340 ft. 5895 m
MASAI STEPPE
PEMBA
ZANZIBAR

INDIAN OCEAN

Equator

Margherita Peak 16,763 ft. 5109 m

CABINDA (Angola)

Kinshasa

TANZANIA
Dar es Salaam

MAFIA ISLAND

Luanda

Kwanza

Lake Mweru
Lubumbashi
Lake Bangweulu

Lake Tanganyika

ASCENSION (U.K.)

ATLANTIC

Cubango

ANGOLA

Cuando

Cuanza

ZAMBIA
Lusaka

Kafue
MALAWI
Lake Nyasa

COMOROS

MAYOTTE

TANJONA BOBAOMBY
Maromokotro 9436 ft. 2876 m

ST. HELENA (U.K.)

NAMIBIA

Cunene

Okavango

Zambezi
Victoria Falls

Lake Kariba
Harare

MOZAMBIQUE

ZIMBABWE

TANJONA VILANANDRO

MADAGASCAR

Antananarivo

OCEAN

Tropic of Capricorn

NAMIB DESERT

Windhoek

BOTSWANA

KALAHARI DESERT

Johannesburg

Vaal
Orange

Limpopo

Save

ÎLE EUROPA (Fr.)

PONTA DA BARRA

Maputo

MOZAMBIQUE CHANNEL

TANJONA VOHIMENA

SWAZILAND

DRAKENSBERG
LESOTHO
Durban

SOUTH AFRICA

GREAT KAROO

Thabana-Ntlenyana 11,424 ft. 3482 m

Cape Town
CAPE OF GOOD HOPE
AGULHAS

© Rand McNally
A-101714-1

0   200   400   600   800   1000   1200 Miles
0   200   400   600   800   1000   1200   1400   1600   1800   2000 Kilometers

Lambert Azimuthal Equal Area Projection
Scale 1:40,000,000
One inch to 640 miles
One cm to 400 km

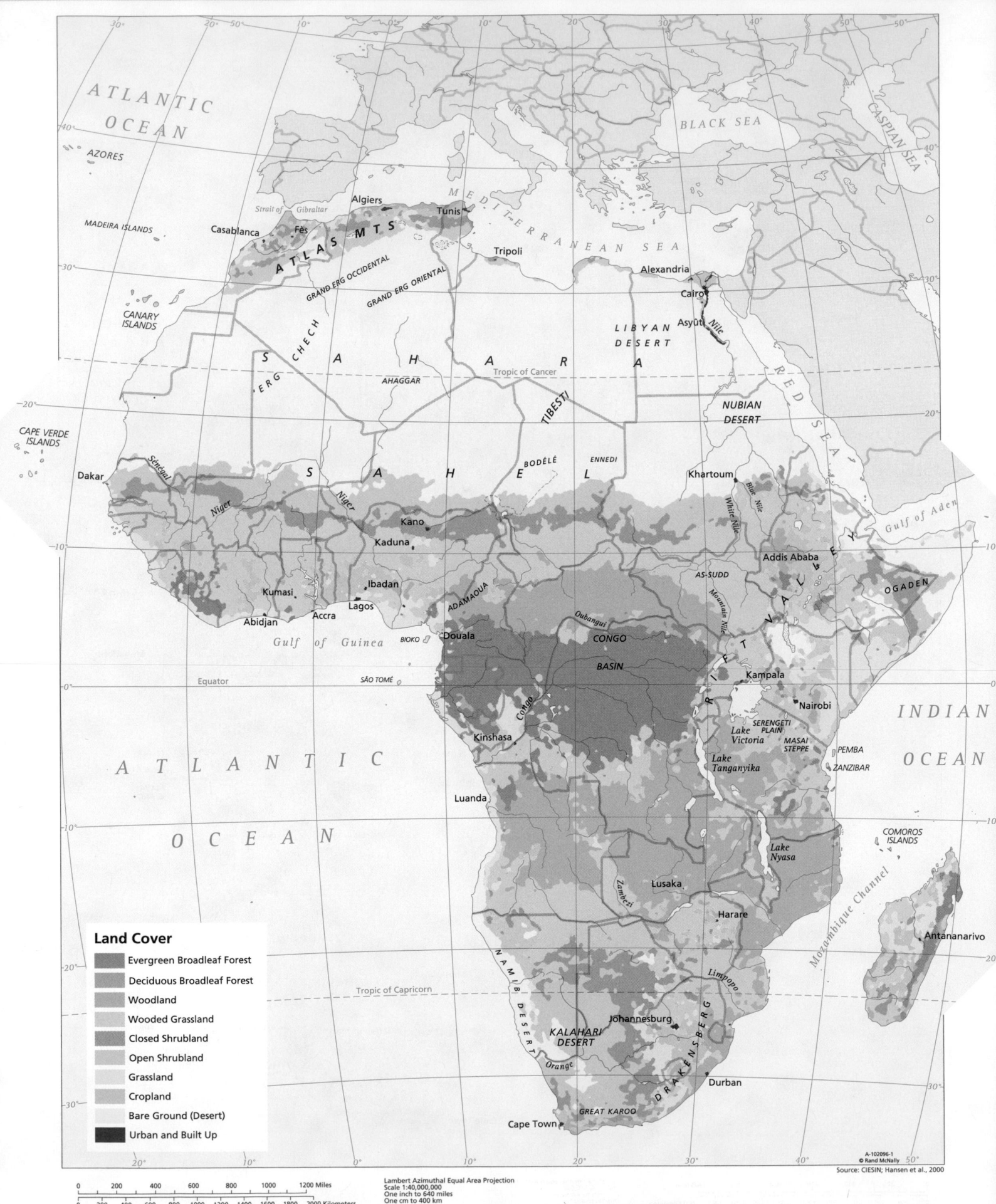

ATLANTIC OCEAN

AZORES

MADEIRA ISLANDS

BLACK SEA

CASPIAN SEA

CANARY ISLANDS

Strait of Gibraltar

Casablanca Fès Algiers Tunis

ATLAS MTS

Tripoli

GRAND ERG OCCIDENTAL GRAND ERG ORIENTAL

MEDITERRANEAN SEA

Alexandria

Cairo

Asyût Nile

LIBYAN DESERT

CAPE VERDE ISLANDS

S A H A R A

ERG CHECH

AHAGGAR

Tropic of Cancer

TIBESTI

NUBIAN DESERT

RED SEA

Dakar

Senégal

Niger

Niger

S A H E L

Kano

Kaduna

BODÉLÉ ENNEDI

Khartoum

White Nile Blue Nile

Gulf of Aden

Kumasi

Ibadan

Addis Ababa

AS-SUDD

OGADEN

Abidjan Accra Lagos

ADAMAOUA

Gulf of Guinea BIOKO Douala

CONGO BASIN

Oubangui

Mountain Nile

RIFT VALLEY

Kampala

INDIAN OCEAN

Equator SÃO TOMÉ

Congo

Nairobi

Kinshasa

Lake Victoria SERENGETI PLAIN MASAI STEPPE

PEMBA ZANZIBAR

ATLANTIC OCEAN

Luanda

Lake Tanganyika

COMOROS ISLANDS

Lake Nyasa

Mozambique Channel

Lusaka

Zambezi

Harare

Antananarivo

Land Cover

Evergreen Broadleaf Forest

Deciduous Broadleaf Forest

Woodland

Wooded Grassland

Closed Shrubland

Open Shrubland

Grassland

Cropland

Bare Ground (Desert)

Urban and Built Up

Tropic of Capricorn

NAMIB DESERT

Limpopo

KALAHARI DESERT

Johannesburg

DRAKENSBERG

Orange

Durban

GREAT KAROO

Cape Town

A-102096-1
© Rand McNally

Source: CIESIN; Hansen et al., 2000

0 200 400 600 800 1000 1200 Miles

0 200 400 600 800 1000 1200 1400 1600 1800 2000 Kilometers

Lambert Azimuthal Equal Area Projection
Scale 1:40,000,000
One inch to 640 miles
One cm to 400 km

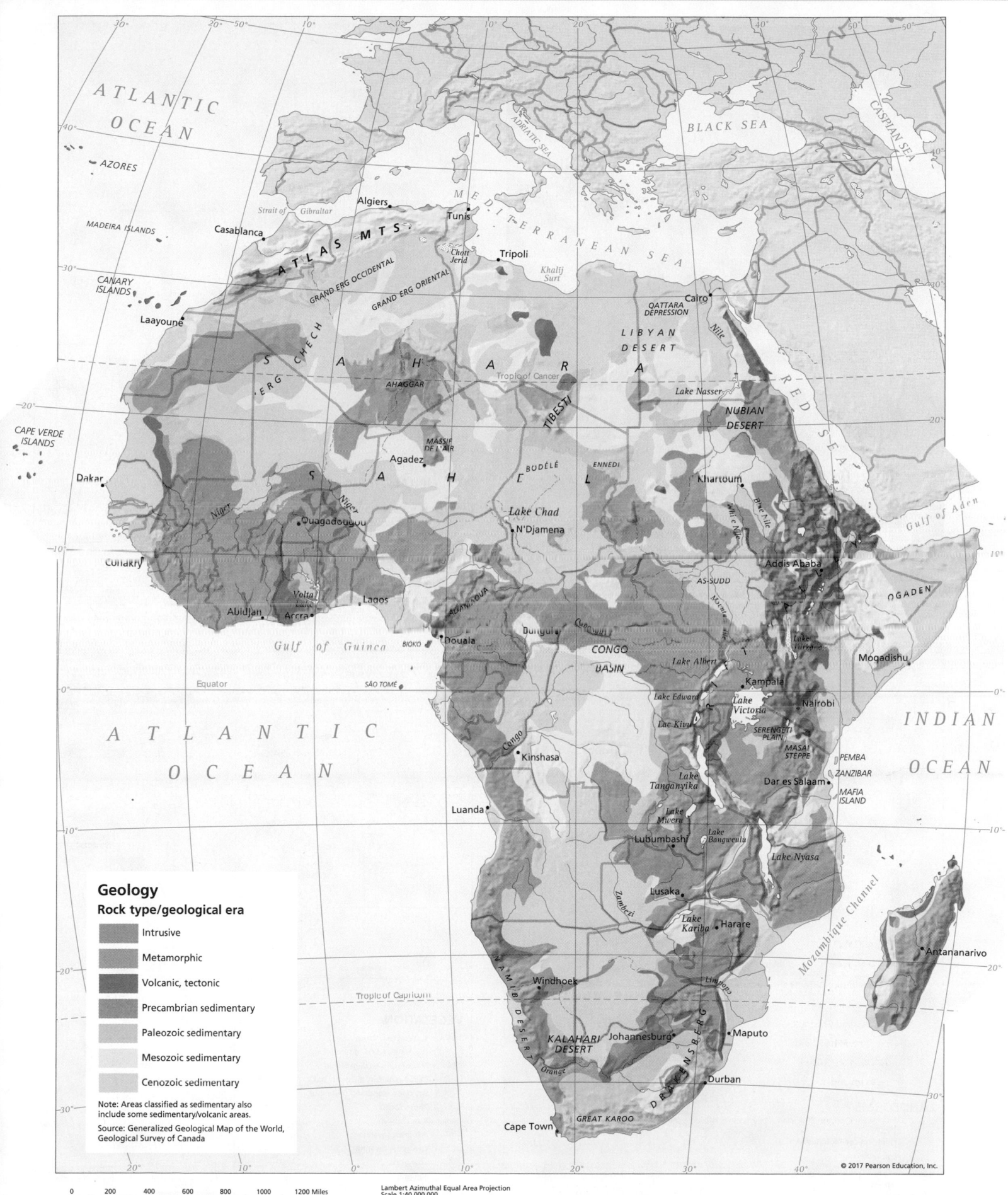

ATLANTIC OCEAN

AZORES

MADEIRA ISLANDS

Casablanca

CANARY ISLANDS

Laayoune

Strait of Gibraltar

Algiers

Tunis

ATLAS MTS

GRAND ERG OCCIDENTAL

GRAND ERG ORIENTAL

Chott Jerid

Tripoli

Khalij Surt

MEDITERRANEAN SEA

BLACK SEA

CASPIAN SEA

ADRIATIC SEA

CAIRO

QATTARA DEPRESSION

LIBYAN DESERT

Lake Nasser

NUBIAN DESERT

RED SEA

CAPE VERDE ISLANDS

Dakar

S A H A R A

ERG CHECH

AHAGGAR

Tropic of Cancer

MASSIF DE L'AIR

Agadez

S A H E L

BODÉLÉ

TIBESTI

ENNEDI

Khartoum

Blue Nile

White Nile

Gulf of Aden

Conakry

Ouagadougou

Niger

Niger

Lake Chad

N'Djamena

AS-SUDD

Addis Ababa

OGADEN

Mogadishu

Volta

Abidjan

Accra

Lagos

Gulf of Guinea

BIOKO

Douala

ADAMAOUA

SÃO TOMÉ

Bangui

Cameroon

CONGO BASIN

Lake Albert

Lake Edward

Lac Kivu

Kampala

Lake Victoria

Nairobi

SERENGETI PLAIN

MASAI STEPPE

Lake Turkana

Equator

ATLANTIC OCEAN

Congo

Kinshasa

Luanda

Lake Tanganyika

Lake Mweru

PEMBA

ZANZIBAR

Dar es Salaam

MAFIA ISLAND

INDIAN OCEAN

Lubumbashi

Lake Bangweulu

Lake Nyasa

Mozambique Channel

Lusaka

Zambezi

Lake Kariba

Harare

Windhoek

NAMIB DESERT

Limpopo

Antananarivo

Tropic of Capricorn

KALAHARI DESERT

Johannesburg

Maputo

Orange

DRAKENSBERG

Durban

Cape Town

GREAT KAROO

Geology
Rock type/geological era

- Intrusive
- Metamorphic
- Volcanic, tectonic
- Precambrian sedimentary
- Paleozoic sedimentary
- Mesozoic sedimentary
- Cenozoic sedimentary

Note: Areas classified as sedimentary also include some sedimentary/volcanic areas.

Source: Generalized Geological Map of the World, Geological Survey of Canada

© 2017 Pearson Education, Inc.

0 200 400 600 800 1000 1200 Miles

0 200 400 600 800 1000 1200 1400 1600 1800 2000 Kilometers

Lambert Azimuthal Equal Area Projection
Scale 1:40,000,000
One inch to 640 miles
One cm to 400 km

Moderate Resolution
Imaging Spectroradiometer (MODIS)
true-color mosaic satellite image

Source: NASA Visible Earth program (http://visibleearth.nasa.gov/)

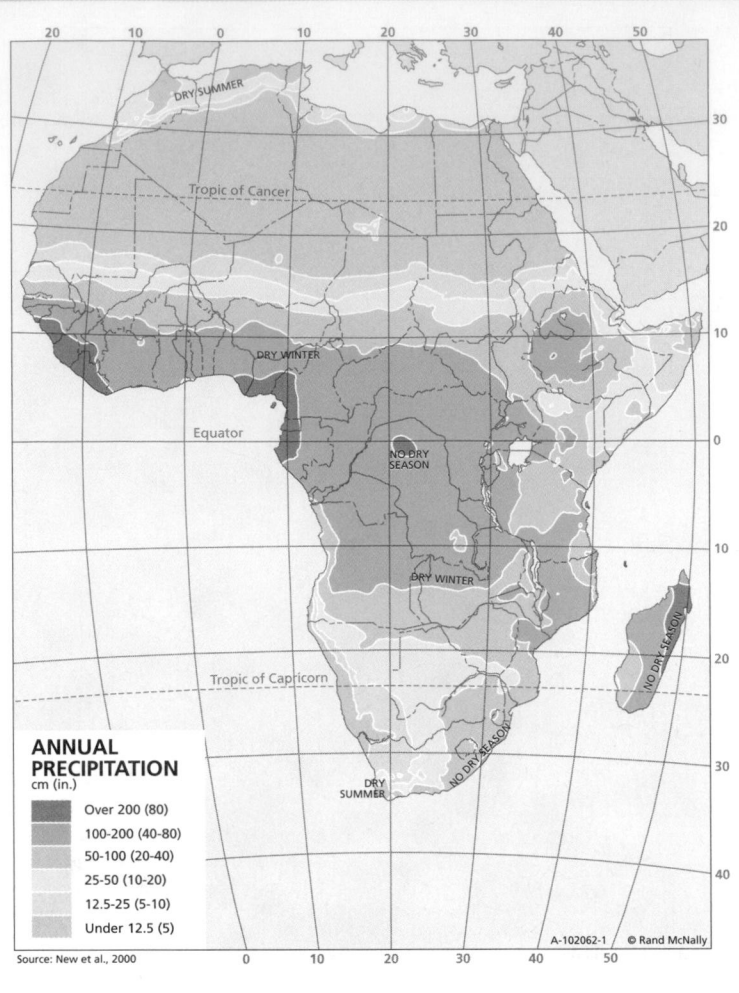

ANNUAL PRECIPITATION
cm (in.)

■	Over 200 (80)
■	100–200 (40–80)
■	50–100 (20–40)
■	25–50 (10–20)
■	12.5–25 (5–10)
□	Under 12.5 (5)

Source: New et al., 2000

A-102062-1   © Rand McNally

DRY SUMMER
Tropic of Cancer
DRY WINTER
Equator
NO DRY SEASON
DRY WINTER
Tropic of Capricorn
NO DRY SEASON
DRY SUMMER
NO DRY SEASON

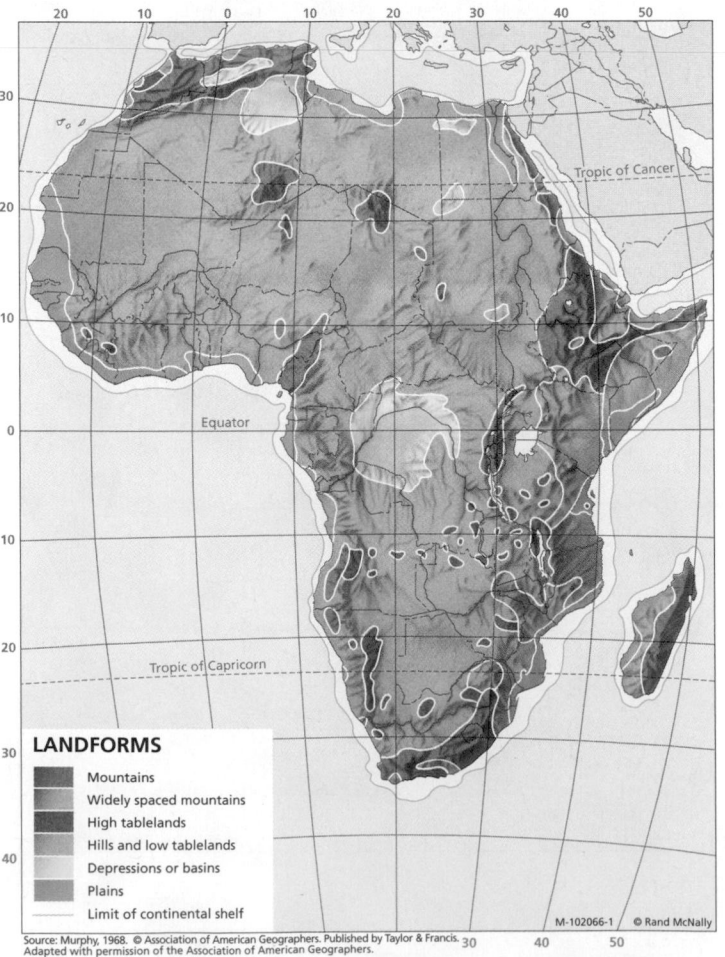

Tropic of Cancer
Equator
Tropic of Capricorn

LANDFORMS

■	Mountains
■	Widely spaced mountains
■	High tablelands
■	Hills and low tablelands
■	Depressions or basins
□	Plains
⎯	Limit of continental shelf

Source: Murphy, 1968. © Association of American Geographers. Published by Taylor & Francis.
Adapted with permission of the Association of American Geographers.

M-102066-1   © Rand McNally

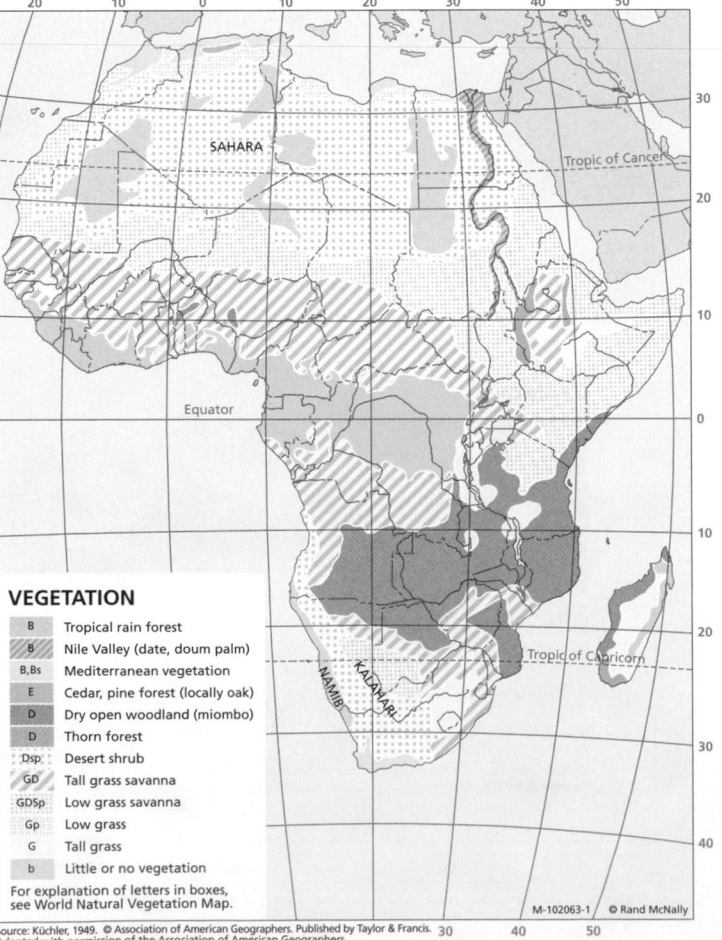

SAHARA
Tropic of Cancer
Equator
Tropic of Capricorn
NAMIB
KALAHARI

VEGETATION

■	B	Tropical rain forest
▨	B	Nile Valley (date, doum palm)
■	B,Bs	Mediterranean vegetation
■	E	Cedar, pine forest (locally oak)
■	D	Dry open woodland (miombo)
■	D	Thorn forest
⠿	Dsp	Desert shrub
▧	GD	Tall grass savanna
⠿	GDsp	Low grass savanna
□	Gp	Low grass
□	G	Tall grass
□	b	Little or no vegetation

For explanation of letters in boxes,
see World Natural Vegetation Map.

Source: Küchler, 1949. © Association of American Geographers. Published by Taylor & Francis.
Adapted with permission of the Association of American Geographers.

M-102063-1   © Rand McNally

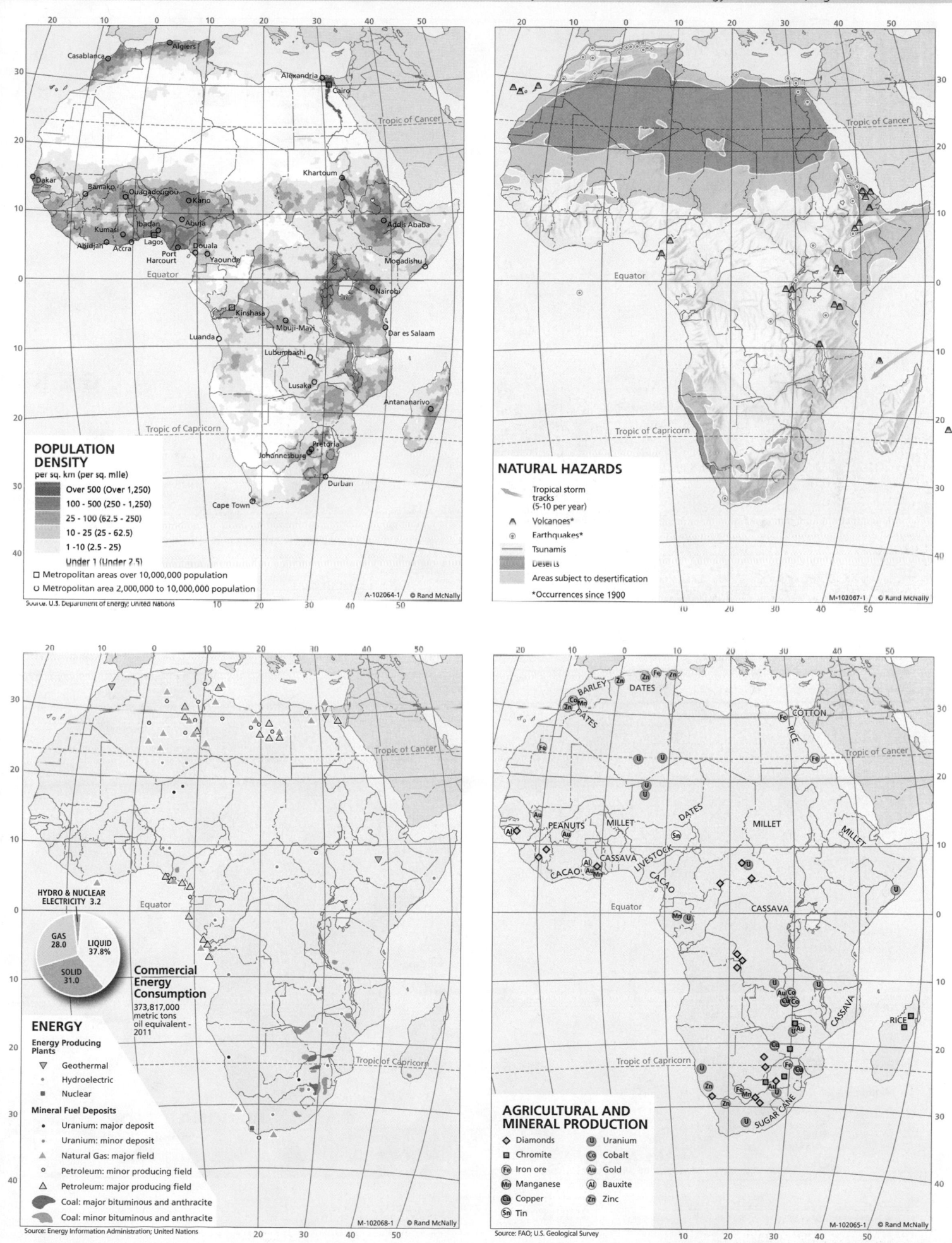

Population Density map (top left):

Casablanca, Algiers, Alexandria, Cairo

Tropic of Cancer

Dakar, Bamako, Ouagadougou, Kano, Khartoum, Kumasi, Ibadan, Abuja, Addis Ababa, Abidjan, Accra, Lagos, Douala, Port Harcourt, Yaounde, Mogadishu

Equator

Kinshasa, Mbuji-Mayi, Nairobi, Luanda, Lubumbashi, Dar es Salaam, Lusaka, Antananarivo

Tropic of Capricorn

Pretoria, Johannesburg, Durban, Cape Town

POPULATION DENSITY
per sq. km (per sq. mile)

- Over 500 (Over 1,250)
- 100 - 500 (250 - 1,250)
- 25 - 100 (62.5 - 250)
- 10 - 25 (25 - 62.5)
- 1 -10 (2.5 - 25)
- Under 1 (Under 2.5)

□ Metropolitan areas over 10,000,000 population
○ Metropolitan area 2,000,000 to 10,000,000 population

Source: U.S. Department of Energy; United Nations
A-102064-1   © Rand McNally

Natural Hazards map (top right):

Tropic of Cancer

Equator

Tropic of Capricorn

NATURAL HAZARDS

- Tropical storm tracks (5-10 per year)
- △ Volcanoes*
- ⊙ Earthquakes*
- Tsunamis
- Deserts
- Areas subject to desertification

*Occurrences since 1900

M=102067-1   © Rand McNally

Energy map (bottom left):

Equator

HYDRO & NUCLEAR ELECTRICITY 3.2

GAS 28.0   LIQUID 37.8%   SOLID 31.0

Commercial Energy Consumption
373,817,000 metric tons oil equivalent - 2011

Tropic of Capricorn

ENERGY

Energy Producing Plants
- ▽ Geothermal
- • Hydroelectric
- ■ Nuclear

Mineral Fuel Deposits
- • Uranium: major deposit
- • Uranium: minor deposit
- ▲ Natural Gas: major field
- ○ Petroleum: minor producing field
- △ Petroleum: major producing field
- Coal: major bituminous and anthracite
- Coal: minor bituminous and anthracite

Source: Energy Information Administration; United Nations
M-102068-1   © Rand McNally

Minerals, Agriculture map (bottom right):

BARLEY, DATES, COTTON, DATES, RICE, PEANUTS, MILLET, DATES, MILLET, MILLET, CACAO, CASSAVA, LIVESTOCK, CACAO, CASSAVA, CASSAVA, RICE, SUGAR CANE

Equator

Tropic of Cancer

Tropic of Capricorn

AGRICULTURAL AND MINERAL PRODUCTION

- ◇ Diamonds
- ■ Chromite
- Fe Iron ore
- Mn Manganese
- Cu Copper
- Sn Tin
- U Uranium
- Co Cobalt
- Au Gold
- Al Bauxite
- Zn Zinc

Source: FAO; U.S. Geological Survey
M-102065-1   © Rand McNally

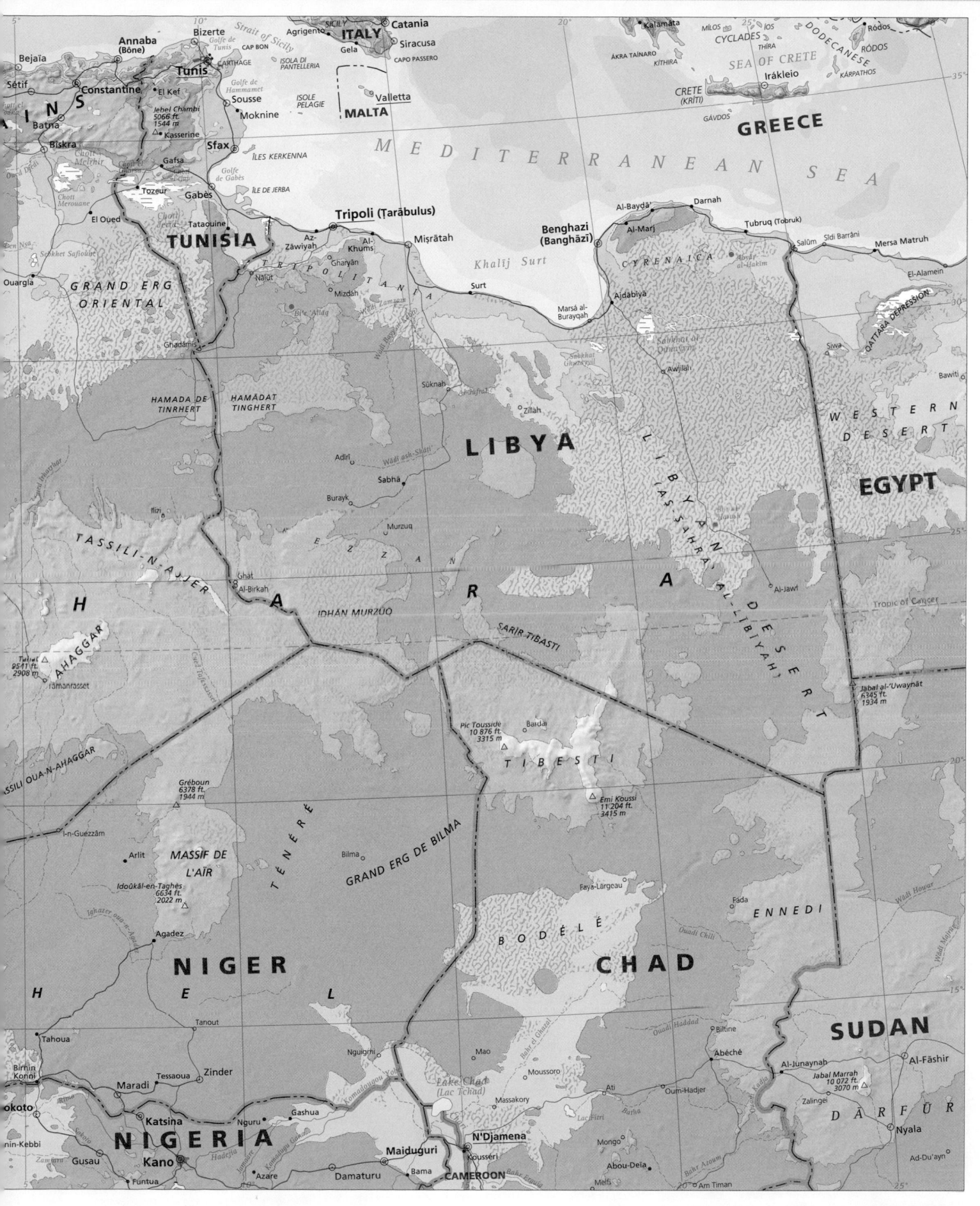

MEDITERRANEAN SEA

GREECE

SEA OF CRETE

CRETE
(KRÍTI)

Irákleio

CYCLADES

DODECANESE

RÓDOS

KÁRPATHOS

MÍLOS
THÍRA
ÍOS
KÍTHIRA
GÁVDOS
ÁKRA TAÍNARO
Kalamáta

SICILY
ITALY
Catania
Siracusa
Gela
Agrigento
CAPO PASSERO
Strait of Sicily
Golfe de Tunis
CAP BON
CARTHAGE
ISOLA DI
PANTELLERIA
ISOLE
PELAGIE
MALTA
Valletta

Annaba
(Bône)
Bizerte
Tunis
Bejaïa
Sétif
Constantine
El Kef
Golfe de
Hammamet
Sousse
Moknine
Sfax
ÎLES KERKENNA
Batna
Jebel Chambi
5066 ft.
1544 m
Kasserine
Biskra
Gafsa
Chott el
Djerid
Golfe de Gabès
ÎLE DE JERBA
Chott
Melrhir
Tozeur
Gabès
Chott
Merouane
El Oued
Tataouine
Az-Zāwiyah
Al-Khums
Mişrātah
Ouargla
Chott
Fejaj
TUNISIA
Gharyān
Naļūt
Mizdāh
Tripoli (Ṭarābulus)
Ghadāmis
TRIPOLITANIA
Surt
Marsā al-
Burayqah
Ajdābiyā
Benghazi
(Banghāzī)
Al-Marj
Al-Baydā'
Darnah
Ṭubruq (Tobruk)
Salūm
Sīdi Barrāni
Mersa Matruh
El-Alamein
CYRENAICA
Abyār
al-Hakīm
QATTARA DEPRESSION
Siwa
Bawiti
EGYPT
WESTERN
DESERT
Khalīj Surt
Sabkhat al
Qunayfidhah
Sabkhat
Ghuzayyil
Awjilah
Zillah
LIBYA
GRAND ERG
ORIENTAL
Sokhet Safioune
Den Nsa
Ouled Djellal
HAMADA DE
TINRHERT
HAMĀDAT
TINGHERT
Bîr Allaq
Wadi Zamzam
Wadi Bayy al-Kabir
Sūknah
Adīrī
Wadi ash-Shāṭiʾ
Sabhā
Burayk
Murzuq
F
E
Z
Z
A
N
Idri
TASSILI-N-AJJER
Ghat
Al-Birkah
IDHĀN MURZŪQ
LIBYAN
SAHARA
AL-LIBIYAH
Al-Jawf
Jabal al-'Uwaynāt
6345 ft.
1934 m
Tropic of Cancer
DESERT
SARĪR TIBASTI
H
A
R
A
AHAGGAR
Tahat
9541 ft.
2908 m
Tamanrasset
TASSILI OUA-N-AHAGGAR
Oued Tafassasset
Pic Toussidé
10 876 ft.
3315 m
Bardaï
TIBESTI
Emi Koussi
11 204 ft.
3415 m
I-n-Guézzâm
Gréboun
6378 ft.
1944 m
MASSIF DE
L'AÏR
Arlit
Idoûkâl-en-Taghès
6634 ft.
2022 m
Agadez
T
É
N
É
R
É
Bilma
GRAND ERG DE BILMA
Faya-Largeau
Fada
ENNEDI
Wadi Howar
BODÉLÉ
Ouadi Chili
Ouadi Haddad
Igharghar
Ighazer oua-n-Agadez
NIGER
H
E
L
L
Tanout
Tahoua
Birnin
Konni
Tessaoua
Zinder
Maradi
okoto
Gusau
Funtua
Katsina
Nguru
Gashua
Damaturu
Azare
Kano
NIGERIA
Bama
Maiduguri
Kousséri
CAMEROON
N'Djamena
Nguigmi
Komadougou Yobé
Hadejia
Mao
Massakory
Lake Chad
(Lac Tchad)
Lac Fitri
Bol
Bahr el Ghazal
Moussoro
Ati
Oum-Hadjer
Biltine
Abéché
Mongo
Abou-Deïa
Melfi
Am Timan
CHAD
Bahr Azoum
Ouadi Kadja
Bahr Keïta
Al-Junaynah
Jabal Marrah
10 072 ft.
3070 m
Zalingei
Nyala
Ad-Du'ayn
Al-Fāshir
D A R F U R
Wadi Magrur
SUDAN
Komadougou Gana
Bahr Araouk
nin-Kebbi
Zamfara
Sokoto

Map labels

Cabo Verde (inset, upper left)
SANTO ANTÃO
Mindelo
SÃO VICENTE
SÃO NICOLAU
SAL
BOA VISTA
CABO VERDE
SANTIAGO
MAIO
BRAVA
FOGO
Praia

© Rand McNally
A-101876-1

Main map
ET TÎDRA
RÂS TIMIRIST
Nouakchott
Tidjikja
Tichît
AOUKÂR
Lac Faguibin
S
MAURITANIA
Boutilimit
Rosso
Bogué
Kiffa
Boû Gâdoûm
Gounda
Saint-Louis
Kaédi
KUMBI SALEH
Adel Bagrou
Louga
Hamoud
Nioro
Lac Débo
Vallée du Ferlo
CAP VERT
Thiès
Kayes
PARC NATIONAL DE LA BOUCLE DU BAOULÉ
Mopti
DAKAR
SENEGAL
MALI
Mbour
Kaolack
Niger
Djenné
Banjul
Georgetown
Tambacounda
Ségou
Banifing
THE GAMBIA
PARC NATIONAL DU NIOKOLO KOBA
Kita
San
Ziguinchor
Kolda
Kédougou
Bamako
Kati
Cacheu
Bafatá
Mali
Koutiala
Gambie
GUINEA-BISSAU
Casamance
Bobo-Dioulasso
Bissau
Koumbia
Labé
Sikasso
ARQUIPÉLAGO DOS BIJAGÓS
Banfora
Fria
GUINEA
Tinkisso
Siguiri
Kindia
Niger
Kankan
Ferkéssédougou
CONAKRY
Kabala
Korhogo
Bintimani 6381 ft 1945 m
Kissidougou
Pic de Tibé 4934 ft 1504 m
PARC NATIONAL DE LA COMOE
Makeni
Koidu-Sefadu
SIERRA LEONE
CÔTE D'IVOIR (IVORY COAST)
Lunsar
Voinjama
Freetown
Bo
Kenema
Nzérékoré
Mont Nimba 5748 ft 1752 m
Man
Bouaké
SHERBRO ISLAND
Gbanga
Daloa
Lac de Kossou
Yamoussoukr
Kakata
LIBERIA
Abengourou
Gagnoa
Divo
Monrovia
Buchanan
Zwedru
Anyama
ABIDJAN
Greenville
Granc Bassar
Harper
GROWA POINT
San-Pédro

Inset map (Nigeria), lower left

a

Oke Odde
Ilorin
Igbaja
Otu
Ogbomosho
Offa
Ajasse
Egbe
Oke-Iseyin
Iganna
Oyan
Ila
Omu-Aran
Oyo
Ikirun
Osi
Ikole
Kabba
Inisa
Oshogbo
Ijero-Ekiti
Okene
BENIN
Fiditi
Ede
Oke-Mesi
Igede-Ekiti
Ikare
Eruwa
Iwo
Ilesha
Ado-Ekiti
Oka
Igbo-Ora
Gbongan
Ife
Ilawe-Ekiti
Ikerre
Igarra
Igbasa-Odo
Lalupon
IBADAN
Ikire
Emure-Ekiti
Ajaokuta
Kétou
Oke-Igbo
Akure
Owo
Idah
Zangnanado
Abeokuta
Ondo
Idanre
Ifon
Auchi
Pobè
Ijebu-Igbo
Ore
Ekpoma
Nsukka
Ilaro
Ishara
Uromi
Aku
Eha-Amufu
Adjohoun
Sakété
Shagamu
Ijebu-Ode
Ubiaja
Amagunze
Ado
Epe
Okitipupa
NIGERIA
Abakaliki
Porto-Novo
Ottan
Ikorodu
Enugu
Ikeja
Badagri
Siluko
Udi
Mbana
Mushin
Benin City
Ogwashi-Uku
Asaba
Awka
Obubra
Cotonou
LAGOS
Lagos Lagoon
Awgu
Okigwi
Afikpo
Ikom
Obiaruku
Onitsha
Nnewi
Ajasá
Ose
Sapele
Abraka
Kwale
Ozubulu
Iala
Olu
Okwe
Bende
OBAN HILLS
Bight of Benin
Ozoro
Oguta
Owerri
Umuahia
Arochukwu
Oban
Warri
Ughelli
Ivorogbo
Omoko
Amaigbo
Ugep
Ikot-Ekpene
Forcados
Burutu
Ahoada
Elele
Aba
Uyo
Calabar
Ikang
Yenagoa
Degema
Abak
Oron
NIGER DELTA
Okrika
Bori
Opobo
Port-Harcourt
Abonnema
Buguma
Eket
Opobo Town
Nembe
Bonny
CAMEROON
Brass
Bight of Biafra
Gulf of Guinea

© Rand McNally
A-101799-1

Inset map a
Lambert Conformal Conic Projection
Scale 1:4,000,000
One inch to 64 miles
One cm to 40 km

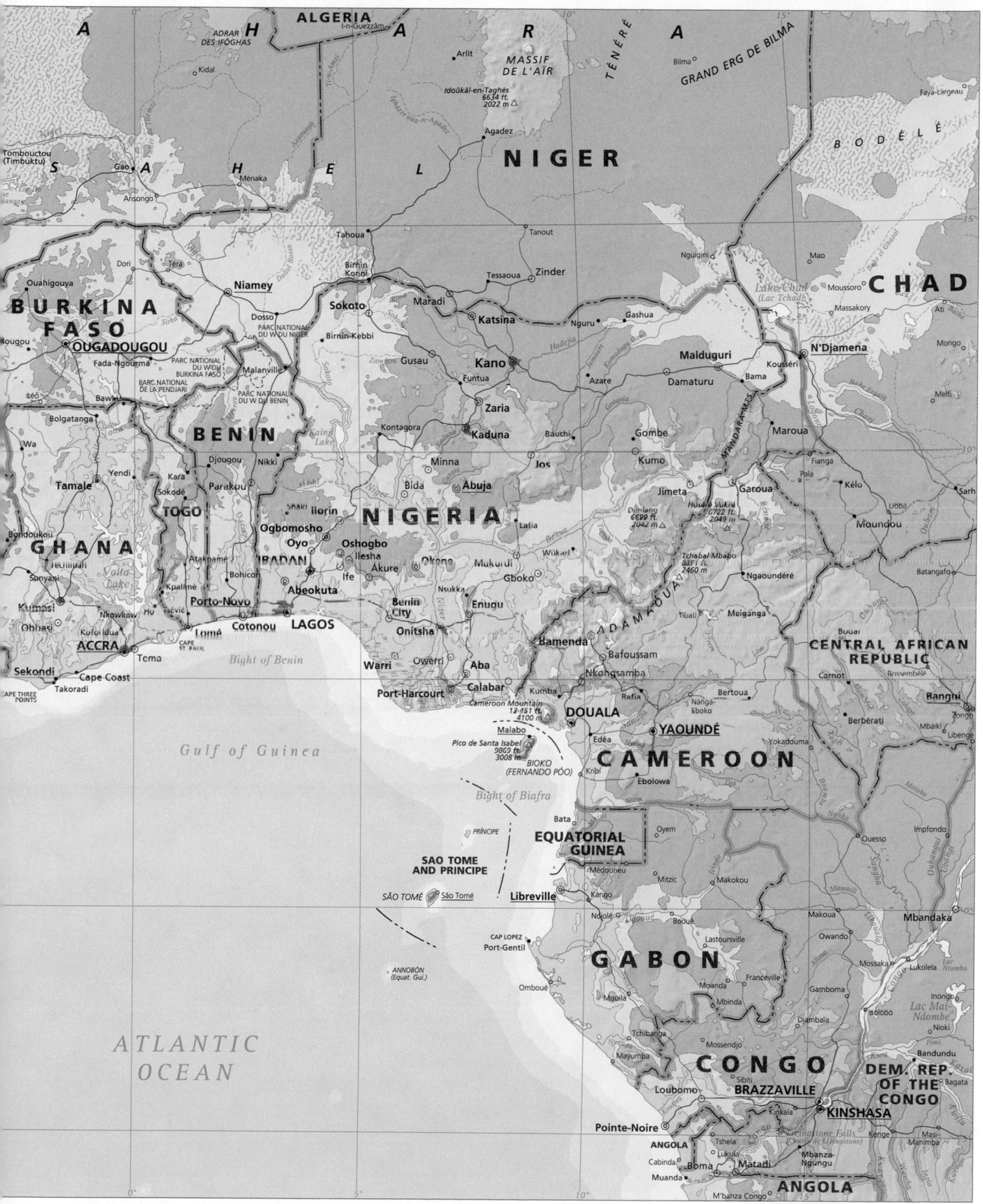

ALGERIA
In-Guezzâm

ADRAR DES IFOGHAS

MASSIF DE L'AÏR

TÉNÉRÉ

GRAND ERG DE BILMA

Arlit

Bilma

Faya-Largeau

Kidal

Idoûkâl-en-Taghès
6634 ft.
2022 m △

Agadez

BODÉLÉ

Tombouctou
(Timbuktu)

Gao

Ménaka

NIGER

Nguigmi

Mao

CHAD

Ansongo

Tahoua

Tanout

Lake Chad
(Lac Tchad)

Moussoro

Massakory

BURKINA FASO

Ouahigouya

Dori

Téra

Niamey

Birnin Konni

Tessaoua

Zinder

Gashua

Nguru

N'Djamena

Mongo

OUGADOUGOU

Fada-Ngourma

Dosso

Sokoto

Maradi

Katsina

Koussèri

Lac Fitri

Malanville

Birnin-Kebbi

Gusau

Kano

Azare

Damaturu

Maiduguri

Bama

Ati

PARC NATIONAL DU W DU NIGER

PARC NATIONAL DU W DU BENIN

Bawku

Funtua

Melfi

Bolgatanga

BENIN

Kontagora

Zaria

Bauchi

Gombe

Maroua

Fianga

Wa

PARC NATIONAL DE LA PENDJARI

Kaduna

Kumo

Pala

Kélo

Sarh

Djougou

Nikki

Minna

Jos

Garoua

Moundou

Tamale

Yendi

Kara

Bida

Abuja

Jimeta

Dimlang
6699 ft.
2042 m △

Husira Vokri
6722 ft.
2049 m △

Uoba

Sokodé

Parakou

NIGERIA

Latia

TOGO

Shaki

Ilorin

Oyo

Oshogbo

Okene

Mukurdi

Wukari

Tchabal Mbabo
8071 ft.
2460 m

Ngaoundéré

Batangafo

GHANA

Bondoukou

Ogbomosho

Ilesha

Ife

Akure

Gboko

Bafoussam

CENTRAL AFRICAN REPUBLIC

Techiman

IBADAN

Abeokuta

Nsukka

Enugu

Bamenda

Tibati

Meiganga

Bossembélé

Sunyani

Volta Lake

Kpalimé

Benin City

Onitsha

ADAMAOUA

Kumasi

Ho

Porto-Novo

Lomé

Cotonou

Owerri

Nkongsamba

Bertoua

Carnot

Bangui

Obuasi

ACCRA

Tema

Warri

Aba

Kumba

Nanga-Eboko

Berbérati

Mbaïki

Libenge

Sekondi

Cape Coast

LAGOS

Port-Harcourt

Calabar

Kumba

Rafia

Zongo

CAPE THREE POINTS

Takoradi

Bight of Benin

Cameroon Mountain
13 451 ft.
4100 m △

DOUALA

Edéa

YAOUNDÉ

Yokadouma

Malabo

Gulf of Guinea

Pico de Santa Isabel
9869 ft.
3008 m △

BIOKO
(FERNANDO PÓO)

Kribi

CAMEROON

Ebolowa

Bight of Biafra

Bata

Oyem

PRÍNCIPE

EQUATORIAL GUINEA

Médouneu

Mitzic

Makokou

Impfondo

SAO TOME AND PRINCIPE

SÃO TOMÉ

São Tomé

Libreville

Kango

Ndjolé

Makoua

Owando

Mbandaka

ANNOBÓN
(Equat. Gui.)

CAP LOPEZ

Port-Gentil

GABON

Lastoursville

Francéville

Mossaka

Lukolela

Lac Mai-Ndombe

ATLANTIC OCEAN

Omboué

Booué

Moanda

Mbinda

Gamboma

Inongo

Nioki

Mouila

CONGO

Djambala

Bolobo

Fimi

Tchibanga

Mossendjo

BRAZZAVILLE

Bandundu

Mayumba

Sibiti

Loubomo

KINSHASA

Bagata

Pointe-Noire

Kinkala

Kenge

Masi-Manimba

ANGOLA

Tshela

Mbanza-Ngungu

Livingstone Falls
(Chutes de Livingstone)

DEM. REP. OF THE CONGO

Cabinda

Boma

Matadi

Lukula

Muanda

M'banza Congo

ANGOLA

60 120 180 240 300 360 Miles
60 120 180 240 300 360 420 480 540 600 Kilometers

Lambert Azimuthal Equal Area Projection
Scale 1:12,000,000
One inch to 190 miles
One cm to 120 km

NIGER

Tanout
Nguigmi
Maradi
Tessaoua
Zinder

Lake Chad
(Lac Tchad)

Mao

CHAD

Ouadi Haddad
Biltine

Abéché

SUDAN

Katsina

Kano

Funtua

Nguru
Gashua

Damaturu

Maiduguri

Massakory

N'Djamena
Kousséri

Moussoro

Ati

Bahai

Al-Fāshir

Mongo

Oum-Hadjer

Al-Junaynah

Jabal Marrah
10 072 ft.
3070 m

An-Nuhūd

DARFŪR

Zaria

Kaduna

Azare
Bama

Lac Fitri

Abou-Deïa

Melfi

Am Timan

Zalingei

Nyala

Ad-Du'ayn

Bauchi

Gombe

Jos
Kumo

NIGERIA

Maroua

MANDARA MTS.

Garoua

Fianga

Pala

Kélo

Moundou

Doba

Sarh

Bahr Salamat

Bahr Aouk

Birao

Raga

Uwayl

Abuja

Lafia

Jimeta

Dimlang
6699 ft.
2042 m

Hoséré Vokré
6772 ft.
2049 m

Ngaoundéré

Batangafo

Kaga
Bandoro

PARC NATIONAL
DU BAMINGUI-
BANGORAN

Ndélé

MASSIF DES BONGO

SOUTH
SUDAN

Makurdi

Wukari

Tchabal Mbabo
8071 ft.
2460 m

ADAMAOUA

Tibati

Meiganga

Bouar

Sibut

Ippy

Bambari

Bria

Wâw

Gboko

Nsukka

Enugu

Bamenda

Bafoussam

Bertoua

Nanga-
Eboko

Carnot

Bossembélé

Berbérati

Mbaïki

Bangui

Zongo

Bangassou

Zémio

Tambura

Aba

Calabar

Kumba

Port-Harcourt

Nkongsamba

Bafia

Cameroon Mountain
13 451 ft.
4100 m

DOUALA

Edéa

YAOUNDÉ

CAMEROON

Yokadouma

Libenge

Gbádolite

Yakoma

Bondo

Dungu

Uele

Bomokandi

Malabo

Pico de Santa Isabel
9869 ft.
3008 m

BIOKO
(FERNANDO PÓO)

Kribi

Ebolowa

Gemena

Businga

Ebola

Aketi

Buta

Bafwasende

Isiro

Nepoko

Bight of Biafra

Gulf of
Guinea

PRÍNCIPE

SAO TOME
AND PRINCIPE

São Tomé

SÃO TOMÉ

Bata

EQUATORIAL
GUINEA

Médouneu

Oyem

Mitzic

Makokou

Quesso

ÎLE SUMBA

Impfondo

Mankanza

Basankusu

Binga

Lisala

ÎLE
ESUMBA

Bumba

Basoko

Yangambi

Mbandaka

Boende

Ubundu

Kisangani

Stanley
Falls

Wamba

Libreville

Kango

Ndjolé

Booué

Makoua

Owando

Ingende

Lukolela

Lac
Ntomba

CAP LOPEZ

Port-Gentil

Lastoursville

GABON

Moanda

Franceville

Gamboma

Mossaka

Lac Mai
Ndombe

Inongo

Nioki

Bolobo

Monkoto

Ikela

Punia

Kalima

Bukavu

Shabunda

Omboué

Mouila

Mbinda

Mayumba

Tchibanga

Mossendjo

Djambala

CONGO

Bandundu

Oshwe

Lodja

Kibombo

DEMOCRATIC REPUBLIC
OF THE CONGO

Kindu

Ugoma
9780 ft.
2981 m

Sibiti

Loubomo

BRAZZAVILLE

KINSHASA

Kinkala

Kenge

Masi-
Manimba

Bagata

Kikwit

Ilebo

Mweka

Lusambo

Kasongo

Kongolo

Pointe-Noire

Tshela

Lukula

Idiofa

Kananga

Kabalo

Nyunzu

ANGOLA

Cabinda

Boma

Mbanze-
Ngungu

Matadi

Bulungu

Mbuji-Mayi

Gandajika

Muanda

M'banza
Congo

Kasongo-Lunda

Tshikapa

Mwene-Ditu

Kaniama

Manono

KATANGA

ATLANTIC

OCEAN

Ambriz

Uíge

Negage

Kahemba

Chitato

Kamina

Mulongo

LUANDA

PONTA DAS PALMEIRINHAS

N'dalatando

Sanza Pombo

Marimba

Caungula

Saurimo

Bukama

Lac Upemba

Lake
Mweru

ANGOLA

Malanje

Kolwezi

Likasi

Kasenga

Gabela

Dilolo

© Rand McNally
A-101812-1

0 60 120 180 240 300 360 Miles

0 60 120 180 240 300 360 420 480 540 600 Kilometers

Lambert Azimuthal Equal Area Projection
Scale 1:12,000,000
One inch to 190 miles
One cm to 120 km

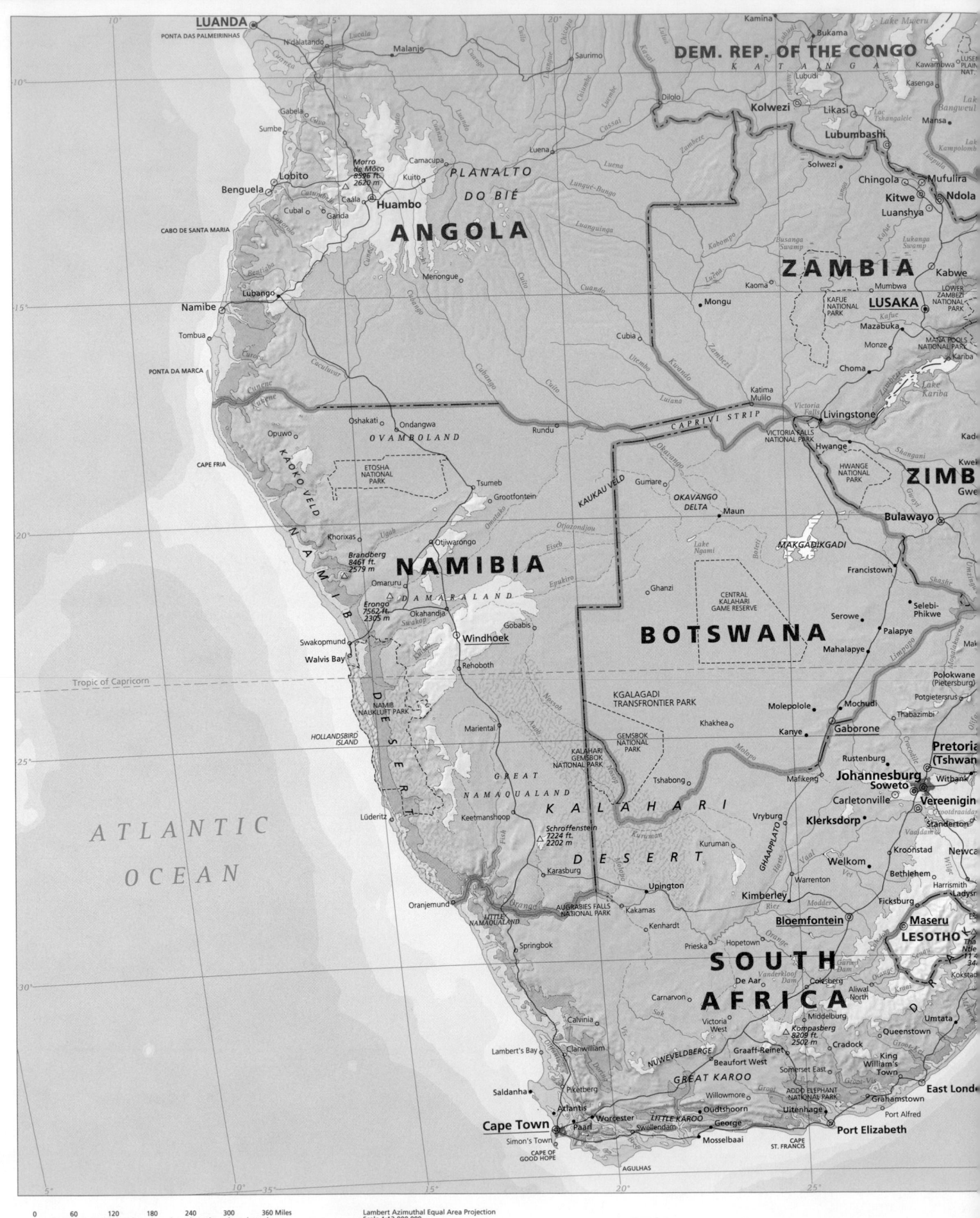

LUANDA
PONTA DAS PALMEIRINHAS
N'dalatando
Malanje
Saurimo
Kamina
Bukama
Lake Mweru
DEM. REP. OF THE CONGO
K A T A N G A
Kawambwa
Kasenga
Mansa
Lake Bangweul

Gabela
Sumbe
Camacupa
Luena
Dilolo
Kolwezi
Likasi
Lubumbashi
Solwezi
Lac Tshangalele
Lake Kampolomb

Benguela
Lobito
Morro
de Môco
8596 ft
2620 m
Kuito
Cuala
PLANALTO
DO BIÉ
Chingola
Kitwe
Luanshya
Mufulira
Ndola

CABO DE SANTA MARIA
Cubal
Ganda
Huambo
ANGOLA
ZAMBIA
Kabwe

Namibe
Lubango
Menongue
Mongu
Kaoma
Mumbwa
KAFUE
NATIONAL
PARK
LUSAKA
LOWER
ZAMBEZI
NATIONAL
PARK

Tombua
Cubia
Mazabuka
MANA POOLS
NATIONAL PARK

PONTA DA MARCA
Monze
Choma
Kariba
Lake
Kariba

CAPE FRIA
Oshakati
Ondangwa
Rundu
Katima
Mulilo
Victoria
Falls
Livingstone
CAPRIVI STRIP

Opuwo
OVAMBOLAND
VICTORIA FALLS
NATIONAL PARK
Hwange

ETOSHA
NATIONAL
PARK
Tsumeb
Gumare
KAUKAU VELD
HWANGE
NATIONAL
PARK
Shangani
ZIMB
Kwel

Khorixas
Grootfontein
OKAVANGO
DELTA
Maun
Gwe

Brandberg
8461 ft
2579 m
Otjiwarongo
MAKGADIKGADI
Bulawayo

NAMIBIA
Omaruru
DAMARALAND
Ghanzi
Lake
Ngami
CENTRAL
KALAHARI
GAME RESERVE
Francistown

Erongo
7562 ft
2305 m
Okahandja
Swakop
Gobabis
Serowe
Selebi-
Phikwe

Swakopmund
Windhoek
Rehoboth
BOTSWANA
Mahalapye
Palapye

Walvis Bay
NAMIB
NAUKLUFT PARK
KGALAGADI
TRANSFRONTIER PARK
Molepolole
Mochudi
Limpopo
Polokwane
(Pietersburg)
Potgietersrus

Tropic of Capricorn
Mariental
DESERT
Thabazimbi

HOLLANDSBIRD
ISLAND
GEMSBOK
NATIONAL
PARK
Khakhea
Kanye
Gaborone
Rustenburg
Pretoria
(Tshwan

KALAHARI
GEMSBOK
NATIONAL
PARK
Tshabong
Mafikeng
Johannesburg Withank
Soweto

Lüderitz
GREAT
Keetmanshoop
K A L A H A R I
Carletonville
Vereenigin

NAMAQUALAND
Schroffenstein
7224 ft
2202 m
Kuruman
D E S E R T
Vryburg
Klerksdorp
Standerton

Fish
Karasburg
Kakamas
Kuruman
Warrenton
Welkom
Kroonstad
Newca

Upington
Kimberley
Bethlehem
Harrismith
Ladysr

ATLANTIC
Oranjemund
Orange
AUGRABIES FALLS
NATIONAL PARK
Kenhardt
Prieska
Hopetown
Bloemfontein
Ficksburg
Maseru

OCEAN
LITTLE
NAMAQUALAND
Springbok
LESOTHO

De Aar
Vanderkloof
Dam
Colesberg
Aliwal
North
Kokstad

Victoria
West
Carnarvon
Middelburg
Queenstown

SOUTH
Calvinia
Kompasberg
8209 ft
2502 m
Cradock
King
William's
Town
Umtata

AFRICA
NUWEVELDBERGE
Beaufort West
Graaff-Reinet
Somerset East
East Lond

Lambert's Bay
Clanwilliam
GREAT KAROO
Willowmore
Oudtshoorn
ADDO ELEPHANT
NATIONAL PARK
Uitenhage
Grahamstown
Port Alfred

Saldanha
Pikelberg
LITTLE KAROO
George
Port Elizabeth

Atlantis
Worcester
Swellendam
Mosselbaai
CAPE
ST. FRANCIS

Cape Town
Paarl
Simon's Town
CAPE OF
GOOD HOPE
AGULHAS

0   60   120   180   240   300   360 Miles
0   60   120 180 240 300 360 420 480 540 600 Kilometers

Lambert Azimuthal Equal Area Projection
Scale 1:12,000,000
One inch to 190 miles
One cm to 120 km

Mbala
Mbeya
Tukuyu
Karonga
Kasama
KIPENGERE RANGE
TANZANIA
Songea
Lindi
Mtwara
Chinsali
Ngoda
8550 ft.
2606 m
Lake Nyasa
Mzuzu
CABO DELGADO
Mocímboa da Praia
Mueda
ILHA QUIRIMBA
GROUPE D'ALDABRA
ASSOMPTION
SEYCHELLES
ATOLL DE COSMOLEDO
ST. PIERRE
ATOLL DE PROVIDENCE
ASTOVE
ATOLL DE FARQUHAR
Bangweulu Swamps
Mpika
Chipata
MALAWI
Kasungu
Lichinga
Montepuez
Pemba
Namapa
NJAZIDJA
Moroni
COMOROS
NZWANI Mutsamudu
MWALI
Mamoudzou
ILES GLORIEUSES
(France; claimed by Madagascar)
ARCHIPEL DES COMORES
TANJONA BOBAOMBY
TANJONA ANORONTANY
Antsirañana
(Diégo-Suarez)
Lilongwe
Mangochi
Lake Malombe
Cuamba
Nacala
MAYOTTE
(France; claimed by Comoros)
NOSY MITSIO
NOSY BE
Ambanja
Maromokotro
9436 ft.
2876 m
Sambava
Zomba
Lake Chilwa
Gurué
Nampula
Ilha de Moçambique
NOSY LAVA
Antsohihy
Antalaha
Blantyre
Sapitwa
9849 ft.
3002 m
Tete
Angoche
ILHA ANGOCHE
Mahajanga
TANJONA ANGONTSY
SAIKANOSY MASOALA
Nsanje
Mocuba
Moma
Marovoay
Farihy Kinkony
Tsaratanana
Farihy Alaotra
Mananara Avaratra
Helodrano Antongila
HARARE
Chitungwiza
Marondera
Rusape
Cantandica
MOZAMBIQUE
Quelimane
Marromeu
TANJONA VILANANDRO
NOSY CHESTERFIELD
Besalampy
Île Juan de Nova
(France; claimed by Madagascar)
Tsiroanomandidy
Amparafaravola
Soanierana Ivongo
NOSY SAINTE MARIE
Toamasina
Ambatondrazaka
WE
Mutare
Inyangani
8504 ft.
2592 m
Chimoio
Dondo
Maintirano
Soavinandriana
ANTANANARIVO
Masvingo
Monte Binga
7995 ft.
2437 m
Beira
NOSY BARREN
Tsiafajavona
8668 ft.
2642 m
Antanifotsy
Mohanoro
GREAT ZIMBABWE
Chiredzi
Save
Belo Tsiribihina
Antsirabe
Ambositra
NOSY VARIKA
GONAREZHOU NATIONAL PARK
Runde
ILHA DO BAZARUTO
PONTA SÃO SEBASTIÃO
Morondava
MADAGASCAR
Mananjary
GREAT LIMPOPO TRANSFRONTIER PARK
Vilankulo
BASSAS DA INDIA
(France; claimed by Madagascar)
TANJONA ANKAOA
Fianarantsoa
Manakara
PARQUE NACIONAL DO LIMPOPO
PONTA DA BARRA FALSA
Murombe
Boby
8720 ft.
2658 m
Manakara
Massinga
ÎLE EUROPA
(France; claimed by Madagascar)
Ankazoabo
Ihosy
Farafangana
Maxixe
PONTA DA BARRA
Toliara
Vangaindrano
KRUGER NATIONAL PARK
Chókwè
Chibuto
Lagoa Poelela
Betioky
Tsromb[e]
Tropic of Capricorn
Berg
8 ft.
1 m
Xai-Xai
Tôlañaro
(Faradofay)
Matola
ILHA DA INHACA
Ambovombe
Mbabane
Maputo
Lobamba
SWAZILAND
Lake Sibayi
TANJONA VOHIMENA
Vryheid
ZULULAND
Lake St. Lucia
© Rand McNally
A-101819-1
Richards Bay
Pietermaritzburg
INDIAN OCEAN
Durban
Port Shepstone
INDIAN OCEAN

a
INDIAN OCEAN
POINTE L'HORTAL
Triolet
Rivière du Rempart
Port Louis
Rose Hill
Curepipe
Mahébourg
MAURITIUS
REUNION
(France)
Saint-Denis
Le Port
Saint-André
Saint-Paul
Saint-Benoît
Piton des Neiges
10 072 ft.
3070 m
Saint-Pierre
Piton de la Fournaise
8635 ft.
2632 m
Saint-Louis
Saint-Joseph
MASCARENE ISLANDS
© Rand McNally
A-101782-1

Inset map a
Lambert Conformal Conic Projection
Scale 1:4,000,000
One inch to 64 miles
One cm to 40 km

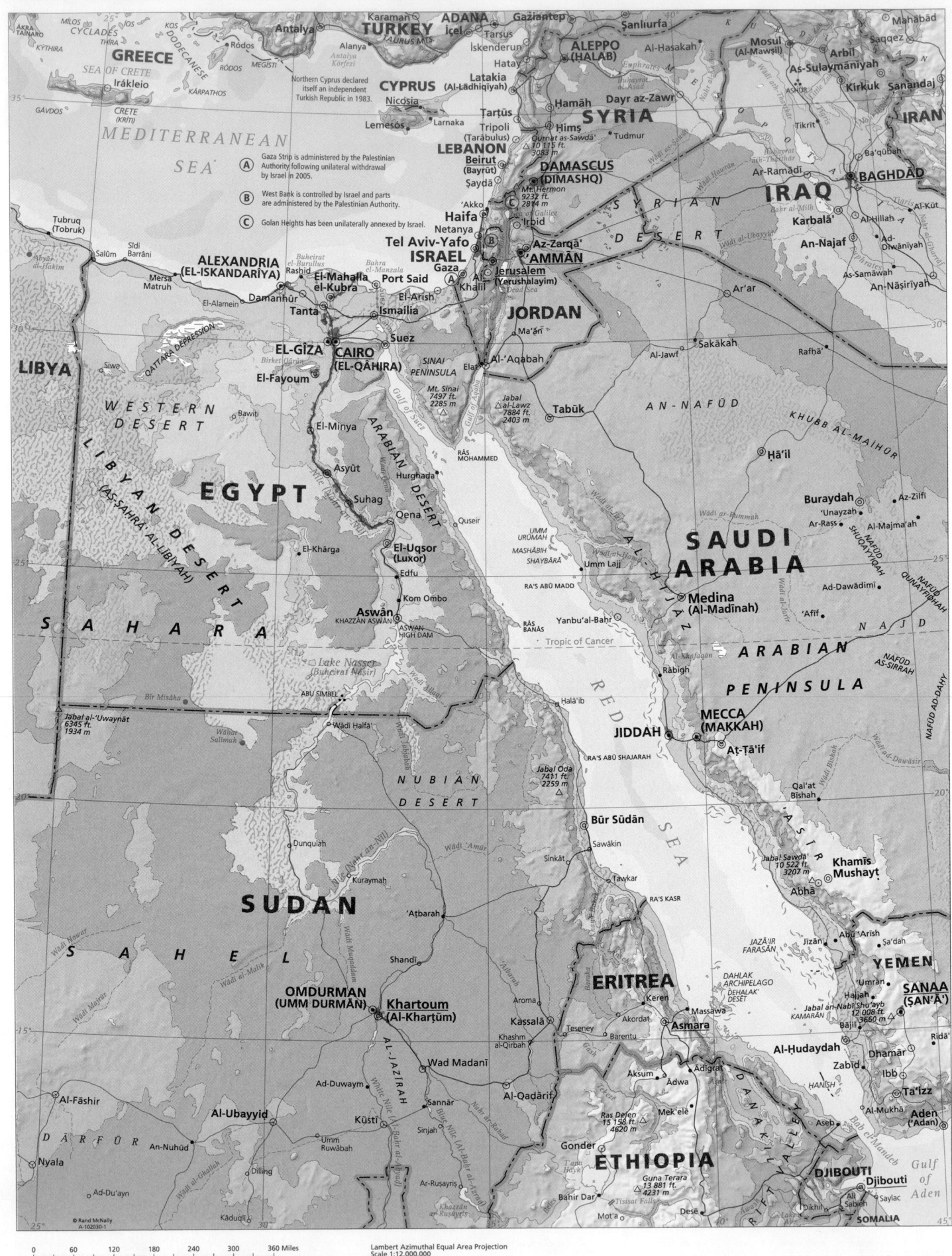

AKRA TAINARO
MILOS
CYCLADES
KOS
THIRA
KYTHIRA
KÁRPATHOS
Karaman
ADANA
İçel
Gaziantep
Mahābād
Antalya
TURKEY
Tarsus
Şanlıurfa
Mosul (Al-Mawşil)
Saqqez
GREECE
Alanya
İskenderun
Al-Hasakah
Arbil
İráklelo
Ródos
TAURUS MTS.
Hatay
ALEPPO (HALAB)
As-Sulaymaniyah
Kirkuk
Sanandaj
DODECANESE
MEGISTI
Antalya Körfezi
Nicosia
Latakia (Al-Ládhiqiyah)
Dayr az-Zawr
Tikrit
IRAN
SEA OF CRETE
CYPRUS
Hamāh
Tudmur
Ba'qūbah
GÁVDOS
Lemesos
Larnaka
Tartūs
Tripoli (Tarâbulus)
Himş
IRAQ
BAGHDĀD
CRETE (KRÍTI)
Northern Cyprus declared itself an independent Turkish Republic in 1983.
SYRIA
Qurnat as-Sawdâ' 10 115 ft 3083 m
Ar-Ramādi
Al-Hillah
Al-Kūt
MEDITERRANEAN
Gaza Strip is administered by the Palestinian Authority following unilateral withdrawal by Israel in 2005.
LEBANON
Beirut (Bayrūt)
Şaydā
DAMASCUS (DIMASHQ)
Mt. Hermon 9232 ft 2814 m
Karbalā'
An-Najaf
Ad-Dwaniyah
SEA
West Bank is controlled by Israel and parts are administered by the Palestinian Authority.
Haifa
Netanya
Akko
Sea of Galilee
Irbid
Az-Zarqā'
AMMĀN
An-Nāşiriyah
As-Samāwah
Tubruq (Tobruk)
Abyar al-Hakim
Golan Heights has been unilaterally annexed by Israel.
Tel Aviv-Yafo
Gaza
Jerusalem (Yerushalayim)
Dead Sea
Ma'ān
Ar'ar
Rafhā'
Sidi Barrâni
Salūm
Mersa Matruh
ALEXANDRIA (EL-ISKANDARIYA)
Rashid
El-Mahalla el-Kubra
Port Said
El-Arish
El Khalil
JORDAN
Sakākah
Al-Jawf
El-Alamein
Damanhûr
Tanta
Ismailia
Suez
Elat
Al-'Aqabah
WESTERN DESERT
Siwa
QATTARA DEPRESSION
Birket Qârûn
Buheirat el-Burullus
Bahra el-Manzala
CAIRO (EL-QÂHIRA)
EL-GÎZA
SINAI PENINSULA
Mt. Sinai 7497 ft 2285 m
Jabal al-Lawz 7884 ft 2403 m
Tabūk
AN-NAFŪD
KHUBB AL-MAIHÛR
El-Fayoum
El-Minya
Gulf of Suez
RÂS MOHAMMED
Hā'il
LIBYA
ARABIAN DESERT
Hurghada
Gulf of Aqaba
Burayḍah
Az-Zilfî
LIBYAN DESERT (AS-SAHRÂ AL-LÎBIYAH)
EGYPT
Asyût
Suhag
Qena
Quseir
UMM URUMAH
UMM MASHÂBIH SHAYBÂRÂ
Umm Lajj
'Unayzah
Ar-Rass
Al-Majma'ah
NAFÚD SHUQAYYIQAH
NAFÚD QUNAYFIDHAH
El-Khârga
El-Uqsor (Luxor)
Edfu
SAUDI ARABIA
Ad-Dawādimî
'Afîf
NAJD
NAFÚD AS-SIRRAH
NAFÚD AD-DAHY
Bawiti
Aswân
KHAZZÂN ASWAN
Kom Ombo
ASWAN HIGH DAM
RA'S ABŪ MADD
RÂS BANĀS
Yanbu'al-Bahr
Medina (Al-Madînah)
SAHARA
Lake Nasser (Buheirat Nasir)
ABU SIMBEL
Tropic of Cancer
Al-Khafaqān
Rābigh
ARABIAN
Al-Khafaqān
Jabal al-'Uwaynât 6345 ft 1934 m
Bîr Misâha
Wahat Salimah
Wādî Halfâ'
Halā'ib
PENINSULA
Jabal Oda 7411 ft 2259 m
MECCA (MAKKAH)
JIDDAH
At-Tā'if
RA'S ABŪ SHAJARAH
NUBIAN DESERT
Dunquiah
Qal'at Bishah
Kuraymah
Būr Sūdān
'ASIR
SUDAN
'Aṭbarah
Sawākin
Jabal Sawda' 10 522 ft 3207 m
Khamis Mushayt
SAHEL
Shandi
Sinkât
Tawkar
RA'S KASR
Abhā
Wādî Howar
Wādî al-Malik
Wādî Muqaddam
RED SEA
Wādî 'Amūr
JAZĀ'IR FARASĀN
Jizān
Abū 'Arîsh
Şa'dah
Al-Fāshir
Al-Ubayyid
An-Nuhûd
OMDURMAN (UMM DURMÂN)
Khartoum (Al-Kharţūm)
Aroma
DAHLAK ARCHIPELAGO (DEHALAK' DESET)
YEMEN
'Umrân
SANAA (ŞAN'Â')
DĀRFŪR
Nyala
Ad-Duwaym
Wad Madanî
AL-JAZÎRA
Kassalā
Khashm al-Qirbah
ERITREA
Keren
Massawa
Barentu
KAMARĀN
Jabal an-Nabi Shu'ayb 12 008 ft 3660 m
Ḥajjah
Bajil
Al-Ḥudaydah
Dhamār
Rida'
Ad-Du'ayn
Dilling
Sannâr
Al-Qadārif
Teseney
Asmara
Aksum
Adwa
Zabid
Ibb
Ta'izz
Kāduqli
Umm Ruwâbah
Kūstî
Sinjah
Akordat
Kassalā
Adigrat
HANISH
Al-Mukhâ
Aden ('Adan)
An-Nuhûd
Ar-Ruşayriş
KHAZZÂN AR-RUŞAYRIŞ
Gonder
Mek'ele
Ras Defen 15 158 ft 4620 m
DANAKIL
Gulf of Aden
Bahir Dar
Tana Hāyk'
ETHIOPIA
Guna Terara 13 881 ft 4231 m
Tisisat Falls
Aseb
Dikhil
'Ali Sabieh
Saylac
DJIBOUTI
Djibouti
Mota
Desê
SOMALIA
Nile (Nahr an-Nīl)
Wādî Ţumilat
Wādî ar-Rummah
Wādî al-Jizl
Wādî al-Hamd
Wādî ad-Dawāsir
Wādî Bîshah
Baraka
Blue Nile (Al-Bahr al-Azraq)
White Nile (Al-Bahr al-Abyaḍ)
Nahr ar-Rahad
Nahr ad-Dinder
Atbarah

© Rand McNally
A-102030-1

0   60   120   180   240   300   360 Miles
0 60 120 180 240 300 360 420 480 540 600 Kilometers

Lambert Azimuthal Equal Area Projection
Scale 1:12,000,000
One inch to 190 miles
One cm to 120 km

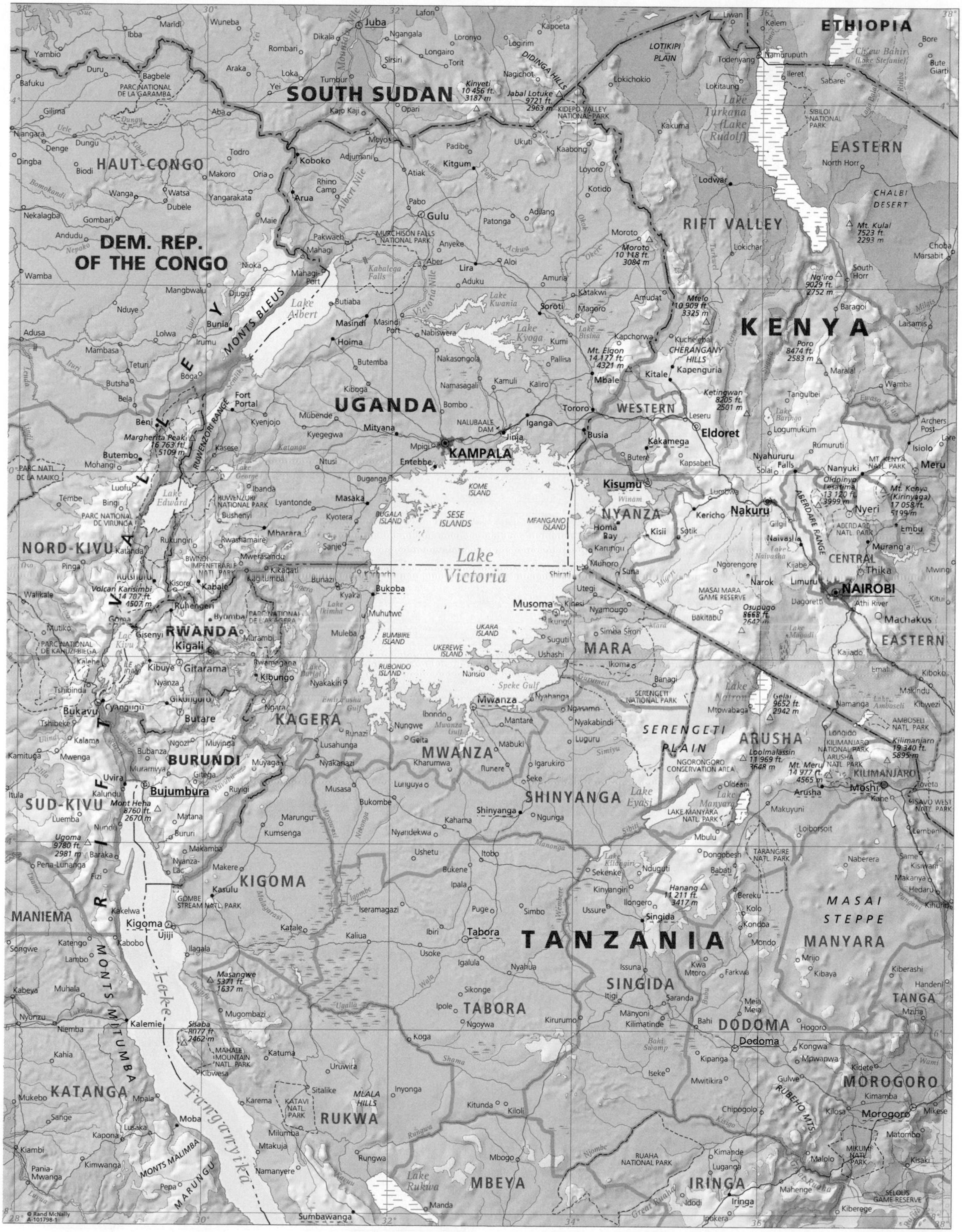

ETHIOPIA

SOUTH SUDAN

DEM. REP. OF THE CONGO

HAUT-CONGO

UGANDA

KENYA

RIFT VALLEY

EASTERN

CHALBI DESERT

WESTERN

NYANZA

NORD-KIVU

RWANDA

CENTRAL

EASTERN

Lake Victoria

SESE ISLANDS

KOME ISLAND

MFANGANO ISLAND

BUGALA ISLAND

MARA

KAGERA

SUD-KIVU

BURUNDI

MWANZA

SHINYANGA

ARUSHA

KILIMANJARO

SERENGETI PLAIN

MASAI STEPPE

MANYARA

KIGOMA

MANIEMA

TANZANIA

SINGIDA

TABORA

DODOMA

TANGA

KATANGA

RUKWA

MOROGORO

MBEYA

IRINGA

Kampala

Nairobi

Juba

Bujumbura

Kigali

Dodoma

Lake Albert

Lake Edward

Lake Kivu

Lake Tanganyika

Lake Turkana (Lake Rudolf)

Lake Kyoga

Lake Natron

Lake Eyasi

Lake Manyara

Mt. Kenya (Kirinyaga) 17 058 ft. 5199 m

Kilimanjaro 19 340 ft. 5895 m

Margherita Peak 16 763 ft. 5109 m

Mt. Elgon 14 177 ft. 4321 m

Volcan Karisimbi 14 787 ft. 4507 m

Lambert Conformal Conic Projection
Scale 1:6,000,000
One inch to 96 miles
One cm to 60 km

© Rand McNally
A-101798-1

0 30 60 90 120 150 180 Miles

0 30 60 90 120 150 180 210 240 270 300 Kilometers

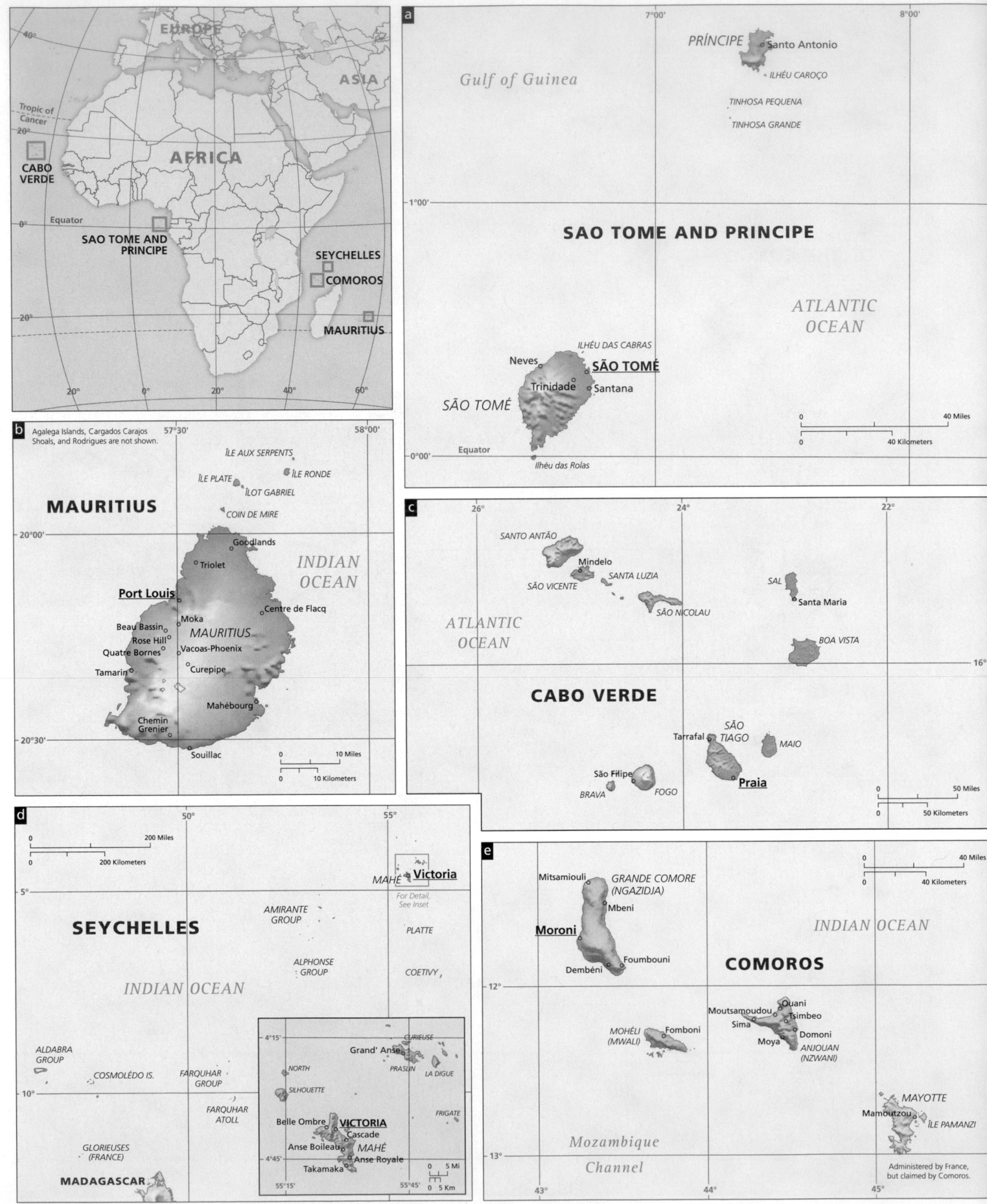

a PRÍNCIPE · Santo Antonio
Gulf of Guinea
· ILHÉU CAROÇO
TINHOSA PEQUENA
TINHOSA GRANDE
7°00'
8°00'

1°00'

SAO TOME AND PRINCIPE

ATLANTIC OCEAN

Neves · ILHÉU DAS CABRAS
SÃO TOMÉ
Trinidade · · Santana
SÃO TOMÉ

0 40 Miles
0 40 Kilometers

Equator
0°00'
Ilhéu das Rolas

EUROPE
ASIA
Tropic of Cancer
40°
20°
CABO VERDE
AFRICA
0° Equator
SAO TOME AND PRINCIPE
SEYCHELLES
COMOROS
20°
MAURITIUS
20° 0° 20° 40° 60°

---

b Agalega Islands, Cargados Carajos Shoals, and Rodrigues are not shown.
57°30'
58°00'
ÎLE AUX SERPENTS
ÎLE PLATE · · ÎLE RONDE
MAURITIUS ÎLOT GABRIEL
· COIN DE MIRE
20°00'
Goodlands
· Triolet INDIAN OCEAN
Port Louis
· Moka Centre de Flacq
Beau Bassin · MAURITIUS
Rose Hill · · Vacoas-Phoenix
Quatre Bornes ·
Tamarin · · Curepipe
· Mahébourg
Chemin Grenier ·
20°30'
Souillac
0 10 Miles
0 10 Kilometers

---

c 26° 24° 22°
SANTO ANTÃO
Mindelo
SÃO VICENTE SANTA LUZIA
SAL
· Santa Maria
SÃO NICOLAU
ATLANTIC OCEAN
BOA VISTA
16°
CABO VERDE
SÃO TIAGO
Tarrafal · · MAIO
São Filipe · **Praia**
BRAVA FOGO
0 50 Miles
0 50 Kilometers

---

d 50° 55°
0 200 Miles
0 200 Kilometers
MAHÉ **Victoria**
For Detail, See Inset
5°
AMIRANTE GROUP
SEYCHELLES PLATTE
ALPHONSE GROUP COETIVY
INDIAN OCEAN
ALDABRA GROUP
COSMOLÉDO IS. FARQUHAR GROUP
10°
FARQUHAR ATOLL
GLORIEUSES (FRANCE)
MADAGASCAR

Inset:
4°15'
CURIEUSE
Grand' Anse
NORTH PRASLIN LA DIGUE
SILHOUETTE
FRIGATE
Belle Ombre · **VICTORIA**
· Cascade
Anse Boileau · MAHÉ
4°45' · Anse Royale
Takamaka
55°15' 55°45'
0 5 Mi
0 5 Km

---

e 0 40 Miles
0 40 Kilometers
Mitsamiouli · GRANDE COMORE (NGAZIDJA)
· Mbeni
Moroni INDIAN OCEAN
· Foumbouni
Dembéni ·
COMOROS
12°
· Ouani
Moutsamoudou · · Tsimbeo
MOHÉLI (MWALI) · Fomboni Sima · · Domoni
Moya · ANJOUAN (NZWANI)
MAYOTTE
Mamoutzou · · ÎLE PAMANZI
Mozambique Channel
13°
Administered by France, but claimed by Comoros.
43° 44° 45°

Lambert Conformal Conic Projection

© 2017 Pearson Education, Inc.

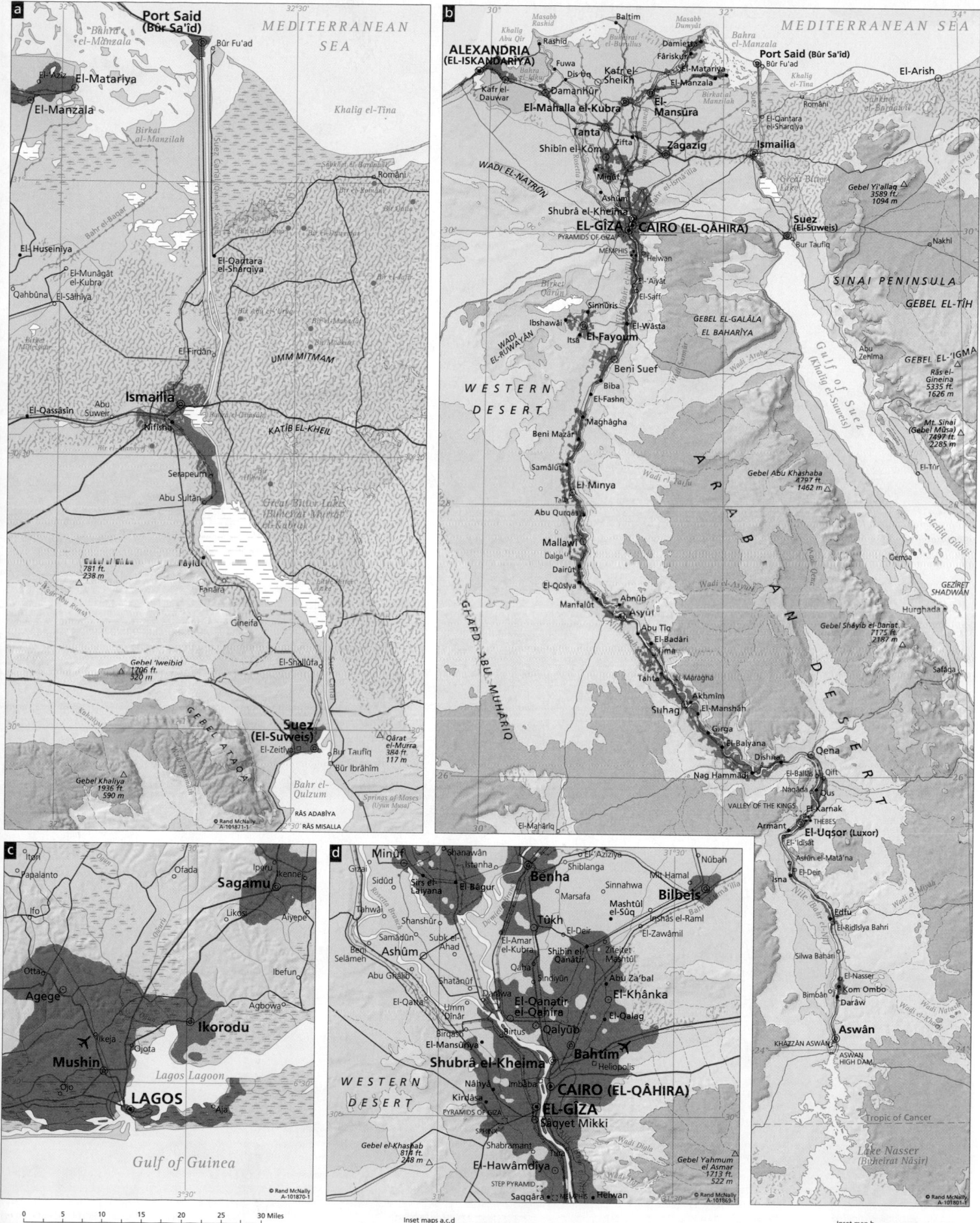

a

Port Said (Bûr Sa'îd)
MEDITERRANEAN SEA
Bûr Fu'ad
Bahra el-Manzala
El-'Azîz
El-Matarîya
El-Manzala
Khalig el-Tîna
Khalig el-Barani
Români
El-Huseinîya
El-Qantara el-Sharqîya
El-Munâgât el-Kubra
Qahbûna
El-Sâlhîya
El-Qassâsîn
El-Firdân
UMM MITMAM
Ismailia
Abu Suweir
Nifisha
KATÎB EL-KHEIL
Serapeum
Abu Sultân
Great Bitter Lake (Buheirat Murrat el-Kabra)
Fâyid
Gebel el-Biâta 781 ft. 238 m
Fanâra
Gineifa
El-Shallûfa
Gebel 'Iweibid 1706 ft. 520 m
GEBEL ATAQA
Suez (El-Suweis)
El-Zeitîya
Bûr Taufîq
Qârat el-Murra 384 ft. 117 m
Bûr Ibrâhîm
Gebel Khaliya 1936 ft. 590 m
Bahr el-Quizum
Springs of Moses ('Uyun Musa)
RÂS ADABÎYA
RÂS MISALLA
© Rand McNally A-101871-1

b

MEDITERRANEAN SEA
Masabb Rashid
Baltim
Masabb Dumyât
Khalig Abu Qîr
Rashîd
Buheirat el-Burullus
Damietta
Fâriskûr
Port Said (Bûr Sa'îd)
Bûr Fu'ad
El-Arîsh
ALEXANDRIA (EL-ISKANDARÎYA)
Fuwa
Dis-ûq
Kafr el-Sheikh
El-Matarîya
Români
Kafr el-Dauwar
Damanhûr
El-Manzala
Khalig el-Tîna
El-Mahalla el-Kubra
El-Mansûra
El-Qantara el-Sharqîya
Tanta
Zifta
Zagazig
Ismailia
Shibîn el-Kôm
WADI EL-NATRÛN
Minûf
Ashûm
Great Bitter Lake
Gebel Yi'allaq 3589 ft. 1094 m
Shubrâ el-Kheima
EL-GÎZA • CAIRO (EL-QÂHIRA)
Suez (El-Suweis)
Bûr Taufîq
Nakhl
PYRAMIDS OF GIZA
MEMPHIS
Helwan
SINAI PENINSULA
GEBEL EL-TÎH
El-'Aiyât
El-Saff
Birket Qârûn
Sinnûris
Ibshawâi
El-Wâsta
GEBEL EL-GALÂLA EL BAHARÎYA
Abu Zenîma
GEBEL EL-'IGMA
Itsa
El-Fayoum
WADI EL-RUWÂYÂN
Beni Suef
Wadi Araba
Râs el-Gineina 5335 ft. 1626 m
WESTERN DESERT
Biba
El-Fashn
Mt. Sinai (Gebel Mûsa) 7497 ft. 2285 m
Maghâgha
Beni Mazâr
Wadi el-Tarfa
El-Tûr
Samâlût
Gebel Abu Khashaba 4797 ft. 1462 m
El-Mînya
Tala
Abu Qurqâs
Mallawi
Dalga
Gebel Shâyib el-Banat 7175 ft. 2187 m
Dairût
El-Qûsîya
Hurghada
Manfalût
Abnûb
Asyût
GEZÎRET SHADWÂN
Abu Tîg
El-Badâri
Tima
Safâga
Tâhta
El-Marâgha
Akhmîm
Suhag
El-Manshâh
Girga
El-Balyana
Dishna
Qena
Nag Hammâdi
El-Ballas
Qift
Naqâda
Qus
VALLEY OF THE KINGS
El-Karnak
Armânt
THEBES
El-Uqsor (Luxor)
El-Idîsât
El-Mahârîq
GHÂRD ABU MUHÂRIQ
A R A B I A N   D E S E R T
© Rand McNally A-101801-1

c

Iton
Ipaní
Ofada
Ikenne
Sagamu
Papalanto
Ifo
Likosi
Aiyepe
Ottà
Agege
Ikeja
Ojota
Mushin
Ojo
LAGOS
Aja
Ikorodu
Lagos Lagoon
Ibefun
Agbowa
Abeokuta
Gulf of Guinea
© Rand McNally A-101870-1

d

Minûf
Shanawân
El-'Azîzîya
Shiblanga
Nûbah
Gizei
Istanha
Sidûd
Sirs el-Laiyana
El-Bâgur
Sinnahwa
Mit Hamal
Benha
Marsafa
Tahwâi
Shanshûr
Samâdûn
Subk el-Ahad
El-Amar el-Kubra
El-Deir
Mashtûl el-Sûq
Inshâs el-Raml
Bilbeis
Tûkh
Shibîn el-Qanâtir
Zifeitet Mashtûl
El-Zawâmil
Ashûm
Qaha
Sindiyûn
Abu Za'bal
Abu Ghâlib
Shatânûf
Darâwa
El-Khânka
Umm Dînâr
El-Qattâ
El-Qanâtir el-Qâhira
El-Qalag
Birgâsh
Qalyûb
El-Mansûrîya
Birtus
Nâhyâ
Kirdâsa
Imbâba
Shubrâ el-Kheima
Bahtîm
Heliopolis
WESTERN DESERT
CAIRO (EL-QÂHIRA)
EL-GÎZA
Sâqyet Mikki
PYRAMIDS OF GIZA
SPHINX
Shabramant
Gebel el-Khashab 818 ft. 248 m
El-Hawâmdiya
Gebel Yahmum el-Asmar 1713 ft. 522 m
Wadi Digla
STEP PYRAMID
Saqqâra
MEMPHIS
Helwan
© Rand McNally A-101869-1

Isna
El-Dein
Edfu
El-Ridîsîya Bahri
Silwa Bahari
El-Nasser
Kom Ombo
Bimbân
Darâw
Aswân
KHAZZÂN ASWÂN
ASWÂN HIGH DAM
Tropic of Cancer
Lake Nasser (Buheirat Nâsir)
© Rand McNally A-101801-1

0 5 10 15 20 25 30 Miles
0 5 10 15 20 25 30 35 40 45 50 Kilometers

Inset maps a,c,d
Lambert Conformal Conic Projection
Scale 1:1,000,000
One inch to 16 miles
One cm to 10 km

Inset map b
Lambert Conformal Conic Projection
Scale 1:4,000,000
One inch to 64 miles
One cm to 40 km

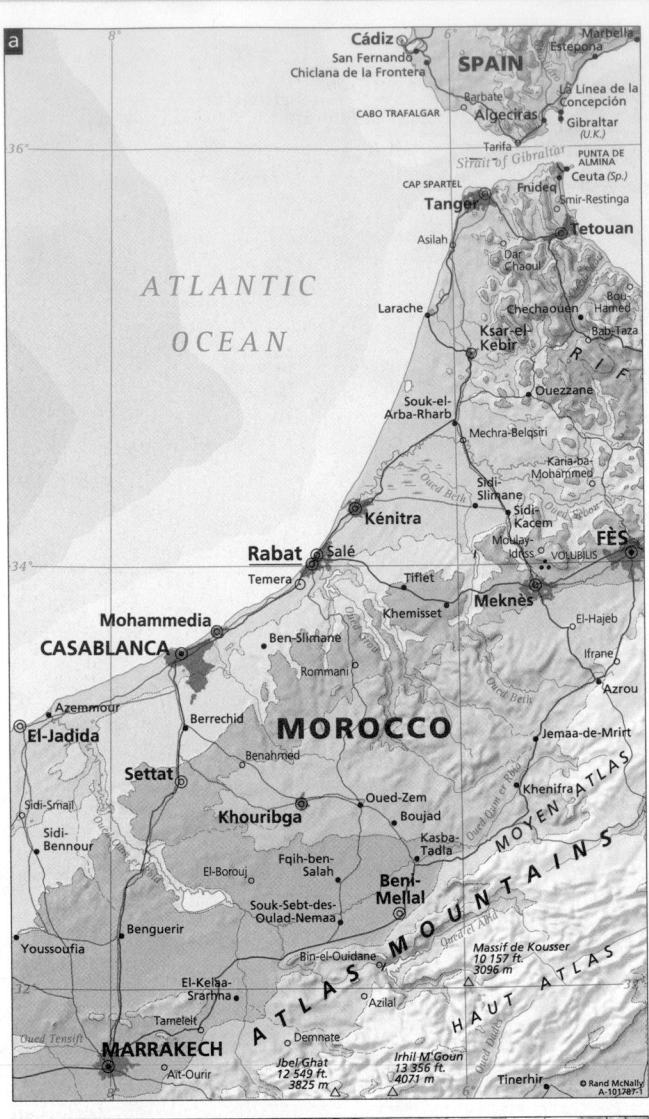

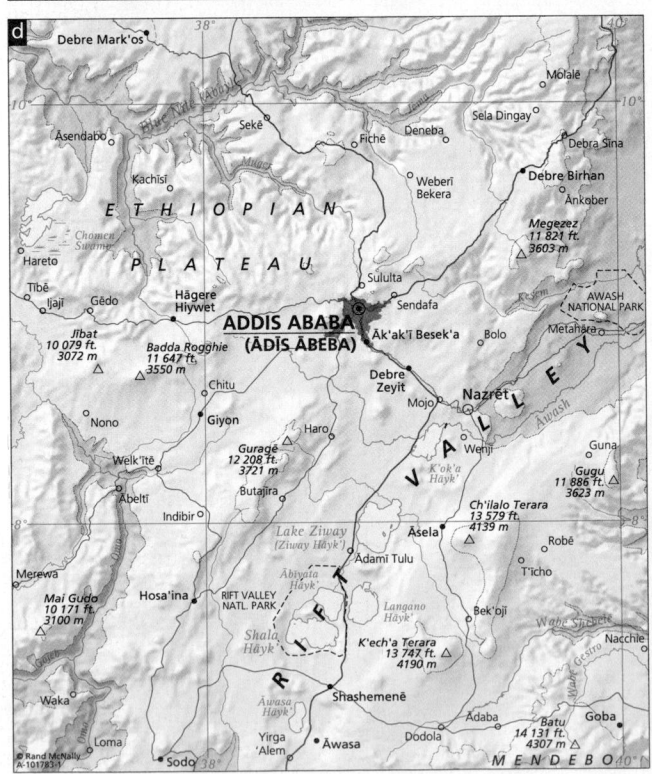

Australia, New Zealand & Oceania

AUSTRALIA,
NEW ZEALAND &
OCEANIA

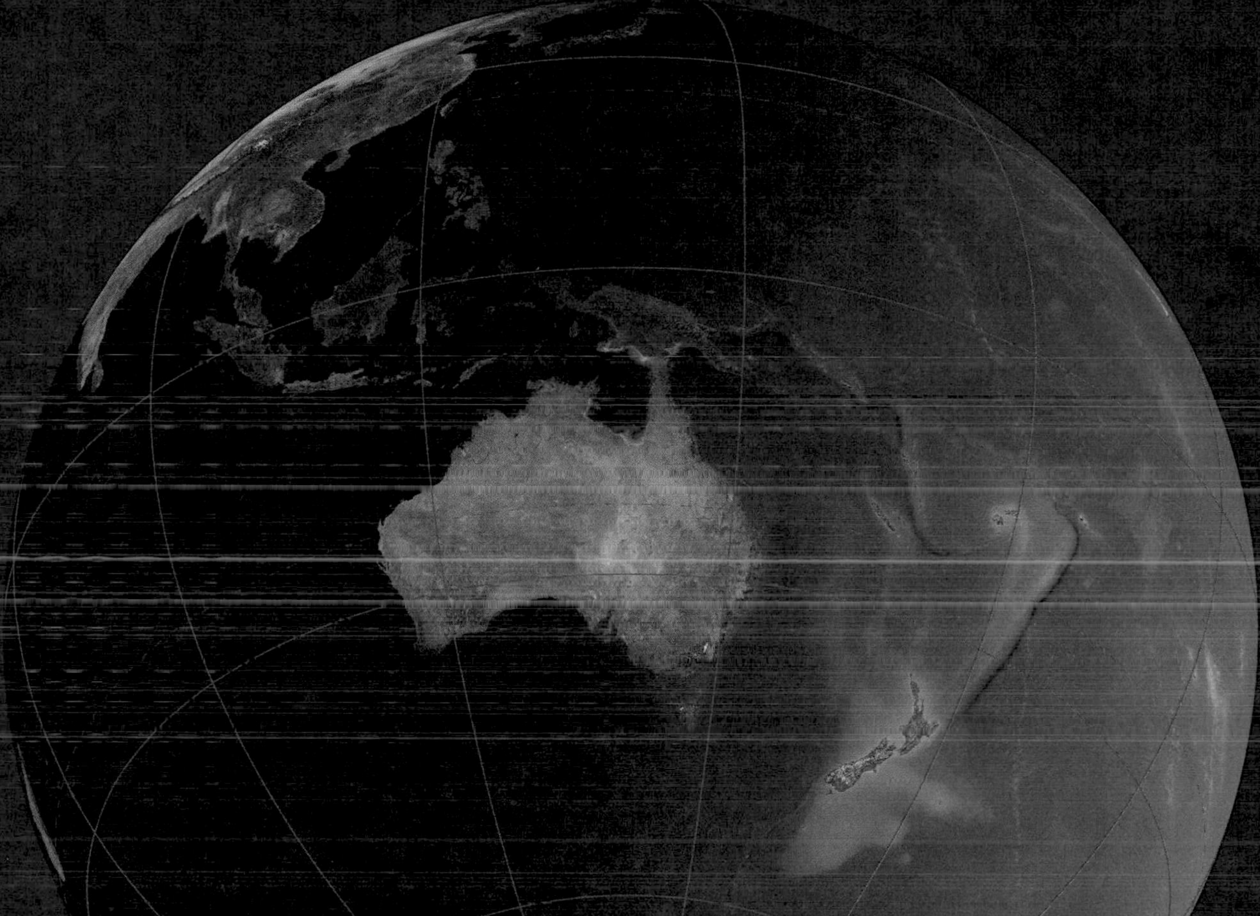

Australia is the world's sixth largest country but also its smallest and flattest continent. Humans first arrived in Australia approximately 45,000 years ago, but the continent remains sparsely populated even today. Dominated by arid and semiarid landscapes, Australia is a leading producer of iron ore, bauxite, gold, lead, uranium, silver, nickel and coal. It also is a principal exporter of wool, accounting for a quarter of the world's supply. Neighboring New Zealand, made up of two major islands, is dominated by mountains, particularly the rugged Southern Alps of South Island. Oceania is made up of thousands of scattered islands in the tropical Pacific Ocean, including Polynesia, Melanesia, and Micronesia.

INDONESIA

JAVA SEA

SEMARANG

SURABAYA

MADURA

Malang

JAVA
(JAWA)

BALI SEA

Denpasar

BALI

LOMBOK

Mataram

SUMBAWA

LESSER SUNDA ISLANDS

FLORES

SUMBA

FLORES
SEA

PULAU
WETAR

SAVU SEA

TIMOR

PULAU
ROTI

Dili

TIMOR-LESTE

PULAU
ALOR

PULAU
BABAR

PULAU
YAMDENA

ARAFURA
SEA

TIMOR SEA

ASHMORE
AND CARTIER
ISLANDS
(Austl.)

INDIAN OCEAN

CAPE LONDONDERRY

Joseph
Bonaparte
Gulf

Beagle
Gulf

MELVILLE
ISLAND

BATHURST
ISLAND

COBOURG
PENINSULA

Van
Diemen
Gulf

CAPE
WESSEL

Darwin

Pine
Creek

ARNHEM
LAND

CAPE
ARNHEM

GROOTE
EYLANDT

GULF OF
CARPENTARIA

Katherine

Daly

Roper

MORNINGTON
ISLAND

CAPE LEVEQUE

Wyndham

Kununurra

Lake
Argyle

Collier
Bay

KIMBERLEY

Fitzroy

Derby

Broome

Fitzroy
Crossing

Halls
Creek

Victoria

Daly
Waters

Borroloola

Burketown

TANAMI

DESERT

Tennant
Creek

Lake
Woods

BARKLY TABLELAND

Camooweal

Port
Hedland

Shay Gap

GREAT SANDY

DESERT

Lake
Gregory

Lake
Wills

Lake
White

NORTHERN

TERRITORY

Mount
Isa

Bou

Karratha

Roebourne

Dr Grey

Marble Bar

Fortescue

Lake
Dora

Lake
Auld

Lake
Mackay

Lake
Macdonald

MACDONNELL
RANGES

AUSTRALIA

Barrow
Creek

Georgina

BARROW
ISLAND

Onslow

NORTH WEST CAPE
Exmouth

Exmouth Gulf

Tom Price

Paraburdoo

Newman

Lake
Disappointment

GIBSON

DESERT

Alice
Springs

Finke

SIMPSON

DESERT

Tropic of Capricorn

Lake
Macleod

Ashburton

WESTERN

AUSTRALIA

Uluru
(Ayers Rock)
△ 2831 ft.
863 m

Mt. Woodroffe △
4708 ft.
1435 m

Birdsville

Carnarvon

Gascoyne

Lake
Carnegie

Macumba

Denham

SHARK
Bay

DIRK HARTOG
ISLAND

Meekatharra

Murchison

Oodnadatta

Cooper

Kalbarri

Cue

Mount Magnet

Laverton

GREAT VICTORIA DESERT

Lake
Eyre
North

Marree

Northampton

Geraldton

Mullewa

Yalgoo

Lake
Barlee

Leonora

Lake
Carey

Lake
Eyre
South

SOUTH
AUSTRALIA

Lake
Frome

Dongara

Lake
Moore

Lake
Ballard

Ooldea

Woomera

Lake
Torrens

Kalgoorlie-
Boulder

Rawlinna

NULLARBOR PLAIN

Hawker

Moora

Coolgardie

Lake
Lefroy

Eucla

Ceduna

Kimba

Lake
Gairdner

Port
Augusta

Quorn

Wanneroo

Perth

Northam

York

Southern
Cross

Lake
Cowan

Norseman

Lake
Dundas

Whyalla

EYRE
PENINSULA

Port
Pirie

Peter-
borou

DARLING RANGE

Beverley

Brookton

Lake
Johnston

Elliston

Spencer
Gulf

Wallaroo

Fremantle

Narrogin

Ravensthorpe

Esperance

GREAT AUSTRALIAN BIGHT

Port
Lincoln

Elizab

Bunbury

Collie

Katanning

Adelaide

CAPE NATURALISTE
Busselton

Bridgetown

Mount Barker

Murray
Bridge

Augusta
CAPE LEEUWIN
Pemberton

Albany

KANGAROO
ISLAND

Gulf St. Vincent

Encounter Bay

Kingston Southe

INDIAN OCEAN

© Rand McNally
A-101715-1

0   80   160   240   320   400   480 Miles

0 80 160 240 320 400 480 560 640 720 800 Kilometers

Lambert Azimuthal Equal Area Projection
Scale 1:16,000,000
One inch to 256 miles
One cm to 160 km

NEW GUINEA

PAPUA NEW GUINEA
Fly
Gulf of Papua
Torres Strait
PRINCE OF WALES ISLAND
CAPE YORK
Port Moresby

SOLOMON SEA
BOUGAINVILLE
CHOISEUL
VELLA LAVELLA
D'ENTRECASTEAUX ISLANDS
MUYUA ISLAND
RENDOVA ISLAND
SANTA ISABEL
NEW GEORGIA
SOLOMON ISLANDS
MALAITA
Honiara
GUADALCANAL
LOUISIADE ARCHIPELAGO
SAN CRISTOBAL
TAGULA ISLAND
RENNELL
SANTA CRUZ ISLANDS

TORRES ISLANDS
BANKS ISLANDS
ESPIRITU SANTO
MALAKULA
AMBRYM
EPI
VANUATU
EFATÉ
Port Vila
ERROMANGO

Weipa
CAPE YORK
CAPE MELVILLE
Karumba
Normanton
Cooktown
Laura
Mareeba
Atherton
Cairns
Innisfail
HINCHINBROOK ISLAND
Ingham
Townsville

CORAL SEA
MELLISH REEF

CORAL SEA ISLANDS TERRITORY (Austl.)
ÎLES CHESTERFIELD
ILE DE SABLE

Cloncurry
Richmond
Charters Towers
Hughenden
Bowen
Collinsville
Mackay
Sarina

GREAT DIVIDING RANGE
GREAT BARRIER REEF

Winton
Aramac
Longreach
Ilfracombe
Clermont
Emerald
Rockhampton
Mount Morgan
Gladstone

WRECK REEF
CATO DE L'OBSERVATOIRE
GIFU ISLAND

TANNA
ANATOM

LIFOU
LOYALTY ISLANDS
MARÉ
NEW CALEDONIA
Nouméa
ILE DES PINS

NEW CALEDONIA (France)

Barcaldine
Blackall
QUEENSLAND
Bundaberg
SANDY CAPE
FRASER ISLAND
Hervey Bay
Maryborough

GREAT ARTESIAN BASIN
Lake Yamma Yamma
Innamincka
Quilpie
Thargomindah
Roma
Dalby
Gympie
Nambour

Tropic of Capricorn

PACIFIC OCEAN

Cunnamulla
Saint George
Toowoomba
BRISBANE
Ipswich
Southport
Warwick
Lismore
Moree
Tenterfield
Grafton

GREY RANGE
Bourke
Narrabri
Inverell
Glen Innes
Coffs Harbour
Armidale
Kempsey

Wilcannia
Cobar
Nyngan
Tamworth
Port Macquarie

LORD HOWE ISLAND (Austl.)

Broken Hill
Dubbo
NEW SOUTH WALES
Orange
Cessnock
Maitland
Newcastle
Gosford

Parramatta
Sydney
Wollongong

Wentworth
Hay
Griffith
Wagga Wagga
Goulburn
JERVIS BAY TERRITORY

Mildura
Swan Hill
Deniliquin
AUSTRALIAN CAPITAL TERRITORY
Canberra
Bordertown
Loxton
Pinnaroo
Shepparton
Albury
Wodonga
Cooma
TASMAN SEA

Mt. Kosciuszko 7313 ft. 2229 m
Bega

Bendigo
Maryborough
Stawell
Ararat
Ballarat
VICTORIA
Hamilton
Geelong
Melbourne
Bairnsdale
CAPE HOWE
Warrnambool
Portland
CAPE OTWAY
KING ISLAND
Bass Strait
FLINDERS ISLAND
FURNEAUX GROUP
CAPE BARREN ISLAND

Smithton
Burnie
Devonport
Launceston
TASMANIA
Mt. Ossa 5305 ft. 1617 m
Queenstown
New Norfolk
Hobart
Port Arthur
SOUTH EAST CAPE

Inset a — NEW ZEALAND
NORTH CAPE
Whangarei
AUCKLAND
Hamilton
Tauranga
EAST CAPE
NORTH ISLAND
Rotorua
New Plymouth
CAPE EGMONT
Gisborne
Napier
Hastings
Wanganui
NEW ZEALAND
TASMAN SEA
CAPE FAREWELL
Palmerston North
Lower Hutt
Wellington
Nelson
Westport
Blenheim
Greymouth
SOUTH ISLAND
SOUTHERN ALPS
Aoraki 12 316 ft. 3754 m
Christchurch
Timaru
Alexandra
PACIFIC OCEAN
WEST CAPE
Dunedin
Invercargill
STEWART ISLAND
BOUNTY TROUGH

© Rand McNally
A-101717-1

Lambert Azimuthal Equal Area Projection
Scale 1:16,000,000
One inch to 256 miles
One cm to 160 km

© Rand McNally
A-101729-1

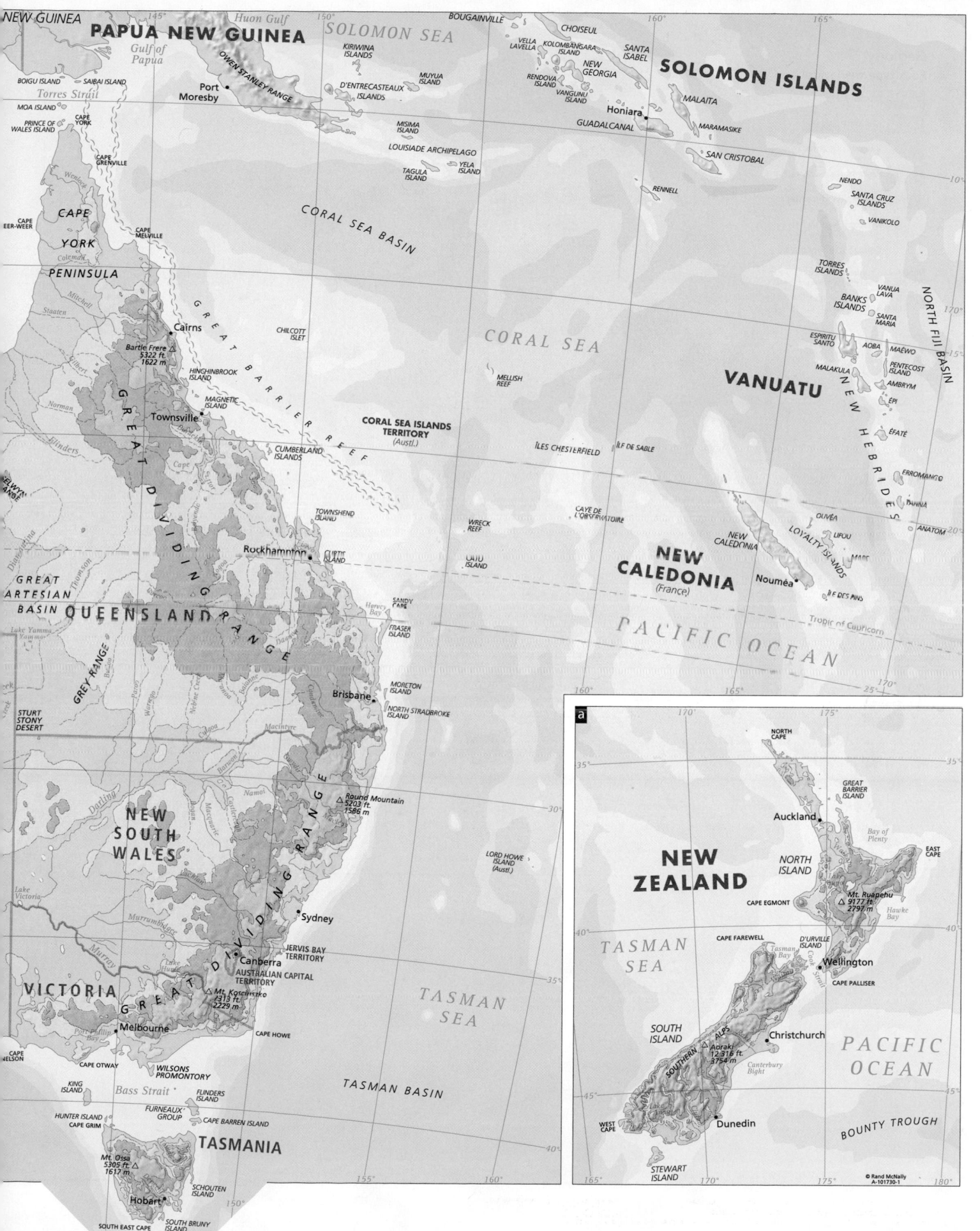

NEW GUINEA

PAPUA NEW GUINEA

Gulf of Papua

Huon Gulf

SOLOMON SEA

BOUGAINVILLE

CHOISEUL

KIRIWINA ISLANDS

VELLA LAVELLA
KOLOMBANGARA ISLAND
NEW GEORGIA
SANTA ISABEL

SOLOMON ISLANDS

Boigu Island
Saibai Island

Port Moresby

OWEN STANLEY RANGE

D'ENTRECASTEAUX ISLANDS

MUYUA ISLAND

RENDOVA ISLAND
VANGUNU ISLAND

MALAITA

Honiara

Torres Strait

MOA ISLAND

PRINCE OF WALES ISLAND

CAPE YORK

MISIMA ISLAND

GUADALCANAL

MARAMASIKE

CAPE GRENVILLE

LOUISIADE ARCHIPELAGO

TAGULA ISLAND
YELA ISLAND

SAN CRISTOBAL

CORAL SEA BASIN

RENNELL

NENDO
SANTA CRUZ ISLANDS

VANIKOLO

CAPE WEER-WEER

CAPE YORK PENINSULA

CAPE MELVILLE

CHILCOTT ISLET

CORAL SEA

TORRES ISLANDS

VANUA LAVA

BANKS ISLANDS

SANTA MARIA

NORTH FIJI BASIN

Cairns

Bartle Frere △ 5322 ft. 1622 m

GREAT BARRIER REEF

MELLISH REEF

ESPIRITU SANTO

AOBA
MAÉWO

VANUATU

MALAKULA
PENTECOST ISLAND
AMBRYM

EPI

HINCHINBROOK ISLAND

MAGNETIC ISLAND

Townsville

CORAL SEA ISLANDS TERRITORY (Austl.)

ÎLES CHESTERFIELD

ÎLE DE SABLE

ÉFATÉ

ERROMANGO

CUMBERLAND ISLANDS

GREAT DIVIDING RANGE

TOWNSHEND ISLAND

WRECK REEF

CAYE DE L'OBSERVATOIRE

TANNA

ANATOM

OUVÉA

LIFOU

LOYALTY ISLANDS

MARÉ

ÎLE DES PINS

Rockhampton

CURTIS ISLAND

GIO ISLAND

GIO ISLAND

NEW CALEDONIA

NEW CALEDONIA (France)

Nouméa

GREAT ARTESIAN BASIN

QUEENSLAND

RANGE

SANDY CAPE

HERVEY BAY

FRASER ISLAND

PACIFIC OCEAN

Tropic of Capricorn

GREY RANGE

Brisbane

MORETON ISLAND
NORTH STRADBROKE ISLAND

STURT STONY DESERT

Round Mountain △ 5203 ft. 1586 m

NEW SOUTH WALES

GREAT DIVIDING RANGE

LORD HOWE ISLAND (Austl.)

Sydney

JERVIS BAY TERRITORY

Canberra

AUSTRALIAN CAPITAL TERRITORY

VICTORIA

△ Mt. Kosciusko 7313 ft. 2229 m

GREAT DIVIDING RANGE

Melbourne

CAPE HOWE

TASMAN SEA

CAPE OTWAY

WILSONS PROMONTORY

CAPE NELSON

KING ISLAND

FLINDERS ISLAND

Bass Strait

TASMAN BASIN

HUNTER ISLAND
CAPE GRIM

FURNEAUX GROUP

CAPE BARREN ISLAND

TASMANIA

Mt. Ossa △ 5305 ft. 1617 m

Hobart

SCHOUTEN ISLAND

SOUTH EAST CAPE

SOUTH BRUNY ISLAND

a

NORTH CAPE

GREAT BARRIER ISLAND

Auckland

Bay of Plenty

EAST CAPE

NEW ZEALAND

NORTH ISLAND

CAPE EGMONT

Mt. Ruapehu △ 9177 ft. 2797 m

Hawke Bay

TASMAN SEA

CAPE FAREWELL

D'URVILLE ISLAND

Tasman Bay

Cook Strait

Wellington

CAPE PALLISER

SOUTH ISLAND

SOUTHERN ALPS

Aoraki △ 12,316 ft. 3754 m

Christchurch

Canterbury Bight

PACIFIC OCEAN

WEST CAPE

Dunedin

BOUNTY TROUGH

STEWART ISLAND

© Rand McNally
A-101730-1

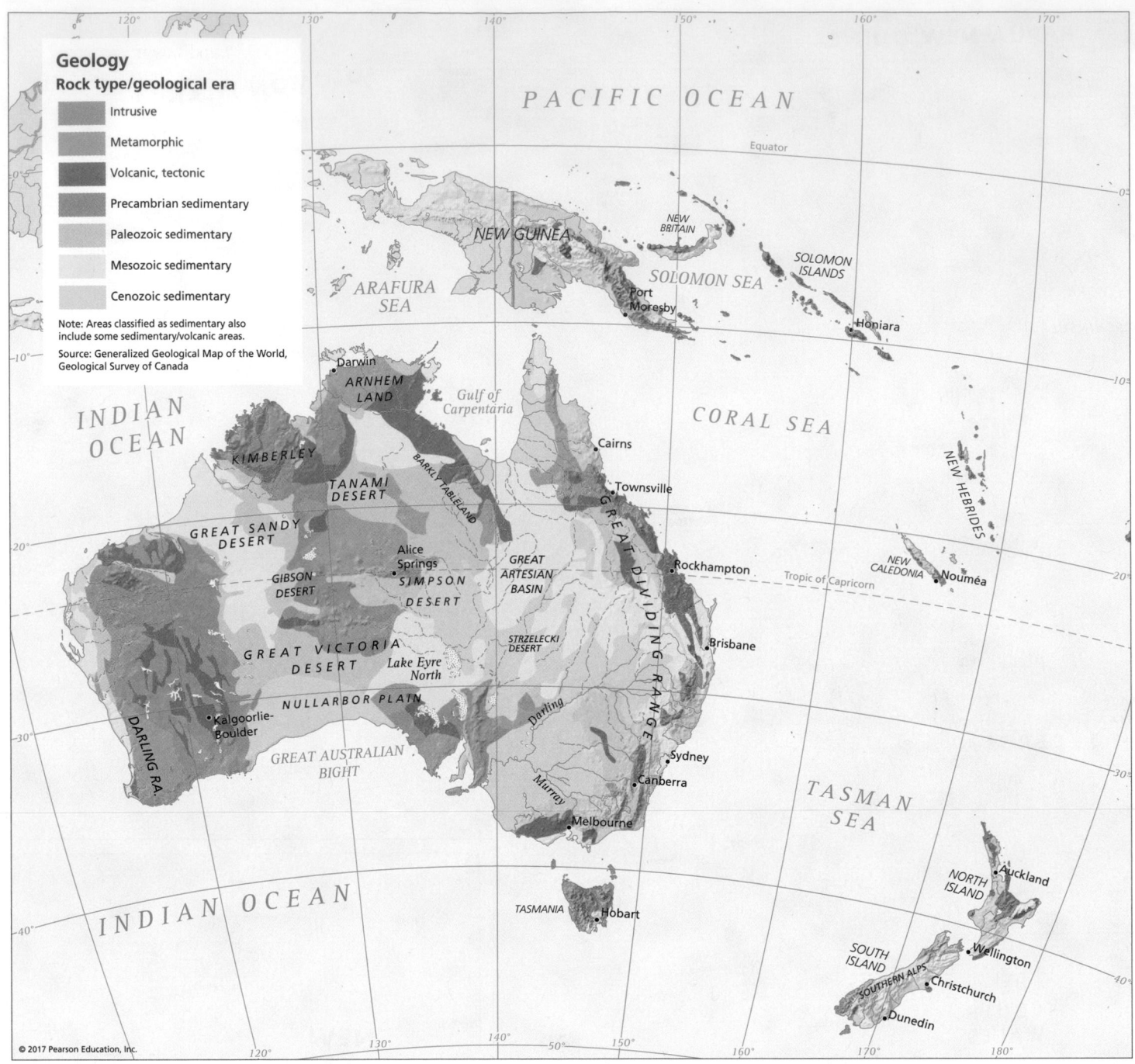

Geology
Rock type/geological era

- Intrusive
- Metamorphic
- Volcanic, tectonic
- Precambrian sedimentary
- Paleozoic sedimentary
- Mesozoic sedimentary
- Cenozoic sedimentary

Note: Areas classified as sedimentary also include some sedimentary/volcanic areas.

Source: Generalized Geological Map of the World, Geological Survey of Canada

PACIFIC OCEAN

Equator

NEW GUINEA

NEW BRITAIN

SOLOMON ISLANDS

SOLOMON SEA

Port Moresby

Honiara

ARAFURA SEA

Gulf of Carpentaria

CORAL SEA

Darwin

ARNHEM LAND

Cairns

Townsville

NEW HEBRIDES

INDIAN OCEAN

KIMBERLEY

TANAMI DESERT

BARKLY TABLELAND

GREAT SANDY DESERT

Alice Springs

GIBSON DESERT

SIMPSON DESERT

GREAT ARTESIAN BASIN

GREAT DIVIDING RANGE

Rockhampton

Tropic of Capricorn

NEW CALEDONIA

Nouméa

GREAT VICTORIA DESERT

Lake Eyre North

STRZELECKI DESERT

Brisbane

NULLARBOR PLAIN

Darling

Kalgoorlie-Boulder

Sydney

DARLING RA.

GREAT AUSTRALIAN BIGHT

Murray

Canberra

TASMAN SEA

Melbourne

NORTH ISLAND

Auckland

INDIAN OCEAN

TASMANIA

Hobart

Wellington

SOUTH ISLAND

SOUTHERN ALPS

Christchurch

Dunedin

Land Cover

- Evergreen Broadleaf Forest
- Mixed Forest
- Woodland
- Wooded Grassland
- Closed Shrubland
- Open Shrubland
- Grassland
- Cropland
- Bare Ground (Desert)
- Urban and Built Up

PACIFIC OCEAN

Equator

NEW GUINEA

NEW BRITAIN

ARAFURA SEA

SOLOMON SEA

SOLOMON ISLANDS

TIMOR SEA

INDIAN OCEAN

CORAL SEA

NEW HEBRIDES

ARNHEM LAND

Gulf of Carpentaria

KIMBERLEY

NEW CALEDONIA

TANAMI DESERT

BARKLY TABLELAND

GREAT SANDY DESERT

Tropic of Capricorn

GIBSON DESERT

SIMPSON DESERT

GREAT ARTESIAN BASIN

GREAT DIVIDING RANGE

GREAT VICTORIA DESERT

STRZELECKI DESERT

Lake Eyre North

Brisbane

NULLARBOR PLAIN

Darling

Perth

DARLING RA.

GREAT AUSTRALIAN BIGHT

Adelaide

Murray

Sydney

TASMAN SEA

Melbourne

INDIAN OCEAN

TASMANIA

NORTH ISLAND

Auckland

SOUTH ISLAND

SOUTHERN ALPS

A-102089-1 © Rand McNally

Source: CIESIN; Hansen et al., 2000

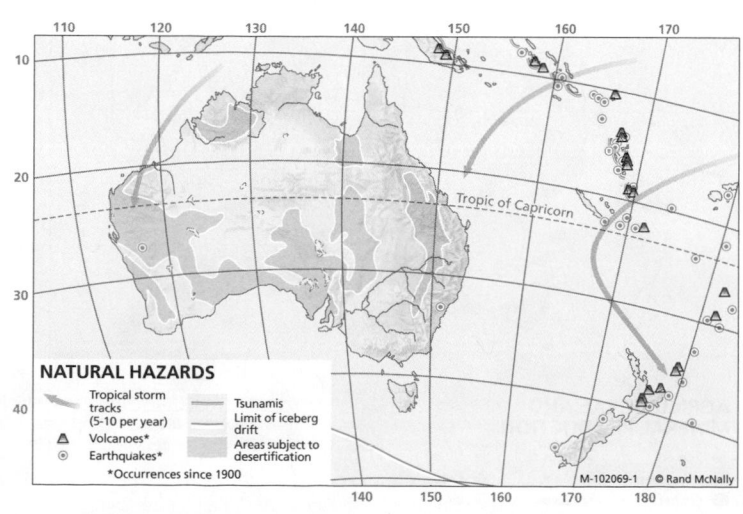

NATURAL HAZARDS

- Tropical storm tracks (5-10 per year)
- ▲ Volcanoes*
- ⊙ Earthquakes*
- Tsunamis
- Limit of iceberg drift
- Areas subject to desertification

*Occurrences since 1900

Tropic of Capricorn

M-102069-1 © Rand McNally

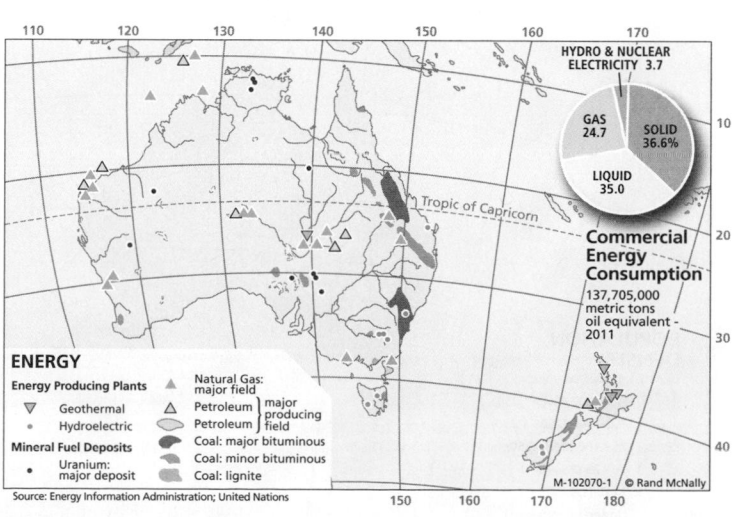

ENERGY

Energy Producing Plants
- ▽ Geothermal
- • Hydroelectric

Mineral Fuel Deposits
- • Uranium: major deposit

- ▲ Natural Gas: major field
- △ Petroleum
- ⬠ Petroleum: major producing field
- ⬮ Coal: major bituminous
- ⬯ Coal: minor bituminous
- ⬭ Coal: lignite

HYDRO & NUCLEAR ELECTRICITY 3.7

GAS 24.7

SOLID 36.6%

LIQUID 35.0

Commercial Energy Consumption

137,705,000 metric tons oil equivalent - 2011

Tropic of Capricorn

Source: Energy Information Administration; United Nations

M-102070-1 © Rand McNally

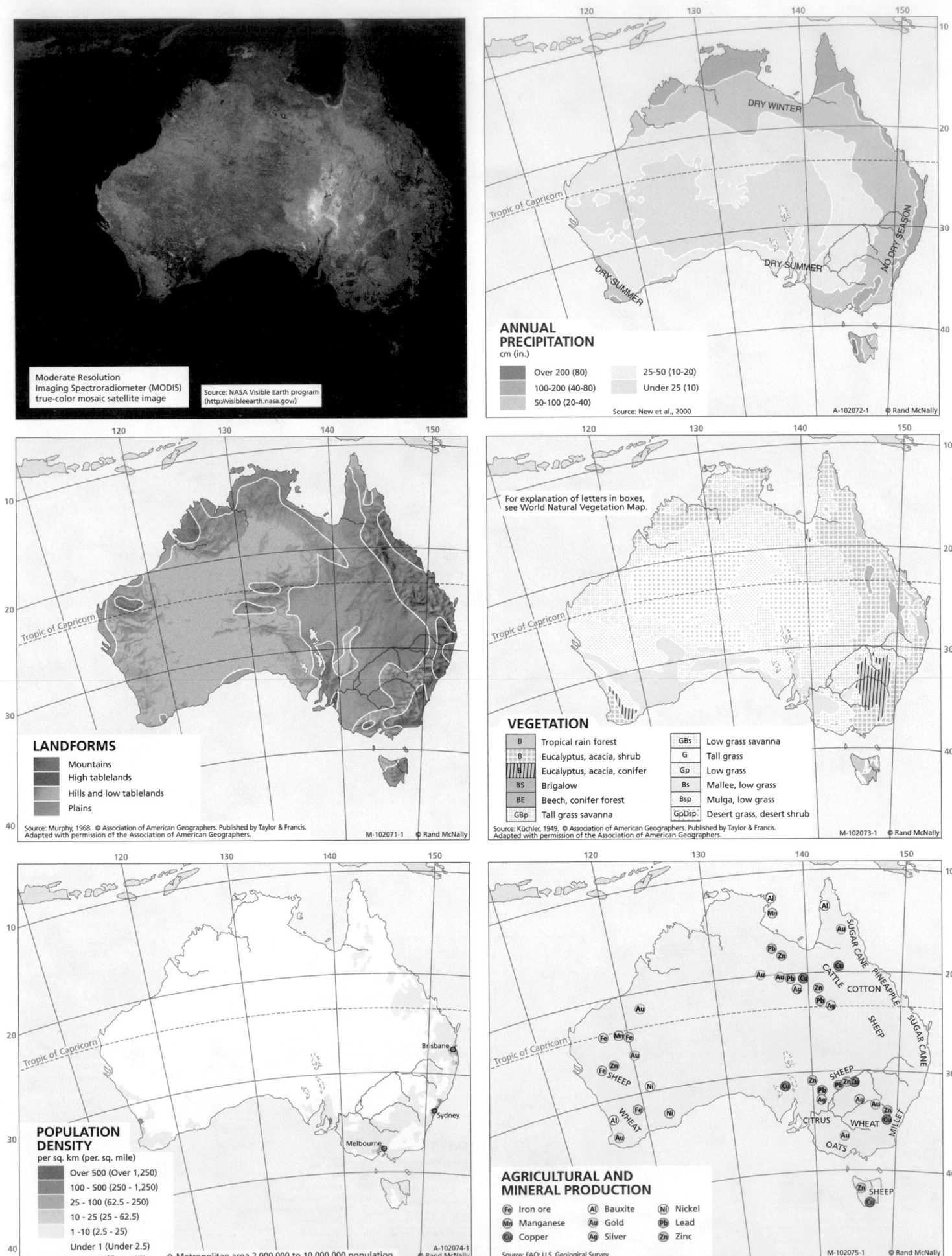

Moderate Resolution
Imaging Spectroradiometer (MODIS)
true-color mosaic satellite image

Source: NASA Visible Earth program
(http://visibleearth.nasa.gov/)

DRY WINTER

Tropic of Capricorn

DRY SUMMER

DRY SUMMER

NO DRY SEASON

**ANNUAL
PRECIPITATION**
cm (in.)

Over 200 (80)
100-200 (40-80)
50-100 (20-40)
25-50 (10-20)
Under 25 (10)

Source: New et al., 2000

A-102072-1   © Rand McNally

Tropic of Capricorn

LANDFORMS

Mountains
High tablelands
Hills and low tablelands
Plains

Source: Murphy, 1968. © Association of American Geographers. Published by Taylor & Francis.
Adapted with permission of the Association of American Geographers.

M-102071-1   © Rand McNally

For explanation of letters in boxes,
see World Natural Vegetation Map.

Tropic of Capricorn

VEGETATION

B	Tropical rain forest		GBs	Low grass savanna
B	Eucalyptus, acacia, shrub		G	Tall grass
	Eucalyptus, acacia, conifer		Gp	Low grass
BS	Brigalow		Bs	Mallee, low grass
BE	Beech, conifer forest		Bsp	Mulga, low grass
GBp	Tall grass savanna		GpDsp	Desert grass, desert shrub

Source: Küchler, 1949. © Association of American Geographers. Published by Taylor & Francis.
Adapted with permission of the Association of American Geographers.

M-102073-1   © Rand McNally

Tropic of Capricorn

Brisbane

Sydney

Melbourne

**POPULATION
DENSITY**
per sq. km (per. sq. mile)

Over 500 (Over 1,250)
100 - 500 (250 - 1,250)
25 - 100 (62.5 - 250)
10 - 25 (25 - 62.5)
1 -10 (2.5 - 25)
Under 1 (Under 2.5)

Source: U.S. Department of Energy; UN

○ Metropolitan area 2,000,000 to 10,000,000 population

A-102074-1   © Rand McNally

Tropic of Capricorn

SUGAR CANE
PINEAPPLE
CATTLE
COTTON
SHEEP
SUGAR CANE
SHEEP
SHEEP
MILLET
CITRUS
WHEAT
OATS
WHEAT
SHEEP

**AGRICULTURAL AND
MINERAL PRODUCTION**

Fe Iron ore   Al Bauxite   Ni Nickel
Mn Manganese   Au Gold   Pb Lead
Cu Copper   Ag Silver   Zn Zinc

Source: FAO; U.S. Geological Survey

M-102075-1   © Rand McNally

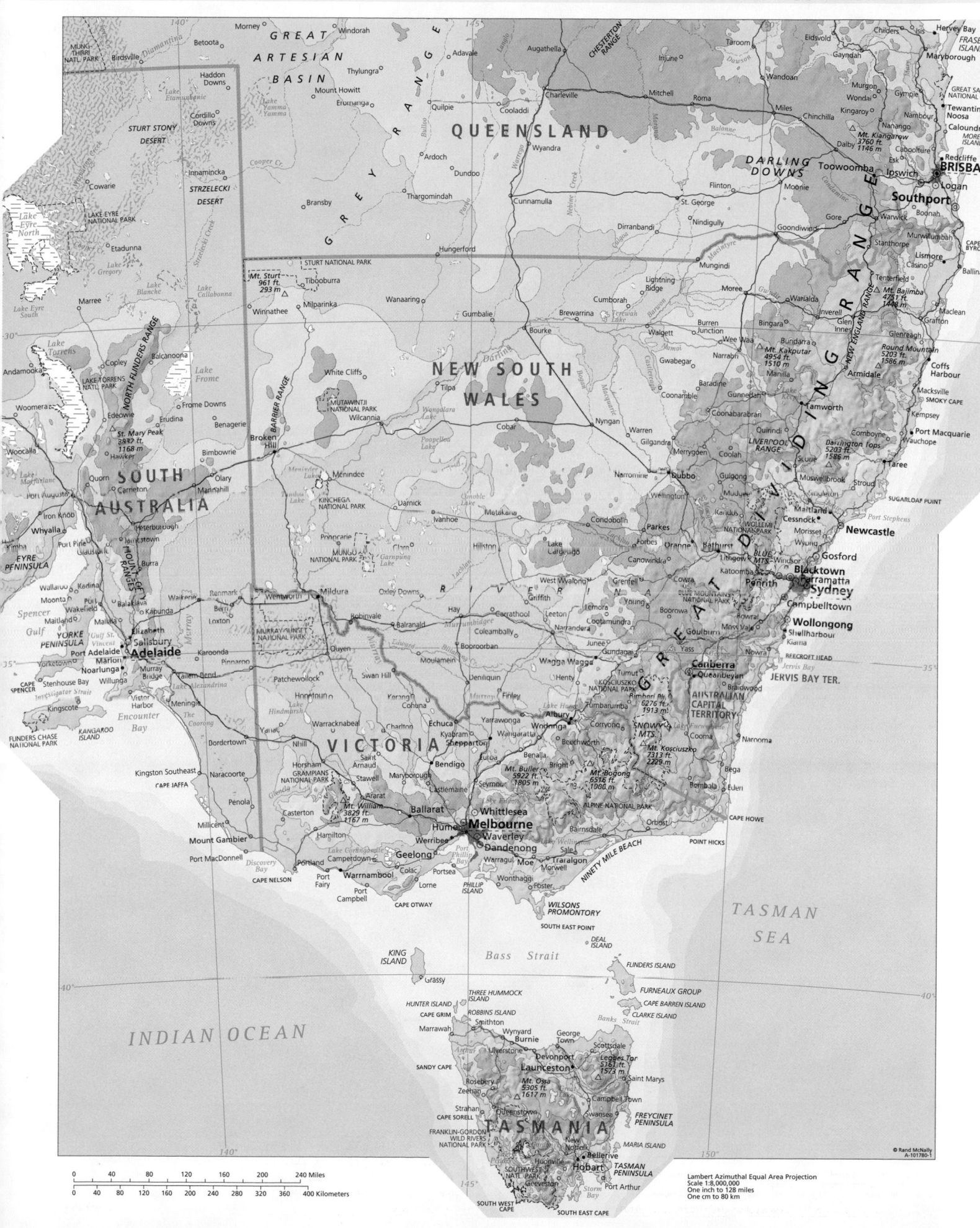

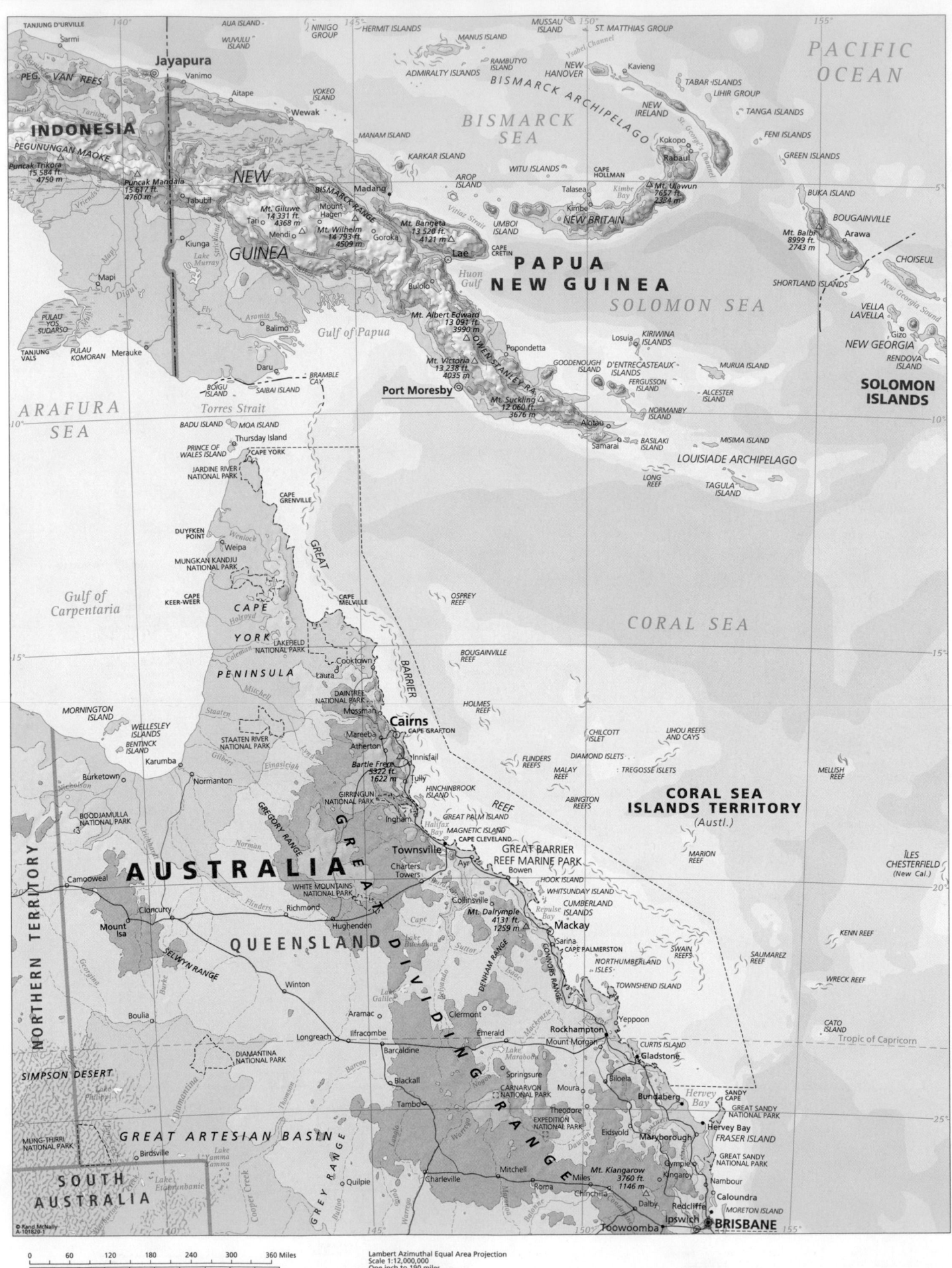

TANJUNG D'URVILLE
PEG. VAN REES
Sarmi
Jayapura
Vanimo
140°
AUA ISLAND ·
WUVULU
ISLAND
NINIGO
GROUP
145°
HERMIT ISLANDS
MUSSAU
ISLAND
150°
ST. MATTHIAS GROUP

PACIFIC
OCEAN
155°

Aitape
Wewak
VOKEO
ISLAND
MANAM ISLAND
ADMIRALTY ISLANDS
MANUS ISLAND
RAMBUTYO
ISLAND
NEW
HANOVER
Kavieng
Ysabel Channel
TABAR ·ISLANDS
LIHIR GROUP

INDONESIA
PEGUNUNGAN MAOKE
Puncak Trikora
15,584 ft.
4750 m
Puncak Mandala
15,617 ft.
4760 m
Tabubil
NEW
Sepik
BISMARCK RANGE
Madang
KARKAR ISLAND
AROP
ISLAND
WITU ISLANDS
CAPE
HOLLMAN
BISMARCK ARCHIPELAGO
NEW
IRELAND
BISMARCK
SEA
Kokopo
Rabaul
TANGA ISLANDS
FENI ISLANDS
GREEN ISLANDS

Kiunga
Tari
GUINEA
Mt. Giluwe
14,331 ft.
4368 m
Mount
Hagen
Mendi
Goroka
Mt. Wilhelm
14,793 ft.
4509 m
Mt. Bangeta
13,520 ft.
4121 m
Talasea
Kimbe
Kimbe
Bay
Mt. Ulawun
7657 ft.
2334 m
BUKA ISLAND
BOUGAINVILLE

Lake
Murray
Stickland
Mapi
Fly
Balimo
Aramia
Lae
CAPE
CRETIN
UMBOI
ISLAND
Vitiaz Strait
NEW BRITAIN
Mt. Balbi
8999 ft.
2743 m
Arawa
CHOISEUL

PULAU
YOS
SUDARSO
Digul
Bulolo
Huon
Gulf
SHORTLAND ISLANDS
VELLA
LAVELLA

TANJUNG
VALS
PULAU
KOMORAN
Merauke
Daru
Mt. Albert Edward
13,091 ft.
3990 m
PAPUA
NEW GUINEA
SOLOMON SEA
NEW GEORGIA
RENDOVA
ISLAND
Gizo
New Georgia Sound

ARAFURA
SEA
BOIGU
ISLAND
SAIBAI ISLAND
BRAMBLE
CAY
Gulf of Papua
Mt. Victoria
13,238 ft.
4035 m
OWEN STANLEY RA.
Popondetta
Losuia
KIRIWINA
ISLANDS
D'ENTRECASTEAUX
ISLANDS
GOODENOUGH
ISLAND
FERGUSSON
ISLAND
MURUA ISLAND
· ALCESTER
ISLAND
SOLOMON
ISLANDS

10°
BADU ISLAND
MOA ISLAND
Thursday Island
Port Moresby
Mt. Suckling
12,060 ft.
3676 m
NORMANBY
ISLAND
BASILAKI
ISLAND
MISIMA ISLAND

Torres Strait
PRINCE OF
WALES ISLAND
CAPE YORK
Alotau
Samarai
LONG
REEF
LOUISIADE ARCHIPELAGO
TAGULA
ISLAND
10°

JARDINE RIVER
NATIONAL PARK
CAPE GRENVILLE

DUYFKEN
POINT
Wenlock
CAPE MELVILLE
OSPREY
REEF

Weipa
MUNGKAN KANDJU
NATIONAL PARK
CAPE
KEER-WEER
CAPE
YORK
Holroyd
GREAT

Gulf of
Carpentaria
PENINSULA
LAKEFIELD
NATIONAL PARK
BOUGAINVILLE
REEF
15°

15°
MORNINGTON
ISLAND
WELLESLEY
ISLANDS
BENTINCK
ISLAND
Coleman
STAATEN RIVER
NATIONAL PARK
Staaten
Mitchell
Cooktown
Laura
DAINTREE
NATIONAL PARK
Mossman
BARRIER
HOLMES
REEF
CHILCOTT
ISLET
LIHOU REEFS
AND CAYS
MELLISH
REEF

Karumba
Burketown
Normanton
Gilbert
Einasleigh
Mareeba
Atherton
Cairns
CAPE GRAFTON
Innisfail
Bartle Frere
5322 ft.
1622 m
Tully
FLINDERS
REEFS
MALAY
REEF
DIAMOND ISLETS
TREGOSSE ISLETS
ABINGTON
REEFS

Nicholson
GIRRINGUN
NATIONAL PARK
HINCHINBROOK
ISLAND
REEF
CORAL SEA
ISLANDS TERRITORY
(Austl.)

BOODJAMULLA
NATIONAL PARK
AUSTRALIA
GREGORY RANGE
GREAT
Ingham
GREAT PALM ISLAND
Halifax
Bay
MAGNETIC ISLAND
CAPE CLEVELAND
MARION
REEF
ÎLES
CHESTERFIELD
(New Cal.)

Camooweal
Norman
Townsville
Ayr
GREAT BARRIER
REEF MARINE PARK
Bowen
20°

20°
Mount
Isa
Cloncurry
Richmond
WHITE MOUNTAINS
NATIONAL PARK
Charters
Towers
Flinders
Collinsville
Mt. Dalrymple
4131 ft.
1259 m
Mackay
WHITSUNDAY ISLAND
CUMBERLAND
ISLANDS
Repulse
Bay
SWAIN
REEFS
KENN REEF

NORTHERN TERRITORY
Georgina
SELWYN RANGE
QUEENSLAND
Hughenden
Cape
Isaac
Burdekin
Sarina
CAPE PALMERSTON
NORTHUMBERLAND
· ISLES
TOWNSHEND ISLAND
SAUMAREZ
REEF
WRECK REEF

Burke
Winton
Lake
Galilee
DENHAM RANGE
DIVIDING
CONNORS RANGE
Mackenzie

Boulia
Diamantina
Aramac
Clermont
Emerald
Rockhampton
Mount Morgan
Yeppoon
CURTIS ISLAND
CATO
ISLAND
Tropic of Capricorn
25°

Longreach
Ilfracombe
Barcaldine
RANGE
Thomson
GREY RANGE
Barcoo
Blackall
Springsure
CARNARVON
NATIONAL PARK
Moura
Theodore
Dawson
Biloela
Gladstone
Bundaberg
Hervey
Bay
SANDY
CAPE
GREAT SANDY
NATIONAL PARK

DIAMANTINA
NATIONAL PARK
Tambo
EXPEDITION
NATIONAL PARK
Eidsvold
Maryborough
FRASER ISLAND

SIMPSON DESERT
GREAT ARTESIAN BASIN
Birdsville
Cooper Creek
Lake
Yamma
Yamma
Barcoo
Mitchell
Charleville
Warrego
Nive
Roma
Miles
Mt. Kiangarow
3760 ft.
1146 m
Chinchilla
Gympie
Kingaroy
Nambour
Caloundra
MORETON ISLAND
25°

MUNG-THIRRI
NATIONAL PARK
SOUTH
AUSTRALIA
Lake
ETTanfbanib
Quilpie
Balonne
GREY RANGE
Condamine
Dalby
Redcliffe
Toowoomba
Ipswich
BRISBANE

© Rand McNally
A-101620-1
140°
145°
150°
155°

0   60   120   180   240   300   360 Miles
0 60 120 180 240 300 360 420 480 540 600 Kilometers

Lambert Azimuthal Equal Area Projection
Scale 1:12,000,000
One inch to 190 miles
One cm to 120 km

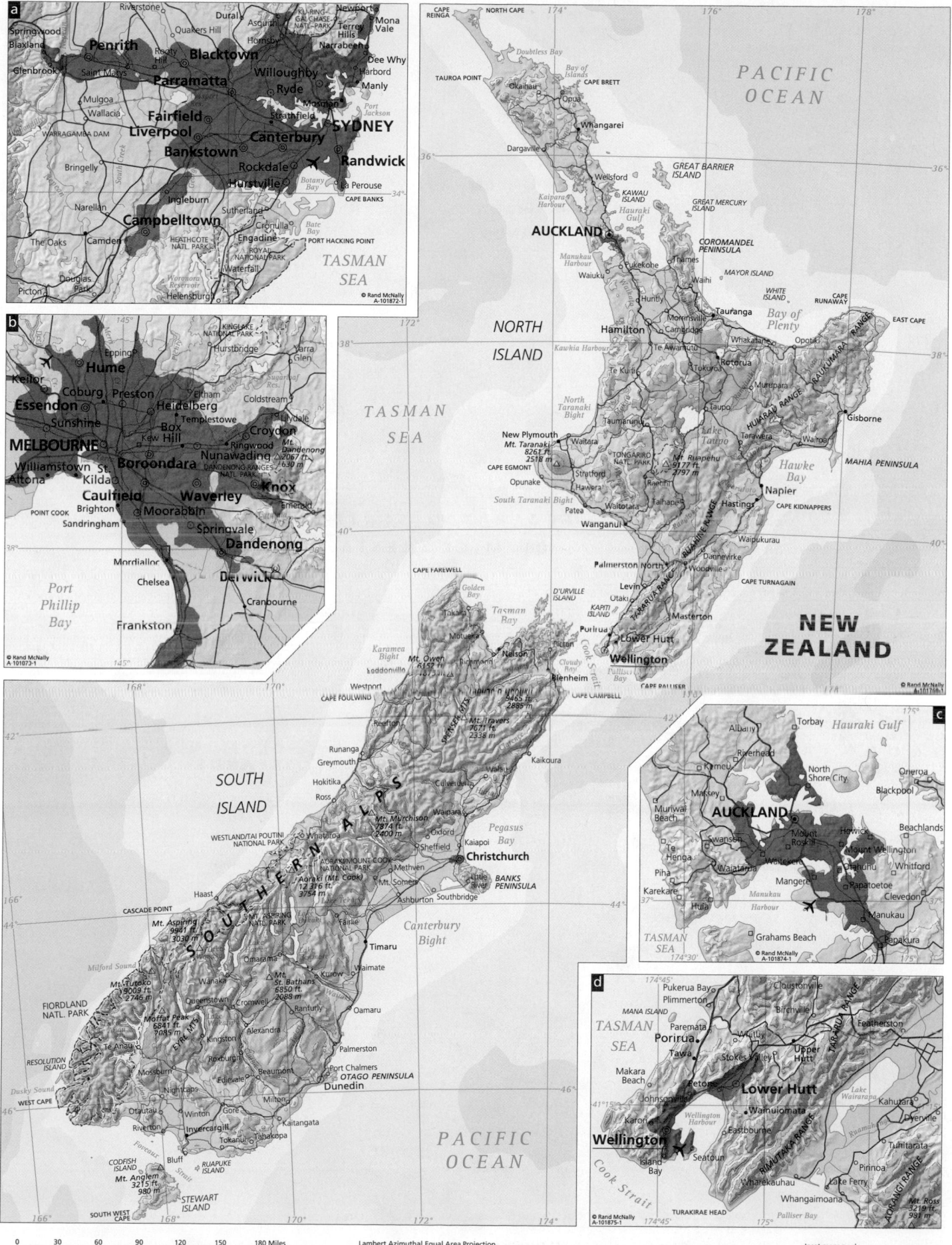

PHILIPPINE BASIN

PHILIPPINE SEA

Tropic of Cancer

MINAMI-DAITŌ-JIMA (Jpn.)
OKINO-DAITŌ-JIMA (Jpn.)

HAHAJIMA-RETTO (Jpn.)
OGASAWARA-SHOTŌ (Jpn.)
KITA-IO-JIMA
KAZAN-RETTO
IWO JIMA
MINAMI-IO-JIMA (Jpn.)

MARCUS ISLAND (Jpn.)

OKINO-TORI-SHIMA (Jpn.)

FARALLON DE PAJAROS

ASUNCION ISLAND
AGRIHAN
PAGAN
ALAMAGAN
GUGUAN
ANATAHAN
SARIGAN
FARALLON DE MEDINILLA
SAIPAN
TINIAN
ROTA

NORTHERN MARIANA ISLANDS (U.S.)

Hagåtña
GUAM (U.S.)

Challenger Deep -35 810 ft. -10 915 m

YAP

ULITHI

GAFERUT

BABELDAOB Melekeok
BELILIOU

SONSOROL ISLANDS

PULO ANNA MERIR

TOBI HELEN ISLAND

PALAU

WEST CAROLINE BASIN

KEPULAUAN MAPIA

C A R O L I N E   I S L A N D S

WOLEAI

EAURIPIK

LAMOTREK

ULUL

HALL ISLANDS

CHUUK (TRUK ISLANDS)

LOSAP ATOLL

OROLUK

NAMOLOK ATOLL

MORTLOCK ISLANDS

SENYAVIN ISLANDS

Palikir
POHNPEI

MWOKIL

PINGELAP

EAST CAROLINE BASIN

NUKUORO

KAPINGAMARANGI

FEDERATED STATES OF MICRONESIA

EAST MARIANA BASIN

M I C R O N E S I A

WAKE ISLAND (U.S.)

M I D   P A C I F I C   M O U N T A I N S

INTERNATIONAL DATE LINE

TAONGI

MARSHALL ISLANDS

ENEWETAK BIKINI BIKAR

RONGELAP UTRIK

UJELANG

WOTHO

KWAJALEIN
LIB

WOTJE
MALOELAP

MAJURO

KOSRAE

KILI

EBON

MILI

BUTARITARI

TARAWA

KURIA ABEMAMA

KIRIBATI

NONOUTI

BERU NIKUNAU

ARORAE

C E N T

P A C I

B A S

PULAU WAIGEO

Equator

NAURU

BANABA

NANUMEA NIUTAO

NIU VAITUPU

Sorong JAZIRAH DOBERAI
Manokwari BIAK
PULAU YAPEN
Fakfak
CERAM
KEPULAUAN KAI
KEPULAUAN ARU

INDONESIA

BANDA SEA

KEPULAUAN TANIMBAR

TANJUNG VALS

Digul

Merauke

ARAFURA SEA

Selat Dampier

TANJUNG D'URVILLE

PEGUNUNGAN VAN REES

Jayapura
Aitape Wewak

PEGUNUNGAN MAOKE

Puncak Jaya 16 503 ft. 5030 m

NEW GUINEA

Mt. Wilhelm 14 793 ft. 4509 m Madang

Mt. Giluwe 14 331 ft. 4368 m

Lae

NINIGO GROUP
WUVULU ISLAND
MANUS ISLAND
MUSSAU ISLAND

ADMIRALTY ISLANDS

BISMARCK ARCHIPELAGO

WITU ISLANDS

NEW HANOVER
Kavieng
TABAR ISLANDS
LIHIR GROUP
NEW IRELAND
Rabaul
Mt. Ulawun 7657 ft. 2334 m

BISMARCK SEA

NEW BRITAIN

BUKA ISLAND

BOUGAINVILLE

PAPUA NEW GUINEA

Popondetta

Gulf of Papua

OWEN STANLEY RANGE

Port Moresby

Samarai

KIRIWINA ISLANDS
MUYUA ISLAND

D'ENTRECASTEAUX ISLANDS

MISIMA ISLAND

LOUISIADE ARCHIPELAGO
TAGULA ISLAND

YELA ISLAND

S O L O M O N   S E A

CHOISEUL

VELLA LAVELLA
NEW GEORGIA
VANGUNU I.

SANTA ISABEL

MALAITA

Honiara
GUADALCANAL

SAN CRISTOBAL

RENNELL

SOLOMON ISLANDS

M E L A N E S I A

TUVALU

NANUMEA NIUTAO

FUNAFUTI

NIULAKITA

ROTUMA

NENDO

SANTA CRUZ ISLANDS

VANIKORO

TORRES ISLANDS

VANUA LAVA BANKS ISLANDS
SANTA MARIA

ESPIRITU SANTO MAÊWO

MALAKULA PENTECÔTE
AMBRYM
EPI
ÉFATÉ
Port Vila
ERROMANGO
TANNA

VANUATU

ANATOM

NEW HEBRIDES

Darwin

MELVILLE ISLAND

COBOURG PENINSULA

CAPE WESSEL

ARNHEM LAND

CAPE ARNHEM

GROOTE EYLANDT

WELLESLEY ISLANDS

Katherine

Birdum

Roper

Gulf of Carpentaria

CAPE YORK

Weipa

CAPE YORK PENINSULA

Cooktown

Normanton

Cairns

Bartle Frere 5322 ft. 1622 m

Townsville

Bowen

Mackay

G R E A T   B A R R I E R   R E E F

Mitchell

CHILCOTT ISLET

CORAL SEA BASIN

MELLISH REEF

ÎLES CHESTERFIELD

CAYE DE L'OBSERVATOIRE

C O R A L   S E A

ÎLE DE SABLE

OUVÉA LOYALTY ISLANDS

LIFOU
MARE

NOUVELLE-CALÉDONIE

Nouméa ÎLE DES PINS

ÎLE HUNTER

NEW CALEDONIA (Fr.)

NORTH FIJI BASIN

VANUA LEVU

VITI LEVU

FIJI

Suva

KANDAVU

SOUTH FIJI BASIN

TANAMI DESERT

GIBSON DESERT

Lake Mackay

Tennant Creek

Mount Isa

Alice Springs

Mt. Woodroffe 4708 ft. 1435 m

GREAT VICTORIA DESERT

SIMPSON DESERT

Birdsville

BARKLY TABLELAND

AUSTRALIA

Charters Towers

Charleville

Longreach

Emerald

GREAT ARTESIAN BASIN

Diamantina

Barcoo

Georgina

Eyre Cr.

Cooper Cr.

Lake Eyre

Tropic of Capricorn

GREY RANGE

GREAT DIVIDING RANGE

Rockhampton

Gladstone

SANDY CAPE

FRASER ISLAND

Harvey Bay

CATO ISLAND

Toowoomba

BRISBANE

Southport

Lismore

Moree

Coffs Harbour

Warrego

Condamine

Flinders

NORFOLK ISLAND (Austl.)

0   150   300   450   600   750 Miles

0   150   300   450   600   750   900   1,050 Kilometers

Lambert Azimuthal Equal Area Projection
Scale 1:27,000,000
One inch to 426 miles
One cm to 270 km

KURE
MIDWAY ISLANDS
(U.S.)

LISIANSKI ISLAND

LAYSAN ISLAND

MARO REEF

H A W A I I A N   I S L A N D S

NECKER ISLAND

FRENCH FRIGATE SHOALS

NIHOA

Tropic of Cancer

UNITED STATES

KAUA'I
NI'IHAU   O'AHU
Honolulu   MOLOKA'I
LĀNA'I   MAUI

Mauna Kea
13 796 ft   Hilo
4205 m   HAWAI'I
KALAE

SCHJETMAN REEF

JOHNSTON ATOLL
(U.S.)

CLARION FRACTURE ZONE

AL

P A C I F I C   O C E A N

IC

N

C H R I S T M A S   R I D G E

CLIPPERTON FRACTURE ZONE

KINGMAN REEF

PALMYRA ATOLL
(U.S.)

TERAINA

TABUAERAN

L I N E   I S L A N D S

KIRIMATI
(CHRISTMAS ISLAND)

HOWLAND ISLAND
(U.S.)
BAKER ISLAND
(U.S.)

JARVIS ISLAND
(U.S.)

Equator

WINSLOW REEF

P

CANTON

BIRNIE   ENDERBURY
RAWAKI

NIKUMARORO
ORONA   MANRA
PHOENIX ISLANDS

MALDEN

K I R I B A T I

STARBUCK

ATAFU
TOKELAU
(N.Z.)
FAKAOFO

PENRHYN

L   I   N   E

S

CAROLINE

SWAINS ISLANDS

NASSAU ISLAND

MANIHIKI

VOSTOK

EIAO
NUKU HIVA
UA POU   **MARQUESAS ISLANDS**
HIVA OA

SAMOA
ÎLES WALLIS
Matā'utu
WALLIS AND FUTUNA
(Fr.)
ÎLE FUTUNA

SAVAI'I   **SAMOA ISLANDS**
Apia
UPOLU   MANUA
TUTUILA   Pago   ISLANDS
Pago

AMERICAN SAMOA
(U.S.)

NORTHERN COOK ISLANDS

SUWARROW

FLINT

FATU HIVA

ÎLES DU ROI GEORGES

ÎLES DU DÉSAPPOINTEMENT

ÎLE TIKEI

TAFAHI

TONGA

VAVA'U

ONO-I-LAU
GROUP

PALMERSTON

COOK ISLANDS
(N.Z.)

MANUAE

MATAIVA

BORA-BORA
RAIATEA
SOCIETY ISLANDS
MOOREA   **Papeete**
TAHITI

MAKATEA

RARAKA   RAROĪA

ANAA   MARUTEA NORD

ÎLES   TUAMOTU

PUKARUA
REAO

NIUE
(N.Z.)

AITUTAKI

MITIARO

SOUTHERN
COOK ISLANDS   ATIU   MAUKE

TONGATAPU
Nuku'alofa   EUA

TONGA ISLANDS

INTERNATIONAL DATE LINE

KERMADEC TRENCH

TONGA TRENCH

RAROTONGA

MANGAIA   ÎLES MARIA

RIMATARA

RURUTU
RAIVAVAE

TUBUAI

FRENCH POLYNESIA
(Fr.)

AHUNUI

ÎLES   TUBUAI

Tropic of Capricorn

TUREIA

MURUROA   MARUTEA SUD

ÎLES GAMBIER

OENO ATOLL   HENDERSON ISLAND

SOUTHWEST PACIFIC BASIN

RAPA   ÎLES MAROTIRI

PITCAIRN ISLAND

PITCAIRN ISLANDS
(U.K.)

KERMADEC ISLANDS
(N.Z.)

© Rand McNally
A-101776-1

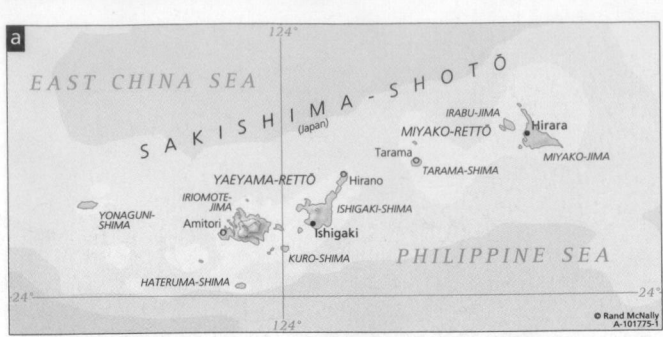

a

EAST CHINA SEA

SAKISHIMA-SHOTŌ
(Japan)

IRABU-JIMA
MIYAKO-RETTŌ
Hirara
MIYAKO-JIMA

Tarama
TARAMA-SHIMA

YAEYAMA-RETTŌ
Hirano
IRIOMOTE-JIMA
ISHIGAKI-SHIMA
Amitori
Ishigaki
KURO-SHIMA
YONAGUNI-SHIMA

HATERUMA-SHIMA

PHILIPPINE SEA

© Rand McNally
A-101775-1

b

PACIFIC OCEAN

CAPE MULINU'U
SAVAI'I
Fagamalo
Falelima
Pu'apu'a
Mauga Silisili
6096 ft.
1858 m
A'opo
SAMOA
Apia
Taga
CAPE ASUISUI
UPOLU
Falelatai
Ti'avea
Lotofaga
Salani

AMERICAN SAMOA
(U.S.)

Ofu OLOSEGA
OFU TAU
Pago Pago Tau
Leone TUTUILA
Nu'uuli MANUA ISLANDS

SAMOA   ISLANDS

© Rand McNally
A-101773-1

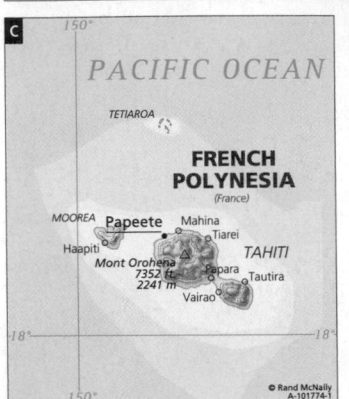

c

PACIFIC OCEAN

TETIAROA

FRENCH POLYNESIA
(France)

MOOREA
Papeete
Mahina
Haapiti
Tiarei
Mont Orohena
7352 ft.
2241 m
Papara
TAHITI
Vairao
Tautira

© Rand McNally
A-101774-1

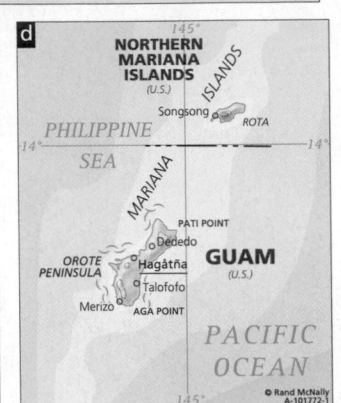

d

NORTHERN MARIANA ISLANDS
(U.S.)

Songsong ROTA

PHILIPPINE SEA

MARIANA ISLANDS

PATI POINT
Dededo
OROTE PENINSULA
Hagåtña
GUAM
(U.S.)
Talofofo
Merizo
AGA POINT

PACIFIC OCEAN

© Rand McNally
A-101772-1

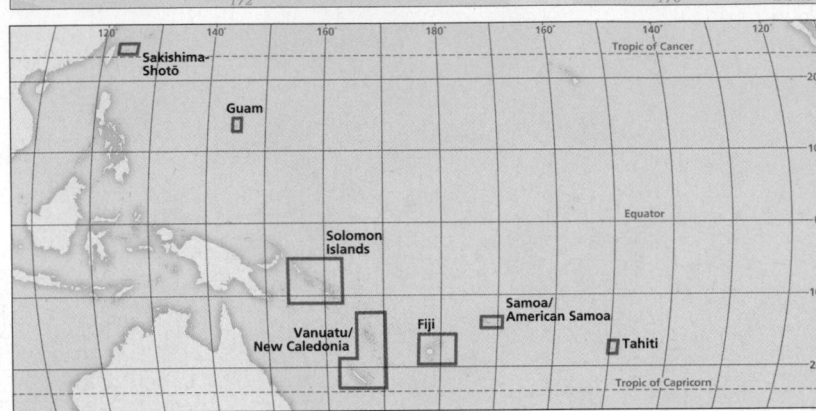

Tropic of Cancer
Sakishima-Shotō
Guam
Solomon Islands
Equator
Vanuatu/ New Caledonia
Fiji
Samoa/ American Samoa
Tahiti
Tropic of Capricorn

e

BUKA ISLAND
Lemankoa
Sohano
PAPUA NEW GUINEA
Dios
Mt. Balbi
8999 ft.
2743 m
Wakunai
Amun
Torokina
Vito
Mt. Takuan
7251 ft.
2210 m
CAPE TOROKINA
Arawa
Empress Augusta Bay
Jaba
Taki
Buin
BOUGAINVILLE
Taro
FAURO I.
Luti
SHORTLAND ISLANDS
MONO ISLAND
SHORTLAND I.

ONTONG JAVA

RONCADOR REEF

PACIFIC OCEAN

BRADLEY REEFS

CHOISEUL
Sasamungga
VAGHENA ISLAND
Kia
SANTA ISABEL
DAI
VELLA LAVELLA
Maravari
KOLOMBANGARA ISLAND
Buala
RANONGGA ISLAND
Gizo
NEW GEORGIA
Susubona
SIMBO ISLAND
VANGUNU ISLAND
Sepi
Dala
Auki
RENDOVA ISLAND
TETEPARE ISLAND
NGGATOKAE I.
SAN JORGE ISLAND
FLORIDA IS.
NGGELA SULE
Bina
MALAITA
Oteotea
MBOROKUA
PAVUVU I.
SAVO I.
NGGELA PILE
Yandina
Visale
Tulaghi
RUSSELL IS.
CAPE ESPERANCE
Tangarare
Honiara
Aola
Mbola
MARAMASIKE
Mt. Makarakomburu
8028 ft.
2447 m
Ayu
ULAWA ISLAND
Inakona
Avu
Kirakira
GUADALCANAL
UKI NI MASI ISLAND
Makira Harbour
SAN CRISTOBAL

SOLOMON ISLANDS

New Georgia Sound

Blanche Channel

Indispensable Strait

SOLOMON SEA

© Rand McNally
A-101766-1

g

PACIFIC OCEAN

HIU
TÈGUA
ÎLES TORRES
URÉPARAPARA
TOGA
MOTA LAVA
VANUA LAVA
BANKS ISLANDS
(ÎLES BANKS)
Sola
SANTA MARIA
Losolava
MÉRÉ LAVA
ESPIRITU SANTO
Nokuku
Big Bay
Malau
Mont Tabwémasana
6165 ft.
1879 m
Longana
VANUATU
Marino
MAÉWO
Wusi
Luganville
AOBA
Nazareth
MALO
PENTECÔTE
Détroit de Bougainville
Lakatoro
AMBRYM
MALAKULA
Lamap
Least
PAAMA
Ringdove
LOPÉVI
ÉPI
Morua
TONGOA
NEW HEBRIDES
(NOUVELLES-HÉBRIDES)
ÉMAÉ
ÎLE NGUNA
Forari
Port Vila
ÉFATÉ

ERROMANGO
Ipota

CORAL SEA

TANNA
Isangel
ANIWA
Walsisi

NEW CALEDONIA
(France)

ÎLE POTT
ÎLES BÉLEP
ÎLE ART
ÎLE BAABA
ÎLE BALABIO
ÎLE YANDÉ
Poum
Pam
Mont Panié
5344 ft.
1629 m
Paagoumène
Hienghène
OUVÉA
Kaala-Gomen
Touho
Mouly
Voh
Ponérihouen
Koné
Houaïlou
WÉ
ÎLE TIGA
Poya
Kouaoua
LIFOU
Bourail
Canala
Rô
MARÉ
La Foa
Thio
NOUVELLE-CALÉDONIE
Boulouparis
Yaté
Xéré
Païta
Le Mont-Dore
Nouméa
ÎLE DES PINS
Vao

LOYALTY ISLANDS
(ÎLES LOYAUTÉ)

Port Patrick
ANATOM
Anelngauhat

© Rand McNally
A-101767-1

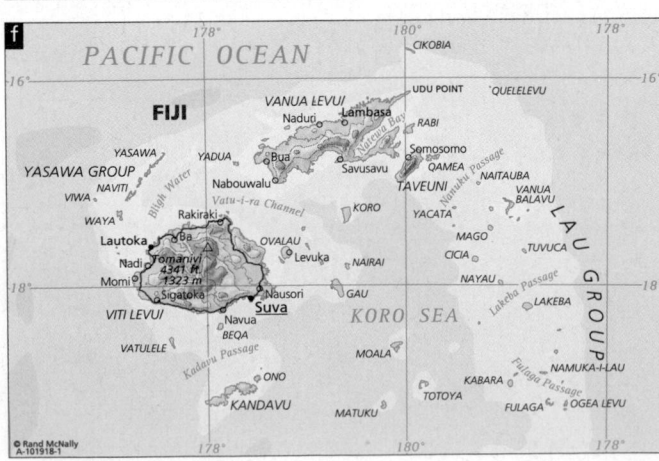

f

PACIFIC OCEAN

CIKOBIA
VANUA LEVU
UDU POINT
QUELELEVU
Naduri
Lambasa
RABI
FIJI
Bua
Somosomo
YASAWA
YADUA
Nabouwalu
Savusavu
QAMEA
NAITAUBA
YASAWA GROUP
Natewa Bay
TAVEUNI
VANUA BALAVU
VIWA
NAVITI
Rakiraki
KORO
YACATA
MAGO
WAYA
Ba
OVALAU
NAIRAI
CICIA
TUVUCA
Lautoka
Levuka
Nadi
Tomanivi
4341 ft.
1323 m
GAU
NAYAU
Momi
NAUSORI
LAKEBA
Sigatoka
VITI LEVU
Navua
BEQA
KORO SEA
VATULELE
Suva
MOALA
NAMUKA-I-LAU
OGEA LEVU
KANDAVU
ONO
TOTOYA
KABARA
FULAGA
MATUKU
LAU GROUP
Nasonini Bay
Vatu-i-ra Channel
Bligh Water
Kadavu Passage
Nanuku Passage
Lakeba Passage
Fulaga Passage

© Rand McNally
A-101918-1

Inset maps a - d
Lambert Azimuthal Equal Area Projection
Scale 1:4,000,000
One inch to 64 miles
One cm to 40 km

Inset maps e - g
Lambert Azimuthal Equal Area Projection
Scale 1:8,000,000
One inch to 128 miles
One cm to 80 km

The Oceans & Polar Regions

Regional Maps: 308–314

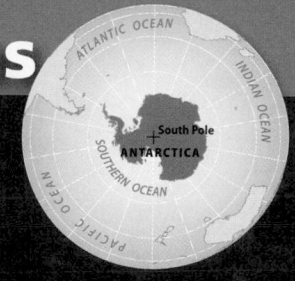

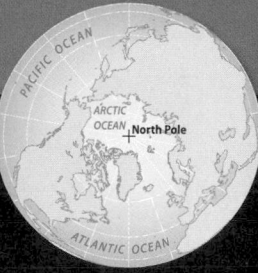

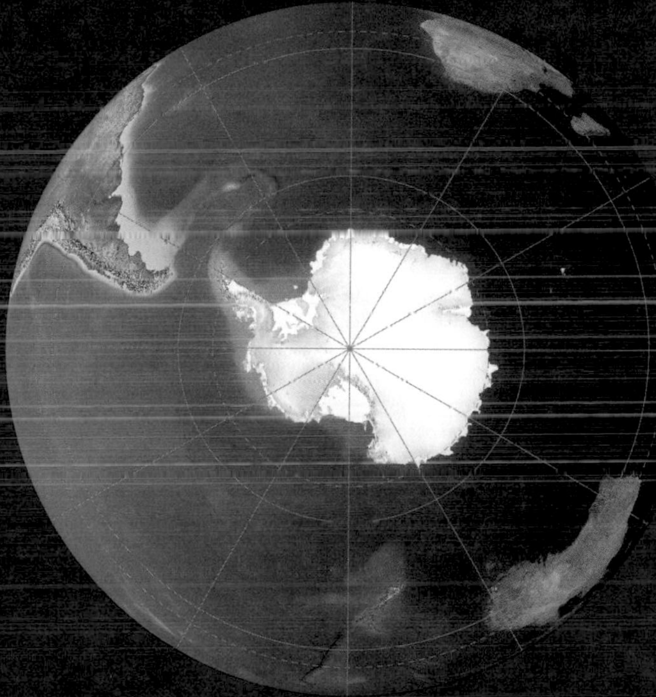

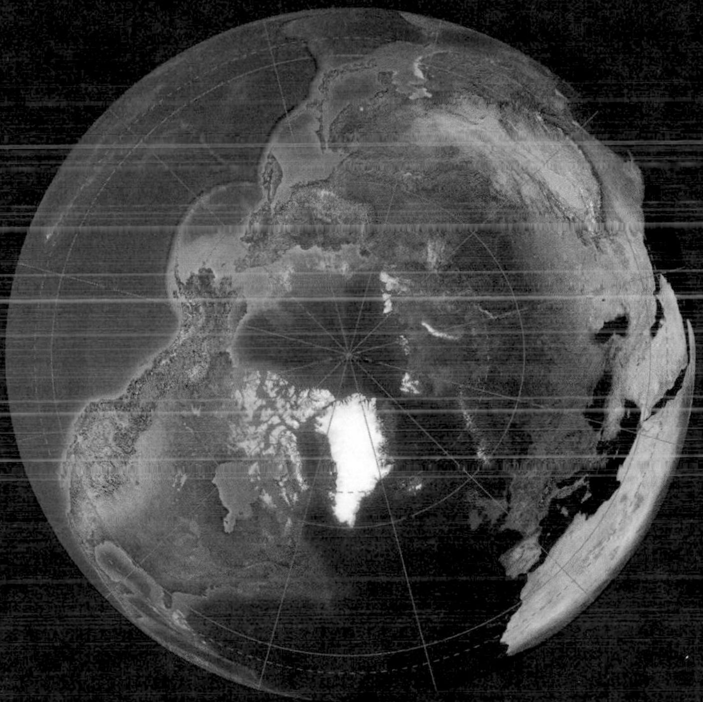

Earth is covered mostly by water. While the continents are huge, they account for only 29 percent of the surface area of the planet. Of Earth's five oceans, the Pacific is the largest. Covering nearly one-third of the planet's surface area, it is larger in area than all of the continents combined. Much of Earth's water is in the form of ice. The polar regions – the Arctic in the north and Antarctica in the south – and covered by sea ice and ice sheets. The Antarctica ice sheet is the largest ice mass on Earth. While the continent receives less than seven inches of precipitation per year on average, snow accumulation over the centuries has resulted in an ice sheet that is more than a mile thick. Due to the great volume of water stored in ice, melting of polar ice due to climate change is of great concern for rising global sea levels.

RUSSIA

CHELYABINSK URAL MTS. YEKATERINBURG OMSK NOVOSIBIRSK

Lake Baikal

SEA OF OKHOTSK

KAMCHATKA PEN.

BERING SEA

ST. LAWRENCE ISLAND

Astana

KOMANDORSKI ISLANDS

Petropavlovsk-Kamchatskiy

ALEUTIAN ISLANDS

KAZAKHSTAN

ALMATY

KYRG.

TASHKENT

TAJIK.

PAK.

ULAANBAATAR

MONGOLIA

ÜRÜMQI

HARBIN

Vladivostok

SHENYANG

SAKHALIN

KURIL ISLANDS

Khabarovsk

SAPPORO HOKKAIDO

NORTH KOREA

BEIJING

SEA OF JAPAN (EAST SEA)

HONSHŪ

JAPAN

HIMALAYAS

CHINA

NANJING

SOUTH KOREA

SEOUL

BUSAN

TŌKYŌ

OSAKA

JAPAN CURRENT

Mt. Everest 29 028 ft. 8848 m

WUHAN

CHONGQING

Hwang

SHANGHAI

SHIKOKU

KYŪSHŪ

EAST CHINA SEA

RYUKYU ISLANDS

BONIN IS.

INTERNATIONAL DATE LINE

NEPAL

KATHMANDU

BHU.

DHAKA

Tropic of Cancer

TAIPEI

INDIA

BNGL.

TAIWAN

KOLKATA

MYANMAR

Ha Noi

HAINAN DAO

SOUTH CHINA SEA

LUZON

PHILIPPINE SEA

MARIANA IS.

MICRONESIA

YANGON

LAOS

Bay of Bengal

ANDAMAN ISLANDS (India)

BANGKOK

THAILAND

CAMBODIA

ANDAMAN SEA

VIETNAM

NORTHERN MARIANA ISLANDS (U.S.)

NORTH EQUATORIAL CURRENT

NICOBAR ISLANDS (India)

Gulf of Thailand

HO CHI MINH CITY

PHILIPPINES

MINDANAO

GUAM (U.S.)

MANILA

MARSHALL ISLANDS

SULU SEA

CAROLINE ISLANDS

MALAYSIA

BRUNEI

KUALA LUMPUR

MALAYSIA

CELEBES SEA

PALAU

FEDERATED STATES OF MICRONESIA

HOWLAND ISLAND (U.S.)

MEDAN

SUMATRA

SINGAPORE

SINGAPORE

BORNEO

Equator

INDONESIA

SULAWESI

MOLUCCAS

BISMARCK SEA

NEW IRELAND

NAURU

BANABA

GILBERT ISLANDS

BAKER ISLAND (U.S.)

KIRIBATI

PHOENIX ISLANDS

POLY

TERA

JAVA SEA

MAKASSAR

BANDA SEA

NEW GUINEA

NEW BRITAIN

MELANESIA

JAKARTA

SURABAYA

JAVA

BALI

FLORES SEA

FLORES

ARAFURA SEA

Lae

PAPUA NEW GUINEA

SOLOMON SEA

SOLOMON ISLANDS

TUVALU

TOKELAU (N.Z.)

CHRISTMAS ISLAND (Austl.)

SAVU SEA

SUMBA

TIMOR SEA

TIMOR-LESTE

TIMOR

Port Moresby

NORTH

COCOS ISLANDS (Austl.)

Darwin

WALLIS AND FUTUNA (Fr.)

SAMOA

AMERICAN SAMOA (U.S.)

Gulf of Carpentaria

CORAL SEA

VANUATU

NEW HEBRIDES

FIJI

TONGA

COCO ISLAN

INDIAN OCEAN

Cairns

NEW CALEDONIA (Fr.)

LOYALTY IS.

NIUE (N.Z.)

Tropic of Capricorn

GREAT DIVIDING RANGE

AUSTRALIA

BRISBANE

EAST AUSTRALIAN CURRENT

NORFOLK ISLAND (Austl.)

WEST AUSTRALIAN CURRENT

Perth

Great Australian Bight

Adelaide

Murray

Sydney

Canberra

NORTH ISLAND

AUCKLAND

ÎLE AMSTERDAM (Fr.)

Melbourne

Mt. Kosciuszko 7313 ft. 2229 m

TASMAN SEA

NEW ZEALAND

ÎLE ST.-PAUL (Fr.)

Bass Strait

TASMANIA

Hobart

SOUTH ISLAND

Wellington

Christchurch

CHATHAM ISLANDS (N.Z.)

Dunedin

STEWART ISLAND

SNARES ISLANDS

BOUNTY IS.

ANTIPODES IS.

ÎLES KERGUÉLEN (Fr.)

AUCKLAND ISLANDS

CAMPBELL ISLAND

HEARD AND MCDONALD ISLANDS (Austl.)

© Rand McNally
A-102151-1

| 0 | 300 | 600 | 900 | 1200 | 1500 Miles |

| 0 | 300 | 600 | 900 | 1200 | 1500 | 1800 | 2100 Kilometers |

Mollweide Projection
Scale 1:55,000,000
One inch to 868 miles
One cm to 550 km

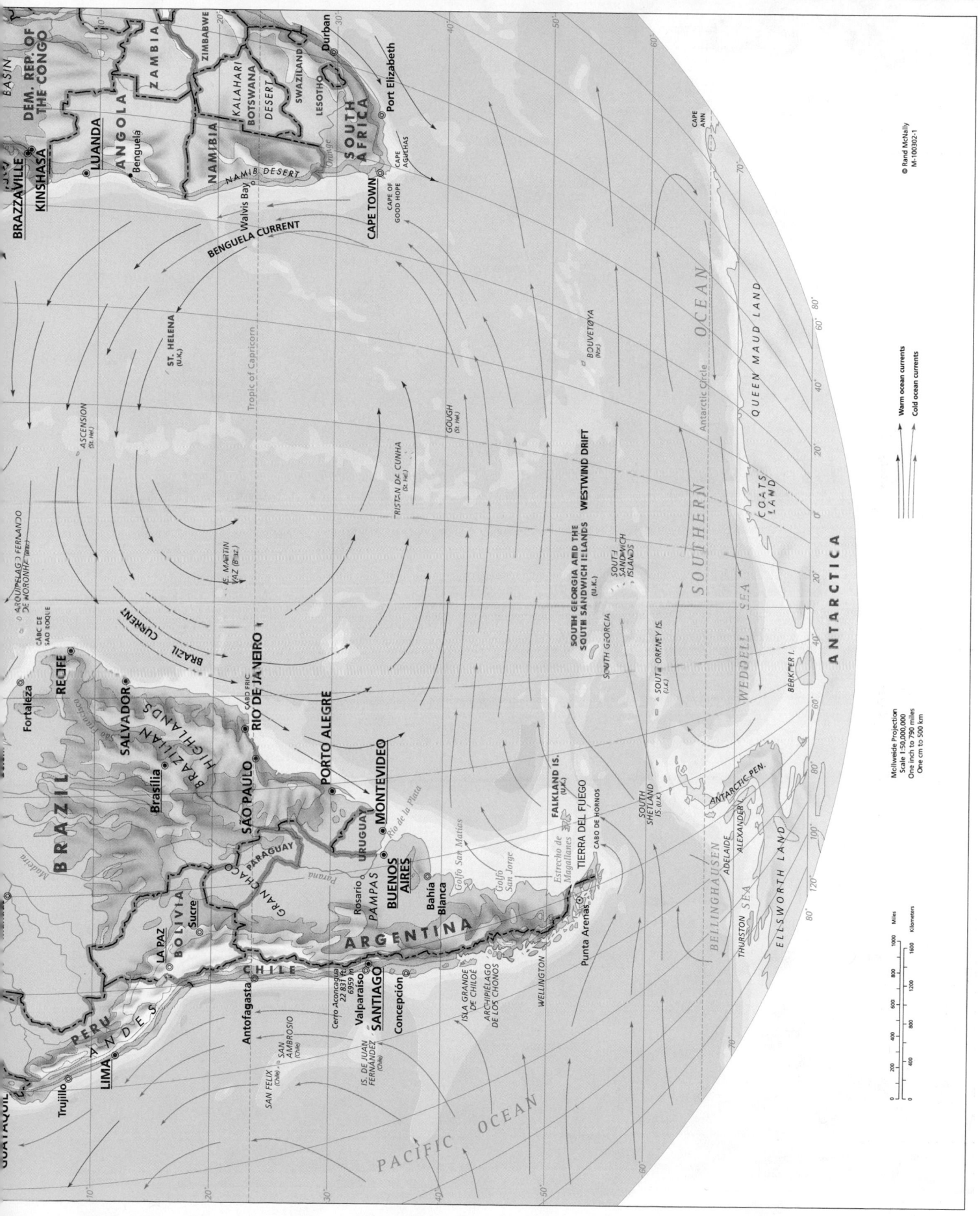

BASIN

DEM. REP. OF
THE CONGO

BRAZZAVILLE
KINSHASA

LUANDA

ANGOLA

Benguela

ZAMBIA

ZIMBABWE

NAMIBIA

KALAHARI
DESERT

BOTSWANA

SWAZILAND

LESOTHO

SOUTH
AFRICA

Durban

Port Elizabeth

CAPE TOWN

CAPE OF
GOOD HOPE

CAPE
AGULHAS

NAMIB DESERT

Walvis Bay

Orange

BENGUELA CURRENT

ST. HELENA
(U.K.)

Tropic of Capricorn

ASCENSION
(St. Hel.)

I.E. MARTIN
VAZ (Braz.)

ARQUIPÉLAGO FERNANDO
DE NORONHA (Braz.)

CABO DE
SÃO ROQUE

RECIFE

Fortaleza

SALVADOR

BRAZILIAN
HIGHLANDS

Brasília

BRAZIL

São Francisco

BOLIVIA

Sucre

LA PAZ

PERU

LIMA

Trujillo

GUAYAQUIL

PACIFIC OCEAN

SAN FELIX
(Chile)

SAN
AMBROSIO
(Chile)

IS. DE JUAN
FERNANDEZ
(Chile)

Antofagasta

Cerro Aconcagua
22,831 ft
6959 m

Valparaíso
SANTIAGO

Concepción

CHILE

ANDES

ARCHIPIÉLAGO
DE LOS CHONOS

ISLA GRANDE
DE CHILOÉ

WELLINGTON

PARAGUAY

GRAN CHACO

Paraná

Rosario

PAMPAS

BUENOS
AIRES

URUGUAY

MONTEVIDEO

Rio de la Plata

SÃO PAULO

PORTO ALEGRE

RIO DE JANEIRO

CABO FRIO

BRAZIL CURRENT

TRISTAN DA CUNHA
(St. Hel.)

GOUGH
(St. Hel.)

Madeira

Bahía
Blanca

ARGENTINA

Golfo San Matías

Golfo
San Jorge

Estrecho de
Magallanes

CABO DE HORNOS

TIERRA DEL FUEGO

Punta Arenas

FALKLAND IS.
(U.K.)

SOUTH GEORGIA AND THE
SOUTH SANDWICH ISLANDS
(U.K.)

SOUTH GEORGIA

SOUTH
SANDWICH
ISLANDS

SOUTH ORKNEY IS.
(U.K.)

WESTWIND DRIFT

BOUVETØYA
(Nor.)

SOUTHERN

OCEAN

CAPE
ANN

QUEEN MAUD LAND

COATS
LAND

WEDDELL
SEA

BERKNER I.

ANTARCTICA

Antarctic Circle

SOUTH
SHETLAND
IS. (U.K.)

ANTARCTIC PEN.

ADELAIDE

ALEXANDER I.

ELLSWORTH LAND

THURSTON I.

BELLINGHAUSEN
SEA

© Rand McNally
M-100302-1

Mollweide Projection
Scale 1:50,000,000
One inch to 790 miles
One cm to 500 km

→ Warm ocean currents
→ Cold ocean currents

Miles

0   200   400   600   800   1000
0   400   800   1200   1600
Kilometers

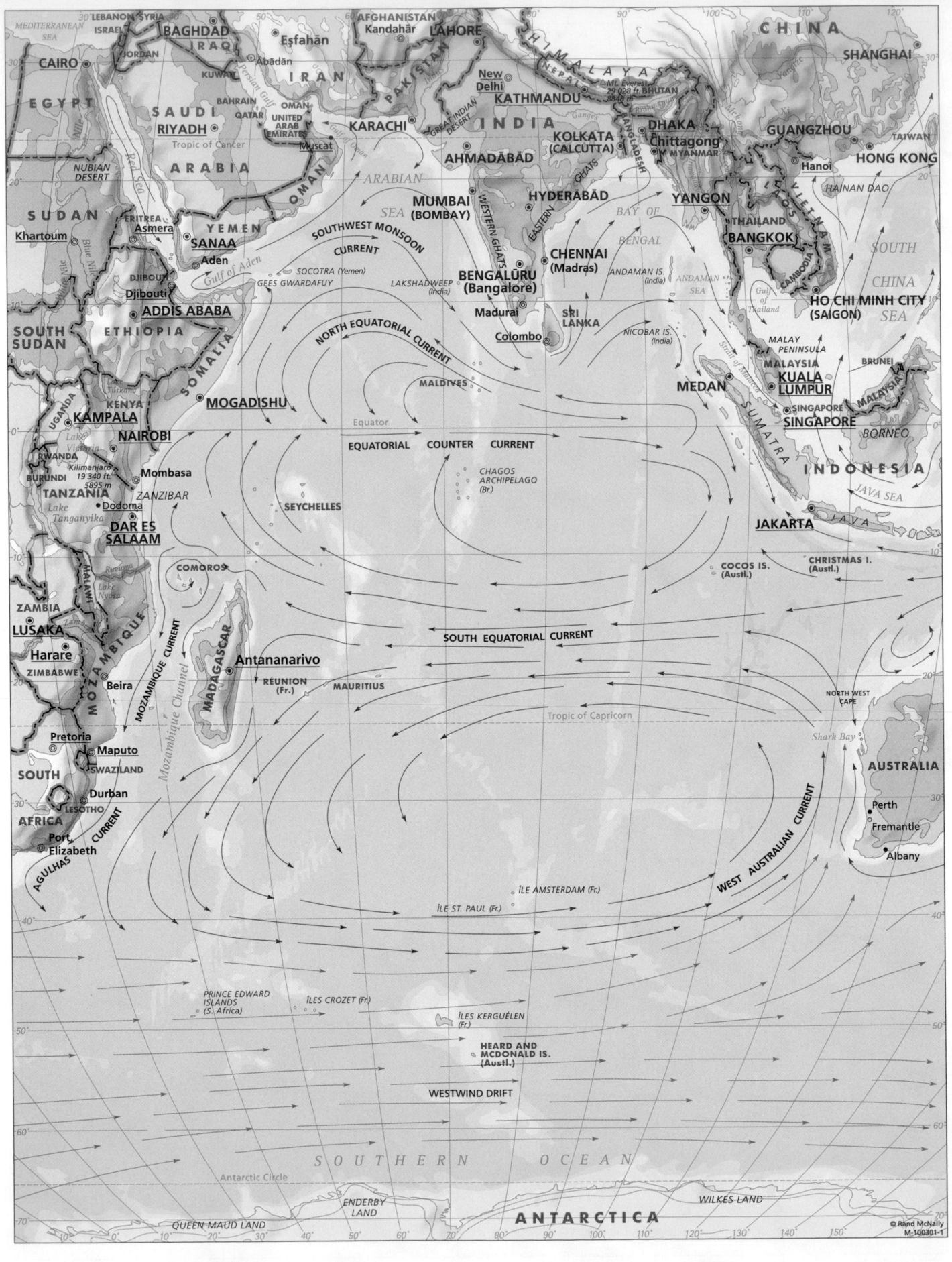

MEDITERRANEAN SEA
LEBANON SYRIA
ISRAEL JORDAN
BAGHDAD
IRAQ
CAIRO
KUWAIT
Esfahān
Kandahār
LAHORE
AFGHANISTAN
New Delhi
KATHMANDU
HIMALAYAS
NEPAL
Mt. Everest 29,028 ft. BHUTAN 8848 m
CHINA
SHANGHAI
EGYPT
Abādān
IRAN
BAHRAIN OMAN
QATAR UNITED ARAB EMIRATES
SAUDI ARABIA
RIYADH
Tropic of Cancer
Muscat
Gulf of Oman
KARACHI
PAKISTAN
INDIA
Great Indian Desert
Ganges
KOLKATA (CALCUTTA)
DHAKA
Chittagong
M. MYANMAR
GUANGZHOU
HONG KONG
TAIWAN
NUBIAN DESERT
SUDAN
ARABIA
AHMADĀBĀD
HYDERĀBĀD
WESTERN GHATS
EASTERN GHATS
YANGON
Hanoi
HAINAN DAO
Khartoum
ERITREA
Asmera
YEMEN
SANAA
Aden
Gulf of Aden
ARABIAN SEA
SOUTHWEST MONSOON CURRENT
MUMBAI (BOMBAY)
SOCOTRA (Yemen)
GEES GWARDAFUY
LAKSHADWEEP (India)
BENGALŪRU (Bangalore)
CHENNAI (Madras)
BAY OF BENGAL
ANDAMAN IS. (India)
ANDAMAN SEA
Gulf of Thailand
THAILAND
BANGKOK
CAMBODIA
SOUTH CHINA SEA
VIETNAM
LAOS
DJIBOUTI
Djibouti
ADDIS ABABA
ETHIOPIA
SOUTH SUDAN
SOMALIA
NORTH EQUATORIAL CURRENT
Madurai
SRI LANKA
Colombo
NICOBAR IS. (India)
HO CHI MINH CITY (SAIGON)
MALAY PENINSULA
MEDAN
STRAIT OF MALACCA
MALAYSIA
KUALA LUMPUR
BRUNEI
MALDIVES
Equator
EQUATORIAL COUNTER CURRENT
CHAGOS ARCHIPELAGO (Br.)
SINGAPORE
SINGAPORE
BORNEO
UGANDA
KAMPALA
KENYA
NAIROBI
Lake Victoria
RWANDA
BURUNDI
Kilimanjaro 19,340 ft. 5895 m
Mombasa
ZANZIBAR
SEYCHELLES
INDONESIA
JAVA SEA
MOGADISHU
TANZANIA
Dodoma
Lake Tanganyika
DAR ES SALAAM
COMOROS
JAKARTA
JAVA
COCOS IS. (Austl.)
CHRISTMAS I. (Austl.)
MALAWI
ZAMBIA
LUSAKA
Lake Nyasa
MOZAMBIQUE CURRENT
MADAGASCAR
Antananarivo
MOZAMBIQUE CHANNEL
SOUTH EQUATORIAL CURRENT
HARARE
Harare
ZIMBABWE
MOZAMBIQUE
Beira
RÉUNION (Fr.)
MAURITIUS
Tropic of Capricorn
NORTH WEST CAPE
Shark Bay
Pretoria
Maputo
SWAZILAND
SOUTH AFRICA
LESOTHO
Durban
AGULHAS CURRENT
Port Elizabeth
AUSTRALIA
Perth
Fremantle
Albany
WEST AUSTRALIAN CURRENT
ÎLE AMSTERDAM (Fr.)
ÎLE ST. PAUL (Fr.)
PRINCE EDWARD ISLANDS (S. Africa)
ÎLES CROZET (Fr.)
ÎLES KERGUÉLEN (Fr.)
HEARD AND McDONALD IS. (Austl.)
WESTWIND DRIFT
SOUTHERN OCEAN
Antarctic Circle
QUEEN MAUD LAND
ENDERBY LAND
WILKES LAND
ANTARCTICA
© Rand McNally
M-300301-1

0 200 400 600 800 1000 Miles
0 400 800 1200 1600 Kilometers

Mollweide Projection
Scale 1:50,000,000
One inch to 790 miles
One cm to 500 km

→ Warm ocean currents
→ Cold ocean currents

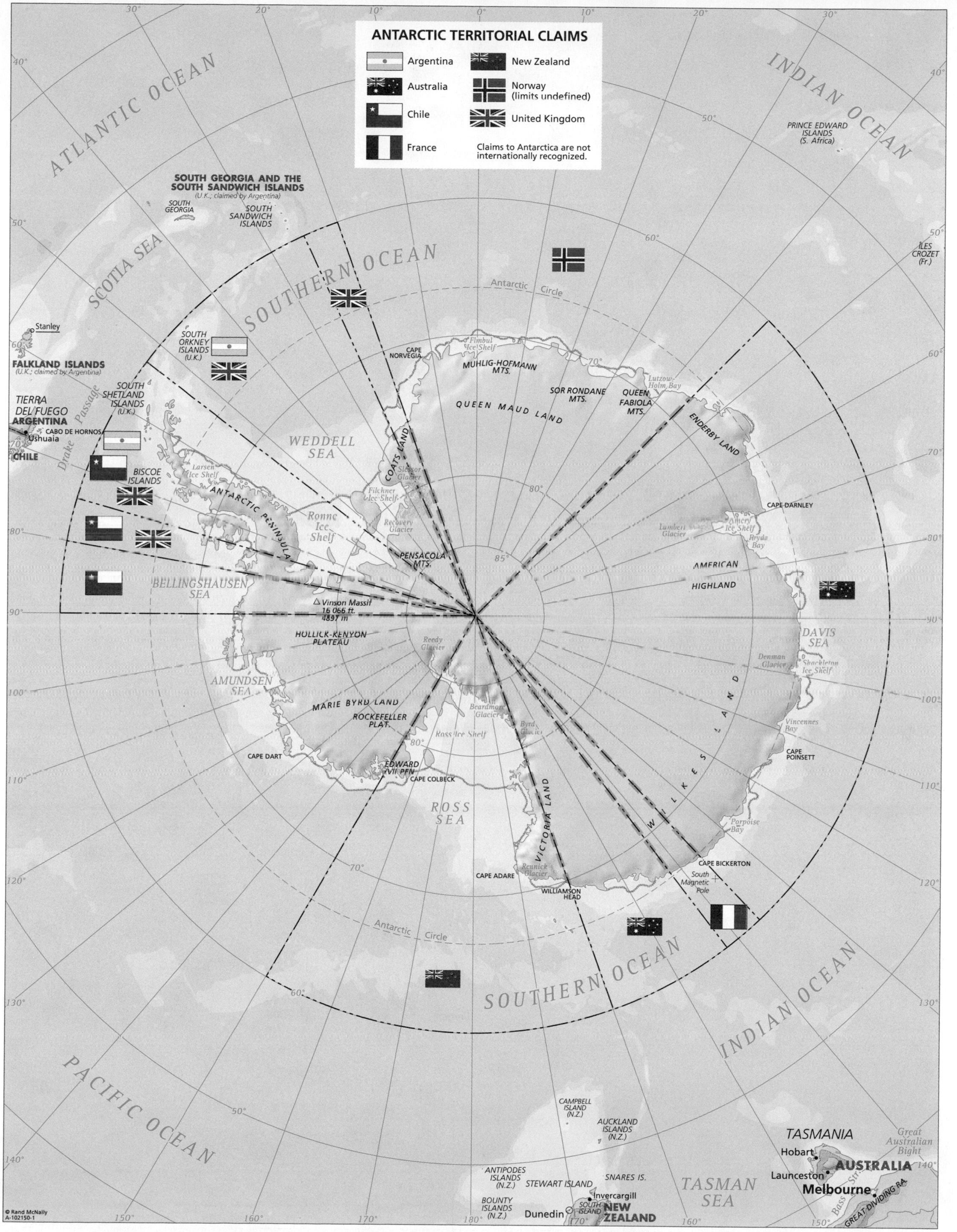

ATLANTIC OCEAN

INDIAN OCEAN

ANTARCTIC TERRITORIAL CLAIMS

Argentina   New Zealand
Australia   Norway (limits undefined)
Chile   United Kingdom
France

Claims to Antarctica are not internationally recognized.

PRINCE EDWARD ISLANDS (S. Africa)

SOUTH GEORGIA AND THE SOUTH SANDWICH ISLANDS
(U.K.; claimed by Argentina)

SOUTH GEORGIA

SOUTH SANDWICH ISLANDS

ÎLES CROZET (Fr.)

SCOTIA SEA

SOUTHERN OCEAN

Antarctic Circle

Stanley

SOUTH ORKNEY ISLANDS (U.K.)

FALKLAND ISLANDS
(U.K.; claimed by Argentina)

CAPE NORVEGIA

Fimbul Ice Shelf

MÜHLIG-HOFMANN MTS.

Lützow-Holm Bay

QUEEN MAUD LAND

SØR RONDANE MTS.

QUEEN FABIOLA MTS.

TIERRA DEL FUEGO

ARGENTINA

SOUTH SHETLAND ISLANDS (U.K.)

CABO DE HORNOS

Ushuaia

CHILE

Drake Passage

ENDERBY LAND

WEDDELL SEA

COATS LAND

Larsen Ice Shelf

BISCOE ISLANDS

ANTARCTIC PENINSULA

Slessor Glacier

Filchner Ice Shelf

Ronne Ice Shelf

Recovery Glacier

CAPE DARNLEY

Amery Ice Shelf

Lambert Glacier

Beaver Bay

PENSACOLA MTS.

AMERICAN HIGHLAND

BELLINGSHAUSEN SEA

△ Vinson Massif 16 066 ft. 4897 m

DAVIS SEA

HOLLICK-KENYON PLATEAU

Reedy Glacier

Denman Glacier

Shackleton Ice Shelf

AMUNDSEN SEA

MARIE BYRD LAND

ROCKEFELLER PLAT.

Beardmore Glacier

Byrd Glacier

Vincennes Bay

CAPE POINSETT

CAPE DART

Ross Ice Shelf

WILKES LAND

Porpoise Bay

ROSS SEA

EDWARD VII PEN.

CAPE COLBECK

VICTORIA LAND

CAPE BICKERTON

South Magnetic Pole

CAPE ADARE

Rennick Glacier

WILLIAMSON HEAD

Antarctic Circle

SOUTHERN OCEAN

INDIAN OCEAN

CAMPBELL ISLAND (N.Z.)

AUCKLAND ISLANDS (N.Z.)

TASMANIA

Great Australian Bight

Hobart

ANTIPODES ISLANDS (N.Z.)

STEWART ISLAND

SNARES IS.

TASMAN SEA

Launceston

AUSTRALIA

Melbourne

BOUNTY ISLANDS (N.Z.)

Invercargill

SOUTH ISLAND

NEW ZEALAND

Dunedin

GREAT DIVIDING RA.

Bass Str.

PACIFIC OCEAN

© Rand McNally
A-102150-1

0 200 400 600 800 1000 1200 Miles

0 200 400 600 800 1000 1200 1400 1600 1800 2000 Kilometers

Lambert Azimuthal Equal Area Projection
Scale 1:40,000,000
One inch to 640 miles
One cm to 400 km

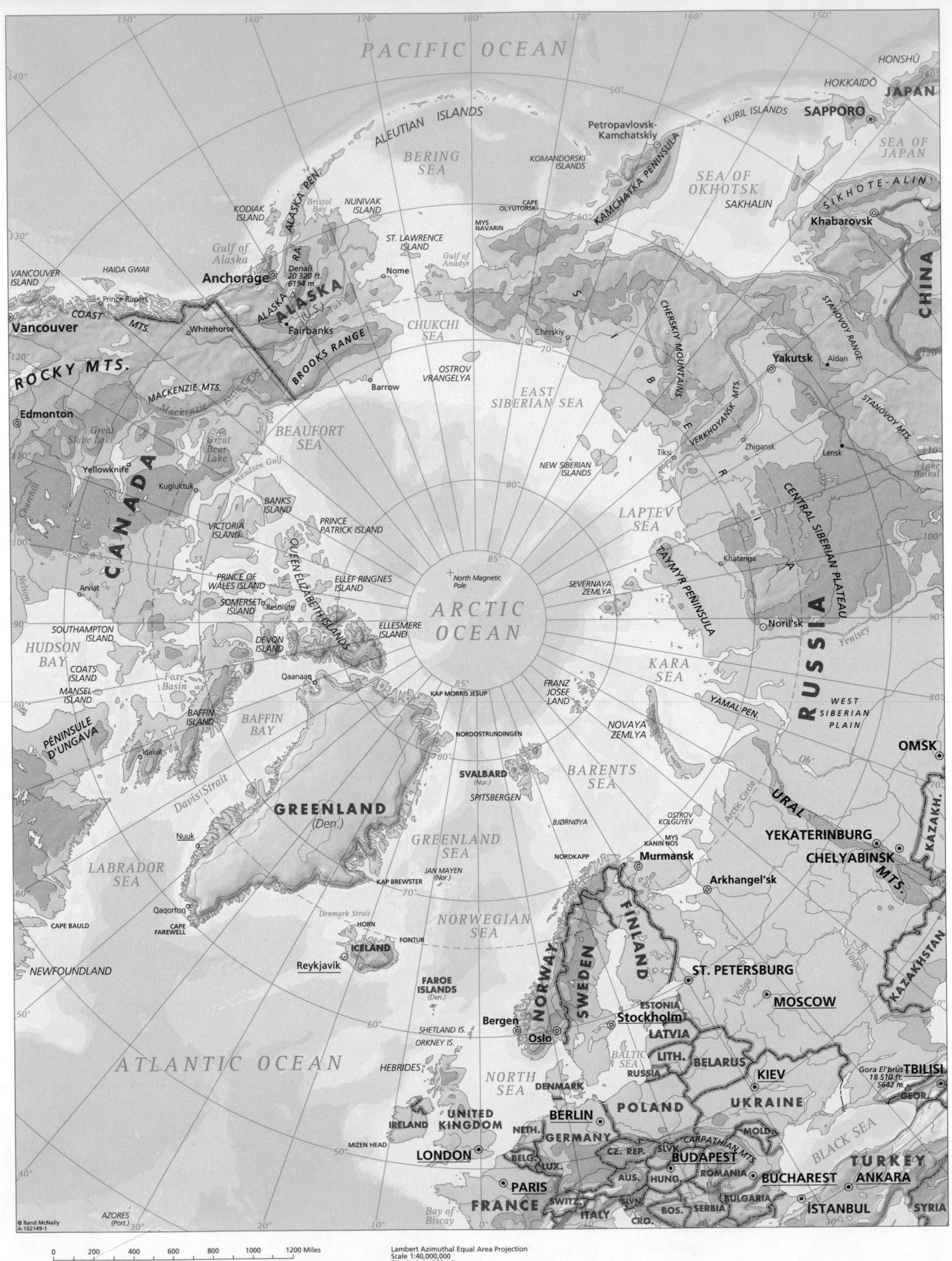

PACIFIC OCEAN

ALEUTIAN ISLANDS

BERING SEA

KODIAK ISLAND
Gulf of Alaska
NUNIVAK ISLAND
ST. LAWRENCE ISLAND
Gulf of Anadyr
CAPE OLYUTORSKI
MYS NAVARIN
KOMANDORSKI ISLANDS
Petropavlovsk-Kamchatskiy
KAMCHATKA PENINSULA
KURIL ISLANDS
SAPPORO
HOKKAIDŌ
HONSHŪ
JAPAN
SEA OF JAPAN
SIKHOTE-ALIN'
Khabarovsk
SEA OF OKHOTSK
SAKHALIN

VANCOUVER ISLAND
HAIDA GWAII
Prince Rupert
COAST MTS.
Anchorage
ALASKA PEN.
Denali 20,320 ft. 6194 m
ALASKA (U.S.)
Whitehorse
Fairbanks
Nome
CHUKCHI SEA
Cherskiy
CHERSKIY MOUNTAINS
STANOVOY RANGE
CHINA
Yakutsk
Aldan

Vancouver
ROCKY MTS.
Edmonton
MACKENZIE MTS.
BROOKS RANGE
Barrow
OSTROV VRANGELYA
EAST SIBERIAN SEA
VERKHOYANSK MTS.
Tiksi
Zhigansk
Lensk
STANOVOY MTS.
Lake Baikal

Yellowknife
Great Slave Lake
Great Bear Lake
Amundsen Gulf
Mackenzie
Arctic Circle
BEAUFORT SEA
NEW SIBERIAN ISLANDS
LAPTEV SEA
Khatanga
CENTRAL SIBERIAN PLATEAU

CANADA
Kugluktuk
BANKS ISLAND
VICTORIA ISLAND
PRINCE PATRICK ISLAND
PRINCE OF WALES ISLAND
QUEEN ELIZABETH ISLANDS
ELLEF RINGNES ISLAND
North Magnetic Pole
ARCTIC OCEAN
SEVERNAYA ZEMLYA
TAYMYR PENINSULA
Noril'sk
RUSSIA
Yenisey

Arviat
SOMERSET ISLAND
Resolute
DEVON ISLAND
ELLESMERE ISLAND
KAP MORRIS JESUP
FRANZ JOSEF LAND
KARA SEA
YAMAL PEN.
Ob
WEST SIBERIAN PLAIN
OMSK

SOUTHAMPTON ISLAND
HUDSON BAY
COATS ISLAND
MANSEL ISLAND
Foxe Basin
Qaanaaq
BAFFIN ISLAND
NORDOSTRUNDINGEN
NOVAYA ZEMLYA
URAL MTS.
YEKATERINBURG
CHELYABINSK
KAZAKH.
KAZAKHSTAN

PÉNINSULE D'UNGAVA
Iqaluit
BAFFIN BAY
SVALBARD (Nor.)
SPITSBERGEN
BARENTS SEA
OSTROV KOLGUYEV
MYS KANIN NOS
Murmansk
Arkhangel'sk
Volga

Churchill
GREENLAND (Den.)
Nuuk
Davis Strait
BJØRNØYA
NORDKAPP
FINLAND
ST. PETERSBURG

LABRADOR SEA
GREENLAND SEA
JAN MAYEN (Nor.)
KAP BREWSTER
NORWAY
SWEDEN
MOSCOW

Qaqortoq
CAPE BAULD
CAPE FAREWELL
Denmark Strait
HORN
ICELAND
FONTUR
NORWEGIAN SEA
Bergen
Oslo
Stockholm
ESTONIA
LATVIA
LITH.
BELARUS
KIEV
Gora El'brus 18,510 ft. 5642 m
TBILISI
GEOR.

NEWFOUNDLAND
Reykjavík
FAROE ISLANDS (Den.)
SHETLAND IS.
ORKNEY IS.
NORTH SEA
BALTIC SEA
RUSSIA
UKRAINE
BLACK SEA
TURKEY
ANKARA

ATLANTIC OCEAN
HEBRIDES
IRELAND
MIZEN HEAD
UNITED KINGDOM
NETH.
DENMARK
BELG.
LUX.
BERLIN
GERMANY
POLAND
CZ. REP.
SLVK.
CARPATHIAN MTS.
ROMANIA
BUDAPEST
BUCHAREST
ISTANBUL
SYRIA

LONDON
PARIS
FRANCE
Bay of Biscay
SWITZ.
AUS.
HUNG.
SLVN.
CRO.
BOS.
SERBIA
ITALY
BULGARIA
BLACK SEA

AZORES (Port.)

© Rand McNally
A-102149-1

0   200   400   600   800   1000   1200 Miles
0   200   400   600   800   1000   1200   1400   1600   1800   2000 Kilometers

Lambert Azimuthal Equal Area Projection
Scale 1:40,000,000
One inch to 640 miles
One cm to 400 km

This table gives the area, population, population density, political status, capital, and predominant languages for every country in the world. The political units listed are categorized by political status in the 'Form of Government and Ruling Power' column of the table, as follows:

A independent countries;

B internally independent political entities which are under the protection of another country in matters of defense and foreign affairs;

C colonies and other dependent political units;

D the major administrative subdivisions of Australia, Canada, China, the United Kingdom, and the United States.

For comparison, the table also includes the continents and the world. All footnotes appear at the end of the table.

The populations are estimates for July 1, 2015, or the latest available data, from United States Census Bureau estimates, official data, and other available information. Area figures include inland water.

Region or political division	Est. Pop. 1/1/09	Area sq. km.	Area sq. mi.	Pop. per sq. km.	Pop. per sq. mi.	Form of Government and Ruling Power	Political Status	Capital	Predominant Languages
† Afghanistan	32,564,000	652,090	251,773	50	129	Transitional	A	Kabul (Kabol)	Dari, Pashto, Uzbek, Turkmen
Africa	1,156,434,000	30,300,000	11,700,000	38	99				
Alabama	4,849,000	135,765	52,419	36	93	State (U.S.)	D	Montgomery	English
Alaska	737,000	1,717,854	663,267	0.4	1.1	State (U.S.)	D	Juneau	English, indigenous
† Albania	3,029,000	28,748	11,100	105	273	Republic	A	Tiranë	Albanian, Greek
Alberta	4,196,000	661,848	255,541	6.3	16	Province (Canada)	D	Edmonton	English
† Algeria	39,542,000	2,381,741	919,595	17	43	Republic	A	Algiers (Alger)	Arabic, Berber dialects, French
American Samoa	54,000	199	77	273	706	Unincorporated territory (U.S.)	C	Pago Pago	Samoan, English
† Andorra	86,000	468	181	183	473	Parliamentary co-principality (Spanish and French)	B	Andorra la Vella	Catalan, Spanish (Castilian), French
† Angola	19,625,000	1,246,700	481,354	16	41	Republic	A	Luanda	Portuguese, indigenous
Anguilla	16,000	96	37	171	444	Overseas territory (U.K. protection)	B	The Valley	English
Anhui	60,298,000	139,000	53,668	434	1,124	Province (China)	D	Hefei	Chinese (Mandarin)
Antarctica	(1)	14,000,000	5,400,000	......	......				
† Antigua and Barbuda	92,000	442	171	209	541	Parliamentary state	A	St. John's	English, local dialects
Aomen see Macau									
† Argentina	43,432,000	2,780,400	1,073,519	16	40	Republic	A	Buenos Aires	Spanish, English, Italian, German, French
Arizona	6,731,000	295,254	113,998	23	59	State (U.S.)	D	Phoenix	English
Arkansas	2,966,000	137,732	53,179	22	56	State (U.S.)	D	Little Rock	English
† Armenia	3,056,000	29,800	11,506	103	266	Republic	A	Yerevan	Armenian, Russian
Aruba	112,000	193	75	581	1,495	Self-governing territory (Netherlands protection)	B	Oranjestad	Dutch, Papiamento, English, Spanish
Ascension	900	88	34	10	26	Dependency (St. Helena)	C	Georgetown	English
Asia	4,343,451,000	44,900,000	17,300,000	97	251				
† Australia	22,751,000	7,692,030	2,969,910	3.0	7.7	Federal parliamentary state	A	Canberra	English, indigenous
Australian Capital Territory	390,000	2,360	911	165	428	Territory (Australia)	D	Canberra	English
† Austria	8,666,000	83,858	32,378	103	268	Federal republic	A	Vienna (Wien)	German
† Azerbaijan	9,781,000	86,600	33,437	113	293	Republic	A	Baku (Bakı)	Azeri, Russian, Armenian
	Bahamas	325,000	13,939	5,382	23	60	Parliamentary State	A	Nassau	English, Creole
† Bahrain	1,347,000	691	267	1,949	5,043	Monarchy	A	Manama (Al-Manāmah)	Arabic, English, Farsi, Urdu
† Bangladesh	168,958,000	147,092	56,598	1,177	3,029	Republic	A	Dhaka	Bangla, English
† Barbados	291,000	430	166	676	1,751	Parliamentary state	A	Bridgetown	English
Beijing	21,140,000	16,000	6,487	1,259	3,260	Autonomous city (China)	D	Beijing	Chinese (Mandarin)
† Belarus	9,590,000	207,600	80,155	46	120	Republic	A	Minsk	Belarussian, Russian
† Belgium	11,324,000	30,528	11,787	371	961	Constitutional monarchy	A	Brussels (Bruxelles)	Dutch (Flemish), French, German
† Belize	347,000	22,966	8,867	15	39	Parliamentary state	A	Belmopan	English, Spanish, Mayan, Garifuna
† Benin	10,449,000	112,622	43,484	93	240	Republic	A	Porto-Novo and Cotonou	French, Fon, Yoruba, indigenous
Bermuda	70,000	54	21	1,300	3,343	Overseas territory (U.K.)	C	Hamilton	English
† Bhutan	742,000	46,500	17,954	16	41	Monarchy (Indian protection)	B	Thimphu	Dzongkha, Tibetan and Nepalese dialects
† Bolivia	10,801,000	1,098,581	424,165	10	25	Republic	A	La Paz and Sucre	Aymara, Quechua, Spanish
† Bosnia and Herzegovina	3,867,000	51,197	19,767	76	196	Republic	A	Sarajevo	Bosnian, Croatian, Serbian
† Botswana	2,183,000	581,730	224,607	3.8	10	Republic	A	Gaborone	English, Tswana
† Brazil	204,260,000	8,547,404	3,300,172	24	62	Federal republic	A	Brasília	Portuguese, Spanish, English, French
British Columbia	4,683,000	944,735	364,764	5.0	13	Province (Canada)	D	Victoria	English
British Indian Ocean Territory	(1)	60	23	......	......	Overseas territory (U.K.)	C		English
British Virgin Islands	33,000	151	58	222	577	Overseas territory (U.K.)	C	Road Town	English
† Brunei	430,000	5,765	2,226	75	193	Monarchy	A	Bandar Seri Begawan	Malay, English, Chinese
† Bulgaria	7,187,000	110,994	42,855	65	168	Republic	A	Sofia (Sofiya)	Bulgarian, Turkish
† Burkina Faso	18,932,000	274,200	105,869	69	179	Republic	A	Ouagadougou	French, indigenous
Burma see Myanmar									
† Burundi	10,742,000	27,830	10,745	386	1,000	Republic	A	Bujumbura	French, Kirundi, Swahili
† Cabo Verde	546,000	4,033	1,557	135	351	Republic	A	Praia	Portuguese, Crioulo
California	38,803,000	423,970	163,696	92	237	State (U.S.)	D	Sacramento	English
† Cambodia	15,709,000	181,035	69,898	87	225	Constitutional monarchy	A	Phnom Penh (Phnum Pénh)	Khmer, French
† Cameroon	23,739,000	475,440	183,568	50	129	Republic	A	Yaoundé	English, French, indigenous
† Canada	35,100,000	9,984,670	3,855,103	3.5	9.1	Federal parliamentary state	A	Ottawa	English, French
Cayman Islands	56,000	264	102	212	550	Overseas territory (U.K.)	C	George Town	English
† Central African Republic	5,392,000	622,984	240,536	8.7	22	Republic	A	Bangui	French, Sango, Arabic, indigenous
† Chad	11,631,000	1,284,000	495,755	9.1	9.1	Republic	A	N'Djamena	Arabic, French, indigenous
Channel Islands	163,000	194	75	842	2,178	Two crown dependencies (U.K. protection)			English, French
† Chile	17,508,000	756,096	291,930	23	60	Republic	A	Santiago	Spanish
† China (incl. Hong Kong and Macau)	1,375,219,000	9,557,172	3,690,045	144	373	Socialist republic	A	Beijing	Chinese dialects
Chongqing	29,700,000	82,400	31,815	360	934	Autonomous city (China)	D	Chongqing	Chinese (Mandarin)
Christmas Island	1,500	135	52	11	29	External territory (Australia)	C	Settlement	English, Chinese, Malay
Cocos (Keeling) Islands	600	14	5.4	43	110	External territory (Australia)	C	West Island	English, Cocos-Malay, Malay
† Colombia	46,737,000	1,138,914	439,737	41	106	Republic	A	Bogotá	Spanish
Colorado	5,356,000	269,601	104,094	20	51	State (U.S.)	D	Denver	English
† Comoros (excl. Mayotte)	781,000	2,235	863	349	905	Federal Islamic republic	A	Moroni	Arabic, French, Comoran
† Congo	4,755,000	342,000	132,047	14	36	Republic	A	Brazzaville	French, Lingala, Kikongo, indigenous
† Congo, Democratic Republic of the	79,375,000	2,345,095	905,446	34	88	Republic	A	Kinshasa	French, Kikongo, Lingala, Swahili, Tshiluba, Kingwana
Connecticut	3,597,000	14,357	5,543	251	649	State (U.S.)	D	Hartford	English
Cook Islands	10,000	236	91	42	108	Self-governing territory (New Zealand protection)	B	Avarua	English, Maori
† Costa Rica	4,814,000	51,100	19,730	94	244	Republic	A	San José	Spanish

Region or political division	Est. Pop. 1/1/09	Area sq. km.	Area sq. mi.	Pop. per sq. km.	Pop. per sq. mi.	Form of Government and Ruling Power	Political Status	Capital	Predominant Languages
† Côte d'Ivoire (Ivory Coast)	23,295,000	322,463	124,504	72	187	Republic	A	Abidjan and Yamoussoukro	French, Dioula and other indigenous
† Croatia	4,465,000	56,538	21,829	79	205	Republic	A	Zagreb	Croatian
† Cuba	11,031,000	110,861	42,804	100	258	Socialist republic	A	Havana (La Habana)	Spanish
Curacao	148,000	444	171	334	868	Self-governing territory (Netherlands protection)	B	Willemstad	Dutch, Papiamento, English
† Cyprus	1,189,000	9,251	3,572	129	129	Republic	A	Nicosia	Greek, Turkish, English
† Czech Republic	10,645,000	78,866	30,450	135	350	Republic	A	Prague (Praha)	Czech, Slovak
Delaware	936,000	6,447	2,489	145	376	State (U.S.)	D	Dover	English
† Denmark	5,582,000	43,096	16,640	130	335	Constitutional monarchy	A	Copenhagen (København)	Danish
District of Columbia	659,000	177	68	3,723	9,690	Federal district (U.S.)	D	Washington	English
† Djibouti	828,000	23,200	8,958	36	92	Republic	A	Djibouti	French, Arabic, Somali, Afar
† Dominica	74,000	751	290	98	254	Republic	A	Roseau	English, French
† Dominican Republic	10,479,000	48,511	18,730	216	559	Republic	A	Santo Domingo	Spanish
East Timor see Timor-Leste									
† Ecuador	15,868,000	283,561	109,484	56	145	Republic	A	Quito	Spanish, Quechua, indigenous
† Egypt	88,487,000	1,001,449	386,662	88	229	Socialist republic	A	Cairo (El Qâhira)	Arabic
† El Salvador	6,141,000	21,041	8,124	292	756	Republic	A	San Salvador	Spanish, Nahua
England	54,317,000	130,422	50,356	416	1,079	Administrative division (U.K.)	D	London	English
† Equatorial Guinea	741,000	28,051	10,831	26	68	Republic	A	Malabo	Spanish, indigenous, English
† Eritrea	6,528,000	117,600	45,406	56	144	Republic	A	Asmera	Tigre, Kunama, Cushitic dialects, Nora Bana, Arabic
† Estonia	1,265,000	45,227	17,462	28	72	Republic	A	Tallinn	Estonian, Latvian, Lithuanian, Russian
† Ethiopia	99,466,000	1,104,300	426,373	90	233	Federal republic	A	Addis Ababa (Ādīs Ābeba)	Amharic, Tigrinya, Orominga, Guaraginga, Somali, Arabic
Europe	745,616,000	9,900,000	3,800,000	75	196				
Falkland Islands (2)	3,400	12,173	4,700	0.3	0.7	Overseas territory (U.K.)	C	Stanley	English
Faroe Islands	50,000	1,399	540	36	93	Self-governing territory (Danish protection)	B	Tórshavn	Danish, Faroese
† Fiji	909,000	18,274	7,056	50	129	Republic	A	Suva	English, Fijian, Hindustani
† Finland	5,477,000	338,145	130,559	16	42	Republic	A	Helsinki	Finnish, Swedish, Lapp, Russian
Florida	19,893,000	170,304	65,755	117	303	State (U.S.)	D	Tallahassee	English
† France (excl. Overseas Departments)	62,814,000	539,965	208,482	116	301	Republic	A	Paris	French
French Guiana	240,000	83,534	32,253	2.9	7.4	Overseas department (France)	C	Cayenne	French
French Polynesia	283,000	4,000	1,544	71	184	Overseas territory (France)	C	Papeete	French, Tahitian
Fujian	37,740,000	120,000	46,332	315	815	Province (China)	D	Fuzhou	Chinese dialects
† Gabon	1,705,000	267,668	103,347	6.4	17	Republic	A	Libreville	French, Fang, indigenous
† Gambia, The	1,968,000	10,689	4,127	184	477	Republic	A	Banjul	English, Malinke, Wolof, Fula, indigenous
Gansu	25,822,000	450,000	173,746	57	149	Province (China)	D	Lanzhou	Chinese (Mandarin), Mongolian, Tibetan dialects
Gaza Strip	1,869,000	360	139	5,192	13,446	Israeli territory with limited self-government			Arabic
Georgia	10,097,000	153,910	59,425	66	170	State (U.S.)	D	Atlanta	English
† Georgia	4,931,000	69,700	26,911	71	183	Republic	A	Tbilisi	Georgian, Russian, Armenian, Azeri
† Germany	80,854,000	357,022	137,847	226	587	Federal republic	A	Berlin	German
† Ghana	26,328,000	238,533	92,098	110	286	Republic	A	Accra	English, Akan and other indigenous
Gibraltar	29,000	6.0	2.3	4,876	12,721	Overseas territory (U.K.)	C	Gibraltar	English, Spanish, Italian, Portuguese
Golan Heights	46,000	1,176	454	39	102	Occupied by Israel			Arabic, Hebrew
Great Britain see United Kingdom									
† Greece	10,776,000	131,957	50,949	82	211	Republic	A	Athens (Athína)	Greek, English, French
Greenland	58,000	2,166,086	836,331	0.03	0.07	Self-governing territory (Danish protection)	B	Nuuk	Danish, Greenlandic, Inuit dialects
† Grenada	111,000	344	133	322	832	Parliamentary state	A	St. George's	English, French
Guadeloupe	403,750	1,628	628	248	642	Overseas department (France)	D	Basse-Terre	French, Creole
Guam	162,000	549	212	295	763	Unincorporated territory (U.S.)	C	Hagåtña	English, Chamorro, Japanese
Guangdong	106,440,000	177,800	68,649	599	1,550	Province (China)	D	Guangzhou	Chinese dialects, Miao-Yao
Guangxi	47,190,000	236,300	91,236	200	517	Autonomous region (China)	D	Nanning	Chinese dialects, Thai, Miao-Yao
† Guatemala	14,919,000	108,889	42,042	137	355	Republic	A	Guatemala	Spanish, Amerindian
Guernsey (incl. Dependencies)	66,000	78	30	847	2,203	Crown dependency (U.K. protection)	B	St. Peter Port	English, French
† Guinea	11,780,000	245,857	94,926	48	124	Republic	A	Conakry	French, indigenous
† Guinea-Bissau	1,726,000	36,125	13,948	48	124	Republic	A	Bissau	Portuguese, Crioulo, indigenous
Guizhou	35,022,000	170,000	65,637	206	534	Province (China)	D	Guiyang	Chinese (Mandarin), Thai, Miao-Yao
† Guyana	735,000	214,969	83,000	3.4	8.9	Republic	A	Georgetown	English, indigenous
Hainan	8,953,000	34,200	13,205	262	678	Province (China)	D	Haikou	Chinese, Min, Tai
† Haiti	10,110,000	27,750	10,714	364	944	Republic	A	Port-au-Prince	Creole, French
Hawaii	1,420,000	28,311	10,931	50	130	State (U.S.)	D	Honolulu	English, Hawaiian, Japanese
Hebei	73,326,000	190,000	73,359	386	1,000	Province (China)	D	Shijiazhuang	Chinese (Mandarin)
Heilongjiang	38,350,000	469,000	181,082	82	212	Province (China)	D	Harbin	Chinese dialects, Mongolian, Tungus
Henan	94,134,000	167,000	64,479	564	1,460	Province (China)	D	Zhengzhou	Chinese (Mandarin)
Holland see Netherlands									
† Honduras	8,747,000	112,088	43,277	78	202	Republic	A	Tegucigalpa	Spanish, indigenous
Hong Kong (Xianggang)	7,141,000	1,100	425	6,492	16,803	Special administrative region (China)	C	Hong Kong (Xianggang)	Chinese (Cantonese), English, Putonghua
Hubei	57,990,000	187,400	72,356	309	801	Province (China)	D	Wuhan	Chinese dialects
Hunan	66,906,000	210,000	81,082	319	825	Province (China)	D	Changsha	Chinese dialects, Miao-Yao
† Hungary	9,898,000	93,030	35,919	106	276	Republic	A	Budapest	Hungarian
† Iceland	332,000	103,000	39,769	3.2	8.3	Republic	A	Reykjavík	Icelandic
Idaho	1,634,000	216,446	83,570	7.6	20	State (U.S.)	D	Boise	English
Illinois	12,881,000	149,998	57,914	86	222	State (U.S.)	D	Springfield	English
† India (incl. part of Jammu and Kashmir)	1,251,696,000	3,166,285	1,222,510	395	1,024	Federal republic	A	New Delhi	English, Hindi, Telugu, Bengali, indigenous
Indiana	6,597,000	94,321	36,418	70	181	State (U.S.)	D	Indianapolis	English
† Indonesia	255,994,000	1,904,443	735,310	134	348	Republic	A	Jakarta	Bahasa Indonesia (Malay), English, Dutch, indigenous
Iowa	3,107,000	145,743	56,272	21	55	State (U.S.)	D	Des Moines	English
† Iran	81,824,000	1,648,195	636,372	50	129	Islamic republic	A	Tehrān	Farsi, Turkish dialects, Kurdish
† Iraq	37,056,000	438,317	169,235	85	219	Republic	A	Baghdād	Arabic, Kurdish, Assyrian, Armenian
† Ireland	4,892,000	70,273	27,133	70	180	Republic	A	Dublin (Baile Átha Cliath)	English, Irish Gaelic
Isle of Man	88,000	572	221	153	396	Crown dependency (U.K. protection)	B	Douglas	English, Manx Gaelic
† Israel (excl. Occupied Areas)	8,049,000	20,770	8,019	388	1,004	Republic	A	Jerusalem (Yerushalayim)	Hebrew, Arabic

Region or political division	Est. Pop. 1/1/09	Area sq. km.	Area sq. mi.	Pop. per sq. km.	Pop. per sq. mi.	Form of Government and Ruling Power	Political Status	Capital	Predominant Languages
† Italy	61,855,000	301,323	116,342	205	532	Republic	A	Rome (Roma)	Italian, German, French, Slovene
Ivory Coast *see Côte d'Ivoire*									
† Jamaica	2,950,000	10,991	4,244	268	695	Parliamentary state	A	Kingston	English, Creole
† Japan	126,920,000	377,750	145,850	336	870	Constitutional monarchy	A	Tōkyō	Japanese
Jersey	97,000	116	45	839	2,162	Crown dependency (U.K. protection)	B	St. Helier	English, French
Jiangsu	79,395,000	102,600	39,614	774	2,004	Province (China)	D	Nanjing	Chinese dialects
Jiangxi	45,222,000	166,600	64,325	271	703	Province (China)	D	Nanchang	Chinese dialects
Jilin	27,513,000	187,000	72,201	147	381	Province (China)	D	Changchun	Chinese (Mandarin), Mongolian, Korean
† Jordan	8,118,000	89,342	34,495	91	235	Constitutional monarchy	A	'Ammān	Arabic
Kansas	2,904,000	213,096	82,277	14	35	State (U.S.)	D	Topeka	English
† Kazakhstan	18,157,000	2,717,300	1,049,156	6.7	17	Republic	A	Astana	Kazakh, Russian
Kentucky	4,413,000	104,659	40,409	42	109	State (U.S.)	D	Frankfort	English
† Kenya	45,925,000	582,646	224,961	79	204	Republic	A	Nairobi	English, Swahili, indigenous
† Kiribati	106,000	811	313	130	338	Republic	A	Bairiki	English, Gilbertese
† Korea, North	24,983,000	120,538	46,540	207	537	Socialist republic	A	P'yŏngyang	Korean
† Korea, South	49,115,000	99,268	38,328	495	1,281	Republic	A	Seoul (Sŏul)	Korean
Kosovo (3)	1,871,000	10,887	4,203	172	445	Republic	A	Priština	Albanian, Serbian
† Kuwait	2,789,000	17,818	6,880	157	405	Constitutional monarchy	A	Kuwait (Al-Kuwayt)	Arabic, English
† Kyrgyzstan	5,665,000	199,900	77,182	28	73	Republic	A	Bishkek	Kirghiz, Russian
† Laos	6,912,000	236,800	91,429	29	76	Socialist republic	A	Vientiane (Viangchan)	Lao, French, English
† Latvia	1,987,000	64,600	24,942	31	80	Republic	A	Rīga	Latvian, Russian, Lithuanian
† Lebanon	6,185,000	10,400	4,016	595	1540	Republic	A	Beirut (Bayrūt)	Arabic, French, Armenian, English
† Lesotho	1,948,000	30,355	11,720	64	166	Constitutional monarchy	A	Maseru	English, Sesotho, Zulu, Xhosa
Liaoning	43,900,000	145,700	56,255	301	780	Province (China)	D	Shenyang	Chinese (Mandarin), Mongolian
† Liberia	4,196,000	111,369	43,000	38	98	Republic	A	Monrovia	English, indigenous
† Libya	6,412,000	1,759,540	679,362	3.6	9.4	Transitional	A	Tripoli (Ṭarābulus)	Arabic
† Liechtenstein	38,000	160	62	235	607	Constitutional monarchy	A	Vaduz	German
† Lithuania	2,884,000	65,300	25,213	44	114	Republic	A	Vilnius	Lithuanian, Polish, Russian
Louisiana	4,650,000	134,264	51,840	35	90	State (U.S.)	D	Baton Rouge	English
† Luxembourg	570,000	2,586	999	221	571	Constitutional monarchy	A	Luxembourg	French, Luxembourgish, German
Macau (Aomen)	593,000	18	6.9	32,930	85,903	Special administrative region (China)	C	Macau (Aomen)	Chinese (Cantonese), Portuguese
† Macedonia	2,096,000	25,713	9,928	82	211	Republic	A	Skopje	Macedonian, Albanian
† Madagascar	23,813,000	587,041	226,658	41	105	Republic	A	Antananarivo	Malagasy, French
Maine	1,330,000	91,646	35,385	15	38	State (U.S.)	D	Augusta	English
† Malawi	17,965,000	118,484	45,747	152	393	Republic	A	Lilongwe	Chichewa, English
† Malaysia	30,514,000	329,758	127,320	93	240	Federal constitutional monarchy	A	Kuala Lumpur and Putrajaya	Malay, Chinese dialects, English, Tamil
† Maldives	393,000	298	115	1,320	3,470	Republic	A	Male'	Divehi
† Mali	16,956,000	1,240,192	478,841	14	35	Republic	A	Bamako	French, Bambara, indigenous
† Malta	414,000	316	122	1,310	3,393	Republic	A	Valletta	English, Maltese
Manitoba	1,293,000	647,797	250,116	2.0	5.2	Province (Canada)	D	Winnipeg	English
† Marshall Islands	72,000	181	70	399	1,031	Republic (U.S. protection)	A	Majuro (island)	English, indigenous, Japanese
Martinique	300,000	1,100	425	363	914	Overseas department (France)	C	Fort-de-France	French, Creole
Maryland	5,976,000	32,133	12,407	186	482	State (U.S.)	D	Annapolis	English
Massachusetts	6,745,000	27,336	10,555	247	639	State (U.S.)	D	Boston	English
† Mauritania	3,597,000	1,030,700	397,956	3.5	9.0	Republic	A	Nouakchott	Arabic, Pular, Soninke, Wolof
† Mauritius (incl. Dependencies)	1,340,000	2,040	788	657	1700	Republic	A	Port Louis	English, Creole, Bhojpuri, French, Hindi, Tamil, others
Mayotte (4)	213,000	374	144	569	1477	Territorial collectivity (France)	C	Mamoudzou	French, Swahili (Mahorian)
† Mexico	121,737,000	1,964,382	758,452	62	161	Federal republic	A	Mexico City (Ciudad de México)	Spanish, indigenous
Michigan	9,910,000	250,494	96,716	40	102	State (U.S.)	D	Lansing	English
† Micronesia, Federated States of	105,000	702	271	150	388	Republic (U.S. protection)	A	Palikir	English, indigenous
Midway Islands	(1)	5.2	2.0			Unincorporated territory (U.S.)	C		English
Minnesota	5,457,000	225,171	86,939	24	63	State (U.S.)	D	St. Paul	English
Mississippi	2,994,000	125,434	48,430	24	62	State (U.S.)	D	Jackson	English
Missouri	6,064,000	180,533	69,704	34	87	State (U.S.)	D	Jefferson City	English
† Moldova	3,547,000	33,851	13,070	105	271	Republic	A	Chişinău	Romanian (Moldovan), Russian
† Monaco	31,000	2.0	0.8	15,268	38,169	Constitutional monarchy	A	Monaco	French, English, Italian, Monegasque
† Mongolia	2,993,000	1,566,500	604,829	1.9	4.9	Republic	A	Ulaanbaatar	Khalkha Mongol, Turkish dialects, Russian, Chinese
Montana	1,024,000	380,838	147,042	2.7	7.0	State (U.S.)	D	Helena	English
† Montenegro	647,000	13,812	5,333	47	121	Republic	A	Podgorica	Serbian, Albanian
Montserrat	5,200	102	39	51	134	Overseas territory (U.K.)	C	Plymouth (abandoned)	English
† Morocco (excl. Western Sahara)	33,323,000	446,550	172,414	75	193	Constitutional monarchy	A	Rabat	Arabic, Berber dialects, French
† Mozambique	25,303,000	801,590	309,496	32	82	Republic	A	Maputo	Portuguese, indigenous
† Myanmar (Burma)	56,320,000	676,578	261,228	83	216	Provisional military government	A	Yangon (Rangoon) and Nay Pyi Taw	Burmese, indigenous
† Namibia	2,212,000	823,144	317,818	2.7	7.0	Republic	A	Windhoek	English, Afrikaans, German, indigenous
† Nauru	10,000	21	8.1	454	1,178	Republic	A	Yaren District	Nauruan, English
Nebraska	1,882,000	200,345	77,354	9.4	24	State (U.S.)	D	Lincoln	English
Nei Mongol (Inner Mongolia)	24,976,000	1,183,000	456,759	21	55	Autonomous region (China)	D	Hohhot	Mongolian
† Nepal	31,551,000	147,181	56,827	214	555	Federal republic	A	Kathmandu (Kāṭhmāṇḍāu)	Nepali, Maithali, Bhojpuri, other indigenous
† Netherlands	16,948,000	41,864	16,164	405	1,048	Constitutional monarchy	A	Amsterdam and The Hague ('s-Gravenhage)	Dutch
Nevada	2,839,000	286,351	110,561	10	26	State (U.S.)	D	Carson City	English
New Brunswick	754,000	72,908	28,150	10	27	Province (Canada)	D	Fredericton	English, French
New Caledonia	272,000	18,575	7,172	15	38	Overseas territory (France)	C	Nouméa	French, indigenous
New Hampshire	1,327,000	24,216	9,350	55	142	State (U.S.)	D	Concord	English
New Jersey	8,938,000	22,588	8,721	396	1,025	State (U.S.)	D	Trenton	English
New Mexico	2,086,000	314,915	121,590	6.6	17	State (U.S.)	D	Santa Fe	English, Spanish
New South Wales	7,597,000	800,640	309,129	9.5	25	State (Australia)	D	Sydney	English
New York	19,746,000	141,299	54,556	140	362	State (U.S.)	D	Albany	English
† New Zealand	4,438,000	270,534	104,454	16	42	Parliamentary state	A	Wellington	English, Maori
Newfoundland and Labrador	528,000	405,212	156,453	1.3	3.4	Province (Canada)	D	St. John's	English
† Nicaragua	5,908,000	129,640	50,054	46	118	Republic	A	Managua	Spanish, English, indigenous
† Niger	18,046,000	1,267,000	489,192	14	37	Provisional military government	A	Niamey	French, Hausa, Djerma, indigenous
† Nigeria	181,562,000	923,768	356,669	197	509	Transitional military government	A	Abuja	English, Hausa, Fulani, Yorbua, Ibo, indigenous
Ningxia	6,542,000	66,400	25,637	99	255	Autonomous region (China)	D	Yinchuan	Chinese (Mandarin)

Region or political division	Est. Pop. 1/1/09	Area sq. km.	Area sq. mi.	Pop. per sq. km.	Pop. per sq. mi.	Form of Government and Ruling Power	Political Status	Capital	Predominant Languages
Niue	1,200	259	100	4.6	12	Self-governing territory (New Zealand protection)	B	Alofi	English, indigenous
Norfolk Island	2,200	36	14	61	158	External territory (Australia)	C	Kingston	English, Norfolk
North America	564,080,000	24,700,000	9,500,000	23	59			...	...
North Carolina	9,944,000	139,389	53,819	71	185	State (U.S.)	D	Raleigh	English
North Dakota	739,000	183,112	70,700	4.0	10	State (U.S.)	D	Bismarck	English
Northern Ireland	1,841,000	13,576	5,242	136	351	Administrative division (U.K.)	D	Belfast	English
Northern Mariana Islands	52,000	464	179	113	292	Commonwealth (U.S. protection)	B	Saipan (island)	English, Chamorro, Carolinian
Northern Territory	244,000	1,349,130	520,902	0.2	0.5	Territory (Australia)	D	Darwin	English, indigenous
Northwest Territories	44,000	1,346,106	519,735	0.03	0.08	Territory (Canada)	D	Yellowknife	English, indigenous
† Norway (incl. Jan Mayen and Svalbard)	5,208,000	323,877	125,050	16	42	Constitutional monarchy	A	Oslo	Norwegian, Lapp, Finnish
Nova Scotia	943,000	55,284	21,345	17	44	Province (Canada)	D	Halifax	English
Nunavut	37,000	2,093,190	808,185	0.02	0.05	Territory (Canada)	D	Iqaluit	English, indigenous
Oceania (incl. Australia)	37,143,000	8,500,000	3,300,000	4.4	11			...	...
Ohio	11,594,000	116,096	44,825	100	259	State (U.S.)	D	Columbus	English
Oklahoma	3,878,000	181,036	69,898	21	55	State (U.S.)	D	Oklahoma City	English
† Oman	3,287,000	309,500	119,499	11	28	Monarchy	A	Muscat (Masqat)	Arabic, English, Baluchi, Urdu, Indian dialects
Ontario	13,792,000	1,076,395	415,599	13	33	Province (Canada)	D	Toronto	English
Oregon	3,970,000	254,805	98,381	16	40	State (U.S.)	D	Salem	English
† Pakistan (incl. part of Jammu and Kashmir)	199,086,000	879,902	339,732	226	586	Federal Islamic republic	A	Islāmābād	English, Urdu, Punjabi, Sindhi, Pashto
† Palau	21,000	487	188	44	113	Republic	A	Melekeok	Angaur, English, Japanese, Palauan, Sonsorolese, Tobi
† Panama	3,657,000	75,517	29,157	48	125	Republic	A	Panamá	Spanish, English
† Papua New Guinea	6,672,000	462,840	178,704	14	37	Parliamentary state	A	Port Moresby	English, Motu, Pidgin, indigenous
† Paraguay	6,783,000	406,752	157,048	17	43	Republic	A	Asunción	Spanish, Guarani
Pennsylvania	12,787,000	119,282	46,055	107	278	State (U.S.)	D	Harrisburg	English
† Peru	30,445,000	1,285,216	496,225	24	61	Republic	A	Lima	Quechua, Spanish, Aymara
† Philippines	100,998,000	300,000	115,831	337	872	Republic	A	Manila	English, Pilipino, Tagalog
Pitcairn Islands (incl. Dependencies)	50	49	19	1.0	2.5	Overseas territory (U.K.)	C	Adamstown	English, Tahitian
† Poland	38,562,000	312,685	120,728	123	319	Republic	A	Warsaw (Warszawa)	Polish
† Portugal	10,825,000	91,985	35,516	118	305	Republic	A	Lisbon (Lisboa)	Portuguese
Prince Edward Island	146,000	5,660	2,185	26	67	Province (Canada)	D	Charlottetown	English
Puerto Rico	3,598,000	9,104	3,515	395	1,024	Commonwealth (U.S. protection)	B	San Juan	Spanish, English
† Qatar	2,195,000	11,427	4,412	192	497	Monarchy	A	Doha (Ad-Dawḩah)	Arabic, English
Qinghai	5,778,000	720,000	277,994	8.0	21	Province (China)	D	Xining	Tibetan dialects, Mongolian, Turkish dialects, Chinese (Mandarin)
Quebec	8,264,000	1,542,056	595,391	5.4	14	Province (Canada)	D	Québec	French, English
Queensland	4,767,000	1,730,650	668,208	2.8	7.1	State (Australia)	D	Brisbane	English
Reunion	834,000	2,510	969	332	861	Overseas department (France)	C	Saint-Denis	French, Creole
Rhode Island	1,055,000	4,002	1,545	264	683	State (U.S.)	D	Providence	English
† Romania	21,666,000	237,500	91,699	91	236	Republic	A	Bucharest (Bucureşti)	Romanian, Hungarian, German
† Russia	142,424,000	17,075,400	6,592,849	8.3	22	Federal republic	A	Moscow (Moskva)	Russian, Tatar, Ukrainian
† Rwanda	12,662,000	26,338	10,169	481	1,245	Republic	A	Kigali	French, Kinyarwanda, Kiswahili
St. Barthélemy	7,237	25	10	289	724	Overseas Collectivity (France)	D	Gustavia	French, English
St. Helena (incl. Dependencies)	7800	314	121	25	64	Overseas territory (U.K.)	C	Jamestown	English
† St. Kitts and Nevis	52,000	261	101	199	514	Parliamentary state	A	Basseterre	English
† St. Lucia	164,000	616	238	266	689	Parliamentary state	A	Castries	English, French
St. Martin	31,754	54	21	588	1,512	Overseas Collectivity (France)	D	Marigot	French, English, Dutch, French Patois
St. Pierre and Miquelon	5,700	242	93	23	61	Territorial collectivity (France)	C	Saint-Pierre	French
† St. Vincent and the Grenadines	103,000	388	150	265	684	Parliamentary state	A	Kingstown	English, French
† Samoa	198,000	2,831	1,093	70	181	Constitutional monarchy	A	Apia	English, Samoan
† San Marino	33,000	61	24	541	1,376	Republic	A	San Marino	Italian
† Sao Tome and Principe	194,000	964	372	201	522	Republic	A	São Tomé	Portuguese, Fang
Saskatchewan	1,134,000	651,036	251,366	1.7	4.5	Province (Canada)	D	Regina	English
† Saudi Arabia	27,752,000	2,149,690	830,000	13	33	Monarchy	A	Riyadh (Ar-Riyāḍ)	Arabic
Scotland	5,348,000	78,133	30,167	68	177	Administrative division (U.K.)	D	Edinburgh	English, Scots Gaelic
† Senegal	13,976,000	196,712	75,951	71	184	Republic	A	Dakar	French, Wolof, Fulani, Serer, indigenous
† Serbia (excl. Kosovo)	7,177,000	77,474	29,913	93	240	Republic	A	Belgrade (Beograd)	Serbian
† Seychelles	92,000	455	176	203	525	Republic	A	Victoria	English, French, Creole
Shaanxi	37,640,000	205,000	79,151	184	476	Province (China)	D	Xi'an	Chinese (Mandarin)
Shandong	97,334,000	153,000	59,074	636	1,648	Province (China)	D	Jinan	Chinese (Mandarin)
Shanghai	24,152,000	6,200	2,394	3,895	10,088	Autonomous city (China)	D	Shanghai	Chinese (Wu)
Shanxi	36,298,000	156,000	60,232	233	603	Province (China)	D	Taiyuan	Chinese (Mandarin)
Sichuan	81,070,000	487,600	188,263	166	431	Province (China)	D	Chengdu	Chinese (Mandarin), Tibetan dialects, Miao-Yao
† Sierra Leone	5,879,000	71,740	27,699	82	212	Transitional military government	A	Freetown	English, Krio, Mende, Temne, indigenous
† Singapore	5,674,000	683	264	8,308	21,494	Republic	A	Singapore	Chinese (Mandarin), English, Malay, Tamil
Sint Maarten	37,132	16	6	2,321	6,189	Constituent country of the Netherlands	B	Philipsburg	English, Spanish, Creole, Dutch
† Slovakia	5,445,000	49,012	18,924	111	288	Republic	A	Bratislava	Slovak, Hungarian
† Slovenia	1,983,000	20,256	7,821	98	254	Republic	A	Ljubljana	Slovenian
† Solomon Islands	622,000	28,370	10,954	22	57	Parliamentary state	A	Honiara	English, indigenous
† Somalia	10,616,000	637,657	246,201	17	43	Transitional	A	Mogadishu (Muqdisho)	Arabic, Somali, English, Italian
† South Africa	53,676,000	1,219,090	470,693	44	114	Republic	A	Pretoria (Tshwane), Cape Town and Bloemfontein	Afrikaans, English, Xhosa, Zulu, other indigenous
South America	409,766,000	17,800,000	6,900,000	23	59				
South Australia	1,696,000	983,480	379,724	1.7	4.5	State (Australia)	D	Adelaide	English
South Carolina	4,832,000	82,932	32,020	58	151	State (U.S.)	D	Columbia	English
South Dakota	853,000	199,731	77,117	4.3	11	State (U.S.)	D	Pierre	English
South Georgia and the South Sandwich Islands (2)	(1)	3,755	1,450	...	...	Overseas territory (U.K.)	C		English
† South Sudan	12,043,000	644,329	248,777	19	48	Republic	A	Juba	English, Arabic, Dinka, Nuer, Bari, Zande, Shilluk, other indigenous
† Spain	48,146,000	504,750	194,885	95	247	Constitutional monarchy	A	Madrid	Spanish (Castilian), Catalan, Galician, Basque
Spanish North Africa (5)	169,000	32	12	5,296	14,123	Five possessions (Spain)	C		Spanish, Arabic, Berber dialects
† Sri Lanka	22,053,000	65,610	25,332	336	871	Socialist republic	A	Colombo and Sri Jayewardenepura Kotte	English, Sinhala, Tamil
† Sudan	36,109,000	1,861,484	718,723	19	50	Provisional military government	A	Khartoum (Al-Kharṭūm)	Arabic, Nubian and other indigenous, English

Region or political division	Est. Pop. 1/1/09	Area sq. km.	Area sq. mi.	Pop. per sq. km.	Pop. per sq. mi.	Form of Government and Ruling Power	Political Status	Capital	Predominant Languages
† Suriname	580,000	163,265	63,037	3.6	9.2	Republic	A	Paramaribo	Dutch, Sranan Tongo, English, Hindustani, Javanese
† Swaziland	1,436,000	17,364	6,704	83	214	Monarchy	A	Mbabane and Lobamba	English, siSwati
† Sweden	9,802,000	449,964	173,732	22	56	Constitutional monarchy	A	Stockholm	Swedish, Lapp, Finnish
† Switzerland	8,122,000	41,293	15,943	197	509	Federal republic	A	Bern	German, French, Italian, Romansch
† Syria	17,065,000	185,180	71,498	92	239	Socialist republic	A	Damascus (Dimashq)	Arabic, Kurdish, Armenian, Aramaic, Circassian
Taiwan	23,415,000	36,002	13,901	650	1,684	Republic	A	T'aipei	Chinese (Mandarin), Taiwanese (Min), Hakka
† Tajikistan	8,192,000	143,100	55,251	57	148	Republic	A	Dushanbe	Tajik, Uzbek, Russian
† Tanzania	51,046,000	945,087	364,900	54	140	Republic	A	Dar es Salaam and Dodoma	English, Swahili, indigenous
Tasmania	516,000	68,400	26,409	7.5	20	State (Australia)	D	Hobart	English
Tennessee	6,549,000	109,151	42,143	60	155	State (U.S.)	D	Nashville	English
Texas	26,957,000	695,621	268,581	39	100	State (U.S.)	D	Austin	English, Spanish
† Thailand	67,976,000	513,115	198,115	132	343	Constitutional monarchy	A	Bangkok (Krung Thep)	Thai, indigenous
Tianjin	14,722,000	11,300	4,363	1,303	3,374	Autonomous city (China)	D	Tianjin	Chinese (Mandarin)
Timor-Leste	1,231,116	14,874	5,743	83	214	Republic	A	Dili	Tetum, Portuguese, Indonesian
† Togo	7,552,000	56,785	21,925	133	344	Provisional military government	A	Lomé	French, Ewe, Mina, Kabye, Dagomba
Tokelau	1,300	12	4.6	111	291	Island territory (New Zealand)	C		English, Tokelauan
† Tonga	107,000	650	251	164	424	Constitutional monarchy	A	Nuku'alofa	Tongan, English
† Trinidad and Tobago	1,222,000	5,128	1,980	238	617	Republic	A	Port of Spain	English, Hindi, French, Spanish
Tristan da Cunha	300	104	40	2.6	6.7	Dependency (St. Helena)	C	Edinburgh of the Seven Seas	English
† Tunisia	11,037,000	163,610	63,170	67	175	Republic	A	Tunis	Arabic, French
† Turkey	79,414,000	783,577	302,541	101	262	Republic	A	Ankara	Turkish, Kurdish, Arabic
† Turkmenistan	5,231,000	488,100	188,457	11	28	Republic	A	Aşgabat	Turkmen, Russian, Uzbek
Turks and Caicos Islands	50,000	430	166	117	303	Overseas territory (U.K.)	C	Grand Turk	English
† Tuvalu	11,000	26	10	418	1,087	Parliamentary state	A	Funafuti	Tuvaluan, English
† Uganda	37,102,000	241,038	93,065	154	399	Republic	A	Kampala	English, Luganda, Swahili, indigenous
† Ukraine	44,429,000	603,700	233,090	74	191	Republic	A	Kiev (Kyiv)	Ukrainian, Russian, Romanian, Polish
† United Arab Emirates	5,780,000	83,600	32,278	69	179	Federation of monarchs	A	Abu Dhabi (Abū Ẓaby)	Arabic, Farsi, English, Hindi, Urdu
† United Kingdom	64,088,000	242,910	93,788	264	683	Parliamentary monarchy	A	London	English, Welsh, Scots Gaelic
† United States	321,369,000	9,826,630	3,794,083	33	85	Federal republic	A	Washington	English, Spanish
† Uruguay	3,342,000	175,016	67,574	19	49	Republic	A	Montevideo	Spanish
Utah	2,943,000	219,887	84,899	13	35	State (U.S.)	D	Salt Lake City	English
† Uzbekistan	29,200,000	447,400	172,742	65	169	Republic	A	Tashkent (Toshkent)	Uzbek, Russian
† Vanuatu	272,000	12,190	4,707	22	58	Republic	A	Port Vila	Bislama, English, French
Vatican City	800	0	0	2,105	4,210	Monarchical-sacerdotal state	A	Vatican City	Italian, Latin, other
† Venezuela	29,275,000	912,050	352,145	32	83	Federal republic	A	Caracas	Spanish, Amerindian
Vermont	627,000	24,901	9,614	25	65	State (U.S.)	D	Montpelier	English
Victoria	5,915,000	227,420	87,807	26	67	State (Australia)	D	Melbourne	English
† Vietnam	94,349,000	331,689	128,066	284	737	Socialist republic	A	Ha Noi	Vietnamese, French, Chinese, English, Khmer, indigenous
Virginia	8,326,000	110,785	42,774	75	195	State (U.S.)	D	Richmond	English
Virgin Islands (U.S.)	104,000	347	134	298	773	Unincorporated territory (U.S.)	C	Charlotte Amalie	English, Spanish, Creole
Wake Island	(1)	8	3.0			Unincorporated territory (U.S.)	C		English
Wales	3,092,000	20,779	8,023	149	385	Administrative division (U.K.)	D	Cardiff	English, Welsh Gaelic
Wallis and Futuna	16,000	255	99	61	158	Overseas territory (France)	C	Mata'Utu	French, Wallisian
Washington	7,062,000	184,665	71,300	38	99	State (U.S.)	D	Olympia	English
West Bank (incl. East Jerusalem)	2,785,000	5,860	2,263	475	1,231	Israeli territory with limited self-government			Arabic, Hebrew
Western Australia	2,587,000	2,529,880	976,792	1.0	2.6	State (Australia)	D	Perth	English
Western Sahara	571,000	266,000	102,703	2.1	5.5	Occupied by Morocco			Arabic
West Virginia	1,850,000	62,755	24,230	29	76	State (U.S.)	D	Charleston	English
Wisconsin	5,758,000	169,639	65,498	34	88	State (U.S.)	D	Madison	English
Wyoming	584,000	253,336	97,814	2.3	6.0	State (U.S.)	D	Cheyenne	English
Xianggang see Hong Kong									
Xinjiang	22,643,000	1,600,000	617,764	14	37	Autonomous region (China)	D	Ürümqi	Turkish dialects, Mongolian, Tungus, English
Xizang (Tibet)	3,120,000	1,220,000	471,045	2.6	6.6	Autonomous region (China)	D	Lhasa	Tibetan dialects
† Yemen	26,737,000	527,968	203,850	51	131	Republic	A	Sanaa (Şan'ā')	Arabic
Yukon	37,000	482,443	186,272	0.08	0.2	Territory (Canada)	D	Whitehorse	English, Inuktitut, indigenous
Yunnan	46,866,000	394,000	152,124	119	308	Province (China)	D	Kunming	Chinese (Mandarin), Tibetan dialects, Khmer, Miao-Yao
† Zambia	15,066,000	752,614	290,586	20	52	Republic	A	Lusaka	English, Tonga, Lozi, other indigenous
Zhejiang	54,980,000	101,800	39,305	540	1,399	Province (China)	D	Hangzhou	Chinese dialects
† Zimbabwe	14,230,000	390,759	150,873	36	94	Republic	A	Harare	English, Shona, Sindebele
WORLD	7,256,490,000	150,100,000	57,900,000	48	125				

(1) No Permanent Population.

(2) Claimed by Argentina.

(3) Kosovo unilaterally declared its independence from Serbia in 2008.

(4) Claimed by Comoros.

(5) Comprises Ceuta, Melilla and several small islands.

† Member of the United Nations (2016)

... None/not applicable.

General Information

Equatorial diameter of Earth 12,756 km (7,926 mi.)
Polar diameter of Earth 12,713 km (7,900 mi.)
Mean diameter of Earth 12,742 km (7,918 mi.)
Equatorial circumference of Earth 40,075 km (24,901 mi.)
Mean distance from Earth to Sun 149,598,000 km (92,955,900 mi.)
Mean distance from Earth to Moon 384,403 km (238,857 mi.)
Total area of Earth 510,100,000 sq. km (197,000,000 sq. mi.)
Highest elevation on Earth's surface, Mt. Everest, Asia 8,848 m (29,028 ft.)
Lowest elevation on Earth's land surface, shores of the Dead Sea, Asia 408 m (1,339 ft.) below sea level
Greatest known depth of the ocean, southwest of Guam, Pacific Ocean 10,924 m (35,840 ft.)
Total land area of Earth (incl. inland water and Antarctica) 150,100,000 sq. km (57,900,000 sq. mi.)
Area of Africa 30,300,000 sq. km (11,700,000 sq. mi.)
Area of Antarctica 14,000,000 sq. km (5,400,000 sq. mi.)
Area of Australia and Oceania 8,500,000 sq. km (3,300,000 sq. mi.)
Area of Asia 44,900,000 sq. km (17,300,000 sq. mi.)
Area of Europe 9,900,000 sq. km (3,800,000 sq. mi.)
Area of North America 24,700,000 sq. km (9,500,000 sq. mi.)
Area of South America 17,800,000 sq. km (6,900,000 sq. mi.)
World Population (est. 7/1/15), 7,256,490,000

Principal Islands Area in sq. km (sq. mi.)

Baffin I., Nu., Can. 507,451 (195,928)
Banks I., N.T., Can. 70,028 (27,038)
Borneo, Asia 748,168 (288,869)
Bougainville, Pap. N. Gui. 9,317 (3,597)
Cape Breton I., N.S., Can. 10,311 (3,981)
Celebes, Indon. 180,680 (69,761)
Ceram, Indon. 17,454 (6,739)
Corsica, Fr. 8,741 (3,375)
Crete, Grc. 8,349 (3,224)
Cuba, Cuba 105,805 (40,852)
Cyprus, Cyp. 9,234 (3,565)
Devon I., Nu., Can. 55,247 (21,331)
Ellesmere I., Nu., Can. 196,236 (75,767)
Flores, Indon. 14,154 (5,465)
Great Britain, U.K. 226,000 (87,259)
Greenland, Green. 2,166,086 (836,330)
Guadalcanal, Sol. Is. 5,352 (2,066)
Hainan Dao, China 33,209 (12,822)
Hawai'i, Hi., U.S. 10,500 (4,054)
Hispaniola, N.A. 73,929 (28,544)
Hokkaidō, Japan 78,719 (30,394)
Honshū, Japan 225,800 (87,182)
Iceland, Ice. 101,826 (39,315)
Ireland, Ire.-U.K. 81,638 (31,521)
Jamaica, Jam. 11,189 (4,320)
Java, Indon. 138,793 (53,588)
Kodiak I., Ak., U.S. 9,578 (3,698)
Kyūshū, Japan 37,437 (14,455)
Leyte, Phil. 7,367 (2,844)
Long I., N.Y., U.S. 3,502 (1,352)
Luzon, Phil. 109,964 (42,457)
Madagascar, Madag. 587,713 (226,917)
Melville I., Can. 42,149 (16,274)
Mindanao, Phil. 97,530 (37,657)
Mindoro, Phil. 10,571 (4,081)
Negros, Phil. 13,074 (5,048)
New Britain, Pap. N. Gui. 35,144 (13,569)
New Caledonia, N. Cal. 16,648 (6,428)
Newfoundland, Nf./L., Can. 108,860 (42,031)
New Guinea, Asia-Oc. 785,753 (303,381)
Nordaustlandet, Nor. 14,247 (5,501)
North I., N.Z. 111,582 (43,082)
Palawan, Phil. 12,188 (4,706)
Panay, Phil. 12,011 (4,637)
Prince of Wales I., Nu., Can. 33,339 (12,872)
Puerto Rico, P.R. 8,733 (3,372)
Sakhalin, Russia 72,492 (27,989)
Samar, Phil. 12,849 (4,961)
Sardinia, Italy 23,949 (9,247)
Shikoku, Japan 18,544 (7,160)
Sicily, Italy 25,662 (9,908)
Southampton I., Nu., Can. 41,214 (15,913)
South I., N.Z. 145,836 (56,308)
Spitsbergen, Nor. 38,980 (15,050)
Sri Lanka, Sri L. 67,654 (26,121)
Sumatra, Indon. 443,065 (171,068)
Taiwan, Tai. 34,506 (13,323)
Tasmania, Austl. 65,519 (25,297)
Tierra del Fuego, S.A. 47,401 (18,302)
Timor, Asia 28,418 (10,972)
Vancouver I., B.C., Can. 31,285 (12,079)
Victoria I., B.C., Can. 217,291 (83,896)
Vrangelya, Ostrov (Wrangel I.), Russia 7,865 (3,037)

Principal Lakes, Oceans and Seas Area in sq. km (sq. mi.)

Arabian Sea, Afr.-Asia 3,864,000 (1,492,000)
Aral Sea, Kaz.-Uzb. fluctuates
Arctic Ocean 14,056,000 (5,400,000)
Athabasca, L., Can. 7,935 (3,064)
Atlantic Ocean 76,762,000 (29,600,000)
Baikal, L., Russia 31,500 (12,162)
Balkhash, L., Kaz. 17,580 (6,788)
Baltic Sea, Eur. 422,000 (163,000)
Bering Sea, Asia-N.A. 2,291,900 (884,900)
Black Sea, Eur.-N.A. 461,000 (178,000)
Caribbean Sea, N.A.-S.A. 10,464 (4,040)
Caspian Sea, Asia-Eur. 371,000 (143,244)
Chad, L., Afr (2001). 1,350 (521)
Erie, L., Can.-U.S. 25,667 (9,910)
Eyre, L., Austl. 9,500 (3,668)
Great Bear Lake, Can. 31,328 (12,096)
Great Salt Lake, U.S. 5,483 (2,117)
Great Slave Lake, Can. 28,568 (11,030)
Hudson Bay, Can. 1,230,000 (475,000)
Huron, L., Can.-U.S. 59,570 (23,000)
Indian Ocean 68,556,000 (26,500,000)
Japan, Sea of, (East Sea) Asia 1,007,800 (389,100)
Kok Nor (Qinghai Hu), China 4,460 (1,722)
Ladoga, L., Russia 16,400 (6,332)
Manitoba, L., Can. 4,624 (1,785)
Maracaibo, L., Ven. 13,010 (5,023)
Mediterranean Sea 2,505,000 (967,000)
Mexico, Gulf of, N.A. 1,500,000 (600,000)
Michigan, L., U.S. 57,757 (22,300)
Nicaragua, Lago de, Nicaragua 8,150 (3,147)
North Sea, Eur. 575,000 (222,000)
Nyasa, L., Afr. 29,500 (11,390)
Onega, L., Russia 9,890 (3,819)
Ontario, L., Can.-U.S. 19,011 (7,340)
Pacific Ocean 155,557,000 (60,100,000)
Red Sea, Afr.-Asia 438,000 (169,000)
Rudolf, L., Eth.-Kenya 6,750 (2,606)
Southern Ocean 20,327,000 (7,800,000)
Superior, L., Can.-U.S. 82,103 (31,700)
Tanganyika. L., Afr. 32,600 (12,587)
Titicaca, Lago, Bol.-Peru 8,372 (3,232)
Torrens, L., Austl. 5,745 (2,218)
Vänern (L.), Swe. 5,648 (2,181)
Van Gölü (L.), Tur. 3,740 (1,444)
Victoria, L., Afr. 68,870 (26,591)
Winnipeg, L., Can. 24,387 (9,416)
Winnipegosis, L., Can. 5,374 (2,075)
Yellow Sea, Asia 1,243,000 (480,000)

Principal Mountains Elevation in m (ft.)

Aconcagua, Cerro, Arg. 6,959 (22,831)
Annapūrṇa, Nepal 8,091 (26,545)
Aoraki (Mt. Cook), N.Z. 3,754 (12,316)
Apo, Mt., Phil. 2,954 (9,692)
Ararat, Mt., Tur. 5,137 (16,854)
Barú, Volcán, Pan. 3,475 (11,401)
Belukha, Mt., Asia 4,506 (14,783)
Bia, Phou, Laos 2,820 (9,252)
Blanc, Mont, Eur. 4,807 (15,771)
Blanca Pk., Co., U.S. 4,372 (14,345)
Bolívar, Pico, Ven. 5,007 (16,427)
Bonete Grande, Cerro, Arg. 6,872 (22,546)
Borah Pk., Id., U.S. 3,859 (12,662)
Boundary Pk., Nv., U.S. 4,006 (13,143)
Cameroon Mtn., Camm. 4,100 (13,451)
Carrauntoohil, Ire. 1,038 (3,406)
Chaltel, Cerro, S.A. 3,340 (10,958)
Chimborazo, Ec. 6,310 (20,702)
Chirripó, Cerro, C.R. 3,819 (12,530)
Colima, Nevado de, Mex. 4,240 (13,911)
Cotopaxi, Ec. 5,897 (19,347)
Cristóbal Colón, Pico, Col. 5,775 (18,947)
Damāvand, Kīh-e, Iran 5,604 (18,386)
Denali (Mt. McKinley), Ak., U.S. 6,194 (20,320)
Dhawalāgiri, Nepal 8,167 (26,795)
Duarte, Pico, Dom. Rep. 3,175 (10,417)
Dufourspitze, Eur. 4,634 (15,203)
Elbert, Mt., Co., U.S. 4,399 (14,433)
El'brus, Gora, Russia 5,642 (18,510)
Elgon, Mt., Afr. 4,321 (14,177)
Erciyes Dağı, Tur. 3,917 (12,851)
Etna, Monte, Italy 3,323 (10,902)
Everest, Mt., Asia 8,848 (29,028)
Fairweather, Mt., N.A. 4,663 (15,300)
Folādī, Koh-e, Afg. 5,135 (16,847)
Fuji-san, Japan 3,776 (12,388)
Galdhøpiggen, Nor. 2,469 (8,100)
Gannett Pk., Wy., U.S. 4,207 (13,804)
Gerlachovský štít, Slvk. 2,655 (8,711)
Giluwe, Mt., Pap. N. Gui. 4,368 (14,331)
Gongga Shan, China 7,590 (24,902)
Grand Teton, Wy., U.S. 4,197 (13,770)
Großglockner, Aus. 3,797 (12,457)
Gunnbjørn Fjeld, Green. 3,700 (12,139)
Hekla, Ice. 1,491 (4,892)
Hkakabo Razi, Mya. 5,881 (19,295)
Hood, Mt., Or., U.S. 3,426 (11,239)
Huascarán, Nevado, Peru 6,746 (22,133)
Huila, Nevado del, Col. 5,750 (18,865)
Hvannadalshnúkur, Ice. 2,119 (6,952)
Illampu, Nevado, Bol. 6,421 (21,066)
Illimani, Nevado de, Bol. 6,457 (21,184)
Imeni Ismail Samani, Pik (Communism Pk.), Taj. 7,495 (24,590)
Inthanon, Doi, Thai. 2,600 (8,530)
Jaya, Puncak, Indon. 5,030 (16,503)
Jungfrau, Switz. 4,158 (13,642)
K2 (Qogir Feng), Asia 8,611 (28,250)
Kāmet, Asia 7,756 (25,446)
Kānchenjunga, Asia 8,598 (28,208)
Karisimbi, Volcan, Afr. 4,507 (14,787)
Kebnekaise, Swe. 2,111 (6,926)
Kenya,Mt.,(Kirinyaga), Kenya 5,199 (17,058)
Kerinci, Gunung, Indon. 3,800 (12,467)
Kilimanjaro, Tan. 5,895 (19,340)
Kinabalu, Gunong, Malay. 4,101 (13,455)
Kinyeti, Sudan 3,187 (10,456)
Klyuchevskaya Sopka, Vulkan, Russia 4,750 (15,584)
Kosciuszko, Mt., Austl. 2,229 (7,313)
Koussi, Emi, Chad 3,415 (11,204)
Kula Kangri, Bhu. 7,554 (24,784)
La Selle, Morne, Haiti 2,674 (8,773)
Lassen Pk., Ca., U.S. 3,187 (10,457)
Llullaillaco, Volcán, S.A. 6,739 (22,110)
Logan, Mt., Yk., Can. 5,959 (19,551)
Longs Pk., Co., U.S. 4,345 (14,255)
Margherita Pk., Afr. 5,109 (16,763)
Maromokotro, Madag. 2,876 (9,436)
Massive, Mt., Co., U.S. 4,396 (14,421)
Matterhorn, Eur. 4,478 (14,692)
Mauna Kea, Hi., U.S. 4,205 (13,796)
Mauna Loa, Hi., U.S. 4,169 (13,677)
Mayon Volcano, Phil. 2,462 (8,077)
Meru, Mt., Tan. 4,565 (14,977)
Misti, Volcán, Peru 5,822 (19,101)
Mitchell, Mt., N.C., U.S. 2,037 (6,684)
Môco, Morro de, Ang. 2,620 (8,596)
Moldoveanu, Vârful, Rom. 2,544 (8,346)
Mulhacén, Spain 3,482 (11,424)
Musala, Blg. 2,925 (9,596)
Muztag, China 7,723 (25,338)
Namjagbarwa Feng, China 7,755 (25,443)
Nanda Devi, India 7,817 (25,645)
Nanga Parbat, Pak. 8,126 (26,660)
Nevis, Ben, Scot., U.K. 1,343 (4,406)
Ojos del Salado, Nevado, S.A. 6,893 (22,615)
Olympus, Mt., Grc. 2,917 (9,570)
Paektu-san, Asia 2,744 (9,003)
Paricutín, Mex. 2,800 (9,186)
Parnassós, Grc. 2,457 (8,061)
Pelée, Montagne, Mart. 1,397 (4,583)
Pico de Orizaba, Volcán, Mex. 5,610 (18,406)
Pidurutalagala, Sri L. 2,524 (8,281)
Pikes Pk., Co., U.S. 4,301 (14,110)
Pinatubo, Mt., Phil. 1,780 (5,840)
Pobeda, Gora Russia 3,147 (10,325)
Popocatépetl, Volcán, Mex. 5,465 (17,930)
Pulog, Mt., Phil. 2,934 (9,626)
Rainier, Mt., Wa., U.S. 4,392 (14,411)
Ramm, Jabal, Jord. 1,754 (5,755)
Ras Dejen, Eth. 4,620 (15,158)
Rinjani, Gunung, Indon. 3,726 (12,224)
Robson, Mt., B.C., Can. 3,954 (12,972)
Roraima, Mt., S.A. 2,875 (9,432)
Ruapehu, Mt., N.Z. 2,797 (9,177)
Ruiz, Nevado del, Col. 5,400 (17,717)
Saint Elias, Mt., N.A. 5,489 (18,009)
Saint Helens, Mt., Wa., U.S. 2,549 (8,364)
Sajama, Nevado, Bol. 6,542 (21,463)
Semeru, Gunung, Indon. 3,676 (12,060)
Shām, Jabal ash-, Oman 3,035 (9,957)
Shasta, Mt., Ca., U.S. 4,317 (14,162)
Snowdon, Wales, U.K. 1,085 (3,560)
Tahat, Alg. 2,908 (9,541)
Tajumulco, Volcán, Guat. 4,220 (13,845)
Tirich Mīr, Pak. 7,690 (25,230)
Toubkal, Jebel, Mor. 4,165 (13,665)
Triglav, Slvn. 2,864 (9,396)
Trikora, Puncak (Wilhelmina Pk.), Indon. 4,750 (15,584)
Tupungato, Cerro, S.A. 6,750 (22,146)
Turquino, Pico, Cuba 1,972 (6,470)
Uluru (Ayers Rock), Austl. 863 (2,831)
Uncompahgre Pk., Co., U.S. 4,361 (14,309)
Vesuvius, Italy 1,281 (4,203)
Vinson Massif, Ant. 4,897 (16,066)
Waddington, Mt., B.C., Can. 4,015 (13,173)
Washington, Mt., N.H., U.S. 1,917 (6,288)
Whitney, Mt., Ca., U.S. 4,418 (14,494)
Wilhelm, Mt., Pap. N. Gui. 4,509 (14,793)
Yü Shan, Tai. 3,997 (13,114)
Zugspitze, Eur. 2,962 (9,718)

Principal Rivers Length in km (mi.)

Albany, N.A. 982 (610)
Aldan, Asia 2,209 (1,373)
Amazonas-Ucayali, S.A. 6,437 (4,000)
Amu Darya, Asia 1,687 (1,048)
Amur, Asia 2,820 (1,752)
Araguaia, S.A. 1,969 (1,223)
Arkansas, N.A. 2,350 (1,460)
Atchafalaya-Red, N.A. 2,285 (1,420)
Athabasca, N.A. 1,231 (765)
Ayeyarwady, Asia 1,573 (977)
Brahmaputra, Asia 3,235 (2,010)
Brazos, N.A. 2,060 (1,280)
Canadian, N.A. 1,458 (906)
Churchill, N.A. 1,609 (1,000)
Colorado, N.A. (U.S.-Mex.) 2,334 (1,450)
Colorado, N.A. (TX) 1,387 (862)
Columbia, N.A. 2,000 (1,243)
Congo, Afr. 4,370 (2,715)
Danube, Eur. 2,860 (1,777)
Darling, Austl. 1,472 (915)
Dnieper, Eur. 2,285 (1,420)
Don, Eur. 1,970 (1,224)
Elbe, Eur. 1,091 (678)
Essequibo, S.A. 970 (603)
Euphrates, Asia 2,412 (1,499)
Fraser, N.A. 1,370 (851)
Ganges, Asia 3,000 (1,864)
Gila, N.A. 1,044 (649)
Godāvari, Asia 1,500 (932)
Huang (Yellow), Asia 4,670 (2,902)
Indigirka, Asia 1,726 (1,072)
Indus, Asia 3,180 (1,976)
Juruá, S.A. 2,758 (1,714)
Kama, Eur. 1,685 (1,047)
Kasai, Afr. 1,968 (1,223)
Kolyma, Asia 2,130 (1,324)
Lena, Asia 4,400 (2,734)
Limpopo, Afr. 1,212 (753)
Loire, Eur. 1,110 (690)
Mackenzie, N.A. 4,241 (2,635)
Madeira, S.A. 3,381 (2,101)
Magdalena, S.A. 1,530 (951)
Marañón, S.A. 1,546 (961)
Mekong, Asia 4,500 (2,796)
Mississippi, N.A. 3,766 (2,340)
Mississippi-Missouri, N.A. 6,420, (3,989)
Missouri, N.A. 4,088 (2,540)
Murray, Austl. 2,375 (1,476)
Negro, S.A. 1,341 (833)
Nelson, N.A. 2,575 (1,600)
Niger, Afr. 4,160 (2,585)
Nile, Afr. 6,650 (4,132)
Ob', Asia 3,650 (2,268)
Oder, Eur. 906 (563)
Ohio, N.A. 2,108 (1,310)
Oka, Eur. 1,304 (810)
Orange, Afr. 2,300 (1,429)
Orinoco, S.A. 2,740 (1,703)
Ottawa, N.A. 1,271 (790)
Paraguay, S.A. 2,297 (1,427)
Paranaíba, S.A. 1,450 (901)
Peace, N.A. 1,923 (1,195)
Pechora, Eur. 1,810 (1,125)
Pecos, N.A. 1,490 (926)
Plata-Paraná, S.A. 4,700 (2,920)
Platte, N.A. 1,593 (990)
Purús, S.A. 2,588 (1,608)
Red, N.A. 2,076 (1,290)
Rhine, Eur. 1,320 (820)
Rhône, Eur 810 (503)
Rio Grande, N.A. 3,058 (1,900)
St. Lawrence, N.A. 3,058 (1,900)
Salado, S.A. 1,156 (718)
São Francisco, S.A. 2,800 (1,740)
Saskatchewan-Bow, N.A. 1,979 (1,230)
Severnaya Dvina (N. Dvina), Eur. 711 (442)
Snake, N.A. 1,674 (1,040)
Songhua (Sungari), Asia 872 (542)
Syr Darya, Asia 1,590 (988)
Tagus, Eur. 1,100 (684)
Tarim, Asia 964 (599)
Tennessee, N.A. 1,426 (886)
Tigris, Asia 1,752 (1,089)
Tisa, Eur. 881 (547)
Tocantins, S.A. 2,124 (1,320)
Ucayali, S.A. 1,484 (922)
Ural, Asia 2,102 (1,306)
Uruguay, S.A. 1,616 (1,004)
Vilyuy, Asia 2,446 (1,520)
Volga, Eur. 3,660 (2,274)
Volta, Afr. 1,600 (994)
Xiang, Asia 934 (580)
Xingu, S.A. 1,883 (1,170)
Yangtze (Chang), Asia 6,301 (3,915)
Yellowstone, N.A. 1,114 (692)
Yenisey, Asia 3,490 (2,169)
Yukon, N.A. 3,187 (1,980)
Zambezi, Afr. 2,660 (1,653)

Abidjan, Cote d'Ivoire (4,150,000) 1,929,079
Abu Dhabi (Abū Z̧aby), United Arab Emirates (879,000) 552,000
Accra, Ghana (2,060,000) 1,594,419
Ad-Dammām, Saudi Arabia 744,321
Addis Ababa (Ādīs Ābeba), Ethiopia (2,919,000) 2,646,000
Ahmadābād, India (6,210,000) 3,520,085
Aleppo (Ḩalab), Syria 4,450,000
Alexandria (El-Iskandarîya), Egypt 4,358,439
Algiers (Alger), Algeria (2,432,000) 1,569,897
Almaty, Kazakhstan 1,507,509
'Ammān, Jordan 2,376,022
Amsterdam, Netherlands (1,057,000) 779,808
Ankara, Turkey (4,166,000) 2,559,471
Antananarivo, Madagascar (2,021,000) 1,015,140
Antwerp (Antwerpen), Belgium (971,000) 498,473
Aşgabat, Turkmenistan 557,600
Asunción, Paraguay (2,040,000) 512,919
Athens (Athína), Greece (3,098,000) 664,046
Atlanta, United States (4,544,000) 420,003
Auckland, New Zealand 1,526,900
Baghdād, Iraq (5,891,000) 3,841,268
Baku (Bakı), Azerbaijan 2,166,355
Bamako, Mali (1,933,000) 658,275
Bandung, Indonesia (2,399,000) 2,394,873
Banghāzī, Libya (678,000) 446,250
Bangkok (Krung Thep), Thailand (8,213,000) 6,355,144
Bangui, Central African Republic (717,000) 451,690
Barcelona, Spain (4,934,000) 1,607,104
Barranquilla, Colombia (1,857,000) 990,547
Beijing, China (16,190,000) 6,690,000
Beirut (Bayrūt), Lebanon (1,990,000) 509,000
Belfast, United Kingdom (596,000) 296,700
Belgrade (Beograd), Serbia 1,357,005
Belo Horizonte, Brazil (5,409,000) 1,433,244
Bengalūru (Bangalore), India (8,275,000) 4,301,326
Berlin, Germany (3,475,000) 3,421,829
Birmingham, United Kingdom (2,429,000) 965,928
Bishkek, Kyrgyzstan 912,454
Bogotá, Colombia (8,506,000) 4,931,796
Bonn, Germany 311,287
Boston, United States (4,185,000) 617,594
Brasília, Brazil (3,710,000) 2,481,272
Brazzaville, Congo (1,574,000) 1,050,000
Brisbane, Australia (2,034,000) 1,041,839
Brussels (Bruxelles), Belgium (1,958,000) 174,383
Bucharest (București), Romania 1,912,515
Budapest, Hungary 1,740,188
Buenos Aires, Argentina (14,246,000) 2,776,138
Bulawayo, Zimbabwe 653,337
Busan, Korea, South 3,416,651
Cairo (El-Qâhira), Egypt (16,899,000) 7,248,671
Calgary, Canada 1,096,833
Cali, Colombia (2,402,000) 1,641,498
Cape Town, South Africa (3,345,000) 987,007
Caracas, Venezuela (2,901,000) 2,104,423
Cardiff, United Kingdom (444,000) 272,129
Casablanca, Morocco (3,405,000) 3,352,399
Changchun, China (3,387,000) 2,470,000
Chelyabinsk, Russia 1,149,829
Chengdu, China (6,234,000) 2,760,000
Chennai (Madras), India (8,523,000) 4,343,645
Chicago, United States (8,616,000) 2,695,598
Chișinău, Moldova 669,694
Chittagong, Bangladesh (4,106,000) 1,566,070
Chongqing, China (11,244,000) 3,870,000
Cincinnati, United States (1,628,000) 296,943
Cleveland, United States (1,781,000) 396,815
Cologne (Köln), Germany 1,034,175
Colombo, Sri Lanka 647,100
Conakry, Guinea (1,666,000) 1,091,500
Copenhagen (København), Denmark (1,192,000) 572,287
Córdoba, Argentina 1,267,521
Cotonou, Benin 862,445
Curitiba, Brazil (3,118,000) 1,751,907
Dakar, Senegal (2,929,000) 1,056,009
Dalian, China (3,862,000) 2,400,000
Dallas, United States (5,149,000) 1,197,816
Damascus (Dimashq), Syria (2,401,000) 1,680,000
Dar es Salaam, Tanzania (3,870,000) 2,497,940
Delhi, India (21,935,000) 9,879,172
Denver, United States (2,385,000) 600,158
Detroit, United States (3,730,000) 713,777
Dhaka, Bangladesh (14,731,000) 3,637,892
Dnipropetrovs'k, Ukraine 987,629
Donets'k, Ukraine 944,552
Douala, Cameroon (2,361,000) 1,907,479
Dubai (Dubayy), United Arab Emirates (1,778,000) 1,171,000
Dublin (Baile Átha Cliath), Ireland (1,100,000) 527,612
Durban, South Africa (2,739,000) 669,242
Dushanbe, Tajikistan 770,027
Düsseldorf, Germany 598,686
Edinburgh, United Kingdom 401,910
Edmonton, Canada (1,137,000) 812,201
El-Gîza, Egypt 3,122,041
Eşfahān, Iran 1,756,126
Essen, Germany 569,884
Faisalābād, Pakistan (3,017,000) 2,008,861
Fortaleza, Brazil (3,520,000) 862,750
Frankfurt am Main, Germany 701,350
Freetown, Sierra Leone 945,423
Fukuoka, Japan (5,556,000) 1,463,743
Glasgow, United Kingdom (1,207,000) 662,954
Goiânia, Brazil (2,049,000) 1,296,324
Guadalajara, Mexico (4,442,000) 1,600,894
Guangzhou (Canton), China (9,620,000) 3,750,000
Guatemala, Guatemala (2,584,000) 1,022,001
Guayaquil, Ecuador (2,492,000) 2,291,158
Hamburg, Germany (1,785,000) 1,746,342
Hannover, Germany 518,386
Ha Noi, Vietnam (2,811,000) 905,939
Harare, Zimbabwe 1,485,231
Harbin, China (4,896,000) 3,120,000
Havana (La Habana), Cuba (2,165,000) 2,121,871
Helsinki, Finland (1,122,000) 608,316
Hiroshima, Japan (2,119,000) 1,173,843
Ho Chi Minh City (Saigon), Vietnam (6,189,000) 2,796,229
Hong Kong (Xianggang), China (7,050,000) 1,250,993
Honolulu, United States (805,000) 337,256
Houston, United States (4,976,000) 2,099,451
Hyderābād, India (7,578,000) 3,637,483
Ibadan, Nigeria (2,814,000) 1,144,000
Islāmābād, Pakistan (1,039,000) 529,180
İstanbul, Turkey (12,703,000) 6,620,241
İzmir, Turkey (2,745,000) 1,757,414
Jaipur, India (3,017,000) 2,322,575
Jakarta, Indonesia (9,630,000) 9,607,787
Jerusalem (Yerushalayim), Israel 822,586
Jiddah, Saudi Arabia (3,452,000) 2,801,481
Jinan, China (3,493,000) 2,150,000
Johannesburg, South Africa (7,992,000) 752,349
Kabul (Kābol), Afghanistan (3,722,000) 3,289,000
Kampala, Uganda (1,594,000) 1,516,210
Kānpur, India (2,904,000) 2,551,337
Kaohsiung, Taiwan (1,514,000) 1,509,510
Karāchi, Pakistan (14,081,000) 9,339,023
Kathmandu (Kāṭhmāṇḍāu), Nepal 1,003,285
Katowice, Poland 308,269
Kharkiv, Ukraine 1,431,461
Khartoum (Al-Kharṭūm), Sudan (4,517,000) 947,483
Kiev (Kyïv), Ukraine (2,806,000) 2,803,716
Kigali, Rwanda (1,044,000) 603,049
Kingston, Jamaica 516,500
Kinshasa, Congo, Dem. Rep. of the (9,382,000) 3,000,000
Kolkata (Calcutta), India (14,283,000) 4,572,876
Kuala Lumpur, Malaysia (5,810,000) 1,698,250
Kuwait (Al-Kuwayt), Kuwait (2,102,000) 544,156
Lagos, Nigeria (10,781,000) 1,213,000
Lahore, Pakistan (7,487,000) 5,143,495
La Paz, Bolivia (1,672,000) 835,361
Leeds, United Kingdom (1,756,000) 424,194
León, Mexico (1,613,000) 1,137,465
Lilongwe, Malawi 669,021
Lima, Peru (8,955,000) 340,422
Lisbon (Lisboa), Portugal (2,812,000) 517,975
Liverpool, United Kingdom (863,000) 481,786
Lomé, Togo 839,566
London, United Kingdom (9,699,000) 7,650,944
Los Angeles, United States (12,160,000) 3,792,621
Luanda, Angola (4,508,000) 1,459,900
Lucknow, India (2,854,000) 2,185,927
Lusaka, Zambia 1,747,152
Lyon, France (1,551,000) 484,344
Madrid, Spain (5,787,000) 3,186,241
Managua, Nicaragua 864,201
Manaus, Brazil 1,792,881
Manchester, United Kingdom (2,538,000) 402,889
Manila, Philippines (11,891,000) 1,652,171
Mannheim, Germany 296,690
Maputo, Mozambique 1,225,868
Maracaibo, Venezuela (2,037,000) 1,339,019
Marrakech, Morocco 978,045
Marseille, France (1,560,000) 850,726
Mashhad, Iran 2,766,258
Mecca (Makkah), Saudi Arabia (1,543,000) 1,294,168
Medan, Indonesia (2,101,000) 2,097,610
Medellín, Colombia (3,510,000) 1,551,160
Melbourne, Australia (3,951,000) 93,625
Mexico City (Ciudad de México), Mexico (20,132,000) 8,720,916
Miami, United States (5,518,000) 399,457
Milan (Milano), Italy (3,056,000) 1,293,135
Milwaukee, United States (1,378,000) 594,833
Minneapolis, United States (2,658,000) 382,578
Minsk, Belarus 1,911,433
Mogadishu (Muqdisho), Somalia 1,212,000
Mombasa, Kenya 915,101
Monrovia, Liberia (1,056,000) 465,000
Monterrey, Mexico (4,113,000) 1,133,070
Montevideo, Uruguay (1,659,000) 1,305,082
Montréal, Canada (3,791,000) 1,649,519
Moscow (Moskva), Russia 11,918,057
Mumbai (Bombay), India (19,422,000) 11,978,450
Munich (München), Germany 1,407,836
Nagoya, Japan (9,165,000) 2,263,894
Nāgpur, India (2,471,000) 2,052,066
Nairobi, Kenya (3,237,000) 3,133,518
Nanjing, China (6,162,000) 2,490,000
Naples (Napoli), Italy (2,216,000) 974,082
N'Djamena, Chad (1,038,000) 451,690
Newcastle upon Tyne, United Kingdom (772,000) 189,150
New Delhi, India 302,363
New York, United States (18,365,000) 8,175,133
Nizhniy Novgorod (Gorky), Russia 1,257,260
Nouakchott, Mauritania 846,871
Novosibirsk, Russia 1,511,369
Nürnberg, Germany 498,876
Odesa, Ukraine 997,189
Omdurman (Umm Durmān), Sudan 1,271,403
Omsk, Russia 1,158,627
Oran (Ouahran), Algeria 705,335
Ōsaka, Japan (19,492,000) 2,665,314
Oslo, Norway (898,000) 634,293
Ottawa, Canada (1,218,000) 883,391
Ouagadougou, Burkina Faso (1,914,000) 1,475,223
Palembang, Indonesia 1,455,284
Panamá, Panama (1,504,000) 465,179
Paris, France (10,460,000) 2,243,833
Perm', Russia 1,007,272
Perth, Australia (1,698,000) 16,714
Philadelphia, United States (5,449,000) 1,526,006
Phnom Penh (Phnum Pénh), Cambodia 1,570,791
Phoenix, United States (3,649,000) 1,445,632
Port-au-Prince, Haiti (2,141,000) 990,558
Portland, United States (1,857,000) 583,776
Port Louis, Mauritius 150,353
Port Moresby, Papua New Guinea 254,158
Porto, Portugal (1,285,000) 224,894
Porto Alegre, Brazil (3,476,000) 1,409,351
Prague (Praha), Czech Republic 1,244,762
Pretoria (Tshwane), South Africa (1,666,000) 692,348
Puebla de Zaragoza, Mexico (2,733,000) 1,399,519
Pune, India (4,951,000) 2,538,473
P'yŏngyang, Korea, North (2,833,000) 2,581,076
Qingdao, China (3,952,000) 2,300,000
Québec, Canada (757,000) 516,622
Quezon City, Philippines 2,761,720
Quito, Ecuador 1,619,146
Rabat, Morocco (1,799,000) 578,644
Recife, Brazil (3,559,000) 1,537,704
Rīga, Latvia 643,492
Rio de Janeiro, Brazil (12,374,000) 6,320,446
Riyadh (Ar-Riyāḍ), Saudi Arabia (5,227,000) 4,087,152
Rome (Roma), Italy (3,592,000) 2,751,082
Rosario, Argentina (1,298,000) 908,163
Rostov-na-Donu, Russia 1,100,091
Rotterdam, Netherlands (996,000) 610,386
Sacramento, United States (1,733,000) 466,488
St. Louis, United States (2,153,000) 319,294
St. Petersburg (Leningrad), Russia 4,990,602
Salvador, Brazil (3,343,000) 2,674,923
Samara, Russia 1,170,381
San Diego, United States (2,964,000) 1,307,402
San Francisco, United States (3,283,000) 805,235
San José, Costa Rica (1,122,000) 288,054
San Juan, Puerto Rico (2,478,000) 374,682
San Salvador, El Salvador (1,086,000) 281,870
Santiago, Chile (6,269,000) 5,128,041
Santo Domingo, Dominican Republic (2,601,000) 1,126,306
São Paulo, Brazil (19,660,000) 11,253,503
Sapporo, Japan (2,591,000) 1,913,545
Saratov, Russia 838,321
Seattle, United States (3,069,000) 608,660
Seoul, Korea, South 10,007,651
Shanghai, China (19,980,000) 8,930,000
Shenyang, China (5,676,000) 4,050,000
Singapore, Singapore 5,469,724
Sofia (Sofiya), Bulgaria 1,210,820
Stockholm, Sweden (1,360,000) 789,024
Stuttgart, Germany 604,297
Surabaya, Indonesia (2,768,000) 2,765,487
Sūrat, India (4,438,000) 2,433,835
Sydney, Australia (4,364,000) 169,505
Tabrīz, Iran 1,494,998
T'aipei, Taiwan (2,654,000) 2,641,856
Tallinn, Estonia 411,063
Tashkent (Toshkent), Uzbekistan (2,209,000) 2,137,218
Tbilisi, Georgia 1,172,700
Tegucigalpa, Honduras 858,437
Tehrān, Iran 8,154,051
Tel Aviv-Yafo, Israel (3,311,000) 416,577
Tianjin, China (9,451,000) 5,000,000
Tijuana, Mexico (1,755,000) 1,286,187
Tōkyō, Japan (36,834,000) 8,945,695
Toronto, Canada (5,499,000) 2,615,060
Tripoli (Ṭarābulus), Libya (1,095,000) 591,062
Tunis, Tunisia (1,916,000) 702,330
Turin (Torino), Italy (1,737,000) 887,114
Ufa, Russia 1,075,007
Ulaanbaatar, Mongolia 1,367,508
Ürümqi, China (2,811,000) 1,130,000
València, Spain 789,364
Vancouver, Canada (2,278,000) 603,502
Vienna (Wien), Austria 1,714,227
Vientiane (Viangchan), Laos (761,000) 464,000
Vilnius, Lithuania 538,430
Volgograd, Russia 1,018,762
Warsaw (Warszawa), Poland 1,711,324
Washington, United States (4,604,000) 601,723
Winnipeg, Canada 663,617
Wuhan, China (7,515,000) 3,870,000
Xi'an, China (5,149,000) 2,410,000
Yangon (Rangoon), Myanmar 5,209,541
Yekaterinburg, Russia 1,386,909
Yerevan, Armenia 1,060,138
Yokohama, Japan 3,688,773
Zagreb, Croatia 790,017
Zürich, Switzerland (1,186,000) 378,884

Values are latest available city populations or recent estimates.
Metropolitan area populations are shown in parentheses.

Annam ... Annamese
Arab ... Arabic
Bantu ... Bantu
Bur ... Burmese
Camb ... Cambodian
Celt ... Celtic
Chn ... Chinese
Czech ... Czech
Dan ... Danish
Du ... Dutch
Fin ... Finnish
Fr ... French
Ger ... German
Gr ... Greek
Hung ... Hungarian
Ice ... Icelandic
India ... India
Indian ... American Indian
Indon ... Indonesian
It ... Italian
Jap ... Japanese
Kor ... Korean
Mal ... Malayan
Mong ... Mongolian
Nor ... Norwegian
Per ... Persian
Pol ... Polish
Port ... Portuguese
Rom ... Romanian
Rus ... Russian
Serb ... Serbian
Siam ... Siamese
So. Slav ... Southern Slavonic
Sp ... Spanish
Swe ... Swedish
Tib ... Tibetan
Tur ... Turkish

å, Nor., Swe. ... brook, river
aa, Dan., Nor ... brook
āb, Per ... water, river
abad, India, Per ... town, city
ada, Tur ... island
adrar, Berber ... mountain
ákra, Gr ... cape
älf, Swe ... river
alp, Ger ... mountain
altipiano, It ... plateau
alto, Sp ... height
archipel, Fr ... archipelago
archipiélago, Sp ... archipelago
arquipélago, Port ... archipelago
arroyo, Sp ... brook, stream
as, Nor., Swe ... ridge
austral, Sp ... southern
baai, Du ... bay
bab, Arab ... gate, port
bach, Ger ... brook, stream
backe, Swe ... Hill
bad, Ger ... bath, spa
bahía, Sp ... bay, gulf
bahr, Arab ... river, sea, lake
baia, It ... bay, gulf
baía, Port ... bay
baie, Fr ... bay, gulf
bajo, Sp ... depression
bak, Indon ... stream
bakke, Dan., Nor ... hill
balkan, Tur ... mountain range
bana, Jap ... point, cape
banco, Sp ... bank
bandao, Chn ... peninsula
bandar, Mal., Per .. town, port, harbor
bang, Siam ... village
bassin, Fr ... basin
batang, Indon., Mal ... river
bei, Chn ... north
ben, Celtic ... mountain, summit
bender, Arab ... harbor, port
bereg, Rus ... coast, shore
berg, Du., Ger., Nor., Swe. ... mountain, hill
bir, Arab ... well
birkat, Arab ... lake, pond, pool
bit, Arab ... house
bjaerg, Dan., Nor ... mountain
bocche, It ... mouth
boğazı, Tur ... strait
bois, Fr ... forest, wood
bolsón, Sp ... flat-floored desert valley
boreal, Sp ... northern
borg, Dan., Nor., Swe ... castle, town
borgo, It ... town, suburb
bosch, Du ... forest, wood
bouche, Fr ... river mouth
bourg, Fr ... town, borough
bro, Dan., Nor., Swe ... bridge
brücke, Ger ... bridge
bucht, Ger ... bay, bight
bugt, Dan., Nor., Swe ... bay, gulf
bulu, Indon ... mountain
burg, Du., Ger ... castle, town
buri, Siam ... town
burun, burnu, Tur ... cape
by, Dan., Nor., Swe ... village
caatinga, Port. (Brazil) ... open brushland
cabezo, Sp ... summit
cabo, Port., Sp. ... cape
campo, It., Port., Sp ... plain, field
campos, Port. (Brazil) ... plains
cañón, Sp ... canyon
cap, Fr ... cape
capo, It ... cape
casa, It., Port., Sp ... house
castello, It., Port ... castle, fort
castillo, Sp ... castle
càte, Fr ... hill
çay, Tur ... stream, river
cayo, Sp ... rock, shoal, islet
cerro, Sp ... mountain, hill
champ, Fr ... field
château, Fr ... castle
chott, Arab ... salt lake
chu, Tib ... water, stream
cidade, Port ... town, city
cima, Sp ... summit, peak
città, It ... town, city
ciudad, Sp ... town, city
cochilha, Port ... ridge
col, Fr ... pass
colina, Sp ... hill
cordillera, Sp ... mountain chain
costa, It., Port., Sp ... coast
côte, Fr ... coast
cuchilla, Sp ... mountain ridge
dağ, Tur ... mountain(s)
dake, Jap ... peak, summit
dal, Dan., Du., Nor., Swe ... valley
dan, Kor ... point, cape
danau, Indon ... lake
dao, Chn ... island
dar, Arab ... house, abode, country
darya, Per ... river, sea
dasht, Per ... plain, desert
deniz, Tur ... sea
désert, Fr ... desert
deserto, It ... desert
desierto, Sp ... desert
détroit, Fr ... strait
dijk, Du ... dam, dike
djebel, Arab ... mountain
do, Kor ... island
dong, Chn ... east
dorf, Ger ... village
dorp, Du ... village
duin, Du ... dune
dzong, Tib fort, administrative capital
eau, Fr ... water
ecuador, Sp ... equator
eiland, Du ... island
elv, Dan., Nor ... river, stream
embalse, Sp ... reservoir
erg, Arab ... dune, sandy desert
est, Fr., It ... east
estado, Sp ... state
este, Port., Sp ... east
estrecho, Sp ... strait
étang, Fr ... pond, lake
état, Fr ... state
eyjar, Ice ... islands
feld, Ger ... field, plain
festung, Ger ... fortress
fiume, It ... river
fjäll, Swe ... mountain
fjärd, Swe ... bay, inlet
fjeld, Nor ... mountain, hill
fjord, Dan., Nor ... fiord, inlet
fjördur, Ice ... fiord, inlet
fleuve, Fr ... river
flod, Dan., Swe ... river
flói, Ice ... bay, marshland
fluss, Ger ... river
foce, It ... river mouth
fontein, Du ... a spring
forêt, Fr ... forest
fors, Swe ... waterfall
forst, Ger ... forest
fos, Dan., Nor ... waterfall
fu, Chn ... town, residence
fuente, Sp ... spring, fountain
fuerte, Sp ... fort
furt, Ger ... ford
gang, Kor ... stream, river
gangri, Tib ... mountain
gat, Dan., Nor ... channel
gàve, Fr ... stream
gawa, Jap ... river
gebergte, Du ... mountain range
gebiet, Ger ... district, territory
gebirge, Ger ... mountains
ghat, India ... pass, mountain range
gobi, Mong ... desert
gol, Mong ... river
göl, gölü, Tur ... lake
golfe, Fr ... gulf, bay
golfo, It., Port., Sp ... gulf, bay
gomba, gompa, ... Tibmonastery
gora, Rus., So. Slav ... mountain
góra, Pol ... mountain
gorod, Rus ... town
grad, Rus., So. Slav ... town
guba, Rus ... bay, gulf
gundung, Indon ... mountain
guntō, Jap ... archipelago
gunung, Mal ... mountain
haf, Swe ... sea, ocean
hafen, Ger ... port, harbor
haff, Ger ... gulf, inland sea
hai, Chn ... sea, lake
hama, Jap ... beach, shore
hamada, Arab ... rocky plateau
hamn, Swe ... harbor
hāmūn, Per ... swampy lake, plain
hantō, Jap ... peninsula
hassi, Arab ... well, spring
haus, Ger ... house
haut, Fr ... summit, top
hav, Dan., Nor ... sea, ocean
havn, Dan., Nor ... harbor, port
havre, Fr ... harbor, port
háza, Hung ... house, dwelling of
heim, Ger ... hamlet, home
hem, Swe ... hamlet, home
higashi, Jap ... east
hisar, Tur ... fortress
hissar, Arab ... fort
ho, Chn ... river
hoek, Du ... cape
hof, Ger ... court, farmhouse
höfn, Ice ... harbor
hoku, Jap ... north
holm, Dan., Nor., Swe ... island
hora, Czech ... mountain
horn, Ger ... peak
hoved, Dan., Nor ... cape
hu, Chn ... lake
huang, Chn ... yellow
hügel, Ger ... hill
huk, Dan., Swe ... point
hus, Dan., Nor., Swe ... house
île, Fr ... island
ilha, Port ... island
indsö, Dan., Nor ... lake
insel, Ger ... island
insjö, Swe ... lake
irmak, irmagi, Tur ... river
isla, Sp ... island
isola, It ... island
istmo, It., Sp ... isthmus
jarvi, jaur, Fin ... lake
jebel, Arab ... mountain
jiang, Chn ... river
jima, Jap ... island
jökel, Nor ... glacier
joki, Fin ... river
jökuli, Ice ... glacier
kaap, Du ... cape
kai, Jap ... bay, gulf, sea
kaikyō, Jap ... channel, strait
kalat, Per ... castle, fortress
kale, Tur ... fort
kali, Mal ... creek, river
kand, Per ... village
kap, Dan., Ger ... cape
kapp, Nor, Swe ... cape
kasr, Arab ... fort, castle
kawa, Jap ... river
kefr, Arab ... village
kei, Jap ... creek, river
ken, Jap ... prefecture
khor, Arab ... bay, inlet
khrebet, Rus ... mountain range
kita, Jap ... north
ko, Jap ... lake
köbstad, Dan ... market-town
kol, Mong ... lake
kólpos, Gr ... gulf
kong, Chn ... river
kopf, Ger ... head, summit, peak
köpstad, Swe ... market town
körfezi, Tur ... gulf
kosa, Rus ... spit
kou, Chn ... river mouth
köy, Tur ... village
kraal, Du. (Africa) ... native village
ksar, Arab ... fortified village
kuala, Mal ... bay, river mouth
kuh, Per ... mountain
kum, Tur ... sand
kuppe, Ger ... summit
küste, Ger ... coast
kyo, Jap ... town, capital
la, Tib ... mountain pass
labuan, Mal ... anchorage, port
lac, Fr ... lake
lago, It., Port., Sp ... lake
lagoa, Port ... lake, bay
laguna, It., Port., Sp ... lagoon, lake
lahti, Fin ... bay, gulf
lan, Swe ... county
landsby, Dan., Nor ... village
liman, Tur ... bay, port
ling, Chn ... pass, ridge, mountain
llanos, Sp ... plains
loch, Celt. (Scotland) ... lake, bay
loma, Sp ... long, low hill
lough, Celt. (Ireland) ... lake, bay
machi, Jap ... town
man, Kor ... bay
mar, Port., Sp ... sea
mare, It., Rom ... sea
marisma, Sp ... marsh, swamp
mark, Ger ... boundary limit
massif, Fr ... block of mountains
mato, Port ... forest, thicket
me, Siam ... river
meer, Du., Ger ... lake, sea
mer, Fr ... sea
mesa, Sp ... flat-topped mountain
meseta, Sp ... plateau
mina, Port., Sp ... mine
minami, Jap ... south
minato, Japan ... harbor, haven
misaki, Jap ... cape, headland
mont, Fr ... mount, mountain
montagna, It ... mountain
montagne, Fr ... mountain
montaña, Sp ... mountain
monte, It., Port., Sp .. mount, mountain
more, Rus., So. Slav ... sea
morro, Port., Sp ... hill, bluff
mühle, Ger ... mill
mund, Ger ... mouth, opening
mündung, Ger ... river mouth
mura, Jap ... township
myit, Bur ... river
mys, Rus ... cape
nada, Jap ... sea
nadi, India ... river, creek
naes, Dan., Nor ... cape
nafud, Arab ... desert of sand dunes
nagar, India ... town, city
nahr, Arab ... river
nam, Siam ... river, water
nan, Chn., Jap ... south
näs, Nor., Swe ... cape
nez, Fr ... point, cape
nishi, nisi, Jap ... west
njarga, Fin ... peninsula
nong, Siam ... marsh
noord, Du ... north
nor, Mong ... lake
nord, Dan., Fr., Ger., It., Nor., Swe .. north
norte, Port., Sp ... north
nos, Rus ... cape
nyasa, Bantu ... lake
ö, Dan., Nor., Swe ... island
occidental, Sp ... western
ocna, Rom ... salt mine
odde, Dan., Nor ... point, cape
oeste, Port., Sp ... west
oka, Jap ... hill
oost, Du ... east
oriental, Sp ... eastern
óros, Gr ... mountain
ost, Ger., Swe ... east
öster, Dan., Nor., Swe ... eastern
ostrov, Rus ... island
oued, Arab ... river, stream
ouest, Fr ... west
ozero, Rus ... lake
pää, Fin ... mountain
padang, Mal ... plain, field
pampas, Sp. (Argentina) .grassy plains
pará, Indian (Brazil) ... river
pas, Fr ... channel, passage
paso, Sp ... mountain pass, passage
passo, It., Port ... mountain pass, passage, strait
patam, India ... city, town
pélagos, Gr ... open sea
pegunungan, Indon ... mountains
peña, Sp ... rock
pendi, Chn ... basin
pertuis, Fr ... strait
pic, Fr ... mountain peak
pico, Port., Sp ... mountain peak
piedra, Sp ... stone, rock
ping, Chn ... plain, flat
planalto, Port ... plateau
planina, Serb ... mountains
playa, Sp ... shore, beach
ploskogor'ye, Rus ... mountains
pnom, Camb ... mountain
pointe, Fr ... point
polder, Du., Ger ... reclaimed marsh
polje, So. Slav ... plain, field
poluostrov, Rus ... peninsula
pont, Fr ... bridge
ponta, Port ... point, headland
ponte, It., Port ... bridge
pore, India ... city, town
porthmós, Gr ... strait
porto, It., Port ... port, harbor
potamós, Gr ... river
prado, Sp ... field, meadow
presqu'île, Fr ... peninsula
proliv, Rus ... strait
pueblo, Sp ... town, village
puerto, Sp ... port, harbor
pulau, Indon ... island
punkt, Ger ... point
punt, Du ... point
punta, It., Sp ... point
pur, India ... city, town
puy, Fr ... peak
qal'a, qal'at, Arab ... fort, village
qasr, Arab ... fort, castle
rann, India ... wasteland
ra's, Arab ... cape, head
reka, Rus., So. Slav ... river
represa, Port ... reservoir
rettō, Jap ... island chain
ría, Sp ... estuary
ribeira, Port ... stream
riberão, Port ... river
rio, It., Port ... stream, river
río, Sp ... river
rivière, Fr ... river
roca, Sp ... rock
rt, Serb ... cape
rīd, Per ... river
saari, Fin ... island
sable, Fr ... sand
sahara, Arab ... desert, plain
saki, Jap ... cape
sal, Sp ... salt
salar, Sp ... salt flat, salt lake
salto, Sp ... waterfall
san, Jap., Kor ... mountain, hill
sat, satul, Rom ... village
schloss, Ger ... castle
sebkha, Arab ... salt marsh
see, Ger ... lake, sea
şehir, Tur ... town, city
selat, Indon ... strait
selvas, Port., (Brazil) .tropical rain forests
seno, Sp ... bay
serra, Port ... mountain chain
serrania, Sp ... mountain ridge
seto, Jap ... strait
severnaya, Rus ... northern
shahr, Per ... town, city
shamo, Chn ... desert
shan, Chn ... mountain, hill, island
shatt, Arab ... river
shi, Jap, Chn ... city
shima, Jap ... island
shōtō, Jap ... archipelago
sierra, Sp ... mountain range
sjö, Nor., Swe ... lake, sea
sö, Dan., Nor ... lake, sea
söder, södra, Swe ... south
song, Annam ... river
sopka, Rus ... peak, volcano
source, Fr ... a spring
spitze, Ger ... summit, point
staat, Ger ... state
stad, Dan., Du., Nor., Swe .. city, town
stadt, Ger ... city, town
stato, It ... state
step', Rus ... treeless plain, steppe
straat, Du ... strait
strand, Dan., Du., Ger., Nor., Swe ... shore, beach
stretto, It ... strait
strom, Ger ... river, stream
ström, Dan., Nor., Swe .. stream, river
stroom, Du ... stream, river
su, suyu, Tur ... water, river
sud, Fr., Sp ... south
süd, Ger ... south
suidō, Jap ... channel
sul, Port ... south
sund, Dan., Nor., Swe ... sound
sungai, sungei, Indon., Mal ... river
sur, Sp ... south
syd, Dan., Nor., Swe ... south
tafelland, Ger. ... plateau
take, Jap ... peak, summit
tal, Ger ... valley
tanjung, tanjong, Mal ... cape
târg, târgul, Rom ... market, town
tell, Arab ... hill
teluk, Indon ... bay, gulf
terra, It ... land
terre, Fr ... earth, land
thal, Ger ... valley
tierra, Sp ... earth, land
tō, Jap ... east; island
tonle, Camb ... river, lake
top, Du ... peak
torp, Swe ... hamlet, cottage
tsangpo, Tib ... river
tso, Tib ... lake
tsu, Jap ... harbor, port
tundra, Rus ... treeless arctic plains
tuz, Tur ... salt
udde, Swe ... cape
ufer, Ger ... shore, riverbank
ujung, Indon ... point, cape
umi, Jap ... sea, gulf
ura, Jap ... bay, coast, creek
ust'ye, Rus ... river mouth
valle, It., Port., Sp ... valley
vallée, Fr ... valley
valli, It ... lake
vár, Hung ... fortress
város, Hung ... town
varoš, So. Slav ... town
veld, Du ... open plain, field
verkh, Rus ... top, summit
ves, Czech ... village
vest, Dan., Nor., Swe ... west
vik, Swe ... cove, bay
vila, Port ... town
villa, Sp ... town
villar, Sp ... village, hamlet
ville, Fr ... town, city
vodokhranilishche, Rus ... reservoir
vostok, Rus ... east
wad, wādī, Arab intermittent stream
wald, Ger ... forest, woodland
wan, Chn., Jap ... bay, gulf
weiler, Ger ... hamlet, village
westersch, Ger ... western
wüste, Ger ... desert
xi, Chn ... west, western
yama, Jap ... mountain
yarimada, Tur ... peninsula
yug, Rus ... south
zaki, Jap ... cape
zaliv, Rus ... bay, gulf
zapad, Rus ... west
zee, Du ... sea
zemlya, Rus ... land
zuid, Du ... south

Abbreviations of Geographic Names and Terms

Ab., Can.	Alberta, Can.
Afg.	Afghanistan
Afr.	Africa
Ak., U.S.	Alaska, U.S.
Al., U.S.	Alabama, U.S.
Alb.	Albania
Alg.	Algeria
Am. Sam.	American Samoa
And.	Andorra
Ang.	Angola
Ant.	Antarctica
Antig.	Antigua and Barbuda
Ar., U.S.	Arkansas, U.S.
Arg.	Argentina
Arm.	Armenia
Aus.	Austria
Austl.	Australia
Az., U.S.	Arizona, U.S.
Azer.	Azerbaijan
b.	Bay, Gulf, Inlet, Lagoon
Bah.	Bahamas
Bahr.	Bahrain
Barb.	Barbados
bas.	Basin
B.C.,	British Columbia, Can.
Bdi.	Burundi
Bel.	Belgium
Bela.	Belarus
Ber.	Bermuda
Bhu.	Bhutan
B.I.O.T.	British Indian Ocean Territory
Blg.	Bulgaria
Bngl.	Bangladesh
Bol.	Bolivia
Bos.	Bosnia and Herzegovina
Bots.	Botswana
Braz.	Brazil
Bru.	Brunei
Br. Vir. Is.	British Virgin Islands
Burkina	Burkina Faso
c.	Cape, Point
Ca., U.S.	California, U.S.
Camb.	Cambodia
Camrn.	Cameroon
can.	Canal
Can.	Canada
C.A.R.	Central African Republic
Cay. Is.	Cayman Islands
C. Iv.	Cote d'Ivoire
clf.	Cliff, Escarpment
co.	County, Parish
Co., U.S.	Colorado, U.S.
Col.	Colombia
Com.	Comoros
cont.	Continent
Cook Is.	Cook Islands
C.R.	Costa Rica
Cro.	Croatia
cst.	Coast, Beach
Ct., U.S.	Connecticut, U.S.
C.V.	Cape Verde
Cyp.	Cyprus
Czech Rep.	Czech Republic
d.	Dam
D.C., U.S.	District of Columbia, U.S.
De., U.S.	Delaware, U.S.
del.	Delta
Den.	Denmark
dep.	Dependency, Colony
depr.	Depression
des.	Desert
Dji.	Djibouti
Dom.	Dominica
Dom. Rep.	Dominican Republic
D.R.C.	Democratic Republic of the Congo
Ec.	Ecuador
El Sal.	El Salvador
Eng., U.K.	England, U.K.
Eq. Gui.	Equatorial Guinea
Erit.	Eritrea
Est.	Estonia
est.	Estuary
Eth.	Ethiopia
E. Timor	East Timor
Eur.	Europe
Falk. Is.	Falkland Islands
Far. Is.	Faroe Islands
Fin.	Finland
Fl., U.S.	Florida, U.S.
for.	Forest, Moor
Fr.	France
Fr. Gu.	French Guiana
Fr. Poly.	French Polynesia
Ga., U.S.	Georgia, U.S.

Gam.	The Gambia
Gaza.	Gaza Strip
Geor.	Georgia
Ger.	Germany
Gib.	Gibraltar
Grc.	Greece
Green.	Greenland
Gren.	Grenada
Guad.	Guadeloupe
Guat.	Guatemala
Guern.	Guernsey
Gui.	Guinea
Gui.-B.	Guinea-Bissau
Guy.	Guyana
Hi., U.S.	Hawaii, U.S.
hist.	Historic Site, Ruins
hist. reg.	Historic Region
Hond.	Honduras
Hung.	Hungary
i.	Island
Ia., U.S.	Iowa, U.S.
ice	Ice Feature, Glacier
Ice.	Iceland
Id., U.S.	Idaho, U.S.
Il., U.S.	Illinois, U.S.
In., U.S.	Indiana, U.S.
Indon.	Indonesia
ind. res.	Indian Reservation
I. of Man	Isle of Man
Ire.	Ireland
is.	Islands
Isr.	Israel
isth.	Isthmus
Jam.	Jamaica
Jord.	Jordan
Kaz.	Kazakhstan
Kir.	Kiribati
Kor., N.	Korea, North
Kor., S.	Korea, South
Kos.	Kosovo
Ks., U.S.	Kansas, U.S.
Kuw.	Kuwait
Ky., U.S.	Kentucky, U.S.
Kyrg.	Kyrgyzstan
La., U.S.	Louisiana, U.S.
Lat.	Latvia
Leb.	Lebanon
Leso.	Lesotho
Lib.	Liberia
Liech.	Liechtenstein
Lith.	Lithuania
lk.	Lake
Lux.	Luxembourg
Ma., U.S.	Massachusetts, U.S.
Mac.	Macedonia
Madag.	Madagascar
Malay.	Malaysia
Mald.	Maldives
Marsh. Is.	Marshall Islands
Mart.	Martinique
Maur.	Mauritania
May.	Mayotte
Mb., Can.	Manitoba, Can.
Md., U.S.	Maryland, U.S.
Me., U.S.	Maine, U.S.
Mex.	Mexico
Mi., U.S.	Michigan, U.S.
Micron.	Micronesia, Federated States of
Mn., U.S.	Minnesota, U.S.
Mo., U.S.	Missouri, U.S.
Mol.	Moldova
Mong.	Mongolia
Mont.	Montenegro
Mor.	Morocco
Moz.	Mozambique
Ms., U.S.	Mississippi, U.S.
Mt., U.S.	Montana, U.S.
mth.	River Mouth or Channel
mtn.	Mountain
mts.	Mountains
Mya.	Myanmar
N.A.	North America
nat. cap.	National Capital
N.B., Can.	New Brunswick, Can.
N.C., U.S.	North Carolina, U.S.
N. Cal.	New Caledonia
N.D., U.S.	North Dakota, U.S.
Ne., U.S.	Nebraska, U.S.
Neth.	Netherlands
Neth. Ant.	Netherlands Antilles
Nf., Can.	Newfoundland, Can.
N.H., U.S.	New Hampshire, U.S.
Nic.	Nicaragua
Nig.	Nigeria

N. Ire., U.K.	Northern Ireland, U.K.
N.J., U.S.	New Jersey, U.S.
N.M., U.S.	New Mexico, U.S.
N. Mar. Is.	Northern Mariana Islands
Nmb.	Namibia
Nor.	Norway
n.p.	National Park or Monument
N.S., Can.	Nova Scotia, Can.
N.T., Can.	Northwest Territories, Can.
Nu., Can.	Nunavut, Can.
Nv., U.S.	Nevada, U.S.
N.Y., U.S.	New York, U.S.
N.Z.	New Zealand
oc.	Ocean
Oc.	Australia and Oceania
Oh., U.S.	Ohio, U.S.
Ok., U.S.	Oklahoma, U.S.
On., Can.	Ontario, Can.
Or., U.S.	Oregon, U.S.
p.	Pass
Pa., U.S.	Pennsylvania, U.S.
Pak.	Pakistan
Pan.	Panama
Pap. N. Gui.	Papua New Guinea
P.E., Can.	Prince Edward I., Can.
Para.	Paraguay
pen.	Peninsula
Phil.	Philippines
Pit.	Pitcairn
pk.	Park, Reserve
pl.	Plain, Flat
plat.	Plateau, Highland
p.o.i.	Point of Interest
Pol.	Poland
Port.	Portugal
P.R.	Puerto Rico
Qc., Can.	Québec, Can.
r.	Rock
rec.	Recreational Site, Park
reg.	Physical Region
res.	Reservoir
Reu.	Reunion
rf.	Reef, Shoal
R.I., U.S.	Rhode Island, U.S.
Rom.	Romania
Rw.	Rwanda
s.	Sea
S.A.	South America
S. Afr.	South Africa
Sau. Ar.	Saudi Arabia
S.C., U.S.	South Carolina, U.S.
Scot., U.K.	Scotland, U.K.
S.D., U.S.	South Dakota, U.S.
Sen.	Senegal
Serb.	Serbia
Sey.	Seychelles
S. Geor.	South Georgia
Sing.	Singapore
Sk., Can.	Saskatchewan, Can.
S.L.	Sierra Leone
Slvk.	Slovakia
Slvn.	Slovenia
S. Mar.	San Marino
Sol. Is.	Solomon Islands
Som.	Somalia
Sp. N. Afr.	Spanish North Africa
Sri L.	Sri Lanka
state	State, Province, Department, Region, etc.
St. Hel.	St. Helena
St. K./N.	St. Kitts and Nevis
St. Luc.	St. Lucia
stm.	River, Creek, Stream
St. P./M.	St. Pierre and Miquelon
S. Tom./P.	Sao Tome and Principe
strt.	Strait, Channel, Sound
St. Vin.	St. Vincent and the Grenadines
Sur.	Suriname
sw.	Swamp, Marsh
Swaz.	Swaziland
Swe.	Sweden
Switz.	Switzerland
Tai.	Taiwan
Taj.	Tajikistan
Tan.	Tanzania
T./C. Is.	Turks and Caicos Islands
Thai.	Thailand
Tn., U.S.	Tennessee, U.S.
Tok.	Tokelau
Trin.	Trinidad and Tobago
Tun.	Tunisia
Tur.	Turkey
Turkmen.	Turkmenistan

Tx., U.S.	Texas, U.S.
U.A.E.	United Arab Emirates
Ug.	Uganda
U.K.	United Kingdom
Ukr.	Ukraine
Ur.	Uruguay
U.S.	United States
Ut., U.S.	Utah, U.S.
Uzb.	Uzbekistan
Va., U.S.	Virginia, U.S.
val.	Valley, Watercourse
Ven.	Venezuela
Viet.	Vietnam
V.I.U.S.	Virgin Islands (U.S.)
vol.	Volcano
Vt., U.S.	Vermont, U.S.
Wa., U.S.	Washington, U.S.
Wal./F.	Wallis and Futuna
W.B.	West Bank
Wi., U.S.	Wisconsin, U.S.
W. Sah.	Western Sahara
wtfl.	Waterfall
W.V., U.S.	West Virginia, U.S.
Wy., U.S.	Wyoming, U.S.
Yk., Can.	Yukon, Can.
Zam.	Zambia
Zimb.	Zimbabwe

Pronunciation of Geographic Names

Key to the sound values of letters and symbols used in the index to indicate pronunciation

ă	ăt; băttle
ǎ	finǎl; appeǎl
ā	rāte; elāte
å	senåte; inanimåte
ä	ärm; cälm
ȧ	ȧsk; bȧth
a	sofa; marine (short neutral or indeterminate sound)
â	fâre; prepâre
ch	choose; church
dh	as th in other; either
ē	bē; ēve
ê	êvent; crêate
ĕ	bĕt; ĕnd
ĕ	recĕnt (short neutral or indeterminate sound)
ē	cratēr; cindēr
g	gō; gāme
gh	guttural g
ĭ	bĭt; wĭll
ĭ	(short neutral or indeterminate sound)
ī	rīde; bīte
κ	gutteral k as ch n German ich
ng	sing
ŋ	bank; linger
N	indicates nasalized
ŏ	nŏd; ŏdd
ŏ	cŏmmit; cŏnnect
ō	ōld; bōld
ô	ôbey; hôtel
ô	ôrder; lông
oi	boil
oo	food; root
o	as oo in foot; wood
ou	out; thou
s	soft; so; sane
sh	dish; finish
th	thin; thick
ū	pūre; cūre
û	ûnite; ûsûrp
û	ûrn; fûr
ŭ	stŭd; ŭp
ŭ	circŭs; sŭbmit
ü	as in French tu
zh	as z in azure
'	indeterminate vowel sound

In many cases the spelling of foreign geographical names does not even remotely indicate the pronunciation to an American, e.g., Słupsk in Poland is pronounced swŏpsk; Jujuy in Argentina is pronounced hōō hwē; La Spezia in Italy is lä-spe'zyä.

This condition is hardly surprising, however, when we consider that in our own language Worcester, Massachusetts, is pronounced wŏs'tēr; Sioux City, Iowa, sōō sĭ'tē; Schuylkill Haven, Pennsylvania, skōōl'kĭl hä-vĕn; Poughkeepsie, New York, pŏ-kĭp'sē.

The indication of pronunciation of geographic names presents several peculiar problems:

1. Many foreign languages use sounds that are not present in the English language and which an American cannot normally articulate. Thus, though the nearest English equivalent sound has been indicated, only approximate results are possible.

2. There are several dialects in each foreign language that cause variation in the local pronunciation of names. This also occurs in identical names in the various divisions of a great language group.

3. Within the United States there are marked differences in pronunciation, not only of local geographic names, but also of common words, indicating that the sound and tone values for letters as well as the placing of the emphasis vary considerably from one part of the country to another.

4. A number of different letters and diacritical combinations could be used to indicate essentially the same or approximate pronunciations.

Some variation in pronunciation other than that indicated in this index may be encountered, but such a difference does not necessarily indicate that either is in error, and in many cases it is a matter of individual choice as to which is preferred. In fact, an exact indication of pronunciation of many foreign names using English letters and diacritical marks is extremely diffiicult and sometimes impossible.

The following sources have been consulted during the process of creating and updating the thematic maps and statistics for the 23nd Edition.

Andreassen, L., M. Beedle, E. Berthier, F. Cawkwell, N. Dickmann, E. Dolgova, A. Fountain, N. Glasser, E. Hansson, U. Haritashya, G. Hartman, C. Helm, L. Iacovelli, H. Jiskoot, G. Kapustin, T. Khromova, J. Kincaid, S. Kutuzov, I. Lavrentiev, X. Li, L. Mabileau, J. Meyer, P. Mool, A. Muravyev, G. Nosenko, F. Paul, A. Racoviteanu, F. Rau, A. Rivera, M. Schnirch, Y. Seliverstov, O. Sigurdsson, S. Taschner, P. Zenteno, and N. Zheltyhina. (2001-2008). *GLIMS Glacier Database*. National Snow and Ice Data Center/World Data Center for Glaciology.

American Wind Energy Association (AWEA). (http://www.awea.org/)

Brinkhoff, Thomas. *City Population*. (http://www.citypopulation.de)

Brohan, P., J.J. Kennedy, I. Harris, S.F.B. Tett, and P.D. Jones. (2006). Uncertainty estimates in regional and global observed temperature changes: A new dataset from 1850. *Journal of Geophysical Research*, 111, D12106, doi:10.1029/2005JD006548. (http://www.cru.uea.ac.uk/cru/data/temperature)

Brown, J., O.J. Ferrians, Jr., J.A. Heginbottom, and E.S. Melnikov. (1998). *Circum-Arctic Map of Permafrost and Ground Ice Conditions*. National Snow and Ice Data Center/World Data Center for Glaciology.

Canadian Wind Energy Association (CanWEA). (http://www.canwea.ca/)

Census of Canada. *Population Counts, for Canada, Provinces and Territories, and Census Divisions by Urban and Rural, 2001 Census*.

Center for International Earth Science Information Network (CIESIN), Columbia University; International Food Policy Research Institute (IFPRI); The World Bank; and Centro Internacional de Agricultural Tropical (CIAT). (2005). *Global Rural-Urban Mapping Project (GRUMP), Alpha Version*. Socioeconomic Data and Applications Center (SEDAC), Columbia University. (http://sedac.ciesin.columbia.edu/gpw/)

Central Intelligence Agency. *The World Factbook*. (https://www.cia.gov/library/publications/the-world- factbook/)

Chorlton, L B. (2007). *Generalized Geology of the World: Bedrock Domains and Major Faults in GIS Format: A Small-Scale World Geology Map with an Extended Geological Attribute Database*. Geological Survey of Canada, Open File 5529.

Christopherson, Robert W. and Ginger Birkeland. (2014). *Geosystems, 9th Edition*. Upper Saddle River, NJ: Pearson.

Farr, T.G., P.A. Rosen, E. Caro, R. Crippen, R. Duren, S. Hensley, M. Kobrick, M. Paller, E. Rodroguez, L. Roth, D. Seal, S. Shaffer, J. D. Alsdorf. (2007). The Shuttle Radar Topography Mission. Reviews of Geophysics, 45, RG2004, doi:10.1029/2005RG000183.

Federal Aviation Administration (FAA). *Aviation Data and Statistics*. (https://www.faa.gov/data_research/aviation_data_statistics/)

Federation of American Scientists. *Status of World Nuclear Forces*. (http://fas.org/issues/nuclear-weapons/status-world-nuclear-forces/)

Fetterer, F. and K. Knowles. (2002). *Sea Ice Index*. National Snow and Ice Data Center.

Food and Agriculture Organization of the United Nations (FAO). *FAOSTAT*. (http://faostat.fao.org)

Fund for Peace. *Fragile States Index*. (http://fsi.fundforpeace.org/)

Geological Survey of Canada. (1995). *Generalized Geological Map of the World and Linked Databases*. (http://dx.doi.org/10.4095/195142)

Global Wind Energy Council. *Global Statistics*. (http://www.gwec.net/global-figures/graphs/)

Halpern, B.S., S. Walbridge, K.A. Selkoe, C.V. Kappel, F. Micheli, C. D'Agrosa, J.F. Bruno, K.S. Casey, C. Ebert, H.E. Fox, R. Fujita, D. Heinemann, H.S. Lenihan, E.M. P. Madin, M.T. Perry, E.R. Selig, M. Spalding, R. Steneck, and R. Watson. (2008). A global map of human impact on marine ecosystems. *Science*, 319(5865), pp. 948-952. doi: 10.1126/science.1149345.

Hansen, M.C., S.V. Stehman, P.V. Potapov, T.R. Loveland, J.R.G. Townshend, R.S. DeFries, K.W. Pittman, F. Stolle, M.K. Steininger, M. Carroll, and C. Dimiceli. (2008). Humid tropical forest clearing from 2000 to 2005 quantified using multi-temporal and multi-resolution remotely sensed data. *PNAS*, 105(27), pp. 9439-9444.

Hijmans, R.J., S.E. Cameron, J.L. Parra, P.G. Jones and A. Jarvis. (2005). Very high resolution interpolated climate surfaces for global land areas. *International Journal of Climatology* 25: 1965-1978.

Hulme, M. (1998). *Global Land Precipitation Dataset, Version 1.0*. Climatic Research Unit, University of East Anglia.

International Center for Prison Studies. *Prison Population Total*. (http://www.prisonstudies.org/highest-to-lowest/prison-population-total)

International Labour Organization, Department of Statistics. (http://www.ilo.org/stat/lang--en/)

Inter-Parliamentary Union. *Women in National Parliaments*. (http://www.ipu.org/wmn-e/classif.htm)

Inter-Parliamentary Union. *Women's Suffrage: A World Chronology of the Recognition of Women's Rights to Vote and to Stand for Election*. (http://www.ipu.org/wmn-e/suffrage.htm)

Küchler, A.W. (1949). A physiognomic classification of vegetation. *Annals of the Association of American Geographers*, 39(3), pp. 201-210.

Lim, Elliot. (2008). Crustal Age Grid. NOAA National Centers for Environmental Information. (https://www.ngdc.noaa.gov/mgg/image/crustalimages.html)

McDaniel, P. (2008). The Twelve Soil Orders. (http://soils.ag.uidaho.edu/soilorders/)

Murphy, R.E. (1968). Annals map supplement number 9. Landforms of the world. Annals of the Association of American Geographers, 58(1), pp. 198-200.

National Aeronautics and Space Administration (NASA), Atmospheric Science Data Center. (http://eosweb.larc.nasa.gov/)

National Aeronautics and Space Administration (NASA), Earth Observatory (http://earthobservatory.nasa.gov/)

National Aeronautics and Space Administration (NASA). (http://www.nasa.gov/)

National Center for Atmospheric Research (NCAR), *Climate Data Guide*. (https://climatedataguide.ucar.edu/)

National Oceanic and Atmospheric Administration, Laboratory for Satellite Altimetry (http://www.star.nesdis.noaa.gov/sod/lsa/)

National Oceanic and Atmospheric Administration, National Hurricane Center. (http://www.nhc.noaa.gov/)

National Oceanic and Atmospheric Administration, National Weather Service, Storm Prediction Center. (http://www.spc.noaa.gov/)

National Snow and Ice Data Center. *Sea Ice Index*. (http://nsidc.org/data/seaice_index/

Natural Resources Canada. *The Atlas of Canada*. (http://www.nrcan.gc.ca/earth-sciences/geography/atlas-canada)

New, M., D. Lister, M. Hulme, and I. Makin. (2000). A high-resolution data set of surface climate over global land areas. *Climate Research*, 21, pp. 1-25.

Olson, D.M., E. Dinerstein, E.D. Wikramanayake, N.D. Burgess, G.V.N. Powell, E.C. Underwood, J.A. D'Amico, I. Itoua, H.E. Strand, J.C. Morrison, C.J. Loucks, T.F. Allnutt, T.H. Ricketts, Y. Kura, J.F. Lamoreux, W.W. Wettengel, P. Hedao, and K.R. Kassem. (2001). Terrestrial ecoregions of the world: A new map of life on earth. BioScience, 51(11), pp. 933-938.

Permanent Service for Mean Sea Level (PSMSL). (http://www.psmsl.org/)

Smithsonian Institution, Global Volcanism Program. (http://volcano.si.edu/)

Statistics Canada. *Air Carrier Traffic at Canadian Airports*. (http://www.statcan.gc.ca/pub/51-203-x/51-203-x2014000-eng.htm)

Statistics Canada. Population and Demography. (http://www5.statcan.gc.ca/subject-sujet/theme-theme.action?pid=3867&lang=eng&more=0)

Stockholm International Peace Research Institute. *SIPRI Military Expenditure Database*. (https://www.sipri.org/databases/milex)

Tapley, B., J. Ries, S. Bettadpur, D. Chambers, M. Cheng, F. Condi, B. Gunter, Z. Kang, P.Nagel, R. Pastor, T. Pekker, S.Poole, and F. Wang, (2005). GGM02 - An improved Earth gravity field model from GRACE. *Journal of Geodesy*, doi 10.1007/s00190-005-0480-z. (http://www.csr.utexas.edu/grace/gravity/)

TeleGeography Research. (http://www.telegeography.com/)

Thornthwaite, C.W. (1944). Report of committee on transpiration and evaporation. American Geophysical Union Transactions, 25(5), pp. 683-693.

Transport Canada. (https://www.tc.gc.ca/eng/menu.htm)

Trewartha, G.T. (1968). *An Introduction to Climate*, 4th Edition. New York: McGraw-Hill Book Company.

U.K. Natural Environment Research Council, National Oceanography Centre. Permanent Service for Mean Sea Level. (http://www.psmsl.org/)

United Nations (UN). *Comtrade Database*. (http://comtrade.un.org/)

United Nations Environment Program (UNEP-WCMC). *Environmental Data Explorer*. (http://geodata.grid.unep.ch/)

United Nations Environment Program-World Conservation Monitoring Center (UNEP-WCMC). (http://www.unep-wcmc.org/)

United Nations Educational, Scientific and Cultural Organization (UNESCO). *UIS.STAT*. (http://data.uis.unesco.org/)

United Nations High Commissioner for Refugees (UNHCR). (http://www.unhcr.org/)

United Nations Peacekeeping. *Current Peacekeeping Operations*. (http://www.un.org/en/peacekeeping/operations/current.shtml)

United Nations, Department of Economic and Social Affairs, Population Division. (http://www.un.org/en/development/desa/population/)

United Nations, Joint Program on HIV/AIDS. *Global Report on the AIDS Epidemic*. (http://www.unaids.org/en/resources/documents/)

United Nations, Office on Drugs and Crime. *World Drug Report*. (http://www.unodc.org/)

United States Bureau of Labor Statistics. *Databases, Tables & Calculators by Subject*. (http://www.bls.gov/data/)

United States Bureau of the Census. https://www.census.gov/

United States Department of Agriculture, National Resources Conservation Service. *Global Desertification Vulnerability Map*. (http://www.nrcs.usda.gov/wps/portal/nrcs/detail/soils/use/?cid=nrcs142p2_054003)

United States Department of Agriculture (USDA). *National Agricultural Statistics Service*. (https://www.nass.usda.gov/Data_and_Statistics/)

United States Department of Energy. *Landscan 2001 High Resolution Global Population Data Set*. © 2003 UT-Battelle, LLC. All rights reserved. Notice: These data were produced by UT-Battelle, LLC under Contract No. DE-AC05-00OR22725 with the Department of Energy. The Government has certain rights in this data. Neither UT-Battelle, LLC nor the United States Department of Energy, nor any of their employees, makes any warranty, express or implied, or assumes any legal liability or responsibility for the accuracy, completeness, or usefulness of any data, apparatus, product, or process disclosed, or represents that its use would not infringe privately owned rights.

United States Department of State, Office to Monitor and Combat Trafficking in Persons. *2015 Trafficking in Persons Report*. (http://www.state.gov/j/tip/rls/tiprpt/2015/)

United States Energy Information Administration (EIA). International Energy Statistics. (http://www.eia.gov/cfapps/ipdbproject/IEDIndex3.cfm)

United States Geological Survey (USGS), National Minerals Information Center. *Commodity Statistics and Information*. (http://minerals.usgs.gov/minerals/pubs/commodity/)

United States Geological Survey (USGS). *The National Map*. (http://www.nationalmap.gov/)

United States Geological Survey (USGS). *Significant Earthquakes Archive*. (http://earthquake.usgs.gov/earthquakes/eqinthenews/)

University of California San Diego, Scripps Institution of Oceanography. The Keeling Curve. (https://scripps.ucsd.edu/programs/keelingcurve/)

Women Suffrage and Beyond. *Confronting the Democratic Deficit*. (http://womensuffrage.org/)

World Bank Group. World Development Indicators. (http://wdi.worldbank.org/)

World Energy Council. *World Energy Resources*. (https://www.worldenergy.org/data/resources/)

World Food Programme. *Food Aid Information System*. (http://www.wfp.org/fais/)

World Health Organization (WHO). *Global Health Observatory*. (http://www.who.int/gho/database/en/)

World Lung Foundation. (http://wouldlungfoundation.org)

Listed below are page references for major topics covered by the thematic maps and graphs, the introductory text, and the tables.

This universal index includes in a single alphabetical list the names of selected features that appear on the reference maps. Each name is followed by a page number and geographical coordinates.

Abbreviation and Capitalization. Abbreviations of names on the maps have been standardized as much as possible. Names that are abbreviated on the maps are generally spelled out in full in the index.

Most initial letters of names are capitalized, except for a few Dutch names such as "s-Gravenhage". Capitalization of non-initial words in a name generally follows local practice.

Alphabetization. Names are alphabetized in the order of the letters of the English alphabet. Spanish *ll* and *ch*, for example, are not treated as separate letters. Furthermore, diacritical marks are disregarded in alphabetization – German or Scandinavian *ä* or *ö* are treated as *a* or *o*.

The names of physical features may appear inverted, since they are always alphabetized under the proper, not the generic, part of the name, thus: "Gibraltar, Strait of", not "Strait of Gibraltar". In this case "Gibraltar" is the proper part of the name and "Strait of" is the generic. Otherwise every entry, whether consisting of one word or more, is alphabetized as a single continuous entity on the basis of the proper part of the name. "Lakeland", for example, appears after "Lake Havasu City" and before "La Luz".

In the case of identical names, towns are listed first, then political divisions, then physical features.

Generic Terms. Except for cities, the names of all features are followed by terms that represent broad classes of features, for example, "Mississippi, stm." or "Alabama, state". A list of all abbreviations used in the index is on page 323.

Country names and the names of features that extend beyond the boundaries of one country are followed by the name of the continent in which each is located. Country designations follow the names of all other places in the index. The locations of places in the United States, Canada and the United Kingdom are further defined by abbreviations that include the state or political division in which each is located.

Pronunciations. Pronunciations are included for many of the names listed. An explanation of the pronunciation system used appears on page 323.

Page References and Geographical Coordinates. The page references and geographical coordinates are found in the last columns of each entry.

If a page contains several maps or insets, a lowercase letter identifies the specific map or inset.

Latitude and longitude coordinates for point features, such as cities and mountain peaks, indicate the location of the symbols. For extensive areal features, such as countries or mountain ranges, the locations are for the approximate center of the feature. For rivers, locations are given for the mouth.

Feature (Pronunciation)	Page	Lat.	Long.
A			
Aachen, Ger. (ä´kěn)	210-11	50°46′N	6°06′E
Aalborg, Den. (ôl´bôr)	208-09	57°02′N	9°55′E
Aalen, Ger. (ä´lěn)	210-11	48°50′N	10°06′E
Aali, Sadd el-, d., Egypt			
see Aswan High Dam	291b	23°59′N	32°53′E
Aarau, Switz. (ärðu)	210-11	47°24′N	8°03′E
Aarhus, Den. (ôr´hōōs)	208-09	56°09′N	10°13′E
Aasiaat, Green.	151	68°43′N	52°52′W
Aba, China	258-59	33°06′N	101°59′E
Aba, Nig.	282a	5°07′N	7°22′E
Abacaxis, stm., Braz.	180-81	3°54′S	58°46′W
Abaco, i., Bah.	154-55	26°28′N	77°05′W
Ābādān, Iran (ä-bä-dän´)	248-49	30°21′N	48°17′E
Abadla, Alg.	204-05	31°01′N	2°41′W
Abaetetuba, Braz. (ä´bä̇ě-tě-tōō´bȧ)	180-81	1°44′S	48°53′W
Abagnar Qi, China *see* Xilinhot	260-61	43°56′N	116°03′E
Abag Qi, China	260-61	43°43′N	114°39′E
Abakan, Russia (ŭ-bá-kän´)	236-37	53°43′N	91°27′E
Abancay, Peru (ä-bän-kä´ė)	184	13°37′S	72°53′W
Abashiri, Japan (ä-bä-shē´rè)	264	44°01′N	144°16′E
Abasolo, Mex. (ä-bä-sō´lò)	158-59	24°04′N	98°22′W
Ābay, stm., Afr. *see* Blue Nile	275	15°38′N	32°30′E
Ābaya Hāyk´, lk., Eth. (ä-bä´yȧ)	284-85	6°18′N	37°52′E
Abbé, Lac, lk., Afr. *see* Abe, Lake	284-85	11°10′N	41°48′E
Abbeville, Fr. (àb-vēl´)	212-13	50°07′N	1°50′E
Abbeville, Al., U.S. (ăb´ě-vĭl)	134-35	31°34′N	85°15′W
Abbeville, Ga., U.S. (ăb´ě-vĭl)	134-35	31°60′N	83°18′W
Abbeville, La., U.S. (ăb´ě-vĭl)	132-33	29°58′N	92°08′W
Abbeville, S.C., U.S. (ăb´ě-vĭl)	134-35	34°11′N	82°23′W
Abbotsford, B.C., Can. (ăb´ŭts-fĕrd)	142-43	49°03′N	122°17′W
Abbottābād, Pak.	252-53	34°09′N	73°13′E
ʿAbd al-Kūrī, i., Yemen (äbd-ĕl-kó´rė)	238-39	12°12′N	52°13′E
Abdulino, Russia (äb-dò-lē´nō)	202-03	53°41′N	53°40′E
Abe, Lake, lk., Afr.	284-85	11°10′N	41°48′E
Abéché, Chad.	280-81	13°50′N	20°50′E
Abemama, at., Kir.	304-05	0°26′N	187°54′E
Abengourou, C. Iv.	282-83	6°44′N	3°29′W
Abeokuta, Nig. (ä-bä́-ò-kōō´tä)	282a	7°09′N	3°21′E
Aberdare, Wales, U.K. (ăb-ĕr-dâr´)	206-07	51°43′N	3°28′W
Aberdare National Park, n.p., Kenya	289	0°30′S	36°45′E
Aberdeen, Scot., U.K. (ab-ĕr-dēn´)	206-07	57°09′N	2°06′W
Aberdeen, Id., U.S. (ăb-ĕr-dēn´)	122-23	42°57′N	112°51′W
Aberdeen, Md., U.S. (ăb-ĕr-dēn´)	126-27	39°31′N	76°10′W
Aberdeen, Ms., U.S. (ăb-ĕr-dēn´)	134-35	33°50′N	88°33′W
Aberdeen, S.D., U.S. (ăb-ĕr-dēn´)	124-25	45°28′N	98°29′W
Aberdeen, Wa., U.S. (ăb-ĕr-dēn´)	122-23	46°59′N	123°49′W
Aberdeen Lake, lk., Nu., Can.	140-41	64°27′N	99°00′W
Aberystwyth, Wales, U.K. (ā-bĕr-ĭst´wĭth)	206-07	52°25′N	4°05′W
Abez´, Russia	202-03	66°32′N	61°44′E
Abhā, Sau. Ar.	288	18°13′N	42°30′E
Abhé Bad, lk., Afr. *see* Abe, Lake	284-85	11°10′N	41°48′E
Ābhē Bid Hāyk´, lk., Afr. *see* Abe, Lake	284-85	11°10′N	41°48′E
Abidjan, nat. cap., C. Iv. (ä-bēd-zhäɴ´)	282-83	5°20′N	4°01′W
Abilene, Ks., U.S. (ăb´ĭ-lēn)	130-31	38°55′N	97°13′W
Abilene, Tx., U.S. (ăb´ĭ-lēn)	130-31	32°27′N	99°44′W
Abingdon, Il., U.S. (ăb´ĭng-dŭn)	124-25	40°48′N	90°23′W
Abingdon, Va., U.S. (ăb´ĭng-dŭn)	134-35	36°43′N	81°59′W
Abitibi, stm., On., Can.	140-41	51°03′N	80°55′W
Abitibi, Lac, lk., Can. (läk äb-ĭ-tĭb´ĭ) *see* Abitibi, Lake	146-47	48°41′N	79°35′W
Abitibi, Lake, lk., Can. (läk äb-ĭ-tĭb´ĭ)	146-47	48°41′N	79°35′W
Ābīyata Hāyk´, lk., Eth.	292d	7°36′N	38°36′E
Abkhazeti Autonomis Respublika, state, Geor. *see* Abkhazia	245	43°10′N	41°00′E
Abkhazia, state, Geor.	245	43°10′N	41°00′E
Åbo, Fin. *see* Turku	208-09	60°27′N	22°16′E
Abou-Deïa, Chad.	280-81	11°27′N	19°17′E
Abou Simbel, hist., Egypt *see* Abu Simbel	288	22°22′N	31°38′E
Abovyan, Arm.	245	40°15′N	44°35′E
Abrantes, Port. (ȧ-brän´tĕs)	214-15	39°28′N	8°12′W
Abra Pampa, Arg.	182-83	22°43′S	65°42′W
Abruka saar, i., Est. (ä-brò´kä-sä´är)	208-09	58°09′N	22°13′E
Abū ʿAlī, i., Sau. Ar.	250-51	27°19′N	49°35′E
Abū ʿArīsh, Sau. Ar. (ä-bōō ä-rēsh´)	288	16°58′N	42°50′E
Abu Dhabi, nat. cap., U.A.E. (ä´bōō dä´bē)	250-51	24°28′N	54°22′E
Abuja, nat. cap., Nig. (ä-bū´jȧ)	282-83	9°12′N	7°11′E
Abū Kamāl, Syria	248-49	34°27′N	40°56′E
Abū Mūsā, i., Asia	250-51	25°52′N	55°02′E
Abū Mūsā, Jazīreh-ye, i., Asia *see* Abū Mūsā	250-51	25°52′N	55°02′E
Abunã, Braz.	180-81	9°41′S	65°22′W
Abuná, stm., S.A.	180-81	9°40′S	65°26′W
Abunã, stm., S.A. (ä-bōō-nä´)	180-81	9°40′S	65°26′W
Ābu Road, India (ä´bōō rōd)	254-55	24°30′N	72°49′E
Abū Shajarah, Ra's, c., Sudan	288	21°05′N	37°13′E
Abu Simbel, hist., Egypt	288	22°22′N	31°38′E
Abū Sunbul, hist., Egypt *see* Abu Simbel	288	22°22′N	31°38′E
Abū Ẓaby, nat. cap., U.A.E. *see* Abu Dhabi	250-51	24°28′N	54°22′E
Abyaḍ, Al-Baḥr al-, stm., Sudan *see* White Nile	275	15°38′N	32°31′E
Abyssinia, nation, Afr. *see* Ethiopia	274	9°0′N	39°00′E
Acacías, Col. (á-ká-sē´äs)	188c	4°00′N	73°45′W
Acadia National Park, n.p., Me., U.S. (ä-kā´dĭ-á näsh´ŭn-ăl pärk)	127a	44°20′N	68°14′W
Acajutla, El Sal. (ä-kä-hōōt´lä)	160	13°35′N	89°50′W
Acámbaro, Mex. (ä-käm´bä-rō)	158-59	20°03′N	100°43′W
Acaponeta, Mex. (ä-kä-pô-nā´tä)	158-59	22°29′N	105°22′W
Acaponeta, stm., Mex. (ä-kä-pô-nä´tä)	158-59	22°23′N	105°38′W
Acapulco de Juárez, Mex.	158-59	16°51′N	99°54′W
Acaraí, Serra, mts., S.A. *see* Acarai Mountains	178-79	1°30′N	58°15′W
Acarai Mountains, mts., S.A.	178-79	1°30′N	58°15′W
Acaraú. Braz.	180-81	2°53′S	40°07′W
Acarigua, Ven. (äkä-rē´gwä)	178-79	9°34′N	69°12′W
Acatlán de Osorio, Mex. (ä-kät-län´dä ō-sō´rē-ō)	158-59	18°13′N	98°03′W
Acayucan, Mex. (ä-kä-yōō´kän)	158-59	17°57′N	94°54′W
Accra, nat. cap., Ghana (ä´krä)	282-83	5°34′N	0°12′W
Acerra, Italy (ä-chě´r-rä)	216-17	40°57′N	14°22′E
Achacachi, Bol. (ä-chä-kä´chě)	182-83	16°02′S	68°41′W
Achalpur, India.	254-55	21°18′N	77°31′E
Acheng, China	260-61	45°32′N	126°59′E
Achinsk, Russia (á-chěnsk´)	236-37	56°16′N	90°30′E
Acireale, Italy (ä-chē-rä-ä´lä)	216-17	37°37′N	15°10′E
Acklins, i., Bah. (ăk´lĭns)	154-55	22°26′N	73°58′W
Acklins, Bight of, b., Bah. (bīt ŭv ăk´lĭns)	154-55	22°32′N	74°08′W

328 Aconcagua, Cerro – Alataw Shan PRONOUNCING INDEX

Feature (Pronunciation)	Page	Lat.	Long.
Aconcagua, Cerro, mtn., Arg. (sĕ´r-rô ä-kôn-kä´gwä)	188e	32°39'S	70°02'W
Açores, is., Port. (ä-zō´rĕs) see Azores	215c	38°30'N	28°00'W
A Coruña, Spain	214-15	43°22'N	8°25'W
Acoyapa, Nic. (ä-kō-yä´pä)	161	11°58'N	85°11'W
Acre, Isr. see 'Akko	248-49	32°55'N	35°06'E
Acre, state, Braz. (ä´krä)	180-81	9°0'S	70°00'W
Acre, stm., S.A. (ä´krä)	184	8°45'S	67°24'W
Actopan, Mex. (äk-tô-pän´)	158-59	20°16'N	98°57'W
Ada, Mn., U.S. (ā´dŭ)	124-25	47°18'N	96°31'W
Ada, Oh., U.S. (ā´dŭ)	126-27	40°46'N	83°49'W
Ada, Ok., U.S. (ā´dŭ)	130-31	34°47'N	96°41'W
Adak Island, i., Ak., U.S. (ā-dăk´ ī´lănd)	136a	51°43'N	176°43'W
Adam, Oman	238-39	22°23'N	57°31'E
Adama, Eth. see Nazrēt	292d	8°32'N	39°16'E
Adamaoua, mts., Afr.	282-83	7°0'N	12°00'E
Adamawa, mts., Afr. see Adamaoua	282-83	7°0'N	12°00'E
Adams, Ma., U.S. (ăd´ămz)	126-27	42°38'N	73°07'W
Adams, Wi., U.S. (ăd´ămz)	126-27	43°57'N	89°49'W
Adams, stm., B.C., Can. (ăd´ămz)	142-43	50°54'N	119°33'W
Adams, Mount, vol., Wa., U.S.	122-23	46°13'N	121°29'W
Adams Lake, lk., B.C., Can.	142-43	51°13'N	119°33'W
'Adan, Yemen see Aden	288	12°49'N	45°02'E
Adana, Tur. (ä´dä-nä)	248-49	37°00'N	35°20'E
Adapazarı, Tur. (ä-dä-pä-zä´rē) see Sakarya	202-03	40°47'N	30°24'E
Adare, Cape, c., Ant.	313	71°20'S	170°08'E
Adavale, Austl.	301	25°55'S	144°36'E
Ad-Dahnā', des., Sau. Ar.	238-39	24°30'N	48°10'E
Ad-Dammām, Sau. Ar.	250-51	26°26'N	50°07'E
Ad-Dawādimī, Sau. Ar.	288	24°28'N	44°18'E
Ad-Dawḥah, nat. cap., Qatar see Doha	250-51	25°17'N	51°32'E
Ad-Dilam, Sau. Ar.	250-51	23°56'N	47°06'E
Addis Ababa, nat. cap., Eth. (ä´dĭs ä´bä-bä)	292d	9°02'N	38°45'E
Ad-Dīwānīyah, Iraq	248-49	31°59'N	44°55'E
Addo Elephant National Park, n.p., S. Afr.	286-87	33°29'S	25°46'E
Ad-Du'ayn, Sudan	288	11°26'N	26°10'E
Ad-Duwaym, Sudan (ad-dō-ām´)	288	13°59'N	32°18'E
Adel, Ga., U.S. (ā-dĕl´)	134-35	31°08'N	83°25'W
Adel, Ia., U.S. (ā-dĕl´)	124-25	41°37'N	94°01'W
Adelaide, Austl. (ăd´ē-lād)	301	34°55'S	138°35'E
Adelaide Peninsula, pen., Nu., Can.	140-41	68°09'N	97°45'W
Aden, Yemen (ä´dĕn)	288	12°49'N	45°02'E
Aden, Gulf of, b., (gŭlf ŭv ä´dĕn)	238-39	12°40'N	48°07'E
Ādīgrat, Eth.	288	14°17'N	39°27'E
Ādilābād, India (ŭ-dĭl-ä-băd´)	254-55	19°41'N	78°33'E
Adīrī, Libya	280-81	27°32'N	13°13'E
Adirondack Mountains, mts., N.Y., U.S. (ăd-ĭ-rŏn´dăk moun´tĭnz)	126-27	44°0'N	74°00'W
Ādīs Ābeba, nat. cap., Eth. (ä-dēs´ ä´bä-bä) see Addis Ababa	292d	9°02'N	38°45'E
Adjud, Rom. (äd´zhŏd)	218-19	46°06'N	27°11'E
Adjuntas, Presa de las, res., Mex. see Vicente Guerrero, Presa	158-59	23°57'N	98°46'W
Admiralty Island National Monument, n.p., U.S. (ăd´mĭ-răl-tē ī´lănd năsh´ŭn-ăl mŏn´ū-mĕnt)	136	57°40'N	134°16'W
Admiralty Islands, is., Pap. N. Gui. (ăd´mĭ-răl-tē ī´lăndz)	302	2°10'S	147°00'E
Adolfo Gonzales Chaves, Arg.	187	38°01'S	60°08'W
Adonara, Pulau, i., Indon.	268-69	8°20'S	123°10'E
Ādoni, India	256	15°38'N	77°16'E
Adra, Spain (ä´drä)	214-15	36°45'N	3°01'W
Adrano, Italy (ä-drä´nō)	216-17	37°40'N	14°50'E
Adrar, Alg.	280-81	27°52'N	0°18'W
Adrâr, reg., Maur.	280-81	20°26'N	12°46'W
Adria, Italy (ä´drē-ä)	216-17	45°03'N	12°04'E
Adrian, Mi., U.S. (ā´drĭ-ăn)	126-27	41°53'N	84°02'W
Adrian, Mn., U.S. (ā´drĭ-ăn)	124-25	43°38'N	95°56'W
Adrianople, Tur. see Edirne	216-17	41°41'N	26°34'E
Adriatico, Mare, s., Eur. see Adriatic Sea	216-17	42°30'N	16°00'E
Adriatic Sea, s., Eur. (ā-drē-ă´tĭc sē)	216-17	42°30'N	16°00'E
Adriatik, Deti, s., Eur. see Adriatic Sea	216-17	42°30'N	16°00'E
Ādwa, Eth.	288	14°11'N	38°53'E
Adycha, stm., Russia (ä´dĭ-chá)	236-37	68°13'N	134°48'E
Adygea, state, Russia see Adygheya	202-03	45°0'N	40°00'E
Adygheya, state, Russia	202-03	45°0'N	40°00'E
Adz'va, stm., Russia (ädz´vá)	202-03	66°36'N	59°24'E
Aegean Sea, s., (ē-jē´ăn sē)	216-17	38°30'N	25°00'E
Affon, stm., Benin see Ouémé	282a	6°27'N	2°33'E
Afghānestān, nation, Asia see Afghanistan	224-25	33°0'N	65°00'E
Afghanistan, nation, Asia (ăf-găn-ĭ-stăn´)	224-25	33°0'N	65°00'E
'Afīf, Sau. Ar.	288	23°55'N	42°56'E
Afikpo, Nig.	282a	5°55'N	7°55'E
Aflou, Alg. (ä-flōō´)	204-05	34°07'N	2°06'E
Afognak Island, i., Ak., U.S. (ä-fŏg-năk´ ī´lănd)	136	58°14'N	152°39'W
Africa, cont., (ăf´rĭ-kă)	275	10°0'N	22°00'E
Afşin, Tur.	247	38°15'N	36°55'E
Afton, Ok., U.S. (ăf´tŭn)	130-31	36°42'N	94°58'W
Afton, Wy., U.S. (ăf´tŭn)	122-23	42°43'N	110°56'W
'Afula, Isr. (ä-fō´lá)	248-49	32°36'N	35°18'E
Afyon, Tur. (ä-fē-ōn)	202-03	38°46'N	30°33'E
Afyonkarahisar, Tur. see Afyon	202-03	38°46'N	30°33'E
Agadez, Niger (ä´gȧ-dĕs)	280-81	16°58'N	7°59'E
Agadir, Mor. (ä-gȧ-dēr)	280-81	30°28'N	9°39'W
Agadyr, Kaz.	244	48°16'N	72°53'E
Agana, nat. cap., Guam see Hagåtña	306c	13°28'N	144°45'E
Āgaro, Eth.	284-85	7°50'N	36°40'E
Agartala, India	254-55	23°50'N	91°16'E
Agate Fossil Beds National Monument, n.p., Ne., U.S.	124-25	42°25'N	103°43'W
Ağdam, Azer. (äg´dăm)	245	39°59'N	46°56'E
Agde, Fr. (ägd)	212-13	43°19'N	3°28'E
Agen, Fr. (á-zhäN´)	212-13	44°12'N	0°38'E
Āghā Jārī, Iran	250-51	30°42'N	49°50'E
Agno, Phil. (äg´nō)	270	16°07'N	119°48'E
Agno, stm., Phil. (äg´nō)	270	16°02'N	120°09'E
Āgra, India (ä´grä)	254-55	27°11'N	78°00'E
Ağrı, Tur.	245	39°43'N	43°04'E
Ağrı Dağı, vol., Tur. see Ararat, Mount	245	39°42'N	44°18'E
Agrigento, Italy	216-17	37°18'N	13°35'E
Agrihan, i., N. Mar. Is.	304-05	18°46'N	145°40'E
Agryz, Russia	202-03	56°31'N	53°01'E
Aguadas, Col. (ä-gwä´dȧs)	188c	5°38'N	75°27'W
Aguadilla, P.R. (ä-gwä-dēl´yä)	154a	18°26'N	67°09'W
Aguadulce, Pan. (ä-gwä-dōōl´sä)	162	8°15'N	80°31'W
Aguán, stm., Hond. (ä-gwá´n)	161	15°58'N	85°44'W
Aguanaval, stm., Mex. (á-guä-nä-väl´)	156-57	25°25'N	102°49'W
Aguanish, stm., Qc., Can.	148-49	50°15'N	62°07'W
Agua Prieta, Mex.	156-57	31°19'N	109°33'W
Aguarico, stm., S.A.	184	0°58'S	75°11'W
Aguascalientes, Mex.	158-59	21°53'N	102°18'W
Aguascalientes, state, Mex. (ä´gwäs-käl-yĕn´tās)	158-59	22°0'N	102°30'W
Água Vermelha, Represa de, res., Braz.	182-83	20°0'S	50°00'W
Águilas, Spain (ä´gē-läs)	214-15	37°25'N	1°35'W
Aguililla, Mex. (ä-gē-lēl-yä)	158-59	18°44'N	102°44'W
Agulhas, c., S. Afr. (ä-gōōl´yäs)	286-87	34°49'S	20°03'E
Agusan, stm., Phil. (ä-gōō´sän)	270	9°01'N	125°31'E
Ahaggar, mts., Alg. (á-há-gär´)	280-81	23°0'N	6°30'E
Ahaggar, Tassili oua-n-, plat., Alg.	280-81	21°0'N	6°00'E
Ahar, Iran	245	38°28'N	47°04'E
Ahlen, Ger. (ä´lĕn)	210-11	51°46'N	7°54'E
Ahmadābād, India (ŭ-mĕd-ä-băd´)	254-55	23°02'N	72°35'E
Ahmadnagar, India (ä´mŭd-nŭ-gûr)	256	19°05'N	74°45'E
Aḥmar, Al-Baḥr al-, s., see Red Sea	288	20°0'N	38°00'E
Ahmar Mountains, mts., Eth.	284-85	9°14'N	41°25'E
Ahoskie, N.C., U.S. (ā-hŏs´kē)	134-35	36°17'N	76°59'W
Ahuacatlán, Mex. (ä-wä-kät-län´)	158-59	21°05'N	104°29'W
Ahumada, Mex.	156-57	30°37'N	106°31'W
Ahunui, at., Fr. Poly.	304-05	19°39'S	140°25'W
Åhus, Swe. (ô´hŏs)	208-09	55°55'N	14°18'E
Ahvāz, Iran	248-49	31°19'N	48°42'E
Ahvenanmaa, is., Fin. (ä´vĕ-nán-mô) see Aland Islands	208-09	60°14'N	19°46'E
Aidar, stm., Eur.	218-19	48°44'N	39°16'E
Aigaíon Pélagos, s., see Aegean Sea	216-17	38°30'N	25°00'E
Aiken, S.C., U.S. (ā´kĕn)	134-35	33°33'N	81°43'W
Ailao Shan, mts., China	258-59	24°13'N	101°20'E
Aimorés, Braz.	186	19°30'S	41°05'W
Aïn Beni Mathar, Mor.	204-05	34°01'N	2°01'W
Aïn Sefra, Alg.	204-05	32°46'N	0°34'W
Ainsworth, Ne., U.S. (ānz´wûrth)	124-25	42°33'N	99°52'W
'Aïn Temouchent, Alg. (ä´ēntĕ-mōō-shaN´)	214-15	35°18'N	1°09'W
Aipe, Col. (ī´pĕ)	188c	3°13'N	75°14'W
Aïr, Massif de l', mts., Niger	280-81	18°0'N	8°30'E
Air Force Island, i., Nu., Can.	140-41	67°55'N	74°10'W
Aïssa, Djebel, mtn., Alg.	280-81	32°51'N	0°30'W
Aitape, Pap. N. Gui. (ä-ē-tä´pā)	302	3°09'S	142°20'E
Aitkin, Mn., U.S. (āt´kĭn)	124-25	46°32'N	93°43'W
Aitutaki, at., Cook Is. (ī-tōō-tä´kē)	304-05	18°52'S	159°45'W
Aiud, Rom. (ä´ē-ŏd)	210-11	46°19'N	23°44'E
Aix-en-Provence, Fr. (ĕks-prô-văns)	212-13	43°32'N	5°27'E
Aix-la-Chapelle, Ger. see Aachen	210-11	50°46'N	6°06'E
Aix-les-Bains, Fr. (ĕks´-lā-baN´)	212-13	45°42'N	5°55'E
Āīzawl, India	254-55	23°44'N	92°43'E
Aizu-wakamatsu, Japan	265	37°30'N	139°56'E
Ajaccio, Fr. (ä-yät´chō)	200-01	41°56'N	8°44'E
Ajdābiyā, Libya	204-05	30°45'N	20°14'E
Ajjer, Tassili-n-, plat., Alg.	280-81	25°41'N	7°29'E
'Ajmān, U.A.E.	250-51	25°24'N	55°28'E
Ajmer, India (ŭj-mēr´)	254-55	26°27'N	74°38'E
Ajo, Az., U.S. (ä´hō)	128-29	32°23'N	112°52'W
Akagera, stm., Afr. see Kagera	289	0°56'S	31°47'E
Akan-kokuritsu-kōen, n.p., Japan	264	43°30'N	144°15'E
Akashi, Japan (ä´kä-shē)	265	34°39'N	134°59'E
Akdeniz, s., see Mediterranean Sea	204-05	35°0'N	20°00'E
Aketi, D.R.C. (ä-kȧ-tē)	284-85	2°45'N	23°46'E
Akhaltsikhe, Geor. (äkȧ´l-tsĭ-kĕ)	245	41°38'N	42°59'E
Akhisar, Tur. (äk-hĭs-sär´)	216-17	38°56'N	27°50'E
Akhtuba, stm., Russia	202-03	46°40'N	48°08'E
Akhtubinsk, Russia	202-03	48°16'N	46°10'E
Akimiski Island, i., Nu., Can. (ä-kĭ-mĭ´skĭ ī´lănd)	140-41	53°0'N	81°20'W
Akita, Japan (ä´kē-tä)	264	39°43'N	140°07'E
Akkerman, Ukr. see Bilhorod-Dnistrovs'kyi	218-19	46°12'N	30°18'E
'Akko, Isr.	248-49	32°55'N	35°06'E
Aklavik, N.T., Can. (äk´lä-vĭk)	138-39	68°15'N	135°06'W
Ākobo, stm., Afr.	284-85	7°47'N	33°03'E
Akola, India (á-kō´lä)	254-55	20°43'N	77°00'E
Akordat, Erit.	288	15°32'N	37°53'E
Akpatok Island, i., Nu., Can. (äk´pȧ-tŏk iī´lănd))	140-41	60°25'N	67°60'W
Akron, Co., U.S. (ăk´rŭn)	130-31	40°10'N	103°13'W
Akron, Oh., U.S. (ăk´rŭn)	126-27	41°05'N	81°31'W
Aksaray, Tur. (äk-sȧ-rī´)	202-03	38°23'N	34°03'E
Akşehir, Tur. (äk´shȧ-hēr)	202-03	38°21'N	31°25'E
Akşehir Gölü, lk., Tur. (äk´shȧ-hēr)	202-03	38°32'N	31°27'E
Aksu, China (ä-kŭ-sōō)	244	41°08'N	80°15'E
Āksum, Eth.	288	14°08'N	38°43'E
Aktyubinsk, Kaz. see Aqtöbe	244	50°18'N	57°10'E
Akūbū, stm., Afr.	284-85	7°47'N	33°03'E
Akune, Japan (ä´kŏ-nȧ)	265	32°01'N	130°12'E
Akure, Nig.	282a	7°15'N	5°11'E
Akureyri, Ice. (ä-kō-rā´rē)	206a	65°39'N	18°07'W
Akyab, Mya. see Sittwe	266-67	20°09'N	92°54'E
Al-'Amārah, Iraq	248-49	31°50'N	47°09'E
Al-'Aqabah, Jord.	248-49	29°32'N	35°01'E
Al-'Arabīyah as-Su'ūdīyah, nation, Asia see Saudi Arabia	224-25	25°0'N	45°00'E
Al-'Ayn, U.A.E.	250-51	24°13'N	55°45'E
Al-'Azīzīyah, Libya	204-05	32°32'N	13°01'E
Al-'Īrāq, nation, Asia see Iraq	224-25	33°0'N	44°00'E
Al-'Uqaylah, Libya	204-05	30°15'N	19°12'E
Alabama, state, U.S. (ăl-ȧ-băm´ȧ)	118-19	32°50'N	87°00'W
Alabama, stm., Al., U.S. (ăl-ȧ-băm´ȧ)	134-35	31°08'N	87°57'W
Alabat Island, i., Phil. (ä-lä-bät´ ī´lănd)	270	14°07'N	122°03'E
Alacant, Spain	214-15	38°21'N	0°30'W
Alagoa Grande, Braz.	188d	7°03'S	35°38'W
Alagoas, state, Braz. (ä-lä-gō´äzh)	188d	9°0'S	36°00'W
Alagoinhas, Braz. (ä-lä-gō-ēn´yäzh)	180-81	12°08'S	38°25'W
Alagón, stm., Spain (ä-lä-gōn´)	214-15	39°45'N	6°52'W
Alajuela, C.R. (ä-lä-hwa´lä)	161	10°01'N	84°13'W
Alajuela, Lago, res., Pan. (lä´gō-ä-lä-hwa´lä)	162	9°15'N	79°35'W
Alaköl köli, lk., Kaz.	244	46°10'N	81°45'E
Alamagan, i., N. Mar. Is.	304-05	17°36'N	145°50'E
Alamein, Egypt see El-Alamein	204-05	30°49'N	28°58'E
Alamo, Nv., U.S. (ä´lȧ-mō)	128-29	37°22'N	115°10'W
Alamogordo, N.M., U.S. (ăl-ȧ-mȧ-gôr´dō)	130-31	32°54'N	105°57'W
Alamosa, Co., U.S. (ăl-ȧ-mō´sȧ)	128-29	37°28'N	105°52'W
Aland Islands, is., Fin. (ô´länd ī´lăndz)	208-09	60°14'N	19°46'E
Alanya, Tur.	248-49	36°33'N	32°01'E
Alaotra, Farihy, lk., Madag. (ä-lä-ō´trȧ)	286-87	17°25'S	48°33'E
Alappuzha, India	256	9°29'N	76°20'E
Alashanyouqi, China	260-61	40°04'N	103°33'E
Alaska, state, U.S. (ȧ-lăs´kȧ)	118-19	65°0'N	153°00'W
Alaska, Gulf of, b., Ak., U.S. (gŭlf ŭv ȧ-lăs´kȧ)	136	58°0'N	146°00'W
Alaska Peninsula, pen., Ak., U.S. (ȧ-lăs´kȧ pĕ-nĭn´sūlȧ)	136	57°0'N	158°00'W
Alaska Range, mts., Ak., U.S. (ȧ-lăs´kȧ rānj)	136	63°26'N	149°07'W
Alataw Shan, mts., Asia	244	45°0'N	81°00'E

PRONOUNCING INDEX
Alatyr' – Altamaha 329

Feature (Pronunciation)	Page	Lat.	Long.
Alatyr', Russia (ä´lä-tür)	202-03	54°51'N	46°34'E
Alausí, Ec.	184	2°13'S	78°51'W
Alayskiy khrebet, mts., Kyrg.	244	39°51'N	72°07'E
Alazeya, stm., Russia	236-37	70°51'N	153°39'E
Alba, Italy (äl´bä)	216-17	44°42'N	8°02'E
Albacete, Spain (äl-bä-thā´tā)	214-15	38°59'N	1°52'W
Al-Baḥrayn, nation, Asia *see* Bahrain	250-51	26°0'N	50°30'E
Albania, nation, Eur. (ăl-bā´nĭ-ȧ)	190-91	41°0'N	20°00'E
Albano Laziale, Italy (äl-bä´nō lät-zē-ä´lā)	216-17	41°44'N	12°39'E
Albany, Austl. (ôl´bȧ-nĭ)	294-95	35°01'S	117°53'E
Albany, Ga., U.S. (ôl´bȧ-nĭ)	134-35	31°34'N	84°09'W
Albany, Ky., U.S. (ôl´bȧ-nĭ)	134-35	36°41'N	85°08'W
Albany, Mo., U.S. (ôl´bȧ-nĭ)	130-31	40°15'N	94°20'W
Albany, N.Y., U.S. (ôl´bȧ-nĭ)	126-27	42°40'N	73°47'W
Albany, Or., U.S. (ôl´bȧ-nĭ)	122-23	44°38'N	123°05'W
Albany, Tx., U.S. (ôl´bȧ-nĭ)	130-31	32°44'N	99°17'W
Albany, stm., On., Can. (ôl´bȧ-nĭ)	140-41	52°17'N	81°32'W
Al-Baṣrah, Iraq *see* Basra	248-49	30°30'N	47°48'E
Al-Batrūn, Leb. (ăl-bä-trōōn´)	248-49	34°15'N	35°40'E
Al-Bayḍā', Libya	204-05	32°45'N	21°37'E
Albemarle, N.C., U.S. (ăl´bē-märl)	134-35	35°14'N	80°12'W
Albemarle Island, i., Ec. *see* Isabela, Isla	184a	0°30'S	91°06'W
Albemarle Sound, strt., N.C., U.S. (ăl´bē-märl sound)	134-35	36°03'N	76°12'W
Albenga, Italy (äl-bĕŋ´gä)	216-17	44°03'N	8°13'E
Alberga Creek, stm., Austl. (ăl-bŭr´gȧ krēk)	296-97	27°07'S	135°30'E
Albert, Fr. (ȧl-bâr´)	212-13	49°60'N	2°39'E
Albert, Lac, lk., Afr. (läk ȧl-bâr´) *see* Albert, Lake	289	1°40'N	31°00'E
Albert, Lake, lk., Afr. (lāk ăl´bĕrt)	289	1°40'N	31°00'E
Alberta, state, Can. (ăl-bûr´tȧ)	138-39	54°0'N	113°00'W
Alberta, Mount, mtn., Ab., Can. (mount ăl-bûr´tȧ)	142-43	52°18'N	117°28'W
Albert Edward, Mount, mtn., Pap. N. Gui. (mount ăl´bĕrt ĕd´wĕrd)	302	8°24'S	147°22'E
Albert Lea, Mn., U.S. (ăl´bĕrt lē´)	124-25	43°39'N	93°22'W
Albert Nile, stm., Ug. (ăl-bâr´ nīl)	289	3°36'N	32°02'E
Alberton, P.E., Can. (ăl´bĕr-tŭn)	148-49	46°49'N	64°04'W
Albertville, Fr. (ȧl-bâr-vēl´)	212-13	45°40'N	6°23'E
Albertville, Al., U.S. (ăl´bĕrt-vĭl)	134-35	34°16'N	86°13'W
Albi, Fr. (äl-bē´)	212-13	43°55'N	2°08'E
Albia, Ia., U.S. (ăl-bĭ-ȧ)	124-25	41°02'N	92°48'W
Albion, Il., U.S. (ăl´bĭ-ŭn)	126-27	38°22'N	88°04'W
Albion, In., U.S. (ăl´bĭ-ŭn)	126-27	41°23'N	85°24'W
Albion, Mi., U.S. (ăl´bĭ ŭn)	126-27	42°15'N	84°45'W
Albion, Ne., U.S. (ăl´bĭ-ŭn)	124-25	41°41'N	98°00'W
Al-Birkah, Libya	280-81	24°52'N	10°12'E
Alborán, Isla de, i., Spain (ē´s-lä-dāl-äl-bō-rä´n)	214-15	35°57'N	3°02'W
Alborz, Reshteh-ye Kūhhā-ye, mts., Iran *see* Elburz Mountains	252-53	36°0'N	53°00'E
Albuquerque, N.M., U.S. (ăl-bů-kûr´kē)	128-29	35°05'N	106°38'W
Albury, Austl. (ôl´bĕr-ē)	301	36°04'S	146°56'E
Alcalá de Henares, Spain (äl-kä-lä´ dā ā-na´räs)	214-15	40°29'N	3°22'W
Alcalá la Real, Spain (äl-kä-lä´lä rā-äl´)	214-15	37°28'N	3°56'W
Alcamo, Italy (äl´kȧ-mō)	216-17	37°59'N	12°58'E
Alcanar, Spain (äl-kä-när´)	214-15	40°33'N	0°29'E
Alcañiz, Spain (äl-kän-yēth´)	214-15	41°03'N	0°08'W
Alcântara, Braz. (äl-kän´tȧ-rȧ)	180-81	2°24'S	44°24'W
Alcázar de San Juan, Spain (äl-kä´thär dā sän hwän´)	214-15	39°23'N	3°12'W
Alcazarquivir, Mor. *see* Er-Rachidia	204-05	31°57'N	4°26'W
Alcazarquivir, Mor. *see* Ksar-el-Kebir	292a	35°01'N	5°54'W
Alcira, Spain (ä-thē´rä) *see* Alzira	214-15	39°09'N	0°26'W
Alcobaça, Braz.	186	17°31'S	39°13'W
Alcobendas, Spain (äl-kō-bĕn´dās)	214-15	40°33'N	3°38'W
Alcoi, Spain	214-15	38°42'N	0°28'W
Alcoy, Spain *see* Alcoi	214-15	38°42'N	0°28'W
Aldabra, Groupe d', is., Sey. (grūp-däl-dä´brä)	286-87	9°24'S	46°27'E
Aldama, Mex. (äl-dä´mä)	158-59	22°55'N	98°04'W
Aldama, Mex. (äl-dä´mä)	132-33	28°51'N	105°54'W
Aldan, Russia	236-37	58°36'N	125°24'E
Aldan, stm., Russia	236-37	63°26'N	129°26'E
Aldan Plateau, plat., Russia (ŭl-dän´) *see* Aldanskoye Nagor'ye	236-37	57°0'N	127°00'E
Aldanskoye Nagor'ye, plat., Russia	236-37	57°0'N	127°00'E
Alderney, i., Guern. (ôl´dĕr-nĭ)	212-13	49°43'N	2°13'W
Aldershot, Eng., U.K. (ôl´dĕr-shŏt)	206-07	51°15'N	0°46'W
Aledo, Il., U.S. (ȧ-lē´dō)	124-25	41°12'N	90°45'W
Alegranza, Isla, i., Spain	215d	29°24'N	13°30'W
Alegrete, Braz. (ä-lå-grā´tā)	187	29°47'S	55°47'W
Aleksandrov, Russia (ä-lyĕk-sän´ drôf)	218-19	56°24'N	38°43'E
Aleksandrov-Gay, Russia	202-03	50°08'N	48°34'E
Aleksandrovsk-Sakhalinskiy, Russia	236-37	50°54'N	142°10'E
Aleksandrów Kujawski, Pol. (ä-lĕk-säh´drōōv kōō-yav´skē)	210-11	52°52'N	18°43'E
Alekseevka, Kaz.	244	52°00'N	70°57'E
Alekseyevka, Russia (ä-lyĕk-sâ-yĕf´kȧ)	218-19	50°38'N	38°41'E
Aleksin, Russia (äb´ĭng-tŭn)	218-19	54°30'N	37°05'E
Além Paraíba, Braz. (ä-lĕ´m-pȧ-rȧē´bȧ)	186	21°52'S	42°40'W
Alençon, Fr. (ȧ-läN-sôN´)	212-13	48°26'N	0°05'E
Alenquer, Braz. (ä-lĕŋ-kĕr´)	180-81	1°56'S	54°46'W
Alep, Syria *see* Aleppo	248-49	36°13'N	37°10'E
Aleppo, Syria (ȧ-lĕp-ō)	248-49	36°13'N	37°10'E
Alert, Nu., Can.	151	82°29'N	62°15'W
Alès, Fr. (ä-lĕs´)	212-13	44°08'N	4°05'E
Alessandria, Italy (ä-lĕs-sän´drē-ä)	216-17	44°55'N	8°37'E
Ålesund, Nor. (ô´lĕ-soon´)	200-01	62°28'N	6°10'E
Aleutian Islands, is., Ak., U.S. (ȧ-lu´shăn ī´lăndz)	136a	52°0'N	176°00'W
Alexander City, Al., U.S. (ăl-ĕg-zăn´dẽr sĭ´tē)	134-35	32°57'N	85°57'W
Alexandra, N.Z.	303	45°15'S	169°23'E
Alexandretta, Tur. *see* İskenderun	248-49	36°35'N	36°11'E
Alexandretta, Gulf of, b., Tur. *see* İskenderun Körfezi	248-49	36°30'N	35°40'E
Alexandria, On., Can. (ăl-ĕg-zăn´drĭ-ȧ)	146-47	45°18'N	74°38'W
Alexandria, Egypt (ăl-ĕg-zăn´drĭ-ȧ)	291b	31°11'N	29°54'E
Alexandria, Rom. (ăl-ĕg-zăn´drĭ-ȧ)	216-17	43°59'N	25°21'E
Alexandria, In., U.S. (ăl-ĕg-zăn´drĭ-ȧ)	126-27	40°15'N	85°40'W
Alexandria, La., U.S. (ăl-ĕg-zăn´drĭ-ȧ)	132-33	31°18'N	92°27'W
Alexandria, Mn., U.S. (ăl-ĕg-zăn´drĭ-ȧ)	124-25	45°53'N	95°23'W
Alexandria, S.D., U.S. (ăl-ĕg-zăn´drĭ-ȧ)	124-25	43°39'N	97°47'W
Alexandria, Va., U.S. (ăl-ĕg-zăn´drĭ-ȧ)	126-27	38°48'N	77°03'W
Alexandria Bay, N.Y., U.S. (ăl-ĕg-zăn´drĭ-ȧ bā)	126-27	44°20'N	75°55'W
Alexandrina, Lake, lk., Austl.	301	35°26'S	139°10'E
Alexandroúpoli, Grc.	216-17	40°52'N	25°53'E
Aleysk, Russia	244	52°29'N	82°46'E
Alfaro, Spain (äl-färō)	214-15	42°11'N	1°45'W
Al-Fāshir, Sudan (äl-fä´shēr)	288	13°38'N	25°21'E
Alfenas, Braz. (äl-fĕ´nȧs)	186	21°27'S	45°57'W
Al-Furāt, stm., Asia *see* Euphrates	226-27	30°60'N	47°27'E
Algeciras, Spain (äl-hä-thē´räs)	214-15	36°08'N	5°27'W
Alger, nat. cap., Alg. *see* Algiers	292b	36°46'N	3°03'E
Algeria, nation, Afr. (ăl-gē´rĭ-ȧ)	274	28°0'N	3°00'E
Algérie, nation, Afr. *see* Algeria	274	28°0'N	3°00'E
Al-Ghayḍah, Yemen	238-39	16°13'N	52°12'E
Alghero, Italy (äl-gâ´rō)	216-17	40°34'N	8°19'E
Algiers, nat. cap., Alg. (ăl-jērs)	292b	36°46'N	3°03'E
Al-Ḥamād, pl., Sau. Ar.	248-49	32°0'N	39°30'E
Al-Ḥasakah, Syria	248-49	36°30'N	40°46'E
Al-Ḥawrah, Yemen	238-39	13°50'N	47°34'E
Al-Ḥayy, Iraq	248-49	32°10'N	46°03'E
Al-Ḥijāz, reg., Sau. Ar.	288	24°30'N	38°30'E
Al-Ḥillah, Iraq	248-49	32°29'N	44°26'E
Al-Hoceima, Mor.	214-15	35°15'N	3°56'W
Al-Ḥudaydah, Yemen	288	14°48'N	42°57'E
Al-Hufūf, Sau. Ar.	250-51	25°22'N	49°34'E
Alicante, Spain *see* Alacant	214-15	38°21'N	0°30'W
Alice, Tx., U.S. (ăl´ĭs)	132-33	27°45'N	98°05'W
Alice Springs, Austl. (ăl´ĭs sprĭngz)	294-95	23°42'S	133°52'E
Alīgarh, India (ä-lē-gŭr´)	254-55	27°54'N	78°04'E
Alima, stm., Congo	284-85	1°31'S	16°40'E
Al-Imārāt al-'Arabīyah al-Muttaḥidah, nation, Asia *see* United Arab Emirates	224-25	24°0'N	54°00'E
Alingsås, Swe. (ä´lĭŋ-sôs)	208-09	57°56'N	12°32'E
'Ali Sabieh, Dji.	288	11°08'N	42°42'E
Aliwal North, S. Afr. (ä-lē-wäl´ nôrth)	286-87	30°42'S	26°43'E
Al-Jawf, Libya	280-81	24°12'N	23°17'E
Al-Jawf, Sau. Ar.	248-49	29°48'N	39°52'E
Al-Jazaïr, nat. cap., Alg. *see* Algiers	292b	36°46'N	3°03'E
Al-Jazīrah, reg., Sudan	288	14°17'N	32°53'E
Aljezur, Port. (äl-zhä-zōōr´)	214-15	37°18'N	8°48'W
Al-Jubayl, Sau. Ar.	250-51	27°01'N	49°40'E
Al-Jufrah, well, Libya	280-81	29°06'N	15°57'E
Al-Junaynah, Sudan	284-85	13°27'N	22°27'E
Al-Karak, Jord. (äl-kĕ-räk´)	248-49	31°11'N	35°42'E
Al-Khābūrah, Oman	250-51	23°58'N	57°06'E
Al-Khalīl, W.B. *see* Hebron	248-49	31°32'N	35°06'E
Al-Kharṭūm, nat. cap., Sudan *see* Khartoum	288	15°35'N	32°32'E
Al-Khaṣab, Oman	250-51	26°12'N	56°15'E
Al-Khubar, Sau. Ar.	250-51	26°17'N	50°12'E
Al-Khums, Libya	204-05	32°39'N	14°16'E
Alkmaar, Neth. (älk-mär´)	206-07	52°38'N	4°45'E
Al-Kūt, Iraq	248-49	32°30'N	45°49'E
Al-Kuwayt, nation, Asia *see* Kuwait	224-25	29°30'N	47°45'E
Al-Kuwayt, nat. cap., Kuw. (äl-kōō-wit) *see* Kuwait	248-49	29°19'N	47°60'E
Al-Lādhiqīyah, Syria *see* Latakia	248-49	35°31'N	35°48'E
Allahābād, India (ŭl-ŭ-hä-bäd´)	254-55	25°26'N	81°51'E
Allakaket, Ak., U.S.	136	66°33'N	152°38'W
'Allāq, Bi'r, well, Libya	280-81	31°05'N	11°58'E
Allegan, Mi., U.S. (ăl´ē-găn)	126-27	42°32'N	85°51'W
Allegheny Plateau, plat., U.S. (ăl-ē-gā´nĭ plă-tō´)	126-27	41°30'N	78°00'W
Allendale, S.C., U.S. (ăl´ĕn-dāl)	134-35	33°00'N	81°19'W
Allende, Mex. (äl-yĕn´då)	132-33	28°20'N	100°50'W
Allende, Mex. (äl-yĕn´då)	132-33	25°17'N	100°01'W
Allenstein, Pol. *see* Olsztyn	210-11	53°47'N	20°29'E
Allentown, Pa., U.S.	126-27	40°37'N	75°29'W
Alleppey, India (ȧ-lĕp´ē) *see* Alappuzha	256	9°29'N	76°20'E
Aller, stm., Ger. (äl´ẽr)	210-11	52°57'N	9°11'E
Alliance, Ne., U.S. (ȧ-lī´ăns)	124-25	42°06'N	102°52'W
Alliance, Oh., U.S. (ȧ-lī´ăns)	126-27	40°55'N	81°06'W
Allier, stm., Fr. (ȧ-lyā´)	212-13	46°58'N	3°04'E
Allinge, Den. (äl´ĭŋ-ĕ)	210-11	55°16'N	14°48'E
Al-Lubnān, nation, Asia *see* Lebanon	224-25	34°0'N	36°00'E
Alma, N.B., Can. (ăl´mȧ)	148-49	45°36'N	64°57'W
Alma, Qc., Can. (ăl´mȧ)	146-47	48°33'N	71°39'W
Alma, Ga., U.S. (ăl´mȧ)	134-35	31°33'N	82°28'W
Alma, Mi., U.S. (ăl´mȧ)	126-27	43°23'N	84°39'W
Alma, Ne., U.S. (ăl´mȧ)	130-31	40°06'N	99°21'W
Alma, Wi., U.S. (ăl´mȧ)	124-25	44°20'N	91°54'W
Almadén, Spain (äl-mä-dhän´)	214-15	38°46'N	4°50'W
Al-Madīnah, Sau. Ar. *see* Medina	288	24°28'N	39°37'E
Al-Maghrib, nation, Afr. *see* Morocco	274	32°0'N	5°00'W
Almagro, Spain (äl-mä´grō)	214-15	38°53'N	3°43'W
Al-Majma'ah, Sau. Ar.	252-53	25°55'N	45°21'E
Al-Makhā', Yemen *see* Mocha	288	13°19'N	43°15'E
Al Manāmah, nat. cap., Bahr. (äl-mă-nă´mȧ)	250-51	26°13'N	50°35'E
Almansa, Spain (äl-män´sä)	214-15	38°52'N	1°06'W
Al-Marj, Libya	204-05	32°30'N	20°53'E
Almas, Pico das, mtn., Braz.	180-81	13°33'S	41°56'W
Almaty, Kaz.	244	43°17'N	76°56'E
Al-Mawṣil, Iraq *see* Mosul	248-49	36°20'N	43°08'E
Almazán, Spain (äl-mä-thän´)	214-15	41°29'N	2°32'W
Almelo, Neth. (äl´mē-lō)	206-07	52°22'N	6°39'E
Almenara, Braz.	182-83	16°12'S	40°41'W
Almendralejo, Spain (äl-mān-drä-lā´hō)	214-15	38°41'N	6°24'W
Almería, Spain (äl-mā-rē´ä)	214-15	36°51'N	2°27'W
Almería, Golfo de, b., Spain (gôl-fô-dĕ-äl-māī-reN´)	214-15	36°46'N	2°30'W
Al'met'yevsk, Russia	202-03	54°54'N	52°19'E
Älmhult, Swe. (ĕlm´hōōlt)	208-09	56°33'N	14°09'E
Almirante, Pan. (äl-mē-rän´tā)	162	9°17'N	82°24'W
Almonte, On., Can. (äl-mŏn´tē)	146-47	45°13'N	76°11'W
Almora, India	254-55	29°36'N	79°40'E
Al-Mubarraz, Sau. Ar.	250-51	25°25'N	49°35'E
Al-Muḥarraq, Bahr.	250-51	26°16'N	50°37'E
Al-Mukallā, Yemen	238-39	14°32'N	49°08'E
Almuñécar, Spain (äl-mōōn-yā´kär)	214-15	36°45'N	3°41'W
Alnön, i., Swe.	208-09	62°25'N	17°26'E
Alor, Pulau, i., Indon. (pōō-lou ä´lôr)	268-69	8°15'S	124°45'E
Alor Setar, Malay. (ä´lôr stär)	266-67	6°07'N	100°23'E
Alpen, mts., Eur. *see* Alps	200-01	46°25'N	10°00'E
Alpena, Mi., U.S. (ăl-pē´nȧ)	126-27	45°03'N	83°26'W
Alpes, mts., Eur. *see* Alps	200-01	46°25'N	10°00'E
Alpi, mts., Eur. *see* Alps	200-01	46°25'N	10°00'E
Alpine, Az., U.S. (ăl´pīn)	128-29	33°50'N	109°08'W
Alpine, Tx., U.S. (ăl´pīn)	132-33	30°21'N	103°40'W
Alpine National Park, n.p., Austl.	301	36°57'S	147°12'E
Alps, mts., Eur. (ălps)	210-11	46°25'N	10°00'E
Al-Qaḍārif, Sudan	288	14°02'N	35°23'E
Al-Qaṭīf, Sau. Ar.	250-51	26°33'N	50°00'E
Al-Quds, nat. cap., Isr. *see* Jerusalem	248-49	31°47'N	35°14'E
Als, i., Den. (äls)	210-11	54°59'N	9°55'E
Alsace, hist. reg., Fr. (äl-sȧ´s)	212-13	48°30'N	7°30'E
Alta Gracia, Arg. (äl´tä grä´sē-a)	187	31°40'S	64°26'W
Altagracia, Ven.	178-79	10°43'N	71°30'W
Altamaha, stm., Ga., U.S. (ôl-tȧ-mȧ-hô´)	134-35	31°19'N	81°18'W

330 Altamira – Angoulême PRONOUNCING INDEX

Feature (Pronunciation)	Page	Lat.	Long.
Altamira, Braz. (äl-tä-mē´rä)	180-81	3°11'S	52°14'W
Altamira, Chile	182-83	25°48'S	69°51'W
Altamura, Italy (äl-tä-mo͞o´rä)	216-17	40°50'N	16°33'E
Altavista, Va., U.S. (äl-tä-vĭs´tȧ)	134-35	37°07'N	79°18'W
Altay, Mong.	260-61	46°24'N	96°15'E
Altay, Mong.	260-61	49°42'N	96°24'E
Altay, state, Russia	244	51°0'N	86°00'E
Altay, mts., Asia *see* Altay Mountains	240-41	48°0'N	90°00'E
Altay Mountains, mts., Asia (äl´tī´ moun´tĭnz)	240-41	48°0'N	90°00'E
Altay Shan, mts., Asia *see* Altay Mountains	240-41	48°0'N	90°00'E
Altenburg, Ger. (äl-tĕn-bo͝orgh)	210-11	50°59'N	12°26'E
Altiplano, plat., S.A. (äl-tē-plȧ´nō)	182-83	18°0'S	68°00'W
Alto Araguaia, Braz.	182-83	17°19'S	53°13'W
Alton, Il., U.S. (ôl´tŭn)	130-31	38°55'N	90°12'W
Altona, Mb., Can.	144-45	49°06'N	97°35'W
Altoona, Ia., U.S. (ăl-to͞o´nȧ)	124-25	41°38'N	93°28'W
Altoona, Pa., U.S. (ăl-to͞o´nȧ)	126-27	40°29'N	78°24'W
Altoona, Wi., U.S. (ăl-to͞o´nȧ)	124-25	44°48'N	91°26'W
Alto Parnaíba, Braz.	180-81	9°07'S	45°57'W
Alto Río Senguer, Arg.	185	45°03'S	70°51'W
Altun Shan, mts., China (äl-tȯn shän)	240-41	38°0'N	88°00'E
Alturas, Ca., U.S. (ăl-to͞o´rȧs)	122-23	41°30'N	120°32'W
Altus, Ok., U.S. (ăl´tŭs)	130-31	34°38'N	99°20'W
Alu, i., Sol. Is. *see* Shortland Island	306e	7°04'S	155°43'E
Al-Ubayyiḍ, Sudan	288	13°11'N	30°13'E
Alūksne, Lat. (ä´lȯks-nĕ)	208-09	57°25'N	27°04'E
Alula, Som. *see* Caluula	284-85	11°57'N	50°46'E
Al-Urdun, nation, Asia *see* Jordan	224-25	31°0'N	36°00'E
Al-Urdunn, stm., Asia *see* Jordan	248-49	31°46'N	35°34'E
Alushta, Ukr. (ä´lshȯ-tȧ)	218-19	44°41'N	34°24'E
Alva, Ok., U.S. (ăl´vȧ)	130-31	36°48'N	98°40'W
Alvarado, Mex. (äl-vä-rä´dhō)	158-59	18°46'N	95°46'W
Älvdalen, Swe. (ĕlv´dä-lĕn)	208-09	61°14'N	14°02'E
Alvear, Arg.	187	29°03'S	56°33'W
Alvesta, Swe. (äl-vĕs´tä)	208-09	56°54'N	14°33'E
Alwar, India (ŭl´wŭr)	254-55	27°34'N	76°37'E
Alxa Zuoqi, China	260-61	38°49'N	105°35'E
Al-Yaman, nation, Asia *see* Yemen	224-25	15°0'N	44°00'E
Alytus, Lith. (ä´le̊-tȯs)	210-11	54°24'N	24°04'E
Alzira, Spain	214-15	39°09'N	0°26'W
Amadjuak Lake, lk., Nu., Can. (ä-mädj´wäk lāk)	140-41	65°0'N	71°00'W
Amahai, Indon.	268-69	3°20'S	128°56'E
Amakuso-Shimo-shima, i., Japan (ämä-ko͞o´sä shē-mō she̊´mä)	265	32°20'N	130°05'E
Åmål, Swe. (o̊´mo̊l)	208-09	59°03'N	12°42'E
Amalfi, Col. (ä´mȧ´l-fē)	178-79	6°55'N	75°05'W
Amambaí, Braz.	182-83	23°07'S	55°13'W
Amami-Ō-shima, i., Japan	264a	28°15'N	129°20'E
Amami-shotō, is., Japan	264a	27°58'N	129°02'E
Amapá, Braz.	180-81	2°02'N	50°46'W
Amapá, state, Braz.	180-81	1°0'N	52°00'W
Amarante, Braz. (ä-mä-rän´tā)	180-81	6°14'S	42°50'W
Amarillo, Tx., U.S. (ăm-ȧ-rĭl´ō)	130-31	35°13'N	101°50'W
Amarkantak, India	254-55	22°40'N	81°46'E
Amaro, Monte, mtn., Italy (mo̊n-tĕ ä-mä´rō)	216-17	42°05'N	14°06'E
Amasya, Tur. (ä-mä´se̊-ȧ)	202-03	40°39'N	35°50'E
Amazon, stm., S.A. (ä´mă-zŏn)	178-79	0°04'S	49°15'W
Amazonas, state, Braz. (ä-mä-thō´näs)	180-81	5°0'S	63°00'W
Amazonas, stm., S.A. (ä-mä-thō´näs) *see* Amazon	178-79	0°04'S	49°15'W
Ambāla, India (ŭm-bä´lŭ)	254-55	30°21'N	76°49'E
Ambanja, Madag.	286-87	13°41'S	48°27'E
Ambargasta, Salinas de, pl., Arg.	182-83	29°15'S	64°30'W
Ambato, Ec. (äm-bä´tō)	184	1°15'S	78°37'W
Ambatondrazaka, Madag.	286-87	17°52'S	48°24'E
Ambelau, Pulau, i., Indon.	268-69	3°51'S	127°12'E
Amberg, Ger. (äm´bĕrgh)	210-11	49°27'N	11°52'E
Ambergris Cay, i., Belize (ăm´bẽr-grēs kā)	160	18°03'N	87°55'W
Ambert, Fr. (äɴ-bĕr´)	212-13	45°33'N	3°45'E
Ambikāpur, India	254-55	23°07'N	83°12'E
Amboina, Indon. *see* Ambon	268-69	3°44'S	128°11'E
Amboise, Fr. (äɴ-bwăz´)	212-13	47°24'N	0°60'E
Ambon, Indon.	268-69	3°44'S	128°11'E
Ambon, Pulau, i., Indon.	268-69	3°40'S	128°05'E
Amboseli National Park, n.p., Kenya	289	2°36'S	37°12'E
Ambositra, Madag. (äɴ-bo̊-sē´trä)	286-87	20°32'S	47°15'E
Ambovombe, Madag.	286-87	25°11'S	46°05'E
Amboy, Il., U.S. (ăm´boi)	126-27	41°43'N	89°20'W
Ambre, Cap d', c., Madag. *see* Bobaomby, Tanjona	286-87	11°58'S	49°15'E
Ambridge, Pa., U.S. (ăm´brĭdj)	126-27	40°36'N	80°13'W
Ambriz, Ang.	284-85	7°51'S	13°10'E
Ambrym, i., Vanuatu	306g	16°15'S	168°10'E
Amchitka Pass, strt., Ak., U.S. (ăm-chĭt´kȧ păs)	136a	51°30'N	179°30'W
Amderma, Russia	202-03	69°45'N	61°39'E
Amdo, China	254-55	32°17'N	91°44'E
Ameca, Mex. (ä-mĕ´kä)	158-59	20°33'N	104°02'W
Amecameca de Juárez, Mex.	158-59	19°07'N	98°46'W
Ameland, i., Neth.	206-07	53°27'N	5°45'E
Amelia Island, i., Fl., U.S.	134-35	30°37'N	81°27'W
American Falls, Id., U.S. (ȧ-mĕr´ĭ-kăn fôlz)	122-23	42°47'N	112°51'W
American Falls Reservoir, res., Id., U.S. (ȧ-mĕr´ĭ-kăn fôlz rĕ´sẽr-vwär)	122-23	42°57'N	112°44'W
American Fork, Ut., U.S. (ȧ-mĕr´ĭ-kăn fôrk)	128-29	40°24'N	111°48'W
American Highland, plat., Ant. (ȧ-mĕr´ĭ-kăn)	313	72°30'S	78°00'E
Americanos, Barra de los, i., Mex.	132-33	24°53'N	97°35'W
American Samoa, dep., Oc. (ȧ-mĕr´ĭ-kăn să-mō´ă)	306b	14°20'S	170°00'W
Americus, Ga., U.S. (ȧ-mĕr´ĭ-kŭs)	134-35	32°05'N	84°14'W
Amerika Samoa, dep., Oc. *see* American Samoa	306b	14°20'S	170°00'W
Amersfoort, Neth. (ä´mĕrz-fōrt)	206-07	52°10'N	5°24'E
Ames, Ia., U.S. (āmz)	124-25	42°01'N	93°37'W
Amesbury, Ma., U.S. (āmz´bẽr-e̊)	126-27	42°52'N	70°56'W
Ámfissa, Grc. (ȧm-fĭ´sȧ)	216-17	38°32'N	22°23'E
Amga, Russia (ŭm-gä´)	236-37	60°54'N	131°58'E
Amga, stm., Russia (ŭm-gä´)	236-37	62°35'N	135°04'E
Amgun, stm., Russia	236-37	52°56'N	139°41'E
Amherst, N.S., Can. (ăm´hẽrst)	148-49	45°50'N	64°12'W
Amherst, Ma., U.S. (ăm´hẽrst)	126-27	42°23'N	72°31'W
Amherst, N.Y., U.S. (ăm´hẽrst)	126-27	42°58'N	78°47'W
Amherst, Va., U.S. (ăm´hẽrst)	126-27	37°35'N	79°04'W
Amiens, Fr. (ä-myäɴ´)	212-13	49°54'N	2°18'E
Amistad, Presa de la, res., N.A. *see* Amistad Reservoir	132-33	29°28'N	101°07'W
Amistad Reservoir, res., N.A.	132-33	29°28'N	101°07'W
Amite, La., U.S. (ă-mēt´)	134-35	30°43'N	90°31'W
Amite, stm., La., U.S. (ă-mēt´)	134-35	30°13'N	90°36'W
Amlia Island, i., Ak., U.S. (ȧ´mle̊ä ī´lȧnd)	136a	52°07'N	187°34'W
'Ammān, nat. cap., Jord. (äm´mȧn)	248-49	31°57'N	35°56'E
Amnok-kang, stm., Asia *see* Yalu	263	39°57'N	124°22'E
Āmol, Iran	252-53	36°28'N	52°21'E
Amorgós, i., Grc. (ä-môr´gōs)	216-17	36°52'N	25°56'E
Amory, Ms., U.S. (āmo̊-rē)	134-35	33°59'N	88°29'W
Amos, Qc., Can. (ā´mŭs)	146-47	48°34'N	78°08'W
Amoy, China *see* Xiamen	243a	24°27'N	118°07'E
Amposta, Spain (äm-pōs´tä)	214-15	40°44'N	0°35'E
Amraoti, India *see* Amrāvati	254-55	20°56'N	77°46'E
Amrāvati, India	254-55	20°56'N	77°46'E
Amreli, India	254-55	21°36'N	71°13'E
Amritsar, India (ŭm-rĭt´sŭr)	254-55	31°38'N	74°52'E
Amroha, India	254-55	28°54'N	78°28'E
Amsterdam, N.Y., U.S. (ăm´stẽr-dăm)	126-27	42°57'N	74°11'W
Amsterdam, nat. cap., Neth. (äm-stẽr-däm´)	206-07	52°22'N	4°54'E
Amstetten, Aus. (äm´stĕt-ĕn)	210-11	48°07'N	14°52'E
Am Timan, Chad (äm´te̊-män´)	280-81	11°02'N	20°17'E
Amu Darya, stm., Asia (ä-mȯ-dä´rēä)	244	44°14'N	59°41'E
Āmū Daryā, stm., Asia *see* Amu Darya	244	44°14'N	59°41'E
Amukta Pass, strt., Ak., U.S. (ä-mo͞ok´tȧ păs)	136a	52°26'N	171°51'W
Amundsen Gulf, b., Can. (ä´mŭn-sĕn-gŭlf´)	95	71°0'N	124°00'W
Amundsen Sea, s., Ant. (ä´mŭn-sĕn sē)	313	72°30'S	112°00'W
Amuntai, Indon.	268-69	2°25'S	115°15'E
Amur, stm., Asia (ä-mo͞or´)	236-37	52°57'N	141°10'E
Anaa, at., Fr. Poly.	304-05	17°26'S	145°31'W
Anabar, stm., Russia (ăn-ȧ-bär´)	236-37	73°13'N	113°32'E
Anaco, Ven. (ä-nȧ´kô)	188b	9°25'N	64°28'W
Anaconda, Mt., U.S. (ăn-ȧ-kŏn´dȧ)	122-23	46°08'N	112°58'W
Anacortes, Wa., U.S. (ăn-ȧ-kôr´tĕz)	122-23	48°30'N	122°37'W
Anadarko, Ok., U.S. (ăn-ȧ-där´kō)	130-31	35°04'N	98°15'W
Anadyr', Russia (ŭ-nȧ-dĭr´)	236-37	64°44'N	177°30'E
Anadyr', stm., Russia (ŭ-nȧ-dĭr´)	236-37	64°52'N	176°15'E
Anadyr, Gulf of, b., Russia (gŭlf ŭv ä-nă-dyĭr´)	236-37	64°0'N	179°00'W
Anadyr Mountains, plat., Russia (ä-nă-dyĭr´ moun´tĭnz) *see* Anadyrskoye Ploskogor'ye	236-37	67°0'N	174°00'E
Anadyrskiy Zaliv, b., Russia *see* Anadyr, Gulf of	236-37	64°0'N	179°00'W
Anadyrskoye Ploskogor'ye, plat., Russia	236-37	67°0'N	174°00'E
Anaheim, Ca., U.S. (ăn´ȧ-hīm)	128-29	33°50'N	117°55'W
Ānai Mudi, mtn., India	256	10°10'N	77°04'E
Anaktuvuk Pass, Ak., U.S.	136	68°09'N	151°43'W
Anambas, Kepulauan, is., Indon. (ä-näm-bäs)	266-67	3°0'N	106°00'E
Anamosa, Ia., U.S. (ăn-ȧ-mō´sȧ)	124-25	42°06'N	91°16'W
Anamur, Tur.	248-49	36°04'N	32°50'E
Anantapur, India	256	14°41'N	77°36'E
Anantnāg, India	254-55	33°45'N	75°08'E
Anapa, Russia (ȧ-nä´pä)	218-19	44°54'N	37°20'E
Anápolis, Braz. (ä-nȧ´pō-le̊s)	182-83	16°21'S	48°57'W
Anatahan, i., N. Mar. Is.	304-05	16°22'N	145°40'E
Anatom, i., Vanuatu	306g	20°12'S	169°48'E
Añatuya, Arg. (ȧ-nyä-to͞o´yȧ)	187	28°27'S	62°49'W
Anauá, stm., Braz.	180-81	0°58'N	61°22'W
Anbanjing, China	258-59	23°57'N	100°54'E
Anchiang, China *see* Qianyang	258-59	27°11'N	110°02'E
Anchorage, Ak., U.S. (ăŋ´kẽr-åj)	136	61°12'N	149°53'W
Ancona, Italy (än-kō´nä)	216-17	43°37'N	13°31'E
Ancud, Chile (äŋ-ko͞odh´)	185	41°53'S	73°49'W
Ancud, Golfo de, b., Chile (go̊l-fô-dĕ-äŋ-ko͞odh´)	185	42°05'S	73°00'W
Anda, China	260-61	46°24'N	125°19'E
Andalgalá, Arg.	182-83	27°35'S	66°19'W
Andalucía, hist. reg., Spain (än-dä-lo͞o-se̊´ä)	214-15	37°15'N	4°30'W
Andalusia, Al., U.S. (ăn-dȧ-lo͞o´zhĭȧ)	134-35	31°19'N	86°29'W
Andaman and Nicobar Islands, state, India	266-67	11°0'N	93°00'E
Andaman Islands, is., India (ăn-dȧ-măn´ ī´lȧndz)	266-67	12°0'N	92°45'E
Andaman Sea, s., Asia (ăn-dȧ-măn´ sē)	266-67	10°0'N	95°00'E
Anderson, Ca., U.S. (ăn´dẽr-sŭn)	122-23	40°29'N	122°22'W
Anderson, In., U.S. (ăn´dẽr-sŭn)	126-27	40°05'N	85°41'W
Anderson, S.C., U.S. (ăn´dẽr-sŭn)	134-35	34°30'N	82°39'W
Anderson, stm., N.T., Can. (ăn´dẽr-sŭn)	140-41	69°42'N	128°54'W
Andes, mts., S.A. (ăn´dēz) (än´dās)	173	20°0'S	67°00'W
Andhra Pradesh, state, India	256	16°0'N	79°00'E
Andijon, Uzb.	252-53	40°47'N	72°21'E
Andkhvoy, Afg.	252-53	36°55'N	65°07'E
Andong, Kor., S. (än´dŭng´)	263	36°34'N	128°43'E
Andorra, nation, Eur. (än-do̊r´rä)	212-13	42°30'N	1°30'E
Andorra la Vella, nat. cap., And.	214-15	42°30'N	1°31'E
Andover, Mn., U.S. (ăn´do̊-vẽr)	124-25	45°14'N	93°17'W
Andøya, i., Nor. (änd-ûe̊)	200-01	69°08'N	15°54'E
Andradina, Braz.	182-83	20°55'S	51°23'W
Andrews, N.C., U.S. (ăn´dro͞oz)	134-35	35°12'N	83°49'W
Andrews, S.C., U.S. (ăn´dro͞oz)	134-35	33°27'N	79°34'W
Andrews, Tx., U.S. (ăn´dro͞oz)	130-31	32°19'N	102°33'W
Andria, Italy (än´dre̊-ä)	216-17	41°13'N	16°17'E
Andros, i., Bah. (ăn´drŏs)	154-55	24°26'N	77°57'W
Ándros, i., Grc. (ȧn´dro̊s)	216-17	37°50'N	24°53'E
Anegada, i., Br. Vir. Is.	155b	18°45'N	64°20'W
Aneto, mtn., Spain (ä-nĕ´to̊)	214-15	42°38'N	0°40'E
Angamos, Punta, c., Chile	182-83	23°02'S	70°31'W
Ang'angxi, China (äŋ-äŋ-shyē)	260-61	47°09'N	123°48'E
Angara, stm., Russia	236-37	58°06'N	93°02'E
Angarsk, Russia	240-41	52°35'N	103°55'E
Ángel, Salto, wtfl., Ven. (säl´tō-ȧ´n-hĕl) *see* Angel Falls	178-79	6°01'N	62°28'W
Ángel de la Guarda, Isla, i., Mex. (e̊´s-lä-ȧ´n-hĕl-dĕ-lä-gwä´r-dä)	156-57	29°22'N	113°28'W
Angeles, Phil. (än´hå-lās)	270	15°08'N	120°36'E
Angel Falls, wtfl., Ven. (ān´jĕl fôlz)	178-79	6°01'N	62°28'W
Ängelholm, Swe. (ĕng´ĕl-hôlm)	208-09	56°15'N	12°53'E
Angers, Fr.	212-13	47°28'N	0°33'W
Angicos, Braz.	188d	5°40'S	36°36'W
Angijak Island, i., Nu., Can.	140-41	65°40'N	62°15'W
Angkor Wat, hist., Camb. (äng´kôr)	266-67	13°26'N	103°52'E
Anglesey, i., Wales, U.K. (ăŋ´g'l-se̊)	206-07	53°17'N	4°22'W
Angleton, Tx., U.S. (aŋ´g'l-tŭn)	132-33	29°10'N	95°26'W
Angoche, Moz.	286-87	16°14'S	39°55'E
Angoche, Ilha, i., Moz. (e̊´lä-ăn-gō´chå)	286-87	16°21'S	39°51'E
Angol, Chile (aŋ-gōl´)	185	37°48'S	72°43'W
Angola, In., U.S. (ăŋ-gō´lȧ)	126-27	41°38'N	84°59'W
Angola, nation, Afr. (ăŋ-gō´lȧ)	274	12°30'S	18°30'E
Angontsy, Tanjona, c., Madag.	286-87	15°13'S	50°27'E
Angora, nat. cap., Tur. *see* Ankara	202-03	39°56'N	32°53'E
Angostura, Ven. *see* Ciudad Bolívar	178-79	8°07'N	63°33'W
Angostura, Presa de la, res., Mex.	156-57	16°02'N	92°22'W
Angoulême, Fr. (äŋ´go͞o-lâm´)	212-13	45°39'N	0°09'E

ăt; finăl; rāte; senåte; ärm; ȧsk; sofȧ; fâre; ch-choose; dh-as th in other; bē; e̊vent; bĕt; recĕnt; cratẽr; g-gō; gh-guttural g; bĭt; ĭ-short neutral; rīde; ᴋ-guttural k as ch in German ich;

PRONOUNCING INDEX
Angra dos Reis – Arctic Ocean 331

Feature (Pronunciation)	Page	Lat.	Long.
Angra dos Reis, Braz. (aŋ´grä dōs rā´ēs)	186	23°01'S	44°19'W
Angren, Uzb.	252-53	41°01'N	70°08'E
Anguilla, dep., N.A. (ăŋ-gwĭl´ȧ)	152-53	18°15'N	63°05'W
Anguilla Cays, is., Bah. (ăŋ-gwĭl´ȧ kēs)	154-55	23°31'N	79°33'W
Anguille, Cape, c., Nf./L., Can. (kăp´-äŋ-gē´yĕ)	148-49	47°55'N	59°24'W
Anholt, i., Den. (än´hōlt)	208-09	56°42'N	11°34'E
Anhui, state, China (än-hwā)	258-59	32°0'N	117°00'E
Anhwei, state, China *see* Anhui	258-59	32°0'N	117°00'E
Aniak, Ak., U.S. (ä-nyà´k)	136	61°35'N	159°33'W
Anina, Rom. (ä-nē´nä)	216-17	45°05'N	21°51'E
Anita, Pa., U.S. (à-nē´ȧ)	126-27	41°0'N	78°58'W
Aniva, Zaliv, b., Russia (zä´lĭf à-nē´và)	264	46°16'N	142°48'E
Anjār, India	254-55	23°07'N	70°02'E
Anjouan, i., Com. *see* Nzwani	286-87	12°15'S	44°25'E
Anju-ŭp, Kor., N.	263	39°37'N	125°40'E
Ankaboa, Tanjona, c., Madag.	286-87	21°55'S	43°18'E
Ankang, China (än-käŋ)	258-59	32°41'N	109°01'E
Ankara, nat. cap., Tur. (än´kȧ-rȧ)	202-03	39°56'N	32°53'E
Ankazoabo, Madag.	286-87	22°18'S	44°31'E
Ānkober, Eth.	292d	9°35'N	39°44'E
Anlong, China (än-loŋ)	258-59	25°07'N	105°28'E
Anlu, China (än´lōō´)	258-59	31°16'N	113°41'E
Anmyeondo, i., Kor., S.	263	36°30'N	126°22'E
Anna, Il., U.S. (ăn´ȧ)	126-27	37°27'N	89°14'W
Annaba, Alg.	200-01	36°54'N	7°46'E
An-Nafūd, des., Sau. Ar.	248-49	28°30'N	41°00'E
An-Najaf, Iraq (än nä-jäf´)	248-49	32°00'N	44°20'E
Annamitique, Chaîne, mts., Asia	266-67	17°0'N	106°00'E
Annapolis, Md., U.S. (ă-năp´ô-lĭs)	126-27	38°58'N	76°31'W
Annapūrṇa, mtn., Nepal	254-55	28°34'N	83°50'E
Ann Arbor, Mi., U.S. (ăn är´bẽr)	126-27	42°16'N	83°43'W
An-Nāṣirīyah, Iraq	248-49	31°03'N	46°15'E
An-Nawfalīyah, Libya	204-05	30°46'N	17°50'E
Annecy, Fr. (àn sē´)	212-13	45°54'N	6°07'E
Annemasse, Fr. (än'mäs´)	212-13	46°12'N	6°14'E
An Nhon, Viet.	266-67	13°54'N	109°05'E
Anniston, Al., U.S. (ăn´ĭs-tŭn)	134-35	33°39'N	85°50'W
Annobón, i., Eq. Gui.	282-83	1°26'S	5°37'E
Annonay, Fr. (àn´ĭs-tsiŭn)	212-13	45°15'N	4°40'E
An-Nuhūd, Sudan	288	12°42'N	28°26'E
Anori, Braz..	180-81	3°45'S	61°42'W
Anorontany, Tanjona, c., Madag.	286-87	12°26'S	48°45'E
Anpu, China (än-pōō)	258-59	21°27'N	110°01'E
Anqing, China	258-59	30°30'N	117°02'E
Ansbach, Ger. (äns´bäk)	210-11	49°10'N	10°35'E
Anse-d'Hainault, Haiti (äns´dĕnō)	154-55	18°30'N	74°26'W
Anserma, Col. (à´n-sĕ´r-mà)	188c	5°14'N	75°48'W
Anshan, China	263	41°08'N	122°60'E
Anshun, China (än-shōōn´)	258-59	26°15'N	105°56'E
Anson, Tx., U.S. (ăn´sŭn)	130-31	32°44'N	99°53'W
Ansongo, Mali	280-81	15°40'N	0°30'E
Antakya, Tur. *see* Antioch	248-49	36°12'N	36°10'E
Antalaha, Madag.	286-87	14°55'S	50°17'E
Antalya, Tur. (än-tä´lê-ä) (ä-dä´lê-ä)	202-03	36°54'N	30°42'E
Antalya, Gulf of, b., Tur. *see* Antalya Körfezi	202-03	36°30'N	30°60'E
Antalya Körfezi, b., Tur.	202-03	36°30'N	30°60'E
Antananarivo, nat. cap., Madag. (än-tä´nä-nä-rē´vō)	286-87	18°55'S	47°32'E
An tAonach, Ire. *see* Nenagh	206-07	52°52'N	8°12'W
Antarctica, cont., (ănt-ärk´tĭ-kȧ)	313	87°0'S	60°00'E
Antarctic Peninsula, pen., Ant.	313	70°15'S	65°55'W
Antequera, Spain (än-tĕ-kĕ´rä)	214-15	37°01'N	4°33'W
Anthony, Tx., U.S. (ăn´thô-nê)	128-29	31°60'N	106°36'W
Anti-Atlas, mts., Mor.	280-81	30°0'N	8°30'W
Antibes, Fr. (äN-tēb´)	212-13	43°35'N	7°07'E
Anticosti, Île d', i., Qc., Can. (ăn-tĭ-kŏs´tê)	148-49	49°30'N	63°00'W
Antigo, Wi., U.S. (ăn´tĭ-gō)	126-27	45°08'N	89°08'W
Antigonish, N.S., Can. (ăn-tĭ-gô-nêsh´)	148-49	45°37'N	61°60'W
Antigua, i., Antig.	155b	17°05'N	61°49'W
Antigua and Barbuda, nation, N.A. (ăn-tē´gwȧ ănd bär-bōō´dȧ)	152-53	17°03'N	61°48'W
Antigua Guatemala, Guat.	160	14°33'N	90°44'W
Antillas, Archipiélago de las, is., *see* West Indies	152-53	19°0'N	70°00'W
Antillas, Mar de las, s., *see* Caribbean Sea	152-53	15°0'N	73°00'W
Antillas Mayores, is., N.A. *see* Greater Antilles	154-55	20°0'N	74°00'W
Antilles, Grandes, is., N.A. *see* Greater Antilles	154-55	20°0'N	74°00'W

Feature (Pronunciation)	Page	Lat.	Long.
Antilles, Mer des, s., *see* Caribbean Sea	152-53	15°0'N	73°00'W
Antilles, Petites, is., *see* Lesser Antilles	155b	15°0'N	61°00'W
Antioch, Tur.	248-49	36°12'N	36°10'E
Antioch, Il., U.S. (ăn´tĭ-ŏk)	126-27	42°29'N	88°06'W
Antioquia, Col. (än-tê-ō´kêä)	188c	6°34'N	75°49'W
Antipodes Islands, is., N.Z.	313	49°40'S	178°47'E
Antlers, Ok., U.S. (ănt´lẽrz)	130-31	34°13'N	95°37'W
Antofagasta, Chile (än-tō-fä-gäs´tä)	182-83	23°39'S	70°23'W
Antón, Pan. (än-tōn´)	162	8°24'N	80°14'W
Antongila, Helodrano, b., Madag.	286-87	15°45'S	49°50'E
António Enes, Moz. (än-to´nyō ĕn´ēs) *see* Angoche	286-87	16°14'S	39°55'E
Antsirabe, Madag. (änt-sê-rä´bā)	286-87	19°52'S	47°02'E
Antsirañana, Madag.	286-87	12°17'S	49°17'E
Antsirane, Madag. *see* Antsirañana	286-87	12°17'S	49°17'E
Antsla, Est. (änt´slȧ)	208-09	57°50'N	26°32'E
Antsohihy, Madag.	286-87	14°49'S	48°03'E
Antung, China *see* Dandong	263	40°07'N	124°21'E
Antwerp, Bel. (ăn´twûrp)	206-07	51°13'N	4°25'E
Antwerpen, Bel. *see* Antwerp	206-07	51°13'N	4°25'E
Anugul, India	254-55	20°51'N	85°06'E
Anūpgarh, India (ŭ-nòp´gŭr)	254-55	29°11'N	73°13'E
Anuradhapura, Sri L. (ŭ-nōō´rä-dŭ-pōō´rŭ)	256	8°21'N	80°24'E
Anvers, Bel. *see* Antwerp	206-07	51°13'N	4°25'E
Anvers Island, i., Ant.	313	64°33'S	63°35'W
Anxi, China (än-shyē)	243a	25°04'N	118°11'E
Anxi, China (än-shyē)	260-61	40°29'N	95°47'E
Anyang, China (än´yäng)	260-61	36°06'N	114°20'E
Anykščiai, Lith. (anïksh-chä´ê)	208-09	55°32'N	25°07'E
Anyuanyi, China *see* Tianzhu	260-61	36°60'N	103°07'E
Anzhero-Sudzhensk, Russia (än´zhâ-rô-sôd´zhĕnsk)	236-37	56°05'N	86°01'E
Anzio, Italy (änt´zê-ō)	216-17	41°27'N	12°37'E
Aoba, i., Vanuatu	306g	15°25'S	167°50'E
Aoga-shima, i., Japan	264	32°28'N	139°46'E
Aomen, China	258-59	22°13'N	113°33'E
Aomori, Japan (äō-mō´rê)	264	40°49'N	140°45'E
Aoraki/Mount Cook National Park, n.p., N.Z.	303	43°35'S	170°15'E
Aoraki, mtn., N.Z.	303	43°36'S	170°10'E
Aôral, Phnum, mtn., Camb.	266-67	12°02'N	104°10'E
Aouk, Bahr, stm., Afr. (bär ä-ök´)	284-85	8°51'N	18°52'E
Aoukâr, reg., Maur.	280-81	18°0'N	9°30'W
Aoulef, Alg.	280-81	26°58'N	1°04'E
Apalachicola, Fl., U.S. (ăp-ȧ-lăch-ĭ-kō´lȧ)	134-35	29°44'N	84°00'W
Apalachicola, stm., Fl., U.S. (ăpȧ-lăch´ĭ-cōlā)	134-35	29°44'N	84°59'W
Apaporis, stm., S.A. (ä-pä-pō´rĭs)	184	1°21'S	69°25'W
Aparri, Phil. (ä-pär´rē)	270	18°20'N	121°40'E
Apatin, Serb. (ŏ´pŏ tĭn)	216-17	45°40'N	18°59'E
Apatity, Russia	200-01	67°34'N	33°23'E
Apeldoorn, Neth. (ä´pĕl-dōōrn)	206-07	52°13'N	5°58'E
Apennines, mts., Italy (ă´-pȧ-nīnz)	216-17	43°0'N	13°00'E
Apía, Col. (á-pē´ä)	188c	5°06'N	75°58'W
Apia, nat. cap., Samoa (ä´-pē-ȧ) (ä-pē´-ȧ)	306b	13°50'S	171°45'W
Apiacás, Serra dos, plat., Braz.	180-81	10°15'S	57°15'W
Apizaco, Mex. (ä-pê-zä´kō)	158-59	19°25'N	98°08'W
Apo, Mount, mtn., Phil. (mount ä´pō)	270	6°59'N	125°16'E
Apolo, Bol.	180-81	14°43'S	68°31'W
Aporé, stm., Braz.	182-83	19°28'S	50°56'W
Apostle Islands, is., Wi., U.S. (ä-pŏs´l ĭ´lăndz)	124-25	46°50'N	90°30'W
Apóstoles, Arg.	187	27°55'S	55°48'W
Apostolove, Ukr.	218-19	47°39'N	33°43'E
Appalaches, Les, mts., N.A. *see* Appalachian Mountains	120-21	41°0'N	77°00'W
Appalachia, Va., U.S. (ăpȧ-lăch´ĭ-ȧ)	134-35	36°54'N	82°48'W
Appalachian Mountains, mts., N.A. (ăp-ȧ-lăch´ĭ-ȧn moun´tĭnz)	120-21	41°0'N	77°00'W
Appennino, mts., Italy (äp-pĕn-nê´nô) *see* Apennines	216-17	43°0'N	13°00'E
Appleton, Mn., U.S. (ăp´l-tŭn)	124-25	45°12'N	96°01'W
Appleton, Wi., U.S. (ăp´l-tŭn)	126-27	44°15'N	88°25'W
Appleton City, Mo., U.S. (ăp´l-tŭn sĭ´tê)	130-31	38°11'N	94°02'W
Apt, Fr. (äpt)	212-13	43°52'N	5°24'E
Apucarana, Braz.	182-83	23°33'S	51°27'W
Apure, stm., Ven. (ä-pōō´rā)	178-79	7°37'N	66°23'W
Apurímac, stm., Peru (ä-pōō-rê-mäk´)	184	11°52'S	73°57'W
Aqaba, Gulf of, b., (gŭlf ŭv ä´kȧ-bȧ)	248-49	29°05'N	34°44'E
Aqmola, nat. cap., Kaz. *see* Astana	244	51°12'N	71°27'E
Aqtaū, Kaz..	202-03	43°38'N	51°11'E

Feature (Pronunciation)	Page	Lat.	Long.
Aqtöbe, Kaz.	244	50°18'N	57°10'E
Aquidauana, Braz. (ä-kē-däwä´nä)	182-83	20°29'S	55°48'W
Aquila, Italy *see* L'Aquila	216-17	42°21'N	13°24'E
Aquin, Haiti (ä-kăn´)	154-55	18°17'N	73°24'W
Ar'ar, Sau. Ar.	248-49	30°56'N	41°04'E
Ara, India	254-55	25°34'N	84°40'E
Ara, stm., Japan (ä-rä)	265	35°40'N	139°51'E
Ara, stm., Japan (ä-rä)	265	38°09'N	139°25'E
'Arab, Baḩr al-, stm., Sudan	284-85	9°02'N	29°28'E
'Arab, Shaṭṭ al-, stm., Asia	248-49	29°57'N	48°33'E
Arabian Desert, des., Egypt (ȧ-rā´bĭ-ăn dĕs´ẽrt)	288	28°0'N	32°00'E
Arabian Gulf, b., Asia *see* Persian Gulf	250-51	27°0'N	51°00'E
Arabian Peninsula, pen., Asia (ȧ-rā´bĭ-ăn pĕ-nĭn´sūlȧ)	238-39	25°0'N	45°00'E
Arabian Sea, s., (ȧ-rā´bĭ-ăn sē)	238-39	15°0'N	65°00'E
Aracaju, Braz. (ä-rä´kä-zhōō´)	188d	10°54'S	37°04'W
Aracati, Braz. (ä-rä´kä-tē´)	180-81	4°34'S	37°46'W
Araçatuba, Braz. (ä-rä-sà-tōō´bä)	182-83	21°12'S	50°27'W
Aracruz, Braz. (ä-rä-krōō´s)	186	19°49'S	40°16'W
Araçuaí, Braz.	182-83	16°53'S	42°04'W
Arad, Rom. (ŏ´rŏd)	210-11	46°11'N	21°19'E
Arafura, Laut, s., *see* Arafura Sea	242-43	9°0'S	133°00'E
Arafura Sea, s., (ä-rä-fōō´rä sē)	242-43	9°0'S	133°00'E
Aragarças, Braz.	182-83	15°55'S	52°15'W
Aragón, hist. reg., Spain (ä-rä-gōn´)	214-15	41°30'N	1°00'W
Araguacema, Braz.	180-81	8°50'S	49°34'W
Aragua de Barcelona, Ven.	188b	9°27'N	64°50'W
Araguaia, stm., Braz. (ä-rä-gwä´yä)	180-81	5°20'S	48°42'W
Araguari, Braz. (ä-rä-gwä´rê)	186	18°39'S	48°12'W
Araguari, stm., Braz.	180-81	1°13'N	50°02'W
Araguatins, Braz. (ä-rä-gwä-tēns)	180-81	5°39'S	48°06'W
Arāk, Iran	252-53	34°05'N	49°41'E
Aral, Kaz.	244	46°48'N	61°40'E
Aral Sea, lk., Asia (ă´-rŭl sē)	244	45°0'N	60°00'E
Aral Tengizi, lk., Asia *see* Aral Sea	244	45°0'N	60°00'E
Aramac, Austl.	302	22°58'S	145°15'E
Aramberri, Mex. (ä-ràm-bĕr-rē´)	158-59	24°06'N	99°49'W
Aranda de Duero, Spain (ä-rän´dä dä dwā´rō)	214-15	41°41'N	3°41'W
Arandas, Mex. (ä-rän´däs)	158-59	20°43'N	102°20'W
Aranjuez, Spain (ä-rän-hwāth´)	214-15	40°02'N	3°37'W
Aransas Pass, Tx., U.S. (ȧ-răn´sȧs pȧs)	132-33	27°54'N	97°09'W
Aranyaprathet, Thai..	266-67	13°41'N	102°31'E
Arapiraca, Braz.	188d	9°45'S	36°39'W
Araranguá, Braz.	186	28°56'S	49°29'W
Araraquara, Braz. (ä-rä-rä-kwà´rä)	186	21°47'S	48°10'W
Ararat, Austl. (ăr ȧrat)	301	37°17'S	142°56'E
Ararat, Mount, vol., Tur. (mount ăr´ȧrăt)	245	39°42'N	44°18'E
Araripe, Chapada do, plat., Braz. (shä-pä´dä dô ä rä rē´pĕ)	180-81	7°23'S	39°49'W
Araruama, Lagoa de, b., Braz. (là-gô´ä-ä-rä-rōō-ä´mä)	186	22°53'S	42°12'W
Aras, stm., Asia (à-räs)	245	40°01'N	48°28'E
Arauca, Col. (ä-rou´kä)	178-79	7°04'N	70°45'W
Arauca, stm., S.A. (ä-rou´kä)	178-79	7°25'N	66°31'W
Arāvalli Range, mts., India (ä-rä´vŭ-lê rānj)	254-55	24°42'N	73°19'E
Araxá, Braz.	186	19°36'S	46°55'W
Araya, Punta de, c., Ven. (pŭn´tä-dĕ-ä-rä´yà)	188b	10°38'N	64°17'W
Araz, stm., Asia.	245	40°01'N	48°28'E
Ārba Minch', Eth..	284-85	6°01'N	37°34'E
Arbīl, Iraq	248-49	36°11'N	44°01'E
Arboga, Swe. (är-bō´gä)	208-09	59°24'N	15°51'E
Arboréa, Italy (är-bō-rĕ´ä)	216-17	39°46'N	8°35'E
Arbroath, Scot., U.K. (är-brōth´)	206-07	56°34'N	2°36'W
Arcachon, Fr. (är-kä-shôn´)	212-13	44°40'N	1°10'W
Arcadia, Fl., U.S. (är-kā´dĭ-ȧ)	135a	27°13'N	81°51'W
Arcadia, La., U.S. (är-kā´dĭ-ȧ)	130-31	32°34'N	92°56'W
Arcadia, Wi., U.S. (är-kā´dĭ-ȧ)	124-25	44°15'N	91°29'W
Arcata, Ca., U.S. (är-kȧ´tȧ)	122-23	40°53'N	124°05'W
Arc Dome, mtn., Nv., U.S. (ärk dōm)	128-29	38°50'N	117°21'W
Arcelia, Mex. (är-sā´lê-ä)	158-59	18°18'N	100°17'W
Archangel, Russia *see* Arkhangel'sk	202-03	64°32'N	40°25'E
Archbald, Pa., U.S. (ärch´bôld)	126-27	41°30'N	75°33'W
Archer Bend National Park, n.p., Austl. *see* Mungkan Kandju National Park	302	13°32'S	142°37'E
Arches National Park, n.p., Ut., U.S. (är´ches năsh´ŭn-ăl pärk)	128-29	38°43'N	109°36'W
Arco, Id., U.S. (är´kô)	122-23	43°39'N	113°18'W
Arcoverde, Braz.	188d	8°25'S	37°04'W
Arctic Ocean, oc., (ärk´tĭk ōshŭn)	314	85°0'N	170°00'E

n-sing; ŋ-baŋk; ɴ-nasalized n; nŏd; cŏmmit; ōld; ôbey; ôrder; oi-boil; fōōd; ŏ-as oo in foot; ou-out; s-soft; sh-dish; th-thin; pūre; ûnite; ûrn; stŭd; circŭs; ü-as in French tu; ´-indeterminate vowel.

332 Arctic Red – Athína PRONOUNCING INDEX

Feature (Pronunciation)	Page	Lat.	Long.
Arctic Red, stm., N.T., Can.	140-41	67°27'N	133°45'W
Arctic Village, Ak., U.S.	136	68°05'N	145°31'W
Ardabīl, Iran	245	38°15'N	48°18'E
Ardahan, Tur. (är-dà-hän′)	245	41°06'N	42°43'E
Ardebil, Iran *see* Ardabīl	245	38°15'N	48°18'E
Ardennen, reg., Eur. *see* Ardennes	206-07	50°10'N	5°45'E
Ardennes, reg., Eur. (àr-dĕn′)	206-07	50°10'N	5°45'E
Ardila, stm., Eur. (är-dē′là)	214-15	38°10'N	7°29'W
Ardmore, Ok., U.S. (ärd′mōr)	130-31	34°10'N	97°09'W
Arecibo, P.R. (ä-rå-sē′bō)	154a	18°28'N	66°43'W
Areia Branca, Braz. (ä-rĕ′yä-brà′n-kä)	188d	4°56'S	37°07'W
Arena, Point, c., Ca., U.S. (point ä-rā′n*á*)	128-29	38°57'N	123°44'W
Arena, Punta, c., Mex.	156-57	23°34'N	109°28'W
Arendal, Nor. (ä′rĕn-däl)	208-09	58°27'N	8°48'E
Arequipa, Peru (ä-rå-kē′pä)	184	16°24'S	71°32'W
Arezzo, Italy (ä-rĕt′sō)	216-17	43°28'N	11°53'E
Argentan, Fr. (àr-zhäN-täN′)	212-13	48°45'N	0°01'W
Argenteuil, Fr. (är-zhäN-tû′y′)	212-13	48°57'N	2°14'E
Argentina, nation, S.A. (är-jĕn-tē′n*á*)	172	34°0'S	64°00'W
Argentino, Lago, lk., Arg. (lä′gô är-κĕn-tē′nō)	185	50°14'S	72°26'W
Argenton-sur-Creuse, Fr. (är-zhäN′tôN-sür-krôs)	212-13	46°35'N	1°31'E
Arghandāb, stm., Afg.	252-53	31°27'N	64°23'E
Argonne, reg., Fr. (ä′r-gôn)	212-13	49°07'N	5°14'E
Árgos, Grc. (är′gōs)	216-17	37°38'N	22°44'E
Arguello, Point, c., Ca., U.S. (point är-gwāl′yō)	128-29	34°35'N	120°38'W
Argun', stm., Asia (år-gōōn′)	236-37	53°19'N	121°27'E
Ariake-kai, b., Japan (ä′rē-ä′kå)	265	33°0'N	130°20'E
Arica, Chile (ä-rē′kä)	182-83	18°29'S	70°19'W
Arica, Col.	178-79	2°07'S	71°44'W
Arīḩā, W.B. *see* Jericho	248-49	31°52'N	35°27'E
Arima, Trin.	155b	10°37'N	61°17'W
Arinos, stm., Braz. (ä-rē′nôzsh)	180-81	10°26'S	58°20'W
Aripuanã, stm., Braz. (á-rê-pwän′y*á*)	180-81	5°07'S	60°23'W
Ariquemes, Braz.	180-81	9°54'S	63°05'W
Aristazabal Island, i., B.C., Can.	142-43	52°38'N	129°07'W
Arizona, Arg.	185	35°43'S	65°19'W
Arizona, state, U.S. (ăr-ĭ-zō′n*á*)	118-19	34°0'N	112°00'W
Arkadelphia, Ar., U.S. (är-k*á*-dĕl′fĭ-*á*)	130-31	34°07'N	93°04'W
Arkansas, state, U.S. (är′kăn-sô) (är-kăn′s*á*s)	118-19	34°50'N	92°30'W
Arkansas, stm., U.S. (är′kăn-sô) (är-kăn′s*á*s)	120-21	33°47'N	91°04'W
Arkansas City, Ks., U.S.	130-31	37°04'N	97°02'W
Arkhangel'sk, Russia (år-κän′gĕlsk)	202-03	64°32'N	40°25'E
Arkhangel'skoye, Russia (är-kän-gĕl′skô-yĕ)	218-19	53°16'N	37°42'E
Arles, Fr. (ärl)	212-13	43°41'N	4°38'E
Arlington, S.D., U.S. (är′lĕng-t*ŭ*n)	124-25	44°22'N	97°08'W
Arlington, Tx., U.S. (år′lĭng-t*ŭ*n)	130-31	32°45'N	97°07'W
Arlington, Va., U.S. (år′lĭng-t*ŭ*n)	126-27	38°52'N	77°07'W
Arlington, Vt., U.S. (år′lĭng-t*ŭ*n)	126-27	43°04'N	73°09'W
Arlington Heights, Il., U.S. (är′lĕng-t*ŭ*n hīts)	126-27	42°05'N	87°59'W
Arlit, Niger	280-81	18°45'N	7°21'E
Armant, Egypt (är-mänt′)	291b	25°37'N	32°32'E
Armavir, Russia (år-må-vĭr′)	202-03	44°59'N	41°07'E
Armenia, Col. (år-mĕ′nē*á*)	188c	4°31'N	75°42'W
Armenia, nation, Asia (är-mē′nē-*á*)	245	40°0'N	45°00'E
Armeniya, nation, Asia *see* Armenia	245	40°0'N	45°00'E
Armentières, Fr. (år-mäN-tyâr′)	212-13	50°41'N	2°53'E
Armidale, Austl. (är′mĭ-dāl)	301	30°31'S	151°40'E
Armour, S.D., U.S. (är′mẽr)	124-25	43°19'N	98°21'W
Armstrong, On., Can.	144-45	50°19'N	89°04'W
Arnaud, stm., Qc., Can.	140-41	59°58'N	69°58'W
Arnedo, Spain (är-nā′dō)	214-15	42°14'N	2°07'W
Arnhem, Neth. (ärn′hĕm)	206-07	51°59'N	5°55'E
Arnhem, Cape, c., Austl. (kāp ärn′hĕm)	296-97	12°22'S	136°57'E
Arnhem Land, reg., Austl. (ärn′hĕm-lănd)	296-97	13°13'S	133°50'E
Arnold, Mn., U.S. (år′nŭld)	124-25	46°53'N	92°05'W
Arnprior, On., Can. (ärn-prī′ẽr)	146-47	45°26'N	76°21'W
Arnsberg, Ger. (ärns′bĕrgh)	210-11	51°24'N	8°04'E
Arnstadt, Ger. (ärn′shtät)	210-11	50°50'N	10°57'E
Aroma, Sudan	288	15°48'N	36°08'E
Aroostook, stm., N.A.	148-49	46°49'N	67°43'W
Arop Island, i., Pap. N. Gui.	302	5°20'S	147°05'E
Arorae, i., Kir.	304-05	2°38'S	176°49'E
Arqalyq, Kaz.	244	50°15'N	66°53'E
Arraias, Braz.	180-81	12°58'S	46°55'W
Ar-Ramādī, Iraq	248-49	33°26'N	43°19'E
Arran, Island of, i., Scot., U.K. (ĭ′l*á*nd ŏv ă′răn)	206-07	55°35'N	5°15'W
Arras, Fr. (à-räs′)	212-13	50°17'N	2°47'E
Ar-Rass, Sau. Ar.	288	25°52'N	43°30'E
Arrecife, Spain	215d	28°57'N	13°33'W
Arrecifes, Arg. (är-rå-sē′fās)	187	34°04'S	60°06'W
Arriaga, Mex. (är-rēä′gä)	158-59	16°14'N	93°53'W
Ar-Riyāḑ, nat. cap., Sau. Ar. *see* Riyadh	250-51	24°38'N	46°43'E
Ar-Rub'al-Khālī, des., Asia *see* Rub'al-Khali	238-39	20°0'N	51°00'E
Ar-Ruṣayriṣ, Sudan	288	11°48'N	34°22'E
Ar-Ruṭbah, Iraq	248-49	33°02'N	40°17'E
Arsen'yev, Russia	264	44°09'N	133°17'E
Art, Île, i., N. Cal.	306g	19°43'S	163°39'E
Árta, Grc. (är′tä)	216-17	39°09'N	20°59'E
Arteaga, Mex. (är-tā-ä′gä)	158-59	18°20'N	102°18'W
Arteaga, Mex. (är-tā-ä′gä)	132-33	25°28'N	100°51'W
Artëm, Russia (*á*r-tyôm′)	264	43°21'N	132°11'E
Artemisa, Cuba (är-tå-mē′sä)	154-55	22°49'N	82°46'W
Artesia, N.M., U.S. (är-tē′sĭ-*á*)	130-31	32°51'N	104°24'W
Artibonite, stm., Haiti (är-tê-bô-nē′tä)	154-55	19°15'N	72°46'W
Artigas, Ur.	187	30°24'S	56°28'W
Artillery Lake, lk., N.T., Can.	140-41	63°09'N	107°52'W
Artvin, Tur.	245	41°10'N	41°50'E
Artyk, Russia	236-37	64°09'N	145°12'E
Aru, Kepulauan, is., Indon.	242-43	6°0'S	134°30'E
Aru, Tanjung, c., Indon.	268-69	2°11'S	116°35'E
Arua, Ug. (ä′rōō-ä)	289	3°01'N	30°55'E
Aruanã, Braz.	180-81	14°54'S	51°05'W
Aruba, dep., N.A. (ä-rōō′b*ă*)	152-53	12°30'N	69°58'W
Arunāchal Pradesh, state, India	254-55	28°30'N	95°00'E
Aruppukkottai, India	256	9°31'N	78°06'E
Arusha, Tan. (å-rōō′shä)	289	3°22'S	36°41'E
Aruwimi, stm., D.R.C.	284-85	1°13'N	23°36'E
Arvayheer, Mong.	260-61	46°15'N	102°48'E
Arviat, Nu., Can.	138-39	61°08'N	94°07'W
Arvidsjaur, Swe.	200-01	65°36'N	19°07'E
Arvika, Swe. (är-vē′kå)	208-09	59°40'N	12°38'E
Arxan, China	260-61	47°11'N	119°57'E
Arys, Kaz.	244	42°26'N	68°48'E
Arzamas, Russia (är-zä-mäs′)	202-03	55°23'N	43°50'E
Aš, Czech Rep. (äsh′)	210-11	50°13'N	12°12'E
Asad, Buḩayrat al-, res., Syria	248-49	36°00'N	38°10'E
Asahi-dake, vol., Japan	264	43°40'N	142°51'E
Asahigawa, Japan *see* Asahikawa	264	43°46'N	142°22'E
Asahikawa, Japan	264	43°46'N	142°22'E
Āsānsol, India	254-55	23°41'N	86°59'E
Asbestos, Qc., Can. (ăs-bĕs′tōs)	146-47	45°46'N	71°57'W
Asbury Park, N.J., U.S. (ăz′bĕr-ĭ pärk)	126-27	40°13'N	74°01'W
Ascensión, Mex. (äs-sĕn-sê-ōn′)	156-57	31°06'N	107°60'W
Ascension, i., St. Hel. (*á*-sĕn′sh*ŭ*n)	275	7°57'S	14°22'W
Aschaffenburg, Ger. (ä-shäf′ĕn-bòrgh)	210-11	49°59'N	9°09'E
Aschersleben, Ger. (äsh′ĕrs-lā-bĕn)	210-11	51°46'N	11°28'E
Ascoli Piceno, Italy (äs′kô-lēpē-chā′nō)	216-17	42°52'N	13°35'E
Aseb, Erit.	288	12°58'N	42°42'E
Āsela, Eth.	292d	7°58'N	39°08'E
Åsele, Swe.	200-01	64°10'N	17°21'E
Aşgabat, nat. cap., Turkmen.	252-53	37°57'N	58°23'E
Asha, Russia (ä′shä)	202-03	55°00'N	57°16'E
Ashburn, Ga., U.S. (ăsh′b*ŭ*rn)	134-35	31°42'N	83°39'W
Ashburton, stm., Austl. (ăsh′bûr-t*ŭ*n)	296-97	21°42'S	114°55'E
Ashdown, Ar., U.S. (ăsh′doun)	130-31	33°41'N	94°08'W
Asheboro, N.C., U.S. (ăsh′bŭr-ô)	134-35	35°42'N	79°49'W
Asheville, N.C., U.S. (ăsh′vĭl)	134-35	35°36'N	82°34'W
Ashgabat, nat. cap., Turkmen. *see* Aşgabat	252-53	37°57'N	58°23'E
Ashikaga, Japan (ä′shē-kä′gå)	265	36°20'N	139°27'E
Ashkhabad, nat. cap., Turkmen. *see* Aşgabat	252-53	37°57'N	58°23'E
Ashland, Ky., U.S. (ăsh′l*á*nd)	126-27	38°28'N	82°39'W
Ashland, Me., U.S. (ăsh′l*á*nd)	127a	46°38'N	68°28'W
Ashland, Ne., U.S. (ăsh′l*á*nd)	124-25	41°02'N	96°22'W
Ashland, Oh., U.S. (ăsh′l*á*nd)	126-27	40°52'N	82°18'W
Ashland, Or., U.S. (ăsh′l*á*nd)	122-23	42°12'N	122°42'W
Ashland, Va., U.S. (ăsh′l*á*nd)	126-27	37°45'N	77°29'W
Ashland, Wi., U.S. (ăsh′l*á*nd)	124-25	46°36'N	90°53'W
Ashley, N.D., U.S. (ăsh′lē)	124-25	46°02'N	99°22'W
Ashmore and Cartier Islands, dep., Oc.	242-43	12°25'S	123°20'E
Ashqelon, Isr. (äsh′kĕ-lōn)	248-49	31°40'N	34°35'E
Ash-Shāriqah, U.A.E. *see* Sharjah	250-51	25°22'N	55°24'E
Ash-Shiḩr, Yemen	238-39	14°46'N	49°37'E
Ashtabula, Oh., U.S. (ăsh-t*á*-bū′l*á*)	126-27	41°51'N	80°48'W
Ashton, Id., U.S. (ăsh′t*ŭ*n)	122-23	44°05'N	111°27'W
Ashur, hist., Iraq	248-49	35°30'N	43°16'E
Asia, cont., (ā′zh*ă*)	226-27	50°0'N	100°00'E
Asinara, Golfo dell', b., Italy (gôl′fō-dĕl-ä-sē-nä′rä)	216-17	41°0'N	8°32'E
Asinara, Isola, i., Italy	216-17	41°04'N	8°16'E
Asino, Russia	236-37	56°60'N	86°08'E
'Asīr, reg., Sau. Ar. (ä-sēr′)	238-39	19°0'N	42°00'E
Askersund, Swe. (äs′kĕr-sònd)	208-09	58°53'N	14°54'E
Asmara, nat. cap., Erit. (äz-mä′-r*á*)	288	15°20'N	38°55'E
Asmera, nat. cap., Erit. (äs-mä′rä) *see* Asmara	288	15°20'N	38°55'E
Asotin, Wa., U.S. (*á*-sō′tĭn)	122-23	46°20'N	117°03'W
Aspen, Co., U.S. (ăs′p*ĕ*n)	128-29	39°12'N	106°49'W
Aspiring, Mount, mtn., N.Z.	303	44°23'S	168°44'E
Assab, Erit. *see* Aseb	288	12°58'N	42°42'E
Assam, state, India (ăs-săm′)	254-55	26°0'N	93°00'E
As-Samāwah, Iraq	248-49	31°19'N	45°17'E
Assateague Island, i., U.S.	126-27	38°05'N	75°12'W
Assens, Den. (äs′s*ĕ*ns)	208-09	55°16'N	9°55'E
Assiniboia, Sk., Can.	144-45	49°38'N	105°59'W
Assiniboine, stm., Can. (ă-sĭn′ĭ-boin)	144-45	49°53'N	97°08'W
Assiniboine, Mount, mtn., Can. (mount ă-sĭn′ĭ-boin)	142-43	50°52'N	115°39'W
Assis, Braz. (ä-sē′s)	182-83	22°40'S	50°26'W
Assomption, i., Sey.	286-87	9°44'S	46°30'E
As-Sūdān, nation, Afr. *see* Sudan	274	15°0'N	30°00'E
As-Sudd, reg., S. Sudan	284-85	8°0'N	31°00'E
As-Sulaymānīyah, Iraq	248-49	35°34'N	45°27'E
As-Sūrīyah, nation, Asia *see* Syria	224-25	35°0'N	38°00'E
As-Suwaydā', Syria	248-49	32°42'N	36°34'E
Astana, nat. cap., Kaz. (ä′stä-nä′)	244	51°12'N	71°27'E
Astara, Azer.	245	38°28'N	48°52'E
Asterābād, Iran *see* Gorgān	252-53	36°51'N	54°26'E
Asti, Italy (äs′tē)	216-17	44°55'N	8°13'E
Astorga, Spain (äs-tôr′gä)	214-15	42°28'N	6°03'W
Astoria, Or., U.S. (ăs-tō′rĭ-*á*)	122-23	46°11'N	123°50'W
Astove, i., Sey.	286-87	10°06'S	47°45'E
Astrakhan', Russia (äs-trà-kän′)	202-03	46°21'N	48°02'E
Asunción, nat. cap., Para. (ä-sōōn-syōn′)	182-83	25°16'S	57°39'W
Asuncion Island, i., N. Mar. Is.	304-05	19°42'N	145°24'E
Aswân, Egypt	291b	24°05'N	32°55'E
Aswan High Dam, d., Egypt	291b	23°59'N	32°53'E
Asyût, Egypt	291b	27°11'N	31°11'E
Atacama, Desierto de, des., Chile (dĕ-syĕ′r-tô-dĕ-ä-tä-kà′mä)	182-83	20°08'S	69°53'W
Atacama, Puna de, plat., S.A. (pōō′nä-dĕ-ä-tä-kà′mä)	182-83	23°46'S	67°45'W
Atacama, Salar de, pl., Chile (sà-lär′dĕ-àtä-kà′mä)	182-83	23°33'S	68°14'W
Atacama Desert, des., Chile (ä-tä-kà′mä) *see* Atacama, Desierto de	182-83	20°08'S	69°53'W
Ataco, Col. (ä-tá′kō)	188c	3°35'N	75°23'W
Atafu, at., Tok.	304-05	8°33'S	186°30'W
Atakpamé, Togo	282-83	7°32'N	1°09'E
Atamyrat, Turkmen.	252-53	37°50'N	65°13'E
Aţar, Maur. (ä-tär′)	280-81	20°32'N	13°02'W
Atascadero, Ca., U.S. (ăt-ăs-k*á*-dâ′rō)	128-29	35°30'N	120°39'W
Atasū, Kaz.	244	48°41'N	71°39'E
Atbara, stm., Afr. *see* 'Aṭbarah	288	17°40'N	33°58'E
'Aṭbarah, Sudan (ät′bá-rä)	288	17°42'N	33°59'E
'Aṭbarah, stm., Afr.	288	17°40'N	33°58'E
Atbasar, Kaz. (ät′bä-sär′)	244	51°48'N	68°21'E
Atchafalaya, stm., La., U.S. (ăch-*á*-f*á*-lī′*á*)	134-35	29°28'N	91°16'W
Atchafalaya Bay, b., La., U.S. (ăch-*á*-f*á*-lī′*á* bā)	134-35	29°27'N	91°23'W
Atchara Autonomis Respublika, state, Geor.	245	41°40'N	42°00'E
Atchison, Ks., U.S. (ăch′ĭ-s*ŭ*n)	130-31	39°34'N	95°07'W
Athabasca, Ab., Can. (ăth-*á*-băs′k*á*)	142-43	54°42'N	113°17'W
Athabasca, stm., Ab., Can. (ăth-*á*-băs′k*á*)	140-41	58°40'N	110°55'W
Athabasca, Lake, lk., Can. (lăk ăth-*á*-băs′k*á*)	140-41	59°07'N	109°59'W
Athens, Al., U.S. (ăth′ĕnz)	134-35	34°48'N	86°58'W
Athens, Ga., U.S. (ăth′ĕnz)	134-35	33°57'N	83°22'W
Athens, Oh., U.S. (ăth′ĕnz)	126-27	39°20'N	82°06'W
Athens, Tn., U.S. (ăth′ĕnz)	134-35	35°27'N	84°36'W
Athens, Tx., U.S. (ăth′ĕnz)	132-33	32°13'N	95°51'W
Athens, nat. cap., Grc. (ăth′ĕnz)	216-17	37°59'N	23°44'E
Atherton, Austl.	302	17°16'S	145°30'E
Athi, stm., Kenya (ä′tē)	284-85	2°58'S	38°31'E
Athína, nat. cap., Grc. (ä-thē′nĕ) *see* Athens	216-17	37°59'N	23°44'E

PRONOUNCING INDEX
Athos, Mount – Bahía Blanca 333

Feature (Pronunciation)	Page	Lat.	Long.
Athos, Mount, mtn., Grc.	216-17	40°09'N	24°19'E
Athy, Ire. (*à*-thī)	206-07	52°60'N	6°59'W
Ati, Chad.	280-81	13°12'N	18°19'E
Atikokan, On., Can.	146-47	48°45'N	91°37'W
Atikonak Lake, lk., Nf./L., Can.	140-41	52°38'N	64°30'W
Atiu, i., Cook Is.	304-05	20°02'S	158°07'W
Atka, Russia	236-37	60°45'N	151°46'E
Atka Island, i., Ak., U.S. (ăt´k*à* ī´l*ă*nd)	136a	52°15'N	174°08'W
Atkarsk, Russia (åt-kärsk´)	202-03	51°53'N	45°00'E
Atkinson, Ne., U.S. (ăt´kĭn-s*ŭ*n)	124-25	42°32'N	98°59'W
Atlanta, Ga., U.S. (ăt-lăn´t*à*)	134-35	33°46'N	84°25'W
Atlanta, Il., U.S. (ăt-lăn´t*à*)	126-27	40°16'N	89°14'W
Atlanta, Tx., U.S. (ăt-lăn´t*à*)	130-31	33°07'N	94°11'W
Atlantic, Ia., U.S. (ăt-lăn´tĭk)	124-25	41°24'N	95°01'W
Atlantic City, N.J., U.S. (ăt-lăn´tĭk sĭ´tê)	126-27	39°21'N	74°26'W
Atlantic Ocean, oc., (ăt-lăn´tĭk ōshŭn)	4-5	5°0'S	25°00'W
Atlantis, S. Afr.	286-87	33°32'S	18°29'E
Atlas Mountains, mts., Afr. (ăt´lăs moun´tĭnz)	280-81	33°0'N	2°00'W
Atlin, B.C., Can.	138-39	59°34'N	133°41'W
Atmore, Al., U.S. (ât´mōr)	134-35	31°01'N	87°30'W
Atoka, Ok., U.S. (á-tō´k*à*)	130-31	34°23'N	96°08'W
Atoui, Khaṭṭ, stm., Afr. (á-tōō-ē´)	280-81	20°03'N	15°58'W
Atoyac, stm., Mex. (ä-tô-yäk´)	158-59	18°07'N	98°44'W
Atoyac de Álvarez, Mex. (ä-tô-yäk´dā äl´vä-rāz)	158-59	17°11'N	100°25'W
Atrak, stm., Asia	252-53	37°26'N	53°53'E
Atrato, stm., Col. (ä-trä´tō)	178-79	8°11'N	76°56'W
Atrek, stm., Asia *see* Atrak	252-53	37°26'N	53°53'E
Aṭ-Ṭā'if, Sau. Ar.	288	21°16'N	40°25'E
Attapu, Laos	266-67	14°48'N	106°51'E
Attawapiskat, On., Can.	138-39	52°56'N	82°25'W
Attawapiskat, stm., On., Can. (ăt´*à*-w*à*-pĭs´kăt)	140-41	52°57'N	82°18'W
Attawapiskat Lake, lk., On., Can.	144-45	52°17'N	87°55'W
Attica, N.Y., U.S. (ăt´ĭ-k*à*)	126-27	42°52'N	78°17'W
Attleboro, Ma., U.S. (ăt´'l-bŭr-ô)	126-27	41°57'N	71°17'W
Attu Island, i., Ak., U.S. (ăt-tōō´ ī´l*ă*nd)	136a	52°55'N	187°00'E
Atuel, stm., Arg.	185	36°16'S	66°51'W
Åtvidaberg, Swe. (ôt-vē´då-bĕrgh)	208-09	58°12'N	16°00'E
Atyraū, Kaz.	202-03	47°07'N	51°55'E
Aubagne, Fr. (ō-bän´y')	212-13	43°17'N	5°34'E
Aubry Lake, lk., N.T., Can.	140-41	67°22'N	126°27'W
Auburn, Al., U.S. (ô´b*ŭ*rn)	134-35	32°37'N	85°29'W
Auburn, Il., U.S. (ô´b*ŭ*rn)	130-31	39°35'N	89°45'W
Auburn, In., U.S. (ô´b*ŭ*rn)	126-27	41°22'N	85°03'W
Auburn, Ma., U.S. (ô´b*ŭ*rn)	126-27	42°12'N	71°50'W
Auburn, Ne., U.S. (ô´b*ŭ*rn)	130-31	40°23'N	95°51'W
Auburn, N.Y., U.S.	126-27	42°56'N	76°34'W
Auburn, Wa., U.S. (ô´b*ŭ*rn)	122-23	47°19'N	122°12'W
Aubusson, Fr. (ō-bü-sôN´)	212-13	45°57'N	2°10'E
Auch, Fr. (ōsh)	212-13	43°39'N	0°35'E
Auckland, N.Z. (ôk´lănd)	303	36°51'S	174°45'E
Auckland Islands, is., N.Z. (ôk´lănd ī´l*ă*ndz)	313	50°46'S	166°12'E
Audubon, Ia., U.S. (ô´dŏ-bŏn)	124-25	41°43'N	94°56'W
Augathella, Austl. (ôr´g*á*´thĕ-l*à*)	301	25°48'S	146°34'E
Augrabies Falls National Park, n.p., S. Afr.	286-87	28°35'S	20°19'E
Augsburg, Ger. (ouks´bŏrgh)	210-11	48°23'N	10°53'E
Augusta, Austl.	294-95	34°19'S	115°10'E
Augusta, Ar., U.S. (ô-gŭs´t*à*)	134-35	35°17'N	91°22'W
Augusta, Ga., U.S. (ô-gŭs´t*à*)	134-35	33°28'N	81°59'W
Augusta, Ky., U.S. (ô-gŭs´t*à*)	126-27	38°46'N	84°00'W
Augusta, Me., U.S.	127a	44°19'N	69°47'W
Augusta, Wi., U.S. (ô-gŭs´t*à*)	124-25	44°41'N	91°07'W
Augustus Island, i., Austl.	296-97	15°21'S	124°31'E
Auob, stm., Afr. (ä´wŏb)	286-87	26°26'S	20°37'E
Aurangābād, India (ou-rŭŋ-gä-bäd´)	254-55	19°53'N	75°20'E
Aurillac, Fr. (ō-rē-yåk´)	212-13	44°55'N	2°26'E
Aurora, On., Can. (ô-rō´r*à*)	146-47	43°60'N	79°28'W
Aurora, Co., U.S. (ô-rō´r*à*)	130-31	39°44'N	104°52'W
Aurora, Il., U.S. (ô-rō´r*à*)	126-27	41°45'N	88°20'W
Aurora, In., U.S. (ô-rō´r*à*)	126-27	39°03'N	84°55'W
Aurora, Mn., U.S. (ô-rō´r*à*)	124-25	47°32'N	92°14'W
Aurora, Mo., U.S. (ô-rō´r*à*)	130-31	36°58'N	93°43'W
Aurora, Ne., U.S. (ô-rō´r*à*)	124-25	40°52'N	98°01'W
Au Sable, stm., Mi., U.S. (ô-să´b'l)	126-27	44°24'N	83°19'W
Aussig, Czech Rep. *see* Ústí nad Labem	210-11	50°40'N	14°02'E
Austin, Mn., U.S. (ôs´tĭn)	124-25	43°40'N	92°58'W
Austin, Nv., U.S. (ôs´tĭn)	128-29	39°31'N	117°07'W
Austin, Tx., U.S. (ôs´tĭn)	132-33	30°16'N	97°42'W
Austin, Lake, lk., Austl.	296-97	27°40'S	118°00'E
Australia, nation, Oc. (ôs-trā´lĭ-*à*)	294-95	25°0'S	135°00'E
Australian Capital Territory, state, Austl. (ôs-trā´lĭ-ăn)	301	35°30'S	149°00'E
Austral Islands, is., Fr. Poly. *see* Tubuaï, Îles.	304-05	23°0'S	150°00'W
Austria, nation, Eur. (ôs´trĭ-*à*)	190-91	47°20'N	13°20'E
Austvågøya, i., Nor.	200-01	68°21'N	14°38'E
Autlán de Navarro, Mex.	158-59	19°47'N	104°22'W
Autun, Fr. (ō-tŭN´)	212-13	46°57'N	4°18'E
Auxerre, Fr. (ō-sâr´)	212-13	47°48'N	3°34'E
Auyán Tepuy, mtn., Ven.	178-79	5°51'N	62°25'W
Auzangate, Nevado, mtn., Peru	184	13°48'S	71°14'W
Ava, Mo., U.S. (ā´v*à*)	130-31	36°57'N	92°40'W
Avallon, Fr. (å-vå-lôN´)	212-13	47°30'N	3°54'E
Avalon, Ca., U.S. (ăv´*à*-lŏn)	128-29	33°20'N	118°19'W
Avaré, Braz.	186	23°07'S	48°55'W
Aveiro, Port. (ä-vā´rŏ)	214-15	40°38'N	8°39'W
Avellaneda, Arg. (ä-vĕl-yä-nā´dhä)	187	29°07'S	59°40'W
Avellaneda, Arg. (ä-vĕl-yä-nā´dhä)	187	34°40'S	58°23'W
Avellino, Italy (ä-vĕl-lē´nō)	216-17	40°55'N	14°47'E
Avesta, Swe. (ä-vĕs´tä)	208-09	60°09'N	16°11'E
Avezzano, Italy (ä-våt-sä´nō)	216-17	42°02'N	13°25'E
Avignon, Fr. (å-vē-nyôN´)	212-13	43°57'N	4°49'E
Ávila, Spain (ä-vē-lä)	214-15	40°40'N	4°42'W
Avilés, Spain (ä-vē-lās´)	214-15	43°34'N	5°54'W
Avon, Ct., U.S. (ā´vŏn)	126-27	41°49'N	72°50'W
Avon, stm., Eng., U.K. (ā´vŭn)	206-07	50°44'N	1°47'W
Avon, stm., Eng., U.K. (ā´vŭn)	206-07	51°59'N	2°11'W
Avon Park, Fl., U.S. (ā´vŏn pärk´)	135a	27°36'N	81°30'W
Avranches, Fr. (å-vräNsh´)	212-13	48°41'N	1°22'W
Awaji-shima, i., Japan	265	34°21'N	134°51'E
Āwasa, Eth.	292d	6°56'N	38°32'E
Āwash, stm., Eth.	284-85	11°09'N	41°41'E
Awjilah, Libya	204-05	29°08'N	21°18'E
Axiós, stm., Eur.	216-17	40°31'N	22°43'E
Ax-les-Thermes, Fr. (äks´lā tĕrm´)	212-13	42°43'N	1°51'E
Ayacucho, Arg. (ä-yä-kōō´chō)	187	37°09'S	58°29'W
Ayacucho, Peru	184	13°08'S	74°14'W
Ayaköz, Kaz.	244	47°57'N	80°26'E
Ayamonte, Spain (ä-yä-mŏ´n-tĕ)	214-15	37°13'N	7°24'W
Ayan, Russia (å-yän´)	236-37	56°26'N	138°13'E
Ayan, stm., Russia	236-37	70°10'N	95°47'E
Ayapel, Col.	178-79	8°19'N	75°09'W
Ayaviri, Peru (ä-yä-vē´rē)	184	14°53'S	70°35'W
Aydar, stm., Eur. (ī-där´) *see* Aidar	218-19	48°44'N	39°16'E
Aydar Kŭl, lk., Uzb.	252-53	40°49'N	67°20'E
Ayden, N.C., U.S. (ā´dĕn)	134-35	35°28'N	77°25'W
Aydın, Tur. (aıy-dĕn)	216-17	37°51'N	27°50'E
Ayers Rock, mtn., Austl. *see* Uluru	296-97	25°20'S	130°60'E
Ayeyarwady, stm., Mya.	238-39	15°51'N	95°05'E
Aylesbury, Eng., U.K. (ālz´bēr-ĭ)	206-07	51°49'N	0°50'W
Aylmer Lake, lk., N.T., Can. (āl´mēr lāk)	140-41	64°05'N	108°30'W
Ayon, Ostrov, i., Russia (ôs-trôf´ ī-ôn´)	236-37	69°47'N	168°41'E
Ayr, Austl.	302	19°34'S	147°24'E
Ayr, Scot., U.K. (âr)	206-07	55°28'N	4°38'W
Ayvalık, Tur. (āīy-wă-lĭk)	216-17	39°20'N	26°42'E
Azaouagh, stm., Afr.	280-81	15°30'N	3°18'E
Azärbaycan, nation, Asia *see* Azerbaijan	245	40°30'N	47°30'E
Azare, Nig.	282-83	11°41'N	10°11'E
Azemmour, Mor. (à-zē-mōōr´)	292a	33°18'N	8°21'W
Azerbaidzhan, nation, Asia *see* Azerbaijan	245	40°30'N	47°30'E
Azerbaijan, nation, Asia (ä´zēr-bä-ē-jän´)	245	40°30'N	47°30'E
Azogues, Ec. (ä-sō´gās)	184	2°44'S	78°50'W
Azores, is., Port. (ā´zōrz) (*à*-zōrz´)	215c	38°30'N	28°00'W
Azov, Russia (á-zôf´) (ä-zôf)	218-19	47°07'N	39°26'E
Azov, Sea of, s., Eur. (sē ŭv ä-zôf´)	218-19	46°0'N	36°00'E
Azovs'ke more, s., Eur. (ä-zôf´skô-yĕ mô´rĕ) *see* Azov, Sea of	218-19	46°0'N	36°00'E
Azovskoye More, s., Eur. *see* Azov, Sea of	218-19	46°0'N	36°00'E
Azraq, Al-Baḥr al-, stm., Afr. *see* Blue Nile	275	15°38'N	32°30'E
Azrou, Mor.	292a	33°26'N	5°12'W
Aztec, N.M., U.S. (ăz´tĕk)	128-29	36°49'N	108°00'W
Azua, Dom. Rep. (ä´swä)	154-55	18°27'N	70°44'W
Azuaga, Spain (ä-thwä´gä)	214-15	38°15'N	5°40'W
Azuero, Península de, pen., Pan.	162	7°40'N	80°35'W
Azul, Arg. (ä-sōōl´)	187	36°47'S	59°52'W
Az-Zahrān, Sau. Ar. *see* Dhahran	250-51	26°18'N	50°08'E
Az-Zarqā', Jord.	248-49	32°03'N	36°05'E
Az-Zāwiyah, Libya	204-05	32°47'N	12°44'E
Az-Zilfī, Sau. Ar.	288	26°18'N	44°49'E

B

Feature (Pronunciation)	Page	Lat.	Long.
Ba'qūbah, Iraq	248-49	33°45'N	44°40'E
Baaba, Île, i., N. Cal.	306g	20°03'S	163°58'E
Babadayhan, Turkmen.	252-53	37°42'N	60°24'E
Babaeski, Tur. (bä´bä-ĕs´kĭ)	216-17	41°26'N	27°06'E
Babahoyo, Ec. (bä-bä-ō´yō)	184	1°48'S	79°32'W
Babar, Pulau, i., Indon. (pōō-lou bä´bär)	242-43	7°55'S	129°45'E
Babeldaob, i., Palau	304-05	7°30'N	134°35'E
Bab el Mandeb, strt.	288	12°44'N	43°21'E
Babelthuap, i., Palau *see* Babeldaob	304-05	7°30'N	134°35'E
Babi, Pulau, i., Indon.	266-67	2°05'N	96°39'E
Babine Lake, lk., B.C., Can. (băb´ēn lāk)	142-43	54°45'N	126°00'W
Bābol, Iran	252-53	36°33'N	52°41'E
Babrujsk, Bela.	218-19	53°08'N	29°14'E
Babuyan Island, i., Phil.	270a	19°32'N	121°57'E
Babuyan Islands, is., Phil. (bä-bōō-yän´ ī´l*ă*ndz)	270a	19°15'N	121°40'E
Bacabal, Braz.	180-81	4°14'S	44°47'W
Bacan, Pulau, i., Indon.	268-69	0°35'S	127°30'E
Bacău, Rom.	218-19	46°34'N	26°55'E
Bac Bo, Vinh, b., Asia *see* Tonkin, Gulf of	266-67	20°0'N	108°00'E
Back, stm., Nu., Can. (băk)	140-41	67°09'N	95°21'W
Bačka Palanka, Serb. (bäch´kä pälän-kä)	216-17	45°15'N	19°24'E
Bac Lieu, Viet.	266-67	9°17'N	105°43'E
Bac Ninh, Viet. (bäk´nên´')	266-67	21°12'N	106°05'E
Baco, Mount, mtn., Phil. (mount bä´kŏ)	270	12°49'N	121°10'E
Bacolod, Phil. (bä-kō´lŏd)	270	10°40'N	122°57'E
Badajoz, Spain (bå-dhä-hōth´)	214-15	38°53'N	6°58'W
Badalona, Spain (bä-dhä-lō´nä)	214-15	41°28'N	2°16'E
Bad Axe, Mi., U.S. (băd´ ăks)	126-27	43°48'N	82°59'W
Baden-Baden, Ger. (bä´dĕn bä´dĕn)	210-11	48°46'N	8°14'E
Bad Hersfeld, Ger. (bät hĕrsh´fĕlt)	210-11	50°52'N	9°42'E
Bad Kissingen, Ger. (bät kĭs´ĭng-ĕn)	210-11	50°12'N	10°05'E
Badlands, hills, U.S. (băd´ lănds)	124-25	46°14'N	103°37'W
Badlands National Park, n.p., S.D., U.S. (băd´ lănds năsh´ŭn-*ă*l pärk)	124-25	43°50'N	102°21'W
Bad Reichenhall, Ger. (bät rī´κĕn-häl)	210-11	47°44'N	12°53'E
Bad River Indian Reservation, ind. res., Wi., U.S. (băd rĭv´ēr ĭn´dĭ-*ă*n rĕ-sēr-vā´shĕn)	124-25	46°33'N	90°40'W
Bad Tölz, Ger. (bät tûltz)	210-11	47°45'N	11°35'E
Badu Island, i., Austl.	302	10°07'S	142°08'E
Baena, Spain (bä-ā´nä)	214-15	37°37'N	4°19'W
Bafatá, Gui.-B.	282-83	12°10'N	14°41'W
Baffin Bay, b., N.A. (băf´ĭn bā)	95	73°0'N	66°00'W
Baffin Bugt, b., N.A. *see* Baffin Bay	95	73°0'N	66°00'W
Baffin Island, i., Nu., Can. (băf´ĭn ī l*ă*nd)	95	68°0'N	70°00'W
Bafia, Camrn.	282-83	4°45'N	11°16'E
Bafing, stm., Afr.	282-83	13°47'N	10°50'W
Bafoussam, Camrn.	282-83	5°29'N	10°25'E
Bāfq, Iran (bäfk)	252-53	31°35'N	55°24'E
Bafra, Tur. (bäf´rä)	202-03	41°34'N	35°53'E
Bafwasende, D.R.C.	284-85	1°06'N	27°16'E
Bagan, hist., Mya.	266-67	21°13'N	94°54'E
Bagansiapiapi, Indon.	266-67	2°09'N	100°48'E
Bagata, D.R.C.	284-85	3°44'S	17°59'E
Bagdad, nat. cap., Iraq *see* Baghdād	248-49	33°21'N	44°25'E
Bagé, Braz.	187	31°19'S	54°06'W
Baghdād, nat. cap., Iraq (bågh-däd´) (băg´dăd)	248-49	33°21'N	44°25'E
Bagheria, Italy (bä-gå-rē´ä)	216-17	38°05'N	13°30'E
Baghlān, Afg.	252-53	36°08'N	68°42'E
Bagley, Mn., U.S. (băg´lê)	124-25	47°30'N	95°23'W
Bagnères-de-Bigorre, Fr. (bän-yâr´dē-bē-gor´)	212-13	43°04'N	0°09'E
Bago, Mya.	266-67	17°20'N	96°29'E
Bagoé, stm., Afr. (bå-gô´å)	282-83	12°35'N	6°34'W
Baguio, Phil. (bä-gē-ō´)	270	16°25'N	120°36'E
Bahama, Canal Viejo de, strt., N.A. *see* Old Bahama Channel	154-55	22°40'N	78°41'W
Bahama Islands, is., Bah.	4-5	24°15'N	76°00'W
Bahamas, nation, N.A. (b*à*-hā´m*à*s)	152-53	24°15'N	76°00'W
Baharampur, India	254-55	24°06'N	88°15'E
Bahāwalpur, Pak. (b*ŭ*-hä´wŭl-pōōr)	252-53	29°23'N	71°40'E
Bäherden, Turkmen.	252-53	38°26'N	57°26'E
Bahia, Braz. (bä-ē´å) *see* Salvador	180-81	12°59'S	38°30'W
Bahia, state, Braz.	180-81	12°0'S	42°00'W
Bahía, Islas de la, is., Hond. (ē´s-läs-dĕ-lä-bä-ē´ä)	161	16°20'N	86°30'W
Bahía Blanca, Arg. (bä-ē´ä blăŋ´kä)	187	38°43'S	62°17'W

n-sing; ŋ-baŋk; N-nasalized n; nŏd; cŏmmit; ōld; ŏbey; ôrder; oi-boil; fōōd; ŏ-as oo in foot; ou-out; s-soft; sh-dish; th-thin; pūre; ûnite; ûrn; stŭd; circŭs; ü-as in French tu; ´-indeterminate vowel.

334 Bahía de Caráquez – Baños PRONOUNCING INDEX

Feature (Pronunciation)	Page	Lat.	Long.
Bahía de Caráquez, Ec. (bä-e´ä dä kä-rä´kĕz)	184	0°37´S	80°26´W
Bahir Dar, Eth.	288	11°35´N	37°24´E
Bahraich, India	254-55	27°35´N	81°36´E
Bahrain, nation, Asia (bä-rān´)	250-51	26°0´N	50°30´E
Baḥrānī, Ḥālat al-, i., U.A.E.	250-51	24°28´N	54°21´E
Baia Mare, Rom. (bä´yä mä´rä)	210-11	47°39´N	23°35´E
Baicheng, China	260-61	45°37´N	122°51´E
Baidoa, Som.	284-85	3°07´N	43°39´E
Baie-Comeau, Qc., Can.	148-49	49°13´N	68°10´W
Baie-Saint-Paul, Qc., Can. (bā´săn´-pôl´)	148-49	47°27´N	70°30´W
Baikal, Lake, lk., Russia (lāk bī-käl´)	236-37	53°0´N	107°40´E
Baile Átha Cliath, nat. cap., Ire. see Dublin	206-07	53°21´N	6°15´W
Bailén, Spain (bä-ê-lān´)	214-15	38°06´N	3°46´W
Băilești, Rom. (bă-ĭ-lĕsh´tê)	216-17	44°02´N	23°21´E
Bailong, stm., China	258-59	32°21´N	105°43´E
Bainbridge, Ga., U.S. (bān´brĭj)	134-35	30°54´N	84°34´W
Bainbridge, Oh., U.S. (bān´brĭj)	126-27	39°14´N	83°16´W
Baiquan, China (bī-chyuäŋ)	260-61	47°36´N	126°05´E
Baird, Tx., U.S. (bârd)	130-31	32°24´N	99°23´W
Bairin Zuoqi, China	260-61	43°59´N	119°23´E
Bairnsdale, Austl. (bârnz´dāl)	301	37°50´S	147°37´E
Baishuijiang, China	258-59	33°29´N	106°02´E
Baitou Shan, mtn., Asia see Paektu-san	263	41°59´N	128°07´E
Baiyin, China	260-61	36°33´N	104°12´E
Baja, Hung. (bŏ´yŏ)	210-11	46°11´N	18°57´E
Baja California, state, Mex. (bä-hä kăl-ĭ-fôr´nĭ-ȧ)	156-57	30°0´N	115°00´W
Baja California, pen., Mex. (bä-hä kăl-ĭ-fôr´nĭ-ȧ)	156-57	27°53´N	113°28´W
Baja California Norte, state, Mex. see Baja California	156-57	30°0´N	115°00´W
Baja California Sur, state, Mex. (bä-hä kăl-ĭ-fôr´nĭ-ȧ so͞or´)	156-57	26°0´N	112°00´W
Bajestān, Iran	252-53	34°31´N	58°11´E
Bājil, Yemen	288	15°04´N	43°17´E
Bajo Boquete, Pan.	162	8°47´N	82°26´W
Baker, Ca., U.S. (bā´kẽr)	128-29	35°16´N	116°04´W
Baker, La., U.S. (bā´kẽr)	134-35	30°35´N	91°10´W
Baker, Mt., U.S. (bā´kẽr)	122-23	46°22´N	104°17´W
Baker, Or., U.S. (bā´kẽr)	122-23	44°47´N	117°50´W
Baker, Mount, vol., Wa., U.S. (mount bā´kẽr)	122-23	48°47´N	121°49´W
Baker Island, dep., Oc. (bā´kẽr ī´lȧnd)	304-05	0°15´N	176°27´W
Baker Island, i., Oc.	304-05	0°12´N	176°29´W
Baker Lake, Nu., Can.	138-39	64°18´N	95°55´W
Baker Lake, lk., Nu., Can. (bā´kẽr lāk)	140-41	64°10´N	95°30´W
Bakersfield, Ca., U.S. (bā´kẽrz-fēld)	128-29	35°22´N	119°01´W
Bakhmach, Ukr. (bȧk-mäch´)	218-19	51°11´N	32°50´E
Bākhtarān, Iran see Kermānshāh	248-49	34°18´N	47°04´E
Bakhtegān, Daryācheh-ye, lk., Iran	250-51	29°20´N	54°05´E
Bakı, nat. cap., Azer. see Baku	245	40°23´N	49°51´E
Bakony, mts., Hung. (bȧ-kōn´y´)	210-11	47°01´N	17°45´E
Bakoy, stm., Afr. (bȧ-kô´ê)	280-81	13°48´N	10°49´W
Baku, nat. cap., Azer. (bȧ-ko͞o´)	245	40°23´N	49°51´E
Bakwanga, D.R.C. see Mbuji-Mayi	284-85	6°08´S	23°39´E
Balabac, Selat, strt., Asia see Balabac Strait	270	7°35´N	117°00´E
Balabac Island, i., Phil. (bä´lä-bäk ī´lȧnd)	270	7°57´N	117°01´E
Balabac Strait, strt., Asia (bä´lä-bäk strāt)	270	7°35´N	117°00´E
Balabanovo, Russia (bȧ-lȧ-bä´nô-vô)	218-19	55°11´N	36°40´E
Balabio, Île, i., N. Cal.	306g	20°07´S	164°11´E
Bālāghāt, India	254-55	21°49´N	80°11´E
Balaguer, Spain (bä-lä-gĕr´)	214-15	41°48´N	0°49´E
Balakovo, Russia (bȧ´lȧ-kô´vŏ)	202-03	52°01´N	47°47´E
Balambangan, Pulau, i., Malay.	268-69	7°16´N	116°55´E
Balāngīr, India	254-55	20°43´N	83°30´E
Balaözen, stm., Eur.	202-03	48°58´N	49°38´E
Balashov, Russia (bá´lȧ-shôf)	202-03	51°32´N	43°10´E
Balasore, India (bä-lȧ-sōr´) see Bāleshwar	254-55	21°29´N	86°57´E
Balassagyarmat, Hung. (bô´lôsh-shô-dyôr´môt)	210-11	48°04´N	19°19´E
Balaton, lk., Hung. (bô´lô-tôn)	210-11	46°50´N	17°45´E
Balayan, Phil. (bä-lä-yän´)	270	13°57´N	120°44´E
Balbina, Represa, res., Braz.	180-81	1°20´S	59°40´W
Balcarce, Arg. (bäl-kär´sā)	187	37°51´S	58°15´W
Baldock Lake, lk., Mb., Can.	144-45	56°33´N	97°57´W
Baldwinsville, N.Y., U.S. (bôld´wĭns-vĭl)	126-27	43°10´N	76°20´W
Baldy Peak, mtn., Az., U.S. (bôl´dê pēk)	128-29	33°55´N	109°35´W

Feature (Pronunciation)	Page	Lat.	Long.
Bâle, Switz. see Basel	210-11	47°33´N	7°36´E
Baleares, Islas, is., Spain see Balearic Islands	214-15	39°29´N	3°01´E
Balearic Islands, is., Spain (bäl-ē-ă´-rĭk ī´lȧndz)	214-15	39°29´N	3°01´E
Balears, Illes, is., Spain see Balearic Islands	214-15	39°29´N	3°01´E
Baler, Phil. (bä-lar´)	270	15°46´N	121°34´E
Bāleshwar, India	254-55	21°29´N	86°57´E
Baley, Russia (bȧl-yā´)	240-41	51°34´N	116°38´E
Bali, i., Indon. (bä´lê)	268-69	8°20´S	115°00´E
Bali, Laut, s., Indon. see Bali Sea	268-69	7°45´S	115°30´E
Balıkesir, Tur. (balĭk´ĭysĭr)	216-17	39°39´N	27°53´E
Balikpapan, Indon. (bä´lêk-pä´pän)	268-69	1°16´S	116°50´E
Balimo, Pap. N. Gui.	302	8°03´S	142°56´E
Balin, China	260-61	48°19´N	122°19´E
Balintang Channel, strt., Phil. (bä-lĭn-täng´ chăn´ĕl)	270a	19°59´N	121°51´E
Bali Sea, s., Indon. (bä´lê sē)	268-69	7°45´S	115°30´E
Balkanabat, Turkmen.	252-53	39°31´N	54°23´E
Balkan Peninsula, pen., Eur. (bôl´kȧn pĕ-nĭn´sūlȧ)	216-17	44°0´N	23°00´E
Balkaria, state, Russia	245	43°30´N	43°30´E
Balkh, Afg. (bälk)	252-53	36°45´N	66°54´E
Balkh, stm., Afg.	252-53	36°38´N	66°56´E
Balkhash, Kaz. see Balqash	244	46°51´N	74°58´E
Balkhash, Lake, lk., Kaz. (lāk bȧl-käsh´)	244	46°0´N	74°00´E
Ballarat, Austl. (băl´ȧ-răt)	301	37°34´S	143°51´E
Ballard, Lake, lk., Austl. (lāk băl´ȧrd)	296-97	29°27´S	120°55´E
Ballia, India	254-55	25°45´N	84°09´E
Ballina, Austl. (băl-ĭ-nä´)	301	28°52´S	153°33´E
Ballinasloe, Ire. (băl´ĭ-nȧ-slō´)	206-07	53°20´N	8°14´W
Ballinger, Tx., U.S. (băl´ĭn-jẽr)	132-33	31°44´N	99°57´W
Ballston Spa, N.Y., U.S. (bôls´tŭn spä´)	126-27	43°00´N	73°51´W
Balonne, stm., Austl. (băl-ōn´)	301	28°37´S	148°10´E
Balqash, Kaz.	244	46°51´N	74°58´E
Balqash köli, lk., Kaz. see Balkhash, Lake	244	46°0´N	74°00´E
Balranald, Austl. (băl´-rȧn-ăld)	301	34°38´S	143°33´E
Balsas, Braz. (bäl´säs)	180-81	7°33´S	46°04´W
Balsas, stm., Braz.	180-81	7°14´S	44°34´W
Balsas, stm., Mex.	158-59	17°54´N	102°11´W
Balta, Ukr. (bál´tȧ)	218-19	47°56´N	29°40´E
Baltasar Brum, Ur.	187	30°42´S	57°19´W
Bălți, Mol.	218-19	47°46´N	27°55´E
Baltic Sea, s., Eur. (bôl´tĭk sē)	208-09	57°0´N	19°00´E
Baltijas jūra, s., Eur. see Baltic Sea	208-09	57°0´N	19°00´E
Baltijos jūra, s., Eur. see Baltic Sea	208-09	57°0´N	19°00´E
Baltim, Egypt (bál-tēm´)	291b	31°34´N	31°05´E
Baltimore, Md., U.S. (bôl´tĭ-môr)	126-27	39°17´N	76°37´W
Baltiysk, Russia (bäl-tēysk´)	210-11	54°39´N	19°55´E
Baltiyskoye More, s., Eur. see Baltic Sea	208-09	57°0´N	19°00´E
Bałtyckie, Morze, s., Eur. see Baltic Sea	208-09	57°0´N	19°00´E
Balūchestān, hist. reg., Asia see Baluchistan	252-53	28°0´N	63°00´E
Baluchistan, hist. reg., Asia	252-53	28°0´N	63°00´E
Baluchistān, hist. reg., Asia (bȧ-lò-chĭ-stän´) see Baluchistan	252-53	28°0´N	63°00´E
Balykchy, Kyrg.	244	42°28´N	76°12´E
Balyqshy, Kaz.	202-03	47°05´N	51°54´E
Bam, Iran	250-51	29°07´N	58°21´E
Bama, Nig.	282-83	11°32´N	13°41´E
Bamako, nat. cap., Mali (bä-mä-kō´)	280-81	12°39´N	7°60´W
Bambari, C.A.R. (bäm-bȧ-rē)	284-85	5°46´N	20°39´E
Bamberg, Ger. (bäm´bĕrgh)	210-11	49°54´N	10°54´E
Bamberg, S.C., U.S. (băm´bûrg)	134-35	33°18´N	81°02´W
Bamenda, Camrn.	282-83	5°58´N	10°09´E
Bāmīān, Afg.	252-53	34°50´N	67°49´E
Bamingui, stm., C.A.R.	284-85	8°34´N	19°04´E
Bamingui-Bangoran, Parc National du, n.p., C.A.R.	284-85	7°54´N	19°42´E
Bampūr, Iran (bŭm-po͞or´)	250-51	27°11´N	60°26´E
Banaba, i., Kir.	304-05	0°52´S	169°33´E
Banaras, India see Vārānasi	254-55	25°20´N	82°59´E
Banās, stm., India (bän-äs´)	254-55	25°55´N	76°44´E
Banâs, Râs, c., Egypt	288	23°54´N	35°47´E
Ban Bat, Viet.	266-67	13°13´N	108°40´E
Bancroft, On., Can. (băn´krŏft)	146-47	45°03´N	77°51´W
Bānda, India (bän´dä)	254-55	25°29´N	80°20´E
Banda, Laut, s., Indon. see Banda Sea	242-43	5°0´S	128°00´E
Banda Aceh, Indon.	266-67	5°33´N	95°19´E
Bandama, stm., C. Iv.	282-83	5°08´N	4°60´W
Bandar, India see Machilīpatnam	256	16°11´N	81°09´E
Bandar 'Abbās, Iran	250-51	27°11´N	56°16´E

Feature (Pronunciation)	Page	Lat.	Long.
Bandar Beheshtī, Iran	250-51	25°18´N	60°38´E
Bandar-e Anzalī, Iran	245	37°28´N	49°28´E
Bandar-e Khomeynī, Iran	248-49	30°26´N	49°06´E
Bandar-e Lengeh, Iran	250-51	26°34´N	54°53´E
Bandar-e Pahlavī, Iran see Bandar-e Anzalī	245	37°28´N	49°28´E
Bandar-e Shāhpūr, Iran see Bandar-e Khomeynī	248-49	30°26´N	49°06´E
Bandar-e Torkeman, Iran	252-53	36°54´N	54°04´E
Bandar Lampung, Indon.	266-67	5°26´S	105°16´E
Bandar Maharani, Malay. (bän-där´ mä-hä-rä´nê) see Muar	266-67	2°02´N	102°34´E
Bandar Penggaram, Malay. see Batu Pahat	266-67	1°51´N	102°56´E
Bandar Seri Begawan, nat. cap., Bru. (bän´där sĕr´ē bŭ´gä-wän)	268-69	4°56´N	114°56´E
Banda Sea, s., Indon. (bän´-dä sē) (băn´-dȧ sē)	242-43	5°0´S	128°00´E
Bandeira, Pico da, mtn., Braz. (pē´kò dä bän dä´rä)	186	20°26´S	41°47´W
Bandelier National Monument, n.p., N.M., U.S. (băn-dĕ-lēr´ năsh´ŭn-ăl mŏn´ū-mĕnt)	128-29	35°45´N	106°20´W
Bandera, Arg.	187	28°53´S	62°16´W
Banderas, Bahía de, b., Mex. (bä-ē´ä dĕ bän-dĕ´räs)	158-59	20°38´N	105°27´W
Bandiantaolehai, China	260-61	41°47´N	104°05´E
Bandırma, Tur. (bän-dĭr´mȧ)	216-17	40°22´N	27°59´E
Ban Don, Thai. see Surat Thani	266-67	9°06´N	99°18´E
Bandon, Or., U.S. (băn´dŭn)	122-23	43°07´N	124°23´W
Bandundu, D.R.C.	284-85	3°16´S	17°21´E
Bandung, Indon.	268-69	6°54´S	107°36´E
Banes, Cuba (bä´nās)	154-55	20°58´N	75°42´W
Banff, Ab., Can. (bămf)	142-43	51°10´N	115°36´W
Banff National Park, n.p., Ab., Can. (bămf năsh´ŭn-ăl pärk)	142-43	51°38´N	116°22´W
Banfora, Burkina	282-83	10°39´N	4°45´W
Bangalore, India (băŋ´gȧ´lôr) see Bengalūru	256	12°59´N	77°36´E
Bangassou, C.A.R. (bän-gȧ-so͞o´)	284-85	4°44´N	22°49´E
Banggai, Indon.	268-69	1°35´S	123°30´E
Banggai, Kepulauan, is., Indon. (bäng-gī´)	268-69	1°30´S	123°15´E
Banggai, Pulau, i., Indon.	268-69	1°37´S	123°33´E
Banggi, Pulau, i., Malay.	268-69	7°16´N	117°10´E
Banggong Co, lk., Asia (bäŋ-gôŋ tswo) see Pangong Tso	254-55	33°45´N	78°42´E
Banghāzī, Libya see Benghazi	280-81	32°07´N	20°04´E
Bangka, Pulau, i., Indon. (po͞o-lou băŋ´kȧ)	266-67	2°15´S	106°00´E
Bangka, Selat, strt., Indon.	266-67	2°20´S	105°45´E
Bangkalan, Indon. (bäng-kȧ-län´)	268-69	7°02´S	112°45´E
Bangkok, nat. cap., Thai. (băɴ´kŏk)	266-67	13°45´N	100°31´E
Bangkulu, Pulau, i., Indon.	268-69	1°50´S	123°06´E
Bangladesh, nation, Asia (bän´-glă-dĕsh´)	224-25	24°0´N	90°00´E
Bangor, N. Ire., U.K. (băŋ´ŏr)	206-07	54°39´N	5°41´W
Bangor, Wales, U.K. (băŋ´ŏr)	206-07	53°14´N	4°09´W
Bangor, Me., U.S.	127a	44°48´N	68°47´W
Bangor, Mi., U.S. (băn´gẽr)	126-27	42°19´N	86°06´W
Bangor, Pa., U.S. (băn´gẽr)	126-27	40°51´N	75°13´W
Bangued, Phil. (bän-gād´)	270	17°36´N	120°37´E
Bangui, nat. cap., C.A.R. (bäɴ-gē´)	284-85	4°22´N	18°33´E
Bangweulu, Lake, lk., Zam. (lāk băng-wē-o͞o´lo͞o)	286-87	11°04´S	29°53´E
Ban Hat Yai, Thai. see Hat Yai	266-67	7°01´N	100°28´E
Ban Houayxay, Laos	266-67	20°15´N	100°24´E
Baní, Dom. Rep. (bä´-nê)	154-55	18°17´N	70°19´W
Banifing, stm., Mali	282-83	14°29´N	4°13´W
Banja Luka, Bos. (bän-yä-lo͞o´kȧ)	216-17	44°46´N	17°12´E
Banjarbaru, Indon.	268-69	3°24´S	114°50´E
Banjarmasin, Indon. (bän-jẽr-mä´sĕn)	268-69	3°20´S	114°36´E
Banjul, nat. cap., Gam. (bôn-jo͞ol´)	282-83	13°27´N	16°36´W
Banks, Îles, is., Vanuatu see Banks Islands	306g	13°25´S	167°42´E
Banks Island, i., B.C., Can. (bănks ī´lȧnd)	142-43	53°25´N	130°10´W
Banks Island, i., N.T., Can. (bănks ī´lȧnd)	95	73°15´N	121°30´W
Banks Islands, is., Vanuatu	306g	13°25´S	167°42´E
Banks Peninsula, pen., N.Z. (bănks pĕ-nĭn´sūlȧ)	303	43°45´S	187°00´E
Banks Strait, strt., Austl. (bănks strāt)	301	40°40´S	148°07´E
Banningville, D.R.C. see Bandundu	284-85	3°16´S	17°21´E
Bannu, Pak.	252-53	32°59´N	70°37´E
Baños, Ec. (bä´-nyôs)	184	1°24´S	78°25´W

ăt; finăl; rāte; senāte; ärm; åsk; sofȧ; fâre; ch-choose; dh-as th in other; bē; ēvent; bĕt; recĕnt; cratẽr; g-gō; gh-guttural g; bĭt; ĭ-short neutral; rīde; ĸ-guttural k as ch in German ich;

PRONOUNCING INDEX

Bānswāra – Bayou Bodcau Reservoir 335

Feature (Pronunciation)	Page	Lat.	Long.
Bānswāra, India	254-55	23°33'N	74°27'E
Bantaeng, Indon.	268-69	5°32'S	119°56'E
Bantayan Island, i., Phil.	270	11°13'N	123°44'E
Bantry, Ire. (băn′trĭ)	206-07	51°41'N	9°27'W
Banyak, Kepulauan, is., Indon.	266-67	2°10'N	97°15'E
Banyuwangi, Indon. (bän-jò-wäŋ′gē)	268-69	8°12'S	114°21'E
Baode, China	260-61	39°01'N	111°05'E
Baoding, China (bou-dĭŋ)	260-61	38°51'N	115°29'E
Baoji, China (bou-jyē)	258-59	34°23'N	107°09'E
Bao Lac, Viet.	266-67	11°33'N	107°47'E
Baoshan, China (bou-shän)	258-59	25°07'N	99°10'E
Baoting, China	258-59	18°38'N	109°47'E
Baotou, China (bou-tō)	260-61	40°35'N	109°58'E
Baoying, China (bou-yĭŋ)	258-59	33°14'N	119°19'E
Baquedano, Chile	182-83	23°20'S	69°50'W
Bar, Mont.	216-17	42°05'N	19°06'E
Baraboo, Wi., U.S. (bâr′ȧ-bōō)	126-27	43°28'N	89°44'W
Baracoa, Cuba (bä-rä-kō′ä)	154-55	20°21'N	74°30'W
Baradero, Arg. (bä-rä-dĕ′ŏ)	187	33°48'S	59°31'W
Baragaon, India *see* Nālanda	254-55	25°08'N	85°24'E
Barahona, Dom. Rep. (bä-rä-ô′nä)	154-55	18°12'N	71°06'W
Baranavičy, Bela.	210-11	53°08'N	26°01'E
Baranof Island, i., Ak., U.S.			
(bä-rä′nôf ī′lȧnd)	136	57°0'N	135°00'W
Barão de Melgaço, Braz.			
(bä-roun-dĕ-mĕl-gä′sô)	182-83	16°13'S	55°58'W
Barat Daya, Kepulauan, is., Indon.	268-69	7°25'S	128°00'E
Baraya, Col. (bä-rȧ′yä)	188c	3°10'N	75°04'W
Barbacena, Braz. (bär-bä-sā′nȧ)	186	21°13'S	43°45'W
Barbacoas, Col. (bär-bä-kō′äs)	178-79	1°41'N	78°09'W
Barbados, nation, N.A. (bär-bā′dōz)	152-53	13°10'N	59°32'W
Barbas, Cap, c., W. Sah.	280-81	22°18'N	16°40'W
Barbastro, Spain (bär-bäs′trō)	214-15	42°02'N	0°08'E
Barberton, Oh., U.S. (bär′bẽr-tŭn)	126-27	41°02'N	81°36'W
Barbosa, Col. (bär-bô′-sȧ)	188c	6°26'N	75°20'W
Barboursville, W.V., U.S.			
(bar′bers-vil)	126-27	38°25'N	82°18'W
Barbuda, i., Antig. (bär-bōō′dȧ)	155b	17°38'N	61°48'W
Barcaldine, Austl. (bar′kol-din)	302	23°34'S	145°18'E
Barce, Libya *see* Al-Marj	204-05	32°30'N	20°53'E
Barcelona, Spain (bär-thȧ-lō′nä)	214-15	41°24'N	2°10'E
Barcelona, Ven. (bär-sȧ-lō′nä)	188b	10°08'N	64°41'W
Barcelos, Braz. (bär-sĕ′lôs)	180-81	0°59'S	62°54'W
Barcelos, Port. (bär-thȧ′lōs)	214-15	41°32'N	8°37'W
Barcoo, stm., Austl.	302	25°12'S	142°50'E
Bardaï, Chad	280-81	21°22'N	16°59'E
Barddhamān, India	254-55	23°14'N	87°52'E
Bardsey Island, i., Wales, U.K.			
(bärd′sē ī′lȧnd)	206-07	52°46'N	4°48'W
Bardstown, Ky., U.S. (bärds′toun)	126-27	37°49'N	85°28'W
Bardwell, Ky., U.S. (bärd′wĕl)	134-35	36°52'N	89°01'W
Bareilly, India	254-55	28°21'N	79°25'E
Barentsevo More, s., Eur.			
see Barents Sea	236-37	74°0'N	36°00'E
Barentshavet, s., Eur. *see* Barents Sea	236-37	74°0'N	36°00'E
Barents Sea, s., Eur. (bä′rĕnts sē)	236-37	74°0'N	36°00'E
Barentu, Erit. (bä-rĕn′tōō)	288	15°07'N	37°35'E
Barfleur, Pointe de, c., Fr.			
(pwănt′ dĕ bȧr-flûr′)	212-13	49°42'N	1°16'W
Barguzin, stm., Russia	236-37	53°25'N	108°59'E
Bar Harbor, Me., U.S. (bär här′bẽr)	127a	44°23'N	68°13'W
Bari, Italy (bä′rē)	216-17	41°07'N	16°52'E
Bariloche, Arg.			
see San Carlos de Bariloche	185	41°09'S	71°18'W
Barinas, Ven. (bä-rē′näs)	178-79	8°38'N	70°13'W
Baring, Cape, c., N.T., Can.			
(kāp bâr′ĭng)	140-41	70°03'N	117°16'W
Bāripada, India	254-55	21°56'N	86°43'E
Barisāl, Bngl.	254-55	22°42'N	90°22'E
Barito, stm., Indon. (bä-rē′tō)	268-69	3°20'S	114°32'E
Barkley, Lake, res., U.S.	134-35	36°44'N	87°57'W
Barkley Sound, strt., B.C., Can.	142-43	48°53'N	125°20'W
Barkly Tableland, plat., Austl.			
(bär′klē tā′-bĕl-lănd)	296-97	18°0'S	136°00'E
Barkol, China (bär-kůl)	260-61	43°33'N	93°02'E
Bar-le-Duc, Fr. (bär-lē-dük′)	212-13	48°47'N	5°10'E
Barlee, Lake, lk., Austl. (lăk bär-lē′)	296-97	29°10'S	119°30'E
Barletta, Italy (bär-lĕt′tä)	216-17	41°19'N	16°17'E
Bārmer, India	254-55	25°44'N	71°24'E
Barnaul, Russia (bär-nä-ōl′)	244	53°22'N	83°45'E
Barnesville, Ga., U.S. (bärnz′vĭl)	134-35	33°03'N	84°10'W
Barnesville, Mn., U.S. (bärnz′vĭl)	124-25	46°39'N	96°25'W
Barnsley, Eng., U.K. (bärnz′lĭ)	206-07	53°34'N	1°29'W
Barnstaple, Eng., U.K. (bärn′stā-p'l)	206-07	51°05'N	4°03'W
Barnwell, S.C., U.S. (bärn′wĕl)	134-35	33°14'N	81°22'W
Baro, stm., Afr.	284-85	8°26'N	33°13'E
Baroda, India (bär-rō′dä)			
see Vadodara	254-55	22°18'N	73°11'E
Barpeta, India	254-55	26°19'N	91°00'E
Barqah, hist. reg., Libya *see* Cyrenaica	280-81	31°0'N	22°30'E
Barquisimeto, Ven. (bär-kē-sē-mā′tō)	178-79	10°05'N	69°19'W
Barra, Braz. (bär′rä)	180-81	11°05'S	43°09'W
Barra, Ponta da, c., Moz.	286-87	23°48'S	35°31'E
Barra do Corda, Braz.			
(bär′rä dò cōr-dä)	180-81	5°31'S	45°15'W
Barra Falsa, Ponta da, c., Moz.	286-87	22°54'S	35°34'E
Barra Mansa, Braz. (bär′rä män′sä)	186	22°33'S	44°10'W
Barranca, Peru	184	10°45'S	77°46'W
Barrancabermeja, Col.			
(bär-räŋ′kä-bēr-mā′hä)	178-79	7°04'N	73°51'W
Barrancas, Ven.	178-79	8°44'N	62°11'W
Barranquilla, Col. (bär-rän-kēl′yä)	178-79	10°59'N	74°48'W
Barras, Braz. (bȧ′r-räs)	180-81	4°15'S	42°18'W
Barre, Vt., U.S. (băr′ē)	126-27	44°12'N	72°30'W
Barreiras, Braz. (bär-rā′räs)	180-81	12°09'S	45°01'W
Barreiro, Port. (bär-rĕ′ē-rò)	214-15	38°39'N	9°04'W
Barreiros, Braz.	188d	8°49'S	35°12'W
Barren, Nosy, is., Madag.	286-87	18°30'S	43°53'E
Barretos, Braz. (bär-rā′tōs)	186	20°34'S	48°34'W
Barrhead, Ab., Can. (băr′ĭd)	142-43	54°07'N	114°24'W
Barrie, On., Can. (băr′ĭ)	146-47	44°23'N	79°41'W
Barrington Tops, mtn., Austl.			
(bă-rĕng-tŏn tŏps)	301	32°0'S	151°28'E
Barron, Wi., U.S. (băr′ŭn)	124-25	45°24'N	91°51'W
Barrow, Ak., U.S. (băr′ō)	136	71°18'N	156°38'W
Barrow, stm., Ire. (bȧ-rȧ)	206-07	52°17'N	7°00'W
Barrow, Point, c., Ak., U.S. (point băr′ō)	136	71°23'N	156°29'W
Barrow Creek, Austl.	294-95	21°31'S	133°55'E
Barrow Island, i., Austl.	296-97	20°48'S	115°23'E
Bārsi, India	256	18°14'N	75°42'E
Barstow, Ca., U.S. (bär′stō)	120-29	34°54'N	117°01'W
Bartica, Guy. (bär′tĭ-kȧ)	178-79	6°24'N	58°37'W
Bartın, Tur. (bär′tĭn)	202-03	41°38'N	32°21'E
Bartle Frere, mtn., Austl. (bärt′'l frēr′)	302	17°23'S	145°49'E
Bartlesville, Ok., U.S. (bär′tlz-vĭl)	130-31	36°45'N	95°59'W
Bartlett, Tn., U.S. (bärt′lĕt)	134-35	35°13'N	89°52'W
Bartlett, Tx., U.S. (bärt′lĕt)	132-33	30°48'N	97°26'W
Bartoszyce, Pol. (bär-tô-shĭ′tsȧ)	210-11	54°15'N	20°49'E
Bartow, Fl., U.S. (bär′tō)	135a	27°54'N	81°50'W
Baru, stm., Afr. *see* Baro	284-85	8°26'N	33°13'E
Barú, Volcán, vol., Pan.	162	8°48'N	82°33'W
Baruun-Urt, Mong.	260-61	46°41'N	113°17'E
Barwon, stm., Austl. (bär′wŭn)	301	30°08'S	147°23'E
Barycz, stm., Pol. (bä′rĭch)	210-11	51°41'N	16°15'E
Barysau, Bela.	208-09	54°14'N	28°31'E
Barysh, Russia	202-03	53°39'N	47°07'E
Basankusu, D.R.C. (bä-sän-kōō′sōō)	284-85	1°13'N	19°48'E
Basarabia, hist. reg., Eur.			
see Bessarabia	218-19	46°53'N	28°44'E
Basco, Phil.	270a	20°27'N	121°58'E
Bascuñán, Cabo, c., Chile	182-83	28°52'S	71°29'W
Basel, Switz. (bä′z'l)	210-11	47°33'N	7°36'E
Basey, Phil.	270	11°18'N	125°04'E
Bashi Channel, strt., Asia			
(bäsh′ē chăn′ĕl)	240-41	22°0'N	121°00'E
Bashkortostan, state, Russia	202-03	54°0'N	56°00'E
Bashtanka, Ukr. (bȧsh-tän′kä)	218-19	47°24'N	32°27'E
Basilaki Island, i., Pap. N. Gui.	302	10°37'S	150°60'E
Basilan, Phil. *see* Isabela	270	6°41'N	121°58'E
Basilan Island, i., Phil.	270	6°34'N	122°03'E
Basin, Wy., U.S. (bā′s'n)	122-23	44°23'N	108°03'W
Basingstoke, Eng., U.K. (bā′zĭng-stōk)	206-07	51°16'N	1°07'W
Başkale, Tur. (bäsh-kä′lĕ)	245	38°03'N	44°01'E
Baskatong, Réservoir, res., Qc., Can.	146-47	46°46'N	75°50'W
Basoko, D.R.C. (bȧ-sō′kō)	284-85	1°14'N	23°36'E
Bas Qafqaz Silsilasi, mts.,			
see Caucasus Mountains	245	42°38'N	45°00'E
Basra, Iraq (bäs′-rä)	248-49	30°30'N	47°48'E
Bassano, Ab., Can. (bäs-sän′ō)	142-43	50°47'N	112°27'W
Bassein, Mya. *see* Pathein	266-67	16°46'N	94°44'E
Basse-Terre, i., Guad. (bȧs′ târ′)	155b	16°10'N	61°40'W
Basse-Terre, nat. cap., Guad. (bȧs′ târ′)	155b	16°00'N	61°43'W
Basseterre, nat. cap., St. K./N.	155b	17°18'N	62°44'W
Bassett, Ne., U.S. (băs′sĕt)	124-25	42°35'N	99°32'W
Bassett, Va., U.S. (băs′sĕt)	134-35	36°46'N	79°59'W
Bass Strait, strt., Austl. (băs strāt)	301	39°20'S	145°30'E
Båstad, Swe. (bô′stät)	208-09	56°25'N	12°52'E
Bastia, Fr. (bäs′tē-ä)	200-01	42°42'N	9°27'E
Bastrop, La., U.S. (băs′trŭp)	130-31	32°47'N	91°55'W
Bastrop, Tx., U.S. (băs′trŭp)	132-33	30°06'N	97°18'W
Basutoland, nation, Afr. *see* Lesotho	274	29°30'S	28°30'E
Bata, Eq. Gui. (bä′tä)	282-83	1°52'N	9°46'E
Batabanó, Golfo de, b., Cuba			
(gôl-fô-dĕ-bä-tä-bȧ-nô′)	154-55	22°15'N	82°30'W
Batagay, Russia	236-37	67°40'N	134°40'E
Batagay-Alyta, Russia	236-37	67°48'N	130°25'E
Batala, India	254-55	31°49'N	75°13'E
Batang, China (bä-täŋ)	258-59	30°02'N	99°11'E
Batangafo, C.A.R.	284-85	7°19'N	18°18'E
Batangas, Phil. (bä-tän′gäs)	270	13°46'N	121°04'E
Batan Island, i., Phil.	270	13°15'N	124°00'E
Batan Island, i., Phil.	270a	20°27'N	121°59'E
Batan Islands, is., Phil. (bä-tän′ ī′lȧndz)	270a	20°30'N	121°50'E
Batanta, Pulau, i., Indon.	242-43	0°52'S	130°39'E
Batavia, Il., U.S. (bȧ-tā′vĭ-ȧ)	126-27	41°51'N	88°19'W
Batavia, Oh., U.S. (bȧ-tā′vĭ-ȧ)	126-27	39°05'N	84°11'W
Batavia, nat. cap., Indon. *see* Jakarta	268-69	6°11'S	106°50'E
Bataysk, Russia (bȧ-tīsk′)	218-19	47°08'N	39°46'E
Bătdâmbâng, Camb. (bȧt-tam-bang′)	266-67	13°06'N	103°12'E
Batesville, Ar., U.S. (bāts′vĭl)	134-35	35°47'N	91°39'W
Batesville, In., U.S. (bāts′vĭl)	126-27	39°18'N	85°13'W
Batesville, Ms., U.S. (bāts′vĭl)	134-35	34°19'N	89°57'W
Bath, N.B., Can. (bȧth)	148-49	46°31'N	67°35'W
Bath, Eng., U.K. (bȧth)	206-07	51°23'N	2°22'W
Bathurst, Austl. (băth′ŭrst)	301	33°25'S	149°35'E
Bathurst, N.B., Can.	148-49	47°36'N	65°39'W
Bathurst, Cape, c., N.T., Can.			
(kāp bath′-ûrst)	140-41	70°35'N	128°00'W
Bathurst Island, i., Austl.			
(bȧth′ŭrst ī′lȧnd)	296-97	11°37'S	130°17'E
Batna, Alg. (bät′nä)	292b	35°34'N	6°11'E
Baton Rouge, La., U.S. (băt′ŭn rōōzh′)	134-35	30°27'N	91°08'W
Battambang, Camb. *see* Bătdâmbâng	266-67	13°06'N	103°12'E
Batticaloa, Sri L.	256	7°43'N	81°42'E
Battle, stm., Can.	140-41	52°42'N	108°15'W
Battle Creek, Mi., U.S. (băt′'l krēk′)	126-27	42°19'N	85°11'W
Battle Creek, Ne., U.S. (băt′'l krēk′)	124-25	42°00'N	97°36'W
Battle Harbour, Nf./L., Can.			
(băt′'l här′bẽr)	138-39	52°16'N	55°35'W
Battle Mountain, Nv., U.S.			
(băt′'l moun′tĭn)	122-23	40°39'N	116°53'W
Batu, mtn., Eth.	292d	6°55'N	39°46'E
Batu, Kepulauan, is., Indon. (bä′tōō)	266-67	0°18'S	98°28'E
Batuata, Pulau, i., Indon.	268-69	6°12'S	122°42'E
Batumi, Geor. (bŭ-tōō′mē)	245	41°39'N	41°39'E
Batu Pahat, Malay.	266-67	1°51'N	102°56'E
Baturaja, Indon.	266-67	4°08'S	104°09'E
Baturité, Braz.	180-81	4°20'S	38°53'W
Baubau, Indon.	268-69	5°28'S	122°37'E
Bauchi, Nig. (bȧ-ōō′chē)	282-83	10°19'N	9°50'E
Bauld, Cape, c., Nf./L., Can.	148-49	51°38'N	55°26'W
Bauru, Braz. (bou-rōō′)	186	22°19'S	49°04'W
Bauska, Lat. (bou′skȧ)	208-09	56°24'N	24°14'E
Bautzen, Ger. (bout′sĕn)	210-11	51°11'N	14°26'E
Bavaria, hist. reg., Ger. (bȧ-vâ-rĭ-ȧ)	210-11	48°30'N	11°30'E
Bawdwin, Mya.	266-67	23°07'N	97°15'E
Bawean, Pulau, i., Indon.			
(pōō-lou bȧ′vē-än)	268-69	5°46'S	112°40'E
Bawiti, Egypt	204-05	28°21'N	28°52'E
Bawku, Ghana	282-83	11°04'N	0°15'W
Baxley, Ga., U.S. (băks′lĭ)	134-35	31°47'N	82°21'W
Bay, Laguna de, lk., Phil.			
(lä-gōō′nä dā bä′ē)	270	14°23'N	121°15'E
Bayamo, Cuba (bä-yä′mō)	154-55	20°23'N	76°38'W
Bayan Har Shan, mts., China	258-59	33°47'N	97°54'E
Bayanhongor, Mong.	260-61	46°10'N	100°42'E
Bayano, Lago, res., Pan.	162	9°12'N	78°44'W
Bayan Obo, China	260-61	41°59'N	110°08'E
Bayard, Ne., U.S. (bā′ẽrd)	124-25	41°46'N	103°20'W
Bayard, N.M., U.S. (bā′ẽrd)	128-29	32°46'N	108°08'W
Bayburt, Tur. (bä′ĭ-bòrt)	245	40°16'N	40°14'E
Bay City, Mi., U.S. (bā sĭ′tē)	126-27	43°36'N	83°53'W
Bay City, Tx., U.S. (bā sĭ′tĭ)	132-33	28°59'N	95°58'W
Baydhabo, Som. *see* Baidoa	284-85	3°07'N	43°39'E
Baydrag, stm., Mong.	260-61	45°37'N	99°15'E
Bayern, hist. reg., Ger. (bī′ẽrn)			
see Bavaria	210-11	48°30'N	11°30'E
Bayeux, Fr. (bȧ-yû′)	212-13	49°17'N	0°42'W
Baykal, Ozero, lk., Russia			
see Baikal, Lake	236-37	53°0'N	107°40'E
Baykit, Russia (bī-kĕt′)	236-37	61°41'N	96°25'E
Baykonur, Kaz. *see* Bayqongyr	244	45°38'N	63°18'E
Bay Minette, Al., U.S. (bā′mĭn-ĕt′)	134-35	30°53'N	87°47'W
Bayombong, Phil. (bä-yôm-bŏng′)	270	16°29'N	121°09'E
Bayonne, Fr. (bȧ-yôn′)	212-13	43°30'N	1°29'W
Bayonne, N.J., U.S. (bā-yōn′)	126-27	40°40'N	74°07'W
Bayou Bodcau Reservoir, res., La., U.S.			
(bī′yōō bŏd′kō rĕ′sẽr-vwär)	130-31	32°48'N	93°27'W

ŋ-sing; ŋ-baŋk; N-nasalized n; nŏd; cŏmmit; ōld; ôbey; ôrder; oi-boil; fōōd; ò-as oo in foot; ou-out; s-soft; sh-dish; th-thin; pūre; ûnite; ûrn; stŭd; circŭs; ü-as in French tu; ′-indeterminate vowel.

336 Bayqongyr – Berea PRONOUNCING INDEX

Feature (Pronunciation)	Page	Lat.	Long.
Bayqongyr, Kaz.	244	45°38'N	63°18'E
Bayram-Ali, Turkmen.	252-53	37°37'N	62°10'E
Bayreuth, Ger. (bī-roit′)	210-11	49°57'N	11°34'E
Bay Roberts, Nf./L., Can. (bā rŏb′ẽrts)	148-49	47°35'N	53°18'W
Bayrūt, nat. cap., Leb. see Beirut	248-49	33°53'N	35°30'E
Bays, Lake of, lk., On., Can. (lāk ŭv bās)	146-47	45°15'N	78°60'W
Bay Saint Louis, Ms., U.S. (bā′ sānt lo͞o′ĭs)	134-35	30°19'N	89°20'W
Bayt Laḩm, W.B. see Bethlehem	248-49	31°43'N	35°12'E
Baytown, Tx., U.S. (bā′town)	132-33	29°44'N	94°59'W
Baza, Spain (bä′thä)	214-15	37°29'N	2°46'W
Bazaruto, Ilha do, i., Moz. (ē′lä-dô-bä-zä-ro͞o′tō)	286-87	21°41'S	35°28'E
Be, Nosy, i., Madag.	286-87	13°20'S	48°15'E
Beach, N.D., U.S. (bēch)	124-25	46°55'N	104°00'W
Beachy Head, c., Eng., U.K. (bēchē hĕd)	206-07	50°45'N	0°15'E
Beacon, N.Y., U.S. (bē′kŭn)	126-27	41°31'N	73°58'W
Beagle Gulf, b., Austl.	296-97	12°0'S	130°20'E
Beardmore, On., Can.	144-45	49°36'N	87°58'W
Beardstown, Il., U.S. (bērds′toun)	130-31	40°01'N	90°26'W
Bear Island, i., Nor. (bâr ī′lănd) see Bjørnøya	236-37	74°27'N	19°02'E
Bear Lake, lk., Mb., Can. (bâr lāk)	144-45	55°08'N	96°00'W
Bear Lake, lk., U.S. (bâr lāk)	122-23	42°0'N	111°20'W
Bear River Range, mts., U.S. (bâr rĭv′ẽr rānj)	122-23	41°29'N	111°41'W
Beata, Cabo, c., Dom. Rep. (kä′bô-bĕ-ä′tä)	154-55	17°37'N	71°25'W
Beata, Isla, i., Dom. Rep.	154-55	17°35'N	71°31'W
Beatrice, Ne., U.S. (bē′ă-trĭs)	130-31	40°16'N	96°45'W
Beatton, stm., B.C., Can.	142-43	56°05'N	120°22'W
Beatty, Nv., U.S. (bēt′ē)	128-29	36°55'N	116°45'W
Beattyville, Ky., U.S. (bēt′ē-vĭl)	126-27	37°34'N	83°43'W
Beaucaire, Fr. (bō-kâr′)	212-13	43°48'N	4°39'E
Beaufort, Malay.	268-69	5°22'N	115°44'E
Beaufort, N.C., U.S. (bō′fŏrt)	134-35	34°43'N	76°40'W
Beaufort, S.C., U.S. (bō′fŏrt)	134-35	32°24'N	80°44'W
Beaufort Sea, s., N.A. (bō′fŏrt sē)	95	73°0'N	140°00'W
Beaufort West, S. Afr.	286-87	32°21'S	22°35'E
Beaumont, Tx., U.S. (bō′mŏnt)	132-33	30°05'N	94°08'W
Beaune, Fr. (bōn)	212-13	47°01'N	4°50'E
Beaupré, Qc., Can.	148-49	47°02'N	70°54'W
Beausejour, Mb., Can.	144-45	50°04'N	96°31'W
Beauvais, Fr. (bō-vĕ′)	212-13	49°26'N	2°05'E
Beaver, Ok., U.S. (bē′vẽr)	130-31	36°49'N	100°31'W
Beaver, Ut., U.S. (bē′vẽr)	128-29	38°17'N	112°38'W
Beaver, stm., Can.	140-41	55°26'N	107°47'W
Beaver Dam, Wi., U.S. (bē′vẽr dăm)	126-27	43°27'N	88°50'W
Beaverhead Mountains, mts., U.S. (bē′vẽr-hĕd moun′tĭnz)	122-23	44°58'N	113°26'W
Beaver Island, i., Mi., U.S. (bē′vẽr ī′lănd)	126-27	45°40'N	85°32'W
Beaverton, Or., U.S. (bē′vẽr-tŭn)	122-23	45°29'N	122°49'W
Beāwar, India	254-55	26°06'N	74°19'E
Bečej, Serb. (bĕc′chā)	216-17	45°37'N	20°03'E
Béchar, Alg.	280-81	31°37'N	2°14'W
Bechuanaland, nation, Afr. see Botswana	274	22°0'S	24°00'E
Beckley, W.V., U.S. (bĕk′lĭ)	126-27	37°48'N	81°11'W
Bedford, Qc., Can. (bĕd′fẽrd)	146-47	45°07'N	72°59'W
Bedford, In., U.S. (bĕd′fẽrd)	126-27	38°51'N	86°29'W
Bedford, Va., U.S. (bĕd′fẽrd)	134-35	37°20'N	79°31'W
Beebe, Ar., U.S. (bē′bē)	134-35	35°04'N	91°53'W
Beecroft Head, c., Austl. (bē′krûft hĕd)	301	35°00'S	150°51'E
Beersheba, Isr.	248-49	31°14'N	34°48'E
Be'er Sheva', Isr. (bēr-shē′bá) see Beersheba	248-49	31°14'N	34°48'E
Beeville, Tx., U.S. (bē′vĭl)	132-33	28°24'N	97°45'W
Bega, Austl. (bā′gă)	301	36°41'S	149°51'E
Beggs, Ok., U.S. (bĕgz)	130-31	35°45'N	96°05'W
Behbahān, Iran	250-51	30°35'N	50°14'E
Behchokǫ̀, N.T., Can.	138-39	62°50'N	116°02'W
Bei, stm., China (bā)	258-59	23°09'N	112°49'E
Bei'an, China (bā-än)	260-61	48°14'N	126°31'E
Beibu Wan, b., Asia see Tonkin, Gulf of	266-67	20°0'N	108°00'E
Beida, Libya see Al-Bayḍā'	204-05	32°45'N	21°37'E
Beihai, China (bā-hī)	258-59	21°27'N	109°05'E
Beijing, state, China	260-61	40°15'N	116°30'E
Beijing, nat. cap., China (bā-jyĭŋ)	260-61	39°55'N	116°22'E
Beipan, stm., China	258-59	25°01'N	106°04'E
Beipiao, China	260-61	41°48'N	120°46'E
Beira, Moz. (bā′rá)	286-87	19°50'S	34°50'E

Feature (Pronunciation)	Page	Lat.	Long.
Beirut, nat. cap., Leb. (bā-ro͞ot′)	248-49	33°53'N	35°30'E
Beitbridge, Zimb.	286-87	22°12'S	30°01'E
Beja, Port. (bā′zhä)	214-15	38°01'N	7°52'W
Béja, Tun.	200-01	36°44'N	9°11'E
Bejaïa, Alg.	292b	36°45'N	5°04'E
Bejuco, Pan. (bĕ-ko͞o′kō)	162	8°36'N	79°53'W
Bekdash, Turkmen. see Karabogaz	202-03	41°32'N	52°35'E
Békés, Hung. (bā′kāsh)	210-11	46°46'N	21°08'E
Békéscsaba, Hung. (bā′kāsh-chô′bô)	210-11	46°40'N	21°05'E
Bekobod, Uzb.	252-53	40°13'N	69°11'E
Bela, Pak.	252-53	26°13'N	66°18'E
Bela-Bela, S. Afr.	292c	24°53'S	28°19'E
Bela Crkva, Serb. (bĕ′lä tsĕrk′vä)	216-17	44°54'N	21°26'E
Belaga, Malay.	268-69	2°43'N	113°47'E
Belagāvi, India	256	15°51'N	74°31'E
Belarus, nation, Eur. (byĭ-lă-ro͞os′) (bĕ-lă-ro͞os′)	190-91	53°50'N	28°00'E
Bela Vista, Braz.	182-83	22°06'S	56°32'W
Belaya, stm., Russia (byĕ′lá-yá)	245	45°06'N	39°29'E
Belaya, stm., Russia (byĕ′lá-yá)	202-03	55°47'N	54°04'E
Belaya Tserkov, Ukr. see Bila Tserkva	218-19	49°48'N	30°08'E
Belcher Islands, is., Nu., Can. (bĕl′chẽr ī′lăndz)	140-41	56°20'N	79°30'W
Belding, Mi., U.S. (bĕl′dĭng)	126-27	43°06'N	85°13'W
Belebey, Russia (byĕ′lĕ-bă′ĭ)	202-03	54°06'N	54°08'E
Belém, Braz. (bå-lĕɴ)	180-81	1°27'S	48°29'W
Belén, Para. (bā-lān′)	182-83	23°28'S	57°15'W
Belen, N.M., U.S. (bĕ-lån′)	128-29	34°40'N	106°46'W
Belëv, Russia (byĕl′yĕf)	218-19	53°48'N	36°09'E
Belfast, S. Afr.	292c	25°43'S	30°04'E
Belfast, N. Ire., U.K. (bĕl′fast)	206-07	54°36'N	5°56'W
Belfort, Fr. (bā-fôr′)	212-13	47°38'N	6°51'E
België, nation, Eur. see Belgium	190-91	50°50'N	4°00'E
Belgique, nation, Eur. see Belgium	190-91	50°50'N	4°00'E
Belgium, nation, Eur. (bĕl′jĭ-ŭm)	190-91	50°50'N	4°00'E
Belgorod, Russia (byĕl′gŭ-rut)	218-19	50°37'N	36°35'E
Belgrade, nat. cap., Serb. (bĕl′grăd)	216-17	44°50'N	20°28'E
Belhaven, N.C., U.S. (bĕl′hā-vĕn)	134-35	35°32'N	76°37'W
Beliliou, i., Palau	304-05	7°00'N	134°15'E
Belitung, i., Indon.	266-67	2°50'S	107°55'E
Belize, nation, N.A. (bĕ-lēz′)	94	17°15'N	88°45'W
Belize, stm., Belize (bĕ-lēz′)	160	17°30'N	88°11'W
Belize City, Belize (bĕ-lēz′ sĭ′tē)	160	17°30'N	88°11'W
Bel'kovskiy, Ostrov, i., Russia (ôs-trôf′ byĕl-kôf′skĭ)	236-37	75°32'N	135°44'E
Bella Bella, B.C., Can.	142-43	52°09'N	128°07'W
Bella Coola, B.C., Can.	142-43	52°21'N	126°46'W
Bellary, India (bĕl-lä′rē)	256	15°09'N	76°55'E
Bella Unión, Ur. (bĕ′l-yä-o͞o-nyô′n)	187	30°15'S	57°35'W
Bella Vista, Arg. (bā′lyà vēs′tá)	182-83	27°02'S	65°18'W
Bella Vista, Arg. (bā′lyà vēs′tá)	187	28°30'S	59°02'W
Bellavista, Peru	184	7°04'S	76°35'W
Belle Bay, b., Nf./L., Can. (bĕl bā)	148-49	47°36'N	55°18'W
Bellefontaine, Oh., U.S. (bel-fŏn′tån)	126-27	40°21'N	83°45'W
Belle Fourche, S.D., U.S. (bĕl′ fo͞orsh′)	124-25	44°40'N	103°51'W
Belle Glade, Fl., U.S. (bĕl glād)	135a	26°42'N	80°40'W
Belle Isle, Strait of, strt., Nf./L., Can.	140-41	51°36'N	56°28'W
Belle Plaine, Ia., U.S. (bĕl plān′)	124-25	41°54'N	92°17'W
Belleville, On., Can. (bĕl′vĭl)	146-47	44°10'N	77°23'W
Belleville, Il., U.S. (bĕl′vĭl)	130-31	38°31'N	89°59'W
Belleville, Ks., U.S. (bĕl′vĭl)	130-31	39°49'N	97°38'W
Bellevue, Ia., U.S. (bĕl′vū)	124-25	42°15'N	90°25'W
Bellevue, Id., U.S. (bĕl′vū)	122-23	43°28'N	114°16'W
Bellevue, Ne., U.S. (bĕl′vū)	124-25	41°09'N	95°55'W
Bellevue, Oh., U.S. (bĕl′vū)	126-27	41°16'N	82°50'W
Bellevue, Wa., U.S. (bĕl′vū)	122-23	47°37'N	122°12'W
Belley, Fr. (bĕ-lē′)	212-13	45°46'N	5°41'E
Bellingham, Wa., U.S. (bĕl′ĭng-hăm)	122-23	48°46'N	122°29'W
Bellingshausen Sea, s., Ant. (bĕl′ĭngz houz′n sē)	313	71°0'S	85°00'W
Bellinzona, Switz. (bĕl-ĭn-tsō′nä)	210-11	46°11'N	9°01'E
Bell Island, i., Nf./L., Can. (bĕl ī′lănd)	148-49	50°44'N	55°35'W
Bello, Col. (bĕ′l-yô)	188c	6°20'N	75°34'W
Bell Peninsula, pen., Nu., Can. (bĕl pĕ-nĭn′sū-lă)	140-41	63°50'N	81°60'W
Belluno, Italy (bĕl-lo͞o′nō)	216-17	46°09'N	12°13'E
Bell Ville, Arg. (bĕl vēl′)	187	32°39'S	62°41'W
Belmond, Ia., U.S. (bĕl′mŏnd)	124-25	42°51'N	93°37'W
Belmonte, Braz. (bĕl-mōn′tä)	182-83	15°54'S	38°53'W
Belmopan, nat. cap., Belize (bĕl-mō-pän′)	160	17°14'N	88°47'W
Belogorsk, Russia	240-41	50°54'N	128°30'E
Belo Horizonte, Braz. (bĕ′lôre-sô′n-tĕ)	186	19°55'S	43°56'W
Beloit, Ks., U.S. (bē-loit′)	130-31	39°27'N	98°06'W
Beloit, Wi., U.S. (bē-loit′)	126-27	42°31'N	89°02'W
Belomorsk, Russia (byĕl-ŏ-môrsk′)	202-03	64°32'N	34°45'E

Feature (Pronunciation)	Page	Lat.	Long.
Belorechensk, Russia	202-03	44°46'N	39°52'E
Beloretsk, Russia (byĕ′lō-rĕtsk)	244	53°58'N	58°24'E
Belorussia, nation, Eur. see Belarus	190-91	53°50'N	28°00'E
Belorussiya, nation, Eur. see Belarus	190-91	53°50'N	28°00'E
Belo Tsiribihina, Madag.	286-87	19°42'S	44°33'E
Belovo, Russia (byĕ′lŭ-vû)	236-37	54°25'N	86°19'E
Beloye, Ozero, lk., Russia	202-03	60°11'N	37°37'E
Beloye More, s., Russia see White Sea	202-03	65°37'N	37°52'E
Beloz'orsk, Russia	202-03	60°02'N	37°48'E
Belton, Mo., U.S. (bĕl′tŭn)	130-31	38°48'N	94°32'W
Belton, Tx., U.S. (bĕl′tŭn)	132-33	31°03'N	97°27'W
Belts, Mol. see Bălți	218-19	47°46'N	27°55'E
Beltsy, Mol. see Bălți	218-19	47°46'N	27°55'E
Belukha, Gora, mtn., Asia see Belukha, Mount	244	49°51'N	86°29'E
Belukha, Mount, mtn., Asia (mount byĭ-lo͞o′-khŭ)	244	49°51'N	86°29'E
Belvidere, Il., U.S. (bĕl-vĕ-dēr′)	126-27	42°15'N	88°49'W
Belzoni, Ms., U.S. (bĕl-zō′nē)	134-35	33°11'N	90°29'W
Bembézar, stm., Spain (bĕm-bā-thär′)	214-15	37°45'N	5°12'W
Bemidji, Mn., U.S. (bē-mĭj′ĭ)	124-25	47°29'N	94°54'W
Benalla, Austl. (bĕn-ăl′à)	301	36°33'S	145°59'E
Benares, India see Vārānasi	254-55	25°20'N	82°59'E
Benavente, Spain (bā-nä-vĕn′tā)	214-15	42°00'N	5°40'W
Bend, Or., U.S. (bĕnd)	122-23	44°04'N	121°18'W
Bender, Mol. see Tighina	218-19	46°50'N	29°29'E
Bender Cassim, Som.	284-85	11°17'N	49°11'E
Bendery, Mol. see Tighina	218-19	46°50'N	29°29'E
Bendigo, Austl. (bĕn′dĭ-gō)	301	36°46'S	144°17'E
Benedito Leite, Braz.	180-81	7°13'S	44°34'W
Benešov, Czech Rep. (bĕn′ĕ-shôf)	210-11	49°47'N	14°43'E
Benevento, Italy (bā-nā-vĕn′tō)	216-17	41°08'N	14°46'E
Bengal, Bay of, b., Asia (bā ŭv bĕn-gôl′)	238-39	15°0'N	90°00'E
Bengalūru, India	256	12°59'N	77°36'E
Bengbu, China (bŭŋ-bo͞o)	258-59	32°57'N	117°21'E
Benghazi, Libya	280-81	32°07'N	20°04'E
Bengkulu, Indon.	266-67	3°48'S	102°16'E
Benguela, Ang. (bĕn-gĕl′à)	286-87	12°35'S	13°25'E
Beni, D.R.C.	289	0°30'N	29°28'E
Beni, stm., Bol. (bā′nē)	180-81	10°59'S	66°07'W
Beni Abbes, Alg.	204-05	30°08'N	2°10'W
Beni Mazâr, Egypt	291b	28°29'N	30°48'E
Beni-Mellal, Mor.	292a	32°21'N	6°22'W
Benin, nation, Afr. (bĕn-ēn′)	274	9°30'N	2°15'E
Benin, stm., Nig. (bĕn-ēn′)	282a	5°45'N	5°04'E
Benin, Bight of, b., Afr. (bīt ŭv bĕn-ēn′)	282-83	5°30'N	3°00'E
Bénin, Golfe de, b., Afr. see Benin, Bight of	282-83	5°30'N	3°00'E
Benin City, Nig. (bĕn-ēn′ sĭ′tē)	282a	6°20'N	5°38'E
Beni Saf, Alg. (bā′nē säf′)	214-15	35°18'N	1°23'W
Beni Suef, Egypt	291b	29°04'N	31°06'E
Benito Juárez, Arg.	187	37°41'S	59°48'W
Benjamín, Isla, i., Chile	185	44°40'S	74°08'W
Benjamin Constant, Braz.	180-81	4°28'S	70°01'W
Benkelman, Ne., U.S. (bĕn-kĕl-măn)	130-31	40°03'N	101°32'W
Bennetta, Ostrov, i., Russia	236-37	76°41'N	149°06'E
Bennettsville, S.C., U.S. (bĕn′ĕts vĭl)	134-35	34°37'N	79°41'W
Benoni, S. Afr. (bĕ-nō′nī)	292c	26°11'S	28°19'E
Bénoué, stm., Afr.	282-83	7°48'N	6°46'E
Benson, Az., U.S. (bĕn-sŭn)	128-29	31°58'N	110°18'W
Benson, Mn., U.S. (bĕn-sŭn)	124-25	45°19'N	95°36'W
Bentinck Island, i., Austl.	302	17°04'S	139°30'E
Benton, Ar., U.S. (bĕn′tŭn)	130-31	34°34'N	92°36'W
Benton, Il., U.S. (bĕn′tŭn)	126-27	37°60'N	88°55'W
Benton, Ky., U.S. (bĕn′tŭn)	134-35	36°52'N	88°21'W
Benton, La., U.S. (bĕn′tŭn)	130-31	32°42'N	93°45'W
Benton Harbor, Mi., U.S. (bĕn′tŭn här′bẽr)	126-27	42°06'N	86°28'W
Bentonville, Ar., U.S. (bĕn′tŭn-vĭl)	130-31	36°22'N	94°12'W
Benue, stm., Afr. (bā′no͞o-å)	282-83	7°48'N	6°46'E
Benxi, China (bŭn-shyē)	263	41°18'N	123°45'E
Beograd, nat. cap., Serb. (bĕ-ō′gråd) see Belgrade	216-17	44°50'N	20°28'E
Beppu, Japan (bĕ′po͞o)	265	33°17'N	131°30'E
Bequia, i., St. Vin. (bĕk-ē′ä)	155b	13°02'N	61°13'W
Berau, Teluk, b., Indon.	242-43	2°30'S	132°30'E
Berbera, Som. (bûr′bûr-á)	284-85	10°26'N	45°01'E
Berbérati, C.A.R.	284-85	4°14'N	15°48'E
Berbice, stm., Guy.	178-79	6°15'N	57°32'W
Berck, Fr. (bĕrk)	212-13	50°25'N	1°35'E
Berdians'k, Ukr.	218-19	46°45'N	36°49'E
Berdigestyakh, Russia	236-37	62°06'N	126°41'E
Berdychiv, Ukr.	218-19	49°54'N	28°37'E
Berea, Ky., U.S. (bē-rē′á)	126-27	37°34'N	84°18'W

ăt; finăl; rāte; senāte; ärm; åsk; sofá; fâre; ch-choose; dh-as th in other; bē; ēvent; bĕt; recĕnt; cratẽr; g-gō; gh-guttural g; bĭt; ĭ-short neutral; rīde; ĸ-guttural k as ch in German ich;

PRONOUNCING INDEX

Berens River – Bissagos 337

Feature (Pronunciation)	Page	Lat.	Long.
Berens River, Mb., Can.			
(bĕrĕnz rĭv′ẽr)	144-45	52°22′N	97°02′W
Beresford, S.D., U.S. (bĕr′ĕs-fẽrd)	124-25	43°05′N	96°47′W
Berettyóújfalu, Hung.			
(bĕ′rĕt-tyō-ōō′y′fô-lōō)	210-11	47°13′N	21°33′E
Berezhany, Ukr. (bĕr-yĕ′zhà-nĕ)	210-11	49°27′N	24°57′E
Berezniki, Russia (bĕr-yôz′nyē-kē)	202-03	59°24′N	56°46′E
Berga, Spain (bĕr′gä)	214-15	42°06′N	1°51′E
Bergama, Tur. (bĕr′gä-mä)	216-17	39°08′N	27°11′E
Bergamo, Italy (bĕr′gä-mō)	216-17	45°42′N	9°41′E
Bergantín, Ven. (bĕr-gän-tē′n)	188b	10°01′N	64°21′W
Bergen, Nor. (bĕr′gĕn)	208-09	60°22′N	5°21′E
Bergerac, Fr. (bĕr-zhē-ràk′)	212-13	44°51′N	0°29′E
Bergville, S. Afr. (bĕrg′vĭl)	292c	28°44′S	29°21′E
Berhala, Selat, strt., Indon.	266-67	0°48′S	104°25′E
Beringa, Ostrov, i., Russia	236-37	54°54′N	166°24′E
Beringovo More, s., *see* Bering Sea	136	60°0′N	175°00′W
Beringov Proliv, strt., *see* Bering Strait	136	65°30′N	169°00′W
Beringovskiy, Russia	236-37	63°04′N	179°22′E
Bering Sea, s., (bē′rĭng sē)	136	60°0′N	175°00′W
Bering Strait, strt., (bē′rĭng strāt)	136	65°30′N	169°00′W
Berja, Spain (bĕr′hä)	214-15	36°51′N	2°57′W
Berkakit, Russia	236-37	56°34′N	124°48′E
Berkane, Mor.	214-15	34°56′N	2°19′W
Berkeley, Ca., U.S. (bûrk′lĭ)	128-29	37°52′N	122°16′W
Berkeley Springs, W.V., U.S.			
(bûrk′lĭ sprĭngz)	126-27	39°37′N	78°15′W
Berlenga, i., Port. (bĕr-lĕn′gäzh)	214-15	39°25′N	9°30′W
Berlin, Md., U.S. (bûr-lĭn)	126-27	38°19′N	75°13′W
Berlin, N.H., U.S. (bûr-lĭn)	126-27	44°28′N	71°10′W
Berlin, Wi., U.S. (bûr-lĭn′)	126-27	43°58′N	88°57′W
Berlin, nat. cap., Ger. (bĕr-lēn′)	210-11	52°31′N	13°26′E
Bermejo, stm., Arg.	182-83	32°17′S	67°22′W
Bermejo, stm., S.A. (bĕr-ma′ho)	182-83	26°52′S	58°22′W
Bermejo, Paso del, p., S.A.	188e	32°49′S	70°05′W
Bermeo, Spain (bĕr-mā′yō)	214-15	43°25′N	2°44′W
Bermuda, dep., N.A. (bûr-myū′-dă)	152-53	32°20′N	64°45′W
Bermuda Islands, is., Ber.	152-53	32°21′N	64°46′W
Bern, nat. cap., Switz. (bĕrn)	210-11	46°57′N	7°26′E
Bernasconi, Arg.	187	37°54′S	63°44′W
Berne, In., U.S. (bûrn)	126-27	40°39′N	84°57′W
Berne, nat. cap., Switz. *see* Bern	210-11	46°57′N	7°26′E
Bernier Island, i., Austl.			
(bĕr-nēr′ ī′lănd)	296-97	24°52′S	113°08′E
Bernina, Piz, mtn., Eur.	210-11	46°22′N	9°51′E
Bernina, Pizzo, mtn., Eur.			
see Bernina, Piz	210-11	46°22′N	9°51′E
Beroun, Czech Rep. (bā′rōn)	210-11	49°58′N	14°04′E
Berryville, Ar., U.S. (bĕr′ĕ-vĭl)	130-31	36°22′N	93°35′W
Bershad', Ukr. (byĕr′shät)	218-19	48°23′N	29°34′E
Bertoua, Camrn.	282-83	4°35′N	13°41′E
Beru, i., Kir.	304-05	1°20′S	176°00′E
Berwick, Pa., U.S. (bûr′wĭk)	126-27	41°04′N	76°15′W
Berwick-upon-Tweed, Eng., U.K.			
(bûr′ĭk-ŭp′ŏn-twēd)	206-07	55°47′N	2°01′W
Besalampy, Madag. (bĕz-à-làm-pē′)	286-87	16°45′S	44°30′E
Besançon, Fr. (bē-säN-sôn)	212-13	47°15′N	6°02′E
Besni, Tur.	247	37°42′N	37°52′E
Bessarabia, hist. reg., Eur.	218-19	46°53′N	28°44′E
Bessemer, Al., U.S. (bĕs′ē-mẽr)	134-35	33°23′N	86°57′W
Bessemer, Mi., U.S. (bĕs′ē-mẽr)	124-25	46°29′N	90°03′W
Bessemer City, N.C., U.S.			
(bĕs′ē-mẽr sĭ′tē)	134-35	35°17′N	81°17′W
Betanzos, Spain (bĕ-tän′thōs)	214-15	43°17′N	8°12′W
Bethal, S. Afr. (bĕth′ăl)	292c	26°27′S	29°28′E
Bethel, Ak., U.S. (bĕth′ĕl)	136	60°48′N	161°46′W
Bethlehem, S. Afr.	292c	28°14′S	28°19′E
Bethlehem, Pa., U.S. (bĕth′lē-hĕm)	126-27	40°38′N	75°23′W
Bethlehem, W.B. (bĕth′lē-hĕm)	248-49	31°43′N	35°12′E
Béthune, Fr. (bā-tün′)	212-13	50°32′N	2°39′E
Betioky, Madag.	286-87	23°43′S	44°23′E
Betpaqdala, des., Kaz.	244	46°0′N	70°00′E
Betsiamites, stm., Qc., Can.	148-49	48°56′N	68°37′W
Betsiboka, stm., Madag.			
(bĕt-sĭ-bō′kà)	286-87	16°03′S	46°35′E
Betūl, India	254-55	21°54′N	77°54′E
Beuthen, Pol. *see* Bytom	210-11	50°21′N	18°55′E
Beverley, Austl.	294-95	32°07′S	116°55′E
Bexhill, Eng., U.K. (bĕks′hĭl)	206-07	50°51′N	0°28′E
Beyneū, Kaz.	244	45°25′N	55°09′E
Beypazarı, Tur. (bā-pà-zä′rĭ)	202-03	40°10′N	31°56′E
Beyşehir, Tur.	246	37°41′N	31°43′E
Beyşehir Gölü, lk., Tur.	202-03	37°40′N	31°30′E
Bezhetsk, Russia (byĕ-zhĕtsk′)	218-19	57°48′N	36°42′E
Béziers, Fr. (bā-zyā′)	212-13	43°21′N	3°13′E
Bezwada, India *see* Vijayawāda	256	16°31′N	80°37′E
Bhadrak, India	254-55	21°03′N	86°30′E
Bhadrāvati, India	256	13°50′N	75°42′E
Bhāgalpur, India (bä′gŭl-pòr)	254-55	25°15′N	86°59′E
Bhaktapur, Nepal	254-55	27°41′N	85°26′E
Bhamo, Mya. (bŭ-mō′)	258-59	24°17′N	97°15′E
Bhandāra, India	254-55	21°09′N	79°39′E
Bharat, nation, Asia *see* India	224-25	20°0′N	77°00′E
Bharatpur, India (bērt′pòr)	254-55	27°13′N	77°29′E
Bharūch, India	254-55	21°43′N	72°60′E
Bhatinda, India (bŭ-tĭn-dà)	254-55	30°12′N	74°57′E
Bhātpāra, India	254-55	22°52′N	88°24′E
Bhāvnagar, India	254-55	21°46′N	72°08′E
Bhawānipatna, India	254-55	19°55′N	83°10′E
Bhilai, India	254-55	21°13′N	81°26′E
Bhilainagar, India *see* Bhilai	254-55	21°13′N	81°26′E
Bhīlwāra, India	254-55	25°21′N	74°38′E
Bhīma, stm., India (bē′mà)	256	16°24′N	77°17′E
Bhind, India	254-55	26°34′N	78°47′E
Bhiwāni, India	254-55	28°47′N	76°08′E
Bhopāl, India (bô-päl)	254-55	23°15′N	77°24′E
Bhubaneshwar, India	254-55	20°14′N	85°50′E
Bhuj, India (bōōj)	254-55	23°15′N	69°40′E
Bhusāwal, India	254-55	21°03′N	75°47′E
Bhutan, nation, Asia (bōō-tän′)	224-25	27°30′N	90°30′E
Bia, Phou, mtn., Laos	266-67	18°59′N	103°09′E
Biafra, Bahía de, b., Afr.			
see Biafra, Bight of	282-83	4°0′N	8°00′E
Biafra, Bight of, b., Afr.	282-83	4°0′N	8°00′E
Biafra, Golfe de, b., Afr.			
see Biafra, Bight of	282-83	4°0′N	8°00′E
Biak, i., Indon. (bē′ăk)	242-43	1°0′S	136°00′E
Biała Podlaska, Pol.			
(byä′wä pōd läs′kä)	210-11	52°02′N	23°00′E
Białystok, Pol. (byä-wĭs′tōk)	210-11	53°08′N	23°09′E
Bianco, Monte, mtn., Eur.			
see Blanc, Mont	212-13	45°50′N	6°52′E
Biarritz, Fr. (byä-rēts′)	212-13	43°29′N	1°33′W
Bickerton, Cape, c., Ant.	313	66°20′S	136°56′E
Bicknell, In., U.S. (bĭk′nĕl)	126-27	38°40′N	87°18′W
Bida, Nig. (bē′dä)	282-83	9°05′N	5°60′E
Bīdar, India	256	17°54′N	77°31′E
Biddeford, Me., U.S. (bĭd′ē-fẽrd)	126-27	43°30′N	70°27′W
Bié, Planalto do, plat., Ang.	286-87	13°30′S	17°02′E
Biebrza, stm., Pol. (byĕb′zha)	210-11	53°13′N	22°26′E
Bielefeld, Ger. (bē′lē-fĕlt)	210-11	52°01′N	8°32′E
Biella, Italy (byĕl′lä)	216-17	45°34′N	8°04′E
Bielsko-Biała, Pol.	210-11	49°49′N	19°03′E
Bielsk Podlaski, Pol.			
(byĕlsk pŭd-lä′skĭ)	210-11	52°46′N	23°12′E
Bien Dong, s., Asia			
see South China Sea	242-43	10°0′N	113°00′E
Bien Hoa, Viet.	266-67	10°57′N	106°50′E
Bienville, Lac, lk., Qc., Can.	140-41	55°05′N	72°40′W
Biga, Tur. (bē′ghà)	216-17	40°14′N	27°15′E
Big Belt Mountains, mts., Mt., U.S.			
(bĭg bĕlt moun′tĭnz)	122-23	46°40′N	111°25′W
Big Bend National Park, n.p., Tx., U.S.			
(bĭg bĕnd năsh′ŭn-ăl pärk)	132-33	29°12′N	103°12′W
Big Black, stm., Ms., U.S. (bĭg blăk)	134-35	32°03′N	91°04′W
Big Cypress Indian Reservation,			
ind. res., Fl., U.S.			
(bĭg sī′prĕs ĭn′dĭ-ăn rĕ-sẽr-vā′shĕn)	135a	26°17′N	80°59′W
Big Cypress Swamp, sw., Fl., U.S.			
(bĭg sī′prĕs swŏmp)	135a	26°10′N	81°38′W
Big Delta, Ak., U.S. (bĭg dĕl′tà)	136	64°09′N	145°47′W
Biggar, Sk., Can.	144-45	52°03′N	107°58′W
Bigge Island, i., Austl.	296-97	14°35′S	125°10′E
Bighorn, stm., U.S. (bĭg hôrn)	120-21	46°09′N	107°29′W
Bighorn Mountains, mts., U.S.			
(bĭg hôrn moun′tĭnz)	122-23	43°59′N	107°04′W
Big Island, i., Nu., Can. (bĭg ī′lănd)	140-41	62°43′N	70°43′W
Big Island, i., On., Can. (bĭg ī′lănd)	144-45	49°09′N	94°37′W
Big Lake, Tx., U.S. (bĭg lāk)	132-33	31°11′N	101°28′W
Big Quill Lake, lk., Sk., Can.	144-45	51°55′N	104°22′W
Big Rapids, Mi., U.S. (bĭg răp′ĭdz)	126-27	43°42′N	85°29′W
Big River, Sk., Can. (bĭg rĭv′ẽr)	144-45	53°51′N	107°01′W
Big Sandy, stm., Wy., U.S. (bĭg sănd′ē)	122-23	41°52′N	109°47′W
Big Sioux, stm., U.S. (bĭg sōō)	124-25	42°29′N	96°28′W
Big Spring, Tx., U.S. (bĭg sprĭng)	130-31	32°15′N	101°29′W
Big Stone Gap, Va., U.S. (bĭg stōn)	134-35	36°52′N	82°47′W
Big Timber, Mt., U.S. (bĭg tĭm′-bẽr)	122-23	45°50′N	109°57′W
Big Trout Lake, lk., On., Can.	144-45	53°44′N	89°57′W
Bihār, state, India (bē-här′)	254-55	25°0′N	86°00′E
Bihar Sharīf, India	254-55	25°12′N	85°32′E
Bijagós, Arquipélago dos, is., Gui.-B.	282-83	11°22′N	16°18′W
Bij-Chem, stm., Russia			
see Bol'shoy Yenisey	236-37	51°44′N	94°28′E
Bijie, China (bē-jyĕ)	258-59	27°18′N	105°17′E
Bīkāner, India (bĭ-kă′nûr)	254-55	28°01′N	73°19′E
Bikar, at., Marsh. Is.	304-05	12°14′N	170°08′E
Bikeqi, China	260-61	40°43′N	111°17′E
Bikin, Russia (bē-kēn′)	264	46°49′N	134°17′E
Bikin, stm., Russia (bē-kēn′)	264	46°51′N	134°02′E
Bikini, at., Marsh. Is.	304-05	11°35′N	165°23′E
Bilāspur, India (bē-läs′pōōr)	254-55	22°05′N	82°10′E
Bilāspur, India (bē-läs′pōōr)	254-55	31°18′N	76°46′E
Bila Tserkva, Ukr.	218-19	49°48′N	30°08′E
Bilauktaung Range, mts., Asia	266-67	13°0′N	99°00′E
Bilbao, Spain (bĭl-bä′ō)	214-15	43°15′N	2°56′W
Bilecik, Tur. (bē-lĕd-zhĕk′)	216-17	40°10′N	29°59′E
Biłgoraj, Pol. (bēw-gō′rī)	210-11	50°33′N	22°42′E
Bilhorod-Dnistrovs'kyi, Ukr.	218-19	46°12′N	30°18′E
Bili, stm., D.R.C.	284-85	4°08′N	22°29′E
Bilibino, Russia	236-37	68°04′N	166°21′E
Biliran Island, i., Phil.	270	11°36′N	124°29′E
Billabong Creek, stm., Austl.			
(bĭl′ă-bŏng krēk)	301	35°06′S	144°02′E
Billings, Mt., U.S. (bĭl′ĭngz)	122-23	45°47′N	108°31′W
Billiton, i., Indon. *see* Belitung	266-67	2°50′S	107°55′E
Bilma, Niger (bēl′mä)	280-81	18°41′N	12°56′E
Biloela, Austl.	302	24°24′S	150°30′E
Biloxi, Ms., U.S. (bĭ-lŏk′sĭ)	134-35	30°24′N	88°53′W
Biltine, Chad	280-81	14°32′N	20°55′E
Bimberi Peak, mtn., Austl.			
(bĭm′bĕrĭ pēk)	301	35°40′S	148°47′E
Bimini Islands, is., Bah.	154-55	25°42′N	79°15′W
Binga, D.R.C.	284-85	2°22′N	20°31′E
Binga, Monte, mtn., Afr.	286-87	19°47′S	33°03′E
Binga, Mount, mtn., Afr.			
see Binga, Monte	286-87	19°47′S	33°03′E
Binghamton, N.Y., U.S. (bĭng′ăm-tŭn)	126-27	42°06′N	75°55′W
Binhai, China	258-59	34°00′N	119°50′E
Binjai, Indon.	266-67	3°36′N	98°30′E
Binongko, Pulau, i., Indon.	268-69	5°57′S	124°02′E
Bintimani, mtn., S.L.	282-83	9°13′N	11°07′W
Bintulu, Malay. (bēn′tōō-lōō)	268-69	3°10′N	113°02′E
Binxian, China (bĭn-shyän)	258-59	35°02′N	108°06′E
Binxian, China (bĭn shyän)	260-61	37°28′N	117°58′E
Binzert, Tun. *see* Bizerte	200-01	37°16′N	9°52′E
Bioko, i., Eq. Gui. (bē-ō′-kō)	282-83	3°30′N	8°40′E
Bīr, India	256	18°60′N	75°46′E
Bira, Russia (bē′rà)	260-61	48°60′N	132°28′E
Bira, stm., Russia (bē′rà)	260-61	48°10′N	133°17′E
Birao, C.A.R.	284-85	10°17′N	22°48′E
Birch Mountains, hills, Ab., Can.			
(bûrch moun′tĭnz)	140-41	57°34′N	113°07′W
Birdsville, Austl. (bûrdz′vĭl)	301	25°54′S	139°21′E
Birecik, Tur. (bē-rĕd-zhĕk′)	248-49	37°03′N	38°03′E
Bīrjand, Iran (bēr′jănd)	252-53	32°53′N	59°13′E
Bîrlad, Rom.	218-19	46°14′N	27°40′E
Birmingham, Eng., U.K.	206-07	52°28′N	1°53′W
Birmingham, Al., U.S. (bûr′mĭng-hăm)	134-35	33°31′N	86°49′W
Bîr Mogreïn, Maur.	280-81	25°14′N	11°35′W
Birnie, at., Kir.	304-05	3°35′S	171°31′W
Birnin-Kebbi, Nig.	282-83	12°28′N	4°12′E
Birnin Konni, Niger	280-81	13°48′N	5°15′E
Birobidzhan, Russia (bē′rô-bē-jän′)	260-61	48°47′N	132°55′E
Birsk, Russia (bĭrsk)	202-03	55°25′N	55°34′E
Biryusa, stm., Russia (bēr-yōō′sä)	236-37	57°43′N	95°27′E
Biržai, Lith. (bēr-zhä′ē)	208-09	56°12′N	24°46′E
Bisbee, Az., U.S. (bĭz′bē)	128-29	31°27′N	109°55′W
Biscay, Bay of, b., Eur. (bĭs′kā′ bā)	212-13	44°0′N	4°00′W
Biscayne Bay, b., Fl., U.S. (bĭs-kān′ bā)	135a	25°33′N	80°15′W
Biscayne National Park, n.p., Fl., U.S.	135a	25°25′N	80°12′W
Biscoe Islands, is., Ant.	313	65°60′S	66°30′W
Bishkek, nat. cap., Kyrg. (bĭsh-kĕk′)	244	42°52′N	74°35′E
Bishop, Ca., U.S. (bĭsh′ŭp)	128-29	37°22′N	118°23′W
Bishop, Tx., U.S. (bĭsh′ŭp)	132-33	27°35′N	97°47′W
Bishop's Falls, Nf./L., Can.	148-49	49°02′N	55°30′W
Bishopville, S.C., U.S. (bĭsh′ŭp-vĭl)	134-35	34°13′N	80°15′W
Biskra, Alg.	292b	34°50′N	5°43′E
Bislig, Phil.	270	8°13′N	126°19′E
Bismarck, N.D., U.S. (bĭz′märk)	124-25	46°48′N	100°48′W
Bismarck Archipelago, is., Pap. N. Gui.			
(bĭz′märk är′kà-pĕ′-à-gō)	302	5°0′S	150°00′E
Bismarck Range, mts., Pap. N. Gui.			
(bĭz′märk rānj)	302	5°30′S	144°45′E
Bismarck Sea, s., Pap. N. Gui.			
(bĭz′märk sē)	302	4°0′S	148°00′E
Bissagos, is., Gui.-B. *see* Bijagós,			
Arquipélago dos	282-83	11°22′N	16°18′W

n-sing; ŋ-baŋk; N-nasalized n; nŏd; cŏmmit; ōld; ôbey; ôrder; oi-boil; fōōd; ȯ-as oo in foot; ou-out; s-soft; sh-dish; th-thin; pūre; ûnite; ûrn; stŭd; circŭs; ü-as in French tu; ′-indeterminate vowel.

338 Bissau – Boorama PRONOUNCING INDEX

Feature (Pronunciation)	Page	Lat.	Long.
Bissau, nat. cap., Gui.-B. (bê-sa´o͞o)	282-83	11°52'N	15°36'W
Bissett, Mb., Can.	144-45	51°02'N	95°40'W
Bistcho Lake, lk., Ab., Can.	140-41	59°44'N	118°46'W
Bistineau, Lake, res., La., U.S. (lăk bĭs-tĭ-nō´)	130-31	32°25'N	93°22'W
Bistrița, Rom. (bĭs-trĭt-sȧ)	210-11	47°08'N	24°30'E
Bistrița, stm., Rom. (bĭs-trĭt-sȧ)	204-05	46°28'N	26°57'E
Bitlis, Tur. (bĭt-lēs´)	245	38°22'N	42°06'E
Bitola, Mac. (bē´tô-lä) (mô´nä-stêr)	216-17	41°02'N	21°20'E
Bitolj, Mac. *see* Bitola	216-17	41°02'N	21°20'E
Bitonto, Italy (bê-tôn´tō)	216-17	41°06'N	16°42'E
Bitra Island, i., India	256	11°36'N	72°11'E
Bitterfeld, Ger. (bĭt´ēr-fĕlt)	210-11	51°37'N	12°19'E
Bitterroot Range, mts., U.S. (bĭt´ēr-o͞ot rānj)	120-21	47°06'N	115°10'W
Bitung, Indon.	268-69	1°26'N	125°08'E
Bityug, stm., Russia (bĭt´yo͞og)	202-03	50°38'N	39°56'E
Biwa-ko, lk., Japan (bē-wä´kō)	265	35°15'N	136°05'E
Biya, stm., Russia (bĭ´yȧ)	236-37	52°26'N	85°00'E
Biysk, Russia (bēsk)	244	52°34'N	85°15'E
Bizerte, Tun. (bê-zĕrt´)	200-01	37°16'N	9°52'E
Bjarèzina, stm., Bela. (bĕr-yĕ´zĕ-nȧ)	218-19	52°33'N	30°14'E
Bjelovar, Cro. (byĕ-lō´vär)	216-17	45°54'N	16°50'E
Björneborg, Fin. *see* Pori	208-09	61°29'N	21°47'E
Bjørnøya, i., Nor.	236-37	74°27'N	19°02'E
Black, stm., Asia (blăk)	266-67	21°15'N	105°21'E
Black, stm., Ar., U.S. (blăk)	130-31	35°38'N	91°19'W
Blackall, Austl. (blăk´ăl)	302	24°26'S	145°28'E
Black Bay, b., On., Can. (blăk bā)	146-47	48°34'N	88°32'W
Blackburn, Eng., U.K. (blăk´bûrn)	206-07	53°45'N	2°29'W
Black Canyon of the Gunnison National Park, n.p., Co., U.S.	128-29	38°34'N	107°44'W
Blackduck, Mn., U.S. (blăk´dŭk)	124-25	47°43'N	94°32'W
Blackfeet Indian Reservation, ind. res., Mt., U.S. (blăk´fēt ĭn´dĭ-ăn rĕ-sēr-vā´shĕn)	122-23	48°40'N	113°00'W
Blackfoot, Id., U.S. (blăk´fo͝ot)	122-23	43°12'N	112°21'W
Black Forest, mts., Ger. *see* Schwarzwald	210-11	48°21'N	8°11'E
Black Hills, mts., U.S. (blăk hĭlz)	124-25	44°0'N	104°00'W
Blackpool, Eng., U.K. (blăk´po͞ol)	206-07	53°50'N	3°02'W
Black Range, mts., N.M., U.S. (blăk rānj)	128-29	33°20'N	107°50'W
Black River Falls, Wi., U.S. (blăk rĭv´ēr fôlz)	124-25	44°18'N	90°50'W
Black Rock Desert, des., Nv., U.S. (blăk rŏk dĕs´ērt)	122-23	41°06'N	118°51'W
Blacksburg, Va., U.S. (blăks´bûrg)	134-35	37°14'N	80°25'W
Black Sea, s., (blăk sē)	202-03	43°0'N	35°00'E
Blackshear, Ga., U.S. (blăk´shĭr)	134-35	31°18'N	82°15'W
Blackstone, Va., U.S. (blăk´stôn)	134-35	37°05'N	77°60'W
Blacktown, Austl. (blăk´toun)	301	33°46'S	150°54'E
Black Volta, stm., Afr. (blăk vôl´tȧ)	282-83	8°41'N	0°60'W
Blackwell, Ok., U.S. (blăk´wĕl)	130-31	36°48'N	97°18'W
Blagodarnoye, Russia (blä´gô-där-nō´yĕ)	202-03	45°06'N	43°25'E
Blagoveshchensk, Russia	260-61	50°17'N	127°33'E
Blaine, Mn., U.S. (blān)	124-25	45°11'N	93°15'W
Blaine, Wa., U.S. (blān)	122-23	48°60'N	122°45'W
Blair, Ne., U.S. (blâr)	124-25	41°33'N	96°08'W
Blairsville, Pa., U.S. (blârs´vĭl)	126-27	40°26'N	79°16'W
Blakely, Ga., U.S. (blăk´lē)	134-35	31°23'N	84°56'W
Blanc, Cap, c., Afr. *see* Nouâdhibou, Râs	280-81	20°47'N	17°03'W
Blanc, Mont, mtn., Eur. (môn blän)	212-13	45°50'N	6°52'E
Blanca, Bahía, b., Arg. (bä-ē´ä-blán´kä)	187	38°55'S	62°10'W
Blanca Peak, mtn., Co., U.S. (blăŋ´kȧ pēk)	130-31	37°35'N	105°29'W
Blanche, Lake, lk., Austl. (lăk blănch)	301	29°15'S	139°39'E
Blanco, Cabo, c., C.R. (kȧ´bô-blăŋ´kō)	161	9°34'N	85°07'W
Blanco, Cape, c., Or., U.S. (kāp blăn´kō)	122-23	42°50'N	124°33'W
Blanquilla, Isla, i., Ven.	178-79	11°51'N	64°37'W
Blantyre, Malawi (blăn-tīyr)	286-87	15°47'S	35°01'E
Blenheim, N.Z.	303	41°31'S	187°58'E
Bleus, Monts, mts., D.R.C.	289	1°37'N	30°28'E
Blida, Alg.	292b	36°29'N	2°49'E
Blind River, On., Can. (blīnd rĭv´ēr)	146-47	46°11'N	82°56'W
Blissfield, Mi., U.S. (blĭs-fēld)	126-27	41°50'N	83°52'W
Blitar, Indon.	268-69	8°06'S	112°10'E
Bloemfontein, nat. cap., S. Afr.	292c	29°07'S	26°12'E
Blois, Fr. (blwä)	212-13	47°35'N	1°20'E
Bloodvein, stm., Can.	144-45	51°48'N	96°53'W
Bloomer, Wi., U.S. (blo͞om´ēr)	124-25	45°06'N	91°29'W
Bloomfield, Ia., U.S. (blo͞om´fēld)	124-25	40°46'N	92°25'W
Bloomfield, In., U.S. (blo͞om´fēld)	126-27	39°01'N	86°56'W
Bloomfield, Mo., U.S. (blo͞om´fēld)	134-35	36°53'N	89°56'W
Bloomfield, Ne., U.S. (blo͞om´fēld)	124-25	42°36'N	97°39'W
Blooming Prairie, Mn., U.S. (blo͞om´ĭng prâ´rĭ)	124-25	43°52'N	93°03'W
Bloomington, Il., U.S. (blo͞om´ĭng-tŭn)	126-27	40°29'N	88°60'W
Bloomington, In., U.S. (blo͞om´ĭng-tŭn)	126-27	39°09'N	86°32'W
Bloomington, Mn., U.S. (blo͞om´ĭng-tŭn)	124-25	44°50'N	93°19'W
Bloomsburg, Pa., U.S. (blo͞omz´bûrg)	126-27	40°60'N	76°27'W
Blossburg, Pa., U.S. (blŏs´bûrg)	126-27	41°41'N	77°05'W
Blountstown, Fl., U.S. (blŭnts´tun)	134-35	30°27'N	85°03'W
Bludenz, Aus. (blo͞o-dĕnts´)	210-11	47°09'N	9°50'E
Blue Earth, Mn., U.S. (blo͞o ûrth)	124-25	43°38'N	94°06'W
Bluefield, W.V., U.S. (blo͞o´fēld)	134-35	37°15'N	81°14'W
Bluefields, Nic. (blo͞o´fēldz)	161	12°01'N	83°46'W
Blue Mountain, mtn., Nf./L., Can. (blo͞o moun´tĭn)	148-49	50°24'N	57°10'W
Blue Mountain Peak, mtn., Jam.	154-55	18°03'N	76°35'W
Blue Mountains, mts., Austl. (blo͞o moun´tĭnz)	301	33°37'S	150°17'E
Blue Mountains, mts., U.S. (blo͞o moun´tĭnz)	122-23	45°16'N	118°42'W
Blue Mountains National Park, n.p., Austl.	301	33°47'S	150°23'E
Blue Nile, stm., Afr. (blo͞o nīl)	275	15°38'N	32°30'E
Bluenose Lake, lk., Nu., Can.	140-41	68°25'N	119°45'W
Blue Ridge, mts., U.S. (blo͞o rĭj)	134-35	37°0'N	82°00'W
Blue River, B.C., Can. (blo͞o rĭv´ēr)	142-43	52°06'N	119°20'W
Bluff, Ut., U.S. (blŭf)	128-29	37°17'N	109°33'W
Bluffton, In., U.S. (blŭf-tŭn)	126-27	40°44'N	85°10'W
Blumenau, Braz. (blo͞o´mĕn-ou)	186	26°56'S	49°05'W
Blyth, Eng., U.K. (blīth)	206-07	55°08'N	1°31'W
Blytheville, Ar., U.S. (blīth´vĭl)	134-35	35°56'N	89°55'W
Bo, S.L.	282-83	7°59'N	11°44'W
Boaco, Nic. (bô-ä´kō)	161	12°29'N	85°39'W
Bo'ai, China (bwo-ī)	260-61	35°10'N	113°04'E
Boano, Pulau, i., Indon.	268-69	2°58'S	127°56'E
Boa Vista, i., C.V. (bō-ä-vēsh´tȧ)	282-83	16°05'N	22°50'W
Bobaomby, Tanjona, c., Madag.	286-87	11°58'S	49°15'E
Bobbili, India	256	18°35'N	83°22'E
Bobo-Dioulasso, Burkina (bō´bô-dyo͞o-lȧs-sō´)	282-83	11°11'N	4°18'W
Bobruysk, Bela. *see* Babrujsk	218-19	53°08'N	29°14'E
Boby, mtn., Madag.	286-87	22°13'S	46°55'E
Boca do Acre, Braz.	180-81	8°45'S	67°23'W
Bocas del Toro, Pan. (bō´käs dĕl tō´rō)	162	9°20'N	82°15'W
Bochnia, Pol. (bôκ´nyä)	210-11	49°58'N	20°25'E
Bocholt, Ger. (bō´κōlt)	210-11	51°50'N	6°37'E
Bodaybo, Russia (bō-dī´bō)	236-37	57°51'N	114°11'E
Bodélé, reg., Chad (bō-dȧ-lā´)	280-81	16°30'N	16°30'E
Boden, Swe.	200-01	65°50'N	21°43'E
Bodh Gaya, India	254-55	24°42'N	84°58'E
Bodmin, Eng., U.K. (bŏd´mĭn)	206-07	50°29'N	4°43'W
Bodø, Nor.	200-01	67°17'N	14°24'E
Bodrum, Tur.	216-17	37°02'N	27°26'E
Boende, D.R.C. (bô-ĕn´dȧ)	284-85	0°14'S	20°52'E
Boerne, Tx., U.S. (bō´ērn)	132-33	29°47'N	98°43'W
Bōfu, Japan (bō´fo͞o) *see* Hōfu	265	34°03'N	131°35'E
Bogal, Lagh, stm., Kenya	284-85	0°46'N	40°50'E
Bogale, Mya.	266-67	16°17'N	95°24'E
Bogalusa, La., U.S. (bō-gȧ-lo͞o´sȧ)	134-35	30°47'N	89°51'W
Bogan, stm., Austl. (bô´gĕn)	301	29°58'S	146°20'E
Bogo, Phil.	270	11°02'N	124°01'E
Bogong, Mount, mtn., Austl.	301	36°44'S	147°18'E
Bogor, Indon.	268-69	6°35'S	106°47'E
Bogoroditsk, Russia (bŏ-gŏ´rŏ-dĭtsk)	218-19	53°47'N	38°08'E
Bogotá, nat. cap., Col. (bō-gō-tä´)	188c	4°37'N	74°06'W
Bogra, Bngl.	254-55	24°50'N	89°22'E
Boguchany, Russia	236-37	58°23'N	97°29'E
Bogué, Maur.	280-81	16°35'N	14°16'W
Bo Hai, b., China	260-61	38°30'N	120°02'E
Bohai Haixia, strt., China (bwo-hī hī-shyä)	260-61	38°15'N	121°00'E
Bohain-en-Vermandois, Fr. (bô-ăɴ-ŏɴ-vâr-män-dwä´)	212-13	49°59'N	3°27'E
Bohea Hills, mts., China *see* Wuyi Shan	258-59	27°42'N	117°09'E
Bohemia, hist. reg., Czech Rep.	210-11	49°50'N	14°00'E
Bohol, i., Phil. (bô-hôl´)	270	9°55'N	123°44'E
Bohol Sea, s., Phil.	270	9°10'N	124°25'E
Boipeba, Ilha de, i., Braz.	180-81	13°38'S	38°56'W
Bois, Lac des, lk., N.T., Can.	140-41	66°46'N	125°08'W
Boise, Id., U.S. (boi´zē)	122-23	43°37'N	116°13'W
Boise City, Ok., U.S. (boi´zē sĭ´tē)	130-31	36°44'N	102°30'W
Boissevain, Mb., Can. (bois´vān)	144-45	49°14'N	100°03'W
Bojeador, Cape, c., Phil.	270	18°30'N	120°35'E
Bojnūrd, Iran	252-53	37°29'N	57°20'E
Boksitogorsk, Russia	202-03	59°28'N	33°52'E
Bokurdak, Turkmen.	252-53	38°46'N	58°29'E
Bolbec, Fr. (bôl-bĕk´)	212-13	49°34'N	0°29'E
Bolgatanga, Ghana	282-83	10°48'N	0°51'W
Boli, China (bwo-lē)	264	45°45'N	130°34'E
Bolívar, Col.	188c	4°21'N	76°10'W
Bolivar, Mo., U.S. (bŏl´ĭ-vȧr)	130-31	37°37'N	93°25'W
Bolivar, Tn., U.S. (bŏl´ĭ-vȧr)	134-35	35°16'N	88°59'W
Bolívar, Cerro, mtn., Ven.	178-79	7°28'N	63°25'W
Bolívar, Pico, mtn., Ven.	178-79	8°33'N	71°01'W
Bolivar Peninsula, pen., Tx., U.S. (bŏl´ĭ-vȧr pĕ-nĭn´sūlȧ)	132-33	29°27'N	94°39'W
Bolivia, nation, S.A. (bô-lĭv´ĭ-ȧ)	172	17°0'S	65°00'W
Bolkhov, Russia (bôl-κôf´)	218-19	53°27'N	36°00'E
Bollnäs, Swe. (bôl´nĕs)	208-09	61°21'N	16°24'E
Bolmen, lk., Swe. (bôl´mĕn)	208-09	56°55'N	13°40'E
Bolobo, D.R.C. (bō´lô-bô)	284-85	2°11'S	16°15'E
Bologna, Italy (bô-lōn´yä)	216-17	44°30'N	11°20'E
Bologoye, Russia (bŏ-lŏ-gô´yĕ)	218-19	57°54'N	34°03'E
Bol'shevik, Ostrov, i., Russia	236-37	78°40'N	102°30'E
Bol'shezemel'skaya Tundra, reg., Russia	202-03	67°30'N	55°60'E
Bol'shoy Begichëv, Ostrov, i., Russia	236-37	74°20'N	112°30'E
Bol'shoy Kavkaz, mts., *see* Caucasus Mountains	245	42°38'N	45°00'E
Bol'shoy Lyakhovskiy, Ostrov, i., Russia	236-37	73°35'N	142°00'E
Bol'shoy Uzen', stm., Eur. *see* Ülkenözen	202-03	48°60'N	49°59'E
Bol'shoy Yenisey, stm., Russia	236-37	51°44'N	94°28'E
Bolvadin, Tur.	246	38°43'N	31°03'E
Bolu, Tur. (bō´lò)	202-03	40°44'N	31°36'E
Bolzano, Italy (bôl-tsä´nō)	216-17	46°30'N	11°21'E
Boma, D.R.C. (bō´mä)	284-85	5°51'S	13°04'E
Bombala, Austl. (bŭm-bä´lă)	301	36°55'S	149°14'E
Bombay, India *see* Mumbai	256	18°57'N	72°50'E
Bom Jesus da Lapa, Braz.	180-81	13°15'S	43°25'W
Bømlo, i., Nor. (bûmlô)	208-09	59°47'N	5°12'E
Bomokandi, stm., D.R.C.	284-85	3°39'N	26°08'E
Bomu, stm., Afr.	284-85	4°09'N	22°29'E
Bon, Cap, c., Tun. (kȧp bôɴ)	280-81	37°05'N	11°03'E
Bonaire, i., N.A. (bô-nâr´)	152a	12°10'N	68°15'W
Bonaventure, Qc., Can.	148-49	48°02'N	65°30'W
Bonavista, Nf./L., Can. (bŏ-nȧ-vĭs´tȧ)	148-49	48°39'N	53°07'W
Bonavista Bay, b., Nf./L., Can. (bŏ-nȧ-vĭs´tȧ bā)	148-49	48°50'N	53°21'W
Bondo, D.R.C. (bôn´dô)	284-85	3°49'N	23°41'E
Bondoc Peninsula, pen., Phil. (bôn-dŏk´ pĕ-nĭn´sūlȧ)	270	13°30'N	122°30'E
Bondoukou, C. Iv. (bŏn-do͞o´ko͞o)	282-83	8°02'N	2°48'W
Bône, Alg. *see* Annaba	200-01	36°54'N	7°46'E
Bone, Indon. *see* Watampone	268-69	4°32'S	120°19'E
Bone, Teluk, b., Indon.	268-69	4°0'S	120°40'E
Bonerate, Pulau, i., Indon.	268-69	7°21'S	121°07'E
Bonete Grande, Cerro, mtn., Arg. (sĕ´r-rô bô´nĕtĕh grän´dĕ)	182-83	27°57'S	68°45'W
Bongo, Massif des, mts., C.A.R.	284-85	8°36'N	22°50'E
Bonham, Tx., U.S. (bŏn´ăm)	130-31	33°35'N	96°11'W
Bonifacio, Bouches de, strt., Eur. *see* Bonifacio, Strait of	216-17	41°18'N	9°15'E
Bonifacio, Strait of, strt., Eur. (străt ŭv bō-nē-fä´chō)	216-17	41°18'N	9°15'E
Bonifay, Fl., U.S. (bŏn-ĭ-fā´)	134-35	30°47'N	85°41'W
Bonin Islands, is., Japan (bō´nĭn ī´lăndz)	308-09	26°58'N	142°14'E
Bonn, Ger. (bôn)	210-11	50°44'N	7°05'E
Bonners Ferry, Id., U.S. (bonērz fĕr´ē)	122-23	48°41'N	116°19'W
Bonne Terre, Mo., U.S. (bŏn târ´)	130-31	37°55'N	90°33'W
Bonny, Nig. (bŏn´ē)	282a	4°26'N	7°10'E
Bonnyville, Ab., Can. (bŏnē-vĭl)	142-43	54°16'N	110°44'W
Bontang, Indon.	268-69	0°08'N	117°30'E
Bontoc, Phil. (bŏn-tŏk´)	270	17°05'N	120°60'E
Boodjamulla National Park, n.p., Austl.	302	18°45'S	138°27'E
Booker T. Washington National Monument, n.p., Va., U.S. (bŏŏk´ēr tē wŏsh´ĭng-tŭn năsh´ŭn-ăl mŏn´ū-mĕnt)	134-35	37°01'N	79°45'W
Boonah, Austl.	301	28°00'S	152°41'E
Boone, Ia., U.S. (bo͞on)	124-25	42°04'N	93°53'W
Boone, N.C., U.S. (bo͞on)	134-35	36°13'N	81°41'W
Booneville, Ar., U.S. (bo͞on´vĭl)	130-31	35°08'N	93°56'W
Booneville, Ms., U.S. (bo͞on´vĭl)	134-35	34°39'N	88°34'W
Boonville, In., U.S. (bo͞on´vĭl)	126-27	38°03'N	87°17'W
Boonville, Mo., U.S. (bo͞on´vĭl)	130-31	38°58'N	92°45'W
Boorama, Som.	284-85	9°58'N	43°09'E

ăt; finăl; rāte; senâte; ärm; åsk; sofȧ; fâre; ch-choose; dh-as th in other; bē; ēvent; bĕt; recĕnt; cratēr; g-gō; gh-guttural g; bĭt; ĭ-short neutral; rīde; κ-guttural k as ch in German ich;

PRONOUNCING INDEX
Boosaaso – Bristol 339

Feature (Pronunciation)	Page	Lat.	Long.
Boosaaso, Som. *see* Bender Cassim	284-85	11°17′N	49°11′E
Boothbay Harbor, Me., U.S. (bōōth′bā här′bẽr)	127a	43°51′N	69°38′W
Boothia, Gulf of, b., Nu., Can. (gŭlf ŭv bōō′thĭ-ȧ)	95	71°0′N	91°00′W
Boothia Peninsula, pen., Nu., Can.	140-41	70°30′N	95°00′W
Booué, Gabon	282-83	0°06′S	11°57′E
Bordoy, i., Far. Is.	206b	62°17′N	6°33′W
Bora-Bora, i., Fr. Poly.	304-05	16°30′S	151°45′W
Borah Peak, mtn., Id., U.S. (bō′rä pēk)	122-23	44°08′N	113°48′W
Borås, Swe. (bò′rōs)	208-09	57°44′N	12°57′E
Borāzjān, Iran (bō-räz-jän′)	250-51	29°16′N	51°12′E
Borba, Braz. (bôr′bä)	180-81	4°23′S	59°35′W
Bordeaux, Fr. (bôr-dō′)	212-13	44°50′N	0°34′W
Bordentown, N.J., U.S. (bôr′dĕn-toun)	126-27	40°08′N	74°44′W
Bordertown, Austl.	301	36°19′S	140°46′E
Bordj Bou Arreridj, Alg. (bôrj-bōō-ȧ-rä-rēj′)	292b	36°04′N	4°46′E
Borgå, Fin. *see* Porvoo	208-09	60°24′N	25°40′E
Borgarnes, Ice.	206a	64°34′N	21°55′W
Borger, Tx., U.S. (bôr′gẽr)	130-31	35°40′N	101°24′W
Borgholm, Swe. (bôrg-hōlm′)	208-09	56°52′N	16°40′E
Borgne, Lake, b., La., U.S. (lāk bôrn′y′)	134-35	30°05′N	89°35′W
Borgomanero, Italy (bôr′gō-mä-nâ′rō)	216-17	45°42′N	8°28′E
Borgo Val di Taro, Italy (bô′r-zhō-väl-dē-tá′rō)	216-17	44°29′N	9°46′E
Borisoglebsk, Russia (bŏ-rē sô-glyĕpsk′)	202-03	51°22′N	42°06′E
Borken, Ger. (bōr′kĕn)	210-11	51°51′N	6°52′E
Borkum, i., Ger. (bōr′kōōm)	210-11	53°36′N	6°42′E
Borlänge, Swe. (bôr-lĕng′gĕ)	208-09	60°29′N	15°27′E
Borneo, i., Asia (bôr′nē-ō)	268-69	0°30′N	114°00′E
Bornholm, i., Den. (bôrn-hôlm)	208-09	55°09′N	14°55′E
Boro, stm., S. Sudan	284-85	8°51′N	26°11′E
Borogontsy, Russia	236-37	62°41′N	131°09′E
Borovichi, Russia (bŏ-rō-vē′chē)	218-19	58°23′N	33°55′E
Borroloola, Austl. (bŏr-rŏ-lōō′lȧ)	294-95	16°05′S	136°17′E
Borūjerd, Iran	248-49	33°53′N	48°45′E
Borzna, Ukr. (bôrz′nȧ)	218-19	51°15′N	32°26′E
Borzya, Russia (bôrz′yȧ)	260-61	50°22′N	116°31′E
Bosa, Italy (bō′sä)	216-17	40°18′N	8°30′E
Bosanska Gradiška, Bos. (bŏ′sän-skä grä-dĭsh′kä)	216-17	45°09′N	17°15′E
Bosanski Novi, Bos. (bŏ′s sän-skĭ nō′vē)	216-17	45°04′N	16°23′E
Boscobel, Wi., U.S. (bŏs′kō-bĕl)	124-25	43°08′N	90°42′W
Bose, China (bwo-sŭ)	258-59	23°55′N	106°38′E
Boshan, China (bwo-shan)	260-61	36°29′N	117°51′E
Bosna, stm., Bos.	216-17	45°04′N	18°28′E
Bosna i Hercegovina, nation, Eur. *see* Bosnia and Herzegovina	190-91	44°15′N	17°50′E
Bosnia and Hercegovina, nation, Eur. *see* Bosnia and Herzegovina	190-91	44°15′N	17°50′E
Bosnia and Herzegovina, nation, Eur. (bŏs′nĭ-ȧ ănd hĕr-tsĕ-gō′vē-nȧ)	190-91	44°15′N	17°50′E
Bosporus, strt., Tur. (bŏs′pȧ-rŭs)	216-17	41°06′N	29°04′E
Bossembélé, C.A.R.	284-85	5°16′N	17°39′E
Bossier City, La., U.S. (bŏsh′ẽr sĭ′tĭ)	130-31	32°31′N	93°44′W
Bosso, Dallol, stm., Niger	280-81	12°24′N	2°52′E
Bosten Hu, lk., China (bwo-stŭn hōō)	240-41	42°0′N	87°00′E
Boston, Ma., U.S. (bôs′tŭn)	126-27	42°22′N	71°03′W
Boston Mountains, mts., Ar., U.S. (bôs′tŭn moun′tĭnz)	130-31	35°50′N	93°20′W
Boteti, stm., Bots.	286-87	20°09′S	23°23′E
Bothaville, S. Afr. (bō′tä-vĭl)	292c	27°24′S	26°37′E
Bothnia, Gulf of, b., Eur. (gŭlf ŭv bŏth′nĭ-ȧ)	200-01	63°0′N	20°00′E
Botoşani, Rom. (bŏ-tŏ-shän′ĭ)	210-11	47°45′N	26°40′E
Botswana, nation, Afr. (bŏtswănă)	274	22°0′S	24°00′E
Bottineau, N.D., U.S. (bŏt-ĭ-nō′)	124-25	48°49′N	100°27′W
Bottniska Viken, b., Eur. *see* Bothnia, Gulf of	200-01	63°0′N	20°00′E
Botucatu, Braz.	186	22°52′S	48°27′W
Botwood, Nf./L., Can. (bŏt′wŏŏd)	148-49	49°09′N	55°22′W
Bouaké, C. Iv.	282-83	7°42′N	5°02′W
Bouar, C.A.R. (bōō-är)	284-85	5°57′N	15°36′E
Boufarik, Alg. (bōō-fä-rēk′)	292b	36°35′N	2°54′E
Bougainville, i., Pap. N. Gui. (bōō-găN-vēl′)	306e	6°0′S	155°00′E
Bougie, Alg. *see* Bejaïa	292b	36°45′N	5°04′E
Bouira, Alg. (boo-ē′rȧ)	292b	36°23′N	3°54′E
Boujdour, Cap, c., W. Sah.	280-81	26°08′N	14°29′W
Boulder, Co., U.S. (bōl′dẽr)	128-29	40°02′N	105°15′W
Boulder, Mt., U.S. (bōl′dẽr)	122-23	46°14′N	112°08′W
Boulder, stm., Mt., U.S. (bōl′dẽr)	122-23	45°52′N	111°57′W
Boulder City, Nv., U.S. (bōl′dẽr sĭ′tē)	128-29	35°59′N	114°50′W
Boulia, Austl.	302	22°55′S	139°55′E
Boulogne-Billancourt, Fr. (bōō-lôn′y′-bē-yäN-kōōr′)	212-13	48°51′N	2°15′E
Boulogne-sur-Mer, Fr. (bōō-lôn′y-sür-mâr′)	212-13	50°43′N	1°36′E
Boundary Peak, mtn., Nv., U.S.	128-29	37°51′N	118°21′W
Bountiful, Ut., U.S. (boun′tĭ-fŏŏl)	122-23	40°53′N	111°53′W
Bounty Islands, is., N.Z.	313	47°42′S	179°04′E
Bourg-en-Bresse, Fr. (bōōr-gĕN-brĕs′)	212-13	46°13′N	5°13′E
Bourges, Fr. (bōōrzh)	212-13	47°05′N	2°24′E
Bourke, Austl. (bŭrk)	301	30°06′S	145°56′E
Bournemouth, Eng., U.K. (bôrn′mŭth)	206-07	50°44′N	1°52′W
Bou Saâda, Alg. (bōō-sä′dä)	292b	35°13′N	4°11′E
Boutilimit, Maur.	280-81	17°33′N	14°42′W
Bouvetøya, i., Afr.	310-11	54°26′S	3°24′E
Bow, stm., Ab., Can. (bō)	142-43	49°56′N	111°42′W
Bowbells, N.D., U.S. (bō′bĕls)	124-25	48°48′N	102°15′W
Bowen, Austl. (bō′ĕn)	302	20°01′S	148°14′E
Bowie, Md., U.S. (bōō′ĭ) (bō′ē)	126-27	39°00′N	76°46′W
Bowie, Tx., U.S. (bōō′ĭ) (bō′ē)	130-31	33°34′N	97°51′W
Bowling Green, Ky., U.S. (bōlĭng grēn)	134-35	36°59′N	86°27′W
Bowling Green, Mo., U.S. (bōlĭng grēn)	130-31	39°21′N	91°12′W
Bowling Green, Oh., U.S. (bōlĭng grēn)	126-27	41°22′N	83°39′W
Bowling Green, Va., U.S. (bōlĭng grēn)	126-27	38°03′N	77°21′W
Bowman, N.D., U.S. (bō′măn)	124-25	46°11′N	103°24′W
Bowral, Austl.	301	34°28′S	150°26′E
Bowron, stm., B.C., Can. (bō′rŭn)	142-43	54°03′N	121°50′W
Boxing, China (bwo-shyĭŋ)	260-61	37°08′N	118°07′E
Boyabat, Tur.	246	41°28′N	34°46′E
Boyang, China (bwo-yän)	258-59	28°60′N	116°40′E
Boyle, Ire. (boil)	206-07	53°59′N	8°18′W
Boyoma Falls, wtfl., D.R.C. *see* Stanley Falls	284-85	0°29′N	25°13′E
Boysun, Uzb.	252-53	38°12′N	67°12′E
Bozeman, Mt., U.S. (bōz′măn)	122-23	45°41′N	111°03′W
Bozen, Italy *see* Bolzano	216-17	46°30′N	11°21′E
Bozhen, China (bwo-jŭn)	260-61	38°05′N	116°33′E
Bozhou, China	258-59	33°52′N	115°46′E
Bozüyük, Tur.	246	39°54′N	30°02′E
Bra, Italy (brä)	216-17	44°41′N	7°51′E
Bracebridge, On., Can. (brās′brĭj)	146-47	45°02′N	79°18′W
Brackettville, Tx., U.S. (brăk′ĕt-vĭl)	132-33	29°18′N	100°25′W
Bradano, stm., Italy (brä-dä′nō)	216-17	40°23′N	16°51′E
Bradenton, Fl., U.S. (brā′dĕn-tŭn)	135a	27°30′N	82°33′W
Bradford, Eng., U.K. (brăd′fẽrd)	206-07	53°48′N	1°45′W
Bradley, Il., U.S. (brăd′lĭ)	126-27	41°00′N	87°51′W
Brady, Tx., U.S. (brā′dĭ)	132-33	31°08′N	99°20′W
Braga, Port. (brä′gä)	214-15	41°33′N	8°26′W
Bragado, Arg. (brä-gä′dō)	187	35°08′S	60°30′W
Bragança, Braz. (brä-gän′sä)	180-81	1°03′S	46°46′W
Bragança, Port.	214-15	41°49′N	6°45′W
Brāhmanbāria, Bngl.	254-55	23°59′N	91°07′E
Brāhmani, stm., India	254-55	20°47′N	87°01′E
Brahmapur, India	256	19°18′N	84°49′E
Brahmaputra, stm., Asia (brä′mȧ-pōō′trȧ)	254-55	24°02′N	91°00′E
Braidwood, Il., U.S. (brād′wŏŏd)	126-27	41°16′N	88°12′W
Brăila, Rom. (brē′ēlȧ)	218-19	45°16′N	27°58′E
Brainerd, Mn., U.S. (brän′ẽrd)	124-25	46°21′N	94°12′W
Brampton, On., Can. (brămp′tŭn)	146-47	43°42′N	79°45′W
Branco, stm., Braz. (bräŋ′kō)	178-79	1°24′S	61°52′W
Brandberg, mtn., Nmb.	286-87	21°10′S	14°33′E
Brandenburg, Ger. (brän′dĕn-bŏŏrgh)	210-11	52°25′N	12°33′E
Brandfort, S. Afr. (brăn′d-fôrt)	292c	28°42′S	26°28′E
Brandon, Mb., Can. (brăn′dŭn)	144-45	49°50′N	99°58′W
Brandon, Ms., U.S. (brăn′dŭn)	134-35	32°16′N	89°59′W
Brandon, S.D., U.S. (brăn′dŭn)	124-25	43°36′N	96°34′W
Brandon, Vt., U.S. (brăn′dŭn)	126-27	43°48′N	73°05′W
Braniewo, Pol. (brä-nyĕ′vō)	210-11	54°23′N	19°50′E
Brantford, On., Can. (brănt′fẽrd)	146-47	43°09′N	80°15′W
Bras d'Or Lake, lk., N.S., Can. (brä-dôr′ lāk)	148-49	45°52′N	60°50′W
Brasil, nation, S.A. *see* Brazil	172	10°0′S	55°00′W
Brasiléia, Braz.	180-81	10°60′S	68°45′W
Brasília, nat. cap., Braz. (brä-sē′lvä)	182-83	15°48′S	47°53′W
Braşov, Rom.	210-11	45°39′N	25°37′E
Brass, Nig. (brăs)	282a	4°19′N	6°15′E
Brassó, Rom. *see* Braşov	210-11	45°39′N	25°37′E
Bratislava, nat. cap., Slvk. (brä′tĭs-lä-vä)	210-11	48°09′N	17°07′E
Bratsk, Russia (brätsk)	236-37	56°08′N	101°39′E
Bratskoye Vodokhranilishche, res., Russia	236-37	55°57′N	101°52′E
Bratsk Reservoir, res., Russia (brätsk rĕ′sẽr-vwär) *see* Bratskoye Vodokhranilishche	236-37	55°57′N	101°52′E
Bratslav, Ukr. (brät′släf)	218-19	48°49′N	28°57′E
Brattleboro, Vt., U.S. (brăt′l-bŭr-ō)	126-27	42°51′N	72°34′W
Braunschweig, Ger. (broun′shvīgh)	210-11	52°16′N	10°31′E
Brava, i., C.V.	282-83	14°52′N	24°43′W
Bravo, stm., N.A. *see* Rio Grande	120-21	25°57′N	97°09′W
Bravo del Norte, stm., N.A. *see* Rio Grande	120-21	25°57′N	97°09′W
Brawley, Ca., U.S. (brô′lĭ)	128-29	32°59′N	115°33′W
Brazeau, stm., Ab., Can.	142-43	52°55′N	115°14′W
Brazeau, Mount, mtn., Ab., Can. (mount brȧ-zō′)	142-43	52°33′N	117°21′W
Brazil, In., U.S. (brȧ-zĭl′)	126-27	39°31′N	87°07′W
Brazil, nation, S.A. (brȧ-zĭl′)	172	10°0′S	55°00′W
Brazos, stm., Tx., U.S. (brä′zōs)	132-33	33°15′N	100°00′W
Brazos, Salt Fork, stm., U.S. (sôlt fôrk)	130-31	33°16′N	100°01′W
Brazzaville, nat. cap., Congo (brȧ-zȧ-vēl′)	284-85	4°16′S	15°17′E
Brčko, Bos. (bĕrch′kō)	216-17	44°52′N	18°49′E
Breckenridge, Mn., U.S. (brĕk′ĕn-rĭj)	124-25	46°15′N	96°34′W
Breckenridge, Tx., U.S. (brĕk′ĕn-rĭj)	130-31	32°45′N	98°55′W
Břeclav, Czech Rep. (brzhĕl′láf)	210-11	48°46′N	16°54′E
Breda, Neth. (brā-dä′)	206-07	51°35′N	4°46′E
Bregenz, Aus. (brā′gĕnts)	210-11	47°30′N	9°46′E
Bregovo, Blg. (brĕ′gō-vō)	216-17	44°10′N	22°39′E
Breiðafjördur, b., Ice.	206a	65°15′N	23°15′W
Brejo, Braz. (brā′zhō)	180-81	3°41′S	42°47′W
Bremen, Ger. (brā-mĕn)	210-11	53°04′N	8°51′E
Bremen, Ga., U.S. (brē′mĕn)	134-35	33°43′N	85°09′W
Bremen, In., U.S. (brē′mĕn)	126-27	41°27′N	86°08′W
Bremerhaven, Ger. (brām ẽr hä′fĕn)	210-11	53°32′N	8°36′E
Bremerton, Wa., U.S. (brĕm′ẽr-tŭn)	122-23	47°34′N	122°39′W
Brenham, Tx., U.S. (brĕn′ăm)	132-33	30°10′N	96°24′W
Brentwood, N.Y., U.S. (brĕnt′wŏŏd)	126-27	40°47′N	73°15′W
Brentwood, Tn., U.S. (brĕnt wŏŏd)	134-35	36°02′N	86°47′W
Brescia, Italy (brā′shä)	216-17	45°33′N	10°13′E
Breslau, Pol. *see* Wrocław	210-11	51°07′N	17°02′E
Bressanone, Italy (brĕs-sä-nō′nā)	216-17	46°43′N	11°39′E
Bressuire, Fr. (grĕ-swēr′)	212-13	46°50′N	0°29′W
Brest, Bela.	210-11	52°07′N	23°42′E
Brest, Fr. (brĕst)	212-13	48°24′N	4°30′W
Bretagne, hist. reg., Fr. (brĕ-tän′yĕ) *see* Brittany	212-13	48°0′N	3°00′W
Breton Sound, strt., La., U.S. (brĕt′ŭn sound)	134-35	29°34′N	89°16′W
Brevard, N.C., U.S. (brē värd′)	134-35	35°14′N	82°44′W
Breves, Braz. (brā′vĕzh)	180-81	1°40′S	50°29′W
Brewarrina, Austl. (brōō-ẽr-rē′nȧ)	301	29°57′S	146°52′E
Brewster, Wa., U.S. (brōō′stẽr)	122-23	48°06′N	119°47′W
Brewster, Kap, c., Green.	95	70°09′N	22°06′W
Brewton, Al., U.S. (brōō′tŭn)	134-35	31°07′N	87°05′W
Brezhnev, Russia *see* Naberezhnye Chelny	202-03	55°42′N	52°19′E
Bria, C.A.R.	284-85	6°33′N	21°58′E
Briançon, Fr. (brē-äN-sôN′)	212-13	44°54′N	6°37′E
Bridgeport, Al., U.S. (brĭj′pôrt)	134-35	34°57′N	85°43′W
Bridgeport, Ca., U.S. (brĭj′pôrt)	128-29	38°16′N	119°14′W
Bridgeport, Ct., U.S. (brĭj′pôrt)	126-27	41°11′N	73°14′W
Bridgeport, Il., U.S. (brĭj′pôrt)	126-27	38°42′N	87°46′W
Bridgeport, Ne., U.S. (brĭj′pôrt)	124-25	41°40′N	103°05′W
Bridgeport, Tx., U.S. (brĭj′pôrt)	130-31	33°13′N	97°46′W
Bridgetown, Austl.	294-95	33°58′S	116°08′E
Bridgetown, N.S., Can. (brĭj′ toun)	148-49	44°52′N	65°16′W
Bridgetown, nat. cap., Barb. (brĭj′ toun)	155b	13°06′N	59°37′W
Bridgewater, N.S., Can.	148-49	44°22′N	64°31′W
Bridgton, Me., U.S. (brĭj′tŭn)	126-27	44°04′N	70°42′W
Bridlington, Eng., U.K. (brĭd′lĭng-tŭn)	206-07	54°05′N	0°12′W
Brig, Switz. (brēg)	210-11	46°19′N	8°00′E
Brigham City, Ut., U.S. (brĭg′ăm sĭ′tē)	122-23	41°31′N	112°01′W
Bright, Austl. (brīt)	301	36°44′S	146°58′E
Brighton, Eng., U.K. (brīt′ŭn)	206-07	50°50′N	0°08′W
Brighton, Co., U.S. (brīt′ŭn)	130-31	39°59′N	104°49′W
Brighton, N.Y., U.S. (brīt′ŭn)	126-27	43°09′N	77°33′W
Brindisi, Italy (brēn′dē-zē)	216-17	40°38′N	17°56′E
Brinkley, Ar., U.S. (brĭŋk′lĭ)	134-35	34°54′N	91°11′W
Brioude, Fr. (brē-ōōd′)	212-13	45°18′N	3°23′E
Brisbane, Austl. (brĭz′băn)	301	27°28′S	153°02′E
Bristol, Eng., U.K.	206-07	51°27′N	2°36′W
Bristol, Ct., U.S. (brĭs′tŭl)	126-27	41°41′N	72°57′W
Bristol, R.I., U.S. (brĭs′tŭl)	126-27	41°40′N	71°16′W

340 Bristol – Buynaksk PRONOUNCING INDEX

Feature (Pronunciation)	Page	Lat.	Long.
Bristol, Tn., U.S. (brĭs´tŭl)	134-35	36°35′N	82°11′W
Bristol, Va., U.S. (brĭs´tŭl)	134-35	36°36′N	82°11′W
Bristol Bay, b., Ak., U.S. (brĭs´tŭl bā)	136	58°0′N	159°00′W
Bristol Channel, strt., U.K.	206-07	51°23′N	4°01′W
Bristow, Ok., U.S. (brĭs´tō)	130-31	35°50′N	96°24′W
British Columbia, state, Can. (brĭt´ĭsh kŏl´ŭm-bĭ-*ȧ*)	138-39	54°0′N	125°00′W
British Guiana, nation, S.A. *see* Guyana	172	5°0′N	59°00′W
British Honduras, nation, N.A. *see* Belize	94	17°15′N	88°45′W
British Indian Ocean Territory, dep., Afr.	224-25	7°0′S	72°00′E
British Solomon Islands, nation, Oc. *see* Solomon Islands	306e	8°0′S	159°00′E
British Virgin Islands, dep., N.A.	152-53	18°30′N	64°30′W
Britt, Ia., U.S. (brĭt)	124-25	43°06′N	93°49′W
Brittany, hist. reg., Fr.	212-13	48°0′N	3°00′W
Britton, S.D., U.S. (brĭt´*ŭ*n)	124-25	45°48′N	97°45′W
Brive-la-Gaillarde, Fr. (brēv-lä-gī-yärd´*ĕ*)	212-13	45°09′N	1°32′E
Brixen, Italy *see* Bressanone	216-17	46°43′N	11°39′E
Brno, Czech Rep. (b'r´nô)	210-11	49°12′N	16°37′E
Brockport, N.Y., U.S. (brŏk´pōrt)	126-27	43°13′N	77°56′W
Brockton, Ma., U.S. (brŏk´t*ŭ*n)	126-27	42°05′N	71°01′W
Brockville, On., Can. (brŏk´vĭl)	146-47	44°36′N	75°41′W
Brodnica, Pol. (brôd´nĭt-sȧ)	210-11	53°15′N	19°24′E
Brody, Ukr. (brô´dĭ)	210-11	50°05′N	25°10′E
Broken Arrow, Ok., U.S. (brō´k*ĕ*n ăr´ō)	130-31	36°03′N	95°47′W
Broken Bow, Ne., U.S. (brō´k*ĕ*n bō)	124-25	41°24′N	99°39′W
Broken Bow, Ok., U.S. (brō´k*ĕ*n bō)	130-31	34°02′N	94°44′W
Broken Hill, Austl. (brōk´*ĕ*n hĭl)	301	31°58′S	141°27′E
Broken Hill, Zam. *see* Kabwe	286-87	14°27′S	28°27′E
Brokopondo, Sur.	178-79	5°04′N	54°59′W
Brokopondo Stuwmeer, res., Sur. *see* W.J. van Blommestein Meer	178-79	4°49′N	55°04′W
Bromberg, Pol. *see* Bydgoszcz	210-11	53°07′N	18°01′E
Bronlund Peak, mtn., B.C., Can.	140-41	57°26′N	126°38′W
Brookfield, Mo., U.S. (brŏŏk´fēld)	130-31	39°48′N	93°05′W
Brookfield, Wi., U.S. (brŏŏk´fēld)	126-27	43°04′N	88°07′W
Brookhaven, Ms., U.S. (brŏŏk´hāv'n)	134-35	31°34′N	90°27′W
Brookings, Or., U.S. (brŏŏk´ĭngs)	122-23	42°04′N	124°17′W
Brookings, S.D., U.S. (brŏŏk´ĭngs)	124-25	44°19′N	96°48′W
Brooklyn Park, Mn., U.S. (brŏŏk´lĭn pärk)	124-25	45°07′N	93°20′W
Brooks, Ab., Can. (brŏŏks)	142-43	50°34′N	111°54′W
Brooks Range, mts., Ak., U.S. (brŏŏks rānj)	136	68°0′N	154°00′W
Brooksville, Fl., U.S. (brŏŏks´vĭl)	135a	28°33′N	82°24′W
Brookton, Austl.	294-95	32°22′S	117°01′E
Broome, Austl. (brōōm)	294-95	18°00′S	122°14′E
Brownfield, Tx., U.S. (broun´fēld)	130-31	33°11′N	102°16′W
Browning, Mt., U.S. (broun´ĭng)	122-23	48°33′N	112°60′W
Brownstown, In., U.S. (brounz´toun)	126-27	38°53′N	86°03′W
Brownsville, Tn., U.S. (brounz´vĭl)	134-35	35°36′N	89°16′W
Brownsville, Tx., U.S. (brounz´vĭl)	132-33	25°56′N	97°29′W
Brownwood, Tx., U.S. (broun´wŏŏd)	132-33	31°43′N	98°59′W
Bruce, Mount, mtn., Austl. (mount brōōs)	296-97	22°35′S	118°08′E
Bruchsal, Ger. (brŏŏk´zäl)	210-11	49°08′N	8°36′E
Bruit, Pulau, i., Malay.	268-69	2°35′N	111°20′E
Bruneau, stm., U.S. (brōō-nō´)	122-23	42°57′N	115°57′W
Brunei, nation, Asia (brŏŏ-nī´)	224-25	4°30′N	114°40′E
Brunei, nat. cap., Bru. *see* Bandar Seri Begawan	268-69	4°56′N	114°56′E
Brünn, Czech Rep. *see* Brno	210-11	49°12′N	16°37′E
Brunswick, Ger. *see* Braunschweig	210-11	52°16′N	10°31′E
Brunswick, Ga., U.S. (brŭnz´wĭk)	134-35	31°11′N	81°30′W
Brunswick, Md., U.S. (brŭnz´wĭk)	126-27	39°19′N	77°38′W
Brunswick, Me., U.S. (brŭnz´wĭk)	127a	43°55′N	69°58′W
Brunswick, Península, pen., Chile	185	53°11′S	71°11′W
Brush, Co., U.S. (brŭsh)	130-31	40°15′N	103°38′W
Brusque, Braz. (brōō´s-kōōĕ)	186	27°07′S	48°56′W
Brussel, nat. cap., Bel. *see* Brussels	206-07	50°50′N	4°22′E
Brussels, nat. cap., Bel. (brŭs´*ĕ*ls)	206-07	50°50′N	4°22′E
Brüx, Czech Rep. *see* Most	210-11	50°31′N	13°39′E
Bruxelles, nat. cap., Bel. (brü-sĕl´) *see* Brussels	206-07	50°50′N	4°22′E
Bryan, Oh., U.S. (brī´*ă*n)	126-27	41°28′N	84°33′W
Bryan, Tx., U.S. (brī´*ă*n)	132-33	30°41′N	96°23′W
Bryansk, Russia	218-19	53°14′N	34°22′E
Bryce Canyon National Park, n.p., Ut., U.S. (brīs kăn´y*ŭ*n năsh´ŭn-*ă*l pärk)	128-29	37°29′N	112°15′W
Bryson City, N.C., U.S. (brĭs´*ŭ*n sĭ´tē)	134-35	35°26′N	83°27′W
Bryukhovetskaya, Russia (b'ryŭk´ô-vyĕt-skä´yä)	218-19	45°49′N	39°00′E
Bua Yai, Thai.	266-67	15°35′N	102°26′E
Būbiyān, i., Kuw.	250-51	29°45′N	48°15′E
Bucak, Tur.	246	37°28′N	30°36′E
Bucaramanga, Col. (bōō-kä´rä-män´gä)	178-79	7°03′N	73°05′W
Buchach, Ukr. (bŏŏ´chȧch)	210-11	49°04′N	25°25′E
Buchanan, Lib. (bŭ-kăn´*ă*n)	282-83	5°53′N	10°02′W
Buchanan, Mi., U.S. (bû-kăn´*ă*n)	126-27	41°49′N	86°22′W
Buchanan, Va., U.S. (bû-kăn´*ă*n)	126-27	37°31′N	79°41′W
Buchanan, Lake, lk., Tx., U.S. (lāk bû-kăn´*ă*n)	132-33	30°48′N	98°25′W
Buchans, Nf./L., Can.	148-49	48°49′N	56°52′W
Bucharest, nat. cap., Rom. (bōō-kȧ-rĕst´)	216-17	44°26′N	26°06′E
Buckhannon, W.V., U.S. (bŭk-hăn´*ŭ*n)	126-27	38°59′N	80°14′W
Buckhaven, Scot., U.K. (bŭk-hā´v'n)	206-07	56°11′N	3°03′W
Buckie, Scot., U.K. (bŭk´ĭ)	206-07	57°40′N	2°59′W
Bucksport, Me., U.S. (bŭks´pôrt)	127a	44°34′N	68°48′W
București, nat. cap., Rom. (bōō-kŏŏ-rĕsh´tĭ) *see* Bucharest	216-17	44°26′N	26°06′E
Bucyrus, Oh., U.S. (bû-sī´r*ŭ*s)	126-27	40°48′N	82°58′W
Budapest, nat. cap., Hung. (bōō´dȧ-pĕsht´)	210-11	47°30′N	19°05′E
Budaun, India	254-55	28°02′N	79°08′E
Budennovsk, Russia	202-03	44°47′N	44°09′E
Budweis, Czech Rep. *see* České Budějovice	210-11	48°59′N	14°28′E
Buena Esperanza, Arg.	185	34°45′S	65°16′W
Buenaventura, Col. (bwā´nä-vĕn-tōō´rȧ)	188c	3°53′N	77°04′W
Buenaventura, Mex.	156-57	29°51′N	107°28′W
Buena Vista, Bol.	182-83	17°27′S	63°40′W
Buena Vista, Co., U.S. (bū´n*ȧ* vĭs´t*ȧ*)	128-29	38°50′N	106°09′W
Buena Vista, Va., U.S. (bū´n*ȧ* vĭs´t*ȧ*)	126-27	37°44′N	79°21′W
Buenos Aires, state, Arg. (bwā´nōs ī´rās)	187	36°0′S	60°00′W
Buenos Aires, nat. cap., Arg. (bwā´nōs ī´rās)	187	34°37′S	58°23′W
Buenos Aires, Lago, lk., S.A. (lä´gō-bwā´nōs ī´rās)	185	46°26′S	71°40′W
Buffalo, Mn., U.S. (buf´*ȧ*-lō)	124-25	45°11′N	93°53′W
Buffalo, Mo., U.S. (buf´*ȧ*-lō)	130-31	37°39′N	93°06′W
Buffalo, N.Y., U.S. (buf´*ȧ*-lō)	126-27	42°53′N	78°52′W
Buffalo, Ok., U.S. (buf´*ȧ*-lō)	130-31	36°50′N	99°38′W
Buffalo, Tx., U.S. (buf´*ȧ*-lō)	132-33	31°27′N	96°04′W
Buffalo, Wy., U.S. (buf´*ȧ*-lō)	122-23	44°22′N	106°42′W
Buffalo, stm., Tn., U.S. (buf´*ȧ*-lō)	134-35	35°60′N	87°50′W
Buffalo Lake, lk., N.T., Can.	140-41	60°10′N	115°30′W
Buford, Ga., U.S. (bū´fērd)	134-35	34°07′N	84°00′W
Buga, Col. (bōō´gä)	188c	3°54′N	76°18′W
Bugojno, Bos. (bŏŏ-gō´ĭ nô)	216-17	44°03′N	17°27′E
Bugsuk Island, i., Phil.	270	8°15′N	117°18′E
Bugt, China	260-61	48°46′N	121°54′E
Bugul'ma, Russia (bŏŏ-gŏŏl´mȧ)	202-03	54°31′N	52°47′E
Buguma, Nig.	282a	4°43′N	6°53′E
Buguruslan, Russia (bŏŏ-gŏŏ-rŏŏs-lán´)	202-03	53°39′N	52°27′E
Buhl, Id., U.S. (būl)	122-23	42°37′N	114°46′W
Buin, Chile (bŏŏ-ēn´)	188e	33°42′S	70°43′W
Buir Nur, lk., Asia (bōō-ēr nōōr)	260-61	47°48′N	117°42′E
Buitenzorg, Indon. *see* Bogor	268-69	6°35′S	106°47′E
Bujumbura, nat. cap., Bdi. (bōō-jŭm-bōō´r*ȧ*)	289	3°23′S	29°22′E
Buka Island, i., Pap. N. Gui.	306e	5°15′S	154°35′E
Bukama, D.R.C. (bōō-kä´mä)	284-85	9°12′S	25°51′E
Bukavu, D.R.C.	289	2°30′S	28°51′E
Bukhara, Uzb. (bŏŏ-kä´rä) *see* Buxoro	252-53	39°46′N	64°26′E
Bukittinggi, Indon.	266-67	0°18′S	100°22′E
Bukoba, Tan.	289	1°19′S	31°48′E
Bulan, Phil.	270	12°40′N	123°53′E
Bulancak, Tur.	247	40°56′N	38°14′E
Bulawayo, Zimb. (bōō-lȧ-wä´yō)	286-87	20°10′S	28°35′E
Bulgan, Mong.	260-61	48°49′N	103°33′E
Bulgaria, nation, Eur. (bŏŏl-gā´rĭ-ă)	190-91	43°0′N	25°00′E
Bŭlgariya, nation, Eur. *see* Bulgaria	190-91	43°0′N	25°00′E
Bulkley Ranges, mts., B.C., Can. (bŭlk´lē rānj*ĕ*z)	142-43	54°30′N	127°30′W
Bulloo, stm., Austl.	296-97	28°40′S	142°31′E
Bull Shoals Lake, res., U.S. (bŏŏl shōlz lāk)	130-31	36°29′N	92°47′W
Bultfontein, S. Afr. (bŏŏlt´fōn-tān´)	292c	28°17′S	26°09′E
Bulungu, D.R.C. (bōō-lŏŏŋ´gōō)	284-85	6°03′S	21°53′E
Bumba, D.R.C. (bŏŏm´bȧ)	284-85	2°11′N	22°28′E
Bumbire Island, i., Tan.	289	1°39′S	31°53′E
Bunbury, Austl. (bŭn´b*ŭ*rĭ)	294-95	33°19′S	115°38′E
Bundaberg, Austl. (bŭn´d*ȧ*-bûrg)	302	24°52′S	152°20′E
Būndi, India	254-55	25°27′N	75°38′E
Bungo-suidō, strt., Japan	265	33°0′N	132°13′E
Bunia, D.R.C.	289	1°32′N	30°15′E
Bunkie, La., U.S. (bŭŋ´kĭ)	134-35	30°57′N	92°11′W
Buntok, Indon.	268-69	1°44′S	114°50′E
Buon Ma Thuot, Viet.	266-67	12°40′N	108°03′E
Buqayq, Sau. Ar.	250-51	25°56′N	49°40′E
Burang, China	254-55	30°14′N	81°11′E
Buraydah, Sau. Ar.	288	26°19′N	43°59′E
Burayk, Libya	280-81	26°37′N	13°07′E
Burbank, Ca., U.S. (bûr´bănk)	128-29	34°11′N	118°19′W
Burco, Som.	284-85	9°32′N	45°33′E
Burdur, Tur. (bōōr-dŏŏr´)	202-03	37°43′N	30°17′E
Bureya, stm., Russia (bŏŏ-rā´yä)	260-61	49°25′N	129°32′E
Burgas, Blg. (bŏŏr-gäs´)	216-17	42°31′N	27°28′E
Burgaw, N.C., U.S. (bûr´gô)	134-35	34°33′N	77°56′W
Burgos, Spain (bōō´r-gôs)	214-15	42°21′N	3°42′W
Burgsvik, Swe. (bŏŏrgs´vĭk)	208-09	57°02′N	18°18′E
Burhānpur, India (bŏŏr´hän-pōōr)	254-55	21°18′N	76°14′E
Burias Island, i., Phil. (bōō´rē-äs ī´l*ă*nd)	270	12°57′N	123°08′E
Burica, Punta, c., N.A. (pōō´n-tä-bōō´rē-kä)	162	8°03′N	82°52′W
Burin, Nf./L., Can. (bûr´ĭn)	148-49	47°02′N	55°11′W
Burkburnett, Tx., U.S. (bûrk-bûr´nĕt)	130-31	34°06′N	98°34′W
Burke, stm., Austl.	302	23°12′S	139°34′E
Burketown, Austl. (bûrk´toun)	302	17°44′S	139°33′E
Burkina Faso, nation, Afr. (bōōr-kē´-nȧ fä´sō)	274	13°0′N	1°30′W
Burley, Id., U.S. (bûr´lĭ)	122-23	42°33′N	113°47′W
Burlington, On., Can. (bûr´lĭng-t*ŭ*n)	146-47	43°19′N	79°48′W
Burlington, Co., U.S. (bûr´lĭng-tŭn)	130-31	39°18′N	102°16′W
Burlington, Ia., U.S. (bûr´lĭng-t*ŭ*n)	124-25	40°48′N	91°06′W
Burlington, N.C., U.S. (bûr´lĭng-t*ŭ*n)	134-35	36°06′N	79°26′W
Burlington, N.D., U.S. (bûr´lĭng-t*ŭ*n)	124-25	48°16′N	101°25′W
Burlington, Vt., U.S.	126-27	44°29′N	73°12′W
Burlington, Wi., U.S. (bûr´lĭng-t*ŭ*n)	126-27	42°41′N	88°16′W
Burma, nation, Asia *see* Myanmar	224-25	22°0′N	98°00′E
Burnie, Austl. (bûr´nē)	301	41°04′S	145°54′E
Burnley, Eng., U.K. (bûrn´lē)	206-07	53°48′N	2°15′W
Burns, Or., U.S. (bûrnz)	122-23	43°36′N	119°03′W
Burnside, stm., Nu., Can.	140-41	66°51′N	108°12′W
Burns Lake, B.C., Can. (bûrnz lāk)	142-43	54°14′N	125°46′W
Burntwood, stm., Mb., Can.	144-45	56°08′N	96°20′W
Burqin, China	244	47°43′N	86°54′E
Burra, Austl.	301	33°40′S	138°55′E
Bursa, Tur. (bōōr´sȧ)	216-17	40°12′N	29°04′E
Bûr Sa'îd, Egypt *see* Port Said	291b	31°16′N	32°18′E
Būr Sūdān, Sudan *see* Port Sudan	288	19°37′N	37°13′E
Burton, Mi., U.S. (bûr´t*ŭ*n)	126-27	43°00′N	83°35′W
Burton upon Trent, Eng., U.K. (bŭr´t*ŭ*n-ŭp´-ŏn-trĕnt)	206-07	52°49′N	1°38′W
Buru, i., Indon.	268-69	3°24′S	126°40′E
Burundi, nation, Afr. (bŭ-rŭn´-dē)	274	3°15′S	30°00′E
Burun-Shibertuy, Gora, mtn., Russia	260-61	49°42′N	109°58′E
Burwell, Ne., U.S. (bûr´w*ĕ*l)	124-25	41°46′N	99°08′W
Buryatia, state, Russia	240-41	53°0′N	109°00′E
Buryatiya, state, Russia *see* Buryatia	240-41	53°0′N	109°00′E
Bury Saint Edmunds, Eng., U.K. (bĕr´ĭ-sānt ĕd´m*ŭ*ndz)	206-07	52°15′N	0°42′E
Busan, Kor., S.	263	35°05′N	129°03′E
Būshehr, Iran	250-51	28°58′N	50°51′E
Bushire, Iran *see* Būshehr	250-51	28°58′N	50°51′E
Bushnell, Il., U.S. (bŏŏsh´nĕl)	130-31	40°33′N	90°30′W
Businga, D.R.C. (bŏŏ-sĭŋ´gȧ)	284-85	3°20′N	20°53′E
Busselton, Austl. (bûs´l-t*ŭ*n)	294-95	33°39′S	115°21′E
Busto Arsizio, Italy (bōōs´tō är-sēd´zē-ō)	216-17	45°37′N	8°51′E
Busuanga Island, i., Phil. (bōō-swäŋ´gä ī´l*ă*nd)	270	12°05′N	120°05′E
Buta, D.R.C. (bōō´tȧ)	284-85	2°49′N	24°45′E
Butare, Rw.	289	2°36′S	29°44′E
Butaritari, at., Kir.	304-05	3°06′N	186°50′E
Bute Inlet, b., B.C., Can.	142-43	50°37′N	124°53′W
Butembo, D.R.C.	289	0°08′N	29°18′E
Butere, Kenya	289	0°13′N	34°30′E
Butha-Buthe, Leso. (bōō-thȧ-bōō´thȧ)	292c	28°45′S	28°16′E
Butha Qi, China *see* Zalantun	260-61	47°60′N	122°45′E
Butler, In., U.S. (bŭt´lēr)	126-27	41°25′N	84°52′W
Butler, Mo., U.S. (bŭt´lēr)	130-31	38°15′N	94°20′W
Butler, Pa., U.S.	126-27	40°51′N	79°54′W
Buton, Pulau, i., Indon.	268-69	5°02′S	122°53′E
Butte, Mt., U.S. (būt)	122-23	45°60′N	112°32′W
Butterworth, Malay.	266-67	5°24′N	100°23′E
Butuan, Phil. (bōō-tōō´än)	270	8°57′N	125°32′E
Buṭwal, Nepal	254-55	27°43′N	83°28′E
Buxoro, Uzb.	252-53	39°46′N	64°26′E
Buy, Russia (bwē)	202-03	58°29′N	41°33′E
Buyant-Uhaa, Mong.	260-61	44°55′N	110°09′E
Buynaksk, Russia	245	42°50′N	47°06′E

ăt; fin*ă*l; rāte; senåte; ärm; åsk; sof*ȧ*; fâre; ch-choose; dh-as th in other; bē; ēvent; bĕt; recĕnt; cratēr; g-gō; gh-guttural g; bĭt; ĭ-short neutral; rīde; K-guttural k as ch in German ich;

PRONOUNCING INDEX Buyr nuur – Canastota 341

Feature (Pronunciation)	Page	Lat.	Long.
Buyr nuur, lk., Asia *see* Buir Nur	**260-61**	47°48′N	117°42′E
Büyük Ağrı Dağı, vol., Tur. *see* Ararat, Mount	**245**	39°42′N	44°18′E
Buzău, Rom. (bo͞o-zĕ′o̊)	**218-19**	45°09′N	26°50′E
Búzi, stm., Moz.	**286-87**	19°53′S	34°45′E
Buzuluk, Russia (bo̊-zo̊-lo̊k′)	**202-03**	52°47′N	52°15′E
Byala Slatina, Blg. (byä′la slä′tēnä)	**216-17**	43°28′N	23°58′E
Byblos, Leb. *see* Jubayl	**248-49**	34°08′N	35°40′E
Bydgoszcz, Pol. (bĭd′go̊shch)	**210-11**	53°07′N	18°01′E
Byelorussia, nation, Eur. *see* Belarus	**190-91**	53°50′N	28°00′E
Bytantay, stm., Russia (byän′täy)	**236-37**	68°45′N	134°27′E
Bytom, Pol. (bĭ′tŭm)	**210-11**	50°21′N	18°55′E
Byumba, Rw.	**289**	1°36′S	30°04′E
Byzantium, Tur. *see* İstanbul	**216-17**	41°02′N	28°59′E
C			
Ca, stm., Asia	**266-67**	18°44′N	105°45′E
Caacupé, Para.	**182-83**	25°22′S	57°08′W
Caála, Ang.	**286-87**	12°51′S	15°33′E
Caazapá, Para.	**182-83**	26°11′S	56°22′W
Cabanatuan, Phil. (kä-bä-nä-twän′)	**270**	15°29′N	120°59′E
Cabano, Qc., Can. (kä-bä-nō′)	**148-49**	47°41′N	68°53′W
Cabedelo, Braz. (kä-bē-dā′lo̊)	**188d**	6°58′S	34°50′W
Cabeza del Buey, Spain (kȧ-bā′thä dĕl bwā′)	**214-15**	38°43′N	5°13′W
Cabimas, Ven. (kä-bē̊′mäs)	**178-79**	10°24′N	71°26′W
Cabinda, Ang. (kä-bĭn′dȧ)	**282-83**	5°33′S	12°12′E
Cabinet Mountains, mts., U.S. (kăb′ĭ-nĕt moun′tĭnz)	**122-23**	48°19′N	116°12′W
Cabo, Braz.	**188d**	8°17′S	35°02′W
Cabo Frio, Braz. (kȧ′bō-frē′ō)	**186**	22°53′S	42°02′W
Cabonga, Réservoir, res., Qc., Can.	**146-47**	47°17′N	76°33′W
Caborca, Mex.	**156-57**	30°43′N	112°09′W
Cabot Strait, strt., Can. (kăb′ŭt strāt)	**148-49**	47°20′N	59°30′W
Cabo Verde, nation, Afr.	**274**	16°0′N	24°00′W
Cabra, Spain (käb′rä)	**214-15**	37°29′N	4°27′W
Cabrera, Illa de, i., Spain	**214-15**	39°09′N	2°57′E
Cabrera, Isla de, i., Spain *see* Cabrera, Illa de	**214-15**	39°09′N	2°57′E
Cabriel, stm., Spain (kä-brē-ĕl′)	**214-15**	39°14′N	1°03′W
Caçador, Braz.	**182-83**	26°47′S	51°01′W
Čačak, Serb. (chä′chȧk)	**216-17**	43°54′N	20°21′E
Cáceres, Braz. (kȧ′sĕ-rĕs)	**182-83**	16°04′S	57°42′W
Cáceres, Spain (kȧ′thȧ-rās)	**214-15**	39°28′N	6°22′W
Cache, stm., Ar., U.S. (kȧsh)	**134-35**	34°42′N	91°20′W
Cache Creek, B.C., Can. (kăsh krēk)	**142-43**	50°49′N	121°19′W
Cachimbo, Serra do, mts., Braz.	**180-81**	8°25′S	55°45′W
Cachoeira do Sul, Braz. (ka-sho̊-a′ra-do̊-soo′l)	**187**	30°02′S	52°54′W
Cachoeiras de Macacu, Braz. (kä-sho̊-ā′räs-dĕ-mä-kȧ′ko͞o)	**186**	22°28′S	42°39′W
Cachoeiro de Itapemirim, Braz.	**186**	20°51′S	41°08′W
Cadereyta Jiménez, Mex. (kä-dä-rā′tä hē-mā′nāz)	**132-33**	25°35′N	99°60′W
Cadillac, Mi., U.S. (kăd′ĭ-lăk)	**126-27**	44°15′N	85°24′W
Cádiz, Spain (kȧ′dēz)	**214-15**	36°31′N	6°17′W
Cadiz, Ky., U.S. (kā′dĭz)	**134-35**	36°52′N	87°50′W
Cadiz, Oh., U.S. (kā′dĭz)	**126-27**	40°16′N	80°60′W
Cádiz, Golfo de, b., Eur. (go̊l-fō-dĕ-kȧ′dēz) *see* Cadiz, Gulf of	**214-15**	36°50′N	7°10′W
Cadiz, Gulf of, b., Eur. (gŭlf ŭv kā′dĭz)	**214-15**	36°50′N	7°10′W
Caen, Fr. (käN)	**212-13**	49°11′N	0°21′W
Caetité, Braz.	**180-81**	14°04′S	42°29′W
Cafayate, Arg.	**182-83**	26°04′S	65°59′W
Cagayan, stm., Phil.	**270**	18°22′N	121°37′E
Cagayan de Oro, Phil.	**270**	8°29′N	124°38′E
Cagayan Islands, is., Phil.	**270**	9°40′N	121°16′E
Cagayan Sulu Island, i., Phil.	**270**	7°01′N	118°30′E
Cagliari, Italy (käl′yä-rē)	**216-17**	39°14′N	9°07′E
Cagliari, Golfo di, b., Italy (go̊l-fō-dē-käl′yä-rē)	**216-17**	39°08′N	9°11′E
Cagua, Ven. (kä′gwä)	**188b**	10°12′N	67°26′W
Caguas, P.R. (kä′gwäs)	**154a**	18°14′N	66°02′W
Cahaba, stm., Al., U.S. (kȧ hä-bä)	**134-35**	32°20′N	87°06′W
Cahors, Fr. (kȧ-o̊r′)	**212-13**	44°27′N	1°26′E
Cahul, Mol.	**218-19**	45°55′N	28°12′E
Caibarién, Cuba (kī-bä-rē-ĕn′)	**154-55**	22°31′N	79°28′W
Caicedonia, Col. (kī-sĕ-do̊-nĕä)	**188c**	4°19′N	75°48′W
Caicó, Braz.	**188d**	6°27′S	37°06′W
Caicos Islands, is., T./C. Is. (kī′kōs ī′lȧndz)	**154-55**	21°42′N	71°54′W
Caicos Passage, strt., N.A. (kī′kōs păs′ĭj)	**154-55**	22°00′N	72°30′W
Caimanera, Cuba (kī-mä-nā′rä)	**154-55**	19°59′N	75°10′W
Cairns, Austl. (kârnz)	**302**	16°56′S	145°45′E
Cairo, Ga., U.S. (kā′rō)	**134-35**	30°53′N	84°13′W
Cairo, Il., U.S. (kā′rō)	**134-35**	37°00′N	89°11′W
Cairo, nat. cap., Egypt (kī′rō)	**291b**	30°03′N	31°14′E
Cajamarca, Peru (kä-hä-mär′kä)	**184**	7°10′S	78°31′W
Cajazeiras, Braz.	**180-81**	6°54′S	38°34′W
Čakovec, Cro. (chȧ′kō-vĕts)	**216-17**	46°23′N	16°26′E
Calabar, Nig. (kăl-ȧ-bär′)	**282a**	4°58′N	8°19′E
Calabozo, Ven. (kä-lä-bō′zō)	**188b**	8°55′N	67°26′W
Calafat, Rom. (kȧ-lȧ-fȧt′)	**216-17**	43°59′N	22°57′E
Calagua Islands, is., Phil. (kä-gä-yän ī′lȧndz)	**270**	14°27′N	122°55′E
Calahorra, Spain (kä-lä-o̊r′rä)	**214-15**	42°18′N	1°58′W
Calais, Fr. (kȧ-lĕ′)	**212-13**	50°58′N	1°51′E
Calais, Me., U.S.	**127a**	45°11′N	67°16′W
Calais, Pas de, strt., Eur. *see* Dover, Strait of	**206-07**	50°59′N	1°31′E
Calama, Chile (kä-lä′mä)	**182-83**	22°27′S	68°55′W
Calamar, Col. (kä-lä-mär′)	**178-79**	1°58′N	72°42′W
Calamian Group, is., Phil. (kä-lä-myän′ gro͞op)	**270**	12°0′N	120°00′E
Calapan, Phil. (kä-lä-pän′)	**270**	13°24′N	121°11′E
Călăraşi, Rom. (kŭ-lŭ-rȧsh′ĭ)	**218-19**	44°13′N	27°19′E
Calatayud, Spain (kä-lä-tä-yo͞odh′)	**214-15**	41°21′N	1°38′W
Calayan Island, i., Phil.	**270a**	19°20′N	121°27′E
Calbayog, Phil.	**270**	12°04′N	124°34′E
Calcasieu, stm., La., U.S. (kăl′kȧ-shū)	**132-33**	30°03′N	93°19′W
Calcasieu Lake, lk., La., U.S. (kăl′kȧ-shū lāk)	**132-33**	29°56′N	93°16′W
Calçoene, Braz.	**180-81**	2°30′N	50°57′W
Calcutta, India (kăl-kŭt′ȧ) *see* Kolkata	**254-55**	22°32′N	88°22′E
Caldas, Col. (kȧ′l-däs)	**188c**	6°04′N	75°38′W
Caldas da Rainha, Port. (käl′däs dä rīn′yă)	**214-15**	39°24′N	9°08′W
Caldera, Chile (käl-dā′rä)	**182-83**	27°04′S	70°50′W
Caldwell, Id., U.S. (kôld′wĕl)	**122-23**	43°40′N	116°41′W
Caldwell, Oh., U.S. (kôld′wĕl)	**126-27**	39°44′N	81°31′W
Caldwell, Tx., U.S. (kôld′wĕl)	**132-33**	30°31′N	96°42′W
Caledonia, Mn., U.S. (kăl-ē-dō′nĭ-ȧ)	**124-25**	43°39′N	91°31′W
Calella, Spain (kä-lĕl′yä)	**214-15**	41°37′N	2°40′E
Calexico, Ca., U.S. (kȧ-lĕk′sĭ-ko)	**128-29**	32°41′N	115°30′W
Calgary, Ab., Can. (kăl′gȧ-rĭ)	**142-43**	51°03′N	114°05′W
Calhoun, Ga., U.S. (kăl-ho͞on′)	**134-35**	34°30′N	84°58′W
Calhoun, Ky., U.S. (kăl-ho͞on′)	**126-27**	37°32′N	87°15′W
Cali, Col. (kä′lē)	**178-79**	3°27′N	76°31′W
Calicut, India *see* Kozhikode	**256**	11°16′N	75°47′E
California, Mo., U.S. (kăl-ĭ-fôr′nĭ-ȧ)	**130-31**	38°38′N	92°34′W
California, state, U.S. (kăl-ĭ-fôr′nĭ-ȧ)	**118-19**	37°30′N	119°30′W
California, Golfo de, b., Mex. (go̊l-fô-dĕ-kä-lē̊-fôr-nyä)	**156-57**	28°0′N	112°00′W
California, Gulf of, b., Mex. (gŭlf ŭv kăl-ĭ-fôr′nĭ-ȧ) *see* California, Golfo de	**156-57**	28°0′N	112°00′W
Calimere, Point, c., India	**256**	10°17′N	79°52′E
Calipatria, Ca., U.S. (kăl-ĭ-pȧt′rĭ-ȧ)	**128-29**	33°08′N	115°31′W
Calkiní, Mex. (käl-kē-nē′)	**160**	20°23′N	90°02′W
Callabonna, Lake, lk., Austl. (lāk călă′bŏnȧ)	**301**	29°41′S	140°03′E
Callao, Peru (käl-yä′ō)	**188a**	12°04′S	77°08′W
Calling Lake, lk., Ab., Can. (kôl′ĭng lāk)	**142-43**	55°13′N	113°15′W
Calmar, Swe. *see* Kalmar	**208-09**	56°40′N	16°22′E
Caloosahatchee, stm., Fl., U.S. (kȧ-loo-sȧ-hăch′ē)	**135a**	26°32′N	82°01′W
Caltagirone, Italy (käl-tä-jē-rō′nȧ)	**216-17**	37°14′N	14°31′E
Caltanissetta, Italy (käl-tä-nē-sĕt′tä)	**216-17**	37°29′N	14°04′E
Caluula, Som.	**284-85**	11°57′N	50°46′E
Calvert Island, i., B.C., Can.	**142-43**	51°33′N	128°02′W
Calvillo, Mex. (käl-vēl′yō)	**158-59**	21°51′N	102°43′W
Calvinia, S. Afr. (kăl-vĭn′ĭ-ȧ)	**286-87**	31°28′S	19°46′E
Camacupa, Ang.	**286-87**	12°01′S	17°28′E
Camagüey, Cuba (kä-mä-gwā′)	**154-55**	21°22′N	77°55′W
Camagüey, state, Cuba (kä-mä-gwā′)	**154-55**	21°30′N	78°00′W
Camaná, Peru	**184**	16°37′S	72°42′W
Camaquã, Braz.	**182-83**	30°51′S	51°49′W
Camará, Braz.	**180-81**	3°55′S	62°44′W
Camarón, Cabo, c., Hond. (kȧ′bô-kä-mä-rōn′)	**161**	15°59′N	85°02′W
Camarones, Arg.	**185**	44°48′S	65°43′W
Camas, Wa., U.S. (kăm′ȧs)	**122-23**	45°35′N	122°24′W
Ca Mau, Viet.	**266-67**	9°11′N	105°09′E
Ca Mau, Mui, c., Viet.	**266-67**	8°37′N	104°43′E
Cambodia, nation, Asia (kăm-bō′dē-ȧ)	**224-25**	13°0′N	105°00′E
Camborne, Eng., U.K. (kăm′bôrn)	**206-07**	50°13′N	5°18′W
Cambrai, Fr. (käN-brĕ′)	**212-13**	50°11′N	3°15′E
Cambrian Mountains, mts., Wales, U.K. (kăm′brĭ-ăn moun′tĭnz)	**206-07**	52°35′N	3°35′W
Cambridge, On., Can. (kām′brĭj)	**146-47**	43°21′N	80°18′W
Cambridge, Eng., U.K. (kām′brĭj)	**206-07**	52°13′N	0°08′E
Cambridge, Il., U.S. (kām′brĭj)	**124-25**	41°18′N	90°11′W
Cambridge, Ma., U.S.	**126-27**	42°22′N	71°06′W
Cambridge, Md., U.S.	**126-27**	38°33′N	76°04′W
Cambridge, Mn., U.S. (kām′brĭj)	**124-25**	45°34′N	93°13′W
Cambridge, Ne., U.S. (kām′brĭj)	**130-31**	40°17′N	100°10′W
Cambridge, Oh., U.S. (kām′brĭj)	**126-27**	40°02′N	81°35′W
Cambridge Bay, Nu., Can.	**138-39**	69°07′N	105°04′W
Cambridge City, In., U.S. (kām′brĭj sĭ′tē)	**126-27**	39°49′N	85°11′W
Camden, Al., U.S. (kăm′dĕn)	**134-35**	31°59′N	87°17′W
Camden, Ar., U.S. (kăm′dĕn)	**130-31**	33°36′N	92°50′W
Camden, Me., U.S. (kăm′dĕn)	**127a**	44°13′N	69°05′W
Camden, N.J., U.S.	**126-27**	39°56′N	75°07′W
Camden, S.C., U.S. (kăm′dĕn)	**134-35**	34°15′N	80°36′W
Cameron, Mo., U.S. (kăm′ēr-ŭn)	**130-31**	39°44′N	94°14′W
Cameron, Tx., U.S. (kăm′ēr-ŭn)	**132-33**	30°51′N	96°59′W
Cameron, Wi., U.S. (kăm′ēr-ŭn)	**124-25**	45°25′N	91°45′W
Cameroon, nation, Afr. (kă′mȧ-ro͞on′)	**274**	6°0′N	12°00′E
Cameroon Mountain, vol., Camrn.	**282-83**	4°12′N	9°11′E
Cameroun, nation, Afr. *see* Cameroon	**274**	6°0′N	12°00′E
Cametá, Braz.	**180-81**	2°15′S	49°31′W
Camiguin Island, i., Phil.	**270a**	18°56′N	121°55′E
Camiling, Phil. (kä-mē̊-lĭng′)	**270**	15°41′N	120°25′E
Camilla, Ga., U.S. (kȧ-mĭl′ȧ)	**134-35**	31°14′N	84°12′W
Caminha, Port. (kä-mĭn′yȧ)	**214-15**	41°52′N	8°49′W
Camiranga, Braz.	**180-81**	1°49′S	46°16′W
Camiri, Bol.	**182-83**	20°03′S	63°31′W
Camocim, Braz. (kä-mô-sēN′)	**180-81**	2°54′S	40°51′W
Camooweal, Austl.	**302**	19°55′S	138°08′E
Campana, Arg. (käm-pä′nä)	**187**	34°10′S	58°57′W
Campana, Isla, i., Chile (ē̊′s-lä-käm-pän′yä)	**185**	48°20′S	75°15′W
Campbell Island, i., N.Z.	**313**	52°33′S	169°08′E
Campbell River, B.C., Can.	**142-43**	50°01′N	125°15′W
Campbellsville, Ky., U.S. (kăm′bĕlz vĭl)	**134-35**	37°21′N	85°21′W
Campbellton, N.B., Can. (kăm′bĕl-tŭn)	**148-49**	47°60′N	66°41′W
Campbelltown, Austl. (kăm′bĕl-toun)	**301**	34°04′S	150°49′E
Campeche, Mex. (käm-pā′chȧ)	**160**	19°50′N	90°31′W
Campeche, state, Mex. (käm-pā′chȧ)	**160**	19°0′N	90°30′W
Campechuela, Cuba (käm-pȧ-chwā′lä)	**154-55**	20°14′N	77°17′W
Cam Pha, Viet.	**266-67**	21°02′N	107°21′E
Campina Grande, Braz. (käm-pē′nä grän′dĕ)	**188d**	7°13′S	35°53′W
Campinas, Braz. (käm-pē′näzh)	**186**	22°55′S	47°05′W
Campo Alegre de Goiás, Braz.	**186**	17°38′S	47°46′W
Campobasso, Italy (käm′pô-bäs′sō)	**216-17**	41°34′N	14°40′E
Campo Belo, Braz.	**186**	20°54′S	45°16′W
Campo de Criptana, Spain (käm′pō dā krēp-tä′nä)	**214-15**	39°24′N	3°07′W
Campo Gallo, Arg.	**187**	26°34′S	62°50′W
Campo Grande, Braz. (käm-po̊ grän′dĕ)	**182-83**	20°28′S	54°38′W
Campo Maior, Braz. (käm-po̊ mä-yôr′)	**180-81**	4°49′S	42°10′W
Campo Mourão, Braz.	**182-83**	24°02′S	52°24′W
Campos, Braz. (kä′m-pôs)	**186**	21°45′S	41°21′W
Camrose, Ab., Can. (kăm-rōz)	**142-43**	53°01′N	112°50′W
Canada, nation, N.A. (kăn′ȧ-dȧ)	**94**	60°0′N	95°00′W
Cañada de Gómez, Arg. (kä-nyä′dä-dĕ-gô′mĕz)	**187**	32°49′S	61°24′W
Canadian, Tx., U.S. (kȧ-nā′dĭ-ăn)	**130-31**	35°55′N	100°23′W
Canadian, stm., U.S. (kȧ-nā′dĭ-ăn)	**120-21**	35°27′N	95°05′W
Canajoharie, N.Y., U.S. (kăn-ȧ-jō-hăr′ē)	**126-27**	42°54′N	74°35′W
Çanakkale, Tur. (chä-näk-kä′lĕ)	**216-17**	40°09′N	26°25′E
Çanakkale Boğazı, strt., Tur. *see* Dardanelles	**216-17**	40°17′N	26°33′E
Canandaigua, N.Y., U.S. (kăn-ăn-dā′gwȧ)	**126-27**	42°53′N	77°17′W
Cananea, Mex. (kä-nä-nĕ′ä)	**156-57**	30°59′N	110°18′W
Canarias, Islas, is., Spain (ē̊′s-läs-kä-nä′ryäs) *see* Canary Islands	**215d**	28°01′N	15°35′W
Canary Islands, is., Spain (kȧ-nâ′-rē ī′lȧndz)	**215d**	28°01′N	15°35′W
Cañas, C.R. (kä′-nyäs)	**161**	10°25′N	85°06′W
Canastota, N.Y., U.S. (kăn-ȧs-tō′tȧ)	**126-27**	43°05′N	75°46′W

n-sing; ŋ-baŋk; N-nasalized n; nŏd; cŏmmit; ōld; ôbey; ôrder; oi-boil; fo͞od; o̊-as oo in foot; ou-out; s-soft; sh-dish; th-thin; pūre; ůnite; ûrn; stŭd; circŭs; ü-as in French tu; ′-indeterminate vowel.

342 Canatlán – Caseyr PRONOUNCING INDEX

Feature (Pronunciation)	Page	Lat.	Long.
Canatlán, Mex. (kä-nät-län′)	158-59	24°31′N	104°46′W
Canaveral, Cape, c., Fl., U.S.	135a	28°27′N	80°32′W
Canavieiras, Braz. (kä-nȧ-vē-ā′räs)	182-83	15°39′S	38°57′W
Canberra, nat. cap., Austl. (kăn′bĕr-*ȧ*)	301	35°17′S	149°08′E
Canby, Mn., U.S. (kăn′bĭ)	124-25	44°43′N	96°17′W
Cancún, Mex.	160	21°08′N	86°51′W
Candala, Som. *see* Qandala	284-85	11°28′N	49°52′E
Candeias, Braz.	180-81	12°40′S	38°32′W
Candelaria, Cuba (kän-dĕ-lä′ryä)	154-55	22°45′N	82°58′W
Candelaria, stm., Mex. (kän-dĕ-lä-ryä)	160	18°38′N	91°17′W
Cando, N.D., U.S. (kăn′dō)	124-25	48°29′N	99°13′W
Candon, Phil. (kän-dōn′)	270	17°11′N	120°27′E
Canea, Grc. *see* Chaniá	216a	35°31′N	24°01′E
Canelones, Ur. (kä-nĕ-lô-nĕs)	187	34°32′S	56°17′W
Cangas, Spain (kän′gäs)	214-15	42°16′N	8°47′W
Cangas de Narcea, Spain (kä′n-gäs-dĕ-när-sĕ-ä)	214-15	43°11′N	6°33′W
Cangkuang, Tanjung, c., Indon.	268-69	6°50′S	105°15′E
Canguçu, Braz.	187	31°21′S	52°37′W
Cangzhou, China (tsäŋ-jō)	260-61	38°18′N	116°52′E
Caniapiscau, stm., Qc., Can.	140-41	57°41′N	69°29′W
Caniapiscau, Réservoir de, res., Qc., Can. *see* Caniapiscau, Lac	140-41	0°0′	0°00′
Caniapiscau, Lac, res., Qc., Can.	140-41	54°09′N	69°51′W
Canicattì, Italy (kä-nē-kät′tē)	216-17	37°21′N	13°51′E
Çankırı, Tur.	202-03	40°36′N	33°37′E
Cannelton, In., U.S. (kăn′*ĕl*-*tŭn*)	126-27	37°55′N	86°45′W
Cannes, Fr. (kȧn)	212-13	43°33′N	7°01′E
Canning, N.S., Can. (kăn′ĭng)	148-49	45°09′N	64°25′W
Canoas, stm., Braz.	182-83	27°37′S	51°26′W
Canon City, Co., U.S. (kăn′*yŭn* sĭ′tē)	130-31	38°27′N	105°15′W
Canonsburg, Pa., U.S. (kăn′*ŭnz*-bûrg)	126-27	40°16′N	80°11′W
Canora, Sk., Can. (k*ȧ*-nōr*ȧ*)	144-45	51°37′N	102°26′W
Canouan, i., St. Vin.	155b	12°43′N	61°20′W
Canso, N.S., Can. (kăn′sō)	148-49	45°20′N	61°00′W
Cantabrian Mountains, mts., Spain (kăn-tā′brē-*ăn* moun′tĭnz) *see* Cantábrica, Cordillera	214-15	43°0′N	5°00′W
Cantábrica, Cordillera, mts., Spain	214-15	43°0′N	5°00′W
Cantandica, Moz.	286-87	18°02′S	33°08′E
Cantanhede, Port. (kän-tän-yā′dā)	214-15	40°21′N	8°36′W
Cantaura, Ven.	188b	9°18′N	64°21′W
Canterbury, Eng., U.K. (kăn′tĕr-bĕr-ē)	206-07	51°17′N	1°05′E
Canterbury Bight, b., N.Z.	303	44°15′S	171°38′E
Can Tho, Viet.	266-67	10°02′N	105°47′E
Canton, China *see* Guangzhou	258-59	23°08′N	113°16′E
Canton, Ms., U.S.	134-35	32°37′N	90°02′W
Canton, Oh., U.S.	126-27	40°48′N	81°23′W
Canton, i., Kir.	304-05	2°49′S	171°41′W
Cañuelas, Arg. (kä-nyoĕ′-läs)	187	35°03′S	58°45′W
Canutama, Braz.	180-81	6°31′S	64°21′W
Canyon, Tx., U.S. (kăn′*yŭn*)	130-31	34°59′N	101°55′W
Canyon de Chelly National Monument, n.p., Az., U.S.	128-29	36°07′N	109°27′W
Canyonlands National Park, n.p., Ut., U.S. (kăn′*yŭn*-lăndz năsh′ŭn-*ăl* pärk)	128-29	38°10′N	110°00′W
Cao Bang, Viet.	266-67	22°40′N	106°15′E
Capanaparo, stm., S.A.	178-79	7°03′N	67°04′W
Cap aux Meules, Île du, i., Qc., Can.	148-49	47°23′N	61°55′W
Cap-Chat, Qc., Can. (kȧp-shä′)	148-49	49°05′N	66°41′W
Cape Barren Island, i., Austl.	301	40°25′S	148°12′E
Cape Breton Highlands National Park, n.p., N.S., Can.	148-49	46°45′N	60°45′W
Cape Breton Island, i., N.S., Can. (kāp brĕt′*ŭn* ī′l*ă*nd)	148-49	46°04′N	60°30′W
Cape Charles, Va., U.S. (kāp chärlz)	134-35	37°16′N	76°01′W
Cape Coast, Ghana	282-83	5°07′N	1°16′W
Cape Dorset, Nu., Can.	138-39	64°14′N	76°33′W
Cape Fear, stm., N.C., U.S. (kāp fēr)	134-35	33°53′N	78°01′W
Cape Girardeau, Mo., U.S. (kāp jē-rär-dō′)	130-31	37°18′N	89°32′W
Cape May, N.J., U.S. (kāp mā)	126-27	38°56′N	74°56′W
Cape Town, nat. cap., S. Afr. (kāp toun)	286-87	33°55′S	18°30′E
Cape Verde, nation, Afr. *see* Cabo Verde	274	16°0′N	24°00′W
Cape York Peninsula, pen., Austl. (kāp yôrk pĕ-nĭn′sū-l*ȧ*)	302	14°0′S	142°30′E
Cap-Haïtien, Haiti (kȧp ȧ-ē-syăn′)	154-55	19°45′N	72°12′W
Capim, stm., Braz.	180-81	1°41′S	47°47′W
Capitol Reef National Park, n.p., Ut., U.S. (kăp′ĭ-tŏl rēf năsh′ŭn-*ăl* pärk)	128-29	38°15′N	111°10′W
Capiz, Phil. *see* Roxas	270	11°35′N	122°45′E
Caprara, Punta, c., Italy (pōō′n-tä-kä-prä′rä)	216-17	41°07′N	8°19′E
Capreol, On., Can.	146-47	46°42′N	80°55′W
Capri, Isola di, i., Italy (ē′-sō-lä-dē-kä′prē)	216-17	40°33′N	14°13′E
Caprivi Strip, hist. reg., Nmb.	286-87	17°59′S	23°00′E
Cap Saint Jacques, Viet. *see* Vung Tau	266-67	10°21′N	107°05′E
Capulin Volcano National Monument, n.p., N.M., U.S. (kȧ-pū′lĭn vŏl-kā′nō năsh′ŭn-*ăl* mŏn′*ū*-m*ĕ*nt)	130-31	36°47′N	103°56′W
Caquetá, stm., S.A.	180-81	3°08′S	64°46′W
Caracal, Rom. (kȧ-rȧ-kȧl′)	216-17	44°07′N	24°22′E
Caracaraí, Braz.	180-81	1°50′N	61°08′W
Caracas, nat. cap., Ven. (kä-rä′käs)	178-79	10°30′N	66°56′W
Caraguatatuba, Braz. (kä-rä-gwä-tȧ-tōō′bä)	186	23°37′S	45°25′W
Caraïbes, Îles des, is., *see* West Indies	152-53	19°0′N	70°00′W
Caraïbes, Mer des, s., *see* Caribbean Sea	152-53	15°0′N	73°00′W
Carajás, Braz.	180-81	6°06′S	50°23′W
Carajás, Serra dos, hills, Braz. (sĕ′r-rä-dôs-kä-rä-zhà′s)	180-81	6°16′S	51°21′W
Carangola, Braz. (kä-rán′gō′lä)	186	20°43′S	42°02′W
Caraquet, N.B., Can. (kă-rȧ-kĕt′)	148-49	47°47′N	64°57′W
Caratasca, Laguna de, b., Hond. (lä-gō′nä-dĕ-kä-rä-täs′kä)	161	15°24′N	83°54′W
Caratinga, Braz.	186	19°47′S	42°09′W
Carauari, Braz.	180-81	4°52′S	66°52′W
Caravelas, Braz. (kȧ-rä-vĕl′äzh)	186	17°44′S	39°15′W
Carazinho, Braz. (kä-rá′zē-ny*ŏ*)	182-83	28°17′S	52°46′W
Carballo, Spain (kär-bäl′yō)	214-15	43°13′N	8°41′W
Carberry, Mb., Can.	144-45	49°52′N	99°21′W
Carbonara, Capo, c., Italy (kä′pō är-bō-nä′rä)	216-17	39°06′N	9°31′E
Carbondale, Il., U.S. (kär′bŏn-dāl)	126-27	37°43′N	89°13′W
Carbondale, Pa., U.S. (kär′bŏn-dāl)	126-27	41°35′N	75°30′W
Carbonear, Nf./L., Can. (kär-bō-nēr′)	148-49	47°45′N	53°14′W
Carbon Hill, Al., U.S. (kär′bŏn hĭl)	134-35	33°54′N	87°32′W
Carcassonne, Fr. (kȧr-kȧ-sôn′)	212-13	43°13′N	2°21′E
Carcross, Yk., Can. (kär′krôs)	138-39	60°11′N	134°42′W
Cárdenas, Cuba (kär′dā-näs)	154-55	23°02′N	81°12′W
Cárdenas, Mex. (ká′r-dĕ-näs)	158-59	18°00′N	93°22′W
Cárdenas, Mex. (ká′r-dĕ-näs)	158-59	21°60′N	99°39′W
Cardiel, Lago, lk., Arg.	185	48°55′S	71°15′W
Cardiff, Wales, U.K. (kär′dĭf)	206-07	51°29′N	3°11′W
Cardigan, Wales, U.K. (kär′dĭ-g*ăn*)	206-07	52°05′N	4°39′W
Cardston, Ab., Can. (kärds′tŭn)	142-43	49°12′N	113°18′W
Carei, Rom. (kä-rĕ′)	210-11	47°41′N	22°28′E
Careiro, Braz.	180-81	3°14′S	59°46′W
Careiro, Ilha do, i., Braz.	180-81	3°09′S	59°48′W
Carey, Oh., U.S. (kā′rē)	126-27	40°57′N	83°23′W
Carey, Lake, lk., Austl. (lāk kâr′ē)	296-97	29°04′S	122°19′E
Caribbean Sea, s., (kăr-ĭ-bē′*ăn* sē)	152-53	15°0′N	73°00′W
Caribe, Mar, s., *see* Caribbean Sea	152-53	15°0′N	73°00′W
Caribische Zee, s., *see* Caribbean Sea	152-53	15°0′N	73°00′W
Cariboo Mountains, mts., B.C., Can. (kă′rĭ-bōō moun′tĭnz)	142-43	53°0′N	121°00′W
Caribou, Me., U.S.	127a	46°51′N	68°00′W
Caribou Mountains, mts., Ab., Can.	140-41	59°06′N	115°10′W
Caricyn, Russia *see* Volgograd	202-03	48°44′N	44°25′E
Carinhanha, Braz. (kä-rĭ-nyän′yä)	180-81	14°19′S	43°48′W
Caripito, Ven.	178-79	10°06′N	63°06′W
Carleton, Mount, mtn., N.B., Can.	148-49	47°23′N	66°53′W
Carleton Place, On., Can. (kärl′tŭn plās)	146-47	45°09′N	76°09′W
Carletonville, S. Afr.	292c	26°21′S	27°24′E
Carlinville, Il., U.S. (kär′lĭn-vĭl)	130-31	39°16′N	89°53′W
Carlisle, Eng., U.K. (kär-līl′)	206-07	54°54′N	2°56′W
Carlos Casares, Arg. (kär-lôs-kä-sá′rĕs)	187	35°38′S	61°21′W
Carlow, Ire. (kär′lō)	206-07	52°50′N	6°55′W
Carlsbad, Czech Rep. *see* Karlovy Vary	210-11	50°14′N	12°53′E
Carlsbad, Ca., U.S. (kärlz′băd)	128-29	33°09′N	117°20′W
Carlsbad, N.M., U.S. (kärlz′băd)	130-31	32°25′N	104°14′W
Carlsbad Caverns National Park, n.p., N.M., U.S. (kärlz′băd kăv′ērnz năsh′ŭn-*ăl* pärk)	130-31	32°08′N	104°35′W
Carlyle, Il., U.S. (kärlīl′)	126-27	38°36′N	89°22′W
Carmacks, Yk., Can.	138-39	62°05′N	136°15′W
Carman, Mb., Can. (kär′mȧn)	144-45	49°31′N	97°59′W
Carmarthen, Wales, U.K. (kär-mär′th*ĕ*n)	206-07	51°52′N	4°19′W
Carmaux, Fr. (kȧr-mō′)	212-13	44°03′N	2°10′E
Carmel, In., U.S. (kär′m*ĕ*l)	126-27	39°58′N	86°07′W
Carmelo, Ur. (kär-mĕ′lo)	187	33°60′S	58°17′W
Carmen, Mex. *see* Ciudad del Carmen	160	18°39′N	91°49′W
Carmen, Isla, i., Mex.	156-57	26°00′N	111°08′W
Carmen, Isla del, i., Mex. (ē′s-lä-dĕl-kȧ′r-mĕn)	160	18°43′N	91°40′W
Carmen de Areco, Arg. (kär′mĕn′ dā ä-rā′kō)	187	34°23′S	59°50′W
Carmi, Il., U.S. (kär′mī)	126-27	38°05′N	88°10′W
Carnarvon, Austl. (kär-när′v*ŭ*n)	294-95	24°52′S	113°40′E
Carnarvon, S. Afr.	286-87	30°58′S	22°08′E
Carnarvon National Park, n.p., Austl.	302	24°42′S	147°55′E
Carnegie, Ok., U.S. (kär-nĕg′ĭ)	130-31	35°07′N	98°35′W
Carnegie, Lake, lk., Austl.	296-97	26°11′S	122°31′E
Carnot, C.A.R.	284-85	4°56′N	15°53′E
Carnsore Point, c., Ire. (kärn′sôr point)	206-07	52°11′N	6°22′W
Caro, Mi., U.S. (kâ′rō)	126-27	43°29′N	83°23′W
Carolina, Braz. (kä-rô-lē′nä)	180-81	7°21′S	47°25′W
Carolina, S. Afr. (kăr-ô-lī′n*ȧ*)	292c	26°04′S	30°08′E
Caroline, at., Kir.	304-05	9°58′S	150°13′W
Caroline Islands, is., Oc. (kă′-rô-līn′ ī′l*ă*ndz)	304-05	8°0′N	147°00′E
Caroní, stm., Ven. (kä-rō′nē)	178-79	8°21′N	62°48′W
Carora, Ven. (kä-rô′rä)	178-79	10°10′N	70°05′W
Carpathian Mountains, mts., Eur. (kär-pā′thĭ-*ăn* moun′tĭnz)	202-03	48°0′N	24°00′E
Carpaţii, mts., Eur. *see* Carpathian Mountains	202-03	48°0′N	24°00′E
Carpaţii Meridionali, mts., Rom. *see* Transylvanian Alps	216-17	45°25′N	23°33′E
Carpentaria, Gulf of, b., Austl. (gŭlf ŭv kär-pĕn-târ′ĭ*ȧ*)	296-97	14°0′S	139°00′E
Carpentras, Fr. (kär-pän-träs′)	212-13	44°04′N	5°03′E
Carranza, Cabo, c., Chile	185	35°36′S	72°38′W
Carrara, Italy (kä-rä′rä)	216-17	44°05′N	10°06′E
Carrauntoohil, mtn., Ire.	206-07	51°59′N	9°45′W
Carreta, Punta, c., Peru (pōō′n-tä-kär-rĕ′tĕ′rȧ)	184	14°11′S	76°17′W
Carriacou, i., Gren.	155b	12°30′N	61°26′W
Carrington, N.D., U.S. (kăr′ĭng-t*ŭ*n)	124-25	47°27′N	99°07′W
Carrizal Bajo, Chile	182-83	28°06′S	71°09′W
Carrizozo, N.M., U.S. (kär-rē-zō′zō)	130-31	33°39′N	105°53′W
Carroll, Ia., U.S. (kăr′*ŭ*l)	124-25	42°04′N	94°52′W
Carrollton, Ga., U.S. (kăr-*ŭ*l-t*ŭ*n)	134-35	33°35′N	85°05′W
Carrollton, Il., U.S. (kăr-*ŭ*l-t*ŭ*n)	130-31	39°18′N	90°24′W
Carrollton, Ky., U.S. (kăr-*ŭ*l-t*ŭ*n)	126-27	38°41′N	85°11′W
Carrollton, Mi., U.S. (kăr-*ŭ*l-t*ŭ*n)	126-27	43°27′N	83°57′W
Carrollton, Mo., U.S. (kăr-*ŭ*l-t*ŭ*n)	130-31	39°22′N	93°30′W
Carrollton, Tx., U.S. (kăr-*ŭ*l-t*ŭ*n)	130-31	32°58′N	96°53′W
Carrot, stm., Can.	144-45	53°50′N	101°19′W
Çarşamba, Tur.	247	41°12′N	36°43′E
Carson City, Nv., U.S.	128-29	39°10′N	119°46′W
Cartagena, Col. (kär-tä-hā′nä)	178-79	10°25′N	75°30′W
Cartagena, Spain (kär-tä-kĕ′nä)	214-15	37°37′N	0°59′W
Cartago, Col. (kär-tä′gō)	188c	4°45′N	75°55′W
Cartago, C.R.	161	9°51′N	83°55′W
Cartersville, Ga., U.S. (kär′tērs-vĭl)	134-35	34°10′N	84°48′W
Carthage, Il., U.S. (kär′thâj)	130-31	40°25′N	91°08′W
Carthage, Mo., U.S. (kär′thâj)	130-31	37°11′N	94°19′W
Carthage, Ms., U.S. (kär′thâj)	134-35	32°44′N	89°32′W
Carthage, N.Y., U.S. (kär′thâj)	126-27	43°59′N	75°37′W
Carthage, Tx., U.S. (kär′thâj)	132-33	32°09′N	94°22′W
Cartwright, Nf./L., Can. (kärt′rīt)	138-39	53°41′N	56°60′W
Caruaru, Braz. (kä-rò-á-rōō′)	188d	8°17′S	35°58′W
Carúpano, Ven. (kä-rōō′pä-nō)	178-79	10°40′N	63°15′W
Carutapera, Braz.	180-81	1°13′S	46°00′W
Caruthersville, Mo., U.S. (k*ȧ*-rŭdh′ērz-vĭl)	134-35	36°11′N	89°40′W
Carvoeiro, Braz.	180-81	1°26′S	61°60′W
Carvoeiro, Cabo, c., Port. (ká′bō-kär-vô-ĕ′y-rō)	214-15	39°21′N	9°24′W
Cary, N.C., U.S. (kă′rē)	134-35	35°47′N	78°47′W
Casablanca, Mor. (kä-sä-bläŋ′kä)	292a	33°36′N	7°36′W
Casa Branca, Braz. (ká′sä-brá′n-kä)	186	21°48′S	47°04′W
Casa Grande, Az., U.S. (kä′sä grän′dā)	128-29	32°53′N	111°45′W
Casa Grande Ruins National Monument, n.p., Az., U.S.	128-29	32°59′N	111°32′W
Casamance, stm., Sen. (kä-sä-mäns′)	282-83	12°33′N	16°45′W
Casanare, stm., Col.	178-79	6°02′N	69°51′W
Cascade Mountains, mts., N.A. (kăs-kād′ moun′tĭnz)	120-21	45°14′N	121°56′W
Cascade Point, c., N.Z. (kăs-kād′ point)	303	44°00′S	168°22′E
Cascade Range, mts., N.A. (kăs-kād′ rānj) *see* Cascade Range	120-21	45°14′N	121°56′W
Cascais, Port. (käs-k*ȧ*-ēzh)	214-15	38°42′N	9°25′W
Cascavel, Braz.	182-83	24°58′S	53°27′W
Caserta, Italy (kä-zĕr′tä)	216-17	41°05′N	14°19′E
Casey, Il., U.S. (kā′sĭ)	126-27	39°18′N	87°60′W
Caseyr, c., Som. *see* Gwardafuy, Gees	284-85	11°50′N	51°17′E

Feature (Pronunciation)	Page	Lat.	Long.
Cashmere, Wa., U.S. (kăsh´mĭr)	122-23	47°31'N	120°27'W
Casilda, Arg. (kä-sē´l-dä)	187	33°03'S	61°11'W
Casino, Austl. (kȧ-sē´nō)	301	28°52'S	153°03'E
Casiquiare, stm., Ven. (kä-sē-kyä´rā)	178-79	1°60'N	67°08'W
Caspe, Spain (käs´pā)	214-15	41°14'N	0°03'W
Casper, Wy., U.S. (kăs´pẽr)	122-23	42°51'N	106°20'W
Caspian Depression, pl., (kăs´pĭ-ăn dĭ-prĕ´shŭn)	202-03	48°0'N	52°00'E
Caspian Sea, lk., (kăs´pĭ-ăn sē)	244	41°18'N	50°59'E
Cass, W.V., U.S. (kăs)	126-27	38°24'N	79°55'W
Cassai, stm., Afr. (kä-sä´ē)	284-85	3°02'S	16°56'E
Cass City, Mi., U.S. (kăs sĭ´tê)	126-27	43°36'N	83°10'W
Casselman, On., Can. (kăs´'l-mȧn)	146-47	45°19'N	75°06'W
Casselton, N.D., U.S. (kăs´'l-tŭn)	124-25	46°54'N	97°13'W
Cássia, Braz. (kȧ´syä)	186	20°33'S	46°56'W
Cassiar, B.C., Can.	138-39	59°16'N	129°43'W
Cassiar Mountains, mts., Can.	140-41	59°0'N	129°00'W
Cassino, Italy (käs-sē´nō)	216-17	41°30'N	13°51'E
Cass Lake, Mn., U.S. (kăs lāk)	124-25	47°22'N	94°37'W
Cassopolis, Mi., U.S. (kăs-ŏ´pô-lĭs)	126-27	41°55'N	86°01'W
Cassville, Mo., U.S. (kăs´vĭl)	130-31	36°41'N	93°52'W
Cassville, Wi., U.S. (kăs´vĭl)	124-25	42°43'N	90°59'W
Castelli, Arg. (kȧs-tĕ´zhē)	187	36°06'S	57°49'W
Castelli, Arg. (kȧs-tĕ´zhē)	182-83	25°57'S	60°37'W
Castelló de la Plana, Spain	214-15	39°59'N	0°02'W
Castellón de la Plana, Spain *see* Castelló de la Plana	214-15	39°59'N	0°02'W
Castelnaudary, Fr. (kȧs´tĕl-nō-dȧ-rē´)	212-13	43°19'N	1°57'E
Castelo, Braz. (käs-tĕ´lô)	186	20°35'S	41°13'W
Castelo Branco, Port.	214-15	39°49'N	7°29'W
Castelsarrasin, Fr. (kȧs´tĕl-sȧ-rȧ-zăN´)	212-13	44°02'N	1°06'E
Castelvetrano, Italy (käs´tĕl-vê-trä´nō)	216-17	37°41'N	12°47'E
Castilla, Peru (kȧs-tē´l-yä)	184	5°12'S	80°37'W
Castillo de San Marcos National Monument, n.p., Fl., U.S. (käs-tē´lyā de-sän mär-kōs năsh´ŭn-ăl mŏn´ū-mĕnt)	134-35	29°55'N	81°19'W
Castillos, Ur.	187	34°13'S	53°50'W
Castle Dale, Ut., U.S. (kȧs´'l dāl)	128-29	39°24'N	110°27'W
Castlegar, B.C., Can. (kȧs´'l-gär)	142-43	49°19'N	117°40'W
Castlemaine, Austl. (käs´'l-mān)	301	37°04'S	144°13'E
Castle Peak, mtn., Co., U.S. (kȧs´'l pēk)	128-29	39°01'N	106°52'W
Castle Peak, mtn., Id., U.S. (kȧs´'l pēk)	122-23	44°02'N	114°35'W
Castlereagh, stm., Austl.	301	30°12'S	147°31'E
Castle Rock, Co., U.S. (kȧs´'l rŏk)	130-31	39°23'N	104°51'W
Castle Rock, Wa., U.S. (kȧs´'l rŏk)	122-23	46°17'N	122°54'W
Castres, Fr. (kȧs´tr')	212-13	43°36'N	2°15'E
Castries, nat. cap., St. Luc. (kȧs-trē´)	155b	14°01'N	60°59'W
Castro, Braz. (käs´trō)	182-83	24°47'S	49°60'W
Castro, Chile (käs´tro)	185	42°29'S	73°46'W
Castro Verde, Port. (käs-trō vĕr´dĕ)	214-15	37°41'N	8°05'W
Castrovillari, Italy (käs´trō-vēl-lyä´rē)	216-17	39°49'N	16°13'E
Catacamas, Hond. (kä-tä-kä´mäs)	161	14°51'N	85°54'W
Catahoula Lake, lk., La., U.S. (kăt-ȧ-hō´lȧ lāk)	134-35	31°30'N	92°08'W
Catalão, Braz. (kä-tä-loun´)	186	18°11'S	47°56'W
Catalina, Chile	182-83	25°13'S	69°44'W
Catalina, i., Ca., U.S. *see* Santa Catalina Island	128-29	33°23'N	118°24'W
Catamarca, state, Arg. (kä-tä-mär´kä)	182-83	27°0'S	67°00'W
Catanduanes Island, i., Phil. (kä-tän-dwä´nĕs ī´lănd)	270	13°45'N	124°15'E
Catanduva, Braz. (kä-tȧn-dōō´vä)	186	21°08'S	48°58'W
Catania, Italy (kä-tä´nyä)	216-17	37°30'N	15°06'E
Catanzaro, Italy (kä-tän-dzä´rō)	216-17	38°54'N	16°36'E
Catarman, Phil.	270	12°30'N	124°38'E
Catbalogan, Phil. (kät-bä-lō´gän)	270	11°50'N	124°51'E
Cathedral Mountain, mtn., Tx., U.S. (kȧ-thē´drăl moun´tĭn)	132-33	30°10'N	103°40'W
Cat Island, i., Bah.	154-55	24°26'N	75°32'W
Catlettsburg, Ky., U.S. (kăt´lĕts-bûrg)	126-27	38°25'N	82°37'W
Catoche, Cabo, c., Mex. (kȧ´bô kä-tô´chĕ)	160	21°36'N	87°06'W
Catonsville, Md., U.S. (kā´tŭnz-vĭl)	126-27	39°17'N	76°44'W
Catorce, Mex. (kä-tôr´sȧ)	158-59	23°42'N	100°56'W
Catriló, Arg.	187	36°24'S	63°25'W
Catrimani, stm., Braz.	180-81	0°28'N	61°43'W
Catskill, N.Y., U.S. (kăts´kĭl)	126-27	42°14'N	73°52'W
Catskill Mountains, mts., N.Y., U.S. (kăts´kĭl moun´tĭnz)	126-27	42°10'N	74°30'W
Cattaraugus Indian Reservation, ind. res., N.Y., U.S. (kăt´tȧ-rȧ-gŭs ĭn´dĭ-ăn rĕ-sẽr-vā´shĕn)	126-27	42°32'N	78°59'W
Catumbela, stm., Ang. (kä´tŏm-bĕl´ȧ)	286-87	12°27'S	13°30'E
Caubvick, Mount, mtn., Can.	140-41	58°53'N	63°43'W
Cauca, stm., Col. (kou´kä)	178-79	8°54'N	74°28'W
Caucasus Mountains, mts., (kô´kȧ-sŭs moun´tĭnz)	245	42°38'N	45°00'E
Caungula, Ang.	284-85	8°26'S	18°38'E
Cauquenes, Chile (kou-kā´nās)	185	35°58'S	72°19'W
Caura, stm., Ven. (kou´rä)	178-79	7°38'N	64°53'W
Caution, Cape, c., B.C., Can. (kăp kô´shŭn)	142-43	51°10'N	127°46'W
Cauto, stm., Cuba (kou´tō)	154-55	20°33'N	77°14'W
Cavalcante, Braz. (kä-väl-kän´tā)	180-81	13°47'S	47°30'W
Cavalier, N.D., U.S. (kăv-ȧ-lēr´)	124-25	48°47'N	97°37'W
Cavalla, stm., Afr.	282-83	4°21'N	7°31'W
Cavally, stm., Afr.	282-83	4°21'N	7°31'W
Cavan, Ire. (kăv´ăn)	206-07	53°59'N	7°22'W
Caviana de Fora, Ilha, i., Braz.	180-81	0°10'N	50°10'W
Cavite, Phil. (kä-vē´tā)	270	14°29'N	120°54'E
Cawnpore, India *see* Kānpur	254-55	26°28'N	80°19'E
Caxias, Braz. (kä´shē-äzh)	180-81	4°50'S	43°21'W
Caxias do Sul, Braz. (kä´shē-äzh-dô-sōō´l)	182-83	29°11'S	51°11'W
Caxito, Ang. (kä-shē´tō)	284-85	8°33'S	13°36'E
Cayambe, Ec. (kȧ-īä´m-bĕ)	184	0°03'N	78°09'W
Cayambe, vol., Ec.	184	0°02'N	77°59'W
Cayenne, nat. cap., Fr. Gu. (kā-ĕn´)	178-79	4°56'N	52°19'W
Cayman Brac, i., Cay. Is. (kā-män´ brăk)	154-55	19°43'N	79°49'W
Cayman Islands, dep., N.A. (kā´măn ī´lăndz) (kī-män´ ī´lăndz)	152-53	19°30'N	80°40'W
Ceará, Braz. *see* Fortaleza	180-81	3°44'S	38°30'W
Ceará, state, Braz.	180-81	5°0'S	40°00'W
Ceará-Mirim, Braz. (sā-ä-rä´mē-rē´N)	188d	5°38'S	35°26'W
Ceatharlach, Ire. *see* Carlow	206-07	52°50'N	6°55'W
Cebaco, Isla de, i., Pan. (ē´s-lä-dĕ-sȧ-bä´kō)	162	7°32'N	81°09'W
Cebu, Phil. (sā-bōō´)	270	10°19'N	123°54'E
Cebu, i., Phil.	270	10°20'N	123°45'E
Čechy, hist. reg., Czech Rep. *see* Bohemia	210-11	49°50'N	14°00'E
Cedar, stm., U.S. (sē´dẽr)	124-25	41°17'N	91°20'W
Cedar Breaks National Monument, n.p., Ut., U.S. (sē´dẽr brāks năsh´ŭn-ăl mŏn´ū-mĕnt)	128-29	37°38'N	112°50'W
Cedarburg, Wi., U.S. (sē´dẽr bûrg)	126-27	43°17'N	87°59'W
Cedar City, Ut., U.S. (sē´dẽr sĭ´tê)	128-29	37°41'N	113°04'W
Cedar Falls, Ia., U.S. (sē´dẽr fôlz)	124-25	42°31'N	92°27'W
Cedar Lake, res., Mb., Can. (sē´dẽr lāk)	144-45	53°15'N	100°10'W
Cedar Rapids, Ia., U.S. (sē´dẽr răp´ĭdz)	124-25	41°58'N	91°40'W
Cedar Springs, Mi., U.S. (sē´dẽr sprĭngz)	126-27	43°13'N	85°33'W
Cedartown, Ga., U.S. (sē´dẽr-toun)	134-35	34°02'N	85°14'W
Cedros, Isla, i., Mex.	156-57	28°11'N	115°13'W
Ceduna, Austl. (sē-dō´nȧ)	294-95	32°07'S	133°41'E
Ceerigaabo, Som.	284-85	10°37'N	47°22'E
Cegléd, Hung. (tsā´glād)	210-11	47°10'N	19°48'E
Celaya, Mex. (sā-lä´yä)	158-59	20°31'N	100°49'W
Celebes, i., Indon. (sĕ´lĕ-bēz)	268-69	2°0'S	121°00'E
Celebes Sea, s., Asia (sĕ´-lĕ-bēz sē)	268-69	3°0'N	122°00'E
Çeleken, Turkmen.	252-53	39°26'N	53°07'E
Celestún, Mex. (sĕ-lĕs-tōō´n)	160	20°52'N	90°23'W
Celina, Oh., U.S. (sĕlī´na)	126-27	40°33'N	84°34'W
Celje, Slvn. (tsĕl´yĕ)	216-17	46°14'N	15°16'E
Celle, Ger. (tsĕl´ē)	210-11	52°37'N	10°05'E
Celtic Sea, s., Eur.	206-07	51°0'N	6°30'W
Cenderawasih, Teluk, b., Indon.	242-43	2°30'S	135°20'E
Center, Tx., U.S. (sĕn´tẽr)	132-33	31°48'N	94°10'W
Centerville, Ia., U.S. (sĕn´tẽr-vĭl)	124-25	40°44'N	92°52'W
Centerville, Pa., U.S. (sĕn´tẽr-vĭl)	126-27	40°03'N	79°59'W
Centerville, S.D., U.S. (sĕn´tẽr-vĭl)	124-25	43°07'N	96°60'W
Central, Cordillera, mts., Phil. (kôr-dēl-yĕ´rä-sĕn´träl)	270	17°02'N	120°53'E
Central, La., U.S. (sĕn´trăl)	134-35	30°33'N	91°02'W
Central, Massif, mts., Fr.	212-13	44°42'N	3°19'E
Central, Sistema, mts., Spain	214-15	40°34'N	4°29'W
Central African Republic, nation, Afr. (sĕn´trăl ăf´rĭ-kăn rē-pŭb´lĭk)	274	7°0'N	21°00'E
Central City, Ky., U.S. (sĕn´trăl sĭ´tĭ)	134-35	37°18'N	87°08'W
Central City, Ne., U.S. (sĕn´trăl sĭ´tê)	124-25	41°07'N	97°60'W
Centralia, Il., U.S. (sĕn-trā´lĭ-ȧ)	126-27	38°32'N	89°08'W
Centralia, Mo., U.S. (sĕn-trā´lĭ-ȧ)	130-31	39°13'N	92°08'W
Centralia, Wa., U.S. (sĕn-trā´lĭ-ȧ)	122-23	46°43'N	122°57'W
Central Kalahari Game Reserve, pk., Bots.	286-87	22°15'S	23°45'E
Central Russian Upland, plat., Russia (sĕn´trăl rŭsh´ăn ŭp´lănd) *see* Srednerusskaya Vozvyshennost'	218-19	52°0'N	38°00'E
Centreville, Al., U.S. (sĕn´tẽr-vĭl)	134-35	32°57'N	87°08'W
Century, Fl., U.S. (sĕn´tū-rĭ)	134-35	30°58'N	87°16'W
Ceram, i., Indon. (sārăm´) (sā´räm)	242-43	3°0'S	129°00'E
Ceram Sea, s., Indon. (sārăm´ sē) (sā´räm sē)	242-43	2°30'S	128°00'E
Cerignola, Italy (chā-rē-nyô´lä)	216-17	41°17'N	15°54'E
Cernăuți, Ukr. *see* Chernivtsi	210-11	48°17'N	25°58'E
Cerralvo, Mex. (sĕr-räl´vō)	132-33	26°06'N	99°36'W
Cerralvo, Isla, i., Mex. (ē´s-lä-sĕr-räl´vō)	156-57	24°14'N	109°52'W
Cerritos, Mex. (sĕr-rē´tōs)	158-59	22°26'N	100°17'W
Cerro de Pasco, Peru (sĕr´rō dā päs´kō)	184	10°41'S	76°16'W
Cervino, mtn., Eur. *see* Matterhorn	210-11	45°59'N	7°43'E
César E. Chávez National Monument, n.p., Ca., U.S.	128-29	35°13'N	118°34'W
Cesena, Italy (chĕ´sĕ-nä)	216-17	44°09'N	12°15'E
Cēsis, Lat. (sā´sĭs)	208-09	57°19'N	25°17'E
Česká Lípa, Czech Rep. (chĕs´kä lē´pa)	210-11	50°41'N	14°32'E
Česká Republika, nation, Eur. *see* Czech Republic	190-91	49°40'N	15°10'E
České Budějovice, Czech Rep. (chĕs´kā bōō´dyĕ-yô-vēt-sĕ)	210-11	48°59'N	14°28'E
Çeşme, Tur. (chĕsh´mĕ)	216-17	38°18'N	26°19'E
Cessnock, Austl.	301	32°50'S	151°21'E
Cestos, stm., Lib.	282-83	5°29'N	9°33'W
Cetatea Albă, Ukr. *see* Bilhorod-Dnistrovs'kyi	218-19	46°12'N	30°18'E
Cetinje, Mont.	216-17	42°24'N	18°56'E
Ceuta, Sp. N. Afr. (thā-ōō´tä)	292a	35°54'N	5°19'W
Cévennes, reg., Fr. (sā-vĕn´)	212-13	44°07'N	3°32'E
Chacabuco, Arg. (chä-kä-bōō´kō)	187	34°38'S	60°29'W
Chachapoyas, Peru (chä-chä-poi´yäs)	184	6°13'S	77°52'W
Chaco, state, Arg. (chä´kō)	187	26°0'S	60°30'W
Chad, nation, Afr. (chăd)	274	15°0'N	19°00'E
Chad, Lake, lk., Afr. (lāk chăd)	280-81	13°03'N	14°33'E
Chadbourn, N.C., U.S. (chăd´bŭrn)	134-35	34°19'N	78°50'W
Chadileuvu, stm., Arg. *see* Salado	185	38°49'S	64°59'W
Chadron, Ne., U.S. (chăd´rŭn)	124-25	42°50'N	103°00'W
Chaffee, Mo., U.S. (chăf´ê)	134-35	37°11'N	89°39'W
Chagos Archipelago, is., B.I.O.T. (chä´-gōs är´kȧ-pĕ´-ȧ-gō)	226-27	6°0'S	72°00'E
Chahanwusu, China *see* Dulan	260-61	36°10'N	98°22'E
Chaiyaphum, Thai.	266-67	15°48'N	102°02'E
Chalatenango, El Sal. (chäl-ä-tĕ-näŋ´gō)	160	14°02'N	88°56'W
Chalbi Desert, des., Kenya	289	3°00'N	37°20'E
Chalcis, Grc. *see* Chalkida	216-17	38°28'N	23°36'E
Chalkída, Grc.	216-17	38°28'N	23°36'E
Chalmette, La., U.S. (shăl mĕt´)	[illegible]	[illegible]	[illegible]
Chaltel, Cerro, mtn., S.A. (sĕ´r-rō-chäl´tĕl)	185	49°17'S	73°05'W
Chālūs, Iran	252-53	36°39'N	51°25'E
Chaman, Pak. (chŭm-än´)	252-53	30°55'N	66°27'E
Chambal, stm., India (chŭm-bäl´)	254-55	26°29'N	79°15'E
Chamberlain, S.D., U.S. (chäm´bẽr-lĭn)	124-25	43°49'N	99°19'W
Chambersburg, Pa., U.S.	126-27	39°56'N	77°40'W
Chambéry, Fr. (shäm-bā-rē´)	212-13	45°35'N	5°55'E
Chambi, Jebel, mtn., Tun.	280-81	35°13'N	8°40'E
Chamical, Arg.	182-83	30°21'S	66°18'W
Ch'amo Häyk', lk., Eth.	284-85	5°50'N	37°34'E
Chamonix-Mont-Blanc, Fr. (shȧ-mō-nē´-môN-bläN)	212-13	45°56'N	6°52'E
Champagne, hist. reg., Fr. (shäm-pän´yĕ)	212-13	49°0'N	4°30'E
Champaign, Il., U.S. (shăm-pān´)	126-27	40°07'N	88°15'W
Champdoré, Lac, lk., Qc., Can.	140-41	55°55'N	65°48'W
Champerico, Guat. (chăm-pȧ-rē´kō)	160	14°17'N	91°55'W
Champlain, Lac, lk., N.A. *see* Champlain, Lake	126-27	44°45'N	73°15'W
Champlain, Lake, lk., N.A. (lāk shăm-plān´)	126-27	44°45'N	73°15'W
Champotón, Mex. (chäm-pō-tōn´)	160	19°21'N	90°43'W
Chañaral, Chile (chän-yä-räl´)	182-83	26°21'S	70°37'W
Chan Chan, hist., Peru	184	8°03'S	79°08'W
Chanchan, Ruinas de, hist., Peru *see* Chan Chan	184	8°03'S	79°08'W
Chandalar, stm., Ak., U.S.	136	66°38'N	146°02'W
Chandeleur Islands, is., La., U.S. (shăn-dĕ-lōōr´ ī´lăndz)	134-35	29°49'N	88°54'W
Chandīgarh, India	254-55	30°44'N	76°54'E
Chandīgarh, state, India	254-55	30°45'N	76°50'E
Chandler, Qc., Can. (chăn´dlẽr)	148-49	48°21'N	64°41'W
Chandler, Az., U.S. (chăn´dlẽr)	128-29	33°18'N	111°53'W
Chandler, Ok., U.S. (chăn´dlẽr)	130-31	35°42'N	96°53'W
Chāndpur, Bngl.	254-55	23°13'N	90°40'E
Chandrapur, India	254-55	19°57'N	79°18'E
Chang, stm., China *see* Yangtze	258-59	31°24'N	121°54'E

n-sing; ŋ-baŋk; N-nasalized n; nŏd; cŏmmit; ōld; ôbey; ôrder; oi-boil; fōōd; ŏŏ-as oo in foot; ou-out; s-soft; sh-dish; th-thin; pūre; ûnite; ûrn; stŭd; circŭs; ü-as in French tu; ´-indeterminate vowel.

Feature (Pronunciation)	Page	Lat.	Long.
Changan, China *see* Xi'an	258-59	34°15′N	108°52′E
Changane, stm., Moz.	286-87	24°44′S	33°32′E
Changbaek-sanjulgi, mts., Asia *see* Changbai Shan	263	41°53′N	128°02′E
Changbai Shan, mts., Asia	263	41°53′N	128°02′E
Chang Cheng, p.o.i., China *see* Great Wall	260-61	40°0′N	112°30′E
Changchun, China (chäŋ-chòn)	260-61	43°53′N	125°19′E
Changde, China (chäŋ-dŭ)	258-59	29°02′N	111°41′E
Changhua, Tai. (chäng'hwä')	243a	24°04′N	120°30′E
Changji, China	240-41	44°01′N	87°18′E
Changjiang, China	258-59	19°16′N	109°02′E
Changkiakow, China *see* Zhangjiakou	260-61	40°49′N	114°53′E
Changli, China (chäŋ-lē)	260-61	39°42′N	119°10′E
Changmar, China	254-55	34°16′N	79°57′E
Changning, China (chäŋ-niŋ)	258-59	24°58′N	99°43′E
Changning, China (chäŋ-niŋ)	258-59	26°19′N	112°21′E
Changning, China (chäŋ-niŋ)	258-59	28°21′N	104°53′E
Changqing, China (chäŋ-chyiŋ)	260-61	36°33′N	116°44′E
Changsha, China	258-59	28°12′N	112°58′E
Changshu, China (chäŋ-shōō)	258-59	31°38′N	120°44′E
Changting, China	258-59	25°50′N	116°21′E
Changyi, China (chäŋ-yē)	260-61	36°52′N	119°24′E
Changzhi, China (chäŋ-jr)	260-61	36°11′N	113°07′E
Changzhou, China (chäŋ-jō)	258-59	31°47′N	119°57′E
Chaniá, Grc.	216a	35°31′N	24°01′E
Chankiang, China *see* Zhanjiang	258-59	21°12′N	110°23′E
Channel Islands, is. Eur (chăn'ĕl ī'lăndz)	212-13	49°20′N	2°20′W
Channel Islands, is., Ca., U.S. (chăn'ĕl ī'lăndz)	128-29	34°0′N	120°00′W
Channel Islands National Park, n.p., Ca., U.S.	128-29	33°28′N	119°02′W
Channel-Port aux Basques, Nf./L., Can.	148-49	47°35′N	59°10′W
Chanthaburi, Thai.	266-67	12°37′N	102°07′E
Chantilly, Fr. (shäɴ-tē-yē')	212-13	49°12′N	2°28′E
Chanute, Ks., U.S. (shȧ-nōōt')	130-31	37°41′N	95°27′W
Chany, Ozero, lk., Russia (ô'zē-rô chä'nê)	244	54°50′N	77°30′E
Chao'an, China (chou-än)	258-59	23°40′N	116°39′E
Chao Hu, lk., China	258-59	31°31′N	117°33′E
Chao Phraya, stm., Thai.	266-67	13°32′N	100°36′E
Chaor, stm., China (chou-r)	260-61	46°48′N	123°35′E
Chaoxian, China (chou shyĕn)	258-59	31°35′N	117°51′E
Chaoyang, China (chou-yäŋ)	258-59	23°16′N	116°35′E
Chaoyang, China (chou-yäŋ)	260-61	41°35′N	120°28′E
Chapala, Mex. (chä-pä'lä)	158-59	20°17′N	103°11′W
Chapala, Laguna de, lk., Mex. (lä-ō'nä-dĕ-chä-pä'lä)	158-59	20°15′N	103°00′W
Chaparral, Col. (chä-pär-rá'l)	188c	3°44′N	75°28′W
Chapayevsk, Russia (châ-pī'ĕfsk)	202-03	52°58′N	49°42′E
Chapecó, Braz.	182-83	27°06′S	52°38′W
Chapel Hill, N.C., U.S. (chăp''l hĭl)	134-35	35°55′N	79°04′W
Chapleau, On., Can. (chăp-lō')	146-47	47°51′N	83°25′W
Chapman, Mount, mtn., B.C., Can. (mount chăp'măn)	142-43	51°50′N	118°20′W
Chappell, Ne., U.S. (chă-pĕl')	124-25	41°06′N	102°28′W
Charadai, Arg.	187	27°39′S	59°52′W
Chār Borjak, Afg.	250-51	30°18′N	62°01′E
Charcas, Mex. (chär'käs)	158-59	23°08′N	101°08′W
Chärdjew, Turkmen. *see* Türkmenabat	252-53	39°05′N	63°35′E
Chardzhou, Turkmen. *see* Türkmenabat	252-53	39°05′N	63°35′E
Chari, stm., Afr. (shä-rē')	284-85	12°56′N	14°34′E
Chārīkār, Afg.	252-53	35°01′N	69°10′E
Chariton, Ia., U.S. (châr'ĭ-tŭn)	124-25	41°01′N	93°19′W
Charkhlik, China *see* Ruoqiang	238-39	39°01′N	88°11′E
Charleroi, Bel. (shȧr-lē-rwä')	206-07	50°25′N	4°26′E
Charleroi, Pa., U.S. (shär'lē-roi)	126-27	40°08′N	79°54′W
Charles, Cape, c., Va., U.S. (kāp chärlz)	134-35	37°08′N	75°58′W
Charles City, Ia., U.S. (chärlz sĭ'tē)	124-25	43°04′N	92°41′W
Charleston, Il., U.S. (chärlz'tŭn)	126-27	39°29′N	88°11′W
Charleston, Mo., U.S. (chärlz'tŭn)	134-35	36°55′N	89°21′W
Charleston, S.C., U.S. (chärlz'tŭn)	134-35	32°47′N	79°56′W
Charleston, W.V., U.S. (chärlz'tŭn)	126-27	38°21′N	81°38′W
Charlestown, In., U.S. (chärlz'toun)	126-27	38°27′N	85°40′W
Charles Young Buffalo Soldiers National Monument, n.p., Oh., U.S.	126-27	39°42′N	83°53′W
Charleville, Austl. (chär'lē-vĭl)	301	26°24′S	146°14′E
Charlevoix, Mi., U.S. (shär'lē-voi)	126-27	45°18′N	85°15′W
Charlotte, Mi., U.S. (shär'lŏt)	126-27	42°33′N	84°50′W
Charlotte, N.C., U.S. (shär'lŏt)	134-35	35°14′N	80°51′W
Charlotte Amalie, nat. cap., V.I.U.S. (shär-lŏt'ĕ ä-mä'lĭ-å)	155b	18°21′N	64°56′W
Charlotte Harbor, b., Fl., U.S. (shär'lŏt här'bẽr)	135a	26°45′N	82°11′W
Charlottenberg, Swe. (shär-lŭt'ĕn-bĕrg)	208-09	59°54′N	12°18′E
Charlottesville, Va., U.S. (shär'lŏtz-vĭl)	126-27	38°02′N	78°29′W
Charlottetown, P.E., Can.	148-49	46°14′N	63°08′W
Charlton Island, i., Nu., Can.	140-41	52°0′N	79°30′W
Chārsadda, Pak. (chŭr-sä'dä)	252-53	34°09′N	71°44′E
Charters Towers, Austl.	302	20°04′S	146°16′E
Chartres, Fr. (shärt'r')	212-13	48°27′N	1°29′E
Chascomús, Arg. (chäs-kō-mōōs')	187	35°35′S	58°01′W
Chase City, Va., U.S. (chās sĭ'tĭ)	134-35	36°47′N	78°28′W
Châteaudun, Fr. (shä-tō-däɴ')	212-13	48°04′N	1°20′E
Château-Gontier, Fr. (chȧ-tō'gôɴ'tyā')	212-13	47°50′N	0°42′W
Châteauguay, Qc., Can. (chȧ-tō-gā)	146-47	45°23′N	73°44′W
Châteauroux, Fr. (shȧ-tō-rōō')	212-13	46°48′N	1°42′E
Château-Thierry, Fr. (shȧ-tō'ty-ĕr-rē')	212-13	49°03′N	3°24′E
Châtellerault, Fr. (shä-tĕl-rō')	212-13	46°49′N	0°33′E
Chatham, On., Can. (chăt'ăm)	146-47	42°24′N	82°11′W
Chatham, Il., U.S. (chăt'ăm)	126-27	39°40′N	89°42′W
Chatham, i., Ec. *see* San Cristóbal, Isla	184a	0°50′S	89°26′W
Chatham, Isla, i., Chile	185	50°38′S	74°27′W
Chatham Sound, strt., B.C., Can. (chăt'ăm sound)	142-43	54°32′N	130°35′W
Chatham Strait, strt., Ak., U.S. (chăt'ăm strāt)	136	57°30′N	134°45′W
Chatrapur, India	256	19°21′N	84°60′E
Chattahoochee, Fl., U.S. (chăt-tȧ-hōō' cheē)	134-35	30°42′N	84°51′W
Chattahoochee, stm., U.S. (chăt-tȧ-hōō' cheē)	120-21	30°46′N	84°52′W
Chattanooga, Tn., U.S. (chăt-ȧ-nōō'gȧ)	134-35	35°03′N	85°18′W
Chaudière, stm., Qc., Can. (shō-dyĕr')	148-49	46°45′N	71°17′W
Chauk, Mya.	266-67	20°54′N	94°49′E
Chaumont, Fr. (shō-môɴ')	212-13	48°06′N	5°08′E
Chauny, Fr. (shō-nē')	212-13	49°37′N	3°13′E
Chaves, Port. (chä'vĕzh)	214-15	41°44′N	7°28′W
Chavin, hist., Peru	184	9°37′S	77°14′W
Chavin de Huantar, hist., Peru *see* Chavin	184	9°37′S	77°14′W
Chaykovskij, Russia	202-03	56°46′N	54°06′E
Cheb, Czech Rep. (κĕb)	210-11	50°05′N	12°22′E
Cheboksary, Russia (chyĕ-bôk-sä'rē)	202-03	56°08′N	47°15′E
Cheboygan, Mi., U.S. (shē-boi'găn)	126-27	45°38′N	84°29′W
Chech, 'Erg, des., Afr.	280-81	24°43′N	2°31′W
Chechen', Ostrov, i., Russia (ôs-trôf' chyĕch'ĕn)	245	43°59′N	47°41′E
Chechnya, state, Russia	245	43°20′N	45°45′E
Checotah, Ok., U.S. (chē-kō'tȧ)	130-31	35°29′N	95°31′W
Chedabucto Bay, b., N.S., Can. (chĕd-ȧ-bŭk-tō bā)	148-49	45°23′N	61°10′W
Cheduba Island, i., Mya.	266-67	18°48′N	93°38′E
Cheektowaga, N.Y., U.S. (chēk-tō-wä'gȧ)	126-27	42°54′N	78°45′W
Chefoo, China *see* Yantai	260-61	37°32′N	121°21′E
Chegdomyn, Russia	236-37	51°08′N	133°05′E
Chehalis, Wa., U.S. (chē-hā'lĭs)	122-23	46°40′N	122°58′W
Cheju, Kor., S. *see* Jeju	260-61	33°30′N	126°32′E
Chekiang, state, China *see* Zhejiang	258-59	29°0′N	120°00′E
Chelan, Wa., U.S. (chē-lăn')	122-23	47°50′N	120°00′W
Chelif, Oued, stm., Alg. (wĕd shā-lēf)	214-15	36°03′N	0°08′E
Chełm, Pol. (κĕlm)	210-11	51°08′N	23°30′E
Chełmno, Pol. (κĕlm'nō)	210-11	53°21′N	18°27′E
Chelmsford, Eng., U.K. (chĕlm's-fẽrd)	206-07	51°44′N	0°28′E
Chelsea, Mi., U.S. (chĕl'sē)	126-27	42°19′N	84°01′W
Chelsea, Ok., U.S. (chĕl'sē)	130-31	36°32′N	95°26′W
Cheltenham, Eng., U.K. (chĕlt'nŭm)	206-07	51°54′N	2°04′W
Chelyabinsk, Russia (chĕl-yä-bĕnsk')	244	55°10′N	61°26′E
Chelyuskin, Mys, c., Russia (mĭs chĕl-yôs'-kĭn)	236-37	77°45′N	104°20′E
Chemnitz, Ger.	210-11	50°50′N	12°56′E
Chemulpo, Kor., S. *see* Inch'ŏn	263	37°28′N	126°38′E
Chenāb, stm., Asia (chē-näb)	252-53	29°21′N	71°02′E
Cheney, Wa., U.S. (chē'nå)	122-23	47°29′N	117°35′W
Chengchow, China *see* Zhengzhou	258-59	34°46′N	113°39′E
Chengde, China (chŭŋ-dŭ)	260-61	40°58′N	117°56′E
Chengdu, China (chŭŋ-dōō)	258-59	30°39′N	104°04′E
Chengshan Jiao, c., China (jyou chŭŋ-shän)	260-61	37°23′N	122°42′E
Chennai, India	256	13°06′N	80°15′E
Chenyang, China *see* Shenyang	263	41°48′N	123°24′E
Chenzhou, China	258-59	25°48′N	112°59′E
Cheonan, Kor., S.	263	36°48′N	127°10′E
Cheongju, Kor., S.	263	36°38′N	127°30′E
Cheongsando, i., Kor., S.	263	34°11′N	126°54′E
Cheorwon, Kor., S.	263	38°17′N	127°14′E
Chepén, Peru (chĕ-pĕ'n)	184	7°14′S	79°25′W
Chepo, Pan. (chā'pō)	162	9°10′N	79°06′W
Cheraw, S.C., U.S. (chē'rô)	134-35	34°42′N	79°54′W
Cherbourg, Fr. (shär-bòr')	212-13	49°39′N	1°38′W
Cheremkhovo, Russia (chĕr'yĕm-kô-vō)	236-37	53°09′N	103°04′E
Cherepanovo, Russia (chĕr'yĕ pä-nô'vō)	244	54°13′N	83°21′E
Cherepovets, Russia (chĕr-yĕ-pô'vyĕtz)	218-19	59°08′N	37°55′E
Chergui, Chott ech, lk., Alg. (chĕr-gē)	280-81	34°13′N	0°26′E
Cherkassy, Ukr. *see* Cherkasy	218-19	49°26′N	32°05′E
Cherkasy, Ukr.	218-19	49°26′N	32°05′E
Cherkessia, state, Russia	245	44°0′N	42°00′E
Cherkessk, Russia	245	44°13′N	42°04′E
Cherlak, Russia (chĭr-läk')	244	54°09′N	74°49′E
Chermoz, Russia (chĕr-môz')	202-03	58°47′N	56°09′E
Chernigov, Ukr. *see* Chernihiv	218-19	51°30′N	31°17′E
Chernihiv, Ukr.	218-19	51°30′N	31°17′E
Chernivtsi, Ukr.	210-11	48°17′N	25°58′E
Chernobyl, *see* Chornobyl', Ukr.	218-19	51°17′N	30°14′E
Cherno More, s., *see* Black Sea	202-03	43°0′N	35°00′E
Chernovtsy, Ukr. *see* Chernivtsi	210-11	48°17′N	25°58′E
Chernoye More, s., *see* Black Sea	202-03	43°0′N	35°00′E
Chernyakhovsk, Russia	210-11	54°38′N	21°49′E
Chernyanka, Russia (chĕrn-yäŋ'kä)	218-19	50°56′N	37°49′E
Cherokee, Ia., U.S. (chĕr-ô-kē')	124-25	42°45′N	95°33′W
Cherokee, Ok., U.S. (chĕr-ô-kē')	130-31	36°45′N	98°21′W
Cherrapunji, India	254-55	25°13′N	91°42′E
Cherryville, N.C., U.S. (chĕr'ĭ-vĭl)	134-35	35°23′N	81°23′W
Cherskiy, Russia	236-37	68°46′N	161°24′E
Cherskiy Mountains, mts., Russia (chĕr'skē moun'tĭnz) *see* Cherskogo, Khrebet	236-37	65°0′N	144°00′E
Cherskogo, Khrebet, mts., Russia	236-37	65°0′N	144°00′E
Cherson, Ukr. *see* Kherson	218-19	46°38′N	32°35′E
Chervonohrad, Ukr.	210-11	50°23′N	24°14′E
Chesaning, Mi., U.S. (chĕs'ȧ-nĭng)	126-27	43°11′N	84°07′W
Chesapeake, Va., U.S. (chĕs'ȧ-pēk)	134-35	36°48′N	76°16′W
Chesapeake Bay, b., U.S. (chĕs'ȧ-pēk bā)	126-27	38°38′N	76°27′W
Chester, Eng., U.K. (chĕs'tẽr)	206-07	53°12′N	2°54′W
Chester, Il., U.S. (chĕs'tẽr)	130-31	37°55′N	89°49′W
Chester, Mt., U.S. (chĕs'tẽr)	122-23	48°32′N	110°57′W
Chester, Pa., U.S. (chĕs'tẽr)	126-27	39°51′N	75°21′W
Chester, S.C., U.S. (chĕs'tẽr)	134-35	34°42′N	81°13′W
Chester, Va., U.S. (chĕs'tẽr)	126-27	37°21′N	77°26′W
Chester, W.V., U.S. (chĕs'tẽr)	126-27	40°36′N	80°33′W
Chesterfield, Eng., U.K. (chĕs'tẽr-fēld)	206-07	53°14′N	1°26′W
Chesterfield, Îles, is., N. Cal.	304-05	19°30′S	158°00′E
Chesterfield, Nosy, i., Madag.	286-87	16°20′S	43°58′E
Chesterfield Inlet, Nu., Can.	138-39	63°21′N	90°43′W
Chetek, Wi., U.S. (chĕ'tĕk)	124-25	45°19′N	91°39′W
Chettlatt Island, i., India	256	11°41′N	72°42′E
Chetumal, Mex.	160	18°30′N	88°18′W
Chetumal, Bahía, b., N.A. (bä-ē-ä-chĕt-ōō-mäl')	160	18°39′N	88°06′W
Chevak, Ak., U.S.	136	61°39′N	165°17′W
Cheviot, Oh., U.S. (shĕv'ĭ-ŭt)	126-27	39°09′N	84°37′W
Ch'ew Bahir, lk., Afr.	289	4°40′N	36°50′E
Chewelah, Wa., U.S. (chē-wē'lä)	122-23	48°17′N	117°43′W
Cheyenne, Wy., U.S. (shī-ĕn')	124-25	41°10′N	104°48′W
Cheyenne, stm., U.S. (shī-ĕn')	120-21	44°47′N	100°44′W
Cheyenne River Indian Reservation, ind. res., S.D., U.S. (shī-ĕn' rĭv'ẽr ĭn'dĭ-ăn rĕ-sẽr-vā'shĕn)	124-25	45°0′N	100°40′W
Cheyenne Wells, Co., U.S. (shī-ĕn' wĕls)	130-31	38°49′N	102°21′W
Chhapra, India	254-55	25°47′N	84°45′E
Chhatarpur, India	254-55	24°55′N	79°36′E
Chhattisgarh, state, India	254-55	21°30′N	82°00′E
Chhindwāra, India	254-55	22°03′N	78°57′E
Chi, stm., Thai.	266-67	15°11′N	104°43′E
Chiai, Tai. (chī'ī')	243a	23°29′N	120°27′E
Chiang Mai, Thai.	266-67	18°48′N	99°00′E
Chiang Rai, Thai.	266-67	19°55′N	99°50′E
Chiapa de Corzo, Mex. (chē-ä'pä dā kôr'zō)	158-59	16°41′N	92°60′W
Chiapas, state, Mex. (chē-ä'päs)	158-59	16°30′N	92°30′W
Chiavari, Italy (kyä-vä'rē)	216-17	44°20′N	9°20′E
Chiba, Japan (chē'bȧ)	265	35°36′N	140°08′E
Chiba, state, Japan (chē'bȧ)	265	35°30′N	140°20′E

ăt; finăl; rāte; senâte; ärm; åsk; sofȧ; fâre; ch-choose; dh-as th in other; bē; ēvent; bĕt; recĕnt; cratẽr; g-gō; gh-guttural g; bĭt; ĭ-short neutral; rīde; κ-guttural k as ch in German ich;

PRONOUNCING INDEX

Chibougamau – Cintalapa 345

Feature (Pronunciation)	Page	Lat.	Long.
Chibougamau, Qc., Can. (chē-bōō´gä-mou)	148-49	49°55'N	74°22'W
Chibougamau, Lac, lk., Qc., Can. (läk chē-bōō´gä-mou)	148-49	49°50'N	74°15'W
Chibuto, Moz.	286-87	24°42'S	33°34'E
Chicago, Il., U.S. (shĭ-kô-gō) (chĭ-kä´gō)	126-27	41°52'N	87°38'W
Chicago Heights, Il., U.S. (shĭ-kô-gō hīts)	126-27	41°30'N	87°39'W
Chicapa, stm., Afr. (chē-kä´pä)	284-85	6°25'S	20°48'E
Chichagof Island, i., Ak., U.S. (chė-chä´gôf ī´lȧnd)	136	57°07'N	135°12'W
Chichén Itzá, hist., Mex.	160	20°40'N	88°35'W
Chichester, Eng., U.K. (chĭch´ĕs-tẽr)	206-07	50°51'N	0°47'W
Chichimilá, Mex. (chē-chē-mė´lä)	160	20°37'N	88°13'W
Chichiriviche, Ven. (chē-chē-rē-vē-chĕ)	188b	10°56'N	68°17'W
Chickamauga, Ga., U.S. (chĭk-ȧ-mô´gȧ)	134-35	34°53'N	85°18'W
Chickasawhay, stm., Ms., U.S. (chĭk-ȧ-sô´wā)	134-35	31°0'N	88°45'W
Chickasha, Ok., U.S. (chĭk´ȧ-shā)	130-31	35°03'N	97°57'W
Chiclana de la Frontera, Spain (chė-klä´nä dĕ-lä-frôn-tĕ´rä)	214-15	36°25'N	6°08'W
Chiclayo, Peru (chė-klä´yō)	184	6°46'S	79°51'W
Chico, Ca., U.S. (chē´kō)	128-29	39°44'N	121°49'W
Chico, stm., Arg.	185	43°49'S	66°26'W
Chico, stm., Arg.	185	49°52'S	68°35'W
Chicopee, Ma., U.S. (chĭk´ỏ-pē)	126-27	42°09'N	72°37'W
Chicoutimi, Qc., Can. (shē-kōō´tē-mē´)	146-47	48°26'N	71°04'W
Chicxulub, Mex. (chēk-sōō-lōō´b)	160	21°08'N	89°31'W
Chidambaram, India	256	11°24'N	79°42'E
Chiefland, Fl., U.S. (chēf´lȧnd)	134-35	29°29'N	82°52'W
Chieri, Italy (kyâ´rē)	216-17	45°01'N	7°49'E
Chieti, Italy (kyĕ´tē)	216-17	42°21'N	14°10'E
Chifeng, China (chr-fŭn)	260-61	42°16'N	118°58'E
Chignecto Bay, b., Can. (shĭg-nĕk´tō bā)	148-49	45°35'N	64°45'W
Chihli, Gulf of, b., China see Bo Hai	260-61	38°30'N	120°02'E
Chihuahua, Mex. (chė-wä´wä)	156-57	28°38'N	106°05'W
Chihuahua, state, Mex.	132-33	28°30'N	106°00'W
Chihuahua, Desierto de, des., N.A. see Chihuahuan Desert	156-57	35°0'N	106°00'W
Chihuahuan Desert, des., N.A.	156-57	35°0'N	106°00'W
Chikkamagalūru, India	256	13°19'N	75°47'E
Chikoy, stm., Asia	260-61	51°02'N	106°38'E
Chilcotin, stm., B.C., Can. (chĭl-kō´tĭn)	142-43	51°44'N	122°24'W
Childers, Austl.	301	25°14'S	152°17'E
Childress, Tx., U.S. (chĭld´rĕs)	130-31	34°26'N	100°13'W
Chile, nation, S.A. (chē´lā)	172	30°0'S	71°00'W
Chile Chico, Chile	185	46°33'S	71°44'W
Chilecito, Arg. (chė-lȧ-sē´tō)	182-83	29°10'S	67°30'W
Chilecito, Arg. (chė-lȧ-sē´tō)	188e	33°53'S	69°04'W
Chilika Lake, lk., India	254-55	19°45'N	85°25'E
Chilko, stm., B.C., Can. (chĭl´kō)	142-43	52°06'N	123°28'W
Chilko Lake, lk., B.C., Can.	142-43	51°17'N	124°04'W
Chillán, Chile (chēl-yän´)	185	36°36'S	72°07'W
Chillicothe, Il., U.S. (chĭl-ĭ-kŏth´ė)	126-27	40°55'N	89°29'W
Chillicothe, Mo., U.S. (chĭl-ĭ-kŏth´ė)	130-31	39°48'N	93°33'W
Chillicothe, Oh., U.S. (chĭl-ĭ-kŏth´ė)	126-27	39°20'N	82°59'W
Chilliwack, B.C., Can. (chĭl´ĭ-wăk)	142-43	49°10'N	121°57'W
Chiloé, Isla Grande de, i., Chile	185	42°30'S	73°55'W
Chilpancingo de los Bravo, Mex.	158-59	17°33'N	99°30'W
Chilton, Wi., U.S. (chĭl´tŭn)	126-27	44°02'N	88°09'W
Chilung, Tai. (chĭ´lung)	243a	25°08'N	121°44'E
Chilwa, Lake, lk., Afr.	286-87	15°20'S	35°43'E
Chimaltenango, Guat. (chė-mäl-tå-näŋ´gō)	160	14°40'N	90°49'W
Chimborazo, vol., Ec. (chėm-bō-rä´zō)	184	1°28'S	78°48'W
Chimbote, Peru (chėm-bō´tå)	184	9°04'S	78°35'W
Chimboy, Uzb.	244	42°56'N	59°47'E
Chimkent, Kaz. see Shymkent	244	42°18'N	69°36'E
Chimoio, Moz.	286-87	19°09'S	33°30'E
China, Mex. (chē´nä)	132-33	25°42'N	99°14'W
China, nation, Asia (chī´nȧ)	224-25	35°0'N	105°00'E
Chinandega, Nic. (chē-nän-dā´gä)	161	12°38'N	87°08'W
China Selatan, Laut, s., Asia see South China Sea	242-43	10°0'N	113°00'E
Chincha Alta, Peru (chĭn´chä äl´tä)	188a	13°26'S	76°08'W
Chinchilla, Austl. (chĭn-chĭl´ȧ)	301	26°45'S	150°38'E
Chindwinn, stm., Mya. (chĭn-dwĭn)	258-59	21°24'N	95°16'E
Chingola, Zam. (chĭng-gōlä)	286-87	12°32'S	27°52'E
Chinhae, Kor., S. see Jinhae	263	35°08'N	128°40'E
Chin Hills, hills, Mya.	254-55	22°30'N	93°30'E
Chinhoyi, Zimb.	286-87	17°22'S	30°11'E
Chiniot, Pak.	252-53	31°43'N	72°59'E
Chinju, Kor., S. see Jinju	263	35°10'N	128°05'E
Chinko, stm., C.A.R. (shĭn´kō)	284-85	4°51'N	23°53'E
Chinmen Tao, i., Tai.	243a	24°27'N	118°23'E
Chinnampo, Kor., N. see Namp'o	263	38°45'N	125°23'E
Chinon, Fr. (shē-nôɴ´)	212-13	47°11'N	0°15'E
Chinook, Mt., U.S. (shĭn-ŏk´)	122-23	48°36'N	109°14'W
Chinsali, Zam.	286-87	10°33'S	32°04'E
Chioggia, Italy (kyôd´jä)	216-17	45°13'N	12°17'E
Chíos, Grc. (kē´ŏs)	216-17	38°23'N	26°09'E
Chíos, i., Grc.	216-17	38°23'N	26°04'E
Chipata, Zam.	286-87	13°37'S	32°38'E
Chipley, Fl., U.S. (chĭp´lĭ)	134-35	30°47'N	85°32'W
Chipman, N.B., Can. (chĭp´mȧn)	148-49	46°10'N	65°53'W
Chippewa, stm., Wi., U.S. (chĭp´ė-wä)	124-25	44°24'N	92°04'W
Chippewa Falls, Wi., U.S. (chĭp´ė-wä fôlz)	124-25	44°56'N	91°24'W
Chiquimula, Guat. (chē-kė-mōō´lä)	160	14°48'N	89°33'W
Chiquimulilla, Guat. (chē-kē-mōō-lē´l-yä)	160	14°05'N	90°23'W
Chiquinquirá, Col. (chė-kēŋ´kė-rä´)	188c	5°37'N	73°48'W
Chirchiq, Uzb.	252-53	41°28'N	69°35'E
Chire, stm., Afr. see Shire	286-87	17°42'S	35°19'E
Chiricahua National Monument, n.p., Az., U.S. (chĭ-rā-cä´hwä năsh´ŭn-ăl mŏn´ū-mĕnt)	128-29	31°59'N	109°22'W
Chiricahua Peak, mtn., Az., U.S.	128-29	31°52'N	109°20'W
Chiriquí Grande, Pan. (chē-rė-kē´ grän´dā)	162	8°57'N	82°08'W
Chirripó, Cerro, mtn., C.R.	161	9°29'N	83°30'W
Chirua, Lago, lk., Afr. see Chilwa, Lake	286-87	15°20'S	35°43'E
Chisasibi, Qc., Can.	138-39	53°48'N	79°02'W
Chishima-rettō, is., Russia see Kuril Islands	236-37	47°14'N	152°18'E
Chisholm, Mn., U.S. (chĭz´ŭm)	124-25	47°29'N	92°53'W
Chisimayu, Som.	284-85	0°22'S	42°32'E
Chişinău, nat. cap., Mol.	218-19	47°02'N	28°50'E
Chistopol, Russia (chĭs-tô´pỏl-y')	202-03	55°22'N	50°37'E
Chistyakovo, Ukr. see Torez	218-19	48°02'N	38°38'E
Chita, Russia (chė-tá´)	240-41	52°02'N	113°29'E
Chitato, Ang.	284-85	7°20'S	20°46'E
Chitradurga, India	256	14°14'N	76°24'E
Chitral, Pak. (chē-trāl´)	252-53	35°52'N	71°49'E
Chitré, Pan.	162	7°58'N	80°26'W
Chittagong, Bngl. (chĭt-á-gông´)	266-67	22°20'N	91°50'E
Chittaurgarh, India	254-55	24°54'N	74°37'E
Chittoor, India	256	13°13'N	79°06'E
Chiumbe, stm., Afr. (chė-ỏm´bå)	284-85	6°59'S	21°11'E
Chivasso, Italy (kė-väs´sō)	216-17	45°12'N	7°54'E
Chivilcoy, Arg. (chē-vėl-koi´)	187	34°54'S	60°02'W
Chkalov, Russia	202-03	51°48'N	55°06'E
Chlef, Alg.	214-15	36°10'N	1°20'E
Chŏâm Khsant, Camb.	266-67	14°13'N	104°56'E
Chochis, Cerro, mtn., Bol.	182-83	18°08'S	59°54'W
Chodzież, Pol. (κōj´yĕsh)	210-11	52°60'N	16°55'E
Choele Choel, Arg. (chỏ-ĕ´lĕ-chỏ-ĕ´l)	185	39°17'S	65°39'W
Choiseul, i., Sol. Is. (shwä-zûl´)	306e	7°05'S	157°00'E
Chojnice, Pol. (κōĭ-nē-tsĕ)	210-11	53°42'N	17°34'E
Chokurdakh, Russia	236-37	70°38'N	147°53'E
Chókwe, Moz.	286-87	24°33'S	32°60'E
Cholet, Fr. (shỏ-lĕ´)	212-13	47°04'N	0°53'W
Choluteca, Hond. (chō-lōō-tā´kä)	161	13°18'N	87°12'W
Choma, Zam.	286-87	16°49'S	26°59'E
Chomutov, Czech Rep. (kō´mỏ-tôf)	210-11	50°28'N	13°25'E
Chona, stm., Russia (chō´nä)	236-37	62°54'N	111°06'E
Chon Buri, Thai.	266-67	13°22'N	101°00'E
Chone, Ec. (chô´nĕ)	184	0°42'S	80°05'W
Ch'ŏngjin, Kor., N. (chŭng-jĭn´)	263	41°47'N	129°48'E
Chongqing, China	258-59	29°34'N	106°34'E
Chongqing, state, China	258-59	30°0'N	108°00'E
Chongzuo, China	258-59	22°24'N	107°22'E
Chonos, Archipiélago de los, is., Chile	185	45°0'S	74°00'W
Chorne more, s., see Black Sea	202-03	43°0'N	35°00'E
Chornobyl', Ukr.	218-19	51°17'N	30°14'E
Chōshi, Japan (chō´shē)	265	35°44'N	140°50'E
Chos Malal, Arg.	185	37°23'S	70°16'W
Chosŏn minjujuŭi-inmīn-konghwaguk, nation, Asia see North Korea	224-25	40°0'N	127°00'E
Choszczno, Pol. (chôsh´chnō)	210-11	53°10'N	15°25'E
Choteau, Mt., U.S. (shō´tō)	122-23	47°49'N	112°12'W
Chouk'ou, China see Shangshui	258-59	33°33'N	114°34'E
Choushan Islands, is., China see Zhoushan Qundao	258-59	30°0'N	122°00'E
Chown, Mount, mtn., Ab., Can. (mount choun)	142-43	53°24'N	119°22'W
Choybalsan, Mong.	260-61	48°04'N	114°32'E
Choyr, Mong.	260-61	46°22'N	108°22'E
Christchurch, N.Z. (krīst´chûrch)	303	43°32'S	186°39'E
Christiansburg, Va., U.S. (krĭs´chănz-bûrg)	134-35	37°08'N	80°25'W
Christina, stm., Ab., Can.	144-45	56°40'N	111°04'W
Christmas Island, dep., Oc. (krĭs´-măs ī´lȧnd)	268-69	10°30'S	105°40'E
Christmas Island, at., Kir. see Kiritimati	304-05	1°48'N	157°19'W
Chrudim, Czech Rep. (κrōō´dyĕm)	210-11	49°57'N	15°48'E
Chrzanów, Pol. (κzhä´nỏf)	210-11	50°07'N	19°26'E
Chu, stm., Asia	266-67	19°53'S	105°45'E
Chubut, state, Arg. (chỏ-bōōt´)	185	44°0'S	69°00'W
Chubut, stm., Arg. (chỏ-bōōt´)	185	43°21'S	65°03'W
Chucunaque, stm., Pan. (chōō-kōō-nä´kå)	162	8°08'N	77°44'W
Chudovo, Russia (chỏ´dỏ-vỏ)	208-09	59°07'N	31°41'E
Chudskoye Ozero, lk., Eur. (chỏt´skỏ-yĕ ô´zĕ-rỏ) see Peipus, Lake	208-09	58°45'N	27°25'E
Chuguyevka, Russia (chỏ-gōō´yĕf-kå)	264	44°09'N	133°52'E
Chukchi Sea, s., (chōōk´chē sē)	136	69°0'N	171°00'W
Chukotskiy, Mys, c., Russia	136	64°16'N	187°07'W
Chukotskiy Poluostrov, pen., Russia see Chukotsk Peninsula	236-37	66°0'N	175°00'W
Chukotskoye More, s., see Chukchi Sea	136	69°0'N	171°00'W
Chukotsk Peninsula, pen., Russia	236-37	66°0'N	175°00'W
Chula Vista, Ca., U.S. (chōō´lȧ vĭs´tȧ)	128-29	32°38'N	117°05'W
Chul'man, Russia	236-37	56°51'N	124°53'E
Chulucanas, Peru	184	5°06'S	80°09'W
Chulym, stm., Russia	236-37	57°42'N	83°52'E
Chumbicha, Arg.	182-83	28°52'S	66°14'W
Chumphon, Thai.	266-67	10°30'N	99°08'E
Chumysh, stm., Russia	244	53°31'N	83°10'E
Chuna, stm., Russia	236-37	57°44'N	95°27'E
Chuncheon, Kor., S.	263	37°52'N	127°44'E
Chungju, Kor., S.	263	36°58'N	127°50'E
Chungking, China see Chongqing	258-59	29°34'N	106°34'E
Chungking, state, China see Chongqing	258-59	30°0'N	108°00'E
Chunya, stm., Russia (chỏn´yä´)	236-37	61°37'N	96°32'E
Chuquicamata, Chile (chōō-kė-kä-mä´tä)	182-83	22°19'S	68°56'W
Chur, Switz. (kōōr)	210-11	46°51'N	9°31'E
Churchill, Mb., Can. (chûrch´ĭl)	138-39	58°47'N	94°10'W
Churchill, stm., Can. (chûrch´ĭl)	140-41	58°49'N	94°11'W
Churchill, Cape, c., Mb., Can. (kăp chûrch´ĭl)	140-41	58°46'N	93°14'W
Churchill Lake, lk., Sk., Can. (chûrch´ĭl lāk)	144-45	55°55'N	108°20'W
Chūru, India	254-55	28°18'N	74°58'E
Chusovaya, stm., Russia (chōō-sô-vä´yȧ)	192-93	58°09'N	57°03'E
Chusovoy, Russia (chōō-sô-vôy´)	202-03	58°18'N	57°49'E
Chuuk, is., Micron.	304-05	7°16'N	151°44'E
Chuvashia, state, Russia	202-03	55°30'N	47°00'E
Chuvashiya, state, Russia see Chuvashia	202-03	55°30'N	47°00'E
Chuxian, China (chōō shyĕn)	258-59	32°19'N	118°18'E
Chuxiong, China (chōō-shyôŋ)	258-59	25°03'N	101°33'E
Chüy, stm., Asia see Shū	244	45°00'N	67°45'E
Cianjur, Indon.	268-69	6°48'S	107°08'E
Cicero, Il., U.S. (sĭs´ẽr-ō)	126-27	41°51'N	87°45'W
Cicia, i., Fiji	306f	17°45'S	179°18'W
Ciechanów, Pol. (tsyĕ-kä´nỏf)	210-11	52°53'N	20°37'E
Ciego de Ávila, Cuba (syā´gō-dĕ-ä´vė-lä)	154-55	21°51'N	78°46'W
Ciego de Ávila, state, Cuba (syā´gō-dĕ-ä´vė-lä)	154-55	22°0'N	78°40'W
Ciempozuelos, Spain (thyĕm-pỏ-thwā´lōs)	214-15	40°10'N	3°37'W
Ciénaga, Col. (syā´nä-gä)	178-79	11°00'N	74°15'W
Cienfuegos, Cuba (syĕn-fwā´gōs)	154-55	22°10'N	80°26'W
Cienfuegos, state, Cuba (syĕn-fwā´gōs)	154-55	22°10'N	80°30'W
Cieszyn, Pol. (tsyĕ´shĕn)	210-11	49°45'N	18°38'E
Cieza, Spain (thyā´thä)	214-15	38°14'N	1°25'W
Cihuatlán, Mex. (sē-wä-tlá´n)	158-59	19°14'N	104°34'W
Cikobia, i., Fiji	306f	15°43'S	179°58'W
Cilacap, Indon.	268-69	7°44'S	109°01'E
Cill Chainnigh, Ire. see Kilkenny	206-07	52°39'N	7°15'W
Cimarron, stm., U.S. (sĭm-ȧ-rōn´)	130-31	36°10'N	96°17'W
Cina Selatan, Laut, s., Asia see South China Sea	242-43	10°0'N	113°00'E
Cincinnati, Oh., U.S. (sĭn-sĭ-năt´ĭ)	126-27	39°11'N	84°28'W
Cintalapa, Mex. (sĕn-tä-lä´pä)	158-59	16°41'N	93°43'W

n-sing; ŋ-baŋk; ɴ-nasalized n; nŏd; cŏmmit; ōld; ỏbey; ôrder; oi-boil; fōōd; ỏ-as oo in foot; ou-out; s-soft; sh-dish; th-thin; pūre; ůnite; ûrn; stŭd; circŭs; ü-as in French tu; ´-indeterminate vowel.

346 Cinto, Monte – Colmar PRONOUNCING INDEX

Feature (Pronunciation)	Page	Lat.	Long.
Cinto, Monte, mtn., Fr. (môɴ chēn´tō)	**200-01**	42°23´N	8°56´E
Cipolletti, Arg.	**185**	38°56´S	67°59´W
Circleville, Oh., U.S. (sûr´k'lvĭl)	**126-27**	39°36´N	82°57´W
Circleville, Ut., U.S. (sûr´k'lvĭl)	**128-29**	38°10´N	112°16´W
Cirebon, Indon.	**268-69**	6°45´S	108°34´E
Cisco, Tx., U.S. (sĭs´kō)	**130-31**	32°24´N	98°59´W
Cisneros, Col. (sês-nĕ´rôs)	**188c**	6°33´N	75°04´W
Cisterna di Latina, Italy (chēs-tĕ´r-nä-dē-lä-tē´nä)	**216-17**	41°36´N	12°49´E
Citlaltépetl, Volcán, vol., Mex. *see* Pico de Orizaba, Volcán	**158-59**	19°01´N	97°16´W
Citronelle, Al., U.S. (cĭt-rō´nĕl)	**134-35**	31°06´N	88°14´W
Città di Castello, Italy (chēt-tä´dē käs-tĕl´lō)	**216-17**	43°28´N	12°15´E
Ciudad Acuña, Mex.	**132-33**	29°19´N	100°56´W
Ciudad Altamirano, Mex. (syōō-dä´d-äl-tä-mē-rä´nō)	**158-59**	18°21´N	100°39´W
Ciudad Bolívar, Ven. (syōō-dhädh´ bô-lē´vär)	**178-79**	8°07´N	63°33´W
Ciudad Camargo, Mex.	**132-33**	27°42´N	105°10´W
Ciudad Cortés, C.R.	**161**	8°58´N	83°32´W
Ciudad Darío, Nic. (syōō-dhädh´dä´rē-ō)	**161**	12°43´N	86°07´W
Ciudad del Carmen, Mex. (syōō-dä´d-dĕl-ká´r-mĕn)	**160**	18°39´N	91°49´W
Ciudad del Maíz, Mex. (syōō-dhädh´del mä-ēz´)	**158-59**	22°24´N	99°36´W
Ciudad de México, nat. cap., Mex. *see* Mexico City	**158-59**	19°24´N	99°09´W
Ciudad de Nutrias, Ven.	**178-79**	8°05´N	69°17´W
Ciudad Guayana, Ven.	**178-79**	8°21´N	62°39´W
Ciudad Hidalgo, Mex. (syōō-dä´d-ē-dä´l-gô)	**158-59**	19°41´N	100°34´W
Ciudad Jiménez, Mex. *see* Jiménez	**132-33**	27°08´N	104°56´W
Ciudad Juárez, Mex. (syōō-dhädh hwä´rāz)	**132-33**	31°45´N	106°28´W
Ciudad Lerdo, Mex. *see* Lerdo	**132-33**	25°32´N	103°31´W
Ciudad Madero, Mex. (syōō-dä´d-mä-dĕ´rô)	**158-59**	22°16´N	97°50´W
Ciudad Mante, Mex. (syōō-dä´d-mán´tĕ)	**158-59**	22°44´N	98°58´W
Ciudad Netzahualcóyotl, Mex.	**158-59**	19°27´N	99°03´W
Ciudad Obregón, Mex. (syōō-dhädh-ô-brĕ-gô´n)	**156-57**	27°29´N	109°57´W
Ciudad Ojeda, Ven.	**178-79**	10°13´N	71°19´W
Ciudad Real, Spain (thyōō-dhädh´rā-äl´)	**214-15**	38°59´N	3°55´W
Ciudad Rodrigo, Spain (thyōō-dhädh´rô-drē´gō)	**214-15**	40°36´N	6°32´W
Ciudad Valles, Mex.	**158-59**	21°59´N	99°00´W
Ciudad Victoria, Mex. (syōō-dhädh´vêk-tō´rē-ä)	**158-59**	23°44´N	99°08´W
Civitavecchia, Italy (chē´vê-tä-vĕk´kyä)	**216-17**	42°06´N	11°48´E
Clairton, Pa., U.S. (klârt*ŭ*n)	**126-27**	40°18´N	79°53´W
Clanton, Al., U.S. (klăn´t*ŭ*n)	**134-35**	32°51´N	86°38´W
Clanwilliam, S. Afr.	**286-87**	32°10´S	18°54´E
Clare, Mi., U.S. (klâr)	**126-27**	43°49´N	84°45´W
Claremont, N.H., U.S. (klâr´mŏnt)	**126-27**	43°22´N	72°20´W
Claremore, Ok., U.S. (klâr´mōr)	**130-31**	36°19´N	95°37´W
Claremorris, Ire. (klâr-mŏr´ĭs)	**206-07**	53°43´N	8°60´W
Clarence, Isla, i., Chile	**185**	54°11´S	71°49´W
Clarence Island, i., Ant.	**313**	61°11´S	54°03´W
Clarence Strait, strt., Austl. (klăr´ĕns strāt)	**296-97**	12°0´S	131°00´E
Clarendon, Tx., U.S. (klâr´ĕn-d*ŭ*n)	**130-31**	34°56´N	100°54´W
Claresholm, Ab., Can. (klâr´ĕs-hōlm)	**142-43**	50°01´N	113°35´W
Clarinda, Ia., U.S. (kl*ȧ*-rĭn´d*ȧ*)	**130-31**	40°44´N	95°02´W
Clarines, Ven. (klä-rē´nĕs)	**188b**	9°58´N	65°09´W
Clarion, Ia., U.S. (klăr´ĭ-*ŭ*n)	**124-25**	42°44´N	93°44´W
Clarion, Pa., U.S. (klăr´ĭ-*ŭ*n)	**126-27**	41°13´N	79°23´W
Clarión, Isla, i., Mex.	**156-57**	18°22´N	114°44´W
Clark, S.D., U.S. (klärk)	**124-25**	44°53´N	97°44´W
Clarke Island, i., Austl.	**301**	40°33´S	148°10´E
Clarksburg, W.V., U.S. (klärkz´bûrg)	**126-27**	39°16´N	80°20´W
Clarksdale, Ms., U.S. (klärks-dāl)	**134-35**	34°12´N	90°34´W
Clark's Harbour, N.S., Can. (klärks här´bẽr)	**148-49**	43°28´N	65°38´W
Clarks Hill Lake, res., U.S. (klärks hĭl lāk) *see* J. Strom Thurmond Reservoir	**134-35**	33°45´N	82°16´W
Clarkston, Wa., U.S. (klärks´t*ŭ*n)	**122-23**	46°25´N	117°04´W
Clarksville, Ar., U.S. (klärks-vĭl)	**130-31**	35°28´N	93°28´W
Clarksville, Tn., U.S. (klärks-vĭl)	**134-35**	36°31´N	87°21´W
Clarksville, Tx., U.S. (kläks-vĭl)	**130-31**	33°37´N	95°04´W
Claxton, Ga., U.S. (kläks´t*ŭ*n)	**134-35**	32°10´N	81°54´W
Clay Center, Ks., U.S. (klā sĕn´tẽr)	**130-31**	39°23´N	97°07´W
Clay City, Ky., U.S. (klā sĭ´tĭ)	**126-27**	37°52´N	83°57´W
Clayton, Ga., U.S. (klā´t*ŭ*n)	**134-35**	34°53´N	83°23´W
Clayton, N.C., U.S. (klā´t*ŭ*n)	**134-35**	35°39´N	78°28´W
Clayton, N.M., U.S. (klā´t*ŭ*n)	**130-31**	36°27´N	103°11´W
Clearfield, Ut., U.S. (klēr-fēld)	**122-23**	41°07´N	112°01´W
Clear Lake, Ia., U.S. (klēr lāk)	**124-25**	43°08´N	93°23´W
Clear Lake, S.D., U.S. (klēr lāk)	**124-25**	44°45´N	96°42´W
Clear Lake, lk., Ca., U.S. (klēr lāk)	**128-29**	39°02´N	122°50´W
Clearwater, Fl., U.S. (klēr-wô´tẽr)	**135a**	27°58´N	82°47´W
Clearwater, stm., Can. (klēr-wô´tẽr)	**144-45**	56°45´N	111°23´W
Clearwater, stm., Ab., Can. (klēr-wô´tẽr)	**142-43**	52°22´N	114°57´W
Clearwater Mountains, mts., Id., U.S. (klēr-wô´tẽr moun´tĭnz)	**122-23**	46°00´N	115°30´W
Cleburne, Tx., U.S. (klē´bŭrn)	**130-31**	32°22´N	97°24´W
Cle Elum, Wa., U.S. (klē ĕl´*ŭ*m)	**122-23**	47°12´N	120°56´W
Clermont, Austl. (klẽr´mŏnt)	**302**	22°49´S	147°40´E
Clermont-Ferrand, Fr. (klĕr-môn´fĕr-răn´)	**212-13**	45°47´N	3°06´E
Cleveland, Ms., U.S. (klēv´l*ă*nd)	**134-35**	33°45´N	90°43´W
Cleveland, Oh., U.S. (klēv´l*ă*nd)	**126-27**	41°29´N	81°42´W
Cleveland, Ok., U.S. (klēv´l*ă*nd)	**130-31**	36°19´N	96°28´W
Cleveland, Tn., U.S. (klēv´l*ă*nd)	**134-35**	35°10´N	84°52´W
Cleveland, Tx., U.S. (klēv´l*ă*nd)	**132-33**	30°21´N	95°05´W
Cleveland, Cape, c., Austl.	**302**	19°13´S	147°02´E
Cleveland Heights, Oh., U.S. (klēv´l*ă*nd hīts)	**126-27**	41°30´N	81°36´W
Cleves, Ger. *see* Kleve	**210-11**	51°47´N	6°09´E
Clewiston, Fl., U.S. (klē´wis-t*ŭ*n)	**135a**	26°45´N	80°54´W
Clifden, Ire. (klĭf´d*ĕ*n)	**206-07**	53°29´N	10°01´W
Clifton, Az., U.S. (klĭf´t*ŭ*n)	**128-29**	33°04´N	109°18´W
Clifton, Il., U.S. (klĭf´t*ŭ*n)	**126-27**	40°56´N	87°56´W
Clifton, N.J., U.S. (klĭf´t*ŭ*n)	**126-27**	40°53´N	74°10´W
Clifton, Tx., U.S. (klĭf´t*ŭ*n)	**132-33**	31°46´N	97°35´W
Clifton Forge, Va., U.S. (klĭf´t*ŭ*n fōrj)	**126-27**	37°49´N	79°50´W
Clinch, stm., U.S. (klĭnch)	**134-35**	35°53´N	84°30´W
Clingmans Dome, mtn., U.S. (klĭng´măns dōm)	**134-35**	35°35´N	83°30´W
Clinton, B.C., Can. (klĭn-´t*ŭ*n)	**142-43**	51°05´N	121°37´W
Clinton, On., Can. (klĭn-´t*ŭ*n)	**146-47**	43°37´N	81°32´W
Clinton, Ar., U.S. (klĭn-´t*ŭ*n)	**130-31**	35°36´N	92°28´W
Clinton, Ia., U.S. (klĭn-´t*ŭ*n)	**124-25**	41°51´N	90°12´W
Clinton, Il., U.S. (klĭn-´t*ŭ*n)	**126-27**	40°09´N	88°57´W
Clinton, In., U.S. (klĭn-´t*ŭ*n)	**126-27**	39°39´N	87°24´W
Clinton, Ky., U.S. (klĭn-´t*ŭ*n)	**134-35**	36°40´N	89°00´W
Clinton, Mo., U.S. (klĭn-´t*ŭ*n)	**130-31**	38°22´N	93°46´W
Clinton, Ms., U.S. (klĭn-´t*ŭ*n)	**134-35**	32°21´N	90°19´W
Clinton, N.C., U.S. (klĭn-´t*ŭ*n)	**134-35**	34°59´N	78°19´W
Clinton, Ok., U.S. (klĭn-´t*ŭ*n)	**130-31**	35°31´N	98°58´W
Clinton, S.C., U.S. (klĭn-´t*ŭ*n)	**134-35**	34°28´N	81°53´W
Clinton, Tn., U.S. (klĭn-´t*ŭ*n)	**134-35**	36°06´N	84°08´W
Clintonville, Wi., U.S. (klĭn´t*ŭ*n-vĭl)	**126-27**	44°36´N	88°46´W
Clio, Mi., U.S. (klē´ō)	**126-27**	43°10´N	83°44´W
Clipperton, Île, at., Oc.	**152-53**	10°18´N	109°13´W
Clipperton Island, at., Oc. *see* Clipperton, Île	**152-53**	10°18´N	109°13´W
Clodomira, Arg.	**187**	27°33´S	64°07´W
Cloncurry, Austl. (klŏn-kŭr´ē)	**302**	20°43´S	140°30´E
Cloquet, Mn., U.S. (klô-kā´)	**124-25**	46°43´N	92°28´W
Cloud Peak, mtn., Wy., U.S. (kloud pēk)	**122-23**	44°25´N	107°10´W
Clover, S.C., U.S. (klō´vẽr)	**134-35**	35°07´N	81°14´W
Cloverdale, Ca., U.S. (klō´vẽr-dāl)	**128-29**	38°48´N	123°01´W
Clovis, Ca., U.S. (klō´vĭs)	**128-29**	36°50´N	119°42´W
Clovis, N.M., U.S. (klō´vĭs)	**130-31**	34°24´N	103°13´W
Cluj-Napoca, Rom.	**210-11**	46°47´N	23°36´E
Cluny, Fr. (klü-nē´)	**212-13**	46°26´N	4°39´E
Clyde, Oh., U.S. (klīd)	**126-27**	41°18´N	82°58´W
Clyde, Tx., U.S. (klīd)	**130-31**	32°25´N	99°29´W
Cnossus, hist., Grc. *see* Knossos	**216a**	35°17´N	25°12´E
Côa, stm., Port. (kô´ä)	**214-15**	41°05´N	7°06´W
Coahuila, state, Mex. (kō-ä-wē´lȧ)	**132-33**	27°20´N	102°00´W
Coalcomán de Matamoros, Mex.	**158-59**	18°47´N	103°09´W
Coaldale, Ab., Can. (kōl´dāl)	**142-43**	49°44´N	112°37´W
Coalgate, Ok., U.S. (kōl´gāt)	**130-31**	34°32´N	96°14´W
Coalinga, Ca., U.S. (kō-*ȧ*-lĭŋ´g*ȧ*)	**128-29**	36°09´N	120°22´W
Coari, Braz. (kō-är´ē)	**180-81**	4°06´S	63°07´W
Coari, stm., Braz.	**180-81**	4°27´S	63°29´W
Coast Mountains, mts., N.A. (kōst moun´tĭnz)	**140-41**	55°0´N	129°00´W
Coast Ranges, mts., U.S. (kōst rānj*ĕ*z)	**120-21**	40°46´N	123°38´W
Coatesville, Pa., U.S. (kōts´vĭl)	**126-27**	39°59´N	75°49´W
Coaticook, Qc., Can. (kō´tĭ-ko̤ok)	**146-47**	45°08´N	71°48´W
Coats Island, i., Nu., Can. (kōts ī´l*ă*nd)	**140-41**	62°30´N	82°60´W
Coats Land, reg., Ant. (kōts lănd)	**313**	77°0´S	28°00´W
Coatzacoalcos, Mex.	**158-59**	18°08´N	94°26´W
Cobá, hist., Mex. (kô´bä)	**160**	20°36´N	87°35´W
Cobalt, On., Can. (kō´bôlt)	**146-47**	47°24´N	79°41´W
Cobán, Guat. (kō-bän´)	**160**	15°28´N	90°22´W
Cobar, Austl.	**301**	31°30´S	145°50´E
Cobija, Bol. (kô-bē´hä)	**180-81**	11°02´S	68°44´W
Coblenz, Ger. *see* Koblenz	**210-11**	50°21´N	7°35´E
Cobourg, On., Can. (kō´bo̤orgh)	**146-47**	43°58´N	78°09´W
Cobourg Peninsula, pen., Austl.	**296-97**	11°22´S	132°17´E
Coburg, Ger. (kō´bo̤org)	**210-11**	50°16´N	10°58´E
Cocanada, India *see* Kākināda	**256**	16°57´N	82°15´E
Cochabamba, Bol.	**182-83**	17°23´S	66°10´W
Cochran, Ga., U.S. (kŏk´răn)	**134-35**	32°23´N	83°21´W
Cochrane, Ab., Can. (kŏk´răn)	**142-43**	51°11´N	114°29´W
Cochrane, On., Can. (kŏk´răn)	**146-47**	49°04´N	81°02´W
Cochrane, Lago, lk., S.A.	**185**	47°21´S	71°56´W
Cockburn Island, i., On., Can. (kŏk-bûrn ī´l*ă*nd)	**146-47**	45°55´N	83°22´W
Cockburn Town, nat. cap., T./C. Is. *see* Grand Turk	**154-55**	21°27´N	71°08´W
Coco, stm., N.A. (kô-kô)	**161**	14°59´N	83°11´W
Coco, Cayo, i., Cuba (kä´-yō-kô´kô)	**154-55**	22°29´N	78°28´W
Coco, Isla del, i., C.R. (ē´s-lä-dĕl-kô-kô)	**95**	5°32´N	87°04´W
Cocoa, Fl., U.S. (kō´kō)	**135a**	28°22´N	80°44´W
Cocoa Beach, Fl., U.S. (kō´kō bēch)	**135a**	28°19´N	80°37´W
Coco Channel, strt., Asia	**266-67**	13°45´N	93°01´E
Coco Islands, is., Mya.	**266-67**	14°09´N	93°25´E
Cocos Islands, dep., Oc. (kō´kôs ī´l*ă*ndz)	**242-43**	12°10´S	96°55´E
Cocula, Mex. (kō-ko̅o̅´lä)	**158-59**	20°23´N	103°50´W
Cod, Cape, pen., Ma., U.S. (kăp kŏd)	**126-27**	41°42´N	70°15´W
Codajás, Braz. (kō-dä-häzh´)	**180-81**	3°49´S	62°05´W
Codera, Cabo, c., Ven. (kȧ´bô-kō-dĕ´rä)	**188b**	10°34´N	66°03´W
Codó, Braz.	**180-81**	4°29´S	43°53´W
Cody, Wy., U.S. (kō´dĭ)	**122-23**	44°32´N	109°03´W
Coeur d'Alene, Id., U.S. (kûr dȧ-lān´)	**122-23**	47°41´N	116°47´W
Coeur d'Alene Indian Reservation, ind. res., Id., U.S. (kûr dȧ-lān´ ĭn´dĭ-*ă*n rĕ-sẽr-vā´shĕn)	**122-23**	47°18´N	116°45´W
Coffeyville, Ks., U.S. (kŏf´ĭ-vĭl)	**130-31**	37°02´N	95°37´W
Coffs Harbour, Austl.	**301**	30°19´S	153°08´E
Cognac, Fr. (kôn-yak´)	**212-13**	45°42´N	0°20´W
Cohoes, N.Y., U.S. (kô-hōz´)	**126-27**	42°47´N	73°42´W
Coiba, Isla de, i., Pan.	**162**	7°27´N	81°45´W
Coig, stm., Arg. (kô´ĕk)	**185**	50°57´S	69°09´W
Coihaique, Chile	**185**	45°34´S	72°04´W
Coimbatore, India (kô-ēm-bȧ-tôr´)	**256**	10°60´N	76°58´E
Coimbra, Port. (kō-ēm´brä)	**214-15**	40°13´N	8°25´W
Coín, Spain (kō-ēn´)	**214-15**	36°40´N	4°46´W
Coire, Switz. *see* Chur	**210-11**	46°51´N	9°31´E
Cojutepeque, El Sal. (kô-hô-tĕ-pā´kå)	**160**	13°43´N	88°56´W
Cokato, Mn., U.S. (kô-kā´tō)	**124-25**	45°04´N	94°12´W
Colac, Austl. (kŏ´lȧc)	**301**	38°21´S	143°35´E
Colatina, Braz. (kô-lä-tē´nä)	**186**	19°32´S	40°39´W
Colbeck, Cape, c., Ant.	**313**	77°25´S	157°33´W
Colby, Ks., U.S. (kōl´bĭ)	**130-31**	39°24´N	101°03´W
Colchester, Eng., U.K. (kōl´chĕs-tẽr)	**206-07**	51°53´N	0°54´E
Cold Lake, Ab., Can.	**142-43**	54°27´N	110°10´W
Cold Lake, lk., Can. (kōld lāk)	**142-43**	54°33´N	110°05´W
Coldwater, Mi., U.S. (kōld´wô-tẽr)	**126-27**	41°57´N	85°00´W
Coleman, Tx., U.S. (kōl´m*ă*n)	**132-33**	31°50´N	99°25´W
Colenso, S. Afr. (kŏ-lĕnz´ō)	**292c**	28°45´S	29°50´E
Coleraine, Mn., U.S. (kōl-rān´)	**124-25**	47°17´N	93°26´W
Colesberg, S. Afr.	**286-87**	30°43´S	25°06´E
Colfax, Ia., U.S. (kōl´făks)	**124-25**	41°41´N	93°15´W
Colfax, La., U.S. (kōl´făks)	**132-33**	31°31´N	92°42´W
Colfax, Wa., U.S. (kōl´făks)	**122-23**	46°53´N	117°22´W
Colhué Huapi, Lago, lk., Arg. (lä´gô kôl-wä´ôȧ´pê)	**185**	45°32´S	68°46´W
Colima, Mex.	**158-59**	19°14´N	103°44´W
Colima, state, Mex. (kōlē´mä)	**158-59**	19°10´N	104°00´W
Colima, Nevado de, vol., Mex. (nĕ-vä´dô-dĕ-kō-lē´mä)	**158-59**	19°33´N	103°38´W
Colinas, Braz.	**180-81**	6°02´S	44°14´W
Coll, i., Scot., U.K. (kōl)	**206-07**	56°38´N	6°34´W
College, Ak., U.S. (kŏl´ĕj)	**136**	64°51´N	147°46´W
College Park, Ga., U.S. (kŏl´ĕj pärk)	**134-35**	33°39´N	84°27´W
Collie, Austl. (kŏl´ē)	**294-95**	33°21´S	116°09´E
Collier Bay, b., Austl. (kŏl-yẽr bā)	**296-97**	16°10´S	124°15´E
Collingwood, On., Can. (kŏl´ĭng-wo̤od)	**146-47**	44°29´N	80°12´W
Collins, Ms., U.S. (kŏl´ĭns)	**134-35**	31°38´N	89°34´W
Collinsville, Austl.	**302**	20°34´S	147°51´E
Collinsville, Il., U.S. (kŏl´ĭnz-vĭl)	**130-31**	38°41´N	89°58´W
Collinsville, Ok., U.S. (kŏl´ĭnz-vĭl)	**130-31**	36°22´N	95°50´W
Collipulli, Chile	**185**	37°57´S	72°26´W
Colmar, Fr. (kōl´mär)	**212-13**	48°05´N	7°22´E

PRONOUNCING INDEX

Colmenar Viejo – Corral 347

Feature (Pronunciation)	Page	Lat.	Long.
Colmenar Viejo, Spain (kôl-mä-när´vyā´hō)	214-15	40°40'N	3°46'W
Cologne, Ger. (kŭ-lōn´)	210-11	50°56'N	6°57'E
Colomb-Béchar, Alg. *see* Béchar	280-81	31°37'N	2°14'W
Colombia, Col. (kô-lôm´bē-ä)	188c	3°24'N	74°48'W
Colombia, nation, S.A. (kô-lôm´bē-ä)	172	4°0'N	72°00'W
Colombo, nat. cap., Sri L. (kô-lôm´bō)	256	6°55'N	79°52'E
Colón, Arg. (kō-lōn´)	187	32°16'S	58°08'W
Colón, Arg. (kō-lōn´)	187	33°55'S	61°04'W
Colón, Cuba (kô-lô´n)	154-55	22°43'N	80°54'W
Colón, Pan. (kō-lō´n)	162	9°22'N	79°54'W
Colón, Archipiélago de, is., Ec.	184a	0°43'N	91°30'W
Colonia Alvear Norte, Arg. *see* General Alvear	185	34°59'S	67°42'W
Colonia del Sacramento, Ur.	187	34°29'S	57°50'W
Colonia Dora, Arg.	187	28°37'S	62°57'W
Colonia Suiza, Ur. (kô-lô´nēä-sōē´zä)	187	34°18'S	57°14'W
Colorado, state, U.S. (kŏl-ô-rä´dō)	118-19	39°0'N	105°30'W
Colorado, stm., Arg.	185	39°52'S	62°09'W
Colorado, stm., N.A. (kŏl-ô-rä´dō)	120-21	31°55'N	114°58'W
Colorado, stm., Tx., U.S. (kŏl-ô-rä´dō)	132-33	28°36'N	95°59'W
Colorado City, Tx., U.S. (kŏl-ô-rä´dō sĭ´tĭ)	130-31	32°23'N	100°52'W
Colorado National Monument, n.p., Co., U.S. (kŏl-ŏ-rä´dō năsh´ŭn-ăl mŏn´ŭ-mĕnt)	128-29	39°03'N	108°41'W
Colorado Plateau, plat., U.S. (kŏl-ô-rä´dō plă-tō´)	128-29	38°0'N	109°00'W
Colorado Springs, Co., U.S. (kŏl-ô-rä´dō sprĭngz)	130-31	38°50'N	104°49'W
Colotepec, stm., Mex. (kô-lô´tĕ-pĕk)	158-59	15°48'N	97°01'W
Colotlán, Mex. (kô-lô-tlän´)	158-59	22°06'N	103°15'W
Colquechaca, Bol. (kôl-kā-chä´kä)	182-83	18°41'S	66°02'W
Colstrip, Mt., U.S. (kōl´strip)	122-23	45°53'N	106°39'W
Columbia, Il., U.S. (kô-lŭm´bĭ-ȧ)	130-31	38°26'N	90°11'W
Columbia, Ky., U.S. (kô-lŭm´bĭ-ȧ)	134-35	37°06'N	85°18'W
Columbia, Md., U.S. (kô-lŭm´bĭ-ȧ)	126-27	39°14'N	76°50'W
Columbia, Mo., U.S. (kô-lŭm´bĭ-ȧ)	130-31	38°57'N	92°21'W
Columbia, Ms., U.S. (kô-lŭm´bĭ-ȧ)	134-35	31°15'N	89°50'W
Columbia, S.C., U.S. (kô-lŭm´bĭ-ȧ)	134-35	34°00'N	81°02'W
Columbia, Tn., U.S. (kô-lŭm´bĭ-ȧ)	134-35	35°37'N	87°02'W
Columbia, stm., N.A. (kô-lŭm´bĭ-ȧ)	140-41	46°14'N	124°06'W
Columbia, Mount, mtn., Ab., Can. (mount kô-lŭm´bĭ-ȧ)	142-43	52°09'N	117°25'W
Columbia City, In., U.S. (kô lŭm´bĭ ȧ sĭ´tê)	126-27	41°09'N	85°28'W
Columbia Icefield, ice, Can. (kô-lŭm´bĭ-ȧ ī´fēld)	142-43	52°08'N	117°27'W
Columbia Mountains, mts., N.A. (kô-lŭm´bĭ-ȧ moun´tĭnz)	140-41	52°0'N	119°00'W
Columbiana, Al., U.S. (kô-lŭm-bĭ-ă´nȧ)	134-35	33°11'N	86°36'W
Columbus, Ga., U.S. (kô-lŭm´bŭs)	134-35	32°28'N	84°58'W
Columbus, In., U.S. (kô-lŭm´bŭs)	126-27	39°12'N	85°55'W
Columbus, Ms., U.S. (kô-lŭm´bŭs)	134-35	33°29'N	88°25'W
Columbus, Mt., U.S. (kô-lŭm´bŭs)	122-23	45°39'N	109°16'W
Columbus, Ne., U.S. (kô-lŭm´bŭs)	124-25	41°26'N	97°22'W
Columbus, N.M., U.S. (kô-lŭm´bŭs)	128-29	31°50'N	107°38'W
Columbus, Oh., U.S.	126-27	39°58'N	82°60'W
Columbus, Tx., U.S. (kô-lŭm´bŭs)	132-33	29°42'N	96°32'W
Columbus, Wi., U.S. (kô-lŭm´bŭs)	126-27	43°21'N	89°01'W
Colusa, Ca., U.S. (kô-lū´sȧ)	128-29	39°12'N	122°01'W
Colville, Wa., U.S. (kŏl´vĭl)	122-23	48°33'N	117°54'W
Colville, stm., Ak., U.S. (kŏl´vĭl)	136	70°27'N	150°18'W
Colville Indian Reservation, ind. res., Wa., U.S. (kŏl´vĭl ĭn´dĭ-ăn rĕ-sĕr-vā´shĕn)	122-23	48°15'N	119°00'W
Comacchio, Italy (kô-mäk´kyō)	216-17	44°43'N	12°11'E
Comala, Mex. (kō-mä-lä´)	158-59	19°19'N	103°45'W
Comalcalco, Mex. (kō-mäl-käl´kō)	158-59	18°16'N	93°12'W
Comanche, Ok., U.S. (kô-măn´chê)	130-31	34°21'N	97°58'W
Comanche, Tx., U.S. (kô-măn´chê)	132-33	31°54'N	98°36'W
Comandante Fontana, Arg.	182-83	25°19'S	59°40'W
Comayagua, Hond. (kō-mä-yä´gwä)	161	14°27'N	87°39'W
Combarbalá, Chile	182-83	31°11'S	71°03'W
Comilla, Bngl. (kô-mĭl´ä)	254-55	23°27'N	91°11'E
Comino, Capo, c., Italy (kä´pō kô-mē´nō)	216-17	40°32'N	9°49'E
Comitán de Domínguez, Mex.	160	16°15'N	92°07'W
Commentry, Fr. (kô-mäN-trē´)	212-13	46°17'N	2°44'E
Commerce, Ga., U.S. (kŏm´ērs)	134-35	34°12'N	83°27'W
Commerce, Ok., U.S. (kŏm´ērs)	130-31	36°56'N	94°52'W
Commerce, Tx., U.S. (kŏm´ērs)	130-31	33°16'N	95°54'W
Committee Bay, b., Nu., Can.	140-41	68°30'N	86°30'W
Communism Peak, mtn., Taj. *see* Imeni Ismail Samani, Pik	244	38°57'N	72°01'E
Como, Italy (kō´mō)	216-17	45°47'N	9°05'E
Comodoro Rivadavia, Arg.	185	45°52'S	67°30'W
Comoé, Parc National de la, n.p., C. Iv.	282-83	9°0'N	3°30'W
Comores, nation, Afr. *see* Comoros	286-87	12°10'S	44°15'E
Comores, Archipel des, is., Afr.	286-87	12°07'S	44°03'E
Comorin, Cape, c., India (kāp kō´mô-rĭn)	256	8°05'N	77°34'E
Comoros, nation, Afr. (kŏm´ō-rōz) (kȧ-mō´-rōz)	286-87	12°10'S	44°15'E
Comox, B.C., Can. (kō´mŏks)	142-43	49°41'N	124°56'W
Compiègne, Fr. (kôN-pyĕn´y')	212-13	49°25'N	2°49'E
Compostela, Mex. (kŏm-pô-stā´lä)	158-59	21°14'N	104°53'W
Conakry, nat. cap., Gui. (kô-nȧ-krē´)	282-83	9°31'N	13°43'W
Concarneau, Fr. (kôN-kär-nō´)	212-13	47°52'N	3°55'W
Conceição do Araguaia, Braz.	180-81	8°15'S	49°19'W
Concepción, Bol. (kôn-sĕp´syōn´)	180-81	11°29'S	66°36'W
Concepción, Bol. (kôn-sĕp´syōn´)	182-83	16°15'S	62°04'W
Concepción, Chile	185	36°49'S	73°05'W
Concepción, Para.	182-83	23°24'S	57°25'W
Concepción del Oro, Mex. (kôn-sĕp-syōn´ dĕl ō´rō)	158-59	24°37'N	101°24'W
Concepción del Uruguay, Arg. (kôn-sĕp-syô´n-dĕl-ōō-rōō-gwī´)	187	32°29'S	58°14'W
Conception Bay, b., Nf./L., Can. (kŏn-sĕp´shŭn bā)	148-49	47°44'N	52°59'W
Conchos, stm., Mex. (kōn´chōs)	156-57	24°56'N	97°38'W
Conchos, stm., Mex. (kōn´chōs)	156-57	29°34'N	104°24'W
Concord, Ca., U.S. (kŏŋ´kŏrd)	128-29	37°59'N	122°02'W
Concord, N.C., U.S. (kŏŋ´kŏrd)	134-35	35°24'N	80°36'W
Concord, N.H., U.S.	126-27	43°12'N	71°32'W
Concordia, Arg. (kŏn-kôr´dĭ-ȧ)	187	31°23'S	58°01'W
Concórdia, Braz.	182-83	27°14'S	52°02'W
Concordia, Mex. (kôn-kô´r-dyä)	158-59	23°16'N	106°04'W
Concordia, Ks., U.S. (kŏn-kô´r-dyä)	130-31	39°34'N	97°40'W
Condega, Nic. (kŏn-dĕ´gä)	161	13°22'N	86°24'W
Condobolin, Austl.	301	33°05'S	147°09'E
Condoto, Col.	188c	5°05'N	76°38'W
Conegliano, Italy (kō-nāl-yä´nō)	216-17	45°53'N	12°18'E
Conghua, China (tsôŋ-hwä)	258-59	23°33'N	113°35'E
Congo, nation, Afr. (kŏn´gō)	274	1°0'S	15°00'E
Congo, stm., Afr. (kŏn´gō)	284-85	5°58'S	12°44'E
Congo, Democratic Republic of the, nation, Afr. (dĕ-mō-kră´tĭc rē-pŭb´lĭk ŭv thĕ kŏn´gō)	274	4°0'S	25°00'E
Congo, République démocratique du, nation, Afr. *see* Congo, Democratic Republic of the	274	4°0'S	25°00'E
Conjeeveram, India *see* Kānchipuram	256	12°50'N	79°43'E
Conneaut, Oh., U.S. (kŏn-ē-ôt´)	126-27	41°57'N	80°34'W
Connecticut, state, U.S. (kŏ-nĕt´ĭ-kŭt)	118-19	41°45'N	72°45'W
Connecticut, stm., U.S.	120-21	41°16'N	72°20'W
Connellsville, Pa., U.S. (kŏn´nĕlz-vĭl)	126-27	40°01'N	79°30'W
Connersville, In., U.S. (kŏn´ērz-vĭl)	126-27	39°39'N	85°08'W
Connors Range, mts., Austl. (kŏn´nŏrs rānj)	302	21°40'S	149°10'E
Conrad, Mt., U.S. (kŏn´răd)	122-23	48°10'N	111°56'W
Conroe, Tx., U.S. (kŏn´rō)	132-33	30°19'N	95°28'W
Conselheiro Lafaiete, Braz.	186	20°40'S	43°47'W
Conshohocken, Pa., U.S. (kŏn-shô-hŏk´ĕn)	126-27	40°05'N	75°18'W
Constance, Ger. *see* Konstanz	210-11	47°40'N	9°10'E
Constance, Lake, lk., Eur. (lāk kŏn´stăns)	210-11	47°39'N	8°54'E
Constanța, Rom. (kōn-stán´tsá)	218-19	44°11'N	28°38'E
Constantine, Alg. (kôn-stäN´tēn´)	292b	36°22'N	6°37'E
Constantine, Mi., U.S. (kŏn´stăn-tēn)	126-27	41°50'N	85°39'W
Constantinople, Tur. *see* İstanbul	216-17	41°02'N	28°59'E
Constitución, Chile (kōn´stĭ-tōō-syōn´)	185	35°19'S	72°25'W
Contreras, Isla, i., Chile	185	51°52'S	74°57'W
Contwoyto Lake, lk., Can.	140-41	65°42'N	110°50'W
Converse, Tx., U.S. (kŏn´vērs)	132-33	29°31'N	98°18'W
Conway, Ar., U.S. (kŏn´wā)	130-31	35°05'N	92°27'W
Conway, N.H., U.S. (kŏn´wā)	126-27	43°59'N	71°07'W
Conway, S.C., U.S. (kŏn´wā)	134-35	33°50'N	79°03'W
Cook, Cape, c., B.C., Can. (kāp kŏŏk)	142-43	50°07'N	127°54'W
Cook, Mount, mtn., N.Z. (mount kŏŏk) *see* Aoraki	303	43°36'S	170°10'E
Cookeville, Tn., U.S. (kŏŏk´vĭl)	134-35	36°10'N	85°31'W
Cook Inlet, b., Ak., U.S. (kŏŏk ĭn´lĕt)	136	60°32'N	151°40'W
Cook Islands, dep., Oc. (kŏŏk ī´lăndz)	304-05	20°0'S	158°00'W
Cook Strait, strt., N.Z. (kŏŏk strāt)	303	41°15'S	174°30'E
Cooktown, Austl. (kŏŏk´toun)	302	15°29'S	145°15'E
Coolgardie, Austl. (kōōl-gär´dê)	294-95	30°57'S	121°10'E
Cooma, Austl. (kōō´mȧ)	301	36°14'S	149°08'E
Coonabarabran, Austl.	301	31°16'S	149°17'E
Coonamble, Austl. (kōō-năm´b'l)	301	30°58'S	148°23'E
Coonoor, India	256	11°20'N	76°48'E
Coon Rapids, Mn., U.S. (kōōn răp´ĭdz)	124-25	45°10'N	93°20'W
Cooper, Tx., U.S. (kōōp´ēr)	130-31	33°23'N	95°41'W
Cooper Creek, stm., Austl.	301	28°18'S	137°29'E
Cooperstown, N.D., U.S. (kōōp´ērs-toun)	124-25	47°27'N	98°07'W
Coosa, stm., U.S. (kōō´sȧ)	134-35	32°30'N	86°16'W
Coos Bay, Or., U.S. (kōōs bā)	122-23	43°22'N	124°13'W
Cootamundra, Austl. (kŏŏtă-mŭnd´ră)	301	34°39'S	148°01'E
Copan, hist., Hond.	161	14°50'N	89°09'W
Copenhagen, nat. cap., Den. (kō´pĕn-hā´gĕn)	208-09	55°41'N	12°34'E
Copiapó, Chile (kō-pyä-pō´)	182-83	27°22'S	70°20'W
Copley, Austl.	301	30°33'S	138°26'E
Copper, stm., Ak., U.S. (kŏp´ēr)	136	60°33'N	144°52'W
Copper Harbor, Mi., U.S. (kŏp´ēr här´bēr)	124-25	47°28'N	87°55'W
Coppermine, Nu., Can. *see* Kugluktuk	138-39	67°47'N	115°11'W
Coppermine, stm., Can. (kŏp´ēr mīn)	140-41	67°48'N	115°08'W
Coquimbo, Chile (kō-kēm´bō)	182-83	29°58'S	71°20'W
Corabia, Rom. (kô-rä´bĭ-ȧ)	216-17	43°47'N	24°31'E
Corail, Mer de, s., Oc. *see* Coral Sea	296-97	20°0'S	158°00'E
Coral Harbour, Nu., Can.	138-39	64°08'N	83°12'W
Coral Sea, s., Oc. (kŏr´ăl sē)	296-97	20°0'S	158°00'E
Corangamite, Lake, lk., Austl. (lāk cŏr-ăŋg´á-mīt)	301	38°10'S	143°25'E
Corato, Italy (kô´rä-tô)	216-17	41°09'N	16°25'E
Corbin, Ky., U.S. (kôr´bĭn)	134-35	36°57'N	84°06'W
Corby, Eng., U.K. (kôr´bĭ)	206-07	52°29'N	0°40'W
Corcaigh, Ire. *see* Cork	206-07	51°54'N	8°28'W
Corcovado, Golfo, b., Chile (gôl-fô-kôr-kô-vä´dhō)	185	43°30'S	73°30'W
Corcovado, Volcán, vol., Chile	185	43°12'S	72°48'W
Cordele, Ga., U.S. (kôr-dēl´)	134-35	31°58'N	83°47'W
Cordell, Ok., U.S. (kôr-dĕl´)	130-31	35°18'N	98°59'W
Córdoba, Arg. (kôr´dô-vä)	187	31°24'S	64°12'W
Córdoba, Mex. (kô´r-dô-bä)	158-59	18°53'N	96°56'W
Córdoba, Spain (kô´r-dô-bä)	214-15	37°54'N	4°47'W
Córdoba, state, Arg. (kôr´dô-vä)	187	32°0'S	64°00'W
Cordova, Spain *see* Córdoba	214-15	37°54'N	4°47'W
Cordova, Ak., U.S. (kôr´dô-vä)	136	60°33'N	145°45'W
Cordova, Al., U.S. (kôr´dô-ȧ)	134-35	33°46'N	87°11'W
Corfu, i., Grc. *see* Kérkyra	216-17	39°40'N	19°45'E
Corinth, Grc.	216-17	37°56'N	22°58'E
Corinth, Ms., U.S. (kŏr´ĭnth)	134-35	34°56'N	88°31'W
Corinto, Braz. (kô-rē´n-tō)	186	18°23'S	44°27'W
Cork, Ire. (kôrk)	206-07	51°54'N	8°28'W
Corleone, Italy (kôr-lă-ō´nā)	216-17	37°49'N	13°18'E
Cornelia, Ga., U.S. (kôr-nē´lyȧ)	134-35	34°31'N	83°32'W
Cornell, Wi., U.S. (kôr-nĕl´)	124-25	45°10'N	91°09'W
Corner Brook, Nf./L., Can. (kôr´nēr brŏŏk)	140-49	40°57'N	57°50'W
Corning, Ar., U.S. (kôr´nĭng)	134-35	36°25'N	90°35'W
Corning, Ia., U.S. (kôr´nĭng)	124-25	40°60'N	94°45'W
Corning, N.Y., U.S. (kôr´nĭng)	126-27	42°09'N	77°03'W
Corno Grande, mtn., Italy (kôr´nō grän´dĕ)	216-17	42°28'N	13°34'E
Cornwall, On., Can. (kôrn´wôl)	146-47	45°02'N	74°44'W
Coro, Ven. (kō´rō)	178-79	11°27'N	69°40'W
Corocoro, Bol. (kō-rō-kō´rō)	182-83	17°11'S	68°27'W
Coromandel Coast, cst., India (kŏr-ô-man´dĕl kōst)	256	13°30'N	80°30'E
Coronado, Ca., U.S. (kŏr-ô-nä´dō)	128-29	32°41'N	117°11'W
Coronado, Bahía de, b., C.R. (bä-ē´ä-dĕ-kô-rō-nä´dō)	161	9°0'N	83°50'W
Coronation Gulf, b., Nu., Can. (kŏr-ô-nā´shŭn gŭlf)	140-41	68°24'N	109°56'W
Coronel, Chile (kō-rō-nĕl´)	185	37°01'S	73°08'W
Coronel Dorrego, Arg. (kŏ-rō-nĕl-dôr-rĕ´gô)	187	38°43'S	61°17'W
Coronel Fabriciano, Braz.	186	19°31'S	42°39'W
Coronel Oviedo, Para. (kŏ-rō-nĕl-ō-vēĕ´dō)	182-83	25°27'S	56°26'W
Coronel Pringles, Arg. (kŏ-rō-nĕl-prēn´glĕs)	187	37°59'S	61°22'W
Coronel Suárez, Arg. (kŏ-rō-nĕl-swä´rās)	187	37°27'S	61°56'W
Coropuna, Nevado, vol., Peru (nĕ-vä-dô-kō-rō-pōō´nä)	184	15°31'S	72°42'W
Corozal, Belize (cŏr-ŏth-ăl´)	160	18°24'N	88°24'W
Corpus Christi, Tx., U.S. (kôr´pŭs krĭstē)	132-33	27°48'N	97°24'W
Corpus Christi, Lake, res., Tx., U.S. (lāk kôr´pŭs krĭstē)	132-33	28°10'N	97°55'W
Corpus Christi Bay, b., Tx., U.S. (kôr´pŭs krĭstē bā)	132-33	27°48'N	97°20'W
Corral, Chile (kō-räl´)	185	39°53'S	73°28'W

n-sing; ŋ-baŋk; N-nasalized n; nŏd; cŏmmit; ōld; ôbey; ôrder; oi-boil; fōōd; ŏŏ-as oo in foot; ou-out; s-soft; sh-dish; th-thin; pūre; ûnite; ûrn; stŭd; circŭs; ü-as in French tu; ´-indeterminate vowel.

348 Corral de Almaguer – Curtis Island PRONOUNCING INDEX

Feature (Pronunciation)	Page	Lat.	Long.
Corral de Almaguer, Spain (kô-räl′dä äl-mä-gâr′)	214-15	39°45′N	3°10′W
Corrente, stm., Braz.	180-81	13°08′S	43°28′W
Correntina, Braz. (kô-rĕn-tē-nȧ)	180-81	13°21′S	44°39′W
Corrientes, Arg. (kō-ryĕn′tās)	187	27°29′S	58°50′W
Corrientes, state, Arg. (kō-ryĕn′tās)	187	29°0′S	58°00′W
Corrientes, stm., S.A.	184	3°44′S	74°33′W
Corrientes, Cabo, c., Col. (kå′bô-kō-ryĕn′tās)	188c	5°30′N	77°32′W
Corrientes, Cabo, c., Cuba (kä′bô-kôr-rē-ĕn′tĕs)	154-55	21°46′N	84°31′W
Corrientes, Cabo, c., Mex.	158-59	20°24′N	105°41′W
Corriverton, Guy.	178-79	5°53′N	57°09′W
Corry, Pa., U.S. (kŏr′ĭ)	126-27	41°55′N	79°38′W
Corse, i., Fr. *see* Corsica	200-01	42°0′N	9°00′E
Corse, Cap, c., Fr. (kȧp kôrs)	200-01	43°01′N	9°25′E
Corsica, i., Fr. (kô′r-sē-kä)	200-01	42°0′N	9°00′E
Corsicana, Tx., U.S. (kôr-sĭ-kăn′ȧ)	132-33	32°06′N	96°29′W
Cortazar, Mex. (kôr-tä-zär)	158-59	20°29′N	100°56′W
Cortés, Mar de, b., Mex. *see* California, Golfo de	156-57	28°0′N	112°00′W
Cortez, Co., U.S.	128-29	37°21′N	108°35′W
Cortez, Sea of, b., Mex. *see* California, Golfo de	156-57	28°0′N	112°00′W
Corubal, stm., Afr.	280-81	11°57′N	15°03′W
Coruche, Port. (kô-rōō′she)	214-15	38°58′N	8°31′W
Çorum, Tur. (chô-rōōm′)	202-03	40°33′N	34°57′E
Corumbá, Braz.	182-83	19°01′S	57°39′W
Corumbá, stm., Braz.	182-83	18°19′S	48°54′W
Corunna, Spain *see* A Coruña	214-15	43°22′N	8°25′W
Corunna, Mi., U.S. (kô-rŭn′ȧ)	126-27	42°59′N	84°07′W
Coruripe, Braz. (kō-rò-rē′pĭ)	188d	10°09′S	36°10′W
Corvallis, Or., U.S. (kŏr-văl′ĭs)	122-23	44°35′N	123°16′W
Corvo, i., Port.	215c	39°42′N	31°06′W
Corydon, In., U.S. (kŏr′ĭ-dŭn)	126-27	38°13′N	86°07′W
Cos, i., Grc. *see* Kos	216-17	36°50′N	27°10′E
Cosenza, Italy (kô-zĕnt′sä)	216-17	39°17′N	16°15′E
Coshocton, Oh., U.S. (kô-shŏk′tŭn)	126-27	40°16′N	81°51′W
Cosmoledo, Atoll de, at., Sey. (kŏs-mô-lā′dō)	286-87	9°42′S	47°31′E
Cosne-sur-Loire, Fr. (kōn-sür-lwär′)	212-13	47°24′N	2°56′E
Costa Rica, nation, N.A. (kŏs′tȧ rē′kȧ)	94	10°0′N	84°00′W
Costermansville, D.R.C. *see* Bukavu	289	2°30′S	28°51′E
Cotabato, Phil. (kō-tä-bä′tō)	270	7°12′N	124°14′E
Côte d'Ivoire, nation, Afr. (kōt-dē-vwär)	274	8°0′N	5°00′W
Cotija de la Paz, Mex. (kô-tē′-ĸä-dĕ-lä-pá′z)	158-59	19°49′N	102°43′W
Cotonou, nat. cap., Benin (kô-tô-nōō′)	282a	6°22′N	2°26′E
Cotopaxi, vol., Ec. (kō-tō-păk′sē)	184	0°41′S	78°27′W
Cotswold Hills, hills, Eng., U.K. (kŭtz′wōld hĭlz)	206-07	51°49′N	1°57′W
Cottage Grove, Or., U.S. (kŏt′āj grōv)	122-23	43°48′N	123°03′W
Cottbus, Ger. (kōtt′bōōs)	210-11	51°45′N	14°19′E
Cotulla, Tx., U.S. (kŏ-tŭl′lȧ)	132-33	28°26′N	99°14′W
Coudersport, Pa., U.S. (kou′dẽrz-port)	126-27	41°46′N	78°01′W
Coudres, Île aux, i., Qc., Can.	148-49	47°24′N	70°23′W
Coulommiers, Fr. (kōō-lô-myā′)	212-13	48°49′N	3°05′E
Council Bluffs, Ia., U.S. (koun′sĭl blŭfs)	124-25	41°16′N	95°51′W
Courtenay, B.C., Can. (cōōrt-nā′)	142-43	49°41′N	124°60′W
Coushatta, La., U.S. (kou-shăt′ȧ)	132-33	32°01′N	93°21′W
Couture, Lac, lk., Qc., Can.	140-41	60°07′N	75°20′W
Coventry, Eng., U.K. (kŭv′ĕn-trĭ)	206-07	52°25′N	1°30′W
Covington, Ga., U.S. (kŭv′ĭng-tŭn)	134-35	33°36′N	83°51′W
Covington, In., U.S. (kŭv′ĭng-tŭn)	126-27	40°08′N	87°23′W
Covington, Ky., U.S. (kŭv′ĭng-tŭn)	126-27	39°05′N	84°31′W
Covington, La., U.S. (kŭv′ĭng-tŭn)	134-35	30°28′N	90°06′W
Covington, Tn., U.S. (kŭv′ĭng-tŭn)	134-35	35°34′N	89°39′W
Covington, Va., U.S. (kŭv′ĭng-tŭn)	126-27	37°48′N	79°60′W
Cowan, Lake, lk., Austl. (lāk kou′ăn)	296-97	31°50′S	121°50′E
Cowes, Eng., U.K. (kouz)	206-07	50°46′N	1°18′W
Cowra, Austl. (kou′rȧ)	301	33°50′S	148°41′E
Coxim, Braz. (kō-shēɴ′)	182-83	18°30′S	54°45′W
Cox's Bāzār, Bngl.	266-67	21°26′N	91°58′E
Coyame, Mex. (kō-yä′mā)	132-33	29°28′N	105°07′W
Coyle, stm., Arg. *see* Coig	185	50°57′S	69°09′W
Coyuca de Benítez, Mex. (kō-yōō′kä dā bā-nē′tāz)	158-59	17°01′N	100°05′W
Coyuca de Catalán, Mex. (kō-yōō′kä dā kä-tä-län′)	158-59	18°19′N	100°42′W
Cozad, Ne., U.S. (kō′zăd)	124-25	40°52′N	99°59′W
Cozumel, Mex. (kŏ-zōō-mĕ′l)	160	20°31′N	86°55′W
Cozumel, Isla, i., Mex. (ē′s-lä-kŏ-zōō-mĕ′l)	160	20°25′N	86°55′W
Cracow, Pol. *see* Kraków	210-11	50°04′N	19°58′E
Cradock, S. Afr. (krä′dŭk)	286-87	32°10′S	25°37′E

Feature (Pronunciation)	Page	Lat.	Long.
Craig, Ak., U.S. (krāg)	136	55°28′N	133°06′W
Craig, Co., U.S. (krāg)	122-23	40°31′N	107°32′W
Craiova, Rom. (krȧ-yō′vȧ)	216-17	44°19′N	23°48′E
Cranbrook, B.C., Can. (krăn′brŏŏk)	142-43	49°31′N	115°46′W
Crandon, Wi., U.S. (krăn′dŭn)	126-27	45°34′N	88°54′W
Cranston, R.I., U.S. (krăns′tŭn)	126-27	41°48′N	71°26′W
Crater Lake, lk., Or., U.S. (krā′tẽr lāk)	122-23	42°56′N	122°06′W
Crater Lake National Park, n.p., Or., U.S. (krā′tẽr lāk năsh′ŭn-ăl pärk)	122-23	42°52′N	122°10′W
Craters of the Moon National Monument and Preserve, n.p., Id., U.S. (krā′tẽrz ŭv thȧ mōōn năsh′ŭn-ăl mŏn′ū-mĕnt ănd prĭ-zûrv)	122-23	43°25′N	113°33′W
Crateús, Braz. (krä-tå-ōōzh′)	180-81	5°10′S	40°40′W
Crato, Braz. (krä′tò)	180-81	7°14′S	39°23′W
Crawford, Ne., U.S. (krô′fẽrd)	124-25	42°41′N	103°25′W
Crawfordsville, In., U.S. (krô′fẽrdz-vĭl)	126-27	40°02′N	86°54′W
Crazy Mountains, mts., Mt., U.S. (krā′zĭ moun′tĭnz)	122-23	46°08′N	110°20′W
Cree, stm., Sk., Can.	140-41	58°55′N	105°46′W
Cree Lake, lk., Sk., Can. (krē lāk)	140-41	57°30′N	106°30′W
Creil, Fr. (krĕ′y′)	212-13	49°16′N	2°29′E
Crema, Italy (krā′mä)	216-17	45°22′N	9°42′E
Cremona, Italy (krā-mō′nä)	216-17	45°09′N	10°01′E
Crépy-en-Valois, Fr. (krā-pē′ĕɴ-vä-lwä′)	212-13	49°14′N	2°54′E
Crescent City, Ca., U.S.	122-23	41°46′N	124°12′W
Cresco, Ia., U.S. (krĕs′kō)	124-25	43°22′N	92°07′W
Crestline, Oh., U.S. (krĕst-līn)	126-27	40°47′N	82°44′W
Creston, B.C., Can. (krĕs′tŭn)	142-43	49°06′N	116°31′W
Creston, Ia., U.S. (krĕs′tŭn)	124-25	41°04′N	94°22′W
Crestview, Fl., U.S. (krĕst′vū)	134-35	30°46′N	86°34′W
Crestwood, Ky., U.S. (krĕst′wòd)	126-27	38°19′N	85°28′W
Crete, Ne., U.S. (krēt)	130-31	40°37′N	96°58′W
Crete, i., Grc. (krēt)	216a	35°13′N	25°00′E
Crete, Sea of, s., Grc. (sē ŭv krēt)	204-05	35°54′N	25°01′E
Crewe, Eng., U.K. (krōō)	206-07	53°06′N	2°27′W
Criciúma, Braz.	186	28°41′S	49°24′W
Crimea, pen., Ukr. *see* Crimean Peninsula	218-19	45°0′N	34°00′E
Crimean Peninsula, pen., Ukr.	218-19	45°0′N	34°00′E
Cripple Creek, Co., U.S. (krĭp′'l krēk)	130-31	38°45′N	105°11′W
Crisfield, Md., U.S. (krĭs-fēld)	126-27	37°59′N	75°51′W
Cristalândia, Braz.	180-81	10°36′S	49°12′W
Cristóbal Colón, Pico, mtn., Col. (pē′kô-krēs-tô′bäl-kō-lôn′)	178-79	10°50′N	73°41′W
Crna Gora, nation, Eur. *see* Montenegro	190-91	42°30′N	19°18′E
Croatia, nation, Eur. (krō-ā′-shȧ)	190-91	45°10′N	15°30′E
Crockett, Tx., U.S. (krŏk′ĕt)	132-33	31°18′N	95°28′W
Crocodile, stm., S. Afr.	292c	24°11′S	26°53′E
Crooked, stm., B.C., Can. (krŏŏk′ĕd)	142-43	54°50′N	122°53′W
Crooked Island, i., Bah.	154-55	22°45′N	74°13′W
Crooked Island Passage, strt., Bah. (krŏŏk′ĕd ī′lȧnd păs′ĭj)	154-55	22°43′N	74°35′W
Crookston, Mn., U.S. (kròks′tŭn)	124-25	47°47′N	96°37′W
Crosby, Mn., U.S. (krôz′bĭ)	124-25	46°29′N	93°58′W
Crosby, N.D., U.S. (krôz′bĭ)	124-25	48°55′N	103°18′W
Cross, stm., Afr. (krôs)	282a	4°49′N	8°15′E
Cross City, Fl., U.S. (krôs sĭ′tĭ)	134-35	29°38′N	83°07′W
Crossett, Ar., U.S. (krôs′ĕt)	130-31	33°08′N	91°58′W
Cross Lake, res., Mb., Can.	144-45	54°45′N	97°30′W
Cross Sound, strt., Ak., U.S. (krŏs sound)	136	58°10′N	136°30′W
Crotone, Italy (krō-tô′nĕ)	216-17	39°05′N	17°07′E
Crow Creek Indian Reservation, ind. res., S.D., U.S. (krō krēk ĭn′dĭ-ăn rĕ-sẽr-vā′shĕn)	124-25	44°11′N	99°30′W
Crow Indian Reservation, ind. res., Mt., U.S. (krō ĭn′dĭ-ăn rĕ-sẽr-vā′shĕn)	122-23	45°27′N	108°00′W
Crowley, La., U.S. (krou′lē)	132-33	30°13′N	92°23′W
Crown Point, N.Y., U.S. (kroun point′)	126-27	43°57′N	73°26′W
Crowsnest Pass, Ab., Can.	142-43	49°37′N	114°25′W
Crozet, Archipel, is., Afr. *see* Crozet, Îles	313	46°0′S	52°00′E
Crozet, Îles, is., Afr. (ēl-krô-zĕ′)	313	46°0′S	52°00′E
Cruz, Cabo, c., Cuba (ká′-bô-krōōz)	154-55	19°51′N	77°44′W
Cruz Alta, Braz. (krōōz äl′tä)	187	28°37′S	53°36′W
Cruz del Eje, Arg. (krōō′s-dĕl-ĕ-kĕ)	182-83	30°43′S	64°49′W
Cruzeiro, Braz. (krōō-zā′rò)	186	22°35′S	44°58′W
Cruzeiro do Sul, Braz. (krōō-zā′rò dò sōōl)	184	7°39′S	72°41′W
Crystal City, Tx., U.S. (krĭs′tăl sĭ′tĭ)	132-33	28°41′N	99°50′W
Crystal Falls, Mi., U.S. (krĭs′tăl fôlz)	124-25	46°06′N	88°19′W
Crystal Lake, Il., U.S. (krĭs′tăl lāk lāk)	126-27	42°14′N	88°17′W

Feature (Pronunciation)	Page	Lat.	Long.
Crystal Springs, Ms., U.S. (krĭs′tăl sprĭngz)	134-35	31°59′N	90°22′W
Csongrád, Hung. (chôn′gräd)	210-11	46°43′N	20°09′E
Cúa, Ven. (kōō′ä)	188b	10°09′N	66°53′W
Cua Lo, Viet.	266-67	18°49′N	105°43′E
Cuamba, Moz.	286-87	14°47′S	36°32′E
Cuando, stm., Afr.	286-87	18°30′S	23°36′E
Cuango, stm., Afr.	284-85	3°13′S	17°23′E
Cuanza, stm., Ang. (kwän′zä)	286-87	9°21′S	13°09′E
Cuatrociénegas, Mex. (kwä′trō syā′nå-gäs)	132-33	26°60′N	102°04′W
Cuauhtémoc, Mex. (kwä-ōō-tĕ-mŏk′)	158-59	22°33′N	98°09′W
Cuauhtémoc, Mex. (kwä-ōō-tĕ-mŏk′)	156-57	28°25′N	106°52′W
Cuautitlán, Mex. (kwä-ōō-têt-län′)	158-59	19°27′N	104°21′W
Cuautla, Mex. (kwä-ōō′tlà)	158-59	18°49′N	98°57′W
Cuba, nation, N.A. (kū′bȧ)	152-53	21°30′N	80°00′W
Cubagua, Isla, i., Ven. (ē′s-lä-kōō-bä′gwä)	188b	10°49′N	64°11′W
Cubal, Ang.	286-87	13°03′S	14°17′E
Cubango, stm., Afr. (kōō-băŋ′gō)	286-87	18°57′S	22°25′E
Cubia, Ang.	286-87	16°01′S	21°43′E
Çubuk, Tur.	246	40°14′N	33°02′E
Cúcuta, Col. (kōō′kōō-tä)	178-79	7°53′N	72°29′W
Cuddalore, India (kŭd ȧ-lōr′)	256	11°45′N	79°46′E
Cuddapah, India (kŭd′ȧ-pä)	256	14°28′N	78°49′E
Cue, Austl. (kū)	294-95	27°26′S	117°54′E
Cuenca, Ec. (kwĕn′kä)	184	2°53′S	79°00′W
Cuenca, Spain	214-15	40°05′N	2°08′W
Cuencamé de Ceniceros, Mex.	132-33	24°52′N	103°42′W
Cuernavaca, Mex. (kwĕr-nä-vä′kä)	158-59	18°55′N	99°14′W
Cuero, Tx., U.S. (kwā′rō)	132-33	29°06′N	97°17′W
Cuiabá, Braz.	182-83	15°36′S	56°05′W
Cuiabá, stm., Braz.	180-81	17°54′S	57°28′W
Cuicatlán, Mex. (kwē-kä-tlän′)	158-59	17°46′N	96°58′W
Cuilapa, Guat. (kò-ē-lä′pä)	160	14°17′N	90°18′W
Cuilo, stm., Afr.	284-85	5°53′S	16°35′E
Cuilo, stm., Afr.	282-83	3°23′S	17°23′E
Cuíto, stm., Ang. (kōō-ē′tō)	286-87	18°01′S	20°47′E
Cuitzeo, Lago de, lk., Mex. (lä′gô-dĕ-kwēt′zā-ō)	158-59	19°55′N	101°05′W
Culebra, Isla de, i., P.R. (ē′s-lä-dĕ-kōō-lā′brä)	154a	18°19′N	65°17′W
Culfa, Azer.	245	38°58′N	45°38′E
Culgoa, stm., Austl. (kŭl-gō′ȧ)	301	29°59′S	146°07′E
Culiacán, Mex. (kōō-lyä-kä′n)	158-59	24°04′N	107°05′W
Culiacán, Mex.	156-57	24°49′N	107°24′W
Culion Island, i., Phil.	270	11°51′N	119°57′E
Cullera, Spain (kōō-lyā′rä)	214-15	39°10′N	0°14′W
Cullinan, S. Afr. (kò′lĭ-năn)	292c	25°40′S	28°33′E
Cullman, Al., U.S. (kŭl′măn)	134-35	34°10′N	86°51′W
Culpeper, Va., U.S. (kŭl′pĕp-ẽr)	126-27	38°27′N	77°60′W
Culuene, stm., Braz.	180-81	12°56′S	52°50′W
Culver, In., U.S. (kŭl′vẽr)	126-27	41°13′N	86°25′W
Cumaná, Ven.	188b	10°27′N	64°11′W
Cumbal, Nevado, vol., Col.	178-79	0°57′N	77°52′W
Cumberland, Md., U.S.	126-27	39°39′N	78°46′W
Cumberland, Wi., U.S. (kŭm′bẽr-lănd)	124-25	45°32′N	92°01′W
Cumberland, stm., U.S. (kŭm′bẽr-lănd)	120-21	37°08′N	88°25′W
Cumberland, Lake, res., Ky., U.S. (lāk kŭm′bẽr-lănd)	134-35	36°57′N	84°55′W
Cumberland Peninsula, pen., Nu., Can. (kŭm′bẽr-lănd pĕ-nĭn′sūlȧ)	140-41	66°32′N	64°13′W
Cumberland Plateau, plat., U.S. (kŭm′bẽr-lănd plă-tō′)	134-35	36°0′N	85°00′W
Cumberland Sound, strt., Nu., Can. (kŭm′bẽr-lănd sound)	140-41	65°10′N	65°30′W
Cumbres de Monterrey, Parque Nacional, n.p., Mex.	132-33	25°30′N	100°25′W
Cunani, Braz.	178-79	2°52′N	51°06′W
Cunco, Chile	185	38°56′S	72°02′W
Cunene, stm., Afr.	286-87	17°15′S	11°45′E
Cunnamulla, Austl. (kŭn-ȧ-mŭl-ȧ)	301	28°04′S	145°41′E
Curaçao, dep., N.A. (kōō-rä-sä′ō)	152a	12°11′N	69°00′W
Curacautín, Chile (kä-rä-käōō-tē′n)	185	38°25′S	71°56′W
Curacó, stm., Arg. *see* Salado	185	38°49′S	64°59′W
Curanilahue, Chile	185	37°28′S	73°21′W
Curaray, stm., S.A.	184	2°26′S	74°04′W
Curepipe, Mauritius	287a	20°19′S	57°31′E
Curicó, Chile	185	34°59′S	71°14′W
Curitiba, Braz. (kōō-rē-tē′bȧ)	186	25°26′S	49°16′W
Currais Novos, Braz. (kōōr-rä′ēs nō-vōs)	188d	6°15′S	36°31′W
Curralinho, Braz.	180-81	1°48′S	49°47′W
Current, stm., U.S. (kûr′ĕnt)	130-31	36°15′N	90°55′W
Curtis Island, i., Austl.	302	23°38′S	151°09′E

ăt; finăl; rāte; senăte; ärm; ȧsk; sofȧ; fâre; ch-choose; dh-as th in other; bē; ēvent; bĕt; recĕnt; cratẽr; g-gō; gh-guttural g; bĭt; ĭ-short neutral; rīde; ĸ-guttural k as ch in German ich;

Feature (Pronunciation)	Page	Lat.	Long.
Curuá, stm., Braz.	**180-81**	1°57'S	55°08'W
Curuá, stm., Braz.	**180-81**	5°21'S	54°28'W
Cururupu, Braz. (ko͞o-rò-pōō')	**180-81**	1°50'S	44°52'W
Curuzú Cuatiá, Arg.	**187**	29°48'S	58°03'W
Curvelo, Braz. (kòr-vĕl'ò)	**186**	18°45'S	44°27'W
Cusco, Peru	**184**	13°31'S	71°59'W
Cushing, Ok., U.S. (kŭsh'ĭng)	**130-31**	35°59'N	96°46'W
Custer, S.D., U.S. (kŭs'tẽr)	**124-25**	43°45'N	103°36'W
Cut, Nuhu, i., Indon.	**242-43**	5°29'S	133°06'E
Cut Bank, Mt., U.S. (kŭt bănk)	**122-23**	48°38'N	112°20'W
Cuthbert, Ga., U.S. (kŭth'bẽrt)	**134-35**	31°46'N	84°47'W
Cuttack, India (kŭ-tăk')	**254-55**	20°30'N	85°50'E
Cutzamalá, stm., Mex. (ko͞o-tzä-mä-lä')	**158-59**	18°23'N	100°41'W
Cuvo, stm., Ang. (ko͞o'vō)	**286-87**	10°52'S	13°48'E
Cuyabá, Braz. *see* Cuiabá	**182-83**	15°36'S	56°05'W
Cuyahoga Falls, Oh., U.S. (kī-*à*-hō'*gà* fôlz)	**126-27**	41°08'N	81°29'W
Cuyo Islands, is., Phil. (ko͞o'yō ī'l*à*ndz)	**270**	11°04'N	120°57'E
Cuyuni, stm., S.A. (ko͞o-yo͞o'nē)	**178-79**	6°23'N	58°40'W
Cyangugu, Rw.	**289**	2°29'S	28°54'E
Cyclades, is., Grc. (sĭ'cl*à*-dēz)	**216-17**	37°30'N	25°00'E
Cynthiana, Ky., U.S. (sĭn-thĭ-ăn'*à*)	**126-27**	38°23'N	84°18'W
Cyprus, nation, Asia (sī'pr*ŭ*s)	**248-49**	35°0'N	33°00'E
Cyrenaica, hist. reg., Libya (sĭr-*à*-nā'ĭ-k*à*)	**280-81**	31°0'N	22°30'E
Czech Republic, nation, Eur. (chĕk rĕ-pŭb'lĭk)	**190-91**	49°40'N	15°10'E
Czernowitz, Ukr. *see* Chernivtsi	**210-11**	48°17'N	25°58'E
Częstochowa, Pol. (chăɴ-stô kô'và)	**210-11**	50°49'N	19°08'E

D

Feature (Pronunciation)	Page	Lat.	Long.
Da, Song, stm., Asia *see* Black	**266-67**	21°15'N	105°21'E
Da'an, China (dä-än)	**260-61**	45°30'N	124°18'E
Dabhoi, India	**254-55**	22°08'N	73°25'E
Dabie Shan, mts., China (dä-bĭĕ shän)	**258-59**	31°06'N	115°56'E
Dacca, nat. cap., Bngl. *see* Dhaka	**254-55**	23°43'N	90°25'E
Dac Glei, Viet.	**266-67**	15°11'N	107°48'E
Dachau, Ger. (dä'κou)	**210-11**	48°15'N	11°27'E
Dade City, Fl., U.S. (dād sĭ'tĭ)	**135a**	28°22'N	82°11'W
Dadeville, Al., U.S. (dād'vĭl)	**134-35**	32°50'N	85°46'W
Dadiangas, Phil. *see* General Santos	**270**	6°07'N	125°10'E
Dādra and Nagar Haveli, state, India	**254-55**	20°05'N	73°00'E
Dādu, Pak.	**252-53**	26°44'N	67°46'E
Dadu, stm., China (dä-do͞o)	**258-59**	29°33'N	103°46'E
Daegu, Kor., S.	**263**	35°52'N	128°35'E
Daejeon, Kor., S.	**263**	36°20'N	127°26'E
Daet, Phil.	**270**	14°07'N	122°57'E
Dagestan, state, Russia (dä-gĕs-tän')	**245**	43°0'N	47°00'E
Dagupan, Phil. (dä-go͞o'pän)	**270**	16°02'N	120°21'E
Da Hinggan Ling, mts., China (dä hĭŋ-gän lĭŋ) *see* Greater Khingan Range	**240-41**	49°0'N	122°00'E
Dahlak Archipelago, is., Erit. (dä-läk' är'kà-pĕ'-å-gō)	**288**	15°45'N	40°30'E
Dahlak Kebir, i., Erit. *see* Dehalak' Desēt	**288**	15°40'N	40°06'E
Dahomey, nation, Afr. *see* Benin	**274**	9°30'N	2°15'E
Daḩy, Nafūd ad-, des., Sau. Ar.	**288**	22°20'N	45°35'E
Daireaux, Arg.	**187**	36°36'S	61°45'W
Dairen, China *see* Dalian	**260-61**	38°54'N	121°34'E
Dairûț, Egypt	**291b**	27°33'N	30°49'E
Dai-sen, vol., Japan (dī'sĕn')	**265**	35°22'N	133°33'E
Dajabón, Dom. Rep. (dä-kä-bô'n)	**154-55**	19°33'N	71°42'W
Dajian Shan, mtn., China	**258-59**	26°42'N	103°34'E
Dakar, nat. cap., Sen. (dȧ-kär')	**282-83**	14°41'N	17°27'W
Dakhla, W. Sah.	**280-81**	23°43'N	15°56'W
Dalälven, stm., Swe.	**208-09**	60°38'N	17°27'E
Dalandzadgad, Mong.	**260-61**	43°35'N	104°25'E
Da Lat, Viet.	**266-67**	11°58'N	108°27'E
Dālbandin, Pak.	**252-53**	28°53'N	64°25'E
Dalby, Austl. (dôl'bē)	**301**	27°11'S	151°16'E
Dale, Nor. (dä'lĕ)	**208-09**	61°22'N	5°25'E
Dale Hollow Lake, res., U.S. (dāl hŏl'ō lāk)	**134-35**	36°37'N	85°20'W
Dalhart, Tx., U.S. (dăl härt)	**130-31**	36°04'N	102°31'W
Dalhousie, N.B., Can. (dăl-ho͞o'zē)	**148-49**	48°04'N	66°24'W
Dalhousie, India	**254-55**	32°34'N	75°58'E
Dalhousie, Cape, c., N.T., Can.	**140-41**	70°14'N	129°42'W
Dali, China (dä-lē)	**258-59**	34°47'N	109°57'E
Dali, China (dä-lē)	**258-59**	25°36'N	100°13'E
Dalian, China (dä-lĭĕn)	**260-61**	38°54'N	121°34'E
Daliang Shan, mts., China	**258-59**	28°0'N	103°00'E
Dallas, Or., U.S. (dăl'l*à*s)	**122-23**	44°55'N	123°18'W
Dallas, Tx., U.S. (dăl'l*à*s)	**130-31**	32°47'N	96°46'W
Dall Island, i., Ak., U.S. (dăl ī'l*ă*nd ī'l*à*nd)	**136**	54°57'N	133°01'W
Dalmā, i., U.A.E.	**250-51**	24°31'N	52°19'E
Dal'negorsk, Russia	**264**	44°33'N	135°34'E
Dalnerechensk, Russia	**264**	45°56'N	133°44'E
Daloa, C. Iv.	**282-83**	6°53'N	6°27'W
Dalrymple, Mount, mtn., Austl. (mount dăl'rĭm-p'l)	**302**	21°02'S	148°38'E
Dāltenganj, India	**254-55**	24°03'N	84°04'E
Dalton, Ga., U.S. (dôl't*ŭ*n)	**134-35**	34°46'N	84°58'W
Dalupiri Island, i., Phil.	**270a**	19°05'N	121°14'E
Daly, stm., Austl. (dā'lĭ)	**296-97**	13°19'S	130°16'E
Daly Waters, Austl.	**294-95**	16°15'S	133°22'E
Damān, India	**254-55**	20°25'N	72°51'E
Damān and Diu, state, India	**254-55**	20°25'N	72°50'E
Damanhûr, Egypt	**291b**	31°02'N	30°28'E
Damar, Pulau, i., Indon.	**268-69**	1°01'S	128°23'E
Damaraland, hist. reg., Nmb. (dă'n*à*-r*à*-lănd)	**286-87**	22°11'S	17°35'E
Damas, nat. cap., Syria *see* Damascus	**248-49**	33°31'N	36°18'E
Damascus, nat. cap., Syria (d*à*-măs'kŭs)	**248-49**	33°31'N	36°18'E
Damāvand, Kūh-e, vol., Iran	**252-53**	35°56'N	52°08'E
Dāmghān, Iran (dām-gān')	**252-53**	36°10'N	54°21'E
Damietta, Egypt	**291b**	31°25'N	31°49'E
Dāmodar, stm., India	**254-55**	22°17'N	88°06'E
Damoh, India	**254-55**	23°51'N	79°27'E
Dampier, Selat, strt., Indon. (så-lät' däm'pēr)	**242-43**	0°40'S	130°40'E
Danakil, reg., Afr.	**288**	13°0'N	41°00'E
Da Nang, Viet.	**266-67**	16°03'N	108°12'E
Danbury, Ct., U.S. (dăn'bẽr-ĭ)	**126-27**	41°23'N	73°27'W
Dandenong, Austl. (dăn'dẻ-nông)	**301**	38°00'S	145°13'E
Dandong, China (dän-dôŋ)	**263**	40°07'N	124°21'E
Dang, stm., China	**260-61**	40°26'N	94°34'E
Dângrêk, Chuŏr Phnum, mts., Asia *see* Phanom Dongrak Range	**266-67**	14°25'N	103°30'E
Dangriga, Belize	**160**	16°58'N	88°13'W
Danilov, Russia (dä'nẻ-lôf)	**218-19**	58°11'N	40°11'E
Dankov, Russia (däŋ'kôf)	**218-19**	53°16'N	39°09'E
Danmark, nation, Eur. *see* Denmark	**190-91**	56°0'N	10°00'E
Danmarksstrædet, strt., *see* Denmark Strait	**95**	67°0'N	25°00'W
Dannemora, N.Y., U.S. (dăn-ẻ-mō'r*à*)	**126-27**	44°43'N	73°43'W
Dannhauser, S. Afr. (dän'hou-zẽr)	**292c**	28°02'S	30°04'E
Dansville, N.Y., U.S. (dănz'vĭl)	**126-27**	42°33'N	77°41'W
Danube, stm., Eur. (dăn'ūb)	**204-05**	45°23'N	29°36'E
Danville, Ar., U.S. (dăn'vĭl)	**130-31**	35°04'N	93°23'W
Danville, Il., U.S. (dăn'vĭl)	**126-27**	40°08'N	87°38'W
Danville, In., U.S. (dăn'vĭl)	**126-27**	39°45'N	86°31'W
Danville, Ky., U.S. (dăn'vĭl)	**126-27**	37°39'N	84°47'W
Danville, Va., U.S. (dăn'vĭl)	**134-35**	36°35'N	79°24'W
Danxian, China (dän shyĕn)	**258-59**	19°31'N	109°33'E
Danzig, Pol. *see* Gdańsk	**210-11**	54°21'N	18°38'E
Daocheng, China	**258-59**	29°04'N	100°36'E
Daoxian, China (dou shyĕn)	**258-59**	25°35'N	111°27'E
Da Qaidam, China	**260-61**	37°53'N	95°19'E
Darabani, Rom. (dă-ră-băn'ĭ)	**210-11**	48°11'N	26°36'E
Darbhanga, India (dŭr-bŭŋ'gä)	**254-55**	26°09'N	85°54'E
Dardanelles, strt., Tur. (där-d*à*-nĕlz')	**216-17**	40°17'N	26°33'E
Dar-el-Beida, Mor. *see* Casablanca	**292a**	33°36'N	7°36'W
Dar es Salaam, nat. cap., Tan. (där ĕs s*à*-läm')	**284-85**	6°49'S	39°16'E
Dārfūr, hist. reg., Sudan (där-fo͞or')	**284-85**	13°30'N	23°30'E
Darganata, Turkmen.	**252-53**	40°29'N	62°10'E
Darhan, Mong.	**260-61**	49°29'N	105°56'E
Darién, Col. (dä-rĭ-ĕn')	**188c**	3°56'N	76°30'W
Darlag, China	**258-59**	33°46'N	99°40'E
Darling, stm., Austl. (där'lĭng)	**301**	34°07'S	141°55'E
Darling Downs, reg., Austl. (där'lĭng dounz)	**301**	27°30'S	150°30'E
Darling Range, mts., Austl. (där'lĭng rānj)	**296-97**	32°31'S	116°21'E
Darlington, Eng., U.K. (där'lĭng-t*ŭ*n)	**206-07**	54°32'N	1°34'W
Darlington, S.C., U.S. (där'lĭng-t*ŭ*n)	**134-35**	34°18'N	79°52'W
Darlington, Wi., U.S. (där'lĭng-t*ŭ*n)	**126-27**	42°41'N	90°07'W
Darmstadt, Ger. (därm'shtät)	**210-11**	49°52'N	8°39'E
Darnah, Libya	**204-05**	32°46'N	22°39'E
Darnley, Cape, c., Ant.	**313**	67°43'S	69°30'E
Daror, Eth.	**284-85**	8°14'N	44°41'E
Dart, Cape, c., Ant.	**313**	73°06'S	126°20'W
Dartmoor, for., Eng., U.K. (därt'mo͞or)	**206-07**	50°35'N	3°55'W
Dartmouth, N.S., Can. (därt'mŭth)	**148-49**	44°41'N	63°34'W
Daru, Pap. N. Gui. (dä'ro͞o)	**302**	9°05'S	143°14'E
Daruvar, Cro. (där'ro͞o-vär)	**216-17**	45°36'N	17°13'E
Darvaza, Turkmen.	**252-53**	40°10'N	58°29'E
Darwin, Austl. (där'wĭn)	**242-43**	12°27'S	130°50'E
Dasht, stm., Pak. (dŭsht)	**250-51**	25°13'N	61°42'E
Daşköpri, Turkmen.	**252-53**	36°17'N	62°38'E
Daşoguz, Turkmen.	**244**	41°50'N	59°58'E
Datia, India	**254-55**	25°40'N	78°27'E
Datong, China (dä-tôŋ)	**260-61**	37°03'N	101°36'E
Datong, China (dä-tôŋ)	**260-61**	40°05'N	113°17'E
Datong, stm., China	**260-61**	36°20'N	102°50'E
Datu Piang, Phil.	**270**	7°02'N	124°29'E
Daua, stm., Afr. *see* Dawa	**284-85**	4°10'N	42°06'E
Daugava, stm., Eur. *see* Western Dvina	**208-09**	57°04'N	24°03'E
Daugavpils, Lat. (dä'ò-gäv-pēls)	**208-09**	55°53'N	26°32'E
Daule, stm., Ec.	**184**	2°18'S	79°50'W
Daung Kyun, i., Mya.	**266-67**	12°14'N	98°05'E
Dauphin, Mb., Can. (dô'fĭn)	**144-45**	51°09'N	100°03'W
Dauphin Lake, lk., Mb., Can. (dô'fĭn lāk)	**144-45**	51°17'N	99°48'W
Dāvangere, India	**256**	14°28'N	75°55'E
Davao, Phil. (dä'vä-ō)	**270**	7°04'N	125°35'E
Davao Gulf, b., Phil.	**270**	6°40'N	125°55'E
Davenport, Ia., U.S. (dăv'ĕn-pōrt)	**124-25**	41°31'N	90°35'W
Davenport, Wa., U.S. (dăv'ĕn-pōrt)	**122-23**	47°39'N	118°08'W
David, Pan. (d*à*-vēdh')	**162**	8°26'N	82°26'W
David City, Ne., U.S. (dā'vĭd sĭ'tē)	**124-25**	41°15'N	97°08'W
Davis, Ca., U.S. (dā'vĭs)	**128-29**	38°33'N	121°44'W
Davis, Ok., U.S. (dā'vĭs)	**130-31**	34°30'N	97°07'W
Davis Mountains, mts., Tx., U.S. (dā'vĭs moun'tĭnz)	**132-33**	30°42'N	104°10'W
Davis Sea, s., Ant.	**313**	66°0'S	92°00'E
Davisstrædet, strt., N.A. *see* Davis Strait	**95**	67°0'N	57°00'W
Davis Strait, strt., N.A. (dā'vĭs strāt)	**95**	67°0'N	57°00'W
Davlekanovo, Russia	**202-03**	54°13'N	55°02'E
Davos, Switz. (dä'vōs)	**210-11**	46°48'N	9°49'E
Dawa, stm., Afr.	**284-85**	4°10'N	42°06'E
Dawāsir, Wādī ad-, stm., Sau. Ar.	**238-39**	20°26'N	45°35'E
Dawei, Mya.	**266-67**	14°05'N	98°13'E
Dawna Range, mts., Mya. (dô na rānj)	**266-67**	16°50'N	98°15'E
Dawson, Yk., Can. (dô's*ŭ*n)	**138-39**	64°03'N	139°24'W
Dawson, Ga., U.S. (dô's*ŭ*n)	**134-35**	31°47'N	84°26'W
Dawson, Mn., U.S. (dô's*ŭ*n)	**124-25**	44°56'N	96°04'W
Dawson, stm., Austl. (dô's*ŭ*n)	**302**	23°38'S	149°46'E
Dawson, Isla, i., Chile	**185**	53°55'S	70°43'W
Dawson Creek, B.C., Can. (dô's*ŭ*n krēk)	**142-43**	55°47'N	120°20'W
Dawson Springs, Ky., U.S. (dô's*ŭ*n sprĭngz)	**134-35**	37°10'N	87°41'W
Dawu, China (dä-wo͞o)	**258-59**	31°33'N	114°07'E
Dax, Fr. (dȧks)	**212-13**	43°43'N	1°03'W
Daxian, China (dä-shyĕn)	**258-59**	31°13'N	107°30'E
Daxue Shan, mts., China	**258-59**	30°30'N	101°46'E
Dayr az-Zawr, Syria (då-ērẻz-zôr')	**248-49**	35°21'N	40°09'E
Dayton, Oh., U.S. (dā't*ŭ*n)	**126-27**	39°45'N	84°12'W
Dayton, Tn., U.S. (dā't*ŭ*n)	**134-35**	35°30'N	85°01'W
Dayton, Tx., U.S. (dā't*ŭ*n)	**132-33**	30°03'N	94°54'W
Dayton, Wa., U.S. (dā't*ŭ*n)	**122-23**	46°19'N	117°58'W
Daytona Beach, Fl., U.S. (dā-tō'n*à* bēch)	**134-35**	29°13'N	81°02'W
Dayu, China (dä-yo͞o)	**258-59**	25°24'N	114°22'E
Da Yunhe, can., China (dä yòn-hŭ) *see* Grand Canal	**260-61**	32°11'N	119°33'E
De Aar, S. Afr. (dē-är')	**286-87**	30°39'S	24°01'E
Dead Sea, lk., Asia (dĕd sē)	**248-49**	31°30'N	35°30'E
Deadwood, S.D., U.S. (dĕd'wòd)	**124-25**	44°23'N	103°44'W
Deal Island, i., Austl.	**301**	39°28'S	147°20'E
Dean, stm., B.C., Can. (dēn)	**142-43**	52°48'N	126°59'W
Dean Channel, strt., B.C., Can. (dēn chăn'ĕl)	**142-43**	52°33'N	127°13'W
Deán Funes, Arg. (dĕ-ȧ'n-fo͞o-nĕs)	**187**	30°25'S	64°21'W
Dearborn, Mi., U.S. (dēr'b*ŭ*rn)	**126-27**	42°19'N	83°10'W
Dearg, Beinn, mtn., Scot., U.K. (bĕn dŭrg)	**206-07**	57°47'N	4°56'W
Dease, stm., B.C., Can.	**140-41**	59°55'N	128°29'W
Death Valley, Ca., U.S. (dĕth văl'ē)	**128-29**	36°19'N	116°25'W
Death Valley, val., Ca., U.S. (dĕth văl'ē)	**128-29**	36°30'N	117°00'W
Death Valley National Park, n.p., U.S. (dĕth văl'ē năsh'ŭn-*ă*l pärk)	**128-29**	36°25'N	116°56'W
Debao, China (dŭ-bou)	**258-59**	23°20'N	106°37'E
Dęblin, Pol. (dăɴ'blĭn)	**210-11**	51°35'N	21°53'E
Dębno, Pol. (dĕb-nô')	**210-11**	52°44'N	14°43'E
Debrecen, Hung. (dĕ'brĕ-tsĕn)	**210-11**	47°32'N	21°38'E
Debre Mark'os, Eth.	**292d**	10°20'N	37°44'E
Decatur, Al., U.S. (dẻ-kā't*ŭ*r)	**134-35**	34°36'N	86°59'W
Decatur, Il., U.S. (dẻ-kā't*ŭ*r)	**126-27**	39°50'N	88°57'W

n-sing; ŋ-baŋk; ɴ-nasalized n; nŏd; c*ŏ*mmit; ōld; ôbey; ôrder; oi-boil; fo͞od; ò-as oo in foot; ou-out; s-soft; sh-dish; th-thin; pūre; ůnite; ûrn; stŭd; circ*ŭ*s; ü-as in French tu; '-indeterminate vowel.

350 Decatur – Djambala PRONOUNCING INDEX

Feature (Pronunciation)	Page	Lat.	Long.
Decatur, In., U.S. (dē-kā′tŭr)	126-27	40°50′N	84°55′W
Decatur, Mi., U.S. (dē-kā′tŭr)	126-27	42°06′N	85°58′W
Decatur, Tx., U.S. (dē-kā′tŭr)	130-31	33°14′N	97°35′W
Decazeville, Fr. (dē-käz′vēl′)	212-13	44°34′N	2°15′E
Deccan, plat., India (dĕk′ăn)	256	17°0′N	78°00′E
Deception Lake, lk., Sk., Can. (dē-sĕp′shŭn lāk)	144-45	56°33′N	104°10′W
Decorah, Ia., U.S. (dē-kō′rȧ)	124-25	43°18′N	91°47′W
Dee, stm., Scot., U.K. (dē)	206-07	57°08′N	2°05′W
Deep River, On., Can. (dēp rĭv′ēr)	146-47	46°06′N	77°30′W
Deerfield, Ma., U.S. (dēr′fēld)	126-27	42°33′N	72°36′W
Deer Lake, Nf./L., Can. (dēr lāk)	148-49	49°11′N	57°26′W
Deer Lake, lk., On., Can. (dēr lāk)	144-45	52°39′N	94°29′W
Deer Lodge, Mt., U.S. (dēr lŏj)	122-23	46°24′N	112°44′W
Deer Park, Wa., U.S. (dēr pärk)	122-23	47°57′N	117°28′W
Deer River, Mn., U.S. (dēr rĭv′ēr)	124-25	47°19′N	93°48′W
Defiance, Oh., U.S. (dē-fī′ăns)	126-27	41°16′N	84°22′W
Defuniak Springs, Fl., U.S. (dē fū′nĭ-ăk sprĭngz)	134-35	30°43′N	86°07′W
Dêgê, China	258-59	31°50′N	98°40′E
Degeh Bur, Eth.	284-85	8°13′N	43°33′E
Deggendorf, Ger. (dĕ′ghĕn-dôrf)	210-11	48°51′N	12°58′E
Dehalak' Desēt, i., Erit.	288	15°40′N	40°06′E
Dehiwala-Mount Lavinia, Sri L.	256	6°52′N	79°52′E
Dehra Dūn, India (dā′rŭ dŭn)	254-55	30°19′N	78°02′E
Dehri, India	254-55	24°54′N	84°11′E
Dehua, China (dŭ-hwä)	243a	25°30′N	118°14′E
Dehui, China	260-61	44°32′N	125°43′E
Dej, Rom. (dāzh)	210-11	47°09′N	23°53′E
De Kalb, Il., U.S. (dē kălb′)	126-27	41°56′N	88°45′W
DeLand, Fl., U.S. (dē lănd′)	134-35	29°02′N	81°18′W
Delano, Ca., U.S. (dĕl′ȧ-nō)	128-29	35°47′N	119°15′W
Delavan, Wi., U.S. (dĕl′ȧ-văn)	126-27	42°38′N	88°39′W
Delaware, Oh., U.S. (dĕl′ȧ-wâr)	126-27	40°18′N	83°04′W
Delaware, state, U.S. (dĕl′ȧ-wâr)	118-19	39°10′N	75°30′W
Delaware, stm., Ks., U.S. (dĕl′ȧ-wâr)	130-31	39°04′N	95°25′W
Delaware, stm., U.S.	126-27	39°20′N	75°25′E
Delaware Bay, b., U.S. (dĕl′ȧ-wâr bā)	126-27	39°05′N	75°15′W
De Leon, Tx., U.S. (dē lē-ŏn′)	132-33	32°07′N	98°33′W
Delft, Neth. (dĕlft)	206-07	52°01′N	4°22′E
Delgado, Cabo, c., Moz. (kȧ′bô-dĕl-gä′dō)	286-87	10°41′S	40°38′E
Delger, stm., Mong.	260-61	49°17′N	100°42′E
Delhi, India (dĕl′hī)	254-55	28°40′N	77°14′E
Delhi, La., U.S. (dĕl′hī)	134-35	32°28′N	91°30′W
Delhi, state, India (dĕl′hī)	254-55	28°37′N	77°10′E
Delicias, Mex.	132-33	28°12′N	105°29′W
Déline, N.T., Can.	138-39	65°11′N	123°25′W
Delingha, China	260-61	37°15′N	97°11′E
Dell Rapids, S.D., U.S. (dĕl răp′ĭdz)	124-25	43°49′N	96°43′W
Delmarva Peninsula, pen., U.S.	126-27	38°30′N	75°30′W
Delmenhorst, Ger. (dĕl′mĕn-hôrst)	210-11	53°04′N	8°38′E
De Long Mountains, mts., Ak., U.S. (dē′lông moun′tĭnz)	136	68°20′N	162°00′W
Delphi, In., U.S. (dĕl′fī)	126-27	40°35′N	86°40′W
Delphos, Oh., U.S. (dĕl′fŏs)	126-27	40°51′N	84°21′W
Del Rio, Tx., U.S. (dĕl rē′ō)	132-33	29°22′N	100°54′W
Delta Junction, Ak., U.S.	136	64°02′N	145°44′W
Dëma, stm., Russia (dyĕm′ä)	192-93	54°44′N	55°54′E
Dembī Dolo, Eth.	284-85	8°32′N	34°48′E
Demidov, Russia (dzyĕ′mē-dô′f)	208-09	55°16′N	31°32′E
Deming, N.M., U.S. (dĕm′ĭng)	128-29	32°16′N	107°45′W
Demini, stm., Braz.	180-81	0°46′S	62°56′W
Demopolis, Al., U.S. (dē-mŏp′ō-lĭs)	134-35	32°31′N	87°50′W
Dempo, Gunung, vol., Indon. (gōō-nŏng dĕm′pô)	266-67	4°00′S	103°09′E
Demyanka, stm., Russia (dyĕm-yän′kä)	236-37	59°31′N	69°05′E
Demyansk, Russia (dyĕm-yänsk′)	208-09	57°38′N	32°28′E
Denakil, reg., Afr. *see* Danakil	288	13°0′N	41°00′E
Denali, mtn., Ak., U.S.	136	63°04′N	151°00′W
Denali National Park and Preserve, n.p., Ak., U.S.	136	63°15′N	150°30′W
Dêngqên, China	258-59	31°27′N	95°27′E
Denham, Austl.	294-95	25°55′S	113°33′E
Denham, Mount, mtn., Jam.	154-55	18°13′N	77°32′W
Denham Range, mts., Austl.	302	21°41′S	147°55′E
Deniliquin, Austl. (dĕ-nĭl′ĭ-kwĭn)	301	35°32′S	144°58′E
Denison, Ia., U.S. (dĕn′ĭ-sŭn)	124-25	42°01′N	95°22′W
Denison, Tx., U.S. (dĕn′ĭ-sŭn)	130-31	33°45′N	96°33′W
Denizli, Tur. (dĕn-ĭz-lē′)	216-17	37°47′N	29°05′E
Denmark, S.C., U.S. (dĕn′märk)	134-35	33°20′N	81°09′W
Denmark, nation, Eur. (dĕn′märk)	190-91	56°0′N	10°00′E
Denmark Strait, strt., (dĕn′märk strāt)	95	67°0′N	25°00′W
Denov, Uzb.	252-53	38°16′N	67°54′E
Denpasar, Indon.	268-69	8°39′S	115°13′E

Feature (Pronunciation)	Page	Lat.	Long.
Denton, Tx., U.S. (dĕn′tŭn)	130-31	33°14′N	97°08′W
D'Entrecasteaux Islands, is., Pap. N. Gui. (däɴ-tr'-kȧs-tō′ ī′lȧndz)	302	9°27′S	150°32′E
Denver, Co., U.S. (dĕn′vēr)	130-31	39°44′N	104°58′W
Deolāli, India	254-55	19°57′N	73°50′E
De Pere, Wi., U.S. (dē pēr′)	126-27	44°26′N	88°03′W
Dêqên, China	258-59	28°38′N	98°52′E
De Queen, Ar., U.S. (dē kwēn′)	130-31	34°03′N	94°21′W
De Quincy, La., U.S. (dē kwĭn′sĭ)	132-33	30°27′N	93°26′W
Dera, Lach, stm., Afr. (dā′rä)	284-85	0°13′N	42°18′E
Dera Ghāzi Khān, Pak. (dā′rŭ gä-zē′ ĸän′)	252-53	30°03′N	70°38′E
Dera Ismāīl Khān, Pak. (dā′rŭ ĭs-mä-ēl′ ĸän′)	252-53	31°49′N	70°55′E
Derbent, Russia (dĕr-bĕnt′)	245	42°03′N	48°17′E
Derby, Austl. (där′bē) (dûr′bē)	294-95	17°20′S	123°38′E
Derby, Eng., U.K. (där′bē)	206-07	52°55′N	1°29′W
Derby, Ks., U.S. (dûr′bē)	130-31	37°32′N	97°15′W
De Ridder, La., U.S. (dē rĭd′ēr)	132-33	30°51′N	93°18′W
Dermott, Ar., U.S. (dûr′mŏt)	134-35	33°32′N	91°26′W
Derry, N. Ire., U.K. *see* Londonderry	206-07	54°59′N	7°20′W
Derry, N.H., U.S. (dâr′ĭ)	126-27	42°53′N	71°20′W
Derventa, Bos. (dĕr′ven-tȧ)	216-17	44°59′N	17°54′E
Derzhavinsk, Kaz.	244	51°06′N	66°19′E
Desaguadero, stm., S.A.	184	18°04′S	67°06′W
Désappointement, Îles du, is., Fr. Poly.	304-05	14°10′S	141°20′W
Deschambault Lake, lk., Sk., Can.	144-45	54°40′N	103°35′W
Deschutes, stm., Or., U.S. (dā-shōōt′)	122-23	45°39′N	120°55′W
Desē, Eth.	288	11°09′N	39°38′E
Deseado, stm., Arg. (dā-sā-ä′dhō)	185	47°46′S	65°53′W
Desengaño, Punta, c., Arg.	185	49°15′S	67°37′W
De Smet, S.D., U.S. (dē smĕt′)	124-25	44°23′N	97°33′W
Des Moines, Ia., U.S. (dē moin′)	124-25	41°34′N	93°36′W
Des Moines, stm., U.S. (dē moin′)	120-21	40°22′N	91°26′W
Desna, stm., Eur. (dyĕs-ná′)	218-19	50°33′N	30°34′E
Desolación, Isla, i., Chile (ē′s-lä-dĕ-sô-lä-syō′n)	185	53°00′S	74°09′W
De Soto, Mo., U.S. (dē sō′tō)	130-31	38°08′N	90°34′W
Dessau, Ger. (dĕsʹou)	210-11	51°50′N	12°14′E
Destruction Bay, Yk., Can.	138-39	61°14′N	138°50′W
Detmold, Ger. (dĕt′mōld)	210-11	51°56′N	8°53′E
Detroit, Mi., U.S. (dē-troit′)	126-27	42°21′N	83°04′W
Detroit Lakes, Mn., U.S. (dē-troit′ lākz)	124-25	46°49′N	95°51′W
Detva, Slvk. (dyĕt′vá)	210-11	48°32′N	19°29′E
Deutschland, nation, Eur. *see* Germany	190-91	51°0′N	10°00′E
Deva, Rom. (dā′vä)	210-11	45°53′N	22°55′E
Deventer, Neth. (dĕv′ĕn-tēr)	206-07	52°16′N	6°10′E
Develi, Tur.	247	38°23′N	35°30′E
Devils, stm., Tx., U.S. (dĕv′'lz)	132-33	29°32′N	100°59′W
Devils Lake, N.D., U.S. (dĕv′'lz lāk)	124-25	48°07′N	98°52′W
Devils Postpile National Monument, n.p., Ca., U.S. (dĕv′'lz pōst-pīl năsh′ŭn-ăl mŏn′ū-mĕnt)	128-29	37°37′N	119°05′W
Devils Tower National Monument, n.p., Wy., U.S. (dĕv′'lz tou′ēr năsh′ŭn-ăl mŏn′ū-mĕnt)	122-23	44°36′N	104°43′W
Devon Island, i., Nu., Can.	95	75°0′N	87°00′W
Devonport, Austl. (dĕv′ŭn-pôrt)	301	41°11′S	146°21′E
Dewās, India	254-55	22°58′N	76°03′E
Dewey, Ok., U.S. (dū′ĭ)	130-31	36°48′N	95°56′W
De Witt, Ar., U.S. (dē wĭt′)	134-35	34°18′N	91°20′W
De Witt, Ia., U.S. (dē wĭt′)	124-25	41°49′N	90°32′W
Dexter, Me., U.S. (dĕks′tēr)	127a	45°02′N	69°17′W
Dexter, Mo., U.S. (dĕks′tēr)	134-35	36°48′N	89°58′W
Dexter, N.M., U.S. (dĕks′tēr)	130-31	33°12′N	104°22′W
Dezfūl, Iran	248-49	32°23′N	48°24′E
Dezhnëva, Mys, c., Russia (mĭs dyĕzh′nyĭf)	136	66°08′N	169°41′W
Dezhou, China (dŭ-jō)	260-61	37°27′N	116°17′E
Dhahran, Sau. Ar.	250-51	26°18′N	50°08′E
Dhaka, nat. cap., Bngl. (dä′kä) (dăk′ȧ)	254-55	23°43′N	90°25′E
Dhamār, Yemen	288	14°33′N	44°24′E
Dhanbād, India	254-55	23°48′N	86°26′E
Dhār, India	254-55	22°36′N	75°18′E
Dhawalāgiri, mtn., Nepal	254-55	28°42′N	83°30′E
Dhuburi, India	254-55	26°01′N	89°59′E
Dhule, India	254-55	20°54′N	74°46′E
Diablo, Pico del, mtn., Mex.	156-57	30°59′N	115°45′W
Diablo Range, mts., Ca., U.S. (dyä′blô rānj)	128-29	37°0′N	121°20′W
Diamante, Arg.	187	32°04′S	60°39′W
Diamantina, Braz.	186	18°15′S	43°37′W
Diamantina, stm., Austl. (dī′man-tē′nȧ)	302	26°58′S	138°49′E

Feature (Pronunciation)	Page	Lat.	Long.
Diamantina National Park, n.p., Austl.	302	23°43′S	141°11′E
Diamantino, Braz. (dē-á-män-tē′no)	180-81	14°25′S	56°27′W
Dian Chi, lk., China (dĭĕn chē)	258-59	24°50′N	102°42′E
Dianópolis, Braz.	180-81	11°38′S	46°50′W
D' Iberville, Mont, mtn., Can. *see* Caubvick, Mount	140-41	58°53′N	63°43′W
Dibrugarh, India	254-55	27°29′N	94°55′E
Dickinson, N.D., U.S. (dĭk′ĭn-sŭn)	124-25	46°53′N	102°47′W
Dickson, Tn., U.S. (dĭk′sŭn)	134-35	36°04′N	87°22′W
Dickson City, Pa., U.S. (dĭk′sŭn sĭ′tē)	126-27	41°28′N	75°37′W
Dicle, stm., Asia *see* Tigris	226-27	30°60′N	47°27′E
Diego de Almagro, Isla, i., Chile	185	51°28′S	75°11′W
Diégo-Suarez, Madag. *see* Antsirañana	286-87	12°17′S	49°17′E
Dien Bien, Viet.	266-67	21°23′N	103°01′E
Dien Bien Phu, Viet. *see* Dien Bien	266-67	21°23′N	103°01′E
Dieppe, N.B., Can. (dē-ĕp′)	148-49	46°06′N	64°44′W
Dieppe, Fr.	212-13	49°56′N	1°05′E
Difuri, i., Mald.	256	5°24′N	73°38′E
Digboi, India	254-55	27°23′N	95°38′E
Digby, N.S., Can. (dĭg′bĭ)	148-49	44°37′N	65°46′W
Digul, stm., Indon.	302	7°10′S	138°41′E
Dijlah, stm., Asia *see* Tigris	226-27	30°60′N	47°27′E
Dijon, Fr. (dē-zhôɴ′)	212-13	47°19′N	5°02′E
Dikhil, Dji.	288	11°06′N	42°22′E
Dikson, Russia (dĭk′sŏn)	236-37	73°30′N	80°33′E
Dīla, Eth.	284-85	6°19′N	38°14′E
Dili, nat. cap., Timor-L. (dĭl′ē)	268-69	8°35′S	125°35′E
Dilling, Sudan	288	12°02′N	29°40′E
Dillingham, Ak., U.S. (dĭl′ĕng-hăm)	136	59°03′N	158°28′W
Dillon, Mt., U.S. (dĭl′ŭn)	122-23	45°13′N	112°38′W
Dillon, S.C., U.S. (dĭl′ŭn)	134-35	34°25′N	79°22′W
Dilolo, D.R.C. (dē-lō′lō)	284-85	10°42′S	22°21′E
Dimāpur, India	254-55	25°55′N	93°44′E
Dimashq, nat. cap., Syria *see* Damascus	248-49	33°31′N	36°18′E
Dimitrovgrad, Blg.	216-17	42°03′N	25°37′E
Dimitrovgrad, Russia	202-03	54°13′N	49°36′E
Dimlang, mtn., Nig.	282-83	8°24′N	11°47′E
Dinagat Island, i., Phil.	270	10°12′N	125°35′E
Dinājpur, Bngl.	254-55	25°38′N	88°38′E
Dinan, Fr. (dē-näɴ′)	212-13	48°28′N	2°03′W
Dinant, Bel. (dē-näɴ′)	206-07	50°16′N	4°55′E
Dinar, Tur.	246	38°04′N	30°10′E
Dinara Planina, mts., Eur. (dē′nä-rä plä′nē-na) *see* Dinaric Alps	216-17	43°55′N	16°38′E
Dinaric Alps, mts., Eur.	216-17	43°55′N	16°38′E
Dinariche, Alpi, mts., Eur. *see* Dinaric Alps	216-17	43°55′N	16°38′E
Dindigul, India	256	10°22′N	77°59′E
Dingalan Bay, b., Phil. (dĭŋ-gä′län bā)	270	15°18′N	121°25′E
Dinggyê, China	254-55	28°35′N	86°37′E
Dinghai, China	258-59	30°01′N	122°06′E
Dingwall, Scot., U.K. (dĭng′wôl)	206-07	57°36′N	4°26′W
Dingxi, China	260-61	35°33′N	104°32′E
Dingxian, China (dĭŋ shyĕn)	260-61	38°31′N	114°60′E
Dingyuan, China (dĭŋ-yůän)	258-59	32°32′N	117°40′E
Dinosaur National Monument, n.p., U.S. (dī′nô-sôr năsh′ŭn-ăl mŏn′ū-mĕnt)	122-23	40°32′N	108°58′W
Dipolog, Phil.	270	8°35′N	123°21′E
Dirē Dawa, Eth.	284-85	9°35′N	41°52′E
Diriamba, Nic. (dēr-yäm′bä)	161	11°51′N	86°15′W
Dirj, Libya	204-05	30°10′N	10°28′E
Dirranbandi, Austl. (dĭ-rȧ-băn′dē)	301	28°35′S	148°14′E
Disappointment Islands, is., Fr. Poly. (dĭs′ȧ-point′ment ī′lȧndz) *see* Désappointement, Îles du	304-05	14°10′S	141°20′W
Dispur, India	254-55	26°08′N	91°48′E
Disraëli, Qc., Can. (dĭs-rā′lĭ)	146-47	45°54′N	71°21′W
District of Columbia, state, U.S.	118-19	38°54′N	77°01′W
Distrito Federal, state, Braz. (dēs-trē′tô-fĕ-dĕ-rä′l)	182-83	15°45′S	47°45′W
Distrito Federal, state, Mex.	158-59	19°15′N	99°10′W
Dis ûq, Egypt	291b	31°08′N	30°39′E
Diu, India (dē′ōō)	254-55	20°43′N	70°59′E
Divinópolis, Braz. (dē-vē-nô′pō-lēs)	186	20°09′S	44°53′W
Divnoye, Russia	202-03	45°55′N	43°22′E
Divo, C. Iv.	282-83	5°50′N	5°22′W
Dixon, Il., U.S. (dĭks′ŭn)	126-27	41°50′N	89°29′W
Dixon Entrance, strt., N.A.	136	54°25′N	132°30′W
Diyarbakır, Tur. (dē-yär-bĕk′ĭr)	245	37°55′N	40°14′E
Dja, stm., Afr.	282-83	2°02′N	15°12′E
Djajapura, Indon. *see* Jayapura	302	2°32′S	140°43′E
Djakarta, nat. cap., Indon. *see* Jakarta	268-69	6°11′S	106°50′E
Djambala, Congo	284-85	2°33′S	14°46′E

PRONOUNCING INDEX
Djedi, Oued – Dunhua 351

Feature (Pronunciation)	Page	Lat.	Long.
Djedi, Oued, stm., Alg.	280-81	34°28'N	6°06'E
Djelfa, Alg.	292b	34°41'N	3°15'E
Djenné, Mali	280-81	13°54'N	4°33'W
Djérem, stm., Camrn.	282-83	5°19'N	13°24'E
Djibouti, nation, Afr. (jē-bōō-tē')	274	11°30'N	43°00'E
Djibouti, nat. cap., Dji. (jē-bōō-tē')	288	11°34'N	43°09'E
Djokjakarta, Indon. *see* Yogyakarta	268-69	7°48'S	110°22'E
Djougou, Benin	282-83	9°42'N	1°40'E
Djugu, D.R.C.	289	1°58'N	30°30'E
Dmitrov, Russia (d'mē'trôf)	218-19	56°21'N	37°31'E
Dnepr, stm., Eur. *see* Dnieper	218-19	46°31'N	32°22'E
Dneprodzerzhinsk, Ukr. *see* Dniprodzerzhyns'k	218-19	48°29'N	34°41'E
Dnestr, stm., Eur. *see* Dniester	204-05	46°19'N	30°17'E
Dnieper, stm., Eur. (nē'pûr)	218-19	46°31'N	32°22'E
Dniester, stm., Eur. (nēs'-tēr)	204-05	46°19'N	30°17'E
Dnipro, stm., Eur. *see* Dnieper	218-19	46°31'N	32°22'E
Dniprodzerzhyns'k, Ukr.	218-19	48°29'N	34°41'E
Dnipropetrovs'k, Ukr.	218-19	48°28'N	34°58'E
Dnister, stm., Eur. *see* Dniester	204-05	46°19'N	30°17'E
Dnjapro, stm., Eur. *see* Dnieper	218-19	46°31'N	32°22'E
Dno, Russia (d'nô')	208-09	57°49'N	29°59'E
Doba, Chad	284-85	8°39'N	16°51'E
Doberai, Jazirah, pen., Indon.	242-43	1°30'S	132°30'E
Doboj, Bos. (dō'boi)	216-17	44°44'N	18°06'E
Dobrich, Blg.	216-17	43°35'N	27°50'E
Dobryanka, Russia (dôb-ryän'kå)	202-03	58°28'N	56°25'E
Doce, stm., Braz. (dō'så)	186	19°39'S	39°49'W
Doctor Arroyo, Mex. (dōk-tōr' är-rō'yō)	158-59	23°42'N	100°11'W
Dodecanese, is., Grc. (dō'dĕ-cä-nēs')	216-17	36°30'N	27°00'E
Dodekanisoy, is., Grc. *see* Dodecanese	216-17	36°30'N	27°00'E
Dodge City, Ks., U.S. (dŏj sĭ'tē)	130-31	37°45'N	100°01'W
Dodgeville, Wi., U.S. (dŏj'vĭl)	126-27	42°58'N	90°08'W
Dodola, Eth.	292d	7°00'N	39°07'E
Dodoma, nat. cap., Tan. (dō'dō-må)	289	6°11'S	35°45'E
Dogai Coring, lk., China	254-55	34°35'N	88°59'E
Dog Island, i., Anguilla	155b	18°17'N	63°15'W
Dog Lake, lk., On., Can. (dŏg lāk)	146-47	48°46'N	89°33'W
Dōgo, i., Japan	265	36°15'N	133°16'E
Doğu Karadeniz Dağları, mts., Tur.	245	40°30'N	40°30'E
Doha, nat. cap., Qatar (dō'hå)	250-51	25°17'N	51°32'E
Dolbeau-Mistassini, Qc., Can.	146-47	48°53'N	72°13'W
Dole, Fr. (dōl)	212-13	47°06'N	5°29'E
Dolgaya Kosa, spit, Russia (dôl-gä'yä kô'sä)	218-19	46°41'N	37°43'E
Dolinsk, Russia (då-lēnsk')	240-41	47°20'N	142°48'E
Dolisie, Congo *see* Loubomo	284-85	4°11'S	12°40'E
Dolores, Arg. (dô-lō'rĕs)	187	36°20'S	57°41'W
Dolores, Ur.	187	33°32'S	58°12'W
Dolores Hidalgo, Mex. (dô-lô'rĕs-ē-dal'gō)	158-59	21°09'N	100°56'W
Dolsando, i., Kor., S.	263	34°38'N	127°45'E
Domažlice, Czech Rep. (dō'mäzh-lē-tsĕ)	210-11	49°27'N	12°56'E
Dombarovskiy, Russia	244	50°46'N	59°32'E
Dombås, Nor.	208-09	62°05'N	9°08'E
Dombóvár, Hung. (dōm'bō-vär)	210-11	46°22'N	18°08'E
Domeyko, Chile	182-83	28°57'S	70°54'W
Domeyko, Cordillera, mts., Chile (kôr-dēl-yĕ'rä-dô-mā'kō)	182-83	24°45'S	69°09'W
Dominica, nation, N.A. (dō-mĭ-nē'kå)	152-53	15°30'N	61°20'W
Dominican Republic, nation, N.A. (dō-mĭn'ĭ-kăn rē-pŭb'lĭk)	152-53	19°0'N	70°40'W
Dominion, Cape, c., Nu., Can.	140-41	66°09'N	74°27'W
Dominique, Canal de la, strt., N.A. *see* Martinique Passage	155b	15°10'N	61°15'W
Domodedovo, Russia (dō-mō-dyĕ'do-vô)	218-19	55°26'N	37°47'E
Dom Pedrito, Braz.	187	30°59'S	54°40'W
Domuyo, Volcán, vol., Arg.	185	36°38'S	70°26'W
Don, stm., Russia (dŏn)	202-03	47°05'N	39°15'E
Don, stm., Scot., U.K. (dŏn)	206-07	57°11'N	2°06'W
Donaldsonville, La., U.S. (dŏn'ăld-sŭn-vĭl)	134-35	30°06'N	90°60'W
Donau, stm., Eur. *see* Danube	204-05	45°23'N	29°36'E
Don Benito, Spain (dōn'bå-nē'tō)	214-15	38°57'N	5°52'W
Doncaster, Eng., U.K. (doŋ'kăs-tēr)	206-07	53°32'N	1°08'W
Dondo, Moz.	286-87	19°35'S	34°44'E
Dondra Head, c., Sri L.	256	5°55'N	80°35'E
Donegal, Ire. (dŏn-ē-gôl')	206-07	54°39'N	8°07'W
Donegal Bay, b., Ire. (dŏn-ē-gôl' bā)	206-07	54°30'N	8°30'W
Donets'k, Ukr.	218-19	47°60'N	37°48'E
Dong, stm., China (dôŋ)	258-59	23°05'N	113°60'E
Dong, stm., China (dôŋ)	260-61	42°17'N	101°06'E
Dongara, Austl. (dôn-gă'rá)	294-95	29°15'S	114°58'E
Dongfang, China (dôŋ-fäŋ)	258-59	19°05'N	108°38'E
Donggala, Indon. (dôn-gä'lä)	268-69	0°41'S	119°44'E
Dongguan, China (dôŋ-gůän)	258-59	23°03'N	113°44'E
Dong Hai, s., Asia *see* East China Sea	240-41	30°0'N	126°00'E
Donghai Dao, i., China	258-59	21°02'N	110°25'E
Dong Hoi, Viet. (dông-hô-ē')	266-67	17°29'N	106°37'E
Dong Nai, stm., Viet.	266-67	10°44'N	106°46'E
Dongola, Sudan *see* Dunqulah	288	19°11'N	30°28'E
Dong San Shen, hist. reg., China *see* Manchuria	260-61	47°0'N	125°00'E
Dongting Hu, lk., China (dôŋ-tĭŋ hōō)	258-59	29°20'N	112°54'E
Dongyang, China	258-59	29°16'N	120°14'E
Dongzhi, China	258-59	30°07'N	117°01'E
Doniphan, Mo., U.S. (dŏn'ĭ-făn)	134-35	36°37'N	90°50'W
Doniphan, Ne., U.S. (dŏn'ĭ-făn)	124-25	40°46'N	98°23'W
Donostia, Spain *see* Donostia-San Sebastián	214-15	43°19'N	1°60'W
Donostia-San Sebastián, Spain	214-15	43°19'N	1°60'W
Door Peninsula, pen., Wi., U.S. (dōr pē-nĭn'sūlă)	126-27	44°55'N	87°20'W
Dorchester, Eng., U.K. (dôr'chĕs-tēr)	206-07	50°43'N	2°27'W
Dorchester, Cape, c., Nu., Can.	140-41	65°27'N	77°26'W
Dordogne, stm., Fr. (dôr-dôn'yĕ)	212-13	45°02'N	0°35'W
Dore Lake, lk., Sk., Can.	144-45	54°46'N	107°17'W
Dores do Indaiá, Braz.	186	19°28'S	45°36'W
Dori, Burkina	282-83	14°02'N	0°02'W
Dornbirn, Aus. (dôrn'bĕrn)	210-11	47°25'N	9°45'E
Dorogobuzh, Russia (dôrôgô'-bōō'zh)	218-19	54°55'N	33°18'E
Dorohoi, Rom. (dō-rō-hoi')	210-11	47°57'N	26°24'E
Dorpat, Est. *see* Tartu	208-09	58°23'N	26°43'E
Dorre Island, i., Austl. (dôr ī'lănd)	296-97	25°07'S	113°06'E
Dortmund, Ger. (dôrt'mŏont)	210-11	51°31'N	7°28'E
Dörtyol, Tur. (dûrt'yōl)	248-49	36°52'N	36°13'E
Dosatuy, Russia	260-61	50°22'N	118°34'E
Dos Bahías, Cabo, c., Arg (kå'bō-dōs-bä-ē'äs)	185	44°56'S	65°33'W
Döşemealtı, Tur	246	37°01'N	30°36'E
Dos Hermanas, Spain (dōs ĕr-mä'näs)	214-15	37°17'N	5°55'W
Dosso, Niger (dôs-sō')	280-81	13°03'N	3°12'E
Dossor, Kaz.	202-03	47°32'N	52°59'E
Dothan, Al., U.S. (dō'thăn)	134-35	31°13'N	85°24'W
Douai, Fr. (doo-å')	212-13	50°22'N	3°05'E
Douala, Camrn. (dōō-ä'lä)	282-83	4°03'N	9°42'E
Douarnenez, Fr. (dōō-är nē-nĕs')	212-13	48°06'N	4°20'W
Douglas, Az., U.S. (dŭg'lăs)	128-29	31°20'N	109°34'W
Douglas, Ga., U.S. (dŭg'lăs)	134-35	31°30'N	82°51'W
Douglas, Mi., U.S. (dŭg'lăs)	126-27	42°39'N	86°12'W
Douglas, Wy., U.S. (dŭg'lăs)	122-23	42°45'N	105°23'W
Douglas, nat. cap., I. of Man (dŭg'lăs)	206-07	54°10'N	4°29'W
Douglas Channel, strt., B.C., Can. (dŭg'lăs chăn'ĕl)	142-43	53°30'N	129°12'W
Douglasville, Ga., U.S. (dŭg'lăs-vĭl)	134-35	33°45'N	84°45'W
Dourada, Serra, plat., Braz. (sĕ'r-rä-dôōō-rá'dä)	180-81	13°10'S	48°34'W
Dourados, Braz.	182-83	22°13'S	54°49'W
Douro, stm., Eur. (dô'ò-rō)	214-15	41°09'N	8°41'W
Dover, Eng., U.K. (dō vēr)	206-07	51°08'N	1°18'E
Dover, De., U.S.	126-27	39°09'N	75°31'W
Dover, N.H., U.S.	126-27	43°12'N	70°53'W
Dover, N.J., U.S. (dō vēr)	126-27	40°54'N	74°32'W
Dover, Oh., U.S. (dō vēr)	126-27	40°31'N	81°28'W
Dover, Strait of, strt., Eur. (strāt ŭv dō vēr)	206-07	50°59'N	1°31'E
Dover-Foxcroft, Me., U.S. (dō'vēr fŏks'krôft)	127a	45°11'N	69°14'W
Dovrefjell, mts., Nor. (dŏv'rĕ fyĕl')	208-09	62°06'N	9°25'E
Dowagiac, Mi., U.S. (dô-wô'jăk)	126-27	41°59'N	86°06'W
Drâa, Hamada du, des., Alg.	280-81	29°0'N	6°45'W
Drâa, Oued, stm., Afr. (wĕd drá)	280-81	28°41'N	11°07'W
Drăgăşani, Rom. (drä-gá-shän'ĭ)	216-17	44°40'N	24°16'E
Draguignan, Fr. (drá-gēn-yäN')	212-13	43°33'N	6°28'E
Drake, Pasaje de, strt., *see* Drake Passage	313	58°0'S	70°00'W
Drakensberg, mts., Afr. (drä'kĕnz-bĕrgh)	286-87	27°0'S	30°00'E
Drake Passage, strt., (drāk păs'ĭj)	313	58°0'S	70°00'W
Dráma, Grc. (drä'mä)	216-17	41°09'N	24°09'E
Drammen, Nor. (dräm'ĕn)	208-09	59°45'N	10°13'E
Drau, stm., Eur. (drou) *see* Drava	216-17	45°33'N	18°56'E
Drava, stm., Eur. (drä'vä)	216-17	45°33'N	18°56'E
Drayton Valley, Ab., Can. (drā'tŭn văl'ē)	142-43	53°14'N	114°59'W
Dresden, Ger. (drās'dĕn)	210-11	51°03'N	13°44'E
Dreux, Fr. (drû)	212-13	48°44'N	1°22'E
Drin, stm., Alb. (drēn)	216-17	41°45'N	19°35'E
Drina, stm., Eur. (drē'nä)	216-17	44°54'N	19°21'E
Drøbak, Nor. (drû'bäk)	208-09	59°39'N	10°39'E
Drohobych, Ukr.	210-11	49°21'N	23°31'E
Druc', stm., Bela. (drōōt)	218-19	53°04'N	30°02'E
Druk-Yul, nation, Asia *see* Bhutan	224-25	27°30'N	90°30'E
Drumheller, Ab., Can. (drŭm-hĕl-ēr)	142-43	51°28'N	112°42'W
Drummond Island, i., Mi., U.S. (drŭm'ŭnd ī'lănd)	126-27	46°0'N	83°40'W
Drummondville, Qc., Can. (drŭm'ŭnd-vĭl)	146-47	45°53'N	72°30'W
Drumright, Ok., U.S. (drŭm'rīt)	130-31	35°60'N	96°36'W
Drwęca, stm., Pol. (d'r-văn'tsä)	210-11	52°60'N	18°41'E
Dryden, On., Can. (drī-dĕn)	144-45	49°47'N	92°51'W
Dry Tortugas, is., Fl., U.S. (drī tôr-tōō'gäz)	135a	24°38'N	82°55'W
Dry Tortugas National Park, n.p., Fl., U.S. (drī tôr-tōō'gäz năsh'ŭn-ăl pärk)	135a	24°37'N	82°54'W
Duala, Camrn. *see* Douala	282-83	4°03'N	9°42'E
Duarte, Pico, mtn., Dom. Rep. (pēcô dŭ'ärtēh)	154-55	19°02'N	70°59'W
Dubai, U.A.E.	250-51	25°16'N	55°19'E
Dubawnt, stm., Can. (dōō-bônt')	140-41	64°31'N	100°05'W
Dubawnt Lake, lk., Can. (dōō-bônt' lāk)	140-41	63°08'N	101°30'W
Dubayy, U.A.E. *see* Dubai	250-51	25°16'N	55°19'E
Dubbo, Austl. (dŭb'ō)	301	32°15'S	148°36'E
Dublin, Ga., U.S. (dŭb'lĭn)	134-35	32°32'N	82°54'W
Dublin, Oh., U.S. (dŭb'lĭn)	126-27	40°06'N	83°07'W
Dublin, Tx., U.S. (dŭb'lĭn)	132-33	32°05'N	98°20'W
Dublin, nat. cap., Ire. (dŭb'lĭn)	206-07	53°21'N	6°15'W
Dubno, Ukr. (dōō'b-nô)	210-11	50°24'N	25°45'E
Du Bois, Pa., U.S. (dò-bois')	126-27	41°07'N	78°46'W
Dubovka, Russia (dò-bôf'kå)	202-03	49°04'N	44°49'E
Dubrovka, Russia (dōō-brôf'kä)	218-19	53°42'N	33°31'E
Dubrovnik, Cro. (dò'brôv-nĕk)	216-17	42°39'N	18°06'E
Dubuque, Ia., U.S. (dò-būk')	124-25	42°30'N	90°41'W
Duchesne, Ut., U.S. (dò-shān')	128-29	40°10'N	110°24'W
Duck Lake, Sk., Can. (dŭk lāk)	144-45	52°49'N	106°14'W
Duck Valley Indian Reservation, ind. res., U.S. (dŭk văl'ē ĭn'dĭ-ăn rĕ-sēr-vā'shēn)	122-23	42°00'N	116°10'W
Dudinka, Russia (dōō-dĭn'kå)	236-37	69°24'N	86°11'E
Dudley, Eng., U.K. (dŭd'lĭ)	206-07	52°30'N	2°05'W
Duero, stm., Eur.	214-15	41°09'N	8°41'W
Dufourspitze, mtn., Eur.	210-11	45°55'N	7°52'E
Dugi Otok, i., Cro. (dōō'gē o'tôk)	216-17	43°59'N	15°04'E
Duisburg, Ger. (dōō'ĭs-bòrgh)	210-11	51°26'N	6°47'E
Duitama, Col.	178-79	5°50'N	73°02'W
Dukhān, Qatar	250-51	25°25'N	50°48'E
Dukhovshchina, Russia (dōō-kôfsh-'chēnà)	218-19	55°12'N	32°25'E
Dulan, China	260-61	36°10'N	98°22'E
Dulce, Golfo, b., C.R. (gōl'fō dōōl'sā)	161	8°32'N	83°14'W
Duluth, Mn., U.S. (dò-lōōth')	124-25	46°46'N	92°09'W
Dumaguete, Phil.	270	9°18'N	123°18'E
Dumai, Indon.	266-67	1°40'N	101°27'E
Dumali Point, c., Phil. (dōō-mä'lē point)	270	13°07'N	121°33'E
Dumaran Island, i., Phil.	270	10°33'N	119°51'E
Dumaring, Indon.	268-69	1°32'N	118°13'E
Dumfries, Scot., U.K. (dŭm-frēs')	206-07	55°04'N	3°37'W
Duna, stm., Eur. *see* Danube	204-05	45°23'N	29°36'E
Dünaburg, Lat. *see* Daugavpils	208-09	55°53'N	26°32'E
Dunai, stm., Eur. *see* Danube	204-05	45°23'N	29°36'E
Dunaj, stm., Eur. *see* Danube	204-05	45°23'N	29°36'E
Dunărea, stm., Eur. *see* Danube	204-05	45°23'N	29°36'E
Dunav, stm., Eur. *see* Danube	204-05	45°23'N	29°36'E
Duncan, B.C., Can. (dŭŋ'kăn)	142-43	48°47'N	123°42'W
Duncan, Ok., U.S. (dŭŋ'kăn)	130-31	34°30'N	97°57'W
Duncan, stm., B.C., Can. (dŭŋ'kăn)	142-43	50°11'N	116°57'W
Duncansby Head, c., Scot., U.K. (dŭn'kănz-bĭ hĕd)	206-07	58°39'N	3°02'W
Dundalk, Ire. (dŭn'kôk)	206-07	54°01'N	6°24'W
Dundas, Lake, lk., Austl. (lāk dŭn-dăs)	296-97	32°35'S	121°50'E
Dundas Island, i., B.C., Can. (dŭn-dăs' ī'lănd)	142-43	54°33'N	130°55'W
Dún Dealgan, Ire. *see* Dundalk	206-07	54°01'N	6°24'W
Dundee, S. Afr.	292c	28°09'S	30°15'E
Dundee, Scot., U.K.	206-07	56°29'N	2°59'W
Dund-Us, Mong.	260-61	47°60'N	91°38'E
Dunedin, N.Z.	303	45°52'S	170°29'E
Dunfermline, Scot., U.K. (dŭn-fērm'lĭn)	206-07	56°04'N	3°29'W
Düngarpur, India	254-55	23°50'N	73°43'E
Dungarvan, Ire. (dŭn-gär'văn)	206-07	52°06'N	7°38'W
Dungu, D.R.C.	289	3°37'N	28°34'E
Dunhua, China (dòn-hwä)	264	43°22'N	128°14'E

n-sing; ŋ-baŋk; N-nasalized n; nŏd; cŏmmit; ōld; ôbey; ôrder; oi-boil; fōōd; ò-as oo in foot; ou-out; s-soft; sh-dish; th-thin; pūre; ûnite; ûrn; stŭd; circŭs; ü-as in French tu; ´-indeterminate vowel.

352 Dunhuang – Eldoret PRONOUNCING INDEX

Feature (Pronunciation)	Page	Lat.	Long.
Dunhuang, China	260-61	40°08'N	94°40'E
Dunkerque, Fr. (dŭn-kĕrk')	212-13	51°03'N	2°23'E
Dunkirk, Fr. *see* Dunkerque	212-13	51°03'N	2°23'E
Dunkirk, In., U.S. (dŭn'kûrk)	126-27	40°22'N	85°12'W
Dunkirk, N.Y., U.S. (dŭn'kûrk)	126-27	42°29'N	79°19'W
Dún Laoghaire, Ire. (dŭn-lâ'rĕ)	206-07	53°17'N	6°08'W
Dunlap, Ia., U.S. (dŭn'lăp)	124-25	41°51'N	95°36'W
Dunlap, Tn., U.S. (dŭn'lăp)	134-35	35°22'N	85°23'W
Dunleary, Ire. *see* Dún Laoghaire	206-07	53°17'N	6°08'W
Dunmore, Pa., U.S. (dŭn'mōr)	126-27	41°27'N	75°38'W
Dunn, N.C., U.S. (dŭn)	134-35	35°18'N	78°37'W
Dunqulah, Sudan	288	19°11'N	30°28'E
Dunsmuir, Ca., U.S. (dŭnz'mūr)	122-23	41°13'N	122°17'W
Duolun, China (dwō-lōōn)	260-61	42°12'N	116°28'E
Duomula, China	254-55	34°07'N	82°30'E
Duque de York, Isla, i., Chile	185	50°40'S	75°20'W
Duquesne, Pa., U.S. (dŏo-kān')	126-27	40°22'N	79°51'W
Du Quoin, Il., U.S. (dŏo-kwoin')	126-27	38°01'N	89°15'W
Durand, Mi., U.S. (dŭ-rănd')	126-27	42°55'N	83°59'W
Durand, Wi., U.S. (dŭ-rănd')	124-25	44°37'N	91°57'W
Durango, Mex. (dōō-rä'n-gō)	158-59	24°02'N	104°41'W
Durango, Co., U.S. (dŏo-răŋ'gō)	128-29	37°17'N	107°52'W
Durango, state, Mex. (dōō-rä'n-gō)	132-33	24°50'N	104°50'W
Durant, Ms., U.S. (dŭ-rănt')	134-35	33°05'N	89°52'W
Durant, Ok., U.S. (dŭ-rănt')	130-31	33°60'N	96°23'W
Durazno, Ur. (dōō-räz'nō)	187	33°25'S	56°30'W
Durazzo, Alb. *see* Durrës	216-17	41°19'N	19°27'E
Durban, S. Afr. (dûr'băn)	286-87	29°55'S	30°56'E
Durbe, Lat. (dōōr'bĕ)	208-09	56°35'N	21°21'E
Düren, Ger. (dü'rĕn)	210-11	50°48'N	6°29'E
Durg, India	254-55	21°11'N	81°17'E
Durham, Eng., U.K. (dûr'ăm)	206-07	54°47'N	1°34'W
Durham, N.C., U.S.	134-35	35°60'N	78°54'W
Durmitor, mtn., Mont.	216-17	43°08'N	19°01'E
Durrës, Alb. (dŏor'ĕs)	216-17	41°19'N	19°27'E
Durrësi, Alb. *see* Durrës	216-17	41°19'N	19°27'E
D'Urville, Tanjung, c., Indon.	302	1°28'S	137°54'E
D'Urville Island, i., N.Z.	303	40°50'S	187°52'E
Dushan, China (dōō-shän)	258-59	25°50'N	107°32'E
Dushanbe, nat. cap., Taj. (dū-shän'-bá) (dū-shän-bä')	252-53	38°34'N	68°47'E
Düsseldorf, Ger. (düs'ĕl-dôrf)	210-11	51°14'N	6°48'E
Duyfken Point, c., Austl.	302	12°34'S	141°38'E
Duyun, China (dōō-yŏon)	258-59	26°16'N	107°30'E
Düziçi, Tur.	247	37°15'N	36°27'E
Dwārka, India	254-55	22°14'N	68°59'E
Dwight, Il., U.S. (dwīt)	126-27	41°05'N	88°25'W
Dyat'kovo, Russia (dyät'kō-vō)	218-19	53°35'N	34°21'E
Dyersburg, Tn., U.S. (dī'ērz-bûrg)	134-35	36°02'N	89°23'W
Dyersville, Ia., U.S. (dī'ērz-vĭl)	124-25	42°29'N	91°07'W
Dzavhan, stm., Mong.	260-61	48°53'N	93°26'E
Dzerzhinsk, Russia	202-03	56°15'N	43°24'E
Dzhambul, Kaz. *see* Taraz	244	42°54'N	71°21'E
Dzhankoi, Ukr.	218-19	45°43'N	34°23'E
Dzharylhach, ostriv, i., Ukr.	218-19	46°02'N	32°55'E
Dzhebariki-Khaya, Russia	236-37	62°11'N	135°46'E
Dzhezkazgan, Kaz. *see* Zhezqazghan	244	47°47'N	67°41'E
Dzhugdzhur, Khrebet, mts., Russia	236-37	58°0'N	136°00'E
Dzhugdzhur Range, mts., Russia (jŏog-jŏor' rănj) *see* Dzhugdzhur, Khrebet	236-37	58°0'N	136°00'E
Dzhungarian Alatau Mountains, mts., Asia *see* Alataw Shan	244	45°0'N	81°00'E
Dzibalchén, Mex. (zē-bäl-chĕ'n)	160	19°28'N	89°44'W
Dzilam González, Mex. (zē-lä'm-gôn-zä'lĕz)	160	21°17'N	88°56'W
Dzitás, Mex. (zē-tä's)	160	20°51'N	88°31'W
Dzuunmod, Mong.	260-61	47°43'N	106°57'E

E

Feature (Pronunciation)	Page	Lat.	Long.
Eagle, Id., U.S. (ē'gl)	122-23	43°41'N	116°19'W
Eagle Grove, Ia., U.S. (ē'gl grōv)	124-25	42°40'N	93°54'W
Eagle Lake, Me., U.S. (ē'gl lāk)	127a	47°02'N	68°36'W
Eagle Lake, Tx., U.S. (ē'gl lāk)	132-33	29°35'N	96°20'W
Eagle Pass, Tx., U.S. (ē'gl pås)	132-33	28°42'N	100°29'W
Earle, Ar., U.S. (ûrl)	134-35	35°16'N	90°28'W
Earlington, Ky., U.S. (ûr'lĭng-tŭn)	134-35	37°16'N	87°31'W
Easley, S.C., U.S. (ēz'lĭ)	134-35	34°50'N	82°35'W
East Angus, Qc., Can. (ēst ăŋ'gŭs)	146-47	45°29'N	71°40'W
Eastbourne, Eng., U.K. (ēst'bôrn)	206-07	50°46'N	0°17'E
East Caicos, i., T./C. Is. (ēst kī'kōs)	154-55	21°42'N	71°29'W
East Cape, c., N.Z.	303	37°41'S	178°33'E
East Chicago, In., U.S. (ēst shĭ-kô'gō)	126-27	41°39'N	87°26'W
East China Sea, s., Asia (ēst chī'ná sē)	240-41	30°0'N	126°00'E
East Detroit, Mi., U.S. (ēst dē-troit') *see* Eastpointe	126-27	42°28'N	82°57'W
Easter Island, i., Chile (ē'stēr ī'lănd) *see* Pascua, Isla de	308-09	27°07'S	109°22'W
Eastern Desert, des., Egypt *see* Arabian Desert	288	28°0'N	32°00'E
Eastern Ghāts, mts., India (ē'stērn ghäts) (ē'stērn ghôts)	256	14°0'N	78°50'E
East Falkland, i., Falk. Is.	185	51°53'S	59°11'W
East Grand Forks, Mn., U.S. (ēst grănd fôrks)	124-25	47°55'N	97°00'W
East Helena, Mt., U.S. (ēst hĕ-lē'ná)	122-23	46°35'N	111°55'W
East Jordan, Mi., U.S. (ēst jôr'dăn)	126-27	45°09'N	85°07'W
Eastland, Tx., U.S. (ēst'lănd)	130-31	32°24'N	98°49'W
East Lansing, Mi., U.S. (ēst lăn'sĭng)	126-27	42°44'N	84°29'W
East Liverpool, Oh., U.S. (ēst lĭv'ēr-pōōl)	126-27	40°38'N	80°35'W
East London, S. Afr. (ēst ŭn'dŭn)	286-87	32°60'S	27°54'E
Eastmain, Qc., Can.	138-39	52°13'N	78°33'W
Eastmain, stm., Qc., Can. (ēst'mān)	140-41	52°15'N	78°35'W
Eastmain-Opinaca, Réservoir, res., Qc., Can.	140-41	52°23'N	76°35'W
Eastman, Ga., U.S. (ēst'măn)	134-35	32°12'N	83°10'W
East Moline, Il., U.S. (ēst mō-lēn')	124-25	41°32'N	90°25'W
East Nishnabotna, stm., Ia., U.S. (ēst nĭsh-ná-bŏt'ná)	130-31	40°39'N	95°38'W
East Orange, N.J., U.S. (ēst ŏr'ĕnj)	126-27	40°46'N	74°12'W
East Pakistan, nation, Asia *see* Bangladesh	224-25	24°0'N	90°00'E
East Peoria, Il., U.S. (ēst pē-ō'rĭ-á)	126-27	40°40'N	89°35'W
Eastpointe, Mi., U.S.	126-27	42°28'N	82°57'W
Eastport, Me., U.S. (ēst'pōrt)	127a	44°54'N	66°60'W
East Providence, R.I., U.S. (ēst prŏv'ĭ-dĕns)	126-27	41°49'N	71°23'W
East Saint Louis, Il., U.S.	130-31	38°37'N	90°09'W
East Sea, s., Asia *see* Japan, Sea of	240-41	40°0'N	135°00'E
East Siberian Sea, s., Russia (ēst sī-bĭr'y'n sē)	236-37	74°0'N	166°00'E
East Stroudsburg, Pa., U.S. (ēst stroudz'bûrg)	126-27	41°00'N	75°11'W
East Tawas, Mi., U.S. (ēst tô'wăs)	126-27	44°17'N	83°28'W
East Timor, nation, Asia *see* Timor-Leste	268-69	8°35'S	126°00'E
Eaton, Oh., U.S. (ē'tŭn)	126-27	39°45'N	84°38'W
Eaton Rapids, Mi., U.S. (ē'tŭn răp'ĭdz)	126-27	42°30'N	84°39'W
Eatonton, Ga., U.S. (ē'tŭn-tŭn)	134-35	33°20'N	83°23'W
Eau Claire, Wi., U.S. (ō klâr')	124-25	44°49'N	91°30'W
Eau Claire, Lac à l', lk., Qc., Can.	140-41	56°11'N	74°26'W
Eauripik, at., Micron.	304-05	6°41'N	143°03'E
Ebinur Hu, lk., China	244	44°56'N	82°52'E
Eboli, Italy (ĕb'ō-lē)	216-17	40°38'N	15°04'E
Ebolowa, Camrn.	282-83	2°55'N	11°09'E
Echoing, stm., Can. (ĕk'ō-ĭng)	144-45	55°51'N	92°04'W
Echuca, Austl. (ê-chŏo'kå)	301	36°08'S	144°45'E
Écija, Spain (ā'thē-hä)	214-15	37°32'N	5°05'W
Ecuador, nation, S.A. (ĕk'wá-dôr)	172	2°0'S	77°30'W
Eddyville, Ia., U.S. (ĕd'ĭ-vĭl)	124-25	41°09'N	92°38'W
Eddyville, Ky., U.S. (ĕd'ĭ-vĭl)	134-35	37°06'N	88°05'W
Ede, Nig.	282a	7°44'N	4°25'E
Edéa, Camrn. (ĕ-dā'ä)	282-83	3°48'N	10°08'E
Eden, Austl.	301	37°04'S	149°54'E
Eden, N.C., U.S. (ē'dĕn)	134-35	36°29'N	79°46'W
Eden, Tx., U.S. (ē'dĕn)	132-33	31°13'N	99°51'W
Eden Prairie, Mn., U.S. (ē'dĕn prâr'ĭ)	124-25	44°51'N	93°28'W
Edenton, N.C., U.S. (ē'dĕn-tŭn)	134-35	36°04'N	76°36'W
Edenville, S. Afr. (ē'd'n-vĭl)	292c	27°34'S	27°41'E
Edfu, Egypt	291b	24°58'N	32°52'E
Edgefield, S.C., U.S. (ĕj'fēld)	134-35	33°47'N	81°56'W
Edgeley, N.D., U.S. (ĕj'lĭ)	124-25	46°21'N	98°43'W
Edgemont, S.D., U.S. (ĕj'mŏnt)	124-25	43°18'N	103°48'W
Edgerton, Wi., U.S. (ĕj'ēr-tŭn)	126-27	42°50'N	89°04'W
Edinburg, Tx., U.S. (ĕd''n-bûrg)	132-33	26°18'N	98°10'W
Edinburgh, Scot., U.K. (ĕd''n-bŭr-ō)	206-07	55°57'N	3°13'W
Edirne, Tur.	216-17	41°41'N	26°34'E
Edmond, Ok., U.S. (ĕd'mŭnd)	130-31	35°39'N	97°29'W
Edmonton, Ab., Can. (ĕd'mŭn-tŭn)	142-43	53°33'N	113°30'W
Edmundston, N.B., Can. (ĕd'mŭn-stŭn)	148-49	47°22'N	68°19'W
Edna, Tx., U.S. (ĕd'ná)	132-33	28°59'N	96°39'W
Édouard, Lac, lk., Afr. *see* Edward, Lake	289	0°23'S	29°36'E
Edremit, Tur. (ĕd-rĕ-mēt')	216-17	39°36'N	27°01'E
Edson, Ab., Can. (ĕd'sŭn)	142-43	53°35'N	116°26'W
Eduardo Castex, Arg.	187	35°55'S	64°18'W
Edward, Lake, lk., Afr. (lāk ĕd'wērd)	289	0°23'S	29°36'E
Edwards Plateau, plat., Tx., U.S.	132-33	30°46'N	100°47'W
Edwardsville, Il., U.S. (ĕd'wērdz-vĭl)	130-31	38°48'N	89°57'W
Edward VII Peninsula, pen., Ant.	313	77°40'S	154°60'W
Eel, stm., Ca., U.S. (ēl)	120-21	40°38'N	124°20'W
Eesti, nation, Eur. *see* Estonia	190-91	59°0'N	26°00'E
Éfaté, i., Vanuatu (å-fä'tā)	306g	17°40'S	168°25'E
Effigy Mounds National Monument, n.p., Ia., U.S. (ĕf'ĭ-jŭ mounds năsh'ŭn-ăl mŏn'ū-mĕnt)	124-25	43°06'N	91°13'W
Effingham, Il., U.S. (ĕf'ĭng-hăm)	126-27	39°07'N	88°33'W
Eg, stm., Mong.	260-61	49°23'N	103°38'E
Egadi, Isole, is., Italy (ĕ'sō-lĕ-ĕ'gä-dē)	216-17	37°58'N	12°16'E
Ege Denizi, s., *see* Aegean Sea	216-17	38°30'N	25°00'E
Eger, Czech Rep. *see* Cheb	210-11	50°05'N	12°22'E
Eger, Hung. (ĕ gĕr)	210-11	47°54'N	20°23'E
Egersund, Nor. (ĕ'ghĕr-sòn')	208-09	58°27'N	6°00'E
Egmont, Cape, c., N.Z. (kāp ĕg'mŏnt)	303	39°17'S	187°45'E
Eğridir Gölü, lk., Tur.	202-03	38°02'N	30°53'E
Egum Atoll, at., Pap. N. Gui.	302	9°25'S	151°55'E
Egvekinot, Russia	236-37	66°18'N	179°11'W
Egypt, nation, Afr. (ē'jĭpt)	274	27°0'N	30°00'E
Eichstätt, Ger. (īk'shtät)	210-11	48°54'N	11°11'E
Eidfjord, Nor. (ĕīd'fyōr)	208-09	60°28'N	7°05'E
Eidsvoll, Nor. (īdhs'vôl)	208-09	60°19'N	11°14'E
Eifel, mts., Ger. (ī'fĕl)	210-11	50°14'N	6°42'E
Eight Degree Channel, strt., Asia	256	8°0'N	73°00'E
Einbeck, Ger. (īn'bĕk)	210-11	51°49'N	9°52'E
Eindhoven, Neth. (īnd'hō-vĕn)	206-07	51°26'N	5°29'E
Éire, nation, Eur. *see* Ireland	190-91	53°0'N	8°00'W
Eirunepé, Braz.	180-81	6°39'S	69°52'W
Eisenach, Ger. (ī'zĕn-äк)	210-11	50°59'N	10°19'E
Eivissa, Spain	214-15	38°55'N	1°25'E
Eivissa, i., Spain	214-15	39°0'N	1°25'E
Ejin Qi, China	260-61	41°52'N	100°56'E
Ejmiatsin, Arm.	245	40°10'N	44°18'E
Ejutla de Crespo, Mex. (å-hōt'lä dā krās'pō)	158-59	16°34'N	96°44'W
Ekenäs, Fin. (ĕ'kĕ-nâs)	208-09	59°58'N	23°26'E
Ekibastuz, Kaz.	244	51°43'N	75°20'E
Ekimchan, Russia	236-37	53°04'N	132°57'E
Eko, Nig. *see* Lagos	282a	6°27'N	3°24'E
Ekwan, stm., On., Can.	140-41	53°12'N	82°14'W
El Aaiún, nat. cap., W. Sah. *see* Laayoune	280-81	27°10'N	13°12'W
El Abiodh Sidi Cheikh, Alg.	204-05	32°53'N	0°32'E
El Affroun, Alg. (ĕl áf-froun')	292b	36°28'N	2°37'E
El-Agheila, Libya *see* Al-'Uqaylah	204-05	30°15'N	19°12'E
El-Alamein, Egypt	204-05	30°49'N	28°58'E
Elands, stm., S. Afr. (ĕlånds)	292c	25°17'S	27°32'E
El-Arish, Egypt	291b	31°08'N	33°50'E
El Asnam, Alg. *see* Chlef	214-15	36°10'N	1°20'E
Elat, Isr.	248-49	29°33'N	34°57'E
Elat, Gulf of, b., *see* Aqaba, Gulf of	248-49	29°05'N	34°44'E
Elath, Isr. *see* Elat	248-49	29°33'N	34°57'E
Elazığ, Tur. (ĕl-ä'zĕz)	202-03	38°41'N	39°15'E
Elba, Al., U.S. (ĕl'bá)	134-35	31°25'N	86°04'W
Elba, Isola d', i., Italy (ê-sō lä-d-ĕl'bá)	216-17	42°46'N	10°17'E
El Banco, Col. (ĕl băn'cō)	178-79	9°01'N	73°58'W
Elbe, stm., Eur. (ĕl'bĕ)	210-11	53°53'N	9°01'E
Elbert, Mount, mtn., Co., U.S. (mount ĕl'bērt)	128-29	39°07'N	106°27'W
Elberton, Ga., U.S. (ĕl'bēr-tŭn)	134-35	34°07'N	82°51'W
Elbeuf, Fr. (ĕl-bûf')	212-13	49°17'N	1°01'E
Elbing, Pol. *see* Elbląg	210-11	54°10'N	19°24'E
Elbistan, Tur. (ēl-bē-stän')	202-03	38°13'N	37°12'E
Elbląg, Pol. (ĕl'blăng)	210-11	54°10'N	19°24'E
El Bonillo, Spain (ĕl bō-nēl'yō)	214-15	38°57'N	2°33'W
Elbow Lake, Mn., U.S. (ĕl'bō lāk)	124-25	45°60'N	95°59'W
El'brus, Gora, mtn., Russia (gå-rä'ĕl'brŏs')	245	43°21'N	42°26'E
Elburz Mountains, mts., Iran (ĕl'bŏrz' moun'tĭnz)	252-53	36°0'N	53°00'E
El Cajon, Ca., U.S.	128-29	32°48'N	116°57'W
El Calafate, Arg.	185	50°21'S	72°17'W
El Campo, Tx., U.S. (ĕl-kăm'pō)	132-33	29°11'N	96°16'W
El Carmen de Bolívar, Col.	178-79	9°43'N	75°07'W
El Centro, Ca., U.S. (ĕl-sĕn'trō)	128-29	32°48'N	115°33'W
Elche, Spain *see* Elx	214-15	38°15'N	0°42'W
Elda, Spain (ĕl'dä)	214-15	38°29'N	0°47'W
El'dikan, Russia	236-37	60°46'N	135°09'E
El Djazaïr, nation, Afr. *see* Algeria	274	28°0'N	3°00'E
El Djazaïr, nat. cap., Alg. *see* Algiers	292b	36°46'N	3°03'E
Eldon, Mo., U.S. (ĕl-dŭn)	130-31	38°21'N	92°35'W
Eldora, Ia., U.S. (ĕl-dō'rá)	124-25	42°22'N	93°06'W
Eldorado, Arg.	187	26°24'S	54°38'W
El Dorado, Ar., U.S. (ĕl dō-rä'dō)	130-31	33°12'N	92°41'W
El Dorado, Ks., U.S. (ĕl dō-rä'dō)	130-31	37°49'N	96°51'W
Eldoret, Kenya (ĕl-dō-rĕt')	289	0°31'N	35°17'E

PRONOUNCING INDEX

Electra – Espírito Santo 353

Feature (Pronunciation)	Page	Lat.	Long.
Electra, Tx., U.S. (ē-lĕk´trá)	**130-31**	34°03′N	98°55′W
Elek, stm., Asia	**202-03**	51°30′N	53°20′E
Elektrostal', Russia (ĕl-yĕk´trō-stál)	**218-19**	55°47′N	38°28′E
El Encanto, Col.	**178-79**	1°42′S	73°14′W
Elephant Butte Reservoir, res., N.M., U.S. (ĕl´ē-fănt būt rĕ´sĕr-vwär)	**128-29**	33°17′N	107°10′W
Elets, Russia	**218-19**	52°37′N	38°30′E
Eleuthera, i., Bah. (ē-lū´thēr-á)	**154-55**	25°11′N	76°13′W
El-Fayoum, Egypt	**291b**	29°19′N	30°50′E
El Ferrol del Caudillo, Spain *see* Ferrol	**214-15**	43°29′N	8°14′W
El Galpón, Arg.	**182-83**	25°24′S	64°38′W
Elgin, Scot., U.K. (ĕl´jĭn)	**206-07**	57°39′N	3°19′W
Elgin, Il., U.S. (ĕl´jĭn)	**126-27**	42°02′N	88°17′W
Elgin, Or., U.S. (ĕl´jĭn)	**122-23**	45°34′N	117°55′W
Elgin, Tx., U.S. (ĕl´jĭn)	**132-33**	30°21′N	97°22′W
El-Gîza, Egypt *see* Giza	**291b**	30°01′N	31°13′E
El Golea, Alg.	**204-05**	30°33′N	2°54′E
Elgon, Mount, mtn., Afr. (mount ĕl´gŏn)	**289**	1°08′N	34°33′E
El Guapo, Ven. (ĕl-gwá´pô)	**188b**	10°09′N	65°58′W
El Ḥank, clf., Afr.	**280-81**	24°23′N	6°36′W
El Hierro, i., Spain	**215d**	27°45′N	18°00′W
Elila, stm., D.R.C. (ĕ-lē´lá)	**284-85**	2°44′S	25°53′E
Elisenvaara, Russia (ā-lē´sĕn-vä´rá)	**208-09**	61°24′N	29°46′E
El-Iskandarîya, Egypt *see* Alexandria	**291b**	31°11′N	29°54′E
Elista, Russia	**202-03**	46°19′N	44°16′E
Elizabeth, Austl.	**301**	34°43′S	138°40′E
Elizabeth City, N.C., U.S. (ē-lĭz´á-bĕth sĭ´tĭ)	**134-35**	36°18′N	76°13′W
Elizabethton, Tn., U.S. (ē-lĭz-á-bĕth´tŭn)	**134-35**	36°21′N	82°14′W
Elizabethtown, Ky., U.S. (ē-lĭz´ă-bĕth-toun)	**126-27**	37°42′N	85°52′W
Elizabethtown, Pa., U.S. (ē-lĭz´ă-bĕth-toun)	**126-27**	40°09′N	76°37′W
El-Jadida, Mor.	**292a**	33°15′N	8°31′W
Elk, stm., B.C., Can. (ĕlk)	**142-43**	49°10′N	115°14′W
Elk City, Ok., U.S. (ĕlk sĭ´tē)	**130-31**	35°25′N	99°25′W
El Kef, Tun. (ĕl-xĕf´)	**200-01**	36°11′N	8°43′E
El-Kharga, Egypt	**288**	25°27′N	30°33′E
Elkhart, In., U.S. (ĕlk´härt)	**126-27**	41°42′N	85°57′W
Elkhorn, Wi., U.S. (ĕlk´hôrn)	**126-27**	42°40′N	88°33′W
Elkin, N.C., U.S. (ĕl´kĭn)	**134-35**	36°15′N	80°51′W
Elk Island National Park, n.p., Ab., Can. (ĕlk ī´lánd năsh´ŭn-ăl pärk)	**142-43**	53°36′N	112°54′W
Elko, Nv., U.S. (ĕl´kō)	**122-23**	40°50′N	115°46′W
Elk Point, S.D., U.S. (ĕlk point)	**124-25**	42°41′N	96°41′W
Elk Rapids, Mi., U.S. (ĕlk răp´ĭdz)	**126-27**	44°53′N	85°24′W
Elk River, Mn., U.S. (ĕlk rĭv´ĕr)	**124-25**	45°18′N	93°35′W
Elkton, Ky., U.S. (ĕlk´tŭn)	**134-35**	36°49′N	87°09′W
Elkton, S.D., U.S. (ĕlk´tŭn)	**124-25**	44°14′N	96°29′W
Ellás, nation, Eur. *see* Greece	**190-91**	39°0′N	22°00′E
Ellendale, N.D., U.S. (ĕl´ĕn-dāl)	**124-25**	46°00′N	98°32′W
Ellensburg, Wa., U.S. (ĕl´ĕnz-bûrg)	**122-23**	46°60′N	120°32′W
Ellesmere Island, i., Nu., Can. (ĕlz´mēr ī´lănd)	**95**	81°0′N	80°00′W
Elliot Lake, On., Can.	**146-47**	46°23′N	82°39′W
Elliston, Austl.	**294-95**	33°39′S	134°54′E
Ellisville, Ms., U.S. (ĕl´ĭs-vĭl)	**134-35**	31°36′N	89°12′W
Ellore, India *see* Elūru	**256**	16°43′N	81°07′E
Ellsworth, Ks., U.S. (ĕlz´wûrth)	**130-31**	38°44′N	98°13′W
Elma, Wa., U.S. (ĕl´má)	**122-23**	47°00′N	123°24′W
Elmadağ, Tur.	**246**	39°55′N	33°14′E
El-Mahalla el-Kubra, Egypt	**291b**	30°58′N	31°10′E
El Malpais National Monument, n.p., N.M., U.S.	**128-29**	34°57′N	107°51′W
El-Mansûra, Egypt	**291b**	31°02′N	31°23′E
Elmhurst, Il., U.S. (ĕlm´hûrst)	**126-27**	41°53′N	87°57′W
El-Minya, Egypt	**291b**	28°06′N	30°45′E
Elmira, N.Y., U.S.	**126-27**	42°05′N	76°48′W
Elmira Heights, N.Y., U.S. (ĕl-mī´rá hīts)	**126-27**	42°09′N	76°50′W
El Nevado, Cerro, mtn., Arg. *see* Nevado, Cerro	**185**	35°34′S	68°28′W
El-Obeid, Sudan *see* Al-Ubayyiḍ	**288**	13°11′N	30°13′E
El Oued, Alg.	**204-05**	33°21′N	6°53′E
El Pao, Ven. (ĕl pá´ō)	**188b**	9°38′N	68°08′W
El Paso, Tx., U.S. (ĕl-păs´ō)	**128-29**	31°48′N	106°27′W
El Paso de Robles, Ca., U.S. *see* Paso Robles	**128-29**	35°38′N	120°41′W
El Pital, Cerro, mtn., N.A.	**160**	14°23′N	89°08′W
El Porvenir, Pan. (ĕl-pôr-vä-nēr´)	**162**	9°33′N	78°59′W
El-Qâhira, nat. cap., Egypt *see* Cairo	**291b**	30°03′N	31°14′E
El Reno, Ok., U.S. (ĕl-rē´nō)	**130-31**	35°32′N	97°57′W
El Salto, Mex. (ĕl-säl´tō)	**158-59**	23°46′N	105°21′W
El Salvador, nation, N.A. (ĕl săl´vä-dôr)	**94**	13°50′N	88°55′W
Elsaß, hist. reg., Fr. *see* Alsace	**212-13**	48°30′N	7°30′E
Elsberry, Mo., U.S. (ĕlz´bĕr-ĭ)	**130-31**	39°10′N	90°48′W
Elsinore, Den. *see* Helsingør	**208-09**	56°02′N	12°37′E
El-Suweis, Egypt *see* Suez	**291b**	29°58′N	32°33′E
El Tajín, hist., Mex.	**158-59**	20°27′N	97°23′W
El Tigre, Ven. (ĕl-tē´grĕ)	**188b**	8°54′N	64°15′W
El-Uqsor, Egypt *see* Luxor	**291b**	25°42′N	32°39′E
Elūru, India	**256**	16°43′N	81°07′E
Elvas, Port. (ĕl´väzh)	**214-15**	38°53′N	7°10′W
Elverum, Nor. (ĕl´vĕ-ròm)	**208-09**	60°53′N	11°34′E
El Viejo, Nic. (ĕl-vyĕ´kō)	**161**	12°40′N	87°10′W
Elwood, In., U.S. (ĕ´wòd)	**126-27**	40°16′N	85°50′W
Elx, Spain	**214-15**	38°15′N	0°42′W
Ely, Mn., U.S. (ē´lĭ)	**124-25**	47°54′N	91°52′W
Ely, Nv., U.S.	**128-29**	39°15′N	114°53′W
Elyria, Oh., U.S. (ē-lĭr´ĭ-á)	**126-27**	41°22′N	82°06′W
Émaé, i., Vanuatu	**306g**	17°04′S	168°22′E
Emāmshahr, Iran *see* Shāhrūd	**252-53**	36°25′N	54°58′E
Embi, Kaz.	**244**	48°50′N	58°09′E
Embira, stm., Braz. *see* Envira	**180-81**	6°42′S	69°48′W
Embu, Kenya	**289**	0°32′S	37°27′E
Emden, Ger. (ĕm´dĕn)	**210-11**	53°22′N	7°12′E
Emerald, Austl.	**302**	23°31′S	148°10′E
Emiliano Zapata, Mex. (ĕ-mē-lyá´nô-zä-pá´tá)	**160**	17°45′N	91°46′W
Emiliano Zapata, Mex. (ĕ-mē-lyá´nô-zä-pá´tá)	**158-59**	18°51′N	99°11′W
Emirdağ, Tur.	**247**	39°01′N	31°09′E
Eminence, Ky., U.S. (ĕm´ĭ-nĕns)	**126-27**	38°22′N	85°11′W
Emmen, Neth. (ĕm´ĕn)	**206-07**	52°47′N	6°54′E
Emmetsburg, Ia., U.S. (ĕm´ĕts-bûrg)	**124-25**	43°07′N	94°41′W
Emmett, Id., U.S. (ĕm´ĕt)	**122-23**	43°53′N	116°30′W
Emory Peak, mtn., Tx., U.S. (ĕ´mŏ-rē pēk)	**132-33**	29°14′N	103°19′W
Empalme, Mex.	**156-57**	27°58′N	110°49′W
Empedrado, Arg.	**187**	27°57′S	58°48′W
Empoli, Italy (ām´pô-lē)	**216-17**	43°43′N	10°57′E
Emporia, Ks., U.S. (ĕm-pō´rĭ-á)	**130-31**	38°24′N	96°11′W
Emporia, Va., U.S. (ĕm-pō´rĭ-á)	**134-35**	36°42′N	77°33′W
Emporium, Pa., U.S. (ĕm-pō´rĭ-ŭm)	**126-27**	41°30′N	78°15′W
Empty Quarter, des., Asia *see* Rub'al-Khali	**238-39**	20°0′N	51°00′E
En, stm., Eur. *see* Inn	**210-11**	48°34′N	13°28′E
Encarnación, Para. (ĕn-kär-nä-syōn´)	**187**	27°20′S	55°52′W
Encinal, Tx., U.S. (ĕn´sĭ-nôl)	**132-33**	28°02′N	99°21′W
Encontrados, Ven. (ĕn-kōn-trä´dōs)	**178-79**	9°03′N	72°14′W
Encounter Bay, b., Austl. (ĕn-koun´tĕr bā)	**201**	35°35′S	138°44′E
Ende, Indon.	**268-69**	8°50′S	121°40′E
Enderbury, at., Kir. (ĕn´dĕr-bûrĭ)	**304-05**	3°08′S	171°05′W
Enderby Land, reg., Ant. (ĕn´dĕr-bĭĭ lănd)	**313**	68°05′S	52°53′E
Enderlin, N.D., U.S. (ĕn´dĕr-lĭn)	**124-25**	46°37′N	97°36′W
Endicott, N.Y., U.S. (ĕn´dĭ-kŏt)	**126-27**	42°07′N	76°04′W
Ene, stm., Peru	**184**	11°10′S	74°15′W
Enewetak, at., Marsh. Is.	**304-05**	11°39′N	162°17′E
Enfield, N.C., U.S. (ĕn´fēld)	**134-35**	36°11′N	77°40′W
Engaño, Cabo, c., Dom. Rep. (ká´-bô- ĕn-gä-nō)	**154-55**	18°37′N	68°20′W
Engel's, Russia (ĕn´gĕls)	**202-03**	51°29′N	46°08′E
Enggano, Pulau, i., Indon. (pōō-lou ĕng-gä´nō)	**266-67**	5°24′S	102°16′E
England, Ar., U.S. (ĭŋ´glănd)	**134-35**	34°33′N	91°58′W
England, state, U.K. (ĭŋ´glănd)	**206-07**	52°30′N	1°30′W
Englehart, On., Can.	**146-47**	47°49′N	79°52′W
Englewood, Co., U.S. (ĕn´g´l-wòd)	**130-31**	39°39′N	104°59′W
English, stm., On., Can. (ĭn´glĭsh)	**144-45**	50°11′N	95°03′W
English Bāzār, India *see* Ingrāj Bāzār	**254-55**	24°60′N	88°09′E
English Channel, strt., Eur. (ĭn´glĭsh chăn´ĕl)	**206-07**	50°13′N	2°20′W
Enguri, stm., Geor. (ĕn-gòr´)	**245**	42°24′N	41°33′E
Enid, Ok., U.S. (ē´nĭd)	**130-31**	36°24′N	97°52′W
Eniwetok, at., Marsh. Is. *see* Enewetak	**304-05**	11°39′N	162°17′E
eNjesuthi, mtn., Afr.	**292c**	29°09′S	29°23′E
Enköping, Swe. (ĕn´kû-pĭng)	**208-09**	59°38′N	17°05′E
Ennedi, plat., Chad (ĕn-nĕd´ē)	**280-81**	17°15′N	22°00′E
Enniskillen, N. Ire., U.K. (ĕn-ĭs-kĭl´ĕn)	**206-07**	54°21′N	7°38′W
Enriquillo, Dom. Rep. (ĕn-rē-kē´l-yō)	**154-55**	17°53′N	71°16′W
Enriquillo, Lago, lk., Dom. Rep. (lä´gô-ĕn-rē-kē´l-yō)	**154-55**	18°29′N	71°38′W
Enschede, Neth. (ĕns´κá-dĕ)	**206-07**	52°13′N	6°54′E
Ensenada, Mex. (ĕn-sĕ-nä´dä)	**156-57**	31°52′N	116°37′W
Enshi, China (ŭn-shr)	**258-59**	30°14′N	109°27′E
Entebbe, Ug.	**289**	0°04′N	32°28′E
Enterprise, Al., U.S. (ĕn´tĕr-prīz)	**134-35**	31°19′N	85°51′W
Enterprise, Or., U.S. (ĕn´tĕr-prīz)	**122-23**	45°26′N	117°17′W
Enugu, Nig. (ĕ-nōō´gōō)	**282a**	6°27′N	7°27′E
Enumclaw, Wa., U.S. (ĕn´ŭm-klô)	**122-23**	47°12′N	121°59′W
Enurmino, Russia	**136**	66°55′N	171°48′W
Envigado, Col. (ĕn-vē-gá´dō)	**188c**	6°10′N	75°35′W
Envira, stm., Braz.	**180-81**	6°42′S	69°48′W
Épernay, Fr. (ā-pĕr-nĕ´)	**212-13**	49°03′N	3°58′E
Ephraim, Ut., U.S. (ē´frā-ĭm)	**128-29**	39°22′N	111°35′W
Ephrata, Pa., U.S. (ĕfrā´tá)	**126-27**	40°11′N	76°11′W
Ephrata, Wa., U.S. (ĕfrā´tá)	**122-23**	47°19′N	119°33′W
Épi, i., Vanuatu	**306g**	16°43′S	168°16′E
Épinal, Fr. (ā-pē-nál´)	**212-13**	48°11′N	6°26′E
Equatorial Guinea, nation, Afr. (ē-kwä-tô´rĭ-ăl gĭn´ē)	**274**	2°0′N	9°00′E
Erbaa, Tur.	**247**	40°40′N	36°34′E
Erciyes Daği, vol., Tur.	**202-03**	38°32′N	35°28′E
Erdenebulgan, Mong.	**260-61**	50°06′N	101°35′E
Erding, Ger. (ĕr´dēng)	**210-11**	48°18′N	11°55′E
Erechim, Braz. (ĕ-rĕ-shĕ´N)	**182-83**	27°38′S	52°17′W
Ereğli, Tur. (ĕ-rā´ĭ-le)	**202-03**	37°31′N	34°04′E
Ereğli, Tur. (ĕ-rā´ĭ-le)	**202-03**	41°17′N	31°26′E
Erenhot, China	**260-61**	43°39′N	111°60′E
Erfoud, Mor.	**204-05**	31°26′N	4°14′W
Erfurt, Ger. (ĕr´fòrt)	**210-11**	50°58′N	11°01′E
Erguig, Bahr, stm., Chad	**282-83**	11°21′N	15°24′E
Ergun, stm., Asia	**260-61**	53°19′N	121°27′E
Er Hai, lk., China	**258-59**	25°48′N	100°11′E
Erick, Ok., U.S. (âr´ĭk)	**130-31**	35°14′N	99°52′W
Erie, Pa., U.S.	**126-27**	42°07′N	80°03′W
Erie, Lake, lk., N.A. (lāk ē´rĭ)	**126-27**	42°15′N	80°60′W
Erimo-misaki, c., Japan (ā´rē-mō mē´sä-kē)	**264**	41°55′N	143°15′E
Eritrea, nation, Afr. (ā-rē-trā´á)	**288**	15°20′N	39°00′E
Erlangen, Ger. (ĕr´läng-ĕn)	**210-11**	49°36′N	11°01′E
Ermelo, S. Afr.	**292c**	26°31′S	29°59′E
Erne, Lower Lough, lk., N. Ire., U.K. (lō´ĕr lŏk ûrn)	**206-07**	54°28′N	7°45′W
Erode, India	**256**	11°21′N	77°44′E
Eromanga, Austl.	**201**	26°40′S	143°16′E
Er-Rachidia, Mor.	**204-05**	31°57′N	4°26′W
Erromango, i., Vanuatu	**306g**	18°45′S	169°05′E
Ertis, Kaz.	**244**	53°20′N	75°28′E
Ertis, stm., Asia *see* Irtysh	**236-37**	61°05′N	68°47′E
Ertix, stm., Asia *see* Irtysh	**236-37**	61°05′N	68°47′E
Ērtra, nation, Afr. *see* Eritrea	**288**	15°20′N	39°00′E
Erwin, Tn., U.S. (ûr´wĭn)	**134-35**	36°08′N	82°25′W
Erzin, Russia	**260-61**	50°15′N	95°09′E
Erzincan, Tur. (ĕr-zĭn-jän´)	**245**	39°44′N	39°31′E
Erzurum, Tur. (ĕrz'rōōm´)	**245**	39°54′N	41°17′E
Esashi, Japan (ĕs´á-shē)	**264**	41°52′N	140°10′E
Esbjerg, Den. (ĕs´byĕrgh)	**208-09**	55°28′N	8°27′E
Escalante, Ut., U.S. (ĕs-kā-lăn´tē)	**128-29**	37°46′N	111°36′W
Escambia, stm., Fl., U.S. (ĕs-kăm´bĭ-á)	**134-35**	30°32′N	87°11′W
Escanaba, Mi., U.S. (ĕs-ká-nô´bá)	**126-27**	45°45′N	87°04′W
Escarpada Point, c., Phil.	**270**	18°31′N	122°13′E
Escondido, Ca., U.S. (ĕs-kŏn-dē´dō)	**128-29**	33°07′N	117°05′W
Escuinapa de Hidalgo, Mex.	**158-59**	22°51′N	105°48′W
Escuintla, Guat. (ĕs-kwēn´tlä)	**160**	14°18′N	90°47′W
Esenguly, Turkmen.	**252-53**	37°27′N	54°01′E
Eşfahān, Iran	**252-53**	32°39′N	51°40′E
Esh-Sham, nat. cap., Syria *see* Damascus	**248-49**	33°31′N	36°18′E
Esil, Kaz.	**244**	51°57′N	66°24′E
Esil, stm., Asia *see* Ishim	**236-37**	57°43′N	71°12′E
Eskifjörður, Ice. (ĕs´kē-fyûr´dōōr)	**206a**	65°04′N	13°57′W
Eskilstuna, Swe. (å´shĕl-stü-na)	**208-09**	59°22′N	16°31′E
Eskimo Point, Nu., Can. *see* Arviat	**138-39**	61°08′N	94°07′W
Eskişehir, Tur. (ĕs-kē-shĕ´h'r)	**202-03**	39°47′N	30°31′E
Esla, stm., Spain (ĕs-lä)	**214-15**	41°29′N	6°03′W
Eslāmshahr, Iran	**252-53**	35°33′N	51°14′E
Eslöv, Swe. (ĕs´lûv)	**208-09**	55°50′N	13°19′E
Esmeralda, Isla, i., Chile	**185**	48°57′S	75°25′W
Esmeraldas, Ec. (ĕs-mä-räl´däs)	**184**	0°57′N	79°39′W
España, nation, Eur. *see* Spain	**190-91**	40°0′N	4°00′W
Espanola, On., Can. (ĕs-pá-nō´lá)	**146-47**	46°16′N	81°47′W
Española, Isla, i., Ec.	**184a**	1°23′S	89°42′W
Esperance, Austl. (ĕs´pē-răns)	**294-95**	33°51′S	121°53′E
Esperanza, Arg.	**187**	31°27′S	60°56′W
Espichel, Cabo, c., Port. (ká´bō-ĕs-pē-shĕl´)	**214-15**	38°25′N	9°13′W
Espinal, Col. (ĕs-pē-näl´)	**188c**	4°09′N	74°53′W
Espinhaço, Serra do, mts., Braz. (sĕ´r-rä-dô-ĕs-pē-ná-sō)	**186**	17°25′S	43°40′W
Espírito Santo, Braz. (ĕs-pē´rē-tō-sän´tō) *see* Vila Velha	**186**	20°20′S	40°17′W

n-sing; ŋ-baŋk; N-nasalized n; nŏd; cŏmmit; ōld; ôbey; ôrder; oi-boil; fōōd; ò-as oo in foot; ou-out; s-soft; sh-dish; th-thin; pūre; ůnite; ûrn; stŭd; circŭs; ü-as in French tu; ´-indeterminate vowel.

354 Espírito Santo – Fear, Cape PRONOUNCING INDEX

Feature (Pronunciation)	Page	Lat.	Long.
Espírito Santo, state, Braz. (ĕs-pē´rē-tô-sän´tô)	186	19°30´S	40°30´W
Espiritu Santo, i., Vanuatu (ĕs-pē´rē-to͞o sän´tō)	306g	15°15´S	166°50´E
Espíritu Santo, Isla del, i., Mex.	156-57	24°29´N	110°21´W
Espita, Mex. (ĕs-pē´tä)	160	21°02´N	88°18´W
Esposende, Port. (ĕs-pō-zĕn´dā)	214-15	41°32´N	8°47´W
Esquel, Arg. (ĕs-kĕ´l)	185	42°54´S	71°19´W
Esquimalt, B.C., Can. (ĕs-kwī´mŏlt)	142-43	48°26´N	123°24´W
Esquina, Arg.	187	30°01´S	59°32´W
Essaouira, Mor.	280-81	31°30´N	9°45´W
Essen, Ger. (ĕs´sĕn)	210-11	51°28´N	7°01´E
Essequibo, stm., Guy. (ĕs-ā-kē´bō)	178-79	7°04´N	58°26´W
Essex, Md., U.S. (ĕs´ĕks)	126-27	39°19´N	76°29´W
Essexville, Mi., U.S. (ĕs´ĕks-vĭl)	126-27	43°37´N	83°50´W
Est, Pointe de l', c., Qc., Can.	148-49	49°08´N	61°41´W
Estación Colonia Alvear Norte, Arg. see General Alvear	185	34°59´S	67°42´W
Estación Foguista J. F. Juárez, Arg. see El Galpón	182-83	25°24´S	64°38´W
Estación Gobernador Vera, Arg. see Vera	187	29°28´S	60°13´W
Estación J. J. Castelli, Arg. see Castelli	182-83	25°57´S	60°37´W
Estados, Isla de los, i., Arg.	185	54°48´S	64°33´W
Estância, Braz. (ĕs-tän´sĭ-ä)	180-81	11°16´S	37°26´W
Estarreja, Port. (ĕ-tär-rā´zhä)	214-15	40°45´N	8°34´W
Estcourt, S. Afr. (ĕst-coort)	292c	29°00´S	29°53´E
Estelí, Nic.	161	13°05´N	86°22´W
Estepona, Spain (ĕs-tå-pō´nä)	214-15	36°26´N	5°08´W
Esterhazy, Sk., Can. (ĕs´tẽr-hä-zē)	144-45	50°39´N	102°05´W
Estevan, Sk., Can. (ĕ-stē´văn)	144-45	49°08´N	103°00´W
Estherville, Ia., U.S. (ĕs´tẽr-vĭl)	124-25	43°25´N	94°50´W
Estill, S.C., U.S. (ĕs´tĭl)	134-35	32°45´N	81°14´W
Eston, Sk., Can.	144-45	51°10´N	108°46´W
Estonia, nation, Eur. (ĕs-tō´nĭ-å)	190-91	59°0´N	26°00´E
Estrela, mtn., Port. (ĕs-trā´lȧ)	214-15	40°19´N	7°37´W
Estremoz, Port. (ĕs-trā-mōzh´)	214-15	38°51´N	7°35´W
Estrondo, Serra do, plat., Braz. (sĕr´-rȧ dò ĕs-trôn´-dò)	180-81	9°0´S	48°45´W
Esumba, Île, i., D.R.C.	284-85	2°0´N	21°12´E
Eszék, Cro. see Osijek	216-17	45°33´N	18°42´E
Étampes, Fr. (ā-tänp´)	212-13	48°26´N	2°10´E
Étaples, Fr. (ā-täp´l´)	212-13	50°31´N	1°38´E
Etāwah, India	254-55	26°46´N	79°01´E
Ethiopia, nation, Afr. (ē-thē-ō´pē-å)	274	9°0´N	39°00´E
Etna, Monte, vol., Italy (mỏn-tå ĕt´nȧ)	216-17	37°45´N	15°00´E
Etolin Strait, strt., Ak., U.S. (ĕt ō lĭn strāt)	136	60°20´N	165°15´W
Etorofu-tō, i., Russia see Iturup, Ostrov	236-37	44°51´N	147°27´E
Etosha National Park, n.p., Nmb.	286-87	18°60´S	15°07´E
Etosha Pan, pl., Nmb. (ĕtō´shä)	286-87	18°45´S	16°15´E
Etowah, Tn., U.S. (ĕt´ô-wä)	134-35	35°19´N	84°32´W
Et Tîdra, i., Maur.	280-81	19°44´N	16°24´W
Eua, i., Tonga	304-05	21°22´S	174°56´W
Euboea, i., Grc. see Évvoia	216-17	38°34´N	23°50´E
Eucla, Austl. (ū´klȧ)	294-95	31°42´S	128°54´E
Euclid, Oh., U.S. (ū´klĭd)	126-27	41°35´N	81°31´W
Eufaula, Al., U.S. (ů-fô´lȧ)	134-35	31°54´N	85°09´W
Eufaula, Ok., U.S. (ů-fô´lȧ)	130-31	35°16´N	95°36´W
Eugene, Or., U.S. (ů-jēn´)	122-23	44°03´N	123°05´W
Eugenia, Punta, c., Mex.	156-57	27°51´N	115°05´W
Eunice, La., U.S. (ū´nĭs)	132-33	30°30´N	92°25´W
Eunice, N.M., U.S. (ū´nĭs)	130-31	32°26´N	103°10´W
Euphrates, stm., Asia (ů-frā´tēz)	226-27	30°60´N	47°27´E
Eureka, Ca., U.S.	122-23	40°47´N	124°09´W
Eureka, Il., U.S. (û-rē´kȧ)	126-27	40°43´N	89°16´W
Eureka, Ks., U.S. (û-rē´kȧ)	130-31	37°50´N	96°18´W
Eureka, Mt., U.S. (û-rē´kȧ)	122-23	48°53´N	115°03´W
Eureka, S.D., U.S. (û-rē´kȧ)	124-25	45°46´N	99°38´W
Eureka Springs, Ar., U.S. (û-rē´kȧ sprĭngz)	130-31	36°24´N	93°45´W
Europa, Île, i., Reu.	286-87	22°20´S	40°21´E
Europa Island, i., Reu. see Europa, Île	286-87	22°20´S	40°21´E
Europe, cont., (ū´rŭp)	4-5	50°0´N	28°00´E
Eustis, Fl., U.S. (ūs´tĭs)	134-35	28°51´N	81°41´W
Eutaw, Al., U.S. (ū-tâ)	134-35	32°50´N	87°53´W
Eutsuk Lake, lk., B.C., Can. (o͞ot´sŭk lăk)	142-43	53°19´N	126°44´W
Evans, Lac, lk., Qc., Can.	140-41	50°55´N	77°00´W
Evanston, Il., U.S. (ĕv´ăn-stŭn)	126-27	42°01´N	87°41´W
Evanston, Wy., U.S. (ĕv´ăn-stŭn)	122-23	41°17´N	110°58´W
Evansville, In., U.S. (ĕv´ănz-vĭl)	126-27	37°59´N	87°35´W
Evansville, Wi., U.S. (ĕv´ănz-vĭl)	126-27	42°47´N	89°17´W
Evansville, Wy., U.S. (ĕv´ănz-vĭl)	122-23	42°50´N	106°15´W
Eva Perón, Arg. see La Plata	187	34°55´S	57°57´W
Eveleth, Mn., U.S. (ĕv´ê-lĕth)	124-25	47°28´N	92°32´W
Evensk, Russia	236-37	61°57´N	159°15´E
Everard, Lake, lk., Austl. (lăk ĕv´ẽr-ȧrd)	296-97	31°25´S	135°06´E
Everest, Mount, mtn., Asia (mount ĕv´ẽr-ĕst)	254-55	27°59´N	86°56´E
Everett, Wa., U.S. (ĕv´ẽr-ĕt)	122-23	47°58´N	122°12´W
Everglades, The, sw., Fl., U.S. (thȧ ĕv´ẽr-glădz swŏmp)	135a	26°0´N	80°40´W
Everglades National Park, n.p., Fl., U.S. (ĕv´ẽr-glādz năsh´ŭn-ăl pärk)	135a	25°27´N	80°53´W
Evergreen, Al., U.S. (ĕv´ẽr-grēn)	134-35	31°26´N	86°57´W
Evergreen, Co., U.S. (ĕv´ẽr-grēn)	128-29	39°38´N	105°19´W
Evergreen, Mt., U.S. (ĕv´ẽr-grēn)	122-23	48°13´N	114°17´W
Évora, Port. (ĕv´ô-rä)	214-15	38°34´N	7°54´W
Evpatoria, Ukr. see Yevpatoriia	218-19	45°12´N	33°22´E
Évreux, Fr. (ā-vrû´)	212-13	49°01´N	1°09´E
Évvoia, i., Grc.	216-17	38°34´N	23°50´E
Exe, stm., Eng., U.K. (ĕks)	206-07	50°42´N	3°29´W
Exeter, Eng., U.K. (ĕk´sê-tẽr)	206-07	50°43´N	3°32´W
Exmoor, plat., Eng., U.K. (ĕks´mŏr)	206-07	51°09´N	3°44´W
Exmouth, Austl.	294-95	21°56´S	114°07´E
Exmouth, Eng., U.K. (ĕks´mŭth)	206-07	50°38´N	3°24´W
Exmouth Gulf, b., Austl.	296-97	22°0´S	114°20´E
Exploits, stm., Nf./L., Can. (ĕks-ploits´)	148-49	49°05´N	55°18´W
Exuma Sound, strt., Bah. (ĕk-so͞o´mä sound)	154-55	24°12´N	76°01´W
Eyasi, Lake, lk., Tan. (lāk å-yä´sê)	289	3°40´S	35°05´E
Eyl, Som.	284-85	7°59´N	49°49´E
Eyre Creek, stm., Austl.	302	26°38´S	138°59´E
Eyre North, Lake, lk., Austl. (lăk âr nôrth)	296-97	28°33´S	137°15´E
Eyre Peninsula, pen., Austl.	296-97	33°15´S	135°48´E
Eyre South, Lake, lk., Austl. (lăk âr south)	301	29°18´S	137°26´E
Eysturoy, i., Far. Is.	206b	62°12´N	6°55´W
Ezequiel Ramos Mexía, Embalse, res., Arg.	185	39°27´S	69°01´W
Ezine, Tur. (å´zĭ-nå)	216-17	39°47´N	26°20´E

F

Feature (Pronunciation)	Page	Lat.	Long.
Faaborg, Den. (fô´bỏrg)	210-11	55°06´N	10°15´E
Fabriano, Italy (fä-brē-ä´nō)	216-17	43°21´N	12°54´E
Fada, Chad (fä´dä)	280-81	17°12´N	21°35´E
Fada-Ngourma, Burkina (fä´dä´n go͞or´mä)	282-83	12°03´N	0°22´E
Faddeyevskiy, Ostrov, i., Russia (ôs-trôf´ fȧd-yä´skĭ)	236-37	75°27´N	144°20´E
Faenza, Italy (fä-ĕnd´zä)	216-17	44°17´N	11°53´E
Færøerne, dep., Eur. see Faroe Islands	190-91	62°0´N	7°00´W
Fafen, stm., Eth.	284-85	5°39´N	44°08´E
Făgăraș, Rom. (fà-gä´räsh)	210-11	45°50´N	24°59´E
Fagernes, Nor.	208-09	60°59´N	9°15´E
Fagnano, Lago, lk., S.A. (lä´gô fäk-nä´nô)	185	54°34´S	67°58´W
Faguibine, Lac, lk., Mali	280-81	16°49´N	3°50´W
Faial, i., Port. (fä-yä´l)	215c	38°34´N	28°42´W
Fairbanks, Ak., U.S. (fâr´bănks)	136	64°51´N	147°42´W
Fairbury, Ne., U.S. (fâr´bĕr-ĭ)	130-31	40°09´N	97°11´W
Fairfax, Mn., U.S. (fâr´făks)	124-25	44°31´N	94°43´W
Fairfax, S.C., U.S. (fâr´făks)	134-35	32°57´N	81°14´W
Fairfield, Ia., U.S. (fâr´fēld)	124-25	41°01´N	91°58´W
Fairfield, Il., U.S. (fâr´fēld)	126-27	38°22´N	88°22´W
Fairfield, Oh., U.S. (fâr´fēld)	126-27	39°21´N	84°34´W
Fairfield, Tx., U.S. (fâr´fēld)	132-33	31°43´N	96°10´W
Fair Haven, Vt., U.S. (fâr ā´vĕn)	126-27	43°36´N	73°16´W
Fair Isle, i., Scot., U.K. (fâr īl)	206c	59°32´N	1°39´W
Fairmont, Mn., U.S. (fâr´mŏnt)	124-25	43°40´N	94°28´W
Fairmont, W.V., U.S. (fâr´mŏnt)	126-27	39°29´N	80°08´W
Fair Ness, c., Nu., Can.	140-41	63°25´N	72°02´W
Fairview, Ok., U.S. (fâr´vū)	130-31	36°16´N	98°29´W
Fairweather, Mount, mtn., N.A. (mount fâr-wĕdh´ẽr)	136	58°54´N	137°32´W
Fairweather Mountain, mtn., N.A. see Fairweather, Mount	136	58°54´N	137°32´W
Faisalābād, Pak.	252-53	31°25´N	73°05´E
Faith, S.D., U.S. (fāth)	124-25	45°01´N	102°01´W
Faiyum, Egypt see El-Fayoum	291b	29°19´N	30°50´E
Faizābād, India	254-55	26°47´N	82°08´E
Fakaofo, at., Tok.	304-05	9°22´S	171°14´W
Fakfak, Indon.	242-43	2°56´S	132°18´E
Faku, China (fä-ko͞o)	263	42°30´N	123°25´E
Falam, Mya.	266-67	22°54´N	93°41´E
Falémé, stm., Afr.	280-81	14°46´N	12°15´W
Falfurrias, Tx., U.S. (făl´fo͞o-rē´ås)	132-33	27°14´N	98°09´W
Falher, Ab., Can. (făl´ẽr)	142-43	55°44´N	117°13´W
Falkenberg, Swe. (fäl´kĕn-bĕrgh)	208-09	56°54´N	12°29´E
Falkensee, Ger. (fäl´kĕn-zā)	210-11	52°34´N	13°04´E
Falkirk, Scot., U.K. (fôl´kûrk)	206-07	56°00´N	3°47´W
Falkland Islands, dep., S.A. (fôk´lȧnd ī´lȧndz)	172	51°45´S	59°00´W
Falkland Islands, is., Falk. Is. (fôk´lȧnd ī´lȧndz)	185	51°41´S	59°08´W
Falkland Sound, strt., Falk. Is.	185	51°45´S	59°25´W
Falköping, Swe. (fäl´chûp-ĭng)	208-09	58°10´N	13°32´E
Fall River, Ma., U.S. (fôl rĭv´ẽr)	126-27	41°42´N	71°10´W
Falls City, Ne., U.S. (fôlz sĭ´tê)	130-31	40°04´N	95°37´W
Falmouth, Jam. (făl´mŭth)	154-55	18°29´N	77°39´W
Falmouth, Eng., U.K. (făl´mŭth)	206-07	50°09´N	5°03´W
Falmouth, Ky., U.S. (făl´mŭth)	126-27	38°41´N	84°20´W
False Divi Point, c., India	256	15°43´N	80°50´E
Falster, i., Den. (fäls´tẽr)	208-09	54°48´N	11°58´E
Fălticeni, Rom. (fŭl-tē-chān´y')	210-11	47°28´N	26°19´E
Falun, Swe. (fä-lo͞on´)	208-09	60°36´N	15°38´E
Famagusta, Cyp. see Gazimağusa	248-49	35°07´N	33°57´E
Fanch'eng, China see Xiangfan	258-59	32°02´N	112°09´E
Fangxian, China (fäŋ-shyĕn)	258-59	32°03´N	110°44´E
Fanning Island, at., Kir. see Tabuaeran	304-05	3°51´N	159°18´W
Fano, Italy (fä´nō)	216-17	43°51´N	13°01´E
Fanø, i., Den. (fän´û)	208-09	55°25´N	8°25´E
Fan Si Pan, mtn., Viet.	266-67	22°15´N	103°46´E
Faradofay, Madag. see Tôlañaro	286-87	25°02´S	47°00´E
Farafangana, Madag. (fä-rä-fäŋ-gä´nä)	286-87	22°49´S	47°50´E
Farāh, Afg. (fä-rä´)	252-53	32°22´N	62°04´E
Farāh, stm., Afg.	252-53	31°27´N	61°27´E
Farallon de Medinilla, i., N. Mar. Is.	304-05	16°01´N	146°04´E
Farallon de Pajaros, i., N. Mar. Is.	304-05	20°32´N	144°54´E
Farasān, Jazā'ir, is., Sau. Ar.	288	16°48´N	41°54´E
Farewell, Cape, c., Green.	95	59°46´N	43°60´W
Farewell, Cape, c., N.Z. (kāp fâr-wĕl´)	303	40°30´S	186°41´E
Farg'ona, Uzb.	252-53	40°23´N	71°48´E
Fargo, N.D., U.S. (fär´gō)	124-25	46°52´N	96°49´W
Faribault, Mn., U.S. (fâ´rĭ-bō)	124-25	44°18´N	93°17´W
Farīdpur, Bngl.	254-55	23°36´N	89°51´E
Farmersburg, In., U.S. (fär´mẽrz-bûrg)	126-27	39°15´N	87°23´W
Farmersville, Tx., U.S. (fär´mẽrz-vĭl)	130-31	33°10´N	96°22´W
Farmington, Il., U.S. (färm-ĭng-tŭn)	124-25	40°42´N	90°00´W
Farmington, Mo., U.S. (färm-ĭng-tŭn)	130-31	37°47´N	90°25´W
Farmington, N.M., U.S. (färm-ĭng-tŭn)	128-29	36°44´N	108°12´W
Farmville, N.C., U.S. (färm-vĭl)	134-35	35°35´N	77°36´W
Farmville, Va., U.S. (färm-vĭl)	134-35	37°18´N	78°24´W
Farnham, Qc., Can. (fär´năm)	146-47	45°17´N	72°59´W
Faro, Braz. (fä´rò)	180-81	2°10´S	56°45´W
Faro, Port. (fä´rò)	214-15	37°01´N	7°56´W
Faro, stm., Afr.	282-83	9°20´N	12°54´E
Faroe Islands, dep., Eur. (fâr´ō ī´lȧndz)	190-91	62°0´N	7°00´W
Fårön, i., Swe.	208-09	57°56´N	19°08´E
Farquhar, Atoll de, at., Sey.	286-87	10°10´S	51°10´E
Farrukhābād, India (fŭ-ròk-hä-bäd´)	254-55	27°24´N	79°35´E
Farsund, Nor. (fär´sòn)	208-09	58°05´N	6°48´E
Fartak, Ra's, c., Yemen	238-39	15°39´N	52°12´E
Farukolu, i., Mald.	256	6°12´N	73°16´E
Farvel, Kap, c., Green. see Farewell, Cape	95	59°46´N	43°60´W
Farwell, Tx., U.S. (fär´wĕl)	130-31	34°24´N	103°01´W
Fasano, Italy (fä-zä´nō)	216-17	40°50´N	17°21´E
Fatehpur Sīkri, India	254-55	27°06´N	77°40´E
Fatsa, Tur.	247	41°02´N	37°30´E
Fauro Island, i., Sol. Is.	306e	6°55´S	156°04´E
Fauske, Nor.	200-01	67°16´N	15°24´E
Fawn, stm., On., Can.	144-45	55°22´N	88°20´W
Faxaflói, b., Ice.	206a	64°25´N	23°00´W
Faya-Largeau, Chad	280-81	17°56´N	19°07´E
Fayette, Al., U.S. (fā-yĕt´)	134-35	33°41´N	87°50´W
Fayette, Mo., U.S. (fā-yĕt´)	130-31	39°09´N	92°41´W
Fayette, Ms., U.S. (fā-yĕt´)	134-35	31°43´N	91°04´W
Fayetteville, Ar., U.S. (fā-yĕt´vĭl)	130-31	36°05´N	94°10´W
Fayetteville, N.C., U.S. (fā-yĕt´vĭl)	134-35	35°03´N	78°53´W
Fayetteville, Tn., U.S. (fā-yĕt´vĭl)	134-35	35°09´N	86°34´W
Fayetteville, W.V., U.S. (fā-yĕt´vĭl)	126-27	38°03´N	81°06´W
Faylakah, i., Kuw.	250-51	29°26´N	48°20´E
Fayyum, Egypt see El-Fayoum	291b	29°19´N	30°50´E
Fazzān, hist. reg., Libya see Fezzan	280-81	26°0´N	14°00´E
Fear, Cape, c., N.C., U.S. (kāp fēr)	134-35	33°57´N	77°56´W

ăt; finăl; rāte; senåte; ärm; åsk; sofȧ; fâre; ch-choose; dh-as th in other; bē; ėvent; bĕt; recĕnt; cratẽr; g-gō; gh-guttural g; bĭt; ĭ-short neutral; rīde; ĸ-guttural k as ch in German ich;

PRONOUNCING INDEX

Fécamp – Fort McPherson 355

Feature (Pronunciation)	Page	Lat.	Long.
Fécamp, Fr. (fā-kän′)	212-13	49°46′N	0°23′E
Fedala, Mor. see Mohammedia	292a	33°42′N	7°23′W
Federal, Arg.	187	30°57′S	58°47′W
Federated States of Micronesia, nation, Oc. see Micronesia, Federated States of	304-05	5°0′N	152°00′E
Feia, Lagoa, b., Braz. (lä′gō-á fĕ′yä)	186	22°0′S	41°20′W
Feijó, Braz.	180-81	8°11′S	70°24′W
Feira de Santana, Braz. (fĕ′á-rä dä sänt-än′ä)	180-81	12°15′S	38°58′W
Feixian, China (fā-shyĕn)	260-61	35°16′N	117°58′E
Fejaj, Chott, lk., Tun.	280-81	33°53′N	9°10′E
Felanitx, Spain (fā-lä-nēch′)	214-15	39°29′N	3°09′E
Feldkirch, Aus. (fĕlt′kĭrk)	210-11	47°14′N	9°36′E
Felipe Carrillo Puerto, Mex.	160	19°35′N	88°02′W
Felix, Cape, c., Nu., Can.	140-41	69°53′N	97°57′W
Feltre, Italy (fĕl′trā)	216-17	46°01′N	11°54′E
Fen, stm., China	260-61	35°28′N	110°34′E
Fengcheng, China (fŭŋ-chŭŋ)	258-59	28°10′N	115°46′E
Fengcheng, China (fŭŋ-chŭŋ)	263	40°27′N	124°04′E
Fengdu, China (fŭŋ-dōō)	258-59	29°58′N	107°46′E
Fengfeng, China	260-61	36°29′N	114°14′E
Fengtien, China see Shenyang	263	41°48′N	123°24′E
Fengxian, China (fŭŋ-shyĕn)	258-59	33°57′N	106°40′E
Fengyang, China (fŭng′yäng′)	258-59	32°52′N	117°33′E
Fengzhen, China (fŭŋ-jŭn)	260-61	40°26′N	113°09′E
Feni Islands, is., Pap. N. Gui.	302	4°04′S	153°38′E
Fenton, Mi., U.S. (fĕn-tŭn)	126-27	42°48′N	83°42′W
Fenyang, China	260-61	37°16′N	111°47′E
Feodosiia, Ukr.	218-19	45°02′N	35°22′E
Ferdows, Iran	252-53	34°01′N	58°10′E
Fergus Falls, Mn., U.S. (fûr′gŭs fôlz)	124-25	46°17′N	96°05′W
Fergusson Island, i., Pap. N. Gui.	302	9°31′S	150°39′E
Ferlo, Vallée du, stm., Sen.	282-83	15°50′N	15°43′W
Fermo, Italy (fĕr′mō)	216-17	43°10′N	13°43′E
Fermoy, Ire. (fûr-moi′)	206-07	52°08′N	8°17′W
Fernandina, Isla, i., Ec.	184a	0°26′S	91°30′W
Fernandina Beach, Fl., U.S. (fûr-nän-dē′ná bēch)	134-35	30°40′N	81°27′W
Fernando de Noronha, Ilha, i., Braz.	173	3°51′S	32°25′W
Fernandópolis, Braz.	182-83	20°16′S	50°15′W
Fernando Póo, i., Eq. Gui. see Bioko	282-83	3°30′N	8°40′E
Fernie, B.C., Can. (fûr′nĭ)	142-43	49°30′N	115°03′W
Ferrara, Italy (fĕr-rä′rä)	216-17	44°51′N	11°36′E
Ferrat, Cap, c., Alg. (kăp fĕr-rät)	214-15	35°55′N	0°23′W
Ferreñafe, Peru (fĕr-rĕn-yà′fĕ)	184	6°38′S	79°47′W
Ferriday, La., U.S. (fĕr′ĭ-dā)	134-35	31°38′N	91°33′W
Ferro, i., Spain see El Hierro	215d	27°45′N	18°00′W
Ferrol, Spain	214-15	43°29′N	8°14′W
Fès, Mor. (fĕs)	292a	34°03′N	5°00′W
Fessenden, N.D., U.S. (fĕs′ĕn-dĕn)	124-25	47°39′N	99°38′W
Festus, Mo., U.S. (fĕst′ŭs)	130-31	38°14′N	90°24′W
Fethiye, Tur. (fĕt-hē′yĕ)	216-17	36°37′N	29°08′E
Feuilles, stm., Qc., Can.	140-41	58°39′N	70°25′W
Feyzābād, Afg.	252-53	37°08′N	70°34′E
Fez, Mor. see Fès	292a	34°03′N	5°00′W
Fezzan, hist. reg., Libya	280-81	26°0′N	14°00′E
Fianarantsoa, Madag. (fyá-nä′rán-tsō′á)	286-87	21°25′S	47°07′E
Fianga, Chad	284-85	9°55′N	15°09′E
Ficksburg, S. Afr. (fĭks′bûrg)	292c	28°52′S	27°53′E
Fife Ness, c., Scot., U.K. (fĭf′nes′)	206-07	56°17′N	2°36′W
Figeac, Fr. (fē-zhäk′)	212-13	44°37′N	2°02′E
Figueira da Foz, Port. (fē-gwĕy-rä-dá-fō′z)	214-15	40°09′N	8°51′W
Figuig, Mor.	204-05	32°08′N	1°13′W
Fiji, nation, Oc. (fē′jē)	306f	18°0′S	178°00′E
Filicudi, Isola, i., Italy (ē′-sō-lä fē′le-kōō′dē)	216-17	38°34′N	14°34′E
Fillmore, Ut., U.S. (fĭl′mōr)	128-29	38°58′N	112°20′W
Fimi, stm., D.R.C.	284-85	3°02′S	16°56′E
Findlay, Oh., U.S. (fĭnd′lā)	126-27	41°02′N	83°38′W
Finisterre, Cabo de, c., Spain see Fisterra, Cabo de	214-15	42°53′N	9°16′W
Finland, nation, Eur. (fĭn′lănd)	190-91	64°0′N	26°00′E
Finland, Gulf of, b., Eur. (gŭlf ŭv fĭn′lănd)	208-09	60°0′N	27°00′E
Finskiy Zaliv, b., Eur. see Finland, Gulf of	208-09	60°0′N	27°00′E
Fiordland National Park, n.p., N.Z.	303	45°30′S	167°20′E
Fırat, stm., Asia see Euphrates	226-27	30°60′N	47°27′E
Firenze, Italy see Florence	216-17	43°47′N	11°14′E
Fīrozābād, India	254-55	27°09′N	78°24′E
Fīrozpur, India	254-55	30°55′N	74°37′E
Fisterra, Cabo de, c., Spain	214-15	42°53′N	9°16′W
Fitchburg, Ma., U.S. (fĭch′bûrg)	126-27	42°35′N	71°48′W
Fitzgerald, Ga., U.S. (fĭts-jĕr′ăld)	134-35	31°43′N	83°15′W
Fitz Roy, Arg.	185	47°03′S	67°14′W
Fitz Roy, Monte, mtn., S.A.	185	49°17′S	73°05′W
Fitzroy Crossing, Austl.	294-95	18°12′S	125°34′E
Fiume, Cro. see Rijeka	216-17	45°20′N	14°27′E
Fizi, D.R.C.	289	4°19′S	28°56′E
Flagstaff, Az., U.S. (flăg-stáf)	128-29	35°11′N	111°39′W
Flaherty Island, i., Nu., Can.	140-41	56°14′N	79°17′W
Flåm, Nor. (flôm)	208-09	60°51′N	7°07′E
Flaming Gorge Reservoir, res., U.S. (flā′mĭng gôrj rĕ′sĕr-vwär)	122-23	41°14′N	109°35′W
Flandreau, S.D., U.S. (flăn′drō)	124-25	44°03′N	96°36′W
Flathead Indian Reservation, ind. res., Mt., U.S. (flăt′hĕd ĭn′dĭ-ăn rĕ-sĕr-vā′shĕn)	122-23	47°30′N	114°25′W
Flathead Lake, lk., Mt., U.S. (flăt′hĕd lāk)	122-23	47°52′N	114°08′W
Flat Rock, Mi., U.S. (flăt rŏk)	126-27	42°06′N	83°17′W
Flattery, Cape, c., Wa., U.S. (kāp flăt′ẽr-ĭ)	122-23	48°23′N	124°43′W
Flekkefjord, Nor. (fläk′kĕ-fyôr)	208-09	58°18′N	6°41′E
Flemingsburg, Ky., U.S. (flĕm′ĭngz-bûrg)	126-27	38°25′N	83°45′W
Flensburg, Ger. (flĕns′bòrgh)	210-11	54°47′N	9°26′E
Flers, Fr. (flĕr)	212-13	48°45′N	0°34′W
Flinders, stm., Austl. (flĭn′dẽrz)	302	17°36′S	140°36′E
Flinders Chase National Park, n.p., Austl.	301	35°58′S	136°44′E
Flinders Island, i., Austl. (flĭn′dẽrz ī′lănd)	301	40°0′S	148°00′E
Flin Flon, Mb., Can. (flĭn flŏn)	144-45	54°46′N	101°53′W
Flint, Mi., U.S. (flĭnt)	126-27	43°00′N	83°41′W
Flint, i., Kir.	304-05	11°25′S	151°48′W
Flint, stm., Ga., U.S. (flĭnt)	134-35	30°46′N	84°48′W
Flora, Il., U.S. (flō′rá)	126-27	38°40′N	88°29′W
Florala, Al., U.S. (flor-ăl′ a)	134-35	31°00′N	86°20′W
Florence, Italy (flŏr′ĕns)	216-17	43°47′N	11°14′E
Florence, Al., U.S. (flŏr′ĕns)	134-35	34°48′N	87°41′W
Florence, Az., U.S. (flŏr′ĕns)	128-29	33°02′N	111°23′W
Florence, Ky., U.S. (flŏr′ĕns)	126-27	38°60′N	84°38′W
Florence, Or., U.S. (flŏr′ĕns)	122-23	43°59′N	124°06′W
Florence, S.C., U.S. (flŏr′ĕns)	134-35	34°11′N	79°46′W
Florencia, Col. (flō-rĕn′sē-á)	178-79	1°36′N	75°36′W
Flores, i., Indon. (flō′rĕs)	268-69	8°38′S	120°56′E
Flores, i., Port.	215c	39°26′N	31°13′W
Flores, Laut, s., Indon. see Flores Sea	268-69	8°0′S	120°00′E
Flores Island, i., B.C., Can.	142-43	49°20′N	126°10′W
Flores Sea, s., Indon. (flō′rĕs sē)	268-69	8°0′S	120°00′E
Floresville, Tx., U.S. (flō′rĕs-vĭl)	132-33	29°08′N	98°09′W
Floriano, Braz. (flō-rà-ä′nò)	180-81	6°47′S	43°01′W
Florianópolis, Braz. (flō-rē-ä-nō′pō-lēs)	186	27°35′S	48°32′W
Florida, Col. (flo-re′da)	188c	3°21′N	76°15′W
Florida, Ur. (flò-rē-dha)	187	34°06′S	56°12′W
Florida, state, U.S. (flŏr′ĭ-dá)	118-19	28°0′N	82°00′W
Florida, Estrecho de la, strt., N.A. see Florida, Straits of	154-55	24°59′N	79°45′W
Florida, Straits of, strt., N.A. (strāts ŭv flŏr′ĭ-dá)	154-55	24°59′N	79°45′W
Florida Bay, b., Fl., U.S. (flŏr′ĭ-dá bā)	135a	24°58′N	80°48′W
Florida Keys, is., Fl., U.S. (flŏr′ĭ-dá kēs)	135a	24°47′N	81°06′W
Florido, stm., Mex. (flō-rē′dō)	156-57	27°43′N	105°11′W
Flórina, Grc. (flô-rē′nä)	216-17	40°47′N	21°24′E
Florissant, Mo., U.S. (flŏr′ĭ-sănt)	130-31	38°48′N	90°20′W
Florø, Nor.	208-09	61°35′N	5°01′E
Floydada, Tx., U.S. (floi-dā′dá)	130-31	33°59′N	101°20′W
Flushing, Mi., U.S. (flûsh′ĭng)	126-27	43°04′N	83°50′W
Fly, stm., (flī)	302	8°14′S	142°09′E
Foča, Bos. (fō′chä)	216-17	43°30′N	18°47′E
Fochville, S. Afr. (fŏk′vĭl)	292c	26°29′S	27°31′E
Focşani, Rom. (fŏk-shä′nē)	218-19	45°42′N	27°12′E
Fogang, China (fwo-gäŋ)	258-59	23°52′N	113°32′E
Foggia, Italy (fōd′jä)	216-17	41°28′N	15°32′E
Fogo, Nf./L., Can. (fō′gō)	148-49	49°43′N	54°18′W
Fogo, i., C.V.	282-83	14°54′N	24°23′W
Fogo Island, i., Nf./L., Can. (fō′gō ī′lănd)	148-49	49°39′N	54°11′W
Foguista J. F. Juárez, Arg. see El Galpón	182-83	25°24′S	64°38′W
Foix, Fr. (fwä)	212-13	42°58′N	1°37′E
Fokino, Russia	218-19	53°26′N	34°26′E
Foladī, Koh-e, mtn., Afg.	252-53	34°38′N	67°32′E
Foley Island, i., Nu., Can.	140-41	68°32′N	75°07′W
Foligno, Italy (fō-lēn′yō)	216-17	42°58′N	12°42′E
Fond-du-Lac, Sk., Can.	138-39	59°20′N	107°10′W
Fond du Lac, Wi., U.S. (fŏn dū lăk′)	126-27	43°46′N	88°27′W
Fond du Lac Indian Reservation, ind. res., Mn., U.S. (fŏn dū lăk′ ĭn′dĭ-ăn rĕ-sĕr-vā′shĕn)	124-25	46°45′N	92°37′W
Fondi, Italy (fōn′dē)	216-17	41°22′N	13°26′E
Fonseca, Golfo de, b., N.A. (gōl-fō-dĕ-fōn-sā′kä)	161	13°10′N	87°40′W
Fontainebleau, Fr. (fôn-tĕn-blō′)	212-13	48°24′N	2°42′E
Fontana, Ca., U.S. (fŏn-tă′ná)	128-29	34°06′N	117°26′W
Fonte Boa, Braz. (fōn′tá bō′ă)	180-81	2°32′S	66°01′W
Fontenay-le-Comte, Fr. (fônt-nĕ′lē-kônt′)	212-13	46°28′N	0°48′W
Fontur, c., Ice.	206a	66°22′N	14°35′W
Foochow, China see Fuzhou	243a	26°06′N	119°17′E
Forbach, Fr. (fōr′bäk)	212-13	49°12′N	6°54′E
Forbes, Austl. (fôrbz)	301	33°23′S	148°00′E
Forchheim, Ger. (fōrk′hīm)	210-11	49°43′N	11°04′E
Fordyce, Ar., U.S. (fôr′dīs)	130-31	33°49′N	92°25′W
Forest, Ms., U.S. (fŏr′ĕst)	134-35	32°22′N	89°28′W
Forest City, Ia., U.S. (fŏr′ĕst sĭ′tē)	124-25	43°16′N	93°39′W
Forest City, N.C., U.S. (fŏr′ĕst sĭ′tĭ)	134-35	35°20′N	81°52′W
Forest City, Pa., U.S. (fŏr′ĕst sĭ′tē)	126-27	41°39′N	75°28′W
Forestville, Qc., Can. (fŏr′ĕst-vĭl)	148-49	48°45′N	69°06′W
Forfar, Scot., U.K. (fôr′fár)	206-07	56°38′N	2°54′W
Forlì, Italy (fōr-lē′)	216-17	44°13′N	12°03′E
Formentera, i., Spain (fōr-mĕn-tā′rä)	214-15	38°42′N	1°28′E
Formiga, Braz. (fōr-mē′gá)	186	20°28′S	45°26′W
Formosa, Arg. (fōr-mō′sä)	182-83	26°10′S	58°12′W
Formosa, Braz.	182-83	15°32′S	47°20′W
Formosa, nation, Asia (fōr-mō′sá) see Taiwan	224-25	23°30′N	121°00′E
Formosa, state, Arg. (fōr-mō′sä)	182-83	25°0′S	60°00′W
Formosa, Serra, plat., Braz. (sĕ′r-rä fōr-mō′sä)	180-81	12°0′S	55°00′W
Formosa Strait, strt., Asia see Taiwan Strait	243a	24°0′N	119°00′E
Føroyar, dep., Eur. see Faroe Islands	190-91	62°0′N	7°00′W
Forrest City, Ar., U.S. (fŏr′ĕst sĭ′tĭ)	134-35	35°00′N	90°48′W
Forsyth, Ga., U.S. (fôr-sīth′)	134-35	33°02′N	83°57′W
Forsyth, Mt., U.S. (fôr-sīth′)	122-23	46°16′N	106°41′W
Fort Albany, On., Can. (fôrt ôl′bá nĭ)	138-39	52°13′N	81°40′W
Fortaleza, Braz. (fôr′tä-lā′zà)	180-81	3°44′S	38°30′W
Fort Archambault, Chad see Sarh	284-85	9°09′N	18°23′E
Fort Atkinson, Wi., U.S. (fôrt ăt′kĭn-sŭn)	126-27	42°55′N	88°51′W
Fort Bayard, China see Zhanjiang	258-59	21°12′N	110°23′E
Fort Benton, Mt., U.S. (fôrt bĕn′tŭn)	122-23	47°49′N	110°41′W
Fort Berthold Indian Reservation, ind. res., N.D., U.S. (fôrt bĕrth′ôld ĭn′dĭ-ăn rĕ-sĕr-vā′shĕn)	124-25	47°40′N	102°25′W
Fort Bragg, Ca., U.S.	128-29	39°27′N	123°48′W
Fort Branch, In., U.S. (fôrt brănch)	126-27	38°15′N	87°35′W
Fort Chipewyan, Ab., Can.	138-39	58°43′N	111°10′W
Fort Collins, Co., U.S. (fôrt kŏl′ĭns)	130-31	40°35′N	105°05′W
Fort-Dauphin, Madag. see Tôlañaro	286-87	25°02′S	47°00′E
Fort-de-France, nat. cap., Mart. (dĕ fräns)	155b	14°36′N	61°04′W
Fort Dodge, Ia., U.S. (fôrt dŏj)	124-25	42°30′N	94°11′W
Fort Edward, N.Y., U.S. (fôrt wẽrd)	126-27	43°16′N	73°35′W
Fortescue, stm., Austl. (fôr′tĕs-kū)	296-97	21°00′S	116°06′E
Fort-Foureau, Camrn. see Kousséri	282-83	12°05′N	15°02′E
Fort Frances, On., Can. (fôrt frăn′sĕs)	144-45	48°37′N	93°24′W
Fort Franklin, N.T., Can. see Déline	138-39	65°11′N	123°25′W
Fort Frederica National Monument, n.p., Ga., U.S. (fôrt frĕd′ĕ-rĭ-ká năsh′ŭn-ăl mŏn′ŭ-mĕnt)	134-35	31°12′N	81°26′W
Fort-George, Qc., Can. see Chisasibi	138-39	53°48′N	79°02′W
Fort Gibson, Ok., U.S. (fôrt gĭb′sŭn)	130-31	35°49′N	95°15′W
Fort Good Hope, N.T., Can. (fôrt good hōp)	138-39	66°15′N	128°37′W
Forth, Firth of, b., Scot., U.K. (fûrth ŏv fôrth)	206-07	56°07′N	2°58′W
Fort Johnston, Malawi see Mangochi	286-87	14°28′S	35°15′E
Fort Kent, Me., U.S. (fôrt kĕnt)	127a	47°15′N	68°35′W
Fort Lauderdale, Fl., U.S. (fôrt lô′dẽr-dāl)	135a	26°07′N	80°09′W
Fort Liard, N.T., Can.	138-39	60°14′N	123°27′W
Fort Lupton, Co., U.S. (fôrt lŭp′tŭn)	130-31	40°05′N	104°49′W
Fort Macleod, Ab., Can. (fôrt má-kloud′)	142-43	49°43′N	113°25′W
Fort Madison, Ia., U.S. (fôrt măd′ĭ-sŭn)	130-31	40°38′N	91°19′W
Fort McMurray, Ab., Can. (fôrt mák-mŭr′ĭ)	144-45	56°44′N	111°25′W
Fort McPherson, N.T., Can. (fôrt mák-fŭr′s′n)	138-39	67°25′N	134°52′W

n-sing; ŋ-baŋk; ɴ-nasalized n; nŏd; cŏmmit; ōld; ôbey; ôrder; oi-boil; fōōd; ò-as oo in foot; ou-out; s-soft; sh-dish; th-thin; pūre; ůnite; ûrn; stŭd; circŭs; ü-as in French tu; ′-indeterminate vowel.

Feature (Pronunciation)	Page	Lat.	Long.
Fort Meade, Fl., U.S. (fōrt mēd)	135a	27°46'N	81°48'W
Fort Mill, S.C., U.S. (fōrt mĭl)	134-35	35°00'N	80°57'W
Fort Mojave Indian Reservation, ind. res., Az., U.S. (fōrt mō-hä'vâ ĭn'dĭ-ăn rĕ-sĕr-vā'shĕn)	128-29	34°55'N	114°35'W
Fort Monroe National Monument, n.p., Va., U.S.	134-35	37°00'N	76°18'W
Fort Morgan, Co., U.S. (fōrt môr'găn)	130-31	40°15'N	103°48'W
Fort Myers, Fl., U.S. (fōrt mī'ērz)	135a	26°38'N	81°52'W
Fort Nelson, B.C., Can. (fōrt nĕl'sŭn)	138-39	58°49'N	122°41'W
Fort Nelson, stm., B.C., Can. (fōrt nĕl'sŭn)	140-41	59°31'N	124°03'W
Fort Norman, N.T., Can. *see* Tulita	138-39	64°54'N	125°34'W
Fort Payne, Al., U.S. (fōrt pān)	134-35	34°27'N	85°43'W
Fort Peck Indian Reservation, ind. res., Mt., U.S. (fōrt pĕk ĭn'dĭ-ăn rĕ-sĕr-vā'shĕn)	122-23	48°22'N	105°40'W
Fort Peck Lake, res., Mt., U.S. (fōrt pĕk lāk)	122-23	47°45'N	106°45'W
Fort Pierce, Fl., U.S. (fōrt pērs)	135a	27°27'N	80°20'W
Fort Portal, Ug. (fōrt pōr'tál)	289	0°40'N	30°17'E
Fort Providence, N.T., Can. (fōrt prŏv'ĭ-dĕns)	138-39	61°21'N	117°35'W
Fort Pulaski National Monument, n.p., Ga., U.S. (fōrt pu-lăs'kĭ năsh'ŭn-ăl mŏn'ū-mĕnt)	134-35	32°01'N	80°55'W
Fort Qu'Appelle, Sk., Can.	144-45	50°46'N	103°48'W
Fort Resolution, N.T., Can. (fōrt rĕz'ō-lū'shŭn)	138-39	61°10'N	113°38'W
Fort Rosebery, Zam. *see* Mansa	286-87	11°12'S	28°53'E
Fort Saint James, B.C., Can. (fōrt sānt jāmz)	142-43	54°28'N	124°16'W
Fort Saint John, B.C., Can. (fōrt sānt jŏn)	142-43	56°17'N	120°54'W
Fort Saskatchewan, Ab., Can. (fōrt săs-kăt'chōō-ân)	142-43	53°43'N	113°14'W
Fort Severn, On., Can. (fōrt sĕv'ērn)	144-45	55°60'N	87°38'W
Fort-Shevchenko, Kaz. (fōrt shĕv-chĕn'kō)	202-03	44°30'N	50°16'E
Fort Simpson, N.T., Can. (fōrt sĭmp'sŭn)	138-39	61°51'N	121°22'W
Fort Smith, N.T., Can. (fōrt smĭth)	138-39	60°01'N	111°54'W
Fort Smith, Ar., U.S. (fōrt smĭth)	130-31	35°23'N	94°25'W
Fort Stanwix National Monument, n.p., N.Y., U.S.	126-27	43°13'N	75°27'W
Fort Stockton, Tx., U.S. (fōrt stŏk'tŭn)	132-33	30°54'N	102°53'W
Fort Sumner, N.M., U.S. (fōrt sŭm'nēr)	130-31	34°29'N	104°14'W
Fort Sumter National Monument, n.p., S.C., U.S. (fōrt sŭm'tēr năsh'ŭn-ăl mŏn'ū-mĕnt)	134-35	32°45'N	79°52'W
Fortuna, Ca., U.S. (fôr-tū'nà)	122-23	40°36'N	124°09'W
Fortune, Nf./L., Can. (fôr'tŭn)	148-49	47°04'N	55°50'W
Fortune Bay, b., Nf./L., Can. (fôr'tŭn bā)	148-49	47°25'N	55°25'W
Fort Union National Monument, n.p., N.M., U.S. (fōrt ūn'yŭn năsh'ŭn-ăl mŏn'ū-mĕnt)	130-31	35°56'N	105°03'W
Fort Valley, Ga., U.S. (fōrt văl'ē)	134-35	32°33'N	83°53'W
Fort Vermilion, Ab., Can. (fōrt vĕr-mĭl'yŭn)	138-39	58°23'N	116°02'W
Fort Walton Beach, Fl., U.S.	134-35	30°25'N	86°36'W
Fort Wayne, In., U.S. (fōrt wān)	126-27	41°04'N	85°07'W
Fort William, Scot., U.K. (fōrt wĭl'yŭm)	206-07	56°49'N	5°06'W
Fort Worth, Tx., U.S. (fōrt wûrth)	130-31	32°45'N	97°21'W
Fort Yukon, Ak., U.S. (fōrt yōō'kŏn)	136	66°34'N	145°15'W
Forūr, Jazīreh-ye, i., Iran	250-51	26°17'N	54°31'E
Foshan, China	258-59	23°03'N	113°07'E
Fossano, Italy (fōs-sä'nō)	216-17	44°33'N	7°44'E
Fossil Butte National Monument, n.p., Wy., U.S.	122-23	41°50'N	110°40'W
Fosston, Mn., U.S. (fôs'tŭn)	124-25	47°34'N	95°45'W
Foster, Austl.	301	38°39'S	146°12'E
Foster, stm., Sk., Can.	144-45	55°47'N	105°48'W
Fostoria, Oh., U.S. (fŏs-tō'rĭ-à)	126-27	41°10'N	83°24'W
Fougères, Fr. (fōō-zhâr')	212-13	48°21'N	1°12'W
Foulwind, Cape, c., N.Z. (kāp foul'wĭnd)	303	41°45'S	171°28'E
Fountain, Co., U.S.	130-31	38°41'N	104°42'W
Fouriesburg, S. Afr. (fō'rēz-bûrg)	292c	28°37'S	28°13'E
Fourmies, Fr. (fōōr-mē')	212-13	50°01'N	4°03'E
Foveaux Strait, strt., N.Z. (fō-vō' strāt)	303	46°35'S	168°00'E
Fowler, In., U.S. (foul'ēr)	126-27	40°37'N	87°19'W
Foxe Basin, b., Nu., Can. (fŏks bā's'n)	140-41	68°25'N	76°60'W
Foxe Peninsula, pen., Nu., Can. (fŏks pĕ-nĭn'sūlă)	140-41	65°0'N	76°00'W
Foz do Iguaçu, Braz.	182-83	25°33'S	54°35'W
Fraga, Spain (frä'gä)	214-15	41°32'N	0°21'E
Franca, Braz. (frä'n-kä)	186	20°32'S	47°24'W
France, nation, Eur. (frǎns)	190-91	46°0'N	2°00'E
Francés Viejo, Cabo, c., Dom. Rep. (kà'bō-frän'sås vyā'hō)	154-55	19°39'N	69°55'W
Franceville, Gabon (fräns-vēl')	282-83	1°38'S	13°35'E
Francis Case, Lake, res., S.D., U.S. (lāk frǎn'sĭs kās)	124-25	43°15'N	98°57'W
Francistown, Bots. (frǎn'sis-toun)	286-87	21°10'S	27°30'E
Francois Lake, lk., B.C., Can.	142-43	54°02'N	125°43'W
Francs Peak, mtn., Wy., U.S.	122-23	43°58'N	109°20'W
Frankfort, S. Afr. (frănk'fŏrt)	292c	27°17'S	28°31'E
Frankfort, In., U.S. (frăŋk'fŭrt)	126-27	40°17'N	86°30'W
Frankfort, Ky., U.S.	126-27	38°12'N	84°50'W
Frankfort, Mi., U.S. (frăŋk'fŭrt)	126-27	44°38'N	86°14'W
Frankfurt, Ger.	210-11	52°21'N	14°32'E
Frankfurt am Main, Ger.	210-11	50°07'N	8°40'E
Franklin, In., U.S. (frănk'lĭn)	126-27	39°28'N	86°03'W
Franklin, Ky., U.S. (frănk'lĭn)	134-35	36°43'N	86°35'W
Franklin, La., U.S. (frănk'lĭn)	134-35	29°47'N	91°30'W
Franklin, N.C., U.S. (frănk'lĭn)	134-35	35°11'N	83°23'W
Franklin, N.H., U.S. (frănk'lĭn)	126-27	43°27'N	71°40'W
Franklin, Tn., U.S. (frănk'lĭn)	134-35	35°56'N	86°52'W
Franklin, Tx., U.S. (frănk'lĭn)	132-33	31°01'N	96°29'W
Franklin, Va., U.S. (frănk'lĭn)	134-35	36°41'N	76°56'W
Franklin, Wi., U.S. (frănk'lĭn)	126-27	42°52'N	87°60'W
Franklin, W.V., U.S. (frănk'lĭn)	126-27	38°39'N	79°21'W
Franklin Mountains, mts., N.T., Can. (frănk'lĭn moun'tĭnz)	140-41	62°59'N	123°43'W
Franklinton, La., U.S. (frăŋk'lĭn-tŭn)	134-35	30°51'N	90°09'W
Frantsa-Iosifa, Zemlya, is., Russia *see* Franz Josef Land	236-37	81°0'N	55°00'E
Franz Josef Land, is., Russia	236-37	81°0'N	55°00'E
Frascati, Italy (fräs-kä'tē)	216-17	41°48'N	12°41'E
Fraser, stm., B.C., Can.	142-43	49°06'N	123°11'W
Fraserburgh, Scot., U.K. (frā'zēr-bûrg)	206-07	57°42'N	2°01'W
Fraser Island, i., Austl.	301	25°15'S	153°10'E
Fraser Plateau, plat., B.C., Can.	142-43	52°0'N	123°00'W
Fray Bentos, Ur.	187	33°08'S	58°18'W
Frazee, Mn., U.S. (frā-zē')	124-25	46°44'N	95°42'W
Fredericia, Den. (frĕdh-ĕ-rē'tsē-à)	208-09	55°35'N	9°46'E
Frederick, Ok., U.S. (frĕd'ēr-ĭk)	130-31	34°23'N	99°01'W
Fredericksburg, Tx., U.S. (frĕd'ēr-ĭkz-bûrg)	132-33	30°16'N	98°52'W
Fredericksburg, Va., U.S. (frĕd'ēr-ĭkz-bûrg)	126-27	38°18'N	77°28'W
Fredericktown, Mo., U.S. (frĕd'ēr-ĭk-toun)	130-31	37°34'N	90°18'W
Fredericton, N.B., Can. (frĕd'-ēr-ĭk-tŭn)	148-49	45°57'N	66°39'W
Frederikshavn, Den. (frĕdh'ĕ-rĕks-houn)	208-09	57°26'N	10°32'E
Fredonia, Col. (frĕ-dō'nyà)	188c	5°56'N	75°40'W
Fredonia, N.Y., U.S. (frê-dō'nĭ-à)	126-27	42°26'N	79°20'W
Fredrikstad, Nor. (frådh'rĕks-ståd)	208-09	59°12'N	10°56'E
Freels, Cape, c., Nf./L., Can. (kāp frēlz)	148-49	49°15'N	53°28'W
Freeport, Bah. (frē'pōrt)	154-55	26°31'N	78°39'W
Freeport, Il., U.S. (frē'pōrt)	126-27	42°18'N	89°37'W
Freeport, N.Y., U.S. (frē'pōrt)	126-27	40°39'N	73°35'W
Freeport, Tx., U.S. (frē'pōrt)	132-33	28°57'N	95°22'W
Freetown, nat. cap., S.L. (frē'toun)	282-83	8°29'N	13°13'W
Freiberg, Ger. (frī'bĕrgh)	210-11	50°55'N	13°21'E
Freirina, Chile (frå-ĭ-rē'nä)	182-83	28°30'S	71°06'W
Freising, Ger. (frī'zĭng)	210-11	48°24'N	11°44'E
Fréjus, Fr. (frā-zhüs')	212-13	43°26'N	6°45'E
Fremantle, Austl. (frē'măn-t'l)	294-95	32°03'S	115°45'E
Fremont, Ca., U.S. (frê-mŏnt')	128-29	37°33'N	121°59'W
Fremont, Mi., U.S. (frê'-mŏnt)	126-27	43°28'N	85°57'W
Fremont, Ne., U.S. (frê'-mŏnt)	124-25	41°27'N	96°30'W
Fremont, Oh., U.S. (frê'-mŏnt)	126-27	41°21'N	83°07'W
French Guiana, dep., S.A. (frĕnch gē-ä'nä)	172	4°0'N	53°00'W
French Lick, In., U.S. (frĕnch lĭk)	126-27	38°33'N	86°37'W
French Polynesia, dep., Oc. (frĕnch pŏl-ĭ-nē'zhă)	304-05	15°0'S	140°00'W
French Somaliland, nation, Afr. *see* Djibouti	274	11°30'N	43°00'E
Fresco, stm., Braz.	180-81	6°40'S	52°00'W
Freshfield, Mount, mtn., Can. (mount frĕsh'fēld)	142-43	51°44'N	116°57'W
Fresnillo, Mex. (frås-nēl'yō)	158-59	23°10'N	102°52'W
Fresno, Col. (frĕs'nō)	188c	5°09'N	75°01'W
Fresno, Ca., U.S.	128-29	36°45'N	119°46'W
Fria, Cape, c., Nmb. (kāp frĭă)	286-87	18°29'S	12°02'E
Frías, Arg. (frē'äs)	182-83	28°38'S	65°07'W
Friedberg, Ger. (frēd'bĕrgh)	210-11	48°22'N	10°59'E
Friedrichshafen, Ger. (frē-drēks-häf'ĕn)	210-11	47°40'N	9°29'E
Friend, Ne., U.S. (frĕnd)	130-31	40°39'N	97°17'W
Friesische Inseln, is., Eur. *see* Frisian Islands	206-07	53°27'N	5°50'E
Frio, Cabo, c., Braz. (kà'bō-frê'ō)	186	22°53'S	42°00'W
Frisian Islands, is., Eur. (frē'zhăn ī'lăndz)	206-07	53°27'N	5°50'E
Frobisher Bay, Nu., Can. *see* Iqaluit	138-39	63°44'N	68°28'W
Frobisher Bay, b., Nu., Can. (frŏb'ĭsh'ēr bā)	140-41	62°30'N	65°60'W
Frobisher Lake, lk., Sk., Can. (frŏb'ĭsh'ēr lāk)	144-45	56°22'N	108°17'W
Frolovo, Russia	202-03	49°47'N	43°39'E
Frome, Lake, lk., Austl. (lāk frōōm)	301	30°42'S	139°48'E
Frontera, Mex. (frôn-tā'rä)	158-59	18°32'N	92°38'W
Frontera, Mex. (frôn-tā'rä)	132-33	26°56'N	101°27'W
Front Royal, Va., U.S. (frŭnt roi'ăl)	126-27	38°55'N	78°12'W
Frosinone, Italy (frō-zē-nō'nå)	216-17	41°39'N	13°21'E
Frostburg, Md., U.S. (frôst'bûrg)	126-27	39°39'N	78°55'W
Frøya, i., Nor.	200-01	63°43'N	8°42'E
Fruita, Co., U.S. (frōōt-ă)	128-29	39°10'N	108°44'W
Fuchun, stm., China (fōō-chòn)	258-59	30°06'N	120°10'E
Fuego, Volcán de, vol., Guat. (vôl-kà'n-dĕ-fwā'gō)	160	14°29'N	90°53'W
Fuente de Cantos, Spain (fwĕn'tå dā kän'tōs)	214-15	38°15'N	6°18'W
Fuerte, stm., Mex. (fōō-ĕ'r-tĕ)	156-57	25°51'N	109°25'W
Fuerte Olimpo, Para. (fwĕr'tå ō-lēm-pō)	182-83	21°05'S	57°52'W
Fuerteventura, i., Spain (fwĕr'tå-vĕn-tōō'rä)	215d	28°20'N	14°00'W
Fuga Island, i., Phil.	270a	18°52'N	121°22'E
Fuji, Japan (fōō'jê)	265	35°09'N	138°40'E
Fuji, stm., Japan (fōō'jê)	265	35°07'N	138°39'E
Fujian, state, China (fōō-jyĕn)	243a	26°0'N	118°00'E
Fujin, China (fōō-jyĭn)	260-61	47°15'N	132°02'E
Fuji-san, vol., Japan (fōō'jê-sän)	265	35°22'N	138°44'E
Fujiyama, vol., Japan *see* Fuji-san	265	35°22'N	138°44'E
Fukien, state, China *see* Fujian	243a	26°0'N	118°00'E
Fukuchiyama, Japan (fō'kò-chê-yä'ma)	265	35°18'N	135°07'E
Fukue-jima, i., Japan (fò-kōō'ā jē'mă)	264	32°40'N	128°45'E
Fukui, Japan (fōō'kōō-ê)	265	36°04'N	136°13'E
Fukuoka, Japan	265	33°35'N	130°25'E
Fukushima, Japan (fōō'kò-shē'mä)	265	37°45'N	140°28'E
Fulaga, i., Fiji	306f	19°09'S	178°36'W
Fulaga Passage, strt., Fiji	306f	18°56'S	178°36'W
Fulda, Ger. (fòl'dä)	210-11	50°33'N	9°41'E
Fuling, China (fōō-lĭŋ)	258-59	29°42'N	107°25'E
Fullerton, Ne., U.S. (fòl'ēr-tŭn)	124-25	41°22'N	97°58'W
Fulton, Il., U.S. (fŭl'tŭn)	124-25	41°52'N	90°10'W
Fulton, Ky., U.S. (fŭl'tŭn)	134-35	36°31'N	88°53'W
Fulton, Mo., U.S. (fŭl'tŭn)	130-31	38°51'N	91°57'W
Fulton, Ms., U.S. (fŭl'tŭn)	134-35	34°14'N	88°24'W
Fulton, N.Y., U.S. (fŭl'tŭn)	126-27	43°19'N	76°25'W
Funafuti, at., Tuvalu	304-05	8°29'S	179°11'E
Funan, China *see* Fushun	263	41°52'N	123°54'E
Funchal, Port. (fòn-shäl')	280-81	32°39'N	16°54'W
Fundación, Col. (fōōn-dä-syō'n)	178-79	10°31'N	74°11'W
Fundy, Bay of, b., Can. (bā ūv fŭn'dĭ)	148-49	45°0'N	66°00'W
Fundy National Park, n.p., N.B., Can. (fŭn'dĭ năsh'ŭn-ăl pärk)	148-49	45°38'N	65°00'W
Fünfkirchen, Hung. *see* Pécs	210-11	46°04'N	18°13'E
Funing, China (fōō-nĭŋ)	258-59	23°34'N	105°37'E
Furnas, Represa de, res., Braz.	186	21°12'S	45°57'W
Furneaux Group, is., Austl. (fûr'nō grōōp)	301	40°10'S	148°05'E
Fürstenwalde, Ger. (für'stĕn-väl-dĕ)	210-11	52°21'N	14°04'E
Fürth, Ger. (fürt)	210-11	49°28'N	10°59'E
Fusan, Kor., S. *see* Pusan	263	35°05'N	129°03'E
Fushun, China (fōō'shōōn')	263	41°52'N	123°54'E
Fusong, China (fōō-soŋ)	263	42°20'N	127°17'E
Fusui, China	258-59	22°38'N	107°55'E
Futuna, Île, i., Wal./F.	304-05	14°18'S	178°09'W
Fuxian, China (fōō shyĕn)	260-61	36°02'N	109°22'E
Fuxian, China (fōō shyĕn) *see* Wafangdian	260-61	39°37'N	122°01'E
Fuxian Hu, lk., China	258-59	24°29'N	102°53'E
Fuxin, China (fōō-shyĭn)	260-61	42°08'N	121°45'E
Fuyang, China (fōō-yäŋ)	258-59	32°54'N	115°49'E
Fuyang, stm., China (fōō-yäŋ)	260-61	38°11'N	116°04'E
Fuyu, China (fōō-yōō)	260-61	45°10'N	124°49'E
Fuyu, China (fōō-yōō)	260-61	47°49'N	124°28'E
Fuzhou, China	243a	26°06'N	119°17'E
Fuzhou, China (fōō-jō)	258-59	28°01'N	116°20'E
Fyn, i., Den. (fü'n)	208-09	55°20'N	10°30'E

ăt; finăl; rāte; senåte; ärm; ȧsk; sofȧ; fâre; ch-choose; dh-as th in other; bē; ėvent; bĕt; recĕnt; cratẽr; g-gō; gh-guttural g; bĭt; ĭ-short neutral; rīde; κ-guttural k as ch in German ich;

PRONOUNCING INDEX
Gaalkacyo – Genil 357

Feature (Pronunciation)	Page	Lat.	Long.
G			
Gaalkacyo, Som.	284-85	6°46'N	47°26'E
Gabela, Ang.	286-87	10°51'S	14°22'E
Gaberones, nat. cap., Bots. *see* Gaborone	286-87	24°40'S	25°56'E
Gabès, Tun. (gä'bĕs)	204-05	33°54'N	10°06'E
Gabès, Golfe de, b., Tun. (gôlf-dĕ-gä'bĕs)	280-81	34°14'N	10°30'E
Gabon, nation, Afr. (gȧ-bôN')	274	1°0'S	11°45'E
Gaborone, nat. cap., Bots. (gä-bō-rō'-nā) (gä'bỏ-rōō-nā)	286-87	24°40'S	25°56'E
Gabrovo, Blg. (gäb'rỏ-vō)	216-17	42°51'N	25°19'E
Gachsārān, Iran	250-51	30°12'N	50°47'E
Gacko, Bos. (gäts'kỏ)	216-17	43°10'N	18°32'E
Gadag, India	256	15°25'N	75°37'E
Gadsden, Al., U.S. (gădz'dĕn)	134-35	34°00'N	86°01'W
Găeşti, Rom. (gä-yĕsh'tē)	216-17	44°43'N	25°20'E
Gaeta, Italy (gä-å'tä)	216-17	41°13'N	13°34'E
Gaferut, i., Micron.	304-05	9°12'N	145°23'E
Gaffney, S.C., U.S. (găf'nĭ)	134-35	35°04'N	81°39'W
Gafsa, Tun. (gäf'sä)	204-05	34°24'N	8°49'E
Gagnoa, C. Iv.	282-83	6°07'N	5°56'W
Gagra, Geor.	245	43°20'N	40°15'E
Gaillimh, Ire. *see* Galway	206-07	53°16'N	9°03'W
Gainesville, Fl., U.S. (gānz'vĭl)	134-35	29°39'N	82°18'W
Gainesville, Ga., U.S. (gānz'vĭl)	134-35	34°18'N	83°49'W
Gainesville, Tx., U.S. (gānz'vĭl)	130-31	33°38'N	97°09'W
Gainesville, Va., U.S. (gānz'vĭl)	126-27	38°47'N	77°38'W
Gairdner, Lake, lk., Austl. (lāk gård'nẽr)	296-97	31°33'S	135°57'E
Gaithersburg, Md., U.S. (gā'thẽrs'bûrg)	126-27	39°08'N	77°12'W
Gaixian, China (gī-shyĕn)	260-61	40°24'N	122°22'E
Galán, Cerro, mtn., Arg.	182-83	25°57'S	66°54'W
Galana, stm., Kenya	284-85	3°09'S	40°08'E
Galapagos Islands, is., Ec. (gä-lä'-pä-gōs ī'lȧndz) *see* Colón, Archipiélago de	184a	0°43'N	91°30'W
Galashiels, Scot., U.K. (găl-ȧ-shēlz)	206-07	55°38'N	2°50'W
Galaţi, Rom.	218-19	45°26'N	28°03'E
Galatina, Italy (gä-lä-tē'nä)	216-17	40°10'N	18°10'E
Galatz, Rom. *see* Galaţi	218-19	45°26'N	28°03'E
Galdhøpiggen, mtn., Nor.	208-09	61°37'N	8°17'E
Galeana, Mex. (gä-lå-ä'nä)	132-33	24°50'N	100°04'W
Galela, Indon.	268-69	1°50'N	127°50'E
Galena, Ak., U.S. (gȧ-lē'nȧ)	136	64°44'N	156°57'W
Galena, Il., U.S. (gȧ-lē'nȧ)	124-25	42°25'N	90°25'W
Galera, Punta, c., Chile	185	39°58'S	73°40'W
Galera, Punta, c., Ec.	184	0°49'N	80°03'W
Galesburg, Il., U.S. (gālz'bûrg)	124-25	40°57'N	90°22'W
Galeton, Pa., U.S. (gāl'tŭn)	126-27	41°44'N	77°39'W
Galich, Russia (gȧl'ĭch)	202-03	58°23'N	42°22'E
Galicia, hist. reg., Eur. (gȧ-lĭsh'ĭ-ȧ)	210-11	49°0'N	22°00'E
Galicja, hist. reg., Eur. *see* Galicia	210-11	49°0'N	22°00'E
Galilee, Sea of, lk., Isr. (sē ŭv găl'ĭ-lē)	248-49	32°48'N	35°35'E
Galion, Oh., U.S. (găl'ĭ-ŭn)	126-27	40°44'N	82°47'W
Galkynyş, Turkmen.	252-53	39°16'N	63°11'E
Gallatin, Mo., U.S. (găl'ȧ-tĭn)	130-31	39°55'N	93°58'W
Gallatin, Tn., U.S. (găl'ȧ-tĭn)	134-35	36°24'N	86°27'W
Galle, Sri L. (gäl)	256	6°02'N	80°13'E
Gallinas, Punta, c., Col. (pōō'n-tä-gä-lyē'näs)	178-79	12°28'N	71°40'W
Gallipoli, Italy (gäl-lē'pỏ-lē)	216-17	40°03'N	17°59'E
Gallipoli, Tur.	216-17	40°26'N	26°41'E
Gallipoli Peninsula, pen., Tur. (găl-lē'pỏ-lē pĕ-nĭn'sūlȧ)	216-17	40°20'N	26°30'E
Gallipolis, Oh., U.S. (găl-ĭ-pỏ-lēs)	126-27	38°49'N	82°12'W
Gällivare, Swe. (yĕl-ĭ-vär'ĕ)	200-01	67°08'N	20°41'E
Gallup, N.M., U.S. (găl'ŭp)	128-29	35°32'N	108°45'W
Galva, Il., U.S. (găl'vȧ)	124-25	41°10'N	90°02'W
Galveston, Tx., U.S. (găl'vĕs-tŭn)	132-33	29°18'N	94°48'W
Galveston Bay, b., Tx., U.S. (găl'vĕs-tŭn bā)	132-33	29°36'N	94°57'W
Galveston Island, i., Tx., U.S. (găl'vĕs-tŭn ī'lȧnd)	132-33	29°13'N	94°55'W
Gálvez, Arg.	187	32°02'S	61°13'W
Galway, Ire. (gôl'wā)	206-07	53°16'N	9°03'W
Gamba, China (gäm-bä)	254-55	28°17'N	88°31'E
Gambell, Ak., U.S.	136	63°47'N	171°44'W
Gambia, The, nation, Afr. (thȧ găm'bē-ȧ)	274	13°30'N	15°30'W
Gambier, Îles, is., Fr. Poly.	304-05	23°08'S	134°57'W
Gamboma, Congo (gäm-bō'mä)	284-85	1°53'S	15°51'E
Gamlakarleby, Fin. *see* Kokkola	200-01	63°50'N	23°09'E
Gamleby, Swe. (gäm'lĕ-bü)	208-09	57°55'N	16°23'E
Gan, stm., China (gän)	240-41	49°11'N	125°10'E
Gananoque, On., Can.	146-47	44°20'N	76°10'W
Gäncä, Azer.	245	40°41'N	46°21'E
Ganda, Ang.	286-87	13°02'S	14°39'E
Gandajika, D.R.C.	284-85	6°44'S	23°57'E
Gander, Nf./L., Can. (găn'dẽr)	148-49	48°57'N	54°35'W
Gander, stm., Nf./L., Can. (găn'dẽr)	148-49	49°29'N	54°24'W
Gander Lake, lk., Nf./L., Can. (găn'dẽr lāk)	148-49	48°57'N	54°39'W
Gāndhinagar, India	254-55	23°13'N	72°40'E
Gandia, Spain	214-15	38°58'N	0°11'W
Ganga, stm., Asia *see* Ganges	254-55	21°58'N	90°57'E
Gangānagar, India	254-55	29°55'N	73°52'E
Gangaw, Mya.	266-67	22°11'N	94°09'E
Gangdisê Shan, mts., China	254-55	31°0'N	82°00'E
Ganges, stm., Asia (găn'jēz)	254-55	21°58'N	90°57'E
Ganges, Mouths of the, mth., Asia (mouthz ŭv thȧ găn'jēz)	254-55	22°0'N	89°00'E
Gangneung, Kor., S.	263	37°46'N	128°54'E
Gangotri, India	254-55	30°60'N	78°59'E
Gangtok, India	254-55	27°19'N	88°38'E
Gangu, China	258-59	34°45'N	105°20'E
Gannan, China (gän-nän)	260-61	47°56'N	123°30'E
Gannett Peak, mtn., Wy., U.S. (găn'ĕt pēk)	122-23	43°11'N	109°39'W
Gansu, state, China (gän-sōō)	260-61	37°0'N	103°00'E
Ganzê, China	258-59	31°38'N	100°01'E
Ganzhou, China (gän-jō)	258-59	25°53'N	114°55'E
Gao, Mali (gä'ō)	280-81	16°16'N	0°02'W
Gao'an, China (gou-än)	258-59	28°26'N	115°23'E
Gaoyi, China (gou-yē)	260-61	37°37'N	114°36'E
Gaoyou, China (gou-yō)	258-59	32°47'N	119°26'E
Gaoyou Hu, lk., China (kä'ō-yōō'hōō)	258-59	32°50'N	119°20'E
Gap, Fr. (gȧp)	212-13	44°34'N	6°05'E
Garabogazköl Aylagy, b., Turkmen. *see* Kara-Bogaz-Gol Gulf	252-53	41°15'N	53°24'E
Garagum, des., Turkmen. *see* Kara Kum	244	39°0'N	60°00'E
Garagum Kanaly, can., Turkmen. *see* Kara-Kum Canal	252-53	37°34'N	65°41'E
Garanhuns, Braz. (gä-rän-yòNsh')	188d	8°54'S	36°29'W
Garber, Ok., U.S. (gȧr'bẽr)	130-31	36°26'N	97°35'W
Garden City, Ga., U.S. (gär'd'n sĭ'tė)	134-35	32°07'N	81°09'W
Garden City, Ks., U.S. (gär'd'n sĭ'tė)	130-31	37°58'N	100°52'W
Gardeyz, Afg.	252-53	33°36'N	69°13'E
Gardiner, Mt., U.S. (gärd'nẽr)	122-23	45°02'N	110°42'W
Gardner, Ks., U.S. (gärd'nẽr)	130-31	38°49'N	94°56'W
Gardner, Ma., U.S. (gärd'nẽr)	126-27	42°35'N	71°60'W
Gardner Canal, b., B.C., Can. (gärd'nẽr kå'năl)	142-43	53°28'N	128°18'W
Gardner Pinnacles, r., Hi., U.S. (gärd'nẽr pĭn'ȧ-k'lz)	137b	25°0'N	167°55'W
Gargždai, Lith. (gärgzh'dī)	208-09	55°43'N	21°24'E
Garibaldi, Mount, vol., B.C., Can. (mount gâr-ĭ-băl'dē)	142-43	49°51'N	122°59'W
Garissa, Kenya	284-85	0°27'S	39°39'E
Garland, Tx., U.S. (gär'lănd)	130-31	32°56'N	96°38'W
Garmisch-Partenkirchen, Ger. (gär'mēsh pär'tĕn-kēr'κĕn)	210-11	47°30'N	11°06'E
Garnett, Ks., U.S. (gär'nĕt)	130-31	38°17'N	95°14'W
Garoua, Camrn. (gär'wä)	282-83	9°19'N	13°23'E
Garqu Yan, China	258-59	33°54'N	92°19'E
Garrett, In., U.S. (găr'ĕt)	126-27	41°21'N	85°07'W
Garrison, N.D., U.S. (găr'ĭ-sŭn)	124-25	47°39'N	101°25'W
Garry Lake, lk., Nu., Can. (găr'ĭ lāk)	140-41	66°0'N	100°00'W
Garut, Indon.	268-69	7°12'S	107°54'E
Garwolin, Pol. (gär-vō'lēn)	210-11	51°54'N	21°38'E
Gary, In., U.S. (gā'rĭ)	126-27	41°36'N	87°21'W
Garyarsa, China	254-55	31°43'N	80°20'E
Garzón, Col. (gär-thōn')	178-79	2°12'N	75°38'W
Gas City, In., U.S. (găs sĭ'tē)	126-27	40°29'N	85°37'W
Gascogne, Golfe de, b., Eur. *see* Biscay, Bay of	212-13	44°0'N	4°00'W
Gash, stm., Afr.	288	16°45'N	35°54'E
Gaspé, Qc., Can.	148-49	48°49'N	64°29'W
Gasteiz, Spain	214-15	42°51'N	2°40'W
Gastonia, N.C., U.S. (găs-tỏ'nĭ-ȧ)	134-35	35°16'N	81°11'W
Gastre, Arg. (gäs-trĕ)	185	42°17'S	69°14'W
Gata, Cabo de, c., Spain (kȧ'bô-dĕ-gä'tä)	214-15	36°44'N	2°11'W
Gata, Sierra de, mts., Spain (syĕr'rȧ dā gä'tä)	214-15	40°16'N	6°44'W
Gatchina, Russia (gä-chē'nȧ)	208-09	59°33'N	30°08'E
Gates of the Arctic National Park and Preserve, n.p., Ak., U.S.	136	67°45'N	153°30'W
Gatesville, Tx., U.S. (gāts'vĭl)	132-33	31°25'N	97°44'W
Gatineau, Qc., Can. (gá'tē-nō)	146-47	45°29'N	75°38'W
Gatineau, stm., Qc., Can. (gá'tē-nō)	146-47	45°27'N	75°42'W
Gauer Lake, lk., Mb., Can.	144-45	57°0'N	97°50'W
Gauja, stm., Eur. (gå'ò-yä)	208-09	57°09'N	24°17'E
Gaustatoppen, mtn., Nor.	208-09	59°50'N	8°35'E
Gávdos, i., Grc. (gäv'dỏs)	216a	34°50'N	24°06'E
Gävle, Swe. (yĕv'lĕ)	208-09	60°40'N	17°10'E
Gavrilov-Yam, Russia (gá'vrê-lôf yäm')	218-19	57°18'N	39°52'E
Gaxun Nur, lk., China	260-61	42°22'N	100°34'E
Gaya, India (gŭ'yä)(gī'ȧ)	254-55	24°48'N	85°00'E
Gaylord, Mi., U.S. (gā'lôrd)	126-27	45°02'N	84°40'W
Gaylord, Mn., U.S. (gā'lôrd)	124-25	44°33'N	94°14'W
Gayndah, Austl. (gān'däh)	301	25°38'S	151°36'E
Gayny, Russia	202-03	60°18'N	54°19'E
Gaza, Gaza (gä'zȧ) (gā'zȧ)	248-49	31°30'N	34°28'E
Gazanjyk, Turkmen.	252-53	39°15'N	55°32'E
Gaziantep, Tur. (gä-zē-än'tĕp)	248-49	37°04'N	37°23'E
Gazimağusa, Cyp.	248-49	35°07'N	33°57'E
Gbadolite, D.R.C.	284-85	4°15'N	21°00'E
Gbanga, Lib.	282-83	7°00'N	9°29'W
Gboko, Nig.	282-83	7°20'N	8°60'E
Gdańsk, Pol. (g'dänsk)	210-11	54°21'N	18°38'E
Gdov, Russia (g'dôf')	208-09	58°45'N	27°49'E
Gdynia, Pol. (g'dĕn'yȧ)	210-11	54°32'N	18°31'E
Geary, Ok., U.S. (gē'rĭ)	130-31	35°38'N	98°19'W
Gediz, stm., Tur.	216-17	38°36'N	26°48'E
Geelong, Austl. (jē-lông')	301	38°08'S	144°21'E
Geeveston, Austl.	301	43°10'S	146°55'E
Gefle, Swe. *see* Gävle	208-09	60°40'N	17°10'E
Geita, Tan.	289	2°52'S	32°10'E
Gejiu, China (gŭ-jĭo)	258-59	23°22'N	103°09'E
Gela, Italy	216-17	37°04'N	14°15'E
Gelasa, Selat, strt., Indon.	266-67	2°55'S	107°13'E
Gelibolu, Tur. (gĕ-lĭb'ỏ-lò) *see* Gallipoli	216-17	40°26'N	26°41'E
Gelibolu Yarımadası, pen., Tur. *see* Gallipoli Peninsula	216-17	40°20'N	26°30'E
Gemena, D.R.C.	284-85	3°14'N	19°47'E
Gemlik, Tur. (gĕm'lĭk)	216-17	40°26'N	29°09'E
Gemsbok National Park, n.p., Bots.	286-87	25°15'S	21°10'E
Gen, stm., China	260-61	50°15'N	119°21'E
Genalē, stm., Afr.	284-85	0°15'S	42°39'E
General Acha, Arg.	187	37°23'S	64°36'W
General Alvear, Arg. (hĕ-nĕ-räl' äl-vĕ-ä'r)	185	34°59'S	67°42'W
General Alvear, Arg. (hĕ-nĕ-räl' äl-vĕ-ä'r)	187	36°01'S	60°01'W
General Belgrano, Arg. (hĕ-nĕ-räl' bĕl-grä'nỏ)	187	35°46'S	58°29'W
General Carrera, Lago, lk., S.A.	185	46°26'S	71°40'W
General Cepeda, Mex. (hĕ-nĕ-räl' sĕ-pĕ'dä)	132-33	25°23'N	101°28'W
General Conesa, Arg. (hĕ-nĕ-räl' kô-nĕ'sä)	185	40°07'S	64°26'W
General Eugenio A. Garay, Para.	182-83	20°30'S	62°11'W
General Guido, Arg. (hĕ-nĕ-räl' gē'dô)	187	36°40'S	57°48'W
General Juan Madariaga, Arg.	187	37°00'S	57°09'W
General La Madrid, Arg.	187	37°15'S	61°17'W
General Lavalle, Arg. (hĕ-nĕ-räl' lä-vä'l-yĕ)	187	36°25'S	56°57'W
General Levalle, Arg.	187	34°01'S	63°55'W
General Manuel Belgrano, Cerro, mtn., Arg.	182-83	29°01'S	67°50'W
General Pico, Arg. (hĕ-nĕ-räl' pē'kô)	187	35°40'S	63°46'W
General Pinedo, Arg.	187	27°19'S	61°17'W
General Roca, Arg. (hĕ-nĕ-räl' rô-kä)	185	39°01'S	67°35'W
General San Martín, Arg. (hĕ-nĕ-räl' sän-mär-tē'n)	187	34°34'S	58°33'W
General Santos, Phil.	270	6°07'N	125°10'E
General Viamonte, Arg. (hĕ-nĕ-räl' veạ'môn-tĕ)	187	34°60'S	61°02'W
General Villegas, Arg.	187	35°02'S	63°01'W
Geneseo, Il., U.S. (jē-nĕsĕō)	124-25	41°27'N	90°09'W
Geneva, Switz. (jĕ-nē'vȧ)	210-11	46°12'N	6°09'E
Geneva, Al., U.S. (jĕ-nē'vȧ)	134-35	31°02'N	85°52'W
Geneva, In., U.S. (jĕ-nē'vȧ)	126-27	40°35'N	84°57'W
Geneva, Ne., U.S. (jĕ-nē'vȧ)	130-31	40°32'N	97°36'W
Geneva, N.Y., U.S. (jĕ-nē'vȧ)	126-27	42°52'N	76°59'W
Geneva, Oh., U.S. (jĕ-nē'vȧ)	126-27	41°48'N	80°56'W
Geneva, Lake, lk., Eur. (lāk jĕ-nē'vȧ)	210-11	46°24'N	6°22'E
Genève, Switz. *see* Geneva	210-11	46°12'N	6°09'E
Genève, Lac de, lk., Eur. *see* Geneva, Lake	210-11	46°24'N	6°22'E
Genf, Switz. *see* Geneva	210-11	46°12'N	6°09'E
Genil, stm., Spain (hå-nēl')	214-15	37°42'N	5°19'W

n-sing; ŋ-baŋk; ɴ-nasalized n; nŏd; cŏmmit; ōld; ôbey; ôrder; oi-boil; fōōd; ŏŏ-as oo in foot; ou-out; s-soft; sh-dish; th-thin; pūre; ûnite; ûrn; stŭd; circŭs; ü-as in French tu; '-indeterminate vowel.

358 Genoa – Goodenough Island PRONOUNCING INDEX

Feature (Pronunciation)	Page	Lat.	Long.
Genoa, Italy (jen´ō-ȧ)	216-17	44°25′N	8°57′E
Genova, Italy see Genoa	216-17	44°25′N	8°57′E
Genova, Golfo di, b., Italy (gôl-fō-dē-jĕn´ō-vä)	216-17	44°10′N	8°55′E
Genovesa, Isla, i., Ec. (ĕ´s-lä-gĕ-nō-vĕ-sä)	184a	0°20′N	89°57′W
Gensan, Kor., N. see Wŏnsan	263	39°09′N	127°26′E
Geogeumdo, i., Kor., S.	263	34°27′N	127°11′E
Geographe Bay, b., Austl. (jē-ô-graf´ bā)	296-97	33°35′S	115°15′E
Geojedo, i., Kor., S.	263	34°52′N	128°37′E
George, S. Afr.	286-87	33°58′S	22°27′E
George, stm., Qc., Can.	140-41	58°46′N	66°08′W
George, Lake, lk., Ug. (lāk jôrg)	289	0°02′N	30°12′E
George, Lake, lk., Fl., U.S. (lāk jôr-ĭj)	134-35	29°17′N	81°36′W
Georgetown, On., Can. (jôr-ĭj-toun)	146-47	43°39′N	79°55′W
Georgetown, P.E., Can. (jôr-ĭj-toun)	148-49	46°11′N	62°32′W
Georgetown, Gam.	282-83	13°33′N	14°46′W
George Town, Malay.	266-67	5°25′N	100°20′E
Georgetown, De., U.S. (jôrg-toun)	126-27	38°41′N	75°23′W
Georgetown, Il., U.S. (jôrg-toun)	126-27	39°58′N	87°38′W
Georgetown, Ky., U.S. (jôrg-toun)	126-27	38°12′N	84°34′W
Georgetown, Oh., U.S. (jôrg-toun)	126-27	38°51′N	83°52′W
Georgetown, S.C., U.S. (jôr-ĭj-toun)	134-35	33°23′N	79°18′W
Georgetown, Tx., U.S. (jôrg-toun)	132-33	30°38′N	97°41′W
George Town, nat. cap., Cay. Is. (jôr-ĭj-toun)	154-55	19°18′N	81°22′W
Georgetown, nat. cap., Guy. (jôrj´toun)	178-79	6°48′N	58°09′W
George Washington Birthplace National Monument, n.p., Va., U.S. (jôrj wŏsh´ĭng-tŭn bûrth´plās nāsh´ŭn-ăl mŏn´ū-mĕnt)	126-27	38°11′N	76°56′W
George Washington Carver National Monument, n.p., Mo., U.S. (jôrg wăsh-ĭng-tŭn kär´vēr nāsh´ŭn-ăl mŏn´ū-mĕnt)	130-31	37°00′N	94°21′W
George West, Tx., U.S. (jôrg wĕst)	132-33	28°20′N	98°07′W
Georgia, nation, Asia (jôr´ji-ă)	245	42°0′N	44°00′E
Georgia, state, U.S. (jôr´ji-ă)	118-19	32°50′N	83°15′W
Georgiana, Al., U.S. (jôr-jē-ăn´ȧ)	134-35	31°38′N	86°45′W
Georgian Bay, b., On., Can.	146-47	45°15′N	80°50′W
Georgian Bay Islands National Park, n.p., On., Can.	146-47	44°52′N	79°52′W
Georgiyevsk, Russia (gyôr-gyĕfsk´)	245	44°09′N	43°29′E
Gera, Ger. (gā´rä)	210-11	50°52′N	12°05′E
Geral, Serra, mts., Braz. (sĕr´rȧ zhā-räl´)	182-83	26°30′S	50°30′W
Geraldton, Austl. (jĕr´ăld-tŭn)	294-95	28°46′S	114°37′E
Geraldton, On., Can.	138-39	49°41′N	86°60′W
Gereshk, Afg.	252-53	31°49′N	64°34′E
Gering, Ne., U.S. (gē´rĭng)	124-25	41°48′N	103°40′W
Gerlachovský štít, mtn., Slvk.	210-11	49°11′N	20°09′E
Germantown, Tn., U.S. (jûr´mȧn-toun)	134-35	35°06′N	89°49′W
Germantown, Wi., U.S. (jûr´mȧn-toun)	126-27	43°14′N	88°07′W
Germany, nation, Eur. (jûr´mȧ-nĭ)	190-91	51°0′N	10°00′E
Germiston, S. Afr. (jûr´mĭs-tŭn)	292c	26°13′S	28°11′E
Gerona, Spain see Girona	214-15	41°59′N	2°49′E
Getafe, Spain (hā-tä´fā)	214-15	40°19′N	3°44′W
Gettysburg, S.D., U.S. (gĕt´ĭs-bûrg)	124-25	45°01′N	99°57′W
Geumgang, stm., Kor., S.	263	35°60′N	126°42′E
Ghaapplato, plat., S. Afr.	286-87	27°29′S	24°19′E
Ghadāmis, Libya	204-05	30°12′N	9°33′E
Ghāghara, stm., Asia	254-55	25°45′N	84°48′E
Ghāghra, stm., Asia see Ghāghara	254-55	25°45′N	84°48′E
Ghana, nation, Afr. (gän´ä)	274	8°0′N	1°00′W
Ghanzi, Bots. (gän´zē)	286-87	21°42′S	21°39′E
Ghardaïa, Alg. (gär-dä´ē-ä)	280-81	32°33′N	3°40′E
Gharm, Taj.	252-53	39°02′N	70°23′E
Gharyān, Libya	204-05	32°10′N	13°01′E
Ghāt, Libya	280-81	24°56′N	10°12′E
Ghawdex, i., Malta see Gozo	216b	36°03′N	14°15′E
Ghazal, Bahr el, stm., Chad (bär ĕl ghä-zäl´)	280-81	13°05′N	15°20′E
Ghāziābād, India	254-55	28°40′N	77°26′E
Ghaznī, Afg.	252-53	33°33′N	68°25′E
Ghazzah, Gaza (gä´ziä) see Gaza	248-49	31°30′N	34°28′E
Ghijduwon, Uzb.	252-53	40°06′N	64°41′E
Ghorīān, Afg.	252-53	34°21′N	61°29′E
Gibara, Cuba (hē-bä´rä)	154-55	21°07′N	76°08′W
Gibraleón, Spain (hē-brä-lå-ōn´)	214-15	37°23′N	6°58′W
Gibraltar, dep., Eur. (gĭ-bräl-tă´r)	190-91	36°08′N	5°21′W
Gibraltar, nat. cap., Gib. (gĭ-bräl-tă´r)	214-15	36°08′N	5°21′W
Gibraltar, Estrecho de, strt., see Gibraltar, Strait of	214-15	35°57′N	5°36′W
Gibraltar, Strait of, strt., (stāt ŭv gĭ-bräl-tä´r)	214-15	35°57′N	5°36′W
Gibson City, Il., U.S. (gĭb´sŭn sĭ´tê)	126-27	40°28′N	88°22′W
Gibson Desert, des., Austl. (gĭb´sŭn dĕs´ērt)	296-97	24°30′S	126°00′E
Giddings, Tx., U.S. (gĭd´ĭngz)	132-33	30°11′N	96°56′W
Gien, Fr. (zhē-ăɴ´)	212-13	47°41′N	2°38′E
Gießen, Ger. (gēs´sĕn)	210-11	50°35′N	8°40′E
Gifu, Japan (gē´fōō)	265	35°25′N	136°45′E
Gijón, Spain (hē-hōn´)	214-15	43°32′N	5°40′W
Gila, stm., U.S. (hē´lȧ)	120-21	32°43′N	114°33′W
Gila Bend, Az., U.S. (hē´lȧ bĕnd)	128-29	32°57′N	112°43′W
Gila Cliff Dwellings National Monument, n.p., N.M., U.S. (hē´lȧ klĭf dwĕl´ĭngz nāsh´ŭn-ăl mŏn´ū-mĕnt)	128-29	33°02′N	108°16′W
Gilbert, Az., U.S. (gĭl´bērt)	128-29	33°21′N	111°47′W
Gilbert, Mn., U.S. (gĭl´bērt)	124-25	47°29′N	92°28′W
Gilbert, Mount, mtn., B.C., Can. (mount gĭl-bērt)	142-43	50°54′N	124°17′W
Gilbert Islands, is., Kir. (gĭl-bērt ī´lȧndz) see Kiribati	304-05	0°30′S	174°00′E
Gilbués, Braz.	180-81	9°50′S	45°21′W
Gilford Island, i., B.C., Can. (gĭl´fērd ī´lȧnd)	142-43	50°45′N	126°20′W
Gilgandra, Austl.	301	31°43′S	148°40′E
Gilgit, Pak. (gĭl´gĭt)	252-53	35°53′N	74°21′E
Gilgit, stm., Pak.	252-53	35°42′N	74°38′E
Gil Island, i., B.C., Can. (gĭl ī´lȧnd)	142-43	53°11′N	129°15′W
Gillam, Mb., Can.	144-45	56°21′N	94°43′W
Gillette, Wy., U.S. (jĭ-lĕt´)	122-23	44°18′N	105°30′W
Gillingham, Eng., U.K. (gĭl´ĭng ăm)	206-07	51°23′N	0°34′E
Gilman, Il., U.S. (gĭl´măn)	126-27	40°46′N	87°59′W
Gilmer, Tx., U.S. (gĭl´mēr)	130-31	32°44′N	94°57′W
Gīlo, stm., Eth.	284-85	8°07′N	33°11′E
Gilroy, Ca., U.S. (gĭl-roi´)	128-29	37°01′N	121°34′W
Giluwe, Mount, mtn., Pap. N. Gui.	302	6°02′S	143°51′E
Gilyuy, stm., Russia	236-37	53°59′N	127°27′E
Gimcheon, Kor., S.	263	36°07′N	128°07′E
Gimli, Mb., Can. (gĭm´lē)	144-45	50°38′N	96°59′W
Gioia del Colle, Italy (jô´yä dĕl kôl´lā)	216-17	40°48′N	16°55′E
Girardot, Col. (hē-rär-dōt´)	188c	4°18′N	74°47′W
Giresun, Tur. (ghĕr´ĕ-sòn´)	202-03	40°55′N	38°24′E
Girga, Egypt	291b	26°20′N	31°53′E
Girīdīh, India (jē-rē-dē)	254-55	24°11′N	86°18′E
Girona, Spain	214-15	41°59′N	2°49′E
Girvan, Scot., U.K. (gûr´văn)	206-07	55°15′N	4°52′W
Gisborne, N.Z. (gĭz´bŭrn)	303	38°40′S	178°01′E
Gisenyi, Rw.	289	1°42′S	29°16′E
Gisors, Fr. (zhē-zôr´)	212-13	49°17′N	1°47′E
Gitarama, Rw.	289	2°04′S	29°44′E
Gitega, Bdi.	289	3°21′S	29°54′E
Giurgiu, Rom. (jòr´jò)	216-17	43°54′N	25°58′E
Givet, Fr. (zhē-vĕ´)	212-13	50°08′N	4°50′E
Giyon, Eth.	292d	8°32′N	37°59′E
Giza, Egypt	291b	30°01′N	31°13′E
Gizo, Sol. Is.	306e	8°06′S	156°50′E
Giżycko, Pol. (gĭ´zhĭ-ko)	210-11	54°02′N	21°46′E
Gjoa Haven, Nu., Can.	138-39	68°39′N	95°55′W
Gjøvik, Nor. (gyû´vĕk)	208-09	60°48′N	10°41′E
Glace Bay, N.S., Can. (gläs bā)	148-49	46°13′N	59°58′W
Glacier Bay National Park and Preserve, n.p., Ak., U.S.	136	59°04′N	136°36′W
Glacier National Park, n.p., B.C., Can. (glā´shēr nāsh´ŭn-ăl pärk)	142-43	51°15′N	117°35′W
Glacier National Park, n.p., Mt., U.S.	122-23	48°35′N	113°40′W
Glacier Peak, vol., Wa., U.S. (glā´shēr pēk) (glā´shēr pēk)	122-23	48°07′N	121°07′W
Gladstone, Austl. (glăd´stōn)	302	23°51′S	151°15′E
Gladstone, Austl. (glăd´stōn)	301	33°17′S	138°21′E
Gladstone, Mi., U.S. (glăd´stōn)	126-27	45°51′N	87°01′W
Gladstone, Mo., U.S. (glăd´stōn)	130-31	39°14′N	94°35′W
Gladwin, Mi., U.S. (glăd´wĭn)	126-27	43°59′N	84°29′W
Glåma, stm., Nor. see Glomma	200-01	59°11′N	10°58′E
Glasgow, Scot., U.K. (glås´gō)	206-07	55°53′N	4°15′W
Glasgow, Ky., U.S.	134-35	37°00′N	85°55′W
Glasgow, Mt., U.S.	122-23	48°12′N	106°38′W
Glauchau, Ger. (glou´ĸou)	210-11	50°49′N	12°33′E
Glazov, Russia (glä´zôf)	202-03	58°08′N	52°39′E
Gleiwitz, Pol. see Gliwice	210-11	50°17′N	18°40′E
Glen Canyon, val., U.S. (glĕn kăn´yŭn)	128-29	37°10′N	110°50′W
Glencoe, S. Afr. (glĕn-cō)	292c	28°12′S	30°06′E
Glendale, Az., U.S. (glĕn´dāl)	128-29	33°32′N	112°12′W
Glendale, Ca., U.S. (glĕn´dāl)	128-29	34°08′N	118°14′W
Glendive, Mt., U.S. (glĕn´dīv)	122-23	47°06′N	104°43′W
Glen Innes, Austl. (glĕn ĭn´ĕs)	301	29°44′S	151°44′E
Glenns Ferry, Id., U.S. (glĕns fĕr´ē)	122-23	42°58′N	115°18′W
Glenrock, Wy., U.S. (glĕn´rŏk)	122-23	42°52′N	105°53′W
Glens Falls, N.Y., U.S. (glĕnz fôlz)	126-27	43°19′N	73°39′W
Glittertinden, mtn., Nor.	208-09	61°39′N	8°33′E
Gliwice, Pol. (gwĭ-wĭt´sĕ)	210-11	50°17′N	18°40′E
Globe, Az., U.S. (glōb)	128-29	33°24′N	110°47′W
Glomma, stm., Nor.	200-01	59°11′N	10°58′E
Glorieuses, Îles, is., Reu.	286-87	11°30′S	47°20′E
Glorioso Islands, is., Reu. see Glorieuses, Îles	286-87	11°30′S	47°20′E
Gloucester, Eng., U.K. (glŏs´tēr)	206-07	51°53′N	2°14′W
Gloversville, N.Y., U.S. (glŭv´ērz-vĭl)	126-27	43°03′N	74°21′W
Glovertown, Nf./L., Can. (glŭv´ēr-toun)	148-49	48°40′N	54°03′W
Glückstadt, Ger. (glük-shtät)	210-11	53°47′N	9°26′E
Gmunden, Aus. (g'mòn´dĕn)	210-11	47°55′N	13°47′E
Gnesen, Pol. see Gniezno	210-11	52°32′N	17°37′E
Gniezno, Pol. (g'nyåz´nō)	210-11	52°32′N	17°37′E
Gnjilane, Kos. (gnyē´lå-nĕ)	216-17	42°28′N	21°29′E
Goa, state, India (gō´ȧ)	256	15°20′N	74°00′E
Goālpāra, India	254-55	26°10′N	90°37′E
Goba, Eth. (gō´bä)	292d	7°00′N	39°59′E
Gobabis, Nmb. (gō-bä´bĭs)	286-87	22°27′S	18°58′E
Gobernador Gregores, Arg.	185	48°46′S	70°15′W
Gobernador Vera, Arg. see Vera	187	29°28′S	60°13′W
Gobi Desert, des., Asia (gō´be dĕs´ērt)	260-61	43°0′N	105°00′E
Goch, Ger. (gôk)	210-11	51°41′N	6°09′E
Godāvari, stm., India (gô-dä´vŭ-rē)	256	16°59′N	81°47′E
Goderich, On., Can. (gōd´rĭch)	146-47	43°45′N	81°42′W
Godfrey, Il., U.S. (gŏd´frē)	130-31	38°57′N	90°11′W
Godhra, India	254-55	22°46′N	73°37′E
Godoy Cruz, Arg.	188e	32°55′S	68°50′W
Gods, stm., Mb., Can. (gŏdz)	144-45	56°23′N	92°51′W
Gods Lake, lk., Mb., Can.	144-45	54°43′N	94°14′W
Godwin Austen, mtn., Asia see K2	252-53	35°53′N	76°30′E
Goeie Hoop, Kaap die, c., S. Afr. see Good Hope, Cape of	286-87	34°21′S	18°28′E
Goiana, Braz.	188d	7°33′S	34°59′W
Goiânia, Braz. (gô-vá´nyä)	182-83	16°40′S	49°16′W
Goiás, Braz. (gô-yá´s)	182-83	15°55′S	50°07′W
Goiás, state, Braz. (gô-yá´s)	180-81	16°0′S	50°00′W
Gökçeada, i., Tur.	216-17	40°10′N	25°50′E
Gökova Körfezi, b., Tur.	216-17	36°54′N	27°51′E
Göksu, stm., Tur. (gûk´sōō´)	248-49	36°19′N	34°03′E
Gol, Nor. (gûl)	208-09	60°42′N	8°57′E
Gölbaşı, Tur.	247	37°47′N	37°38′E
Gold Coast, Austl. see Southport	301	27°58′S	153°25′E
Golden, B.C., Can. (gōl´dĕn)	142-43	51°18′N	116°58′W
Golden, Co., U.S. (gōl´dĕn)	128-29	39°45′N	105°13′W
Goldendale, Wa., U.S. (gōl´dĕn-dāl)	122-23	45°49′N	120°50′W
Golden Hinde, mtn., B.C., Can. (gōl´dĕn hīnd)	142-43	49°40′N	125°45′W
Goldsboro, N.C., U.S. (gōldz-bûr´ō)	134-35	35°23′N	77°60′W
Goldthwaite, Tx., U.S. (gōld´thwāt)	132-33	31°27′N	98°34′W
Golfito, C.R. (gôl-fē´tō)	161	8°38′N	83°10′W
Goliad, Tx., U.S. (gō-lĭ-ăd´)	132-33	28°40′N	97°23′W
Golmud, China	260-61	36°25′N	94°54′E
Goma, D.R.C.	289	1°41′S	29°13′E
Gombe, Nig.	282-83	10°17′N	11°10′E
Gomel', Bela. see Homel'	218-19	52°26′N	30°59′E
Gómez Palacio, Mex. (gō´mĕz pä-lä´syō)	132-33	25°35′N	103°30′W
Gonābād, Iran	252-53	34°21′N	58°41′E
Gonaïves, Haiti (gō-nȧ-ēv´)	154-55	19°27′N	72°41′W
Gonam, stm., Russia	236-37	57°19′N	131°15′E
Gonarezhou National Park, n.p., Zimb.	286-87	21°34′S	31°56′E
Gonâve, Île de la, i., Haiti (ēl-dē-lȧ-gō-nȧv´)	154-55	18°51′N	73°03′W
Gonbad-e Kāvūs, Iran	252-53	37°15′N	55°10′E
Gonda, India	254-55	27°08′N	81°58′E
Gondar, Eth. see Gonder	288	12°37′N	37°28′E
Gonder, Eth.	288	12°37′N	37°28′E
Gondia, India	254-55	21°28′N	80°12′E
Gongbo'gyamda, China	258-59	29°55′N	93°26′E
Gongga Shan, mtn., China (gôŋ-gä shän)	258-59	29°35′N	101°51′E
Gongola, stm., Nig.	282-83	9°29′N	12°03′E
Gongxian, China	260-61	34°48′N	113°03′E
Gongzhuling, China	260-61	43°30′N	124°49′E
Gonzales, La., U.S. (gŏn-ză´lĕz)	134-35	30°14′N	90°55′W
Gonzales, Tx., U.S. (gŏn-ză´lĕz)	132-33	29°30′N	97°27′W
Goodenough Island, i., Pap. N. Gui.	302	9°20′S	150°15′E

ăt; finăl; rāte; senåte; ärm; åsk; sofȧ; fâre; ch-choose; dh-as th in other; bē; ēvent; bĕt; recĕnt; cratēr; g-gō; gh-guttural g; bĭt; ĭ-short neutral; rīde; ĸ-guttural k as ch in German ich;

PRONOUNCING INDEX
Good Hope, Cape of – Great Grimsby 359

Feature (Pronunciation)	Page	Lat.	Long.
Good Hope, Cape of, c., S. Afr. (kāp ŏv go͝od hōp)	286-87	34°21'S	18°28'E
Good Hope Mountain, mtn., B.C., Can. (go͝od hōp moun´tĭn)	142-43	51°09'N	124°10'W
Gooding, Id., U.S. (go͝od´ĭng)	122-23	42°57'N	114°43'W
Goodland, Ks., U.S. (go͝od´lănd)	130-31	39°20'N	101°43'W
Goole, Eng., U.K. (go͞ol)	206-07	53°42'N	0°53'W
Goondiwindi, Austl.	301	28°32'S	150°19'E
Goose Lake, lk., U.S. (go͞os lāk)	122-23	41°57'N	120°25'W
Goqên, China	258-59	29°09'N	97°14'E
Gorakhpur, India (gō´rŭk-po͞or)	254-55	26°46'N	83°22'E
Gorda, Punta, c., Cuba (po͞o´n-tä-gôr-dä)	154-55	22°23'N	82°09'W
Gorgān, Iran	252-53	36°51'N	54°26'E
Gorgona, Isla, i., Col.	178-79	2°58'N	78°11'W
Gorgona, Isola di, i., Italy (gôr-gō´nä)	216-17	43°26'N	9°54'E
Gori, Geor. (gō´rē)	245	41°59'N	44°06'E
Gorica, Italy *see* Gorizia	216-17	45°57'N	13°38'E
Gorinchem, Neth. (gō´rĭn-κĕm)	206-07	51°50'N	5°01'E
Gorizia, Italy (gô-rē´tsē-yä)	216-17	45°57'N	13°38'E
Gorkhā, Nepal	254-55	28°0'N	84°37'E
Gorky, Russia *see* Nizhniy Novgorod	202-03	56°19'N	44°01'E
Gorki Reservoir, res., Russia (gôr´kē rĕ´sĕr-vwär) *see* Gor'kovskoye Vodokhranilishche	202-03	57°02'N	43°10'E
Gor'kovskoye Vodokhranilishche, res., Russia	202-03	57°02'N	43°10'E
Gorlice, Pol. (gôr-lē´tsĕ)	210-11	49°39'N	21°10'E
Görlitz, Ger. (gûr´lĭts)	210-11	51°09'N	14°59'E
Gorlovka, Ukr. *see* Horlivka	218-19	48°20'N	38°03'E
Gorna Oryakhovitsa, Blg. (gôr´nä-ôr-yĕk´ô-vē-tsȧ)	216-17	43°08'N	25°42'E
Gornji Milanovac, Serb. (gôrn´yē-mē´la-nô-väts)	216-17	44°01'N	20°27'E
Gorno-Altaysk, Russia (gôr´nŭ´ŭl-tīsk´)	244	51°58'N	85°51'E
Gornozavodsk, Russia	264	46°33'N	141°51'E
Gorodets, Russia	202-03	56°39'N	43°28'E
Goroka, Pap. N. Gui.	302	6°05'S	145°24'E
Gorontalo, Indon. (gō-rōn-tä´lō)	268-69	0°32'N	123°04'E
Görz, Italy *see* Gorizia	216-17	45°57'N	13°38'E
Gorzów Wielkopolski, Pol. (gō-zho͞ov´vyĕl-ko-pōl´skē)	210-11	52°44'N	15°14'E
Gosford, Austl.	301	33°25'S	151°21'E
Goshen, In., U.S. (gō´shĕn)	126-27	41°35'N	85°50'W
Goslar, Ger. (gŏs´lär)	210-11	51°55'N	10°26'E
Gostivar, Mac. (gos´tē-vär)	216-17	41°48'N	20°55'E
Gostynin, Pol. (gôs tē´nĭn)	210-11	52°26'N	19°28'E
Göta, stm., Swe. (gōĕtä)	208-09	57°41'N	11°53'E
Göteborg, Swe.	208-09	57°43'N	11°58'E
Gotha, Ger. (gō´tȧ)	210-11	50°57'N	10°42'E
Gothenburg, Swe. *see* Göteborg	208-09	57°43'N	11°58'E
Gothenburg, Ne., U.S. (gŏth´ĕn-bûrg)	124-25	40°56'N	100°10'W
Gotland, i., Swe.	208-09	57°30'N	18°33'E
Gotō-rettō, is., Japan	264	32°50'N	129°00'E
Gotska Sandön, i., Swe.	208-09	58°22'N	19°16'E
Göttingen, Ger. (gût´ĭng-ĕn)	210-11	51°32'N	9°56'E
Gouda, Neth. (gou´dä)	206-07	52°01'N	4°42'E
Gouin, Réservoir, res., Qc., Can.	146-47	48°37'N	74°55'W
Goulburn, Austl. (gōl´bŭrn)	301	34°45'S	149°43'E
Goundam, Mali (go͞on-däɴ´)	280-81	16°25'N	3°40'W
Gouverneur, N.Y., U.S. (gŭv-ẽr-no͞or´)	126-27	44°20'N	75°28'W
Governador Valadares, Braz. (gô-vĕr-nä-dō-´r vä-lä-dä´rĕs)	186	18°53'S	41°58'W
Goya, Arg. (gō´yä)	187	29°09'S	59°15'W
Goyania, Braz. *see* Goiânia	182-83	16°40'S	49°16'W
Göyçay, Azer. (gĕ-ôk´chī)	245	40°39'N	47°45'E
Gozo, i., Malta	216b	36°03'N	14°15'E
Graaff-Reinet, S. Afr. (gräf´rī´nĕt)	286-87	32°16'S	24°33'E
Gračac, Cro. (grä´chäts)	216-17	44°18'N	15°50'E
Graceville, Fl., U.S. (grās´vĭl)	134-35	30°57'N	85°31'W
Gracias a Dios, Cabo, c., N.A.	161	14°60'N	83°10'W
Graciosa, i., Port. (grä-syō´sä)	215c	39°04'N	28°00'W
Gradačac, Bos. (gra-dä´chats)	216-17	44°53'N	18°26'E
Gradaús, Braz.	180-81	7°43'S	51°10'W
Grænlandshav, s., *see* Greenland Sea	314	77°0'N	1°00'W
Grænlandssund, strt., *see* Denmark Strait	95	67°0'N	25°00'W
Grafton, Austl. (graf´tŭn)	301	29°42'S	152°56'E
Grafton, N.D., U.S. (graf´tŭn)	124-25	48°25'N	97°25'W
Grafton, W.V., U.S. (graf´tŭn)	126-27	39°20'N	80°01'W
Grafton, Cape, c., Austl.	302	16°53'S	145°56'E
Graham, N.C., U.S. (grā´ăm)	134-35	36°04'N	79°24'W
Graham, Tx., U.S. (grā´ăm)	130-31	33°07'N	98°35'W
Graham Island, i., B.C., Can. (grā´ăm ī´lănd)	142-43	53°47'N	132°34'W
Grahamstad, S. Afr. *see* Grahamstown	286-87	33°18'S	26°31'E
Grahamstown, S. Afr. (grā´ăms´toun)	286-87	33°18'S	26°31'E
Grajaú, Braz.	180-81	5°47'S	46°07'W
Grajaú, stm., Braz.	180-81	3°41'S	44°49'W
Grajewo, Pol. (grä-yā´vo)	210-11	53°39'N	22°28'E
Grampian Mountains, mts., Scot., U.K. (grăm´pĭ-ăn moun´tĭnz)	206-07	56°55'N	4°00'W
Grampians National Park, n.p., Austl.	301	37°15'S	142°25'E
Granada, Nic. (grä-nä´dhä)	161	11°56'N	85°58'W
Granada, Spain (grä-nä´dä)	214-15	37°11'N	3°36'W
Granbury, Tx., U.S. (grăn´bĕr-ĭ)	130-31	32°27'N	97°48'W
Granby, Qc., Can. (grăn´bĭ)	146-47	45°24'N	72°43'W
Granby, Co., U.S. (grăn´bĭ)	128-29	40°06'N	105°57'W
Granby, Mo., U.S. (grăn´bĭ)	130-31	36°55'N	94°15'W
Gran Canaria, i., Spain (grän´kä-nä´rē-ä)	215d	27°52'N	15°37'W
Gran Chaco, reg., S.A. (grän´chä´kō)	182-83	23°0'S	60°00'W
Grand, stm., On., Can. (grănd)	146-47	42°52'N	79°34'W
Grand, stm., Mi., U.S. (grănd)	126-27	43°04'N	86°14'W
Grand Bahama, i., Bah. (grănd bȧ-hä´mȧ)	154-55	26°38'N	78°25'W
Grand Bank, Nf./L., Can. (grănd băngk)	148-49	47°06'N	55°45'W
Grand-Bassam, C. Iv. (gräɴ bȧ-säɴ´)	282-83	5°13'N	3°45'W
Grand-Bourg, Guad. (gräɴ bo͞or´)	155b	15°54'N	61°19'W
Grand Canal, can., China (grănd kȧ´năl)	260-61	32°11'N	119°33'E
Grand Canyon, Az., U.S. (grănd kăn´yŭn)	128-29	36°02'N	112°10'W
Grand Canyon, val., Az., U.S. (grănd kăn´yŭn)	128-29	36°22'N	112°30'W
Grand Canyon National Park, n.p., Az., U.S. (grănd kăn´yŭn năsh´ŭn-ăl pärk)	128-29	36°20'N	112°53'W
Grand Canyon-Parashant National Monument, n.p., Az., U.S.	128-29	36°20'N	113°44'W
Grand Cayman, i., Cay. Is. (grănd kā´măn) (grănd kī-män´)	154-55	19°20'N	81°15'W
Grand Coulee Dam, d., Wa., U.S. (grănd ko͞o´lē dăm)	122-23	47°57'N	119°01'W
Grande, stm., Bol. (grän´dĕ)	182-83	15°50'S	64°47'W
Grande, stm., Braz. (grän´dĕ)	180-81	11°05'S	43°09'W
Grande, stm., Braz. (grän´dĕ)	182-83	20°08'S	51°00'W
Grande, Bahía, b., Arg. (bä ē´ä grän´dĕ)	185	51°15'S	68°31'W
Grande, Cuchilla, mts., Ur. (ko͞o-chē´l-yä grän´dĕ)	187	33°25'S	55°06'W
Grande, Ilha, i., Braz. (ē´lä-grän´dĕ)	186	23°09'S	44°14'W
Grande, Rio, stm., N.A. (rē´ō grän´dā) *see* Rio Grande	120-21	25°57'N	97°09'W
Grande Cayemite, i., Haiti	154-55	18°37'N	73°45'W
Grande Comore, i., Com. *see* Njazidja	286-87	11°35'S	43°20'E
Grande de Santiago, stm., Mex. (grä´n-dĕ-dĕ-sän-tya´gō)	158-59	21°37'N	105°28'W
Grande do Gurupá, Ilha, i., Braz.	180-81	1°0'S	51°30'W
Grande Prairie, Ab., Can. (grănd prâr´ĭ)	142-43	55°10'N	118°48'W
Grand Erg de Bilma, des., Niger	280-81	18°30'N	14°00'E
Grand Erg Occidental, des., Alg.	280-81	30°56'N	1°35'E
Grand Erg Oriental, des., Alg.	280-81	30°30'N	7°00'E
Grandes, Salinas, pl., Arg.	182-83	30°06'S	65°14'W
Grandes Antillas, Islas, is., N.A. *see* Greater Antilles	154-55	20°0'N	74°00'W
Grande-Terre, i., Guad.	155b	16°19'N	61°22'W
Grand Falls, N.B., Can. (grănd fôlz)	148-49	47°03'N	67°44'W
Grand Falls-Windsor, Nf./L., Can.	148-49	48°56'N	55°39'W
Grandfather Mountain, mtn., N.C., U.S. (grănd-fä-thẽr moun´tĭn)	134-35	36°07'N	81°48'W
Grandfield, Ok., U.S. (grănd´fēld)	130-31	34°14'N	98°41'W
Grand Forks, B.C., Can. (grănd fôrks)	142-43	49°02'N	118°27'W
Grand Forks, N.D., U.S. (grănd fôrks)	124-25	47°55'N	97°03'W
Grand Haven, Mi., U.S. (grănd hā´v'n)	126-27	43°04'N	86°13'W
Grand Island, Ne., U.S. (grănd ī´lănd)	124-25	40°55'N	98°21'W
Grand Island, i., Mi., U.S. (grănd ī´lănd)	124-25	46°30'N	86°40'W
Grand Junction, Co., U.S. (grănd jŭngk´shŭn)	128-29	39°04'N	108°34'W
Grand Lake, lk., N.B., Can. (grănd lāk)	148-49	45°53'N	66°03'W
Grand Lake, lk., Nf./L., Can. (grănd lăk)	148-49	48°59'N	57°22'W
Grand Lake, lk., La., U.S. (grănd lāk)	132-33	29°53'N	92°45'W
Grand Ledge, Mi., U.S. (grănd lĕj)	126-27	42°45'N	84°44'W
Grand Manan Island, i., N.B., Can. (grănd mȧ-năn ī´lănd)	148-49	44°43'N	66°49'W
Grândola, Port. (grăɴ´dō-lȧ)	214-15	38°10'N	8°34'W
Grand Portage Indian Reservation, ind. res., Mn., U.S. (grănd pōr´tĭj ĭn´dĭ-ăn rĕ-sẽr-vā´shĕn)	124-25	47°58'N	89°47'W
Grand Rapids, Mb., Can. (grănd răp´ĭdz)	144-45	53°12'N	99°17'W
Grand Rapids, Mi., U.S. (grănd răp´ĭdz)	126-27	42°58'N	85°40'W
Grand Rapids, Mn., U.S. (grănd răp´ĭdz)	124-25	47°14'N	93°31'W
Grand-Sault, N.B., Can. *see* Grand Falls	148-49	47°03'N	67°44'W
Grand Staircase-Escalante National Monument, n.p., Ut., U.S.	128-29	37°30'N	111°30'W
Grand Teton, mtn., Wy., U.S. (grănd tē´tŏn)	122-23	43°44'N	110°48'W
Grand Teton National Park, n.p., Wy., U.S. (grănd tē´tŏn năsh´ŭn-ăl pärk)	122-23	43°56'N	110°46'W
Grand Traverse Bay, b., Mi., U.S. (grănd trăv´ẽrs bā)	126-27	45°02'N	85°30'W
Grand Turk, nat. cap., T./C. Is. (grănd tûrk)	154-55	21°27'N	71°08'W
Grandview, Mb., Can.	144-45	51°10'N	100°42'W
Grandview, Wa., U.S. (grănd´vyo͞o)	122-23	46°15'N	119°54'W
Grangeville, Id., U.S. (grānj´vĭl)	122-23	45°56'N	116°07'W
Granite City, Il., U.S. (grăn´ĭt sĭ´tē)	130-31	38°42'N	90°09'W
Granite Falls, Mn., U.S. (grăn´ĭt fôlz)	124-25	44°49'N	95°33'W
Granite Falls, N.C., U.S. (grăn´ĭt fôlz)	134-35	35°48'N	81°26'W
Granite Peak, mtn., Mt., U.S.	122-23	45°10'N	109°48'W
Gränna, Swe. (grĕn´ȧ)	208-09	58°00'N	14°28'E
Granollers, Spain (grä-nôl-yĕrs´)	214-15	41°37'N	2°17'E
Grantham, Eng., U.K. (grăn´tăm)	206-07	52°55'N	0°39'W
Grants, N.M., U.S.	128-29	35°10'N	107°51'W
Grants Pass, Or., U.S. (grȧnts păs)	122-23	42°26'N	123°19'W
Granville, Fr. (gräɴ-vēl´)	212-13	48°51'N	1°35'W
Granville, N.Y., U.S. (grăn´vĭl)	126-27	43°24'N	73°16'W
Granville Lake, lk., Mb., Can.	144-45	56°17'N	100°30'W
Gräsö, i., Swe.	208-09	60°24'N	18°25'E
Grasse, Fr. (gräs)	212-13	43°40'N	6°55'E
Grasslands National Park, n.p., Sk., Can.	144-45	49°04'N	106°58'W
Grates Point, c., Nf./L., Can. (grāts point)	148-49	48°10'N	52°57'W
Graudenz, Pol. *see* Grudziądz	210-11	53°29'N	18°44'E
Gravatá, Braz.	188d	8°12'S	35°34'W
Gravelbourg, Sk., Can. (grăv´ĕl-bôrg)	144-45	49°52'N	106°34'W
Gravenhage, 's-, nat. cap., Neth. *see* Hague, The	206-07	52°05'N	4°18'E
Gray, Fr. (grȧ)	212-13	47°26'N	5°35'E
Grayling, Mi., U.S. (grā´lĭng)	126-27	44°39'N	84°42'W
Grays Peak, mtn., Co., U.S. (grāz pēk)	128-29	39°37'N	105°45'W
Graz, Aus. (gräts)	210-11	47°05'N	15°27'E
Great Artesian Basin, bas., Austl. (grāt är-tēzh-ăn bā´s'n)	296-97	25°0'S	143°00'E
Great Australian Bight, b., Austl. (grāt ôs-trā´lĭ-ăn bĭt)	296-97	35°0'S	130°00'E
Great Barrier Island, i., N.Z. (grāt băr´ĭ-ẽr ī´lănd)	303	36°10'S	175°25'E
Great Barrier Reef, rf., Austl.	302	18°0'S	146°50'E
Great Barrier Reef Marine Park, n.p., Austl.	302	18°0'S	146°50'E
Great Basin, bas., U.S. (grāt bā´s'n)	120-21	40°0'N	117°00'W
Great Basin National Park, n.p., Nv., U.S.	128-29	38°55'N	114°14'W
Great Bear Lake, lk., N.T., Can. (grāt bâr lāk)	140-41	66°0'N	120°00'W
Great Bend, Ks., U.S. (grāt bĕnd)	130-31	38°22'N	98°45'W
Great Britain, nation, Eur. *see* United Kingdom	190-91	54°0'N	2°00'W
Great Britain, i., U.K. (grāt brĭt´'n)	206-07	54°0'N	2°00'W
Great Channel, strt., Asia	266-67	6°25'N	94°20'E
Great Dismal Swamp, sw., U.S. (grāt dĭz´măl swŏmp)	134-35	36°28'N	76°28'W
Great Divide Basin, bas., Wy., U.S. (grāt dĭ-vīd´ bā´s'n)	122-23	42°0'N	108°10'W
Great Dividing Range, mts., Austl. (grāt dĭ-vī-dĭng rānj)	294-95	25°0'S	147°00'E
Greater Antilles, is., N.A. (grāt´ẽr ăn-tĭ´lēz)	154-55	20°0'N	74°00'W
Greater Khingan Range, mts., China (grāt´ẽr hĭŋ-gän rānj)	240-41	49°0'N	122°00'E
Greater Sunda Islands, is., Asia (grāt´ẽr so͞on´dȧ ī´lăndz)	268-69	2°0'S	110°00'E
Great Exuma, i., Bah. (grāt ĕk-so͞o´mä)	154-55	23°32'N	75°50'W
Great Falls, Mt., U.S. (grāt fôlz)	122-23	47°30'N	111°18'W
Great Falls, S.C., U.S. (grāt fôlz)	134-35	34°34'N	80°54'W
Great Grimsby, Eng., U.K. *see* Grimsby	206-07	53°35'N	0°05'W

n-sing; ŋ-baŋk; ɴ-nasalized n; nŏd; cŏmmit; ōld; ôbey; ôrder; oi-boil; fo͞od; ô-as oo in foot; ou-out; s-soft; sh-dish; th-thin; pūre; ůnite; ûrn; stŭd; circŭs; ü-as in French tu; ´-indeterminate vowel.

360 Great Guana Cay – Guinea PRONOUNCING INDEX

Feature (Pronunciation)	Page	Lat.	Long.
Great Guana Cay, i., Bah. (grāt gwä´nä kē)	**154-55**	24°0'N	76°20'W
Great Inagua, i., Bah. (grāt ē-nä´gwä)	**154-55**	21°05'N	73°18'W
Great Indian Desert, des., Asia (grāt ĭn´dĭ-*ă*n dĕs´ẽrt)	**252-53**	27°0'N	71°00'E
Great Karoo, plat., S. Afr.	**286-87**	32°47'S	22°32'E
Great Limpopo Transfrontier Park, n.p., Afr.	**286-87**	23°0'S	31°30'E
Great Namaqualand, hist. reg., Nmb.	**286-87**	25°0'S	17°00'E
Great Nicobar, i., India (grāt nĭk-ô-bär´)	**266-67**	7°0'N	93°50'E
Great Palm Island, i., Austl.	**302**	18°43'S	146°37'E
Great Pee Dee, stm., S.C., U.S. (grāt pē-dē´)	**134-35**	33°18'N	79°17'W
Great Plains, pl., U.S. (grāt plāns)	**120-21**	42°0'N	100°00'W
Great Ruaha, stm., Tan.	**289**	7°56'S	37°48'E
Great Salt Lake, lk., Ut., U.S. (grāt sôlt lāk)	**122-23**	41°10'N	112°30'W
Great Salt Lake Desert, des., Ut., U.S. (grāt sôlt lāk dĕs´ẽrt)	**120-21**	40°40'N	113°30'W
Great Sand Dunes National Park and Preserve, n.p., Co., U.S.	**128-29**	37°46'N	105°33'W
Great Sandy Desert, des., Austl. (grāt săn´dē dĕs´ẽrt)	**296-97**	21°30'S	125°00'E
Great Sandy National Park, n.p., Austl.	**302**	24°55'S	153°16'E
Great Slave Lake, lk., N.T., Can. (grāt slāv lāk)	**140-41**	61°30'N	114°00'W
Great Smoky Mountains National Park, n.p., U.S. (grāt smōk-ē moun´tĭnz năsh´ŭn-*ă*l pärk)	**134-35**	35°39'N	83°30'W
Great Victoria Desert, des., Austl. (grāt vĭk-tō´rĭ-*à* dĕs´ẽrt)	**296-97**	28°30'S	127°45'E
Great Wall, p.o.i., China	**260-61**	40°0'N	112°30'E
Gréboun, mtn., Niger	**280-81**	20°0'N	8°35'E
Gredos, Sierra de, mts., Spain (syĕr´rä dā grā´dōs)	**214-15**	40°20'N	4°51'W
Greece, nation, Eur. (grēs)	**190-91**	39°0'N	22°00'E
Greeley, Co., U.S. (grē´lĭ)	**124-25**	40°24'N	104°41'W
Greeley, Ne., U.S. (grē´lĭ)	**124-25**	41°33'N	98°32'W
Green, stm., U.S. (grēn)	**120-21**	38°11'N	109°53'W
Green, stm., Ky., U.S. (grēn)	**126-27**	37°54'N	87°31'W
Green Bay, Wi., U.S. (grēn bā)	**126-27**	44°30'N	87°60'W
Green Bay, b., U.S. (grēn bā)	**126-27**	44°58'N	87°35'W
Greencastle, In., U.S. (grēn-kȧs´'l)	**126-27**	39°38'N	86°51'W
Green Cove Springs, Fl., U.S. (grēn kōv sprĭngz)	**134-35**	29°60'N	81°42'W
Greenfield, Ca., U.S. (grēn´fēld)	**128-29**	36°19'N	121°15'W
Greenfield, Ia., U.S. (grēn´fēld)	**124-25**	41°18'N	94°28'W
Greenfield, In., U.S. (grēn´fēld)	**126-27**	39°47'N	85°46'W
Greenfield, Oh., U.S. (grēn´fēld)	**126-27**	39°20'N	83°23'W
Greenfield, Tn., U.S. (grēn´fēld)	**134-35**	36°09'N	88°48'W
Greenfield, Wi., U.S. (grēn´fēld)	**126-27**	42°57'N	88°01'W
Green Islands, is., Pap. N. Gui.	**302**	4°30'S	154°10'E
Greenland, dep., N.A. (grēn´l*ă*nd)	**94**	70°0'N	40°00'W
Greenland, i., Green. (grēn´l*ă*nd)	**95**	70°0'N	40°00'W
Greenland Sea, s., (grēn´l*ă*nd sē)	**314**	77°0'N	1°00'W
Green Mountains, mts., N.A. (grēn moun´tĭnz)	**126-27**	43°45'N	72°45'W
Greenock, Scot., U.K. (grēn´*ŭ*k)	**206-07**	55°57'N	4°45'W
Green River, Ut., U.S. (grēn rĭv´ẽr)	**128-29**	38°60'N	110°09'W
Green River, Wy., U.S. (grēn rĭv´ẽr)	**122-23**	41°32'N	109°28'W
Greensboro, Al., U.S. (grēnz´bŭro)	**134-35**	32°42'N	87°36'W
Greensboro, Ga., U.S. (grēns-bûr´ō)	**134-35**	33°35'N	83°11'W
Greensboro, N.C., U.S. (grēns-bûr´ō)	**134-35**	36°04'N	79°48'W
Greensburg, In., U.S. (grēnz´bûrg)	**126-27**	39°20'N	85°29'W
Greensburg, Ky., U.S. (grēns-bûrg)	**134-35**	37°16'N	85°30'W
Greenville, Lib. (grēn´vĭl)	**282-83**	5°02'N	9°03'W
Greenville, Al., U.S. (grēn´vĭl)	**134-35**	31°50'N	86°37'W
Greenville, Il., U.S. (grēn´vĭl)	**126-27**	38°53'N	89°25'W
Greenville, Ky., U.S. (grēn´vĭl)	**134-35**	37°12'N	87°11'W
Greenville, Me., U.S. (grēn´vĭl)	**127a**	45°27'N	69°34'W
Greenville, Mi., U.S. (grēn´vĭl)	**126-27**	43°11'N	85°15'W
Greenville, Ms., U.S. (grēn´vĭl)	**134-35**	33°25'N	91°03'W
Greenville, N.C., U.S. (grēn´vĭl)	**134-35**	35°37'N	77°22'W
Greenville, Oh., U.S. (grēn´vĭl)	**126-27**	40°06'N	84°38'W
Greenville, Pa., U.S. (grēn´vĭl)	**126-27**	41°24'N	80°23'W
Greenville, S.C., U.S. (grēn´vĭl)	**134-35**	34°51'N	82°24'W
Greenville, Tx., U.S. (grēn´vĭl)	**130-31**	33°09'N	96°07'W
Greenwood, Ar., U.S. (grēn-wo͝od)	**130-31**	35°12'N	94°16'W
Greenwood, In., U.S. (grēn-wo͝od)	**126-27**	39°36'N	86°05'W
Greenwood, La., U.S. (grēn-wo͝od)	**130-31**	32°27'N	93°58'W
Greenwood, Ms., U.S. (grēn-wo͝od)	**134-35**	33°31'N	90°11'W
Greenwood, S.C., U.S. (grēn-wo͝od)	**134-35**	34°12'N	82°09'W
Greer, S.C., U.S. (grēr)	**134-35**	34°56'N	82°14'W
Gregory, S.D., U.S. (grĕg´ō-rĭ)	**124-25**	43°14'N	99°26'W
Gregory, Lake, lk., Austl. (lāk grĕg´ō-rē)	**301**	28°55'S	139°00'E
Gregory Range, mts., Austl.	**302**	19°0'S	143°05'E
Greifswald, Ger. (grīfs´vält)	**210-11**	54°05'N	13°23'E
Greiz, Ger. (grīts)	**210-11**	50°39'N	12°13'E
Gremyachinsk, Russia (grâ´myȧ-chĭnsk)	**202-03**	58°35'N	57°51'E
Grenada, Ms., U.S. (grē-nä´dȧ)	**134-35**	33°46'N	89°49'W
Grenada, nation, N.A. (grĕ-nā´-d*à*)	**152-53**	12°07'N	61°40'W
Grenadines, is., N.A. (grĕn´*à*-dēnz)	**155b**	12°40'N	61°15'W
Grenoble, Fr. (grē-nô´bl')	**212-13**	45°11'N	5°42'E
Grenville, Cape, c., Austl.	**302**	11°58'S	143°14'E
Gresham, Or., U.S. (grĕsh´*ă*m)	**122-23**	45°29'N	122°26'W
Gresik, Indon.	**268-69**	7°10'S	112°39'E
Gretna, La., U.S. (grĕt´n*à*)	**134-35**	29°55'N	90°04'W
Gretna, Ne., U.S. (grĕt´n*à*)	**124-25**	41°08'N	96°15'W
Grey, stm., Nf./L., Can. (grā)	**148-49**	47°33'N	57°08'W
Greybull, Wy., U.S. (grā´bo͝ol)	**122-23**	44°30'N	108°03'W
Grey Islands, is., Nf./L., Can.	**148-49**	50°50'N	55°37'W
Greymouth, N.Z. (grā´mouth)	**303**	42°28'S	171°13'E
Grey Range, mts., Austl. (grā rānj)	**301**	27°0'S	143°35'E
Greytown, S. Afr. (grā´toun)	**292c**	29°04'S	30°35'E
Gribanovskiy, Russia	**202-03**	51°27'N	41°58'E
Gribbel Island, i., B.C., Can.	**142-43**	53°25'N	129°00'W
Griffin, Ga., U.S. (grĭf´ĭn)	**134-35**	33°15'N	84°16'W
Griffith, Austl. (grĭf-ĭth)	**301**	34°17'S	146°03'E
Grim, Cape, c., Austl. (kāp grĭm)	**301**	40°39'S	144°43'E
Grimsby, Eng., U.K.	**206-07**	53°35'N	0°05'W
Grímsey, i., Ice. (grĭms´â)	**206a**	66°33'N	18°01'W
Grimstad, Nor. (grĭm-städh)	**208-09**	58°20'N	8°36'E
Groesbeck, Tx., U.S. (grōs´bĕk)	**132-33**	31°31'N	96°32'W
Gronau, Ger. (grō´nou)	**210-11**	52°13'N	7°01'E
Groningen, Neth. (grō´nĭng-*ĕ*n)	**206-07**	53°13'N	6°34'E
Grønland, dep., N.A. *see* Greenland	**94**	70°0'N	40°00'W
Grønland, i., Green. *see* Greenland	**95**	70°0'N	40°00'W
Grønlandshavet, s., *see* Greenland Sea	**314**	77°0'N	1°00'W
Groote Eylandt, i., Austl. (grō´tē ī´länt)	**296-97**	13°60'S	136°38'E
Grootfontein, Nmb. (grōt´fōn-tān´)	**286-87**	19°34'S	18°06'E
Groot Karroo, plat., S. Afr. *see* Great Karoo	**286-87**	32°47'S	22°32'E
Groot Namaland, hist. reg., Nmb. *see* Great Namaqualand	**286-87**	25°0'S	17°00'E
Gros Morne, mtn., Nf./L., Can. (grō môrn´)	**148-49**	49°36'N	57°48'W
Gros Morne National Park, n.p., Nf./L., Can. (grō môrn´ năsh´ŭn-*ă*l pärk)	**148-49**	49°40'N	57°45'W
Grosseto, Italy (grôs-sā´tō)	**216-17**	42°46'N	11°07'E
Großglockner, mtn., Aus.	**210-11**	47°04'N	12°42'E
Grosswardein, Rom. *see* Oradea	**210-11**	47°04'N	21°56'E
Groton, Ct., U.S. (grŏt´*ŭ*n)	**126-27**	41°21'N	72°05'W
Groton, S.D., U.S. (grŏt´*ŭ*n)	**124-25**	45°27'N	98°06'W
Grouard Mission, Ab., Can.	**142-43**	55°32'N	116°09'W
Groveton, N.H., U.S. (grōv´t*ŭ*n)	**126-27**	44°36'N	71°31'W
Growa Point, c., Lib.	**282-83**	4°21'N	7°36'W
Groznyy, Russia (grôz´nĭ)	**245**	43°19'N	45°41'E
Grudziądz, Pol. (gro͝o´jyŏnts)	**210-11**	53°29'N	18°44'E
Grünberg, Pol. *see* Zielona Góra	**210-11**	51°56'N	15°31'E
Grundy Center, Ia., U.S. (grŭn´dĭ sĕn´tẽr)	**124-25**	42°21'N	92°47'W
Gruziya, nation, Asia *see* Georgia	**245**	42°0'N	44°00'E
Gryazi, Russia (gryä´zĭ)	**202-03**	52°29'N	39°57'E
Gryazovets, Russia (gryä´zō-vĕts)	**202-03**	58°53'N	40°15'E
Gryfice, Pol. (grĭ´fĭ-tsĕ)	**210-11**	53°55'N	15°12'E
Guacanayabo, Golfo de, b., Cuba (gôl-fô-dĕ-gwä-kä-nä-yä´bō)	**154-55**	20°28'N	77°30'W
Guacara, Ven. (gwȧ´kä-rä)	**188b**	10°14'N	67°53'W
Guadalajara, Mex. (gwä-dhä-lä-hä´rä)	**158-59**	20°40'N	103°20'W
Guadalajara, Spain (gwä-dä-lä-kä´rä)	**214-15**	40°38'N	3°10'W
Guadalcanal, i., Sol. Is. (gwä-dhäl-kä-näl´)	**306e**	9°32'S	160°12'E
Guadalquivir, stm., Spain (gwä-dhäl-kē-vēr´)	**214-15**	36°47'N	6°24'W
Guadalupe, Mex.	**132-33**	25°41'N	100°15'W
Guadalupe, stm., Tx., U.S. (gwä-dhä-lo͞o´pȧ)	**132-33**	28°27'N	96°49'W
Guadalupe, Isla, i., Mex.	**152-53**	29°03'N	118°21'W
Guadalupe, Sierra de, mts., Spain (syĕr´rä dā gwä-dhä-lo͞o´pä)	**214-15**	39°29'N	5°28'W
Guadalupe Mountains, mts., U.S. (gwä-dhä-lo͞o´pȧ moun´tĭnz)	**132-33**	32°24'N	105°04'W
Guadalupe Mountains National Park, n.p., Tx., U.S.	**132-33**	31°55'N	104°55'W
Guadalupe Peak, mtn., Tx., U.S. (gwä-dhä-lo͞o´pȧ pēk)	**132-33**	31°50'N	104°52'W
Guadarrama, Sierra de, mts., Spain (syĕr´rä dā gwä-dhär-rä´mä)	**214-15**	40°51'N	4°01'W
Guadeloupe, dep., N.A. (gwä-dē-lo͞op)	**152-53**	16°15'N	61°35'W
Guadiana, stm., Eur. (gwä-dvä´nä)	**214-15**	37°10'N	7°24'W
Guadix, Spain (gwä-dēsh´)	**214-15**	37°18'N	3°08'W
Guafo, Isla, i., Chile	**185**	43°36'S	74°43'W
Guaíra, Braz. (gwä-ē-rä)	**186**	20°19'S	48°18'W
Guaíra, Braz. (gwä-ē-rä)	**182-83**	24°06'S	54°15'W
Guajaba, Cayo, i., Cuba (kä´yō-gwä-hä´bä)	**154-55**	21°50'N	77°30'W
Guajará-Mirim, Braz. (gwä-zhä-rä´mē-rēN´)	**180-81**	10°48'S	65°22'W
Gualeguay, Arg. (gwä-lĕ-gwä´y)	**187**	33°09'S	59°20'W
Gualeguay, stm., Arg. (gwä-lĕ-gwä´y)	**187**	33°19'S	59°39'W
Gualeguaychú, Arg.	**187**	33°01'S	58°31'W
Guam, dep., Oc. (gwäm)	**306c**	13°28'N	144°47'E
Guam, i., Guam (gwäm)	**306c**	13°28'N	144°47'E
Guaminí, Arg.	**187**	37°01'S	62°25'W
Guamo, Col. (gwȧ´mō)	**188c**	4°02'N	74°58'W
Guanaja, Isla de, i., Hond.	**161**	16°29'N	85°53'W
Guanajuato, Mex. (gwä-nä-hwä´tō)	**158-59**	21°01'N	101°16'W
Guanajuato, state, Mex. (gwä-nä-hwä´tō)	**158-59**	21°0'N	101°00'W
Guanambi, Braz.	**180-81**	14°14'S	42°47'W
Guanare, Ven. (gwä-nä´rå)	**178-79**	9°03'N	69°46'W
Guandacol, Arg.	**182-83**	29°31'S	68°32'W
Guane, Cuba (gwä´nå)	**154-55**	22°12'N	84°05'W
Guang'an, China	**258-59**	30°28'N	106°38'E
Guangchang, China (gůäŋ-chäŋ)	**258-59**	26°51'N	116°19'E
Guangdong, state, China (gůäŋ-dôŋ)	**313**	23°0'N	113°00'E
Guanghua, China *see* Laohekou	**258-59**	32°25'N	111°36'E
Guangnan, China	**258-59**	24°10'N	105°06'E
Guangxi, state, China	**258-59**	24°0'N	109°00'E
Guangyuan, China	**258-59**	32°25'N	105°49'E
Guangzhou, China	**258-59**	23°08'N	113°16'E
Guanta, Ven. (gwän´tä)	**188b**	10°14'N	64°36'W
Guantánamo, Cuba (gwän-tä´nä-mô)	**154-55**	20°08'N	75°13'W
Guantánamo, state, Cuba (gwän-tä´nä-mô)	**154-55**	20°20'N	75°00'W
Guanxian, China (gůän-shyĕn)	**258-59**	31°00'N	103°36'E
Guanxian, China (gůän-shyĕn)	**260-61**	36°28'N	115°26'E
Guapí, Col.	**178-79**	2°36'N	77°54'W
Guápiles, C.R. (gwä-pē-lĕs)	**161**	10°12'N	83°47'W
Guaporé, stm., S.A. (gwä-pô-rā´)	**180-81**	11°55'S	65°00'W
Guarabira, Braz. (gwä-rä-bē´r*à*)	**188d**	6°51'S	35°29'W
Guaranda, Ec. (gwä-rán´dä)	**184**	1°36'S	78°60'W
Guarapari, Braz. (gwä-rä-pä´rē)	**186**	20°40'S	40°30'W
Guarapuava, Braz. (gwä-rä-pwä´v*à*)	**182-83**	25°23'S	51°29'W
Guarda, Port. (gwär´dä)	**214-15**	40°33'N	7°15'W
Guarulhos, Braz. (gwȧ-ro͞o´l-yôs)	**186**	23°29'S	46°32'W
Guasave, Mex.	**156-57**	25°34'N	108°28'W
Guasdualito, Ven.	**178-79**	7°15'N	70°45'W
Guasipati, Ven. (gwä-sē-pä´tē)	**178-79**	7°29'N	61°53'W
Guastalla, Italy (gwäs-täl´lä)	**216-17**	44°55'N	10°39'E
Guatemala, nation, N.A. (guä-tȧ-mä´lä)	**94**	15°30'N	90°15'W
Guatemala, nat. cap., Guat. (guä-tȧ-mä´lä)	**160**	14°38'N	90°32'W
Guaviare, stm., Col.	**178-79**	4°04'N	67°43'W
Guaxupé, Braz.	**186**	21°18'S	46°42'W
Guayama, P.R. (gwä-yä´mä)	**154a**	17°59'N	66°07'W
Guayana, Ven. *see* Ciudad Guayana	**178-79**	8°21'N	62°39'W
Guayana, nation, S.A. *see* Guyana	**172**	5°0'N	59°00'W
Guayaquil, Ec. (gwī-ä-kēl´)	**184**	2°12'S	79°54'W
Guayaquil, Golfo de, b., S.A. (gôl-fô-dĕ gwī-ä-kēl´)	**184**	2°57'S	80°36'W
Guaymas, Mex. (gwȧ´y-mäs)	**156-57**	27°55'N	110°55'W
Gûbâi, Madîq, strt., Egypt	**291b**	27°40'N	33°55'E
Gubakha, Russia (go͞o-bä´kå)	**202-03**	58°52'N	57°33'E
Gubbio, Italy (go͞ob´byô)	**216-17**	43°21'N	12°34'E
Gubkin, Russia	**218-19**	51°17'N	37°33'E
Gucheng, China (go͞o-chŭŋ)	**258-59**	32°18'N	111°35'E
Gudermes, Russia	**245**	43°21'N	46°06'E
Guebwiller, Fr. (gĕb-vē-lâr´)	**212-13**	47°55'N	7°12'E
Guelph, On., Can. (gwĕlf)	**146-47**	43°33'N	80°15'W
Guercif, Mor.	**200-01**	34°14'N	3°20'W
Guéret, Fr. (gā-rĕ´)	**212-13**	46°10'N	1°52'E
Guernesey, dep., Eur. *see* Guernsey	**212-13**	49°28'N	2°35'W
Guernsey, dep., Eur. (gûrn´zĭ)	**212-13**	49°28'N	2°35'W
Guerrero, Mex. (gĕr-rā´rō)	**156-57**	28°33'N	107°30'W
Guerrero, state, Mex. (gĕr-rā´rō)	**158-59**	17°40'N	100°00'W
Gugē, mtn., Eth.	**284-85**	6°11'N	37°24'E
Guguan, i., N. Mar. Is.	**304-05**	17°19'N	145°51'E
Guide, China	**260-61**	36°01'N	101°27'E
Guilin, China (gwā-lĭn)	**258-59**	25°17'N	110°17'E
Guimarães, Port. (gē-mä-răNsh´)	**214-15**	41°27'N	8°18'W
Guimaras Island, i., Phil.	**270**	10°35'N	122°37'E
Guiné, Golfo da, b., Afr. *see* Guinea, Gulf of	**282-83**	2°0'N	2°30'E
Guinea, nation, Afr. (gĭn´ē)	**274**	11°0'N	10°00'W

ăt; finăl; rāte; senåte; ärm; åsk; sofȧ; fâre; ch-choose; dh-as th in other; bē; ēvent; bĕt; recĕnt; cratẽr; g-gō; gh-guttural g; bĭt; ĭ-short neutral; rīde; K-guttural k as ch in German ich;

PRONOUNCING INDEX
Guinea, Golfo de – Hanjiang 361

Feature (Pronunciation) Page Lat. Long.

Guinea, Golfo de, b., Afr.
see Guinea, Gulf of ..... **282-83** 2°0'N 2°30'E
Guinea, Gulf of, b., Afr.
(gŭlf ŭv gĭn'ē) ..... **282-83** 2°0'N 2°30'E
Guinea-Bissau, nation, Afr.
(gĭn'ē bē-sa'ōō) ..... **274** 12°0'N 15°00'W
Guinea Ecuatorial, nation, Afr.
see Equatorial Guinea ..... **274** 2°0'N 9°00'E
Guiné-Bissau, nation, Afr.
see Guinea-Bissau ..... **274** 12°0'N 15°00'W
Guinée, nation, Afr. *see* Guinea ..... **274** 11°0'N 10°00'W
Guinée, Golfe de, b., Afr.
see Guinea, Gulf of ..... **282-83** 2°0'N 2°30'E
Güines, Cuba ..... **154-55** 22°51'N 82°02'W
Guingamp, Fr. (găɴ-găɴ') ..... **212-13** 48°34'N 3°09'W
Guiping, China ..... **258-59** 23°23'N 110°04'E
Güira de Melena, Cuba
(gwē'rä dĕ må-lā'nä) ..... **154-55** 22°48'N 82°30'W
Guiratinga, Braz. ..... **182-83** 16°21'S 53°45'W
Güiria, Ven. (gwē-rē'ä) ..... **155b** 10°35'N 62°18'W
Guiuan, Phil. ..... **270** 11°02'N 125°44'E
Guixian, China ..... **258-59** 23°06'N 109°39'E
Guiyang, China ..... **258-59** 26°35'N 106°43'E
Guizhou, state, China (gwā-jō) ..... **258-59** 27°0'N 107°00'E
Gujarāt, state, India ..... **254-55** 22°0'N 72°00'E
Gujrānwāla, Pak. ..... **252-53** 32°09'N 74°11'E
Gujrāt, Pak. ..... **252-53** 32°34'N 74°05'E
Gulbarga, India (gool-bûr'gȧ) ..... **256** 17°20'N 76°50'E
Gulbene, Lat. (gool-bă'nĕ) ..... **208-09** 57°10'N 26°46'E
Gulfport, Ms., U.S. (gŭlf'pōrt) ..... **134-35** 30°22'N 89°06'W
Gulian, China ..... **240-41** 52°56'N 122°19'E
Guliston, Uzb. ..... **252-53** 40°30'N 68°46'E
Gulja, China *see* Yining ..... **244** 43°55'N 81°18'E
Gull Lake, Sk., Can. (gŭl lāk) ..... **144-45** 50°08'N 108°27'W
Gull Lake, lk., Ab., Can. (gŭl lāk) ..... **142-43** 52°33'N 114°02'W
Gulu, Ug. ..... **289** 2°47'N 32°18'E
Gumaca, Phil. (gōō-mä-kä') ..... **270** 13°55'N 122°06'E
Gumdag, Turkmen. ..... **252-53** 39°12'N 54°36'E
Gummersbach, Ger. (gōōm'ĕrs-bäk) ..... **210-11** 51°02'N 7°34'E
Gumti, stm., India ..... **254-55** 25°31'N 83°10'E
Gümüşhane, Tur. ..... **245** 40°28'N 39°28'E
Guna, India ..... **254-55** 24°39'N 77°18'E
Gundagai, Austl. ..... **301** 35°04'S 148°06'E
Gunisao, stm., Mb., Can. (gŭn *i*-sā'ō) ..... **144-45** 53°53'N 97°60'W
Gunisao Lake, lk., Mb., Can.
(gŭn-*i*-sā'ō lāk) ..... **144-45** 53°33'N 96°15'W
Gunnbjørn Fjeld, mtn., Green. ..... **95** 68°53'N 30°00'W
Gunnedah, Austl. (gŭ'nē-dä) ..... **301** 30°59'S 150°15'E
Gunnison, Co., U.S. (gŭn'ĭ-sŭn) ..... **128-29** 38°33'N 106°55'W
Gunnison, Ut., U.S. (gŭn'ĭ-sŭn) ..... **128-29** 39°09'N 111°49'W
Gunsan, Kor., S. ..... **263** 35°59'N 126°43'E
Guntersville, Al., U.S. (gŭn'tẽrz-vĭl) ..... **134-35** 34°22'N 86°18'W
Guntūr, India (goon'tōōr) ..... **256** 16°18'N 80°27'E
Gunungsitoli, Indon. ..... **266-67** 1°15'N 97°37'E
Guoyang, China (gwō-yäŋ) ..... **258-59** 33°30'N 116°12'E
Gurara, stm., Nig. ..... **282-83** 8°11'N 6°42'E
Gurdon, Ar., U.S. (gûr'dŭn) ..... **130-31** 33°55'N 93°10'W
Guri, Embalse de, res., Ven. ..... **178-79** 7°30'N 62°50'W
Gurnee, Il., U.S. (gûr'nē) ..... **126-27** 42°22'N 87°54'W
Gurué, Moz. ..... **286-87** 15°27'S 36°59'E
Gurupá, Braz. ..... **180-81** 1°25'S 51°39'W
Gurupi, Braz. ..... **180-81** 11°43'S 49°02'W
Gurupi, stm., Braz. ..... **180-81** 1°16'S 46°09'W
Guryev, Kaz. *see* Atyraū ..... **202-03** 47°07'N 51°55'E
Gusau, Nig. (gōō-zä'ōō) ..... **282-83** 12°10'N 6°40'E
Guşgy, Turkmen. ..... **252-53** 35°16'N 62°21'E
Gushi, China (gōō-shr) ..... **258-59** 32°11'N 115°41'E
Gusinoozërsk, Russia ..... **260-61** 51°17'N 106°31'E
Gus'-Khrustal'nyy, Russia
(gōōs-ĸrōō-stäl'ny') ..... **202-03** 55°37'N 40°40'E
Gütersloh, Ger. (gü'tĕrs-lo) ..... **210-11** 51°54'N 8°23'E
Guthrie, Ok., U.S. (gŭth'rĭ) ..... **130-31** 35°53'N 97°26'W
Gutian, China ..... **258-59** 26°36'N 118°46'E
Gutiérrez Zamora, Mex.
(gōō-tē-âr'räz zä-mō'rä) ..... **158-59** 20°27'N 97°05'W
Guttenberg, Ia., U.S. (gŭt'ĕn-bûrg) ..... **124-25** 42°47'N 91°06'W
Guwāhāti, India ..... **254-55** 26°11'N 91°44'E
Guyana, nation, S.A. (gŭy'ănă) ..... **172** 5°0'N 59°00'W
Guyane, dep., S.A. *see* French Guiana ..... **172** 4°0'N 53°00'W
Guyang, China (gōō-yäŋ) ..... **260-61** 41°02'N 110°04'E
Guymon, Ok., U.S. (gī'mŏn) ..... **130-31** 36°41'N 101°29'W
Guyuan, China ..... **260-61** 35°59'N 106°18'E
Guzhen, China (gōō-jŭn) ..... **258-59** 33°19'N 117°19'E
Guzmán, Mex. ..... **158-59** 19°42'N 103°28'W
G'uzor, Uzb. ..... **252-53** 38°36'N 66°15'E
Gvardeysk, Russia (gvår-dĕysk') ..... **210-11** 54°39'N 21°05'E

Gwādar, Pak. (gwä'dŭr) ..... **250-51** 25°08'N 62°20'E
Gwalior, India ..... **254-55** 26°13'N 78°09'E
Gwangju, Kor., S. ..... **263** 35°09'N 126°54'E
Gwardafuy, Gees, c., Som. ..... **284-85** 11°50'N 51°17'E
Gweru, Zimb. ..... **286-87** 19°27'S 29°49'E
Gwinn, Mi., U.S. (gwĭn) ..... **124-25** 46°17'N 87°25'W
Gyandzha, Azer. *see* Gäncä ..... **245** 40°41'N 46°21'E
Gyangzê, China ..... **254-55** 28°56'N 89°34'E
Gyaring Co, lk., China ..... **254-55** 31°05'N 88°24'E
Gyaring Hu, lk., China ..... **260-61** 34°54'N 97°15'E
Gyeongju, Kor., S. ..... **263** 35°51'N 129°13'E
Gympie, Austl. (gĭm'pē) ..... **301** 26°12'S 152°40'E
Gyöngyös, Hung. (dyûn'dyûsh) ..... **210-11** 47°47'N 19°56'E
Győr, Hung. (dyûr) ..... **210-11** 47°41'N 17°39'E
Gyula, Hung. (dyō'lä) ..... **210-11** 46°39'N 21°20'E
Gyumri, Arm. ..... **245** 40°47'N 43°51'E
Gyzylarbat, Turkmen. ..... **252-53** 38°59'N 56°16'E
Gyzyletrek, Turkmen. ..... **252-53** 37°36'N 54°47'E

H

Haapamäki, Fin. (häp'ä-mĕ-kē) ..... **208-09** 62°15'N 24°27'E
Haapsalu, Est. (häp'sä-lo) ..... **208-09** 58°56'N 23°33'E
Haar, Ger. (här) ..... **210-11** 48°06'N 11°44'E
Haarlem, Neth. (här'lĕm) ..... **206-07** 52°23'N 4°38'E
Hachijō-jima, i., Japan ..... **264** 33°05'N 139°48'E
Hachinohe, Japan (hä'chē-nō'hå) ..... **264** 40°30'N 141°29'E
Ḥadd, Ra's al-, c., Oman ..... **250-51** 22°32'N 59°48'E
Hadejia, stm., Nig. ..... **282-83** 12°50'N 10°51'E
Hadera, Isr. (κå-dĕ'rȧ) ..... **248-49** 32°27'N 34°55'E
Haderslev, Den. (hä'dhĕrs-lĕv) ..... **208-09** 55°15'N 9°30'E
Hadībū, Yemen ..... **238-39** 12°39'N 54°02'E
Ḥadīthah, Iraq ..... **248-49** 34°07'N 42°22'E
Ḥaḍramawt, reg., Yemen ..... **238-39** 15°0'N 50°00'E
Hadyai, Thai. *see* Hat Yai ..... **266-67** 7°01'N 100°28'E
Haeju, Kor., N. (hä'ĕ-jŭ) ..... **263** 38°03'N 125°43'E
Haft Gel, Iran ..... **252-53** 31°20'N 49°32'E
Hagåtña, nat. cap., Guam ..... **306c** 13°28'N 144°45'E
Hagen, Ger. (hä'gĕn) ..... **210-11** 51°22'N 7°28'E
Hagerstown, Md., U.S. (hā'gẽrz-toun) ..... **126-27** 39°39'N 77°43'W
Haggin, Mount, mtn., Mt., U.S. ..... **122-23** 46°05'N 113°05'W
Hagi, Japan (hä'gĭ) ..... **265** 34°24'N 131°24'E
Hague, Cap de la, c., Fr.
(kȧp dē lȧ ȧg') ..... **212-13** 49°43'N 1°56'W
Hague, The, nat. cap., Neth. ..... **206-07** 52°06'N 4°18'E
Haguenau, Fr. (ȧg'nō') ..... **212-13** 48°49'N 7°47'E
Hahajima-rettō, is., Japan ..... **304-05** 26°37'N 142°10'E
Haicheng, China (hī-chŭŋ) ..... **260-61** 40°51'N 122°46'E
Haida Gwaii, is., B.C.,
Can. (häl'dä gwī) ..... **142-43** 53°0'N 132°00'W
Haifa, Isr. (hä'ē-fa) ..... **248-49** 32°49'N 35°00'E
Haifeng, China (hä'ē-fĕng') ..... **258-59** 22°58'N 115°20'E
Haikang, China ..... **258-59** 20°55'N 110°05'E
Haikou, China ..... **258-59** 20°03'N 110°22'E
Ḥā'il, Sau. Ar. ..... **288** 27°31'N 41°42'E
Hailar, China ..... **260-61** 49°11'N 119°44'E
Hailar, stm., China ..... **260-61** 49°30'N 117°51'E
Hailey, Id., U.S. (hā'lĭ) ..... **122-23** 43°32'N 114°19'W
Hailun, China (hä'ē-lōōn') ..... **260-61** 47°27'N 126°58'E
Hailuoto, i., Fin. ..... **200-01** 65°02'N 24°42'E
Hainan, state, China (hī'-nän') ..... **258-59** 19°0'N 109°30'E
Hainan Dao, i., China (hī-nän dou) ..... **258-59** 19°0'N 109°30'E
Hainan Strait, strt., China (hī'-nän' strāt)
see Qiongzhou Haixia ..... **258-59** 20°10'N 110°15'E
Haines, Ak., U.S. (hānz) ..... **136** 59°14'N 135°27'W
Haines City, Fl., U.S. (hānz sĭ'tĭ) ..... **135a** 28°07'N 81°38'W
Haines Junction, Yk., Can. ..... **138-39** 60°45'N 137°28'W
Hai Ninh, Viet. ..... **266-67** 21°32'N 107°56'E
Hai Phong, Viet.
(hī fŏng')(hä'ēp-hŏng) ..... **266-67** 20°52'N 106°41'E
Haiti, nation, N.A. (hā'tĭ) ..... **152-53** 19°0'N 72°25'W
Haïti, i., N.A. *see* Hispaniola ..... **154-55** 19°0'N 71°00'W
Haizhou, China ..... **258-59** 34°35'N 119°08'E
Hajdúböszörmény, Hung.
(hôl'dō-bû'sûr-mān') ..... **210-11** 47°40'N 21°31'E
Hajdúnánás, Hung. (hô'ĭ-dō-nä'näsh) ..... **210-11** 47°50'N 21°27'E
Ḥajjah, Yemen ..... **288** 15°42'N 43°36'E
Ḥakīm, Abyār al-, well, Libya ..... **280-81** 31°36'N 23°29'E
Hakodate, Japan (hä-kō-dä't å) ..... **264** 41°45'N 140°43'E
Haku-san, vol., Japan (hä'kōō-sän') ..... **265** 36°09'N 136°46'E
Ḥalab, Syria *see* Aleppo ..... **248-49** 36°13'N 37°10'E
Ḥalā'ib, Sudan ..... **288** 22°13'N 36°38'E
Halberstadt, Ger. (häl'bĕr-shtät) ..... **210-11** 51°53'N 11°03'E

Halcon, Mount, mtn., Phil.
(mount häl-kōn') ..... **270** 13°16'N 121°00'E
Halden, Nor. (häl'dĕn) ..... **208-09** 59°08'N 11°23'E
Haleakalā National Park, n.p., Hi., U.S.
(hä'lå-ä'kä-lä năsh'ŭn-*ă*l pärk) ..... **137a** 20°44'N 156°13'W
Haleyville, Al., U.S. (hā'lĭ-vĭl) ..... **134-35** 34°14'N 87°38'W
Halfway, stm., B.C., Can. ..... **142-43** 56°13'N 121°26'W
Halifax, N.S., Can. (hăl'ĭ-făks) ..... **148-49** 44°39'N 63°36'W
Halifax Bay, b., Austl. (hăl'ĭ-făx bā) ..... **302** 18°50'S 146°30'E
Hallasan, mtn., Kor., S. (häl'lå-sän') ..... **260-61** 33°22'N 126°32'E
Halle, Ger. ..... **210-11** 51°29'N 11°58'E
Hallettsville, Tx., U.S. (hăl'ĕts-vĭl) ..... **132-33** 29°27'N 96°56'W
Hall Islands, is., Micron. ..... **304-05** 8°35'N 151°59'E
Hallock, Mn., U.S. (hăl'*ŭ*k) ..... **124-25** 48°46'N 96°57'W
Hall Peninsula, pen., Nu., Can.
(hôl pĕ-nĭn'sū*lă*) ..... **140-41** 63°30'N 66°00'W
Hallsberg, Swe. (häls'bĕrgh) ..... **208-09** 59°05'N 15°08'E
Halls Creek, Austl. (hôlz krēk) ..... **294-95** 18°15'S 127°40'E
Halmahera, i., Indon. (häl-mä-hā'rä) ..... **268-69** 1°0'N 128°00'E
Halmahera, Laut, s., Indon.
see Halmahera Sea ..... **242-43** 1°0'S 129°00'E
Halmahera Sea, s., Indon.
(häl-mä-hā'rä sē) ..... **242-43** 1°0'S 129°00'E
Halmstad, Swe. (hälm'städ) ..... **208-09** 56°40'N 12°53'E
Hälsingborg, Swe. *see* Helsingborg ..... **208-09** 56°03'N 12°42'E
Haltern, Ger. (häl'tĕrn) ..... **210-11** 51°45'N 7°11'E
Haltiatunturi, mtn., Eur. ..... **200-01** 69°18'N 21°16'E
Halton Hills, On., Can.
see Georgetown ..... **146-47** 43°39'N 79°55'W
Halys, stm., Tur. *see* Kızılırmak ..... **202-03** 41°44'N 35°58'E
Hamada, Japan ..... **265** 34°53'N 132°05'E
Hamadān, Iran (hŭ-mŭ-dän') ..... **248-49** 34°48'N 48°31'E
Ḥamāh, Syria (hä'mä) ..... **248-49** 35°08'N 36°45'E
Hamamatsu, Japan (hä'mä-mät'sō) ..... **265** 34°43'N 137°42'E
Hamar, Nor. (hä'mär) ..... **208-09** 60°48'N 11°05'E
Hamburg, Ger. (häm'bōōrgh) ..... **210-11** 53°33'N 9°59'E
Hamburg, Ar., U.S. (hăm'bûrg) ..... **134-35** 33°14'N 91°48'W
Hamburg, N.Y., U.S. (hăm'bûrg) ..... **126-27** 42°43'N 78°50'W
Hamden, Ct., U.S. (hăm'dĕn) ..... **126-27** 41°24'N 72°54'W
Hämeenlinna, Fin. (hĕ'măn-lĭn-na) ..... **208-09** 60°58'N 24°31'E
HaMelah, Yam, lk., Asia
see Dead Sea ..... **248-49** 31°30'N 35°30'E
Hameln, Ger. (hä'mĕln) ..... **210-11** 52°06'N 9°22'E
Hamersley Range, mts., Austl.
(hăm'ẽrz-lē rānj) ..... **296-97** 22°24'S 117°34'E
Hamhŭng, Kor., N. (häm'hoong') ..... **263** 39°55'N 127°32'E
Hami, China (hä-me) ..... **260-61** 42°50'N 93°31'E
Hamilton, Austl. (hăm'ĭl-t*ŭ*n) ..... **301** 37°45'S 142°01'E
Hamilton, On., Can. (hăm'ĭl-t*ŭ*n) ..... **146-47** 43°15'N 79°51'W
Hamilton, N.Z. (hăm'ĭl-t*ŭ*n) ..... **303** 37°47'S 175°17'E
Hamilton, Al., U.S. (hăm'ĭl-t*ŭ*n) ..... **134-35** 34°08'N 87°59'W
Hamilton, Mo., U.S. (hăm'ĭl-t*ŭ*n) ..... **130-31** 39°45'N 94°00'W
Hamilton, Mt., U.S. (hăm'ĭl-t*ŭ*n) ..... **122-23** 46°14'N 114°10'W
Hamilton, Oh., U.S. (hăm'ĭl-t*ŭ*n) ..... **126-27** 39°24'N 84°34'W
Hamilton, Tx., U.S. (hăm'ĭl-t*ŭ*n) ..... **132-33** 31°41'N 98°08'W
Hamilton, nat. cap., Ber. (hăm'ĭl-t*ŭ*n) ..... **152-53** 32°18'N 64°48'W
Hamina, Fin. (hä'mē-nȧ) ..... **208-09** 60°34'N 27°19'E
Hamlet, N.C., U.S. (hăm'lĕt) ..... **134-35** 34°54'N 79°42'W
Hamlin, Tx., U.S. (hăm'lĭn) ..... **130-31** 32°53'N 100°08'W
Hamm, Ger. (häm) ..... **210-11** 51°41'N 7°49'E
Hammamet, Tun. ..... **200-01** 36°24'N 10°37'E
Hammamet, Golfe de, b., Tun. ..... **280-81** 36°05'N 10°40'E
Hammerfest, Nor. (hä'mĕr-fĕst) ..... **200-01** 70°40'N 23°42'E
Hammond, In., U.S. (hăm'*ŭ*nd) ..... **126-27** 41°35'N 87°30'W
Hammond, La., U.S. (hăm'*ŭ*nd) ..... **134-35** 30°30'N 90°28'W
Hammonton, N.J., U.S. (hăm'ŭn-t*ŭ*n) ..... **126-27** 39°38'N 74°48'W
Hampton, N.B., Can. (hămp't*ŭ*n) ..... **148-49** 45°32'N 65°51'W
Hampton, Ia., U.S. (hămp't*ŭ*n) ..... **124-25** 42°45'N 93°12'W
Hampton, S.C., U.S. (hămp't*ŭ*n) ..... **134-35** 32°52'N 81°07'W
Hampton, Va., U.S. (hămp't*ŭ*n) ..... **134-35** 37°02'N 76°21'W
Ḥamrā', Al-Ḥamādah al-, des., Libya ..... **204-05** 30°0'N 12°00'E
Hāmūn, Daryācheh-ye, lk., Iran ..... **250-51** 30°43'N 61°07'E
Han, stm., China (hän) ..... **258-59** 23°41'N 116°38'E
Han, stm., China (hän) ..... **258-59** 30°34'N 114°17'E
Hāna, Hi., U.S. (hä'nä) ..... **137a** 20°45'N 155°59'W
Hancheng, China ..... **260-61** 35°29'N 110°25'E
Hancock, Mi., U.S. (hăn'kŏk) ..... **124-25** 47°08'N 88°36'W
Handan, China (hän-dän) ..... **260-61** 36°37'N 114°28'E
Hanford, Ca., U.S. (hăn'fẽrd) ..... **128-29** 36°20'N 119°38'W
Hangayn nuruu, mts., Mong. ..... **260-61** 47°32'N 98°42'E
Hangchow, China *see* Hangzhou ..... **258-59** 30°15'N 120°10'E
Hanggin Houqi, China ..... **260-61** 40°57'N 107°14'E
Hanggin Qi, China ..... **260-61** 39°55'N 108°52'E
Hangö, Fin. (häŋ'gû) *see* Hanko ..... **208-09** 59°50'N 22°58'E
Hangzhou, China (häng'chō') ..... **258-59** 30°15'N 120°10'E
Hanjiang, China ..... **243a** 25°30'N 119°06'E

n-sing; ŋ-baŋk; ɴ-nasalized n; nŏd; c*ŏ*mmit; ōld; ôbey; ôrder; oi-boil; fōōd; ŏ-as oo in foot; ou-out; s-soft; sh-dish; th-thin; pūre; ûnite; ûrn; stŭd; circ*ŭ*s; ü-as in French tu; '-indeterminate vowel.

362 Hankinson – Hengxian PRONOUNCING INDEX

Feature (Pronunciation)	Page	Lat.	Long.
Hankinson, N.D., U.S. (hăŋ´kĭn-sŭn)	124-25	46°04´N	96°55´W
Hanko, Fin.	208-09	59°50´N	22°58´E
Hankow, China *see* Wuhan	258-59	30°34´N	114°17´E
Hanna, Ab., Can. (hăn´á)	142-43	51°39´N	111°56´W
Hannibal, Mo., U.S. (hăn´ĭ băl)	130-31	39°42´N	91°22´W
Hannover, Ger. (hän-ō´vĕr)	210-11	52°24´N	9°44´E
Ha Noi, nat. cap., Viet. (hä-noi´)	266-67	21°02´N	105°50´E
Hanover, On., Can. (hăn´ô-vĕr)	146-47	44°09´N	81°01´W
Hanover, Ger. *see* Hannover	210-11	52°24´N	9°44´E
Hanover, N.H., U.S. (hăn´ô-vĕr)	126-27	43°42´N	72°17´W
Hanover, Pa., U.S. (hăn´ô-vĕr)	126-27	39°48´N	76°59´W
Hanover, Va., U.S. (hăn´ô-vĕr)	126-27	37°46´N	77°23´W
Hanover, Isla, i., Chile	185	50°58´S	74°45´W
Hantsport, N.S., Can. (hănts´pōrt)	148-49	45°04´N	64°12´W
Hanuy, stm., Mong.	260-61	49°21´N	102°22´E
Hanzhong, China (hän-jôŋ)	258-59	33°04´N	107°02´E
Hāora, India	254-55	22°35´N	88°20´E
Haparanda, Swe. (hä-pa-rän´dä)	200-01	65°50´N	24°06´E
Happy Valley-Goose Bay, Nf./L., Can.	138-39	53°20´N	60°25´W
Hāpur, India	254-55	28°44´N	77°47´E
Harare, nat. cap., Zimb. (hä-rä´-rē)	286-87	17°50´S	31°03´E
Ḩarash, Bi'r al-, well, Libya	280-81	25°39´N	22°08´E
Harbin, China	260-61	45°45´N	126°38´E
Harbor Beach, Mi., U.S. (här´bẽr bēch)	126-27	43°51´N	82°39´W
Harbour Breton, Nf./L., Can. (här´bẽr brĕt´ŭn) (brĕ-tôN´)	148-49	47°30´N	55°49´W
Harbour Grace, Nf./L., Can. (här´bẽr grās)	148-49	47°44´N	53°15´W
Hardangerfjorden, b., Nor.	208-09	60°10´N	6°00´E
Hardin, Mt., U.S. (här´dĭn)	122-23	45°44´N	107°37´W
Hardoi, India	254-55	27°23´N	80°09´E
Hare Bay, b., Nf./L., Can. (hâr bā)	148-49	51°16´N	55°51´W
Hareidlandet, i., Nor.	200-01	62°21´N	5°57´E
Hārer, Eth.	284-85	9°18´N	42°08´E
Hargeysa, Som. (här-gā´ē-sá)	284-85	9°34´N	44°04´E
Har Horin, hist., Mong. *see* Karakorum	260-61	47°14´N	102°50´E
Har Hu, lk., China	260-61	38°15´N	97°40´E
Hari, stm., Indon.	266-67	1°04´S	104°12´E
Haridwār, India	254-55	29°56´N	78°07´E
Harīrūd, stm., Asia	252-53	37°24´N	60°31´E
Harlan, Ia., U.S. (här´lăn)	124-25	41°39´N	95°20´W
Harlan, Ky., U.S. (här´lăn)	134-35	36°51´N	83°19´W
Harlem, Mt., U.S. (hár´lĕm)	122-23	48°32´N	108°47´W
Harlingen, Neth. (här´lĭng-ĕn)	206-07	53°10´N	5°26´E
Harlingen, Tx., U.S. (här´lĭng-ĕn)	132-33	26°12´N	97°42´W
Harlow, Eng., U.K. (här´lō)	206-07	51°47´N	0°07´E
Harlowton, Mt., U.S. (hár´lō-tun)	122-23	46°26´N	109°50´W
Harney Basin, bas., Or., U.S. (här´nĭ bā´s'n)	122-23	43°15´N	119°00´W
Harney Peak, mtn., S.D., U.S. (här´nĭ pēk)	124-25	43°52´N	103°32´W
Härnösand, Swe. (hĕr-nû-sänd)	200-01	62°38´N	17°56´E
Haro, Spain (ä´rō)	214-15	42°36´N	2°52´W
Hārot, stm., Afg.	252-53	31°29´N	61°16´E
Harper, Lib. (här´pẽr)	282-83	4°23´N	7°43´W
Harpers Ferry, W.V., U.S. (här´pẽrz fĕr´ē)	126-27	39°19´N	77°45´W
Harricana, stm., Can.	140-41	51°10´N	79°47´W
Harriet Tubman Underground Railroad National Monument, n.p., Md., U.S.	126-27	38°27´N	76°06´W
Harriman, Tn., U.S. (hă´ĭ-măn)	134-35	35°56´N	84°33´W
Harrington, De., U.S. (hăr´ĭng-tŭn)	126-27	38°56´N	75°34´W
Harris, Lake, lk., Austl.	296-97	31°06´S	135°11´E
Harrisburg, Il., U.S. (hăr´ĭs-bûrg)	126-27	37°44´N	88°32´W
Harrisburg, Pa., U.S.	126-27	40°16´N	76°54´W
Harrismith, S. Afr. (hă-rĭs´mĭth)	292c	28°17´S	29°08´E
Harrison, Ar., U.S. (hăr´ĭ-sŭn)	130-31	36°14´N	93°07´W
Harrison, Mi., U.S. (hăr´ĭ-sŭn)	126-27	44°01´N	84°48´W
Harrison, Cape, c., Nf./L., Can.	140-41	54°55´N	57°56´W
Harrisonburg, Va., U.S. (hăr´ĭ-sŭn-bûrg)	126-27	38°27´N	78°52´W
Harrison Lake, lk., B.C., Can. (hăr´ĭ-sŭn lāk)	142-43	49°33´N	121°52´W
Harrisonville, Mo., U.S. (hăr-ĭ-sŭn-vĭl)	130-31	38°39´N	94°21´W
Harrisville, Mi., U.S. (hăr´ĭs-vĭl)	126-27	44°39´N	83°18´W
Harrisville, W.V., U.S. (hăr´ĭs-vĭl)	126-27	39°13´N	81°03´W
Harrodsburg, Ky., U.S. (hăr´ŭdz-bûrg)	126-27	37°46´N	84°51´W
Harstad, Nor. (här´städh)	200-01	68°47´N	16°34´E
Hart, Mi., U.S. (härt)	126-27	43°42´N	86°22´W
Hartford, Ct., U.S. (härt´fẽrd)	126-27	41°46´N	72°41´W
Hartford, Ky., U.S. (härt´fẽrd)	126-27	37°27´N	86°54´W
Hartford, Mi., U.S. (härt´fẽrd)	126-27	42°12´N	86°10´W
Hartford, S.D., U.S. (härt´fẽrd)	124-25	43°38´N	96°58´W
Hartford City, In., U.S. (härt´fẽrd sĭ´tē)	126-27	40°27´N	85°22´W
Hartlepool, Eng., U.K. (här´t'l-pōōl)	206-07	54°42´N	1°12´W
Hart Mountain, mtn., Mb., Can. (härt moun´tĭn)	144-45	52°29´N	101°25´W
Harts, stm., S. Afr.	286-87	28°24´S	24°17´E
Hartselle, Al., U.S. (härt´sĕl)	134-35	34°27´N	86°56´W
Hartshorne, Ok., U.S. (härts´hôrn)	130-31	34°51´N	95°34´W
Hartsville, S.C., U.S. (härts´vĭl)	134-35	34°22´N	80°05´W
Hartwell, Ga., U.S. (härt´wĕl)	134-35	34°21´N	82°55´W
Hartwell Lake, res., U.S. (härt´wĕl lāk)	134-35	34°28´N	82°51´W
Har Us nuur, lk., Mong.	260-61	48°0´N	92°10´E
Harvard, Il., U.S. (här´vărd)	126-27	42°25´N	88°37´W
Harvey, N.D., U.S.	124-25	47°46´N	99°55´W
Harwich, Eng., U.K. (här´wĭch)	206-07	51°57´N	1°17´E
Haryāna, state, India	254-55	29°20´N	76°20´E
Harz, mts., Ger. (härts)	210-11	51°45´N	10°30´E
Haskell, Tx., U.S. (hăs´kĕl)	130-31	33°09´N	99°44´W
Hassan, India	256	12°60´N	76°06´E
Hassi Messaoud, Alg.	204-05	31°41´N	6°04´E
Hässleholm, Swe. (hăs´lĕ-hōlm)	208-09	56°09´N	13°46´E
Hastings, N.Z. (hās´tĭngz)	303	39°38´S	176°51´E
Hastings, Eng., U.K. (hās´tĭngz)	206-07	50°52´N	0°35´E
Hastings, Mi., U.S. (hās´tĭngz)	126-27	42°39´N	85°17´W
Hastings, Mn., U.S. (hās´tĭngz)	124-25	44°44´N	92°51´W
Hastings, Ne., U.S. (hās´tĭngz)	130-31	40°35´N	98°24´W
Hatay, Tur. *see* Antioch	248-49	36°12´N	36°10´E
Haţeg, Rom. (kät-säg´)	216-17	45°36´N	22°57´E
Hāthras, India	254-55	27°36´N	78°03´E
Ha Tinh, Viet.	266-67	18°20´N	105°54´E
Hatteras, Cape, c., N.C., U.S. (kāp hăt´ẽr-ăs)	134-35	35°13´N	75°32´W
Hatteras Island, i., N.C., U.S.	134-35	35°25´N	75°29´W
Hattiesburg, Ms., U.S. (hăt´ĭz-bûrg)	134-35	31°20´N	89°17´W
Hatvan, Hung. (hôt´vôn)	210-11	47°40´N	19°41´E
Hat Yai, Thai.	266-67	7°01´N	100°28´E
Haugesund, Nor. (hou´gẽ-soon´)	208-09	59°25´N	5°18´E
Hauraki Gulf, b., N.Z. (hä-ōō-rä´kē gŭlf)	303	36°35´S	175°05´E
Haut Atlas, mts., Mor.	280-81	31°47´N	6°04´W
Haute-Volta, nation, Afr. *see* Burkina Faso	274	13°0´N	1°30´W
Havana, Il., U.S. (hȧ-vă´nȧ)	130-31	40°18´N	90°03´W
Havana, nat. cap., Cuba (hȧ-vă´nȧ)	154-55	23°06´N	82°27´W
Havel, stm., Ger. (hä´fĕl)	210-11	52°53´N	12°01´E
Haverhill, Ma., U.S. (hā´vẽr-hĭl)	126-27	42°47´N	71°05´W
Havre, Fr. *see* Le Havre	212-13	49°29´N	0°08´E
Havre, Mt., U.S. (hăv´ẽr)	122-23	48°33´N	109°41´W
Havre Aubert, Île du, i., Qc., Can.	148-49	47°14´N	61°57´W
Havre de Grace, Md., U.S. (hăv´ẽr dē grās´)	126-27	39°33´N	76°06´W
Havre-Saint-Pierre, Qc., Can.	148-49	50°15´N	63°36´W
Hawai'i, i., Hi., U.S. (häw wī´ē)	137a	19°29´N	155°30´W
Hawai'ian Islands, is., Hi., U.S. (hä-wī´ăn ī´lăndz)	137b	24°0´N	157°00´W
Hawai'i Volcanoes National Park, n.p., Hi., U.S.	137a	19°23´N	155°17´W
Hawaii, state, U.S. (häw wī´ē)	118-19	20°0´N	157°45´W
Hawi, Hi., U.S. (hä´wē)	137a	20°15´N	155°50´W
Hawick, Scot., U.K. (hô´ĭk)	206-07	55°25´N	2°47´W
Hawke Bay, b., N.Z. (hôk bā)	303	39°20´S	177°30´E
Hawker, Austl. (hô´kẽr)	301	31°53´S	138°25´E
Hawkesbury, On., Can. (hôks´bẽr-ĭ)	146-47	45°37´N	74°36´W
Hawkesbury Island, i., B.C., Can.	142-43	53°38´N	129°00´W
Hawkinsville, Ga., U.S. (hô´kĭnz-vĭl)	134-35	32°17´N	83°28´W
Hawley, Mn., U.S. (hô´lĭ)	124-25	46°53´N	96°19´W
Hawthorne, Nv., U.S.	128-29	38°32´N	118°37´W
Haxtun, Co., U.S. (hăks´tŭn)	124-25	40°38´N	102°38´W
Hay, Austl.	301	34°31´S	144°50´E
Hay, stm., Can. (hā)	140-41	60°52´N	115°44´W
HaYarden, stm., Asia *see* Jordan	248-49	31°46´N	35°34´E
Hayastan, nation, Asia *see* Armenia	245	40°0´N	45°00´E
Hayden, Id., U.S. (hā´dĕn)	122-23	47°46´N	116°47´W
Hayes, stm., Mb., Can. (hāz)	144-45	57°03´N	92°14´W
Hayes, Mount, mtn., Ak., U.S. (mount hāz)	136	63°37´N	146°43´W
Haynesville, La., U.S. (hānz´vĭl)	130-31	32°58´N	93°08´W
Hay River, N.T., Can. (hā rĭv´ẽr)	138-39	60°49´N	115°48´W
Hays, Ks., U.S. (hāz)	130-31	38°52´N	99°18´W
Hazard, Ky., U.S. (hăz´ărd)	134-35	37°15´N	83°12´W
Hazārībāg, India	254-55	23°60´N	85°22´E
Hazelton, B.C., Can. (hā´z'l-tŭn)	142-43	55°15´N	127°42´W
Hazelton Mountains, mts., B.C., Can. (hā´z'l-tŭn moun´tĭnz)	142-43	54°51´N	128°00´W
Hazleton, Pa., U.S. (hā´z'l-tŭn)	126-27	40°57´N	75°59´W
Headland, Al., U.S. (hĕd´lănd)	134-35	31°21´N	85°21´W
Healdsburg, Ca., U.S. (hēldz´bûrg)	128-29	38°37´N	122°52´W
Healdton, Ok., U.S. (hēld´tŭn)	130-31	34°14´N	97°29´W
Heard and McDonald Islands, dep., Oc.	308-09	53°05´S	73°00´E
Heard Island, i., Austl. (hûrd ī´lănd)	312	53°06´S	73°30´E
Hearne, Tx., U.S. (hûrn)	132-33	30°53´N	96°36´W
Hearst, On., Can. (hûrst)	138-39	49°41´N	83°42´W
Heavener, Ok., U.S. (hēv´nẽr)	130-31	34°54´N	94°36´W
Hebbronville, Tx., U.S. (hē´brŭn-vĭl)	132-33	27°18´N	98°41´W
Hebei, state, China (hŭ-bā)	260-61	38°0´N	116°00´E
Heber City, Ut., U.S. (hē´bẽr sĭ´tē)	128-29	40°31´N	111°25´W
Heber Springs, Ar., U.S. (hē´bẽr sprĭngz)	134-35	35°29´N	92°02´W
Hebi, China	260-61	35°58´N	114°09´E
Hebrides, is., Scot., U.K.	206-07	57°0´N	6°30´W
Hebron, Nf./L., Can. (hĕb´rŭn)	138-39	58°12´N	62°38´W
Hebron, N.D., U.S. (hĕb´rŭn)	124-25	46°54´N	102°03´W
Hebron, Ne., U.S. (hĕb´rŭn)	130-31	40°09´N	97°35´W
Hebron, W.B.	248-49	31°32´N	35°06´E
Hecate Strait, strt., B.C., Can. (hĕk´ă-tē strāt)	142-43	53°0´N	131°00´W
Hecelchakán, Mex. (ā-sĕl-chä-kän´)	160	20°10´N	90°08´W
Hechi, China (hŭ-chr)	258-59	24°42´N	108°02´E
Hechuan, China (hŭ-chyuän)	258-59	29°60´N	106°16´E
Hedemora, Swe. (hĭ-dĕ-mō´rä)	208-09	60°17´N	15°59´E
Ḩefa, Isr. *see* Haifa	248-49	32°49´N	35°00´E
Hefei, China (hŭ-fā)	258-59	31°51´N	117°17´E
Heflin, Al., U.S. (hĕf´lĭn)	134-35	33°39´N	85°35´W
Hegang, China	260-61	47°19´N	130°16´E
Heho, Mya.	266-67	20°43´N	96°49´E
Heidelberg, Ger. (hīdĕl-bĕrgh)	210-11	49°25´N	8°42´E
Heihe, China	260-61	50°14´N	127°30´E
Heijō, nat. cap., Kor., N. *see* P'yŏngyang	263	39°01´N	125°44´E
Heilbron, S. Afr. (hīl´brōn)	292c	27°17´S	27°59´E
Heilbronn, Ger. (hīl´brōn)	210-11	49°08´N	9°12´E
Heilong, stm., Asia *see* Amur	236-37	52°57´N	141°10´E
Heilongjiang, state, China (hā-lôŋ-jyäŋ)	260-61	48°0´N	128°00´E
Heilongjiang, stm., Asia *see* Amur	236-37	52°57´N	141°10´E
Heilungkiang, state, China *see* Heilongjiang	260-61	48°0´N	128°00´E
Heinola, Fin. (hå-nō´lä)	208-09	61°12´N	26°03´E
Hejaz, reg., Sau. Ar. (hĕ-jäz´) (hē-jäz´) *see* Al-Hijāz	288	24°30´N	38°30´E
Hejian, China (hŭ-jyĕn)	260-61	38°26´N	116°05´E
Hekla, vol., Ice.	206a	64°0´N	19°39´W
Hel, Pol. (hāl)	210-11	54°37´N	18°48´E
Helagsfjället, mtn., Swe.	200-01	62°55´N	12°27´E
Helena, Ar., U.S. (hē-lē´nȧ)	134-35	34°32´N	90°36´W
Helena, Mt., U.S.	122-23	46°36´N	112°02´W
Helen Island, i., Palau	304-05	2°58´N	131°49´E
Helgoland, i., Ger. (hĕl´gô-länd)	210-11	54°11´N	7°52´E
Hellín, Spain (ĕl-yēn´)	214-15	38°30´N	1°42´W
Hells Canyon, val., U.S. (hĕls kăn´yŭn)	122-23	45°17´N	116°40´W
Helmand, stm., Asia (hĕl´mŭnd)	252-53	31°19´N	61°29´E
Helmond, Neth. (hĕl´mônt) (ĕl´môN´)	206-07	51°29´N	5°40´E
Helmstedt, Ger. (hĕlm´shtĕt)	210-11	52°13´N	11°01´E
Helsingborg, Swe. (hĕl´sĭng-bôrgh)	208-09	56°03´N	12°42´E
Helsingfors, nat. cap., Fin. *see* Helsinki	208-09	60°10´N	24°57´E
Helsingør, Den. (hĕl-sĭng-ûr´)	208-09	56°02´N	12°37´E
Helsinki, nat. cap., Fin. (hĕl´sēn-kē)	208-09	60°10´N	24°57´E
Helvetia, nation, Eur. *see* Switzerland	190-91	47°0´N	8°00´E
Hemingford, Ne., U.S. (hĕm´ĭng-fẽrd)	124-25	42°19´N	103°05´W
Hempstead, N.Y., U.S. (hĕmp´stĕd)	126-27	40°42´N	73°41´W
Hempstead, Tx., U.S. (hĕmp´stĕd)	132-33	30°05´N	96°05´W
Hemse, Swe. (hĕm´sĕ)	208-09	57°14´N	18°23´E
Henan, state, China (hŭ-nän)	258-59	34°0´N	114°00´E
Henderson, Ky., U.S. (hĕn´dẽr-sŭn)	126-27	37°50´N	87°35´W
Henderson, N.C., U.S. (hĕn´dẽr-sŭn)	134-35	36°20´N	78°24´W
Henderson, Nv., U.S. (hĕn´dẽr-sŭn)	128-29	36°02´N	114°59´W
Henderson, Tn., U.S. (hĕn´dẽr-sŭn)	134-35	35°26´N	88°38´W
Henderson, Tx., U.S. (hĕn´dẽr-sŭn)	132-33	32°09´N	94°48´W
Henderson Island, i., Pit.	304-05	24°22´S	128°19´W
Hendersonville, N.C., U.S. (hĕn´dẽr-sŭn-vĭl)	134-35	35°19´N	82°28´W
Hendersonville, Tn., U.S. (hĕn´dẽr-sŭn-vĭl)	134-35	36°18´N	86°37´W
Hendorābī, Jazīreh-ye, i., Iran	250-51	26°41´N	53°38´E
Hendrina, S. Afr. (hĕn-drē´nȧ)	292c	26°10´S	29°44´E
Hendū Kosh, mts., Asia *see* Hindu Kush	252-53	36°0´N	71°30´E
Hengām, Jazīreh-ye, i., Iran	250-51	26°39´N	55°53´E
Hengelo, Neth. (hĕŋgĕ-lō)	206-07	52°16´N	6°48´E
Hengshan, China (hĕng´shän´)	258-59	27°15´N	112°51´E
Hengshan, China (hĕng´shän´)	260-61	37°57´N	109°18´E
Hengshui, China (hĕng´shōō-ē´)	260-61	37°44´N	115°42´E
Hengxian, China (hŭŋ shyĕn)	258-59	22°41´N	109°12´E

ăt; finăl; rāte; senâte; ärm; ȧsk; sofȧ; fâre; ch-choose; dh-as th in other; bē; ėvent; bĕt; recĕnt; cratẽr; g-gō; gh-guttural g; bĭt; ĭ-short neutral; rīde; ĸ-guttural k as ch in German ich;

PRONOUNCING INDEX
Hengyang – Horton 363

Feature (Pronunciation)	Page	Lat.	Long.
Hengyang, China	**258-59**	26°54'N	112°36'E
Henlopen, Cape, c., De., U.S. (kāp hĕn-lō´pĕn)	**126-27**	38°47'N	75°06'W
Hennebont, Fr. (ĕn-bôɴ´)	**212-13**	47°48'N	3°16'W
Hennessey, Ok., U.S. (hĕn´ĕ-sĭ)	**130-31**	36°07'N	97°54'W
Henrietta, Tx., U.S. (hen-rĭ-ĕ´tȧ)	**130-31**	33°49'N	98°12'W
Henrietta Maria, Cape, c., On., Can. (kāp hĕn-rĭ-ĕt´ȧ mȧ-rē´ȧ)	**140-41**	55°08'N	82°20'W
Henzada, Mya.	**266-67**	17°38'N	95°28'E
Hepu, China (hū-pōō)	**258-59**	21°41'N	109°11'E
Herāt, Afg. (hĕ-rät´)	**252-53**	34°21'N	62°12'E
Heredia, C.R. (ā-rā´dhē-ä)	**161**	9°59'N	84°07'W
Hereford, Eng., U.K. (hĕrĕ´fērd)	**206-07**	52°04'N	2°43'W
Hereford, Tx., U.S. (hĕr´ĕ-fērd)	**130-31**	34°50'N	102°24'W
Herford, Ger. (hĕr´fôrt)	**210-11**	52°07'N	8°40'E
Herkimer, N.Y., U.S. (hûr´kĭ-mēr)	**126-27**	43°02'N	74°58'W
Herlen, stm., Asia *see* Kerulen	**260-61**	48°44'N	117°03'E
Hermannstadt, Rom. *see* Sibiu	**210-11**	45°47'N	24°09'E
Hermansville, Mi., U.S. (hûr´măns-vĭl)	**126-27**	45°42'N	87°36'W
Hermit Islands, is., Pap. N. Gui. (hûr´mĭt ī´lăndz)	**302**	1°30'S	145°05'E
Hermon, Mount, mtn., Asia	**248-49**	33°25'N	35°51'E
Hermosillo, Mex. (ĕr-mô-sē´l-yō)	**156-57**	29°05'N	110°58'W
Hermus, stm., Tur. *see* Gediz	**216-17**	38°36'N	26°48'E
Herning, Den. (hĕr´nĭng)	**208-09**	56°08'N	8°59'E
Herrin, Il., U.S. (hĕr´ĭn)	**126-27**	37°48'N	89°02'W
Herstal, Bel. (hĕr´stäl)	**206-07**	50°40'N	5°38'E
Hertford, N.C., U.S. (hûrt´fērd)	**134-35**	36°11'N	76°29'W
Hervey Bay, Austl.	**301**	25°17'S	152°50'E
Hervey Bay, b., Austl.	**296-97**	25°0'S	153°00'E
Hesperia, Ca., U.S.	**128-29**	34°25'N	117°18'W
Hessen, hist. reg., Ger. (hĕs´ĕn)	**210-11**	50°30'N	9°15'E
Hettinger, N.D., U.S. (hĕt´ĭn-jēr)	**124-25**	46°00'N	102°38'W
Hexian, China (hŭ shyĕn)	**258-59**	24°18'N	111°39'E
Heyuan, China (hŭ-yůän)	**258-59**	23°43'N	114°42'E
Heze, China (hŭ-dzŭ)	**260-61**	35°15'N	115°27'E
Hialeah, Fl., U.S. (hī-ȧ-lē´ăh)	**135a**	25°51'N	80°17'W
Hiawatha, Ia., U.S. (hī-ȧ-wô´thȧ)	**124-25**	42°02'N	91°41'W
Hibbing, Mn., U.S. (hĭb´ĭng)	**124-25**	47°25'N	92°56'W
Hickman, Ky., U.S. (hĭk´mȧn)	**134-35**	36°34'N	89°11'W
Hickory, N.C., U.S. (hĭk´ō-rĭ)	**134-35**	35°44'N	81°21'W
Hicks, Point, c., Austl.	**301**	37°48'S	149°16'E
Hidalgo, Mex. (ē-dhäl´gō)	**158-59**	24°15'N	99°26'W
Hidalgo, Mex. (ē-dhäl´gō)	**132-33**	25°58'N	100°27'W
Hidalgo, state, Mex. (ē-dhäl´gō)	**158-59**	20°30'N	99°00'W
Hidalgo del Parral, Mex. (ē-dä´l-gō-dĕl-pär-rä´l)	**132-33**	26°57'N	105°40'W
Higashi Shina Kai, s., Asia *see* East China Sea	**240-41**	30°0'N	126°00'E
Higginsville, Mo., U.S. (hĭg´ĭnz-vĭl)	**130-31**	39°05'N	93°43'W
Highland, Il., U.S. (hī´lănd)	**130-31**	38°44'N	89°41'W
Highland Park, Il., U.S. (hī´lănd pärk)	**126-27**	42°11'N	87°48'W
Highland Park, Mi., U.S. (hī´lănd pärk)	**126-27**	42°24'N	83°07'W
High Level, Ab., Can.	**138-39**	58°31'N	117°08'W
Highmore, S.D., U.S. (hī´mōr)	**124-25**	44°31'N	99°26'W
High Point, N.C., U.S. (hī point)	**134-35**	35°57'N	80°01'W
High Prairie, Ab., Can. (hī prā´rĭ)	**142-43**	55°26'N	116°29'W
High River, Ab., Can. (hī rĭv´ēr)	**142-43**	50°35'N	113°52'W
Hightstown, N.J., U.S. (hīts-toun)	**126-27**	40°17'N	74°32'W
High Wycombe, Eng., U.K. (hī wĭ-kŭm)	**206-07**	51°38'N	0°45'W
Higüey, Dom. Rep. (ē-gwĕ´y)	**154-55**	18°37'N	68°43'W
Hiiumaa, i., Est. (hē´ŏm-ô)	**208-09**	58°55'N	22°38'E
Hikone, Japan (hē´kō-nĕ)	**265**	35°15'N	136°15'E
Hildesheim, Ger. (hĭl´dĕs-hīm)	**210-11**	52°09'N	9°57'E
Hilla, Iraq *see* Al-Ḥillah	**248-49**	32°29'N	44°26'E
Hillaby, Mount, mtn., Barb. (mount hĭl´ȧ-bĭ)	**155b**	13°12'N	59°35'W
Hillerød, Den. (hē´lĕ-rûdh hĭl)	**208-09**	55°56'N	12°19'E
Hillsboro, Il., U.S. (hĭlz´bŭr-ō)	**126-27**	39°09'N	89°29'W
Hillsboro, N.D., U.S. (hĭlz´bŭr-ō)	**124-25**	47°24'N	97°04'W
Hillsboro, Oh., U.S. (hĭlz´bŭr-ō)	**126-27**	39°12'N	83°37'W
Hillsboro, Or., U.S. (hĭlz´bŭr-ō)	**122-23**	45°31'N	122°59'W
Hillsboro, Tx., U.S. (hĭlz´bŭr-ō)	**132-33**	32°01'N	97°08'W
Hillsboro, W.V., U.S. (hĭlz´bŭr-ō)	**126-27**	38°08'N	80°13'W
Hillsdale, Mi., U.S. (hĭls-dāl hĭlz)	**126-27**	41°55'N	84°38'W
Hillston, Austl.	**301**	33°29'S	145°33'E
Hilo, Hi., U.S. (hē´lō)	**137a**	19°43'N	155°05'W
Himāchal Pradesh, state, India	**254-55**	32°0'N	77°00'E
Himalayas, mts., Asia (hĭ-mä´lȧ-yȧz) (hĭ-mȧ-lā´-yȧz)	**254-55**	28°0'N	84°00'E
Himalaya Shan, mts., Asia *see* Himalayas	**254-55**	28°0'N	84°00'E
Himatnagar, India	**254-55**	23°35'N	72°58'E
Himeji, Japan (hē´mȧ-jē)	**265**	34°50'N	134°42'E
Ḥimṣ, Syria	**248-49**	34°45'N	36°44'E
Hinche, Haiti (hēn´chȧ) (äɴsh)	**154-55**	19°09'N	72°00'W
Hinchinbrook Island, i., Austl. (hĭn-chĭn-brŏŏk ī´lȧnd)	**302**	18°23'S	146°17'E
Hindenburg, Pol. *see* Zabrze	**210-11**	50°18'N	18°46'E
Hindu Kush, mts., Asia (hĭn´dōō kōōsh´)	**252-53**	36°0'N	71°30'E
Hindupur, India (hĭn´dōō-pōōr)	**256**	13°49'N	77°30'E
Hinnøya, i., Nor.	**200-01**	68°32'N	15°59'E
Hinton, Ab., Can. (hĭn´tŭn)	**142-43**	53°25'N	117°34'W
Hinton, W.V., U.S. (hĭn´tŭn)	**126-27**	37°40'N	80°53'W
Hirado-shima, i., Japan (hē´rä-dō shē´mä)	**265**	33°20'N	129°30'E
Hirara, Japan	**306a**	24°47'N	125°17'E
Hīrmand, stm., Asia *see* Helmand	**252-53**	31°19'N	61°29'E
Hirosaki, Japan (hē´rō-sä´kē)	**264**	40°36'N	140°29'E
Hiroshima, Japan (hē-rō-shē´mȧ)	**265**	34°24'N	132°28'E
Hirosima, Japan *see* Hiroshima	**265**	34°24'N	132°28'E
Hirschberg, Pol. *see* Jelenia Góra	**210-11**	50°54'N	15°44'E
Hirson, Fr. (ēr-sôɴ´)	**212-13**	49°56'N	4°05'E
Hisār, India	**254-55**	29°09'N	75°44'E
Hispaniola, i., N.A. (hĭ´spăn-ĭ-ō-lȧ)	**154-55**	19°0'N	71°00'W
Hitachi, Japan (hē-tä´chē)	**265**	36°36'N	140°39'E
Hitoyoshi, Japan (hē´tō-yō´shē)	**265**	32°12'N	130°46'E
Hitra, i., Nor. (hĭträ)	**200-01**	63°33'N	8°45'E
Hiu, i., Vanuatu	**306g**	13°08'S	166°33'E
Hiva Oa, i., Fr. Poly.	**304-05**	9°45'S	139°00'W
Hjo, Swe. (yō)	**208-09**	58°18'N	14°17'E
Hjørring, Den. (jûr´ĭng)	**208-09**	57°28'N	9°59'E
Hkakabo Razi, mtn., Mya.	**258-59**	28°17'N	97°46'E
Hlohovec, Slvk. (hlō´ho-vĕts)	**210-11**	48°26'N	17°48'E
Ho, Ghana	**282-83**	6°36'N	0°28'E
Hoa Binh, Viet.	**266-67**	20°50'N	105°20'E
Hobart, Austl. (hō´bȧrt)	**301**	42°52'S	147°18'E
Hobart, Ok., U.S. (hō´bȧrt)	**130-31**	35°02'N	99°06'W
Hobbs, N.M., U.S. (hŏbs)	**130-31**	32°42'N	103°08'W
Hobro, Den. (hō-brō´)	**208-09**	56°38'N	9°48'E
Ho Chi Minh City, Viet. (hō-chē-mĭn sĭ´tē)	**266-67**	10°45'N	106°40'E
Hodeida, Yemen *see* Al-Ḥudaydah	**288**	14°48'N	42°57'E
Hodgenville, Ky., U.S. (hŏj´ĕn-vĭl)	**126-27**	37°34'N	85°44'W
Hódmezővásárhely, Hung. (hōd´mĕ-zû-vō´shor-hĕl-y´)	**210-11**	46°25'N	20°20'E
Hodna, Chott el, lk., Alg.	**292b**	35°25'N	4°45'E
Hodonín, Czech Rep. (hē´dō-nēn)	**210-11**	48°51'N	17°08'E
Hoei, Bel. *see* Huy	**206-07**	50°31'N	5°14'E
Hof, Ger. (hōf)	**210-11**	50°18'N	11°55'E
Hofsjökull, Ice, Ice. (hôfs´yû´kōōl)	**206a**	64°50'N	18°54'W
Hōfu, Japan	**265**	34°03'N	131°35'E
Hofuf, Sau. Ar. *see* Al-Hufūf	**250-51**	25°22'N	49°34'E
Hogansville, Ga., U.S. (hō´gănz-vĭl)	**134-35**	33°11'N	84°55'W
Hoggar, mts., Alg. *see* Ahaggar	**280-81**	23°0'N	6°30'E
Hohensalza, Pol. *see* Inowrocław	**210-11**	52°48'N	18°15'E
Hohe Tauern, mts., Aus. (hō´ĕ tou´ērn)	**210-11**	47°06'N	12°56'E
Hohhot, China (hŭ-hō-tŭ)	**260-61**	40°49'N	111°39'E
Hoihow, China *see* Haikou	**258-59**	20°03'N	110°22'E
Hoisington, Ks., U.S. (hoi´zĭng-tŭn)	**130-31**	38°31'N	98°46'W
Hokitika, N.Z. (hō-kĭ-tē´kä)	**303**	42°44'S	170°58'E
Hokkaidō, i., Japan (hōk´kī-dō)	**264**	44°0'N	143°00'E
Hola, Kenya	**284-85**	1°30'S	40°01'E
Holbrook, Az., U.S. (hōl´brŏŏk)	**128-29**	34°54'N	110°10'W
Holden, Mo., U.S. (hōl´dĕn)	**130-31**	38°43'N	93°60'W
Holden, W.V., U.S. (hōl´dĕn)	**126-27**	37°49'N	82°04'W
Holdenville, Ok., U.S. (hōl´dĕn-vĭl)	**130-31**	35°05'N	96°24'W
Holdrege, Ne., U.S. (hōl´drĕj)	**130-31**	40°27'N	99°22'W
Holguín, Cuba (ôl-gēn´)	**154-55**	20°53'N	76°15'W
Holguín, state, Cuba (ôl-gēn´)	**154-55**	20°55'N	75°50'W
Holland, Mi., U.S. (hŏl´ănd)	**126-27**	42°47'N	86°06'W
Holland, nation, Eur. *see* Netherlands	**190-91**	52°15'N	5°30'E
Hollandia, Indon. *see* Jayapura	**302**	2°32'S	140°43'E
Hollandsbird Island, i., Nmb.	**286-87**	24°39'S	14°32'E
Hollick-Kenyon Plateau, plat., Ant.	**313**	79°0'S	97°00'W
Hollis, Ok., U.S. (hŏl´ĭs)	**130-31**	34°41'N	99°55'W
Hollister, Mo., U.S. (hŏl´ĭs-tēr)	**130-31**	36°37'N	93°13'W
Holly Springs, Ms., U.S. (hŏl´ĭ sprĭngz)	**134-35**	34°46'N	89°27'W
Hollywood, Fl., U.S. (hŏl´ē-wŏŏd)	**135a**	26°01'N	80°09'W
Holmestrand, Nor. (hôl´mĕ-strän)	**208-09**	59°29'N	10°18'E
Holstebro, Den. (hôl´stĕ-brō)	**208-09**	56°22'N	8°38'E
Holyhead, Wales, U.K. (hŏl´ē-hĕd)	**206-07**	53°19'N	4°38'W
Holyoke, Co., U.S. (hōl´yōk)	**124-25**	40°35'N	102°18'W
Holyoke, Ma., U.S. (hōl´yōk)	**126-27**	42°12'N	72°37'W
Homa Bay, Kenya	**289**	0°31'S	34°27'E
Homalin, Mya.	**258-59**	24°51'N	94°56'E
Homel', Bela.	**218-19**	52°26'N	30°59'E
Homer, Ak., U.S. (hō´mēr)	**136**	59°39'N	151°31'W
Homer, La., U.S. (hō´mēr)	**130-31**	32°48'N	93°04'W
Homestead, Fl., U.S. (hōm´stĕd)	**135a**	25°29'N	80°28'W
Homestead National Monument of America, n.p., Ne., U.S. (hōm´stĕd)	**130-31**	40°16'N	96°48'W
Homewood, Al., U.S. (hōm´wŏŏd)	**134-35**	33°28'N	86°48'W
Hominy, Ok., U.S. (hŏm´ĭ-nĭ)	**130-31**	36°25'N	96°24'W
Homs, Libya *see* Al-Khums	**204-05**	32°39'N	14°16'E
Homs, Syria *see* Ḥimṣ	**248-49**	34°45'N	36°44'E
Honan, state, China *see* Henan	**258-59**	34°0'N	114°00'E
Honda, Col. (hōn´dȧ)	**188c**	5°13'N	74°45'W
Hondo, stm., N.A. (hon-dō´)	**160**	18°29'N	88°18'W
Honduras, nation, N.A. (hŏn-dōō´rȧs)	**94**	15°0'N	86°30'W
Honduras, Golfo de, b., N.A. *see* Honduras, Gulf of	**160**	16°05'N	87°58'W
Honduras, Gulf of, b., N.A. (gŭlf ŭv hŏn-dōō´rȧs)	**160**	16°05'N	87°58'W
Hønefoss, Nor. (hē´nē-fōs)	**208-09**	60°11'N	10°15'E
Honesdale, Pa., U.S. (hōnz´dāl)	**126-27**	41°34'N	75°16'W
Honfleur, Fr. (ôɴ-flûr´)	**212-13**	49°25'N	0°14'E
Hong, Song, stm., Asia *see* Red	**258-59**	20°18'N	106°32'E
Hon Gai, Viet.	**266-67**	21°03'N	107°04'E
Hongjiang, China	**258-59**	27°04'N	109°58'E
Hong Kong, China (hŏng kŏng) (hŏng kŏng)	**258-59**	22°16'N	114°10'E
Hongliuyuan, China	**260-61**	41°02'N	95°25'E
Hongshui, stm., China (hôŋ-shwā)	**258-59**	23°48'N	109°32'E
Hongtong, China	**260-61**	36°17'N	111°40'E
Honguedo, Détroit d', strt., Qc., Can.	**148-49**	49°15'N	64°00'W
Hongze Hu, lk., China	**258-59**	33°16'N	118°34'E
Honiara, nat. cap., Sol. Is. (hō-nē-ä´-rä)	**306e**	9°26'S	159°57'E
Honiton, Eng., U.K. (hŏn´ĭ-tŏn)	**206-07**	50°48'N	3°12'W
Honolulu, Hi., U.S. (hŏn-ō-lōō´lōō)	**137a**	21°19'N	157°52'W
Honouliuli National Monument, n.p., Hi., U.S	**137a**	21°23'N	158°04'W
Honshū, i., Japan (hŏn´-shōō)	**264**	36°0'N	138°00'E
Hood, i., Ec. *see* Española, Isla	**184a**	1°23'S	89°42'W
Hood, Mount, vol., Or., U.S. (mount hŏŏd)	**122-23**	45°23'N	121°42'W
Hood River, Or., U.S. (hŏŏd rĭv´ēr)	**122-23**	45°42'N	121°31'W
Hooker, Ok., U.S. (hŏŏk´ēr)	**130-31**	36°52'N	101°13'W
Hook Island, i., Austl.	**302**	20°08'S	148°55'E
Hoonah, Ak., U.S. (hōō´nä)	**136**	58°07'N	135°26'W
Hooper Bay, Ak., U.S. (hŏŏp´ēr bā)	**136**	61°31'N	166°06'W
Hoopeston, Il., U.S. (hōōps´tŭn)	**126-27**	40°28'N	87°39'W
Hoosick Falls, N.Y., U.S. (hōō´sĭk fôlz)	**126-27**	42°54'N	73°21'W
Hoover Dam, d., U.S. (hōō´vēr dăm)	**128-29**	36°02'N	114°43'W
Hope, B.C., Can.	**142-43**	49°23'N	121°26'W
Hope, Ar., U.S. (hōp)	**130-31**	33°40'N	93°35'W
Hope, Ben, mtn., Scot., U.K. (bĕn hōp)	**206-07**	58°24'N	4°37'W
Hopedale, Nf./L., Can. (hōp´dāl)	**138-39**	55°28'N	60°13'W
Hopeh, state, China *see* Hebei	**260-61**	38°0'N	116°00'E
Hopelchén, Mex. (ō-pĕl-chĕ´n)	**160**	19°46'N	89°51'W
Hopes Advance, Cap, c., Qc., Can. (kȧp hōps ăd-vans´)	**140-41**	61°04'N	69°34'W
Hopetoun, Austl. (hōp´toun)	**301**	35°44'S	142°22'E
Hopetown, S. Afr. (hōp´toun)	**286-87**	29°35'S	24°04'E
Hopewell, Va., U.S. (hōp´wĕl)	**134-35**	37°18'N	77°17'W
Hopi Indian Reservation, ind. res., Az., U.S. (hō´pē ĭn´dĭ-ăn rĕ-sēr-vā´shĕn)	**128-29**	35°45'N	110°35'W
Hopkinsville, Ky., U.S. (hŏp´kĭns-vĭl)	**134-35**	36°52'N	87°29'W
Hoquiam, Wa., U.S. (hō´kwĭ-ăm)	**122-23**	46°59'N	123°53'W
Horicon, Wi., U.S. (hŏr´ĭ-kŏn)	**126-27**	43°27'N	88°37'W
Horlivka, Ukr.	**218-19**	48°20'N	38°03'E
Hormoz, Jazīreh-ye, i., Iran	**250-51**	27°04'N	56°28'E
Hormuz, Strait of, strt., Asia (strāt ŭv hôr´mŭz´)	**250-51**	26°34'N	56°15'E
Horn, c., Ice.	**206a**	66°28'N	22°28'W
Horn, Cape, c., Chile (kāp hôrn)	**185**	55°59'S	67°16'W
Hornavan, lk., Swe.	**200-01**	66°10'N	17°46'E
Hornell, N.Y., U.S. (hôr-nĕl´)	**126-27**	42°20'N	77°39'W
Hornepayne, On., Can.	**146-47**	49°12'N	84°47'W
Horn Island, i., Ms., U.S.	**134-35**	30°15'N	88°43'W
Hornos, Cabo de, c., Chile *see* Horn, Cape	**185**	55°59'S	67°16'W
Horn Plateau, plat., N.T., Can.	**140-41**	62°08'N	120°16'W
Horqin Youyi Qianqi, China *see* Ulanhot	**260-61**	46°04'N	122°04'E
Horqueta, Para. (ŏr-kĕ´tä)	**182-83**	23°20'S	57°03'W
Horse Islands, is., Nf./L., Can. (hôrs ī´lăndz)	**148-49**	50°13'N	55°45'W
Horsens, Den. (hôrs´ĕns)	**208-09**	55°52'N	9°52'E
Horsham, Austl. (hôr´shăm) (hôrs´ăm)	**301**	36°43'S	142°12'E
Horten, Nor. (hôr´tĕn)	**208-09**	59°25'N	10°29'E
Horton, stm., N.T., Can.	**140-41**	69°55'N	127°02'W

n-sing; ŋ-baŋk; ɴ-nasalized n; nŏd; cŏmmit; ōld; ōbey; ôrder; oi-boil; fōōd; ŏŏ-as oo in foot; ou-out; s-soft; sh-dish; th-thin; pūre; ûnite; ûrn; stŭd; circŭs; ü-as in French tu; ´-indeterminate vowel.

Feature (Pronunciation)	Page	Lat.	Long.
Hosa'ina, Eth.	292d	7°37'N	37°56'E
Hosapete, India	256	15°16'N	76°23'E
Hoséré Vokré, mtn., Camrn.	282-83	8°20'N	13°15'E
Hoshangābād, India	254-55	22°45'N	77°43'E
Hoshiārpur, India	254-55	31°32'N	75°55'E
Hoste, Isla, i., Chile (ê's-lä-ôs'tā)	185	55°05'S	69°15'W
Hotan, China (hwô-tän)	244	37°07'N	79°55'E
Hotan, stm., China (hwô-tän)	244	40°30'N	80°56'E
Hot Springs, Ar., U.S. (hŏt sprĭngz)	130-31	34°30'N	93°04'W
Hot Springs, N.M., U.S. (hŏt sprĭngz) *see* Truth or Consequences	128-29	33°08'N	107°15'W
Hot Springs, S.D., U.S. (hŏt sprĭngz)	124-25	43°26'N	103°29'W
Hot Springs, Va., U.S. (hŏt sprĭngz)	126-27	37°60'N	79°49'W
Hot Springs National Park, n.p., Ar., U.S. (hŏt sprĭngz năsh'ŭn-ăl pärk)	130-31	34°31'N	93°02'W
Hottah Lake, lk., N.T., Can.	140-41	65°04'N	118°29'W
Houghton, Mi., U.S. (hŏ'tŭn)	124-25	47°06'N	88°36'W
Houghton Lake, lk., Mi., U.S. (hŏ'tŭn lăk)	126-27	44°20'N	84°45'W
Houlton, Me., U.S. (hōl'tŭn)	127a	46°08'N	67°50'W
Houma, China	260-61	35°37'N	111°21'E
Houma, La., U.S. (hōō'mà)	134-35	29°35'N	90°43'W
Houston, Ms., U.S. (hūs'tŭn)	134-35	33°54'N	88°60'W
Houston, Tx., U.S. (hūs'tŭn)	132-33	29°45'N	95°22'W
Hovd, Mong. *see* Dund-Us	260-61	47°60'N	91°38'E
Hovd, stm., Mong.	236-37	48°05'N	92°13'E
Hövsgöl nuur, lk., Mong.	260-61	51°0'N	100°30'E
Howard, S.D., U.S. (hou'ărd)	124-25	44°00'N	97°32'W
Howe, Cape, c., Austl. (kāp hou)	301	37°30'S	149°58'E
Howell, Mi., U.S. (hou'ĕl)	126-27	42°36'N	83°56'W
Howe Sound, strt., B.C., Can. (hou sound)	142-43	49°22'N	123°18'W
Howland Island, dep., Oc. (hou'lănd ī'lănd)	304-05	0°51'N	176°38'W
Howland Island, i., Oc. (hou'lănd ī'lănd)	304-05	0°48'N	176°38'W
Hoxie, Ar., U.S. (hŏk'sĭ)	134-35	36°03'N	90°59'W
Hradec Králové, Czech Rep.	210-11	50°12'N	15°50'E
Hranice, Czech Rep. (hrän'yĕ-tsĕ)	210-11	49°33'N	17°45'E
Hrodna, Bela.	210-11	53°41'N	23°50'E
Hrubieszów, Pol. (hrōō-byā'shŏŏf)	210-11	50°49'N	23°56'E
Hrvatska, nation, Eur. (hr-väts'kä) *see* Croatia	190-91	45°10'N	15°30'E
Hsinchu, Tai. (hsĭn'chōō')	243a	24°48'N	120°58'E
Hsinhailien, China *see* Lianyungang	258-59	34°37'N	119°11'E
Hsipaw, Mya.	266-67	22°37'N	97°18'E
Huacho, Peru	184	11°08'S	77°37'W
Huadian, China (hwä-dĭĕn)	260-61	42°58'N	126°45'E
Hua Hin, Thai.	266-67	12°35'N	99°57'E
Huai'an, China (hwī-än)	258-59	33°31'N	119°08'E
Huai'an, China (hwī-än)	260-61	40°40'N	114°25'E
Huaicheng, China *see* Huai'an	258-59	33°31'N	119°08'E
Huaide, China *see* Gongzhuling	260-61	43°30'N	124°49'E
Huailai, China	260-61	40°23'N	115°34'E
Huainan, China	258-59	32°40'N	117°01'E
Huaiyang, China (hōōäī'yang)	258-59	33°44'N	114°53'E
Huajuapan de León, Mex. (wäj-wä'päm dā lā-ōn')	158-59	17°49'N	97°45'W
Hualfín, Arg.	182-83	27°14'S	66°50'W
Hualien, Tai. (hwä'lyĕn')	243a	23°58'N	121°35'E
Huallaga, stm., Peru (wäl-yä'gä)	184	5°06'S	75°36'W
Huallanca, Peru	184	8°49'S	77°52'W
Huambo, Ang.	286-87	12°46'S	15°44'E
Huancavelica, Peru (wän'kä-vā-lē'kä)	188a	12°47'S	75°01'W
Huancayo, Peru (wän-kä'yō)	188a	12°05'S	75°13'W
Huang, stm., China (hůäŋ)	240-41	37°49'N	118°53'E
Huangchuan, China (hůäŋ-chůän)	258-59	32°08'N	115°03'E
Huang Hai, s., Asia *see* Yellow Sea	240-41	36°0'N	123°00'E
Huangho, stm., China *see* Huang	240-41	37°49'N	118°53'E
Huanghua, China (hůäŋ-hwä)	260-61	38°22'N	117°21'E
Huangshan, China	258-59	29°45'N	118°18'E
Huangshi, China	258-59	30°13'N	115°05'E
Huangyuan, China (hůäŋ-yůän)	260-61	36°41'N	101°16'E
Huanren, China (hůän-rŭn)	263	41°13'N	125°20'E
Huánuco, Peru (wä-nōō'kō)	184	9°56'S	76°15'W
Huanuni, Bol. (wä-nōō'nē)	182-83	18°17'S	66°50'W
Huaral, Peru (wä-rä'l)	188a	11°30'S	77°12'W
Huaraz, Peru	184	9°32'S	77°33'W
Huascarán, Nevado, mtn., Peru (nĕ-vä'dō wäs-kä-rän')	184	9°07'S	77°37'W
Huasco, Chile (wäs'kō)	182-83	28°28'S	71°15'W
Huatabampo, Mex.	156-57	26°50'N	109°38'W
Huauchinango, Mex. (wä-ōō-chē-näŋ'gō)	158-59	20°11'N	98°03'W
Huautla, Mex. (wä-ōō'tlä)	158-59	18°08'N	96°50'W
Huaxian, China (hwä shyĕn)	258-59	23°23'N	113°12'E
Huaynamota, stm., Mex. (wäy-nä-mō'tä)	158-59	21°57'N	104°32'W
Hubballi, India	256	15°21'N	75°09'E
Hubbard, Tx., U.S. (hŭb'ẽrd)	132-33	31°50'N	96°48'W
Hubbard Creek Reservoir, lk., Tx., U.S. (hŭb'ẽrd krēk rĕ'zẽr-vwär)	130-31	32°48'N	99°01'W
Hubei, state, China (hōō-bā)	258-59	31°0'N	112°00'E
Huddersfield, Eng., U.K. (hŭd'ẽrz-fēld)	206-07	53°39'N	1°47'W
Hudiksvall, Swe. (hōō'dĭks-väl)	208-09	61°44'N	17°07'E
Hudson, Mi., U.S. (hŭd'sŭn)	126-27	41°51'N	84°21'W
Hudson, Wi., U.S. (hŭd'sŭn)	124-25	44°59'N	92°45'W
Hudson, stm., U.S. (hŭd'sŭn)	120-21	40°41'N	74°02'W
Hudson, Détroit d', strt., Can. *see* Hudson Strait	140-41	62°30'N	71°60'W
Hudson Bay, Sk., Can. (hŭd'sŭn bā)	144-45	52°52'N	102°23'W
Hudson Bay, b., Can. (hŭd'sŭn bā)	140-41	60°0'N	86°00'W
Hudson Falls, N.Y., U.S. (hŭd'sŭn fôlz)	126-27	43°19'N	73°35'W
Hudson Strait, strt., Can. (hŭd'sŭn strāt)	140-41	62°30'N	71°60'W
Hue, Viet. (ü-ā')	266-67	16°28'N	107°35'E
Huehuetenango, Guat. (wā-wå-tå-näŋ'gō)	160	15°21'N	91°27'W
Huejuquilla El Alto, Mex. (wā-hōō-kēl'yä ĕl äl'tō)	158-59	22°38'N	103°54'W
Huelva, Spain (wĕl'vä)	214-15	37°16'N	6°57'W
Huesca, Spain (wĕs-kä)	214-15	42°08'N	0°25'W
Huéscar, Spain (wās'kär)	214-15	37°48'N	2°33'W
Huetamo de Núñez, Mex.	158-59	18°35'N	100°53'W
Hughenden, Austl. (hū'ĕn-dĕn)	302	20°51'S	144°13'E
Hugli, stm., India (hōōg'lĭ)	254-55	21°36'N	87°60'E
Hugo, Ok., U.S. (hū'gō)	130-31	34°01'N	95°31'W
Hugoton, Ks., U.S. (hū'gō-tŭn)	130-31	37°11'N	101°21'W
Huhehot, China *see* Hohhot	260-61	40°49'N	111°39'E
Huichapan, Mex. (wē-chä-pän')	158-59	20°22'N	99°40'W
Hŭich'ŏn, Kor., N.	263	40°10'N	126°17'E
Huila, Nevado del, vol., Col. (nĕ-vä-dō-del-wē'lä)	188c	2°59'N	75°58'W
Huili, China	258-59	26°40'N	102°14'E
Huimin, China (hōōĭ mĭn)	260-61	37°29'N	117°32'E
Huinan, China	263	42°41'N	126°02'E
Huixtla, Mex.	160	15°08'N	92°27'W
Huize, China	258-59	26°25'N	103°18'E
Huizhou, China	258-59	23°05'N	114°24'E
Hujirt, Mong.	260-61	48°53'N	101°14'E
Hukuoka, Japan *see* Fukuoka	265	33°35'N	130°25'E
Hulan, China (hōō'län')	260-61	45°59'N	126°36'E
Hulan Ergi, China	260-61	47°12'N	123°38'E
Hulin, China (hōō'lĭn')	264	45°46'N	132°59'E
Hulun, China *see* Hailar	260-61	49°11'N	119°44'E
Hulun Nur, lk., China (hōō-lòn nòr)	260-61	49°01'N	117°32'E
Humacao, P.R. (ōō-mä-kä'ō)	154a	18°09'N	65°49'W
Humahuaca, Arg.	182-83	23°12'S	65°21'W
Humaitá, Braz.	180-81	7°30'S	63°02'W
Humaitá, Para.	187	27°05'S	58°32'W
Humble, Tx., U.S. (hŭm'b'l)	132-33	29°59'N	95°16'W
Humboldt, Sk., Can. (hŭm'bōlt)	144-45	52°11'N	105°07'W
Humboldt, Ia., U.S. (hŭm'bōlt)	124-25	42°43'N	94°13'W
Humboldt, Tn., U.S. (hŭm'bōlt)	134-35	35°50'N	88°55'W
Humboldt, stm., Nv., U.S. (hŭm'bōlt)	120-21	40°01'N	118°33'W
Hume, Lake, res., Austl.	301	36°08'S	147°02'E
Humphreys Peak, mtn., Az., U.S. (hŭm'frĭs pēk)	128-29	35°20'N	111°40'W
Húnaflói, b., Ice. (hōō'nä-flō'ĭ)	206a	65°50'N	20°50'W
Hunan, state, China (hōō'nän')	258-59	28°0'N	111°00'E
Hunchun, China (hòn-chŭn)	263	42°52'N	130°22'E
Hunedoara, Rom. (kōō'nĕd-wä'rà)	216-17	45°46'N	22°55'E
Hungary, nation, Eur. (hŭŋ'gå-rĭ)	190-91	47°0'N	20°00'E
Hŭngdŏki-dong, Kor., N.	263	39°50'N	127°38'E
Hungerford, Austl. (hŭn'gẽr-fẽrd)	301	28°59'S	144°24'E
Hŭngnam, Kor., N. *see* Hŭngdŏki-dong	263	39°50'N	127°38'E
Hunjiang, China	263	41°57'N	126°28'E
Hunsrück, mts., Ger. (hōōns'rûk)	210-11	49°46'N	7°08'E
Hunter, Île, i., N. Cal.	304-05	22°24'S	186°06'E
Hunter Island, i., Austl.	301	40°32'S	144°45'E
Hunter Island, i., B.C., Can.	142-43	51°54'N	128°03'W
Huntingburg, In., U.S. (hŭnt'ĭng-bûrg)	126-27	38°18'N	86°57'W
Huntingdon, Qc., Can. (hŭnt'ĭng-dŭn)	146-47	45°06'N	74°10'W
Huntingdon, Tn., U.S. (hŭnt'ĭng-dŭn)	134-35	36°00'N	88°26'W
Huntington, In., U.S. (hŭnt'ĭng-tŭn)	126-27	40°52'N	85°28'W
Huntington, Ut., U.S. (hŭnt'ĭng-tŭn)	128-29	39°20'N	110°58'W
Huntington, W.V., U.S. (hŭnt'ĭng-tŭn)	126-27	38°25'N	82°26'W
Huntington Beach, Ca., U.S. (hŭnt'ĭng-tŭn bēch)	128-29	33°39'N	117°59'W
Huntsville, On., Can. (hŭnts'vĭl)	146-47	45°19'N	79°12'W
Huntsville, Al., U.S. (hŭnts'vĭl)	134-35	34°43'N	86°36'W
Huntsville, Mo., U.S. (hŭnts'vĭl)	130-31	39°26'N	92°33'W
Huntsville, Tx., U.S. (hŭnts'vĭl)	132-33	30°43'N	95°33'W
Hunucmá, Mex.	160	21°01'N	89°52'W
Hunyuan, China	260-61	39°42'N	113°41'E
Huong Thuy, Viet.	266-67	16°25'N	107°40'E
Huon Gulf, b., Pap. N. Gui.	302	7°10'S	147°25'E
Huonville, Austl.	301	43°01'S	147°02'E
Huoqiu, China (hwô-chyō)	258-59	32°20'N	116°16'E
Huoshan, China (hwô-shän)	258-59	31°24'N	116°20'E
Hurd, Cape, c., On., Can. (kāp hûrd)	146-47	45°14'N	81°42'W
Hurghada, Egypt	291b	27°14'N	33°50'E
Hurley, Wi., U.S. (hûr'lĭ)	124-25	46°27'N	90°11'W
Huron, S.D., U.S. (hū'rŏn)	124-25	44°22'N	98°13'W
Huron, Lake, lk., N.A. (lāk hū'rŏn)	126-27	44°30'N	82°15'W
Hurricane, Ut., U.S. (hŭr'ĭ-kān)	128-29	37°11'N	113°17'W
Húsavík, Ice.	206a	66°03'N	17°19'W
Huşi, Rom. (kósh')	218-19	46°41'N	28°04'E
Husum, Ger. (hōō'zòm)	210-11	54°28'N	9°04'E
Hutchinson, Ks., U.S. (hŭch'ĭn-sŭn)	130-31	38°03'N	97°55'W
Hutchinson, Mn., U.S. (hŭch'ĭn-sŭn)	124-25	44°53'N	94°23'W
Huy, Bel. (û-ē') (hû'ē)	206-07	50°31'N	5°14'E
Huzhou, China	258-59	30°52'N	120°06'E
Hvannadalshnúkur, mtn., Ice.	206a	64°01'N	16°41'W
Hvar, Otok, i., Cro. (ŏ'tŏk кhvär)	216-17	43°09'N	16°45'E
Hwaining, China *see* Anqing	258-59	30°30'N	117°02'E
Hwange, Zimb.	286-87	18°22'S	26°30'E
Hwange National Park, n.p., Zimb.	286-87	19°0'S	26°35'E
Hwang-hae, s., Asia *see* Yellow Sea	240-41	36°0'N	123°00'E
Hyargas nuur, lk., Mong.	260-61	49°12'N	93°24'E
Hyde Park, Guy.	178-79	6°30'N	58°16'W
Hyde Park, N.Y., U.S. (hīd pärk)	126-27	41°47'N	73°56'W
Hyderābād, India (hī-dẽr-å-băd')	256	17°23'N	78°29'E
Hyderābād, Pak. (hī-dẽr-å-băd')	252-53	25°23'N	68°21'E
Hyères, Fr. (ē-år')	212-13	43°08'N	6°08'E
Hyesan, Kor., N.	263	41°24'N	128°10'E
Hyndman Peak, mtn., Id., U.S. (hīnd'măn pēk)	122-23	43°45'N	114°08'W
Hyōgo, state, Japan (hĭyō'gō)	265	35°0'N	135°00'E

I

Feature (Pronunciation)	Page	Lat.	Long.
Iaco, stm., S.A.	184	9°02'S	68°35'W
Iaşi, Rom. (yä'shē)	218-19	47°10'N	27°36'E
Iba, Phil. (ē'bä)	270	15°20'N	119°58'E
Ibadan, Nig. (ē-bä'dän)	282a	7°23'N	3°54'E
Ibagué, Col.	188c	4°27'N	75°15'W
Iban, Pegunungan, mts., Asia *see* Iran Mountains	268-69	2°05'N	114°55'E
Ibarra, Ec. (ē-bär'rä)	184	0°22'N	78°08'W
Ibb, Yemen	288	13°58'N	44°11'E
Iberian Peninsula, pen., Eur. (ī-bēr'ĕ-ăn pĕ-nĭn'sū-lă)	192-93	40°0'N	5°00'W
Ibérica, Península, pen., Eur. *see* Iberian Peninsula	192-93	40°0'N	5°00'W
Ibiá, Braz.	186	19°29'S	46°32'W
Ibicaraí, Braz.	180-81	14°51'S	39°37'W
Ibicuí, stm., Braz.	187	29°25'S	56°47'W
Ibiza, Spain *see* Eivissa	214-15	38°55'N	1°25'E
Ibiza, i., Spain (ē-bē'thä) *see* Eivissa	214-15	39°0'N	1°25'E
Ica, Peru (ē'kà)	184	14°04'S	75°45'W
Içá, stm., S.A.	184	3°07'S	67°56'W
Içana, Braz. (ē-sä'nä)	180-81	0°21'N	67°19'W
İçel, Tur.	248-49	36°49'N	34°38'E
Iceland, nation, Eur. (īs'lănd)	190-91	65°0'N	18°00'W
Ichalkaranji, India	256	16°41'N	74°28'E
Ichilo, stm., Bol.	182-83	15°50'S	64°47'W
Icó, Braz.	180-81	6°24'S	38°51'W
Idabel, Ok., U.S. (ī'dȧ-bĕl)	130-31	33°53'N	94°49'W
Ida Grove, Ia., U.S. (ī'dȧ-grōv)	124-25	42°21'N	95°28'W
Idah, Nig. (ē'dä)	282a	7°07'N	6°45'E
Idaho, state, U.S. (ī'dȧ-hō)	118-19	45°0'N	115°00'W
Idaho Falls, Id., U.S. (ī'dȧ-hō fôlz)	122-23	43°30'N	112°03'W
Idaho Springs, Co., U.S. (ī'dȧ-hō sprĭngz)	128-29	39°44'N	105°31'W
Ider, stm., Mong.	260-61	49°16'N	100°41'E
Idi, Indon. (ē'dē)	266-67	4°57'N	97°46'E
Idi Amin Dada, Lac, lk., Afr. *see* Edward, Lake	289	0°23'S	29°36'E
Idiofa, D.R.C.	284-85	5°01'S	19°35'E
Ídi Óros, mtn., Grc.	216a	35°18'N	24°43'E
Idlib, Syria	248-49	35°56'N	36°39'E
Idoûkâl-en-Taghès, mtn., Niger	280-81	17°43'N	8°45'E
Iesi, Italy (yā'sē) *see* Jesi	216-17	43°31'N	13°14'E

ăt; finăl; rāte; senåte; ärm; åsk; sofȧ; fâre; ch-choose; dh-as th in other; bē; ėvent; bĕt; recĕnt; cratẽr; g-gō; gh-guttural g; bĭt; ĭ-short neutral; rīde; к-guttural k as ch in German ich;

Feature (Pronunciation)	Page	Lat.	Long.
Ife, Nig.	**282a**	7°28'N	4°33'E
Ifôghas, Adrar des, mts., Afr.	**280-81**	20°0'N	2°00'E
Igarka, Russia (ē-gär'kȧ)	**236-37**	67°28'N	86°38'E
Iglesias, Italy (ē-lĕ'syôs)	**216-17**	39°19'N	8°32'E
Igloolik, Nu., Can.	**138-39**	69°23'N	81°48'W
Igluligaarjuk, Nu., Can.			
see Chesterfield Inlet	**138-39**	63°21'N	90°43'W
Iglulik, Nu., Can. *see* Igloolik	**138-39**	69°23'N	81°48'W
Igombe, stm., Tan.	**289**	4°43'S	31°23'E
Iguaçu, stm., S.A. (ē-gwä-sōō')	**182-83**	25°36'S	54°36'W
Iguaçu, Cataratas do, wtfl., S.A.			
see Iguassu Falls	**182-83**	25°42'S	54°27'W
Iguaçu, Saltos do, wtfl., S.A.			
see Iguassu Falls	**182-83**	25°42'S	54°27'W
Iguala, Mex. (ė-gwä'lä)	**158-59**	18°21'N	99°32'W
Igualada, Spain (ē-gwä-lä'dä)	**214-15**	41°35'N	1°39'E
Iguape, Braz.	**186**	24°43'S	47°34'W
Iguassu Falls, wtfl., S.A.			
(ē-gwä-sōō' fôlz)	**182-83**	25°42'S	54°27'W
Iguatu, Braz. (ē-gwä-tōō')	**180-81**	6°22'S	39°18'W
Iguazú, stm., S.A. *see* Iguaçu	**182-83**	25°36'S	54°36'W
Iguazú, Cataratas del, wtfl., S.A.			
see Iguassu Falls	**182-83**	25°42'S	54°27'W
Ihosy, Madag.	**286-87**	22°24'S	46°07'E
Iida, Japan (ē'ē-dȧ)	**265**	35°31'N	137°50'E
Iisalmi, Fin.	**200-01**	63°32'N	27°17'E
Iizuka, Japan (ē'ē-zō-kȧ)	**265**	33°38'N	130°41'E
Ijâfene, des., Maur.	**280-81**	22°04'N	7°42'W
Ijebu-Ode, Nig. (ė-jĕ'bōō ōdå)	**282a**	6°49'N	3°56'E
IJsselmeer, lk., Neth. (ī'sĕl-mār)	**206-07**	52°45'N	5°25'E
Ijuí, Braz.	**187**	28°23'S	53°55'W
Ikaalinen, Fin. (ē'kä-lī-nĕn)	**208-09**	61°45'N	23°04'E
Ikaría, i., Grc. (ē-kä'ryȧ)	**216-17**	37°36'N	26°09'E
Ikela, D.R.C.	**284-85**	1°11'S	23°17'E
Ikhtiman, Blg. (ėk'tē-män)	**216-17**	42°26'N	23°50'E
Iki, i., Japan (ē'kē)	**265**	33°47'N	129°43'E
Ikom, Nig.	**282a**	5°58'N	8°42'E
Ikoma, Tan. (ė-kō'mä)	**289**	2°05'S	34°38'E
Ikopa, stm., Madag.	**206-07**	17°00'S	46°45'E
Iksan, Kor., S.	**263**	35°56'N	126°57'E
Ilagan, Phil.	**270**	17°08'N	121°53'E
Ilam, nation, Asia *see* Sri Lanka	**224-25**	7°0'N	81°00'E
Ilan, Tai. (ē'län')	**243a**	24°46'N	121°45'E
Ile, stm., Asia	**244**	45°20'N	74°05'E
Île-à-la-Crosse, Sk., Can.	**144-45**	55°26'N	107°55'W
Île-à-la-Crosse, Lac, lk., Sk., Can.	**144-45**	55°42'N	107°41'W
Ilebo, D.R.C.	**284-85**	4°20'S	20°36'E
Ilek, stm., Asia *see* Elek	**202-03**	51°30'N	53°20'E
Ilesha, Nig.	**282a**	7°38'N	4°45'E
Ilfracombe, Austl.	**302**	23°29'S	144°30'E
Ilfracombe, Eng., U.K. (ĭl-frȧ-kōōm')	**206-07**	51°12'N	4°08'W
Ilgın, Tur.	**246**	38°17'N	31°55'E
Ilha de Moçambique, Moz.	**286-87**	15°02'S	40°41'E
Ilha Grande, Baía da, b., Braz.			
(bäē'ä dĕ ēl'yȧ grän'dĕ)	**186**	23°09'S	44°30'W
Ílhavo, Port. (ēl'yȧ-vô)	**214-15**	40°36'N	8°40'W
Ilhéos, Braz. *see* Ilhéus	**180-81**	14°47'S	39°03'W
Ilhéus, Braz. (ē-lĕ'ōōs)	**180-81**	14°47'S	39°03'W
Ili, stm., Asia *see* Ile	**244**	45°20'N	74°05'E
Iliamna Lake, lk., Ak., U.S.			
(ē-lē-ăm'nȧ lăk)	**136**	59°37'N	154°49'W
Iligan, Phil.	**270**	8°15'N	124°16'E
Ilion, N.Y., U.S. (ĭl'ĭ-ŭn)	**126-27**	43°01'N	75°03'W
Ilizi, Alg.	**280-81**	26°29'N	8°29'E
Illampu, Nevado, mtn., Bol.			
(nĕ-vá'dô-ēl-yäm-pōō')	**182-83**	15°50'S	68°34'W
Illapel, Chile (ē-zhä-pĕ'l)	**182-83**	31°37'S	71°09'W
Illimani, Nevado de, mtn., Bol.			
(nĕ-vá'dô-dĕ-ēl-yē-mä'nē)	**182-83**	16°50'S	67°54'W
Illinois, state, U.S.			
(ĭl-ĭ-noi') (ĭl-ĭ-noiz')	**118-19**	40°0'N	89°00'W
Illinois, stm., Il., U.S.			
(ĭl-ĭ-noi') (ĭl-ĭ-noiz')	**120-21**	38°58'N	90°25'W
Il'men', Ozero, lk., Russia			
(ô'zĕ-rô el''men'') (ĭl'mĕn)	**208-09**	58°17'N	31°20'E
Ilo, Peru	**184**	17°38'S	71°20'W
Iloilo, Phil. (ē-lô-ē'lō)	**270**	10°42'N	122°34'E
Ilorin, Nig. (ē-lô-rēn')	**282a**	8°30'N	4°33'E
Ilulissat, Green.	**151**	69°13'N	51°06'W
Ilwaki, Indon.	**268-69**	7°55'S	126°25'E
Imabari, Japan (ē'mä-bä'rē)	**265**	34°04'N	133°00'E
Imandra, Ozero, lk., Russia			
(ô'zĕ-rô ē-män'drȧ)	**202-03**	67°33'N	33°00'E
Imatra, Fin.	**208-09**	61°10'N	28°46'E
Imbituba, Braz.	**186**	28°14'S	48°41'W
Imeni Ismail Samani, Pik, mtn., Taj.	**244**	38°57'N	72°01'E
Imlay City, Mi., U.S. (ĭm'lā sĭ'tē)	**126-27**	43°01'N	83°05'W
Imola, Italy (ē'mô-lä)	**216-17**	44°21'N	11°43'E
Imotski, Cro. (ē-môts'kē)	**216-17**	43°27'N	17°13'E
Imperatriz, Braz.	**180-81**	5°31'S	47°28'W
Imperia, Italy (ēm-pā'rē-ä)	**216-17**	43°54'N	8°03'E
Imperial, Ne., U.S. (ĭm-pē'rĭ-ăl)	**130-31**	40°31'N	101°39'W
Impfondo, Congo (ĭmp-fōn'dô)	**284-85**	1°37'N	18°04'E
Imphāl, India (ĭmp'hŭl)	**254-55**	24°47'N	93°57'E
Inari, Fin.	**200-01**	68°54'N	26°60'E
Inari, lk., Fin. *see* Inarijärvi	**202-03**	69°0'N	28°00'E
Inarijärvi, lk., Fin.	**202-03**	69°0'N	28°00'E
Inca, Spain (ēŋ'kä)	**214-15**	39°43'N	2°55'E
Inca de Oro, Chile	**182-83**	26°45'S	69°54'W
İnce Burun, c., Tur. (ĭn'jå)	**202-03**	42°05'N	34°58'E
Incheon, Kor., S.	**263**	37°28'N	126°38'E
Indefatigable, i., Ec.			
see Santa Cruz, Isla	**184a**	0°38'S	90°23'W
Independence, Ia., U.S.			
(ĭn-dē-pĕn'dĕns)	**124-25**	42°29'N	91°54'W
Independence, Ks., U.S.			
(ĭn-dē-pĕn'dĕns)	**130-31**	37°13'N	95°42'W
Independence, Ky., U.S.			
(ĭn-dē-pĕn'dĕns)	**126-27**	38°56'N	84°32'W
Independence, Mo., U.S.			
(ĭn-dē-pĕn'dĕns)	**130-31**	39°05'N	94°25'W
Independence Mountains, mts., Nv.,			
U.S. (ĭn-dē-pĕn'dĕns moun'tĭnz)	**122-23**	41°18'N	116°00'W
Inderbor, Kaz.	**202-03**	48°32'N	51°42'E
India, nation, Asia (ĭn'dĭ-ȧ)	**224-25**	20°0'N	77°00'E
Indiana, Pa., U.S. (ĭn-dĭ-än'ȧ)	**126-27**	40°35'N	79°09'W
Indiana, state, U.S. (ĭn-dĭ-än'ȧ)	**118-19**	40°0'N	86°15'W
Indianapolis, In., U.S.			
(ĭn-dĭ-ăn-ăp'ô-lĭs)	**126-27**	39°46'N	86°08'W
Indian Head, Sk., Can. (ĭn'dĭ-ăn hĕd)	**144-45**	50°32'N	103°40'W
Indian Ocean, oc., (ĭn'dĭ-ăn ōshŭn)	**4-5**	10°0'S	70°00'E
Indianola, Ia., U.S. (ĭn-dĭ-ăn-ō'lȧ)	**124-25**	41°21'N	93°33'W
Indianola, Ms., U.S. (ĭn-dĭ-ăn-ō'lȧ)	**134-35**	33°27'N	90°39'W
Indian Springs, Nv., U.S.			
(ĭn'dĭ-ăn sprĭngz)	**128-29**	36°34'N	115°41'W
Indigirka, stm., Russia (ēn-dē-gēr'kȧ)	**236-37**	70°49'N	148°54'E
Indio, Ca., U.S.	**128-29**	33°43'N	116°13'W
Indochina, reg., Asia (ĭn-dō-chī'nȧ)	**266-67**	16°0'N	107°00'E
Indonesia, nation, Asia (ĭn'dô-nē-zhȧ)	**224-25**	5°0'S	120°00'E
Indore, India (ĭn-dōr')	**254-55**	22°43'N	75°52'E
Indragiri, stm., Indon. (ĭn-drȧ-jē'rē)	**266-67**	0°22'S	103°26'E
Indrāvati, stm., India (ĭn-drŭ-vä'tē)	**256**	18°44'N	80°17'E
Indus, stm., Asia (ĭn'dŭs)	**226-27**	24°60'N	68°16'E
Inferior, Laguna, b., Mex.			
(lä-gō'nä-ēn-fĕr-rôr)	**158-59**	16°17'N	94°40'W
Infiernillo, Presa del, res., Mex.	**158-59**	18°37'N	101°46'W
Ingende, D.R.C.	**284-85**	0°13'S	18°58'E
Ingersoll, On., Can. (ĭn'gĕr-sŏl)	**146-47**	43°02'N	80°53'W
Ingham, Austl. (ĭng'ăm)	**302**	18°39'S	146°09'E
Ingoda, stm., Russia (ēn-gō'dȧ)	**240-41**	51°43'N	115°48'E
Ingolstadt, Ger. (ĭŋ'gôl-shtät)	**210-11**	48°46'N	11°26'E
Ingrāj Bāzār, India	**254-55**	24°60'N	88°09'E
I-n-Guezzâm, Alg.	**280-81**	19°27'N	5°48'E
Ingushetia, state, Russia	**245**	43°15'N	45°00'E
Ingushetiya, state, Russia *see* Ingushetia	**245**	43°15'N	45°00'E
Inhaca, Ilha da, i., Moz.	**286-87**	26°01'S	32°57'E
Inhambupe, Braz. (ēn-yäm-bōō'pä)	**180-81**	11°49'S	38°20'W
Inírida, stm., Col. (ē-nē-rē'dä)	**178-79**	3°55'N	67°51'W
Injune, Austl. (ĭn'jōn)	**301**	25°51'S	148°34'E
Inland Sea, s., Japan (ĭn'lănd sē)			
see Seto-naikai	**265**	34°22'N	133°37'E
Inn, stm., Eur. (ĭn)	**210-11**	48°34'N	13°28'E
Innamincka, Austl. (ĭnn-ȧ'mĭn-kȧ)	**301**	27°45'S	140°44'E
Inner Mongolia, state, China			
see Nei Mongol	**260-61**	43°0'N	115°00'E
Innisfail, Austl.	**302**	17°33'S	146°02'E
Innisfail, Ab., Can.	**142-43**	52°02'N	113°58'W
Innoko, stm., Ak., U.S.	**136**	62°11'N	159°44'W
Innsbruck, Aus. (ĭns'bro͝ok)	**210-11**	47°16'N	11°24'E
Inongo, D.R.C. (ē-nôŋ'gō)	**284-85**	1°55'S	18°18'E
Inowrocław, Pol. (ē-nō-vrôts'läf)	**210-11**	52°48'N	18°15'E
I-n-Salah, Alg.	**280-81**	27°11'N	2°29'E
Inta, Russia	**202-03**	66°02'N	60°09'E
International Falls, Mn., U.S.			
(ĭn'tĕr-năsh'ŭn-ăl fôlz)	**124-25**	48°36'N	93°25'W
Inthanon, Doi, mtn., Thai.	**266-67**	18°35'N	98°29'E
Intiyaco, Arg.	**187**	28°40'S	60°04'W
Inukjuak, Qc., Can.	**138-39**	58°28'N	78°06'W
Inuvik, N.T., Can.	**138-39**	68°20'N	133°39'W
Invercargill, N.Z.	**303**	46°25'S	168°22'E
Inverell, Austl.	**301**	29°47'S	151°07'E
Inverness, Scot., U.K. (ĭn-vĕr-nĕs')	**206-07**	57°28'N	4°15'W
Inverness, Fl., U.S. (ĭn-vĕr-nĕs')	**134-35**	28°50'N	82°20'W
Investigator Strait, strt., Austl.			
(ĭn-vĕst'ĭ'gā-tôr strāt)	**301**	35°25'S	137°10'E
Inyangani, mtn., Zimb. (ēn-yän-gä'nē)	**286-87**	18°17'S	32°50'E
Inza, Russia	**202-03**	53°51'N	46°22'E
Ioánnina, Grc. (yô-ä'nē-nȧ)	**216-17**	39°39'N	20°51'E
Iō-jima, i., Japan *see* Iwo Jima	**304-05**	24°47'N	141°20'E
Ionia, Mi., U.S. (ī-ō'nĭ-ȧ)	**126-27**	42°59'N	85°04'W
Ionian Islands, is., Grc.			
(ī-ō'nĭ-ăn ī'lăndz)	**216-17**	38°30'N	20°30'E
Ionian Sea, s., Eur. (ī-ō'nĭ-ăn sē)	**216-17**	39°0'N	19°00'E
Ionio, Mar, s., Eur. *see* Ionian Sea	**216-17**	39°0'N	19°00'E
Iónioi Nísoi, is., Grc.			
see Ionian Islands	**216-17**	38°30'N	20°30'E
Iónion Pélagos, s., Eur.			
see Ionian Sea	**216-17**	39°0'N	19°00'E
Íos, i., Grc. (ī'ŏs)	**216-17**	36°43'N	25°20'E
Iō-tō, i., Japan *see* Iwo Jima	**304-05**	24°47'N	141°20'E
Iowa, state, U.S. (ī'ô-wȧ)	**118-19**	42°15'N	93°15'W
Iowa, stm., Ia., U.S. (ī'ô-wȧ)	**124-25**	41°10'N	91°01'W
Iowa City, Ia., U.S. (ī'ô-wȧ sĭ'tē)	**124-25**	41°40'N	91°32'W
Iowa Falls, Ia., U.S. (ī'ô-wȧ fôlz)	**124-25**	42°32'N	93°16'W
Iowa Park, Tx., U.S. (ī'ô-wȧ pärk)	**130-31**	33°58'N	98°40'W
Ipameri, Braz.	**186**	17°43'S	48°09'W
Ipiales, Col. (ē-pē-ä'lās)	**178-79**	0°50'N	77°38'W
Ipiaú, Braz.	**180-81**	14°08'S	39°44'W
Ipoh, Malay.	**266-67**	4°36'N	101°04'E
Iporã, Braz.	**182-83**	16°27'S	51°07'W
Ippy, C.A.R.	**284-85**	6°16'N	21°12'E
Ipswich, Austl. (ĭps'wĭch)	**301**	27°37'S	152°47'E
Ipswich, Eng., U.K. (ĭps'wĭch)	**206-07**	52°04'N	1°09'E
Ipswich, S.D., U.S. (ĭps'wĭch)	**124-25**	45°27'N	99°02'W
Ipu, Braz. (ē-pōō)	**180-81**	4°20'S	40°42'W
Iqalult, Nu., Can.	**138-39**	63°44'N	68°28'W
Iquique, Chile (ē-kē'kĕ)	**182-83**	20°13'S	70°09'W
Iquitos, Peru (ė-ke'tos)	**184**	3°47'S	73°15'W
Irákleio, Grc.	**216a**	35°20'N	25°08'E
Iran, nation, Asia (ē-rän')	**224-25**	32°0'N	53°00'E
Īrān, nation, Asia *see* Iran	**224-25**	32°0'N	53°00'E
Iran, Pergunungan, mts., Asia			
see Iran Mountains	**268-69**	2°05'N	114°55'E
Iran Mountains, mts., Asia			
(ē-rän' moun'tĭnz)	**268-69**	2°05'N	114°55'E
Īrānshahr, Iran	**250-51**	27°13'N	60°42'E
Irapuato, Mex. (ē-rä-pwä'tō)	**158-59**	20°41'N	101°21'W
Iraq, nation, Asia (ē-räk')	**224-25**	33°0'N	44°00'E
Irazú, Volcán, vol., C.R.			
(vōl-kä'n ē-rä-zōō')	**161**	9°58'N	83°53'W
Irbid, Jord. (ėr-bēd')	**248-49**	32°34'N	35°51'E
Irbīl, Iraq *see* Arbīl	**248-49**	36°11'N	44°01'E
Ireland, nation, Eur. (īr-lănd)	**190-91**	53°0'N	8°00'W
Irian, i., *see* New Guinea	**302**	5°0'S	140°00'E
Iringa, Tan. (ė-rĭŋ'gä)	**289**	7°47'S	35°42'E
Iriomote-jima, i., Japan			
(ērē'-ō-mō-tä jē'mȧ)	**306a**	24°20'N	123°50'E
Iriri, stm., Braz.	**180-81**	3°48'S	52°36'W
Irish Sea, s., Eur. (ī'rĭsh sē)	**206-07**	53°30'N	5°20'W
Irkutsk, Russia (ĭr-ko͝otsk')	**240-41**	52°18'N	104°17'E
Iron Knob, Austl. (ī-ăn nŏb)	**301**	32°44'S	137°08'E
Iron Mountain, Mi., U.S.			
(ī'ĕrn moun'tĭn)	**126-27**	45°49'N	88°03'W
Iron River, Mi., U.S. (ī'ĕrn rĭv'ĕr)	**124-25**	46°06'N	88°39'W
Ironton, Oh., U.S. (ī'ĕrn-tŭn)	**126-27**	38°32'N	82°41'W
Ironwood, Mi., U.S. (ī'ĕrn-wo͝od)	**124-25**	46°27'N	90°10'W
Ironwood Forest National			
Monument, n.p., Az., U.S.	**128-29**	32°27'N	111°30'W
Iroquois Falls, On., Can.			
(ĭr'ô-kwoi fôlz)	**146-47**	48°46'N	80°40'W
Irrawaddy, stm., Mya. (ĭr-ȧ-wäd'ē)			
see Ayeyarwady	**238-39**	15°51'N	95°05'E
Irtysh, stm., Asia (ĭr-tĭsh')	**236-37**	61°05'N	68°47'E
Irumu, D.R.C. (ē-rō'mōō)	**289**	1°27'N	29°52'E
Irún, Spain (ē-rōōn')	**214-15**	43°21'N	1°48'W
Iruña, Spain *see* Pamplona	**214-15**	42°49'N	1°39'W
Irvine, Ky., U.S. (ûr'vĭn)	**126-27**	37°42'N	83°58'W
Irving, Tx., U.S. (ûr'vēng)	**130-31**	32°49'N	96°57'W
Isabela, Phil.	**270**	6°41'N	121°58'E
Isabela, Cabo, c., Dom. Rep.			
(kȧ'bô-ē-sä-bĕ'lä)	**154-55**	19°55'N	71°01'W
Isabela, Isla, i., Ec. (ē's-lä-ē-sä-bä'lä)	**184a**	0°30'S	91°06'W
Isabelia, Cordillera, mts., Nic.			
(kôr-dēl-yĕ'rä-ē-sä-bĕlyä)	**161**	13°30'N	85°32'W
Isabella Indian Reservation, ind. res.,			
Mi., U.S. (ĭs-ȧ-bĕl'-lȧ			
ĭn'dĭ-ăn rĕ-sĕr-vā'shĕn)	**126-27**	43°41'N	84°48'W
Ísafjörður, Ice. (ēs'ȧ-fÿr-dôr)	**206a**	66°03'N	23°07'W

n-sing; ŋ-baŋk; ɴ-nasalized n; nŏd; cŏmmit; ōld; ôbey; ôrder; oi-boil; fo͞od; o͝o-as oo in foot; ou-out; s-soft; sh-dish; th-thin; pūre; ůnite; ûrn; stŭd; circŭs; ü-as in French tu; ′-indeterminate vowel.

366 Ischia, Isola d' – Jarocin PRONOUNCING INDEX

Feature (Pronunciation)	Page	Lat.	Long.
Ischia, Isola d', i., Italy (ē´-sō-lä-dē´sh-kyä)	216-17	40°43'N	13°54'E
Ise, Japan (ĭs´hĕ) (û´gē-yä´mä´dȧ)	265	34°30'N	136°42'E
Isernia, Italy (ē-zĕr´nyä)	216-17	41°36'N	14°14'E
Iset', stm., Russia	236-37	56°36'N	66°17'E
Ise-wan, b., Japan (ē´sĕ wän)	265	34°43'N	136°43'E
Iseyin, Nig.	282a	7°58'N	3°36'E
Isfahan, Iran *see* Eşfahān	252-53	32°39'N	51°40'E
Ishigaki-shima, i., Japan	306a	24°24'N	124°12'E
Ishikari, stm., Japan	264	43°16'N	141°23'E
Ishikari-wan, b., Japan (ē´shē-kä-rē wän)	264	43°25'N	141°01'E
Ishim, Russia (ĭsh-ĕm´)	236-37	56°07'N	69°30'E
Ishim, stm., Asia (ĭsh-ĕm´)	236-37	57°43'N	71°12'E
Ishimskaya Ravnina, pl., Asia	244	55°0'N	70°00'E
Ishinomaki, Japan (ĭsh-nō-mä´kē)	264	38°25'N	141°18'E
Ishpeming, Mi., U.S. (ĭsh´pē-mĭng)	124-25	46°29'N	87°39'W
Isil'kul', Russia	244	54°54'N	71°16'E
Isiolo, Kenya	289	0°21'N	37°35'E
Isiro, D.R.C.	284-85	2°46'N	27°37'E
İskenderun, Tur. (ĭs-kĕn´dĕr-ōōn)	248-49	36°35'N	36°11'E
İskenderun Körfezi, b., Tur.	248-49	36°30'N	35°40'E
Iskitim, Russia	244	54°39'N	83°18'E
Iskŭr, stm., Blg. (ĭs´k'r)	216-17	43°45'N	24°26'E
Isla Cristina, Spain (ī´lä-krē-stē´nä)	214-15	37°12'N	7°19'W
Islāmābād, nat. cap., Pak. (ĭs´lä-mä-bäd´) (ĭs-lä´-mä-bäd´)	252-53	33°39'N	73°05'E
Isla Mujeres, Mex. (ē´s-lä-mōō-kĕ´rĕs)	160	21°12'N	86°43'W
Ísland, nation, Eur. *see* Iceland	190-91	65°0'N	18°00'W
Island Lake, lk., Mb., Can. (ī´lănd lāk ī´lănd)	144-45	53°47'N	94°25'W
Islands, Bay of, b., Nf./L., Can. (bā ŭv ī´lăndz)	148-49	49°10'N	58°15'W
Íslandshaf, s., Eur. *see* Norwegian Sea	200-01	70°0'N	2°00'E
Islay, i., Scot., U.K. (ī´lā)	206-07	55°49'N	6°17'W
Isle of Man, dep., Eur. (īl ŭv măn)	206-07	54°15'N	4°30'W
Isle Royale National Park, n.p., Mi., U.S. (īl´roi-ăl´ năsh´ŭn-ăl pärk)	124-25	47°58'N	88°55'W
Ismailia, Egypt (ēs-mä-ēl´ēȧ)	291b	30°36'N	32°16'E
Isparta, Tur. (ē-spär´tȧ)	202-03	37°46'N	30°33'E
Israel, nation, Asia (ĭz´rē-ĕl)	224-25	31°30'N	34°45'E
Isrā'īl, nation, Asia *see* Israel	224-25	31°30'N	34°45'E
Issoire, Fr. (ē-swȧr´)	212-13	45°33'N	3°15'E
Issoudun, Fr. (ē-sōō-dăɴ´)	212-13	46°57'N	2°00'E
Issyk-Kul, Lake, lk., Kyrg. (lăk ē´-sĭk-kōōl´)	244	42°25'N	77°15'E
İstanbul, Tur. (ē-stän-bōōl´)	216-17	41°02'N	28°59'E
İstanbul Boğazı, strt., Tur. *see* Bosporus	216-17	41°06'N	29°04'E
Istaravshan, Taj.	252-53	39°54'N	69°00'E
Istiaía, Grc. (ĭs-tyī´yä)	216-17	38°57'N	23°09'E
Istmina, Col. (ēst-mē´nä)	188c	5°09'N	76°41'W
Istra, pen., Eur. (ē-strä)	216-17	45°17'N	13°57'E
Istria, pen., Eur. *see* Istra	216-17	45°17'N	13°57'E
Itabaiana, Braz. (ē-tä-bä-yȧ-nä)	188d	7°20'S	35°20'W
Itabaiana, Braz. (ē-tä-bä-yȧ-nä)	180-81	10°41'S	37°26'W
Itabapoana, Braz. (ē-tä´-bä-pôä´nä)	186	21°18'S	40°59'W
Itaberaí, Braz.	182-83	16°01'S	49°48'W
Itabira, Braz.	186	19°38'S	43°14'W
Itabuna, Braz. (ē-tä-bōō´nȧ)	180-81	14°47'S	39°17'W
Itacoatiara, Braz. (ē-tȧ-kwȧ-tyä´rȧ)	180-81	3°08'S	58°26'W
Itagüí, Col. (ē-tä´gwē)	188c	6°10'N	75°38'W
Itaipu, Represa de, res., S.A.	182-83	24°56'S	54°26'W
Itaipu Reservoir, res., S.A. (ē-tī´pōō rĕ´sĕr-vwär) *see* Itaipu, Represa de	182-83	24°56'S	54°26'W
Itaituba, Braz. (ē-tä´ī-tōō´bȧ)	180-81	4°15'S	55°59'W
Itajaí, Braz. (ē-tä-zhī´)	186	26°54'S	48°40'W
Itajubá, Braz.	186	22°26'S	45°27'W
Italia, nation, Eur. *see* Italy	190-91	43°0'N	13°00'E
Italy, nation, Eur. (ĭt´ȧ-lē)	190-91	43°0'N	13°00'E
Itämeri, s., Eur. *see* Baltic Sea	208-09	57°0'N	19°00'E
Itānagar, India	254-55	27°09'N	93°33'E
Itaparica, Ilha de, i., Braz.	180-81	13°0'S	38°42'W
Itapecuru-Mirim, Braz. (ē-tä-pĕ´kōō-rōō-mē-rēɴ´)	180-81	3°24'S	44°20'W
Itapemirim, Braz.	186	21°01'S	40°49'W
Itaperuna, Braz. (ē-tȧ´pā-rōō´nä)	186	21°12'S	41°54'W
Itapetinga, Braz.	182-83	15°15'S	40°16'W
Itapetininga, Braz. (ē-tä-pĕ-tē-nē´ɴ-gä)	186	23°35'S	48°02'W
Itapicuru, stm., Braz.	180-81	2°51'S	44°12'W
Itapicuru, stm., Braz.	180-81	11°45'S	37°31'W
Itaquari, Braz.	186	20°20'S	40°23'W
Itaqui, Braz.	187	29°08'S	56°32'W
Itararé, Braz.	186	24°07'S	49°21'W
Itārsi, India	254-55	22°36'N	77°46'E
Itasca, Tx., U.S. (ī-tăs´kȧ)	132-33	32°10'N	97°09'W
Itaúna, Braz. (ē-tä-ōō´nä)	186	20°04'S	44°34'W
Itbayat Island, i., Phil.	270a	20°46'N	121°50'E
Itenes, stm., S.A. *see* Iténez	180-81	11°55'S	65°00'W
Iténez, stm., S.A.	180-81	11°55'S	65°00'W
Ithaca, Mi., U.S. (ĭth´ȧ-kȧ)	126-27	43°17'N	84°36'W
Ithaca, N.Y., U.S. (ĭth´ȧ-kȧ)	126-27	42°26'N	76°30'W
Itu, Braz. (ē-tōō´)	186	23°16'S	47°18'W
Ituango, Col. (ē-twäŋ´gō)	178-79	7°06'N	75°44'W
Itui, stm., Braz.	178-79	4°39'S	70°15'W
Ituiutaba, Braz. (ē-tōō-ēōō-tä´bä)	182-83	18°58'S	49°27'W
Itumbiara, Braz.	186	18°25'S	49°12'W
Iturbide, Mex. (ē´tōōr-bē´dhā)	160	19°38'N	89°36'W
Ituri, stm., D.R.C.	284-85	1°40'N	27°02'E
Iturup, Ostrov, i., Russia (ŏs-trôf ē-tōō-rōōp´)	236-37	44°51'N	147°27'E
Ituxi, stm., Braz.	180-81	7°18'S	64°51'W
Ituzaingó, Arg. (ē-tōō-zä-ē´n-gō)	187	27°36'S	56°40'W
Ītyop'iya, nation, Afr. *see* Ethiopia	274	9°0'N	39°00'E
Iuka, Ms., U.S. (ī-ū´kȧ)	134-35	34°49'N	88°11'W
Iul'tin, Russia	236-37	67°43'N	178°51'W
Ivaí, stm., Braz.	182-83	23°18'S	53°44'W
Ivalo, Fin.	200-01	68°40'N	27°32'E
Ivanhoe, Austl. (ĭv´ăn-hō)	301	32°55'S	144°19'E
Ivano-Frankivs'k, Ukr.	210-11	48°55'N	24°44'E
Ivanovo, Russia (ē-vä´nō-vō)	202-03	57°01'N	40°59'E
Ivanovo-Voznesensk, Russia *see* Ivanovo	202-03	57°01'N	40°59'E
Ivdel', Russia (ĭv´dyĕl)	202-03	60°41'N	60°27'E
Iviza, Spain *see* Eivissa	214-15	38°55'N	1°25'E
Ivory Coast, nation, Afr. *see* Côte d'Ivoire	274	8°0'N	5°00'W
Ivrea, Italy (ē-vrĕ´ä)	216-17	45°28'N	7°53'E
Ivujivik, Qc., Can.	138-39	62°23'N	77°55'W
Iwaki, Japan	265	37°03'N	140°55'E
Iwo, Nig.	282a	7°38'N	4°11'E
Iwo Jima, i., Japan (ē´wō jē´mȧ)	304-05	24°47'N	141°20'E
Ixmiquilpan, Mex. (ēs-mē-kēl´pän)	158-59	20°29'N	99°13'W
Ixtepec, Mex. (ēks-tĕ´pĕk)	158-59	16°32'N	95°05'W
Ixtlán de Juárez, Mex. (ēs-tlän´ dā hwä´rāz)	158-59	17°20'N	96°30'W
Ixtlán del Río, Mex. (ēs-tlän´dĕl rē´ō)	158-59	21°02'N	104°22'W
Iyo-nada, s., Japan (ē´yō nä-dä)	265	33°40'N	132°20'E
Izabal, Lago de, lk., Guat. (lä´gō-dĕ-ē´zä-bäl´)	160	15°30'N	89°10'W
Izamal, Mex. (ē-zä-mä´l)	160	20°56'N	89°01'W
Izberbash, Russia	245	42°33'N	47°52'E
Izhevsk, Russia (ē-zhyĕfsk´)	202-03	56°50'N	53°12'E
Izhma, stm., Russia	202-03	65°19'N	52°55'E
Izium, Ukr.	218-19	49°13'N	37°17'E
Izmaïl, Ukr.	216-17	45°21'N	28°50'E
İzmir, Tur. (ĭz-mēr´)	216-17	38°26'N	27°09'E
İzmit, Tur. (ĭz-mĕt´)	216-17	40°47'N	29°57'E
Izuhara, Japan (ē´zōō-hä´rä)	263	34°12'N	129°17'E
Izumo, Japan (ē´zōō-mō)	265	35°22'N	132°46'E
Izu-shotō, is., Japan	264	32°0'N	140°00'E

J

Feature (Pronunciation)	Page	Lat.	Long.
Jabal, Baḩr al-, stm., S. Sudan *see* Mountain Nile	284-85	9°30'N	30°30'E
Jabalpur, India	254-55	23°10'N	79°56'E
Jaboatão, Braz. (zhä-bô-ȧ-touɴ)	188d	8°07'S	35°01'W
Jaca, Spain (hä´kä)	214-15	42°35'N	0°34'W
Jacala, Mex. (hä-kȧ´lä)	158-59	21°01'N	99°11'W
Jacaltenango, Guat. (hä-kȧl-tĕ-nȧŋ´gō)	160	15°40'N	91°44'W
Jacareí, Braz.	186	23°19'S	45°58'W
Jacarezinho, Braz. (zhä-kä-rĕ´zē-nyȯ)	182-83	23°10'S	49°59'W
Jacksboro, Tx., U.S. (jăks´bŭr-ō)	130-31	33°13'N	98°09'W
Jackson, Al., U.S. (jăk´sŭn)	134-35	31°31'N	87°54'W
Jackson, Ga., U.S. (jăk´sŭn)	134-35	33°18'N	83°58'W
Jackson, Ky., U.S. (jăk´sŭn)	126-27	37°33'N	83°24'W
Jackson, La., U.S. (jăk´sŭn)	134-35	30°50'N	91°13'W
Jackson, Mi., U.S. (jăk´sŭn)	126-27	42°15'N	84°24'W
Jackson, Mn., U.S. (jăk´sŭn)	124-25	43°37'N	94°59'W
Jackson, Mo., U.S. (jăk´sŭn)	130-31	37°23'N	89°40'W
Jackson, Ms., U.S. (jăk´sŭn)	134-35	32°19'N	90°11'W
Jackson, Oh., U.S. (jăk´sŭn)	126-27	39°03'N	82°39'W
Jackson, Tn., U.S. (jăk´sŭn)	134-35	35°37'N	88°49'W
Jackson, Wy., U.S. (jăk´sŭn)	122-23	43°29'N	110°45'W
Jackson Lake, lk., Wy., U.S. (jăk´sŭn lāk)	122-23	43°55'N	110°40'W
Jacksonville, Al., U.S. (jăk´sŭn-vĭl)	134-35	33°49'N	85°46'W
Jacksonville, Ar., U.S. (jăk´sŭn-vĭl)	130-31	34°52'N	92°07'W
Jacksonville, Fl., U.S. (jăk´sŭn-vĭl)	134-35	30°21'N	81°39'W
Jacksonville, Il., U.S. (jăk´sŭn-vĭl)	130-31	39°44'N	90°14'W
Jacksonville, N.C., U.S. (jăk´sŭn-vĭl)	134-35	34°45'N	77°25'W
Jacksonville, Tx., U.S. (jăk´sŭn-vĭl)	132-33	31°58'N	95°16'W
Jacksonville Beach, Fl., U.S. (jăk´sŭn-vĭl bēch)	134-35	30°17'N	81°24'W
Jacmel, Haiti (zhȧk-mĕl´)	154-55	18°14'N	72°32'W
Jacobābād, Pak.	252-53	28°17'N	68°26'E
Jacobina, Braz. (zhä-kô-bē´nȧ)	180-81	11°11'S	40°31'W
Jacques-Cartier, Détroit de, strt., Qc., Can.	148-49	49°53'N	62°45'W
Jacques-Cartier, Mont, mtn., Qc., Can.	148-49	48°59'N	65°57'W
Jacquet River, N.B., Can. (zhȧ-kĕ´ rĭv´ĕr) (jăk´ĕt rĭv´ĕr)	148-49	47°55'N	66°01'W
Jacuí, stm., Braz.	182-83	30°02'S	51°15'W
Jadransko more, s., Eur. *see* Adriatic Sea	216-17	42°30'N	16°00'E
Jadransko morje, s., Eur. *see* Adriatic Sea	216-17	42°30'N	16°00'E
Jaén, Peru (кä-ĕ´n)	184	5°43'S	78°47'W
Jaén, Spain	214-15	37°46'N	3°48'W
Jaffa, Cape, c., Austl. (kāp jăf´ȧ)	301	36°58'S	139°40'E
Jaffna, Sri L. (jäf´nȧ)	256	9°40'N	80°01'E
Jagādhri, India	254-55	30°10'N	77°18'E
Jagdalpur, India	256	19°05'N	82°02'E
Jägerndorf, Czech Rep. *see* Krnov	210-11	50°05'N	17°42'E
Jaguarão, Braz.	187	32°34'S	53°23'W
Jaguariaíva, Braz.	182-83	24°15'S	49°42'W
Jaguaribe, stm., Braz.	180-81	4°25'S	37°46'W
Jagüey Grande, Cuba (hä´gwā grän´dā)	154-55	22°32'N	81°08'W
Jahrom, Iran	250-51	28°29'N	53°33'E
Jaipur, India	254-55	26°55'N	75°48'E
Jaisalmer, India	254-55	26°55'N	70°55'E
Jajce, Bos. (yī´tsĕ)	216-17	44°20'N	17°17'E
Jājpur, India	254-55	20°51'N	86°20'E
Jakarta, nat. cap., Indon. (yä-kär´tä)	268-69	6°11'S	106°50'E
Jakobstad, Fin. (yȧ´kôb-städh)	200-01	63°41'N	22°43'E
Jalālābād, Afg. (jŭ-lä-lȧ-bäd)	252-53	34°26'N	70°27'E
Jalal-Abad, Kyrg.	244	40°56'N	73°00'E
Jalandhar, India	254-55	31°19'N	75°35'E
Jalapa, Guat. (hä-lä´pȧ)	160	14°38'N	89°59'W
Jalapa, Mex. *see* Xalapa	158-59	19°32'N	96°55'W
Jālgaon, India	254-55	21°01'N	75°34'E
Jalisco, state, Mex. (hä-lēs´kō)	158-59	20°20'N	103°40'W
Jālna, India	254-55	19°51'N	75°54'E
Jalón, stm., Spain (hä-lōn´)	214-15	41°47'N	1°03'W
Jālor, India	254-55	25°21'N	72°37'E
Jalostotitlán, Mex. (hä-lōs-tē-tlȧn´)	158-59	21°11'N	102°27'W
Jalpa, Mex. (häl´pä)	158-59	21°38'N	102°60'W
Jalpāiguri, India	254-55	26°31'N	88°42'E
Jamaame, Som.	284-85	0°01'N	42°42'E
Jamaica, nation, N.A. (jȧ-mā´kȧ)	152-53	18°15'N	77°30'W
Jamanxim, stm., Braz.	180-81	4°45'S	56°27'W
Jambi, Indon. (mäm´bē)	266-67	1°37'S	103°36'E
Jambongan, Pulau, i., Malay.	268-69	6°40'N	117°27'E
James, stm., U.S. (jāmz)	120-21	42°52'N	97°19'W
James, stm., Va., U.S. (jāmz)	126-27	36°56'N	76°26'W
James Bay, b., Can. (jāmz bā)	140-41	53°30'N	80°30'W
Jamestown, Austl.	301	33°12'S	138°36'E
Jamestown, Ky., U.S. (jāmz´toun)	134-35	36°59'N	85°04'W
Jamestown, N.D., U.S. (jāmz´toun)	124-25	46°54'N	98°42'W
Jamestown, N.Y., U.S. (jāmz´toun)	126-27	42°06'N	79°14'W
Jammu, India (jŭm´ōō)	254-55	32°43'N	74°51'E
Jammu and Kashmir, state, India (jŭm´ōō ănd kȧsh-mēr´)	254-55	34°0'N	76°00'E
Jammu and Kashmir, hist. reg., Asia	254-55	34°0'N	76°00'E
Jamnagar, India (jäm-nû´gŭr)	254-55	22°28'N	70°04'E
Jamshedpur, India (jäm´shēd-pōōr)	254-55	22°48'N	86°11'E
Jamuna, stm., Bngl.	254-55	23°43'N	89°49'E
Janaucu, Ilha, i., Braz.	180-81	0°30'N	50°10'W
Janesville, Ca., U.S. (jānz´vĭl)	128-29	40°18'N	120°32'W
Janesville, Wi., U.S. (jānz´vĭl)	126-27	42°41'N	89°02'W
Jangīpur, India	254-55	24°28'N	88°04'E
Jan Mayen, dep., Eur. (yän mī´ĕn)	314	71°02'N	8°19'W
Jan Mayen, i., Nor. (yän mī´ĕn)	314	71°03'N	8°19'W
Januária, Braz. (zhä-nwä´rē-ä)	182-83	15°29'S	44°22'W
Japan, nation, Asia (jȧ-păn´)	224-25	36°0'N	138°00'E
Japan, Sea of, s., Asia (sē ŭv jȧ-păn´)	240-41	40°0'N	135°00'E
Japurá, stm., S.A.	180-81	3°08'S	64°46'W
Jaraguá do Sul, Braz.	186	26°29'S	49°05'W
Jarama, stm., Spain (hä-rä´mä)	214-15	40°02'N	3°39'W
Jari, stm., Braz. (zhä-rē)	180-81	1°09'S	51°53'W
Jarkand, China *see* Shache	244	38°25'N	77°15'E
Jarocin, Pol. (yä-rō´tsyēn)	210-11	51°58'N	17°30'E

ăt; fĭnăl; rāte; senåte; ärm; åsk; sofȧ; fâre; ch-choose; dh-as th in other; bē; ēvent; bĕt; recĕnt; cratẽr; g-gō; gh-guttural g; bĭt; ĭ-short neutral; rīde; ĸ-guttural k as ch in German ich;

PRONOUNCING INDEX
Jarosław – Juanjuí 367

Feature (Pronunciation)	Page	Lat.	Long.
Jarosław, Pol. (yȧ-rôs-wȧf)	210-11	50°01′N	22°41′E
Jarud Qi, China (jya-lo͞o-tŭ shyē)	260-61	44°34′N	120°54′E
Jarvis Island, dep., Oc.	304-05	0°19′S	160°01′W
Jarvis Island, i., Oc.	304-05	0°23′S	160°01′W
Jāsk, Iran (jäsk)	250-51	25°39′N	57°47′E
Jasło, Pol. (yäs′wō)	210-11	49°45′N	21°28′E
Jason Islands, is., Falk. Is.	185	51°09′S	60°54′W
Jasper, Ab., Can. (jăs′pēr)	142-43	52°52′N	118°05′W
Jasper, Al., U.S. (jăs′pēr)	134-35	33°50′N	87°17′W
Jasper, Fl., U.S. (jăs′pēr)	134-35	30°31′N	82°57′W
Jasper, Ga., U.S. (jăs′pēr)	134-35	34°28′N	84°25′W
Jasper, In., U.S. (jăs′pēr)	126-27	38°23′N	86°56′W
Jasper, Tx., U.S. (jăs′pēr)	132-33	30°54′N	94°00′W
Jasper National Park, n.p., Ab., Can. (jăs′pēr năsh′ŭn-ăl pärk)	142-43	52°53′N	118°03′W
Jassy, Rom. *see* Iași	218-19	47°10′N	27°36′E
Jataí, Braz.	182-83	17°53′S	51°45′W
Jaú, Braz.	186	22°18′S	48°33′W
Jauja, Peru (ḳä-ō′ḳ)	188a	11°47′S	75°29′W
Jaumave, Mex. (hou-mä′vå)	158-59	23°25′N	99°23′W
Jaunpur, India	254-55	25°44′N	82°41′E
Java, i., Indon. (jä′v*ȧ*)	268-69	7°30′S	109°59′E
Javari, stm., S.A. (ḳȧ-vä-rē)	184	4°21′S	70°02′W
Java Sea, s., Indon. (jä′v*ȧ* sē)	268-69	5°0′S	110°00′E
Javhlant, Mong. *see* Uliastay	260-61	47°44′N	96°51′E
Jawa, i., Indon. *see* Java	268-69	7°30′S	109°59′E
Jawa, Laut, s., Indon. *see* Java Sea	268-69	5°0′S	110°00′E
Jawhar, Som.	284-85	2°47′N	45°31′E
Jaworzno, Pol. (yä-vôzh′nô)	210-11	50°12′N	19°15′E
Jaya, Puncak, mtn., Indon.	242-43	4°05′S	137°11′E
Jayapura, Indon.	302	2°32′S	140°43′E
Jaz Mūrīān, Hāmūn-e, lk., Iran	250-51	27°14′N	58°49′E
Jeanerette, La., U.S. (jĕn-ẽr-et′) (zhän-rĕt′)	134-35	29°55′N	91°40′W
Jeddah, Sau. Ar. *see* Jiddah	288	21°30′N	39°12′E
Jędrzejów, Pol. (yăɴ-dzhā′yôf)	210-11	50°39′N	20°19′E
Jefferson, Ia., U.S. (jĕf′ẽr-s*ŭ*n)	124-25	42°01′N	94°23′W
Jefferson, Oh., U.S. (jĕf′ẽr-s*ŭ*n)	126-27	41°44′N	80°46′W
Jefferson, Tx., U.S. (jĕf′ẽr-s*ŭ*n)	130-31	32°46′N	94°21′W
Jefferson, Wi., U.S. (jĕf′ẽr-s*ŭ*n)	126-27	43°00′N	88°48′W
Jefferson, Mount, mtn., Nv., U.S. (mount jĕf′ẽr-s*ŭ*n)	128-29	38°46′N	116°55′W
Jefferson City, Mo., U.S.	130-31	38°33′N	92°10′W
Jefferson City, Tn., U.S. (jĕf′ẽr-s*ŭ*n sĭ′tẻ)	134-35	36°07′N	83°30′W
Jeffersontown, Ky., U.S. (jĕf′ẽr-s*ŭ*n-toun)	126-27	38°13′N	85°35′W
Jeffersonville, In., U.S. (jĕf′ẽr-s*ŭ*n-vĭl)	126-27	38°17′N	85°44′W
Jeju, Kor., S. (jĕ′joo)	260-61	33°30′N	126°32′E
Jejudo, i., Kor., S. (jĕ′jo͞o dō)	260-61	33°22′N	126°30′E
Jēkabpils, Lat. (yĕk′ȧb-pĭls)	208-09	56°30′N	25°52′E
Jelenia Góra, Pol. (yĕ-lĕn′yȧ gŏ′rȧ)	210-11	50°54′N	15°44′E
Jelgava, Lat.	208-09	56°39′N	23°44′E
Jellico, Tn., U.S. (jĕl′ĭ-kō)	134-35	36°35′N	84°08′W
Jemaja, Pulau, i., Indon.	266-67	2°55′N	105°45′E
Jember, Indon.	268-69	8°10′S	113°42′E
Jena, Ger. (yā′nä)	210-11	50°56′N	11°35′E
Jengish Chokusu, mtn., Asia	244	42°02′N	80°05′E
Jenkins, Ky., U.S. (jĕŋ′kĭnz)	134-35	37°10′N	82°39′W
Jennings, La., U.S. (jĕn′ĭngz)	132-33	30°13′N	92°39′W
Jeonju, Kor., S.	263	35°49′N	127°09′E
Jequié, Braz.	180-81	13°52′S	40°05′W
Jequitinhonha, stm., Braz. (zhĕ-kē-tēŋ-ō′n-yä)	182-83	15°51′S	38°53′W
Jerada, Mor.	200-01	34°19′N	2°10′W
Jerba, Île de, i., Tun.	280-81	33°48′N	10°54′E
Jérémie, Haiti (zhā-rā-mē′)	154-55	18°39′N	74°07′W
Jeremoabo, Braz. (zhĕ-rā-mō-ȧ′bō)	180-81	10°06′S	38°19′W
Jerevan, nat. cap., Arm. *see* Yerevan	245	40°11′N	44°30′E
Jerez de la Frontera, Spain	214-15	36°42′N	6°08′W
Jericho, W.B.	248-49	31°52′N	35°27′E
Jerid, Chott, lk., Tun. (shôt jĕr′ĭd)	280-81	33°42′N	8°26′E
Jerome, Id., U.S. (jẻ-rōm′)	122-23	42°44′N	114°31′W
Jersey, dep., Eur. (jûr′zĭ)	212-13	49°15′N	2°10′W
Jersey City, N.J., U.S. (jûr′zĭ sĭ′tẻ)	126-27	40°44′N	74°04′W
Jersey Shore, Pa., U.S. (jûr′zĭ shōr)	126-27	41°12′N	77°15′W
Jerseyville, Il., U.S. (jẽr′zẻ-vĭl)	130-31	39°07′N	90°20′W
Jerusalem, nat. cap., Isr. (jẻ-ro͞o′sȧ-lĕm)	248-49	31°47′N	35°14′E
Jesi, Italy	216-17	43°31′N	13°14′E
Jesselton, Malay. *see* Kota Kinabalu	268-69	5°58′N	116°05′E
Jessore, Bngl.	254-55	23°10′N	89°13′E
Jesup, Ga., U.S. (jĕs′*ŭ*p)	134-35	31°36′N	81°53′W
Jesús Carranza, Mex. (hē-so͞o′s-kär-rá′n-zä)	158-59	17°24′N	95°02′W
Jesús María, Arg.	187	30°59′S	64°05′W
Jewel Cave National Monument, n.p., S.D., U.S. (jū′ĕl kāv)	124-25	43°45′N	103°51′W
Jhālāwār, India	254-55	24°36′N	76°10′E
Jhang Sadar, Pak.	252-53	31°16′N	72°19′E
Jhānsi, India (jän′sẻ)	254-55	25°27′N	78°35′E
Jharkhand, state, India	254-55	23°30′N	85°00′E
Jhelum, Pak.	252-53	32°56′N	73°43′E
Jhelum, stm., Asia (jā′lŭm)	252-53	31°12′N	72°08′E
Jhunjhunūn, India	254-55	28°08′N	75°24′E
Jiading, China (jyä-dĭŋ)	258-59	31°23′N	121°14′E
Jiali, China	258-59	30°45′N	93°20′E
Jialing, China *see* Guangyuan	258-59	32°25′N	105°49′E
Jialing, stm., China (jyä-lĭŋ)	258-59	29°34′N	106°35′E
Jiamusi, China	264	46°48′N	130°22′E
Ji'an, China (jyē-än)	258-59	27°07′N	114°59′E
Ji'an, China (jyē-än)	263	41°06′N	126°10′E
Jianchuan, China	258-59	26°34′N	99°53′E
Jiangjin, China	258-59	29°17′N	106°15′E
Jiangkou, China	258-59	23°35′N	110°11′E
Jiangling, China (jyäŋ-lĭŋ)	258-59	30°19′N	112°12′E
Jiangmen, China	258-59	22°34′N	113°05′E
Jiangsu, state, China (jyäŋ-so͞o)	258-59	33°0′N	120°00′E
Jiangxi, state, China (jyäŋ-shyē)	258-59	28°0′N	116°00′E
Jiangyin, China (jyäŋ-yĭn)	258-59	31°54′N	120°15′E
Jianli, China (jyĕn-lē)	258-59	29°49′N	112°54′E
Jianning, China (jyĕn-nĭŋ)	258-59	26°50′N	116°49′E
Jian'ou, China (jyĕn-ō)	258-59	27°02′N	118°19′E
Jianshi, China (jyĕn-shr)	258-59	30°36′N	109°44′E
Jianshui, China	258-59	23°37′N	102°49′E
Jiaohe, China (jyou-hŭ)	260-61	43°43′N	127°20′E
Jiaoxian, China (jyou shyĕn)	260-61	36°17′N	119°60′E
Jiaozuo, China (jyou-dzwō)	260-61	35°15′N	113°14′E
Jiashan, China (jyä-shän)	258-59	32°46′N	117°59′E
Jiashun Hu, lk., China	254-55	34°24′N	85°47′E
Jiaxing, China (jyä-shyĭn)	258-59	30°46′N	120°45′E
Jiayu, China (jyä-yo͞o)	258-59	29°58′N	113°55′E
Jibuti, nat. cap., Dji. *see* Djibouti	288	11°34′N	43°09′E
Jicarilla Apache Indian Reservation, ind. res., N.M., U.S. (ḳē-kȧ-rēl′yȧ ĭn′dĭ *ă*n rĕ zẽr vā′shŭn)	178-79	36°40′N	107°00′W
Jicarón, Isla, i., Pan. (ē′s-lä-ḳē-kä-rōn′)	162	7°16′N	81°49′W
Jiddah, Sau. Ar.	288	21°30′N	39°12′E
Jieyang, China (jyĕ yäŋ)	258-59	23°33′N	116°21′E
Jiguaní, Cuba (ḳē-gwä-nē′)	154-55	20°22′N	76°25′W
Jijiga, Eth.	284-85	9°21′N	42°48′E
Jilin, China (jyē-lĭn)	260-61	43°51′N	126°33′E
Jilin, state, China	260-61	44°0′N	126°00′E
Jīma, Eth.	284-85	7°38′N	36°50′E
Jiménez, Mex. (ḳẻ-mā′nåz)	132-33	27°08′N	104°56′W
Jiménez, Mex. (ḳẻ-mā′nåz)	132-33	29°02′N	100°41′W
Jiménez del Téul, Mex. (ḳē-mā′nåz dĕl tē-o͞o′l)	158-59	23°14′N	103°49′W
Jimeta, Nig.	282-83	9°16′N	12°26′E
Jim Thorpe, Pa., U.S. (jĭm′ thôrp′)	126-27	40°52′N	75°44′W
Jinan, China (jyē-nän)	260-61	36°40′N	116°59′E
Jincheng, China (jyĭn-chŭŋ)	260-61	35°30′N	112°50′E
Jindo, i., Kor., S.	263	34°27′N	126°15′E
Jindřichuv Hradec, Czech Rep. (yĕn′d'r-zhĭ-ko͞of hrä′dĕts)	210-11	49°09′N	15°01′E
Jing, stm., China (jyĭŋ)	258-59	34°28′N	109°05′E
Jingdezhen, China (jyĭn-dŭ-jŭn)	258-59	29°17′N	117°12′E
Jinggangshan, China	258-59	26°36′N	114°05′E
Jinghong, China	258-59	21°59′N	100°49′E
Jingning, China (jyĭŋ-nĭŋ)	260-61	35°32′N	105°44′E
Jingxian, China (jyĭŋ shyĕn)	258-59	26°40′N	109°25′E
Jingxian, China (jyĭŋ shyĕn)	258-59	30°41′N	118°24′E
Jingxian, China (jyĭŋ shyĕn)	260-61	37°41′N	116°16′E
Jinhae, Kor., S.	263	35°08′N	128°40′E
Jinhua, China (jyĭn-hwä)	258-59	29°07′N	119°39′E
Jining, China (jyē-nĭŋ)	260-61	35°24′N	116°34′E
Jining, China (jyē-nĭŋ)	260-61	41°02′N	113°06′E
Jinja, Ug. (jĭn′jä)	289	0°26′N	33°13′E
Jinju, Kor., S.	263	35°10′N	128°05′E
Jinmu Jiao, c., China	258-59	18°11′N	109°35′E
Jinning, China	258-59	24°40′N	102°35′E
Jinotega, Nic. (kē-nô-tā′gä)	161	13°05′N	85°60′W
Jinotepe, Nic. (kē-nô-tā′pā)	161	11°51′N	86°12′W
Jinsen, Kor., S. *see* Inch'ŏn	263	37°28′N	126°38′E
Jinsha, stm., China *see* Yangtze	258-59	31°24′N	121°54′E
Jinshi, China	258-59	29°38′N	111°52′E
Jinta, China (jyĭn-tä)	260-61	40°00′N	98°53′E
Jinxi, China	260-61	40°45′N	120°50′E
Jinyun, China (jyĭn-yön)	258-59	28°40′N	120°03′E
Jinzhai, China (jyĭn-jī)	258-59	31°45′N	115°55′E
Jinzhou, China (jyĭn-jō)	260-61	39°06′N	121°43′E
Jinzhou, China (jyĭn-jō)	260-61	41°07′N	121°08′E
Ji-Paraná, Braz.	180-81	10°52′S	61°57′W
Jiparaná, stm., Braz. *see* Machado	180-81	8°02′S	62°53′W
Jipijapa, Ec. (ḳē-pē-hä′pä)	184	1°21′S	80°35′W
Jirisan, mtn., Kor., S.	263	35°20′N	127°44′E
Jiujiang, China (jyō-jyän)	258-59	29°43′N	115°59′E
Jiulian Shan, mts., China	258-59	24°17′N	114°36′E
Jiuling Shan, mts., China	258-59	28°46′N	114°45′E
Jiuquan, China (jyō-chyän)	260-61	39°45′N	98°30′E
Jiutai, China	260-61	44°09′N	125°50′E
Jixi, China	264	45°17′N	130°58′E
Jixian, China (jyē shyĕn)	264	46°43′N	131°08′E
Jixian, China (jyē shyĕn)	260-61	35°25′N	114°04′E
Jixian, China (jyē shyĕn)	260-61	40°02′N	117°24′E
Jīzān, Sau. Ar.	288	16°54′N	42°36′E
Jizzax, Uzb.	252-53	40°08′N	67°51′E
J. J. Castelli, Arg. *see* Castelli	182-83	25°57′S	60°37′W
João Belo, Moz. *see* Xai-Xai	286-87	25°03′S	33°39′E
João Pessoa, Braz.	188d	7°07′S	34°52′W
Joaquín V. González, Arg.	182-83	25°05′S	64°09′W
Jódar, Spain (hō′där)	214-15	37°50′N	3°21′W
Jodhpur, India (hōd′po͝or)	254-55	26°17′N	73°01′E
Joensuu, Fin. (yô-ĕn′so͞o)	200-01	62°36′N	29°47′E
Joetsu, Japan	265	37°09′N	138°15′E
Joffre, Mount, mtn., Can. (mount jŏ′f'r)	142-43	50°32′N	115°13′W
Jõgeva, Est. (yû′gĕ-vȧ)	208-09	58°45′N	26°24′E
Jogjakarta, Indon. *see* Yogyakarta	268-69	7°48′S	110°22′E
Johannesburg, S. Afr. (yô-hän′ĕs-bŏrgh)	292c	26°12′S	28°05′E
John Day, stm., Or., U.S. (jŏn′ dā)	122-23	45°44′N	120°39′W
Johnsonburg, Pa., U.S. (jŏn′sŭn-bûrg)	126-27	41°29′N	78°41′W
Johnson City, N.Y., U.S. (jŏn′s*ŭ*n sĭ′tẻ)	126-27	42°07′N	75°58′W
Johnson City, Tn., U.S. (jŏn′s*ŭ*n sĭ′tẻ)	134-35	36°19′N	82°22′W
Johnson City, Tx., U.S. (jŏn′s*ŭ*n sĭ′tẻ)	132-33	30°16′N	98°24′W
Johnston, Lake, lk., Austl.	296-97	32°18′S	120°46′E
Johnston Atoll, dep., Oc.	304-05	16°45′N	169°32′W
Johnston Atoll, at., Oc. (jŏn′s*ŭ*n ȧ′tŏl)	304-05	16°45′N	169°32′W
Johnstown, Pa., U.S. (jonz′toun)	126-27	40°19′N	78°55′W
Johor Bahru, Malay.	266-67	1°28′N	103°45′E
Joigny, Fr. (zhwȧn-yē′)	212-13	47°59′N	3°24′E
Joinville, Braz.	186	26°18′S	48°50′W
Jokkmokk, Swe.	200-01	66°37′N	19°50′E
Joliet, Il., U.S. (jō-lĭ-ĕt′)	126-27	41°31′N	88°04′W
Joliette, Qc., Can. (zhō-lyĕt′)	146-47	46°02′N	73°25′W
Jolo, Phil. (hō-lō)	270	6°02′N	120°60′E
Jolo Group, is., Phil.	270	6°01′N	121°18′E
Jolo Island, i., Phil. (hō-lō ī′l*ȧ*nd)	270	5°58′N	121°06′E
Jomda, China	258-59	31°27′N	98°15′E
Jon, Deti, s., Eur. *see* Ionian Sea	216-17	39°0′N	19°00′E
Jonava, Lith. (yō-nä′vȧ)	208-09	55°05′N	24°17′E
Jonesboro, Ar., U.S. (jōnz′bûro)	134-35	35°51′N	90°42′W
Jonesboro, La., U.S. (jōnz′bûro)	130-31	32°14′N	92°43′W
Jonesville, La., U.S. (jōnz′vĭl)	134-35	31°37′N	91°49′W
Joniškis, Lith. (yō′nĭsh-kĭs)	208-09	56°14′N	23°38′E
Jönköping, Swe. (yûn′chû-pĭng)	208-09	57°47′N	14°11′E
Jonquière, Qc., Can. (zhôɴ-kyâr′)	146-47	48°26′N	71°11′W
Jonuta, Mex. (hō-no͞o′tä)	160	18°06′N	92°07′W
Joplin, Mo., U.S. (jŏp′lĭn)	130-31	37°05′N	94°31′W
Jordan, Mt., U.S.	122-23	47°20′N	106°57′W
Jordan, nation, Asia (jôr′d*ă*n)	224-25	31°0′N	36°00′E
Jordan, stm., Asia (jôr′d*ă*n)	248-49	31°46′N	35°34′E
Jorhāt, India (jôr-hät′)	254-55	26°46′N	94°13′E
Jos, Nig.	282-83	9°56′N	8°53′E
José Batlle y Ordóñez, Ur.	187	33°29′S	55°08′W
José de San Martín, Arg.	185	44°02′S	70°29′W
Joseph Bonaparte Gulf, b., Austl. (jō′sĕf bō′nȧ-pärt gŭlf)	296-97	14°15′S	128°30′E
Joshua Tree National Park, n.p., Ca., U.S. (jō′shū-*ȧ* trē năsh′ŭn-*ă*l pärk)	128-29	33°55′N	116°00′W
Jostedalsbreen, ice, Nor.	208-09	61°40′N	7°00′E
Jovellanos, Cuba (hō-vĕl-yä′nōs)	154-55	22°48′N	81°11′W
J. Strom Thurmond Reservoir, res., U.S.	134-35	33°45′N	82°16′W
Juan Aldama, Mex. (ḳöȧ′n-äl-dȧ′mä)	158-59	24°19′N	103°19′W
Juan de Fuca, Strait of, strt., N.A. (strāt ŭv hwän′ dā fo͞o′kä)	122-23	48°18′N	124°00′W
Juan de Fuca Strait, strt., N.A. *see* Juan de Fuca, Strait of	122-23	48°18′N	124°00′W
Juan Fernández, Archipiélago, is., Chile	173	33°0′S	80°00′W
Juanjuí, Peru	184	7°10′S	76°45′W

n-sing; ŋ-baŋk; ɴ-nasalized n; nŏd; c*ŏ*mmit; ōld; ôbey; ôrder; oi-boil; fo͞od; o͝o-as oo in foot; ou-out; s-soft; sh-dish; th-thin; pūre; ûnite; ûrn; stŭd; circ*ŭ*s; ü-as in French tu; ′-indeterminate vowel.

Feature (Pronunciation)	Page	Lat.	Long.
Juárez, Arg. (hōōá'rĕz) see Benito Juárez	187	37°41'S	59°48'W
Juazeiro, Braz. (zhōōá'zā'rō)	180-81	9°25'S	40°30'W
Juazeiro do Norte, Braz. (zhōōá'zā'rō-dō-nôr-tĕ)	180-81	7°12'S	39°20'W
Juba, nat. cap., S. Sudan	289	4°51'N	31°37'E
Jubal, Strait of, strt., Egypt see Gûbâi, Madîq	291b	27°40'N	33°55'E
Jubayl, Leb. (jōō-bīl')	248-49	34°08'N	35°40'E
Jubba, stm., Afr.	284-85	0°15'S	42°39'E
Juby, Cap, c., Mor. (kăp yōō'bē)	280-81	27°57'N	12°55'W
Júcaro, Cuba (hōō'kà-rō)	154-55	21°38'N	78°51'W
Juchipila, Mex. (hōō-chē-pē'lä)	158-59	21°24'N	103°07'W
Juchitán de Zaragoza, Mex.	158-59	16°26'N	95°01'W
Juchitlán, Mex. (hōō-chē-tlän)	158-59	20°05'N	104°06'W
Juddah, Sau. Ar. see Jiddah	288	21°30'N	39°12'E
Juidongshan, China	258-59	23°44'N	117°30'E
Juigalpa, Nic. (hwē-gäl'pä)	161	12°06'N	85°22'W
Juiz de Fora, Braz. (zhò-ēzh' dā fō'rä)	186	21°45'S	43°22'W
Jujuy, Arg. (hōō-hwē') see San Salvador de Jujuy	182-83	24°12'S	65°18'W
Jujuy, state, Arg. (hōō-hwē')	182-83	23°0'S	66°00'W
Julesburg, Co., U.S. (jōōlz'bûrg)	124-25	40°59'N	102°17'W
Juliaca, Peru (hōō-lē-ä'kä)	184	15°30'S	70°08'W
Juliana Top, mtn., Sur.	178-79	3°39'N	56°32'W
Jumentos Cays, is., Bah. (hōō-mĕn'tōs kēs)	154-55	22°42'N	75°55'W
Jumilla, Spain (hōō-mēl'yä)	214-15	38°28'N	1°20'W
Jūnāgadh, India (jò-nä'gŭd)	254-55	21°31'N	70°27'E
Junction, Tx., U.S. (jŭngk'shŭn)	132-33	30°29'N	99°47'W
Junction City, Ks., U.S. (jŭngk'shŭn sĭ'tē)	130-31	39°02'N	96°50'W
Junction City, Or., U.S. (jŭngk'shŭn sĭ'tē)	122-23	44°14'N	123°11'W
Jundiaí, Braz.	186	23°11'S	46°53'W
Juneau, Ak., U.S. (jōō'nō)	136	58°20'N	134°25'W
Junee, Austl.	301	34°52'S	147°35'E
Jungar Qi, China	260-61	39°49'N	111°10'E
Jungfrau, mtn., Switz. (yòng'frou)	210-11	46°32'N	7°58'E
Junín, Arg. (hōō-nē'n)	187	34°36'S	60°58'W
Junín de los Andes, Arg.	185	39°55'S	71°05'W
Jūniyah, Leb. (jōō-nē'ē)	248-49	33°60'N	35°39'E
Junxian, China	258-59	32°32'N	111°31'E
Juquiá, Braz.	186	24°19'S	47°38'W
Jur, stm., S. Sudan (jòr)	284-85	8°39'N	29°17'E
Jura, i., Scot., U.K. (jōō'rà)	206-07	56°01'N	5°56'W
Jura, mts., Eur. (zhü-rà')	210-11	47°06'N	6°50'E
Jurbarkas, Lith. (yōōr-bär'käs)	208-09	55°05'N	22°47'E
Jūrmala, Lat.	208-09	56°58'N	23°42'E
Juruá, stm., S.A.	180-81	2°35'S	65°47'W
Juruena, stm., Braz. (zhōō-rōōĕ'nä)	180-81	7°21'S	58°09'W
Justo Daract, Arg.	185	33°52'S	65°11'W
Jutaí, stm., Braz.	180-81	2°44'S	66°48'W
Jutiapa, Guat. (hōō-tē-ä'pä)	160	14°18'N	89°54'W
Juticalpa, Hond. (hōō-tē-käl'pä)	161	14°40'N	86°13'W
Jutland, reg., Den. see Jylland	208-09	56°0'N	9°15'E
Juventud, Isla de la, i., Cuba	154-55	21°40'N	82°50'W
Jyekundo, China see Yushu	258-59	33°00'N	97°00'E
Jylland, reg., Den.	208-09	56°0'N	9°15'E
Jyväskylä, Fin.	208-09	62°15'N	25°45'E

K

Feature (Pronunciation)	Page	Lat.	Long.
K2, mtn., Asia (kā-tōō)	252-53	35°53'N	76°30'E
Ka'ena Point, c., Hi., U.S. (kä'ā-nä point)	137a	21°35'N	158°17'W
Kaapstad, nat. cap., S. Afr. see Cape Town	286-87	33°55'S	18°30'E
Kaarlela, Fin. see Kokkola	200-01	63°50'N	23°09'E
Kaba, stm., Afr. see Little Scarcies	282-83	8°51'N	13°07'W
Kabaena, Pulau, i., Indon. (pōō-lou kä-bà-ā'nà)	268-69	5°15'S	121°55'E
Kabala, S.L. (kà-bà'là)	282-83	9°35'N	11°33'W
Kabale, Ug.	289	1°11'S	29°56'E
Kabalega Falls, wtfl., Ug.	289	2°17'N	31°42'E
Kabalo, D.R.C. (kä-bä'lō)	284-85	6°03'S	26°55'E
Kabara, i., Fiji	306f	18°57'S	178°57'W
Kabardin-Balkaria, state, Russia see Balkaria	245	43°30'N	43°30'E
Kabardino-Balkariya, state, Russia see Balkaria	245	43°30'N	43°30'E
Kabīr Kūh, mts., Iran	248-49	33°36'N	46°12'E
Kābol, nat. cap., Afg. (kä'bōōl) see Kabul	252-53	34°32'N	69°10'E
Kābol, stm., Asia	252-53	33°55'N	72°14'E
Kabompo, stm., Zam. (kà-bôm'pō)	286-87	14°12'S	23°11'E
Kabul, nat. cap., Afg. (kä'bōōl)	252-53	34°32'N	69°10'E
Kabul, stm., Asia (kä'bòl) see Kābol	252-53	33°55'N	72°14'E
Kaburuang, Pulau, i., Indon.	268-69	3°48'N	126°48'E
Kabwe, Zam.	286-87	14°27'S	28°27'E
Kachchh, Gulf of, b., India	254-55	22°37'N	69°30'E
Kachchh, Rann of, reg., Asia see Kutch, Rann of	254-55	24°15'N	70°46'E
Kachul, Mol. see Cahul	218-19	45°55'N	28°12'E
Kadamatt Island, i., India	256	11°13'N	72°47'E
Kadan Kyun, i., Mya.	266-67	12°30'N	98°22'E
Kadéï, stm., Afr.	282-83	3°31'N	16°03'E
Kadina, Austl.	301	33°58'S	137°43'E
Kadirli, Tur.	247	37°22'N	36°06'E
Kadiyevka, Ukr. see Stakhanov	218-19	48°34'N	38°40'E
Kadoma, Zimb.	286-87	18°21'S	29°54'E
Kaduna, Nig. (kä-dōō'nä)	282-83	10°32'N	7°25'E
Kaduna, stm., Nig. (kä-dōō'nä)	282-83	8°45'N	5°48'E
Kāduqlī, Sudan	288	10°60'N	29°43'E
Kadzherom, Russia	202-03	64°41'N	55°55'E
Kaédi, Maur. (kä-ā-dē')	280-81	16°10'N	13°29'W
Kaesŏng, Kor., N. (kä'ĕ-sŭng) (kĭ'jō)	263	37°59'N	126°34'E
Kafue, stm., Zam. (kä'fōō)	286-87	15°56'S	28°57'E
Kafue National Park, n.p., Zam. (kä'fōō năsh'ŭn-ăl pärk)	286-87	15°22'S	25°25'E
Kaga Bandoro, C.A.R.	284-85	6°59'N	19°12'E
Kagera, stm., Afr. (kä-gā'rà)	289	0°56'S	31°47'E
Kagoshima, Japan (kä'gô-shē'mà)	265	31°36'N	130°33'E
Kagoshima-wan, b., Japan (kä'gô-shē'mä wän)	265	31°24'N	130°38'E
Kahama, Tan.	289	3°49'S	32°36'E
Kahayan, stm., Indon.	268-69	3°16'S	114°06'E
Kahemba, D.R.C.	284-85	7°18'S	18°59'E
Kaho'olawe, i., Hi., U.S. (kä-hōō-lä'wē)	137a	20°33'N	156°37'W
Kahoka, Mo., U.S. (kà-hō'kà)	130-31	40°26'N	91°43'W
Kahramanmaraş, Tur.	202-03	37°35'N	36°57'E
Kahuku Point, c., Hi., U.S. (kä-hōō'kōō point)	137a	21°43'N	157°59'W
Kai, Kepulauan, is., Indon.	242-43	5°35'S	132°45'E
Kaibab Indian Reservation, ind. res., Az., U.S. (kä'ē-bàb ĭn'dĭ-ăn rĕ-sĕr-vā'shĕn)	128-29	36°55'N	112°40'W
Kaidu, stm., China (kī-dōō)	240-41	41°58'N	86°44'E
Kaifeng, China (kī-fŭng)	258-59	34°47'N	114°21'E
Kaijo, Kor., N. see Kaesŏng	263	37°59'N	126°34'E
Kai Kecil, i., Indon.	242-43	5°46'S	132°43'E
Kailas, mtn., China see Kangrinboqê Feng	254-55	31°04'N	81°18'E
Kailas Range, mts., China (kī-läs rānj) see Gangdisê Shan	254-55	31°0'N	82°00'E
Kailu, China	260-61	43°36'N	121°19'E
Kailua, Hi., U.S. (kä'ē-lōō'ä)	137a	21°24'N	157°45'W
Kailua, Hi., U.S. (kä'ē-lōō'ä)	137a	19°39'N	155°58'W
Kailua Kona, Hi., U.S. see Kailua	137a	19°39'N	155°58'W
Kaiping, China	258-59	22°22'N	112°37'E
Kairouan, Tun.	200-01	35°41'N	10°07'E
Kaiserslautern, Ger. (kī-zĕrs-lou'tĕrn)	210-11	49°26'N	7°45'E
Kaiyuan, China (kū-yuän)	263	42°32'N	124°02'E
Kaiyuan, China (kū-yuän)	258-59	23°42'N	103°14'E
Kajaani, Fin. (kä'yà-nĕ)	200-01	64°14'N	27°45'E
Kaka, Turkmen.	252-53	37°20'N	59°37'E
Kakabia, Pulau, i., Indon.	268-69	6°54'S	122°13'E
Kakamas, S. Afr.	286-87	28°46'S	20°36'E
Kakamega, Kenya	289	0°17'N	34°45'E
Kakhovka, Ukr. (kä-kôf'kà)	218-19	46°49'N	33°30'E
Kakhovka Reservoir, res., Ukr see Kakhovs'ke vodoskhovyshche	218-19	47°28'N	34°06'E
Kakhovs'ke vodoskhovyshche, res., Ukr.	218-19	47°28'N	34°06'E
Kakhul, Mol. see Cahul	218-19	45°55'N	28°12'E
Kākināda, India	256	16°57'N	82°15'E
Kakshaal-Too, mts., Asia	244	41°0'N	78°00'E
Kaktovik, Ak., U.S. (kăk-tō'vĭk)	136	70°08'N	143°38'W
Kakuma, Kenya	289	3°42'N	34°52'E
Kalaallit Nunaat, dep., N.A. see Greenland	94	70°0'N	40°00'W
Kalabahi, Indon.	268-69	8°15'S	124°32'E
Kalach, Russia (kà-lách')	202-03	50°25'N	41°00'E
Kalachinsk, Russia	244	55°02'N	74°35'E
Kalach-na-Donu, Russia	202-03	48°43'N	43°28'E
Kalae, c., Hi., U.S.	137a	18°55'N	155°41'W
Kalahari Desert, des., Afr. (kä-lä-hä'rē dĕs'ērt)	286-87	24°0'S	21°30'E
Kalahari Gemsbok National Park, n.p., S. Afr.	286-87	25°30'S	20°30'E
Kalama, Wa., U.S. (kà-lăm'à)	122-23	46°01'N	122°50'W
Kalamáta, Grc.	216-17	37°03'N	22°07'E
Kalamazoo, Mi., U.S. (kăl-à-mà-zōō')	126-27	42°17'N	85°35'W
Kalamazoo, stm., Mi., U.S. (kăl-à-mà-zōō')	126-27	42°40'N	86°12'W
Kalanchak, Ukr. (kä-län-chäk')	218-19	46°15'N	33°18'E
Kalao, Pulau, i., Indon.	268-69	7°18'S	120°58'E
Kalaotoa, Pulau, i., Indon.	268-69	7°22'S	121°47'E
Kalāt, Pak. (kŭ-lät')	252-53	29°02'N	66°35'E
Kalaw, Mya.	266-67	20°38'N	96°34'E
Kalbarri, Austl.	294-95	27°42'S	114°10'E
Kaledupa, Pulau, i., Indon.	268-69	5°32'S	123°47'E
Kalemie, D.R.C.	289	5°55'S	29°11'E
Kalemyo, Mya.	266-67	23°13'N	94°07'E
Kalevala, Russia	202-03	65°11'N	31°11'E
Kalewa, Mya.	266-67	23°12'N	94°18'E
Kalgoorlie-Boulder, Austl. (kăl-gōōr'lē-bōld'ĕr)	294-95	30°44'S	121°27'E
Kalibo, Phil.	270	11°43'N	122°23'E
Kalima, D.R.C.	284-85	2°36'S	26°37'E
Kalimantan, i., Asia see Borneo	268-69	0°30'N	114°00'E
Kālimpang, India	254-55	27°04'N	88°28'E
Kaliningrad, Russia (kä-lē-nēn'grät)	210-11	54°43'N	20°30'E
Kalisch, Pol. see Kalisz	210-11	51°46'N	18°06'E
Kalispell, Mt., U.S. (kăl'ĭ-spĕl)	122-23	48°12'N	114°19'W
Kalisz, Pol. (kä'lēsh)	210-11	51°46'N	18°06'E
Kalixälven, stm., Swe.	200-01	65°53'N	23°03'E
Kalmar, Swe. (käl'mär)	208-09	56°40'N	16°22'E
Kalmarsund, strt., Swe. (käl'mär)	208-09	56°40'N	16°25'E
Kal'mius, stm., Ukr. (käl'myōōs)	218-19	47°05'N	37°34'E
Kalmykia, state, Russia	202-03	46°30'N	45°30'E
Kalmykiya, state, Russia see Kalmykia	202-03	46°30'N	45°30'E
Kalpeni Island, i., India	256	10°05'N	73°38'E
Kalsūbai, mtn., India	254-55	19°36'N	73°43'E
Kaluga, Russia (kà-lò'gä)	218-19	54°32'N	36°17'E
Kalundborg, Den. (kà-lòn''bôr')	208-09	55°41'N	11°07'E
Kalush, Ukr. (kä'lòsh)	210-11	49°02'N	24°22'E
Kalyān, India	256	19°16'N	73°08'E
Kalyazin, Russia (käl-yà'zĕn)	218-19	57°14'N	37°54'E
Kálymnos, i., Grc.	216-17	37°0'N	27°00'E
Kama, stm., Russia (kä'mä)	202-03	55°35'N	51°29'E
Kamaishi, Japan (kä'mä-ē'shē)	264	39°16'N	141°53'E
Kamakura, Japan (kä'mä-kōō'rä)	265	35°19'N	139°33'E
Kama Reservoir, res., Russia (kä'mä rĕ'sĕr-vwär) see Kamskoye Vodokhranilishche	202-03	58°52'N	56°15'E
Kambarka, Russia	202-03	56°15'N	54°13'E
Kamchatka, Poluostrov, pen., Russia see Kamchatka Peninsula	236-37	56°0'N	160°00'E
Kamchatka Peninsula, pen., Russia (kăm-chăt-kà pĕ-nĭn'sūlà)	236-37	56°0'N	160°00'E
Kamenjak, Rt, c., Cro. (kä'mĕ-nyäk)	216-17	44°46'N	13°55'E
Kamenka, Russia	202-03	53°11'N	44°03'E
Kamen'-na-Obi, Russia (kä-mĭny'nŭ ô'bē)	244	53°48'N	81°20'E
Kāmet, mtn., Asia	254-55	30°54'N	79°37'E
Kam'ianets'-Podil's'kyi, Ukr.	210-11	48°40'N	26°36'E
Kamina, D.R.C.	284-85	8°44'S	25°00'E
Kaminak Lake, lk., Nu., Can.	140-41	62°09'N	95°07'W
Kamino-shima, i., Japan	263	34°35'N	129°25'E
Kaminuriak Lake, lk., Nu., Can.	140-41	62°59'N	95°35'W
Kamituga, D.R.C.	289	3°02'S	28°14'E
Kamloops, B.C., Can. (kăm'lōōps)	142-43	50°40'N	120°20'W
Kampala, nat. cap., Ug. (käm-pä'lä)	289	0°19'N	32°34'E
Kampar, stm., Indon. (käm'pär)	266-67	0°14'N	102°42'E
Kamphaeng Phet, Thai.	266-67	16°28'N	99°32'E
Kâmpóng Cham, Camb.	266-67	12°0'N	105°27'E
Kâmpóng Chhnăng, Camb.	266-67	12°15'N	104°40'E
Kâmpóng Saôm, Camb.	266-67	10°38'N	103°31'E
Kâmpóng Saôm, Chhâk, b., Camb.	266-67	10°50'N	103°32'E
Kâmpóng Thum, Camb. (kŏm'pŏng-tŏm)	266-67	12°42'N	104°54'E
Kâmpôt, Camb. (käm'pōt)	266-67	10°37'N	104°11'E
Kampuchea, nation, Asia see Cambodia	224-25	13°0'N	105°00'E
Kamsack, Sk., Can. (kăm'săk)	144-45	51°34'N	101°54'W
Kamskoye Vodokhranilishche, res., Russia	202-03	58°52'N	56°15'E
Kamuela, Hi., U.S. see Waimea	137a	20°02'N	155°40'W
Kámuk, Cerro, mtn., C.R. (sĕ'r-rô-kä-mōō'k)	161	9°17'N	83°01'W
Kamyshin, Russia (kä-mwēsh'ĭn)	202-03	50°07'N	45°24'E
Kanaaupscow, stm., Qc., Can.	140-41	53°40'N	76°44'W
Kanab, Ut., U.S. (kăn'ăb)	128-29	37°03'N	112°32'W

ăt; finăl; rāte; senāte; ärm; ásk; sofà; fâre; ch-choose; dh-as th in other; bē; ēvent; bĕt; recĕnt; cratĕr; g-gō; gh-guttural g; bĭt; ĭ-short neutral; rīde; κ-guttural k as ch in German ich;

PRONOUNCING INDEX
Kanab Plateau – Kavkasioni 369

Feature (Pronunciation)	Page	Lat.	Long.
Kanab Plateau, plat., U.S. (kăn'ăb plă-tō')	**128-29**	36°36'N	112°45'W
Kanagawa, state, Japan (kä'nä-gä'wä)	**265**	35°30'N	139°15'E
Kananga, D.R.C.	**284-85**	5°54'S	22°25'E
Kanash, Russia	**202-03**	55°31'N	47°29'E
Kanawha, stm., W.V., U.S. (kȧ-nô'wȧ)	**126-27**	38°50'N	82°09'W
Kanazawa, Japan (kä'nȧ-zä'wȧ)	**265**	36°34'N	136°39'E
Kanchanjanggā, mtn., Asia *see* Kānchenjunga	**254-55**	27°41'N	88°10'E
Kānchenjunga, mtn., Asia (kĭn-chĭn-jŏŋ'gȧ)	**254-55**	27°41'N	88°10'E
Kānchipuram, India	**256**	12°50'N	79°43'E
Kandahār, Afg.	**252-53**	31°37'N	65°43'E
Kandalaksha, Russia (kån-dȧ-lȧk'shȧ)	**200-01**	67°09'N	32°24'E
Kandangan, Indon.	**268-69**	2°48'S	115°16'E
Kandavu, i., Fiji	**306f**	19°00'S	178°11'E
Kandy, Sri L. (kän'dē)	**256**	7°18'N	80°38'E
Kane, Pa., U.S. (kān)	**126-27**	41°40'N	78°48'W
Kāne'ohe, Hi., U.S. (kä-nä-ō'hā)	**137a**	21°25'N	157°48'W
Kanevskaya, Russia (kȧ-nyĕf'skȧ)	**218-19**	46°05'N	38°58'E
Kangar, Malay.	**266-67**	6°27'N	100°12'E
Kangaroo Island, i., Austl. (kăŋ-gȧ-rōō' ī'lȧnd)	**301**	35°50'S	137°05'E
Kangāvar, Iran (kŭŋ'gä-vär)	**248-49**	34°30'N	47°58'E
Kangding, China	**258-59**	30°04'N	102°01'E
Kangean, Kepulauan, is., Indon. (käŋ'gē-än)	**268-69**	6°55'S	115°30'E
Kangean, Pulau, i., Indon.	**268-69**	6°54'S	115°20'E
Kanggye, Kor., N. (käng'gyĕ)	**263**	40°58'N	126°36'E
Kangiqsliniq, Nu., Can. *see* Rankin Inlet	**138-39**	62°49'N	92°10'W
Kangiqsualujjuaq, Qc., Can.	**138-39**	58°42'N	65°59'W
Kangiqsujuaq, Qc., Can.	**138-39**	61°35'N	71°58'W
Kangirsuk, Qc., Can.	**138-39**	60°02'N	70°01'W
Kangnŭng, Kor., S. *see* Gangneung	**263**	37°46'N	128°54'E
Kango, Gabon (käN-gō)	**282-83**	0°11'N	10°05'E
Kangrinboqê Feng, mtn., China	**254-55**	31°04'N	81°18'E
Kangto, mtn., Asia	**254-55**	27°52'N	92°30'E
Kanhsien, China *see* Ganzhou	**258-59**	25°53'N	114°55'E
Kaniama, D.R.C.	**284-85**	7°33'S	24°10'E
Kanin, Poluostrov, pen., Russia	**202-03**	68°0'N	45°00'E
Kanin Nos, Mys, c., Russia	**202-03**	68°39'N	43°17'E
Kankakee, Il., U.S. (kăŋ-kȧ-kē')	**126-27**	41°07'N	87°51'W
Kankan, Gui. (käN-käN) (kän-kän')	**282-83**	10°23'N	9°18'W
Kankō, Kor., N. *see* Hamhŭng	**263**	39°55'N	127°32'E
Kanmaw Kyun, i., Mya.	**266-67**	11°40'N	98°28'E
Kannapolis, N.C., U.S. (kăn-ăp'ô-lĭs)	**134-35**	35°29'N	80°37'W
Kannur, India	**256**	11°52'N	75°22'E
Kano, Nig. (kä'nō)	**282-83**	12°01'N	8°30'E
Kānpur, India (kän'pûr)	**254-55**	26°28'N	80°19'E
Kansas, state, U.S. (kăn'zȧs)	**118-19**	38°45'N	98°15'W
Kansas City, Ks., U.S. (kăn'zȧs sĭ'tē)	**130-31**	39°07'N	94°38'W
Kansas City, Mo., U.S. (kăn'zȧs sĭ'tē)	**130-31**	39°06'N	94°34'W
Kansk, Russia	**236-37**	56°12'N	95°43'E
Kansu, state, China *see* Gansu	**260-61**	37°0'N	103°00'E
Kantang, Thai. (kän'täng')	**266-67**	7°24'N	99°32'E
Kanton, i., Kir. *see* Canton	**304-05**	2°49'S	171°41'W
Kantunilkin, Mex. (kän-tōō-nēl-kē'n)	**160**	21°06'N	87°29'W
Kanye, Bots.	**286-87**	24°59'S	25°19'E
Kaohsiung, Tai. (kä-ô-syŏng')	**243a**	22°38'N	120°17'E
Kaoko Veld, plat., Nmb.	**286-87**	20°0'S	14°00'E
Kaolack, Sen.	**282-83**	14°09'N	16°04'W
Kaoma, Zam.	**286-87**	14°47'S	24°48'E
Kapenguria, Kenya	**289**	1°09'N	35°01'E
Kapfenberg, Aus. (käp'fĕn-bĕrgh)	**210-11**	47°27'N	15°17'E
Kapingamarangi, at., Micron.	**304-05**	1°04'N	154°46'E
Kapit, Malay.	**268-69**	2°00'N	112°56'E
Kapoeta, S. Sudan	**289**	4°47'N	33°35'E
Kaposvár, Hung. (kô'pōsh-vär)	**210-11**	46°22'N	17°48'E
Kapuas, stm., Indon.	**268-69**	0°09'S	109°08'E
Kapuas Hulu, Pegunungan, mts., Asia *see* Upper Kapuas Mountains	**268-69**	1°15'N	113°30'E
Kapuas Hulu, Pergunungan, mts., Asia *see* Upper Kapuas Mountains	**268-69**	1°15'N	113°30'E
Kapuskasing, On., Can.	**146-47**	49°25'N	82°25'W
Kapuskasing, stm., On., Can.	**146-47**	49°38'N	82°16'W
Kara, Togo	**282-83**	9°33'N	1°12'E
Kara, stm., Russia (kärȧ)	**202-03**	69°07'N	64°45'E
Kara-Balta, Kyrg.	**244**	42°48'N	73°51'E
Karabogaz, Turkmen.	**202-03**	41°32'N	52°35'E
Kara-Bogaz-Gol Gulf, b., Turkmen. (kȧ-rä' bŭ-gäs' gôl gŭlf)	**252-53**	41°15'N	53°24'E
Karabük, Tur.	**202-03**	41°13'N	32°37'E
Karachay, state, Russia *see* Cherkessia	**245**	44°0'N	42°00'E
Karachay-Cherkessia, state, Russia *see* Cherkessia	**245**	44°0'N	42°00'E
Karachayevo-Cherkesiya, state, Russia *see* Cherkessia	**245**	44°0'N	42°00'E
Karachev, Russia (kȧ-rȧ-chôf')	**218-19**	53°07'N	34°59'E
Karāchi, Pak. (kȧ-rä'chē)	**252-53**	24°54'N	67°01'E
Kara Deniz, s., *see* Black Sea	**202-03**	43°0'N	35°00'E
Karaginskiy, Ostrov, i., Russia	**236-37**	58°50'N	164°00'E
Karaginskiy Zaliv, b., Russia	**236-37**	58°50'N	164°00'E
Karaj, Iran	**252-53**	35°50'N	50°59'E
Karakax, stm., China	**244**	38°03'N	80°32'E
Karakelong, Pulau, i., Indon.	**268-69**	4°16'N	126°49'E
Karakol, Kyrg.	**244**	42°29'N	78°23'E
Karakoram Range, mts., Asia (kä'rä kō'rȯm rānj)	**254-55**	35°30'N	77°00'E
Karakorum, hist., Mong.	**260-61**	47°14'N	102°50'E
Karakorum Shan, mts., Asia *see* Karakoram Range	**254-55**	35°30'N	77°00'E
Kara Kum, des., Turkmen. (kärä-kōōm')	**244**	39°0'N	60°00'E
Kara-Kum Canal, can., Turkmen. (kärä-kōōm' kȧ'năl)	**252-53**	37°34'N	65°41'E
Karakumy, des., Turkmen. *see* Kara Kum	**244**	39°0'N	60°00'E
Karaman, Tur. (kä-rä-män')	**202-03**	37°11'N	33°13'E
Karamay, China (kär-äm-ä)	**244**	45°36'N	84°51'E
Karamea Bight, b., N.Z. (kȧ-rȧ-mē'ȧ bīt)	**303**	41°30'S	171°40'E
Karasburg, Nmb.	**286-87**	28°01'S	18°45'E
Kara Sea, s., Russia (kärȧ sē)	**236-37**	76°0'N	80°00'E
Karasuk, Russia	**244**	53°43'N	78°03'E
Karatau Range, mts., Kaz.	**244**	43°36'N	68°52'E
Karatsu, Japan (kä'rȧ-tsōō)	**265**	33°26'N	129°59'E
Karawang, Indon.	**268-69**	6°18'S	107°18'E
Karbalā', Iraq	**248-49**	32°37'N	44°02'E
Karcag, Hung. (kär'tsäg)	**210-11**	47°19'N	20°56'E
Kardeljevo, Cro.	**216-17**	43°04'N	17°26'E
Kärdla, Est. (kĕrd'lȧ)	**208-09**	58°60'N	22°45'E
Kargasok, Russia	**236-37**	59°04'N	80°50'E
Kargopol', Russia (kär-gō-pōl'')	**202-03**	61°30'N	38°58'E
Kariba, Zimb.	**286-87**	16°31'S	28°48'E
Kariba, Lake, res., Afr.	**286-87**	17°0'S	28°00'E
Karimata, Kepulauan, is., Indon. (kä-rē-mä'tä)	**266-67**	1°25'S	109°05'E
Karimata, Pulau, i., Indon.	**266-67**	1°36'S	108°55'E
Karimata, Selat, strt., Indon.	**268-69**	2°05'S	108°40'E
Karīmnagar, India	**256**	18°26'N	79°09'E
Karimunjawa, Kepulauan, is., Indon. (kä'rē-mōōn-yä'vä)	**268-69**	5°50'S	110°25'E
Karisimbi, Volcan, vol., Afr.	**289**	1°30'S	29°27'E
Karkar Island, i., Pap. N. Gui. (kär'kär ī'lȧnd)	**302**	4°40'S	146°00'E
Karkūk, Iraq *see* Kirkuk	**248-49**	35°28'N	44°24'E
Karleby, Fin. *see* Kokkola	**200-01**	63°50'N	23°09'E
Karl-Marx-Stadt, Ger. *see* Chemnitz	**210-11**	50°50'N	12°56'E
Karlovac, Cro. (kär'lō-väts)	**216-17**	45°29'N	15°33'E
Karlovo, Blg. (kär'lō-vō)	**216-17**	42°39'N	24°48'E
Karlovy Vary, Czech Rep. (kär'lō-vē vä'rē)	**210-11**	50°14'N	12°53'E
Karlshamn, Swe. (kärls'häm)	**208-09**	56°09'N	14°51'E
Karlskrona, Swe. (kärls'krō-nä)	**208-09**	56°10'N	15°36'E
Karlsruhe, Ger. (kärls'rōō-ĕ)	**210-11**	49°01'N	8°23'E
Karlstad, Swe. (kärl'städ)	**208-09**	59°23'N	13°31'E
Karmøy, i., Nor. (kärm-ûe)	**208-09**	59°15'N	5°15'E
Karnāl, India	**254-55**	29°41'N	76°59'E
Karnātaka, state, India	**256**	14°0'N	76°00'E
Karonga, Malawi (kȧ-rōŋ'gȧ)	**286-87**	9°55'S	33°56'E
Kárpathos, i., Grc.	**204-05**	35°41'N	27°09'E
Karpaty, mts., Eur. *see* Carpathian Mountains	**202-03**	48°0'N	24°00'E
Karpinsk, Russia (kär'pĭnsk)	**202-03**	59°46'N	60°00'E
Karpogory, Russia	**202-03**	64°00'N	44°23'E
Karratha, Austl.	**294-95**	20°43'S	116°48'E
Kars, Tur. (kärs)	**245**	40°36'N	43°05'E
Karshi, Uzb. (kär'shē) *see* Qarshi	**252-53**	38°52'N	65°48'E
Karskoye More, s., Russia *see* Kara Sea	**236-37**	76°0'N	80°00'E
Kartaly, Russia (kȧr'tȧ lē)	**244**	53°03'N	60°39'E
Karumba, Austl.	**302**	17°28'S	140°51'E
Karūr, India	**256**	10°57'N	78°05'E
Kārwār, India	**256**	14°48'N	74°08'E
Kasai, stm., Afr.	**284-85**	3°02'S	16°56'E
Kasama, Zam. (kȧ-sä'mȧ)	**286-87**	10°12'S	31°11'E
Kasar, Ras, c., Afr. *see* Kasr, Ra's	**288**	18°01'N	38°34'E
Kasba Lake, lk., Can.	**140-41**	60°18'N	102°07'W
Kasba-Tadla, Mor. (käs'bȧ-täd'lȧ)	**292a**	32°37'N	6°16'W
Kaschau, Slvk. *see* Košice	**210-11**	48°43'N	21°16'E
Kasenga, D.R.C. (kȧ-seŋ'gȧ)	**284-85**	10°22'S	28°37'E
Kasese, Ug.	**289**	0°10'N	30°05'E
Kāshān, Iran (kä-shän')	**252-53**	33°59'N	51°26'E
Kashgar, China (käsh-gär) *see* Kashi	**244**	39°28'N	75°59'E
Kashi, China (kä-shr)	**244**	39°28'N	75°59'E
Kashihara, Japan (kä'shē-hä'rä)	**265**	34°30'N	135°48'E
Kashin, Russia (kä-shēn')	**218-19**	57°22'N	37°37'E
Kashira, Russia (kä-shē'rȧ)	**218-19**	54°51'N	38°10'E
Kashiwazaki, Japan (kä'shē-wä-zä'kē)	**265**	37°22'N	138°33'E
Kāshmar, Iran	**252-53**	35°13'N	58°28'E
Kasia, India	**254-55**	26°45'N	83°55'E
Kasimov, Russia (kȧ-sē'môf)	**202-03**	54°56'N	41°23'E
Kaskaskia, stm., Il., U.S. (kăs-kăs'kĭ-ȧ)	**130-31**	37°58'N	89°57'W
Kaskattama, stm., Mb., Can. (kăs-kȧ-tä'mȧ)	**144-45**	57°03'N	90°05'W
Kasongo, D.R.C. (kä-sôŋ'gō)	**284-85**	4°27'S	26°40'E
Kasongo-Lunda, D.R.C.	**284-85**	6°29'S	16°50'E
Kaspīy Mangy oypaty, pl., *see* Caspian Depression	**202-03**	48°0'N	52°00'E
Kaspiysk, Russia	**245**	42°53'N	47°38'E
Kaspiyskiy, Russia	**202-03**	45°24'N	47°21'E
Kaspiyskoye More, lk., *see* Caspian Sea	**244**	41°18'N	50°59'E
Kasr, Ra's, c., Afr.	**288**	18°01'N	38°34'E
Kassa, Slvk. *see* Košice	**210-11**	48°43'N	21°16'E
Kassalā, Sudan	**288**	15°27'N	36°23'E
Kassel, Ger. (käs'ĕl)	**210-11**	51°19'N	9°29'E
Kasserine, Tun.	**200-01**	35°09'N	8°50'E
Kasson, Mn., U.S. (kăs'ŭn)	**124-25**	44°02'N	92°46'W
Kastamonu, Tur. (kä-stȧ-mō'nōō)	**202-03**	41°23'N	33°47'E
Kastellorizo, i., Grc. *see* Megísti	**204-05**	36°08'N	29°36'E
Kastoría, Grc. (kȧs-tō'rĭ-ȧ)	**216-17**	40°32'N	21°17'E
Kasulu, Tan.	**289**	4°34'S	30°06'E
Kasungu, Malawi	**286-87**	13°03'S	33°28'E
Kasūr, Pak.	**252-53**	31°07'N	74°27'E
Katahdin, Mount, mtn., Me., U.S. (mount kȧ-tä'dĭn)	**127a**	45°55'N	68°55'W
Katanda, D.R.C.	**289**	0°50'S	29°22'E
Katanga, hist. reg., D.R.C. (kȧ-täŋ'gȧ)	**284-85**	10°0'S	26°00'E
Katanga, stm., Russia	**236-37**	60°00'N	102°11'E
Katanning, Austl. (kȧ-tăn'ĭng)	**294-95**	33°42'S	117°33'E
Katchall Island, i., India	**266-67**	7°55'N	93°23'E
Katha, Mya.	**258-59**	24°10'N	96°20'E
Katherine, Austl. (kăth'ẽr-ĭn)	**294-95**	14°29'S	132°16'E
Kāthiāwār Peninsula, pen., India (kä'tyȧ-wär' pĕ-nĭn'sūlă)	**254-55**	22°0'N	71°00'E
Kāṭhmāṇḍāu, nat. cap., Nepal *see* Kathmandu	**254-55**	27°42'N	85°19'E
Kathmandu, nat. cap., Nepal (kät-män-dōō')	**254-55**	27°42'N	85°19'E
Katihār, India	**254-55**	25°33'N	87°34'E
Katima Mulilo, Nmb.	**286-87**	17°30'S	24°16'E
Ka Tiriti o te Moana, mts., N.Z. *see* Southern Alps	**303**	43°30'S	170°30'E
Katmai National Park and Preserve, n.p., Ak., U.S. (kăt'mī năsh'ŭn-ăl pärk ănd prĭ-zûrv')	**136**	58°30'N	155°05'W
Kātmāndu, nat. cap., Nepal *see* Kathmandu	**254-55**	27°42'N	85°19'E
Katni, India *see* Murwāra	**254-55**	23°50'N	80°24'E
Katoomba, Austl.	**301**	33°43'S	150°18'E
Katowice, Pol.	**210-11**	50°16'N	19°01'E
Katrineholm, Swe. (kä-trē'nĕ-hōlm)	**208-09**	58°59'N	16°12'E
Katsina, Nig. (kät'sē-nȧ)	**282-83**	12°60'N	7°36'E
Kattaqo'rg'on, Uzb.	**252-53**	39°55'N	66°16'E
Kattegat, strt., Eur. (kăt'ĕ-gät)	**208-09**	57°0'N	11°00'E
Kattegatt, strt., Eur. *see* Kattegat	**208-09**	57°0'N	11°00'E
Kattowitz, Pol. *see* Katowice	**210-11**	50°16'N	19°01'E
Katun', stm., Russia (kȧ-tōn')	**244**	52°26'N	85°00'E
Kaua'i, i., Hi., U.S.	**137a**	22°0'N	159°30'W
Kaufbeuren, Ger. (kouf'boi-rĕn)	**210-11**	47°53'N	10°37'E
Kaufman, Tx., U.S. (kôf'măn)	**130-31**	32°35'N	96°20'W
Kaukauna, Wi., U.S. (kô-kô'nȧ)	**126-27**	44°16'N	88°16'W
Kaukau Veld, plat., Afr.	**286-87**	19°30'S	20°30'E
Kaunakakai, Hi., U.S. (kä'ōō-nȧ-kä'kī)	**137a**	21°06'N	157°01'W
Kaunas, Lith. (kou'nȧs) (kôv'nō)	**208-09**	54°54'N	23°54'E
Kauriālā, stm., Asia *see* Ghāghara	**254-55**	25°45'N	84°48'E
Kau-ye Kyun, i., Mya.	**266-67**	10°60'N	98°31'E
Kavála, Grc. (kä-vä'lä)	**216-17**	40°57'N	24°24'E
Kavalerovo, Russia	**264**	44°16'N	135°03'E
Kavaratti Island, i., India	**256**	10°34'N	72°38'E
Kavieng, Pap. N. Gui. (kä-vē-ĕng')	**302**	2°34'S	150°48'E
Kavīr, Dasht-e, des., Iran (dŭsht-ē-ka-vēr')	**252-53**	34°40'N	54°30'E
Kavkasioni, mts., *see* Caucasus Mountains	**245**	42°38'N	45°00'E

n-sing; ŋ-baŋk; N-nasalized n; nŏd; cŏmmit; ōld; ôbey; ôrder; oi-boil; fōōd; ŏŏ-as oo in foot; ou-out; s-soft; sh-dish; th-thin; pūre; ûnite; ûrn; stŭd; circŭs; ü-as in French tu; '-indeterminate vowel.

370 Kawaguchi – Kiangarow, Mount PRONOUNCING INDEX

Feature (Pronunciation)	Page	Lat.	Long.
Kawaguchi, Japan (kä-wä-gōō-chē)	265	35°48'N	139°43'E
Kawambwa, Zam.	286-87	9°48'S	29°05'E
Kawasaki, Japan (kä-wä-sä'kė)	265	35°32'N	139°42'E
Kaxgar, stm., China	244	39°25'N	76°26'E
Kayak Island, i., Ak., U.S.	136	59°54'N	144°27'W
Kayan, stm., Indon.	268-69	2°55'N	117°35'E
Kaycee, Wy., U.S. (kā-sē')	122-23	43°43'N	106°40'W
Kayes, Mali (kāz)	280-81	14°27'N	11°26'W
Kayoa, Pulau, i., Indon.	268-69	0°04'S	127°24'E
Kayseri, Tur. (kī'sĕ-rē)	202-03	38°44'N	35°29'E
Kayuagung, Indon.	266-67	3°23'S	104°50'E
Kazakh Hills, hills, Kaz. (kä-zäk' hĭlz)	244	49°0'N	72°00'E
Kazakhstan, nation, Asia (kä-zäk-stän')	224-25	47°0'N	76°00'E
Kazan', Russia (kȧ-zän')	202-03	55°51'N	49°04'E
Kazan, stm., Can.	140-41	64°10'N	95°22'W
Kazanka, Ukr. (kȧ-zän'kȧ)	218-19	47°50'N	32°50'E
Kazanlŭk, Blg. (kȧ'zȧn-lėk)	216-17	42°37'N	25°24'E
Kazan-rettō, is., Japan	304-05	25°0'N	141°00'E
Kazbek, Gora, vol., (gȧ-rä' kȧz-bĕk')	245	42°42'N	44°31'E
Kāzerūn, Iran	250-51	29°37'N	51°39'E
Kazincbarcika, Hung. (kô'zĭnts-bôr-tsĭ-ko)	210-11	48°15'N	20°39'E
Kazvin, Iran *see* Qazvīn	252-53	36°16'N	49°58'E
Kazym, stm., Russia (kä-zēm')	236-37	63°53'N	65°53'E
Kearney, Ne., U.S. (kär'nĭ)	124-25	40°42'N	99°05'W
Keban Barajı, res., Tur.	202-03	38°56'N	38°55'E
Kebnekaise, mtn., Swe. (kĕp'nĕ-kȧ-ĕs'ĕ)	200-01	67°55'N	18°35'E
K'ebrī Dehar, Eth.	284-85	6°45'N	44°17'E
Kech, stm., Pak.	252-53	25°59'N	62°44'E
Kecskemét, Hung. (kĕch'kĕ-māt)	210-11	46°54'N	19°42'E
Kėdainiai, Lith. (kē-dī'nĭ-ī)	208-09	55°18'N	23°59'E
Kedgwick, N.B., Can. (kĕdj'wĭk)	148-49	47°38'N	67°23'W
Kediri, Indon.	268-69	7°49'S	112°01'E
Kédougou, Sen.	282-83	12°33'N	12°11'W
Keele Peak, mtn., Yk., Can.	140-41	63°26'N	130°19'W
Keeling Islands, dep., Oc. (kē'lĭng ī'lăndz) *see* Cocos Islands	242-43	12°10'S	96°55'E
Keelung, Tai. *see* Chilung	243a	25°08'N	121°44'E
Keene, N.H., U.S.	126-27	42°56'N	72°17'W
Keer-Weer, Cape, c., Austl.	302	13°51'S	141°29'E
Keetmanshoop, Nmb. (kāt'mȧns-hōp)	286-87	26°35'S	18°09'E
Keewatin, Mn., U.S. (kē-wä'tĭn)	124-25	47°24'N	93°04'W
Keflavík, Ice.	206a	64°01'N	22°35'W
Ke-hsi Mānsām, Mya.	266-67	21°56'N	97°51'E
Keijō, nat. cap., Kor., S. *see* Seoul	263	37°33'N	127°01'E
Keila, Est. (kā'lȧ)	208-09	59°18'N	24°26'E
Kelang, Pulau, i., Indon.	268-69	3°12'S	127°44'E
Kellogg, Id., U.S. (kĕl'ŏg)	122-23	47°32'N	116°08'W
Kelmė, Lith. (kĕl-mȧ)	208-09	55°38'N	22°55'E
Kélo, Chad	284-85	9°19'N	15°49'E
Kelowna, B.C., Can.	142-43	49°53'N	119°29'W
Keluang, Malay.	266-67	2°02'N	103°20'E
Kem', Russia (kĕm)	202-03	64°57'N	34°36'E
Kemano, B.C., Can.	142-43	53°32'N	127°57'W
Kemerovo, Russia	236-37	55°23'N	86°03'E
Kemi, Fin. (kā'mė)	200-01	65°46'N	24°34'E
Kemijärvi, Fin. (kā'mė-yĕr-vē)	200-01	66°43'N	27°24'E
Kemijoki, stm., Fin. (kā'mė-yô'kė)	200-01	65°47'N	24°30'E
Kemmerer, Wy., U.S. (kĕm'ēr-ēr)	122-23	41°48'N	110°33'W
Kemper, Fr. *see* Quimper	212-13	47°60'N	4°06'W
Kempsey, Austl. (kĕmp'sė)	301	31°06'S	152°50'E
Kempt, Lac, lk., Qc., Can. (läk kĕmpt)	146-47	47°25'N	74°15'W
Kemul, Kong, mtn., Indon.	268-69	1°52'N	116°13'E
Kenadsa, Alg.	204-05	31°33'N	2°25'W
Kenai, Ak., U.S. (kĕ-nī')	136	60°34'N	151°13'W
Kenai Fjords National Park, n.p., Ak., U.S. (kĕ-nī' fē-ôrdz' năsh'ŭn-ăl pärk)	136	59°50'N	150°09'W
Kenai Peninsula, pen., Ak., U.S. (kĕ-nī' pĕ-nĭn'sū-lă)	136	60°10'N	150°00'W
Kendal, Eng., U.K. (kĕn'dăl)	206-07	54°20'N	2°44'W
Kendall, Cape, c., Nu., Can.	140-41	63°36'N	87°09'W
Kendallville, In., U.S. (kĕn'dăl-vĭl)	126-27	41°26'N	85°15'W
Kendari, Indon.	268-69	3°57'S	122°36'E
Kenedy, Tx., U.S. (kĕn'ĕ-dĭ)	132-33	28°49'N	97°50'W
Kenema, S.L.	282-83	7°52'N	11°11'W
Kenge, D.R.C.	284-85	4°52'S	16°56'E
Kēng Tung, Mya.	266-67	21°17'N	99°36'E
Kenhardt, S. Afr.	286-87	29°20'S	21°09'E
Kénitra, Mor. (kĕ-nē'trȧ)	292a	34°16'N	6°35'W
Kenmare, N.D., U.S. (kĕn-mâr')	124-25	48°42'N	102°05'W
Kennebec, stm., Me., U.S. (kĕn-ė-bĕk')	127a	43°45'N	69°46'W
Kennebunk, Me., U.S. (kĕn-ė-bŭnk')	126-27	43°23'N	70°33'W
Kennedy, Cape, c., Fl., U.S. *see* Canaveral, Cape	135a	28°27'N	80°32'W
Kenner, La., U.S. (kĕn'ēr)	134-35	29°59'N	90°15'W
Kennett, Mo., U.S. (kĕn'ĕt)	134-35	36°14'N	90°02'W
Kennewick, Wa., U.S. (kĕn'ė-wĭk)	122-23	46°12'N	119°08'W
Kenney Dam, d., B.C., Can.	142-43	53°37'N	124°58'W
Kenora, On., Can. (kė-nō'rȧ)	144-45	49°46'N	94°29'W
Kenosha, Wi., U.S. (kĕ-nō'shȧ)	126-27	42°34'N	87°50'W
Kenova, W.V., U.S. (kė-nō'vȧ)	126-27	38°24'N	82°35'W
Kent, Oh., U.S. (kĕnt)	126-27	41°09'N	81°21'W
Kent, Wa., U.S. (kĕnt)	122-23	47°23'N	122°12'W
Kentaū, Kaz.	244	43°31'N	68°31'E
Kentland, In., U.S. (kĕnt'lănd)	126-27	40°46'N	87°27'W
Kenton, Oh., U.S. (kĕn'tŭn)	126-27	40°39'N	83°36'W
Kent Peninsula, pen., Nu., Can. (kĕnt pĕ-nĭn'sū-lă)	140-41	68°30'N	107°00'W
Kentucky, state, U.S. (kĕn-tŭk'ĭ)	118-19	37°30'N	85°15'W
Kentucky, stm., Ky., U.S. (kĕn-tŭk'ĭ)	126-27	38°41'N	85°11'W
Kentucky Lake, res., U.S. (kĕn-tŭk'ĭ lāk)	134-35	36°41'N	88°04'W
Kentville, N.S., Can.	148-49	45°05'N	64°29'W
Kentwood, La., U.S. (kĕnt'wŏŏd)	134-35	30°56'N	90°31'W
Kentwood, Mi., U.S. (kĕnt'wŏŏd)	126-27	42°55'N	85°35'W
Kenya, nation, Afr. (kēn'yȧ)	274	1°0'N	38°00'E
Kenya, Mount, mtn., Kenya (mount kēn'yȧ)	289	0°09'S	37°19'E
Kenyon, Mn., U.S. (kĕn'yŭn)	124-25	44°17'N	93°00'W
Keokuk, Ia., U.S. (kē'ô-kŭk)	130-31	40°24'N	91°23'W
Kępno, Pol. (kăn'pnō)	210-11	51°17'N	18°00'E
Kerala, state, India	256	10°0'N	76°30'E
Keramian, Pulau, i., Indon.	268-69	5°05'S	114°36'E
Kerang, Austl. (kė-răng')	301	35°44'S	143°55'E
Kerbela, Iraq *see* Karbalā'	248-49	32°37'N	44°02'E
Kerch, Ukr.	218-19	45°21'N	36°28'E
Kerchens'ka protoka, strt., Eur. *see* Kerch Strait	218-19	45°23'N	36°41'E
Kerchenskiy Proliv, strt., Eur. (kĕr-chĕn'skĭ prô'lĭf) *see* Kerch Strait	218-19	45°23'N	36°41'E
Kerch Strait, strt., Eur.	218-19	45°23'N	36°41'E
Keren, Erit.	288	15°46'N	38°27'E
Kerguelen, Îles, i., Afr.	312	49°20'S	69°16'E
Kerguélen, Îles, is., Afr. (ēl-kĕr'gȧ-lĕn)	312	49°15'S	69°10'E
Kericho, Kenya	289	0°22'S	35°16'E
Kerinci, Gunung, vol., Indon.	266-67	1°42'S	101°16'E
Keriya, stm., China (kē'rė-yä)	254-55	35°17'N	81°34'E
Kerkenna, Îles, is., Tun. (ēl-dĕ-kĕr'kĕn-nä)	280-81	34°44'N	11°12'E
Kerki, Turkmen. (kĕr'kė) *see* Atamyrat	252-53	37°50'N	65°13'E
Kérkyra, Grc.	216-17	39°37'N	19°55'E
Kérkyra, i., Grc.	216-17	39°40'N	19°45'E
Kermadec Islands, is., N.Z. (kĕr-măd'ĕk ī'lăndz)	304-05	30°10'S	178°15'W
Kermān, Iran (kĕr-män')	250-51	30°17'N	57°04'E
Kermānshāh, Iran	248-49	34°18'N	47°04'E
Kerme, Gulf of, b., Tur. *see* Gökova Körfezi	216-17	36°54'N	27°51'E
Kerrobert, Sk., Can.	144-45	51°55'N	109°08'W
Kerrville, Tx., U.S. (kûr'vĭl)	132-33	30°03'N	99°08'W
Kerulen, stm., Asia (kĕr'ōō-lĕn)	260-61	48°44'N	117°03'E
Keşan, Tur. (kĕ'shȧn)	216-17	40°52'N	26°39'E
Kesennuma, Japan	264	38°54'N	141°35'E
Keshan, China (kŭ-shän)	260-61	48°01'N	125°52'E
Keshod, India	254-55	21°18'N	70°14'E
Kestell, S. Afr. (kĕs'tĕl)	292c	28°18'S	28°38'E
Keszthely, Hung. (kĕst'hĕl-lĭ)	210-11	46°46'N	17°15'E
Ket', stm., Russia (kyĕt)	236-37	58°55'N	81°32'E
Keta, Ozero, lk., Russia	236-37	68°44'N	90°00'E
Ketapang, Indon. (kė-tȧ-päng')	268-69	1°52'S	109°58'E
Ketchikan, Ak., U.S. (kĕch-ĭ-kăn')	136	55°20'N	131°35'W
Ketchum, Id., U.S.	122-23	43°41'N	114°23'W
Ketoy, Ostrov, i., Russia	236-37	47°20'N	152°28'E
Kettering, Eng., U.K. (kĕt'ēr-ĭng)	206-07	52°24'N	0°45'W
Kettering, Oh., U.S. (kĕt'ēr-ĭng)	126-27	39°42'N	84°10'W
Kewanee, Il., U.S. (kė-wä'nė)	126-27	41°14'N	89°55'W
Kewaunee, Wi., U.S. (kė-wô'nė)	126-27	44°27'N	87°30'W
Keweenaw Peninsula, pen., Mi., U.S. (kē'wė-nô pĕ-nĭn'sū-lă)	124-25	47°12'N	88°25'W
Keweenaw Point, c., Mi., U.S.	124-25	47°27'N	87°50'W
Key Largo, i., Fl., U.S.	135a	25°11'N	80°22'W
Keyser, W.V., U.S. (kī'sēr)	126-27	39°26'N	78°59'W
Key West, Fl., U.S. (kē wĕst')	135a	24°33'N	81°47'W
Kgalagadi Transfrontier Park, n.p., Afr.	286-87	25°23'S	20°44'E
Khabarovsk, Russia (ĸä-bä'rôfsk)	260-61	48°26'N	135°08'E
Khairpur, Pak.	252-53	27°32'N	68°45'E
Khajurāho, India	254-55	24°50'N	79°58'E
Khakhea, Bots.	286-87	24°45'S	23°31'E
Khal'mer-Yu, Russia (kŭl-myĕr'-yōō')	202-03	67°57'N	64°45'E
Khambhāt, India	254-55	22°19'N	72°37'E
Khambhāt, Gulf of, b., India	254-55	20°57'N	72°26'E
Khamīs Mushayţ, Sau. Ar.	288	18°18'N	42°44'E
Khammam, India	256	17°15'N	80°09'E
Khānābād, Afg.	252-53	36°41'N	69°07'E
Khandwa, India	254-55	21°49'N	76°21'E
Khandyga, Russia	236-37	62°40'N	135°32'E
Khānewāl, Pak.	252-53	30°18'N	71°56'E
Khanka, Lake, lk., Asia (lāk ĸän'kȧ)	264	45°11'N	132°25'E
Khanka, Ozero, lk., Asia *see* Khanka, Lake	264	45°11'N	132°25'E
Khānpur, Pak.	252-53	28°38'N	70°40'E
Khantayskoye Vodokhranilishche, res., Russia	236-37	68°0'N	88°00'E
Khanty-Mansiysk, Russia (ĸŭn-te'mŭn-sēsk')	236-37	60°59'N	69°01'E
Khao Laem Reservoir, res., Thai.	266-67	14°55'N	98°33'E
Khapcheranga, Russia	260-61	49°42'N	112°23'E
Kharagpur, India (kŭ-rŭg'pór)	254-55	22°20'N	87°20'E
Kharg Island, i., Iran *see* Khārk, Jazīreh-ye	250-51	29°15'N	50°19'E
Khargon, India	254-55	21°50'N	75°36'E
Khārk, Jazīreh-ye, i., Iran	250-51	29°15'N	50°19'E
Kharkiv, Ukr.	218-19	49°60'N	36°14'E
Kharkov, Ukr. *see* Kharkiv	218-19	49°60'N	36°14'E
Kharmanli, Blg. (ĸȧr-män'lė)	216-17	41°56'N	25°55'E
Khartoum, nat. cap., Sudan (kär-tōōm')	288	15°35'N	32°32'E
Khasavyurt, Russia	245	43°15'N	46°35'E
Khāsh, stm., Afg.	252-53	30°48'N	61°46'E
Khashm al-Qirbah, Sudan	288	14°58'N	35°55'E
Khaskovo, Blg. (ĸás'kô-vô)	216-17	41°56'N	25°34'E
Khatanga, Russia (ĸȧ-tän'gȧ)	236-37	71°58'N	102°30'E
Khatangskiy Zaliv, b., Russia (kä-tän'g-skė zä'lĭf)	226-27	73°35'N	109°45'E
Khatt, Oued al, stm., W. Sah.	280-81	26°55'N	13°03'W
Khaybar, Kowtal-e, p., Asia *see* Khyber Pass	252-53	34°06'N	71°07'E
Khazar, Daryā-ye, lk., *see* Caspian Sea	244	41°18'N	50°59'E
Khemis Miliana, Alg.	292b	36°16'N	2°13'E
Khersān, stm., Iran	252-53	31°34'N	50°22'E
Kherson, Ukr. (ĸĕr-sôn')	218-19	46°38'N	32°35'E
Kherson, co., Ukr. (ĸĕr-sôn')	218-19	46°45'N	33°30'E
Kheta, stm., Russia	236-37	71°55'N	102°06'E
Khilok, Russia	260-61	51°21'N	110°27'E
Khilok, stm., Russia	260-61	51°19'N	106°59'E
Khimki, Russia (ĸĕm'kĭ)	218-19	55°54'N	37°26'E
Khiwa, Uzb.	252-53	41°24'N	60°22'E
Khmel'nyts'kyi, Ukr.	210-11	49°26'N	27°01'E
Khodzheyli, Uzb. *see* Khŭjayli	244	42°48'N	59°25'E
Kholm, Afg.	252-53	36°41'N	67°42'E
Kholm, Russia (ĸôlm)	208-09	57°09'N	31°11'E
Kholmsk, Russia (kŭlmsk)	264	47°03'N	142°03'E
Khomeynīshahr, Iran	252-53	32°41'N	51°32'E
Khong, stm., Asia *see* Mekong	266-67	10°33'N	105°27'E
Khong, stm., Asia *see* Salween	226-27	16°33'N	97°40'E
Salween, stm., Asia	226-27	16°33'N	97°40'E
Khon Kaen, Thai.	266-67	16°27'N	102°50'E
Khoper, stm., Russia (ĸô'pēr)	202-03	49°37'N	42°19'E
Khor, stm., Russia (ĸôr')	260-61	47°49'N	134°41'E
Khorixas, Nmb.	286-87	20°21'S	14°59'E
Khorol, Ukr. (ĸô'rôl)	218-19	49°47'N	33°17'E
Khorol, stm., Ukr. (ĸô'rôl)	218-19	49°28'N	33°47'E
Khorramābād, Iran	248-49	33°29'N	48°21'E
Khorramshahr, Iran (kô-ram'shär)	248-49	30°25'N	48°11'E
Khorugh, Taj.	252-53	37°29'N	71°33'E
Khouribga, Mor.	292a	32°53'N	6°55'W
Khromtaū, Kaz.	244	50°15'N	58°27'E
Khudzhand, Taj. *see* Khujand	252-53	40°17'N	69°39'E
Khujand, Taj.	252-53	40°17'N	69°39'E
Khŭjayli, Uzb.	244	42°48'N	59°25'E
Khulna, Bngl.	254-55	22°49'N	89°34'E
Khunjerab Pass, p., Asia	252-53	36°52'N	75°28'E
Khust, Ukr. (kóst)	210-11	48°11'N	23°18'E
Khuzdār, Pak.	252-53	27°48'N	66°37'E
Khvalynsk, Russia (ĸvá-lĭnsk')	202-03	52°29'N	48°05'E
Khvoy, Iran	245	38°33'N	44°58'E
Khyber Pass, p., Asia (kī'bēr pȧs)	252-53	34°06'N	71°07'E
Kiamba, Phil.	270	5°60'N	124°37'E
Kiambi, D.R.C. (kyäm'bė)	289	7°19'S	28°01'E
Kiamichi, stm., Ok., U.S. (kyȧ-mē'chė)	130-31	33°57'N	95°14'W
Kiamusze, China *see* Jiamusi	264	46°48'N	130°22'E
Kiangarow, Mount, mtn., Austl.	301	26°50'S	151°32'E

ăt; finăl; rāte; senāte; ärm; ȧsk; sofȧ; fâre; ch-choose; dh-as th in other; bē; ēvent; bĕt; recĕnt; cratēr; g-gō; gh-guttural g; bĭt; ĭ-short neutral; rīde; ĸ-guttural k as ch in German ich;

Feature (Pronunciation)	Page	Lat.	Long.
Kiangsi, state, China *see* Jiangxi	258-59	28°0′N	116°00′E
Kiangsu, state, China *see* Jiangsu	258-59	33°0′N	120°00′E
Kiantajärvi, lk., Fin. (kyàn′tȧ-yĕr-vē)	202-03	65°03′N	29°07′E
Kibombo, D.R.C.	284-85	3°54′S	25°55′E
Kibre Mengist, Eth.	284-85	5°52′N	39°00′E
Kıbrıs, nation, Asia *see* Cyprus	248-49	35°0′N	33°00′E
Kičevo, Mac. (kē′chĕ-vô)	216-17	41°32′N	20°57′E
Kıcık Qafqaz daqları, mts., Asia			
see Lesser Caucasus	245	40°60′N	44°35′E
Kicking Horse Pass, p., Can.	142-43	51°27′N	116°20′W
Kidal, Mali (kē-dál′)	280-81	18°26′N	1°24′E
Kiel, Ger. (kēl)	210-11	54°19′N	10°07′E
Kiel Canal, can., Ger. (kēl kä-näl′)	210-11	53°54′N	9°10′E
Kielce, Pol. (kyĕl′tsĕ)	210-11	50°53′N	20°38′E
Kiev, nat. cap., Ukr. (kē′ĕf) (kē′ĕv)	218-19	50°26′N	30°30′E
Kiev Reservoir, res., Ukr.			
(kē′ĕf rĕ′sĕr-vwär) (kē′ĕv rĕ′sĕr-vwär)			
see Kyïvs'ke vodoskhovyshche	218-19	50°51′N	30°32′E
Kiffa, Maur. (kēf′ȧ)	280-81	16°37′N	11°24′W
Kigali, nat. cap., Rw. (kē-gä′lē)	289	1°56′S	30°04′E
Kigoma, Tan. (kē-gō′mä)	289	4°53′S	29°37′E
Kiirun, Tai. *see* Chilung	243a	25°08′N	121°44′E
Kii-suidō, strt., Japan (kē sōō-ē′dō)	265	33°55′N	134°55′E
Kikládes, is., Grc. *see* Cyclades	216-17	37°30′N	25°00′E
Kikori, stm., Pap. N. Gui.	302	7°22′S	144°14′E
Kikwit, D.R.C. (kē′kwēt)	284-85	5°02′S	18°49′E
Kil, Swe. (kēl)	208-09	59°31′N	13°19′E
Kīlauea, Hi., U.S. (kē-lä-ōō-ā′ä)	137a	22°13′N	159°25′W
Kili, i., Marsh. Is.	304-05	5°39′N	169°07′E
Kilimanjaro, mtn., Tan. (kyl-ê-măn-jä′rô)	289	3°04′S	37°22′E
Kilimatinde, Tan. (kĭl-ê-mä-tĭn′då)	289	5°52′S	34°58′E
Killingi-Nõmme, Est.			
(kē′lĭn-gê-nôm′mĕ)	208-09	58°09′N	24°58′E
Kilis, Tur. (kē′lēs)	248-49	36°43′N	37°07′E
Kilkenny, Ire. (kĭl-kĕn-ĭ)	206-07	52°39′N	7°15′W
Kilkís, Grc. (kĭl′kĭs)	216-17	40°59′N	22°53′E
Killala, Ire. (kĭ lä′lȧ)	206-07	54°13′N	9°13′W
Killarney, Mb., Can.	144-45	49°12′N	99°42′W
Killeen, Tx., U.S.	132-33	31°06′N	97°42′W
Kilmarnock, Scot., U.K. (kĭl-mär′nŭk)	206-07	55°36′N	4°30′W
Kilombero, stm., Tan.	284-85	8°31′S	37°22′E
Kilosa, Tan.	289	6°50′S	36°59′E
Kilrush, Ire. (kĭl′rŭsh)	206-07	52°39′N	9°29′W
Kilttān Island, i., India	256	11°29′N	73°00′E
Kimamba, Tan.	289	6°46′S	37°08′E
Kimba, Austl. (kĭm′bȧ)	301	33°07′S	136°26′E
Kimball, Ne., U.S. (kĭm-bȧl)	124-25	41°14′N	103°40′W
Kimball, S.D., U.S. (kĭm-bȧl)	124-25	43°45′N	98°57′W
Kimberley, B.C., Can. (kĭm′bẽr-lĭ)	142-43	49°41′N	115°59′W
Kimberley, S. Afr. (kĭm′bẽr-lĭ)	286-87	28°44′S	24°45′E
Kimberley, plat., Austl.	296-97	17°0′S	127°00′E
Kimberley, reg., Austl.	296-97	16°0′S	127°00′E
Kimch'aek, Kor., N.	263	40°41′N	129°12′E
Kimch'ŏn, Kor., S. *see* Gimcheon	263	36°07′N	128°07′E
Kimmirut, Nu., Can.	138-39	62°51′N	69°53′W
Kimovsk, Russia	218-19	53°58′N	38°32′E
Kimry, Russia (kĭm′rê)	218-19	56°52′N	37°22′E
Kinabalu, Gunong, mtn., Malay.	268-69	6°05′N	116°33′E
Kincardine, On., Can. (kĭn-kär′dĭn)	146-47	44°10′N	81°38′W
Kincolith, B.C., Can.	142-43	55°00′N	129°57′W
Kinder, La., U.S. (kĭn′dẽr)	132-33	30°29′N	92°51′W
Kindersley, Sk., Can. (kĭn′dẽrz-lê)	144-45	51°29′N	109°10′W
Kindia, Gui. (kĭn′dê-ȧ)	282-83	10°04′N	12°51′W
Kindu, D.R.C.	284-85	2°57′S	25°55′E
Kinel', Russia	202-03	53°14′N	50°38′E
Kineshma, Russia (kê-nĕsh′mȧ)	202-03	57°27′N	42°08′E
Kingaroy, Austl. (kĭŋ′gă-roi)	301	26°33′S	151°51′E
King City, Ca., U.S. (kĭng sĭ′tĭ)	128-29	36°12′N	121°08′W
Kingfisher, Ok., U.S. (kĭng′fĭsh-ẽr)	130-31	35°52′N	97°56′W
Kingisepp, Russia (kĭŋ-gê-sep′)	208-09	59°22′N	28°37′E
King Island, i., Austl. (kĭng ī′lănd)	301	39°50′S	144°00′E
King Island, i., B.C., Can.	142-43	52°12′N	127°42′W
Kingman, Az., U.S. (kĭng′măn)	128-29	35°12′N	114°02′W
Kingman, Ks., U.S. (kĭng′măn)	130-31	37°39′N	98°07′W
Kings Canyon National Park,			
n.p., Ca., U.S. (kĭngz kăn′yŭn			
năsh′ŭn-ăl pärk)	128-29	36°56′N	118°35′W
Kingscote, Austl. (kĭngz′kŭt)	301	35°39′S	137°37′E
King's Lynn, Eng., U.K. (kĭngz lĭn′)	206-07	52°46′N	0°24′E
Kings Mountain, N.C., U.S.			
(kĭngz moun′tĭn)	134-35	35°15′N	81°20′W
Kings Peak, mtn., Ut., U.S. (kĭngz pēk)	122-23	40°46′N	110°22′W
Kingsport, Tn., U.S. (kĭngz′pōrt)	134-35	36°33′N	82°34′W
Kingston, On., Can. (kĭngz′tŭn)	146-47	44°14′N	76°30′W
Kingston, N.Y., U.S.	126-27	41°56′N	74°00′W
Kingston, Pa., U.S. (kĭngz′tŭn)	126-27	41°15′N	75°54′W
Kingston, nat. cap., Jam. (kĭngz′tŭn)	154-55	18°00′N	76°48′W
Kingston Southeast, Austl.	301	36°50′S	139°51′E
Kingston upon Hull, Eng., U.K.	206-07	53°45′N	0°20′W
Kingstown, Ire. *see* Dún Laoghaire	206-07	53°17′N	6°08′W
Kingstown, nat. cap., St. Vin.			
(kĭngz′toun)	155b	13°09′N	61°14′W
Kingstree, S.C., U.S. (kĭngz′trē)	134-35	33°40′N	79°50′W
Kingsville, Tx., U.S. (kĭngz′vĭl)	132-33	27°31′N	97°51′W
King William Island, i., Nu., Can.			
(kĭng wĭl′yăm ī′lănd)	140-41	69°0′N	97°30′W
King William's Town, S. Afr.			
(kĭng-wĭl′-yŭmz-toun)	286-87	32°51′S	27°22′E
Kinkala, Congo	284-85	4°22′S	14°46′E
Kinkony, Farihy, lk., Madag.	286-87	16°08′S	45°50′E
Kinnaird Head, c., Scot., U.K.			
(kĭn-ârd′hĕd)	206-07	57°42′N	2°01′W
Kinneret, Yam, lk., Isr.			
see Galilee, Sea of	248-49	32°48′N	35°35′E
Kinngait, Nu., Can. *see* Cape Dorset	138-39	64°14′N	76°33′W
Kinsale, Old Head of, c., Ire.			
(ōld hĕd ŏv kĭn-sāl)	206-07	51°37′N	8°32′W
Kinshasa, nat. cap., D.R.C.			
(kĭn-shä′sä)	284-85	4°21′S	15°18′E
Kinsley, Ks., U.S. (kĭnz′lĭ)	130-31	37°56′N	99°24′W
Kinston, N.C., U.S. (kĭnz′tŭn)	134-35	35°16′N	77°35′W
Kinyeti, mtn., S. Sudan	289	3°57′N	32°54′E
Kipawa, Lac, res., Qc., Can.	146-47	46°54′N	78°59′W
Kipengere Range, mts., Tan.	286-87	9°23′S	34°26′E
Kípros, nation, Asia *see* Cyprus	248-49	35°0′N	33°00′E
Kirby, Tx., U.S. (kŭr′bĭ)	132-33	29°29′N	98°22′W
Kirbyville, Tx., U.S. (kŭr′bĭ-vĭl)	132-33	30°40′N	93°54′W
Kirenga, stm., Russia (kê-rĕŋ′gȧ)	236-37	57°46′N	108°06′E
Kirensk, Russia (kē-rĕnsk′)	236-37	57°48′N	108°10′E
Kirghizia, nation, Asia *see* Kyrgyzstan	244	41°30′N	75°00′E
Kirgiziya, nation, Asia *see* Kyrgyzstan	244	41°30′N	75°00′E
Kirgiz Range, mts., Asia	244	42°29′N	73°50′E
Kiribati, nation, Oc. (kē rä bäs)	304-05	5°0′S	170°00′W
Kiribati, is., Kir. (kē-rä-bäs)	304-05	0°30′S	174°00′E
Kırıkkale, Tur.	202-03	39°51′N	33°31′E
Kirin, state, China *see* Jilin	200-01	44°0′N	126°00′E
Kirinyaga, mtn., Kenya			
see Kenya, Mount	289	0°09′S	37°19′E
Kiritimati, at., Kir.	304-05	1°48′N	157°19′W
Kiriwina Islands, is., Pap. N. Gui.	302	8°35′S	151°05′E
Kirkcaldy, Scot., U.K. (kẽr-kô′dĭ)	206-07	56°07′N	3°10′W
Kirkenes, Nor.	200-01	69°43′N	30°02′E
Kirkland, Wa., U.S. (kûrk′lănd)	122-23	47°40′N	122°12′W
Kirkland Lake, On., Can.	146-47	48°10′N	80°01′W
Kırklareli, Tur. (kẻrk′lär-ĕ′lẻ)	216-17	41°45′N	27°14′E
Kirksville, Mo., U.S. (kûrks′vĭl)	130-31	40°12′N	92°34′W
Kirkuk, Iraq	248-49	35°28′N	44°24′E
Kirkwall, Scot., U.K. (kûrk′wôl)	206c	58°59′N	2°58′W
Kirov, Russia	218-19	54°04′N	34°19′E
Kirov, Russia	202-03	58°36′N	49°40′E
Kirov Bay, b., Azer. (kē′rŭf bā)			
see Qızılağac körfäzi	245	39°05′N	49°01′E
Kirovohrad, Ukr.	218-19	48°31′N	32°16′E
Kirovsk, Russia (kê-rôfsk′)	200-01	67°37′N	33°40′E
Kirs, Russia	202-03	59°21′N	52°15′E
Kirsanov, Russia (kêr-sȧ′nôf)	202-03	52°39′N	42°45′E
Kırşehir, Tur. (kẻr-shĕ′hĕr)	202-03	39°09′N	34°10′E
Kīrthar Range, mts., Pak. (kĭr-tŭr rănj)	252-53	27°0′N	67°10′E
Kiruna, Swe. (kē-rōō′nä)	200-01	67°51′N	20°16′E
Kirzhach, Russia (kêr-zhäk′)	218-19	56°09′N	38°52′E
Kisaki, Tan. (kē-sä′kē)	289	7°27′S	37°37′E
Kisangani, D.R.C.	284-85	0°32′N	25°12′E
Kisar, Pulau, i., Indon.	268-69	8°05′S	127°10′E
Kīsh, Jazīreh-ye, i., Iran	250-51	26°32′N	53°56′E
Kishinev, nat. cap., Mol. *see* Chișinău	218-19	47°02′N	28°50′E
Kisii, Kenya	289	0°40′S	34°46′E
Kiska Island, i., Ak., U.S. (kĭs′kȧ ī′lănd)	136a	51°60′N	177°31′E
Kiskitto Lake, lk., Mb., Can.			
(kĭs-kĭ′tō lăk)	144-45	54°16′N	98°34′W
Kiskunfélegyháza, Hung.			
(kĭsh′kòn-fā′lĕd-y′hä′zô)	210-11	46°43′N	19°50′E
Kiskunhalas, Hung.			
(kĭsh′kòn-hô′lôsh)	210-11	46°25′N	19°30′E
Kislovodsk, Russia	245	43°55′N	42°44′E
Kismaayo, Som. *see* Chisimayu	284-85	0°22′S	42°32′E
Kiso, stm., Japan (kē′sō)	265	35°04′N	136°44′E
Kissidougou, Gui. (kē′sê-dōō′gōō)	282-83	9°11′N	10°06′W
Kissimmee, Fl., U.S. (kĭ-sĭm′ê)	135a	28°18′N	81°24′W
Kissimmee, stm., Fl., U.S. (kĭ-sĭm′ê)	135a	27°08′N	80°52′W
Kisumu, Kenya (kē′sōō-mōō)	289	0°05′S	34°46′E
Kita, Mali (kē′tȧ)	280-81	13°02′N	9°30′W
Kita-Daitō-jima, i., Japan	240-41	25°57′N	131°18′E
Kita-Iō-jima, i., Japan	304-05	25°26′N	141°17′E
Kitakyūshū, Japan	265	33°54′N	130°51′E
Kitale, Kenya	289	1°01′N	34°60′E
Kitami, Japan	264	43°48′N	143°54′E
Kitchener, On., Can. (kĭch′ĕ-nẽr)	146-47	43°27′N	80°29′W
Kitega, Bdi. *see* Gitega	289	3°21′S	29°54′E
Kitgum, Ug. (kĭt′gòm)	289	3°17′N	32°52′E
Kitimat, B.C., Can. (kĭ′tĭ-măt)	142-43	54°01′N	128°42′W
Kitimat Ranges, mts., B.C., Can.			
(kĭ′tĭ-măt rānjĕz)	142-43	53°30′N	128°50′W
Kittanning, Pa., U.S. (kĭ-tăn′ĭng)	126-27	40°49′N	79°31′W
Kittery, Me., U.S. (kĭt′ẽr-ĭ)	126-27	43°06′N	70°45′W
Kitty Hawk, N.C., U.S. (kĭt′tê hôk)	134-35	36°04′N	75°44′W
Kitui, Kenya	289	1°22′S	38°01′E
Kitwe, Zam.	286-87	12°49′S	28°13′E
Kivalina, Ak., U.S.	136	67°44′N	164°32′W
Kivu, Lac, lk., Afr.	289	2°03′S	28°54′E
Kiyev, nat. cap., Ukr. *see* Kiev	218-19	50°26′N	30°30′E
Kizel, Russia (kē′zĕl)	202-03	59°04′N	57°39′E
Kızılırmak, stm., Tur.	202-03	41°44′N	35°58′E
Kizlyar, Russia (kĭz-lyär′)	245	43°51′N	46°42′E
Kladno, Czech Rep. (kläd′nō)	210-11	50°08′N	14°06′E
Klagenfurt, Aus. (klä′gĕn-fōrt)	210-11	46°38′N	14°19′E
Klaipėda, Lith. (klī′pā-dȧ)	208-09	55°43′N	21°08′E
Klamath, stm., U.S. (klăm′ăth)	120-21	41°33′N	124°06′W
Klamath Falls, Or., U.S.			
(klăm′ăth fôlz)	122-23	42°14′N	121°48′W
Klamath Mountains, mts., U.S.			
(klăm′ăth moun′tĭnz)	122-23	41°31′N	123°14′W
Klang, Malay.	266-67	3°03′N	101°27′E
Klatovy, Czech Rep. (klȧ′tô-vê)	210-11	49°24′N	13°18′E
Klausenburg, Rom.			
see Cluj-Napoca	210-11	46°47′N	23°36′E
Klein Karroo, plat., S. Afr.			
see Little Karoo	286-87	33°45′S	21°30′E
Klerksdorp, S. Afr. (klĕrks′dôrp)	292c	26°52′S	26°39′E
Kletnya, Russia (klyĕt′nyȧ)	218-19	53°23′N	33°13′E
Kleve, Ger. (klĕ′fĕ)	210-11	51°47′N	6°09′E
Klimovsk, Russia (klĭ′môfsk)	218-19	55°22′N	37°32′E
Klin, Russia (klēn)	218-19	56°20′N	36°43′E
Klintehamn, Swe. (klēn′tē-häm)	208-09	57°24′N	18°12′E
Klintsy, Russia (klĭn′tsĭ)	218-19	52°45′N	32°15′E
Klip, stm., S. Afr. (klĭp)	292c	27°03′S	29°04′E
Klosterneuburg, Aus.			
(klōs-tĕr-noi′bōōrgh)	210-11	48°18′N	16°21′E
Kluane National Park and Reserve,			
n.p., Yk., Can.	138-39	60°45′N	139°30′W
Kluczbork, Pol. (klōōch′bōrk)	210-11	50°58′N	18°14′E
Klyazma, stm., Russia (klyäz′mȧ)	202-03	56°10′N	42°58′E
Klyuchevskaya Sopka, Vulkan,			
vol., Russia (klyōō-chĕfskä′yä)	236-37	56°04′N	160°38′E
Klyuchi, Russia (klyōō′chĭ)	236-37	56°19′N	160°51′E
Knee Lake, lk., Mb., Can.	144-45	55°06′N	94°36′W
Knight Inlet, b., B.C., Can. (nīt ĭn′lĕt)	142-43	50°42′N	125°43′W
Knin, Cro. (knēn)	216-17	44°02′N	16°12′E
Knossos, hist., Grc.	216a	35°17′N	25°12′E
Knox, In., U.S. (nŏks)	126-27	41°17′N	86°37′W
Knox, Cape, c., B.C., Can.	142-43	54°11′N	133°04′W
Knoxville, Ia., U.S. (nŏks′vĭl)	124-25	41°19′N	93°06′W
Knoxville, Il., U.S. (nŏks′vĭl)	124-25	40°54′N	90°17′W
Knoxville, Tn., U.S. (nŏks′vĭl)	134-35	35°58′N	83°55′W
Kōbe, Japan (kō′bĕ)	265	34°41′N	135°10′E
København, nat. cap., Den.			
(kû-b'n-houn′) *see* Copenhagen	208-09	55°41′N	12°34′E
Koblenz, Ger. (kō′blĕntz)	210-11	50°21′N	7°35′E
Kobroor, Pulau, i., Indon.	242-43	6°12′S	134°32′E
Kobryn, Bela. (kō′brĕn′)	210-11	52°13′N	24°21′E
Kobuk, stm., Ak., U.S. (kō′bŭk)	136	66°34′N	161°33′W
Kobuk Valley National Park, n.p.,			
Ak., U.S. (kō′bŭk văl′ê			
năsh′ŭn-ăl pärk)	136	67°20′N	159°00′W
Kobuleti, Geor. (kô-bò-lyā′tê)	245	41°49′N	41°48′E
Kocaeli, Tur. *see* İzmit	216-17	40°47′N	29°57′E
Kočevje, Slvn. (kô′chāv-ye)	216-17	45°38′N	14°52′E
Koch Bihār, India	254-55	26°19′N	89°27′E
Kochechum, stm., Russia	236-37	64°17′N	100°11′E
Kochi, India	256	9°56′N	76°15′E
Kōchi, Japan	265	33°33′N	133°32′E
Kodiak, Ak., U.S. (kō′dyăk)	136	57°49′N	152°22′W
Kodiak Island, i., Ak., U.S.			
(kō′dyăk ī′lănd)	136	57°30′N	153°30′W
Koforidua, Ghana (kō fō-rĭ-dōō′ȧ)	282-83	6°05′N	0°17′W
Kōfu, Japan	265	35°39′N	138°34′E
Køge, Den. (kû′gĕ)	208-09	55°27′N	12°11′E
Kogon, Uzb.	252-53	39°43′N	64°33′E
Kohāt, Pak.	252-53	33°35′N	71°27′E

n-sing; ŋ-baŋk; ɴ-nasalized n; nŏd; cŏmmit; ōld; ôbey; ôrder; oi-boil; fōōd; ŏŏ-as oo in foot; ou-out; s-soft; sh-dish; th-thin; pūre; ůnite; ûrn; stŭd; circŭs; ü-as in French tu; ′-indeterminate vowel.

ăt; fīnăl; rāte; senåte; ärm; ȧsk; sofȧ; fâre; ch-choose; dh-as th in other; bē; ĕvent; bĕt; recĕnt; cratēr; g-gō; gh-guttural g; bĭt; ī-short neutral; rīde; ᴋ-guttural k as ch in German ich;

n-sing; ŋ-baŋk; N-nasalized n; nŏd; cŏmmit; ōld; ôbey; ôrder; oi-boil; fōōd; ȯ-as oo in foot; ou-out; s-soft; sh-dish; th-thin; pūre; ūnite; ûrn; stŭd; circŭs; ü-as in French tu; ´-indeterminate vowel.

Feature (Pronunciation)	Page	Lat.	Long.
Lafia, Nig.	282-83	8°30'N	8°31'E
La Flèche, Fr. (lȧ flāsh´)	212-13	47°42'N	0°04'W
Lagan, stm., Swe.	208-09	56°33'N	12°56'E
Lågen, stm., Nor. (lô´ghěn)	208-09	59°02'N	10°04'E
Lages, Braz.	182-83	27°49'S	50°18'W
Laghouat, Alg. (lä-gwät´)	204-05	33°51'N	2°51'E
Lagoa da Prata, Braz. (lȧ-gô´ä-dȧ-prä´tȧ)	186	20°02'S	45°33'W
La Gomera, i., Spain	215d	28°07'N	17°11'W
Lagos, Nig. (lä´gōs)	282a	6°27'N	3°24'E
Lagos, Port. (lä´gōzh)	214-15	37°06'N	8°40'W
Lagos de Moreno, Mex. (lä´gōs dā mô-rā´nō)	158-59	21°22'N	101°54'W
La Grand'Combe, Fr. (lȧ grän kanb´)	212-13	44°13'N	4°01'E
La Grande, Or., U.S. (lȧ grănd´)	122-23	45°20'N	118°05'W
La Grande Deux, Réservoir, res., Qc., Can.	140-41	53°40'N	76°55'W
La Grande Quatre, Réservoir, res., Qc., Can.	140-41	54°0'N	73°15'W
Lagrange, Ga., U.S.	134-35	33°02'N	85°02'W
La Grange, Ky., U.S. (lä grānj)	126-27	38°24'N	85°23'W
La Gran Sabana, pl., Ven.	178-79	5°21'N	62°04'W
La Guajira, Península de, pen., S.A.	178-79	12°0'N	71°40'W
Laguna, Braz. (lä-gōō´nä)	186	28°28'S	48°47'W
Lagunillas, Bol. (lä-gōō-nēl´yäs)	182-83	19°38'S	63°43'W
La Habana, nat. cap., Cuba (lä-ä-bȧ´ nä) *see* Havana	154-55	23°06'N	82°27'W
Lahad Datu, Malay.	270	5°02'N	118°20'E
Lahaina, Hi., U.S. (lä-hä´ē-nä)	137a	20°53'N	156°40'W
Lahat, Indon.	266-67	3°48'S	103°32'E
Lāhījān, Iran	252-53	37°12'N	50°00'E
Laholm, Swe. (lä´hôlm)	208-09	56°31'N	13°03'E
Lahore, Pak. (lä-hōr´)	252-53	31°35'N	74°20'E
Lahr, Ger. (lär)	210-11	48°21'N	7°52'E
Lahti, Fin. (lä´tê)	208-09	60°59'N	25°40'E
Laibach, nat. cap., Slvn. *see* Ljubljana	216-17	46°03'N	14°31'E
Laibin, China (lī-bĭn)	258-59	23°42'N	109°14'E
Laichow Bay, b., China *see* Laizhou Wan	260-61	37°20'N	119°19'E
L'Aigle, Fr. (lĕ´gl')	212-13	48°46'N	0°38'E
Laiwui, Indon.	268-69	1°21'S	127°39'E
Laiyang, China (läī´yäng)	260-61	36°58'N	120°43'E
Laizhou Bay, b., China (lī-jō bä) *see* Laizhou Wan	260-61	37°20'N	119°19'E
Laizhou Wan, b., China (lī-jō wän)	260-61	37°20'N	119°19'E
Lajeado, Braz. (lä-zhěȧ´dô)	182-83	29°24'S	51°57'W
Lajes, Braz. (lȧ´zhěs)	188d	5°41'S	36°14'W
Lajinha, Braz. (lä-zhē´nyä)	186	20°09'S	41°37'W
La Junta, Co., U.S. (lȧ hōōn´tȧ)	130-31	37°59'N	103°33'W
Lake Arthur, La., U.S. (lāk är´thŭr)	132-33	30°05'N	92°41'W
Lakeba, i., Fiji	306f	18°13'S	178°47'W
Lakeba Passage, strt., Fiji	306f	17°55'S	178°45'W
Lake Cargelligo, Austl.	301	33°19'S	146°22'E
Lake Charles, La., U.S. (lāk chärlz´)	132-33	30°14'N	93°13'W
Lake City, Fl., U.S. (lāk sĭ´tĭ)	134-35	30°12'N	82°38'W
Lake City, Mn., U.S. (lāk sĭ´tê)	124-25	44°27'N	92°17'W
Lake City, S.C., U.S. (lāk sĭ´tĭ)	134-35	33°52'N	79°45'W
Lake Cowichan, B.C., Can. (lāk kou´ĭ-chȧn)	142-43	48°50'N	124°03'W
Lake Crystal, Mn., U.S. (lāk krĭs´tȧl)	124-25	44°07'N	94°13'W
Lake Geneva, Wi., U.S. (lāk jě-nē´vȧ)	126-27	42°36'N	88°26'W
Lake Harbour, Nu., Can. *see* Kimmirut	138-39	62°51'N	69°53'W
Lake Havasu City, Az., U.S. (lāk hăv´ȧ-sōō sĭ´tê)	128-29	34°29'N	114°21'W
Lakeland, Fl., U.S. (lāk wûrth´)	135a	28°03'N	81°58'W
Lake Linden, Mi., U.S. (lāk lĭn´děn)	124-25	47°12'N	88°24'W
Lake Louise, Ab., Can. (lāk lōō-ēz´)	142-43	51°27'N	116°13'W
Lake Mills, Ia., U.S. (lāk mĭlz´)	124-25	43°25'N	93°32'W
Lake Oswego, Or., U.S. (lāk ŏs-wē´go)	122-23	45°25'N	122°43'W
Lake Placid, N.Y., U.S. (lāk plăs´ĭd)	126-27	44°17'N	73°59'W
Lake Preston, S.D., U.S. (lāk prĕs´tŭn)	124-25	44°22'N	97°23'W
Lake Providence, La., U.S. (lāk prŏv´ĭ-děns)	134-35	32°49'N	91°11'W
Lakeview, Or., U.S.	122-23	42°12'N	120°21'W
Lake Village, Ar., U.S. (lāk vĭl´ȧj)	134-35	33°19'N	91°17'W
Lake Wales, Fl., U.S. (lāk wālz´)	135a	27°54'N	81°35'W
Lakewood, Co., U.S. (lāk´wŏŏd)	130-31	39°44'N	105°07'W
Lakewood, N.J., U.S. (lāk´wŏŏd)	126-27	40°07'N	74°14'W
Lakewood, Oh., U.S. (lāk´wŏŏd)	126-27	41°29'N	81°48'W
Lakewood, Wa., U.S. (lāk´wŏŏd)	122-23	47°11'N	122°31'W
Lake Worth, Fl., U.S. (lāk wûrth´)	135a	26°37'N	80°03'W
Lakhdenpokh'ya, Russia (l´äk-dĭe´npōkyȧ)	208-09	61°31'N	30°12'E
Lakhīmpur, India	254-55	27°57'N	80°47'E
Lakota, N.D., U.S. (lȧ-kō´tȧ)	124-25	48°02'N	98°21'W
Lakshadweep, state, India	256	10°0'N	73°00'E
Lakshadweep, is., India	256	10°0'N	73°00'E
Lakshadweep Sea, s., Asia	256	7°0'N	76°00'E
La Libertad, Guat. (lä lē-běr-tädh´)	160	16°47'N	90°07'W
La Ligua, Chile (lä lē´gwä)	188e	32°27'S	71°15'W
Lalitpur, India	254-55	24°41'N	78°25'E
Lalitpur, Nepal	254-55	27°40'N	85°19'E
La Loche, Sk., Can.	144-45	56°29'N	109°26'W
La Louvière, Bel. (lȧ lōō-vyâr´)	206-07	50°29'N	4°12'E
La Luz, Mex. (lä lōōz´)	158-59	24°12'N	97°52'W
Lama, Ozero, lk., Russia	236-37	69°32'N	90°27'E
La Madrid, Arg.	182-83	27°39'S	65°15'W
La Malbaie, Qc., Can. (lȧ mäl-bā´)	148-49	47°40'N	70°09'W
La Mancha, reg., Spain (lä män´chä)	214-15	39°21'N	2°28'W
La Manche, strt., Eur. *see* English Channel	206-07	50°13'N	2°20'W
Lamar, Co., U.S. (lȧ-mär´)	130-31	38°05'N	102°37'W
Lamar, Mo., U.S. (lȧ-mär´)	130-31	37°30'N	94°16'W
La Marmora, Punta, mtn., Italy (pŏ´n-tä-lä-mä´r-mô-rä)	216-17	39°59'N	9°20'E
Lamas, Peru (lä´mȧs)	184	6°25'S	76°35'W
Lamballe, Fr. (län-bäl´)	212-13	48°28'N	2°32'W
Lambayeque, Peru (läm-bä-yā´kȧ)	184	6°41'S	79°54'W
Lambertsbaai, S. Afr. *see* Lambert's Bay	286-87	32°06'S	18°19'E
Lambert's Bay, S. Afr.	286-87	32°06'S	18°19'E
Lame Deer, Mt., U.S. (lām dēr´)	122-23	45°39'N	106°41'W
La Méditerranée, s., *see* Mediterranean Sea	204-05	35°0'N	20°00'E
Lamego, Port. (lä-mā´gō)	214-15	41°06'N	7°49'W
Lamesa, Tx., U.S.	130-31	32°45'N	101°58'W
Lamía, Grc. (lȧ-mē´ȧ)	216-17	38°54'N	22°26'E
Lamon Bay, b., Phil. (lä-mōn´ bā)	270	14°28'N	122°01'E
Lamoni, Ia., U.S.	130-31	40°38'N	93°56'W
Lamotrek, at., Micron.	304-05	7°21'N	146°20'E
La Moure, N.D., U.S. (lȧ mōōr´)	124-25	46°20'N	98°17'W
Lampang, Thai.	266-67	18°17'N	99°29'E
Lampasas, Tx., U.S. (lăm-păs´ȧs)	132-33	31°04'N	98°11'W
Lampazos de Naranjo, Mex.	132-33	27°02'N	100°31'W
Lamphun, Thai.	266-67	18°35'N	99°01'E
Lamu, Kenya (lä´mōō)	284-85	2°17'S	40°53'E
Lan', stm., Bela. (lán')	210-11	52°09'N	27°17'E
Lāna'i, i., Hi., U.S. (lä-nä´ê)	137a	20°50'N	156°55'W
Lanark, Scot., U.K. (lăn´ȧrk)	206-07	55°41'N	3°47'W
Lancang, stm., Asia *see* Mekong	266-67	10°33'N	105°27'E
Lancaster, Eng., U.K.	206-07	54°03'N	2°50'W
Lancaster, Ca., U.S. (lăŋ´kȧs-tē)	128-29	34°42'N	118°08'W
Lancaster, Ky., U.S. (lăŋ´kȧs-tē)	126-27	37°37'N	84°35'W
Lancaster, Oh., U.S. (lăŋ´kȧs-tē)	126-27	39°43'N	82°36'W
Lancaster, Pa., U.S.	126-27	40°03'N	76°19'W
Lancaster, S.C., U.S. (lăŋ´kȧs-tē)	134-35	34°43'N	80°46'W
Lancaster, Wi., U.S. (lăŋ´kȧs-tē)	124-25	42°51'N	90°43'W
Lanchow, China *see* Lanzhou	260-61	36°04'N	103°43'E
Lander, Wy., U.S. (lăn´dēr)	122-23	42°50'N	108°44'W
Landerneau, Fr. (län-děr-nō´)	212-13	48°27'N	4°16'W
Landes, reg., Fr. (länd)	212-13	44°10'N	0°52'W
Landsberg, Pol. *see* Gorzów Wielkopolski	210-11	52°44'N	15°14'E
Landsberg an der Warthe, Pol. *see* Gorzów Wielkopolski	210-11	52°44'N	15°14'E
Land's End, c., Eng., U.K. (lăndz ěnd)	206-07	50°03'N	5°44'W
Landshut, Ger. (länts´hōōt)	210-11	48°33'N	12°09'E
Landskrona, Swe. (láns-krŏ´nȧ)	208-09	55°52'N	12°50'E
Lanett, Al., U.S. (lȧ-nět´)	134-35	32°52'N	85°11'W
La'nga Co, lk., China (läŋ-lä tswo)	254-55	30°43'N	81°13'E
Langano Hāyk', lk., Eth.	292d	7°36'N	38°46'E
Langdon, N.D., U.S.	124-25	48°46'N	98°23'W
Langeland, i., Den.	208-09	55°00'N	10°51'E
Langeoog, i., Ger.	210-11	53°45'N	7°29'E
Langjökull, ice, Ice. (läng-yû´kōōl)	206a	64°42'N	20°12'W
Langkawi, Pulau, i., Malay.	266-67	6°24'N	99°50'E
Langley, B.C., Can. (lăng´lĭ)	142-43	49°06'N	122°39'W
Langon, Fr. (län-gôn´)	212-13	44°33'N	0°15'W
Langøya, i., Nor.	200-01	68°44'N	14°50'E
Langqên, stm., Asia *see* Sutlej	254-55	29°21'N	71°02'E
Langres, Fr. (län´gr')	212-13	47°52'N	5°19'E
Langsa, Indon. (läng´sȧ)	266-67	4°28'N	97°58'E
Lang Son, Viet. (läng´sōn´)	266-67	21°51'N	106°45'E
Langzhong, China (läŋ-jōŋ)	258-59	31°34'N	105°59'E
Lanigan, Sk., Can. (lăn´ĭ-gȧn)	144-45	51°52'N	105°02'W
Länkäran, Azer. (lěn-kô-rän´)	245	38°45'N	48°51'E
Lansdale, Pa., U.S. (lănz´dāl)	126-27	40°15'N	75°17'W
L'Anse, Mi., U.S. (láns)	124-25	46°45'N	88°26'W
Lansing, Mi., U.S.	126-27	42°45'N	84°33'W
Lanta Yai, Ko, i., Thai.	266-67	7°34'N	99°03'E
Lanxi, China	258-59	29°12'N	119°28'E
Lanzarote, i., Spain (län-zȧ-rō´tä)	215d	29°0'N	13°40'W
Lanzhou, China (län-jō)	260-61	36°04'N	103°43'E
Lao, nation, Asia *see* Laos	224-25	18°0'N	105°00'E
Laoag, Phil. (lä-wäg´)	270	18°12'N	120°36'E
Laoang, Phil.	270	12°35'N	125°02'E
Lao Cai, Viet.	266-67	22°30'N	103°58'E
Laoha, stm., China	260-61	43°25'N	120°45'E
Laohekou, China	258-59	32°25'N	111°36'E
Laon, Fr. (län)	212-13	49°34'N	3°39'E
La Orchila, Isla, i., Ven.	178-79	11°48'N	66°09'W
La Oroya, Peru (lä-ô-rō´yä)	188a	11°30'S	75°56'W
Laos, nation, Asia (lā-ōs) (lȧ-ōs´)	224-25	18°0'N	105°00'E
La Palma, Pan. (lä-päl´mä)	162	7°42'N	80°11'W
La Palma, Pan.	162	8°24'N	78°09'W
La Palma, i., Spain (lä-päl´mä)	215d	28°40'N	17°52'W
La Paloma, Ur.	187	34°40'S	54°10'W
La Paragua, Ven.	178-79	6°51'N	63°19'W
La Paz, Arg. (lä päz´)	187	30°44'S	59°38'W
La Paz, Arg. (lä päz´)	185	33°27'S	67°34'W
La Paz, Hond. (lä-pȧ´z)	161	14°19'N	87°41'W
La Paz, Mex. (lä-pȧ´z)	158-59	23°41'N	100°43'W
La Paz, Mex.	156-57	24°10'N	110°18'W
La Paz, nat. cap., Bol. (lä-pȧ´z)	182-83	16°30'S	68°09'W
Lapeer, Mi., U.S. (lȧ-pēr´)	126-27	43°03'N	83°18'W
Lapland, reg., Eur. (lăp´lȧnd)	200-01	68°0'N	25°00'E
La Plata, Arg. (lä plä´tä)	187	34°55'S	57°57'W
La Plata, Mo., U.S. (lȧ plā´tȧ)	130-31	40°02'N	92°29'W
La Pocatière, Qc., Can. (lȧ pô-kȧ-tyâr´)	148-49	47°22'N	70°02'W
La Porte, In., U.S. (lȧ pōrt´)	126-27	41°37'N	86°42'W
La Porte City, Ia., U.S. (lȧ pōrt´ sĭ´tê)	124-25	42°19'N	92°11'W
Lappland, reg., Eur. *see* Lapland	200-01	68°0'N	25°00'E
Laptev Sea, s., Russia (läp´tyĭf sē)	226-27	76°0'N	126°00'E
Laptevykh, More, s., Russia *see* Laptev Sea	226-27	76°0'N	126°00'E
La Quiaca, Arg. (lä kē-ä´kä)	182-83	22°07'S	65°36'W
L'Aquila, Italy (lä´kē-lä)	216-17	42°21'N	13°24'E
Lār, Iran (lär)	250-51	27°40'N	54°20'E
Larache, Mor. (lä-räsh´)	292a	35°12'N	6°09'W
Lārak, Jazīreh-ye, i., Iran	250-51	26°52'N	56°22'E
Laramie, Wy., U.S. (lăr´ȧ-mĭ)	122-23	41°19'N	105°35'W
Larantuka, Indon.	268-69	8°19'S	122°58'E
Larat, Pulau, i., Indon.	242-43	7°08'S	131°50'E
Laredo, Spain (lȧ-rā´dhō)	214-15	43°25'N	3°25'W
Laredo, Tx., U.S. (lȧ-rā´dhō)	132-33	27°31'N	99°28'W
Largo, Cayo, i., Cuba (kä´yō-lär´gō)	154-55	21°38'N	81°28'W
Larimore, N.D., U.S. (lăr´ĭ-môr)	124-25	47°54'N	97°38'W
La Rioja, Arg. (lä rē-ōhä)	182-83	29°25'S	66°51'W
La Rioja, state, Arg. (lä-rē-ō´kä)	182-83	30°0'S	67°30'W
Lárisa, Grc. (lä´rē-sȧ)	216-17	39°38'N	22°25'E
Larissa, Grc. *see* Lárisa	216-17	39°38'N	22°25'E
Lārkāna, Pak.	252-53	27°33'N	68°13'E
Larnaca, Cyp. *see* Larnaka	248-49	34°55'N	33°38'E
Larnaka, Cyp.	248-49	34°55'N	33°38'E
Larned, Ks., U.S. (lär´něd)	130-31	38°11'N	99°05'W
La Rochelle, Fr. (lȧ rô-shěl´)	212-13	46°10'N	1°10'W
La Roche-sur-Yon, Fr. (lȧ rôsh´sûr-yôn´)	212-13	46°40'N	1°26'W
La Roda, Spain (lä rō´dä)	214-15	39°12'N	2°09'W
La Ronge, Sk., Can.	144-45	55°06'N	105°17'W
La Rubia, Arg.	187	30°08'S	61°48'W
Larvik, Nor. (lär´vēk)	208-09	59°04'N	10°01'E
La Salle, Il., U.S. (lȧ săl´)	126-27	41°20'N	89°06'W
La Sarre, Qc., Can.	146-47	48°48'N	79°12'W
Las Aves, Isla, i., Ven.	155b	15°41'N	63°37'W
Lascano, Ur.	187	33°40'S	54°13'W
Las Cruces, N.M., U.S. (läs-krōō´sěs)	128-29	32°19'N	106°47'W
La Selle, Morne, mtn., Haiti (mōrn lä´sěl´)	154-55	18°22'N	71°59'W
La Serena, Chile (lä-sě-rě´nä)	182-83	29°54'S	71°15'W
Las Flores, Arg. (läs flo´rěs)	187	36°01'S	59°06'W
Las Heras, Arg.	185	46°31'S	68°56'W
Lashio, Mya. (lăsh´ê-ō)	266-67	22°57'N	97°45'E
Lashkar, India *see* Gwalior	254-55	26°13'N	78°09'E
Las Lajas, Arg.	185	38°30'S	70°22'W
Las Lomitas, Arg.	182-83	24°43'S	60°36'W
Las Minas, Cerro, mtn., Hond.	161	14°33'N	88°39'W
La Solana, Spain (lä-sō-lä-nä)	214-15	38°57'N	3°14'W
Las Palmas de Gran Canaria, Spain (läs päl´mäs)	215d	28°07'N	15°26'W
La Spezia, Italy (lä-spě´zyä)	216-17	44°07'N	9°50'E
Las Piedras, Ur. (läs-pyě´dräs)	187	34°44'S	56°13'W
Las Piedras, stm., Peru	184	12°31'S	69°14'W
Las Plumas, Arg.	185	43°43'S	67°14'W
Las Rosas, Mex. (läs rō thäs)	158-59	16°22'N	92°22'W

ăt; finăl; rāte; senåte; ärm; ȧsk; sofȧ; fâre; ch-choose; dh-as th in other; bē; ēvent; bět; recěnt; cratēr; g-gō; gh-guttural g; bĭt; ĭ-short neutral; rīde; κ-guttural k as ch in German ich;

PRONOUNCING INDEX Lassen Peak – Lexington 375

Feature (Pronunciation)	Page	Lat.	Long.
Lassen Peak, vol., Ca., U.S.			
(lăs′ĕn pēk)	**122-23**	40°29′N	121°31′W
Lassen Volcanic National Park, n.p.,			
Ca., U.S. (lăs′ĕn vŏl-kăn′ĭk			
năsh′ŭn-ăl pärk)	**122-23**	40°30′N	121°27′W
Las Tablas, Pan. (läs tä′bläs)	**162**	7°46′N	80°17′W
Last Mountain Lake, lk., Sk., Can.			
(låst moun′tĭn lāk)	**144-45**	51°06′N	105°15′W
Las Tórtolas, Cerro, mtn., S.A.	**182-83**	29°57′S	69°53′W
Lastoursville, Gabon (lȧs-tōōr-vēl′)	**282-83**	0°48′S	12°42′E
Las Tunas, Cuba	**154-55**	20°58′N	76°57′W
Las Varillas, Arg.	**187**	31°52′S	62°42′W
Las Vegas, N.M., U.S. (läs vā′gäs)	**130-31**	35°36′N	105°13′W
Las Vegas, Nv., U.S. (läs vā′gäs)	**128-29**	36°11′N	115°08′W
Latacunga, Ec. (lä-tä-kōŋ′gä)	**184**	0°56′S	78°36′W
Latakia, Syria	**248-49**	35°31′N	35°48′E
La Teste-de-Buch, Fr. (lä-tĕst-dĕ-büsh)	**212-13**	44°38′N	1°09′W
Lathrop, Mo., U.S. (lā′thrŭp)	**130-31**	39°33′N	94°20′W
La Tortuga, Isla, i., Ven.			
(ē′s-lä-lä-tôr-tōō′gä)	**188b**	10°56′N	65°20′W
Latouche Treville, Cape, c., Austl.	**296-97**	18°28′S	121°50′E
La Tremblade, Fr. (lä-trĕN-bläd′)	**212-13**	45°46′N	1°08′W
Latrobe, Pa., U.S. (lȧ-trōb′)	**126-27**	40°18′N	79°22′W
La Tuque, Qc., Can. (lȧ′tük′)	**146-47**	47°26′N	72°47′W
Lātūr, India (lä-tōōr′)	**256**	18°24′N	76°35′E
Latvia, nation, Eur. (lăt′vē-ȧ)	**190-91**	57°0′N	25°00′E
Latvija, nation, Eur. *see* Latvia	**190-91**	57°0′N	25°00′E
Lauenburg, Pol. *see* Lębork	**210-11**	54°32′N	17°46′E
Lau Group, is., Fiji	**306f**	18°20′S	178°30′W
Lauis, Switz. *see* Lugano	**210-11**	46°01′N	8°57′E
Launceston, Austl. (lôn′sĕs-tŭn)	**301**	41°25′S	147°08′E
La Unión, Chile (lä-ōō-nyô′n)	**185**	40°18′S	73°05′W
La Unión, El Sal.	**160**	13°20′N	87°51′W
La Unión, Mex. (lä ōōn-nyōn′)	**158-59**	17°58′N	101°49′W
Laura, Austl. (lôrȧ)	**302**	15°33′S	144°26′E
Laurel, De., U.S. (lô′rĕl)	**126-27**	38°33′N	75°34′W
Laurel, Md., U.S. (lô′rĕl)	**126-27**	39°06′N	76°51′W
Laurel, Ms., U.S. (lô′rĕl)	**134-35**	31°42′N	89°08′W
Laurel, Mt., U.S. (lô′rĕl)	**122-23**	45°40′N	108°46′W
Laurel, Ne., U.S. (lô′rĕl)	**124-25**	42°26′N	97°00′W
Laurens, S.C., U.S. (lô′rĕnz)	**134-35**	34°30′N	82°01′W
Laurentides, Les, plat., Qc., Can.	**140-41**	48°0′N	71°00′W
Laurinburg, N.C., U.S. (lô′rĭn-bûrg)	**134-35**	34°47′N	79°28′W
Laurium, Mi., U.S. (lô′rĭ-ŭm)	**124-25**	47°14′N	88°26′W
Lausanne, Switz. (lō-zȧn′)	**210-11**	46°31′N	6°38′E
Lausitzer Neiße, stm., Eur. *see* Neisse	**210-11**	52°04′N	14°46′E
Laut, Pulau, i., Indon.	**268-69**	3°40′S	116°10′E
Lautaro, Chile (lou-tä′rō)	**185**	38°31′S	72°26′W
Laut Kecil, Kepulauan, is., Indon.	**268-69**	4°49′S	115°44′E
Lava, Nosy, i., Madag.	**286-87**	14°33′S	47°36′E
Lava Beds National Monument, n.p.,			
Ca., U.S. (lä′vȧ bĕds			
năsh′ŭn-ăl mŏn′ū-mĕnt)	**122-23**	41°45′N	121°32′W
Laval, Qc., Can.	**146-47**	45°33′N	73°44′W
Laval, Fr. (lä-väl′)	**212-13**	48°04′N	0°46′W
Lāvān, Jazīreh-ye, i., Iran	**250-51**	26°49′N	53°15′E
Lavapié, Punta, c., Chile	**185**	37°09′S	73°35′W
La Vega, Dom. Rep. (lä-vĕ′gä)	**154-55**	19°13′N	70°31′W
Laverton, Austl. (lā′vēr-tŭn)	**294-95**	28°37′S	122°24′E
La Victoria, Ven. (lä vēk-tō′rē-ä)	**188b**	10°13′N	67°20′W
Lavras, Braz. (lä′vräzh)	**186**	21°14′S	45°00′W
Lavrentiya, Russia	**136**	65°35′N	171°01′W
Lawas, Malay.	**268-69**	4°51′N	115°24′E
Lawn Hill National Park, n.p., Austl.			
see Boodjamulla National Park	**302**	18°45′S	138°27′E
Lawrence, In., U.S. (lô′rĕns)	**126-27**	39°50′N	86°01′W
Lawrence, Ks., U.S. (lô′rĕns)	**130-31**	38°57′N	95°15′W
Lawrence, Ma., U.S. (lô′rĕns)	**126-27**	42°42′N	71°10′W
Lawrenceburg, In., U.S. (lô′rĕns-bûrg)	**126-27**	39°05′N	84°52′W
Lawrenceburg, Ky., U.S. (lô′rĕns-bûrg)	**126-27**	38°03′N	84°54′W
Lawrenceburg, Tn., U.S. (lô′rĕns-bûrg)	**134-35**	35°15′N	87°20′W
Lawrenceville, Ga., U.S. (lô-rĕns-vĭl)	**134-35**	33°58′N	83°59′W
Lawrenceville, Il., U.S. (lô-rĕns-vĭl)	**126-27**	38°43′N	87°41′W
Lawrenceville, Va., U.S. (lô-rĕns-vĭl)	**134-35**	36°46′N	77°51′W
Lawton, Ok., U.S. (lô′tŭn)	**130-31**	34°36′N	98°24′W
Lawz, Jabal al-, mtn., Sau. Ar.	**248-49**	28°40′N	35°18′E
La'youn, nat. cap., W. Sah.			
see Laayoune	**280-81**	27°10′N	13°12′W
Laysan Island, i., Hi., U.S.	**137b**	25°50′N	171°50′W
Layton, Ut., U.S. (lā′tŭn)	**122-23**	41°05′N	111°58′W
Lazarev, Russia	**236-37**	52°12′N	141°30′E
Lázaro Cárdenas, Mex.	**158-59**	17°57′N	102°12′W
Lazdijai, Lith. (läzh′dē-yī′)	**210-11**	54°14′N	23°32′E
Lead, S.D., U.S. (lēd)	**124-25**	44°21′N	103°46′W
Leader, Sk., Can.	**144-45**	50°53′N	109°31′W
Leadville, Co., U.S. (lĕd′vĭl)	**128-29**	39°15′N	106°18′W
Leaf, stm., Ms., U.S. (lēf)	**134-35**	31°0′N	88°45′W
League City, Tx., U.S. (lēg sĭ′tĭ)	**132-33**	29°30′N	95°06′W
Leamington, On., Can. (lĕm′ĭng-tŭn)	**146-47**	42°02′N	82°36′W
Leavenworth, Ks., U.S. (lĕv′ĕn-wûrth)	**130-31**	39°18′N	94°56′W
Leavenworth, Wa., U.S.			
(lĕv′ĕn-wûrth)	**122-23**	47°36′N	120°40′W
Łeba, Pol. (lā′bä)	**210-11**	54°45′N	17°33′E
Lebak, Phil.	**270**	6°31′N	124°02′E
Lebanon, In., U.S. (lĕb′ȧ-nŭn)	**126-27**	40°03′N	86°28′W
Lebanon, Ky., U.S. (lĕb′ȧ-nŭn)	**126-27**	37°34′N	85°15′W
Lebanon, Mo., U.S. (lĕb′ȧ-nŭn)	**130-31**	37°41′N	92°40′W
Lebanon, N.H., U.S. (lĕb′ȧ-nŭn)	**126-27**	43°39′N	72°15′W
Lebanon, Oh., U.S. (lĕb′ȧ-nŭn)	**126-27**	39°26′N	84°13′W
Lebanon, Or., U.S. (lĕb′ȧ-nŭn)	**122-23**	44°32′N	122°54′W
Lebanon, Tn., U.S. (lĕb′ȧ-nŭn)	**134-35**	36°13′N	86°17′W
Lebanon, Va., U.S. (lĕb′ȧ-nŭn)	**134-35**	36°54′N	82°05′W
Lebanon, nation, Asia (lĕb′ȧ-nŭn)	**224-25**	34°0′N	36°00′E
Lebedyan', Russia (lyĕ′bĕ-dyän′)	**218-19**	53°01′N	39°08′E
Lębork, Pol. (län-bôrk′)	**210-11**	54°32′N	17°46′E
Lebrija, Spain (lȧ-brē′hä)	**214-15**	36°56′N	6°04′W
Lebu, Chile	**185**	37°37′S	73°39′W
Lecce, Italy (lĕt′chā)	**216-17**	40°21′N	18°10′E
Lecco, Italy (lĕk′kō)	**216-17**	45°51′N	9°23′E
Le Creusot, Fr. (lĕkrû-zô)	**212-13**	46°48′N	4°26′E
Ledo, India	**254-55**	27°17′N	95°44′E
Ledu, China	**260-61**	36°28′N	102°24′E
Leduc, Ab., Can. (lĕ-dōōk′)	**142-43**	53°15′N	113°32′W
Ledyanaya, Gora, mtn., Russia	**236-37**	61°53′N	171°09′E
Ledyard Bay, b., Ak., U.S.	**136**	69°14′N	164°31′W
Leech Lake, lk., Mn., U.S. (lēch lāk)	**124-25**	47°09′N	94°23′W
Leeds, Eng., U.K. (lēdz)	**206-07**	53°50′N	1°35′W
Leeds, N.D., U.S. (lēdz)	**124-25**	48°17′N	99°27′W
Leesburg, Fl., U.S. (lēz′bûrg)	**134-35**	28°49′N	81°53′W
Leesburg, Va., U.S. (lēz′bûrg)	**126-27**	39°06′N	77°34′W
Leesville, La., U.S. (lēz′vĭl)	**132-33**	31°08′N	93°16′W
Leeton, Austl.	**301**	34°33′S	146°24′E
Leeuwarden, Neth. (lā′wär-dĕn)	**206-07**	53°12′N	5°47′E
Leeuwin, Cape, c., Austl. (kāp ōō′wĭn)	**296-97**	34°23′S	115°08′E
Leeward Islands, is., N.A.			
(lē′wẽrd ī′lȧndz)	**155b**	17°0′N	63°00′W
Lefkáda, i., Grc.	**216-17**	38°42′N	20°39′E
Lefkoşa, nat. cap., Cyp. *see* Nicosia	**248-49**	35°10′N	33°22′E
Lefroy, Lake, lk., Austl. (lăk lē-troi′)	**296-97**	31°15′S	121°40′E
Leganés, Spain (lå-gä′nås)	**214-15**	40°20′N	3°46′W
Legaspi, Phil.	**270**	13°08′N	123°45′E
Leghorn, Italy *see* Livorno	**216-17**	43°34′N	10°19′E
Legnica, Pol. (lĕk-nĭt′sȧ)	**210-11**	51°13′N	16°10′E
Leh, India (lā)	**254-55**	34°10′N	77°35′E
Le Havre, Fr. (lĕ åv′r′)	**212-13**	49°29′N	0°08′E
Lehi, Ut., U.S.	**128-29**	40°24′N	111°51′W
Leicester, Eng., U.K. (lĕs′tẽr)	**206-07**	52°39′N	1°08′W
Leikanger, Nor. (lī′käŋ′gẽr)	**208-09**	61°12′N	6°50′E
Leine, stm., Ger. (lī′nĕ)	**210-11**	52°43′N	9°36′E
Leipzig, Ger. (līp′tsĭk)	**210-11**	51°20′N	12°23′E
Leiria, Port. (lā-rē′ä)	**214-15**	39°45′N	8°48′W
Leitchfield, Ky., U.S. (lēch′fēld)	**126-27**	37°29′N	86°18′W
Leizhou Bandao, pen., China			
(lā-jō bän-dou)	**258-59**	20°47′N	110°05′E
Leksand, Swe. (lĕk′sänd)	**208-09**	60°43′N	15°01′E
Leland, Mi., U.S. (lē′lănd)	**126-27**	45°01′N	85°46′W
Leland, Ms., U.S. (lē′lănd)	**134-35**	33°25′N	90°54′W
Leli Shan, mtn., China	**254-55**	33°26′N	81°42′E
Le Maire, Estrecho de, strt., Arg.			
(ĕs-trĕ′chô-dĕ-lĕ-mī′rĕ)	**185**	54°50′S	64°60′W
Léman, Lac, lk., Eur.			
see Geneva, Lake	**210-11**	46°24′N	6°22′E
Le Mans, Fr. (lĕ mäN′)	**212-13**	48°00′N	0°12′E
Le Mars, Ia., U.S. (lē märz′)	**124-25**	42°48′N	96°10′W
Lemesós, Cyp.	**248-49**	34°41′N	33°03′E
Lemhi Range, mts., Id., U.S.			
(lĕm′hī rānj)	**122-23**	44°33′N	113°36′W
Lemmon, S.D., U.S. (lĕm′ŭn)	**124-25**	45°56′N	102°10′W
Lemnos, i., Grc. *see* Límnos	**216-17**	39°55′N	25°18′E
Lempa, stm., N.A. (lĕm′pä)	**160**	13°15′N	88°49′W
Lena, stm., Russia (lē′nȧ) (lyĕ′nŭ)	**236-37**	72°25′N	126°40′E
Lençóis, Braz.	**180-81**	12°34′S	41°23′W
Lenexa, Ks., U.S. (lē′nĕx-ă)	**130-31**	38°58′N	94°44′W
Lenghu, China	**238-39**	38°50′N	93°26′E
Lenin, Qullai, mtn., Asia			
see Lenin Peak	**252-53**	39°20′N	72°55′E
Lenina, Pik, mtn., Asia *see* Lenin Peak	**252-53**	39°20′N	72°55′E
Lenin Atyndagy Choku, mtn., Asia			
see Lenin Peak	**252-53**	39°20′N	72°55′E
Leningrad, Russia			
see Saint Petersburg	**208-09**	59°57′N	30°15′E
Leningradskaya, Russia			
(lyĕ-nĭn-gräd′skȧ-yȧ)	**218-19**	46°19′N	39°23′E
Leninogor, Kaz. *see* Ridder	**244**	50°21′N	83°30′E
Leninogorsk, Russia	**202-03**	54°35′N	52°29′E
Lenin Peak, mtn., Asia	**252-53**	39°20′N	72°55′E
Leninsk, Kaz. *see* Bayqongyr	**244**	45°38′N	63°18′E
Leninsk-Kuznetskiy, Russia	**236-37**	54°41′N	86°12′E
Leninskoye, Russia	**260-61**	47°60′N	132°38′E
Lennox, S.D., U.S. (lĕn′ŭks)	**124-25**	43°21′N	96°55′W
Lennox, Isla, i., Chile	**185**	55°18′S	66°50′W
Lenoir, N.C., U.S. (lē-nōr′)	**134-35**	35°55′N	81°32′W
Lensk, Russia	**236-37**	60°44′N	114°56′E
Léo, Burkina	**282-83**	11°06′N	2°06′W
Leoben, Aus. (lå-ō′bĕn)	**210-11**	47°22′N	15°06′E
Léogâne, Haiti (lā-ō-gan′)	**154-55**	18°31′N	72°38′W
Leominster, Ma., U.S. (lĕm′ĭn-stẽr)	**126-27**	42°32′N	71°45′W
León, Mex. (lå-ōn′)	**158-59**	21°07′N	101°42′W
León, Nic. (lĕ-ô′n)	**161**	12°26′N	86°52′W
León, Spain (lĕ-ô′n)	**214-15**	42°36′N	5°34′W
Leon, Ia., U.S. (lē′ŏn)	**124-25**	40°45′N	93°44′W
León, hist. reg., Spain (lĕ-ô′n)	**214-15**	42°0′N	6°00′W
Leon, stm., Tx., U.S. (lē′ŏn)	**132-33**	30°59′N	97°24′W
León de los Aldamas, Mex. *see* León	**158-59**	21°07′N	101°42′W
Leonforte, Italy (lā-ôn-fôr′tā)	**216-17**	37°38′N	14°23′E
Leonora, Austl.	**294-95**	28°53′S	121°20′E
Léopold II, Lac, lk., D.R.C.			
see Mai-Ndombe, Lac	**282-83**	2°25′S	18°18′E
Leopoldina, Braz. (lā-ô-pôl-dē′nä)	**186**	21°32′S	42°38′W
Lepe, Spain (lā′pā)	**214-15**	37°15′N	7°12′W
Leping, China (lŭ-pĭŋ)	**258-59**	28°57′N	117°06′E
Le Port, Reu.	**287a**	20°55′S	55°18′E
Le Puy, Fr. (lē pwē′)	**212-13**	45°03′N	3°53′E
Lerdo, Mex. (lĕr′dō)	**132-33**	25°32′N	103°31′W
Lérida, Spain *see* Lleida	**214-15**	41°37′N	0°38′E
Lerma, stm., Mex. (lĕr′mä)	**158-59**	20°13′N	102°41′W
Le Roy, N.Y., U.S. (lĕ roi′)	**126-27**	42°58′N	77°59′W
Lerwick, Scot., U.K. (lĕr′ĭk) (lûr′wĭk)	**206c**	60°09′N	1°09′W
Lesbos, i., Grc. *see* Lésvos	**216-17**	39°10′N	26°20′E
Les Cayes, Haiti	**154-55**	18°12′N	73°45′W
Leshan, China (lŭ-shän)	**238-39**	29°34′N	103°45′E
Leshukonskoye, Russia	**202-03**	64°53′N	45°42′E
Leskovac, Serb. (lĕs′kô-väts)	**216-17**	43°01′N	21°57′E
Leslie, Mi., U.S. (lĕz′lĭ)	**126-27**	42°27′N	84°25′W
Lesosibirsk, Russia	**236-37**	58°14′N	92°29′E
Lesotho, nation, Afr. (lĕsō′thô)	**274**	29°30′S	28°30′E
Lesozavodsk, Russia (lyĕ-sô-zȧ-vôdsk′)	**264**	45°28′N	133°24′E
Les Sables-d'Olonne, Fr.			
(lā sȧ′bl′dô-lŭn′)	**212-13**	46°30′N	1°47′W
Les Saintes, is., Guad. (lā-săNt′)	**155b**	15°52′N	61°37′W
Lesser Antilles, is., (lĕs′ẽr ăn-tĭ′lēz)	**155b**	15°0′N	61°00′W
Lesser Caucasus, mts., Asia	**245**	40°60′N	44°35′E
Lesser Khingan Range, mts., China	**260-61**	48°45′N	127°00′E
Lesser Slave, stm., Ab., Can.			
(lĕs′ẽr slāv)	**142-43**	55°10′N	114°03′W
Lesser Slave Lake, lk., Ab., Can.			
(lĕs′ẽr slāv lāk)	**142-43**	55°29′N	115°10′W
Lesser Sunda Islands, is., Asia			
(lĕs′ẽr sōōn′dȧ ī′lȧndz)	**268-69**	9°0′S	120°00′E
Le Sueur, Mn., U.S. (lē sōōr′)	**124-25**	44°28′N	93°55′W
Lésvos, i., Grc.	**216-17**	39°10′N	26°20′E
Leszno, Pol. (lĕsh′nô)	**210-11**	51°51′N	16°35′E
Lethbridge, Ab., Can. (lĕth′brĭj)	**142-43**	49°42′N	112°49′W
Lethem, Guy.	**178-79**	3°23′N	59°48′W
Leti, Kepulauan, is., Indon.	**268-69**	8°13′S	127°50′E
Leticia, Col. (lĕ-tē′syȧ)	**178-79**	4°10′S	69°56′W
Leucas, i., Grc. *see* Lefkáda	**216-17**	38°42′N	20°39′E
Levanger, Nor. (lĕ-väng′ẽr)	**200-01**	63°45′N	11°18′E
Leveque, Cape, c., Austl. (kāp lĕ-vĕk′)	**296-97**	16°26′S	122°56′E
Leverkusen, Ger. (lĕ′fĕr-kōō-zĕn)	**210-11**	51°03′N	6°59′E
Levice, Slvk. (lā′vĕt-sĕ)	**210-11**	48°13′N	18°37′E
Le Vigan, Fr. (lĕ vē-gäN′)	**212-13**	43°59′N	3°35′E
Lévis, Qc., Can. (lā-vē′) (lē′vĭs)	**146-47**	46°48′N	71°11′W
Levittown, Pa., U.S. (lĕ′vĭt-toun)	**126-27**	40°09′N	74°51′W
Levkosía, nat. cap., Cyp. *see* Nicosia	**248-49**	35°10′N	33°22′E
Lewes, De., U.S. (lōō′ĭs)	**126-27**	38°46′N	75°08′W
Lewis, Isle of, i., Scot., U.K.			
(īl ŏv lōō′ĭs)	**206-07**	58°08′N	6°45′W
Lewisburg, Tn., U.S. (lū′ĭs-bûrg)	**134-35**	35°27′N	86°48′W
Lewisburg, W.V., U.S. (lū′ĭs-bûrg)	**126-27**	37°48′N	80°27′W
Lewisporte, Nf./L., Can. (lū′ĭs-pōrt)	**148-49**	49°16′N	55°05′W
Lewiston, Id., U.S. (lū′ĭs-tŭn)	**122-23**	46°24′N	117°00′W
Lewiston, Me., U.S. (lū′ĭs-tŭn)	**126-27**	44°06′N	70°13′W
Lewistown, Il., U.S. (lū′ĭs-toun)	**130-31**	40°23′N	90°09′W
Lewistown, Mt., U.S. (lū′ĭs-toun)	**122-23**	47°04′N	109°26′W
Lexington, Il., U.S. (lĕk′sĭng-tŭn)	**126-27**	40°38′N	88°46′W
Lexington, Ky., U.S. (lĕk′sĭng-tŭn)	**126-27**	38°03′N	84°31′W

n-sing; ŋ-baŋk; N-nasalized n; nŏd; cŏmmit; ōld; ôbey; ôrder; oi-boil; fōōd; ȯ-as oo in foot; ou-out; s-soft; sh-dish; th-thin; pūre; ûnite; ûrn; stŭd; circŭs; ü-as in French tu; ′-indeterminate vowel.

376 Lexington – Livingstone PRONOUNCING INDEX

Feature (Pronunciation)	Page	Lat.	Long.
Lexington, Ma., U.S. (lĕk´sĭng-tŭn)	126-27	42°27'N	71°14'W
Lexington, Mo., U.S. (lĕk´sĭng-tŭn)	130-31	39°11'N	93°53'W
Lexington, N.C., U.S. (lĕk´sĭng-tŭn)	134-35	35°49'N	80°15'W
Lexington, Ne., U.S. (lĕk´sĭng-tŭn)	124-25	40°47'N	99°44'W
Lexington, Oh., U.S. (lĕk´sĭng-tŭn)	126-27	40°41'N	82°34'W
Lexington, Tn., U.S. (lĕk´sĭng-tŭn)	134-35	35°39'N	88°24'W
Lexington, Va., U.S. (lĕk´sĭng-tŭn)	126-27	37°46'N	79°27'W
Leyte, i., Phil. (lā´tā)	270	10°50'N	124°50'E
Leyte Gulf, b., Phil.	270	10°50'N	125°25'E
Leżajsk, Pol. (lĕ´zhä-ĭsk)	210-11	50°16'N	22°26'E
L'gov, Russia (lgôf)	218-19	51°39'N	35°16'E
Lhasa, China (läs´ä)	254-55	29°39'N	91°08'E
Lhasa, stm., China	254-55	29°20'N	90°46'E
Lhokseumawe, Indon.	266-67	5°11'N	97°08'E
Lhorong, China	258-59	30°47'N	95°51'E
Li, stm., China	258-59	29°12'N	112°11'E
Liangzhou, China *see* Wuwei	260-61	37°56'N	102°38'E
Lianjiang, China (lĭĕn-jyäŋ)	243a	26°12'N	119°31'E
Lianxian, China	258-59	24°47'N	112°21'E
Lianyungang, China (lĭĕn-yòn-gäŋ)	258-59	34°37'N	119°11'E
Lianzhou, China *see* Hepu	258-59	21°41'N	109°11'E
Liao, stm., China	260-61	40°41'N	122°09'E
Liaocheng, China (lĭou-chŭŋ)	260-61	36°27'N	115°59'E
Liaodong Bandao, pen., China (lĭou-dôŋ bän-dou)	260-61	39°55'N	122°19'E
Liaodong Wan, b., China (lĭou-dôŋ wäŋ)	260-61	40°30'N	121°30'E
Liaoning, state, China	263	41°0'N	123°00'E
Liaotung, Gulf of, b., China *see* Liaodong Wan	260-61	40°30'N	121°30'E
Liaotung Peninsula, pen., China *see* Liaodong Bandao	260-61	39°55'N	122°19'E
Liaoyang, China (lyä´ō-yäng´)	263	41°16'N	123°10'E
Liaoyuan, China (lĭou-yůän)	260-61	42°55'N	125°08'E
Liard, stm., Can. (lē-är´)	140-41	61°51'N	121°19'W
Lib, i., Marsh. Is.	304-05	8°19'N	167°25'E
Libagon, Phil.	270	10°18'N	125°03'E
Líbano, Col. (lē´bȧ-nō)	188c	4°56'N	75°04'W
Libau, Lat. *see* Liepāja	208-09	56°31'N	21°01'E
Libby, Mt., U.S. (lĭb´ē)	122-23	48°23'N	115°33'W
Libenge, D.R.C. (lē-bĕŋ´gā)	284-85	3°39'N	18°38'E
Liberal, Ks., U.S. (lĭb´ẽr-ăl)	130-31	37°02'N	100°56'W
Liberec, Czech Rep. (lē´bĕr-ĕts)	210-11	50°46'N	15°04'E
Liberia, C.R.	161	10°37'N	85°26'W
Liberia, nation, Afr. (lī-bē´rĭ-ȧ)	274	6°30'N	9°30'W
Liberty, Ky., U.S. (lĭb´ẽr-tĭ)	134-35	37°19'N	84°56'W
Liberty, Mo., U.S. (lĭb´ẽr-tĭ)	130-31	39°15'N	94°25'W
Liberty, N.Y., U.S. (lĭb´ẽr-tĭ)	126-27	41°48'N	74°44'W
Liberty, S.C., U.S. (lĭb´ẽr-tĭ)	134-35	34°47'N	82°42'W
Liberty, Tx., U.S. (lĭb´ẽr-tĭ)	132-33	30°04'N	94°48'W
Lībiyā, nation, Afr. *see* Libya	274	27°0'N	17°00'E
Lībīyah, Aṣ-Ṣaḥrā' al-, des., Afr. *see* Libyan Desert	275	24°0'N	25°00'E
Libourne, Fr. (lē-bōōrn´)	212-13	44°55'N	0°14'W
Libres, Mex. (lē´bräs)	158-59	19°28'N	97°41'W
Libreville, nat. cap., Gabon (lē-br'vēl´)	282-83	0°24'N	9°28'E
Libya, nation, Afr. (lĭb´ē-ä)	274	27°0'N	17°00'E
Libyan Desert, des., Afr. (lĭb´ē-ăn dĕs´ẽrt)	275	24°0'N	25°00'E
Licancábur, Volcán, vol., S.A.	182-83	22°50'S	67°50'W
Licantén, Chile (lē-kän-tĕ´n)	185	34°59'S	72°06'W
Lichinga, Moz.	286-87	13°17'S	35°15'E
Lichtenburg, S. Afr. (lĭk´tĕn-bĕrgh)	292c	26°10'S	26°10'E
Licking, stm., Ky., U.S. (lĭk´ĭng)	126-27	39°05'N	84°31'W
Licungo, stm., Moz.	286-87	17°38'S	37°22'E
Lida, Bela. (lē´dȧ)	210-11	53°54'N	25°18'E
Lidköping, Swe. (lēt´chû-pĭng)	208-09	58°30'N	13°11'E
Lidzbark, Pol. (lĭts´bärk)	210-11	53°16'N	19°50'E
Liechtenstein, nation, Eur. (lēk´tĕn-shtīn)	210-11	47°09'N	9°35'E
Liège, Bel.	206-07	50°38'N	5°34'E
Liegnitz, Pol. *see* Legnica	210-11	51°13'N	16°10'E
Lienz, Aus. (lē-ĕnts´)	210-11	46°50'N	12°46'E
Liepāja, Lat. (le´pä-yä´)	208-09	56°31'N	21°01'E
Lietuva, nation, Eur. *see* Latvia	190-91	56°0'N	24°00'E
Lièvre, stm., Qc., Can.	146-47	45°31'N	75°26'W
Lifou, i., N. Cal.	306g	20°44'S	167°14'E
Ligao, Phil. (lē-gä´ō)	270	13°14'N	123°34'E
Ligonha, stm., Moz. (lē-gō´nyȧ)	286-87	16°53'S	39°08'E
Ligonier, In., U.S. (lĭg-ō-nēr)	126-27	41°28'N	85°34'W
Lihir Group, is., Pap. N. Gui.	302	2°55'S	152°36'E
Līhu'e, Hi., U.S. (lē-hōō´ā)	137a	21°59'N	159°22'W
Liivi laht, b., Eur. *see* Riga, Gulf of	208-09	57°30'N	23°35'E
Lijiang, China (lē-jyäŋ)	258-59	26°52'N	100°14'E
Likasi, D.R.C.	284-85	10°59'S	26°43'E
Likhoslavl, Russia (lyē-kôsläv´'l)	218-19	57°07'N	35°28'E
Likouala, stm., Congo	284-85	1°12'S	16°49'E
Lille, Fr. (lēl)	212-13	50°39'N	3°07'E
Lillehammer, Nor. (lēl´ĕ-häm´mẽr)	208-09	61°07'N	10°28'E
Lillesand, Nor. (lēl´ĕ-sän')	208-09	58°15'N	8°24'E
Lillestrøm, Nor. (lēl´ĕ-strûm)	208-09	59°58'N	11°04'E
Lillooet, B.C., Can. (lĭ´lōō-ĕt)	142-43	50°42'N	121°56'W
Lillooet, stm., B.C., Can. (lĭ´lōō-ĕt)	142-43	49°45'N	122°08'W
Lilongwe, nat. cap., Malawi (lē-lō-än)	286-87	13°59'S	33°44'E
Liloy, Phil.	270	8°07'N	122°40'E
Lima, Oh., U.S. (lī´mȧ)	126-27	40°45'N	84°07'W
Lima, nat. cap., Peru (lē´mä)	184	12°04'S	77°03'W
Limassol, Cyp. *see* Lemesós	248-49	34°41'N	33°03'E
Limay, stm., Arg. (lē-mä´ē)	185	38°59'S	68°00'W
Limbaži, Lat. (lēm´bä-zĭ)	208-09	57°31'N	24°43'E
Limeira, Braz. (lē-mā´rä)	186	22°34'S	47°24'W
Limerick, Ire. (lĭm´nȧk)	206-07	52°40'N	8°38'W
Límnos, i., Grc.	216-17	39°55'N	25°18'E
Limoges, Fr.	212-13	45°50'N	1°15'E
Limón, Hond. (lē-mō´n)	161	15°51'N	85°31'W
Limon, Co., U.S. (lī´mŏn)	130-31	39°16'N	103°42'W
Limoux, Fr. (lē-mōō´)	212-13	43°04'N	2°12'E
Limpopo, stm., Afr. (lĭm-pō´pō)	286-87	25°12'S	33°31'E
Limpopo, Grande Parque Transfronteiriço do, n.p., Afr. *see* Great Limpopo Transfrontier Park	286-87	23°0'S	31°30'E
Limpopo, Parque Nacional do, n.p., Moz.	286-87	23°21'S	31°54'E
Linapacan Island, i., Phil.	270	11°27'N	119°49'E
Linares, Chile (lē-nä´räs)	185	35°50'S	71°36'W
Linares, Mex.	132-33	24°51'N	99°34'W
Linares, Spain (lē-nä´rĕs)	214-15	38°06'N	3°38'W
Lincoln, Arg. (lĭŋ´kŭn)	187	34°52'S	61°31'W
Lincoln, Eng., U.K. (lĭŋ´kŭn)	206-07	53°14'N	0°33'W
Lincoln, Il., U.S. (lĭŋ´kŭn)	126-27	40°09'N	89°22'W
Lincoln, Ks., U.S. (lĭŋ´kŭn)	130-31	39°02'N	98°09'W
Lincoln, Me., U.S. (lĭŋ´kŭn)	127a	45°22'N	68°31'W
Lincoln, Ne., U.S.	124-25	40°48'N	96°43'W
Lincoln, Mount, mtn., Co., U.S. (mount lĭŋ´kŭn)	128-29	39°21'N	106°07'W
Lincolnton, N.C., U.S. (lĭŋ´kŭn-tŭn)	134-35	35°28'N	81°15'W
Lindale, Ga., U.S. (lĭn´dāl)	134-35	34°12'N	85°11'W
Linden, Guy.	178-79	6°05'N	58°17'W
Linden, Al., U.S. (lĭn´dĕn)	134-35	32°19'N	87°48'W
Linden, Tx., U.S. (lĭn´dĕn)	130-31	33°01'N	94°22'W
Lindesberg, Swe. (lĭn´dĕs-bĕrgh)	208-09	59°36'N	15°13'E
Lindesnes, c., Nor. (lĭn´ĕs-nĕs)	208-09	58°0'N	7°02'E
Lindi, Tan. (lĭn´dē)	286-87	9°60'S	39°42'E
Lindi, stm., D.R.C.	284-85	0°33'N	25°05'E
Lindian, China (lĭn-dĭĕn)	260-61	47°11'N	124°52'E
Lindley, S. Afr. (lĭnd´lē)	292c	27°53'S	27°56'E
Lindsay, On., Can. (lĭn´zē)	146-47	44°21'N	78°44'W
Lindsay, Ok., U.S. (lĭn´zē)	130-31	34°50'N	97°37'W
Line Islands, is., Oc. (līn ī´lȧndz)	304-05	0°05'N	157°00'W
Linfen, China	260-61	36°05'N	111°31'E
Lingao, China (lĭn-gou)	258-59	19°54'N	109°40'E
Lingayen, Phil. (lĭŋ´gä-yän´)	270	16°01'N	120°14'E
Lingayen Gulf, b., Phil.	270	16°15'N	120°14'E
Lingen, Ger. (lĭŋ´gĕn)	210-11	52°31'N	7°19'E
Lingga, Kepulauan, is., Indon.	266-67	0°05'S	104°35'E
Lingling, China (lĭŋ-lĭŋ) *see* Yongzhou	258-59	26°13'N	111°37'E
Lingyuan, China (lĭŋ-yůän)	260-61	41°15'N	119°16'E
Linh, Ngoc, mtn., Viet.	266-67	15°04'N	107°59'E
Linhai, China	258-59	28°51'N	121°07'E
Linhe, China (lĭn-hŭ)	260-61	40°49'N	107°30'E
Linjiang, China (lĭn-jyäŋ)	263	41°49'N	126°55'E
Linköping, Swe. (lĭn´chû-pĭng)	208-09	58°25'N	15°37'E
Linkou, China	264	45°19'N	130°16'E
Linqing, China (lĭn-chyĭŋ)	260-61	36°51'N	115°42'E
Linqu, China (lĭn-chyōō)	260-61	36°31'N	118°32'E
Linru, China	258-59	34°10'N	112°50'E
Lins, Braz. (lē´ns)	182-83	21°40'S	49°45'W
Lintao, China	260-61	35°22'N	103°51'E
Linton, In., U.S. (lĭn´tŭn)	126-27	39°02'N	87°10'W
Linton, N.D., U.S. (lĭn´tŭn)	124-25	46°16'N	100°14'W
Linxi, China (lĭn-shyē)	260-61	43°36'N	118°03'E
Linxia, China	260-61	35°36'N	103°13'E
Linyi, China (lĭn-yē)	260-61	35°04'N	118°22'E
Linyi, China (lĭn-yē)	260-61	37°11'N	116°52'E
Linz, Aus. (lĭnts)	210-11	48°18'N	14°18'E
Lion, Golfe du, b., Fr.	212-13	43°0'N	4°00'E
Lipa, Phil. (lē-pä´)	270	13°56'N	121°10'E
Lipari, Isola, i., Italy (ē´-sō-lä-lē´pä-rē)	216-17	38°29'N	14°56'E
Lipetsk, Russia (lyē´pĕtsk)	218-19	52°37'N	39°37'E
Lípez, Cerro, mtn., Bol.	182-83	21°55'S	66°53'W
Liping, China (lē-pĭŋ)	258-59	26°17'N	108°60'E
Lippe, stm., Ger. (lĭp´ĕ)	210-11	51°39'N	6°37'E
Lippstadt, Ger. (lĭp´shtät)	210-11	51°40'N	8°20'E
Lipu, China (lē-pōō)	258-59	24°25'N	110°29'E
Lira, Ug.	289	2°15'N	32°54'E
Lisakovsk, Kaz.	244	52°32'N	62°33'E
Lisala, D.R.C. (lē-sä´lä)	284-85	2°09'N	21°31'E
Lisboa, nat. cap., Port. (lēzh-bō´ä) *see* Lisbon	214-15	38°43'N	9°08'W
Lisbon, N.D., U.S. (lĭz´bŭn)	124-25	46°26'N	97°41'W
Lisbon, Oh., U.S. (lĭz´bŭn)	126-27	40°46'N	80°46'W
Lisbon, nat. cap., Port. (lĭz´bŭn)	214-15	38°43'N	9°08'W
Lisbon Falls, Me., U.S. (lĭz´bŭn fôlz)	126-27	44°00'N	70°03'W
Lisburn, N. Ire., U.K. (lĭs´bŭrn)	206-07	54°31'N	6°03'W
Lisburne, Cape, c., Ak., U.S.	136	68°52'N	166°14'W
Lishui, China (lĭ´shwĭ´)	258-59	28°27'N	119°54'E
Lisianski Island, i., Hi., U.S.	137b	26°02'N	174°00'W
Lisichansk, Ukr. *see* Lysychans'k	218-19	48°55'N	38°26'E
Lisieux, Fr. (lē-zyû´)	212-13	49°09'N	0°14'E
Liski, Russia (lyēs´kē)	218-19	50°59'N	39°31'E
Lismore, Austl. (lĭz´môr)	301	28°49'S	153°17'E
Litang, China	258-59	23°12'N	109°09'E
Litang, China	258-59	29°60'N	100°16'E
Litang, stm., China	258-59	28°03'N	101°32'E
Litchfield, Il., U.S. (lĭch´fēld)	126-27	39°10'N	89°39'W
Litchfield, Mn., U.S. (lĭch´fēld)	124-25	45°08'N	94°32'W
Lithgow, Austl. (lĭth´gō)	301	33°30'S	150°09'E
Lithuania, nation, Eur. (lĭth-ů-ā-´nĭ-ȧ)	190-91	56°0'N	24°00'E
Litoměřice, Czech Rep. (lē´tô-myĕr´zhĭ-tsĕ)	210-11	50°33'N	14°08'E
Litovko, Russia	260-61	49°15'N	135°10'E
Little Abaco, i., Bah. (lĭt´'l ä´bä-kō)	154-55	26°54'N	77°43'W
Little Andaman, i., India (lĭt´'l ăn-dȧ-măn´)	266-67	10°45'N	92°30'E
Little Belt Mountains, mts., Mt., U.S. (lĭt´'l bĕlt moun´tĭnz)	122-23	46°45'N	110°35'W
Little Bighorn, stm., U.S. (lĭt´'l bĭg-hôrn)	122-23	45°44'N	107°34'W
Little Bighorn Battlefield National Monument, n.p., Mt., U.S. (lĭt´'l bĭg-hôrn băt´'l-fēld năsh´ŭn-ăl mŏn´ū-mĕnt)	122-23	45°32'N	107°20'W
Little Cayman, i., Cay. Is. (lĭt´'l kā´mȧn) (lĭt´'l kī-män´)	154-55	19°42'N	80°02'W
Little Current, On., Can.	146-47	45°58'N	81°55'W
Little Exuma, i., Bah. (lĭt´'l ĕk-sōō´mä)	154-55	23°27'N	75°37'W
Little Falls, Mn., U.S. (lĭt´'l fôlz)	124-25	45°59'N	94°22'W
Little Falls, N.Y., U.S. (lĭt´'l fôlz)	126-27	43°03'N	74°52'W
Littlefield, Tx., U.S. (lĭt´'l-fēld)	130-31	33°55'N	102°20'W
Little Inagua, i., Bah. (lĭt´'l ē-nä´gwä)	154-55	21°30'N	72°60'W
Little Karoo, plat., S. Afr.	286-87	33°45'S	21°30'E
Little Karroo, plat., S. Afr. (lĭt´'l kä-rōō) *see* Little Karoo	286-87	33°45'S	21°30'E
Little Missouri, stm., U.S. (lĭt´'l mĭ-sōō´rĭ)	120-21	47°36'N	102°17'W
Little Nicobar, i., India	266-67	7°20'N	93°40'E
Little Powder, stm., U.S. (lĭt´'l pou´dẽr)	122-23	45°28'N	105°21'W
Little Rock, Ar., U.S.	130-31	34°43'N	92°19'W
Little Scarcies, stm., Afr.	282-83	8°51'N	13°07'W
Little Sioux, stm., U.S. (lĭt´'l sōō)	124-25	41°49'N	96°06'W
Little Smoky, stm., Ab., Can. (lĭt´'l smōk´ĭ)	142-43	55°40'N	117°38'W
Littleton, Co., U.S. (lĭt´'l-tŭn)	130-31	39°35'N	105°01'W
Littleton, N.H., U.S. (lĭt´'l-tŭn)	126-27	44°18'N	71°46'W
Litzmannstadt, Pol. *see* Łódź	210-11	51°47'N	19°31'E
Liuaniua, at., Sol. Is. *see* Ontong Java	306e	5°19'S	159°16'E
Liubliana, nat. cap., Slvn. *see* Ljubljana	216-17	46°03'N	14°31'E
Liubotyn, Ukr.	218-19	49°56'N	35°57'E
Liuchow, China *see* Liuzhou	258-59	24°19'N	109°23'E
Liuyang, China (lyōō´yäng´)	258-59	28°08'N	113°38'E
Liuzhou, China (lĭō-jō)	258-59	24°19'N	109°23'E
Live Oak, Fl., U.S. (lĭv ōk)	134-35	30°18'N	82°59'W
Livermore, Ca., U.S. (lĭv´ẽr-mōr)	128-29	37°41'N	121°46'W
Livermore, Ky., U.S. (lĭv´ẽr-mōr)	126-27	37°29'N	87°08'W
Liverpool, N.S., Can. (lĭv´ẽr-pōōl)	148-49	44°03'N	64°43'W
Liverpool, Eng., U.K. (lĭv´ẽr-pōōl)	206-07	53°25'N	2°57'W
Liverpool Range, mts., Austl. (lĭv´ẽr-pōōl rānj)	301	31°51'S	150°18'E
Livingston, Guat.	160	15°50'N	88°46'W
Livingston, Al., U.S. (lĭv´ĭng-stŭn)	134-35	32°36'N	88°12'W
Livingston, Mt., U.S. (lĭv´ĭng-stŭn)	122-23	45°40'N	110°34'W
Livingston, Tn., U.S. (lĭv´ĭng-stŭn)	134-35	36°23'N	85°19'W
Livingston, Tx., U.S. (lĭv´ĭng-stŭn)	132-33	30°42'N	94°57'W
Livingston, Lake, res., Tx., U.S.	132-33	30°43'N	95°08'W
Livingstone, Zam. (lĭv-ĭng-stōn)	286-87	17°52'S	25°51'E

ăt; finăl; rāte; senāte; ärm; ȧsk; sofȧ; fâre; ch-choose; dh-as th in other; bē; ēvent; bĕt; recĕnt; cratẽr; g-gō; gh-guttural g; bĭt; ĭ-short neutral; rīde; ĸ-guttural k as ch in German ich;

PRONOUNCING INDEX Livingstone, Chutes de – Lübeck 377

Feature (Pronunciation)	Page	Lat.	Long.
Livingstone, Chutes de, wtfl., Afr.			
see Livingstone Falls	282-83	4°51′S	14°29′E
Livingstone Falls, wtfl., Afr.	282-83	4°51′S	14°29′E
Livno, Bos. (lēv′nô)	216-17	43°50′N	17°00′E
Livny, Russia (lēv′nė)	218-19	52°25′N	37°37′E
Livorno, Italy (lē-vôr′nō) (lĕg′hôrn)	216-17	43°34′N	10°19′E
Livramento, Braz. (lē-vrȧ-mĕ′n-tô)			
see Santana do Livramento	187	30°53′S	55°31′W
Lixi, China	258-59	29°15′N	114°47′E
Lixian, China (lē shyĕn)	258-59	29°30′N	111°38′E
Lixian, China (lē shyĕn)	258-59	34°09′N	105°07′E
Lixian, stm., Asia *see* Black	266-67	21°15′N	105°21′E
Lizard Point, c., Eng., U.K.			
(lĭz′ȧrd point)	206-07	49°58′N	5°13′W
Ljubljana, nat. cap., Slvn.			
(lyōō′blyä′na)	216-17	46°03′N	14°31′E
Ljungby, Swe. (lyòng′bü)	208-09	56°50′N	13°56′E
Ljusdal, Swe. (lyōōs′däl)	208-09	61°50′N	16°06′E
Ljusnan, stm., Swe.	200-01	61°09′N	17°10′E
Llandudno, Wales, U.K. (lăn-düd′nō)	206-07	53°19′N	3°50′W
Llanelli, Wales, U.K. (lȧ-nĕl′ĭ)	206-07	51°41′N	4°09′W
Llanes, Spain (lyä′nås)	214-15	43°25′N	4°45′W
Llano, Tx., U.S. (lă′nō) (lyä′nō)	132-33	30°46′N	98°40′W
Llano, stm., Tx., U.S. (lă′nō) (lyä′nō)	132-33	30°39′N	98°25′W
Llanos, pl., S.A. (lyä′nōs)	178-79	5°0′N	70°00′W
Lleida, Spain	214-15	41°37′N	0°38′E
Lloydminster, Sk., Can.	142-43	53°17′N	110°01′W
Llullaillaco, Cerro, vol., S.A.			
see Llullaillaco, Volcán	182-83	24°43′S	68°33′W
Llullaillaco, Volcán, vol., S.A.			
(vôl-kȧ′n lyōō-lyī-lyä′kō)	182-83	24°43′S	68°33′W
Loa, stm., Chile	182-83	21°26′S	70°03′W
Loanda, Braz.	182-83	23°00′S	53°11′W
Loanda, nat. cap., Ang. *see* Luanda	286-87	8°49′S	13°14′E
Loange, stm., Afr. (lô-äŋ′gā)	284-85	4°17′S	20°02′E
Lobamba, nat. cap., Swaz.			
(lōō′-bäm-bä) (lō-bäm′-bä)	286-87	26°27′S	31°12′E
Lobaye, stm., C.A.R.	284-85	3°41′N	18°35′E
Lobería, Arg. (lô-bĕ′rē′ä)	187	38°09′S	58°47′W
Lobito, Ang. (lô bē′tō)	286-87	12°21′S	13°33′E
Lobos, Arg. (lō′bôs)	187	35°11′S	59°06′W
Loches, Fr. (lôsh)	212-13	47°08′N	0°60′E
Lockhart, Tx., U.S. (lŏk′härt)	132-33	29°53′N	97°40′W
Lock Haven, Pa., U.S. (lŏk′hā-vĕn)	126-27	41°07′N	77°27′W
Loc Ninh, Viet. (lōk′nĭng′)	266-67	11°50′N	106°37′E
Lodève, Fr. (lô-dĕv′)	212-13	43°43′N	3°19′E
Lodeynoye Pole, Russia			
(lô-dĕy-nô′yĕ)	208-09	60°44′N	33°34′E
Lodi, Italy (lô′dē)	216-17	45°19′N	9°30′E
Lodi, Ca., U.S. (lō′dī)	128-29	38°08′N	121°16′W
Lodi, Wi., U.S. (lō′dī)	126-27	43°19′N	89°31′W
Lodja, D.R.C.	284-85	3°26′S	23°27′E
Lodsch, Pol. *see* Łódź	210-11	51°47′N	19°31′E
Lodwar, Kenya	289	3°08′N	35°38′E
Łódź, Pol.	210-11	51°47′N	19°31′E
Loei, Thai.	266-67	17°27′N	101°31′E
Lofa, stm., Afr.	282-83	6°39′N	11°04′W
Loffa, stm., Afr. *see* Lofa	282-83	6°39′N	11°04′W
Lofoten, is., Nor. (lō′fō-tĕn)	200-01	68°08′N	14°10′E
Logan, N.M., U.S. (lō′gȧn)	130-31	35°22′N	103°25′W
Logan, Oh., U.S. (lō′gȧn)	126-27	39°32′N	82°25′W
Logan, Ut., U.S. (lō′gȧn)	122-23	41°45′N	111°50′W
Logan, W.V., U.S. (lō′gȧn)	126-27	37°51′N	81°60′W
Logan, Mount, mtn., Yk., Can.			
(mount lō′gȧn)	140-41	60°34′N	140°24′W
Logansport, In., U.S. (lō′gănz-pōrt)	126-27	40°45′N	86°21′W
Logone, stm., Afr. (lō-gō′nā) (lô-gôn′)	282-83	12°05′N	15°02′E
Logroño, Spain (lô-grō′nyō)	214-15	42°28′N	2°27′W
Løgstør, Den. (lügh-stûr′)	208-09	56°58′N	9°16′E
Loi-kaw, Mya.	266-67	19°40′N	97°13′E
Loire, stm., Fr. (lwâr)	212-13	47°18′N	2°00′W
Loja, Ec. (lō′hä)	184	3°59′S	79°12′W
Loja, Spain (lō′-kä)	214-15	37°10′N	4°09′W
Lokoro, stm., D.R.C.	284-85	1°43′S	18°22′E
Lol, stm., S. Sudan (lōl)	284-85	9°13′N	28°59′E
Lolland, i., Den. (lôl′än′)	208-09	54°47′N	11°16′E
Lom, Blg. (lōm)	216-17	43°50′N	23°15′E
Lom, stm., Afr.	282-83	5°19′N	13°24′E
Lomami, stm., D.R.C.	284-85	0°47′N	24°17′E
Lomas de Zamora, Arg.			
(lō′mäs dā zä-mō′rä)	187	34°46′S	58°24′W
Lomblen, Pulau, i., Indon.			
(pōō-lou lōm-blĕn′)	268-69	8°25′S	123°30′E
Lombok, i., Indon. (lŏm-bŏk′)	268-69	8°45′S	116°30′E
Lomé, nat. cap., Togo (lō′mē)	282-83	6°08′N	1°13′E
Lomela, stm., D.R.C. (lō-mā′lȧ)	284-85	0°18′S	20°45′E
Lomond, Loch, lk., Scot., U.K.			
(lôk lō′mŭnd)	206-07	56°06′N	4°37′W
Lomonosov, Russia (lô-mô′nô-sof)	208-09	59°55′N	29°48′E
Lompoc, Ca., U.S. (lŏm-pŏk′)	128-29	34°39′N	120°27′W
Łomża, Pol. (lôm′zhȧ)	210-11	53°11′N	22°05′E
Lonaconing, Md., U.S. (lō-nȧ-kō′nĭng)	126-27	39°34′N	78°58′W
Loncoche, Chile	185	39°22′S	72°38′W
London, On., Can. (lŭn′dŭn)	146-47	42°59′N	81°14′W
London, Ky., U.S. (lŭn′dŭn)	134-35	37°07′N	84°05′W
London, Oh., U.S. (lŭn′dŭn)	126-27	39°53′N	83°27′W
London, nat. cap., Eng., U.K. (lŭn′dŭn)	206-07	51°30′N	0°10′W
Londonderry, N. Ire., U.K.	206-07	54°59′N	7°20′W
Londonderry, Cape, c., Austl.	296-97	13°45′S	126°56′E
Londonderry, Isla, i., Chile	185	55°03′S	70°35′W
Londrina, Braz. (lôn-drē′nä)	182-83	23°19′S	51°10′W
Long Beach, Ca., U.S. (lông bēch)	128-29	33°46′N	118°12′W
Long Beach, Ms., U.S. (lông bēch)	134-35	30°21′N	89°10′W
Long Beach, Wa., U.S. (lông bēch)	122-23	46°22′N	124°03′W
Long Branch, N.J., U.S. (lông brȧnch)	126-27	40°18′N	73°60′W
Long Cay, i., Bah. (lông kē)	154-55	22°35′N	74°22′W
Longchang, China	258-59	29°21′N	105°17′E
Long Eaton, Eng., U.K. (lông ē′tŭn)	206-07	52°53′N	1°16′W
Longford, Ire. (lŏng′fērd)	206-07	53°44′N	7°48′W
Long Island, i., Bah.	154-55	23°15′N	75°07′W
Long Island, i., N.S., Can. (lông ī′lȧnd)	148-49	44°20′N	66°16′W
Long Island, i., Nu., Can.	140-41	54°52′N	79°21′W
Long Island, i., Pap. N. Gui.			
(lông ī′lȧnd) *see* Arop Island	302	5°20′S	147°05′E
Long Island, i., N.Y., U.S. (lông ī′lȧnd)	126-27	40°47′N	73°17′W
Long Island Sound, strt., U.S.			
(lông ī′lȧnd sound)	126-27	41°05′N	72°58′W
Longjiang, China	260-61	47°20′N	123°11′E
Longkou, China (lôŋ-kō)	260-61	37°39′N	120°21′E
Long Lake, lk., On., Can. (lông lāk)	146-47	49°30′N	86°50′W
Longli, China	258-59	26°28′N	106°58′E
Longmont, Co., U.S. (lông′mŏnt)	130-31	40°10′N	105°06′W
Longnawan, Indon.	268-69	1°48′N	114°53′E
Long Point, c., Nf./L., Can. (lông point)	148-49	48°47′N	58°47′W
Long Point Bay, b., On., Can.			
(lông point bā)	146-47	42°40′N	80°14′W
Longquan, China	258-59	28°03′N	119°06′E
Long Range Mountains, mts., Nf./L., Can.			
(lông rānj moun′tĭnz)	148-49	49°20′N	57°39′W
Longreach, Austl. (lông′rēch)	302	23°27′S	144°15′E
Longs Peak, mtn., Co., U.S. (lôngz pēk)	122-23	40°18′N	105°37′W
Longueuil, Qc., Can. (lôɴ-gû′y′)	146-47	45°32′N	73°30′W
Longview, Tx., U.S. (lông-vū)	130-31	32°30′N	94°44′W
Longview, Wa., U.S. (lông-vū)	122-23	46°08′N	122°57′W
Longwy, Fr. (lôɴ-wē′)	212-13	49°31′N	5°47′E
Longxi, China (lôŋ-shyē)	258-59	34°57′N	104°42′E
Long Xuyen, Viet. (loung′ sōō′yĕn)	266-67	10°23′N	105°26′E
Longzhou, China (lôŋ-jō)	258-59	22°21′N	106°51′E
Lonoke, Ar., U.S. (lō′nōk)	134-35	34°48′N	91°54′W
Lons-le-Saunier, Fr. (lôɴ-lē-sō-nyā′)	212-13	46°41′N	5°32′E
Lookout, Cape, c., N.C., U.S.			
(kāp căp lòk′out)	134-35	34°36′N	76°32′W
Loop Head, c., Ire. (lōōp hĕd)	206-07	52°34′N	9°56′W
Lopatka, Mys, c., Russia			
(mĭs lô-pät′kȧ)	236-37	50°53′N	156°40′E
Lop Buri, Thai.	266-67	14°48′N	100°37′E
Lopévi, i., Vanuatu	306g	16°31′S	168°20′E
Lopez, Cap, c., Gabon	282-83	0°38′S	8°42′E
Lop Nor, lk., China *see* Lop Nur	240-41	40°29′N	90°16′E
Lop Nur, lk., China	240-41	40°29′N	90°16′E
Lopori, stm., D.R.C. (lō-pō′rē)	284-85	1°14′N	19°49′E
Lora, Hāmūn-i-, lk., Asia	252-53	29°17′N	64°47′E
Lorain, Oh., U.S. (lô-rān′)	126-27	41°28′N	82°11′W
Loralai, Pak. (lō-rŭ-lī′)	252-53	30°22′N	68°36′E
Lorca, Spain (lôr′kä)	214-15	37°41′N	1°41′W
Lord Howe Island, i., Austl.			
(lôrd hou ī′lȧnd)	296-97	31°34′S	159°06′E
Lordsburg, N.M., U.S. (lôrdz′bûrg)	128-29	32°21′N	108°42′W
Loreto, Mex.	156-57	26°01′N	111°21′W
Lorica, Col. (lô-rē′kä)	178-79	9°14′N	75°49′W
Lorient, Fr. (lô-rē′äɴ′)	212-13	47°45′N	3°22′W
Lörrach, Ger. (lûr′äk)	210-11	47°37′N	7°40′E
Lorraine, hist. reg., Fr.	212-13	49°0′N	6°00′E
Los Alamos, N.M., U.S. (lōs ăl-ȧ-mōs′)	128-29	35°53′N	106°18′W
Los Andes, Chile (lôs ȧn′dĕs)	188e	32°50′S	70°36′W
Los Angeles, Chile (lôs äŋ′hå-lās)	185	37°27′S	72°19′W
Los Angeles, Ca., U.S.			
(lōs ăŋ′gĕl-s)	128-29	34°03′N	118°14′W
Losap Atoll, at., Micron.	304-05	6°58′N	152°39′E
Los Gatos, Ca., U.S. (lōs gä′tōs)	128-29	37°13′N	121°59′W
Los Lagos, Chile	185	39°52′S	72°48′W
Los Mochis, Mex.	156-57	25°47′N	108°60′W
Los Roques, Islas, is., Ven.	178-79	11°50′N	66°45′W
Los Teques, Ven. (lōs tĕ′kĕs)	188b	10°21′N	67°02′W
Lost River Range, mts., Id., U.S.			
(lŏst rĭv′ēr rānj)	122-23	44°10′N	113°35′W
Losuia, Pap. N. Gui.	302	8°33′S	151°03′E
Los Vilos, Chile (lôs vē′lôs)	182-83	31°53′S	71°29′W
Lota, Chile (lō′tä)	185	37°05′S	73°09′W
Lothringen, hist. reg., Fr. *see* Lorraine	212-13	49°0′N	6°00′E
Lotung, Tai.	243a	24°41′N	121°46′E
Louangphrabang, Laos			
(lōō-ang′prä-bäng′)	266-67	19°52′N	102°08′E
Loubomo, Congo	284-85	4°11′S	12°40′E
Loudon, Tn., U.S. (lou′dŭn)	134-35	35°44′N	84°21′W
Louga, Sen.	282-83	15°37′N	16°13′W
Louisa, Ky., U.S. (lōō′ēz-ȧ)	126-27	38°06′N	82°37′W
Louisa, Va., U.S. (lōō′ēz-ȧ)	126-27	38°01′N	78°00′W
Louise Island, i., B.C., Can.	142-43	52°58′N	131°50′W
Louisiade Archipelago, is., Pap. N. Gui.	302	11°0′S	153°00′E
Louisiana, Mo., U.S. (lōō-ē-zē-ăn′ȧ)	130-31	39°27′N	91°04′W
Louisiana, state, U.S. (lōō-ē-zē-ăn′ȧ)	118-19	31°15′N	92°15′W
Louis Trichardt, S. Afr. (lōō′ĭs trĭchårt)			
see Makhado	286-87	23°03′S	29°55′E
Louisville, Ga., U.S. (lōō′ē-vĭl)	134-35	32°60′N	82°24′W
Louisville, Il., U.S. (lōō′ē-vĭl)	126-27	38°46′N	88°31′W
Louisville, Ky., U.S. (lōō′ē-vĭl)	126-27	38°15′N	85°46′W
Louisville, Ms., U.S. (lōō′ē-vĭl)	134-35	33°07′N	89°03′W
Louis-XIV, Pointe, c., Qc., Can.	140-41	54°38′N	79°45′W
Loukhi, Russia	202-03	66°04′N	33°03′E
Louny, Czech Rep. (lō′nė)	210-11	50°21′N	13°48′E
Lourdes, Fr. (lōōrd)	212-13	43°06′N	0°03′W
Louviers, Fr. (lōō-vyā′)	212-13	49°13′N	1°10′E
Lovat′, stm., Russia	208-09	58°13′N	31°27′E
Lovech, Blg. (lō′vĕts)	216-17	43°08′N	24°43′E
Loveland, Co., U.S. (lŭv′lănd)	130-31	40°25′N	105°05′W
Lovell, Wy., U.S. (lŭv′ĕl)	122-23	44°50′N	108°24′W
Lovelock, Nv., U.S. (lŭv′lŏk)	128-29	40°11′N	118°28′W
Loviisa, Fin. (lô′vē-sȧ)	208-09	60°27′N	26°13′E
Lovisa, Fin. *see* Loviisa	208-09	60°27′N	26°13′E
Low, Cape, c., Nu., Can. (kāp lō)	140-41	63°07′N	85°18′W
Lowa, stm., D.R.C. (lō′wȧ)	284-85	1°25′S	25°52′E
Lowell, Ma., U.S.	126-27	42°38′N	71°19′W
Lower Brule Indian Reservation,			
ind. res., S.D., U.S. (lō′ẽr brü′lā			
ĭn′dĭ-ȧn rĕ-sẽr-vā′shĕn)	124-25	44°05′N	99°54′W
Lower California, pen., Mex.			
see Baja California	156-57	27°53′N	113°28′W
Lower Hutt, N.Z. (lō′ẽr hŭt)	303	41°13′S	174°56′E
Lower Post, B.C., Can.	138-39	59°56′N	128°29′W
Lower Red Lake, lk., Mn., U.S.			
(lō′ẽr rĕd lāk)	124-25	47°57′N	95°01′W
Lower Zambezi National Park,			
n.p., Zam.	286-87	15°32′S	29°56′E
Lowestoft, Eng., U.K. (lō′stŏf)	206-07	52°29′N	1°44′E
Łowicz, Pol. (lô′vĭch)	210-11	52°06′N	19°57′E
Loxton, Austl. (lôks′tŭn)	301	34°27′S	140°34′E
Loyalty Islands, is., N. Cal.	306g	21°0′S	167°00′E
Loyauté, Îles, is., N. Cal.			
see Loyalty Islands	306g	21°0′S	167°00′E
Loznica, Serb. (lôz′nē-tsȧ)	216-17	44°32′N	19°14′E
Lualaba, stm., D.R.C. (lōō-ȧ-lä′bȧ)	284-85	0°22′N	25°21′E
Luama, stm., D.R.C. (lōō′ä-mȧ)	284-85	4°46′S	26°53′E
Lu′an, China (lōō-än)	258-59	31°44′N	116°29′E
Luan, stm., China	260-61	39°24′N	119°17′E
Luanda, nat. cap., Ang. (lōō-än′dä)	286-87	8°49′S	13°14′E
Luando, stm., Ang.	286-87	10°21′S	16°27′E
Luanginga, stm., Afr.			
see Luanguinga	286-87	15°12′S	22°55′E
Luang Prabang, Laos			
see Louangphrabang	266-67	19°52′N	102°08′E
Luangue, stm., Afr.	284-85	4°17′S	20°02′E
Luanguinga, stm., Afr.			
(lōō-ä-gĭn′gȧ)	286-87	15°12′S	22°55′E
Luangwa, Zam.	286-87	15°37′S	30°25′E
Luanshya, Zam.	286-87	13°08′S	28°24′E
Luapula, stm., Afr.	286-87	9°24′S	28°31′E
Lubaczów, Pol. (lōō-bä′chóf)	210-11	50°10′N	23°07′E
Lubań, Pol. (lōō′bän′)	210-11	51°07′N	15°18′E
Lubang, Phil. (lōō-bäng′)	270	13°51′N	120°07′E
Lubang Island, i., Phil.	270	13°46′N	120°11′E
Lubang Islands, is., Phil.			
(lōō-bäng′ ī′lȧndz)	270	13°45′N	120°17′E
Lubango, Ang.	286-87	14°55′S	13°30′E
Lubāns, lk., Lat. (lōō-bä′nȧs)	208-09	56°46′N	26°53′E
Lubartów, Pol. (lōō-bär′tòf)	210-11	51°28′N	22°37′E
Lubbock, Tx., U.S. (lŭb′ŭk)	130-31	33°34′N	101°51′W
Lübeck, Ger. (lü′bĕk)	210-11	53°52′N	10°40′E

n-sing; ŋ-baŋk; ɴ-nasalized n; nŏd; cŏmmit; ōld; ôbey; ôrder; oi-boil; fōōd; ò-as oo in foot; ou-out; s-soft; sh-dish; th-thin; pūre; ůnite; ûrn; stŭd; circŭs; ü-as in French tu; ′-indeterminate vowel.

378 Lubiana – Magallanes, Estrecho de PRONOUNCING INDEX

Feature (Pronunciation)	Page	Lat.	Long.
Lubiana, nat. cap., Slvn. *see* Ljubljana	216-17	46°03'N	14°31'E
Lubilash, stm., D.R.C. (lo͞o-bē-lȧsh´)	284-85	6°03'S	23°45'E
Lublin, Pol. (lyö´blēn´)	210-11	51°14'N	22°35'E
Lubnān, nation, Asia *see* Lebanon	224-25	34°0'N	36°00'E
Lubny, Ukr. (lòb´nê)	218-19	50°01'N	33°00'E
Lubuagan, Phil. (lò-bwä-gä´n)	270	17°22'N	121°10'E
Lubudi, D.R.C. (lò-bò´dē)	284-85	9°57'S	25°58'E
Lubudi, stm., D.R.C. (lò-bò´dē)	284-85	4°02'S	21°23'E
Lubudi, stm., D.R.C. (lò-bò´dē)	284-85	9°13'S	25°38'E
Lubumbashi, D.R.C. (lo͞o-bŭm-bä´shē)	284-85	11°41'S	27°28'E
Lucca, Italy (lo͞ok´kä)	216-17	43°51'N	10°30'E
Lucena, Phil. (lo͞o-sā´nä)	270	13°56'N	121°37'E
Lucena, Spain (lo͞o-thā´nä)	214-15	37°24'N	4°29'W
Lučenec, Slvk. (lo͞o´châ-nyĕts)	210-11	48°20'N	19°40'E
Lucera, Italy (lo͞o-châ´rä)	216-17	41°31'N	15°20'E
Lucerne, Switz. *see* Luzern	210-11	47°03'N	8°19'E
Lucipara, Kepulauan, is., Indon.	268-69	5°33'S	127°27'E
Lucknow, India (lŭk´nou)	254-55	26°52'N	80°55'E
Luçon, Fr. (lü-sôɴ´)	212-13	46°27'N	1°10'W
Lüda, China *see* Dalian	260-61	38°54'N	121°34'E
Lüderitz, Nmb. (lü´dēr-ĭts) (lü´dĕ-rĭts)	286-87	26°39'S	15°09'E
Ludhiāna, India	254-55	30°54'N	75°51'E
Ludington, Mi., U.S. (lŭd´ĭng-tŭn)	126-27	43°57'N	86°26'W
Ludlow, Eng., U.K. (lŭd´lō)	206-07	52°22'N	2°43'W
Ludvika, Swe. (loodh-vē´kȧ)	208-09	60°08'N	15°11'E
Ludza, Lat. (lo͞od´zȧ)	208-09	56°32'N	27°44'E
Luena, Ang.	286-87	11°47'S	19°54'E
Luena, stm., Ang.	286-87	12°30'S	22°34'E
Lufeng, China	258-59	22°56'N	115°37'E
Lufira, stm., D.R.C. (lo͞o-fē´rȧ)	284-85	8°21'S	26°26'E
Lufkin, Tx., U.S. (lŭf´kĭn)	132-33	31°20'N	94°43'W
Luga, Russia (lo͞o´gȧ)	208-09	58°44'N	29°52'E
Luga, stm., Russia (lo͞o´gȧ)	208-09	59°40'N	28°18'E
Lugano, Switz. (lo͞o-gä´nō)	210-11	46°01'N	8°57'E
Lugenda, stm., Moz.	286-87	11°25'S	38°29'E
Lugo, Italy (lo͞o´gô)	216-17	44°25'N	11°55'E
Lugo, Spain (lo͞o´gō)	214-15	43°01'N	7°33'W
Luhans'k, Ukr.	218-19	48°34'N	39°20'E
Luik, Bel. *see* Liège	206-07	50°38'N	5°34'E
Luimneach, Ire. *see* Limerick	206-07	52°40'N	8°38'W
Lukanga Swamp, sw., Zam. (lo͞o-käŋ´gȧ swŏmp)	286-87	14°25'S	27°45'E
Lukenie, stm., D.R.C. (lo͞o-kā´ynȧ)	284-85	2°44'S	18°10'E
Lukolela, D.R.C.	284-85	1°04'S	17°11'E
Łuków, Pol. (wó´kóf)	210-11	51°56'N	22°23'E
Lukuga, stm., D.R.C. (lo͞o-ko͞o´gȧ)	284-85	5°40'S	26°55'E
Lukula, D.R.C.	284-85	5°22'S	12°57'E
Lulaka, stm., D.R.C.	284-85	0°53'S	20°11'E
Luleå, Swe.	200-01	65°36'N	22°10'E
Lüleburgaz, Tur. (lü´lĕ-bòr-gäs´)	216-17	41°25'N	27°22'E
Lüliang Shan, mts., China	260-61	37°25'N	111°20'E
Luling, Tx., U.S. (lū´lĭng)	132-33	29°41'N	97°39'W
Lulonga, stm., D.R.C.	284-85	0°38'N	18°21'E
Lulua, stm., D.R.C.	284-85	5°02'S	21°06'E
Lumberton, Ms., U.S. (lŭm´bēr-tŭn)	134-35	31°00'N	89°30'W
Lumberton, N.C., U.S. (lŭm´bēr-tŭn)	134-35	34°38'N	79°01'W
Lumberton, Tx., U.S. (lŭm´bēr-tŭn)	132-33	30°14'N	94°12'W
Lund, Swe. (lŭnd)	208-09	55°42'N	13°11'E
Lüneburg, Ger. (lü´nē-bòrgh)	210-11	53°16'N	10°25'E
Lunel, Fr. (lü-nĕl´)	212-13	43°40'N	4°08'E
Lunenburg, N.S., Can. (lo͞o´nĕn-bûrg)	148-49	44°23'N	64°19'W
Lunéville, Fr. (lü-nå-vel´)	212-13	48°36'N	6°30'E
Lunga, stm., Zam.	286-87	14°34'S	26°26'E
Lūni, stm., India	254-55	24°37'N	71°17'E
Lunsar, S.L.	282-83	8°41'N	12°32'W
Luo, stm., China	258-59	34°41'N	110°08'E
Luoding, China (lwô-dĭŋ)	258-59	22°47'N	111°33'E
Luohe, China (lwô-hŭ)	258-59	33°34'N	114°02'E
Luoyang, China (lwô-yäŋ)	258-59	34°41'N	112°27'E
Luqu, China	258-59	34°38'N	102°14'E
Luray, Va., U.S. (lū-rā´)	126-27	38°39'N	78°28'W
Lurgan, N. Ire., U.K. (lûr´gǎn)	206-07	54°28'N	6°20'W
Lurín, Peru	188a	12°17'S	76°52'W
Lúrio, stm., Moz.	286-87	13°30'S	40°32'E
Lusaka, nat. cap., Zam. (lò-sä´kȧ)	286-87	15°24'S	28°17'E
Lusambo, D.R.C. (lo͞o-säm´bō)	284-85	4°57'S	23°30'E
Lushan, China	258-59	30°15'N	102°58'E
Lu Shan, mtn., China	258-59	29°31'N	115°58'E
Lüshun, China (lü-shŭn)	260-61	38°49'N	121°15'E
Lusk, Wy., U.S. (lŭsk)	124-25	42°46'N	104°27'W
Lūt, Dasht-e, des., Iran (dă´sht-ē-lo͞ot)	252-53	32°0'N	58°00'E
Lutherstadt Wittenberg, Ger.	210-11	51°52'N	12°39'E
Luton, Eng., U.K. (lū´tŭn)	206-07	51°53'N	0°25'W
Lutong, Malay.	268-69	4°28'N	113°60'E
Luts'k, Ukr.	210-11	50°45'N	25°20'E
Lutzow-Holm Bay, b., Ant.	313	69°10'S	37°30'E
Luverne, Al., U.S. (lū-vûn´)	134-35	31°43'N	86°16'W
Luverne, Mn., U.S. (lū-vûn´)	124-25	43°39'N	96°13'W
Luvua, stm., D.R.C.	284-85	6°45'S	26°57'E
Luwegu, stm., Tan.	284-85	8°31'S	37°23'E
Luwuk, Indon.	268-69	0°56'S	122°47'E
Luxembourg, nation, Eur. *see* Luxemburg	206-07	49°45'N	6°05'E
Luxembourg, nat. cap., Lux.	206-07	49°37'N	6°07'E
Luxemburg, nation, Eur. (lŭk´-sŭm-bûrg)	206-07	49°45'N	6°05'E
Luxi, China	258-59	24°21'N	98°23'E
Luxor, Egypt	291b	25°42'N	32°39'E
Luza, Russia	202-03	60°37'N	47°16'E
Luzern, Switz. (lò-tsĕrn)	210-11	47°03'N	8°19'E
Luzhou, China (lo͞o-jō)	258-59	28°53'N	105°27'E
Luziânia, Braz. (lo͞o-zyȧ´nēȧ)	182-83	16°15'S	47°55'W
Lužická Nisa, stm., Eur. *see* Neisse	210-11	52°04'N	14°46'E
Luzon, i., Phil. (lo͞o-zŏn´)	270	16°0'N	121°00'E
Luzon Strait, strt., Asia (lo͞o-zŏn´ strāt)	258-59	20°30'N	121°00'E
L'viv, Ukr.	210-11	49°51'N	24°02'E
Lwów, Ukr. *see* L'viv	210-11	49°51'N	24°02'E
Lyallpur, Pak. *see* Faisalābād	252-53	31°25'N	73°05'E
Lycksele, Swe.	200-01	64°36'N	18°41'E
Lydenburg, S. Afr. (lī´dĕn-bûrg)	292c	25°08'S	30°27'E
Lykens, Pa., U.S. (lī´kĕnz)	126-27	40°34'N	76°43'W
Lynchburg, Va., U.S. (lĭnch´bûrg)	134-35	37°25'N	79°09'W
Lyndonville, Vt., U.S. (lĭn´dŭn-vĭl)	126-27	44°32'N	72°00'W
Lynn, Ma., U.S. (lĭn)	126-27	42°28'N	70°57'W
Lynn Lake, Mb., Can. (lĭn lāk)	144-45	56°51'N	101°00'W
Lyon, Fr. (lē-ôn´)	212-13	45°45'N	4°49'E
Lyons, Ga., U.S. (lī´ŭnz)	134-35	32°12'N	82°19'W
Lyons, Ks., U.S. (lī´ŭnz)	130-31	38°21'N	98°12'W
Lyons, Ne., U.S. (lī´ŭnz)	124-25	41°56'N	96°29'W
Lysekil, Swe. (lü´sĕ-kêl)	208-09	58°17'N	11°27'E
Lys'va, Russia (lĭs´vȧ)	202-03	58°06'N	57°48'E
Lysychans'k, Ukr.	218-19	48°55'N	38°26'E
Lyuban', Russia (lyo͞o´bȧn)	208-09	59°21'N	31°15'E
Lyubertsy, Russia (lyo͞o´bĕr-tsē)	218-19	55°41'N	37°53'E
Lyudinovo, Russia (lū-dē´novō)	218-19	53°52'N	34°28'E

M

Feature (Pronunciation)	Page	Lat.	Long.
Ma, stm., Asia	266-67	19°47'N	105°52'E
Ma'ān, Jord. (mä-än´)	248-49	30°12'N	35°44'E
Ma'anshan, China	258-59	31°42'N	118°30'E
Maastricht, Neth. (mäs´trĭĸt)	206-07	50°52'N	5°42'E
Mabank, Tx., U.S. (mā´bănk)	130-31	32°23'N	96°06'W
Macaé, Braz.	186	22°24'S	41°47'W
MacAlpine Lake, lk., Nu., Can.	140-41	66°38'N	102°51'W
Macapá, Braz.	180-81	0°03'N	51°03'W
Macará, Ec.	184	4°22'S	79°56'W
Macau, Braz. (mä-kȧ´ò)	188d	5°07'S	36°38'W
Macclesfield, Eng., U.K. (măk´'lz-fēld)	206-07	53°16'N	2°08'W
MacDonnell Ranges, mts., Austl. (măk-dŏn´ĕl rānjĕz)	296-97	23°52'S	133°14'E
MacDowell Lake, lk., On., Can. (măk-dou ĕl lāk)	144-45	52°15'N	92°45'W
Macdui, Ben, mtn., Scot., U.K. (bĕn măk-do͞o´ē)	206-07	57°05'N	3°39'W
Macedonia, nation, Eur. (măs-ē-dō´nĭ-ȧ)	190-91	41°50'N	22°00'E
Macedonia, hist. reg., Eur. (măs-ē-dō´nĭ-ȧ)	216-17	41°0'N	23°00'E
Maceió, Braz.	188d	9°40'S	35°43'W
Macerata, Italy (mä-chå-rä´tä)	216-17	43°18'N	13°27'E
Macfarlane, Lake, lk., Austl. (lāk măc´fär-lān)	301	31°58'S	136°43'E
Machado, stm., Braz.	180-81	8°02'S	62°53'W
Machagai, Arg.	187	26°56'S	60°02'W
Machakos, Kenya	289	1°31'S	37°16'E
Machala, Ec. (mä-chá´lä)	184	3°16'S	79°57'W
Machilīpatnam, India	256	16°11'N	81°09'E
Machiques, Ven.	178-79	10°04'N	72°32'W
Machu Picchu, hist., Peru	184	13°07'S	72°34'W
Măcin, Rom. (mȧ-chēn´)	218-19	45°15'N	28°08'E
Mackay, Austl. (mă-kī´)	302	21°10'S	149°12'E
MacKay Lake, lk., N.T., Can. (măk-kā´ lāk)	140-41	63°54'N	110°23'W
Mackenzie, stm., N.T., Can. (mȧ-kĕn´zĭ)	140-41	58°60'N	111°25'W
Mackenzie Bay, b., Can. (mȧ-kĕn´zĭ bā)	136	69°0'N	136°30'W
Mackenzie Mountains, mts., Can. (mȧ-kĕn´zĭ moun´tĭnz)	140-41	64°0'N	130°00'W
Mackinaw City, Mi., U.S. (măk´ĭ-nô sĭ´tē)	126-27	45°46'N	84°43'W
Maclean, Austl.	301	29°28'S	153°13'E
Macleod, Lake, lk., Austl.	296-97	24°04'S	113°42'E
Macomb, Il., U.S. (mȧ-ko͞om´)	130-31	40°28'N	90°40'W
Mâcon, Fr. (mä-kôɴ)	212-13	46°19'N	4°50'E
Macon, Ga., U.S. (mā´kŏn)	134-35	32°50'N	83°38'W
Macon, Mo., U.S. (mā´kŏn)	130-31	39°44'N	92°28'W
Macon, Ms., U.S. (mā´kŏn)	134-35	33°06'N	88°34'W
Macquarie, stm., Austl. (mȧ-kwŏr´ē)	301	30°08'S	147°23'E
Mada, stm., Nig.	282-83	7°59'N	7°58'E
Madagascar, nation, Afr. (măd-ȧ-găs´kȧr)	274	19°0'S	46°00'E
Madagasikara, nation, Afr. *see* Madagascar	274	19°0'S	46°00'E
Madame, Isle, i., N.S., Can. (īl mȧ-dȧm´)	148-49	45°33'N	61°02'W
Madang, Pap. N. Gui. (mä-däng´)	302	5°17'S	145°45'E
Madawaska, stm., On., Can. (măd-ȧ-wôs´kȧ)	146-47	45°27'N	76°21'W
Madeira, i., Port. (mä-dā´rä)	280-81	32°44'N	17°00'W
Madeira, stm., S.A. (mä-dā´-rȧ)	180-81	3°22'S	58°45'W
Madeira, Arquipélago da, is., Port. (är-kē-pĕ´lä-gō-dä-mädĕy´-rä) *see* Madeira Islands	280-81	32°40'N	16°45'W
Madeira Islands, is., Port. (mä-dā´rä ī´lȧndz)	280-81	32°40'N	16°45'W
Madeleine, Îles de la, is., Qc., Can.	148-49	47°30'N	61°45'W
Madelia, Mn., U.S. (mȧ-dē´lĭ-ȧ)	124-25	44°03'N	94°25'W
Madhya Pradesh, state, India (mŭd´vŭ prŭ-dāsh´)	254-55	23°0'N	79°00'E
Madidi, stm., Bol.	180-81	12°31'S	66°58'W
Madikeri, India	256	12°25'N	75°45'E
Madill, Ok., U.S. (mȧ-dĭl´)	130-31	34°05'N	96°47'W
Madison, Al., U.S. (măd´ĭ-sŭn)	134-35	34°41'N	86°45'W
Madison, Fl., U.S. (măd´ĭ-sŭn)	134-35	30°28'N	83°25'W
Madison, Ga., U.S. (măd´ĭ-sŭn)	134-35	33°36'N	83°28'W
Madison, In., U.S. (măd´ĭ-sŭn)	126-27	38°44'N	85°23'W
Madison, Me., U.S. (măd´ĭ-sŭn)	126-27	44°48'N	69°53'W
Madison, Mn., U.S. (măd´ĭ-sŭn)	124-25	45°01'N	96°12'W
Madison, Ms., U.S. (măd´ĭ-sŭn)	134-35	32°28'N	90°07'W
Madison, N.C., U.S. (măd´ĭ-sŭn)	134-35	36°23'N	79°58'W
Madison, Ne., U.S. (măd´ĭ-sŭn)	124-25	41°50'N	97°27'W
Madison, S.D., U.S. (măd´ĭ-sŭn)	124-25	44°00'N	97°07'W
Madison, Wi., U.S. (măd´ĭ-sŭn)	126-27	43°05'N	89°22'W
Madison, W.V., U.S. (măd´ĭ-sŭn)	126-27	38°04'N	81°49'W
Madisonville, Ky., U.S. (măd´ĭ-sŭn-vĭl)	134-35	37°20'N	87°30'W
Madisonville, Tx., U.S. (măd´ĭ-sŭn-vĭl)	132-33	30°56'N	95°55'W
Madiun, Indon.	268-69	7°37'S	111°31'E
Madoi, China	260-61	34°55'N	98°12'E
Madona, Lat. (mȧ´dō´nȧ)	208-09	56°51'N	26°14'E
Madras, India *see* Chennai	256	13°06'N	80°15'E
Madras, state, India *see* Tamil Nādu	256	11°0'N	78°15'E
Madre, Laguna, b., Mex. (lä-ò´nä mä´drå)	132-33	25°01'N	97°40'W
Madre, Laguna, b., Tx., U.S.	132-33	26°58'N	97°26'W
Madre, Sierra, mts., Phil. (sē-ĕ´r-rä-mȧ´drĕ)	270	16°20'N	122°00'E
Madre de Dios, stm., S.A. (mä´drå dä dē-ōs´)	180-81	10°24'S	65°24'W
Madre de Dios, Isla, i., Chile (ē´s-lä-mȧ´drå dä dē-ōs´)	185	50°15'S	75°05'W
Madre del Sur, Sierra, mts., Mex. (sē-ĕ´r-rä-mä´drå dĕl-so͞or´)	158-59	17°0'N	100°00'W
Madre Occidental, Sierra, mts., Mex. (sē-ĕ´r-rä-mȧ´drĕ-äk-sĭ-dĕn´-tl)	156-57	25°0'N	105°00'W
Madre Oriental, Sierra, mts., Mex. (sē-ĕ´r-rä-mȧ´drĕ ō-rē-ĕn-täl´)	156-57	21°26'N	99°50'W
Madrid, nat. cap., Spain (mä-drē´d)	214-15	40°24'N	3°41'W
Madridejos, Spain (mä-dhrē-dhä´hōs)	214-15	39°28'N	3°32'W
Madurai, India (mä-do͞o´rä)	256	9°55'N	78°08'E
Maebashi, Japan (mä-ĕ-bä´shē)	265	36°23'N	139°05'E
Mae Hong Son, Thai.	266-67	19°16'N	97°57'E
Mae Klong, stm., Thai.	266-67	13°22'N	99°60'E
Mae Sot, Thai.	266-67	16°43'N	98°35'E
Maestra, Sierra, mts., Cuba (sē-ĕ´r-rȧ-mä-ās´trä)	154-55	20°06'N	76°24'W
Maéwo, i., Vanuatu	306g	15°10'S	168°10'E
Mafeking, S. Afr. (mȧf´ĕ´kĭng) *see* Mafikeng	286-87	25°53'S	25°39'E
Mafia Island, i., Tan.	284-85	7°50'S	39°50'E
Mafikeng, S. Afr.	286-87	25°53'S	25°39'E
Mafra, Braz. (mä´frä)	182-83	26°08'S	49°49'W
Mafra, Port. (mäf´rȧ)	214-15	38°57'N	9°19'W
Magadan, Russia (mȧ-gȧ-dän´)	236-37	59°35'N	150°50'E
Magallanes, Chile *see* Punta Arenas	185	53°09'S	70°55'W
Magallanes, Estrecho de, strt., S.A.	185	54°0'S	71°00'W

PRONOUNCING INDEX
Magangué – Mandsaur 379

Feature (Pronunciation)	Page	Lat.	Long.
Magangué, Col.	**178-79**	9°18'N	74°48'W
Magat, stm., Phil. (mä-gät′)	**270**	17°02'N	121°50'E
Magdagachi, Russia	**240-41**	53°27'N	125°49'E
Magdalena, Bol. (mäg-dä-lā′nä)	**180-81**	13°20'S	64°08'W
Magdalena, Mex. (mäg-dä-lā′nä)	**158-59**	20°54'N	103°57'W
Magdalena, N.M., U.S. (mäg-dä-lā′nä)	**128-29**	34°07'N	107°15'W
Magdalena, stm., Col. (mäg-dä-lā′nä)	**178-79**	11°06'N	74°51'W
Magdalena, Bahía, b., Mex.			
(bä-ē′ä-mäg-dä-lā′nä)	**156-57**	24°35'N	112°00'W
Magdalena, Isla, i., Chile			
(ē′s-lä-mäg-dä-lā′nä)	**185**	44°40'S	73°10'W
Magdalena de Kino, Mex.	**156-57**	30°38'N	110°58'W
Magdeburg, Ger. (mäg′dĕ-bòrgh)	**210-11**	52°08'N	11°38'E
Magelang, Indon.	**268-69**	7°28'S	110°13'E
Magellan, Strait of, strt., S.A.			
(strāt ŭv mȧ-gĕl′-ăn)			
see Magallanes, Estrecho de	**185**	54°0'S	71°00'W
Magerøya, i., Nor.	**200-01**	71°02'N	25°42'E
Magnesia, Tur. *see* Manisa	**216-17**	38°37'N	27°26'E
Magnetic Island, i., Austl.	**302**	19°08'S	146°50'E
Magnitogorsk, Russia			
(mäg-nyē′tô-gôrsk)	**244**	53°26'N	59°04'E
Magnolia, Ar., U.S. (măg-nō′lĭ-ȧ)	**130-31**	33°16'N	93°15'W
Magnolia, Ms., U.S. (măg-nō′lĭ-ȧ)	**134-35**	31°09'N	90°28'W
Mago, i., Fiji	**306f**	17°27'S	179°09'W
Magog, Qc., Can. (mȧ-gŏg′)	**146-47**	45°16'N	72°09'W
Magpie, stm., On., Can. (Măg′pī)	**146-47**	47°56'N	84°50'W
Magpie, stm., Qc., Can. (Măg′pī)	**148-49**	50°19'N	64°27'W
Magpie, Lac, lk., Qc., Can.			
(läk măg′pī)	**148-49**	51°0'N	64°41'W
Maguari, Cabo, c., Braz.	**180-81**	0°18'S	48°22'W
Magway, Mya.	**266-67**	20°30'N	94°30'E
Magyarország, nation, Eur.			
see Hungary	**190-91**	47°0'N	20°00'E
Mahābād, Iran	**248-49**	36°46'N	45°44'E
Mahagi, D.R.C.	**289**	2°19'N	31°01'E
Mahajanga, Madag.	**286-87**	15°43'S	46°19'E
Mahakam, stm., Indon.	**268-69**	0°35'S	117°17'E
Mahalapye, Bots.	**286-87**	23°06'S	26°50'E
Mahalla el-Kubra, Egypt			
see El-Mahalla el-Kubra	**291b**	30°58'N	31°10'E
Mahānadi, stm., India	**254-55**	20°19'N	86°47'E
Mahanoro, Madag. (mȧ-hȧ-nō′rō)	**286-87**	19°55'S	48°48'E
Mahārāshtra, state, India	**256**	19°0'N	76°00'E
Maha Sarakham, Thai.	**266-67**	16°11'N	103°10'E
Mahbūbnagar, India	**256**	16°44'N	77°59'E
Mahe, India (mä-ā′)	**256**	11°42'N	75°32'E
Mahébourg, Mauritius	**287a**	20°24'S	57°42'E
Mahendra Giri, mtn., India	**256**	18°58'N	84°21'E
Mahendranagar, Nepal	**254-55**	28°58'N	80°10'E
Mahenge, Tan. (mä-hĕŋ′gȧ)	**289**	7°38'S	36°16'E
Mahesāna, India	**254-55**	23°36'N	72°23'E
Mahilëu, Bela.	**208-09**	53°56'N	30°21'E
Mahnomen, Mn., U.S. (mô-nō′mĕn)	**124-25**	47°19'N	95°59'W
Mahón, Spain *see* Maó	**214-15**	39°53'N	4°16'E
Mahone Bay, b., N.S., Can.			
(mȧ-hōn′ bā)	**148-49**	44°30'N	64°15'W
Maicuru, stm., Braz.	**180-81**	2°12'S	54°18'W
Maiduguri, Nig. (mä′ē-dä-gōō′rē)	**282-83**	11°51'N	13°09'E
Maiko, stm., D.R.C.	**284-85**	0°11'N	25°32'E
Maikop, Russia *see* Maykop	**202-03**	44°36'N	40°06'E
Mai-Ndombe, Lac, lk., D.R.C.	**282-83**	2°25'S	18°18'E
Maine, state, U.S. (mān)	**118-19**	45°15'N	69°15'W
Maine, Gulf of, b., N.A.	**148-49**	43°0'N	68°00'W
Mainland, i., Scot., U.K. (mān-lănd)	**206c**	60°16'N	1°16'W
Maintenon, Fr. (măn-tē-nôN′)	**212-13**	48°35'N	1°35'E
Maintirano, Madag. (mä′ēn-tē-rä′nō)	**286-87**	18°03'S	44°02'E
Mainz, Ger. (mīnts)	**210-11**	50°00'N	8°16'E
Maio, i., C.V. (mä′yo)	**282-83**	15°11'N	23°10'W
Maipo, stm., Chile (mī′pȯ)	**188e**	33°37'S	71°38'W
Maipo, Volcán, vol., S.A.			
(vōl-kä′n mī′pȯ)	**188e**	34°10'S	69°50'W
Maipú, Arg.	**187**	36°52'S	57°54'W
Maiquetía, Ven. (mī-kĕ-tē′ä)	**188b**	10°36'N	66°58'W
Maitland, Austl. (māt′lănd)	**301**	32°44'S	151°33'E
Maitland, Austl. (māt′lănd)	**301**	34°23'S	137°40'E
Maíz, Islas del, is., Nic.	**161**	12°15'N	83°00'W
Maizuru, Japan (mä-ĭ′zōō-rōō)	**265**	35°28'N	135°24'E
Majene, Indon.	**268-69**	3°32'S	118°57'E
Majī, Eth.	**284-85**	6°11'N	35°35'E
Majorca, i., Spain (mȧ-jôr′-kȧ)			
see Mallorca	**214-15**	39°30'N	3°00'E
Majuro, at., Marsh. Is.	**304-05**	7°05'N	171°09'E
Makanya, Tan. (mä-kän′yä)	**289**	4°21'S	37°50'E
Makarov, Russia	**240-41**	48°38'N	142°46'E
Makarska, Cro. (mä′kär-skȧ)	**216-17**	43°17'N	17°01'E

Feature (Pronunciation)	Page	Lat.	Long.
Makasar, Selat, strt., Indon.			
see Makassar Strait	**268-69**	2°0'S	117°30'E
Makassar, Indon. (mȧ-kä′sûr)	**268-69**	5°08'S	119°25'E
Makassar Strait, strt., Indon.			
(mȧ-kä′-sûr strāt)	**268-69**	2°0'S	117°30'E
Makatea, i., Fr. Poly.	**304-05**	15°50'S	148°16'W
Makedonija, nation, Eur.			
see Macedonia	**190-91**	41°50'N	22°00'E
Makedonija, hist. reg., Eur.			
see Macedonia	**216-17**	41°0'N	23°00'E
Makeni, S.L.	**282-83**	8°53'N	12°03'W
Makeyevka, Ukr. *see* Makiïvka	**218-19**	48°02'N	37°58'E
Makgadikgadi, pl., Bots.	**286-87**	20°17'S	25°43'E
Makhachkala, Russia (mäκ′äch-kä′lä)	**245**	42°59'N	47°30'E
Makhado, S. Afr.	**286-87**	23°03'S	29°55'E
Makiïvka, Ukr.	**218-19**	48°02'N	37°58'E
Makindu, Kenya	**289**	2°16'S	37°50'E
Makinsk, Kaz.	**244**	52°39'N	70°25'E
Makkah, Sau. Ar. *see* Mecca	**288**	21°27'N	39°51'E
Makokou, Gabon (mȧ-kȯ-kōō′)	**282-83**	0°35'N	12°51'E
Makona, stm., Afr. *see* Moa	**282-83**	6°60'N	11°34'W
Makoua, Congo	**284-85**	0°00'S	15°38'E
Makung, Tai.	**243a**	23°34'N	119°34'E
Makurdi, Nig.	**282-83**	7°44'N	8°31'E
Mala, Punta, c., Pan. (pò′n-tä-mä′lä)	**162**	7°28'N	80°01'W
Malabang, Phil.	**270**	7°38'N	124°04'E
Malabar Coast, cst., India			
(măl′ȧ-bär kōst)	**256**	11°0'N	75°00'E
Malabo, nat. cap., Eq. Gui. (mä-lă′bō)	**282-83**	3°45'N	8°47'E
Malacca, Malay. *see* Kota Kinabalu	**266-67**	2°12'N	102°16'E
Malacca, Strait of, strt., Asia			
(strāt ŭv mȧ-lăk′ȧ)	**266-67**	2°30'N	101°20'E
Malad City, Id., U.S. (mȧ-läd′ sĭ′tē)	**122-23**	42°12'N	112°15'W
Maladzečna, Bela.	**210-11**	54°19'N	26°52'E
Málaga, Col. (mȧ′lä-gȧ)	**178-79**	6°42'N	72°44'W
Málaga, Spain (mȧ′lä-gȧ)	**214-15**	36°44'N	4°25'W
Malagasy Republic, nation, Afr.			
see Madagascar	**274**	19°0'S	46°00'E
Malaita, i., Sol. Is. (mȧ-lä′ē-tȧ)	**306e**	9°0'S	161°00'E
Malaka, Malay. *see* Kota Kinabalu	**266-67**	2°12'N	102°16'E
Malaka, Selat, strt., Asia			
see Malacca, Strait of	**266-67**	2°30'N	101°20'E
Malakāl, S. Sudan (mȧ-lȧ-käl′)	**284-85**	9°31'N	31°39'E
Malakula, i., Vanuatu (mä-lä-kōō′lä)	**306g**	16°15'S	167°30'E
Malang, Indon.	**268-69**	7°59'S	112°38'E
Malanje, Ang. (mä-läŋ-gȧ)	**286-87**	9°32'S	16°20'E
Malanville, Benin	**282-83**	11°52'N	3°23'E
Mälaren, lk., Swe.	**208-09**	59°30'N	17°12'E
Malargüe, Arg.	**185**	35°28'S	69°35'W
Malartic, Qc., Can.	**146-47**	48°09'N	78°07'W
Malatya, Tur. (mȧ-lä′tyȧ)	**202-03**	38°21'N	38°18'E
Malawi, nation, Afr. (mä-lä′-wē)	**274**	13°30'S	34°00'E
Malawi, Lake, lk., Afr. (lāk mä-lä′-wē)			
see Nyasa, Lake	**286-87**	12°0'S	34°30'E
Malaya Vishera, Russia	**208-09**	58°51'N	32°14'E
Malaybalay, Phil.	**270**	8°09'N	125°08'E
Malay Peninsula, pen., Asia			
(mȧ-lā′ pĕ-nĭn′sūlȧ) (mā′lā)	**266-67**	6°0'N	101°00'E
Malaysia, nation, Asia (mȧ-lā′zhȧ)	**224-25**	2°30'N	112°30'E
Malbork, Pol. (mäl′bôrk)	**210-11**	54°02'N	19°02'E
Malden, Mo., U.S. (môl′dĕn)	**134-35**	36°34'N	89°58'W
Malden, i., Kir. (môl′dĕn)	**304-05**	4°03'S	154°59'W
Maldive Islands, nation, Asia			
see Maldives	**224-25**	3°15'N	73°00'E
Maldives, nation, Asia			
(măl′dīvz) (môl′dēvz)	**224-25**	3°15'N	73°00'E
Maldonado, Ur. (mäl-dŏ-nä′dȯ)	**187**	34°55'S	54°57'W
Male', nat. cap., Mald. (mä-lā′)	**256**	4°10'N	73°30'E
Maléas, Ákra, c., Grc.	**216-17**	36°26'N	23°12'E
Male Atoll, at., Mald.	**256**	4°25'N	73°30'E
Mālegaon, India	**254-55**	20°33'N	74°32'E
Malheur Lake, lk., Or., U.S.			
(mȧ-lōōr′ lāk)	**122-23**	43°20'N	118°45'W
Mali, nation, Afr. (mä′-lē)	**274**	17°0'N	4°00'W
Mali, stm., Mya.	**258-59**	25°43'N	97°31'E
Malik, Wādī al-, stm., Sudan	**288**	18°03'N	30°58'E
Mali Kyun, i., Mya.	**266-67**	13°06'N	98°16'E
Malinaltepec, Mex.			
(mä-lē-näl-tȧ-pĕk′)	**158-59**	17°05'N	98°39'W
Malindi, Kenya (mä-lēn′dē)	**284-85**	3°13'S	40°06'E
Malino, Bukit, mtn., Indon.	**268-69**	0°42'N	120°51'E
Malkara, Tur. (mäl′κȧ-rȧ)	**216-17**	40°52'N	26°55'E
Malko Tŭrnovo, Blg.			
(mäl′kō-t′r′nō-vȧ)	**216-17**	41°59'N	27°32'E
Mallawi, Egypt	**291b**	27°44'N	30°51'E
Mallery Lake, lk., Nu., Can.	**140-41**	63°55'N	98°25'W

Feature (Pronunciation)	Page	Lat.	Long.
Mallorca, i., Spain	**214-15**	39°30'N	3°00'E
Malmö, Swe.	**208-09**	55°36'N	13°01'E
Maloelap, at., Marsh. Is.	**304-05**	8°45'N	171°03'E
Malolos, Phil. (mä-lō′lōs)	**270**	14°51'N	120°49'E
Maloshuyka, Russia	**202-03**	63°44'N	37°25'E
Måløy, Nor.	**200-01**	61°56'N	5°08'E
Maloyaroslavets, Russia			
(mä′lō-yä-rō-slä-vyĕts)	**218-19**	55°01'N	36°28'E
Malpelo, Isla de, i., Col.			
(ē′s-lä-dĕ-mäl-pā′lō)	**178-79**	3°59'N	81°35'W
Malpeque Bay, b., P.E., Can.			
(môl-pĕk′ bā)	**148-49**	46°30'N	63°47'W
Malta, Mt., U.S. (môl′tȧ)	**122-23**	48°21'N	107°52'W
Malta, nation, Eur. (môl′tȧ)	**190-91**	35°50'N	14°35'E
Malta, i., Malta	**216b**	35°53'N	14°27'E
Maluku, is., Indon. *see* Moluccas	**268-69**	2°0'S	128°00'E
Maluku, Laut, s., Indon.			
see Molucca Sea	**268-69**	0°13'N	125°10'E
Malvern, Ar., U.S. (măl′vērn)	**130-31**	34°22'N	92°49'W
Malyy Anyuy, stm., Russia	**236-37**	68°31'N	160°55'E
Malyye Derbety, Russia	**202-03**	47°58'N	44°43'E
Malyy Kavkaz, mts., Asia			
see Lesser Caucasus	**245**	40°60'N	44°35'E
Malyy Shantar, Ostrov, i., Russia	**236-37**	54°30'N	137°36'E
Malyy Taymyr, Ostrov, i., Russia	**236-37**	78°08'N	107°12'E
Malyy Uzen', stm., Eur. *see* Balaözen	**202-03**	48°58'N	49°38'E
Mamberamo, stm., Indon.	**302**	1°35'S	137°52'E
Mambéré, stm., C.A.R.	**284-85**	3°32'N	16°03'E
Mamburao, Phil. (mäm-bōō′rä-ō)	**270**	13°15'N	120°35'E
Mammoth Cave National Park,			
n.p., Ky., U.S. (măm′ŏth kāv			
năsh′ŭn-ăl pärk)	**134-35**	37°11'N	86°08'W
Mamoré, stm., S.A.	**180-81**	10°24'S	65°23'W
Mamoudzou, nat. cap., May.	**286-87**	12°47'S	45°14'E
Mamry, Jezioro, lk., Pol. (mäm′rī)	**210-11**	54°07'N	21°44'E
Man, C. Iv.	**282-83**	7°24'N	7°33'W
Man, Isle of, dep., Eur.			
see Isle of Man	**206-07**	54°15'N	4°30'W
Manacapuru, Braz.	**180-81**	3°17'S	60°36'W
Manacor, Spain (mä-nä-kōr′)	**214-15**	39°34'N	3°12'E
Manado, Indon.	**268-69**	1°29'N	124°51'E
Managua, nat. cap., Nic. (mä-nä′gwä)	**161**	12°09'N	86°17'W
Managua, Lago de, lk., Nic.			
(lȧ′gô-dĕ-mä-nä′gwä)	**161**	12°20'N	86°20'W
Manakara, Madag. (mä-nä-kä′rŭ)	**286-87**	22°09'S	48°01'E
Manāli, India	**254-55**	32°16'N	77°09'E
Manama, nat. cap., Bahr. (mă-nă′mȧ)			
see Al-Manāmah	**250-51**	26°13'N	50°35'E
Manam Island, i., Pap. N. Gui.	**302**	4°05'S	145°02'E
Mananara, stm., Madag.			
(mä-nä-nä′rŭ)	**286-87**	23°21'S	47°42'E
Mananara Avaratra, Madag.	**286-87**	16°10'S	49°46'E
Mananjary, Madag. (mä-nän-zhä′rē)	**286-87**	21°14'S	48°21'E
Manáos, Braz. *see* Manaus	**180-81**	3°07'S	60°01'W
Mana Pools National Park, n.p., Zimb.	**286-87**	15°52'S	29°15'E
Manas Hu, lk., China	**244**	45°43'N	85°54'E
Manassas, Va., U.S. (mȧ-năs′ȧs)	**126-27**	38°45'N	77°28'W
Manaus, Braz. (mä-nä′ōōzh)	**180-81**	3°07'S	60°01'W
Mancelona, Mi., U.S. (măn-sē-lō′nȧ)	**126-27**	44°54'N	85°04'W
Manchester, Eng., U.K. (măn′chĕs-tēr)	**206-07**	53°27'N	2°15'W
Manchester, Ct., U.S. (măn′chĕs-tēr)	**126-27**	41°47'N	72°31'W
Manchester, Ga., U.S. (măn′chĕs-tēr)	**134-35**	32°51'N	84°37'W
Manchester, Ia., U.S. (măn′chĕs-tēr)	**124-25**	42°29'N	91°28'W
Manchester, Ky., U.S. (măn′chĕs-tēr)	**134-35**	37°09'N	83°47'W
Manchester, N.H., U.S.	**126-27**	42°59'N	71°28'W
Manchester, Tn., U.S. (măn′chĕs-tēr)	**134-35**	35°29'N	86°05'W
Manchuria, hist. reg., China			
(măn-chōō′rē-ȧ)	**260-61**	47°0'N	125°00'E
Mand, stm., Iran	**250-51**	28°09'N	51°16'E
Manda Island, i., Kenya	**284-85**	2°15'S	40°57'E
Mandal, Nor. (män′däl)	**208-09**	58°02'N	7°27'E
Mandala, Puncak, mtn., Indon.	**302**	4°43'S	140°18'E
Mandalay, Mya. (măn′dȧ-lā)	**266-67**	21°58'N	96°05'E
Mandalgovĭ, Mong.	**260-61**	45°46'N	106°16'E
Mandalī, Iraq	**248-49**	33°45'N	45°32'E
Mandan, N.D., U.S. (măn′dăn)	**124-25**	46°49'N	100°55'W
Mandara, Monts, mts., Afr.			
see Mandara Mountains	**282-83**	10°45'N	13°40'E
Mandara Mountains, mts., Afr.			
(män-dä′rä moun′tĭnz)	**282-83**	10°45'N	13°40'E
Mandeb, Bab el, strt., (băb′ĕl män-dĕb′)			
see Bab el Mandeb	**288**	12°44'N	43°21'E
Mandera, Kenya	**284-85**	3°56'N	41°52'E
Mandioli, Pulau, i., Indon.	**268-69**	0°44'S	127°14'E
Mandla, India	**254-55**	22°35'N	80°23'E
Mandsaur, India	**254-55**	24°03'N	75°05'E

380 Manduria – Marquette PRONOUNCING INDEX

Feature (Pronunciation)	Page	Lat.	Long.
Manduria, Italy (män-dōō′rē-ä)	216-17	40°24′N	17°38′E
Māndvi, India (mŭnd′vē)	254-55	22°51′N	69°22′E
Manfalût, Egypt	291b	27°19′N	30°58′E
Manfredonia, Italy (män-frå-dô′nyä)	216-17	41°38′N	15°55′E
Mangabeiras, Chapada das, hills, Braz.	180-81	9°55′S	46°32′W
Mangaia, i., Cook Is.	304-05	21°55′S	157°54′W
Mangalore, India *see* Mangalūru	256	12°52′N	74°51′E
Mangalūru, India	256	12°52′N	74°51′E
Mangchang, China	258-59	25°08′N	107°31′E
Mangkalihat, Tanjung, c., Indon.	268-69	1°02′N	118°59′E
Mangochi, Malawi	286-87	14°28′S	35°15′E
Mangoky, stm., Madag. (män-gō′kē)	286-87	21°20′S	43°32′E
Mangole, Pulau, i., Indon.	268-69	1°51′S	125°51′E
Mangshi, China *see* Luxi	258-59	24°21′N	98°23′E
Mangueira, Lagoa, b., Braz.	187	33°06′S	52°48′W
Mangum, Ok., U.S. (măŋ′gŭm)	130-31	34°53′N	99°30′W
Mangya, China	238-39	37°40′N	90°50′E
Manhattan, Ks., U.S. (măn-hăt′ăn)	130-31	39°11′N	96°34′W
Manhattan, Mt., U.S. (măn-hăt′ăn)	122-23	45°51′N	111°20′W
Manhuaçu, Braz. (män-ŏŏá′sōō)	186	20°15′S	42°02′W
Manicoré, Braz.	180-81	5°49′S	61°16′W
Manicouagan, stm., Qc., Can.	148-49	49°10′N	68°09′W
Manicouagan, Réservoir, res., Qc., Can.	140-41	51°22′N	68°44′W
Manihiki, at., Cook Is. (mä′nē-hē′kē)	304-05	10°24′S	161°01′W
Manila, nat. cap., Phil. (mȧ-nĭl′ȧ)	270	14°35′N	120°60′E
Manila Bay, b., Phil. (mȧ-nĭl′ȧ bā)	270	14°30′N	120°45′E
Manipa, Pulau, i., Indon.	268-69	3°18′S	127°33′E
Manipur, state, India	254-55	25°0′N	94°00′E
Manisa, Tur. (mä′nē-sȧ)	216-17	38°37′N	27°26′E
Manistee, Mi., U.S. (măn-ĭs-tē′)	126-27	44°15′N	86°19′W
Manistique, Mi., U.S. (măn-ĭs-tēk′)	126-27	45°58′N	86°14′W
Manitoba, state, Can. (măn-ĭ-tō′bȧ)	138-39	54°0′N	97°00′W
Manitoba, Lake, lk., Mb., Can. (lāk măn-ĭ-tō′bȧ)	144-45	50°47′N	98°43′W
Manitoulin Island, i., On., Can. (măn-ĭ-tōō′lĭn ī′lȧnd)	146-47	45°47′N	82°20′W
Manitou Springs, Co., U.S. (măn′ĭ-tōō sprĭngz)	130-31	38°52′N	104°54′W
Manitowoc, Wi., U.S. (măn-ĭ-tô-wŏk′)	126-27	44°06′N	87°39′W
Maniwaki, Qc., Can.	146-47	46°23′N	75°59′W
Manizales, Col. (mä-nê-zä′lås)	188c	5°04′N	75°31′W
Mānjra, stm., India	256	18°49′N	77°52′E
Mankanza, D.R.C.	284-85	1°33′N	19°04′E
Mankato, Ks., U.S. (măn-kā′tō)	130-31	39°47′N	98°13′W
Mankato, Mn., U.S. (măn-kā′tō)	124-25	44°10′N	93°59′W
Manlleu, Spain (män-lyā′ōō)	214-15	42°0′N	2°17′E
Manna, Indon.	266-67	4°28′S	102°55′E
Mannar, Sri L. (mȧ-när′)	256	8°59′N	79°55′E
Mannar, Gulf of, b., Asia	256	8°30′N	79°00′E
Mannar Island, i., Sri L.	256	9°03′N	79°50′E
Mannheim, Ger. (män′hīm)	210-11	49°30′N	8°28′E
Manning, S.C., U.S. (măn′ĭng)	134-35	33°42′N	80°13′W
Mannington, W.V., U.S. (măn′ĭng-tŭn)	126-27	39°31′N	80°23′W
Manokwari, Indon. (mȧ-nŏk-wä′rê)	242-43	0°51′S	134°05′E
Manono, D.R.C.	284-85	7°18′S	27°25′E
Manosque, Fr. (mȧ-nôsh′)	212-13	43°50′N	5°47′E
Manouane, Lac, res., Qc., Can.	148-49	50°42′N	70°46′W
Manra, at., Kir.	304-05	4°27′S	171°15′W
Manresa, Spain (män-rā′sä)	214-15	41°44′N	1°49′E
Mansa, Zam.	286-87	11°12′S	28°53′E
Mansel Island, i., Nu., Can. (măn′sĕl ī′lȧnd)	140-41	61°60′N	79°50′W
Mansfield, Eng., U.K. (mănz′fēld)	206-07	53°09′N	1°12′W
Mansfield, La., U.S. (mănz′fēld)	132-33	32°02′N	93°43′W
Mansfield, Mo., U.S. (mănz′fēld)	130-31	37°06′N	92°35′W
Mansfield, Oh., U.S. (mănz′fēld)	126-27	40°45′N	82°31′W
Mansfield, Pa., U.S. (mănz′fēld)	126-27	41°48′N	77°04′W
Mansura, Egypt *see* El-Mansûra	291b	31°02′N	31°23′E
Manta, Ec. (män′tä)	184	0°57′S	80°43′W
Mantes-la-Jolie, Fr. (mänt-ĕ-lä-zhō-lē′)	212-13	48°59′N	1°43′E
Mantiqueira, Serra da, mts., Braz.	186	22°14′S	44°53′W
Manturovo, Russia	202-03	58°20′N	44°47′E
Manuae, at., Cook Is.	304-05	19°21′S	158°56′W
Manuae, at., Fr. Poly.	304-05	16°30′S	154°40′W
Manua Islands, is., Am. Sam.	306b	14°13′S	169°35′W
Manuel Rodríguez, Isla, i., Chile	185	52°34′S	73°51′W
Manui, Pulau, i., Indon. (pōō-lou mä-nōō′ē)	268-69	3°35′S	123°08′E
Manus Island, i., Pap. N. Gui. (mä′nōōs ī′lȧnd)	302	2°05′S	147°00′E
Manyame, stm., Afr.	286-87	15°37′S	30°39′E
Manyara, Lake, lk., Tan.	289	3°35′S	35°50′E
Manych, stm., Russia	202-03	47°15′N	40°15′E
Manyoni, Tan.	289	5°45′S	34°50′E
Manzanillo, Cuba (män′zä-nēl′yō)	154-55	20°21′N	77°07′W
Manzanillo, Mex.	158-59	19°03′N	104°20′W
Manzhouli, China (män-jō-lē)	260-61	49°35′N	117°27′E
Mao, Chad (mä′ō)	280-81	14°07′N	15°19′E
Maó, Spain	214-15	39°53′N	4°16′E
Maoke, Pegunungan, mts., Indon.	302	4°0′S	138°00′E
Maoming, China	258-59	21°41′N	110°51′E
Mapastepec, Mex. (ma-päs-tå-pĕk′)	158-59	15°26′N	92°54′W
Mapi, Indon.	302	7°05′S	139°24′E
Mapimí, Mex. (mä-pê-mē′)	132-33	25°49′N	103°51′W
Mapimí, Bolsón de, des., Mex. (bôl-sŏ′n-dĕ-mä-pē′mē)	132-33	26°30′N	104°00′W
Maple Creek, Sk., Can. (mā′p'l crēk krēk)	144-45	49°55′N	109°29′W
Maplewood, Mn., U.S. (mā′p'l wŏŏd)	124-25	45°01′N	93°04′W
Mapuera, stm., Braz.	180-81	1°05′S	57°03′W
Maputo, nat. cap., Moz. (mä-pōō′-tō)	286-87	25°58′S	32°35′E
Maqat, Kaz.	202-03	47°39′N	53°22′E
Maquan, stm., China	254-55	29°33′N	84°07′E
Maquinchao, Arg.	185	41°15′S	68°41′W
Maquoketa, Ia., U.S. (mȧ-kō-kē-tȧ)	124-25	42°04′N	90°40′W
Mar, Serra do, mts., Braz. (sĕr′rȧ dō mär′)	186	23°30′S	45°30′W
Mara, stm., Afr.	289	1°32′S	33°59′E
Marabá, Braz.	180-81	5°21′S	49°06′W
Maracá, Ilha de, i., Braz.	180-81	2°05′N	50°25′W
Maracaibo, Ven. (mä-rä-kī′bō)	178-79	10°40′N	71°38′W
Maracaibo, Lago de, lk., Ven. (lä′gô-dĕ-mä-rä-kī′bō)	178-79	9°43′N	71°50′W
Maracaibo, Lake, lk., Ven. (lāk mä-rä-kī′bō) *see* Maracaibo, Lago de	178-79	9°43′N	71°50′W
Maracay, Ven. (mä-rä-kāy′)	188b	10°16′N	67°37′W
Marādah, Libya	204-05	29°13′N	19°12′E
Maradi, Niger (mȧ-rȧ-dē′)	280-81	13°29′N	7°06′E
Marāgheh, Iran	248-49	37°23′N	46°15′E
Maragogipe, Braz.	180-81	12°46′S	38°55′W
Marahuaca, Cerro, mtn., Ven.	178-79	3°35′N	65°27′W
Marajó, Baía de, b., Braz.	180-81	1°0′S	48°30′W
Maralal, Kenya	289	1°05′N	36°41′E
Maramasike, i., Sol. Is.	306e	9°32′S	161°27′E
Marand, Iran	245	38°26′N	45°46′E
Maranhão, state, Braz. (mä-rän-youɴ)	180-81	5°0′S	45°00′W
Maranoa, stm., Austl. (mä-rä-nō′ä)	296-97	27°44′S	148°44′E
Marañón, stm., Peru (mä-rä-nyŏn′)	184	4°29′S	73°30′W
Maraş, Tur. *see* Kahramanmaraş	202-03	37°35′N	36°57′E
Marathon, On., Can. (măr′ȧ-thŏn)	146-47	48°43′N	86°23′W
Marathon, Fl., U.S. (măr′ȧ-thŏn)	135a	24°42′N	81°06′W
Marathon, N.Y., U.S. (măr′ȧ-thŏn)	126-27	42°27′N	76°02′W
Marawwaḩ, i., U.A.E.	250-51	24°17′N	53°15′E
Marble Bar, Austl. (märb′'l bär)	294-95	21°10′S	119°45′E
Marble Hall, S. Afr.	292c	24°57′S	29°14′E
Marburg an der Drau, Slvn. *see* Maribor	216-17	46°33′N	15°39′E
Marca, Ponta da, c., Ang.	286-87	16°31′S	11°42′E
Marceline, Mo., U.S. (mär-sĕ-lēn′)	130-31	39°43′N	92°57′W
Marchena, Isla, i., Ec. (ê′s-lä-mär-chĕ′nä)	184a	0°21′N	90°29′W
Mar Chiquita, Laguna, lk., Arg. (lä-gōō′nä-mär-chê-kē′tä)	187	30°42′S	62°36′W
Marcus Island, i., Japan (mär′kŭs ī′lȧnd)	304-05	24°18′N	153°58′E
Marcy, Mount, mtn., N.Y., U.S. (mount mär′sê)	126-27	44°07′N	73°56′W
Mardān, Pak.	252-53	34°12′N	72°03′E
Mar del Plata, Arg. (mär dĕl- plä′ta)	187	37°60′S	57°34′W
Mardin, Tur. (mär-dēn′)	248-49	37°18′N	40°45′E
Maré, i., N. Cal. (mȧ-rā′)	306g	21°30′S	167°59′E
Mareeba, Austl.	302	16°60′S	145°24′E
Marengo, Il., U.S. (mȧ-rĕŋ′gō)	126-27	42°15′N	88°37′W
Marfa, Tx., U.S. (mär′fȧ)	132-33	30°19′N	104°02′W
Margarita, Isla de, i., Ven. (ê′s-lä dĕ mä-gȧ-rē′tä)	188b	11°0′N	64°00′W
Margate, Eng., U.K. (mär′gāt)	206-07	51°23′N	1°25′E
Margelan, Uzb. *see* Marghilon	252-53	40°28′N	71°44′E
Margherita, Som. *see* Jamaame	284-85	0°01′N	42°42′E
Margherita Peak, mtn., Afr.	289	0°22′N	29°51′E
Marghilon, Uzb.	252-53	40°28′N	71°44′E
Mārgow, Dasht-e, des., Afg.	252-53	30°45′N	63°10′E
Marguerite, Pic, mtn., Afr. *see* Margherita Peak	289	0°22′N	29°51′E
Marhanets', Ukr.	218-19	47°39′N	34°38′E
Maria, Îles, is., Fr. Poly.	304-05	21°44′S	154°38′W
María Cleofas, Isla, i., Mex.	158-59	21°18′N	106°15′W
María Elena, Chile	182-83	22°20′S	69°40′W
Maria Island, i., Austl.	301	42°39′S	148°04′E
María Madre, Isla, i., Mex.	158-59	21°37′N	106°35′W
María Magdalena, Isla, i., Mex.	158-59	21°27′N	106°26′W
Mariana Islands, is., Oc. (mä-ryä′nä ī′lȧndz)	304-05	15°60′N	145°44′E
Marianna, Ar., U.S. (mā-rĭ-ăn′ȧ)	134-35	34°46′N	90°46′W
Marianna, Fl., U.S. (mā-rĭ-ăn′ȧ)	134-35	30°46′N	85°15′W
Mariánské Lázně, Czech Rep. (mär′yȧn-skĕ′läz′nyĕ)	210-11	49°58′N	12°42′E
Maria Theresiopel, Serb. *see* Subotica	216-17	46°06′N	19°41′E
Mariato, Punta, c., Pan.	162	7°13′N	80°53′W
Maribo, Den. (mä′rē-bô)	210-11	54°46′N	11°31′E
Maribor, Slvn. (mä′re-bôr)	216-17	46°33′N	15°39′E
Marīdī, S. Sudan	289	4°55′N	29°28′E
Marie Byrd Land, reg., Ant.	313	80°0′S	120°00′W
Marie-Galante, i., Guad. (mȧ-rē′ gȧ-länt′)	155b	15°56′N	61°16′W
Mari El, state, Russia	202-03	56°30′N	48°00′E
Marienbad, Czech Rep. *see* Mariánské Lázně	210-11	49°58′N	12°42′E
Marienburg, Pol. *see* Malbork	210-11	54°02′N	19°02′E
Mariental, Nmb.	286-87	24°37′S	17°58′E
Mariestad, Swe. (mä-rē′ĕ-städ′)	208-09	58°43′N	13°51′E
Marietta, Ga., U.S. (mā-rĭ-ĕt′ȧ)	134-35	33°57′N	84°33′W
Marietta, Oh., U.S. (mā-rĭ-ĕt′ȧ)	126-27	39°25′N	81°28′W
Marietta, Ok., U.S. (mā-rĭ-ĕt′ȧ)	130-31	33°56′N	97°07′W
Marília, Braz. (mä-rē′lyȧ)	182-83	22°13′S	49°57′W
Marimba, Ang.	284-85	8°22′S	16°59′E
Marinduque, i., Phil. (mä-rên-dōō′kā)	270	13°24′N	121°58′E
Marine City, Mi., U.S. (mȧ-rēn′ sĭ′tē)	126-27	42°43′N	82°29′W
Marinette, Wi., U.S. (măr-ĭ-nĕt′)	126-27	45°06′N	87°37′W
Maringá, Braz.	182-83	23°25′S	51°56′W
Maringa, stm., D.R.C. (mä-rĭŋ′gä)	284-85	1°13′N	19°50′E
Marinha Grande, Port. (mä-rēn′yȧ grän′dĕ)	214-15	39°45′N	8°56′W
Marion, Al., U.S. (măr′ĭ-ŭn)	134-35	32°38′N	87°19′W
Marion, Ar., U.S. (măr′ĭ-ŭn)	134-35	35°13′N	90°12′W
Marion, Ia., U.S. (măr′ĭ-ŭn)	124-25	42°02′N	91°36′W
Marion, Il., U.S. (măr′ĭ-ŭn)	126-27	37°44′N	88°56′W
Marion, In., U.S. (măr′ĭ-ŭn)	126-27	40°33′N	85°39′W
Marion, Ky., U.S. (măr′ĭ-ŭn)	134-35	37°20′N	88°05′W
Marion, N.C., U.S. (măr′ĭ-ŭn)	134-35	35°41′N	82°01′W
Marion, Oh., U.S. (măr′ĭ-ŭn)	126-27	40°35′N	83°07′W
Marion, S.C., U.S. (măr′ĭ-ŭn)	134-35	34°11′N	79°24′W
Marion, Va., U.S. (măr′ĭ-ŭn)	134-35	36°50′N	81°31′W
Marion, Lake, res., S.C., U.S. (lāk măr′ĭ-ŭn)	134-35	33°32′N	80°29′W
Mariquita, Col. (mä-rē-kē′tä)	188c	5°11′N	74°54′W
Mariscal Estigarribia, Para.	182-83	22°02′S	60°37′W
Maritzburg, S. Afr. *see* Pietermaritzburg	286-87	29°36′S	30°23′E
Mariupol', Ukr.	218-19	47°06′N	37°34′E
Mariy-El, state, Russia *see* Mari El	202-03	56°30′N	48°00′E
Marka, Som.	284-85	1°43′N	44°46′E
Markaryd, Swe. (mär′kä-rüd)	208-09	56°28′N	13°35′E
Marked Tree, Ar., U.S. (märkt trē)	134-35	35°32′N	90°25′W
Markha, Russia	236-37	60°36′N	123°19′E
Markha, stm., Russia	236-37	63°27′N	118°53′E
Markham, On., Can. (märk′ȧm)	146-47	43°52′N	79°16′W
Markovo, Russia (mär′kô-vô)	236-37	64°40′N	170°27′E
Marks, Russia	202-03	51°42′N	46°44′E
Marksville, La., U.S. (märks′vĭl)	134-35	31°07′N	92°05′W
Marlette, Mi., U.S. (mär-lĕt′)	126-27	43°20′N	83°04′W
Marlin, Tx., U.S. (mär′lĭn)	132-33	31°17′N	96°53′W
Marlinton, W.V., U.S. (mär′lĭn-tŭn)	126-27	38°14′N	80°06′W
Marlow, Ok., U.S. (mär′lō)	130-31	34°38′N	97°58′W
Marmande, Fr. (mȧr-mäɴd′)	212-13	44°30′N	0°10′E
Marmara, Sea of, s., Tur. (mär′mȧ-rȧ)	216-17	40°40′N	28°15′E
Marmara Denizi, s., Tur. *see* Marmara, Sea of	216-17	40°40′N	28°15′E
Marmarth, N.D., U.S. (mär′märth)	124-25	46°18′N	103°55′W
Marmelos, stm., Braz.	180-81	6°05′S	61°46′W
Maroa, Ven. (mä-rō′ä)	178-79	2°44′N	67°33′W
Maromokotro, mtn., Madag.	286-87	14°01′S	48°58′E
Marondera, Zimb.	286-87	18°10′S	31°32′E
Marosvásárhely, Rom. *see* Târgu Mureș	210-11	46°33′N	24°34′E
Marotiri, Îles, is., Fr. Poly.	304-05	27°53′S	143°21′W
Maroua, Camrn. (mär′wä)	282-83	10°37′N	14°19′E
Marovoay, Madag.	286-87	16°07′S	46°39′E
Marquesas Islands, is., Fr. Poly. (mär-kĕ′säs ī′lȧndz)	304-05	8°59′S	139°31′W
Marquesas Keys, is., Fl., U.S. (mȧr-kē′zȧs kēs)	135a	24°34′N	82°08′W
Marquette, Mi., U.S.	124-25	46°32′N	87°23′W

ăt; finăl; rāte; senāte; ärm; åsk; sofȧ; fâre; ch-choose; dh-as th in other; bē; ēvent; bĕt; recĕnt; cratẽr; g-gō; gh-guttural g; bĭt; ĭ-short neutral; rīde; ĸ-guttural k as ch in German ich;

PRONOUNCING INDEX
Marquises, Îles – Meadow Lake 381

Feature (Pronunciation)	Page	Lat.	Long.
Marquises, Îles, is., Fr. Poly. *see* Marquesas Islands	**304-05**	8°59'S	139°31'W
Marrah, Jabal, vol., Sudan (jĕb´ĕl mär´ä)	**284-85**	13°03'N	24°21'E
Marrakech, Mor. (mȧr-rä´kĕsh)	**292a**	31°38'N	8°01'W
Marrakesh, Mor. *see* Marrakech	**292a**	31°38'N	8°01'W
Marree, Austl. (măr´rē)	**301**	29°39'S	138°04'E
Marromeu, Moz.	**286-87**	18°16'S	35°52'E
Marsá al-Burayqah, Libya	**204-05**	30°23'N	19°36'E
Marsabit, Kenya	**289**	2°20'N	37°60'E
Marsala, Italy (mär-sä´lä)	**216-17**	37°48'N	12°26'E
Marseille, Fr. (mȧr-sâ´y')	**212-13**	43°18'N	5°24'E
Marseilles, Il., U.S. (mär-sĕlz´)	**126-27**	41°20'N	88°42'W
Marshall, Il., U.S. (mär´shăl)	**126-27**	39°23'N	87°42'W
Marshall, Mi., U.S. (mär´shăl)	**126-27**	42°16'N	84°57'W
Marshall, Mn., U.S. (mär´shăl)	**124-25**	44°27'N	95°48'W
Marshall, Mo., U.S. (mär´shăl)	**130-31**	39°07'N	93°12'W
Marshall, Tx., U.S. (mär´shăl)	**130-31**	32°33'N	94°22'W
Marshall Islands, nation, Oc. (mär´shăl ī´lăndz)	**304-05**	11°0'N	168°00'E
Marshalltown, Ia., U.S. (mär´shăl-toun)	**124-25**	42°03'N	92°54'W
Marshfield, Mo., U.S. (märsh´fēld)	**130-31**	37°20'N	92°54'W
Marshfield, Wi., U.S. (märsh´fēld)	**126-27**	44°40'N	90°10'W
Marsh Harbour, Bah. (märsh´ här´bẽr)	**154-55**	26°32'N	77°04'W
Marsh Island, i., La., U.S.	**134-35**	29°35'N	91°53'W
Mart, Tx., U.S. (märt)	**132-33**	31°32'N	96°50'W
Martaban, Gulf of, b., Mya. (gŭlf ŭv mär-tŭ-bän´)	**266-67**	16°46'N	97°01'E
Martha's Vineyard, i., Ma., U.S. (mär´thȧz vĭn´yȧrd)	**126-27**	41°24'N	70°38'W
Martigny, Switz. (mȧr-tė-nyē´)	**210-11**	46°06'N	7°04'E
Martin, S.D., U.S. (mär´tĭn)	**124-25**	43°10'N	101°44'W
Martin, Tn., U.S. (mär´tĭn)	**134-35**	36°21'N	88°51'W
Martina Franca, Italy (mar-te´na fraŋ´ka)	**216-17**	40°42'N	17°20'E
Martinez, Ga., U.S. (mär-tē´nĕz)	**134-35**	33°31'N	82°05'W
Martinique, dep., N.A. (mȧr-tė-nēk´)	**152-53**	14°40'N	61°00'W
Martinique Passage, strt., N.A.	**155b**	15°10'N	61°15'W
Martinsburg, W.V., U.S. (mär´tĭnz-bûrg)	**126-27**	39°27'N	77°57'W
Martinsville, In., U.S. (mär´tĭnz-vĭl)	**126-27**	39°25'N	86°25'W
Martinsville, Va., U.S. (mär´tĭnz-vĭl)	**134-35**	36°41'N	79°52'W
Martin Vaz, Ilhas, is., Braz.	**173**	20°30'S	28°51'W
Martos, Spain (mär´tōs)	**214-15**	37°43'N	3°58'W
Martre, Lac la, lk., N.T., Can. (läk lä märtr)	**140-41**	63°15'N	117°55'W
Marungu, mts., D.R.C.	**289**	7°42'S	30°01'E
Mary, Turkmen. (mä´rė)	**252-53**	37°35'N	61°49'E
Maryborough, Austl. (mā´rĭ-bŭr-ȯ)	**301**	25°32'S	152°42'E
Maryborough, Austl. (mā´rĭ-bŭr-ȯ)	**301**	37°03'S	143°44'E
Maryland, state, U.S. (mĕr´ĭ-lănd)	**118-19**	39°0'N	76°45'W
Marystown, Nf./L., Can. (mâr´ĭz-toun)	**148-49**	47°11'N	55°10'W
Marysville, Ca., U.S.	**128-29**	39°09'N	121°35'W
Marysville, Ks., U.S. (mā´rĭz-vĭl)	**130-31**	39°50'N	96°39'W
Marysville, Oh., U.S. (mā´rĭz-vĭl)	**126-27**	40°14'N	83°22'W
Marysville, Wa., U.S. (mā´rĭz-vĭl)	**122-23**	48°04'N	122°10'W
Maryville, Mo., U.S. (mā´rĭ-vĭl)	**130-31**	40°21'N	94°52'W
Maryville, Tn., U.S. (mā´rĭ-vĭl)	**134-35**	35°46'N	83°58'W
Masai Mara Game Reserve, pk., Kenya	**289**	1°15'S	35°15'E
Masai Steppe, plat., Tan.	**289**	4°45'S	37°00'E
Masaka, Ug.	**289**	0°20'S	31°44'E
Masalembu Besar, Pulau, i., Indon.	**268-69**	5°34'S	114°26'E
Masan, Kor., S. (mä-sän´)	**263**	35°12'N	128°34'E
Masatepe, Nic. (mä-sä-tĕ´pĕ)	**161**	11°54'N	86°09'W
Masaya, Nic. (mä-sä´yä)	**161**	11°58'N	86°06'W
Masbate, Phil. (mäs-bä´tä)	**270**	12°22'N	123°38'E
Masbate, i., Phil. (mäs-bä´tä)	**270**	12°15'N	123°30'E
Mascara, Alg.	**214-15**	35°23'N	0°08'E
Mascareignes, Îles, is., Afr.	**287a**	21°0'S	57°00'E
Mascarene Islands, is., Afr. *see* Mascareignes, Îles	**287a**	21°0'S	57°00'E
Mascota, Mex. (mäs-kō´tä)	**158-59**	20°31'N	104°47'W
Mascoutah, Il., U.S. (mäs-kū´tä)	**130-31**	38°29'N	89°48'W
Maseru, nat. cap., Leso. (măz´ẽr-ōō)	**292c**	29°19'S	27°29'E
Mashābih, i., Sau. Ar.	**288**	25°38'N	36°31'E
Mashhad, Iran	**252-53**	36°17'N	59°36'E
Māshkel, Hāmūn-i-, lk., Pak. (hā-mōōn´ē māsh-kĕl´)	**252-53**	28°15'N	63°00'E
Masi-Manimba, D.R.C.	**284-85**	4°46'S	17°57'E
Masindi, Ug. (mä-sēn´dė)	**289**	1°41'N	31°43'E
Masira, Gulf of, b., Oman *see* Maṣīrah, Khalīj	**238-39**	20°10'N	58°15'E
Maṣīrah, i., Oman	**238-39**	20°27'N	58°48'E
Maṣīrah, Khalīj, b., Oman	**238-39**	20°10'N	58°15'E
Masjed-e Soleymān, Iran	**252-53**	31°58'N	49°18'E

Feature (Pronunciation)	Page	Lat.	Long.
Masoala, Saikanosy, pen., Madag.	**286-87**	15°26'S	50°04'E
Mason, Mi., U.S. (mā´sŭn)	**126-27**	42°35'N	84°26'W
Mason, Tx., U.S. (mā´sŭn)	**132-33**	30°45'N	99°15'W
Mason City, Ia., U.S. (mā´sŭn sĭ´tĭ)	**124-25**	43°09'N	93°12'W
Masqaṭ, nat. cap., Oman *see* Muscat	**250-51**	23°36'N	58°32'E
Massa, Italy (mäs´sä)	**216-17**	44°03'N	10°09'E
Massachusetts, state, U.S. (măs-ȧ-chōō´sĕts)	**118-19**	42°15'N	71°50'W
Massafra, Italy (mäs-sä´frä)	**216-17**	40°35'N	17°08'E
Massakory, Chad	**280-81**	12°60'N	15°44'E
Massawa, Erit.	**288**	15°37'N	39°26'E
Massena, N.Y., U.S. (mă-sē´nȧ)	**126-27**	44°56'N	74°53'W
Masset, B.C., Can. (măs´ĕt)	**142-43**	54°02'N	132°08'W
Massillon, Oh., U.S. (măs´ĭ-lŏn)	**126-27**	40°48'N	81°31'W
Massinga, Moz. (mä-sĭn´gä)	**286-87**	23°20'S	35°24'E
Massive, Mount, mtn., Co., U.S. (mount mȧs´ĭv)	**128-29**	39°12'N	106°28'W
Maştağa, Azer.	**245**	40°32'N	49°59'E
Mastung, Pak.	**252-53**	29°48'N	66°52'E
Masuda, Japan (mä-sōō´dȧ)	**265**	34°41'N	131°51'E
Masulipatam, India *see* Machilīpatnam	**256**	16°11'N	81°09'E
Masvingo, Zimb.	**286-87**	20°04'S	30°49'E
Matadi, D.R.C. (mȧ-tä´dė)	**284-85**	5°49'S	13°29'E
Matagalpa, Nic. (mä-tä-gäl´pä)	**161**	12°60'N	85°44'W
Matagami, Qc., Can.	**138-39**	49°45'N	77°39'W
Matagorda Island, i., Tx., U.S.	**132-33**	28°15'N	96°37'W
Mataiva, at., Fr. Poly.	**304-05**	14°53'S	148°40'W
Matamoros, Mex. (mä-tä-mō´rōs)	**132-33**	25°52'N	97°30'W
Matamoros, Mex. (mä-tä-mō´rōs)	**132-33**	25°32'N	103°14'W
Matandu, stm., Tan.	**284-85**	8°43'S	39°22'E
Matane, Qc., Can. (mȧ-tån´)	**148-49**	48°50'N	67°31'W
Matanzas, Cuba (mä-tän´zäs)	**154-55**	23°03'N	81°34'W
Matanzas, state, Cuba (mä-tän´zäs)	**154-55**	22°40'N	81°20'W
Matapalo, Cabo, c., C.R. (kä´bō-mä-tä-pä´lō)	**161**	8°23'N	83°17'W
Matapan, Cape, c., Grc. *see* Taínaro, Ákra	**216-17**	36°23'N	22°29'E
Matapédia, Qc., Can. (mă-tȧ-pā´dē-ȧ)	**148-49**	47°58'N	66°56'W
Matapédia, Lac, lk., Qc., Can. (läk mă-tȧ-pā´dē-ȧ)	**148-49**	48°33'N	67°33'W
Matara, Sri L. (mä-tä´rä)	**256**	5°57'N	80°34'E
Mataram, Indon.	**268-69**	8°35'S	116°07'E
Mataró, Spain	**214-15**	41°32'N	2°26'E
Matasiri, Pulau, i., Indon.	**268-69**	4°48'S	115°49'E
Matâ'utu, nat. cap., Wal./F.	**304-05**	13°17'S	176°09'W
Matehuala, Mex. (mä-tă-wä´lä)	**158-59**	23°40'N	100°38'W
Matera, Italy (mä-tå´rä)	**216-17**	40°41'N	16°36'E
Mathura, India (mu-tȯ´rŭ)	**254-55**	27°30'N	77°41'E
Mathurai, India *see* Madurai	**256**	9°55'N	78°08'E
Matias Barbosa, Braz. (mä-tē´äs-bȧr-bô-sä)	**186**	21°53'S	43°19'W
Mato, Cerro, mtn., Ven.	**178-79**	7°16'N	65°15'W
Mato Grosso, state, Braz. (mät´ō grōs´ō)	**180-81**	12°0'S	57°00'W
Mato Grosso, Planalto do, plat., Braz. (plä-nȧl´tō-dō mät´ō grōs´ō)	**180-81**	14°59'S	53°37'W
Mato Grosso do Sul, state, Braz.	**182-83**	20°0'S	55°00'W
Matola, Moz.	**286-87**	25°49'S	32°27'E
Matosinhos, Port.	**214-15**	41°11'N	8°41'W
Maṭraḥ, Oman (mȧ-trä´)	**250-51**	23°37'N	58°31'E
Matsue, Japan (mät´sȯ-ĕ)	**265**	35°28'N	133°04'E
Matsumoto, Japan (mät´sȯ-mō´tȯ)	**265**	36°14'N	137°58'E
Matsu Tao, i., Tai.	**243a**	26°09'N	119°56'E
Matsuyama, Japan (mät´sȯ-yä´mä)	**265**	33°50'N	132°46'E
Mattawa, On., Can. (măt´ȧ-wȧ)	**146-47**	46°18'N	78°41'W
Matterhorn, mtn., Eur. (mät´ẽr-hôrn)	**210-11**	45°59'N	7°43'E
Matthew Town, Bah. (măth´ū toun)	**154-55**	21°01'N	73°42'W
Mattoon, Il., U.S. (mă-tōōn´)	**126-27**	39°29'N	88°23'W
Maturín, Ven. (mä-tōō-rēn´)	**178-79**	9°44'N	63°11'W
Maubeuge, Fr. (mô-bûzh´)	**212-13**	50°17'N	3°58'E
Maués, Braz. (mȧ-wĕ´s)	**180-81**	3°22'S	57°43'W
Maui, i., Hi., U.S. (mä´ōō-ē)	**137a**	20°45'N	156°15'W
Maumee, Oh., U.S. (mô-mē´)	**126-27**	41°34'N	83°39'W
Maumee, stm., U.S. (mô-mē´)	**126-27**	41°42'N	83°27'W
Maun, Bots. (mä-ōn´)	**286-87**	19°60'S	23°25'E
Mauna Kea, vol., Hi., U.S. (mä´ō-näkä´ä)	**137a**	19°50'N	155°28'W
Mauna Loa, vol., Hi., U.S. (mä´ō-nälō´ȧ)	**137a**	19°29'N	155°36'W
Maunoir, Lac, lk., N.T., Can.	**140-41**	67°30'N	125°00'W
Maurepas, Lake, lk., La., U.S. (lāk mō-rē-pä´)	**134-35**	30°15'N	90°30'W
Mauritania, nation, Afr. (mô-rē-tā´nĭ-ȧ)	**274**	20°0'N	12°00'W
Mauritanie, nation, Afr. *see* Mauritania	**274**	20°0'N	12°00'W
Mauritius, nation, Afr. (mô-rĭsh´ĭ-ŭs)	**274**	20°17'S	57°33'E
Mauston, Wi., U.S. (môs´tŭn)	**126-27**	43°47'N	90°04'W

Feature (Pronunciation)	Page	Lat.	Long.
Mawlamyaing, Mya. *see* Mawlamyine	**266-67**	16°30'N	97°38'E
Mawlamyine, Mya.	**266-67**	16°30'N	97°38'E
Maxixe, Moz.	**286-87**	23°52'S	35°21'E
Maya, stm., Russia (mä´yä)	**236-37**	60°25'N	134°34'E
Mayaguana, i., Bah.	**154-55**	22°23'N	72°57'W
Mayagüez, P.R. (mä-yä-gwāz´)	**154a**	18°12'N	67°09'W
Mayfield, Ky., U.S. (mā´fēld)	**134-35**	36°45'N	88°38'W
Maykop, Russia	**202-03**	44°36'N	40°06'E
Maymyo, Mya. (mī´myō)	**266-67**	22°02'N	96°28'E
Mayo, Yk., Can. (mā-yō´)	**138-39**	63°36'N	135°51'W
Mayodan, N.C., U.S. (mā-yō´dăn)	**134-35**	36°24'N	79°59'W
Mayon Volcano, vol., Phil. (mä-yōn´ vōl-kā´nō)	**270**	13°15'N	123°41'E
Mayotte, dep., Afr. (mȧ-yŏt´)	**286-87**	12°50'S	45°10'E
Maysville, Ky., U.S. (māz´vĭl)	**126-27**	38°38'N	83°46'W
Mayumba, Gabon	**282-83**	3°22'S	10°40'E
Māyūram, India	**256**	11°06'N	79°39'E
Mayville, N.D., U.S. (mā´vĭl)	**124-25**	47°30'N	97°19'W
Mayville, Wi., U.S. (mā´vĭl)	**126-27**	43°30'N	88°32'W
Mayyit, Al-Baḥr al-, lk., Asia *see* Dead Sea	**248-49**	31°30'N	35°30'E
Maza, Arg.	**187**	36°48'S	63°20'W
Mazabuka, Zam. (mä-zä-bōō´kä)	**286-87**	15°51'S	27°46'E
Mazagan, Mor. *see* El-Jadida	**292a**	33°15'N	8°31'W
Mazagão, Braz. (mä-zȧ-gou´ɴ)	**180-81**	0°07'S	51°17'W
Mazara del Vallo, Italy (mät-sä´rä dĕl väl´lō)	**216-17**	37°39'N	12°36'E
Mazār-e Sharīf, Afg.	**252-53**	36°42'N	67°07'E
Mazarrón, Spain (mä-zär-rô´n)	**214-15**	37°36'N	1°19'W
Mazaruni, stm., Guy.	**178-79**	6°26'N	58°36'W
Mazatenango, Guat. (mä-zä-tå-näŋ´gō)	**160**	14°32'N	91°30'W
Mazatlán, Mex. (mä-zä-tlän´)	**158-59**	23°13'N	106°25'W
Mażeikiai, Lith. (mȧ-zhā´kė-ī)	**208-09**	56°19'N	22°21'E
Mazoe, stm., Afr. *see* Mazowe	**286-87**	16°32'S	33°26'E
Mazowe, stm., Afr.	**286-87**	16°32'S	33°26'E
Mazyr, Bela.	**218-19**	52°03'N	29°16'E
Mbabane, nat. cap., Swaz. (m'bä-bä´nē)	**286-87**	26°20'S	31°09'E
Mbaïki, C.A.R. (m'bȧ-ē´kė)	**284-85**	3°52'N	17°60'E
Mbala, Zam.	**286-87**	8°51'S	31°22'E
Mbale, Ug.	**289**	1°05'N	34°10'E
Mbandaka, D.R.C.	**284-85**	0°02'N	18°15'E
M'banza Congo, Ang.	**284-85**	6°16'S	14°15'E
Mbanza-Ngungu, D.R.C.	**284-85**	5°14'S	14°53'E
Mbarara, Ug.	**289**	0°36'S	30°38'E
Mbari, stm., C.A.R.	**284-85**	4°36'N	22°44'E
Mbeya, Tan.	**286-87**	8°54'S	33°30'E
Mbinda, Congo	**284-85**	2°07'S	12°53'E
Mbini, stm., Afr.	**282-83**	1°35'N	9°38'E
Mbomou, stm., Afr. (m'bō´mōō)	**284-85**	4°09'N	22°29'E
Mbour, Sen.	**282-83**	14°25'N	16°58'W
Mbuji-Mayi, D.R.C.	**284-85**	6°08'S	23°39'E
Mbuji-Mayi, stm., D.R.C.	**284-85**	6°02'S	23°44'E
McAdam, N.B., Can. (măk-ăd´ăm)	**148-49**	45°35'N	67°20'W
McAlester, Ok., U.S. (măk ăl´ĕs-tẽr)	**130-31**	34°56'N	95°46'W
McAllen, Tx., U.S. (măk-ăl´ĕn)	**132-33**	26°12'N	98°14'W
McBride, B.C., Can. (măk-brīd´)	**142-43**	53°18'N	120°10'W
McCamey, Tx., U.S. (mă-kā´mĭ)	**132-33**	31°08'N	102°13'W
McCauley Island, i., B.C., Can.	**142-43**	53°40'N	130°15'W
McColl, S.C., U.S.	**134-35**	34°40'N	79°33'W
McComb, Ms., U.S. (mȧ-kōm´)	**134-35**	31°14'N	90°27'W
McCook, Ne., U.S. (mȧ-ko͝ok´)	**130-31**	40°12'N	100°37'W
McGehee, Ar., U.S. (mȧ-gē´)	**134-35**	33°38'N	91°24'W
McGill, Nv., U.S. (mȧ-gĭl´)	**128-29**	39°25'N	114°49'W
McGrath, Ak., U.S. (măk-gräth)	**136**	62°58'N	155°38'W
McGregor, Tx., U.S. (măk-grĕg´ẽr)	**132-33**	31°26'N	97°24'W
McGregor, stm., B.C., Can. (măk-grĕg´ẽr)	**142-43**	54°10'N	122°01'W
McKeesport, Pa., U.S. (mȧ-kez´pōrt)	**126-27**	40°21'N	79°52'W
McKenzie, Tn., U.S. (mȧ-kĕn´zĭ)	**134-35**	36°08'N	88°31'W
McKinley, Mount, mtn., Ak., U.S. *see* Denali	**136**	63°04'N	151°00'W
McKinney, Tx., U.S. (mȧ-kĭn´ĭ)	**130-31**	33°12'N	96°37'W
McLaughlin, S.D., U.S. (mȧk-lŏf´lĭn)	**124-25**	45°49'N	100°48'W
McLennan, Ab., Can. (măk-lĭn´nȧn)	**142-43**	55°41'N	116°52'W
McLeod, stm., Ab., Can.	**142-43**	54°09'N	115°42'W
McLoughlin, Mount, mtn., Or., U.S. (mount mȧk-lŏk´lĭn)	**122-23**	42°27'N	122°19'W
McMinnville, Or., U.S. (măk-mĭn´vĭl)	**122-23**	45°13'N	123°11'W
McMinnville, Tn., U.S. (măk-mĭn´vĭl)	**134-35**	35°41'N	85°47'W
McPherson, Ks., U.S. (mȧk-fŭr´s'n)	**130-31**	38°22'N	97°40'W
McRae, Ga., U.S. (mȧk-rā´)	**134-35**	32°04'N	82°54'W
Mead, Lake, res., U.S. (lāk mēd)	**128-29**	36°08'N	114°26'W
Meade, stm., Ak., U.S.	**136**	70°55'N	156°00'W
Meadow Lake, Sk., Can. (mĕd´ō lāk)	**144-45**	54°08'N	108°26'W

n-sing; ŋ-baŋk; ɴ-nasalized n; nŏd; cŏmmit; ōld; ôbey; ôrder; oi-boil; fo͞od; ȯ-as oo in foot; ou-out; s-soft; sh-dish; th-thin; pūre; ûnite; ûrn; stŭd; circŭs; ü-as in French tu; ´-indeterminate vowel.

382 Meadville – Michalovce PRONOUNCING INDEX

Feature (Pronunciation)	Page	Lat.	Long.
Meadville, Pa., U.S.	126-27	41°39′N	80°09′W
Meaford, On., Can. (mē′fẽrd)	146-47	44°36′N	80°35′W
Meaux, Fr. (mō)	212-13	48°58′N	2°53′E
Mecca, Sau. Ar. (mĕk′á)	288	21°27′N	39°51′E
Mechanic Falls, Me., U.S. (mē-kăn′ĭk fôlz)	126-27	44°07′N	70°24′W
Mechanicsburg, Pa., U.S. (mē-kăn′ĭks-bûrg)	126-27	40°12′N	77°01′W
Mechanicsville, Va., U.S. (mē-kăn′ĭks-vĭl)	126-27	37°36′N	77°22′W
Mecubúri, stm., Moz.	286-87	14°10′S	40°32′E
Medan, Indon. (må-dän′)	266-67	3°35′N	98°41′E
Medanosa, Punta, c., Arg. (pōō′n-tä-mĕ-dä-nô′sä)	185	48°06′S	65°55′W
Médéa, Alg.	292b	36°12′N	2°51′E
Medellín, Col. (må-dhĕl-yēn′)	178-79	6°15′N	75°35′W
Medenine, Tun. (mā-dē-nēn′)	204-05	33°20′N	10°30′E
Medford, Ok., U.S. (mĕd′fẽrd)	130-31	36°49′N	97°43′W
Medford, Or., U.S. (mĕd′fẽrd)	122-23	42°20′N	122°52′W
Medford, Wi., U.S. (mĕd′fẽrd)	126-27	45°08′N	90°20′W
Medgyes, Rom. *see* Mediaș	210-11	46°10′N	24°22′E
Mediaș, Rom. (mĕd-yäsh′)	210-11	46°10′N	24°22′E
Medical Lake, Wa., U.S. (mĕd′ĭ-kǎl lāk)	122-23	47°37′N	117°43′W
Medicine Hat, Ab., Can. (mĕd′ĭ-sĭn hăt)	142-43	50°03′N	110°41′W
Medicine Lodge, Ks., U.S. (mĕd′ĭ-sĭn lŏj)	130-31	37°17′N	98°35′W
Medina, Sau. Ar. (må-dē′ná)	288	24°28′N	39°37′E
Medina, N.Y., U.S. (mē-dī′ná)	126-27	43°13′N	78°23′W
Medina, Oh., U.S. (mē-dī′ná)	126-27	41°08′N	81°51′W
Medina del Campo, Spain (må-dē′nä dĕl käm′pō)	214-15	41°19′N	4°55′W
Medina de Ríoseco, Spain (må-dē′nä dā rē-ō-sā′kō)	214-15	41°52′N	5°02′W
Medinīpur, India	254-55	22°26′N	87°20′E
Medio, Punta, c., Chile	182-83	27°07′S	70°56′W
Mediterranean Sea, s., (mĕd-ĭ-tĕr-ā′nē-ăn sē)	204-05	35°0′N	20°00′E
Méditerranée, Mer, s., *see* Mediterranean Sea	204-05	35°0′N	20°00′E
Mediterráneo, Mar, s., *see* Mediterranean Sea	204-05	35°0′N	20°00′E
Mediterraneo, Mar, s., *see* Mediterranean Sea	204-05	35°0′N	20°00′E
Mediterrània, Mar, s., *see* Mediterranean Sea	204-05	35°0′N	20°00′E
Mednogorsk, Russia	244	51°25′N	57°35′E
Médouneu, Gabon	282-83	0°59′N	10°55′E
Medveditsa, stm., Russia (mĕd-vyĕ′dē tsá)	202-03	49°35′N	42°39′E
Medvezhyegorsk, Russia	202-03	62°55′N	34°28′E
Medyn', Russia (mĕ-dēn′)	218-19	54°57′N	35°53′E
Meekatharra, Austl. (mē-kǎ-thăr′á)	294-95	26°35′S	118°30′E
Meeker, Co., U.S. (mēk′ẽr)	128-29	40°03′N	107°55′W
Meelpaeg Lake, res., Nf./L., Can. (mēl′pá-ĕg lāk)	148-49	48°16′N	56°35′W
Meerut, India (mē′rŏt)	254-55	28°59′N	77°42′E
Meghālaya, state, India	254-55	25°30′N	91°15′E
Meghna, stm., Bngl.	254-55	22°50′N	90°42′E
Megísti, i., Grc.	204-05	36°08′N	29°36′E
Mehun-sur-Yèvre, Fr. (mē-ŭN-sür-yĕvr′)	212-13	47°09′N	2°13′E
Meiganga, Camrn.	282-83	6°34′N	14°07′E
Meiktila, Mya.	266-67	20°52′N	95°52′E
Meixian, China *see* Meizhou	258-59	24°20′N	116°07′E
Meizhou, China	258-59	24°20′N	116°07′E
Mejillones, Chile (må-ĸē-lyō′nås)	182-83	23°06′S	70°27′W
Mek'elē, Eth.	288	13°30′N	39°28′E
Meknès, Mor. (mĕk′nĕs) (mĕk-nĕs′)	292a	33°54′N	5°33′W
Mekong, stm., Asia (mā-kông′)	266-67	10°33′N	105°27′E
Mékôngk, stm., Asia *see* Mekong	266-67	10°33′N	105°27′E
Mékrou, stm., Afr.	282-83	12°24′N	2°50′E
Melaka, Malay.	266-67	2°12′N	102°16′E
Melaka, Selat, strt., Asia *see* Malacca, Strait of	266-67	2°30′N	101°20′E
Melanesia, is., Oc. (mĕl-á-nē′-zhá)	304-05	13°0′S	164°00′E
Mélanésie, is., Oc. *see* Melanesia	304-05	13°0′S	164°00′E
Melawi, stm., Indon.	268-69	0°05′N	111°29′E
Melbourne, Austl. (mĕl′bŭrn)	301	37°49′S	144°57′E
Melbourne, Fl., U.S. (mĕl′bŭrn)	135a	28°05′N	80°37′W
Melbourne Island, i., Nu., Can.	140-41	68°30′N	104°45′W
Melchor, Isla, i., Chile	185	45°08′S	73°57′W
Melekeok, nat. cap., Palau	304-05	7°29′N	134°37′E
Meleuz, Russia	202-03	52°58′N	55°56′E
Mélèzes, stm., Qc., Can.	140-41	57°41′N	69°29′W
Melfi, Chad	280-81	11°03′N	17°56′E
Melfort, Sk., Can. (mĕl′fôrt)	144-45	52°52′N	104°36′W
Melilla, Sp. N. Afr. (mā-lēl′yä)	214-15	35°18′N	2°57′W
Melipilla, Chile (mā-lē-pē′lyä)	188e	33°41′S	71°13′W
Melita, Mb., Can.	144-45	49°16′N	100°59′W
Melitopol', Ukr. (mā-lē-tô′pōl-y′)	218-19	46°51′N	35°21′E
Mellen, Wi., U.S. (mĕl′ĕn)	124-25	46°20′N	90°40′W
Mellerud, Swe. (mâl′ĕ-rōōdh)	208-09	58°42′N	12°28′E
Melo, Ur. (mā′lō)	187	32°22′S	54°11′W
Melos, i., Grc. (mē′lŏs) *see* Mílos	216-17	36°41′N	24°28′E
Melrhir, Chott, lk., Alg.	280-81	34°18′N	6°17′E
Melrose, Mn., U.S. (mĕl′rōz)	124-25	45°40′N	94°49′W
Melton Mowbray, Eng., U.K. (mĕl′tŭn mō′brâ)	206-07	52°46′N	0°53′W
Melun, Fr. (mē-lŭN′)	212-13	48°32′N	2°40′E
Melville, Sk., Can. (mĕl′vĭl)	144-45	50°55′N	102°48′W
Melville, Cape, c., Austl. (kāp mĕl′vĭl)	302	14°11′S	144°30′E
Melville, Lake, lk., Nf./L., Can. (lāk mĕl′vĭl)	140-41	53°40′N	59°44′W
Melville Island, i., Austl. (mĕl′vĭl ī′lánd)	296-97	11°40′S	131°00′E
Melville Island, i., Can.	95	75°15′N	109°59′W
Melville Peninsula, pen., Nu., Can. (mĕl′vĭl pĕ-nĭn′sūlá)	140-41	68°0′N	84°00′W
Memel, Lith. *see* Klaipėda	208-09	55°43′N	21°08′E
Memel, S. Afr. (mĕ′mĕl)	292c	27°41′S	29°34′E
Memmingen, Ger. (mĕm′ĭng-ĕn)	210-11	47°59′N	10°11′E
Mempawah, Indon.	266-67	0°20′N	108°58′E
Memphis, Mo., U.S. (mĕm′fĭs)	130-31	40°28′N	92°10′W
Memphis, Tn., U.S. (mĕm′fĭs)	134-35	35°09′N	90°03′W
Memphis, Tx., U.S. (mĕm′fĭs)	130-31	34°44′N	100°33′W
Mena, Ukr. (mē-ná′)	218-19	51°31′N	32°14′E
Mena, Ar., U.S. (mē′ná)	130-31	34°35′N	94°15′W
Menado, Indon. *see* Manado	268-69	1°29′N	124°51′E
Ménaka, Mali	280-81	15°55′N	2°24′E
Menard, Tx., U.S. (mē-närd′)	132-33	30°55′N	99°47′W
Menasha, Wi., U.S. (mē-năsh′á)	126-27	44°12′N	88°26′W
Mendawai, stm., Indon.	268-69	3°14′S	113°19′E
Mende, Fr. (mänd)	212-13	44°30′N	3°30′E
Mendi, Pap. N. Gui.	302	6°10′S	143°40′E
Mendocino, Cape, c., Ca., U.S. (kāp mĕn′dô-sē′nō)	122-23	40°25′N	124°23′W
Mendota, Ca., U.S. (mĕn-dō′tá)	128-29	36°46′N	120°23′W
Mendota, Il., U.S. (mĕn-dō′tá)	126-27	41°33′N	89°07′W
Mendoza, Arg. (mĕn-dō′sä)	188e	32°53′S	68°49′W
Mendoza, state, Arg. (mĕn-dō′sä)	188e	34°30′S	68°30′W
Mengcheng, China (mŭŋ-chŭŋ)	258-59	33°16′N	116°33′E
Menggala, Indon.	266-67	4°29′S	105°15′E
Menghai, China	258-59	21°59′N	100°27′E
Menindee, Austl. (mĕ-nĭn-dē)	301	32°24′S	142°26′E
Menominee, Mi., U.S. (mē-nŏm′ĭ-nē)	126-27	45°08′N	87°37′W
Menominee, stm., U.S. (mē-nŏm′ĭ-nē)	124-25	45°06′N	87°36′W
Menongue, Ang.	286-87	14°39′S	17°41′E
Menorca, i., Spain (mĕ-nŏ′r-kä)	214-15	40°0′N	4°00′E
Mentawai, Selat, strt., Indon.	266-67	1°45′S	100°00′E
Menzel Bourguiba, Tun.	200-01	37°10′N	9°48′E
Meoqui, Mex.	132-33	28°16′N	105°29′W
Meppel, Neth. (mĕp′ĕl)	206-07	52°42′N	6°12′E
Meppen, Ger. (mĕp′ĕn)	210-11	52°42′N	7°18′E
Merauke, Indon. (må-rou′kä)	302	8°30′S	140°24′E
Merca, Som. *see* Marka	284-85	1°43′N	44°46′E
Merced, Ca., U.S. (mẽr-sĕd′)	128-29	37°18′N	120°29′W
Mercedario, Cerro, mtn., Arg. (sĕ′r-rō mĕr-så-dhä′rē-ō)	182-83	31°59′S	70°08′W
Mercedes, Arg. (mĕr-sā′dhās)	187	29°11′S	58°03′W
Mercedes, Arg. (mĕr-sā′dhās)	187	34°40′S	59°26′W
Mercedes, Ur.	187	33°15′S	58°02′W
Mercy, Cape, c., Nu., Can.	140-41	64°54′N	63°35′W
Merefa, Ukr. (må-rĕf′á)	218-19	49°51′N	36°05′E
Mergui, Mya. (mẽr-gē′)	266-67	12°26′N	98°37′E
Mergui Archipelago, is., Mya. (mẽr-gē′ är′kå-pĕ′-å-gō)	266-67	12°0′N	98°00′E
Mérida, Mex.	160	20°59′N	89°37′W
Mérida, Spain	214-15	38°55′N	6°20′W
Mérida, Ven. (mĕ′rē-dhä)	178-79	8°37′N	71°09′W
Meriden, Ct., U.S. (mĕr′ĭ-dĕn)	126-27	41°32′N	72°48′W
Meridian, Id., U.S. (mē-rĭd-ĭ-ăn)	122-23	43°36′N	116°21′W
Meridian, Ms., U.S. (mē-rĭd-ĭ-ăn)	134-35	32°22′N	88°42′W
Meridian, Tx., U.S. (mē-rĭd-ĭ-ăn)	132-33	31°55′N	97°40′W
Merikarvia, Fin. (mā′rē-kár′vē-á)	208-09	61°51′N	21°30′E
Merín, Laguna, b., S.A. *see* Mirim, Lagoa	187	32°45′S	52°50′W
Merir, i., Palau	304-05	4°19′N	132°19′E
Merkel, Tx., U.S. (mûr′kĕl)	130-31	32°28′N	100°01′W
Merrill, Mi., U.S. (mĕr′ĭl)	126-27	43°25′N	84°20′W
Merrill, Wi., U.S. (mĕr′ĭl)	126-27	45°11′N	89°41′W
Merritt, B.C., Can. (mĕr′ĭt)	142-43	50°06′N	120°46′W
Merryville, La., U.S. (mĕr′ĭ-vĭl)	132-33	30°45′N	93°33′W
Mersa Matruh, Egypt	204-05	31°21′N	27°14′E
Merseburg, Ger. (mĕr′zĕ-bōōrgh)	210-11	51°21′N	11°60′E
Mersin, Tur. *see* İçel	248-49	36°49′N	34°38′E
Merthyr Tydfil, Wales, U.K. (mûr′thẽr tĭd′vĭl)	206-07	51°46′N	3°23′W
Méru, Fr. (mā-rü′)	212-13	49°14′N	2°08′E
Meru, Kenya (mā′rōō)	289	0°03′N	37°39′E
Meru, Mount, vol., Tan.	289	3°14′S	36°45′E
Merzifon, Tur. (mĕr′ze-fōn)	202-03	40°52′N	35°27′E
Mesa, Az., U.S. (mā′sá)	128-29	33°24′N	111°49′W
Mesabi Range, hills, Mn., U.S. (mā-sŏb′bē rānj)	124-25	47°30′N	92°50′W
Mesagne, Italy (mā-sän′yå)	216-17	40°34′N	17°49′E
Mesa Verde National Park, n.p., Co., U.S. (mā′sá vĕr′dē năsh′ŭn-ăl pärk)	128-29	37°15′N	108°26′W
Mescalero Apache Indian Reservation, ind. res., N.M., U.S. (mĕs-kä-lā′rō ă-păch′ē ĭn′dĭ-ăn rĕ-sẽr-vā′shĕn)	130-31	33°12′N	105°40′W
Mesewa, Erit. *see* Massawa	288	15°37′N	39°26′E
Meshchovsk, Russia (myĕsh′chĕfsk)	218-19	54°19′N	35°17′E
Meshed, Iran *see* Mashhad	252-53	36°17′N	59°36′E
Mesogéios Thálassa, s., *see* Mediterranean Sea	204-05	35°0′N	20°00′E
Mesopotamia, hist. reg., Asia	248-49	34°0′N	44°00′E
Mesoyéios Thálassa, s., *see* Mediterranean Sea	204-05	35°0′N	20°00′E
Messalo, stm., Moz.	286-87	11°41′S	40°26′E
Messina, Italy (mĕ-sē′ná)	216-17	38°11′N	15°33′E
Messina, Stretto di, strt., Italy (stĕ′t-tô dē mĕ-sē′ná)	216-17	38°09′N	15°35′E
Meta, stm., S.A.	178-79	6°11′N	67°28′W
Métabetchouane, stm., Qc., Can. (mĕ-tá-bĕt-chōō-än′)	146-47	48°26′N	71°58′W
Meta Incognita Peninsula, pen., Nu., Can.	140-41	62°45′N	68°30′W
Metán, Arg. (mĕ-tá′n)	182-83	25°30′S	64°57′W
Metapán, El Sal. (må-täpän′)	160	14°20′N	89°26′W
Metković, Cro. (mĕt′kō-vĭch)	216-17	43°03′N	17°39′E
Metlakatla, Ak., U.S. (mĕt-lá-kăt′lá)	136	55°07′N	131°35′W
Metropolis, Il., U.S. (mē-trŏp′ō-lĭs)	134-35	37°09′N	88°44′W
Metter, Ga., U.S. (mĕt′ẽr)	134-35	32°24′N	82°04′W
Metz, Fr. (mĕtz)	212-13	49°08′N	6°10′E
Meulaboh, Indon.	266-67	4°09′N	96°08′E
Mexia, Tx., U.S. (må-hē′ä)	132-33	31°40′N	96°29′W
Mexiana, Ilha, i., Braz.	180-81	0°02′S	49°35′W
Mexicali, Mex. (måk-sē-kä′lē)	156-57	32°39′N	115°30′W
Mexicana, Altiplanicie, plat., Mex.	4-5	25°29′N	104°00′W
Mexican Hat, Ut., U.S. (mĕk′sĭ-kăn hăt)	128-29	37°12′N	109°52′W
Mexico, Me., U.S. (mĕk′sĭ-kō)	126-27	44°34′N	70°33′W
Mexico, Mo., U.S. (mĕk′sĭ-kō)	130-31	39°10′N	91°53′W
Mexico, nation, N.A. (mĕk′sĭ-kō)	94	23°0′N	102°00′W
México, state, Mex.	158-59	19°20′N	99°45′W
México, Golfo de, b., N.A. *see* Mexico, Gulf of	152-53	25°0′N	90°00′W
Mexico, Gulf of, b., N.A. (gŭlf ŭv mĕk′sĭ-kō)	152-53	25°0′N	90°00′W
Mexico City, nat. cap., Mex. (mĕk′sĭ-kō sĭ′tē)	158-59	19°24′N	99°09′W
Meyersdale, Pa., U.S. (mī′ẽrz-dāl)	126-27	39°49′N	79°02′W
Meymaneh, Afg.	252-53	35°56′N	64°48′E
Mezen', Russia	202-03	65°50′N	44°15′E
Mezen', stm., Russia	202-03	65°53′N	44°09′E
Mézenc, Mont, mtn., Fr. (mŏN-mā-zĕN′)	212-13	44°55′N	4°11′E
Mezha, stm., Russia (myā′zhá)	218-19	55°43′N	31°31′E
Mezhdurechensk, Russia	244	53°41′N	88°07′E
Mezőkövesd, Hung. (mĕ′zû-kû′vĕsht)	210-11	47°48′N	20°35′E
Mezőtúr, Hung. (mĕ′zû-tōōr)	210-11	47°00′N	20°37′E
Mezquital, Mex. (måz-kē-täl′)	158-59	23°29′N	104°22′W
Mfangano Island, i., Kenya	289	0°28′S	34°01′E
M'Goun, Irhil, mtn., Mor.	280-81	31°31′N	6°25′W
Miahuatlán de Porfirio Díaz, Mex.	158-59	16°19′N	96°36′W
Miajadas, Spain (mē-ä-hä′dās)	214-15	39°09′N	5°54′W
Miami, Fl., U.S. (mī-ă′-mē)	135a	25°47′N	80°13′W
Miami Beach, Fl., U.S.	135a	25°48′N	80°08′W
Miāneh, Iran	248-49	37°26′N	47°42′E
Mianyang, China	258-59	31°28′N	104°44′E
Miaoli, Tai. (mē-ou′lĭ)	243a	24°33′N	120°49′E
Miass, Russia (mī-äs′)	244	54°59′N	60°06′E
Michalovce, Slvk. (mē′ĸä-lôf′tsĕ)	210-11	48°46′N	21°56′E

ăt; finăl; rāte; senåte; ärm; åsk; sofá; fâre; ch-choose; dh-as th in other; bē; ēvent; bĕt; recĕnt; cratẽr; g-gō; gh-guttural g; bĭt; ĭ-short neutral; rīde; ĸ-guttural k as ch in German ich;

PRONOUNCING INDEX Michelson, Mount – Mobridge 383

Feature (Pronunciation)	Page	Lat.	Long.
Michelson, Mount, mtn., Ak., U.S. (mount mĭch´ĕl-sŭn)	**136**	69°19'N	144°17'W
Michigan, state, U.S. (mĭsh-´ĭ-găn)	**118-19**	44°0'N	85°00'W
Michigan, Lake, lk., U.S. (lāk mĭsh-´ĭ-găn)	**126-27**	44°0'N	87°00'W
Michigan City, In., U.S. (mĭsh-´ĭ-găn sĭ´tė)	**126-27**	41°43'N	86°53'W
Michipicoten Island, i., On., Can.	**146-47**	47°45'N	85°45'W
Michoacán, state, Mex.	**158-59**	19°10'N	101°50'W
Michurinsk, Russia (mĭ-chōō-rĭnsk´)	**202-03**	52°54'N	40°29'E
Micronesia, is., Oc. (mī-krō-nē´zhá)	**304-05**	11°0'N	159°00'E
Micronesia, Federated States of, nation, Oc. (fĕ´ĕr-ā´ĕd stāts ŭv mī-krō-nē´zhá)	**304-05**	5°0'N	152°00'E
Middelburg, S. Afr.	**286-87**	31°30'S	25°00'E
Middelfart, Den. (mĕd´'l-färt)	**208-09**	55°30'N	9°45'E
Middle, stm., B.C., Can. (mĕd´'l)	**142-43**	54°52'N	125°08'W
Middle Andaman, i., India (mĕd´'l ăn-dá-măn´)	**266-67**	12°30'N	92°50'E
Middle Caicos, i., T./C. Is.	**154-55**	21°48'N	71°47'W
Middlesboro, Ky., U.S. (mĭd´'lz-bŭr-ô)	**134-35**	36°36'N	83°43'W
Middlesbrough, Eng., U.K. (mĭd´'lz-brŭ)	**206-07**	54°34'N	1°14'W
Middleton, N.S., Can. (mĭd´'l-tŭn)	**148-49**	44°57'N	65°04'W
Middleton Island, i., Ak., U.S.	**136**	59°26'N	146°19'W
Middletown, Oh., U.S.	**126-27**	39°31'N	84°23'W
Midland, On., Can. (mĭd´lănd)	**146-47**	44°45'N	79°52'W
Midland, Mi., U.S.	**126-27**	43°36'N	84°14'W
Midland, Tx., U.S.	**132-33**	32°00'N	102°05'W
Midway, Ky., U.S. (mĭd´wā)	**126-27**	38°08'N	84°42'W
Midway Islands, dep., Oc. (mĭd´wā ī´lándz)	**304-05**	28°13'N	177°22'W
Międzyrzecz, Pol. (myăn-dzû´zhĕch)	**210-11**	52°27'N	15°35'E
Mier, Mex. (myâr)	**132-33**	26°26'N	99°09'W
Mieres, Spain (myā´rās)	**214-15**	43°16'N	5°46'W
Mier y Noriega, Mex. (myâr´ē nô-rē-ā´gä)	**158-59**	23°25'N	100°08'W
Miguel Alemán, Presa, res., Mex. (prā'sä-mē-gāl´-ä-lĕ-mà´n)	**158-59**	18°13'N	96°32'W
Mikhaylov, Russia (mē-käy´lôf)	**218-19**	54°14'N	39°02'E
Mikhaylovka, Russia	**202-03**	50°04'N	43°15'E
Mikkeli, Fin. (mēk´ĕ-lĭ)	**208-09**	61°42'N	27°16'E
Mikun', Russia	**202-03**	62°22'N	50°05'E
Mikura-jima, i., Japan (mē´kōō-rá jē´má)	**264**	33°52'N	139°36'E
Milaca, Mn., U.S. (mė-lak´a)	**124-25**	45°45'N	93°39'W
Milagro, Arg.	**182-83**	31°01'S	65°60'W
Milagro, Ec.	**184**	2°08'S	79°36'W
Milan, Italy (mē-län´)	**216-17**	45°28'N	9°12'E
Milan, Mi., U.S. (mī´lăn)	**126-27**	42°05'N	83°41'W
Milan, Mo., U.S. (mī´lăn)	**130-31**	40°12'N	93°07'W
Milan, Oh., U.S. (mī´lăn)	**126-27**	41°10'N	82°37'W
Milan, Tn., U.S. (mī´lăn)	**134-35**	35°55'N	88°46'W
Milano, Italy (mė-lä´nō) see Milan	**216-17**	45°28'N	9°12'E
Milâs, Tur. (mė´läs)	**216-17**	37°19'N	27°47'E
Milbank, S.D., U.S. (mĭl´băŋk)	**124-25**	45°13'N	96°38'W
Mildura, Austl. (mĭl-dū´rá)	**301**	34°12'S	142°10'E
Mile, China	**258-59**	24°26'N	103°27'E
Miles, Austl.	**301**	26°40'S	150°11'E
Miles City, Mt., U.S. (mīlz sĭ´tė)	**122-23**	46°25'N	105°50'W
Milford, Ct., U.S. (mĭl´fērd)	**126-27**	41°13'N	73°04'W
Milford, De., U.S. (mĭl´fērd)	**126-27**	38°55'N	75°26'W
Milford, Ne., U.S. (mĭl´fērd)	**124-25**	40°46'N	97°03'W
Milford, Ut., U.S. (mĭl´fērd)	**128-29**	38°24'N	113°01'W
Milford Sound, b., N.Z.	**303**	44°31'S	167°48'E
Milk, stm., N.A.	**122-23**	48°03'N	106°19'W
Mil'kovo, Russia	**236-37**	54°42'N	158°38'E
Millau, Fr. (mē-yō´)	**212-13**	44°06'N	3°05'E
Milledgeville, Ga., U.S. (mĭl´ĕj-vĭl)	**134-35**	33°05'N	83°14'W
Mille Lacs, Lac des, lk., On., Can. (läk dĕ mēl läks)	**146-47**	48°50'N	90°30'W
Mille Lacs Lake, lk., Mn., U.S.	**124-25**	46°15'N	93°40'W
Millen, Ga., U.S. (mĭl´ĕn)	**134-35**	32°48'N	81°56'W
Millenium, at., Kir. see Caroline	**304-05**	9°58'S	150°13'W
Miller, S.D., U.S. (mĭl´ēr)	**124-25**	44°31'N	98°59'W
Millerovo, Russia (mĭl´ĕ-rô-vô)	**202-03**	48°56'N	40°24'E
Millersburg, Ky., U.S. (mĭl´ĕrz-bûrg)	**126-27**	38°18'N	84°09'W
Millersburg, Oh., U.S. (mĭl´ĕrz-bûrg)	**126-27**	40°33'N	81°54'W
Millicent, Austl. (mĭl-ĭ-sĕnt)	**301**	37°36'S	140°20'E
Millinocket, Me., U.S. (mĭl-ĭ-nŏk´ĕt)	**127a**	45°40'N	68°42'W
Mills Lake, lk., N.T., Can.	**140-41**	61°30'N	118°10'W
Mílos, i., Grc. (mē´lŏs)	**216-17**	36°41'N	24°28'E
Milton, On., Can. (mĭl´tŭn)	**146-47**	43°31'N	79°53'W
Milton, Fl., U.S. (mĭl´tŭn)	**134-35**	30°38'N	87°02'W
Milton, Pa., U.S. (mĭl´tŭn)	**126-27**	41°01'N	76°51'W
Milton, Wi., U.S. (mĭl´tŭn)	**126-27**	42°47'N	88°56'W
Milwaukee, Wi., U.S. (mĭl-wô´kė)	**126-27**	43°01'N	87°56'W
Min, stm., China (mĕn)	**240-41**	26°04'N	119°33'E
Min, stm., China (mĕn)	**258-59**	28°46'N	104°38'E
Minami-Daitō-jima, i., Japan	**240-41**	25°50'N	131°15'E
Minami-Iō-jima, i., Japan	**304-05**	24°14'N	141°28'E
Minami-Tori-shima, i., Japan see Marcus Island	**304-05**	24°18'N	153°58'E
Minas, Cuba (mē´näs)	**154-55**	21°29'N	77°36'W
Minas, Ur. (mē´näs)	**187**	34°23'S	55°14'W
Minas Basin, b., N.S., Can. (mī´nás bā´s'n)	**148-49**	45°20'N	64°00'W
Minas Channel, strt., N.S., Can. (mī´nás chăn´ĕl)	**148-49**	45°15'N	64°45'W
Minas de Oro, Hond. (mē´näs-dĕ-dĕ-ō-rô)	**161**	14°46'N	87°20'W
Minas Gerais, state, Braz.	**186**	18°0'S	44°00'W
Minas Novas, Braz. (mē´näzh nō´väzh)	**186**	17°15'S	42°36'W
Minatitlán, Mex. (mē-nä-tė-tlän´)	**158-59**	17°59'N	94°32'W
Mindanao, i., Phil. (mĭn-dä-nou´)	**270**	8°0'N	125°00'E
Mindanao Sea, s., Phil. (mĭn-dä-nou´ sē) see Bohol Sea	**270**	9°10'N	124°25'E
Mindelo, C.V.	**282-83**	16°52'N	24°60'W
Minden, Ger. (mĭn´dĕn)	**210-11**	52°18'N	8°55'E
Mindoro, i., Phil. (mĭn-dō´rō)	**270**	12°50'N	121°05'E
Mindoro Strait, strt., Phil. (mĭn-dō´rō strāt)	**270**	12°20'N	120°40'E
Mineiros, Braz.	**182-83**	17°34'S	52°34'W
Mineola, Tx., U.S. (mĭn-ē-ō´lá)	**130-31**	32°40'N	95°29'W
Mineral'nyye Vody, Russia	**245**	44°12'N	43°08'E
Mineral Point, Wi., U.S. (mĭn´ēr-ăl point)	**126-27**	42°52'N	90°10'W
Mineral Wells, Tx., U.S. (mĭn´ēr-ăl wĕlz)	**130-31**	32°48'N	98°07'W
Minfeng, China	**244**	37°04'N	82°39'E
Mingäçevir, Azer.	**245**	40°46'N	47°02'E
Mingäçevir su anbarı, res., Azer.	**245**	40°55'N	46°48'E
Mingāora, Pak.	**252-53**	34°49'N	72°21'E
Mingechaur, Azer. see Mingäçevir	**245**	40°46'N	47°02'E
Mingechaur Reservoir, res., Azer. see Mingäçevir su anbarı	**245**	40°55'N	46°48'E
Minicoy Island, i., India	**256**	8°16'N	73°03'E
Minigwal, Lake, lk., Austl.	**296-97**	29°35'S	123°12'E
Minle, China	**260-61**	38°28'N	100°56'E
Minna, Nig. (mĭn´á)	**282-83**	9°37'N	6°33'E
Minneapolis, Mn., U.S. (mĭn-ē-ăp´ō-lĭs)	**124-25**	44°59'N	93°17'W
Minnedosa, Mb., Can. (mĭn-ē-dō´sá)	**144-45**	50°14'N	99°49'W
Minneota, Mn., U.S. (mĭn-ē-ō´tá)	**124-25**	44°34'N	96°00'W
Minnesota, state, U.S. (mĭn-ē-sō´tá)	**118-19**	46°0'N	94°15'W
Minnesota, stm., Mn., U.S.	**124-25**	44°54'N	93°11'W
Minnitaki Lake, lk., On., Can. (mĭ´nĭ tä´kē lāk)	**144-45**	49°58'N	92°00'W
Minonk, Il., U.S. (mī´nŏnk)	**126-27**	40°54'N	89°02'W
Minorca, i., Spain see Menorca	**214-15**	40°0'N	4°00'E
Minot, N.D., U.S.	**124-25**	48°14'N	101°18'W
Minsk, state, Bela. (mênsk)	**210-11**	53°45'N	27°45'E
Minsk, nat. cap., Bela. (mênsk)	**210-11**	53°54'N	27°33'E
Mińsk Mazowiecki, Pol. (mên´sk mä-zô-vyĕt´skĭ)	**210-11**	52°11'N	21°34'E
Minto, Lac, lk., Qc., Can.	**140-41**	57°12'N	74°58'W
Minturno, Italy (mēn-tōōr´nō)	**216-17**	41°16'N	13°45'E
Minxian, China	**258-59**	34°26'N	104°02'E
Minya, Egypt see El-Minya	**291b**	28°06'N	30°45'E
Minya Konka, mtn., China see Gongga Shan	**258-59**	29°35'N	101°51'E
Min'yar, Russia	**202-03**	55°03'N	57°33'E
Miracema do Tocantins, Braz.	**180-81**	9°33'S	48°24'W
Mirador, Braz. (mē-rá-dōr´)	**180-81**	6°22'S	44°22'W
Miraflores, Col. (mē-rä-flō´rās)	**178-79**	1°25'N	72°17'W
Miramar, Arg.	**187**	38°16'S	57°51'W
Miramichi, N.B., Can.	**148-49**	47°02'N	65°28'W
Miramichi Bay, b., N.B., Can. (mĭr´á-mē´shē bā)	**148-49**	47°08'N	65°08'W
Miranda, stm., Braz.	**182-83**	19°25'S	57°20'W
Miranda de Ebro, Spain (mē-rä´n-dä-dĕ-ĕ´brō)	**214-15**	42°42'N	2°56'W
Miranda do Douro, Port. (mē-rän´dä dô-dwĕ´rô)	**214-15**	41°30'N	6°16'W
Mirandela, Port. (mē-rän-dā´lá)	**214-15**	41°29'N	7°11'W
Mirecourt, Fr. (mēr-kōōr´)	**212-13**	48°18'N	6°08'E
Miri, Malay. (mē´rē)	**268-69**	4°23'N	113°59'E
Mirim, Lagoa, b., S.A. (lá-gô´ä-mė-rēɴ´)	**187**	32°45'S	52°50'W
Mirnyy, Russia	**236-37**	62°31'N	113°59'E
Mīrpur Khās, Pak. (mēr´pōōr кäs)	**252-53**	25°31'N	69°01'E
Mirzāpur, India (mēr´zä-pōōr)	**254-55**	25°08'N	82°34'E
Misâha, Bîr, well, Egypt	**288**	22°12'N	27°57'E
Misantla, Mex. (mė-sän´tlä)	**158-59**	19°56'N	96°50'W
Miscou Island, i., N.B., Can. (mĭs´kō ī´lánd)	**148-49**	47°57'N	64°32'W
Mishan, China (mĭ´shän)	**264**	45°32'N	131°52'E
Mishawaka, In., U.S. (mĭsh-á-wôk´á)	**126-27**	41°40'N	86°10'W
Mishmi Hills, hills, Asia	**258-59**	29°0'N	96°00'E
Misima Island, i., Pap. N. Gui.	**302**	10°42'S	152°45'E
Misiones, state, Arg. (mē-syō´nās)	**187**	27°0'S	55°00'W
Miskitos, Cayos, is., Nic.	**161**	14°23'N	82°46'W
Miskolc, Hung. (mĭsh´kôlts)	**210-11**	48°06'N	20°47'E
Misool, Pulau, i., Indon. (pōō-lou mė-sôl´)	**242-43**	1°52'S	130°10'E
Mişr, nation, Afr. see Egypt	**274**	27°0'N	30°00'E
Mişrātah, Libya	**204-05**	32°22'N	15°06'E
Missinaibi, stm., On., Can. (mĭs´ĭn-ä´ė-bė)	**140-41**	50°45'N	81°31'W
Missinaibi Lake, lk., On., Can. (mĭs´ĭn-ä´ė-bė lāk)	**146-47**	48°21'N	83°43'W
Mission, S.D., U.S. (mĭsh´ŭn)	**124-25**	43°18'N	100°38'W
Mission, Tx., U.S. (mĭsh´ŭn)	**132-33**	26°13'N	98°19'W
Mississippi, state, U.S. (mĭs-ĭ-sĭp´ė)	**118-19**	32°50'N	89°30'W
Mississippi, stm., U.S. (mĭs-ĭ-sĭp´ė)	**120-21**	28°60'N	89°08'W
Mississippi River Delta, del., La., U.S.	**120-21**	29°10'N	89°15'W
Mississippi Sound, strt., U.S. (mĭs-ĭ-sĭp´ė sound)	**134-35**	30°15'N	88°40'W
Missoula, Mt., U.S. (mĭ-zōō´lá)	**122-23**	46°52'N	114°00'W
Missouri, state, U.S. (mĭ-sōō´rė)	**118-19**	38°30'N	93°30'W
Missouri, stm., U.S. (mĭ-sōō´rė)	**120-21**	38°49'N	90°07'W
Missouri City, Tx., U.S. (mĭ-sōō´rė sĭ´tĭ)	**132-33**	29°37'N	95°31'W
Missouri Valley, Ia., U.S. (mĭ-sōō´rė văl´ė)	**124-25**	41°33'N	95°54'W
Mistassibi, stm., Qc., Can.	**146-47**	48°53'N	72°14'W
Mistassini, Lac, lk., Qc., Can. (läk mĭs-tá-sĭ´nė)	**140-41**	51°0'N	73°37'W
Misti, Volcán, vol., Peru	**184**	16°18'S	71°24'W
Mita, Punta de, c., Mex. (pōō´n-tä-dĕ-mē´tä)	**158-59**	20°47'N	105°32'W
Mitau, Lat. see Jelgava	**208-09**	56°39'N	23°44'E
Mitchell, Austl.	**301**	26°29'S	147°58'E
Mitchell, In., U.S. (mĭch´ĕl)	**126-27**	38°44'N	86°29'W
Mitchell, Ne., U.S. (mĭch´ĕl)	**124-25**	41°57'N	103°48'W
Mitchell, S.D., U.S. (mĭch´ĕl)	**124-25**	43°43'N	98°02'W
Mitchell, Mount, mtn., N.C., U.S. (mount mĭch´ĕl)	**134-35**	35°46'N	82°16'W
Mitiaro, i., Cook Is.	**304-05**	19°48'S	157°43'W
Mito, Japan	**265**	36°22'N	140°29'E
Mitsio, Nosy, i., Madag.	**286-87**	12°54'S	48°36'E
Mitsiwa, Erit. see Massawa	**288**	15°37'N	39°26'E
Mittellandkanal, can., Ger. (mĭt´ĕl-lănd kä-näl´)	**210-11**	52°14'N	11°43'E
Mitú, Col.	**178-79**	1°08'N	70°03'W
Mitumba, Monts, mts., D.R.C.	**289**	6°0'S	29°00'E
Mitzic, Gabon	**282-83**	0°47'N	11°34'E
Miyake-jima, i., Japan (mė´yä-kå jē´má)	**264**	34°05'N	139°32'E
Miyako, Japan	**264**	39°38'N	141°57'E
Miyako-jima, i., Japan	**306a**	24°47'N	125°20'E
Miyakonojō, Japan	**265**	31°43'N	131°04'E
Miyazaki, Japan	**265**	31°54'N	131°26'E
Miyazu, Japan	**265**	35°32'N	135°11'E
Miyoshi, Japan (mė-yō´shė´)	**265**	34°49'N	132°51'E
Miyun, China	**260-61**	40°22'N	116°50'E
Mizdah, Libya (mĕz´dä)	**204-05**	31°26'N	12°59'E
Mizen Head, c., Ire.	**206-07**	51°27'N	9°49'W
Mizil, Rom. (mē´zēl)	**216-17**	44°59'N	26°27'E
Mizoram, state, India	**266-67**	23°30'N	93°00'E
Mizque, Bol.	**182-83**	17°57'S	65°20'W
Mjölby, Swe. (myûl´bü)	**208-09**	58°20'N	15°09'E
Mjøsa, lk., Nor. (myûsä)	**208-09**	60°40'N	11°00'E
Mkinvartsveri, Mt'a, vol., see Kazbek, Gora	**245**	42°42'N	44°31'E
Mladá Boleslav, Czech Rep. (mlä´dä bô´lĕ-slåf)	**210-11**	50°25'N	14°54'E
Mlanje Peak, mtn., Malawi see Sapitwa	**286-87**	15°57'S	35°36'E
Mława, Pol. (mwä´vá)	**210-11**	53°07'N	20°22'E
Moa, stm., Afr.	**282-83**	6°60'N	11°34'W
Moa, Pulau, i., Indon.	**268-69**	8°10'S	127°56'E
Moab, Ut., U.S. (mō´ăb)	**128-29**	38°35'N	109°33'W
Moa Island, i., Austl.	**302**	10°12'S	142°16'E
Moala, i., Fiji	**306f**	18°36'S	179°53'E
Moanda, Gabon	**282-83**	1°34'S	13°13'E
Moba, D.R.C.	**289**	7°04'S	29°44'E
Moberly, Mo., U.S. (mō´bēr-lĭ)	**130-31**	39°25'N	92°26'W
Mobile, Al., U.S. (mô-bēl´)	**134-35**	30°41'N	88°03'W
Mobile Bay, b., Al., U.S. (mô-bēl´ bā)	**134-35**	30°34'N	87°60'W
Mobridge, S.D., U.S. (mō´brĭj)	**124-25**	45°32'N	100°26'W

n-sing; ŋ-baŋk; ɴ-nasalized n; nŏd; cŏmmit; ōld; ôbey; ôrder; oi-boil; fōōd; ŏ-as oo in foot; ou-out; s-soft; sh-dish; th-thin; pūre; ûnite; ûrn; stŭd; circŭs; ü-as in French tu; ´-indeterminate vowel.

384 Mobutu Sese Seko, Lac – Moorea PRONOUNCING INDEX

Feature (Pronunciation)	Page	Lat.	Long.
Mobutu Sese Seko, Lac, lk., Afr. see Albert, Lake	289	1°40'N	31°00'E
Moca, Dom. Rep. (mô'kä)	154-55	19°24'N	70°31'W
Moçambique, Moz. (mō-säN-bē'kĕ) see Ilha de Moçambique	286-87	15°02'S	40°41'E
Moçambique, nation, Afr. see Mozambique	274	18°15'S	35°00'E
Moçambique, Canal de, strt., Afr. see Mozambique Channel	286-87	19°0'S	41°00'E
Moçâmedes, Ang. (mô-zà-mĕ-dĕs) see Namibe	286-87	15°12'S	12°10'E
Mocha, Yemen	288	13°19'N	43°15'E
Mocha, Isla, i., Chile	185	38°22'S	73°55'W
Mochudi, Bots. (mō-chōō'dê)	292c	24°23'S	26°09'E
Mocímboa da Praia, Moz. (mô-sē'êm-bô-à prä'ēà)	286-87	11°20'S	40°22'E
Môco, Morro de, mtn., Ang.	286-87	12°28'S	15°10'E
Mococa, Braz. (mô-kô'kà)	186	21°28'S	46°60'W
Mocorito, Mex.	156-57	25°29'N	107°55'W
Moctezuma, Mex. (môk'tå-zōō'mä)	156-57	29°48'N	109°42'W
Mocuba, Moz.	286-87	16°51'S	36°60'E
Modder, stm., S. Afr.	286-87	29°03'S	24°38'E
Modena, Italy (mô'dĕ-nä)	216-17	44°39'N	10°55'E
Modesto, Ca., U.S.	128-29	37°39'N	120°60'W
Modimolle, S. Afr.	292c	24°42'S	28°25'E
Mödling, Aus. (mûd'lĭng)	210-11	48°05'N	16°18'E
Moe, Austl.	301	38°11'S	146°15'E
Moengo, Sur.	178-79	5°38'N	54°24'W
Moeris, Lake, lk., Egypt see Qârûn, Birket	291b	29°28'N	30°39'E
Moero, Lac, lk., Afr. see Mweru, Lake	284-85	9°0'S	28°45'E
Mogadishu, nat. cap., Som. (mŏg'à-dĭ'shōō)	284-85	2°03'N	45°20'E
Mogador, Mor. see Essaouira	280-81	31°30'N	9°45'W
Mogaung, Mya. (mō-gä'òng)	258-59	25°18'N	96°56'E
Mogilno, Pol. (mô-gēl'nô)	210-11	52°40'N	17°59'E
Mogocha, Russia	236-37	53°44'N	119°45'E
Mogok, Mya. (mō-gōk')	266-67	22°56'N	96°31'E
Mogotón, mtn., N.A.	161	13°45'N	86°23'W
Moguer, Spain (mō-gĕr')	214-15	37°16'N	6°50'W
Mohács, Hung. (mô'häch)	210-11	46°00'N	18°41'E
Mohall, N.D., U.S. (mō'hôl)	124-25	48°46'N	101°30'W
Mohammedia, Mor.	292a	33°42'N	7°23'W
Mohe, China (mwo-hŭ)	240-41	53°29'N	122°20'E
Mohéli, i., Com. see Mwali	286-87	12°18'S	43°42'E
Mohyliv-Podil's'kyi, Ukr.	218-19	48°28'N	27°47'E
Mo i Rana, Nor.	200-01	66°19'N	14°10'E
Moisie, stm., Qc., Can.	148-49	50°15'N	66°05'W
Moissac, Fr. (mwä-sàk')	212-13	44°06'N	1°05'E
Mojave, Ca., U.S. (mô-hä'vā)	128-29	35°04'N	118°10'W
Mojave Desert, des., Ca., U.S. (mô-hä'vā dĕs'ērt)	128-29	35°0'N	117°00'W
Mojiguaçu, stm., Braz. (mô-gê-gwà'sōō)	186	20°54'S	48°11'W
Moknine, Tun.	200-01	35°38'N	10°54'E
Mokpo, Kor., S.	263	34°48'N	126°24'E
Moksha, stm., Russia	202-03	54°45'N	41°53'E
Moldavia, nation, Eur. see Moldova	190-91	47°0'N	29°00'E
Molde, Nor. (môl'dĕ)	200-01	62°45'N	7°11'E
Moldova, nation, Eur. (mäl-dō'và)	190-91	47°0'N	29°00'E
Moldoveanu, Vârful, mtn., Rom.	216-17	45°36'N	24°44'E
Molepolole, Bots. (mō-lå-pô-lō'lå)	286-87	24°25'S	25°31'E
Molfetta, Italy (môl-fĕt'tä)	216-17	41°12'N	16°36'E
Molina, Chile (mō-lē'nä)	185	35°07'S	71°17'W
Molina de Aragón, Spain (mô-lē'nä dĕ ä-rä-gô'n)	214-15	40°51'N	1°53'W
Molina de Segura, Spain (mô-lē'nä dĕ sĕ-gōō'rä)	214-15	38°03'N	1°13'W
Moline, Il., U.S. (mô-lēn')	124-25	41°30'N	90°29'W
Mollendo, Peru (mô-lyĕn'dō)	184	17°01'S	72°02'W
Mölndal, Swe. (mûln'däl)	208-09	57°41'N	11°56'E
Moloka'i, i., Hi., U.S. (mō-lô kä'ē)	137a	21°07'N	157°00'W
Molopo, stm., Afr. (mō-lō-pô)	286-87	28°31'S	20°13'E
Molson Lake, lk., Mb., Can. (mōl'sŭn lāk)	144-45	54°12'N	96°45'W
Moluccas, is., Indon. (mō-lŭk'ŭz)	268-69	2°0'S	128°00'E
Molucca Sea, s., Indon. (mō-lŭk'à sē)	268-69	0°13'N	125°10'E
Moma, Moz.	286-87	16°50'S	39°09'E
Mombasa, Kenya (mōm-bä'sä)	284-85	4°03'S	39°40'E
Mombetsu, Japan (môm'bĕt-sōō')	264	44°21'N	143°21'E
Mompós, Col. (mōm-pōs')	178-79	9°12'N	74°25'W
Møn, i., Den. (mûn)	208-09	55°00'N	12°20'E
Mona, Canal de la, strt., N.A. see Mona Passage	155b	18°30'N	67°45'W
Mona, Isla de, i., P.R.	155b	18°05'N	67°54'W
Monaco, nation, Eur. (mŏn'à-kō)	212-13	43°45'N	7°25'E
Mona Passage, strt., N.A. (mō'nă päs'ĭj)	155b	18°30'N	67°45'W
Monarch Mountain, mtn., B.C., Can. (mŏn'ērk moun'tĭn)	142-43	51°54'N	125°53'W
Monastir, Mac. see Bitola	216-17	41°02'N	21°20'E
Monastyrshchina, Russia (mô-nás-tērsh'chĭ-nà)	218-19	54°21'N	31°51'E
Monchegorsk, Russia (mōn'chĕ-gôrsk)	200-01	67°55'N	32°50'E
Monclova, Mex. (môn-klō'vä)	132-33	26°54'N	101°25'W
Moncton, N.B., Can. (mŭngk'tŭn)	148-49	46°06'N	64°48'W
Mondego, stm., Port. (mōn-dĕ'gō)	214-15	40°08'N	8°45'W
Mondego, Cabo, c., Port. (kà'bō môn-dā'gò)	214-15	40°11'N	8°54'W
Mondovi, Wi., U.S. (mŏn-dō'vĭ)	124-25	44°34'N	91°39'W
Monett, Mo., U.S. (mô-nĕt')	130-31	36°55'N	93°56'W
Monforte de Lemos, Spain (môn-fôr'tä dĕ lĕ'môs)	214-15	42°32'N	7°30'W
Mongala, stm., D.R.C. (mōn-gàl'à)	284-85	1°53'N	19°50'E
Möng Hsat, Mya.	266-67	20°31'N	99°13'E
Mongibello, vol., Italy see Etna, Monte	216-17	37°45'N	15°00'E
Mongo, Chad	280-81	12°11'N	18°42'E
Mongol Altayn nuruu, mts., Asia	240-41	46°30'N	93°00'E
Mongol Ard Uls, nation, Asia see Mongolia	224-25	46°0'N	105°00'E
Mongolia, nation, Asia (mŏŋ-gō'lĭ-à)	224-25	46°0'N	105°00'E
Mongu, Zam. (mŏŋ-gōō')	286-87	15°17'S	23°08'E
Monkoto, D.R.C. (môn-kō'tô)	284-85	1°37'S	20°40'E
Monmouth, Il., U.S. (mŏn'mŭth)(mŏn'mouth)	124-25	40°54'N	90°39'W
Monmouth, Or., U.S. (mŏn'mŭth)(mŏn'mouth)	122-23	44°51'N	123°14'W
Monmouth Mountain, mtn., B.C., Can. (mŏn'mŭth moun'tĭn)	142-43	51°0'N	123°47'W
Mono, stm., Afr.	282-83	6°16'N	1°49'E
Mono Island, i., Sol. Is.	306e	7°22'S	155°33'E
Mono Lake, lk., Ca., U.S. (mō'nō lāk)	128-29	38°0'N	119°00'W
Monon, In., U.S. (mō'nŏn)	126-27	40°52'N	86°52'W
Monongahela, Pa., U.S. (mô-nŏn-gà-hē'là)	126-27	40°11'N	79°55'W
Monopoli, Italy (mô-nô'pô-lē)	216-17	40°57'N	17°18'E
Monroe, Ga., U.S. (mŭn-rō')	134-35	33°48'N	83°43'W
Monroe, La., U.S. (mŭn-rō')	130-31	32°31'N	92°07'W
Monroe, Mi., U.S. (mŭn-rō')	126-27	41°55'N	83°25'W
Monroe, N.C., U.S. (mŭn-rō')	134-35	34°59'N	80°33'W
Monroe, Ut., U.S. (mŭn-rō')	128-29	38°38'N	112°07'W
Monroe, Wi., U.S. (mŭn-rō')	126-27	42°36'N	89°38'W
Monroe City, Mo., U.S. (mŭn-rō' sĭ'tê)	130-31	39°39'N	91°44'W
Monroeville, Al., U.S. (mŭn-rō'vĭl)	134-35	31°31'N	87°20'W
Monrovia, nat. cap., Lib. (mŏn-rō'vĭ-à)	282-83	6°19'N	10°47'W
Mönsterås, Swe. (mŭn'stĕr-ôs)	208-09	57°02'N	16°27'E
Montague, P.E., Can. (mŏn'tà-gū)	148-49	46°10'N	62°39'W
Montague, Ca., U.S. (mŏn'tà-gū)	122-23	41°44'N	122°31'W
Montague, Mi., U.S. (mŏn'tà-gū)	126-27	43°25'N	86°21'W
Montague, Isla, i., Mex.	156-57	31°43'N	114°44'W
Montague Island, i., Ak., U.S. (mŏn'tà-gū ī'lànd)	136	60°10'N	147°18'W
Montana, state, U.S. (mŏn-tăn'à)	118-19	47°0'N	110°00'W
Montargis, Fr. (môn-tàr-zhē')	212-13	48°00'N	2°44'E
Montauban, Fr.	212-13	44°01'N	1°21'E
Montauk, N.Y., U.S. (mŏn-tôk')	126-27	41°03'N	71°57'W
Montauk Point, c., N.Y., U.S. (mŏn-tôk' point)	126-27	41°04'N	71°52'W
Montbard, Fr. (môn-bàr')	212-13	47°38'N	4°20'E
Montbéliard, Fr. (môN-bā-lyàr')	212-13	47°31'N	6°46'E
Montbrison, Fr. (môn-brē-zon')	212-13	45°37'N	4°04'E
Mont-de-Marsan, Fr. (môN-dē-màr-säN')	212-13	43°53'N	0°30'W
Montdidier, Fr. (môN-dē-dyā')	212-13	49°39'N	2°34'E
Monte Alegre, Braz.	180-81	2°00'S	54°05'W
Monte Azul, Braz.	182-83	15°09'S	42°52'W
Monte Caseros, Arg. (mô'n-tĕ-kä-sĕ'rôs)	187	30°15'S	57°39'W
Monte Comán, Arg.	188e	34°35'S	67°53'W
Monte Cristi, Dom. Rep. (mô'n-tĕ-krê's-tē)	154-55	19°51'N	71°38'W
Monte Escobedo, Mex. (mōn'tå ĕs-kô-bā'dhō)	158-59	22°18'N	103°32'W
Montego Bay, Jam. (mŏn-tē'gō bā)	154-55	18°28'N	77°55'W
Montélimar, Fr. (môN-tā-lē-mär')	212-13	44°34'N	4°45'E
Monte Lindo, stm., Para.	182-83	23°54'S	57°17'W
Montello, Wi., U.S. (mŏn-tĕl'ō)	126-27	43°48'N	89°20'W
Montemorelos, Mex. (mōn'tå-mō-rā'lōs)	132-33	25°11'N	99°50'W
Montemor-o-Novo, Port. (mōN-tĕ-môr'ō-nô'vò)	214-15	38°38'N	8°13'W
Montenegro, nation, Eur. (môn-tà-nā'grō) (mōn-tê-nē'grō)	190-91	42°30'N	19°18'E
Montepuez, Moz.	286-87	13°07'S	38°60'E
Montepulciano, Italy (môn'tå-pōōl-chä'nō)	216-17	43°06'N	11°47'E
Monte Quemado, Arg.	182-83	25°48'S	62°49'W
Montereau-Faut-Yonne, Fr. (môN-t'rō'fō-yôn')	212-13	48°23'N	2°57'E
Monterey, Ca., U.S. (mŏn-tĕ-rā')	128-29	36°36'N	121°54'W
Monterey, Tn., U.S. (mŏn-tĕ-rā')	134-35	36°09'N	85°16'W
Monterey, Va., U.S. (mŏn-tĕ-rā')	126-27	38°24'N	79°35'W
Monterey Bay, b., Ca., U.S. (mŏn-tĕ-rā' bā)	128-29	36°48'N	121°55'W
Montería, Col. (mōn-tā-rä'ä)	178-79	8°45'N	75°53'W
Monteros, Arg. (môn-tĕ'rôs)	182-83	27°10'S	65°30'W
Monterotondo, Italy (môn-tĕ-rô-tô'n-dō)	216-17	42°03'N	12°36'E
Monterrey, Mex. (mōn-tĕ-rā')	132-33	25°41'N	100°19'W
Montesano, Wa., U.S. (mŏn-tê-sä'nō)	122-23	46°59'N	123°35'W
Monte Sant'Angelo, Italy (mô'n-tĕ sän ä'n-gzhĕ-lô)	216-17	41°43'N	15°57'E
Montes Claros, Braz. (môn-tĕs-klä'rôs)	182-83	16°44'S	43°51'W
Montevallo, Al., U.S. (môn-tê-văl'ō)	134-35	33°06'N	86°51'W
Montevarchi, Italy (mōn-tå-vär'kē)	216-17	43°32'N	11°35'E
Montevideo, Mn., U.S. (mŏn'tå-vê-dhā'ō)	124-25	44°57'N	95°43'W
Montevideo, nat. cap., Ur. (mōn'tå-vê-dhā'ō)	187	34°54'S	56°11'W
Monte Vista, Co., U.S. (mŏn'tê vĭs'tà)	128-29	37°35'N	106°09'W
Montezuma, Ga., U.S. (mŏn-tē-zōō'mà)	134-35	32°18'N	84°03'W
Montgomery, Pak. see Sāhīwāl	252-53	30°40'N	73°06'E
Montgomery, Al., U.S. (mŏnt-gŭm'ēr-ĭ)	134-35	32°23'N	86°18'W
Monticello, Ar., U.S. (mŏn-tĭ-sĕl'ō)	134-35	33°38'N	91°47'W
Monticello, Fl., U.S. (mŏn-tĭ-sĕl'ō)	134-35	30°32'N	83°52'W
Monticello, Ga., U.S. (mŏn-tĭ-sĕl'ō)	134-35	33°18'N	83°41'W
Monticello, Ia., U.S. (mŏn-tĭ-sĕl'ō)	124-25	42°14'N	91°12'W
Monticello, Il., U.S. (mŏn-tĭ-sĕl'ō)	126-27	40°00'N	88°35'W
Monticello, In., U.S. (mŏn-tĭ-sĕl'ō)	126-27	40°45'N	86°45'W
Monticello, Ky., U.S. (mŏn-tĭ-sĕl'ō)	134-35	36°50'N	84°52'W
Monticello, Mn., U.S. (mŏn-tĭ-sĕl'ō)	124-25	45°18'N	93°48'W
Monticello, Ut., U.S. (mŏn-tĭ-sĕl'ō)	128-29	37°53'N	109°21'W
Montijo, Port. (môn-tē'zhō)	214-15	38°42'N	8°58'W
Montijo, Spain (môn-tē'hō)	214-15	38°55'N	6°37'W
Montijo, Golfo de, b., Pan. (gôl-fô-dĕ-môn-tē'hō)	162	7°40'N	81°07'W
Mont-Joli, Qc., Can. (môN zhô-lē')	148-49	48°37'N	68°07'W
Mont-Laurier, Qc., Can.	146-47	46°32'N	75°30'W
Montluçon, Fr. (môN-lü-sôN')	212-13	46°20'N	2°36'E
Montmagny, Qc., Can. (môN-màn-yē')	148-49	46°59'N	70°33'W
Montmorillon, Fr. (môN'mô-rê-yôN')	212-13	46°26'N	0°52'E
Montpelier, Id., U.S. (mŏnt-pēl'yēr)	122-23	42°20'N	111°18'W
Montpelier, Oh., U.S. (mŏnt-pēl'yēr)	126-27	41°34'N	84°36'W
Montpelier, Vt., U.S.	126-27	44°16'N	72°35'W
Montpellier, Fr. (môN-pĕ-lyā')	212-13	43°37'N	3°52'E
Montréal, Qc., Can. (mŏn-trê-ôl')	146-47	45°29'N	73°34'W
Montreal, stm., On., Can. (mŏn-trê-ôl')	146-47	47°08'N	79°26'W
Montreal, stm., On., Can. (mŏn-trê-ôl')	146-47	47°15'N	84°39'W
Montreal Lake, lk., Sk., Can. (mŏn-trê-ôl' lāk)	144-45	54°20'N	105°40'W
Montreux, Switz. (môn-trû')	210-11	46°26'N	6°55'E
Montrose, Scot., U.K. (mŏnt-rōz')	206-07	56°43'N	2°28'W
Montrose, Co., U.S. (mŏn-trōz')	128-29	38°29'N	107°53'W
Monts, Pointe des, c., Qc., Can. (pwăNt' dā môN')	148-49	49°20'N	67°23'W
Montserrat, dep., N.A. (mŏnt-sĕ-răt')	152-53	16°45'N	62°12'W
Monywa, Mya. (mŏn'yōō-wà)	266-67	22°06'N	95°08'E
Monza, Italy (mōn'tsä)	216-17	45°35'N	9°17'E
Monze, Zam.	286-87	16°17'S	27°29'E
Monzón, Spain (mōn-thōn')	214-15	41°55'N	0°12'E
Moody, Tx., U.S. (mōō'dĭ)	132-33	31°18'N	97°21'W
Mooi, stm., S. Afr. (mōō'ĭ)	292c	26°52'S	26°57'E
Mooi, stm., S. Afr. (mōō'ĭ)	292c	28°46'S	30°34'E
Moon, Mountains of the, mts., Afr. see Ruwenzori Range	289	0°20'N	29°53'E
Moonta, Austl. (mōōn'tà)	301	34°04'S	137°35'E
Moora, Austl. (mòr'à)	294-95	30°38'S	116°00'E
Moore, Lake, lk., Austl. (lāk mōr)	296-97	29°44'S	117°32'E
Moorea, i., Fr. Poly.	306d	17°32'S	149°50'W

ăt; finăl; rāte; senåte; ärm; åsk; sofà; fâre; ch-choose; dh-as th in other; bē; êvent; bĕt; recĕnt; cratēr; g-gō; gh-guttural g; bĭt; ĭ-short neutral; rīde; κ-guttural k as ch in German ich;

PRONOUNCING INDEX Mooresville – Muang Pak-Lay 385

Feature (Pronunciation)	Page	Lat.	Long.
Mooresville, In., U.S. (mōrz´vĭl)	126-27	39°36′N	86°22′W
Mooresville, N.C., U.S. (mōrz´vĭl)	134-35	35°35′N	80°49′W
Moorhead, Mn., U.S. (mōr´hĕd)	124-25	46°53′N	96°45′W
Moorhead, Ms., U.S. (mōr´hĕd)	134-35	33°28′N	90°31′W
Moosehead Lake, lk., Me., U.S.	127a	45°38′N	69°39′W
Moose Jaw, Sk., Can. (mōōs jô)	144-45	50°23′N	105°32′W
Moose Jaw, stm., Sk., Can. (mōōs jô)	144-45	50°34′N	105°17′W
Moose Lake, Mb., Can. (mōōs lāk)	144-45	53°42′N	100°21′W
Moosomin, Sk., Can. (mōō´sô-mĭn)	144-45	50°08′N	101°41′W
Moosonee, On., Can. (mōō´sô-nē)	138-39	51°17′N	80°40′W
Moppo, Kor., S. *see* Mokpo	263	34°48′N	126°24′E
Mopti, Mali (mŏp´tē)	280-81	14°29′N	4°12′W
Moquegua, Peru (mô-kā´gwä)	184	17°12′S	70°57′W
Mora, Spain (mô-rä)	214-15	39°41′N	3°46′W
Mora, Swe. (mō´rä)	208-09	61°00′N	14°35′E
Mora, Mn., U.S. (mō´rȧ)	124-25	45°53′N	93°18′W
Mora, N.M., U.S. (mō´rȧ)	130-31	35°58′N	105°22′W
Morādābād, India (mô-rä-dä-bäd´)	254-55	28°50′N	78°47′E
Moraleda, Canal, strt., Chile	185	44°30′S	73°30′W
Morant Cays, is., Jam.	154-55	17°24′N	75°59′W
Morant Point, c., Jam. (mô-rănt´ point)	154-55	17°55′N	76°11′W
Moratuwa, Sri L.	256	6°48′N	79°53′E
Morava, hist. reg., Czech Rep.	210-11	49°30′N	16°60′E
Moravská Ostrava, Czech Rep. *see* Ostrava	210-11	49°50′N	18°17′E
Morawhanna, Guy. (mô-rá-hwä´ná)	178-79	8°17′N	59°44′W
Moray Firth, b., Scot., U.K. (mŭr´å fûrth)	206-07	58°02′N	3°05′W
Morbi, India	254-55	22°49′N	70°50′E
Morden, Mb., Can. (môr´dĕn)	144-45	49°11′N	98°05′W
Moreau, stm., S.D., U.S. (mô-rō´)	124-25	45°19′N	100°20′W
Moree, Austl. (mō´rē)	301	29°28′S	149°51′E
Morehead, Ky., U.S. (mōr´hĕd)	126-27	38°11′N	83°27′W
Morehead City, N.C., U.S. (mōr´hĕd sĭ´tĭ)	134-35	34°43′N	76°45′W
Morelia, Mex. (mô-rā´lyä)	158-59	19°42′N	101°12′W
Morella, Spain (mô-räl´yä)	214-15	40°37′N	0°07′W
Morelos, Mex. (mô-rā´lōs)	132-33	28°25′N	100°53′W
Morelos, state, Mex.	158-59	18°45′N	99°00′W
Morena, Sierra, mts., Spain (syĕr´rä mô-rā´nä)	214-15	38°0′N	5°00′W
Morenci, Mi., U.S. (mô-rĕn´sĭ)	126-27	41°43′N	84°13′W
Moresby Island, i., B.C., Can. (mōrz´bĭ ī´lănd)	142-43	52°50′N	131°55′W
Moreton Island, i., Austl. (môr´tŭn ī´lȧnd)	301	27°11′S	153°24′E
Morgan City, La., U.S. (môr´găn sĭ´tĭ)	134-35	29°42′N	91°12′W
Morganfield, Ky., U.S. (môr´găn-fēld)	126-27	37°41′N	87°55′W
Morganton, N.C., U.S. (môr´găn-tŭn)	134-35	35°45′N	81°41′W
Morgantown, Ky., U.S. (môr´găn-toun)	134-35	37°13′N	86°41′W
Morgantown, W.V., U.S. (môr´găn-toun)	126-27	39°38′N	79°57′W
Morgenzon, S. Afr. (môr´gănt-sŏn)	292c	26°44′S	29°37′E
Morghāb, stm., Asia	252-53	38°38′N	61°10′E
Morioka, Japan (mō´rē-ō´kȧ)	264	39°42′N	141°09′E
Morkoka, stm., Russia (môr-kô´kȧ)	236-37	65°11′N	115°51′E
Morlaix, Fr. (mŏr-lĕ´)	212-13	48°35′N	3°50′W
Mornington, Isla, i., Chile	185	49°45′S	75°23′W
Mornington Island, i., Austl.	302	16°33′S	139°24′E
Morocco, nation, Afr. (mô-rŏk´ō)	274	32°0′N	5°00′W
Morogoro, Tan. (mō-rô-gō´rō)	289	6°49′S	37°40′E
Moro Gulf, b., Phil.	270	6°51′N	123°00′E
Moroleón, Mex. (mô-rō-lā-ōn´)	158-59	20°08′N	101°12′W
Morombe, Madag. (mōō-rōōm´bā)	286-87	21°45′S	43°22′E
Morón, Arg. (mo-rō´n)	187	34°39′S	58°37′W
Morón, Cuba (mô-rōn´)	154-55	22°07′N	78°38′W
Mörön, Mong.	260-61	49°38′N	100°10′E
Morón, Ven. (mô-rō´n)	188b	10°29′N	68°12′W
Morona, stm., Peru	184	4°45′S	77°04′W
Morondava, Madag. (mô-rōn-dȧ´vȧ)	286-87	20°18′S	44°17′E
Morón de la Frontera, Spain (mô-rōn´dā läf rōn-tā´rä)	214-15	37°07′N	5°27′W
Moroni, nat. cap., Com.	286-87	11°42′S	43°15′E
Moron Us, stm., China	254-55	34°40′N	94°56′E
Morozovsk, Russia	202-03	48°21′N	41°50′E
Morrill, Ne., U.S. (mŏr´ĭl)	124-25	41°57′N	103°57′W
Morrilton, Ar., U.S. (mŏr´ĭl-tŭn)	130-31	35°09′N	92°45′W
Morrinhos, Braz. (mô-rēn´yōzh)	186	17°44′S	49°06′W
Morris, Mb., Can. (mŏr´ĭs)	144-45	49°21′N	97°23′W
Morris, Il., U.S. (mŏr´ĭs)	126-27	41°22′N	88°26′W
Morris, Mn., U.S. (mŏr´ĭs)	124-25	45°35′N	95°55′W
Morris, stm., Mb., Can. (mŏr´ĭs)	144-45	49°21′N	97°21′W
Morrison, Il., U.S. (mŏr´ĭ-sŭn)	126-27	41°49′N	89°57′W
Morristown, Tn., U.S. (mŏr´rĭs-toun)	134-35	36°13′N	83°17′W
Morro do Chapéu, Braz. (môr-ò dò-shä-pĕ´ōō)	180-81	11°33′S	41°09′W
Morshansk, Russia (môr-shansk´)	202-03	53°26′N	41°49′E
Morskoy araly, i., Kaz.	202-03	44°59′N	50°18′E
Morteros, Arg. (môr-tĕ´tôs)	187	30°42′S	62°00′W
Mortes, stm., Braz.	180-81	11°43′S	50°43′W
Mortlock Islands, is., Micron.	304-05	5°28′N	153°41′E
Morwell, Austl.	301	38°14′S	146°24′E
Mosal'sk, Russia (mō-zálsk´)	218-19	54°30′N	34°58′E
Moscow, Id., U.S. (mŏs´kō)	122-23	46°44′N	117°00′W
Moscow, nat. cap., Russia (mŏs´kō)	218-19	55°45′N	37°38′E
Moscow, stm., Russia (mŏs´kō) *see* Moskva	218-19	55°04′N	38°51′E
Mosel, stm., Eur. (mō´sĕl) (mô-zĕl)	210-11	50°22′N	7°37′E
Moselle, stm., Eur.	210-11	50°22′N	7°37′E
Moshi, Tan. (mō´shē)	289	3°20′S	37°20′E
Mosjøen, Nor.	200-01	65°50′N	13°12′E
Moskenesøya, i., Nor.	200-01	67°60′N	13°06′E
Moskva, nat. cap., Russia (mŏs-kvä´) *see* Moscow	218-19	55°45′N	37°38′E
Moskva, stm., Russia (mŏs-kvä´)	218-19	55°04′N	38°51′E
Mosquera, Col.	178-79	2°30′N	78°26′W
Mosquito Coast, hist. reg., Nic. *see* Mosquitos, Costa de	161	13°0′N	83°45′W
Mosquitos, Costa de, hist. reg., Nic. (kôs-tä-dĕ-mōs-kē´tō)	161	13°0′N	83°45′W
Mosquitos, Golfo de los, b., Pan.	162	9°0′N	81°15′W
Moss, Nor. (môs)	208-09	59°26′N	10°42′E
Mossaka, Congo	284-85	1°09′S	16°50′E
Mosselbaai, S. Afr. (mô´sŭl bā)	286-87	34°11′S	22°08′E
Mossel Bay, S. Afr. *see* Mosselbaai	286-87	34°11′S	22°08′E
Mossendjo, Congo	284-85	2°53′S	12°40′E
Mossoró, Braz.	188d	5°11′S	37°20′W
Moss Point, Ms., U.S. (môs point)	134-35	30°23′N	88°33′W
Most, Czech Rep. (môst)	210-11	50°31′N	13°39′E
Mostaganem, Alg.	214-15	35°56′N	0°05′E
Mostar, Bos. (mōs´tär)	216-17	43°20′N	17°40′E
Mostardas, Braz.	182-83	31°07′S	50°57′W
Mosul, Iraq (mo´ul) (mosool´)	248-49	36°20′N	43°08′E
Mot'a, Eth.	288	11°04′N	37°53′E
Motagua, stm., N.A. (mô-tä´gwä)	160	15°43′N	88°13′W
Motala, Swe. (mô-tô´lä)	208-09	58°32′N	15°04′E
Motherwell, Scot., U.K. (mŭdh´ēr-wĕl)	206-07	55°48′N	3°60′W
Motril, Spain (mô-trēl´)	214-15	36°45′N	3°31′W
Motygino, Russia	236-37	58°12′N	94°39′E
Mouhoun, stm., Afr. *see* Black Volta	282-83	8°41′N	0°60′W
Mouila, Gabon	282-83	1°52′S	11°00′E
Moulins, Fr. (mōō-lăn´)	212-13	46°34′N	3°20′E
Moulmein, Mya. *see* Mawlamyine	266-67	16°30′N	97°38′E
Moulouya, Oued, stm., Mor. (wēd mōō-lōō´yȧ)	280-81	35°08′N	2°21′W
Moultrie, Ga., U.S. (mōl´trĭ)	134-35	31°11′N	83°47′W
Mound City, Il., U.S. (mound sĭ´tē)	134-35	37°05′N	89°10′W
Mound City, Mo., U.S. (mound sĭ´tē)	130-31	40°08′N	95°14′W
Moundou, Chad	284-85	8°34′N	16°05′E
Moundsville, W.V., U.S. (moundz´vĭl)	126-27	39°55′N	80°44′W
Mountain Brook, Al., U.S. (moun´tĭn brŏŏk)	134-35	33°29′N	86°42′W
Mountain Grove, Mo., U.S. (moun´tĭn grōv)	130-31	37°08′N	92°16′W
Mountain Home, Ar., U.S. (moun´tĭn hōm)	130-31	36°20′N	92°23′W
Mountain Home, Id., U.S. (moun´tĭn hōm)	122-23	43°08′N	115°41′W
Mountain Nile, stm., S. Sudan (moun´tĭn nīl)	284-85	9°30′N	30°30′E
Mountain View, Mo., U.S. (moun´tĭn vū)	134-35	36°60′N	91°42′W
Mountain Village, Ak., U.S.	136	62°05′N	163°44′W
Mount Airy, N.C., U.S. (mount âr´ĭ)	134-35	36°30′N	80°37′W
Mount Ayr, Ia., U.S. (mount âr)	130-31	40°43′N	94°14′W
Mount Barker, Austl.	294-95	34°38′S	117°40′E
Mount Carmel, Il., U.S. (mount kär´mĕl)	126-27	38°24′N	87°46′W
Mount Carmel, Pa., U.S. (mount kär´mĕl)	126-27	40°48′N	76°25′W
Mount Cook National Park, n.p., N.Z. *see* Aoraki/Mount Cook National Park	303	43°35′S	170°15′E
Mount Desert Island, i., Me., U.S. (mount dē-zûrt´ ī´lȧnd)	127a	44°20′N	68°20′W
Mount Dora, Fl., U.S. (mount dō´rȧ)	134-35	28°48′N	81°38′W
Mount Forest, On., Can. (mount fŏr´ĕst)	146-47	43°59′N	80°44′W
Mount Gambier, Austl. (mount găm´bēr)	301	37°50′S	140°47′E
Mount Gilead, Oh., U.S. (mount gĭl´ĕăd)	126-27	40°33′N	82°49′W
Mount Hagen, Pap. N. Gui.	302	5°52′S	144°14′E
Mount Isa, Austl. (mount ī´zȧ)	302	20°44′S	139°29′E
Mount Kenya National Park, n.p., Kenya	289	0°09′S	37°19′E
Mount Magnet, Austl. (mount măg-nĕt)	294-95	28°04′S	117°51′E
Mount McKinley National Park, n.p., Ak., U.S. *see* Denali National Park and Preserve	136	63°15′N	150°30′W
Mount Morgan, Austl. (mount môr-găn)	302	23°39′S	150°23′E
Mount Morris, Mi., U.S. (mount mŏr´ĭs)	126-27	43°07′N	83°42′W
Mount Morris, N.Y., U.S. (mount mŏr´ĭs)	126-27	42°44′N	77°52′W
Mount Olive, N.C., U.S. (mount ŏl´ĭv)	134-35	35°12′N	78°04′W
Mount Pleasant, Ia., U.S. (mount plĕz´ănnt)	124-25	40°58′N	91°32′W
Mount Pleasant, Mi., U.S. (mount plĕz´ănnt)	126-27	43°36′N	84°46′W
Mount Pleasant, S.C., U.S. (mount plĕz´ănnt)	134-35	32°47′N	79°52′W
Mount Pleasant, Tn., U.S. (mount plĕz´ănnt)	134-35	35°32′N	87°12′W
Mount Pleasant, Tx., U.S. (mount plĕz´ănnt)	130-31	33°09′N	94°58′W
Mount Pleasant, Ut., U.S. (mount plĕz´ănnt)	128-29	39°33′N	111°27′W
Mount Rainier National Park, n.p., Wa., U.S. (mount rȧ-nēr´ năsh´ŭn-ăl pärk)	122-23	46°52′N	121°43′W
Mount Shasta, Ca., U.S. (mount shăs´tȧ)	122-23	41°19′N	122°18′W
Mount Sterling, Il., U.S. (mount stûr´lĭng)	130-31	39°59′N	90°46′W
Mount Sterling, Ky., U.S. (mount stûr´lĭng)	126-27	38°03′N	83°57′W
Mount Stewart, P.E., Can. (mount stū´ărt)	148-49	46°22′N	62°53′W
Mount Vernon, Il., U.S. (mount vûr´nŭn)	126-27	38°19′N	88°55′W
Mount Vernon, In., U.S. (mount vûr´nŭn)	126-27	37°56′N	87°54′W
Mount Vernon, Ky., U.S. (mount vûr´nŭn)	134-35	37°21′N	84°22′W
Mount Vernon, Mo., U.S. (mount vûr´nŭn)	130-31	37°06′N	93°49′W
Mount Vernon, N.Y., U.S. (mount vûr´nŭn)	126-27	40°55′N	73°50′W
Mount Vernon, Oh., U.S. (mount vûr´nŭn)	126-27	40°23′N	82°29′W
Mount Vernon, Wa., U.S. (mount vûr´nŭn)	122-23	48°25′N	122°20′W
Moura, Braz. (mō´rȧ)	180-81	1°29′S	61°37′W
Mourne Mountains, mts., N. Ire., U.K. (môrn moun´tĭnz)	206-07	54°10′N	6°04′W
Moussoro, Chad	280-81	13°38′N	16°30′E
Moûtiers, Fr. (mōō-tyâr´)	212-13	45°29′N	6°31′E
Moutong, Indon.	268-69	0°29′N	121°14′E
Moyahua, Mex. (mô-yä´wä)	158-59	21°16′N	103°10′W
Moyale, Kenya (mô-yä´lä)	284-85	3°32′N	39°03′E
Moyen Atlas, mts., Mor.	280-81	33°30′N	5°00′W
Moyero, stm., Russia	236-37	68°44′N	103°38′E
Moyo, Pulau, i., Indon.	268-69	8°15′S	117°34′E
Moyobamba, Peru (mō-yô-bäm´bä)	184	6°04′S	76°56′W
Mozambique, nation, Afr. (mō-zăm-bēk´)	274	18°15′S	35°00′E
Mozambique, Canal du, strt., Afr. *see* Mozambique Channel	286-87	19°0′S	41°00′E
Mozambique Channel, strt., Afr. (mō-zăm-bek´ chăn´ĕl)	286-87	19°0′S	41°00′E
Mozdok, Russia (môz-dôk´)	245	43°44′N	44°39′E
Mozhaysk, Russia (mô-zhäysk´)	218-19	55°30′N	36°01′E
Mozhga, Russia	202-03	56°27′N	52°12′E
Mozyr, Bela. *see* Mazyr	218-19	52°03′N	29°16′E
Mpika, Zam.	286-87	11°51′S	31°27′E
Mpwapwa, Tan. ('m-pwä´pwä)	289	6°19′S	36°26′E
M'Sila, Alg. (m'sē´lȧ)	292b	35°42′N	4°33′E
Msta, stm., Russia (m'stá´)	218-19	58°29′N	31°27′E
Mtkvari, stm., Asia *see* Kür	245	39°17′N	49°26′E
Mtsensk, Russia (m'tsĕnsk)	218-19	53°17′N	36°35′E
Mtwara, Tan.	286-87	10°21′S	40°15′E
Muanda, D.R.C.	284-85	5°57′S	12°22′E
Muang Khammouan, Laos	266-67	17°25′N	104°49′E
Muang Không, Laos	266-67	14°07′N	105°51′E
Muang Ngoy, Laos	266-67	20°42′N	102°40′E
Muang Pak-Lay, Laos	266-67	18°13′N	101°24′E

n-sing; ŋ-baŋk; ɴ-nasalized n; nŏd; cŏmmit; ōld; ôbey; ôrder; oi-boil; fōōd; ŏŏ-as oo in foot; ou-out; s-soft; sh-dish; th-thin; pūre; ûnite; ûrn; stŭd; circŭs; ü-as in French tu; ´-indeterminate vowel.

386 Muang Pakxan – Nakhon Phanom PRONOUNCING INDEX

Feature (Pronunciation)	Page	Lat.	Long.
Muang Pakxan, Laos	266-67	18°25'N	103°39'E
Muang Sing, Laos	266-67	21°11'N	101°09'E
Muang Vangviang, Laos	266-67	18°55'N	102°26'E
Muang Xaignabouri, Laos	266-67	19°17'N	101°43'E
Muar, Malay.	266-67	2°02'N	102°34'E
Muaratewe, Indon.	268-69	0°56'S	114°52'E
Mubende, Ug.	289	0°35'N	31°24'E
Mucajaí, stm., Braz.	178-79	2°24'N	60°50'W
Muchinga Mountains, mts., Zam.	286-87	11°40'S	31°44'E
Mucuri, stm., Braz.	186	18°05'S	39°34'W
Mudan, stm., China (mo͞o-dän)	264	46°18'N	129°32'E
Mudanjiang, China (mo͞o-dän-jyäŋ)	264	44°35'N	129°36'E
Mudgee, Austl. (mŭ-jē)	301	32°36'S	149°35'E
Mueda, Moz.	286-87	11°40'S	39°34'E
Mufulira, Zam.	286-87	12°33'S	28°14'E
Muğla, Tur. (mo͞og'lä)	216-17	37°13'N	28°22'E
Mühlhausen, Ger. (mül'hou-zĕn)	210-11	51°13'N	10°28'E
Muhlig-Hofmann Mountains, mts., Ant.	313	72°10'S	4°53'E
Muhu, i., Est. (mo͞o'ho͞o)	208-09	58°37'N	23°13'E
Mukacheve, Ukr.	210-11	48°26'N	22°45'E
Mukah, Malay.	268-69	2°54'N	112°06'E
Mukalla, Yemen *see* Al-Mukallā	238-39	14°32'N	49°08'E
Mukden, China *see* Shenyang	263	41°48'N	123°24'E
Mukry, Turkmen.	252-53	37°36'N	65°43'E
Mula, Spain (mo͞o'lä)	214-15	38°03'N	1°30'W
Muladu, i., Mald.	256	7°01'N	72°59'E
Mulchatna, stm., Ak., U.S.	136	59°39'N	157°08'W
Mulhacén, mtn., Spain	214-15	37°03'N	3°19'W
Mulhouse, Fr. (mü-lo͞oz')	212-13	47°45'N	7°20'E
Muling, China (mo͞o-lĭŋ)	264	44°31'N	130°16'E
Muling, China (mo͞o-lĭŋ)	264	44°56'N	130°32'E
Muling, stm., China (mo͞o-lĭŋ)	264	45°52'N	133°30'E
Mull, Island of, i., Scot., U.K. (ī'lănd ŏv mŭl)	206-07	56°27'N	6°00'W
Mullan, Id., U.S. (mŭl'ăn)	122-23	47°28'N	115°48'W
Muller, Pegunungan, mts., Indon. (mŭl'ēr)	268-69	0°40'N	113°50'E
Mullewa, Austl.	294-95	28°33'S	115°31'E
Mullins, S.C., U.S. (mŭl'ĭnz)	134-35	34°12'N	79°15'W
Mulongo, D.R.C.	284-85	7°49'S	26°60'E
Multān, Pak. (mo͝ol-tän')	252-53	30°11'N	71°27'E
Mulvane, Ks., U.S. (mŭl-vān')	130-31	37°28'N	97°15'W
Mumbai, India	256	18°57'N	72°50'E
Mumbwa, Zam. (mo͝om'bwä)	286-87	14°59'S	27°04'E
Mun, stm., Thai.	266-67	15°19'N	105°31'E
Muna, Mex. (mo͞o'nå)	160	20°29'N	89°43'W
Muna, stm., Russia	236-37	67°53'N	123°05'E
Muna, Pulau, i., Indon.	268-69	4°53'S	122°27'E
München, Ger. *see* Munich	210-11	48°08'N	11°35'E
Muncie, In., U.S. (mŭn'sĭ)	126-27	40°11'N	85°22'W
Munger, India	254-55	25°23'N	86°28'E
Mungindi, Austl. (mŭn-gĭn'dē)	301	28°59'S	148°59'E
Mungkan Kandju National Park, n.p., Austl.	302	13°32'S	142°37'E
Munich, Ger. (mū'nĭk)	210-11	48°08'N	11°35'E
Munising, Mi., U.S. (mū'nĭ-sĭng)	124-25	46°24'N	86°39'W
Munkács, Ukr. *see* Mukacheve	210-11	48°26'N	22°45'E
Münster, Ger.	210-11	51°57'N	7°37'E
Muntok, Indon. (mo͝on-tŏk')	266-67	2°04'S	105°10'E
Muonio, Fin.	200-01	67°58'N	23°40'E
Muqayshiṭ, i., U.A.E.	250-51	24°10'N	53°45'E
Muqdisho, nat. cap., Som. *see* Mogadishu	284-85	2°03'N	45°20'E
Muradiye, Tur. (mo͞o-rä'dē-yĕ)	245	38°59'N	43°50'E
Murashi, Russia	202-03	59°24'N	48°58'E
Murat, stm., Tur. (mo͞o-rät')	245	38°40'N	39°53'E
Murchison, stm., Austl. (mûr'chĭ-sŭn)	296-97	27°42'S	114°08'E
Murchison Falls, wtfl., Ug. *see* Kabalega Falls	289	2°17'N	31°42'E
Murchison Falls National Park, n.p., Ug.	289	2°15'N	31°50'E
Murcia, Spain (mo͞or'thyä)	214-15	37°59'N	1°08'W
Mur-de-Barrez, Fr.	212-13	44°51'N	2°39'E
Murdo, S.D., U.S. (mûr'dō)	124-25	43°53'N	100°41'W
Muret, Fr. (mü-rĕ')	212-13	43°28'N	1°19'E
Murfreesboro, N.C., U.S. (mûr'frēz-bŭr-ō)	134-35	36°27'N	77°06'W
Murfreesboro, Tn., U.S. (mûr'frēz-bŭr-ō)	134-35	35°50'N	86°23'W
Murgap, stm., Asia (mo͞or-gäp')	252-53	38°38'N	61°10'E
Murgon, Austl.	301	26°15'S	151°57'E
Mūrītāniyā, nation, Afr. *see* Mauritania	274	20°0'N	12°00'W
Murmansk, Russia (mo͞or-mänsk')	200-01	68°58'N	33°05'E
Murom, Russia (mo͞o'rŏm)	202-03	55°34'N	42°02'E
Muroran, Japan (mo͞o'rō-rän)	264	42°19'N	140°59'E
Muros, Spain (mo͞o'rōs)	214-15	42°47'N	9°04'W
Murphy, N.C., U.S. (mŭr'fĭ)	134-35	35°05'N	84°02'W
Murphysboro, Il., U.S. (mŭr'fĭz-bŭr-ō)	126-27	37°46'N	89°20'W
Murray, Ky., U.S. (mŭr'ĭ)	134-35	36°37'N	88°19'W
Murray, Ut., U.S. (mŭr'ĭ)	122-23	40°39'N	111°54'W
Murray, stm., Austl. (mŭr'ĭ)	301	35°22'S	139°21'E
Murray, stm., B.C., Can. (mŭr'ĭ)	142-43	55°43'N	121°13'W
Murray, Lake, lk., Pap. N. Gui.	302	7°0'S	141°30'E
Murray Bridge, Austl. (mŭr'ĭ brĭj)	301	35°08'S	139°16'E
Murray Harbour, P.E., Can. (mŭr'ĭ här'bēr)	148-49	45°60'N	62°32'W
Murray-Sunset National Park, n.p., Austl.	301	34°45'S	141°30'E
Murrumbidgee, stm., Austl. (mŭr-ŭm-bĭd'jē)	301	34°42'S	143°08'E
Murska Sobota, Slvn. (mo͞or'skä sô'bô-tä)	216-17	46°40'N	16°10'E
Murua Island, i., Pap. N. Gui.	302	9°06'S	152°45'E
Murud, Gunong, mtn., Malay.	268-69	3°52'N	115°30'E
Mururoa, at., Fr. Poly.	304-05	21°52'S	138°55'W
Murwāra, India	254-55	23°50'N	80°24'E
Murwillumbah, Austl. (mŭr-wĭl'lŭm-bŭ)	301	28°21'S	153°24'E
Murzuq, Libya	280-81	25°56'N	13°55'E
Murzūq, Idhān, des., Libya	280-81	24°30'N	13°00'E
Mürzzuschlag, Aus. (mürts'tso͞o-shlägh)	210-11	47°36'N	15°41'E
Muş, Tur. (mo͞osh)	245	38°43'N	41°29'E
Musala, mtn., Blg.	216-17	42°11'N	23°34'E
Musay'īd, Qatar	250-51	24°59'N	51°33'E
Muscat, nat. cap., Oman (mŭs-kăt')	250-51	23°36'N	58°32'E
Muscat and Oman, nation, Asia *see* Oman	224-25	22°0'N	58°00'E
Muscatine, Ia., U.S. (mŭs-kå-tēn)	124-25	41°25'N	91°02'W
Muscle Shoals, Al., U.S. (mŭs''l shōlz)	134-35	34°44'N	87°40'W
Mushin, Nig.	282a	6°31'N	3°21'E
Musi, stm., Indon. (mo͞o'sē)	266-67	2°22'S	104°55'E
Muskegon, Mi., U.S. (mŭs-kē'gŭn)	126-27	43°14'N	86°15'W
Muskegon, stm., Mi., U.S. (mŭs-kē'gŭn)	126-27	43°13'N	86°19'W
Muskegon Heights, Mi., U.S. (mŭs-kē'gŭn hīts)	126-27	43°12'N	86°14'W
Muskingum, stm., Oh., U.S. (mŭs-kĭŋ'gŭm)	126-27	39°24'N	81°28'W
Muskogee, Ok., U.S. (mŭs-kō'gē)	130-31	35°44'N	95°22'W
Muskoka, Lake, lk., On., Can. (lāk mŭs-kō'kå)	146-47	45°02'N	79°25'W
Musoma, Tan.	289	1°30'S	33°48'E
Mussau Island, i., Pap. N. Gui. (mo͞o-sä'o͞o ī'lănd)	302	1°27'S	149°37'E
Musselshell, stm., Mt., U.S. (mŭs''l-shĕl)	122-23	47°27'N	107°55'W
Mustvee, Est. (mo͞ost'vĕ-ĕ)	208-09	58°51'N	26°56'E
Musu-dan, c., Kor., N. (mo͞o'so͞o dän)	263	40°51'N	129°43'E
Muswellbrook, Austl. (mŭs'wŭnl-bro͝ok)	301	32°16'S	150°54'E
Mutare, Zimb.	286-87	18°58'S	32°40'E
Mutsamudu, Com.	286-87	12°08'S	44°26'E
Mutsu, Japan	264	41°17'N	141°10'E
Mutsu-wan, b., Japan (mo͞ot'so͞o wän)	264	41°05'N	140°55'E
Mutton Bay, Qc., Can. (mŭt''n bā)	148-49	50°47'N	59°02'W
Muttra, India *see* Mathura	254-55	27°30'N	77°41'E
Mutum, Braz. (mo͞o-to͞o'm)	186	19°48'S	41°27'W
Muynak, Uzb. *see* Mŭynoq	244	43°46'N	59°02'E
Mŭynoq, Uzb.	244	43°46'N	59°02'E
Muyua Island, i., Pap. N. Gui. *see* Murua Island	302	9°06'S	152°45'E
Muzaffarnagar, India	254-55	29°28'N	77°42'E
Muzaffarpur, India	254-55	26°07'N	85°23'E
Muztag, mtn., China	254-55	36°03'N	80°07'E
Muztag, mtn., China	254-55	36°25'N	87°25'E
Muztaū biīgi, mtn., Asia *see* Belukha, Mount	244	49°51'N	86°29'E
Mwali, i., Com.	286-87	12°18'S	43°42'E
Mwanza, Tan. (mwän'zä)	289	2°31'S	32°54'E
Mweka, D.R.C.	284-85	4°51'S	21°34'E
Mwene-Ditu, D.R.C.	284-85	7°03'S	23°27'E
Mweru, Lake, lk., Afr. (lāk mwā'ro͞o)	284-85	9°0'S	28°45'E
Mweru Wantipa, Lake, lk., Zam.	284-85	8°45'S	29°40'E
Mwokil, at., Micron.	304-05	6°39'N	159°47'E
Myanaung, Mya.	266-67	18°17'N	95°19'E
Myanmar, nation, Asia (myän-mär)	224-25	22°0'N	98°00'E
Myaundzha, Russia	236-37	63°03'N	147°11'E
Myaungmya, Mya.	266-67	16°35'N	94°55'E
Myingyan, Mya. (myĭng-yŭn')	266-67	21°27'N	95°23'E
Myitkyinā, Mya. (myĭ'chē-nå)	258-59	25°23'N	97°24'E
Mykolaïv, Ukr.	218-19	46°58'N	31°59'E
Mymensingh, Bngl.	254-55	24°45'N	90°24'E
Myohyang-san, mtn., Kor., N. (myō'hyang-sän')	263	40°01'N	126°21'E
Mýrdalsjökull, ice, Ice. (mür'däls-yû'kŏl)	206a	63°40'N	19°05'W
Myrtle Beach, S.C., U.S. (mûr't'l bēch)	134-35	33°42'N	78°54'W
Mysore, India *see* Mysūru	256	12°18'N	76°39'E
Mysore, state, India *see* Karnātaka	256	14°0'N	76°00'E
Mys Shmidta, Russia	136	68°52'N	179°37'W
Mysūru, India	256	12°18'N	76°39'E
My Tho, Viet.	266-67	10°22'N	106°22'E
Mytilíni, Grc.	216-17	39°06'N	26°33'E
Mytishchi, Russia (mĕ-tēsh'chi)	218-19	55°55'N	37°46'E
Mzuzu, Malawi	286-87	11°24'S	33°57'E

N

Feature (Pronunciation)	Page	Lat.	Long.
Naantali, Fin. (nän'tå-lē)	208-09	60°30'N	22°04'E
Naberezhnye Chelny, Russia	202-03	55°42'N	52°19'E
Nabeul, Tun. (nā-bûl')	200-01	36°27'N	10°46'E
Nabī Shu'ayb, Jabal an-, mtn., Yemen	238-39	15°17'N	43°59'E
Nābulus, W.B.	248-49	32°14'N	35°17'E
Nacala, Moz. (nă-ká'lá)	286-87	14°33'S	40°40'E
Náchod, Czech Rep. (näk'ôt)	210-11	50°25'N	16°11'E
Nacogdoches, Tx., U.S. (năk'ō-dō'chĕz)	132-33	31°35'N	94°39'W
Nacozari de García, Mex.	156-57	30°24'N	109°39'W
Nadadores, Mex. (nä-dä-dō'rās)	132-33	27°02'N	101°35'W
Nadiād, India	254-55	22°41'N	72°52'E
Nador, Mor.	214-15	35°11'N	2°56'W
Nadym, Russia	236-37	65°35'N	72°39'E
Nadym, stm., Russia (ná'dĭm)	236-37	66°13'N	72°00'E
Næstved, Den. (nĕst'vĭdh)	210-11	55°14'N	11°46'E
Naga, Phil. (nä'gä)	270	13°38'N	123°11'E
Nāgāland, state, India	254-55	26°0'N	95°00'E
Nagano, Japan (nä'gä-nō)	265	36°39'N	138°12'E
Nagaoka, Japan (nä'gá-ō'ká)	265	37°27'N	138°51'E
Nagaon, India	254-55	26°21'N	92°41'E
Nāgappattinam, India	256	10°46'N	79°51'E
Nagarote, Nic. (nä-gä-rô'tĕ)	161	12°16'N	86°34'W
Nagasaki, Japan (nä'gá-sä'kē)	265	32°45'N	129°53'E
Nāgaur, India	254-55	27°12'N	73°44'E
Nāgercoil, India	256	8°10'N	77°26'E
Nagorno-Karabakh, hist. reg., Azer. (nu-gôr'nŭ-kŭ-rŭ-bäk')	245	40°00'N	46°40'E
Nagoya, Japan	265	35°10'N	136°55'E
Nāgpur, India (näg'po͞or)	254-55	21°09'N	79°05'E
Nagqu, China	254-55	31°31'N	92°05'E
Nagua, Dom. Rep. (ná'gwä)	154-55	19°23'N	69°51'W
Naguna, Île, i., Vanuatu *see* Nguna, Île	306g	17°27'S	168°21'E
Nagybánya, Rom. *see* Baia Mare	210-11	47°39'N	23°35'E
Nagykanizsa, Hung. (nôd'y'kô'nĕ-shô)	210-11	46°27'N	16°60'E
Nagykőrös, Hung. (nôd'y'kû-rŭsh)	210-11	47°02'N	19°46'E
Nagyvarad, Rom. *see* Oradea	210-11	47°04'N	21°56'E
Naha, Japan (nä'hä)	264a	26°13'N	127°42'E
Nāhan, India	254-55	30°32'N	77°17'E
Nahanni National Park Reserve, n.p., N.T., Can.	138-39	61°35'N	125°45'W
Nahe, China	260-61	48°29'N	124°53'E
Nahr al-Urdunn, stm., Asia *see* Jordan	248-49	31°46'N	35°34'E
Nahuel Huapi, Lago, lk., Arg. (lä'gō nä'wĕl wä'pē)	185	40°58'S	71°30'W
Naica, Mex. (nä-ē'kä)	132-33	27°51'N	105°30'W
Nain, Nf./L., Can. (nīn)	138-39	56°33'N	61°43'W
Nā'īn, Iran	252-53	32°52'N	53°05'E
Naini Tāl, India	254-55	29°24'N	79°26'E
Nairn, Scot., U.K. (nârn)	206-07	57°35'N	3°53'W
Nairobi, nat. cap., Kenya (nī-rō'bē)	289	1°16'S	36°49'E
Naitauba, i., Fiji	306f	17°01'S	179°16'W
Naivasha, Kenya (nī-vä'shá)	289	0°45'S	36°26'E
Najafābād, Iran	252-53	32°38'N	51°22'E
Najasa, stm., Cuba (nä-hä'sä)	154-55	20°43'N	77°59'W
Najd, hist. reg., Sau. Ar.	288	26°07'N	44°40'E
Najin, Kor., N. (nä'jĭn)	263	42°15'N	130°18'E
Naju, Kor., S. (nä'jo͞o')	263	35°02'N	126°43'E
Nakambé, stm., Afr. *see* White Volta	282-83	8°57'N	1°10'W
Nakanbe, stm., Afr. *see* White Volta	282-83	8°57'N	1°10'W
Nakano-shima, i., Japan	264	29°50'N	129°52'E
Nakhichevan, Azer. *see* Naxçıvan	245	39°13'N	45°25'E
Nakhodka, Russia (nŭ-kôt'kŭ)	264	42°49'N	132°53'E
Nakhon Pathom, Thai.	266-67	13°49'N	100°04'E
Nakhon Phanom, Thai.	266-67	17°24'N	104°47'E

ăt; finăl; rāte; senåte; ärm; åsk; sofả; fâre; ch-choose; dh-as th in other; bē; ėvent; bĕt; recĕnt; cratēr; g-gō; gh-guttural g; bĭt; ĭ-short neutral; rīde; κ-guttural k as ch in German ich;

PRONOUNCING INDEX
Nakhon Ratchasima – Negro 387

Feature (Pronunciation)	Page	Lat.	Long.
Nakhon Ratchasima, Thai.	266-67	14°58'N	102°06'E
Nakhon Sawan, Thai.	266-67	15°42'N	100°06'E
Nakhon Si Thammarat, Thai.	266-67	8°26'N	99°58'E
Nakskov, Den. (näk´skou)	210-11	54°50'N	11°08'E
Nakuru, Kenya	289	0°17'S	36°04'E
Nālanda, India	254-55	25°08'N	85°24'E
Nalchik, Russia (nȧl-chēk´)	245	43°29'N	43°37'E
Nalgonda, India	256	17°03'N	79°16'E
Nalubaale Dam, d., Ug.	289	0°27'N	33°11'E
Nālūt, Libya (nä-lo͞ot´)	204-05	31°53'N	10°60'E
Namak, Daryācheh-ye, lk., Iran	252-53	34°30'N	51°50'E
Namangan, Uzb. (nȧ-mȧn-gän´)	252-53	40°60'N	71°40'E
Namapa, Moz.	286-87	13°42'S	39°49'E
Nambour, Austl. (năm´bŏr)	301	26°38'S	152°58'E
Namcha Barwa, mtn., China *see* Namjagbarwa Feng	258-59	29°38'N	95°04'E
Nam Co, lk., China (näm tswo)	254-55	30°41'N	90°32'E
Nam Dinh, Viet. (näm dĕnk´)	266-67	20°26'N	106°10'E
Namhaedo, i., Kor., S.	263	34°48ÿN	127°57ÿE
Namhkam, Mya.	258-59	23°50'N	97°41'E
Namib Desert, des., Nmb. (nä-mēb´ dĕs´ĕrt)	286-87	23°0'S	15°00'E
Namibe, Ang.	286-87	15°12'S	12°10'E
Namibia, nation, Afr. (nä-mĭ´-bē-ȧ)	274	22°0'S	17°00'E
Namib Naukluft Park, pk., Nmb.	286-87	24°40'S	15°17'E
Namjagbarwa Feng, mtn., China	258-59	29°38'N	95°04'E
Namlea, Indon.	268-69	3°16'S	127°06'E
Nam Ngum Reservoir, res., Laos	266-67	18°33'N	102°37'E
Namoi, stm., Austl. (nămȯi)	301	30°00'S	148°04'E
Namolok Atoll, at., Micron.	304-05	5°55'N	153°08'E
Nampa, Id., U.S. (năm´pȧ)	122-23	43°35'N	116°33'W
Namp'o, Kor., N.	263	38°45'N	125°23'E
Nampula, Moz.	286-87	15°07'S	39°16'E
Namsang, Mya.	266-67	20°53'N	97°43'E
Namsos, Nor. (näm´sȯs)	200-01	64°28'N	11°32'E
Namu, B.C., Can.	142-43	51°50'N	127°52'W
Namuka-i-Lau, i., Fiji	306f	18°51'S	178°38'W
Namyit Island, i., Asia	242-43	10°24'N	114°27'E
Nan, Thai.	266-67	18°46'N	100°46'E
Nan, stm., Thai.	266-67	15°42'N	100°09'E
Nanaimo, B.C., Can. (nȧ-nī´mō)	142-43	49°10'N	123°57'W
Nanam, Kor., N. (nä´nän´)	263	41°43'N	129°42'E
Nanango, Austl.	301	26°41'S	152°00'E
Nanao, Japan (nä´nä-ō)	265	37°03'N	136°58'E
Nanchang, China	258-59	28°41'N	115°53'E
Nancheng, China (nän-chäŋ) *see* Hanzhong	258-59	33°04'N	107°02'E
Nancheng, China (nän-chäŋ)	258-59	27°34'N	116°39'E
Nanchong, China (nän-chôŋ)	258-59	30°47'N	106°05'E
Nancy, Fr. (näɴ-sē´)	212-13	48°41'N	6°10'E
Nanda Devi, mtn., India (nän´dä dä´vē)	254-55	30°23'N	79°59'E
Nānded, India	256	19°09'N	77°18'E
Nanga-Eboko, Camrn.	282-83	4°40'N	12°22'E
Nanga Parbat, mtn., Pak.	252-53	35°15'N	74°36'E
Nangis, Fr. (näɴ-zhē´)	212-13	48°34'N	3°01'E
Nanhai, China *see* Foshan	258-59	23°03'N	113°07'E
Nan Hai, s., Asia *see* South China Sea	242-43	10°0'N	113°00'E
Nanjing, China (nän-jyĭŋ)	258-59	24°31'N	117°23'E
Nanjing, China (nän-jyĭŋ)	258-59	32°03'N	118°47'E
Nanking, China *see* Nanjing	258-59	32°03'N	118°47'E
Nan Ling, mts., China	258-59	25°0'N	112°00'E
Nanliu, stm., China (nän-lĭō)	258-59	21°40'N	109°05'E
Nanning, China (nän´nĭng´)	258-59	22°48'N	108°20'E
Nanpan, stm., China (nän-pän)	258-59	24°57'N	106°08'E
Nanping, China (nän-pĭŋ)	258-59	26°38'N	118°10'E
Nansei-shotō, is., Japan *see* Ryukyu Islands	264a	25°44'N	126°58'E
Nanshan Island, i., Asia	242-43	10°44'N	115°49'E
Nansio, Tan.	289	2°06'S	33°03'E
Nantai-zan, vol., Japan (nän-täē-zän)	265	36°46'N	139°30'E
Nantes, Fr. (näɴt´)	212-13	47°14'N	1°33'W
Nanticoke, Pa., U.S. (năn´tĭ-kōk)	126-27	41°12'N	76°00'W
Nantong, China (nän-tôŋ)	258-59	32°01'N	120°51'E
Nantucket Island, i., Ma., U.S. (năn-tŭk´ĕt ī´lȧnd)	126-27	41°16'N	70°03'W
Nanumea, at., Tuvalu	304-05	5°42'S	176°09'E
Nanuque, Braz.	186	17°50'S	40°20'W
Nanxiong, China (nän-shôŋ)	258-59	25°07'N	114°20'E
Nanyang, China (nän-yäŋ)	258-59	33°00'N	112°32'E
Nanyuki, Kenya	289	0°01'N	37°05'E
Nao, Cabo de la, c., Spain *see* Nao, Cap de la	214-15	38°44'N	0°14'E
Nao, Cap de la, c., Spain	214-15	38°44'N	0°14'E
Náousa, Grc. (nä´o͞o-sȧ)	216-17	40°37'N	22°03'E
Napa, Ca., U.S. (năp´ȧ)	128-29	38°18'N	122°17'W

Feature (Pronunciation)	Page	Lat.	Long.
Napaktulik Lake, lk., Nu., Can.	140-41	66°15'N	113°05'W
Napanee, On., Can. (năp´ȧ-nē)	146-47	44°15'N	76°57'W
Naperville, Il., U.S. (nā´pĕr-vĭl)	126-27	41°46'N	88°09'W
Napier, N.Z. (nā´pĭ-ĕr)	303	39°29'S	176°54'E
Naples, Italy (nā´p'lz)	216-17	40°51'N	14°17'E
Naples, Fl., U.S. (nā´p'lz)	135a	26°09'N	81°48'W
Napo, stm., S.A. (nä´pō)	184	3°29'S	72°38'W
Napoleon, Oh., U.S. (nȧ-pō´lē-ŭn)	126-27	41°23'N	84°08'W
Napoli, Italy (nä´pē-lē) *see* Naples	216-17	40°51'N	14°17'E
Nappanee, In., U.S. (năp´ȧ-nē)	126-27	41°26'N	85°59'W
Nara, Japan (nä´rä)	265	34°41'N	135°50'E
Nara, state, Japan (nä´rä)	265	34°30'N	135°50'E
Naracoorte, Austl. (nȧ-rȧ-ko͝on´tē)	301	36°58'S	140°44'E
Naraụ, stm., Eur. *see* Narew	210-11	52°31'N	21°05'E
Nārāyanganj, Bngl.	254-55	23°37'N	90°30'E
Narbonne, Fr. (nȧr-bŏn´)	212-13	43°11'N	2°60'E
Narborough, i., Ec. *see* Fernandina, Isla	184a	0°26'S	91°30'W
Nares Stræde, strt., N.A. *see* Nares Strait	95	80°30'N	68°00'W
Nares Strait, strt., N.A.	95	80°30'N	68°00'W
Narew, stm., Eur. (när´ĕf)	210-11	52°31'N	21°05'E
Narmada, stm., India	254-55	21°41'N	72°45'E
Nārnaul, India	254-55	28°03'N	76°06'E
Narodnaya, Gora, mtn., Russia (gȧ-rä´ nȧ-rôd´nȧ-yȧ)	202-03	65°04'N	60°09'E
Naro-Fominsk, Russia (nä´rô-mĕnsk´)	218-19	55°23'N	36°44'E
Narok, Kenya	289	1°05'S	35°52'E
Narrabri, Austl.	301	30°20'S	149°47'E
Narrandera, Austl. (nȧ-rȧn-dē´rȧ)	301	34°45'S	146°33'E
Narrogin, Austl. (năr´ô-gĭn)	294-95	32°56'S	117°11'E
Narromine, Austl.	301	32°14'S	148°14'E
Narsimhapur, India	254-55	22°57'N	79°12'E
Narva, Est. (nȧr´vȧ)	208-09	59°23'N	28°12'E
Narva laht, b., Eur.	208-09	59°30'N	27°40'E
Narvik, Nor. (när´vĕk)	200-01	68°26'N	17°25'E
Narvskiy Zaliv, b., Eur. (nȧr´vskĭ zä´lĭf) *see* Narva laht	208-09	59°30'N	27°40'E
Naryan-Mar, Russia (när´yȧn mär´)	202-03	67°37'N	52°60'E
Naryn, Kyrg.	244	41°26'N	75°59'E
Naryn, stm., Asia (nŭ-rĭn´)	244	41°46'N	73°10'E
Nasca, Peru	184	14°50'S	74°57'W
Nāshik, India	254-55	20°00'N	73°47'E
Nashua, Ia., U.S. (năsh´ŭ-ȧ)	124-25	42°57'N	92°33'W
Nashua, N.H., U.S.	126-27	42°46'N	71°28'W
Nashville, Ar., U.S. (năsh´vĭl)	130-31	33°57'N	93°51'W
Nashville, Ga., U.S. (năsh´vĭl)	134-35	31°12'N	83°15'W
Nashville, Il., U.S. (năsh´vĭl)	126-27	38°21'N	89°23'W
Nashville, Mi., U.S. (năsh´vĭl)	126-27	42°36'N	85°06'W
Nashville, Tn., U.S. (năsh´vĭl)	134-35	36°09'N	86°47'W
Našice, Cro. (nä´shĕ-tsĕ)	216-17	45°29'N	18°04'E
Nāṣir, Buḥayrat, res., Afr. *see* Nasser, Lake	288	22°40'N	32°00'E
Nâsir, Buheirat, res., Afr. *see* Nasser, Lake	288	22°40'N	32°00'E
Nasīrābād, Bngl. *see* Mymensingh	254-55	24°45'N	90°24'E
Nass, stm., B.C., Can. (năs)	142-43	54°59'N	129°40'W
Nassau, nat. cap., Bah. (năs´ô)	154-55	25°04'N	77°20'W
Nassau Island, i., Cook Is.	304-05	11°34'S	165°24'W
Nasser, Lake, res., Afr. (lăk nä-sĕr)	288	22°40'N	32°00'E
Natagaima, Col. (nä-tä-gī´mä)	188c	3°38'N	75°06'W
Natal, Braz. (nä-täl´)	188d	5°47'S	35°13'W
Natal, Indon.	266-67	0°34'N	99°07'E
Natashquan, Qc., Can. (nä-tăsh´kwăn)	148-49	50°12'N	61°49'W
Natashquan, stm., Can.	148-49	50°11'N	61°35'W
Natchez, Ms., U.S. (năch´ĕz)	134-35	31°34'N	91°24'W
Natchitoches, La., U.S. (năk´ĭ-tŏsh)(nȧch-ĭ-tŏsh´)	132-33	31°46'N	93°06'W
National City, Ca., U.S. (năsh´ŭn-ăl sĭ´tĭ)	128-29	32°41'N	117°06'W
Natividade, Braz. (nä-tē-vē-dä´dĕ)	180-81	11°42'S	47°47'W
Natron, Lake, lk., Afr. (lăk nā´trŏn)	289	2°25'S	35°60'E
Natuna Besar, Kepulauan, is., Indon.	266-67	4°40'N	108°00'E
Natuna Selatan, Kepulauan, is., Indon.	266-67	2°45'N	109°00'E
Natural Bridges National Monument, n.p., Ut., U.S. (năt´ů-răl brĭj´ĕs năsh´ŭn-ăl mŏn´ŭ-mĕnt)	128-29	37°37'N	109°59'W
Naturaliste, Cape, c., Austl. (kăp năt-ů-rȧ-lĭst´)	296-97	33°33'S	115°01'E
Naugatuck, Ct., U.S. (nô´gȧ-tŭk)	126-27	41°29'N	73°03'W
Naujat, Nu., Can. *see* Repulse Bay	138-39	66°32'N	86°14'W
Naumburg, Ger. (noum´bŏrgh)	210-11	51°10'N	11°48'E
Nauru, nation, Oc. (nä-o͞o´-ro͞o)	304-05	0°32'S	166°55'E
Nautla, Mex. (nä-o͞ot´lä)	158-59	20°13'N	96°47'W
Nava, Mex. (nä´vä)	132-33	28°25'N	100°46'W
Navadwīp, India	254-55	23°25'N	88°22'E

Feature (Pronunciation)	Page	Lat.	Long.
Navahermosa, Spain (nä-vä-ĕr-mō´sä)	214-15	39°38'N	4°28'W
Navajo Indian Reservation, ind. res., U.S. (năv´ȧ-hō ĭn´dĭ-ăn rĕ-sĕr-vā´shĕn)	128-29	36°39'N	109°46'W
Navajo National Monument, n.p., Az., U.S. (năv´ȧ-hō năsh´ŭn-ăl mŏn´ŭ-mĕnt)	128-29	36°44'N	110°29'W
Navanagar, India *see* Jamnagar	254-55	22°28'N	70°04'E
Navarin, Mys, c., Russia	236-37	62°17'N	179°06'E
Navarino, Isla, i., Chile (ĕ´s-lä-nȧ-vä-rē´nō)	185	55°05'S	67°49'W
Navasota, Tx., U.S. (năv-ȧ-sō´tȧ)	132-33	30°23'N	96°06'W
Navasota, stm., Tx., U.S. (năv-ȧ-sō´tȧ)	132-33	30°20'N	96°09'W
Navassa Island, dep., N.A.	152-53	18°24'N	75°01'W
Navassa Island, i., N.A. (nȧ-văs´ȧ ī´lȧnd)	154-55	18°24'N	75°01'W
Navidad, Chile (nä-vē-dä´d)	188e	33°56'S	71°50'W
Naviti, i., Fiji	306f	17°07'S	177°15'E
Navoiy, Uzb.	252-53	40°07'N	65°23'E
Navojoa, Mex. (nä-vô-kô´ä)	156-57	27°05'N	109°27'W
Navsāri, India	254-55	20°57'N	72°56'E
Nawa, Japan *see* Naha	264a	26°13'N	127°42'E
Nawābshāh, Pak. (nȧ-wäb´shä)	252-53	26°14'N	68°24'E
Naxçıvan, Azer.	245	39°13'N	45°25'E
Naxçıvan Muxtar Respublikası, state, Azer.	245	39°20'N	45°30'E
Náxos, i., Grc. (nȧk´sȯs)	216-17	37°03'N	25°31'E
Nayarit, state, Mex. (nä-yä-rēt´)	158-59	22°0'N	105°00'W
Nayau, i., Fiji	306f	17°58'S	179°03'W
Nayoro, Japan	264	44°21'N	142°28'E
Nay Pyi Taw, nat. cap., Mya.	266-67	19°45'N	96°07'E
Nazaré, Port.	214-15	39°36'N	9°04'W
Nazaré da Mata, Braz. (nä-zä-rĕ´ dä-mä-tä)	188d	7°44'S	35°14'W
Nazas, Mex. (nä´zäs)	132-33	25°14'N	104°08'W
Nazas, stm., Mex. (nä´zäs)	132-33	25°35'N	105°03'W
Naze, Japan	264a	28°22'N	129°30'E
Naze, The, c., Nor. *see* Lindesnes	208-09	58°0'N	7°02'E
Nazilli, Tur. (nä-zĭ-lē´)	216-17	37°55'N	28°20'E
Nazrēt, Eth.	292d	8°32'N	39°16'E
N'dalatando, Ang.	286-87	9°18'S	14°54'E
Ndélé, C.A.R.	284-85	8°25'N	20°39'E
N'Djamena, nat. cap., Chad (ŭn-jä-mē-nä´)	280-81	12°07'N	15°03'E
Ndjolé, Gabon	282-83	0°08'S	10°45'E
Ndola, Zam. (n'dō´lä)	286-87	12°57'S	28°38'E
Neagh, Lough, lk., N. Ire., U.K. (lŏk nā)	206-07	54°37'N	6°23'W
Neagră, Marea, s., *see* Black Sea	202-03	43°0'N	35°00'E
Near Islands, is., Ak., U.S. (nēr ī´lȧndz)	136a	52°37'N	187°03'E
Neath, Wales, U.K. (nēth)	206-07	51°40'N	3°48'W
Nebine Creek, stm., Austl. (nĕ-bēnē´ krēk)	301	29°21'S	146°45'E
Neblina, Cerro de la, mtn., S.A. *see* Neblina, Pico da	178-79	0°50'N	65°59'W
Neblina, Pico da, mtn., S.A.	178-79	0°50'N	65°59'W
Nebraska, state, U.S. (nē-brăs´kȧ)	118-19	41°30'N	100°00'W
Nebraska City, Ne., U.S. (nē-brăs´kȧ sĭ´tē)	130-31	40°41'N	95°52'W
Nechako, stm., B.C., Can.	142-43	53°55'N	122°43'W
Nechako Plateau, plat., B.C., Can. (nĭ-chă´kō plä-tō´)	142-43	54°0'N	124°30'W
Nechako Range, mts., B.C., Can. (nĭ-chă´kō rānj)	142-43	53°21'N	124°37'W
Nechako Reservoir, res., B.C., Can. (nĭ-chă´kō rĕ´sĕr-vwär)	142-43	53°33'N	124°53'W
Neches, stm., Tx., U.S. (nĕch´ĕz)	132-33	29°59'N	93°52'W
Necker Island, i., Hi., U.S.	137b	23°35'N	164°42'W
Necochea, Arg. (nā-kô-chā´ä)	187	38°34'S	58°44'W
Nederland, nation, Eur. *see* Netherlands	190-91	52°15'N	5°30'E
Nêdong, China	254-55	29°13'N	91°47'E
Needles, Ca., U.S. (nē´d'lz)	128-29	34°50'N	114°36'W
Neenah, Wi., U.S. (nē´nȧ)	126-27	44°11'N	88°28'W
Neepawa, Mb., Can.	144-45	50°14'N	99°28'W
Neftçala, Azer.	245	39°24'N	49°15'E
Negage, Ang.	284-85	7°45'S	15°17'E
Negapatam, India *see* Nāgappattinam	256	10°46'N	79°51'E
Negaunee, Mi., U.S. (nē-gô´nē)	124-25	46°30'N	87°36'W
Negēlē, Eth.	284-85	5°20'N	39°35'E
Negombo, Sri L.	256	7°13'N	79°51'E
Negotin, Serb. (nĕ´gō-tēn)	216-17	44°13'N	22°33'E
Negra, Punta, c., Peru	184	6°05'S	81°06'W
Negritos, Peru	184	4°40'S	81°17'W
Negro, stm., Arg.	185	41°02'S	62°47'W

n-sing; ŋ-baŋk; ɴ-nasalized n; nŏd; cŏmmit; ōld; ôbey; ôrder; oi-boil; fo͞od; ȯ-as oo in foot; ou-out; s-soft; sh-dish; th-thin; pūre; ûnite; ûrn; stŭd; circŭs; ü-as in French tu; ´-indeterminate vowel.

388 Negro – Nicoya, Golfo de PRONOUNCING INDEX

Feature (Pronunciation)	Page	Lat.	Long.
Negro, stm., S.A. (nā'grō)	**178-79**	3°08'S	59°55'W
Negro, stm., S.A. (nā'grō)	**187**	33°26'S	58°27'W
Negros, i., Phil. (nā'grōs)	**270**	10°0'N	123°00'E
Nehbandān, Iran	**252-53**	31°32'N	60°02'E
Neiba, Dom. Rep. (nå-ē'bä)	**154-55**	18°29'N	71°25'W
Neijiang, China (nā-jyäŋ)	**258-59**	29°35'N	105°03'E
Neillsville, Wi., U.S. (nēlz'vĭl)	**124-25**	44°34'N	90°35'W
Nei Mongol, state, China	**260-61**	43°0'N	115°00'E
Neiqiu, China (nā-chyō)	**260-61**	37°17'N	114°30'E
Neira, Col. (nā'rä)	**188c**	5°09'N	75°31'W
Neisse, stm., Eur. (nēs)	**210-11**	52°04'N	14°46'E
Neiva, Col. (nå-ē'vä)(nā'vä)	**178-79**	2°56'N	75°17'W
Nejd, hist. reg., Sau. Ar. *see* Najd	**288**	26°07'N	44°40'E
Nek'emtē, Eth.	**284-85**	9°02'N	36°29'E
Nekoosa, Wi., U.S. (nē-kōō'sá)	**126-27**	44°18'N	89°54'W
Nelidovo, Russia	**218-19**	56°13'N	32°47'E
Neligh, Ne., U.S. (nē'-lē)	**124-25**	42°08'N	98°02'W
Nellore, India (nĕl-lōr')	**256**	14°27'N	79°59'E
Nelson, B.C., Can. (nĕl'sŭn)	**142-43**	49°29'N	117°18'W
Nelson, N.Z. (nĕl'sŭn)	**303**	41°18'S	187°15'E
Nelson, stm., Mb., Can. (nĕl'sŭn)	**140-41**	57°08'N	92°21'W
Nelson, Cape, c., Austl. (kāp nĕl'sŭn)	**301**	38°25'S	141°32'E
Nelspruit, S. Afr.	**286-87**	25°28'S	30°59'E
Neman, Russia (ŋĕ'-mán)	**208-09**	55°02'N	22°02'E
Neman, stm., Eur. (ŋĕ'-mán)	**208-09**	55°21'N	21°16'E
Nëman, stm., Eur.	**208-09**	55°21'N	21°16'E
Nemunas, stm., Eur.	**208-09**	55°21'N	21°16'E
Nemuro, Japan (nā'mō-rō)	**264**	43°20'N	145°35'E
Nen, stm., China (nŭn)	**260-61**	45°26'N	124°39'E
Nenagh, Ire. (nē'ná)	**206-07**	52°52'N	8°12'W
Nendo, i., Sol. Is.	**296-97**	10°45'S	165°54'E
Neosho, Mo., U.S. (nē-ō'shō)	**130-31**	36°52'N	94°23'W
Neosho, stm., U.S. (nē-ō'shō)	**130-31**	35°48'N	95°18'W
Nepal, nation, Asia (nĕ-pôl')	**224-25**	28°0'N	84°00'E
Nepāl, nation, Asia *see* Nepal	**224-25**	28°0'N	84°00'E
Nepālgañj, Nepal	**254-55**	28°04'N	81°37'E
Nephi, Ut., U.S. (nē'fī)	**128-29**	39°43'N	111°50'W
Nercha, stm., Russia	**236-37**	51°56'N	116°39'E
Nerchinsk, Russia (nyĕr' chĕnsk)	**240-41**	51°59'N	116°35'E
Nerekhta, Russia (nyĕ-rĕk'tá)	**218-19**	57°28'N	40°34'E
Nerja, Spain (nĕr'hä)	**214-15**	36°45'N	3°52'W
Neskaupstadur, Ice.	**206a**	65°09'N	13°42'W
Ness, Loch, lk., Scot., U.K. (lŏk nĕs)	**206-07**	57°17'N	4°29'W
Ness City, Ks., U.S. (nĕs sĭ'tē)	**130-31**	38°27'N	99°54'W
Nesterov, Russia (nyĕs-tă'rôf)	**208-09**	54°38'N	22°35'E
Netanya, Isr.	**248-49**	32°21'N	34°52'E
Netherlands, nation, Eur. (nĕdh'ĕr-lăndz)	**190-91**	52°15'N	5°30'E
Netherlands Guiana, nation, S.A. *see* Suriname	**172**	4°0'N	56°00'W
Nettuno, Italy (nĕt-tōō'nō)	**216-17**	41°28'N	12°39'E
Neubrandenburg, Ger. (noi-brän'dĕn-bòrgh)	**210-11**	53°33'N	13°15'E
Neudamm, Pol. *see* Dębno	**210-11**	52°44'N	14°43'E
Neufchâtel-en-Bray, Fr. (nû-shä-tĕl'ĕn-brā')	**212-13**	49°44'N	1°27'E
Neumarkt, Rom. *see* Târgu Mureş	**210-11**	46°33'N	24°34'E
Neumünster, Ger. (noi'münstĕr)	**210-11**	54°04'N	9°59'E
Neunkirchen, Aus. (noin'kĭrk-ĕn)	**210-11**	47°44'N	16°06'E
Neuquén, Arg. (nĕ-ò-kān')	**185**	38°57'S	68°04'W
Neuquén, state, Arg. (nĕ-ò-kān')	**185**	39°0'S	70°00'W
Neuquén, stm., Arg. (nĕ-ò-kān')	**185**	38°59'S	68°00'W
Neuruppin, Ger. (noi'rōō-pēn)	**210-11**	52°55'N	12°49'E
Neusatz, Serb. *see* Novi Sad	**216-17**	45°15'N	19°50'E
Neuse, stm., N.C., U.S. (nūz)	**134-35**	35°09'N	76°31'W
Neustrelitz, Ger. (noi-strā'lĭts)	**210-11**	53°22'N	13°04'E
Neuwied, Ger. (noi'vēdt)	**210-11**	50°26'N	7°28'E
Nevada, Ia., U.S. (nē-vä'dá)	**124-25**	42°02'N	93°27'W
Nevada, Mo., U.S. (nē-vä'dá)	**130-31**	37°51'N	94°21'W
Nevada, state, U.S. (nĕ-vá'dá)	**118-19**	39°0'N	117°00'W
Nevada, Sierra, mts., Spain (syĕr'rä nā-vä'dhä)	**214-15**	37°05'N	3°10'W
Nevada, Sierra, mts., Ca., U.S. (sē-ĕ'r-rä nē-vä'dá)	**128-29**	38°0'N	119°15'W
Nevado, Cerro, mtn., Arg.	**185**	35°34'S	68°28'W
Nevado, Cerro, mtn., Col. (sĕ'r-rō-nĕ-vä'dō)	**188c**	3°59'N	74°04'W
Nevel', Russia (nyĕ'vĕl)	**208-09**	56°01'N	29°56'E
Nevel'sk, Russia	**264**	46°40'N	141°52'E
Nevers, Fr. (nē-vâr')	**212-13**	46°60'N	3°09'E
Nevinnomyssk, Russia	**202-03**	44°38'N	41°56'E
Nevis, i., St. K./N. (nē'vĭs)	**155b**	17°10'N	62°34'W
Nevis, Ben, mtn., Scot., U.K. (bĕn nē'vĭs)	**206-07**	56°48'N	5°01'W
Nevşehir, Tur. (nĕv-shĕ'hĕr)	**202-03**	38°37'N	34°43'E
New, stm., U.S. (nū)	**120-21**	38°09'N	81°12'W
New Albany, In., U.S. (nū ôl'bá-nĭ)	**126-27**	38°17'N	85°50'W
New Albany, Ms., U.S. (nū ôl'bá-nĭ)	**134-35**	34°30'N	89°01'W
New Amsterdam, Guy. (nū ăm'stĕr-dăm)	**178-79**	6°15'N	57°30'W
Newark, De., U.S. (nōō'ärk)	**126-27**	39°41'N	75°45'W
Newark, N.J., U.S.	**126-27**	40°43'N	74°10'W
Newark, N.Y., U.S. (nū'ērk)	**126-27**	43°03'N	77°05'W
Newark, Oh., U.S. (nōō'ûrk)	**126-27**	40°03'N	82°24'W
Newaygo, Mi., U.S. (nū'wā-go)	**126-27**	43°25'N	85°48'W
New Bedford, Ma., U.S. (nū bĕd'fērd)	**126-27**	41°38'N	70°56'W
Newberg, Or., U.S. (nū'bûrg)	**122-23**	45°18'N	122°58'W
New Bern, N.C., U.S. (nū bûrn)	**134-35**	35°06'N	77°04'W
Newberry, Mi., U.S. (nū'bĕr-ĭ)	**124-25**	46°22'N	85°28'W
Newberry, S.C., U.S. (nū'bĕr-ĭ)	**134-35**	34°17'N	81°37'W
New Boston, Oh., U.S. (nū bôs'tŭn)	**126-27**	38°45'N	82°56'W
New Boston, Tx., U.S. (nū bôs'tŭn)	**130-31**	33°28'N	94°25'W
New Braunfels, Tx., U.S. (nū broun'fĕls)	**132-33**	29°42'N	98°07'W
New Britain, Ct., U.S. (nū brĭt''n)	**126-27**	41°40'N	72°46'W
New Britain, i., Pap. N. Gui. (nū brĭt''n)	**302**	6°0'S	150°00'E
New Brunswick, state, Can. (nū brŭnz'wĭk)	**138-39**	46°30'N	66°15'W
Newburgh, N.Y., U.S.	**126-27**	41°30'N	74°02'W
Newbury, Eng., U.K. (nū'bĕr-ĭ)	**206-07**	51°24'N	1°19'W
Newburyport, Ma., U.S. (nū'bĕr-ĭ-pōrt)	**126-27**	42°48'N	70°52'W
New Caledonia, dep., Oc. (nū kăl-ē-dō'nĭ-á)	**306g**	21°30'S	165°30'E
New Caledonia, i., N. Cal. (nū kăl-ē-dō'nĭ-á) *see* Nouvelle-Calédonie	**306g**	21°33'S	165°42'E
New Carlisle, Qc., Can. (nū kär-līl')	**148-49**	48°01'N	65°21'W
Newcastle, Austl. (nū-kås''l)	**301**	32°56'S	151°45'E
Newcastle, S. Afr.	**292c**	27°46'S	29°55'E
New Castle, De., U.S. (nū kås''l)	**126-27**	39°40'N	75°33'W
New Castle, In., U.S. (nū kås''l)	**126-27**	39°55'N	85°22'W
Newcastle, Ok., U.S. (nū-kås''l)	**130-31**	35°14'N	97°36'W
New Castle, Pa., U.S.	**126-27**	40°60'N	80°21'W
Newcastle, Wy., U.S. (nū-kås''l)	**124-25**	43°51'N	104°13'W
Newcastle upon Tyne, Eng., U.K.	**206-07**	54°59'N	1°40'W
New Delhi, nat. cap., India (nū dĕl'hī)	**254-55**	28°36'N	77°13'E
Newell, S.D., U.S. (nū'ĕl)	**124-25**	44°43'N	103°25'W
New England Range, mts., Austl. (nů ĭŋ'glănd rānj)	**301**	29°52'S	151°44'E
Newenham, Cape, c., Ak., U.S. (kāp ū-ĕn-hăm)	**136**	58°39'N	162°10'W
Newfoundland, i., Nf./L., Can. (nū-fŭn'lănd') (nū'fĭnd-lănd) (nū'found-lănd')	**148-49**	48°30'N	56°00'W
Newfoundland and Labrador, state, Can.	**138-39**	52°0'N	56°00'W
New Georgia, i., Sol. Is. (nū ôr'jĭ-á)	**306e**	8°09'S	157°26'E
New Glasgow, N.S., Can. (nū glás'gō)	**148-49**	45°36'N	62°38'W
New Guinea, i., (nū gĭne)	**302**	5°0'S	140°00'E
New Hampshire, state, U.S. (nū hămp'shĭr)	**118-19**	43°35'N	71°40'W
New Hampton, Ia., U.S. (nū hămp'tŭn)	**124-25**	43°03'N	92°19'W
New Hanover, S. Afr. (nū hăn'ôvĕr)	**292c**	29°21'S	30°31'E
New Hanover, i., Pap. N. Gui. (nū hăn'ôvĕr)	**302**	2°30'S	150°15'E
New Harmony, In., U.S. (nū här'mō-nĭ)	**126-27**	38°08'N	87°56'W
New Haven, Ct., U.S. (nū hā'vĕn)	**126-27**	41°19'N	72°56'W
New Haven, In., U.S. (nū hāv''n)	**126-27**	41°04'N	85°02'W
New Hazelton, B.C., Can.	**142-43**	55°15'N	127°35'W
New Hebrides, is., Vanuatu (nū hĕ'brĭ-dēz)	**306g**	16°0'S	167°00'E
New Iberia, La., U.S. (nū jĭ-bē'rĭ-á)	**134-35**	30°00'N	91°49'W
New Jersey, state, U.S. (nū jûr'zĭ)	**118-19**	40°15'N	74°30'W
Newkirk, Ok., U.S. (nū'kûrk)	**130-31**	36°53'N	97°03'W
New Kowloon, China *see* Xinjiulong	**258-59**	22°21'N	114°10'E
New Lexington, Oh., U.S. (nū lĕk'sĭng-tŭn)	**126-27**	39°42'N	82°13'W
New Lisbon, Wi., U.S. (nū lĭz'bŭn)	**126-27**	43°52'N	90°10'W
New Liskeard, On., Can.	**146-47**	47°31'N	79°40'W
New London, Ct., U.S. (nū lŭn'dŭn)	**126-27**	41°21'N	72°07'W
New London, Wi., U.S. (nū lŭn'dŭn)	**126-27**	44°23'N	88°44'W
New Madrid, Mo., U.S. (nū măd'rĭd)	**134-35**	36°35'N	89°32'W
Newmarket, On., Can. (nū'mär-kĕt)	**146-47**	44°03'N	79°27'W
New Martinsville, W.V., U.S. (nū mär'tĭnz-vĭl)	**126-27**	39°38'N	80°52'W
New Mexico, state, U.S. (nū mĕk'sĭ-kō)	**118-19**	34°30'N	106°00'W
Newnan, Ga., U.S. (nū'năn)	**134-35**	33°23'N	84°48'W
New Norfolk, Austl. (nū nôr'fŏk)	**301**	42°47'S	147°03'E
New Orleans, La., U.S. (nū ôr'lănz)	**134-35**	29°59'N	90°05'W
New Philadelphia, Oh., U.S. (nū fĭl-á-dĕl'fĭ-á)	**126-27**	40°30'N	81°27'W
New Plymouth, N.Z. (nū plĭm'ŭth)	**303**	39°04'S	174°05'E
Newport, Eng., U.K. (nū-pôrt)	**206-07**	50°42'N	1°18'W
Newport, Wales, U.K. (nū-pôrt)	**206-07**	51°35'N	3°00'W
Newport, Ar., U.S. (nū'pōrt)	**134-35**	35°37'N	91°17'W
Newport, In., U.S. (nū'pōrt)	**126-27**	39°53'N	87°25'W
Newport, Or., U.S. (nū'pōrt)	**122-23**	44°39'N	124°03'W
Newport, R.I., U.S. (nū'pōrt)	**126-27**	41°29'N	71°19'W
Newport, Tn., U.S. (nū'pōrt)	**134-35**	35°58'N	83°11'W
Newport, Wa., U.S. (nū'pōrt)	**122-23**	48°11'N	117°07'W
Newport News, Va., U.S. (nū'pōrt nūz)	**134-35**	36°59'N	76°25'W
New Providence, i., Bah. (nū prŏv'ĭ-dĕns)	**154-55**	25°02'N	77°24'W
New Richmond, Wi., U.S. (nū rĭch'mŭnd)	**124-25**	45°07'N	92°32'W
New Roads, La., U.S. (nū rōds)	**134-35**	30°42'N	91°27'W
New Rochelle, N.Y., U.S. (nū rū-shĕl')	**126-27**	40°54'N	73°49'W
New Rockford, N.D., U.S. (nū rŏk'fērd)	**124-25**	47°41'N	99°09'W
New Siberian Islands, is., Russia (nū sī-bĭr'y'n ī'lándz)	**236-37**	75°0'N	142°00'E
New Smyrna Beach, Fl., U.S. (nū smûr'ná bēch)	**134-35**	29°02'N	80°56'W
New South Wales, state, Austl. (nū south wālz)	**301**	33°0'S	146°00'E
Newton, Ia., U.S. (nū'tŭn)	**124-25**	41°42'N	93°03'W
Newton, Il., U.S. (nū'tŭn)	**126-27**	38°59'N	88°10'W
Newton, Ks., U.S. (nū'tŭn)	**130-31**	38°03'N	97°21'W
Newton, Ma., U.S. (nū'tŭn)	**126-27**	42°20'N	71°13'W
Newton, Ms., U.S. (nū'tŭn)	**134-35**	32°20'N	89°10'W
Newton, N.C., U.S. (nū'tŭn)	**134-35**	35°40'N	81°13'W
Newton, Tx., U.S. (nū'tŭn)	**132-33**	30°50'N	93°46'W
New Ulm, Mn., U.S. (nū ŭlm)	**124-25**	44°19'N	94°28'W
New Waterford, N.S., Can. (nū wô'tēr-fērd)	**148-49**	46°15'N	60°06'W
New York, N.Y., U.S. (nū yôrk)	**126-27**	40°43'N	74°01'W
New York, state, U.S. (nū yôrk)	**118-19**	43°0'N	75°00'W
New Zealand, nation, Oc. (nū zē'lánd)	**303**	41°0'S	174°00'E
Neyshābūr, Iran	**252-53**	36°11'N	58°52'E
Nezahualcóyotl, Presa, res., Mex.	**158-59**	17°10'N	93°40'W
Nez Perce Indian Reservation, ind. res., Id., U.S. (nĕz' pûrs' ĭn'dĭ-ăn rĕ-sĕr-vā'shĕn)	**122-23**	46°20'N	116°30'W
Ngami, Lake, lk., Bots. (lāk n'gä'mē)	**286-87**	20°29'S	22°46'E
Ngangla Ringco, lk., China (ŋäŋ-lä rĭŋ-tswo)	**254-55**	31°34'N	83°01'E
Ngaoundéré, Camrn.	**282-83**	7°19'N	13°35'E
Ng'iro, Ewaso, stm., Kenya	**284-85**	0°28'N	39°55'E
Ngoring Hu, lk., China	**260-61**	34°53'N	97°41'E
Ngorongoro Conservation Area, pk., Tan.	**289**	3°0'S	35°30'E
Ngozi, Bdi.	**289**	2°54'S	29°53'E
Nguigmi, Niger ('n-gĕg'mē)	**280-81**	14°15'N	13°07'E
Nguna, Île, i., Vanuatu	**306g**	17°27'S	168°21'E
Nguru, Nig. ('n-gōō'rōō)	**282-83**	12°52'N	10°27'E
Nhamundá, stm., Braz.	**180-81**	1°58'S	56°58'W
Nha Trang, Viet. (nyä-träng')	**266-67**	12°16'N	109°12'E
Ni'ihau, i., Hi., U.S. (nē'ē-ha'ōō)	**137a**	21°54'N	160°09'W
Niagara, Wi., U.S. (nī-ăg'á-rá)	**126-27**	45°46'N	88°01'W
Niagara Falls, On., Can. (nī-ăg'á-rá fôlz)	**146-47**	43°05'N	79°02'W
Niah, Malay.	**268-69**	3°52'N	113°43'E
Niamey, nat. cap., Niger (nē-ä-mā')	**280-81**	13°31'N	2°07'E
Niangara, D.R.C. (nē-äŋ-gá'rá)	**289**	3°42'N	27°54'E
Nias, Pulau, i., Indon. (pōō-lou nē'äs')	**266-67**	1°05'N	97°35'E
Niassa, Lago, lk., Afr. *see* Nyasa, Lake	**286-87**	12°0'S	34°30'E
Nicaragua, nation, N.A. (nĭk-á-rä'gwá)	**94**	13°0'N	85°00'W
Nicaragua, Lago de, lk., Nic. (lä'gô dĕ-nĭk-á-rä'gwä)	**161**	11°39'N	85°26'W
Nicaragua, Lake, lk., Nic. (lāk nĭk-á-rä'gwä) *see* Nicaragua, Lago de	**161**	11°39'N	85°26'W
Nice, Fr. (nēs)	**212-13**	43°43'N	7°16'E
Nichinan, Japan	**265**	31°36'N	131°23'E
Nicholas Channel, strt., N.A. (nĭk'ō-lás chăn'ĕl)	**154-55**	23°21'N	80°21'W
Nicholasville, Ky., U.S. (nĭk'ō-lás-vĭl)	**126-27**	37°53'N	84°35'W
Nicobar Islands, is., India (nĭk-ō-bär' ī'lándz)	**266-67**	8°0'N	93°30'E
Nicomedia, Tur. *see* İzmit	**216-17**	40°47'N	29°57'E
Nicosia, nat. cap., Cyp. (nē-kô-sē'á)	**248-49**	35°10'N	33°22'E
Nicosia, nat. cap., Cyp. (nē-kô-zē'ă)	**248-49**	35°10'N	33°22'E
Nicoya, C.R. (nē-kō'yä)	**161**	10°09'N	85°27'W
Nicoya, Golfo de, b., C.R. (gôl-fô dĕ nē-kō'yä)	**161**	9°47'N	84°48'W

ăt; finăl; rāte; senåte; ärm; åsk; sofá; fâre; ch-choose; dh-as th in other; bē; ēvent; bĕt; recĕnt; cratēr; g-gō; gh-guttural g; bĭt; ĭ-short neutral; rīde; κ-guttural k as ch in German ich;

PRONOUNCING INDEX
Nicoya, Península de – Norwalk 389

Feature (Pronunciation)	Page	Lat.	Long.
Nicoya, Península de, pen., C.R.	161	10°01′N	85°25′W
Nictheroy, Braz. *see* Niterói	186	22°54′S	43°07′W
Nidzica, Pol. (nê-jēt′sȧ)	210-11	53°22′N	20°26′E
Nienburg, Ger. (nē′ĕn-bōrgh)	210-11	52°38′N	9°13′E
Nieuw Nickerie, Sur. (nē′ū nê-nē′kĕ-rē′)	178-79	5°56′N	56°60′W
Niğde, Tur. (nĭg′dĕ)	202-03	37°60′N	34°44′E
Nigel, S. Afr. (nī′jĕl)	292c	26°26′S	28°28′E
Niger, nation, Afr. (nī′jẽr)	274	16°0′N	8°00′E
Niger, stm., Afr. (nī′jẽr) (nē-zhâr′)	282-83	4°17′N	6°04′E
Nigeria, nation, Afr. (nī-jē′rĭ-ȧ)	274	10°0′N	8°00′E
Nihoa, i., Hi., U.S.	137b	23°03′N	161°56′W
Nihon, nation, Asia *see* Japan	224-25	36°0′N	138°00′E
Nihon-kai, s., Asia *see* Japan, Sea of	240-41	40°0′N	135°00′E
Niigata, Japan (nē′ê-gä′tȧ)	265	37°55′N	139°04′E
Nii-jima, i., Japan (nē jē′mȧ)	265	34°22′N	139°16′E
Nijmegen, Neth. (nī′mā-gĕn)	206-07	51°50′N	5°50′E
Nikel', Russia	200-01	69°25′N	30°15′E
Nikkō, Japan	265	36°45′N	139°37′E
Nikolayev, Ukr. *see* Mykolaïv	218-19	46°58′N	31°59′E
Nikolayevsk-na-Amure, Russia	236-37	53°09′N	140°44′E
Nikol'sk, Russia (nê-kôlsk′)	202-03	53°42′N	46°05′E
Nikol'sk, Russia (nê-kôlsk′)	202-03	59°32′N	45°27′E
Nikopol', Ukr.	218-19	47°34′N	34°24′E
Niksar, Tur.	247	40°36′N	36°57′E
Nikumaroro, at., Kir.	304-05	4°40′S	174°32′W
Nikunau, i., Kir.	304-05	1°22′S	176°27′E
Nīl, Bahr el-, stm., Afr. *see* Nile	288	30°10′N	31°07′E
Nīl, Nahr an-, stm., Afr. *see* Nile	288	30°10′N	31°07′E
Nile, stm., Afr. (nīl)	288	30°10′N	31°07′E
Niles, Mi., U.S. (nīlz)	126-27	41°50′N	86°15′W
Niles, Oh., U.S. (nīlz)	126-27	41°11′N	80°44′W
Nimba, Mont, mtn., Afr. (môɴ nĭm′bȧ)	282-83	7°37′N	8°25′W
Nimba, Mount, mtn., Afr. *see* Nimba, Mont	282-83	7°37′N	8°25′W
Nîmes, Fr. (nēm)	212-13	43°51′N	4°22′E
Nine Degree Channel, strt., India	256	9°0′N	73°00′E
Ninety Mile Beach, cst., Austl.	301	38°13′S	147°23′E
Ning'an, China (nĭŋ-än)	264	44°20′N	129°28′E
Ningbo, China (nĭŋ-bwo)	258-59	29°53′N	121°32′E
Ningcheng, China	260-61	41°33′N	119°20′E
Ningde, China (nĭŋ-dŭ)	258-59	26°43′N	119°33′E
Ningdu, China	258-59	26°31′N	115°58′E
Ningming, China	258-59	22°08′N	107°05′E
Ningshan, China	258-59	33°19′N	108°19′E
Ningwu, China (nĭng′wōō′)	260-61	39°03′N	112°12′E
Ningxia, state, China (nĭŋ-shyä)	260-61	37°0′N	106°00′E
Ninh Binh, Viet. (nĭn bĭnh′)	266-67	20°15′N	105°59′E
Ninigo Group, is., Pap. N. Gui.	242-43	1°15′S	144°15′E
Ninnescah, stm., Ks., U.S. (nĭn′ĕs-kä)	130-31	37°20′N	97°10′W
Nioaque, Braz. (nēô-ȧ′-kĕ)	182-83	21°09′S	55°50′W
Niobrara, stm., U.S. (nī-ô-brâr′ȧ)	120-21	42°46′N	98°03′W
Nioki, D.R.C.	284-85	2°39′S	17°42′E
Nioro, Mali	280-81	15°14′N	9°35′W
Nipawin, Sk., Can.	144-45	53°22′N	104°00′W
Nipe, Bahía de, b., Cuba (bä-ē′ä-dĕ-nē′pā)	154-55	20°47′N	75°42′W
Nipigon, On., Can. (nĭp′ĭ-gŏn)	146-47	49°01′N	88°15′W
Nipigon, Lake, res., On., Can. (lāk nĭp′ĭ-gŏn)	144-45	49°40′N	88°34′W
Nipigon Bay, b., On., Can. (nĭp′ĭ-gŏn bā)	146-47	48°54′N	87°56′W
Nipissing, Lake, lk., On., Can. (lāk nĭp′ĭ-sĭng)	146-47	46°15′N	79°42′W
Niquero, Cuba (nē-kā′rō)	154-55	20°03′N	77°35′W
Niš, Serb.	216-17	43°19′N	21°54′E
Nisa, Port. (nē′sȧ)	214-15	39°30′N	7°39′W
Nish, Serb. *see* Niš	216-17	43°19′N	21°54′E
Nishapur, Iran *see* Neyshābūr	252-53	36°11′N	58°52′E
Nisser, lk., Nor. (nĭs′ẽr)	208-09	59°10′N	8°30′E
Nistru, stm., Eur. *see* Dniester	204-05	46°19′N	30°17′E
Niterói, Braz. (nē tĕ rô′ĭ)	186	22°54′S	43°07′W
Nitra, Slvk. (nē′trȧ)	210-11	48°19′N	18°06′E
Nitro, W.V., U.S. (nī′trō)	126-27	38°25′N	81°51′W
Niue, dep., Oc. (nĭ′ò)	304-05	19°02′S	169°52′W
Niulakita, i., Tuvalu	304-05	10°45′S	179°30′E
Niut, Gunung, mtn., Indon.	268-69	1°0′N	109°55′E
Niutao, i., Tuvalu	304-05	6°07′S	177°19′E
Nixon, Tx., U.S. (nĭk′sŭn)	132-33	29°16′N	97°46′W
Nizāmābād, India	256	18°40′N	78°06′E
Nizhnekamsk, Russia	202-03	55°33′N	51°58′E
Nizhnekamskoye Vodokhranilishche, res., Russia	202-03	55°50′N	53°00′E
Nizhneudinsk, Russia (nĕzh′nyĭ-ōōdĕnsk′)	236-37	54°54′N	99°02′E
Nizhnevartovsk, Russia	236-37	60°56′N	76°34′E
Nizhniy Novgorod, Russia	202-03	56°19′N	44°01′E
Nizhniy Tagil, Russia (nyēzh′-nyē tŭgēl′)	236-37	57°55′N	59°59′E
Nizhnyaya Tunguska, stm., Russia	236-37	65°46′N	87°54′E
Nizhyn, Ukr.	218-19	51°02′N	31°54′E
Njazidja, i., Com.	286-87	11°35′S	43°20′E
Njombe, stm., Tan.	289	6°56′S	35°06′E
Nkawkaw, Ghana	282-83	6°33′N	0°47′W
Nkongsamba, Camrn.	282-83	4°57′N	9°56′E
Nmai, stm., Mya.	258-59	25°43′N	97°31′E
Noākhāli, Bngl.	254-55	22°49′N	91°06′E
Noatak, stm., Ak., U.S. (nō-ȧ′tȧk)	136	67°00′N	162°30′W
Nobeoka, Japan (nō-bâ-ō′kȧ)	265	32°35′N	131°41′E
Noblesville, In., U.S. (nō′bl′z-vĭl)	126-27	40°02′N	86°00′W
Nochistlán, Mex. (nō-chēs-tlän′)	158-59	21°22′N	102°51′W
Nogales, Mex. (nō-gä′lĕs)	156-57	31°19′N	110°56′W
Nogales, Az., U.S. (nō-gä′lĕs)	128-29	31°21′N	110°56′W
Nogent-le-Rotrou, Fr. (nō-zhŏɴ-lĕ-rō-trōō′)	212-13	48°19′N	0°49′E
Noginsk, Russia (nô-gēnsk′)	218-19	55°52′N	38°28′E
Nogoyá, Arg.	187	32°24′S	59°48′W
Noir, Isla, i., Chile	185	54°29′S	73°01′W
Noirmoutier, Île de, i., Fr. (ēl-dē-nwȧr-mōō-tyā′)	212-13	47°0′N	2°15′W
Nokomis, Il., U.S. (nô-kō′mĭs)	126-27	39°18′N	89°17′W
Nolinsk, Russia (nô-lênsk′)	202-03	57°33′N	49°57′E
Nombre de Dios, Mex. (nôm-brĕ-dĕ-dyô′s)	158-59	23°50′N	104°14′W
Nombre de Dios, Pan. (nô′m-brĕ dĕ-dyô′s)	162	9°35′N	79°28′W
Nome, Ak., U.S. (nōm)	136	64°30′N	165°24′W
Nonacho Lake, lk., N.T., Can.	140-41	61°42′N	109°40′W
Nondalton, Ak., U.S.	136	60°01′N	154°49′W
Nong'an, China (nôŋ-än)	260-61	44°26′N	125°11′E
Nong Khai, Thai.	266-67	17°52′N	102°45′E
Nonouti, at., Kir.	304-05	0°38′S	174°26′E
Noord Zee, s., Eur. *see* North Sea	200-01	56°0′N	3°00′E
Noorvik, Ak., U.S.	136	66°53′N	160°59′W
Nootka Island, i., B.C., Can. (nōōt′kȧ ī′lȧnd)	142-43	49°44′N	126°46′W
Nordegg, Ab., Can. (nûr′dĕg)	142-43	52°28′N	116°05′W
Norderney, i., Ger. (nōr′dĕr-nĕy)	210-11	53°43′N	7°11′E
Nordhausen, Ger. (nôrt′hau-zĕn)	210-11	51°30′N	10°48′E
Nordhorn, Ger. (nôrt′hôrn)	210-11	52°26′N	7°04′E
Nordkapp, c., Nor.	200-01	71°10′N	25°47′E
Nord-Ostsee-Kanal, can., Ger. (nōrd-ûst-zā kä-näl′) *see* Kiel Canal	210-11	53°54′N	9°10′E
Nordsee, s., Eur. *see* North Sea	200-01	56°0′N	3°00′E
Nordsjøen, s., Eur. *see* North Sea	200-01	56°0′N	3°00′E
Norfolk, Ne., U.S. (nôr′fŏk)	124-25	42°02′N	97°25′W
Norfolk, Va., U.S. (nôr′fŏk)	134-35	36°51′N	76°16′W
Norfolk Island, dep., Oc. (nôr-fŭk ī′lȧnd)	304-05	29°02′S	167°57′E
Norge, nation, Eur. *see* Norway	190-91	62°0′N	10°00′E
Noril'sk, Russia (nô rēlsk′)	236-37	69°19′N	88°14′E
Norin, stm., Asia *see* Naryn	244	41°46′N	73°13′E
Normal, Il., U.S. (nôr′mȧl)	126-27	40°30′N	88°59′W
Norman, Ok., U.S.	130-31	35°13′N	97°26′W
Norman, stm., Austl. (nôr′mȧn)	302	17°28′S	140°50′E
Normanby Island, i., Pap. N. Gui.	302	10°05′S	151°05′E
Normandie, hist. reg., Fr. (nŏr-mäɴ-dē′) *see* Normandy	212-13	49°0′N	0°05′W
Normandy, hist. reg., Fr. (nŏr-män-dē′)	212-13	49°0′N	0°05′W
Normanton, Austl. (nôr′mȧn-tŭn)	302	17°40′S	141°05′E
Norman Wells, N.T., Can.	138-39	65°17′N	126°42′W
Ñorquinco, Arg.	185	41°51′S	70°55′W
Norristown, Pa., U.S. (nŏr′ĭs-toun)	126-27	40°07′N	75°21′W
Norrköping, Swe. (nôr′chûp′ĭng)	208-09	58°36′N	16°11′E
Norrtälje, Swe. (nôr-tĕl′yĕ)	208-09	59°46′N	18°44′E
Norseman, Austl. (nôrs′mȧn)	294-95	32°12′S	121°48′E
Norskehavet, s., Eur. *see* Norwegian Sea	200-01	70°0′N	2°00′E
Norte, Serra do, plat., Braz. (sĕ′r-rä-dô-nôr′te)	180-81	11°20′S	59°00′W
North, Cape, c., N.S., Can.	148-49	47°02′N	60°24′W
North Adams, Ma., U.S. (nôrth ăd′ămz)	126-27	42°42′N	73°07′W
Northam, Austl. (nôr-dhăm)	294-95	31°39′S	116°40′E
Northam, S. Afr. (nôr′thăm)	292c	25°03′S	27°11′E
North America, cont., (nôrth ȧ-mĕr′ĭ-kȧ)	4-5	45°0′N	100°00′W
Northampton, Austl. (nôr-thămp′tŭn)	294-95	28°21′S	114°38′E
Northampton, Eng., U.K. (nôrth-ămp′tŭn)	206-07	52°15′N	0°54′W
North Andaman, i., India (nôrth ăn-dȧ-măn′)	266-67	13°15′N	92°55′E
North Battleford, Sk., Can. (nôrth ăt′'l-fērd)	144-45	52°46′N	108°16′W
North Bay, On., Can.	146-47	46°19′N	79°26′W
North Bend, Or., U.S. (nôrth bĕnd)	122-23	43°24′N	124°13′W
North Caicos, i., T./C. Is. (nôrth kī′kôs)	154-55	21°56′N	71°59′W
North Cape, c., N.Z. (nôrth kāp)	303	34°24′S	187°02′E
North Caribou Lake, lk., On., Can.	144-45	52°50′N	90°40′W
North Carolina, state, U.S. (nôrth kăr-ô-lī′nȧ)	118-19	35°30′N	80°00′W
North Cascades National Park, n.p., Wa., U.S.	122-23	48°30′N	121°00′W
North Channel, strt., On., Can.	146-47	46°02′N	82°50′W
North Channel, strt., U.K.	206-07	55°10′N	5°40′W
North Charleston, S.C., U.S. (nôrth chärlz′tŭn)	134-35	32°53′N	79°60′W
North Chicago, Il., U.S. (nôrth shĭ-kô′gō)	126-27	42°18′N	87°52′W
North Dakota, state, U.S. (nôrth dȧ-kō′tȧ)	118-19	47°30′N	100°15′W
Northeast Providence Channel, strt., Bah. (nôrth-ēst′ prŏv′ĭ-dĕns chăn′ĕl)	154-55	25°40′N	77°09′W
Northeim, Ger. (nôrt′hīm)	210-11	51°42′N	10°00′E
Northern Cook Islands, is., Cook Is.	304-05	10°0′S	161°00′W
Northern Donets, stm., Eur. (nôrth′ẽrn dŏn-ĕts′)	202-03	47°36′N	40°54′E
Northern Indian Lake, lk., Mb., Can.	140-41	57°21′N	97°19′W
Northern Ireland, state, U.K. (nôrth′ẽrn īr′lănd)	206-07	54°40′N	6°45′W
Northern Mariana Islands, dep., Oc. (nôrth′ẽrn mä-rē-ä′nȧ ī′lȧndz)	304-05	16°0′N	149°00′E
Northern Sporades, is., Grc. *see* Vóreioi Sporades	216-17	39°15′N	23°55′E
Northern Territory, state, Austl.	294-95	20°0′S	134°00′E
Northfield, Mn., U.S. (nôrth′fēld)	124-25	44°28′N	93°10′W
North Island, i., N.Z. (nôrth ī′lȧnd)	303	39°0′S	176°00′E
North Judson, In., U.S. (nôrth jŭd′sŭn)	126-27	41°13′N	86°46′W
North Korea, nation, Asia (nôrth kô-rē′-ȧ)	224-25	40°0′N	127°00′E
North Lakhimpur, India	254-55	27°14′N	94°07′E
North Little Rock, Ar., U.S. (nôrth lĭt′'l rŏk)	130-31	34°46′N	92°18′W
North Magnetic Pole, p.o.i.,	314	86°04′N	153°21′W
North Manchester, In., U.S. (nôrth măn′chĕs-tẽr)	126-27	40°60′N	85°46′W
North Ogden, Ut., U.S. (nôrth ŏg′dĕn)	122-23	41°19′N	111°58′W
North Ossetia, state, Russia	245	43°0′N	44°15′E
North Platte, Ne., U.S. (nôrth plăt)	124-25	41°08′N	100°45′W
North Platte, stm., U.S. (nôrth plăt)	120-21	41°07′N	100°42′W
Northport, Al., U.S. (nôrth′pōrt)	134-35	33°14′N	87°34′W
North Saskatchewan, stm., Can. (nôrth săn-kăch′ê-wän)	140-41	53°14′N	105°05′W
North Sea, s., Eur. (nôrth sē)	200-01	56°0′N	3°00′E
North Siberian Lowland, pl., Russia (nôrth sī-bĭr′y′n lō′lȧnd) *see* Severo-Sibirskaya Nizmennost'	236-37	73°0′N	100°00′E
North Sydney, N.S., Can. (nôrth sĭd′nê)	148-49	46°13′N	60°16′W
North Thompson, stm., B.C., Can.	142-43	50°41′N	120°20′W
North Tonawanda, N.Y., U.S. (nôrth tŏn-ȧ-wŏn′dȧ)	126-27	43°02′N	78°52′W
Northumberland Strait, strt., Can. (nôr thŭm′bẽr-lȧnd strāt)	148-49	46°0′N	63°30′W
North Vancouver, B.C., Can. (nôrth văn-kōō′vẽr)	142-43	49°19′N	123°04′W
North Vernon, In., U.S. (nôrth vûr′nŭn)	126-27	39°00′N	85°38′W
North West Cape, c., Austl. (nôrth wĕst kāp)	296-97	21°48′S	114°10′E
Northwest Providence Channel, strt., Bah. (nôrth-wĕst′ prŏv′ĭ-dĕns chăn′ĕl)	154-55	26°10′N	78°20′W
Northwest Territories, state, Can. (nôrth′wĕst tĕr′ĭ-tō′rĭs)	138-39	65°0′N	120°00′W
North Wilkesboro, N.C., U.S. (nôrth wĭlks′bŭrô)	134-35	36°11′N	81°09′W
Northwood, Ia., U.S. (nôrth′wòd)	124-25	43°27′N	93°13′W
Northwood, N.D., U.S. (nôrth′wòd)	124-25	47°44′N	97°34′W
Norton, Ks., U.S. (nôr′tŭn)	130-31	39°50′N	99°53′W
Norton, Va., U.S. (nôr′tŭn)	134-35	36°56′N	82°38′W
Norton Sound, strt., Ak., U.S. (nôr′tŭn sound)	136	63°50′N	164°00′W
Norvegia, Cape, c., Ant.	313	71°25′S	12°18′W
Norwalk, Ct., U.S. (nôr′wôk)	126-27	41°07′N	73°25′W

n-sing; ŋ-baŋk; ɴ-nasalized n; nŏd; cŏmmit; ōld; ôbey; ôrder; oi-boil; fōōd; ò-as oo in foot; ou-out; s-soft; sh-dish; th-thin; pūre; ûnite; ûrn; stŭd; circŭs; ü-as in French tu; ′-indeterminate vowel.

390 Norwalk – Odemira PRONOUNCING INDEX

Feature (Pronunciation)	Page	Lat.	Long.
Norwalk, Oh., U.S. (nôr´wôk)	126-27	41°15´N	82°36´W
Norway, Me., U.S. (nôr´wôk)	126-27	44°13´N	70°32´W
Norway, nation, Eur. (nôr´wā)	190-91	62°0´N	10°00´E
Norway House, Mb., Can. (nôr´wā hous)	144-45	53°59´N	97°48´W
Norwegian Sea, s., Eur. (nôr-wē´jăn sē sē)	200-01	70°0´N	2°00´E
Norwich, Eng., U.K.	206-07	52°38´N	1°17´E
Norwich, Ct., U.S. (nôr´wĭch)	126-27	41°32´N	72°05´W
Norwood, Ma., U.S. (nôr´wŏŏd)	126-27	42°11´N	71°12´W
Norwood, Oh., U.S. (nôr´wŏŏd)	126-27	39°10´N	84°27´W
Noshiro, Japan (nō´shē-rô)	264	40°12´N	140°02´E
Nosivka, Ukr. (nô´sôf-kȧ)	218-19	50°56´N	31°36´E
Nosop, stm., Afr.	286-87	26°53´S	20°41´E
Nossob, stm., Afr. (nŏ´sŏb)	286-87	26°53´S	20°41´E
Nosy-Varika, Madag.	286-87	20°35´S	48°32´E
Noteć, stm., Pol. (nô´tĕcn)	210-11	52°44´N	15°25´E
Notodden, Nor. (nōt´ôd'n)	208-09	59°34´N	9°16´E
Noto-hantō, pen., Japan	265	37°20´N	137°00´E
Notozero, Ozero, lk., Russia	202-03	66°28´N	32°05´E
Notre-Dame, Monts, mts., Qc., Can.	148-49	48°10´N	68°00´W
Notre Dame Bay, b., Nf./L., Can. (nō´t'r dȧm´ bā)	148-49	49°46´N	55°15´W
Nottawasaga Bay, b., On., Can. (nŏt´ȧ-wȧ-sā´gȧ bā)	146-47	44°35´N	80°15´W
Nottingham, Eng., U.K. (nŏt´ĭng-ăm)	206-07	52°57´N	1°07´W
Nottingham Island, i., Nu., Can.	140-41	63°20´N	77°55´W
Nouâdhibou, Maur.	280-81	20°55´N	17°02´W
Nouâdhibou, Râs, c., Afr.	280-81	20°47´N	17°03´W
Nouakchott, nat. cap., Maur. (nü-äk´-shôt)	280-81	18°06´N	15°58´W
Nouméa, nat. cap., N. Cal. (nōō-mā´ä)	306g	22°17´S	166°27´E
Nouvelle, Qc., Can. (nōō-vĕl´)	148-49	48°08´N	66°19´W
Nouvelle-Calédonie, dep., Oc. *see* New Caledonia	306g	21°30´S	165°30´E
Nouvelle-Calédonie, i., N. Cal.	306g	21°33´S	165°42´E
Nouvelle-France, Cap de, c., Qc., Can.	140-41	62°27´N	73°42´W
Nouvelles-Hébrides, nation, Oc. *see* Vanuatu	306g	16°0´S	167°00´E
Nouvelles-Hébrides, is., Vanuatu *see* New Hebrides	306g	16°0´S	167°00´E
Nova Freixo, Moz. *see* Cuamba	286-87	14°47´S	36°32´E
Nova Friburgo, Braz. (nō´vȧ frē-bōōr´gō)	186	22°16´S	42°32´W
Nova Goa, India *see* Panaji	256	15°30´N	73°50´E
Nova Iguaçu, Braz. (nō´vä-ē-gwä-sōō´)	186	22°45´S	43°27´W
Nova Kakhovka, Ukr.	218-19	46°45´N	33°25´E
Nova Lima, Braz. (nō´vȧ lē´mä)	186	19°60´S	43°51´W
Novara, Italy (nô-vä´rä)	216-17	45°27´N	8°37´E
Nova Scotia, state, Can. (nō´vȧ skō´shȧ)	138-39	45°0´N	63°00´W
Novaya Ladoga, Russia (nô´vȧ-ya lȧ-dō-gȧ)	208-09	60°05´N	32°15´E
Novaya Sibir', Ostrov, i., Russia (ôs-trôf´ nô´vȧ-ya sē-bēr´)	236-37	75°0´N	149°00´E
Novaya Zemlya, is., Russia (nô´vȧ-ya zĕm-lyȧ´)	236-37	74°0´N	57°00´E
Nova Zagora, Blg. (nô´vä zä´gô-rȧ)	216-17	42°30´N	26°01´E
Novelda, Spain (nō-vĕl´dȧ)	214-15	38°23´N	0°46´W
Nové Zámky, Slvk. (nô´vĕ zȧm´kē)	210-11	47°60´N	18°11´E
Novgorod, Russia (nôv´gō-rōt)	208-09	58°32´N	31°18´E
Novi Pazar, Blg. (nō´vĭ pȧ-zär´)	216-17	43°21´N	27°12´E
Novi Pazar, Serb. (nō´vĭ pȧ-zär´)	216-17	43°08´N	20°31´E
Novi Sad, Serb. (nō´vĭ säd´)	216-17	45°15´N	19°50´E
Novoanninskiy, Russia	202-03	50°32´N	42°41´E
Novo Aripuanã, Braz.	180-81	5°08´S	60°21´W
Novocherkassk, Russia (nô´vô-chĕr-kȧsk´)	202-03	47°25´N	40°06´E
Novodvinsk, Russia	202-03	64°25´N	40°49´E
Novohrad-Volyns'kyi, Ukr.	210-11	50°36´N	27°38´E
Novokuybyshevsk, Russia	202-03	53°06´N	49°56´E
Novokuznetsk, Russia (nō´vô-kō´z-nyĕ´tsk)	244	53°45´N	87°07´E
Novo Mesto, Slvn. (nôvô mäs´tô)	216-17	45°48´N	15°10´E
Novomoskovsk, Russia (nô´vô-môs-kôfsk´)	218-19	54°05´N	38°13´E
Novomoskovs'k, Ukr. (nō´vô-kō´z-nyĕ´tsk)	218-19	48°38´N	35°12´E
Novorossiysk, Russia (nô´vô-rô-sēsk´)	218-19	44°43´N	37°46´E
Novorzhev, Russia (nô´vô-rzhĕv´)	208-09	57°02´N	29°20´E
Novoshakhtinsk, Russia	218-19	47°48´N	39°54´E
Novosibirsk, Russia (nô´vô-sē-bērsk´)	236-37	55°01´N	82°53´E
Novosibirskiye Ostrova, is., Russia *see* New Siberian Islands	236-37	75°0´N	142°00´E
Novosibirskoye Vodokhranilishche, res., Russia	244	54°35´N	82°35´E
Novosil', Russia (nô´vô-sĭl)	218-19	52°58´N	37°03´E

Feature (Pronunciation)	Page	Lat.	Long.
Novosokol'niki, Russia (nô´vô-sô-kôl´nē-kē)	208-09	56°21´N	30°10´E
Novouzensk, Russia (nô-vô-ô-zĕnsk´)	202-03	50°29´N	48°10´E
Novovolyns'k, Ukr.	210-11	50°44´N	24°08´E
Novozybkov, Russia (nô´vô-zēp´kôf)	218-19	52°32´N	31°56´E
Nový Jičín, Czech Rep. (nô´vē yē´chēn)	210-11	49°36´N	18°01´E
Novyy Oskol, Russia (nô´vē ôs-kôl´)	218-19	50°46´N	37°53´E
Novyy Uzen, Kaz. *see* Zhangaözen	202-03	43°19´N	52°47´E
Nowata, Ok., U.S. (nô-wä´tȧ)	130-31	36°42´N	95°38´W
Nowra, Austl. (nou´rȧ)	301	34°53´S	150°36´E
Nowshera, Pak.	252-53	34°01´N	71°59´E
Nowy Dwór Mazowiecki, Pol. (nō´vĭ dvōōr mä-zo-vyĕts´ke)	210-11	52°26´N	20°43´E
Nowy Targ, Pol. (nō´vē tärk´)	210-11	49°28´N	20°03´E
Noxubee, stm., U.S. (nŏks´û-bē)	134-35	32°50´N	88°10´W
Nsanje, Malawi	286-87	16°58´S	35°12´E
Nsukka, Nig.	282a	6°51´N	7°24´E
Ntem, stm., Afr.	282-83	2°20´N	9°50´E
Ntomba, Lac, lk., D.R.C.	284-85	0°48´S	18°03´E
Nu, stm., Asia (nōō) *see* Salween	226-27	16°33´N	97°40´E
Nubian Desert, des., Sudan (nōō´bĭ-ăn dĕs´ērt)	288	20°30´N	33°00´E
Nueces, stm., Tx., U.S. (nů-ā´sȧs)	132-33	27°50´N	97°22´W
Nueltin Lake, lk., Can. (nwĕl´tin lăk)	140-41	60°19´N	99°40´W
Nueva, Isla, i., Chile	185	55°14´S	66°32´W
Nueva Gerona, Cuba (nwā´vä ĸĕ-rô´nä)	154-55	21°53´N	82°48´W
Nueva Imperial, Chile	185	38°44´S	72°57´W
Nueva Palmira, Ur. (nwā´vä päl-mē´rä)	187	33°52´S	58°23´W
Nueva Rosita, Mex. (nòĕ´vä rô-sē´tä)	132-33	27°57´N	101°13´W
Nueva Toltén, Chile	185	39°12´S	73°13´W
Nueve de Julio, Arg. (nwā´vȧ dā hōō´lyō)	187	35°27´S	60°53´W
Nuevitas, Cuba (nwā-vē´tās)	154-55	21°33´N	77°16´W
Nuevo, Cayo, i., Mex.	158-59	21°51´N	92°06´W
Nuevo, Golfo, b., Arg.	185	42°42´S	64°36´W
Nuevo Casas Grandes, Mex.	156-57	30°25´N	107°55´W
Nuevo Laredo, Mex. (nwā´vô lä-rā´dhō)	132-33	27°28´N	99°31´W
Nuevo León, state, Mex. (nwā´vô lȧ-ōn´)	132-33	25°40´N	100°00´W
Nuguria Islands, is., Pap. N. Gui.	302	3°21´S	154°41´E
Nui, at., Tuvalu	304-05	7°15´S	177°10´E
Nukha, Azer. *see* Şeki	245	41°10´N	47°10´E
Nuku'alofa, nat. cap., Tonga (nōō´-kōō-ä-lō´-fȧ)	304-05	21°08´S	175°13´W
Nukuoro, at., Micron.	304-05	3°51´N	154°58´E
Nukus, Uzb.	244	42°28´N	59°36´E
Nullarbor Plain, pl., Austl. (nů-lär´bôr plān)	296-97	31°0´S	129°00´E
Numara, i., Mald.	256	6°25´N	73°04´E
Numazu, Japan (nōō´mä-zōō)	265	35°06´N	138°52´E
Numedalslågen, stm., Nor. *see* Lågen	208-09	59°02´N	10°04´E
Numfoor, Pulau, i., Indon.	242-43	1°03´S	134°54´E
Nunavut, state, Can.	138-39	70°0´N	95°00´W
Nunivak Island, i., Ak., U.S. (nōō´nĭ-văk ī´lȧnd)	136	60°00´N	166°29´W
Nunjiang, China	260-61	49°10´N	125°14´E
Nuomin, stm., China	260-61	48°13´N	124°31´E
Nuoro, Italy (nwō´rō)	216-17	40°20´N	9°20´E
Nüra, stm., Kaz.	244	50°22´N	69°15´E
Nuremberg, Ger. *see* Nürnberg	210-11	49°27´N	11°04´E
Nürnberg, Ger. (nürn´bĕrgh)	210-11	49°27´N	11°04´E
Nushagak, stm., Ak., U.S. (nū-shă-găk´)	136	59°03´N	158°24´W
Nu Shan, mts., China	258-59	27°0´N	99°00´E
Nushki, Pak. (nŭsh´kē)	252-53	29°35´N	66°04´E
Nuuk, nat. cap., Green.	310-11	64°11´N	51°44´W
Nuweveldberge, mts., S. Afr.	286-87	32°14´S	21°48´E
Nyahururu Falls, Kenya	289	0°02´N	36°22´E
Nyainqêntanglha Shan, mts., China (nyä-ĭn-chyŭn-tăŋ-lä shän)	254-55	30°0´N	90°00´E
Nyala, Sudan	288	12°03´N	24°54´E
Nyandoma, Russia	202-03	61°40´N	40°13´E
Nyanza, Rw.	289	2°21´S	29°45´E
Nyasa, Lake, lk., Afr. (lăk nyä´sä)	286-87	12°0´S	34°30´E
Nyasaland, nation, Afr. *see* Malawi	274	13°30´S	34°00´E
Nyborg, Den. (nü´bôr'')	208-09	55°19´N	10°47´E
Nybro, Swe. (nü´brō)	208-09	56°45´N	15°55´E
Nyeri, Kenya	289	0°25´S	36°57´E
Nyíregyháza, Hung. (nyē´rĕd-y'hä´zä)	210-11	47°57´N	21°43´E
Nyköping, Swe. (nü´chû-pĭng)	208-09	58°45´N	16°60´E
Nylstroom, S. Afr. (nĭl´strōm) *see* Modimolle	292c	24°42´S	28°25´E
Nynäshamn, Swe. (nü-nĕs-hȧm'n)	208-09	58°55´N	17°57´E
Nyngan, Austl. (nĭŋ´gȧn)	301	31°33´S	147°10´E

Feature (Pronunciation)	Page	Lat.	Long.
Nyong, stm., Camrn. (nyông)	282-83	3°16´N	9°55´E
Nysa Łużycka, stm., Eur. *see* Neisse	210-11	52°04´N	14°46´E
Nyslott, Fin. *see* Savonlinna	208-09	61°52´N	28°54´E
Nytva, Russia	202-03	57°56´N	55°20´E
Nyunzu, D.R.C.	289	5°58´S	28°02´E
Nyurba, Russia	236-37	63°17´N	118°20´E
Nyuvchim, Russia	202-03	61°23´N	50°36´E
Nyuya, stm., Russia (nyōō´yȧ)	236-37	60°32´N	116°18´E
Nzérékoré, Gui.	282-83	7°45´N	8°49´W
Nzwani, i., Com. (äɴ-zhwäɴ)	286-87	12°15´S	44°25´E

O

Feature (Pronunciation)	Page	Lat.	Long.
O'ahu, i., Hi., U.S. (ō-ä´hōō) (ō-ä´hü)	137a	21°30´N	158°00´W
Oahe, Lake, res., U.S.	124-25	45°29´N	100°20´W
Oak Bay, B.C., Can. (ōk bā)	142-43	48°27´N	123°18´W
Oak Creek, Wi., U.S. (ōk krēk´)	126-27	42°52´N	87°54´W
Oakdale, La., U.S. (ōk´dāl)	132-33	30°49´N	92°40´W
Oakes, N.D., U.S. (ōks)	124-25	46°08´N	98°06´W
Oak Grove, Ky., U.S. (ōk grōv)	134-35	36°40´N	87°26´W
Oak Harbor, Wa., U.S. (ōk här´bēr)	122-23	48°18´N	122°40´W
Oakland, Ca., U.S. (ōk´lănd)	128-29	37°48´N	122°17´W
Oakland, Md., U.S. (ōk´lănd)	126-27	39°24´N	79°24´W
Oakland, Ne., U.S. (ōk´lănd)	124-25	41°50´N	96°28´W
Oak Lawn, Il., U.S. (ōk lôn)	126-27	41°43´N	87°45´W
Oakley, Ks., U.S. (ōk´lĭ)	130-31	39°08´N	100°51´W
Oak Ridge, Tn., U.S. (ōk rĭj)	134-35	36°01´N	84°15´W
Oakville, On., Can. (ōk´vĭl)	146-47	43°27´N	79°40´W
Oakville, Mo., U.S. (ōk´vĭl)	130-31	38°28´N	90°19´W
Oaxaca, state, Mex. (wä-hä´kä)	158-59	17°0´N	96°30´W
Oaxaca de Juárez, Mex.	158-59	17°03´N	96°43´W
Ob', stm., Russia (ōb)	236-37	66°47´N	68°56´E
Oban, Scot., U.K. (ō´băn)	206-07	56°25´N	5°28´W
Oberlin, Ks., U.S. (o´bēr-lĭn)	130-31	39°49´N	100°32´W
Oberlin, Oh., U.S. (o´bēr-lĭn)	126-27	41°17´N	82°13´W
Obi, Kepulauan, is., Indon. (ō´bē)	268-69	1°27´S	127°38´E
Obi, Pulau, i., Indon.	268-69	1°30´S	127°45´E
Óbidos, Braz. (ō´bē-dōzh)	180-81	1°54´S	55°31´W
Obihiro, Japan (ō´bē-hē´rō)	264	42°55´N	143°12´E
Obluchye, Russia	260-61	49°01´N	131°04´E
Obninsk, Russia	218-19	55°06´N	36°37´E
Oboyan, Russia (ô-bô-yän´)	218-19	51°13´N	36°17´E
Observatoire, Caye de l', i., N. Cal.	296-97	21°25´S	158°50´E
Obsgchiy Syrt, mts., Eur. *see* Zhalpy Syrt	202-03	52°0´N	51°30´E
Obskaya Guba, b., Russia	236-37	69°0´N	73°00´E
Obuasi, Ghana	282-83	6°13´N	1°41´W
Ocala, Fl., U.S. (ô-kä´lȧ)	134-35	29°11´N	82°08´W
Ocampo, Mex. (ô-käm´pō)	156-57	28°11´N	108°23´W
Ocaña, Col. (ô-kän´yä)	178-79	8°14´N	73°21´W
Ocaña, Spain (ô-kä´n-yä)	214-15	39°57´N	3°30´W
Occidental, Cordillera, mts., Col.	178-79	5°0´N	76°00´W
Ocean City, Md., U.S. (ō´shăn sĭ´tē)	126-27	38°20´N	75°05´W
Ocean City, N.J., U.S. (ō´shăn sĭ´tē)	126-27	39°17´N	74°35´W
Ocean Falls, B.C., Can. (ō´shăn fôlz)	142-43	52°21´N	127°41´W
Ocean Grove, N.J., U.S. (ō´shăn grōv)	126-27	40°13´N	74°00´W
Ocean Island, i., Kir. *see* Banaba	304-05	0°52´S	169°33´E
Oceanside, Ca., U.S. (ō´shăn-sīd)	128-29	33°12´N	117°22´W
Ocean Springs, Ms., U.S. (ō´shăn sprĭngs sprĭngz)	134-35	30°26´N	88°50´W
Ochlockonee, stm., U.S. (ŏk-lô-kō´nē)	134-35	29°59´N	84°26´W
Ocilla, Ga., U.S. (ô-sĭl´ȧ)	134-35	31°36´N	83°15´W
Ockelbo, Swe. (ôk´ĕl-bô)	208-09	60°54´N	16°44´E
Ocmulgee National Monument, n.p., Ga., U.S. (ôk-mŭl´gē năsh´ŭn-ăl mŏn´ū-mĕnt)	134-35	32°43´N	83°38´W
Oconee, stm., Ga., U.S. (ô-kō´nē)	134-35	31°58´N	82°32´W
Oconomowoc, Wi., U.S. (ô-kŏn´ô-mô-wŏk´)	126-27	43°06´N	88°29´W
Oconto, Wi., U.S. (ô-kŏn´tō)	126-27	44°54´N	87°52´W
Oconto Falls, Wi., U.S. (ô-kŏn´tō fôlz)	126-27	44°52´N	88°08´W
Ocosingo, Mex.	160	16°55´N	92°06´W
Ocotal, Nic. (ō-kô-täl´)	161	13°38´N	86°28´W
Ocotlán, Mex. (ō-kô-tlän´)	158-59	20°21´N	102°47´W
Ocotlán de Morelos, Mex. (ō-kô-tlän´ dā mô-rā´lōs)	158-59	16°47´N	96°40´W
Ocracoke Island, i., N.C., U.S.	134-35	35°06´N	75°59´W
October Revolution Island, i., Russia *see* Oktyabr'skoy Revolyutsii, Ostrov	236-37	79°30´N	96°60´E
Ocumare del Tuy, Ven. (ō-kōō-mä´ra del twē´)	188b	10°07´N	66°46´W
Oda, Jabal, mtn., Sudan	288	20°21´N	36°39´E
Odda, Nor. (ôdh-ȧ)	208-09	60°04´N	6°32´E
Odemira, Port. (ō-dä-mē´rȧ)	214-15	37°35´N	8°38´W

ăt; finăl; rāte; senâte; ärm; ȧsk; sofȧ; fâre; ch-choose; dh-as th in other; bē; ēvent; bĕt; recĕnt; cratēr; g-gō; gh-guttural g; bĭt; ĭ-short neutral; rīde; ĸ-guttural k as ch in German ich;

Feature (Pronunciation)	Page	Lat.	Long.
Ödemiş, Tur. (û′dĕ-mēsh)	216-17	38°14′N	27°59′E
Odendaalsrus, S. Afr. (ō′dĕn-däls-rûs′)	292c	27°52′S	26°42′E
Odense, Den. (ō′dhĕn-sĕ)	208-09	55°24′N	10°23′E
Oder, stm., Eur. (ō′dẽr)	210-11	53°55′N	14°17′E
Odesa, Ukr.	218-19	46°29′N	30°42′E
Odessa, De., U.S. (ô-dĕs′á)	126-27	39°27′N	75°40′W
Odessa, Tx., U.S. (ô-dĕs′á)	132-33	31°51′N	102°22′W
Odin, Mount, mtn., B.C., Can.	142-43	50°33′N	118°08′W
Odintsovo, Russia (ô-dēn′tsô-vô)	218-19	55°40′N	37°16′E
Odisha, state, India (o′dēsh′ä)	254-55	20°0′N	84°00′E
Odra, stm., Eur. (ô′drà) *see* Oder	210-11	53°55′N	14°17′E
Odrzywół, Pol.	210-11	51°32′N	20°33′E
Oeiras, Braz. (wå-ē-räzh′)	180-81	7°01′S	42°08′W
Oelwein, Ia., U.S. (ōl′wīn)	124-25	42°40′N	91°55′W
Oeno Atoll, at., Pit.	304-05	23°55′S	130°44′W
O'Fallon, Mo., U.S. (ō-făl′ŭn)	130-31	38°49′N	90°42′W
Offenburg, Ger. (ôf′ĕn-bòrgh)	210-11	48°28′N	7°57′E
Ofu, i., Am. Sam.	306b	14°10′S	169°40′W
Ogaadeen, reg., Afr. *see* Ogaden	284-85	8°0′N	44°00′E
Ogaden, reg., Afr. (ō-gä′dĕn)	284-85	8°0′N	44°00′E
Ogallala, Ne., U.S. (ō-gä-lä′lä)	124-25	41°08′N	101°43′W
Ogasawara-guntō, is., Japan			
see Bonin Islands	308-09	26°58′N	142°14′E
Ogasawara-shotō, is., Japan	304-05	26°0′N	142°00′E
Ogbomosho, Nig. (ŏg-bō-mō′shō)	282a	8°08′N	4°15′E
Ogden, Ia., U.S. (ŏg′dĕn)	124-25	42°03′N	94°02′W
Ogden, Ut., U.S. (ŏg′dĕn)	122-23	41°14′N	111°57′W
Ogdensburg, N.Y., U.S.			
(ŏg′dĕnz-bûrg)	126-27	44°42′N	75°30′W
Ogea Levu, i., Fiji	306f	19°08′S	178°24′W
Ogeechee, stm., Ga., U.S. (ō-gē′chē)	134-35	31°51′N	81°05′W
Ogilvie Mountains, mts., Yk., Can.			
(ō′g'l-vĭ moun′tĭnz)	140-41	65°03′N	139°29′W
Ogooué, stm., Afr.	282-83	0°49′S	9°00′E
Ogulin, Cro. (ō-gōō-lēn′)	216-17	45°16′N	15°14′E
Ogurja Ada, i., Turkmen.	244	38°57′N	53°03′E
O'Higgins, Lago, lk., S.A.	185	48°53′S	72°39′W
Ohio, state, U.S. (ô′hī′ō)	118-19	40°15′N	82°45′W
Ohio, stm., U.S. (ô′hī′ō)	120-21	36°59′N	89°09′W
Ohlau, Pol. *see* Oława	210-11	50°56′N	17°19′E
Ohōtuku-kai, s., Asia			
see Okhotsk, Sea of	236-37	53°0′N	150°00′E
Ohrid, Mac. (ō′кrēd)	216-17	41°07′N	20°49′E
Oiapoque, Braz.	178-79	3°51′N	51°49′W
Oiapoque, stm., S.A.	178-79	4°10′N	51°37′W
Oil City, Pa., U.S. (oil sĭ′tĭ sĭ′tê)	126-27	41°26′N	79°42′W
Ōita, Japan (ō′ē-tä)	265	33°14′N	131°37′E
Ojinaga, Mex. (ō-кē-nä′gä)	132-33	29°33′N	104°25′W
Ojocaliente, Mex. (ô-кō-kä-lyĕ′n-tĕ)	158-59	22°33′N	102°13′W
Ojos del Salado, Cerro, mtn., S.A.			
see Ojos del Salado, Nevado	182-83	27°06′S	68°32′W
Ojos del Salado, Nevado, mtn., S.A.	182-83	27°06′S	68°32′W
Oka, stm., Russia (ô-kä′)	236-37	55°16′N	102°18′E
Oka, stm., Russia (ô-kä′)	202-03	56°20′N	43°59′E
Okahandja, Nmb.	286-87	21°58′S	16°54′E
Okanagan, stm., N.A. (ō′kȧ-näg′ȧn)			
see Okanogan	142-43	48°06′N	119°44′W
Okanagan Lake, lk., B.C., Can.			
(ō′kȧ-näg′ȧn lāk)	142-43	49°55′N	119°31′W
Okanogan, stm., N.A.	142-43	48°06′N	119°44′W
Okāra, Pak.	252-53	30°48′N	73°27′E
Okavango, stm., Afr. (ō-kȧ-vän′gō)	286-87	18°57′S	22°25′E
Okavango Delta, del., Bots.	286-87	19°29′S	22°32′E
Okavango Swamp, del., Bots.			
see Okavango Delta	286-87	19°29′S	22°32′E
Okaya, Japan (ō′kȧ-yȧ)	265	36°04′N	138°03′E
Okayama, Japan (ō′kȧ-yä′mȧ)	265	34°40′N	133°55′E
Okazaki, Japan (ō′kȧ-zä′kē)	265	34°57′N	137°10′E
Okeechobee, Fl., U.S. (ō-kē-chō′bē)	135a	27°15′N	80°50′W
Okeechobee, Lake, lk., Fl., U.S.			
(lāk ō-kē-chō′bē)	135a	26°55′N	80°45′W
Okefenokee Swamp, sw., U.S.			
(ō′kē-fē-no′kē swômp)	134-35	30°42′N	82°20′W
Okemah, Ok., U.S. (ô-kē′mä)	130-31	35°26′N	96°18′W
Okene, Nig.	282a	7°29′N	6°15′E
Okha, Russia (ŭ-кä′)	236-37	53°35′N	142°57′E
Okhota, stm., Russia	236-37	59°20′N	143°04′E
Okhotsk, Russia (ô-кôtsk′)	236-37	59°22′N	143°18′E
Okhotsk, Sea of, s., Asia			
(sē ŭv ô-kôtsk′)	236-37	53°0′N	150°00′E
Okhotskoye More, s., Asia			
see Okhotsk, Sea of	236-37	53°0′N	150°00′E
Okhtyrka, Ukr.	218-19	50°18′N	34°54′E
Okinawa-jima, i., Japan	264a	26°32′N	127°60′E
Okino-Daitō-jima, i., Japan	240-41	24°28′N	131°11′E
Okino-Tori-shima, i., Japan	240-41	20°27′N	136°04′E
Oki-shotō, is., Japan	265	36°11′N	133°11′E
Oklahoma, state, U.S. (ô-klȧ-hō′mȧ)	118-19	35°30′N	98°00′W
Oklahoma City, Ok., U.S.			
(ô-klȧ-hō′mȧ sĭ′tĭ)	130-31	35°29′N	97°29′W
Okmulgee, Ok., U.S. (ôk-mŭl′gē)	130-31	35°36′N	95°58′W
Okolona, Ms., U.S. (ō-kô-lō′nȧ)	134-35	33°60′N	88°45′W
Okotoks, Ab., Can.	142-43	50°44′N	113°59′W
Oktyabr'sk, Kaz.	244	49°28′N	57°25′E
Oktyabrskiy, Russia	202-03	54°29′N	53°29′E
Oktyabr'skiy, Russia	236-37	52°41′N	156°13′E
Oktyabr'skoy Revolyutsii, Ostrov, i.,			
Russia	236-37	79°30′N	96°60′E
Okushiri-tō, i., Japan (ō′koo-shē′rē tō)	264	42°10′N	139°27′E
Ola, Russia	236-37	59°37′N	151°20′E
Olanchito, Hond. (ō′län-chē′tô)	161	15°29′N	86°34′W
Öland, i., Swe. (û-länd′)	208-09	56°45′N	16°38′E
Olary, Austl.	301	32°17′S	140°19′E
Olathe, Ks., U.S. (ô-lā′thē)	130-31	38°53′N	94°49′W
Olavarría, Arg. (ō-lä-vär-rē′ä)	187	36°54′S	60°19′W
Oława, Pol. (ô-lä′vȧ)	210-11	50°56′N	17°19′E
Olbia, Italy (ô′l-byä)	216-17	40°56′N	9°30′E
Old Bahama Channel, strt., N.A.			
(ōld bȧ-hā′mȧ chăn′ĕl)	154-55	22°40′N	78°41′W
Old Crow, Yk., Can. (ōld crō)	138-39	67°36′N	139°49′W
Oldenburg, Ger. (ôl′dĕn-bòrgh)	210-11	53°09′N	8°13′E
Old Forge, Pa., U.S. (ōld fôrj)	126-27	41°22′N	75°44′W
Olds, Ab., Can. (ōldz)	142-43	51°47′N	114°05′W
Old Wives Lake, lk., Sk., Can.			
(ōld wīvz lāk)	144-45	50°06′N	106°00′W
Olean, N.Y., U.S. (ō-lē-ăn′)	126-27	42°05′N	78°26′W
Olekma, stm., Russia (ô-lyĕk-má′)	236-37	60°23′N	120°41′E
Olëkminsk, Russia (ô-lyĕk-mênsk′)	236-37	60°22′N	120°26′E
Oleksandriia, Ukr.	218-19	48°40′N	33°07′E
Olenegorsk, Russia	200-01	68°09′N	33°14′E
Olenëk, stm., Russia (ô-lyĕ-nyôk′)	226-27	73°00′N	119°45′E
Olga, Russia (ôl′gȧ)	264	43°44′N	135°17′E
Ölgiy, Mong.	240-41	48°58′N	89°58′E
Olhão, Port. (ôl-youɴ′)	214-15	37°02′N	7°51′W
Ólimbos, mtn., Cyp.	248-49	34°56′N	32°51′E
Olímpia, Braz.	186	20°44′S	48°55′W
Olinda, Braz. (ô-lê′n-dä)	188d	8°01′S	34°51′W
Oliva, Arg.	187	32°03′S	63°34′W
Oliva, Spain (ô-lê′vä)	214-15	38°55′N	0°07′W
Olive Hill, Ky., U.S. (ŏl′ĭv hĭl)	126-27	38°18′N	83°11′W
Oliveira, Braz. (ô-lē-vā′rȧ)	186	20°42′S	44°49′W
Oliver, B.C., Can. (ŏ′lĭ-vẽr)	142-43	49°11′N	119°33′W
Olivia, Mn., U.S. (ō-lĭv′ē-ȧ)	124-25	44°47′N	94°59′W
Ollagüe, Chile (ô-lyä′gå)	182-83	21°13′S	68°16′W
Olmaliq, Uzb.	252-53	40°51′N	69°35′E
Olmos, Peru	184	5°56′S	79°46′W
Olmütz, Czech Rep. *see* Olomouc	210-11	49°36′N	17°16′E
Olney, Il., U.S. (ŏl′nĭ)	126-27	38°43′N	88°06′W
Olney, Tx., U.S. (ŏl′nē)	130-31	33°22′N	98°45′W
Olomane, stm., Qc., Can. (ō′lô mȧ′nĕ)	148-49	50°14′N	60°38′W
Olomouc, Czech Rep. (ô′lô-mōts)	210-11	49°36′N	17°16′E
Olonets, Russia (ô-lô′nĕts)	208-09	60°59′N	32°59′E
Olongapo, Phil.	270	14°52′N	120°17′E
Oloron-Sainte-Marie, Fr.			
(ō-lô-rônt′sănt mȧ-rē′)	212-13	43°11′N	0°36′W
Olot, Spain (ô-lōt′)	214-15	42°11′N	2°29′E
Olovyannaya, Russia	260-61	50°57′N	115°34′E
Olsztyn, Pol. (ôl′shtĭn)	210-11	53°47′N	20°29′E
Olt, stm., Rom.	202-03	43°43′N	24°48′E
Olten, Switz. (ôl′tĕn)	210-11	47°21′N	7°54′E
Oltenița, Rom. (ôl-tā′nĭ-tsȧ)	216-17	44°05′N	26°38′E
Olutanga Island, i., Phil.	270	7°22′N	122°52′E
Olvera, Spain (ôl-vĕ′rä)	214-15	36°56′N	5°16′W
Olympia, Wa., U.S.	122-23	47°02′N	122°53′W
Olympic Mountains, mts., Wa., U.S.			
(ô-lĭm′pĭk moun′tĭnz)	122-23	47°50′N	123°45′W
Olympic National Park, n.p., Wa., U.S.			
(ô-lĭm′pĭk năsh′ŭn-ăl pärk)	122-23	47°51′N	123°44′W
Ólympos, mtn., Grc.			
see Olympus, Mount	216-17	40°05′N	22°21′E
Olympus, mtn., Cyp. *see* Ólimbos	248-49	34°56′N	32°51′E
Olympus, Mount, mtn., Grc.			
(mount ô-lĭm′pŭs)	216-17	40°05′N	22°21′E
Olympus, Mount, mtn., Wa., U.S.			
(mount ô-lĭm′pŭs)	122-23	47°48′N	123°43′W
Olyutorski, Cape, c., Russia			
(kāp ŭl-yōō′tôr-skē)			
see Olyutorskiy, Mys	236-37	59°57′N	170°22′E
Olyutorskiy, Mys, c., Russia			
(mĭs ŭl-yōō′tŏr-skē)	236-37	59°57′N	170°22′E
Olyutorskiy Zaliv, b., Russia	236-37	60°15′N	168°30′E
Om', stm., Russia	236-37	54°59′N	73°22′E
Omagh, N. Ire., U.K. (ō′mä)	206-07	54°36′N	7°18′W
Omaha, Ne., U.S. (ō′mȧ-hä)	124-25	41°15′N	95°56′W
Omaha Indian Reservation, ind. res.,			
Ne., U.S. (ō′mȧ-hä			
ĭn′dĭ-ăn rĕ-sẽr-vā′shĕn)	124-25	42°07′N	96°32′W
Omak, Wa., U.S.	122-23	48°25′N	119°31′W
Oman, nation, Asia (ō-män′)	224-25	22°0′N	58°00′E
'Omān, Daryā-ye, b., Asia			
see Oman, Gulf of	250-51	24°30′N	58°30′E
Oman, Gulf of, b., Asia			
(gŭlf ŭv ō-män′)	250-51	24°30′N	58°30′E
Omaruru, Nmb. (ō-mä-rōō′rōō)	286-87	21°26′S	15°57′E
Omatako, stm., Nmb.	286-87	17°57′S	20°28′E
Omboué, Gabon	282-83	1°37′S	9°16′E
Omdurman, Sudan (ŏm-dûr-măn′)	288	15°39′N	32°29′E
Ometepe, Isla de, i., Nic.			
(ē′s-lä-dĕ-ō-mĕ-tā′pā)	161	11°30′N	85°35′W
Ometepec, Mex. (ô-mĕ-tȧ-pĕk′)	158-59	16°41′N	98°24′W
Ōminato, Japan *see* Mutsu	264	41°17′N	141°10′E
Omineca, stm., B.C., Can.			
(ô-mĭ-nĕk′ȧ)	142-43	56°07′N	124°28′W
Omineca Mountains, mts., B.C., Can.	140-41	56°0′N	125°00′W
Omo, stm., Afr. (ō′mō)	284-85	4°31′N	36°03′E
Omolon, stm., Russia (ō′mō)	236-37	68°42′N	158°43′E
Omro, Wi., U.S. (ŏm′rō)	126-27	44°02′N	88°45′W
Omsk, Russia (ômsk)	244	54°57′N	73°23′E
Omsukchan, Russia	236-37	62°30′N	155°46′E
Ōmura, Japan (ō′mōō-rȧ)	265	32°55′N	129°58′E
Ōmuta, Japan (ō-mò-tȧ)	265	33°01′N	130°27′E
Omutninsk, Russia (ô′mōō-tnēnsk)	202-03	58°39′N	52°11′E
Onawa, Ia., U.S. (ŏn-ȧ-wȧ)	124-25	42°02′N	96°06′W
Onda, Spain (ōn′dä)	214-15	39°58′N	0°15′W
Ondangwa, Nmb.	286-87	17°56′S	16°00′E
Ondo, Nig.	282a	7°06′N	4°50′E
Öndörhaan, Mong.	260-61	47°20′N	110°40′E
Onega, Russia (ô-nyĕ′gȧ)	202-03	63°55′N	38°06′E
Onega, stm., Russia (ô-nyĕ′gȧ)	202-03	63°57′N	37°57′E
Onega, Lake, lk., Russia			
(lāk ô-nyĕ′-gȧ)	202-03	61°30′N	35°45′E
Oneida, N.Y., U.S. (ô-nī′dȧ)	126-27	43°06′N	75°39′W
O'Neill, Ne., U.S. (ō-nēl′)	124-25	42°28′N	98°39′W
Onekotan, Ostrov, i., Russia	236-37	49°21′N	154°42′E
Oneonta, Al., U.S. (ō-nē-ŏn′tȧ)	134-35	33°57′N	86°28′W
Oneonta, N.Y., U.S. (ō-nē-ŏn′tȧ)	126-27	42°28′N	75°04′W
Onezhskoye Ozero, lk., Russia			
see Onega, Lake	202-03	61°30′N	35°45′E
Ongi, stm., Mong.	260-61	44°31′N	103°40′E
Onitsha, Nig. (ô-nĭt′shȧ)	282a	6°09′N	6°47′E
Ono-i-Lau, i., Fiji	304-05	20°39′S	178°42′W
Onomichi, Japan (ō′nô-mē′chê)	265	34°25′N	133°12′E
Onon, stm., Asia (ô′nôn)	240-41	51°42′N	115°49′E
Onoto, Ven. (ô-nŏ′tô)	188b	9°36′N	65°11′W
Onotoa, at., Kir.	304-05	1°53′S	175°34′E
Onslow, Austl. (ŏnz′lō)	294-95	21°39′S	115°07′E
Onslow Bay, b., N.C., U.S. (ônz′lō bā)	134-35	34°20′N	77°20′W
Ontake-san, vol., Japan (ôn′tä-kå-sän)	265	35°53′N	137°29′E
Ontario, Or., U.S. (ŏn-tā′rĭ-ō)	122-23	44°02′N	116°57′W
Ontario, state, Can. (ŏn-tā′rĭ-ō)	138-39	51°0′N	85°00′W
Ontario, Lake, lk., N.A. (lāk ŏn-tā′rĭ-ō)	126-27	43°45′N	78°00′W
Ontonagon, Mi., U.S. (ŏn-tô-năg′ŏn)	124-25	46°52′N	89°19′W
Ontong Java, at., Sol. Is.	306e	5°19′S	159°16′E
Onverwacht, Sur.	178-79	5°36′N	55°12′W
Oodnadatta, Austl. (ōōd′nȧ-dȧ′tȧ)	294-95	27°33′S	135°27′E
Ooldea, Austl.	294-95	30°28′S	131°51′E
Oos-Londen, S. Afr. *see* East London	286-87	32°60′S	27°54′E
Oostende, Bel. (ōst-ĕn′dē)	206-07	51°14′N	2°55′E
Opalaca, Cordillera, mts., Hond.			
(kôr-dēl-yĕ′rä-ô-pä-lä′kä)	161	14°30′N	88°20′W
Oparino, Russia	202-03	59°51′N	48°17′E
Opasquia, On., Can. (ō-păs′kwē-ȧ)	144-45	53°16′N	93°35′W
Opelika, Al., U.S. (ŏp-ē-lī′kȧ)	134-35	32°38′N	85°23′W
Opelousas, La., U.S. (ŏp-ē-lōō′sȧs)	134-35	30°32′N	92°05′W
Opeongo Lake, lk., On., Can.			
(ŏp-ē-ôŋ′gō lāk)	146-47	45°42′N	78°24′W
Opobo, Nig.	282a	4°35′N	7°34′E
Opochka, Russia (ō-pôch′kȧ)	208-09	56°43′N	28°40′E
Opoczno, Pol. (ô-pôch′nô)	210-11	51°23′N	20°18′E
Opole, Pol. (ô-pôl′á)	210-11	50°40′N	17°57′E
Oporto, Port. *see* Porto	214-15	41°09′N	8°37′W
Opp, Al., U.S. (ŏp)	134-35	31°17′N	86°15′W
Oppeln, Pol. *see* Opole	210-11	50°40′N	17°57′E
Opuwo, Nmb.	286-87	18°02′S	13°41′E
Oqsuqtooq, Nu., Can.			
see Gjoa Haven	138-39	68°39′N	95°55′W
Oradea, Rom. (ô-räd′yä)	210-11	47°04′N	21°56′E
Orai, India	254-55	25°59′N	79°28′E
Oral, Kaz.	202-03	51°13′N	51°22′E

n-sing; ŋ-baŋk; ɴ-nasalized n; nŏd; cŏmmit; ōld; ôbey; ôrder; oi-boil; fōōd; ȯ-as oo in foot; ou-out; s-soft; sh-dish; th-thin; pūre; ûnite; ûrn; stŭd; circŭs; ü-as in French tu; ′-indeterminate vowel.

P

ăt; finăl; rāte; senăte; ärm; ásk; sofá; fāre; ch-choose; dh-as th in other; bē; ĕvent; bĕt; recĕnt; cratēr; g-gō; gh-guttural g; bīt; ĭ-short neutral; rīde; κ-guttural k as ch in German ich;

PRONOUNCING INDEX Pābna – Paris 393

Feature (Pronunciation)	Page	Lat.	Long.
Pābna, Bngl.	254-55	24°00'N	89°14'E
Pacaraima, Serra, mts., S.A. (sĕr´rȧ pä-kä-rä-ē´mȧ) *see* Pakaraima Mountains	178-79	5°06'N	60°39'W
Pacaraima, Sierra de, mts., S.A. *see* Pakaraima Mountains	178-79	5°06'N	60°39'W
Pacasmayo, Peru (pä-käs-mä´yō)	184	7°24'S	79°33'W
Pachmarhi, India	254-55	22°28'N	78°26'E
Pachuca de Soto, Mex.	158-59	20°06'N	98°45'W
Pacific Ocean, oc., (pȧ-sĭf´ĭk ōshŭn)	4-5	10°0'S	150°00'W
Pacific Ranges, mts., B.C., Can. (pȧ-sĭf´ĭk rānjĕz)	142-43	51°11'N	125°33'W
Pacific Rim National Park Reserve, n.p., B.C., Can. (pȧ-sĭf´ĭk rĭm năsh´ŭn-ăl pärk rĭ-zûrv´)	142-43	48°45'N	125°06'W
Padang, Indon.	266-67	0°57'S	100°22'E
Padang, Indon. (pä-däng´)	266-67	1°39'S	108°55'E
Padangsidempuan, Indon.	266-67	1°23'N	99°16'E
Paden City, W.V., U.S. (pā´dĕn sĭ´tĭ)	126-27	39°37'N	80°51'W
Paderborn, Ger. (pä-dĕr-bôrn´)	210-11	51°43'N	8°45'E
Padma, stm., Asia *see* Ganges	254-55	21°58'N	90°57'E
Padova, Italy (pä´dô-vä)	216-17	45°24'N	11°52'E
Padre Island, i., Tx., U.S. (pä´drā ī´lȧnd)	132-33	27°01'N	97°23'W
Padua, Italy (păd´ū-ȧ) *see* Padova	216-17	45°24'N	11°52'E
Paducah, Ky., U.S.	134-35	37°05'N	88°37'W
Paektu-san, mtn., Asia (pâk´tōō-sän´)	263	41°59'N	128°07'E
Pagadian, Phil.	270	7°50'N	123°25'E
Pagalu, i., Eq. Gui. *see* Annobón	282-83	1°26'S	5°37'E
Pagan, i., N. Mar. Is.	304-05	18°07'N	145°46'E
Pago Pago, nat. cap., Am. Sam. (pän´-gō pän´-gō)	306b	14°16'S	170°42'W
Pagosa Springs, Co., U.S. (pȧ-gō´sȧ sprĭngz)	128-29	37°16'N	107°02'W
Pāhala, Hi., U.S. (pä-hä´lä)	137a	19°12'N	155°28'W
Pahang, stm., Malay.	266-67	3°30'N	103°24'E
Pahlevi, Iran *see* Bandar-e Anzalī	245	37°28'N	49°28'E
Palde, Est. (pı´dĕ)	208-09	58°54'N	25°35'E
Päijänne, lk., Fin. (pĕ´ē-yĕn-nĕ)	208-09	61°35'N	25°30'E
Painesville, Oh., U.S. (pānz´vĭl)	126-27	41°43'N	81°15'W
Painted Desert, des., Az., U.S. (pānt´ĕd dĕs´ẽrt)	128-29	35°45'N	111°07'W
Paintsville, Ky., U.S. (pānts´vĭl)	126-27	37°48'N	82°49'W
Paisley, Scot., U.K. (pāz´lĭ)	206-07	55°51'N	4°25'W
Paita, Peru (pä-ē´tä)	184	5°06'S	81°06'W
Pajala, Swe.	200-01	67°13'N	23°23'E
Pakaraima Mountains, mts., S.A.	178-79	5°06'N	60°39'W
Pakistan, nation, Asia (pă´-kĭ-stăn)	224-25	30°0'N	70°00'E
Pakistan, East, nation, Asia *see* Bangladesh	224-25	24°0'N	90°00'E
Pakokku, Mya. (pȧ-kŏk´kȯ)	266-67	21°19'N	95°06'E
Paks, Hung. (pôksh)	210-11	46°39'N	18°53'E
Pak Sane, Laos *see* Muang Pakxan	266-67	18°25'N	103°39'E
Pakxé, Laos	266-67	15°08'N	105°48'E
Pala, Chad	284-85	9°21'N	14°54'E
Palacios, Tx., U.S. (pä-lä´syōs)	132-33	28°42'N	96°13'W
Palaiseau, Fr. (pȧ-lĕ-zō´)	212-13	48°43'N	2°16'E
Palakkad, India	256	10°46'N	76°39'E
Palana, Russia	236-37	59°05'N	159°59'E
Palangkaraya, Indon.	268-69	2°10'S	113°54'E
Palani, India	256	10°27'N	77°31'E
Pālanpur, India (pä´lŭn-pōōr)	254-55	24°10'N	72°27'E
Palapye, Bots. (pȧ-läp´yĕ)	286-87	22°34'S	27°07'E
Palatka, Russia	236-37	60°06'N	150°57'E
Palatka, Fl., U.S. (pȧ-lăt´kȧ)	134-35	29°39'N	81°39'W
Palau, nation, Oc. (pä-lä´ō)	304-05	5°0'N	137°00'E
Palauig, Phil. (pȧ-lou´ēg)	270	15°26'N	119°56'E
Palawan, i., Phil. (pä-lä´wan)	270	9°30'N	118°30'E
Paldiski, Est. (päl´dĭ-skĭ)	208-09	59°20'N	24°06'E
Palembang, Indon. (pä-lĕm-bäng´)	266-67	2°58'S	104°46'E
Palencia, Spain (pä-lĕ´n-syä)	214-15	42°01'N	4°32'W
Palenque, Mex. (pä-lĕŋ´kå)	160	17°31'N	91°57'W
Palenque, hist., Mex.	160	17°30'N	91°60'W
Palenque, Punta, c., Dom. Rep. (pōō´n-tä pä-lĕŋ´kå)	154-55	18°15'N	70°09'W
Palermo, Italy (pä-lĕr´mô)	216-17	38°07'N	13°21'E
Palesse, reg., Eur. *see* Pripet Marshes	210-11	52°0'N	27°00'E
Palestine, Tx., U.S. (păl´ĕs-tīn)	132-33	31°45'N	95°38'W
Paletwa, Mya. (pŭ-lĕt´wä)	266-67	21°18'N	92°51'E
Pāli, India	254-55	25°47'N	73°20'E
Palikir, nat. cap., Micron.	304-05	6°58'N	158°13'E
Pālitāna, India	254-55	21°31'N	71°49'E
Palizada, Mex. (pä-lē-zä´dä)	160	18°15'N	92°05'W
Palk Strait, strt., Asia (pôk strāt)	256	10°0'N	79°45'E
Palliser, Cape, c., N.Z.	303	41°37'S	175°17'E
Palma de Mallorca, Spain	214-15	39°34'N	2°39'E
Palmares, Braz. (päl-mȧ´rĕs)	188d	8°41'S	35°36'W
Palmas, Braz. (päl´mäs)	182-83	26°30'S	52°01'W
Palmas, Braz.	180-81	10°06'S	48°20'W
Palma Soriano, Cuba (päl´mä-sô-rē-ä´nō)	154-55	20°13'N	75°59'W
Palmeira dos Índios, Braz. (pä-mā´rä-dôs-ē´n-dyôs)	188d	9°25'S	36°37'W
Palmeirinhas, Ponta das, c., Ang.	286-87	9°05'S	12°60'E
Palmer, Ak., U.S. (päm´ẽr)	136	61°32'N	149°05'W
Palmerston, at., Cook Is.	304-05	18°03'S	163°10'W
Palmerston, Cape, c., Austl.	302	21°33'S	149°28'E
Palmerston North, N.Z. (päm´ẽr-stŭn nôrth)	303	40°21'S	175°37'E
Palmetto, Fl., U.S. (pȧl-mĕt´ō)	135a	27°31'N	82°35'W
Palmi, Italy (päl´mē)	216-17	38°21'N	15°51'E
Palmira, Col. (päl-mē´rä)	188c	3°33'N	76°18'W
Palm Springs, Ca., U.S.	128-29	33°50'N	116°32'W
Palmyra, Syria *see* Tudmur	248-49	34°33'N	38°17'E
Palmyra, Mo., U.S. (păl-mī´rȧ)	130-31	39°48'N	91°31'W
Palmyra, N.Y., U.S. (păl-mī´rȧ)	126-27	43°03'N	77°14'W
Palmyra Atoll, at., Oc.	304-05	5°51'N	162°05'W
Palo Alto, Ca., U.S. (pä´lō äl´tō)	128-29	37°26'N	122°08'W
Paloe, Pulau, i., Indon.	268-69	8°20'S	121°43'E
Palopo, Indon.	268-69	3°00'S	120°11'E
Palos, Cabo de, c., Spain (kȧ´bô-dĕ-pä´lôs)	214-15	37°38'N	0°41'W
Palu, Indon.	268-69	0°54'S	119°52'E
Palu, Tur. (pä-loo´)	245	38°41'N	39°60'E
Paluan, Phil. (pä-lōō´än)	270	13°26'N	120°27'E
Pāmban Island, i., India	256	9°16'N	79°19'E
Pamekasan, Indon.	268-69	7°10'S	113°29'E
Pamiers, Fr. (pȧ-myā´)	212-13	43°07'N	1°36'E
Pamir, mts., Asia *see* Pamirs	252-53	38°0'N	73°00'E
Pāmīr, Daryā-ye, mts., Asia *see* Pamirs	252-53	38°0'N	73°00'E
Pamirs, mts., Asia (pä-mērz)	252-53	38°0'N	73°00'E
Pamlico Sound, strt., N.C., U.S. (păm´lĭ-kō sound)	134-35	35°20'N	75°55'W
Pampa, Tx., U.S. (păm´pȧ)	130-31	35°32'N	100°58'W
Pampa, reg., Arg. (päm´pȧ) *see* Pampas	187	35°0'S	63°00'W
Pampanga, stm., Phil. (päm-päŋ´gä)	270	14°46'N	120°39'E
Pampas, reg., Arg. (päm´päs)	187	35°0'S	63°00'W
Pampas, stm., Peru	184	13°25'S	73°13'W
Pampeluna, Spain *see* Pamplona	214-15	42°49'N	1°39'W
Pamplona, Col. (päm-plō´nä)	178-79	7°22'N	72°38'W
Pamplona, Spain (päm-plô´nä)	214-15	42°49'N	1°39'W
Pana, Il., U.S. (pā´nȧ)	126-27	39°23'N	89°05'W
Panagyurishte, Blg. (pȧ-nȧ-gyoo´rĕsh-tĕ)	216-17	42°30'N	24°12'E
Panaitan, Pulau, i., Indon.	268-69	6°36'S	105°12'E
Panaji, India	256	15°30'N	73°50'E
Panama, nation, N.A. (păn-ȧ-mä´ sĭ´tĭ)	94	9°0'N	80°00'W
Panamá, nat. cap., Pan. (păn-ȧ-mä´)	162	8°58'N	79°32'W
Panamá, Golfo de, b., Pan.	162	8°0'N	79°30'W
Panama, Gulf of, b., Pan. *see* Panamá, Golfo de	162	8°0'N	79°30'W
Panama, Isthmus of, isth., Pan. *see* Panamá, Istmo de	162	9°0'N	80°00'W
Panamá, Istmo de, isth., Pan.	162	9°0'N	80°00'W
Panama Canal, can., Pan.	162	9°23'N	79°56'W
Panama City, Fl., U.S. (păn-ȧ-mä´ sĭ´tĭ)	134-35	30°10'N	85°40'W
Panay, i., Phil. (pä-nī´)	270	11°15'N	122°30'E
Panay Gulf, b., Phil.	270	10°15'N	122°15'E
Pančevo, Serb. (pȧn´chĕ-vô)	216-17	44°53'N	20°40'E
Panevėžys, Lith. (pä´nyĕ-vāzh´ēs)	208-09	55°44'N	24°23'E
Pangani, stm., Tan. (pän-gä´nē)	284-85	5°24'S	38°57'E
Pangkalanbuun, Indon.	268-69	2°42'S	111°38'E
Pangkalpinang, Indon. (päng-käl´pē-näng´)	266-67	2°08'S	106°06'E
Pangnirtung, Nu., Can.	138-39	66°08'N	65°43'W
Pangong Tso, lk., Asia	254-55	33°45'N	78°42'E
Panguitch, Ut., U.S. (păn´gwĭch)	128-29	37°50'N	112°26'W
Pangutaran Group, is., Phil.	270	6°14'N	120°39'E
Panhame, stm., Afr. *see* Manyame	286-87	15°37'S	30°39'E
Pānīpat, India	254-55	29°23'N	76°58'E
Panj, stm., Asia	252-53	37°00'N	68°16'E
Panjgūr, Pak.	252-53	26°58'N	64°05'E
Panjim, India *see* Panaji	256	15°30'N	73°50'E
Panna, India	254-55	24°43'N	80°11'E
Pannirtuuq, Nu., Can. *see* Pangnirtung	138-39	66°08'N	65°43'W
Pantar, Pulau, i., Indon. (pōō-lou pän´tär)	268-69	8°25'S	124°07'E
Pantelleria, Isola di, i., Italy (ē´sō-lä-dē-pän-tĕl-lå-rē´ä)	216-17	36°47'N	12°00'E
Pante Makasar, Timor-L.	268-69	9°13'S	124°21'E
Pánuco, Mex. (pä´nōō-kô)	158-59	22°02'N	98°11'W
Pánuco, stm., Mex. (pä´nōō-kô)	158-59	22°16'N	97°47'W
Panxian, China	258-59	25°49'N	104°35'E
Panzós, Guat. (pä-zōs´)	160	15°24'N	89°39'W
Paoli, In., U.S. (pȧ-ō´lĭ)	126-27	38°33'N	86°28'W
Pápa, Hung. (pä´pô)	210-11	47°20'N	17°28'E
Papagayo, Golfo de, b., C.R. (gôl-fô-dĕ-pä-pä-gä´yō)	161	10°42'N	85°50'W
Papantla de Olarte, Mex. (pä-pän´tlä dā-ô-lä´r-tĕ)	158-59	20°27'N	97°19'W
Papeete, nat. cap., Fr. Poly. (pä-pē´-tē)	306d	17°32'S	149°34'W
Papenburg, Ger. (päp´ĕn-bōrgh)	210-11	53°06'N	7°24'E
Papua, Gulf of, b., Pap. N. Gui. (gŭlf ŭv păp-ōō-ȧ)	302	8°30'S	145°00'E
Papua New Guinea, nation, Oc. (păp-ōō-ȧ nū gĭne)	302	6°0'S	147°00'E
Papudo, Chile (pä-pōō´dō)	188e	32°31'S	71°28'W
Papun, Mya.	266-67	18°04'N	97°27'E
Pará, Braz. *see* Belém	180-81	1°27'S	48°29'W
Pará, state, Braz.	180-81	4°0'S	53°00'W
Pará, stm., Braz.	180-81	1°29'S	48°49'W
Paraburdoo, Austl.	294-95	23°12'S	117°44'E
Paracatu, Braz. (pä-rä-kä-tōō´)	186	17°14'S	46°52'W
Paracatu, stm., Braz.	182-83	16°35'S	45°06'W
Paracel Islands, is., China	242-43	15°46'N	112°17'E
Paraćin, Serb. (pá´rä-chēn)	216-17	43°52'N	21°25'E
Pāradwīp, India	254-55	20°17'N	86°41'E
Paragould, Ar., U.S. (păr´ȧ-gōōld)	134-35	36°04'N	90°30'W
Paraguá, stm., Bol.	180-81	13°32'S	61°49'W
Paragua, stm., Ven.	178-79	6°56'N	62°55'W
Paraguaçu, stm., Braz. (pä-rä-gwä-zōō´)	180-81	12°50'S	38°48'W
Paraguay, stm., S.A. (pä-rä-gwä´y)	187	27°19'S	58°36'W
Paraguai, stm., S.A. *see* Paraguay	187	27°19'S	58°36'W
Paraguaná, Península de, pen., Ven.	178-79	11°56'N	70°03'W
Paraguarí, Para.	182-83	25°37'S	57°09'W
Paraguay, nation, S.A. (păr´ȧ-gwā)	172	23°0'S	58°00'W
Parahyba, Braz. *see* João Pessoa	188d	7°07'S	34°52'W
Paraíba, Braz. *see* João Pessoa	188d	7°07'S	34°52'W
Paraíba, state, Braz. (pä-rä-ē´bä)	188d	7°15'S	36°30'W
Paraíba do Sul, stm., Braz.	186	21°37'S	41°02'W
Paraíso, Mex.	158-59	18°23'N	93°14'W
Paraiso, Pan. (pä-rä-ē´sō)	162	9°03'N	79°38'W
Parakou, Benin (pȧ-rȧ-kōō´)	282-83	9°20'N	2°37'E
Paramaribo, nat. cap., Sur. (pȧ-rȧ-mȧ´rĕ-bō)	178-79	5°49'N	55°10'W
Paramirim, Braz.	180-81	13°27'S	42°14'W
Paramushir, Ostrov, i., Russia	236-37	50°25'N	155°50'E
Paraná, Arg. (pä-ä-nä´)	187	31°44'S	60°31'W
Paranã, Braz.	180-81	12°33'S	47°52'W
Paraná, state, Braz.	182-83	24°0'S	51°00'W
Paranã, stm., Braz.	180-81	12°30'S	48°14'W
Paraná, stm., S.A. (pä-ä-nä´)	182-83	33°48'S	59°14'W
Paranaguá, Braz.	186	25°31'S	48°31'W
Paranaguá, Baía de, b., Braz.	186	25°27'S	48°22'W
Paranaíba, Braz. (pä-rä-nȧ-ē´bȧ)	182-83	19°41'S	51°11'W
Paranaíba, stm., Braz. (pä-rä-nȧ-ē´bȧ)	182-83	20°08'S	51°00'W
Paranapanema, stm., Braz. (pä-rä´nȧ´pä-nĕ-mä)	182-83	22°42'S	53°10'W
Paranavaí, Braz.	182-83	23°04'S	52°29'W
Parapara, Ven. (pä-rä-pä-rä)	188b	9°44'N	67°17'W
Paray-le-Monial, Fr. (pȧ-rĕ´lē-mō-nyäl´)	212-13	46°27'N	4°07'E
Pārbat, stm., India	254-55	25°51'N	76°33'E
Parbhani, India	256	19°16'N	76°46'E
Pardo, stm., Braz. (pär´dō)	180-81	15°39'S	38°57'W
Pardo, stm., Braz. (pär´dō)	186	20°09'S	48°37'W
Pardubice, Czech Rep. (pär´dȯ-bĭt-sĕ)	210-11	50°02'N	15°46'E
Parece Vela, i., Japan *see* Okino-Tori-shima	240-41	20°27'N	136°04'E
Parent, Qc., Can.	146-47	47°56'N	74°37'W
Parepare, Indon.	268-69	4°01'S	119°38'E
Paria, Golfo de, b., (gôl-fô-dĕ-br-pä-rē-ä) *see* Paria, Gulf of	152-53	10°20'N	62°00'W
Paria, Gulf of, b.,	152-53	10°20'N	62°00'W
Paricutín, vol., Mex.	158-59	19°28'N	102°15'W
Parima, Serra, mts., S.A. (sĕr´rȧ pä-rē´mȧ) *see* Parima, Sierra	178-79	3°24'N	64°10'W
Parima, Sierra, mts., S.A.	178-79	3°24'N	64°10'W
Pariñas, Punta, c., Peru (pōō´n-tä-pä-rē´n-yäs)	184	4°40'S	81°20'W
Parintins, Braz. (pä-rĭn-tĭnzh´)	180-81	2°37'S	56°45'W
Paris, Ar., U.S. (păr´ĭs)	130-31	35°18'N	93°44'W
Paris, Il., U.S. (păr´ĭs)	126-27	39°37'N	87°42'W
Paris, Ky., U.S. (păr´ĭs)	126-27	38°12'N	84°16'W

n-sing; ŋ-baŋk; ɴ-nasalized n; nŏd; cŏmmit; ōld; ôbey; ôrder; oi-boil; fōōd; ȯ-as oo in foot; ou-out; s-soft; sh-dish; th-thin; pūre; ûnite; ûrn; stŭd; circŭs; ü-as in French tu; ´-indeterminate vowel.

Feature (Pronunciation)	Page	Lat.	Long.
Paris, Mo., U.S. (păr´ĭs)	**130-31**	39°29'N	92°00'W
Paris, Tn., U.S. (păr´ĭs)	**134-35**	36°18'N	88°20'W
Paris, Tx., U.S. (păr´ĭs)	**130-31**	33°40'N	95°33'W
Paris, nat. cap., Fr. (på-rē´)	**212-13**	48°52'N	2°21'E
Parita, Bahía de, b., Pan. (bä-ē´ä-dĕ-pä-rē´tä)	**162**	8°08'N	80°24'W
Park City, Ks., U.S. (pärk sĭ´tē)	**130-31**	37°48'N	97°18'W
Parker, Co., U.S. (pär´kẽr pärk)	**130-31**	39°31'N	104°46'W
Parker, S.D., U.S. (pär´kẽr pärk)	**124-25**	43°24'N	97°08'W
Parkersburg, W.V., U.S. (pär´kẽrz-bûrg)	**126-27**	39°15'N	81°33'W
Parkes, Austl. (pärks)	**301**	33°09'S	148°10'E
Park Falls, Wi., U.S. (pärk fôlz)	**124-25**	45°56'N	90°26'W
Park Range, mts., Co., U.S. (pärk rānj)	**122-23**	40°40'N	106°40'W
Park Rapids, Mn., U.S. (pärk răp´ĭdz)	**124-25**	46°55'N	95°04'W
Park River, N.D., U.S. (pärk rĭv´ẽr)	**124-25**	48°24'N	97°45'W
Parkston, S.D., U.S. (pärks´tŭn)	**124-25**	43°24'N	97°59'W
Parla, Spain (pär´lä)	**214-15**	40°14'N	3°46'W
Parlākimidi, India	**256**	18°47'N	84°06'E
Parma, Italy (pär´mä)	**216-17**	44°49'N	10°20'E
Parnaguá, Braz.	**180-81**	10°13'S	44°38'W
Parnaíba, Braz. (pär-nä-ē´bä)	**180-81**	2°54'S	41°47'W
Parnaíba, stm., Braz. (pär-nä-ē´bä)	**180-81**	2°46'S	41°50'W
Parnassós, mtn., Grc.	**216-17**	38°32'N	22°35'E
Pärnu, Est. (pĕr´no͞o)	**208-09**	58°22'N	24°33'E
Paroo, stm., Austl. (pä´ro͞o)	**301**	30°23'S	143°59'E
Parowan, Ut., U.S. (păr´ô-wăn)	**128-29**	37°51'N	112°50'W
Parral, Chile (pär-rä´l)	**185**	36°09'S	71°50'W
Parramatta, Austl.	**301**	33°49'S	151°00'E
Parras de la Fuente, Mex.	**132-33**	25°27'N	102°10'W
Parrsboro, N.S., Can. (pärz´bŭr-ô)	**148-49**	45°25'N	64°20'W
Parry, Cape, c., N.T., Can.	**140-41**	70°08'N	124°24'W
Parry, Mount, mtn., B.C., Can. (mount păr´ĭ)	**142-43**	52°53'N	128°45'W
Parry Sound, On., Can. (păr´ĭ sound)	**146-47**	45°20'N	80°02'W
Parsnip, stm., B.C., Can. (pärs´nĭp)	**142-43**	55°10'N	123°02'W
Parsons, Ks., U.S. (pär´s'nz)	**130-31**	37°20'N	95°16'W
Parsons, W.V., U.S. (pär´s'nz)	**126-27**	39°06'N	79°41'W
Parthenay, Fr. (pår-t'nĕ´)	**212-13**	46°39'N	0°15'W
Partinico, Italy (pär-tē´nē-kô)	**216-17**	38°03'N	13°07'E
Paru, stm., Braz.	**180-81**	1°35'S	52°31'W
Parys, S. Afr. (på-rīs´)	**292c**	26°54'S	27°28'E
Pasadena, Ca., U.S. (păs-å-dē´nå)	**128-29**	34°09'N	118°09'W
Pasadena, Tx., U.S. (păs-å-dē´nå)	**132-33**	29°41'N	95°13'W
Pasaje, Ec.	**184**	3°20'S	79°48'W
Pa Sak, stm., Thai.	**266-67**	14°21'N	100°35'E
Pascagoula, Ms., U.S. (păs-kå-go͞o´lå)	**134-35**	30°22'N	88°33'W
Pascagoula, stm., Ms., U.S. (păs-kå-go͞o´lå)	**134-35**	30°22'N	88°37'W
Paşcani, Rom. (päsh-kän´)	**210-11**	47°15'N	26°44'E
Pasco, Wa., U.S. (păs´kō)	**122-23**	46°14'N	119°05'W
Pascua, Isla de, i., Chile	**308-09**	27°07'S	109°22'W
Pasni, Pak.	**252-53**	25°16'N	63°27'E
Paso de Indios, Arg.	**185**	43°51'S	68°56'W
Paso de los Libres, Arg. (pä-sô-dĕ-lôs-lē´brĕs)	**187**	29°42'S	57°09'W
Paso de los Toros, Ur. (pä-sô-dĕ-lôs tō´rôs)	**187**	32°49'S	56°31'W
Paso Robles, Ca., U.S. (pă´sō rō´blĕs)	**128-29**	35°38'N	120°41'W
Passaic, N.J., U.S. (pă-sā´ĭk)	**126-27**	40°52'N	74°08'W
Passau, Ger. (päsou)	**210-11**	48°34'N	13°27'E
Passero, Capo, c., Italy (kä´pō päs-sĕ´rô)	**216-17**	36°40'N	15°09'E
Passo Fundo, Braz. (pä´sŏ fōn´dō)	**182-83**	28°15'S	52°25'W
Passos, Braz. (pä´sōs)	**186**	20°43'S	46°37'W
Pastaza, stm., S.A. (päs-tä´zä)	**184**	4°55'S	76°24'W
Pasto, Col. (päs´tō)	**178-79**	1°12'N	77°16'W
Pasuruan, Indon.	**268-69**	7°38'S	112°54'E
Pasvalys, Lith. (päs-vä-lēs´)	**208-09**	56°04'N	24°24'E
Patagonia, reg., Arg. (păt-å-gō´nĭ-å)	**185**	44°0'S	68°00'W
Pātan, India	**254-55**	23°51'N	72°07'E
Pate Island, i., Kenya	**284-85**	2°06'S	41°03'E
Paterson, N.J., U.S. (păt´ẽr-sŭn)	**126-27**	40°55'N	74°10'W
Pathānkot, India	**254-55**	32°16'N	75°39'E
Pathein, Mya.	**266-67**	16°46'N	94°44'E
Pathfinder Reservoir, res., Wy., U.S. (păth´fīn-dẽr rĕ´sẽr-vwär)	**122-23**	42°25'N	106°55'W
Patiāla, India (pŭt-ē-ä´lŭ)	**254-55**	30°19'N	76°23'E
Pātkai Range, mts., Asia	**258-59**	27°0'N	96°00'E
Patna, India	**254-55**	25°36'N	85°07'E
Patnanongan Island, i., Phil. (pät-nä-nôŋ´gän ī´lånd)	**270**	14°48'N	122°11'E
Pato Branco, Braz.	**182-83**	26°14'S	52°41'W
Patos, Braz. (pä´tōzh)	**188d**	7°01'S	37°16'W
Patos, Lagoa dos, b., Braz. (lä´gō-å dozh pä´tōzh)	**182-83**	31°06'S	51°15'W
Patos de Minas, Braz. (pä´tōzh dĕ-mē´näzh)	**186**	18°35'S	46°31'W
Patquía, Arg.	**182-83**	30°02'S	66°52'W
Pátra, Grc.	**216-17**	38°14'N	21°44'E
Patricio Lynch, Isla, i., Chile	**185**	48°37'S	75°26'W
Patrocínio, Braz. (pä-trō-sē´nē-ō)	**186**	18°56'S	46°60'W
Pattani, Thai. (pät´å-nē)	**266-67**	6°52'N	101°15'E
Patten, Me., U.S. (păt´'n)	**127a**	45°59'N	68°27'W
Patterson, La., U.S. (păt´ẽr-sŭn)	**134-35**	29°42'N	91°18'W
Patuca, stm., Hond.	**161**	15°48'N	84°18'W
Patuca, Punta, c., Hond. (po͞o´n-tä-pä-to͞o´kä)	**161**	15°49'N	84°18'W
Pátzcuaro, Mex. (päts´kwä-rô)	**158-59**	19°31'N	101°37'W
Pau, Fr. (pō)	**212-13**	43°18'N	0°22'W
Pauini, stm., Braz.	**180-81**	7°47'S	67°05'W
Pauk, Mya.	**266-67**	21°27'N	94°28'E
Paulding, Oh., U.S. (pôl´dĭng)	**126-27**	41°08'N	84°35'W
Paulis, D.R.C. *see* Isiro	**284-85**	2°46'N	27°37'E
Paulistana, Braz.	**180-81**	8°09'S	41°09'W
Paulo Afonso, Braz.	**180-81**	9°21'S	38°14'W
Paul Roux, S. Afr. (pôrl ro͞o)	**292c**	28°18'S	27°58'E
Pauls Valley, Ok., U.S. (pôlz văl´ē)	**130-31**	34°44'N	97°13'W
Paungde, Mya.	**266-67**	18°29'N	95°30'E
Pavia, Italy (pä-vē´ä)	**216-17**	45°12'N	9°10'E
Pavlodar, Kaz. (páv-lô-dár´)	**244**	52°17'N	76°59'E
Pavlovo, Russia	**202-03**	55°57'N	43°04'E
Pavuvu Island, i., Sol. Is.	**306e**	9°03'S	159°06'E
Pawan, stm., Indon.	**268-69**	1°51'S	109°56'E
Pawhuska, Ok., U.S. (pô-hŭs´kå)	**130-31**	36°40'N	96°20'W
Pawnee, Ok., U.S. (pô-nē´)	**130-31**	36°20'N	96°48'W
Pawnee, stm., Ks., U.S. (pô-nē´)	**130-31**	38°10'N	99°06'W
Pawnee City, Ne., U.S. (pô-nē´ sĭ´tē)	**130-31**	40°07'N	96°09'W
Paw Paw, Mi., U.S. (pô pô)	**126-27**	42°13'N	85°53'W
Pawtucket, R.I., U.S. (pô-tŭk´ĕt)	**126-27**	41°53'N	71°23'W
Paxton, Il., U.S. (păks´tŭn)	**126-27**	40°27'N	88°05'W
Payakumbuh, Indon.	**266-67**	0°14'S	100°38'E
Payette, Id., U.S. (på-ĕt´)	**122-23**	44°05'N	116°56'W
Pay-Khey, Khrebet, mts., Russia	**202-03**	69°0'N	63°00'E
Paynesville, Mn., U.S. (pānz´vĭl)	**124-25**	45°23'N	94°43'W
Paysandú, Ur. (pī-sän-do͞o´)	**187**	32°20'S	58°05'W
Payson, Az., U.S. (pā´s'n)	**128-29**	34°10'N	111°19'W
Pazardzhik, Blg. (pä-zår-dzhek´)	**216-17**	42°12'N	24°20'E
Peabody, Ks., U.S. (pē´bŏd-ĭ)	**130-31**	38°10'N	97°06'W
Peace, stm., Can. (pēs)	**140-41**	58°60'N	111°25'W
Peace, stm., Fl., U.S. (pēs)	**135a**	26°58'N	82°01'W
Peace River, Ab., Can. (pēs rĭv´ẽr)	**142-43**	56°15'N	117°16'W
Pearl, stm., U.S. (pûrl)	**134-35**	30°11'N	89°32'W
Pearland, Tx., U.S. (pûrl´ănd)	**132-33**	29°33'N	95°17'W
Pearl and Hermes Atoll, at., Hi., U.S.	**137b**	27°55'N	175°45'W
Pearl Harbor, b., Hi., U.S. (pûrl här´bẽr)	**137a**	21°22'N	157°59'W
Pearsall, Tx., U.S. (pēr´sôl)	**132-33**	28°54'N	99°06'W
Pebble Island, i., Falk. Is.	**185**	51°20'S	59°34'W
Peçanha, Braz. (på-kän´yå)	**186**	18°32'S	42°34'W
Pechenga, Russia (pyĕ´chĕŋ-gå)	**200-01**	69°34'N	31°14'E
Pechora, Russia	**202-03**	65°08'N	57°09'E
Pechora, stm., Russia (pyĕ-chô´rå)	**202-03**	67°59'N	53°56'E
Pechorskoye More, s., Russia	**202-03**	70°0'N	54°00'E
Pecos, Tx., U.S. (pā´kōs)	**132-33**	31°25'N	103°30'W
Pecos, stm., U.S. (pā´kōs)	**120-21**	29°42'N	101°22'W
Pécs, Hung. (pāch)	**210-11**	46°04'N	18°13'E
Pedernales, Ven.	**155b**	9°57'N	62°15'W
Pedra Azul, Braz. (pā´drä-zo͞o´l)	**182-83**	15°60'S	41°17'W
Pedreiras, Braz. (pĕ-drā´räs)	**180-81**	4°34'S	44°39'W
Pedro Afonso, Braz.	**180-81**	8°60'S	48°10'W
Pedro II, Braz. (pā´drō så-gôn´dō)	**180-81**	4°25'S	41°28'W
Pedro Juan Caballero, Para. (pĕ´drō hòá´n-kä-bäl-yĕ´rō)	**182-83**	22°33'S	55°45'W
Peebles, Scot., U.K. (pē´b'lz)	**206-07**	55°39'N	3°12'W
Peekskill, N.Y., U.S. (pēks´kĭl)	**126-27**	41°17'N	73°55'W
Peel, stm., Can.	**140-41**	67°42'N	134°31'W
Pegasus Bay, b., N.Z. (pĕg´å-sŭs bā)	**303**	43°20'S	187°00'E
Pegu, Mya. *see* Bago	**266-67**	17°20'N	96°29'E
Peiching, nat. cap., China *see* Beijing	**260-61**	39°55'N	116°22'E
Peipsi järv, lk., Eur. *see* Peipus, Lake	**208-09**	58°45'N	27°25'E
Peipus, Lake, lk., Eur. (lāk pī´pŭs)	**208-09**	58°45'N	27°25'E
Peiraiás, Grc.	**216-17**	37°57'N	23°39'E
Peixe, stm., Braz.	**182-83**	21°30'S	51°57'W
Pekalongan, Indon.	**268-69**	6°53'S	109°40'E
Pekanbaru, Indon.	**266-67**	0°31'N	101°27'E
Pekin, Il., U.S. (pē´kĭn)	**126-27**	40°34'N	89°39'W
Peking, nat. cap., China *see* Beijing	**260-61**	39°55'N	116°22'E
Pelagie, Isole, is., Italy	**200-01**	35°40'N	12°40'E
Pelat, Mont, mtn., Fr. (môɴ pē-lá´)	**212-13**	44°16'N	6°42'E
Peleduy, Russia (pyĕl-yĭ-do͞o´ē)	**236-37**	59°39'N	112°44'E
Pelée, Montagne, vol., Mart. (môɴ-pē-lā´ pȧ-lē´)	**155b**	14°48'N	61°10'W
Pelee Island, i., On., Can. (pē´lē ī´lånd)	**146-47**	41°46'N	82°39'W
Peleliu, i., Palau *see* Beliliou	**304-05**	7°00'N	134°15'E
Peleng, Pulau, i., Indon.	**268-69**	1°15'S	123°08'E
Pelham, Ga., U.S. (pĕl´hăm)	**134-35**	31°08'N	84°09'W
Pelican Rapids, Mn., U.S. (pĕl´ĭ-kăn răp´ĭdz)	**124-25**	46°35'N	96°04'W
Pella, Ia., U.S. (pĕl´å)	**124-25**	41°25'N	92°55'W
Pellworm, i., Ger. (pĕl´vôrm)	**210-11**	54°31'N	8°38'E
Pelly, stm., Yk., Can. (pĕl´ĭ)	**140-41**	62°46'N	137°20'W
Pelly Crossing, Yk., Can.	**138-39**	62°50'N	136°35'W
Pelly Mountains, mts., Yk., Can. (pĕl´ĭ moun´tĭnz)	**140-41**	62°0'N	133°00'W
Peloponnesus, pen., Grc.	**216-17**	37°30'N	22°00'E
Pelopónnisos, pen., Grc. *see* Peloponnesus	**216-17**	37°30'N	22°00'E
Pelotas, Braz. (på-lō´tåzh)	**187**	31°45'S	52°19'W
Pelotas, stm., Braz.	**182-83**	27°28'S	51°54'W
Pematangsiantar, Indon.	**266-67**	2°57'N	99°04'E
Pemba, Moz. (pĕm´bå)	**286-87**	13°01'S	40°32'E
Pemba, i., Tan. (pĕm´bå)	**284-85**	5°10'S	39°48'E
Pemberton, Austl.	**294-95**	34°27'S	116°01'E
Pembina, N.D., U.S. (pĕm´bĭ-nå)	**124-25**	48°58'N	97°15'W
Pembina, stm., Ab., Can. (pĕm´bĭ-nå)	**142-43**	54°45'N	114°17'W
Pembroke, On., Can. (pĕm´brōk)	**146-47**	45°49'N	77°07'W
Pembroke, Cape, c., Nu., Can.	**140-41**	62°56'N	81°56'W
Pembuang, stm., Indon.	**268-69**	3°21'S	112°33'E
Peñalara, Pico de, mtn., Spain (pē´kō-dĕ-pā-nyä-lä´rä)	**214-15**	40°51'N	3°57'W
Penang, Malay. *see* George Town	**266-67**	5°25'N	100°20'E
Peñarroya-Pueblonuevo, Spain (pĕn-yär-rô´yä-pwĕ´blô-nwĕ´vô)	**214-15**	38°18'N	5°16'W
Peñas, Cabo de, c., Spain (kå´bô-dĕ-pā´nyäs)	**214-15**	43°39'N	5°51'W
Penas, Golfo de, b., Chile (gôl-fô-dĕ-pĕ´n-äs)	**185**	47°22'S	74°50'W
Pender, Ne., U.S. (pĕn´dẽr)	**124-25**	42°07'N	96°43'W
Pendjari, stm., Afr.	**282-83**	10°55'N	0°50'E
Pendleton, Or., U.S. (pĕn´d'l-tŭn)	**122-23**	45°40'N	118°48'W
Pend Oreille, Lake, lk., Id., U.S. (lăk pŏn-dô-rā´) (lāk pĕn-dô-rĕl´)	**122-23**	48°10'N	116°17'W
Penedo, Braz. (på-nā´dô)	**188d**	10°16'S	36°35'W
Penetanguishene, On., Can. (pĕn´ē-tăŋ-gĭ-shēn´)	**146-47**	44°46'N	79°56'W
Penganga, stm., India	**256**	19°54'N	79°10'E
P'enghu, Tai. *see* Makung	**243a**	23°34'N	119°34'E
P'enghu Ch'üntao, is., Tai.	**243a**	23°30'N	119°30'E
Penglai, China (pŭŋ-lī)	**260-61**	37°48'N	120°43'E
Pengshui, China	**258-59**	29°18'N	108°09'E
Pengxian, China	**258-59**	30°59'N	103°56'E
Peniche, Port. (pĕ-nē´chå)	**214-15**	39°21'N	9°22'W
Penida, Nusa, i., Indon.	**268-69**	8°44'S	115°32'E
Pennines, mts., Eng., U.K. (pĕn-īn')	**206-07**	54°11'N	2°02'W
Pennsylvania, state, U.S. (pĕn-sĭl-vā´nĭ-å)	**118-19**	40°45'N	77°30'W
Penn Yan, N.Y., U.S. (pĕn yăn´)	**126-27**	42°40'N	77°03'W
Penobscot, stm., Me., U.S.	**127a**	44°29'N	68°48'W
Penobscot Bay, b., Me., U.S. (pē-nŏb´skŏt bā)	**127a**	44°15'N	68°52'W
Penola, Austl.	**301**	37°23'S	140°50'E
Penonomé, Pan.	**162**	8°31'N	80°22'W
Penrhyn, at., Cook Is.	**304-05**	9°0'S	158°00'W
Pensacola, Fl., U.S. (pĕn-så-kō´lå)	**134-35**	30°25'N	87°13'W
Pensacola Mountains, mts., Ant.	**313**	84°21'S	47°02'W
Pensilvania, Col. (pĕn-sēl-vá´nyä)	**188c**	5°32'N	75°03'W
Pentecost Island, i., Vanuatu (pĕn´tē-kŏst ī´lånd) *see* Pentecôte	**306g**	15°42'S	168°10'E
Pentecôte, i., Vanuatu	**306g**	15°42'S	168°10'E
Penticton, B.C., Can.	**142-43**	49°30'N	119°35'W
Pentland Firth, strt., Scot., U.K. (pĕnt´lănd fûrth)	**206-07**	58°44'N	3°07'W
Penyu, Kepulauan, is., Indon.	**268-69**	5°22'S	127°46'E
Penza, Russia (pĕn´zå)	**202-03**	53°12'N	45°00'E
Penzance, Eng., U.K. (pĕn-zăns´)	**206-07**	50°07'N	5°33'W
Penzhina, stm., Russia (pyĭn-zē-nŭ)	**236-37**	62°29'N	165°15'E
People's Democratic Republic of Korea, nation, Asia *see* Korea, North	**224-25**	40°0'N	127°00'E
Peoria, Il., U.S. (pē-ō´rĭ-å)	**126-27**	40°41'N	89°36'W
Peotone, Il., U.S. (pē´ŏ-tōn)	**126-27**	41°20'N	87°47'W
Pequeñas Antillas, is., *see* Lesser Antilles	**155b**	15°0'N	61°00'W
Perabumulih, Indon.	**266-67**	3°27'S	104°15'E
Perak, stm., Malay.	**266-67**	3°58'N	100°53'E
Perdido, Monte, mtn., Spain (mōn-tå-pĕr-dē´dō)	**214-15**	42°40'N	0°05'E
Pereira, Col. (på-rā´rä)	**188c**	4°50'N	75°42'W

ăt; finăl; rāte; senåte; ärm; åsk; sofȧ; fâre; ch-choose; dh-as th in other; bē; ēvent; bĕt; recĕnt; cratẽr; g-gō; gh-guttural g; bĭt; ĭ-short neutral; rīde; ĸ-guttural k as ch in German ich;

PRONOUNCING INDEX Pereslavl'-Zalesskiy – Pins, Île des 395

Feature (Pronunciation)	Page	Lat.	Long.
Pereslavl'-Zalesskiy, Russia (pâ-râ-slâv´'l zá-lyĕs´kĭ)	**218-19**	56°44′N	38°51′E
Pergamino, Arg. (pĕr-gä-mē´nō)	**187**	33°54′S	60°35′W
Perham, Mn., U.S. (pĕr´hăm)	**124-25**	46°36′N	95°35′W
Péribonka, stm., Qc., Can.	**140-41**	48°46′N	72°03′W
Périgueux, Fr. (pā-rē-gû´)	**212-13**	45°11′N	0°43′E
Perito Moreno, Arg.	**185**	46°36′S	70°55′W
Perlas, Laguna de, b., Nic. (lä-gö´nä-dĕ-läs-pĕr´läs)	**161**	12°30′N	83°40′W
Perleberg, Ger. (pĕr´lĕ-bĕrg)	**210-11**	53°05′N	11°52′E
Perm', Russia (pĕrm)	**202-03**	58°00′N	56°16′E
Pernambuco, Braz. *see* Recife	**188d**	8°03′S	34°54′W
Pernambuco, state, Braz. (pĕr-näm-bo͞o´kō)	**180-81**	8°0′S	37°00′W
Pernik, Blg. (pĕr-nēk´)	**216-17**	42°37′N	23°03′E
Péronne, Fr. (pā-rôn´)	**212-13**	49°56′N	2°56′E
Perote, Mex. (pĕ-rō´tĕ)	**158-59**	19°34′N	97°15′W
Perpignan, Fr. (pĕr-pē-nyäɴ´)	**212-13**	42°42′N	2°53′E
Perros, Bahía de, strt., Cuba (bä-ē´ä-dĕ-pā´rōs)	**154-55**	22°21′N	78°31′W
Perry, Fl., U.S. (pĕr´ĭ)	**134-35**	30°07′N	83°35′W
Perry, Ga., U.S. (pĕr´ĭ)	**134-35**	32°28′N	83°44′W
Perry, Ia., U.S. (pĕr´ĭ)	**124-25**	41°51′N	94°07′W
Perry, N.Y., U.S. (pĕr´ĭ)	**126-27**	42°43′N	78°01′W
Perry, Ok., U.S. (pĕr´ĭ)	**130-31**	36°17′N	97°17′W
Perrysburg, Oh., U.S. (pĕr´ĭz-bûrg)	**126-27**	41°34′N	83°37′W
Perryton, Tx., U.S. (pĕr´ĭ-tŭn)	**130-31**	36°23′N	100°49′W
Perryville, Mo., U.S. (pĕr-ĭ-vĭl)	**130-31**	37°43′N	89°52′W
Persepolis, hist., Iran (pẽr-sĕpȯ-lĭs)	**250-51**	29°57′N	52°52′E
Persia, nation, Asia *see* Iran	**224-25**	32°0′N	53°00′E
Persian Gulf, b., Asia (pûr´zh*ă*n gŭlf)	**250-51**	27°0′N	51°00′E
Perth, Austl. (pûrth)	**294-95**	31°57′S	115°51′E
Perth, On., Can. (pûrth)	**146-47**	44°55′N	76°15′W
Perth, Scot., U.K. (pûrth)	**206-07**	56°24′N	3°27′W
Perth Amboy, N.J., U.S. (pûrth ăm´boi)	**126-27**	40°31′N	74°14′W
Pertuis, Fr. (pĕr-tüē´)	**212-13**	43°42′N	5°30′E
Peru, Il., U.S. (pė-ro͞o´)	**126-27**	41°20′N	89°07′W
Peru, In., U.S. (pė-ro͞o´)	**126-27**	40°45′N	86°03′W
Peru, nation, S.A. (pė-ro͞o´)	**172**	10°0′S	76°00′W
Perugia, Italy (pā-ro͞o´jä)	**216-17**	43°07′N	12°22′E
Pervomais'k, Ukr.	**218-19**	48°03′N	30°51′E
Pervouralsk, Russia (pĕr-vô-ô-rälsk´)	**236-37**	56°55′N	59°57′E
Pesaro, Italy (pā´zä rō)	**216-17**	43°55′N	12°55′E
Pescadores, is., Tai. *see* P'enghu Ch'üntao	**243a**	23°30′N	119°30′E
Pescara, Italy (pās-kä´rä)	**216-17**	42°29′N	14°12′E
Peshāwar, Pak. (pĕ-shä´wŭr)	**252-53**	33°60′N	71°33′E
Peshtigo, Wi., U.S. (pĕsh´tė-gō)	**126-27**	45°03′N	87°44′W
Pesqueira, Braz.	**188d**	8°22′S	36°42′W
Pesyakov, Ostrov, i., Russia	**202-03**	68°45′N	57°41′E
Petacalco, Bahía, b., Mex. (bä-ē´ä-dĕ-pĕ-tä-käl´kȯ)	**158-59**	17°56′N	101°57′W
Petah Tiqwa, Isr.	**248-49**	32°06′N	34°54′E
Petaluma, Ca., U.S. (pĕt-*ȧ*-lö´m*ȧ*)	**128-29**	38°14′N	122°38′W
Petare, Ven. (pĕ-tä´rĕ)	**188b**	10°29′N	66°49′W
Petatlán, Mex. (pā-tä-tlän´)	**158-59**	17°31′N	101°16′W
Peterborough, Austl.	**301**	32°58′S	138°50′E
Peterborough, On., Can. (pē´tẽr-bŭr-ȯ)	**146-47**	44°18′N	78°20′W
Peterhead, Scot., U.K. (pē-tẽr-hĕd´)	**206-07**	57°31′N	1°47′W
Peter Pond Lake, lk., Sk., Can. (pē´tẽr pŏnd lāk)	**144-45**	56°06′N	109°06′W
Petersburg, Ak., U.S. (pē´tẽrz-bûrg)	**136**	56°48′N	132°57′W
Petersburg, Il., U.S. (pē´tẽrz-bûrg)	**130-31**	40°00′N	89°50′W
Petersburg, In., U.S. (pē´tẽrz-bûrg)	**126-27**	38°29′N	87°17′W
Petersburg, Va., U.S. (pē´tẽrz-bûrg)	**134-35**	37°14′N	77°24′W
Petersburg, W.V., U.S. (pē´tẽrz-bûrg)	**126-27**	38°59′N	79°08′W
Peter the Great Bay, b., Russia *see* Petra Velikogo, Zaliv	**264**	42°40′N	132°00′E
Petitcodiac, N.B., Can. (pē-tē-kō-dyȧk´)	**148-49**	45°56′N	65°11′W
Petit-Goâve, Haiti (pē-tē´ gō-äv´)	**154-55**	18°26′N	72°51′W
Petlalcingo, Mex. (pĕ-tläl-sĕŋ´gȯ)	**158-59**	18°04′N	97°56′W
Peto, Mex. (pĕ´tȯ)	**160**	20°08′N	88°54′W
Petorca, Chile (pā-tōr´kȧ)	**188e**	32°15′S	70°57′W
Petoskey, Mi., U.S. (pė-tŏs-kĭ)	**126-27**	45°22′N	84°57′W
Petra Velikogo, Zaliv, b., Russia	**264**	42°40′N	132°00′E
Petrich, Blg. (pā´trĭch)	**216-17**	41°24′N	23°13′E
Petrified Forest National Park, n.p., Az., U.S. (pĕt´rĭ-fīd fŏr´ĕst năsh´ŭn-*ă*l pärk)	**128-29**	34°54′N	109°47′W
Petrinja, Cro. (pā´trēn-yȧ)	**216-17**	45°27′N	16°16′E
Petrodvorets, Russia (pyĕ-trô-dvô-ryĕts´)	**208-09**	59°53′N	29°52′E
Petrolia, On., Can. (pĕ-trō´lĭ-*ȧ*)	**146-47**	42°53′N	82°09′W

Feature (Pronunciation)	Page	Lat.	Long.
Petrolina, Braz. (pĕ-trō-lē´n*ȧ*)	**180-81**	9°24′S	40°30′W
Petropavlovsk, Kaz. (pyĕ-trô-päv´lôvsk)	**244**	54°52′N	69°09′E
Petropavlovsk-Kamchatskiy, Russia (pyĕ-trô-päv´lôvsk käm-chät´skĭ)	**236-37**	53°01′N	158°41′E
Petrópolis, Braz. (pâ-trô-pô-lēzh´)	**186**	22°31′S	43°10′W
Petroșani, Rom.	**216-17**	45°25′N	23°23′E
Petrovgrad, Serb. *see* Zhovti Vody	**216-17**	45°23′N	20°24′E
Petrovsk, Russia (pyĕ-trôfsk´)	**202-03**	52°19′N	45°23′E
Petrovsk-Zabaykal'skiy, Russia (pyĕ-trôfskzä-bī-käl´skĭ)	**260-61**	51°16′N	108°50′E
Petrozavodsk, Russia (pyā´trô-zȧ-vôtsk´)	**202-03**	61°47′N	34°21′E
Petrozsény, Rom. *see* Petroșani	**216-17**	45°25′N	23°23′E
Petukhovo, Russia	**244**	55°04′N	67°54′E
Pevek, Russia	**236-37**	69°41′N	170°21′E
Peza, stm., Russia (pyā´zȧ)	**202-03**	65°36′N	44°37′E
Pézenas, Fr. (pā-zē-nä´)	**212-13**	43°27′N	3°26′E
Pforzheim, Ger. (pfôrts´hīm)	**210-11**	48°54′N	8°42′E
Pha-an, Mya.	**266-67**	16°53′N	97°38′E
Phangan, Ko, i., Thai.	**266-67**	9°45′N	100°01′E
Phangnga, Thai.	**266-67**	8°27′N	98°32′E
Phanom Dong Rak, Thiu Khao, mts., Asia *see* Phanom Dongrak Range	**266-67**	14°25′N	103°30′E
Phanom Dongrak Range, mts., Asia	**266-67**	14°25′N	103°30′E
Phan Rang, Viet.	**266-67**	11°34′N	108°60′E
Phan Si Pan, mtn., Viet. *see* Fan Si Pan	**266-67**	22°15′N	103°46′E
Phan Thiet, Viet.	**266-67**	10°56′N	108°06′E
Phenix City, Al., U.S. (fē´nĭks sĭ´tĭ)	**134-35**	32°28′N	85°01′W
Phetchabun, Thiu Khao, mts., Thai.	**266-67**	16°32′N	100°55′E
Philadelphia, Ms., U.S. (fĭl-*ȧ*-dĕl´phĭ-*ȧ*)	**134-35**	32°46′N	89°07′W
Philadelphia, Pa., U.S. (fĭl *ȧ* dĕl´phĭ *ȧ*)	**126-27**	39°57′N	75°10′W
Philip, S.D., U.S. (fĭl´ĭp)	**124-25**	44°02′N	101°39′W
Philippeville, Alg. *see* Skikda	**292b**	36°53′N	6°55′E
Philippines, nation, Asia (fĭl´ĭ-pēnz)	**224-25**	13°0′N	122°00′E
Philippine Sea, s., (fĭl´ĭ-pēn sē)	**240-41**	20°0′N	135°00′E
Philipsburg, Mt., U.S. (fĭl´lĭps-bẽrg)	**122-23**	46°20′N	113°18′W
Phillip Island, i., Austl. (fĭl´ĭp ī´l*ă*nd)	**301**	38°29′S	145°14′E
Phillips, Wi., U.S. (fĭl´ĭps)	**126-27**	45°42′N	90°24′W
Phillipsburg, Ks., U.S. (fĭl´lĭps-bẽrg)	**130-31**	39°45′N	99°19′W
Phillipsburg, N.J., U.S. (fĭl´lĭps-bẽrg)	**126-27**	40°42′N	75°11′W
Phitsanulok, Thai.	**266-67**	16°50′N	100°16′E
Phnom Penh, nat. cap., Camb. (nŏm´pĕn´)	**266-67**	11°34′N	104°54′E
Phnum Pénh, nat. cap., Camb. (nŏm´pĕn´) *see* Phnom Penh	**266-67**	11°34′N	104°54′E
Phoenix, Az., U.S. (fē´nĭks)	**128-29**	33°26′N	112°03′W
Phoenix Islands, is., Kir. (fē´nĭks ī´l*ă*ndz)	**304-05**	4°0′S	186°00′W
Phoenixville, Pa., U.S. (fē´nĭks-vĭl)	**126-27**	40°08′N	75°31′W
Phôngsali, Laos	**266-67**	21°43′N	102°07′E
Phra Chedi Sam Ong, p., Asia *see* Three Pagodas Pass	**266-67**	15°18′N	98°22′E
Phrae, Thai.	**266-67**	18°08′N	100°09′E
Phra Nakhon, nat. cap., Thai. *see* Bangkok	**266-67**	13°45′N	100°31′E
Phra Nakhon Si Ayutthaya, Thai.	**266-67**	14°21′N	100°34′E
Phuket, Thai.	**266-67**	7°52′N	98°23′E
Phuket, Ko, i., Thai.	**266-67**	8°0′N	98°22′E
Phu Ly, Viet.	**266-67**	20°31′N	105°56′E
Phu Quoc, Dao, i., Viet.	**266-67**	10°12′N	104°00′E
Piacenza, Italy (pyä-chĕnt´sä)	**216-17**	45°03′N	9°42′E
Piatra-Neamț, Rom.	**210-11**	46°57′N	26°24′E
Piauí, state, Braz.	**180-81**	7°0′S	43°00′W
Piazza Armerina, Italy (pyät´sä är-må-rē´nä)	**216-17**	37°23′N	14°22′E
Pic, stm., On., Can. (pĕk)	**146-47**	48°36′N	86°18′W
Picayune, Ms., U.S. (pĭk´á yo͞on)	**134-35**	30°32′N	89°42′W
Pichanal, Arg.	**182-83**	23°18′S	64°14′W
Pichilemu, Chile (pē-chē-lĕ´mo͞o)	**188e**	34°23′S	72°00′W
Pichucalco, Mex. (pē-cho͞o-käl´kȯ)	**158-59**	17°30′N	93°10′W
Pickle Lake, On., Can.	**144-45**	51°30′N	90°04′W
Pico, i., Port. (pē´kȯ)	**215c**	38°28′N	28°20′W
Pico de Orizaba, Volcán, vol., Mex. (vȯl-kȧ´n-pē´kō-dĕ-ō-rē-zä´bä)	**158-59**	19°01′N	97°16′W
Picos, Braz. (pē´kōzh)	**180-81**	7°05′S	41°28′W
Picton, On., Can. (pĭk´tŭn)	**146-47**	43°60′N	77°08′W
Picton, Isla, i., Chile	**185**	55°03′S	66°55′W
Pictou, N.S., Can. (pĭk-to͞o´)	**148-49**	45°41′N	62°42′W
Pidurutalagala, mtn., Sri L. (pē´dȯ-rȯ-tä´lȧ-gä´lȧ)	**256**	6°60′N	80°46′E
Piedmont, Al., U.S. (pēd´mŏnt)	**134-35**	33°55′N	85°37′W
Piedmont, Mo., U.S. (pēd´mŏnt)	**134-35**	37°09′N	90°42′W
Piedra del Águila, Arg.	**185**	40°03′S	70°03′W

Feature (Pronunciation)	Page	Lat.	Long.
Piedras, Punta, c., Arg. (po͞o´n-tä-pyĕ´dräs)	**187**	35°26′S	57°07′W
Piedras Negras, Mex. (pyā´dräs nā´gräs)	**132-33**	28°42′N	100°31′W
Pierce, Ne., U.S. (pērs)	**124-25**	42°12′N	97°32′W
Pierre, S.D., U.S. (pēr)	**124-25**	44°22′N	100°21′W
Pietarsaari, Fin. *see* Jakobstad	**200-01**	63°41′N	22°43′E
Pietermaritzburg, S. Afr. (pē-tẽr-mȧ-rĭts-bûrg)	**286-87**	29°36′S	30°23′E
Pietersburg, S. Afr. (pē´tẽrz-bûrg) *see* Polokwane	**292c**	23°53′S	29°26′E
Pigeon Lake, lk., Ab., Can. (pĭj´*ŭ*n lāk)	**142-43**	53°0′N	114°00′W
Pigeon Lake, lk., On., Can. (pĭj´*ŭ*n lāk)	**146-47**	44°30′N	78°30′W
Piggott, Ar., U.S. (pĭg-ŭt)	**134-35**	36°23′N	90°12′W
Pigüé, Arg.	**187**	37°37′S	62°25′W
Pihkva järv, lk., Eur. *see* Pskov, Lake	**208-09**	58°0′N	28°00′E
Pijijiapan, Mex. (pēkē-kē-ä´pän)	**158-59**	15°42′N	93°13′W
Pikalëvo, Russia	**202-03**	59°31′N	34°11′E
Pikes Peak, mtn., Co., U.S. (pīks pēk)	**130-31**	38°51′N	105°03′W
Piketberg, S. Afr.	**286-87**	32°55′S	18°46′E
Pikeville, Ky., U.S. (pīk´vĭl)	**126-27**	37°30′N	82°33′W
Piła, Pol. (pē´lȧ)	**210-11**	53°09′N	16°44′E
Pilanesberg, hill, S. Afr. (pē´*ă*ns´bûrg)	**292c**	25°12′S	27°05′E
Pilar, Arg. (pē´lär)	**187**	31°26′S	61°16′W
Pilar, Para.	**187**	26°54′S	58°19′W
Pilcomayo, stm., S.A. (pēl-cō-mī´ȯ)	**182-83**	25°17′S	57°40′W
Pili, Phil. (pē´lė)	**270**	13°32′N	123°17′E
Pīlībhīt, India	**254-55**	28°38′N	79°48′E
Pilica, stm., Pol. (pė-lēt´sä)	**210-11**	51°52′N	21°17′E
Pilipinas, nation, Asia *see* Philippines	**224-25**	13°0′N	122°00′E
Pilsen, Czech Rep. *see* Plzeň	**210-11**	49°45′N	13°23′E
Pimpri-Chinchwad, India	**256**	18°38′N	73°47′E
Pinamalayan, Phil. (pē-nä-mä-lä´yän)	**270**	13°02′N	121°29′E
Pinang, Malay. *see* George Town	**266-67**	5°25′N	100°20′E
Pinar del Río, Cuba (pē-när´ dĕl rē´ȯ)	**154-55**	22°25′N	83°41′W
Pinar del Río, state, Cuba (pē-när´ dĕl rē´ȯ)	**154-55**	22°30′N	83°45′W
Pinatubo, Mount, vol., Phil. (mount pē-nä-to͞o´bȯ)	**270**	15°00′N	120°21′E
Pincher Creek, Ab., Can. (pĭn´chẽr krēk)	**142-43**	49°29′N	113°57′W
Pinckneyville, Il., U.S. (pĭnk´nĭ-vĭl)	**126-27**	38°05′N	89°23′W
Pindaré, stm., Braz.	**180-81**	3°18′S	44°47′W
Píndos Óros, mts., Grc.	**216-17**	39°49′N	21°14′E
Pindus Mountains, mts., Grc. (pĭn´d*ŭ*s moun´tĭnz) *see* Píndos Óros	**216-17**	39°49′N	21°14′E
Pine, stm., B.C., Can. (pīn)	**142-43**	56°09′N	120°44′W
Pine Bluff, Ar., U.S. (pīn blŭf)	**134-35**	34°14′N	92°02′W
Pine City, Mn., U.S. (pīn sĭ´tė)	**124-25**	45°50′N	92°58′W
Pine Creek, Austl. (pīn crēk krēk)	**294-95**	13°48′S	131°50′E
Pine Falls, Mb., Can. (pīn fôlz)	**144-45**	50°34′N	96°14′W
Pinega, stm., Russia (pė-nyĕ´gȧ)	**202-03**	64°08′N	41°54′E
Pinehouse Lake, lk., Sk., Can.	**144-45**	55°34′N	106°31′W
Pine Ridge, S.D., U.S. (pīn rĭj)	**124-25**	43°01′N	102°33′W
Pinerolo, Italy (pē-nå-rô´lō)	**216-17**	44°54′N	7°20′E
Pines, Isle of, i., Cuba (īl ŭv pīnz) *see* Juventud, Isla de la	**154-55**	21°40′N	82°50′W
Pineville, Ky., U.S. (pīn´vĭl)	**134-35**	36°45′N	83°42′W
Pineville, La., U.S. (pīn´vĭl)	**132-33**	31°20′N	92°26′W
Ping, stm., Thai.	**266-67**	15°42′N	100°09′E
Pingdingshan, China	**258-59**	33°45′N	113°18′E
Pingdu, China (pĭŋ-do͞o)	**260-61**	36°47′N	119°56′E
Pingelap, at., Micron.	**304-05**	6°13′N	160°42′E
Pingjiang, China	**258-59**	28°42′N	113°35′E
Pingle, China (pĭŋ-lŭ)	**258-59**	24°38′N	110°40′E
Pingliang, China (pĭng´lyäng´)	**260-61**	35°33′N	106°42′E
Pingquan, China (pĭŋ-chyůän)	**260-61**	40°59′N	118°39′E
Pingtan, China (pĭŋ-tän)	**243a**	25°31′N	119°47′E
Pingtan Dao, i., China (pĭŋ-tän dou)	**243a**	25°33′N	119°48′E
Pingtung, Tai.	**243a**	22°40′N	120°29′E
Pingwu, China (pĭŋ-wo͞o)	**258-59**	32°25′N	104°33′E
Pingxiang, China (pĭŋ-shyäŋ)	**258-59**	22°08′N	106°44′E
Pingxiang, China (pĭŋ-shyäŋ)	**258-59**	27°38′N	113°50′E
Pingyao, China	**260-61**	37°16′N	112°14′E
Pingyi, China (pĭŋ-yē)	**260-61**	35°30′N	117°38′E
Pingyuan, China (pĭŋ-yůän)	**258-59**	24°36′N	115°55′E
Pinheiro, Braz.	**180-81**	2°31′S	45°05′W
Pinnacles National Park, n.p., Ca., U.S. (pĭn´*ȧ*-k'lz năsh´ŭn-*ă*l pärk)	**128-29**	36°30′N	121°11′W
Pinnaroo, Austl.	**301**	35°16′S	140°54′E
Pinos, Isla de, i., Cuba *see* Juventud, Isla de la	**154-55**	21°40′N	82°50′W
Pinrang, Indon.	**268-69**	3°48′S	119°39′E
Pins, Île des, i., N. Cal.	**306g**	22°37′S	167°28′E

n-sing; ŋ-baŋk; ɴ-nasalized n; nŏd; c*ŏ*mmit; ōld; ôbey; ôrder; oi-boil; fo͞od; ȯ-as oo in foot; ou-out; s-soft; sh-dish; th-thin; pūre; ůnite; ûrn; stŭd; circ*ŭ*s; ü-as in French tu; ´-indeterminate vowel.

396 Pinsk – Portage PRONOUNCING INDEX

Feature (Pronunciation)	Page	Lat.	Long.
Pinsk, Bela. (pēn'sk)	210-11	52°07'N	26°07'E
Pinsk Marshes, reg., Eur. see Pripet Marshes	210-11	52°0'N	27°00'E
Pinta, Isla, i., Ec.	184a	0°35'N	90°44'W
Pinyug, Russia	202-03	60°15'N	47°47'E
Piombino, Italy (pyôm-bē'nō)	216-17	42°56'N	10°32'E
Pioneer Mountains, mts., Mt., U.S. (pī'ô-nēr' moun'tĭnz)	122-23	45°31'N	112°60'W
Pioner, Ostrov, i., Russia	236-37	79°50'N	92°30'E
Piorini, stm., Braz.	180-81	3°23'S	63°30'W
Piotrków Trybunalski, Pol. (pyŏtr'ko͞ov trĭ-bo͞o-nal'skē)	210-11	51°24'N	19°42'E
Pipe Spring National Monument, n.p., Az., U.S. (pīp spring năsh'ŭn-ăl mŏn'ū-mĕnt)	128-29	36°50'N	112°49'W
Pipestone, Mn., U.S. (pīp'stōn)	124-25	43°60'N	96°19'W
Pipestone, stm., On., Can.	144-45	52°54'N	89°15'W
Pipestone National Monument, n.p., Mn., U.S. (pīp'stōn năsh'ŭn-ăl mŏn'ū-mĕnt)	124-25	44°0'N	96°18'W
Pipinas, Arg.	187	35°32'S	57°19'W
Pipmuacan, Réservoir, res., Qc., Can. (pĭp-mä-kän')	148-49	49°37'N	70°27'W
Piqua, Oh., U.S. (pĭk'wȧ)	126-27	40°09'N	84°15'W
Piracicaba, Braz. (pē-rä-sē-kä'bä)	186	22°43'S	47°36'W
Piraeus, Grc. see Peiraiás	216-17	37°57'N	23°39'E
Piran, Slvn. (pē-rå'n)	216-17	45°32'N	13°34'E
Pirané, Arg.	182-83	25°43'S	59°05'W
Pirapora, Braz. (pē-rȧ-pō'rȧ)	186	17°21'S	44°56'W
Pires do Rio, Braz.	186	17°18'S	48°17'W
Piriápolis, Ur.	187	34°52'S	55°16'W
Pirineos, mts., Eur. see Pyrenees	212-13	42°40'N	1°00'E
Pirna, Ger. (pĭr'nä)	210-11	50°58'N	13°57'E
Pirot, Serb. (pē'rōt)	216-17	43°10'N	22°35'E
Piru, Indon. (pē-ro͞o')	268-69	3°03'S	128°11'E
Pisa, Italy (pē'sä)	216-17	43°44'N	10°24'E
Pisagua, Chile (pē-sä'gwä)	182-83	19°34'S	70°12'W
Pisco, Peru (pēs'kō)	184	13°42'S	76°12'W
Písek, Czech Rep. (pē'sĕk)	210-11	49°19'N	14°09'E
Pishan, China	244	37°37'N	78°16'E
Pisticci, Italy (pēs-tē'chē)	216-17	40°23'N	16°34'E
Pistoia, Italy (pēs-tô'yä)	216-17	43°56'N	10°55'E
Pisuerga, stm., Spain (pē-swĕr'gä)	214-15	41°33'N	4°52'W
Pitalito, Col. (pē-tä-lē'tō)	178-79	1°52'N	76°01'W
Pitanga, Braz.	182-83	24°44'S	51°45'W
Pitcairn Island, i., Pit.	304-05	25°04'S	130°06'W
Pitcairn Islands, dep., Oc. (pĭt'kârn ī'lăndz)	304-05	25°04'S	130°05'W
Piteå, Swe.	200-01	65°19'N	21°29'E
Piteälven, stm., Swe.	200-01	65°23'N	21°19'E
Pitești, Rom. (pē-tĕsht'')	216-17	44°51'N	24°52'E
Pithiviers, Fr. (pē-tē-vyā')	212-13	48°10'N	2°15'E
Pitti Island, i., India	256	10°50'N	72°37'E
Pitt Island, i., B.C., Can. (pĭt ī'lănd)	142-43	53°35'N	129°45'W
Pittsburg, Ks., U.S. (pĭts'bûrg)	130-31	37°25'N	94°42'W
Pittsburg, Tx., U.S. (pĭts'bûrg)	130-31	33°00'N	94°58'W
Pittsburgh, Pa., U.S. (pĭts'bûrg)	126-27	40°27'N	80°01'W
Pittsfield, Il., U.S. (pĭts'fēld)	130-31	39°36'N	90°48'W
Pittston, Pa., U.S. (pĭts'tŭn)	126-27	41°20'N	75°47'W
Pium, Braz.	180-81	10°27'S	49°11'W
Piura, Peru (pē-o͞o'rä)	184	5°11'S	80°38'W
Pivdennyi Buh, stm., Ukr. see Southern Bug	218-19	46°39'N	31°56'E
Placentia Bay, b., Nf./L., Can.	148-49	47°15'N	54°30'W
Placerville, Ca., U.S. (plăs'ēr-vĭl)	128-29	38°44'N	120°48'W
Placetas, Cuba (plä-thā'täs)	154-55	22°19'N	79°39'W
Plainview, Mn., U.S. (plān'vū)	124-25	44°10'N	92°09'W
Plainview, Ne., U.S. (plān'vū)	124-25	42°21'N	97°47'W
Plainview, Tx., U.S. (plān'vū)	130-31	34°11'N	101°42'W
Plainwell, Mi., U.S. (plan'wĕl)	126-27	42°26'N	85°38'W
Plano, Il., U.S. (plā'nō)	126-27	41°40'N	88°32'W
Plano, Tx., U.S. (plā'nō)	130-31	33°03'N	96°41'W
Plant City, Fl., U.S. (plȧnt sĭ'tĭ)	135a	28°01'N	82°07'W
Plaquemine, La., U.S. (plăk'mēn')	134-35	30°17'N	91°14'W
Plasencia, Spain (plä-sĕn'thē-ä)	214-15	40°02'N	6°05'W
Plaster Rock, N.B., Can. (plȧs'tēr rŏk)	148-49	46°55'N	67°24'W
Plastun, Russia (plȧs-to͞on')	264	44°45'N	136°18'E
Plata, Río de la, est., S.A. (rē'ō dālä plä'tä)	187	35°0'S	57°00'W
Platinum, Ak., U.S. (plăt'ĭ-nŭm)	136	59°01'N	161°49'W
Plato, Col. (plä'tō)	178-79	9°48'N	74°47'W
Platte, S.D., U.S. (plăt)	124-25	43°23'N	98°51'W
Platte, stm., U.S. (plăt)	130-31	39°16'N	94°51'W
Platte, stm., Ne., U.S. (plăt)	120-21	41°04'N	95°53'W
Platteville, Wi., U.S. (plăt'vĭl)	124-25	42°44'N	90°29'W
Plattsburg, Mo., U.S. (plăts'bûrg)	130-31	39°34'N	94°27'W
Plattsburgh, N.Y., U.S.	126-27	44°42'N	73°28'W
Plattsmouth, Ne., U.S. (plăts'mŭth)	124-25	41°01'N	95°54'W
Plauen, Ger. (plou'ĕn)	210-11	50°30'N	12°08'E
Playa Vicente, Mex. (plä-yä vē-sĕn'tā)	158-59	17°48'N	95°49'W
Play Ku, Viet.	266-67	13°59'N	108°01'E
Pleasanton, Tx., U.S. (plĕz'ăn-tŭn)	132-33	28°58'N	98°29'W
Pleiku, Viet. see Play Ku	266-67	13°59'N	108°01'E
Plenty, Bay of, b., N.Z. (bā ŭv plĕn'tē)	303	37°40'S	177°00'E
Plentywood, Mt., U.S. (plĕn'tē-wo͝od)	122-23	48°46'N	104°32'W
Plesetsk, Russia	202-03	62°43'N	40°18'E
Plessisville, Qc., Can. (plĕ-sē'vēl')	146-47	46°13'N	71°46'W
Pleszew, Pol. (plĕ'zhĕf)	210-11	51°54'N	17°47'E
Płock, Pol. (pwôtsk)	210-11	52°33'N	19°42'E
Ploërmel, Fr. (plô-ĕr-mĕl')	212-13	47°56'N	2°24'W
Ploești, Rom. see Ploiești	216-17	44°57'N	26°02'E
Ploiești, Rom. (plô-yĕsht'')	216-17	44°57'N	26°02'E
Plomb du Cantal, mtn., Fr. (plôɴ'dü-käɴ-täl')	212-13	45°04'N	2°45'E
Plonge, Lac la, lk., Sk., Can. (läk lä plōɴzh)	144-45	55°08'N	107°17'W
Plovdiv, Blg. (plôv'dĭf) (fĭl-ĭp-ŏp'ō-lĭs)	216-17	42°09'N	24°45'E
Plungė, Lith. (plŏn'gā)	208-09	55°54'N	21°52'E
Plyeven, Blg.	216-17	43°25'N	24°37'E
Plymouth, Eng., U.K. (plĭm'ŭth)	206-07	50°23'N	4°10'W
Plymouth, In., U.S. (plĭm'ŭth)	126-27	41°20'N	86°18'W
Plymouth, N.C., U.S. (plĭm'ŭth)	134-35	35°52'N	76°45'W
Plymouth, N.H., U.S. (plĭm'ŭth)	126-27	43°46'N	71°41'W
Plymouth, Pa., U.S. (plĭm'ŭth)	126-27	41°14'N	75°58'W
Plymouth, Vt., U.S. (plĭm'ŭth)	126-27	43°33'N	72°44'W
Plymouth, Wi., U.S. (plĭm'ŭth)	126-27	43°46'N	87°59'W
Plzeň, Czech Rep.	210-11	49°45'N	13°23'E
Po, stm., Italy	210-11	44°59'N	12°03'E
Pobeda, Gora, mtn., Russia	236-37	65°12'N	146°12'E
Pobedy, Pik, mtn., Asia see Jengish Chokusu	244	42°02'N	80°05'E
Pocahontas, Ar., U.S. (pō-kȧ-hŏn'tȧs)	134-35	36°16'N	90°58'W
Pocahontas, Ia., U.S. (pō-kȧ-hŏn'tȧs)	124-25	42°44'N	94°40'W
Pocatello, Id., U.S. (pō-kȧ-tĕl'ō)	122-23	42°53'N	112°29'W
Pochëp, Russia (pô-chĕp')	218-19	52°55'N	33°29'E
Pocomoke City, Md., U.S. (pō-kō-mōk' sĭ'tē)	126-27	38°04'N	75°33'W
Poços de Caldas, Braz. (pō-sôs-dĕ-käl'dȧs)	186	21°47'S	46°34'W
Podgorica, nat. cap., Mont.	216-17	42°27'N	19°16'E
Podkamennaya Tunguska, stm., Russia	236-37	61°36'N	90°09'E
Podolsk, Russia (pô-dôl''sk)	218-19	55°26'N	37°34'E
Podporozhye, Russia	208-09	60°55'N	34°10'E
Poggibonsi, Italy (pôd-jē-bôn'sē)	216-17	43°28'N	11°09'E
Pogranichnyy, Russia	264	44°24'N	131°23'E
Pohang, Kor., S.	263	36°03'N	129°22'E
Pohjanlahti, b., Eur. see Bothnia, Gulf of	200-01	63°0'N	20°00'E
Pohnpei, i., Micron.	304-05	6°55'N	158°15'E
Poinsett, Cape, c., Ant.	313	65°48'S	113°10'E
Point Au Fer Island, i., La., U.S.	134-35	29°15'N	91°15'W
Pointe-à-Pitre, Guad. (pwănt' ȧ pē-tr')	155b	16°15'N	61°32'W
Pointe-des-Galets, Reu. see Le Port	287a	20°55'S	55°18'E
Pointe-Noire, Congo	284-85	4°48'S	11°52'E
Point Hope, Ak., U.S. (point hōp)	136	68°21'N	166°41'W
Point Pleasant, Oh., U.S. (point plĕz'ănt)	126-27	38°54'N	84°14'W
Point Pleasant, W.V., U.S. (point plĕz'ănt)	126-27	38°52'N	82°08'W
Poitiers, Fr. (pwȧ-tyā')	212-13	46°35'N	0°20'E
Pokharā, Nepal	254-55	28°13'N	83°60'E
Pokhvistnevo, Russia	202-03	53°39'N	52°08'E
Pokrovsk, Russia	236-37	61°31'N	129°11'E
Pokrovskoye, Russia (pô-krôf'skô-yĕ)	218-19	52°37'N	36°51'E
Polack, Bela.	208-09	55°30'N	28°47'E
Poland, nation, Eur. (pō'lănd)	190-91	52°0'N	19°00'E
Polatlı, Tur.	202-03	39°36'N	32°10'E
Polessk, Russia (pô'lĕsk)	208-09	54°52'N	21°05'E
Polesye, reg., Eur. see Pripet Marshes	210-11	52°0'N	27°00'E
Polillo Island, i., Phil.	270	14°50'N	121°57'E
Polillo Islands, is., Phil. (pô-lēl'yō ī'lăndz)	270	14°50'N	122°05'E
Polissya, reg., Eur. see Pripet Marshes	210-11	52°0'N	27°00'E
Pollāchi, India	256	10°39'N	77°01'E
Polokwane, S. Afr.	292c	23°53'S	29°26'E
Polonnaruwa, Sri L.	256	7°56'N	81°01'E
Polotsk, Bela. see Polack	208-09	55°30'N	28°47'E
Polska, nation, Eur. see Poland	190-91	52°0'N	19°00'E
Polson, Mt., U.S. (pōl'sŭn)	122-23	47°41'N	114°09'W
Poltava, Ukr. (pôl-tä'vä)	218-19	49°36'N	34°31'E
Põltsamaa, Est. (pŏlt'sȧ-mä)	208-09	58°39'N	25°58'E
Polunochnoye, Russia (pô-lo͞o-nô'ch-nô'yĕ)	202-03	60°52'N	60°26'E
Polyarnyy, Russia (pŭl-yär'nē)	236-37	69°06'N	178°39'E
Polyarnyy, Russia (pŭl-yär'nē)	200-01	69°11'N	33°28'E
Polynesia, is., Oc. (pŏl-ĭ-nē'zhȧ)	304-05	4°0'S	156°00'W
Polynésie, is., Oc. see Polynesia	304-05	4°0'S	156°00'W
Polynésie française, dep., Oc. see French Polynesia	304-05	15°0'S	140°00'W
Pomerania, hist. reg., Eur. (pŏm-ē-rā'nĭ-ȧ)	210-11	54°0'N	16°00'E
Pomeroy, Oh., U.S. (pŏm'ēr-oi)	126-27	39°02'N	82°01'W
Pomeroy, Wa., U.S. (pŏm'ēr-oi)	122-23	46°28'N	117°35'W
Pommern, hist. reg., Eur. see Pomerania	210-11	54°0'N	16°00'E
Pomona, Ca., U.S. (pō-mō'nȧ)	128-29	34°03'N	117°45'W
Pomorze, hist. reg., Eur. see Pomerania	210-11	54°0'N	16°00'E
Pompano Beach, Fl., U.S. (pŏm'pȧ-nō bēch)	135a	26°14'N	80°08'W
Pompei, hist., Italy	216-17	40°45'N	14°30'E
Pompeii, hist., Italy see Pompei	216-17	40°45'N	14°30'E
Ponape, i., Micron. see Pohnpei	304-05	6°55'N	158°15'E
Ponca, Ne., U.S. (pŏn'kȧ)	124-25	42°33'N	96°43'W
Ponca City, Ok., U.S. (pŏn'kȧ sĭ'tĭ)	130-31	36°42'N	97°05'W
Ponce, P.R. (pōn'sā)	154a	18°01'N	66°37'W
Ponferrada, Spain (pôn-fĕr-rä'dhä)	214-15	42°33'N	6°35'W
Pongolo, stm., S. Afr.	286-87	26°52'S	32°21'E
Ponoka, Ab., Can. (pŏ-nō'kȧ)	142-43	52°40'N	113°35'W
Ponoy, stm., Russia	202-03	66°60'N	41°16'E
Ponta Delgada, Port. (pôn'tȧ dĕl-gä'dȧ)	215c	37°45'N	25°40'W
Ponta Grossa, Braz. (pôn'tä grō'sȧ)	182-83	25°05'S	50°10'W
Ponta Porã, Braz.	182-83	22°33'S	55°42'W
Pontarlier, Fr. (pôɴ-tȧr-lyā')	212-13	46°54'N	6°22'E
Pont-Audemer, Fr. (pôɴ-tōd'mâr')	212-13	49°21'N	0°31'E
Pontchartrain, Lake, lk., La., U.S. (lāk pôɴ-shȧr-trăɴ')	134-35	30°10'N	90°10'W
Pontedera, Italy (pōn-tā-dā'rä)	216-17	43°40'N	10°38'E
Ponte Nova, Braz. (pô'n-tĕ-nô'vȧ)	186	20°25'S	42°54'W
Pontevedra, Spain (pôn-tĕ-vĕ-drä)	214-15	42°25'N	8°38'W
Pontiac, Il., U.S. (pŏn'tĭ-ăk)	126-27	40°53'N	88°37'W
Pontiac, Mi., U.S. (pŏn'tĭ-ăk)	126-27	42°38'N	83°18'W
Pontianak, Indon. (pôn-tē-ä'nȧk)	266-67	0°01'S	109°20'E
Pontine, Isole, is., Italy see Ponziane, Isole	216-17	40°55'N	12°57'E
Pontine Islands, is., Italy see Ponziane, Isole	216-17	40°55'N	12°57'E
Pontivy, Fr. (pôɴ-tē-vē')	212-13	48°04'N	2°58'W
Pontoise, Fr. (pôɴ-twȧz')	212-13	49°03'N	2°05'E
Pontotoc, Ms., U.S. (pŏn-tô-tŏk')	134-35	34°15'N	88°60'W
Pontremoli, Italy (pôn-trĕm'ô-lē)	216-17	44°23'N	9°53'E
Pontus Mountains, mts., Tur. see Doğu Karadeniz Dağları	245	40°30'N	40°30'E
Ponziane, Isole, is., Italy	216-17	40°55'N	12°57'E
Poole, Eng., U.K. (po͞ol)	206-07	50°44'N	1°59'W
Poona, India see Pune	256	18°32'N	73°52'E
Poopó, Lago, lk., Bol.	182-83	18°47'S	67°05'W
Popayán, Col. (pō-pä-yän')	178-79	2°27'N	76°36'W
Popigay, stm., Russia	236-37	72°57'N	106°10'E
Poplar, Mt., U.S. (pŏp'lēr)	122-23	48°06'N	105°13'W
Poplar, stm., Can.	144-45	53°02'N	97°27'W
Poplar Bluff, Mo., U.S. (pŏp'lēr blŭf)	134-35	36°45'N	90°24'W
Poplar Plains, Ky., U.S. (pŏp'lēr plānz)	126-27	38°22'N	83°40'W
Poplarville, Ms., U.S. (pŏp'lēr-vĭl)	134-35	30°51'N	89°34'W
Popocatépetl, Volcán, vol., Mex.	158-59	19°01'N	98°37'W
Popondetta, Pap. N. Gui.	302	8°47'S	148°13'E
Popovo, Blg. (pô'pô-vō)	216-17	43°21'N	26°14'E
Poprad, Slvk.	210-11	49°03'N	20°19'E
Porangatu, Braz.	180-81	13°26'S	49°12'W
Porbandar, India (pōr-bŭn'dŭr)	254-55	21°39'N	69°37'E
Porcher Island, i., B.C., Can. (pôr'kēr ī'lănd)	142-43	53°57'N	130°30'W
Porcupine, stm., N.A.	136	66°34'N	145°21'W
Pordenone, Italy (pōr-dā-nō'nā)	216-17	45°58'N	12°39'E
Pori, Fin. (pô'rē)	208-09	61°29'N	21°47'E
Porkhov, Russia (pôr'kôf)	208-09	57°46'N	29°34'E
Porlamar, Ven. (pôr-lä-mär')	178-79	10°57'N	63°51'W
Pornic, Fr. (pôr-nēk')	212-13	47°07'N	2°06'W
Poronaysk, Russia (pô'rô-nīsk)	236-37	49°14'N	143°06'E
Porpoise Bay, b., Ant.	313	66°30'S	128°30'E
Porsgrunn, Nor. (pôrs'grŏn')	208-09	59°09'N	9°40'E
Port, Reu. see Le Port	287a	20°55'S	55°18'E
Portachuelo, Bol. (pôrt-ä-chwä'lō)	182-83	17°21'S	63°24'W
Portage, In., U.S. (pôr'tåj)	126-27	41°35'N	87°11'W
Portage, Mi., U.S. (pôr'tåj)	126-27	42°12'N	85°36'W

PRONOUNCING INDEX
Portage – Prince of Wales Island 397

Feature (Pronunciation)	Page	Lat.	Long.
Portage, Pa., U.S. (pôr'tȧj)	126-27	40°23'N	78°40'W
Portage, Wi., U.S. (pôr'tȧj)	126-27	43°33'N	89°28'W
Portage la Prairie, Mb., Can. (pōr'tĭj lä-prâ'rī)	144-45	49°58'N	98°18'W
Port Alberni, B.C., Can. (pōr äl-bĕr-nē')	142-43	49°15'N	124°48'W
Portalegre, Port. (pôr-tä-lā'grĕ)	214-15	39°17'N	7°25'W
Portales, N.M., U.S. (pôr-tä'lĕs)	130-31	34°11'N	103°20'W
Port Alfred, S. Afr.	286-87	33°36'S	26°54'E
Port Alice, B.C., Can. (pōrt ăl'ĭs)	142-43	50°23'N	127°26'W
Port Allegany, Pa., U.S. (pōrt ăl-ē-gā'nĭ)	126-27	41°49'N	78°17'W
Port Angeles, Wa., U.S. (pōrt ăn'jē-lĕs)	122-23	48°07'N	123°26'W
Port Antonio, Jam.	154-55	18°10'N	76°27'W
Port Arthur, Austl.	301	43°09'S	147°50'E
Port Arthur, China *see* Lüshun	260-61	38°49'N	121°15'E
Port Arthur, Tx., U.S.	132-33	29°54'N	93°56'W
Port Augusta, Austl. (pōrt ô-gŭs'tă)	301	32°30'S	137°46'E
Port au Port Bay, b., Nf./L., Can. (pôr'tō pōr' bā)	148-49	48°40'N	58°45'W
Port-au-Prince, nat. cap., Haiti (pôr'tō präNs')	154-55	18°32'N	72°20'W
Port Austin, Mi., U.S. (pōrt ôs'tĭn)	126-27	44°02'N	83°00'W
Port Blair, India (pōrt blâr)	266-67	11°39'N	92°45'E
Port Borden, P.E., Can. (pōrt bôr'dĕn)	148-49	46°15'N	63°42'W
Port-Cartier, Qc., Can.	148-49	50°02'N	66°52'W
Port Clinton, Oh., U.S. (pōrt klĭn'tŭn)	126-27	41°30'N	82°58'W
Port-de-Paix, Haiti (pōrt dē pĕ)	154-55	19°56'N	72°49'W
Port Dickson, Malay. (pōrt dĭk'sŭn)	266-67	2°31'N	101°49'E
Port Edward, China *see* Weihai	260-61	37°30'N	122°07'E
Portel, Braz.	180-81	1°58'S	50°48'W
Port Elgin, N.B., Can. (pōrt ĕl'jĭn)	148-49	46°03'N	64°06'W
Port Elgin, On., Can. (pōrt ĕl'jĭn)	146-47	44°26'N	81°23'W
Port Elizabeth, S. Afr. (pōrt ē-lĭz'ȧ-bĕth)	286-87	33°58'S	25°34'E
Porterville, Ca., U.S. (pōr'tēr-vĭl)	128-29	36°05'N	119°02'W
Port Fairy, Austl.	301	38°23'S	142°14'E
Port-Francqui, D.R.C. *see* Ilebo	284-85	4°20'S	20°36'E
Port-Gentil, Gabon (pōr-zhäN-tē')	282-83	0°43'S	8°47'E
Port-Harcourt, Nig. (pōrt här'kŭrt)	282a	4°47'N	7°01'E
Port Hardy, B.C., Can. (pōrt här'dĭ)	142-43	50°43'N	127°30'W
Port Hedland, Austl. (pōrt hĕd'lănd)	294-95	20°19'S	118°36'E
Port Hood, N.S., Can. (pōrt hȯd)	148-49	46°01'N	61°32'W
Port Hope, On., Can. (pōrt hōp)	146-47	43°57'N	78°17'W
Port Huron, Mi., U.S. (pōrt hū'rŏn)	126-27	42°58'N	82°26'W
Portimao, Port. (pôr-tē-mouN)	214-15	37°08'N	8°32'W
Port Jervis, N.Y., U.S. (pōrt jûr'vĭs)	126-27	41°23'N	74°42'W
Port Lairge, Ire. *see* Waterford	206-07	52°15'N	7°06'W
Portland, Austl. (pōrt'lănd)	301	38°21'S	141°36'E
Portland, In., U.S. (pōrt'lănd)	126-27	40°26'N	84°59'W
Portland, Me., U.S. (pōrt'lănd)	126-27	43°40'N	70°17'W
Portland, Mi., U.S. (pōrt'lănd)	126-27	42°52'N	84°54'W
Portland, Or., U.S. (pōrt'lănd)	122-23	45°32'N	122°40'W
Portland, Tn., U.S. (pōrt'lănd)	134-35	36°35'N	86°31'W
Portland, Tx., U.S. (pōrt'lănd)	132-33	27°53'N	97°19'W
Portland Bight, b., Jam. (pōrt'lănd bīt)	154-55	17°50'N	77°06'W
Portland Inlet, b., B.C., Can. (pōrt'lănd ĭn'lĕt)	142-43	54°50'N	130°14'W
Portland Point, c., Jam. (pōrt'lănd point)	154-55	17°43'N	77°11'W
Port Lavaca, Tx., U.S. (pōrt lȧ-vä'kȧ)	132-33	28°37'N	96°38'W
Port Lincoln, Austl. (pōrt lĭŋ-kŭn)	294-95	34°44'S	135°52'E
Port Louis, nat. cap., Mauritius	287a	20°10'S	57°30'E
Port-Lyautey, Mor. *see* Kénitra	292a	34°16'N	6°35'W
Port Macquarie, Austl. (pōrt mȧ-kwŏ'rĭ)	301	31°27'S	152°55'E
Port Moresby, nat. cap., Pap. N. Gui. (pōrt mōrz'bē)	302	9°28'S	147°12'E
Port Neches, Tx., U.S. (pōrt nĕch'ĕz)	132-33	29°59'N	93°57'W
Porto, Port. (pōr'tȯ)	214-15	41°09'N	8°37'W
Porto Alegre, Braz. (pōr'tō ä-lā'grĕ)	182-83	30°03'S	51°12'W
Porto Amélia, Moz. *see* Pemba	286-87	13°01'S	40°32'E
Portobelo, Pan. (pôr'tȯ-bā'lȯ)	162	9°33'N	79°39'W
Porto de Moz, Braz.	180-81	1°44'S	52°14'W
Porto de Pedras, Braz. (pōr'tō pā'drăzh)	188d	9°09'S	35°17'W
Porto Esperança, Braz.	182-83	19°37'S	57°27'W
Porto Esperidião, Braz.	182-83	15°51'S	58°28'W
Portoferraio, Italy (pōr'tȯ-fĕr-rä'yō)	216-17	42°49'N	10°19'E
Port of Spain, nat. cap., Trin. (pōrt ŭv spān')	155b	10°39'N	61°30'W
Portogruaro, Italy (pōr'tȯ-grȯ-ä'rō)	216-17	45°47'N	12°50'E
Porto Murtinho, Braz. (pōr'tō mȯr-tēn'yȯ)	182-83	21°42'S	57°52'W
Porto Nacional, Braz. (pōr'tō ná-syȯ-näl')	180-81	10°42'S	48°25'W
Porto-Novo, nat. cap., Benin (pōr'tȯ-nō'vō)	282a	6°29'N	2°37'E
Porto Santo, i., Port. (pōr'tȯ sän'tȯ)	280-81	33°04'N	16°20'W
Porto Seguro, Braz. (pōr'tō sā-gōō'rȯ)	182-83	16°26'S	39°04'W
Porto Torres, Italy (pōr'tō tôr'rĕs)	216-17	40°50'N	8°24'E
Porto União, Braz.	182-83	26°15'S	51°04'W
Porto-Vecchio, Fr. (pōr'tô-vĕk'ē-ō)	200-01	41°36'N	9°16'E
Porto Velho, Braz. (pōr'tō väl'yȯ)	180-81	8°46'S	63°54'W
Portoviejo, Ec. (pôr-tō-vyā'hō)	184	1°03'S	80°27'W
Port Phillip Bay, b., Austl. (pōrt fĭl'ĭp bā)	301	38°07'S	144°48'E
Port Pirie, Austl. (pōrt pĭ'rē)	301	33°12'S	138°00'E
Port Said, Egypt (pōrt sä-ēd')	291b	31°16'N	32°18'E
Port Saint Lucie, Fl., U.S. (pôrt sānt lū'sē)	135a	27°20'N	80°20'W
Port Shepstone, S. Afr. (pōrt hĕps'tŭn)	286-87	30°45'S	30°25'E
Portsmouth, Eng., U.K. (pôrts'mŭth)	206-07	50°48'N	1°05'W
Portsmouth, N.H., U.S. (pôrts'mŭth)	126-27	43°04'N	70°46'W
Portsmouth, Oh., U.S. (pôrts'mŭth)	126-27	38°44'N	82°60'W
Portsmouth, Va., U.S. (pôrts'mŭth)	134-35	36°50'N	76°19'W
Port Stanley, nat. cap., Falk. Is. *see* Stanley	185	51°43'S	57°49'W
Port Sudan, Sudan (pōrt sōō-dän')	288	19°37'N	37°13'E
Port Sulphur, La., U.S. (pōrt sŭl'fēr)	134-35	29°29'N	89°42'W
Port Townsend, Wa., U.S. (pōrt tounz'ĕnd)	122-23	48°07'N	122°46'W
Portugal, nation, Eur. (pōr'tu-gȧl)	190-91	39°30'N	8°00'W
Portugalete, Spain (pōr-tōō-gä-lā'tā)	214-15	43°19'N	3°01'W
Portuguese Guinea, nation, Afr. *see* Guinea-Bissau	274	12°0'N	15°00'W
Port Vila, nat. cap., Vanuatu (port vē'lȧ)	306g	17°45'S	168°19'E
Port Wakefield, Austl. (pōrt wāk'fēld)	301	34°11'S	138°01'E
Port Washington, Wi., U.S. (pōrt wŏsh'ĭng tŭn)	126-27	43°23'N	87°53'W
Porvenir, Chile	185	53°18'S	70°22'W
Porvoo, Fln.	208-09	60°24'N	25°40'E
Posadas, Arg. (pō-sä'dhäs)	187	27°22'S	55°54'W
Posen, Pol. *see* Poznań	210-11	52°24'N	16°54'E
Poso, Indon.	268-69	1°23'S	120°46'E
Poso, Danau, lk., Indon. (pō'sō)	268-69	1°52'S	120°35'E
Posse, Braz.	180-81	14°06'S	46°22'W
Post, Tx., U.S. (pōst)	130-31	33°11'N	101°23'W
Postojna, Slvn. (pōs-tōyná)	216-17	45°47'N	14°13'E
Pos'yet, Russia (pos-yĕt')	263	42°39'N	130°49'E
Potawatomi Indian Reservation, ind. res., Ks., U.S. (pŏt-å-wä'tō-mē ĭn'dĭ-ăn rĕ-sēr-vā'shēn)	130-31	39°20'N	95°50'W
Potchefstroom, S. Afr. (pŏch'ĕf-strōm)	292c	26°43'S	27°07'E
Poteau, Ok., U.S. (pȯ-tō')	130-31	35°04'N	94°38'W
Poteet, Tx., U.S. (pȯ-tēt)	132-33	29°02'N	98°34'W
Potenza, Italy (pȯ-tĕnt'sä)	216-17	40°39'N	15°48'E
Potgietersrus, S. Afr. (pŏt-ĸē'tērs-rûs)	292c	24°11'S	29°01'E
Poti, Geor. (pô'tē)	245	42°09'N	41°40'E
Potomac, stm., U.S. (pȯ-tō'măk)	120-21	37°59'N	76°18'W
Potosí, Bol.	182-83	19°35'S	65°45'W
Potrerillos, Chile	182-83	26°26'S	69°29'W
Potsdam, Ger. (pôts'däm)	210-11	52°24'N	13°04'E
Potsdam, N.Y., U.S. (pŏts'dăm)	126-27	44°40'N	74°59'W
Pott, Île, i., N. Cal.	306g	19°35'S	163°35'E
Pottstown, Pa., U.S. (pŏts'toun)	126-27	40°16'N	75°39'W
Pottsville, Pa., U.S. (pôts'vĭl)	126-27	40°41'N	76°12'W
Poughkeepsie, N.Y., U.S. (pŏ-kĭp'sē)	126-27	41°42'N	73°56'W
Pouso Alegre, Braz. (pō'zȯ ä-lā'grĕ)	186	22°14'S	45°56'W
Poŭthĭsăt, Camb.	266-67	12°32'N	103°56'E
Poverty Point National Monument, n.p., La., U.S.	130-31	32°39'N	91°25'W
Póvoa de Varzim, Port. (pō-vō'ȧ dā vär'zēN)	214-15	41°23'N	8°45'W
Povorino, Russia	202-03	51°12'N	42°16'E
Povungnituk, Qc., Can.	138-39	60°03'N	77°19'W
Povungnituk, stm., Qc., Can.	140-41	60°01'N	77°23'W
Powder, stm., U.S. (pou'dēr)	120-21	46°44'N	105°27'W
Powell, Wy., U.S. (pou'ĕl)	122-23	44°45'N	108°45'W
Powell, Lake, res., U.S. (lāk pou'ĕl)	128-29	37°29'N	110°44'W
Powell Lake, lk., B.C., Can. (pou'ĕl lāk)	142-43	50°11'N	124°24'W
Powell River, B.C., Can. (pou'ĕl rĭv'ēr)	142-43	49°53'N	124°33'W
Poxoréu, Braz.	182-83	15°50'S	54°23'W
Poyang Hu, lk., China	258-59	29°0'N	116°25'E
Poyarkovo, Russia	260-61	49°38'N	128°39'E
Požarevac, Serb. (pô'zhá'rĕ-váts)	216-17	44°38'N	21°11'E
Poznań, Pol.	210-11	52°24'N	16°54'E
Pozoblanco, Spain (pō-thō-bläŋ'kō)	214-15	38°23'N	4°51'W
Pozsony, nat. cap., Slvk. *see* Bratislava	210-11	48°09'N	17°07'E
Pozuelo de Alarcón, Spain (pō-thwā'lō dā ä-lär-kōn')	214-15	40°27'N	3°46'W
Pra, stm., Ghana (prȧ)	282-83	5°01'N	1°38'W
Prachin Buri, Thai. (prä'chēn)	266-67	14°03'N	101°23'E
Prachuap Khiri Khan, Thai.	266-67	11°49'N	99°48'E
Pradera, Col. (prä-dĕ'rä)	188c	3°25'N	76°15'W
Prades, Fr. (pråd)	212-13	42°37'N	2°26'E
Prado, Braz.	186	17°18'S	39°15'W
Prague, nat. cap., Czech Rep. (präg)	210-11	50°05'N	14°26'E
Praha, nat. cap., Czech Rep. (prä'hȧ) *see* Prague	210-11	50°05'N	14°26'E
Praia, nat. cap., C.V. (prä'yȧ)	282-83	14°55'N	23°31'W
Prainha Nova, Braz.	180-81	7°29'S	60°38'W
Prairie du Chien, Wi., U.S. (prā'rī dȯ shēn')	124-25	43°03'N	91°08'W
Pratas Island, i., Tai.	240-41	20°42'N	116°43'E
Prathet Thai, nation, Asia *see* Thailand	224-25	15°0'N	100°00'E
Prato, Italy (prä'tō)	216-17	43°53'N	11°06'E
Pratt, Ks., U.S. (prăt)	130-31	37°39'N	98°44'W
Prattville, Al., U.S. (prăt'vĭl)	134-35	32°28'N	86°28'W
Praya, Indon.	268-69	8°43'S	116°17'E
Pregolya, stm., Russia (prĕ-gô'lă)	210-11	54°41'N	20°23'E
Premont, Tx., U.S. (prē-mŏnt')	132-33	27°22'N	98°07'W
Prenzlau, Ger. (prĕnts'lou)	210-11	53°19'N	13°52'E
Preparis Island, i., Mya.	266-67	14°53'N	93°41'E
Preparis North Channel, strt., Mya.	266-67	15°32'N	94°06'E
Preparis South Channel, strt., Mya.	266-67	14°37'N	93°53'E
Přerov, Czech Rep. (przhě'rôf)	210-11	49°27'N	17°28'E
Prescott, On., Can. (prĕs'kŭt)	146-47	44°43'N	75°31'W
Prescott, Ar., U.S. (prĕs'kŏt)	130-31	33°49'N	93°23'W
Prescott, Az., U.S. (prĕs'kŏt)	128-29	34°33'N	112°28'W
Presho, S.D., U.S. (prĕsh'ȯ)	124-25	43°55'N	100°04'W
Presidencia Roque Sáenz Peña, Arg.	187	26°47'S	60°26'W
Presidente Epitácio, Braz. (prā-sē dĕn'tĕ å-pē-tä'syȯ)	182-83	21°46'S	52°07'W
Presidente Prudente, Braz.	182-83	22°07'S	51°24'W
Presidio, Tx., U.S. (prĕ-sĭ'dĭ-ȯ)	132-33	29°34'N	104°23'W
Presidio, stm., Mex. (prĕ-sē'dyô)	158-59	23°05'N	106°17'W
Prešov, Slvk. (prĕ'shôf)	210-11	48°59'N	21°15'E
Presque Isle, Me., U.S. (prĕsk-ēl')	127a	46°41'N	68°01'W
Proßburg, nat. cap., Slvk. *see* Bratislava	210-11	48°09'N	17°07'E
Preston, Eng., U.K. (prĕs'tŭn)	206-07	53°46'N	2°42'W
Preston, Id., U.S. (prĕs'tŭn)	122-23	42°06'N	111°53'W
Preto, stm., Braz.	180-81	11°21'S	43°52'W
Pretoria, nat. cap., S. Afr. (prē-tō'rĭ-ȧ)	292c	25°45'S	28°11'E
Préveza, Grc. (prĕ'vȧ-zȧ)	216-17	38°57'N	20°45'E
Priboj, Serb. (prē'boi)	216-17	43°35'N	19°32'E
Price, Ut., U.S. (prīs)	128-29	39°36'N	110°48'W
Price Island, i., B.C., Can.	142-43	52°23'N	128°41'W
Prichard, Al., U.S. (prĭt'chârd)	134-35	30°44'N	88°05'W
Prienai, Lith. (prē-ĕn'ī)	208-09	54°38'N	23°57'E
Prieska, S. Afr. (prē-ĕs'kȧ)	286-87	29°40'S	22°44'E
Prijedor, Bos. (prē'yĕ-dôr)	216-17	44°59'N	16°42'E
Prijepolje, Serb. (prĕ'yĕ-pô'lyĕ)	216-17	43°24'N	19°39'E
Prikaspiyskaya Nizmennost', pl., *see* Caspian Depression	202-03	48°0'N	52°00'E
Prilep, Mac. (prē'lĕp)	216-17	41°21'N	21°34'E
Primorsk, Russia (prē-môrsk')	208-09	60°22'N	28°38'E
Primrose Lake, lk., Can.	142-43	54°55'N	109°45'W
Prince Albert, Sk., Can. (prĭns ăl'bērt)	144-45	53°13'N	105°45'W
Prince Albert National Park, n.p., Sk., Can. (prĭns ăl'bērt năsh'ŭn-ăl pärk)	144-45	54°0'N	106°25'W
Prince Albert Sound, b., N.T., Can. (prĭns ăl'bērt sound)	140-41	70°27'N	114°50'W
Prince Charles Island, i., Nu., Can. (prĭns chärlz ī'lănd)	140-41	67°47'N	76°06'W
Prince Edward Island, state, Can. (prĭns ĕd'wērd ī'lănd)	138-39	46°20'N	63°20'W
Prince Edward Island, i., P.E., Can.	140-41	46°20'N	63°20'W
Prince Edward Island National Park, n.p., P.E., Can.	148-49	46°30'N	63°25'W
Prince Edward Islands, is., S. Afr. (prĭns ĕd'wērd ī'lăndz)	313	46°45'S	37°49'E
Prince George, B.C., Can. (prĭns jôrj)	142-43	53°54'N	122°46'W
Prince of Wales, Cape, c., Ak., U.S. (kāp prĭns ŭv wālz)	136	65°40'N	168°07'W
Prince of Wales Island, i., Austl. (prĭns ŭv wālz ī'lănd)	302	10°40'S	142°10'E
Prince of Wales Island, i., Nu., Can. (prĭns ŭv wālz ī'lănd)	314	72°40'N	99°00'W

n-sing; ŋ-baŋk; N-nasalized n; nŏd; cŏmmit; ōld; ōbey; ôrder; oi-boil; fōōd; ȯ-as oo in foot; ou-out; s-soft; sh-dish; th-thin; pūre; ûnite; ûrn; stŭd; circŭs; ü-as in French tu; '-indeterminate vowel.

398 Prince of Wales Island – Qaraton PRONOUNCING INDEX

Feature (Pronunciation)	Page	Lat.	Long.
Prince of Wales Island, i., Ak., U.S. (prĭns ŭv wālz ī´lănd)	136	55°47′N	132°50′W
Prince Rupert, B.C., Can. (prĭns roo´pĕrt)	142-43	54°19′N	130°17′W
Princess Royal Island, i., B.C., Can.	142-43	52°57′N	128°49′W
Princeton, B.C., Can. (prĭns´tŭn)	142-43	49°27′N	120°31′W
Princeton, Il., U.S. (prĭns´tŭn)	126-27	41°22′N	89°28′W
Princeton, In., U.S. (prĭns´tŭn)	126-27	38°21′N	87°35′W
Princeton, Ky., U.S. (prĭns´tŭn)	134-35	37°07′N	87°53′W
Princeton, Mn., U.S. (prĭns´tŭn)	124-25	45°34′N	93°35′W
Princeton, Mo., U.S. (prĭns´tŭn)	130-31	40°24′N	93°35′W
Princeton, N.J., U.S. (prĭns´tŭn)	126-27	40°22′N	74°39′W
Princeton, W.V., U.S. (prĭns´tŭn)	134-35	37°22′N	81°06′W
Prince William Sound, strt., Ak., U.S. (prĭns wĭl´yăm sound)	136	60°42′N	147°07′W
Príncipe, i., S. Tom./P. (prēn´sē-pě)	282-83	1°37′N	7°25′E
Principe Channel, strt., B.C., Can. (prĭn´sĭ-pē chăn´ĕl)	142-43	53°28′N	130°00′W
Príncipe da Beira, Braz.	180-81	12°25′S	64°25′W
Prineville, Or., U.S. (prīn´vĭl)	122-23	44°18′N	120°51′W
Prinzapolka, stm., Nic. (prēn-zä-pōl´kä)	161	13°24′N	83°34′W
Priozërsk, Russia (prĭ-ô´zĕrsk)	208-09	61°02′N	30°09′E
Pripet Marshes, reg., Eur.	210-11	52°0′N	27°00′E
Priština, nat. cap., Kos. (prēsh´tĭ-nä)	216-17	42°40′N	21°10′E
Pritzwalk, Ger. (prēts´välk)	210-11	53°09′N	12°10′E
Privas, Fr. (prē-väs´)	212-13	44°44′N	4°37′E
Privolzhskaya Vozvyshennost', plat., Russia	202-03	52°0′N	46°00′E
Privolzhskiy, Russia	202-03	51°24′N	46°02′E
Priyutovo, Russia	202-03	53°53′N	53°56′E
Prizren, Kos. (prē´zrĕn)	216-17	42°13′N	20°45′E
Probolinggo, Indon.	268-69	7°45′S	113°13′E
Proctor, Mn., U.S. (prŏk´tẽr)	124-25	46°44′N	92°14′W
Proddatūr, India	256	14°45′N	78°33′E
Progreso, Mex. (prô-grā´sō)	160	21°16′N	89°39′W
Prokopyevsk, Russia	244	53°54′N	86°44′E
Prokuplje, Serb. (prô´kòp´l-yě)	216-17	43°14′N	21°36′E
Prome, Mya.	266-67	18°49′N	95°13′E
Pronja, stm., Bela. (prô´nyä)	208-09	53°27′N	31°01′E
Propriá, Braz.	188d	10°13′S	36°50′W
Prosser, Wa., U.S. (prŏs´ẽr)	122-23	46°12′N	119°46′W
Prostějov, Czech Rep. (prôs´tyě-yôf)	210-11	49°29′N	17°07′E
Protoka, stm., Russia (prôt´ô-kȧ)	218-19	45°44′N	37°47′E
Providence, R.I., U.S. (prŏv´ĭ-dĕns)	126-27	41°50′N	71°25′W
Providence, Atoll de, at., Sey.	286-87	9°14′S	51°03′E
Providencia, Isla de, i., Col.	178-79	13°21′N	81°22′W
Providenciales, i., T./C. Is.	154-55	21°47′N	72°17′W
Provideniya, Russia (prô-vĭ-dă´nĭ-yȧ)	136	64°23′N	187°18′W
Provo, Ut., U.S. (prō´vō)	128-29	40°13′N	111°38′W
Prudhoe Bay, b., Ak., U.S.	136	70°21′N	148°22′W
Prudnik, Pol. (prŏd´nĭk)	210-11	50°19′N	17°35′E
Pruszków, Pol. (prosh´kòf)	210-11	52°10′N	20°49′E
Prut, stm., Eur. (proot)	202-03	45°28′N	28°13′E
Prydz Bay, b., Ant.	313	69°0′S	76°00′E
Pryluky, Ukr.	218-19	50°36′N	32°23′E
Pryor, Ok., U.S. (prī´ẽr)	130-31	36°19′N	95°19′W
Przemyśl, Pol. (pzhě´mĭsh´l)	210-11	49°47′N	22°47′E
Przhevalsk, Kyrg. (p'r-zhĭ-välsk´) *see* Karakol	244	42°29′N	78°23′E
Pskov, Russia (pskôf)	208-09	57°49′N	28°22′E
Pskov, Lake, lk., Eur. (lāk pskôf)	208-09	58°0′N	28°00′E
Pskovskoye Ozero, lk., Eur. (p'skôv´skô´yě ôzě-rô) *see* Pskov, Lake	208-09	58°0′N	28°00′E
Ptuj, Slvn. (ptoo´ē)	216-17	46°26′N	15°52′E
Pucallpa, Peru	184	8°23′S	74°32′W
Pucheng, China (poo-chŭŋ)	258-59	27°55′N	118°32′E
Pucheng, China (poo-chŭŋ)	258-59	34°58′N	109°35′E
Puck, Pol. (pòtsk)	210-11	54°43′N	18°24′E
Pudozh, Russia (poo´dôzh)	202-03	61°48′N	36°34′E
Puducherry, state, India	256	11°56′N	79°50′E
Pudukkottai, India	256	10°23′N	78°49′E
Puebla, state, Mex. (pwā´blä)	158-59	18°50′N	98°00′W
Puebla de Zaragoza, Mex.	158-59	19°03′N	98°12′W
Pueblo, Co., U.S. (pwā´blō)	130-31	38°16′N	104°38′W
Puente Genil, Spain (pwĕn´tå-hå-nēl´)	214-15	37°24′N	4°47′W
Puerto Aisén, Chile (pwĕ´r-tō ä´y-sĕ´n)	185	45°15′S	72°15′W
Puerto Ángel, Mex. (pwĕ´r-tō äŋ´hål)	158-59	15°40′N	96°29′W
Puerto Armuelles, Pan. (pwe´r-tō är-moo-ā´lyās)	162	8°17′N	82°52′W
Puerto Asís, Col.	178-79	0°31′N	76°31′W
Puerto Ayacucho, Ven.	178-79	5°40′N	67°38′W
Puerto Baquerizo Moreno, Ec.	184a	0°54′S	89°36′W
Puerto Barrios, Guat. (pwĕ´r-tô bär´rê-ôs)	160	15°43′N	88°35′W
Puerto Bermúdez, Peru (pwĕ´r-tō bĕr-moo´dāz)	184	10°20′S	74°54′W
Puerto Berrío, Col. (pwĕ´r-tō bĕr-rē´ō)	188c	6°28′N	74°26′W
Puerto Cabello, Ven. (pwĕ´r-tō kä-bĕl´yō)	188b	10°28′N	68°01′W
Puerto Cabezas, Nic. (pwĕ´r-tô kä-bā´zäs)	161	14°01′N	83°23′W
Puerto Carreño, Col.	178-79	6°11′N	67°30′W
Puerto Chicama, Peru (pwĕ´r-tō chē-kä´mä)	184	7°42′S	79°25′W
Puerto Cortés, Hond. (pwĕ´r-tô kôr-tās´)	161	15°51′N	87°57′W
Puerto Cumarebo, Ven. (pwĕ´r-tō koo-mä-rĕ´bô)	178-79	11°29′N	69°21′W
Puerto de la Cruz, Spain	215d	28°23′N	16°33′W
Puerto Deseado, Arg. (pwĕ´r-tō dā-så-ä´dhō)	185	47°44′S	65°54′W
Puerto Juárez, Mex.	160	21°10′N	86°49′W
Puerto la Cruz, Ven. (pwĕ´r-tō lä kroo´z)	188b	10°13′N	64°38′W
Puerto Leguízamo, Col.	178-79	0°11′S	74°46′W
Puerto Libertad, Mex.	156-57	29°55′N	112°41′W
Puerto Limón, C.R.	161	9°59′N	83°02′W
Puertollano, Spain (pwĕ-tôl-yä´nō)	214-15	38°41′N	4°06′W
Puerto Madryn, Arg. (pwĕ´r-tō mä-drĕn´)	185	42°46′S	65°03′W
Puerto Maldonado, Peru (pwĕ´r-tō mäl-dō-nä´dô)	184	12°36′S	69°12′W
Puerto Montt, Chile (pwĕ´r-tō mô´nt)	185	41°28′S	72°57′W
Puerto Morazán, Nic.	161	12°50′N	87°11′W
Puerto Natales, Chile (pwĕ´r-tō nä-tá´lĕs)	185	51°42′S	72°29′W
Puerto Padre, Cuba (pwĕ´r-tō pä´drā)	154-55	21°12′N	76°36′W
Puerto Peñasco, Mex. (pwĕ´r-tô pĕn-yä´s-kô)	156-57	31°19′N	113°32′W
Puerto Pinasco, Para. (pwĕ´r-tō pē-ná´s-kô)	182-83	22°37′S	57°49′W
Puerto Pirámides, Arg.	185	42°34′S	64°15′W
Puerto Píritu, Ven. (pwĕ´r-tō pē´rē-too)	188b	10°02′N	65°02′W
Puerto Plata, Dom. Rep. (pwĕ´r-tō plä´tä)	154-55	19°45′N	70°39′W
Puerto Princesa, Phil.	270	9°44′N	118°45′E
Puerto Rico, Bol.	180-81	11°06′S	67°32′W
Puerto Rico, dep., N.A. (pwĕr´tô rē´kō)	152-53	18°15′N	66°30′W
Puerto Rico, i., P.R. (pwĕr´tō rē´kō)	154a	18°15′N	66°30′W
Puerto Salgar, Col. (pwĕ´r-tō säl-gär´)	188c	5°28′N	74°39′W
Puerto San José, Guat.	160	13°56′N	90°49′W
Puerto San Julián, Arg.	185	49°18′S	67°43′W
Puerto Santa Cruz, Arg. (pwĕ´r-tō sän´tä krooz´)	185	50°01′S	68°34′W
Puerto Sastre, Para.	182-83	22°02′S	58°01′W
Puerto Suárez, Bol. (pwĕ´r-tō swä´râz)	182-83	18°57′S	57°51′W
Puerto Tejada, Col. (pwĕ´r-tō tĕ-kä´dä)	188c	3°14′N	76°25′W
Puerto Vallarta, Mex. (pwĕ´r-tō väl-yär´tä)	158-59	20°37′N	105°14′W
Puerto Varas, Chile (pwĕ´r-tō vä´räs)	185	41°20′S	72°58′W
Puerto Villamil, Ec.	184a	0°56′S	91°01′W
Puerto Wilches, Col. (pwĕ´r-tō vĕl´c-hĕs)	178-79	7°20′N	73°54′W
Pueyrredón, Lago, lk., S.A.	185	47°21′S	71°56′W
Puget Sound, b. Wa., U.S.	122-23	47°49′N	122°27′W
Pugachev, Russia (poo´gȧ-chyôf)	202-03	52°02′N	48°49′E
Puhi-waero, c., N.Z. *see* South West Cape	303	47°17′S	167°28′E
Pukch'ŏng-ŭp, Kor., N.	263	40°14′N	128°19′E
Pukou, China	258-59	32°06′N	118°43′E
Pula, Cro. (poo´lä)	216-17	44°52′N	13°51′E
Pulacayo, Bol. (poo-lä-kä´yō)	182-83	20°23′S	66°42′W
Pulaski, N.Y., U.S. (pů-lăs´kĭ)	126-27	43°34′N	76°07′W
Pulaski, Tn., U.S. (pů-lăs´kĭ)	134-35	35°12′N	87°02′W
Pulaski, Va., U.S. (pů-lăs´kĭ)	134-35	37°03′N	80°47′W
Puławy, Pol. (pò-wä´vě)	210-11	51°25′N	21°59′E
Pullman, Wa., U.S. (pòl´măn)	122-23	46°44′N	117°10′W
Pulo Anna, i., Palau	304-05	4°40′N	131°58′E
Pulog, Mount, mtn., Phil. (mount poo´lôg)	270	16°36′N	120°54′E
Puma Yumco, lk., China (poo-mä yoom-tswo)	254-55	28°33′N	90°24′E
Puná, Isla, i., Ec.	184	2°47′S	80°08′W
Punakha, Bhu. (poo-nŭk´ŭ)	254-55	27°37′N	89°52′E
Punata, Bol. (poo-nä´tä)	182-83	17°33′S	65°50′W
Pune, India	256	18°32′N	73°52′E
Punia, D.R.C.	284-85	1°28′S	26°27′E
Punjab, state, India (pŭn´jäb´)	254-55	31°0′N	75°30′E
Puno, Peru (poo´nô)	184	15°51′S	70°02′W
Punta Alta, Arg.	187	38°53′S	62°04′W
Punta Arenas, Chile (poo´n-tä-rĕ´näs)	185	53°09′S	70°55′W
Punta de Piedras, Ven. (poo´n-tä dĕ pyĕ´dräs)	188b	10°54′N	64°06′W
Punta Gorda, Belize (pòn´tä gôr´dä)	160	16°06′N	88°48′W
Punta Gorda, Fl., U.S. (pŭn´tȧ gôr´dȧ)	135a	26°56′N	82°03′W
Puntarenas, C.R. (pònt-ä-rā´näs)	161	9°58′N	84°50′W
Punto Fijo, Ven. (poo´n-tô fē´κô)	178-79	11°43′N	70°12′W
Punxsutawney, Pa., U.S. (pŭnk-sŭ-tô´nē)	126-27	40°56′N	78°58′W
Puqi, China	258-59	29°43′N	113°53′E
Puquio, Peru (poo´kyô)	184	14°42′S	74°09′W
Pur, stm., Russia	236-37	67°21′N	77°55′E
Purcell, Ok., U.S. (pûr-sĕl´)	130-31	35°02′N	97°22′W
Puri, India (pò´rē)	254-55	19°48′N	85°51′E
Purificación, Col. (poo-rē-fē-kä-syōn´)	188c	3°51′N	74°55′W
Purificación, Mex. (poo-rē-fē-kä-syô´n)	158-59	19°43′N	104°36′W
Pūrnia, India	254-55	25°47′N	87°29′E
Pursat, Camb. *see* Poŭthĭsăt	266-67	12°32′N	103°56′E
Purus, stm., S.A.	180-81	3°41′S	61°28′W
Purús, stm., S.A. (poo-roo´s)	180-81	3°41′S	61°28′W
Purwokerto, Indon.	268-69	7°25′S	109°14′E
Pusan, Kor., S. *see* Busan	263	35°05′N	129°03′E
Pushkin, Russia (pòsh´kĭn)	208-09	59°43′N	30°26′E
Pustoshka, Russia (pûs-tôsh´kȧ)	208-09	56°20′N	29°22′E
Putaendo, Chile (poo-tä-ĕn-dô)	188e	32°37′S	70°44′W
Putao, Mya.	258-59	27°21′N	97°24′E
Putian, China (poo-tĭĕn)	243a	25°26′N	119°00′E
Puting, Tanjung, c., Indon.	268-69	3°32′S	111°49′E
Putla de Guerrero, Mex. (poo´tlä-dĕ-gĕr-rĕ´rō)	158-59	17°00′N	97°54′W
Putnam, Ct., U.S. (pŭt´năm)	126-27	41°55′N	71°54′W
Putorana, Gory, plat., Russia	236-37	69°0′N	95°00′E
Putrajaya, nat. cap., Malay.	266-67	2°56′N	101°43′E
Puttalam, Sri L.	256	8°01′N	79°51′E
Putumayo, stm., S.A. (pò-too-mä´yō)	184	3°07′S	67°56′W
Puyallup, Wa., U.S. (pū-ăl´ŭp)	122-23	47°11′N	122°17′W
Puyang, China (poo-yäŋ)	260-61	35°42′N	115°00′E
Puyo, Ec.	184	1°29′S	77°59′W
Pweto, D.R.C. (pwā´tō)	284-85	8°28′S	28°54′E
Pyakupur, stm., Russia	236-37	64°56′N	77°44′E
Pyandzh, stm., Asia *see* Panj	252-53	37°00′N	68°16′E
Pyasina, stm., Russia (pyä-sē´nȧ)	236-37	73°52′N	87°09′E
Pyasino, Ozero, lk., Russia	236-37	69°45′N	87°45′E
Pyatigorsk, Russia (pyȧ-tē-gôrsk´)	245	44°04′N	43°04′E
Pyè, Mya. *see* Prome	266-67	18°49′N	95°13′E
Pyinmana, Mya. (pyĕn-mä´nŭ)	266-67	19°44′N	96°13′E
P'yŏngyang, nat. cap., Kor., N. (pyŭng´gäng´)	263	39°01′N	125°44′E
Pyramid Lake, lk., Nv., U.S. (pĭ´rȧ-mĭd lāk)	128-29	40°01′N	119°35′W
Pyramid Lake Indian Reservation, ind. res., Nv., U.S. (pĭ´rȧ-mĭd lāk ĭn´dĭ-ăn rĕ-sĕr-vā´shĕn)	128-29	40°13′N	119°36′W
Pyrenees, mts., Eur. (pĭr-e-nēz´)	212-13	42°40′N	1°00′E
Pýrgos, Grc.	216-17	37°40′N	21°27′E
Pyritz, Pol. *see* Pyrzyce	210-11	53°09′N	14°53′E
Pyrzyce, Pol. (pĕzhĭ´tsĕ)	210-11	53°09′N	14°53′E

Q

Feature (Pronunciation)	Page	Lat.	Long.
Qaanaaq, Green.	288	77°29′N	69°20′W
Qā'en, Iran	252-53	33°44′N	59°10′E
Qaidam, stm., China	260-61	36°52′N	95°57′E
Qaidam Pendi, bas., China	240-41	37°0′N	95°00′E
Qal'at Bīshah, Sau. Ar.	288	19°60′N	42°36′E
Qalāt, Afg.	252-53	32°07′N	66°54′E
Qamani'tuaq, Nu., Can. *see* Baker Lake	138-39	64°18′N	95°55′W
Qamar, Ghubbat al-, b., Yemen	238-39	16°0′N	52°30′E
Qamdo, China (chyäm-dwō)	258-59	31°10′N	97°09′E
Qamea, i., Fiji	306f	16°46′S	179°46′W
Qandahār, Afg. *see* Kandahār	252-53	31°37′N	65°43′E
Qandala, Som.	284-85	11°28′N	49°52′E
Qapshaghay, Kaz.	244	43°52′N	77°04′E
Qapshaghay bögeni, res., Kaz.	244	43°49′N	77°42′E
Qaqortoq, Green.	310-11	60°44′N	46°02′W
Qaraghandy, Kaz.	244	49°53′N	73°10′E
Qarataū, Kaz.	244	43°10′N	70°28′E
Qarataū zhotasy, mts., Kaz. *see* Karatau Range	244	43°36′N	68°52′E
Qaraton, Kaz.	202-03	46°26′N	53°31′E

ăt; finăl; rāte; senåte; ärm; åsk; sofȧ; fâre; ch-choose; dh-as th in other; bē; ēvent; bĕt; recĕnt; cratẽr; g-gō; gh-guttural g; bĭt; ĭ-short neutral; rīde; κ-guttural k as ch in German ich;

Feature (Pronunciation)	Page	Lat.	Long.
Qarazhal, Kaz.	**244**	48°01′N	70°49′E
Qarqan, stm., China	**240-41**	39°26′N	88°22′E
Qarshi, Uzb.	**252-53**	38°52′N	65°48′E
Qârûn, Birket, lk., Egypt	**291b**	29°28′N	30°39′E
Qāsh, Nahr al-, stm., Afr. *see* Gash	**288**	16°45′N	35°54′E
Qatar, nation, Asia (kä′tär)	**224-25**	25°0′N	51°10′E
Qattâra, Munkhafad el-, depr., Egypt			
see Qattara Depression	**288**	30°0′N	27°30′E
Qattara Depression, depr., Egypt			
(kä-tä′rȧ dĭ-prĕ′shŭn)	**288**	30°0′N	27°30′E
Qazaqstan, nation, Asia			
see Kazakhstan	**224-25**	47°0′N	76°00′E
Qazaqtyng usaqshoqylyghy, hills, Kaz.			
see Kazakh Hills	**244**	49°0′N	72°00′E
Qazımämmäd, Azer.	**245**	40°02′N	48°56′E
Qazvīn, Iran	**252-53**	36°16′N	49°58′E
Qena, Egypt	**291b**	26°11′N	32°43′E
Qeqertarsuaq, Green.	**310-11**	69°15′N	53°33′W
Qeshm, Jazīreh-ye, i., Iran	**250-51**	26°45′N	55°45′E
Qezel Owzan, stm., Iran	**248-49**	36°47′N	49°09′E
Qianyang, China	**258-59**	27°11′N	110°02′E
Qiemo, China	**238-39**	38°10′N	85°30′E
Qijiang, China (chyē-jyäŋ)	**258-59**	29°02′N	106°39′E
Qilian Shan, mts., China			
(chyē-lĭĕn shän)	**260-61**	39°06′N	98°40′E
Qing'an, China (chyĭŋ-än)	**260-61**	46°52′N	127°30′E
Qingdao, China (chyĭn-dou)	**260-61**	36°05′N	120°20′E
Qinghai, state, China (chyĭŋ-hī)	**260-61**	36°0′N	96°00′E
Qinghai Hu, lk., China (chyĭŋ-hī hōō)	**260-61**	36°48′N	100°06′E
Qingjiang, China (chyĭŋ-jyäŋ)	**258-59**	33°36′N	119°01′E
Qingshui, stm., China	**258-59**	27°08′N	109°37′E
Qingshui, stm., China	**260-61**	37°28′N	105°32′E
Qingtang, China	**258-59**	24°12′N	113°51′E
Qingyang, China (chyĭŋ-yäŋ)	**258-59**	30°38′N	117°51′E
Qingyang, China (chyĭŋ-yäŋ)	**260-61**	36°01′N	107°52′E
Qingyuan, China (chyĭŋ-yōän)			
see Baoding	**260-61**	38°51′N	115°29′E
Qingyuan, China (chyĭŋ-yōän)	**258-59**	23°42′N	113°02′E
Qingyuan, China (chyĭŋ-yōän)	**258-59**	27°37′N	119°06′E
Qingyuan, China (chyĭŋ-yōän)	**263**	42°07′N	124°50′E
Qingyuan, China (chyĭŋ-yōän)	**258-59**	24°35′N	108°45′E
Qing Zang Gaoyuan, plat., China			
see Tibet, Plateau of	**240-41**	33°0′N	92°00′E
Qinhuangdao, China (chyĭn-huaŋ-dou)	**260-61**	39°56′N	119°36′E
Qin Ling, mts., China (chyĭn lĭŋ)	**258-59**	34°0′N	108°00′E
Qinzhou, China (chyĭn-jō)	**258-59**	21°58′N	108°37′E
Qionghai, China (chyōŋ-hī)	**258-59**	19°16′N	110°28′E
Qionglai, China	**258-59**	30°25′N	103°28′E
Qiongzhong, China	**258-59**	19°02′N	109°48′E
Qiongzhou Haixia, strt., China	**258-59**	20°10′N	110°15′E
Qiqihar, China (chyē-chyē-här)	**260-61**	47°20′N	123°58′E
Qitai, China (chyē-tī)	**240-41**	44°01′N	89°35′E
Qixian, China (chyē-shyĕn)	**258-59**	34°33′N	114°47′E
Qiyang, China (chyē-yäŋ)	**258-59**	26°29′N	111°43′E
Qızılağac körfäzi, b., Azer.	**245**	39°05′N	49°01′E
Qizilqum, des., Asia	**244**	42°0′N	64°00′E
Qo′qon, Uzb.	**252-53**	40°32′N	70°56′E
Qogir Feng, mtn., Asia *see* K2	**252-53**	35°53′N	76°30′E
Qom, Iran	**252-53**	34°39′N	50°53′E
Qomolangma Feng, mtn., Asia			
see Everest, Mount	**254-55**	27°59′N	86°56′E
Qomsheh, Iran	**252-53**	32°00′N	51°52′E
Qondūz, Afg. *see* Kondoz	**252-53**	36°44′N	68°51′E
Qorako'l, Uzb.	**252-53**	39°32′N	63°55′E
Qosshaghyl, Kaz.	**202-03**	46°51′N	53°48′E
Qostanay, Kaz.	**244**	53°12′N	63°37′E
Quadra Island, i., B.C., Can.	**142-43**	50°12′N	125°16′W
Quakertown, Pa., U.S. (kwā′kẽr-toun)	**126-27**	40°27′N	75°21′W
Quanah, Tx., U.S. (kwä′nȧ)	**130-31**	34°18′N	99°45′W
Quang Ngai, Viet. (kwäng n'gä′ē)	**266-67**	15°07′N	108°47′E
Quan Long, Viet. *see* Ca Mau	**266-67**	9°11′N	105°09′E
Quanzhou, China (chyuän-jō)	**243a**	24°55′N	118°35′E
Qu'Appelle, stm., Can.	**144-45**	50°26′N	101°20′W
Quartu Sant'Elena, Italy			
(kwär-tōō′ sänt a′lå-nä)	**216-17**	39°15′N	9°11′E
Quatsino Sound, strt., B.C., Can.			
(kwŏt-sē′nō sound)	**142-43**	50°26′N	127°59′W
Quba, Azer. (kōō′bä)	**245**	41°22′N	48°31′E
Qūchān, Iran	**252-53**	37°06′N	58°31′E
Queanbeyan, Austl.	**301**	35°21′S	149°14′E
Québec, Qc., Can.			
(kwē-bĕk′) (kā-bĕk′)	**146-47**	46°49′N	71°13′W
Québec, state, Can.			
(kwē-bĕk′) (kē-bĕk′)	**138-39**	52°0′N	72°00′W
Quedlinburg, Ger. (kvĕd′lēn-bōōrgh)	**210-11**	51°47′N	11°09′E
Queen Charlotte, B.C., Can.	**142-43**	53°16′N	132°05′W
Queen Charlotte Mountains, mts., B.C.,			
Can. (kwēn shär′lŏt moun′tĭnz)	**142-43**	53°0′N	132°00′W
Queen Charlotte Sound, strt., B.C., Can.			
(kwēn shär′lŏt sound)	**142-43**	51°30′N	129°30′W
Queen Charlotte Strait, strt., B.C., Can.			
(kwēn shär′lŏt strāt)	**142-43**	50°50′N	127°25′W
Queen Elizabeth Islands, is., Can.	**150**	77°00′S	100°00′W
Queen Fabiola Mountains, mts., Ant.	**313**	71°30′S	35°40′E
Queen Maud Gulf, b., Nu., Can.			
(kwēn mäd gŭlf)	**140-41**	68°25′N	102°30′W
Queen Maud Land, reg., Ant.			
(kwēn mäd lănd)	**313**	74°59′S	15°51′E
Queensland, state, Austl. (kwēnz′lănd)	**302**	22°0′S	145°00′E
Queenstown, Austl. (kwēnz′toun)	**301**	42°04′S	145°33′E
Queenstown, S. Afr.	**286-87**	31°54′S	26°53′E
Quelelevu, i., Fiji	**306f**	16°05′S	179°09′W
Quelimane, Moz. (kā-lē-mä′nĕ)	**286-87**	17°53′S	36°53′E
Quelpart Island, i., Kor., S.			
see Jejudo	**260-61**	33°22′N	126°30′E
Quemado de Güines, Cuba			
(kā-mä′dhä-dĕ-gwē′nĕs)	**154-55**	22°48′N	80°15′W
Quemoy, i., Tai. *see* Chinmen Tao	**243a**	24°27′N	118°23′E
Querétaro, Mex. (kå-rā′tä-rō)	**158-59**	20°35′N	100°23′W
Querétaro, state, Mex. (kå-rā′tä-rō)	**158-59**	21°0′N	99°55′W
Quesnel, B.C., Can. (kā-nĕl′)	**142-43**	52°58′N	122°29′W
Quesnel, stm., B.C., Can. (kā-nĕl′)	**142-43**	52°58′N	122°30′W
Quesnel Lake, lk., B.C., Can.			
(kā-nĕl′ lāk)	**142-43**	52°32′N	121°05′W
Quetta, Pak. (kwĕt′ä)	**252-53**	30°13′N	67°01′E
Quetzaltenango, Guat.	**160**	14°50′N	91°31′W
Quevedo, Ec.	**184**	1°01′S	79°27′W
Quezon City, Phil. (kā-zōn sĭ′tē)	**270**	14°38′N	121°03′E
Qufu, China (chyōō-fōō)	**260-61**	35°36′N	117°02′E
Quibdó, Col. (kēb′dō)	**188c**	5°42′N	76°39′W
Quila, Mex.	**158-59**	24°25′N	107°13′W
Quillacollo, Bol.	**182-83**	17°26′S	66°17′W
Quillota, Chile (kēl-yō′tä)	**188e**	32°51′S	71°14′W
Quilon, India (kwē-lōn′) *see* Kollam	**256**	8°53′N	76°35′E
Quilpie, Austl. (kwĭl′pē)	**301**	26°37′S	144°16′E
Quimbaya, Col. (kēm-bä′yä)	**188c**	4°30′N	75°40′W
Quimilí, Arg.	**187**	27°38′S	62°25′W
Quimper, Fr. (kăɴ-pĕr′)	**212-13**	47°60′N	4°06′W
Quince Mil, Peru	**184**	13°15′S	70°37′W
Quincy, Fl., U.S. (kwĭn′sē)	**134-35**	30°35′N	84°35′W
Quincy, Il., U.S. (kwĭn′sē)	**130-31**	39°56′N	91°24′W
Quincy, Ma., U.S. (kwĭn′sē)	**126-27**	42°15′N	71°02′W
Quincy, Mi., U.S. (kwĭn′sē)	**126-27**	41°57′N	84°53′W
Quincy, Wa., U.S. (kwĭn′sē)	**122-23**	47°14′N	119°51′W
Quines, Arg.	**182-83**	32°14′S	65°47′W
Quintana Roo, state, Mex.			
(kēn-tä-nä rō′ō)	**160**	19°40′N	88°30′W
Quintero, Chile (kēn-tĕ′rō)	**188e**	32°47′S	71°32′W
Quirihue, Chile	**185**	36°17′S	72°33′W
Quirimba, Ilha, i., Moz.	**286-87**	12°20′S	40°36′E
Quiroga, Mex. (kē-rō′gä)	**158-59**	19°40′N	101°32′W
Quitman, Ga., U.S. (kwĭt′măn)	**134-35**	30°47′N	83°34′W
Quitman, Ms., U.S. (kwĭt′măn)	**134-35**	32°02′N	88°44′W
Quito, nat. cap., Ec. (kē′tō)	**184**	0°12′S	78°30′W
Quixadá, Braz.	**180-81**	4°58′S	39°01′W
Qujing, China	**258-59**	25°35′N	103°50′E
Qulyndy Zhazyghy, pl., Asia	**244**	53°0′N	79°00′E
Qum, Iran *see* Qom	**252-53**	34°39′N	50°53′E
Qumarlêb, China	**258-59**	34°30′N	95°21′E
Qŭnghirot, Uzb.	**244**	43°03′N	58°51′E
Quorn, Austl. (kwôrn)	**301**	32°21′S	138°03′E
Qŭrghonteppa, Taj.	**252-53**	37°50′N	68°47′E
Quseir, Egypt	**288**	26°07′N	34°17′E
Quxian, China (chyōō-shyĕn)	**258-59**	30°51′N	106°58′E
Quy Nhon, Viet.	**266-67**	13°46′N	109°15′E
Quzhou, China (chyoō-jō)	**258-59**	28°57′N	118°52′E
Qyrghyz zhotasy, mts., Asia			
see Kirgiz Range	**244**	42°29′N	73°50′E
Qyzylorda, Kaz. (kzĕl-ôr′dȧ)	**244**	44°51′N	65°30′E
Qyzylqum, des., Asia	**244**	42°0′N	64°00′E

R

Feature (Pronunciation)	Page	Lat.	Long.
Raab, Hung. *see* Győr	**210-11**	47°41′N	17°39′E
Raahe, Fin. (rä′ĕ)	**200-01**	64°41′N	24°31′E
Raba, Indon.	**268-69**	8°29′S	118°45′E
Rabat, nat. cap., Mor. (rȧ-bät′)	**292a**	34°01′N	6°50′W
Rabaul, Pap. N. Gui. (rä′boul)	**302**	4°12′S	152°11′E
Rabi, i., Fiji	**306f**	16°30′S	179°58′W
Rābigh, Sau. Ar.	**288**	22°48′N	39°02′E
Race, Cape, c., Nf./L., Can. (kāp rās)	**148-49**	46°40′N	53°06′W
Rach Gia, Viet.	**266-67**	10°01′N	105°06′E
Racine, Wi., U.S. (rȧ-sēn′)	**126-27**	42°43′N	87°47′W
Radford, Va., U.S. (răd′fẽrd)	**134-35**	37°07′N	80°35′W
Radom, Pol. (rä′dôm)	**210-11**	51°24′N	21°09′E
Radomsko, Pol. (rä-dôm′skô)	**210-11**	51°04′N	19°27′E
Radomyshl', Ukr. (rä-dô-mēsh′'l)	**218-19**	50°30′N	29°15′E
Radviliškis, Lith. (räd′vē-lēsh′kēs)	**208-09**	55°49′N	23°33′E
Radzyń Podlaski, Pol.			
(räd′zēn-y′ pŭd-lä′skĭ)	**210-11**	51°47′N	22°36′E
Rāe Bareli, India	**254-55**	26°13′N	81°14′E
Raeford, N.C., U.S. (rā′fẽrd)	**134-35**	34°58′N	79°16′W
Rafaela, Arg. (rä-fä-å′lä)	**187**	31°15′S	61°29′W
Rafḩā', Sau. Ar.	**248-49**	29°41′N	43°28′E
Rafsanjān, Iran	**250-51**	30°24′N	55°59′E
Raga, S. Sudan	**284-85**	8°28′N	25°41′E
Ragay Gulf, b., Phil.	**270**	13°30′N	122°45′E
Ragged Island, i., Bah.	**154-55**	22°14′N	75°44′W
Ragged Island Range, is., Bah.	**154-55**	22°34′N	75°52′W
Ragusa, Italy (rä-gōō′sä)	**216-17**	36°55′N	14°44′E
Rāichūr, India (rä′ē-chōōr′)	**256**	16°12′N	77°21′E
Raigarh, India (rī′gŭr)	**254-55**	21°54′N	83°24′E
Rainbow Bridge National Monument,			
n.p., Ut., U.S. (rān′bō brĭj			
năsh′ŭn-ăl mŏn′ū-mĕnt)	**128-29**	37°06′N	110°57′W
Rainier, Mount, vol., Wa., U.S.			
(mount rā-nēr′)	**122-23**	46°51′N	121°45′W
Rainy Lake, lk., N.A. (rān′ē lāk)	**144-45**	48°39′N	93°17′W
Rainy River, On., Can. (rān′ē rĭv′ẽr)	**144-45**	48°44′N	94°34′W
Raipur, India	**254-55**	21°15′N	81°39′E
Raivavae, i., Fr. Poly.	**304-05**	23°52′S	147°40′W
Rājahmundry, India (räj-ŭ-mŭn′drē)	**256**	17°01′N	81°47′E
Rajang, stm., Malay.	**268-69**	2°10′N	111°21′E
Rājapālaiyam, India	**256**	9°27′N	77°34′E
Rājasthān, state, India (rä′jŭs-tän)	**254-55**	27°0′N	74°00′E
Rājgarh, India	**254-55**	23°55′N	76°55′E
Rājkot, India (räj′kōt)	**254-55**	22°18′N	70°48′E
Rājshāhi, Bngl.	**254-55**	24°22′N	88°36′E
Rakaposhi, mtn., Pak.	**252-53**	36°10′N	74°30′E
Rakata, Pulau, i., Indon. *see* Krakatoa	**268-69**	6°10′S	105°26′E
Rakiura, i., N.Z. *see* Stewart Island	**303**	47°0′S	167°50′E
Rakvere, Est. (rȧk′vĕ-rĕ)	**208-09**	59°21′N	26°22′E
Raleigh, N.C., U.S.	**134-35**	35°47′N	78°39′W
Rambouillet, Fr. (räɴ-bōō-yĕ′)	**212-13**	48°39′N	1°50′E
Rambutyo Island, i., Pap. N. Gui.	**302**	2°18′S	147°49′E
Rāmeswaram, India	**256**	9°17′N	79°19′E
Ramm, Jabal, mtn., Jord.	**248-49**	29°35′N	35°24′E
Ramos, Mex. (rä′mōs)	**158-59**	22°50′N	101°55′W
Ramos Arizpe, Mex. (rä′mōs ä-rēz′på)	**132-33**	25°32′N	100°57′W
Rāmpur, India (räm′pōōr)	**254-55**	20°40′N	79°01′E
Rampur Boalia, Bngl. *see* Rājshāhi	**254-55**	24°22′N	88°36′E
Ramree Island, i., Mya.			
(räm′rē′ ī′lănd)	**266-67**	19°06′N	93°48′E
Ramsey Lake, lk., On., Can.			
(răm′zē lāk)	**146-47**	47°15′N	82°16′W
Ramsgate, Eng., U.K. (rămz′′gāt)	**206-07**	51°20′N	1°24′E
Ramu, stm., Pap. N. Gui. (rä′mōō)	**302**	4°03′S	144°40′E
Ranau, Malay.	**268-69**	5°57′N	116°41′E
Rancagua, Chile (rän-kä′gwä)	**188e**	34°10′S	70°44′W
Rānchi, India	**254-55**	23°21′N	85°20′E
Randers, Den. (rän′ĕrs)	**208-09**	56°28′N	10°03′E
Randleman, N.C., U.S. (răn′d'l-măn)	**134-35**	35°49′N	79°49′W
Randolph, Ne., U.S. (răn′dôlf)	**124-25**	42°23′N	97°21′W
Random Island, i., Nf./L., Can.			
(răn′dŭm ī′lănd)	**148-49**	48°08′N	53°45′W
Rāngāmāti, Bngl.	**266-67**	22°42′N	92°08′E
Rangeley, Me., U.S. (rānj′lē rānj)	**126-27**	44°58′N	70°39′W
Ranger, Tx., U.S. (rān′jẽr rānj)	**130-31**	32°29′N	98°41′W
Rangoon, nat. cap., Mya. (ráŋ-gōōn′)			
see Yangon	**266-67**	16°47′N	96°12′E
Rangpur, Bngl. (rŭng′pōōr)	**254-55**	25°45′N	89°16′E
Rānīganj, India (rä-nē-gŭnj′)	**254-55**	23°36′N	87°07′E
Rankin Inlet, Nu., Can.	**138-39**	62°49′N	92°10′W
Rann of Kutch, reg., Asia			
see Kutch, Rann of	**254-55**	24°15′N	70°46′E
Ranongga Island, i., Sol. Is.	**306e**	8°05′S	156°34′E
Rantauprapat, Indon.	**266-67**	2°06′N	99°49′E
Rantekombola, Bulu, mtn., Indon.	**268-69**	3°24′S	120°02′E
Rantoul, Il., U.S. (răn-tōōl′)	**126-27**	40°18′N	88°09′W
Raoul Island, i., N.Z.	**304-05**	29°16′S	177°55′W
Rapa, i., Fr. Poly.	**304-05**	27°36′S	144°20′W
Rapallo, Italy (rä-päl′lō)	**216-17**	44°21′N	9°14′E
Rapel, stm., Chile (rä-pĕl′)	**188e**	33°54′S	71°50′W
Rapid City, S.D., U.S. (răp′ĭd sĭ′tē)	**124-25**	44°04′N	103°13′W
Rapla, Est. (räp′lȧ)	**208-09**	59°00′N	24°47′E

400 Rappahannock – Rimavská Sobota PRONOUNCING INDEX

Feature (Pronunciation)	Page	Lat.	Long.
Rappahannock, stm., Va., U.S. (răp'ă-hăn'ŭk)	126-27	37°35'N	76°18'W
Rapu Rapu Island, i., Phil.	270	13°12'N	124°09'E
Raraka, at., Fr. Poly.	304-05	16°10'S	144°54'W
Raroïa, i., Fr. Poly.	304-05	16°01'S	142°27'W
Rarotonga, i., Cook Is.	304-05	21°14'S	159°46'W
Ras Dashen Terara, mtn., Eth. (räs dä-shän') see Ras Dejen	288	13°16'N	38°24'E
Ras Dejen, mtn., Eth.	288	13°16'N	38°24'E
Raseiniai, Lith. (rä-syā'nyī)	208-09	55°23'N	23°08'E
Rashid, Egypt see Rosetta	291b	31°24'N	30°25'E
Rashin, Kor., N. see Najin	263	42°15'N	130°18'E
Rasht, Iran	248-49	37°17'N	49°35'E
Rasshua, Ostrov, i., Russia	236-37	47°45'N	153°01'E
Rasskazovo, Russia (räs-kä'sô-vô)	202-03	52°40'N	41°53'E
Rastatt, Ger. (rä-shtät)	210-11	48°51'N	8°12'E
Ratangarh, India (rŭ-tŭn'gŭr)	254-55	28°05'N	74°37'E
Rathenow, Ger. (rä'tĕ-nō)	210-11	52°36'N	12°20'E
Rat Islands, is., Ak., U.S. (răt ī'lăndz)	136a	52°0'N	178°00'E
Ratlām, India	254-55	23°20'N	75°02'E
Ratnāgiri, India	256	16°59'N	73°18'E
Raton, N.M., U.S. (rȧ-tōn')	130-31	36°55'N	104°26'W
Rättvik, Swe. (rĕt'vēk)	208-09	60°53'N	15°07'E
Rauch, Arg. (rä'ōōch)	187	36°47'S	59°06'W
Rauma, Fin. (rä'ô-mȧ)	208-09	61°08'N	21°30'E
Raurkela, India	254-55	22°13'N	84°52'E
Ravenna, Italy (rä-vĕn'nä)	216-17	44°25'N	12°12'E
Ravenna, Ne., U.S. (rȧ-vĕn'ȧ)	124-25	41°02'N	98°55'W
Ravensburg, Ger. (rä'vĕns-bōōrgh)	210-11	47°47'N	9°37'E
Ravensthorpe, Austl. (rā'vĕns-thôrp)	294-95	33°35'S	120°03'E
Ravenswood, W.V., U.S. (rā'vĕnz-wŏŏd)	126-27	38°57'N	81°46'W
Rāvi, stm., Asia	252-53	30°37'N	71°53'E
Rawaki, at., Kir.	304-05	3°43'S	170°43'W
Rāwalpindi, Pak. (rä-wŭl-pēn'dē)	252-53	33°36'N	73°04'E
Rawicz, Pol. (rä'vēch)	210-11	51°37'N	16°52'E
Rawlinna, Austl.	294-95	31°02'S	125°18'E
Rawlins, Wy., U.S. (rô'lĭnz)	122-23	41°47'N	107°14'W
Rawson, Arg.	185	43°19'S	65°06'W
Raxaul, India	254-55	26°59'N	84°50'E
Ray, Cape, c., Nf./L., Can. (kāp rā)	148-49	47°38'N	59°18'W
Raya, Bukit, mtn., Indon.	268-69	0°40'S	112°41'E
Raychikhinsk, Russia	260-61	49°48'N	129°24'E
Raymond, N.H., U.S. (rā'mŭnd)	126-27	43°02'N	71°11'W
Raymond, Wa., U.S. (rā'mŭnd)	122-23	46°41'N	123°44'W
Raymondville, Tx., U.S. (rā'mŭnd-vĭl)	132-33	26°29'N	97°46'W
Rayne, La., U.S. (rān)	132-33	30°14'N	92°16'W
Raytown, Mo., U.S. (rā'toun)	130-31	38°59'N	94°28'W
Rayville, La., U.S. (rā-vĭl)	134-35	32°29'N	91°46'W
Raz, Pointe du, c., Fr. (pwănt dü rä)	212-13	48°03'N	4°44'W
Razdol'noye, Russia (räz-dôl'nô-yĕ)	264	43°30'N	131°49'E
Razlog, Blg. (räz'lôk)	216-17	41°53'N	23°29'E
Razorback Mountain, mtn., B.C., Can. (rā'zēr-băk moun'tĭn)	142-43	51°35'N	124°42'W
Ré, Île de, i., Fr.	212-13	46°12'N	1°24'W
Reading, Eng., U.K. (rĕd'ĭng)	206-07	51°28'N	0°59'W
Reading, Pa., U.S.	126-27	40°20'N	75°56'W
Real, Cordillera, mts., S.A.	182-83	16°50'S	66°34'W
Realicó, Arg.	187	35°02'S	64°14'W
Rebun-tō, i., Japan (rē'bōōn tō)	264	45°23'N	141°02'E
Recife, Braz. (rå-sē'fĕ)	180-81	8°03'S	34°54'W
Reconquista, Arg. (rā-kôn-kēs'tä)	187	29°09'S	59°38'W
Recreo, Arg.	182-83	29°17'S	65°04'W
Rector, Ar., U.S. (rĕk'tēr)	134-35	36°16'N	90°18'W
Rècyča, Bela.	218-19	52°22'N	30°25'E
Red, stm., Asia (rĕd)	258-59	20°18'N	106°32'E
Red, stm., N.A. (rĕd)	120-21	50°25'N	96°47'W
Red, stm., U.S. (rĕd)	120-21	29°49'N	91°23'W
Red, stm., Ky., U.S. (rĕd)	126-27	37°50'N	84°06'W
Redang, Pulau, i., Malay.	266-67	5°46'N	103°01'E
Red Bank, Tn., U.S. (rĕd băngk)	134-35	35°07'N	85°17'W
Red Bluff, Ca., U.S.	128-29	40°11'N	122°14'W
Red Bluff Reservoir, res., U.S. (rĕd blŭf rĕ'sēr-vwär)	132-33	31°57'N	103°56'W
Redcliff, Ab., Can. (rĕd'clĭf)	142-43	50°05'N	110°47'W
Redcliffe, Austl. (rĕd'clĭf)	301	27°14'S	153°07'E
Red Cloud, Ne., U.S. (rĕd kloud)	130-31	40°05'N	98°31'W
Red Deer, Ab., Can. (rĕd dēr)	142-43	52°16'N	113°49'W
Red Deer, stm., Can. (rĕd dēr)	142-43	50°55'N	109°53'W
Red Deer, stm., Can. (rĕd dēr)	144-45	52°59'N	100°52'W
Red Deer Lake, lk., Mb., Can. (rĕd dēr lāk)	144-45	52°56'N	101°20'W
Redding, Ca., U.S.	122-23	40°35'N	122°23'W
Redfield, S.D., U.S. (rĕd'fēld)	124-25	44°52'N	98°31'W
Red Indian Lake, lk., Nf./L., Can. (rĕd ĭn'dĭ-ăn lāk)	148-49	48°39'N	56°50'W
Red Lake, On., Can. (rĕd lāk)	144-45	51°01'N	93°49'W
Red Lake, lk., On., Can.	144-45	51°01'N	94°05'W
Red Lake Falls, Mn., U.S. (rĕd lāk fôlz)	124-25	47°53'N	96°16'W
Red Lake Indian Reservation, ind. res., Mn., U.S. (rĕd lāk ĭn'dĭ-ăn rĕ-sēr-vā'shĕn)	124-25	48°03'N	94°59'W
Red Lion, Pa., U.S. (rĕd lī'ŭn)	126-27	39°54'N	76°36'W
Redmond, Or., U.S. (rĕd'mŭnd)	122-23	44°17'N	121°10'W
Redmond, Wa., U.S. (rĕd'mŭnd)	122-23	47°40'N	122°07'W
Red Oak, Ia., U.S. (rĕd ōk)	124-25	41°01'N	95°14'W
Redon, Fr. (rē-dôn')	212-13	47°39'N	2°05'W
Redonda, i., Antig. (rĕ-dŏn'dȧ)	155b	16°56'N	62°21'W
Red Sea, s., (rĕd sē)	288	20°0'N	38°00'E
Red Sucker Lake, lk., Mb., Can. (rĕd sŭk'ēr lāk)	144-45	54°09'N	93°40'W
Red Wing, Mn., U.S.	124-25	44°34'N	92°32'W
Redwood Falls, Mn., U.S. (rĕd'wŏŏd fôlz)	124-25	44°32'N	95°07'W
Redwood National Park, n.p., Ca., U.S. (rĕd'wŏŏd năsh'ŭn-ăl pärk)	122-23	41°20'N	124°02'W
Reed City, Mi., U.S. (rēd sĭ'tē)	126-27	43°53'N	85°32'W
Reed Lake, lk., Mb., Can. (rēd lāk)	144-45	54°38'N	100°30'W
Reedley, Ca., U.S. (rēd'lē)	128-29	36°35'N	119°26'W
Reedsburg, Wi., U.S. (rēdz'bûrg)	126-27	43°32'N	89°60'W
Reedsport, Or., U.S. (rēdz'pôrt)	122-23	43°42'N	124°06'W
Reform, Al., U.S. (rē-fôrm')	134-35	33°23'N	88°01'W
Refugio, Tx., U.S. (rå-fōō'hyō) (rĕ-fū'jō)	132-33	28°18'N	97°17'W
Rega, stm., Pol. (rĕ-gä)	210-11	54°09'N	15°17'E
Regensburg, Ger. (rā'ghĕns-bôrgh)	210-11	49°01'N	12°06'E
Reggio di Calabria, Italy (rĕ'jô dē kä-lä'brē-ä)	216-17	38°07'N	15°39'E
Reghin, Rom. (rå-gēn')	210-11	46°47'N	24°43'E
Regina, Sk., Can. (rē-jī'nȧ)	144-45	50°27'N	104°38'W
Registan, reg., Afg. see Rīgestān	252-53	31°0'N	65°00'E
Rehoboth, Nmb.	286-87	23°19'S	17°05'E
Reẖovot, Isr.	248-49	31°54'N	34°49'E
Reidsville, N.C., U.S. (rēdz'vĭl)	134-35	36°21'N	79°40'W
Reims, Fr. (răɴs)	212-13	49°15'N	4°02'E
Reindeer Lake, lk., Can. (rān'dēr lāk)	144-45	57°16'N	102°15'W
Reinosa, Spain (rå-ē-nō'sä)	214-15	42°60'N	4°08'W
Remada, Tun.	204-05	32°19'N	10°23'E
Remanso, Braz.	180-81	9°37'S	42°07'W
Remedios, Pan. (rē-mĕ'dyōs)	162	8°13'N	81°50'W
Remiremont, Fr. (rē-mēr-môɴ')	212-13	48°01'N	6°36'E
Rendova Island, i., Sol. Is. (rĕn'dô-vȧ ī'lănd)	306e	8°32'S	157°20'E
Rendsburg, Ger. (rĕnts'bôrgh)	210-11	54°18'N	9°40'E
Renfrew, On., Can. (rĕn'frōō)	146-47	45°29'N	76°42'W
Rengo, Chile (rĕn'gō)	188e	34°29'S	70°53'W
Reni, Ukr. (ran')	216-17	45°28'N	28°17'E
Renmark, Austl. (rĕn'märk)	301	34°11'S	140°45'E
Rennell, i., Sol. Is. (rĕn-nĕl')	296-97	11°33'S	160°05'E
Rennes, Fr. (rĕn)	212-13	48°07'N	1°41'W
Reno, Nv., U.S. (rē'nō)	128-29	39°32'N	119°49'W
Reno, Tx., U.S. (rē'nō)	130-31	32°56'N	97°35'W
Renovo, Pa., U.S. (rē-nō'vō)	126-27	41°20'N	77°45'W
Rensselaer, In., U.S. (rĕn'sē-lâr)	126-27	40°57'N	87°09'W
Rensselaer, N.Y., U.S. (rĕn'sē-lâr)	126-27	42°40'N	73°45'W
Renton, Wa., U.S. (rĕn'tŭn)	122-23	47°30'N	122°11'W
Reo, Indon.	268-69	8°19'S	120°29'E
Repetek, Turkmen.	252-53	38°34'N	63°11'E
Republic, Mo., U.S. (rē-pŭb'lĭk)	130-31	37°08'N	93°29'W
República Dominicana, nation, N.A. see Dominican Republic	152-53	19°0'N	70°40'W
Republican, stm., U.S. (rē-pŭb'lĭ-kăn)	120-21	39°03'N	96°48'W
Republican, South Fork, stm., U.S. (south fôrk rē-pŭb'lĭ-kăn)	130-31	40°04'N	101°31'W
Republic of Korea, nation, Asia	224-25	36°30'N	128°00'E
République centrafricaine, nation, Afr. see Central African Republic	274	7°0'N	21°00'E
Repulse Bay, Nu., Can.	138-39	66°32'N	86°14'W
Repulse Bay, b., Austl. (rē-pŭls' bā)	302	20°36'S	148°43'E
Requena, Spain (rå-kā'nä)	214-15	39°29'N	1°06'W
Resht, Iran see Rasht	248-49	37°17'N	49°35'E
Resistencia, Arg. (rā-sēs-tĕn'syä)	187	27°27'S	59°00'W
Reşiţa, Rom. (rå'shē-tȧ)	216-17	45°18'N	21°53'E
Resolution Island, i., Nu., Can. (rĕz-ō-lū'shŭn ī'lănd)	140-41	61°30'N	65°00'W
Resolution Island, i., N.Z. (rĕz-ŏl-ûshûn ī'lănd)	303	45°40'S	166°40'E
Resolute, Nu., Can.	150	74°41'N	94°54'W
Restrepo, Col. (rĕs-trĕ'pô)	188c	3°48'N	76°31'W
Retalhuleu, Guat. (rā-täl-ōō-lān')	160	14°32'N	91°41'W
Rethel, Fr. (r-tĕl')	212-13	49°31'N	4°22'E
Reunion, dep., Afr. (rā-ü-nyôn')	287a	21°06'S	55°36'E
Réunion, dep., Afr. see Reunion	287a	21°06'S	55°36'E
Reus, Spain (rā'ōōs)	214-15	41°09'N	1°07'E
Reutlingen, Ger. (roit'lĭng-ĕn)	210-11	48°30'N	9°12'E
Reval, nat. cap., Est. see Tallinn	208-09	59°26'N	24°48'E
Revda, Russia (ryâv'dȧ)	202-03	67°58'N	34°34'E
Revelstoke, B.C., Can. (rĕv'ĕl-stōk)	142-43	50°59'N	118°11'W
Revillagigedo, Islas, is., Mex. (ĕ's-läs-rĕ-vēl-yä-hĕ'gĕ-dô)	156-57	18°48'N	112°06'W
Revin, Fr. (rē-văɴ)	212-13	49°56'N	4°39'E
Rewa, India (rā'wä)	254-55	24°32'N	81°18'E
Rexburg, Id., U.S. (rĕks'bûrg)	122-23	43°50'N	111°47'W
Rey, Isla del, i., Pan. (ĕ's-lä-dĕl-rā'ē)	162	8°22'N	78°55'W
Rey, Laguna del, lk., Mex. (lä-gò'nä-dĕl-rā)	132-33	27°01'N	103°24'W
Reyes, Bol. (rā'yĕs)	180-81	14°19'S	67°22'W
Reyes, Point, c., Ca., U.S. (point rā'yĕs)	128-29	38°0'N	123°01'W
Reykjanes, pen., Ice. (rā'kyȧ-nĕs)	206a	63°49'N	22°43'W
Reykjavík, nat. cap., Ice. (rā'kyȧ-vēk)	206a	64°08'N	21°56'W
Reynosa, Mex. (rā-ē-nō'sä)	132-33	26°05'N	98°17'W
Reẕā'īyeh, Iran see Orūmīyeh	248-49	37°32'N	45°05'E
Rēzekne, Lat. (rā'zĕk-nĕ)	208-09	56°30'N	27°20'E
Rheims, Fr. see Reims	212-13	49°15'N	4°02'E
Rhein, stm., Eur. see Rhine	210-11	51°53'N	6°02'E
Rheine, Ger. (rī'nĕ)	210-11	52°17'N	7°27'E
Rhin, stm., Eur. see Rhine	210-11	51°53'N	6°02'E
Rhine, stm., Eur. (rīn)	210-11	51°53'N	6°02'E
Rhinelander, Wi., U.S. (rīn'lăn-dēr)	126-27	45°38'N	89°24'W
Rhir, Cap, c., Mor.	280-81	30°38'N	9°53'W
Rhode Island, state, U.S. (rōd ī'lănd)	118-19	41°40'N	71°30'W
Rhodes, Grc. see Ródos	216-17	36°26'N	28°14'E
Rhodes, i., Grc. (rōdz) see Ródos	216-17	36°10'N	28°00'E
Rhône, stm., Eur. (rōn)	212-13	43°53'N	4°39'E
Riachão, Braz. (rē-ä-chouɴ')	180-81	7°22'S	46°39'W
Riau, Kepulauan, is., Indon.	266-67	1°0'N	104°30'E
Ribe, Den. (rē'bĕ)	208-09	55°20'N	8°46'E
Ribeirão Preto, Braz. (rē-bā-rouɴ-prĕ'tô)	186	21°10'S	47°48'W
Riberalta, Bol. (rē-bȧ-räl'tä)	180-81	11°00'S	66°05'W
Rib Lake, Wi., U.S. (rĭb lāk)	126-27	45°19'N	90°12'W
Rice Lake, Wi., U.S. (rīs lāk)	124-25	45°30'N	91°44'W
Rice Lake, lk., On., Can. (rīs lāk)	146-47	44°08'N	78°13'W
Richards Bay, S. Afr.	286-87	28°47'S	32°05'E
Richardson, Tx., U.S. (rĭch'ērd-sŭn)	130-31	32°58'N	96°44'W
Richardson Mountains, mts., Can. (rĭch'ērd-sŭn moun'tĭnz)	140-41	67°22'N	136°04'W
Richfield, Ut., U.S. (rĭch'fēld)	128-29	38°46'N	112°05'W
Rich Hill, Mo., U.S. (rĭch hĭl)	130-31	38°06'N	94°22'W
Richland, Ga., U.S. (rĭch'lănd)	134-35	32°05'N	84°40'W
Richland, Wa., U.S. (rĭch'lănd)	122-23	46°16'N	119°17'W
Richland Center, Wi., U.S. (rĭch'lănd sĕn'tēr)	124-25	43°20'N	90°23'W
Richmond, Austl. (rĭch'mŭnd)	302	20°44'S	143°08'E
Richmond, B.C., Can. (rĭch'mŭnd)	142-43	49°09'N	123°10'W
Richmond, Qc., Can. (rĭch'mŭnd)	146-47	45°40'N	72°09'W
Richmond, In., U.S. (rĭch'mŭnd)	126-27	39°49'N	84°54'W
Richmond, Ky., U.S. (rĭch'mŭnd)	126-27	37°45'N	84°18'W
Richmond, Mi., U.S. (rĭch'mŭnd)	126-27	42°49'N	82°45'W
Richmond, Mo., U.S. (rĭch'mŭnd)	130-31	39°17'N	93°59'W
Richmond, Va., U.S. (rĭch'mŭnd)	126-27	37°33'N	77°27'W
Richmond Hill, On., Can. (rĭch'mŭnd hĭl)	146-47	43°52'N	79°26'W
Richwood, La., U.S. (rĭch'wŏŏd)	130-31	32°27'N	92°06'W
Richwood, W.V., U.S. (rĭch'wŏŏd)	126-27	38°13'N	80°34'W
Ridā', Yemen	288	14°38'N	44°54'E
Ridder, Kaz.	244	50°21'N	83°30'E
Riding Mountain National Park, n.p., Mb., Can. (rĭd'ĭng moun'tĭn năsh'ŭn-ăl pärk)	144-45	50°55'N	100°25'W
Riesa, Ger. (rē'zȧ)	210-11	51°18'N	13°18'E
Riesco, Isla, i., Chile	185	52°59'S	72°38'W
Rieti, Italy (rē-ā'tē)	216-17	42°24'N	12°52'E
Rif, mts., Mor.	280-81	35°0'N	4°00'W
Rift Valley, val., Afr. (rĭft văl'ē)	275	3°0'S	29°00'E
Riga, nat. cap., Lat. (rē'gȧ)	208-09	56°57'N	24°06'E
Riga, Gulf of, b., Eur. (gŭlf ŭv rē'gȧ)	208-09	57°30'N	23°35'E
Rīgas jūras līcis, b., Eur. see Riga, Gulf of	208-09	57°30'N	23°35'E
Rigby, Id., U.S. (rĭg'bē)	122-23	43°40'N	111°56'W
Rīgestān, reg., Afg.	252-53	31°0'N	65°00'E
Rijeka, Cro. (rĭ-yĕ'kä)	216-17	45°20'N	14°27'E
Rijn, stm., Eur. see Rhine	210-11	51°53'N	6°02'E
Rima, stm., Nig.	282-83	13°04'N	5°07'E
Rimatara, i., Fr. Poly.	304-05	22°38'S	152°51'W
Rimavská Sobota, Slvk. (rē'mȧf-skȧ sŏ'bô-tȧ)	210-11	48°23'N	20°05'E

ăt; finăl; rāte; senāte; ärm; åsk; sofȧ; fâre; ch-choose; dh-as th in other; bē; ēvent; bĕt; recĕnt; cratēr; g-gō; gh-guttural g; bĭt; ĭ-short neutral; rīde; κ-guttural k as ch in German ich;

Feature (Pronunciation)	Page	Lat.	Long.
Rimbo, Swe. (rēm´bō)	208-09	59°45'N	18°22'E
Rimini, Italy (rē´mē-nē)	216-17	44°04'N	12°35'E
Rimouski, Qc., Can. (rē-mōōs´kē)	148-49	48°27'N	68°33'W
Rincón del Bonete, Lago Artificial de, res., Ur.	187	32°43'S	56°01'W
Rincón de Romos, Mex. (rēn-kōn dā rō-mōs´)	158-59	22°14'N	102°18'W
Ringkøbing, Den. (rĭng´kûb-ĭng)	208-09	56°05'N	8°15'E
Ringsted, Den. (rĭng´stĕdh)	208-09	55°27'N	11°50'E
Ringvassøya, i., Nor. (rĭng´väs-ûē)	200-01	69°55'N	19°15'E
Rinjani, Gunung, vol., Indon.	268-69	8°24'S	116°28'E
Riobamba, Ec. (rē´ō-bäm-bä)	184	1°40'S	78°39'W
Rio Branco, Braz. (rē´ō brän´kō)	180-81	9°58'S	67°48'W
Río Branco, Ur. (rīō brănсō)	187	32°36'S	53°23'W
Rio Casca, Braz. (rē´ō-kà´s-kä)	186	20°14'S	42°39'W
Río Chico, Ven. (rē´ō chē´kŏ)	188b	10°18'N	65°59'W
Rio Claro, Braz. (rē´ō klä´rō)	186	22°26'S	47°33'W
Río Colorado, Arg.	187	38°60'S	64°07'W
Río Cuarto, Arg. (rē´ō kwär´tō)	187	33°08'S	64°21'W
Rio de Janeiro, Braz. (rē´ō dā zhä-nâ´ē-rō)	186	22°54'S	43°14'W
Rio de Janeiro, state, Braz. (rē´ō dā zhä-nâ´ē-rō)	186	22°0'S	42°30'W
Rio do Sul, Braz.	182-83	27°13'S	49°39'W
Río Gallegos, Arg. (rē´ō gä-lā´gōs)	185	51°38'S	69°13'W
Río Grande, Arg.	185	53°49'S	67°47'W
Rio Grande, Braz. (rē´ō grän´dĕ)	187	32°02'S	52°06'W
Río Grande, Mex. (rē´ō grän´dā)	158-59	15°59'N	97°27'W
Río Grande, Mex. (rē´ō grän´dā)	158-59	23°50'N	103°03'W
Rio Grande, stm., N.A. (rē´ō grän´dā)	120-21	25°57'N	97°09'W
Rio Grande do Sul, Braz. *see* Rio Grande	187	32°02'S	52°06'W
Rio Grande do Sul, state, Braz. (rē´ō grän´dĕ-dō-sōō´l)	187	30°0'S	54°00'W
Ríohacha, Col. (rē´ō-ä´chä)	178-79	11°33'N	72°55'W
Río Hato, Pan. (rē´ō-ä´tō)	162	8°23'N	80°10'W
Rio Largo, Braz.	188d	9°29'S	35°51'W
Riom, Fr. (rē-ôɴ´)	212-13	45°54'N	3°07'E
Río Mayo, Arg.	185	45°41'S	70°14'W
Rio Negro, Braz.	182-83	26°06'S	49°47'W
Ríonegro, Col. (rē´ō-nĕ´grō)	188c	6°08'N	75°23'W
Río Negro, state, Arg. (rē´ō nā´grō)	185	40°0'S	67°00'W
Rio Pardo, Braz.	182-83	29°59'S	52°22'W
Rio Pardo de Minas, Braz. (rē´ō pär´dō-dē-mē´näs)	182-83	15°37'S	42°33'W
Ríosucio, Col. (rē´ō-sōō´syō)	188c	5°25'N	75°42'W
Ríosucio, Col. (rē´ō-sōō´syō)	178-79	7°25'N	77°06'W
Río Tercero, Arg. (rē´ō dĕr-sĕ´rō)	187	32°11'S	64°07'W
Rio Tinto, Braz.	188d	6°48'S	35°05'W
Rio Verde, Braz. (rē´ō vĕr´dĕ)	182-83	17°47'S	50°55'W
Rioverde, Mex. (rē´ō-vĕr´dā)	158-59	21°56'N	99°59'W
Ripley, Ms., U.S. (rĭp´lē)	134-35	34°45'N	88°57'W
Ripley, Tn., U.S. (rĭp´lē)	134-35	35°45'N	89°32'W
Ripley, W.V., U.S. (rĭp´lē)	126-27	38°48'N	81°44'W
Ripoll, Spain (rē-pōl´')	214-15	42°12'N	2°12'E
Ripon, Wi., U.S. (rĭp´ŏn)	126-27	43°51'N	88°50'W
Rishiri-tō, i., Japan (rē-shē´rē tō)	264	45°11'N	141°15'E
Rising Sun, In., U.S. (rīz´ĭng sŭn)	126-27	38°57'N	84°52'W
Risør, Nor. (rēs´ûr)	208-09	58°43'N	9°14'E
Rittman, Oh., U.S. (rĭt´năn)	126-27	40°58'N	81°47'W
Ritzville, Wa., U.S. (rĭts´vĭl)	122-23	47°07'N	118°22'W
Rivas, Nic. (rē´väs)	161	11°27'N	85°52'W
Rivera, Ur. (rē-vā´rä)	187	30°54'S	55°33'W
River Falls, Wi., U.S. (rĭv´ẽr fôlz)	124-25	44°52'N	92°37'W
Riverhead, N.Y., U.S. (rĭv´ẽr hĕd)	126-27	40°55'N	72°40'W
Rivers, Mb., Can. (rĭv´ẽrz)	144-45	50°02'N	100°14'W
Riverside, Ca., U.S.	128-29	33°58'N	117°21'W
Rivers Inlet, B.C., Can.	142-43	51°42'N	127°15'W
Rivesaltes, Fr. (rēv'zält')	212-13	42°46'N	2°52'E
Riviera Beach, Fl., U.S. (rĭv-ĭ-ẽr´à bēch)	135a	26°46'N	80°04'W
Rivière-du-Loup, Qc., Can. (rē-vyâr´ dü lōō´)	148-49	47°50'N	69°32'W
Rivne, Ukr.	210-11	50°37'N	26°14'E
Riyadh, nat. cap., Sau. Ar. (rī-äd´)	250-51	24°38'N	46°43'E
Rize, Tur. (rē´zĕ)	245	41°01'N	40°31'E
Rjukan, Nor. (ryōō´kän)	208-09	59°53'N	8°35'E
Road Town, nat. cap., Br. Vir. Is. (rōd toun)	155b	18°26'N	64°37'W
Roanne, Fr. (rō-än´)	212-13	46°02'N	4°04'E
Roanoke, Al., U.S. (rō´à-nōk)	134-35	33°09'N	85°22'W
Roanoke, Va., U.S. (rō´à-nōk)	134-35	37°16'N	79°57'W
Roanoke, stm., U.S. (rō´à-nōk)	134-35	35°57'N	76°43'W
Roanoke Rapids, N.C., U.S. (rō´à-nōk răp´ĭdz)	134-35	36°28'N	77°39'W
Roan Plateau, plat., U.S. (rōn plä-tō´)	128-29	39°30'N	109°40'W
Roatán, Hond. (rō-ä-tän´)	161	16°20'N	86°32'W
Roatán, Isla de, i., Hond.	161	16°22'N	86°29'W
Roberval, Qc., Can. (rŏb´ẽr-văl) (rō-bĕr-vál´)	146-47	48°31'N	72°14'W
Robinson, Il., U.S. (rŏb´ĭn-sŭn)	126-27	39°01'N	87°45'W
Robinvale, Austl. (rŏb-ĭn´văl)	301	34°36'S	142°46'E
Roblin, Mb., Can.	144-45	51°15'N	101°23'W
Roboré, Bol.	182-83	18°20'S	59°45'W
Robson, Mount, mtn., B.C., Can. (mount rŏb´sŭn)	142-43	53°07'N	119°09'W
Robstown, Tx., U.S. (rŏbz´toun)	132-33	27°47'N	97°40'W
Roca, Cabo da, c., Port. (kà´bō-dä-rō´kă)	214-15	38°47'N	9°29'W
Roca Partida, Isla, i., Mex.	156-57	19°00'N	112°04'W
Rocha, Ur. (rō´chás)	187	34°30'S	54°19'W
Rochefort, Fr. (rôsh-fôr´)	212-13	45°57'N	0°58'W
Rochelle, Il., U.S. (rō-shĕl´)	126-27	41°55'N	89°04'W
Rochester, In., U.S. (rŏch´ĕs-tẽr)	126-27	41°04'N	86°12'W
Rochester, Mn., U.S. (rŏch´ĕs-tẽr)	124-25	44°00'N	92°29'W
Rochester, N.H., U.S. (rŏch´ĕs-tẽr)	126-27	43°18'N	70°59'W
Rochester, N.Y., U.S.	126-27	43°09'N	77°36'W
Rock, stm., U.S. (rŏk)	124-25	41°29'N	90°38'W
Rockdale, Tx., U.S. (rŏk´dāl)	132-33	30°39'N	97°00'W
Rockefeller Plateau, plat., Ant.	313	80°0'S	135°00'W
Rock Falls, Il., U.S. (rŏk fôlz)	126-27	41°47'N	89°41'W
Rockford, Il., U.S. (rŏk´fẽrd)	126-27	42°16'N	89°05'W
Rockford, Mi., U.S. (rŏk´fẽrd)	126-27	43°07'N	85°34'W
Rockhampton, Austl. (rŏk-hămp´tŭn)	302	23°23'S	150°31'E
Rock Hill, S.C., U.S. (rŏk hĭl)	134-35	34°56'N	81°02'W
Rockingham, N.C., U.S. (rŏk´ĭng-hăm)	134-35	34°56'N	79°46'W
Rock Island, Il., U.S. (rŏk ī´lănd)	124-25	41°30'N	90°34'W
Rockland, On., Can. (rŏk´lănd)	146-47	45°33'N	75°17'W
Rockland, Me., U.S.	127a	44°07'N	69°07'W
Rockport, In., U.S. (rŏk´pōrt)	126-27	37°53'N	87°03'W
Rockport, Tx., U.S. (rŏk´pōrt)	132-33	28°01'N	97°03'W
Rock Rapids, Ia., U.S. (rŏk răp´ĭdz)	124-25	43°26'N	96°10'W
Rocksprings, Tx., U.S. (rŏk sprĭngs)	132-33	30°01'N	100°12'W
Rock Springs, Wy., U.S. (rŏk sprĭngz)	122-23	41°35'N	109°13'W
Rockstone, Guy. (rŏk´stōn)	178-79	5°59'N	58°32'W
Rock Valley, Ia., U.S. (rŏk văl´ĭ văl´ē)	124-25	43°12'N	96°18'W
Rockville, In., U.S. (rŏk´vĭl)	126-27	39°45'N	87°14'W
Rockwell City, Ia., U.S. (rŏk´wĕl sĭ´tē)	124-25	42°24'N	94°38'W
Rockwood, Me., U.S. (rŏk-wŏŏd)	126-27	45°40'N	69°45'W
Rockwood, Tn., U.S. (rŏk-wŏŏd)	134-35	35°52'N	84°41'W
Rocky Ford, Co., U.S. (rŏk´-ē fôrd)	130-31	38°03'N	103°43'W
Rocky Island Lake, res., On., Can. (rŏk´-ē ī´lănd lāk)	146-47	46°56'N	82°57'W
Rocky Mount, N.C., U.S. (rŏk´-ē mount)	134-35	35°57'N	77°48'W
Rocky Mount, Va., U.S. (rŏk´-ē mount)	134-35	37°00'N	79°54'W
Rocky Mountain House, Ab., Can. (rŏk´-ē moun´tĭn hous)	142-43	52°23'N	114°56'W
Rocky Mountain National Park, n.p., Co., U.S. (rŏk´-e moun´tĭn năsh´ŭn-ăl pärk)	122-23	40°21'N	105°42'W
Rocky Mountains, mts., N.A. (rŏk´-ē moun´tĭnz)	95	48°0'N	116°00'W
Rodeo, Arg.	182-83	30°12'S	69°06'W
Rodeo, Mex. (rō-dā´ō)	132-33	25°11'N	104°34'W
Rodez, Fr. (rō-dĕz´)	212-13	44°21'N	2°34'E
Rodniki, Russia (rôd´nē-kē)	202-03	57°06'N	41°44'E
Ródos, Grc.	216-17	36°26'N	28°14'E
Ródos, i., Grc.	216-17	36°10'N	28°00'E
Roebourne, Austl. (rō´bŭrn)	294-95	20°46'S	117°10'E
Rogagua, Laguna, lk., Bol.	180-81	13°42'S	67°07'W
Rogaguado, Laguna, lk., Bol. (rō´gō-ä-gwä-dō)	180-81	12°52'S	65°43'W
Rogers, Ar., U.S. (rŏj-ẽrz)	130-31	36°20'N	94°07'W
Rogers, Mount, mtn., Va., U.S.	134-35	36°39'N	81°33'W
Rogers City, Mi., U.S. (rŏj-ẽrz sĭ´tē)	126-27	45°25'N	83°49'W
Rohtak, India	254-55	28°53'N	76°36'E
Roi Georges, Îles du, is., Fr. Poly.	304-05	14°32'S	145°08'W
Rojas, Arg. (rō´häs)	187	34°12'S	60°44'W
Rojo, Cabo, c., Mex. (kà´bō rō´hō)	158-59	21°33'N	97°20'W
Rojo, Cabo, c., P.R. (kà´bō rō´hō)	154a	17°56'N	67°11'W
Rokan, stm., Indon.	266-67	1°50'N	100°55'E
Rokeby National Park, n.p., Austl. *see* Mungkan Kandju National Park	302	13°32'S	142°37'E
Rokycany, Czech Rep. (rô´kĭ´tsa-nĭ)	210-11	49°45'N	13°36'E
Rolândia, Braz.	182-83	23°18'S	51°23'W
Roldanillo, Col. (rôl-dä-nē´l-yō)	188c	4°24'N	76°09'W
Rolla, Mo., U.S.	130-31	37°57'N	91°46'W
Roma, Austl. (rō´mà)	301	26°35'S	148°47'E
Roma, nat. cap., Italy (rō´mä) *see* Rome	216-17	41°54'N	12°29'E
Romaine, stm., Can. (rō-mĕn´)	148-49	50°18'N	63°48'W
Roman, Rom. (rō´män)	210-11	46°56'N	26°57'E
Romang, Pulau, i., Indon.	268-69	7°34'S	127°26'E
Romania, nation, Eur. (rō-mā´nē-à)	190-91	46°0'N	25°30'E
Roman-Kosh, hora, mtn., Ukr.	218-19	44°37'N	34°15'E
Romano, Cape, c., Fl., U.S. (kāp rō-mä´nō)	135a	25°50'N	81°41'W
Romano, Cayo, i., Cuba (kä´yō-rō-mà´nō)	154-55	22°04'N	77°50'W
Romblon, Phil. (rōm-blōn´)	270	12°34'N	122°16'E
Rome, Ga., U.S. (rōm)	134-35	34°16'N	85°10'W
Rome, N.Y., U.S. (rōm)	126-27	43°13'N	75°28'W
Rome, nat. cap., Italy (rōm)	216-17	41°54'N	12°29'E
Romeo, Mi., U.S. (rō´mē-ō)	126-27	42°48'N	83°00'W
Romilly-sur-Seine, Fr. (rō-mē-yē´sür-sān´)	212-13	48°31'N	3°44'E
Romny, Ukr. (rôm´nĭ)	218-19	50°45'N	33°29'E
Rømø, i., Den. (rûm´û)	208-09	55°08'N	8°31'E
Romorantin-Lanthenay, Fr. (rō-mō-räɴ-tăɴ´)	212-13	47°22'N	1°44'E
Rona, i., Scot., U.K.	200-01	59°07'N	5°49'W
Ronan, Mt., U.S. (rō´năn)	122-23	47°31'N	114°06'W
Roncador, Serra do, plat., Braz. (sĕr´ra dō rōn-kä-dōr´)	180-81	12°0'S	52°00'W
Ronda, Spain (rōn´dä)	214-15	36°44'N	5°10'W
Rondônia, state, Braz.	180-81	11°0'S	63°00'W
Rondonópolis, Braz.	182-83	16°28'S	54°38'W
Ronge, Lac la, lk., Sk., Can. (läk lä rōnzh)	144-45	55°10'N	105°00'W
Rongelap, at., Marsh. Is.	304-05	11°20'N	166°50'E
Rongjiang, China (rōŋ-jyäŋ)	258-59	25°51'N	108°35'E
Rønne, Den. (rûn´ĕ)	210-11	55°06'N	14°42'E
Ronneby, Swe. (rōn´ĕ-bü)	208-09	56°12'N	15°18'E
Ronuro, stm., Braz.	180-81	11°56'S	53°33'W
Roorkee, India	254-55	29°52'N	77°53'E
Roosendaal, Neth. (rō´zĕn-däl)	206-07	51°32'N	4°28'E
Roosevelt, Ut., U.S. (rōz´'vĕlt)	128-29	40°19'N	109°59'W
Roosevelt, stm., Braz. (rō´sĕ-vĕlt)	180-81	7°34'S	60°41'W
Roper, stm., Austl. (rōp´ẽr)	296-97	14°44'S	135°23'E
Roque Pérez, Arg. (rō´kĕ-pĕ´rĕz)	187	35°25'S	59°20'W
Roraima, state, Braz. (rō´rī-mà)	180-81	1°0'N	61°00'W
Roraima, Monte, mtn., S.A. *see* Roraima, Mount	178-79	5°13'N	60°44'W
Roraima, Mount, mtn., S.A. (mount rō-rä-ē´mä)	178-79	5°13'N	60°44'W
Røros, Nor. (rûr'ôs)	200-01	62°35'N	11°23'E
Ros', stm., Ukr. (rôs)	218-19	49°41'N	31°36'E
Rosales, Mex. (rō-zä´lās)	132-33	28°12'N	105°33'W
Rosamorada, Mex. (rō´zä-mō-rä´dhä)	158-59	22°06'N	105°12'W
Rosario, Arg. (rō-zä´rē-ō)	187	32°57'S	60°40'W
Rosário, Braz. (rō-zä´rē-ō)	180-81	2°57'S	44°14'W
Rosario, Mex. (rō-zä´rē-ō)	158-59	23°00'N	105°52'W
Rosario, Para.	182-83	24°25'S	57°06'W
Rosario, Ur. (rō-zä´rē-ō)	187	34°19'S	57°21'W
Rosario de la Frontera, Arg.	182-83	25°48'S	64°58'W
Rosario de Lerma, Arg.	182-83	24°59'S	65°35'W
Rosário do Sul, Braz. (rō-zä´rē-ō-dō-sōō´l)	187	30°15'S	54°56'W
Rosário Oeste, Braz. (rō-zä´rē-ō ō´ĕst´ĕ)	180-81	14°50'S	56°25'W
Roscoe, Tx., U.S. (rōs´kō)	130-31	32°27'N	100°33'W
Roseau, Mn., U.S. (rō-zō´)	124-25	48°51'N	95°46'W
Roseau, nat. cap., Dom.	155b	15°18'N	61°23'W
Rosebud, stm., Ab., Can. (rōz´bŭd)	142-43	51°25'N	112°37'W
Rosebud Indian Reservation, ind. res., S.D., U.S. (rōz´bŭd ĭn´dĭ-ăn rĕ-sẽr-vā´shĕn)	124-25	43°08'N	100°33'W
Roseburg, Or., U.S.	122-23	43°14'N	123°20'W
Rosenheim, Ger. (rō´zĕn-hīm)	210-11	47°52'N	12°08'E
Rosetown, Sk., Can. (rōz´toun)	144-45	51°32'N	108°01'W
Rosetta, Egypt	291b	31°24'N	30°25'E
Roseville, Ca., U.S. (rōz´vĭl)	128-29	38°45'N	121°17'W
Roseville, Mn., U.S. (rōz´vĭl)	124-25	45°01'N	93°10'W
Roşiori de Vede, Rom. (rō-shôr´ē dĕ vĕ-dĕ)	216-17	44°07'N	24°60'E
Roskilde, Den. (rôs´kēl-dĕ)	208-09	55°39'N	12°06'E
Roslavl', Russia (rôs´läv'l)	218-19	53°57'N	32°52'E
Rossano, Italy (rō-sä´nō)	216-17	39°35'N	16°39'E
Rossiya, nation, Eur. *see* Russia	190-91	60°0'N	100°00'E
Rossland, B.C., Can. (rôs´lănd)	142-43	49°05'N	117°48'W
Rosso, Maur.	280-81	16°31'N	15°48'W
Rossosh', Russia (rôs´sŭsh)	202-03	50°12'N	39°35'E
Ross River, Yk., Can.	138-39	62°00'N	132°26'W
Ross Sea, s., Ant. (rôs sē)	313	76°0'S	175°00'W
Rossville, Ga., U.S. (rôs´vĭl)	134-35	34°59'N	85°18'W
Rosthern, Sk., Can.	144-45	52°40'N	106°20'W
Rostock, Ger. (rôs´tŭk)	210-11	54°05'N	12°07'E

n-sing; ŋ-baŋk; ɴ-nasalized n; nŏd; cŏmmit; ōld; ôbey; ôrder; oi-boil; fōōd; ŏŏ-as oo in foot; ou-out; s-soft; sh-dish; th-thin; pūre; ûnite; ûrn; stŭd; circŭs; ü-as in French tu; ´-indeterminate vowel.

402 Rostov – Saint Christopher PRONOUNCING INDEX

Feature (Pronunciation)	Page	Lat.	Long.
Rostov, Russia (rôstôv)	218-19	57°11'N	39°25'E
Rostov-na-Donu, Russia (rôstôv-nȧ-dô-nōō)	218-19	47°13'N	39°43'E
Roswell, Ga., U.S. (rŏz'wĕl)	134-35	34°02'N	84°21'W
Roswell, N.M., U.S. (rŏz'wĕl)	130-31	33°24'N	104°33'W
Rota, i., N. Mar. Is.	306c	14°10'N	145°12'E
Rotherham, Eng., U.K. (rŏdh'ēr-ăm)	206-07	53°25'N	1°23'W
Rothesay, Scot., U.K. (rŏth'sā)	206-07	55°50'N	5°03'W
Roti, Pulau, i., Indon. (pōō-lou rō'tē)	268-69	10°45'S	123°10'E
Rotorua, N.Z.	303	38°09'S	176°14'E
Rotterdam, Neth. (rŏt'ēr-däm')	206-07	51°55'N	4°28'E
Rottweil, Ger. (rōt'vīl)	210-11	48°10'N	8°38'E
Rotuma, i., Fiji	304-05	12°30'S	177°05'E
Roubaix, Fr. (rōō-bĕ')	212-13	50°41'N	3°10'E
Rouen, Fr. (rōō-äN')	212-13	49°27'N	1°07'E
Rouge, stm., Qc., Can. (rōōzh)	146-47	45°38'N	74°42'W
Round Mountain, mtn., Austl.	301	30°27'S	152°14'E
Round Rock, Tx., U.S. (round rŏk)	132-33	30°30'N	97°41'W
Roundup, Mt., U.S. (round'ŭp)	122-23	46°27'N	108°33'W
Rouyn-Noranda, Qc., Can.	146-47	48°14'N	79°01'W
Rovaniemi, Fin. (rô'vȧ-nyĕ'mĭ)	200-01	66°30'N	25°42'E
Rovereto, Italy (rō-vā-rā'tō)	216-17	45°54'N	11°02'E
Rovigo, Italy (rō-vē'gō)	216-17	45°05'N	11°47'E
Rovinj, Cro. (rō'ēn')	216-17	45°05'N	13°38'E
Rovira, Col. (rō-vē'rä)	188c	4°14'N	75°15'W
Rovno, Ukr. *see* Rivne	210-11	50°37'N	26°14'E
Rovuma, stm., Afr.	286-87	10°31'S	40°24'E
Rowley Island, i., Nu., Can.	140-41	69°05'N	78°52'W
Roxas, Phil.	270	11°35'N	122°45'E
Roy, Ut., U.S. (roi)	122-23	41°10'N	112°01'W
Royale, Isle, i., Mi., U.S.	124-25	48°0'N	89°00'W
Royal Oak, Mi., U.S. (roi'ăl ōk)	126-27	42°30'N	83°08'W
Royal Tunbridge Wells, Eng., U.K.	206-07	51°08'N	0°16'E
Royan, Fr. (rwä-yäN')	212-13	45°38'N	1°01'W
Rožňava, Slvk. (rôzh'nyȧ-vȧ)	210-11	48°40'N	20°33'E
Rtishchevo, Russia ('r-tĭsh'chĕ-vô)	202-03	52°16'N	43°47'E
Ruaha National Park, n.p., Tan.	289	7°30'S	34°40'E
Ruapehu, Mount, vol., N.Z. (mount r'oo-ȧ-pā'hōō)	303	39°17'S	175°34'E
Rub'al-Khali, des., Asia	238-39	20°0'N	51°00'E
Rubizhne, Ukr.	218-19	49°01'N	38°23'E
Rubondo Island, i., Tan.	289	2°20'S	31°52'E
Rubtsovsk, Russia	244	51°31'N	81°12'E
Ruby Mountains, mts., Nv., U.S. (rōō'bē moun'tĭnz)	128-29	40°25'N	115°31'W
Rudkøbing, Den. (rōōdh'kûb-ĭng)	210-11	54°56'N	10°44'E
Rūdnyy, Kaz.	244	52°59'N	63°07'E
Rudolf, Lake, lk., Afr. (lāk rōō'dōlf)	289	3°30'N	36°00'E
Rudolf Hāyk', lk., Afr. *see* Rudolf, Lake	289	3°30'N	36°00'E
Ruffec, Fr. (rü-fĕk')	212-13	46°01'N	0°12'E
Rufiji, stm., Tan. (rō-fē'jē)	284-85	7°58'S	39°25'E
Rufino, Arg.	187	34°16'S	62°42'W
Rugao, China (rōō-gou)	258-59	32°24'N	120°33'E
Rugby, Eng., U.K. (rŭg'bē)	206-07	52°23'N	1°16'W
Rugby, N.D., U.S.	124-25	48°22'N	99°60'W
Rügen, i., Ger. (rü'ghĕn)	210-11	54°25'N	13°24'E
Rugufu, stm., Tan.	289	5°30'S	30°01'E
Ruhengeri, Rw.	289	1°30'S	29°38'E
Rui'an, China (rwā-än)	258-59	27°50'N	120°35'E
Ruijin, China	258-59	25°52'N	116°00'E
Ruiz, Mex. (rōē'z)	158-59	21°57'N	105°09'W
Ruiz, Nevado del, vol., Col. (nĕ-vä'dō-dĕl-rōōē'z)	188c	4°53'N	75°20'W
Rūjiena, Lat. (rō'yĭ-ă-nȧ)	208-09	57°54'N	25°20'E
Rukwa, Lake, lk., Tan. (lāk rōōk-wä')	289	8°0'S	32°25'E
Ruma, Serb. (rōō'mȧ)	216-17	45°00'N	19°49'E
Rumbek, S. Sudan (rŭm'bĕk)	284-85	6°48'N	29°41'E
Rum Cay, i., Bah. (rŭm kē)	154-55	23°41'N	74°53'W
Rumford, Me., U.S. (rŭm'fērd)	126-27	44°33'N	70°33'W
Rumoi, Japan	264	43°56'N	141°39'E
Runan, China (rōō-nän)	258-59	33°00'N	114°21'E
Runde, stm., Zimb.	286-87	21°18'S	32°24'E
Rundu, Nmb.	286-87	17°55'S	19°45'E
Rŭng, Kaôh, i., Camb.	266-67	10°44'N	103°14'E
Rungwa, stm., Tan.	289	7°37'S	31°49'E
Ruo, stm., China (rwō)	260-61	41°04'N	100°20'E
Ruoqiang, China	238-39	39°01'N	88°11'E
Rupat, Pulau, i., Indon. (pōō-lou rōō'pät)	266-67	1°50'N	101°35'E
Rupert, Id., U.S. (rōō'pērt)	122-23	42°38'N	113°41'W
Rurrenabaque, Bol.	180-81	14°28'S	67°30'W
Rurutu, i., Fr. Poly.	304-05	22°26'S	151°20'W
Rusape, Zimb.	286-87	18°32'S	32°08'E
Ruse, Blg. (rōō'sĕ) (rō'sĕ)	216-17	43°51'N	25°57'E
Rushville, Il., U.S. (rŭsh'vĭl)	130-31	40°07'N	90°33'W
Rushville, In., U.S. (rŭsh'vĭl)	126-27	39°36'N	85°27'W
Rushville, Ne., U.S. (rŭsh'vĭl)	124-25	42°43'N	102°28'W

Feature (Pronunciation)	Page	Lat.	Long.
Rusk, Tx., U.S. (rŭsk)	132-33	31°48'N	95°09'W
Russas, Braz. (rōō's-säs)	180-81	4°56'S	37°58'W
Russell, Mb., Can. (rŭs'ĕl)	144-45	50°47'N	101°15'W
Russell, Ks., U.S. (rŭs'ĕl)	130-31	38°50'N	98°50'W
Russell, Ky., U.S. (rŭs'ĕl)	126-27	38°31'N	82°42'W
Russell Lake, lk., Mb., Can. (rŭs'ĕl lāk)	144-45	56°15'N	101°32'W
Russellville, Al., U.S. (rŭs'ĕl-vĭl)	134-35	34°30'N	87°44'W
Russellville, Ar., U.S. (rŭs'ĕl-vĭl)	130-31	35°17'N	93°09'W
Russellville, Ky., U.S. (rŭs'ĕl-vĭl)	134-35	36°51'N	86°53'W
Russia, nation, Eur. (rŭ'shȧ)	236-37	60°0'N	100°00'E
Rustavi, Geor.	245	41°32'N	45°02'E
Rustenburg, S. Afr. (rŭs'tĕn-bûrg)	292c	25°40'S	27°15'E
Ruston, La., U.S. (rŭs'tŭn)	130-31	32°32'N	92°38'W
Ruteng, Indon.	268-69	8°36'S	120°29'E
Rutherfordton, N.C., U.S. (rŭdh'ēr-fērd-tŭn)	134-35	35°22'N	81°58'W
Rutland, Vt., U.S.	126-27	43°37'N	72°59'W
Rutog, China	254-55	33°26'N	79°42'E
Rutshuru, D.R.C. (rōōt-shōō'rōō)	289	1°11'S	29°27'E
Ruvuma, stm., Afr.	286-87	10°31'S	40°24'E
Ruwenzori Range, mts., Afr.	289	0°20'N	29°53'E
Ruzayevka, Russia	202-03	54°04'N	44°57'E
Rwanda, nation, Afr. (rü-än'-dȧ)	274	2°0'S	30°00'E
Ryazan', Russia (ryä-zän'')	218-19	54°38'N	39°44'E
Ryazhsk, Russia (ryäzh'sk')	202-03	53°42'N	40°05'E
Rybachiy, Poluostrov, pen., Russia	200-01	69°42'N	32°36'E
Rybachye, Kyrg. *see* Balykchy	244	42°28'N	76°12'E
Rybinsk, Russia	218-19	58°03'N	38°52'E
Rybnik, Pol. (rĭb'nĕk)	210-11	50°06'N	18°33'E
Ryde, Eng., U.K. (rīd)	206-07	50°44'N	1°10'W
Ryeosu, Kor., S. *see* Yŏsu	263	34°44'N	127°44'E
Rylsk, Russia (rēl''sk)	218-19	51°34'N	34°42'E
Ryojun, China *see* Lüshun	260-61	38°49'N	121°15'E
Ryōtsu, Japan (ryōt'sōō)	265	38°05'N	138°26'E
Ryukyu Islands, is., Japan (rü-kü ĭ'lăndz)	264a	25°44'N	126°58'E
Rzeszów, Pol. (zhä-shōf)	210-11	50°03'N	22°01'E
Rzhev, Russia ('r-zhĕf)	218-19	56°17'N	34°19'E

S

Feature (Pronunciation)	Page	Lat.	Long.
Saale, stm., Ger. (sä-lĕ)	210-11	51°57'N	11°55'E
Saalfeld, Ger. (säl'fĕlt)	210-11	50°39'N	11°22'E
Saarbrücken, Ger. (zähr'brü-kĕn)	210-11	49°14'N	6°60'E
Saaremaa, i., Est.	208-09	58°25'N	22°30'E
Saavedra, Arg. (sä-ȧ-vä'drä)	187	37°46'S	62°21'W
Saba, i., N.A. (sä'bä)	155b	17°38'N	63°14'W
Šabac, Serb. (shä'bȧts)	216-17	44°46'N	19°42'E
Sabadell, Spain (sä-bä-dhāl')	214-15	41°33'N	2°06'E
Sabah, hist. reg., Malay.	268-69	5°20'N	117°10'E
Sabanagrande, Hond. (sä-bä'nä-grä'n-dĕ)	161	13°49'N	87°17'W
Sabanalarga, Col. (sä-bá'nä-lär'gä)	178-79	10°38'N	74°55'W
Sabancuy, Mex. (sä-bäŋ-kwē')	160	18°58'N	91°11'W
Sabang, Indon. (sä'bäng)	268-69	0°13'N	119°53'E
Sabang, Indon. (sä'bäng)	266-67	5°53'N	95°20'E
Şāberī, Hāmūn-e, lk., Asia	252-53	31°30'N	61°20'E
Sabhā, Libya	280-81	27°01'N	14°28'E
Sabi, stm., Afr. (sä'bē) *see* Save	286-87	20°58'S	35°04'E
Sabinal, Cayo, i., Cuba (kä'yō sä-bē-näl')	154-55	21°40'N	77°18'W
Sabinas, Mex.	132-33	27°51'N	101°07'W
Sabinas, stm., Mex. (sä-bē'näs)	132-33	26°51'N	99°35'W
Sabinas, stm., Mex. (sä-bē'näs)	132-33	27°29'N	100°40'W
Sabinas Hidalgo, Mex. (sä-bē'näs ē-däl'gō)	132-33	26°30'N	100°10'W
Sabine, stm., U.S.	120-21	30°00'N	93°46'W
Sable, Cape, c., N.S., Can. (kāp sā'b'l)	148-49	43°25'N	65°37'W
Sable, Cape, pen., Fl., U.S. (kāp sā'b'l)	135a	25°12'N	81°05'W
Sable, Île de, i., N. Cal.	296-97	19°15'S	159°56'E
Sable Island, i., N.S., Can.	148-49	43°56'N	59°56'W
Sablé-sur-Sarthe, Fr. (säb-lā-sür-särt')	212-13	47°50'N	0°20'W
Sabor, stm., Port. (sä-bōr')	214-15	41°11'N	7°07'W
Sabzevār, Iran	252-53	36°13'N	57°40'E
Sac, stm., Mo., U.S. (sŏk)	130-31	38°01'N	93°44'W
Sac City, Ia., U.S. (sŏk sĭ'tē)	124-25	42°25'N	94°60'W
Sacheon, Kor., S.	263	34°56'N	128°05'E
Sachigo, stm., On., Can.	144-45	55°04'N	88°59'W
Sachigo Lake, lk., On., Can. (săch'ĭ-gō lāk)	144-45	53°49'N	92°08'W
Sachsen, hist. reg., Ger. (zäk'sĕn) *see* Saxony	210-11	52°45'N	9°30'E

Feature (Pronunciation)	Page	Lat.	Long.
Sackville, N.B., Can. (săk'vĭl)	148-49	45°54'N	64°22'W
Saco, Me., U.S. (sô'kô)	126-27	43°30'N	70°27'W
Sacramento, Ca., U.S. (săk-rȧ-mĕn'tō)	128-29	38°35'N	121°29'W
Sacramento, stm., Ca., U.S. (săk-rȧ-mĕn'tō)	120-21	38°03'N	121°53'W
Sacramento Mountains, mts., N.M., U.S.	130-31	32°42'N	105°37'W
Şa'dah, Yemen	288	16°49'N	43°48'E
Sadiya, India (sŭ-dē'yä)	254-55	27°50'N	95°40'E
Sado, i., Japan (sä'dō)	265	38°0'N	138°25'E
Saeki, Japan (sä'ȧ-kē) *see* Saiki	265	32°58'N	131°55'E
Safâga, Egypt	291b	26°45'N	33°56'E
Safford, Az., U.S. (săf'fērd)	128-29	32°50'N	109°43'W
Safi, Mor. (sä'fē) (äs'fē)	280-81	32°18'N	9°13'W
Safīd Koh, Selseleh-ye, mts., Afg.	252-53	34°30'N	63°30'E
Safonovo, Russia	218-19	55°07'N	33°15'E
Safranbolu, Tur.	246	41°15'N	32°42'E
Saga, China	254-55	29°29'N	85°09'E
Saga, Japan	265	33°15'N	130°18'E
Sagaing, Mya.	266-67	21°53'N	95°59'E
Sagami-nada, b., Japan (sä'gä'mē nä-dä)	265	34°60'N	139°30'E
Saganaga Lake, lk., N.A. (să-gȧ-nä'gȧ lāk)	144-45	48°14'N	90°52'W
Sāgar, India	254-55	23°50'N	78°45'E
Sagarmāthā, mtn., Asia *see* Everest, Mount	254-55	27°59'N	86°56'E
Sagavanirktok, stm., Ak., U.S.	136	70°21'N	148°11'W
Saginaw, Mi., U.S. (săg'ĭ-nô)	126-27	43°26'N	83°58'W
Saginaw Bay, b., Mi., U.S. (săg'ĭ-nô bā)	126-27	43°50'N	83°40'W
Sagua de Tánamo, Cuba (sä-gwä dĕ tä'nä-mō)	154-55	20°35'N	75°14'W
Sagua la Grande, Cuba (sä-gwä lä grä'n-dĕ)	154-55	22°49'N	80°04'W
Saguaro National Park, n.p., Az., U.S. (säg-wä'rō năsh'ŭn-ăl pärk)	128-29	32°16'N	111°12'W
Saguenay, stm., Qc., Can. (săg-ē-nā')	148-49	48°08'N	69°41'W
Sagunt, Spain	214-15	39°41'N	0°16'W
Sagunto, Spain (sä-gōn'tō) *see* Sagunt	214-15	39°41'N	0°16'W
Sa'gya, China	254-55	28°54'N	88°04'E
Sahara, des., Afr. (sȧ-hä'rȧ)	280-81	26°0'N	13°00'E
Sahāranpur, India (sŭ-hä'rŭn-pōōr')	254-55	29°58'N	77°33'E
Sahel, reg., Afr.	280-81	12°0'N	17°00'E
Sāḩil, reg., Afr. *see* Sahel	280-81	12°0'N	17°00'E
Sāhīwāl, Pak.	252-53	30°40'N	73°06'E
Şaḩrā', des., Afr. *see* Sahara	280-81	26°0'N	13°00'E
Saïda, Alg.	200-01	34°50'N	0°09'E
Saidpur, Bngl.	254-55	25°47'N	88°54'E
Saigon, Viet. *see* Ho Chi Minh City	266-67	10°45'N	106°40'E
Saiki, Japan	265	32°58'N	131°55'E
Saimaa, lk., Fin. (sä'ĭ-mä)	208-09	61°15'N	28°15'E
Saín Alto, Mex. (sä-ēn' äl'tō)	158-59	23°35'N	103°13'W
Saint Albans, Eng., U.K. (sănt ôl'bănz)	206-07	51°45'N	0°21'W
Saint Albans, Vt., U.S. (sănt ôl'bănz)	126-27	44°49'N	73°05'W
Saint Albans, W.V., U.S. (sănt ôl'bănz)	126-27	38°23'N	81°50'W
Saint Albert, Ab., Can. (sănt ăl'bērt)	142-43	53°38'N	113°38'W
Saint-Amand-Mont-Rond, Fr. (săN't ȧ-mäN' môN-rôN')	212-13	46°43'N	2°30'E
Saint-André, Cap, c., Madag. *see* Vilanandro, Tanjona	286-87	16°12'S	44°28'E
Saint Andrews, Scot., U.K.	206-07	56°20'N	2°48'W
Saint-Anselme, Qc., Can. (săN' tăN-sĕlm')	148-49	46°37'N	70°57'W
Saint Anthony, Nf./L., Can. (săN ăn'thô-nē)	148-49	51°22'N	55°37'W
Saint Anthony, Id., U.S. (sănt ăn'thô-nē)	122-23	43°58'N	111°41'W
Saint-Augustin, Qc., Can.	148-49	51°14'N	58°38'W
Saint Augustine, Fl., U.S. (sănt ô'gŭs-tēn)	134-35	29°54'N	81°19'W
Saint-Barthélemy, i., Guad.	155b	17°54'N	62°50'W
Saint Bees Head, c., Eng., U.K. (sănt bēz'hĕd)	206-07	54°31'N	3°38'W
Saint Bride, Mount, mtn., Ab., Can. (mount sănt brīd)	142-43	51°31'N	115°57'W
Saint-Brieuc, Fr. (săN' brēs')	212-13	48°31'N	2°45'W
Saint Catharines, On., Can. (sănt kăth'ȧ-rĭnz)	146-47	43°10'N	79°14'W
Saint-Chamond, Fr. (săN' shȧ-môN')	212-13	45°29'N	4°31'E
Saint Charles, Il., U.S. (sănt chärlz')	126-27	41°55'N	88°19'W
Saint Charles, Md., U.S. (sănt chärlz')	126-27	38°35'N	76°57'W
Saint Charles, Mi., U.S. (sănt chärlz')	126-27	43°18'N	84°08'W
Saint Charles, Mn., U.S. (sănt chärlz')	124-25	43°58'N	92°03'W
Saint Charles, Mo., U.S. (sănt chärlz')	130-31	38°48'N	90°29'W
Saint Christopher, i., St. K./N.	155b	17°20'N	62°45'W

ăt; finăl; rāte; senāte; ärm; ȧsk; sofȧ; fâre; ch-choose; dh-as th in other; bē; ēvent; bĕt; recĕnt; cratẽr; g-gō; gh-guttural g; bĭt; ĭ-short neutral; rīde; K-guttural k as ch in German ich;

PRONOUNCING INDEX

Saint Christopher and Nevis – Salina, Isola 403

Feature (Pronunciation)	Page	Lat.	Long.
Saint Christopher and Nevis, nation, N.A. *see* Saint Kitts and Nevis	**152-53**	17°20'N	62°45'W
Saint Clair, Mi., U.S. (sānt klâr)	**126-27**	42°50'N	82°29'W
Saint Clair, Mo., U.S. (sānt klâr)	**130-31**	38°21'N	90°59'W
Saint-Claude, Fr. (săn´ klōd´)	**212-13**	46°23'N	5°51'E
Saint Cloud, Fl., U.S. (sānt kloud´)	**135a**	28°15'N	81°17'W
Saint Cloud, Mn., U.S. (sānt kloud)	**124-25**	45°33'N	94°10'W
Saint Croix, i., V.I.U.S. (sānt kroi´)	**155b**	17°45'N	64°45'W
Saint Croix, stm., N.A. (sānt kroi´)	**148-49**	45°10'N	67°09'W
Saint Croix, stm., U.S. (sānt kroi´)	**124-25**	44°45'N	92°48'W
Saint-Denis, Fr. (săn´dē-nē´)	**212-13**	48°57'N	2°21'E
Saint-Denis, nat. cap., Reu. (săn´dē-nē´)	**287a**	20°52'S	55°28'E
Saint-Dizier, Fr. (săn dē-zyā´)	**212-13**	48°39'N	4°57'E
Sainte-Agathe-des-Monts, Qc., Can.	**146-47**	46°03'N	74°17'W
Sainte-Foy, Qc., Can. (săn fwä)	**146-47**	46°47'N	71°17'W
Sainte Genevieve, Mo., U.S. (sānt jĕn´ĕ-vēv)	**130-31**	37°59'N	90°03'W
Saint Elias, Mount, mtn., N.A. (mount sānt ē-lī´*ă*s)	**136**	60°18'N	140°55'W
Saint-Élie, Fr. Gu.	**178-79**	4°50'N	53°17'W
Sainte-Lucie, Canal de, strt., N.A. *see* Saint Lucia Channel	**155b**	14°09'N	60°57'W
Sainte-Marguerite, stm., Qc., Can.	**148-49**	50°09'N	66°36'W
Sainte-Marie, Cap, c., Madag. *see* Vohimena, Tanjona	**286-87**	25°36'S	45°09'E
Sainte Marie, Nosy, i., Madag.	**286-87**	16°50'S	49°57'E
Saint-Étienne, Fr.	**212-13**	45°26'N	4°24'E
Saint-Eustache, Qc., Can. (săn´ tû-stásh´)	**146-47**	45°34'N	73°55'W
Saint-Félicien, Qc., Can. (săn fā-lē-syăn´)	**146-47**	48°39'N	72°27'W
Saint-Florent-sur-Cher, Fr. (săn´ flô-rän´sür-shâr´)	**212-13**	46°59'N	2°15'E
Saint-Flour, Fr. (săn flōōr´)	**212-13**	45°02'N	3°05'E
Saint Francis, Cape, c., S. Afr.	**286-87**	34°11'S	24°50'E
Saint Gaudens, Fr. (săn gō-däns´)	**212-13**	43°07'N	0°44'E
Saint George, Austl. (sånt jôrj´)	**301**	28°03'S	148°35'E
Saint George, N.B., Can. (săn jôrj´)	**148-49**	45°08'N	66°49'W
Saint George, S.C., U.S. (sānt jôrj´)	**134-35**	33°11'N	80°35'W
Saint George, Ut., U.S. (sānt jôrj´)	**128-29**	37°06'N	113°34'W
Saint George, Cape, c., Nf./L., Can. (kāp sånt jôr-jĕz)	**148-49**	48°29'N	59°15'W
Saint George, Cape, c., Fl., U.S. (kāp sānt jôr-jĕz)	**134-35**	29°35'N	85°04'W
Saint George Island, i., Fl., U.S.	**134-35**	29°39'N	84°53'W
Saint-Georges, Fr. Gu.	**178-79**	3°57'N	51°48'W
Saint George's, nat. cap., Gren. (sånt jôrj´ĕs)	**155b**	12°04'N	61°45'W
Saint George's Bay, b., Nf./L., Can. (sånt jôr-jĕz bā)	**148-49**	48°20'N	59°00'W
Saint Georges Bay, b., N.S., Can. (sånt jôr-jĕz bā)	**148-49**	45°50'N	61°45'W
Saint George's Channel, strt., Eur. (sånt jôr-jĕz chăn´ĕl)	**206-07**	52°0'N	6°00'W
Saint-Girons, Fr. (săn zhē-rôn´)	**212-13**	42°59'N	1°09'E
Saint Helena, dep., Afr. (sānt hĕ-lē´n*á*)	**274**	15°57'S	5°42'W
Saint Helena, i., St. Hel. (sānt hĕ-lē´n*á*)	**275**	15°57'S	5°43'W
Saint Helens, Or., U.S. (sānt hĕl´ĕnz)	**122-23**	45°52'N	122°48'W
Saint Helens, Mount, vol., Wa., U.S. (mount sānt hĕl´ĕnz)	**122-23**	46°12'N	122°11'W
Saint Helier, nat. cap., Jersey (sånt hyĕl´yēr)	**212-13**	49°12'N	2°07'W
Saint-Hyacinthe, Qc., Can.	**146-47**	45°38'N	72°57'W
Saint Ignace, Mi., U.S. (sānt ĭg´nås)	**126-27**	45°52'N	84°44'W
Saint Ignace Island, i., On., Can. (sånt ĭg´nås ī´l*ă*nd)	**146-47**	48°48'N	87°56'W
Saint James, Mn., U.S. (sānt jāmz´)	**124-25**	43°59'N	94°38'W
Saint James, Mo., U.S. (sānt jāmz´)	**130-31**	37°60'N	91°37'W
Saint James, Cape, c., B.C., Can. (kāp sånt jāmz´)	**142-43**	51°56'N	131°01'W
Saint-Jean, Lac, res., Qc., Can. (läk săn´ zhän´)	**146-47**	48°35'N	72°05'W
Saint-Jean-d'Angély, Fr. (săn-zhän´-dän-zhå-lē´)	**212-13**	45°57'N	0°31'W
Saint-Jean-de-Luz, Fr. (săn-zhän´ dē lüz´)	**212-13**	43°24'N	1°39'W
Saint-Jean-sur-Richelieu, Qc., Can.	**146-47**	45°19'N	73°16'W
Saint-Jérôme, Qc., Can. (săn zhā-rōm´)	**146-47**	45°47'N	74°00'W
Saint John, N.B., Can. (sånt jŏn)	**148-49**	45°17'N	66°04'W
Saint John, i., V.I.U.S. (sånt jŏn)	**155b**	18°20'N	64°45'W
Saint John, stm., N.A. (sånt jŏn)	**148-49**	45°16'N	66°04'W
Saint John, Cape, c., Nf./L., Can. (kāp sånt jŏn)	**148-49**	49°59'N	55°32'W
Saint John's, Nf./L., Can. (sånt jŏns)	**148-49**	47°34'N	52°43'W
Saint Johns, Az., U.S. (sānt jŏnz)	**128-29**	34°30'N	109°22'W
Saint Johns, Mi., U.S. (sānt jŏnz)	**126-27**	42°60'N	84°33'W
Saint John's, nat. cap., Antig. (sånt jŏnz)	**155b**	17°07'N	61°51'W
Saint Johns, stm., Fl., U.S. (sānt jŏnz)	**135a**	30°24'N	81°23'W
Saint Johnsbury, Vt., U.S. (sānt jŏnz´bĕr-ē)	**126-27**	44°26'N	72°01'W
Saint Joseph, Mi., U.S. (sānt jō´sĕf)	**126-27**	42°05'N	86°29'W
Saint Joseph, Mo., U.S. (sānt jō´sĕf)	**130-31**	39°46'N	94°50'W
Saint Joseph, stm., U.S. (sānt jō´sĕf)	**126-27**	42°06'N	86°29'W
Saint Joseph, Lake, lk., On., Can.	**144-45**	51°03'N	90°52'W
Saint-Joseph-de-Beauce, Qc., Can. (sĕn zhō-zĕf´dĕ bōs)	**148-49**	46°18'N	70°52'W
Saint-Junien, Fr. (săn´zhü-nyăn´)	**212-13**	45°53'N	0°54'E
Saint Kilda, i., Scot., U.K. (sånt kĭl´d*á*)	**206-07**	57°49'N	8°36'W
Saint Kitts, i., St. K./N. (sånt kĭtts) *see* Saint Christopher	**155b**	17°20'N	62°45'W
Saint Kitts and Nevis, nation, N.A. (sånt kĭts ănd nē´v*ŭ*s)	**152-53**	17°20'N	62°45'W
Saint-Laurent, stm., N.A. *see* Saint Lawrence	**95**	49°14'N	67°01'W
Saint-Laurent, Golfe du, b., Can. *see* Saint Lawrence, Gulf of	**148-49**	48°0'N	62°00'W
Saint-Laurent du Maroni, Fr. Gu.	**178-79**	5°28'N	54°02'W
Saint Lawrence, Nf./L., Can. (sånt lô´rĕns)	**148-49**	46°56'N	55°24'W
Saint Lawrence, stm., N.A. (sånt lô´rĕns)	**95**	49°14'N	67°01'W
Saint Lawrence, Gulf of, b., Can. (gŭlf ŭv sånt lô´rĕns)	**148-49**	48°0'N	62°00'W
Saint Lawrence Island, i., Ak., U.S. (sānt lô´rĕns ī´l*ă*nd)	**136**	63°30'N	170°30'W
Saint-Louis, Sen.	**282-83**	16°01'N	16°29'W
Saint Louis, Mi., U.S. (sānt lōō´ĭs)	**126-27**	43°24'N	84°36'W
Saint Louis, Mo., U.S. (sānt lōō´ĭs) (lōō´ē)	**130-31**	38°39'N	90°13'W
Saint Lucia, nation, N.A. (sānt lōō´-sh*ă*)	**152-53**	13°53'N	60°58'W
Saint Lucia, Lake, lk., S. Afr.	**286-87**	28°04'S	32°28'E
Saint Lucia Channel, strt., N.A. (sānt lū´shĭ-*á* chăn´ĕl)	**155b**	14°09'N	60°57'W
Saint-Malo, Fr. (săn´ mä-lō´)	**212-13**	48°39'N	2°01'W
Saint-Marc, Haiti (săn´ mårk´)	**154-55**	19°07'N	72°41'W
Saint Maries, Id., U.S. (sānt mā´rēs)	**122-23**	47°19'N	116°34'W
Saint-Martin, i., N.A. (săn-mär´tĭn)	**155b**	18°04'N	63°04'W
Saint Martinville, La., U.S. (sānt mär´tĭn-vĭl)	**134-35**	30°08'N	91°50'W
Saint Marys, Austl. (sānt mā´rēz)	**301**	41°35'S	148°11'E
Saint Marys, Ga., U.S. (sānt mā´rēz)	**134-35**	30°44'N	81°33'W
Saint Marys, Oh., U.S. (sānt mā´rēz)	**126-27**	40°33'N	84°24'W
Saint Marys, Pa., U.S. (sānt mā´rēz)	**126-27**	41°25'N	78°35'W
Saint Marys, W.V., U.S. (sānt mā´rēz)	**126-27**	39°23'N	81°12'W
Saint Mary's, Cape, c., Nf./L., Can.	**148-49**	46°50'N	54°12'W
Saint Mary's Bay, b., Nf./L., Can.	**148-49**	46°50'N	53°47'W
Saint Matthew Island, i., Ak., U.S. (sånt măth´ū ī´l*ă*nd)	**236-37**	60°29'N	186°53'W
Saint Matthews, S.C., U.S. (sånt măth´ūz)	**134-35**	33°40'N	80°47'W
Saint Matthias Group, is., Pap. N. Gui.	**302**	1°36'S	149°47'E
Saint-Maurice, stm., Qc., Can. (săn´ mō-rēs´) (sånt mô´rĭs)	**146-47**	46°21'N	72°31'W
Saint Michael, Ak., U.S. (sānt mī´k*ĕ*l)	**136**	63°29'N	162°02'W
Saint-Mihiel, Fr. (săn´ mē-yĕl´)	**212-13**	48°54'N	5°32'E
Saint-Nazaire, Fr. (săn´ná-zâr´)	**212-13**	47°17'N	2°13'W
Saint-Omer, Fr. (săn´tô-mâr´)	**212-13**	50°45'N	2°16'E
Saint Paul, Ab., Can. (sånt pôl´)	**142-43**	53°60'N	111°17'W
Saint-Paul, Reu.	**287a**	21°0'S	55°16'E
Saint Paul, Mn., U.S. (sānt pôl)	**124-25**	44°57'N	93°06'W
Saint Paul, Ne., U.S. (sānt pôl)	**124-25**	41°13'N	98°28'W
Saint Paul, stm., Lib.	**282-83**	6°25'N	10°44'W
Saint Pauls, N.C., U.S. (sānt pôls)	**134-35**	34°49'N	78°58'W
Saint Peter, Mn., U.S. (sānt pē´tēr)	**124-25**	44°20'N	93°58'W
Saint Peter Port, nat. cap., Guern. (sånt pē´tēr pōrt)	**212-13**	49°28'N	2°33'W
Saint Petersburg, Russia (sånt pē´tērz-bûrg)	**208-09**	59°57'N	30°15'E
Saint Petersburg, Fl., U.S. (sānt pē´tērz-bûrg)	**135a**	27°46'N	82°40'W
Saint-Pierre, Reu.	**287a**	21°19'S	55°29'E
Saint-Pierre, i., Sey.	**286-87**	9°19'S	50°43'E
Saint-Pierre, nat. cap., St. P./M. (săn´pyâr´)	**148-49**	46°47'N	56°12'W
Saint Pierre and Miquelon, dep., N.A. (sånt pē-âr´ ănd mĭk-ē-lôn´)	**148-49**	46°55'N	56°20'W
Saint-Pierre-et-Miquelon, dep., N.A. *see* Saint Pierre and Miquelon	**148-49**	46°55'N	56°20'W
Saint-Pol-de-Léon, Fr. (săn-pô´dĕ-lā-ôn´)	**212-13**	48°41'N	3°59'W
Saint-Quentin, Fr. (săn´kän-tăn´)	**212-13**	49°51'N	3°18'E
Saint-Sébastien, Cap, c., Madag. *see* Anorontany, Tanjona	**286-87**	12°26'S	48°45'E
Saint Stephen, N.B., Can. (sånt stē´vĕn)	**148-49**	45°12'N	67°17'W
Saint Thomas, On., Can. (sånt tŏm´*ă*s)	**146-47**	42°47'N	81°11'W
Saint Thomas, i., V.I.U.S.	**155b**	18°21'N	64°55'W
Saint-Tropez, Fr. (săn trô-pĕ´)	**212-13**	43°16'N	6°38'E
Saint Vincent, i., St. Vin.	**155b**	13°15'N	61°12'W
Saint-Vincent, Cap, c., Madag. *see* Ankaboa, Tanjona	**286-87**	21°55'S	43°18'E
Saint Vincent, Gulf, b., Austl. (gŭlf vĭn´sĕnt)	**301**	34°47'S	138°06'E
Saint Vincent and the Grenadines, nation, N.A. (sānt vĭn´sĕnt ănd th*á* grĕn´*ă*-dēnz)	**152-53**	13°15'N	61°12'W
Saipan, i., N. Mar. Is.	**304-05**	15°12'N	145°45'E
Saitama, state, Japan (sī´tä-mä)	**265**	36°0'N	139°30'E
Sajama, Nevado, mtn., Bol. (nĕ-vá´dō-sä-há´mä)	**182-83**	18°06'S	68°54'W
Sak, stm., S. Afr.	**286-87**	30°06'S	20°42'E
Sakai, Japan (sä´kä-ē)	**265**	34°35'N	135°29'E
Sakākah, Sau. Ar.	**248-49**	29°58'N	40°13'E
Sakakawea, Lake, res., N.D., U.S.	**124-25**	47°44'N	102°18'W
Sakami, Lac, lk., Qc., Can.	**140-41**	53°15'N	76°45'W
Sakart'velo, nation, Asia *see* Georgia	**245**	42°0'N	44°00'E
Sakarya, Tur.	**202-03**	40°47'N	30°24'E
Sakarya, stm., Tur. (sȧ-kär´yȧ)	**202-03**	41°07'N	30°39'E
Sakata, Japan (sä´kä-tä)	**264**	38°55'N	139°51'E
Sakha, state, Russia *see* Yakutia	**236-37**	67°0'N	125°00'E
Sakhalin, i., Russia (sȧ-kȧ-lēn´)	**236-37**	51°0'N	143°00'E
Šakiai, Lith. (shä´kĭ-ī)	**208-09**	54°58'N	23°04'E
Sakishima-shotō, is., Japan (sä kē-shē mä gōn tō´)	**306a**	24°33'N	124°26'E
Sal, i., C.V. (säal)	**282-83**	16°49'N	22°57'W
Sal, stm., Russia (sȧl)	**202-03**	47°31'N	40°44'E
Sal, Cay, i., Bah. (kē säl)	**154-55**	23°43'N	80°25'W
Sala, Swe. (sô´lä)	**208-09**	59°56'N	16°37'E
Salaberry-de-Valleyfield, Qc., Can.	**146-47**	45°15'N	74°08'W
Sala Consilina, Italy (sä´lä kōn-sē-lē´nä)	**216-17**	40°25'N	15°34'E
Salada, Laguna, lk., Mex. (lä gō´nä sä lä´dä)	**128-29**	32°20'N	115°40'W
Saladas, Arg.	**187**	28°14'S	58°39'W
Saladillo, Arg. (sä-lä-dēl´yō)	**187**	35°38'S	59°47'W
Salado, stm., Arg. (sä-lä´dō)	**182-83**	31°41'S	60°44'W
Salado, stm., Arg. (sä-lä´dō)	**187**	35°45'S	57°23'W
Salado, stm., Arg. (sä-lä´dō)	**186**	38°49'S	64°50'W
Salado, stm., Mex. (sä-lä´dô)	**132-33**	26°52'N	99°19'W
Şalālah, Oman	**238-39**	17°01'N	54°06'E
Salamanca, Chile (sä-lä-mä´n-kä)	**182-83**	31°46'S	70°59'W
Salamanca, Mex. (sä-lä-mä´n-kä)	**158-59**	20°34'N	101°12'W
Salamanca, Spain (sä-lä-mä´n-kȧ)	**214-15**	40°58'N	5°39'W
Salamanca, N.Y., U.S. (săl-*ȧ*-măŋ´k*ȧ*)	**126-27**	42°10'N	78°43'W
Salamat, Bahr, stm., Chad (bär sä-lä-mät´)	**284-85**	9°27'N	18°06'E
Salamina, Col. (sä-lä-mē´-nä)	**188c**	5°25'N	75°29'W
Salatiga, Indon.	**268-69**	7°20'S	110°31'E
Salavat, Russia	**202-03**	53°22'N	55°56'E
Salaverry, Peru (sä-lä-vă´rē)	**184**	8°14'S	78°58'W
Salawati, i., Indon. (sä-lä-wä´tē)	**242-43**	1°07'S	130°52'E
Sala y Gómez, Isla, i., Chile	**308-09**	26°26'S	105°26'W
Saldanha, S. Afr.	**286-87**	32°60'S	17°57'E
Saldus, Lat. (sȧl´dŏs)	**208-09**	56°40'N	22°30'E
Sale, Austl. (sāl)	**301**	38°07'S	147°04'E
Salé, Mor.	**292a**	34°03'N	6°48'W
Salebabu, Pulau, i., Indon.	**268-69**	3°56'N	126°42'E
Salekhard, Russia (sŭ-lyĭ-kärt)	**236-37**	66°32'N	66°37'E
Salem, India (sä´lĕm)	**256**	11°39'N	78°10'E
Salem, Il., U.S. (sā´lĕm)	**126-27**	38°37'N	88°57'W
Salem, In., U.S. (sā´lĕm)	**126-27**	38°36'N	86°06'W
Salem, Mo., U.S. (sā´lĕm)	**130-31**	37°39'N	91°32'W
Salem, Oh., U.S. (sā´lĕm)	**126-27**	40°54'N	80°51'W
Salem, Or., U.S.	**122-23**	44°56'N	123°01'W
Salem, S.D., U.S. (sā´lĕm)	**124-25**	43°44'N	97°23'W
Salem, Va., U.S. (sā´lĕm)	**134-35**	37°18'N	80°03'W
Salem, W.V., U.S. (sā´lĕm)	**126-27**	39°17'N	80°34'W
Salerno, Italy (sä-lĕr´nō)	**216-17**	40°41'N	14°47'E
Salerno, Golfo di, b., Italy (gôl-fō-dē-sä-lĕr´nō)	**216-17**	40°32'N	14°42'E
Salgótarján, Hung. (shôl´gō-tôr-yän)	**210-11**	48°06'N	19°50'E
Salida, Co., U.S. (sȧ-lī´dȧ)	**128-29**	38°32'N	105°60'W
Salīmah, Wāḥat, well, Sudan	**288**	21°22'N	29°19'E
Salina, Ks., U.S. (s*ȧ*-lī´n*ȧ*)	**130-31**	38°50'N	97°36'W
Salina, Ut., U.S. (s*ȧ*-lī´n*ȧ*)	**128-29**	38°58'N	111°52'W
Salina, Isola, i., Italy (ē´-sō-lä-sä-lē´nä)	**216-17**	38°34'N	14°50'E

n-sing; ŋ-baŋk; ɴ-nasalized n; nŏd; c*ŏ*mmit; ōld; ŏbey; ôrder; oi-boil; fōōd; ȯ-as oo in foot; ou-out; s-soft; sh-dish; th-thin; pūre; ůnite; ûrn; stŭd; circ*ŭ*s; ü-as in French tu; ´-indeterminate vowel.

404 Salina Cruz – San Ildefonso ó la Granja PRONOUNCING INDEX

Feature (Pronunciation)	Page	Lat.	Long.
Salina Cruz, Mex. (sä-lē´nä krōōz´)	158-59	16°11′N	95°11′W
Salinas, Ec.	184	2°13′S	80°57′W
Salinas, Ca., U.S. (sȧ-lē´nȧs)	128-29	36°41′N	121°40′W
Salinas de Hidalgo, Mex.	158-59	22°38′N	101°44′W
Saline, stm., Ar., U.S. (sȧ-lēn´)	130-31	33°09′N	92°08′W
Salisbury, Md., U.S.	126-27	38°22′N	75°36′W
Salisbury, Mo., U.S. (sôlz´bē-rē)	130-31	39°26′N	92°48′W
Salisbury, N.C., U.S. (sôlz´bē-rē)	134-35	35°40′N	80°28′W
Salisbury Island, i., Nu., Can.	140-41	63°30′N	76°60′W
Salliq, Nu., Can. *see* Coral Harbour	138-39	64°08′N	83°12′W
Sallisaw, Ok., U.S. (săl´ĭ-sô)	130-31	35°28′N	94°48′W
Salluit, Qc., Can.	138-39	62°13′N	75°36′W
Salmon, Id., U.S. (săm´ŭn)	122-23	45°11′N	113°54′W
Salmon, stm., B.C., Can. (săm´ŭn)	142-43	54°04′N	122°33′W
Salmon, stm., N.B., Can. (săm´ŭn)	148-49	46°04′N	65°55′W
Salmon, stm., Id., U.S. (săm´ŭn)	122-23	45°51′N	116°47′W
Salmon Arm, B.C., Can. (săm´ŭn ärm)	142-43	50°42′N	119°19′W
Salmon River Mountains, mts., Id., U.S. (săm´ŭn rĭv´ẽr moun´tĭnz)	122-23	44°58′N	114°52′W
Salon-de-Provence, Fr. (sȧ-lôN-dẽ-prô-väNs´)	212-13	43°39′N	5°05′E
Salonika, Grc. *see* Thessaloníki	216-17	40°38′N	22°59′E
Salsk, Russia (sälsk)	202-03	46°28′N	41°33′E
Salt, stm., Az., U.S. (sôlt)	128-29	33°23′N	112°17′W
Salta, Arg. (säl´tä)	182-83	24°48′S	65°25′W
Salta, state, Arg. (säl´tä)	182-83	25°0′S	64°30′W
Saltillo, Mex. (säl-tēl´yō)	132-33	25°26′N	101°00′W
Salt Lake City, Ut., U.S. (sôlt lāk sĭ´tĭ sĭ´tē)	122-23	40°47′N	111°54′W
Salto, Arg. (säl´tō)	187	34°18′S	60°15′W
Salto, Ur.	187	31°23′S	57°58′W
Salto Grande, Embalse, res., S.A.	187	30°55′S	57°54′W
Salto Grande, Embalse de, res., S.A. *see* Salto Grande, Embalse	187	30°55′S	57°54′W
Salton Sea, lk., Ca., U.S. (sôlt´ŭn sē)	128-29	33°19′N	115°50′W
Saltville, Va., U.S. (sôlt´vĭl)	134-35	36°53′N	81°46′W
Saluda, S.C., U.S. (sȧ-lōō´dȧ)	134-35	34°00′N	81°47′W
Salûm, Egypt	204-05	31°34′N	25°09′E
Saluzzo, Italy (sä-lōōt´sō)	216-17	44°39′N	7°29′E
Salvador, Braz. (säl-vä-dōr´)	180-81	12°59′S	38°30′W
Salvador, El, nation, N.A. *see* El Salvador	94	13°50′N	88°55′W
Salvador, Lake, lk., La., U.S. (lāk säl´-vä-dōr lāk)	134-35	29°45′N	90°15′W
Salvatierra, Mex. (säl-vä-tyĕr´rä)	158-59	20°13′N	100°54′W
Salyan, Azer.	245	39°35′N	48°58′E
Salzburg, Aus. (sälts´bôrgh)	210-11	47°49′N	13°03′E
Salzwedel, Ger. (sälts-vā´dĕl)	210-11	52°51′N	11°09′E
Samâlût, Egypt (sȧ-mä-lōōt´)	291b	28°18′N	30°42′E
Samana Cay, i., Bah.	154-55	23°05′N	73°44′W
Samar, i., Phil. (sä´mär)	270	12°0′N	125°00′E
Samara, Russia (sȧ-mä´rȧ)	202-03	53°11′N	50°07′E
Samara, stm., Russia (sȧ-mä´rȧ)	202-03	53°10′N	50°04′E
Samara, stm., Ukr. (sȧ-mä´rȧ)	218-19	48°28′N	35°06′E
Samarai, Pap. N. Gui. (sä-mä-rä´ē)	302	10°36′S	150°42′E
Samarinda, Indon.	268-69	0°30′S	117°09′E
Samarqand, Uzb.	252-53	39°40′N	66°56′E
Sāmarrā', Iraq	248-49	34°11′N	43°53′E
Samaúna, Braz.	180-81	7°56′S	60°01′W
Sambalpur, India (sŭm´bŭl-pŏr)	254-55	21°28′N	83°59′E
Sambas, Indon.	266-67	1°19′N	109°16′E
Sambava, Madag.	286-87	14°16′S	50°09′E
Sambhal, India	254-55	28°35′N	78°34′E
Sāmbhar, India	254-55	26°54′N	75°13′E
Sambir, Ukr.	210-11	49°31′N	23°13′E
Samborombón, Bahía, b., Arg. (bä-ē´ä-säm-bô-rôm-bô´n)	187	36°0′S	57°12′W
Samcheok, Kor., S.	263	37°27′N	129°10′E
Same, Tan.	289	4°04′S	37°44′E
Samoa, nation, Oc. (sȧ-mō´ȧ)	306b	13°55′S	186°00′W
Samoa Islands, is., Oc. (sȧ-mō´ȧ ī´lȧndz)	306b	14°0′S	171°00′W
Samoded, Russia	202-03	63°37′N	40°30′E
Samokov, Blg. (sä´mô-kôf)	216-17	42°20′N	23°34′E
Sámos, i., Grc. (să´mŏs)	216-17	37°42′N	26°50′E
Samothrace, i., Grc. *see* Samothráki	216-17	40°29′N	25°36′E
Samothráki, i., Grc.	216-17	40°29′N	25°36′E
Sampit, Indon.	268-69	2°33′S	112°57′E
Sam Rayburn Reservoir, res., Tx., U.S.	132-33	31°13′N	94°17′W
Samsun, Tur. (sȧm´sōōn´)	202-03	41°17′N	36°20′E
Samtredia, Geor. (sȧm´trĕ-dē)	245	42°10′N	42°21′E
Samui, Ko, i., Thai.	266-67	9°32′N	100°01′E
San, Mali (sän)	280-81	13°18′N	4°54′W
Sanđoy, i., Far. Is.	206b	61°50′N	6°45′W
Şan'ā', nat. cap., Yemen (sän´ä) *see* Sanaa	288	15°21′N	44°12′E
Sanaa, nat. cap., Yemen (sän´ä)	288	15°21′N	44°12′E
Sanaga, stm., Camrn. (sä-nä´gä)	282-83	3°33′N	9°39′E
San Agustin, Cape, c., Phil.	270	6°18′N	126°12′E
Sanana, Pulau, i., Indon.	268-69	2°12′S	125°55′E
Sanandaj, Iran	248-49	35°19′N	47°00′E
San Andreas, Ca., U.S. (săn ăn´drē-ăs)	128-29	38°12′N	120°41′W
San Andrés, Col.	162	12°33′N	81°42′W
San Andrés, Isla de, i., Col. (ē´s-lä-dĕ-sän-än-drĕ´s)	178-79	12°33′N	81°43′W
San Andres Mountains, mts., N.M., U.S. (săn ăn´drē-ăs moun´tĭnz)	128-29	32°59′N	106°36′W
San Andrés Tuxtla, Mex. (sän-än-drā´s-tōōs´tlä)	158-59	18°26′N	95°13′W
San Angelo, Tx., U.S. (săn ăn-jĕ-lō)	132-33	31°29′N	100°26′W
San Antonio, Chile (sän-ȧn-tô´nyō)	188e	33°36′S	71°36′W
San Antonio, Col. (sän-ȧn-tô´nyō)	188c	3°55′N	75°29′W
San Antonio, Tx., U.S. (săn ăn-tō´nē-ō)	132-33	29°25′N	98°29′W
San Antonio, stm., Tx., U.S. (săn ăn-tō´nē-ō)	132-33	28°30′N	96°53′W
San Antonio, Cabo, c., Arg.	187	36°40′S	56°42′W
San Antonio, Cabo de, c., Cuba (ká´bô-dĕ-sän-ăn-tô´nyô)	154-55	21°52′N	84°57′W
San Antonio Bay, b., Tx., U.S. (săn ăn-tō´nē-ô bā)	132-33	28°20′N	96°45′W
San Antonio de los Cobres, Arg. (sän-ȧn-tô´nyō då lōs kō´brås)	182-83	24°13′S	66°19′W
San Antonio Oeste, Arg. (sän-nä-tō´nyô ô-ĕs´tå)	185	40°45′S	64°58′W
San Augustine, Tx., U.S. (săn ô´gŭs-tēn)	132-33	31°31′N	94°07′W
San Benedetto del Tronto, Italy (sän bā´nâ-dĕt´tô dĕl trōn´tô)	216-17	42°58′N	13°53′E
San Benedicto, Isla, i., Mex.	156-57	19°19′N	110°49′W
San Benito, Guat.	160	16°55′N	89°54′W
San Benito, Tx., U.S. (săn bĕ-nē´tō)	132-33	26°08′N	97°38′W
San Bernardino, Ca., U.S. (săn bûr-när-dē´nō)	128-29	34°06′N	117°17′W
San Bernardino Strait, strt., Phil.	270	12°32′N	124°10′E
San Bernardo, Chile (sän bĕr-när´dô)	188e	33°36′S	70°42′W
San Blas, Mex. (sän bläs´)	158-59	21°33′N	105°17′W
San Blas, Mex. (sän bläs´)	156-57	26°05′N	108°46′W
San Blas, Cape, c., Fl., U.S. (kāp săn bläs´)	134-35	29°40′N	85°22′W
San Borja, Bol.	180-81	14°49′S	66°51′W
San Buenaventura, Mex. (sän bwā´nȧ-vĕn-tōō´rȧ)	132-33	27°04′N	101°33′W
San Buenaventura, Ca., U.S. *see* Ventura	128-29	34°17′N	119°17′W
San Carlos, Chile (sän-ká´r-lŏs)	185	36°26′S	71°57′W
San Carlos, Mex. (sän kär´lōs)	132-33	29°01′N	100°51′W
San Carlos, Nic. (sän-kä´r-lôs)	161	11°07′N	84°47′W
San Carlos, Phil.	270	10°30′N	123°25′E
San Carlos, Phil.	270	15°56′N	120°21′E
San Carlos, Az., U.S. (săn kär´lōs)	128-29	33°21′N	110°27′W
San Carlos, Ven.	178-79	9°40′N	68°35′W
San Carlos, stm., C.R. (sän kär´lōs)	161	10°47′N	84°12′W
San Carlos de Bariloche, Arg.	185	41°09′S	71°18′W
San Carlos de Bolívar, Arg.	187	36°13′S	61°07′W
San Carlos del Zulia, Ven.	178-79	9°02′N	71°56′W
San Carlos de Río Negro, Ven.	178-79	1°55′N	67°04′W
San Carlos Indian Reservation, ind. res., Az., U.S. (săn kär´lōs ĭn´dĭ-ăn rĕ-sẽr-vā´shŭn)	128-29	33°23′N	110°09′W
San Cataldo, Italy (sän kä-täl´dō)	216-17	37°29′N	13°59′E
Sánchez, Dom. Rep. (sän´chĕz)	154-55	19°14′N	69°37′W
San Clemente, Monte, mtn., Chile *see* San Valentín, Monte	185	46°36′S	73°20′W
San Clemente Island, i., Ca., U.S. (săn klå-mĕn´tå ī´lȧnd)	128-29	32°54′N	118°29′W
San Cristóbal, Arg.	187	30°19′S	61°13′W
San Cristóbal, Dom. Rep. (sän krēs-tō´bäl)	154-55	18°25′N	70°06′W
San Cristóbal, Ven. (sän krēs-tô´bäl)	178-79	7°45′N	72°13′W
San Cristobal, i., Sol. Is.	306e	10°36′S	161°45′E
San Cristóbal, Isla, i., Ec.	184a	0°50′S	89°26′W
San Cristóbal de las Casas, Mex.	158-59	16°45′N	92°38′W
Sancti Spíritus, Cuba (säŋk´tē spē´rē-tōōs)	154-55	21°56′N	79°27′W
Sancti Spíritus, state, Cuba (säŋk´tē spē´rē-tōōs)	154-55	22°0′N	79°20′W
Sancy, Puy de, mtn., Fr. (pwē-dĕ-säN-sē´)	212-13	45°32′N	2°49′E
Sandakan, Malay. (sȧn-dä´kȧn)	270	5°51′N	118°06′E
Sandefjord, Nor. (sän´dĕ-fyôr´)	208-09	59°08′N	10°14′E
Sanders, Az., U.S. (săn´dẽrz)	128-29	35°14′N	109°20′W
Sanderson, Tx., U.S. (săn´dẽr-sŭn)	132-33	30°09′N	102°24′W
Sandersville, Ga., U.S. (săn´dẽrz-vĭl)	134-35	32°59′N	82°49′W
Sand Hills, hills, Ne., U.S. (sănd hĭlz)	124-25	42°0′N	101°00′W
Sandia, Peru	184	14°16′S	69°27′W
San Diego, Ca., U.S. (săn dē-ā´gô)	128-29	32°43′N	117°08′W
San Diego, Tx., U.S. (săn dē-ā´gô)	132-33	27°46′N	98°14′W
San Diego, Cabo, c., Arg.	185	54°39′S	65°08′W
San Diego de la Unión, Mex. (sän dē-å-gô då lä ōō-nyōn´)	158-59	21°28′N	100°52′W
Sandıklı, Tur.	246	38°28′N	30°16′E
Sandnes, Nor. (sänd´nĕs)	208-09	58°51′N	5°44′E
Sandomierz, Pol. (sȧn-dô´myĕzh)	210-11	50°41′N	21°46′E
San Donà di Piave, Italy (sän dô ná´ dē pyä´vĕ)	216-17	45°38′N	12°34′E
Sandoway, Mya. (sän-dô-wī´)	266-67	18°28′N	94°22′E
Sandpoint, Id., U.S. (sănd point)	122-23	48°17′N	116°33′W
Sand Springs, Ok., U.S. (sănd sprĭngz)	130-31	36°08′N	96°07′W
Sandstone, Mn., U.S. (sănd´stōn)	124-25	46°08′N	92°52′W
Sandusky, Mi., U.S. (săn-dŭs´kē)	126-27	43°25′N	82°49′W
Sandusky, Oh., U.S. (săn-dŭs´kē)	126-27	41°27′N	82°42′W
Sandwich, Il., U.S. (sănd´wĭch)	126-27	41°39′N	88°37′W
Sandy, Ut., U.S. (sănd´ē)	122-23	40°37′N	111°54′W
Sandy Cape, c., Austl.	302	24°42′S	153°16′E
Sandykgaçy, Turkmen.	252-53	36°33′N	62°33′E
Sandy Lake, lk., Nf./L., Can. (sănd´ē lāk)	148-49	49°16′N	57°00′W
Sandy Lake, lk., On., Can. (sănd´ē lāk)	144-45	53°02′N	93°00′W
Sandy Springs, Ga., U.S. (sănd´ē sprĭngz)	134-35	33°56′N	84°23′W
San Estanislao, Para. (sän ĕs-tä-nēs-lå´ô)	182-83	24°39′S	56°29′W
San Felipe, Chile (sän fĕ-lē´på)	188e	32°45′S	70°43′W
San Felipe, Mex. (sän fĕ-lē´pĕ)	158-59	21°29′N	101°13′W
San Felipe, Mex. (sän fĕ-lē´pĕ)	156-57	31°02′N	114°51′W
San Felipe, Ven. (sän fĕ-lē´pĕ)	178-79	10°20′N	68°44′W
San Felipe, Cayos de, is., Cuba (kä´yōs-dĕ-sän-fĕ-lē´pĕ)	154-55	21°58′N	83°30′W
San Félix, Isla, i., Chile (ē´s-lä-dĕ-sän fā-lēks´)	173	26°17′S	80°06′W
San Fernando, Chile	188e	34°35′S	70°59′W
San Fernando, Mex. (sän fĕr-nän´dô)	132-33	24°51′N	98°10′W
San Fernando, Phil.	270	15°01′N	120°41′E
San Fernando, Phil.	270	16°37′N	120°19′E
San Fernando, Trin.	155b	10°17′N	61°27′W
San Fernando de Apure, Ven. (sän-fĕr-nä´n-dō-dĕ-ä-pōō´rå)	178-79	7°53′N	67°27′W
San Fernando de Atabapo, Ven. (sän-fĕr-nä´n-dō-dĕ-ä-tä-bä´pô)	178-79	4°02′N	67°41′W
San Fernando del Valle de Catamarca, Arg.	182-83	28°28′S	65°47′W
Sanford, Fl., U.S. (săn´fôrd)	134-35	28°47′N	81°17′W
Sanford, Me., U.S. (săn´fẽrd)	126-27	43°27′N	70°47′W
Sanford, N.C., U.S. (săn´fẽrd)	134-35	35°29′N	79°11′W
San Francisco, Arg. (sän frăn´sĭs´kô)	187	31°26′S	62°05′W
San Francisco, Ca., U.S. (săn frăn-sĭs´kô)	128-29	37°47′N	122°25′W
San Francisco del Oro, Mex. (sän frăn´sĭs´kô-dĕl ō´rō)	132-33	26°52′N	105°51′W
San Francisco del Rincón, Mex. (sän frăn´sĭs´kô-dĕl rēn-kōn´)	158-59	21°01′N	101°52′W
San Francisco de Macorís, Dom. Rep. (sän frăn´sĭs´kô-dä-mä-kō´rēs)	154-55	19°18′N	70°15′W
San Gabriel Chilac, Mex. (sän-gä-brē-ĕl-chē-läk´)	158-59	18°20′N	97°21′W
Sangar, Russia	236-37	63°55′N	127°29′E
Sangarius, stm., Tur. *see* Sakarya	202-03	41°07′N	30°39′E
Sangay, vol., Ec.	184	2°0′S	78°20′W
Sangeang, Pulau, i., Indon.	268-69	8°12′S	119°04′E
Sangerhausen, Ger. (säng´ĕr-hou-zĕn)	210-11	51°28′N	11°18′E
Sanggan, stm., China	260-61	40°21′N	115°25′E
Sanggau, Indon.	268-69	0°07′N	110°35′E
Sangha, stm., Afr.	282-83	1°12′S	16°50′E
Sangihe, Kepulauan, is., Indon.	268-69	3°0′N	125°30′E
Sangihe, Pulau, i., Indon.	268-69	3°35′N	125°32′E
San Gil, Col. (sän-кē´l)	178-79	6°33′N	73°08′W
San Giovanni in Fiore, Italy (sän jô-vän´nē ēn fyō´rā)	216-17	39°15′N	16°42′E
Sangju, Kor., S. (säng´jōō´)	263	36°25′N	128°10′E
Sāngli, India	256	16°52′N	74°34′E
San Gregorio, Ur.	187	32°38′S	55°50′W
Sangue, stm., Braz.	180-81	10°57′S	58°20′W
Sanibel Island, i., Fl., U.S. (săn´ĭ-bĕl ī´lȧnd)	135a	26°27′N	82°08′W
San Ignacio, Arg.	187	27°16′S	55°33′W
San Ignacio, Mex.	156-57	27°17′N	112°54′W
San Ignacio de Moxo, Bol.	180-81	14°56′S	65°37′W
San Ignacio de Velasco, Bol.	182-83	16°23′S	60°57′W
San Ildefonso, Cape, c., Phil. (kāp sän-ēl-dĕ-fōn-sō)	270	16°02′N	122°00′E
San Ildefonso ó la Granja, Spain (sän-ēl-dĕ-fōn-sō ō lä grän´khä)	214-15	40°54′N	4°00′W

ăt; finăl; rāte; senâte; ärm; ȧsk; sofȧ; fâre; ch-choose; dh-as th in other; bē; ēvent; bĕt; recĕnt; cratẽr; g-gō; gh-guttural g; bĭt; ĭ-short neutral; rīde; к-guttural k as ch in German ich;

PRONOUNCING INDEX

San Isidro – Santiaguillo, Laguna 405

Feature (Pronunciation)	Page	Lat.	Long.
San Isidro, Arg. (sän ē-sē´drô)	187	34°28′S	58°31′W
San Jacinto, Phil. (sän hä-sēn´tō)	270	12°34′N	123°44′E
San Javier, Arg.	187	30°34′S	59°56′W
San Javier, Bol.	182-83	16°20′S	62°38′W
San Joaquín, Bol.	180-81	13°04′S	64°49′W
San Joaquín, stm., Bol.	180-81	13°08′S	63°41′W
San Joaquin Valley, val., Ca., U.S.	128-29	36°55′N	120°29′W
San Jorge, Golfo, b., Arg. (gôl-fô-sän-ĸô´r-ĸě)	185	46°0′S	67°00′W
San Jorge Island, i., Sol. Is.	306e	8°27′S	159°35′E
San Jose, Phil.	270	12°21′N	121°04′E
San Jose, Ca., U.S.	128-29	37°21′N	121°54′W
San José, nat. cap., C.R. (sän hô-sā´)	161	9°56′N	84°05′W
San José, Isla, i., Mex. (ē´s-lä-sän ĸô-sě´)	156-57	25°00′N	110°38′W
San José, Isla, i., Pan. (ē´s-lä-sän hô-sā´)	162	8°15′N	79°07′W
San José de Chiquitos, Bol.	182-83	17°50′S	60°44′W
San José de Feliciano, Arg. (sän ĸô-sě´ då lä ěs-kē´nå)	187	30°23′S	58°45′W
San José de Jáchal, Arg.	182-83	30°14′S	68°45′W
San José del Cabo, Mex.	156-57	23°03′N	109°41′W
San José del Guaviare, Col.	178-79	2°34′N	72°38′W
San José de Mayo, Ur.	187	34°21′S	56°42′W
San Jose Island, i., Tx., U.S.	132-33	28°02′N	96°55′W
San Juan, Arg. (sän hwän´)	182-83	31°32′S	68°32′W
San Juan, state, Arg. (sän hwän´)	182-83	31°0′S	69°00′W
San Juan, nat. cap., P.R. (sän hwän´)	154a	18°28′N	66°07′W
San Juan, stm., Arg.	182-83	32°17′S	67°22′W
San Juan, stm., Mex. (sän-ho͞o-än´)	132-33	26°22′N	98°51′W
San Juan, stm., N.A.	161	10°56′N	83°43′W
San Juan, stm., U.S. (sän hwän´)	120-21	37°11′N	110°43′W
San Juan, Pico, mtn., Cuba (pē´kô-sän-kôä´n)	154-55	21°59′N	80°09′W
San Juan Bautista, Para. (sän hwän´ bou-tēs´tä)	187	26°53′S	57°01′W
San Juan de la Maguana, Dom. Rep.	154-55	18°48′N	71°13′W
San Juan del Norte, Nic.	161	10°55′N	83°42′W
San Juan de los Morros, Ven. (sän-ho͞o-än´dĕ-lôs-mô´r-rôs)	188b	9°55′N	67°21′W
San Juan del Río, Mex. (sän hwän del rē´ô)	158-59	20°23′N	100°00′W
San Juan del Río, Mex. (sän hwän del rē´ô)	132-33	24°48′N	104°27′W
San Juan del Sur, Nic. (sän hwän děl so͞or)	161	11°15′N	85°52′W
San Juan Evangelista, Mex. (sän-ho͞o-ä´n-å-väŋ-kå-lěs´ta´)	158-59	17°54′N	95°07′W
San Juanito, Isla, i., Mex.	158-59	21°46′N	106°41′W
San Juan Mountains, mts., Co., U.S. (san hwän´ moun´tĭnz)	128-29	37°32′N	107°31′W
San Justo, Arg. (sän ho͞os´tô)	187	30°47′S	60°35′W
Sankt Michel, Fin. *see* Mikkeli	208-09	61°42′N	27°16′E
Sankt Pölten, Aus. (zäŋkt-pûl´těn)	210-11	48°12′N	15°37′E
Sankuru, stm., D.R.C. (sän-ko͞o´ro͞o)	284-85	4°17′S	20°24′E
San Lázaro, Cabo, c., Mex. (kå´bô sän-lå´zä-rō)	156-57	24°48′N	112°18′W
Şanlıurfa, Tur.	248-49	37°10′N	38°48′E
San Lorenzo, Arg. (sän lô-rěn´zô)	187	32°44′S	60°45′W
San Lorenzo, Ec.	184	1°15′N	78°50′W
San Lorenzo, Cabo, c., Ec.	184	1°04′S	80°54′W
San Lorenzo, Isla, i., Peru	188a	12°05′S	77°14′W
Sanlúcar de Barrameda, Spain (sän-lo͞o´kär)	214-15	36°47′N	6°21′W
San Lucas, Bol. (sän lo͞o´kås)	182-83	20°06′S	65°08′W
San Lucas, Cabo, c., Mex.	156-57	22°52′N	109°54′W
San Luis, Arg. (sän lô-ēs´)	185	33°18′S	66°21′W
San Luis, Guat. (sän lô-ēs´)	160	16°13′N	89°27′W
San Luis, state, Arg. (sän lô-ēs´)	185	34°0′S	66°00′W
San Luís, Laguna, lk., Bol.	180-81	13°45′S	64°00′W
San Luis de la Paz, Mex. (sän lô-ēs´ då lä päz´)	158-59	21°18′N	100°31′W
San Luis Obispo, Ca., U.S. (sån lo-ēes´ ô-bĭs´po)	128-29	35°17′N	120°40′W
San Luis Potosí, Mex.	158-59	22°09′N	100°59′W
San Luis Potosí, state, Mex.	158-59	22°30′N	100°30′W
San Luis Río Colorado, Mex.	156-57	32°28′N	114°46′W
San Marcos, Mex. (sän mär´kôs)	158-59	16°49′N	99°23′W
San Marcos, Tx., U.S. (sän mär´kôs)	132-33	29°53′N	97°56′W
San Marcos de Colón, Hond. (sän-má´r-kōs-dě-kô-lô´n)	161	13°26′N	86°49′W
San Marino, nation, Eur. (sän mē-rē´nô)	216-17	43°56′N	12°25′E
San Martín, Arg.	188e	33°05′S	68°29′W
San Martín, Col. (sän mär-tē´n)	178-79	3°42′N	73°42′W
San Martín, stm., Bol.	180-81	13°08′S	63°47′W
San Martín, Lago, lk., S.A. (lä´gô sän mär-tē´n)	185	48°53′S	72°39′W
San Martín de los Andes, Arg.	185	40°10′S	71°22′W
San Mateo, Ca., U.S. (săn mä-tā´ô)	128-29	37°34′N	122°19′W
San Mateo, Ven. (sän má-tě´ô)	188b	9°45′N	64°33′W
San Matías, Golfo, b., Arg. (gôl-fô-sän-mä-tē´äs)	185	41°30′S	64°15′W
Sanmenxia, China	258-59	34°47′N	111°12′E
San Miguel, El Sal. (sän mē-gāl´)	160	13°28′N	88°11′W
San Miguel, Mex. (sän mē-gāl´)	132-33	29°10′N	101°28′W
San Miguel, Pan. (sän mē-gāl´)	162	8°27′N	78°56′W
San Miguel, stm., Bol. (sän-mē-gěl´)	180-81	13°53′S	63°54′W
San Miguel, Golfo de, b., Pan. (gôl-fô-dě-sän mē-gāl´)	162	8°22′N	78°17′W
San Miguel del Monte, Arg.	187	35°27′S	58°49′W
San Miguel de Tucumán, Arg.	182-83	26°49′S	65°13′W
San Miguel El Alto, Mex. (sän mē-gāl´ ĕl äl´tô)	158-59	21°01′N	102°19′W
Sannār, Sudan	288	13°34′N	33°33′E
San Nicolas, Phil. (sän nē-kô-läs´)	270	18°10′N	120°36′E
San Nicolás, stm., Mex. (sän nē-kô-lá´s)	158-59	19°38′N	105°13′W
San Nicolás, Canal de, strt., N.A. *see* Nicholas Channel	154-55	23°21′N	80°21′W
San Nicolás de los Arroyos, Arg.	187	33°20′S	60°14′W
Sanok, Pol. (sä´nôk)	210-11	49°34′N	22°13′E
San Pablo, Phil. (sän-pä-blô)	270	14°04′N	121°19′E
San Pedro, Arg. (sän pā´drô)	182-83	24°15′S	64°52′W
San Pedro, Arg. (sän pā´drô)	187	33°41′S	59°41′W
San Pedro, Chile (sän pě´drô)	188e	33°54′S	71°26′W
San-Pédro, C. Iv.	282-83	4°45′N	6°38′W
San Pedro, Punta, c., Chile	182-83	25°31′S	70°38′W
San Pedro, Volcán, vol., Chile	182-83	21°53′S	68°25′W
San Pedro de Jujuy, Arg. *see* San Pedro	182-83	24°15′S	64°52′W
San Pedro de las Colonias, Mex. (sän pā´drô dě-läs-kô-lô´nyäs)	132-33	25°46′N	102°59′W
San Pedro de Macorís, Dom. Rep. (sän-pě´drô-då mä-kô-rēs´)	154-55	18°28′N	69°18′W
San Pedro de Ycuamandiyú, Para.	182-83	24°05′S	57°08′W
San Pedro Sula, Hond. (sän pā´drô so͞o´lä)	161	15°30′N	88°02′W
San Pietro, Isola di, i., Italy (ē´sō-lä-dē-sän pyā´trô)	216-17	39°08′N	8°16′E
San Quintín, Cabo, c., Mex.	156-57	30°22′N	115°60′W
San Rafael, Arg. (sän rä-fä-āl´)	188e	34°37′S	68°20′W
San Ramón de la Nueva Orán, Arg.	182-83	23°09′S	64°20′W
San Remo, Italy (sän rā´mô)	216-17	43°50′N	7°46′E
San Roque, Punta, c., Mex.	156-57	27°11′N	114°25′W
San Saba, Tx., U.S. (săn sä´bå)	132-33	31°12′N	98°43′W
San Saba, stm., Tx., U.S. (săn sä´bå)	132-33	31°15′N	98°36′W
San Salvador, i., Bah. (săn săl´vå-dôr)	154-55	24°02′N	74°27′W
San Salvador, nat. cap., El Sal. (sän säl-vä-dōr´)	160	13°40′N	89°13′W
San Salvador, Isla, i., Ec. (ē´s-lä-sän säl-vä-dōr´) *see* Santiago, Isla	184a	0°14′S	90°45′W
San Salvador de Jujuy, Arg.	182-83	24°12′S	65°18′W
San Sebastián, Spain (sän så-bås-tyän´) *see* Donostia-San Sebastián	214-15	43°19′N	1°60′W
San Severo, Italy (sän sě-vå´rō)	216-17	41°41′N	15°23′E
Sanshui, China (sän-shwā)	258-59	23°11′N	112°53′E
San Simon, stm., Az., U.S. (sån sī-mōn´)	122-23	32°52′N	109°33′W
Santa, stm., Peru	184	9°01′S	78°38′W
Santa Ana, Bol.	180-81	13°45′S	65°35′W
Santa Ana, El Sal. (sän´tä ä´nä)	160	13°59′N	89°34′W
Santa Ana, Mex. (sän´tä ä´nä)	158-59	24°04′N	100°30′W
Santa Ana, Mex. (sän´tä ä´nä)	156-57	30°32′N	111°07′W
Santa Ana, Ca., U.S. (săn´tå ăn´å)	128-29	33°45′N	117°53′W
Santa Anna, Tx., U.S. (săn´tå ăn´å)	132-33	31°44′N	99°19′W
Santa Bárbara, Hond. (sän-tä-bä´r-bä-rä)	161	14°55′N	88°14′W
Santa Bárbara, Mex. (sän-tä-bä´r-bä-rä)	132-33	26°49′N	105°48′W
Santa Barbara, Ca., U.S. (săn-tå-bär´bå-rå)	128-29	34°25′N	119°42′W
Santa Catalina Island, i., Ca., U.S. (săn´tå kă-tå-lē´nå ī´lånd)	128-29	33°23′N	118°24′W
Santa Catarina, Mex. (sän´tä kä-tä-rē´nä)	132-33	25°41′N	100°28′W
Santa Catarina, state, Braz. (sän-tä-kä-tä-rē´nå)	182-83	27°0′S	50°00′W
Santa Catarina, Ilha de, i., Braz.	186	27°36′S	48°30′W
Santa Clara, Cuba (sän´tå klä´rä)	154-55	22°25′N	79°58′W
Santa Clarita, Ca., U.S.	128-29	34°24′N	118°33′W
Santa Cruz, Braz. (sän-tä-kro͞o´s)	188d	6°13′S	36°01′W
Santa Cruz, Ca., U.S. (săn´tå kro͞oz´)	128-29	36°59′N	122°02′W
Santa Cruz, stm., Arg. (sän´tå kro͞oz´)	185	50°08′S	68°21′W
Santa Cruz, Isla, i., Ec. (ē´s-lä-sän-tä-kro͞o´z)	184a	0°38′S	90°23′W
Santa Cruz de la Palma, Spain	215d	28°41′N	17°46′W
Santa Cruz de la Sierra, Bol.	182-83	17°48′S	63°10′W
Santa Cruz del Sur, Cuba (sän-tä-kro͞o´s-děl-sö´r)	154-55	20°43′N	77°59′W
Santa Cruz de Tenerife, Spain (sän´tä kro͞oz då tā-nå-rē´fā)	215d	28°28′N	16°15′W
Santa Cruz do Sul, Braz.	182-83	29°43′S	52°26′W
Santa Cruz Islands, is., Sol. Is.	296-97	10°60′S	166°15′E
Santa Fe, Arg. (sän´tä fā´)	187	31°38′S	60°42′W
Santa Fe, Spain (sän´tä-fā´)	214-15	37°11′N	3°43′W
Santa Fe, N.M., U.S. (săn´tå fā´)	128-29	35°41′N	105°59′W
Santa Fe, state, Arg. (sän´tä fā´)	187	31°0′S	61°00′W
Santa Fe de Bogotá, nat. cap., Col. *see* Bogotá	178-79	4°37′N	74°06′W
Santa Fé do Sul, Braz.	182-83	20°13′S	50°56′W
Santai, China (san-tī)	258-59	31°09′N	105°01′E
Santa Inés, Isla, i., Chile (ē´s-lä-sän´tä ē-nās´)	185	53°46′S	72°44′W
Santa Isabel, Arg.	185	36°15′S	66°56′W
Santa Isabel, i., Sol. Is.	306e	8°0′S	159°00′E
Santa Isabel, nat. cap., Eq. Gui. *see* Malabo	282-83	3°45′N	8°47′E
Santa Magdalena, Isla, i., Mex.	156-57	24°54′N	112°13′W
Santa Margarita, Isla, i., Mex. (ē´s-lä-sän´tå mär-gå-rē´tå)	156-57	24°27′N	111°51′W
Santa Maria, Braz. (sän´tå mä-rē´å)	187	29°41′S	53°49′W
Santa Maria, Ca., U.S. (săn-tå må-rē´å)	128-29	34°57′N	120°26′W
Santa Maria, i., Port. (sän-tä-mä-rē´ä)	215c	36°58′N	25°06′W
Santa Maria, i., Vanuatu	306g	14°14′S	167°28′E
Santa María, stm., Mex. (sän´tå mä-rē´å)	158-59	21°48′N	99°10′W
Santa Maria, Cabo de, c., Ang.	286-87	13°25′S	12°32′E
Santa Maria, Cabo de, c., Port. (kå´bō-dě-sän-tä-mä-rē´ä)	214-15	36°58′N	7°54′W
Santa María, Isla, i., Ec.	184a	1°17′S	90°26′W
Santa María del Oro, Mex. (sän´tä-mä-rē´a-děl-ô-rô)	132-33	25°56′N	105°23′W
Santa Marta, Col. (săn´tå mär´tä)	178-79	11°15′N	74°12′W
Santa Monica, Ca., U.S. (săn´tå mŏn´ĭ-kå)	128-29	34°01′N	118°29′W
Santana do Livramento, Braz.	187	30°53′S	55°31′W
Santander, Col. (sän-tän-děr´)	188c	3°03′N	76°29′W
Santander, Phil.	270	9°25′N	123°20′E
Santander, Spain (sän-tän-dår´)	214-15	43°28′N	3°48′W
Sant'Antioco, Isola di, i., Italy (ē´sō-lä-dē-sän-än-tyŏ´kô)	216-17	39°02′N	8°25′E
Santarém, Braz. (sän-tä-rěn´)	180-81	2°26′S	54°42′W
Santarém, Port.	214-15	39°14′N	8°41′W
Santaren Channel, strt., Bah. (săn-tå-rěn´ chăn´ěl)	154-55	24°0′N	79°30′W
Santa Rita, Hond.	161	15°10′N	87°54′W
Santa Rosa, Arg.	187	36°37′S	64°17′W
Santa Rosa, Braz.	187	27°52′S	54°26′W
Santa Rosa, Ec.	184	3°27′S	79°57′W
Santa Rosa, Ca., U.S.	128-29	38°26′N	122°43′W
Santa Rosa, N.M., U.S. (săn´tå rō´så)	130-31	34°56′N	104°41′W
Santa Rosa de Copán, Hond.	161	14°46′N	88°47′W
Santa Rosalía, Mex. (sän´tå rô-zä´lē-å)	156-57	27°20′N	112°17′W
Santa Rosa Range, mts., Nv., U.S. (săn´tå rō´zå rānj)	122-23	41°35′N	117°40′W
Santa Sylvina, Arg.	187	27°50′S	61°08′W
Santa Vitória do Palmar, Braz. (sän-tä-vē-tô´ryä-dô-päl-mår)	187	33°31′S	53°22′W
Santee, Ca., U.S. (săn tē´)	128-29	32°50′N	116°57′W
Santee, stm., S.C., U.S. (săn tē´)	134-35	33°14′N	79°28′W
Santiago, Braz. (sän-tyá´gô)	187	29°11′S	54°52′W
Santiago, Pan. (sän-tyá´gô)	162	8°06′N	80°58′W
Santiago, i., C.V.	282-83	15°02′N	23°39′W
Santiago, nat. cap., Chile (sän-tē-á´gô)	185	33°27′S	70°40′W
Santiago, Isla, i., Ec.	184a	0°14′S	90°45′W
Santiago de Compostela, Spain	214-15	42°53′N	8°32′W
Santiago de Cuba, Cuba (sän-tyá´gô-dě ko͞o´bå)	154-55	20°02′N	75°49′W
Santiago de Cuba, state, Cuba (sän-tyá´gô-dě ko͞o´bå)	154-55	20°10′N	75°55′W
Santiago del Estero, Arg. (sän-tē-á´gô-děl ěs-tā-rô)	182-83	27°47′S	64°16′W
Santiago del Estero, state, Arg. (sän-tē-á´gô-děl ěs-tā-rô)	187	28°0′S	63°30′W
Santiago de los Caballeros, Dom. Rep.	154-55	19°27′N	70°42′W
Santiago Jamiltepec, Mex.	158-59	16°18′N	97°50′W
Santiago Papasquiaro, Mex.	132-33	25°03′N	105°25′W
Santiaguillo, Laguna, lk., Mex. (lä-oō´nä-sän-tē-a-gēl´yô)	132-33	24°45′N	104°48′W

n-sing; ŋ-baŋk; ɴ-nasalized n; nŏd; cŏmmit; ōld; ôbey; ôrder; oi-boil; fo͞od; ô-as oo in foot; ou-out; s-soft; sh-dish; th-thin; pūre; ûnite; ûrn; stŭd; circŭs; ü-as in French tu; ´-indeterminate vowel.

406 Santo Amaro – Schwarzwald PRONOUNCING INDEX

Feature (Pronunciation)	Page	Lat.	Long.
Santo Amaro, Braz. (sän´tô ä-mä´rô)	180-81	12°32'S	38°42'W
Santo André, Braz.	186	23°40'S	46°31'W
Santo Ângelo, Braz. (sän-tô-ä´n-zhĕ-lô)	187	28°16'S	54°16'W
Santo Antão, i., C.V. (sän´tô ä´n-zhĕ-lô)	282-83	17°03'N	25°07'W
Santo Antônio de Jesus, Braz.	180-81	12°57'S	39°14'W
Santo Antônio do Içá, Braz.	180-81	3°04'S	67°56'W
Santo Domingo, Nic. (sän-tô-dô-mē´n-gō)	161	12°16'N	85°05'W
Santo Domingo, i., N.A. *see* Hispaniola	154-55	19°0'N	71°00'W
Santo Domingo, nat. cap., Dom. Rep. (sän´tô dô-mĭn´gô)	154-55	18°30'N	69°53'W
Santoña, Spain (sän-tō´nyä)	214-15	43°26'N	3°27'W
Santorini, i., Grc. *see* Thíra	216-17	36°26'N	25°27'E
Santos, Braz. (sän´tozh)	186	23°56'S	46°20'W
Santos Dumont, Braz. (sän´tôs-dò-mô´nt)	186	21°28'S	43°33'W
Santo Tomé, Arg.	187	28°33'S	56°02'W
Santo Tomé de Guayana, Ven. *see* Ciudad Guayana	178-79	8°21'N	62°39'W
San Valentín, Monte, mtn., Chile (mô´n-tĕ-sän-vä-lĕn-tē´n)	185	46°36'S	73°20'W
San Vicente, El Sal. (sän vê-sĕn´tâ)	160	13°38'N	88°47'W
San Vicente de Cañete, Peru	188a	13°05'S	76°24'W
San Vicente del Caguán, Col.	178-79	2°07'N	74°47'W
San Xavier Indian Reservation, ind. res., Az., U.S. (sän x-ā´vēr ĭn´dĭ-ăn rĕ-sĕr-vā´shĕn)	128-29	32°02'N	111°08'W
Sanya, China	258-59	18°14'N	109°30'E
Sanyuan, China	258-59	34°37'N	108°55'E
Sanza Pombo, Ang.	284-85	7°19'S	15°60'E
São Bento, Braz.	180-81	2°42'S	44°50'W
São Borja, Braz. (soun-bôr-zhä)	187	28°39'S	56°01'W
São Carlos, Braz. (soun kär´lōzh)	186	22°02'S	47°54'W
São Cristóvão, Braz. (soun-krēs-tō-voun)	188d	11°01'S	37°12'W
São Domingos, Braz.	180-81	13°24'S	46°21'W
São Francisco, Braz. (soun frän-sēsh´kò)	182-83	15°57'S	44°52'W
São Francisco, stm., Braz. (soun frän-sēsh´kò)	173	10°30'S	36°24'W
São Francisco, Ilha de, i., Braz.	186	26°18'S	48°37'W
São Francisco do Sul, Braz. (soun frän-sēsh´kò-dô-sōō´l)	186	26°15'S	48°37'W
São Gabriel, Braz. (soun´gä-brê-ĕl´)	187	30°20'S	54°19'W
São João da Barra, Braz. (soun-zhôun-dä-bä´rä)	186	21°38'S	41°02'W
São João da Boa Vista, Braz. (soun-zhôun-dä-bôä-vē´s-tä)	186	21°59'S	46°48'W
São João Del Rei, Braz. (soun zhô-oun´dĕl-rā)	186	21°08'S	44°15'W
São Jorge, i., Port. (soun zhôr´zhĕ)	215c	38°38'N	28°03'W
São José do Rio Preto, Braz. (soun zhô-zĕ´dô-re´ō-prĕ-tō)	182-83	20°49'S	49°23'W
São José dos Campos, Braz. (soun zhô-zā´dōzh kän pōzh´)	186	23°11'S	45°53'W
São Leopoldo, Braz. (soun-lĕ-ô-pôl´dô)	182-83	29°46'S	51°08'W
São Lourenço do Sul, Braz.	187	31°22'S	51°58'W
São Luís, Braz.	180-81	2°31'S	44°16'W
São Luís Gonzaga, Braz.	187	28°24'S	54°57'W
São Manuel, stm., Braz.	180-81	7°21'S	58°08'W
São Mateus, Braz. (soun mä-tā´ozh)	186	18°44'S	39°52'W
São Miguel, i., Port.	215c	37°47'N	25°30'W
Saona, Isla, i., Dom. Rep. (ē´s-lä-sä-ō´nä)	154-55	18°09'N	68°40'W
Saône, stm., Fr. (sōn)	212-13	45°43'N	4°50'E
São Nicolau, i., C.V. (soun´ nê-kô-loun´)	282-83	16°36'N	24°11'W
São Paulo, Braz. (soun´ pou´lò)	186	23°33'S	46°38'W
São Paulo, state, Braz. (soun´ pou´lò)	186	22°0'S	49°00'W
São Paulo de Olivença, Braz. (soun´pou´lòdå ô-lē-vĕn´så)	180-81	3°28'S	68°57'W
São Raimundo Nonato, Braz. (soun´ rī-mò´n-do nô-nä´tò)	180-81	9°01'S	42°42'W
São Roque, Braz. (soun´ rō´kĕ)	186	23°32'S	47°08'W
São Roque, Cabo de, c., Braz. (kå´bō-dĕ-soun´ rō´kĕ)	188d	5°29'S	35°16'W
São Salvador, Braz. *see* Salvador	180-81	12°59'S	38°30'W
São Sebastião, Braz. (soun så-bäs-tê-oun´)	186	23°49'S	45°25'W
São Sebastião, Ilha de, i., Braz.	186	23°51'S	45°20'W
São Sebastião, Ponta, c., Moz.	286-87	22°08'S	35°29'E
São Simão, Braz. (soun-sē-moun)	186	21°29'S	47°33'W
São Simão, Represa de, res., Braz.	182-83	18°37'S	49°59'W
São Tiago, i., C.V. (soun tê-ä´gò) *see* Santiago	282-83	15°02'N	23°39'W
São Tomé, i., S. Tom./P.	282-83	0°12'N	6°36'E
São Tomé, nat. cap., S. Tom./P.	282-83	0°20'N	6°44'E
São Tomé, Cabo de, c., Braz.	186	21°59'S	40°59'W
Sao Tome and Principe, nation, Afr. (soun tòmā änd prēn´sê-pĕ)	282-83	1°0'N	7°00'E
São Tomé e Príncipe, nation, Afr. *see* Sao Tome and Principe	282-83	1°0'N	7°00'E
Saoura, Oued, stm., Alg.	280-81	29°01'N	0°57'W
São Vicente, Braz. (soun ve-se´n-tĕ)	186	23°58'S	46°22'W
São Vicente, i., C.V. (soun vê-sĕn´tå)	282-83	16°49'N	24°55'W
São Vicente, Cabo de, c., Port. (kå´bō-dĕ-sän-vê-sĕ´n-tĕ)	214-15	37°01'N	8°59'W
Sap, Tonle, lk., Camb. (tôn´lā säp´)	266-67	13°0'N	104°00'E
Sapé, Braz.	188d	7°06'S	35°13'W
Sapele, Nig. (sä-pā´lā)	282a	5°54'N	5°40'E
Sapitwa, mtn., Malawi	286-87	15°57'S	35°36'E
Sapporo, Japan (säp-pô´rô)	264	43°04'N	141°21'E
Sapulpa, Ok., U.S. (så-pŭl´på)	130-31	36°00'N	96°06'W
Saqqez, Iran	248-49	36°14'N	46°18'E
Sarāb, Iran	245	37°56'N	47°31'E
Saragossa, Spain *see* Zaragoza	214-15	41°39'N	0°53'W
Sarajevo, nat. cap., Bos. (sä-rá-yĕv´ō) (sä-rä´ya-vô)	216-17	43°52'N	18°25'E
Sarakhs, Iran	252-53	36°32'N	61°10'E
Saranac Lake, N.Y., U.S. (săr´á-năk lāk)	126-27	44°20'N	74°08'W
Sarandí Grande, Ur. (sä-rän´dē-grän´dĕ)	187	33°45'S	56°20'W
Sarang, Kaz.	244	49°46'N	72°52'E
Sarangani Islands, is., Phil.	270	5°25'N	125°26'E
Saransk, Russia (sá-ränsk´)	202-03	54°11'N	45°09'E
Sarapul, Russia (sä-räpòl')	202-03	56°28'N	53°48'E
Sarasota, Fl., U.S. (săr-*á*-sōt*á*)	135a	27°20'N	82°32'W
Saratoga, Wy., U.S. (săr-*á*-tō´g*á*)	122-23	41°28'N	106°48'W
Saratoga Springs, N.Y., U.S. (săr-*á*-tō´g*á* sprĭngz)	126-27	43°05'N	73°47'W
Saratov, Russia (sá rä´tôf)	202-03	51°34'N	45°60'E
Saratovskoye Vodokhranilishche, res., Russia	202-03	52°47'N	48°26'E
Saravan, Laos	266-67	15°43'N	106°25'E
Sarawak, hist. reg., Malay. (sá-rä´wäk)	268-69	2°30'N	113°30'E
Sarayevo, nat. cap., Bos. *see* Sarajevo	216-17	43°52'N	18°25'E
Sardegna, i., Italy *see* Sardinia	216-17	40°0'N	9°00'E
Sardinia, i., Italy (sär-dĭn´ĭ*á*)	216-17	40°0'N	9°00'E
Sardis Lake, res., Ms., U.S. (sär´dĭs lāk)	134-35	34°27'N	89°43'W
Sardis Lake, res., Ok., U.S. (sär´dĭs lāk)	130-31	34°42'N	95°21'W
Sargent, Ne., U.S. (sär´jĕnt)	124-25	41°38'N	99°22'W
Sargodha, Pak.	252-53	32°05'N	72°40'E
Sarh, Chad (är-chan-bō´)	284-85	9°09'N	18°23'E
Sārī, Iran	252-53	36°34'N	53°04'E
Sarina, Austl.	302	21°26'S	149°14'E
Sariqamish Kuli, lk., Asia *see* Sarygamysh köli	244	41°56'N	57°25'E
Sariwŏn, Kor., N.	263	38°30'N	125°46'E
Şarköy, Tur. (shär´kû-ē)	216-17	40°38'N	27°07'E
Sarmi, Indon.	302	1°51'S	138°42'E
Sarmiento, Arg.	185	45°35'S	69°05'W
Sarmiento de Gambia, Cerro, mtn., Chile	185	54°27'S	70°50'W
Särna, Swe.	208-09	61°41'N	13°08'E
Sarnia, On., Can. (sär´nê-*á*)	146-47	42°58'N	82°24'W
Sarny, Ukr. (sär´nê)	210-11	51°20'N	26°37'E
Sarpsborg, Nor. (särps´bôrg)	208-09	59°16'N	11°09'E
Sarrebourg, Fr. (sär-bōōr´)	212-13	48°44'N	7°03'E
Sarrebruck, Ger. *see* Saarbrücken	210-11	49°14'N	6°60'E
Sarreguemines, Fr. (sär-gē-mēn´)	212-13	49°07'N	7°04'E
Sartang, stm., Russia	236-37	67°27'N	133°15'E
Sárvár, Hung. (shär´vär)	210-11	47°15'N	16°56'E
Sarych, mys, c., Ukr. (mĭs sá-rēch´)	218-19	44°25'N	33°45'E
Sarygamysh köli, lk., Asia	244	41°56'N	57°25'E
Sarysū, stm., Kaz. (sä´rê-sōō)	244	45°11'N	66°30'E
Sary-Tash, Kyrg.	252-53	39°44'N	73°15'E
Sāsarām, India (sŭs-ŭ-räm´)	254-55	24°57'N	84°01'E
Sasebo, Japan (sä´sá-bô)	265	33°10'N	129°43'E
Saskatchewan, state, Can. (săs-kăch´ê-wän)	138-39	54°0'N	105°00'W
Saskatchewan, stm., Can. (săs-kăch´ê-wän)	144-45	53°16'N	98°50'W
Saskatoon, Sk., Can. (săs-k*á*-tōōn´)	144-45	52°08'N	106°39'W
Saskylakh, Russia	224-25	71°52'N	114°06'E
Sasovo, Russia (sás´ô-vô)	202-03	54°20'N	41°57'E
Sassandra, stm., C. Iv. (sás-sän´drá)	282-83	4°58'N	6°04'W
Sassari, Italy (säs´sä-rē)	216-17	40°43'N	8°33'E
Sata-misaki, c., Japan	265	30°60'N	130°40'E
Sātāra, India	256	17°41'N	74°00'E
Säter, Swe. (sĕ´tĕr)	208-09	60°21'N	15°45'E
Satilla, stm., Ga., U.S. (sá-tĭl´*á*)	134-35	30°59'N	81°29'W
Satīt, stm., Afr. *see* Tekezē	288	14°20'N	35°51'E
Satluj, stm., Asia *see* Sutlej	254-55	29°21'N	71°02'E
Satna, India	254-55	24°34'N	80°50'E
Sátoraljaújhely, Hung. (shä´tô-rô-lyô-ōō´yĕl´)	210-11	48°24'N	21°41'E
Sattahip, Thai.	266-67	12°40'N	100°54'E
Satu Mare, Rom. (sá´tōō-má´rĕ)	210-11	47°47'N	22°53'E
Sauđárkrókur, Ice.	206a	65°45'N	19°41'W
Sauce, Arg.	187	30°05'S	58°47'W
Saucillo, Mex.	132-33	28°01'N	105°16'W
Saudi Arabia, nation, Asia (sá-ó´dĭ á-rä´bĭ-á)	224-25	25°0'N	45°00'E
Saugatuck, Mi., U.S. (sô´g*á*-tŭk)	126-27	42°40'N	86°11'W
Saujbulagh, Iran *see* Mahābād	248-49	36°46'N	45°44'E
Sauk Centre, Mn., U.S. (sôk sĕn´tĕr)	124-25	45°44'N	94°57'W
Sauk City, Wi., U.S. (sôk sĭ´tê)	126-27	43°16'N	89°43'W
Sauk Rapids, Mn., U.S. (sôk răp´ĭdz)	124-25	45°36'N	94°10'W
Saül, Fr. Gu.	178-79	3°38'N	53°12'W
Sault Sainte Marie, On., Can. (sōō sånt må-rē´)	146-47	46°31'N	84°20'W
Sault Sainte Marie, Mi., U.S. (sōō sānt må-rē´)	124-25	46°29'N	84°21'W
Saunders Island, i., Falk. Is.	185	51°23'S	60°13'W
Saurimo, Ang.	286-87	9°40'S	20°23'E
Sava, stm., Eur. (sä´vä)	216-17	44°50'N	20°27'E
Savai'i, i., Samoa	306b	13°35'S	186°25'W
Savanna, Il., U.S. (s*á*-văn´*á*)	124-25	42°05'N	90°08'W
Savannah, Ga., U.S. (s*á*-văn´*á*)	134-35	32°03'N	81°06'W
Savannah, Mo., U.S. (s*á*-văn´*á*)	130-31	39°56'N	94°50'W
Savannah, Tn., U.S. (s*á*-văn´*á*)	134-35	35°14'N	88°14'W
Savannah, stm., U.S. (s*á*-văn´*á*)	120-21	32°01'N	80°53'W
Savannakhét, Laos	266-67	16°34'N	104°45'E
Savanna-la-Mar, Jam. (sá-vän´á lá mär´)	154-55	18°14'N	78°08'W
Savè, Benin	282a	8°01'N	2°25'E
Save, stm., Afr. (sä´vĕ)	286-87	20°58'S	35°04'E
Sāveh, Iran	252-53	35°01'N	50°21'E
Saverne, Fr. (sá-vĕrn´)	212-13	48°45'N	7°22'E
Savo Island, i., Sol. Is.	306e	9°08'S	159°49'E
Savona, Italy (sä-nō´nä)	216-17	44°19'N	8°28'E
Savonlinna, Fin. (sá´vôn-lēn´nà)	208-09	61°52'N	28°54'E
Savran', Ukr. (säv-rän´)	218-19	48°08'N	30°06'E
Savu Sea, s., Indon. (sä-vōō sē)	268-69	9°40'S	122°00'E
Sawahlunto, Indon.	266-67	0°40'S	100°46'E
Sawāi Mādhopur, India	254-55	25°59'N	76°22'E
Sawākin, Sudan	288	19°06'N	37°20'E
Sawdā', Jabal, mtn., Sau. Ar.	288	18°18'N	42°22'E
Sawdā', Qurnat as-, mtn., Leb.	248-49	34°18'N	36°07'E
Sawu, Laut, s., Indon. *see* Savu Sea	268-69	9°40'S	122°00'E
Sawu, Pulau, i., Indon.	268-69	10°30'S	121°54'E
Saxony, hist. reg., Ger.	210-11	52°45'N	9°30'E
Sayan Mountains, mts., Asia (sŭ-hän´ moun´tĭnz)	236-37	53°32'N	94°50'E
Sayanogorsk, Russia	236-37	53°06'N	91°24'E
Sayany, mts., Asia *see* Sayan Mountains	236-37	53°32'N	94°50'E
Şaydā, Leb. *see* Sidon	248-49	33°34'N	35°23'E
Saylac, Som.	288	11°20'N	43°28'E
Saynshand, Mong. *see* Buyant-Uhaa	260-61	44°55'N	110°09'E
Sayre, Ok., U.S. (sā´ĕr)	130-31	35°18'N	99°38'W
Sayre, Pa., U.S. (sā´ĕr)	126-27	41°59'N	76°31'W
Sayula, Mex. (sä-yōō´lä)	158-59	19°53'N	103°36'W
Saywūn, Yemen	238-39	15°56'N	48°45'E
Scandinavia, reg., Eur.	200-01	62°30'N	15°00'E
Scappoose, Or., U.S. (skă-pōōs´)	122-23	45°45'N	122°52'W
Scarborough, Trin.	155b	11°11'N	60°44'W
Scarborough, Eng., U.K. (skär´bŭr-ô)	206-07	54°17'N	0°25'W
Schässburg, Rom. *see* Sighişoara	210-11	46°14'N	24°48'E
Schefferville, Qc., Can.	138-39	54°48'N	66°50'W
Schenectady, N.Y., U.S. (skĕ-nĕk´tá-dê)	126-27	42°48'N	73°56'W
Schiermonnikoog, i., Neth.	206-07	53°29'N	6°11'E
Schio, Italy (skē´ô)	216-17	45°43'N	11°21'E
Schleswig, Ger. (shĕls´vêgh)	210-11	54°32'N	9°33'E
Schneidemühl, Pol. *see* Piła	210-11	53°09'N	16°44'E
Schofield, Wi., U.S. (skō´fēld)	126-27	44°53'N	89°36'W
Schreiber, On., Can.	146-47	48°48'N	87°15'W
Schuyler, Ne., U.S. (slī´ler)	124-25	41°27'N	97°04'W
Schuylkill Haven, Pa., U.S. (skōōl´kĭl hā-vĕn)	126-27	40°38'N	76°10'W
Schwabach, Ger. (shvä´bäk)	210-11	49°20'N	11°02'E
Schwäbisch Hall, Ger. (shvă´bēsh häl)	210-11	49°07'N	9°45'E
Schwaner, Pegunungan, mts., Indon. (skvän´ĕr)	268-69	0°40'S	112°40'E
Schwarzwald, mts., Ger. (shvärts´väld)	210-11	48°21'N	8°11'E

PRONOUNCING INDEX

Schwechat – Shangzhi 407

Feature (Pronunciation)	Page	Lat.	Long.
Schwechat, Aus. (shvĕk´ät)	210-11	48°08'N	16°29'E
Schwedt, Ger. (shvĕt)	210-11	53°04'N	14°17'E
Schweinfurt, Ger. (shvīn´fōrt)	210-11	50°03'N	10°13'E
Schweiz, nation, Eur. *see* Switzerland	190-91	47°0'N	8°00'E
Schwerin, Ger. (shvĕ-rēn´)	210-11	53°38'N	11°25'E
Sciacca, Italy (shė-äk´kä)	216-17	37°31'N	13°03'E
Scilly, Isles of, is., Eng., U.K. (īls ŏv sĭl´ė)	206-07	49°55'N	6°20'W
Scobey, Mt., U.S. (skō´bė)	122-23	48°48'N	105°25'W
Scotland, state, U.K. (skŏt´*lă*nd)	206-07	57°0'N	4°00'W
Scotland Neck, N.C., U.S. (skŏt´*lă*nd nĕk)	134-35	36°07'N	77°25'W
Scotstown, Qc., Can. (skŏts´toun)	148-49	45°31'N	71°16'W
Scott, Cape, c., B.C., Can. (kāp skŏt)	142-43	50°47'N	128°25'W
Scott City, Ks., U.S. (skŏt sĭ´tė)	130-31	38°29'N	100°54'W
Scottsbluff, Ne., U.S. (skŏts´blŭf)	124-25	41°52'N	103°40'W
Scottsboro, Al., U.S. (skŏts´bŭro)	134-35	34°41'N	86°01'W
Scottsburg, In., U.S. (skŏts´bûrg)	126-27	38°41'N	85°46'W
Scottsdale, Austl. (skŏts´dāl)	301	41°10'S	147°31'E
Scottsdale, Az., U.S.	128-29	33°35'N	111°52'W
Scottsville, Ky., U.S. (skŏts´vĭl)	134-35	36°45'N	86°11'W
Scottville, Mi., U.S. (skŏt´vĭl)	126-27	43°57'N	86°17'W
Scranton, Pa., U.S.	126-27	41°26'N	75°39'W
Scugog, Lake, lk., On., Can. (lāk skū´gôg)	146-47	44°10'N	78°50'W
Scunthorpe, Eng., U.K. (skŭn´thôrp)	206-07	53°36'N	0°40'W
Scutari, Alb. *see* Shkodër	216-17	42°04'N	19°31'E
Seaford, De., U.S. (sē´fērd)	126-27	38°39'N	75°37'W
Sea Islands, is., U.S. (sē ī´*lă*ndz)	134-35	31°20'N	81°20'W
Seal Cays, is., T./C. Is. (sēl kēs)	154-55	21°10'N	71°38'W
Sealy, Tx., U.S. (sē´lė)	132-33	29°47'N	96°09'W
Searcy, Ar., U.S. (sûr´sė)	134-35	35°15'N	91°45'W
Seaside, Or., U.S. (sē´sīd)	122-23	45°59'N	123°55'W
Seattle, Wa., U.S. (sė-ăt´'l)	122-23	47°36'N	122°20'W
Sébaco, Nic. (sĕ-bä´kô)	161	12°51'N	86°06'W
Sebastián Vizcaíno, Bahía, b., Mex.	156-57	28°0'N	114°30'W
Sebatik, Pulau, i., Asia	268-69	4°10'N	117°45'E
Sebewaing, Mi., U.S. (se´bė-wăng)	126-27	43°44'N	83°26'W
Sebree, Ky., U.S. (sė-brē´)	126-27	37°37'N	87°32'W
Sebring, Fl., U.S. (sē´brĭng)	135a	27°30'N	81°26'W
Sebuku, Pulau, i., Indon.	268-69	3°30'S	116°22'E
Sechura, Bahía de, b., Peru	184	5°39'S	81°01'W
Seda, China	258-59	32°20'N	100°41'E
Sedalia, Mo., U.S.	130-31	38°42'N	93°14'W
Sedan, Fr. (sē-dän)	212-13	49°42'N	4°56'E
Sedro-Woolley, Wa., U.S. (sē´drō-wŏl´ė)	122-23	48°30'N	122°14'W
Segama, stm., Malay.	268-69	5°31'N	118°48'E
Segamat, Malay. (sā´gà-màt)	266-67	2°30'N	102°49'E
Segesvár, Rom. *see* Sighișoara	210-11	46°14'N	24°48'E
Segezha, Russia	202-03	63°44'N	34°18'E
Ségou, Mali (sā-gōō´)	280-81	13°26'N	6°16'W
Segovia, Spain (så-gō´vė-ä)	214-15	40°57'N	4°07'W
Segre, stm., Eur. (så´grā)	214-15	41°22'N	0°18'E
Seguin, Tx., U.S. (sĕ-gēn´)	132-33	29°34'N	97°58'W
Segura, stm., Spain (så-gōō´rä)	214-15	38°07'N	0°39'W
Segura, Sierra de, mts., Spain (sē-ĕ´r-rä-dĕ sē-gū´là)	214-15	38°0'N	2°43'W
Seiland, i., Nor.	200-01	70°25'N	23°15'E
Seinäjoki, Fin. (så´ē-nĕ-yô´kė)	200-01	62°47'N	22°51'E
Seine, stm., On., Can. (sån)	144-45	48°38'N	92°58'W
Seine, stm., Fr.	212-13	49°29'N	0°29'E
Seishin, Kor., N. *see* Ch'ŏngjin	263	41°47'N	129°48'E
Seixas, Ponta do, c., Braz.	188d	7°09'S	34°47'W
Şeki, Azer.	245	41°10'N	47°10'E
Sekondi, Ghana	282-83	4°59'N	1°43'W
Selaru, Pulau, i., Indon.	242-43	8°10'S	130°59'E
Selatan, Tanjung, c., Indon. (tän´jông så-lä´tän)	268-69	4°10'S	114°38'E
Selawik, Ak., U.S. (sē-lå-wĭk)	136	66°40'N	160°07'W
Selawik Lake, lk., Ak., U.S.	136	66°30'N	160°40'W
Selayar, Pulau, i., Indon.	268-69	6°05'S	120°30'E
Selemdzha, stm., Russia (sâ-lĕmt-zhä´)	236-37	51°44'N	128°53'E
Selenga, stm., Asia (sĕ-lĕŋ-gä´) *see* Selenge	260-61	52°17'N	106°16'E
Selenge, stm., Asia	260-61	52°17'N	106°16'E
Selennyakh, stm., Russia (sĕl-yĭn-yäk)	236-37	67°51'N	144°53'E
Sélestat, Fr. (sē-lē-stä´)	212-13	48°15'N	7°27'E
Seliger, Ozero, lk., Russia (ô´zē-rô sĕl´lē-gĕr)	218-19	57°13'N	33°03'E
Selizharovo, Russia (så´lė-zhä´rô-vô)	218-19	56°51'N	33°28'E
Selkirk, Mb., Can. (sĕl´kûrk)	144-45	50°09'N	96°52'W
Selma, Al., U.S. (sĕl´*mȧ*)	134-35	32°24'N	87°01'W
Selma, Ca., U.S. (sĕl´*mȧ*)	128-29	36°35'N	119°37'W
Selma, N.C., U.S. (sĕl´*mȧ*)	134-35	35°32'N	78°17'W
Selva, Arg.	187	29°46'S	62°03'W
Selvas, reg., Braz.	180-81	5°0'S	68°00'W
Selwyn Lake, lk., Can. (sĕl´wĭn lāk)	140-41	59°55'N	104°22'W
Selwyn Mountains, mts., Can. (sĕl´wĭn moun´tĭnz)	140-41	63°10'N	130°20'W
Selwyn Range, mts., Austl.	302	21°35'S	140°35'E
Semara, W. Sah.	280-81	26°44'N	11°41'W
Semarang, Indon. (sĕ-mä´räng)	268-69	6°58'S	110°25'E
Semënov, Russia	202-03	56°47'N	44°29'E
Semeru, Gunung, vol., Indon.	268-69	8°06'S	112°55'E
Semey, Kaz.	244	50°24'N	80°14'E
Seminole, Ok., U.S. (sĕm´ĭ-nōl)	130-31	35°13'N	96°40'W
Seminole, Tx., U.S. (sĕm´ĭ-nōl)	130-31	32°43'N	102°39'W
Seminole, Lake, res., U.S. (lāk sĕm´ĭ-nōl)	134-35	30°46'N	84°50'W
Semliki, stm., Afr. (sĕm´lė-kē)	289	1°12'N	30°31'E
Semnān, Iran	252-53	35°34'N	53°24'E
Semporna, Malay.	268-69	4°28'N	118°36'E
Senador Pompeu, Braz. (sĕ-nä-dōr-pôm-pĕ´ŏ)	180-81	5°35'S	39°22'W
Sena Madureira, Braz.	180-81	9°04'S	68°40'W
Senatobia, Ms., U.S. (sĕ-nȧ-tō´bė-*ȧ*)	134-35	34°38'N	89°58'W
Sendai, Japan (sĕn-dī´)	265	31°49'N	130°19'E
Sendai, Japan	265	38°15'N	140°53'E
Seneca, Ks., U.S. (sĕn´ĕ-*kȧ*)	130-31	39°50'N	96°04'W
Seneca, S.C., U.S. (sĕn´ĕ-*kȧ*)	134-35	34°41'N	82°57'W
Seneca Falls, N.Y., U.S. (sĕn´ĕ-*kȧ* fôlz)	126-27	42°54'N	76°48'W
Senegal, nation, Afr. (sĕn-ĕ-gôl´)	274	14°0'N	14°00'W
Sénégal, stm., Afr.	280-81	15°48'N	16°32'W
Senekal, S. Afr. (sĕn´ē-*kă*l)	292c	28°19'S	27°38'E
Senftenberg, Ger. (zĕnf´tĕn-bĕrgh)	210-11	51°32'N	14°00'E
Senhor do Bonfim, Braz. (sĕn-yôr dô bôn-fē´n)	180-81	10°27'S	40°11'W
Senigallia, Italy (så-nė-gäl´lyä)	216-17	43°43'N	13°13'E
Senj, Cro. (sĕn´)	216-17	44°59'N	14°55'E
Senja, i., Nor. (sĕnyä)	200-01	69°20'N	17°30'E
Senneterre, Qc., Can.	146-47	48°23'N	77°14'W
Senqu, stm., Afr. *see* Orange	286-87	28°35'S	16°28'E
Sens, Fr. (säns)	212-13	48°12'N	3°17'E
Senta, Serb. (sĕn´tä)	216-17	45°56'N	20°06'E
Senyavin Islands, is., Micron.	304-05	6°54'N	158°04'E
Seoni, India	254-55	22°05'N	79°33'E
Seoul, nat. cap., Kor., S. (sōl)	263	37°33'N	127°01'E
Sepanjang, Pulau, i., Indon.	268-69	7°11'S	115°50'E
Sepetiba, Baía de, b., Braz. (bäe´ä dĕ så-på-te´b*a*)	186	23°0'S	43°48'W
Sepik, stm., (sĕp-ēk´)	302	3°53'S	144°28'E
Sept-Îles, Qc., Can. (sĕ-tēl´)	148-49	50°12'N	66°22'W
Sequoia National Park, n.p., Ca., U.S. (sė-kwoi´*ȧ* năsh´ŭn-*ă*l pärk)	128-29	36°31'N	118°34'W
Serafimovich, Russia	202-03	49°35'N	42°45'E
Seram, i., Indon. *see* Ceram	242-43	3°0'S	129°00'E
Seram, Laut, s., Indon. *see* Ceram Sea	242-43	2°30'S	128°00'E
Serang, Indon. (så-räng´)	268-69	6°07'S	106°09'E
Sarajevo, nat. cap., Bos. *see* Sarajevo	216-17	43°52'N	18°25'E
Serbia, nation, Eur. (sĕr´bė-*ă*)	190-91	44°0'N	21°00'E
Serdobsk, Russia (sĕr-dôpsk´)	202-03	52°27'N	44°13'E
Şereflikoçhisar, Tur.	246	38°57'N	33°33'E
Seremban, Malay. (sĕr-ĕm-bän´)	266-67	2°43'N	101°57'E
Serengeti National Park, n.p., Tan.	289	2°20'S	34°50'E
Serengeti Plain, pl., Tan.	289	2°55'S	35°00'E
Sergeyevka, Russia	264	43°20'N	133°21'E
Sergipe, state, Braz. (sĕr-zhē´pĕ)	180-81	10°30'S	37°30'W
Sergiyev Posad, Russia	218-19	56°19'N	38°09'E
Serian, Malay.	268-69	1°10'N	110°33'E
Sérifos, i., Grc.	216-17	37°10'N	24°29'E
Serik, Tur.	246	36°55'N	31°06'E
Sermata, Pulau, i., Indon.	268-69	8°13'S	128°55'E
Serov, Russia (syĕ-rôf´)	236-37	59°36'N	60°35'E
Serowe, Bots. (sĕ-rō´wĕ)	286-87	22°23'S	26°43'E
Serpukhov, Russia (syĕr´pò-kôf)	218-19	54°55'N	37°26'E
Serra do Navio, Braz.	180-81	0°55'N	52°01'W
Serra Talhada, Braz.	180-81	7°59'S	38°18'W
Sérres, Grc. (sĕr´rĕ) (sĕr´ĕs)	216-17	41°05'N	23°33'E
Serrinha, Braz. (sĕr-rēn´yȧ)	180-81	11°39'S	38°60'W
Sertã, Port. (sĕr´tȧ)	214-15	39°48'N	8°06'W
Sertânia, Braz. (sĕr-tȧ´nyä)	188d	8°05'S	37°16'W
Serutu, Pulau, i., Indon.	266-67	1°42'S	108°45'E
Sêrxü, China	258-59	33°08'N	97°55'E
Sesayap Lama, Indon.	268-69	3°35'N	116°60'E
Sese Islands, is., Ug.	289	0°20'S	32°20'E
Sesimbra, Port. (sĕ-sē´m-brä)	214-15	38°26'N	9°05'W
Sestri Levante, Italy (sĕs´trė lå-vän´tå)	216-17	44°16'N	9°24'E
Sestroretsk, Russia (sĕs-trô-rĕtsk)	208-09	60°06'N	29°58'E
Sète, Fr. (sĕt)	212-13	43°25'N	3°42'E
Sete Lagoas, Braz. (sĕ-tĕ lä-gô´äs)	186	19°28'S	44°15'W
Sétif, Alg.	292b	36°11'N	5°25'E
Seto-naikai, s., Japan (sĕ´tô nī´kī)	265	34°22'N	133°37'E
Settat, Mor. (sĕt-ät´) (sĕ-tá´)	292a	32°59'N	7°36'W
Settlers, S. Afr. (sĕt´lērs)	292c	25°01'S	28°28'E
Setúbal, Port. (så-tōō´bäl)	214-15	38°31'N	8°53'W
Setúbal, Baía de, b., Port. (bä-ē´ä-dĕ-så-tōō´bäl)	214-15	38°18'N	9°04'W
Seul, Lac, lk., On., Can. (läk sûl)	144-45	50°36'N	91°49'W
Sevana Lich, lk., Arm. (syĭ-văn´)	245	40°18'N	45°19'E
Sevastopol', Ukr. (syĕ-vás-tô´pôl'´)	218-19	44°36'N	33°32'E
Severn, stm., On., Can. (sĕv´ērn)	144-45	55°59'N	87°36'W
Severna Park, Md., U.S. (sĕv´ērn-å pärk)	126-27	39°04'N	76°33'W
Severnaya Dvina, stm., Russia	202-03	64°40'N	39°51'E
Severnaya Osetiya-Alaniya, state, Russia *see* North Ossetia	245	43°0'N	44°15'E
Severnaya Sos'va, stm., Russia	236-37	64°10'N	65°27'E
Severnaya Zemlya, is., Russia (sĕ-vyĭr-nīŭ zĭ-m'lyä´)	236-37	79°30'N	98°00'E
Severnyye Uvaly, hills, Russia	202-03	59°28'N	48°13'E
Severodvinsk, Russia	202-03	64°34'N	39°50'E
Severo-Kuril'sk, Russia	236-37	50°42'N	156°07'E
Severomorsk, Russia	200-01	69°04'N	33°28'E
Severo-Sibirskaya Nizmennost', pl., Russia	236-37	73°0'N	100°00'E
Severouralsk, Russia (sĕ-vyĭ-rŭ-ōō-rälsk´)	202-03	60°10'N	59°58'E
Severskiy Donets, stm., Eur. *see* Northern Donets	202-03	47°36'N	40°54'E
Sevier, stm., Ut., U.S. (sė-vēr´)	128-29	39°02'N	113°06'W
Sevier Lake, lk., Ut., U.S. (sė-vēr´ lāk)	128-29	38°56'N	113°09'W
Sevilla, Col. (sĕ-vė´l-yä)	188c	4°15'N	75°56'W
Sevilla, Spain (så-vēl´yä)	214-15	37°23'N	5°59'W
Seville, Spain *see* Sevilla	214-15	37°23'N	5°59'W
Sevsk, Russia (syĕfsk)	218-19	52°09'N	34°31'E
Seward, Ak., U.S. (sū´ård)	136	60°07'N	149°27'W
Seward, Ne., U.S. (sū´ård)	124-25	40°54'N	97°06'W
Seward Peninsula, pen., Ak., U.S. (sū´ård pĕ-nĭn´sūl*ă*)	136	65°0'N	164°00'W
Sewell, Chile (sē´ŏ-ĕl)	188e	34°05'S	70°23'W
Seydişehir, Tur.	246	37°25'N	31°51'E
Seyđisfjörđur, Ice. (sā´dės-fyûr-dòr)	206a	65°15'N	14°01'W
Seybaplaya, Mex. (sā-ē-bä-plä´yä)	160	19°40'N	90°40'W
Seychelles, nation, Afr. (sā-shĕl´)	274	4°35'S	55°40'E
Seymchan, Russia	236-37	62°54'N	152°24'E
Seymour, In., U.S. (sē´mōr)	126-27	38°58'N	85°53'W
Seymour, Tx., U.S. (sē´mōr)	130-31	33°36'N	99°16'W
Sfax, Tun. (sfäks)	200-01	34°45'N	10°46'E
's-Gravenhage, nat. cap., Neth. ('s ĸrä´vĕn hä´gĕ) (hag) *see* Hague, The	206-07	52°06'N	4°18'E
Sha, stm., China (shä)	258-59	33°37'N	114°38'E
Shabeelle, stm., Afr. (shä´bå-lē)	284-85	0°10'N	42°46'E
Shabunda, D.R.C.	284-85	2°42'S	27°21'E
Shache, China (shä-chŭ)	244	38°25'N	77°15'E
Shagonar, Russia	240-41	51°32'N	92°49'E
Shahdol, India	254-55	23°18'N	81°22'E
Shāhjahānpur, India (shä-jŭ-hän´pōōr)	254-55	27°53'N	79°54'E
Shahrisabz, Uzb.	252-53	39°03'N	66°50'E
Shāhrūd, Iran	252-53	36°25'N	54°58'E
Shaker Heights, Oh., U.S. (shā´kēr hīts)	126-27	41°29'N	81°36'W
Shakhty, Russia (shäĸ´tė)	202-03	47°42'N	40°13'E
Shakhunya, Russia	202-03	57°40'N	46°38'E
Shaki, Nig.	282-83	8°40'N	3°24'E
Shaktoolik, Ak., U.S.	136	64°20'N	161°09'W
Shala Hāyk', lk., Eth. (shä´là)	292d	7°28'N	38°31'E
Shalqar, Kaz.	244	47°50'N	59°37'E
Shām, Bādiyat ash-, des., Asia *see* Syrian Desert	248-49	32°0'N	40°00'E
Shām, Jabal ash-, mtn., Oman	250-51	23°13'N	57°16'E
Shamokin, Pa., U.S. (shȧ-mō´kĭn)	126-27	40°47'N	76°35'W
Shamrock, Tx., U.S. (shăm´rŏk)	130-31	35°13'N	100°15'W
Shandī, Sudan	288	16°41'N	33°26'E
Shandong, state, China (shän-dôŋ)	260-61	36°0'N	118°00'E
Shandong Bandao, pen., China (shän-dôŋ bän-dou)	260-61	37°0'N	121°00'E
Shand uul, mtn., Mong.	260-61	43°28'N	104°03'E
Shangani, stm., Zimb.	286-87	18°31'S	27°12'E
Shangcheng, China (shäŋ-chŭŋ)	258-59	31°48'N	115°24'E
Shangdu, China (shäŋ-dōō)	260-61	41°34'N	113°31'E
Shanghai, China (shäng´hī´)	258-59	31°14'N	121°28'E
Shanghai, state, China	258-59	31°10'N	121°30'E
Shanglin, China (shäŋ-lĭn)	258-59	23°30'N	108°32'E
Shangqiu, China (shäŋ-chyŏ)	258-59	34°27'N	115°39'E
Shangrao, China (shäŋ-rou)	258-59	28°26'N	117°58'E
Shangshui, China	258-59	33°33'N	114°34'E
Shangxian, China	258-59	33°52'N	109°56'E
Shangzhi, China (shäŋ-jr)	264	45°13'N	127°59'E

n-sing; ŋ-baŋk; ɴ-nasalized n; nŏd; c*ŏ*mmit; ōld; ôbey; ôrder; oi-boil; fōōd; ò-as oo in foot; ou-out; s-soft; sh-dish; th-thin; pūre; ůnite; ûrn; stŭd; circ*ŭ*s; ü-as in French tu; ´-indeterminate vowel.

408 Shanhaiguan – Sífnos PRONOUNCING INDEX

Feature (Pronunciation)	Page	Lat.	Long.
Shanhaiguan, China	260-61	40°01'N	119°45'E
Shannon, stm., Ire. (shăn´ŏn)	206-07	52°35'N	9°41'W
Shansi, state, China *see* Shanxi	260-61	37°0'N	112°00'E
Shantarskiye Ostrova, is., Russia (shän´tär-skyě ôs-trôf´)	236-37	54°54'N	137°33'E
Shantou, China (shän-tō)	258-59	23°21'N	116°40'E
Shantung, state, China *see* Shandong	260-61	36°0'N	118°00'E
Shantung Peninsula, pen., China *see* Shandong Bandao	260-61	37°0'N	121°00'E
Shanxi, state, China (shän-shyē)	260-61	37°0'N	112°00'E
Shanxian, China (shän shyĕn)	258-59	34°48'N	116°05'E
Shanyin, China	260-61	39°31'N	112°50'E
Shaoguan, China (shou-gůän)	258-59	24°49'N	113°36'E
Shaowu, China	258-59	27°19'N	117°30'E
Shaoxing, China (shou-shyĭŋ)	258-59	29°59'N	120°34'E
Shaoyang, China	258-59	27°15'N	111°28'E
Shar, Kaz.	244	49°35'N	81°03'E
Sharjah, U.A.E.	250-51	25°22'N	55°24'E
Shark Bay, b., Austl. (shärk bā)	296-97	25°30'S	113°30'E
Sharktooth Mountain, mtn., B.C., Can.	140-41	58°35'N	127°57'W
Sharon, Pa., U.S. (shăr´ŏn)	126-27	41°13'N	80°30'W
Sharon Springs, Ks., U.S. (shăr´ŏn sprĭngz)	130-31	38°54'N	101°45'W
Sharonville, Oh., U.S. (shăr´ŏn vĭl)	126-27	39°16'N	84°25'W
Sharpsburg, Md., U.S. (shärps´bûrg)	126-27	39°27'N	77°45'W
Sharqīyah, Aṣ-Ṣaḥrā' ash-, des., Egypt *see* Arabian Desert	288	28°0'N	32°00'E
Sharya, Russia	202-03	58°22'N	45°31'E
Shashe, stm., Afr.	286-87	22°11'S	29°21'E
Shashi, China (shä-shē)	258-59	30°19'N	112°14'E
Shasta, Mount, vol., Ca., U.S. (mount shăs´tȧ)	122-23	41°25'N	122°13'W
Shasta Lake, res., Ca., U.S. (shăs´tȧ lāk)	122-23	40°46'N	122°22'W
Shatt al-Arab, stm., Asia *see* 'Arab, Shaṭṭ al-	248-49	29°57'N	48°33'E
Shattuck, Ok., U.S. (shăt´ŭk)	130-31	36°17'N	99°53'W
Shatura, Russia	218-19	55°34'N	39°32'E
Shaunavon, Sk., Can.	144-45	49°39'N	108°24'W
Shaw, Ms., U.S. (shô)	134-35	33°37'N	90°46'W
Shawano, Wi., U.S. (shȧ-wô´nô)	126-27	44°46'N	88°36'W
Shawinigan, Qc., Can.	146-47	46°33'N	72°45'W
Shawnee, Ks., U.S. (shô-nē´)	130-31	39°01'N	94°44'W
Shawnee, Ok., U.S. (shô-nē´)	130-31	35°20'N	96°55'W
Shawneetown, Il., U.S. (shô´nē-toun)	126-27	37°42'N	88°11'W
Shaybārā, i., Sau. Ar.	288	25°26'N	36°50'E
Shay Gap, Austl.	294-95	20°30'S	120°05'E
Shaykh, Jabal ash-, mtn., Asia *see* Hermon, Mount	248-49	33°25'N	35°51'E
Shchekino, Russia	218-19	54°01'N	37°31'E
Shchelkovo, Russia (shchĕl´kô-vô)	218-19	55°54'N	38°01'E
Shchigry, Russia (shchē´grē)	218-19	51°52'N	36°55'E
Shchors, Ukr. (shchôrs)	218-19	51°49'N	31°57'E
Shchūchīnsk, Kaz.	244	52°56'N	70°11'E
Sheberghān, Afg.	252-53	36°40'N	65°45'E
Sheboygan, Wi., U.S. (shė-boi´gǎn)	126-27	43°45'N	87°43'W
Sheboygan Falls, Wi., U.S. (shė-boi´gǎn fôlz)	126-27	43°44'N	87°48'W
Shediac, N.B., Can. (shĕ´dė-ăk)	148-49	46°12'N	64°34'W
Shedin Peak, mtn., B.C., Can. (shĕd´ĭn pēk)	142-43	55°55'N	127°32'W
Sheenjek, stm., Ak., U.S.	136	66°45'N	144°34'W
Sheffield, Eng., U.K.	206-07	53°23'N	1°28'W
Sheffield, Al., U.S. (shĕf´fēld)	134-35	34°45'N	87°42'W
Shekhūpura, Pak.	252-53	31°42'N	73°59'E
Sheki, Azer. *see* Şeki	245	41°10'N	47°10'E
Shelagyote Peak, mtn., B.C., Can.	142-43	55°58'N	127°12'W
Shelbina, Mo., U.S. (shĕl-bī´nȧ)	130-31	39°41'N	92°03'W
Shelburn, In., U.S. (shĕl´bŭrn)	126-27	39°11'N	87°24'W
Shelburne, N.S., Can.	148-49	43°46'N	65°19'W
Shelby, Mi., U.S. (shĕl´bė)	126-27	43°37'N	86°22'W
Shelby, Ms., U.S. (shĕl´bė)	134-35	33°57'N	90°46'W
Shelby, Mt., U.S. (shĕl´bė)	122-23	48°30'N	111°51'W
Shelby, N.C., U.S. (shĕl´bė)	134-35	35°18'N	81°32'W
Shelby, Oh., U.S. (shĕl´bė)	126-27	40°53'N	82°39'W
Shelbyville, Il., U.S. (shĕl´bė-vĭl)	126-27	39°24'N	88°48'W
Shelbyville, In., U.S. (shĕl´bė-vĭl)	126-27	39°31'N	85°46'W
Shelbyville, Ky., U.S. (shĕl´bė-vĭl)	126-27	38°13'N	85°13'W
Shelbyville, Tn., U.S. (shĕl´bė-vĭl)	134-35	35°29'N	86°27'W
Shelbyville, Lake, res., Il., U.S. (lāk shĕl´bė-vĭl)	126-27	39°30'N	88°43'W
Sheldon, Ia., U.S. (shĕl´dǔn)	124-25	43°11'N	95°51'W
Shelikhova, Zaliv, b., Russia	236-37	60°0'N	158°00'E
Shelikof Strait, strt., Ak., U.S. (shĕ´lė-kôf strāt)	136	57°18'N	155°41'W

Feature (Pronunciation)	Page	Lat.	Long.
Shellbrook, Sk., Can.	144-45	53°14'N	106°23'W
Shelley, Id., U.S. (shĕl´lė)	122-23	43°23'N	112°07'W
Shelton, Ct., U.S. (shĕl´tǔn)	126-27	41°19'N	73°05'W
Shelton, Wa., U.S. (shĕl´tǔn)	122-23	47°12'N	123°06'W
Shemonaīkha, Kaz.	244	50°39'N	81°54'E
Shenandoah, Ia., U.S. (shĕn-ǎn-dō´ȧ)	124-25	40°46'N	95°23'W
Shenandoah, Pa., U.S. (shĕn-ǎn-dō´ȧ)	126-27	40°49'N	76°12'W
Shenandoah, Va., U.S. (shĕn-ǎn-dō´ȧ)	126-27	38°29'N	78°37'W
Shenandoah National Park, n.p., Va., U.S. (shĕn-ǎn-dō´ȧ năsh´ŭn-ǎl pärk)	126-27	38°34'N	78°20'W
Sheng-li Feng, mtn., Asia *see* Jengish Chokusu	244	42°02'N	80°05'E
Shenkursk, Russia (shĕn-ko͞orsk´)	202-03	62°07'N	42°53'E
Shenxian, China (shŭn shyĕn)	260-61	38°01'N	115°33'E
Shenyang, China (shŭn-yäŋ)	263	41°48'N	123°24'E
Shenzhen, China	258-59	22°34'N	114°07'E
Shepetivka, Ukr.	210-11	50°11'N	27°04'E
Shepparton, Austl. (shĕp´ăr-tǔn)	301	36°23'S	145°25'E
Sherbro Island, i., S.L.	282-83	7°34'N	12°43'W
Sherbrooke, Qc., Can.	146-47	45°24'N	71°54'W
Sheridan, Ar., U.S. (shĕr´ĭ-dǎn)	130-31	34°18'N	92°24'W
Sheridan, Wy., U.S. (shĕr´ĭ-dǎn)	122-23	44°48'N	106°58'W
Sherlovaya Gora, Russia	260-61	50°32'N	116°18'E
Sherman, Tx., U.S. (shĕr´măn)	130-31	33°35'N	96°36'W
Sherridon, Mb., Can.	144-45	55°07'N	101°05'W
Sherwood Park, Ab., Can.	142-43	53°31'N	113°18'W
Shetland Islands, is., Scot., U.K. (shĕt´lǎnd ī´lǎndz)	206c	60°25'N	1°39'W
Shexian, China (shŭ shyĕn)	258-59	29°53'N	118°26'E
Sheyenne, stm., N.D., U.S. (shī-ĕn´)	124-25	47°01'N	96°50'W
Shiashkotan, Ostrov, i., Russia	236-37	48°52'N	154°10'E
Shibām, Yemen (shē´bäm)	238-39	15°54'N	48°40'E
Shidao, China	260-61	36°54'N	122°24'E
Shīeli, Kaz.	244	44°10'N	66°44'E
Shijiazhuang, China (shr-jyä-jůäŋ)	260-61	38°02'N	114°29'E
Shikārpur, Pak.	252-53	27°57'N	68°39'E
Shikoku, i., Japan (shē´kō´ko͞o)	265	33°45'N	133°30'E
Shikotan, Ostrov, i., Russia	236-37	43°47'N	146°45'E
Shikotan-tō, i., Russia *see* Shikotan, Ostrov	236-37	43°47'N	146°45'E
Shiliguri, India	254-55	26°43'N	88°26'E
Shilka, Russia	240-41	51°52'N	116°02'E
Shilka, stm., Russia (shĭl´kȧ)	240-41	53°21'N	121°27'E
Shillong, India (shėl-lông´)	254-55	25°34'N	91°53'E
Shimanovsk, Russia	240-41	52°00'N	127°41'E
Shimber Berris, mtn., Som. *see* Shimbiris	284-85	10°44'N	47°15'E
Shimbiris, mtn., Som.	284-85	10°44'N	47°15'E
Shimian, China	258-59	29°16'N	102°17'E
Shimla, India	254-55	31°06'N	77°10'E
Shimonoseki, Japan	265	33°58'N	130°56'E
Shimono-shima, i., Japan	263	34°12'N	129°15'E
Shinano, stm., Japan (shė-nä´nō)	265	37°57'N	139°04'E
Shīndand, Afg.	252-53	33°18'N	62°08'E
Shingishū, Kor., N. *see* Sinŭiju	263	40°06'N	124°24'E
Shingū, Japan	265	33°43'N	136°00'E
Shinyanga, Tan. (shĭn-yäŋ´gä)	289	3°40'S	33°26'E
Shiono-misaki, c., Japan (shė-ō´nō mė´sä-kė)	265	33°26'N	135°46'E
Shiqizhen, China *see* Zhongshan	258-59	22°31'N	113°22'E
Shirati, Tan. (shė-rä´tė)	289	1°07'S	33°60'E
Shīrāz, Iran (shē-räz´)	250-51	29°36'N	52°32'E
Shire, stm., Afr. (shē´rȧ)	286-87	17°42'S	35°19'E
Shiretoko-misaki, c., Japan	264	44°21'N	145°20'E
Shishaldin Volcano, vol., Ak., U.S. (shī-shăl´dĭn vŏl-kā´nō)	136	54°45'N	163°57'W
Shivamogga, India	256	13°56'N	75°35'E
Shively, Ky., U.S. (shīv´lė)	126-27	38°12'N	85°49'W
Shivpuri, India	254-55	25°25'N	77°39'E
Shizuoka, Japan (shė´zo͞o´ōkȧ)	265	34°58'N	138°23'E
Shkodër, Alb. (shkô´dûr) (sko͞o´tarė)	216-17	42°04'N	19°31'E
Shkodra, Alb. *see* Shkodër	216-17	42°04'N	19°31'E
Shmidta, Ostrov, i., Russia	226-27	81°08'N	90°48'E
Shoal Lake, lk., Can. (shōl lāk)	144-45	49°32'N	95°00'W
Shoals, In., U.S. (shōlz)	126-27	38°40'N	86°47'W
Shōdo-shima, i., Japan (shō´dō shė´mä)	265	34°30'N	134°17'E
Shortland Island, i., Sol. Is.	306e	7°04'S	155°43'E
Shoshone, Id., U.S. (shô-shōn´tė)	122-23	42°56'N	114°25'W
Shostka, Ukr. (shôst´kȧ)	218-19	51°52'N	33°29'E
Shouguang, China (shō-gůäŋ)	260-61	36°53'N	118°44'E
Shouxian, China (shō shyĕn)	258-59	32°34'N	116°46'E
Shpola, Ukr. (shpô´lȧ)	218-19	49°00'N	31°24'E
Shqipëria, nation, Eur. *see* Albania	190-91	41°0'N	20°00'E
Shreveport, La., U.S. (shrēv´pōrt)	130-31	32°30'N	93°45'W
Shrewsbury, Eng., U.K. (shro͞oz´bēr-ĭ)	206-07	52°43'N	2°45'W
Shū, Kaz.	244	43°36'N	73°45'E

Feature (Pronunciation)	Page	Lat.	Long.
Shū, stm., Asia	244	45°00'N	67°45'E
Shuajingsi, China	258-59	32°00'N	103°17'E
Shuangcheng, China (shůäŋ-chŭŋ)	260-61	45°22'N	126°19'E
Shuangliao, China	260-61	43°30'N	123°30'E
Shuangyashan, China	264	46°35'N	131°19'E
Shubrâ el-Kheima, Egypt	291b	30°06'N	31°15'E
Shumagin Islands, is., Ak., U.S. (sho͞o´mȧ-gėn ī´lǎndz)	136	55°06'N	159°43'W
Shumen, Blg.	216-17	43°16'N	26°57'E
Shumerlya, Russia	202-03	55°29'N	46°25'E
Shunde, China (shòn-dŭ)	258-59	22°50'N	113°15'E
Shuqayyiqah, Nafūd, sand, Sau. Ar.	288	25°45'N	43°55'E
Shūshtar, Iran (sho͞osh´tŭr)	248-49	32°03'N	48°51'E
Shuswap Lake, lk., B.C., Can. (sho͞os´wŏp lāk)	142-43	50°57'N	119°15'W
Shuya, Russia (sho͞o´yȧ)	202-03	56°51'N	41°23'E
Shuyang, China (sho͞o yäng)	258-59	34°08'N	118°47'E
Shwangliao, China *see* Liaoyuan	260-61	42°55'N	125°08'E
Shwebo, Mya.	266-67	22°34'N	95°42'E
Shymkent, Kaz.	244	42°18'N	69°36'E
Shyok, stm., Asia	252-53	35°14'N	75°55'E
Siālkot, Pak. (sė-äl´kōt)	252-53	32°31'N	74°33'E
Siam, nation, Asia *see* Thailand	224-25	15°0'N	100°00'E
Siam, Gulf of, b., Asia *see* Thailand, Gulf of	266-67	10°0'N	101°00'E
Sian, China *see* Xi'an	258-59	34°15'N	108°52'E
Siargao Island, i., Phil.	270	9°53'N	126°02'E
Siasi Island, i., Phil.	270	5°33'N	120°51'E
Siau, Pulau, i., Indon.	268-69	2°46'N	125°23'E
Šiauliai, Lith. (shė-ou´lė-ī)	208-09	55°56'N	23°20'E
Sibay, Russia (sē´bāy)	244	52°42'N	58°40'E
Šibenik, Cro. (shė-bä´nĕk)	216-17	43°44'N	15°54'E
Siberia, reg., Russia (sī-bĭr´ė-ȧ)	236-37	65°0'N	110°00'E
Sibi, Pak.	252-53	29°33'N	67°53'E
Sibir', reg., Russia *see* Siberia	236-37	65°0'N	110°00'E
Sibiryakova, Ostrov, i., Russia	236-37	72°50'N	79°00'E
Sibiti, Congo (sė-bė-tē´)	284-85	3°41'S	13°21'E
Sibiu, Rom. (sė-bĭ-o͞o´)	210-11	45°47'N	24°09'E
Sibley, Ia., U.S. (sĭb´lė)	124-25	43°24'N	95°45'W
Sibolga, Indon. (sė-bō´gä)	266-67	1°45'N	98°47'E
Sibsāgar, India (sėb-sŭ´gŭr)	254-55	26°59'N	94°39'E
Sibu, Malay.	268-69	2°18'N	111°50'E
Sibut, C.A.R.	284-85	5°44'N	19°05'E
Sibutu Island, i., Phil.	270	4°46'N	119°29'E
Sibuyan Island, i., Phil. (sē-bo͞o-yän´ ī´lǎnd)	270	12°27'N	122°34'E
Sibuyan Sea, s., Phil. (sē-bo͞o-yän´ sē)	270	12°50'N	122°40'E
Sichuan, state, China (sz-chůän)	258-59	31°0'N	105°00'E
Sicilia, i., Italy *see* Sicily	216-17	37°30'N	14°00'E
Sicily, i., Italy (sĭs´ĭ-lė)	216-17	37°30'N	14°00'E
Sico Tinto, stm., Hond. (sė-kō tēn´tō)	161	15°50'N	85°03'W
Sicuani, Peru	184	14°16'S	71°13'W
Sidhi, India	254-55	24°24'N	81°53'E
Sîdi Barrâni, Egypt	204-05	31°37'N	25°56'E
Sidi Bel Abbès, Alg. (sē´dē-bĕl á-bĕs´)	214-15	35°12'N	0°11'W
Sidi-Bennour, Mor.	292a	32°39'N	8°25'W
Sidikalang, Indon.	266-67	2°44'N	98°20'E
Sidney, B.C., Can. (sĭd´nė)	142-43	48°39'N	123°24'W
Sidney, Mt., U.S. (sĭd´nė)	122-23	47°43'N	104°09'W
Sidney, Ne., U.S. (sĭd´nė)	124-25	41°09'N	102°59'W
Sidney, N.Y., U.S. (sĭd´nė)	126-27	42°19'N	75°23'W
Sidney, Oh., U.S. (sĭd´nė)	126-27	40°17'N	84°10'W
Sidney Lanier, Lake, res., Ga., U.S. (lāk sĭd´nė lăn´yēr)	134-35	34°15'N	83°57'W
Sidon, Leb.	248-49	33°34'N	35°23'E
Sidra, Gulf of, b., Libya (gŭlf ŭv sĭ´drȧ) *see* Surt, Khalīj	280-81	31°30'N	18°00'E
Siedlce, Pol. (syĕd´'l-tsĕ)	210-11	52°10'N	22°17'E
Siegburg, Ger. (zēg´bo͞orgh)	210-11	50°47'N	7°12'E
Siegen, Ger. (zē´ghĕn)	210-11	50°52'N	8°01'E
Siemiatycze, Pol. (syĕm´yä´tė-chĕ)	210-11	52°26'N	22°52'E
Siĕmréab, Camb.	266-67	13°22'N	103°51'E
Siena, Italy (sė-ĕn´ä)	216-17	43°19'N	11°20'E
Sienyang, China *see* Xianyang	258-59	34°20'N	108°42'E
Sieradz, Pol. (syĕ´rädz)	210-11	51°36'N	18°45'E
Sierpc, Pol. (syĕrpts)	210-11	52°51'N	19°40'E
Sierra Blanca, Tx., U.S. (sē-ĕ´rȧ blaŋ-kä)	132-33	31°11'N	105°21'W
Sierra Blanca Peak, mtn., N.M., U.S. (sē-ĕ´rä blăn´kȧ pēk)	130-31	33°23'N	105°48'W
Sierra Colorada, Arg.	185	40°36'S	67°45'W
Sierra Leone, nation, Afr. (sė-ĕr´rä lā-ō´nā)	274	8°30'N	11°30'W
Sierra Nevada, mts., Ca., U.S. *see* Nevada, Sierra	128-29	38°0'N	119°15'W
Sífnos, i., Grc.	216-17	36°58'N	24°43'E

ăt; finǎl; rāte; senāte; ärm; åsk; sofȧ; fâre; ch-choose; dh-as th in other; bē; ėvent; bĕt; recĕnt; cratẽr; g-gō; gh-guttural g; bĭt; ĭ-short neutral; rīde; ĸ-guttural k as ch in German ich;

n-sing; ŋ-baŋk; ɴ-nasalized n; nŏd; cŏmmit; ōld; ôbey; ôrder; oi-boil; fōōd; ò-as oo in foot; ou-out; s-soft; sh-dish; th-thin; pūre; ûnite; ûrn; stŭd; circŭs; ü-as in French tu; ′-indeterminate vowel.

410 Society Islands – Spezia PRONOUNCING INDEX

Feature (Pronunciation)	Page	Lat.	Long.
Society Islands, is., Fr. Poly. (sō-sī´ĕ-tē ī´lándz)	304-05	17°0'S	150°00'W
Socoltenango, Mex. (sō-kōl-tĕ-näŋ´gō)	160	16°12'N	92°14'W
Socorro, Col. (sô-kôr´rō)	178-79	6°29'N	73°16'W
Socorro, N.M., U.S. (sō-kō´r-rō)	128-29	34°04'N	106°54'W
Socorro, Tx., U.S. (sō-kō´r-rō)	132-33	31°39'N	106°18'W
Socorro, Isla, i., Mex.	156-57	18°45'N	110°58'W
Socotra, i., Yemen (sō-kō´trá) *see* Suquṭrā	238-39	12°31'N	53°54'E
Soc Trang, Viet.	266-67	9°36'N	105°58'E
Socuéllamos, Spain (sō-kōō-āl´yä-mōs)	214-15	39°17'N	2°47'W
Sodankylä, Fin.	200-01	67°25'N	26°34'E
Soda Springs, Id., U.S. (sō´dá sprĭngz)	122-23	42°40'N	111°36'W
Söderhamn, Swe. (sû-dĕr-häm´'n)	208-09	61°19'N	17°05'E
Södertälje, Swe. (sû-dĕr-tĕl´yĕ)	208-09	59°12'N	17°37'E
Sodo, Eth.	292d	6°52'N	37°46'E
Soe, Indon.	268-69	9°52'S	124°17'E
Soerabaja, Indon. *see* Surabaya	268-69	7°15'S	112°45'E
Sofia, nat. cap., Blg. (sô´fē-á)	216-17	42°42'N	23°19'E
Sofia, stm., Madag.	286-87	15°25'S	47°14'E
Sofiya, nat. cap., Blg. (sô´fē-á) *see* Sofia	216-17	42°42'N	23°19'E
Sogamoso, Col. (sô-gä-mô´sō)	178-79	5°44'N	72°56'W
Sognefjorden, b., Nor.	208-09	61°06'N	5°10'E
Sogo Nur, lk., China	260-61	42°17'N	101°14'E
Sog Xian, China	258-59	31°50'N	93°47'E
Sōhu Gan, i., Japan	264	29°49'N	140°21'E
Soissons, Fr. (swä-sôn´)	212-13	49°23'N	3°20'E
Sŏjosŏn-man, b., Kor., N.	263	39°20'N	124°50'E
Sokal', Ukr. (sô´käl')	210-11	50°29'N	24°17'E
Sokcho, Kor., S.	263	38°11'N	128°34'E
Söke, Tur. (sû´kĕ)	216-17	37°45'N	27°24'E
Sokhumi, Geor.	245	43°00'N	41°00'E
Sokodé, Togo	282-83	8°59'N	1°09'E
Sokol, Russia	202-03	59°28'N	40°07'E
Sokółka, Pol. (sô-kōl´kå)	210-11	53°24'N	23°31'E
Sokołów Podlaski, Pol. (sô-kô-wôf´ pŭd-lä´skĭ)	210-11	52°24'N	22°15'E
Sokoto, Nig. (sō´kô-tō)	282-83	13°04'N	5°15'E
Sokoto, stm., Nig.	282-83	11°24'N	4°08'E
Solano, Phil. (sō-lä´nō)	270	16°31'N	121°11'E
Solāpur, India	256	17°41'N	75°54'E
Soledad, Col. (sō-lĕ-dä´d)	178-79	10°56'N	74°46'W
Soledad Díez Gutiérrez, Mex.	158-59	22°12'N	100°56'W
Solikamsk, Russia (sô-lē-kamsk´)	202-03	59°40'N	56°46'E
Sol'-Iletsk, Russia	202-03	51°09'N	55°00'E
Solimões, stm., S.A. *see* Amazon	178-79	0°04'S	49°15'W
Solingen, Ger. (zō´lĭng-ĕn)	210-11	51°10'N	7°05'E
Sollefteå, Swe.	200-01	63°11'N	17°16'E
Solnechnogorsk, Russia	218-19	56°11'N	36°59'E
Solo, Indon. *see* Surakarta	268-69	7°34'S	110°50'E
Solomon Islands, nation, Oc. (sŏ´lō-mŭn ī´lándz)	306e	8°0'S	159°00'E
Solomon Sea, s., Oc. (sŏ´lō-mŭn sē)	302	8°0'S	155°00'E
Solon, China (swo-lōōn)	260-61	46°36'N	121°13'E
Solor, Pulau, i., Indon.	268-69	8°28'S	122°59'E
Solov'yëvsk, Russia	260-61	49°55'N	115°42'E
Soltau, Ger. (sōl´tou)	210-11	52°59'N	9°50'E
Sol'tsy, Russia (sôl´tsė)	208-09	58°07'N	30°20'E
Solvay, N.Y., U.S. (sŏl´vā)	126-27	43°04'N	76°14'W
Sölvesborg, Swe. (sûl´vĕs-bôrg)	208-09	56°03'N	14°35'E
Sol'vychegodsk, Russia (sôl´vė-chĕ-gôtsk´)	202-03	61°20'N	46°55'E
Solway Firth, b., U.K. (sŏl´wā fûrth´)	206-07	54°50'N	3°35'W
Solwezi, Zam.	286-87	12°11'S	26°25'E
Somalia, nation, Afr. (sō-mä´lē-ā)	274	6°0'N	48°00'E
Somaliland, nation, Afr. *see* Somalia	274	6°0'N	48°00'E
Somali Republic, nation, Afr. *see* Somalia	274	6°0'N	48°00'E
Sombor, Serb. (sôm´bôr)	216-17	45°47'N	19°07'E
Sombrerete, Mex. (sōm-brå-rā´tå)	158-59	23°41'N	103°39'W
Sombrero, i., St. K./N.	155b	18°36'N	63°26'W
Somerset, Ky., U.S. (sŭm´ēr-sĕt)	134-35	37°05'N	84°37'W
Somerset, Oh., U.S. (sŭm´ēr-sĕt)	126-27	39°48'N	82°18'W
Somerset East, S. Afr. (sŭm´ēr-sĕt ēst)	286-87	32°44'S	25°35'E
Somersworth, N.H., U.S. (sŭm´ērz-wûrth)	126-27	43°16'N	70°52'W
Somerville, Tn., U.S. (sŭm´ēr-vĭl)	134-35	35°15'N	89°21'W
Somerville, Tx., U.S. (sŭm´ēr-vĭl)	132-33	30°20'N	96°32'W
Somoto, Nic. (sô-mō´tō)	161	13°28'N	86°35'W
Son, stm., India (sōn)	254-55	25°42'N	84°52'E
Sønderborg, Den. (sûn´'er-bôrgh)	210-11	54°55'N	9°48'E
Sonepur, India	254-55	20°49'N	83°54'E
Song Da, stm., Asia *see* Black	266-67	21°15'N	105°21'E
Songea, Tan. (sôn-gä´å)	286-87	10°41'S	35°39'E
Songhua, stm., China	260-61	47°43'N	132°31'E

Feature (Pronunciation)	Page	Lat.	Long.
Songhua Hu, res., China	260-61	43°25'N	127°10'E
Songjiang, China	258-59	31°01'N	121°14'E
Sŏngjin, Kor., N. (sŭng´jĭn´) *see* Kimch'aek	263	40°41'N	129°12'E
Songkhla, Thai. (sông´klä´)	266-67	7°12'N	100°36'E
Songnim, Kor., N.	263	38°44'N	125°38'E
Songo, Moz.	286-87	15°39'S	32°43'E
Sonid Youqi, China	260-61	42°44'N	112°40'E
Sonmiāni Bay, b., Pak.	252-53	25°15'N	66°30'E
Sonneberg, Ger. (sôn´ē-bĕrgh)	210-11	50°21'N	11°11'E
Sonora, Ca., U.S. (sō-nō´rá)	128-29	37°59'N	120°22'W
Sonora, Tx., U.S. (sō-nō´rá)	132-33	30°34'N	100°39'W
Sonora, state, Mex. (sō-nō´rá)	156-57	29°20'N	110°40'W
Sonora, stm., Mex. (sō-nō´rá)	156-57	29°05'N	110°54'W
Sonora, Desierto de, des., N.A. *see* Sonoran Desert	156-57	30°0'N	113°00'W
Sonoran Desert, des., N.A.	156-57	30°0'N	113°00'W
Sonsón, Col. (sôn-sôn´)	188c	5°43'N	75°18'W
Sonsonate, El Sal. (sōn-sō-nä´tå)	160	13°43'N	89°43'W
Sonsorol Islands, is., Palau (sŏn-sō-rōl´ ī´lándz)	304-05	5°20'N	132°13'E
Son Tay, Viet.	266-67	21°08'N	105°30'E
Soomaaliya, nation, Afr. *see* Somalia	274	6°0'N	48°00'E
Soome laht, b., Eur. *see* Finland, Gulf of	208-09	60°0'N	27°00'E
Sora, Italy (sô´rä)	216-17	41°44'N	13°37'E
Sorell, Cape, c., Austl.	301	42°12'S	145°10'E
Sorel-Tracy, Qc., Can.	146-47	46°03'N	73°05'W
Sorgun, Tur.	247	39°49'N	35°11'E
Soria, Spain (sō´rē-ä)	214-15	41°46'N	2°28'W
Sorocaba, Braz. (sō-rô-kä´bá)	186	23°30'S	47°28'W
Sorochinsk, Russia	202-03	52°26'N	53°10'E
Sorong, Indon. (sō-rông´)	242-43	0°53'S	131°15'E
Soroti, Ug. (sō-rō´tē)	289	1°43'N	33°36'E
Sørøya, i., Nor.	200-01	70°34'N	22°22'E
Sorrento, Italy (sôr-rĕn´tō)	216-17	40°38'N	14°22'E
Sor Rondane Mountains, mts., Ant.	313	72°0'S	25°00'E
Sorsogon, Phil. (sôr-sôgōn´)	270	12°59'N	124°01'E
Sortavala, Russia (sôr´tä-vä-lä)	208-09	61°42'N	30°40'E
Sosna, stm., Russia (sôs´nå)	218-19	52°42'N	38°55'E
Sosnogorsk, Russia	202-03	63°36'N	53°53'E
Sosnowiec, Pol. (sôs-nô´vyĕts)	210-11	50°18'N	19°08'E
Sos'va, stm., Russia (sôs´vå)	236-37	59°33'N	62°20'E
Soto la Marina, Barra, i., Mex.	158-59	24°10'N	97°44'W
Soufrière, vol., Guad. (sōō-frē-âr´)	155b	16°04'N	61°40'W
Sŏul, nat. cap., Kor., S. *see* Seoul	263	37°33'N	127°01'E
Sounding Creek, stm., Ab., Can. (soun´dĭng krēk)	142-43	52°06'N	110°28'W
Souris, Mb., Can. (sōō´rē´)	144-45	49°38'N	100°16'W
Souris, P.E., Can. (sōō´rē´)	148-49	46°21'N	62°15'W
Souris, stm., N.A. (sōō´rē´)	140-41	49°40'N	99°35'W
Sousa, Braz.	180-81	6°45'S	38°14'W
Sousse, Tun. (sōōs)	200-01	35°49'N	10°38'E
South, stm., N.C., U.S. (south)	134-35	34°35'N	78°16'W
South Africa, nation, Afr. (south ăf´rĭ-kă)	274	30°0'S	26°00'E
South America, cont., (south á-mĕr´ĭ-kă)	173	15°0'S	60°00'W
Southampton, Eng., U.K. (south-ămp´tŭn)	206-07	50°55'N	1°24'W
Southampton Island, i., Nu., Can.	140-41	64°20'N	84°40'W
South Andaman, i., India (south ăn-dá-măn´)	266-67	11°48'N	92°44'E
South Australia, state, Austl. (south ôs-trā´lĭ-á)	294-95	30°0'S	135°00'E
South Bend, In., U.S. (south bĕnd)	126-27	41°41'N	86°14'W
South Bend, Wa., U.S. (south bĕnd)	122-23	46°40'N	123°46'W
South Boston, Va., U.S. (south bôs´tŭn)	134-35	36°42'N	78°54'W
Southbridge, Ma., U.S. (south´brĭj)	126-27	42°05'N	72°03'W
South Carolina, state, U.S. (south kăr-ō-lī´ná)	118-19	34°0'N	81°00'W
South China Sea, s., Asia (south chī´ná sē)	242-43	10°0'N	113°00'E
South Dakota, state, U.S. (south dá-kō´tá)	118-19	44°15'N	100°00'W
South East Cape, c., Austl.	301	43°38'S	146°52'E
South East Point, c., Austl.	301	39°08'S	146°25'E
Southend-on-Sea, Eng., U.K. (south-ĕnd´-ŏn-sē)	206-07	51°33'N	0°45'E
Southern Alps, mts., N.Z. (sŭ-thûrn ălps)	303	43°30'S	170°30'E
Southern Bug, stm., Ukr. (sŭ-thûrn bōōg)	218-19	46°39'N	31°56'E
Southern Cook Islands, is., Cook Is.	304-05	20°0'S	159°00'W
Southern Cross, Austl.	294-95	31°14'S	119°19'E
Southern Indian Lake, lk., Mb., Can. (sŭth´ērn ĭn´dĭ-ăn lāk)	144-45	57°13'N	98°21'W

Feature (Pronunciation)	Page	Lat.	Long.
Southern Ocean, oc., (sŭ-thûrn ōshŭn)	4-5	50°0'S	135°00'E
Southern Pines, N.C., U.S. (sŭth´ērn pīnz)	134-35	35°10'N	79°24'W
Southern Ute Indian Reservation, ind. res., Co., U.S. (sŭth´ērn ūt ĭn´dĭ-ăn rĕ-sĕr-vā´shĕn)	128-29	37°05'N	107°45'W
South Georgia, i., S. Geor. (south jôr´já)	313	54°15'S	36°45'W
South Georgia and the South Sandwich Islands, dep., S.A.	313	54°0'S	38°00'W
South Haven, Mi., U.S. (south hāv´'n)	126-27	42°24'N	86°16'W
South Henik Lake, lk., Nu., Can.	140-41	61°30'N	97°27'W
South Indian Lake, Mb., Can.	144-45	56°48'N	98°57'W
Southington, Ct., U.S. (sŭdh´ĭng-tŭn)	126-27	41°36'N	72°53'W
South Island, i., India	256	10°03'N	72°17'E
South Island, i., N.Z. (south ī´lánd)	303	43°0'S	171°00'E
South Korea, nation, Asia (south kô-rē´-á)	224-25	36°30'N	128°00'E
South Luangwa National Park, n.p., Zam.	286-87	12°56'S	31°38'E
South Magnetic Pole, p.o.i.,	313	64°18'S	136°44'E
South Nahanni, stm., N.T., Can.	140-41	61°03'N	123°20'W
South Negril Point, c., Jam. (south nå-grēl´ point)	154-55	18°15'N	78°22'W
South Ogden, Ut., U.S. (south ŏg´dĕn)	122-23	41°12'N	111°59'W
South Orkney Islands, is., Ant.	313	60°35'S	44°07'W
South Paris, Me., U.S. (south păr´ĭs)	126-27	44°13'N	70°31'W
South Pittsburg, Tn., U.S. (south pĭs´bûrg)	134-35	35°01'N	85°43'W
South Platte, stm., U.S. (south plăt)	120-21	41°07'N	100°42'W
Southport, Austl. (south´pōrt)	301	27°58'S	153°25'E
Southport, Eng., U.K. (south´pôrt)	206-07	53°39'N	3°01'W
South River, On., Can.	146-47	45°50'N	79°22'W
South Sandwich Islands, is., S. Geor. (south sănd´wĭch ī´lándz)	313	57°31'S	26°37'W
South Saskatchewan, stm., Can. (south sás-kach´ĕ-wän)	140-41	53°14'N	105°04'W
South Shetland Islands, is., Ant.	313	62°0'S	58°00'W
South Shields, Eng., U.K. (south shēldz)	206-07	54°60'N	1°25'W
South Sioux City, Ne., U.S. (south sōō sĭt´ē sĭ´tē)	124-25	42°27'N	96°25'W
South Sudan, nation, Afr.	274	6°00'N	30°00'E
South Thompson, stm., B.C., Can. (south tŏmp´sŭn)	142-43	50°41'N	120°20'W
South West Cape, c., Austl.	301	43°33'S	146°04'E
South West Cape, c., N.Z.	303	47°17'S	167°28'E
Southwest Miramichi, stm., N.B., Can. (south-wĕst´ mĭr á-mė´shē)	148-49	46°58'N	65°34'W
Southwest National Park, n.p., Austl.	301	43°05'S	146°09'E
Sovetsk, Russia (sô-vyĕtsk´)	208-09	55°05'N	21°53'E
Sovetsk, Russia (sô-vyĕtsk´)	202-03	57°35'N	48°58'E
Sovetskaya Gavan, Russia (sŭ-vyĕt´skĭ-u gä´vŭn')	240-41	48°58'N	140°18'E
Soweto, S. Afr.	292c	26°17'S	27°51'E
Sozopol, Blg. (sôz´ô-pôl')	216-17	42°25'N	27°42'E
Spa, Bel. (spä)	206-07	50°29'N	5°52'E
Spain, nation, Eur. (spān)	190-91	40°0'N	4°00'W
Spanish Fork, Ut., U.S. (spăn´ĭsh fôrk)	128-29	40°08'N	111°39'W
Spanish Town, Jam.	154-55	17°60'N	76°58'W
Sparks, Nv., U.S. (spärks)	128-29	39°32'N	119°44'W
Sparta, Grc. (spär´tá)	216-17	37°05'N	22°26'E
Sparta, Tn., U.S. (spär´tá)	134-35	35°56'N	85°28'W
Sparta, Wi., U.S. (spär´tá)	124-25	43°56'N	90°48'W
Spartanburg, S.C., U.S. (spär´tăn-bûrg)	134-35	34°57'N	81°56'W
Spartel, Cap, c., Mor. (kăp spär-tĕl´)	292a	35°48'N	5°55'W
Spárti, Grc. *see* Sparta	216-17	37°05'N	22°26'E
Spartivento, Capo, c., Italy (kä´pō spär-tė-vĕn´tô)	216-17	37°55'N	16°04'E
Spartivento, Capo, c., Italy (kä´pō spär-tė-vĕn´tô)	216-17	38°53'N	8°50'E
Spas-Demensk, Russia (spås dyĕ-mĕnsk´)	218-19	54°25'N	34°02'E
Spas-Klepiki, Russia (spås klĕp´ē-kė)	218-19	55°08'N	40°12'E
Spassk-Dal'niy, Russia (spŭsk´däl´nyē)	264	44°36'N	132°50'E
Spear, Cape, c., Nf./L., Can. (kăp spēr)	148-49	47°31'N	52°39'W
Spearfish, S.D., U.S. (spēr´fĭsh)	124-25	44°30'N	103°52'W
Speedway, In., U.S. (spēd´wā)	126-27	39°47'N	86°13'W
Spence Bay, Nu., Can. *see* Taloyoak	138-39	69°32'N	93°31'W
Spencer, Ia., U.S. (spĕn´sēr)	124-25	43°09'N	95°09'W
Spencer, In., U.S. (spĕn´sēr)	126-27	39°17'N	86°46'W
Spencer, W.V., U.S. (spĕn´sēr)	126-27	38°47'N	81°22'W
Spencer, Cape, c., Austl.	301	35°18'S	136°53'E
Spencer Gulf, b., Austl. (spĕn´sēr gŭlf)	296-97	34°0'S	137°00'E
Speyer, Ger. (shpī´ēr)	210-11	49°20'N	8°26'E
Spezia, Italy *see* La Spezia	216-17	44°07'N	9°50'E

ăt; finăl; rāte; senāte; ärm; ȧsk; sofȧ; fâre; ch-choose; dh-as th in other; bē; ėvent; bĕt; recĕnt; cratẽr; g-gō; gh-guttural g; bĭt; ĭ-short neutral; rīde; ĸ-guttural k as ch in German ich;

PRONOUNCING INDEX

Spinazzola – Sukabumi 411

Feature (Pronunciation)	Page	Lat.	Long.
Spinazzola, Italy (spē-nät´zô-lä)	216-17	40°58'N	16°05'E
Spires, Ger. *see* Speyer	210-11	49°20'N	8°26'E
Spirit Lake, Ia., U.S. (spĭr´ĭt lāk)	124-25	43°26'N	95°06'W
Spirit Lake, Id., U.S. (spĭr´ĭt lāk)	122-23	47°58'N	116°53'W
Spišská Nová Ves, Slvk. (spēsh´skä nō´vä vĕs)	210-11	48°57'N	20°34'E
Spitsbergen, i., Nor. (spĭts´bûr-gĕn)	236-37	78°45'N	16°00'E
Split, Cro. (splēt)	216-17	43°30'N	16°26'E
Split Lake, res., Mb., Can.	144-45	56°08'N	96°15'W
Spokane, Wa., U.S. (spō-kăn´)	122-23	47°39'N	117°24'W
Spokane Indian Reservation, ind. res., Wa., U.S. (spōkăn´ ĭn´dĭ-ăn rĕ-sēr-vā´shĕn)	122-23	47°55'N	118°00'W
Spokane Valley, Wa., U.S. (spōkăn vălē)	122-23	47°40'N	117°13'W
Spoleto, Italy (spô-lā´tō)	216-17	42°44'N	12°44'E
Spooner, Wi., U.S. (spōōn´ẽr)	124-25	45°49'N	91°53'W
Spratly Islands, is., Asia	242-43	10°0'N	114°00'E
Springbok, S. Afr. (sprĭng´bŏk)	286-87	29°43'S	17°55'E
Springdale, Nf./L., Can. (sprĭng´dāl)	148-49	49°31'N	56°04'W
Springdale, Ar., U.S. (sprĭng´dāl)	130-31	36°11'N	94°09'W
Springer, N.M., U.S. (sprĭng´ẽr)	130-31	36°22'N	104°36'W
Springfield, Co., U.S. (sprĭng´fēld)	130-31	37°24'N	102°37'W
Springfield, Fl., U.S. (sprĭng´fēld)	134-35	30°12'N	85°37'W
Springfield, Il., U.S.	126-27	39°48'N	89°39'W
Springfield, Ky., U.S. (sprĭng´fēld)	126-27	37°41'N	85°13'W
Springfield, Ma., U.S.	126-27	42°07'N	72°35'W
Springfield, Mn., U.S. (sprĭng´fēld)	124-25	44°14'N	94°59'W
Springfield, Mo., U.S. (sprĭng´fēld)	130-31	37°13'N	93°17'W
Springfield, Oh., U.S. (sprĭng´fēld)	126-27	39°55'N	83°49'W
Springfield, Or., U.S. (sprĭng´fēld)	122-23	44°03'N	123°01'W
Springfield, Tn., U.S. (sprĭng´fēld)	134-35	36°30'N	86°53'W
Springfield, Vt., U.S. (sprĭng´fēld)	126-27	43°18'N	72°29'W
Springhill, N.S., Can. (sprĭng-hĭl´)	148-49	45°39'N	64°03'W
Springs, S. Afr. (sprĭngs)	292c	26°15'S	28°26'E
Springsure, Austl.	302	24°07'S	148°05'E
Spring Valley, Il., U.S. (sprĭng crēk văl´ē)	126-27	41°19'N	89°12'W
Spring Valley, Mn., U.S. (sprĭng văl´ē)	124-25	43°41'N	92°25'W
Spruce Grove, Ab., Can. (sprōōs grōv)	142-43	53°32'N	113°55'W
Spruce Knob, mtn., W.V., U.S.	126-27	38°42'N	79°32'W
Squamish, B.C., Can. (skwŏ´mĭsh)	142-43	49°42'N	123°08'W
Squamish, stm., B.C., Can. (skwŏ´mĭsh)	142-43	49°39'N	123°14'W
Srbija, nation, Eur. (sr bē-yä) *see* Serbia	190-91	44°0'N	21°00'E
Sredinny Khrebet, mts., Russia	236-37	56°0'N	158°00'E
Srednekolymsk, Russia (s'rĕd´nyĕ kô-lêmsk´)	236-37	67°27'N	153°40'E
Srednerusskaya Vozvyshennost', plat., Russia	218-19	52°0'N	38°00'E
Śrem, Pol. (shrĕm)	210-11	52°05'N	17°02'E
Sremska Mitrovica, Serb. (srĕm´skä mē´trô-vē-tsä´)	216-17	44°59'N	19°37'E
Sri Aman, Malay.	268-69	1°13'N	111°28'E
Sri Jayewardenepura Kotte, nat. cap., Sri L.	256	6°54'N	79°54'E
Srīkākulam, India	256	18°18'N	83°54'E
Sri Lanka, nation, Asia (shrē´-län-kȧ) (srē´-län-kȧ)	224-25	7°0'N	81°00'E
Srīnagar, India	254-55	34°05'N	74°48'E
Staaten River National Park, n.p., Austl.	302	16°40'S	143°00'E
Stafford, Eng., U.K. (stăf´fẽrd)	206-07	52°48'N	2°07'W
Stafford, Va., U.S. (stăf´fẽrd)	126-27	38°25'N	77°24'W
Stakhanov, Ukr.	218-19	48°34'N	38°40'E
Stalin, Rom. *see* Brașov	210-11	45°39'N	25°37'E
Stambaugh, Mi., U.S. (stăm´bô)	124-25	46°05'N	88°38'W
Stamford, Eng., U.K. (stăm´fẽrd)	206-07	52°39'N	0°29'W
Stamford, Ct., U.S. (stăm´fẽrd)	126-27	41°03'N	73°33'W
Stamford, Tx., U.S. (stăm´fẽrd)	130-31	32°55'N	99°49'W
Stamps, Ar., U.S. (stămps)	130-31	33°22'N	93°30'W
Standerton, S. Afr. (stăn´dẽr-tŭn)	292c	26°57'S	29°15'E
Standing Rock Indian Reservation, ind. res., U.S. (stănd´ĭng rŏk ĭn´dĭ-ăn rĕ-sēr-vā´shĕn)	124-25	45°50'N	101°10'W
Stanford, Ky., U.S. (stăn´fẽrd)	126-27	37°31'N	84°41'W
Stanisławów, Ukr. *see* Ivano-Frankivs'k	210-11	48°55'N	24°44'E
Stanley, N.D., U.S. (stăn´lē)	124-25	48°19'N	102°24'W
Stanley, nat. cap., Falk. Is. (stăn´lē)	185	51°43'S	57°49'W
Stanley Falls, wtfl., D.R.C.	284-85	0°29'N	25°13'E
Stanovoye Nagor'ye, mts., Russia	236-37	56°0'N	114°00'E
Stanovoy Khrebet, mts., Russia (stŭn-à-voi´)	236-37	55°48'N	125°34'E
Stanovoy Mountains, mts., Russia (stŭn-à-voi´ moun´tĭnz) *see* Stanovoye Nagor'ye	236-37	56°0'N	114°00'E
Stanovoy Range, mts., Russia (stŭn-à-voi´ rānj) *see* Stanovoy Khrebet	236-37	55°48'N	125°34'E
Stanthorpe, Austl.	301	28°40'S	151°56'E
Stanton, Ky., U.S. (stăn´tŭn)	126-27	37°50'N	83°55'W
Stanton, Tx., U.S. (stăn´tŭn)	132-33	32°08'N	101°47'W
Staples, Mn., U.S. (stā´p'lz)	124-25	46°21'N	94°48'W
Staraya Russa, Russia (stä´rá-yá rōōsȧ)	208-09	57°60'N	31°21'E
Stara Zagora, Blg. (stä´rä zä´gô-rȧ)	216-17	42°26'N	25°39'E
Starbuck, Mb., Can. (stär´bŭk)	144-45	49°46'N	97°37'W
Starbuck, i., Kir.	304-05	5°37'S	155°53'W
Staritsa, Russia (stä´rē-tsȧ)	218-19	56°30'N	34°56'E
Starke, Fl., U.S. (stärk)	134-35	29°57'N	82°07'W
Starkville, Ms., U.S. (stärk´vĭl)	134-35	33°27'N	88°49'W
Starodub, Russia (stä-rô-drŏp´)	218-19	52°35'N	32°46'E
Starominskaya, Russia (stä´rô mĭn´skȧ-yȧ)	218-19	46°32'N	39°02'E
Start Point, c., Eng., U.K. (stärt point)	206-07	50°14'N	3°39'W
Staryy Oskol, Russia (stä´rē ôs-kôl´)	218-19	51°18'N	37°51'E
Staszów, Pol. (stä´shòf)	210-11	50°34'N	21°20'E
State College, Pa., U.S. (stāt kŏl´ĕj)	126-27	40°47'N	77°52'W
Staten Island, i., Arg. *see* Estados, Isla de los	185	54°48'S	64°33'W
Statesboro, Ga., U.S. (stāts´bŭr-ô)	134-35	32°27'N	81°47'W
Statesville, N.C., U.S. (stās´vĭl)	134-35	35°47'N	80°54'W
Staunton, Il., U.S. (stôn´tŭn)	130-31	39°01'N	89°47'W
Staunton, Va., U.S. (stôn´tŭn)	126-27	38°09'N	79°05'W
Staunton, stm., U.S. *see* Roanoke	134-35	35°57'N	76°43'W
Stavanger, Nor. (stä´väng´ẽr)	208-09	58°58'N	5°45'E
Stavropol', Russia	202-03	45°02'N	41°59'E
Stawell, Austl.	301	37°04'S	142°46'E
Steamboat Springs, Co., U.S. (stēm´bōt´ sprĭngz)	122-23	40°29'N	106°50'W
Steel, stm., On., Can. (stēl)	146-47	48°47'N	86°54'W
Steens Mountain, mts., Or., U.S. (stēnz moun´tĭn)	122-23	42°35'N	118°40'W
Stefanie, Lake, lk., Afr. *see* Ch'ew Bahir	289	4°40'N	36°50'E
Steinamanger, Hung. *see* Szombathely	210-11	47°14'N	16°38'E
Steinbach, Mb., Can.	144-45	49°31'N	96°41'W
Steinkjer, Nor. (stĕĭn-kyĕr)	200-01	64°01'N	11°29'E
Stellarton, N.S., Can. (stĕl´ăr-tŭn)	148-49	45°33'N	62°39'W
Stendal, Ger. (shtĕn´däl)	210-11	52°36'N	11°51'E
Stephens Island, i., B.C., Can.	142-43	54°10'N	130°45'W
Stephens Lake, res., Mb., Can.	144-45	56°26'N	95°07'W
Stephenville, Nf./L., Can. (stē´vĕn-vĭl)	148-49	48°33'N	58°37'W
Sterling, Ak., U.S. (stûr´lĭng)	136	60°32'N	150°48'W
Sterling, Co., U.S. (stûr´lĭng)	124-25	40°37'N	103°13'W
Sterling, Il., U.S. (stûr´lĭng)	126-27	41°48'N	89°42'W
Sterlitamak, Russia (styĕr´lĕ-ta-mȧk´)	202-03	53°37'N	55°58'E
Šternberk, Czech Rep. (shtĕrn´bĕrk)	210-11	49°44'N	17°18'E
Stettin, Pol. *see* Szczecin	210-11	53°26'N	14°32'E
Steubenville, Oh., U.S. (stū´bĕn-vĭl)	126-27	40°22'N	80°38'W
Stevens Point, Wi., U.S.	126-27	44°31'N	89°34'W
Stevensville, Mt., U.S. (stē´vĕnz-vĭl)	122-23	46°30'N	114°05'W
Stewart, B.C., Can.	142-43	55°56'N	129°58'W
Stewart, Isla, i., Chile	185	54°52'S	71°12'W
Stewart Island, i., N.Z.	303	47°0'S	167°50'E
Steynsrus, S. Afr. (stīns´rōōs)	292c	27°57'S	27°34'E
Steyr, Aus. (shtīr)	210-11	48°03'N	14°25'E
Stikine, stm., N.A. (stĭ-kēn´)	140-41	56°41'N	132°14'W
Stillwater, Ok., U.S. (stĭl´wô-tẽr)	130-31	36°08'N	97°05'W
Stillwater Range, mts., Nv., U.S. (stĭl´wô-tẽr rānj)	128-29	39°53'N	118°06'W
Štip, Mac. (shtĭp)	216-17	41°45'N	22°12'E
Stirling, Scot., U.K. (stûr´lĭng)	206-07	56°07'N	3°56'W
Stjørdalshalsen, Nor. (styûr-däls-hälsĕn)	200-01	63°29'N	10°56'E
Stockholm, nat. cap., Swe. (stŏk´hôlm)	208-09	59°20'N	18°03'E
Stockport, Eng., U.K. (stŏk´pôrt)	206-07	53°25'N	2°10'W
Stockton, Ca., U.S.	128-29	37°57'N	121°17'W
Stockton, Ks., U.S. (stŏk´tŭn)	130-31	39°26'N	99°16'W
Stockton Plateau, plat., Tx., U.S. (stŏk´tŭn plă-tō´)	132-33	30°30'N	102°30'W
Stœng Tréng, Camb. (stông´trĕng´)	266-67	13°31'N	105°58'E
Stoke-on-Trent, Eng., U.K. (stōk-ŏn-trĕnt)	206-07	52°60'N	2°10'W
Stolbovoy, Ostrov, i., Russia	236-37	74°05'N	136°00'E
Stolin, Bela. (stô´lēn)	210-11	51°54'N	26°52'E
Stolp, Pol. *see* Słupsk	210-11	54°28'N	17°02'E
Stonehaven, Scot., U.K. (stōn´hā-v'n)	206-07	56°58'N	2°13'W
Stonewall, Mb., Can. (stōn´wôl)	144-45	50°08'N	97°19'W
Storm Bay, b., Austl.	301	43°10'S	147°32'E
Storm Lake, Ia., U.S.	124-25	42°39'N	95°13'W
Stornoway, Scot., U.K. (stôr´nô-wā)	206-07	58°13'N	6°24'W
Storsjøen, lk., Nor. (stòr-syûĕn)	208-09	60°21'N	11°41'E
Storsjön, lk., Swe.	200-01	63°12'N	14°18'E
Storuman, Swe.	200-01	65°05'N	17°05'E
Stosch, Isla, i., Chile	185	49°09'S	75°26'W
Strabane, N. Ire., U.K. (stră-băn´)	206-07	54°50'N	7°27'W
Strahan, Austl. (strā´ăn)	301	42°09'S	145°19'E
Strakonice, Czech Rep. (strä´kô-nyē-tsĕ)	210-11	49°15'N	13°55'E
Stralsund, Ger. (shräl´sȯnt)	210-11	54°19'N	13°05'E
Stranraer, Scot., U.K. (străn-rär´)	206-07	54°54'N	5°02'W
Strasbourg, Fr. (strás-bōōr´)	212-13	48°35'N	7°45'E
Stratford, On., Can. (străt´fẽrd)	146-47	43°22'N	80°58'W
Stratford, Ct., U.S. (străt´fẽrd)	126-27	41°13'N	73°08'W
Stratford, Tx., U.S. (străt´fẽrd)	130-31	36°20'N	102°04'W
Straubing, Ger. (strou´bĭng)	210-11	48°53'N	12°35'E
Strausberg, Ger. (strous´bĕrgh)	210-11	52°34'N	13°53'E
Streator, Il., U.S. (strē´tẽr)	126-27	41°08'N	88°49'W
Strehaia, Rom. (strĕ-кä´yȧ)	216-17	44°38'N	23°13'E
Streymoy, i., Far. Is.	206b	62°08'N	7°00'W
Strickland, stm., Pap. N. Gui. (strĭk´lănd)	302	7°35'S	141°23'E
Strongsville, Oh., U.S. (strŏngz´vĭl)	126-27	41°18'N	81°50'W
Stronsay, i., Scot., U.K. (strŏn´sā)	206c	59°06'N	2°36'W
Stroudsburg, Pa., U.S. (stroudz´bûrg)	126-27	40°59'N	75°12'W
Strugi-Krasnyye, Russia (strōō´gĭ krä´s-ny´yĕ)	208-09	58°16'N	29°07'E
Strumica, Mac. (strōō´mĭ-tsȧ)	216-17	41°26'N	22°37'E
Stryi, Ukr.	210-11	49°15'N	23°51'E
Strzelce Opolskie, Pol. (stzhĕl´tsĕ o-pŏl´skyĕ)	210-11	50°31'N	18°19'E
Strzelecki Creek, stm., Austl.	301	29°21'S	139°48'E
Stuart, Fl., U.S. (stū´ẽrt)	135a	27°12'N	80°15'W
Stuart, Ia., U.S. (stū´ẽrt)	124-25	41°30'N	94°20'W
Stuart, stm., B.C., Can.	142-43	53°59'N	123°33'W
Stuart Island, i., Ak., U.S. (stū´ẽrt ī´lănd)	136	63°35'N	162°31'W
Stuart Lake, lk., B.C., Can. (stū´ẽrt lāk)	142-43	54°32'N	124°35'W
Stuhlweissenburg, Hung. *see* Székesfehérvár	210-11	47°12'N	18°25'E
Stupino, Russia	218-19	54°54'N	38°05'E
Sturgeon, stm., On., Can. (stûr´jŭn)	146-47	46°16'N	79°56'W
Sturgeon Bay, Wi., U.S. (stûr´jŭn bā)	126-27	44°50'N	87°22'W
Sturgeon Bay, b., Mb., Can. (stûr´jŭn bā)	144-45	52°0'N	97°50'W
Sturgeon Falls, On., Can. (stûr´jŭn fôlz)	146-47	46°22'N	79°55'W
Sturt Stony Desert, des., Austl.	301	28°30'S	141°00'E
Stuttgart, Ger. (shtōōt´gärt)	210-11	48°48'N	9°11'E
Stuttgart, Ar., U.S. (stŭt´gart)	134-35	34°30'N	91°33'W
Styr, stm., Eur. (stẻr)	210-11	52°07'N	26°36'E
Suduroy, i., Far. Is.	206b	61°32'N	6°50'W
Suao, Tai. (sōōŏu)	243a	24°36'N	121°50'E
Subansiri, stm., Asia	258-59	26°46'N	93°45'E
Subarnarekha, stm., India	254-55	21°34'N	87°23'E
Sūbāṭ, stm., S. Sudan	284-85	9°22'N	31°33'E
Subotica, Serb. (sōō´bô´tē-tsȧ)	216-17	46°06'N	19°41'E
Suceava, Rom. (sōō-chȧ-ä´vȧ)	210-11	47°40'N	26°17'E
Sucre, nat. cap., Bol. (sōō´krā)	182-83	19°02'S	65°16'W
Sudan, nation, Afr. (sōō-dän´)	274	15°0'N	30°00'E
Sudan, reg., Afr. (sōō-dän´) *see* Sahel	280-81	12°0'N	17°00'E
Sudbury, On., Can. (sŭd´bẽr-ē)	146-47	46°29'N	80°59'W
Sudd, reg., Sudan *see* As-Sudd	284-85	8°0'N	31°00'E
Sudost', stm., Eur. (sò-dôst´)	218-19	52°20'N	33°23'E
Sudzha, Russia (sòd´zhȧ)	218-19	51°11'N	35°18'E
Sue, stm., S. Sudan	284-85	7°40'N	28°02'E
Sueca, Spain (swā´kä)	214-15	39°12'N	0°19'W
Suez, Egypt (sōō-ĕz´)	291b	29°58'N	32°33'E
Suez, Gulf of, b., Egypt (gŭlf ŭv sōō-ĕz´)	291b	29°0'N	32°50'E
Suez Canal, can., Egypt (sōō-ĕz´ kȧ´năl)	291b	29°57'N	32°35'E
Suffolk, Va., U.S. (sŭf´ŭk)	134-35	36°44'N	76°35'W
Suhag, Egypt	291b	26°33'N	31°42'E
Şuḩār, Oman	250-51	24°20'N	56°44'E
Sühbaatar, Mong.	260-61	50°13'N	106°12'E
Suhl, Ger. (zōōl)	210-11	50°36'N	10°41'E
Suid-Afrika, nation, Afr. *see* South Africa	274	30°0'S	26°00'E
Suide, China (swā-dŭ)	260-61	37°31'N	110°15'E
Suifenhe, China (swā-fŭn-hŭ)	264	44°24'N	131°08'E
Suihua, China	260-61	46°39'N	126°59'E
Suining, China (sōō´ē-nĭng´)	258-59	30°30'N	105°35'E
Suipacha, Arg. (swē-pä´chä)	187	34°47'S	59°42'W
Suisse, nation, Eur. *see* Switzerland	190-91	47°0'N	8°00'E
Suixian, China (swā shyĕn)	258-59	34°26'N	115°04'E
Suizhong, China (swā-jōŋ)	260-61	40°20'N	120°20'E
Suizhou, China	258-59	31°42'N	113°22'E
Sukabumi, Indon.	268-69	6°55'S	106°55'E

n-sing; ŋ-baŋk; ɴ-nasalized n; nŏd; cŏmmit; ōld; ôbey; ôrder; oi-boil; fōōd; ȯ-as oo in foot; ou-out; s-soft; sh-dish; th-thin; pūre; ûnite; ûrn; stŭd; circŭs; ü-as in French tu; ´-indeterminate vowel.

412 Sukagawa – Tafassasset, Oued PRONOUNCING INDEX

Feature (Pronunciation)	Page	Lat.	Long.
Sukagawa, Japan (sōō´kä-gä´wä)	265	37°17'N	140°23'E
Sukarnapura, Indon. *see* Jayapura	302	2°32'S	140°43'E
Sukarno, Pegunungan, mtn., Indon. *see* Jaya, Puncak	242-43	4°05'S	137°11'E
Sukhinichi, Russia (sōō´kē´nē-chē)	218-19	54°07'N	35°22'E
Sukhona, stm., Russia (sò-kô´nà)	202-03	60°45'N	46°18'E
Sukhothai, Thai.	266-67	17°01'N	99°49'E
Sukhumi, Geor. (sò-kòm´) *see* Sokhumi	245	43°00'N	41°00'E
Sukkozero, Russia	200-01	63°14'N	32°18'E
Sukkur, Pak. (sŭk´ŭr)	252-53	27°42'N	68°52'E
Sūknah, Libya	204-05	29°04'N	15°47'E
Sukumo, Japan (sōō´kò-mô)	265	32°56'N	132°44'E
Sula, i., Nor.	208-09	61°08'N	4°55'E
Sula, stm., Ukr. (sōō-lá´)	218-19	49°38'N	32°43'E
Sula, Kepulauan, is., Indon.	268-69	1°52'S	125°22'E
Sulaimaniya, Iraq *see* As-Sulaymānīyah	248-49	35°34'N	45°27'E
Sulaimān Range, mts., Pak. (sò-lä-ē-män´ rānj)	252-53	30°30'N	70°10'E
Sulawesi, i., Indon. *see* Celebes	268-69	2°0'S	121°00'E
Sulawesi, Laut, s., Asia *see* Celebes Sea	268-69	3°0'N	122°00'E
Sulina, Rom. (sōō-lē´ná)	218-19	45°09'N	29°40'E
Sulitelma, mtn., Eur. (sōō-lē-tyĕl´mȧ)	200-01	67°08'N	16°24'E
Sulitjelma, mtn., Eur. *see* Sulitelma	200-01	67°08'N	16°24'E
Sullana, Peru (sōō-lyä´nä)	184	4°54'S	80°41'W
Sulligent, Al., U.S. (sŭl´ĭ-jĕnt)	134-35	33°54'N	88°08'W
Sullivan, Il., U.S. (sŭl´ĭ-văn)	126-27	39°36'N	88°37'W
Sullivan, In., U.S. (sŭl´ĭ-văn)	126-27	39°05'N	87°24'W
Sullivan, Mo., U.S. (sŭl´ĭ-văn)	130-31	38°12'N	91°10'W
Sulmona, Italy (sōōl-mō´nä)	216-17	42°04'N	13°55'E
Sulphur, La., U.S. (sŭl´fŭr)	132-33	30°14'N	93°22'W
Sulphur, Ok., U.S. (sŭl´fŭr)	130-31	34°30'N	96°58'W
Sulphur Springs, Tx., U.S. (sŭl´fŭr sprĭngz)	130-31	33°09'N	95°36'W
Sultanabad, Iran *see* Arāk	252-53	34°05'N	49°41'E
Sulu, Laut, s., Asia *see* Sulu Sea	270	8°0'N	120°00'E
Sulu Archipelago, is., Phil. (sōō´lōō är´kȧ-pĕ´-å-gō)	270	6°0'N	121°00'E
Sulūq, Libya	204-05	31°40'N	20°15'E
Suluova, Tur.	247	40°50'N	35°39'E
Sulu Sea, s., Asia (sōō´lōō sē)	270	8°0'N	120°00'E
Sumatera, i., Indon. *see* Sumatra	266-67	0°05'S	102°00'E
Sumatra, i., Indon. (sò-mä-trȧ)	266-67	0°05'S	102°00'E
Sumba, i., Indon. (sŭm´bȧ)	268-69	10°0'S	120°00'E
Sumba, Île, i., D.R.C.	284-85	1°44'N	19°32'E
Sumbawa, i., Indon. (sòm-bä´wä)	268-69	8°49'S	117°56'E
Sumbawa Besar, Indon.	268-69	8°30'S	117°24'E
Sumbawanga, Tan.	289	7°60'S	31°38'E
Sumbe, Ang.	286-87	11°14'S	13°51'E
Sumenep, Indon.	268-69	7°00'S	113°52'E
Summerland, B.C., Can. (sŭ´mĕr-lănd)	142-43	49°36'N	119°41'W
Summerside, P.E., Can. (sŭm´ēr-sīd)	148-49	46°24'N	63°47'W
Summerville, Ga., U.S. (sŭm´ēr-vĭl)	134-35	34°29'N	85°21'W
Summerville, S.C., U.S. (sŭm´ēr-vĭl)	134-35	33°01'N	80°11'W
Summit Lake, B.C., Can.	142-43	54°17'N	122°37'W
Summit Peak, mtn., Co., U.S. (sŭm´mĭt pēk)	128-29	37°21'N	106°42'W
Šumperk, Czech Rep. (shòm´pĕrk)	210-11	49°58'N	16°59'E
Sumqayıt, Azer.	245	40°35'N	49°38'E
Sumter, S.C., U.S. (sŭm´tēr)	134-35	33°55'N	80°21'W
Sumy, Ukr. (sōō´mĭ)	218-19	50°55'N	34°48'E
Sumzom, China	258-59	29°44'N	96°08'E
Sunchales, Arg.	187	30°56'S	61°34'W
Suncheon, Kor., S.	263	34°57'N	127°30'E
Sunda, Selat, strt., Indon. *see* Sunda Strait	268-69	6°00'S	105°46'E
Sundance, Wy., U.S. (sŭn´dăns)	124-25	44°24'N	104°23'W
Sunda Strait, strt., Indon. (sōōn´dȧ strāt)	268-69	6°00'S	105°46'E
Sunderland, Eng., U.K. (sŭn´dēr-lănd)	206-07	54°55'N	1°23'W
Sundsvall, Swe. (sòonds´väl)	208-09	62°23'N	17°19'E
Sungaipenuh, Indon.	266-67	2°04'S	101°24'E
Sungari, stm., China *see* Songhua	260-61	47°43'N	132°31'E
Sungari Reservoir, res., China *see* Songhua Hu	260-61	43°25'N	127°10'E
Sunne, Swe. (sōōn´ĕ)	208-09	59°50'N	13°10'E
Sunnyvale, Ca., U.S. (sŭn-nē-văl)	128-29	37°22'N	122°01'W
Sunset Crater Volcano National Monument, n.p., Az., U.S. (sŭn-sĕt krā´tēr vŏl-kā´nō năsh´ŭn-ăl mŏn´ŭ-mĕnt)	128-29	35°22'N	111°31'W
Suntar, Russia (sòn-tär´)	236-37	62°10'N	117°38'E
Sun Valley, Id., U.S.	122-23	43°43'N	114°23'W
Sunyani, Ghana	282-83	7°21'N	2°20'W

Feature (Pronunciation)	Page	Lat.	Long.
Suomenlahti, b., Eur. *see* Finland, Gulf of	208-09	60°0'N	27°00'E
Suomi, nation, Eur. *see* Finland	190-91	64°0'N	26°00'E
Suomussalmi, Fin.	200-01	64°53'N	29°02'E
Superior, Az., U.S. (su-pē´rĭ-ēr)	128-29	33°18'N	111°06'W
Superior, Ne., U.S. (su-pē´rĭ-ēr)	130-31	40°01'N	98°04'W
Superior, Wi., U.S. (su-pē´rĭ-ēr)	124-25	46°43'N	92°05'W
Superior, Laguna, b., Mex. (lä-gò´nä sōō-pä-rē-ōr´)	158-59	16°21'N	94°55'W
Superior, Lake, lk., N.A. (lăk su-pē´rĭ-ēr)	124-25	48°0'N	88°00'W
Suphan Buri, Thai.	266-67	14°28'N	100°08'E
Suqian, China (sōō-chyĕn)	258-59	33°57'N	118°18'E
Suquṭrā, i., Yemen	238-39	12°31'N	53°54'E
Şūr, Oman	250-51	22°34'N	59°30'E
Sura, stm., Russia	202-03	55°37'N	46°02'E
Surabaja, Indon. *see* Surabaya	268-69	7°15'S	112°45'E
Surabaya, Indon.	268-69	7°15'S	112°45'E
Surakarta, Indon.	268-69	7°34'S	110°50'E
Sūrat, India (sò´rŭt)	254-55	21°12'N	72°50'E
Surat Thani, Thai.	266-67	9°06'N	99°18'E
Surazh, Russia (sōō-räzh´)	218-19	53°01'N	32°25'E
Surendranagar, India	254-55	22°43'N	71°38'E
Surgut, Russia (sòr-gòt´)	236-37	61°16'N	73°12'E
Surigao, Phil.	270	9°46'N	125°29'E
Suriname, nation, S.A. (sōō-rē-näm´)	172	4°0'N	56°00'W
Sūrīyah, nation, Asia *see* Syria	224-25	35°0'N	38°00'E
Sūrmaq, Iran	250-51	31°05'N	52°48'E
Surprise, Az., U.S.	128-29	33°37'N	112°20'W
Surt, Libya	204-05	31°12'N	16°35'E
Surt, Khalīj, b., Libya	280-81	31°30'N	18°00'E
Suruga-wan, b., Japan (sōō´rōō-gä wän)	265	34°51'N	138°33'E
Susanville, Ca., U.S.	128-29	40°25'N	120°39'W
Susong, China (sōō-sôŋ)	258-59	30°09'N	116°07'E
Susquehanna, Pa., U.S. (sŭs´kwē-hăn´ȧ)	126-27	41°56'N	75°36'W
Susquehanna, stm., U.S. (sŭs´kwē-hăn´ȧ)	126-27	39°32'N	76°05'W
Susques, Arg.	182-83	23°25'S	66°30'W
Sussex, N.B., Can. (sŭs´ĕks)	148-49	45°43'N	65°31'W
Susuman, Russia	236-37	62°46'N	148°10'E
Sutlej, stm., Asia (sŭt´lĕj)	254-55	29°21'N	71°02'E
Sutton, W.V., U.S. (sut´'n)	126-27	38°39'N	80°45'W
Sutton, Monts, mts., N.A. *see* Green Mountains	126-27	43°45'N	72°45'W
Suva, nat. cap., Fiji (sōō-vȧ)	306f	18°07'S	178°27'E
Suwałki, Pol. (sò-vou´kē)	210-11	54°06'N	22°56'E
Suwanose-jima, i., Japan	264	29°38'N	129°43'E
Suwarrow, at., Cook Is.	304-05	13°15'S	163°05'W
Suweis, Khalîg el-, b., Egypt *see* Suez, Gulf of	291b	29°0'N	32°50'E
Suweis, Qanâ el-, can., Egypt *see* Suez Canal	291b	29°57'N	32°35'E
Suwon, Kor., S.	263	37°16'N	127°01'E
Suzdal', Russia (sōōz´dȧl)	218-19	56°25'N	40°26'E
Suzhou, China (sōō-jō)	258-59	33°38'N	116°59'E
Suzhou, China (sōō-jō)	258-59	31°18'N	120°37'E
Svalbard, dep., Eur. (sväl´bärt)	226-27	78°0'N	17°00'E
Svay Riĕng, Camb.	266-67	11°04'N	105°49'E
Svelvik, Nor. (svĕl´vĕk)	208-09	59°37'N	10°24'E
Svendborg, Den. (svĕn-bôrgh)	210-11	55°04'N	10°37'E
Sverdlovs'k, Ukr.	218-19	48°05'N	39°39'E
Sverige, nation, Eur. *see* Sweden	190-91	62°0'N	15°00'E
Svetlaya, Russia (svyĕt´lá-yȧ)	264	46°34'N	138°20'E
Svetlograd, Russia	202-03	45°20'N	42°50'E
Svilengrad, Blg. (svĕl´ĕn-grȧt)	216-17	41°46'N	26°13'E
Svir', stm., Russia	202-03	60°30'N	32°48'E
Svishtov, Blg. (svēsh´tôf)	216-17	43°37'N	25°21'E
Svizzera, nation, Eur. *see* Switzerland	190-91	47°0'N	8°00'E
Svobodnyy, Russia (svô-bôd´nĭ)	240-41	51°23'N	128°08'E
Svolvær, Nor. (svôl´vĕr)	200-01	68°15'N	14°33'E
Swainsboro, Ga., U.S. (swānz´bŭr-ô)	134-35	32°36'N	82°20'W
Swakop, stm., Nmb.	286-87	22°41'S	14°32'E
Swakopmund, Nmb. (svä´kôp-mònt) (swȧ´kôp-mŏnd)	286-87	22°40'S	14°32'E
Swan, stm., Can. (swŏn)	144-45	52°34'N	100°45'W
Swan Hill, Austl. (swŏn hĭl)	301	35°21'S	143°33'E
Swan Lake, lk., Mb., Can. (swŏn lăk)	144-45	52°31'N	100°45'W
Swan Range, mts., Mt., U.S. (swŏn rānj)	122-23	47°50'N	113°40'W
Swan River, Mb., Can. (swŏn rĭv´ēr)	144-45	52°05'N	101°16'W
Swansea, Wales, U.K. (swŏn´sē)	206-07	51°38'N	3°58'W
Swaziland, nation, Afr. (swä´zē-lănd)	274	26°30'S	31°30'E
Sweden, nation, Eur. (swē´dĕn)	190-91	62°0'N	15°00'E
Sweetwater, Tn., U.S. (swēt´wô-tēr)	134-35	35°36'N	84°28'W
Sweetwater, Tx., U.S. (swēt´wô-tēr)	130-31	32°28'N	100°24'W
Swellendam, S. Afr.	286-87	34°01'S	20°26'E

Feature (Pronunciation)	Page	Lat.	Long.
Świecie, Pol. (shvyăn´tsyĕ)	210-11	53°24'N	18°27'E
Swift Current, Sk., Can. (swĭft kûr´ĕnt)	144-45	50°17'N	107°47'W
Swindon, Eng., U.K. (swĭn´dŭn)	206-07	51°34'N	1°47'W
Swinemünde, Pol. *see* Świnoujście	210-11	53°54'N	14°15'E
Świnoujście, Pol. (shvĭ-nĭ-ô-wēsh´chyĕ)	210-11	53°54'N	14°15'E
Switzerland, nation, Eur. (swĭt´zēr-lănd)	190-91	47°0'N	8°00'E
Sycamore, Il., U.S. (sĭk´ȧ-mōr)	126-27	41°59'N	88°41'W
Sychëvka, Russia (sē-chôf´kȧ)	218-19	55°49'N	34°17'E
Sydney, Austl. (sĭd´nē)	301	33°52'S	151°13'E
Sydney, N.S., Can. (sĭd´nē)	148-49	46°09'N	60°12'W
Sydney Mines, N.S., Can. (sĭd´nē mīns)	148-49	46°15'N	60°15'W
Syktyvkar, Russia (sük-tüf´kär)	202-03	61°39'N	50°49'E
Sylacauga, Al., U.S. (sĭl-ȧ-kô´gȧ)	134-35	33°10'N	86°15'W
Sylhet, Bngl.	254-55	24°54'N	91°52'E
Sylvania, Ga., U.S. (sĭl-vā´nĭ-ȧ)	134-35	32°45'N	81°38'W
Sylvester, Ga., U.S. (sĭl-vĕs´tēr)	134-35	31°32'N	83°50'W
Syracuse, Ks., U.S. (sĭr´ȧ-kūs)	130-31	37°59'N	101°45'W
Syracuse, Ne., U.S. (sĭr´ȧ-kūs)	130-31	40°39'N	96°11'W
Syracuse, N.Y., U.S.	126-27	43°03'N	76°09'W
Syrdarīya, stm., Asia *see* Syr Darya	244	46°04'N	60°04'E
Syr Darya, stm., Asia (sĭr-dä´rē-ä)	244	46°04'N	60°04'E
Syria, nation, Asia (sĭr´ĭ-ȧ)	224-25	35°0'N	38°00'E
Syriam, Mya.	266-67	16°46'N	96°15'E
Syrian Desert, des., Asia	248-49	32°0'N	40°00'E
Sýros, i., Grc.	216-17	37°26'N	24°55'E
Syzran', Russia (sĕz-rän´)	202-03	53°09'N	48°26'E
Szabadka, Serb. *see* Subotica	216-17	46°06'N	19°41'E
Szamotuły, Pol. (shȧ-mô-tōō´wē)	210-11	52°37'N	16°35'E
Szatmárnémeti, Rom. *see* Satu Mare	210-11	47°47'N	22°53'E
Szczecin, Pol. (shchĕ´tsĭn)	210-11	53°26'N	14°32'E
Szechwan, state, China *see* Sichuan	258-59	31°0'N	105°00'E
Szeged, Hung. (sĕ´gĕd)	210-11	46°16'N	20°10'E
Székesfehérvár, Hung. (sā´kĕsh-fĕ´hār-vär)	210-11	47°12'N	18°25'E
Szekszárd, Hung. (sĕk´särd)	210-11	46°21'N	18°43'E
Szentes, Hung. (sĕn´tĕsh)	210-11	46°39'N	20°16'E
Szolnok, Hung.	210-11	47°11'N	20°12'E
Szombathely, Hung. (sôm´bôt-hĕl')	210-11	47°14'N	16°38'E
Szydłowiec, Pol. (shid-wô´vyets)	210-11	51°14'N	20°52'E

T

Feature (Pronunciation)	Page	Lat.	Long.
Ta'izz, Yemen	288	13°35'N	44°01'E
Taal, Lake, lk., Phil. (lăk tä-äl´)	270	13°60'N	121°01'E
Tabaco, Phil. (tä-bä´kō)	270	13°23'N	123°43'E
Tabar Islands, is., Pap. N. Gui.	302	2°45'S	151°57'E
Ţabas, Iran	252-53	33°36'N	56°55'E
Tabasco, state, Mex. (tä-bäs´kô)	158-59	18°15'N	93°00'W
Taber, Ab., Can.	142-43	49°47'N	112°09'W
Tablas Island, i., Phil. (tä´bläs ī´lȧnd)	270	12°23'N	122°02'E
Tábor, Czech Rep. (tä´bôr)	210-11	49°25'N	14°41'E
Tabora, Tan. (tä-bō´rä)	289	5°01'S	32°50'E
Tabrīz, Iran (tȧ-brēz´)	245	38°05'N	46°17'E
Tabuaeran, at., Kir.	304-05	3°51'N	159°18'W
Tabūk, Sau. Ar.	248-49	28°23'N	36°35'E
Tacheng, China (tä-chŭŋ)	244	46°45'N	82°58'E
Tacloban, Phil. (tä-klō´bän)	270	11°14'N	124°60'E
Tacna, Peru (täk´nä)	184	18°01'S	70°15'W
Tacoma, Wa., U.S. (tȧ-kō´mȧ)	122-23	47°15'N	122°26'W
Taconic Range, mts., U.S. (tȧ-kŏn´ĭk rānj)	126-27	42°30'N	73°20'W
Tacotalpa, stm., Mex. (tä-kô-täl´pä)	158-59	17°48'N	92°51'W
Tacuarembó, Ur.	187	31°42'S	55°59'W
Tademaït, Plateau du, plat., Alg. (plä-tō´ dü tä-dē-mä´ēt)	280-81	28°20'N	2°47'E
Tadoussac, Qc., Can. (tȧ-dōō-sȧk´)	148-49	48°10'N	69°42'W
Tādpatri, India	256	14°54'N	78°00'E
Tadzhikistan, nation, Asia *see* Tajikistan	252-53	39°0'N	71°00'E
T'aebaek-sanmaek, mts., Asia (tī-bĭk´ sän-mĭk´)	263	37°30'N	128°31'E
Taedong-gang, stm., Kor., N. (tī-dŏng gäng´)	263	38°43'N	125°07'E
Taegu, Kor., S. *see* Daegu	263	35°52'N	128°35'E
Taehan-min'guk, nation, Asia *see* South Korea	224-25	36°30'N	128°00'E
Taejŏn, Kor., S. *see* Daejeon	263	36°20'N	127°26'E
Tafalla, Spain (tä-fäl´yä)	214-15	42°31'N	1°40'W
Tafassasset, Oued, stm., Afr.	280-81	21°52'N	9°59'E

ăt; finăl; rāte; senâte; ärm; àsk; sofȧ; fâre; ch-choose; dh-as th in other; bē; ēvent; bĕt; recĕnt; cratēr; g-gō; gh-guttural g; bĭt; ĭ-short neutral; rīde; κ-guttural k as ch in German ich;

PRONOUNCING INDEX
Taft – Tarragona 413

Feature (Pronunciation)	Page	Lat.	Long.
Taft, Ca., U.S. (tȧft)	128-29	35°08′N	119°26′W
Taganrog, Russia (tȧ-gȧn-rôk′)	218-19	47°14′N	38°54′E
Taganrogskiy Zaliv, b., Eur. (tȧ-gȧn-rôk′skĭ zä′lĭf)	218-19	47°0′N	38°23′E
Tagbilaran, Phil.	270	9°40′N	123°52′E
Tagdempt, Alg. *see* Tiaret	214-15	35°28′N	1°21′E
Tagtabazar, Turkmen.	252-53	35°58′N	62°55′E
Taguatinga, Braz.	180-81	12°24′S	46°27′W
Taguke, China	254-55	32°06′N	84°43′E
Tagula Island, i., Pap. N. Gui. (tä′gōō-lȧ ī′lănd)	302	11°30′S	153°30′E
Tagus, stm., Eur. (tā′gŭs)	214-15	38°51′N	8°57′W
Tahan, Gunong, mtn., Malay.	266-67	4°38′N	102°14′E
Tahanroz's'ka zatoka, b., Eur. *see* Taganrogskiy Zaliv	218-19	47°0′N	38°23′E
Tahat, mtn., Alg. (tä-hät′)	280-81	23°17′N	5°32′E
Tahiti, i., Fr. Poly. (tä-hē′tē) (tä′ē-tē′)	306d	17°37′S	149°27′W
Tahlequah, Ok., U.S. (tă-lĕ-kwä′)	130-31	35°55′N	94°58′W
Tahoe, Lake, lk., U.S. (lāk tä′hō)	128-29	39°07′N	120°03′W
Tahoua, Niger (tä′ōō-ä)	280-81	14°53′N	5°16′E
Tahta, Egypt	291b	26°46′N	31°30′E
Tahulandang, Pulau, i., Indon.	268-69	2°20′N	125°25′E
Tahuna, Indon.	268-69	3°36′N	125°30′E
Tai'an, China (tī-än)	260-61	36°11′N	117°07′E
Taibai Shan, mtn., China (tī-bī shän)	258-59	33°54′N	107°46′E
Taibus Qi, China (tī-bōō-sz chyē)	260-61	41°53′N	115°17′E
Taichung, Tai. (tī′chŏng)	243a	24°09′N	120°41′E
Taiden, Kor., S. *see* Taejŏn	263	36°20′N	127°26′E
Taif, Sau. Ar. *see* Aṭ-Ṭā'if	288	21°16′N	40°25′E
Taigu, China (tī-gōō)	260-61	37°25′N	112°33′E
Taihang Shan, mts., China (tī-häŋ shän)	260-61	38°0′N	114°00′E
Taihe, China (tī-hŭ)	258-59	26°49′N	114°54′E
Taihoku, nat. cap., Tai. *see* Taipei	243a	25°03′N	121°30′E
Tai Hu, lk., China (tī hōō)	258-59	31°13′N	120°11′E
Taikyu, Kor., S. *see* Taegu	263	35°52′N	128°35′E
Tailai, China (tī-lī)	260-61	46°23′N	123°25′E
Tailem Bend, Austl. (tā-lĕm bĕnd)	301	35°16′S	139°27′E
Tainan, Tai. (tī′nän′)	243a	23°0′N	120°12′E
Taínaro, Ákra, c., Grc.	216-17	36°23′N	22°29′E
Taining, China (tī′nĭng′)	258-59	26°54′N	117°09′E
Taipei, nat. cap., Tai. (tī′pā′)	243a	25°03′N	121°30′E
Taiping, Malay.	266-67	4°51′N	100°44′E
Taira, Japan *see* Iwaki	265	37°03′N	140°55′E
Taishun, China	258-59	27°33′N	119°43′E
Taitao, Península de, pen., Chile	185	46°30′S	74°25′W
Taitung, Tai. (tī′tōōng′)	243a	22°45′N	121°08′E
Taiwan, nation, Asia (tī-wän)	224-25	23°30′N	121°00′E
T'ai-wan Hai-hsia, strt., Asia *see* Taiwan Strait	243a	24°0′N	119°00′E
Taiwan Haixia, strt., Asia *see* Taiwan Strait	243a	24°0′N	119°00′E
Taiwan Strait, strt., Asia (tī-wän strāt strāt)	243a	24°0′N	119°00′E
Taiyuan, China (tī-yůän)	260-61	37°52′N	112°33′E
Taizhao, China	258-59	30°02′N	92°57′E
Taizhou, China (tī-jō)	258-59	32°30′N	119°55′E
Tajikistan, nation, Asia (tä-jēk′-ĭ-stän′) (tä-jĭk′-ĭ-stăn′)	252-53	39°0′N	71°00′E
Tajo, stm., Eur. *see* Tagus	214-15	38°51′N	8°57′W
Tajumulco, Volcán, vol., Guat. (vôl-kä′n tä-hōō-mōōl′kō)	160	15°02′N	91°55′W
Tajuña, stm., Spain (tä-ᴋōō′n-yä)	214-15	40°07′N	3°35′W
Tak, Thai.	266-67	16°53′N	99°09′E
Takamatsu, Japan	265	34°20′N	134°03′E
Takao, Tai. *see* Kaohsiung	243a	22°38′N	120°17′E
Takaoka, Japan (tä′kä′ō-kä′)	265	36°45′N	137°01′E
Takasaki, Japan (tä′kät′sōō-kē′)	265	36°19′N	139°01′E
Takatsuki, Japan (tä′kät′sōō-kē′)	265	34°51′N	135°38′E
Takayama, Japan (tä′kä′yä′mä)	265	36°08′N	137°15′E
Takefu, Japan (tä′kĕ-fōō)	265	35°54′N	136°10′E
Takengon, Indon.	266-67	4°37′N	96°51′E
Takêv, Camb.	266-67	10°59′N	104°45′E
Takhli, Thai.	266-67	15°16′N	100°21′E
Takht-e Jamshīd, hist., Iran *see* Persepolis	250-51	29°57′N	52°52′E
Takla Lake, lk., B.C., Can.	142-43	55°25′N	125°54′W
Takla Makan Desert, des., China	240-41	39°0′N	83°00′E
Taklimakan Shamo, des., China *see* Takla Makan Desert	240-41	39°0′N	83°00′E
Tala, Mex. (tä′lä)	158-59	20°39′N	103°43′W
Talagante, Chile (tä-lä-gä′n-tĕ)	188e	33°39′S	70°55′W
Talara, Peru (tä-lä′rä)	184	4°35′S	81°16′W
Talas, Kyrg.	244	42°32′N	72°15′E
Talasea, Pap. N. Gui. (tä-lä-sā′ä)	302	5°18′S	150°02′E
Talaud, Kepulauan, is., Indon. (tä-lout′)	268-69	4°20′N	126°50′E
Talavera de la Reina, Spain	214-15	39°58′N	4°49′W
Talca, Chile (täl′kä)	185	35°25′S	71°39′W
Talcahuano, Chile (täl-kä-wä′nō)	185	36°42′S	73°07′W
Taldom, Russia (täl-dôm)	218-19	56°44′N	37°32′E
Taldyqorghan, Kaz.	244	45°01′N	78°23′E
Talghar, Kaz.	244	43°18′N	77°14′E
Talkeetna, Ak., U.S. (tȧl-kēt′nȧ)	136	62°20′N	150°07′W
Tall 'Afar, Iraq	248-49	36°22′N	42°27′E
Talladega, Al., U.S. (tăl-ȧ-dē′gȧ)	134-35	33°26′N	86°06′W
Tallahassee, Fl., U.S. (tăl-ȧ-hăs′ē)	134-35	30°26′N	84°17′W
Tallapoosa, stm., U.S. (tăl-ȧ-pōō′sȧ)	134-35	32°30′N	86°16′W
Tallassee, Al., U.S. (tăl′ȧ-sē)	134-35	32°33′N	85°55′W
Tallinn, nat. cap., Est. (tȧl′lĕn) (rä′väl)	208-09	59°26′N	24°48′E
Tallulah, La., U.S. (tă-lōō′lȧ)	134-35	32°25′N	91°12′W
Talo, mtn., Eth.	284-85	10°44′N	37°54′E
Talok, Indon.	268-69	1°03′N	118°49′E
Taloyoak, Nu., Can.	138-39	69°32′N	93°31′W
Talpa de Allende, Mex. (täl′pä dā äl-yĕn′dâ)	158-59	20°23′N	104°50′W
Talsi, Lat. (tal′sĭ)	208-09	57°15′N	22°37′E
Taltal, Chile (täl-täl′)	182-83	25°24′S	70°29′W
Talurqjuak, Nu., Can. *see* Taloyoak	138-39	69°32′N	93°31′W
Tama, Ia., U.S. (tā′mä)	124-25	41°58′N	92°35′W
Tamale, Ghana (tä-mä′lå)	282-83	9°24′N	0°50′W
Taman', Russia (tȧ-män′′)	218-19	45°12′N	36°43′E
Tamanrasset, Alg.	280-81	22°48′N	5°31′E
Tamaqua, Pa., U.S. (tȧ-mô′kwă)	126-27	40°48′N	75°58′W
Tamaulipas, state, Mex. (tä-mä-ōō-lē′päs′)	158-59	24°0′N	98°45′W
Tamazunchale, Mex. (tä-mä-zȯn-chä′lå)	158-59	21°16′N	98°47′W
Tambacounda, Sen. (täm-bä-kōōn′dȧ)	282-83	13°47′N	13°40′W
Tambelan, Kepulauan, is., Indon. (täm-bå-län′)	266-67	1°0′N	107°30′E
Tambo, Austl. (tăm′bō)	302	24°53′S	146°15′E
Tambo, stm., Peru	184	17°09′S	71°49′W
Tambora, Gunung, vol., Indon.	268-69	8°14′S	117°55′E
Tambov, Russia (tȧm-bôf′)	202-03	52°43′N	41°25′E
Tambura, S. Sudan (täm-bōō′rä)	284-85	5°36′N	27°28′E
Tame, Col.	178-79	6°28′N	71°44′W
Tamiahua, Mex. (tä myä wä)	158-59	21°16′N	97°27′W
Tamiahua, Laguna de, lk., Mex. (lä-gō′nä-dĕ-tä-myä-wä)	158-59	21°35′N	97°35′W
Tamiami Canal, can., Fl., U.S. (tă-mī-ăm′ĭ kȧ′năl)	135a	25°46′N	80°11′W
Tamil Nādu, state, India	256	11°0′N	78°15′E
Tam Ky, Viet.	266-67	15°33′N	108°30′E
Tammerfors, Fin. *see* Tampere	208-09	61°30′N	23°46′E
Tammisaari, Fin. *see* Ekenäs	208-09	59°58′N	23°26′E
Tampa, Fl., U.S. (tăm′pȧ)	135a	27°58′N	82°27′W
Tampa Bay, b., Fl., U.S. (tăm′pȧ bā)	135a	27°45′N	82°35′W
Tampere, Fin. (tȧm′pĕ-rĕ)	208-09	61°30′N	23°46′E
Tampico, Mex. (täm-pē′kō)	158-59	22°13′N	97°51′W
Tamworth, Austl. (tăm′wûrth)	301	31°06′S	150°55′E
Tana, stm., Eur.	200-01	70°26′N	28°16′E
Tana, stm., Kenya (tä′nä)	284-85	2°31′S	40°31′E
Tanabe, Japan (tä-nä′bā)	265	33°44′N	135°24′E
T'ana Hāyk', lk., Eth.	284-85	11°57′N	37°17′E
Tanahjampea, Pulau, i., Indon.	268-69	7°05′S	120°42′E
Tanami Desert, des., Austl.	296-97	20°0′S	129°30′E
Tanana, Ak., U.S. (tă′nȧ-nô)	136	65°11′N	152°05′W
Tanana, stm., Ak., U.S. (tă′nȧ-nô)	136	65°09′N	152°04′W
Tananarive, nat. cap., Madag. *see* Antananarivo	286-87	18°55′S	47°32′E
Tanch'ŏn-ŭp, Kor., N.	263	40°27′N	128°54′E
Tancítaro, Pico de, mtn., Mex. (pē′kô-dĕ tän-sē′tä-rō)	158-59	19°23′N	102°13′W
Tandag, Phil.	270	9°04′N	126°12′E
Tandil, Arg. (tän-dēl′)	187	37°19′S	59°08′W
Tanega-shima, i., Japan (tä′nå-gä′ shē′mä)	264	30°40′N	131°00′E
Tang, stm., China (täŋ)	258-59	33°18′N	117°46′E
Tanga, Tan. (tăŋ′gȧ)	284-85	5°04′S	39°06′E
Tanga Islands, is., Pap. N. Gui.	302	3°29′S	153°13′E
Tanganika, Lac, lk., Afr. *see* Tanganyika, Lake	284-85	6°0′S	29°30′E
Tanganyika, nation, Afr. *see* Tanzania	274	6°0′S	35°00′E
Tanganyika, Lac, lk., Afr. *see* Tanganyika, Lake	284-85	6°0′S	29°30′E
Tanganyika, Lake, lk., Afr. (lăk tăn′gȧn-yē′kȧ)	284-85	6°0′S	29°30′E
Tanger, Mor. (tăn-jēr′) *see* Tangier	292a	35°47′N	5°48′W
Tanggu, China (täŋ-gōō)	260-61	39°01′N	117°40′E
Tanggula Shan, mts., China (täŋ-gōō-lä shän)	258-59	33°0′N	92°00′E
Tangier, Mor. (tăn-jēr′)	292a	35°47′N	5°48′W
Tangipahoa, stm., U.S. (tăn′jē-pȧ-hō′ȧ)	134-35	30°20′N	90°16′W
Tangra Yumco, lk., China (täŋ-rä yōōm-tswo)	254-55	31°01′N	86°34′E
Tangshan, China	260-61	39°37′N	118°12′E
Tanimbar, Kepulauan, is., Indon.	242-43	7°30′S	131°30′E
Tanjore, India *see* Thanjāvūr	256	10°47′N	79°09′E
Tanjungbalai, Indon. (tän′jông-bä′lå)	266-67	2°58′N	99°48′E
Tanjungkarang-Telukbetung, Indon. *see* Bandar Lampung	266-67	5°26′S	105°16′E
Tanjungpandan, Indon.	266-67	2°44′S	107°39′E
Tanjungpinang, Indon. (tän′jŏng-pē′näng)	266-67	0°55′N	104°28′E
Tanjungselor, Indon.	268-69	2°50′N	117°21′E
Tanna, i., Vanuatu	306g	19°30′S	169°20′E
Tannūrah, Ra's, c., Sau. Ar.	250-51	26°38′N	50°10′E
Tanout, Niger	280-81	14°58′N	8°53′E
Tanta, Egypt	291b	30°47′N	31°00′E
Tan-Tan, Mor.	280-81	28°26′N	11°06′W
Tantoyuca, Mex. (tän-tō-yōō′kä)	158-59	21°21′N	98°14′W
Tanzania, nation, Afr. (tăn-zä-nē′ȧ)	274	6°0′S	35°00′E
Tao, stm., China (tou)	260-61	35°55′N	103°20′E
Tao'er, stm., China (tou-är)	260-61	45°41′N	123°49′E
Taongi, at., Marsh. Is.	304-05	14°37′N	168°58′E
Taormina, Italy (tä-ôr-mē′nä)	216-17	37°51′N	15°17′E
Taos, N.M., U.S. (tä′ôs)	130-31	36°25′N	105°35′W
Taoyüan, Tai.	243a	24°59′N	121°18′E
Tapa, Est. (tȧ′pȧ)	208-09	59°16′N	25°58′E
Tapachula, Mex.	160	14°54′N	92°17′W
Tapajós, stm., Braz. (tä-pä-zhô′s)	180-81	2°27′S	54°38′W
Tapalqué, Arg. (tä-päl-kĕ′)	187	36°21′S	60°02′W
Tapanahoni, stm., Sur.	178-79	4°21′N	54°26′W
Tapauá, stm., Braz.	180-81	5°47′S	64°24′W
Tāpi, stm., India	254-55	21°09′N	72°44′E
Tapuruquara, Braz.	180-81	0°23′S	65°05′W
Taquari Novo, stm., Braz.	182-83	19°15′S	57°14′W
Tara, Russia (tä′rȧ)	236-37	56°54′N	74°22′E
Tara, stm., Russia (tä′rȧ)	236-37	56°42′N	74°36′E
Taraba, stm., Nig.	282-83	8°33′N	10°14′E
Ṭarābulus, Leb. (tȧ-rä′bȯ-lōōs) *see* Tripoli	248-49	34°26′N	35°51′E
Ṭarābulus, hist. reg., Libya *see* Tripolitania	280-81	31°0′N	15°00′E
Ṭarābulus, nat. cap., Libya *see* Tripoli	280-81	32°52′N	13°10′E
Tarakan, Indon.	268-69	3°19′N	117°35′E
Tarancón, Spain (tä-rän-kōn′)	214-15	40°01′N	3°00′W
Taranto, Italy (tä′rän-tô)	216-17	40°28′N	17°15′E
Taranto, Golfo di, b., Italy (gôl-fô-dē tä′rän-tô)	216-17	40°10′N	17°20′E
Tarapoto, Peru (tä-rä-pô′tō)	184	6°30′S	76°24′W
Taraquá, Braz.	180-81	0°06′N	68°24′W
Tarare, Fr. (tȧ-rär′)	212-13	45°54′N	4°26′E
Tarashcha, Ukr. (tä′rȧsh-chȧ)	218-19	49°34′N	30°31′E
Tarauacá, stm., Braz.	180-81	7°29′S	70°04′W
Tarawa, at., Kir.	304-05	1°21′N	187°08′E
Taraz, Kaz.	244	42°54′N	71°21′E
Tarazona, Spain (tä-rä-thō′nä)	214-15	41°55′N	1°43′W
Tarbagatai Shan, mts., Asia *see* Tarbagatay, khrebet	244	47°12′N	83°00′E
Tarbagatay, khrebet, mts., Asia	244	47°12′N	83°00′E
Tarbes, Fr. (tȧrb)	212-13	43°14′N	0°05′E
Tarboro, N.C., U.S. (tär′bŭr-ō)	134-35	35°54′N	77°33′W
Taree, Austl. (tă-rē′)	301	31°54′S	152°28′E
Tarfaya, Mor.	280-81	27°57′N	12°55′W
Târgu Mureș, Rom.	210-11	46°33′N	24°34′E
Tari, Pap. N. Gui.	302	5°51′S	142°59′E
Tarija, Bol. (tär-rē′hä)	182-83	21°33′S	64°43′W
Tarim, stm., China (tȧ-rĭm′)	240-41	39°32′N	88°26′E
Tarim Pendi, bas., China	240-41	39°0′N	83°00′E
Taritatu, stm., Indon.	302	2°55′S	138°28′E
Tarkhankut, mys, c., Ukr. (mĭs tȧr-ᴋän′kȯt)	218-19	45°21′N	32°30′E
Tarkio, Mo., U.S. (tär′kĭ-ō)	130-31	40°26′N	95°23′W
Tarko-Sale, Russia	236-37	64°56′N	77°47′E
Tarlac, Phil. (tär′läk)	270	15°29′N	120°36′E
Tarma, Peru (tär′mä)	188a	11°24′S	75°44′W
Tarnopol, Ukr. *see* Ternopil'	210-11	49°33′N	25°37′E
Tarnów, Pol. (tär′nōf)	210-11	50°01′N	21°01′E
Taroom, Austl.	301	25°39′S	149°48′E
Tarpon Springs, Fl., U.S. (tär′pŏn sprĭngz)	135a	28°09′N	82°46′W
Tarquinia, Italy (tär-kwē′nē-ä)	216-17	42°15′N	11°45′E
Tarragona, Spain (tär-rä-gō′nä)	214-15	41°07′N	1°14′E

414 Tàrrega – Thanlwin PRONOUNCING INDEX

Feature (Pronunciation)	Page	Lat.	Long.
Tàrrega, Spain	214-15	41°39'N	1°09'E
Tárrega, Spain (tä rå-gä) see Tàrrega	214-15	41°39'N	1°09'E
Tarsus, Tur. (tår'sōs) (tär'sŭs)	248-49	36°54'N	34°55'E
Tartagal, Arg. (tär-tä-gá'l)	182-83	22°33'S	63°50'W
Tartu, Est. (tär'tōō)	208-09	58°23'N	26°43'E
Ṭarṭūs, Syria	248-49	34°53'N	35°54'E
Tarutao, Ko, i., Thai.	266-67	6°35'N	99°40'E
Tarutung, Indon.	266-67	2°01'N	98°58'E
Taseyeva, stm., Russia	236-37	58°05'N	94°01'E
Tashauz, Turkmen. see Daşoguz	244	41°50'N	59°58'E
Ṭashk, Daryācheh-ye, lk., Iran	250-51	29°45'N	53°30'E
Tashkent, nat. cap., Uzb. (tásh'kěnt)	252-53	41°19'N	69°17'E
Tāshkurghān, Afg. see Kholm	252-53	36°41'N	67°42'E
Tashtagol, Russia	244	52°46'N	87°53'E
Tasiilaq, Green.	310-11	65°35'N	37°50'W
Tasikmalaya, Indon.	268-69	7°20'S	108°13'E
Tasman Bay, b., N.Z. (tăz'măn bā)	303	41°0'S	187°20'E
Tasmania, state, Austl.	301	43°0'S	147°00'E
Tasman Peninsula, pen., Austl. (tăz'măn pě-nĭn'sūlă)	301	43°05'S	147°50'E
Tasman Sea, s., Oc. (tăz'măn sē)	308-09	40°0'S	163°00'E
Tatabánya, Hung.	210-11	47°34'N	18°26'E
Tataouine, Tun.	204-05	32°55'N	10°28'E
Tatarskiy Proliv, strt., Russia	236-37	50°0'N	141°15'E
Tatar Strait, strt., Russia (tá-tär' strāt) see Tatarskiy Proliv	236-37	50°0'N	141°15'E
Tateyama, Japan (tä'tě-yä'mä)	265	34°59'N	139°52'E
Tathlina Lake, lk., N.T., Can.	140-41	60°32'N	117°32'W
Tatnam, Cape, c., Mb., Can.	144-45	57°14'N	90°54'W
Tatta, Pak.	252-53	24°45'N	67°56'E
Tatvan, Tur.	245	38°31'N	42°18'E
Tau, i., Am. Sam.	306b	14°15'S	169°29'W
Taunggyi, Mya.	266-67	20°47'N	97°02'E
Taupo, N.Z.	303	38°41'S	176°06'E
Taupo, Lake, lk., N.Z. (lăk tä'ōō-pō)	303	38°49'S	175°55'E
Tauragė, Lith. (tou'rá-gā)	208-09	55°15'N	22°17'E
Tauranga, N.Z.	303	37°42'S	176°09'E
Tauroa Point, c., N.Z.	303	35°10'S	187°04'E
Taurus Mountains, mts., Tur. (tôr'ŭs moun'tĭnz)	202-03	37°0'N	33°00'E
Tavastehus, Fin. see Hämeenlinna	208-09	60°58'N	24°31'E
Tavda, stm., Russia (táv-dá')	236-37	57°48'N	67°15'E
Tavira, Port. (tä-vē'rá)	214-15	37°07'N	7°39'W
Tavoy, Mya. see Dawei	266-67	14°05'N	98°13'E
Tavşanlı, Tur. (táv'shän-lĭ)	216-17	39°32'N	29°29'E
Tawas City, Mi., U.S. (tô'wǎs sĭ'tĭ)	126-27	44°16'N	83°31'W
Tawau, Malay.	268-69	4°16'N	117°53'E
Tawitawi Island, i., Phil.	270	5°11'N	119°60'E
Ṭawkar, Sudan	288	18°25'N	37°44'E
Taxco de Alarcón, Mex. (täs'kō dě ä-lär-kô'n)	158-59	18°33'N	99°36'W
Taxkorgan Tajik Zizhixian, China	254-55	37°47'N	75°14'E
Tayabas Bay, b., Phil. (tä-yä'bäs bā)	270	13°45'N	121°45'E
Taylor, Tx., U.S. (tā'lẽr)	132-33	30°34'N	97°24'W
Taylorville, Il., U.S. (tā'lẽr-vĭl)	126-27	39°33'N	89°18'W
Taymura, stm., Russia	236-37	63°46'N	98°07'E
Taymyr, Ozero, lk., Russia (ô'zě-rô tī-mĭr')	226-27	74°36'N	102°24'E
Taymyr, Poluostrov, pen., Russia (tī-mĭr')	236-37	76°0'N	104°00'E
Taymyr Peninsula, pen., Russia (tī-mĭr' pě-nĭn'sūlă) see Taymyr, Poluostrov	236-37	76°0'N	104°00'E
Tayshet, Russia (tī-shět')	236-37	55°56'N	98°00'E
Taytay, Phil.	270	10°49'N	119°31'E
Taz, stm., Russia (táz)	236-37	67°30'N	78°44'E
Taza, Mor. (tä'zä)	200-01	34°14'N	4°01'W
Tazovskiy, Russia	236-37	67°29'N	78°42'E
Tbilisi, nat. cap., Geor. ('tbĭl-yē'sē)	245	41°44'N	44°47'E
Tchad, nation, Afr. see Chad	274	15°0'N	19°00'E
Tchad, Lac, lk., Afr. see Chad, Lake	280-81	13°03'N	14°33'E
Tchibanga, Gabon (chē-băŋ'gä)	282-83	2°51'S	11°01'E
Teapa, Mex. (tā-ä'pä)	158-59	17°34'N	92°58'W
Tébessa, Alg.	200-01	35°24'N	8°07'E
Tebingtinggi, Indon.	266-67	3°19'N	99°10'E
Tecalitlán, Mex. (tā-kä-lē-tlän')	158-59	19°28'N	103°17'W
Techiman, Ghana	282-83	7°36'N	1°56'W
Tecka, Arg.	185	43°28'S	70°50'W
Tecomán, Mex. (tā-kō-män')	158-59	18°55'N	103°52'W
Tecpan de Galeana, Mex. (těk-pän' dā gä-lā-ä'nä)	158-59	17°13'N	100°36'W
Tecuala, Mex. (tě-kwä-lä)	158-59	22°24'N	105°28'W
Tecuci, Rom. (ta-kōch')	218-19	45°51'N	27°26'E
Tecumseh, Mi., U.S. (tē-kŭm'sĕ)	126-27	41°60'N	83°56'W
Tecumseh, Ne., U.S. (tē-kŭm'sĕ)	130-31	40°22'N	96°12'W
Tecumseh, Ok., U.S. (tē-kŭm'sĕ)	130-31	35°15'N	96°56'W
Tees, stm., Eng., U.K. (tēz)	206-07	54°36'N	1°15'W
Tefé, Braz.	180-81	3°23'S	64°43'W
Tefé, stm., Braz.	180-81	3°31'S	64°57'W
Tegal, Indon.	268-69	6°52'S	109°08'E
Tégua, i., Vanuatu	306g	13°15'S	166°37'E
Tegucigalpa, nat. cap., Hond. (tā-gōō-sē-gäl'pä)	161	14°05'N	87°13'W
Tehek Lake, lk., Nu., Can.	140-41	64°56'N	95°37'W
Teheran, nat. cap., Iran see Tehrān	252-53	35°40'N	51°25'E
Tehrān, nat. cap., Iran (tě-hrän')	252-53	35°40'N	51°25'E
Tehuacán, Mex. (tā-wä-kän')	158-59	18°27'N	97°24'W
Tehuantepec, Golfo de, b., Mex. (gôl-fô dě tå-wän-tå-pěk')	158-59	15°60'N	94°50'W
Tehuantepec, Istmo de, isth., Mex. (ē'st-mô dě tå-wän-tå-pěk')	158-59	17°0'N	95°00'W
Teide, Pico del, mtn., Spain	215d	28°16'N	16°38'W
Tejen, Turkmen.	252-53	37°22'N	60°31'E
Tejen, stm., Asia	252-53	37°24'N	60°31'E
Tejo, stm., Eur. see Tagus	214-15	38°51'N	8°57'W
Tejupan, Punta, c., Mex. (pōō'n-tä-tě-κōō-pä'n)	158-59	18°20'N	103°30'W
Tejupilco de Hidalgo, Mex. (tå-hōō-pēl'kô dā ē-dhäl'gō)	158-59	18°54'N	100°09'W
Tekamah, Ne., U.S. (tê-kā'mȧ)	124-25	41°47'N	96°13'W
Tekeli, Kaz.	244	44°48'N	78°51'E
Tekezē, stm., Afr.	288	14°20'N	35°51'E
Tekirdağ, Tur.	216-17	40°60'N	27°31'E
Tekit, Mex. (tě-kē't)	160	20°32'N	89°20'W
Tekkeköy, Tur.	247	41°13'S	36°28'E
Tela, Hond. (tā'lä)	161	15°46'N	87°28'W
Telangana, state, India	256	17°15'N	79°00'E
Tel Aviv-Jaffa, Isr. see Tel Aviv-Yafo	248-49	32°03'N	34°47'E
Tel Aviv-Yafo, Isr. (těl-ä-vēv'já'já'fá)	248-49	32°03'N	34°47'E
Telegraph Creek, B.C., Can. (těl'ê-gráf krēk)	138-39	57°55'N	131°10'W
Telén, Arg.	185	36°16'S	65°31'W
Telen, stm., Indon.	268-69	0°09'S	116°41'E
Telescope Peak, mtn., Ca., U.S. (těl'ê-skōp pēk)	128-29	36°10'N	117°05'W
Teletskoye Ozero, lk., Russia	244	51°38'N	87°40'E
Tell City, In., U.S. (těl sĭ'tê)	126-27	37°57'N	86°46'W
Tello, Col. (tě'l-yô)	188c	3°04'N	75°08'W
Telluride, Co., U.S. (těl'ū-rīd)	128-29	37°57'N	107°48'W
Teloloapan, Mex. (tā'lô-lô-ä'pän)	158-59	18°21'N	99°52'W
Telos, i., Grc. see Tílos	216-17	36°26'N	27°23'E
Telsen, Arg.	185	42°27'S	66°58'W
Telšiai, Lith. (těl'sha'ê)	208-09	55°59'N	22°15'E
Teluk Intan, Malay.	266-67	4°01'N	101°02'E
Tema, Ghana	282-83	5°38'N	0°01'E
Temagami, Lake, lk., On., Can.	146-47	47°0'N	80°05'W
Temax, Mex. (tě'mäx)	160	21°09'N	88°56'W
Tembenchi, stm., Russia	236-37	64°37'N	99°56'E
Tembesi, stm., Indon.	266-67	1°42'S	103°06'E
Tembilahan, Indon.	266-67	0°16'S	103°13'E
Temecula, Ca., U.S.	128-29	33°30'N	117°09'W
Temesvár, Rom. see Timișoara	216-17	45°45'N	21°13'E
Temirtaū, Kaz.	244	50°03'N	72°57'E
Tempe, Az., U.S.	128-29	33°24'N	111°55'W
Tempio Pausania, Italy (těm'pê-ō pou-sä'nê-ä)	216-17	40°54'N	9°06'E
Temple, Tx., U.S. (těm'p'l)	132-33	31°06'N	97°21'W
Tempoal, stm., Mex. (těm-pô-ä'l)	158-59	21°46'N	98°27'W
Temryuk, Russia (tyěm-ryòk')	218-19	45°16'N	37°22'E
Temuco, Chile (tå-mōō'kō)	185	38°44'S	72°36'W
Tena, Ec.	184	0°59'S	77°49'W
Tenāli, India	256	16°15'N	80°35'E
Tenasserim, Mya. (těn-ăs'ẽr-ĭm)	266-67	12°05'N	99°01'E
Ten Degree Channel, strt., India	266-67	10°0'N	93°00'E
Tendrara, Mor.	204-05	33°04'N	2°00'W
Ténéré, des., Niger	280-81	18°43'N	10°51'E
Tenerife, i., Spain (tå-nå-rē'fä) (těn-ẽr-ĭf')	215d	28°19'N	16°34'W
Ténès, Alg. (tā-nĕs')	214-15	36°30'N	1°18'E
Tengchong, China	258-59	25°01'N	98°30'E
Tenggara, Nusa, is., Asia see Lesser Sunda Islands	268-69	9°0'S	120°00'E
Tengiz köli, lk., Kaz.	244	50°22'N	68°56'E
Tengxian, China (tŭŋ shyěn)	258-59	23°20'N	110°53'E
Tengxian, China (tŭŋ shyěn)	260-61	35°05'N	117°09'E
Tennant Creek, Austl. (těn'ănt krēk)	294-95	19°39'S	134°11'E
Tennessee, state, U.S. (těn-ě-sē')	118-19	35°50'N	85°30'W
Tennessee, stm., U.S. (těn-ě-sē')	120-21	37°05'N	88°34'W
Teno, stm., Eur.	200-01	70°26'N	28°16'E
Tenom, Malay.	268-69	5°07'N	115°56'E
Tenosique, Mex. (tā-nō-sē'kå)	160	17°29'N	91°26'W
Tenryū, stm., Japan (těn'ryōō')	265	34°40'N	137°48'E
Tensas, stm., La., U.S. (těn'sô)	134-35	31°37'N	91°48'W
Tenterfield, Austl. (těn'tẽr-fēld)	301	29°04'S	152°01'E
Teocaltiche, Mex. (tā'ô-käl-tē'chå)	158-59	21°26'N	102°34'W
Teófilo Otoni, Braz. (tě-ô'fē-lō-tô'nê)	186	17°53'S	41°31'W
Teotihuacán, hist., Mex.	158-59	19°44'N	98°50'W
Tepalcatepec, Mex. (tā'päl-kä-tå'pěk)	158-59	19°11'N	102°51'W
Tepatitlán de Morelos, Mex. (tā-pä-tê-tlän' dā mô-rä'lôs)	158-59	20°48'N	102°44'W
Tepeaca, Mex. (tā-på-ä'kä)	158-59	18°58'N	97°54'W
Tepic, Mex. (tā-pēk')	158-59	21°30'N	104°54'W
Tequila, Mex. (tå-kē'lä)	158-59	20°52'N	103°50'W
Tequisquiapan, Mex. (tå-kēs-kê-ä'pän)	158-59	20°32'N	99°54'W
Téra, Niger	280-81	14°00'N	0°46'E
Teraina, i., Kir.	304-05	4°42'N	160°45'W
Teramo, Italy (tā'rä-mô)	216-17	42°40'N	13°42'E
Terceira, i., Port. (těr-sā'rä)	215c	38°43'N	24°13'W
Tercero, stm., Arg.	187	32°55'S	62°20'W
Terek, stm., Russia	245	43°44'N	46°33'E
Teresina, Braz. (těr-å-sē'ná)	180-81	5°05'S	42°49'W
Teresópolis, Braz. (těr-ā-sô'pō-lēzh)	186	22°26'S	42°59'W
Tergüün Bogd uul, mtn., Mong.	260-61	44°57'N	100°15'E
Teriběrka, Russia (tyěr-ê-byôr'kå)	202-03	69°07'N	35°08'E
Terme, Tur.	247	41°13'N	36°58'E
Términos, Laguna de, b., Mex. (lä-gò'nä dě ě'r-mē-nôs)	160	18°36'N	91°34'W
Termiz, Uzb.	252-53	37°14'N	67°16'E
Termoli, Italy (těr'mô-lê)	216-17	42°00'N	14°60'E
Ternate, Indon. (těr-nä'tā)	268-69	0°49'N	127°18'E
Terney, Russia	264	45°03'N	136°36'E
Terni, Italy (těr'nê)	216-17	42°34'N	12°39'E
Ternopil', Ukr.	210-11	49°33'N	25°37'E
Terpeniya, Mys, c., Russia (mĭs těr-pă'nĭ-yá)	236-37	48°39'N	144°44'E
Terpeniya, Zaliv, b., Russia (zä'lĭf těr-pă'nĭ-yá)	236-37	49°0'N	143°30'E
Terrace, B.C., Can. (těr'ĭs)	142-43	54°32'N	128°35'W
Terracina, Italy (těr-rä-chē'nä)	216-17	41°17'N	13°15'E
Terranova di Sicilia, Italy see Gela	216-17	37°04'N	14°15'E
Terra Nova National Park, n.p., Nf./L., Can.	148-49	48°31'N	53°56'W
Terrebonne, Qc., Can. (těr-bōn')	146-47	45°42'N	73°37'W
Terre Haute, In., U.S. (těr-ê hōt')	126-27	39°27'N	87°25'W
Teruel, Spain (tå-rōō-ěl')	214-15	40°21'N	1°06'W
Tes, stm., Asia	240-41	50°29'N	93°03'E
Teseney, Erit.	288	15°08'N	36°42'E
Tes-Khem, stm., Asia see Tes	240-41	50°29'N	93°03'E
Teslin, Yk., Can. (těs-lĭn)	138-39	60°11'N	132°43'W
Teslin, stm., Can. (těs-lĭn)	140-41	61°34'N	134°53'W
Teslin Lake, lk., Can. (těs-lĭn lăk)	140-41	60°15'N	132°57'W
Tessalit, Mali	280-81	20°12'N	1°00'E
Tessaoua, Niger (těs-sä'ô-ä)	280-81	13°45'N	7°59'E
Tete, Moz. (tā'tě)	286-87	16°09'S	33°36'E
Tetepare Island, i., Sol. Is.	306e	8°43'S	157°33'E
Teterow, Ger. (tā'tě-rō)	210-11	53°46'N	12°34'E
Tetiaroa, at., Fr. Poly.	306d	17°00'S	149°34'W
Tetouan, Mor.	292a	35°35'N	5°22'W
Tetovo, Mac. (tā'tô-vô)	216-17	42°00'N	20°59'E
Teuco, stm., Arg.	182-83	25°39'S	60°10'W
Tevere, stm., Italy see Tiber	216-17	41°45'N	12°14'E
Texarkana, Ar., U.S. (těk-sär-kän'ȧ)	130-31	33°26'N	94°03'W
Texarkana, Tx., U.S. (těk-sär-kän'ȧ)	130-31	33°26'N	94°04'W
Texas, state, U.S. (těk'sŭs)	118-19	31°30'N	99°00'W
Texas City, Tx., U.S. (těk'sŭs sĭ'tĭ)	132-33	29°23'N	94°54'W
Texoma, Lake, res., U.S. (lăk těk'ō-mä)	130-31	33°54'N	96°37'W
Teykovo, Russia (těy-kô-vô)	218-19	56°51'N	40°33'E
Teziutlán, Mex. (tå-zē-ōō-tlän')	158-59	19°49'N	97°21'W
Tezpur, India	254-55	26°37'N	92°48'E
Thabana-Ntlenyana, mtn., Leso.	286-87	29°28'S	29°16'E
Thabazimbi, S. Afr.	292c	24°37'S	27°24'E
Thai Lan, Vinh, b., Asia see Thailand, Gulf of	266-67	10°0'N	101°00'E
Thailand, nation, Asia (tī'ănd)	224-25	15°0'N	100°00'E
Thailand, Gulf of, b., Asia (gŭlf ŭv tī'ănd)	266-67	10°0'N	101°00'E
Thai Nguyen, Viet.	266-67	21°36'N	105°50'E
Thakhek, Laos see Muang Khammouan	266-67	17°25'N	104°49'E
Thal, Pak.	252-53	33°22'N	70°33'E
Thames, stm., On., Can. (těmz)	146-47	42°19'N	82°27'W
Thames, stm., Eng., U.K. (těmz)	206-07	51°27'N	0°21'E
Thāne, India	256	19°14'N	72°59'E
Thanh Hoa, Viet. (tän'hō'á)	266-67	19°48'N	105°46'E
Thanh Pho Ho Chi Minh, Viet. see Ho Chi Minh City	266-67	10°45'N	106°40'E
Thanjāvūr, India	256	10°47'N	79°09'E
Thanlwin, stm., Asia see Salween	226-27	16°33'N	97°40'E

ăt; finăl; rāte; senâte; ärm; åsk; sofȧ; fâre; ch-choose; dh-as th in other; bē; ēvent; bět; recěnt; cratẽr; g-gō; gh-guttural g; bĭt; ĭ-short neutral; rīde; κ-guttural k as ch in German ich;

Feature (Pronunciation)	Page	Lat.	Long.
Thann, Fr. (tän)	212-13	47°49'N	7°05'E
Thar Desert, des., Asia (tär dĕs´ĕrt)	252-53	27°0'N	71°00'E
Thargomindah, Austl. (thár´gō-mĭn´då)	301	28°0'S	143°49'E
Tharrawaddy, Mya.	266-67	17°39'N	95°47'E
Thásos, i., Grc. (thă´sôs)	216-17	40°39'N	24°40'E
Thayer, Mo., U.S. (thâ´ĕr)	134-35	36°31'N	91°33'W
Thayetmyo, Mya.	266-67	19°19'N	95°11'E
Thazi, Mya.	266-67	20°51'N	96°04'E
Thebes, Grc. (thēbz) *see* Thíva	216-17	38°20'N	23°19'E
Thebes, hist., Egypt (thēbz)	291b	25°42'N	32°39'E
The Coorong, b., Austl. (thė kö´rŏng)	301	35°46'S	139°15'E
The Dalles, Or., U.S. (thė dălz)	122-23	45°36'N	121°11'W
The Hague, nat. cap., Neth. (thė hāg) *see* Hague, The	206-07	52°06'N	4°18'E
The Minch, strt., Scot., U.K.	206-07	58°10'N	5°50'W
Theodore, Austl. (thēō´dôr)	302	24°57'S	150°05'E
Theodore Roosevelt National Park (North Unit), n.p., N.D., U.S. (thē-ô-dôr rōō-sȧ-vĕlt năsh´ŭn-ăl pärk)	124-25	47°34'N	103°24'W
Theodore Roosevelt National Park (South Unit), n.p., N.D., U.S. (thē-ô-dôr rōō-sȧ-vĕlt năsh´ŭn-ăl pärk)	124-25	46°58'N	103°25'W
The Pas, Mb., Can. (thė pä)	144-45	53°49'N	101°13'W
Thermopolis, Wy., U.S. (thĕr-mŏp´ô-lĭs)	122-23	43°39'N	108°13'W
The Snares, is., N.Z. *see* Snares Islands	313	48°0'S	166°30'E
Thessaloníki, Grc. (thĕs-sá-lô-nē´kė)	216-17	40°38'N	22°59'E
Thetford Mines, Qc., Can. (thĕt´fĕrd mīns)	146-47	46°05'N	71°18'W
The Valley, nat. cap., Anguilla	155b	18°13'N	63°04'W
Thibodaux, La., U.S. (tė-bô-dō´)	134-35	29°48'N	90°49'W
Thief River Falls, Mn., U.S. (thēf rĭv´ĕr fôlz)	124-25	48°07'N	96°11'W
Thiers, Fr. (tyâr)	212-13	45°51'N	3°32'E
Thiès, Sen. (tė-ĕs´)	282-83	14°48'N	16°56'W
Thika, Kenya	289	1°03'S	37°04'E
Thimphu, nat. cap., Bhu. (tĭm-pōō´)	254-55	27°28'N	89°39'E
Thingvellir, Ice.	206a	64°16'N	21°07'W
Thionville, Fr. (tyôɴ-vēl´)	212-13	49°22'N	6°10'E
Thíra, i., Grc.	216-17	36°26'N	25°27'E
Thiruvananthapuram, India	256	8°31'N	76°57'E
Thisted, Den. (tēs´tĕdh)	208-09	56°57'N	8°42'E
Thíva, Grc.	216-17	38°20'N	23°19'E
Thjórsá, stm., Ice. (tyûr´så)	206a	63°55'N	20°40'W
Thomas, Ok., U.S. (tŏm´ăs)	130-31	35°45'N	98°45'W
Thomaston, Ga., U.S. (tŏm´ăs-tŭn)	134-35	32°53'N	84°19'W
Thomasville, Al., U.S. (tŏm´ăs-vĭl)	134-35	31°54'N	87°45'W
Thomasville, Ga., U.S. (tŏm´ăs-vĭl)	134-35	30°50'N	83°59'W
Thomasville, N.C., U.S. (tŏm´ăs-vĭl)	134-35	35°53'N	80°05'W
Thompson, Mb., Can. (tŏm-sŏn)	144-45	55°44'N	97°51'W
Thompson, stm., B.C., Can. (tŏm-sŏn)	142-43	50°14'N	121°35'W
Thompson, stm., U.S. (tŏm-sŏn)	130-31	39°45'N	93°37'W
Thompson Falls, Mt., U.S. (tŏm-sŏn fôlz)	122-23	47°35'N	115°21'W
Thomson, stm., Austl. (tŏm-sŏn)	302	25°11'S	142°50'E
Thomson's Falls, Kenya (tŏm-sŏns fôlz) *see* Nyahururu Falls	289	0°02'N	36°22'E
Thonon-les-Bains, Fr. (tô-nôɴ´lå-băɴ´)	212-13	46°23'N	6°29'E
Thorn, Pol. *see* Toruń	210-11	53°01'N	18°37'E
Thorshavn, nat. cap., Far. Is. *see* Tórshavn	206b	62°01'N	6°46'W
Thouars, Fr. (tōō-är´)	212-13	46°59'N	0°13'W
Thousand Islands National Park, n.p., On., Can.	146-47	44°25'N	75°52'W
Thrace, hist. reg., Eur. (thrās)	216-17	41°20'N	26°45'E
Thráki, hist. reg., Eur. *see* Thrace	216-17	41°20'N	26°45'E
Three Forks, Mt., U.S. (thrē fôrks)	122-23	45°53'N	111°34'W
Three Gorges Reservoir, res., China	258-59	31°0'N	110°30'E
Three Hummock Island, i., Austl.	301	40°26'S	144°55'E
Three Oaks, Mi., U.S. (thrē ōks)	126-27	41°48'N	86°36'W
Three Pagodas Pass, p., Asia	266-67	15°18'N	98°22'E
Three Points, Cape, c., Ghana	282-83	4°45'N	2°05'W
Three Rivers, Mi., U.S. (thrē rĭv´ĕrz)	126-27	41°56'N	85°37'W
Thrissur, India	256	10°31'N	76°13'E
Thu, Cu Lao, i., Viet.	266-67	10°32'N	108°57'E
Thunder Bay, On., Can. (thŭn´dĕr bā)	146-47	48°24'N	89°15'W
Thunder Bay, b., On., Can. (thŭn´dĕr bā)	146-47	48°24'N	89°00'W
Thursday Island, Austl.	302	10°36'S	142°15'E
Thurso, Scot., U.K.	206-07	58°35'N	3°32'W
Thysville, D.R.C. *see* Mbanza-Ngungu	284-85	5°14'S	14°53'E
Tiandong, China (tĭĕn-dôŋ)	258-59	23°36'N	107°08'E
Tianjin, China (tĭĕn-jyĭ)	260-61	39°08'N	117°11'E
Tianjin, state, China	260-61	39°30'N	117°15'E
Tianjun, China	260-61	37°20'N	98°57'E
Tianmen, China (tĭĕn-mŭn)	258-59	30°39'N	113°10'E
Tian Shan, mts., Asia (tĭĕn shän) *see* Tien Shan	244	42°0'N	80°00'E
Tianshui, China (tĭĕn-shwā)	258-59	34°32'N	105°54'E
Tiantai, China	258-59	29°08'N	121°00'E
Tianzhu, China	260-61	36°60'N	103°07'E
Tiaret, Alg.	214-15	35°28'N	1°21'E
Tibasti, Sarīr, des., Libya	280-81	24°0'N	17°00'E
Tibati, Camrn.	282-83	6°27'N	12°37'E
Tiber, stm., Italy (Itī´bẽr)	216-17	41°45'N	12°14'E
Tiberias, Lake, lk., Isr. *see* Galilee, Sea of	248-49	32°48'N	35°35'E
Tibesti, mts., Afr. (tĭ-bĕs´-tē)	280-81	21°30'N	17°30'E
Tibet, state, China (tĭ-bĕt´)	254-55	32°0'N	88°00'E
Tibet, Plateau of, plat., China (plä-tō´ ŭv tĭ-bĕt´)	240-41	33°0'N	92°00'E
Tibooburra, Austl.	301	29°26'S	142°01'E
Tiburón, Isla, i., Mex.	156-57	29°0'N	112°23'W
Tīchīt, Maur.	280-81	18°29'N	9°28'W
Ticonderoga, N.Y., U.S. (tī-kŏn-dĕr-ō´gȧ)	126-27	43°51'N	73°26'W
Ticul, Mex. (tē-kōō´l)	160	20°24'N	89°32'W
Tidaholm, Swe. (tē´då-hōlm)	208-09	58°11'N	13°58'E
Tidjikja, Maur.	280-81	18°33'N	11°25'W
Tidore, Indon.	268-69	0°38'N	127°24'E
Tieli, China	260-61	46°59'N	128°04'E
Tieling, China (tĭĕ-lĭŋ)	263	42°18'N	123°51'E
Tien Giang, stm., Asia *see* Mekong	266-67	10°33'N	105°27'E
Tien Shan, mts., Asia (tĭĕn shän)	244	42°0'N	80°00'E
Tientsin, China *see* Tianjin	260-61	39°08'N	117°11'E
Tientsin, state, China *see* Tianjin	260-61	39°30'N	117°15'E
Tierp, Swe. (tyĕrp)	208-09	60°20'N	17°31'E
Tierra Blanca, Mex. (tyĕ´r-rä-blä´n-kä)	158-59	18°26'N	96°21'W
Tierra del Fuego, i., S.A. (tyĕr´rä dĕl fwā´gô)	185	54°0'S	69°00'W
Tietê, stm., Braz.	182-83	20°37'S	51°34'W
Tiffin, Oh., U.S. (tĭf´in)	126-27	41°06'N	83°10'W
Tifton, Ga., U.S. (tĭf´tŭn)	134-35	31°27'N	83°30'W
Tiga, Île, i., N. Cal.	306g	21°08'S	167°48'E
Tighina, Mol.	218-19	46°50'N	29°29'E
Tigil', Russia	236-37	57°47'N	158°42'E
Tignish, P.E., Can. (tĭg´nĭsh)	148-49	46°58'N	64°02'W
Tigre, stm., Peru	184	4°29'S	74°05'W
Tigris, stm., Asia (tī-grĭs)	226-27	30°60'N	47°27'E
Tihuatlán, Mex. (tė-wä-tlän´)	158-59	20°44'N	97°34'W
Tijuana, Mex. (tė-hwä´nä)	156-57	32°32'N	117°01'W
Tikal, hist., Guat. (tė-käl´)	160	17°15'N	89°39'W
Tikei, Île, i., Fr. Poly.	304-05	14°58'S	144°33'W
Tikhoretsk, Russia (tė-κôr-yĕtsk´)	202-03	45°51'N	40°08'E
Tikhvin, Russia (têκ-vēn´)	208-09	59°39'N	33°32'E
Tikrit, Iraq	240-41	34°36'N	43°42'E
Tiksi, Russia (têk-sē´)	236-37	71°39'N	128°48'E
Tilburg, Neth. (tĭl´bûrg)	206-07	51°34'N	5°05'E
Tîlemsi, Vallée du, stm., Mali	280-81	16°18'N	0°01'E
Tillamook, Or., U.S. (tĭl´ȧ-mŏk)	122-23	45°27'N	123°50'W
Tillsonburg, On., Can. (tĭl´sŭn-bûrg)	146-47	42°52'N	80°43'W
Tílos, i., Grc.	216-17	36°26'N	27°23'E
Tilpa, Austl.	301	30°57'S	144°24'E
Tim, Russia (têm)	218-19	51°38'N	37°07'E
Timan Ridge, hills, Russia *see* Timanskiy Kryazh	202-03	65°0'N	51°00'E
Timanskiy Kryazh, hills, Russia	202-03	65°0'N	51°00'E
Timaru, N.Z. (tĭm´å-rōō)	303	44°24'S	171°14'E
Timbalier Bay, b., La., U.S. (tĭm´bå-lĕr bā)	134-35	29°10'N	90°20'W
Timbuktu, Mali *see* Tombouctou	280-81	16°47'N	3°01'W
Timimoun, Alg. (tē-mė-mōōn´)	204-05	29°14'N	0°16'E
Timirist, Râs, c., Maur.	280-81	19°23'N	16°32'W
Timișoara, Rom.	216-17	45°45'N	21°13'E
Timmins, On., Can. (tĭm´ĭnz)	146-47	48°29'N	81°21'W
Timor, i., Asia (tė-mōr´)	268-69	9°0'S	125°00'E
Timor, Laut, s., *see* Timor Sea	296-97	11°0'S	128°00'E
Timor-Leste, nation, Asia	268-69	8°35'S	126°00'E
Timor Sea, s., (tė-mōr´ sē)	296-97	11°0'S	128°00'E
Timpanogos Cave National Monument, n.p., Ut., U.S. (tĭ-măn´ō-gōz kāv năsh´ŭn-ăl mŏn´ū-mĕnt)	128-29	40°26'N	111°44'W
Tinaca Point, c., Phil.	270	5°34'N	125°20'E
Tindouf, Alg. (tėn-dōōf´)	280-81	27°49'N	8°08'W
Tinghert, Ḩamādat, plat., Afr.	280-81	29°0'N	9°00'E
Tingo María, Peru (tē´ngô-mä-rē´ä)	184	9°10'S	75°56'W
Tingri, China *see* Dinggyê	254-55	28°35'N	86°37'E
Tingsryd, Swe. (tĭngs´rüd)	208-09	56°32'N	14°59'E
Tinian, i., N. Mar. Is.	304-05	15°00'N	145°38'E
Tinkisso, stm., Gui.	282-83	11°21'N	9°11'W
Tinogasta, Arg. (tē-nô-gäs´tä)	182-83	28°03'S	67°34'W
Tínos, i., Grc.	216-17	37°36'N	25°10'E
Tinrhert, Hamada de, plat., Afr.	280-81	29°0'N	9°00'E
Tintina, Arg.	187	27°02'S	62°43'W
Tioman, Pulau, i., Malay.	266-67	2°48'N	104°10'E
Tipitapa, Nic. (tē-pė-tä´pä)	161	12°12'N	86°06'W
Tip Top Mountain, mtn., On., Can.	146-47	48°16'N	85°59'W
Tīrān, i., Sau. Ar.	248-49	27°57'N	34°33'E
Tīrân, Madīq, strt., *see* Tiran, Strait of	248-49	27°58'N	34°28'E
Tiran, Strait of, strt.,	248-49	27°58'N	34°28'E
Tirana, nat. cap., Alb. *see* Tiranë	216-17	41°20'N	19°50'E
Tiranë, nat. cap., Alb. (tė-rä´nä)	216-17	41°20'N	19°50'E
Tirano, Italy (tė-rä´nō)	216-17	46°13'N	10°11'E
Tiraspol, Mol.	218-19	46°51'N	29°38'E
Tire, Tur. (tē´rĕ)	216-17	38°06'N	27°45'E
Tiree, i., Scot., U.K. (tī-rē´)	206-07	56°31'N	6°52'W
Tîrgu Mureș, Rom. *see* Târgu Mureș	210-11	46°33'N	24°34'E
Tirich Mīr, mtn., Pak.	252-53	36°15'N	71°50'E
Tirreno, Mar, s., Eur. *see* Tyrrhenian Sea	216-17	40°0'N	12°00'E
Tiruchchirāppalli, India (tĭr´ô-chĭ-rä´på-lĭ)	256	10°49'N	78°42'E
Tirunelveli, India	256	8°44'N	77°41'E
Tiruppur, India	256	11°06'N	77°21'E
Tisa, stm., Eur.	216-17	45°08'N	20°17'E
Tisdale, Sk., Can. (tĭz´dāl)	144-45	52°51'N	104°03'W
Tisisat Falls, wtfl., Eth.	288	11°29'N	37°35'E
T'īs Isat Fwafwatē, wtfl., Eth. *see* Tisisat Falls	288	11°29'N	37°35'E
Tīsta, stm., Asia	254-55	25°31'N	89°42'E
Tisza, stm., Eur. (tė´sä)	216-17	45°08'N	20°17'E
Titicaca, Lago, lk., S.A. (lä´gô-tē-tė-kä´kä)	182-83	15°50'S	69°20'W
Titograd, nat. cap., Mont. *see* Podgorica	216-17	42°27'N	19°16'E
Titov Veles, Mac. (tė´tôv vĕ´lĕs)	216-17	41°43'N	21°47'E
Titusville, Fl., U.S. (tī´tŭs-vĭl)	135a	28°37'N	80°49'W
Titusville, Pa., U.S. (tī´tŭs-vĭl)	126-27	41°38'N	79°41'W
Tivoli, Italy (tē´vô-lė)	216-17	41°58'N	12°48'E
Tiwanaku, hist., Bol.	182-83	16°33'S	68°41'W
Tizimín, Mex. (tē-zē-mė´n)	160	21°10'N	88°10'W
Tizi Ouzou, Alg. (tē´zė-ōō-zōō´)	292b	36°48'N	4°02'E
Tiznit, Mor. (tĕz-nēt)	280-81	29°42'N	9°43'W
Tjörn, i., Swe.	208-09	58°0'N	11°38'E
Tlacotalpan, Mex. (tlä-kô-täl´pän)	158-59	18°38'N	95°40'W
Tlacotepec, Mex. (tlä-kô-tå-pĕ´k)	158-59	17°47'N	99°59'W
Tlahualilo de Zaragoza, Mex.	132-33	26°06'N	103°26'W
Tlalnepantla, Mex. (tläl-nå-pàn´tlä)	158-59	19°32'N	99°12'W
Tlaquepaque, Mex. (tlä-kĕ-pä´kĕ)	158-59	20°39'N	103°19'W
Tlaxcala, state, Mex.	158-59	19°25'N	98°10'W
Tlaxcala de Xicohténcatl, Mex.	158-59	19°19'N	98°14'W
Tlemcen, Alg.	214-15	34°53'N	1°18'W
Toamasina, Madag.	286-87	18°09'S	49°24'E
Tobago, i., Trin. (tô-bā´gō)	155b	11°15'N	60°40'W
Tobejuba, Isla, i., Ven.	155b	9°20'N	60°52'W
Tobelo, Indon.	268-69	1°44'N	128°00'E
Tobi, i., Palau	304-05	3°0'N	131°10'E
Tobol, stm., Asia (tô-bôl´)	236-37	58°09'N	68°13'E
Tobol'sk, Russia (tô-bôlsk´)	236-37	58°11'N	68°15'E
Tobruk, Libya	280-81	32°05'N	23°57'E
Tobyl, stm., Asia	236-37	58°09'N	68°13'E
Tocantinópolis, Braz. (tō-kän-tē-nō´pō-lēs)	180-81	6°19'S	47°25'W
Tocantins, state, Braz. (tō-kän-tēns´)	180-81	10°0'S	48°00'W
Tocantins, stm., Braz. (tō-kän-tēɴs´)	180-81	1°45'S	49°12'W
Toccoa, Ga., U.S. (tŏk´ô-ȧ)	134-35	34°35'N	83°20'W
Tocoa, Hond. (tō-kô´ä)	161	15°38'N	86°01'W
Tocopilla, Chile (tō-kô-pēl´yä)	182-83	22°06'S	70°11'W
Tocuyo de la Costa, Ven. (tô-kōō´yō-dĕ-lä-kôs´tä)	188b	11°02'N	68°22'W
Todos Santos, Bol.	182-83	16°48'S	65°08'W
Toemoek Hoemak Gebergte, mts., S.A. *see* Tumuc-Humac Mountains	178-79	2°19'N	54°35'W
Tofino, B.C., Can. (tō-fē´nō)	142-43	49°08'N	125°54'W
Toga, i., Vanuatu	306g	13°25'S	166°41'E
Togian, Kepulauan, is., Indon.	268-69	0°20'S	122°00'E
Togliatti, Russia	202-03	53°32'N	49°26'E
Togo, nation, Afr. (tō´gō)	274	8°0'N	1°10'E
Tok, Ak., U.S.	136	63°20'N	143°00'W
Tokachi, stm., Japan (tō-kä´chė)	264	42°44'N	143°43'E
Tokat, Tur. (tô-kät´)	202-03	40°19'N	36°34'E
Tokelau, dep., Oc. (tō-kė-lä´ô)	304-05	9°0'S	171°45'W
Tokmok, Kyrg.	244	42°50'N	75°18'E
Tokushima, Japan (tō´kŏ´shē-mä)	265	34°04'N	134°34'E
Tokuyama, Japan (tō´kŏ´yä-mä)	265	34°03'N	131°48'E
Tōkyō, nat. cap., Japan (tō´kė-ō)	265	35°42'N	139°47'E

416 Tôlañaro – Trinity PRONOUNCING INDEX

Feature (Pronunciation)	Page	Lat.	Long.
Tôlañaro, Madag.	286-87	25°02'S	47°00'E
Toledo, Spain (tô-lĕ'dô)	214-15	39°53'N	4°03'W
Toledo, Ia., U.S. (tô-lē'dō)	124-25	41°60'N	92°35'W
Toledo, Oh., U.S. (tô-lē'dō)	126-27	41°39'N	83°33'W
Toledo, Or., U.S. (tô-lē'dō)	122-23	44°38'N	123°56'W
Toledo, Montes de, mts., Spain (mŏ'n-tĕs-dĕ-tô-lĕ'dô)	214-15	39°33'N	4°20'W
Toledo Bend Reservoir, res., U.S. (tô-lē'dō bĕnd rĕ'sĕr-vwär)	132-33	31°30'N	93°45'W
Toliara, Madag.	286-87	23°22'S	43°40'E
Tolima, Nevado del, vol., Col. (nĕ-vä-dô-dĕl-tô-lē'mä)	188c	4°40'N	75°19'W
Tolitoli, Indon.	268-69	1°02'N	120°49'E
Tolmezzo, Italy (tôl-mĕt'zô)	216-17	46°25'N	13°01'E
Tolo, Teluk, b., Indon. (tō'lō)	268-69	2°0'S	122°30'E
Tolosa, Spain (tô-lō'sä)	214-15	43°09'N	2°05'W
Toluca, Il., U.S. (tô-lōō'kȧ)	126-27	40°60'N	89°08'W
Toluca, Nevado de, vol., Mex. (nĕ-vä-dô-dĕ-tô-lōō'kä)	158-59	19°05'N	99°44'W
Toluca de Lerdo, Mex.	158-59	19°17'N	99°39'W
Tolyatti, Russia *see* Togliatti	202-03	53°32'N	49°26'E
Tom', stm., Russia	236-37	56°53'N	84°27'E
Tomah, Wi., U.S. (tō'mȧ)	124-25	43°59'N	90°30'W
Tomahawk, Wi., U.S. (tŏm'ȧ-hôk)	126-27	45°28'N	89°44'W
Tomakomai, Japan	264	42°38'N	141°36'E
Tomar, Port. (tō-mär')	214-15	39°36'N	8°25'W
Tomaszów Lubelski, Pol. (tô-mä'shòf lōō-bĕl'skĭ)	210-11	50°27'N	23°25'E
Tomaszów Mazowiecki, Pol. (tô-mä'shòf mä-zô'vyĕt-skĭ)	210-11	51°33'N	20°01'E
Tomatlán, Mex. (tô-mä-tlä'n)	158-59	19°55'N	105°14'W
Tombador, Serra do, plat., Braz. (sĕr'rȧ dò tōm-bä-dôr')	180-81	12°0'S	57°40'W
Tombigbee, stm., U.S. (tŏm-bĭg'bē)	120-21	31°04'N	87°58'W
Tombouctou, Mali	280-81	16°47'N	3°01'W
Tombstone, Az., U.S. (tōōm'stōn)	128-29	31°43'N	110°04'W
Tombstone Mountain, mtn., Yk., Can.	140-41	64°25'N	138°30'W
Tombua, Ang. (ȧ-lĕ-zhän'drĕ)	286-87	15°48'S	11°49'E
Tomé, Chile	185	36°36'S	72°57'W
Tomea, Pulau, i., Indon.	268-69	5°45'S	123°56'E
Tomelilla, Swe. (tô'mĕ-lĕl-lä)	208-09	55°33'N	13°57'E
Tomelloso, Spain (tō-mâl-lyō'sō)	214-15	39°09'N	3°01'W
Tomini, Indon.	268-69	0°32'N	120°32'E
Tomini, Teluk, b., Indon.	268-69	0°20'S	121°00'E
Tommot, Russia (tŏm-mŏt')	236-37	58°58'N	126°18'E
Tomo, stm., Col.	178-79	5°19'N	67°50'W
Tom Price, Austl.	294-95	22°41'S	117°48'E
Tomsk, Russia (tŏmsk)	236-37	56°30'N	84°58'E
Tonalá, Mex.	158-59	16°05'N	93°45'W
Tondano, Indon. (tŏn-dä'nō)	268-69	1°18'N	124°55'E
Tønder, Den. (tûn'nĕr)	210-11	54°56'N	8°52'E
Tone, stm., Japan (tō'nĕ)	265	35°45'N	140°51'E
Tonga, nation, Oc. (tŏŋ'gȧ)	304-05	20°0'S	175°00'W
Tonga Islands, is., Tonga (tŏŋ'gȧ ī'lȧndz)	304-05	20°0'S	175°00'W
Tong'an, China (tŏŋ-än)	243a	24°44'N	118°09'E
Tongatapu, i., Tonga	304-05	21°10'S	175°10'W
Tongbei, China (tŏŋ-bā)	260-61	47°46'N	126°46'E
Tongcheng, China	258-59	31°03'N	116°57'E
Tongchuan, China	258-59	35°04'N	109°04'E
Tongguan, China (tŏŋ-gūän)	258-59	34°36'N	110°17'E
Tonghai, China	258-59	24°07'N	102°47'E
Tonghe, China (tŏŋ-hŭ)	264	45°58'N	128°45'E
Tonghua, China (tŏŋ-hwä)	263	41°43'N	125°56'E
Tongjiang, China (tŏŋ-jyäŋ)	258-59	31°56'N	107°14'E
Tongjiang, China (tŏŋ-jyäŋ)	260-61	47°38'N	132°30'E
Tongjosŏn-man, b., Kor., N.	263	39°30'N	128°00'E
Tongliao, China (tŏŋ-lĭou)	260-61	43°37'N	122°17'E
Tongoy, Chile (tōn-goi')	182-83	30°16'S	71°29'W
Tongren, China (tŏŋ-rŭn)	258-59	27°43'N	109°11'E
Tongsa Dzong, Bhu.	254-55	27°31'N	90°30'E
Tongxian, China (tŏŋ shyĕn)	260-61	39°54'N	116°39'E
Tongyu, China	260-61	44°48'N	123°05'E
Tongzi, China	258-59	28°09'N	106°49'E
Tonk, India (Tŏŋk)	254-55	26°10'N	75°48'E
Tonkawa, Ok., U.S. (tŏŋ kȧ-wô)	130-31	36°41'N	97°19'W
Tonkin, Gulf of, b., Asia (gŭlf ŭv tôn-kăn')	266-67	20°0'N	108°00'E
Tônlé Sab, Bœng, lk., Camb. *see* Sap, Tonle	266-67	13°0'N	104°00'E
Tonneins, Fr. (tô-năɴ')	212-13	44°23'N	0°19'E
Tonopah, Nv., U.S. (tō-nô-pä')	128-29	38°05'N	117°13'W
Tønsberg, Nor. (tûns'bĕrgh)	208-09	59°16'N	10°26'E
Tonto National Monument, n.p., Az., U.S. (tôn'tô)	128-29	33°34'N	111°02'W
Tooele, Ut., U.S. (tò-ĕl ē)	128-29	40°32'N	112°18'W

Feature (Pronunciation)	Page	Lat.	Long.
Toowoomba, Austl. (tò wōōm'bȧ)	301	27°34'S	151°57'E
Topeka, Ks., U.S.	130-31	39°02'N	95°41'W
Topol'čany, Slvk. (tô-pôl'chä-nü)	210-11	48°34'N	18°10'E
Topolobampo, Mex. (tō-pō-lô-bä'm-pô)	156-57	25°36'N	109°03'W
Topozero, Ozero, lk., Russia	202-03	65°40'N	32°00'E
Toppenish, Wa., U.S. (tôp'ĕn-ĭsh)	122-23	46°23'N	120°19'W
Torawitan, Tanjung, c., Indon.	268-69	1°45'N	124°60'E
Torbat-e Ḥeydarīyeh, Iran	252-53	35°17'N	59°13'E
Torbat-e Jām, Iran	252-53	35°15'N	60°38'E
Torbay, Nf./L., Can. (tôr-bā')	148-49	47°40'N	52°45'W
Torbay, Eng., U.K. *see* Torquay	206-07	50°28'N	3°32'W
Torch, stm., Sk., Can.	144-45	53°52'N	103°06'W
Torch Lake, lk., Mi., U.S. (tôrch lāk)	126-27	45°03'N	85°20'W
Torda, Rom. *see* Turda	210-11	46°34'N	23°47'E
Torez, Ukr.	218-19	48°02'N	38°38'E
Torghay, stm., Kaz.	244	48°02'N	62°34'E
Torghay üstirti, plat., Kaz.	244	51°0'N	64°00'E
Torino, Italy *see* Turin	216-17	45°03'N	7°41'E
Tori Sima, i., Japan	264	30°29'N	140°19'E
Torit, S. Sudan	289	4°24'N	32°34'E
Tormes, stm., Spain (tōr'mās)	214-15	41°18'N	6°27'W
Torneälven, stm., Eur.	200-01	65°49'N	24°09'E
Torneträsk, lk., Swe. (tôr'nĕ trĕsk)	200-01	68°20'N	19°23'E
Torngat, Monts, mts., Can. *see* Torngat Mountains	140-41	59°0'N	64°00'W
Torngat Mountains, mts., Can.	140-41	59°0'N	64°00'W
Torngat Mountains National Park, n.p., Nf./L., Can.	138-39	59°11'N	64°00'W
Tornio, Fin. (tôr'nĭ-ô)	200-01	65°51'N	24°10'E
Tornionjoki, stm., Eur.	200-01	65°49'N	24°09'E
Tornquist, Arg.	187	38°06'S	62°13'W
Toronto, On., Can. (tô-rŏn'tō)	146-47	43°38'N	79°24'W
Toropets, Russia (tô'rô-pyĕts)	208-09	56°30'N	31°40'E
Tororo, Ug.	289	0°42'N	34°11'E
Toros Dağları, mts., Tur. *see* Taurus Mountains	202-03	37°0'N	33°00'E
Torquay, Eng., U.K. (tôr-kē')	206-07	50°28'N	3°32'W
Torrance, Ca., U.S. (tŏr'rănc)	128-29	33°51'N	118°20'W
Torrelavega, Spain (tôr-rā'lä-vā'gä)	214-15	43°21'N	4°03'W
Torremaggiore, Italy (tôr'rā mäd-jō'rā)	216-17	41°41'N	15°17'E
Torrens, Lake, lk., Austl. (lăk tōr-ĕns)	301	31°03'S	137°51'E
Torreón, Mex. (tôr-rå-ōn')	132-33	25°33'N	103°26'W
Torres, Îles, is., Vanuatu	306g	13°17'S	166°39'E
Torres Islands, is., Vanuatu (tôr'rĕs ī'lȧndz) (tŏr'ĕz ī'lȧndz) *see* Torres, Îles	306g	13°17'S	166°39'E
Torres Novas, Port. (tôr'rĕzh nō'väzh)	214-15	39°28'N	8°32'W
Torres Strait, strt., Oc. (tôr'rĕs strāt)	302	10°25'S	142°10'E
Torres Vedras, Port. (tôr'rĕsh vă'dräzh)	214-15	39°05'N	9°15'W
Torrevella, Spain	214-15	37°59'N	0°41'W
Torrevieja, Spain (tôr-rā-vyā'hä) *see* Torrevella	214-15	37°59'N	0°41'W
Torrington, Ct., U.S. (tŏr'ĭng-tŭn)	126-27	41°48'N	73°07'W
Torrington, Wy., U.S. (tŏr'ĭng-tŭn)	124-25	42°05'N	104°12'W
Torsby, Swe. (tôrs'bü)	208-09	60°08'N	13°01'E
Tórshavn, nat. cap., Far. Is. (tôrs-houn')	206b	62°01'N	6°46'W
Tortola, i., Br. Vir. Is. (tôr-tō'lä)	155b	18°27'N	64°36'W
Tórtolas, Cerro de las, mtn., S.A. *see* Las Tórtolas, Cerro	182-83	29°57'S	69°53'W
Tortona, Italy (tôr-tō'nä)	216-17	44°54'N	8°52'E
Tortosa, Spain (tôr-tō'sä)	214-15	40°48'N	0°31'E
Tortue, Île de la, i., Haiti (ēl-dē-lȧ-tôr-tü') *see* Tortuga Island	154-55	20°03'N	72°47'W
Tortuga Island, i., Haiti *see* Tortue, Île de la	154-55	20°03'N	72°47'W
Toruń, Pol.	210-11	53°01'N	18°37'E
Tõrva, Est. (t'r'vȧ)	208-09	58°01'N	25°56'E
Torzhok, Russia (tôr'zhôk)	218-19	57°03'N	34°58'E
Toscana, hist. reg., Italy (tôs-kä'nä) *see* Tuscany	216-17	43°25'N	11°00'E
Toshkent, nat. cap., Uzb. *see* Tashkent	252-53	41°19'N	69°17'E
Tosno, Russia (tôs'nô)	208-09	59°33'N	30°52'E
Tostado, Arg. (tôs-tä'dô)	187	29°14'S	61°46'W
Totana, Spain (tô-tä-nä)	214-15	37°46'N	1°30'W
Tot'ma, Russia (tôt'mȧ)	202-03	59°58'N	42°45'E
Totoras, Arg. (tô-tô'räs)	187	32°36'S	61°10'W
Totoya, i., Fiji	306f	18°56'S	179°51'W
Tottori, Japan (tô'tô-rē)	265	35°30'N	134°14'E
Toubkal, Jebel, mtn., Mor.	280-81	31°05'N	7°55'W
Toûil, Oued, stm., Alg. (wĕd tōō-ēl')	292b	35°33'N	2°36'E
Toul, Fr. (tōōl)	212-13	48°41'N	5°53'E
Toulnustouc, stm., Qc., Can.	148-49	49°35'N	68°25'W
Toulon, Fr. (tōō-lôɴ')	212-13	43°08'N	5°56'E

Feature (Pronunciation)	Page	Lat.	Long.
Toulouse, Fr. (tōō-lōōz')	212-13	43°36'N	1°27'E
Toungoo, Mya. (tō-òŋ-gōō')	266-67	18°56'N	96°26'E
Tourane, Viet. *see* Da Nang	266-67	16°03'N	108°12'E
Tourcoing, Fr. (tòr-kwaɴ')	212-13	50°43'N	3°09'E
Tours, Fr. (tōōr)	212-13	47°24'N	0°43'E
Toussidé, Pic, vol., Chad (pĭk tōō-sē-dā')	280-81	21°02'N	16°28'E
Towner, N.D., U.S. (tou'nẽr)	124-25	48°21'N	100°24'W
Townsend, Mt., U.S. (toun'zĕnd)	122-23	46°19'N	111°31'W
Townshend Island, i., Austl.	302	22°15'S	150°30'E
Townsville, Austl. (tounz'vĭl)	302	19°16'S	146°48'E
Towson, Md., U.S. (tou'sŭn)	126-27	39°24'N	76°36'W
Towuti, Danau, lk., Indon. (tô-wōō'tē)	268-69	2°45'S	121°32'E
Toxkan, stm., China	244	41°07'N	80°12'E
Toyama, Japan (tō'yä-mä)	265	36°41'N	137°13'E
Toyohashi, Japan (tō'yô-hä'shê)	265	34°46'N	137°23'E
Tozeur, Tun. (tô-zûr')	204-05	33°55'N	8°08'E
Trabzon, Tur. (trȧb'zôn)	245	40°60'N	39°44'E
Tracy, Ca., U.S. (trā'sē)	128-29	37°44'N	121°26'W
Tracy, Mn., U.S. (trā'sē)	124-25	44°14'N	95°37'W
Trafalgar, Cabo, c., Spain (kȧ'bô-trä-fäl-gä'r)	214-15	36°11'N	6°02'W
Trail, B.C., Can. (trāl)	142-43	49°06'N	117°42'W
Trakiya, hist. reg., Eur. *see* Thrace	216-17	41°20'N	26°45'E
Tranås, Swe. (trän'ôs)	208-09	58°03'N	14°59'E
Trancas, Arg.	182-83	26°13'S	65°17'W
Trang, Thai.	266-67	7°33'N	99°36'E
Trangan, Pulau, i., Indon. (pōō-lou träŋ'gän)	242-43	6°35'S	134°20'E
Trani, Italy (trä'nē)	216-17	41°16'N	16°25'E
Transylvania, hist. reg., Rom. (trăn-sĭl-vā'nĭ-ȧ)	210-11	46°44'N	23°37'E
Transylvanian Alps, mts., Rom. (trăn-sĭl-vā'nĭ-ȧn ălps)	216-17	45°25'N	23°33'E
Trapani, Italy	216-17	38°01'N	12°31'E
Traralgon, Austl. (trä'rǎl-gŏn)	301	38°12'S	146°32'E
Traverse City, Mi., U.S. (träv'ẽrs sĭ'tē)	126-27	44°45'N	85°37'W
Travnik, Bos. (träv'nēk)	216-17	44°14'N	17°40'E
Trebinje, Bos. (trå'bēn-yĕ)	216-17	42°43'N	18°23'E
Trebišov, Slvk. (trĕ'bē-shôf)	210-11	48°38'N	21°44'E
Trebizond, Tur. *see* Trabzon	245	40°60'N	39°44'E
Treinta y Tres, Ur. (trå-ēn'tä ē träs')	187	33°14'S	54°23'W
Trelew, Arg. (trē'lū)	185	43°15'S	65°18'W
Trelleborg, Swe.	208-09	55°23'N	13°11'E
Tremblant, Mont, mtn., Qc., Can.	146-47	46°16'N	74°35'W
Trenčín, Slvk.	210-11	48°54'N	18°04'E
Trenque Lauquen, Arg. (trĕn'kĕ-lä'ō-kĕ'n)	187	35°58'S	62°45'W
Trent, Italy *see* Trento	216-17	46°04'N	11°08'E
Trent, stm., On., Can. (trĕnt)	146-47	44°06'N	77°34'W
Trento, Italy (trĕn'tô)	216-17	46°04'N	11°08'E
Trenton, N.S., Can. (trĕn'tŭn)	148-49	45°37'N	62°38'W
Trenton, On., Can. (trĕn'tŭn)	146-47	44°06'N	77°35'W
Trenton, Mo., U.S. (trĕn'tŭn)	130-31	40°05'N	93°37'W
Trenton, N.J., U.S.	126-27	40°13'N	74°45'W
Trenton, Tn., U.S. (trĕn'tŭn)	134-35	35°59'N	88°57'W
Tres Arroyos, Arg. (träs'är-rō'yōs)	187	38°22'S	60°16'W
Três Corações, Braz. (trĕ's kō-rä-zô'ĕs)	186	21°42'S	45°15'W
Tres Esquinas, Col.	178-79	0°44'N	75°14'W
Tres Lagoas, Braz. (trĕ's lä-gô'ás)	182-83	20°47'S	51°43'W
Tres Marías, Islas, is., Mex.	158-59	21°32'N	106°32'W
Três Marias, Represa de, res., Braz.	186	18°14'S	45°16'W
Tres Picos, Cerro, mtn., Arg.	187	38°09'S	61°57'W
Tres Puntas, Cabo, c., Arg.	185	47°06'S	65°53'W
Três Rios, Braz. (trĕ's rē'ōs)	186	22°07'S	43°12'W
Treviglio, Italy (trā-vē'lyô)	216-17	45°32'N	9°36'E
Treviso, Italy (trĕ-vê'sō)	216-17	45°40'N	12°14'E
Trichardt, S. Afr. (trĭ-kärt')	292c	26°30'S	29°14'E
Trichinopoly, India *see* Tiruchchirāppalli	256	10°49'N	78°42'E
Trichūr, India *see* Thrissur	256	10°31'N	76°13'E
Trieste, Italy (trē-ĕs'tä)	216-17	45°40'N	13°46'E
Triglav, mtn., Slvn.	216-17	46°23'N	13°50'E
Trikora, Puncak, mtn., Indon.	302	4°18'S	138°40'E
Trincomalee, Sri L. (trĭŋ-kô-mȧ-lē')	256	8°34'N	81°14'E
Trinidad, Bol. (trē-nē-dhädh')	180-81	14°49'S	64°54'W
Trinidad, Col.	178-79	5°25'N	71°40'W
Trinidad, Cuba (trē-nē-dhädh')	154-55	21°48'N	79°59'W
Trinidad, Co., U.S. (trĭn'ĭdăd)	130-31	37°10'N	104°30'W
Trinidad, Ur.	187	33°32'S	56°54'W
Trinidad, i., Trin. (trĭn'ĭ-dăd)	155b	10°30'N	61°15'W
Trinidad, Isla, i., Arg.	187	39°10'S	61°57'W
Trinidad and Tobago, nation, N.A. (trĭn'ĭ-dăd ănd tô-bā'gō)	152-53	11°0'N	61°00'W
Trinity, Tx., U.S. (trĭn'ĭ-tē)	132-33	30°56'N	95°23'W
Trinity, stm., Ca., U.S. (trĭn'ĭ-tē)	122-23	41°11'N	123°42'W
Trinity, stm., Tx., U.S. (trĭn'ĭ-tē)	132-33	29°46'N	94°41'W

ăt; finăl; rāte; senāte; ärm; åsk; sofȧ; fâre; ch-choose; dh-as th in other; bē; ēvent; bĕt; recĕnt; cratẽr; g-gō; gh-guttural g; bĭt; ĭ-short neutral; rīde; ᴋ-guttural k as ch in German ich;

418 Ţuwayq, Jabal – Urundi

PRONOUNCING INDEX

Feature (Pronunciation)	Page	Lat.	Long.
Ţuwayq, Jabal, mts., Sau. Ar.	238-39	23°0'N	46°00'E
Tuxpan, Mex. (tōōs'pän)	158-59	21°56'N	105°17'W
Tuxpan de Rodríguez Cano, Mex.	158-59	20°58'N	97°24'W
Tuxtepec, Mex. (tós-tå-pĕk')	158-59	18°05'N	96°07'W
Tuxtla Gutiérrez, Mex. (tós'tlä gōō-tyår'rĕs)	158-59	16°45'N	93°06'W
Tuyen Quang, Viet.	266-67	21°50'N	105°11'E
Tuy Hoa, Viet.	266-67	13°05'N	109°19'E
Tuymazy, Russia	202-03	54°36'N	53°43'E
Tuz Gölü, lk., Tur.	202-03	38°45'N	33°25'E
Tuzla, Bos. (tōōz'lá)	216-17	44°33'N	18°40'E
Tver', Russia	218-19	56°52'N	35°55'E
Tvertsa, stm., Russia (tvĕr'tsá)	218-19	56°52'N	35°56'E
Tweed, stm., U.K. (twēd)	206-07	55°46'N	1°60'W
Tweeling, S. Afr. (twē'lĭng)	292c	27°39'S	28°30'E
Twin Falls, Id., U.S. (twĭn fôlz)	122-23	42°34'N	114°28'W
Two Rivers, Wi., U.S. (tōō rĭv'ẽrz)	126-27	44°09'N	87°34'W
Tyan' Shan', mts., Asia *see* Tien Shan	244	42°0'N	80°00'E
Tyler, Mn., U.S. (tī'lẽr)	124-25	44°17'N	96°08'W
Tyler, Tx., U.S. (tī'lẽr)	130-31	32°21'N	95°19'W
Tylertown, Ms., U.S. (tī'lẽr-toun)	134-35	31°07'N	90°09'W
Tym, stm., Russia	236-37	59°26'N	80°01'E
Tymovskoye, Russia	236-37	50°51'N	142°38'E
Tynda, Russia	236-37	55°09'N	124°43'E
Tyndall, S.D., U.S. (tĭn'dál)	124-25	42°60'N	97°52'W
Tyrma, Russia	260-61	50°03'N	132°10'E
Tyrrhenian Sea, s., Eur. (tĭr-rē'nĭ-ăn sē)	216-17	40°0'N	12°00'E
Tyrrhénienne, Mer, s., Eur. *see* Tyrrhenian Sea	216-17	40°0'N	12°00'E
Tyul'gan, Russia	202-03	52°24'N	56°14'E
Tyumen', Russia (tyōō-mĕn')	236-37	57°10'N	65°33'E
Tyung, stm., Russia	236-37	63°46'N	121°32'E
Tyva, state, Russia	240-41	52°0'N	95°00'E
Tzaneen, S. Afr.	292c	23°49'S	30°10'E
Tzeliutsing, China *see* Zigong	258-59	29°22'N	104°45'E
Tzucacab, Mex. (tzōō-kä-kä'b)	160	20°04'N	89°02'W
Tzupo, China *see* Boshan	260-61	36°29'N	117°51'E

U

Feature (Pronunciation)	Page	Lat.	Long.
Uatumã, stm., Braz.	180-81	2°24'S	57°33'W
Uaupés, stm., S.A.	178-79	0°02'N	67°15'W
Ubá, Braz.	186	21°07'S	42°56'W
Ubangi, stm., Afr. (ōō-bäŋ'gē)	284-85	0°25'S	17°47'E
Ubatuba, Braz. (ōō-bå-tōō'bá)	186	23°26'S	45°04'W
Ube, Japan	265	33°57'N	131°15'E
Úbeda, Spain (ōō'bå-dä)	214-15	38°01'N	3°22'W
Uberaba, Braz. (ōō-bå-rä'bȧ)	186	19°46'S	47°56'W
Uberlândia, Braz. (ōō-bĕr-lá'n-dyä)	186	18°54'S	48°15'W
Ubon Ratchathani, Thai. (ōō'bŭn rä'chätä-nē)	266-67	15°14'N	104°52'E
Ubrique, Spain (ōō-brē'kå)	214-15	36°41'N	5°27'W
Ubsu-Nur, Ozero, lk., Asia *see* Uvs Lake	240-41	50°20'N	92°45'E
Ubundu, D.R.C.	284-85	0°21'S	25°25'E
Ucayali, stm., Peru (ōō'kä-yä'lē)	184	4°30'S	73°30'W
Uchaly, Russia (û-chä'lĭ)	244	54°18'N	59°27'E
Uchiura-wan, b., Japan (ōō'chē-ōō'rä wän)	264	42°20'N	140°40'E
Uchiza, Peru	184	8°25'S	76°25'W
Uchur, stm., Russia (ö-chör')	236-37	58°47'N	130°36'E
Uda, stm., Russia (ö'dá)	240-41	51°49'N	107°34'E
Uda, stm., Russia (ö'dá)	236-37	54°43'N	135°18'E
Uda, stm., Russia (ö'dá)	236-37	56°02'N	99°38'E
Udachnyy, Russia	236-37	66°29'N	112°15'E
Udagamandalam, India	256	11°25'N	76°42'E
Udai, stm., Ukr. (ö'dá)	218-19	50°04'N	33°07'E
Udaipur, India (ö-dŭ'ē-pōōr)	254-55	24°35'N	73°42'E
Uddevalla, Swe. (ōōd'dĕ-väl-á)	208-09	58°21'N	11°55'E
Udine, Italy (ōō'dē-nå)	216-17	46°04'N	13°15'E
Udmurtia, state, Russia	202-03	57°0'N	53°00'E
Udmurtiya, state, Russia *see* Udmurtia	202-03	57°0'N	53°00'E
Udon Thani, Thai.	266-67	17°24'N	102°47'E
Ueda, Japan (wā'dä)	265	36°24'N	138°15'E
Uele, stm., D.R.C. (wā'lå)	284-85	4°07'N	22°26'E
Uelen, Russia	136	66°09'N	169°48'W
Uelzen, Ger. (ült'sĕn)	210-11	52°58'N	10°34'E
Uere, stm., D.R.C.	284-85	3°33'N	25°15'E
Ufa, Russia (ö'fa)	202-03	54°42'N	55°58'E
Ufa, stm., Russia (ö'fa)	202-03	54°41'N	56°02'E
Ugab, stm., Nmb. (ōō'gäb)	286-87	21°11'S	13°38'E
Ugalla, stm., Tan. (ōō-gä'lä)	289	5°17'S	30°58'E
Uganda, nation, Afr. (ōō-gän'dä) (û-găn'dȧ)	274	1°0'N	32°00'E
Uglegorsk, Russia (ōō-glĕ-gôrsk)	236-37	49°04'N	142°03'E
Uglich, Russia (ōōg-lĕch')	218-19	57°32'N	38°20'E
Ugoma, mtn., D.R.C.	289	4°0'S	28°45'E
Uhrichsville, Oh., U.S. (ū'rĭks-vĭl)	126-27	40°24'N	81°21'W
Uíge, Ang.	284-85	7°38'S	15°04'E
Uina, stm., Camrn. *see* Vina	282-83	7°52'N	15°46'E
Uitenhage, S. Afr.	286-87	33°40'S	25°27'E
Uji, Japan (ōō'jē)	265	34°54'N	135°49'E
Ujiji, Tan. (ōō-jē'jē)	289	4°55'S	29°41'E
Uji-yamada, Japan *see* Ise	265	34°30'N	136°42'E
Ujjain, India (ōō-jŭēn)	254-55	23°11'N	75°47'E
Ujungpandang, Indon. *see* Makassar	268-69	5°08'S	119°25'E
Újvidék, Serb. *see* Novi Sad	216-17	45°15'N	19°50'E
Ukara Island, i., Tan.	289	1°50'S	33°03'E
Ukerewe Island, i., Tan.	289	2°03'S	33°00'E
Ukhta, Russia (ōōk'tá)	202-03	63°34'N	53°44'E
Ukiah, Ca., U.S.	128-29	39°09'N	123°12'W
Ukiah, Or., U.S. (ū-kī'á)	122-23	45°09'N	118°56'W
Ukkusiksalik National Park, n.p., Nu., Can.	138-39	66°0'N	90°00'W
Ukmergė, Lith. (ók'mĕr-ghå)	208-09	55°15'N	24°46'E
Ukraïna, nation, Eur. *see* Ukraine	190-91	49°0'N	32°00'E
Ukraine, nation, Eur. (yōō-krān')	190-91	49°0'N	32°00'E
Ukyr, Russia	260-61	49°28'N	108°52'E
Ulaanbaatar, nat. cap., Mong. (ōō'län-bä'tôr)	260-61	47°55'N	106°56'E
Ulaangom, Mong.	260-61	49°59'N	92°04'E
Ulaan-Uul, Mong.	260-61	44°23'N	111°12'E
Ulan Bator, nat. cap., Mong. *see* Ulaanbaatar	260-61	47°55'N	106°56'E
Ulanhad, China *see* Chifeng	260-61	42°16'N	118°58'E
Ulanhot, China	260-61	46°04'N	122°04'E
Ulan-Ude, Russia (ōō'län ōō'då)	240-41	51°50'N	107°36'E
Ulawa Island, i., Sol. Is.	306e	9°47'S	161°57'E
Uldz, stm., Asia	260-61	49°55'N	115°33'E
Uldza, stm., Asia *see* Uldz	260-61	49°55'N	115°33'E
Uleåborg, Fin. *see* Oulu	200-01	65°01'N	25°28'E
Ulhāsnagar, India	256	19°14'N	73°08'E
Uliast, Mong.	260-61	48°57'N	91°09'E
Uliastay, Mong.	260-61	47°44'	96°51'E
Ulindi, stm., D.R.C. (ōō-lĭn'dē)	284-85	1°40'S	25°52'E
Ulithi, at., Micron.	304-05	9°55'N	139°42'E
Uljin, Kor., S.	263	36°59'N	129°23'E
Ülkenözen, stm., Eur.	202-03	48°60'N	49°59'E
Ulm, Ger. (ōlm)	210-11	48°24'N	9°59'E
Ulónguè, Moz.	286-87	14°37'S	34°19'E
Ulricehamn, Swe. (ōl-rē'sĕ-häm)	208-09	57°48'N	13°25'E
Ulsan, Kor., S. (ōōl'sän')	263	35°33'N	129°19'E
Ulúa, stm., Hond. (ōō-lōō'á)	161	15°52'N	87°44'W
Ulul, i., Micron.	304-05	8°36'N	149°40'E
Ulungur, stm., China (ōō-lōōn-gŭr)	240-41	46°59'N	87°26'E
Ulungur Hu, lk., China	240-41	47°13'N	87°16'E
Uluru, mtn., Austl.	296-97	25°20'S	130°60'E
Ulverstone, Austl. (ŭl'vẽr-stŭn)	301	41°10'S	146°11'E
Ulyanovsk, Russia (ōō-lyä'nôfsk)	202-03	54°19'N	48°22'E
Ulysses, Ks., U.S. (ū-lĭs'ēz)	130-31	37°35'N	101°21'W
Umán, Mex. (ōō-män')	160	20°53'N	89°45'W
Uman', Ukr. (ö-mán')	218-19	48°44'N	30°14'E
'Umān, nation, Asia *see* Oman	224-25	22°0'N	58°00'E
'Umān, Khalīj, b., Asia *see* Oman, Gulf of	250-51	24°30'N	58°30'E
Umarkot, Pak.	252-53	25°22'N	69°45'E
Umatilla Indian Reservation, ind. res., Or., U.S. (ū-má-tĭl'á ĭn'dĭ-ăn rĕ-sẽr-vā'shĕn)	122-23	45°41'N	118°31'W
Umba, Russia	202-03	66°41'N	34°18'E
Umboi Island, i., Pap. N. Gui.	302	5°36'S	147°53'E
Umeå, Swe.	200-01	63°50'N	20°16'E
Umeälven, stm., Swe.	200-01	63°47'N	20°19'E
Umm Durmān, Sudan *see* Omdurman	288	15°39'N	32°29'E
Umm Lajj, Sau. Ar.	288	25°07'N	37°16'E
Umm Ruwābah, Sudan	288	12°54'N	31°12'E
Umm Urūmah, i., Sau. Ar.	288	25°46'N	36°33'E
Umnak Island, i., Ak., U.S. (ōōm'nák ī'lȧnd)	136a	53°25'N	168°10'W
Umpqua, stm., Or., U.S. (ŭmp'kwá)	122-23	43°42'N	124°05'W
'Umrān, Yemen	288	15°58'N	43°58'E
Umtata, S. Afr. (ōm-tä'tä)	286-87	31°35'S	28°47'E
Umuarama, Braz.	182-83	23°46'S	53°19'W
Unalakleet, Ak., U.S. (ū-ná-lák'lēt)	136	63°53'N	160°47'W
Unalaska, Ak., U.S. (ū-nä-lás'ká)	136a	53°52'N	166°32'W
Unalaska Island, i., Ak., U.S.	136a	53°45'N	166°45'W
Unauna, Pulau, i., Indon.	268-69	0°10'S	121°35'E
'Unayzah, Sau. Ar.	288	26°05'N	43°59'E
Uncia, Bol. (ōōn'sē-ä)	182-83	18°26'S	66°35'W
Uncompahgre Peak, mtn., Co., U.S. (ŭn-kŭm-pä'grē pēk)	128-29	38°04'N	107°28'W
Unecha, Russia (ö-nĕ'chá)	218-19	52°51'N	32°42'E
Ungava, Baie d', b., Can. *see* Ungava Bay	140-41	59°30'N	67°30'W
Ungava, Péninsule d', pen., Qc., Can.	140-41	60°0'N	74°00'W
Ungava Bay, b., Can. (ŭŋ-gá'vá bā)	140-41	59°30'N	67°30'W
Ungava Peninsula, pen., Qc., Can. (ŭŋ-gá'vá pĕ-nĭn'sūlȧ) *see* Ungava, Péninsule d'	140-41	60°0'N	74°00'W
Ungvár, Ukr. *see* Uzhhorod	210-11	48°37'N	22°19'E
União, Braz.	180-81	4°35'S	42°52'W
União dos Palmares, Braz.	188d	9°10'S	36°02'W
Unimak Island, i., Ak., U.S. (ōō-nē-mák' ī'lȧnd)	136	54°43'N	164°27'W
Unini, stm., Braz.	180-81	1°41'S	61°31'W
Union, Mo., U.S. (ūn'yŭn)	130-31	38°27'N	91°01'W
Union, S.C., U.S. (ūn'yŭn)	134-35	34°43'N	81°37'W
Union City, In., U.S. (ūn'yŭn sĭ'tē)	126-27	40°11'N	85°00'W
Union City, Mi., U.S. (ūn'yŭn sĭ'tē)	126-27	42°04'N	85°09'W
Union City, Pa., U.S. (ūn'yŭn sĭ'tē)	126-27	41°54'N	79°50'W
Union City, Tn., U.S. (ūn'yŭn sĭ'tē)	134-35	36°25'N	89°03'W
Union Springs, Al., U.S. (ūn'yŭn sprĭngz)	134-35	32°09'N	85°43'W
Uniontown, Al., U.S. (ūn'yŭn-toun)	134-35	32°27'N	87°31'W
Uniontown, Pa., U.S.	126-27	39°54'N	79°44'W
Unionville, Mo., U.S. (ūn'yŭn-vĭl)	130-31	40°29'N	93°01'W
United Arab Emirates, nation, Asia (ū-nī'tĕd âr'ăb ĕ'mĕr-ĕts)	224-25	24°0'N	54°00'E
United Arab Republic, nation, Afr. *see* Egypt	274	27°0'N	30°00'E
United Kingdom, nation, Eur. (ū-nī'tĕd kĭng'dŭm)	190-91	54°0'N	2°00'W
United States, nation, N.A. (ū-nī'tĕd stāts)	94	38°0'N	97°00'W
Unity, Sk., Can.	144-45	52°27'N	109°07'W
Ünye, Tur.	247	41°08'N	37°17'W
Upa, stm., Russia (ö'pá)	218-19	54°02'N	36°21'E
Upata, Ven. (ōō-pä'tä)	178-79	8°01'N	62°24'W
Upemba, Lac, lk., D.R.C.	284-85	8°36'S	26°26'E
Upington, S. Afr. (ŭp'ĭng-tŭn)	286-87	28°27'S	21°14'E
Upland, In., U.S. (ŭp'lȧnd)	126-27	40°28'N	85°29'W
Upolu, i., Samoa	306b	13°55'S	171°45'W
Upolu Point, c., Hi., U.S. (ōō-pō'lōō point)	137a	20°16'N	155°51'W
Upper Arrow Lake, lk., B.C., Can. (ŭp'ẽr ăr'ō lāk)	142-43	50°31'N	117°56'W
Upper Kapuas Mountains, mts., Asia	268-69	1°15'N	113°30'E
Upper Klamath Lake, lk., Or., U.S. (ŭp'ẽr klăm'ăth lāk)	122-23	42°24'N	121°54'W
Upper Red Lake, lk., Mn., U.S. (ŭp'ẽr rĕd lāk)	124-25	48°10'N	94°40'W
Upper Sandusky, Oh., U.S. (ŭp'ẽr săn-dŭs'kē)	126-27	40°49'N	83°17'W
Uppsala, Swe. (ōōp'sá-lä)	208-09	59°52'N	17°38'E
Upsala, Swe. *see* Uppsala	208-09	59°52'N	17°38'E
Ural, stm., (ö-räl') (ū-rôl)	192-93	46°50'N	51°33'E
Ural Mountains, mts., Russia (ö-räl'' moun'tĭnz) (ū-rôl moun'tĭnz)	236-37	60°0'N	60°00'E
Ural'skiye Gory, mts., Russia *see* Ural Mountains	236-37	60°0'N	60°00'E
Ura-Tyube, Taj. *see* Istaravshan	252-53	39°54'N	69°00'E
Urbana, Il., U.S. (ûr-băn'á)	126-27	40°07'N	88°13'W
Urbana, Oh., U.S. (ûr-băn'á)	126-27	40°07'N	83°45'W
Urbino, Italy (ōōr-bē'nō)	216-17	43°44'N	12°39'E
Urdinarrain, Arg. (ōōr-dē-när-rάē'n)	187	32°41'S	58°54'W
Urfa, Tur. *see* Şanlıurfa	248-49	37°10'N	38°48'E
Urganch, Uzb.	252-53	41°33'N	60°38'E
Urgench, Uzb. *see* Urganch	252-53	41°33'N	60°38'E
Urla, Tur. (ör'lá)	216-17	38°20'N	26°46'E
Urmi, stm., Russia (ör'mē)	260-61	48°36'N	135°01'E
Urmia, Iran *see* Orūmīyeh	248-49	37°32'N	45°05'E
Urmia, Lake, lk., Iran (lāk ör'mēȧ) *see* Orūmīyeh, Daryācheh-ye	245	37°40'N	45°30'E
Urrao, Col. (ōōr-rá'ō)	188c	6°20'N	76°08'W
Uruapan del Progreso, Mex.	158-59	19°25'N	102°04'W
Urubamba, stm., Peru (ōō-rōō-bäm'bä)	184	10°44'S	73°44'W
Uruguai, stm., S.A.	187	34°10'S	58°18'W
Uruguaiana, Braz.	187	29°46'S	57°04'W
Uruguay, nation, S.A. (ōō-rōō-gwī') (ū'rōō-gwā)	172	33°0'S	56°00'W
Uruguay, stm., S.A. (ōō-rōō-gwī')	187	34°10'S	58°18'W
Urumchi, China *see* Ürümqi	240-41	43°48'N	87°35'E
Ürümqi, China (ū-rŭm-chyē)	240-41	43°48'N	87°35'E
Urundi, nation, Afr. *see* Burundi	274	3°15'S	30°00'E

ăt; finȧl; rāte; senåte; ärm; ȧsk; sofȧ; fâre; ch-choose; dh-as th in other; bē; ēvent; bĕt; recĕnt; cratẽr; g-gō; gh-guttural g; bĭt; ĭ-short neutral; rīde; ĸ-guttural k as ch in German ich;

PRONOUNCING INDEX
Urup, Ostrov – Venezuela, Golfo de 419

Feature (Pronunciation)	Page	Lat.	Long.
Urup, Ostrov, i., Russia (ôs-trôf′ ŏ′rŏp′)	236-37	46°0′N	150°00′E
Uryupinsk, Russia (ŏr′yŏ-pēn-sk′)	202-03	50°48′N	42°01′E
Urzhum, Russia	202-03	57°07′N	50°01′E
Urziceni, Rom. (ŏ-zē-chĕn′′)	218-19	44°43′N	26°40′E
Usa, stm., Russia (ŏ′sȧ)	202-03	65°58′N	56°57′E
Uşak, Tur. (o͞o′shȧk)	216-17	38°41′N	29°24′E
Usborne, Mount, mtn., Falk. Is.	185	51°41′S	58°50′W
Üshtöbe, Kaz.	244	45°14′N	77°58′E
Ushuaia, Arg. (o͞o-sho͞o-ī′ä)	185	54°47′S	68°19′W
Usinsk, Russia	202-03	65°57′N	57°24′E
Üsküb, nat. cap., Mac. *see* Skopje	216-17	42°00′N	21°28′E
Usman', Russia (o͞os-mȧn′)	218-19	52°03′N	39°45′E
Uspanapa, stm., Mex. (o͞os-pä-nä′pä)	158-59	17°56′N	94°28′W
Ussel, Fr. (üs′ĕl)	212-13	45°33′N	2°18′E
Ussuri, stm., Asia (o͞o-so͞o′rė)	260-61	48°27′N	135°04′E
Ussuriysk, Russia	264	43°48′N	131°59′E
Ust'-Barguzin, Russia	236-37	53°25′N	109°02′E
Ust-Bolsheretsk, Russia	236-37	52°49′N	156°17′E
Ust'-Ilimsk, Russia	236-37	58°00′N	102°40′E
Ústí nad Labem, Czech Rep.	210-11	50°40′N	14°02′E
Ustinov, Russia *see* Izhevsk	202-03	56°50′N	53°12′E
Üstirt, plat., Asia *see* Ust-Urt Plateau	244	43°0′N	56°00′E
Ust-Kamchatsk, Russia	236-37	56°13′N	162°29′E
Ust'-Kut, Russia	236-37	56°47′N	105°43′E
Ust'-Kuyga, Russia	236-37	69°60′N	135°35′E
Ust-Maya, Russia (ŏst mȧ′yȧ)	236-37	60°25′N	134°30′E
Ust'-Nera, Russia	236-37	64°34′N	143°18′E
Ust'-Omchug, Russia	236-37	61°08′N	149°37′E
Ust-Tsilma, Russia (ŏst tsĭl′mȧ)	202-03	65°25′N	52°05′E
Ust-Urt Plateau, plat., Asia (o͞ost-o͞ort plä-tō′)	244	43°0′N	56°00′E
Ustyurt Platosi, plat., Asia *see* Ust-Urt Plateau	244	43°0′N	56°00′E
Ustyuzhna, Russia (yo͞ozh′nȧ)	218-19	58°51′N	36°28′E
Usu, China (ů-so͞o)	244	44°26′N	84°41′E
Usulután, El Sal. (o͞o-so͞o-lä-tän′)	160	13°20′N	88°26′W
Usumacinta, stm., N.A. (o͞o′so͞o-mä-sēn′tō)	160	18°23′N	92°39′W
Usumbura, nat. cap., Bdi. *see* Bujumbura	289	3°23′S	29°22′E
Utah, state, U.S. (ū′taw)	110-19	39°30′N	111°30′W
Utah Lake, lk., Ut., U.S. (ū′taw lāk)	128-29	40°13′N	111°49′W
Utembo, stm., Ang.	286-87	17°04′S	21°58′E
Utena, Lith. (o͞o′tā-nä)	208-09	55°30′N	25°37′E
Uthai Thani, Thai.	266-67	15°23′N	100°02′E
Utiariti, Braz.	180-81	13°02′S	58°17′W
Utica, N.Y., U.S.	126-27	43°06′N	75°15′W
Utiel, Spain (o͞o-tyâl′)	214-15	39°34′N	1°13′W
Utila, Isla de, i., Hond. (ē′s-lä-dĕ-o͞o-tē′lä)	161	16°06′N	86°56′W
Uto, Japan (o͞o′tō′)	265	32°41′N	130°40′E
Utrecht, Neth. (ü′trĕκt) (ū′trĕkt)	206-07	52°05′N	5°08′E
Utrecht, S. Afr.	292c	27°40′S	30°19′E
Utrera, Spain (o͞o-trā′rä)	214-15	37°11′N	5°47′W
Utsunomiya, Japan (o͞ot′sŏ-nŏ-mē-yȧ′)	265	36°33′N	139°54′E
Uttaradit, Thai.	266-67	17°38′N	100°06′E
Uttarakhand, state, India	254-55	30°0′N	79°30′E
Uttaranchal, state, India *see* Uttarakhand	254-55	30°0′N	79°30′E
Uttar Pradesh, state, India (ŏt-tär-prä-dĕsh)	254-55	27°0′N	80°00′E
Uummannarsuaq, c., Green. *see* Farewell, Cape	95	59°46′N	43°60′W
Uvá, stm., Col.	178-79	3°56′N	68°34′W
Uvalde, Tx., U.S. (ū-văl′dē)	132-33	29°13′N	99°47′W
Uvira, D.R.C. (o͞o-vē′rä)	289	3°23′S	29°09′E
Uvs Lake, lk., Asia	240-41	50°20′N	92°45′E
Uvs nuur, lk., Asia *see* Uvs Lake	240-41	50°20′N	92°45′E
Uwajima, Japan (o͞o-wä′jē-mä)	265	33°13′N	132°34′E
Uwayl, S. Sudan	284-85	8°46′N	27°24′E
'Uwaynāt, Jabal al-, mtn., Afr.	280-81	21°53′N	25°02′E
Uxmal, hist., Mex. (o͞o′x-mä′l)	160	20°22′N	89°46′W
Uyuni, Bol. (o͞o-yo͞o′nē)	182-83	20°28′S	66°50′W
Uyuni, Salar de, pl., Bol. (sä-lär-dĕ o͞o-yo͞o′nē)	182-83	20°17′S	68°07′W
Uzbekistan, nation, Asia (o͞oz-bĕk′-ĕ--stăn′)	252-53	41°0′N	64°00′E
Ŭzbekiston, nation, Asia *see* Uzbekistan	252-53	41°0′N	64°00′E
Uzh, stm., Ukr. (ōzh)	218-19	51°15′N	30°15′E
Uzhhorod, Ukr.	210-11	48°37′N	22°19′E
Užice, Serb. (o͞o′zhē-tsĕ)	216-17	43°52′N	19°51′E
Uzlovaya, Russia	218-19	53°59′N	38°11′E

V

Feature (Pronunciation)	Page	Lat.	Long.
Vaal, stm., S. Afr. (väl)	286-87	29°04′S	23°38′E
Vaasa, Fin. (vä′sȧ)	200-01	63°06′N	21°37′E
Vache, Île à, i., Haiti	154-55	18°04′N	73°38′W
Vadodara, India	254-55	22°18′N	73°11′E
Vaduz, nat. cap., Liech. (vä′dŏts)	210-11	47°09′N	9°32′E
Vaga, stm., Russia (va′gȧ)	202-03	62°49′N	42°53′E
Vágar, i., Far. Is.	206b	62°05′N	7°17′W
Vaghena Island, i., Sol. Is.	306e	7°26′S	157°46′E
Váh, stm., Slvk. (väκ)	210-11	47°45′N	18°09′E
Vaitupu, i., Tuvalu	304-05	7°28′S	178°41′E
Vakh, stm., Russia (väκ)	236-37	60°49′N	76°48′E
Vākhān, hist. reg., Afg.	252-53	37°0′N	73°00′E
Vakhsh, stm., Taj.	252-53	37°07′N	68°19′E
Valcheta, Arg.	185	40°42′S	66°09′W
Valdai Hills, hills, Russia (vȧl-dī′ hĭlz) *see* Valdayskaya Vozvyshennost'	218-19	57°0′N	33°30′E
Valday, Russia (vȧl-dī′)	218-19	57°59′N	33°15′E
Valdayskaya Vozvyshennost', hills, Russia *see* Valdai Hills	218-19	57°0′N	33°30′E
Valdepeñas, Spain (väl-då-pān′yäs)	214-15	38°46′N	3°23′W
Valdés, Península, pen., Arg. (pĕ-nē′n-so͞o-lä vȧl-dĕ′s)	185	42°30′S	64°00′W
Valdez, Ak., U.S. (văl′dĕz)	136	61°08′N	146°20′W
Valdivia, Chile (väl-dē′vä)	185	39°49′S	73°13′W
Val-d'Or, Qc., Can.	146-47	48°06′N	77°46′W
Valdosta, Ga., U.S. (văl-dŏs′tȧ)	134-35	30°50′N	83°16′W
Vale, Or., U.S. (vāl)	122-23	43°59′N	117°15′W
Valença, Braz. (vä-lĕN′sȧ)	180-81	13°22′S	39°05′W
Valença, Braz. (vä-lĕN′sȧ)	186	22°15′S	43°42′W
Valence, Fr.	212-13	44°56′N	4°54′E
València, Spain	214-15	39°28′N	0°22′W
Valencia, Spain *see* València	214-15	39°28′N	0°22′W
Valencia, Ven. (vä-lĕn′syä)	188b	10°11′N	68°00′W
Valenciennes, Fr. (vȧ-läN-syĕn′)	212-13	50°21′N	3°31′E
Valentine, Ne., U.S. (vȧ läN-tė-nyē′)	124-25	42°53′N	100°33′W
Valera, Ven. (vä-lĕ′rä)	178-79	9°19′N	70°37′W
Valga, Est. (väl′gȧ)	208-09	57°47′N	26°03′E
Valjevo, Serb. (väl′yå-vŏ)	216-17	44°16′N	19°54′E
Valladolid, Mex. (väl-yä-dhŏ-lēdh′)	160	20°41′N	88°12′W
Valladolid, Spain (väl-yä-dhŏ-lēdh′)	214-15	41°39′N	4°43′W
Valle de Guanape, Ven. (vä′l-yĕ-dĕ-gwä-nä′pĕ)	188b	9°54′N	65°41′W
Valle de la Pascua, Ven. (väl′yā dā lä-pä′s-ko͞oä)	188b	9°13′N	66°00′W
Valle de Santiago, Mex. (väl′yā dā sän-tē-ä′gŏ)	158-59	20°24′N	101°12′W
Valledupar, Col. (väl′yā-do͞o pär′)	178-79	10°28′N	73°15′W
Vallegrande, Bol. (väl′yā grän′dā)	182-83	18°29′S	64°06′W
Vallejo, Ca., U.S.	128-29	38°07′N	122°15′W
Vallenar, Chile (väl-yå-när′)	182-83	28°34′S	70°46′W
Valletta, nat. cap., Malta (väl-lĕt′ä)	216b	35°54′N	14°31′E
Valley City, N.D., U.S. (văl′ė sĭ′tĭ)	124-25	46°56′N	98°00′W
Valleyview, Ab., Can.	142-43	55°05′N	117°17′W
Vallimanca, Arroyo, stm., Arg. (är-rŏ′yŏ väl-yē-mä′n-kä)	187	35°44′S	60°08′W
Valls, Spain (väls)	214-15	41°17′N	1°15′E
Valmiera, Lat. (vȧl′myĕ-rȧ)	208-09	57°32′N	25°26′E
Valognes, Fr. (vȧ-lôn′y′)	212-13	49°31′N	1°28′W
Valona, Alb. *see* Vlorë	216-17	40°29′N	19°30′E
Vālpārai, India	256	10°19′N	76°54′E
Valparaíso, Chile (väl′pä-rä-ē′sŏ)	188e	33°03′S	71°37′W
Valparaíso, Mex.	158-59	22°46′N	103°35′W
Valparaiso, Fl., U.S. (văl-pȧ-rā′zō)	134-35	30°31′N	86°30′W
Valparaiso, In., U.S. (văl-pȧ-rā′zō)	126-27	41°28′N	87°03′W
Valréas, Fr. (väl-rȧ-ä′)	212-13	44°23′N	4°59′E
Vals, Tanjung, c., Indon.	302	8°24′S	137°38′E
Valuyki, Russia (vȧ-lŏ-ē′kė)	218-19	50°12′N	38°08′E
Valverde del Camino, Spain (väl-vĕr-dĕ-dĕl-kä-mē′nō)	214-15	37°34′N	6°45′W
Van, Tur. (vàn)	245	38°30′N	43°24′E
Van, Lake, lk., Tur. *see* Van Gölü	245	38°33′N	42°46′E
Vanadzor, Arm.	245	40°48′N	44°29′E
Vanavara, Russia	236-37	60°21′N	102°16′E
Van Buren, Ar., U.S. (văn bū′rĕn)	130-31	35°27′N	94°22′W
Van Buren, Me., U.S. (văn bū′rĕn)	127a	47°10′N	67°57′W
Vanceburg, Ky., U.S. (văns′bûrg)	126-27	38°36′N	83°19′W
Vancouver, B.C., Can. (văn-ko͞o′vẽr)	142-43	49°17′N	123°07′W
Vancouver, Wa., U.S. (văn-ko͞o′vẽr)	122-23	45°38′N	122°39′W
Vancouver Island, i., B.C., Can. (văn-ko͞o′vẽr ī′lȧnd)	140-41	49°45′N	126°00′W
Vancouver Island Ranges, mts., B.C., Can. (văn-ko͞o′vẽr ī′lȧnd rānjĕz)	142-43	49°25′N	125°25′W
Vandalia, Il., U.S. (văn-dā′lĭ-ȧ)	126-27	38°58′N	89°06′W
Vandalia, Mo., U.S. (văn-dā′lĭ-ȧ)	130-31	39°19′N	91°30′W
Vandalia, Oh., U.S. (văn-dā′lĭ-ȧ)	126-27	39°54′N	84°12′W
Vanderbijlpark, S. Afr.	292c	26°42′S	27°50′E
Vanderhoof, B.C., Can.	142-43	54°01′N	124°06′W
Vanderlin Island, i., Austl.	296-97	15°44′S	137°02′E
Van Diemen Gulf, b., Austl. (văn dē′mĕn gŭlf)	296-97	11°50′S	132°00′E
Vanegas, Mex. (vä-nĕ′gäs)	158-59	23°53′N	100°56′W
Vänern, lk., Swe.	208-09	58°55′N	13°30′E
Vänersborg, Swe. (vĕ′nĕrs-bôr′)	208-09	58°23′N	12°19′E
Vangaindrano, Madag.	286-87	23°21′S	47°36′E
Van Gölü, lk., Tur.	245	38°33′N	42°46′E
Vanikolo, i., Sol. Is.	296-97	11°37′S	166°52′E
Vanimo, Pap. N. Gui.	302	2°44′S	141°20′E
Vankarem, Russia	136	67°50′N	175°50′W
Van Lear, Ky., U.S. (văn lēr′)	126-27	37°47′N	82°48′W
Vannes, Fr. (vȧn)	212-13	47°39′N	2°46′W
Van Rees, Pegunungan, mts., Indon.	302	2°35′S	138°15′E
Vanua Balavu, i., Fiji	306f	17°14′S	178°57′W
Vanua Lava, i., Vanuatu	306g	13°45′S	167°28′E
Vanua Levu, i., Fiji	306f	16°33′S	179°15′E
Vanuatu, nation, Oc. (vä-no͞o-ä′-to͞o)	306g	16°0′S	167°00′E
Van Wert, Oh., U.S. (văn wûrt′)	126-27	40°52′N	84°35′W
Vārānasi, India	254-55	25°20′N	82°59′E
Varaždin, Cro. (vä′räzh′dĭn)	216-17	46°18′N	16°20′E
Varberg, Swe. (vär′bĕrg)	208-09	57°06′N	12°16′E
Vardar, stm., Eur. (vär′där) *see* Axiós	216-17	40°31′N	22°43′E
Vardø, Nor.	200-01	70°21′N	31°01′E
Varėna, Lith. (vä-rä′nȧ)	210-11	54°13′N	24°35′E
Vareš, Bos. (vä′rĕsh)	216-17	44°09′N	18°19′E
Varese, Italy (vä-rā′så)	216-17	45°49′N	8°50′E
Varginha, Braz. (vär-zhė′n-yä)	186	21°34′S	45°26′W
Varkaus, Fin. (vär′kous)	200-01	62°19′N	27°54′E
Varna, Blg. (vär′nȧ)	216-17	43°13′N	27°54′E
Värnamo, Swe. (vĕr′nȧ-mŏ)	208-09	57°11′N	14°03′E
Vasa, Fin. *see* Vaasa	200-01	63°06′N	21°37′E
Vaslui, Rom. (vȧs-lo͞o′ė)	218-19	46°38′N	27°45′E
Vassar, Mi., U.S. (văs′ẽr)	126-27	43°22′N	83°35′W
Västerås, Swe. (vĕs′tĕr ôs)	208-09	59°37′N	16°33′E
Västervik, Swe. (vĕs′tĕr-vēk)	208-09	57°45′N	16°39′E
Vasto, Italy (väs′tô)	216-17	42°07′N	14°43′E
Vasyugan, stm., Russia (vȧs-yo͞o-gȧn′)	236-37	59°07′N	80°46′E
Vasyugan'ye, sw., Russia	236-37	58°0′N	77°00′E
Vatican City, nation, Eur. (văt′ĭkăn sĭ′tė)	216-17	41°54′N	12°27′E
Vaticano, Città del, nation, Eur. *see* Vatican City	216-17	41°54′N	12°27′E
Vatnajökull, ice, Ice. (vät′nȧ-yû-kŏl)	206a	64°25′N	16°50′W
Vatra Dornei, Rom. (vät′rȧ dôr′nă′)	210-11	47°21′N	25°22′E
Vättern, lk., Swe.	208-09	58°24′N	14°36′E
Vatu-i-ra Channel, strt., Fiji	306f	17°17′S	178°31′E
Vaughn, N.M., U.S.	130-31	34°36′N	105°15′W
Vaupés, stm., S.A. (vȧ′o͞o-pĕ′s)	178-79	0°02′N	67°15′W
Växjö, Swe. (vĕks′shû)	208-09	56°53′N	14°49′E
Vaygach, Ostrov, i., Russia (ôs-trôf′ vī-gȧch′)	202-03	70°0′N	59°30′E
Vedea, stm., Rom. (vå′dyȧ)	216-17	43°43′N	25°33′E
Vedia, Arg. (vĕ′dyä)	187	34°30′S	61°33′W
Vega, i., Nor.	200-01	65°39′N	11°50′E
Vegreville, Ab., Can.	142-43	53°30′N	112°03′W
Vejle, Den. (vī′lĕ)	208-09	55°42′N	9°32′E
Velebit, mts., Cro. (vā′lĕ-bēt)	216-17	44°38′N	15°03′E
Vélez-Málaga, Spain (vā′lāth-mä′lä-gä)	214-15	36°47′N	4°06′W
Velhas, stm., Braz.	186	17°13′S	44°49′W
Velikaya, stm., Russia (vå-lē′kȧ-yȧ)	208-09	57°52′N	28°09′E
Velikaya, stm., Russia (vå-lē′kȧ-yȧ)	236-37	64°35′N	176°12′E
Veliki Bečkerek, Serb. *see* Zhovti Vody	216-17	45°23′N	20°24′E
Velikiye Luki, Russia (vyē-lē′-kyĕ lo͞o′ke)	208-09	56°20′N	30°33′E
Velikiy Ustyug, Russia (vå-lē′kĭ o͞os-tyŏg′)	202-03	60°45′N	46°19′E
Veliko Tŭrnovo, Blg.	216-17	43°04′N	25°38′E
Velizh, Russia (vā′lĕzh)	208-09	55°36′N	31°12′E
Vella Lavella, i., Sol. Is.	306e	7°45′S	156°40′E
Velletri, Italy (vĕl-lā′trē)	216-17	41°41′N	12°46′E
Vellore, India (vĕl-lōr′)	256	12°55′N	79°08′E
Vel'sk, Russia (vĕlsk)	202-03	61°04′N	42°06′E
Venadillo, Col. (vĕ-nä-dē′l-yŏ)	188c	4°43′N	74°55′W
Venado Tuerto, Arg. (vĕ-nä′dō-to͞oĕ′r-tō)	187	33°45′S	61°58′W
Vendôme, Fr. (väN-dōm′)	212-13	47°47′N	1°04′E
Venëv, Russia (vĕn-ĕf′)	218-19	54°21′N	38°16′E
Venezia, Italy *see* Venice	216-17	45°26′N	12°20′E
Venezuela, nation, S.A. (vĕn-ē-zwē′lȧ)	172	8°0′N	66°00′W
Venezuela, Golfo de, b., S.A. (gôl-fô-dĕ vĕn-ĕ-zwē′lȧ)	178-79	11°30′N	71°00′W

n-sing; ŋ-baŋk; N-nasalized n; nŏd; cŏmmit; ōld; ôbey; ôrder; oi-boil; fo͞od; ŏ-as oo in foot; ou-out; s-soft; sh-dish; th-thin; pūre; ůnite; ûrn; stŭd; circŭs; ü-as in French tu; ′-indeterminate vowel.

420 Venezuela, Gulf of – Virudunagar PRONOUNCING INDEX

Feature (Pronunciation)	Page	Lat.	Long.
Venezuela, Gulf of, b., S.A. (gŭlf ŭv vĕn-ē-zwē´l*ȧ*) see Venezuela, Golfo de	**178-79**	11°30′N	71°00′W
Venice, Italy (vĕn´ĭs)	**216-17**	45°26′N	12°20′E
Venice, Fl., U.S. (vĕn´ĭs)	**135a**	27°06′N	82°27′W
Venta, stm., Eur. (vĕn´tȧ)	**208-09**	57°24′N	21°34′E
Ventersburg, S. Afr. (vĕn-tērs´bûrg)	**292c**	28°06′S	27°09′E
Ventersdorp, S. Afr. (vĕn-tērs´dôrp)	**292c**	26°19′S	26°51′E
Ventimiglia, Italy (vĕn-tē-mēl´yä)	**216-17**	43°48′N	7°36′E
Ventspils, Lat. (vĕnt´spēls)	**208-09**	57°24′N	21°35′E
Ventuari, stm., Ven. (vĕn-tōōȧ´rē)	**178-79**	3°58′N	67°03′W
Ventura, Ca., U.S. (vĕn-tōō´r*ȧ*)	**128-29**	34°17′N	119°17′W
Venustiano Carranza, Mex. (vĕ-nōōs-tyä´nō-kär-rä´n-zä)	**158-59**	16°22′N	92°34′W
Venustiano Carranza, Mex. (vĕ-nōōs-tyä´nō-kär-rä´n-zä)	**158-59**	19°45′N	103°46′W
Vera, Arg. (vĕ-rä)	**187**	29°28′S	60°13′W
Vera, Spain (vā´rä)	**214-15**	37°15′N	1°52′W
Veracruz, Mex.	**158-59**	19°12′N	96°08′W
Veracruz, state, Mex. (vā-rä-krōōz´)	**158-59**	19°20′N	96°40′W
Verāval, India (vĕr´vŭ-väl)	**254-55**	20°55′N	70°22′E
Vercelli, Italy (vĕr-chĕl´lē)	**216-17**	45°20′N	8°25′E
Verde, stm., Braz.	**182-83**	21°12′S	51°53′W
Verde, Cape, c., Sen. see Vert, Cap	**282-83**	14°44′N	17°30′W
Verden, Ger. (fĕr´dĕn)	**210-11**	52°55′N	9°14′E
Verdun-sur-Meuse, Fr.	**212-13**	49°10′N	5°23′E
Vereeniging, S. Afr. (vĕ-rā´nĭ-gĭng)	**292c**	26°40′S	27°57′E
Vereshchagino, Russia	**202-03**	58°05′N	54°39′E
Vereya, Russia (vĕ-rå´yȧ)	**218-19**	55°21′N	36°12′E
Verín, Spain (vå-rēn´)	**214-15**	41°57′N	7°26′W
Verkhnetulomskoye Vodokhranilishche, res., Russia	**202-03**	68°30′N	31°05′E
Verkhneudinsk, Russia see Ulan-Ude	**240-41**	51°50′N	107°36′E
Verkhniy Baskunchak, Russia	**202-03**	48°14′N	46°43′E
Verkhnyaya Inta, Russia	**202-03**	65°59′N	60°20′E
Verkhoyansk, Russia (vyĕr-kô-yänsk´)	**236-37**	67°32′N	133°25′E
Verkhoyanskiy Khrebet, mts., Russia	**236-37**	67°0′N	129°00′E
Verkhoyansk Mountains, mts., Russia (vyĕr-kô-yänsk´ moun´tĭnz) see Verkhoyanskiy Khrebet	**236-37**	67°0′N	129°00′E
Vermilion, Ab., Can. (vẽr-mĭl´y*ŭn*)	**142-43**	53°21′N	110°51′W
Vermilion, stm., Ab., Can. (vẽr-mĭl´y*ŭn*)	**142-43**	53°39′N	110°20′W
Vermilion Bay, b., La., U.S. (vẽr-mĭl´y*ŭn* bā)	**134-35**	29°40′N	92°00′W
Vermilion Lake, lk., Mn., U.S. (vẽr-mĭl´y*ŭn* lāk)	**124-25**	47°53′N	92°25′W
Vermillion, S.D., U.S. (vẽr-mĭl´y*ŭn*)	**124-25**	42°47′N	96°56′W
Vermillion, stm., S.D., U.S. (vẽr-mĭl´y*ŭn*)	**124-25**	42°44′N	96°53′W
Vermont, state, U.S. (vẽr-mŏnt´)	**118-19**	43°50′N	72°45′W
Vernal, Ut., U.S. (vûr´n*ă*l)	**122-23**	40°28′N	109°33′W
Vernon, B.C., Can. (vẽr-nôɴ´)	**142-43**	50°16′N	119°16′W
Vernon, Tx., U.S. (vûr´n*ŭn*)	**130-31**	34°10′N	99°17′W
Vero Beach, Fl., U.S. (vē´rō bēch)	**135a**	27°38′N	80°24′W
Verona, Italy (vā-rō´nä)	**216-17**	45°27′N	10°60′E
Versailles, Fr. (vĕr-sī´y′)	**212-13**	48°48′N	2°08′E
Versailles, In., U.S. (vẽr-sālz´)	**126-27**	39°04′N	85°15′W
Versailles, Ky., U.S. (vẽr-sālz´)	**126-27**	38°03′N	84°44′W
Versec, Serb. see Vršac	**216-17**	45°07′N	21°18′E
Vershino-Shakhtaminskiy, Russia	**260-61**	51°18′N	117°52′E
Vert, Cap, c., Sen.	**282-83**	14°44′N	17°30′W
Verviers, Bel. (vĕr-vyā´)	**206-07**	50°35′N	5°52′E
Vesoul, Fr. (vē-sōōl´)	**212-13**	47°37′N	6°10′E
Vesterålen, is., Nor. (vĕs´tẽr ô´lĕn)	**200-01**	68°40′N	15°33′E
Vesterhavet, s., Eur. see North Sea	**200-01**	56°0′N	3°00′E
Vestfjorden, b., Nor.	**200-01**	68°08′N	15°00′E
Vestmannaeyjar, Ice. (vĕst´män-ä-ā´yär)	**206a**	63°26′N	20°17′W
Vestvågøya, i., Nor.	**200-01**	68°13′N	13°42′E
Vesuvio, vol., Italy (vĕ-sōō´vyä) see Vesuvius	**216-17**	40°49′N	14°26′E
Vesuvius, vol., Italy (vĕ-sōō´vy-*ŭs*)	**216-17**	40°49′N	14°26′E
Ves'yegonsk, Russia (vĕ-syĕ-gônsk´)	**218-19**	58°40′N	37°16′E
Veszprém, Hung. (vĕs´prām)	**210-11**	47°06′N	17°54′E
Vet, stm., S. Afr. (vĕt)	**286-87**	27°41′S	25°39′E
Vetlanda, Swe. (vĕt-län´dä)	**208-09**	57°26′N	15°05′E
Vetluga, Russia (vyĕt-lōō´gȧ)	**202-03**	57°51′N	45°47′E
Vevay, In., U.S. (vē´vā)	**126-27**	38°45′N	85°04′W
Vezirköprü, Tur.	**247**	41°09′N	35°27′E
Viacha, Bol. (vēȧ´chȧ)	**182-83**	16°39′S	68°18′W
Vian, Ok., U.S. (vī´*ă*n)	**130-31**	35°30′N	94°58′W
Viana, Braz. (vē-ä´nä)	**180-81**	3°13′S	45°01′W
Viana do Castelo, Port. (vē-ä´n*ȧ* dò käs-tā´lò)	**214-15**	41°42′N	8°50′W
Viangchan, nat. cap., Laos see Vientiane	**266-67**	17°57′N	102°37′E
Viareggio, Italy (vē-ä-rĕd´jô)	**216-17**	43°53′N	10°15′E
Viborg, Den. (vē´bôr)	**208-09**	56°27′N	9°25′E
Viborg, Russia	**208-09**	60°43′N	28°46′E
Vibo Valentia, Italy (vē´bô-vä-lĕ´n-tyä)	**216-17**	38°40′N	16°06′E
Vicebsk, Bela.	**208-09**	55°12′N	30°12′E
Vicente Guerrero, Presa, res., Mex.	**158-59**	23°57′N	98°46′W
Vicenza, Italy (vē-chĕnt´sä)	**216-17**	45°33′N	11°33′E
Vichada, stm., Col.	**178-79**	4°56′N	67°50′W
Vichadero, Ur.	**187**	31°48′S	54°42′W
Vichuga, Russia (vē-chōō´gȧ)	**202-03**	57°13′N	41°55′E
Vichy, Fr. (vē-shē´)	**212-13**	46°08′N	3°26′E
Vicksburg, Ms., U.S. (vĭks´bûrg)	**134-35**	32°22′N	90°52′W
Viçosa, Braz. (vē-sô´sä)	**188d**	9°24′S	36°14′W
Viçosa, Braz. (vē-sô´sä)	**186**	20°46′S	42°52′W
Victor Harbor, Austl.	**301**	35°33′S	138°37′E
Victoria, Arg. (vēk-tô´rēä)	**187**	32°36′S	60°09′W
Victoria, Braz. see Vitória	**186**	20°19′S	40°21′W
Victoria, B.C., Can.	**142-43**	48°26′N	123°22′W
Victoria, Chile (vēk-tô-rēä)	**185**	38°14′S	72°20′W
Victoria, China see Hong Kong	**258-59**	22°16′N	114°10′E
Victoria, Malay. see Labuan	**268-69**	5°17′N	115°15′E
Victoria, Tx., U.S. (vĭk-tō´rĭ-*ȧ*)	**132-33**	28°48′N	96°60′W
Victoria, Va., U.S. (vĭk-tō´rĭ-*ȧ*)	**134-35**	36°59′N	78°14′W
Victoria, state, Austl.	**301**	38°0′S	145°00′E
Victoria, stm., Austl.	**296-97**	15°07′S	129°40′E
Victoria, Lake, lk., Afr. (lāk vĭk-tō´rĭ-*ȧ*)	**289**	1°0′S	33°00′E
Victoria, Mount, mtn., Pap. N. Gui.	**302**	8°54′S	147°32′E
Victoria Falls, wtfl., Afr. (vĭk-tō´rĭ-*ȧ* fôlz)	**286-87**	17°55′S	25°51′E
Victoria Falls National Park, n.p., Zimb.	**286-87**	17°55′S	25°40′E
Victoria Island, i., Can. (vĭk-tō´rĭ-*ȧ* ī´l*ă*nd)	**95**	71°0′N	110°00′W
Victoria Lake, res., Nf./L., Can. (vĭk-tō´rĭ-*ȧ* lāk)	**148-49**	48°19′N	57°28′W
Victoria Land, reg., Ant. (vĭk-tō´rĭ-*ȧ* lănd)	**313**	75°0′S	163°00′E
Victoria Nile, stm., Ug. (vĭk-tō´rĭ-*ȧ* nīl)	**289**	2°14′N	31°26′E
Victoria Peak, mtn., Belize (vēk-tōrĭ´*ȧ* pēk)	**160**	16°48′N	88°37′W
Victoria Peak, mtn., B.C., Can. (vĭk-tō´rĭ-*ȧ* pēk)	**142-43**	50°03′N	126°06′W
Victoriaville, Qc., Can. (vĭk-tō´rĭ-*ȧ*-vĭl)	**146-47**	46°03′N	71°57′W
Victoria West, S. Afr.	**286-87**	31°24′S	23°07′E
Vicuña Mackenna, Arg.	**187**	33°54′S	64°24′W
Vidalia, Ga., U.S. (vĭ-dā´lĭ-*ȧ*)	**134-35**	32°13′N	82°25′W
Vidalia, La., U.S. (vĭ-dā´lĭ-*ȧ*)	**134-35**	31°33′N	91°26′W
Vidin, Blg. (vĭ´dĭn)	**216-17**	43°60′N	22°53′E
Vidisha, India	**254-55**	23°31′N	77°49′E
Vidzy, Bela. (vĕ´dzĭ)	**208-09**	55°25′N	26°38′E
Viedma, Arg. (vyād´mä)	**185**	40°49′S	62°60′W
Viedma, Lago, lk., Arg.	**185**	49°35′S	72°35′W
Vienna, Ga., U.S. (vē-ĕn´*ȧ*)	**134-35**	32°05′N	83°47′W
Vienna, Il., U.S. (vē-ĕn´*ȧ*)	**126-27**	37°25′N	88°54′W
Vienna, W.V., U.S. (vē-ĕn´*ȧ*)	**126-27**	39°19′N	81°33′W
Vienna, nat. cap., Aus. (vē-ĕn´*ȧ*)	**210-11**	48°13′N	16°20′E
Vienne, Fr. (vyĕn´)	**212-13**	45°32′N	4°52′E
Vientiane, nat. cap., Laos (vyĕn´tän)	**266-67**	17°57′N	102°37′E
Vieques, Isla de, i., P.R. (ē´s-lä-dĕ-vyā´kås)	**154a**	18°08′N	65°25′W
Vierfontein, S. Afr. (vēr´fōn-tān)	**292c**	27°05′S	26°45′E
Vierzon, Fr. (vyâr-zôɴ´)	**212-13**	47°13′N	2°05′E
Viesca, Mex. (vē-ās´kä)	**132-33**	25°21′N	102°48′W
Vieste, Italy (vyĕs´tā)	**216-17**	41°53′N	16°11′E
Viet Nam, nation, Asia (vyĕt´näm´)	**224-25**	16°0′N	108°00′E
Vigan, Phil. (vēgän)	**270**	17°35′N	120°23′E
Vigevano, Italy (vē-jå-vä´nô)	**216-17**	45°19′N	8°52′E
Vigo, Spain (vē´gō)	**214-15**	42°14′N	8°43′W
Vihti, Fin. (vē´tĭ)	**208-09**	60°25′N	24°19′E
Vijayawāda, India	**256**	16°31′N	80°37′E
Vikna, i., Nor.	**200-01**	64°57′N	10°58′E
Vila Cabral, Moz. see Lichinga	**286-87**	13°17′S	35°15′E
Vila Coutinho, Moz. see Ulónguè	**286-87**	14°37′S	34°19′E
Vila do Conde, Port. (vē´lȧ dò kôn´d*ĕ*)	**214-15**	41°21′N	8°45′W
Vila Franca de Xira, Port. (vē´l*ȧ*-frän´kȧ dā shē´rȧ)	**214-15**	38°56′N	8°60′W
Vila Gouveia, Moz. see Cantandica	**286-87**	18°02′S	33°08′E
Vilanandro, Tanjona, c., Madag.	**286-87**	16°12′S	44°28′E
Viļāni, Lat. (vē´lá-nĭ)	**208-09**	56°34′N	26°57′E
Vilankulo, Moz.	**286-87**	21°60′S	35°19′E
Vila Nova de Gaia, Port. (vē´l*ȧ* nō´vȧ dā gä´yä)	**214-15**	41°08′N	8°37′W
Vila Pery, Moz. see Chimoio	**286-87**	19°09′S	33°30′E
Vila Real, Port. (vē´lȧ rā-äl´)	**214-15**	41°18′N	7°45′W
Vila Velha, Braz.	**186**	20°20′S	40°17′W
Vila Viçosa, Port. (vē´l*ȧ*-vē-sō´zȧ)	**214-15**	38°47′N	8°13′W
Vilejka, Bela.	**208-09**	54°30′N	26°55′E
Vileyka, Bela. (vē-lā´ē-kȧ) see Vilejka	**208-09**	54°30′N	26°55′E
Vilhelmina, Swe.	**200-01**	64°38′N	16°39′E
Vilhena, Braz.	**180-81**	12°43′S	60°07′W
Viljandi, Est. (vēl´yän-dē)	**208-09**	58°22′N	25°36′E
Vilkaviškis, Lith. (vēl-kȧ-vēsh´kēs)	**208-09**	54°39′N	23°02′E
Villa Ángela, Arg. (vē´l-yä á´n-kĕ-lä)	**187**	27°35′S	60°43′W
Villa Bella, Bol. (vē´l-yä-bĕ´l-yä)	**180-81**	10°26′S	65°24′W
Villacañas, Spain (vēl-yä-kän´yäs)	**214-15**	39°37′N	3°20′W
Villach, Aus. (fē´läĸ)	**210-11**	46°36′N	13°50′E
Villacidro, Italy (vē-lä-chē´drô)	**216-17**	39°28′N	8°45′E
Villa Constitución, Arg. (vēl´yä-kōn-stē-tōō-syōn´)	**187**	33°14′S	60°20′W
Villa de Cura, Ven. (vēl´yä-ē-kōō´rä)	**188b**	10°02′N	67°29′W
Villa Dolores, Arg. (vēl´yä dō-lō´rās)	**182-83**	31°56′S	65°11′W
Villa Flores, Mex. (vēl´yä-flō´rās)	**158-59**	16°14′N	93°14′W
Villa Grove, Il., U.S. (vĭl´*ȧ* grōv´)	**126-27**	39°51′N	88°10′W
Villaguay, Arg. (vē´l-yä-gwī)	**187**	31°52′S	59°02′W
Villa Hayes, Para. (vēl´yä äyås)(hāz)	**182-83**	25°05′S	57°34′W
Villahermosa, Mex. (vēl´yä-ĕr-mō´sä)	**158-59**	17°59′N	92°55′W
Villa Hidalgo, Mex. (vēl´yäē-däl´gō)	**158-59**	21°40′N	102°36′W
Villaldama, Mex. (vēl-yäl-dä´mä)	**132-33**	26°30′N	100°25′W
Villa María, Arg. (vē´l-yä-mä-rē´ä)	**187**	32°25′S	63°14′W
Villa Mercedes, Arg. (vēl´yä-mĕr-sā´dås)	**185**	33°40′S	65°28′W
Villa Montes, Bol. (vē´l-yä-mô´n-tĕs)	**182-83**	21°15′S	63°29′W
Villanueva, Mex. (vēl´yä-noē´vä)	**158-59**	22°20′N	102°52′W
Villanueva de la Serena, Spain (vēl-yä-nwĕ´vä-dä lä sā-rā´nä)	**214-15**	38°58′N	5°48′W
Villa Ocampo, Arg.	**187**	28°29′S	59°21′W
Villa Ocampo, Mex. (vēl´yä-ô-käm´pō)	**132-33**	26°28′N	105°33′W
Villa Regina, Arg.	**185**	39°06′S	67°05′W
Villarrica, Para. (vēl-yä-rē´kä)	**182-83**	25°47′S	56°28′W
Villarrobledo, Spain (vēl-yär-rô-blā´dhō)	**214-15**	39°16′N	2°36′W
Villa Unión, Arg.	**187**	29°24′S	62°47′W
Villa Unión, Mex. (vēl´yä-ōō-nyōn´)	**158-59**	23°10′N	106°12′W
Villavicencio, Col. (vē´l-yä-vē-sĕ´n-syō)	**178-79**	4°10′N	73°38′W
Villazón, Bol. (vē´l-yä-zô´n)	**182-83**	22°05′S	65°35′W
Villena, Spain (vē-lyā´nȧ)	**214-15**	38°38′N	0°52′W
Villeneuve-sur-Lot, Fr. (vēl´nûv´sür-lō´)	**212-13**	44°25′N	0°42′E
Ville Platte, La., U.S. (vēl plȧt´)	**132-33**	30°41′N	92°16′W
Villers-Cotterêts, Fr. (vē-ār´kô-trā´)	**212-13**	49°15′N	3°06′E
Villeta, Col. (vē´l-yĕ´tȧ)	**188c**	4°60′N	74°30′W
Villeurbanne, Fr. (vēl-ûr-bän´)	**212-13**	45°45′N	4°52′E
Villiers, S. Afr. (vĭl´ĭ-ērs)	**292c**	27°02′S	28°36′E
Vilnius, nat. cap., Lith. (vĭl´nē-ōs)	**208-09**	54°40′N	25°17′E
Vilyuy, stm., Russia (vēl´yī)	**236-37**	64°23′N	126°26′E
Vilyuysk, Russia (vē-lyōō´ĭsk′)	**236-37**	63°45′N	121°37′E
Vilyuyskoye Vodokhranilishche, res., Russia	**236-37**	62°34′N	111°13′E
Vimmerby, Swe. (vĭm´ēr-bü)	**208-09**	57°40′N	15°52′E
Vina, stm., Camrn.	**282-83**	7°52′N	15°46′E
Viña del Mar, Chile (vē´nyä dĕl mär´)	**188e**	33°01′S	71°33′W
Vincennes, In., U.S. (vĭn-zĕnz´)	**126-27**	38°41′N	87°32′W
Vincennes Bay, b., Ant.	**313**	66°18′S	108°49′E
Vindhya Range, mts., India (vĭnd´yä rănj)	**254-55**	23°0′N	77°00′E
Vineland, N.J., U.S. (vīn´l*ă*nd)	**126-27**	39°29′N	75°01′W
Vinh, Viet. (vĕn´y′)	**266-67**	18°40′N	105°41′E
Vinh Long, Viet.	**266-67**	10°14′N	105°59′E
Vinita, Ok., U.S. (vĭ-nē´t*ȧ*)	**130-31**	36°38′N	95°09′W
Vinkovci, Cro. (vēn´kôv-tsē)	**216-17**	45°17′N	18°49′E
Vinnitsa, Ukr. see Vinnytsia	**218-19**	49°14′N	28°32′E
Vinnytsia, Ukr.	**218-19**	49°14′N	28°32′E
Vinson Massif, mtn., Ant. (vĭn´sŭn mă-sēf)	**313**	78°32′S	85°14′W
Vinton, Ia., U.S. (vĭn´tŭn)	**124-25**	42°10′N	92°01′W
Vinton, La., U.S. (vĭn´tŭn)	**132-33**	30°11′N	93°35′W
Virac, Phil. (vē-räk´)	**270**	13°36′N	124°14′E
Virden, Mb., Can. (vûr´d*ĕ*n)	**144-45**	49°50′N	100°55′W
Virgin Gorda, i., Br. Vir. Is.	**155b**	18°30′N	64°24′W
Virginia, S. Afr. (vẽr-jĭn´y*ȧ*)	**292c**	28°05′S	26°53′E
Virginia, Mn., U.S. (vẽr-jĭn´y*ȧ*)	**124-25**	47°31′N	92°32′W
Virginia, state, U.S. (vẽr-jĭn´y*ȧ*)	**118-19**	37°30′N	78°45′W
Virginia Beach, Va., U.S. (vẽr-jĭn´y*ȧ* bēch)	**134-35**	36°52′N	75°59′W
Virginia City, Nv., U.S.	**128-29**	39°19′N	119°39′W
Virgin Islands, dep., N.A. (vûr´jĭn ī´l*ă*ndz)	**152-53**	18°20′N	64°50′W
Viroqua, Wi., U.S. (vĭ-rō´kw*ȧ*)	**124-25**	43°34′N	90°53′W
Virovitica, Cro. (vē-rō-vē´tē-tsȧ)	**216-17**	45°50′N	17°24′E
Virrat, Fin. (vĭr´ät)	**208-09**	62°15′N	23°45′E
Virserum, Swe. (vĭr´sĕ-ròm)	**208-09**	57°19′N	15°36′E
Virudunagar, India	**256**	9°35′N	77°57′E

PRONOUNCING INDEX
Vis, Otok – Warm Baths 421

Feature (Pronunciation)	Page	Lat.	Long.
Vis, Otok, i., Cro.	216-17	43°02'N	16°11'E
Visalia, Ca., U.S. (vĭ-sā'lĭ-ȧ)	128-29	36°20'N	119°18'W
Visayan Sea, s., Phil.	270	11°35'N	123°51'E
Visby, Swe. (vĭs'bü)	208-09	57°38'N	18°19'E
Viscount Melville Sound, strt., Can.	95	74°10'N	108°00'W
Višegrad, Bos. (vē'shĕ-gräd)	216-17	43°47'N	19°18'E
Vishākhapatnam, India	256	17°43'N	83°19'E
Vishera, stm., Russia (vĭ'shĕ-rȧ)	202-03	59°54'N	56°26'E
Visoko, Bos. (vē'sô-kô)	216-17	43°59'N	18°11'E
Vistula, stm., Pol. (vĭs'tů-lȧ)	210-11	54°21'N	18°56'E
Vitarte, Peru	188a	12°02'S	76°56'W
Viterbo, Italy (vē-tĕr'bō)	216-17	42°25'N	12°06'E
Viti, nation, Oc. *see* Fiji	306f	18°0'S	178°00'E
Viti, nation, Oc. *see* Fiji	306f	18°0'S	178°00'E
Viti Levu, i., Fiji	306f	18°0'S	178°00'E
Vitim, stm., Russia (vē'tēm)	236-37	59°28'N	112°35'E
Vitória, Braz. (vē-tō'rē-ä)	180-81	2°53'S	52°00'W
Vitória, Braz.	186	20°19'S	40°21'W
Vitoria, Spain (vē-tô-ryä) *see* Gasteiz	214-15	42°51'N	2°40'W
Vitória da Conquista, Braz.	180-81	14°51'S	40°51'W
Vitry-le-François, Fr. (vē-trē'lĕ-fräN-swȧ')	212-13	48°44'N	4°36'E
Vivian, La., U.S. (vĭv'ĭ-ăn)	130-31	32°53'N	93°59'W
Vizcaya, Golfo de, b., Eur. *see* Biscay, Bay of	212-13	44°0'N	4°00'W
Vize, Ostrov, i., Russia	236-37	79°33'N	76°50'E
Vizianagaram, India	256	18°07'N	83°25'E
Vladikavkaz, Russia	245	43°03'N	44°39'E
Vladimir, Russia (vlȧ-dyē'mēr)	218-19	56°08'N	40°24'E
Vladivostok, Russia (vlȧ-dē-vôs-tôk')	260-61	43°08'N	131°56'E
Vlonë, Alb. *see* Vlorë	216-17	40°29'N	19°30'E
Vlora, Alb. *see* Vlorë	216-17	40°29'N	19°30'E
Vlorë, Alb.	216-17	40°29'N	19°30'E
Vogel Peak, mtn., Nig. *see* Dimlang	282-83	8°24'N	11°47'E
Voghera, Italy (vô-gā'rä)	216-17	44°60'N	9°01'E
Vohimena, Tanjona, c., Madag.	286-87	25°36'S	45°09'E
Voi, Kenya	284-85	3°23'S	38°34'E
Voinjama, Lib.	282-83	8°25'N	9°45'W
Voiron, Fr. (vwȧ-rôN')	212-13	45°22'N	5°35'E
Volcano Islands, is., Japan (vŏl-kā'nō ī'lăndz) *see* Kazan-rettō	304-05	25°0'N	141°00'E
Volga, stm., Russia (vôl'gä)	202-03	45°45'N	47°56'E
Volga Upland, plat., Russia (vôl'gä ŭp'lănd) *see* Privolzhskaya Vozvyshennost'	202-03	52°0'N	46°00'E
Volgodonsk, Russia	202-03	47°31'N	42°08'E
Volgograd, Russia (vŏl-gō-gră't)	202-03	48°44'N	44°25'E
Volgograd Reservoir, res., Russia (vŏl-gō-gră't rĕ'sēr-vwär) *see* Volgogradskoye Vodokhranilishche	202-03	50°18'N	45°49'E
Volgogradskoye Vodokhranilishche, res., Russia	202-03	50°18'N	45°49'E
Volkhov, Russia (vôl'kôf)	208-09	59°55'N	32°19'E
Volksrust, S. Afr.	292c	27°22'S	29°54'E
Vologda, Russia (vô'lôg-dȧ)	202-03	59°14'N	39°55'E
Volokolamsk, Russia (vô-lô-kô'lȧmsk)	218-19	56°02'N	35°58'E
Vólos, Grc.	216-17	39°22'N	22°57'E
Vol'sk, Russia (vôl'sk)	202-03	52°03'N	47°22'E
Volta, stm., Ghana (vôl'tȧ)	282-83	5°46'N	0°40'E
Volta Blanche, stm., Afr. (vôl'tȧ blänsh) *see* White Volta	282-83	8°57'N	1°10'W
Volta Lake, res., Ghana (vôl'tȧ lāk)	282-83	7°30'N	0°07'E
Volta Noire, stm., Afr. (vôl'tȧ nwär) *see* Black Volta	282-83	8°41'N	0°60'W
Volta Redonda, Braz. (vōl'tä-rā-dōn'dä)	186	22°32'S	44°07'W
Volzhsk, Russia	202-03	55°52'N	48°21'E
Volzhskiy, Russia	202-03	48°50'N	44°45'E
Vordingborg, Den. (vôr'dĭng-bôr)	210-11	55°01'N	11°55'E
Vóreioi Sporades, is., Grc.	216-17	39°15'N	23°55'E
Vorgashor, Russia	202-03	67°32'N	64°05'E
Vorkuta, Russia (vôr-kōō'tȧ)	202-03	67°29'N	64°03'E
Vormsi, i., Est. (vôrm'sĭ)	208-09	59°00'N	23°15'E
Vorona, stm., Russia (vô-rô'na)	202-03	51°21'N	42°02'E
Voronezh, Russia (vô-rô'nyĕzh)	218-19	51°40'N	39°10'E
Voronezh, stm., Russia (vô-rô'nyĕzh)	202-03	51°32'N	39°06'E
Voronya, stm., Russia (vô-rô'nyȧ)	202-03	69°00'N	35°42'E
Voroshilov, Russia *see* Ussuriysk	264	43°48'N	131°59'E
Voroshilovsk, Russia *see* Stavropol'	202-03	45°02'N	41°59'E
Võru, Est. (vô'rû)	208-09	57°50'N	27°01'E
Voskresensk, Russia (vôs-krē-sĕnsk')	218-19	55°19'N	38°42'E
Voss, Nor. (vôs)	208-09	60°38'N	6°26'E
Vostochno-Sibirskoye More, s., Russia *see* East Siberian Sea	236-37	74°0'N	166°00'E
Vostok, i., Kir.	304-05	10°06'S	152°23'W
Votkinsk, Russia (vôt-kēnsk')	202-03	57°03'N	53°59'E
Votuporanga, Braz.	182-83	20°26'S	49°58'W

Feature (Pronunciation)	Page	Lat.	Long.
Voyageurs National Park, n.p., Mn., U.S.	124-25	48°30'N	93°00'W
Voznesens'k, Ukr.	218-19	47°34'N	31°20'E
Vrangelya, Ostrov, i., Russia	236-37	71°14'N	179°21'W
Vranje, Serb. (vrän'yĕ)	216-17	42°34'N	21°55'E
Vratsa, Blg. (vrät'tsȧ)	216-17	43°12'N	23°34'E
Vrbas, Serb. (v'r'bäs)	216-17	45°34'N	19°39'E
Vrchlabí, Czech Rep. (v'r'chlä-bē)	210-11	50°37'N	15°37'E
Vrede, S. Afr. (vrĭ'dĕ)(vrēd)	292c	27°26'S	29°10'E
Vredefort, S. Afr. (vrĭ'dĕ-fôrt)(vrēd'fōrt)	292c	27°00'S	27°22'E
Vrindāvan, India	254-55	27°35'N	77°42'E
Vršac, Serb. (v'r'shäts)	216-17	45°07'N	21°18'E
Vryburg, S. Afr. (vrī'bûrg)	286-87	26°58'S	24°44'E
Vryheid, S. Afr. (vrī'hīt)	286-87	27°46'S	30°48'E
Vsetín, Czech Rep. (fsĕt'yēn)	210-11	49°20'N	18°00'E
Vukovar, Cro. (vô'kô-vär)	216-17	45°21'N	19°00'E
Vulcano, Isola, i., Italy (ē'-sō-lä-vōōl-kä'nô)	216-17	38°24'N	14°58'E
Vung Tau, Viet.	266-67	10°21'N	107°05'E
Vyatka, Russia *see* Kirov	202-03	58°36'N	49°40'E
Vyatka, stm., Russia (vyät'kȧ)	202-03	55°35'N	51°29'E
Vyatskiye Polyany, Russia	202-03	56°14'N	51°05'E
Vyazemskiy, Russia (vyȧ-zĕm'skĭ)	260-61	47°33'N	134°46'E
Vyazma, Russia (vyȧz'mȧ)	218-19	55°12'N	34°17'E
Vyazniki, Russia (vyȧz'nē-kē)	202-03	56°15'N	42°08'E
Vychegda, stm., Russia (vē'chĕg-dȧ)	202-03	61°17'N	46°37'E
Vygozero, Ozero, lk., Russia	202-03	63°35'N	34°42'E
Vym', stm., Russia (vwēm)	202-03	62°13'N	50°24'E
Vyritsa, Russia (vē'rĭ-tsȧ)	208-09	59°24'N	30°20'E
Vyshniy Volochëk, Russia (vêsh'nyĭ vôl-ô-chĕk')	218-19	57°35'N	34°34'E
Vyškov, Czech Rep. (vêsh'kôf)	210-11	49°17'N	16°60'E
Vysokogornyy, Russia	236-37	50°06'N	139°09'E
Vysokovsk, Russia (vĭ-sô'kôfsk)	218-19	56°19'N	36°34'E
Vytegra, Russia (vû'tĕg-rȧ)	202-03	61°00'N	36°28'E

W

Feature (Pronunciation)	Page	Lat.	Long.
Wa, Ghana	282-83	10°03'N	2°30'W
Wabana, Nf./L., Can.	148-49	47°39'N	52°57'W
Wabasca, stm., Ab., Can.	140-41	58°21'N	115°20'W
Wabasca-Desmarais, Ab., Can.	142-43	55°58'N	113°52'W
Wabash, In., U.S. (wô'băsh)	126-27	40°48'N	85°49'W
Wabash, stm., U.S. (wô'băsh)	120-21	37°48'N	88°01'W
Wabasha, Mn., U.S. (wä'bȧ-shô)	124-25	44°23'N	92°02'W
Wabē Gestro, stm., Eth.	284-85	4°17'N	42°03'E
Wabē Shebelē, stm., Afr.	284-85	0°10'N	42°46'E
Wabowden, Mb., Can. (wă-bō'd'n)	144-45	54°54'N	98°37'W
W.A.C. Bennett Dam, d., B.C., Can.	142-43	56°01'N	122°10'W
Waccasassa Bay, b., Fl., U.S. (wă-kȧ-să'sȧ bā)	134-35	29°06'N	82°52'W
Waco, Tx., U.S. (wā'kō)	132-33	31°33'N	97°09'W
Waco Mammoth National Monument, n.p., Tx., U.S.	132-33	31°36'N	97°11'W
Waddān, Libya	204-05	29°10'N	16°10'E
Waddeneilanden, is., Eur. *see* Frisian Islands	206-07	53°27'N	5°50'E
Waddington, Mount, mtn., B.C., Can. (mount wŏd'dĭng-tŭn)	142-43	51°22'N	125°16'W
Wadena, Sk., Can.	144-45	51°56'N	103°47'W
Wadena, Mn., U.S. (wŏ-dē'nȧ)	124-25	46°26'N	95°09'W
Wadesboro, N.C., U.S. (wādz'bŭr-ô)	134-35	34°58'N	80°05'W
Wādī Ḥalfā', Sudan	288	21°48'N	31°20'E
Wadley, Ga., U.S. (wŭd'lē)	134-35	32°52'N	82°24'W
Wad Madanī, Sudan (wäd mē-dä'nē)	288	14°23'N	33°31'E
Wadsworth, Nv., U.S. (wŏdz'wûrth)	128-29	39°38'N	119°17'W
Wafangdian, China	260-61	39°37'N	122°01'E
Wagadugu, nat. cap., Burkina *see* Ouagadougou	282-83	12°23'N	1°32'W
Wager Bay, b., Nu., Can. (wā'jēr bā)	140-41	65°26'N	88°40'W
Wagga Wagga, Austl. (wôg'ȧ wôg'ȧ)	301	35°07'S	147°21'E
Wagoner, Ok., U.S. (wăg'ŭn-ēr)	130-31	35°58'N	95°23'W
Wągrowiec, Pol. (vôn-grô'vyĕts)	210-11	52°49'N	17°13'E
Wāh Cantonment, Pak.	252-53	33°48'N	72°41'E
Wahoo, Ne., U.S. (wä-hōō')	124-25	41°13'N	96°37'W
Wahpeton, N.D., U.S. (wô'pē-tŭn)	124-25	46°16'N	96°36'W
Wahrān, Alg. *see* Oran	214-15	35°41'N	0°39'W
Wai'anae, Hi., U.S. (wä'ē-ȧ-nä'ā)	137a	21°27'N	158°11'W
Waigeo, Pulau, i., Indon. (pōō-lou wä-ē-gā'ô)	242-43	0°10'S	130°55'E
Waikabubak, Indon.	268-69	9°38'S	119°25'E
Waikato, stm., N.Z. (wä'ē-kä'to)	303	37°23'S	174°43'E

Feature (Pronunciation)	Page	Lat.	Long.
Waikerie, Austl. (wā'kēr-ē)	301	34°11'S	139°59'E
Wailuku, Hi., U.S. (wä'ē-lōō'kōō)	137a	20°54'N	156°30'W
Waimea, Hi., U.S. (wä-ē-mā'ä)	137a	21°58'N	159°40'W
Waimea, Hi., U.S. (wä-ē-mā'ä)	137a	20°02'N	155°40'W
Waimea, Hi., U.S. (wä-ē-mā'ä)	137a	20°02'N	155°40'W
Wainganga, stm., India (wä-ēn-gŭŋ'gä)	254-55	19°37'N	79°48'E
Waingapu, Indon.	268-69	9°40'S	120°16'E
Wainwright, Ab., Can. (wān-rīt)	142-43	52°51'N	110°51'W
Wainwright, Ak., U.S. (wān-rīt)	136	70°39'N	159°59'W
Waitsburg, Wa., U.S. (wāts'bûrg)	122-23	46°16'N	118°09'W
Wajima, Japan (wä'jē-mȧ)	265	37°24'N	136°54'E
Wajir, Kenya	284-85	1°45'N	40°04'E
Waka, Eth.	292d	7°10'N	37°21'E
Wakamatsu, Japan *see* Aizu-wakamatsu	265	37°30'N	139°56'E
Wakasa-wan, b., Japan (wä'kä-sä wän)	265	35°45'N	135°40'E
Wakatipu, Lake, lk., N.Z. (lāk wä-kä-tē'pōō)	303	45°05'S	168°34'E
Wakayama, Japan (wä-kä'yä-mä)	265	34°13'N	135°10'E
WaKeeney, Ks., U.S. (wô-kē'nē)	130-31	39°01'N	99°53'W
Wakefield, Ne., U.S. (wāk-fēld)	124-25	42°16'N	96°52'W
Wake Forest, N.C., U.S. (wāk fŏr'ĕst)	134-35	35°58'N	78°31'W
Wake Island, dep., Oc. (wāk ī'lȧnd)	304-05	19°17'N	166°36'E
Wakhān, hist. reg., Afg. *see* Vākhān	252-53	37°0'N	73°00'E
Wakkanai, Japan (wä'kä-nä'ē)	264	45°24'N	141°41'E
Wakkerstroom, S. Afr. (vȧk'ēr-strōm)(wăk'ēr-strōōm)	292c	27°20'S	30°08'E
Wałbrzych, Pol. (väl'bzhûk)	210-11	50°46'N	16°17'E
Waldenburg, Pol. *see* Wałbrzych	210-11	50°46'N	16°17'E
Waldorf, Md., U.S. (wăl'dōrf)	126-27	38°37'N	76°54'W
Wales, Ak., U.S. (wālz)	136	65°36'N	168°04'W
Wales, state, U.K. (wālz)	206-07	52°30'N	3°30'W
Wales Island, i., Nu., Can.	140-41	68°0'N	86°43'W
Walgett, Austl. (wôl'gĕt)	301	30°02'S	148°07'E
Walhalla, N.D., U.S. (wŭl-hăl'ȧ)	124-25	48°56'N	97°55'W
Walhalla, S.C., U.S. (wŭl-hăl'ȧ)	134-35	34°46'N	83°04'W
Walikale, D.R.C.	289	1°25'S	28°03'E
Walker, Mi., U.S. (wôk'ēr)	126-27	42°59'N	85°45'W
Walker, Mn., U.S. (wôk'ēr)	124-25	47°06'N	94°35'W
Walker Lake, lk., Mb., Can. (wôk'ēr lāk)	144-45	54°42'N	96°57'W
Walker Lake, lk., Nv., U.S.	128-29	38°44'N	118°43'W
Wallaceburg, On., Can.	146-47	42°36'N	82°23'W
Wallaroo, Austl. (wŏl-ȧ-rōō)	301	33°56'S	137°37'E
Walla Walla, Wa., U.S. (wŏl'ȧ wŏl'ȧ)	122-23	46°04'N	118°19'W
Wallis, Îles, is., Wal./F.	304-05	13°18'S	176°10'W
Wallis and Futuna, dep., Oc. (wŏl'ĭs ănd fōō-tōō'nȧ)	304-05	14°0'S	177°00'W
Wallis et Futuna, dep., Oc. *see* Wallis and Futuna	304-05	14°0'S	177°00'W
Wallowa Mountains, mts., Or., U.S. (wôl'ô-wa moun tinz)	122-23	45°16'N	117°21'W
Walnut Creek, stm., Ks., U.S. (wôl'nŭt krēk)	130-31	38°21'N	98°41'W
Walnut Ridge, Ar., U.S. (wôl'nŭt rĭj)	134-35	36°04'N	90°58'W
Walsall, Eng., U.K. (wôl-sôl)	206-07	52°36'N	1°59'W
Walsenburg, Co., U.S. (wôl'sĕn-bûrg)	130-31	37°37'N	104°47'W
Walters, Ok., U.S. (wôl'tērz)	130-31	34°21'N	98°19'W
Walvisbaai, Nmb. *see* Walvis Bay	286-87	22°57'S	14°31'E
Walvis Bay, Nmb. (wôl'vĭs bā)	286-87	22°57'S	14°31'E
Walworth, Wi., U.S. (wôl'wûrth)	126-27	42°32'N	88°37'W
Wamba, D.R.C.	289	2°09'N	28°00'E
Wamego, Ks., U.S. (wŏ-mē'gō)	130-31	39°12'N	96°18'W
Wami, stm., Tan. (wä'mē)	284-85	6°15'S	38°51'E
Wanfoxia, China	260-61	40°05'N	95°55'E
Wanganui, N.Z. (wôŋ'gȧ-nōō'ē)	303	39°56'S	175°02'E
Wangaratta, Austl. (wôŋ'gȧ-răt'ȧ)	301	36°22'S	146°19'E
Wangiwangi, Pulau, i., Indon.	268-69	5°20'S	123°35'E
Wangpan Yang, b., China	258-59	30°30'N	121°46'E
Wanneroo, Austl.	294-95	31°45'S	115°48'E
Wanxian, China (wän-shyĕn)	258-59	30°49'N	108°22'E
Wanzai, China (wän-dzī)	258-59	28°05'N	114°27'E
Wapakoneta, Oh., U.S. (wä'pȧ-kô-nĕt'ȧ)	126-27	40°34'N	84°12'W
Wapawekka Lake, lk., Sk., Can. (wŏ'pä-wĕ'kä lăk)	144-45	54°55'N	104°40'W
Wapello, Ia., U.S. (wŏ-pĕl'ō)	124-25	41°10'N	91°12'W
Wapiti, stm., Can.	142-43	55°08'N	118°18'W
Wapusk National Park, n.p., Can.	138-39	58°0'N	93°30'W
Warangal, India (wŭ'rȧŋ-gȧl)	256	18°00'N	79°35'E
Warburton Creek, stm., Austl.	301	27°59'S	137°25'E
Warden, S. Afr. (wôr'dĕn)	292c	27°51'S	28°58'E
Wardha, India (wŭr'dä)	254-55	20°45'N	78°37'E
Wardha, stm., India	254-55	19°36'N	79°47'E
Warialda, Austl.	301	29°33'S	150°35'E
Warmbad, S. Afr. *see* Bela-Bela	292c	24°53'S	28°19'E
Warm Baths, S. Afr. *see* Bela-Bela	292c	24°53'S	28°19'E

n-sing; ŋ-baŋk; N-nasalized n; nŏd; cŏmmit; ōld; ôbey; ôrder; oi-boil; fōōd; ŏŏ-as oo in foot; ou-out; s-soft; sh-dish; th-thin; pūre; ûnite; ûrn; stŭd; circŭs; ü-as in French tu; '-indeterminate vowel.

422 Warm Springs Indian Reservation – Wetaskiwin PRONOUNCING INDEX

Feature (Pronunciation)	Page	Lat.	Long.
Warm Springs Indian Reservation, ind. res., Or., U.S. (wôrm springz ĭn´dĭ-ăn rĕ-sĕr-vā´shĕn)	122-23	44°53'N	121°23'W
Warrego, stm., Austl. (wôr´ē-gō)	294-95	30°25'S	145°21'E
Warren, Ar., U.S. (wŏr´ĕn)	130-31	33°37'N	92°04'W
Warren, Mi., U.S. (wŏr´ĕn)	126-27	42°30'N	83°02'W
Warren, Mn., U.S. (wŏr´ĕn)	124-25	48°12'N	96°47'W
Warren, Oh., U.S. (wŏr´ĕn)	126-27	41°14'N	80°49'W
Warrensburg, Mo., U.S. (wŏr´ĕnz-bûrg)	130-31	38°46'N	93°44'W
Warrenton, S. Afr.	286-87	28°07'S	24°51'E
Warrenton, Mo., U.S. (wŏr´ĕn-tŭn)	130-31	38°49'N	91°09'W
Warrenton, Or., U.S. (wŏr´ĕn-tŭn)	122-23	46°10'N	123°55'W
Warrenton, Va., U.S. (wŏr´ĕn-tŭn)	126-27	38°43'N	77°48'W
Warri, Nig. (wär´ē)	282a	5°31'N	5°46'E
Warrnambool, Austl. (wôr´năm-bool)	301	38°23'S	142°29'E
Warroad, Mn., U.S. (wôr´rōd)	124-25	48°54'N	95°19'W
Warsaw, In., U.S. (wôr´sô)	126-27	41°14'N	85°51'W
Warsaw, Ky., U.S. (wôr´sô)	126-27	38°47'N	84°54'W
Warsaw, Mo., U.S. (wôr´sô)	130-31	38°15'N	93°23'W
Warsaw, N.C., U.S. (wôr´sô)	134-35	34°59'N	78°05'W
Warsaw, Va., U.S. (wôr´sô)	126-27	37°57'N	76°46'W
Warsaw, nat. cap., Pol. (wôr´sô)	210-11	52°15'N	21°00'E
Warszawa, nat. cap., Pol. (vàr-shä´và) *see* Warsaw	210-11	52°15'N	21°00'E
Warwick, Austl. (wŏr´ĭk)	301	28°13'S	152°02'E
Warwick, R.I., U.S. (wŏr´ĭk)	126-27	41°43'N	71°23'W
Wasco, Ca., U.S. (wăs´kō)	128-29	35°36'N	119°20'W
Waseca, Mn., U.S. (wô-sē´kȧ)	124-25	44°05'N	93°30'W
Washburn, N.D., U.S. (wŏsh´bŭrn)	124-25	47°17'N	101°02'W
Washburn, Wi., U.S. (wŏsh´bŭrn)	124-25	46°40'N	90°54'W
Washington, Ga., U.S. (wŏsh´ĭng-tŭn)	134-35	33°44'N	82°44'W
Washington, Ia., U.S. (wŏsh´ĭng-tŭn)	124-25	41°18'N	91°41'W
Washington, Il., U.S. (wŏsh´ĭng-tŭn)	126-27	40°42'N	89°25'W
Washington, In., U.S. (wŏsh´ĭng-tŭn)	126-27	38°39'N	87°10'W
Washington, Mo., U.S. (wŏsh´ĭng-tŭn)	130-31	38°33'N	91°01'W
Washington, N.C., U.S. (wŏsh´ĭng-tŭn)	134-35	35°33'N	77°04'W
Washington, state, U.S. (wŏsh´ĭng-tŭn)	118-19	47°30'N	120°30'W
Washington, nat. cap., D.C., U.S. (wŏsh´ĭng-tŭn)	126-27	38°53'N	77°02'W
Washington, Mount, mtn., N.H., U.S. (mount wŏsh´ĭng-tŭn)	126-27	44°16'N	71°18'W
Washington Island, i., Kir. *see* Teraina	304-05	4°42'N	160°45'W
Washington Island, i., Wi., U.S. (wŏsh´ĭng-tŭn ī´lȧnd)	126-27	45°22'N	86°54'W
Waskaganish, Qc., Can.	138-39	51°29'N	78°45'W
Waskaiowaka Lake, lk., Mb., Can. (wŏ´skă-yō´wŏ-kă lāk)	144-45	56°31'N	96°18'W
Watampone, Indon.	268-69	4°32'S	120°19'E
Wataru, i., Mald.	256	5°43'N	73°23'E
Waterberge, mts., S. Afr. (wôrtĕr´bûrg)	292c	24°28'S	27°58'E
Waterbury, Ct., U.S. (wô´tĕr-bĕr-ē)	126-27	41°33'N	73°02'W
Waterbury, Vt., U.S. (wô´tĕr-bĕr-ē)	126-27	44°20'N	72°45'W
Waterford, Ire. (wô´tĕr-fĕrd)	206-07	52°15'N	7°06'W
Waterhen Lake, lk., Mb., Can.	144-45	52°06'N	99°34'W
Waterloo, Bel. (wô-tĕr-loo´)	206-07	50°43'N	4°24'E
Waterloo, On., Can. (wô-tĕr-loo´)	146-47	43°28'N	80°30'W
Waterloo, Qc., Can. (wô-tĕr-loo´)	146-47	45°21'N	72°30'W
Waterloo, Ia., U.S. (wô-tĕr-loo´)	124-25	42°30'N	92°21'W
Waterloo, Il., U.S. (wô-tĕr-loo´)	130-31	38°20'N	90°09'W
Waterton Lakes National Park, n.p., Ab., Can.	142-43	49°06'N	114°01'W
Watertown, N.Y., U.S.	126-27	43°59'N	75°55'W
Watertown, S.D., U.S. (wô´tĕr-toun)	124-25	44°54'N	97°07'W
Watertown, Wi., U.S. (wô´tĕr-toun)	126-27	43°11'N	88°43'W
Water Valley, Ms., U.S. (văl´ē văl´ē)	134-35	34°10'N	89°38'W
Waterville, Me., U.S.	127a	44°33'N	69°38'W
Watervliet, N.Y., U.S. (wô´tĕr-vlēt´)	126-27	42°44'N	73°43'W
Watford, Eng., U.K. (wŏt´fŏrd)	206-07	51°40'N	0°25'W
Watford City, N.D., U.S.	124-25	47°48'N	103°18'W
Watling Island, i., Bah. *see* San Salvador	154-55	24°02'N	74°27'W
Watonga, Ok., U.S. (wŏ-tôŋ´gȧ)	130-31	35°51'N	98°25'W
Watrous, Sk., Can.	144-45	51°41'N	105°27'W
Watsa, D.R.C. (wät´sä)	289	3°02'N	29°32'E
Watseka, Il., U.S. (wŏt-sē´kȧ)	126-27	40°46'N	87°44'W
Watson Lake, Yk., Can. (wŏt´sŭn lāk)	138-39	60°04'N	128°44'W
Watsonville, Ca., U.S. (wŏt´sŭn-vĭl)	128-29	36°55'N	121°45'W
Wauchula, Fl., U.S. (wô-choo´lȧ)	135a	27°33'N	81°49'W
Waukegan, Il., U.S. (wŏ-kē´găn)	126-27	42°21'N	87°51'W
Waukesha, Wi., U.S. (wô´kē-shô)	126-27	43°01'N	88°14'W
Waukon, Ia., U.S. (wô kŏn)	124-25	43°16'N	91°29'W
Waupaca, Wi., U.S. (wô-păk´ȧ)	126-27	44°21'N	89°05'W
Waupun, Wi., U.S. (wô-pŭn´)	126-27	43°38'N	88°44'W
Waurika, Ok., U.S. (wô-rē´kȧ)	130-31	34°10'N	97°60'W
Wausau, Wi., U.S. (wô´sô)	126-27	44°57'N	89°37'W
Wausaukee, Wi., U.S. (wô-sô´kē)	126-27	45°23'N	87°59'W
Wauseon, Oh., U.S. (wô´sē-ŏn)	126-27	41°33'N	84°08'W
Wautoma, Wi., U.S. (wô-tō´mȧ)	126-27	44°04'N	89°18'W
Waverly, Ia., U.S. (wā´vĕr-lē)	124-25	42°44'N	92°29'W
Waverly, Ne., U.S. (wā´vĕr-lē)	124-25	40°55'N	96°32'W
Waverly, Oh., U.S. (wā´vĕr-lē)	126-27	39°07'N	82°59'W
Waverly, Tn., U.S. (wā´vĕr-lē)	134-35	36°05'N	87°48'W
Wāw, S. Sudan	284-85	7°41'N	27°59'E
Wawa, On., Can.	146-47	47°59'N	84°47'W
Waxahachie, Tx., U.S. (wăk-sȧ-hăch´ē)	130-31	32°24'N	96°51'W
Waya, i., Fiji	306f	17°18'S	177°08'E
Wayabula, Indon.	268-69	2°18'N	128°12'E
Waycross, Ga., U.S.	134-35	31°13'N	82°22'W
Wayne, Ne., U.S. (wān)	124-25	42°14'N	97°01'W
Wayne, W.V., U.S. (wān)	126-27	38°13'N	82°27'W
Waynesboro, Ga., U.S. (wānz´bŭr-ô)	134-35	33°05'N	82°01'W
Waynesboro, Ms., U.S. (wānz´bŭr-ô)	134-35	31°40'N	88°39'W
Waynesboro, Pa., U.S. (wānz´bŭr-ô)	126-27	39°45'N	77°35'W
Waynesboro, Va., U.S. (wānz´bŭr-ô)	126-27	38°04'N	78°53'W
Waynesville, Mo., U.S. (wānz´vĭl)	130-31	37°50'N	92°12'W
Waynesville, N.C., U.S. (wānz´vĭl)	134-35	35°29'N	82°60'W
Waynoka, Ok., U.S. (wā-nō´kȧ)	130-31	36°35'N	98°53'W
Weagamow Lake, lk., On., Can. (wē´ăg-ă-mou lāk)	144-45	52°53'N	91°22'W
Weatherford, Ok., U.S. (wĕ-dhĕr-fĕrd)	130-31	35°32'N	98°43'W
Weatherford, Tx., U.S. (wĕ-dhĕr-fĕrd)	130-31	32°46'N	97°48'W
Weddell Island, i., Falk. Is.	185	51°53'S	61°05'W
Weddell Sea, s., Ant. (wĕd´ĕl sē)	313	72°0'S	45°00'W
Wedgeport, N.S., Can. (wĕj´pōrt)	148-49	43°45'N	65°60'W
Weed, Ca., U.S. (wēd)	122-23	41°25'N	122°23'W
Weenen, S. Afr. (vā´nĕn)	292c	28°51'S	30°05'E
Wei, stm., China (wā)	258-59	34°37'N	110°17'E
Wei, stm., China (wā)	260-61	36°49'N	115°41'E
Weichang, China (wā-chăŋ)	260-61	42°00'N	117°40'E
Weifang, China	260-61	36°42'N	119°06'E
Weihai, China (wa´hāī´)	260-61	37°30'N	122°07'E
Weilheim, Ger. (vīl´hīm´)	210-11	47°50'N	11°09'E
Weimar, Ger. (vī´mår)	210-11	50°59'N	11°19'E
Weinan, China	258-59	34°29'N	109°29'E
Weipa, Austl.	302	12°42'S	141°56'E
Weiser, Id., U.S. (wē´zĕr)	122-23	44°15'N	116°58'W
Weißenfels, Ger. (vī´sĕn-fĕlz)	210-11	51°12'N	11°58'E
Weixi, China (wā-shyē)	258-59	27°11'N	99°17'E
Welch, W.V., U.S. (wĕlch)	134-35	37°26'N	81°35'W
Welkom, S. Afr. (wĕl´kŏm)	292c	27°58'S	26°44'E
Welland, On., Can. (wĕl´ănd)	146-47	42°59'N	79°15'W
Wellesley Islands, is., Austl.	302	16°42'S	139°30'E
Wellington, Austl. (wĕl´lĭng-tŭn)	301	32°34'S	148°57'E
Wellington, Co., U.S. (wĕl´lĭng-tŭn)	130-31	40°42'N	104°60'W
Wellington, Ks., U.S. (wĕl´lĭng-tŭn)	130-31	37°16'N	97°24'W
Wellington, Oh., U.S. (wĕl´lĭng-tŭn)	126-27	41°10'N	82°13'W
Wellington, Tx., U.S. (wĕl´lĭng-tŭn)	130-31	34°51'N	100°13'W
Wellington, nat. cap., N.Z. (wĕl´lĭng-tŭn)	303	41°18'S	174°46'E
Wellington, Isla, i., Chile (ĕ´s-lä ŏĕ´lĕng-tŏn)	185	49°20'S	74°40'W
Wells, Mn., U.S. (wĕlz)	124-25	43°45'N	93°44'W
Wells, Nv., U.S. (wĕlz)	122-23	41°07'N	114°58'W
Wells, Lake, lk., Austl. (lāk wĕlz)	296-97	26°41'S	123°11'E
Wellsboro, Pa., U.S. (wĕlz´bŭ-rô)	126-27	41°45'N	77°17'W
Wellsburg, W.V., U.S. (wĕlz´bûrg)	126-27	40°16'N	80°36'W
Wellston, Oh., U.S. (wĕlz´tŭn)	126-27	39°07'N	82°32'W
Wellsville, N.Y., U.S. (wĕlz´vĭl)	126-27	42°07'N	77°56'W
Wellsville, Oh., U.S. (wĕlz´vĭl)	126-27	40°36'N	80°39'W
Wellsville, Ut., U.S. (wĕlz´vĭl)	122-23	41°39'N	111°56'W
Wels, Aus. (vĕls)	210-11	48°10'N	14°01'E
Welshpool, Wales, U.K. (wĕlsh´pool)	206-07	52°40'N	3°09'W
Wembere, stm., Tan.	289	4°09'S	34°11'E
Wenatchee, Wa., U.S. (wê-năch´ē)	122-23	47°25'N	120°19'W
Wenatchee Mountains, mts., Wa., U.S. (wê-năch´ē moun´tĭnz)	122-23	47°20'N	120°45'W
Wenchang, China (wŭn-chăŋ)	258-59	19°33'N	110°45'E
Wenchow, China *see* Wenzhou	258-59	28°01'N	120°38'E
Wendover, Ut., U.S.	122-23	40°44'N	114°02'W
Wenlock, stm., Austl.	302	12°15'S	141°56'E
Wenshan, China	258-59	23°30'N	104°28'E
Wentworth, Austl. (wĕnt´wûrth)	301	34°06'S	141°55'E
Wenzhou, China (wŭn-jō)	258-59	28°01'N	120°38'E
Werdēr, Eth.	284-85	6°58'N	45°21'E
Wesel, Ger. (vā´zĕl)	210-11	51°40'N	6°38'E
Weser, stm., Ger. (vā´zĕr)	210-11	53°32'N	8°34'E
Wesermünde, Ger. *see* Bremerhaven	210-11	53°32'N	8°36'E
Weslaco, Tx., U.S. (wĕs-lā´kō)	132-33	26°10'N	97°59'W
Wessel, Cape, c., Austl.	296-97	11°02'S	136°45'E
Wessington Springs, S.D., U.S. (wĕs´ĭng-tŭn springz)	124-25	44°05'N	98°34'W
West Allis, Wi., U.S. (wĕst ăl´ĭs)	126-27	43°01'N	88°01'W
West Bend, Ia., U.S. (wĕst bĕnd)	124-25	42°57'N	94°26'W
West Bend, Wi., U.S. (wĕst bĕnd)	126-27	43°25'N	88°10'W
West Bengal, state, India (wĕst bĕn-gôl´)	254-55	24°0'N	88°00'E
West Branch, Ia., U.S. (wĕst brȧnch)	124-25	41°40'N	91°21'W
West Branch, Mi., U.S. (wĕst brȧnch)	126-27	44°16'N	84°14'W
Westbrook, Me., U.S. (wĕst´brŏk)	126-27	43°41'N	70°21'W
West Caicos, i., T./C. Is. (wĕst kī´kōs) (wĕst kăē´kō)	154-55	21°39'N	72°28'W
West Cape, c., N.Z. (wĕst kāp)	303	45°55'S	166°25'E
West Chester, Pa., U.S. (wĕst chĕs´tĕr)	126-27	39°58'N	75°36'W
West Columbia, S.C., U.S. (wĕst cŏl´ŭm-bē-ȧ)	134-35	33°59'N	81°05'W
West Columbia, Tx., U.S. (wĕst cŏl´ŭm-bē-ȧ)	132-33	29°09'N	95°39'W
West Des Moines, Ia., U.S. (wĕst dē moin´)	124-25	41°34'N	93°44'W
West End, Bah. (wĕst ĕnd)	154-55	26°42'N	78°59'W
Westerly, R.I., U.S. (wĕs´tĕr-lē)	126-27	41°23'N	71°50'W
Western Australia, state, Austl. (wĕst´tĕrn ôs-trā´lĭ-ȧ)	294-95	25°0'S	122°00'E
Western Desert, des., Egypt (wĕst´tĕrn dĕs´ĕrt)	288	27°0'N	27°00'E
Western Dvina, stm., Eur.	208-09	57°04'N	24°03'E
Western Ghāts, mts., India (wĕst´tĕrn ghäts) (wĕst´tĕrn ghôts)	256	14°0'N	75°00'E
Westernport, Md., U.S. (wĕs´tĕrn pōrt)	126-27	39°29'N	79°03'W
Western Sahara, dep., Afr. (wĕst´tĕrn sȧ-hä´rȧ)	274	24°30'N	13°00'W
Western Samoa, nation, Oc. *see* Samoa	306b	13°55'S	186°00'W
Westerville, Oh., U.S. (wĕs´tĕr-vĭl)	126-27	40°07'N	82°55'W
West Falkland, i., Falk. Is.	185	51°50'S	59°60'W
Westfield, Ma., U.S. (wĕst´fēld)	126-27	42°08'N	72°45'W
Westfield, N.Y., U.S. (wĕst´fēld)	126-27	42°19'N	79°35'W
Westfield, Wi., U.S. (wĕst´fēld)	126-27	43°53'N	89°30'W
West Frankfort, Il., U.S. (wĕst frăŋk´fŭrt)	126-27	37°54'N	88°56'W
West Helena, Ar., U.S. (wĕst hĕl´ĕn-ȧ)	134-35	34°33'N	90°39'W
West Indies, is., (wĕst ĭn´dēz)	152-53	19°0'N	70°00'W
West Lafayette, In., U.S. (wĕst lä-fâ-yĕt´)	126-27	40°25'N	86°54'W
West Liberty, Ia., U.S. (wĕst lĭb´ĕr-tĭ)	124-25	41°34'N	91°16'W
West Liberty, Oh., U.S. (wĕst lĭb´ĕr-tĭ)	126-27	40°15'N	83°47'W
Westlock, Ab., Can. (wĕst´lŏk)	142-43	54°09'N	113°52'W
Westminster, Co., U.S. (wĕst´min-stĕr)	130-31	39°51'N	105°04'W
West Nishnabotna, stm., Ia., U.S. (wĕst nĭsh-nȧ-bŏt´nȧ)	130-31	40°30'N	95°42'W
Weston, W.V., U.S. (wĕs´tŭn)	126-27	39°02'N	80°29'W
Weston-super-Mare, Eng., U.K. (wĕs´tŭn sū´pĕr-mā´rē)	206-07	51°21'N	2°58'W
West Palm Beach, Fl., U.S. (wĕst pälm bēch)	135a	26°44'N	80°08'W
West Pensacola, Fl., U.S. (wĕst pĕn-sȧ-kō´lȧ)	134-35	30°25'N	87°16'W
West Plains, Mo., U.S. (wĕst-plānz´)	134-35	36°44'N	91°52'W
West Point, Ga., U.S. (wĕst point)	134-35	32°54'N	85°09'W
West Point, Ms., U.S. (wĕst point)	134-35	33°36'N	88°39'W
West Point, Ne., U.S. (wĕst point)	124-25	41°51'N	96°43'W
West Point, N.Y., U.S. (wĕst point)	126-27	41°24'N	73°58'W
West Point, Va., U.S. (wĕst point)	126-27	37°32'N	76°48'W
West Point, c., Austl.	296-97	35°01'S	135°57'E
West Point Lake, res., U.S. (wĕst point lāk)	134-35	32°60'N	85°12'W
Westport, N.Z.	303	41°45'S	171°36'E
Westport, Wa., U.S. (wĕst´pôrt)	122-23	46°54'N	124°06'W
West Road, stm., B.C., Can. (wĕst rōd)	142-43	53°18'N	122°53'W
West Siberian Plain, pl., Russia (wĕst sī-bĭr´y´n plān)	236-37	60°0'N	75°00'E
West Union, Ia., U.S. (wĕst ūn´yŭn)	124-25	42°57'N	91°49'W
West Union, Oh., U.S. (wĕst ūn´yŭn)	126-27	38°47'N	83°33'W
West Valley City, Ut., U.S.	122-23	40°42'N	112°00'W
Westville, N.S., Can. (wĕst´vĭl)	148-49	45°34'N	62°42'W
West Virginia, state, U.S. (wĕst vĕr-jĭn´ĭ-ȧ)	118-19	38°45'N	80°30'W
West Warwick, R.I., U.S. (wĕst wŏr´ĭk)	126-27	41°42'N	71°32'W
West Wyalong, Austl. (wĕst wīălông)	301	33°55'S	147°12'E
Wetar, Pulau, i., Indon. (pōō-lou wĕt´ár)	268-69	7°48'S	126°18'E
Wetaskiwin, Ab., Can. (wĕ-tăs´kē-wŏn)	142-43	52°58'N	113°22'W

ăt; finȧl; rāte; senâte; ärm; ȧsk; sofȧ; fâre; ch-choose; dh-as th in other; bē; ēvent; bĕt; recĕnt; cratĕr; g-gō; gh-guttural g; bĭt; ĭ-short neutral; rīde; ĸ-guttural k as ch in German ich;

Feature (Pronunciation)	Page	Lat.	Long.
Wete, Tan.	**284-85**	5°02'S	39°44'E
Wetumpka, Al., U.S. (wē-tŭmp′kȧ)	**134-35**	32°33'N	86°12'W
Wetzlar, Ger. (vets′lär)	**210-11**	50°33'N	8°29'E
Wewak, Pap. N. Gui. (wā-wäk′)	**302**	3°35'S	143°39'E
Wewoka, Ok., U.S. (wē-wō′kȧ)	**130-31**	35°10'N	96°30'W
Weyburn, Sk., Can. (wā′bûrn)	**144-45**	49°40'N	103°51'W
Weymouth, Eng., U.K. (wā′mŭth)	**206-07**	50°37'N	2°28'W
Whangarei, N.Z.	**303**	35°43'S	174°19'E
Whapmagoostui, Qc., Can.	**138-39**	55°17'N	77°45'W
Wharton, Tx., U.S. (hwôr′tŭn)	**132-33**	29°19'N	96°06'W
Wharton Lake, lk., Nu., Can.	**140-41**	64°01'N	99°52'W
Wheatland, Wy., U.S. (hwēt′lănd)	**122-23**	42°03'N	104°57'W
Wheaton, Md., U.S. (hwē′tŭn)	**126-27**	39°02'N	77°03'W
Wheaton, Mn., U.S. (hwē′tŭn)	**124-25**	45°48'N	96°30'W
Wheeler Peak, mtn., N.M., U.S.	**130-31**	36°34'N	105°25'W
Wheeling, W.V., U.S. (hwēl′ĭng)	**126-27**	40°04'N	80°41'W
Whitby, On., Can. (hwĭt′bē)	**146-47**	43°53'N	78°56'W
White, stm., On., Can. (hwīt)	**146-47**	48°33'N	86°16'W
White, stm., N.A.	**136**	63°11'N	139°35'W
White, stm., U.S. (hwīt)	**120-21**	33°53'N	91°04'W
White, stm., U.S. (hwīt)	**124-25**	43°42'N	99°27'W
White, stm., In., U.S. (hwīt)	**126-27**	38°25'N	87°45'W
White, East Fork, stm., In., U.S. (ēst fôrk hwīt)	**126-27**	38°32'N	87°14'W
White Bay, b., Nf./L., Can. (hwīt bā)	**148-49**	50°07'N	56°27'W
White Cloud, Mi., U.S. (hwīt kloud)	**126-27**	43°33'N	85°46'W
Whitecourt, Ab., Can. (hwīt′côrt)	**142-43**	54°08'N	115°43'W
White Earth Indian Reservation, ind. res., Mn., U.S. (hwīt ûrth ĭn′dĭ-ăn rĕ-sĕr-vā′shĕn)	**124-25**	47°18'N	95°50'W
Whitefish, Mt., U.S.	**122-23**	48°25'N	114°20'W
Whitefish Lake, lk., N.T., Can.	**140-41**	62°41'N	106°48'W
White Hall, Ar., U.S. (hwīt hôl)	**130-31**	34°16'N	92°07'W
Whitehall, Mi., U.S. (hwīt hôl)	**126-27**	43°24'N	86°20'W
Whitehall, Mt., U.S. (hwīt′hôl)	**122-23**	45°52'N	112°06'W
Whitehall, N.Y., U.S. (hwīt′hôl)	**126-27**	43°33'N	73°24'W
Whitehaven, Eng., U.K. (hwīt′hā-vĕn)	**206-07**	54°33'N	3°35'W
Whitehorse, Yk., Can. (hwīt′hôrs)	**138-39**	60°43'N	135°08'W
White Lake, lk., On., Can. (hwīt lāk)	**146-47**	45°17'N	76°32'W
White Lake, lk., La., U.S. (hwīt lāk)	**132-33**	29°43'N	92°28'W
White Mountain Peak, mtn., Ca., U.S. (hwīt moun′tĭn pēk)	**128-29**	37°38'N	118°16'W
White Mountains, mts., N.H., U.S. (hwīt moun′tĭnz)	**126-27**	44°10'N	71°35'W
White Nile, stm., Africa (hwīt nīl)	**275**	15°38'N	32°31'E
White Pass, p., N.A. (hwīt păs)	**136**	59°38'N	135°10'W
White Russia, nation, Eur. *see* Belarus	**190-91**	53°50'N	28°00'E
Whitesail Lake, res., B.C., Can. (hwīt′sāl lāk)	**142-43**	53°30'N	127°00'W
White Sands National Monument, n.p., N.M., U.S.	**128-29**	32°46'N	106°20'W
White Sea, s., Russia (hwīt sē sē)	**202-03**	65°37'N	37°52'E
White Sulphur Springs, Mt., U.S. (hwīt sŭl′fŭr sprĭngz)	**122-23**	46°33'N	110°55'W
White Sulphur Springs, W.V., U.S. (hwīt sŭl′fŭr sprĭngz)	**126-27**	37°48'N	80°18'W
Whiteville, N.C., U.S. (hwīt′vĭl)	**134-35**	34°20'N	78°42'W
White Volta, stm., Afr. (hwīt vôl′tȧ)	**282-83**	8°57'N	1°10'W
Whitewater, Wi., U.S. (hwīt-wŏt′ĕr)	**126-27**	42°50'N	88°43'W
Whitewater Lake, lk., Mb., Can. (hwīt-wŏt′ĕr lāk)	**144-45**	49°15'N	100°20'W
Whitewright, Tx., U.S. (hwīt′rīt)	**130-31**	33°31'N	96°24'W
Whitney, Lake, res., Tx., U.S. (lāk hwĭt′nē)	**132-33**	31°56'N	97°26'W
Whitney, Mount, mtn., Ca., U.S. (mount hwĭt′nē)	**128-29**	36°35'N	118°18'W
Whitsunday Island, i., Austl. (hwĭt′s'n-dā ī′lȧnd)	**302**	20°15'S	148°59'E
Whittle, Cap, c., Qc., Can.	**148-49**	50°11'N	60°10'W
Wholdaia Lake, lk., N.T., Can.	**140-41**	60°41'N	104°18'W
Whyalla, Austl. (hwī-ăl′ȧ)	**301**	33°02'S	137°34'E
Wiarton, On., Can. (wī′ȧr-tŭn)	**146-47**	44°44'N	81°08'W
Wichita, Ks., U.S. (wĭch′ĭ-tô)	**130-31**	37°41'N	97°19'W
Wichita Falls, Tx., U.S. (wĭch′ĭ-tô fôlz)	**130-31**	33°55'N	98°30'W
Wichita Mountains, mts., Ok., U.S. (wĭch′ĭ-tô moun′tĭnz)	**130-31**	34°45'N	98°40'W
Wick, Scot., U.K. (wĭk)	**206-07**	58°26'N	3°06'W
Wicklow Mountains, mts., Ire. (wĭk′lō moun′tĭnz)	**206-07**	53°02'N	6°24'W
Wieliczka, Pol. (vyĕ-lēch′kȧ)	**210-11**	49°59'N	20°04'E
Wien, nat. cap., Aus. (vēn) *see* Vienna	**210-11**	48°13'N	16°20'E
Wiener Neustadt, Aus. (vē′nĕr noi′shtät)	**210-11**	47°49'N	16°15'E
Wieprz, stm., Pol. (vyĕpzh)	**210-11**	51°33'N	21°50'E
Wiesbaden, Ger. (vēs′bä-dĕn)	**210-11**	50°05'N	8°14'E
Wiggins, Ms., U.S. (wĭg′ĭnz)	**134-35**	30°51'N	89°09'W
Wight, Isle of, i., Eng., U.K. (īl ŏv wīt)	**206-07**	50°40'N	1°20'W
Wilber, Ne., U.S. (wĭl′bĕr)	**130-31**	40°29'N	96°58'W
Wilburton, Ok., U.S. (wĭl′bĕr-tŭn)	**130-31**	34°56'N	95°19'W
Wilcannia, Austl. (wĭl-căn-ĭȧ)	**301**	31°34'S	143°23'E
Wildcat Hill, mtn., Sk., Can. (wīld′kăt hĭl)	**144-45**	53°17'N	102°30'W
Wildhay, stm., Ab., Can. (wīld′hā)	**142-43**	53°59'N	117°17'W
Wilge, stm., S. Afr. (wĭl′jĕ)	**292c**	27°02'S	28°21'E
Wilhelm, Mount, mtn., Pap. N. Gui.	**302**	5°47'S	145°00'E
Wilhelmina Peak, mtn., Indon. *see* Trikora, Puncak	**302**	4°18'S	138°40'E
Wilhelmshaven, Ger. (vēl-hĕlms-hä′fĕn)	**210-11**	53°32'N	8°07'E
Wilkes-Barre, Pa., U.S. (wĭlks′băr-ē)	**126-27**	41°15'N	75°53'W
Wilkes Land, reg., Ant.	**313**	69°0'S	120°00'E
Wilkie, Sk., Can. (wĭlk′ē)	**144-45**	52°25'N	108°40'W
Willamette, stm., Or., U.S.	**122-23**	45°39'N	122°46'W
Willard, Oh., U.S. (wĭl′ȧrd)	**126-27**	41°03'N	82°43'W
Willcox, Az., U.S. (wĭl′kŏks)	**128-29**	32°15'N	109°50'W
Willemstad, nat. cap., Curaçao (vĭ′lĕm-stät)	**152a**	12°06'N	68°56'W
Williams, Az., U.S. (wĭl′yămz)	**128-29**	35°15'N	112°12'W
Williamsburg, Ia., U.S. (wĭl′yămz-bŭrg)	**124-25**	41°40'N	92°01'W
Williamsburg, Ky., U.S. (wĭl′yămz-bŭrg)	**134-35**	36°44'N	84°10'W
Williamsburg, Va., U.S. (wĭl′yămz-bŭrg)	**134-35**	37°16'N	76°43'W
Williams Lake, B.C., Can. (wĭl′yămz lāk)	**142-43**	52°07'N	122°09'W
Williamson, W.V., U.S. (wĭl′yăm-sŭn)	**126-27**	37°40'N	82°17'W
Williamson Head, c., Ant.	**313**	69°12'S	157°47'E
Williamsport, Pa., U.S.	**126-27**	41°14'N	77°01'W
Williamston, N.C., U.S. (wĭl′yămz-tŭn)	**134-35**	35°51'N	77°04'W
Williamston, S.C., U.S. (wĭl′yămz-tŭn)	**134-35**	34°37'N	82°29'W
Williamstown, Ky., U.S. (wĭl′yămz-toun)	**126-27**	38°38'N	84°34'W
Williamstown, W.V., U.S. (wĭl′yămz-toun)	**126-27**	39°24'N	81°27'W
Willimantic, Ct., U.S. (wĭl-ĭ-măn′tĭk)	**126-27**	41°43'N	72°12'W
Willis, Tx., U.S. (wĭl′ĭs)	**132-33**	30°25'N	95°29'W
Williston, Fl., U.S. (wĭl′ĭs-tŭn)	**134-35**	29°23'N	82°27'W
Williston, N.D., U.S. (wĭl′ĭs-tŭn)	**124-25**	48°09'N	103°37'W
Williston, S.C., U.S. (wĭl′ĭs-tŭn)	**134-35**	33°24'N	81°26'W
Williston Lake, res., B.C., Can. (wĭl′ĭs-tŭn lāk)	**140-41**	56°00'N	123°54'W
Willmar, Mn., U.S. (wĭl′mär)	**124-25**	45°07'N	95°03'W
Willow, Ak., U.S. (wĭl′ō)	**136**	61°45'N	150°03'W
Willowmore, S. Afr. (wĭl′ō-môr)	**286-87**	33°18'S	23°29'E
Willows, Ca., U.S. (wĭl′ōz)	**128-29**	39°31'N	122°12'W
Willow Springs, Mo., U.S. (wĭl′ō sprĭngz)	**134-35**	36°60'N	91°58'W
Wills Point, Tx., U.S. (wĭlz point)	**130-31**	32°43'N	96°01'W
Wilmington, De., U.S.	**126-27**	39°45'N	75°32'W
Wilmington, N.C., U.S. (wĭl′mĭng-tŭn)	**134-35**	34°13'N	77°57'W
Wilmington, Oh., U.S. (wĭl′mĭng-tŭn)	**126-27**	39°26'N	83°50'W
Wilmore, Ky., U.S. (wĭl′mōr)	**126-27**	37°46'N	84°37'W
Wilna, nat. cap., Lith. *see* Vilnius	**208-09**	54°40'N	25°17'E
Wilson, N.C., U.S. (wĭl′sŭn)	**134-35**	35°44'N	77°55'W
Wilson, Ok., U.S. (wĭl′sŭn)	**130-31**	34°10'N	97°25'W
Wilson, Cape, c., Nu., Can.	**140-41**	67°01'N	81°30'W
Wilson, Mount, mtn., Ca., U.S. (mount wĭl′sŭn)	**128-29**	34°14'N	118°04'W
Wilson, Mount, mtn., Co., U.S. (mount wĭl′sŭn)	**128-29**	37°51'N	107°59'W
Wilsons Promontory, pen., Austl.	**301**	38°55'S	146°20'E
Wilton, N.D., U.S. (wĭl′tŭn)	**124-25**	47°09'N	100°47'W
Winamac, In., U.S. (wĭn′ȧ măk)	**126-27**	41°03'N	86°36'W
Winburg, S. Afr. (wĭm-bûrg)	**292c**	28°31'S	27°01'E
Winchester, In., U.S. (wĭn′chĕs-tĕr)	**126-27**	40°10'N	84°59'W
Winchester, Ky., U.S. (wĭn′chĕs-tĕr)	**126-27**	37°60'N	84°11'W
Winchester, Tn., U.S. (wĭn′chĕs-tĕr)	**134-35**	35°11'N	86°06'W
Winchester, Va., U.S. (wĭn′chĕs-tĕr)	**126-27**	39°11'N	78°10'W
Windau, Lat. *see* Ventspils	**208-09**	57°24'N	21°35'E
Wind Cave National Park, n.p., S.D., U.S. (wĭnd kāv năsh′ŭn-ăl pärk)	**124-25**	43°34'N	103°29'W
Windhoek, nat. cap., Nmb. (vĭnt′hŏk)	**286-87**	22°34'S	17°05'E
Windom, Mn., U.S. (wĭn′dŭm)	**124-25**	43°52'N	95°07'W
Windorah, Austl.	**301**	25°25'S	142°39'E
Wind River Indian Reservation, ind. res., Wy., U.S. (wĭnd rĭv′ĕr ĭn′dĭ-ăn rĕ-sĕr-vā′shĕn)	**122-23**	43°26'N	109°00'W
Wind River Range, mts., Wy., U.S. (wĭnd rĭv′ĕr rānj)	**122-23**	43°05'N	109°25'W
Windsor, Austl. (wĭn′zĕr)	**301**	33°37'S	150°49'E
Windsor, N.S., Can. (wĭn′zĕr)	**148-49**	44°59'N	64°08'W
Windsor, On., Can. (wĭn′zĕr)	**146-47**	42°18'N	83°02'W
Windsor, Eng., U.K. (wĭn′zĕr)	**206-07**	51°29'N	0°36'W
Windsor, Co., U.S. (wĭn′zĕr)	**130-31**	40°28'N	104°53'W
Windsor, Mo., U.S. (wĭn′zĕr)	**130-31**	38°32'N	93°32'W
Windsor, N.C., U.S. (wĭn′zĕr)	**134-35**	35°60'N	76°57'W
Windsor, Vt., U.S. (wĭn′zĕr)	**126-27**	43°29'N	72°23'W
Windward Islands, is., N.A. (wĭnd′wĕrd ī′lȧndz)	**155b**	13°0'N	61°00'W
Windward Passage, strt., N.A. (wĭnd′wĕrd păs′ĭj)	**154-55**	19°56'N	73°52'W
Winfield, Ks., U.S.	**130-31**	37°14'N	96°60'W
Wingham, On., Can.	**146-47**	43°53'N	81°18'W
Winisk, stm., On., Can.	**140-41**	55°16'N	85°07'W
Winisk Lake, lk., On., Can.	**144-45**	52°55'N	87°22'W
Winkler, Mb., Can. (wĭnk′lĕr)	**144-45**	49°11'N	97°56'W
Winnebago, Mn., U.S. (wĭn′ē-bā′gō)	**124-25**	43°46'N	94°11'W
Winnebago, Lake, lk., Wi., U.S. (lāk wĭn′ē-bā′gō)	**126-27**	44°0'N	88°25'W
Winnebago Indian Reservation, ind. res., Ne., U.S. (wĭn′ē-bā′gō ĭn′dĭ-ăn rĕ-sĕr-vā′shĕn)	**124-25**	42°15'N	96°31'W
Winnemucca, Nv., U.S. (wĭn-ē-mŭk′ȧ)	**122-23**	40°58'N	117°44'W
Winner, S.D., U.S. (wĭn′ĕr)	**124-25**	43°23'N	99°52'W
Winnfield, La., U.S. (wĭn′fēld)	**132-33**	31°55'N	92°38'W
Winnipeg, Mb., Can. (wĭn′ĭ-pĕg)	**144-45**	49°52'N	97°10'W
Winnipeg, stm., Can. (wĭn′ĭ-pĕg)	**144-45**	50°41'N	96°24'W
Winnipeg, Lake, lk., Mb., Can. (lāk wĭn′ĭ-pĕg)	**144-45**	52°0'N	97°00'W
Winnipegosis, Lake, lk., Mb., Can.	**144-45**	52°30'N	99°59'W
Winnsboro, La., U.S. (wĭnz′bŭr′ō)	**134-35**	32°10'N	91°43'W
Winnsboro, S.C., U.S. (wĭnz′bŭr′ō)	**134-35**	34°23'N	81°05'W
Winnsboro, Tx., U.S. (wĭnz′bŭr′ō)	**130-31**	32°58'N	95°17'W
Winona, Mi., U.S. (wĭ-nō′nȧ)	**124-25**	46°52'N	88°55'W
Winona, Mn., U.S. (wĭ-nō′nȧ)	**124-25**	44°03'N	91°38'W
Winona, Ms., U.S. (wĭ-nō′nȧ)	**134-35**	33°29'N	89°44'W
Winslow, Az., U.S. (wĭnz′lō)	**128-29**	35°02'N	110°42'W
Winsted, Ct., U.S. (wĭn′stĕd)	**126-27**	41°56'N	73°04'W
Winston-Salem, N.C., U.S. (wĭn stŭn-sā′lĕm)	**134-35**	36°06'N	80°14'W
Winter Garden, Fl., U.S. (wĭn′tĕr gär′d'n)	**135a**	28°34'N	81°35'W
Winter Haven, Fl., U.S. (wĭn′tĕr hā′vĕn)	**135a**	28°02'N	81°44'W
Winters, Tx., U.S. (wĭn′tĕrz)	**132-33**	31°57'N	99°58'W
Winterset, Ia., U.S. (wĭn′tĕr-sĕt)	**124-25**	41°20'N	94°01'W
Winterthur, Switz. (vĭn′tĕr-tōōr)	**210-11**	47°30'N	8°43'E
Winthrop, Me., U.S. (wĭn′thrŭp)	**126-27**	44°18'N	69°59'W
Winton, Austl. (wĭn-tŭn)	**302**	22°23'S	143°02'E
Wirāṭnagar, Nepal	**254-55**	26°28'N	87°17'E
Wīrgañj, Nepal	**254-55**	27°01'N	84°52'E
Wisby, Swe. *see* Visby	**208-09**	57°38'N	18°19'E
Wisconsin, state, U.S. (wĭs-kŏn′sĭn)	**118-19**	44°45'N	89°30'W
Wisconsin, stm., Wi., U.S. (wĭs-kŏn′sĭn)	**124-25**	42°59'N	91°09'W
Wisconsin Dells, Wi., U.S. (wĭs-kŏn′sĭn dĕlz)	**126-27**	43°38'N	89°45'W
Wisconsin Rapids, Wi., U.S. (wĭs-kŏn′sĭn răp′ĭdz)	**126-27**	44°23'N	89°49'W
Wishek, N.D., U.S. (wĭsh′ĕk)	**124-25**	46°16'N	99°33'W
Wisła, stm., Pol. (vēs′wȧ) *see* Vistula	**210-11**	54°21'N	18°56'E
Wisłoka, stm., Pol. (vēs-wô′kȧ)	**210-11**	50°27'N	21°24'E
Wismar, Ger. (vĭs′mär)	**210-11**	53°53'N	11°28'E
Wisner, Ne., U.S. (wĭz′nĕr)	**124-25**	41°59'N	96°56'W
Wissembourg, Fr. (vē-sän-bōōr′)	**212-13**	49°02'N	7°57'E
Witbank, S. Afr. (wĭt-băŋk)	**292c**	25°53'S	29°14'E
Withlacoochee, stm., U.S. (wĭth-lȧ-kōō′chē)	**134-35**	30°23'N	83°10'W
Witu Islands, is., Pap. N. Gui.	**302**	4°45'S	149°19'E
W.J. van Blommestein Meer, res., Sur.	**178-79**	4°49'N	55°04'W
Wkra, stm., Pol. (f′krȧ)	**210-11**	52°27'N	20°45'E
Włocławek, Pol. (vwô-tswä′vĕk)	**210-11**	52°39'N	19°04'E
Włodawa, Pol. (vwō-dä′vä)	**210-11**	51°33'N	23°34'E
Włoszczowa, Pol. (vwôsh-chô′vä)	**210-11**	50°51'N	19°58'E
Wodonga, Austl.	**301**	36°08'S	146°53'E
Wokam, Pulau, i., Indon.	**242-43**	5°37'S	134°30'E
Woleai, at., Micron.	**304-05**	7°21'N	143°53'E
Woleu, stm., Afr. *see* Mbini	**282-83**	1°35'N	9°38'E
Wolf, Volcán, vol., Ec.	**184a**	0°00'S	91°20'W
Wolf Point, Mt., U.S. (wŏlf point)	**122-23**	48°05'N	105°37'W

n-sing; ŋ-baŋk; ɴ-nasalized n; nŏd; cŏmmit; ōld; ōbey; ôrder; oi-boil; fōōd; ŏ-as oo in foot; ou-out; s-soft; sh-dish; th-thin; pūre; ûnite; ûrn; stŭd; circŭs; ü-as in French tu; ′-indeterminate vowel.

424 Wolfsburg – Yamal, Poluostrov PRONOUNCING INDEX

Feature (Pronunciation)	Page	Lat.	Long.
Wolfsburg, Ger. (vôlfs′bo͞orgh)	210-11	52°26′N	10°47′E
Wolfville, N.S., Can. (wo͝olf′vĭl)	148-49	45°05′N	64°22′W
Wollaston, Islas, is., Chile	185	55°45′S	67°37′W
Wollaston Lake, lk., Sk., Can. (wo͝ol′ȧs-tŭn lāk)	140-41	58°15′N	103°20′W
Wollaston Peninsula, pen., Can. (wo͝ol′ȧs-tŭn pĕ-nĭn′sūlȧ)	140-41	70°0′N	115°00′W
Wollongong, Austl. (wo͝ol′ŭn-gông)	301	34°25′S	150°54′E
Wołomin, Pol. (vô-wō′mēn)	210-11	52°20′N	21°15′E
Wolseley, Sk., Can.	144-45	50°25′N	103°16′W
Wolverhampton, Eng., U.K. (wo͝ol′vēr-hămp-tŭn)	206-07	52°35′N	2°08′W
Wondai, Austl.	301	26°19′S	151°53′E
Wonju, Kor., S.	263	37°21′N	127°57′E
Wŏnsan, Kor., N. (wŭn′sän′)	263	39°09′N	127°26′E
Wonthaggi, Austl. (wônt-hăg′ê)	301	38°37′S	145°35′E
Woodbine, Ia., U.S. (wo͝od′bīn)	124-25	41°44′N	95°43′W
Woodbridge, Va., U.S. (wo͝od′brĭj′)	126-27	38°39′N	77°15′W
Wood Buffalo National Park, n.p., Can. (wo͝od bŭf′ȧ-lō năsh′ŭn-ăl pärk)	138-39	59°06′N	112°58′W
Woodburn, Or., U.S. (wo͝od′bûrn)	122-23	45°08′N	122°51′W
Woodlark, i., Pap. N. Gui. (wo͝od′lärk) *see* Murua Island	302	9°06′S	152°45′E
Woodroffe, Mount, mtn., Austl. (mount wo͝od′rŭf)	296-97	26°20′S	131°45′E
Woodruff, S.C., U.S. (wo͝od′rŭf)	134-35	34°45′N	82°02′W
Woods, Lake of the, lk., N.A. (lāk ŭv thȧ wo͝odz)	144-45	49°15′N	94°45′W
Woodsfield, Oh., U.S. (wo͝odz-fēld)	126-27	39°46′N	81°07′W
Woodstock, N.B., Can. (wo͝od′stŏk)	148-49	46°09′N	67°34′W
Woodstock, On., Can. (wo͝od′stŏk)	146-47	43°08′N	80°45′W
Woodstock, Il., U.S. (wo͝od′stŏk)	126-27	42°19′N	88°26′W
Woodstock, N.Y., U.S. (wo͝od′stŏk)	126-27	42°03′N	74°07′W
Woodstock, Va., U.S. (wo͝od′stŏk)	126-27	38°53′N	78°30′W
Woodsville, N.H., U.S. (wo͝odz′vĭl)	126-27	44°09′N	72°02′W
Woodville, Ms., U.S. (wo͝od′vĭl)	134-35	31°07′N	91°18′W
Woodville, Tx., U.S. (wo͝od′vĭl)	132-33	30°46′N	94°25′W
Woodward, Ok., U.S. (wo͝od′wôrd)	130-31	36°27′N	99°23′W
Woomera, Austl. (wo͝om′ērȧ)	301	31°12′S	136°50′E
Woonsocket, R.I., U.S. (wo͞on-sŏk′ĕt)	126-27	42°01′N	71°31′W
Wooramel, stm., Austl.	296-97	25°51′S	114°16′E
Wooster, Oh., U.S. (wo͝os′tēr)	126-27	40°48′N	81°56′W
Worcester, S. Afr. (wo͝os′tēr)	286-87	33°39′S	19°27′E
Worcester, Eng., U.K. (wo͝o′stēr)	206-07	52°11′N	2°14′W
Worcester, Ma., U.S.	126-27	42°16′N	71°48′W
Worden, Mt., U.S. (wôr′dĕn)	122-23	45°58′N	108°13′W
Workington, Eng., U.K. (wûr′kĭng-tŭn)	206-07	54°38′N	3°34′W
Worksop, Eng., U.K. (wûrk′sŏp) (wûr′sŭp)	206-07	53°19′N	1°07′W
Worland, Wy., U.S. (wûr′lănd)	122-23	44°01′N	107°57′W
World War II Valor in the Pacilc National Monument, n.p., Hi., U.S.	137	21°22′N	157°57′W
Worms, Ger. (vōrms)	210-11	49°38′N	8°21′E
Worthing, Eng., U.K. (wûr′dhĭng)	206-07	50°49′N	0°23′W
Worthington, In., U.S. (wûr′dhĭng-tŭn)	126-27	39°07′N	86°59′W
Worthington, Mn., U.S. (wûr′dhĭng-tŭn)	124-25	43°38′N	95°36′W
Wotho, at., Marsh. Is.	304-05	10°06′N	166°01′E
Wotje, at., Marsh. Is.	304-05	9°27′N	170°02′E
Wowoni, Pulau, i., Indon. (po͞o-lou wō-wō′nē)	268-69	4°08′S	123°06′E
Wrangel Island, i., Russia (răn′gĕl ī′lȧnd) *see* Vrangelya, Ostrov	236-37	71°14′N	179°21′W
Wrangell, Ak., U.S. (răn′gĕl)	136	56°29′N	132°22′W
Wrangell, Cape, c., Ak., U.S. (kāp răn′gĕl)	136a	52°55′N	186°30′E
Wrangell, Mount, mtn., Ak., U.S. (mount răn′gĕl)	136	62°0′N	144°06′W
Wrangell Mountains, mts., Ak., U.S. (răn′gĕl moun′tĭnz)	136	62°0′N	143°00′W
Wrangell-Saint Elias National Park and Preserve, n.p., Ak., U.S.	136	61°37′N	142°57′W
Wrath, Cape, c., Scot., U.K. (kāp răth)	206-07	58°38′N	4°60′W
Wray, Co., U.S. (rā)	130-31	40°05′N	102°14′W
Wrens, Ga., U.S. (rĕnz)	134-35	33°12′N	82°20′W
Wrexham, Wales, U.K. (rĕk′săm)	206-07	53°03′N	2°60′W
Wrightsville, Ga., U.S. (rīts′vĭl)	134-35	32°44′N	82°43′W
Wrigley, N.T., Can.	138-39	63°16′N	123°38′W
Wrocław, Pol. (vrôtsläv) (brĕs′lou)	210-11	51°07′N	17°02′E
Września, Pol. (vzhăsh′nyȧ)	210-11	52°20′N	17°35′E
Wu, stm., China (wo͞o′)	258-59	24°49′N	113°35′E
Wu, stm., China (wo͞o′)	258-59	27°11′N	109°48′E
Wuchang, China (wo͞o-chäŋ)	260-61	44°55′N	127°10′E
Wuchang, China *see* Wuhan	258-59	30°34′N	114°17′E
Wudaoliang, China	254-55	35°12′N	93°05′E
Wudu, China	258-59	33°25′N	104°51′E
Wugang, China	258-59	26°44′N	110°38′E
Wugong Shan, mts., China	258-59	27°21′N	113°50′E
Wuhai, China	260-61	39°40′N	106°48′E
Wuhan, China (wo͞o-hän′)	258-59	30°34′N	114°17′E
Wuhu, China (wo͞o′ho͞o)	258-59	31°21′N	118°22′E
Wŭjang, China	254-55	33°37′N	79°48′E
Wukari, Nig.	282-83	7°53′N	9°47′E
Wuliang Shan, mts., China	258-59	24°29′N	100°39′E
Wuliaru, Pulau, i., Indon.	242-43	7°27′S	131°04′E
Wunnummin Lake, lk., On., Can.	144-45	52°55′N	89°10′W
Wupatki National Monument, n.p., Az., U.S.	128-29	35°32′N	111°26′W
Wuppertal, Ger. (vo͝op′ēr-täl)	210-11	51°17′N	7°11′E
Würzburg, Ger. (vürts′bo͝orgh)	210-11	49°48′N	9°56′E
Wushan, China	258-59	31°06′N	109°50′E
Wushenqi, China	260-61	38°58′N	109°01′E
Wusuli, stm., Asia *see* Ussuri	260-61	48°27′N	135°04′E
Wutai, China	260-61	38°44′N	113°21′E
Wutai Shan, mtn., China	260-61	39°04′N	113°35′E
Wutongqiao, China	258-59	29°24′N	103°49′E
Wutsin, China *see* Changzhou	258-59	31°47′N	119°57′E
Wuvulu Island, i., Pap. N. Gui.	302	1°45′S	142°50′E
Wuwei, China (wo͞o′wā′)	258-59	31°18′N	117°54′E
Wuwei, China (wo͞o′wā′)	260-61	37°56′N	102°38′E
Wuxi, China (wo͞o-shyē)	258-59	31°22′N	109°33′E
Wuxi, China (wo͞o-shyē)	258-59	31°35′N	120°18′E
Wuxing, China (wo͞o-shyĭŋ) *see* Huzhou	258-59	30°52′N	120°06′E
Wuyi Shan, mts., China (wo͞o-yē shän)	258-59	27°42′N	117°09′E
Wuyuan, China	260-61	41°03′N	108°22′E
Wuzhong, China	260-61	37°59′N	106°12′E
Wuzhou, China (wo͞o-jō)	258-59	23°30′N	111°21′E
Wyandotte, Mi., U.S. (wī′ăn-dŏt)	126-27	42°13′N	83°09′W
Wyandra, Austl.	301	27°16′S	145°59′E
Wymore, Ne., U.S. (wī′mōr)	130-31	40°07′N	96°40′W
Wyndham, Austl. (wĭnd′ăm)	294-95	15°29′S	128°07′E
Wynne, Ar., U.S. (wĭn)	134-35	35°13′N	90°48′W
Wynnewood, Ok., U.S. (wĭn′wo͝od)	130-31	34°39′N	97°10′W
Wynyard, Sk., Can. (wĭn′yērd)	144-45	51°47′N	104°10′W
Wyoming, Mi., U.S. (wī-ō′mĭng)	126-27	42°55′N	85°43′W
Wyoming, state, U.S. (wī-ō′mĭng)	118-19	43°0′N	107°30′W
Wyong, Austl.	301	33°17′S	151°25′E
Wyszków, Pol. (vêsh′ko͝of)	210-11	52°36′N	21°28′E
Wytheville, Va., U.S. (wĭth′vĭl)	134-35	36°57′N	81°06′W

X

Feature (Pronunciation)	Page	Lat.	Long.
Xaafuun, Raas, c., Som.	284-85	10°26′N	51°25′E
Xaidulla, China	254-55	36°26′N	77°58′E
Xainza, China	254-55	30°55′N	88°40′E
Xai-Xai, Moz.	286-87	25°03′S	33°39′E
Xalapa, Mex.	158-59	19°32′N	96°55′W
Xam, stm., Asia *see* Chu	266-67	19°53′N	105°45′E
Xam Nua, Laos	266-67	20°25′N	104°03′E
Xankändi, Azer.	245	39°49′N	46°45′E
Xapuri, Braz.	180-81	10°39′S	68°31′W
Xar Moron, stm., China	260-61	43°25′N	120°45′E
Xau, Lake, pl., Bots.	286-87	21°18′S	24°44′E
Xäzär Dänizi, lk., *see* Caspian Sea	244	41°18′N	50°59′E
Xcalak, Mex. (sä-lä′k)	160	18°16′N	87°50′W
Xenia, Oh., U.S. (zē′nĭ-ȧ)	126-27	39°41′N	83°56′W
Xeres, Spain *see* Jerez de la Frontera	214-15	36°42′N	6°08′W
Xi, stm., China (shyē)	258-59	22°20′N	113°18′E
Xi, stm., China (shyē)	260-61	42°25′N	100°55′E
Xiaguan, China *see* Dali	258-59	25°36′N	100°13′E
Xiahe, China	260-61	35°24′N	102°32′E
Xiamen, China	243a	24°27′N	118°07′E
Xi'an, China (shyē-än)	258-59	34°15′N	108°52′E
Xiangfan, China	258-59	32°02′N	112°09′E
Xianggang, China *see* Hong Kong	258-59	22°16′N	114°10′E
Xiangkhoang, Laos	266-67	19°20′N	103°22′E
Xiangquan, stm., Asia *see* Sutlej	254-55	29°21′N	71°02′E
Xiangride, China	260-61	35°60′N	97°59′E
Xiangtan, China (shyäŋ-tän)	258-59	27°51′N	112°54′E
Xiantao, China	258-59	30°22′N	113°27′E
Xianyang, China (shyĕn-yäŋ)	258-59	34°20′N	108°42′E
Xianyou, China	243a	25°22′N	118°40′E
Xiaogan, China	258-59	30°55′N	113°54′E
Xiao Hinggan Ling, mts., China *see* Lesser Khingan Range	260-61	48°45′N	127°00′E
Xiapu, China (shyä-po͞o)	258-59	26°52′N	120°01′E
Xibaxa, stm., Asia *see* Subansiri	258-59	26°46′N	93°45′E
Xichang, China	258-59	27°54′N	102°16′E
Xicoténcatl, Mex. (sē-kô-tĕn-kät′l)	158-59	23°00′N	98°56′W
Xifeng, China (shyē-fŭŋ)	260-61	42°44′N	124°43′E
Xigazê, China	254-55	29°16′N	88°54′E
Xilinhot, China	260-61	43°56′N	116°03′E
Ximiao, China	260-61	41°07′N	100°17′E
Xinchang, China (shyĭn-chäŋ)	258-59	29°31′N	120°53′E
Xing'an, China (shyĭŋ-än)	258-59	25°37′N	110°31′E
Xinghua, China (shyĭŋ-hwä)	258-59	32°56′N	119°50′E
Xingkai Hu, lk., Asia *see* Khanka, Lake	264	45°11′N	132°25′E
Xingtai, China (shyĭŋ-tī)	260-61	37°04′N	114°30′E
Xingu, stm., Braz. (zhĕŋ-go͞o′)	180-81	1°30′S	51°50′W
Xingyi, China	258-59	25°05′N	104°54′E
Xinhua, China (shyĭn-hwä)	258-59	27°37′N	111°02′E
Xining, China (shyē-nĭŋ)	260-61	36°38′N	101°50′E
Xinjiang, China	260-61	35°37′N	111°13′E
Xinjiang, state, China (shyĭn-jyäŋ)	238-39	40°0′N	85°00′E
Xinjiulong, China	258-59	22°21′N	114°10′E
Xinmin, China (shyĭn-mĭn)	263	41°59′N	122°50′E
Xinpu, China *see* Lianyungang	258-59	34°37′N	119°11′E
Xintai, China (shyĭn-tī)	260-61	35°54′N	117°46′E
Xinxian, China (shyĭn shyĕn)	260-61	38°24′N	112°44′E
Xinxiang, China (shyĭn-shyäŋ)	260-61	35°18′N	113°52′E
Xinyang, China (shyĭn-yäŋ)	258-59	32°07′N	114°04′E
Xinye, China (shyĭn-yŭ)	258-59	32°33′N	112°21′E
Xiping, China (shyē-pĭŋ)	258-59	33°23′N	114°01′E
Xique-Xique, Braz.	180-81	10°50′S	42°43′W
Xırdalan, Azer.	245	40°28′N	49°46′E
Xisha Qundao, is., China *see* Paracel Islands	242-43	15°46′N	112°17′E
Xishui, China (shyē-shwä)	258-59	30°28′N	115°15′E
Xixian, China (shyē shyĕn)	258-59	32°21′N	114°44′E
Xizang, state, China (shyē-dzäŋ) *see* Tibet	254-55	32°0′N	88°00′E
Xongka, stm., Asia *see* Ca	266-67	18°44′N	105°45′E
Xuancheng, China (shyůän-chŭŋ)	258-59	30°57′N	118°45′E
Xuanhua, China (shyůän-hwä)	260-61	40°36′N	115°02′E
Xuchang, China (shyo͞o-chäŋ)	258-59	34°02′N	113°49′E
Xun, stm., China (shyòn)	258-59	23°26′N	111°30′E
Xuwen, China	258-59	20°20′N	110°11′E
Xuyong, China	258-59	28°10′N	105°25′E
Xuzhou, China	258-59	34°16′N	117°11′E

Y

Feature (Pronunciation)	Page	Lat.	Long.
Yaan, China (yä-än)	258-59	30°01′N	103°04′E
Yablonovy Range, mts., Russia (yȧ-blô-nô-vē′ fănj) *see* Yablonovyy Khrebet	236-37	53°30′N	115°00′E
Yablonovyy Khrebet, mts., Russia	236-37	53°30′N	115°00′E
Yaco, stm., S.A. *see* Iaco	184	9°02′S	68°35′W
Yacuiba, Bol. (yá-ko͞o-ē′bȧ)	182-83	22°02′S	63°42′W
Yacyretá, Isla, i., Para.	187	27°25′S	56°30′W
Yadong, China	254-55	27°29′N	88°54′E
Yafran, Libya	204-05	32°04′N	12°31′E
Yagodnoye, Russia	236-37	62°32′N	149°37′E
Yaguajay, Cuba (yä-guä-hä′ē)	154-55	22°20′N	79°14′W
Yahualica, Mex. (yä-wä-lē′kä)	158-59	21°09′N	102°51′W
Yaitopya, nation, Afr. *see* Ethiopia	274	9°0′N	39°00′E
Yakima, Wa., U.S. (yăk′ĭmȧ)	122-23	46°36′N	120°30′W
Yakoma, D.R.C.	284-85	4°04′N	22°26′E
Yaku-shima, i., Japan (yä′ko͞o shē′mä)	264	30°20′N	130°30′E
Yakutat, Ak., U.S. (yăk′ô-tăt)	136	59°32′N	139°43′W
Yakutat Bay, b., Ak., U.S. (yo͞o-kū-tăt′ bā)	136	59°40′N	140°00′W
Yakutia, state, Russia	236-37	67°0′N	125°00′E
Yakutsk, Russia (yȧ-ko͝otsk′)	236-37	62°02′N	129°42′E
Yala, Thai.	266-67	6°33′N	101°17′E
Yalgoo, Austl.	294-95	28°21′S	116°41′E
Yalong, stm., China (yä-lôŋ)	258-59	26°36′N	101°48′E
Yalta, Ukr. (yäl′tȧ)	218-19	44°30′N	34°10′E
Yalu, stm., Asia	263	39°57′N	124°22′E
Yamagata, Japan	265	38°15′N	140°20′E
Yamaguchi, Japan	265	34°11′N	131°29′E
Yamal, Poluostrov, pen., Russia (yä-mäl′)	236-37	70°0′N	70°00′E

PRONOUNCING INDEX
Yamal Peninsula – Yü Shan 425

Feature (Pronunciation)	Page	Lat.	Long.
Yamal Peninsula, pen., Russia (yŭ-mäl´ pĕ-nĭn´sū*lȧ*) *see* Yamal, Poluostrov	**236-37**	70°0'N	70°00'E
Yamantau, Gora, mtn., Russia (gȧ-rä´ yä´man-tȧw)	**244**	54°15'N	58°06'E
Yamarovka, Russia	**260-61**	50°34'N	110°25'E
Yambio, S. Sudan	**289**	4°34'N	28°24'E
Yambol, Blg. (yȧm´bôl)	**216-17**	42°29'N	26°31'E
Yamdena, Pulau, i., Indon.	**242-43**	7°36'S	131°25'E
Yamethin, Mya. (yŭ-mē´thĕn)	**266-67**	20°28'N	96°09'E
Yamma Yamma, Lake, lk., Austl. (lāk yăm´ȧ yăm´ȧ)	**301**	26°20'S	141°25'E
Yamoussoukro, nat. cap., C. Iv. (yä-mo͞o-so͞o´-krō)	**282-83**	6°49'N	5°17'W
Yamuna, stm., India	**254-55**	25°26'N	81°54'E
Yamzho Yumco, lk., China (yäm-jwo yo͞om-tswo)	**254-55**	28°58'N	90°45'E
Yana, stm., Russia (yä´nȧ)	**236-37**	71°32'N	136°38'E
Yanac, Austl. (yăn´*ă*k)	**301**	36°09'S	141°25'E
Yanam, India (yŭnŭm´)	**256**	16°44'N	82°15'E
Yan'an, China (yän-än)	**260-61**	36°35'N	109°29'E
Yanbu'al-Baḥr, Sau. Ar.	**288**	24°05'N	38°05'E
Yanchang, China	**260-61**	36°35'N	110°01'E
Yancheng, China (yän-chŭŋ)	**258-59**	33°24'N	120°09'E
Yanchi, China	**260-61**	37°47'N	107°23'E
Yandé, Île, i., N. Cal.	**306g**	20°03'S	163°48'E
Yangambi, D.R.C.	**284-85**	0°46'N	24°27'E
Yangchun, China (yäŋ-chȯn)	**258-59**	22°11'N	111°47'E
Yanggu, China (yäŋ-go͞o)	**260-61**	36°07'N	115°46'E
Yangiyŭl, Uzb.	**252-53**	41°09'N	69°07'E
Yangjiang, China (yäŋ-jyäŋ)	**258-59**	21°53'N	111°58'E
Yangon, nat. cap., Mya. (yăn´gŏn´)	**266-67**	16°47'N	96°12'E
Yangquan, China (yäŋ-chyůän)	**260-61**	37°51'N	113°34'E
Yangtze, stm., China (yäng´tse)	**258-59**	31°24'N	121°54'E
Yangxin, China (yäŋ-shyĭn)	**258-59**	29°50'N	115°13'E
Yangzhou, China (yäŋ-jō)	**258-59**	32°24'N	119°25'E
Yanji, China (yän-jyē)	**263**	42°46'N	129°26'E
Yanji, China (yän-jyē)	**263**	42°54'N	129°30'E
Yankton, S.D., U.S. (yănk´t*ŭ*n)	**124-25**	42°52'N	97°24'W
Yanqi, China	**240-41**	42°03'N	86°34'E
Yanshan, China (yän-shän)	**258-59**	23°37'N	104°20'E
Yanshou, China (yän-shō)	**264**	45°26'N	128°21'E
Yantai, China	**260-61**	37°32'N	121°21'E
Yanzhou, China (yäŋ-jō)	**260-61**	35°33'N	116°49'E
Yaoundé, nat. cap., Camrn. (yȧ´-o͞on-dā´)	**282-83**	3°52'N	11°31'E
Yap, i., Micron. (yăp)	**304-05**	9°31'N	138°06'E
Yapacaní, Bol.	**182-83**	16°45'S	64°18'W
Yapen, Pulau, i., Indon.	**242-43**	1°45'S	136°15'E
Yaponskoye More, s., Asia *see* Japan, Sea of	**240-41**	40°0'N	135°00'E
Yaque del Norte, stm., Dom. Rep. (yä´kȧ dĕl nôr´tȧ)	**154-55**	19°50'N	71°41'W
Yaqui, stm., Mex. (yä´kē)	**156-57**	27°40'N	110°38'W
Yaransk, Russia (yä-ränsk´)	**202-03**	57°18'N	47°54'E
Yarensk, Russia	**202-03**	62°09'N	49°02'E
Yarí, stm., Col.	**178-79**	0°19'S	72°21'W
Yarkand, stm., China *see* Yarkant	**244**	40°28'N	80°51'E
Yarkant, China *see* Shache	**244**	38°25'N	77°15'E
Yarkant, stm., China	**244**	40°28'N	80°51'E
Yarlung, stm., Asia *see* Brahmaputra	**254-55**	24°02'N	91°00'E
Yarmouth, N.S., Can. (yär´m*ŭ*th)	**148-49**	43°50'N	66°06'W
Yaroslavl', Russia (yȧ-rô-släv´'l)	**218-19**	57°37'N	39°52'E
Yar-Sale, Russia	**236-37**	66°51'N	70°53'E
Yartsevo, Russia (yär´tsyĕ-vô)	**218-19**	55°04'N	32°42'E
Yarumal, Col. (yä-ro͞o-mäl´)	**178-79**	6°58'N	75°24'W
Yasawa, i., Fiji	**306f**	16°47'S	177°31'E
Yashiro-jima, i., Japan	**265**	33°55'N	132°15'E
Yass, Austl.	**301**	34°51'S	148°55'E
Yassy, Rom. *see* Iași	**218-19**	47°10'N	27°36'E
Yata, stm., Bol.	**180-81**	10°29'S	65°26'W
Yathkyed Lake, lk., Nu., Can. (yȧth-kī-ĕd´ lāk)	**140-41**	62°41'N	98°00'W
Yatsuga-take, mtn., Japan (yät´so͞o-gä-dä´kä)	**265**	35°59'N	138°23'E
Yatsushiro, Japan (yät´so͞o´shē-rō)	**265**	32°30'N	130°36'E
Yaundé, nat. cap., Camrn. *see* Yaoundé	**282-83**	3°52'N	11°31'E
Yautepec, Mex. (yä-o͞o-tȧ-pĕk´)	**158-59**	18°52'N	99°03'W
Yavarí, stm., S.A.	**184**	4°21'S	70°02'W
Yaví, Cerro, mtn., Ven.	**178-79**	5°32'N	65°59'W
Yawata, Japan (yä´wä-tä) *see* Kitakyūshū	**265**	33°54'N	130°51'E
Yawatahama, Japan (yä´wä´tä´hä-mä)	**265**	33°27'N	132°26'E
Yaxian, China (yä shyĕn) *see* Sanya	**258-59**	18°14'N	109°30'E
Yazd, Iran	**252-53**	31°54'N	54°22'E
Yazoo, stm., Ms., U.S. (yă´zo͞o)	**134-35**	32°23'N	91°00'W
Yazoo City, Ms., U.S. (yă´zo͞o sĭ´tĭ)	**134-35**	32°51'N	90°25'W
Ye, Mya. (yā)	**266-67**	15°15'N	97°52'E
Yecla, Spain (yā´klä)	**214-15**	38°37'N	1°07'W
Yefremov, Russia (yĕ-frä´môf)	**218-19**	53°09'N	38°07'E
Yegor'yevsk, Russia (yĕ-gôr´yĕfsk)	**218-19**	55°23'N	39°02'E
Yei, S. Sudan	**289**	4°06'N	30°40'E
Yei, stm., S. Sudan	**284-85**	6°15'N	30°13'E
Yekaterinburg, Russia	**236-37**	56°51'N	60°36'E
Yekaterinoslav, Ukr. *see* Dnipropetrovs'k	**218-19**	48°28'N	34°58'E
Yelabuga, Russia (yĕ-lä´bȯ-gȧ)	**202-03**	55°46'N	52°05'E
Yelizavety, Mys, c., Russia (mĭs yĕ-lyē-sȧ-vyĕ´tĭ)	**236-37**	54°24'N	142°42'E
Yellow, stm., China *see* Huang	**240-41**	37°49'N	118°53'E
Yellowhead Pass, p., Can. (yĕl´ō-hĕd păs)	**142-43**	52°54'N	118°22'W
Yellowknife, N.T., Can. (yĕl´ō-nīf)	**138-39**	62°27'N	114°21'W
Yellow Sea, s., Asia (yĕl´ō sē)	**240-41**	36°0'N	123°00'E
Yellowstone, stm., U.S. (yĕl´ō-stōn)	**120-21**	47°59'N	103°60'W
Yellowstone Lake, lk., Wy., U.S.	**122-23**	44°28'N	110°23'W
Yellowstone National Park, n.p., U.S. (yĕl´ō-stōn năsh´ŭn-*ă*l pärk)	**122-23**	44°30'N	110°35'W
Yel'nya, Russia (yĕl´nyȧ)	**218-19**	54°34'N	33°11'E
Yemen, nation, Asia (yĕm´*ĕ*n)	**224-25**	15°0'N	44°00'E
Yenangyaung, Mya. (yā´nän-d oung)	**266-67**	20°27'N	94°53'E
Yen Bai, Viet.	**266-67**	21°42'N	104°52'E
Yendi, Ghana (yĕn´dẻ)	**282-83**	9°26'N	0°00'W
Yenisey, stm., Russia (yĕ-nẻ-sĕ´ẻ)	**236-37**	71°54'N	82°20'E
Yeo Lake, lk., Austl. (yō lāk)	**296-97**	27°58'S	124°25'E
Yeongdeok, Kor., S.	**263**	36°26'N	129°23'E
Yeosu, Kor., S.	**263**	34°44'N	127°44'E
Yeppoon, Austl.	**302**	23°08'S	150°45'E
Yerevan, nat. cap., Arm. (yĕ-rĕ-vän´)	**245**	40°11'N	44°30'E
Yergeni, hills, Russia	**202-03**	47°0'N	44°00'E
Yerington, Nv., U.S. (yẻ´rĭng-t*ŭ*n)	**128-29**	38°59'N	119°09'W
Yershov, Russia	**202-03**	51°21'N	48°16'E
Yerupaja, Nevado, mtn., Peru	**184**	10°16'S	76°54'W
Yerushalayim, nat. cap., Isr. *see* Jerusalem	**248-49**	31°47'N	35°14'E
Yessentuki, Russia	**245**	44°02'N	42°51'E
Ye-u, Mya.	**266-67**	22°40'N	95°20'E
Yeu, Île d', i., Fr. (ēl dyû)	**212-13**	46°42'N	2°20'W
Yevlax, Azer.	**245**	40°37'N	47°09'E
Yevpatoriia, Ukr.	**218-19**	45°12'N	33°22'E
Yeya, stm., Russia (yă´yȧ)	**202-03**	46°40'N	38°36'E
Yeysk, Russia (yĕysk)	**218-19**	46°42'N	38°10'E
Yezd, Iran *see* Yazd	**252-53**	31°54'N	54°22'E
Ygatimí, Para.	**182-83**	24°05'S	55°24'W
Yi'an, China	**260-61**	47°53'N	125°18'E
Yibin, China (yē-bĭn)	**258-59**	28°46'N	104°37'E
Yichang, China (yē-chäŋ)	**258-59**	30°42'N	111°17'E
Yichun, China	**258-59**	27°48'N	114°23'E
Yichun, China	**260-61**	47°43'N	128°55'E
Yidu, China (yē-do͞o)	**258-59**	30°24'N	111°26'E
Yilan, China (yē-län)	**264**	46°18'N	129°32'E
Yiliang, China	**258-59**	24°57'N	103°08'E
Yinchuan, China (yĭn-chůän)	**260-61**	38°28'N	106°16'E
Ying, stm., China	**258-59**	32°30'N	116°31'E
Yingkou, China (yĭŋ-kō)	**260-61**	40°40'N	122°14'E
Yingtan, China	**258-59**	28°14'N	117°02'E
Yining, China (yē-nĭŋ)	**244**	43°55'N	81°18'E
Yirga 'Alem, Eth.	**292d**	6°52'N	38°24'E
Yirol, S. Sudan	**284-85**	6°33'N	30°30'E
Yishui, China (yē-shwā)	**260-61**	35°47'N	118°38'E
Yisra'el, nation, Asia *see* Israel	**224-25**	31°30'N	34°45'E
Yitulihe, China	**260-61**	50°37'N	121°33'E
Yiyang, China (yē-yäŋ)	**258-59**	28°24'N	117°26'E
Yiyang, China (yē-yän)	**258-59**	28°35'N	112°20'E
Yoakum, Tx., U.S. (yō´k*ŭ*m)	**132-33**	29°17'N	97°09'W
Yockanookany, stm., Ms., U.S. (yŏk´ȧ-no͞o-kă-nī)	**134-35**	32°40'N	89°41'W
Yog Point, c., Phil. (yōg point)	**270**	14°06'N	124°12'E
Yogyakarta, Indon. (yŏg-yȧ-kär´tȧ)	**268-69**	7°48'S	110°22'E
Yokadouma, Camrn.	**282-83**	3°31'N	15°03'E
Yokkaichi, Japan (yō´kä´ē-chē)	**265**	34°58'N	136°39'E
Yokohama, Japan (yō´kō-hä´mä)	**265**	35°27'N	139°37'E
Yokosuka, Japan (yō-kō´sȯ-kä)	**265**	35°16'N	139°40'E
Yolöten, Turkmen.	**252-53**	37°17'N	62°22'E
Yom, stm., Thai.	**266-67**	15°52'N	100°16'E
Yonago, Japan (yō´nä-gō)	**265**	35°26'N	133°20'E
Yonezawa, Japan (yō´nĕ´zȧ-wä)	**265**	37°55'N	140°07'E
Yong'an, China (yȯŋ-än)	**258-59**	25°58'N	117°22'E
Yongdeng, China	**260-61**	36°43'N	103°16'E
Yongding, China	**258-59**	24°44'N	116°44'E
Yongding, stm., China (yȯŋ-dĭŋ)	**260-61**	39°16'N	117°04'E
Yongfeng, China	**258-59**	27°19'N	115°24'E
Yongnian, China (yȯŋ-nĭĕn)	**260-61**	36°47'N	114°29'E
Yongren, China	**258-59**	26°06'N	101°48'E
Yongshan, China	**258-59**	28°09'N	103°32'E
Yongshun, China (yȯŋ-shȯn)	**258-59**	29°00'N	109°51'E
Yongzhou, China	**258-59**	26°13'N	111°37'E
Yonkers, N.Y., U.S. (yŏŋ´kẽrz)	**126-27**	40°56'N	73°54'W
York, Austl.	**294-95**	31°53'S	116°46'E
York, Eng., U.K.	**206-07**	53°58'N	1°05'W
York, Al., U.S. (yôrk)	**134-35**	32°30'N	88°18'W
York, Ne., U.S. (yôrk)	**124-25**	40°52'N	97°35'W
York, Pa., U.S.	**126-27**	39°58'N	76°44'W
York, S.C., U.S. (yôrk)	**134-35**	34°59'N	81°15'W
York, Cape, c., Austl. (kāp yôrk)	**302**	10°42'S	142°32'E
York Factory, Mb., Can.	**144-45**	57°03'N	92°15'W
Yorkton, Sk., Can. (yôrk´t*ŭ*n)	**144-45**	51°13'N	102°28'W
Yorktown, Tx., U.S. (yôrk´toun)	**132-33**	28°59'N	97°30'W
Yorktown, Va., U.S. (yôrk´toun)	**134-35**	37°14'N	76°31'W
Yoro, Hond. (yō´rō)	**161**	15°09'N	87°08'W
Yosemite National Park, n.p., Ca., U.S. (yȯ-sĕm´ĭ-tẻ năsh´ŭn-*ă*l pärk)	**128-29**	37°56'N	119°36'W
Yoshino, stm., Japan (yō´shē-nō)	**265**	34°04'N	134°37'E
Yoshkar-Ola, Russia (yôsh-kär´ō-lä´)	**202-03**	56°39'N	47°52'E
Yösöbulag, Mong. *see* Altay	**260-61**	46°24'N	96°15'E
Yos Sudarso, Pulau, i., Indon.	**302**	7°50'S	138°30'E
You, stm., China (yō)	**258-59**	22°50'N	108°06'E
You, stm., China (yō)	**243a**	26°24'N	118°27'E
You, stm., China (yō)	**258-59**	28°27'N	110°23'E
Youghal, Ire. (yo͞o´ôl) (yôl)	**206-07**	51°57'N	7°51'W
Young, Austl. (yŭng)	**301**	34°19'S	148°18'E
Young, Ur. (yô-o͞o´ng)	**187**	32°42'S	57°38'W
Youngstown, Oh., U.S.	**126-27**	41°06'N	80°39'W
Youssoufia, Mor.	**292a**	32°15'N	8°32'W
Youyang, China	**258-59**	28°58'N	108°41'E
Yozgat, Tur. (yŏz´gȧd)	**202-03**	39°49'N	34°48'E
Ypsilanti, Mi., U.S. (ĭp-sĭ-lăn´tĭ)	**126-27**	42°15'N	83°37'W
Yreka, Ca., U.S. (wī-rē´k*ȧ*)	**122-23**	41°44'N	122°38'W
Ystad, Swe.	**208-09**	55°26'N	13°50'E
Ysyk-Köl, lk., Kyrg. *see* Issyk-Kul, Lake	**244**	42°25'N	77°15'E
Ytyk-Kyuyël', Russia	**236-37**	62°21'N	133°33'E
Yu, stm., China	**258-59**	23°24'N	110°06'E
Yuan, stm., Asia *see* Red	**258-59**	20°18'N	106°32'E
Yuan, stm., China (yůän)	**258-59**	28°45'N	111°20'E
Yuan'an, China (yůän-än)	**258-59**	31°04'N	111°25'E
Yuanling, China (yůän-lĭŋ)	**258-59**	28°20'N	110°16'E
Yuanmou, China	**258-59**	25°43'N	101°52'E
Yuba City, Ca., U.S. (yo͞o´b*ȧ* sĭ´tĭ)	**128-29**	39°08'N	121°37'W
Yucatán, state, Mex. (yo͞o-kä-tän´)	**160**	20°50'N	89°00'W
Yucatán, Canal de, strt., N.A. *see* Yucatan Channel	**152-53**	21°42'N	86°04'W
Yucatán, Península de, pen., N.A. *see* Yucatan Peninsula	**160**	19°30'N	89°00'W
Yucatan Channel, strt., N.A. (yo͞o-kä-tän´ chăn´ĕl)	**152-53**	21°42'N	86°04'W
Yucatan Peninsula, pen., N.A. (yo͞o-kä-tän´ pĕ-nĭn´sū*lȧ*)	**160**	19°30'N	89°00'W
Yuci, China (yo͞o-tsz)	**260-61**	37°41'N	112°44'E
Yudoma, stm., Russia (yo͞o-dō´mȧ)	**236-37**	59°10'N	135°14'E
Yueqing, China (yůĕ-chyĭn)	**258-59**	28°07'N	120°57'E
Yueyang, China (yůĕ-yäŋ)	**258-59**	29°22'N	113°06'E
Yug, stm., Russia (yȯg)	**202-03**	60°43'N	46°19'E
Yukagirskoye Ploskogor'ye, plat., Russia	**236-37**	66°0'N	155°00'E
Yukhnov, Russia (yȯk´nof)	**218-19**	54°44'N	35°15'E
Yukon, state, Can. (yo͞o´kŏn)	**138-39**	64°0'N	135°00'W
Yukon, stm., N.A. (yo͞o´kŏn)	**136**	62°36'N	164°49'W
Yulin, China (yo͞o-lĭn)	**258-59**	22°38'N	110°07'E
Yulin, China (yo͞o-lĭn)	**260-61**	38°17'N	109°45'E
Yuma, Az., U.S. (yo͞o´mä)	**128-29**	32°43'N	114°37'W
Yuma, Co., U.S. (yo͞o´mä)	**130-31**	40°07'N	102°44'W
Yumen, China (yo͞o-mŭn)	**260-61**	39°50'N	97°34'E
Yuncheng, China (yȯn-chŭŋ)	**258-59**	35°01'N	110°59'E
Yuncheng, China (yȯn-chŭŋ)	**260-61**	35°36'N	115°56'E
Yunnan, state, China (yun´nän´)	**258-59**	24°0'N	101°00'E
Yunxian, China (yȯn shyĕn)	**258-59**	24°31'N	100°02'E
Yunxian, China (yȯn shyĕn)	**258-59**	32°49'N	110°49'E
Yunxiao, China (yȯn-shyou)	**258-59**	24°04'N	117°19'E
Yurga, Russia	**236-37**	55°43'N	84°55'E
Yurimaguas, Peru (yo͞o-rē-mä´gwäs)	**184**	5°54'S	76°06'W
Yuruá, stm., S.A. *see* Juruá	**180-81**	2°35'S	65°47'W
Yuryev, Est. *see* Tartu	**208-09**	58°23'N	26°43'E
Yuscarán, Hond. (yo͞os-kä-rän´)	**161**	13°56'N	86°50'W
Yü Shan, mtn., Tai.	**243a**	23°28'N	120°57'E

n-sing; ŋ-baŋk; ɴ-nasalized n; nŏd; c*ŏ*mmit; ōld; ôbey; ôrder; oi-boil; fo͞od; ȯ-as oo in foot; ou-out; s-soft; sh-dish; th-thin; pūre; ůnite; ûrn; stŭd; circ*ŭ*s; ü-as in French tu; ´-indeterminate vowel.

Feature (Pronunciation)	Page	Lat.	Long.
Yushu, China (yōō-shōō)	258-59	33°00′N	97°00′E
Yushu, China (yōō-shōō)	260-61	44°50′N	126°34′E
Yutian, China (yōō-tĭĕn) (kü-r-yä)	254-55	36°51′N	81°40′E
Yuty, Para. (yōō-tē′)	187	26°38′S	56°11′W
Yuxian, China (yōō shyĕn)	258-59	34°10′N	113°28′E
Yuxian, China (yōō shyĕn)	260-61	38°06′N	113°24′E
Yuzha, Russia (yōō′zhä)	202-03	56°35′N	42°01′E
Yuzhno-Sakhalinsk, Russia (yōōzh′nô-sä-кä-lĭnsk′)	264	46°58′N	142°43′E
Yuzovka, Ukr. see Donets′k	218-19	47°60′N	37°48′E
Yvetot, Fr. (ēv-tō′)	212-13	49°37′N	0°46′E

Z

Feature (Pronunciation)	Page	Lat.	Long.
Zaandam, Neth. (zän′däm)	206-07	52°26′N	4°49′E
Zabaykal′sk, Russia	260-61	49°39′N	117°19′E
Zabīd, Yemen	288	14°11′N	43°19′E
Zābol, Iran	250-51	31°02′N	61°30′E
Zabrze, Pol. (zäb′zhĕ)	210-11	50°18′N	18°46′E
Zacapa, Guat. (sä-kä′pä)	160	14°58′N	89°32′W
Zacatecas, Mex. (sä-kä-tä′käs)	158-59	22°46′N	102°34′W
Zacatecas, state, Mex.	158-59	23°0′N	103°00′W
Zacatlán, Mex. (sä-kä-tlän′)	158-59	19°56′N	97°58′W
Zacoalco de Torres, Mex. (sä-kô-äl′kô dä tōr′rěs)	158-59	20°13′N	103°34′W
Zacualtipan, Mex. (sä-kô-äl-tē-pän′)	158-59	20°38′N	98°39′W
Zadar, Cro. (zä′där)	216-17	44°07′N	15°14′E
Zagazig, Egypt	291b	30°35′N	31°31′E
Zagreb, nat. cap., Cro. (zä′grěb)	216-17	45°49′N	15°58′E
Zāgros, Kūhhā-ye, mts., Iran see Zagros Mountains	238-39	33°03′N	48°33′E
Zagros Mountains, mts., Iran (zä′grŏs moun′tĭnz)	238-39	33°03′N	48°33′E
Za′gya, stm., China	254-55	31°56′N	88°60′E
Zāhedān, Iran (zä′hå-dän)	250-51	29°30′N	60°52′E
Zaḥlah, Leb. (zä′lä′)	248-49	33°51′N	35°55′E
Zaḥodnjaja Dzvina, stm., Eur. see Western Dvina	208-09	57°04′N	24°03′E
Zaire, stm., Afr. see Congo	284-85	5°58′S	12°44′E
Zaječar, Serb. (zä′yě-chär′)	216-17	43°55′N	22°17′E
Zakamensk, Russia	260-61	50°23′N	103°16′E
Zakopane, Pol. (zá-kô-pä′ně)	210-11	49°18′N	19°58′E
Zákynthos, i., Grc.	216-17	37°47′N	20°47′E
Zalaegerszeg, Hung. (zŏ′lŏ-ě′gěr-sěg)	210-11	46°51′N	16°51′E
Zalantun, China	260-61	47°60′N	122°45′E
Zalingei, Sudan	12°55′N	23°29′E
Zambeze, stm., Afr. see Zambezi	286-87	18°49′S	36°15′E
Zambezi, stm., Afr. (zäm-bā′zě)	286-87	18°49′S	36°15′E
Zambia, nation, Afr. (zäm′bě-à)	274	14°30′S	27°30′E
Zamboanga, Phil. (säm-bô-aṅ′gä)	270	6°54′N	122°04′E
Zamboanga Peninsula, pen., Phil.	270	7°32′N	122°16′E
Zambrów, Pol. (zäm′brôf)	210-11	52°58′N	22°15′E
Zamfara, stm., Nig.	282-83	12°02′N	4°03′E
Zamora, Spain (thä-mō′rä)	214-15	41°31′N	5°45′W
Zamora de Hidalgo, Mex.	158-59	19°59′N	102°17′W
Zanesville, Oh., U.S. (zānz′vĭl)	126-27	39°56′N	82°01′W
Zanjān, Iran	248-49	36°40′N	48°29′E
Zante, i., Grc. see Zákynthos	216-17	37°47′N	20°47′E
Zanzibar, Tan. (zän′zĭ-bär)	284-85	6°10′S	39°12′E
Zanzibar, i., Tan. (zän′zĭ-bär)	284-85	6°10′S	39°20′E
Zaozhuang, China (dzou-jǔäṅ)	258-59	34°52′N	117°33′E
Zapadnaya Dvina, stm., Eur. see Western Dvina	208-09	57°04′N	24°03′E
Zapadno-Sibirskaya Ravnina, pl., Russia see West Siberian Plain	236-37	60°0′N	75°00′E
Zapala, Arg. (zä-pä′lä)	185	38°54′S	70°03′W
Zapata, Tx., U.S. (sä-pä′tä)	132-33	26°55′N	99°16′W
Zapata, Península de, pen., Cuba (pě-nē′n-sōō-lä-dě-zä-pá′tä)	154-55	22°20′N	81°35′W
Zapopan, Mex. (sä-pō′pän)	158-59	20°43′N	103°24′W
Zaporizhzhia, Ukr.	218-19	47°51′N	35°10′E
Zaporozhye, Ukr. see Zaporizhzhia	218-19	47°51′N	35°10′E
Zara, Cro. see Zadar	216-17	44°07′N	15°14′E
Zaragoza, Mex. (sä-rä-gō′sä)	158-59	23°56′N	99°46′W
Zaragoza, Mex. (sä-rä-gō′sä)	132-33	28°29′N	100°55′W
Zaragoza, Mex. (sä-rä-gō′sä)	132-33	31°43′N	106°23′W
Zaragoza, Spain (thä-rä-gō′thä)	214-15	41°39′N	0°53′W
Zaranj, Afg.	252-53	31°07′N	61°53′E
Zarasai, Lith. (zä-rä-sī′)	208-09	55°44′N	26°16′E
Zárate, Arg. (zä-rä′tā)	187	34°06′S	59°02′W
Zaraysk, Russia (zä-rä′ěsk)	218-19	54°45′N	38°53′E
Zard Kūh, mtn., Iran	252-53	32°22′N	50°04′E
Zarghon Shahr, Afg.	252-53	32°51′N	68°27′E
Zaria, Nig. (zä′rě-ä)	282-83	11°07′N	7°43′E
Zarzal, Col. (zär-zá′l)	188c	4°23′N	76°03′W
Žatec, Czech Rep. (zhä′těts)	210-11	50°20′N	13°33′E
Zavitinsk, Russia	260-61	50°07′N	129°27′E
Zawiercie, Pol. (zá-vyěr′tsyě)	210-11	50°30′N	19°26′E
Zaysan, Kaz. (zī′sán)	244	47°28′N	84°52′E
Zaza, stm., Cuba (zá′zä)	154-55	21°39′N	79°33′W
Zbarazh, Ukr. (zbä-räzh′)	210-11	49°40′N	25°48′E
Zbruch, stm., Ukr. (zbrŏch)	210-11	48°32′N	26°27′E
Zduńska Wola, Pol. (zdōōn′′skä vō′lä)	210-11	51°36′N	18°58′E
Zeehan, Austl.	301	41°53′S	145°20′E
Zeeland, Mi., U.S. (zē′lănd)	126-27	42°49′N	86°01′W
Zeerust, S. Afr.	292c	25°33′S	26°06′E
Zelënodol′sk, Russia	202-03	55°50′N	48°32′E
Zelenogorsk, Russia (zě-lä′nô-gôrsk)	208-09	60°12′N	29°43′E
Zémio, C.A.R. (za-myō′)	284-85	5°02′N	25°08′E
Zenica, Bos. (zě′nět-sä)	216-17	44°12′N	17°54′E
Zernograd, Russia	202-03	46°50′N	40°18′E
Zeya, Russia (zá′yä)	236-37	53°44′N	127°15′E
Zeya, stm., Russia	236-37	50°14′N	127°36′E
Zeyskoye Vodokhranilishche, res., Russia	236-37	54°28′N	127°45′E
Zêzere, stm., Port. (zě′zá-rě′)	214-15	39°28′N	8°20′W
Zgierz, Pol. (zgyězh)	210-11	51°52′N	19°25′E
Zhalpy Syrt, mts., Eur.	202-03	52°0′N	51°30′E
Zhambyl, Kaz. see Taraz	244	42°54′N	71°21′E
Zhangaözen, Kaz.	202-03	43°19′N	52°47′E
Zhangaqazaly, Kaz.	244	45°51′N	62°09′E
Zhangatas, Kaz.	244	43°34′N	69°44′E
Zhangdian, China see Zibo	260-61	36°47′N	118°03′E
Zhangjiakou, China	260-61	40°49′N	114°53′E
Zhangping, China	258-59	25°18′N	117°26′E
Zhangye, China (jäṅ-yu)	260-61	38°56′N	100°27′E
Zhangzhou, China (jäṅ-jō)	258-59	24°31′N	117°40′E
Zhänibek, Kaz.	202-03	49°26′N	46°50′E
Zhanjiang, China (jän-jyäṅ)	258-59	21°12′N	110°23′E
Zhao′an, China (jou-än)	258-59	23°44′N	117°09′E
Zhaodong, China (jou-dôṅ)	260-61	46°04′N	125°59′E
Zhaoqing, China	258-59	23°04′N	112°28′E
Zhaotong, China (jou-tôṅ)	258-59	27°21′N	103°43′E
Zhaoyuan, China (jou-yuän)	260-61	37°22′N	120°24′E
Zhaoyuan, China (jou-yuän)	260-61	45°31′N	125°09′E
Zharkent, Kaz.	244	44°10′N	79°60′E
Zhaxigang, China	254-55	32°31′N	79°41′E
Zhaysang köli, lk., Kaz.	244	48°0′N	84°00′E
Zhayyq, stm., see Ural.	192-93	46°50′N	51°33′E
Zhdanov, Ukr. see Mariupol′	218-19	47°06′N	37°34′E
Zhejiang, state, China (jü-jyäṅ)	258-59	29°0′N	120°00′E
Zheleznogorsk-Ilimskiy, Russia	236-37	56°35′N	104°08′E
Zheltye Vody, Ukr. see Zhovti Vody	218-19	48°21′N	33°31′E
Zhem, stm., Kaz.	244	46°43′N	53°08′E
Zhengding, China (jǔṅ-dĭṅ)	260-61	38°09′N	114°34′E
Zhengyang, China (jǔṅ-yäṅ)	258-59	32°36′N	114°23′E
Zhengzhou, China (jǔṅ-jō)	258-59	34°46′N	113°39′E
Zhenjiang, China (jǔṅ-jyäṅ)	258-59	32°12′N	119°25′E
Zhenping, China	258-59	33°08′N	112°19′E
Zhenyuan, China (jǔṅ-yüän)	258-59	26°52′N	108°19′E
Zhetiqara, Kaz.	244	52°11′N	61°12′E
Zhezqazghan, Kaz.	244	47°47′N	67°41′E
Zhidoi, China	258-59	33°08′N	94°50′E
Zhigansk, Russia (zhē-gánsk′)	236-37	66°48′N	123°21′E
Zhijiang, China (jr-jyäṅ)	258-59	27°26′N	109°40′E
Zhitomir, Ukr. see Zhytomyr	218-19	50°16′N	28°41′E
Zhob, Pak.	252-53	31°21′N	69°27′E
Zhob, stm., Pak.	252-53	32°03′N	69°49′E
Zhongba, China	254-55	29°54′N	83°40′E
Zhongdian, China	258-59	27°49′N	99°42′E
Zhongghar Alataū zhotasy, mts., Asia see Alataw Shan	244	45°0′N	81°00′E
Zhongguo, nation, Asia see China	224-25	35°0′N	105°00′E
Zhongning, China	260-61	37°28′N	105°39′E
Zhongshan, China	258-59	22°31′N	113°22′E
Zhongwei, China (jôṅ-wä)	260-61	37°31′N	105°11′E
Zhongxian, China (jôṅ shyěn)	258-59	30°29′N	108°05′E
Zhongxiang, China	258-59	31°10′N	112°35′E
Zhosaly, Kaz.	244	45°29′N	64°06′E
Zhoucun, China (jō-tsōōn)	260-61	36°48′N	117°50′E
Zhoushan Dao, i., China	258-59	30°03′N	122°08′E
Zhoushan Qundao, is., China (jō-shän-chyón-dou) see Choushan Islands	258-59	30°0′N	122°00′E
Zhovti Vody, Ukr.	218-19	48°21′N	33°31′E
Zhuanghe, China (jǔäṅ-hǔ)	263	39°42′N	122°58′E
Zhucheng, China (jōō-chǔṅ)	260-61	36°0′N	119°24′E
Zhuji, China (jōō-jyē) see Shangqiu.	260-61	34°27′N	115°39′E
Zhuji, China (jōō-jyē)	258-59	29°43′N	120°15′E
Zhujiang Kou, est., Asia (jōō-jyäṅ kō)	258-59	22°36′N	113°44′E
Zhukovskiy, Russia (zhô-kôf′skī)	218-19	55°36′N	38°09′E
Zhuozhou, China	260-61	39°29′N	115°58′E
Zhushan, China	258-59	32°10′N	110°19′E
Zhuzhou, China	258-59	27°50′N	113°09′E
Zhytomyr, Ukr.	218-19	50°16′N	28°41′E
Zi, stm., China (dzě)	258-59	28°44′N	112°32′E
Zi, stm., China (dzě)	258-59	31°58′N	96°57′E
Zibo, China (dzē-bwo)	260-61	36°47′N	118°03′E
Zielona Góra, Pol. (zhyě-lô′nä gōō′rä)	210-11	51°56′N	15°31′E
Zigong, China	258-59	29°22′N	104°45′E
Ziguinchor, Sen.	282-83	12°34′N	16°17′W
Zihuatanejo, Mex.	158-59	17°39′N	101°33′W
Zile, Tur. (zě-lě′)	202-03	40°18′N	35°54′E
Žilina, Slvk. (zhě′lĭ-nä)	210-11	49°13′N	18°44′E
Zillah, Libya	280-81	28°34′N	17°34′E
Zima, Russia (zě′má)	236-37	53°55′N	102°04′E
Zimapán, Mex. (sē-mä′pän)	158-59	20°44′N	99°23′W
Zimbabwe, nation, Afr. (zĭm-bäb′-wä) (zĭm-bäb′-wē)	274	20°0′S	30°00′E
Zimbabwe Ruins, hist., Zimb.	286-87	20°17′S	30°57′E
Zimnicea, Rom. (zěm-nē′chá)	216-17	43°40′N	25°23′E
Zinder, Niger (zĭn′děr)	280-81	13°48′N	8°59′E
Zion, Il., U.S. (zī′ŭn)	126-27	42°27′N	87°50′W
Zion National Park, n.p., Ut., U.S.	128-29	37°18′N	113°02′W
Zipaquirá, Col.	188c	5°00′N	74°01′W
Zittau, Ger. (tsě′tou)	210-11	50°54′N	14°48′E
Ziway, Lake, lk., Eth.	292d	7°59′N	38°50′E
Ziway Hāyk′, lk., Eth. see Ziway, Lake	292d	7°59′N	38°50′E
Ziya, stm., China (dzě-yä)	260-61	39°09′N	117°10′E
Ziyun, China	258-59	25°43′N	106°05′E
Zizhong, China	258-59	29°47′N	104°50′E
Zlatoust, Russia (zlä-tô-òst′)	244	55°10′N	59°40′E
Zlynka, Russia (zlěn′ká)	218-19	52°25′N	31°44′E
Znaim, Czech Rep. see Znojmo.	210-11	48°51′N	16°03′E
Znamenka, Ukr. see Znam′ianka	218-19	48°43′N	32°40′E
Znam′ianka, Ukr.	218-19	48°43′N	32°40′E
Znojmo, Czech Rep. (znoi′mô)	210-11	48°51′N	16°03′E
Zolotonosha, Ukr. (zŏ′lô-tŏ-nô′shá)	218-19	49°40′N	32°02′E
Zomba, Malawi (zŏm′bá)	286-87	15°23′S	35°19′E
Zongo, D.R.C. (zŏṅ′gô)	284-85	4°21′N	18°37′E
Zonguldak, Tur. (zŏn′gōōl′dàk)	202-03	41°28′N	31°49′E
Zouérat, Maur.	280-81	22°40′N	12°27′W
Zrenjanin, Serb.	216-17	45°23′N	20°24′E
Zubtsov, Russia (zŏp-tsôf′)	218-19	56°10′N	34°35′E
Zugdidi, Geor.	245	42°30′N	41°52′E
Zugspitze, mtn., Eur.	210-11	47°25′N	10°59′E
Zújar, stm., Spain (zōō′кär)	214-15	39°01′N	5°47′W
Zululand, hist. reg., S. Afr.	286-87	28°10′S	32°00′E
Zumbrota, Mn., U.S. (zŭm-brō′tá)	124-25	44°18′N	92°41′W
Zunhua, China (dzón-hwä)	260-61	40°11′N	117°57′E
Zunyi, China	258-59	27°42′N	106°56′E
Zürich, Switz. (tsü′rĭk)	210-11	47°23′N	8°32′E
Zuyevka, Russia	202-03	58°24′N	51°09′E
Zvishavane, Zimb.	286-87	20°20′S	30°03′E
Zvolen, Slvk. (zvô′lěn)	210-11	48°35′N	19°08′E
Zvornik, Bos. (zvôr′něk)	216-17	44°23′N	19°07′E
Zwedru, Lib.	282-83	6°04′N	8°07′W
Zweibrücken, Ger. (tsvī-brük′ěn)	210-11	49°15′N	7°22′E
Zwickau, Ger. (tsvĭkŏu)	210-11	50°43′N	12°30′E
Zwolle, Neth. (zvŏl′ě)	206-07	52°31′N	6°06′E
Zyryanka, Russia (zě-ryän′ká)	236-37	65°45′N	150°51′E
Zyryanovsk, Kaz.	244	49°44′N	84°18′E

ăt; finăl; rāte; senâte; ärm; åsk; sofá; fâre; ch-choose; dh-as th in other; bē; ěvent; bĕt; recĕnt; cratēr; g-gō; gh-guttural g; bĭt; ĭ-short neutral; rīde; к-guttural k as ch in German ich;